Metals Handbook® Ninth Edition

Volume 11
Failure Analysis and Prevention

Prepared under the direction of
the ASM Handbook Committee

Gordon W. Powell, Coordinator
Salah E. Mahmoud, Coordinator

Kathleen Mills, Manager of Editorial Operations
Joseph R. Davis, Senior Technical Editor
James D. Destefani, Technical Editor
Deborah A. Dieterich, Production Editor
George M. Crankovic, Assistant Editor
Heather J. Frissell, Assistant Editor
Diane M. Jenkins, Word Processing Specialist
Karen Lynn O'Keefe, Word Processing Specialist

Robert L. Stedfeld, Director of Reference Publications

Editorial Assistance
Robert T. Kiepura
Bonnie R. Sanders

AMERICAN SOCIETY FOR METALS
METALS PARK, OHIO 44073

Metals Handbook is a collective effort involving thousands of technical specialists. It brings together in one book a wealth of information from world-wide sources to help scientists, engineers, and technicians solve current and long-range problems.

Great care is taken in the compilation and production of this volume, but it should be made clear that no warranties, express or implied, are given in connection with the accuracy or completeness of this publication, and no responsibility can be taken for any claims that may arise.

Nothing contained in the Metals Handbook shall be construed as a grant of any right of manufacture, sale, use, or reproduction, in connection with any method, process, apparatus, product, composition, or system, whether or not covered by letters patent, copyright, or trademark, and nothing contained in the Metals Handbook shall be construed as a defense against any alleged infringement of letters patent, copyright, or trademark, or as a defense against liability for such infringement.

Comments, criticisms, and suggestions are invited, and should be forwarded to the American Society for Metals.

Library of Congress Cataloging in Publication Data

American Society for Metals

Metals handbook.

Includes bibliographies and indexes.
Contents: v. 1. Properties and selection—[etc.]—
v. 9. Metallography and Microstructures—[etc.]—
v. 11. Failure analysis and prevention.

1. Metals—Handbooks, manuals, etc. I. American
Society for Metals. Handbook Committee.
TA459.M43   1978   669   78-14934
ISBN 0-87170-007-7 (v. 1)
SAN 204-7586

Printed in the United States of America

Foreword

Volume 11 of the 9th Edition of *Metals Handbook* is a tribute to the science of failure analysis, a science that benefits all of us every day. Although the word "failure" conjures up negative images—anything from a snapped lawnmower bolt to a major airline disaster—the analysis of failures is the most positive of disciplines. By systematically identifying, exploring, understanding, and finally solving problems, the failure analyst rewards us with improved product reliability and safety.

This Volume continues the *Metals Handbook* tradition of providing its readers with practical information that can be used on the job. The hundreds of case histories described in this Handbook guide you through detailed, step-by-step analyses of actual service failures—concluding with recommendations for preventing similar failures in the future.

It should be noted that in this Handbook the term "failure" is used as a technical term, as are many other terms relating to failure. As a technical term, "failure" means cessation of function or usefulness. Failures may stem from many causes, but the term carries no implication of negligence or malfeasance.

Handbook Coordinators Gordon Powell and Salah Mahmoud are to be congratulated for recruiting a fine array of knowledgeable contributors and for successfully pulling a large undertaking together. Thanks to the many authors and reviewers who volunteered their expertise to this project. We also appreciate the assistance of ASM's Handbook Committee and Failure Analysis Committee (a special nod to former chairman Ray Fessler and member Bruce Christ) and the hard work of the ASM editorial staff. As always, it is the combined efforts of dedicated individuals that make *Metals Handbook* a valued reference source.

Raymond F. Decker
President, ASM

Edward L. Langer
Managing Director, ASM

The Ninth Edition of Metals Handbook
is dedicated to the memory of
TAYLOR LYMAN, A.B. (Eng.), S.M., Ph.D.
(1917–1973)
Editor, Metals Handbook, 1945–1973

Preface

Failure Analysis and Prevention was published initially in 1975 as Volume 10 of the 8th Edition of *Metals Handbook*. That Volume has served as the best single source of information on the failure modes and service failures of metallic materials. This new Volume is an updated and expanded version of its predecessor.

The present Handbook essentially maintains the format used in the 8th Edition—beginning with a review of the engineering aspects of failure, following with in-depth studies of various failure mechanisms and related environmental factors, and concluding with extensive analyses of hundreds of actual service failures. Nearly all of the articles have been significantly revised and expanded, and brand-new articles on solid metal induced embrittlement, failures in sour gas environments, and failures of bridge components, pipelines, and locomotives axles have been included.

This edition also includes, for the first time, articles on the failure analysis of ceramics, polymers, and continuous fiber reinforced composites; this change is in keeping with ASM's commitment to being a society for materials. Another article that represents a step forward for the Handbook in terms of increased coverage is concerned with failure analysis of integrated circuits.

On behalf of ASM, we express our appreciation to the authors for their time and effort expended in the preparation of this Volume and for their willingness to share with its readers their knowledge and the lessons derived from practical experience. Most of the contributors have established national and international reputations in their respective fields; several are authors of noteworthy textbooks. Each article has also undergone review by members of the ASM Handbook Committee and by other experts from industry and universities. We extend thanks to all of them.

Dr. Gordon W. Powell
The Ohio State University

Dr. Salah E. Mahmoud
Exxon Production Research Company

Previous Chairmen of the ASM Handbook Committee

R.S. Archer
(1940-1942) (Member, 1937-1942)

L.B. Case;
(1931-1933) (Member, 1927-1933)

T.D. Cooper
(1984-1986) (Member, 1981-1986)

E.O. Dixon
(1952-1954) (Member, 1947-1955)

R.L. Dowdell
(1938-1939) (Member, 1935-1939)

J.P. Gill
(1937) (Member, 1934-1937)

J.D. Graham
(1966-1968) (Member, 1961-1970)

J.F. Harper
(1923-1926) (Member, 1923-1926)

C.H. Herty, Jr.
(1934-1936) (Member, 1930-1936)

J.B. Johnson
(1948-1951) (Member, 1944-1951)

L.J. Korb
(1983) (Member, 1978-1983)

R.W.E. Leiter
(1962-1963) (Member, 1955-1958, 1960-1964)

G.V. Luerssen
(1943-1947) (Member, 1942-1947)

G.N. Maniar
(1979-1980) (Member, 1974-1980)

J.L. McCall
(1982) (Member, 1977-1982)

W.J. Merten
(1927-1930) (Member, 1923-1933)

N.E. Promisel
(1955-1961) (Member, 1954-1963)

G.J. Shubat
(1973-1975) (Member, 1966-1975)

W.A. Stadtler
(1969-1972) (Member, 1962-1972)

R. Ward
(1976-1978) (Member, 1972-1978)

M.G.H. Wells
(1981) (Member, 1976-1981)

D.J. Wright
(1964-1965) (Member, 1959-1967)

ASM Failure Analysis Committee

Charles Morin
Chairman
Packer Engineering Associates, Inc.

Raymond R. Fessler
Former Chairman
Battelle Columbus Laboratories

Robert C. Anderson
Anderson & Associates, Inc.

James Bennett
AT&T Bell Laboratories

Dale H. Breen
Packer Engineering Associates, Inc.

Herman Burghard
Southwest Research Institute

Bruce W. Christ
National Bureau of Standards

Dwight R. Diercks
Argonne National Laboratory

Robert J. Gray
Consultant

Jan M. Hoegfeldt
Honeywell, Inc.

Steve Hopkins
Failure Analysis Associates

Salah E. Mahmoud
Exxon Production Research Company

Michael Marx
National Transportation Safety Board

Regis M. Pelloux
Massachusetts Institute of Technology

Henry Piehler
Carnegie-Mellon University

Gordon W. Powell
The Ohio State University

M.H. Rafiee
Combustion Engineering, Inc.

John E. Roland
General Motors Company

L.E. Samuels
Australian Department of Defense

Helmut Thielsch
Thielsch Engineering Associates, Inc.

William R. Warke
Amoco Corporation

Paul Weihrauch
Massachusetts Materials Research

Donald J. Wulpi
Consultant

John H. Zirnhelt
Canadian Welding Bureau

Policy on Units of Measure

By a resolution of its Board of Trustees, the American Society for Metals has adopted the practice of publishing data in both metric and customary U.S. units of measure. In preparing this Handbook, the editors have attempted to present data primarily in metric units based on Système International d'Unités (SI), with secondary mention of the corresponding values in customary U.S. units. The decision to use SI as the primary system of units was based on the aforementioned resolution of the Board of Trustees, the widespread use of metric units throughout the world, and the expectation that the use of metric units in the United States will increase substantially during the anticipated lifetime of this Handbook.

For the most part, numerical engineering data in the text and in tables are presented in SI-based units with the customary U.S. equivalents in parentheses (text) or adjoining columns (tables). For example, pressure, stress, and strength are shown both in SI units, which are pascals (Pa) with a suitable prefix, and in customary U.S. units, which are pounds per square inch (psi). To save space, large values of psi have been converted to kips per square inch (ksi), where 1 kip = 1000 lb. Some strictly scientific data are presented in SI units only.

To clarify some illustrations that depict machine parts described in the text, only one set of dimensions is presented on artwork. References in the accompanying text to dimensions in the illustrations are presented in both SI-based and customary U.S. units.

On graphs and charts, grids correspond to SI-based units, which appear along the left and bottom edges; where appropriate, corresponding customary U.S. units appear along the top and right edges. Some illustrations previously published in the 8th Edition, particularly graphs illustrating mechanical property values, have been reproduced in their original form. In some instances, only customary U.S. units have been used.

Data pertaining to a specification published by a specification-writing group may be given in only the units used in that specification or in dual units, depending on the nature of the data. For example, the typical yield strength of aluminum sheet made to a specification written in customary U.S. units would be presented in dual units, but the thickness specified in that specification might be presented only in inches.

Data obtained according to standardized test methods for which the standard recommends a particular system of units are presented in the units of that system. Wherever feasible, equivalent units are also presented.

Conversions and rounding have been done in accordance with ASTM Standard E 380, with careful attention to the number of significant digits in the original data. For example, an annealing temperature of 1575 °F contains three significant digits. In this instance, the equivalent temperature would be given as 855 °C; the exact conversion to 857.22 °C would not be appropriate. For an invariant physical phenomenon that occurs at a precise temperature (such as the melting of pure silver), it would be appropriate to report the temperature as 961.93 °C or 1763.5 °F. In many instances (especially in tables and data compilations), temperature values in °C and °F are alternatives rather than conversions.

The policy on units of measure in this Handbook contains several exceptions to strict conformance to ASTM E 380; in each instance, the exception has been made to improve the clarity of the Handbook. The most notable exception is the use of $MPa\sqrt{m}$ rather than $MN \cdot m^{-3/2}$ or $MPa \cdot m^{0.5}$ as the SI unit of measure for fracture toughness. Other examples of such exceptions are the use of ''L'' rather than ''l'' as the abbreviation for liter and the use of g/cm^3 rather than kg/m^3 as the unit of measure for density (mass per unit volume).

SI practice requires that only one virgule (diagonal) appear in units formed by combination of several basic units. Therefore, all of the units preceding the virgule are in the numerator and all units following the virgule are in the denominator of the expression; no parentheses are required to prevent ambiguity.

Authors and Reviewers

Lester E. Alban
Fairfield Manufacturing Company, Inc.
Steven D. Antolovich
Georgia Institute of Technology
William G. Ashbaugh
Cortest Engineering Services Inc.
Aziz I. Asphahani
Cabot Corporation
J.G. Baetz
Atlantic Research Corporation
C.M. Bailey, Jr.
AT&T Bell Laboratories (retired)
Joan Barna
Babcock & Wilcox
Leo Barnard
Scanning Electron Analysis Laboratories
Inc.
John M. Barsom
U.S. Steel Corporation
William F. Bates
Lockheed-Georgia Company
Clive P. Bosnyak
Dow Chemical U.S.A.
David Broek
FractuREsearch Inc.
I. Brough
University of Manchester Institute of
Science and Technology
T.V. Bruno
Metallurgical Consultants, Inc.
Jon Bryant
Otis Engineering Corporation
J.H. Bucher
Lukens Steel
Kenneth G. Budinski
Eastman Kodak Company
Dwight Buford
Colorado School of Mines
M.E. Burnett
The Timken Company
H.I. Burrier
The Timken Company
D.P. Chesire
AT&T Bell Laboratories
Robin K. Churchill
ESCO Corporation
Russ Cippola
Aptech Engineering Services
B. Cohen
Air Force Wright Aeronautical Laboratories

Vito J. Colangelo
Benet Weapons Laboratory
Watervliet Arsenal
Jeffery A. Colwell
Battelle Columbus Laboratories
R.D. Corneliussen
Drexel University
Dwight L. Crook
Intel Corporation
Jim Curiel
Scanning Electron Analysis Laboratories
Inc.
C.V. Darragh
The Timken Company
T.J. Davies
University of Manchester Institute of
Science and Technology
John R. Dillon
ESCO Corporation
Bill Dobbs
Air Force Wright Aeronautical Laboratories
Robert J. Eiber
Battelle Columbus Laboratories
J. Epperson
National Transportation Safety Board
David Eylofson
ESCO Corporation
D.P. Fairchild
Exxon Production Research Company
Frank Fechek
Air Force Wright Aeronautical Laboratories
John W. Fisher
Lehigh University
Anna C. Fraker
National Bureau of Standards
Robert J. Franco
Exxon Company – USA
Karl H. Frank
The University of Texas at Austin
Robert Franklin
Exxon Research and Engineering
David N. French
David N. French, Inc., Metallurgists
William E. Fuller
ESCO Corporation
Richard I. Garbor
Exxon Production Research Company
Alan G. Glover
Welding Institute of Canada
Barry M. Gordon
General Electric Company

Ray A. Grove
Boeing Commercial Airplane Company
Paul S. Gupton
Bell & Associates
A.M. Hall
Materials Technology Institute
Frederick G. Hammitt
University of Michigan
Stephen Harris
Scanning Electron Analysis Laboratories
Inc.
Daniel Hauser
Edison Welding Institute
Harold J. Henning
Consultant
Donald H. Hensler
AT&T Bell Laboratories
Frank J. Heymann
Westinghouse Electric Corporation
Thomas Hikido
Pyromet
Richard G. Hoagland
Ohio State University
J.L. Hughes
University of Mississippi
Richard Hughes
Reynolds Aluminum
F.R. Hutchings
British Engine Insurance Ltd.
S. Jacobs
Intel Corporation
F.L. Jamieson
Stelco, Inc. (retired)
Walter L. Jensen
Lockheed-Georgia Company
J. Wayne Jones
University of Michigan
Michael H. Kamdar
Benet Weapons Laboratory
Watervliet Arsenal
Lawrence L. Kashar
Scanning Electron Analysis Laboratories
Inc.
J.F. Kiefner
Battelle Columbus Laboratories
C.D. Kim
U.S. Steel Corporation
Roy T. King
Cabot Corporation
Henry P. Kirchner
Ceramic Finishing Company

Contents

Glossary of Terms

A

abrasion. The process of grinding or wearing away through the use of abrasives; a roughening or scratching of a surface due to *abrasive wear*.

abrasive wear. The removal of material from a surface when hard particles slide or roll across the surface under pressure. The particles may be loose or may be part of another surface in contact with the surface being abraded. Compare with *adhesive wear*.

adhesive wear. The removal or displacement of material from a surface by the welding together and subsequent shearing of minute areas of two surfaces that slide across each other under pressure. Compare with *abrasive wear*.

alligatoring. The longitudinal splitting of flat slabs in a plane parallel to the rolled surface. Also called fishmouthing.

alligator skin. See *orange peel*.

ambient. Surrounding; usually used in relation to temperature, as ''ambient temperature'' surrounding a certain part or assembly.

annealing twin. A *twin* formed in a crystal during recrystallization.

anode. The electrode of an electrolytic cell at which oxidation occurs. Contrast with *cathode*.

arrest lines (marks). See *beach marks*.

asperity. In *tribology*, a protuberance in the small-scale topographical irregularities of a solid surface.

axial. Longitudinal, or parallel to the axis or centerline of a part. Usually refers to axial compression or axial tension.

axial strain. Increase (or decrease) in length resulting from a stress acting parallel to the longitudinal axis of a test specimen.

B

banded structure. A segregated structure consisting of alternating, nearly parallel bands of different composition, typically aligned in the direction of primary hot working.

beach marks. Macroscopic (visible) progression marks on a fracture surface that indicate successive position of the advancing crack front. The classic appearance is of irregular elliptical or semielliptical rings, radiating outward from one or more origins. Beach marks (also known as clamshell marks, tide marks, or arrest marks) are typically found on service fractures where the part is loaded randomly, intermittently, or with periodic variations in mean stress or alternating stress. Not to be confused with *striations*, which are microscopic and form differently.

breaking stress. See *rupture stress*.

Brinell hardness number, HB. A number related to the applied load and to the surface area of the permanent impression made by a ball indenter computed from:

$$HB = \frac{2P}{\pi D(D - \sqrt{D^2 - d^2})}$$

where P is applied load, kgf; D is diameter of ball, mm; and d is mean diameter of the impression, mm.

Brinell hardness test. A test for determining the hardness of a material by forcing a hard steel or carbide ball of specified diameter into it under a specified load. The result is expressed as the *Brinell hardness number*.

brinelling. Damage to a solid bearing surface characterized by one or more plastically formed indentations brought about by overload. This term is often applied in the case of rolling-element bearings. See also *false brinelling*.

brittle. Permitting little or no plastic (permanent) deformation prior to fracture.

brittle crack propagation. A very sudden propagation of a crack with the absorption of no energy except that stored elastically in the body. Microscopic examination may reveal some deformation not noticeable to the unaided eye. Contrast with *ductile crack propagation*.

brittle erosion behavior. Erosion behavior having characteristic properties (e.g., little or no plastic flow, the formation of cracks) that can be associated with *brittle fracture* of the exposed surface. The maximum volume removal occurs at an angle near 90°, in contrast to approximately 25° for *ductile erosion behavior*.

brittle fracture. Separation of a solid accompanied by little or no macroscopic plastic deformation. Typically, brittle fracture occurs by rapid crack propagation with less expenditure of energy than for *ductile fracture*.

brittleness. The tendency of a material to fracture without first undergoing significant *plastic deformation*. Contrast with *ductility*.

buckle. (1) An indented valley in the surface of a sand casting due to expansion of the molding sand. (2) A local waviness in metal bar or sheet, usually transverse to the direction of rolling.

buckling. A compression phenomenon that occurs when, after some critical level of load, a bulge, bend, bow, kink, or other wavy condition is produced in a beam, column, plate, bar, or sheet product form.

bulk modulus. See *bulk modulus of elasticity*.

bulk modulus of elasticity, K. The measure of resistance to change in volume; the ratio of hydrostatic stress to the corresponding unit change in volume. This elastic constant can be expressed by:

$$K = \frac{\sigma_m}{\Delta} = \frac{-p}{\Delta} = \frac{1}{\beta}$$

where K is bulk modulus of elasticity, σ_m is hydrostatic or mean stress tensor, p is hydrostatic pressure, and β is compressibility. Also known as bulk modulus, compression modulus, hydrostatic modulus, and volumetric modulus of elasticity.

C

carbon flotation. Segregation in which free graphite has separated from the molten iron. This defect tends to occur at the upper surfaces of the cope of the castings.

casting shrinkage. See *liquid shrinkage, shrinkage cavity, solidification shrinkage,* and *solid shrinkage*.

catastrophic wear. Rapidly occurring or accelerating surface damage, deterioration, or change of shape caused by wear to such a degree that the service life of a part is appreciably shortened or its function is destroyed.

caustic cracking. A form of *stress-corrosion cracking* most frequently encountered in carbon steels or iron-chromium-nickel alloys that are exposed to concentrated hydroxide solutions at temperatures of 200 to 250 °C (400 to 480 °F). Also known as caustic embrittlement.

caustic embrittlement. See *caustic cracking*.

cavitation. The formation and rapid collapse within a liquid of cavities or bubbles that contain vapor or gas or both. Cavitation caused by severe turbulent flow often leads to *cavitation damage*.

cavitation damage. The degradation of a solid body resulting from its exposure to cavitation. This may include loss of material, surface deformation, or changes in properties or appearance.

cavitation erosion. See *cavitation damage*.

centerline shrinkage. Shrinkage or porosity occurring along the central plane or axis of a cast metal section.

chafing fatigue. Fatigue initiated in a surface damaged by rubbing against another body. See also *fretting*.

Charpy test. An impact test in which a V-notched, keyhole-notched, or U-notched specimen, supported at both ends, is struck behind the notch by a striker mounted at the lower end of a bar that can swing as a pendulum. The energy that is absorbed in fracture is calculated from the height to which the striker would have risen had there been no specimen and the height to which it actually rises after fracture of the specimen. Contrast with *Izod test*.

chevron pattern. A fractographic pattern of radial marks (shear ledges) that looks like nested letters ''V''; sometimes called a herringbone pattern. Chevron patterns are typically found on brittle fracture surfaces in parts whose widths are considerably greater than their thicknesses. The points of the chevrons can be traced back to the fracture origin.

chill. A white iron structure that is produced by rapid solidification.

chord modulus. The slope of the chord drawn between any two specific points on a stress-strain curve. See also *modulus of elasticity*.

clamshell marks. See *beach marks*.

cleavage. (1) Fracture of a crystal by crack propagation across a crystallographic plane of low index. (2) The tendency to cleave or split along definite crystallographic planes.

cleavage crack. A crack that extends along a plane of easy *cleavage* in a crystalline material.

cleavage fracture. A fracture, usually of a polycrystalline metal, in which most of the grains have failed by *cleavage*, resulting in bright reflecting facets. It is one type of *crystalline fracture* and is associated with *low-energy brittle fracture*. Contrast with *shear fracture*.

cleavage plane. A characteristic crystallographic plane or set of planes in a crystal on which *cleavage fracture* occurs easily.

cold shot. A small globule of metal that solidified prematurely and is embedded in but not entirely fused with the surface of the casting.

cold shut. A discontinuity on or immediately beneath the surface of a casting, caused by the meeting of two streams of liquid metal that failed to merge. A cold shut may have the appearance of a crack or seam with smooth, rounded edges.

columnar structure. A coarse structure of parallel, elongated grains formed by unidirectional growth that is most often observed in castings. This results from diffusional growth accompanied by a solid-state transformation.

composite material. A heterogeneous, solid structural material consisting of two or more distinct components that are mechanically or metallurgically bonded together, such as a wire or filament of a high-melting substance embedded in a metal or nonmetal matrix.

compression modulus. See *bulk modulus of elasticity*.

compressive. Pertaining to forces on a body or part of a body that tend to crush, or compress, the body.

compressive strength. The maximum *compressive stress* a material is capable of developing. With a brittle material that fails in compression by fracturing, the compressive strength has a definite value. In the case of ductile, malleable, or semiviscous materials (which do not fail in compression by a shattering fracture), the value obtained for compressive stength is an arbitrary value dependent on the degree of distortion that is regarded as effective failure of the material.

compressive stress. A stress that causes an elastic body to deform (shorten) in the direction of the applied load. Contrast with *tensile stress*.

contact fatigue. Cracking and subsequent pitting of a surface subjected to alternating Hertzian stresses such as those produced under rolling contact or combined rolling and sliding. The phenomenon of contact fatigue is encountered most often in rolling-element bearings or in gears, where the surface stresses are high due to the concentrated loads and are repeated many times during normal operation.

corrosion. The chemical or electrochemical reaction between a material, usually a metal, and its environment that produces a deterioration of the material and its properties. See also *corrosion fatigue, crevice corrosion, denickelification, dezincification, erosion-corrosion, exfoliation, filiform corrosion, fretting corrosion, galvanic corrosion, general corrosion, graphitic corrosion, impingement attack, interdendritic corrosion, intergranular corrosion, internal oxidation, oxidation, parting, pitting, poultice corrosion, rust, selective leaching, stray-current corrosion, stress-corrosion cracking,* and *sulfide stress cracking*.

corrosion fatigue. Cracking produced by the combined action of repeated or fluctuating stress and a corrosive environment at lower stress levels or fewer cycles than would be required in the absence of a corrosive environment.

corrosive wear. *Wear* in which chemical or electrochemical reaction with the environment is significant.

crack extension, Δa. An increase in crack size. See also *crack length, effective crack size, original crack size,* and *physical crack size*.

crack length (depth), a. In *fatigue* and *stress-corrosion cracking*, the *physical crack size* used to determine the crack growth rate and the *stress-intensity factor*. For a compact-type specimen, crack length is measured from the line connecting the bearing points of load application. For a center-crack tension specimen, crack length is measured from the perpendicular bisector of the central crack. See also *crack size*.

crack mouth opening displacement (CMOD). See *crack opening displacement*.

crack opening displacement (COD). On a K_{Ic} specimen, the opening displacement of the notch surfaces at the notch and in the direction perpendicular to the plane of the notch and the crack. The displacement at the tip is called the crack tip opening displacement (CTOD); at the mouth, it is called the crack mouth opening displacement (CMOD).

crack plane orientation. An identification of the plane and direction of a fracture in relation to product geometry. This identification is designated by a hyphenated code, the first letter(s) representing the direction normal to the crack plane and the second letter(s) designating the expected direction of crack propagation.

crack size, a. A lineal measure of a principal planar dimension of a crack. This measure is commonly used in the calculation of quantities descriptive of the stress and displacement fields. In practice, the value of crack size is obtained from procedures for measurement of *physical crack size, original crack size,* or *effective crack size,* as appropriate to the situation under consideration. See also *crack length (depth)*.

crack tip opening displacement (CTOD). See *crack opening displacement*.

crack-tip plane strain. A stress-strain field near a crack tip that approaches *plane strain* to the degree required by an empirical criterion.

creep. Time-dependent strain occurring under stress. The *creep strain* occurring at a diminishing rate is called primary or transient creep; that occurring at a minimum and almost constant rate, secondary or steady-rate creep; that occurring at an accelerating rate, tertiary creep.

creep rate. The slope of the creep-time curve at a given time determined from a Cartesian plot.

creep-rupture strength. The stress that will cause fracture in a creep test at a given time in a specified constant environment. Also known as stress-rupture strength.

creep strain. The time-dependent total strain (extension plus initial gage length) produced by applied stress during a creep test.

creep strength. The stress that will cause a given *creep strain* in a creep test at a given time in a specified constant environment.

creep stress. The constant load divided by the original cross-sectional area of the specimen.

crevice corrosion. Localized *corrosion* of a metal surface at, or immediately adjacent to, an area that is shielded from full exposure to the environment because of close proximity between the metal and the surface of another material.

cross direction. See *transverse direction*.

crush. An indentation in a casting surface due to displacement of sand into the mold cavity when the mold is closed.

crystalline fracture. A pattern of brightly reflecting crystal facets on the fracture surface of a polycrystalline metal, resulting from *cleavage fracture* of many individual crystals. Contrast with *fibrous fracture* and *silky fracture*; see also *granular fracture*.

cup fracture (cup-and-cone fracture). A mixed-mode fracture, often seen in tension test specimens of a ductile material, where the central portion undergoes plane-strain fracture and the surrounding region undergoes plane-stress fracture. One of the mating fracture surfaces looks like a miniature cup; it has a central depressed flat-face region surrounded by a shear lip. The other fracture surface looks like a miniature truncated cone.

cupping. The condition sometimes occurring in heavily cold worked rods and wires, in which the outside fibers are still intact and the central zone has failed in a series of cup-and-cone fractures.

cut. A raised rough surface on a casting due to erosion by the metal stream of part of the sand mold or core.

cycle, *N*. In fatigue, one complete sequence of values of applied load that is repeated periodically.

cyclic load. (1) Repetitive loading, as with regularly recurring stresses on a part, that sometimes leads to fatigue fracture. (2) Loads that change value by following a regular repeating sequence of change.

cyclic stressing. See *cyclic load*.

D

decarburization. Loss of carbon from the surface layer of a carbon-containing alloy due to reaction with one or more chemical substances in a medium that contacts the surface.

deformation. A change in the form of a body due to stress, thermal change, change in moisture, or other causes. Measured in units of length.

deformation bands. Bands produced within individual grains during cold working which differ variably in orientation from the matrix.

deformation curve. See *stress-strain diagram*.

dendrite. A crystal with a treelike branching pattern. Dendrites are most evident in cast metals slowly cooled through the solidification range.

denickelification. *Corrosion* in which nickel is selectively leached from nickel-containing alloys. Most commonly observed in copper-nickel alloys after extended service in fresh water. See also *selective leaching*.

depletion. Selective removal of one component of an alloy, usually from the surface or preferentially from grain-boundary regions. See also *selective leaching*.

deposit attack. See *poultice corrosion*.

deposit corrosion. See *poultice corrosion*.

dezincification. *Corrosion* in which zinc is selectively leached from zinc-containing alloys. Most commonly found in copper-zinc alloys containing less than 85% Cu after extended service in water containing dissolved oxygen. See also *selective leaching*.

diamond pyramid hardness test. See *Vickers hardness test*.

dimpled rupture fracture. A fractographic term describing *ductile fracture* that occurs through the formation and coalescence of microvoids (dimples) along the fracture path. The fracture surface of such a ductile fracture appears dimpled when observed at high magnification and usually is most clearly resolved when viewed in a scanning electron microscope.

distortion. Any deviation from an original size, shape, or contour that occurs because of the application of *stress* or the release of *residual stress*.

ductile crack propagation. Slow crack propagation that is accompanied by noticeable *plastic deformation* and requires energy to be supplied from outside the body. Contrast with *brittle crack propagation*.

ductile erosion behavior. Erosion behavior having characteristic properties (i.e., considerable *plastic deformation*) that can be associated with *ductile fracture* of the exposed solid surface. A characteristic ripple pattern forms on the exposed surface at low values of angle of attack. Contrast with *brittle erosion behavior*.

ductile fracture. Fracture characterized by tearing of metal accompanied by appreciable gross *plastic deformation* and expenditure of considerable energy. Contrast with *brittle fracture*.

ductility. The ability of a material to deform plastically before fracturing. Measured by *elongation* or *reduction of area* in a tension test, by height of *cupping* in a cupping test, or by the radius or angle of bend in a bend test. Contrast with *brittleness*; see also *plastic deformation*.

dynamic. Moving, or having high velocity. Frequently used with high strain rate (>0.1 s^{-1}) testing of metal specimens. Contrast with *static*.

E

effective crack size, a_e. The *physical crack size* augmented for the effects of crack-tip plastic deformation. Sometimes the effective crack size is calculated from a measured value of a physical crack size plus a calculated value of a plastic-zone adjustment. A preferred method for calculation of effective crack size compares compliance from the secant of a load-deflection trace with the elastic compliance from a calibration for the type of specimen.

elastic constants. The factors of proportionality that relate elastic displacement of a material to applied forces. See also *bulk modulus of elasticity, modulus of elasticity, Poisson's ratio,* and *shear modulus*.

elastic deformation. A change in dimensions directly proportional to and in phase with an increase or decrease in applied force.

elastic limit. The maximum *stress* a material is capable of sustaining without any permanent *strain* (deformation) remaining upon complete release of the stress. See also *proportional limit*.

elastic strain. See *elastic deformation*.

elasticity. The property of a material by virtue of which deformation caused by *stress* disappears upon removal of the stress. A perfectly elastic body completely recovers its original shape and dimensions after release of stress.

elongation. A term used in mechanical testing to describe the amount of extension of a test piece when stressed. See also *elongation, percent* and *stress*.

elongation, percent. The extension of a uniform section of a specimen expressed as percentage of the original gage length:

$$\text{Elongation, \%} = \frac{L_f - L_o}{L_o} \times 100$$

where L_o is original gage length and L_f is final gage length. See also *elongation*.

embrittlement. The severe loss of ductility and/or toughness of a material, usually a metal or alloy.

endurance limit. The maximum *stress* below which a material can presumably endure an infinite number of *stress cycles*. If the stress is not completely reversed, the value of the mean stress, the minimum stress, or the *stress ratio* also should be stated. Compare with *fatigue limit*.

erosion. Destruction of materials by the abrasive action of moving fluids, usually accelerated by the presence of solid particles carried with the fluid. See also *erosion-corrosion*.

erosion-corrosion. A conjoint action involving *corrosion* and *erosion* in the presence of a moving corrosive fluid, leading to the accelerated loss of material.

exfoliation. *Corrosion* that proceeds laterally from the sites of initiation along planes parallel to the surface, generally at grain boundaries, forming corrosion products that force metal away from the body of the material, giving rise to a layered appearance.

F

failure. A general term used to imply that a part in service (1) has become completely inoperable, (2) is still operable but is incapable of satisfactorily performing its intended function, or (3) has deteriorated seriously, to the point that it has become unreliable or unsafe for continued use.

false brinelling. Damage to a solid bearing surface characterized by indentations not caused by *plastic deformation*, resulting from overload but thought to be due to other causes such as *fretting corrosion*. See also *brinelling*.

fatigue. The phenomenon leading to *fracture* under repeated or fluctuating stresses having a maximum value less than the ultimate tensile strength of the material. See also *fatigue failure, high-cycle fatigue, low-cycle fatigue,* and *ultimate strength*.

fatigue crack growth rate, da/dN. The rate of crack extension caused by constant-amplitude fatigue loading, expressed in terms of crack extension per cycle of load application.

fatigue failure. Failure that occurs when a specimen undergoing fatigue completely fractures into two parts or has softened or been otherwise significantly reduced in stiffness by thermal heating or cracking. Fatigue failure generally occurs at loads which applied statically would produce little perceptible effect. Fatigue failures are progressive, beginning as minute cracks that grow under the action of the fluctuating stress.

fatigue life. The number of *stress cycles* that can be sustained prior to failure under a stated test condition.

fatigue limit. The maximum *stress* that presumably leads to fatigue fracture in a specified number of *stress cycles*. If the stress is not completely reversed, the value of the mean stress, the minimum stress, or the *stress ratio* also should be stated. Compare with *endurance limit*.

fatigue notch factor, K_f. The ratio of the *fatigue strength* of an unnotched specimen to the fatigue strength of a notched specimen of the same material and condition; both strengths are determined at the same number of *stress cycles*.

fatigue notch sensitivity, q. An estimate of the effect of a notch or hole of a given size and shape on the fatigue properties of a material; measured by $q = (K_f - 1)/(K_t - 1)$, where K_f is the *fatigue notch factor* and K_t is the *stress-concentration factor*. A material is said to be fully notch sensitive if q approaches a value of 1.0; it is not notch sensitive if the ratio approaches 0.

fatigue ratio. The *fatigue limit* under completely reversed flexural stress divided by the *tensile strength* for the same alloy and condition.

fatigue strength. The maximum *stress* that can be sustained for a specified number of *stress cycles* without failure, the stress being completely reversed within each cycle unless otherwise stated.

fatigue striation. See *striation*.

fatigue wear. *Wear* of a solid surface caused by *fracture* arising from material fatigue.

fiber. (1) The characteristic of wrought metal that indicates directional properties. It is revealed by etching of a longitudinal section or is manifested by the fibrous or woody appearance of a fracture. It is caused chiefly by extension of the constituents of the metal, both metallic and nonmetallic, in the direction of working. (2) The pattern of *preferred orientation* of metal crystals after a given deformation process, usually wiredrawing. See also *texture*.

fiber-reinforced composite. A material consisting of two or more discrete physical phases, in which a fibrous phase is dispersed in a continuous matrix phase. The fibrous phase may be macro-, micro-, or submicroscopic, but it must retain its physical identity so that it could conceivably be removed from the matrix intact.

fiber stress. Local *stress* through a small area (a point or line) on a section where the stress is not uniform, as in a beam under a bending load.

fibrous fracture. A gray and amorphous *fracture* that results when a metal is sufficiently ductile for the crystals to elongate before fracture occurs. When a fibrous fracture is obtained in an impact test, it may be regarded as definite evidence of toughness of the metal. See also *crystalline fracture* and *silky fracture*.

fibrous structure. (1) In forgings, a structure revealed as laminations, not necessarily detrimental, on an etched section or as a ropy appearance on a *fracture*. (2) In wrought iron, a structure consisting of slag fibers embedded in ferrite. (3) In rolled steel plate stock, a uniform, lamination-free, fine-grained structure on a fractured surface.

filiform corrosion. *Corrosion* that occurs under some coatings in the form of randomly distributed threadlike filaments.

fisheye. A discontinuity found on the fracture surface of a weld in steel that consists of a small pore or inclusion surrounded by an approximately round, bright area.

fishmouthing. See *alligatoring*.

flake. A short, discontinuous internal crack in ferrous metals attributed to stresses produced by localized transformation and hydrogen-solubility effects during cooling after hot working. In fracture surfaces, flakes appear as bright, silvery areas with a coarse texture. In deep acid-etched transverse sections, they appear as discontinuities that are usually in the midway to center location of the section. Also termed hairline cracks and shatter cracks.

flow. Movement (slipping or sliding) of essentially parallel planes within an element of a material in parallel directions; occurs under the action of *shear stress*. Continuous action in this manner, at constant volume and without disintegration of the material, is termed *yield, creep,* or *plastic deformation*.

flow lines. Texture showing the direction of metal flow during hot or cold working. Flow lines often can be revealed by etching the surface or a section of a metal part.

fluting. A type of *pitting* in which cavities occur in a regular pattern, forming grooves or flutes. Fluting is caused by *fretting* or by electric arcing.

fold. A defect in metal, usually on or near the surface, caused by continued fabrication of overlapping surfaces.

fractography. Descriptive explanation of a fracture process, especially in metals, with specific reference to photographs of the fracture surface. Macrofractography involves low magnification ($<25\times$); microfractography, at high magnification ($>25\times$).

fracture. The irregular surface produced when a piece of metal is broken. See also *crystalline fracture, fibrous fracture, granular fracture, intergranular fracture, silky fracture,* and *transgranular fracture.*

fracture mechanics. See *linear elastic fracture mechanics.*

fracture stress. See *rupture stress.*

fracture test. Test in which a specimen is broken and its fracture surface is examined with the unaided eye or with a low-power microscope to determine such factors as composition, grain size, case depth, or discontinuities.

fracture toughness. A generic term for measures of resistance to extension of a crack. The term is sometimes restricted to results of fracture mechanics tests, which are directly applicable in fracture control. However, the term commonly includes results from simple tests of notched or precracked specimens not based on fracture mechanics analysis. Results from tests of the latter type are often useful for fracture control, based on either service experience or empirical correlations with fracture mechanics tests. See also *stress-intensity factor.*

fretting. *Wear* that occurs between tight-fitting surfaces subjected to oscillation at very small amplitude. This type of wear can be a combination of *oxidative wear* and *abrasive wear*. See also *fretting corrosion.*

fretting corrosion. The deterioration at the interface between contacting surfaces as the result of *corrosion* and slight oscillatory slip between the two surfaces.

fretting fatigue. Fatigue fracture that initiates at a surface area where *fretting* has occurred.

G

galling. A condition whereby excessive friction between high spots results in localized welding with subsequent *spalling* and a further roughening of the rubbing surfaces of one or both of two mating parts.

galvanic corrosion. Accelerated *corrosion* of a metal because of an electrical contact with a more noble metal or nonmetallic conductor in a corrosive electrolyte.

gas hole. A hole in a casting or weld formed by gas escaping from molten metal as it solidifies. Gas holes may occur individually or in clusters, or may be distributed throughout the solidified metal.

gas porosity. Fine holes or pores within a metal that are caused by entrapped gas or by evolution of dissolved gas during solidification.

general corrosion. A form of deterioration that is distributed more or less uniformly over a surface. See also *corrosion.*

glide. See *slip.*

grain. An individual crystal in a polycrystalline metal or alloy, including twinned regions or *subgrains* if present.

grain boundary. An interface separating two grains at which the orientation of the lattice changes from that of one grain to that of the other. When the orientation change is very small the boundary is sometimes referred to as a *sub-boundary structure.*

grain-boundary corrosion. Same as *intergranular corrosion*; see also *corrosion* and *interdendritic corrosion.*

grain flow. Fiberlike lines on polished and etched sections of forgings caused by orientation of the constituents of the metal in the direction of working during forging. Grain flow produced by proper die design can improve required *mechanical properties* of forgings.

granular fracture. A type of irregular surface produced when metal is broken that is characterized by a rough, grainlike appearance, rather than a smooth or fibrous one. It can be subclassified as *transgranular* or *intergranular*. This type of fracture is frequently called *crystalline fracture*; however, the inference that the metal broke because it "crystallized" is not justified, because all metals are crystalline in the solid state. See also *fibrous fracture* and *silky fracture.*

graphitic corrosion. Deterioration of gray cast iron in which the metallic constituents are selectively leached or converted to corrosion products, leaving the graphite intact; it occurs in relatively mild aqueous solutions and in buried pipe and fittings. The term "graphitization" is commonly used to identify this form of *corrosion*, but is not recommended because of its use in metallurgy for the decomposition of carbide to graphite.

H

hairline crack. See *flake.*

hardness. A measure of the resistance of a material to surface indentation or abrasion; may be thought of as a function of the *stress* required to produce some specified type of surface deformation. There is no absolute scale for hardness; therefore, to express hardness quantitatively, each type of test has its own scale of arbitrarily defined hardness. Indentation hardness can be measured by *Brinell, Knoop, Rockwell, Scleroscope,* and *Vickers hardness tests.*

Hartmann lines. See *Lüders lines.*

heat-affected zone. That portion of the base metal that was not melted during brazing, cutting, or welding, but whose *microstructure* and *mechanical properties* were altered by the heat.

herringbone pattern. See *chevron pattern.*

high-cycle fatigue. *Fatigue* that occurs at relatively large numbers of cycles. The arbitrary, but commonly accepted, dividing line between high-cycle fatigue and *low-cycle fatigue* is considered to be about 10^4 to 10^5 cycles. In practice, this distinction is made by determining whether the dominant component of the *strain* imposed during cyclic loading is elastic (high cycle) or plastic (low cycle), which in turn depends on the properties of the metal and on the magnitude of the nominal *stress.*

Hooke's law. A material in which *stress* is linearly proportional to *strain* is said to obey Hooke's law. This law is valid only up to the *proportional limit*, or the end of the straight-line portion of the *stress-strain diagram*. See also *modulus of elasticity.*

hot crack. See *solidification shrinkage crack.*

hot tear. A crack or fracture formed before completion of solidification because of hindered contraction. A hot tear is frequently open to the surface of the casting and thus exposed to the atmosphere. This may result in *oxidation, decarburization,* or other metal-atmosphere reactions at the tear surface.

hydrogen blistering. The formation of blisters on or below a metal surface from excessive internal hydrogen pressure. Hydrogen may be formed during cleaning, plating, corrosion, etc.

hydrogen damage. A general term for the embrittlement, cracking, blistering, and hydride formation that can occur when hydrogen is present in some metals.

hydrogen embrittlement. A condition of low ductility or hydrogen-induced cracking in metals resulting from the absorption of hydrogen. See also *hydrogen-induced delayed cracking.*

hydrogen-induced delayed cracking. A term sometimes used to identify a form of *hydrogen embrittlement* in which a metal appears to fracture spontaneously under a steady stress less than the *yield stress*. There is usually a delay between the application of stress (or exposure of the stressed metal to hydrogen) and the onset of cracking. Also referred to as static fatigue.

hydrostatic modulus. See *bulk modulus of elasticity.*

I

impact energy. The amount of energy required to fracture a material, usually measured by means of an *Izod test* or *Charpy test*. The type of specimen and test conditions affect the values and therefore should be specified.

impact load. An especially severe shock load such as that caused by instantaneous arrest of a falling mass, by shock meeting of two parts (in a mechanical hammer, for example), or by explosive impact, in which there can be an exceptionally rapid buildup of stress.

impact strength. See *impact energy*.

impingement attack. *Corrosion* associated with turbulent flow of liquid. May be accelerated by entrained gas bubbles. See also *erosion-corrosion*.

inclusion. A particle of foreign material in a metallic matrix. The particle is usually a compound (such as an oxide, sulfide, or silicate), but may be of any substance that is foreign to (and essentially insoluble in) the matrix. Inclusions are usually considered undesirable, although in some cases—such as in free-machining metals—manganese sulfides, phosphorus, selenium, or tellurium may be deliberately introduced to improve machinability.

intercrystalline. See *intergranular*.

intercrystalline cracking. See *intergranular cracking*.

intercrystalline corrosion. See *intergranular corrosion*.

interdendritic corrosion. Corrosive attack that progresses preferentially along interdendritic paths. This type of attack results from local differences in composition commonly encountered in alloy castings. See also *corrosion*.

interface. The boundary between two contacting parts or regions of parts.

intergranular. Between crystals or grains. Also termed intercrystalline. Contrast with *transgranular*.

intergranular corrosion. *Corrosion* occurring preferentially at grain boundaries, usually with slight or negligible attack on the adjacent grains. See also *interdendritic corrosion*.

intergranular cracking. Cracking or fracturing that occurs between the grains or crystals in a polycrystalline aggregate. Contrast with *transgranular cracking*.

intergranular fracture. Brittle fracture of a metal in which the fracture is between the grains, or crystals, that form the metal. Contrast with *transgranular fracture*.

intergranular stress-corrosion cracking. *Stress-corrosion cracking* in which the cracking occurs along grain boundaries.

internal oxidation. (1) The formation of isolated particles of *corrosion* products beneath the metal surface. This occurs as the result of preferential oxidation of certain alloy constituents by inward diffusion of oxygen, nitrogen, sulfur, etc. Also called subsurface cor-rosion. (2) Preferential *in situ* oxidation of certain components of phases within the bulk of a solid alloy accomplished by diffusion of oxygen into the body. This is commonly used to prepare electrical contact materials.

intracrystalline. See *transgranular*.

intracrystalline cracking. See *transgranular cracking*.

Izod test. A type of impact test in which a V-notched specimen, mounted vertically, is subjected to a sudden blow delivered by the weight at the end of a pendulum arm. The energy required to break off the free end is a measure of the impact strength or toughness of the material. Contrast with *Charpy test*.

J

J-integral. A mathematical expression; a line or surface integral that encloses the crack front from one crack surface to the other, used to characterize the *fracture toughness* of a material having appreciable plasticity before fracture. The J-integral eliminates the need to describe the behavior of the material near the crack tip by considering the local stress-strain field around the crack front; J_{Ic} is the critical value of the J-integral required to initiate growth of a pre-existing crack.

K

Knoop hardness number, HK. A number related to the applied load and to the projected area of the permanent impression made by a rhombic-based pyramidal diamond indenter having included edge angles of 172° 30′ and 130° 0′ computed from the equation:

$$HK = \frac{P}{0.07028\, d^2}$$

where P is applied load, kgf; and d is long diagonal of the impression, mm. In reporting Knoop hardness numbers, the test load is stated.

Knoop hardness test. An indentation hardness test using calibrated machines to force a rhombic-based pyramidal diamond indenter having specified edge angles, under specified conditions, into the surface of the material under test and to measure the long diagonal after removal of the load.

L

lamination. (1) A type of discontinuity with separation or weakness generally aligned parallel to the worked surface of a metal. (2) In electrical components such as motors, a blanked piece of electrical sheet that is stacked up with several other identical pieces to make a stator or rotor.

lap. (1) A surface imperfection on worked metal caused by folding over a fin overfill or similar surface condition, then impressing this into the surface by subsequent working without welding it. (2) A flat surface that holds an abrasive for polishing operations.

leaching. See *selective leaching*.

linear-elastic fracture mechanics. A method of fracture analysis that can determine the *stress* (or load) required to induce fracture instability in a structure containing a crack-like flaw of known size and shape. See also *stress-intensity factor*.

liquid metal embrittlement. The decrease in *ductility* of a metal caused by contact with a liquid metal.

liquid shrinkage. The reduction in volume of liquid metal as it cools to the liquidus.

longitudinal direction. That direction parallel to the direction of maximum elongation in a worked material. See also *normal direction* and *transverse direction*.

low-cycle fatigue. *Fatigue* that occurs at relatively small numbers of cycles ($<10^4$ cycles). Low-cycle fatigue may be accompanied by some plastic, or permanent, deformation. Compare with *high-cycle fatigue*.

Lüders lines. Elongated surface markings or depressions, often visible to the unaided eye, that form along the length of a tension specimen at an angle of approximately 45° to the loading axis. Caused by localized *plastic deformation*, they result from discontinuous (inhomogeneous) yielding. Also known as Lüders bands, Hartmann lines, Piobert lines, or stretcher strains.

M

macroscopic. Visible at magnifications at or below $25\times$.

macroshrinkage. Isolated, clustered, or interconnected voids in a casting that are detectable macroscopically. Such voids are usually associated with abrupt changes in section size and are caused by feeding that is insufficient to compensate for *solidification shrinkage*.

macrostructure. The structure of metals as revealed by macroscopic examination of a specimen. The examination may be carried out using an as-polished or a polished and etched specimen.

magnification. The ratio of the length of a line in the image plane (for example, ground glass or a photographic plate) to the length of the same line in the object. Magnifications are usually expressed in linear terms and in units called diameters.

malleability. The characteristic of metals that permits *plastic deformation* in compression without fracture. See also *ductility*.

matrix. The continuous or principal phase in which another constituent is dispersed.

maximum strength. See *ultimate strength.*

mechanical (cold) crack. A crack or fracture in a casting resulting from rough handling or from thermal shock, such as may occur at shakeout or during heat treatment.

mechanical properties. The properties of a material that reveal its elastic and inelastic (plastic) behavior when force is applied, thereby indicating its suitability for mechanical (load-bearing) applications. Examples are *elongation, fatigue limit, hardness, modulus of elasticity, tensile strength,* and *yield strength.* Compare with *physical properties.*

mechanical twin. A *twin* formed in a crystal by simple shear under external loading.

metal penetration. An imperfection on the surface of a casting caused by the penetration of molten metal into voids between refractory particles of the mold.

microcrack. A crack of microscopic proportions. Also termed microfissure.

microfissure. See *microcrack.*

microporosity. Extremely fine *porosity* in castings.

microscopic. Visible only at magnifications above $25 \times$.

microshrinkage. A casting imperfection consisting of interdendritic voids. Microshrinkage results from contraction during solidification where the opportunity to supply filler material is inadequate to compensate for shrinkage. Alloys with wide ranges in solidification temperature are particularly susceptible.

microstructure. The structure of metals and alloys as revealed after polishing and etching a specimen, at magnifications greater than $25 \times$.

misrun. A casting not fully formed because of solidification of metal before the mold is filled.

mode. One of the three classes of crack (surface) displacements adjacent to the crack tip. These displacement modes are associated with stress-strain fields around the crack tip and are designated I, II, and III. See also *crack-tip plane strain* and *crack opening displacement.*

modulus of elasticity, *E*. (1) The measure of rigidity or stiffness of a metal; the ratio of stress, below the *proportional limit,* to the corresponding strain. In terms of the *stress-strain diagram,* the modulus of elasticity is the slope of the stress-strain curve in the range of linear proportionality of stress to strain. Also known as Young's modulus. (2) For materials that do not conform to *Hooke's law* throughout the elastic range, the slope of either the tangent to the stress-strain curve at the origin or at low stress, the secant drawn from the origin to any specified point on the

stress-strain curve, or the chord connecting any two specific points on the stress-strain curve is usually taken to be the modulus of elasticity. In these cases, the modulus is referred to as the *tangent modulus, secant modulus,* or *chord modulus,* respectively.

modulus of rigidity. See *shear modulus.*

modulus of rupture. Nominal stress at fracture in a bend test or torsion test. In bending, modulus of rupture is the bending moment at fracture divided by the section modulus. In torsion, modulus of rupture is the torque at fracture divided by the polar section modulus. See also *modulus of rupture in bending* and *modulus of rupture in torsion.*

modulus of rupture in bending, S_b. The value of maximum tensile or compressive stress (whichever causes failure) in the extreme fiber on a beam loaded to failure in bending, computed from:

$$S_b = \frac{Mc}{I}$$

where M is maximum bending moment, computed from the maximum load and the original moment arm; c is initial distance from the neutral axis to the extreme fiber where failure occurs; and I is initial moment of inertia of the cross section about the neutral axis. See also *modulus of rupture.*

modulus of rupture in torsion, S_s. The value of maximum shear stress in the extreme fiber of a member of circular cross section loaded to failure in torsion, computed from:

$$S_s = \frac{Tr}{J}$$

where T is maximum twisting moment, r is original outer radius, and J is polar moment of inertia of the original cross section. See also *modulus of rupture.*

N

necking. (1) Reduction of the cross-sectional area of metal in a localized area by stretching. (2) Reduction in the diameter of a portion of the length of a cylindrical shell or tube.

neutron embrittlement. Embrittlement resulting from bombardment with neutrons, usually encountered in metals that have been exposed to a neutron flux in the core of a reactor. In steels, neutron embrittlement is evidenced by a rise in the ductile-to-brittle transition temperature. See also *radiation damage.*

nominal strength. See *ultimate strength.*

normal direction. That direction perpendicular to the plane of working in a worked material.

See also *longitudinal direction* and *transverse direction.*

notch. See *stress concentration.*

notch acuity. Relates to the severity of the *stress concentration* produced by a given notch in a particular structure. If the depth of the notch is very small compared with the width (or diameter) of the narrowest cross section, acuity may be expressed as the ratio of the notch depth to the notch root radius. Otherwise, acuity is defined as the ratio of one-half the width (or diameter) of the narrowest cross section to the notch root radius.

notch brittleness. Susceptibility of a material to *brittle fracture* at points of *stress concentration.* For example, in a notch tension test, the material is said to be notch brittle if the notch strength is less than the tensile strength of an unnotched specimen. Otherwise, it is said to be notch ductile.

notch depth. The distance from the surface of a test specimen to the bottom of the notch. In a cylindrical test specimen, the percentage of the original cross-sectional area removed by machining an annular groove.

notch rupture strength. The ratio of applied load to original area of the minimum cross section in a stress-rupture test of a notched specimen.

notch sensitivity. A measure of the reduction in strength of a metal caused by the presence of *stress concentration.* Values can be obtained from static, impact, or fatigue tests.

notch strength. The maximum load on a notched tension-test specimen divided by the minimum cross-sectional area (the area at the root of the notch). Also called notch tensile strength.

notch tensile strength. See *notch strength.*

O

orange peel. A surface roughening in the form of a pebble-grained pattern where a metal of unusually coarse grain is stressed beyond its elastic limit. Also known as pebbles and alligator skin.

original crack size, a_o. The *physical crack size* at the start of testing.

oxidation. (1) A reaction in which there is an increase in valence resulting from a loss of electrons. Contrast with *reduction.* (2) A *corrosion* reaction in which the corroded metal forms an oxide; usually applied to reaction with a gas containing elemental oxygen, such as air.

oxidative wear. A type of *wear* resulting from the sliding action between two metallic components that generates oxide films on the metal surfaces. These oxide films prevent the formation of a metallic bond between the sliding surfaces, resulting in fine wear debris and low wear rates.

P

parting. The selective *corrosion* of one or more components of a solid-solution alloy.

pebbles. See *orange peel*.

physical crack size, a_p. The distance from a reference plane to the observed crack front. This distance may represent an average of several measurements along the crack front. The reference plane depends on the specimen form, and it is normally taken to be either the boundary or a plane containing either the load line or the centerline of a specimen or plate.

physical properties. Properties of a metal or alloy that are relatively insensitive to structure and can be measured without the application of force; for example, density, electrical conductivity, coefficient of thermal expansion, magnetic permeability, and lattice parameter. Does not include chemical reactivity. Compare with *mechanical properties*.

pinhole. A small, rounded hole just below the surface of a casting, sometimes visible only after machining. Such holes, often localized, have bright interior surfaces.

Piobert lines. See *Lüders lines*.

pitting. (1) *Corrosion* of a metal surface, confined to a point or small area, that takes the form of cavities. (2) In *tribology*, a type of wear characterized by the presence of surface cavities formed by processes such as fatigue, local adhesion, or cavitation.

plane strain. The stress condition in *linear elastic fracture mechanics* in which there is zero strain in a direction normal to both the axis of applied *tensile stress* and the direction of crack growth (i.e., parallel to the crack front); most nearly achieved in loading thick plates along a direction parallel to the plate surface. Under plane-strain conditions, the plane of fracture instability is normal to the axis of the principal tensile stress.

plane-strain fracture toughness, K_{Ic}. The crack extension resistance under conditions of *crack-tip plane strain*. See also *stress-intensity factor*.

plane stress. The stress condition in *linear elastic fracture mechanics* in which the stress in the thickness direction is zero; most nearly achieved in loading very thin sheet along a direction parallel to the surface of the sheet. Under plane-stress conditions, the plane of fracture instability is inclined 45° to the axis of the principal *tensile stress*.

plane-stress fracture toughness, K_c. The value of the crack-extension resistance at the instability condition determined from the tangency between the *R-curve* and the critical crack-extension force curve of the specimen. See also *stress-intensity factor*.

plastic deformation. The permanent (inelastic) distortion of metals under applied stresses that strain the material beyond its *elastic limit*.

plowing. In *tribology*, the formation of grooves by *plastic deformation* of the softer of two surfaces in relative motion.

Poisson's ratio. The absolute value of the ratio of the transverse strain to the corresponding axial strain, in a body subjected to uniaxial stress; usually applied to elastic conditions.

polycrystalline. Comprising an aggregate of more than one crystal and usually a large number of crystals.

pore. (1) A small void in the body of a metal. (2) A minute cavity in a powder metallurgy compact, sometimes intentional. (3) A minute perforation in an electroplated coating.

porosity. Fine holes or pores within a metal.

poultice corrosion. A term used in the automotive industry to describe the *corrosion* of vehicle body parts due to the collection of road salts and debris on ledges and in pockets that are kept moist by weather and washing. Also called deposit attack or deposit corrosion.

preferred orientation. A condition of a polycrystalline aggregate in which the crystal orientations are not random, but rather exhibit a tendency for alignment with a specific direction in the bulk material, commonly related to the direction of working. See also *fiber* and *texture*.

primary creep. The first, or initial, stage of *creep*, or time-dependent deformation.

principal stress (normal). The maximum or minimum value of the normal stress at a point in a plane considered with respect to all possible orientations of the considered plane. On such principal planes the shear stress is zero. There are three principal stresses on three mutually perpendicular planes. The state of stress at a point may be: (1) uniaxial, a state of stress in which two of the three principal stresses are zero; (2) biaxial, a state of stress in which only one of the three principal stresses is zero; or (3) triaxial, a state of stress in which none of the principal stresses is zero. Multiaxial stress refers to either biaxial or triaxial stress.

proportional limit. The maximum *stress* at which *strain* remains directly proportional to stress; the upper end of the straight-line portion of the stress-strain or load-elongation curve. See also *elastic limit*.

Q

quasi-cleavage fracture. A fracture mode that combines the characteristics of *cleavage fracture* and *dimpled rupture fracture*. An intermediate type of fracture found in certain high-strength metals.

quenching crack. A crack formed as a result of thermal stresses produced by rapid cooling from a high temperature.

R

R-curve. A plot of crack-extension resistance as a function of stable crack extension, which is the difference between either the *physical crack size* or the *effective crack size* and the *original crack size*. *R*-curves normally depend on specimen thickness and, for some materials, on temperature and strain rate.

radial marks. Lines on a fracture surface that radiate from the fracture origin and are visible to the unaided eye or at low magnification. Radial marks result from the intersection and connection of brittle fractures propagating at different levels. Also known as shear ledges. See also *chevron pattern*.

radiation damage. A general term for the alteration of properties of a material arising from exposure to ionizing radiation (penetrating radiation), such as x-rays, gamma rays, neutrons, heavy-particle radiation, or fission fragments in nuclear fuel material. See also *neutron embrittlement*.

ratchet marks. Lines on a fatigue fracture surface that result from the intersection and connection of fatigue fractures propagating from multiple origins. Ratchet marks are parallel to the overall direction of crack propagation and are visible to the unaided eye or at low magnification.

rattail. A shallow, indented, and irregular line on a casting surface due to sand expansion.

reduction. (1) In cupping and deep drawing, a measure of the percentage decrease from blank diameter to cup diameter, or of diameter reduction in redrawing. (2) In forging, rolling, and drawing, either the ratio of the original to final cross-sectional area or the percentage decrease in cross-sectional area. (3) A reaction in which there is a decrease in valence resulting from a gain in electrons. Contrast with *oxidation*.

reduction in area. See *reduction of area*.

reduction of area. The difference between the original cross-sectional area of a tension specimen and the smallest area at or after fracture as specified for the material being tested. Also known as reduction in area.

residual stress. Stress present in a body that is free of external forces or thermal gradients.

river pattern. A characteristic pattern of cleavage steps running parallel to the local direction of crack propagation on the fracture surfaces of grains that have separated by *cleavage*.

rock candy fracture. A fracture that exhibits separated-grain facets; most often used to describe an *intergranular fracture* in a large-grained metal.

Rockwell hardness number, HR. A number derived from the net increase in the depth of impression as the load on an indenter is increased from a fixed minor load to a major load and then returned to the minor load. Rockwell hardness numbers are always quoted with a scale symbol representing the penetrator, load, and dial used.

Rockwell hardness test. An indentation hardness test using a calibrated machine that utilizes the depth of indentation, under constant load, as a measure of hardness. Either a 120° diamond cone with a slightly rounded point or a 1.6- or 3.2-mm ($\frac{1}{16}$- or $\frac{1}{8}$-in.) diam steel ball is used as the indenter.

Rockwell superficial hardness number. See *Rockwell hardness number* and *Rockwell superficial hardness test*.

Rockwell superficial hardness test. Same as *Rockwell hardness test*, except that smaller minor and major loads are used.

rupture stress. The stress at failure. Also known as breaking stress or fracture stress.

rust. A *corrosion* product consisting primarily of hydrated iron oxide. A term properly applied only to ferrous alloys.

S

sand hole. A pit in the surface of a sand casting resulting from a deposit of loose sand on the surface of the mold.

scab. A raised and rough area on the surface of a casting due to sand being dislodged from the surface of the mold.

Scleroscope hardness number, HSc or HSd. A number related to the height of rebound of a diamond-tipped hammer dropped on the material being tested. It is measured on a scale determined by dividing into 100 units the average rebound of the hammer from a quenched (to maximum hardness) and untempered AISI W-5 tool steel test block.

Scleroscope hardness test. A dynamic indentation hardness test using a calibrated instrument that drops a diamond-tipped hammer from a fixed height onto the surface of the material being tested. The height of rebound of the hammer is a measure of the hardness of the material.

scoring. In *tribology*, a severe form of *wear* characterized by the formation of extensive grooves and scratches in the direction of sliding.

scratching. In *tribology*, the mechanical removal and/or displacement of material from a surface by the action of abrasive particles or protuberances sliding across the surfaces. See also *plowing*.

scuffing. A form of *adhesive wear* that produces superficial scratches or a high polish on the rubbing surfaces. It is observed most often on inadequately lubricated parts.

seam. An unfused fold or lap that appears as a crack on a metal surface.

secant modulus. The slope of the secant drawn from the origin to any specified point on a stress-strain curve. See also *modulus of elasticity*.

season cracking. Cracking resulting from the combined effects of corrosion and internal stress. A term usually applied to *stress-corrosion cracking* of brass.

secondary creep. See *creep*.

segregation. Nonuniform distribution of alloying elements, impurities, or phases.

selective leaching. *Corrosion* in which one element is preferentially removed from an alloy, leaving a residue (often porous) of the elements that are more resistant to the particular environment. See also *decarburization*, *denickelification*, *dezincification*, and *graphitic corrosion*.

sensitization. In austenitic stainless steels, the precipitation of chromium carbides, usually at grain boundaries, on exposure to temperatures of about 550 to 850 °C (1000 to 1550 °F), leaving the grain boundaries depleted of chromium and therefore susceptible to preferential attack by a corroding (oxidizing) medium.

shatter crack. See *flake*.

shear bands. Bands in which *deformation* has been concentrated inhomogeneously in sheets that extend across regional groups of grains. Only one system is usually present in each regional group of grains, different systems being present in adjoining groups. The bands are noncrystallographic and form on planes of maximum *shear stress* (55° to the compression direction). They carry most of the deformation at large strains.

shear fracture. A *ductile fracture* in which a crystal (or a polycrystalline mass) has separated by sliding or tearing under the action of shear stresses. See also *shear stress*.

shear ledges. See *radial marks*.

shear lip. A narrow, slanting ridge along the edge of a fracture surface. The term sometimes also denotes a narrow, often crescent-shaped, fibrous region at the edge of a fracture that is otherwise of the cleavage type, even though this fibrous region is in the same plane as the rest of the fracture surface.

shear modulus, G. The ratio of *shear stress* to the corresponding *shear strain* for shear stresses below the *proportional limit* of the material. Values of shear modulus are usually determined by torsion testing. Also known as modulus of rigidity.

shear strength. The maximum *shear stress* that a material is capable of sustaining. Shear strength is calculated from the maximum load during a shear or torsion test and is based on the original dimensions of the cross section of the specimen.

shear stress. (1) A *stress* that exists when parallel planes in metal crystals slide across each other. (2) The stress component tangential to the plane on which the forces act. Also known as tangential stress.

shear strain. The tangent of the angular change, due to force, between two lines originally perpendicular to each other through a point in a body.

shock load. The sudden application of an external force that results in a very rapid build-up of *stress*—for example, piston loading in internal combustion engines.

shrinkage. See *casting shrinkage*.

shrinkage cavity. A void left in cast metals as a result of *solidification shrinkage*. Shrinkage cavities occur in the last metal to solidify after casting.

silky fracture. A metal fracture in which the broken metal surface has a fine texture, usually dull in appearance. Characteristic of tough and strong metals. Contrast with *crystalline fracture* and *granular fracture*.

slant fracture. A type of fracture appearance, typical of *plane-stress* fractures, in which the plane of metal separation is inclined at an angle (usually about 45°) to the axis of the applied stress.

slip. *Plastic deformation* by the irreversible shear displacement (translation) of one part of a crystal relative to another in a definite crystallographic direction and usually on a specific crystallographic plane. Sometimes called glide. See also *flow*.

S-N curve. A plot of stress (S) against the number of cycles to failure (N). The stress can be the maximum stress (S_{max}) or the alternating stress amplitude (S_a). The stress values are usually nominal stress; i.e., there is no adjustment for stress concentration. The diagram indicates the S-N relationship for a specified value of the mean stress (S_m) or the stress ratio (A or R) and a specified probability of survival. For N a log scale is almost always used. For S a linear scale is used most often, but a log scale is sometimes used. Also known as S-N diagram.

S-N diagram. See *S-N curve*.

solidification shrinkage. The reduction in volume of metal from beginning to end of solidification.

solidification shrinkage crack. A crack that forms, usually at elevated temperature, because of the internal (shrinkage) stresses that develop during solidification of a metal casting. Also termed hot crack.

solid shrinkage. The reduction in volume of metal from the solidus to room temperature.

spalling. The cracking and flaking of particles out of a surface.

static. Stationary or very slow. Frequently used in connection with routine tension testing of metal specimens. Contrast with *dynamic*.

static fatigue. See *hydrogen-induced delayed cracking*.

steady-rate creep. See *creep*.

strain. The unit of change in the size or shape of a body due to force.

strain hardening. An increase in hardness and strength caused by *plastic deformation* at temperatures below the recrystallization range. Also known as work hardening.

stray-current corrosion. *Corrosion* caused by electric current from a source external to the intended electrical circuit, for example, extraneous current in the earth.

stress. The intensity of the internally distributed forces or components of forces that resist a change in the volume or shape of a material that is or has been subjected to external forces. Stress is expressed in force per unit area and is calculated on the basis of the original dimensions of the cross section of the specimen. Stress can be either direct (tension or compression) or shear. Usually expressed in pounds per square inch (psi) or megapascals (MPa).

stress amplitude. One-half the algebraic difference between the maximum and minimum *stress* in one cycle of a repetitively varying stress.

stress concentration. A change in contour or a discontinuity that causes local increases in *stress* in materials under load. Typical are sharp-cornered grooves or notches, threads, fillets, holes, etc. Also called stress raiser.

stress-concentration factor, K_t. A multiplying factor for applied *stress* that allows for the presence of a structural discontinuity such as a notch or hole; K_t equals the ratio of the greatest stress in the region of the discontinuity to the nominal stress for the entire section. Also known as theoretical stress-concentration factor.

stress-corrosion cracking, SCC. A cracking process that requires the simultaneous action of a corrodent and sustained *tensile stress*. This excludes corrosion-reduced sections that fail by fast fracture. It also excludes intergranular or transgranular corrosion, which can disintegrate an alloy without applied or residual stress. See also *corrosion*.

stress cycle. The smallest segment of the stress-time function that is repeated periodically.

stress-intensity factor, K. A scaling factor used in *linear-elastic fracture mechanics* to describe the intensification of applied *stress* at the tip of a crack of known size and shape. At the onset of rapid crack propagation in any structure containing a crack, the factor is called the critical stress-intensity factor, or the *fracture toughness*. Various subscripts are used to denote different loading conditions or fracture toughnesses:

K_c. Plane-stress fracture toughness. The value of stress intensity at which crack propagation becomes rapid in sections thinner than those in which plane-strain conditions prevail.

K_I. Stress-intensity factor for a loading condition that displaces the crack faces in a direction normal to the crack plane. Also known as the opening mode of deformation.

K_{Ic}. Plane-strain fracture toughness. The minimum value of K_c for any given material and condition, which is attained when rapid crack propagation in the opening mode is governed by plane-strain conditions.

K_{Id}. Dynamic fracture toughness. The fracture toughness determined under dynamic loading conditions; it is used as an approximation of K_{Ic} for very tough materials.

K_{Iscc}. Threshold stress intensity for stress-corrosion cracking when loading conditions meet plane-strain requirements.

K_Q. Provisional value for plane-strain fracture toughness.

K_{th}. Threshold stress intensity for stress-corrosion cracking. A value of stress intensity characteristic of a specific combination of material, material condition, and corrosive environment above which stress-corrosion crack propagation occurs and below which the material is immune from stress-corrosion cracking.

ΔK. The range of the stress-intensity factor during a fatigue cycle.

stress ratio, A or R. The algebraic ratio of two specified *stress* values in a *stress cycle*. Two commonly used stress ratios are (1) the ratio of the alternating stress amplitude to the mean stress, $A = S_a/S_m$, and (2) the ratio of the minimum stress to the maximum stress, $R = S_{min}/S_{max}$.

stress raiser. See *stress concentration*.

stress-rupture strength. See *creep-rupture strength*.

stress-strain curve. See *stress-strain diagram*.

stress-strain diagram. A graph in which corresponding values of *stress* and *strain* are plotted against each other. Values of stress are usually plotted vertically (ordinate or *y* axis) and values of strain horizontally (abscissa or *x* axis). Also known as deformation curve and stress-strain curve.

stretcher strains. See *Lüders lines*.

striation. A fatigue fracture feature often observed in electron micrographs that indicates the position of the crack front after each succeeding cycle of *stress*. The distance between striations indicates the advance of the crack front across that crystal during one *stress cycle*, and a line normal to the striation indicates the direction of local crack propagation. Not to be confused with *beach marks*, which are much larger (macroscopic) and form differently.

stringer. In wrought materials, an elongated configuration of microconstituents or foreign material aligned in the direction of working. The term is commonly associated with elongated oxide or sulfide inclusions in steel.

sub-boundary structure (subgrain structure). A network of low-angle boundaries, usually with misorientations less than 1° within the main grains of a microstructure.

subgrain. A portion of a crystal or *grain*, with an orientation slightly different from the orientation of neighboring portions of the same crystal.

subsurface corrosion. See *internal oxidation*.

sulfidation. The reaction of a metal or alloy with a sulfur-containing species to produce a sulfur compound that forms on or beneath the surface of the metal or alloy.

sulfide stress cracking, SSC. Brittle failure by cracking under the combined action of *tensile stress* and *corrosion* in the presence of water and hydrogen sulfide.

T

tangential stress. See *shear stress*.

tangent modulus. The slope of the stress-strain curve at any specified *stress* or *strain*. See also *modulus of elasticity*.

tangential stress. See *shear stress*.

temper brittleness. Brittleness that results when certain steels are held within, or are cooled slowly through, a certain range of temperature below the transformation range. The brittleness is manifested as an upward shift in ductile-to-brittle transition temperature, but only rarely produces a low value of *reduction of area* in a smooth-bar tension test of the embrittled material.

tensile strength. In *tension testing*, the ratio of maximum load to the original cross-sectional area. See also *ultimate strength*; compare with *yield strength*.

tensile stress. A *stress* that causes two parts of an elastic body, on either side of a typical stress plane, to pull apart. Contrast with *compressive stress*.

tensile testing. See *tension testing*.

tension. The force or load that produces *elongation*.

tension testing. A method of determining the behavior of materials subjected to uniaxial loading, which tends to stretch the metal. A longitudinal specimen of known length and diameter is gripped at both ends and stretched at a slow, controlled rate until rupture occurs. Also known as tensile testing.

tertiary creep. See *creep*.

texture. In a polycrystalline aggregate, the state of distribution of crystal orientations. In the usual sense, it is synonymous with *preferred orientation*, in which the distribution is not random. See also *fiber*.

theoretical stress-concentration factor. See *stress-concentration factor*.

thermal fatigue. Fracture resulting from the presence of temperature gradients that vary with time in such a manner as to produce cyclic stresses in a structure.

thermal stresses. Stresses in metal resulting from nonuniform temperature distribution.

thermal shock. The development of a steep temperature gradient and accompanying high stresses within a structure.

tide marks. See *beach marks*.

torsion. A twisting action applied to a shaftlike or cylindrical member. The twisting may be either reversed (back and forth) or unidirectional (one way).

torsional stress. The *shear stress* on a transverse cross section resulting from a twisting action.

transcrystalline. See *transgranular*.

transcrystalline cracking. See *transgranular cracking*.

transgranular. Through or across crystals or grains. Also called intracrystalline or transcrystalline.

transgranular cracking. Cracking or fracturing that occurs through or across a crystal or grain. Also called transcrystalline cracking. Contrast with *intergranular cracking*.

transgranular fracture. Fracture through or across the crystals or grains of a metal. Also called transcrystalline fracture or intracrystalline fracture. Contrast with *intergranular fracture*.

transient creep. See *creep* and *primary creep*.

transverse direction. Literally, "across," usually signifying a direction or plane perpendicular to the direction of working. In rolled plate or sheet, the direction across the width is often called long transverse; the direction through the thickness, short transverse.

tribology. The science concerned with the design, friction, lubrication, and wear of contacting surfaces that move relative to each other.

tuberculation. The formation of localized *corrosion* products that appear on a surface as knoblike prominences (tubercules).

twin. Two portions of a crystal with a definite orientation relationship; one may be regarded as the parent, the other as the twin. The orientation of the twin is a mirror image of the orientation of the parent across a twinning plane or an orientation that can be derived by rotating the twin portion about a twinning axis. See also *annealing twin* and *mechanical twin*.

twin bands. Bands across a crystal grain, observed on a polished and etched section, where crystallographic orientations have a mirror-image relationship to the orientation of the matrix grain across a composition plane that is usually parallel to the sides of the band.

U

ultimate strength. The maximum *stress* (tensile, compressive, or shear) a material can sustain without fracture, determined by dividing maximum load by the original cross-sectional area of the specimen. Also known as nominal strength or maximum strength.

V

Vickers hardness number, HV. A number related to the applied load and the surface area of the permanent impression made by a square-based pyramidal diamond indenter having included face angles of 136°, computed from:

$$\text{HV} = 2P \sin\frac{\alpha/2}{d^2} = \frac{1.8544P}{d^2}$$

where P is applied load (kgf), d is mean diagonal of the impression (mm), and α is the face angle of the indenter (136°).

Vickers hardness test. An indentation hardness test employing a 136° diamond pyramid indenter (Vickers) and variable loads, enabling the use of one hardness scale for all ranges of hardness—from very soft lead to tungsten carbide. Also known as diamond pyramid hardness test.

volumetric modulus of elasticity. See *bulk modulus of elasticity*.

W

Wallner lines. A distinct pattern of intersecting sets of parallel lines, usually producing a set of V-shaped lines, sometimes observed when viewing brittle fracture surfaces at high magnification in an electron microscope. Wallner lines are attributed to interaction between a shock wave and a brittle crack front propagating at high velocity. Sometimes Wallner lines are misinterpreted as fatigue striations.

wear. Damage to a solid surface, generally involving progressive loss of material, due to relative motion between that surface and a contacting surface or substance.

wear rate. The rate of material removal or dimensional change due to *wear* per unit of exposure parameter—for example, quantity of material removed (mass, volume, thickness) in unit distance of sliding or unit time.

whiskers. (1) Metallic filamentary growths, often microscopic in size, that attain very high strengths. (2) Oxide whiskers, such as sapphire, which because of their strength and inertness at high temperatures are used as reinforcements in metal-matrix composites.

work hardening. See *strain hardening*.

Y

yield. Evidence of *plastic deformation* in structural materials. See also *creep* and *flow*.

yield point. The first *stress* in a material, usually less than the maximum attainable stress, at which an increase in *strain* occurs without an increase in stress. Only certain metals—those that exhibit a localized, heterogeneous type of transition from elastic to plastic deformation—produce a yield point. If there is a decrease in stress after yielding, a distinction may be made between upper and lower yield points. The load at which a sudden drop in the flow curve occurs is called the upper yield point. The constant load shown on the flow curve is the lower yield point.

yield strength. The stress at which a material exhibits a specified deviation from proportionality of *stress* and *strain*. The specified deviation is usually 0.2% for most metals. Compare with *tensile strength*.

yield stress. The stress level of highly ductile materials, such as structural steels, at which large *strains* take place without further increase in *stress*.

Young's modulus. See *modulus of elasticity*.

SELECTED REFERENCES

- A.D. Merriman, *A Dictionary of Metallurgy*, Pitman Publishing, 1958
- "Standard Definitions of Terms Relating to Corrosion and Corrosion Testing," G 15, *Annual Book of ASTM Standards*, Vol 03.02, ASTM, Philadelphia, 1984, p 133-137
- "Standard Definitions of Terms Relating to Fatigue Testing and the Statistical Analysis of Fatigue Data," E 206, *Annual Book of ASTM Standards*, Vol 03.01, ASTM, Philadelphia, 1984, p 340-345
- "Standard Definitions of Terms Relating to Methods of Mechanical Testing," E 6, *Annual Book of ASTM Standards*, Vol 03.01, ASTM, Philadelphia, 1984, p 119-129
- "Standard Terminology Relating to Erosion and Wear," G 40, *Annual Book of ASTM Standards*, Vol 03.02, ASTM, Philadelphia, 1984, p 239-246
- "Standard Terminology Relating to Fracture Testing," E 616, *Annual Book of ASTM Standards*, Vol 03.01, ASTM, Philadelphia, 1984, p 671-684

Engineering Aspects of Failure and Failure Analysis

General Practice in Failure Analysis

D.A. Ryder, Consultant
T.J. Davies and I. Brough, Department of Metallurgy, University of Manchester Institute of Science and Technology
F.R. Hutchings, British Engine Insurance Ltd.

THE GENERAL PROCEDURES, techniques, and precautions employed in the investigation and analysis of metallurgical failures that occur in service will be discussed in this article. The stages of investigation will be discussed, and the various features of the more common causes of failure indicated. Types of failure characteristics will be described, and where appropriate, several of the fundamental mechanisms involved will be explained. Information on procedures and techniques specific to the analysis of failures by various mechanisms and related environmental factors, failures of principal product forms, and failures of manufactured components and assemblies is provided in individual articles in the following four Sections of this Volume.

Objectives of Failure Investigation

A failure investigation and subsequent analysis should determine the primary cause of a failure, and based on the determination, corrective action should be initiated that will prevent similar failures. Frequently, the importance of contributory causes to the failure must be assessed; new experimental techniques may have to be developed, or an unfamiliar field of engineering or science explored. A complex accident investigation, such as investigation into aircraft accidents, usually requires the services of experts in several branches of engineering and the physical sciences, as well as metallurgy.

Stages of an Analysis

Although the sequence is subject to variation, depending upon the nature of a specific failure, the principal stages that comprise the investigation and analysis of a failure are:

- Collection of background data and selection of samples
- Preliminary examination of the failed part (visual examination and record keeping)
- Nondestructive testing
- Mechanical testing (including hardness and toughness testing)
- Selection, identification, preservation, and/or cleaning of all specimens

- Macroscopic examination and analysis (fracture surfaces, secondary cracks, and other surface phenomena)
- Microscopic examination and analysis
- Selection and preparation of metallographic sections
- Examination and analysis of metallographic sections
- Determination of failure mechanism
- Chemical analyses (bulk, local, surface corrosion products, deposits or coatings, and electron microprobe analysis)
- Analysis of fracture mechanics
- Testing under simulated service conditions (special tests)
- Analysis of all the evidence, formulation of conclusions, and writing the report (including recommendations)

The time involved in ascertaining all the circumstances of a failure is time well spent. When a broken component is received for examination, the investigator is sometimes inclined to prepare specimens immediately without devising an investigation procedure. Such lack of forethought should be avoided, because in the end a large amount of time and effort may be wasted; however, by first carefully considering the background of the failure and studying the general features, a more informative procedure will be indicated.

In the investigation of failures of some components, it may be impractical or impossible for the failure analyst to visit the failure site. Under these circumstances, data and samples may be collected by field engineers or by other personnel at the site. A field failure report sheet or check list can be used to ensure that all pertinent information regarding the failure is recorded.

Collection of Background Data and Selection of Samples

Initially, the failure investigation should be directed toward gaining an acquaintance with all pertinent details relating to the failure, collecting the available information regarding the manufacturing, processing, and service his-

tories of the failed component or structure, and reconstructing insofar as possible the sequence of events leading to the failure. The collection of background data on the manufacturing and fabricating history of a component should begin with obtaining specifications and drawings and should encompass all the design aspects of the component. Data relating to manufacturing and fabrication may be grouped into mechanical processing, which may include cold forming, stretching, bending, machining, grinding, and polishing; thermal processing, which may include details of hot forming, heat treating, welding, brazing, or soldering; and chemical processing, which may provide details of cleaning, electroplating, and application of coatings by chemical alloying or diffusion.

Service History. Obtaining a complete service history depends a great deal upon how detailed and thorough the recordkeeping was before the failure. The availability of complete service records greatly simplifies the assignment of the failure analyst. In collecting service histories, special attention should be given to environmental details, such as normal and abnormal loading, accidental overloads, cyclic loads, variations in pressure and temperature, pressure and temperature gradients, and operation in a corrosive environment, including the concentration and/or flow of a liquid environment. In most cases, however, complete service records are not available, forcing the analyst to work from fragmentary service information. When service data are sparse, the analyst must deduce the service conditions. Much depends upon the skill and judgment of the analyst because a misleading deduction can be more harmful than the absence of information.

Photographic Records. The analyst should decide if photographs of the failed component or structure are required. A failure that appears almost inconsequential in a preliminary investigation may later be found to have serious consequences; thus, a complete photographic record of the investigation can be important. If the photographs are to be provided to the analyst from another source, the analyst should be certain that these will be suitable for his purpose, that is, that they will adequately detail the characteristics of the failure.

For the failure analyst who chooses to do his own photography, a single lens reflex 35-mm camera with a variety of lenses and an extension bellows and with a battery-operated flash can produce excellent results. It is desirable to supplement the 35-mm equipment with a Polaroid camera with close-up and portrait lenses. The quality of Polaroid prints will generally be lower than that of prints made from 35-mm film, yet Polaroid prints may be quite adequate for the intended purpose. When accurate color rendition is required, the subject should be photographed against a gray background, and a sample of the actual background should be provided to the photographic studio for use as a guide in developing and printing.

Selection of samples should be done before starting the examination proper, especially if the investigation is to be lengthy or involved. As with photographs, the analyst is responsible for ensuring that the samples will be suitable for the intended purpose and that they represent the characteristics of the failure adequately. It is advisable to look for additional evidence of damage beyond that which is immediately apparent.

It is often necessary to compare failed components with similar components that did not fail to determine whether the failure was brought about by service conditions or was the result of an error in manufacture. For example, if a boiler tube fails and overheating is suspected to be the cause and if investigation reveals the spheroidized structure in the boiler tube, which indicates overheating in service, then comparison with another tube, remote from the region exposed to high temperature, will determine if the tubes were supplied in the spheroidized condition.

As another example, assume that examination of a bolt shows a fatigue fracture that is typical of the type caused by repeated application of excessive bending stresses. Loss of clamping force is the major reason for fatigue fractures of bolts. Generally, it is also necessary to examine nuts or other components associated with the bolt because errors in machining, or wear, of associated components can result in nonaxial loading in service, which could not be established from an examination of the bolt alone. Also, in failures involving corrosion, stress corrosion, or corrosion fatigue, a sample of the fluid that has been in contact with the metal or a sample of any deposits that have been formed will often be required for analysis. Information on sample preparation can be found in the section "Selection, Preservation, and Cleaning of Fracture Surfaces" in this article.

Abnormal Conditions. In addition to developing a general history of the failed component or structure, it is also advisable to determine if any abnormal conditions prevailed or if events occurred in service that may have contributed to the cause of failure and to determine

if any recent repairs or overhauls have been carried out and why. It is also necessary to inquire if the failure under investigation is an isolated example or if others have occurred, either in the component under consideration or in another of a similar design. In routine examination of a brittle fracture, it is important to know if at the time of the accident or failure the prevailing temperature was low and if some measure of shock loading was involved. When dealing with failures of crankshafts or other shafts, it is generally desirable to ascertain the conditions of the bearings and to determine whether any misalignment existed within the machine concerned or between the driving and driven components.

Preliminary Examination of the Failed Part

The failed part, including all of its fragments, should be subjected to a thorough visual examination before any cleaning is undertaken. Soils and debris found on the part often provide useful evidence in establishing the cause of failure or in determining a sequence of events leading to the failure. For example, traces of paint found on a portion of a fracture surface may provide evidence that a crack, into which some paint seeped, was present in the surface for some time before through-fracture occurred. Such evidence should be noted and recorded.

Visual Inspection. The preliminary examination should begin with unaided visual inspection. The unaided eye has exceptional depth of focus and has the ability to examine large areas rapidly and to detect subtle changes of color and texture. Some of these advantages are lost when any optical or electron-optical device is used. Particular attention should be given to the surfaces of fractures and to the paths of cracks. The significance of any indications of abnormal conditions or abuse in service should be observed and assessed, and a general assessment of the basic design and workmanship of the part should be made. All important features, including dimensions, should be recorded in writing or by sketches or photographs.

It cannot be emphasized too strongly that the examination should be performed as searchingly and effectively as possible because clues to the cause of breakdown are often present and may be missed if the observer is not vigilant enough to notice them. In this connection, a low-power microscope (about 6 to 25×), preferably of a binocular type, will be invaluable.

Photographing Fractures. Where fractures are involved, the next step in preliminary examination should be general photography of the entire fractured part, including broken pieces, to record their size and condition and to show how the fracture is related to the components of the part. This should be followed by careful examination of the fracture surface at various angles and magnifications. The examination should begin with the use of direct

lighting and should proceed using various angles of oblique lighting and dark-field illumination to assess how the fracture characteristics can best be delineated and emphasized. This should also assist in determining which areas of the fracture are of prime interest and which magnification should be used (for a given picture size) to bring out fine details. When this evaluation has been completed, it is appropriate to proceed with photography of the fracture, recording what each photograph shows, its magnification, and how it relates to the other photographs.

Nondestructive Testing

Several nondestructive tests are extremely useful in failure investigation and analysis, particularly magnetic-particle inspection of ferrous metals, liquid-penetrant inspection, ultrasonic inspection, and eddy-current inspection of materials that conduct electricity. All of these tests are used to detect surface cracks and discontinuities. Other nondestructive tests used are radiography (mainly for internal examination), acoustic-emission inspection, and experimental stress analysis (for determining machine loads and component stresses that can cause failure).

Magnetic-particle inspection uses magnetic fields to locate surface and subsurface discontinuities in ferromagnetic materials. When the material or part to be tested is magnetized, discontinuities that lie generally transverse to the direction of the magnetic field will cause a leakage field to be formed at and above the surface of the part. This leakage field and therefore the presence of the discontinuity are detected by means of fine ferromagnetic particles applied over the surface; some of these particles are gathered and held by the leakage field. The magnetically held collection of particles forms an outline of the discontinuity and indicates its size, shape, and extent. Frequently, a fluorescent material is combined with the particles so that discontinuities can be readily detected visually under ultraviolet light. Magnetic lines of force (or flux) can be set up by passing a large current of electricity through the component to be inspected, by use of a magnetizing yoke, and by use of a magnetizing coil. Following magnetic-particle inspection, the component is demagnetized.

The advantages of magnetic-particle inspection include:

- It is the best and most reliable method available for detecting surface cracks, especially very fine and shallow cracks and cracks filled with foreign matter
- Techniques are easy to learn, and the process is rapid, simple, and inexpensive to perform
- Indications are produced directly on the surface of the part and are a magnetic picture of the actual discontinuity. There is no electric circuitry or electronic readout to be

calibrated or kept in proper operating condition

- There is little or no limitation on size or shape of the part to be tested
- No elaborate precleaning is ordinarily necessary, and the process will work well through a thin coating of paint or other nonmetallic coverings

Limitations of magnetic-particle inspection are as follows:

- It is not completely reliable for locating discontinuities that lie entirely below the surface
- The magnetic field must be in a direction that will intercept the principal plane of the discontinuity
- Care is required to avoid local heating and burning of surfaces at the points of electrical contact

Liquid-penetrant inspection is used to detect surface flaws in materials. It is used mainly, but not exclusively, with nonmagnetic materials, on which magnetic-particle inspection cannot be used. The technique of liquid-penetrant inspection involves the spreading of a liquid penetrant on the sample. Liquid penetrants can seep into small cracks and flaws (as fine as 1 μm) in the surface of the sample by capillary action. The excess liquid is wiped from the surface, and a developer is applied that causes the liquid to be drawn from the cracks or flaws that are open at the surface. The liquid itself is usually a very bright color or contains fluorescent particles that, under ultraviolet light, cause discontinuities in the material to stand out.

The main advantages of the liquid-penetrant method are its ability to be used on nonmagnetic materials, its low cost, its portability, and the ease with which results can be interpreted. The principal limitations of the liquid-penetrant method are:

- Discontinuities must be open to the surface
- Testpieces must be cleaned before and after testing because the liquid penetrant may corrode the metal
- Surface films may prevent detection of discontinuities
- The process is generally not suited to inspection of low-density powder metallurgy parts or other porous materials

Eddy-current inspection can be used on all materials that conduct electricity. If a coil conducting an alternating current is placed around or near the surface of the sample, it will set up eddy currents within the material by electromagnetic induction. These eddy currents affect the impedance in the exciting coil or any other pickup coil that is nearby. Cracks or flaws within the sample will cause distortions in the eddy current, which in turn cause distortion in the impedance of the coil. The resulting change

in impedance can be detected by attaching the appropriate electrical circuits and a meter. Flaws or cracks will show up as some deflection or fluctuation on the meter.

The advantages of electromagnetic inspection are:

- Both surface and subsurface defects are detectable
- No special operator skills are required
- The process is adaptable to continuous monitoring
- The process may be substantially automated and is capable of high speeds
- No probe contact is needed

Limitations of electromagnetic inspection include:

- Depth of penetration is shallow
- Materials to be inspected must be electrically conductive
- Indications are influenced by more than one variable
- Reference standards are required

Ultrasonic inspection is a nondestructive method in which beams of high-frequency acoustic energy are introduced into the material under evaluation to detect surface and subsurface flaws and to measure the thickness of the material or the distance to a flaw. An ultrasonic beam will travel through a material until it strikes an interface or discontinuity, such as a flaw. Interfaces and flaws interrupt the beam and reflect a portion of the incident acoustic energy. The amount of energy reflected is a function of the nature and orientation of the interface or flaw as well as the acoustic impedance of such a reflector. Energy reflected from various interfaces or flaws can be used to define the presence and locations of flaws, the thickness of the material, or the depth of a flaw beneath a surface.

The advantages of ultrasonic tests are as follows:

- High sensitivity, which permits the detection of minute cracks
- Great penetrating power, which allows the examination of extremely thick sections
- Accuracy in measurement of flaw position and estimation of flaw size

Ultrasonic tests have the following limitations:

- Size-contour complexity and unfavorable discontinuity orientation can pose problems in interpretation of the echo pattern
- Undesirable internal structure—for example, grain size, structure, porosity, inclusion content, or fine dispersed precipitates—can similarly hinder interpretation
- Reference standards are required

Radiography is a nondestructive inspection method that is based on differential absorption

of penetrating radiation—either electromagnetic radiation of very short wavelength or particulate radiation—by the part or testpiece being inspected. Because of differences in density and variations in thickness of the part or because of differences in absorption characteristics caused by variations in composition, different portions of a testpiece absorb different amounts of penetrating radiation. Unabsorbed radiation passing through the part can be recorded on film or photosensitive paper, viewed on a fluorescent screen, or monitored by various types of radiation detectors.

The term radiography usually implies a radiographic process that produces a permanent image on film (conventional radiography) or paper (paper radiography or xeroradiography), although in a broad sense it refers to all forms of radiographic inspection. When inspection involves viewing of a real-time image on a fluorescent screen or image intensifier, the radiographic process is termed real-time inspection. When electronic nonimaging instruments are used to measure the intensity of radiation, the process is termed radiation gaging. Tomography, a radiation inspection method adapted from the medical computerized axial tomography (CAT) scanner, provides a cross-sectional view of a testpiece. All the above terms are mainly used in connection with inspection that involves penetrating electromagnetic radiation in the form of x-rays or γ-rays. Neutron radiography refers to radiographic inspection using neutrons rather than electromagnetic radiation.

Radiographic inspection is used extensively on castings and weldments, particularly where there is a critical need to ensure freedom from internal flaws. For example, radiography is often specified for inspection of thick-wall castings and weldments for steam-power equipment (boiler and turbine components and assemblies) and other high-pressure systems. Radiography can also be used on forgings and mechanical assemblies. It is also well suited to the inspection of semiconductor devices for cracks, broken wires, unsoldered connections, foreign material, and misplaced components.

Compared with other nondestructive methods of inspection, radiography is expensive. Relatively large capital costs and space allocations are required for a radiographic laboratory or a real-time inspection station. On the other hand, when portable x-ray or γ-ray sources are used, capital costs can be relatively low.

Certain types of flaws are difficult to detect by radiography. Laminar defects, such as cracks, present problems unless they are essentially parallel to the radiation beam. Tight, meandering cracks in thick sections usually cannot be detected even when properly oriented. Minute discontinuities, such as inclusions in wrought material, flakes, microporosity, and microfissures, cannot be detected unless they are sufficiently segregated to yield a detectable gross effect. Laminations normally are not detectable by radiography because of their unfavorable orientation—usually parallel to the surface. Laminations seldom yield differences in absorption that enable laminated areas to be distinguished from lamination-free areas.

Acoustic-Emission Inspection. Acoustic emission is defined as the high-frequency stress waves generated by the rapid release of strain energy that occurs within a material during such processes as crack growth, plastic deformation, and phase transformation. This energy may originate from stored elastic energy, as in crack propagation, or from stored chemical-free energy, as in phase transformation.

Sources of acoustic emission that generate stress waves in material include local dynamic movements, such as the initiation and propagation of cracks, twinning, slip, sudden reorientation of grain boundaries, bubble formation during boiling, or martensitic phase transformations. The stresses in a metallic system may be well below the elastic design limits, yet the region near a flaw or crack tip may undergo plastic deformation and fracture from locally high stresses, ultimately resulting in premature or catastrophic failure under service conditions.

Acoustic-emission inspection detects and analyzes minute acoustic-emission signals generated by discontinuities in materials under applied stress. Proper analysis of these signals can provide information concerning the location and structural significance of the detected discontinuities.

Some of the significant applications of acoustic-emission inspection are:

- Continuous surveillance of pressure vessels and nuclear primary-pressure boundaries for the detection and location of active flaws
- Detection of incipient fatigue fracture in aircraft structures
- Monitoring of both fusion and resistance weldments during welding and cooling
- Determination of the onset of stress-corrosion cracking (SCC) and hydrogen damage in susceptible structures
- Use as a study tool for the investigation of fracture mechanisms and of behavior of materials
- Periodic inspection of tanks and aerial-device booms made of composite materials

Experimental stress analysis can be done by several methods, all of which may be valuable in determining machine loads and component stresses that can cause failures. Stress-coating can be used effectively for locating small areas of high strains, determining the directions of the principal strains, and measuring the approximate magnitude of tensile and compressive strains. Gages can then be placed at the high-strain areas and in the principal-strain directions to measure the strain accurately on gage lengths 0.5 to 150 mm (0.02 to 6 in.). Although there are many mechanical, optical, and electrical devices capable of accurate strain measurements, the bonded electrical-resistance strain gage has become the standard tool for general laboratory and field use.

Photoelastic coatings have also been used for laboratory stress measurements. For this technique, a birefringent coating of controlled thickness is bonded to the testpiece with a reflective cement. Optical analysis is similar to conventional analysis, but requires special equipment. The analysis may be recorded on color film with single-frame or movie cameras.

X-ray diffraction is the only available method for direct nondestructive measurement of surface residual stresses in crystalline materials. Stresses are determined by measuring the angle by which the stressed material crystal diffracts an x-ray beam. For detailed information, see the article "X-Ray Diffraction Residual Stress Techniques" in Volume 10 of the 9th Edition of *Metals Handbook*.

Mechanical Testing

Hardness testing is the simplest of the mechanical tests and is often the most versatile tool available to the failure analyst. Among its many applications, hardness testing can be used to assist in evaluating heat treatment (comparing the hardness of the failed component with that prescribed by specification), to provide an approximation of the tensile strength of steel, and to detect work hardening or to detect softening or hardening caused by overheating, decarburization, or carbon or nitrogen pickup. Hardness testing is also essentially nondestructive, except when preparation of a special hardness-test specimen is required, as in microhardness testing.

Other mechanical tests are useful in confirming that the failed component conforms to specification or in evaluating the effects of surface conditions on mechanical properties. Where appropriate, tensile, fatigue, or impact tests should be carried out provided sufficient material for the fabrication of test specimens is available. The determination of plane-strain fracture-toughness values may also be justifiable. It may be necessary to make some tests either at slightly elevated or at low temperatures to simulate service conditions. Also, it may be helpful to test specimens after they have been subjected to particular heat treatments that simulate the thermal treatment of the failed component in service in order to determine how this treatment has modified mechanical properties. For example, treating a steel at a temperature in the embrittling range for about 1 h before impact testing will indicate any tendency to strain-age embrittlement, and the determination of the ductile-to-brittle transition temperature may be useful in investigating brittle fracture of a low-carbon steel.

The failure analyst should exercise care in interpreting mechanical-test results. For example, the fact that a material has a tensile strength 5 to 10% below the minimum specified value does not mean that this is the prime cause of its failure in service. Also, it should be understood that laboratory tests on small specimens may not adequately represent the behavior of a much

larger structure or component in service. For example, it is possible for brittle fracture of a large structure to occur at or near ordinary temperature, while subsequent laboratory tests of Charpy or Izod specimens show a transition temperature well below −18 °C (0 °F). The effects of size in fatigue, stress-corrosion, and hydrogen-embrittlement testing are not well understood. However, on the basis of the limited evidence available, it appears that resistance to these failure processes decreases as specimen size increases. Detailed information on various mechanical test procedures is provided in Volume 8 of the 9th Edition of *Metals Handbook*.

Limitations of Tensile Tests. In most service-failure investigations, the tensile test does not provide much useful information, because relatively few failures result from the use of a material that is deficient in tensile strength or from a tensile load that is applied until rupture occurs. Furthermore, samples cut from components that have failed in a brittle manner generally show adequate ductility under the conditions imposed during a tensile test.

Tensile tests are essential during production in order to ascertain if the material conforms to specification requirements. There is also some justification for tensile testing of components that have failed in service in order to eliminate poor-quality material as a possible cause of failure. These tensile tests for determining material quality are often carried out by manufacturers and suppliers when examining defective components that have been returned to them for analysis.

Preparation of specimens for tensile tests is expensive and time-consuming; the amount of material available may also be insufficient even if only substandard test specimens are to be prepared. The usual care must be exercised in identifying the position and the direction from which the samples have been taken. Altering the properties of the sample by excessive heating during cutting and machining should be avoided.

These remarks are applicable to other types of testpieces, such as those used for fatigue testing, impact testing, and fracture-toughness testing. The properties of the bulk of the material are often required, and its shape and size will dictate the type of testpiece to be used.

A simple bend test usually shows whether or not a material possesses adequate ductility. The amount of ductility, as shown by the elongation value in a tensile test, is not related directly to the behavior of metals in service. For example, a crankshaft failure could not be attributed to the fact that the crankshaft showed only 20% elongation on a standard sample as opposed to the 26% required by the specification.

It is usually sufficient during the initial stages of an investigation to determine whether the material is essentially ductile or brittle, and this can be ascertained from a simple bend test. The reduction of area, thought by some to be related to the ability of the material to deform plastically at a notch and so to relieve stress intensification, gives a much more realistic measure of ductility, but its measurement is often omitted from a tensile test. Testing of cast irons and most nonferrous alloys for tensile strength is almost essential, although some indication of the tensile strength is obtainable from hardness tests.

Results of tensile tests on specimens from components that have failed in service sometimes show that the material is slightly inferior in strength and ductility compared to results of acceptance tests done when the components were made. However, acceptance tests are often carried out on test specimens provided specifically for the purpose. Consequently, some discrepancy is to be expected because of differences in the amount of forging or cold work to which the component and the test specimens have been subjected during manufacture or because of a difference in section thickness has resulted in variations in cooling rate either at the time of casting or during heat treatment. Therefore, such disparities in results should not necessarily be interpreted as an indication that the particular properties of the material have deteriorated in service.

The role of directionality in tensile testing should also be considered. Specimens cut transversely to the longitudinal axis of a component such as a shaft should be expected to give lower yield-strength values and lower elongation values than those cut along the longitudinal axis, because of the marked directionality and the resulting anisotropy produced during rolling or forging. Detailed information on tensile testing can be found in the Section ''Tension Testing'' in Volume 8 of the 9th Edition of *Metals Handbook*.

Selection, Preservation, and Cleaning of Fracture Surfaces

The proper selection, preservation, and cleaning of fracture surfaces are vital to prevent important evidence from being destroyed or obscured. Surfaces of fractures may suffer mechanical or chemical damage. Mechanical damage may arise from several sources, including the striking of the fracture surface by other objects. This can occur during actual fracture in service or when removing or transporting a fractured part for analysis. Information regarding the selection and preservation of surfaces that have been degraded by wear or corrosion is provided in the articles ''Wear Failures'' and ''Corrosion Failures'' in this Volume.

Usually, the surface of a fracture can be protected during shipment by a covering of cloth or cotton, but this may remove some closely adhering material, which often contains the primary clue to the cause of the fracture. The surface of a fracture should not be touched or rubbed with the fingers. Also, no attempt should be made to fit together the sections of a fractured part by placing them in contact. This generally accomplishes nothing and almost always causes damage to the fracture surface.

Chemical (corrosion) damage to a fracture specimen can be prevented in several ways. For example, because the identification of foreign material present on a fracture surface may be important in the overall interpretation of the cause of the fracture, many laboratories prefer not to use corrosion-preventive coatings on a fracture specimen. When possible, it is best to dry the fracture specimen, preferably with a jet of dry compressed air, which will also blow extraneous foreign material from the surface. The specimen should then be placed in a desiccator or packed with a suitable desiccant.

Whenever possible, washing of the fracture surface with water should be avoided. However, specimens contaminated with seawater or with fire-extinguishing fluids require thorough washing, usually with water, followed by rinsing with acetone or alcohol before storage in a desiccator or coating with a desiccant.

Cleaning. Surfaces of fractures should be cleaned only when absolutely necessary. Cleaning may be required to remove obliterating debris and dirt or to prepare for electron microscope examination. Cleaning procedures include use of a dry-air blast or a soft-hair artist's brush, treating with inorganic solvents by immersion or jet, treating with mild acid or alkaline solutions (depending upon the metal) that will attack deposits but to which the base metal is essentially inert (for example, a 6 N solution of hydrochloric acid inhibited with 2 g/L hexamethylene tetramine has been used successfully for steel), ultrasonic cleaning, and application and stripping of plastic replicas.

Cleaning with a cellulose acetate replica is one of the most useful methods, particularly when the surface of a fracture has been affected by corrosion. A strip of acetate sheet about 1 mm (0.04 in.) thick and of suitable size is softened by immersion in acetone and placed on the surface of the fracture. The initial strip is backed by a piece of unsoftened acetate, and the replica is then pressed hard onto the surface of the fracture using a vise or suitable clamps. The drying time will depend upon the extent to which the replicating material was softened, and this will in turn be governed by the texture of the surface of the fracture. Drying times of not less than 1 h are recommended, and overnight drying is desirable if time permits. The dry replica is lifted from the fracture using a scalpel or tweezers. The replicating procedure can be repeated several times if the fracture is badly contaminated. When a clean and uncontaminated replica is obtained, the process is complete. An advantage of this method is that the debris removed from the fracture is preserved for any subsequent examination that may be necessary for identification of the type of debris.

Sectioning. Because examination tools, including hardness testers and optical and electron microscopes, are limited regarding the size of specimen they can accept, it is often necessary to remove from a failed component a fracture-containing portion or section that is of a size convenient to handle and examine. It is important that records, either sketches or photographs, be kept to show the locations of the cuts made during sectioning.

Before cutting or sectioning, the fracture area should be carefully protected. All cutting should be done so that surfaces of fractures and areas adjacent to them are not damaged or altered; this includes keeping the fracture surface dry whenever possible. For large parts, the common method of removing specimens is flame cutting. Cutting must be done at a sufficient distance from the fracture site to avoid altering the microstructure of the metal underlying the surface of the fracture by the heat of the flame and to avoid depositing the molten metal from flame cutting onto the surface of the fracture.

Jewelers' saws, hacksaws, and abrasive cutoff wheels can be used to cut a wide range of part sizes. Dry cutting is preferable because coolants may corrode the fracture site or may wash away foreign matter from the surface of the fracture. A coolant may be required, however, if a dry cut cannot be made at a sufficient distance from the fracture site to avoid heat damage to the area of the fracture.

Brittle substances can be sampled by inducing fractures with a sharp blow, but this must be administered at some distance from the area to be examined. Large unwieldy pieces of cast iron, for example, may be reduced to a manageable size by fracturing in this way. Trepanning is another way in which samples can be removed without undue disturbance of the structure.

Opening Secondary Cracks. When the primary fracture has been damaged or corroded to such a degree that most of the information on the cause of fracture is obliterated, it is desirable to open any secondary cracks to expose their fracture surfaces for examination and study. These cracks may provide more information than the primary fracture. If the cracks are tightly closed, they may have been protected from corrosive conditions, and if they have existed for less time than the primary fracture, they may have corroded less. Also, primary cracks that have not been propagated to total fracture may have to be opened.

In opening cracks for examination, care must be exercised to prevent damage, primarily mechanical, to the surface of the fracture. This can usually be accomplished if opening is done such that the two surfaces of the fracture are moved in opposite directions, normal to the fracture plane. Generally, a saw cut can be made from the back of the fractured part to a point near the tip of the crack, using extreme care to avoid actually reaching the tip of the crack. This saw cut will reduce the amount of solid metal that must be broken. The final breaking of the specimen can be done in several ways: (1) by clamping the two sides of the fractured part in a tensile-testing machine if the shape permits, and pulling, (2) by placing the specimen in a vise and bending one half away from the other by striking it with a hammer in a way that will avoid damage to the surfaces of the crack, or (3) by gripping the halves of the fracture in pliers or vise grips and bending or pulling them apart.

It is desirable to be able to distinguish between a fracture surface produced during opening of a primary or secondary crack and the surface produced by primary or secondary cracking. This can be accomplished by making sure that a different fracture mechanism is active in making the new break, such as by opening the crack at a very low temperature. During opening at low temperature, care should be taken to avoid condensation of water because this could cause corrosion of the fracture surface.

It is recommended that crack separations and crack lengths be measured before opening. Often, the amount of strain that occurred in the specimen can be determined from a measurement of the separation between the adjacent halves of a fracture. This should be done before preparation for opening a secondary crack has begun. The lengths of cracks may also be important for analyses of fatigue fractures or for consideration of fracture mechanics. Information on crack-length measurements for fatigue investigations can be found in the article ''Fatigue Crack Propagation'' in Volume 8 of the 9th Edition of *Metals Handbook*.

Macroscopic Examination of Fracture Surfaces

The detailed examination of fracture surfaces at magnifications ranging from 1 to 100× may be done with the unaided eye, a hand lens, or a low-power optical microscope. Occasionally, it may also be advantageous to use a scanning electron microscope at low magnification. Photography of specimens requires a high-quality camera for magnifications up to 20× and a metallograph with macro objectives and illuminating systems for magnifications from 20 to 50×. The ordinary incident (vertical and oblique) light system and the objectives used in standard metallography are generally best for magnifications of 50 to 100× (microscope components and light systems associated with optical microscopy are discussed in the article ''Optical Microscopy'' in Volume 9 of the 9th Edition of *Metals Handbook*.

Frequently, a specimen may be too large or too heavy for the stage of the metallograph, and cutting or sectioning the specimen may be difficult or undesirable. In these cases, excellent results can be achieved by examining and, where appropriate, photographing replicas made by the method for cleaning fractures (see the section ''Cleaning'' in this article). These replicas can be coated with a thin layer (about 20 nm thick) of vacuum-evaporated gold or aluminum to improve their reflectivity, or they may be shadowed at an angle to increase the contrast of fine detail. The replicas may be examined by incident-light or transmitted-light microscopy. Because they are electrically conductive, the replicas may also be examined by scanning electron microscopy (SEM).

The amount of information that can be obtained from examination of a fracture surface at low-power magnification is surprisingly extensive. Consideration of the configuration of the fracture surfaces may give an indication of the stress system that produced failure. Failure in monotonic tension produces a flat (square) fracture normal to the maximum tensile stress under plane-strain conditions and a slant (shear) fracture at about 45° if plane-stress conditions prevail. Because pure plane-strain and pure plane-stress conditions are ideal situations that seldom occur in service, many fractures are flat at the center, but surrounded by a picture frame of slant fracture. The slant fracture occurs because conditions approximating plane strain operate at the center of the specimen but relax toward plane stress near free surfaces. An example of this behavior is found in the familiar cup-and-cone tensile fracture.

In thin sheets or small-diameter rods, full-slant fracture may occur because through-thickness stresses are relaxed by plastic deformation and a stress state approximating plane strain cannot develop. The term shear lip is often used to describe an area of slant fracture between a flat area and a free surface. This term should be avoided because slant fractures are seldom the result of pure shear. The term 45° fracture for a slant fracture is somewhat misleading, because the angle between principal axis and fracture surface may vary several degrees from this value; in addition, the fracture surface may be a curved plane. Torsional stresses may produce fractures having spiral surfaces, especially if they are generated by fatigue.

Macroscopic examination can usually determine the direction of crack growth and therefore the origin of failure. With brittle flat fractures, determination depends largely upon the fracture surface exhibiting chevron marks of the type shown in Fig. 1. The direction of crack growth is almost always away from the tips of the chevrons.

Chevron marks occur because nearly all cracks are stepped at an early stage in their development, and as the crack front expands, the traces of the steps form chevron marks. In plate and sheet, chevron marks may result from the nucleation of new cracks ahead of a main crack front.

Occasionally, chevron marks may not follow the general pattern, and their tips may point to the last region to fail rather than to the origin, as in the fracture surfaces shown in Fig. 2; in this fracture, the conditions resulted from fatigue-crack initiation along the entire length of a drilled hole due to fretting. When the crack front contracts in the later stages of fracture (rather than expanding, as is usually the case), the chevron marks, being normal to the crack front at any given position, indicate the region of final fracture. This behavior is unusual; however, chevron marks of this type should be looked for when determining crack-growth directions.

Where fracture surfaces show both flat and slant fractures, it may be generally concluded that the flat fracture occurred first. Crack extension, often with crack-front tunneling, relaxes the plane-strain state of stress so that final fracture occurs by slant fracture under plane-stress conditions. Conversely, if a fracture has begun at a free surface, the fracture-origin area is usually characterized by a total absence of slant fracture or shear lip.

Low-power examination of fracture surfaces often reveals regions having a texture different from the region of final fracture. Fatigue, stress-corrosion, and hydrogen-embrittlement fractures may all show these differences.

Figure 3(a) shows the fracture surface of a steel tube and is an excellent example of the type of information that can be obtained by macroscopic examination. In Fig. 3(a), the chevron marks clearly indicate that the fracture origin is at the point marked by the arrow. This region, unlike the rest of the fracture, has no shear lip. The flat fracture surface suggests that the stress causing the failure was tension parallel to the length of the tube. The origin of the fracture as seen at higher magnification in Fig. 3(b) shows several small fracture nuclei with a texture different from that of the remainder of the fracture surface.

Microscopic Examination of Fracture Surfaces

The microscopic examination of fractured surfaces can be carried out using an optical (light) microscope, a transmission electron microscope, and/or a scanning electron microscope. The principles, instrumentation, and applications associated with these instruments are discussed in Volumes 9 and 10 of the 9th Edition of *Metals Handbook*. In addition, the use of these instruments in fractography will be addressed in Volume 12, which follows the current Volume in the *Metals Handbook* series.

Optical Microscopy. Although used extensively for low-magnification (<100×) fractography, the optical microscope has far less applicability for microfractography. This is due to its limited resolution (~10^{-6} m) and depth of field (cannot focus on rough surfaces). Information on the resolution and depth of field of the optical microscope can be found in the article ''Optical Microscopy'' in Volume 9 of the 9th Edition of *Metals Handbook*.

Transmission electron microscopes became commercially available in the 1950s, and it was soon realized that these instruments could be used in applied and fundamental research in materials science and physical metallurgy. These applications are reviewed in the

Fig. 1 Fracture surface on the flange side of a broken SAE 1050 steel axle shaft showing an example of chevron marks

Chevron marks consistently point toward the origin of fracture. Here, the origin is indicated by the arrow, which is located at the fine-structured induction-hardened case. 3×. Courtesy of General Motors Corporation

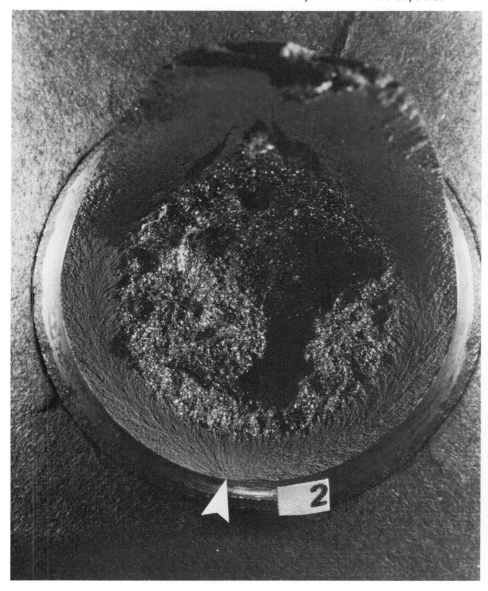

Fig. 2 Surfaces of a fatigue fracture that initiated in the vicinity of a drilled hole and progressed to final overstress fracture

The advancing crack front and the chevron marks that point to the region of final fracture are shown.

Fig. 3 Macroscopic examination of a steel tube

(a) Fracture surface showing point of crack initiation (arrow), chevron marks, and development of shear lips. Approximately actual size. (b) Fracture-origin area. 5×. Note that fracture nuclei differ in texture from the main fracture surface.

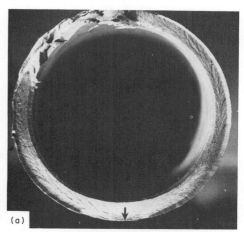

(a)

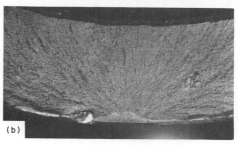

(b)

article "Analytical Transmission Electron Microscopy" in Volume 10 of the 9th Edition of *Metals Handbook*.

Because specimens for transmission electron microscopy (TEM) must be thin enough to permit the transmission of electron beams, replicas of the fracture surface must be obtained. The role of the replica has changed dramatically since its origin in the 1940s. Until the 1970s, a replica was primarily a method of reproducing surface topographic detail for high-magnification fractography. Many precise and sophisticated single- and two-stage techniques were developed for this purpose. A number of these techniques were described in the article

"The Transmission Electron Microscope and its Application to Fractography" in Volume 9 of the 8th Edition of *Metals Handbook*.

The use of scanning electron microscopes with resolutions below 3 nm has considerably diminished the need for replication techniques. However, direct replication is still used in materials science for special problems, such as examining the surface of a large component without cutting it, studying radioactive material that cannot be placed in an ordinary unshielded microscope, or studying extremely fine fatigue striations.

However, a second type of replica that utilizes extraction techniques remains in use, and

interest is increasing with the spread of analytical TEM. Many problems involving the determination of the composition, crystal structure, or orientation of small second-phase particles are simplified if the particles are extracted from their matrix, then supported in the microscope

Fig. 4 Typical dimpled-rupture fracture surface of a ductile fracture

Scanning electron micrograph taken at an angle of approximately 40 to 50° to the fracture surface. 1500×. Courtesy of D.J. Wulpi, Consultant

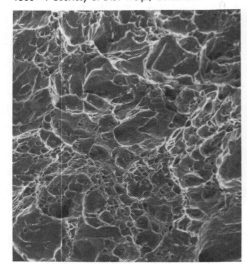

Fig. 6 Tongues in a second area of cleavage of the 1040 steel specimen of Fig. 5 at the same 40° tilt

Grain A shows two well-defined sets of feathered tongues that are approximately orthogonal. A small carbide particle, C, at the boundary of grains A and B initiated the local cleavage crack through the surrounding grains. Secondary cracks are at D and E. Note succession of cleavage steps in grain B. (a) 1560×. (b) 3900×

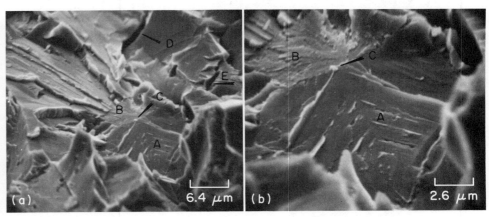

using a replica. Extraction replicas can preserve the relative positions and orientation of second-phase particles if they are small enough to be supported by a carbon film less than 1 μm thick.

The scanning electron microscope is one of the most versatile instruments for investigating the microstructure of metallic materials. Compared to the optical microscope, it expands the resolution range by more than one order of magnitude to approximately 10 nm in routine instruments, with ultimate values below 3 nm. Useful magnification thus extends beyond 10 000× up to 60 000×, closing the gap between the optical and the transmission electron microscope. Compared to optical microscopy, the depth of focus, ranging from 1 μm at 10 000× to 2 mm (0.08 in.) at 10×, is larger by more than two orders of magnitude due to the very small beam aperture.

Fractography is probably the most popular field of SEM. The large depth of focus, the possibility of changing magnification over a wide range, very simple nondestructive specimen preparation with direct inspection, and the three-dimensional appearance of SEM fractographs make the scanning electron microscope an indispensable tool in failure studies and fracture research. Two excellent monographs on the principles, instrumentation, and applications of SEM are provided in Volumes 9 and 10 of the 9th Edition of *Metals Handbook*.

Although the interpretation of microfractographs requires practice and understanding of fracture mechanisms, there are only a small number of basic features that are clearly recognizable and indicative of a particular mode of failure:

- Dimpled fracture, typical of overstress failures of ductile metals and alloys (Fig. 4)
- Cleavage facets, typical of transgranular brittle fracture of body-centered cubic (bcc) and hexagonal close-packed (hcp) metals and alloys (Fig. 5 to 7)
- Brittle intergranular fracture typical of temper-embrittled steel, in which fracture is due to segregation of an embrittling species to grain boundaries, to intergranular SCC (Fig. 8), or to hydrogen embrittlement
- Stage II striations, typical of fatigue failure (Fig. 9)

Additional information on fractography can be found in Ref 1 to 8.

Fig. 5 Cleavage fracture in a notched impact specimen of hot-rolled 1040 steel broken at −196 °C (−321 °F)

The specimen was tilted in the scanning electron microscope at an angle of 40° to the electron beam. The cleavage planes followed by the crack show various alignments, as influenced by the orientations of the individual grains. Grain A, at the center in (a), shows two sets of tongues (arrowheads in b) as the result of local cleavage along the |112| planes of microtwins created by plastic deformation at the tip of the main crack on |100| planes. Grain B and many other facets show the cleavage steps of river patterns. The junctions of the steps point in the direction of crack propagation from grain A through grain B at an angle of about 22° to the horizontal plane. The details of these forks are clear in (c). (a) 1560×. (b) 3900×. (c) 7800×. See also Fig. 6 and 7.

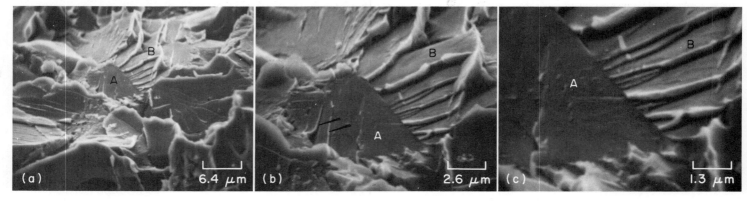

Fig. 7 A site of cleavage-crack nucleation in the fracture surface of the 1040 steel specimen in Fig. 5 and 6 at the same 40° tilt to the electron beam

Cleavage spread in several directions from fractured carbide particle, A, at junction of four grains. Cleavage features differ from grain to grain, with river patterns visible in some and tongues in others. River-pattern lines in grain B are attributed to interaction of local screw dislocations with a cleavage plane. Height of steps increases in direction of crack propagation, to the left and downward. (a) 790×. (b) 1590×

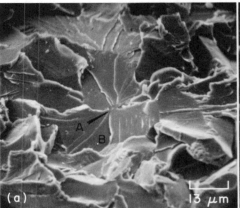 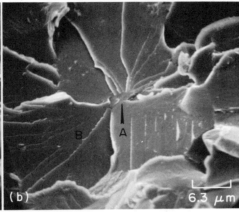

Fig. 9 Fractograph of a fracture surface showing stage II striations typical of fatigue failure

3500×. Source: R.J. Forsyth

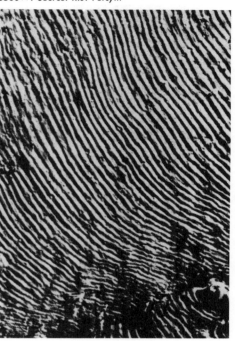

Selection and Preparation of Metallographic Sections

Metallographic examination of polished and polished-and-etched sections by optical microscopy and electron-optical techniques is a vital part of failure investigation and should be carried out as a routine procedure. Metallographic examination provides the investigator with a good indication of the class of material involved and whether it has the desired structure. If abnormalities are present, they may not be associated with undesirable characteristics that predispose early failure. It is sometimes possible to relate them to an unsuitable composition or to the effects of service, such as the aging in low-carbon steel that has caused precipitation of iron nitride. The microscope may also provide information regarding the method of manufacture of the part under investigation and the heat treatment to which it has been subjected, either intentionally during manufacture or accidentally during service. Other service effects, such as corrosion, oxidation, and severe work hardening of surfaces, also are revealed, and their extent can be investigated. Also, the characteristics of any cracks that may be present, particularly their mode of propagation, provide information regarding the factors responsible for their initiation and development.

Only a few general directions can be given as to the best location from which to take specimens for microscopic examination, because

Fig. 8 17-4 PH stainless steel main landing-gear deflection yoke that failed due to intergranular SCC

(a) Fracture surface. The box indicates the area where intergranular attack was noted. 14×. (b) Intergranular attack from area of box in (a). 170×. Courtesy of W.L. Jensen, Lockheed Georgia Company

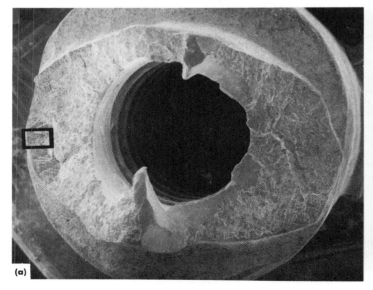 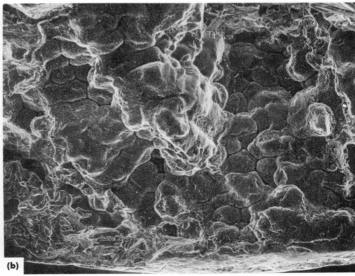

Fig. 10 Metallographic specimen on which nickel plate (smooth light-gray areas along top and right side) was used to protect edges during preparation

Specimen shows profile of a stress-corrosion crack in welded low-carbon steel pipe. Fracture surface is at top; weld, at right; pipe surface, far right. Etched with picral. 500×

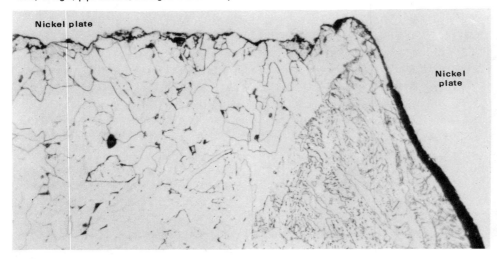

almost every failure has individual features that must be taken into account. In most examinations, however, it must be determined whether the structure of a specimen taken adjacent to a fracture surface or a region at which a service defect has developed is representative of the component as a whole. This can be done only by the examination of specimens taken from other locations, and in general, it is recommended that the number selected for examination should be too many rather than too few. For example, in the case of ruptured or bulged boiler tubes in which failure is usually restricted to one portion only, it is desirable to examine specimens taken from both sides of the fracture, from a location opposite the affected zone, and from an area as remote from the failure as the size of sample permits in order to determine whether the failure has been due to a defect in the material or to overheating and to determine whether the overheating was of a general or localized nature. In investigations involving general overheating, the original condition of the material can sometimes be ascertained only from a sample cut from a part of the tube many feet away from the affected zone.

When examining cracks microscopically, the most valuable information may sometimes be gained from a study of specimens that include the extremities of the cracks. In general, the parts of cracks that are visible to the unaided eye are too wide, ragged, or corroded for their paths to be revealed with certainty under the microscope; however, at their extremities they are finer, and examination of specimens from these regions generally enables the paths to be positively identified as either intergranular or transgranular.

In the investigation of fatigue cracks, it is advisable to take a specimen from the region where the fracture originated to ascertain if the initial development was associated with an abnormality, such as a weld defect, a decarburized surface, a zone rich in inclusions or, in castings, a zone containing severe porosity. Where multiple-origin cracks are concerned, however, such a procedure is not practicable; in these instances, it is most unlikely that the cracks were due to local inhomogeneities. Multiple fatigue-crack initiation is very typical of fretting and corrosion fatigue. Similarly, with surface marks, where their origin is impossible to identify from outward appearances, a microscopic examination will show whether they occurred in rolling or arose from ingot defects, such as scabs, laps, or seams. In brittle fractures, it is useful to examine a specimen cut from where the failure originated, if this can be located with certainty, because failures by brittle fracture are frequently associated with locally work-hardened surfaces, particularly if the steel is of the strain-aging type.

It is sometimes necessary to plate the fracture surface of a specimen with a metal, such as nickel, before mounting and sectioning so that the fracture edge is supported and can be included in the examination. Figure 10, which shows a section through a welded low-carbon steel pipe, is an example of the plating technique. The section is bordered at the top by the surface of a stress-corrosion crack and at the right by the outside surface of the pipe (both shown in profile). Both of these surfaces were nickel plated to prevent rounding of the edges during polishing. This section shows the intergranular nature of the crack and the location of the crack relative to the end of a weld deposit (at right), which suggests that the heat of welding caused the stress; the source of corrosion is not known.

Examination and Analysis of Metallographic Sections

The examination of metallographic sections with a microscope is standard practice in most failure analyses because of the outstanding capability of the microscope of revealing material imperfections caused during processing and of detecting the results in various in-service operating conditions and environments that may have contributed to failure. Inclusions, microstructural segregation, decarburization, carbon pickup, improper heat treatment, untempered white martensite, and intergranular corrosion are among the many metallurgical imperfections and undesirable conditions that can be detected and analyzed by microscopic examination of metallographic sections.

Volume 9 of the 9th Edition of *Metals Handbook* presents a comprehensive selection of micrographs of wrought, cast, and powder metallurgy ferrous and nonferrous alloys. Hundreds of undesirable or abnormal microstructures, defects or significant surface conditions, and microstructures after service or exposure supplement the many pertinent failure-related micrographs presented in this Volume.

Even in the absence of a specific metallurgical imperfection, examination of metallographic sections is invaluable to the investigator in the measurement of parameters, such as case depth, thickness of plated coatings, grain size, and heat-affected zone (HAZ)—all of which may have a bearing on the cause of failure. Metallographic sections are also useful when quantitative metallographic techniques or electron probe microanalysis are used in failure analysis. Additional information is provided in the articles "Quantitative Metallography" and "Electron Probe X-Ray Microanalysis" in Volumes 9 and 10, respectively, of the 9th Edition of *Metals Handbook*.

Determination of Fracture Type

To use the information obtained from examination of the failure region, the fracture surfaces, and metallographic sections to determine the cause of fracture, it is usually necessary to determine the fracture type. However, a satisfactory logical classification of failures involving fracture does not exist. For example, the extensive elongation of a low-carbon steel specimen followed by cleavage might be classified as either brittle or ductile fracture. The low-energy catastrophic fracture of a high-strength aluminum alloy by microvoid coalescence is also difficult to classify because, although the fracture energy is low and failure initiates by fracture or decohesion of brittle particles, the growth and coalescence of the microvoids occurs by plastic deformation. Another difficulty is that cleavage fracture may be initiated by dislocation interactions that by definition involve plasticity.

For the purposes of this article, fractures will be classified in terms of their growth mechanism, and crack initiation will not be considered. Thus, cleavage-crack extension is brittle regardless of the plastic deformation that may have accompanied or preceded crack initiation, and any fracture mainly by microvoid coalescence will be regarded as ductile because the mechanism of crack extension necessarily involves plastic deformation. The fracture classifications used in this article are described below. Further discussion of failure classification is provided in the article ''Identification of Types of Failures'' in this Volume.

Ductile Fracture

Overload fractures of many metals and alloys occur by ductile fracture. Overloading in tension is perhaps the least complex of the overload fractures, although essentially the same processes operate in bending and torsion as well as under the complex states of stress that may have produced a given service failure. Classically, ductile-tensile fracture of a cylindrical specimen involves plastic extension, initially without necking. During this extension, cracking of included particles, which are present in even the purest metals, or decohesion of particle/matrix interfaces occurs, creating microvoids. When the ability of the material to work harden is exhausted, necking begins, and triaxial stresses are set up that cause lateral extension of the microvoids, which coalesce to form a central crack. Fracture of the remaining section to produce an annulus of slant fracture, often incorrectly called shear, is less well understood, but probably occurs by a crack growing circumferentially around the specimen under plane-stress conditions. The total absence of second-phase particles would result in fracture by 100% reduction of area, but this rarely occurs in service failures.

Sheet specimens fracture by similar mechanisms, and if they are thick enough for plane-strain conditions to operate, a flat fracture with shear lips is produced. Full-slant fracture may occur in thin sections, that is, under plane-stress conditions. Exceptions occur in materials that show marked discontinuous yielding accompanied by the generation of Lüders bands across the whole cross section of the specimen. It is possible for fracture to occur by microvoid coalescence within a Lüders band, producing a full-slant fracture in thick sections, that is, under nominally plane-strain conditions (Ref 9).

Fractography of Ductile Fractures. Fractographic examination of flat ductile-fracture surfaces usually reveals approximately equiaxed dimples, generally with evidence of the particles that originated the fracture (Fig. 4). Slant fracture, or ductile fracture involving shear-stress components (such as torsion), generates elongated dimples (Fig. 11). When the elongated dimples are produced by a shear component, the dimples in the mating fracture surfaces point in opposite directions. In ductile-

Fig. 11 Elongated dimples on the fracture surface of a quenched-and-tempered 4140 steel

TEM replica. 10 000×. Courtesy of I. Le May, Metallurgical Consulting Services Ltd.

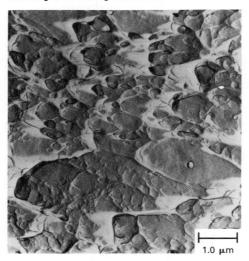

1.0 μm

fracture surfaces produced by tearing, the crack produces elongated dimples on mating surfaces that are mirror images.

Fractures that are ductile as seen macroscopically are usually transgranular, but electron fractography has shown dimple patterns on tensile-fracture surfaces of aluminum-copper alloys when metallographic sections have shown that the fracture path was apparently intergranular and nominally brittle (Ref 9). Isolated dimples on brittle intergranular-fracture surfaces of aluminum-zinc-magnesium alloys have been observed (Ref 9); this situation is generally confined to precipitation-hardened alloys with grain-boundary zones that are precipitate-free.

Transgranular Brittle Fracture

Transgranular cleavage of iron and low-carbon steel is the most commonly encountered process of brittle fracture—so common, in fact, that the term brittle fracture is sometimes misinterpreted as meaning only transgranular cleavage of iron and low-carbon steel. Transgranular cleavage can also occur in several other bcc metals and their alloys—for example, tungsten, molybdenum, and chromium—and some hcp metals—for example, zinc, magnesium, and beryllium. However, face-centered cubic (fcc) metals and alloys are usually regarded as immune from this mechanism of fracture. Iron and low-carbon steel show a ductile-to-brittle transition with decreasing temperature that arises from a strong dependence of the yield stress to temperature (see the article ''Ductile-to-Brittle Fracture Transition'' in this Volume).

Transition temperature is not really a physical constant, but it depends upon several physical factors, including the shape and size of the specimen and strain rate. Thus, a component or

structure that has given satisfactory service may fracture unexpectedly; the catastrophic brittle fracture of ships in heavy seas and the failure of bridges on unusually cold days are examples. Metallurgical changes, especially strain aging, may cause the brittle fracture of such items as crane hooks and chain links after long periods of satisfactory operation.

Cleavage fracture is not difficult to diagnose, because the fracture path is by definition crystallographic and usually, but not invariably, occurs on {100} cube planes in bcc metals and alloys and on {0001} basal planes in hcp metals and alloys. In polycrystalline specimens, this often produces a pattern of brightly reflecting crystal facets, and such fractures are often described as crystalline. The general plane of fracture is approximately normal to the axis of maximum tensile stress, and a shear lip is often present as a picture frame around the fracture. The local absence of a shear lip or slant fracture suggests a possible location for initiation of the fracture.

Fractography of Transgranular Brittle Fractures. The fractography of cleavage fracture in low-carbon steels, iron, zinc, and other single-phase bcc metals and alloys is fairly well established. In polycrystalline specimens, numerous fan-shaped cleavage plateaus, usually showing a high degree of geometric perfection, are present (Fig. 5 to 7). The most characteristic feature of these plateaus is the presence of a pattern of river marks, which consist of cleavage steps and indicate the local direction of crack growth. The rule is, if the tributaries are regarded as flowing into the main stream, then the direction of crack growth is downstream. This is in contrast to macroscopic chevron marks, for which the direction of crack growth, using the river analogy, would be up-stream. Figures 5 and 7 show the generation of river marks on a fracture surface.

Other fractographic features that may be observed include the presence of cleavage on conjugate planes, ductile tears joining cleavage planes at different levels, and tongues, which result from fracture in mechanical twins formed ahead of the advancing crack (Fig. 5 and 7). Cleavage fracture in pearlitic and martensitic steels is less easily interpreted because microstructure tends to modify the fracture surface.

Intergranular Brittle Fracture

Intergranular brittle fracture can usually be easily recognized, but determining the primary cause of the fracture may be difficult. Fractographic examinations can readily identify the presence of large fractions of second-phase particles at grain boundaries. Unfortunately, the segregation of a layer of few atoms thick of some element or compound that produces intergranular fracture often cannot be detected by fractography. Some causes of intergranular brittle fracture are given below. Although not exhaustive, these three causes indicate some of the possibilities that need to be considered, eliminated, or confirmed as contributing to the fracture.

The first cause of intergranular brittle fracture is the absence of sufficient deformation systems to satisfy the Taylor-von Mises criterion, which states that five independent systems (slip or slip-plus-twinning) are necessary for a grain to deform to an arbitrary shape imposed by its neighbors. The fracture of polycrystalline ceramics at low temperature is a good example, but this mechanism of fracture is not usual in fcc metals and alloys. The second cause is the presence at a grain boundary of a large area of second-phase particles, such as carbides in iron-nickel-chromium alloys. The third cause is the segregation of a specific element or compound to a grain boundary where a layer a few atoms thick is sufficient to cause embrittlement. Embrittlement caused by the presence of oxygen in high-purity iron, oxygen in nickel, or antimony in copper and the temper embrittlement of certain steels are examples of intergranular embrittlement in which detection of a second phase at grain boundaries is difficult.

The conditions under which a slowly growing crack may follow an intergranular path before final overload fracture occurs include fatigue fracture, SCC (Fig. 8), liquid-metal embrittlement, hydrogen embrittlement, and creep and stress-rupture failures. These are discussed below.

Fatigue Fracture

Fatigue fracture results from the application of repeated or cyclic stresses, each of which may be substantially below the nominal yield strength of the material (see the article "Fatigue Failures" in this Volume). Because the laboratory fatigue behavior of many metals and alloys is well established, it is perhaps surprising that so many service failures still occur by this mechanism. The difficulty is that there are a great many variables that influence fatigue behavior; these include the magnitude and frequency of application of the fluctuating stress, the presence of a mean stress, temperature, environment, specimen size and shape, state of stress, the presence of residual stresses, surface finish, microstructure, and the presence of fretting damage. This list is not comprehensive, and an additional problem is that one variable may be more important with respect to one material than another. For example, the noble metals are insensitive to most corrosive environments, titanium alloys are especially susceptible to fretting, and high-strength low-toughness materials, such as high-strength steels, are more susceptible to the effect of surface finish than are low-strength tougher alloys.

General Features of Fatigue Fractures. Because most of the surface area of a fatigue crack is generated by a process that is tensile stress dependent, the stress system responsible for fracture can often be deduced from the configuration of the fracture. The article "Fatigue Failures" in this Volume describes the

Fig. 12 Progression marks (beach marks) that indicate successive positions of the advancing fatigue-crack front

Source: J. Schijve

way in which fatigue-fracture configuration is influenced by the stress system, the magnitude of stress, and part shape.

The most noticeable macroscopic features of classic fatigue-fracture surfaces are the progression marks—also known as beach marks, clamshell marks, or tide marks—that indicate successive positions of the advancing crack front (Fig. 12). Fatigue-fracture surfaces are smooth textured near the origins and generally show slight roughening as the crack grows. There is little macroscopic ductility associated with fatigue fracture, and there may be some evidence that the crack has followed specific crystal planes during early growth, thus giving a faceted appearance. Unfortunately, a great many fatigue fractures do not show the classic progression marks.

Most fatigue cracks are transcrystalline without marked branching, although intercrystalline fatigue is not particularly uncommon. Corrosion fatigue in most materials is also transcrystalline; its most striking feature is usually the multiplicity of crack origins, only one of which extends catastrophically. Fatigue initiated by fretting has similar characteristics and is generally diagnosed by the presence of a fretting product filling the multiplicity of cracks and by the presence of a fretting product on the surface of the component. On aluminum alloys, the fretting product is often a hard black deposit; on steels, considerable quantities of a substance

Fig. 13 Brittle fracture of a WC-3Co cemented carbide

The fracture was partly intergranular (smooth grains) and partly transgranular (see, for example, the trapezoidal grain exhibiting Wallner lines indicated by arrow). TEM Formvar replica. Etched with 5% HCl. 12 000×. Courtesy of S.B. Luyckx, University of the Witwatersrand Johannesburg, R.S.A.

resembling cocoa are produced. Fretting product appears to be a mixture of finely divided particles of the base metal, its oxides, and its hydrated oxides.

Microscopically, surfaces of fatigue fractures are characterized by the presence of striations, each of which is produced by a single cycle of stress (Ref 10). It is not true, however, that every cycle of stress produces a striation; in fact, the complete absence of striations does not rule out fatigue fracture. A number of fractographic features may be confused with fatigue striations, particularly Wallner lines, which are produced by shock-wave/crack-front interactions (Fig. 13); tire tracks; and rub marks.

Although Wallner lines resemble fatigue striations, there are at least two identifying characteristics that separate these fracture features:

- Wallner lines are usually found only in very brittle materials or phases, where fatigue striations are seldom observed
- Fatigue striations may propagate in different directions, but they never cross each other, as Wallner lines do (compare Fig. 9 and 13)

Tire tracks are another feature that seems to be associated with high-stress low-cycle fatigue. These marks are thought to result from the relative motion between two closely mating fracture surfaces under the action of a repeated high stress. Typical tire tracks are shown in Fig. 14.

Rubbing and abrasion can also produce striationlike linear indications that are commonly referred to as rub marks (Fig. 15). The failure analyst must learn to distinguish these

Fig. 14 Tire tracks on 4140 steel quenched and tempered at 700 °C (1290 °F)

TEM replica. 2500×. Courtesy of I. LeMay, Metallurgical Consulting Services Ltd.

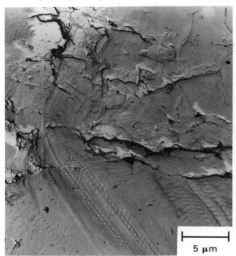

Fig. 15 Striationlike rub marks produced by abrasion

(a) 500×. (b) 1000×. Courtesy of B. Gabriel, Packer Engineering Associates, Inc.

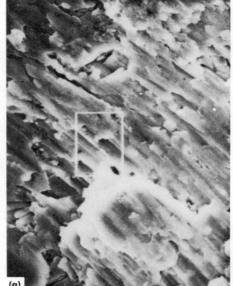 

(a)   (b)

features from striations by examining several areas across the diameter of the fracture surface.

Stress-Corrosion Cracking

Stress-corrosion cracking is a mechanical-environmental failure process in which mechanical stress and chemical attack combine in the initiation and propagation of fracture in a metal part (see the article "Stress-Corrosion Cracking" in this Volume). It is produced by the synergistic action of a sustained tensile stress and a specific corrosive environment, causing failure in less time than the sum of the separate effects of the stress and the corrosive environment would.

Failure by SCC is frequently caused by exposure to a seemingly mild chemical environment while subject to a tensile stress that is well below the yield strength of the metal. Under such conditions fine cracks can penetrate deeply into the part, although the surface may show only apparently insignificant amounts of corrosion. Therefore, there may be no macroscopic indications of impending failure. The most common instances of failure by SCC in service are probably those associated with the following metals and alloys:

- High-strength aluminum alloys, especially of the aluminum-zinc-magnesium type (7*xxx* series) under atmospheric-corrosion conditions. Internal and assembly stresses are often important
- Austenitic stainless steels and nickel alloys of the Inconel type in the presence of very low concentrations of chloride ions
- Low-carbon structural steels, usually in the presence of hot concentrated nitrate of caustic alkali solutions

- High-strength steels (tensile strengths of 1240 MPa, or 180 ksi, and above) in a variety of environments, probably with hydrogen embrittlement playing a dominant part
- Copper alloys, notably 70Cu-30Zn cartridge brass in ammoniacal environments, usually in the presence of internal stresses

General Features of SCC. Stress-corrosion cracks may be intergranular (Fig. 8), transgranular, or a combination of both. In aluminum alloys and low-carbon steels, intergranular fracture is common, although the fracture path may be immediately adjacent to the grain boundary rather than precisely along it. High-strength steels and α-brasses also usually show grain-boundary fracture, with some cracking along matrix/twin interfaces in α-brasses. Transgranular fractures showing extensive branching are typical of SCC in austenitic stainless steels of the 18Cr-8Ni type (Fig. 16), and similar transgranular cracks with branches that follow crystallographic planes have been observed in magnesium alloys.

When SCC is transgranular, deviations may occur on a microscopic scale so that the crack may follow microstructural features, such as grain and twin boundaries or specific crystal planes. When SCC is intergranular, the presence of flat elongated grains means that there is an easy stress-corrosion path normal to the short-transverse direction, which produces "woody" stress-corrosion fracture surfaces. This behavior is typical of extrusions of high-strength aluminum alloys in which solution treatment does not cause recrystallization.

Some stress-corrosion fractures show progression marks and alternating regions of SCC

and overload fracture, with changes of shape of the crack front. The progression marks in the fracture surface shown in Fig. 17 could very easily be confused with fatigue. This fracture occurred in a forging that had not been used in service, and the fracture was due entirely to residual internal stress and atmospheric corrosion. Other features observed on SCC fractures include striations (Fig. 18), cleavage facets, and tongues, which can be easily confused with similar features on cleavage and fatigue fractures.

When intergranular fracture occurs with only superficial corrosion, a rock-candy appearance of the fracture surface is typical (Fig. 19). Unfortunately, only in the high-strength aluminum alloys does this clearly define stress-corrosion fracture, because in high-strength steels, this pattern is also characteristic of hydrogen-induced slow crack growth.

Liquid-Metal Embrittlement

Liquid-metal embrittlement results in either a loss in ductility of a solid metal or its fracture below the normal yield stress under the circumstances that its surface is wetted by some lower-melting liquid metal. Thus, for example, a 70-30 brass wetted by mercury will fracture at a stress near but below that for yielding under simple tensile- or bending-test conditions. Although the ductility of 70-30 brass is very high and tensile rupture of this material exhibits substantial necking, the fracture associated with mercury embrittlement is flat intercrystalline cleavage under SEM examination, and the gage length shows no measurable elongation.

Susceptibility to liquid-metal embrittlement is unique to specific metals. Liquid mercury

Fig. 16 Transverse section showing transgranular cracks in a type 316 stainless steel vent line in a coal liquefaction plant assembly
Failure was due to chloride SCC. Crack branching is evident. 200×. Courtesy of J.R. Keiser and A.R. Olsen, Oak Ridge National Laboratory

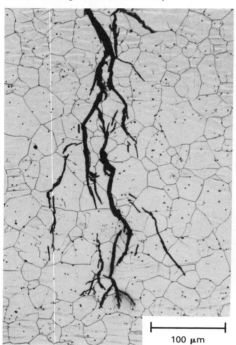

100 μm

Fig. 17 Fracture surface of a high-strength aluminum alloy forging that failed from stress corrosion
Progression marks similar to those observed in fatigue fractures are evident. Source: Ref 10

Fig. 18 SCC striations on the fracture surface of 316L stainless steel
The arrow indicates the crack-growth direction. 10 000×. Courtesy of I. Le May, Metallurgical Consulting Services Ltd.

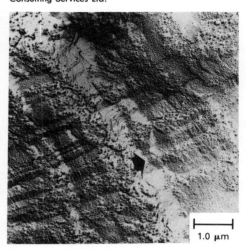

1.0 μm

Fig. 19 Fracture surface with a rock-candy appearance caused by stress corrosion
TEM fractograph of a germanium-shadowed plastic-carbon replica of a fracture surface of an aluminum alloy specimen. 4000×

embrittles copper and aluminum alloys but not simple carbon or alloy-carbon steels. Molten lithium, zinc, indium, and cadmium, and alloys containing these elements can embrittle carbon steels, but have smaller effects or no effects at all on aluminum and copper alloys. Molten bismuth embrittles copper alloys, but has no effect on aluminum or on carbon steels.

There is a specificity in the identities of liquid metals that exert large, small, or no embrittling effects on any given solid alloy under conditions of tensile stress and wetting. Liquid alloying complicates the definition of specificity. Liquid lead, for example, has no embrittling effect on carbon steels, but small percentages of dissolved tin, cadmium, and nickel create a potential for cracking under tensile stresses that are either simple stresses or components of complex stress systems.

Commercially pure metals are usually not, in any practical sense, subject to liquid-metal embrittlement. Two exceptions are zinc, which is embrittled by mercury, and copper, which is embrittled by molten bismuth. Thus, copper and aluminum are not embrittled by liquid mercury, whereas their alloys are very much in hazard.

There is a transition from brittle to ductile behavior with increasing temperature. For example, the temperature range for embrittlement of carbon steels by liquid lead alloys is brack-

eted on the low side by the melting temperature of the lead alloy and by an upper limit of temperature that is only about 300 °C (570 °F) higher. Above this upper limit, the steel has undiminished ductility. Detailed information is provided in the article "Liquid-Metal Embrittlement" in this Volume. In addition, embrittlement of metallic couples that occurs below the melting point of the liquid is described in the article "Embrittlement by Solid-Metal Environments" in this Volume.

Hydrogen Embrittlement

Hydrogen embrittles several metals and alloys, but its deleterious effect on steels, particularly when the strength of the steel exceeds about 1240 MPa (180 ksi) is most important. A few parts per million of hydrogen dissolved in steel can cause hairline cracking and loss of tensile ductility. Even when the quantity of gas in solution is too small to reduce ductility, hydrogen-induced delayed fracture, sometimes called static fatigue, may occur. Gaseous environments containing hydrogen are also damaging. Hairline cracking usually follows prioraustenite grain boundaries and seems to occur when the damaging effect of dissolved hydrogen is superimposed on the stresses that accompany the austenite-to-martensite transformation. Affected areas are recognized on fracture surfaces by their brittle appearance and high reflectivity, which usually contrasts with the matte appearance of surrounding regions of ductile fracture. This has led to such areas being described as flakes or fisheyes.

Hairline cracking is readily recognizable metallographically and is most common near the center of fairly bulky components where constraint of plastic deformation is high, but its incidence may be minimized by modification of steelmaking and heat-treating practices. Hairline cracking is important with respect to service failures because such a crack may extend by fatigue and thus initiate catastrophic fracture.

Fractography of Hydrogen-Embrittled Steels. Hydrogen-embrittled steels, especially those that have suffered delayed fracture, show fracture surfaces very similar to those typical of SCC fracture in aluminum alloys and high-strength steels. In delayed fracture, there is

Fig. 20 Intergranular fracture in a cadmium-plated AISI 8740 steel aircraft wing nut

Hydrogen embrittlement was the cause of failure. The wing nuts had not been sufficiently baked to release the hydrogen picked up during plating. 900×.
Courtesy of W.L. Jensen, Lockheed Georgia Company

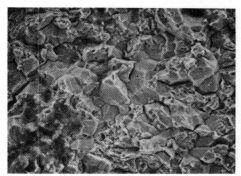

always a region of fracture surface produced by hydrogen-assisted slow crack growth; this crack growth typically follows austenite grain boundaries. Intergranular fracture due to hydrogen embrittlement in an AISI 8740 steel aircraft wing nut is shown in Fig. 20. Out-of-plane branch cracking along such boundaries is also common. In some steels, hydrogen may promote cleavage fracture. Positive identification is often difficult, and it is frequently impossible to differentiate between hydrogen-induced delayed fracture and SCC. Hydrogen embrittlement is discussed in greater detail in the article "Hydrogen-Damage Failures" in this Volume.

Creep and Stress-Rupture Failures

Creep is the change in dimension of a metal or alloy under an applied stress at a temperature exceeding about 0.5 T_M, where T_M is the melting point measured on the absolute scale. Thus, lead, tin, and superpure aluminum may deform by creep at room temperature or a little above, whereas temperatures near 1000 °C (1830 °F) may be necessary to permit creep in refractory bcc metals, such as tungsten and molybdenum, and in nickel-base superalloys used in gas turbines. Clearly, creep strain may produce sufficiently large changes in the dimensions of a component to render it useless for further service before fracture occurs. In other situations, the creep strain may lead to fracture; this type of failure is called stress rupture.

Creep and stress-rupture failures are generally easy to identify. They can often be recognized by the local ductility and multiplicity of intergranular cracks that are usually present (Fig. 21). Stress-rupture failure can often be identified by optical examination of microsections because there is generally a multiplicity of creep voids adjacent to the main fracture (Fig. 22).

Creep and stress-rupture failures are best understood by considering the two general types of creep processes that occur. In the first type, grain-boundary sliding is thought to generate a stress concentration at a triple point that cannot be relieved by plastic deformation in an adjacent grain. This produces a wedge-shaped grain-boundary crack. The second type involves the initiation of voids at grain boundaries, especially those grain boundaries oriented transversely to a tensile stress, and the growth of the voids by the migration and precipitation of vacancies. This process is called cavitation creep. Stress-rupture fracture due to cavitation creep produces voids that are detectable by fractography, but use of fractography is seldom required for identification of stress-rupture failures. Failures due to creep and stress-rupture are discussed in the article "Elevated-Temperature Failures" in this Volume.

Complex Failures

Occasionally, a service failure may occur by the sequential operation of two quite different fracture mechanisms. When conducting a failure analysis, this possibility should always be considered. Figure 23 shows an example of two types of fracture mechanisms occurring together. The fractures originated on the inside surface of a diametrically drilled hole in an aluminum alloy lug at the points indicated by arrows A and B in Fig. 23.

Fractographic examination revealed that the initial cracking was by stress corrosion (arrows A and B, Fig. 23) but that crack extension by this mechanism stopped, probably because of internal stress relief. Crack propagation continued by fatigue, as evidenced by beach marks in a band on one of the fracture surfaces (arrow C, Fig. 23), until catastrophic fracture of the remaining section occurred.

Chemical Analysis

In a failure investigation, routine analysis is recommended to ensure that the material is the one that was specified. Slight deviations from specified compositions are not likely to be of major importance in failure analysis. In fact, because only a minority of service failures result from unsuitable or defective material, the results of chemical analysis rarely disclose the reason for failure. In specific investigations, particularly where corrosion and stress corrosion are involved, chemical analysis of any deposit, scale or corrosion product, or the medium with which the affected material has been in contact is required to assist in establishing the primary cause of failure.

Where analysis shows that the content of a particular element is slightly greater than that required in the specifications, it should not be concluded that such deviation is responsible for the failure. It is often doubtful whether such a deviation has played even a contributory role in failure. For example, sulfur and phosphorus

Fig. 21 Typical creep deformation and intergranular cracking in a jet-engine turbine blade

Courtesy of J. Schijve

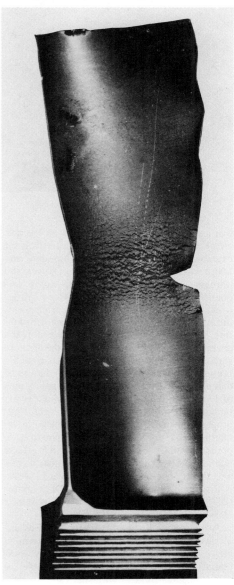

contents in structural steels are limited to 0.04% in many specifications, but rarely can a failure in service be attributed to a sulfur content slightly in excess of 0.04%. Within limits, the distribution of the microstructural constituents in a material is of more importance than their exact proportions. A chemical analysis, except a spectrographic analysis restricted to a limited region of the surface, is usually made on drillings that represent a considerable volume of material and therefore provide no indication of possible local deviations due to segregation and similar effects.

Also, certain gaseous elements, or interstitials, that are normally not reported in a chem-

Fig. 22 Creep-induced failure of a boiler plate

(a) A polished cross section of the plate that shows necking, a feature of short-term creep. 2×. (b) Intergranular voids (dark areas) in an area near the fracture surface. Courtesy of B. Gabriel, Packer Engineering Associates, Inc.

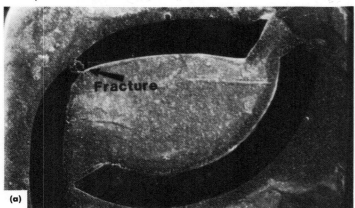

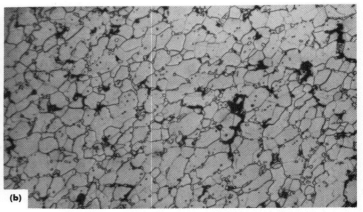

ical analysis have profound effects on the mechanical properties of some metals. In steel, for example, the effects of oxygen, nitrogen, and hydrogen are of major importance. Oxygen and nitrogen may give rise to strain aging and quench aging. Hydrogen may induce brittleness, particularly when absorbed during welding, cathodic cleaning, electroplating, or pickling. Hydrogen is also responsible for the characteristic halos or fisheyes on the fracture surfaces of welds in steels, in which case the presence of hydrogen is often due to the use of damp electrodes. These fisheyes are indications of local rupture that has taken place under the bursting stresses induced by the molecular hydrogen, which diffuses through the metal in the atomic state and collects under pressure in pores and other discontinuities. Various effects due to gas absorption are found in other metals and alloys. The formation of fisheyes in welds is discussed in the article ''Weld Discontinuities'' in Volume 6 of the 9th Edition of *Metals Handbook*.

Analysis of Bulk Materials. Various analytical techniques can be used to determine elemental concentrations and to identify compounds in alloys, bulky deposits and samples of environmental fluids, lubricants, and suspensions. A guide for the selection of applicable analytical techniques for qualitative, semiquantitative, and quantitative chemical analysis of materials is provided in the article ''How to Use the Handbook'' in Volume 10 of the 9th Edition of *Metals Handbook*.

Techniques such as emission spectroscopy, atomic absorption spectroscopy, inductively coupled plasma atomic emission spectroscopy, and classical wet analytical chemistry can be used to determine dissolved metals. Combustion methods such as high-temperature combustion and inert-gas fusion are used for determining the concentration of carbon, sulfur, nitrogen, hydrogen, and oxygen. X-ray diffraction methods identify crystalline compounds either on the metal surface or as a mass of particles and can

be used to analyze corrosion products and other surface deposits. X-ray fluorescence analysis can be used to analyze both crystalline and amorphous solids, as well as liquids and gases. Infrared and ultraviolet spectroscopy are used in analyzing organic materials. When the organic materials, such as solvents, oils, greases, rubber, and plastics, are present in a complex mixture, the mixture is first separated into its individual components by gas chromatography.

Each of the above techniques, as well as many other applicable methods, is discussed in Volume 10 of the 9th Edition of *Metals Handbook* in the Sections ''Optical and X-Ray Spectroscopy,'' ''Classical, Electrochemical, and Radiochemical Analysis,'' and ''Diffraction Methods.''

Analysis of Surfaces and Deposits. Concurrent with the many developments in the analysis and interpretation of fractures by examination of fracture-surface topographies have been the development and application of analytical techniques for providing information regarding the chemical composition of surface constituents. Energy-dispersive and wavelength-dispersive x-ray spectrometers have been used for this purpose. They are employed as accessories to scanning electron microscopes and permit simultaneous viewing and chemical analysis of a surface. If it is desirable to detect the elements in extremely thin surface layers, Auger electron spectroscopy, Mössbauer spectroscopy, secondary ion mass spectroscopy (SIMS), and low-energy ion-scattering spectroscopy (LEISS) are useful.

Auger electron spectroscopy can provide qualitative and semiquantitative determinations of elements with atomic numbers of 3 or more (lithium). The size of the area examined varies greatly with the test conditions; it may be from 1 to 50 μm in diameter.

Mössbauer spectroscopy is primarily applicable to identifying corrosion products on steel. It examines a depth of 300 nm or less when measuring emitted γ-rays.

For chemical analysis of surface areas as small as 1 μm in diameter, the electron probe microanalyzer is widely used. The electron probe microanalyzer can determine the concentration of all but the low atomic number elements ($Z < 11$) with a threshold sensitivity for elemental detection of approximately 0.02%.

In-depth composition profiles of oxide surface layers and corrosion films are possible using SIMS. Hydrogen concentrations in embrittled metals and alloys can also be determined.

The composition of oxidation, corrosion, and other contaminating films (oils, greases, and so on) can also be obtained using LEISS. Elements with atomic numbers of 2 or more are detectable. Additional information on the above techniques can be found in Volume 10 of the 9th Edition of *Metals Handbook* in the articles ''Auger Electron Spectroscopy,'' ''Mössbauer Spectroscopy,'' ''Secondary Ion Mass Spectroscopy,'' and ''Low-Energy Ion-Scattering Spectroscopy.''

Spot tests are relatively simple, qualitative chemical tests that can be used to identify the metal in the failed part, the alloying elements present, deposits, corrosion products, and soil. These tests require little equipment—none of which is complicated or expensive—and can be performed quickly. Spot tests can be performed both in the laboratory and in the field; they do not require extensive training in analytical chemistry. The only requirement is that the substance to be tested be dissolvable; hydrochloric acid or even aqua regia may be used to dissolve the substance.

Spot tests for metallic elements, such as chromium, nickel, cobalt, iron, and molybdenum, are usually carried out by dissolving a small amount of the alloy in acid and mixing a drop of the resulting solution with a drop of a specific reagent on absorbent paper or a porcelain plate. Spot colorings produced in this way indicate the presence or absence of the metallic radical under test. Samples may be removed

Fig. 23 Fracture surfaces of an aluminum alloy lug

Fractures originated by SCC on the surface of a diametrical hole (arrows A and B). The crack propagated by fatigue, as evidenced by the presence of beach marks (arrow C).

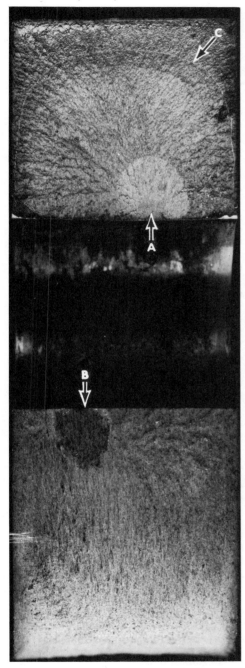

from gross surfaces by spotting the specimen with a suitable acid, allowing time for solution, and collecting the acid spot with an eyedropper. Methods for detecting and identifying both metallic and nonmetallic elements and radicals are described in "Chemical Spot Testing" in the Desk Edition of *Metals Handbook*.

Application of Fracture Mechanics

The mechanics of fracture in metal parts and specimens under load and the application of fracture mechanics concepts to the design and prediction of service life of parts and components are becoming increasingly important in the investigation of failures due to fracture and to the formulation of corrective measures that will prevent similar failures. The concepts of fracture mechanics are useful in measuring fracture toughness and other toughness parameters and in providing a quantitative framework for evaluating structural reliability.

Fracture mechanics concepts and applications are dealt with in the article "Failure Analysis and Fracture Mechanics" in this Volume. The article also provides detailed consideration of notch effects and of toughness testing and evaluation. Among the subjects dealt with in connection with notch effects are stress concentration, triaxiality, plastic constraint, and local strain rate. Test methods that are considered in connection with toughness testing and evaluation are the plane-strain fracture-toughness test, the dynamic tear test, the crack-opening displacement test, and instrumented impact testing.

Fracture mechanics concepts are also used in fatigue, SCC, and hydrogen-embrittlement studies. These concepts are addressed in Volume 8 of the 9th Edition of *Metals Handbook* in the articles "Fatigue Crack Propagation," "Environmental Effects on Fatigue Crack Propagation," "Tests for Stress-Corrosion Cracking," and "Tests for Hydrogen Embrittlement." In addition to these articles, an entire Section in Volume 8 is devoted to the subject of fracture mechanics and its use in determining macroscopic and microscopic fracture behavior.

Simulated-Service Testing

During the concluding stages of an investigation, it may be necessary to conduct tests that attempt to simulate the conditions under which failure is believed to have occurred. Often, simulated-service testing is not practicable because elaborate equipment is required, and even when practicable, it is possible that not all of the service conditions are fully known or understood. Corrosion failures, for example, are difficult to reproduce in a laboratory, and many attempts to reproduce them have given misleading results. Serious errors can arise when attempts are made to reduce the time required for a test by artificially increasing the severity of one of the factors, such as the corrosive medium or the operating temperature.

On the other hand, when its limitations are clearly understood, the simulated testing of the effects of certain selected variables encountered in service may be helpful in planning corrective action that will avoid similar failure or, at least, extend service life. The evaluation of the effi-

ciency of special additives to lubricants to counteract wear is an example of the successful application of simulated-service testing using a selected number of service variables. The aircraft industry has made successful use of such devices as the wind tunnel to simulate some of the conditions encountered in flight, and naval architects have implemented tank tests to evaluate hull modifications, power requirements, steerage, and other variables that might forestall component failure or promote safety at sea.

Taken singly, most of the metallurgical phenomena involved in failures can be satisfactorily reproduced on a laboratory scale, and the information derived from such experiments can be helpful to the investigator provided the limitations of the tests are fully recognized. Such limitations are discussed for a variety of test procedures in Volume 8 of the 9th Edition of *Metals Handbook*.

Analyzing the Evidence, Formulating Conclusions, and Writing the Report

At a certain stage in every investigation, the evidence revealed by examinations and tests that are outlined in this article is analyzed and collated, and preliminary conclusions are formulated. Obviously, many investigations will not involve a series of clear-cut stages. If the probable cause of failure is apparent early in the examination, the pattern and extent of subsequent investigation will be directed toward confirmation of the probable cause and the elimination of other possibilities. Other investigations will follow a logical series of stages, as outlined in this article, and the findings at each stage will determine the manner in which the investigation proceeds. As new facts modify first impressions, different hypotheses of failure will develop and will be retained or abandoned as dictated by the findings. Where extensive laboratory facilities are available to the investigator, maximum effort will be devoted to amassing the results of mechanical tests, chemical analysis, fractography, and microscopy before the formulation of preliminary conclusions is attempted. Finally, in those investigations in which the cause of failure is particularly elusive, a search through published reports of similar instances may be required to suggest possible clues.

Some of the work performed during the course of an investigation may be thought to be unnecessary. It is important, however, to distinguish between work that is unnecessary and that which does not produce useful results. During an examination, it is to be expected that some of the work done will not assist directly in determining the cause of failure; nevertheless, some negative evidence may be helpful in dismissing some causes of failure from consideration.

On the other hand, any tendency to curtail work essential to an investigation should be guarded against. In some cases, it is possible to form an opinion regarding the cause of failure from a single aspect of the investigation, such as a visual examination of a fracture surface or examination of a single metallographic specimen. However, before final conclusions are reached, supplementary data confirming the original opinion, if available, should be sought. Total dependence on the conclusions that can be drawn from a single specimen, such as a metallographic section, may be readily challenged unless a history of similar failures can be drawn upon.

The following check list, which is in the form of a series of questions, has been proposed as an aid in analyzing the evidence derived from examinations and tests and in formulating conclusions (Ref 11). The questions are also helpful in calling attention to details of the overall investigation that may have been overlooked.

- Has failure sequence been established?
- If failure involved cracking or fracture, have the initiation sites been determined?
- Did cracks initiate at the surface or below the surface?
- Was cracking associated with a stress concentrator?
- How long was the crack present?
- What was the intensity of the load?
- What was the type of loading—static, cyclic, or intermittent?
- How were the stresses oriented?
- What was the failure mechanism?
- What was the approximate service temperature at the time of failure?
- Did temperature contribute to failure?
- Did wear contribute to failure?
- Did corrosion contribute to failure? What type of corrosion?
- Was the proper material used? Is a better material required?
- Was the cross section adequate for class of service?
- Was the quality of the material acceptable in accordance with specification?
- Were the mechanical properties of the material acceptable in accordance with specification?
- Was the component that failed properly heat treated?
- Was the component that failed properly fabricated?
- Was the component properly assembled or installed?
- Was the component repaired during service? If so, was the repair performed correctly?
- Was the component properly run in?
- Was the component properly maintained? Properly lubricated?
- Was failure related to abuse in service?
- Can the design of the component be improved to prevent similar failures?
- Are failures likely to occur in similar components now in service? What can be done to prevent their failure?

In general, the answers to these questions will be derived from a combination of records and the examinations and tests previously outlined in this article. However, the cause or causes of failure cannot always be determined with certainty. In this case, the investigation should determine the most probable cause or causes of failure, distinguishing findings based on demonstrated fact from conclusions based on conjecture.

Writing the Report. The failure analysis report should be written clearly, concisely, and logically. One experienced investigator has proposed that the report be divided into the following principal sections:

- Description of the failed component
- Service conditions at time of failure
- Prior service history
- Manufacturing and processing history of component
- Mechanical and metallurgical study of failure
- Metallurgical evaluation of quality
- Summary of mechanisms that caused failure
- Recommendations for prevention of similar failure or for correction of similar components in service

Obviously, not every report will require coverage under every one of these sections. Lengthy reports should begin with an abstract. Because readers of failure analysis reports are often purchasing, operating, and accounting personnel, the avoidance of technical jargon wherever possible is highly desirable. A glossary of terms may also be helpful. The use of appendices, containing detailed calculations, equations, and tables of chemical and metallurgical data can serve to keep the body of the report clear and uncluttered.

Appendix: Use of Microanalytical Techniques in Failure Analysis and Problem Solving

By the ASM Committee on Failure Analysis by Microanalytical Techniques*

RECENT ADVANCES in instrumentation capabilities have brought microanalytical techniques out of the research environment into

*Lawrence Kashar, *Chairman*, Arun Kumar, Shahram Sheybany, Leo Barnard, Michael Neff, Stephen Harris, and Jim Curiel, Scanning Electron Analysis Laboratories, Inc.

everyday problem solving and failure analysis. These advances include the detection of light elements (those having atomic numbers ≤ 11) and the development of accurate quantitative methods with the energy-dispersive x-ray analyzer available with scanning electron microscopes, scanning Auger microscopes with greatly improved spatial resolution, trace element analysis using either secondary ion mass spectroscopy or laser microprobe mass analysis, and small spot x-ray photoelectron spectroscopy.

Unfortunately, in the investigation of many failures or materials problems, the fracture characteristics and/or microstructure may not provide enough information for a complete determination of the cause. Corrosion, oxidation, or mechanical damage may have obliterated the fracture characteristics, or the failure may have been caused by impurities or contaminant films. In some of these difficult cases and in many other analyses, the chemical data provided by microanalytical techniques can be crucial to completing the failure analysis. In this Appendix, the development, characteristics, and application of several microanalytical techniques that have been useful in performing failure analyses will be discussed. These techniques include electron probe microanalysis, scanning Auger electron spectroscopy, secondary ion mass spectroscopy, x-ray photoelectron spectroscopy, and laser microprobe mass analysis. Detailed descriptions of the basic theory, equipment, analysis characteristics, and applications of the first four of these microanalytical techniques are presented in Volume 10 of the 9th Edition of *Metals Handbook*.

Background and Historical Development

Electron Probe Microanalysis (EPMA). The practical development of techniques for analyzing microquantities of material can be traced to the work of R. Castaing, who, as a student of A. Guinier at the University of Paris in 1948, demonstrated that the technique now known as EPMA was possible. In EPMA, a beam of electrons is focused on the sample area of interest; the interaction between the beam and the atoms in that sample volume results in the emission of x-rays and other signals (Fig. 24) with energies that are characteristic of the elements present in the sample volume.

In modern x-ray microanalysis, the analyst has two choices for the x-ray spectrometer: the wavelength-dispersive spectrometer or the energy-dispersive spectrometer. The main difference between the two electron microprobe techniques concerns the method of detecting the x-rays and determining their energies or wavelengths. The original design by Castaing utilized a wavelength-dispersive spectrometer and was the model for the first commercial systems. These systems consisted of an x-ray detector and single-crystal diffractometers that moved synchronously around the sample to detect the

Fig. 24 Schematic representation of the interaction of an electron beam with a specimen

Shown are the various types of signals that result and the information that can be obtained.

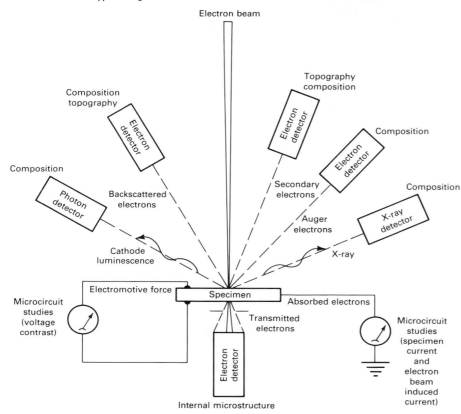

Fig. 25 Schematic diagram of the components of a wavelength-dispersive x-ray spectrometer

Courtesy of Cameca Instruments

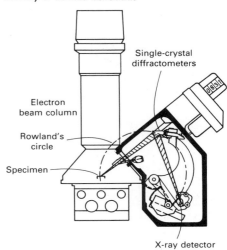

various wavelength x-rays (Fig. 25). This configuration is excellent for the quantitative analysis of flat samples, but is awkward and time-consuming to set up and is difficult to use on irregular topography, such as a fracture surface.

With the advent of semiconductor devices, lithium-drifted silicon detectors were developed that, with no moving parts, could simultaneously count and determine the energy of the x-rays (Fig. 26). These detectors are commonly referred to as energy-dispersive x-ray spectrometry (EDX) detectors, and they are usually mounted on a scanning electron microscope, a combination that permits high-resolution photography of the specimen and the ability to identify easily the elemental composition of selected areas of the specimen. Detailed information on the principles, instrumentation, and applications of EPMA can be found in the article ''Electron Probe X-Ray Microanalysis'' in Volume 10 of the 9th Edition of *Metals Handbook*.

Auger Electron Spectroscopy (AES). The acceptance of the transmission electron microscope, the electron microprobe, and the scanning electron microscope as research and problem-solving tools kept equipment manufacturers involved in making these instruments even more useful. The improvements that were

made in the electron optics, detectors, and vacuum systems for these techniques assisted development of practical Auger analysis systems, which proved to be another microanalytical technique of significant utility in materials science. In AES, an electron beam is focused on the sample area of interest, and the weak Auger electrons, which result from a secondary reaction of the x-rays with the electron orbitals, are detected.

Discovered by Pierre Auger in the 1920s, these electrons have specific energies that are characteristic of the emitting elements. In 1953, Lander suggested the use of AES for identification of surface impurities, but it was not until the late 1960s that Weber and Peria demonstrated a useful Auger electron spectrometer, and Harris showed the high sensitivity of the technique by using the now familiar differentiated energy spectrum, $dN(E)/dE$, to make small peaks visually more apparent (Fig. 27).

The advantage of AES over EPMA is in its surface sensitivity: the electron microprobe analyzes a layer of a pure element that can be from 300 to 3000 nm thick, depending on the sample and the operating conditions, whereas AES analyzes a layer that ranges from 1 to 10 nm thick, depending on the electron energy being analyzed (Fig. 28). The surface sensitivity of Auger anal-

ysis results from the low energy of the Auger electrons (30 to 2000 eV). Because of these low energies, only electrons generated near the surface will be able to escape unaltered from the sample. A brief discussion of the analysis volumes of various techniques is presented in the section ''Comparison of Microanalytical Techniques'' in this article.

Because of the extraordinary sensitivity of AES to surface layer composition, it was useful to be able to compare the composition of the surface with that of the layers just below the surface. To facilitate this type of depth profile analysis, equipment was developed to remove material uniformly and controllably from the sample surface by using an ion beam, usually of argon ions. With such an addition to an Auger system, the variation in composition with depth below the surface can be determined by ion etching or ion sputtering away the original surface and performing Auger analyses on the successively exposed new surfaces. Detailed information on the principles, instrumentation, and applications of AES can be found in the article ''Auger Electron Spectroscopy'' in Volume 10 of the 9th Edition of *Metals Handbook*.

Secondary Ion Mass Spectroscopy (SIMS). Almost simultaneously with the development of commercial Auger systems with depth-profiling capability was the development of systems for detecting the secondary ions created by the interaction of the ion beam with materials. Castaing and Slodzian proposed such a technique in 1962, calling it secondary ion emission microanalysis. It is now usually referred to as SIMS. The advantages of the SIMS techniques are:

• Detection of elements present in part per million quantities

Fig. 26 Schematic of a complete energy-dispersive x-ray spectrometer used in EPMA

Various pulse processing functions and the multichannel analyzer are shown. FET, field effect transistor

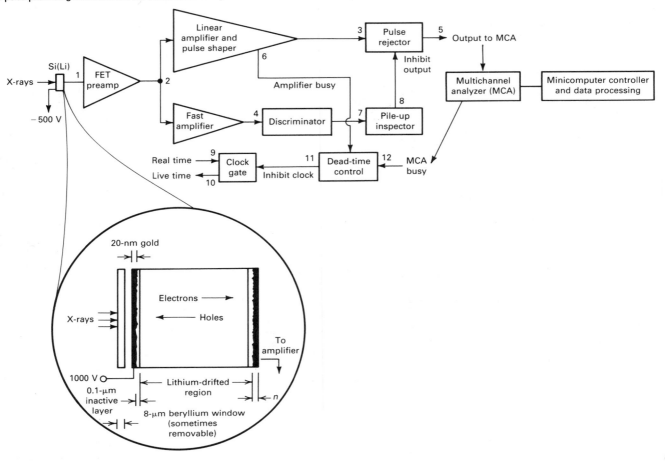

Fig. 27 AES spectrum of pure silver

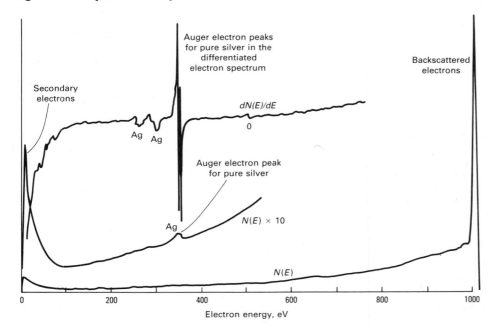

Fig. 28 Electron mean free path versus kinetic energy in metals

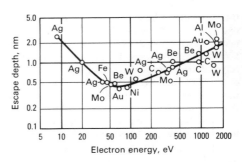

- Detection of all elements, including hydrogen
- Determination of isotopic ratios
- Detection of molecular fragments generated by the sputtering process

The major disadvantages of SIMS are the high variation in detection sensitivity for various elements, which extends over three orders of magnitude; the effect of matrix variations on this detection sensitivity; and, as a result, the very poor quantification of results in all but the most carefully controlled analytical samples.

The mass spectrometers used to detect the secondary ions can be either magnetic or quadrupole instruments. Quadrupole mass spectrometers are simpler, smaller, and less expensive than magnetic systems and are therefore preferred as a modular addition to an existing surface analysis system. Such an addition to an Auger system equipped with an ion etching gun permits simultaneous depth profile studies by AES and SIMS (Fig. 29). There are several different designs of stand-alone dedicated SIMS systems. These systems have several advantages over the modular add-on systems, including smaller ion beam diameters and greatly improved detection sensitivities (Table 1). Detailed information on the principles, instrumentation, and applications of SIMS can be found in the article "Secondary Ion Mass Spectroscopy" in Volume 10 of the 9th Edition of *Metals Handbook*.

X-Ray Photoelectron Spectroscopy (XPS). Although the photoelectric effect was described by Einstein in 1905 and work was done on the transitions by Auger in the 1920s, the commercial x-ray photoelectron spectrometer did not become available until 1969.

Siegbahn's group in Sweden during the 1950s and 1960s used XPS to study the orbital electron energies. They showed that energy shifts occurred and were related to changes in the chemical bonding of the elements, and they systematically measured the binding energies of many elements and found that the chemical environment affected the core-level binding energies. This led to an alternate name for the technique—electron spectroscopy for chemical analysis (ESCA). Because the energies of the emitted photoelectrons are relatively low, the technique is very sensitive to surface chemistry. For reasons analogous to AES, the thickness of the analyzed layer depends on the energy of the emitted photoelectron.

In XPS, a collimated beam of soft x-rays that is directed onto the surface of a sample results in the generation of photoelectrons whose energies are determined by the difference between the x-ray energy and the binding energy of the electron. Historically, one of the drawbacks to XPS derived from the difficulty of focusing an x-ray beam, which resulted in a large area of analysis. Recent developments have successfully reduced the size of the analyzed area to as small as 150 μm in diameter, which is much more suitable for failure analysis and materials problem solving. Detailed information on the principles, instrumentation, and applications of XPS can be found in the article "X-Ray Photoelectron Spectroscopy" in Volume 10 of the 9th Edition of *Metals Handbook*.

Laser microprobe mass analysis (LAMMA), which was developed during the past decade, uses a finely collimated laser beam to vaporize and ionize a microvolume of the sample in the area of interest. The ions

Fig. 29 Various parts of the vacuum chamber on a scanning Auger microscope equipped with an ion gun and a quadrupole mass spectrometer

This combination permits Auger, SIMS, and depth profile analyses with AES and SIMS. 1, quadrupole mass spectrometer; 2, argon ion beam etching gun; 3, electron beam column; 4, Auger electron detector; 5, specimen stage manipulator; 6, UHV analysis chamber; 7, specimen exchange airlock

Table 1   Comparison of microanalytical technique characteristics

Analysis Characteristic	EPMA Window-less(c)	EPMA With Be window	AES Optimum	AES Normal	SIMS Stand alone Optimum	SIMS Stand alone Normal	SIMS Add-on quadrupole	XPS(a) Optimum	XPS(a) Normal	LAMMA(b)
Range of elements analyzed	B to U	F to U	Li to U	Li to U	H to U	H to U	H to U	Li to U	Li to U	H to U
Sensitivity of elemental detection	1.0(d)	0.2%	0.5%	1.0%	1 ppt	10 ppt	1 ppm	0.5%	1.0%	1 ppm
Quantification reliability (10 highest) ..	6(d)	10	7	4	5	3	2	9	8	1
Areal distribution of elements ..	Yes	Yes	Yes	Yes	Yes	Yes	Yes	No	No	No
Compound or molecular information ..	No	No	Some-times	No	Some-times	Some-times	Some-times	Yes	Yes	Yes
Depth of analyzed volume	~300 nm	~10 000 nm	1 nm	1 nm	1 nm	1 nm	1 nm	10 nm	10 nm	100 nm
Diameter of analyzed volume	~300 nm	~10 000 nm	50 nm	300 nm	50 nm	1000 nm	35 μm	150 μm	2000 μm	~2 μm

(a) XPS, X-ray photoelectron spectroscopy. (b) LAMMA, laser microprobe mass analysis. (c) Operating with a 5-kV electron beam. (d) For elements below sodium in periodic table

Fig. 30 Laser microprobe mass analyzer

From left to right: the data display, the electronics control panel, and the specimen chamber that is mounted in front of the laser bench and time-of-flight mass spectrometer

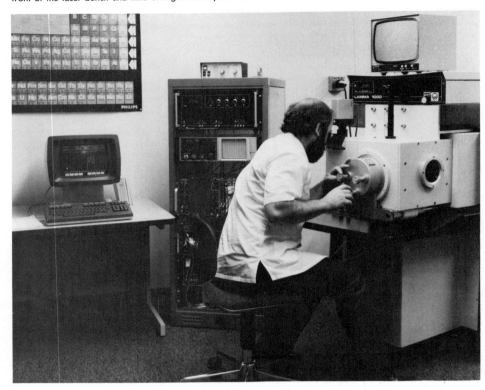

created by the laser pulse are drawn into and analyzed by a time-of-flight mass spectrometer. All elements of the periodic table and their isotopes can be detected with a laser microprobe. In addition, the mass spectrum can contain molecular fragments that permit the identification of organic, inorganic, and metallo-organic compounds in a manner analogous to SIMS. A LAMMA device is shown in Fig. 30.

Sample Considerations and Limitations

All of the microanalytical techniques that have been described in the previous section are carried out with the sample in a vacuum chamber. The necessity of this vacuum arises from two problems: the operation of the equipment itself and the prevention of sample contamination. Without the use of a vacuum, the emitted electrons, ions, or x-rays would be absorbed by air molecules. Such parts of the system as the electron gun, the ion gun, and the detectors also require high vacuum for proper functioning. In thin-film analysis, the environment of the analysis chamber must not react with the sample surface. In such thin-film (surface) analyses as AES, SIMS, and XPS, ultrahigh-vacuum (UHV) systems (10^{-9} torr) are used; in these systems, diffusion pumps are not normally used, because stringent precautions must be taken to prevent backstreaming of the pump

fluid vapors into the analysis chamber. To maintain a satisfactory vacuum, the samples to be analyzed must have a low vapor pressure and must not outgas. Except in rare cases, this restricts the samples to solid materials.

Samples that are to be analyzed by the surface-sensitive techniques should be protected from accidental contamination. This includes fingerprints, dust, atmospheric moisture, and hydrocarbons. The analyst should be aware that hydrocarbon contamination of the specimen surface occurs in most scanning electron microscopes and should take this into account when reviewing the results of surface analyses performed after SEM examinations.

The electrical conductivity of the sample has an appreciable effect on the quality of analysis results. In analyses using EPMA, SIMS, and AES, optimum results are achieved on conductive materials because the probe beam carries a charge and the sample must be sufficiently conductive to dissipate the charge by conducting it to ground. If this does not occur, the sample becomes charged and can adversely affect both the probe beam and the emitted signal. For EPMA analyses, thin conductive coatings can be evaporated onto nonconductive samples; however, these coatings affect the results of a surface analysis.

If the analysis is to be performed in a vacuum chamber, the sample must fit into the chamber. Further, the configuration of the

sample must be amenable to the geometry of the technique. Simply put, the area to be analyzed must be viewable to both the analyzing (probe) beam and the emitted signal detector. A common configuration that is difficult (if not impossible) to analyze is the inside of a deep hole.

Comparison of Microanalytical Techniques

If it is assumed for a particular failure analysis that facilities are available to conduct all of the microanalytical techniques that have been discussed, the choice of technique is determined by the type of information required, the type and size of the sample, and the characteristics of the techniques themselves. The important technique characteristics that should be considered include:

- Analysis area
- Analysis depth
- Elemental range detected
- Compositional range detected
- Elemental or molecular information
- Ease and reliability of quantification
- Spatial distribution of elements
- Probe beam—charged or neutral particles
- Sample damage by probe beam

Some of the characteristics of these microanalytical techniques are compared in Table 1. For most of the techniques, two sets of characteristics are listed: the optimum available on the latest systems and the values generally attained by more standard systems. In selecting which techniques should be implemented to obtain the desired data, the capabilities of the particular equipment available are more important than an academic evaluation of the state of the art possible with that microanalytical technique.

The choice of technique has some secondary implications that are often overlooked. One of these is the time necessary to accomplish the required analyses. Not only do the various techniques require different times simply to acquire a spectrum, but the quality of the data required (sensitivity, resolution, and so on) affects the acquisition time. For example, it may require only a few seconds to obtain a general idea of the composition of a sample using EPMA or LAMMA, but it may take 30 min to obtain a complete XPS analysis of a thin-film surface layer with sufficient counting statistics to obtain reliable quantitative and binding energy information.

If depth profiles are required, then the time taken to ion etch the crater must be added to the spectra acquisition time. To obtain accurate depth profiles, the surface analyzed must be flat and parallel to the original surface. Because the intensity of the ion beams used for etching during the depth profiles varies across their cross section, it is necessary to raster the ion beam to achieve a flat-bottomed crater (Fig.

31). If the analysis desired involves a relatively large surface area, the ion beam has to be rastered over an even larger area, which again increases the time required to complete the analysis.

The outstanding aspects of the microanalytical techniques that are normally used in failure analysis and materials problem solving can be summarized as follows:

- *EPMA*: Identification (quantitative) of elements (above beryllium) in base material, thick films, and particles
- *AES*: Identification (semiquantitative) of elements (above helium) in thin films of conductors, semiconductors, and some insulators
- *XPS*: Identification (quantitative) of elements and compounds in thin films of conductors, semiconductors, and insulators
- *SIMS*: Identification of elements, isotopes, and molecular fragments in thin films of metals and semiconductors
- *LAMMA*: Identification of elements, isotopes, and molecular fragments in metals, nonmetals, organics, and metallo-organics

The following examples have been selected to illustrate the selection of microanalysis techniques and, more importantly, the use of these techniques in failure analysis and problem solving.

Examples

Example 1: Use of EPMA to Identify Microconstituents in a Failed Extrusion Press.
The energy-dispersive x-ray analyzer, in conjunction with the scanning electron microscope, is valuable to the practicing failure analyst. These techniques will resolve the distinct fractographic information necessary to determine the fracture mode and mechanism and will determine whether location variations in chemical composition of the bulk material have resulted in any unusual microconstituents or contaminants at the origin.

In this example, EPMA was used to determine the composition of the fatigue fracture-origin site that was not a normal origin site based on stress analysis or stress concentrations. An unusually high concentration of inclusions was present in the fracture-origin area, which by cyclic crack growth finally led to the catastrophic failure of an extrusion press.

Investigation. The 230-mm (9-in.) thick casing of a 450-Mg (500-ton) extrusion press failed after 27 years of service. The casing, shown schematically in Fig. 32, ruptured along the full longitudinal axis during normal operating conditions. The casing had been fabricated from ASTM 235-55 low-carbon steel. Initial visual examination revealed an area that exhibited multiple origins and classic beach marks radiating out approximately 75 mm (3 in.) from the origin along the wall of a hydraulic-oil bleed hole (Fig. 33). The rest of the fracture surface was very flat and in the same longitudinal plane. The focus of the beach marks was not at the corners of the bleed hole on the inside diameter of the hydraulic cylinder, which a stress analysis revealed as the most likely origin for a fatigue crack. Because of the unusual origin site, material deficiencies had to be considered.

Results. Examination of the origin site with the scanning electron microscope revealed corrosion pits along the bleed hole wall. Microfractographic details, such as fatigue striations, could not be resolved in the electron microscope because of the excessive oxidation/corrosion present on the fracture surfaces.

A portion of the bleed hole was vacuum encapsulated in epoxy and sectioned parallel to the fracture surface. Metallographic examination of this section revealed a microstructure that generally consisted of a matrix of pearlite and ferrite with bands of slightly finer pearlite. A large concentration of inclusion stringers was observed in the area corresponding to the fracture origin. Both the grain and stringer orientations were transverse to the longitudinal axis of the bleed hole and parallel to the major axis of

the extrusion press housing, that is, in the same plane and direction as the fracture.

Analysis of the stringers using the SEM/EDX system revealed high concentrations of sulfur and manganese, indicating the presence of manganese sulfide inclusions (Fig. 34a to c). No other unusual microconstituents were observed.

Conclusions. The fracture appeared to have resulted from corrosion-assisted fatigue. The fatigue initiated at the corrosion pits associated with a concentration of manganese sulfide stringers that were exposed on the inside diameter bleed hole. Crack propagation then proceeded because of cyclic loading of the press in operation; the stress concentration was caused by the inclusions/corrosion pits. In addition, the presence of the local concentration of inclusions would be expected to result in poor mechanical properties of the steel in the region. The final overload fracture of the remaining material was related to its low fracture toughness. Although the bulk chemical analysis conformed to the specification requirements for this alloy, the inclusion concentration in the fracture-initiated area indicated that the chemical-composition limits for sulfur and manganese would have greatly exceeded the specification.

Recommendations. A higher-quality steel was recommended for the replacement unit to lessen the possibility of such gross inclusion segregation and to improve the fracture toughness of the cylinder.

Fig. 31 Schematic of the effects of depth profiling

(a) Cratering effect. Because the ion beam intensity decreases with distance from its center, the edges are etched away more slowly than the center region. This results in contributions from the surface layer (A) that are thought to come from the matrix. (b) The cratering effect can be greatly reduced by rastering the ion beam and electronically gating the detector so that signals are accepted by the detector when the ion beam is away from the crater edges.

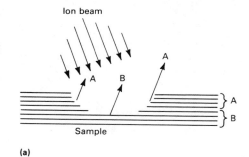

Fig. 32 Schematic of extrusion press casing

Cross-sectional view

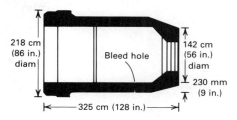

Fig. 33 Segment of fractured casing after bisection of the bleed hole

Note multiple origins.

Fig. 34 EPMA analysis of inclusion stringers in failed extrusion press

(a) SEM micrograph of metallographic section near bleed hole. Note large inclusions. 490×. (b) EDX dot map of manganese. (c) EDX dot map of sulfur

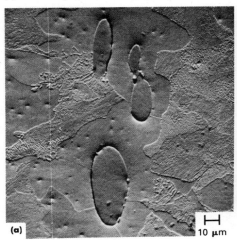
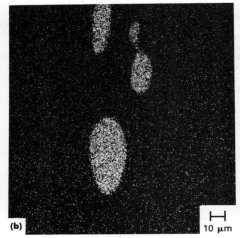
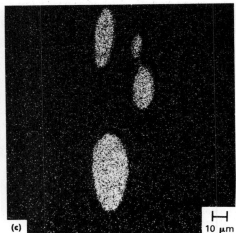

Fig. 35 Percentage of x-rays generated by a sample that are detected by a typical EDX detector

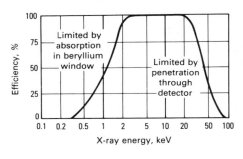

Fig. 36 Overall view of the inside of the bottom of the failed melting pot

Light-Element Detection Using EPMA.

A typical EDX detector can detect a limited range of x-ray energies (Fig. 35). At the low-energy end, the limitation is caused by the absorption of x-rays by the beryllium window (Fig. 26) placed in front of the detector to protect it from contamination. At the high-energy end of the spectrum, the limitation is caused by the increasing probability of the x-rays passing directly through the detector without being absorbed. The elements with high-energy x-rays that might pass directly through the detector (K lines) have other lower-energy characteristic x-rays that can be detected (L or M lines). However, low atomic number elements emit only low-energy characteristic x-rays. If these x-rays do not have sufficient energy to pass through the beryllium window, the element will not be detected.

In the last few years, two improvements have enabled the use of EDX detectors without beryllium windows. First, the vacuum capabilities of the scanning electron microscope have improved, reducing the contamination rate of the unprotected detector. Second, the electronics of the detectors have improved: low-end noise has been reduced so that the low-energy

x-rays can be detected with good signal-to-noise ratios, and the electronics have been made more stable, permitting cycling to room temperature to allow decontamination of the detector surface.

The development of EDX detectors in which the beryllium window can be removed permits a more complete metallurgical evaluation with the ability to identify the presence of fluorine, oxygen, nitrogen, carbon, and (in some of the most recent detectors) boron. The following three examples of the use of windowless detectors demonstrate the usefulness of this capability in failure analysis and problem solving.

Example 2: Identification of Iron Oxide Inclusions. A steel pot used to melt magnesium alloys leaked (Fig. 36), releasing about 35 kg (80 lb) of molten magnesium onto the foundry floor and causing an extensive fire. After the fire, the hole at the bottom of the pot was irregular and several inches across; the hole appeared to have been increased in size by erosion from the molten magnesium and ensuing fire. No evidence of the original leaking hole remained for analysis.

In an attempt to obtain some data that would explain the failure, samples were removed from various areas of the failed pot and examined. Cross sections of the pot wall were polished and etched and, in general, were found to be composed of ferrite and pearlite mixtures, as would be expected for hot-rolled low-carbon steel. However, the sample taken from a location about 75 mm (3 in.) from the hole contained a cluster of unusually large inclusions (Fig. 37).

By removing the beryllium window from in front of the detector, EPMA spectra were obtained from the inclusions and from the steel matrix (Fig. 38). The inclusion spectrum (Fig. 38a) contained primarily iron and oxygen, whereas the matrix spectrum (Fig. 38b) contained primarily iron. X-ray maps were made to show the distribution of iron and oxygen (Fig. 39a to c). These results indicated that the inclusions were iron oxide. A similar inclusion at the failure site in the melting pot may have reacted violently with the molten magnesium, causing the leak.

Example 3: Identification of Carbonitrides. The identification of complex inclusions in steels and superalloys is often critical to the understanding of the steelmaking practice and its adequacy. Titanium carbonitride inclusions are usually identified by their cubic shape and pinkish-orange color (Fig. 40). The EPMA spectra obtained using a typical EDX detector is shown in Fig. 41(a); only an x-ray peak for titanium can be seen. With a windowless EDX detector, the spectrum showed titanium, carbon, and nitrogen (Fig. 41b). It should be noted that titanium emits an x-ray at about the same energy as nitrogen; therefore, both nitrogen and titanium contributed to the intensity of this peak.

Example 4: Combined EDX/AES Analysis of Failed Inconel 600 Steam Line Bellows. Within the first few months of operation of a 8-km (5-mile) long 455-mm (18-in.) diam high-pressure steam line between a coal-fired electricity-generating plant and a paper

Fig. 37 Polished-and-etched cross section of a section of the failed melting pot shown in Fig. 36

The ferrite and pearlite constituents normally found in hot-rolled carbon steel and large anomalous inclusions are evident.

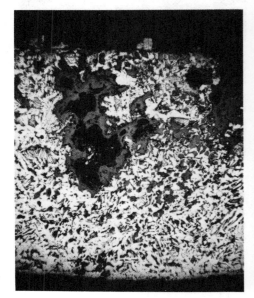

Fig. 38 EPMA spectra obtained with a windowless detector from the two areas indicated in Fig. 39(a)

(a) From the inclusion area showing the presence of iron, oxygen, and some carbon. (b) From the steel matrix showing the presence of only iron with a trace of carbon

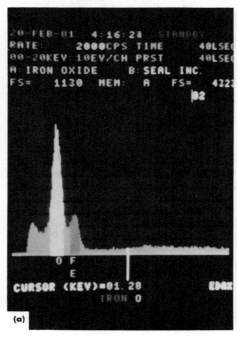

(a)

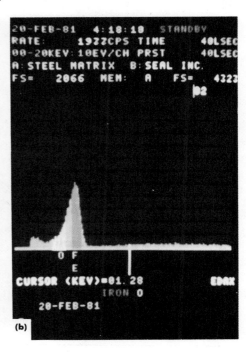

(b)

mill, several of the Inconel 600 bellows failed. The steam line operated at 6030 kPa (875 psi) and 420 °C (790 °F). Each bellows was of a three-ply design; each ply contained two longitudinal gas tungsten arc welds and was cold formed with 13 convolutions. The welds had been inspected with dye penetrant. The Inconel 600 had been received as 1.5-mm (0.060-in.) thick sheets, hot rolled to final thickness at 900 °C (1650 °F) and cooled to room temperature, resulting in a hardness ranging from 78 to 86 HRB. After assembly, each set of bellows was tested to ensure that each of the elements stroked uniformly. The bellows design resulted in the maximum stress being a circumferential

membrane stress of 108 MPa (15.7 ksi), which is well below the ASME-assigned design limit of 138 MPa (20 ksi) for Inconel 600 at 420 °C (790 °F).

Because of the high-pressure high-temperature application for the Inconel 600, which was specified by the customer, both the material supplier and the bellows manufacturer expressed concern about the purity of the steam, especially with regard to alkali contents. Analyses provided by the power plant generating the steam indicated that the steam purity would not pose a problem to the Inconel 600.

In operation, the high-pressure steam was in contact with the outside surface of the bellows, and condensate was drained from the expansion joint to an adjacent 205-mm (8-in.) diam condensate return line through a steam trap.

Investigation and Results. Sections of the failed bellows and sections of bellows contaminated with a white material were submitted for analysis. The outer ply of the failed bellows had fractured circumferentially along the outer ridges of several of the convolutions, whereas the middle ply had cracked on only two convolutions (Fig. 42). The inner ply had only one crack in it. None of the fractures emanated from

Fig. 39 X-ray maps from a section of the failed melting pot shown in Fig. 36

(a) Scanning electron micrograph of a polished cross section. EPMA spectra of regions A and B (arrows) are shown in Fig. 38(a) and (b), respectively. (b) X-ray dot map showing distribution of oxygen. (c) X-ray dot map showing distribution of iron

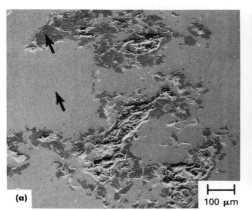

(a)

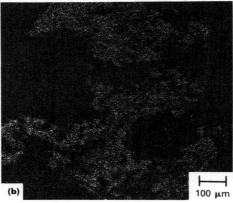

(b)

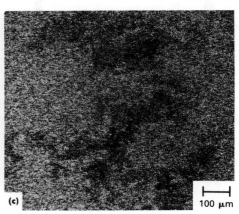

(c)

Fig. 40 Type 321 stainless steel containing titanium carbonitride blocky inclusions

720×

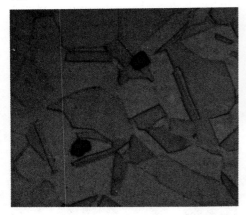

Fig. 41 EPMA analysis obtained on the inclusion shown in Fig. 40 using a 20-kV electron beam voltage

(a) Spectrum obtained with a typical EDX detector showing only the presence of titanium. (b) Spectrum obtained with a windowless EDX detector showing the presence of carbon, nitrogen, and titanium

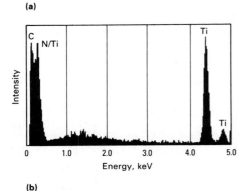

(a)

(b)

the longitudinal weld; in fact, several of the cracks terminated at the weld zone.

Metallographic sections through the fracture showed the intergranular nature of the failure (Fig. 43). Careful metallographic and electron microprobe examinations of the grain boundaries of the Inconel 600 bellows sheets did not

Fig. 42 Photograph of the failed Inconel 600 bellows

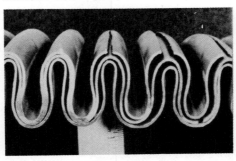

Fig. 43 Metallographic cross section of the failed bellows shown in Fig. 42

Note the intergranular nature of the failure.

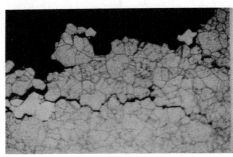

reveal any abnormal conditions. Chemical analysis and tensile tests conducted on samples taken from the failed bellows met all of the specifications for the material.

The fracture surfaces were discolored, apparently by the formation of oxide films after the cracks had been generated. The color of these films varied from blue to brown, which was probably due to variations in thickness. In most areas, however, the film thickness was not sufficient to obscure the intergranular nature of the fracture (Fig. 44). Energy-dispersive x-ray spectra obtained from most of the fracture surfaces showed only the major alloying elements of Inconel 600 and oxygen, but on one fracture area, small x-ray peaks for sodium, silicon, chlorine, and potassium were also present. In addition, on an area that was so heavily contaminated with foreign material that the intergranular fracture was almost completely obscured, the EDX spectrum included carbon, sodium, magnesium, aluminum, silicon, phosphorus, sulfur, and calcium in quantities exceeding the levels expected for the alloy (Fig. 45).

These results indicated that a more careful analysis should be conducted on the apparently clean fracture areas. This was done using the scanning Auger microscope on several portions of the various fractures. The Auger spectra obtained from one fracture, as received in the laboratory and after argon ion etching for 12 min (equal to about 120 nm), showed that the

Fig. 44 SEM fractograph of the failed Inconel 600 bellows

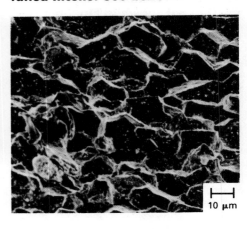

10 μm

Fig. 45 EPMA spectrum obtained from a contaminated intergranular fracture area on the bend of a failed Inconel 600 bellows

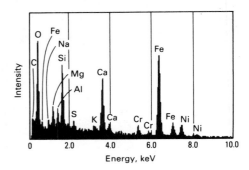

entire oxide thickness was contaminated with relatively large amounts of sodium, calcium, potassium, aluminum, and sulfur. All of the Auger analyses that were performed gave similar indications of alkali, alkali earth, and other contaminants that completely permeated even the thin oxides on the fracture surfaces.

Conclusions. Because the unusual contaminants were found completely through the oxide layers on the fracture surfaces, it was concluded that the operating environment of the bellows must have contained detrimental contaminants far in excess of the expected levels. Alkali-induced failure of Inconel 600 at temperatures in the range of 420 °C (790 °F) have occurred previously and were the cause of the concern raised by the manufacturers of the materials and the bellows. This type of failure might be classified as an elevated-temperature stress-corrosion failure. Considering the forewarning of this type of problem and the careful analysis of the steam products by the customer, it was surprising that these failures had occurred. A careful analysis of the steam and the operating procedures that had been used was therefore undertaken. This verified that the steam normally met the purity levels specified, but during boiler-cleaning operations, the generating plant had allowed contamination to get into the steam line.

Example 5: Quantitative Thin-Film Analyses by EPMA. Modern quantitative EPMA is useful in various applications that have in the past been limited to wet chemical analysis techniques. The limitations of chemical analysis by quantitative EPMA can be largely overcome by using known standards for comparison. However, the easiest and most efficient method for quantitative EPMA is by using a no-standards computer program that corrects the measured x-ray intensities with known atomic number, *Z*, absorption, *A*, and fluorescence, *F*, correction factors (*ZAF* correction). Most current quantitative EPMA systems perform these corrections in the same manner, and the accuracy of the result is then limited only by the quality of the EDX detector, the sample geometry and homogeneity, and the microscope conditions used. The accuracy of a typical no-standards analysis is illustrated in this example (Ref 12). Detailed information on quantitative EPMA can be found in the article ''Electron Probe X-Ray Microanalysis'' in Volume 10 of the 9th Edition of *Metals Handbook*.

Discussion. Phosphorus is added to the glassivation on semiconductor devices to improve mechanical and thermal properties. The phosphorus content is normally maintained between 3 and 7%. Insufficient phosphorus increases the tendency of the glass to crack, whereas excess phosphorus can react with moisture, forming phosphoric acid, which will attack the aluminum metallization on the semiconductor devices. Quantitative EPMA has been found to be a cost-effective production-control technique as compared with wet analysis, which requires the dissolution of the glassivation, a slow destructive technique. In the past, x-ray fluorescence (XRF) analysis using standards of known thickness has been implemented to determine phosphorus in glass, but variations in sample thickness can introduce errors, especially when x-rays from the silicon substrate contribute to the overall spectrum. This signal from the substrate depends on the thickness of the glassivation layer and, for most commercial glassivation thicknesses, cannot be discounted from the XRF analysis. Electron probe microanalysis is concentrated in a much thinner surface layer, and the spatial resolution is much better than XRF analysis. The depth of penetration for electrons in a silicon dioxide (SiO_2) sample varies from approximately 500 nm at 5 kV to 8000 nm at 30 kV.

Use of EPMA permits the variation of electron beam voltage, which determines the depth of electron penetration (Fig. 46), and permits analysis of varying glassivation thicknesses, with the advantages of analysis of small areas either on wafers or assembled devices. This capability is especially useful when analyzing small, thin layers of light elements. For typical glassivation thicknesses, the 5-kV beam volt-

Fig. 46 Electron penetration depth as a function of accelerating voltage and sample density

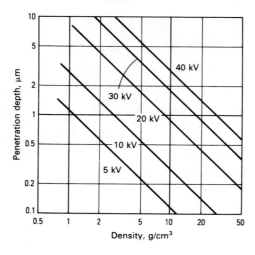

age is required to prevent excitation of the silicon substrate, which would contribute silicon x-rays to the analysis, reducing the percentage of phosphorus reported. Figures 47(a) and (b) illustrate the difference in the actual analyses obtained for a 498.7-nm thick layer of 3.55% P glass on silicon due to electron penetration through the thin film at different accelerating voltages.

Investigation. Quantitative analysis of phosphorus was performed at 5 kV on seven sample glassivation layers on silicon wafers and a pure quartz sample, using a no-standards *ZAF*-corrected computer analysis of spectra obtained with a 10-μm beryllium window in front of the EDX detector. The actual composition of the seven samples was determined by wet chemical analysis (Table 2).

One hundred second acquisitions were obtained in quadruplicate in four different areas of each sample. Except for the 11.84% P sample, all of the samples were found to be extremely homogeneous. The 11.84% P sample varied considerably, probably as a result of interaction with moisture in the atmosphere, and was not used for any of the correlations.

Conclusions. The correlation between the actual phosphorus contents and the analysis achieved with the no-standards computer analysis of data is shown in Fig. 48. The error bars on Fig. 48 encompass the maximum and minimum results for 16 acquisitions. With this technique, the maximum error encountered in the 112 analyses conducted was 0.4% P, or 11% of the actual phosphorus content. The average error was less than one-half of these maximum errors, or less than 0.2% P, indicating the usefulness of the no-standards quantitative EPMA technique for analyzing phosphorus in thin glassivation layers. This technique is now routinely used in production control and in failure analysis.

Fig. 47 EPMA spectra obtained on a 498.7-nm layer of 3.55% P doped silica on a silicon substrate using different electron beam voltages

(a) Spectrum obtained at 20 kV. (b) Spectrum obtained at 5 kV

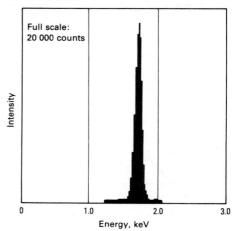

(a)

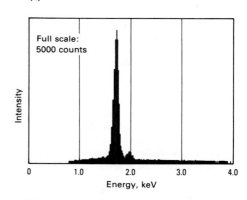

(b)

Table 2 Phosphorus content of seven glassivation layer samples and a pure quartz sample as determined by classical wet chemistry analysis

Phosphorus, %	Oxide thickness, nm
0 (pure quartz sample)	Infinite
1.56	1520.9
3.55	498.7
3.63	1350.0
5.76	1500.0
8.88	896.2
10.25	727.7
11.84	903.0

Example 6: Analysis of Contaminants on Grain-Boundary Fractures. Since electron microscopy for the evaluation of fracture surfaces became common in the late 1960s, SEM and TEM (using replicas) have been the techniques of choice for determining the type, mode, and sometimes, the cause of the fracture

Fig. 48 Correlation between the calculated and the actual phosphorus contents of seven thin-film standards and a quartz sample as determined by EPMA

Data were obtained using a typical EDX detector and a no-standards background-subtracted *ZAF*-corrected computer program.

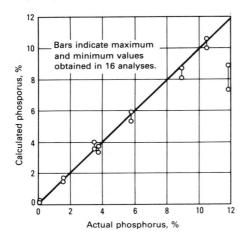

Fig. 49 Combined SEM/AES analysis of failed niobium alloy rocket-nozzle component

(a) Intergranular-fracture surface with contaminant particles on grain facets. (b) AES spectra of contaminated fracture surface

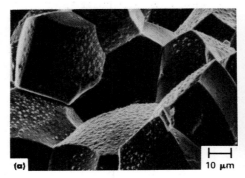

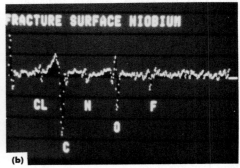

of metals. Although care and experience must be exercised in making such judgments, these techniques provide the trained failure analyst with the best and most direct information on the cause of the fracture. In most cases, if the fracture surfaces have not been altered by mechanical damage or corrosion, the fractographic interpretation is almost unambiguous.

However, if the fracture surface is intergranular, the type and mode of the fracture is not unambiguous. Several causes of intergranular fracture are listed below:

- Overload fracture due to a brittle grain-boundary film
- Overload fracture due to a precipitate-free grain-boundary zone
- Fracture caused by hydrogen embrittlement
- Fracture caused by SCC
- Fracture caused by liquid-metal embrittlement
- Elevated-temperature fracture resulting from the formation of grain-boundary voids and grain-boundary slip

With knowledge of the operating history of the part, some of the possibilities can be eliminated. In many cases, microanalytical techniques provide information that is crucial to understanding the failure cause and sequence. Auger analysis has proved to be the most useful in this regard because it can detect thin-film residues, which may include traces of the stress-corrosion environment, and because it has high spatial resolution for documenting the fractographic features and selecting areas for microanalysis.

Investigation. Cracks were observed in a niobium alloy (Nb-106) part when it was pulled

from storage for assembly into a rocket nozzle. The part had been in protective storage for several months after heat treating. No other evidence of damage was observed.

Results. The part was carefully sectioned to permit opening of the crack and was examined in a scanning electron microscope. The crack surfaces were found to be intergranular, with contaminant particles on a large number of the grain facets (Fig. 49a). Energy-dispersive x-ray analysis of several of these particles showed them to consist of niobium and fluorine.

To determine how these elements were combined in these small particles, plastic replicas were applied and pulled from the fracture surface, removing many of the particles. These replicas were prepared by standard TEM techniques and were analyzed with selected-area electron diffraction (additional information on selected-area diffraction is provided in the article "Analytical Transmission Electron Microscopy" in Volume 10 of the 9th Edition of *Metals Handbook*). The diffraction pattern obtained most closely matched the pattern for niobium tetrafluoride.

Auger analyses were conducted on several grain facet areas that did not have any contaminant particles. Without removing any of the contaminant layer, the Auger electron spectra were similar to that shown in Fig. 49(b), with peaks representing carbon, oxygen, nitrogen, fluorine, and chlorine.

Conclusions. With these results, the processing history of the part was searched. It was quickly learned that, although the part had been heat treated in a vacuum, the affinity of niobium for oxygen is so great that a tenacious oxide film forms. This film was removed by using a combination of strong acids: nitric, hydrochloric, hydrofluoric, and lactic. The contaminants found on the fracture surface by AES analysis represent contributions from each of these: fluorine from the hydrofluoric, chlorine from the hydrochloric, nitrogen from the nitric, and carbon and oxygen from the lactic.

Based on these findings, it was concluded that residues of the cleaning acid on the part had

caused SCC during storage. The tensile stresses necessary to generate SCC were not explicitly identified, but were assumed to have been residual stresses from the heat treatment.

Recommendations. To prevent the recurrence of this problem, more stringent cleaning procedures were instituted to remove any trace of the cleaning acids.

Example 7: Analysis of Nitrided Steel Cases. Nitrogen enrichment of the surfaces of metals, especially steels, has been a very successful method of obtaining a hard wear-resistant material. The nitrogen content of these layers is rarely analyzed, because of the difficulty of obtaining satisfactory analyses and the relative ease of using metallography and microhardness to evaluate the quality of the nitriding. Electron microprobe analysis has shown as much as 8.2% N in the undesirable white layer that is sometimes present on the surface of nitrided parts and as much as 1% N in the hardened cases (Ref 13). These results were obtained with a wavelength-dispersive spectrometer, which is limited in sensitivity for this analysis because of the low x-ray emissivity of nitrogen. Nitrogen analysis using AES has lower detection limits, but this technique requires UHV operation, which means that the polished cross section must be removed from the mounting material because of out-gassing. Laser microprobe mass analysis was found to give the very high spatial resolution and the elemental sensitivity needed to relate the hardness variations with the nitrogen content variations.

Investigation and Results. The microhardness traverse through a nitrided Nitralloy part did not follow the normal decreasing pattern, but had an undesirable steplike pattern (Fig. 50). Examination of the etched cross section showed that a 0.005-mm (0.0002-in.) thick white layer did exist on the part, but in the region in which the step occurred in the hardness traverse, no microstructural anomaly was observed. Laser microprobe mass analyses were obtained from areas adjacent to each of the microhardness indentations. Typical

Fig. 50 Polished (top) and etched (bottom) cross sections of a nitrided Nitralloy part

The microhardness traverse indentations and their associated laser pulse craters are shown.

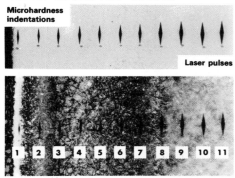

Fig. 51 Laser microprobe spectra obtained near the surface (top) and in the matrix (bottom) of the sample shown in Fig. 50

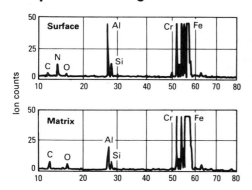

LAMMA spectra from the hardened case and from the Nitralloy matrix are illustrated in Fig. 51. Shown on these spectra are peaks at atomic masses corresponding to carbon, nitrogen, oxygen, aluminum, silicon, chromium, and iron. When the intensity of the nitrogen peaks from each spectra was plotted on the same graph as the hardness profile (Fig. 52), a step in the nitrogen intensity was found that corresponded to the step in the hardness traverse.

Conclusions. The anomaly in the hardness traverse was found to be related to a similar anomaly in the nitrogen concentration profile. This probably resulted from an unusual nitriding cycle, but it was not possible to determine the precise heat-treating lot for this part. Laser microprobe mass analysis is a powerful technique for investigating processing variables in nitriding.

Example 8: Adhesion Failures Caused by Thin-Film Contaminants.
The most widely used test for adhesion is the peel test, in which a layer of material is pulled from its base material and the force required to peel it away is measured. A variation of this test is the tape test, which is used to test surface finish layer adhesion. The microanalytical techniques com-

Fig. 52 The decrease in nitrogen concentration (counts at mass 14) and hardness (converted from HK) along the traverse shown in Fig. 50

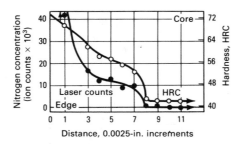

Fig. 53 Optical photograph of a peel test sample partially peeled to reveal the underside of the conversion-coated nickel-phosphorus laminate

monly used to characterize adhesion problems include AES, XPS, SIMS, and LAMMA. These techniques vary greatly in spatial resolution, sensitivity, speed, and the type of information obtained, but all surface-analysis techniques are similar in that the information is obtained from a very thin surface layer. Because many adhesion problems are related to contaminants less than 10 nm thick, surface analysis is often necessary to solve adhesion problems. In this example, it was necessary not only to identify the contaminant but also to determine whether the contamination occurred before or after a chromate conversion surface treatment.

Investigation. A batch of bimetal foil/epoxy laminates was rejected because of poor peel strength. The laminates were manufactured by sintering a nickel/phosphorus powder layer to a copper foil, cleaning, then chromate conversion coating the nickel-phosphorus surface, and laminating the nickel-phosphorus side of the clad bimetal onto an epoxy film, so that the end product contained nickel-phosphorus sandwiched between copper and epoxy, with a chromate conversion layer on the epoxy side of the nickel-phosphorus. Peel test samples were

prepared by masking and etching the copper (Fig. 53). Testing revealed abnormally low adhesion strength for the bad batch of peel test samples. These samples were compared with normal-strength samples using XPS by analyzing the surface chemistry of the exposed underside of the conversion-coated nickel-phosphorus layer after it was peeled away from the epoxy for both good and bad samples. X-ray photoelectron spectroscopy was used because the results would not be affected by the nonconducting epoxy.

Results of the XPS analyses indicated an 8.8% Na concentration on the surface of the bad sample; the good sample contained less than 1% Na on the surface (Fig. 54a and b). Depth profiling was used to determine whether the sodium was present on the surface of the chromate conversion coating or whether it was present before the conversion coating was applied. After 15 min of argon ion etching, corresponding to approximately 45 nm of material removed, high concentrations of sodium were still evident, indicating that the sodium was present before the chromate conversion treatment was performed. A review of the manufacturing procedures showed that sodium hydroxide was used as a cleaning agent before the chromate conversion coating. Apparently, the sodium hydroxide had not been properly removed during water rinsing. This result determined the exact stage in the process to be modified, thus eliminating unnecessary expense.

Example 9: Analysis of a Plating Peeling Problem on an Integrated Circuit Leadframe by XPS.
Peeling of a 90Sn-10Pb solder coating from a 1.25-μm (50-μin.) thick copper underplating on an Alloy 42 (Fe-42Ni) leadframe was observed during the lead-bending operation. The plating separation was at the solder/copper interface, and the exposed copper plating was dark and stained.

Investigation and Results. The leadframe was bent to peel the plating, and the freshly exposed surfaces, that is, exposed copper plating and the underside of the solder plating, were analyzed for surface chemistries by XPS. Both surfaces were analyzed in the as-received condition and after 10 min of argon ion etching to a depth level of approximately 20 nm below the original surface. The XPS spectra are shown in Fig. 55(a) to 55(d).

The surface of the exposed copper plating on the leadframe revealed oxidized copper along with carbon, nitrogen, sulfur, potassium, tin, and lead, in order of decreasing concentration. At approximately 20 nm below the original surface, the carbon and oxygen levels were reduced considerably, nitrogen and sulfur were not present, and metallic copper, with small amounts of potassium, tin, and lead, was observed.

The underside of the solder plating revealed carbon, oxygen, nitrogen, copper, and potassium, in order of decreasing concentration. The sodium and potassium levels were much higher

Fig. 54 Results of XPS analyses of bimetal foil laminates

(a) XPS spectra of the underside of the nickel-phosphorus layer in the low peel strength sample. Note the height of the sodium peak at 1069 eV as compared with (b), which shows the XPS spectra from a high peel strength sample.

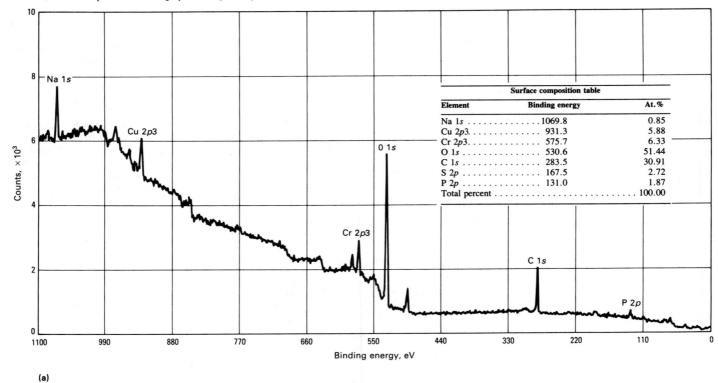

Surface composition table		
Element	Binding energy	At.%
Na 1s	1069.8	0.85
Cu 2p3	931.3	5.88
Cr 2p3	575.7	6.33
O 1s	530.6	51.44
C 1s	283.5	30.91
S 2p	167.5	2.72
P 2p	131.0	1.87
Total percent		100.00

(a)

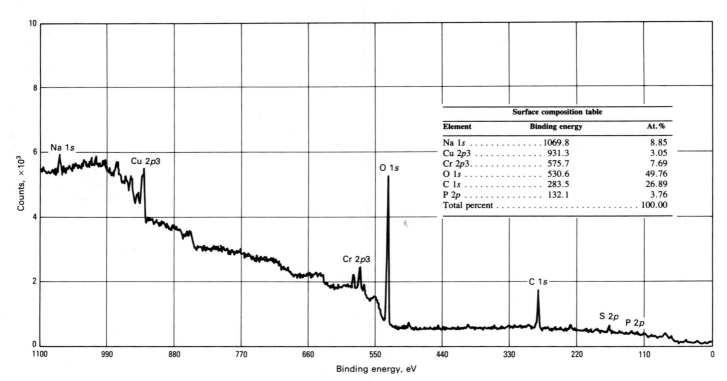

Surface composition table		
Element	Binding energy	At.%
Na 1s	1069.8	8.85
Cu 2p3	931.3	3.05
Cr 2p3	575.7	7.69
O 1s	530.6	49.76
C 1s	283.5	26.89
P 2p	132.1	3.76
Total percent		100.00

(b)

Fig. 55 X-ray photoelectron spectrum obtained during analysis of an integrated circuit leadframe

Spectrum obtained on the as-received surface of the exposed copper underplating on the leadframe.
(b) Spectrum obtained on the surface of the exposed copper underplating on the leadframe after a 10-min argon ion etch. (c) Spectrum obtained on the as-received surface on the underside of the peeled solder plating.
(d) Spectrum obtained on the underside of the peeled solder plating after a 10-min argon ion etch. (A) represents Auger spectrum.

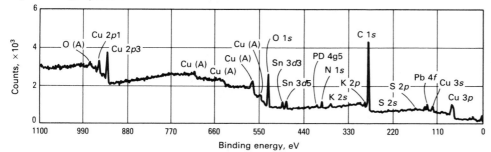

(a)

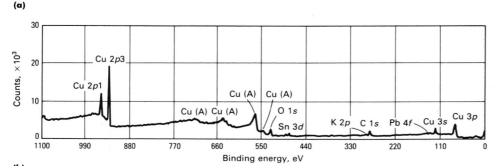

(b)

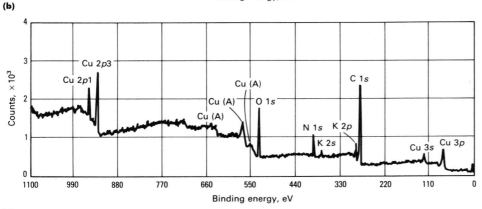

(c)

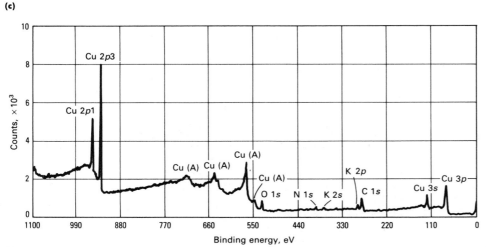

(d)

than those observed on the mating copper underplating surface. Analysis by XPS at a depth of approximately 20 nm below the surface indicated an increase in the amount of metallic copper and a decrease in carbon, oxygen, nitrogen, and potassium. Further ion etching for 30 min revealed only metallic copper. No lead or tin from the solder was observed on the underside of the peeled solder plating at the failed surface, indicating that a copper-rich surface layer from the copper underplating had separated with the 90Sn-10Pb.

The copper underplating was performed using a copper cyanide bath containing potassium cyanide, carbonate and hydroxide, Rochelle salt (potassium-sodium tartarate), and proprietary organic brighteners. Therefore, the presence of potassium, carbon, nitrogen, and oxygen at the failed interface can be explained as an accumulation of plating salts. The solder plating was fused at elevated temperature, which caused countercurrent diffusion of lead and tin into the copper plating and diffusion of copper into the 90Sn-10Pb solder plating.

Conclusions. The 90Sn-10Pb solder plating peeled from the copper underplated leadframe. Excessive accumulation of plating salts and organic brighteners, as well as associated oxidation of copper on the surface of the copper underplating by these agents (by moisture absorption), caused the plating peeling problem.

Recommendations. Excessive buildup of plating salts and organic brighteners in the copper plating bath should be eliminated, and the leadframe surface should be properly cleaned after the copper underplating.

Example 10: Use of SIMS to Study Hydrogen Embrittlement of Gold-Palladium Metallization Layer in Lead Attachments of Resistor Networks. Solder-plated copper leads were attached to a thick-film gold-palladium conductor layer on an alumina substrate in a resistor network with a 63Sn-37Pb solder. Lifting of the leads with the solder and gold-palladium metallization were observed at the metallization/substrate interface after the resistor network had been treated in a hydrogen atmosphere for 48 h at 1205 kPa (175 psig) and at 85 °C (185 °F). When the leads from resistor networks that were not exposed to this hydrogen treatment were pulled, fractures were observed at the expected location, that is, in the solder.

Investigation and Results. Scanning electron microscopy of the separated interfaces (underside of the lead joints) revealed gold-palladium metallization on the hydrogen-treated samples, whereas the untreated control samples revealed a ductile overload fracture in the 63Sn-37Pb solder.

Metallography and EPMA of the cross-section mounts did not reveal any differences between the hydrogen-exposed and the unexposed control samples. Therefore, the underside of the pulled leads were examined with SIMS to detect the differences in composition of the failed interfaces between the hydrogen-treated and the control samples. The SIMS

Fig. 56 Depth profile of SIMS spectra for positive ions obtained on the underside of fractured lead

(a) Control sample. (b) Hydrogen-treated sample

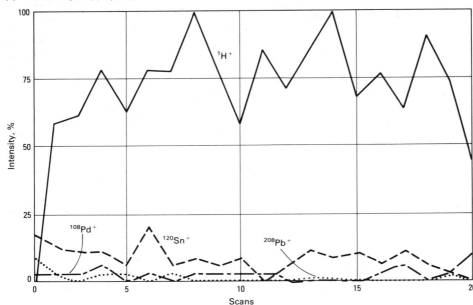

(a)

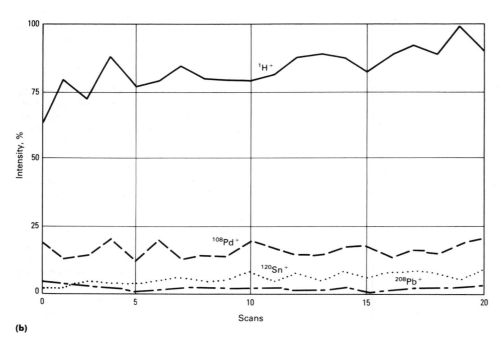

(b)

positive ion depth profiles for $^1H^+$, $^{108}Pd^+$, $^{120}Sn^+$, and $^{208}Pb^+$ were plotted for 20 scans (2 min/scan) at an etch rate of approximately 10 nm/min set on an SiO_2 standard (Fig. 56). The hydrogen concentration was found to be approximately eight to ten times higher for the hydrogen-treated sample as compared to the control sample (comparing the full scales of Fig. 56a and b). In addition, palladium was

associated with hydrogen in the hydrogen-treated sample.

It should be noted that the solubility of hydrogen in palladium at 1205 kPa (175 psig) and at 85 °C (185 °F) was in the 25 to 30% range. The hydrogen atoms dissolved interstitially in the palladium lattice, causing volume expansion, phase change, and distortion, without loss of the metallic character.

Conclusions. The resistor network exposed to the hydrogen atmosphere dissolved a large amount of hydrogen in the gold-palladium metallization layer, causing volume expansion, distortion, and eventual lifting of the lead. The higher palladium content on the hydrogen-embrittled fracture was caused by the fracture propagating through the palladium-rich embrittled phase.

Recommendations. Resistor networks or other devices containing more than 25 ppm Pd in the metallization layer should not be used in a hydrogen atmosphere. Other metallization systems containing silver, gold, and platinum are recommended for use in a hydrogen atmosphere.

REFERENCES

1. *Fractography in Failure Analysis*, STP 645, ASTM, Philadelphia, 1978
2. *Fractography—Microscopic Cracking Process*, STP 600, ASTM, Philadelphia, 1976
3. G. Henry and D. Horstmann, *Fractography and Microfractography*, Vol 5, Verlag Stahleisen, 1979
4. G.G. Garrett and D.L. Marriott, Ed., *Engineering Applications of Fracture Analysis*, Pergamon Press, 1979
5. A. Phillips, V. Kerlins, R.A. Rawe, and B.V. Whiteson, Ed., *Electron Fractography Handbook*, Metals and Ceramics Information Center, Battelle Columbus Laboratories, Columbus, OH, June 1976
6. R. Henderson, A. Phillips, V. Kerlins, G.F. Pittinato, and M.A. Russo, Ed., *SEM/TEM Fractography Handbook*, Metals and Ceramics Information Center, Battelle Columbus Laboratories, Columbus, OH, Dec 1975
7. L. Engel and H. Klingele, *An Atlas of Metal Damage*, Wolfe Publishing Ltd. in association with Carl Hanser Verlag, 1981
8. S. Bhattacharyya, V.E. Johnson, S. Agarwal, and M.A.H. Jones, Ed., *Fracture Handbook—Failure Analysis of Metallic Materials by Scanning Electron Microscopy*, Illinois Institute of Technology Research Institute, Chicago, 1979
9. D.A. Ryder, and A. Smale, *Fracture of Solids*, Interscience, 1963
10. P.J. Forsyth and D.A. Ryder, *Metallurgia*, Vol 63, 1961, p 117
11. G.F. Vander Voort, Conducting the Failure Examination, *Met. Eng. Quart.*, May 1975
12. J. Patterson, L. Kashar, and M. Smith, Reliability of SEM/EDX Analysis of P in SiO_2, in *Proceedings from the International Symposium for Testing and Failure Analysis*, Oct 1982
13. T. Bell, Ferritic Nitrocarburizing, *Met. Eng. Quart.*, May 1976; reprinted in *Source Book on Nitriding*, American Society for Metals, 1977, p 266-282

Failure Analysis and Fracture Mechanics

David Broek, FractuREsearch Inc.

FRACTURE MECHANICS has developed into a useful tool in the design of crack-tolerant structures and in fracture control; it also has a place in failure analysis. Fracture mechanics can provide helpful quantitative information on the circumstances that led to the failure, and it can be used to substantiate preventive measures to avoid the recurrence of failures in similar components. This article will illustrate the role fracture mechanics can play in failure analysis. The concepts of fracture mechanics will be explained briefly. More detailed discussions of these concepts are provided in Ref 1 to 3 and in the article "Fracture Mechanics" in Volume 8 of the 9th Edition of *Metals Handbook*.

Fracture mechanics is the mathematical analysis of the mechanical processes that lead to fracture failure. It includes the process of crack growth by such mechanisms as fatigue and stress corrosion as well as the final fracture event by cleavage or dimple rupture. The analysis is based on established procedures used generally in solid mechanics, for example, the theory of elasticity. The analysis concepts employ the stress and strain fields in a cracked body and the changes in strain energy that take place during cracking and fracture.

Fracture mechanics analysis based on the theory of elasticity is called linear elastic fracture mechanics (LEFM). Linear elastic fracture mechanics can be used as long as there is only a limited amount of plastic deformation occurring at or near the crack tip. When plastic deformation is more excessive, the errors due to the assumption of elastic behavior become too large to be acceptable. In this case, elastic-plastic fracture mechanics (EPFM) is used. In actuality, EPFM is still based on elasticity concepts, but it assumes that the stress-strain curve of the material is nonlinear. Thus, a better term is nonlinear elastic fracture mechanics (NLEFM).

Fracture mechanics analysis dealing with cracking (as opposed to fracture) is sometimes referred to as subcritical fracture mechanics (SCFM), the final fracture event by cleavage or dimple rupture being considered the critical condition. In general, the amount of plastic deformation involved in cracking is very small so that crack-growth analysis uses LEFM concepts. An exception is creep cracking, for which EPFM concepts must be used.

Fracture Mechanics Concepts

Figure 1 defines three modes of loading. Fracture mechanics concepts are essentially the same for each of the modes if the modes occur separately. Combined-mode loading is more difficult to deal with, but if the modes are in phase (and remain proportional), the crack in a very early stage of development will turn into a direction in which it experiences only Mode I, unless it is prevented from doing so due to geometrical confinement. (When the modes are independent and/or out of phase, combined-mode loading presents a problem for which no accepted solution is available.) In view of this, the great majority of all practical cracking and fracture cases are Mode I problems. For this reason, fracture mechanics is generally confined to Mode I.

Linear Elastic Fracture Mechanics

Using the conventional theory of elasticity, it is possible to calculate the stress field at the tip of a crack in an arbitrary body with an arbitrary crack under arbitrary Mode I loading. Using the coordinate system shown in Fig. 2, the crack-tip stresses are:

$$\sigma_{ij} = \frac{K_I}{\sqrt{\pi 2r}} f_{ij}(\theta) + C_1 r^0 + C_2 r^{1/2} + \ldots \quad \text{(Eq 1)}$$

If r is very small, the first term of the solution is very large (infinite for $r = 0$); therefore, the other terms can be neglected. Because all cracking and fracturing take place at or very near the crack tip (where $r \geq 0$), it is justifiable to use only the first term of the solution to describe the stress field in the area of interest. Because the subscripts i and j can stand for either x and y, all stresses are described for any r and θ. For the stress in the y-direction along the plane $\theta = 0$, the function $f_{yy}(\theta) = 1$, so that:

$$\sigma_{yy} = \frac{K_I}{\sqrt{2\pi r}} \quad \text{for } \theta = 0 \quad \text{(Eq 2)}$$

where the subscript I denotes Mode I loading. The following arguments will be based on the stress σ_{yy}, but these do not change if the entire stress field is taken into account.

Equation 2 shows that the crack-tip stress depends on the distance r from the crack tip. By prescribing any r, the stresses can be obtained from Eq 2. The stress also depends on K_I.

Fig. 2 Arbitrary body and coordinate system

See text for explanation of symbols.

Fig. 1 Modes of loading

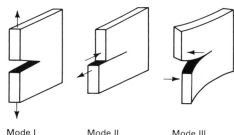

Mode I
Opening mode, tensile mode

Mode II
Sliding mode, shear mode

Mode III
Tearing mode

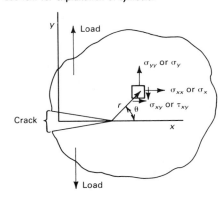

However, K_I remains undefined, because the solution is for an arbitrary case that is undefined by itself. The importance of Eq 2 is that it is possible to obtain a solution for a case in which nothing is defined. This means that Eq 2 is a general solution that applies to all cracked bodies regardless of the loading (as long as it is Mode I) and the geometry and crack size.

All the effects of loading, geometry, and crack size are incorporated into one parameter, K_I. For different geometries, crack sizes, and loading, K_I will have a different value, but apart from the value of K, the stress field will be the same. Equation 2 is the general solution for all crack problems. The parameter K governs the field, and K is the only significant parameter for all crack problems (K is called the stress-intensity factor). Once K is known, the near-crack-tip stress field is known in its entirety.

As the general solution, Eq 2 may be applied to the problem of Fig. 3, an infinite plate subjected to uniform tension, σ, with a crack length of $2a$. The crack shown in Fig. 3 is defined as $2a$ because of the following convention: any crack with two tips is defined as $2a$, and any crack with one tip is defined as a (Fig. 4). There is no objection to defining the crack in Fig. 3 as a, but all quantitative expressions will then differ by a factor of $\sqrt{2}$.

As long as the stresses are elastic, the stress at any point is proportional to the applied stress. This is the case for the crack-tip stress in Fig. 3; therefore, $\sigma_{yy} \propto \sigma$. It must be expected that the crack-tip stress will also depend on the crack size. Because σ_{yy} depends on $1/\sqrt{r}$ according to Eq 2, it is inevitable that it depends on $\sqrt{a}$; otherwise, the dimensions would be wrong. Hence, a simple argument shows that:

$$\sigma_{yy} \propto \frac{\sqrt{a}}{\sqrt{2\pi r}} \qquad \text{(Eq 3)}$$

Dimensional analysis does not provide for any dimensionless constants that may enter the problem. Thus, Eq 3 must use a proportionality sign instead of an equals sign. However, by including a dimensionless constant b, the equation can be expressed with an equals sign as:

$$\sigma_{yy} = \frac{b\sigma\sqrt{a}}{\sqrt{2\pi r}} = \frac{\sigma\sqrt{\pi a}}{\sqrt{2\pi r}} \quad \text{for } b = \sqrt{\pi} \qquad \text{(Eq 4)}$$

A formal algebraic analysis is necessary to obtain the value of b, which in this case equals $\sqrt{\pi}$, as indicated in Eq 4. A comparison of Eq 2 and 4 shows that the stress intensity, K, apparently equals:

$$K_I = \sigma\sqrt{\pi a} \quad (W \text{ large}) \qquad \text{(Eq 5)}$$

Consider the case of Fig. 3 for a plate of finite width W. The above arguments still hold, and it follows that the crack-tip stress must be:

Fig. 3 A center crack in a plate subjected to uniform tension

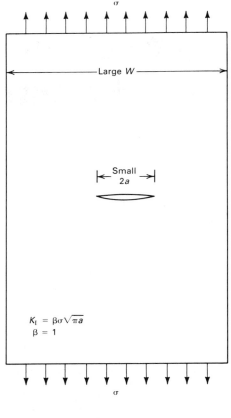

$$\sigma_{yy} = \frac{\beta\sigma\sqrt{\pi a}}{\sqrt{2\pi r}} \qquad \text{(Eq 6)}$$

where β is a dimensionless constant. The crack-tip stresses will be higher when W is smaller. Thus, β must depend on W. It is known that β must be dimensionless, yet β cannot be dimensionless and depend on W at the same time, unless β depends on W/a or a/W, that is, $\beta = f(a/W)$. Comparison of Eq 2 and 6, then, shows that:

$$K_I = \beta\left(\frac{a}{W}\right)\sigma\sqrt{\pi a} \qquad \text{(Eq 7)}$$

If the crack-tip stress is affected by other geometric parameters—for example, if a crack emanates from a hole, the crack-tip stress will depend on the size of the hole—the only effect on the stress-intensity factor (and the crack-tip stresses) will be in β. Consequently, β will be a function of all geometric factors affecting the crack-tip stress: $\beta = \beta(a/W, a/L, a/D)$. Crack-tip stresses are always given by Eq 2; the value of K in Eq 2 is always given by Eq 7. All effects of geometry are reflected in one geometry parameter β. This geometric parameter has been calculated for many generic geometries, and the results have been compiled in various handbooks (Ref 4-7).

Fig. 4 An edge crack in a plate subjected to uniform tension

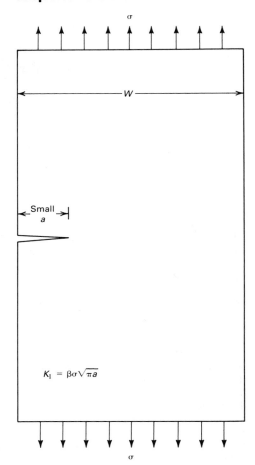

Fracture occurs when the crack-tip stress field exceeds a critical state. According to Eq 2, for the stress field to exceed a critical state, K_I must exceed a critical value. Let this critical value be denoted as K_c. Then, the condition for fracture is:

$$K_I = K_c \qquad \text{(Eq 8)}$$

The critical value must be obtained from a test. Assume that a test is performed on a panel of the configuration of Fig. 3 with very large W and a crack of $2a = 0.1$ m (4 in.) where $a = 0.05$ m (2 in.). The stress at fracture is measured at 137.8 MPa (20 ksi). Comparing Eq 5 and 7, it must be concluded that for this situation with large W the value of $\beta = 1$. Then, it follows that at the time of fracture the stress intensity was $K = 137.8\sqrt{\pi 0.0508} = 55$ MPa$\sqrt{m}$ ($20\sqrt{\pi \times 2} = 50.1$ ksi $\sqrt{in.}$). At a value of $K = 55$ MPa$\sqrt{m}$ (50.1 ksi$\sqrt{in.}$) in Eq 2, the stress field at the crack tip was more intense than the material could bear. The critical value for this material is then $K_c = 55$ MPa$\sqrt{m}$ (50.1 ksi$\sqrt{in.}$). This critical value, K_c, is called the toughness of the material.

Any time the value of K equals K_c in any structural component made of this material, the crack-tip stress field will be the same as at the time of fracture in the test panel: All structures made of this material will show fracture if $K = 55$ MPa$\sqrt{m}$ (50.1 ksi$\sqrt{in.}$). This means that fracture in all structures of this material can be predicted. Because $K = \beta\sigma\sqrt{\pi a}$, fracture will occur when:

$$\beta\sigma\sqrt{\pi a} = K_c \qquad \text{(Eq 9)}$$

If β is known for the geometry at hand, for example, from handbooks, the fracture stress of any structure with a crack of any size can be predicted. For example, assume that a crack of $2a = 0.1524$ m (6 in.) exists in a large panel of the configuration of Fig. 3. As shown, $\beta = 1$ for this case. Fracture occurs in accordance with Eq 9. With the values given, the fracture condition is $\sigma\sqrt{\pi \times 0.0762} = 55$ MPa$\sqrt{m}$ ($\sigma\sqrt{\pi \times 3} = 50.1$ ksi$\sqrt{in.}$); that is, at a stress of $\sigma = 55/\sqrt{\pi \times 0.0762} = 112.4$ MPa ($50.1/\sqrt{\pi \times 3} = 16.33$ ksi), the panel will fail.

The stress intensity K_I represents the mechanical side of the equation; the toughness, K_c, is the material's side. Fracture occurs when $K_I = K_c$. This is analogous to the statement that yielding will occur when the stress, σ, equals the yield stress, σ_{YS}, where σ is the mechanical side of the equation, and σ_{YS} the material's side. The condition for yield in its simplest form is $\sigma = \sigma_{YS}$. The similarity is further illustrated in Fig. 5.

According to Eq 2, the crack-tip stress is infinite for $r = 0$ regardless of the value of K. This is a direct consequence of the use of the theory of elasticity, which states that σ is linearly proportional to strain ϵ without any limitations and which permits the stress to become infinite. In reality, a material will exhibit plastic deformation that limits the stress. It can be shown that the size of the so-called plastic zone at the crack tip (Fig. 6) is determined by the value of K (Ref 1). Consequently, for equal values of K, the plastic zones are equal, and the stresses and

Fig. 6 Crack-tip stress distribution
(a) Elastic. (b) Elastic-plastic

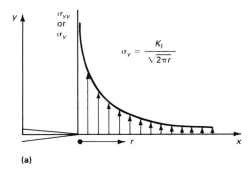

(a)

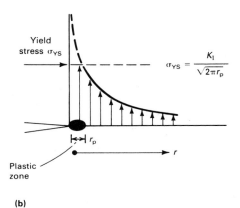

(b)

strains acting on the boundary of the plastic zone are equal. This being the case, equal behavior will take place inside the plastic zones (similitude for equal K). Thus, if the material exhibits a fracture for a certain value of K, it will always exhibit fracture at that value of K.

The arguments presented above remain valid even if some plastic deformation occurs. However, one condition must be satisfied: the plastic zone must be so small that its size is determined fully by K and by K only. This will be the case if the plastic zone does not extend beyond a value of r, at which the first term in

Eq 1 is still much larger than all other terms; otherwise, the constants C_1, C_2, and so on, in Eq 1 will become significant. In general, this will not occur until the value of K_I/σ_{YS} exceeds about 2 (this also depends on geometry). If it does occur, LEFM is no longer valid, and EPFM must be used, although the use of LEFM can be stretched further with simple approximations (Ref 1).

Elastic-Plastic Fracture Mechanics

Most stress analysis problems can be solved using the law of conservation of energy, and the fracture problem is not an exception. At the time of fracture, three energy terms are involved: the strain energy contained in the body, the work done by the external load, and the energy to produce a fracture. The energy to produce the fracture (the actual breaking of material) must be delivered. If the applied load does no work, the only source of energy is the strain energy. If the load does do work, the source of energy is the difference between the work done by the load and the change in strain energy. Because the energy resource can be shown to be equal in the two cases, only one of these cases need be considered (Ref 1).

The strain energy in a plate with a crack of size a equals the area under the load-displacement curve (Fig. 7). If the crack is larger, that is, $a + da$, the stiffness of the body is lower, the load-displacement curve will have a lesser slope, and the strain energy will therefore be less for the same displacement.

It follows that if the crack would grow from a to $a + da$ under constant-end displacement of the body, a certain amount of strain energy would be released equal to the area of the small triangle 0AB shown in Fig. 7(d). If the displacement is constant, the load does no work. Denoting the strain energy by S, the strain energy released in the process is dS/da. It can be shown that dS/da equals K_I^2/E, where E is Young's modulus (Ref 1).

If the crack would grow from a to $a + da$ under constant load, the load would do work, and the strain energy would increase. The difference would be the energy available for fracture. Because this difference is again equal to K_I^2/E, the so-called energy-release rate, dU/da, is always equal to K_I^2/E, regardless of the circumstances, if the crack extends.

Crack extension by fracture over da will occur when sufficient energy is released to deliver the energy for the breaking of the material over a distance da. Let W be the fracture energy and dW/da the energy for growth over da. Then, the condition for fracture is:

$$\frac{dU}{da} = \frac{dW}{da} \qquad \text{(Eq 10)}$$

Usually, dU/da and dW/da are given a different notation, and the condition is written as $G = R$, where $G = dU/da$ and $R = dW/da$.

Fig. 5 Fracture analysis sequence

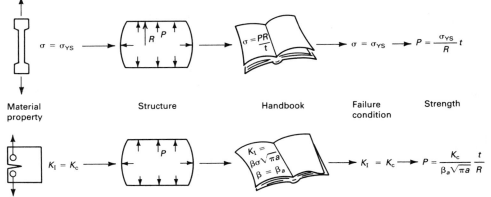

Material property   Structure   Handbook   Failure condition   Strength

Fig. 7 Elastic energy release

(a) Plate with crack a under load P; displacement δ. (b) Load-displacement record. (c) P-δ record for small and larger crack. (d) Crack extension at constant displacement

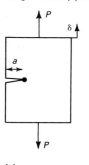

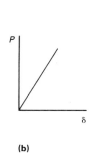

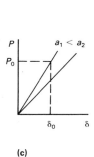

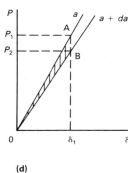

(a)   (b)   (c)   (d)

As discussed above, dU/da equals K_I^2/E, so that Eq 10 becomes:

$$\frac{K_I^2}{E} = R$$

$$K_I = \sqrt{ER} \qquad \text{(Eq 11)}$$

where

$$\frac{\beta^2 \sigma^2 \pi a}{E} = R \qquad \text{(Eq 12)}$$

The condition for fracture is that K must exceed a critical value $\sqrt{ER}$. This condition is the same as obtained previously on the basis of crack-tip stresses. The critical value of K is K_c; therefore, it follows that $\sqrt{ER}$ is nothing but the toughness, and the fracture energy R is then K_c^2/E. The energy criterion and the stress criterion provide the same answer, or one of the criteria would be incorrect.

The energy criterion is used as the fracture condition in EPFM. The criterion of Eq 10 should be valid independent of whether the material has a linear or a nonlinear stress-strain curve. It is somewhat confusing that in EPFM, due to historical developments, the energy-release rate is denoted as J and the fracture work J_R, while in LEFM G and R are used. The condition for crack extension by fracture is then written as $J = J_R$.

In LEFM, an expression for dU/da is obtained as K_I^2/E. An expression for dU/da in EPFM can be derived using an equation to describe the nonlinear stress-strain curve. A curve-fitting equation known as the Ramberg-Osgood equation is often used:

$$\epsilon = \frac{\sigma}{E} + \frac{\sigma^n}{F} = \epsilon_{\text{elastic}} + \epsilon_{\text{plastic}} \qquad \text{(Eq 13)}$$

where the first term is Hooke's law, and the second term describes the plastic strain. It is known that for the elastic term the energy-release rate is K_I^2/E. Therefore, only the contribution of the plastic term needs further evaluation:

$$\epsilon_{\text{plastic}} = \frac{\sigma^n}{F} \qquad \text{(Eq 14)}$$

Equation 14 assumes that a plot of log σ versus log $\epsilon_{\text{plastic}}$ is a straight line, which is approximately the case for many materials. One obtains n (strain-hardening exponent) and F as the slope and intercept of this line.

It is somewhat unfortunate and certainly confusing that Eq 14 was modified by the introduction of an arbitrary number. This was done as follows: one defines an arbitrary reference stress as σ_0 and a reference strain ϵ_0, which must be related by $\epsilon_0 = \sigma_0/E$. Dividing both sides of Eq 14 by ϵ_0 and multiplying the right side by $1 = (\sigma_0/\sigma_0)^n$ yields:

$$\frac{\epsilon}{\epsilon_0} = \frac{1}{\epsilon_0}\frac{1}{F}\left(\frac{\sigma_0}{\sigma_0}\right)^n \sigma^n = \frac{\sigma_0^n}{\epsilon_0 F}\left(\frac{\sigma}{\sigma_0}\right)^n \qquad \text{(Eq 15)}$$

or

$$\frac{\epsilon}{\epsilon_0} = \alpha\left(\frac{\sigma}{\sigma_0}\right)^n \quad \text{with } \alpha = \frac{\sigma_0^n}{\epsilon_0 F} \qquad \text{(Eq 16)}$$

Note that σ_0 is arbitrary, and as long as $\epsilon_0 = \sigma_0/E$, any value of σ_0 can be used with α given by Eq 16. The reference stress σ_0 is called the flow stress, which is misleading because σ_0 has no physical significance as it is arbitrary.

Using Eq 16, the following expression for dU/da (or J) can be obtained:

$$J_{\text{plastic}} = \frac{dU}{da}_{\text{plastic}} = \alpha\epsilon_0\sigma_0 ch\left(\frac{P}{P_0}\right)^{n+1} \qquad \text{(Eq 17)}$$

where c is the uncracked ligament, P is the load, and P_0 is the so-called limit load for the (hypothetical) case that σ_0 is the limit stress. The function h is a dimensionless function, depending on geometry and n. Equation 17 can be simplified by recognizing that:

$$P = k \times \sigma$$

$$P_0 = g \times \sigma_0$$

and

$$c = \left(\frac{c}{a}\right)a = f \times a$$

Substitution in Eq 17 provides:

$$\frac{dU}{da}_{\text{plastic}} = \alpha\epsilon_0\sigma_0 fha\left(\frac{k}{g}\right)^{n+1}\left(\frac{\sigma}{\sigma_0}\right)^{n+1} \qquad \text{(Eq 18)}$$

where k/g and f are dimensionless functions of geometry, and h is a dimensionless function of geometry and n. All these functions can be combined into one function, H, of geometry and n so that:

$$\frac{dU}{da}_{\text{plastic}} = \alpha\epsilon_0\frac{\sigma_0}{\sigma_0^{n+1}}H\sigma^{n+1}a \qquad \text{(Eq 19)}$$

Substitution of α from Eq 16 yields:

$$\frac{dU}{da}_{\text{plastic}} = \frac{H\sigma^{n+1}a}{F} \qquad \text{(Eq 20)}$$

which is the contribution of the plastic part of the stress-strain curve to dU/da. The contribution of the elastic part (see Eq 10 and 12) was:

$$\frac{dU}{da}_{\text{plastic}} = \frac{\pi\beta\sigma^2 a}{E} \qquad \text{(Eq 21)}$$

Clearly, Eq 20 and 21 have a similar form. For an elastic material, $n = 1$, and $F = E$, substitution of which in Eq 20 immediately leads to Eq 21.

The expression for the plastic part of dU/da can be evaluated if H is known; the elastic part can be evaluated if β is known. Handbooks for β for different geometries exist (Ref 4-6), and handbooks for H for different geometries and different n are being developed (Ref 7, 8). Instead of using Eq 20 and providing H, these handbooks are based on Eq 17 and provide h, k, and g. This is cumbersome and suggests that the limit load plays a role. It is clear, however, from the above derivation that the limit load will be divided out. In fact, one can easily rearrange Eq 21 for the elastic part in the same manner and give the false impression that the elastic part depends on the limit load.

From a physical point of view, the true limit load, if one exists, will indeed be the highest possible fracture load. However, use of Eq 13 does not predict such a dependence, and the artificial introduction of a limit load does not change this fact.

The total energy-release rate dU/da equals $G + J$. Hence, the condition for the beginning of fracture becomes:

$$G + J_{\text{plastic}} = \frac{dU}{da}$$

or

$$G + J_{\text{plastic}} = J_R \qquad \text{(Eq 22)}$$

so that the fracture stress follows from:

$$\frac{\pi \beta^2 \sigma^2 a}{E} + \frac{H \sigma^{n+1} a}{F} = J_R \qquad \text{(Eq 23)}$$

If H and β are known, the stress for the commencement of fracture can be calculated from Eq 23 for any crack size a. The solution is obtained by iteration.

Due to the manner in which J is measured, the fracture energy often appears to increase when the fracture is in progress. For low-toughness materials, this increase is generally small and can then be ignored (usually in LEFM). For high-toughness materials, the increase can be considerable. In this case, the fracture may first be stable, and a further increase in load is necessary to cause an uncontrollable (unstable) fracture (Ref 1, 3, 7). Because of historical development, J is usually referred to as the J-integral, and it can be expressed in the form of an integral (Ref 9), which is especially useful for determining the value of J in finite element analysis, and from this the value of H or h.

In EPFM, the toughness is J_R. Because it is expressed in terms of the energy criterion, it has different units from the toughness K_c in LEFM. If the LEFM toughness is expressed in terms of an energy criterion, it equals K_c^2/E so that the dimension will be energy per unit area, such as kJ/m², N · m/mm², or in. · lb/in.² Clearly, the toughness J_R has the same dimension.

Plane Stress and Plane Strain

The stresses at the crack tip are very high (Fig. 6). The strains are also high. High strains in the x- and y-direction will be associated with a contraction in the z-direction (Fig. 8). Because the crack itself is stress free, no contraction occurs behind the crack front. Further away from the crack, the stresses are low; therefore, the contraction is small. Thus, there is a thin cylindrical-like zone of material at the crack tip wanting to contract a great deal, while the surrounding material does not need to contract. If this zone is long and thin, its contraction will be prevented because it is attached to the surrounding material. The surrounding material will constrain the contraction by exerting a tension stress upon the cylindrical zone in the z-direction so as to keep $\epsilon_0 = 0$. The required σ_z follows from Hooke's law:

$$\epsilon_z = \frac{\sigma_z}{E} - \nu \frac{\sigma_x + \sigma_y}{E} = 0$$

or

$$\sigma_z = \nu(\sigma_x + \sigma_y) \qquad \text{(Eq 24)}$$

where ν is Poisson's ratio. If plasticity occurs, the value of ν approaches 0.5.

If the plate is thin, the cylindrical zone will be very short (Fig. 9), and contraction will take place unhindered, in which case:

$$\epsilon_z = -\nu \frac{\sigma_x + \sigma_y}{E} \text{ and } \sigma_z = 0 \qquad \text{(Eq 25)}$$

The condition of Eq 24 is known as plane strain, that of Eq 25 as plane stress.

In plane strain, the crack-tip conditions are more severe, and fracture will occur at lower values of K_I; the toughness is lower. In this case, the toughness is denoted by K_{Ic}, the plane-strain fracture toughness. If constraint is small so that there is either plane stress or an intermediate state of stress, the toughness is higher. The toughness is then denoted by K_c.

The amount of constraint depends on the length-to-diameter ratio of the cylindrical zone of material at the crack tip that needs to undergo large contractions. If the zone is long, there is plane strain; if the zone is short, there is plane stress. For the through-thickness crack shown in Fig. 8, the length of the zone equals the thickness. Therefore, for this situation, the thickness determines whether there is plane strain or plane stress: large thicknesses will give plane strain, and small thicknesses plane stress. Therefore, the toughness will depend on thickness, as shown schematically in Fig. 10. Once there is complete constraint, the situation cannot worsen so that the toughness does not decrease further once it reaches K_{Ic}.

The thickness, B_p, at which K_{Ic} is reached has been determined experimentally (Ref 10) to be of the order of:

$$B_p = 2.5 \left(\frac{K_{Ic}}{\sigma_{YS}} \right)^2 \qquad \text{(Eq 26)}$$

It should be noted again that constraint is determined by the length-to-diameter ratio of the cylindrical zone defined above. For through-thickness cracks, the length of the zone equals the thickness. This is the reason thickness enters into the problem. However, for part-through cracks, such as in Fig. 11, the length of the zone does not depend on the thickness. In the case of part-through cracks (surface flaws and corner cracks), the state of stress is always plane strain (unless the flaw is almost completely through). Hence, in the case of part-through cracks, one should use K_{Ic}, not K_c, regardless of thickness.

The constraint of contraction occurs also in EPFM, although the effect is often ignored. For the case in which $n = 1$, $J = G = K_I^2/E$. Hence, the condition for plane strain would be:

$$B_p = \frac{2.5EJ}{\sigma_{YS}^2} \qquad \text{(Eq 27)}$$

which can be generalized to:

$$B_p = \frac{2.5FJ}{\sigma_{YS}^{n+1}} \qquad \text{(Eq 28)}$$

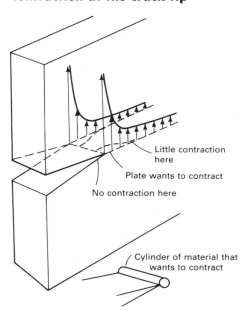

Fig. 8 Schematic showing contraction at the crack tip

Little contraction here

Plate wants to contract

No contraction here

Cylinder of material that wants to contract

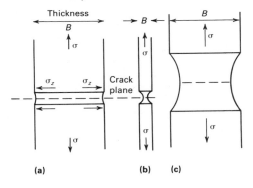

Fig. 9 Comparison of plane strain and plane stress

(a) Low-stress thick plate, thin cylinder no contraction plane strain. (b) Low-stress thin plate, free contraction plane stress. (c) High-stress thick plate, thick cylinder free contraction plane stress

Thickness

B

σ

σ_z σ_z

σ

Crack plane

B

σ

σ

B

σ

σ

(a) (b) (c)

Fig. 10 Dependence of toughness on thickness

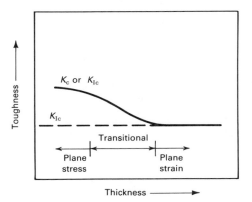

Toughness

K_c or K_{Ic}

K_{Ic}

Transitional

Plane stress

Plane strain

Thickness

Other conditions have been proposed, but they are much less stringent than Eq 28. Until a thorough experimental investigation has been performed, as was done for LEFM (Ref 10), the conditions for plane strain will remain uncertain.

Subcritical Fracture Mechanics

The growth of the crack that eventually leads to fracture occurs by mechanisms entirely different from the fracture itself. During most of the cracking process, the crack is much smaller than the one that would cause fracture at the prevailing stress. Therefore, the crack-tip stress field is less severe, and the size of the plastic zone smaller than at the time of fracture. Due to this small plastic zone, SCFM can often be approached using elastic concepts. If the crack size is close to critical (fracture), this may no longer be true, but because by far the longest time is spent in the growth of much smaller cracks, it is justifiable to use LEFM to obtain the time for crack growth with good accuracy. A notable exception is creep crack growth.

Fatigue. In the case of cyclic loading, crack growth occurs by fatigue. The amount of growth in one cycle depends on the cracktip stress field, which is described by K. During a load cycle, K varies from a minimum $K_{min} = \sigma_{min} \beta\sqrt{\pi a}$ to a maximum $K_{max} = \sigma_{max} \beta\sqrt{\pi a}$, over a range $\Delta K = \Delta\sigma\beta\sqrt{\pi a}$, where σ_{min} is the minimum stress in the cycle, σ_{max} is the maximum stress in the cycle, and $\Delta\sigma$ is the stress range $\Delta\sigma = \sigma_{max} - \sigma_{min}$.

It must be expected that the amount of growth per cycle depends on ΔK and K_{max}. If one defines a stress ratio R as $R = K_{min}/K_{max}$, the above statement is equivalent to the assertion that the amount of growth per cycle depends on ΔK and R. Use of ΔK and R is often more convenient because $R = K_{min}/K_{max} = \sigma_{min} \beta\sqrt{\pi a}/\sigma_{max} \beta\sqrt{\pi a} = \sigma_{min}/\sigma_{max}$. Hence, in the case of constant-amplitude loading where σ_{min} and σ_{max} do not change from cycle to cycle, R remains constant, while K_{max} would depend on crack size.

The amount of growth per cycle is the rate of growth. This means that the above statement is equivalent to:

$$\frac{da}{dN} = \text{dependent on } \Delta K \text{ and } R \qquad \text{(Eq 29)}$$

where N is the number of cycles, and da/dN is the rate of growth.

It cannot be known *a priori* how da/dN depends on ΔK and R; this information must be provided by a test of the material. By measuring the rate of growth in a test and plotting the growth rates as a function of ΔK (calculated from $\Delta K = \beta\Delta\sigma\sqrt{\pi a}$), the dependence on ΔK is obtained, and in a similar fashion, tests at different R stress ratios provide the dependence on R.

Figure 12 shows typical fatigue-crack growth rate test results. Every time the same values of

Fig. 11 Stress intensity of surface flaw
(a) Surface flaw. (b) Φ^2 versus $a/2c$. (c) M_k versus $a/2c$

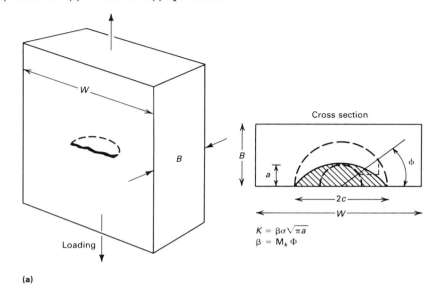

(a)

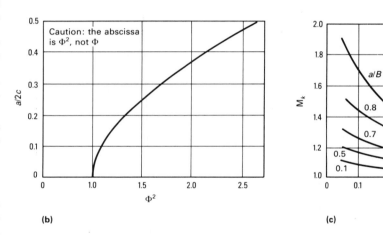

$$K = \beta\sigma\sqrt{\pi a}$$
$$\beta = M_k \Phi$$

(b)

(c)

ΔK and R occur, the material will respond with the same growth rate because the crack-tip stress conditions are the same (i.e., similitude). Thus, the data shown in Fig. 12 can be used to predict crack growth in a structure. Cyclic loading of the structure provides $\Delta K = \beta\Delta\sigma\sqrt{\pi a}$ and R for the structure. If β is known for the structural configuration (handbooks), and the stress is known, the rate of growth follows from the diagram. Crack size as a function of time (number of cycles) is obtained by integration:

$$N = \int_{a_1}^{a_2} \frac{da}{da/dN} \qquad \text{(Eq 30)}$$

The integration is typically performed numerically using a computer.

Integration would be more convenient if the data shown in Fig. 12 could be represented by a simple equation. Figure 12 suggests that the

data for a fixed value of R fall nearly on a straight line on a log-log scale. In this case, the rate data can be represented by the Paris equation (Ref 11):

$$\frac{da}{dN} = C\Delta K^m \qquad \text{(Eq 31)}$$

where C and m are empirical constants.

Unfortunately, the Paris equation covers only one R value. If the lines for different R values are parallel, one could use the modified equation:

$$\frac{da}{dN} = \frac{C_0}{(1 - R)^n} \Delta K^m \qquad \text{(Eq 32)}$$

where n and C_0 are also empirical constants. By noting that $R = K_{min}/K_{max} = (K_{max} - \Delta K)/K_{max}$ so that $K_{max} = \Delta K/(1 - R)$, Eq 32 becomes:

$$\frac{da}{dN} = C_0\frac{\Delta K^n}{(1 - R)^n} \Delta K^{m-n} \qquad \text{(Eq 33)}$$

or (letting $p = m - n$):

$$\frac{da}{dN} = C_0 K^n_{\max}\Delta K^p \qquad \text{(Eq 34)}$$

Equation 34, as first proposed in Ref 12, is now known as the Walker equation. Its advantage over the Paris equation is that it covers all values of R.

Many other equations are used. It should be pointed out, however, that no equation has any physical meaning: all are curve-fitting equations. There is no objection to their use if they fit the data. However, if the numerical integration must be done by a computer, use of the original data in tabular form is as convenient as use of a sometimes poor fitting equation.

Stress-Corrosion Cracking. In the case of stress-corrosion cracking, the rate of growth will depend on K, by the same arguments used for fatigue. In this case, the rate of growth depends directly on time so that the growth rate da/dt is:

$$\frac{da}{dt} = f(k) \qquad \text{(Eq 35)}$$

Again, the rates must be obtained from a test. The results will be shown schematically, as in Fig. 13. Integration of the rate data for the loading and β of a structure will provide the crack size as a function of time.

For combined stress-corrosion and fatigue cracking, one obtains da/dN from a cyclic-loading test in the appropriate environment. Schematically, the results will have the form shown in Fig. 14. Again, a numerical integration will provide the crack-growth curve for the structure regardless of the shape of the da/dN-ΔK data. Complications are due to load interaction and load-environment interaction, as detailed in Ref 1.

In the case of stress corrosion at sustained load, the stress intensity must exceed a certain minimum value for crack growth to occur. This minimum value is known as the stress-corrosion threshold, denoted by K_{Iscc}. The threshold value is best determined by performing tests on cracked specimens and by measuring the time to fracture. A plot of the stress intensity applied as a function of time to failure will show an asymptote (Fig. 15). During each test, the stress intensity will increase due to increasing crack growth. This is not reflected in Fig. 15, because the plot is made for the initial K at the start of the test. If growth does occur, it will continue to occur because K increases, and a failure will result. Conversely, if no failure occurs, no crack growth occurs; therefore, the stress intensity at the asymptote is indeed the threshold. In Fig. 13(b), the thresh-

old could be obtained from a vertical asymptote at the lower end of the curve, but the procedure illustrated in Fig. 15 is more reliable (and more costly).

Tests and Data

Test data for fracture toughness and crack growth can be obtained from any type of

cracked specimen provided β (or H in the case of EPFM) is known for the geometry at hand. If this were not true, the data could not be applied to structural geometries, but the data can be applied on the basis of the similitude of stress fields as provided by K (and indirectly by J).

Standards for toughness testing and crack-growth testing have been established by the American Society for Testing and Materials

Fig. 12 Fatigue-crack growth rate

(a) Crack-growth test. (b) Measured crack-growth curve. (c) Growth-rate curve. (d) Effect of stress ratio, R

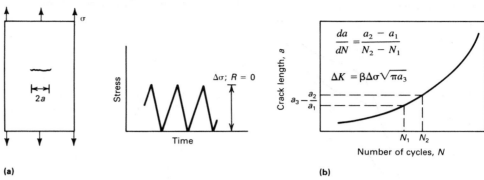

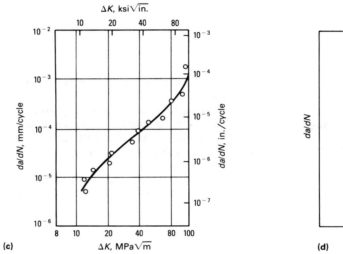

Fig. 13 Stress-corrosion cracking

(a) Crack growth as a function of time. (b) Growth rate as a function of K

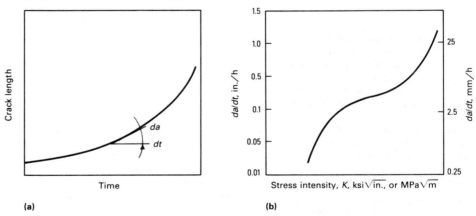

Fig. 14 Crack growth in a pipeline steel in saltwater and in air

cps, cycles per second

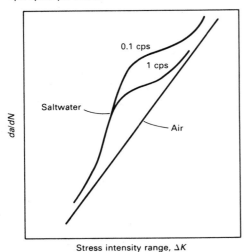

Fig. 15 Stress-corrosion test data

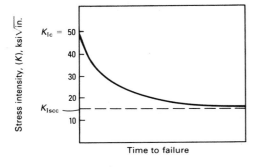

(ASTM). Test standards are discussed in Appendix 2 in this article. Other tests providing information on fracture properties are discussed in Appendix 1, although most of the latter do not provide data that can be used in the application of fracture mechanics concepts as discussed above.

Toughness, whether in terms of J or K, and crack-growth behavior depend strongly on a variety of metallurgical, environmental, and mechanical parameters. The most influential parameters (not in order of importance) include:

- The yield or ultimate strength as affected by heat treatment, cold work, and so on, of an alloy (Fig. 16).
- Alloy composition, intermetallics, and inclusions
- Temperature (Fig. 17)
- Loading rate (Fig. 18)
- Environment (Fig. 12 and 14)
- Thickness (state of stress as discussed) (Fig. 19)

In any application of fracture mechanics, data for the relevant conditions must be available or

Fig. 16 Variation of plane-strain fracture toughness, K_{Ic}, with tensile strength for three steels

Source: Ref 13

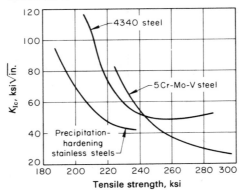

Fig. 17 Comparison of static (K_{Ic}), dynamic (K_{Id}), and dynamic-instrumented (K_{Idi}) impact fracture toughness of precracked specimens of ASTM A533, grade B, steel, as a function of test temperature

The stress-intensity rate, $\dot{K}$, was about 1.098×10^4 MPa$\sqrt{m} \cdot s^{-1}$ (10^4 ksi$\sqrt{in.} \cdot s^{-1}$) for the dynamic tests and about 1.098×10^6 MPa$\sqrt{m} \cdot s^{-1}$ (10^6 ksi$\sqrt{in.} \cdot s^{-1}$) for the dynamic-instrumented tests. Source: Ref 14

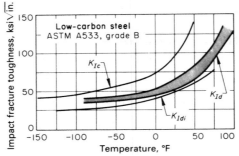

obtained. Because fracture mechanics has been applied in the aerospace industry for a number of years, considerable data on aerospace alloys are available. The enormous cost of data generation (due to the many parameters affecting the properties) cannot be assimilated by one institution or company. Realizing this, the aerospace industry has pooled data, which are compiled in a four-volume data handbook (Ref 17). Data bases for materials used in the nuclear industry are currently being generated under the auspices of the Electric Power Research Institute (EPRI) and the Nuclear Regulatory Commission (NRC). Data for other materials are scarce or nonexistent. Without concerted efforts for the pooling of data, the application of fracture mechanics to various common alloys will be hampered for many years.

Fig. 18 Effect of stress-intensity rate, $\dot{K}$, on plane-strain fracture toughness, K_{Ic}, of two heats of cast 1Cr-4Ni-0.2Mo high-strength steel austenitized at 830 °C (1525 °F), oil quenched, and tempered as noted.

Source: Ref 15

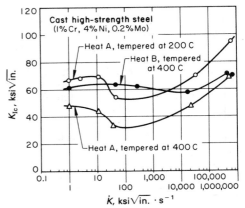

Fig. 19 Effects of section thickness of fracture-toughness specimens of vacuum-melted 18% Ni maraging steel (2070 MPa, or 300 ksi, minimum yield strength) on measured values of critical stress-intensity factor, K_{Ic}

Data were obtained with three types of specimens: center-cracked (pop-in) rectangular, surface-flawed rectangular, and circumferentially notched cylindrical specimens. At thicknesses greater than about 15 mm (0.6 in.), the stress-intensity factor, K_c, drops to the critical plane-strain stress-intensity factor, K_{Ic}. Source: Ref 16

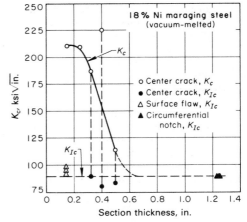

Correlation Between Toughness and Impact Properties. With the scarcity of toughness data for common alloys, the question arises whether impact properties, such as Charpy-V values, can be used to estimate toughness. The Charpy value is in a sense a measure of the fracture energy, and as shown above, K_c^2/E and J_R are also measures of the fracture energy. However, there are important differences between a Charpy test and a toughness test.

First, a Charpy specimen contains a notch, while a toughness specimen contains a crack. This leads to a different state of stress in the two specimens. As discussed, the state of stress has a considerable influence on the fracture condition. This problem could be solved by using cracked instead of notched Charpy specimens, but most available data are for notched Charpy specimens.

Second, a Charpy test measures the total energy required to fracture a specimen. The measured energy includes the plastic energy to "fold" the specimen. At the upper shelf, where fracture often does not occur, this folding energy is the only component measured.

Third, the Charpy test, if it measures a fracture energy provides the energy to fail the entire uncracked ligament, while in the energy concept of fracture mechanics, the fracture energy dW/da is of interest, that is, the energy for a small crack extension da. If W were independent of Δa, then dW/da could still be obtained from the Charpy result, but in general, dW/da depends strongly on Δa. Fourth, because the Charpy test is an impact test, strain rates in the test are higher than those in a normal toughness test.

In view of the above differences, there is no direct physical relationship between a Charpy value and toughness. Nevertheless, one may search for an empirical relationship by simply plotting Charpy data versus available toughness data for the same materials and by curve fitting the plot if it shows any trends. On the basis of such efforts, certain empirical relationships have been established (Ref 18, 19). From a practical point of view, there can be no argument about the usefulness of such empirical relationships provided they are used within the boundaries for which they were established and are interpreted as estimates of toughness only.

Software

Although fracture and crack-growth analysis can be performed manually, most problems are not easily solved in this manner. A computer analysis is usually necessary. Software for all of the above analyses can be developed using the equations given in this article. However, in the foregoing discussions, many complications have been disregarded in view of space limitations. Therefore, software development should not be attempted without consultation of more detailed fracture mechanics texts (Ref 1-3). Software is commercially available. For example, the American Society for Metals markets a software package known as The Fracture Mechanic, which includes:

- Calculation of fatigue-crack growth in structures

- Critical stress or critical crack-size calculation with LEFM
- Critical stress and critical crack-size calculation with EPFM using the J-integral

Accuracy

The results of fracture mechanics analysis often have limited accuracy. Although fracture mechanics can certainly be improved, most of the inaccuracies are due to scatter in the data of the material and inaccuracies in input data for stress, β, and stress history. Yield and tensile stress are bulk properties that are determined as the average properties of all the material (bulk) of a tensile bar, yet scatter is considerable. Fracture and crack-growth properties reflect only the response of a small amount of material that happens to be at the crack tip. As such, the measured properties are local properties and therefore subject to larger scatter. The larger the scatter, the less accurate the analysis will be. Inaccuracies due to scatter cannot be attributed to fracture mechanics. They are troublesome and may adversely affect the usefulness of fracture mechanics, but no improvement in fracture mechanics could change this.

The results of fracture mechanics analysis also depend on how well stresses and cyclic-stress history are known. Because β and H follow from stress analysis, these parameters also contain errors. Lastly, the cyclic-stress history is usually known only as a vague projection. Thus, the stress-intensity factor contains an accumulation of errors. In the case of fatigue, the rate of crack growth depends on the stress intensity to the 4th or 5th power. Then, an error of 10% in K will lead to a difference of $(1.1)^5 = 1.6$, that is, an error of 60% in calculated life. Again, the input, not fracture mechanics, is the source of the problem. Fracture mechanics techniques can certainly be improved, but no improvement will change the inaccuracies discussed above.

Use of Fracture Mechanics

Principles. The strength of a structural component decreases with crack size. The residual strength (rather than the original design strength) at any crack size can be calculated using LEFM or EPFM. Calculations of the residual strength as a function of crack size leads to a curve, as shown schematically in Fig. 20.

Let the maximum service stress be σ_s. The new structure has a strength σ_N, which means there is a safety factor $j = \sigma_N/\sigma_s$. When the strength decreases due to crack growth, the safety factor decreases. If one desires that the safety factor never be less than f, then the minimum permissible residual strength is $\sigma_p = f\sigma_s$. The residual-strength diagram shows that the crack size should then not exceed a_p, where a_p is the maximum permissible crack size.

Subcritical fracture mechanics permits calculation of the crack-propagation curve as in

Fig. 20 Residual strength as a function of crack size

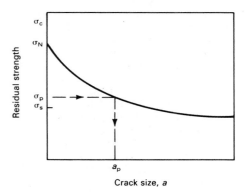

Fig. 21 Subcritical crack-growth curve

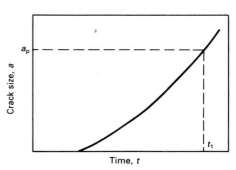

Fig. 21. Given that the maximum permissible crack size is a_p as obtained from the residual-strength diagram, it follows from the crack-propagation curve that the crack must be eliminated before time (cycles) t_1, as shown schematically in Fig. 21. This can be done by retiring the structure (which also eliminates the crack) or by inspecting the structure and repairing the crack.

Fracture control can be exercised only if both the residual-strength diagram and the crack-growth curve are available. If either is missing, nothing worthwhile can be said about the fracture safety of the structure. For example, if a crack is detected, one may be able to show that it can be sustained under the prevailing service loading. However, because the crack may be much longer the next day, it is not possible to keep the structure in operation, unless it is known how long it will take for the crack to grow to a size that cannot be sustained. The crack-growth curve must be known as well.

Use in Failure Analysis. Failure analysis, as discussed in this Volume, will show the mechanism by which the failure took place (fatigue, stress corrosion, hydrogen-assisted growth, creep, or combinations of these), the mechanism by which the final fracture took place (dimple rupture or cleavage), and whether or not material defects (inclusions, intermetallics, porosity, and so on) or design and manufacturing defects (notches, scratches, and so on) were involved.

Quantitative fractography, for example, striation counts, can to some degree provide information on time to failure and the prevailing stresses. With the use of fracture mechanics, more quantitative information can be extracted. Quantitative fractography will be discussed at length in Volume 12 of the 9th Edition of *Metals Handbook*, which follows the current volume in the *Metals Handbook* series.

Final Crack Size. Fractography generally will provide the crack size at which the final fracture took place. Because fracture occurs in accordance with Eq 9 or 23, equations are available for using the fractographic results in various ways.

If the operation stresses are known, equations can be used to calculate the toughness, which, if compared to the anticipated toughness, will show whether or not the material had a toughness higher or lower than expected. If the material specification indicated a toughness value, the calculation will show whether or not the material was in accordance with the specification.

Conversely, if the fracture stress is unknown, the above equations can be used to determine the stress at the time of fracture. In this case, a reasonably accurate (sometimes estimated) toughness value must be used. With the stress at fracture known, it can be determined whether the fracture took place due to an excessively high load or whether in general the operating stress levels were too high.

This information can often be corroborated by measuring the size of the stretched zone from the fractographs, provided stereomicroscopy is used (Fig. 22). The size of the stretched zone is a measure of the crack-tip opening displacement (CTOD) at the time of fracture. It can be shown that the CTOD at fracture is related to the toughness as $CTOD = K_c^2/E\sigma_{YS}$ or $CTOD = J/\sigma_{YS}$; therefore, the toughness can be calculated from the measured stretched zone size. Figure 23 shows that careful measurement can give reliable toughness values. With the toughness known, fracture stresses can be calculated as discussed above.

Crack Growth. Fatigue-striation counts (if possible) generally provide a reasonable account of the crack-growth rates and crack-growth curve. If the crack-growth-rate behavior of the material and the stresses are known, the stress intensity can be calculated, and a comparison can be made between actual and anticipated properties for a conclusion about the adequacy of the material. Conversely, if the stresses are not known, the measured rates and the rate properties can be used to estimate what the acting stresses were. This procedure will at least provide the magnitude of the stresses.

From the amount of crack growth (crack size at fracture), known stresses, and growth-rate properties, a reasonable insight can always be obtained regarding the question of misuse, for example, continuous overloading. The time to failure and final crack size are determined using fracture mechanics, as discussed above. When the results are not in accordance with the

Fig. 23 Correlation between CTOD and toughness

(a) Stretched-zone depth versus CTOD. (b) Stretched-zone width versus CTOD. (c) Stretched-zone width versus depth. Source: Ref 20

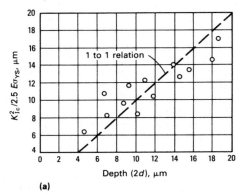

(a)

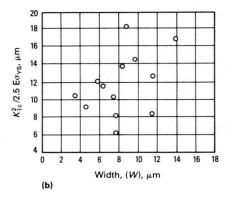

(b)

Fig. 22 Measurement of the stretched zone in a fractured specimen

(a) Crack-tip blunting and the stretched zone. (b) Topography of stretched zone

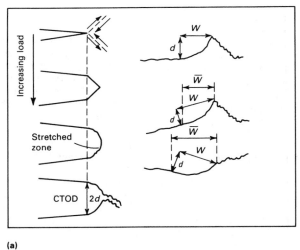

(a)

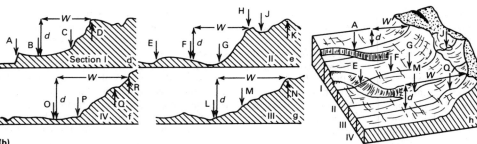

(b)

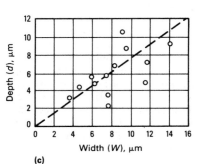

(c)

observations, the analysis can be repeated to determine how much higher (or lower) the stresses would have had to have been to produce the cracking time and extent of cracking as observed.

Any change of loading or change in environment during the cracking process is likely to leave its mark. Such changes produce a change in crack-growth rate, which is usually associated with a change in microfracture topography (surface roughness). Because the roughness affects the reflection of light, a change in roughness will appear as a line (beach mark). Any beach mark is an indication of a change in circumstances during cracking. The beach marks on fatigue-crack surfaces are well known. Similar marks may occur on stress-corrosion-crack surfaces due to changes in loading or environment.

A beach mark is clearly the delineation of the crack front at some point during the cracking process. Thus, the crack size at the time of the change is known. If any information on the nature of possible changes is known, the crack size at which they occurred can be used to obtain information on rate properties or stresses in the manner discussed above. If there is no information about the nature of the changes, such information (when it is a change in loading in particular) can be obtained from known growth-rate properties: the time for growth between beach marks can be calculated, and the stress required to produce the observed crack sizes estimated.

Initial Defects. If fractography shows the presence of initial defects (mechanical or metallurgical), crack-growth analysis can be used, starting from the initial defect. Again, depending on the objective, information can be obtained on stresses (known material properties) or material properties (known stresses). Thus, one can obtain information on whether the design stresses were too high or on whether the material was inadequate.

Residual, Contact, and Thermal Stresses. In general, the stress intensities due to different stress systems are additive as long as all stress systems are of the same mode, for example, Mode I. The stress intensity due to combined bending and tension is the sum of those due to tension and bending. Residual stresses due to welding, cold deformation, surface treatments, or heat treatments give rise to a stress intensity. If this stress intensity is of the same mode as that due to external loads, the two are additive. Although this is very useful in principle, the stress intensity due to residual stresses usually cannot be determined simply because the residual stresses are unknown.

For a fracture problem, it may be assumed that local yielding at the crack tip is so extensive that residual stresses will be annihilated and can be ignored for the fracture analysis. In the case of fatigue-crack growth, the residual stress is static as opposed to the cyclic applied stress. As such, the effect of the residual stress may be interpreted as a simple change in R-ratio. It can be accounted for if the magnitude of the residual stress is known. In the case of sustained-load stress corrosion, the stress intensities are additive. Again, the residual stress could be accounted for if its magnitude is known.

Residual stresses occur only when plastic deformation occurs very locally. Upon release of the elastic strains of the surrounding material, the plastically (permanently) deformed region does not fit in the surroundings. The surrounding elastic material will force it to fit and exert stresses to do so. These stresses are called residual stresses because they occur without the presence of an external load. The surrounding elastic material will essentially return to zero strain, which forces the permanently deformed material nearly back to zero strain. Just as the original plastic strain required yielding, the return to zero strain will require reverse yielding. This leads to the conclusion that residual stresses will always equal the yield stress. However, if the material is at elevated temperature where the yield is lower, the residual stress may be equal to this lower yield stress only.

These guidelines can be used to evaluate the effect of residual stress on K. It should be noted, however, that K is an elastic parameter, and if the residual stress equals the yield, the applicability of K is doubtful. Accounting for residual stresses in EPFM is equally debatable. However, application of this rule can at least lead to an estimate of the effect. Cautious engineering judgment is certainly required. The same arguments hold more or less for thermal stresses and contact stresses. Stress intensities are additive, but the problem is in knowing what to add.

Appendix 1: Notch-Toughness Testing and Evaluation

This Appendix describes and gives references to a variety of tests for determination of notch toughness—both with impact loading and with slow loading.

The Charpy and Izod Tests. The Charpy machine has a total available striking energy of 300 J (220 ft · lb). The test specimen, supported at both ends, is broken by a single blow of the pendulum applied at the middle of the specimen on its unnotched side. The specimen breaks at the notch, the two halves fly away, and the pendulum passes between the two parts of the anvil. Height of fall minus height of rise gives the amount of energy absorption involved in deforming and breaking the specimen. To this is added frictional and other losses amounting to 1.5 or 3 J (1 or 2 ft · lb). The instru-

ment is calibrated to record directly the energy absorbed by the test specimen.

The Izod machine is similar in principle, but the test specimen is held rigidly in a vise as a cantilever beam, with the center of the notch coinciding with the upper face of the jaws. The capacity of the machine is usually 300 J (220 ft · lb). Half the test specimen is broken off, and the impact value is determined as with the Charpy test.

Dimensional details for standard Charpy and Izod specimens are given in ASTM E 23 (Ref 21). The Charpy V-notch specimen is widely used for structural steels, as in the keyhole specimen. The Izod test can record higher values than the Charpy, but is difficult to use for testing at reduced temperature. For testing heat-treated steels, both the V-notch specimen and the keyhole specimen are used, with most users of the test preferring the V-notch. Comparisons of results with Charpy V and keyhole notches for the same steels, tested over a range of temperature, are shown in Fig. 24. More detailed information on the standard Charpy impact test, instrumented Charpy impact tests, and precracked Charpy tests can be found in the article "Dynamic Fracture Testing" in Volume 8 of the 9th Edition of *Metals Handbook*. Related information on interpretation of load-time records (impact-response curves) obtained in instrumented impact tests with precracked Charpy specimens can be found in the same article.

The Schnadt specimen, details of which are shown in Fig. 25, has been used primarily in Europe for testing ship plate (Ref 23). In the Schnadt test, five test pieces are used with different notch radii, ranging from no notch to a severe notch made by pressing a sharp knife into the bottom of a milled groove. A hardened steel pin is inserted in a hole parallel to and behind the notch, replacing the material normally under compression in the Charpy or Izod tests. The specimen is broken by impact as a three-point-loaded beam.

The drop-weight test (DWT) specimen and procedure are shown in Fig. 26(a). The crack inducer is a bead of hard-facing metal about 75 mm (3 in.) long. The specimen, 89 by 356 by 25 mm (3½ by 14 by 1 in.), is placed, weld down, on rounded end supports and is struck by a 27-kg (60-lb) falling weight with sufficient energy to bend the specimen about 5°. A cleavage crack forms in the bead as soon as incipient yield occurs (at about 3° deflection), forming the sharpest possible notch, a cleavage crack in the test specimen. A series of specimens is tested over a range of temperatures to find the nil-ductility transition temperature. This is the temperature below which steel, in the presence of a cleavage crack, will not deform plastically before fracturing, but will fracture at the moment of yielding.

This is a "go, no-go" test in that the specimen will either break or fail to break. Procedures followed and test specimens used in

Fig. 24 Comparison of energy-temperature curves obtained by Charpy V-notch and Charpy keyhole tests of steel specimens
Source: Ref 22

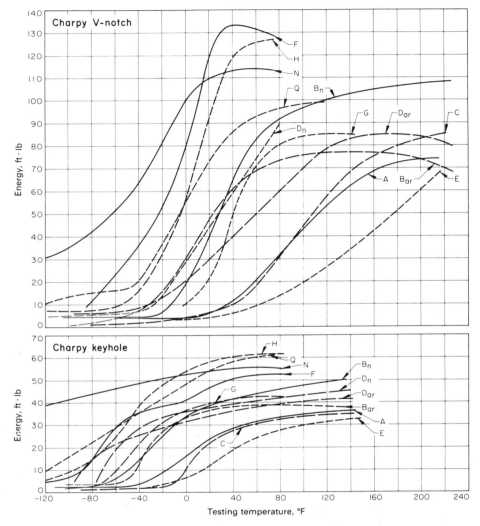

Fig. 25 Details of the Schnadt notched-bar impact-test specimen
Dimensions given in inches. Source: Ref 23

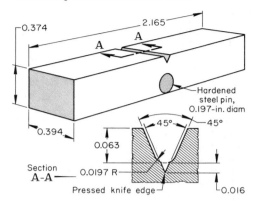

Fig. 26 Drop-weight (a) and explosion-bulge (b) methods of testing

The drop-weight test is used for determining the nil-ductility transition temperature by testing a series of specimens over a range of temperatures. In the explosion-bulge test, steel does not deform before fracturing below this temperature. Dimensions given in inches. See also Fig. 27.

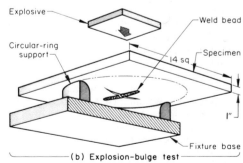

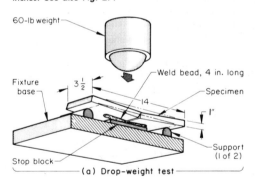

Steel	_____ Composition, % _____											
	C	Mn	Si	P	S	Ni	Al	Cu	Cr	Mo	Sn	N
A	0.26	0.50	0.03	0.012	0.039	0.20	0.012	0.03	0.03	0.006	0.003	0.004
B	0.18	0.73	0.07	0.008	0.030	0.05	0.015	0.07	0.03	0.006	0.012	0.005
C	0.24	0.48	0.05	0.012	0.026	0.02	0.016	0.03	0.03	0.005	0.003	0.009
D	0.22	0.55	0.21	0.013	0.024	0.16	0.020	0.22	0.12	0.022	0.023	0.005
E	0.20	0.33	0.01	0.013	0.020	0.15	0.009	0.18	0.09	0.018	0.024	0.005
F	0.18	0.82	0.15	0.012	0.031	0.04	0.054	0.05	0.03	0.008	0.021	0.006
G	0.20	0.86	0.19	0.020	0.020	0.08	0.045	0.15	0.04	0.018	0.012	0.006
H	0.18	0.76	0.16	0.012	0.019	0.05	0.053	0.09	0.04	0.006	0.004	0.004
N	0.17	0.53	0.25	0.011	0.020	3.39	0.077	0.19	0.06	0.025	0.017	0.005
Q	0.22	1.13	0.05	0.011	0.030	0.05	...	0.13	0.03	0.006	0.018	0.006

Steel	Condition	Yield point		Tensile strength		Elongation in 50 mm, %	Elongation in 200 mm, %	Reduction of area, %	Rockwell B hardness
		MPa	ksi	MPa	ksi				
A	As rolled	248	36.0	407	59.0	41	34	58	60
B$_{ar}$ ...	As rolled	228	33.0	396	57.5	44	34	64	61
B$_n$ ...	Normalized	248	36.0	393	57.0	44	34	63	60
C	As rolled	248	36.0	448	65.0	39	30	56	67
D$_{ar}$ ...	As rolled	258	37.5	448	65.0	...	30	54	...
D$_n$ ...	Normalized	241	35.0	414	60.0	...	32	59	...
E$_{ar}$ ...	As rolled	207	30.0	393	57.0	...	32	56	...
E$_n$ ...	Normalized	241	35.0	396	57.5	...	31	56	...
F	As rolled	234	34.0	420	61.0	...	31	62	...
G	As rolled	286	41.5	483	70.0	...	28	56	...
H	As rolled	248	36.0	438	63.5	42	30	63	70
N	As rolled	400	58.0	552	80.0	35	26	65	84
Q	Quenched and tempered	317	46.0	496	72.0	45	23	62	81

Steels A, B, and C, semikilled; D, F, G, and H, fully deoxidized; E, rimmed

drop-weight testing are outlined in ASTM E 208 (Ref 24).

The explosion-bulge test is shown in Fig. 26(b). The specimen is 356 mm (14 in.) square by about 25 mm (1 in.) thick. A crack-starter weld similar to the one in the DWT is applied to the lower surface of the specimen. The specimen is placed over a die

Fig. 27 Illustrations of plates tested at various temperatures by the explosion-bulge method and their relation to type of fracture in tests on the same material by drop-weight testing and by tension testing

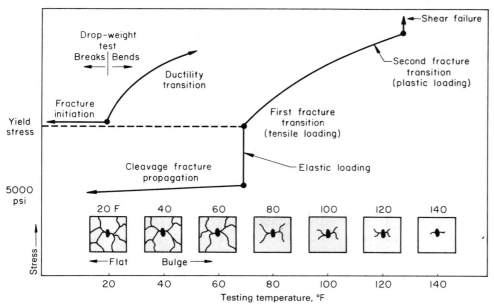

Fig. 28 Notched slow-bend test specimen (Lehigh test specimen) with longitudinal weld bead
Dimensions given in inches unless indicated

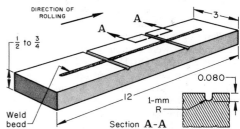

The **Lehigh bend test,** another slow-bend test, is more widely used. It may be applied to specimens containing a longitudinal weld bead or with no weld deposit. The Lehigh test specimen (Fig. 28) is 75 mm (3 in.) wide by 305 mm (12 in.) long by 12.7 to 19 mm (½ to ¾ in.) thick and is notched across its entire width. The notch may be a standard V-notch or a less severe notch (1 mm, or 0.04 in., radius and 2 mm, or 0.080 in., deep). The specimen is bent slowly until fracture occurs, at which time contraction in width just below the notch, percentage of fibrous fracture, and bend angle at maximum load are measured. Ductility transition temperature is usually defined as that point at which a lateral contraction of 1% is obtained by plotting lateral contraction against temperature. As in the Charpy specimen, the fracture appearance transition is defined as a temperature corresponding to 50% shear fracture.

The **Robertson test** utilizes a large rectangular specimen, usually 305 mm (12 in.) wide. This test measures the temperature at which a brittle crack, initiated by impact, stops in a plate under uniform stress with a superimposed temperature gradient. The specimen has a knob through which a hole is drilled

with the weld bead on the tension side and is loaded by the force of a controlled explosion.

Typical results on a steel specimen are shown in Fig. 27. At temperatures lower than the nil-ductility transition temperature, determined by the DWT (about −7 °C, or 20 °F, for this steel, as shown at the upper left corner of Fig. 27), the steel does not deform before fracturing. The specimens break flat under explosive impact, because fracture is readily initiated. Fractures also propagate through the very lightly loaded edge regions supported by the die.

Just above the nil-ductility transition temperature, considerable plastic deformation (bulging) precedes fracture. At this point, it is more difficult to initiate fracture, but cracks propagate to the edges when fracture occurs. As the testing temperature is raised to where fracture is partly in shear, cracks are confined to the area of bulging (forcing); they do not run through the lightly loaded edge regions. Some shearing is evident along the edges of the fracture. At still higher temperatures, brittle fractures are no longer possible, and all fractures occur in shear.

Notched Slow-Bend Test. Opinion varies as to whether the behavior of steel structures can be predicted from impact tests involving very high strain rates, which seldom correspond to service conditions. Consequently, notched slow-bend test have been proposed for investigating brittle fracture of steel structures. In such tests, specimens are geometrically similar to the standard Charpy V-notch specimen and may

range in size from 9.5 mm (⅜ in.) square by 50 mm (2 in.) long to 230 mm (9 in.) square by 1.8 m (6 ft) long. The notch radius is kept constant for each series of specimens. Specimens are bent in a tension-testing machine at very low strain rates with the notch in tension. Load and deflection measurements are recorded. The load-carrying ability (strength) of the material can be calculated from the breaking load. The deflection serves as a measure of relative ductility.

This test has been used primarily for testing at room temperature. Brittle or ductile behavior depends on the size of the specimen; small pieces bend, larger bars of the same material break.

Fig. 29 Robertson specimen (a) the Esso, or Feely, specimen (b), and the Navy tear-test, or Kahn, specimen (c)
The Navy tear-test specimen utilizes full plate thickness, t. Dimensions given in inches

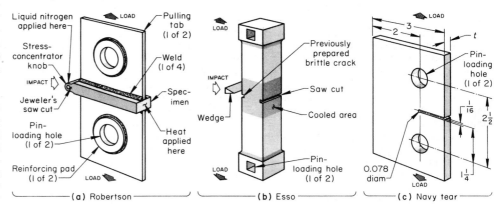

and a jeweler's saw cut is made. The cut acts as a crack initiator. A tensile load is applied, as shown in Fig. 29(a), while a stress gradient is established on the specimen by use of a liquid-nitrogen cup on one side and a strip heater on the opposite side. Impact force is then imposed on the knob.

Tests of this general type using a temperature gradient were carried out in the laboratory of the Standard Oil Development Co. However, to prevent stress-wave reflection due to impact, the specimens were longer than the typical Robertson specimen, and they contained a wire cut to obtain a sharper initial crack. Results of tests using this type of specimen as well as another modification that eliminated the Robertson knob were difficult to evaluate, because there was some doubt as to the stress level existing at the time the crack stopped. In view of this doubt, it was concluded that "go, no-go" tests at constant temperature and stress were subject to fewer uncertainties. This led to the development of the Esso (Feely) test.

The Esso test (Feely) specimen is also made from a slab of steel with end connections suitable for stressing in a tension-testing machine (Fig. 29b). Length of the specimen in the direction of applied tension varies from 0.9 to 1.8 m (3 to 6 ft); width, from 405 mm to 1.8 m (16 in. to 6 ft). Saw cuts are made at the middle of each side, with a subsequent cut on one side using a fine wire and grinding compound. The specimen is then loaded as a simple beam to create a stress in the outer fiber of the wire-cut side by imposing a transverse load on the opposite side. The area near the saw cuts is then cooled by liquid nitrogen.

When the temperature of this area is about that of the liquid nitrogen, a wedge is driven (by impact from a small slug shot at high velocity from a power gun) into the saw cut in which the wire cut has been made. This induces a fine crack at the base of the wire cut that acts as a severe stress concentrator. The saw cut on the opposite side from the wire cut serves merely to improve the distribution of stress.

The specimen, now in the precracked stage, is loaded axially with an insulated box around the central portion. Dry nitrogen cooled in a coil immersed in liquid nitrogen is injected into the box on both sides of the specimen. When the desired temperature is reached (as indicated by thermocouple), a predetermined tensile load is applied to the specimen, and the power gun is again used to drive a wedge into the previously prepared crack. If failure does not occur, the stress is raised, and the specimen is subjected to impact at successively higher stress levels until brittle fracture occurs.

Below a certain critical temperature, reproducible results are obtained for the stress level at which failure will occur. This is true regardless of whether the specimen is subjected to a number of impacts at progressively increasing stress or whether the breaking stress is applied

at the first impact. This situation does not exist at higher temperatures, and more specimens must be used to determine the lowest stress at which a specimen will fail from the first impact.

The Navy tear-test (Kahn) specimen is flame cut from a full plate and is machined in the manner shown in Fig. 29(c). Actually, only the edge opposite the notch is a machined surface. The specimen is supported at the pinholes on pins that are mounted on shackles. In this test, a series of specimens is subjected to slow tensile loading at various temperatures. The maximum load, the energy input required to initiate fracture, and the input required to complete fracture are measured. Transition temperature is the temperature corresponding to 50% shear fracture or the point of abrupt energy change at which tearing begins after initial fracture.

Correlations. Although used to measure notch toughness or transition temperature, slow-bend-test specimens differ markedly in severity of notch, strain rate, and size. The correlation of test results obtained with various specimens is poor, because the same property of the steel may not be measured in each test. Energy measurements derived with one type of specimen should not be compared with ductility measurements derived with a second specimen or with the fracture appearance of a third. When seeking a correlation among different test methods, appearance of the fracture has proved to be the best criterion. Nevertheless, notch toughness of two steels should be compared only if the same type of specimen and criteria are used. Test results are relative, not absolute.

Comparisons of energy-temperature curves obtained by Charpy keyhole and Charpy V-notch tests of the same 12 steels were shown in Fig. 24. Transition temperatures determined

by nine different methods on six steels are plotted in Fig. 30. Included in Fig. 30 are transition temperatures determined by several special tests that are not described in this Appendix. Details of these tests are provided in Ref 22.

Appendix 2: Fracture-Toughness Testing and Evaluation

This Appendix describes details of and gives references to a variety of tests for determining fracture toughness—both in plane-strain situations and in situations to which plane-strain fracture toughness is not applicable.

Plane-Strain Fracture-Toughness Test. Several approaches are currently being taken to develop procedures for evaluating toughness in more ductile metals. For materials having generally high strength and low toughness, the plane-strain toughness parameter, K_{Ic}, is the one most commonly determined. ASTM E 399 (Ref 25) defines K_{Ic} as the material-toughness property measured in terms of the stress-intensity factory, K_{I}, by a specified procedure. The procedure is designed to determine the stress intensity at which unstable crack propagation begins. According to ASTM E 399, K_{I} is "a measure of the stress-field intensity near the tip of an ideal crack in a linear-elastic medium when deformed so that the crack faces are displaced apart, normal to the crack plane (opening mode or Mode I deformation)," as illustrated in Fig. 1. Currently, there are no

Fig. 30 Comparison of transition temperatures determined by nine different methods on six steels

Letter designations correspond to the steel compositions similarly identified in the table that accompanies Fig. 24. For metric unit equivalents, see the "Metric Conversion Guide" in this Volume.

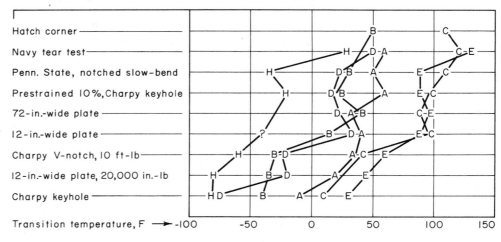

Fig. 31 Two standard specimens for determination of plane-strain fracture toughness

Details on specimen size, configuration, and preparation can be found in ASTM E 399 (Ref 25).

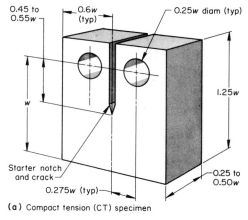

(a) Compact tension (CT) specimen

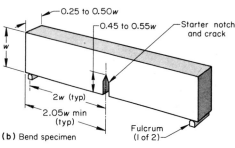

(b) Bend specimen

Fig. 32 Two standard specimens for determination of sharp-notch tensile strength

Dimensions given in inches. Alternate specimen dimensions are given in ASTM 338 (Ref 28).

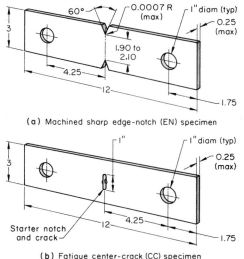

(a) Machined sharp edge-notch (EN) specimen

(b) Fatigue center-crack (CC) specimen

Fig. 33 Two standard specimen sizes for the NRL DT test

In the specimens, a crack with a sharp tip is produced by making a brittle electron-beam weld or by pressing with a knife edge. With either method for providing the crack tip, and with either size of specimen, maximum constraint conditions are attained. Dimensions given in inches

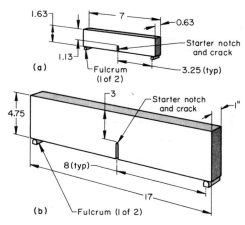

standard procedures for measuring fracture toughness with deformation modes II (shear perpendicular to the crack tip with plane-strain or plane-stress conditions) or III (shear parallel to the crack tip with anti-plane-strain conditions). For K_{Ic} measurements, ASTM E 399 describes procedures using specimens like those shown in Fig. 31. The crack-tip plastic region is small compared to crack length and to specimen dimension in the constraint direction. From a record of load versus crack opening and from previously determined relations of crack configuration to stress intensity, plane-strain fracture toughness can be measured accurately provided all the criteria for a valid test are met.

Using the same type of relation that permits calculation of a K_{Ic} value from load required to initiate extension of a given crack, it is possible to calculate the load that would initiate extension of a crack of a different configuration or to calculate the maximum size of crack that could be present without inducing fracture in a part under a certain load. From such calculations, it is possible to determine how rigorous inspection must be if all cracks that might lead to fracture are to be detected.

Numerous other specimen designs for the determination of K_I values have been proposed, and K_{Ic} values derived from them have been reported in the literature. In Ref 26, which

provides detailed discussion of the theoretical background of fracture processes, there is a discussion of single-edge-crack, double-edge-crack and center-crack rectangular specimens, and circumferentially cracked cylindrical specimens, all loaded in tension (Ref 27). There is interest in developing a specimen in which a surface crack will grow with an elliptical contour, propagating both along the surface and into the material.

For tougher, more ductile steels, specimens large enough to ensure loading under plane-strain conditions are often larger than the structures to be built. Consequently, other toughness tests and procedures are in use. One such test applicable to high-strength sheet metals is described in ASTM E 338 (Ref 28). Specimens such as those shown in Fig. 32 are used to determine the tensile strength of a sharply notched or precracked specimen, which can then be compared with the tensile strength of an unnotched (or uncracked) specimen. Other tests are designed to allow approximation of K_{Ic} values of tough metals that crack with plastic zones too large for valid measurement by the methods in ASTM E 399 (Ref 25). The use of specimens of K_{Ic} configuration but smaller than required for valid K_{Ic} tests may be possible if tested by loading at high speeds, generating K_{Id} values (where d designates dynamic). For large specimens or very high speeds, such testing involves advanced research techniques, as described in the article "Dynamic Fracture Testing" in Volume 8 of the 9th Edition of *Metals Handbook*. For many steels of technical interest, however, Charpy-type specimens can be used effectively with the root of the V-notch

extended by fatigue cracking. As discussed below, testing of precracked Charpy specimens can yield values that are strongly related to K_{Ic} values.

Dynamic Tear (DT) Test. Many metals and alloys, especially at lower strength levels are too tough and too ductile to fracture under plane-strain conditions in the sizes normally used in structures. In an effort to obtain reliable values of fracture toughness of ductile metals and alloys, the Naval Research Laboratory introduced the DT test. This test is intended to evaluate metals and alloys over a wider range of fracture toughness than can other fracture-toughness tests. Correlation of DT toughness and K_{Ic} toughness has been published (Ref 29).

The standard DT-test specimen, two sizes of which are illustrated in Fig. 33, is similar to the Charpy specimen, but has greater depth and has a proportionately deeper notch, which is sharpened by a pressed knife edge. The DT-test specimen is broken by impact opposite the notch in a manner similar to the Charpy specimen, and the energy not absorbed is measured by the swing of the pendulum, if this type of machine is used, or by deformation of lead or aluminum plates if a drop-weight machine is used. Details on DT testing of metallic materials can be found in ASTM E 604 (Ref 30) and in Ref 31.

The DT test is a 1964 modification of the NRL drop-weight tear test (DWTT), which originally had a deep, sharp crack introduced by an electron-beam weld, rather than a notch. A different modification of the DWTT was introduced in 1963 by Battelle Memorial Institute. The Battelle DWTT, described in ASTM E 436 (Ref 32), uses a shallow notch pressed by a sharp chisel edge, rather than a

Fig. 34 RAD for steels, as prepared for trade-off analyses for components 25 mm (1.0 in.) thick
Data were determined using specimens 12.5 to 75 mm (0.5 to 3.0 in.) thick. Source: Ref 33

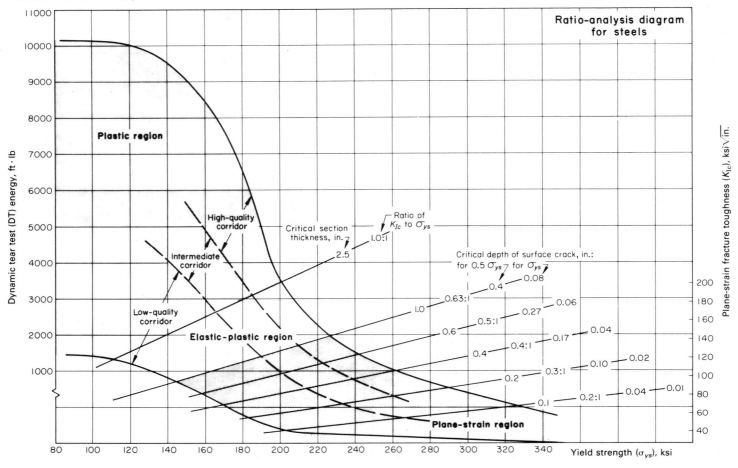

deep crack or notch. In the Battelle DWTT, only the fracture appearance is recorded, while in the DT test, the energy to fracture the specimen is recorded.

Ratio-Analysis Diagram (RAD). From consideration of fracture mechanics and plastic-flow properties, and from numerous measurements, an RAD can be constructed to represent many of the conditions relating to material selection for fracture control. The conception and development of this tool has been discussed (Ref 33). Simplified versions of RADs for steels, aluminum alloys, and titanium alloys originally presented in Ref 33 are given here in Fig. 34 to 36. On the ordinate scales are values of DT energy and values of K_{Ic} that correspond to the DT values. The K_{Ic} scale is extended to the point corresponding to a 1-to-1 ratio of K_{Ic} to σ_{YS} (yield strength) for the best alloy of those to be considered. Sloping lines representing various ratios of K_{Ic} to σ_{YS} are drawn. By the equations of fracture mechanics, these ratios can be related to the critical flaw (crack) depth expected to initiate fracture in a plate of a particular thickness. Each sloping line is applicable to a certain

critical thickness. Such a line can be constructed by plotting values of toughness versus yield strength for each alloy being considered for a design. If the thickness in the design is less than the critical thickness denoted by the sloping line, crack tips in the part will not be stressed under plane-strain conditions and either elastic-plastic or gross-yielding behavior can be expected. If design thickness is greater than critical thickness, the part can sustain even greater crack-tip stress intensity before crack extension will begin.

Because the modulus of elasticity is implicit in the relations between K_{Ic} and crack size, different diagrams are required for steel, aluminum alloys, and titanium alloys. Furthermore, at a given yield strength, different alloys have different fracture-toughness levels. Consequently, the diagrams may be divided into quality corridors based on test results for alloys with high, intermediate, and low toughness. With an RAD, it is then possible to guide the selection of alloys and working yield strengths so as to place the metal in either the ductile regime or the fracture mechanics regime, as may be appropriate for the cost or risk

involved. Results obtained in DT tests extend fracture analysis beyond the linear-elastic fracture mechanics range.

Crack-Opening-Displacement (COD) Test. A procedure that has been utilized, particularly in Great Britain, for fracture control of materials too tough for rigorous plane-strain testing is the COD test. Although this test was designed to measure crack-tip displacement as the load on the testpiece is increased, it is so difficult to make actual measurements of this displacement that measurements of surface-crack openings in specimens resembling plane-strain bend specimens have had to suffice. For COD specimens, size requirements are not so rigorous as for K_{Ic} specimens, and for beam specimens, the angle of bending is sometimes measured rather than the crack opening. Many aspects of this test are described in Ref 34 and 35. Methods for measuring COD and factors affecting measurement accuracy can be found in the article "Fatigue Crack Propagation" in Volume 8 of the 9th Edition of *Metals Handbook*.

J Testing. In EPFM, the toughness is *J*, expressed as the energy for crack growth by

Fig. 35 RAD for aluminum alloys, as prepared for trade-off analyses for components 25 mm (1.0 in.) thick

Data were determined using specimens 25 to 100 mm (1.0 to 4.0 in.) thick. Source: Ref 33

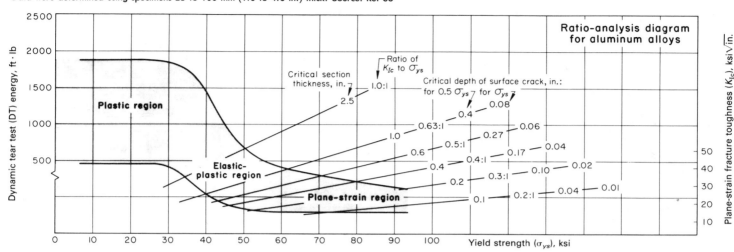

Fig. 36 RAD for titanium alloys, as prepared for trade-off analyses for components 25 mm (1.0 in.) thick

Data were determined using specimens 25 to 100 mm (1.0 to 4.0 in.) thick. Source: Ref 33

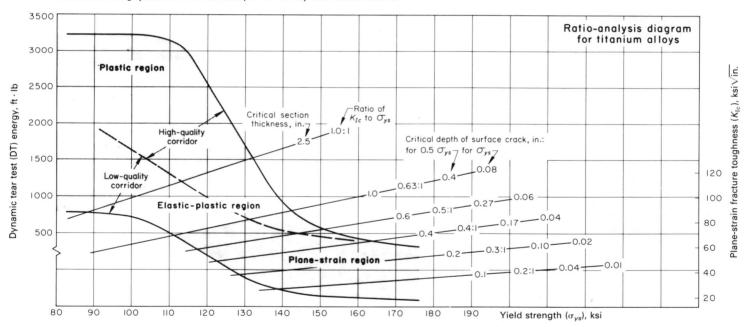

fracture. As discussed in the section "Elastic-Plastic Fracture Mechanics" in this article, the fracture energy increases during the fracturing process. This may lead to a certain amount of stable crack followed by an instability (uncontrollable fracture). Stable crack growth and subcritical crack growth by, for example, fatigue or stress corrosion should not be confused. Stable crack growth is fracture in progress. The amount of stable crack growth before instability depends on the stiffness of the entire loading system and the steepness of the rise in fracture energy.

ASTM standard E 813 (Ref 36) describes a procedure to measure J as a function of the amount of crack extension by fracture. A compact-tension (CT) specimen similar to the one used for K testing is used. During the test, the area under the load displacement curve is measured. In this case, the load-displacement curve will be nonlinear.

The area under the load-displacement curve represents an energy. In the case of a CT specimen, most of this energy will be expended in the cross section of the crack, because the remainder of the specimen undergoes negligible deformation due to the proximity of the loading points and the crack. In such a case, the above-mentioned energy is a measure of the fracture energy J. It can be demonstrated that:

$$J_R = \frac{2A}{bB}$$

where A is the area under the load-displacement curve, b is the uncracked ligament, and B is the thickness.

As fracture proceeds during the test, the crack size will increase so that each point of the

Fig. 37 J_R curves from tests

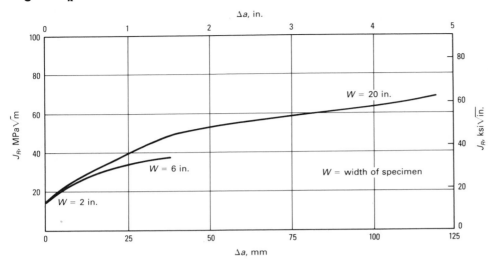

load-displacement curve is associated with a different crack size. The crack sizes can be indicated on the load-displacement curve. Determining the area A under the curve up to crack size a provides J for crack size a from the above equation in which b is the uncracked ligament at crack size a. If the initial crack size was a_0, the value of J at $\Delta a = a - a_0$ is then known. The procedure is repeated for a number of crack sizes. A plot of the resulting J values for the associated a values provides the J curve. An example is shown in Fig. 37.

Instrumented-Impact Test. A recent modification of the Charpy test (and other impact tests), the instrumented-impact test utilizes a strain-gaged tup that strikes the specimen so that a record is obtained of load and energy versus time, which correlates with deflection. From the record of load versus deflection, it is possible to determine the elastic portion of the stress-strain curve, the onset of crack extension, the energy for crack initiation, and the energy for crack propagation (Ref 37). With use of the instrumented impact test on precracked specimens, values of K_{Id} may be obtained under the same restrictions and in a similar manner as valid K_{Ic} values. Information on instrumented-impact tests can be found in the article "Dynamic Fracture Testing" in Volume 8 of the 9th Edition of *Metals Handbook*.

R-Curve Analysis. In testing tougher materials, in which plane-strain conditions do not develop at the crack tip for specimen thicknesses of interest, or in testing product forms that are limited in thickness, such as sheet materials, fracture-toughness properties can be evaluated by means of fracture-extension resistance, or *R*-curve, analysis, in which crack-growth resistance, *R*, per unit of crack extension, Δa, is determined (Ref 38). The crack-growth resistance may be expressed in

terms of either the strain-energy-release rate, *G*, or the fracture mechanics stress-intensity factor, *K*. Investigations at the Naval Research Laboratory have shown that *R*-curve analysis can be used to characterize metals of all fracture states—plane-strain, elastic-plastic, and plastic (Ref 39)—and that DT-test *R*-curve characterizations can be directly translated to design criteria by correlation with results of larger-scale tests that model generic types of configurations and loadings. These investigations also have shown that the energy to fracture a DT-test specimen is related to specimen dimensions and crack extension by:

$$E = R_p(\Delta a)^2 B^{1/2} \qquad \text{(Eq 36)}$$

where *E* is the DT-test energy, Δa is the crack extension, *B* is the specimen thickness, and R_p is a constant that for any given material is related to the *R*-curve slope and is a measure of the inherent resistance of the material to crack extension (Ref 39). This ductile-fracture equation applies equally well for steels at temperatures corresponding to upper-shelf (100% ductile) fracture conditions and for aluminum alloys, but is not applicable for titanium alloys because of temperature-transition effects. The use of the R_p parameter in association with the RAD permits an independent analysis of metallurgical and mechanical aspects of the fracture properties of materials for a wide range of section sizes.

Another approach to the determination of *R*-curves involves expressing the *R*-curve as a plot of stress-intensity factor, *K*, versus crack length, *a*, utilizing a modified wedge-opening-loading (WOL) or CT fracture-toughness specimen (Ref 40). This approach is directed primarily toward higher-strength sheet materials.

REFERENCES

1. D. Broek, *Elementary Engineering Fracture Mechanics*, 3rd ed., Nijhoff, 1981
2. J.F. Knott, *Fundamentals of Fracture Mechanics*, Butterworths, 1973
3. M.F. Kanninen and C. Popelar, *Advanced Fracture Mechanics*, Oxford University Press, 1985
4. H. Tada, P.C. Paris, and G.R. Irwin, *Stress Analysis of Cracks Handbook*, Del Research Corp., Bethlehem, PA, 1973
5. G.C. Sih, *Handbook of Stress Intensity Factors*, Lehigh University, Bethlehem, PA, 1973
6. D.P. Rooke and D.J. Cartwright, *Compendium of Stress Intensity Factors*, Her Majesty's Stationery Office, London, 1976
7. V. Kumar *et al.*, *An Engineering Approach for Elastic-Plastic Fracture Analysis*, EPRI NP-1931, Electric Power Research Institute, Palo Alto, CA, 1981
8. V. Kumar *et al.*, *Advances in Elastic-Plastic Fracture Analysis*, EPRI NP-3607, Electric Power Research Institute, Palo Alto, CA, 1984
9. J.R. Rice, A Path Independent Integral and the Approximate Analysis of Strain Concentrations by Notches and Cracks, *J. Appl. Mech.*, 1968, p 379-386
10. W.F. Brown and J.E. Srawley, *Plane Strain Crack Toughness Testing of High Strength Metallic Materials*, STP 410, ASTM, Philadelphia, 1966
11. P.C. Paris *et al.*, A Rational Analytic Theory of Fatigue, *Trend Eng.*, Vol 13, 1961, p 9-14
12. D. Broek and J. Schijve, *The Influence of the Mean Stress on the Propagation of Fatigue Cracks*, TR-M-2111, National Aerospace Institute, Amsterdam, 1963
13. E.A. Steigerwald, Crack Toughness Measurements of High-Strength Steels, in *Review of Developments in Plane Strain Fracture Toughness Testing*, STP 463, ASTM, Philadelphia, 1970, p 102-123
14. R.A. Wullaert, R.D. Ireland, and A.S. Tetelman, Use of the Precracked Charpy Specimen in Fracture Toughness Testing, in *Fracture Prevention and Control*, American Society for Metals, 1974, p 255-282
15. A. H. Priest and M.J. May, Effect of Loading Rate on the Fracture Toughness of Several High-Strength Steels, *Fracture Toughness of High-Strength Materials: Theory and Practice*, ISI 120, Institute for Scientific Information, Philadelphia, 1970, p 16-23
16. C.F. Tiffany and J.N. Masters, Applied Fracture Mechanics in *Fracture Toughness Testing and Its Applications*, STP 381, ASTM, Philadelphia, 1965, p 249-277

17. H. Hucek, Ed., *Damage Tolerant Design Handbook*, 4 Volumes, Metals and Ceramics Information Center, Columbus, OH, 1983, to be revised in 1986
18. S.T. Rolfe and J.M. Barsom, *Fracture Control of Structures*, Prentice-Hall, 1977
19. B. Marandet and G. Sanz, Evaluation of the Toughness of Thick Medium Strength Steels by Using Linear Elastic Fracture Mechanics and Correlations, in *STP 631*, ASTM, Philadelphia, 1977, p 72
20. D. Broek, *Correlation Between Fracture Toughness and Stretched Zone Size*, International Congress on Fracture, Vol III, 1973, p 422
21. "Standard Metals for Notched Bar Impact Testing of Metallic Materials," E 23, *Annual Book of ASTM Standards*, Vol 03.01, ASTM, Philadelphia, 1984, p 210-233
22. A. Boodberg, H.E. Davis, E.R. Parker, and G.E. Troxell, Causes of Cleavage Fracture in Ship Plate—Tests of Wide Notched Plates, *Weld. J*, April 1948, p 186s-199s
23. E.R. Parker, *Brittle Behavior of Engineering Structures*, John Wiley & Sons, 1957
24. "Standard Method for Conducting Drop Weight Test to Determine Nil-Ductility Transition Temperature of Ferritic Steels," E 208, *Annual Book of ASTM Standards*, Vol 03.01, ASTM, Philadelphia, 1984, p 346-365

25. "Standard Test Method for Plane-Strain Fracture Toughness of Metallic Materials," E 399, *Annual Book of ASTM Standards*, Vol 03.01, ASTM, Philadelphia, 1984, p 519-554
26. H. Liebowitz, Ed., *Fracture*, Vol I to VII, Academic Press, 1968-1972
27. J.E. Srawley, Plane Strain Fracture Toughness, in *Fracture*, Vol IV, H. Liebowitz, Ed., Academic Press, 1969, p 45-68
28. "Standard Method of Sharp-Notch Tension Testing of High-Strength Sheet Materials," E 338, *Annual Book of ASTM Standards*, Vol 03.01, ASTM, Philadelphia, 1984, p 483-490
29. W.S. Pellini, *Evolution of Principles for Fracture-Safe Design of Steel Structures*, NRL 6957, U.S. Naval Research Laboratory, Washington, Sept 1969
30. "Standard Test Method for Dynamic Tear Testing of Metallic Materials," E 604, *Annual Book of ASTM Standards*, Vol 03.01, ASTM, Philadelphia, 1984, p 641-652
31. E.A. Lange, P.P. Puzak, and L.A. Cooley, *Standard Method for the 5/8" Dynamic Tear Test*, NRL 7159, U.S. Naval Research Laboratory, Washington, July 1970
32. "Standard Method for Drop-Weight Tear Tests of Ferritic Steels," E 436, *Annual Book of ASTM Standards*, Vol 03.01, ASTM, Philadelphia, 1984, p 555-560

33. W.S. Pellini, *Criteria for Fracture Control Plans*, NRL 7406, U.S. Naval Research Laboratory, Washington, May 1970
34. M.O. Dobson, Ed., *Practical Fracture Mechanics for Structural Steel*, Chapman & Hall, 1969
35. A.A. Wells, Fracture Control: Past, Present and Future, *Exp. Mech.*, Oct 1973, p 401-410
36. "Standard Test Method for J_{Ic}, a Measure of Fracture Toughness," E 813, *Annual Book of ASTM Standards*, Vol 03.01, ASTM, Philadelphia, 1984, p 763-781
37. C.E. Turner, Measurements of Fracture Toughness by Instrumented Impact Test, in *Impact Testing of Metals*, STP 466, ASTM, Philadelphia, 1970, p 93-114
38. *Fracture Toughness Evaluation by R Curve Methods*, STP 527, ASTM, Philadelphia, 1973
39. R. Judy, Jr., and R. Goode, R-Curve Characterization and Analysis of Fractures in High-Strength Structural Metals, *Met. Eng. Quart.*, Vol 13 (No. 4), Nov 1973, p 27-34
40. D.E. McCabe and R.H. Heyer, R-Curve Determination Using a Crack-Line-Wedge-Loaded (CLWL) Specimen, in *Fracture Toughness Evaluation by R-Curve Methods*, STP 527, ASTM, Philadelphia, 1973, p 17-35

Ductile-to-Brittle Fracture Transition

A.R. Rosenfield and C.W. Marschall, Battelle Memorial Institute

THE DUCTILE-TO-BRITTLE FRACTURE TRANSITION is a marked change in fracture resistance of metal with changes in one or more test variables. It occurs only in certain metals within ranges that depend on the metal. Temperature, stress state, and strain rate are among the variables that can give rise to the fracture transition. Although changing any one of the three can produce a transition, it is customary to speak in terms of a metal having a transition temperature (the temperature dividing ductile behavior from brittle behavior), rather than having a transition stress state or a transition strain rate. The same practice is followed in this article.

Metals that exhibit a fracture transition are those in which the yield strength increases sharply with decreasing temperature and which are capable of fracturing by cleavage or an intergranular mode with very little accompanying plastic deformation. Metals having the body-centered cubic (bcc) crystal structure, notably nonaustenitic steels, are among those materials. When tested below the transition temperature, that is, where cleavage or the intergranular mode prevails, bcc metals behave in a brittle, glasslike manner. Above the transition temperature, however, they display significant plasticity and absorb more energy in the fracture process. Figure 1 shows fracture-transition behavior. To avoid brittle behavior, materials that exhibit such a transition should not be used at temperatures below the fracture-transition zone. Indeed, for many critical applications, the minimum service temperature may need to be toward the upper temperature end of the transition zone.

When considering the role of the ductile-to-brittle transition in defining safe operating conditions for steel structures, two factors are important: (1) the temperature at which the transition occurs relative to the service temperatures and (2) the level of fracture resistance displayed at temperatures well above the transition temperature (known as upper-shelf behavior). Experience has shown that only the first aspect need be considered for the vast majority of low-strength (<275 MPa, or 40 ksi) and medium-strength (275 to 415 MPa, or 40 to 60 ksi) structural steels. For these materials, merely being above the transition temperature usually ensures that the structure will behave as designed. However, for steels of very high strength (>690 MPa, or 100 ksi), the level of fracture resistance on the upper shelf may be more important than the transition temperature. Typically, these higher strength steels derive their strength by quench-and-temper treatments that produce martensitic microstructures having transition temperatures below most service temperatures. Therefore, for the majority of applications involving high-strength quenched-and-tempered steels, the fracture mode will be ductile. Nonetheless, the limited ductility associated with very high strength levels may permit a low-energy shear fracture to occur. Even for medium-strength steels, modifications intended to lower the transition temperature may lower the upper-shelf energy to an unacceptable level.

This article will focus on low- and medium-strength steels with sufficient resistance to ductile fracture that transition temperature is of primary importance. Three questions pertaining to the fracture transition will be examined: (1) how are transition temperatures and safe operating temperatures determined experimentally, (2) how can high fracture resistance and low transition temperatures be promoted through control of the microstructure, and (3) what service environments can cause an elevation of the ductile-to-brittle transition temperature?

Practical Evaluation of Transition Temperature

Although it is customary to speak of a transition temperature for a particular steel, no single clearly defined temperature can be assigned to this term; it depends on the test method used and the definition employed. One of the most straightforward test methods is the uniaxial tensile test, which is described in the Section "Tension Testing" in Volume 8 of the 9th Edition of *Metals Handbook*. Figure 2 shows how the ductility and fracture appearance of tensile specimens of a particular steel are influenced by test temperature. From the tensile tests alone, a variety of transition temperatures could be defined, including:

- Temperature above which fracture is entirely fibrous: −10 °C (15 °F)
- Temperature at which fracture is half fibrous and half brittle: −35 °C (−30 °F)

Fig. 1 Schematic of the effect of temperature on toughness of metals that exhibit a ductile-to-brittle fracture transition

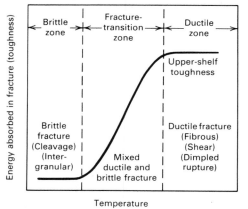

Fig. 2 Ductile-to-brittle fracture transition in tensile tests on a ferritic steel

RT, room temperature. Source: Ref 1

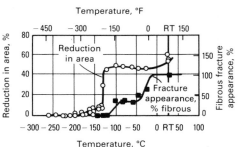

- Temperature below which fracture is entirely brittle (0% fibrous): −120 °C (−185 °F)
- Temperature at which fracture ductility changes abruptly: −130 °C (−200 °F)
- Temperature below which fracture ductility is zero: −200 °C (−330 °F)

Thus, depending on the definition chosen, the transition temperature could be as low as −200 °C (−330 °F) or as high as −10 °C (15 °F), a range of 190 °C (345 °F).

In practice, it is not customary to define transition temperatures on the basis of tensile tests, because three-point-bend testing is experimentally more simple. The most common practice is to employ a bend-test method that is likely to produce more conservative results. As noted above, both the strain rate and the state of stress can influence the fracture transition. Increasing the strain rate by using impact loading and changing from a uniaxial-tension to a multiaxial-tension stress state by introducing a sharp notch both tend to cause an upward shift in the experimentally measured transition temperature. Thus, it is common in transition-temperature studies to use specimens that contain a sharp notch (multiaxial stress state) and to test them under impact (high strain rate). The Charpy V-notch impact test, described in the articles "Failure Analysis and Fracture Mechanics" in this Volume and "Dynamic Fracture Testing" in Volume 8 of the 9th Edition of *Metals Handbook*, embodies both features and is the most commonly used international test for studying fracture-transition behavior. Other methods that employ notched specimens are also described in the article "Failure Analysis and Fracture Mechanics."

As noted above in connection with fracture transition in tensile tests, numerous definitions of transition temperature can be chosen for each test method, including the Charpy V-notch impact test. Figure 3 shows Charpy-test results for an ASTM A533 type B structural steel used in nuclear pressure vessels. Table 1 indicates some of the transition temperatures that could be defined on the basis of Fig. 3. They cover a range of about 145 °C (260 °F).

The limited data presented in Fig. 2 and 3 demonstrate clearly that the transition temperature depends strongly on the definition used. The transition temperatures that are most useful for a particular service component can be determined only by experience involving empirical correlations. For example, it was found that steel plates used in World War II Liberty Ships possessed adequate resistance to fracture if the temperature corresponding to the 20-J (15-ft · lb) energy-absorption level was below the lowest service temperature. In the absence of empirical correlations, a more conservative definition of transition temperature would probably be recommended, such as the temperature corresponding to 50% shear area. The penalty of conservatism is of course increased

costs, resulting from the need to use higher quality steels or more expensive heat treatments.

In one respect, the transition temperature approach suffers because it does not address the question of severe operating stress. This drawback has been addressed over the last 20 to 25 years by the advent of the widespread use of fracture mechanics. In this approach, additional conservatism has been introduced by using specimens in which even more severe multiaxial-tension stress states are developed during testing. This condition is accomplished by the use of very sharp notches or cracks and greater specimen thicknesses. In addition, side grooving of the specimen is sometimes used to increase constraint further and to enhance multiaxial effects. Testing of these fracture mechanics specimens yields values of K_{Ic}, the plane-strain fracture toughness. Fracture-toughness testing and evaluation are discussed in the articles "Failure Analysis and Fracture Mechanics" in this Volume and "Fracture Mechanics," "Fracture Toughness Testing Us-

Fig. 3 Fracture-transition data from Charpy V-notch impact tests on an ASTM A533, type B, steel
Source: Ref 2

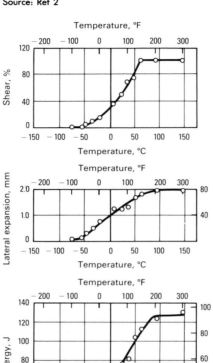

Table 1   Examples of transition temperatures obtained from Charpy V-notch impact tests on ASTM A533, type B, steel

Transition-temperature definition	Transition temperature	
	°C	°F
Upper-shelf energy	90	194
100% shear	65	149
50% shear	25	77
68 J (50 ft · lb) energy	15	59
40 J (30 ft · lb) energy	−15	5
0% shear	−55	−67

ing Chevron-Notched Specimens," and "Dynamic Fracture Testing" in Volume 8 of the 9th Edition of *Metals Handbook*. At present, K_{Ic} as a function of temperature can be determined only at temperatures in the lower part of the fracture-transition zone, while J_{Ic} and crack opening displacement, other measures of fracture toughness, can be obtained only at temperatures at which fracture is completely ductile. No agreed-upon technique exists for the intermediate region. Even when carrying out the fracture mechanics tests, it is customary to include the simpler and less costly Charpy V-notch impact tests so that empirical correlations between the various fracture parameters can be developed.

Not all of the tests that involve cracks and greater specimen thicknesses are used to develop fracture mechanics parameters. Exceptions are the dynamic tear (DT) test and the drop-weight tear test (DWTT), both of which use three-point-bend specimens and impact loading. These tests are also described in the article "Failure Analysis and Fracture Mechanics" in this Volume. The DT test employs specimens of two different sizes, depending on the material form. The smaller specimen has about 6 to 20 times the volume of a Charpy specimen (depending on thickness), and the larger specimen has more than 200 times the volume. In either size, a notch with a very sharp tip is introduced by pressing with a knife edge. The energy to fracture the specimen under impact loading is obtained over a range of temperatures that includes the ductile-to-brittle transition. Thus, this test differs from the Charpy V-notch test only in specimen size and notch sharpness. It too provides results that can be applied to service components only through empirical correlations. The DWTT is an even simpler test in that only fracture appearance is observed as a function of test temperature in the ductile-to-brittle transition range. Specimens are 76.2 × 305 mm (3 × 12 in.) by full-plate thickness and contain a sharp-pressed notch. Application of the DWTT is limited to ferritic steels having yield strength no greater than 825 MPa (120 ksi).

Each test described to this point can yield a variety of transition temperatures, depending on the definition chosen. However, one test, known as the drop-weight test (DWT), yields a single well-defined transition temperature that is widely used in a number of applications. The DWT bend specimen contains a notched weld bead on the tensile surface. The temperature defined by the DWT is called the nil-ductility-transition (NDT) temperature. Below the NDT temperature, it is believed that the steel will display no significant plastic deformation before fracturing in the presence of a cleavage crack and under impact loading. The DWT specimen and test procedure are described in the article "Failure Analysis and Fracture Mechanics" in this Volume.

Another temperature closely related to the NDT temperature is known as the reference nil-ductility temperature, RT_{NDT} (Ref 3). It is based on the results of both DWT and Charpy V-notch tests and is an important reference temperature in nuclear pressure vessel steels. The RT_{NDT} is the highest of three temperatures: the NDT temperature, $T_{CV50} - 60$ °F, and $T_{LE35} - 60$ °F, where T_{CV50} and T_{LE35} are the temperatures corresponding to 50 ft · lb (68 J) of energy absorption and 0.89 mm (0.035 in.) of lateral expansion, respectively, in Charpy V-notch tests.

In summary, the user of ferritic steels must always be aware of the ductile-to-brittle transition phenomenon to ensure avoidance of catastrophic fractures. Numerous tests are available for measuring the transition temperature, but the temperature so determined depends on both the test method and the definition used. Without user experience or empirical correlations, conservative measures may be called for—tests that employ high constraint (stress triaxiality) and rapid loading—as well as a more conservative definition of transition temperature. User experience may permit various degrees of relaxation of these conservatisms.

Factors Influencing the Ductile-to-Brittle Transition Temperature of Structural Steels

Compositional, Heat Treatment, and Processing Effects. Advances in basic research have led to a theoretical picture of the ductile-to-brittle transition that reflects the practical advances in alloy and processing development. The theory envisions hard particles in the microstructure (particularly large carbides) being cracked by slip bands associated with plastic deformation. In turn, these particle-cracks can develop further in one of two ways: the cracks can blunt and become voids, leading to ductile fracture, or the cracks can spread into the surrounding ferrite, leading to cleavage fracture. Considering these points, the keys to preventing cleavage (and the attendant low toughness) are to prevent particle cracking and to promote crack blunting.

When viewed on the microscale, the cracking is caused by the high stresses that are set up when a slip band encounters a particle. Analy-

ses in the scientific literature show that the local stresses are small when the free-slip distance (either the grain size or particle spacing) is small. Thus, fine grain size is beneficial (Fig. 4). The figure also shows the extent of strengthening due to grain refinement, a rare example of a single metallurgical change benefiting two often incompatible properties.

Heat treatments that break up pearlite colonies into fine carbide dispersions also limit slip-line lengths and inhibit cleavage. Small particle sizes are beneficial because they affect the crack-propagation stage by inhibiting the spread of particle cracks into ferrite by cleavage.

As noted above, the theory is consistent with practice. For example, normalizing is a useful method of grain refinement that lowers the transition temperature. Other methods are discussed in Ref 5, which contains a summary of the major trends in the progression toward higher strengths and lower ductile-to-brittle transition temperatures. Several themes run through this progression. Chiefly, the better combination of properties available in modern steels has been achieved by microstructural refinement, specifically:

- Finer grain sizes are achieved by higher manganese/carbon ratios, grain-refining additions, and controlled rolling
- Higher strengths and lower transition temperatures are achieved by adding carbide formers, such as niobium, which produce fine carbides in place of pearlite. Because carbide refinement has been a sufficiently effective method of strengthening, lower carbon levels can be used to achieve desired strength levels, improving weldability over traditional steels
- Sulfide-inclusion control, either by shape control and/or lowering the sulfur content, benefits upper-shelf toughness, particularly in the transverse directions. By increasing the shear-fracture contribution to impact energy, inclusion control may have an indirect beneficial effect on transition temperature. Lack of inclusion control gives rise to the texture effect (Fig. 5)

Nickel has a unique effect because it is the only practical solid-solution strengthener that lowers the transition temperature. The effect is small (5 to 10 °C, or 9 to 18 °F, per percent nickel), and no precise explanation is available. One possibility is a reduction in the microstress concentrating power of slip bands. Nickel also inhibits hot shortness (poor ductility at hot-working temperature) due to copper segregation at grain boundaries. As a result, it has been possible to strengthen steel by a fine dispersion of copper precipitates and retain a low transition temperature. However, these benefits are not available for steels exposed to neutron irradiation, as will be discussed in the section "Radiation Embrittlement" in this article.

Effects of Microstructure. At a given strength level, the transition temperature of a

Fig. 4 Dependence of the ductile-to-brittle transition temperature (T_c) and the lower yield stress at T_c on the average grain diameter d of ferrite in a low-carbon (0.15% C) steel
Source: Ref 4

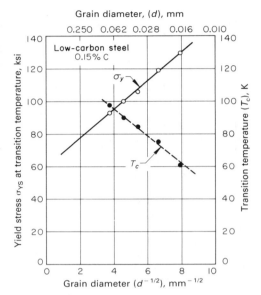

Fig. 5 Effect of specimen orientation on Charpy V-notch impact energy of as-rolled low-carbon (0.12% C) steel plate
Source: Ref 6

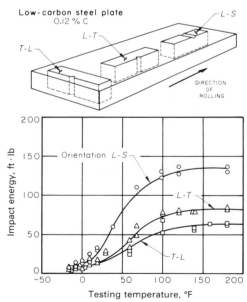

steel is determined by its microstructure. For example, of the major microstructural constituents found in steels, ferrite displays the highest transition temperature, followed by pearlite, upper bainite, and, finally, lower bainite and tempered martensite. The transition temperature of each of these constituents varies with the temperature at which the constituent formed and, where applicable, the temperature at which the steel was tempered. In practice, the cooling or quenching rate and the time-temperature transformation characteristics of a steel (including its conventional hardenability) determine the resulting microstructure or combination of microstructures. The transformation characteristics are in turn controlled by the alloy composition, austenitizing temperature, and austenite grain size.

Isothermally transformed lower bainite has a slightly lower transition temperature than tempered martensite of the same strength. However, mixed structures, which result from incomplete bainitic treatments causing partial transformation to martensite, have much higher transition temperatures than either 100% tempered martensite or 100% lower bainite. Thus, it is important that bainitic treatments be carried to completion to avoid the adverse effects of mixed microstructure.

Finally, retained austenite can inhibit the fast propagation of cleavage fracture in some ferritic and martensitic steels, improving fracture toughness significantly. However, strain-transformable austenite, which occurs in some steels, can be detrimental to toughness (Ref 7).

Deterioration in Service

Two sources of embrittlement in service occur in steels of moderate strength and are usually described in terms of the ductile-to-brittle transition temperature: temper embrittlement and radiation embrittlement.

Temper Embrittlement. Low-alloy steels that are either cooled slowly through the temperature range of 350 to 600 °C (600 to 1100 °F) or exposed to prolonged service in that range may be subject to intergranular failure accompanied by a large increase in transition temperature. The mechanism involves the segregation of tramp elements, such as phosphorus, arsenic, antimony, and tin. Methods of avoiding intergranular failure include using steels of higher purity and reheating at about 600 °C (1100 °F), followed by quenching. Additions of a small percentage of a strong carbide former, such as 0.5% Mo, can also have a beneficial effect, but will not eliminate temper embrittlement entirely. Figure 6 illustrates the characteristics of temper embrittlement, including a large increase in transition temperature due to phosphorus and a further increase due to tempering in the critical range.

Radiation Embrittlement. Neutron irradiation leads to an increase in the transition temperature and a decrease in the upper-shelf energy (Fig. 7). Although the mechanism is controversial, the transition-temperature shift is known to be caused by copper and phosphorus (and probably nickel). Reference 10 provides methods for conservatively estimating tran-

Fig. 6 Transition-temperature shift resulting from tempering in the temper-embrittlement range

Results for doped and undoped low-alloy steel. Scatter band represents phosphorus-doped specimens tempered at 480 °C (900 °F) for 5 to 200 h. Source: Ref 8

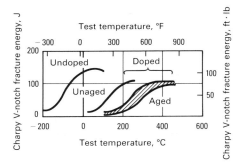

sition-temperature shift (as well as the lowering of the upper shelf). Radiation embrittlement can easily be prevented by limiting these elements to lower levels. However, the steels in many nuclear reactors in service contain relatively high levels of those elements, which may lead to early decommissioning or to the need to anneal out the damage. Either option is extremely costly.

Service Failures Resulting From Brittle Fracture

Example 1: Brittle Fracture of a Support Arm for a Front-End Loader. A support arm on a front-end loader failed in a brittle manner while lifting a load. The arm had a cross section of 50 × 200 mm (2 × 8 in.). Material used for the arm was hot-rolled ASTM A572, grade 42 (type 1), steel, which exhibited poor impact properties in the as-rolled condition and had a ductile-to-brittle transition temperature exceeding 93 °C (200 °F). This transition temperature was much too high for the application.

It was recommended that a modified ASTM A572, grade 42 (0.15% C max), type 1 or 2, steel be used (type 1, which contains niobium, may be needed to meet strength requirements). The steel should be specified to be killed, fine-grained, and normalized, with Charpy V-notch impact-energy values of 20 J (15 ft · lb) at −46 °C (−50 °F) in the longitudinal direction and 20 J (15 ft · lb) at −29 °C (−20 °F) in the transverse direction.

Example 2: Brittle Fracture of a Clamp-Strap Assembly. The clamp-strap assembly shown in Fig. 8, which was used for securing the caging mechanism on a star-tracking telescope, fractured during installation. In installation, the clamp ends were separated manually to 106 mm (4.187 in.) to fit over the telescope. The clamp strap was specified to be type 410 stainless steel—austenitized at 955 to 1010 °C (1750 to 1850

Fig. 7 Effect of neutron irradiation at 288 °C (550 °F) on Charpy V-notch impact properties of a submerged arc weld from a nuclear pressure vessel

Copper content = 0.23%, neutron fluence = 1.5 × 10¹⁹ neutrons/cm². FATT, fracture appearance transition temperature. Source: Ref 9

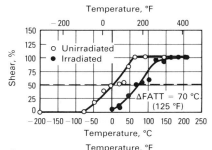

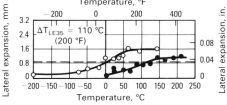

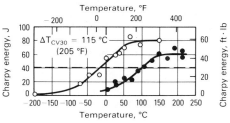

°F), oil quenched, and tempered at 565 °C (1050 °F) for 2 h to achieve a hardness of 30 to 35 HRC.

Investigation. Visual examination showed that the strap had fractured transversely across two rivet holes closest to one edge of the pin retainer (Fig. 8) in a completely brittle manner. Microscopic examination of a section transverse to the fracture showed the fracture to be predominantly transgranular with some grain-boundary separation. The grain size was determined to be as large as ASTM 2 to 3 (Ref 11). This coarse grain size indicated that the material was at an excessively high temperature during austenitizing.

The hardness of the failed strap was found to be 29 to 30.5 HRC (specified hardness was 30 to 35 HRC), and spectrographic analysis confirmed that the material was type 410 stainless steel. For comparison, similar information was determined for a clamp that had not failed. This clamp had a hardness of 27.5 to 28 HRC and a structure that consisted of tempered martensite. The grain size observed was predominantly ASTM 7. Spectrographic analysis showed that the material was type 410 stainless steel.

Fig. 8 Stainless steel clamp-strap assembly that failed in service because of brittleness due to coarse grain size

Dimensions given in inches

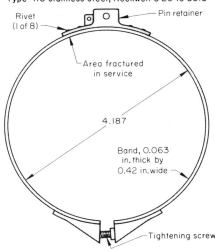

Type 410 stainless steel, Rockwell C 29 to 30.5

Rivet (1 of 8)
Pin retainer
Area fractured in service
4.187
Band, 0.063 in. thick by 0.42 in. wide
Tightening screw

Slow-bend tests were performed on full-width specimens cut from the failed strap and the unfailed strap at locations remote from the rivets. Cracking occurred with an audible click in the specimen from the broken strap at a bend angle of approximately 20°. Fracture occurred by transgranular cleavage with a slight amount of grain-boundary separation. The specimen from the unfailed strap was bent through an angle of approximately 180° with no cracking and only slight orange peel on the outside of the bend.

Additional full-width specimens were removed from both straps for reheat treatment. The specimen from the broken strap was austenitized at 970 °C (1775 °F), which was within the specified range of 955 to 1010 °C (1750 to 1850 °F), oil quenched, and tempered at 565 °C (1050 °F) for 2 h. The specimen from the unfailed ductile strap was austenitized at an excessively high temperature of 1055 °C (1935 °F), oil quenched, and also tempered at 565 °C (1050 °F) for 2 h. Slow-bend tests and metallographic examination were performed on both specimens. The specimen from the failed strap exhibited restored bend ductility and a refined grain size of ASTM 5 to 6. The specimen from the unfailed strap, originally a ductile material, fractured by cleavage and exhibited a coarsened grain size of ASTM 4.

Conclusions. Transgranular-cleavage (brittle) fracture occurred across rivet holes in the clamp strap closest to one edge of the pin retainer. Coarse grain size (ASTM 2 to 3) was responsible for the brittle fracture. Excessively high temperature during austenitizing caused the large grain size.

The fact that the hardness of the strap that failed was lower than the specified hardness of 30 to 35 HRC had no effect on the failure. The actual hardness of 29 to 30.5 HRC was considered sufficient for satisfactory performance of a properly treated strap, and the hardness of the strap that did not fail was even lower.

Recommendation. The strap should be heat treated as specified to maintain the required ductility and grain size.

Example 3: Brittle Fracture of a Rephosphorized, Resulfurized Steel Check-Valve Poppet.

The poppet shown in Fig. 9(a) was used in a check valve to control fluid flow.

The maximum operating pressure was 24 MPa (3.5 ksi).

Specifications required that the part be made of 1213 or 1215 rephosphorized and resulfurized steel for good machinability. The poppet was specified to be case hardened to 55 to 60 HRC, with a case depth of 0.6 to 0.9 mm (0.025 to 0.035 in.); the hardness of the mating valve seat was 40 HRC.

After about two weeks in service, the poppet broke. Fracture occurred through the two 8-mm (0.313-in.) diam holes at the narrowest section of the poppet (Fig. 9a). The valve continued to operate after it broke, which resulted in extensive loss of metal between the holes. Other poppets reportedly made of the same material by the same manufacturing procedures withstood laboratory test-stand runs of 1000 h at 24 MPa (3.5 ksi). The failed poppet and a poppet that had withstood a 1000-h test were submitted to the laboratory for metallurgical evaluation on a production-stoppage basis.

Chemical analysis of the metal in the failed poppet indicated that it was made of 1213 or 1215 steel. However, the part that withstood the 1000-h test contained 0.68% Mn, 0.005% P, 0.048% S, and 0.01% Si, indicating that the material was a low-carbon steel with 0.05% max sulfur.

Metallurgical Examination. Surface-hardness measurements were taken at 12 places along the length of each poppet. Hardness of the failed poppet ranged from 61 to 65 HRC, with an average of 63 HRC. The unfailed poppet had a hardness range of 57 to 62 HRC, with an average of 59.5 HRC.

Both of the parts were sectioned longitudinally for microscopic examination. Figure 9(b) is a micrograph of the failed poppet, which shows cracks extending across nonmetallic in

Fig. 9 Check-valve poppet that was redesigned to eliminate breakage in service caused by the pressure of nonmetallic inclusions in a rephosphorized, resulfurized steel

(a) Original design of the poppet showing origin of fracture and subsequent complete fracture between two opposing holes in the narrowest section of the poppet. (b) Micrograph of an unetched longitudinal section taken through the failed poppet showing cracks extending across nonmetallic inclusions. 80×. (c) Macrograph of a 5% nital etched longitudinal section taken through a poppet that did not fail showing the carburized case extending to the center of the cross section. 4×. (d) Improved design of the poppet. Dimensions given in inches

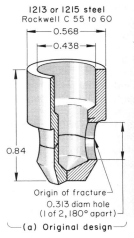

1213 or 1215 steel
Rockwell C 55 to 60
0.568
0.438
0.84
Origin of fracture
0.313 diam hole (1 of 2, 180° apart)
(a) Original design

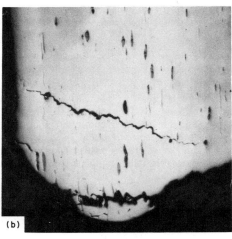

(b)

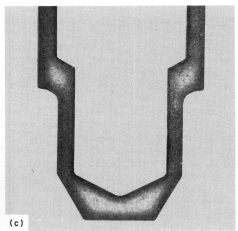

(c)

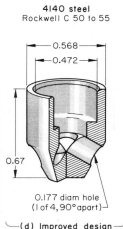

4140 steel
Rockwell C 50 to 55
0.568
0.472
0.67
0.177 diam hole (1 of 4, 90° apart)
(d) Improved design

clusions in the rephosphorized, resulfurized 1215 steel. The unfailed poppet had a much cleaner microstructure and exhibited no evidence of incipient cracking. Figure 9(c), a macrograph of a longitudinal section through the unfailed poppet, shows the carburized case extending to the center of the cross section. The failed part was in a similar condition.

Hardness surveys were taken across the transition section between the two diameters of the poppet body on both failed and unfailed parts. The hardness values were as shown in the following tabulation:

Location	Rockwell C hardness	
	Failed part	Unfailed part
Outer surface	63.5	59.5
Center of section	29–30	50–51
Inner surface	65.5	56

Several poppets were taken from stock and inspected by magnetic-particle and liquid-penetrant methods. None of the parts inspected had crack indications.

Conclusions. The check-valve poppet failed by brittle fracture. Surface hardness was excessive—61 to 65 HRC instead of the specified 55 to 60 HRC. Cracking occurred across nonmetallic inclusions. The inclusions inherent in rephosphorized and resulfurized steel resulted in low resistance to brittle fracture, especially at the specified hardness levels.

Corrective Measures. The valve was redesigned as shown in Fig. 9(d). The material was changed to 4140 steel, hardened, and tempered to 50 to 55 HRC. The improved design and the change in material eliminated failures of this type of poppet.

REFERENCES

1. G.T. Hahn, B.L. Averbach, W.S. Owen, and M. Cohen, in *Fracture*, B.L. Averbach, D.K. Felbeck, G.T. Hahn, and D.A. Thomas, Ed., The Technology Press of MIT and John Wiley & Sons, 1959, p 91-114
2. T.R. Mager and C.W. Marschall, *Development of a Crack Arrest Data Bank Irradiated Reactor Pressure Vessel Steels*, Vol 2, EPRI NP-3616, Electric Power Research Institute, Palo Alto, CA, July 1984
3. Pressure Vessel Research Committee, Ad-Hoc Group on Toughness Requirements, PVRC Recommendations on Toughness Requirements for Ferritic Materials, *Weld. Res. Counc. Bull.*, No. 175, Aug 1972
4. F. deKazinczy, W.A. Backofen, and B. Kapadia, discussion in N.J. Petch, The Ductile-Cleavage Transition in Alpha Iron, in *Fracture*, B.L. Averbach, D.K. Felbeck, G.T. Hahn, and D.A. Thomas, Ed., The Technology Press of MIT and John Wiley & Sons, 1959, p 54-64
5. F.B. Pickering, in *Physical Metallurgy and the Design of Steels*, Applied Science Publishing, 1978, p 50-62
6. P.P. Puzak, E.W. Eschbacher, and W.S. Pellini, Initiation and Propagation of Brittle Fracture in Structural Steels, *Weld. Res. Supp.*, Dec 1952, p 569-s
7. W.C. Leslie, *The Physical Metallurgy of Steels*, Hemisphere Publishing, 1982
8. S. Takayama, T. Ogura, S-C Fu, and C.J. McMahon, Jr., The Calculation of Transition Temperature Changes in Steels Due to Temper Embrittlement, *Metall. Trans. A*, Vol 11, 1980, p 1513-1530
9. C.W. Marschall, A.R. Rosenfield, M.P. Landow, T.R. Mager, R.G. Lott, and S.W. Tagart, Jr., Crack Arrest Behavior of Pressure Vessel Steels and Weldments as Influenced by Radiation and Copper Content, in *STP 870*, ASTM, Philadelphia, 1985, p 1059-1083
10. ''Effect of Residual Elements on Predicted Radiation Damage to Reactor Vessel Materials,'' Regulatory Guide 1.99, Revision 1, Nuclear Regulatory Commission, Washington, April 1977 (revision in preparation, 1986)
11. ''Standard Methods for Determining Average Grain Size,'' E 112, *Annual Book of ASTM Standards*, Vol 02.01, ASTM, Philadelphia, 1984

Failure Mechanisms and Related Environmental Factors

Identification of Types of Failures

Revised by Gordon W. Powell, Ohio State University

ANALYSIS OF A FAILURE of a metal structure or part usually requires identification of the type of failure. Failure can occur by one or more of several mechanisms, including surface damage (such as corrosion or wear), elastic or plastic distortion, and fracture. The articles that follow in this Section cover these various types of failure. For example, surface damage by corrosion is discussed in "Corrosion Failures" and "Elevated-Temperature Failures." Wear is discussed in "Wear Failures" and "Liquid-Erosion Failures." Elastic and plastic distortion are discussed in "Distortion Failures," and time-dependent plastic distortion (creep) is discussed in "Elevated-Temperature Failures." Because there are several types of failure by fracture, including those affected by chemical or thermal environment, the various methods of describing and classifying fractures will be discussed in this article.

Classification of Fractures

Many elements of fracture have been used to describe and categorize the types of fracture encountered in the laboratory and in service. These elements include loading conditions, rate of crack growth, and macro- and microscopic appearance of fracture surfaces.

The failure analyst often finds it useful to classify fractures on a macroscopic scale as ductile fractures, brittle fractures, fatigue fractures, and fractures resulting from the combined effects of stress and environment. The last group includes stress-corrosion cracking, liquid-metal embrittlement, hydrogen embrittlement, corrosion fatigue, and stress rupture. Fractures by combined or mixed mechanisms occur frequently.

Loading Conditions. A fracture that has resulted from loading being increased at a low or moderate rate to the breaking point of the material is often called an overload fracture; when the load is increased at a high rate, the resulting fracture is often called an impact fracture. Stress-rupture fracture is produced by sustained application of a fairly steady load.

Fatigue fracture is produced by repeated or cyclic application of load. The stress system acting on a material can be described by such terms as uniaxial, biaxial, and triaxial tension; compression; bending; torsion; direct shear (bolted or riveted joints); and, in the case of structural or machine components, such as gears, being pressed against one another, contact stresses. Additional terms, such as stress ratio and mean stress, are used to describe the stress system associated with cyclic (fatigue) loading.

Crack-Growth Rate. Cracks that lead to ductile fracture grow at a low rate, generally less than 6 m/s (20 ft/s). This slow cracking, known as stable crack growth, proceeds only while external loading is applied.

Unstable crack growth can proceed at a rate as high as several thousand feet per second. This fast cracking can proceed under internal elastic stressing, without the need for a continuing externally applied load, and can lead to the catastrophic brittle failure of a structure. The rate of crack growth may diminish significantly with increasing temperature so that, at some temperatures, fracture may proceed by a mixture of plastic deformation and brittle cracking or may occur as ductile fracture.

In addition to being affected by temperature, crack-growth rate is affected by the chemical composition, crystal structure, microstructure and grain size of the alloy, the size of the part, the direction of loading, the strain rate, and the chemical environment. The effects of these various factors are discussed in the articles that follow in this Section.

Macroscopic Examination. The macroscopic appearance of a fracture surface is described in terms of light reflection (bright or gray) and in terms of texture (smooth or rough, crystalline or silky, granular or fibrous). Use of the terms brittle and ductile to describe qualitatively the amount of macroscopic strain that occurs before final separation, as well as the terms flat-face and shear-face to describe the macroscopic orientation of the fracture surface is discussed below. The fracture mode depends on the stress system, with brittle flat-face fracture promoted by a plane-strain condi-

tion and shear-face fracture promoted by plane stress.

Microscopic Examination. The topography of a ductile fracture surface is characterized by innumerable dimples that originate by the nucleation and growth of microvoids at second-phase particles in the microstructure of the material. Brittle fracture on the microscopic level is characterized by cleavage facets whose extent is limited by the grain size of the material. In both of these cases, that is, dimple rupture and cleavage, the fracture path is transgranular. Fatigue fractures also follow a transgranular path; the fracture surface is distinguished by the presence of striations.

Brittle fracture induced by temper embrittlement is intergranular, and the topography of the fracture surface is readily identified by the grain boundaries that separate contiguous grains. Intergranular fracture can also occur in steels that have been overheated or in castings; occasionally, the grain size is so large that the intergranular character of the fracture surface can be ascertained by visual examination. The term rock-candy fracture is used to describe this situation. Rock-candy fracture is associated with a high density of second-phase particles (manganese sulfides, carbides, carbonitrides, aluminum nitrides) at the grain boundaries; consequently, the intergranular fracture surface will be dimpled.

Regarding a crystallographic aspect of ductile and brittle fractures, hexagonal close-packed (hcp) metals can undergo slip and cleavage on the same plane, whereas these processes occur on different planes in body-centered cubic (bcc) metals. Cleavage does not occur in face-centered cubic (fcc) metals under normal conditions, but cleavagelike features have been observed in stressed single crystals of nickel embrittled by exposure to liquid mercury.

Ductile Fractures

Ductile fractures are characterized by tearing of metal accompanied by appreciable gross plastic deformation and expenditure of considerable energy. Ductile tensile fractures in most materials have a gray, fibrous appearance and are classified on a macroscopic scale as either flat-face (square) or shear-face (slant-shear) fractures.

Flat-face tensile fractures in ductile materials are produced by a triaxial tensile-stress system that corresponds to a plane-strain condition in the case of thick plate. Necking precedes fracture, and the plane of the flat-face fracture is normal to the direction of the largest tensile stress. The flat-face fracture area is terminated by a shear lip that extends to the free surface (Fig. 1), and the ratio of the area of the flat-face region to the area of the shear lip usually increases with section thickness. Microscopic examination of flat-face tensile fractures in ductile materials at magnifications of about 100× and greater will reveal equiaxed dimples that are formed by microvoid coalescence (Fig. 2).

Shear-face tensile fractures in ductile materials are produced under plane-stress conditions (that is, in thin sections or near free surfaces), with or without necking, and typically occur at angles of about 45° to the surface of the part. Figure 3 shows an alloy steel bolt that fractured in a ductile manner because of overloading caused by the fatigue fracture of another portion of the assembly. The angle of the full-slant fracture surface (approximately 45°) and the fine, gray, silky appearance of the fracture are both characteristic of shear-face tensile fractures in ductile materials. Necking is faintly visible in the threaded area near the fracture surface.

Microscopic examination of shear-face tensile fractures, as well as the shear lips of flat-face tensile fractures, in ductile materials at magnifications of about 100× and greater reveals elongated dimples with their long axes in the direction of the shear force (Fig. 4). The elongated dimples produced by tensile shearing point in opposite directions on mating fracture surfaces. Elongated dimples are also produced by tensile tearing, but such dimples point in the same direction on mating fracture surfaces, which distinguishes them from shear dimples (see the Appendix to the article "Ductile and Brittle Fractures" in this Volume).

Fig. 1 Flat-face tensile fracture in an aluminum alloy plate pulled to rupture in a universal testing machine

(a) A view of a portion of the fracture surface. (b) A profile view of the fracture. The flat-face region is clearly visible across the center of the fracture; above and below it are shear lips, which are at a 45° slant to the flat-face region. 2×

Fig. 2 Equiaxed dimples on flat-face fracture surfaces

(a) 1020 steel specimen broken in tension. Note large dimples that contain smaller dimples. SEM fractograph. 100×. (b) Equiaxed dimples on fracture surface of quenched-and-tempered 4140 steel. Note the carbides at which local fracture initiated (arrow). TEM replica. 10 000×. (Courtesy of I. LeMay)

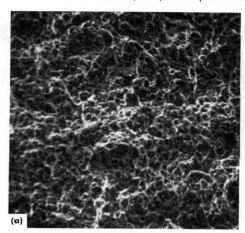
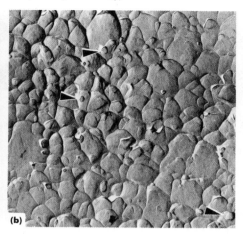

├──┤
1.0 μm

Fig. 3 Shear-face tensile fracture in an alloy steel bolt that broke by overload when another portion of the assembly fractured by fatigue

Actual size

Brittle Fractures

Brittle fractures are characterized by rapid crack propagation with less expenditure of energy than with ductile fractures and without appreciable gross plastic deformation. Brittle tensile fractures have a bright, granular appearance, are of the flat-face type, and are produced under plane-strain conditions with little or no necking. The visual appearance of the fracture surface of steel plate, for example, varies with the temperature at which fracture occurs (Fig. 5). A chevron pattern may be present over a limited range of temperatures; however, at lower temperatures, the chevron pattern becomes less distinct, and the fracture surface may be essentially featureless if the temperature is sufficiently low. Figure 6 shows the fracture surface of a fragment from a titanium alloy pressure vessel that failed catastrophically by brittle tensile fracture under a service load. A well-defined chevron pattern points to the origin of the fracture, which was

Fig. 4 Shear dimples in the shear-lip zone of a Charpy impact fracture in hot-rolled 1040 steel

The free edge of the shear-lip zone is the vertical margin at the right. The elongated dimples produced by tensile shearing point in opposite directions on mating fracture surfaces. SEM fractograph. 400×

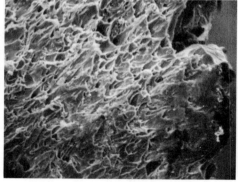

initiated by small cracks produced when the serial number was electrically engraved on the vessel.

Microscopic examination of brittle fractures reveals intergranular or transgranular facets. Intergranular facets are grain surfaces that have been exposed by crack propagation along grain boundaries (Fig. 7). The transgranular facets observed on brittle fractures are produced by cleavage along numerous parallel crystallographic planes, thus creating a terraced fracture surface. The individual levels of the terraced surface are connected by cleavage steps and/or tear ridges. As the cleavage crack propagates, the individual steps or ridges join together to produce steps or ridges of greater height and thereby reduce the number of planes over

Fig. 5 Chevron patterns in low-alloy steel ship-plate samples broken over a range of temperatures

Each fracture began at the notch (top). Mating fracture halves are shown for each temperature. (Courtesy of G.F. Vander Voort)

Fig. 8 River patterns on the surface of a brittle transgranular fracture in an Alnico alloy specimen

The white arrows point to a cleavage step; the black arrow shows the direction of crack propagation. TEM replica. 4900 ×

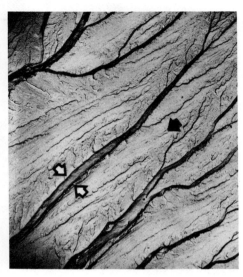

Fig. 6 Brittle tensile fracture in a titanium alloy pressure vessel that broke in a catastrophic manner under a service load

Fracture was initiated by small cracks created by electric engraving of a serial number on the vessel. ¾ ×

Fig. 7 Intergranular brittle fracture in 4340 steel showing facets that are the surfaces of equiaxed grains that have separated without microvoid coalescence

This is an example of a rock candy fracture surface. TEM replica. 1700 ×

Fig. 9 Fatigue fracture in an alloy steel lift pin from a large crane

The pin had insufficient fatigue strength because of improper heat treatment. Two fatigue zones are evident (top and bottom), each containing a set of beach marks. Final fracture was by ductile shear, in the horizontal band near the midsection of the pin. Actual size

which cleavage is occurring. The convergence of the steps and/or ridges produces a river pattern, with the crack propagating in the downstream direction (Fig. 8).

Fatigue Fractures

Fatigue fractures result from cyclic loading and appear brittle on a macroscopic scale. They are characterized by incremental propagation of cracks until the cross section has been reduced to where it can no longer support the maximum applied load, and fast fracture ensues. Frequently, the progress of a service-induced fatigue crack is indicated by a series of macroscopic crescents, or beach marks, progressing from the origin of the crack. Figure 9 illustrates a fatigue fracture caused by reversed bending of an alloy steel lift pin from a large crane. Cracks initiated at the top and bottom of the pin; rachet marks, which indicate multiple crack initiation, can be seen along the top edge of the pin. The horizontal band in the central

region of the fracture surface is the area of final fracture, which occurred in a ductile manner. The failure was caused by an improper heat treatment that did not provide the pin with adequate strength.

Beach marks are also observable in the forged/air-cooled diesel truck crankshaft shown in Fig. 10. This component failed due to

fatigue after more than 1.6×10^6 km (1×10^6 miles). Final fracture was done in a tension-testing machine.

Figure 11 shows the surface of a fatigue fracture in a forged steel trailer towbar that broke after a comparatively low number of load cycles. Examination revealed the towbar to be made of coarse-grain pearlitic steel with a heavily decarburized outer surface. The decarburized surface had greatly reduced tensile strength and caused fracture to initiate under cyclic tensile stress. The fracture surface

Fig. 10 Fatigue fracture of an AISI 1048 diesel truck crankshaft

Note large number of beach marks due to in-service fatigue. ¾×. (Courtesy of A. Burge and S. Vahlstrom)

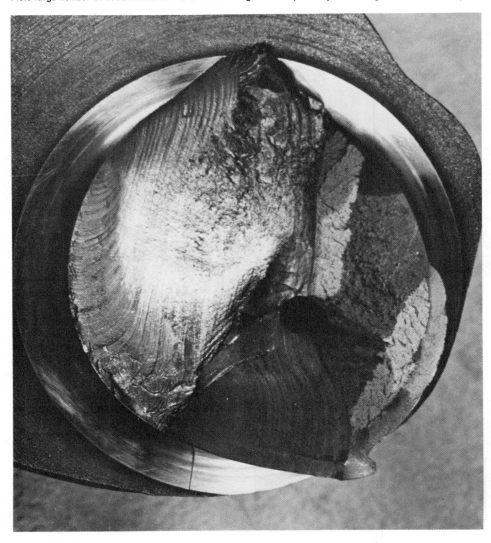

Fig. 11 Surface of a fatigue fracture in a forged steel trailer towbar that broke after a comparatively low number of load cycles

At the bottom zone are several crack origins on slightly different planes, forming ratchet marks; the cracks from these origins combined to form a single crack front, which progressed through the next, rougher zone (bracket A) by low-cycle fatigue; the large, brittle zone at top was formed during final fracture by a single load cycle. Actual size

Fig. 12 High-cycle fatigue striations in a fracture surface of aluminum alloy 6061-T6

The striations are closely spaced and propagate on flat plateaus joined by shear steps (white arrows). Black arrow indicates direction of crack propagation. TEM replica. 4900×

consists of three distinct zones—a fairly smooth, multiple-origin fatigue zone containing ratchet marks, a low-cycle, rougher fatigue zone, and a single-cycle final-fracture zone. The last zone can be described as brittle because it is perpendicular to the principal tensile stress and shows no visible plastic deformation. This is an example of the type of fracture in which the fatigue-cracking mechanism accounts for only a small portion of the entire fracture surface. In highly notch-sensitive materials, the fatigue crack that initiates final fast fracture may be so small as to be nearly invisible to the unaided eye.

Microscopic examination of fatigue fractures often reveals characteristic striations. Each striation is the result of a single cycle of loading; however, a recognizable striation is not necessarily produced by each cycle. Striations formed by high-cycle fatigue in aluminum alloy 6061-T6 are shown in Fig. 12.

Environmentally Affected Fractures

Figure 13 shows a fracture that was caused by the combined effects of stress and corrosive environment. The fracture occurred in a C-ring specimen that was cut from aluminum alloy 7039-T6 plate. The specimen fractured by stress-corrosion cracking under applied tensile stress while immersed in a salt solution. At the low magnification of ¾× neither the ring surface nor the fracture surface exhibits evidence of appreciable corrosive attack. The flat-face fracture surface, with no visible evidence of plastic deformation, is typical of stress-corrosion fractures. Stress-corrosion cracks are often branched, and the crack path may be transgranular, intergranular, or a combination of both. A microscopic feature that often is found on the facets of intergranular fractures caused by stress-corrosion cracking or hydro-

gen embrittlement is the fine tear ridge called a hairline indication (arrows A, Fig. 14).

Liquid metals may also constitute an environment that can embrittle an alloy. Figure 15 shows the fracture surfaces of two aluminum alloy 2024-T4 plates that failed by liquid-metal embrittlement. The plates were inoculated with liquid mercury and then loaded to fracture in tension. The plates broke at a stress well below the nominal yield strength. Rapid fracture occurred under these conditions, producing a

Fig. 13 Stress-corrosion-cracking fracture in a C-ring specimen cut from an aluminum alloy 7039-T6 plate

Fracture occurred under applied tensile stress during immersion in a salt solution. No appreciable evidence of corrosive attack is visible on the ring surface or on the fracture surface. ¾×

Fig. 14 Transgranular-cleavage fracture showing hairline indications (fine tear ridges) at A, together with tongues at B, shallow dimples at C, and secondary cracks at D

Fracture shown was produced in a specimen of 4315 steel by stress-corrosion cracking in a 3.5% NaCl solution. TEM replica. 2680×

Fig. 15 Liquid-metal-embrittlement fractures in two aluminum alloy 2024-T4 plates that were wetted with liquid mercury and then loaded to fracture in tension

Fracture occurred rapidly at a stress well below the nominal yield strength of the plates. Visible on each fracture surface is a flat mercury-damaged region (arrow, upper fracture) surrounded by a shear region. The region nearest the wetting (at left) consists of a dark mass of grain fragments. 1⅔×

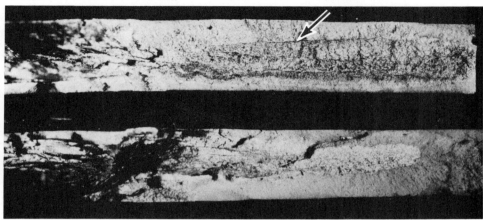

Fig. 16 Circumferential corrosion-fatigue cracks in a low-alloy steel superheater tube (at right)

The corrosion-fatigue cracks are adjacent to a weld joining the superheater tube to a stainless steel tube (at left). At this junction, the temperature normally encountered is approximately 595 to 620 °C (1100 to 1150 °F). The fatigue stress was caused by the tube vibrations inherent in normal operation of the boiler. ⅞×

Fig. 17 Cross section through a plain carbon steel boiler tube showing parallel corrosion-fatigue cracks in the initial stages of development

Note the gray corrosion product that fills the cracks. The fatigue stress was caused by tube vibration. Etched with nital. 100×

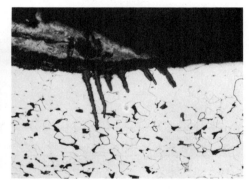

Fig. 18 A transverse fracture in a 1075 steel railroad rail

Fracture nucleus (small dark area near top of railhead) is a hydrogen-induced flake, or fisheye, which initiated a fatigue crack (large light zone surrounding the nucleus).

fracture surface consisting of a flat, mercury-damaged region surrounded by a shear region. The extent of the mercury-damaged region always depends on the amount of mercury available at the site of inoculation, but the surrounding fracture surface may be either ductile or brittle, depending on the notch sensitivity of the attacked metal. In Fig. 15, the mercury-damaged region is clearly delineated on the fracture surface. The inoculation region is on the left side of the plate areas shown. Near the inoculation region, the aluminum alloy is reduced to a mass of grain fragments.

Figures 16 and 17 illustrate the multiple cracking that is typical of corrosion fatigue.

The parallel cracks shown in Fig. 17 are filled with corrosion product. The fracture surface of a steel rail head that broke by fatigue at a subsurface hydrogen-induced flake is shown in Fig. 18. Flake formation in steel is one of the types of damage that can result from exposure to hydrogen. Subsurface initiation is common in fractures of this type.

Determination of Fracture Type

Several analytical procedures are available for distinguishing among the various types of

Table 1 Fracture mode identification chart

Method	Instantaneous failure mode(a)		Progressive failure mode(b)			
	Ductile overload	Brittle overload	Fatigue	Corrosion	Wear	Creep
Visual, 1 to 50× (fracture surface)	• Necking or distortion in direction consistent with applied loads • Dull, fibrous fracture • Shear lips	• Little or no distortion • Flat fracture • Bright or coarse texture, crystalline, grainy • Rays or chevrons point to origin	• Flat progressive zone with beach marks • Overload zone consistent with applied loading direction • Ratchet marks where origins join	• General wastage, roughening, pitting, or trenching • Stress-corrosion and hydrogen damage may create multiple cracks that appear brittle	• Gouging, abrasion, polishing, or erosion • Galling or storing in direction of motion • Roughened areas with compacted powdered debris (fretting) • Smooth gradual transitions in wastage	• Multiple brittle appearing fissures • External surface and internal fissures contain reaction scale coatings • Fracture after limited dimensional change
Scanning electron microscopy, 20 to 10 000× (fracture surface)	• Microvoids (dimples) elongated in direction of loading • Single crack with no branching • Surface slip band emergence	• Cleavage or intergranular fracture • Origin area may contain an imperfection or stress concentrator	• Progressive zone: worn appearance, flat, may show striations at magnifications above 500× • Overload zone: may be either ductile or brittle	• Path of penetration may be irregular, intergranular, or a selective phase attacked • EDS(c) may help identify corrodent	• Wear debris and/or abrasive can be characterized as to morphology and composition • Rolling contact fatigue appears like wear in early stages	• Multiple intergranular fissures covered with reaction scale • Grain faces may show porosity
Metallographic inspection, 50 to 1000× (cross section)	• Grain distortion and flow near fracture • Irregular, transgranular fracture	• Little distortion evident • Intergranular or transgranular • May relate to notches at surface or brittle phases internally	• Progressive zone: usually transgranular with little apparent distortion • Overload zone: may be either ductile or brittle	• General or localized surface attack (pitting, cracking) • Selective phase attack • Thickness and morphology of corrosion scales	• May show localized distortion at surface consistent with direction of motion • Identify embedded particles	• Microstructural change typical of overheating • Multiple intergranular cracks • Voids formed on grain boundaries or wedge shaped cracks at grain triple points • Reaction scales or internal precipitation • Some cold flow in last stages of failure
Contributing factors	• Load exceeded the strength of the part • Check for proper alloy and processing by hardness check or destructive testing, chemical analysis • Loading direction may show failure was secondary • Short-term, high-temperature, high-stress rupture has ductile appearance (see creep)	• Load exceeded the dynamic strength of the part • Check for proper alloy and processing as well as proper toughness, grain size • Loading direction may show failure was secondary or impact induced • Low temperatures	• Cyclic stress exceeded the endurance limit of the material • Check for proper strength, surface finish, assembly, and operation • Prior damage by mechanical or corrosion modes may have initiated cracking • Alignment, vibration, balance • High cycle low stress: large fatigue zone; low cycle high stress: small fatigue zone	• Attack morphology and alloy type must be evaluated • Severity of exposure conditions may be excessive; check: pH, temperature, flow rate, dissolved oxidants, electrical current, metal coupling, aggressive agents • Check bulk composition and contaminants	• For gouging or abrasive wear: check source of abrasives • Evaluate effectiveness of lubricants • Seals or filters may have failed • Fretting induced by slight looseness in clamped joints subject to vibration • Bearing or materials engineering design may reduce or eliminate problem • Water contamination • High velocities or uneven flow distribution, cavitation	• Mild overheating and/or mild overstressing at elevated temperature • Unstable microstructures and small grain size increase creep rates • Ruptures occur after long exposure times • Verify proper alloy

(a) Failure at the time of load application without prior weakening. (b) Failure after a period of time where the strength has degraded due to the formation of cracks, internal defects, or wastage. (c) EDS, energy-dispersive spectroscopy
Compiled by C.R. Morin, S.L. Meiley, and Z.B. Flanders, Packer Engineering Associates, Inc.

---

fractures. For example, the presence or absence of plastic macrodeformation can be determined with the unaided eye or by use of a steel scale, a machinist's micrometer, or a machinist's or measuring microscope. Differences in some dimensional attribute of parts (such as width or thickness) at and well away from the fracture can serve to define macrodeformation after assurance that both points of measurement had the same dimension before fracture.

Fracture-surface matching is also used to determine the presence or absence of plastic deformation. It is very important, however, to resist the temptation to fit the matching fracture surfaces together, because this almost always destroys (smears) microscopic features. The fracture surfaces should never actually touch during fracture-surface matching.

The origin of a fracture may be indicated by a discoloration or by the topography of the fracture surface. A discolored area on a fracture surface may be produced by a pre-existing crack whose surfaces have been corroded or oxidized. For example, the surfaces of a quench crack can be oxidized during a subsequent tempering heat treatment; the oxide film gives a bluish-black color to the surfaces of the crack. Topographical features that often reveal the origin of a fracture are either chevron or river patterns or a set of diverging ledges. If the fracture surface is essentially featureless, the presence of a shear lip can be used to locate, within limits, the origin of a fracture. For example, a shear lip is not formed at the origin of a stress-corrosion crack, but when the crack begins to propagate rapidly, a shear lip is

formed wherever the crack front exits from the interior to the free surface. Beach marks (Fig. 10), which are associated with fatigue-initiated fractures, also provide a definite indication of the crack origin; it should be noted that fracture surfaces having an appearance similar to that of the beach-mark pattern can be produced by stress corrosion.

Generally, cyclic loading produces only a single crack, which is usually located at a site of stress concentration or of a metallurgical defect, whereas additional cracks, formed independently of the main crack and at a distance from it, may be observed on the surface of a structural or machine component subjected to corrosion fatigue or stress corrosion.

On the microscopic level, striations on the fracture surface are unique to fatigue, and the crack path, although normally transgranular, can be intergranular. For example, intergranular fatigue cracking can occur in the case of a carburized steel or in a material that has a high density of second-phase particles at the grain boundaries.

Corrosion-fatigue and stress-corrosion cracks may propagate transgranularly, in-tergranularly, or by a combination of both modes. A distinguishing feature of stress corrosion is the branching of the main crack. If corrosion pits or corrosion products are found only on the slow-growth region of a fracture surface, the environment was in all probability sufficiently corrosive to affect the fracture mechanism. However, if evidence of corrosion is found on both the slow-growth and fast-growth areas, some corrosion took place subsequent to fracture, and the environment may or may not have influenced fracture. Table 1 is a concise summary of the visual and microscopic aspects and the contributory factors associated with the major fracture modes of metallic materials.

Ductile and Brittle Fractures

DUCTILE AND BRITTLE are terms that describe the amount of macroscopic plastic deformation that precedes fracture. Ductile fractures are characterized by tearing of metal accompanied by appreciable gross plastic deformation and expenditure of considerable energy. Ductile tensile fractures in most materials have a gray, fibrous appearance and are classified on a macroscopic scale as either flat (perpendicular to the maximum tensile stress) or shear (at a 45° slant to the maximum tensile stress) fractures.

Brittle fractures are characterized by rapid crack propagation with less expenditure of energy than with ductile fractures and without appreciable gross plastic deformation. Brittle tensile fractures have a bright, granular appearance and exhibit little or no necking. They are generally of the flat type, that is, normal (perpendicular) to the direction of the maximum tensile stress. A chevron pattern may be present on the fracture surface, pointing toward the origin of the crack, especially in brittle fractures in flat platelike components.

It must be pointed out, however, that these terms can also be applied, and are applied, to fracture on a microscopic level. Ductile fractures are those that occur by microvoid formation and coalescence, whereas brittle fractures may occur by either transgranular (cleavage or quasi-cleavage) or intergranular cracking. Intergranular fractures are specific to certain conditions that induce embrittlement. These include embrittlement by thermal treatment or elevated-temperature service and embrittlement by the synergistic effect of stress and environmental conditions. Both types are discussed later in this article.

Clearly, the cup-and-cone fracture shown in Fig. 1(a) has occurred as a result of appreciable plastic deformation and thus is a ductile fracture, whereas the fracture shown in Fig. 1(b) is a brittle fracture. The sequence of events that culminates in a cup-and-cone fracture is illustrated in Fig. 2, which shows the development of voids within the necked region (triaxial tensile stresses) of a tensile specimen and the coalescence of the voids to produce an internal crack by normal rupture. Final separation of the cross section occurs by shear rupture, which produces the wall of the cup. Figure 3 shows scanning electron microscopy (SEM) fractographs of the bottom and the sidewall of the cup. On the microscopic level, a crack is formed by coalescence of microvoids that form as a result of particle-matrix decohesion or cracking of second-phase particles; the microvoids and the associated particles are shown at high magnification in Fig. 4. The process of microvoid formation and coalescence involves considerable localized plastic deformation and requires the expenditure of a large amount of energy, which is the basis of selection of a material with good fracture toughness. The reduction of area of ultrahigh-purity aluminum and copper approaches 100% because of the absence within these materials of void-nucleating particles. In their visual appearance, ductile fractures have a matte or silky texture.

Regarding the brittle fracture shown in Fig. 1, it will be noted that the fracture surface is characterized by radial ridges that emanate from the center of the fracture surface. The ridges run parallel to the direction of crack propagation, and a ridge is produced when two cracks that are not coplanar become connected by tearing of the intermediate material. The cracks, which propagate predominantly by quasi-cleavage, move rapidly toward the periphery of the specimen cross section and, as shown in Fig. 1, penetrate the external surface

Fig. 1 Appearance of ductile (a) and brittle (b) tensile fractures
Source: Ref 1

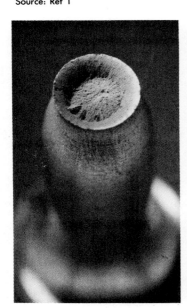

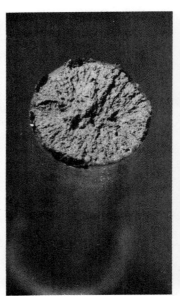
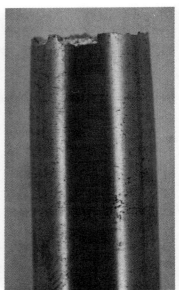

(a)

(b)

Fig. 2 Sections of a tensile specimen at various stages of formation during development of a cup-and-cone fracture

Note that the fracture is initiating internally. 7×. Source: Ref 2

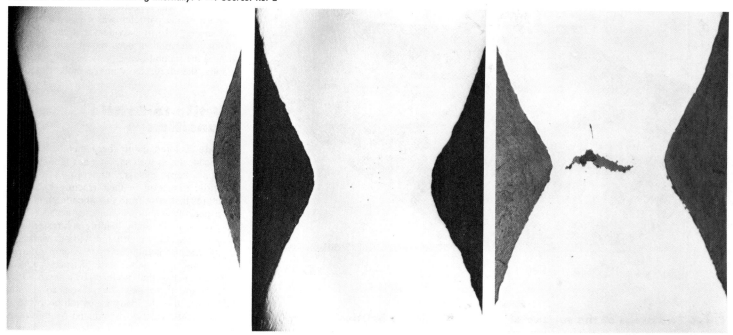

Fig. 3 Fractographs of a ductile cup-and-cone fracture surface

(a) Bottom of the cup. (b) Sidewall of the cup. SEM. 650×. Source: Ref 2

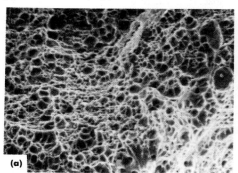
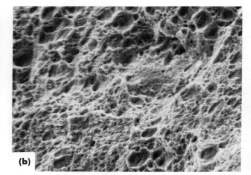

Fig. 4 Large and small sulfide inclusions in a ductile dimple fracture

SEM. 5000×. Courtesy of R.D. Buchheit, Battelle Columbus Laboratories

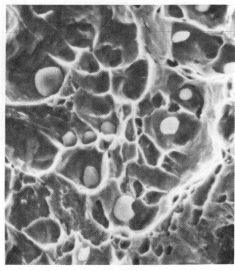

of the specimen by shear rupture along a relatively small shear lip. The shear lip develops as a result of the change in the state of stress from one of triaxial tension to one of plane stress. The extent or width of the shear lip depends on the temperature at which fracture occurs, formation of a shear lip being favored by higher temperatures.

The radial pattern and the shear lip provide useful information relative to the origin and direction of propagation of a crack. For example, the radial pattern on the fracture surface of the machine component shown in Fig. 5 indicates that fracture initiated on the upper edge of the fracture surface, the actual initiation site being designated by the arrow. The ridge pattern shown in Fig. 5 is commonly referred to as a river pattern and is formed by the convergence and joining of tear ridges into a single

tear ridge as the crack front advances. The extent to which a distinctive ridge pattern is developed depends on the temperature at which fracture occurs; higher temperatures (relative to the nil-ductility transition temperature) promote the formation of a readily visible river pattern (this matter is considered again below). Therefore, the absence or presence of a ridge pattern on the fracture surface of a brittle fracture can be used to provide a qualitative estimate of the fracture temperature relative to the nil-ductility transition temperature of the steel.

As noted above, the brittle fracture shown in Fig. 1 terminates with a very small shear lip. This fact can prove helpful when attempting to determine the origin of a brittle fracture. The origin of the fracture is invariably characterized by the absence of a shear lip, whereas a shear lip is expected to be present along the periphery

of the fracture surface where the crack emerges from the interior of the material. Consequently, the periphery of a fracture surface should be examined with these facts in mind. If fracture occurs at a low temperature, then a shear lip may not be formed.

Fracture by Mixed Mechanisms. A fracture surface that exhibits intermingled features

Fig. 5 Brittle fracture in a large steel component

Note the pronounced radial marks indicating the fracture direction. Source: Ref 3

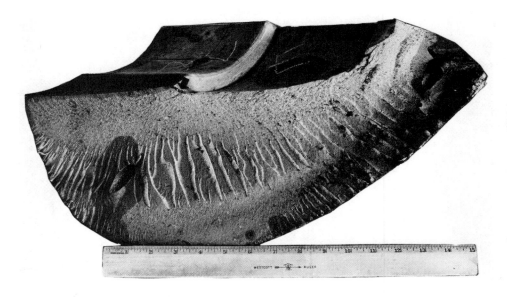

Fig. 6 Two areas of the surface of a macroscopically brittle fracture caused by mixed mechanisms

Fractographs are of an aluminum alloy 7075-T6 specimen containing a large concentration of relatively large second-phase particles. In both fracture-surface areas shown, fine dimples are visible in regions where there are no particles. Fracture of the particles visible in (a) was by cleavage. Considerable plastic deformation occurred in the material around the particles visible in (b). SEM. Both 1000×

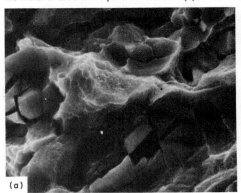

(a)

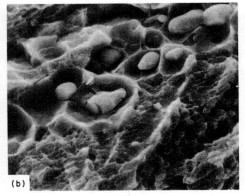

(b)

of two or more mechanisms of fracture in a given area is generally labeled a mixed-mode fracture. This is not to be confused with a fracture surface having features that suggest successive operation of different fracture mechanisms as cracking proceeds across a section. In the latter case, the individual fracture mechanisms can be analyzed sequentially and therefore require no special discussion.

The occurrence of fracture by mixed mechanisms often indicates that the usual factors that determine the operative mechanism, such as state of stress, loading history, microstructure, and environment, favor both mechanisms and that the fracture mechanisms in adjacent microscopic or submicroscopic regions are determined by such factors as grain orientation or microstructure. For example, in an alloy in which second-phase particles are present, two effects are to be anticipated. First, because most of the second-phase particles are considerably harder and stronger than the matrix, their presence will undoubtedly contribute to the overall strength of the alloy. Second, the fracture mechanisms operating in the region where these hard particles are located may be different from that operating in the regions where the particles are not located.

An example of the effects of hard second-phase particles is illustrated in Fig. 6, which shows SEM fractographs of two areas in an aluminum alloy 7075-T6 specimen that broke in a macroscopically brittle manner. In both fractographs, fine dimples are visible in the region where there are no particles, whereas in the regions where there are relatively large second-phase particles, the operative fracture mechanism was different. In Fig. 6(a), the fracture of the particles was clearly by cleavage; in Fig. 6(b), it is evident that considerable plastic deformation occurred in the material around the second-phase particles. Both fractographs, therefore, show mixed-mode fracture.

Ductile-to-Brittle Transition

Body-centered cubic (bcc) metals, such as ferritic steels, as well as some hexagonal close-packed (hcp) metals, undergo a ductile-to-brittle transition in their fracture behavior. Fractures that arise from this phenomenon have an especially insidious character because they may occur under static loading at stresses well below the yield strength and without warning. Brittle fracture is induced by low temperatures, high strain rates, and a state of triaxial tensile stresses, such as that produced by a notch. Only two of the three need be present to initiate brittle fracture. In many cases (ships, storage tanks), brittle fracture is induced by low temperature and a state of triaxial tensile stresses at a critical location in the structure.

The ductile-to-brittle transition can be demonstrated quite readily by the Charpy V-notch impact test. The results of such tests conducted on a low-carbon steel are shown in Fig. 7. The transition from ductile to brittle fracture is manifested geometrically by a decrease in the amount of lateral expansion on the compression side of the specimen and by elimination of the shear lip as the test temperature decreases. In addition, the usual appearance of the fracture surface changes from that of a rough, ductile fracture to a smooth, planar brittle fracture. On a microscopic level, the ductile fracture is characterized by microvoid coalescence, dimples of various orientations, and transgranular cracking, whereas the low-temperature brittle fracture consists of transgranular quasi-cleavage. At temperatures within the transition range, crack propagation occurs by both microvoid coalescence and quasi-cleavage, with the latter being confined predominantly to the central region of the fracture surface.

Face-centered cubic alloys, such as AISI type 316 stainless steel, do not exhibit a ductile-to-brittle transition, as demonstrated in Fig. 8 by SEM fractographs of fractured Charpy specimens. These fracture surfaces were produced by microvoid coalescence and contain dimples of various orientations. The existence of a ductile-to-brittle transition in low-carbon steel is related to the structure and mobility of dislocations in the bcc ferrite; the strong temperature dependence of the flow stress and dislocation-interstitial atom interactions are specific aspects of this phenomenon. Detailed information on the evaluation of transition temperature and factors that influence it can be

Fig. 7 SEM fractographs of ductile (D) and brittle (B) fractures in Charpy V-notch impact specimens shown at top
Both 400 ×

Temperature, °C (°F)	25 (75)	65 (150)	95 (200)
Energy, J (ft · lb)	34 (25)	134 (99)	152 (112)
Lateral expansion, mm (in.)	0.81 (0.032)	1.85 (0.073)	1.85 (0.073)
% fibrous	65	95	100

Temperature, °C (°F)	−18 (0)	−4 (25)	10 (50)
Energy, J (ft · lb)	5.5 (4)	13.5 (10)	23 (17)
Lateral expansion, mm (in.)	0.15 (0.006)	0.35 (0.014)	0.53 (0.021)
% fibrous	15	20	40

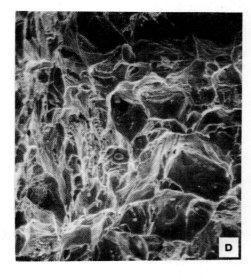
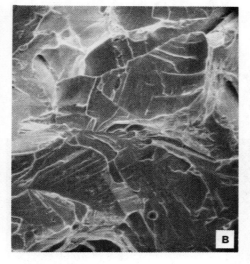

found in the article ''Ductile-to-Brittle Fracture Transition'' in this Volume.

Causes of Brittle Fractures

Brittle fractures that occur in service are invariably initiated by defects that are initially present in the manufactured product or fabricated structure or by defects that develop during service. The defects are essentially stress concentrators and may take any one of the following five forms.

First, notches, which are discontinuities caused by abrupt changes in the direction of a free surface, are often fracture initiators. Among the common intentional notches are sharp fillets and corners (Fig. 9), holes (Fig. 10), threads, splines, and keyways. Notches can also be produced accidentally by mechanical damage, such as from dents, gouges, or scratches (Fig. 11). When fracture occurs it always initiates at the notch, regardless of whether the presence of a notch is intentional or accidental.

Second, laps, folds, flakes (Fig. 12), large inclusions (Fig. 13), forging bursts, laminations in sheet and plate, and undesirable grain flow introduced during working operations also contribute to failure. These types of defects will be discussed in the section ''Failures From Improper Fabrication'' in this article. For a discussion of material defects caused during forging, see the article ''Failures of Forgings'' in this Volume. Failures caused by seams are also discussed in the article ''Failures of Springs'' in this Volume.

Third, segregation, inclusions, undesirable microstructures, porosity, tears, cracks or surface discontinuities introduced during melting, deoxidation, grain refining, and casting operations may or may not lead to failure. For example, a number of defects present in castings and ingots are of minor importance and seldom lead to failure. However, some discontinuities, such as cracks, can have serious consequences. For an extensive discussion of failures caused by casting defects, see the articles ''Failures of Iron Castings'' and ''Failures of Steel Castings'' in this Volume.

Fourth, cracks resulting from machining, quenching, fatigue, hydrogen embrittlement, liquid-metal embrittlement, or stress corrosion also lead to brittle fracture. In fact, the single most prevalent initiator of brittle fracture is the fatigue crack, which conservatively accounts for at least 50% of all brittle fractures in manufactured products (Ref 2). Figure 14 shows an example of a brittle fracture that was initiated by a fatigue crack. Articles on fatigue failures and environmentally induced cracking can be found elsewhere in this Volume. Quench cracking and failures resulting from machining are discussed later in this article.

Lastly, residual stresses, although not defects in a geometric sense, can be an important factor in the initiation of brittle fractures. Figure 15 shows a brittle fracture that originated from a hole punched in the steel plate; the chevron pattern is evident. The surface of the punched hole is plastically deformed, and the strain gradient produces residual stresses that, in conjunction with the applied stresses, initiated a small crack at the surface of the hole. In such situations, moisture may have had access to the hole and contributed to formation of the initial crack. Additional information on failures resulting from residual stress can be found later in this article.

Causes of Ductile Fractures

Service failures that occur solely by ductile fracture are relatively infrequent, but when they do occur, the structure or machine component is often stated to have failed as a result of an overload. Overload failures may be the result of the part having been underdesigned (a term that includes the selection and heat treatment of the materials) for a specific set of service conditions, improperly fabricated, or fabricated from defective materials or may be the result of the part having been abused, that is, subjected to conditions of load and environment that exceeded those of the intended use. Figure 16 shows a simple example of ductile fracture of a machine component. A super-duty rear axle that had been fabricated from an AISI type S7 tool steel failed as a result of a torsional overload; the smaller-diameter section of the

Fig. 8 SEM fractographs of Charpy specimens of type 316 stainless steel

Specimens broken at (a) −185 °C (−300 °F), (b) −130 °C (−200 °F), (c) −75 °C (−100 °F), and (d) −18 °C (0 °F). All 650×

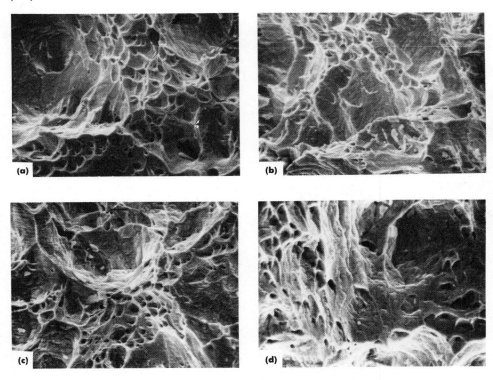

Fig. 9 Fracture surface exhibiting chevron pattern (left) pointing toward fracture origin, at a sharp corner

Fracture initiated at the location indicated by the arrow, where the corner of a snap-ring slot was specified to have a zero minimum radius. Fracture surface is that of a forging of AMS 6434 (vanadium-modified 4335) steel that was heat treated to a yield strength of 196 MPa (190 ksi).

shaft has undergone considerable twisting. The axle was intended to have a quenched-and-tempered microstructure, but it was actually underaustenitized (or not austenitized at all) during heat treatment, as indicated by its spheroidized microstructure and hardness (22 to 27 HRC).

Example 1: Ductile Overload Fracture of an Extension Ladder Made From 6061-T6 Aluminum Alloy Extrusions.

A two-section aluminum extension ladder, owned by the fire department of a large city, broke in service after having been used at the sites of several fires. Fire department officials noticed that one of the structural members of the ladder section that broke had a pattern of cracks along its edges. The ladder was sent to a laboratory for determination of the cause of failure.

The ladder was constructed of aluminum alloy extrusions and stampings that were riveted together at each rung location and at the ends of side rails. Each side rail consisted of two extruded T-sections (made from aluminum alloy 6061-T6) that were held in position by rivets through stamped rung supports (Fig. 17). An end cap, consisting of a curved T-section and two flat plates, was riveted to each end of the side rails (not shown in Fig. 17) in a manner quite similar to the rung supports. The rungs, which were hollow cylindrical extrusions with shallow longitudinal grooves along the outer surface, were spaced about 0.3 m (1 ft) apart and were upset at each end to lock them into the rung supports.

Investigation. Both side rails in the lower section of the ladder, called the bed section, had broken immediately below the fifth rung from the top. Figure 17 shows the right-hand side rail in the area of the fracture and shows that the upper T-section (T-section A, the one that faced away from the burning structure) fractured at a rivet holding the rung support in place.

An extensive amount of macroscopic plastic deformation was associated with the fracture in the lower T-section (B, Fig. 17), whereas the deformation in the upper T-section (A, Fig. 17) was more localized in the fracture region. T-section A also had kinked between the fifth and sixth rungs from the top of the bed section. The upper T-section of the left-hand side rail, not shown in Fig. 17 but designated here as T-section C, also had bent between the fifth and sixth rungs; this location roughly coincided with the location of the bottom end of the upper section (or fly) of the ladder at the moment of failure.

In the left-hand side rail, only T-section C had broken. The lower T-section, also not shown in Fig. 17 but designated here as T-section D, had bent almost at a right angle to its original position. This severe bend also occurred immediately below the fifth rung, at approximately the same location in relation to the rung as the fracture in T-section B in the right-hand side rail. All four T-sections had suffered other damage of a lesser degree, mostly caused by their being bent and/or twisted as the ladder collapsed.

The fracture surfaces were examined visually and by optical (light) stereomicroscopy. From the general fracture appearance and the degree of deformation, it was concluded that all fractures occurred by overload.

The composition of a sample taken from T-section B was found to be within the specified limits for aluminum alloy 6061. Microscopic examination did not disclose any significant differences in microstructure or grain size among the four T-sections.

Thickness measurements at various locations in undeformed areas indicated that among the four T-sections the flanges and stems were of equal thickness within 0.10 mm (0.004 in.). This is well within standard industry tolerances for aluminum extrusions in this size range.

The Rockwell B hardness of all four T-sections was measured at several widely spaced locations; ranges were as follows:

Fig. 10 Fracture surface exhibiting crack-arrest lines (beach marks) that are characteristic of stable (slow) crack growth

Cracking initiated at two rivet holes (arrows A) on the web of a structural-support beam of aluminum alloy 7075-T6, propagated slowly by fatigue to points B on the flange, then changed to unstable crack growth (rapid fracture). Light fractograph. Approximately ¾ ×

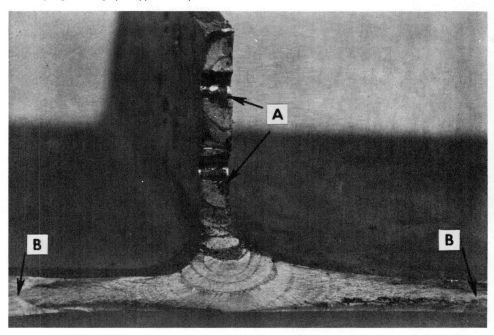

Fig. 11 Fracture surface exhibiting radial marks that converge at fracture origin

Fracture was initiated by a surface scratch, which can be located at lower edge of surface by tracing radial marks to their point of convergence. Fracture surface is that of an A2 tool steel, heat treated to 54 to 56 HRC, that broke upon being loaded to 860 MPa (125 ksi) for a second time.

T-section	Hardness, HRB
A (upper, right side rail)	20-32
B (lower, right side rail)	48-51
C (upper, left side rail)	47-55
D (lower, left side rail)	19-25

These data indicated that T-sections B and C had been properly aged, because their hardnesses were in the lower end of the range of 6061-T6 (acceptable hardness, 47 to 72 HRB). On the other hand, T-sections A and D had hardnesses that were considerably lower than the acceptable hardness for the T6 temper and were within the range for 6061-T4 (acceptable hardness, 19 to 45 HRB). This indicated that T-sections A and D had been naturally aged at room temperature after solution heat treatment.

The four T-sections were visually examined for edge cracking. The edges of both flanges and the edge of the stem of T-section A had numerous small cracks over a substantial length in the general area of the fracture. About 1.5 m (5 ft) above the fracture, cracks were observed on both flanges but not on the stem, whereas cracks were present on only one flange about 1.5 m (5 ft) below the fracture. T-section B exhibited edge cracks on both flanges in the general area of the fracture and on only one flange about 1.5 m (5 ft) above the fracture. No cracks were found on the edge of the stem in T-section B or anywhere on T-sections C and D. Although one of the major fractures passed through an edge crack, it was concluded that

the edge cracks, which were caused by too high a temperature in the workpiece or excessive speed in the extrusion process, had no causative relation to the failure.

The exact load on the ladder and the position of the load at the instant of failure were not reported. However, when this type of ladder is in use, the lower T-section in each side rail (T-section B or D) is generally stressed in tension, whereas the upper T-section (A or C) is generally stressed in compression. Because aluminum alloy 6061-T4 has about half the yield strength of 6061-T6, failure probably occurred because T-section D yielded locally under load (T-section D was in the T4 temper and was stressed in tension). After T-section D had yielded, the load distribution on the side rail changed in the region of the yielding, placing the stem portion of T-section C in tension (and perhaps also affecting the load distribution in the right-hand side rail to put the stem of T-section A in tension as well). T-section C became overloaded and bent, then fracture originated at the edge of the stem near a stress concentration at a rivet hole and propagated toward the flange. The rapid failure of the left-hand side rail caused the right-hand side rail to become overloaded and to fail by ductile fracture as the ladder collapsed.

Three regions of T-section A were examined and all were found to be soft. If T-section A had been softened by exposure to the heat of a fire, T-section B, which would have been closer to the fire, also would have been softened.

However, examination of T-section B showed that it had not been softened.

Conclusions. The fire ladder broke because two of the four extruded aluminum alloy T-sections (A and D) in the side rails had a low yield strength; they had been naturally aged after solution heat treatment, rather than artificially aged. Solution treatment could not have occurred as a result of exposure to the heat of a fire in prior usage because one T-section in each side rail had properties that were in agreement with published values for artificially aged aluminum alloy 6061.

Bending, caused by overload, was the mechanism of failure in one T-section. The three others then failed by ductile fracture, also caused by overload. Edge cracking in two of the T-sections was the result of improper conditions during extrusion of the T-sections; however, this condition was not a primary cause of failure.

Failures From Improper Fabrication

Imperfections produced during forming, machining, joining, plating, and other fabrication processes can lead to ductile or brittle fracture. Examples of some of these imperfections and resulting part failures are presented in this section.

Forming

Forming operations, such as cold heading, cold extrusion, stamping, drawing, bending, and straightening, can produce severe imperfections if proper processing and material controls are not used. For example, depending on the type of alloy, speed cracks can develop on an extrusion if the extrusion speed is above a critical level. Figure 18(a) shows an extreme example of this type of cracking. When the critical extrusion speed is exceeded, die friction

Fig. 12 Hydrogen flakes (arrows) in a 405-mm (16-in.) diam forging

Hairline crack that opened (see inset) shows characteristic shape of hydrogen flake. Some staining (arrows in inset) resulted from cold etching. Source: Ref 1

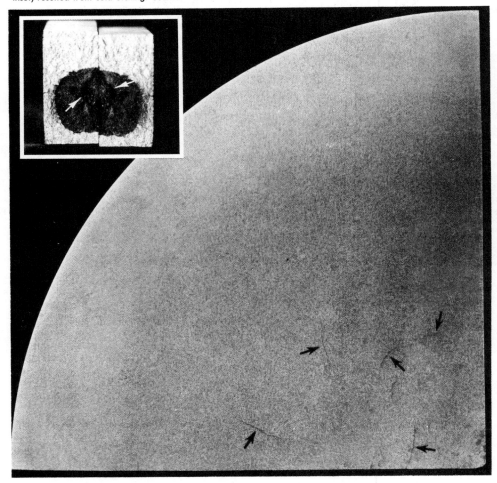

raises the surface temperature to the point at which the grain boundaries can no longer withstand the longitudinal surface stresses produced during extrusion, and transverse intergranular cracks result (Fig. 18b).

Forward cold extrusion can produce internal imperfections known as central bursts, or chevrons. Their occurrence is nearly always restricted to isolated lots of material and usually to only a small percentage of the pieces extruded in any particular production run. Some of the factors that can contribute to the formation of chevrons are incorrect die angles; either too great or too small a reduction of cross-sectional area; incomplete annealing of slug material; excessive work hardenability of the slug material; the presence of an excessive amount of seams and other imperfections in the slug material; segregation in a steel slug that results in hard martensitic particles in the center of the slug, which act as barriers to material flow; and insufficient die lubrication.

Blanking. Burrs produced by blanking can sometimes lead to failure. For example, examination of a spring washer of 17-7 PH stainless steel that fractured in a simulated-service test revealed that fracture originated from the burr side of the washer, which was also the tension side. By processing so that burrs are on the compression side of the washer, no additional fractures have been experienced.

Hot- or cold-forming operations, such as drawing, bending, and straightening, can produce grain deformation, orange peel, thinned corners, local regions of severe work hardening, and cracks. These effects, which may lead to failure, are discussed in the article "Failures of Cold Formed Parts" in this Volume. In the following example, cracks that were produced during a straightening operation led to fracture of an aluminum alloy aircraft fitting.

Example 2: Fracture of an Aluminum Alloy 2014-T6 Catapult-Hook Attachment Fitting for Naval Aircraft. The forged aluminum alloy 2014-T6 catapult-hook attachment fitting shown in Fig. 19 was from a naval aircraft. The fitting broke in service. The surface of the fitting had been anodized by the chromic acid process to protect it from corrosion.

Investigation. The fitting was analyzed spectrographically and found to be within the required composition limits for aluminum alloy 2014. Visual examination revealed a brown stain on the fracture surface. The stain was analyzed and found to be a residue of chromic acid, evidently from the anodizing treatment. Minute cracks were discovered on the inside surface of the bearing hole, and small areas of pitting corrosion were visible on the exterior surface of the fitting.

Microscopic examination revealed a small number of rosettes, suggestive of eutectic melting, in an otherwise normal structure. They did not appear to be associated with the cracks or to have contributed to failure.

Tensile-test coupons were cut from the fitting and were found to have the following mechanical properties:

Tensile strength	447 MPa (64.9 ksi)
Yield strength	406 MPa (58.9 ksi)
Elongation in 50 mm (2 in.)	11%

Minimum longitudinal properties of an aluminum alloy 2014-T6 die forging with a section thickness of 100 mm (4 in.) or less (which apply to this attachment fitting) are:

Tensile strength	448 MPa (65 ksi)
Yield strength	379 MPa (55 ksi)
Elongation in 50 mm (2 in.)	7%

Conclusions. The presence of chromic acid stain on the fracture surface proved that the forging had cracked before anodizing. This suggests that the crack initiated during straightening, either after machining or after heat treatment.

The structure and composition of the alloy appear to have been acceptable. Although the tensile and yield strengths were barely acceptable for aluminum alloy 2014-T6 die forging of 100-mm (4-in.) thickness or less, this fact is not considered to have affected the failure, because the ductility was acceptable. The rosettes found in the microstructure are believed to have been nondamaging. Had they contributed to the failure, the ductility would have been very low.

Recommendations. To avoid a recurrence of this type of failure, the catapult-hook attachment fittings in stock should be inspected for cracks. In addition, the manufacturing process should be revised to include a fluorescent liquid-penetrant inspection before anodizing, because chromic acid destroys the penetrant. This inspection would reduce the possibility of cracked parts being used in service.

Fig. 13 Circular spall that began at a large subsurface inclusion in a hardened steel roll

Source: Ref 2

Fig. 14 Fracture surface of an ASTM A36 structural steel member

Failure began by fatigue (arrows), but progressed only a short distance before brittle fracture occurred. Source: Ref 1

Machining

Parts often contain machining marks when completed. The location and severity of such marks are important in determining how much influence, if any, they have on fracture. Stress raisers consisting of sharp depressions on the surface of a part are generally produced by machining. However, corners and edges resulting from missed machining or deburring operations are more serious stress raisers.

Another type of stress concentrator that can be produced during machining is embedded tool chips in the surface of the workpiece. In one case, during nondestructive testing of a stainless steel part by radiography, a high-density inclusion was detected. The inclusion was identified by SEM to be a chip from a tungsten carbide tool bit. If the part had not been thoroughly inspected, the chip may have caused the part to break.

Grinding. Many steel parts are designed to have smooth hard surfaces for resistance to wear in service. These parts include journals on shafts, piston pins, and bearing surfaces. These types of parts are normally machined, hardened by heat treatment, and ground to extremely close tolerances.

Cracks can result from grinding burns, or scorches, caused by localized frictional heating during grinding. Grinding burns vary in severity from a lightly tempered condition to fresh, untempered martensite. Sometimes, only a shallow layer, 0.0025 to 0.0075 mm (0.0001 to 0.0003 in.) deep, is affected, but heavy grinding produces much deeper burned layers. Heavy grinding causes localized heating sufficient to raise the surface temperature to the austenitizing temperature of 760 to 870 °C (1400 to 1600 °F). Subsequent rapid quenching by the grinding coolant results in brittle surface layers of untempered martensite. Also, if the surface contains residual compressive stresses, such as those developed during carburizing, heating the surface to its austenitizing temperature will relieve these beneficial compressive stresses.

Mild grinding burns consist of shallow layers that have only been tempered; maximum surface temperatures do not exceed Ac_1. In many respects, mild grinding burns resemble heat-affected zones (HAZ) in steel weldments. A similar shallow, tempered layer exists beneath the untempered layer of a severe grinding burn. These tempered layers, although less brittle than untempered layers, are nevertheless regions where material properties differ from those in the interior of the part. Tempering can induce changes of surface stresses or formation of cracks or both. Because the heated area is localized, its ability to contract is restricted by surrounding harder and stronger martensitic regions that have not been heated (tempered) during grinding. The tempered regions subsequently contain residual tensile stress. If this residual tensile stress is great enough to exceed the strength of the tempered material, cracks form.

Fig. 15 Brittle fracture that began at a punched hole in a steel plate

Mating fracture halves show chevrons pointing back toward the punched hole. Source: Ref 2

Fig. 16 305-mm (12-in.) long axle shaft made of AISI S7 tool steel that failed in torsion

Note torsion marks on axle shaft. Improper hardening of the shaft resulted in overload failure. Source: Ref 1

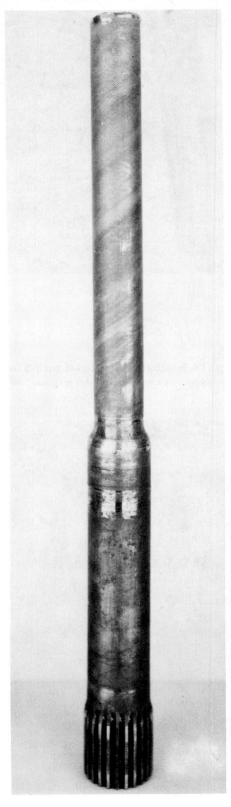

Hardened surfaces containing patches of retained austenite are exceedingly sensitive to grinding operations. If the heat or stress generated in grinding is sufficient to induce transformation of the retained austenite to fresh martensite (normally referred to as untempered martensite), cracks may form in the surface or just below it. If cracks do not occur immediately, the newly transformed regions often crack or spall under the additional stresses encountered under load in service. Severe grinding can also develop grinding cracks in unhardened parts.

Grinding burns can be revealed by etching the ground surface in an aqueous solution of 5 to 10% nitric acid. A surface that contains grinding burns will exhibit a series of dark-gray streaks on a light-gray background. The streaks will follow the grinding direction, and the severity will vary from a few very small streaks to an almost completely dark-gray surface.

Example 3: Brittle Fracture of a Case-Hardened Component Because of Low Impact Resistance and Grinding Burns. The latch tip on the main-clutch stop arm shown in Fig. 20(a) fractured during normal operation in a business machine. In operation, the latch tip was subjected to intermittent impact loading. Three stop arms that failed in service were sent to the metallurgical laboratory for determination of the cause of failure.

The stop arms were machined from 8620 steel. The latch tips were carburized and then induction hardened to a minimum surface hardness of 62 HRC.

Investigation. Examination of two stop arms indicated that fracture of the latch tip (Original design, Detail A, Fig. 20a) occurred at the hardness-transition zone created by induction hardening of the carburized case on the tips. This transition zone, because of its inhomogeneous structure, decreased the overall strength of the parts. In addition to the transition zone, a sharp ground relief step at the location of failure contributed to the low impact resistance of the

arm. The hardness-transition zone usually coincided with, or was slightly outward (radially) from, the relief step.

The fracture surface (Fig. 20b) had a typical brittle appearance, with no necking or shear lips. The fracture surface had a fine-grain appearance in the hardened case (at right, Fig. 20b), but was somewhat coarser throughout the rest of the fracture.

Examination of an etched section through another fractured latch tip disclosed several small cracks that extended almost through the hardened case to the core; two of these cracks are indicated by arrows at right in Fig. 20(c). Fracture of the parts may have occurred through similar cracks. Also observed was a burned layer approximately 0.075 mm (0.003 in.) deep on the latch surface. This layer (arrow at left, Fig. 20c) was caused by improper grinding, and the cracks present in the part were apparently the result of this grinding. Hardness at a depth of 0.025 mm (0.001 in.) in this layer was 52 HRC (a minimum of 55 HRC was specified).

In an attempt to relocate the transition zone and as a method of salvaging finished parts that had the transition zone coinciding with the relief step, several stop arms were induction hardened a second time. Erratic results obtained when samples were impact tested indicated that the parts were not usable.

Conclusions. The induction-hardened stop arms failed by brittle fracture in the hardness-transition zone as the result of excessive impact loading. The presence of several cracks through the hardened case near the region of failure on one of the latch tips indicated that fracture apparently occurred through a similar crack. A burned layer on the surface of the latch indicated that the cracks had been caused by improper grinding after hardening.

Corrective Measures. To strengthen the stop-arm latch tip, the size of the backing web that reinforced the tip was increased (Improved design, Detail A, Fig. 20a). Also, the radius at

Fig. 17 Extruded right-hand side rail of aluminum alloy 6061-T6 that failed from ductile overload fracture

Dimensions given in inches

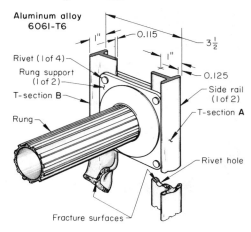

the relief step was increased to 1.5 ± 0.5 mm (0.06 ± 0.02 in.), proper grinding techniques were used, and the hardened zone was specified to extend a minimum of 1.5 mm (0.06 in.) beyond the step. No failures of parts made to those specifications have been reported.

Electric discharge machining (EDM) is a technique for machining electrically conductive materials by the erosive action of spark discharge; the workpiece is the anode, and the electrode is the cathode. Each spark erodes a small amount of metal. Electrical discharge machining has found considerable application in the machining of dies because the process can machine intricate cavities economically.

However, surfaces machined by the EDM technique exhibit some damage. The depth of this damage is related to the EDM procedures used. The rate of metal removal depends on the volume of the crater formed by each spark and the frequency of the sparking. The crater depth establishes the degree of surface roughness. The surface often has a splattered or a wavelike appearance due to the resolidification of molten droplets that were not removed by the spark action during EDM processing.

Metallurgical examination indicates that surface heating is due to the melting action of the spark. The microstructure at the surface of electric discharge machined dies often resembles the microstructure of welds; the surface frequently exhibits an as-solidified appearance. In highly alloyed die steels, the surface layer usually consists of fresh martensite, retained austenite, and eutectic-type carbides. Below this white layer, the steel has been gradiently heated and contains a range of microstructures, hardnesses, and properties. Zones heated above the Ac_3 or the Ac_{cm} temperature of the steel usually contain fresh martensite and retained austenite (Ac_3 represents the temperature at which transformation of ferrite to austenite is completed during heating, whereas Ac_{cm} represents the temperature at which solution of cementite in austenite is completed during heating). Below these zones, the steel has been heated into the two-phase region and may contain fresh martensite, annealed microstructures, and carbides, depending on the steel composition. Below this two-phase zone, the heat of the operation tempers the original microstructure, with the degree of tempering decreasing with depth until the base hardness is obtained.

The depth of the white layer and the gradiently tempered zones vary considerably, depending on the EDM procedures employed. In addition to the commonly observed surface condition discussed above, other surface abnormalities, such as carburization or decarburization, have occasionally been observed on electric discharge machined surfaces. Cracks are also occasionally observed in these surface layers. Residual stresses are also created during cooling after EDM.

Electric discharge machined surfaces frequently exhibit greater wear resistance than that normally achieved after conventional quenching and tempering. This is due to the high hardnesses associated with the white layer. The

Fig. 19 Catapult-hook attachment fitting forged from aluminum alloy 2014-T6

The component cracked during straightening, then fractured in service.

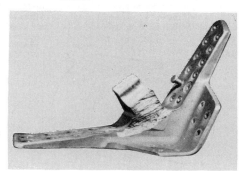

metal at an electric discharge machined surface generally exhibits a lower endurance limit than obtained with the same steel after conventional processing. Therefore, tools or die that have been produced by EDM usually do not perform as well in situations involving fatigue. Lapping of the surface after the EDM operation will restore the original fatigue life. Electric discharge machined surfaces should be removed if the tool or die is subjected to high tensile stresses, high impact loads, or cyclic stressing.

Problems associated with EDM can be avoided if the nature of these problems is understood and the proper steps are taken. Manufacturers of EDM equipment can usually recommend procedures for handling different types of tool steels. The safest procedure to follow is to:

- Perform EDM operations on tool steels in the annealed condition only
- Stress relieve below the Ac_1 temperature of the steel
- Remove the electric discharge machined surface by grinding or lapping
- Quench and temper using the appropriate procedures

Fig. 18 Cracking due to excessive extrusion speed

(a) Aluminum alloy 7004 extrusion with speed cracks on two edges. Actual size. (b) Micrograph of a section through a speed crack that propagated along grain boundaries when critical extrusion speed was exceeded. Etched with Keller's reagent. 350×

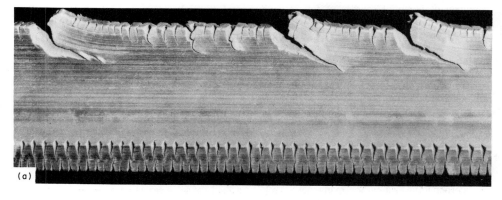

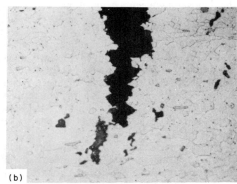

Fig. 20 Main-clutch stop arm of 8620 steel on which the case-hardened latch tip failed in service from brittle fracture because of low impact resistance and grinding burns

(a) View of stop arm showing location of fracture in latch tip, and detail showing original and improved designs of latch tip. Dimensions given in inches. (b) Fractograph of a typical fracture surface showing brittle appearance; note fine-grain appearance of the hardened case (at right). 9×. (c) Micrograph of a section through a fractured latch tip that shows two of several small cracks (arrows at right) that extended through the hardened case to the core, and the burned layer on latch surface (dark band indicated by arrow at left) that resulted from grinding burns

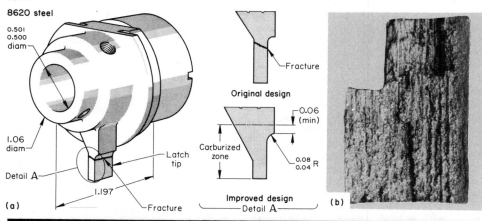

Fig. 21 Lamellar tearing in the HAZ of a carbon-manganese steel corner joint

Etched with 2% nital. Courtesy of R. Dolby, The Welding Institute, United Kingdom

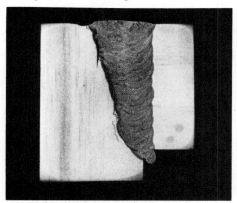

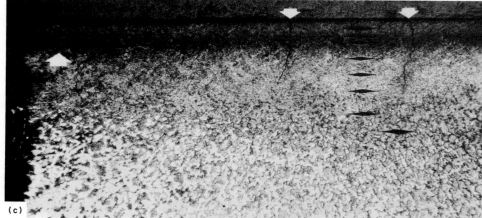

If the part is in the hardened condition before EDM processing, it should be stress relieved (tempered) immediately after machining, using the highest temperature consistent with the desired hardness. Removal of the surface layer by grinding or lapping will help to prevent premature failure. This surface layer may also be removed by electrochemical machining.

Welding and Brazing

Welding and brazing techniques are often used to join metals. Occasionally, welded joints are primary origins of failures. These failures often occur in a catastrophic manner. When failure occurs in a weldment, the failure analyst should make every effort to pinpoint as closely as possible the exact origin and mechanism of the failure.

Prime consideration to several factors, such as stress concentration, should be given early in the design of a weld. Welded and brazed joints are often placed in a design configuration that introduces the possibility of notch-type stress concentrations. The failure analyst should be aware of the possibility of such stress-concentration effects whenever a weld is suspected as the origin of a component failure. See the articles "Failures of Weldments" and "Failures of Brazed Joints" in this Volume for a review of potential sources of failures and methods of failure prevention.

Weld Imperfections. One of the most common factors associated with weld failures is the introduction of weld imperfections, such as porosity, incomplete fusion, inclusions, arc strikes, and hard spots, which act as crack sources and initiation points and often provide ready paths for subsequent crack propagation in a brittle manner. Detailed information on defects associated with weldments can be found in the article "Weld Discontinuities" in Volume 6 of the 9th Edition of *Metals Handbook*.

Figure 21 shows lamellar tearing in the HAZ of a corner joint in a carbon-manganese steel.

Lamellar tearing is a form of base-metal cold cracking that occurs in steels parallel to the plate surface adjacent to the welds. Lamellar tearing is associated with nonmetallic inclusions, such as oxides, sulfides, and silicates, that are elongated in the direction of rolling. The net result of these inclusions is a decrease in the through-thickness ductility. Other factors that affect lamellar tearing are design details, plate thickness, magnitude, sign and distribution of induced stresses, type of edge preparation and properties, and base-metal fabrication.

Figure 22 shows a hydrogen-induced root crack in the HAZ of a high-strength low-alloy steel weldment. Another form of hydrogen embrittlement of welds is known as fisheyes, which are cracks surrounding porosity or slag in the microstructure of a weld. Fisheyes are caused by the presence of hydrogen that collects at these locations and that embrittles the surrounding metal. Use of low-hydrogen electrodes lessens the chances of hydrogen embrittlement in welds. See the article "Hydrogen-Damage Failures" in this Volume for additional information on the deleterious effects of hydrogen.

A void caused by incomplete fusion is illustrated in Fig. 23, which shows a gas tungsten arc weld between a nickel alloy and an alloy steel. Special care must be taken when welding with nickel alloy filler metal in order to avoid this problem of incomplete fusion. The primary factor is weld-joint design. Molten nickel alloy weld metal is sluggish and does not spread or flow as well as steel weld metal. Therefore, joint accessibility must be sufficient to allow proper control of the arc.

The other important factor in welding a nickel alloy is the welding procedure. The inherent lower penetration of nickel alloys requires smaller weld beads and strict guidance of the arc. An increase in amperage will not significantly increase the penetration, but will instead cause overheating and puddling of the

Fig. 22 Hydrogen-induced root cracking in the HAZ of a Fe-0.2C-1.43Mn-0.01Si-0.015Nb steel

Etched with 2% nital. 8×. Courtesy of R. Dolby, The Welding Institute, United Kingdom

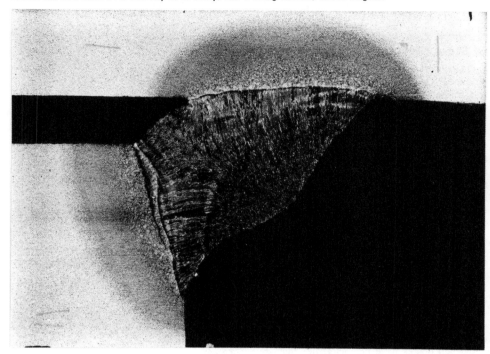

Fig. 23 Section through an automatic gas tungsten arc weld containing voids caused by incomplete fusion

(a) Base metal at left is Incoloy 800 nickel alloy, that at right is 2.25Cr-1.0Mo alloy steel. Filler metal was ERNiCr-3, used with cold wire feed. Macrograph. 1½×. (b) Micrograph of the area circled in (a) showing void. 75×. Both etched with 2% nital

(a)

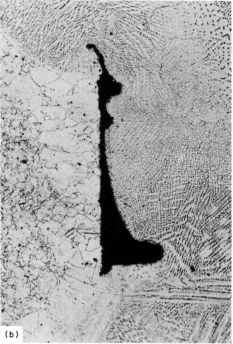

(b)

molten weld metal with resultant loss of deoxidizers and rolling of the weld metal, leading to unsound welds. The void shown in Fig. 23, which was found by radiographic and ultrasonic inspection but could have led to failure in service, was caused by misguidance of the welding arc.

Because weld imperfections are one of the most common factors associated with weld failures, it is good practice, whenever possible, to inspect joints by appropriate nondestructive methods for detection of imperfections. If a weld failure has been attributed to a weld imperfection, the cause of the imperfection must be ascertained and a solution provided. In this solution, the importance of metal and weld cleanness is often stressed. Unfortunately, overzealous joint preparation by an operator who is not aware of metallurgical consideration has also resulted in failures in weldments and brazements. Heavy, rough grinding of certain ferrous alloys is particularly damaging and introduces flaws that produce subsequent failure.

Example 4: Low-Temperature Brittle Fracture in a Steel Tank Car Because of Weld Imperfections. A railway tank car developed a fracture in the region of the sill and shell attachment during operation at −34 °C (−30 °F). On either side of the sill-support member, cracking initiated at the weld between a 6.4-mm (¼-in.) thick frontal cover plate and a 1.6-mm (⅝-in.) thick side support plate. The crack then propagated in a brittle manner upward through the side plate, through the welds

attaching the side plate to the 25-mm (1-in.) thick shell plate, and continued for several millimeters in the shell plate before terminating. The fracture surfaces displayed chevron marks pointing back to the weld (arrow, Fig. 24) between the frontal cover and side support plates at the point of fracture origin.

Investigation. Examination of the weld at the fracture origin revealed imperfections, including small root cracks, unfused regions, and small hard spots on the surfaces of the weld and adjacent plates. Micrographs and hardness values (up to 45 HRC) showed that the hard spots had been rapidly cooled, but why they formed was not ascertained. The presence of these hard spots indicated that the assembly had not been adequately stress relieved after structural modifications that involved welding in the region of the sill and shell attachment.

The shell plate met the chemical-composition requirements of ASTM A212, grade B, steel. Other plates involved were not positively identified, but were generally classified as semi-killed carbon steels.

The toughness properties of the shell and side support plates were measured and were found to be inadequate for the service conditions in the presence of weld imperfections. Charpy V-notch testing gave a 20-J (15-ft · lb) transition temperature of −7 to 5 °C (20 to 40 °F) for the shell plate and −1 to 5 °C (30 to 40 °F) for the side support plates. Drop-weight tests on the shell plate gave a nil-ductility temperature of 5 to 10 °C (40 to 50 °F).

Conclusions. The fracture was initiated by weld imperfections and propagated in a brittle manner as a result of service stresses acting on plate having inadequate toughness at the low temperatures encountered in service.

Recommendations. On the basis of fracture-toughness studies, it was suggested that the specifications for the steel plates be modified to include a toughness requirement; the Charpy V-notch requirement of 20 J (15 ft · lb) at −46 °C (−50 °F), which is specified in ASTM A 300 for ASTM A212, grade B, steel, should be adequate for this application. In addition, the welding and inspection practices should be improved to reduce the incidence of weld imperfections greatly. It should be noted that

Fig. 24 Brittle fracture in a carbon steel weldment

Failure originated at a weld (arrow) between a frontal cover plate (bottom) and a side support plate (center) for a tank car. Chevron marks in both plates point to fracture origin. A side plate is at top.

ASTM A212, grade B, steel has been superseded by ASTM A516, grade 70, steel, which has a low-temperature requirement of 20 J (15 ft · lb) at −34 °C (−30 °F) as specified in ASTM A 20/A 20M (Ref 4).

Postwelding operations, such as stress relieving, should not be overlooked as a possible source of problems associated with joint failures, especially in view of the large thermal stresses and residual stresses caused by welding some metal structures. The role of residual stresses in failures of welds is discussed in the section ''Failures From Residual Stresses'' in this article and in the article ''Residual Stresses and Distortion'' in Volume 6 of the 9th Edition of *Metals Handbook*.

Failures From Improper Thermal Treatment

Imperfections and abnormalities can be produced during any of the thermal treatments associated with hot or cold reduction or shaping, such as homogenization, preheating, and annealing. Overheating and burning, both of which reduce strength and ductility, are two examples that are discussed in the article on ''Failures of Forgings'' in this Volume. Case hardening of parts is another important source of imperfections and abnormalities that can lead to failure. For example, fatigue cracks often initiate at the case/core interface (where the strength gradient is steep) in case-hardened parts subjected to cyclic bending or torsional loads.

Unless precautions are taken, case hardening may jeopardize toughness properties gained by proper material selection. For example, a sharp notch in a steel part requiring good toughness should not be carburized or nitrided. If case hardening of the part is necessary, the notched area should be masked off, or the case formed in this area should be removed by machining or grinding.

Final heat treatment and tempering to the required strength, ductility, and toughness lev-

els is another type of thermal treatment that can cause deficiencies that can lead to failure by brittle or ductile fracture. Quench cracks or embrittlement can occur during these various thermal treatments, as discussed below.

Quench Cracking of Steel (Ref 5)

Quench cracks in steel result from stresses produced during the austenite-to-martensite transformation, which is accompanied by an increase in volume. As-quenched martensite is hard and exhibits almost no ductility.

When a component made of fully hardenable alloy steel is quenched, martensite forms first at the outermost surfaces, which are first to reach the martensite start, M_s, temperature. The martensitic expansion works the softer austenite below and is almost unrestricted in its growth at the outer surfaces. As cooling progresses and the material near the center of the section reaches the M_s temperature, the expansion accompanying the newly formed martensite is restricted by the outer layers of martensite formed earlier. This results in internal stress that places the surface in tension. Cracking occurs when enough martensite has formed to set up an internal stress sufficient to exceed the tensile strength of the as-quenched martensite at the outer surfaces of the component.

Quench cracks have characteristics that are easily recognized. First, the fracture generally runs from the surface toward the center of mass in a relatively straight line (Fig. 25); the crack is also likely to open or spread and may exhibit a shear lip at the extreme surface. Second, because quench cracking occurs at relatively low temperatures, the crack will not exhibit any decarburization when examined macroscopically or microscopically (Fig. 26). Lastly, the fracture surfaces will exhibit a fine crystalline texture. When tempered after quenching, the fracture surfaces may be blackened by oxidation (Fig. 28, Example 5). Microscopic examination of a quench crack tempered at higher temperatures and under oxidizing conditions will disclose tempering scale (Fig. 27).

Factors Controlling Cracking. Any condition that concentrates the stresses encountered in quenching will promote the formation of quench cracks. Whenever possible, sharp changes in section, such as rectangular keyways or holes, should be avoided or plugged during quenching.

Cold stamping marks used to identify parts have also been known to nucleate quench cracks. The distribution of mass and the lack of uniform or concentric cooling of the parts also influence the promotion of quench cracking.

Faster quenching media promote deeper hardening. The selection of a suitable quenching medium is often a significant factor in eliminating quench cracks. The quenchants most commonly used are caustic solutions, brine, water, oil, and air; the fastest quench can be obtained in caustic solutions and the slowest in still air. Molten salts or metals with low

Fig. 25 Quench crack in a round steel bar

Extent of lighter-etching outer region indicates depth of martensitic transformation. Note that the crack is confined to the hardened region and extends in a nearly straight line toward the center of the bar. Etched with nital. Actual size

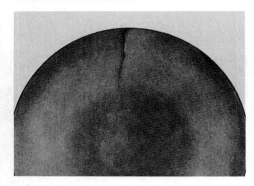

Fig. 26 Quench cracks in steel, which are characteristically free of decarburization

Etched with nital. 100×

melting points are also used for more complicated hardening treatments, such as martempering and austempering.

The selection of heat-treating schedules, quenching media, and tempering schedules is governed by the following factors:

- The hardness and mechanical properties required for the part to give the best service life; these may determine a specific microstructure
- The alloy steel selected for the part, which inherently determines the nature of the treatment to yield the desired properties
- The design of the part in relation to hardening and processing after heat treatment

Fig. 27 Quench crack in steel containing tempering scale

Presence of the scale indicates that crack surfaces were exposed to oxidizing conditions in the tempering furnace. Etched in 4% picral. 500×

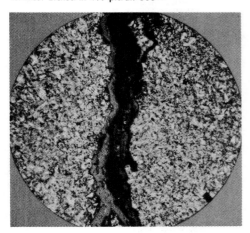

- The equipment available in a specific shop to do the job
- The production economics of the operations in relation to specific shop conditions.

Because the most important and universally accepted method of heat treating steel is quenching and tempering, which yields a product that differs materially from a normalized or annealed product, the basic factors considered in selecting a specific heat-treating schedule and quenching medium to produce the required microstructure should be understood. Failure to obtain the desired properties will result in poor service life in most cases. Ideally, the quench-ing medium should be selected for its quenching rate, that is, the ability of the quench to effect a cooling rate sufficient to develop the desired microstructure. After quenching, it is important that the hardened piece be tempered as soon as possible to relieve the internal stresses formed in quenching. A good rule of thumb is to draw the work from the quenching medium while it is still warm, about 65 to 95 °C (150 to 200 °F), and transfer to a tempering furnace. When oil quenching, the part should be smoking lazily during transfer from the quench to the tempering furnace.

Delayed Cracking. A common misconception is that quench cracks can occur only while the work is in the quenching medium. Actually, quench cracks can occur an hour, a day, or a week after quenching if the work is allowed to stand after hardening without tempering, as illustrated in the following example.

Example 5: Brittle Fracture of Rocket-Motor Case That Originated at Delayed Quench Cracks. The rocket-motor case shown in Fig. 28 failed during proof-pressure testing. Specifications required hydrostatic testing to a pressure of 5780 kPa (838 psig), but the case burst at 3075 kPa (443 psi).

Fabrication. The motor case was made of consumable-electrode vacuum arc remelted D-6ac alloy steel with the following composition:

Element	Composition, %
Carbon	0.47
Manganese	0.75
Silicon	0.22
Chromium	1.05
Nickel	0.55
Molybdenum	1.00
Vanadium	0.11
Iron	rem

The forward dome and aft adapter were made from machined forgings. The five cylinders were made by cold shear spinning of forged, machined, spheroidized preforms. The cylinders were annealed after spinning and before welding. All the components were joined by girth welds using the automatic gas tungsten arc process. There were no longitudinal seams. The rocket-motor case was 7.6 m (25 ft) in length and 1.5 m (5 ft) in diameter and had a stock thickness of 4.6 mm (0.180 in.) at the girth welds and 3.8 mm (0.150 in.) elsewhere.

After being assembled, the motor cases were heat treated as follows: preheat to 480 °C (900 °F), stabilize at 705 °C (1300 °F) for 4 h, austenitize at 900 °C (1650 °F) for 30 min, furnace cool to 845 °C (1550 °F) and hold for 30 min, quench in salt bath at 205 °C (400 °F) for 20 min, air cool to 50 °C (125 °F), wash in hot water, air dry, snap temper at 315 °C (600 °F) for 4 h, air cool, final temper at 550 °C (1025 °F) for 4 h, and air cool.

A 6-h time interval occurred between the start of the quench, out of a gantry furnace, and the start of the 315-°C (600-°F) snap temper in an air-atmosphere pit furnace. The time was consumed in completing the modified marquench in 205-°C (400-°F) salt, cooling in air, rinsing to remove encrusted salt, and drying. For about 4½ of the 6 h, the case was left standing on its aft end, drying and waiting for the pit furnace to be available.

The 315-°C (600-°F) snap temper was used to impart sufficient ductility to the motor case so that it could be fixtured with a series of adjustable spiders and shoes that rounded the casing to final dimensions. The 550-°C (1025-°F) temper served a dual purpose—it developed the required material properties and acted as a

Fig. 28 D-6ac alloy steel rocket-motor case that failed from brittle fracture because of delayed quench cracks that originated in the forward dome and partly penetrated the wall of the case

(a) Configurations and dimensions (given in inches). (b) Fractograph showing oxidized quench crack (arrow A) that was the primary origin of fracture, and adjacent quench crack (arrow B) that was a secondary origin. Approximately 3×. (c) Photograph of the fracture region showing quench-crack patterns (arrows) on both sides of the main fracture. Approximately actual size

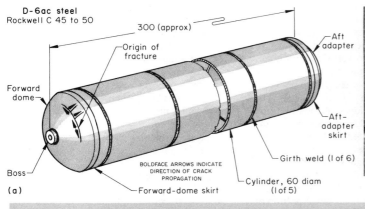

temper-straightening operation. Attempts to eliminate the 315-°C (600-°F) snap temper in favor of a 220-°C (425-°F) temper in the same salt bath used for quenching had been tried unsuccessfully on other parts. Although the 220-°C (425-°F) temper provided some ductility, it was not enough to prevent cracking during installation of the temper-straightening fixtures.

The final hardness of the rocket-motor case was 45 to 50 HRC. Temper scale was removed by light abrasive blasting.

Visual examination of the failed motor case revealed that brittle fracture originated in the elliptical portion of the forward dome and propagated circumferentially, then propagated radially into the immediately adjacent cylinder (Fig. 28a). The late stages of crack propagation were in macroscopic shear, a ductile-fracture mode.

The primary-fracture origin was identified from radial marks and was coincident with an oxidized prior crack (dark region at arrow A, Fig. 28b) that had initiated at the outer surface of the dome and had partly penetrated the dome wall. Several secondary origins were identified at other pre-existing surface cracks, all of which contained oxide scale (dark region at arrow B, Fig. 28b). Close visual examination disclosed an extensive pattern of cracks with a maximum length of 3.8 mm (1.5 in.) and a maximum depth of 1.5 mm (0.06 in.) in the fractured dome. These cracks (arrows, Fig. 28c) followed a chordal path about 90 cm (36 in.) long and 100 mm (4 in.) wide, passing within 305 mm (12 in.) of the boss on the dome.

Magnetic-particle inspection of the entire motor case revealed an extensive pattern of short circumferential cracks on the outer surface encircling the forward dome skirt and the first cylinder. Cracks were found in a pattern that extended about one-half the circumference of the aft-adapter skirt. There were no cracks in the remaining four cylinders.

Microhardness surveys were taken on specimens from the forward dome; one survey was adjacent and parallel to the fracture surface, and one was parallel to the outer surface. The first survey showed partial, but acceptable, decarburization of the outer surface, which occurred during austenitizing. The latter survey showed a uniform hardness and no decarburization of the fracture surface.

These data indicated that cracking occurred after the austenitizing stage of heat treatment, which would decarburize exposed surfaces. Thus, the pre-existing cracks occurred after austenitizing but before tempering.

Microscopic examination of polished sections from the dome, first cylinder, and aft adapter identified the scale in the cracks as temper scale, similar to that which would form on the casing surface during final tempering at 550 °C (1025 °F). This indicated that the cracks appeared before or during the final tempering phase of the heat treatment. A section was taken across one of the cracks in the motor case, and the specimen was given a sensitizing heat treatment to reveal prior-austenite grain boundaries. Carbide precipitation at prior-austenite boundaries was accomplished by heating the specimen to 455 °C (850 °F) and holding at that temperature for 5 h. The specimen was prepared for microscopic examination and etched in ethereal picral (a mixture of picric acid in water, zephiran chloride solution, and anhydrous ether, from which the ether layer is decanted for etching) until the prior-austenite boundaries became visible. Cracks in the etched specimen were intergranular, which is typical of cracks that occur in as-quenched steel. These results indicated that cracking had occurred while the steel was in the as-quenched condition, before the 315-°C (600-°F) snap temper.

X-ray diffraction studies of material from the fractured dome showed a very low level of retained austenite (about 1%). Specimens from the dome that were heat treated in the laboratory to the as-quenched condition had about 3 to 5% retained austenite. This indicated that transformation of retained austenite had occurred during the extended delay between quenching and snap tempering.

Chemical analysis of the metal in the three components that cracked showed a carbon content of 0.47 to 0.50%, compared with 0.44 to 0.46% carbon in the components that did not crack. There was no correlation between cracking and the amount of other alloying elements in the material. Thus, cracking appeared to be related only to the carbon content of the material. Components with higher carbon contents were more susceptible to cracking than those with lower carbon contents.

Bend Tests. Specimens for slow-bend tests were machined from samples taken from the failed dome. Specimens for static-bend tests were machined from a forged ring of D-6ac steel containing 0.50% C and from a ring containing 0.42% C. The former were heat treated in the laboratory to simulate various stages of the production process, then tested. Results showed that only the as-quenched condition produced brittle cracking along prior-austenite grain boundaries. Static loading of as-quenched specimens to various stress levels produced delayed cracking that resembled the processing cracks in the casing in both high-carbon and low-carbon D-6ac steels. The greater probability for delayed quench cracking in higher carbon material, demonstrated in the static-bend tests, correlated with observations of cracking in the three components in the failed case that had the highest carbon contents.

Conclusions. Brittle fracture of the rocket-motor case during hydrostatic testing originated at pre-existing cracks in the elliptical portion of the forward dome. The patterns of pre-existing cracks along a chordal path in the dome and in circumferential bands in the first cylinder and aft adapter were the result of delayed quench cracking. The most likely mechanism of de-

Fig. 29 Quench cracks in steel, which are associated with coarse austenitic grain size
Note that the cracks and crack tributaries follow the coarse prior-austenite grain boundaries. Etched in 4% picral and Vilella's reagent. 60×

layed quench cracking was isothermal transformation of retained austenite to martensite under the influence of residual quenching stresses.

Corrective Measures. To reduce the probability of delayed quench cracking, the quenching portion of the heat-treating cycle was modified as follows: quench in a salt bath at 205 to 220 °C (400 to 425 °F) for 10 min; cool to 40 to 50 °C (100 to 125 °F) in air; as soon as the motor case reaches 40 to 50 °C (100 to 125 °F), return it to the salt bath and hold at 205 to 220 °C (400 to 425 °F) for 2 h, then cool to room temperature.

Tempering in the salt pot used for quenching, immediately after quenching and before the case temperature dropped below 40 °C (100 °F), was successful in eliminating the problem. No other instances of this type of failure were experienced.

Causes of Cracking. Some of the more common causes of quench cracks in steel are the following:

- Overheating during austenitizing so that normally fine-grain steels are likely to coarsen. Coarse-grain steels are deeper hardening and are inherently more susceptible to quench cracking than fine-grain steels (Fig. 29)
- Improper selection of quenching medium—for example, the use of water, brine, or caustic solution when oil is the proper quenching medium for the specific part and type of steel
- Improper selection of steel
- Time delays between quenching and tempering
- Allowing temperature of component to drop too low before tempering (applicable mainly to hypereutectoid high-alloy and tool steels)
- Improper design of keyways, holes, sharp changes in section, and other stress raisers
- Improper entry of the part into the quenching medium relative to its shape, which results in nonuniform or eccentric cooling

Depending on the grade of steel, certain of the items listed above will be more important than others in determining whether a given component is likely to crack upon quenching. Although they may not quench crack, parts treated to high strength levels (high hardness) that contain localized high concentrations of residual stress may fracture instantaneously in service if the residual-stress pattern is acting in the same direction as the applied load.

Failures From Improper Electroplating

One of the most serious problems encountered during electroplating steel parts is the absorption of hydrogen, which can lead to the formation of flakes or to general embrittlement. Hydrogen may also develop during acid pickling or during service (from such sources as hydrogen-bearing fluids and certain corrosion products). Hydrogen embrittlement is discussed in further detail in the article "Hydrogen-Damage Failures" in this Volume.

It should also be noted that certain hard electrodeposits, especially hard chromium, frequently contain microcracks. These microcracks can be detrimental to parts that require good toughness. If plating is required for corrosion resistance, a soft metal, such as zinc or cadmium, is often preferred.

Arc striking is another problem that can occur during electroplating, although arc striking can also occur during electrocleaning, magnetic-particle inspection, and tests requiring electrical contact, as well as welding.

Example 6: Brittle Fracture of a Cadmium-Plated 4140 Steel Retaining Ring at a Hard Spot Caused by an Arc Strike. The retaining ring shown in Fig. 30 was used to hold components of a segmented fitting in place under a constant load. The ring was made of 4140 steel tubing and heat treated to 36 to 40 HRC, then cadmium plated. Several rings that broke after less than 30 days in service were examined to determine the cause of failure.

Investigation. Metallographic examination of a section adjacent to one fracture surface showed that the microstructure was tempered martensite and the inclusion content was low. The fracture was brittle in appearance, and there were no shear lips.

Examination also revealed a pit or burned spot on the outer surface of the ring (Section A-A, Fig. 30a and b). The pit appeared as a semicircular defect in the fracture surface at the outer edge of the ring (Fig. 30c). The defect was approximately 0.18 mm (0.007 in.) deep and 0.5 mm (0.020 in.) in diameter and had a hardness of 58 to 60 HRC. The base metal adjacent to the defect had a hardness of 36 to 40 HRC. At higher magnifications, small cracks or fissures were evident within the defect.

Two unbroken rings were visually examined and found to have defects similar in appearance to those on the rings that fractured in service.

Fig. 30 Retaining ring of cadmium-plated 4140 steel that failed by brittle fracture at a hard spot caused by an arc strike during plating

(a) View of retaining ring, and section showing location of arc strike. Dimensions given in inches.
(b) Photograph of the retaining ring showing pit or burned spot on the outer surface. 12×.
(c) Micrograph showing the pit as a semicircular defect in the fracture surface at the outer edge of the ring. Etched with nital. 100×

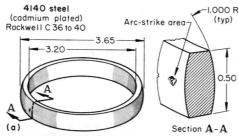

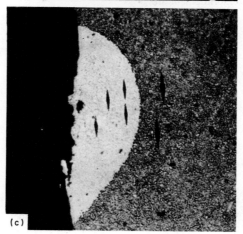

These two rings were fractured for further examination. Each fracture surface contained a small hard spot similar to those on the rings that failed; however, a shear lip at the inside diameter indicated an inherently ductile material.

The difference in hardness, etching characteristics, and surface appearance of other parts of the ring indicated that the small hard spot was untempered martensite that formed as the result of an arc strike during the cadmium-plating operation. Arcing at the point of contact between the ring and a finger on the plating

rack or arcing between rings while on the rack caused local overheating above the critical temperature of 735 °C (1350 °F). This arcing occurred either as the ring entered the plating bath or when it was in the bath. The part was quenched by the bath, resulting in transformation to hard, brittle martensite at the locations of arcing.

Conclusions. The rings failed in brittle fracture as the result of an arc strike (or burn) on the surface of the ring. At the site of the arc strike, a small region of hard, brittle untempered martensite was formed. Fracture occurred readily when the ring was stressed.

Failures From Residual Stresses

Residual stresses are stresses that exist in a part independent of external force or restraint. Nearly every manufacturing operation affects the residual-stress pattern in varying degrees. Cold heading, cold extruding, drawing, bending, straightening, machining, surface rolling, shot blasting, shot peening, and polishing all are mechanical processes that produce residual stresses by plastic deformation. Thermal processes that introduce residual stresses include hot rolling, welding, torch cutting, and heat treating (through thermal expansion and contraction or allotropic transformation) and carburizing and nitriding (through increase in case volume).

Heat-treating processes that usually produce compressive residual stresses at the surfaces of parts are surface hardening, carburizing, and nitriding. Some mechanical procedures (shot peening, surface rolling, cold drawing through a die, extruding, and polishing) produce the same effect. Nearly all other manufacturing operations produce residual tensile stresses at the surface of parts. Grinding, improper heat treatment, straightening, and welding generally produce the most severe residual tensile stresses.

Residual stresses can be of great help if the stress patterns are favorable. In general, residual stresses are beneficial when they are parallel to the direction of the applied load and of opposite sense, that is, a compressive residual stress and a tensile applied load or a tensile residual stress and a compressive applied load. The magnitude of residual stresses is important, particularly when service stresses are fluctuating in nature; residual tensile stresses raise the mean stress, thereby reducing fatigue life. Also, the residual-stress pattern may be altered by the applied stress; that is, localized yielding can occur when the sum of residual and service stresses exceeds the yield stress of the material. When the service stress is removed, the residual-stress pattern will be different from what it was before the service stress was applied.

Welding Stresses. Figure 31 shows a 142-cm (56-in.) diam steel backup roll that fractured catastrophically during shipment. The

Fig. 31 Failed steel backup roll

Component fractured during winter shipment because of shrinkage stresses from low temperatures plus residual stresses from submerged arc welding of a 25-mm (1-in.) thick surface overlay.

backup roll had been salvaged from previous operations in a steel mill, and a 25-mm (1-in.) thick overlay had been applied by submerged arc welding to rebuild the surface. Ultrasonic inspection after welding indicated the weld to be perfectly sound. However, during shipping in a railroad car in winter, the shrinkage stresses caused by low temperatures in combination with residual welding stresses triggered the fracture. The excessive residual hoop stress after welding was due to an inadequate stress-relieving heat treatment. Thorough stress relief after welding would have prevented failure.

Heat-Treating Stresses. Heat treating generally produces appreciable residual stresses. Although stresses resulting from proper heat treatment are generally favorable, such factors as material, case depth, and quench rate influence the nature of the stresses produced. If processing of carburized parts having high core hardness, deep cases, or both induces high residual tensile stresses at the surface, the results can be particularly damaging.

Stress Relieving. Residual stresses can be removed from most parts by the proper thermal treatment. The severity and type of the residual stresses determine the necessary thermal treatment required. Each part must be evaluated as a separate component, with consideration given to its function in service. An effective stress-relief treatment may require an optimum or compromise temperature, which relieves the detrimental stresses but keeps distortion and loss of strength to a minimum. For example, a flange for a steering shaft was made from 1008 steel strip by stamping in a multiple-stage progressive die, stress relieving at 425 °C (800 °F) for 1 h, and broaching of serrations in a center hole. The steering shaft was then staked to the flange. The flanges fractured through the serrated-hole section on impact

loading. The parts were found to have high residual tensile stress in this section after the 425-°C (800-°F) stress relief. The stress-relief temperature was increased to 495 °C (925 °F), which removed the forming stresses and enabled the part to meet impact-test requirements.

Embrittlement of Steels

Many forms of embrittlement of steel parts can lead to brittle fracture. At least nine forms can occur during thermal treatment or elevated-temperature service. These forms of embrittlement (and the types of steel that some forms specifically affect) are:

- Strain-age embrittlement (low-carbon steel)
- Quench-age embrittlement (low-carbon steel)
- Blue brittleness
- Temper embrittlement (alloy steels)
- Tempered-martensite embrittlement (alloy steels)
- Stress-relief embrittlement
- 400- to 500-°C (750- to 930-°F) embrittlement (ferritic stainless steels)
- σ-phase embrittlement
- Graphitization (carbon and low-alloy steels)
- Intermetallic-compound embrittlement (galvanized steel)

In addition, steels (and other metals and alloys) can be embrittled by environmental conditions. The four forms of environmental embrittlement are:

- Neutron embrittlement
- Hydrogen embrittlement
- Stress-corrosion cracking (SCC)
- Liquid-metal embrittlement

Thermally Induced Embrittlement

Strain-Age Embrittlement. If low-carbon steel is deformed, its hardness and strength will increase upon aging at room or slightly elevated temperature, but with a concurrent loss of ductility. Rimmed or capped sheet steels are particularly susceptible to strain-age embrittlement, although strain aging has also been encountered in plate steels and weld HAZs. These steels are often temper rolled to suppress the yield point. The return of the yield point or the presence of Lüders strain in the stress-strain curve is evidence that strain-age embrittlement has occurred. The degree of embrittlement is a function of the amount of cold work, the aging temperature, and the time at temperature. Room-temperature aging may require from a few hours to a year. However, as the aging temperature is increased, the required time decreases, with embrittlement occurring in a matter of minutes at about 200 °C (400 °F).

Quench-Age Embrittlement. If low-carbon steels are rapidly cooled from temperatures slightly below the lower critical temperature, Ac_1, of the steel, the hardness of the steel increases, with a resultant loss of ductility upon

aging at room temperature. As with strain aging, quench-age embrittlement is a function of time at the aging temperature until the maximum degree of embrittlement is reached. An aging period of several weeks at room temperature is required for maximum embrittlement.

A decrease in the quenching temperature decreases the extent of the embrittlement. Quenching from temperatures of 560 °C (1040 °F) and below does not produce quench-age embrittlement. Steels with carbon contents of 0.04 to 0.12% appear most susceptible to quench-age embrittlement; increasing the carbon content above 0.12% reduces the effect. Quench-age embrittlement results from precipitation of solute carbon at existing dislocations and from precipitation hardening because of differences in the solid solubility of carbon in ferrite at different temperatures.

Blue Brittleness. When plain carbon steels and some alloy steels are heated between 230 and 370 °C (450 and 700 °F), there is an increase in strength and a marked decrease in ductility and impact strength. This embrittling phenomenon is known as blue brittleness because it occurs in the blue-heat range. Blue brittleness is an accelerated form of strain-age embrittlement. The increase in strength and decrease in ductility are caused by precipitation hardening within the critical-temperature range. Deformation while the steel is heated in the blue-heat range results in even higher hardness and tensile strength after cooling to room temperature. If the strain rate is increased, the blue-brittle temperature range increases. Use of susceptible steels that have been heated in the blue-brittleness range should be avoided, especially if the steels are subjected to impact loads, because the toughness of these materials will be considerably less than optimum.

Stress-relief embrittlement results in the loss of toughness within the HAZ and/or the weld metal as a result of stress relieving of a welded structure. Stress-relief cracking is also thought to be caused by the same mechanism and leads to intergranular cracking within the weld zone upon stress relieving. This phenomenon is also referred to as postweld heat-treat cracking and reheat cracking.

Both phenomena have been observed only in those alloy systems that undergo precipitation hardening. These systems include low-alloy structural and pressure vessel steels, ferritic creep-resisting steels, austenitic stainless steels, and some nickel-base alloys. In addition, a tendency for creep embrittlement, notch weakening, and poor creep ductility during short-term elevated-temperature tests usually occurs in these materials.

During welding, the HAZ is exposed to high temperatures, ranging up to the melting point of the alloy. At these temperatures, existing precipitates in the base metal (in steels, carbides, and nitrides) are taken into solution, and grain coarsening occurs. During cooling, some precipitation takes place at grain boundaries or within the grains, but the majority of the precipitates remain in solution. Subsequent exposure at stress-relieving temperatures causes precipitation in the HAZ, leading to significant strengthening. This results in the loss of toughness in the HAZ. Residual stresses in the structure are relieved through creep deformation. However, the strengthening of the precipitates of the grain interiors tends to concentrate creep strain at grain boundaries, leading to intergranular cracking.

Temper Embrittlement. The technical importance of temper brittleness arises from the fact that many of the common low-alloy steels exhibit an increase in their ductile-to-brittle transition temperatures after being heated in the range from 375 to 575 °C (700 to 1070 °F) or slowly cooled through this temperature range. Plain carbon steels containing less than 0.30% Mn are not susceptible to temper brittleness, although any steel containing appreciable amounts of manganese, nickel, or chromium will be susceptible provided the steel also contains one or more of the impurities antimony, phosphorus, tin, and arsenic. Thus, a combination of an alloying element and an impurity is required to cause temper embrittlement.

Detection. Temper embrittlement is usually detected by an upward shift in the ductile-to-brittle transition temperature in a notched-bar test, such as the Charpy V-notch impact test. For common commercial levels of phosphorus and tin in nickel-chromium steels, the transition temperature may increase as much as 200 to 300 °C (360 to 540 °F) for steel held at the temperature of most rapid embrittlement for 1000 h. Where large amounts of antimony (about 0.05%) have been deliberately added in experimental materials, shifts of as much as 500 to 600 °C (900 to 1080 °F) have been found. Generally, there is no detectable change in smooth-bar tension-test behavior at room temperature (either in the yield strength or the flow curve or in the elongation or reduction of area) accompanying such large-scale changes in notched-bar behavior. In cases of extreme embrittlement, there may be a detectable drop in reduction of area, which becomes more marked at cryogenic temperatures.

An embrittling treatment called step cooling is sometimes used to estimate the effects of long-term exposures. Step cooling consists of cooling the material through the embrittling range in a series of steps from about 600 to 300 °C (1100 to 570 °F), with the time at each step increasing as the temperature is lowered.

Embrittlement Mechanisms. The increase in transition temperature due to temper embrittlement is accompanied by a gradual change in brittle fracture mode (below the transition temperature) from completely transcrystalline to completely intercrystalline fracture. The fracture path generally follows the prior-austenite grain boundaries in embrittled material. Upon de-embrittlement, the brittle fracture mode re-

verts to transcrystalline. There is no detectable phase change accompanying this change in fracture mode; rather, the increase in transition temperature and change in fracture mode result from segregation of the responsible impurity along prior-austenite grain boundaries.

Control of Temper Embrittlement. Steels that have become embrittled can be restored to their original toughness by heating to about 600 °C (1100 °F) or above, then cooling rapidly to below about 300 °C (570 °F). Alternatively, embrittlement can be forestalled by reducing susceptibility. The principal means of reducing susceptibility is to reduce embrittling impurities as much as possible by control of raw materials and melting practice.

Tempered-martensite embrittlement of quenched-and-tempered high-strength low-alloy steels occurs over a temperature range of approximately 200 to 370 °C (400 to 700 °F). It takes place mainly in steels that have been heat treated to a microstructure of tempered martensite. Steels with microstructures of tempered lower bainite are also susceptible to tempered-martensite embrittlement, but steels with pearlitic microstructures and other bainitic steels are not.

Tempered-martensite embrittlement is evaluated by measuring the effect of tempering temperature on room-temperature impact energy. This is in contrast to temper embrittlement, which is evaluated by measuring the effect of tempering temperature on the ductile-to-brittle transition temperature.

Tempered-martensite embrittlement is believed to be caused by ferrite networks resulting from precipitation of cementite platelets along prior-austenite grain boundaries. However, some investigators believe that the precipitation of grain-boundary cementite platelets as such is responsible for tempered-martensite embrittlement.

Steels containing substantial amounts of chromium or manganese are highly susceptible to this form of embrittlement. Aluminum contents above 0.04% reduce embrittlement, and additions of 0.1% aluminum usually eliminate the problem. Some degree of embrittlement has been observed when phosphorus, antimony, arsenic, tin, silicon, manganese, or nitrogen were added to high-purity steels. Additions of nitrogen produced intergranular fractures, but the other embrittling agents did not (commercial grades of steel that are subjected to tempered-martensite embrittlement treatments fracture intergranularly).

Embrittlement in low-alloy steels heat treated to high strength levels can be minimized by:

- Developing special steels with retarded martensite-tempering characteristics
- Developing steels with faster rates of martensite tempering
- Using steels capable of transformation to 100% upper bainite at the desired strength level and section size

- Avoiding tempering in the region of susceptibility
- Using the lowest possible carbon content consistent with the desired strength level

400- to 500-°C Embrittlement. Fine-grain high-chromium stainless steels normally possess good ductility. However, if they are held for long periods of time at temperatures of 400 to 500 °C (750 to 930 °F), they will become harder, but embrittled. Embrittled high-chromium ferritic stainless steels contain two ferrites, one rich in iron and one rich in chromium.

Susceptibility to 400- to 500-°C embrittlement increases with increasing chromium content, with the highest degree of embrittlement occurring at chromium contents exceeding 19%. At least 15% chromium is necessary for embrittlement to occur. The effect of carbon content on embrittlement is minimal. High chromium steels that contain at least 1% Ti are more susceptible to embrittlement than similar steels with lower titanium contents. The embrittlement caused by prolonged soaking within the 400- to 500-°C range can be removed by soaking at somewhat higher temperatures for several hours.

Sigma-Phase Embrittlement. The formation of σ phase in ferritic and austenitic stainless steels during long periods of exposure to temperatures between approximately 560 and 980 °C (1050 and 1800 °F) results in considerable embrittlement after cooling to room temperature. Sigma phase, an iron-chromium compound approximately equivalent to FeCr, can be formed by either slow cooling from temperatures of 1040 to 1150 °C (1900 to 2100 °F) or by water quenching from 1040 to 1150 °C, followed by heating at 560 to 980 °C, with heating at 850 °C (1560 °F) producing the greatest effect. The embrittlement is most detrimental after the steel has cooled to temperatures below 260 °C (500 °F). At higher temperatures, stainless steels containing σ phase can usually withstand normal design stresses. However, cooling to 260 °C (500 °F) or below results in essentially complete loss of toughness.

The presence of σ phase greatly increases notch sensitivity, particularly in ferritic stainless steels and austenitic alloys that also contain some ferrite. The hardness and tensile strengths are usually not significantly affected by the presence of σ phase, but the impact strength is greatly affected. Sigma phase exerts a strengthening effect at high temperatures; however, the impact strength at high temperatures of an alloy containing σ phase is lower than the impact strength at room temperature of an alloy without σ phase.

Graphitization of carbon and carbon-molybdenum steel piping during service at temperatures above 425 °C (800 °F) has caused numerous failures in steam power plants and refineries. Graphite formation generally occurs in a narrow region in the HAZ of a weld where the metal has been briefly heated above the lower critical temperature. The graphitization

Fig. 32 Orchard heater of galvanized low-carbon steel that broke in a brittle manner because of an iron-zinc intermetallic compound along the grain boundaries

Dimensions given in inches. View A-A: micrograph of an etched section that shows the microstructure of the steel sheet. 400×

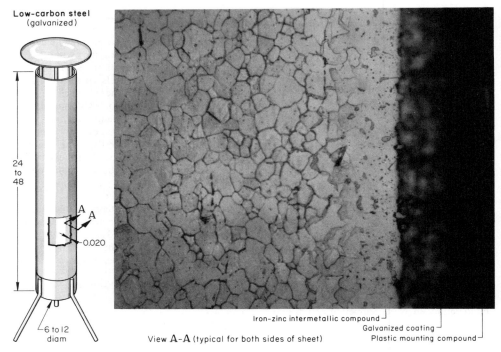

Low-carbon steel (galvanized)

24 to 48

6 to 12 diam

A A

0.020

Iron-zinc intermetallic compound
Galvanized coating
Plastic mounting compound

View A-A (typical for both sides of sheet)

tendency of carbon and carbon-molybdenum steels is increased when the aluminum content exceeds about 0.025%. Steels deoxidized with silicon may also be susceptible to graphitization. Deoxidation with titanium will usually produce good resistance to graphitization. Carbon-molybdenum steels exhibit greater resistance to graphitization than do carbon steels.

The degree of embrittlement depends on the distribution, size, and shape of the graphite. The severity of graphitization is frequently evaluated by bend testing. If graphitization is detected in its early stages, the material can often be rehabilitated by normalizing and tempering just below the lower critical temperature. Steel that has undergone more severe graphitization cannot be salvaged in this manner; the defective region must be cut out and rewelded, or the section must be replaced. Carbon and carbon-molybdenum steels can be rendered less susceptible to graphitization by tempering just below the lower critical temperature.

Intermetallic-Compound Embrittlement. Embrittlement of galvanized steel can result from long periods of exposure at elevated temperatures below the melting point of the zinc in the coating. In this type of embrittlement, zinc diffuses from the galvanized coating to grain boundaries in the steel, resulting in the formation of a brittle intergranular network of iron-zinc intermetallic compound. The presence of this compound can lead to brittle fracture.

Example 7: Brittle Fracture of Galvanized Steel Heater Shells Because of Embrittlement by Intermetallic Compounds. Afer only a short time in service, oil-fired or gas-fired orchard heaters, such as the one shown in Fig. 32, were easily broken, particularly by impact during handling and storage. Failure occurred in the portion of the heater shell that normally reached the highest temperature in service.

The orchard heaters were pipes, generally 150 to 305 mm (6 to 12 in.) in diameter and 60 to 120 cm (24 to 48 in.) in length. When in use, fuel was burned inside a pipe, which served as a windguard and a heat sink, and heat was transferred to nearby trees by radiation and convection. The heaters were used only a few nights a year and were stored outdoors when not in use. To minimize cost and deterioration from atmospheric corrosion, the heater shells were made of galvanized low-carbon steel sheet about 0.5 mm (0.020 in.) thick.

Investigation. Visual examination of a broken heater revealed that the area around the fracture was not shiny like the rest of the pipe, but was gray in color. Metal from these areas was very brittle and easily broken.

Pieces of metal from the failure areas were prepared for metallographic examination. A brittle and somewhat porous metallic layer about 0.025 mm (0.001 in.) thick was observed on both surfaces of the sheet. Next to this was

an apparently single-phase region nearly 0.05 mm (0.002 in.) in thickness (View A-A, Fig. 32). This phase was also present along the grain boundaries of the steel beneath this second layer and became progressively less observable nearer the center of the sheet. The region of observable penetration was about 0.075 mm (0.003 in.) thick, but this phase probably extended to the center of the sheets as a thin band that could not be resolved by optical microscopy.

Conclusions. Prolonged heating of the galvanized steel heater shells caused the zinc-rich surface to become alloyed with iron and reduce the number of layers. Also, heating caused zinc to diffuse along grain boundaries toward the center of the sheet. Zinc in the grain boundaries reacted with iron to form the brittle intergranular phase (View A-A, Fig. 32). The result was failure by brittle fracture at low impact loads during handling and storage.

Corrective Measures. Galvanized steel sheet was replaced by aluminized steel sheet for the combustion chamber. The increase in service life more than compensated for the more costly and difficult-to-fabricate aluminized sheet metal.

Environmentally Assisted Embrittlement

Neutron Embrittlement. Neutron irradiation of steel components in nuclear reactors usually results in a significant rise in the ductile-to-brittle transition temperature of steel. The transition temperature, as determined by Charpy V-notch impact tests, may be substantially increased, depending on such factors as the neutron dose, neutron spectrum, irradiation temperature, and steel composition. The amount by which the transition temperature increases because of neutron irradiation is usually at least 17 °C (30 °F) but less than 200 °C (360 °F).

The increase in the ductile-to-brittle transition temperature or in the nil-ductility transition temperature has been determined for many steels used for these applications under various conditions of irradiation. High-strength steels, which have lower initial nil-ductility transition temperatures than low-strength steels, are generally less susceptible to radiation embrittlement. Steels with low initial nil-ductility transition temperatures, fine-grain microstructures, and high dislocation densities generally offer greater resistance to neutron embrittlement. Neutron embrittlement renders the steel more susceptible to intergranular fracture.

Heat-treatment practice greatly affects the susceptibility of a steel to neutron embrittlement. Steels with tempered-martensite microstructures are less susceptible than those with tempered upper bainite or ferritic microstructures. Vacuum degassing and control of such elements as copper and phosphorus help reduce susceptibility to neutron embrittlement.

Hydrogen embrittlement has been a long-time problem. Quantitative knowledge regarding the influence of hydrogen on metals is difficult to obtain; however, from a qualitative standpoint, the effects of hydrogen have been well documented in the literature and in the article ''Hydrogen-Damage Failures'' in this Volume.

Absorption of hydrogen results in a general loss of ductility. Historically, hydrogen-embrittlement effects have been evaluated by reversed-bend tests, single-bend tests, and fatigue tests. Reduction-of-area and elongation values determined by standard tensile tests also show the effect of hydrogen embrittlement. Impact tests are generally not a good method of detecting hydrogen embrittlement.

The degree of hydrogen embrittlement depends greatly on the strength level of the steel. Resistance to hydrogen embrittlement decreases as the strength level is increased.

Hydrogen embrittlement of steel may occur during pickling to remove scale and rust. The rate of hydrogen pickup depends on the type and concentration of acid, the temperature of the solution, the pickling time, and the presence and concentration of inhibitors. Strongly ionized acids, such as hydrochloric, sulfuric, and hydrofluoric, cause severe embrittlement.

Hydrogen embrittlement of steel may occur during electroplating. The coating itself may become embrittled if hydrogen becomes trapped in the coating during plating. Embrittled coatings frequently develop blisters; ruptures of the plating may result if sufficient hydrogen pressure develops in a blister.

The effects of hydrogen embrittlement decrease as the strain rate increases. These effects are most pronounced at intermediate service temperatures and disappear at high and low temperatures. Because the solubility of hydrogen is greater at high temperatures than at low temperatures, the concentration of hydrogen can exceed solubility limits at room temperature upon rapid cooling from high temperatures, such as after forging and heat treatment. Any excess hydrogen present, therefore, must diffuse to the surface of the steel or form discontinuities in the interior of the part. Pockets of hydrogen in these discontinuities may

develop enough pressure to form hairline cracks or to shatter the steel.

A form of hydrogen embrittlement known as sulfide-stress cracking is encountered in sour environments typical of deep oil and gas wells. Corrosion of steel in aqueous environments containing hydrogen sulfide or in moist atmospheres containing gaseous hydrogen sulfide releases atomic hydrogen that is then absorbed into the steel, where it may cause embrittlement. This problem seriously limits the use of hardenable high-strength steels for well drilling and oil- and gas-production equipment. For example, alloy steels heat treated to yield strengths above 560 MPa (80 ksi) have been known to fracture in a very short time.

The best countermeasures developed to date have been to increase well temperatures to at least 150 °C (300 °F) or to increase the pH of the aqueous environment to 6.0 or above. Austenitic steels, nickel alloys, and cobalt alloys perform better than alloy steels in sour environments, but even these alloys are not totally resistant to cracking. Cracking of alloy steels appears to be greater when they have been cold worked or when the operating conditions include high-pressure hydrogen sulfide, low pH, or temperatures between −45 and 65 °C (−50 and 150 °F). Detailed information on hydrogen sulfide assisted cracking can be found in the article ''Failures in Sour Gas Environments.''

Stress-corrosion cracking is a mechanical-environmental failure process in which mechanical stress and chemical attack combine to initiate and propagate fracture in a metal part. Stress-corrosion cracking is produced by the synergistic action of sustained tensile stress and a specific corrosive environment; this action causes failure to occur more rapidly than it would if the separate effects of the stress and the corrosive environment were simply added together.

Failure by SCC is frequently caused by simultaneous exposure to a seemingly mild chemical environment and to a tensile stress well below the yield strength of the metal. Under such conditions, fine cracks can pene-

trate deeply into the part while the surface exhibits only insignificant amounts of corrosion. Therefore, there may be no macroscopic indications of an impending failure. For detailed information, see the article ''Stress-Corrosion Cracking'' in this Volume.

Liquid-metal embrittlement can cause cracking and fracture in stressed parts of many metals. Not all combinations of solid and liquid metals produce embrittlement. For example, aluminum is embrittled by liquid gallium, sodium, and tin, and steel is embrittled by liquid cadmium and lithium (liquid-metal embrittlement of steel also has been reported to be caused by liquid copper, brass, aluminum bronze, antimony, and tellurium). Both aluminum and steel are embrittled by liquid indium, zinc, and mercury.

The melting temperature and chemical reactivity of a liquid metal are not deciding factors as to whether it will cause embrittlement or not. Instead, most instances of embrittlement are accompanied by low intersolubilities and absence of intermetallic-compound formation. Detailed information is provided in the article ''Liquid-Metal Embrittlement'' in this Volume.

REFERENCES

1. G.F. Vander Voort, Ductile and Brittle Fractures, *Met. Eng. Quart.*, Vol 16 (No. 3), 1976, p 32-57
2. G. Powell and G.F. Vander Voort, Ductile and Brittle Fractures, in *Metals Handbook*, Desk Edition, American Society for Metals, 1985, p 32-9 to 32-13
3. G.F. Vander Voort, Macroscopic Examination Procedures, in *Metallography in Failure Analysis*, Plenum Press, 1978
4. ''Standard Specification for General Requirements for Steel Plates for Pressure Vessels,'' A 20/A 20M, *Annual Book of ASTM Standards*, Vol 01.04, ASTM, Philadelphia, 1984, p 76-105
5. *Republic Alloy Steels*, Republic Steel Corp., Cleveland, 1968, p 312-317

Fatigue Failures

Revised by Stephen D. Antolovich and Ashok Saxena, Georgia Institute of Technology

FATIGUE damage results in progressive localized permanent structural change and occurs in materials subjected to fluctuating stresses and strains. It may result in cracks or fracture after a sufficient number of fluctuations. Fatigue fractures are caused by the simultaneous action of cyclic stress, tensile stress, and plastic strain. If any one of these three is not present, fatigue cracks will not initiate and propagate. The cyclic stress and strain starts the crack; the tensile stress produces crack growth (propagation). Although compressive stress will not cause fatigue cracks to propagate, compression loads may do so. Both tensile and compressive stresses can lead to fatigue damage.

The process of fatigue consists of three stages:

- Initial fatigue damage leading to crack nucleation and crack initiation
- Progressive cyclic growth of a crack (crack propagation) until the remaining uncracked cross section of a part becomes too weak to sustain the loads imposed
- Final, sudden fracture of the remaining cross section

Fatigue cracking normally results from cyclic stresses that are well below the static yield strength of the material. In low-cycle fatigue, however, or if the material has an appreciable work-hardening rate, the stresses may also be above the static yield strength.

Fatigue cracks initiate and propagate in regions where the strain is most severe. Because most engineering materials contain defects and thus regions of stress concentration that intensify strain, most fatigue cracks initiate and grow from structural defects. Under the action of cyclic loading, a plastic zone, or region of deformation, develops at the defect tip. This zone of high deformation becomes an initiation site for a fatigue crack. The crack propagates under the applied stress through the material until complete fracture results. On the microscopic scale, the most important feature of the fatigue process is nucleation of one or more cracks under the influence of reversed stresses that exceed the flow stress, followed by development of cracks at persistent slip bands or at grain boundaries.

Prediction of Fatigue Life

The fatigue life of any specimen or structure is the number of stress (strain) cycles required to cause failure. This number is a function of many variables, including stress level, stress state, cyclic wave form, fatigue environment, and the metallurgical condition of the material. Small changes in the specimen or test conditions can significantly affect fatigue behavior, making analytical prediction of fatigue life difficult. Therefore, the designer may rely on experience with similar components in service rather than on laboratory evaluation of mechanical test specimens. Laboratory tests, however, are essential in understanding fatigue behavior, and current studies with fracture mechanics test specimens are beginning to provide satisfactory design criteria.

Laboratory fatigue tests can be classified as crack initiation or crack propagation. In crack-initiation testing, specimens or components are subjected to the number of stress cycles required for a fatigue crack to initiate and subsequently to grow large enough to produce failure.

In crack propagation testing, fracture mechanics methods are used to determine the crack-growth rates of pre-existing cracks under cyclic loading. Fatigue-crack propagation may be caused by cyclic stresses in a benign environment or by the combined effects of cyclic stresses and an aggressive environment (corrosion fatigue).

The above test methodologies are described in the articles "Fatigue Crack Initiation," "Fatigue Crack Propagation," and "Environmental Effects on Fatigue Crack Propagation" in Volume 8 of the 9th Edition of *Metals Handbook*.

Fatigue-Crack Initiation

Most laboratory fatigue testing is done either with axial loading or in bending, thus producing only tensile and compressive stresses. The stress is usually cycled either between a maximum and a minimum tensile stress or between a maximum tensile stress and a maximum compressive stress. The latter is considered a negative tensile stress, is given an algebraic minus sign, and therefore is known as the minimum stress.

The stress ratio is the algebraic ratio of two specified stress values in a stress cycle. Two commonly used stress ratios are: the ratio, A, of the alternating stress amplitude to the mean stress ($A = S_a/S_m$) and the ratio, R, of the minimum stress to the maximum stress ($R = S_{min}/S_{max}$).

If the stresses are fully reversed, the stress ratio R becomes -1; if the stresses are partially reversed, R becomes a negative number less than 1. If the stress is cycled between a maximum stress and no load, the stress ratio R becomes 0. If the stress is cycled between two tensile stresses, the stress ratio R becomes a positive number less than 1. A stress ratio R of 1 indicates no variation in stress, making the test a sustained-load creep test rather than a fatigue test.

Applied stresses are described by three parameters. The mean stress, S_m, is the algebraic average of the maximum and minimum stresses in one cycle, $S_m = (S_{max} + S_{min})/2$. In the completely reversed test, the mean stress is 0. The range of stress, S_r, is the algebraic difference between the maximum and minimum stresses in one cycle, $S_r = S_{max} - S_{min}$. The stress amplitude, S_a, is one-half the range of stress, $S_a = S_r/2 = (S_{max} - S_{min})/2$.

During a fatigue test, the stress cycle is usually maintained constant, so that the applied stress conditions can be written $S_m \pm S_a$, where S_m is the static or mean stress, and S_a is the alternating stress, which is equal to half the stress range. Figure 1 shows the nomenclature used to describe test parameters involved in cyclic-stress testing.

S-N Curves. The results of fatigue-crack initiation tests are usually plotted as maximum

Fig. 1 Schematic showing test parameters for cyclic-stress testing

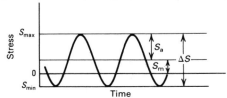

Fig. 2 Typical _S-N_ curves for constant amplitude and sinusoidal loading

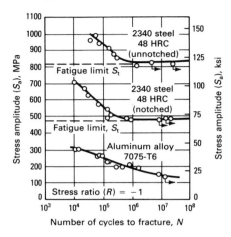

Fig. 3 Typical plot of strain range versus cycles-to-failure for low-cycle fatigue

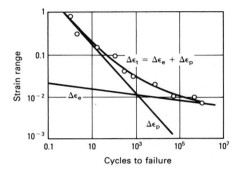

Fig. 4 Fatigue-crack propagation rate data for aluminum alloy 7075-T6

$R < 0$

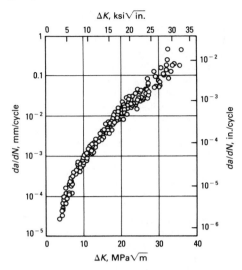

stress, minimum stress, or stress amplitude to number of cycles, _N_, to failure using a logarithmic scale for the number of cycles. Stress is plotted on either a linear or a logarithmic scale. The resulting plot of data is an _S-N_ curve. Three typical _S-N_ curves are shown in Fig. 2.

The number of cycles of stress that a metal can endure before failure increases with decreasing stress. For some engineering materials, such as steel (Fig. 2) and titanium, the _S-N_ curve becomes horizontal at a certain limiting stress. Below this limiting stress, known as the fatigue limit or endurance limit, the material can endure an infinite number of cycles without failure. The shape of the _S-N_ curve and the endurance limit depend on the surface condition of the specimen or component. Statistical characterization of the _S-N_ curve and the techniques for defining a mean fatigue curve and evaluating scatter or variability about that mean are discussed in the article "Fatigue Data Analysis" in Volume 8 of the 9th Edition of _Metals Handbook_.

Fatigue Limit and Fatigue Strength. The horizontal portion of an _S-N_ curve represents the maximum stress that the metal can withstand for an infinitely large number of cycles with 50% probability of failure and is known as the fatigue (endurance) limit, S_f. Most nonferrous metals do not exhibit a fatigue limit. Instead, their _S-N_ curves continue to drop at a slow rate at high numbers of cycles, as shown by the curve for aluminum alloy 7075-T6 in Fig. 2. For these metals, fatigue strength rather than fatigue limit is reported, which is the stress to which the metal can be subjected for a specified number of cycles. Because there is no standard number of cycles, each table of fatigue strengths must specify the number of cycles for which the strengths are reported. The fatigue strength of nonferrous metals at 10^8 or at 5×10^8 cycles is erroneously called the fatigue limit.

Low-Cycle Fatigue. For the low-cycle fatigue region ($N < 10^4$ cycles), tests are con-

ducted with controlled cycles of elastic plus plastic strain, rather than with controlled load or stress cycles. Under controlled strain testing, fatigue-life behavior is represented by a log-log plot of the total strain range, $\Delta\epsilon_t$, versus the number of cycles to failure (Fig. 3).

The total strain range is separated into elastic and plastic components. For many metals and alloys, the elastic strain range, $\Delta\epsilon_e$, is equal to the stress range divided by the modulus of elasticity. The plastic strain range, $\Delta\epsilon_p$, is the difference between the total strain range and the elastic strain range. Additional information on low-cycle fatigue is provided in the article "Fatigue Crack Initiation" in Volume 8 of the 9th Edition of _Metals Handbook_.

Stress-Concentration Factor. Stress is concentrated in a metal by structural discontinuities, such as notches, holes, or scratches, which act as stress raisers. The stress-concentration factor, K_t, is the ratio of the area test stress in the region of the notch (or other stress concentrators) to the corresponding nominal stress. For determination of K_t, the greatest stress in the region of the notch is calculated from the theory of elasticity, or equivalent values are derived experimentally.

The fatigue-notch factor, K_f, is the ratio of the fatigue strength of a smooth (unnotched) specimen to the fatigue strength of a notched specimen at the same number of cycles.

Fatigue-notch sensitivity, q, for a material is determined by comparing the fatigue-notch factor, K_f, and the stress-concentration factor, K_t, for a specimen of a given size containing a stress concentrator of a given shape and size. A common definition of fatigue-notch sensitivity is $q = (K_f - 1)/(K_t - 1)$, in which q may vary between 0 (where $K_f = 1$) and 1 (where $K_f = K_t$). This value may also be stated as a percentage.

Fatigue-Crack Propagation

In large structural components, the existence of a crack does not necessarily imply imminent

failure of the part. Significant structural life may remain in the cyclic growth of the crack to a size at which a critical failure occurs. The objective of fatigue-crack propagation testing is to determine the rates at which subcritical cracks grow under cyclic loadings before reaching a size that is critical for fracture. For an in-depth discussion, see the article "Fatigue Crack Propagation" in Volume 8 of the 9th Edition of _Metals Handbook_.

The growth or extension of a fatigue crack under cyclic loading is principally controlled by maximum load and stress ratio. However, as in crack initiation, a number of additional factors may exert a strong influence, including environment, frequency, temperature, grain direction, and other microstructural factors. Fatigue-crack propagation testing usually involves constant-load-amplitude cycling of notched specimens that have been precracked in fatigue. Crack length is measured as a function of elapsed cycles, and these data are subjected to numerical analysis to establish the rate of crack growth, _da/dN_. Methods for numerically determining the crack-growth rate can be found in the article "Fatigue Data Analysis" in Volume 8 of the 9th Edition of _Metals Handbook_.

Crack-growth rates are expressed as a function of the crack-tip stress-intensity factor range, ΔK. The stress-intensity factor is calculated from expressions based on linear elastic stress analysis and is a function of crack size, load range, and cracked specimen geometry. Detailed information on the expressions used to define stress intensity from stress field analysis is provided in the article "Fracture Mechanics" in Volume 8 of the 9th Edition of _Metals Handbook_. Fatigue-crack growth data are typically presented in a log-log plot of _da/dN_ versus ΔK (Fig. 4).

Stages of Fracture From Fatigue

The fracture surface that results from fatigue failure has a characteristic appearance that can be divided into three zones or progressive stages of fracture.

Stage I is the initiation of cracks and their propagation by slip-plane fracture, extending inward from the surface at approximately 45° to the stress axis. A stage I fracture never extends over more than about two to five grains around the origin. In each grain, the fracture surface is along a well-defined crystallographic plane, which should not be confused with a cleavage plane, although it has the same brittle appearance. There are usually no fatigue striations associated with a stage I fracture surface. In some cases, depending on the material, environment, and stress level, a stage I fracture may not be discernible.

Stage II. The transition from stage I to stage II fatigue fracture is the change of orientation of the main fracture plane in each grain from one or two shear planes to many parallel plateaus separated by longitudinal ridges. The plateaus are usually normal to the direction of maximum tensile stress.

A transition from stage I to stage II in a coarse-grain specimen of aluminum alloy 2024-T3 is shown in Fig. 5. The presence of inclusions rich in iron and silicon did not affect the fracture path markedly. The inclusions, which were fractured, ranged from 5 to 25 μm in diameter. In Fig. 5, the stage II area shows a large number of approximately parallel fatigue patches containing very fine fatigue striations that are not resolved at the magnification used. Fine striations are typical in stage II, but are usually seen only under high magnification.

Stage III occurs during the last stress cycle when the cross section cannot sustain the applied load. The final fracture, which is the result of a single overload, can be brittle, ductile, or a combination of the two.

Fracture Characteristics Revealed by Macroscopy

Examination of fatigue-fracture surfaces usually begins visually or by low-magnification optical (light) microscopy. Macroscopic examination of fracture surfaces involves relatively simple techniques; it can often be done at the failure site, requires little or no preparation of the specimen, requires minimal and relatively simple equipment, and does not destroy the specimen or alter the fracture surfaces. Macroscopic examination is particularly useful in correlating fracture-surface characteristics with part size and shape and with loading conditions. The crack origin can best be found by visual examination or by viewing the fracture surface at low magnifications (25 to 100×).

Beach Marks. The most characteristic feature usually found on fatigue-fracture surfaces is beach marks, which are centered around a

Fig. 5 Transition from stage I to stage II of a fatigue fracture in a coarse-grain specimen of aluminum alloy 2024-T3

common point that corresponds to the fatigue-crack origin. Also called clamshell, conchoidal and arrest marks, beach marks are perhaps the most important characteristic feature in identifying fatigue failures. Beach marks can occur as a result of changes in loading or frequency or by oxidation of the fracture surface during periods of crack arrest from intermittent service of the part or component.

Figure 6 shows examples of fatigue-fracture surfaces containing beach marks. Figure 6(a) shows the fracture surface of an aluminum alloy 7075-T6 plate that was fractured in a laboratory by spectrum-load fatigue testing. The beach marks were produced by changes in crack growth as a result of variations in applied load levels. The final-fracture region (stage III) covers about 40% of the fracture surface. Figure 6(b) shows the surface of a fatigue fracture in a 4130 steel shaft that failed in service; this surface exhibits beach marks produced by oxidation of the fracture when the shaft was idle. The beach marks shown in Fig. 6(c) were produced by a combination of variations in loading and periods of rest normally experienced by an automotive crankshaft—in this case, a ductile iron crankshaft. Fracture originated at a notch in the root of a weld. The final-fracture area was very small and contained shear lips, indicating that plane-stress conditions prevailed as the crack neared the surface.

Many fatigue fractures produced under conditions of uninterrupted crack growth and without load variations do not exhibit beach marks. Figure 7 shows the fracture surface of a specimen tested in a laboratory where the load was repeated at the same intensity and without interruption through final fracture. The fracture surface produced by fatigue contains no beach marks.

Final-Fracture Zone. The final-fracture zone of a fatigue-fracture surface is often fibrous, resembling the fracture surfaces of impact or fracture-toughness test specimens of the

Fig. 6 Beach marks in the fatigue region on fracture surfaces of three different metals

See text for discussion of (a) to (c).

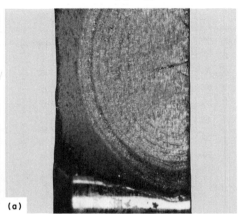

(a)

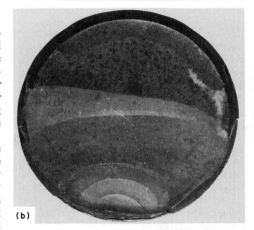

(b)

(c)

Fig. 7 Fracture surface on a laboratory fatigue-test specimen made of titanium alloy Ti-6Al-6V-2Sn

Fatigue zone (A) contains no beach marks, because load was not varied during the test.

Fig. 8 Surface of a fatigue fracture in a 4330V steel part

Chevron marks point to origin of fatigue in lower left corner. Arrows identify shear rupture along the periphery.

same material. The size of the final-fracture zone depends on the magnitude of the loads, and its shape depends on the shape, size and direction of loading of the fractured part. In tough materials, with thick or round sections, the final-fracture zone will consist of a fracture by two distinct modes: (1) tensile fracture (plane-strain mode) extending from the fatigue zone and in the same plane and (2) shear fracture (plane-stress mode) at 45° to the surface of the part bordering the tensile fracture. These two modes are illustrated in the surface of a fatigue fracture through a thick section shown in Fig. 8. In Fig. 8, two features in the final-fracture zone aid in determining the origin of fracture:

- Fatigue usually originates at the surface; therefore, the fatigue origin is not included in the shear-lip fracture
- The presence of characteristic chevron marks in the tensile fracture that point back to the origin of fracture

In thin sheet-metal pieces having sufficient toughness, final fracture occurs somewhat differently. As the crack propagates from the fatigue zone, the fracture plane rotates around an axis in the direction of crack propagation until it forms an angle of about 45° with the loading direction and the surface of the sheet. The fracture plane, inclined 45° to the load direction, can occur on either a single-shear or a double-shear plane (Fig. 9).

Fracture Characteristics Revealed by Microscopy

Examination of fatigue fractures by optical microscopy is often difficult because the height of features on the fracture surface may exceed the depth of field of the microscope, especially at high magnifications. Metallographic examinations of cross sections through suspected fatigue failures typically show that the crack path was transgranular. A cross section through a crack that has not grown enough to cause separation of the component is often useful in showing deficiencies in design or manufacture or in showing the result of unanticipated service conditions.

Striations. In the electron microscope examination of fatigue-fracture surfaces, the most prominent features found are patches of finely spaced parallel marks, called fatigue striations. The fatigue striations are oriented perpendicular to the microscopic direction of crack propagation and, with uniform loading, generally increase in spacing as they progress from the origin of fatigue. Each striation is the result of a single cycle of stress, but every stress cycle does not necessarily produce a striation; striation spacing depends strongly on the level of applied loading. The clarity of the striations depends on the ductility of the material. Striations are more visible at stress levels higher than the fatigue limit, and they

Fig. 9 Fracture planes that are 45° to the direction of loading

(a) Single-shear plane. (b) Double-shear plane

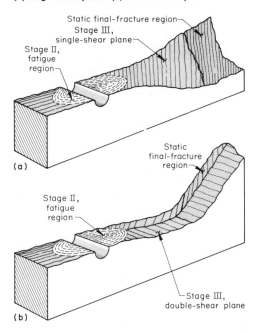

are more readily visible in ductile materials. Thus, patches of fatigue striations in high-strength steel are less visible than in an aluminum alloy. Also, the striations are more visible at high stress levels than they are near the fatigue limit.

At high rates of crack growth (0.0025 mm, or 10^{-4} in., per cycle or more), the striations become wavy and develop a rough front. A large plastic zone exists in front of the crack, which may cause extensive secondary cracking. Each secondary crack propagates as a fatigue crack, creating a network of secondary striations. The local crack direction may differ markedly from the overall direction of crack propagation because of the many changes in direction of the local fracture path.

In steels, fatigue striations that are formed at ordinary crack-growth rates are not always as well defined as they are in aluminum alloys. The striations that are formed in aluminum alloys at very low crack-growth rates (less than 1.3×10^{-4} mm, or 5×10^{-6} in., per cycle) are difficult to resolve and often cannot be distinguished from the network of slip lines and slip bands associated with plastic deformation at and near the crack front as it propagates through the section.

Fatigue Cracking

Fatigue cracking normally results from cyclic stresses that are well below the static yield strength of the material. (In low-cycle fatigue, however, or if the material has an appreciable work-hardening rate, the stresses may also be above the static yield strength.) A fatigue crack generally initiates in a highly stressed region of a component subjected to cyclic stresses of sufficient magnitude. The crack propagates under the applied stress through the material until complete fracture results. On the microscopic scale, the most important feature of the fatigue process is the nucleation of one or more cracks under the influence of reversed stresses that exceed the flow stress, followed by the development of cracks at persistent slip bands or at grain boundaries. Subsequently, fatigue cracks propagate by a series of opening and closing motions at the tip of the crack that produce, within the grains, striations that are parallel to the crack front.

Crack Initiation

Fatigue cracks form at the point or points of maximum local stress and minimum local strength. The local stress pattern is determined by the shape of the part, including such local features as surface and metallurgical imperfections that concentrate macroscopic stress, and by the type and magnitude of the loading. Strength is determined by the material itself, including all discontinuities, anisotropies, and inhomogeneities present. Local surface imperfections, such as scratches, mars, burrs, and other fabrication flaws, are the most obvious flaws at which fatigue cracks start. Surface and subsurface material discontinuities in critical locations will also influence crack initiation. Inclusions of foreign material, hard precipitated particles, and crystal discontinuities such as grain boundaries and twin boundaries are examples of microscopic stress concentrators in the material matrix. On the submicroscopic scale, dislocation density, lattice defects, and the orientation of cross-slip planes control the formation of persistent slip bands, intrusions and extrusions, and dislocation cells, which ultimately control nucleation mechanisms.

In the roots of small fatigue-crack notches, the local stress state is triaxial (plane strain). This reduces the local apparent ductility of the material and helps control the orientation of the crack as long as the crack is small. On a microscopic level, the fracture surface at the origin consists of the crystallographic planes in individual grains that are most favorably oriented for slip.

Crack Nucleation. A variety of crystallographic features have been observed to nucleate fatigue cracks. In pure metals, tubular holes that develop in persistent slip bands, extrusion-intrusion pairs at free surfaces, and twin boundaries are common sites for crack initiation. Cracking also initiates at grain boundaries in polycrystalline materials, even in the absence of inherent grain-boundary weakness; at high strain rates, this seems to be the preferred site for crack nucleation. Nucleation at a grain boundary appears to be purely a geometrical effect related to plastic incompatibility, whereas nucleation at a twin boundary is associated with active slip on crystallographic planes immediately adjacent and parallel to the twin boundary.

The above processes also occur in alloys and heterogeneous materials. However, alloying and commercial production practices introduce segregation, inclusions, second-phase particles, and other features that disturb the structure, and these have a dominant effect on the crack-nucleation process. In general, alloying that enhances cross slip, that enhances twinning, or that increases the rate of work hardening will retard crack nucleation. On the other hand, alloying usually raises the flow stress of a metal, which speeds up crack propagation, thus at least partly offsetting the potentially beneficial effect on fatigue-crack nucleation.

Relation to Environment. In observing locations of crack nucleation, the possibility of environment-related mechanisms, including pitting corrosion, stress-corrosion cracking, and other effects of a hostile environment, must be considered. For example, a great number of fatigue failures in otherwise lightly loaded structures originate in fretted areas. In any structure having joints with some relative motion, fretting provides a possible failure origin. Environmental effects are discussed in more detail in the article "Corrosion-Fatigue Failures" in this Volume.

Crack Propagation

Once a fatigue crack has been nucleated, its rate and direction of growth are controlled by localized stresses and by the structure of the material at the tip of the crack.

Initial Propagation. After a short period at Stage I, macroscopic crack propagation usually occurs perpendicular to the maximum tensile stress. On a microscopic level, the local directions of propagation are controlled to some extent by crystallographic planes and may form on or may be parallel to slip bands in grains near the surface. In interior grains, cleavage cracks often form at the intersection of slip bands with grain boundaries. In high-strength materials containing spheroidal second-phase particles, secondary crack nucleation ahead of the main crack front (slipless fatigue) occurs around such particles. In slipless fatigue, cracks form along lattice planes that are unfavorably located relative to the maximum tensile stress.

Crack Enlargement. After a crack has nucleated and propagated to a finite size, it becomes a macroscopic stress raiser and can be more influential than any stress raiser already in the part. At this point, the crack tip will take over control of the fracture direction. Subsequently, the orientation of the crack surface will depend on the stress field at the crack tip and will often follow a series of void coalescences in advance of the crack front.

On the macroscopic scale, early crack extension occurs under plane-strain conditions. This gives a typical fine-grain, flat-face surface that, when produced under random loading or sequences of high and low stress amplitudes, exhibits characteristic beach marks.

Fig. 10 Case-hardened motorcycle-transmission shaft that fractured in service from high-cycle fatigue

(a) View of shaft showing location of fracture. Shaft was made of carburized-and-hardened 1010 steel (case hardness, 40 HRC; core, 85 HRB). (b) Fracture surface of the shaft showing crack-initiation sites (arrows A) and curved crack-arrest marks (arrows B) that indicate slower crack propagation in the soft core than in the hardened case. 4×

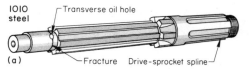

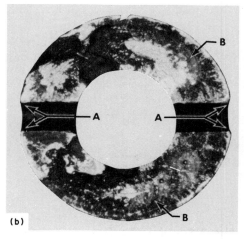

Crack-propagation variations in anisotropic or inhomogeneous materials occasionally produce beach marks that are difficult to interpret. Figure 10(a) shows a case-hardened steel transmission shaft, with a transverse oil hole, that fractured after extended service in a racing motorcycle. Fracture initiated in the hardened case at the outer-circumference corners of the transverse hole (arrows A, Fig. 10b) and progressed around the shaft in both directions. During periods of low operating stress, there was no crack progress, resulting in crack-arrest marks (arrows B, Fig. 10b). During cycles of high stress, crack progress was reinitiated. The crack front in the soft core was curved backward against the propagation direction, indicating greater ductility, lower notch sensitivity, and lower stress over much of the rotation cycle in the core material than in the hardened case. The back curvature of the crack front may also have been the result of the severe stress-concentration effect of the cracked case.

In highly anisotropic materials, such as spring wire, a fatigue-fracture surface may exhibit regions where the crack propagated in typical fatigue and other regions where it propagated preferentially in another mode along planes of weakness (see the article "Failures of Springs" in this Volume).

Final Propagation. After a crack has grown to the size at which it significantly changes the load-carrying capacity of the part, a

Fig. 11 Relationship between fatigue-crack growth rate and applied stress intensity for two values of constant load ratio, R

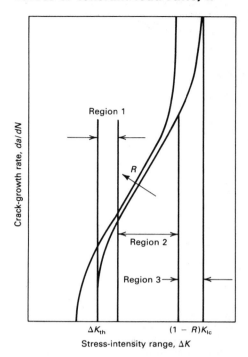

change in the direction of crack growth usually follows. Fractures in sheet metal exhibit a shear lip at approximately 45° to the initial flat-face fracture.

In cylindrical parts subjected to low operating stresses, the fracture may still appear planar, but the surface appearance becomes more fibrous and shows a greater ductility (larger size of plastic zone at the crack front). This indicates a change from crack extension to final fast fracture.

As a general rule, low-stress high-cycle fatigue produces flat-face (plane-strain) fractures. The fracture surface appears fine grained and lightly polished near the crack-nucleation site, where the stress intensification is least. The surface becomes progressively rougher and more fibrous as the crack grows and the intensity of stress increases. On high-stress low-cycle fatigue surfaces, found in certain areas of all complete fatigue fractures, the surface is fibrous, rough, and more typical of plane-stress loading conditions, where the general fracture direction is at 45° to the main tensile load.

Analytical Treatment of Fatigue-Crack Propagation

The crack-tip stress-intensity factor, K, uniquely characterizes the elastic stress field near a crack. Its magnitude depends on the applied nominal stress, geometry, and the characteristic crack dimension, a. In fracture mechanics, K is used to represent the crack driving force that controls the onset of brittle fracture and the rate of subcritical crack growth in structures. Methods for determining K for various load/crack configurations have been derived and are listed in various publications.

The rate of subcritical crack growth from fatigue loading, da/dN, is characterized by the stress-intensity range, $\Delta K = K_{max} - K_{min}$. Figure 11 shows the relationship between da/dN and ΔK at constant load (or stress) ratio $R = K_{min}/K_{max} = S_{min}/S_{max}$. The three regions extend from ΔK levels associated with almost zero crack-growth rate to conditions approaching fast fracture.

The stress-intensity range below which the crack-growth rate becomes diminishingly small is called the fatigue-crack growth threshold, ΔK_{th}. This is an important design parameter for such applications as rotating shafts involving low-stress high-frequency fatigue loading where no crack extension during service can be permitted. In Region 1, crack growth is negligible for an almost unlimited number of cycles. Region 2 behavior is relevant to design situations involving a finite number of fatigue cycles. Region 3 behavior for rapid crack growth is significant only for applications that may experience very few (of the order of ten or fewer) load and unload cycles—for example, pressure vessels in which pressure may be discharged only a few times during the service life.

Because predicting service fatigue life often involves integrating crack growth rates over a range of ΔK values, the da/dN versus ΔK relationship can be represented by:

$$\left(\frac{da}{dN}\right)^{-1} = \frac{A_1}{(\Delta K)^{n_1}} +$$
$$A_2 \left[\frac{1}{(\Delta K)^{n_2}} - \frac{1}{[(1 - R)K_c]^{n_2}} \right] \quad \text{(Eq 1)}$$

where n_1, n_2, A_1, A_2, and K_c are best fit material constants determined by regression analysis. Equation 1 is general and applies to all three regions of crack growth.

Variables that influence fatigue-crack growth behavior are stress ratio, loading frequency (not a factor in benign environments), material chemistry, heat treatment, test temperature, and environment. In a fatigue-crack growth test to determine the material constants, these variables are controlled to simulate actual service conditions. The fracture mechanics approach for characterizing fatigue-crack growth can be used in design applications to estimate maximum flaw sizes that allow a part to reach its design life. This approach is also very useful for conducting failure analyses. Examples 1 and 12 in this article demonstrate the use of these concepts in a design and failure analysis application, respectively.

Example 1: Developing Inspection Criteria for Fan Shafts. Fan shafts are typically forged from medium-carbon steels. Primary fatigue stresses are generated from gravity bending and are fully reversed during each revolution at rates of 900 to 1800 rpm or 15 to 30 Hz. Maximum allowable alternating stress on the fan shaft is typically ±52 MPa (±7.5 ksi). Before shipping, shaft forgings are subjected to various types of nondestructive inspections to reveal inclusions and forging bursts. Fatigue-crack propagation analysis can be used to evaluate such defects.

Investigation. Specimens for mechanical testing were obtained by slicing a 580-mm (23-in.) diam shaft forging into 75-mm (3-in.) thick disks. Tensile specimens and Charpy impact specimens were located at various radial distances from the center of the shaft cross section. Compact-type specimens, 25.4 and 6.25 mm (1 and 0.25 in.) thick, were used for fracture-toughness and fatigue-crack growth rate testing. In addition, 12.5- × 50.8-mm (0.5- × 2-in.) center-crack-tension specimens were also machined from the central region of another disk from the same forging for characterizing fatigue-crack growth at $R = -1.0$.

Test results at several radial distances from the shaft center indicated that tensile strength increases slightly from the peripheral region of the shaft toward the center. Fatigue-crack growth testing was conducted on specimens taken from two extreme radial locations of the shaft, but no trend could be attributed to specimen location. Figure 12 shows the wide range of fatigue-crack growth-rate behavior of the shaft material for load ratios of −1.0, 0.1, and 0.5, which represent a spectrum of typical load histories encountered in service. For a load ratio of −1.0, only the positive half of the load cycle was used to compute ΔK. These data were used to determine the material constants A_1, A_2, n_1, and n_2 using a standard multiple-linear regression computer program.

Discussion. The number of fatigue cycles that accumulate on the fan shaft during the projected 40-year service life can be about 10^{11} cycles, and little crack extension during service is acceptable. Thus, the ΔK_{th} of the shaft material for completely reversed loading ($R = -1.0$) is selected as the maximum allowable value of ΔK.

Conclusions. A growth rate of 10^{-15} m per cycle (4×10^{-14} in. per cycle) can be considered acceptable for this application because it implies a crack extension of 0.1 mm (0.004 in.) during the 10^{11} fatigue cycles. The ΔK corresponding to this growth rate can be determined from Eq 1 by using the appropriate fitting constants at $R = -1.0$. This value is 6.4 MPa√m (5.8 ksi√in.), and a safety factor of 20% on the ΔK_{th} value reduces the allowable K to 5.1 MPa√m (4.6 ksi√in.).

Two common defect geometries are shown in Fig. 13 with the flaw shape factor, Q, for various values of the flaw aspect ratio, $a/2c$, and ratios of applied stress to the yield strength of the material. The value of Q influences the magnitude of K according to:

Fig. 12 Room-temperature fatigue-crack growth behavior of ASTM grade A293 steel

Levels of R represent spectrum of typical in-service load history.

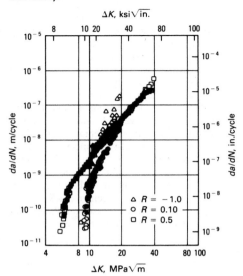

Fig. 13 Flaw shape parameter curves for surface and internal cracks

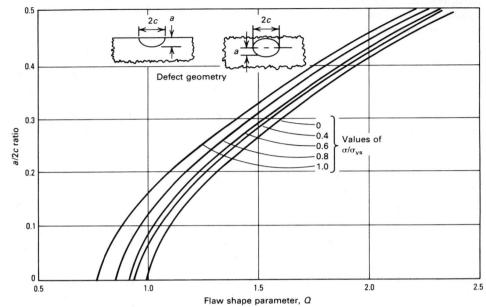

Defect geometry

Values of σ/σ_{ys}
0
0.4
0.6
0.8
1.0

Flaw shape parameter, Q

$$K = M_T \sigma \sqrt{\frac{\pi a}{Q}}$$

For fatigue, this relationship can be rewritten as:

$$\Delta K = M_T \Delta\sigma \sqrt{\frac{\pi a}{Q}}$$

where $\Delta\sigma$ is the applied stress range, and M_T is the finite width magnification factor. Substituting the allowable value of ΔK yields the critical flaw size, a_{cr}.

$$a_{cr} = \left(\frac{5.1}{M_T \Delta\sigma}\right)^2 \frac{Q}{\pi}$$

For deeply embedded internal flaws, $M_T = 1.0$, and for small surface defects in large shafts, $M_T = 1.1$. These values are used to calculate critical flaw sizes of various aspect ratios for several fatigue stress levels (Fig. 14). For surface flaws, critical flaw sizes are a factor of 1.1 lower for equivalent $a/2c$ and stress values.

Effect of Type of Loading and Part Shape

Nucleation and growth of a fatigue crack and the features on the fracture surface are all strongly affected by the shape of the part and the type and magnitude of loading exerted on the part in service, as well as metallurgical and environmental factors. In a beam of simple uniform cross section subjected to fluctuating or alternating stresses, fracture analysis is relatively simple, and the appearance of fracture surfaces in test specimens is predictable.

Fig. 14 Critical sizes of internal flaws at various stress levels

Surface flaws are a factor of 1.1 lower.

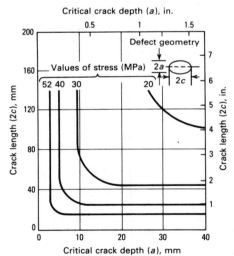

Unidirectional Bending. A beam with a uniform cross section subjected to pure, fluctuating, unidirectional bending has a bending moment that is uniform along the length of the beam; the tensile fiber stress is also uniform along the length of the beam. Therefore, a fatigue crack may be initiated at any point along the beam; in fact, several fatigue cracks may have been active before one of them became large enough to cause the final fracture.

A related form of loading is cantilever loading, in which the bending moment and therefore the tensile fiber stress vary along the length of the beam. Fracture will initiate at the point of highest stress, adjacent to the rigid mounting. Because of the shear-stress component characteristic of cantilever loading, the direction of the maximum tensile stress makes a small angle with the axis of the beam. Consequently, a fatigue crack will usually propagate into that portion of the beam within the fixed mounting.

Alternating Bending. If the loading of a beam is alternating instead of fluctuating, fatigue cracks may initiate on both sides of the beam. In pure bending, the cracks on opposite sides of the beam are not necessarily in the same plane; therefore, final fracture may occur at some angle other than 90° to the axis of the beam. If the same load is applied to the beam in both directions, the two fatigue cracks should be symmetrical. The size of the final-fracture zone is indicative of the relative loading of the beam.

There is one significant difference between the appearance of fatigue cracks formed under fluctuating load and those formed under alternating load. With an alternating load, each crack is opened in one half-cycle and closed and pressed together during the other half-cycle, which rubs and polishes the high points on opposite sides of the fatigue crack. Under fluctuating load, the rubbing is less pronounced. Consequently, beach marks are usually more readily observed after alternating loading than after fluctuating loading.

Rotational Bending. A machine component that is commonly subjected to a bending load is a rotating round shaft. A unique feature of rotational-bending loading is that during one revolution of the shaft both maximum and minimum loading are exerted around the entire circumference of the shaft in the region of maximum bending moment.

Fig. 15 4150 steel pump shaft that fractured in service from reversed-bending and torsional fatigue

The shaft was used in a piston-type pump. Dimensions given in inches. View A-A shows beach marks over a large area of a fracture surface; oval region near bottom center is the final-fracture area.

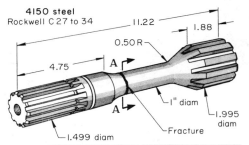

Fig. 16 Schematic of the initiation of torsional-fatigue cracks in a shaft subjected to longitudinal shear (a) or transverse shear (b)

Dashed lines indicate other cracks that can appear when torsional stresses are reversed.

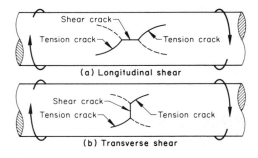

Because the loading is axially symmetrical, a fatigue crack can be initiated at any point, or at several points, around the periphery of the shaft. Multiple cracks do not necessarily occur in the same plane and are separated from each other by ridges called ratchet marks. The presence of multiple cracks normally indicates a relatively high applied load and rotational bending.

Under low or moderate overloads, a rotating shaft may fail as a result of a single fatigue crack. When the shaft is always rotated in the same direction, the crack usually advances asymmetrically; that is, the apparent center of the area of beach marks shifts in a direction opposite to that of shaft rotation (second row from the bottom, Fig. 18). Asymmetric fatigue-crack growth indicates rotational bending and the direction of rotation. Under periodically reversing rotating-bending loads, fatigue cracks grow symmetrically (Fig. 15 and 18). Even without reversing the direction of rotation, a rotating-bending fatigue crack may advance symmetrically. Examples of fatigue failures of shafts can be found in the article "Failures of Shafts" in this Volume.

Example 2: Bending-Fatigue Fracture of a 4150 Steel Pump Shaft. Shafts used in piston-type pumps failed after the pumps had been intermittently used for about 12 months. Reports showed that 4 of 200 shafts in service had failed. Two shafts were examined; one is shown in Fig. 15. The material specified for the shafts was 4150 steel, heat treated to a hardness of 27 to 34 HRC and a minimum tensile strength of 890 MPa (129 ksi). The pumps operated at a speed of 1585 rpm, a pressure of 13.8 MPa (2 ksi), a temperature of 57 °C (135 °F), and a capacity of 172 cm^3 (10.5 in.^{3}) per revolution.

Investigation. Visual and macroscopic examination of the fracture surfaces of the two shafts disclosed beach marks over a large part of each surface. The appearance of the fracture surfaces suggested fatigue resulting from reversed bending and rotational loading with a high stress concentration but a low overstress. In one shaft, the final-fracture area was oval and encircled with a gray area (view A-A, Fig. 15), and it consisted of 10 to 15% of the cross-sectional area. The other shaft exhibited a final-fracture area of similar shape near the center of the shaft.

Metallographic examination of a section through a shaft revealed light and dark bands. The dark bands (alloy-rich areas) had a hardness of 44.5 to 45.5 HRC; the light bands (alloy-lean areas), 29 to 30 HRC. Surface hardness of the shafts from end to end ranged from Rockwell C 29 to 33; surface hardness in the fracture area was about 30 HRC.

Conclusions. The shafts failed in reversed-bending and torsional fatigue that resulted from high stress concentration at a low overstress.

Corrective Measure. The shaft diameter in the failure area was increased from 25 to 28.5 mm (1 to 1.125 in.) for subsequent shafts, which reduced the nominal stresses by about 25% and was expected to provide a safe operating condition.

Torsional Loading. Under torsional loading of a shaft, the maximum local tensile stresses are at 45° to the axis of the shaft. Under a fluctuating torsional load, fatigue cracks may develop normal to the tensile stresses. Under alternating (reversed) torsional loading ($S_m = 0$, $R = -1$), two sets of fatigue cracks, per-

pendicular to each other, may develop. Torsional fatigue cracks can begin in longitudinal shear (Fig. 16a) or transverse shear (Fig. 16b); the relatively equal length of the cracks in each pair in Fig. 16 indicates that the equal and opposite stresses have occurred during loading. At later stages of fatigue-crack growth, one crack of a pair usually grows much faster than the other and eventually causes rupture of the shaft. Beams that are subjected to fluctuating torsion ($S_m < S_a$, $R > 0$) will typically show fatigue cracks in only one direction; the presence of perpendicular fatigue cracks in a component subjected to fluctuating torsion probably indicates the presence of torsional vibration.

Axial Loading. In each of the previously discussed types of loading there are stress gradients within the beam; stresses are greatest at the surface of the beam, which increases the normal likelihood for fatigue cracks to initiate at the surface. However, in pure axial loading of a simple, uniform beam, the stress is constant across the cross section of the beam, and a fatigue crack may initiate at a discontinuity within the member rather than at the surface. The appearance of fatigue cracks caused by fluctuating tensile loads is often similar to that described for bending loads. The stress states within beams subjected to the two types of loading are similar, although purely axial loading is rarely found in service.

Loading of flat components differs from that of long, somewhat cylindrical components. In flat components, biaxial tension is more common, and torsion less so. Under conditions of biaxial tension, fatigue life depends on the maximum shear stress rather than the principal tensile stresses. Thus, fatigue failure may not occur when the loading conditions result in low shear stresses but high tensile stresses.

In sheet or plate materials, the fatigue-crack front may extend under plane-strain conditions to give a wholly flat-face fracture (Fig. 17a). Fatigue fractures in very thin sheet subjected to high stress intensities may shift from flat face (plane-strain conditions) to shear face (plane-stress conditions), as shown in Fig. 17(b).

Effects of Overstress and Stress Concentration

The magnitude of the nominal stress on a cyclically loaded component is often measured by the amount of overstress, that is, the amount by which the nominal stress exceeds the fatigue limit or the long-life fatigue strength of the material used in the component. The number of load cycles that a component under low overstress can endure is high; thus, the term high-cycle fatigue is often applied. Increasing the magnitude of the nominal stress has the following effects:

- Initiation of multiple cracks is more likely
- Striation spacing is increased
- The region of final fast fracture is increased in size

Fig. 17 Fatigue-fracture zones in aluminum alloy 7075-T6 plates

(a) Fatigue crack that grew as a flat-face fracture with a shallow convex crack front. (b) Change in orientation of fatigue fracture from plane strain (arrow A) to plane stress (arrow B)

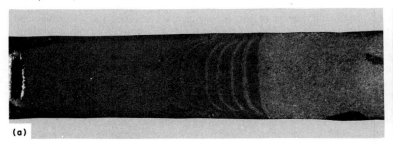

(a)

(b)

With very high overstress, low-cycle fatigue fractures are produced. The arbitrary but commonly accepted dividing line between high-cycle and low-cycle fatigue is considered to be about 10^5 cycles. A more basic distinction is made by determining whether the dominant component of the strain imposed during cyclic loading is elastic (high cycle) or plastic (low cycle), which in turn depends on the properties of the metal as well as the magnitude of the nominal stress. In extreme conditions, the dividing line between high-cycle and low-cycle fatigue may be less than 100 cycles.

Stress Concentrations. Notches, grooves, holes, fillets, threads, keyways, and splines are common design features. All such sectional discontinuities increase the local stress level above that estimated on the basis of minimum cross-sectional area. In addition to the reduction of fatigue strength or fatigue life, increasing the severity of stress concentration has the following effects on fatigue-crack features:

- Initiation of multiple cracks is more likely
- Beach marks usually become convex toward the point of crack origin
- Under rotational loading, the beach marks may completely surround the final-fracture zone
- Combined stress states may be introduced, thereby influencing the direction of crack growth

Figure 18 shows how magnitude of nominal stress, severity of stress concentration, and type of loading affect the appearance of fatigue-fracture surfaces of components with round, square, and rectangular cross sections and those of thick plates. The chart shown in Fig. 18 is intended for use only as a guide. Deviations from this chart will be found for various material, test, and service conditions.

Figure 18 is based on a number of principles. First, the number of active crack nuclei or initiation sites increases with increasing local stress in regions of potential crack nucleation. Therefore, at high overstress or in the presence of a severe stress concentration, multiple crack origins will be seen. In most cases, the cracks from these origins will eventually unite to form

a single crack front. Before the single crack front is formed, the individual microcracks will be separated by small, vertical ledges called ratchet marks. Alternatively, at just above the fatigue limit, or minimum stress for fracture, a single origin will occur, and the entire fracture will emanate from that point.

Second, in the absence of stress concentrations at the surface, cracks propagate more rapidly near the center of a section than at the surface. This occurs because deformation constraints cause the stresses to be triaxial and more severe away from the surface. However, when there is a stress-concentrating notch at the surface, such as a thread with a sharp root, the stress near this notch may be more severe than it is farther below the surface. Under conditions of severe notching, W-shape crack fronts will sometimes be observed.

Third, for a given material, the size of the region of catastrophic fracture (or final fast fracture) relative to the size of the region of subcritical crack propagation will increase with nominal stress. Under an overload that is slightly more than adequate to cause fracture, the region of final fast fracture will be relatively small; under a much higher applied stress, this region will be relatively large.

Fourth, in a fracture caused by rotating bending, the final-fracture region will often be rotated, or offset, toward the origin in a direction opposite to the direction of rotation. Also, all other conditions being the same, the region of final fracture will move toward the center of the section as the nominal stress increases.

Finally, fracture initiation usually occurs at or near the surface, because in most engineering situations, such as in bending or when stress concentrations are present, the surface is subjected to the greatest stress. Subsurface origins have been observed in tension-tension or tension-compression fatigue or in Hertzian-stress situations, such as rolling-contact fatigue of bearings and gear teeth, if there is a large inclusion or imperfection below the surface in the interior of the specimen or part, but this is unusual. Crack initiation at corners or at the ends of drilled holes may result from the presence of burrs remaining at these locations after machining.

Effect of Frequency of Loading

Exclusive of environmental effects, there are no distinguishing surface features of a fatigue fracture produced at high frequency that differentiate it from other types of fatigue fractures during visual or optical microscope examination. At best, when examined with an optical microscope, a fracture surface created at a high frequency will have a brittle appearance, showing mostly a platelike structure throughout the fatigue zone. Beach marks may or may not be present, depending on whether crack growth was steady or intermittent or if load variations occurred.

The fracture surface shown in Fig. 19(a) was produced by subjecting a 0.5-mm (0.020-in.) thick panel of aluminum alloy 7075-T6 to high-frequency (200 Hz) vibration with superimposed loading at lower frequency (1 Hz). Figure 19(b) shows the characteristic platelike structure and well-defined striations that were created as a result of the superimposed low-frequency loads. Examination of the fracture surface with a scanning electron microscope revealed finely spaced (0.13 to 0.15 μm, or 5 to 6 μin.) striations attributable to crack growth resulting from the stresses induced during high-frequency vibration (Fig. 19c). The narrowly spaced striations are expected at high-frequency loading, because in each cycle the length of time at peak loading is very short, and thus the increment of crack growth per cycle is correspondingly small. This is true especially when the temperature is high and time-dependent damage processes are operative.

Effect of Stress on Fatigue Strength

Fatigue cracks generally form preferentially at the surface, because the level of stress is generally higher at the surface. Experimental results indicate that fatigue can occur under high vacuum and at low temperature, suggesting that the primary mechanism of fatigue need not involve corrosive attack or thermal activation, although both may contribute to final failure.

Mean Stress. A series of fatigue tests can be conducted at various mean stresses and the results plotted as a series of *S-N* curves. For design purposes, it is more useful to know how the mean stress affects the permissible alternating stress amplitude for a given life (number of cycles). This is usually accomplished by plotting the allowable stress amplitude for a specific number of cycles as a function of the associated mean stress. At zero mean stress, the allowable stress amplitude is the effective fatigue limit for a specified number of cycles. As the mean stress increases, the permissible amplitudes steadily decrease until, at a mean stress equal to the ultimate tensile strength of the material, the permissible amplitude is zero.

The two straight lines and the curve shown in Fig. 20 represent the three most widely used empirical relations. The straight line joining the alternating fatigue strength to the tensile strength is the modified Goodman law (Eq 2). Goodman's original law, which is no longer used, included the assumption that the alternating fatigue limit was equal to one-third of the tensile strength; this has since been modified to the relation shown in Fig. 20, using the alternating fatigue strength determined experimentally. Gerber found that the early experiments of Wöhler fitted closely to a parabolic relation, and this is known as Gerber's parabola (curve, Fig. 20). Gerber's law is given in Eq 3. The third relation, known as Soderberg's law (Eq 4), is given in Fig. 20 by the straight line from the alternating fatigue strength to the static yield strength. For many purposes it is essential that the static yield strength not be exceeded, and this relation is intended to fulfill the conditions that neither fatigue failure nor yielding occurs. The relations may be written mathematically as:

$$S_a = S\left[1 - \left(\frac{S_m}{S_u}\right)\right] \qquad (Eq\ 2)$$

$$S_a = S\left[1 - \left(\frac{S_m}{S_u}\right)^2\right] \qquad (Eq\ 3)$$

$$S_a = S\left[1 - \left(\frac{S_m}{S_y}\right)\right] \qquad (Eq\ 4)$$

where S_a is the alternating stress associated with a mean stress S_m, S is the alternating fatigue strength, S_u is the tensile strength, and S_y is the yield strength.

An understanding of the Goodman, or constant-life, diagram has resulted in many varied and useful treatments for improving fatigue life. According to a constant-life diagram, increased tension decreases the fatigue life and increased compression increases it. Because most cracks originate at the surface of the part, placing the surface under compressive stress should be beneficial. Recognition of this has resulted in development of such surface treatments as nitriding, carburizing, shot peening, surface rolling, and overstressing. When these treatments are properly applied, the sur-

Fig. 18 Schematic of marks on surfaces of fatigue fractures produced in smooth and notched components with round, square, and rectangular cross sections and in thick plates under various loading conditions at high and low nominal stress

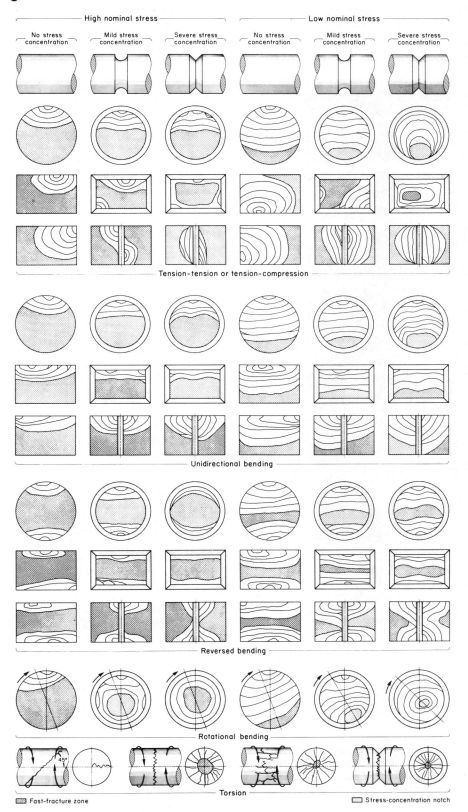

Fig. 19 Fatigue-fracture surface of an aluminum alloy 7075-T6 panel produced by high-frequency vibrations with superimposed loadings at lower frequency

(a) and (b) Light fractographs of part of the fracture surface at 20× and 800×, respectively; (b) shows platelike structure with striations created by low-frequency loading. (c) SEM fractograph of part of the fracture surface showing fine striations between coarse ones. 5000×

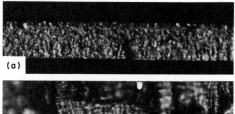

(a)

(b)

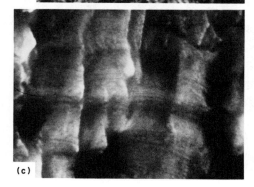

(c)

Fig. 20 Effect of mean stress on the alternating stress amplitude, as shown by the modified Goodman line, Gerber's parabola, and Soderberg line

See text for discussion.

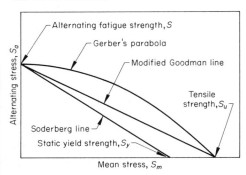

face is in a state of residual compression. However, if not properly applied, they can have a detrimental effect on fatigue life.

Stress Amplitude. Because stress amplitude will vary widely under actual loading conditions, it is necessary to predict fatigue life under various stress amplitudes. The most widely used method of estimating fatigue under complex loading is provided by the linear-damage law. This is a hypothesis first suggested by Palmgren and restated by Miner, and is sometimes known as Miner's law (Ref 1). Application of n_i cycles at a stress amplitude S_i, for which the average number of cycles to failure is N_i, is assumed to cause an amount of fatigue damage that is measured by the cumulative cycle ratio n_i/N_i, and failure is assumed to occur when $\Sigma(n_i/N_i) = 1$.

This method is not applicable in all cases, and numerous alternative theories of cumulative linear damage have been suggested (Ref 2). Some considerations of redistribution of stresses have been clarified, but there is as yet no approach that seems satisfactory in all situations. Miner's law assumes that damage is independent of the stress or strain level, which may or may not be true, depending on the material.

The effect of varying the stress amplitude (linear damage) may be evaluated experimentally by means of a test in which a given number of stress cycles are applied to a testpiece at a one stress amplitude and the test is continued to fracture at a different amplitude. Alternatively, the stress may be changed from one stress amplitude to another at regular intervals; such tests are known as block, or interval, tests. These tests do not simulate service conditions, but are useful in assessing the linear-damage law and indicating its limitations. Results of tests by several investigators have shown that initial overstressing reduces both the fatigue limit and the subsequent fatigue life at stresses above the fatigue limit. The results also show that a slight overstress does not markedly reduce the fatigue limit, even if continued for a large proportion of the normal life of a material. However, this is not true for a high overstress.

Residual Stress. Fatigue fractures generally propagate from the surface. Processing operations, such as grinding, polishing, and machining, that work harden or increase residual stress on the surface can influence the fatigue strength, although there is no generalized formulation that will predict the extent of improved fatigue strength that can be derived from work hardening and residual stress. Compressive residual surface stresses generally increase the fatigue strength, but tensile residual surface stresses do not. There may be a gradual decrease in residual stress if the cyclic stresses cause some plastic deformation. Compressive residual surface stress provides greater improvements in the fatigue strength of harder materials, such as alloy

spring steel; in softer materials, such as low-carbon steel, work hardening effectively improves fatigue strength. This is because the harder material can sustain a high level of residual elastic surface stress, and the tensile strength (and thus the fatigue limit) of the softer material is improved by work hardening.

In a notched high-strength steel, the beneficial effect of prestretching and the detrimental effect of precompression are much greater than in a plain carbon steel because of the type of residual stress present at the notch. A compressive residual stress introduced during quenching from a tempering temperature will increase the fatigue strength, particularly in notched specimens.

In general, residual stresses are introduced by misfit of structural parts, a change in the specific volume of a metal accompanying phase changes, a change in shape following plastic deformation, or thermal stresses resulting from rapid temperature changes such as occur in quenching. The influence of residual stress on fatigue strength is, in principle, similar to that of an externally applied static stress. A static compressive surface stress increases the fatigue strength, and static tensile surface stress reduces it.

Complex Stresses. The criteria of static failure have been applied to fatigue failure. If S_1, S_2, and S_3 are the amplitudes of the principal alternating stresses where $S_1 \geq S_2 \geq S_3$, and if S is the alternating uniaxial fatigue strength, then the following criteria apply:

- Maximum principal-stress criterion $S_1 = S$
- Maximum shear-stress criterion $S_1 - S_3 = S$
- Shear-strain energy $(S_1 - S_2)^2 + (S_2 - S_3)^2 + (S_3 - S_1)^2 = S^2$
- Maximum principal-strain criterion $S_1 = \mu(S_2 + S_3) = S$, where μ is Poisson's ratio

Because fatigue cracks usually propagate from the surface, where one of the principal stresses is zero, only biaxial stresses need to be considered. When the two principal stresses are of the same algebraic sign, the criteria of maximum principal stress and maximum shear stress give the same relationship. If the principal stresses are of opposite algebraic sign, then all criteria give a different relationship. It is often convenient to determine the suitability of the various criteria by comparing the fatigue strength in torsion t and bending b. No single criterion adequately describes the general behavior of the stresses; for ductile materials, the closest correlation with the experimental results of (t/b) is provided by the shear-strain-energy criterion (Ref 3).

Frequency. The frequency range of 500 to 10 000 cycles/min is generally used in a fatigue test. In this range, the fatigue strength of most materials, based on a given number of cycles to fracture, is little affected by frequency. In general, there is a slight decrease in fatigue strength with a decrease in frequency, because the fatigue limit may be related to the amount of

Fig. 21 Effect of stress raisers on stress concentration and distribution of stress at several changes of form in components
See text for discussion.

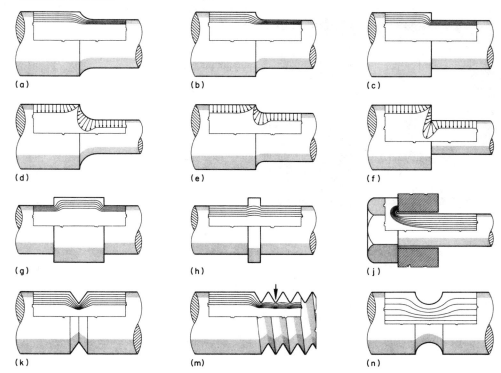

(a)   (b)   (c)

(d)   (e)   (f)

(g)   (h)   (j)

(k)   (m)   (n)

plastic deformation that occurs during the stress cycle. For example, at high frequency there is less relaxation time during each stress cycle for deformation to occur, which results in less damage. For steel, the fatigue limit is not affected between 200 and 5000 cycles/min. However, at frequencies up to 100 000 cycles/min, steels that are 100% ferritic show a marked increase in fatigue strength. In nonferrous metals, the fatigue strength increases continuously with frequency; in plain carbon steel, the fatigue strength reaches a maximum value, then decreases with an increase in frequency.

Effect of Stress Concentrations on Fatigue Strength

Fatigue cracks usually initiate at some region of stress concentration resulting from the presence of surface discontinuities (stress raisers), such as a step or shoulder, a screw thread, an oil hole or bolt hole, or a surface flaw. The stress-concentration factor, K_t, of the discontinuity is a measure of intensity of stress that occurs. In some situations, values of K_t can be calculated using this theory of elasticity or can be measured using photoelastic plastic models. Many of these values are reported in standard references (Ref 4-7).

The mathematical theory of elasticity is based on an ideal isotropic material free of any internal discontinuity, a strictly accurate pro-

file, and an increase in stress concentration due solely to the presence of the surface discontinuity. In actual parts, the stress intensification is affected not only by the surface discontinuity but also, to an undetermined extent, by the size of the part, by local readjustments of stress because of plastic yielding, by surface roughness, and by the heterogeneous structure of the material itself, including anisotropy and inherent internal discontinuities. Therefore, the deleterious effect of a stress raiser on a part is usually determined experimentally and expressed in terms of a fatigue notch factor, K_f. This is the ratio of the fatigue strength without stress concentration to fatigue strength with stress concentration. In general, experimentally determined values of K_f are somewhat less than the values of K_t calculated for the same specimens.

Under static loading conditions in a ductile material, a stress raiser has little or no effect in most situations. If the local stress exceeds the yield strength of the material, plastic deformation occurs and stress is redistributed. Provided the amount of plastic deformation is not excessive, no adverse effect need be anticipated. However, if the part is subjected to fluctuating or alternating stresses and if a fatigue crack nucleates at a stress below the yield strength of the material, stress redistribution by plastic deformation will not occur, and the material will be subject to the full effect of the stress raiser.

Stress concentrations affect the fatigue behavior of different materials differently. For example, relatively brittle materials, such as quenched-and-tempered steels, are more susceptible to the effects of stress raisers than ductile materials, such as normalized or annealed steels. In addition, cast irons, containing innumerable internal stress raisers, show little further adverse effects from externally introduced stress raisers.

Distribution of Stress. To visualize the distribution of stress at a change in section size or shape, it is helpful to consider the part in terms of electricity flowing through a conductor of similar cross section. In diagrammatic form, stress flow can be represented as a series of parallel lines, the stress being inversely proportional to the distance between the lines; that is, the lines bunch together in regions of high stress. Figure 21 shows the flow of stress associated with several of the stress raisers typically found in parts in service.

Progressive increases in stress with decreasing fillet radii are shown in Fig. 21 (a) to (c), and the relative magnitude and distribution of stress resulting from uniform loading of these parts is indicated in Fig. 21(d) to (f). Stress caused by the presence of an integral collar of considerable width is shown in Fig. 21(g); Fig. 21(h) shows the decrease in stress concentration that accompanies a decrease in collar width. Stress conditions are very similar when collars or similar parts are pressed or shrunk into position. The stress flow at the junction of a bolt head and a shank is as represented in Fig. 21(j).

A single notch introduces a considerably greater stress-concentration effect than does a continuous thread: the reason for this is clear when the stress flow is considered. The stress-concentration effect of a single sharp notch is as shown in Fig. 21(k). The stress concentration at the right of the arrow in Fig. 21(m) is very similar to that in the narrow collar in Fig. 21(h) because of the mutual relief afforded by adjacent threads. To the left of the arrow, however, the last thread is relieved from one side only; consequently, there is a considerable stress concentration, similar to that of the single notch in Fig. 21(k). This is why bolts so frequently fracture through the last full thread.

The effect of a groove or gouge on stress concentration (Fig. 21n) is less severe than that of a sharp notch. A series of grooves will have an effect similar to that shown in Fig. 21(m).

Example 3: Fatigue Fracture of a Cast Stainless Steel Lever. A main fuel control was returned to the factory for examination after service on a test aircraft engine that had experienced high vibrations. There were no apparent problems with the fuel control, but it was removed for examination to ensure that no problems had developed as a result of the vibrations. When the fuel control was disassembled, the lever shown in Fig. 22(a) was found to be cracked. The lever was cast from AMS 5350 (AISI type 410) stainless steel that

Fig. 22 Cast stainless steel fuel-control lever that failed when the aircraft engine in which it was mounted experienced excessive vibration
(a) Configuration and dimensions (given in inches). (b) View of a fracture surface; arrow indicates region of fracture origin

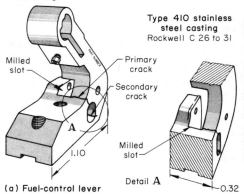

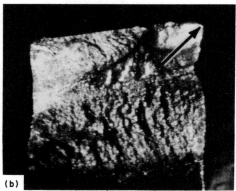

Fig. 23 Aluminum alloy 7075-T73 landing-gear torque-arm assembly that was redesigned to eliminate fatigue fracture at a lubrication hole
(a) Configuration and dimensions (given in inches). (b) Fracture surface showing fatigue beach marks. Approximately 2 ×

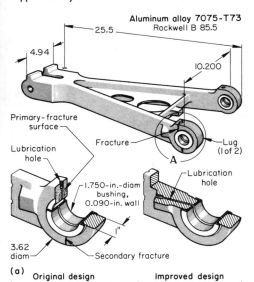

was through-hardened to 26 to 32 HRC and passivated.

Investigation. There were no corrosion products or stains on the lever; however, a slight bluish cast, probably caused during chemical etching of the part number, was found on the slotted arm. The general condition of several holes and surfaces of the lever was good, although the sides of the rectangular slot in the base were abraded from rubbing against the mating part. The sides of the elongated hole in the slotted arm were polished by motion of the mating member. There was negligible wear in the round hole and in the elongated holes where fracture occurred.

The crack initiated at the sharp corner of the milled slot and propagated across to the outer wall (detail A, Fig. 22). The surfaces of the part at the crack were slightly offset, indicating some plastic deformation. This deformation could have occurred after the failure and during disassembly of the lever from the fuel control.

To examine the fracture surface, the sections were spread apart about 30°, resulting in a slight tearing in one corner of the section surrounding the elongated hole. Bending of the part was easily done, indicating that the general ductility of the part was satisfactory. The bent section was subsequently broken off, and another crack was observed in the casting; however, it was remote from and unrelated to the fracture under investigation. The lever had been magnetic-particle inspected during manufacture, and the crack was of a size that should have been detected.

The fracture surface under investigation (Fig. 22b) had beach marks initiating at the sharp corner along the milled slot. Changes in frequency or amplitude of vibration caused different rates of propagation, resulting in a change in pattern. Motion of the part after failure occurred caused some obliteration of the fracture characteristics.

Conclusions. The lever failed in fatigue as a result of excessive vibration of the fuel control

on the test engine. Cracking initiated in a sharp corner of a milled slot.

Corrective Measures. Conditions on the test stand were more severe than those the engine was expected to survive in service without major damage and were solely responsible for failure of the lever. If any corrective measure at all were necessary on the part, it would be to specify a large radius in the corner where cracking originated. This would reduce the stress-concentration factor significantly, thus minimizing the susceptibility of the lever to fatigue.

Example 4: Fatigue Fracture of an Aluminum Alloy 7075-T73 Landing-Gear Torque Arm. The torque-arm assembly for an aircraft nose landing gear (Fig. 23a) failed after 22 779 simulated flights. The part, made from an aluminum alloy 7075-T73 forging, had an expected life of 100 000 simulated flights.

Two cadmium-plated flanged bushings, made of copper alloy C63000 (aluminum bronze), were press-fitted into each bored hole in the lug. A space between the bushings provided an annular groove for a lubricant. A lubrication hole extended from the outer surface to the bore of the lug (Original design, Detail A, Fig. 23).

Investigation. Initial study of the fracture surfaces indicated that the primary fracture initiated from multiple origins on both sides of the lubrication hole (arrows, Fig. 23b). Beach marks on the fracture surface indicate that cracking was initiated and propagated by fatigue until an overload stage was reached. A fracture on the opposite side of the lug (Secondary fracture, Detail A, Fig. 23) appeared to be typical for a static overload failure.

A section was taken normal to the fracture surface and in the region contacted by the bushing flange, and was prepared for metallographic examination. Results showed small fatigue-type cracks in the hole adjacent to the origin of primary fracture. There were no indications of inclusions or discontinuities in the

plane of the section that could contribute to the initiation of lug fracture.

Hardness of the material at the face of the lug adjacent to the fracture surface was 85.5 HRB. Electrical conductivity of the material was 39% IACS, typical for aluminum alloy 7075.

Conclusions. The arm failed in fatigue cracking that initiated on each side of the lubrication hole. No material defects were found at the failure origin.

Corrective Measures. The location of the lubrication hole was changed as shown in the "Improved design" view in Detail A in Fig. 23. The faces of the lug were shot peened for

Fig. 24 Comparison of the distribution of torsional stresses in a shaft at and away from a keyway

See text for discussion.

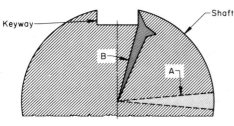

Fig. 25 Stress distribution produced in a notched specimen under various conditions

Stress distribution in a specimen (a) subjected to a load $L = 1$, (b) subjected to a load $L = 1.5$, (c) after the load was decreased to $L = 1$, (d) after removal of the external load (residual stresses), (e) after being shot peened, and (f) subjected to a load after shot peening

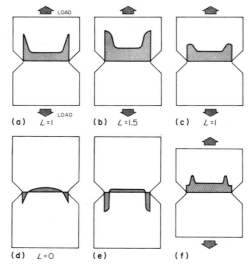

added resistance to fatigue failure. Also, the forging material was changed to aluminum alloy 7175-T736 for its higher mechanical properties.

Keyways in components frequently act as stress raisers. When the stress is predominantly torsional, fatigue cracks usually initiate in the fillets of the keyway, which is to be expected from the disturbance of the stress flow caused by the presence of the keyway. The transition of torsional stress as it reaches a keyway is shown by a comparison of shaded zones A and B in Fig. 24. The distribution of torsional stress in regions away from the keyway (zone A) is smooth and wedge shaped (stress is zero at center and maximum at the surface of the shaft). In regions near the keyway (zone B), the stress distribution is distorted by the stresses caused by contact between the key and keyway, resulting in maximum torsional stress below the surface. When the end of a keyway is formed with a sharp step, a further concentration of stress occurs, especially in the presence of bending stresses, and cracking often begins at this point. This further stress concentration can be reduced considerably by using a sled-runner type of keyway, in which the transition at the end of the keyway is gradual.

Reduction of Stresses. The detrimental effect of stress concentrations can be reduced or eliminated by the use of induced stresses. These stresses may be introduced by plastic deformation of the surface, such as in cold working and shot peening; by phase transformation, such as in case hardening; or by proof loading with a load high enough to cause local plastic flow in the notch but low enough to avoid general yielding.

Extreme or incorrect cold working may have adverse effects under fatigue conditions, because it may give rise to minute cracks in the surface of the material or at least make the surface prone to such cracks. If the correct amount of cold working is applied, however, improvement in the fatigue strength is appreciable. For example, a screw thread produced by rolling is more resistant to fatigue failure than one that has been cut. Surface rolling and shot peening, especially of springs, are other examples of the successful application of cold

working. Apart from the increase in fatigue strength of the surface of the material arising from work hardening, the stress distribution is modified considerably.

Figure 25(a) shows the stress distribution in a specimen notched by machining and then subjected to a tensile load $L = 1$. This load just starts to produce yielding at the root of the notch, which has a stress-concentration factor of 3.0. Figure 25(b) shows the same specimen loaded to $L = 1.5$, which produces yielding to a greater depth. The result of decreasing the load to $L = 1$ is shown in Fig. 25(c). The beneficial effect of the overload is indicated by the reduction of peak stress. The residual-stress distribution after removal of the external load ($L = 0$) is shown in Fig. 25(d). The stress distribution produced in a specimen in which the notches have been shot peened is shown in Fig. 25(e). The stress in the shot-peened specimen after application of load is shown in Fig. 25(f).

If the amount of cold working is so adjusted that the residual compressive stress in the part effectively reduces the applied peak tensile stress below the fatigue limit of the material, fatigue failure will not occur. It is difficult in practice, however, to ensure the correct amount of cold working. Cold working intended to improve the endurance of a part should be chosen with appropriate consideration of the affected stress system. Residual compressive surface stresses improve fatigue strength only under conditions involving bending and torsional stresses. Where the stress is purely ten-

sile (although infrequent in real-life engineering applications), the fatigue strength may be reduced, because the compressive stress at the surface is balanced by a region of tensile stress just below the surface. In tensile loading, the stress is uniformly distributed across the section and is augmented by the residual tensile stress, and failure is likely to take place by cracking just beneath the surface.

Other methods of reducing the effect of stress concentration are to raise the fatigue strength of the material in the region of high stress and to produce compressive residual stresses by flame hardening, carburizing, or nitriding. If performed correctly, any of the three processes is beneficial in raising the fatigue strength, but the extent of the treated region needs to be carefully chosen; otherwise, failure is likely to occur at one of the junctions between the treated and untreated regions.

Influence of Design on Fatigue Strength

Mechanical and structural design encompasses two categories: configuration and material properties. Selection of material is often influenced by requirements other than fatigue characteristics, such as resistance to corrosion or to elevated temperatures. Regardless of the controlling property, the material used in a part designed for fatigue resistance must possess an acceptable combination of many properties. Not all designs necessarily require a value for all possible properties, and data may not be available for all specific materials, conditions and forms.

Simple rules for fatigue-resistance design cannot be stated because of the diversity in function, loads, stresses, materials, and environments. However, a number of useful practices can be defined that, when applied with good engineering judgment, can be expected to result in a significant reduction in the probability of fatigue failure.

Correction Factors for Test Data. The available fatigue data are normally for a specific type of loading, specimen size, and surface roughness. For instance, the Moore rotating-beam fatigue-test machine uses a 7.6-mm (0.3-in.) diam specimen that is free of any stress concentrations because of specimen shape, is polished to a mirror finish, and is subjected to completely reversed bending stresses. It has been suggested that the fatigue-life data used in design calculations be corrected by multiplying the number of cycles (life), N_i', determined in a fatigue test by three factors that account for the variation in the type of loading, part diameter, and surface roughness (Ref 8). Thus, the design life $N_i = K_1 K_d K_s N_i'$, where K_1 is the correction factor for the type of loading, K_d for the part diameter, and K_s for the surface roughness. Values of the three factors for correcting standard Moore fatigue-test data on steel are given in Fig. 26.

Fig. 26 Correction factors for surface roughness (K_s), type of loading (K_l), and part diameter (K_d) for fatigue life of steel parts

See text for application.

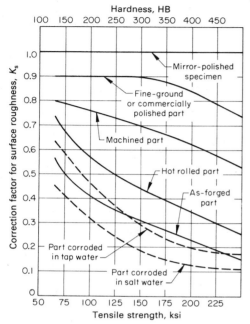

Factor	Type of load		
	Bending	Torsion	Axial
K_l......................	1.0	0.58	0.9(a)
K_d, where:			
d ≤ 0.4 in.............	1.0	1.0	1.0
0.4 in. < d ≤ 2 in.	0.9	0.9	1.0
K_s......................	— From chart above —		

(a) A lower value (0.6 to 0.85) may be used to account for known or suspected undetermined bending because of load eccentricity.

Interference at Line of Contact. The following two design guides can minimize interference at the line of contact between mating parts:

- Provide generous fillets and radii
- Design parts for minimum mismatch at installation for lowest residual and preload tensile strains

In some cases, generous fillet radii can cause interference if the corner radii on the mating parts are too small. Therefore, a large corner radius would result in minimum mismatch at installation.

Example 5: Fatigue Fracture of a D-6ac Steel Structural Member at the Line of Contact With Another Member. A structure had been undergoing fatigue testing for several months when a postlike member ruptured. Fracture occurred in the fillet of the post that contacted the edge of a carry-through box bolted to the member (Fig. 27). At failure, the part was receiving the second set of loads up to 103.6% of design load. The post was made of D-6ac steel and was heat treated to a tensile strength of 1517 to 1655 MPa (220 to 240 ksi).

Fig. 27 Structural member (post) of D-6ac steel that failed by fatigue cracking

The cracking was initiated by rubbing and galling from a mating carry-through box that was bolted to the post.

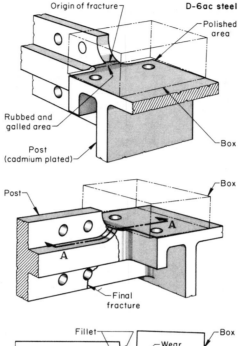

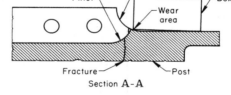

Investigation. The edge of the carry-through box along the line of contact with the post was permanently deformed and galled by rubbing, and adjacent contact points on the post were polished. The line of fracture in the post coincided with the top edge of the rubbed and galled area on the box. A fillet on the riblike surface of the post in the stressed area was tangent to the forward lower surface edge of the box. Rubbing and galling occurred across the entire width of the fillet where it was in contact with the edge of the box (Section A-A, Fig. 27). Also seen was light fretting surrounding many of the boltholes and at several other high spots on the faying surfaces.

Microscopic examination of the fracture surface indicated that the origin of the crack was between 1.3 and 2.5 mm (0.05 and 0.10 in.) from the edge of the post. Also, several beach marks on the fracture surface indicated that various static and dynamic loads had been applied to the structure before final fracture. Chevron marks on the fracture surface showed that the crack propagated toward the bolthole and that the last area to fail was between the bolthole and the back edge of the post. Inspec-

tion of the edge of the fracture surface revealed numerous indications of small secondary cracks.

Metallographic examination of sections through the fracture surface revealed evidence of cold work and secondary cracking in the rubbed and galled area. The grain size and microstructure were satisfactory, and no evidence of decarburization or oxidation was found on the surfaces. The post had apparently been shot peened and cadmium plated.

Electron fractography confirmed that cracking had initiated at a region of tearing and that the cracks had propagated by fatigue. Some intergranular features observed near the origins were associated with the tearing.

Standard-size round tensile specimens were cut from the thick sections of the post on each side of the fracture and prepared. The axis of each specimen was parallel to the long dimension of the post. The mechanical properties of all specimens exceeded the minimum values specified for the post.

Conclusions. Fatigue was the primary cause of failure. A fillet on the post created an area of interference with the edge of the carry-through box bolted to the post. Rubbing of the faying surfaces worked the interference area on the post until small tears developed. These small tears became stress-concentration points that nucleated fatigue cracks and ultimately resulted in fracture.

Recommendations. The edge of the box in the area of contact with the post should be rounded to ensure a tangency fit.

Joint Design. The importance of the effect of joint design on fatigue behavior cannot be overemphasized. The conventional approach to designing a multimember structure for fatigue resistance is to consider the members as first in importance, and the joints as inevitable complications. Of equal importance is the shape of the parts within the joint, because the shape may transform the external loading to a different mode, causing possible stress concentrations.

Example 6: Fatigue Fracture of a Forged 4150 Steel Drive Axle in an Overhead Crane. A stepped drive axle used in a high-speed electric overhead crane broke after 15 months of service. The axle (Fig. 28a) was made from a hardened-and-tempered resulfurized 4150 steel forging. The overhead crane was rated at 6800 kg (7½ tons) and handled about 220 lifts per day, each lift averaging 3625 to 5440 kg (4 to 6 tons).

Investigation. Fracture occurred approximately 50 mm (2 in.) from the driven end of the large-diameter keywayed section on the stepped axle and approximately 38 mm (1½ in.) from one end of the keyway where the crane wheel was keyed to the axle (Fig. 28a). There were visual indications that the crane wheel had moved during operation to a position approximately 12.5 mm (½ in.) farther away from the driven end of the axle. The interference fit between the wheel and the shaft was insufficient to prevent radial and axial movement

Fig. 28 4150 steel drive axle for an overhead-crane wheel that fractured from fatigue in service due to insufficient interference fit between the wheel and the axle

(a) Original design showing location of fracture. (b) Improved axle design that used a narrow shoulder at stress area to prevent shifting. Dimensions given in inches. (c) Fracture surface showing regions of bending fatigue (region A), combined bending and torsional fatigue (region B), and final fast fracture (region C)

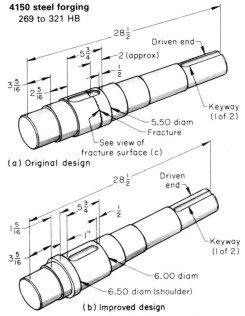

(a) Original design

(b) Improved design

(c)

during operation. The axial movement of the crane wheel allowed the torsional moment to be displaced toward the location of the maximum bending moment, increasing the total effective stress. The edges of the portion of the keyway in the broken axle were battered and misshapen (Fig. 28c), apparently from repeated impact against the key during the later stages of failure.

Visual examination of the fracture surface revealed three fracture regions as shown in Fig. 28(c): a region of bending fatigue (at A); a region of combined bending and torsional fa-

tigue (at B); and a region of final fast fracture (at C). The dark, convex bending-fatigue region accounted for less than 5% of the fracture surface. The greater part of the fracture surface, more than 80%, was the region of combined bending and torsional fatigue—the flat, relatively featureless area at B in Fig. 28(c) that covers the upper half of the fracture surface and looks progressively less smooth as it approaches the rough final-fracture region (at C) adjacent to the bending-fatigue region (at A).

Beach marks in the bending-fatigue region indicated that cracking started at the surface approximately at the center of the bending-fatigue region. Cracking in the region of combined bending and torsional fatigue had multiple origins at the surface. Several of the origins can be seen immediately to the left of the bending-fatigue region (A) in Fig. 26(c). Final fracture was by mixed ductile and brittle fracture, as indicated by a chevron pattern that followed a generally elliptical contour in the final-fracture region (C).

The hardness of the axle varied from 305 HB near the surface to 277 HB at the center. This was within the specified hardness range of 269 to 331 HB.

Macroscopic and microscopic examination revealed indications of slight porosity near the center of the axle, but this had no causative relationship to the fracture. The microstructure was acicular tempered martensite with moderately numerous elongated manganese sulfide inclusions.

The chemical composition was within the normal range for 4150 steel, except for manganese, which was slightly above the high limit for 4150 steel, and sulfur, which was intentionally added to improve machinability of the hardened and tempered material.

Conclusions. Cracking initiated in bending fatigue at a location approximately opposite the keyway. This was followed by torsional-fatigue cracking from multiple origins and by final fracture due to mixed ductile and brittle fracture.

Axial shift in position of the crane wheel during operation, because of insufficient interference fit, was the major cause of fatigue cracking. This shift significantly increased the bending stresses on the shaft.

Corrective Measures. The axle was redesigned as shown in Fig. 28(b). The critical diameter was increased from 140 to 150 mm (5.50 to 6 in.), and a narrow shoulder was added to keep the drive wheel from shifting during operation. Metallurgical changes were not considered necessary, because the metallurgical condition did not appear to be a factor in this failure.

Welded Joints. The fatigue strength of good as-welded joints depends on the stress state in the load path through the weldment, the stress-concentration factor at the toe of the weld, the condition of the material in the heat-affected zone (HAZ), and the presence or absence of inclusions and other defects in the

Fig. 29 Welded stainless steel elbow assembly that, as originally designed, cracked at the root of the weld under cyclic loading

The improved design moved the weld out of the high-stress area. Dimensions given in inches

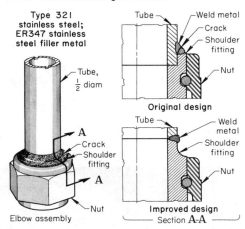

weld. Fatigue strength, at a reasonable life, of as-welded joints in carbon and low-alloy steels is independent of the steel used.

The fatigue strength of as-welded joints can be increased by cold working, because this inhibits the growth of cracks. Stress relieving after welding does not have much effect on the repeated tension-fatigue strength, but because of the lower maximum tensile stress in the loading cycle, it can increase zero-mean-load values.

Example 7: Fatigue Cracking of a Stainless Steel Elbow Assembly at a Welded Joint in a High-Stress Region. The welded elbow assembly shown in Fig. 29 was part of a hydraulic-pump pressure line for a jet aircraft. The other end of the tube was attached to a flexible metal hose, which provided no support and offered no resistance to vibration. The components of the elbow were made of AISI type 321 stainless steel and were joined with ER347 stainless steel filler metal by gas tungsten arc welding. The line was leaking hydraulic fluid at the nut end of the elbow, and the assembly was returned to the manufacturer to determine the cause of failure.

Investigation. Visual examination of the elbow assembly revealed a crack in the fillet weld that joined the tube to the shoulder fitting. In assembly, the tube was inserted into a counterbore in the shoulder fitting (Original design, Section A-A, Fig. 29), then welded by the gas tungsten arc process.

The fitting was separated from the tubing so that the surfaces of the cracked area could be examined. Stains on the fracture surface indicated that fatigue cracks had initiated at the root of the weld, at a notch inherent in the design of the joint. Cracking in the weld metal had progressed partway around the joint, thus pro-

Fig. 30 Duct assembly of medium-carbon steels in which welded bellows liners of type 321 stainless steel fractured in fatigue

(a) Configuration and dimensions (given in inches). (b) Light fractograph showing fracture origin (top edge) and fatigue striations (bottom right). 30×

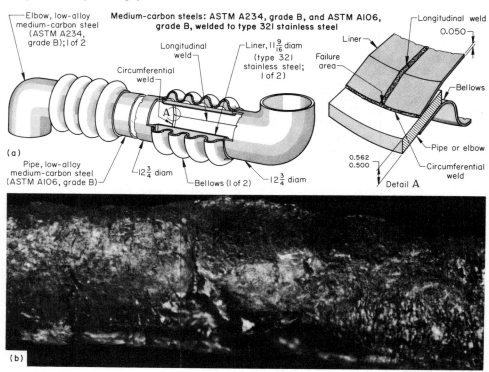

ducing a leakage path through which pressurized hydraulic fluid had escaped.

Conclusion. Failure was by fatigue cracking initiated at a notch at the root of the weld and was propagated by cyclic loading of the tubing as the result of vibration and inadequate support of the hose assembly.

Corrective Measures. The joint design was changed from a cylindrical lap joint (tube inserted into a counterbore) to a square-groove butt joint (Improved design, Section A-A, Fig. 29). The new joint design provided a more flexible joint by eliminating the notch and by moving the weld out of the high-stress area. An additional support was provided for the hose assembly to minimize vibration at the elbow.

Example 8: Fatigue Fracture of Welded Type 321 Stainless Steel Liners for a Bellows-Type Expansion Joint. The liners for bellows-type expansion joints in a duct assembly (Fig. 30a) failed in service. The duct assembly, used in a low-pressure nitrogen-gas system, consisted of two expansion joints (bellows) connected by a 32.4-cm (12¾-in.) OD pipe of ASTM A106, grade B, steel. Elbows of ASTM A234, grade B, steel, 180° to each other, were attached to each end of the assembly. A liner with an outside diameter of 29.4 cm (11⁹⁄₁₆ in.) was welded inside each expansion joint. The liners were 1.3 mm (0.050 in.) thick and made of AISI type 321 stainless steel. The upstream end of one liner was

welded to an elbow, and the upstream end of the other liner was welded to the pipe; this allowed the downstream ends of the liners to remain free and permitted the components to move with expansion and contraction of the bellows.

Investigation. In field inspection, portions of the two failed liners were found downstream in the duct assembly. These were removed and sent to the laboratory for failure analysis. Many of the fracture surfaces had undergone extreme deformation, making the exact fracture interpretation extremely difficult. However, portions of the fracture surfaces were undamaged, and the origin of failure and characteristic fracture pattern could be seen.

Laboratory inspection showed that the origin of failure in the portion of the fracture surface studied was in an area where two welds intersected—the longitudinal or seam weld forming the liner and the circumferential weld attaching the liner to the pipe or elbow. Figure 30(b) shows a fractograph of the fracture surface at the weld intersection indicating the fracture origin and fatigue striations.

Metallographic examination of a cross section containing the longitudinal weld revealed that the microstructure of the weld area contained no irregularities. Metallurgical evaluation of the cross section in the transverse plane of the fractured piece containing the intersection of the longitudinal and circumferential welds exhibited

extreme banding within the microstructure. The banding was the result of chromium and nickel dilution in one area and the enrichment of the same elements in areas adjacent to that area. In addition, the intersection of the two welds contained cracks similar in appearance to those normally formed when variations in thermal expansion and contraction are encountered when welding dissimilar metals.

A microhardness traverse across the weld area revealed a base-metal average hardness of 89 HRB. The hardness of the HAZ on both sides of the weld was 87 HRB, and the weld metal had a hardness of 93 to 98.5 HRB.

Conclusions. The liners failed in fatigue that was initiated at the intersection of the longitudinal weld forming the liner and the circumferential weld that joined the liner to the bellows assembly.

Because the liners were welded at one end and the other end was free, the components moved with expansion and contraction of the joints and were subjected to vibrational stresses imposed by the transfer of nitrogen gas. The stresses, created by cyclic loading, were concentrated mainly at the metallurgical notches formed at the weld metal/base metal interface. Differences in composition caused a variation in strength at the critical region of high residual stresses. Vibrational stresses were concentrated at the circumferential weld metal/base metal interface and at the intersection of the circumferential and longitudinal welds. The origin of the failure was at the intersection of the two welds—the region containing the most severe stress raisers. The design allowed the two liners to be cantilevered in a dynamic system subjected to both thermal stresses and vibrations from gaseous flow.

Corrective Measure. The thickness of the liners was increased from 1.3 to 1.9 mm (0.050 to 0.075 in.), which successfully damped some of the stress-producing vibrations.

Effect of Material Conditions on Fatigue Strength

In fatigue fracture, localized plastic deformation is responsible for crack propagation. The microstructure of the material can influence this crack growth by inhibiting or modifying the plastic-deformation process. Sometimes, the nature of the cracking process is changed from ductile to brittle. Transition of the cracking process from one involving plastic deformation to another involving cleavage is largely determined by the microstructure of the material.

Grain Size

Under high cyclic strain (low-cycle fatigue), the fatigue life of many metals is independent of grain size. In contrast, under low cyclic strain (high-cycle fatigue), the fatigue life of many metals is increased when grain size is reduced. However, the effect of grain size on high-cycle fatigue properties is difficult to assess, because these properties may be altered by the same treatments that alter grain size.

In some alloys, an improvement in resistance to high-cycle fatigue brought about by a decrease in grain size may be partly offset by a deleterious effect on another property. For example, a decrease in grain size is thought to raise the smooth-bar fatigue limit in some steels. However, small grain size increases notch sensitivity in these steels; therefore, the net result may constitute no improvement in resistance to notched-bar fatigue. As another example, fine grain size in high-temperature alloys (which may be subjected to both fatigue and creep) may result in high fatigue life but low stress-rupture life at normal service temperatures; thus, an intermediate grain size may afford the longest service life.

Alloying

The influence of chemical composition on fatigue strength is approximately proportional to its influence on tensile strength. The fatigue limit of plain carbon steel generally increases with carbon content. Molybdenum, chromium, and nickel have a similar effect.

The fatigue limit of high-strength steel with tensile strength in the range of 1379 MPa (200 ksi) can be increased by the addition of copper. Although the phosphorus content of steel is generally kept at a minimum to prevent brittleness, steels with high phosphorus contents have greater fatigue strength. A sulfur content of 0.01% has no effect on fatigue limit. Austenitic steels containing nickel and chromium have a high fatigue limit, together with low notch sensitivity and high resistance to corrosion fatigue.

The fatigue strengths of titanium alloys are higher than those of steel. Some titanium alloys maintain considerable strength up to 500 °C (930 °F) and have high resistance to corrosion fatigue. Even though titanium alloys are particularly susceptible to hydrogen embrittlement, the presence of hydrogen does not affect fatigue properties.

Solid-Solution Strengthening. Aluminum alloys that are solid-solution strengthened show an increase in fatigue strength to about the same degree as the corresponding increase in tensile strength. If the strength of magnesium alloys is increased by solute addition, the fatigue strength is also increased in proportion to the increase in tensile strength.

Second phases that are often present in metallurgical systems affect crack propagation on the basis of three factors:

- The lattice strain caused by the presence of the second phase
- The stress concentration determined by the size, shape, and distribution of the second phase
- The nature of the bond between the second phase and the matrix

Second phases have a marked influence on the mechanism and kinetics of crack nucleation and propagation because they can accelerate or inhibit the crack-propagation rate under various circumstances. For example, investigations on aluminum alloys containing a large number of particles of precipitated, intermetallic second phase showed that the particles acted as raisers from which fatigue cracks nucleated.

The grain boundaries of age-hardened alloys are generally free of precipitate particles and consequently are relatively soft. Therefore, stress relaxation takes place along the boundaries, resulting in localized strain concentration, and crack nucleation, at grain-boundary triple points.

The instability of coherent precipitate particles is considered to be the most important factor responsible for the low fatigue life of high-strength aluminum alloys. Experimental results indicate that reversion takes place during the early stages of cyclic loading. Reversion under cyclic straining involves the passage of dislocations back and forth through the precipitate particles, which disintegrates the particles to subcritical size. The solute then goes back into solution or is distributed along the dislocation network, thereby softening the slip bands and bringing about crack nucleation.

Work Hardening

Work-hardened copper-aluminum alloys (75 wt% Al) do not soften under cyclic stresses. This results in lower rates of crack propagation in work-hardened alloys, thereby indicating a small amount of deformation at the crack tip during fatigue. Experimental results on annealed and cold-worked α-brass suggest that a work-hardened alloy is harder at the crack tip than an annealed alloy.

In alloys that are not strengthened by heat treatment (some copper alloys, aluminum alloys, and stainless steels), fatigue strength can be increased by cold working. However, no improvement in fatigue strength is obtained by cold working of age-hardenable alloys. Alloys that are hardened by transformation processes—for example, martensitic steels—exhibit a lesser degree of improvement by cold working than the nonhardenable alloys.

Heat Treatment

Fatigue strength is generally increased by any heat treatment that increases tensile strength. In steels, a tempered-martensite structure has the best fatigue properties. The lower fatigue resistance of mixed structures is generally the result of metallurgical notches such as are formed by coarse pearlite, free ferrite, retained austenite, and carbide segregation.

In copper alloys, particularly in copper-zinc and copper-tin alloys, the fatigue strength can be improved by solution heat treatment. The fatigue strength of age-hardenable aluminum alloys can also be raised by solution heat treatment; however, subsequent aging, which occurs at room temperature in some alloys, has no further beneficial effect and may even decrease fatigue strength.

Effect of Discontinuities on Fatigue Strength

Discontinuities within a metal, either at the surface or subsurface, can adversely affect fatigue strength. These discontinuities may arise from melting practices or primary or secondary working of the material, or may be a characteristic of a particular alloy system.

Surface Discontinuities

All wrought metal is subjected to thermal-mechanical processing that shapes the metal by plastic deformation, generally by rolling, hammering, squeezing, or drawing. The movement of metal during these processes, whether performed at room temperature or at elevated temperatures, makes them common sources of surface discontinuities, such as laps, seams, and cold shuts. Oxides, slivers or chips of the base material, or foreign material can be embedded into the surface by rolling or forging.

These surface imperfections produce a notch of unknown severity that acts as a stress raiser under load and adversely affects fatigue strength. The imperfections can also serve as sites for crack initiation during fabrication or in service.

Laps and seams are surface discontinuities that are caused by folding over of metal without fusion. They are usually filled with scale and, on steel components, are enclosed by a layer of decarburized metal.

Example 9: Fatigue Fracture of a 15B41 Steel Connecting-Rod Cap. A connecting cap (Fig. 31a) from a truck engine fractured after 65 200 km (40 500 miles) of service. The cap was made from a 15B41 steel forging and was hardened to 29 to 35 HRC. There were no known unusual operating conditions that would have caused fracture.

Investigation. In the laboratory, visual examination of the fracture surface disclosed an open forging defect across one of the outer corners of the cap. The defect extended approximately 9.5 mm (³⁄₈ in.) along the top surface and 16 mm (⁵⁄₈ in.) along the side of the cap. The fracture surface exhibited beach marks typical of fatigue (View A-A, Fig. 31). The surface of the defect was stained, indicating that oxidation occurred either in heat treatment or in heating during forging. Deep etching of the fracture surface revealed grain flow normal for this type of forging, but no visible defects.

Metallographic examination of a section through the fracture surface showed that the microstructure was an acceptable tempered martensite. However, oxide inclusions were present at the fracture surface (Fig. 31b).

Conclusions. Fatigue fracture initiated at a corner of the cap from a forging defect that extended to the surface. Fatigue cracking was propagated by cyclic loading inherent in the part.

Recommendation. Failure of the caps could have been minimized by more careful fluores-

Fig. 31 Forged 15B41 steel connecting-rod cap that fractured from fatigue

Cracking originated at an open forging defect.
(a) Configuration and dimensions (given in inches).
(b) Section through the fracture surface showing oxide inclusions. 400×

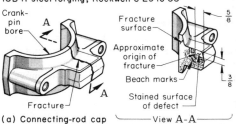

(a) Connecting-rod cap — View A-A

(b)

Fig. 32 Fracture surface of a forged steel rocker arm

Fatigue (arrows) is evident nucleating from a coarse intergranular surface typical of burning.

Fig. 33 Fracture surface of a hardened steel connecting rod

Arrows indicate large inclusions from which fatigue cracking initiated.

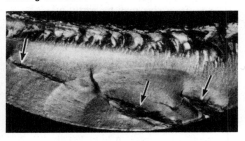

Fig. 34 Fracture surface of a hardened steel valve spring that failed in torsional fatigue

Arrow indicates fracture origin at a subsurface nonmetallic inclusion.

cent magnetic-particle inspection of the forged surfaces before machining and before putting the part into service.

Burning is the result of heating a metal to a temperature so close to its melting point as to damage the metal permanently by intergranular oxidation or incipient melting. Burned steel usually contains both oxide films and voids or cracks at grain boundaries that act as nuclei for fatigue cracking. Burning most often occurs in forging, either during preheating or during the forging operation itself. The fatigue fracture of the forged steel rocker arm shown in Fig. 32 was initiated at a coarse intergranular area of burned metal (see arrows) incurred during forging.

Subsurface Discontinuities

Subsurface and core discontinuities originate in the as-cast ingot. Voids in cast materials from gas porosity, shrinkage porosity, and improper metal fill are common. In ingots that are subsequently hot and cold reduced, the portion dominated by voids is removed and discarded. The remaining internal voids normally weld shut from the temperature and pressure used to reduce the ingot, which results in a continuous, homogeneous product. When the surfaces of the voids are oxidized or otherwise contaminated, the opposing surfaces do not weld together, and the defect is retained in the wrought product.

Gas Porosity or Shrinkage Porosity. Gas porosity or gas holes are rounded cavities (spherical, flattened, elongated, or partially collapsed) that are caused by the generation or accumulation of gas bubbles in the molten metal as it solidifies. Shrinkage cavities result from varying rates of contraction while the metal is changing from the molten to the solid state. Shrinkage porosity is mainly characterized by jagged holes or spongy areas lined with dendrites. The fatigue strength of cast alloys is only slightly reduced by the presence of shrinkage or gas porosity, but is much reduced when the porosity extends to the surface, regardless of severity, because of the notch effect.

Inclusions. The presence of nonmetallic inclusions at or close to the surface is detrimental, because the inclusions act as stress raisers and form points for the initiation of fatigue cracks. The larger the number of discontinuities, the greater the possibility of fatigue failure. However, there does seem to be a limit to the number of inclusions that could cause fatigue. If discontinuities such as graphite flakes in cast iron or sulfide inclusions in free-cutting steels are very numerous, mutual stress relief occurs, based on the principle of a

single groove versus a continuous thread, and the material tends to become less prone to fatigue.

Inclusions in ferrous alloys are usually oxides, sulfides, and silicates. Many inclusions, however, are of a more complex intermediate composition. Every commercial metal has its own characteristic inclusions. In cast metals, the shape of the inclusions is about the same in all polished sections; in wrought metals, the shape depends on the orientation of the polished surface. Deformation in mechanical working causes inclusions to deform plastically to elongated shapes and to appear in longitudinal sections as stringers or streaks, although in transverse section the shape is globular or flat. Hard, refractory, and very small inclusions, such as alumina inclusions in steel, are not deformed by mechanical work.

Fatigue properties of high-strength alloys are degraded by inclusions, with a more marked effect being apparent on the transverse fatigue properties of wrought alloys than on the longitudinal properties. In general, however, the effect of inclusions depends on the size and shape of the inclusions, on their resistance to deformation and their orientation relative to the stress, and on the tensile strength of the alloy. Soft steels, for example, are much less affected by inclusions than are hard steels.

Fig. 35 Fracture surface of a carburized-and-hardened steel roller

As a result of banded alloy segregation, circumferential fatigue fracture initiated at a subsurface origin near the case/core interface (arrow).

The fracture surface of a hardened steel connecting rod is shown in Fig. 33; large inclusions that intersect the surface are indicated by arrows. Loads on this rod were transverse to the forging grain flow. Inclusions, like most discontinuities, are most damaging when they intersect the surface; however, subsurface inclusions can also be responsible for premature fatigue failure when the stress level at the inclusions is high enough to initiate a crack. Figure 34 shows the surface of a torsional-fatigue fracture that initiated at a large, subsurface nonmetallic inclusion in a hardened steel valve spring.

Internal bursts in rolled and forged metals result from the use of equipment that has insufficient capacity to work the metal throughout its cross section. If the working force is not sufficient, the outer layers of the metal will be deformed more than the inside metal, sometimes causing wholly internal, intergranular fissures that can act as stress concentrators, from which fatigue cracks may propagate under tensile, bending, or torsional loading.

Alloy Segregation

Distribution of alloying elements in industrial alloys is not always uniform. Localized deviations from the average composition originate from specific conditions during solidification of the alloys. Hot working and soaking tend to equalize the compositional differences, but these differences sometimes persist into the wrought product. Inhomogeneous distribution of alloying elements in heat-treated alloys is particularly objectionable, because it may lead to the formation of thermal cracks caused by uneven contraction or expansion in heating and cooling.

Banding. When segregation in an alloy occurs in layers, or bands, the alloy is said to have a banded structure. Banding can lead to discontinuities that can in turn cause premature fatigue failure. Figure 35 shows the fatigue fracture of a carburized-and-hardened steel roller. Banded alloy segregation in the metal

used for the rollers resulted in heavy, banded retained austenite, particularly in the carburized case, after heat treatment. When the roller was subjected to service loads, the delayed transformation of the retained austenite to martensite caused microcracks near the case/core interface. These internal microcracks nucleated a fatigue fracture that progressed around the circumference of the roller, following the interface between case and core.

Flakes are internal fissures in ferrous metals. They are attributed to stresses produced by localized transformation and decreased solubility of hydrogen during cooling after hot working. For alloy steels, flakes are often associated with high hydrogen content in the steel; chemical-element segregation, producing regions of high alloy content; and rapid cooling from the hot-working temperature.

Effect of Heat Treatment on Fatigue Strength

Most metals and alloys used in highly stressed components undergo some form of heat treating to improve properties. Heat treatment always involves controlled heating and cooling operations. Imperfections may arise during heat treatment as a result of improper temperatures or furnace atmospheres, improper or uneven rates of heat application or heat removal, and improper preparation of surfaces or shapes before heat treatment.

The fatigue resistance of a component may be increased if the surface layer is made more fatigue resistant. This can be done by case carburizing, nitriding, or carbonitriding, provided the treatment is carried out properly so as to avoid gross structural discontinuities, such as carbide networks (carburizing), excessive white layer (nitriding), unsatisfactory carbon or nitrogen gradient (carbonitriding), or quenching cracks.

Example 10: Fatigue Fracture of an 8617 Steel Pilot-Valve Bushing. The pilot-valve bushing shown in Fig. 36 fractured after only a few hours of service. In operation, the bushing was subjected to torsional stresses with possible slight bending stresses. A slight misalignment occurred in the assembly before fracture.

The bushing was made of 8617 steel and was case hardened to a depth of 0.13 to 0.4 mm (0.005 to 0.015 in.) by carbonitriding. Specifications required that the part be carbonitrided, cooled, rehardened by quenching from 790 °C (1450 °F), then tempered at about 175 °C (350 °F). This heat-treating procedure would produce a part having a tough, fine-grain structure.

Investigation. The mating pieces of the bushing showed that no gross deformation occurred during fracture. A small area on the fracture surface of the large portion of the fractured bushing had been ground flat, etched, and tested for hardness. A spot of red rust on the fracture surface adjacent to the cross hole near the cracks appeared to be the result of the

Fig. 36 Carbonitrided 8617 steel pilot-valve bushing that fractured in fatigue because of improper heat treatment

Dimensions given in inches

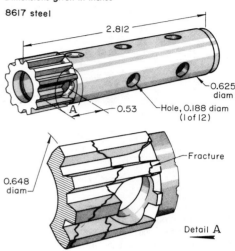

etching. Except for the rust in the etched area, no corrosion products were present, and general wear on the spline teeth was negligible.

Only one spline tooth showed any ductility associated with the fracture. This tooth was apparently in the area of final fracture. Metallographic examination of the fracture surface revealed a smooth pattern and beach marks that indicated propagation of the crack in fatigue. Several cracks initiated at the hole in the spline (Fig. 36), but it was impossible to determine which crack started first. Microstructural examination revealed an unsatisfactory carbonitrided case structure resulting from improper heat treatment, as well as many fine nonmetallic stringer inclusions in the core material.

Conclusions. The bushing fractured in fatigue because of a highly stressed case-hardened surface of unsatisfactory microstructure and subsurface nonmetallic inclusions. Cracks initiated at the highly stressed surface and propagated across the section as a result of cyclic loading. The precise cause of the unsatisfactory microstructure of the carbonitrided case could not be determined, but it was apparent that heat-treating specifications had not been closely followed.

Recommendations. It was recommended that inspection procedures be modified to avoid the use of steel containing nonmetallic stringer inclusions and that specifications for carbonitriding, hardening, and tempering be rigorously observed.

Overheating. In the heat treatment of metals, high temperatures generally cause large grains to develop; depending on the metal and application, large grains may have undesirable attributes. In most metals, large grains will generally reduce fatigue strength. The properties of coarse-grain metals are impaired by the size of the grains and by changes that occur at

Fig. 37 Effect of surface roughness on the fatigue strength of a steel specimen

The test specimen had a tensile strength of 1145 MPa (166 ksi).

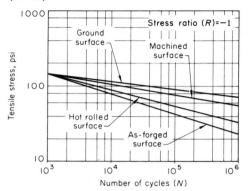

Fig. 38 Drive shaft that fractured from fatigue in the spline area because of sharp fillets and machining marks at spline roots

Dimensions given in inches

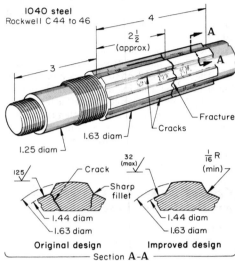

the grain boundaries, such as the precipitation of solid impurities, which may form weak, continuous grain-boundary films. Damage from overheating is especially evident in high-carbon steels, in which all the harmful effects of coarse grains on properties are combined with an increased probability of cracking during quenching from the hardening temperature.

Eutectic Melting. In aluminum and other age-hardenable alloys, solution heat treatment improves mechanical properties by developing the maximum practical concentration of the hardening constituents in solid solution. The solubilities of these constituents increase markedly with temperature, especially just below the eutectic melting temperature. Consequently, the most favorable temperature for effecting maximum solution is very near that at which melting occurs. Melting in age-hardenable alloys produces an intergranular network of nonductile eutectic products and intragranular circular spots (rosettes), both of which reduce ductility and fatigue strength. Similar effects of melting have been observed in high-temperature alloys, such as the Stellites, and in high-speed tool steels.

Quench cracks are a frequent cause of failure in hardened steel parts. The origin of quench cracks in steel is attributed to sudden volume changes that occur in hardening. The transformation of austenite into martensite is always accompanied by expansion, which, under unfavorable conditions, may result in cracking. Conditions that promote the formation of quench cracks are a quenching medium that is too severe, sharp edges and rough finishes, and hardening temperatures that are too high. Sometimes, the cracks do not appear immediately, but are delayed and take time to become visible. Delayed quench cracks are the result of additional transformation of retained austenite in steel. In highly alloyed steels, delayed cracking can be avoided by tempering immediately after quenching.

Quench cracks are generally intergranular. If a quench crack is open to the surface during

tempering, the walls of the crack may be covered with scale and may be decarburized.

Decarburization is a loss of carbon from the surface of a ferrous alloy as a result of heating in a medium that reacts with carbon. Unless special precautions are taken, the risk of losing carbon from the surface of steel is always present in any heating to high temperatures in an oxidizing atmosphere. A marked reduction in fatigue strength is noted in steels with decarburized surfaces. The effect of decarburization is much greater on high-tensile-strength steels than on steels with low tensile strength.

Influence of Manufacturing Practices on Fatigue Strength

Manufacturing practices influence fatigue strength by affecting the intrinsic fatigue strength of the material near the surface, by introducing or removing residual stresses in the surface layers, and by introducing or removing irregularities on the surface that act as stress raisers.

Machining. Most components subjected to fatigue conditions have been machined. Heavy cuts and residual tool marks from rough machining can promote fatigue failure in a component.

Surface irregularities produced by rough machining act as stress raisers. However, a series of parallel grooves, such as results from turning operations, is less severe in its effect than an isolated groove, because parallel grooves provide mutual stress relief. Rough machining also damages the metal to an appreciable depth. A machining tool shears off metal, rather than cutting it, with the result that

the surface is torn and is work hardened to an extent that depends largely on the depth of cut, the type and shape of the tool, and the characteristics of the metal. Components intended to withstand fatigue conditions should be finished with a fine cut or, preferably, ground, and the direction of the final cut or grind should be parallel to that of the principal tensile load whenever practical.

The effect of surface roughness on fatigue strength is shown in Fig. 37. Several surface finishes are compared, such as a smooth ground surface, a rougher machined surface, and as-forged and hot-rolled surfaces resulting from hot working.

Most mechanically finished metal parts have a shallow surface layer in residual compression. Apart from the effect it has on surface roughness, the final finishing process will be beneficial to fatigue life when it increases the depth and intensity of the compressively stressed layer and will be detrimental when it decreases or removes the layer. Processes such as electrolytic polishing and chemical and electrochemical machining, which remove metal without plastic deformation, may reduce fatigue properties. Electric discharge machining can be detrimental to fatigue properties without proper control and subsequent processing because of surface and subsurface microstructural changes. Improperly controlled grinding can have similar effects on fatigue properties.

Occasionally, during machining of a component, a tool may scratch or groove the surface. If the part is highly stressed in service, the result can be a premature fatigue fracture nucleating at the tool mark.

Stress raisers frequently occur at a change in section, such as the shoulder between two shaft sections of different diameters. Rough-machining marks, as well as steps resulting from improper blending of fillets with shaft surfaces, frequently serve as initiation sites for fatigue cracks.

Example 11: Fatigue Fracture of a 1040 Steel Splined Shaft. The splined shaft shown in Fig. 38 was from a front-end loader used in a salt-handling area. The shaft broke after having been in service approximately 2 weeks while operating at temperatures near −18 °C (0 °F). During the summer months, similar shafts had a service life of 5 to 8 months. The shaft was made of 1040 steel and was heat treated to a hardness of 44 to 46 HRC and a tensile strength of approximately 1448 MPa (210 ksi).

Investigation. Visual examination of the splines disclosed heavy chatter marks at the root of the spline, with burrs and tears at the fillet area. Surface finish at the root of the splines was 3.2 μm (125 μin.).

Examination of the fracture surface showed that fatigue cracks propagated through about 75% of the shaft before final fracture occurred in a brittle manner. There were indications that fatigue cracks initiated in sharp fillets at the root of the splines.

Conclusions. The shaft failed in fatigue as the result of stress concentrations in the sharp fillets and rough surfaces at the root of the splines. Failure occurred sooner in cold weather than in hot weather because the ductile-to-brittle transition temperature of the 1040 steel shaft was too high.

Corrective Measures. Premature cold-weather failures of these shafts were prevented by a number of changes in design and material. The fillet radius was increased to a minimum of 1.6 mm (1/16 in.), and the surface finish in the spline area was changed to a maximum of 0.8 μm (32 μin.). The material for the shafts was changed to a low-nickel alloy steel with the following nominal composition:

Element	Composition, %
Carbon	0.42
Manganese	0.78
Phosphorus	0.025 max
Sulfur	0.025 max
Silicon	0.26
Chromium	0.87
Nickel	2.08
Molybdenum	0.28
Vanadium	0.25

The ductile-to-brittle transition temperature of this steel was approximately −73 °C (−100 °F). The new shafts were heat treated to a hardness of 28 to 32 HRC before machining and were used in the as-machined condition.

Drilling. The fatigue strength of components can be reduced merely by the presence of a drilled hole; it is further reduced by failure to remove burrs (incurred during drilling) from the hole edges. Fractures originating at drilled holes are common in complex parts containing internal, intersecting machined passages because of the difficulty and expense of providing adequate break-edge radii at such locations.

Example 12: Fracture Mechanics Analysis of Fatigue Failures in Crankshafts With Drilled Holes and Surface Compression. Several crankshaft failures occurred in equipment that was being used in logging operations in subzero temperatures. Failure usually initiated at a cracked pin oil hole, and the failure origin was approximately 7.6 mm (0.3 in.) from the shaft surface. The holes were produced by gun drilling, giving rise to surface defects. The fracture surface was characteristic of fatigue in that it was flat, relatively shiny, and exhibited beach marks. The crack surface was at 45° to the axis of the shaft, indicating dominant tensile stresses. Possible causes of failure included:

- Poor-quality material with low fracture and fatigue resistance
- Poor machining, which initiates cracks that propagate to failure by a fatigue mechanism in service
- Excessive stresses associated with abusive use or failure of other components in the system to operate according to design

Fig. 39 Schematic stress distribution showing a maximum stress at 7.6 mm (0.3 in.) from the shaft surface

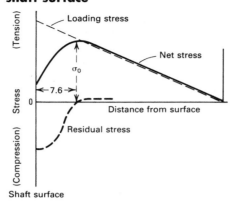

Fig. 40 Schematic stress distribution on plane normal to the oil hole 7.6 mm (0.3 in.) from the shaft surface

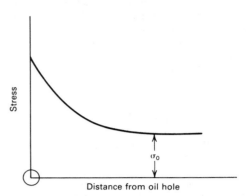

Fig. 41 Macrograph of fractured shaft

Note the elliptical shape of the propagating fatigue crack. The scale markings are 0.1 in.

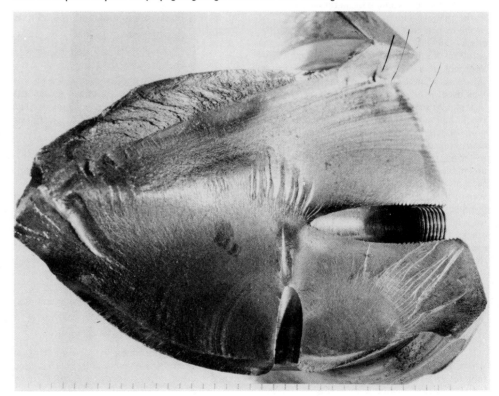

The material was the French designation AFNOR 38CD4 (similar to AISI type 4140H) and was in the quenched-and-tempered condition, with a yield strength of about 760 MPa (110 ksi). It was treated to have compressive surface stresses, and the prior-austenite grain size was ASTM 8. The investigation sought to analyze the failure, provide an estimate of operative stresses, and suggest ways to avoid such failures in the future.

Investigation. Although the location of the origin would normally be at the surface, because the stresses are at a maximum there, the fact that the origin was approximately 7.6 mm (0.3 in.) below the main surface was indicative of compressive stresses on the surface. These stresses are caused by heat treatment and are clearly beneficial.

The picture that develops is that the tensile stress due to torsional loading decreases linearly from a maximum at the surface to zero at the center. The residual surface compression associated with heat treatment is large but decreases rather rapidly. The net stress represents a balance between these two competing factors and is shown schematically in Fig. 39. To be consistent with observation, the maximum occurs in the vicinity of 7.6 mm (0.3 in.) below the surface. At the point of initiation, the stress was amplified by the oil hole. It is a maximum at the oil hole surface and decreases on a plane normal to the hole, as shown schematically in Fig. 40. The crack surface is shown in Fig. 41. It can be seen that the crack size at the point of final failure was large compared to the size of the oil hole. In addition, it can be seen from the beach marks that the crack had an elliptical shape. These observations were useful in estimating the size of the remote stress field 7.6 mm (0.3 in.) from the main surface.

Fracture Mechanics Approach. In principle, the stresses can be computed by measuring the final crack size and shape and noting that at fracture the stress intensity parameter is equal to the plane-strain fracture toughness K_{Ic}. The problem was that the oil holes from which the crack did not originate interfered with the solutions for the stress intensity parameter. This approach therefore had an element of uncertainty associated with it, and another approach was used.

It can be seen quite clearly in Fig. 41 that the crack had an original elliptical shape at the point of origin. The situation is shown schematically in Fig. 42; it was extremely well defined in terms of geometry and stress, and was thus amenable to analysis. Defects around the surface of the oil hole were observed whose size was of the order of 0.076 mm (0.003 in.). For purposes of analysis, these defects were considered to be elliptical starter cracks. For the cracks to propagate, the stress intensity must exceed the threshold stress intensity for fatigue (Ref 9):

$$\hat{K}^* = \frac{1.1\,\sigma\sqrt{\pi a}}{\phi_0}$$

$$\left[\sin^2\beta + \left(\frac{a}{b}\right)^2\cos^2\beta\right]^{1.4} F_\lambda(s) \quad \text{(Eq 5)}$$

where $\hat{K}^*$ is the threshold stress intensity for fatigue; σ is remote stress; ϕ_0 is an elliptic integral; β is a position locator (Fig. 42); a and b are the semi-minor and semi-major axes, respectively; $s = a/R + a$, where R is hole radius; and $F_\lambda(s)$ is a correction factor to account for the stress-concentrating effect of the oil hole (Ref 10). The geometry and correction factors are defined in Fig. 43. In this case, $\lambda = -1$, because torsional loading is equivalent to biaxial tension and compression (Ref 11). Also, because $\beta = \pi/2$, Eq 5 reduces to:

$$\hat{K}^* = \frac{1.1\sigma\sqrt{\pi a}}{\phi_0}[2F_0(s) - F_1(s)] \quad \text{(Eq 6)}$$

From measurement:

Fig. 42 Schematic of elliptical cracks in the origin region

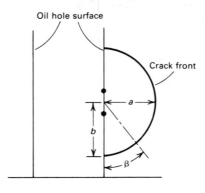

$$s = 0.024$$

From this:

$$F_\lambda(0.024) = 2(3.3) - 2.2 = 4.4$$

To a very good approximation, the elliptic integral is:

$$\phi_0 = 1 + 0.5707\left(\frac{a}{b}\right)^{1.42}$$

The aspect ratio of the elliptical crack was measured as a function of crack depth a. The results are shown in Fig. 44. The appropriate equation is:

$$\left(\frac{a}{b}\right) = ^-0.1a + 0.82$$

where a is in millimeters. For a crack depth of 0.08 mm (0.003 in.):

$$\left(\frac{a}{b}\right) = 0.81$$

$$\phi_0 = 1.74$$

Using the computed and measured quantities, $\hat{K}^*$ can be expressed in terms of stress:

$$\hat{K}^* = 0.27\sigma \quad \text{(for } a = 0.08 \text{ mm)}$$

The threshold stress intensity for type 4140 at the 690-MPa (100-ksi) yield level is, to a good approximation, of the order of 6.6 MPa$\sqrt{\text{m}}$ (6 ksi$\sqrt{\text{in.}}$). Thus, the computed stress is about 154 MPa (22 ksi). It should be noted that this represents the minimum stress necessary for cracks to propagate. The stresses could be higher but certainly not lower. Also, this calculation does not distinguish between any residual stresses that might be associated with machining the hole and loading stresses. The one certain fact is that there is a net stress of at least 154 MPa (22 ksi).

It is of interest to inquire as to the effects of improving the machining operation. For example, assume that the defect zone size is halved. In this case, the following quantities apply:

Fig. 43 Correction factor for cracks emanating from a hole subjected to biaxial stress

Source: Ref 10

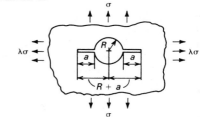

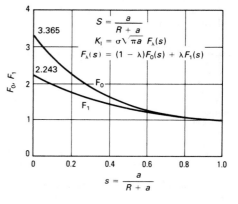

Fig. 44 Aspect ratio as a function of crack depth

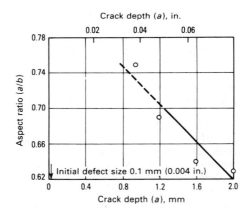

$$a = 0.038 \text{ mm}$$

$$\frac{a}{b} = 0.82$$

$$\phi_0 = 1.75$$

$$F_\lambda(s) = 4.48$$

Using these values and substituting into Eq 6 gives:

$$\hat{K}^* = 0.193\sigma \quad \text{(for } a = 0.038 \text{ mm)}$$

In this case the minimum stress for crack propagation would be 214 MPa (31 ksi).

Fig. 45 Aluminum alloy propeller blade that fractured in fatigue after being deformed in service and then cold straightened

(a) Fracture surface showing brittle appearance, and final-fracture areas (arrows A) at leading and trailing edges. (b) Crack-initiation region showing brittle appearance of fracture surface. 12.5×. (c) Electron fractograph showing crack-progression bands in the fatigue zone. 2300×

Finally, an upper-bound calculation can be made. The object here is to compute the crack size that would not propagate even if the stress were just below yield. It is recognized that the stresses will not be this high. However, if it is feasible to machine the critical oil hole so that the defects are smaller than those that will propagate at this stress, then such an operation is clearly indicated, because any fatigue crack propagation problem could certainly be eliminated. In this case, the following values apply:

$$\sigma = 760 \text{ MPa (110 ksi)}$$

$$\frac{a}{b} = 0.82$$

$$\phi_0 = 1.75$$

$$F_\lambda(s) = 4.48$$

$$a = \frac{1}{1.21\pi}\left(\frac{\hat{K}^* \, \phi_0}{\sigma F_\lambda(s)}\right)^2$$

$$= 3 \times 10^{-3} \text{ mm} \quad \text{(for } \sigma = 760 \text{ MPa)}$$

Any machining operation that would reduce the defect size to below approximately 0.025 mm (0.001 in.) would eliminate any fatigue problem at the oil holes.

Conclusions. The stress 7.6 mm (0.3 in.) from the surface was at least 154 MPa (22 ksi). This included any stresses that could be present as a result of machining. For the most severe

conditions that can be envisioned, improving the machining operation such that machining-related defects were reduced to 0.025 mm (0.001 in.) would eliminate the problem of fatigue crack propagation from the oil holes.

Corrective Measures. The oil holes were drilled by a technique that essentially eliminated surface defects. No further field failures were encountered.

Grinding. Proper grinding practice produces a smooth surface that is essentially free of induced residual stresses or sites for the nucleation of fatigue cracks. However, abusive grinding, particularly in steels heat treated to high hardness, is a common cause of reduced fatigue strength and failure that results from severe induced tensile stresses, intense localized heating, or both. Intense localized heating results in overtempering, formation of untempered martensite, burning, or formation of tight, shallow surface cracks usually referred to as grinding cracks.

Straightening. Components may be unintentionally plastically deformed during manufacture, during shipping, or in service without cracking. These parts can be straightened manually, by heating, in presses, or in roll straighteners. The initial deformation and subsequent working operations can introduce residual stresses or stress raisers. Nicks, scratches, or locally work-hardened surfaces are potential stress raisers.

Aircraft propeller blades are frequently damaged by deformation in minor ground accidents and repaired by cold straightening, either manually or in a press. The straightening operations are restricted to bends of less than a specified angle, varying with the location along the blade.

Example 13: Fatigue Fracture of a Cold-Straightened Aluminum Alloy Propeller Blade. An aluminum alloy propeller blade that had been cold straightened to correct deformation incurred in service fractured soon after being returned to service. A fracture surface of the blade is shown in Fig. 45(a).

Investigation. Examination of the fracture surface revealed that crack initiation occurred at the camber surface (top surface, Fig. 45a) in an area containing numerous surface pits. The macroscopic appearance of the surface was of brittle fracture (Fig. 45b); however, crack-progression bands and fine, poorly resolved fatigue striations were visible in replica electron micrographs (Fig. 45c). A small amount of ductile-overload fracture was found at the leading and trailing edges of the blade (arrows A, Fig. 45a), suggesting that low stress levels were involved in high-cycle fatigue-crack propagation. No evidence was found of a prior crack resulting from the initial deformation or the straightening operation.

A stress analysis using an x-ray method did not detect any residual stress in the camber surface of the propeller blade adjacent to the fracture. However, a spanwise tensile stress of approximately 51 MPa (7.4 ksi) was indicated in the same surface of the unfailed mating blade at the location of the initial bend.

Conclusions. The residual stress may have originated with the straightening operation, and the apparent absence of stress in the fractured blade was the result of relaxation through fracture. Because no prior crack damage could be attributed to the initial deformation or to straightening, the rapid fracture may have been induced by residual stresses contributing to the normal spectrum of cyclic stresses. Also, the presence of surface pits acting as stress raisers would compound this problem. The origin of the pits was not known.

Corrective Measures. Stress-relief annealing after cold straightening, plus refinishing of the surface, reduced fracturing of propeller blades that were cold straightened to correct deformation experienced in service.

Surface Compression. The beneficial effects of compressive residual stresses can be obtained by coining (such as around holes), surface rolling, or shot peening. These processes often produce visible marks on the surface, such as burnished areas and peening dimples. The presence or absence of such marks is not a conclusive indication of the magnitude of stresses produced by these processes.

Residual stresses can be evaluated using x-ray diffraction techniques. However, because the stresses will be dissipated in the immediate

Fig. 46 Helicopter-blade spindle that fractured in fatigue because it was incompletely shot peened before being plated

(a) Spindle showing fracture region (at fillet between shank and fork). Dimensions given in inches. (b) Longitudinal score mark (region A) and circumferential scratches (arrow B) on shank that had been peened over. A galled area (arrow C) is also visible.

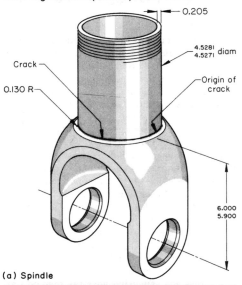

(a) Spindle

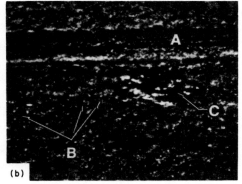

(b)

vicinity of the fracture, stresses should be checked at a location somewhat removed from the fracture.

Example 14: Fatigue Fracture of a Spindle for a Helicopter Blade. The spindle of a helicopter-rotor blade (Fig. 46a) fractured after 7383 h of flight service, causing the aircraft to crash. These spindles were overhauled and inspected at intervals of 1200 h or less. Between the sixth overhaul and the failure, 464 flight hours had been accumulated.

At every overhaul, the spindles were fluorescent-magnetic-particle inspected, and the surface was carefully examined for evidence of wear. If a spindle showed wear, it was reworked by grinding the shank to 0.1 mm (0.004 in.) under the finished diameter. The spindle was then shot peened with S170 shot to an Almen intensity of 0.010 to 0.012 A. Following shot peening, the shank was nickel

sulfamate plated to 0.05 mm (0.002 in.) over the finished diameter, ground to finished size, and cadmium plated. The spindle that failed had been overhauled six times and reworked twice. More information on shot peening and methods for measuring the intensity of peening is available in the article "Shot Peening" in Volume 5 of the 9th Edition of *Metals Handbook*.

Investigation. After the crash, four of the five rotor blades were found in the main impact area. The fifth blade was about 0.4 km (¼ mile) away. The broken spindle was with this blade. The spindle had broken in the shank adjacent to the shoulder at the inboard end of the shank. Visual examination of the fracture surface revealed that a fatigue crack had propagated through approximately 72% of the shank cross section before final fracture.

Examination with a stereomicroscope showed the fatigue origin to be near the line where the cylindrical shank met the 3.3-mm (0.130-in.) radius fillet at the junction of the shank and fork of the spindle. Hardness of the metal near the crack origin was, in some spots, as low as 28 HRC; the specified hardness was 34 to 38 HRC. A banded microstructure was found near the low-hardness areas; it had interfered with the desired response to heat treatment.

On the surface near the fillet at the junction of the shank and fork, faint grinding marks and circumferential grooves could be seen. These marks and grooves would have been peened over and covered with peening dimples if peening had been done properly. Unlike the fillet area, the cylindrical portion of the shank exhibited the typical dimpled surface of peened parts (see Fig. 46b).

Life tests conducted after the accident with 30 spindles, new and used, peened and unpeened, showed that two spindles that had not been shot peened broke after 10% and 20% of the mean life of peened spindles.

Conclusions. The spindle failed in fatigue that originated near the junction of the shank and fork. The nonuniformity of the shot-peened effect on the shank and fillet portions of the spindle resulted from a minimum of attention being afforded the fillet during peening. The fracture was of the low-stress high-cycle type, initiated by stresses well below the gross yield strength and propagated by thousands of load cycles.

Plating on metal surfaces can be detrimental to fatigue strength of the plated parts. Carbon and low-alloy steels, particularly steels of high hardness, are susceptible to hydrogen damage due to absorption of hydrogen during the plating cycle and the associated acid or alkaline cleaning cycles. A soft plating material, such as cadmium, may inhibit the escape of hydrogen from the steel and lead to hydrogen damage. However, the harmful effects of cadmium plating can be significantly minimized by following recommended procedures, which include plating at high current densities (6 to 8 A/dm², or

0.4 to 0.5 A/in.²), followed by baking at about 190 °C (375 °F) for 8 to 24 h. A daily procedure for ensuring that the plating bath will give low hydrogen pickup is highly desirable. Apparatus for such a procedure is available in the form of an electronic-electrochemical instrument that measures the relative amount of hydrogen generated during electroplating.

Hard plating materials, such as chromium, are usually in a state of tensile stress after they have been deposited on the base metal. Cracks develop in the plating material, act as stress raisers, and produce cracking in the base metal. Crack growth in the base metal can be prevented by introducing a compressive residual stress such as that produced by shot peening. Government specifications QQ-C-320 and MIL-G-26074A specify shot peening the part surface before plating with chromium or electroless nickel for all steel parts designed for long life under fatigue loading. Other recommended procedures for chromium plating of steel are given in ASTM B 177. Several effective procedures for decreasing the residual tensile stress imposed by electroless nickel plating have also been developed, including the addition of saccharin and other organic compounds containing sulfur to the plating bath (Ref 12).

Cleaning is sometimes necessary for the removal of oil, grease, or other contaminants from the surface of a workpiece. For production work, vapor degreasing or a dip into a solvent or a simple chemical cleaning solution may be employed. Many alkaline solutions that will remove grease and oil from aluminum are not satisfactory for most cleaning purposes, because they attack the surface. However, alkaline solutions that are suitably inhibited to eliminate the corrosive action can be used successfully. The cleaning treatment should be followed by thorough rinsing in clean water, then drying.

Example 15: Fatigue Fracture of Aluminum Alloy 7178-T6 Aircraft Fuel-Tank Floors. The floors of the fuel tanks in two aircraft failed almost identically after 1076 and 1323 h of service, respectively. The floors had been fabricated from aluminum alloy 7178-T6 sheet, with portions of the sheet chemically milled to reduce thickness as a savings in weight. Failure in both tanks occurred in the rear chemically milled section of the floor (Fig. 47a).

The reduced sections were about 0.5 mm (0.020 in.) thick and were blended into the surrounding lands with a radius of about 0.9 mm (0.035 in.). Land thickness at the central surface was about 1.3 mm (0.050 in.); at the edge, about 2 mm (0.080 in.). An alkaline etchtype cleaner was used on the panels before chemical milling and before painting.

Visual Inspection. The fractures began near the center of the floor close to the radius between the chemically milled bay and the central land surface. Figure 47(a) shows the extent of fracture in the fuel-tank floor. The fractures followed the radius in both directions

Fig. 47 Aluminum alloy 7178-T6 floor of an aircraft fuel tank that failed by fatigue because of alkaline cleaning of the metal before painting

(a) Floor of the fuel tank showing extent of fracture. Dimensions given in inches. (b) Fracture surface showing fatigue marks and dimples indicating a ductile-overload fracture. 200×. (c) Secondary crack initiating at a surface pit. 500×. (d) Surface pits on chemically milled surface. 60×. (e) Pits on load surface. 60×

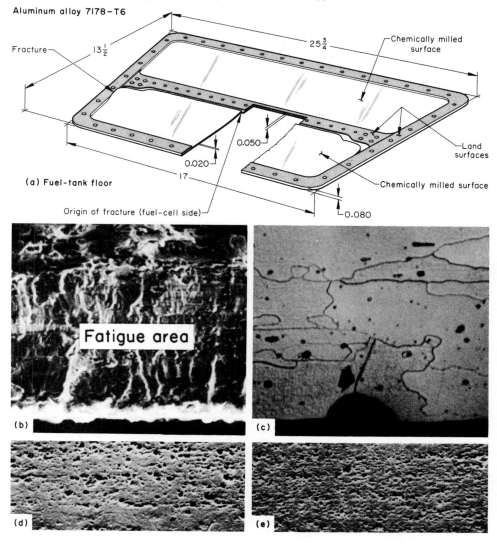

fatigue areas of the fracture surface initiated from corrosion pits. Also observed were secondary fatigue cracks originating in pits on the chemically milled surface and parallel to the fracture surface.

A section of each floor was stripped of all coatings to reveal the degrees and severity of pitting on the chemically milled surfaces and on the land surfaces, which were not chemically milled. Optical microscopic examination of the stripped sections revealed extensive pitting on both sides of the floors, including the chemically milled radii, which were mechanically machined after chemical milling. The pitted surface of a chemically milled area is shown in Fig. 47(d). The land surface (Fig. 47e), which had not been chemically milled, had more but slightly smaller pits.

Previous experience indicated that the alkaline etch-type cleaner used for cleaning the panels before chemical milling and before painting can cause pitting in aluminum alloys.

Chemical analyses, measurements of electrical conductivity, and hardness tests all indicated that the aluminum alloy used for the fuel-tank floors conformed to the specifications for 7178-T6.

Conclusions. The floors failed by fatigue cracking that initiated near the center of the fuel-tank floor and ultimately propagated as rapid ductile-overload fractures. The fatigue cracks originated in pits on the fuel-cell side (surface opposite the chemically milled areas) of the tank floors. Adjacent land surfaces were similarly pitted, but having appreciably thicker walls, were subjected to a lower stress. The pits were attributed to attack caused by the alkaline-etch cleaning process used to prepare the surface for painting.

Corrective Measures. To avoid further fatigue failures from pitting, instructions were issued for critical monitoring of the alkaline-etch cleaning to avoid the formation of pits. Careful inspection following alkaline-etch cleaning was scheduled before release of the floor panels for painting.

Welding practices can have an effect on the fatigue strength of a metal at and below the surface. Surface defects resulting from poor welding practices provide stress raisers at which bending-fatigue and torsional-fatigue cracks can originate. Craters, underbead cracks, and arc strikes are typical surface defects. Subsurface defects, such as flux inclusions, incomplete fusion, and inadequate joint penetration, can originate cracking in parts loaded in tension, bending, and torsion.

Example 16: Fatigue Fracture of a Highway Tractor-Trailer Steel Drawbar. A drawbar connecting the two tank-type trailers of a highway gasoline rig broke while the rig was on an exit ramp of an interstate highway, causing the tractor, semi-trailer, and full trailer to overturn. The driver's description of the way

across the floor, then propagated toward the rear along both sides.

Examination of the fracture surface at low-power magnification revealed small, bright beach marks in the central 20 cm (8 in.) of the fracture. The beach marks initiated on the inside (fuel-cell side, which was opposite the chemically milled surface) edges of the fractures, indicating that failure was caused by fatigue cracks propagating through the thickness of the floor. The fatigue characteristics of both fractures were restricted to the central 20-cm (8-in.) section, with the remaining portions of the fracture surfaces showing features that are characteristic of overload fractures. No physical damage to the floors that could have caused the failures was found near the fractures.

Metallurgical Investigation. Fractographs taken with a scanning electron microscope confirmed that the beach marks observed at low-power magnification contained the parallel striations characteristic of fatigue. Beyond the fatigue zones, the surface had the dimpled features typical of ductile-overload failure (Fig. 47b).

The microstructure of the metal in the fracture region was normal for aluminum alloy 7178-T6. The fracture profile had no unusual features, but a number of pits were on the inside surface of the floor. (The chemically milled surface was the outer surface of the floor.) Examination at 500× revealed that many of the pits contained secondary fatigue cracks, as shown in Fig. 47(c). The electron microscope study also established that the cracks in the

Fig. 48 Highway tractor-trailer steel drawbar that fractured by fatigue at the cast connection at right because of a weld defect

(a) Drawbar showing components, and fracture in the right cast connection at arrow A; connection at left failed by brittle fracture. (b) Fracture surface of right case connection showing fatigue region (arrow B) and fracture origin in weld (arrow C). (c) TEM fractograph showing fatigue striations at region B in fracture surface. 20 000×. (d) TEM fractograph showing corrosion products (black dots) at region C. 6000×

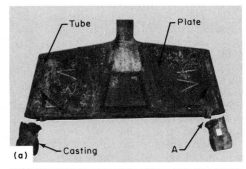

(a)

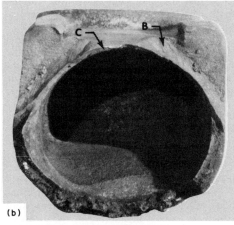

(b)

(c)

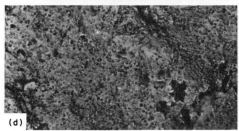

(d)

in which the rig behaved during the accident, together with physical evidence at the scene, suggested that the full trailer pulled to the right and overturned the rig.

Investigation. The drawbar was a weldment of steel plates, tubes, and castings. As shown in Fig. 48(a), the left and right cast connections to the full trailer broke in an area where welds joined the connections to the welded tube and plate.

The fracture surface on the right side of the drawbar (A in Fig. 48a) is shown in Fig. 48(b). The upper half of the fracture is through a weld, and the lower half is through the casting. No conclusive evidence of the fracture mechanism was discernible by light fractography. However, a TEM fractograph at 20 000× (Fig. 48c) revealed fatigue striations in the region at arrow B in Fig. 48(b). Corrosion products on the fracture surface at the region at arrow C in Fig. 46 (d), which are seen as black dots in Fig. 46(d), indicated that this area was probably the site of fracture origin and that it had cracked before the accident happened.

The casting on the right side of the drawbar contained large voids and a significant amount of porosity (lower portion, Fig. 48b). Electron fractography established that the cast connection on the left side failed by brittle fracture.

Metallographic examination showed that both castings and the tubing had microstructures consisting of pearlite in a ferrite matrix. These microstructures were considered suitable for the application. However, weld quality was poor in the casting-to-tube joint, with evidence of incomplete fusion and inadequate penetration into the base metal.

Conclusions. The drawbar fractured in fatigue, which originated in the weld joining the cast connector to the right side of the drawbar assembly. The crack initiated in a region of poor weld quality. A contributing factor to fracture of both connectors was the presence of voids and porosity in the castings.

Corrective Measures. The welding procedures were revised, and receiving inspection of the connection castings was instituted.

Magnetic-Particle Inspection. Localized regions of untempered martensite can be produced by arc burns that result when the probe used in magnetic-particle inspection touches the surface of the workpiece. Fatigue cracking of the pin described in the following example occurred in an area contacted by a conductor (probe) used during magnetic-particle inspection.

Example 17: Fatigue Fracture of AMS 6470 Steel Knuckle Pins. Within about 1 month, several knuckle pins, similar to the one shown in Fig. 49(a), cracked in service. The pins, used in engines, failed over a range of 218 to 463 h in operation.

Specifications for the pins required that they be made of AMS 6470 steel (a nitriding steel containing 1.6% Cr, 0.35% Mo, 1.15% Al, 0.38 to 0.43% C, rem Fe) and have a minimum case hardness of 92 HR15N, a case depth of 0.4 to 0.5 mm (0.017 to 0.022 in.), and a core hardness of 285 to 341 HB. Five knuckle pins, all from the same production lot and averaging 295 h in service, were selected for failure analysis.

Investigation. Visual examination of the fracture surface revealed beach marks typical of fatigue cracks that had nucleated at the base of the longitudinal oil hole (Fig. 49b). Examination showed that the fractures had been nucleated by circular cracks that apparently existed on the pins before their installation in the engine. Enlarged views of the bottom of the oil hole in a fractured pin and in an unused pin are shown in Fig. 49(c) and (d), respectively. The circular crack in the fractured pin (arrows at top left and center right, Fig. 49c) nucleated the fatigue fracture. A similar circular crack was found at the base of the oil hole in an unused pin (arrows, Fig. 49d).

Micrographs of sections through the cracked areas of both unused and failed knuckle pins revealed a remelt zone and an area of untempered martensite within the region circumscribed by the cracks shown in Fig. 49(c) and (d). These findings indicated that a severe local application of heat, sufficient to cause partial melting of metal, had occurred. Such local areas of overheating, or burning, are attributable to electric-arc burning. The thermal stresses accompanying the burning caused the circular cracks that nucleated fracture.

A section was taken through the cracked areas in an unused pin for metallographic study. The cracks were about 0.4 mm (0.017 in.) deep, and the thermal transformation area was about 0.8 mm (0.030 in.) deep. Tests showed that the failed knuckle pins were satisfactory with respect to case and core hardness, case depth, microstructure, and inclusion content.

A check of the inspection procedures disclosed that the pins had been magnetic-particle inspected by inserting a probe into the longitudinal hole. Therefore, it was assumed that arc burning occurred during magnetic-particle inspection as a result of electric arcing between the probe and the knuckle-pin material. Cracking at the bottom of the oil hole by the central-conductor magnetic-particle inspection method was done in the laboratory, thus confirming that cracking can occur during magnetic-particle inspection of the pin. Although arc burning does not always cause cracking, the thermal transformation of metal that accompanies arc burning can cause a severe metallurgical notch.

Conclusions. The knuckle pins failed by fatigue fracture nucleated from small circular cracks at the base of the longitudinal oil hole. The circular cracks were the result of arc burning, attributable to improper technique in magnetic-particle inspection. In addition, thermal transformation of the metal, which accompanies arc burning, causes a stress concentration that may lead to fatigue failure, even though cracking resulting from overheating or burning does not occur.

Recommendations. The conductor should be insulated to prevent arc burning at the base of the longitudinal oil hole. A borescope or metal monitor may be used to inspect the hole for

Fig. 49 AMS 6470 steel knuckle pin that failed in service by fatigue cracking

Cracking originated at an arc burn that had been made at the bottom of a longitudinal oil hole during magnetic-particle inspection. (a) Knuckle pin showing crack. Dimensions given in inches. (b) Fracture surface showing origin of fracture (arrow) and beach marks. 2.8×. (c) Bottom of oil hole in fractured pin showing circular crack around the arc burn (arrow at top left, arrow at right). 20×. Arrow at bottom left indicates fatigue fracture that nucleated from the circular crack. (d) Bottom of an oil hole in an unused pin showing circular crack (arrows) around an arc burn that was made during magnetic-particle inspection. 20×

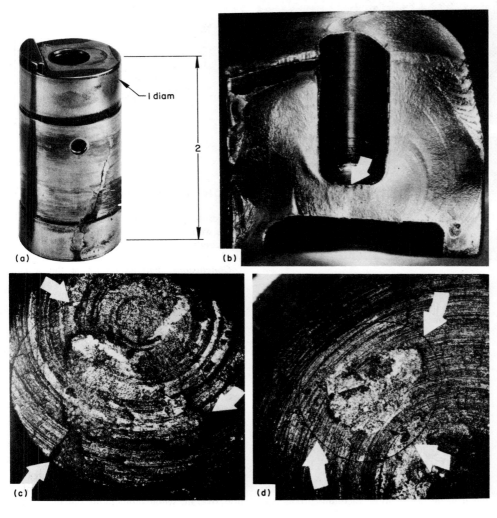

Fig. 50 Integral coupling and gear of chromium-molybdenum steel that failed in fatigue because the part was magnetized and retained metal chips at tooth roots

Dimensions given in inches

evidence of arc burning from magnetic-particle inspection.

Example 18: Fatigue Fracture of a Chromium-Molybdenum Steel Integral Coupling and Gear. Figure 50 shows an integral coupling and gear, used on a turbine-driven main boiler-feed pump, that was removed from service after 1 year of operation because of excessive vibration. Oil lines to the coupling were reportedly not operating shortly after the coupling was installed. Also, chips in the oil were reported on previous inspections of the coupling, but their nature was unknown.

Investigation. Visual examination revealed that the hub of the coupling was cracked, that teeth on the coupling gear were severely damaged, and that a paste of gritty material was pressed into the gear teeth. Severe wear and broken teeth were also observed. A crack was found that apparently started in the keyway and continued circumferentially around the coupling (Details A and B, Fig. 50).

Spectrographic analysis of the coupling material revealed that it was made of a chromium-molybdenum steel, similar in analysis to 4130 steel. The gritty material in the gear teeth was found to be of the same composition as the metal in the coupling.

Surface hardness of the gear teeth was 56 to 58 HRC, and core hardness was 27 to 30 HRC; both hardness ranges were satisfactory for the work metal and the application. Metallographic examination revealed a desirable microstructure with no unusual features.

A section of the coupling containing gear teeth and part of the keyway was removed to permit examination of the fracture surface. Beach marks and evidence of cold work that are typical of fatigue failure were found on the surface.

When the cut was made on the coupling, chips remained in the cut and were difficult to remove, indicating a strong magnetic field in the part. A gauss-meter probe was inserted into the cut, and a reading of 2000 G was obtained, which is a high residual flux density for steel (for comparison, magnetic steels have a flux density of about 9000 G). A field test around the boiler-feed pump indicated virtually no electric forces that could induce magnetism in the gears. Therefore, the gear must have been magnetized before being put into service.

A logical source of magnetism in the gear teeth would be magnetic-particle inspection on the gear before it was put into service. Normally, workpieces are demagnetized after magnetic-particle inspection. However, in this case, the coupling may have been only partly demagnetized or overlooked completely.

Conclusions. Failure of the coupling was by fatigue, as evidenced by the beach marks and signs of cold work. Incomplete demagnetization of the coupling following magnetic-particle inspection caused retention of metal chips in the roots of the teeth. These chips contributed to cyclic loading.

Fig. 51 ASTM A186 steel double-flange trailer wheel, for a coke-oven pusher car that failed in fatigue

Cracking initiated at heavily indented and improperly placed stamp marks. (a) Car wheel showing position of stamp marks and fractures in the rim and web. Dimensions given in inches. (b) Stamp marks showing heavy impression, and fracture extending along the base of the lower row of numbers. (c) Notches (arrows) created from the heavily indented stamp marks, and from which cracks initiated, along the top at the fracture surface

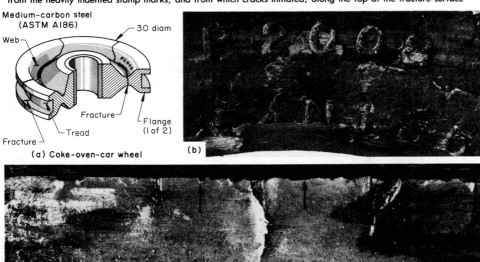

Fig. 52 Fatigue life of a specimen of N-155 alloy subjected to various temperatures and reversed-bending stress

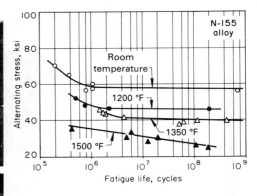

Fig. 51c) were approximately 3.2 mm (⅛ in.) deep.

Microscopic examination showed the metal in the web, rim, and tread to be in the normalized condition. Cleanness and soundness of the steel were satisfactory for the application.

Conclusions. Fatigue failure of the wheel was the result of heavy stamp marks that acted as stress raisers in the weaker web section. Because this was a double-flange wheel, considerable side thrust was applied to the wheel, causing stress concentration at the web.

Recommendation. The ASTM specification A 504 regarding location of stamped identification numbers should be followed.

Fatigue Failure at Elevated Temperatures

Failure by fatigue can usually occur at any temperature below the melting point of a metal and still maintain the characteristic features of fatigue fractures, usually with little deformation, over the whole temperature range. At high temperatures, however, both the fatigue strength and the static strength of metal generally decrease as the operational temperature increases. Figure 52 shows typical *S-N* curves for reversed-bending fatigue tests conducted on a structural metal alloy at various temperatures. The fatigue limit is clearly lower at the higher temperatures. Mechanical-property data on most alloys at high temperatures also show that, just as at room temperature, the fatigue strength is closely related to the tensile strength, unless the temperature is high enough for the fatigue strength to be affected by creep phenomena. Data from tests in which the load is completely reversed during each cycle usually can be interpreted as being uncomplicated by creep. Under actual service conditions, however, this is rarely the case.

At high temperatures, application of a constant load to a metal component produces con-

---

Improper lubrication caused gear teeth to overheat and spall, producing chips that eventually overstressed the gear, causing failure. Because the oil-circulation system was not operating properly, metal chips were not removed from the coupling.

Recommendations. To prevent damage to the replacement coupling, it should be checked for residual magnetism and should be demagnetized if any is present. The lubrication oil should be changed or filtered to remove any debris from previous damage to the gears. The oil-circulation system should be put into operating conditions.

Identification Marking. Excessive stresses may be introduced into components by identification marks (manufacturing date or lot number, steel heat number, size or part number). The location and the method of applying the marks can be important. They should not be in areas of high tensile, bending, or torsional stresses.

Raised numerals or letters are preferable to those that are indented, and hot-forging methods are recommended over cold forging, stamping, or coining. For marks that are made on machined surfaces, use of a marking ink is generally preferable to use of an electric etching pencil or vibrating mechanical engraver, but either of the latter two is less abusive than using a cold steel stamp. However, using steel stamps for numbers in a nonstressed area may be preferable to some other type of identification mark in a more highly stressed region. Characters with straight-line portions have the greatest tendency to cause cracks, although characters with rounded contours can also cause cracking.

If steel stamps must be used, low-stress stamps (stamps with the sharp edges removed from the characters) or dull stamps will cause the least trouble, particularly if light impressions are made. The stamping should be located in known low-stress areas, a practice that was not followed in the following example.

Example 19: Fatigue Fracture of Steel Wheels for a Coke-Oven Car. The double-flange trailer wheel shown in Fig. 51 had been in service on a coke-oven pusher car for about 5 years when it broke. Specifications called for rolled steel track wheels in conformity with ASTM A 186 (since reclassified as A 504), with no grade indicated.

Investigation. Chemical analysis showed the metal in the wheel to be medium-carbon steel within the ranges given in ASTM A 186. Hardness of the tread was 302 HB and of the web was 255 HB.

Visual examination of the broken wheel revealed that cracking had extended circumferentially around the web before complete failure occurred (Fig. 51a). The cracks ran parallel with the base of the lower row of numbers stamped with heavy indentation on the web section (Fig. 51b). The ASTM specification states that the marks identifying the wheels must be stamped on the back face of the rim not less than 3.2 mm (⅛ in.) from the inner edge of the rim.

The fracture surface had an appearance typical of fatigue failure. However, the beach marks usually associated with this type of crack had been mostly removed by the constant rubbing of the fracture faces. The stamp marks from which the cracks initiated (arrows,

Fig. 53 Effect of temperature on the fatigue life of S-816 alloy tested under a fluctuating axial load at a frequency of 216 000 cycles per hour

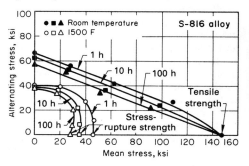

Fig. 54 Surface of a high-temperature fatigue fracture in a nickel alloy jet-engine turbine blade

Arrow indicates a fatigue pattern nucleating on the convex side of the airfoil.

Fig. 55 Fracture surface of a steel turbine disk for a jet engine that failed from fatigue

Fracture was nucleated by stress-rupture cracks in tenons (arrows). High mean stress was a major factor in this high-temperature failure.

tinuous deformation or creep, which will eventually lead to fracture if the load is maintained for a sufficient length of time. The stress-rupture stength is defined as the stress that a metal can withstand for a given time, at a given temperature, without breaking. With increases in temperature, stress-rupture strength decreases rapidly to values that may be considerably lower than fatigue strength. Therefore, the primary requirement of a metal that will be subjected to high temperatures is that it have adequate stress-rupture strength. Many alloys that possess good creep resistance are also resistant to fatigue; however, the condition of an alloy that will provide maximum stress-rupture strength is not necessarily the condition that provides maximum fatigue strength. In practice, it is necessary to design against failure by fatigue and against excessive distortion or fracture by creep, just as it is necessary to consider combined tensile and fatigue loads at room temperature.

At room temperature, and except at very high frequencies, the frequency at which cyclic loads are applied has little effect on the fatigue strength of most metals. The effect, however, becomes much greater as the temperature increases and creep becomes more of a factor. At high temperatures, the fatigue strength often depends on the total time the stress is applied rather than solely on the number of cycles. This behavior occurs because of continuous deformation under load at high temperatures. Under fluctuating stress, the cyclic frequency affects both the fatigue life and the amount of creep. This is shown in Fig. 53, a typical constant-life diagram that illustrates the temperature behavior of S-816 alloy tested under a fluctuating axial load. At room temperature, the curves converge at the tensile strength, plotted along the mean-stress axis. At high temperature, the curves terminate at the stress-rupture strength, which, being a time-dependent property, results in termination at a series of end points along the mean-stress axis.

Fractures resulting from fluctuating stresses at high temperatures may be similar in appearance to fatigue fractures, stress-rupture fractures, or a mixture of the two, depending on the relative magnitudes of the mean and alternating stresses. Figure 54 shows a high-temperature fatigue fracture in a jet-engine turbine blade; alternating stresses were primarily responsible for this fracture, and static stresses played a minor role. A fatigue fracture through the tenons of a turbine disk is shown in Fig. 55. In this fracture, mean stresses played a major role, with fatigue being nucleated by stress-rupture cracking. Generally, the amount of deformation in the region of high-temperature fatigue fractures decreases as the ratio of alternating stress to mean stress increases.

An important requirement for fatigue resistance at high temperatures is that the component have resistance to oxidation and other forms of high-temperature corrosion. Fatigue strength at high temperature can be seriously reduced by surface attack from fuel ash containing vanadium pentoxide, from leaded fuels, and from other contaminants. In general,

however, materials are less notch sensitive at high temperatures than at room temperature.

Extended operation at high temperatures may result in metallurgical changes in the alloy structure, which may also play a role in reducing fatigue strength. Generally, however, brief exposure to high temperatures, which does not result in such metallurgical changes as recrystallization, tempering, phase changes, coarsening, precipitation, melting, or diffusion, will not have a serious effect on fatigue life upon return to normal operating temperatures.

Example 20: Intergranular Fatigue Cracking of a Stainless Steel Expansion Joint. A type 321 stainless steel bellows expansion joint on a 17-cm (6¾-in.) OD inlet line in a gas-turbine test facility cracked during operation. Cracking occurred in welded joints and in unwelded portions of the bellows.

The line consisted of two bellows and a pipe (Fig. 56). The bellows were of type 321 stainless steel with a 2.4-mm (0.093-in.) wall thickness; the pipe was type 347 stainless steel, 15.4-cm (6.065-in.) ID, and 8.6-mm (0.340-in.) nominal wall thickness. The bellows were made by forming the convolution halves from stainless steel sheet, then welding the convolutions together (Detail B, Fig. 56).

The line carried high-purity nitrogen gas at 1034 kPa (150 psi) with a flow rate of 5.4 to 8.2 kg/s (12 to 18 lb/s). During the 800-h operation of the line, the nitrogen gas became contaminated because oil and oil vapor leaked into the system. The nominal gas temperature was 650 °C (1200 °F), although the maximum inlet temperature to the turbine occasionally exceeded 675 °C (1250 °F).

Before the line was put in service, 47 open-cycle tests were conducted on the piping. In the tests, the piping was pressurized with air at 276 to 310 kPa (40 to 45 psi) and subjected to temperatures of 620 to 650 °C (1150 to 1200 °F) for an approximate total exposure time of 100 h at temperature. Closed-cycle operation with high-purity nitrogen gas started approximately 13 months after the piping was installed, then failed after about 23 months. After field repairs, the piping operated about 6 weeks before a second failure occurred. There were 130 thermal cycles; the average time to reach temperature was 1.5 h, and the average cooling time was 1.2 h. A new line was installed, and samples of the cracked line were sent to a laboratory for failure analysis.

Visual examination indicated that the bellows unit nearest the turbine contained the most cracking; therefore, only this bellow section was selected for analysis. The bellows were split in half, the reflector sleeve was removed, and the bellows was vapor blasted and liquid-penetrant inspected to reveal all visible cracking.

The most severe crack occurred in the welded joint between the first and second convolutions from the downstream end of the bellows. The crack appeared to initiate along

Fig. 56 Stainless steel pipeline for carrying hot nitrogen gas from heater to turbine that failed in the bellows section because of grain-boundary embrittlement

Dimensions given in inches

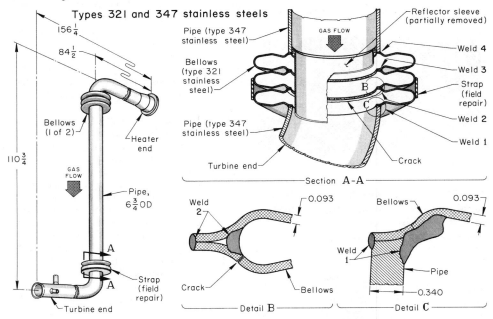

the surface of the inner wall opposite the external weld, then propagated toward the outer surface of the convolution.

No cracks were observed in the welded joint between the second and third convolutions. Cracks were found in the welded joints on each side of the bellows where it was joined to the pipe. A stainless steel strap had been field welded around the top of the first two downstream convolutions to repair the area where leakage occurred.

Heavy deposits of carbon were found on the inner walls of the convolutions and on the back side of the reflector sleeve. The reflector was intended to prevent contaminants from reaching the inner walls of the convolutions and to streamline gas flow across the convolutions.

Chemical analyses were made of the piping and of samples from welds 1 and 2 (Fig. 56). Although some minor deviations from normal composition were found (particularly involving molybdenum, niobium, and titanium contents), these deviations did not affect material properties.

Microhardness measurements were taken on three sections of pipe and on two weld samples. Hardness readings taken within the carburized zone at the inner surface of the pipe wall averaged 36 HRC; near the center of the wall, 90 HRB; and at the outer surface, 95 HRB. The hardness gradient was believed to be the result of oxygen penetration from the outer surface, which was exposed to air, and carbon diffusion from a carburized zone on the inner surface.

The weld samples showed slightly higher hardness values than normally occur in type 321 stainless steel, which was attributed to the presence of a hard grain-boundary precipitate. Hardnesses of the weld metal and adjacent base metal ranged from 95 to 99 HRB.

Metallographic examination showed the piping to be sound, although thin carburized layers occurred on the inner surfaces. Carburization appeared to have been caused by a buildup of carbon, which was deposited during service through breakdown of an oil contaminant in the nitrogen gas. This condition was observed in all of the piping.

The microstructure was normal for annealed type 347 stainless steel. Large particles, probably niobium carbides (NbC) and complex carbonitrides (NbC,N), were randomly dispersed through the structure. These particles were believed to have been present in the as-received material. Fine carbide precipitates noted in the pipe structures were not detrimental, because they showed no preference for the grain boundaries but were well dispersed throughout the austenite grains.

The weld metal appeared to be of good quality and did not indicate the presence of inclusions or show many signs of overheating. No excessive grain growth occurred near the weld metal/base metal interface. Intergranular carbide precipitation was found in both the weld metal and the HAZ of the base metal of weld 3. This was attributed to local overheating during the welding process that had vaporized most of the titanium stabilizer. Carbides then formed with subsequent heating within the

sensitizing-temperature range. Extensive intergranular cracking and intergranular oxidation were found in weld 1. Carbonaceous material was found in unusually large crevices at the head of the cracks.

Intergranular cracking was found in the HAZ in all four of the welded joints in the bellows. At normal operating temperatures for this gas-inlet line, intergranular fatigue failure in austenitic stainless steels usually occurs only in the presence of grain-boundary embrittlement. This type of embrittlement is usually the result of nitriding, intergranular oxidation, carbide precipitation, or the formation of σ or χ phase.

There was no metallographic evidence to indicate that nitriding had occurred. Intergranular oxidation was present, but because failure did not occur until long after air testing was concluded, its presence was not considered to be the mechanism causing the embrittlement, even though it undoubtedly weakened the bellows.

A light-colored precipitate occurred in small amounts in all of the welded joints of the bellows convolutions and was suspected of containing either complex carbides or σ phase. The sample from weld 3 was etched with Vilella's reagent, specifically to reveal carbides and σ phase, then polished lightly to remove the etchant and re-etched with Murakami's reagent, which reveals only carbides. The two etching operations revealed that significant quantities of both complex carbides and σ phase were present.

The grain-boundary precipitate was identified as consisting largely of a continuous σ-phase network. The amount of σ phase formed was small; however, because the phase was in a continuous grain-boundary network, the amount formed was considered sufficient to cause the grain boundaries to be highly embrittled. Because the roots of the convolutions were points of maximum flexure, and because of the loss of plasticity resulting from σ embrittlement, the structure could not withstand the fluctuating thermal and mechanical stresses encountered during service.

Cracking occurred in the weld HAZs, as well as the roots of the convolutions that were areas of maximum flexure. These areas were highly strained, which would accelerate carbon penetration in the cracked areas and would account for carburization at these points, even though carburization was not found in other areas of the samples examined.

Intergranular cracking occurred at the welded joints of the convolutions and originated at the root of crevices. The crevices were caused by surface oxidation during open-cycle testing and acted as open stress raisers. Open-cycle testing also caused stress oxidation within the convolution roots, which accounted for the intergranular oxygen penetration found in the cracks on all of the welded joints examined. In general, intergranular oxygen penetration increases with increasing time and temperature at any given stress above a specific minimum stress.

Fig. 57 Specimen of carburized steel showing surface-origin pitting that resulted from contact-fatigue testing

Arrows indicate direction of rotation. (a) Macrograph showing shell-like surface pit. 4×. (b) Section through the surface fatigue crack. Crack A initiated at the surface and propagated from left to right. Crack B, a large subsidiary crack, is almost normal to the surface. Note small subsidiary cracks and grain-boundary oxide network. Approximately 380×. (c) Surface of the specimen showing apex of pit (at bottom) encircled by the crack. Approximately 55×. (d) Apex of the pit spalled away and crack encircling a larger area. Approximately 55×

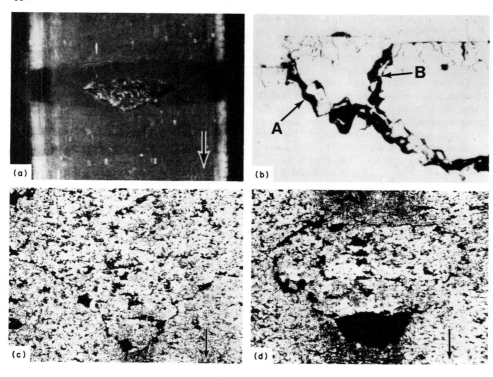

Conclusions. Failure of the bellows occurred by intergranular fatigue cracking along a continuous grain-boundary network of σ phase, which had precipitated under normal operating conditions. Fatigue cracking originated at crevices caused by surface oxidation during open-cycle testing before normal service.

Secondary degrading effects on the piping and bellows included grain-boundary carbide precipitation, intergranular oxidation on the outside surface, and carburization of the inside surface. Quality of the piping and of the welds had no bearing on the cause of failure.

Recommendations. Type 321 stainless steel is satisfactory for the bellows convolutions, provided that open-cycle testing does not result in surface oxidation and crevices. However, type 347 stainless steel would be better, because it has greater stability during welding, even though, like type 321, it is susceptible to σ-phase formation. Inconel 600 would be an even better choice for the bellows because of its improved stress-oxidation properties and because it is not susceptible to σ-phase formation.

Welds in a type 321 bellows should be stress relieved by solution treatment at 950 to 980 °C (1750 to 1800 °F) before open-cycle testing. Reheating at 870 to 900 °C (1600 to 1650 °F) for 1 to 2 h after annealing would precipitate the greater part of the dissolved carbon as titanium carbide, thus at least partially stabilizing the type 321 stainless steel bellows and welds. Prevention of oil leakage into the system would minimize carburization of the piping and bellows.

Thermal Fatigue

Thermal-fatigue failure is the result of temperature cycling, as opposed to fatigue at high temperatures caused by strain cycling. Two conditions necessary for thermal fatigue are some form of mechanical constraint and a temperature change. Thermal expansion or contraction caused by a temperature change acting against a constraint causes thermal stress. Constraint may be external—for example, constraint imposed by rigid mountings for pipes—or it may be internal, in which case it is set up by a temperature gradient within the part. In thick sections, temperature gradients are likely to occur both along and through the material, causing highly triaxial stresses and reducing material ductility, even though the uniaxial ductility often increases with increasing temperature. Reduction in the ductility of the material gives rise to fractures that have a brittle appearance, often with many cleavagelike facets in evidence.

Identifying features of low-cycle thermal-fatigue failures are multiple initiation sites that join randomly by edge sliding to form the main crack, transverse fractures, an oxide wedge filling the crack, and transgranular fracture. Cracks having similar characteristics but distinguished by intergranular fracture are caused by stress-rupture phenomena (long periods at elevated temperature under high static tensile load). The primary failure mechanism involved in stress rupture is grain-boundary sliding. In thermal-fatigue cracks, slip processes and cleavage operate much as they do in failure at normal temperature, but the evidence is often destroyed by oxide formation, flame polishing, and melting processes.

True thermal fatigue occurs in such components as internal-combustion engines, in which thick-section cast materials are used, and in applications such as heat exchangers, in which thin wrought material is used. Another prominent example occurs in jet engine turbine blades. In cast materials, uniform sections, mild strain gradients, and short-flake graphite are desirable design features. On oil-fired and gas-fired furnace heat exchangers, the thermal cycle is important because it controls temperature gradients, and in thin sections, external constraints are of minor importance.

Under certain circumstances, thermal-fatigue and stress-rupture failures blend into each other. Thermal fatigue is the basic mechanism in failures that occur because of numerous, short heating and cooling cycles. Stress rupture becomes an important consideration as the cycle times increase and therefore is primarily a long-term rate process. Most thermal-fatigue fractures are of the low-cycle, high-strain type; the fracture surfaces are rough and faceted at or near the initiation sites and are more fibrous and with shear lips at 45° angles in the final-fracture area. For additional discussion of thermal fatigue, see the article "Elevated-Temperature Failures" in this Volume.

Contact Fatigue

Elements that roll, or roll and slide, against each other under high contact pressure are subject to the development of surface pits or spalls after many repetitions of load. Pitting is a manifestation of metal fatigue from imposed cyclic contact stresses. Factors that govern pitting fatigue are the contact stress, the material properties and metallurgy, and the physical and chemical characteristics of the contacting surfaces, including the oil film lubricating the surfaces.

The magnitude and distribution of stresses at and below the surface of contact have been described (Ref 13-18). The significant stress in rolling-contact fatigue is the maximum alternating shear stress that undergoes a reversal in direction during rolling. In pure rolling, this shear stress occurs on a plane slightly below the surface and can lead to initiation of fatigue cracks in the subsurface material. As these cracks propagate under the repeated loads, they reach the surface and produce small pits.

When sliding is imposed on rolling, the tangential forces and thermal gradient caused by friction alter the magnitude and distribution of stresses in and below the contact area. The alternating shear stress increases in magnitude and is moved nearer to the surface by sliding forces. Thus, initiation of contact-fatigue cracks in gear teeth, which are subjected to significant amounts of sliding adjacent to the pitchline, is found to be in the surface material. These cracks propagate at a shallow angle to the surface, and pits result when the cracks are connected to the surface by secondary cracks. If pitting is severe, the bending strength of the tooth may be decreased to the point at which fracturing can occur.

Surface-Pitting Fatigue. Test specimens of carburized steel examined during and after contact-fatigue testing disclose the mechanism of surface pitting. Figure 57(a) shows a surface pit that has a shell-like appearance, with the apex of the "V" of the pit pointing in the direction of rotation.

A crack begins at the surface and propagates at an acute angle to the surface in the same direction that the loading wheel rolls over the surface opposite to the direction of rotation. The crack and resulting pit are confined to the outer portion of the carburized case. Several stages of pitting development are shown in Fig. 57(b). At the apex, a small volume of metal has been released and is ready to spall off. The small subsidiary cracks shown in Fig. 57(b) will eventually join and release a much larger volume of metal. This progression of failure, viewed on the surface of the specimen, appears as shown in Fig. 57(c) and (d). The apex of the pit is encircled by a crack in Fig. 57(c); in Fig. 57(d), the apex has spalled away and the crack has encircled a larger area. Figures 57(b) to (d) illustrate the sequence of events leading to surface pitting, which is considered to be the typical mode of failure when sliding and rolling are present between contacting surfaces.

The specimen shown in Fig. 57(a) has a single pit, but there is no evidence of other surface damage in the contact path. This is typical of specimens tested at high contact stresses and run for relatively small numbers of cycles (less than 10^7) before developing the failure pit. On the other hand, specimens tested at lower contact stresses and run for many more cycles (greater than 2×10^8) develop scores of small pits that result in a severely worn contact path.

Small pits begin the same way as the large pit in Fig. 57(a); however, the fatigue crack seems to lack the driving force to penetrate very far below the surface and instead quickly returns to the surface and results in a small, shallow pit.

The continuing passage of the loading roller over these small pits tends to obliterate their characteristic V shape.

Subsurface Cracking. Another form of contact-fatigue failure initiates in the subsurface, but results in severe pitting or spalling on the surface. In pure rolling, or where subsurface stress concentrations arise because of rolling and sliding, cracking originates below the contact surfaces, frequently originating at an inclusion, and propagates parallel to the surface. In the early stages, subsidiary cracks may appear on the surface; ultimately, large areas spall away as the main subsurface crack spreads. In case-hardened parts, when insufficient case depth and/or core hardness exists, the cracking frequently occurs at the case/core interface, causing severe spalling; this is termed subcase fatigue. For more discussion of contact fatigue, see the articles "Failures of Rolling-Element Bearings" and "Wear Failures" in this Volume.

Corrosion Fatigue

Corrosion fatigue is associated with alternating or fluctuating stresses that occur in a corrosive environment and cause accelerated crack initiation and propagation at a location where neither the environment nor the stress acting alone would be sufficient to produce a crack. Fatigue cracking in a corrosive environment is identified by the presence of numerous small cracks adjacent to the fracture and of compacted corrosion product on the fracture surface, which may damage and obscure fine surface detail of the fracture.

The corrosive environment usually introduces stress raisers on the surface. The irregular surface that results is detrimental to the fatigue properties of the part in a mechanical or geometric sense. For parts susceptible to embrittlement by hydrogen or for parts exposed to a fairly continuous corrosive environment with intermittent applications of loading, the cracking mechanism may be somewhat more complex.

An important feature of corrosion fatigue is that the stress range required to cause fracture diminishes progressively as the time and number of stress cycles increase. It is, therefore, impractical and uneconomical to attempt solely to design against corrosion fatigue. Although different alloys show differing performance under a given corrosion-fatigue environment, it is customary to protect the surface to achieve adequate performance at low cost. For more information, see the article "Corrosion-Fatigue Failures" in this Volume.

Inspection Schedules and Techniques

Fatigue failures can be reduced by maintaining routine inspection schedules that include nondestructive testing. If a crack or discontinuity is found in a vital component that could fail and cause serious damage, the component should be replaced or repaired as soon as possible. Where failure is not so critical, the part containing a discontinuity can be inspected at regular intervals until the component is considered to be near its failure point before it is replaced. Critical components that are exchanged in assemblies during routine maintenance should be inspected to ensure that the removed part is usable as a spare.

Records of failures should be maintained so that recurring instances of failures of certain components can be noted. Components that fail repeatedly indicate that perhaps a different design and/or material should be used. Where it may be unreasonable to change the design or material, examination of the records should give a rough life expectancy. The analytical procedures described previously in this article may be included in this examination. Knowing this, the part can be replaced before failure results. However, most preventive-maintenance systems assume that all parts are exactly the same, which is not necessarily true; therefore, such systems should be used only to give some rough guidelines and should be supplemented by nondestructive inspection.

Visual inspection can sometimes reveal fatigue cracks. The cracks are frequently located at obvious points of stress concentration, such as section changes, sharp fillets, last thread in threaded components, toe of welds, and keyways. These stress concentrations may be the result of design-fabrication defects or accidental notches. Once fatigue cracks become visible to the naked eye, they usually propagate at a rate such that the remaining life will be only a small percentage of the total life of the part.

Nondestructive Testing. Liquid-penetrant inspection, electromagnetic inspection, and magnetic-particle inspection (of ferrous metals) are reliable methods of detecting surface cracks and discontinuities. However, the number of load applications required to produce fatigue failure usually cannot be predicted from test indications. This information must be used with an analytical procedure to estimate the residual life.

Ultrasonic inspection and radiography are mainly used for internal examination. For crack detection, ultrasonics is much more reliable than radiography. It is not unusual to monitor fatigue cracks by ultrasonics until crack growth is considered to have reached a critical size.

Stresscoating, strain gages, photoelastic coatings, and x-ray diffraction generally are not used for fractured parts, but rather for unfailed mating or similar parts to study the residual stresses in a part or the induced stresses imparted by specific loadings. The results of these tests can be used to analyze the stresses in parts that failed by fatigue. More information on nondestructive testing is presented in the article "General Practice in Failure Analysis" in this Volume.

Determination of Fatigue Damage and Life

Fatigue causes more than 80% of the operating failures of machine elements, and in many of these the stress cycles may be very complex with occasional high peaks—for example, the gust loading of aircraft wings. For satisfactory correlation with service behavior, full-size or large-scale specimens must be tested under conditions as close as possible to those existing in service. This method is expensive, but it does provide valuable data. A less costly testing procedure is simplified laboratory testing. By using the fatigue information obtained from the testing of standard specimens or by using models and applying the proper correction factors for configuration, surface finish, environment, and various other parameters, an approximation of the lifetime of the component can be determined. (For an example of test-data correction factors, see Fig. 26 and the corresponding discussion in this article.) It should be emphasized, however, that this is just an approximation, and without full-scale tests such as those mentioned previously, this information could be almost useless.

For improving production designs, the target load/life test method of evaluating the design of a part has been used by automotive, farm equipment, construction equipment, and aircraft companies. In this method, numerous failures from actual operating machines are observed. From these observations, conclusions can be drawn as to the type of loading that causes the failures. A fatigue test is then set up to apply the appropriate loading. After obtaining failures during the test, certain adjustments can be made in the magnitude and method of applying the loads until the same type of failure is developed in the test as those experienced by the operating machines.

Target lives for the parts in the fatigue test can be established, based on the time-to-failure on the machines in service, the time-to-failure in the fatigue tests, and the desired life in service. For example, if a certain part fails in one-fourth of the desired life of the machine, and if these same parts fail in 25 000 cycles during the fatigue tests, one would not consider any revised parts satisfactory unless they exceeded 100 000 cycles without failing. Once the appropriate loading and number of cycles are established for a part on a production machine, the same test can be used to ensure that similar parts on new machines will be satisfactory.

One application in which this method has been used is the rear axles of automobiles. Maximum tractive torque is applied to the rear axle for 100 000 cycles. If the axle gears withstand this loading, they are considered satisfactory for customer use. The advantages of this procedure are that it makes use of the experience gained from many machines already in use and that it is rather rapid. Its disadvantage is that sufficient machines must be placed in customers' hands to develop the failed parts for study and evaluation of test methods.

The differences in the operating conditions, environment, fatigue properties of the parts, and processing methods between the production design and new design must be small to produce accurate results. Therefore, this technique is primarily used when there is ample experience, as in improving parts that have a history of failure in service or in evaluating machines that are similar to existing production models.

REFERENCES

1. M.A. Miner, Cumulative Damage in Fatigue, *Trans. ASME*, Vol 67, 1945, p A159
2. H.J. Grover, "Fatigue of Aircraft Structures," NAVAIR 01-1A-13, Naval Air Systems Command, U.S. Department of the Navy, 1966
3. P.G. Forrest, *Fatigue of Metals*, Pergamon Press, 1962, p 113
4. R.E. Peterson, *Stress Concentration Design Factors*, John Wiley & Sons, 1974
5. H. Neuber, "Theory of Notch Stresses: Principles for Exact Calculation of Strength With Reference to Structural Form and Material," AEC-Tr-4547, Springer Publishing, 1958; available through NTIS, U.S. Department of Commerce
6. R.J. Roark, *Formulas for Stress and Strain*, 4th ed., McGraw-Hill, 1965
7. T. Topper, R. Wetzel, and J. Morrow, Neuber's Rule Applied to Fatigue of Notched Specimens, *J. Mater.*, Vol 4 (No. 1), March 1969
8. R.C. Juvinall, *Engineering Considerations of Stress, Strain and Strength*, McGraw-Hill, 1967
9. P.C. Paris and G.C. Sih, in *STP 381*, ASTM, Philadelphia, 1965, p 30-83
10. H. Tada, P. Paris, and G. Irwin, *The Stress Analysis of Cracks Handbook*, Del Research, Hellertown, PA, 1973
11. W.J. McGregor Tegart, *Elements of Mechanical Metallurgy*, Macmillan, 1966, chap. 3
12. Electroless Nickel Plating—A Review, *Met. Finish.*, Jan 1975, p 38-44
13. H.R. Hertz, Miscellaneous Papers, Macmillan, 1896
14. J.O. Smith and C.K. Liu, Stresses Due to Tangential and Normal Loads on Elastic Solids With Application to Some Contact Stress Problems, *J. Appl. Mech.*, Vol 20 (No. 2), June 1953
15. R.E. Denning and S.L. Rice, "Surface Fatigue Research With the Geared Roller Test Machine," SAE Paper 620 B, Society of Automotive Engineers, New York, 1963
16. G.J. Moyar and J.D. Morrow, "Surface Failure of Bearings and Other Rolling Elements," Engineering Experiment Station Bulletin No. 468, University of Illinois, 1964
17. W.E. Littman and R.L. Widner, "Propagation of Contact Fatigue From Surface and Subsurface Origins," ASME Paper No. 63-WA/CF-2, American Society of Mechanical Engineers, New York
18. J.P. Sheehan and M.A. Howes, "The Role of Surface Finish in Pitting Fatigue of Carburized Steel," SAE Paper 730580, Society of Automotive Engineers, New York, 1973

SELECTED REFERENCES

- J.E. Campbell, W.W. Gerberich, and J.H. Underwood, Ed., *Application of Fracture Mechanics of Metallic Structural Materials*, American Society for Metals, 1982
- *Damage Tolerance in Aircraft Structures*, STP 486, ASTM, Philadelphia, 1971
- *Fatigue at High Temperature*, STP 459, ASTM, Philadelphia, 1969
- C. Laird and D.J. Duquette, Mechanisms of Fatigue Crack Nucleation, in *Corrosion Fatigue: Chemistry, Mechanics and Microstructure*, National Association of Corrosion Engineers, Houston, 1972
- A.F. Madayag, *Metal Fatigue: Theory and Design*, John Wiley & Sons, 1969
- *Metal Fatigue Damage—Mechanism, Detection, Avoidance and Repair*, STP 495, ASTM, Philadelphia, 1971
- C.C. Osgood, *Fatigue Design*, Wiley-Interscience, 1970
- G. Sines and J.L. Waisman, *Metal Fatigue*, McGraw-Hill, 1959
- *Structural Fatigue in Aircraft*, STP 404, ASTM, Philadelphia, 1966

Distortion Failures

Revised by Donald J. Wulpi, Metallurgical Consultant

DISTORTION FAILURE occurs when a structure or component is deformed so that it no longer can support the load it was intended to carry, is incapable of performing its intended function, or interferes with the operation of another component. Distortion failures can be plastic or elastic and may or may not be accompanied by fracture. There are two main types of distortion: size distortion, which refers to a change in volume (growth or shrinkage), and shape distortion (bending or warping), which refers to a change in geometrical form (Ref 1).

Distortion failures are ordinarily considered to be self-evident, for example, damage of a car body in a collision or bending of a nail being driven into hard wood. However, the failure analyst is often faced with more subtle situations. For example, the immediate cause of distortion (bending) of an automobile-engine valve stem is contact of the valve head with the piston, but the failure analyst must go beyond this immediate cause in order to recommend proper corrective measures. The valve may have stuck open because of faulty lubrication; the valve spring may have broken because corrosion had weakened it; the spring may have had insufficient strength and taken a set, allowing the valve to drop into the path of the piston; or the engine may have been raced beyond its rpm limit many times, causing coil clash and subsequent fatigue fracture of the spring. Without careful consideration of all the evidence, the failure analyst may overlook the true cause of a distortion failure. In this article, several common aspects of failure by distortion will be discussed, and suitable examples of distortion failures will be presented for illustration.

Overloading

Every structure has a load limit beyond which it is considered unsafe or unreliable. Applied loads that exceed this limit are known as overloads and sometimes result in distortion or fracture of one or more structural members. Estimation of load limits is one of the most important aspects of design and is commonly computed by one of two methods—classical design or limit analysis.

Classical Design. The conservative, classical method of design assumes that failure occurs whenever the stress at any point in a structure exceeds the yield strength of the material. Except for members that are loaded in pure tension, the fact that yielding occurs at some point in a structure has little influence on the ability of the structure to support the load. However, yielding has long been considered a prelude to structural collapse or fracture and is therefore a reasonable basis for limiting applied loads.

Classical design keeps allowable stresses entirely within the elastic region and is used routinely in the design of parts. Allowable stresses for static service are generally set at one-half the yield strength for ductile materials and one-sixth for brittle materials, although other fractions may be more suitable for specific applications. The reason for using such low fractions of yield strength is to allow for such factors as possible errors in computational assumptions, accidental overload, introduction of residual stress during processing, temperature effects, variations in material quality (including imperfections), degradation (for example, from corrosion), and inadvertent local increases in applied stress resulting from notch effects.

Classical design is also used for setting allowable stresses in other applications, for example, where fracture can occur by fatigue or stress rupture. In these instances, fatigue strength or stress-rupture strength is substituted for yield strength as a point of reference.

Limit Analysis. The upper limit in design is defined as the load at which a structure will break or collapse under a single application of force. This load can be calculated by a method known as limit analysis (Ref 2, 3). With limit analysis, it is unnecessary to estimate stress distributions, which makes stress analysis much simpler by this method than by classical design. However, limit analysis is based on the concept of tolerance to yielding in the most highly stressed regions of the structure and therefore cannot be used in designing for resistance to fatigue or elastic buckling or in designing flaw-tolerant structures.

Limit analysis assumes an idealized material—one that behaves elastically up to a certain yield strength, then does not work harden but undergoes an indefinite amount of plastic deformation with no change in stress. The inher-ent safety of a structure is more realistically estimated by limit analysis in those instances when the structure will tolerate some plastic deformation before it collapses. Because low-carbon steel, one of the most common materials used in structural members, behaves somewhat like the idealized material, limit analysis is very useful to the designer, especially in the analysis of statically indeterminate structures.

Figure 1 illustrates the relative stress-strain behavior of a low-carbon steel, a strain-hardening material, and an idealized material—all with the same yield strength (the upper yield point for the low-carbon steel, and the stress at 0.2% offset for the strain-hardening material). Load limits for parts made of materials that strain harden significantly when stressed in the plastic region can be estimated by limit analysis, as can those for parts made of other materials whose stress-strain behavior differs from that of the idealized material. In these situations, the designer bases his design calculations on an assumed strength that may actually lie well within the plastic region for the material.

Fig. 1 Schematic comparison of the conventional stress-strain behavior of a low-carbon steel, a strain-hardening material, and the idealized material assumed in limit analysis

All have the same yield strength.

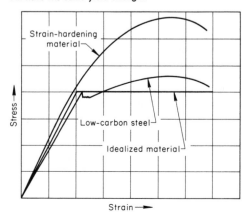

Buckling. Collapse due to instability under compressive stress, or buckling, may or may not be permanent deformation, depending on whether or not the yield strength was exceeded. Long, slender, straight bars, tubes, or columns under axial compressive forces will buckle when the buckling load is exceeded. Buckling failure may also be encountered on the compressive sides of tubes, I-beams, channels, and angles under bending forces. Tubes may also buckle due to torsional forces, causing waves, or folds, generally perpendicular to the direction of the compressive-stress component.

The buckling load depends only upon the dimensions of the part and the modulus of elasticity of the material. Therefore, buckling cannot be prevented by changing the strength or hardness of the metal. The modulus of elasticity of a given metal is affected only by temperature, increasing at lower temperature and decreasing at higher temperature. Buckling can be prevented only by changing the size or shape of the part with respect to the load imposed on it (Ref 4).

Safety Factors. In both classical design and limit analysis, yielding is assumed to be the criterion for calculating safe loads on statically loaded structures. For a given applied load, the two methods differ in that the safety factor (the ratio of the theoretical capacity of a structural member to the maximum allowable load) is generally higher when calculated by limit analysis. For example, classical design limits the capacity of a rectangular beam to the bending moment that will produce tensile yielding in the regions farthest from the neutral axis; limit analysis predicts that complete collapse will occur at a bending moment 1.5 times the limiting bending moment determined by classical design. It is important that the designer be able to relate the actual behavior of a structure to its assumed behavior because, for a given applied load and selected safety factor, a structure designed by limit analysis will usually have thinner sections than a structure designed by classical methods.

Safety factors are important design considerations because they allow for factors that cannot be computed in advance. Overload failure can occur either when the applied stress is increased above the design value or when the material strength is degraded. If either situation is a characteristic of the fabricated structure, the design must be changed to allow for these factors more realistically.

Example 1: Collapse of Extension Ladders by Overloading of Side Rails. Several aluminum alloy extension ladders of the same size and type collapsed in service in the same manner; the extruded aluminum alloy 6063-T6 side rails buckled, but the rungs and hardware remained firmly in place. The ladders had a maximum extended length of 6.4 m (21 ft), and the recommended maximum angle of inclination to the vertical was 15°.

Investigation. Visual examination disclosed that the side-rail extrusions, which had the

Fig. 2 Aluminum alloy 6063-T6 extension-ladder side-rail extrusion that failed by plastic deformation and subsequent buckling

(a) Configuration and dimensions (given in inches).
(b) Relation of maximum applied load to the section thickness of the flanges and web of the side-rail extrusion

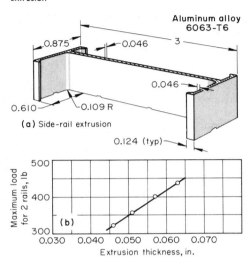

I-beam shape shown in Fig. 2(a), had failed by plastic buckling, with only slight surface cracking in the most severely deformed areas. There were no visible defects in materials or workmanship, and all dimensions of the side rails were within specified tolerances.

Hardness tests using a portable hardness tester, metallographic examination, and tensile tests of specimens from the buckled side rails were conducted. All results agreed with the typical properties reported for aluminum alloy 6063-T6 extrusions.

Stress analysis of the design of the ladder, using actual dimensions, indicated that the side-rail extrusions had been designed with a thickness that would provide a safety factor of 1.2 at ideal loading conditions (15° maximum inclination) and that use of the ladders under other conditions could subject the side rails to stresses beyond the yield strength of the material. Once yielding occurred, buckling would continue until the ladder collapsed.

The stress analysis was extended to include an evaluation of the relation of maximum applied load to section thickness with the ladder at its maximum extension of 6.4 m (21 ft) and at an inclination of 15°. This relation is shown in Fig. 2(b).

Conclusion. The side rails of the ladders buckled when subjected to loads that produced stresses beyond the yield strength of the alloy. Failure was by plastic deformation, with only slight tearing in the most severely deformed regions.

Corrective Measure. The flange and web of the side-rail extrusion were increased in thick-

Fig. 3 Relation of distortion ratio to stress ratio for two steel cantilever beams of rectangular cross section

Distortion ratio is permanent deflection, measured at a distance from the support ten times the beam thickness, divided by beam thickness. Stress ratio is maximum stress, calculated from applied load and original beam dimensions, divided by yield strength.

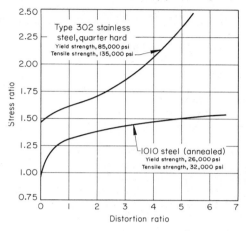

ness from 1.2 to 1.4 mm (0.046 to 0.057 in.). This increased the safety factor from 1.2 to 1.56. After this change, no further failures were reported.

Amount of Distortion. When designing structures using limit analysis, the designer does not always consider the amount of distortion that will be encountered. A rough illustration of the distortion that resulted from overloading of small cantilever beams is given in Fig. 3. Known loads were applied to rectangular-section beams of low-carbon steel and of stainless steel, and the permanent deflection at the loading point was measured. Maximum fiber stresses were calculated from the applied load and original specimen dimensions.

This type of test provides a simplistic but useful concept of distortion by showing how much distortion occurs at strains beyond the yielding point. As shown in Fig. 3, the beam made of low-carbon steel, which strain hardens only slightly, exhibited no distortion when the calculated maximum fiber stress was equal to the yield strength (at a stress ratio of 1.00). However, this beam collapsed at a load equivalent to a fiber stress just above the tensile strength, as shown in Fig. 3 where the lower curve became essentially horizontal. This collapse load agrees with the limit-analysis collapse load of 1.5 times the load at yield. The beam made of stainless steel, which strain hardens at a rather high rate, showed no distortion at fiber stresses up to 1.47 times the yield strength. When the calculated stress equaled the tensile strength (at a stress ratio of 1.59), distortion was 0.7 times the beam thickness, and the beam supported a calculated stress of 1.5 times the tensile strength without collapse.

When loads increase gradually, distortion is gradual, and design can be based on knowledge of the amount of distortion that can be tolerated. Thus, simple bench tests of full-size or scaled-down models can often be used in estimating the loads required to produce various amounts of distortion.

When rapid or impulse loads are applied, as in impact, shock loading, or vibration, the amount of distortion that can occur without fracture is considerably less predictable. For most structural materials, measured values of strength are higher under impulse loading, and values of ductility are lower, than the values measured under static loading. Tensile and yield strengths as much as 20% higher than the slow-tension-test values have been measured under very high rates of loading. In addition, the variation, or scatter, among replicate tests of mechanical properties is greater when strain rates are high than it is when strain rates are low. The amount of distortion that can occur at high rates of loading is difficult to analyze or predict because (1) the crystallographic processes involved in deformation and fracture are influenced by strain rate and temperature, (2) impulse loading creates an adiabatic condition that causes a local increase in temperature, and (3) impulse loading involves the propagation of high-velocity stress waves through the structure.

Effect of Temperature. Distortion failures caused by overload can occur at any temperature at which the flow strength of the material is less than the fracture strength. In this discussion, flow strength is defined as the average true stress required to produce detectable plastic deformation caused by a relatively slow, continuously increasing application of load; fracture strength is the average true stress at fracture caused by a relatively slow, continuously increasing application of load. The flow strength and fracture strength of a material are temperature dependent, as is the elastic modulus (Young's modulus, bulk modulus, or shear modulus).

Figure 4 illustrates this temperature dependence schematically for polycrystalline materials that do not undergo a solid-state transformation. Two flow strengths are shown: one for a material that does not have a ductile-to-brittle transition in fracture behavior, such as metal with a face-centered cubic (fcc) crystal structure, and one for a body-centered cubic (bcc) material that exhibits a ductile-to-brittle transition.

As shown in Fig. 4, the flow strength, fracture strength, and elastic modulus of a material generally decrease as temperature increases. If a structure can carry a certain load at 20 °C (70 °F), it can carry the same load without deforming at lower temperatures. Stressed members made of materials having a ductile-to-brittle fracture transition will sometimes fracture spontaneously if the temperature is lowered to a value below the transition temperature. A more detailed discussion of this

Fig. 4 Schematic diagram of the temperature dependence of elastic, plastic, and fracture behavior of polycrystalline materials that do not exhibit a solid-state transformation

T is the instantaneous absolute temperature, and T_M is the absolute melting temperature of the material.

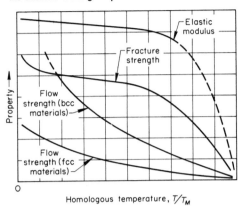

Homologous temperature, T/T_M

phenomenon is provided in the article "Ductile-to-Brittle Fracture Transition" in this Volume.

If the temperature is increased so that the flow strength becomes lower than the applied stress, a structure may deform spontaneously with no increase in load. A change in temperature may also cause an elastic-distortion failure because of a change in modulus, as might occur in a control device whose accuracy depends on a predictable elastic deflection of a control element or a sensing element. For most structural materials, the curve defining the temperature dependence of elastic and plastic properties is relatively flat at temperatures near 20 °C (70 °F).

In fcc materials, and in bcc materials at temperatures above the transition temperature, distortion (gross yielding) always accompanies overload fracture in a section that does not contain a severe stress raiser. In addition, gross yielding is one of the criteria that determine whether fracture in a section containing a stress raiser is a ductile overload (plane-stress) fracture. Localized distortion also accompanies brittle crack extension in ductile materials (plane-strain fracture). In inherently brittle materials, where the fracture stress is less than the flow stress, no gross or localized distortion accompanies fracture.

At temperatures higher than about one-half the absolute melting temperature, phenomena such as creep may cause distortion failure. Creep, or time-dependent strain, is a relatively long-term phenomenon and can be distinguished from overload distortion by relating the length of time at temperature to the amount of distortion, as discussed in the article "Elevated-Temperature Failures" in this Volume.

Changes in operating temperature can affect the properties of a structure in other ways. For example, if a martensitic steel is tempered at a given temperature and then encounters a higher temperature in service, yield strength and tensile strength will decrease because of overtempering. Long-time exposure to moderately elevated temperatures may cause overaging in a precipitation-hardening alloy, with a corresponding loss in strength. The volume change accompanying the transformation of retained austenite in a martensitic steel on exposure to cryogenic temperatures may cause a distortion failure (dimensional growth or warpage) in a close-tolerance assembly, such as a precision bearing. When the temperature is changed, different coefficients of thermal expansion for different materials in a heterogeneous structure can cause interference between structural members (or can produce permanent distortion because of thermally induced stresses if the members are joined together). The failure analyst must understand the effect of temperature on properties of the specific materials involved when analyzing failures that have occurred at temperatures substantially above or below the design or fabrication temperature.

Incorrect Specifications

Large errors in specification of material or method of processing for a part can lead to distortion failures. These errors are often the result of faulty or incomplete information being available to the designer. In such instances, the designer has to make assumptions concerning the conditions of service.

Example 2: Distortion Failure of an Automotive Valve Spring. The engine of an automobile lost power and compression and emitted an uneven exhaust sound after several thousand miles of operation. When the engine was dismantled, it was found that the outer spring on one of the exhaust valves was too short to function properly. The short steel spring and an outer spring taken from another cylinder in the same engine (both shown in Fig. 5) were examined in the laboratory to determine why one had distorted and the other had not.

Investigation. The failed outer spring (at left, Fig. 5) had decreased in length to about the same free length as that of its companion inner spring. Most of the distortion had occurred in the first active coil (at top, Fig. 5), and a surface residue of baked-on oil present on this end of the spring indicated that a temperature of 175 to 205 °C (350 to 400 °F) had been reached. Temperatures lower than 120 °C (250 °F) usually do not cause relaxation (or set) in high-carbon steel springs.

The load required to compress each outer spring to a length of 2.5 cm (1 in.) was measured. The distorted spring needed only 30 kg (67 lb), whereas the longer spring needed 41 kg (90 lb). The distorted spring had suffered

Fig. 5 Valve springs made from patented and drawn high-carbon steel wire

Distorted outer spring (left) exhibited about 25% set because of proeutectoid ferrite in the microstructure and high operating temperature. Outer spring (right) is satisfactory.

25% set, which was the immediate cause of the engine malfunction.

The microstructure of both springs was primarily heavily cold-drawn fine pearlite, but the microstructure of the distorted spring contained small amounts of proeutectoid ferrite. Although the composition of the spring alloy was unknown, the microstructure indicated that the material was patented and cold-drawn high-carbon steel wire. The distorted spring had a hardness of 43 HRC, and the longer spring had a hardness of 46 HRC. Both hardness and microstructure indicated that the material in the deformed spring had 10% lower yield strength than material in the undeformed spring. The estimates of yield strength were considered valid because of two factors: the accuracy of the hardness testing, and characteristically consistent ratios of yield strength to tensile strength for the grades of steel commonly used in spring wire.

Conclusions. The engine malfunctioned because one of the exhaust-valve springs had taken a 25% set in service. Relaxation in the spring material occurred because of the combined effect of improper microstructure (proeutectoid ferrite) plus a relatively high operating temperature. The undeformed spring exhibited little or no set because the tensile strength and corresponding yield strength of the material (estimated from hardness measurements) were about 10% higher than those of the material in the deformed spring.

Recommendations. A higher yield strength and a higher ratio of yield strength to tensile strength can be achieved in the springs by using quenched-and-tempered steel instead of patented and cold-drawn steel. An alternative would be to use a more expensive chromium-vanadium alloy steel instead of plain carbon steel; the chromium-vanadium steel should be quenched and tempered. Regardless of material or processing specifications, if springs are

Fig. 6 Comparison of longitudinal profiles of an 1138 steel shotgun barrel before and after testing

1000 rounds of a new type of ammunition were fired in the test. (a) Inside diameter. (b) Outside diameter

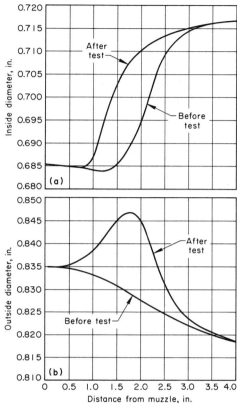

stressed close to the yield point of the material, close control of material and processing plus stringent inspection are needed to ensure satisfactory performance.

Service conditions are sometimes changed, invalidating certain assumptions that were made when the part was originally designed. Such changes include an increase in operating temperature to one at which the material no longer has the required strength, an increase in the load rating of an associated component, which the user may interpret as an increase in the allowable load on the structure as a whole, and an arbitrary increase in applied load by the user on the assumption that the component has a high enough safety factor to accommodate the added load.

Example 3: Bulging of a Shotgun Barrel Caused by a Change From Lead Shot to Iron Shot. A standard commercial shotgun barrel fabricated from 1138 steel deformed during a test that was made with a new type of ammunition. Use of the new ammunition, which contained soft iron shot with a hardness of about 72 HB, was intended to reduce toxicity; the old ammunition had

contained traditional lead shot with a hardness of 30 to 40 HB.

Investigation. The shotgun barrel was of uniform inside diameter from the breech to a point 7.5 cm (3 in.) from the muzzle; at this point, the inside diameter began to decrease ("Before test" curve, Fig. 6a). This taper, or integral choke, which is intended to concentrate the shot pattern, ended about 3.8 cm (1½ in.) from the muzzle, and the final portion of the barrel had a relatively uniform inside diameter.

After a test in which 1000 rounds of ammunition containing soft iron shot were fired, the shotgun barrel had a longitudinal profile of inside diameter as shown in the "After test" curve in Fig. 6(a). Comparison of this curve with the profile before the test shows that the effect of firing soft iron shot was to deform the gun barrel so that the choke taper was shifted toward the muzzle. After the test, there was a bulge on the outside surface of the barrel, shown in comparison to the longitudinal profile of the outside diameter before the test in Fig. 6(b). Deformation of the barrel had been detected after the first 100 rounds of iron-shot ammunition had been fired, and the bulge grew progressively larger as the test continued.

Apparently, the bore of the failed barrel was not concentric with the outside surface, because the wall thickness at a given distance from the breech varied widely among different points around the circumference. For example, at a distance of 5 mm (0.2 in.) from the muzzle, the wall thickness varied from 1.3 to 2 mm (0.051 to 0.080 in.).

The microstructure of the barrel material was a mixture of ferrite and coarse pearlite. The alloy had a hardness of 163 to 198 HB (converted from Vickers hardness measurements).

Based on previous tests, in which the hoop stress in shotgun barrels had been measured when lead-shot ammunition was fired, the safety factor had been estimated at 2.0. In this instance, it was concluded that wall thickness variations had reduced the safety factor to about 1.3 for lead-shot ammunition. Previous tests had also shown that lead shot was deformed extensively by impact with the bore in the choke zone of this type of gun barrel.

Analysis. The major stresses in the choke zone are produced by impact of shot pellets against the bore. When lead shot is used, the lead absorbs a considerable amount of the impact energy as it deforms. Soft iron shot, on the other hand, is much harder than lead and does not deform significantly. More of the impact energy is absorbed by the barrel when iron shot is used, producing higher stresses.

In this instance, had the gun barrel been of more uniform wall thickness around its circumference, it might not have deformed. However, it was believed that conversion to iron-shot ammunition would increase stresses in the barrel enough to warrant an increase in the strength of this type of barrel.

Conclusions. The shotgun barrel deformed because a change to iron-shot ammunition increased stresses in the choke zone of the barrel. Bulging was enhanced by a lack of uniformity in wall thickness.

Recommendations. Three alternative solutions to this problem were proposed, all involving changes in specifications:

- The barrel could be made of steel with a higher yield strength
- The barrel could be made with a greater and more uniform wall thickness
- An alternative nontoxic metal shot with a hardness of about 30 to 40 HB could be developed for use in the ammunition

Failure to Meet Specifications

Parts sometimes do not perform to expectations because the material or processing does not conform to requirements, leaving the part with insufficient strength. For instance, a part can be damaged by decarburization, as discussed below for a spiral power spring.

Figure 7 shows two spiral power springs that were designed to counterbalance a textile-machine beam. The spring at left in Fig. 7 was satisfactory and took a normal set when loaded to solid deflection in a presetting operation. The spring at right in Fig. 7, after having been intentionally overstressed in the same manner as the satisfactory spring, exhibited 15% less reaction force than was required at 180° angular deflection because it had taken a set that was 30° in excess of the normal set.

The material in the satisfactory spring had a hardness of 45 HRC, while the material in the spring that failed had a hardness of 41.5 HRC. This is a 10% disparity in tensile and yield strengths between the two springs. The spring that failed had a 0.08-mm (0.003-in.) thick surface layer of partial decarburization that further weakened the region of the cross section where maximum stresses are developed in the spring under load. The decarburized layer, which had a lower yield strength than the bulk material, yielded excessively during presetting; therefore the spring did not attain its specified shape in the free state following this operation.

Another material deficiency that can lead to deformation failure is variability in response to heat treatment among parts in a given production lot. Certain alloys, particularly hardenable low-alloy steels and some precipitation-hardening alloys, can vary in their response to a specified heat treatment because of slight compositional variations from lot to lot or within a given lot. This can result in some parts having too low a strength for the application even though they were properly heat treated according to specification.

Remedies for variability in response to heat treatment usually involve changes in the heat-treating process, ranging in complexity from tailoring the heat-treating conditions for each lot or sublot to making a small adjustment in the heat-treatment specification. Experiments on each lot or sublot are almost always needed to establish parameters when heat-treating conditions are tailored.

An adjustment in heat-treating conditions was successful in avoiding variation in properties among sublots of heat-treated AISI type 631 (17-7 PH) stainless steel Belleville washers. Two of these washers—one of which was from an acceptable sublot and the other from a deficient sublot—were subjected to examination. The washer from the acceptable sublot had developed the required hardness upon solution heat treating at 955 °C (1750 °F), followed by refrigeration at −75 °C (−100 °F) and aging. The other washer was soft after an identical heat treatment and yielded under load (flattened). The microstructure of the acceptable washer was a mixture of austenite and martensite; the structure of the washer that flattened consisted almost entirely of austenite.

Previous experience with 17-7 PH stainless steel indicated that some alloy segregation was not unusual and that relatively minor variations in composition could affect response to heat treatment, perhaps by depressing the range of martensite-transformation temperatures to a variable degree. As noted in Ref 5, the solution-treating temperature has a marked effect on the martensite start (M_s) temperature in the precipitation-hardening stainless steels that are austenitic as solution annealed and martensitic as aged (17-7 PH, AM-350, AM-355 and PH15-7Mo). Consequently, although it never was clearly established whether temperature variations inside the solution-treating furnace or minor variations in composition were responsible for the observed variability in properties of the 17-7 PH Belleville washers, all sublots attained the required strength when the solution-treating temperature was lowered to 870 °C (1600 °F).

Faulty Heat Treatment. Mistakes made in heat treating hardenable alloys are among the most common causes of premature failure. Temperatures that are either too high or too low can result in the development of inadequate or undesirable mechanical properties. Quenching a steel part too fast can crack it; quenching too slowly can fail to produce the required strength or toughness. If parts are shielded from a heating or cooling medium, they can respond poorly to heat treatment, as discussed below.

Two hold-down clamps, both from the same lot, are shown in Fig. 8. Both clamps were bowed to the same degree after fabrication, as intended, but the clamp at the bottom flattened when it was installed. A small percentage of the clamps, all of which were made from hardened-and-tempered 1070 steel, deformed when a bolt was inserted through the hole and tightened. The clamp at top in Fig. 8 was acceptable, with a microstructure of tempered martensite and a hardness of 46 HRC; the clamp at bottom, which deformed, had a mixed structure of ferrite, coarse pearlite and tempered martensite, and a hardness of only 28 HRC.

Analysis of the heat-treating process revealed that the parts were stacked so that occasional groupings were slack quenched as a result of shielding; this promoted the formation of softer upper transformation products. When the loading practice was changed to ensure more uniform quenching (so that transformation to 100% martensite was accomplished on all parts), the problem was solved.

Example 4: Bending of an Aircraft-Wing Slat Track. A curved member called a slat track (Fig. 9), which supported the extendable portion of the leading edge of the wing on a military aircraft, failed by bending at one end after very short service. It was estimated that the slat track, fabricated from heat-treated 4140 steel, had undergone only one high-load cycle.

Fig. 7 Two spiral power springs from a textile machine

Spring at left is an acceptable part, whereas spring at right took an excessive set (the inner end of the spiral is 30° out of position) because of insufficient yield strength and a decarburized surface layer.

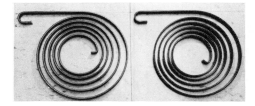

Fig. 8 Two hardened-and-tempered 1070 steel hold-down clamps

The clamp at top was acceptable. The clamp at bottom was slack quenched because of faulty loading practice (stacking), and it failed by distortion (flattening) because of the resultant mixed microstructure.

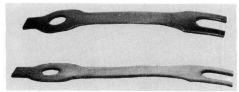

Fig. 9 4140 steel slat track from a military aircraft wing

The track bent because one end did not become fully austenitic during heat treatment, producing a low-strength structure of ferrite and tempered martensite.

Investigation. Hardness measurements were taken at various points along the length of the track. The end that bent (at right, Fig. 9) had a hardness of 30 HRC, compared with a hardness of 41 HRC for the remainder of the part.

Metallographic examination of specimens from both ends of the slat track revealed that the microstructure of the end that bent contained a large number of ferrite islands in a matrix of tempered martensite. The microstructure of the opposite end contained no ferrite.

Conclusions. Bending had occurred in a portion of the slat track because service stresses had exceeded the strength of the material in a region of mixed martensite and ferrite. It was determined that the most likely cause of the mixed structure was nonuniform austenitization during heat treatment. The end that bent never became fully austenitic, because the furnace temperature was locally too low or because the soaking time was too short or both. It was decided that material or design changes were not warranted, considering the nature of the failure and the probable cause of the mixed microstructure in one end of the slat track.

Corrective Measures. Steps were taken to improve control of temperature of parts during austenitization.

Warping during heat treatment or during stress-relief annealing is also a common type of distortion failure. Warping can result from nonuniform residual stress or from thermal or transformational stresses that are introduced during heating or cooling. When relief of residual stress causes distortion, the amount of distortion is proportional to the decrease of the residual stress. When distortion is caused by thermal or transformational stress, the extent of the distortion is greater for parts that have complex configuration or large differences in section thickness and for faster heating or cooling rates. Warping can also be caused by inadequate support in the heat-treating furnace, leading to sagging due to the weight of the part.

Most warping is the result of plastic deformation that occurred in some region of the part at elevated temperature or during a change in temperature. Dimensional changes accompanying stress relief are the result of readjustments involving both elastic and plastic strains. Distortion that occurs during other types of heat treatment involves mainly plastic strain and generally results in high levels of residual stress in the warped part. The magnitude and distribution of residual stresses are determined by the composition, shape, size, and heat-treating conditions of a given part.

Warping is most often severe in heat treatments that involve quenching. In hardenable steels, the principal cause of warping on quenching is nonuniform rates of transformation. The effect of transformational stresses may be intensified if a nonuniform composition exists; this nonuniformity may be the result of segregation, or it may be the result of processing, as in a carburized part. Such inhomogenei-

ties may produce a variation in transformation temperature at locations that are geometrically equivalent and that cool at the same rate. Nonuniform transformational stresses that result from inhomogeneity also can occur during tempering.

Warping can often be minimized by modifying the heat-treating conditions. For example, slow heating and cooling rates are less likely to cause warping because local variances in temperature and in rates of temperature change are minimized. Preheating before austenitizing is often used as a means of minimizing warping of some tool steels and heavy sections because preheating reduces the temperature gradient between the surface and the interior of the part. Induction hardening and nitriding have been used to minimize warping when surface hardness is of primary importance to the performance of a part.

In heat treatments requiring rapid cooling or quenching, excessive warping can usually be reduced by changing the quenching conditions. In many instances, the orientation of a part as it enters the quenchant will influence the amount of distortion that occurs. Quenching in special fixtures (quench presses) is widely used in certain industries to minimize distortion by providing different, controlled cooling rates at different locations in a given part. Martempering has also been used to minimize distortion because in this process transformation rates upon cooling to the M_s temperature are equalized throughout the part.

Faulty Case Hardening. Carburizing, which both increases the surface hardness of a part and provides resistance to wear and indentation, can, if improperly controlled, produce a case that has too low or too high a carbon content. With too low a carbon content, the surface may not be hard enough to withstand normal service loads. This condition may be accompanied by shallow case depth, which aggravates the problem. With an excessively high carbon content, which generally is the result of a high carbon potential or improper diffusion during the carburizing cycle, large amounts of retained austenite may be present in the carburized zone after heat treatment, depending on the composition of the steel. Retained austenite softens the surface and, under certain conditions, may transform to martensite in service. When transformation in service occurs, the resulting untempered martensite may crack and thus promote early failure by surface fatigue, or a distortion failure in a close-fitting assembly may occur because of the volume change that accompanies transformation.

Example 5: Seizing of a Spool-Type Hydraulic Valve. Occasional failures were experienced in spool-type valves used in a hydraulic system. When a valve would fail, the close-fitting rotary valve would seize, causing loss of flow control of the hydraulic oil. The rotating spool in the valve was made of 8620 steel and was gas carburized. The cylinder in

which the spool fitted was made of 1117 steel, also gas carburized.

Investigation. Low-magnification visual examination of the spool and cylinder from a failed valve revealed some burnishing, apparently the result of contact between the spool and the inside cylinder wall. Measurement of the surface profile of the spool showed no evidence of wear or galling. When this profile was compared with that of a spool from a valve that operated satisfactorily, no significant difference was found.

Metallographic sections were made of the cylinder from the failed valve and from the one that operated satisfactorily. The microstructure of the carburized case on the cylinder from the satisfactory valve was well-defined martensite interspersed with some austenite (white-etching areas, Fig. 10a). On the other hand, the microstructure of the case from the failed valve contained a much greater amount of retained austenite, especially near the surface (Fig. 10b). In addition, some patches of untempered martensite were found close to the surface upon overetching the section of the failed valve cylinder.

Microhardness traverses of the two sections revealed that, in general, the hardness of the case from the failed part was about 100 Knoop points lower than the case hardness of the unfailed part. At the surface, the case hardness of the failed part was 300 Knoop points lower.

Conclusions. Momentary sliding contact between the spool and the cylinder wall (probably during valve opening) caused unstable retained austenite in the failed cylinder to transform to martensite. The increase in volume resulted in sufficient size distortion (growth) to cause interference between the cylinder and the spool, seizing, and loss of flow control. The failed parts had been carburized in a process in which the carbon potential was too high, which resulted in a microstructure having excessive retained austenite after heat treatment.

Corrective Measure. The composition of the carburizing atmosphere was modified to yield carburized parts that did not retain significant amounts of austenite when they were heat treated. After this change was made, occasional valve failure by seizing ceased.

Faulty Repairs. Products are often repaired to correct deficiencies that are found in new parts during quality-control inspections or in used parts after they have deteriorated in service. Repair welding and brazing are generally recognized as potential sources of unwanted alterations of the properties of heat-treatable alloys. Parts can be made softer or more brittle by careless repair, depending on the alloy and the conditions under which the repair was made.

Substitution of a part, particularly a fastener, whose properties do not match the properties of the part it replaces can lead to failure of the substitute part, failure of another part, or both, as in the following instance.

Fig. 10 Cross sections through the carburized 1117 steel cylinders from two spool-type hydraulic valves

The cylinder of the valve that operated satisfactorily (a) had little retained austenite in the case, whereas the cylinder of the seized valve (b) had much retained austenite that transformed to martensite in service, resulting in size distortion (growth). 500×

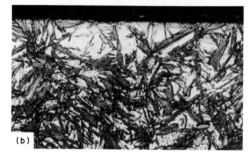

A carrying handle of complex configuration was secured to a heavy portable device by means of a ring clamp at one end of the handle and a special ¼-20 hardened bolt through a small flange at the other end. For some unknown reason, the special slotted hexagonal-head bolt was replaced by a standard commercial hexagonal-head cap screw whose hardness was 93 HRB instead of the specified hardness of 28 HRC.

The commercial cap screw was distorted in service (Fig. 11a), causing the handle to become loose at one end. This in turn caused eccentric and excessive loading on the small flange of the handle with the result that the flange bent and then broke. Figure 11(b) shows the broken flange; the bright area next to the hole is where the loose flange chafed against the distorted bolt and is an indication of the eccentric load distribution on the flange. A fastener of the correct hardness and length is shown in Fig. 11(c) for comparison with the distorted commercial cap screw.

Analyzing Distortion Failures

Distortion failures are often considered to be relatively simple phenomena that are easy to analyze because deformation can occur only when the applied stress exceeds the flow

Fig. 11 Part substitution that resulted in a distortion failure

(a) Distorted commercial cap screw that was used as a replacement for a hardened bolt. (b) Carry-handle flange that broke because the cap screw bent. (c) Correct replacement part

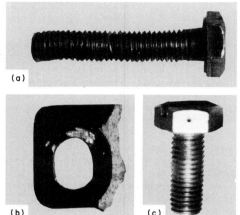

strength of the material. On the contrary, distortion is not always the result of simple overload or use of an improperly processed part. Analysis of a distortion failure often must be exceptionally thorough and rigorous to determine the cause of failure and, more importantly, to specify proper corrective action. The analyst must consider factors that may not have been anticipated in design of the part, such as material substitutions or process changes during manufacture and the misuse, abuse, or occurrence of complex stress fields in service.

A seemingly innocent substitution of material resulted in distortion of small volute springs made of cold-worked spring-temper Inconel. Normally, the material was purchased as cold-flattened wire, but one lot of springs was formed from cold-rolled and slit strip because flattened wire could not be obtained in time to meet the delivery schedule. After a presetting operation, the strip springs were consistently out of tolerance; they had taken an excessive set. The Inconel strip had a hardness of 360 HV, compared to 390 HV for the wire that normally was used. This represents a difference of about 10% in strength and accounts for the observed distortion. Had this problem gone undetected during manufacture, it might well have resulted in distortion failures in service.

Analytical Procedure. The article "General Practice in Failure Analysis" in this Volume gives a general procedure that can be followed for any failure analysis. That article also gives suggestions concerning methods of analysis and precautions that will increase the validity of the analysis. The following ten steps are adapted from the general procedure and are suggested specifically for analysis of a distortion failure:

- Define the effect of the failure on the structure or assembly, and define the desired results of corrective action
- Obtain all available design and service information
- Examine the distorted part, making a record of observations, including a sketch or a photograph of the distorted part along with an undistorted part for comparison. Also, note all pertinent measurements of dimensions. It is usually helpful to enter these measurements, which should be made with at least the same precision as in a quality-control inspection, alongside the design dimensions on a blueprint of the part
- Perform laboratory tests as necessary to confirm the composition, structure and other chemical or metallurgical characteristics of the distorted part
- Trace the failed part through all manufacturing processes to discover whether process deviations occurred during production
- Compare the actual conditions of service with design assumptions
- Compare the actual material properties with design specifications
- Determine whether any differences found in the preceding two steps fully account for the distortion observed in the failed structure. If the differences do not fully account for the observed distortion, the information obtained in the second or fourth step is incorrect or incomplete
- Prepare alternative courses of action to correct the variant factors that caused the observed distortion, and select the course that seems most likely to produce the desired result, which was defined in the first step
- Test the selected course of corrective action to verify its effectiveness. Evaluate side effects of the corrective action, such as its effects on cost or on ease of implementation

Example 6: Deformation of a Gas-Nitrided Drive-Gear Assembly. Slipping of components in the left-side final drive train of a tracked military vehicle was detected after the vehicle had been driven 13 700 km (8500 miles) in combined highway and rough-terrain service. No abnormal service conditions were reported in the history of the vehicle. The slipping was traced to the mating surfaces of the final drive gear and the adjacent splined coupling sleeve (Fig. 12). Failure analysis was conducted to determine the cause of the malfunction and to recommend corrective measures that would prevent similar failures in other vehicles.

Material and Fabrication. Specifications required that the gear and coupling be made from 4140 steel bar oil-quenched and tempered to a hardness of 265 to 290 HB (equivalent to 27 to 31 HRC), and that the finish-machined parts be single-stage gas nitrided to produce a total case depth of 0.5 mm (0.020 in.) and a minimum surface hardness equivalent to 58 HRC.

Fig. 12 Gas-nitrided 4140 steel drive-gear assembly in which gear teeth deformed because of faulty design and low core hardness

Details A and B show deformed areas on drive-gear teeth and mating internal splines. Dimensions given in inches

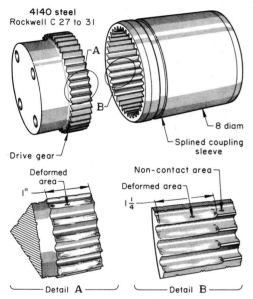

Visual Examination. Low-magnification examination of the drive gear and coupling showed that the teeth of the gear and the mating contact surfaces of the internal splines on the coupling were almost completely worn away. The spline surfaces had been damaged mainly by severe indentation and plastic deformation. No spalling, cracking, or other indications of fatigue damage were visible on the splines, nor was there any indication that abrasive wear had been the mechanism of failure.

Metal on the drive side of the gear teeth had been plastically deformed and subsequently removed (Details A and B, Fig. 12). The damaged areas on the splines were wider (axial dimension) than the gear teeth, indicating that there had been excessive lateral play between the components.

Examination of the surfaces of the internal splines that contacted a gear on the opposite end of the coupling (not illustrated) revealed a much smaller amount of deterioration than on the failed end. For example, the machined flats on the crests of the splines showed no damage, the follower sides of the splines had very shallow, almost imperceptible wear areas that were only about 0.1 mm (0.004 in.) wide, and the drive side of each spline had a gently radiused, concave area of damage approximately 6.4 × 25 mm (¼ × 1 in.) and 0.13 mm (0.005 in.) deep at its center. However, the amount of damage on the drive side of these splines was considered to be excessive for only 13 700 km (8500 miles) of service.

Conformance to Material Specifications. Spectrographic analysis of the two components confirmed that composition was within the range specified for 4140 steel. Core hardness was in the required range of 27 to 31 HRC (264 to 294 HB). Total case depth, as determined by the microscopic method on polished specimens etched in 2% nital, was satisfactory. Surface hardness, as measured using a Knoop indenter, was equivalent to 50 HRC, substantially lower than the required value of 58 HRC.

Microstructure of the cases and cores of the two components was examined at 500× on polished sections etched in 2% nital. There was a white layer (nitrogen-rich iron nitride, Fe_2N) about 0.025 mm (0.001 in.) thick on the surfaces of the gear teeth and the splines, and grain-boundary networks of iron nitride were present to a slight degree near the surface. The microstructure of the core consisted mainly of tempered martensite, but contained large amounts of blocky ferrite.

Gear and Spline Configuration. Measurements of the gear teeth and splines in areas showing little damage established that the parts had been manufactured in conformance with the engineering drawings. The splines were straight axially and convex radially. However, the gear teeth that failed were convex in both directions in the vicinity of the pitch line. This design, which was intended to facilitate alignment and adjustment, provided extremely small contact areas between tooth and spline surfaces and hence very heavy localized loading. The gear that engaged the other end of the internally splined coupling (not shown in Fig. 12) was designed to provide larger contact areas with the splines in the vicinity of the pitch line, providing lighter and more uniform local loading.

Conclusions. The premature failure occurred by crushing, or cracking, of the case as a result of several factors, which are listed below in order of importance:

- Design that produced excessively high localized stresses at the pitch line of the mating components
- Specification of a core hardness (27 to 31 HRC) too low to provide adequate support for a 0.5-mm (0.020-in.) thick case or to permit attainment of the specified surface hardness of 58 HRC after nitriding; actual surface hardness was 50 HRC
- The presence of large amounts of blocky ferrite in the core (a microstructure conducive to case crushing) as a result of faulty heat treatment before nitriding
- The presence of a white nitride layer about 0.025 mm (0.001 in.) thick at the surface and of nitride networks in the case

Recommendations. Measures to correct the first two deficiencies listed above were recommended as the most important steps in obtaining adequate drive-train performance. First, the excessively high local stresses at the pitch line

should be reduced to an acceptable level by modifying the gear-tooth contour in the vicinity of the pitch line to provide a wider and longer initial contact area than in the original design. Second, a core hardness of 35 to 40 HRC should be specified to provide adequate support for the case and to permit attainment of the specified surface hardness of 58 HRC. Closer control of heat treating, which would be necessary to produce the recommended higher core hardness consistently, would also eliminate the presence of blocky ferrite in the core.

For maximum service life, consideration should also be given to controlling single-stage gas nitriding to minimize the thickness of the white layer and the extent of nitride networks in the case or to use double-stage gas nitriding to provide a diffused nitride layer and specifying final lapping or honing to remove the white layer.

Special Types of Distortion Failure

Analysis of distortion failures can be particularly difficult when there is no apparent permanent deformation of the part or when complex stress fields are involved. In this section, three types of distortion failure are discussed, which may provide useful insights into the problems of analyzing unusual mechanisms of distortion.

Elastic Distortion. A distortion failure does not necessarily involve yielding under a single application of load. Most parts deflect elastically under load. For example, if a part ordinarily made of a high-modulus alloy is made of a low-modulus alloy, it will deflect more under a given load than if it were made of the high-modulus alloy. If this greater amount of deflection places the part in the path of another part in an assembly, it could be said to have failed by elastic distortion. As mentioned earlier, a change in the modulus of a material because of a change in temperature can cause an elastic-distortion failure. Elastic buckling of a long, slender column is another type of distortion failure in which the yield strength of the material is not exceeded (unless the structure collapses).

Ratcheting. Cyclic strain accumulation, or ratcheting, requires that a part be stressed by steady-state loading, either uniaxial or multiaxial, and that a cyclically varying strain in a direction other than the direction of principal stress be superimposed on the part. In ratcheting, an oscillating load or a cyclic variation of temperature strains the material beyond the yield point on alternate sides of a single member, or on alternate members of a structure, during each half-cycle. With succeeding cycles, plastic strain accumulates, with the result that one or more of the overall dimensions of the member or the structure change relatively uniformly along the direction of steady-state stress. Deformation produced by a cyclic variation in load is known as isothermal ratcheting (even though a temperature change may occur simultaneously with the load variation). Progressive growth due to plastic strain incurred during a change in temperature is called thermal ratcheting. Ratcheting may ultimately result in ductile fracture or in failure by low-cycle fatigue.

As an illustration of isothermal ratcheting, consider a hollow cylinder that is stressed elastically in tension along its longitudinal axis. If a cyclic torsional load of sufficient magnitude to cause plastic straining is superimposed on the longitudinal load, the cylinder can increase in length by as much as 20% before local instabilities disrupt uniform strain accumulation (Ref 6, 7).

At elevated temperatures, ratcheting must be distinguished from creep or stress relaxation. Ratcheting is solely a strain-dependent phenomenon, whereas creep and stress relaxation are time-dependent phenomena. Exposure to elevated temperature for an extended length of time is necessary for creep or stress relaxation to occur, but extensive deformation by ratcheting can occur in short periods of time—sometimes only minutes. Ratcheting can appear to be time dependent when the cyclic strains are imposed at regular intervals. However, the factor that distinguishes ratcheting is the occurrence of plastic strain during both halves of the cyclic variation.

In general, the proper corrective action for failures by ratcheting involves changing the design of the part or the conditions of service to reduce the magnitude of the service stresses or specifying a material with a higher yield strength for the application.

Inelastic Cyclic Buckling. Some materials exhibit cyclic strain softening—a continuous decrease of elastic limit or tangent modulus that occurs with imposition of alternating stresses whose magnitude lies between the proportional limit and the yield strength. Columns made from materials that exhibit this behavior can fail by lateral displacement at the midspan (buckling) under stresses much lower than those predicted by classical design.

Table 1 presents the results of a test in which cylindrical specimens of cold-worked 1020 steel, resembling tensile specimens with threaded ends, were stressed by alternating tensile and compressive loads of equal magnitude (Ref 8). Buckling occurred at 60 to 90% of the number of cycles to failure at stresses below the 0.2% offset yield strength but above the proportional limit of the material, corresponding to the inelastic portion of the stress-strain curve. When an aluminum alloy was tested in the same manner, buckling did not occur in the range of stresses between the proportional limit and the 0.2% offset yield strength. Aluminum alloys are among those that do not exhibit cyclic strain softening.

Table 1   Inelastic cycle buckling of cylindrical specimens of cold-worked 1020 steel(a)

| Peak stress | | Stress cycles | | Failure |
MPa	ksi	At buckling	At failure	mode
600–648	87–94	¼	¼	(b)
552	80	14	20	(b)
490	71	130	143	(b)
441	64	380	691	(c)
421	61	1900	2377	(c)
400	58	...	4730	(d)

(a) Tensile strength of the steel: 690 MPa (100 ksi); yield strength, 620 MPa (90 ksi); proportional limit, 345 MPa (50 ksi). (b) Buckling, with or without fracture. (c) Buckling followed by buckling-induced fracture. (d) Low-cycle fatigue fracture. Source: Ref 8

SELECTED REFERENCES

1. B.S. Lement, *Distortion in Tool Steels*, American Society for Metals, 1959
2. J.W. Jones, Limit Analysis, *Mach. Des.*, Vol 45 (No. 23), Sept 20, 1973, p 146-151
3. D. Goldner, Plastic Bending in Tubular Beams, *Mach. Des.*, Vol 45 (No. 24), Oct 4, 1973, p 152-155
4. D.J. Wulpi, *Understanding How Components Fail*, American Society for Metals, 1985, p 13-19
5. A.J. Lena, Precipitation Reactions in Iron-Base Alloys, in *Precipitation From Solid Solution*, American Society for Metals, 1959 p 224-327
6. D. Burgreen, Review of Thermal Ratcheting, in *Fatigue at Elevated Temperatures*, STP 520, ASTM, Philadelphia, 1973, p 535-551
7. K.D. Shimmin, Cyclic Strain Accumulation Under Complex Multiaxial Loading, RTD-TDR-63-4120, Dec 1963
8. C.R. Preschmann and R.I. Stephens, Inelastic Cyclic Buckling, *Exp. Mech.*, Vol 12 (No. 9), Sept 1972, p 426-428

Wear Failures

Revised by R.C. Tucker, Jr., Union Carbide Corporation

WEAR, friction, and lubrication are complex, interwoven subjects that may all affect the propensity of a component to failure. While all three are important factors, the major emphasis in this article will be on wear. In general, wear may be defined as damage to a solid surface caused by the removal or displacement of material by the mechanical action of a contacting solid, liquid, or gas. Gradual deterioration is often implied, and the effects are for the most part surface-related phenomena; but these restrictions should not be rigorously applied in failure analysis. Neither should the assumption that wear is entirely mechanical be accepted, because chemical corrosion may combine with other wear factors.

Friction is the resistance to motion when two bodies in contact are forced to move relative to each other. It is closely associated with any wear mechanisms that may be operating and with any lubricant and/or surface films that may be present, as well as the surface topographies. The heat generated as a result of the dissipation of frictional interaction may affect the performance of lubricants, may change the properties of the contacting materials and/or their surface films, and, in some cases, may change the properties of the product being processed. Any of these results of frictional heating can cause severe safety problems because of the danger of mechanical failure of components due to structural weakening, severe wear (for example, seizure), or fire and explosion. In addition, high friction results in higher energy consumption in most moving machinery.

The preceding suggests that low values of friction are the most desirable, but it should be kept in mind that intermediate or high values are often desirable in, for example, braking or wrapping applications or in the control of tension on fibers, belts, and so on. Thus, accurate control of friction is required, not just a universal lowering of it.

Lubrication implies the intentional use of a substance that reduces friction between contacting surfaces and is usually a mitigating factor in wear. Failure of lubricants or lubrication systems may lead to a change in wear mechanisms and to sudden failure of a component. Because of their beneficial effects, lubricants are almost always used whenever possible, and nonlubri-cated wear is generally associated only with those types of applications in which use of a lubricant is either impossible or economically infeasible.

Except in the case of severe galling leading to seizure, wear is normally a relatively gradual process. Therefore, problems associated with wear differ from those associated with outright breakage of a component, and defining failure is not always obvious; the performance of equipment may slowly degrade due to wear rather than cease suddenly. Therefore, selecting the time to replace worn components (or selecting the best materials initially to combat wear) involves consideration of cost in light of the cost of improvement in process productivity or product quality and the cost of downtime.

For example, all mechanical components that undergo sliding or rolling contact are subject to some degree of wear. Typical of such components are bearings, gears, seals, guides, piston rings, splines, brakes, and clutches. Wear of these components may range from mild polishing-type attrition to rapid and severe removal of material with accompanying surface roughening. Whether or not wear constitutes failure of a component depends upon whether the wear deleteriously affects the ability of the component to function. Even mild polishing-type wear of a close-fitting spool in a hydraulic valve may cause excessive leakage and thus constitute failure, even though the surface of the spool is smooth and apparently undamaged. On the other hand, a hammer in a rock crusher can continue to function satisfactorily in spite of severe denting, gouging, and the removal of as much as several inches of surface metal.

Types of Wear

Wear has been categorized in various ways. The phenomenological approach is based on a macroscopic description of the appearance of worn surfaces, for example, scuffing, rubbing, and fretting. The utility of such a system is limited because it does not focus on mechanisms of wear and therefore must rely almost entirely on empirical solutions to wear problems. Another possibility is to categorize wear on the basis of the fundamental mechanism that is operating. Unfortunately, this approach is complicated by the fact that more than one mechanism may be operating at a time and by the lack of sufficient information. Both approaches have been described and summarized in several publications (Ref 1-6). A somewhat different approach involves describing sliding wear mechanisms on the basis of the shape and size of the wear debris particles generated (Ref 7).

While the terminology of wear is unsettled and basic definitions have not been standardized, it is now fairly widely accepted that there are three primary types of wear: adhesive wear, abrasive wear, and erosive wear. These categories will be briefly described (additional information is provided in the References in this article). The delamination theory of wear incorporates elements of adhesion and abrasion (Ref 8). In addition, there are other types of wear that, although not regarded as primary, are sometimes afforded separate status. These include surface fatigue, fretting, and cavitation erosion. Frequently, more than one mechanism is operating simultaneously, and it is difficult to separate the effects of one from the other.

Adhesive Wear

Adhesive wear has been identified, with varying degrees of accuracy, by the terms scoring, galling, seizing, and scuffing. It was defined by the Organization for Economic Cooperation and Development (OECD) as wear by transference of material from one surface to another during relative motion due to a process of solid-phase welding; particles that are removed from one surface are either permanently or temporarily attached to the other surface (Ref 9). The phenomenon is usually described as a sequence of events similar to the following.

Theory. Solid surfaces are almost never perfectly smooth, but rather consist of micro- or macroscopic asperities of various shapes. When two such surfaces are brought into contact under a load normal to the general planes of the surfaces, the asperities come into contact and elastically or plastically deform until the real area of contact is sufficient to carry the load. A bond may then occur between the two surfaces that is stronger than the intrinsic strength of the weaker of the two materials in contact. When relative motion between the two surfaces occurs, the weaker of the two materials fails, and material is transferred to the contacting surface.

In subsequent interactions, this transferred material may be retransferred to the original surface (probably at a different location) or may become totally separated as a wear debris particle of an irregular morphology (Ref 7). Formulas that have been proposed (Ref 10, 11) to describe this phenomenon are of the form:

$$V_{ad} = \frac{kSL}{3H} \quad \text{or} \quad V = \frac{kSL}{H} \qquad \text{(Eq 1)}$$

where V is the wear scar volume, S is the distance of sliding, L is the load, H is the indentation yield strength (hardness) of the softer surface, and k is a probability factor that a given area contact will fracture within the weaker material rather than at the original interface.

Formulations similar to Eq 1 have been shown to describe adhesive wear over fairly wide ranges of sliding distances, under a variety of conditions, over limited ranges of load, and over limited ranges of hardness when the same classes of material were compared (Ref 12, 13). While initial theoretical considerations assumed bare metal-to-metal contact, later work assumed that oxide films, adsorbed films, and/or lubricant effects could be accounted for by changing k or by using more complex formulations (Ref 14). It has also been proposed that true metal-to-metal adhesive wear occurs at some time after motion is initiated when surface films or contaminants are worn away. Presumably, therefore, more than one adhesive wear mechanism could be operating at any given time, depending upon the presence or absence of various surface films in local areas. Changes in the apparent value of k or k/H as a function of load may be the result of penetration of such films at sufficiently high load or the generation of new films as a result of frictional heating.

The wear coefficient, k, has been determined experimentally for a large number of materials couples under various test conditions and geometries. The values found range from about 10^{-3} to 10^{-8} (Ref 7). For example, representative values of k for the end of a cylinder sliding against the flat surface of a ring at 1.8 m/s (6 ft/s) under a 400-g load are given for various combinations of cylinder and ring materials in Table 1. In many laboratory experiments, a stationary specimen with a small surface area rubs against a moving specimen with a large area. This frequently leads to a much higher wear rate on the smaller specimen than on the larger because of the constant contact and associated heating of the smaller specimen. This relative area effect may influence the wear mechanisms operating and may not be representative of field use. For most practical applications, volume loss, as predicted by Eq 1, must be converted to a linear value representing penetration or decrease in length, for example, increase in diameter of a journal bearing bushing, reduction in shaft

diameter, or reduction in the length of brush in an electric motor.

Abrasive Wear

Abrasive wear, or abrasion, is caused by the displacement of material from a solid surface due to hard particles or protuberances sliding along the surface. The OECD divided abrasion into scouring abrasion and abrasive erosion (Ref 15). The latter, being ascribed to solid particles entrained in a fluid moving nearly parallel to a solid surface, will be discussed in the section "Erosion" in this article. Scouring abrasion, or simply abrasion, may be due to loose particles entrapped between two solid surfaces in relative motion or to particles fixed to one of the surfaces as a result of embedment or adherence. It may also be due to relatively hard protuberances or asperities on one of the surfaces abrading the other. The source of particles may be either foreign or wear debris (from either adhesive or abrasive wear).

In one system of common terminology, abrasive wear is classified as gouging abrasion, high-stress (or grinding) abrasion, or low-stress (or scratching) abrasion. In gouging abrasion, large particles are removed from the surface, leaving deep grooves and/or pits. In this system, machine shop grinding can may be classified as gouging abrasion. High-stress, or grinding, abrasion is accompanied by the fracture of the abrasive particle (the source of the term high stress). The worn surface may exhibit varying degrees of scratching with plastic flow of sufficiently ductile phases or fracture of brittle phases. Debris may be formed after repeated plastic flow by a fatiguelike mechanism or by chipping. Low-stress, or scratching, abrasion occurs when the load is low enough that the abrasive particles are not fractured. In machine shop terminology, this may be classified as polishing. A worn surface usually exhibits fine scratches.

Another system of classification divides abrasion into two-body or three-body abrasion. As implied above, in two-body abrasion, the abrading protuberance is fixed to one of two surfaces in relative motion; in three-body abrasion, a loose abrasive particle is trapped between two surfaces in relative motion. The rate

of wear may be much higher in two-body than in three-body abrasion because in three-body abrasion the particle may roll a high percentage of the time, causing little wear (Ref 16). In machine shop terminology, grinding wheels and sand paper act predominantly as two-body systems, while free-abrasive grinding and lapping are predominantly three-body systems. Two- or three-body abrasion may operate in either the high- or low-stress regimes, while two-body may be in the gouging regime as well. If a loose particle becomes embedded on one surface, the system becomes two-body instead of three-body. In many, if not most, situations, more than one mechanism of wear is operating; therefore, more than one category or classification applies.

Theory. There have been many investigations of abrasive wear. Some of the earlier studies can be found in Ref 17 to 23. Generally, an abrasive particle or protuberance moving across a surface under load indents the surface, then either creates a groove as a result of plowing (plastic flow to the sides of the groove) without direct material removal or creates a groove with material loss as a result of chip formation ahead of the moving particle or the fracture of the material plowed to the side. Material plowed to the side may fracture on the first pass of an abrasive particle or only after repeated deformation by more than one particle.

A common expression for the volume rate of abrasive wear per unit length of sliding, V_{ab}, is:

$$V_{ab} = \frac{dQ}{dl} \qquad \text{(Eq 2)}$$

where Q is the volume swept out by abrasive, and l is the sliding distance. Because the volume Q is a function of the depth of indentation of the particle:

$$V_{ab} \propto \frac{W}{H} \qquad \text{(Eq 3)}$$

where W is the load, and H is the hardness of the surface being abraded. An alternative expression has been described (Ref 24) for V, the volume of wear per unit area per unit sliding distance:

Table 1   Wear coefficients for various combinations of materials under conditions of dry sliding(a)

Sliding combination		Wear coefficient, k	Hardness of softer member, 10^6 g/cm^2
Cylinder material	Ring material		
Low-carbon steel	Low-carbon steel	7.0×10^{-3}	18.6
60-40 brass	Hardened steel	6.0×10^{-4}	9.5
PTFE	Hardened steel	2.5×10^{-5}	0.5
Bakelite	Hardened steel	7.5×10^{-6}	2.5
Beryllium copper	Hardened steel	3.7×10^{-5}	21.0
Tool steel	Hardened steel	1.3×10^{-4}	85.0
Stellite	Hardened steel	5.5×10^{-5}	69.0
Tungsten carbide	Low-carbon steel	4.0×10^{-6}	18.6
Tungsten carbide	Tungsten carbide	1.0×10^{-6}	130.0

(a) Wear coefficients given are for the end of a cylinder sliding against the flat surface of a ring at 1.8 m/s (6 ft/s) under a 400-g load.

$$V_{ab} = \frac{K_1 K_2 K_3 \sigma}{H} \qquad \text{(Eq 4)}$$

where K_1 is the probability of wear debris formation, K_2 is the mean proportion of groove volume removed when debris formation occurs, K_3 is a constant that depends on the shape of the abrading particle or protuberance, σ is the applied load per unit area, and H is the hardness of the surface being abraded. Equations 2 to 4 assume that the abrasive particle is harder than the surface and is either rigid or has a shape after penetrating that is accounted for by K_3.

Extensive strain hardening around the grooves occurs during the abrasive wear of most metals; therefore, prior work hardening has little effect on improving wear resistance. This was demonstrated for austenitic manganese steels (Ref 25). Surface hardening should not be confused with the bulk hardness of the material used in Eq 4. A recent review of the effects of surface hardening can be found in Ref 24. The effects of high strain, high strain rates, and localized high transient temperature increases, all associated with abrasive wear, are complex, not completely understood, and beyond the scope of this article, but are undoubtedly crucial to a satisfactory theoretical explanation of abrasive wear.

Equation 4 and the considerations above apply primarily to ductile materials where K_2 probably has a value between 0 and 1 because some of the material removed from the groove is simply plastically pushed to the side without forming wear debris directly. For inherently brittle materials or heavily deformed materials that have reached their strain-hardening limit, debris may form directly with chips greater in size than the nominal volume of the groove, and K_2 may thus be greater than 1 (Ref 16-19, 26).

For brittle materials, indentation or scratching may result in fracture if the critical radii of the particle at which the transition from elastic-plastic to purely elastic behavior is exceeded and the Hertzian fracture stress is exceeded (Ref 27). The critical size may be significantly lower for scratching than for simple indentation because of the tensile stresses associated with sliding. Critical size decreases as the hardness of the surface increases and as fracture toughness decreases. A relationship has been developed (Ref 28) for the upper limit for the volume of wear per unit sliding area per unit sliding distance V_{ab-b}:

$$V_{ab-b} = K_5^{-1/4} \, \sigma^{5/4} \, d^{1/2} \, K_c^{-3/4} \, H^{-1/2} \qquad \text{(Eq 5)}$$

where K_5 depends upon particle shape and distribution, d is the mean particle diameter, and K_c is fracture toughness.

Reference 24 compares the difference in performance between ductile and brittle materials by using data from previous work (Ref 29, 30) involving the abrasion of various materials with 250- and 84-μm SiC bonded abrasives (Fig. 1). The sharp increase in wear for brittle

Fig. 1 Wear on 250-μm SiC versus wear on 84-μm SiC for ductile and brittle materials
Applied load was 1 MPa (6.9 ksi).

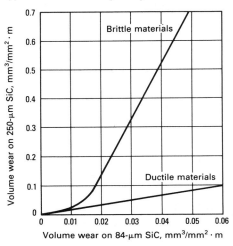

Fig. 2 Wear of various materials on 84-μm SiC bonded abrasive
Experimental data are shown by solid lines; theoretical models, broken lines.

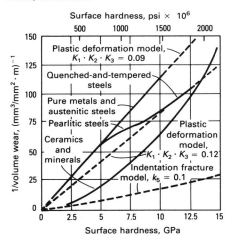

materials is associated with the transition from subcritical to supercritical indentation using 250-μm grit.

Although Eq 5 might be expected to provide some guidance in defining the bounds of wear behavior for ductile and brittle materials, the variety of conditions found in actual service confounds this expectation to some degree. Performance of various materials when worn by 84- and 250-μm SiC and 84-μm flint-bonded abrasives is shown in Fig. 2 to 4 as compared to the predictions of Eq 4 and 5. The value of the product $K_1 \cdot K_2 \cdot K_3$ was determined by a least squares fit, and the value of K_5 was assumed (Ref 24, 29-31).

Equations 2 to 5 are not necessarily valid for high-stress abrasion.

Fig. 3 Wear of various materials on 250-μm SiC bonded abrasive
Experimental data are indicated by solid lines; theoretical models, broken lines.

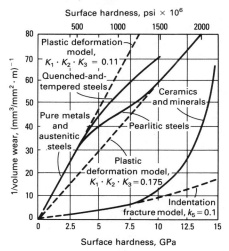

Fig. 4 Wear of various materials on 84-μm flint-bonded abrasive
Experimental data are indicated by solid lines; theoretical models, broken lines.

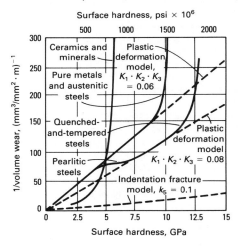

for the wear rate in high-stress abrasion is not available. High-stress abrasion is common in ball and rod milling. In ball milling, the abrasive is caught between adjacent balls or between ball and liner, and the ball weight crushes the hard particles (Ref 27). Although impact may occur in ball mills, it is not the important crushing force. Grinding can be produced without impact. In fact, the most efficient ball mills are those that simply roll and tumble the balls. Because the abrasive may have little opportunity to cut before it is crushed, the abraded surfaces are subjected primarily to repetitive concentrated compressive stress over a period of time. The abraded surface, therefore, is subjected to an attack similar to that resulting from sand blasting with the blast directed perpendicular to the surface. This is substantiated by the fact that there seems to be a correlation between results of sand-blasting wear tests and results of ball-mill wear tests.

Delamination Theory of Wear

A theory has been developed to describe the wear of surfaces in sliding contact based on the delamination of the surface (Ref 8, 32). The theory is strongly tied to frictional behavior and postulates a sequence of events leading to the loss of material in the form of thin sheets. Initially, contact between two surfaces occurs at asperities that are presumed to deform and/or fracture easily, yielding a relatively smooth surface. As a result of surface traction, plastic deformation of the surface layer of these smooth surfaces leads to the generation of subsurface cracks. These cracks are nucleated below the surface because the triaxial state of compressive stress at the surface prevents crack nucleation. The cracks usually nucleate only after repeated (cyclic) loading. They may initiate at second phases or at particulate inclusions, if present; nucleation in a homogeneous material is not as well understood. The cracks propagate parallel to the surface for some distance before eventually branching to the surface and forming loose debris.

The depth of nucleation and distance of propagation are a function of the material properties as well as the load and frictional characteristics of the surface. Whether crack nucleation or propagation is rate controlling depends upon various material characteristics. Crack nucleation can be expected to be controlling in materials that deform plastically at low stress levels or have rapid crack-propagation rates. Crack propagation usually controls in materials with pre-existing cracks or that easily nucleate cracks. It should be noted that new asperities may be formed as a result of intersecting areas of delamination.

Fretting

Fretting is a wear phenomenon that occurs between two mating surfaces; initially, it is adhesive in nature, and vibration or small-amplitude oscillation is an essential causative factor. Fretting is frequently accompanied by corrosion. In general, fretting occurs between two tight-fitting surfaces that are subjected to a cyclic, relative motion of extremely small amplitude.

Fretting generally occurs at contacting surfaces that are intended to be fixed in relation to each other but that actually undergo minute alternating relative motion that is usually produced by vibration. There are exceptions, however, such as contact between balls and raceways in bearings and between mating surfaces in oscillating bearings and flexible couplings. More information on fretting in bearings is available in the articles "Failures of Sliding Bearings" and "Failures of Rolling-Element Bearings" in this Volume. Fretting further differs from ordinary wear in that the bulk of the debris produced is retained at the site of fretting. In ferrous materials, the fretting process creates a mass of reddish oxide particles. Fretting also occurs in nonoxidizing materials, such as gold, platinum, and cupric oxide.

Common sites for fretting are in joints that are bolted, keyed, pinned, press fitted, or riveted; in oscillating bearings, splines, couplings, clutches, spindles, and seals; in press fits on shafts; and in universal joints, base-plates, shackles, and orthopedic implants (see the article "Failures of Metallic Orthopedic Implants" in this Volume). One additional problem with fretting is that it may initiate fatigue cracks, which, in highly stressed components, often result in fatigue fracture.

Mechanism of Fretting

Although certain aspects of the mechanism of fretting are still not thoroughly understood, the fretting process is generally divided into the following three stages: initial adhesion, oscillation accompanied by the generation of oxidized debris, and fatigue and wear in the region of contact.

Initial Adhesion. Measurements of electrical resistance have established that intimate intermetallic contact occurs during the very early stages of fretting. Adhesion between fretting surfaces is developed by the formation of bonded junctions between asperities of the mating surfaces. For relatively high amplitudes of fretting motion, adhesion points may be created and destroyed several times during motion in one direction. Fretting has occurred at amplitudes as small as 0.025 μm (1 μin.). However, if the relative motion is small enough to be absorbed by elastic deformation at the asperities of the surfaces, no fretting damage is produced.

Initially, an oxide film a few angstroms thick, which forms on the surfaces, prevents metallic contact. For the metals to adhere, the oxide film must first be disrupted to permit metal-to-metal contact. If the fretting couple is of identical material, deformation of the material underlying the oxide film will occur at both surfaces equally, and the oxide films on both surfaces will be disrupted. If the fretting couple consists of dissimilar metals, the softer metal will deform the greatest amount; therefore, the oxide film on the softer metal will be disrupted but that on the harder metal will stay intact. Hard metals of the same crystal structure generally produce adhesion coefficients of low values, whereas soft metals produce adhesion coefficients of high values.

Generation of Debris. Disagreement exists among researchers as to whether the metal is oxidized before or after it is removed from the surface. Possibly both occur, with the relative contribution of each being controlled by the conditions of fretting. For example, the major component of the debris produced by low-carbon steel or iron when fretted in air is ferric oxide (α-Fe_2O_3), which is reddish brown and highly abrasive.

The debris formed by many nonferrous metals is largely unoxidized and is larger in particle size than that from ferrous metals. On the other hand, in hard materials, such as tool steel and chromium, the initial wear particles are very small, with much oxide present. If fretting occurs in a protective or inert atmosphere, little debris is produced, although surface damage can be extensive.

If one of the metals in a fretting couple is soft, hard oxide fragments may become embedded in the softer metal and thus reduce the wear rate. Therefore, the oxides formed in fretting can reduce the wear rate if they adhere to the surfaces or can increase the wear rate if they remain loose.

Fatigue and Fretting Wear. Under fretting conditions, fatigue cracks are initiated at very low stresses, well below the fatigue limit of nonfretted specimens. The initiation of fatigue cracks in fretted regions depends mainly upon the state of stress in the surface and particularly upon the stresses superimposed on the cyclic stress. The direction of growth of the fatigue cracks is associated with the direction of contact stresses and takes place in a direction perpendicular to the maximum principal stress in the fretting area. For this reason, fatigue strength that is based on crack initiation decreases linearly with increasing contact pressure.

A phenomenon peculiar to fretting is that some of the fatigue cracks do not propagate, because the effect of contact stress extends only to a very shallow depth below the fretted surface. At this point, favorable compressive residual stresses retard or completely halt crack propagation. Prevention of crack propagation in components that carry considerable stresses, such as axles or shafts, is vitally important because the usual mode of failure of such components is fatigue initiated by fretting.

Friction

Friction is the resistance to motion that is experienced when two surfaces in contact are forced to slide relative to each other. More formally, friction is the resisting force tangential to the common boundary between two bodies when, under the action of an external force, one body moves or tends to move relative to the surface of the other (Ref 33). The influence of friction on wear behavior is complex and involves at least the topography of the surfaces, their composition, and any surface films. More extensive reviews on friction and its effects on wear can be found in Ref 1 and 4 to 6.

DaVinci, in 1519, first expounded the two basic laws of friction: (1) friction is proportional to the normal force between surfaces, and (2) friction is independent of the apparent area of contact. These were restated by Amontons (1699) and experimentally verified by Coulomb

(1785). For the most part, they still hold today. Coulomb's third law, that friction is nearly independent of velocity, is not always found to be valid.

An explanation of the observation that friction is independent of the apparent area of contact begins with the recognition that real surfaces are not perfectly smooth and that contact between two surfaces is only through asperities. For example, when a solid cube is placed on a flat surface, one face of the cube appears to be in intimate contact with the flat surface. However, real surfaces have a certain amount of roughness and waviness; therefore, the actual area of contact between the cube face and the flat surface (true-contact area) is the sum of a very large number of minute areas where high points on opposing surfaces contact one another. Some estimates indicate that individual areas of true contact are about 25×10^{-3} to 25×10^{-5} mm (10^{-3} to 10^{-5} in.) in diameter and are randomly distributed over the apparent-contact area.

The surface roughness and waviness of manufactured parts have distinctive geometrical patterns characteristic of the process that produced the parts. The surface of a turned shaft consists of ridges and furrows; the surface of a ground ball-bearing raceway consists of shallow, parallel, U-shape troughs with thin ridges between them; the surface of a gold-plated electrical contact is normally a distribution of small convex surfaces, resembling a mass of close-packed bubbles. The size and configuration of these fine-scale surface features determine the actual conditions of contact between opposing surfaces, and what occurs at these points of contact has a significant influence on friction and wear.

Microscopic stress conditions at points of contact can be described most simply by the analogy of a smooth sphere supported by a smooth planar surface. The resulting stress at the point of tangency is sufficiently high to cause elastic deformation of both the sphere and the flat surface, and the minute contact area spreads out until the stress is reduced to slightly below the elastic limit. Upon application of an external load in a direction perpendicular to the planar surface, the area of contact will increase roughly in proportion to the increase in load taken to the two-thirds power. As load is increased, the elastic limit of the planar surface, which is related to hardness, is eventually exceeded. When the elastic limit is exceeded, a permanent dent in the surface occurs (this is comparable to what occurs in a hardness test, in which the penetrator is harder than the surface being tested).

The independence of friction on the apparent area of contact is thus due to the load being carried on small areas (asperities) of real contact, which increase in direct proportion to the load. Estimation of individual areas can be made using Hertzian calculations, but in reality, there will be a distribution of sizes and shapes of individual contact areas. Nonetheless, for the total load normal to the surface, L:

$$L = PA \qquad \text{(Eq 6)}$$

where P is the pressure that asperities can support (found to be approximately constant across a surface for most metals), and A is the total area of real contact, that is, the total area of the deformed asperities in contact.

The adhesion theory of friction, accepted for many years, assumed that strong adhesive bonds developed in areas of contact and had to be sheared when the surfaces moved relative to each other. If S is the shear strength of this bond, the friction force, F, required for movement is:

$$F = SA \qquad \text{(Eq 7)}$$

Because $A = L/P$, then $F = SL/P$, or $f = F/L = S/P$ where f is the coefficient of friction. Thus, $f =$ shear strength/yield strength.

The yield strength is often equated with the indentation hardness, H, of the material:

$$f = \frac{S}{H} \qquad \text{(Eq 8)}$$

The coefficient of friction f is also frequently designated as μ.

Equation 8 is conceptually useful, but is difficult to use in practice because of the difficulty in measuring S and because it ignores the work done in plowing and the effects of surface contaminants. It has been proposed (Ref 32) that f is the sum of several effects:

$$f = f_d + f_p + f_a \qquad \text{(Eq 9)}$$

where f_d is due to asperity deformation, f_a is due to adhesion between flats, and f_p is due to plowing by wear particles and/or hard surface asperities. The relative contributions of these components depend upon the condition of the sliding interface, which is affected by the history of sliding, the specific materials used, the surface topography and the environment (Ref 32). It was suggested that f_d should vary from 0.42 to 1, f_a from 0 to 0.4, and f_p from 0 to 1.

It may be noted from the equation for calculating the coefficient of friction, $f = S/P$, that the coefficient is lowest when the hardness of the softer member of the contacting pair is high relative to its shear strength. Because the usual treatments that increase the hardness of a metal or alloy also increase its shear strength, a high ratio of hardness to shear strength is seldom found in a single metal or alloy. A high ratio can be achieved, however, by using a composite material. In wear-resistant electrical contacts, for example, a low coefficient of friction has been achieved by plating a hard substrate with a very thin layer of gold.

On real surfaces, the factor S (shear strength) in Eq 8 probably represents a complex parameter not only associated with the yield properties of the weaker member of the sliding pair but also dependent upon the condition of the sliding surfaces in contact. If a bonded junction is weak and shearing takes place at the bond and not through the subsurface material, the condition of the surface has a decided effect on the friction.

There are many ways of weakening the junction. In fact, a natural surface is so contaminated with adsorbed gases and solids that contact in air generally inhibits bonding of contacting asperities. Problems encountered with sliding surfaces in space missions have demonstrated the effect of truly clean surfaces kept clean in a gasless environment. Under these conditions, in which naturally occurring oxides and adsorbed films have been removed by heat or ion bombardment, metal surfaces bond together. In such an environment, in which conventional lubricants would evaporate, low shear strength, low vapor pressure solids, such as metal sulfides, selenides, and tellurides, have been applied as bonded coatings to reduce friction.

Even a few molecular layers of an organic material on a metal surface can result in a large decrease in its coefficient of friction. Therefore, published friction-coefficient values are often highly questionable because of the effect of surface films.

Naturally occurring metal oxides generally reduce adhesion forces of surfaces. Some oxides are better lubricants than others. For example, the friction produced by hardened steel sliding against hardened steel can have a threefold variation as the partial pressure of oxygen on the surface is varied. The lower oxides of iron (Fe_3O_4 or FeO) exist at low partial pressures of oxygen and have better lubricating properties than Fe_2O_3, the iron oxide normally found in air.

It is possible to encounter high friction in sliding contact when a combination of high speed and heavy load is encountered. This combination produces frictional heating, softening of surface layers, and breakthrough of protective oxide films, bringing clean, active metal surfaces into contact. This process is the basis for friction welding, which can be used to join metals of different chemical compositions.

Other factors that have been found to be influential in controlling frictional forces include surface roughness and crystal structure. Generally, friction is highest when a surface is very rough, and interlocking of jagged high points occurs or when the contacting surfaces are very smooth, allowing surface tension of fluid films or molecular attraction between the surfaces to become significant.

Studies of sliding friction using single-crystal materials in high vacuum have shown that materials with hexagonal close-packed (hcp) crystal structures exhibit lower friction than materials with body-centered cubic (bcc), face-centered cubic (fcc), or tetragonal crystal structures. It has also been found that the coefficient

Fig. 5 Step-by-step development of a hydrodynamic lubricating film in a unidirectionally loaded journal bearing

See text for details.

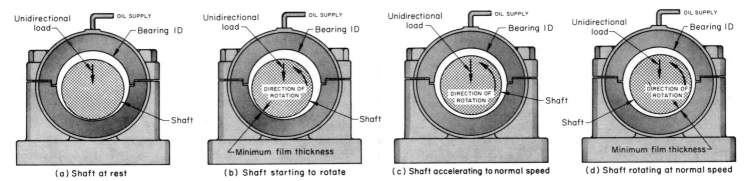

(a) Shaft at rest   (b) Shaft starting to rotate   (c) Shaft accelerating to normal speed   (d) Shaft rotating at normal speed

of friction varies significantly, depending upon the crystallographic direction of sliding. Sliding in the close-packed direction in any crystal system produces the lowest friction. However, regardless of crystallographic direction, the coefficient of friction is high for clean surfaces of crystals; consequently, there are few applications in which the crystallographic direction of sliding is important.

Graphite is an outstanding solid lubricant that has very low friction, apparently because of its hexagonal crystal structure and easy slip in close-packed $\langle 0001 \rangle$ directions. However, the low friction of graphite is actually a consequence of adsorbed water vapor on exposed basal planes. Without adsorbed water vapor, the friction of graphite is high.

Lubricated Wear

One important means of reducing wear is lubrication. Lubrication not only reduces the power consumption needed to overcome friction but also protects rolling and sliding contact surfaces from excessive wear. Even with lubrication, however, wear still occurs.

On lubricated surfaces, the wear process is usually mild and generates fine debris of a particle size as small as 1 or 2 μm. Abrasive wear or delamination wear predominates under lubricated conditions. Electron microscope examination of worn surfaces from lubricated assemblies frequently reveals a multitude of fine scratches oriented in the direction of relative motion. The fine debris generated by abrasion becomes suspended in the oil or grease. In devices using circulating-oil lubrication, advantage has been taken of the fact that wear debris can be analyzed by spectroscopy and that deterioration of the device by wear can be diagnosed from these results. This technique is used to monitor the condition of vital components in aircraft and locomotive engines.

Modes of Lubrication

There are several basic modes of lubrication. In all modes, contact surfaces are separated by a lubricating medium, which may be a solid, a semisolid, or a pressurized liquid or gaseous film.

Hydrodynamic lubrication is a system in which the shape and relative motion of the sliding surfaces cause the formation of a fluid film having sufficient pressure to separate the surfaces. Hydrostatic lubrication is a system in which the lubricant is supplied under sufficient external pressure to separate the opposing surfaces by a fluid film. Elastohydrodynamic lubrication is a system in which the friction and film thickness between the two bodies in relative motion are determined by the elastic properties of the bodies in combination with the viscous properties of the lubricant at the prevailing pressure, temperature, and rate of shear. Dry-film (solid-film) lubrication is a system in which a coating of solid lubricant separates the opposing surfaces and the lubricant itself wears away. Boundary lubrication and thin-film lubrication are two modes in which friction and wear are affected by properties of the contacting surfaces as well as by the properties of the lubricant. In boundary lubrication, each surface is covered by a chemically bonded fluid or semisolid film, which may or may not separate opposing surfaces, and viscosity of the lubricant is not a factor affecting friction and wear. In thin-film lubrication, the lubricant usually is not bonded to the surfaces, it does separate opposing surfaces, and lubricant viscosity affects friction and wear.

Mechanical devices often operate under several lubrication modes simultaneously or alternately. For example, when a hydrodynamic journal bearing starts turning from rest, it operates under boundary lubrication and then thin-film lubrication for a short time until a stable, thick oil film develops and the solid surfaces separate. The process is reversed when rotation is slowed or stopped. Wear occurs during the initial and final boundary-lubricated periods. Gears experience both elastohydrodynamic and boundary lubrication at the same time. For example, during meshing of one tooth of a spur gear with a tooth of a mating gear, initial contact is sliding contact, which results

in wear and scuffing at the tips and roots of the teeth. Contact along the pitchline, however, is essentially rolling contact, and elastohydrodynamic conditions prevail. Pitchline damage takes the form of pitting or spalling and is similar to rolling-contact fatigue found in ball and roller bearings.

Hydrodynamic Lubrication. The step-by-step development of a hydrodynamic fluid film is illustrated in Fig. 5 for a full journal bearing under unidirectional loading. In Fig. 5(a), the machine is at rest. The oil supply is shut off, and most of the oil has leaked from the normally full clearance space (greatly exaggerated in the illustration). The remaining film on the bearing and journal surfaces is extremely thin, and there is probably some metal-to-metal contact between asperities on the mating surfaces at the bottom where the journal surface rests on the bearing.

In Fig. 5(b), the machine has been started, and the shaft has begun to rotate. The oil supply has been turned on, and oil has filled the clearance space. At the start, friction is momentarily high, and the shaft tends to climb up the side of the bearing in a direction opposite to the direction of rotation. As it does so, it rolls onto a thicker oil film, friction is reduced, and the tendency to climb is balanced by a tendency to slip back on the thicker oil film.

As the journal gains speed (Fig. 5c), oil is drawn into the wedge-shaped clearance space at the lower left. A fluid pressure is developed in this region of the film, which pushes the journal to the right and lifts it.

Finally, at full speed, the journal is supported on a thick film and assumes the position shown in Fig. 5(d)—on the opposite side of the bearing from the position at start-up (compare with Fig. 5b). The converging wedge-shaped oil film has moved to a position under the journal, and the point of nearest approach of the journal and bearing, that is, the point of minimum film thickness, is slightly to the right of a vertical line through the center of the bearing.

Fluid dynamics can be used to define and predict load capacity, friction, and heat generation in a fluid film when hydrodynamic lubrication, hydrostatic lubrication, or elastohydrodynamic lubrication prevails. The viscosity of the lubricant is important in determining operating characteristics and whether or not wear can be anticipated. The pressure generated in the liquid-lubricant film by the shearing process supports the load and keeps the solid surfaces separated. As the load increases, film temperature increases, shear rate decreases, film thickness diminishes, and the solid surfaces approach each other. When the thickness of the lubricant film approaches the dimensions of surface roughness, asperity contact begins, and evidence of wear can be detected. For hydrodynamic conditions, the film thickness, t, or closest approach of the bearing surfaces, depends upon:

$$t \approx \left(\frac{\mu v}{W}\right)^{1/2} \qquad \text{(Eq 10)}$$

where μ is the bulk viscosity of the lubricant at bearing temperature, v is the sliding velocity, and W is the bearing load.

A journal bearing operating under hydrodynamic conditions is somewhat self-regulating. That is, with an established thick lubricating film, an increase in velocity will increase the shear rate of the oil film, and the resulting increase in energy input will increase the oil-film temperature, resulting in a decrease in viscosity. In turn, the decrease in viscosity will diminish the effect of the velocity increase on film thickness. In addition, when the correct bearing materials are selected, a wear-in process will decrease the peaks of the surface asperities and allow operation on a thinner and thinner film without metal-to-metal contact and wear as the bearing continues to operate. Many sliding surfaces in machinery do not benefit from this seemingly self-regulating infinite-life lubrication system. Because of misalignment, vibration, limited lubricant supply, and repeated start-stop operation, many bearings continually operate in the boundary regime where asperity contact and wear occur.

Hydrostatic lubrication, often used in high-speed precision bearings, is similar to the thick-film stage of hydrodynamic lubrication in that opposing surfaces slide on a relatively thick film of lubricant. However, in hydrostatic lubrication, the film is maintained by fluid pressure from an external source and by a fixed or controlled rate of leakage from between the surfaces. Because most hydrostatic bearings are designed to have fluid-pressure peaks equally spaced around the bearing surface, the shaft is positioned more nearly in the center of the bearing than in hydrodynamic lubrication, and film thickness is about the same for any two points around the bearing. One of the main advantages of hydrostatic lubrication over hydrodynamic lubrication is that with hydrostatic lubrication the shaft is supported by a full oil film at any speed. Thus, thin-film and boundary lubrication on start-up and shutdown, with the

attendant increased friction and wear, are avoided. In addition, thick-film lubrication can be maintained with low-viscosity lubricants that in a hydrodynamic bearing would not be able to develop enough film pressure to support the shaft load.

Elastohydrodynamic Lubrication. Under elastohydrodynamic rolling-contact conditions, typical of ball and roller bearings, minimum lubricant-film thickness t_{min} follows approximately the relationship expressed in:

$$t_{min} \approx (\mu\alpha)^{0.7}\left(\frac{N^{0.7}}{W^{0.9}}\right) \qquad \text{(Eq 11)}$$

where μ is the bulk viscosity of the lubricant at bearing temperature, α is the coefficient of viscosity increase with pressure, N is rotational speed, and W is bearing load.

Under rolling-contact conditions, film thickness is not as sensitive to load as it is to rotational speed. An increase in load produces an increase in elastic deflection in the contact area and distributes the contact pressure over a larger area. Because rolling contact involves initial line, or point, contact, very large localized contact stress results, necessitating the use of high yield strength bearing materials (heat-treated bearing steels). The extremely thin lubricant films involved (as small as ten millionths of an inch) also require very smooth surface finishes to ensure true elastohydrodynamic lubrication. Nevertheless, wear does occur in ball and roller bearings.

Rolling-contact wear can be insidious, progressing with an improvement in surface finish and with no loss in sphericity, but with sufficient loss of material to cause loss of vital preload, such as can occur in miniature precision bearings. Spalling or pitting is another, more dramatic type of wear that can occur in rolling-contact applications. This is a self-aggravating type of surface damage that causes the performance of rolling-element bearings to become increasingly rough and that ultimately may result in fracture of rolling elements. In recent studies of the role of lubrication in determining the life of ball and roller bearings, it has been found that the ratio of minimum film thickness to combined surface roughness of two opposing surfaces provides a fairly good indicator of useful life. The greater the incidence of asperity contact through the lubricant film, the sooner the onset of spalling.

Boundary lubrication occurs in a large number of mechanical devices because the conditions required for full-film lubrication (or even thin-film lubrication) using a fluid substance often cannot be attained without using a complex and expensive lubrication system. For example, a grease-lubricated bearing subjected to an intermittent oscillating motion under heavy load operates almost exclusively under boundary lubrication. Under these conditions, high points or asperities on surfaces come into contact, but bonding is prevented by very thin, soft, solid films. These films shear easily and

prevent metal removal or heavy scoring of the surfaces.

Boundary films have a wide variety of forms and compositions. Experiments have shown that a single monolayer of stearic acid can lubricate and prevent asperity adhesion. Under practical operating conditions, however, the boundary films active in machinery are complex reaction products of the lubricant, the atmosphere, and the constituents in the bearing surface. Although full-film conditions of lubrication do not exist, with the addition of lubricant the wear rate of contacting surfaces can be reduced to as little as 5% of the rate for nonlubricated wear.

Surface temperature has perhaps the greatest influence on the effectiveness of boundary lubrication. Frictional energy produces heat on sliding surfaces. With boundary lubrication, there generally is not sufficient lubricant flow to carry away the heat; in contrast, hydrodynamic lubrication is quite effective in removing frictional heat. There are several possible consequences of frictional heating during boundary lubrication. Under extreme-pressure (EP) lubrication, chemical additives in the lubricant react with metal surfaces to form soft, solid reaction products, which are presumably the agents that prevent metal adhesion and surface damage. Because heat increases the reaction rate, at asperity contacts, where the load surface temperatures are highest, the reaction rate is greatest, and a solid lubricant is provided in the spots where the potential for adhesion is greatest. Thus, by chemical attack modified by the localized surface temperature, the occurrence of severe wear is prevented by substitution of mild corrosion.

Lubricants

Almost any surface film can act as a lubricant, preventing cold welding of asperities on opposing surfaces or allowing opposing surfaces to slide across one another at a lower frictional force than would prevail if the film were not present. Lubricants may be either liquid or solid (in some cases, gas films may act as lubricants). One of the functions of a lubricant is to carry away heat generated by two surfaces sliding under contact pressure. Liquid lubricants can dissipate heat better than solid or semifluid lubricants, but in all types, the shear properties of the lubricant are critical to its performance.

Properties. Liquid lubricants maintain separation of opposing surfaces by pressure within the film, which opposes the contact force. This pressure may be generated within the film, usually as a result of the shape of the opposing surfaces, or the liquid may be forced between the opposing surfaces by pressure from an external source. Regardless of the means of creating pressure within the film, the opposing surfaces slide on a film of liquid. Friction and wear are directly influenced by the thickness and shear properties (viscosity) of the liquid. Where appropriate, the use of a high-viscosity

Fig. 6 Schematic showing the relation of surface roughness to film thickness

Shown are conditions of full-film, thin-film, and boundary lubrication.

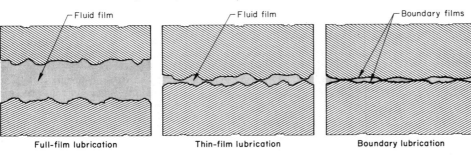

Full-film lubrication   Thin-film lubrication   Boundary lubrication

Fig. 7 Schematic of the polar bonding and orientation of straight-chain fatty-acid molecules in a boundary lubricant between sliding surfaces

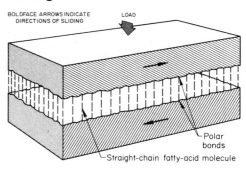

lubricant usually results in a relatively thick film and a low wear rate. However, high sliding speeds cannot be accommodated by a viscous film, because excessive heat generated within the film causes it to become less viscous and to decompose chemically.

Full-film (thick-film) lubrication, such as occurs under hydrostatic or hydrodynamic conditions, effectively separates asperities on opposing surfaces, whereas thin-film and boundary lubrication allow asperity contact. The differences among these three conditions of liquid lubrication are illustrated schematically in Fig. 6.

Some special types of boundary lubricants, most notably the EP lubricants, react with a metallic surface, often at high temperatures, to produce a monomolecular film on the surface. This very thin film contaminates the mating surfaces and prevents metal-to-metal contact or adhesion. Extreme-pressure lubricants often contain extremely reactive constituents that re-form the film instantly if it is scraped off one of the surfaces. Film formation of this type is, in effect, corrosion; when it is uncontrolled or when the film is repeatedly scraped off and re-formed, deterioration of the surface can result.

Solid-film lubricants must be adherent to be effective, or they allow metal-to-metal contact or introduce unwanted particles that roll and slide within the joint. When they can be kept within the joint, graphite and molybdenum disulfide make good lubricants because they shear easily in certain crystallographic directions. Hard, adherent oxide films, such as Fe_3O_4 on steel or anodized Al_2O_3 on aluminum, withstand wear because they resist penetration and do not bond with most mating surfaces.

Lubricating oils are relatively free-flowing organic substances that are used to lower the coefficient of friction in mechanical devices. They are available in a broad range of viscosities, and many are blended or contain additives to make them suitable for specific uses. In general, lubricating substances that are fluid at 20 °C (70 °F) are termed oils; lubricating substances that are solid or semifluid at 20 °C (70 °F) are termed greases or fats.

Oils are derived from petroleum (mineral oils) or from plants or animals (fixed oils). Mineral oils are classified according to source (type of crude), refining process (distillate or residual), and commercial use. The commercial mineral oil base products consist mainly of saturated hydrocarbons (even though naphthene-base crudes are predominantly unsaturated) in the form of chain or ring molecules that are chemically inactive and do not have polar heads. These commercial products may or may not contain waxes, volatile compounds, fixed oils, and special-purpose additives. Fixed oils and fats differ from mineral oils in that they consist of an alcohol radical and a fatty-acid radical, can be reacted with an alkali (sodium hydroxide or potassium hydroxide, for example) to form glycerin or soap, cannot be distilled without decomposing, and contain 9 to 12.5% oxygen. All fixed oils are insoluble in water and, except for castor oil, are insoluble in alcohol at room temperature.

Fixed oils are generally considered to have greater oiliness than mineral oils. Oiliness is a term that describes the relative ability of any lubricant to act as a boundary lubricant. Electron-diffraction experiments have shown that molecules of the effective lubricating agent—a long-chain fatty acid of high molecular weight, such as stearic acid or oleic acid—are attached to a metallic surface by polar bonding and stand up much like individual strands in a pile carpet (Fig. 7). This results in a surface layer with high adhesion, high resistance to contact stress, and low resistance to lateral shear along the surface.

Lubricating grease, as defined by ASTM, is a solid to semifluid product consisting of a dispersion of a thickening agent in a liquid lubricant. In more practical terms, most greases are stabilized mixtures of mineral oil and metallic soap. The soap is usually a calcium, sodium, or lithium compound and is present in the form of fibers (Fig. 8) whose size and configuration are characteristic of the metallic radical in the soap compound.

Solid lubricants, which are solids with lubricating properties, can be maintained between two moving surfaces to reduce friction and wear. Numerous solid inorganic and or-ganic compounds, as well as certain metals and composite materials, may be classified as solid lubricants. Molybdenum disulfide, graphite, and polytetrafluoroethylene (PTFE) are the solid lubricants most commonly used. Several hundred different compounds and mixtures have been described as potential solid lubricants.

Solid lubricants have been used as thin films, structural sections of bearing assemblies, reinforced laminates, and inserts. Figure 9 shows three bearing designs using different means for solid lubrication: a rolling-element bearing having films of solid lubricant bonded to the surfaces of the raceways and the retainer (Fig. 9a), a plain spherical bearing having a liner of resin-bonded PTFE fibers between the spherical inner ring and the outer ring (Fig. 9b), and a journal bearing having a spring-loaded solid insert of molybdenum disulfide in the housing to maintain a supply of lubricant to the shaft (Fig. 9c).

Although solid lubricants may be applied to achieve design simplification or weight reduction, they are usually adopted because of their good stability at elevated temperatures, in chemically active environments, and when exposed to nuclear radiation. Solid lubricants also provide certain advantages in high-vacuum, aerospace, or cryogenic applications in which liquids would evaporate or congeal.

Lubricant Failures Leading to Wear

In devices that depend upon lubricants to combat friction and avoid deterioration by wear, failure of the lubricant can be disastrous. Most lubricant failures occur by chemical decomposition, contamination, changes in properties caused by excessive heat, or outright loss from, or inadequate flow of a pressurized fluid into, lubricated areas. Lubricating oils and greases can fail by any one of the above processes alone. However, in most cases, chemical decomposition, contamination, and temperature are all involved and are interre-

Fig. 8 Scanning electron micrographs showing the fibrous appearance of soap particles in a lithium-base grease
(a) 15 000×. (b) 24 000×

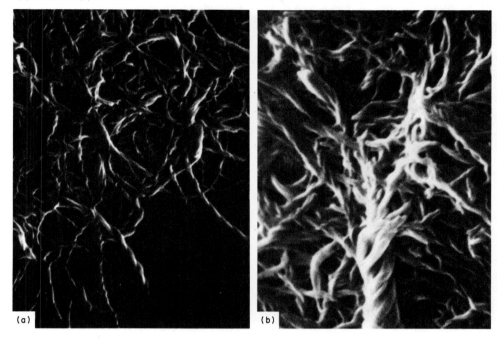

Fig. 9 Bearing designs using three types of solid lubrication
(a) Rolling-element bearing. (b) Spherical bearing. (c) Journal bearing

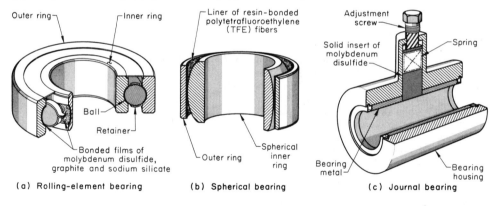

(a) Rolling-element bearing   (b) Spherical bearing   (c) Journal bearing

lated. For example, when oil is heated in the presence of air, oxidation occurs. Oxidation increases the viscosity and organic acid concentration of mineral oils with the result that varnish and lacquer deposits may form on hot metal surfaces. Under severe conditions, the deposits may be converted to hard carbonaceous substances. Fixed oils absorb oxygen more readily than mineral oils do, and some may dry, thicken, and form elastic solids. Certain fixed oils (castor, olive, sperm, and lard oils) oxidize more slowly than others. These fixed oils are widely used in blended oils because of their nondrying characteristics. Temperature affects oxidation rates; in mineral and blended oils, for example, the rate doubles with each 10 °C (18 °F) increase in temperature. Oxidation rates

are also higher when the oil is agitated or foams or when catalysts, such as copper or acids, are present.

In general, solid-film lubricants fail by mechanical removal of microscopically thin layers. Wear debris, which consists primarily of lubricant particles, is generated by the sliding action of a sharp edge against the bonded film on a contact surface. The sharp edge shears a layer of the film (and sometimes the entire film) from the substrate. Contact between a rolling element and a sharp ridge can chip the film, which often initiates more extensive failure. This process eventually results in a lack of dynamic stability (as would result from excessive clearance in a bearing) or in galling and seizing of metallic contact surfaces.

Many bonded solid lubricants derive their adhesion from binders, which are incorporated into the film in quantities up to about 20% by volume. Metals, oxides, silicates, or other ceramics are the most common binders. When wear debris contains binder particles, it abrades the remaining film more rapidly than when it consists solely of particles of the lubricating substance.

Pressurized lubricating systems involving circulation of fluid lubricant are susceptible to certain types of failure that affect the ability of the system to provide the required flow rate or pressure at the point of lubricant injection. Decomposition of wax-containing mineral oils or contamination of oil with certain chemical substances can cause formation of sludge, which may clog flow passages, resulting in loss of flow or pressure or both. Severe agitation of the oil sometimes results in entrapment of air in the form of tiny bubbles (foaming). The presence of water, many additives, and particulate foreign matter (debris) in oils increases the likelihood of their forming stable foams. Foaming causes pump-inlet starvation, loss of circulation, and sponginess in control systems and can cause an oil reservoir to overflow because of the volumetric increase resulting from air entrapment. Increases in viscosity caused by oil decomposition, for example, oxidation, or by excessive cooling of the oil fed to the pump can cause reduced flow of the circulating oil and starvation of bearing surfaces. A decrease in viscosity because of excessively high operating temperatures can cause a reduction in film thickness in a hydrostatic or hydrodynamic bearing, resulting in increased wear or seizure.

Contamination of the lubricant with water or reactive chemical substances can lead to lubricant decomposition, corrosion of contact surfaces, or both. Contamination with abrasive substances or debris can cause abrasive wear, especially when the size of the contaminant particles is about the same as the thickness of the lubricating film.

In internal-combustion engines, water, halide acids, sulfur acids, and products of partial combustion of fuel hydrocarbons are picked up by the lubricating oil. These contaminants can cause various undesirable chemical reactions with oil and metallic surfaces, resulting in the formation of varnish deposits, sludges, or viscous emulsions in the oil or in corrosive wear of engine components.

Because of their chemical nature, fixed oils are particularly susceptible to chemical alteration by alkalis. Alkalis cause saponification (formation of soap) by direct chemical reaction with the fatty acids in fixed oils. This reaction alters the nature of the lubricant and consequently its lubricating properties.

Viscosity of mineral oils, fixed oils, and greases is affected by both temperature and pressure. An increase in pressure causes an increase in viscosity, although the effect is generally not significant except at very high pressures. Conversely, any change in temperature has a very significant effect on viscosity.

Decreasing the temperature of mineral oil from 100 °C (212 °F) to 0 °C (32 °F) can increase viscosity by 100 times or more, making the oil much less free flowing. The viscosity of fixed oils is affected to a lesser degree by temperature; for example, the viscosity of lard oil is increased by only 30 times when the temperature is decreased from 100 °C (212 °F) to 0 °C (32 °F).

Greases that are solid or semisolid at room temperature gradually soften with increasing temperature and can become fluid and free flowing at operating temperatures from 95 to 205 °C (200 to 400 °F), depending upon the type of grease. Every grease has a dropping point, which is defined as the temperature at which the first drop of liquid grease falls in a standard test. This temperature is the lower boundary of the melting range of the grease.

Oils containing substantial quantities of volatile compounds may lose these components by evaporation when operating temperatures are too high. This process not only alters the viscosity but also upsets the chemical nature of the oil, thus changing other properties. In some cases, volatilization can take place within the lubricating film, for example, when frictional heat is ineffectively removed by the circulating fluid. Bubble formation within the film reduces the load-carrying capacity of the film, leading to adhesive or abrasive wear or, in severe cases, cavitation erosion of opposing surfaces.

Transition Temperature. When boundary lubrication is provided by soft metallic soaps, for example, iron stearate, an increase in surface temperature can result in a marked increase in the coefficient of friction and a sudden change in wear rate from mild to severe. The temperature at which this change occurs (transition temperature) is the point at which the soap desorbs from the metal surface and no longer provides a bonded, continuous surface film. Transition temperatures are generally within 120 to 205 °C (250 to 400 °F), depending upon the lubricant and the chemical composition of the metal substrate. Extreme-pressure lubricants function by reaction with the metal surface rather than by adsorption of components in the lubricant and are often used as substitutes for soap-type boundary lubricants when operating temperatures exceed the transition temperature.

Prevention of Lubricant Failures

Lubricant failure can often be traced to the selection of an inappropriate lubricant. Petroleum lubricating oils are available in a wide variety of formulations, with an equally wide variety of special properties. When these properties cannot be obtained by conventional refining techniques or are obtainable only at a very high cost by refining, they are imparted to the lubricant by additives.

Oil additives may serve to improve one or more of the properties of the base oil, to impart to it entirely new performance characteristics, or to reduce the rate at which undesirable changes in the oil take place during service. Some of the more common additives include:

- *Viscosity-index improvers*: These substances decrease the effect that temperature has on viscosity, making the oil more viscous at high temperature than it would be without the additive
- *Pour-point depressants*: These substances make a wax-containing oil less viscous at low temperatures by inhibiting the growth and coalescence of wax crystals suspended in the oil
- *Defoamants*: These additives promote the coalescence of tiny entrapped air bubbles into larger bubbles, which can rise to the surface and collapse
- *Wetting agents and emulsifiers*: These additives enable the oil to displace water from metal surfaces or to absorb the water as a stable emulsion, thus promoting oil-film formation on a metal surface
- *Oxidation inhibitors*: These combat oxidation of the oil itself by interrupting the chain of chemical reactions leading to deterioration or by deactivating catalytic metallic surfaces
- *Detergents and dispersants*: Widely used in lubricants for internal-combustion engines, these additives combat the formation of sludge and varnish
- *Corrosion inhibitors*: These additives reduce or prevent corrosion of lubricated surfaces by contaminants in the oil, such as oxygen, water, acids, and combustion products
- *Lubrication-property improvers*: This category includes a variety of additives that reduce friction (especially under boundary lubrication), speed up a wearing-in process, enhance film strength, or provide lubrication under high contact pressures

Mechanical Design. Certain types of lubricant failure can be prevented by changes in the design of the device itself or in the design of the lubricating system. Starvation of a bearing caused by inadequate lubricant flow or by clogging of oil passages can sometimes be corrected by increasing the size of the passages. An increase or decrease in clearance between sliding surfaces will often enable the lubricant to function more effectively. Shields, covers, and seals can sometimes be used to prevent lubricant contamination from external sources. In other cases, filtration or absorption devices can be incorporated into the system to remove unwanted contaminants.

Effect of Lubrication on Surface Features

The features on a surface that has been worn under lubricated conditions are different from those found on a surface that has undergone nonlubricated wear. Examination of sliding surfaces by high-magnification electron microscopy reveals that, when a lubricant is present, wear occurs by deformation of the highest surface asperities rather than by galling and tearing, which predominate with nonlubricated wear. For example, when a ground surface is subjected to mild lubricated wear, the microscopic surface ridges resulting from the abrasive action of hard particles in the grinding wheel come into contact with the opposing surface and flatten out. The highest ridges come into contact first and are subjected to large contact stresses, which cause plastic flow. When the tops of the ridges deform, they often develop thin tongues of extruded metal that subsequently break off, forming very fine particles of wear debris. Thus, the surface is gradually leveled or smoothed out as more and more ridges come into contact.

On polished or lapped surfaces, lubricated wear produces an extremely fine pattern of microscratches that are often invisible except by electron microscopy. Examination of these microscratches reveals that they are caused by plastic deformation and not by plowing or micromachining. The deformation appears to be the result of contact by hard asperities in the mating surface or by fine debris. Each scratch is usually a shallow trough with a flat bottom and steep sides. The ridges produced by this scratching process are worn, which produces fine debris in much the same way as wear debris is generated from ground surfaces. Additional information on lubricated wear, including examples of lubricated wear in service is provided in the articles "Failures of Sliding Bearings," "Failures of Rolling-Element Bearings," and "Failures of Gears" in this Volume.

Nonlubricated Wear

Metal adhesion and cold welding characterize the process of wear in the absence of a lubricant. The conditions of nonlubricated wear are difficult to define because, in most practical situations, there is some kind of lubricant on any sliding or rolling surfaces. In addition to the naturally occurring oxide on most metals, the atmosphere and its industrial contaminants provide a wide variety of adsorbing organic and inorganic molecules. These surface contaminants protect contacting surfaces in much the same way as boundary lubricants in that they prevent intimate contact between chemically active metal surfaces. Only when metal surfaces are kept in an ultrahigh-vacuum environment and are cleaned by an electron beam, by an electric arc, or by sputtering are they truly nonlubricated. Under these conditions, cold welding of the surfaces can take place immediately upon contact.

Contaminating films on metal surfaces can be penetrated under high contact stresses, resulting in cold welding of asperity contacts. If the asperity junction is stronger than the weaker of the two metals in contact, sliding motion will cause subsurface shear of the junction, and a particle larger than the junction may be torn out of the surface. It is also possible that the junction will not shear off but will grow by subsurface shear until a critical size is reached and the heavily worked junction breaks off. This process, known as prow formation, is found most often under point-contact conditions involving a hard metal sliding on a soft metal. More information on the processes associated with nonlubricated wear is available in the section "Adhesive Wear" in this article.

Erosion

Erosion, or erosive wear, is the loss of material from a solid surface due to relative motion in contact with a fluid that contains solid particles (Ref 34). The term abrasive erosion is sometimes used to describe erosion in which the solid particles move nearly parallel to the solid surface; the term impingement, or impact, erosion is used to describe erosion in which the relative motion of solid particles is nearly normal to the solid surface (cavitation erosion, fluid erosion, and electrical pitting will not be discussed; see the articles "Liquid-Erosion Failures" and "Failures of Rolling-Element Bearings" in this Volume). The detrimental effects of erosion have become an increasingly important problem, particularly on the airfoils and shrouds in various fans, in compressors and turbines, on helicopter blades, in centrifugal pumps, on valve components, and in pipe joints and bends. Beneficial effects of erosion, however, also exist, for example, in grit blasting for cleaning purposes, in drilling for mining and tunneling, and in the cutting of some composites.

Erosion occurs as the result of a number of different mechanisms, depending upon the composition, size, and shape of the eroding particles, their velocity and angle of impact, and the composition and microstructure of the surface being eroded. The sensitivity of brittle and ductile materials to the angle of impact generally follows the behavior shown schematically in Fig. 10, with ductile materials experiencing the maximum rate at about 20 to 30° and brittle materials at about 90°. Erosion of a given material as a function of time often follows a pattern consisting of an incubation period with little or no material removal, followed by an increasing rate, and finally a steady state. Some materials exhibit peaks with decelerating rate with or without a final steady state (Fig. 11). Most mechanistic models are concerned only with the steady state.

The erosion of materials has been attributed to a number of mechanisms, including cutting, plowing, extrusion, fragmentation, elastic fracture, elastic-plastic fracture, and melting. An even greater number of analytical models have been proposed, but none has been completely satisfactory. It is perhaps easiest to consider the erosion of ductile and brittle materials sepa-

Fig. 10 Effect of impact angle on aluminum and glass for 300-μm iron spheres

The velocity of the spheres was 10 m/s (33 ft/s).
Source: Ref 35

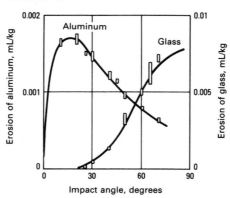

Fig. 11 Characteristic volume-loss rate versus time curves

(a) Type I, (b) Type II, (c) Type III, and (d) Type IV are generalized types of material behavior. Source: Ref 36

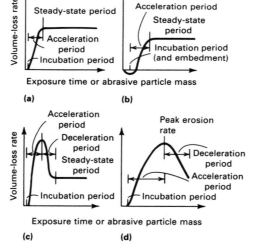

rately, but it must be understood that in reality there is a continuum of materials. In addition, some materials, such as composites and cermets, have both brittle and ductile components and exhibit a mixed behavior. Models have been proposed to explain ductile and brittle behavior simultaneously. For example, (Ref 37):

$$\epsilon = A' \cos^2 \alpha \sin n\alpha + B' \sin^2 \alpha \qquad \text{(Eq 12)}$$

where ϵ is the erosion rate, A' is a constant equal to zero if the material is fully brittle, B' is a constant equal to zero if the material is fully ductile, and n is equal to $\pi/2\alpha$ for $\alpha > \alpha_0$ (α is the angle of impact and α_0 is a constant). Another attempt has also been made to describe

both brittle and ductile erosion in one equation (Ref 35). The scope of these equations, however, appears limited in describing the actual performance of materials.

The erosion of ductile materials at relatively low angles of impingement may occur in one or more stages, the first occurring by cutting or plowing. Erosion by cutting can be considered similar to chip formation in machining in that pieces of material are removed by the impact of a sharp-edged particle. With high-speed photography of single particle impacts, plowing (plastic deformation) was shown to occur by the impact of either spherically shaped particles or angular particles with rake angles of 0 to −17° (Ref 38). Although direct material removal can occur under any of these conditions, a significant portion of the displaced material frequently remains adhered to the surface in the form of a lip of heavily deformed material at the ends and sides of the particle trajectory. This material can then be removed by subsequent impacts in a second stage of erosion.

One of the earlier formulations (Ref 39-41) based on rigid angular particles eroding a ductile material predicted that the volume of wear W due to particles of total mass M with velocity V impacting at an angle α between the velocity vector and the surface and horizontal flow pressure p would be:

$$W \sim \frac{nV^2 f(\alpha)}{p} \qquad \text{(Eq 13)}$$

Although Eq 13 presumably described some features of ductile metal erosion at angles less than 45°, most studies have shown the exponent of V to be greater than 2, usually between 2 and 3. More detailed equations accounted for both particles that left the surface while still cutting and those that did not. Similar equations based on an energy balance have been derived (Ref 35).

Erosion by an extrusion and fragmentation mechanism is also possible (Ref 42). In the primary stage of erosion by this mechanism, impacting particles first produce indentations, with lips of deformed material that are removed by fragments from subsequent shattered impact particles scouring the surface. Equations based on this approach predict a total erosion rate with the correct exponent for velocity, but they rely on empirically determined constants.

The preceding applies primarily to angles of impingement of 45° or less. Erosion of ductile materials at a high angle of impingement may also be significant, but specific mechanisms are more difficult to demonstrate. Work hardening with embrittlement, delamination, low-cycle fatigue, and melting, in addition to extrusion or lip formation and fragmentation with scouring as discussed above, have all been proposed (Ref 43). Of these, most studies seem to favor an extrusion mechanism with subsequent lip removal or a delamination or fatigue mechanism. The formation of a relatively stable hilly topography after an incubation period has been reported on iron and copper surfaces as a result of 90° impingement of glass spheres, with the predominant mechanism of erosion being flaking due to subsurface cracking and hillside plowing (Ref 44).

The erosion of brittle materials has been attributed to cracking as a result of Hertzian stresses occurring during impact (Ref 45) and to work hardening to the point where impact stresses exceed the strength of the material (Ref 35). Equation 14 has been proposed (Ref 35) for this type of deformation wear:

$$W = \frac{M(V \sin \alpha - K)^2}{2e} \qquad \text{(Eq 14)}$$

At least under some circumstances, erosion of glass has been shown to occur by the generation and intersection of ring fractures (Ref 46).

Effects of Temperature and Corrosion on Erosion. Increasing the temperature usually increases the erosion rate, at least for ductile materials (Ref 47). A substantial amount of high-temperature erosion testing has been done in recent years in support of the gas turbine and coal gasification/fluidization industries, most of which supports this temperature effect as exemplified by the work described in Ref 48.

Corrosion may increase or decrease the apparent erosion rate, depending upon the rate of oxide or other corrosion-product formation and the resistance of the product to erosion as compared to the normal surface material. If the corrosion product is slow growing and more erosion resistant than the substrate, it may effectively protect the substrate from erosion. Similarly, deposition of some process material on the surface may also protect it. These protective mechanisms have been observed on some boiler heat-exchanger tubes, but are probably more the exception than the rule. More often, oxidation of the substrate with rapid removal of the fragile and brittle oxide (particularly at high angles of impact) accelerates erosion.

Analyzing Wear Failures

At least three sources of evidence are desirable for an accurate analysis of a wear failure: the worn surface, the operating environment, and any wear debris.

Surface damage can range from polishing or burnishing to removal of a relatively large volume of material. Examination of the worn surface can provide a substantial amount of information: the amount of material removed, the type of damage (scratching, gouging, plowing, erosion, fretting, adhesion, pitting, corrosion, or spalling), the existence and character of surface films, whether certain constituents are being attacked preferentially, the direction of relative motion between a worn surface and abrading particles, or whether abrading particles have become embedded in the surface.

Environmental conditions have such a profound effect on the mechanism and rate of metal removal that detailed knowledge of these conditions should always be sought. For example, a limestone crusher sustained erratic wear, with greater wear occurring when rock from one side of the quarry was processed. The rock from that side looked the same as rock from the rest of the quarry, but geological analysis revealed that it contained the silicified remains of a coral reef with sponges that were much harder and more abrasive than the surrounding limestone.

Wear environments may be corrosive, may have been altered during service (such as by breakdown of a lubricant), may provide inadequate lubrication, or may differ from the assumed environment on which the original material selection was made.

Wear debris, whether found between worn surfaces, embedded in a surface, suspended in the lubricant, or beside the worn part, can provide clues to the wear mechanism. A wear particle that consists of a metallic center with an oxide covering is probably a particle that was detached from the worn surface by abrasive or adhesive wear and was subsequently oxidized by exposure to the environment. On the other hand, a small wear particle that consists solely of oxide may be the result of corrosion on the worn surface with subsequent mechanical removal of the corrosion product.

Procedure for Wear Analysis. Generally, the steps entailed in analyzing a wear failure are as follows:

- Identify the actual materials in the worn part, environment, abrasive erodent, wear debris, and lubricant
- Identify the mechanism, or combination of mechanisms, of wear: adhesive, abrasive, corrosive, surface fatigue, or erosive
- Determine the surface configuration of the worn surface and the original surface
- Determine the relative motions in the system, including direction and velocity
- Determine the force or pressure between mating surfaces or between the worn surface and the wear environment on both the macroscopic and microscopic scales
- Determine the wear rate
- Determine the coefficient of friction
- Determine the effectiveness and type of lubricant: oil, grease, surface film, naturally occurring oxide layer, adsorbed film, or other
- Establish whether the observed wear is normal or abnormal for the particular application
- Devise a solution, if required

Laboratory Examination of Worn Parts

Analysis of a wear failure depends to a large degree upon knowledge of the service conditions under which the wear occurred. However,

as with other failures, proper analysis of a wear failure depends upon consideration of many factors and upon careful examination, both macroscopic and microscopic.

Wear failures are generally the result of relatively long-term exposure, yet certain information obtained at the time the failure is discovered can be useful in establishing cause. For example, analysis of samples of the environment, especially the lubricant or analysis of sludge from the lubricating system or from an oil filter can reveal the nature and amount of wear debris or abrasive in the system.

Physical measurements can define the amount and location of wear damage, but they can seldom provide enough information to establish either the mechanism or the cause of the damage. Examination of a worn part generally begins with visual observation and measurement of dimensions, which usually involves the use of micrometers, calipers, and standard or special gages. The amount and character of surface damage must often be observed on a microscopic scale. An optical comparator, toolmaker's microscope, recording profilometer or other fine-scale measuring equipment may be required for adequate assessment of the amount of damage that has occurred.

Weighing a worn component or assembly and comparing its weight with that of an unused part can help define the amount of material lost, as in abrasive wear, or the amount of material transferred to an opposing surface, as in adhesive wear. Weight-loss estimates can also help define relative wear rates for two opposing surfaces that may be made of different materials or that may have been worn by different mechanisms.

Screening of abrasives or wear debris to determine the particle sizes, as well as the weight percentage of particles of each size, is often helpful. The combination of determination of particle size with chemical analysis of the various screenings can provide useful information—for example, when one component in an abrasive mixture is the primary cause of wear or when wear debris and an abrasive coexist in the wear environment. The combination of screening with microscopy can often reveal such details as progressive alteration of the size and shape of abrasive particles with time, as might occur in a ball mill.

Microscopy, particularly scanning electron microscopy (SEM), is used to study features of the worn surface, including the configuration, distribution, and direction of scratches or gouges as well as indications of the preferential removal of specific constituents of the microstructure. Abrasive particles or wear debris can be viewed under the microscope to study their shape and the configuration of their edges (sharp or rounded) and to establish whether or not they have fractured during the wear process.

Direct observation at magnifications greater than about 50× is difficult when the part does not fit on the stage of a metallurgical micro-

Fig. 12 Scoring damage caused by chipping of chromium plating on a 4340 steel cylinder

(a) Scoring on the cylinder. (b) Scoring on a mating cast aluminum alloy B850-T5 bearing adapter. (c) Cross section through a deep score mark in the aluminum alloy adapter revealing a large embedded particle of chromium. 100×

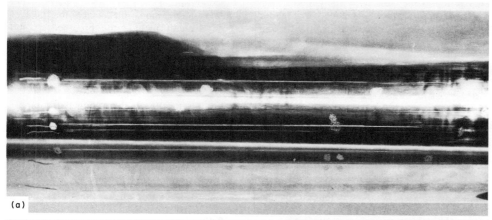

(a)

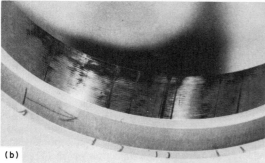

(b)

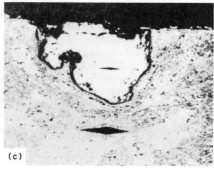

(c)

scope. Sectioning to remove a portion of the worn surface for direct observation precludes repair and reuse of the part. Replication can also be used for optical microscope or SEM observations of the worn surfaces of large parts. Replication using plastic films or harder cast materials offers the additional advantage that a reproduction of the surface can be obtained at a remote site and carried back to the laboratory for detailed study. Hard replicas can be used for physical measurements, such as with a profilometer.

Metallography. Examination of the microstructure of a worn part can reveal such information as whether the initial microstructure was proper or improper, the existence of a localized phase transformation, the existence of a cold-worked surface layer, the presence of an adherent surface film, or, as in the following example, the presence of embedded abrasive particles.

Example 1: Scoring Damage Caused by Chipping of Chromium Plating on a Cylinder. Several large chromium-plated 4340 steel cylinders were removed from service because of deep longitudinal score marks in the plating. One of the damaged cylinders and a mating cast aluminum alloy B850-T5 bearing adapter that also exhibited deep longitudinal

score marks were submitted for examination. In service, the bearing adapter slid along the surface of the cylinder. Scoring on the cylinder and on the bearing adapter are shown in Fig. 12(a) and (b), respectively.

The grooves or score marks on the cylinder were about 25 cm (10 in.) long and were in a band with a width of about 1/12 the circumference of the cylinder. The deeper grooves had completely penetrated the chromium plating, exposing the base steel. The score marks on the aluminum alloy adapter were also in a 1/12-circumference band, but they were deeper than those on the cylinder and were filled with shiny particles. In addition, the adapter showed evidence of heavy localized burnishing in the general area in which scoring occurred.

Investigation. The hardness and adherence of the chromium plating on the cylinder were checked with a steel file and a vibratory etching tool and were found to be satisfactory. No lifting or peeling of the plate was noted in the areas tested.

Microscopic examination of a cross section through one of the deep score marks in the aluminum alloy adapter revealed a large particle of chromium embedded in the groove (Fig. 12c). A hardness test performed on the particle indicated a hardness of 850 HK (66 HRC); the

hardness of the aluminum alloy adjacent to the particle was 186 HK (86 HRB).

Conclusions. It was concluded that high localized loads on the cylinder had resulted in chipping of the chromium plating, particles of which became embedded in the aluminum alloy adapter. The sliding action of the adapter with embedded hard particles resulted in scoring of both the cylinder and the adapter. If the cylinder alone had been available for examination, it might have been concluded that the scoring had been caused by entrapped sand or debris from an external source.

Special Techniques. For valid failure analysis, such techniques as taper sectioning are sometimes needed to allow metallographic observations or microhardness measurements of very thin surface layers. It is almost always necessary to use special materials that support the edge of a specimen in a metallographic mount, for example, nickel plating on the specimen or powdered glass in the mounting material, and to polish the mounted specimen carefully so that the edge is not rounded.

Etchants, in addition to preparing a specimen for the examination of microstructure, can also be used to reveal characteristics of the worn surface. Two examples of features that can be revealed by etching a worn surface are transferred material caused by localized adhesion to an opposing surface and the phase transformations resulting from overheating caused by excessive friction, such as the white layer (untempered martensite) that sometimes develops on steel or cast iron under conditions of heavy sliding contact. Etching a worn surface can also help in detecting the selective removal of specific constituents of the microstructure. The worn surface should be examined under the microscope and photographed before it is etched because some topographical features may be easier to observe on the unetched surface.

Macroscopic and microscopic hardness testing can provide an indication of the resistance of a material to abrasive wear. Because harder materials are likely to cut or scratch softer materials, comparative hardness of two sliding surfaces may be important. Microhardness measurements on martensitic steels may indicate that frictional heat has overtempered the steel and, when used in conjunction with a tempering curve (a plot of hardness versus tempering temperature), can allow a rough estimate of surface temperature. Hardness measurements can also indicate whether or not a worn part was correctly heat treated.

X-ray and electron diffraction analyses can disclose the structure of a crystalline solid. These techniques are particularly valuable for analyzing abrasives, wear debris, or surface films because they can identify compounds, not merely elements. Microstructural features, such as retained austenite, cannot always be seen in microscopic examination of an etched specimen; quantitative diffraction analysis can reveal the relative amounts of such unresolved constit-

Fig. 13 Evidence of galling, or adhesive wear, on the inner surface of a carburized 4720 steel inner cone of a roller bearing

Galling was confirmed by the use of electron probe x-ray microanalysis.

uents in the microstructure. Information on these techniques is available in the articles "X-Ray Powder Diffraction" and "Low-Energy Electron Diffraction" in Volume 10 of the 9th Edition of *Metals Handbook*.

Chemical and Geological Analysis. One or more of the various techniques of chemical analysis—wet analysis, spectroscopy, calorimetry, x-ray spectrometry, atomic absorption spectrometry, or electron microprobe analysis—may be needed for proper analysis of wear failures (see the articles "Classical Wet Analytical Chemistry," "X-Ray Spectrometry," "Atomic Absorption Spectrometry," and "Electron Probe X-Ray Microanalysis" in Volume 10 of the 9th Edition of *Metals Handbook*). The actual compositions of the worn material, the wear debris, the abrasive, and the surface film may be needed in order to devise solutions to a wear problem.

Example 2: Galling Wear on a Steel Inner Cone of a Roller-Bearing Assembly. When a roller-bearing assembly was removed from an aircraft for inspection after a short time in service, several areas of apparent galling were noticed around the inside surface of the inner cone of the bearing. These areas were roughly circular spots of built-up metal, such as the spot shown in Fig. 13. The bearing had not seized, and there was no evidence of heat discoloration in the galled areas.

The inner cone, made of modified 4720 steel and carburized for wear resistance, rode on an AISI type 630 (17-4 PH) stainless steel spacer. Consequently, it was desirable to determine whether the galled spots contained any stainless steel from the spacer. Other items for investigation were the nature of the bond between the galled spot and the inner cone and any evidence of overtempering or rehardening resulting from localized overheating.

Investigation. Electron probe x-ray microanalysis of a cross section through the largest galled spot verified the composition of the spot as type 630 stainless steel. Microscopic exam-

ination revealed that the built-up metal was welded to the cone. Microhardness readings on the unaffected carburized case and on the case beneath the galled area indicated that the heat generated by galling had resulted in localized tempering of the case under the galled spot.

Conclusion. It was concluded that galling had been caused by a combination of local overload and abnormal vibration of mating parts of the roller-bearing assembly.

Importance of Service History in Failure Analysis

One of the early steps in wear-failure analysis is the identification of the type of wear or, if more than one type can be recognized, evaluation of the relative importance of each type as quantitatively as possible. This identification of the type or types of wear requires a detailed description of the service conditions based on close observation and on adequate experience. A casual and superficial description of service conditions is not likely to be of much value.

Descriptions of service conditions are usually incomplete, thus imposing a serious handicap on the failure analyst, especially if he is working in a laboratory remote from the service site. For example, assume that an analyst must study the problem of a badly seized engine cylinder—obviously an instance of adhesive metal-to-metal wear (or lubricated wear, because use of a suitable engine oil is implied). Furthermore, assume that during an oil change the system had been flushed with a solvent, such as kerosene, to rinse out the old oil and had been inadvertently left filled with solvent instead of new oil. Also assume that a slow leak, resulting in loss of the solvent, was not detected during the operating period immediately preceding seizure. The analyst probably would receive the damaged parts (cylinder block and pistons) after they had been removed from the engine, cleaned, and packed. If evidence of the solvent could not be clearly established, determination of the cause of failure would be extremely difficult or perhaps impossible.

Similarly, incomplete descriptions of service conditions can be misleading in analysis of abrasive wear. For example, in describing the source of abrasion that produces wear of mining and ore-handling equipment, generalized references to the ore, such as copper ore, are not uncommon. Such descriptions are too vague to be meaningful; the mineral being extracted usually has little effect on the abrasiveness of the mixture, whereas the bulk rock, or gangue, is the principal source of abrasive particles. Unless the gangue minerals are studied qualitatively and quantitatively, a valid assessment of wear, whether normal or abnormal, is not possible.

In the analysis of conventional lubricated wear, detailed description of the lubricant is essential and often must be supplemented by data regarding pressures applied to mating sur-

faces, operating temperatures, and surface conditions. When corrosion is a factor in lubricated wear, it may be difficult to determine the temperature, degree of aeration, pH and velocity of the lubricant, and the composition and concentration of the corrodent in the lubricant. Other complications that make the analysis more difficult include the presence of substances that inhibit or accelerate corrosion.

Effect of Material Properties on Wear

Various means of identifying the compositions of the worn part, wear debris, surface film, abrasive, and wear environment have been discussed in the section "Laboratory Examination of Worn Parts" in this article. Failure analysis is often successful when the properties of each component in the system and the effects of these properties on the wear process are fully understood.

Adhesive wear is likely to be severe when similar metals rub with little or no lubrication. Metallic particles will be torn from one or both surfaces. Under light contact loads, the particles are likely to be very fine and, if air is present, will probably react completely with oxygen to form oxide wear debris. Under heavy loads, the particles may be somewhat larger, and the wear debris may be mainly metallic, even when air is present. If the metals are dissimilar, they are more likely to be mutually insoluble and thus less susceptible to adhesive wear. Frequently, materials used in sliding bearings are deliberately chosen to be insoluble in the mating material. However, complete insolubility is rare; therefore, satisfactory performance may depend upon a third factor, such as a lubricant or surface film.

Rubbing of a metal against a nonmetal, such as a plastic, sometimes causes reactions similar to metal-to-metal adhesive wear. Particles of the plastic may adhere to the metal and become torn away, eventually resulting in destruction of the plastic component. Plastics have poor thermal conductivity and cannot readily dissipate heat at a junction. For this reason, thin films of certain plastics, such as nylon or PTFE superimposed on a metal base are often better than thicker layers of plastic alone because the metal substrate acts as a heat sink to keep the plastic cool.

Abrasive wear can occur on sliding surfaces that operate in a contaminated environment, as often occurs with journal bearings. Dirt particles may enter the thin space between the journal and the bearing surface and become partly embedded in the soft bearing metal. The resulting projections can cut the journal like many tiny tool bits. Consequently, an important property of many bearing materials is their capacity to embed foreign particles deeply enough to avoid damage to the shaft. Similarly, abrasive particles can be trapped in soft packing material and cause wear of shafts or pump plungers.

Fig. 14 Hard-faced austenitic stainless steel pump sleeve used to pump river water to a brine plant

The sleeve at left, coated with a fused nickel-base hard-facing alloy, shows severe abrasive wear by river-water silt after 3387 h of service. Sleeve at right, coated with plasma-deposited chromium oxide, shows little evidence of wear after 5190 h of service.

Abrasive wear can occur in the dry state or in the presence of a liquid. In the following example, abrasive wear was caused by silt carried by water pumped from a river.

Example 3: Failure of a Hard-Faced Stainless Steel Pump Sleeve Because of Abrasive Wear by River-Water Silt. Whenever river water is used in a manufacturing process, the presence of abrasive silt in the water can be expected to result in wear problems. A typical wear problem was encountered in a brine plant when river water was pumped into the plant by a battery of vertical pumps, each operating at 3600 rpm and at a discharge pressure of 827 kPa (120 psi). The pumps were lubricated by means of controlled leakage. The 3.8-cm (1½-in.) outside-diameter pump sleeves were made of an austenitic stainless steel and were hard faced with a fused nickel-base hard-facing alloy (approximately 58 HRC). Packing for the pumps consisted of a braided PTFE-asbestos material.

After several weeks of operation, the pumps began to leak and to spray water over the platforms on which they were mounted at the edge of the river. In addition to decreasing pumping efficiency, the leaks resulted in the formation of ice on the platforms in cold weather, thus creating a safety hazard. The leaks were caused by excessive sleeve wear that resulted from the presence of fine, abrasive silt in the river water. The silt, which contained hard particles of silica, could not be filtered out of the inlet water effectively. A severely worn sleeve that was removed from a pump after only 3387 h of service is shown at left in Fig. 14. Maximum depth of wear on this sleeve was about 1.6 mm (¹⁄₁₆ in.).

Corrective Measures. To prevent excessive sleeve wear, the nickel-base hard-facing alloy coating on the sleeves was replaced with a plasma-deposited chromium oxide coating (~1300 HV). The coating was ground and lapped to a thickness of 0.1 to 0.13 mm (0.004 to 0.005 in.) and a surface finish of 0.15 to 0.2 μm (6 to 8 μin.). Wear resistance in service was extremely favorable, as shown by the sleeve at right in Fig. 14, which was removed from a pump after 5190 h of service.

Combined Wear Mechanisms

More than one mechanism can be responsible for the wear observed on a particular part. For example, an agricultural tool used in an acidic soil can undergo simultaneous abrasive and corrosive wear. Analysis of wear involving combined mechanisms is very difficult and demands rigorous attention to every detail.

Interaction between wear mechanisms can complicate analysis. For example, in erosion-corrosion, the rate of deterioration by corrosive action can increase by one order of magnitude or more when erosion also occurs. Many corrosion-resistant alloys, such as stainless steels, are relatively stable in a corrosive medium, because they form a thin, tightly adherent surface film that inhibits further corrosion. If the film is removed by abrasive or erosive action, corrosion can proceed on the newly exposed metal surface. Chemical action may reestablish the film, but if erosion removes it as fast as it forms, stability in the corroding medium can no longer exist.

Selection of material for wear application is often based on arbitrary screening tests involving an artificial wear environment. Differences between this artificial environment and the actual conditions of service cannot be overlooked in a failure analysis, and if screening tests are to be used as the basis for corrective action, they should duplicate as closely as possible the actual mechanism or mechanisms of wear in service.

Surface Configuration

Because wear is a surface phenomenon, the original surface configuration of the components in contact influences wear by influencing resistance to relative motion. In bearings, for example, mechanical wear will increase with an increase in surface roughness or out-of-round-ness—factors that increase resistance to rolling or sliding. Cutting tools will wear abnormally if their cutting edges are not sharp or if their cutting angles and clearance angles are incorrect for the application.

Changes in surface configuration that occur during wear affect subsequent stages of wear. In some situations, the process of wearing in, which involves progressive reduction in surface roughness by adhesive or abrasive wear of opposing surfaces, is followed by a period of relatively little wear. The initial smoothing-out of asperities, particularly in lubricated systems that operate under boundary lubrication, reduces the microscopic hills and valleys on the surface to a height that is about the same as the thickness of the lubricant film. The surfaces then ride on each other with no interference between peaks on the opposing surfaces, and wear essentially ceases.

In other cases, particularly if the initial surfaces are somewhat rougher or if boundary lubrication is ineffective, adhesive wear may result in progressive surface roughening and eventual failure. If this process releases wear debris into the joint and if this debris has a particle size that exceeds the thickness of the lubricating film, combined adhesive and abrasive wear between the opposing surfaces and the wear debris can result in rapid deterioration.

Direction of Relative Motion. When only unidirectional sliding is involved, scratches or gouges produced on the worn surface are aligned with the direction of relative motion. In a sleeve bearing, for example, the scratches should run circumferentially on the inner surface of the bearing and on the mating shaft. Scratches that are oriented in other directions indicate such factors as misalignment, vibration, or looseness. These factors can contribute to the severity of the wear.

In devices that undergo combined rolling and sliding, knowledge of the relative velocities and directions of rolling and sliding is necessary to determine the wear mechanism. The direction of rolling is defined as the direction in which the point of contact moves; the direction of rolling is always opposite to the direction of rotation of a rolling element. On a given surface, a condition of positive sliding exists if the direction of sliding is the same as the direction of rolling. Negative sliding occurs on the mating surface, where the directions of rolling and sliding are opposite to each other. Most surface-fatigue failures originate in regions of negative sliding because the shear stresses there are usually more severe than in regions of positive sliding. Negative sliding occurs on the dedenda of gear teeth, on the cam follower riding on a cam, and, in other devices, on the part that has the lower surface velocity in a rolling-sliding system.

Environmental Effects

Service environment influences wear mainly by affecting the chemical reactions at the wear

Fig. 15 Worn 1095 steel yarn eyelet from a textile plant

Service life was improved by changing eyelet material to M2 high-speed tool steel, which contains spheroidal carbides in a martensite matrix.

surface and the chemical and physical stability of materials in the wear system. Environmental effects can be subtle. Even relatively obscure variations, such as differences in atmospheric humidity, have been found to have large effects on friction and wear.

Chemical reactions of surfaces with the environment are involved in corrosive wear and in erosion-corrosion. Some controlled corrosion is often desired, as when EP lubricants are used. However, if the corrosive additives in an EP lubricant are overactive, the contacting surfaces can be damaged by excessive corrosion. It is important to understand the chemical reactions that can occur in the system under study and the inhibiting or enhancing effects that specific environmental constituents can have on these reactions.

Temperature affects the chemical, physical, and mechanical properties of materials. The bulk temperature of a mechanical device or system controls chemical-reaction rates in the environment or at surfaces, solubility of elements or compounds in the environment (or, at very high temperatures, the solubility of phases in a heterogeneous microstructure), and the physical and mechanical properties of contacting materials or the environment. Variations in temperature, such as those that can occur at surfaces because of frictional heat during start-up and shutdown, can cause different effects to prevail in different parts of the system or to prevail at different times. For example, bulk

temperature may indicate that the materials in contacting surfaces are stable; however, overheating caused by excessive friction can produce localized microstructural changes, which can alter the wear resistance of the surface material.

Effects of Microstructure

The microstructural heterogeneity of a wear surface influences the wear process because such constituents as carbides, inclusions, intermetallics, and dispersed phases have properties different from those of the matrix. Hard microconstituents, such as carbides, can make a metal extremely resistant to abrasive wear if they are closely spaced in a relatively hard matrix. For example, the yarn eyelet shown in Fig. 15 was made of 1095 steel hardened and tempered to 60 HRC; it was used in a textile machine to guide highly abrasive synthetic yarn and was severely worn in service. The yarn, which was drawn through the eyelet at high speed, changed direction at the exit of the eyelet, riding on the inside corner. Service life was increased significantly when the material was changed to heat-treated M2 high-speed tool steel, which usually exhibits a microstructure of closely spaced spheroidal carbides in a matrix of ferrite (annealed structure) or tempered martensite (hardened structure). Even greater increases in service life have been achieved in similar applications using chemical vapor deposited or other surface coatings.

Matrix hardness is important to wear resistance. If hard microconstituents are widely dispersed in a matrix that is not hard enough to have good wear resistance of its own, the matrix may wear away rapidly, leaving the hard particles protruding from the surface, where they can cut into a mating surface. For this reason, under dry sliding conditions, fine pearlite exhibits better wear resistance than coarse pearlite or a mixture of ferrite and pearlite.

Example 4: Wear Failure of a 4140 Steel Bolt Because of Low Hardness.

Figure 16 shows a bolt and a thimble used to connect a wire rope to a crane hanger bracket. The bolt, which attached the thimble to the hanger bracket, was worn excessively. Two worn bolts, one new bolt, and a new thimble were examined.

Specifications required the bolts to be made of 4140 steel heat treated to a hardness of 277 to 321 HB. Thimbles were to be made of cast 8625 steel, but no heat treatment or hardness were specified.

Investigation. Transverse sections were taken through the worn areas of the used bolts and through the corresponding area of the new bolt. The thimble was not sectioned, but hardness readings were taken on the outer surface adjacent to the top of the hole that normally

Fig. 16 Bolt and thimble assembly used for connection of a wire rope to a crane hanger bracket showing the worn area of the bolt

Dimensions given in inches

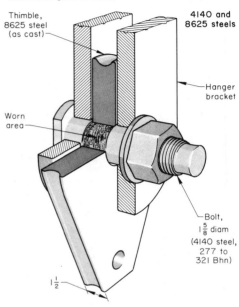

wears under service conditions. Hardness values for the bolt sections and the thimble were:

Sample	Wear surface	Opposite surface	Center
		Hardness, HB	
Used bolt........	327	324	331
Used bolt........	319	327	340
New bolt........	454	...	376
Thimble........	253	231	...

Hardness of the used bolts was within or only slightly above the specified range; the hardness of the new bolt was well above the range. As noted earlier, no hardness was specified for the thimble.

Microstructure of the bolts was tempered martensite. The microstructures of the two used bolts were similar in appearance; however, the microstructure of the new bolt appeared slightly different because it had been tempered at a lower temperature than the used bolts. The thimble showed a bainitic microstructure typical of as-cast steel rather than of heat-treated steel.

Discussion. Mating of the hard bolt surface with the soft thimble surface was conducive to wear of both surfaces. The wearability of the bolt-thimble arrangement could be improved only by equalizing and increasing the hardness of the mating surfaces. Although the bolt was heat treated, the specified hardness of 277 to 321 HB did not produce a surface hardness that yielded good wearability.

Case hardening of the bolt and thimble hole to equal hardness values would produce the best wear characteristics. However, loading of case-hardened bolts could cause the case to spall or fracture from the base metal; also, case hardening the bolt would not improve its impact strength. The wearability and the impact properties of the bolt would be improved by tempering after through hardening.

Corrective Measures. The bolts were through hardened and tempered to the hardness range of 375 to 430 HB. The thimbles were heat treated to a similar microstructure and the same hardness range as those of the bolt. Molybdenum disulfide lubricant was liberally applied during the initial installation of the bolts. A maintenance lubrication program was not suggested, but galling could be reduced by periodic application of a solid lubricant. No further failures were reported after these measures were incorporated.

Carbide-containing steels with a hardness greater than about 0.6 times the hardness of a contacting abrasive have significantly greater abrasion resistance than softer carbide-containing steels. In the softer steels, carbides have little effect on abrasion resistance, probably because the matrix does not provide the necessary support for the dispersed phase. In hypereutectoid steels, maximum hardness is not necessarily optimum for wear resistance. A grain-boundary network of carbides, which may be encountered in many hypereutectoid steels, causes brittle behavior and can result in chipping of the wear surface. Therefore, hypereutectoid steels must be processed so that the carbides are redistributed as a dispersed phase; this usually does not produce maximum hardness.

Austenitic manganese steels are extremely tough nonmagnetic alloys in which the usual hardening transformation has been suppressed by a combination of high carbon and manganese contents and rapid cooling from a high temperature. Manganese steels exceed even the austenitic stainless steels in ability to work harden and probably have no equal in this respect. This property makes these alloys exceptionally resistant to wear accompanied by heavy impact, such as in ore-crushing and earthmoving equipment and in railway frogs and crossings.

It is sometimes believed that unless manganese steel has been work hardened it has poor resistance to wear. This generalization is invalid. The misunderstanding has probably developed because, where much impact and attendant work hardening are present, 12% Mn steel is so clearly superior to other metals that its performance is attributed to the surface hardening. However, controlled-abrasion tests have indicated that in some circumstances the abrasion resistance of austenitic manganese steel is modified little by work hardening, while in others it will outwear harder pearlitic white cast irons without work hardening.

As with other steels, quenching suppresses austenite transformation, with the difference that the critical cooling velocity is quite low below Ar′ (grain-boundary carbide precipitation

Fig. 17 Schematic showing how free lead in a bearing alloy provides boundary lubrication

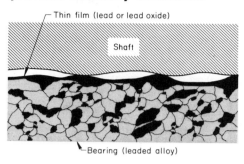

between Ae_{cm} and Ar′ is the primary cause of impaired properties associated with slow cooling). Isothermal transformation between 260 °C (500 °F) and the upper Ae_1 temperature will develop acicular or pearlitic structures, which appear slowly and require very long times for attainment of equilibrium. As with mixed structures in other steels, the mechanical properties of partially transformed austenite are poor, particularly when carbides or other brittle constituents form in large flat plates parallel to crystallographic planes. The mechanical properties are also low when pearlite, which seems to nucleate most readily at grain boundaries, develops in sufficient quantity to form an envelope around each grain. These structures account for the low ductility and reduced strength of austenitic manganese steel after reheating.

Tin bronzes are used widely in plain bearings that operate under boundary lubrication because they will not gall a steel shaft if the lubricant film breaks down. Tin in bronze also strengthens adsorption bonds of the active agent in many boundary lubricants. However, a tin content exceeding 12% can form a hard $Cu_{31}Sn_8$ intermetallic phase that will scratch a steel shaft.

Lead, which is used in bearing bronzes in amounts up to 10%, exists as a separate phase and, under dry sliding conditions, will smear over the surface (Fig. 17) to act as a boundary lubricant, either as metallic lead or as lead oxide. Lead in bearing bronzes can reduce the coefficient of friction in dry sliding against steel by as much as 50%. Lead can also enhance lubrication by influencing the chemical processes that produce semisolid films on the bearing surface in an oil-lubricated system.

Cast irons generally have good wear resistance. Although little is known about the specific roles of the various constituents, particularly graphite, it is generally agreed that abrasion-resistant gray irons should have a microstructure consisting of free graphite in a fine pearlitic matrix with little or no free ferrite. White cast irons can be very resistant to abrasion if the composition is controlled to produce a microstructure of carbides (Fe_3C or Cr_7C_3) in martensite with a small amount of retained

austenite for toughness. Martensitic abrasion-resistant cast irons typically contain chromium, in amounts up to 35%. Low-chromium martensitic irons are alloyed with 4 to 5% Ni for maximum hardness; larger amounts promote austenite retention. Small amounts of molybdenum or copper are sometimes added to enhance hardenability in thick sections.

Coatings. Frequently, many of the effects of wear can be best improved by the use of coatings. A wide variety of coating compositions and deposition methods exist. Deposition methods include thermal spray (detonation gun, plasma, and flame spray), electroplating, electroless plating, chemical vapor deposition (CVD), and physical vapor deposition (PVD). Additional information is available in Volume 5 of the 9th Edition of *Metals Handbook* and in Ref 49.

Thermal spray coatings, particularly detonation gun and plasma, offer perhaps the widest variety of material combinations useful for various types of wear resistance. Virtually any material that can be melted without decomposing can be used to produce coatings. Thus, relatively soft materials, such as aluminum bronze, can be used as a bearing surface with good embeddability, compatibility, conformability, and wear resistance (much greater than its wrought counterpart), while hard tungsten carbide-base cermets and ceramics can be used as adhesion-, abrasion-, and erosion-resistant surfaces. Use of coatings allows the designer to optimize the wear-resistant characteristics of a surface while relying on more common and less costly wrought or cast alloys for structural requirements.

Electroplated chromium coatings are well known for their wear resistance. Electroless coatings are frequently used for corrosion resistance and wear resistance. Extreme care must be taken to ensure proper surface preparation before deposition and strict adherence to deposition procedures to ensure the best quality of either type of coating.

Carbide and nitride coatings produced by CVD and PVD, although usually applied as very thin coatings, have extraordinarily high hardness and provide excellent adhesive wear resistance, abrasive wear under light loads, and erosion resistance in some situations. For example, chemical vapor and physical vapor deposited titanium carbide and titanium nitride are widely used to extend the life of such components as drills, taps, hobs, and burrs.

Laboratory Wear Tests

Although laboratory wear tests rarely duplicate service conditions, they frequently provide useful ranking of materials to facilitate selection for a given application and/or to help analyze wear problems. Various committees of ASTM have developed recommended practices for a variety of wear tests, including those for adhesive wear, abrasive wear, and erosive wear (Ref 50). While these tests may be very useful in relatively ranking materials, caution should be used in trying to extrapolate to actual wear rates in service from laboratory results. Unless previous experience indicates a good correlation between service life and a specific wear test, no attempt should be made to extrapolate from one to the other.

REFERENCES

1. D.A. Rigney, Ed., *Fundamentals of Friction and Wear of Materials*, American Society for Metals, 1981
2. J. Halling, Ed., *Principles of Tribology*, MacMillan, 1975
3. M.B. Peterson and W.O. Winer, Ed., *Wear Control Handbook*, American Society of Mechanical Engineers, New York, 1980
4. D. Scott, Ed., *Wear, Vol 13, Treatise on Materials Science and Technology*, Academic Press, 1979
5. A.Z. Szeri, Ed., *Tribology*, McGraw-Hill, 1980
6. E. Rabinowicz, *Friction and Wear of Materials*, John Wiley & Sons, 1965
7. L.E. Samuels *et al.*, Sliding Wear Mechanisms in *Fundamentals of Friction and Wear of Materials*, D.A. Rigney, Ed., American Society for Metals, 1981, p 13-42
8. N.P. Suh *et al,*. The Delamination Theory of Wear, *Wear*, Vol 44, 1977, p 1-162
9. *Friction, Wear and Lubrication*, Organization for Economic Co-Operation and Development, Paris, 1969, p 14
10. F.P. Bowden and D. Tabor, *The Friction and Lubrication of Solids*, Vol 1, 2, Clarendon Press, 1950, 1966
11. J.F. Archard, Contact and Rubbing of Flat Surfaces, *J. Appl. Phys.*, Vol 24, 1953, p 981-988
12. M.M. Kruschov, Resistance of Metals to Wear by Abrasion; Related to Hardness, in *Institute of Mechanical Engineers Conference: Lubrication and Wear*, 1975, p 655-659
13. R.C. Tucker, Jr., Plasma and Detonation Gun Deposition Techniques, in *Deposition Technologies for Films and Coatings*, R.F. Bunshah *et al.*, Ed., Noyes Publications, 1982, p 454-489
14. C.N. Rowe, Some Aspects of the Heat of Absorption in the Function of a Boundary Lubricant, *Trans. ASLE*, Vol 9, 1966
15. *Friction, Wear and Lubrication*, Organization for Economic Co-Operation and Development, Paris, 1969, p 13
16. E. Rabinowicz, L.A. Dunn, and P.G. Russell, A Study of Abrasive Wear Under Three-Body Conditions, *Wear*, Vol 4, 1961, p 345
17. E. Rabinowicz, L.A. Dunn, and P.G. Russell, The Abrasive Wear Resistance of Some Bearing Steels, *Lubr. Eng.*, Vol 17, 1961, p 587
18. T.O. Mulhern and L.E. Samuels, The Abrasion of Metals: A Model of the Process, *Wear*, Vol 5, 1962, p 478
19. J. Goddard and H. Wilman, A Theory of Friction and Wear During the Abrasion of Metals, *Wear*, Vol 5, 1962, p 114
20. M.M. Kruschov, Resistance of Metals to Wear by Abrasion, as Related to Hardness, in *Proceedings of the Conference on Lubrication Wear*, 1957
21. B.W.E. Avient, J. Goddard, and H. Wilman, An Experimental Study of Friction and Wear During Abrasion of Metals, *Proc. R. Soc. (London) A*, Vol 258, 1960, p 159
22. A.J. Sedricks and T.O. Mulhearn, Mechanics of Cutting and Rubbing in Simulated Abrasive Processes, *Wear*, Vol 6, 1963, p 457
23. M.F. Stroud and H. Wilman, The Proportion of the Groove Volume Removed as Wear in Abrasion of Metals, *Br. J. Appl. Phys.*, Vol 13, 1962, p 173
24. M.A. Moore, Abrasive Wear, in *Fundamentals of Friction and Wear of Materials*, D.A. Rigney, Ed., American Society for Metals, 1981, p 73-118
25. H.S. Avery, Work Hardening in Relation to Abrasion Resistance, in *Materials for the Mining Industry Symposium*, Climax Molybdenum Company, 1974, p 43-77
26. J.N. King and H. Wilman, The Friction and Wear Properties, During Abrasion, of Compressed Graphite Powder Compacts and Commercial Graphitized Carbons, *Wear*, Vol 5, 1962, p 213
27. H.S. Avery, The Measurement of Wear Resistance, *Wear*, Vol 4, 1961, p 427
28. A.G. Evans and T.R. Wilshaw, Quasi-Static Particle Damage in Brittle Solids—I. Observation, Analysis and Implications, *Acta Metall.*, Vol 24, 1976, p 939-956
29. M.A. Moore and F.S. King, Abrasive Wear of Brittle Solids, *Wear*, Vol 60, 1980, p 123-140
30. R.C.D. Richardson, The Wear of Metals by Relatively Hard Abrasives, *Wear*, Vol 10, 1967, p 291-309
31. R.C.D. Richardson, The Wear of Metals by Relatively Soft Abrasives, *Wear*, Vol 11, 1968, p 245-275
32. N.P. Suh, Update on the Delamination Theory of Wear, in *Fundamentals of Friction and Wear of Materials*, D.A. Rigney, Ed., American Society for Metals, 1981, p 43-71
33. *Friction, Wear and Lubrication*, Organization for Economic Co-Operation and Development, Paris, 1969, p 35
34. *Friction, Wear and Lubrication*, Organization for Economic Co-Operation and Development, Paris, 1969, p 31
35. J.G.A. Bitter, A Study of Erosion Phenomena—Part IV, *Wear*, Vol 5-21, 1963, p 169-190
36. P.V. Rao and D.H. Buckley, Time Effect of Erosion by Solid Particle Impingement on Ductile Materials, in *Proceedings of the Sixth International Conference on Erosion by Liquid and Solid Impact*, J.E. Field and N.S. Corney, Ed., University of Cambridge, England, 1983, p 38-1 to 38-10
37. G.P. Tilly, Erosion Caused by Impact of Solid Particles, in *Treatise on Materials Science and Technology*, Vol 13, D. Scott, Ed., Academic Press, 1979, p 287-319
38. I.M. Hutchins, Mechanisms of the Erosion of Metals by Solid Particles, in *Erosion Prevention and Useful Applications*, W.F. Adler, Ed., STP 66, ASTM, Philadelphia, 1979, p 59-76
39. I. Finnie, *Proceedings of the Third U.S. National Congress of Applied Mechanics*, 1958, p 527-532
40. I. Finnie, Erosion by Solid Particles in a Fluid Stream, in *Symposium on Erosion and Cavitation*, STP 307, ASTM, Philadelphia, 1962, p 70-82
41. I. Finnie, Some Observations on the Erosion of Ductile Metals, *Wear*, Vol 19, 1972, p 81-90
42. G.P. Tilly, A Two Stage Mechanism of Ductile Erosion, *Wear*, Vol 23, 1973, p 87-96
43. I. Finnie, A. Levy, and D.H. McFadden, Fundamental Mechanisms of the Erosive Wear of Ductile Metals by Solid Particles, in *Erosion: Prevention and Useful Applications*, W.F. Adler, Ed., STP 664, ASTM, Philadelphia, 1979, p 36-58
44. R. Brown, E. Jin Jun, and J.W. Edington, Wear of Materials—1981, in *Proceedings of the Third International Conference on Wear of Materials*, American Society of Mechanical Engineers, New York, 1981, p 583-591
45. I. Finnie, Erosion of Surfaces by Solid Particles, *Wear*, Vol 3, 1960, p 87-103
46. W.F. Adler, Analytical Modeling of Multiple Particle Impacts on Brittle Materials, *Wear*, Vol 37, 1976, p 353-364
47. G.P. Tilly, Erosion Caused by Airborne Particles, *Wear*, Vol 14, 1969, p 63-79
48. W. Tabakoff, High Temperature Erosion Study of Inco-600 Metal and Fluid Effects on Specimen Sizes, in *Proceedings of the Sixth International Conference on Erosion by Liquid and Solid Impact*, J.E. Field and H.S. Corney, Ed., University of Cambridge, England, 1983, p 55-1 to 55-10
49. R.F. Bunshah *et al.*, *Deposition Technologies for Films and Coatings*, Noyes Publications, 1982
50. *Annual Book of ASTM Standards*, ASTM, Philadelphia, 1986

Liquid-Erosion Failures

Frederick G. Hammitt, Department of Mechanical Engineering and
Applied Mechanics, University of Michigan
Frank J. Heymann, Westinghouse Electric Corporation

EROSION of a solid surface can take place in a liquid medium even without the presence of solid abrasive particles in that medium. Cavitation, one mechanism of liquid erosion, involves the formation and subsequent collapse of bubbles within the liquid. The process by which material is removed from a surface is called cavitation erosion, and the resulting damage is termed cavitation damage. The collision at high speed of liquid droplets with a solid surface results in a form of liquid erosion called liquid-impingement erosion.

Cavitation damage has been observed on ship propellers and hydrofoils; on dams, spillways, gates, tunnels, and other hydraulic structures; and in hydraulic pumps and turbines. High-speed flow of liquid in these devices causes local hydrodynamic pressures to vary widely and rapidly. In mechanical devices, severe restrictions in fluid passages have produced cavitation damage downstream of orifices and in valves, seals, bearings, heat-exchanger tubes, and venturis. Cavitation erosion has also damaged water-cooled diesel-engine cylinder liners.

Liquid-impingement erosion has been observed on many components exposed to high-velocity steam containing moisture droplets, such as blades in the low-pressure end of large steam turbines. Rain erosion, one form of liquid-impingement erosion, frequently damages the aerodynamic surfaces of aircraft and missiles when they fly through rainstorms at high subsonic or supersonic speeds. Liquid-impingement and cavitation erosion are of concern in nuclear-power systems, which operate at lower steam quality than conventional steam systems, and in systems using liquid metals as the working fluid, where the corrosiveness of the liquid metal can promote rapid erosion of components.

Liquid erosion involves the progressive removal of material from a surface by repeated impulse loading at microscopically small areas. Liquid dynamics is of major importance in producing damage, although corrosion also plays a role in the damage process, at least with certain fluid-material combinations. The process of liquid erosion is not as well understood as most other failure processes. It is difficult to define the hydrodynamic conditions that produce erosion and the metallurgical processes by which particles are detached from the surface. Evidently, both cavitation and liquid impingement exert similar hydrodynamic forces on a solid surface. In any event, the appearance of damaged surfaces and the relative resistance of materials to damage are similar for both liquid-impingement and cavitation erosion.

Cavitation

When the local pressure in a liquid is reduced without a change in temperature, a condition may eventually be reached where gas-filled bubbles (or cavities) nucleate and grow within the body of liquid. The gas in the bubbles may be vapor or molecules of a substance that was formerly dissolved in the liquid. If a bubble is formed by vaporization, bubble growth will occur rapidly, but if gas dissolution is required for bubble formation, growth will occur more slowly. Growth of gas-filled bubbles (as opposed to vapor-filled bubbles) depends on the diffusion of dissolved gas to the cavity or on the rate of gas expansion due to pressure reduction. If cavities formed in a low-pressure region pass into a region of higher pressure, their growth will be reversed, and they will collapse and disappear as the vapor condenses or the gas is redissolved in the liquid. A vapor-filled cavity will implode, collapsing very rapidly (perhaps within a few milliseconds); a gas-filled cavity will collapse more slowly—both being the exact or nearly exact reverse of the bubble-growth process (the liquid dynamics of bubble growth and collapse are covered in Ref 1 and 2).

The collapse of cavities (bubbles) produces the damage to materials. The exact mechanism by which cavity collapse transmits severe localized forces to a surface is not fully understood. However, it most likely involves either waves produced by the collapse and immediate reformation of a cavity, a process known as rebound (Fig. 1), or impingement of a microjet of liquid through the collapsing cavity onto the surface being damaged due to nonsymmetrical cavity collapse (Fig. 2). Both rebound and nonsymmetrical collapse with formation of a microjet have been observed experimentally and partly computed analytically (Ref 1, 2).

Collapse pressures were first estimated by Lord Rayleigh in 1917 and have since been estimated by many others using modifications of Rayleigh's theory. Rayleigh found that, for an empty cavity collapsing with spherical symmetry in an incompressible inviscid fluid, the velocity of the collapsing cavity wall and the pressure at the instant of complete collapse were infinitely large. Later analyses, many of which were based on the assumption of adiabatic compression of gas in a collapsing cavity in a compressible fluid, predicted collapse pressures in the range of 30 to 223 MPa (300 to 2200 atm). Although more recent analyses predict wall velocities approaching infinity for a spherical empty cavity at the instant of complete collapse, the presence of gas within the cavity results in wall velocities that rise to a very high value immediately before complete collapse, then fall rapidly to zero at the instant of collapse. Later computer analyses (Ref 2) show more

Fig. 1 The mechanics of cavity growth, collapse, and rebound

(a) Schematic representation of successive stages of growth, collapse, and rebound of a single traveling cavity. (b) Graph of cavity diameter as a function of time for the cavity in (a). Source: Ref 1

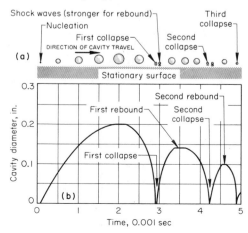

Fig. 2 Schematic representation of successive stages of nonsymmetrical cavity collapse with microjet impingement against a metallic surface

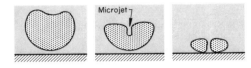

realistic results for symmetrical and non-symmetrical collapses with real fluid parameters.

Actual collapses near a surface do not preserve spherical symmetry very far into the collapse; thus, the Rayleigh model is largely voided. Actual collapses form microjets of liquid (Fig. 2), which probably attain velocities from 100 to 500 m/s (330 to 1640 ft/s). Thus, the actual damaging process may be quite similar to that of liquid impingement, except the jet is much smaller (a few microns in diameter).

One aspect of the collapse of a gas-filled cavity is important: in order to retard collapse significantly and thereby reduce the amount of resulting damage, the gas must be capable of storing much of the thermodynamic work involved in collapsing the cavity. Cavitation usually occurs in a liquid of low vapor pressure and low concentration of dissolved gas when the contents of a cavity are incapable of absorbing any significant amount of the work. Thus, almost all of the energy of collapse will be used to compress the surrounding liquid. The contents of a collapsing cavity have a significant retarding effect on the cavity collapse and the damage that results from it only when the vapor pressure is high compared to ambient pressure or when the dissolved-gas content is high. This behavior is called the thermodynamic effect (Ref 1, 2).

Liquid-Impingement Erosion

The high-velocity impact of a drop of liquid against a solid surface produces two effects that result in damage to the surface: high pressure, which is generated in the area of the impact, and liquid flow along the surface at high speed radially from the area of impact, which occurs as the initial pressure pulse subsides. A first approximation of the average impact pressure, before radial outflow, is the idealized water-hammer pressure that would be generated from the impact between a flat-fronted liquid body and a flat rigid surface. Its value is ρCV, where ρ is the liquid density, C is the acoustic velocity of the liquid, and V is the impact velocity. For example, for water impacting at 480 m/s (1570 ft/s), this pressure is about 1100 MPa (160 ksi)—considerably above the yield strength of many alloys. This value is somewhat reduced by the compressibility of the surface.

In the impact of a spherical drop against a flat surface, the liquid/solid interaction is considerably more complicated; not yet fully understood, it has been a subject of controversy. However, there is now ample analytical and experimental confirmation (Ref 2) that the maximum pressure is developed not at the central point of impact but in a ring around it, and that this maximum pressure is close to twice the idealized water-hammer pressure referred to above (Ref 3). More specifically, as the instantaneous contact area between the impacting drop and the surface increases, the pressure at the perimeter of this area also increases until finally relieved by gross radial outflow of liquid from the impact area. Microscopic observation of damage caused by single impacts has revealed an annular zone of deformation, and sometimes tearing or cracking, which has been attributed to outward-flowing liquid with high superimposed pressures. A central depression is also observed in very ductile metals.

In liquid impingement, each collision between a drop of liquid and a surface can produce damage. On the other hand, it has been established that the collapse of only 1 in 30 000 cavitation bubbles results in visible surface damage (Ref 1).

One model for liquid-impingement erosion processes is illustrated in Fig. 3. The impact-pressure history (and resulting stress waves) can produce circumferential cracks or deformation patterns around the initial area of impact (Fig. 3a), depending on the properties of the surface material and the energy of the impact. Following impact, the liquid flows away radially at high velocity. The spreading liquid may hit nearby surface asperities or the surface steps resulting from plastic deformation caused by the initial impact pressure. The force of this impact stresses the asperity or surface step at its base and may produce a crack (Fig. 3b). Subsequent impacts by other drops may widen the crack or detach a particle entirely (Fig. 3c). Direct hits on existing cracks, pits, or other deep depressions can produce accelerated damage by a microjet-impingement mechanism (Fig. 3d). Eventually, the pits and secondary cracks intersect, and larger pieces of the surface become detached.

Characteristics of Erosion Damage

Materials may be damaged by deformation, ductile fracture, brittle fracture, or fatigue. Corrosion was once thought to play an essential role in cavitation erosion, but some experiments (most notably, tests of plastics in water and of aluminum in toluene) strongly indicate that damage can occur even with the complete lack of corrosion. This does not mean that corrosion does not influence damage in situations where corrosion is known to occur, but rather that corrosion is not a necessary factor in producing damage. However, it may increase total damage considerably (up to tenfold) compared to the mechanical components alone (Ref 2).

Fig. 3 Processes by which a material is damaged by liquid-impingement erosion

(a) Solid surface showing initial impact of a drop of liquid that produces circumferential cracks in the area of impact or produces shallow craters in very ductile materials. (b) High-velocity radial flow of liquid away from the impact area is arrested by a nearby surface asperity, which cracks at its base. (c) Subsequent impact by another drop of liquid breaks the asperity. (d) Direct hit on a deep pit results in accelerated damage, because shock waves bouncing off the sides of the pit cause the formation of a high-energy microjet within the pit.

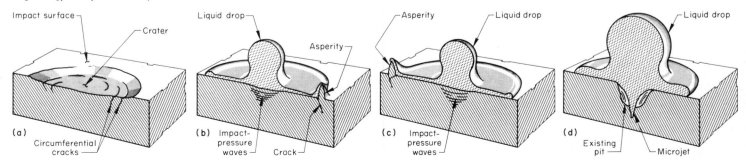

In ductile materials, liquid erosion often occurs by the formation of microscopic craters under the impact of cavitation shock waves, drop impingement, or microjet impingement, as shown in Fig. 4(a) for polycrystalline nickel exposed to intense cavitation in a standard vibratory cavitation test for 5 s at 20 kHz. Longer exposure times result in more widespread damage and a deepening of previously formed shallow pits (Fig. 4b) and eventually in fracture of extruded ridges between adjacent pits (Fig. 4c). Metallographic examination and x-ray diffraction studies have shown that plastic deformation—in the form of both slip and mechanical twinning—can occur in a layer about 30 to 300 μm below the surface during the initial stages of damage. This layer remains fairly constant in thickness throughout the subsequent stages of material removal. Apparently, material is lost by ductile fracture of asperities in the early stages of the erosion process, with fracture of work-hardened surface material and of ridges between erosion pits predominating in later stages. Brittle materials are eroded mainly by fracture and chipping of microscopic particles from the surface.

Experimental evidence has convincingly shown that some erosion damage is the result of single events. Pits observed on erosion-test specimens after short exposures are often essentially unchanged after much longer exposures (Ref 1, 2). However, because fatigue striations are occasionally found on a damaged surface, fatigue cannot be dismissed as a possible damage mechanism. Figure 5 illustrates damage on a stainless steel pump component; erosion occurred in cavitating mercury (Ref 1, 2). Large individual craters are scattered over a background of typical small-scale pitting. Regions of fatigue also were found on this component.

Erosion Rates. The rate of cavitation or liquid-impingement erosion, commonly measured as weight or volume loss per unit of time, often follows one of the patterns shown in Fig. 6. Erosion damage of most materials is not observable as a weight loss until after an incubation period. The erosion rate then usually rises rapidly to a maximum, which may persist for some time, as shown by the dashed curve in Fig. 6, then it usually decreases to a lower value, which either may remain relatively steady or may fluctuate unpredictably. The length of the incubation period, the maximum damage rate, and the shape of subsequent portions of the erosion-rate curve depend on the intensity of cavitation or liquid impingement, the properties of the material, and (to a minor extent) the original surface condition. A smoother surface often increases the incubation period, but does not affect maximum damage rate. At low hydrodynamic intensities, chemical activities of the material and the environment may also influence the rate of damage.

The incubation period appears to coincide with the time needed to develop the subsurface

Fig. 4 Scanning electron micrographs of a surface of polycrystalline nickel damaged by exposure to intense cavitation in a vibratory test at 20 kHz

(a) Shallow craters that formed on the surface after exposure for 5 s. (b) More widespread and deeper attack after exposure for 10 min. (c) Fracture of ridges between deep pits after exposure for 2 h. Source: Ref 4

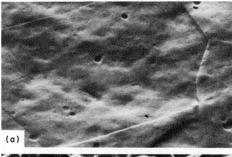
(a)

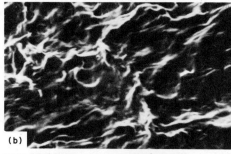
(b)

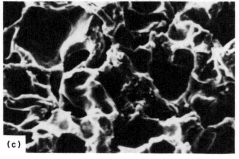
(c)

work-hardened layer mentioned previously. During this time, only random surface pitting occurs with a very slight loss of microscopic particles from widely separated locations on the surface. As detectable weight loss begins, the characteristics of the surface change, with fracture, deep pitting, and fatigue becoming more evident. The exact mechanisms vary with the properties of the material and with the hydrodynamic intensity. Reduction of the damage rate appears to occur when the surface has become so rough that the intensity of individual impacts is reduced by the presence of liquid trapped between deformation ridges or is reduced by the ridges themselves. Advanced stages of liquid erosion produce a characteristic honeycomblike damaged surface, and varia-

Fig. 5 Pitted surface of a stainless steel pump component damaged by exposure to cavitating mercury

Source: Ref 1, 2

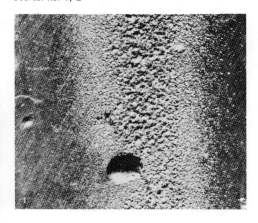

Fig. 6 Schematic representation of typical variation of liquid-erosion rate with exposure time

See text for discussion.

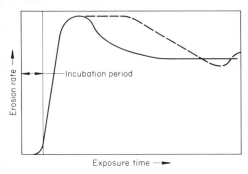

tions in the erosion rate may occur because of gross changes in part contour.

Effect of Flow Velocity. There have been several attempts to correlate the rate of erosion with flow properties of the fluid stream (for cavitation) and with impact velocity of drops (for liquid impingement). Although the results have not been adequately explained by any theory proposed to date, the maximum rate of volume loss per unit of area (sometimes expressed as the mean depth of penetration rate, MDPR) usually varies approximately with an exponential function of the relative velocity between surface and fluid.

In cavitation erosion, the exponent six is most commonly found, but exponents as high as ten and as low as two have been found for the dependence of MDPR on relative velocity (Ref 1, 2). The value of the exponent in a given instance is undoubtedly affected by such factors as cavitation intensity, fluid pressure, type of cavitation, material, and specimen configuration. A strong dependence of damage rate on flow velocity is thus indicated.

In liquid-impingement erosion, the rate of volume loss usually varies with the fifth or sixth power of relative velocity. However, exponents as high as eight to ten have been determined for rain-erosion tests of some nonmetallic materials. Regardless of whether damage occurs by cavitation erosion or liquid impingement, the life of a component in a given erosive situation can be profoundly affected by small changes in relative velocity between the component and the eroding fluid.

Effect of Droplet or Jet Size. Erosion is also dependent on the size of impacting droplets or jets. For a given total mass of liquid impinging, the erosion will be less with smaller drops, even though this results in a larger number of impacts. However, a complete functional dependence and physical explanation is still lacking.

Damage Resistance of Metals

The resistance of specific metals or other materials to liquid erosion, which is commonly evaluated by ASTM G 32 (Ref 5), does not depend on any one property, although many attempts have been made to correlate erosion damage with different intrinsic properties. Various properties, such as hardness, ultimate resilience (one-half the square of ultimate strength divided by the modulus of elasticity), true stress at fracture, strain energy to fracture, corrosion-fatigue strength, and work-hardening rate, appear to be measures of resistance to erosion damage for certain metals or limited classes of alloys; ultimate resilience and hardness appear to be best (Ref 1, 2). However, most such correlations break down when attempts are made to extend them to a wide variety of alloys or to metallic and nonmetallic materials (Ref 6). Even elaborate correlations are often in error by as much as 300%, and for untested materials, they may predict erosion rates that are in error by an order of magnitude or more from the actual rate determined by subsequent testing. Brinell or Vickers hardness appears to be as good a correlating factor as any; its utility is enhanced by its widespread use as a measure of material strength. For many alloys, MDPR varies inversely with HB^n, where the exponent n has a value of about 2. This corresponds to ultimate resilience (Ref 1), which is consistent with an energy-to-brittle fracture model.

Part of the uncertainty involved in developing meaningful correlations is due to the uncertain definition of exposure conditions that produce damage in various laboratory-test mechanisms. Even more influential is that different mechanisms of metal removal appear to exist, depending on the intensity of the cavitation or impingement and the relative importance of corrosion. Intense hydrodynamic conditions seem to favor single-event processes, but conditions that produce impacts of a lesser magnitude are more conducive to fatigue or to

corrosion enhanced by the mechanical removal of protective films of corrosion products.

Effect of Hardness. Hardness is usually a good index of erosion resistance when the same alloy or very similar alloys are considered at different hardness levels. However, erosion resistance of different types of alloys at the same hardness level may vary by an order of magnitude or more.

In some instances, work hardening can increase erosion resistance, especially under mild erosive conditions. However, for long exposures or for intense exposure conditions, erosion resistance may be reduced, probably because work hardening by the eroding medium precedes loss of material by fatigue or fracture. Surface treatments such as shot peening are generally not very effective, because they duplicate the processes that occur during the incubation period.

Thermal treatments, especially those that increase toughness as well as hardness, usually improve erosion resistance. Generally, a ductile and work-hardenable metal of a given hardness will resist erosion better than a brittle metal of the same hardness.

Laboratory experiments and service experience universally confirm that the Stellites, a family of cobalt-chromium-tungsten alloys, are the most resistant of all the structural alloys to liquid erosion. Although the erosion resistance of Stellite alloys is approached by that of some high-strength ausformed or maraging steels and may be equaled by that of some very hard tool steels, the Stellites achieve outstanding erosion resistance with lower hardness and greater resistance to corrosion and stress-corrosion cracking than either high-strength steel or tool steel. Relative to their hardnesses, titanium alloys and the Inconel nickel-base alloys exhibit above average erosion resistance.

Effect of Microstructure. Small grain size and fine dispersions of hard second-phase particles enhance erosion resistance. These characteristics, particularly the latter, appear to give the Stellites and some tool steels their superior erosion resistance.

Some investigations have shown that cavitation impacts induce phase transformations in certain highly erosion-resistant materials, including Stellites and some work-hardening chromium-manganese steels. It has been proposed, but not proven, that the energy absorbed in the transformations contributes to their high erosion resistance.

Ranking for erosion resistance in a given situation is made difficult by the complications of defining both the fluid conditions that result in damage and the metal properties that influence erosion resistance. This is true for laboratory tests and for field evaluations. Even as late as 1960, attempts to rank materials for cavitation resistance produced only a qualitative comparison, because results from different sources varied widely in cavitation conditions and in amount of damage for the same material.

A ranking system that is at least semiquantitative has been developed (Ref 6, 7). In this system, the value of a normalized erosion resistance, defined as the maximum rate of volume loss of a reference material divided by the maximum rate of volume loss for the material being evaluated, is computed. This allows comparison of materials that have been tested under different sets of conditions, provided that the reference material has been tested under each of the different sets of conditions. Figure 7 is a summary of normalized erosion resistance for a wide variety of alloys tested at different conditions, using 18Cr-8Ni austenitic stainless steel with a hardness of 170 HV as the reference material. Figure 7 shows that the most resistant alloys (tool steels, Stellite alloys, and maraging steels) have greater erosion resistance than the reference material, by an order of magnitude or more, and the range of normalized erosion resistance spans almost four orders of magnitude. This range is far greater than any range of intrinsic material properties. For these reasons, the present lack of precision in erosion prediction is not surprising.

Prediction of Erosion Rates. In cavitation, the hydrodynamic conditions are so difficult to describe that no quantitative erosion prediction equation, based on independently measurable parameters, exists. In liquid impingement, however, it is sometimes possible to predict erosion rates based on known values of the amount of liquid impinging, the impact angle and velocity, and perhaps the droplet size. Although theoretically based analytical models for erosion rates have been published, none has yet succeeded in predicting all the empirically observed relationships; some yield erosion rates that are in error by a factor of 10 000 or more.

At the current state of knowledge, better predictions can be made by purely empirical equations derived from compilations of test data. One such equation is given in Ref 6. A development of the same approach, using data from an interlaboratory test program sponsored by ASTM Committee G-2, is given in Ref 7 and summarized below. Equation 1 predicts the maximum erosion rate (the peak in Fig. 6) and cannot be used for long-term extrapolations; it applies to impingement by water droplets or jets:

$$\text{Log } R_e = 4.8 \log V_0 + 0.67 \log d + 0.57J$$
$$- 0.22K - 16.65 - \log \text{NER} \qquad \text{(Eq 1)}$$

where R_e is the rationalized erosion rate, or volume of material removed per unit volume of water impinged, on same area; V_0 is the normal (to surface) component of impact velocity in meters per second; d is the typical droplet or jet diameter in millimeters; J is 0 for impact by droplets, or 1 for lateral impact by cylindrical jets; K is 0 for flat target surfaces, or 1 for curved or cylindrical target surfaces; and NER is the erosion resistance number, essentially identical to the normalized erosion resistance

Fig. 7 Classification of 22 alloys or alloy groups according to their normalized erosion resistances relative to 18Cr-8Ni austenitic stainless steel having a hardness of 170 HV

Source: Ref 6

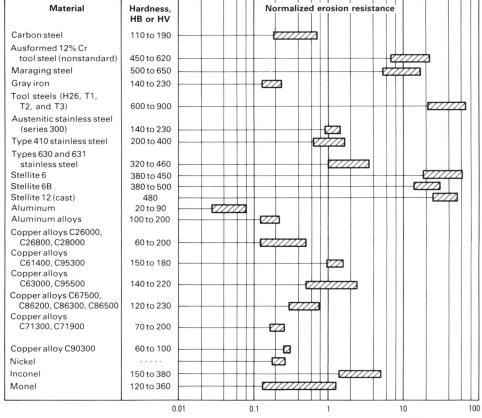

Material	Hardness, HB or HV	Normalized erosion resistance
Carbon steel	110 to 190	
Ausformed 12% Cr tool steel (nonstandard)	450 to 620	
Maraging steel	500 to 650	
Gray iron	140 to 230	
Tool steels (H26, T1, T2, and T3)	600 to 900	
Austenitic stainless steel (series 300)	140 to 230	
Type 410 stainless steel	200 to 400	
Types 630 and 631 stainless steel	320 to 460	
Stellite 6	380 to 450	
Stellite 6B	380 to 500	
Stellite 12 (cast)	480	
Aluminum	20 to 90	
Aluminum alloys	100 to 200	
Copper alloys C26000, C26800, C28000	60 to 200	
Copper alloys C61400, C95300	150 to 180	
Copper alloys C63000, C95500	140 to 220	
Copper alloys C67500, C86200, C86300, C86500	120 to 230	
Copper alloys C71300, C71900	70 to 200	
Copper alloy C90300	60 to 100	
Nickel	. . . . .	
Inconel	150 to 380	
Monel	120 to 360	

0.01   0.1   1   10   100

Normalized erosion resistance relative to 18Cr-8Ni austenitic stainless steel at 170 HV

Fig. 8 Leading edge of a series 400 stainless steel impeller for a boiler feed pump showing deep local damage caused by cavitation erosion

Source: Ref 1, 2

Analysis of Liquid-Erosion Failures

Erosion damage typically appears as a pitted or honeycomblike region, as shown in Fig. 8 on a stainless steel impeller blade from a boiler feed pump. In some instances, erosion damage results in an appreciable loss of metal. For example, the propellers of high-speed ocean liners have sustained sufficient loss of metal in a single crossing of the Atlantic to require replacement.

In hydraulic components, the damaged area will rarely be associated with the region of lowest static pressure. If low static pressure is the cause of cavitation, damage will be downstream of the low-pressure region where vapor pockets cannot be sustained and bubbles implode. It is a common misconception that cavitation damage can occur only in low-pressure regions. Cavitation damage often occurs in relatively high-pressure regions; this is particularly true if sufficiently high flow velocity also occurs. A complete determination of the local cavitation parameter for the flow, K_f, and an estimation of the probable value of the local parameter for incipient cavitation, K_i, can be valuable in establishing whether cavitation is responsible for observed damage (calculation of these parameters is discussed in Ref 1). Regardless of other environmental effects, the existence of drop impingement or hydraulic conditions conducive to cavitation should be positively ascertained before the damage is ascribed to liquid erosion.

Example 1: Failure of a Bronze Pump Impeller by Cavitation Damage. Figure 9(a) shows the impeller from one of two water pumps that were taken out of service because of greatly reduced output. Both impellers showed

of Ref 6 and Fig. 7. There is some evidence that for liquids other than water, an additional factor of 2 log G, where G is the specific gravity of the liquid, should be added. The expected or mean prediction error for Eq 1 is about a factor of three; in some instances, the actual error will be an order of magnitude or more.

Effect of Corrosion

Liquid erosion is known to occur in the absence of any direct evidence of corrosion, yet corrosion can markedly influence the erosion process. Liquid erosion was once thought to be exclusively a corrosion-enhanced process. According to the corrosion theory, collapsing cavitation bubbles cause the mechanical removal of protective surface films. The newly exposed metal surface immediately begins to corrode, forming another film. Repeated removal and re-formation of the film of corrosion products produces the characteristic pitting attack. Although largely rejected as the basic mechanism of liquid erosion, the erosion-corrosion process described above can

drastically accelerate erosive attack, particularly at low hydrodynamic intensities in aggressive environments. At very high hydrodynamic intensities, corrosion is rarely a significant factor, even in aggressive environments.

Because corrosion was once thought to be basic to liquid erosion, cathodic protection was investigated as a means of reducing erosive damage, revealing that cathodic protection did reduce damage. However, it now appears that damage was reduced only when the applied current density was sufficient to generate a layer of hydrogen bubbles on the tested surface. Thus, damage may have been reduced primarily because the layer of hydrogen bubbles cushioned the surface against the hydrodynamic forces of bubble collapse rather than because the cathodic current provided galvanic protection. Cathodic protection will of course reduce total weight loss when corrosion is a significant factor in producing damage. Nevertheless, recent experiments have shown that damage is also sometimes reduced when an anodic current of sufficient density to evolve gas at the tested surface is applied.

considerable material loss over all the interior and exterior surfaces. The pumps drew water from an open tank through a standpipe. Several similar water pumps were operating under almost the same conditions with no observed failures.

The impellers were 25.4 cm (10 in.) in diameter and 1.3 cm (0.5 in.) wide. They were made from a cast bronze alloy and were contained in a cast iron pump casing.

Investigation. Visual examination of the impellers disclosed that the interior surfaces were extremely clean but were pockmarked over the entire area. The flange face on the suction side and the surfaces adjacent to those where material was missing showed evidence of cold work.

Micrographs of sections through the damaged surfaces showed a layer of distorted metal grains (Fig. 9b). At higher magnification, slip lines were visible, indicating that severe cold work had occurred at the surface. There was no evidence of intergranular attack or dezincification.

The clean, pockmarked, severely eroded surfaces of the impellers are characteristic of cavitation damage. In this instance, cavitation damage could have been the result either of a turbulent flow pattern caused by the movement of the impellers in the liquid or of excessive air in the system because the water in the supply tank was low or because air had been drawn through a pump seal. Further investigation revealed that considerable quantities of air were

being drawn into the system when the water in the supply tank was allowed to drop to a low level.

Conclusions. Cavitation erosion caused the metal removal and the microstructural damage evidenced by a shallow layer of severely worked grains at the damaged surface. Cavitation was induced when a low level of water in the supply tank allowed large quantities of air to be drawn into the standpipe along with the water. However, air injection into the bubble-collapse region is sometimes used to reduce damage by cushioning bubble collapse (Ref 1, 2).

Corrective Measures. A water-level control was added to the piping system to maintain a sufficient head of water at the standpipe, and air was excluded from the pump inlet. No further failures occurred.

Example 2: Cavitation Erosion of a Water-Cooled Aluminum Alloy 6061-T6 Combustion Chamber.
Equipment in which an assembly of in-line cylindrical components rotated in water at 1040 rpm displayed excessive vibration after less than 1 h of operation. The malfunction was traced to an aluminum alloy 6061-T6 combustion chamber (Fig. 10a) that was part of the rotating assembly.

The combustion chamber consisted of three hollow cylindrical sections having diameters of 7.5 cm (3 in.), 7.3 cm (2.875 in.), and 3.0 cm (1.1875 in.), respectively (left to right, Fig. 10a).

Investigation. Preliminary examination of the combustion chamber showed pitting on the water-cooled exterior surface in two bands approximately 0.64 cm (0.25 in.) wide that extended completely around the circumference of the chamber at axial locations of 4.8 cm (1.875 in.) and 9 cm (3.5625 in.) from the right-hand end of the 7.3-cm (2.875-in.) diam section of the chamber as shown in Fig. 10(a). The pitting was more severe in the band at the 4.8-cm (1.875-in.) location (particularly over about 180° of the circumference) than in the band at the 9-cm (3.5625-in.) location.

Also, a circumferential groove about 1.3 cm (0.5 in.) wide and having a maximum depth of about 0.25 mm (0.010 in.) had been abraded on the 7.5-cm (3-in.) diam section of the chamber along an arc of approximately 180° at the left edge in Fig. 10(a). At the point at which this wear was observed, the combustion chamber was designed to have a nominal clearance from a concentric housing around it, with cooling water flowing through the intervening annular space. The region of maximum wear was on the same side of the chamber as the region of severest pitting.

In operation, gases in the combustion chamber reached a very high temperature. The high thermal conductivity of the aluminum alloy, the rotation of the chamber, and axial flow of cooling water that was initially at room temperature provided efficient cooling of the chamber.

The 3.0-cm (1.1875-in.) outer-diameter shank served as the fuel inlet, and ignition took place within the main portion of the chamber. Accordingly, the shank was the coolest portion of the chamber and was not expected to be exposed to temperatures above about 95 °C (200 °F), even near the interior surface, on the basis of test data and calculations. Metal temperatures above about 175 °C (350 °F) were expected to be reached only to a very shallow depth on the interior surface in the hottest portions of the main body of the chamber, because of the high heat-transfer rate across the 8-mm (0.3125-in.) thick wall.

Spectrographic analysis showed that the material of the chamber corresponded in composition to aluminum alloy 6061, as specified. Tests also showed that the chamber had been anodized.

Hardness measurements taken at intervals all around the circumference of the chamber near the more severe band of pitting averaged 83 HB, with the lowest reading at 75 HB. The average hardness on the exterior of the shank was 83 HB. These hardnesses were substantially lower than the typical hardness of aluminum alloy 6061-T6, which is 95 HB.

Three cross-sectional specimens were taken for metallographic examination. Specimen 1 was taken through the most severely pitted area, and a portion of this specimen is shown at two magnifications in Fig. 10(b) and (c). This region was generally eroded to a depth of about 0.02 mm (0.001 in.), and some pits (not shown) were several thousandths of an inch

Fig. 9 Water-pump impeller with considerable loss of material from cavitation damage

(a) Photograph of cast bronze impeller. (b) Micrographs of an etched section from the impeller showing a layer of distorted metal grains at the surface subjected to cavitation. 100×

(a)

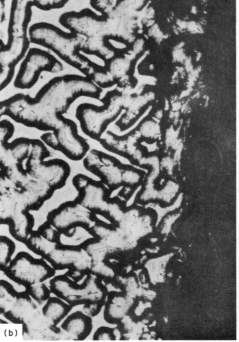

(b)

Fig. 10 Aluminum alloy 6061-T6 combustion chamber damaged by cavitation erosion

The chamber rotated in water at moderate speed. (a) Overall view of the chamber. (b) and (c) Micrographs of cross sections of the chamber wall showing typical cavitation damage. 100 and 500×, respectively

(a)

(b)

(c)

deep. This was also the area where the highest surface temperature on the chamber wall would be expected. Specimen 2 was taken through the most severely abraded region on the 7.6-cm (3-in.) outer diameter section of the part. Specimen 3 was taken through the shank, which was not damaged.

Examination of the three metallographic specimens at a magnification of 800× showed the structure to be essentially the same on each specimen and to contain a fairly dense distribution of a very fine precipitate of magnesium silicide (Mg_2Si) throughout the material. This constituent would be visible only if aluminum alloy 6061 had been heated to temperatures above about 175 °C (350 °F) or if it had been improperly heat treated.

Conclusions. As a result of improper heat treatment, the combustion-chamber material was too soft for successful use in this application. Because even the external surface of the shank, which could not be heated above about 95 °C (200 °F) in use, was just as soft and showed the same distribution of Mg_2Si as the hottest portion of the combustion chamber, overheating in service was eliminated as a possible cause of the observed low hardness.

Misalignment of the combustion chamber and one or both of the mating parts, to which the softness of the chamber material could have been a contributory factor, resulted in eccentric rotation and the excessive vibration that caused malfunction of the assembly. Contact against a surrounding member then caused the exten-

sive abrasion shown at the left edge of Fig. 10(a). The pitting (which showed maximum severity on the same side of the chamber on which there was mechanical abrasion) was produced by cavitation erosion resulting from the combined effects of low hardness of the metal, cyclic pressure variation associated with the eccentric rotation (which induced the low pressures necessary for cavitation bubbles to form in the first place), and metal-surface temperatures near the boiling point of water at the hottest regions of the combustion-chamber exterior.

The operating characteristics of the defective combustion chamber were not sufficiently understood to explain the mechanism by which the cavitation erosion was concentrated at the two bands observed. Irregularities in the housing around the combustion chamber and temperature variation relating to the combustion pattern in the chamber were considered to be possible contributing factors to localization of the cavitation erosion.

Recommendations. The adoption of inspection procedures to ensure that the specified properties of aluminum alloy 6061-T6 were obtained and that the combustion chamber and adjacent components were aligned within specified tolerances was recommended to prevent future occurrences of this type of failure on these assemblies. In a similar situation, consideration should also be given to raising the pressure in the coolant in order to suppress the formation of cavitation bubbles.

Example 3: Liquid Erosion of Hydraulic Dynamometer Stator Vanes. Figure 11 shows severely eroded stator vanes of a hydraulic dynamometer for a steam-turbine test facility. The stator was cast from a copper-manganese-aluminum alloy. This type of dynamometer acts somewhat like a stalled hydraulic coupling: Vanes in the rotor impart tangential momentum to the water recirculating in a spiral path in the toroidal working compartment, and this momentum is killed by the fixed stator vanes. This results in dissipation of mechanical energy, which is converted into heat. The heat is carried away by a continuous through-flow of water superimposed on the recirculating flow. The water enters the working compartment through some of the holes that are visible in the stator vanes and exits through a circumferential slot between the rotor and stator at the outer diameter. The other stator holes vent the center of the recirculating vortex to the atmosphere. The resisting torque developed depends on the fill ratio, the thickness of the recirculating water film. The rate of water through-flow must be sufficient to maintain the temperature rise of the water within acceptable limits.

This dynamometer, designed to absorb up to 51 MW (69 000 hp) at 3670 rpm, constituted an extrapolation of previous design practices and experience. It was subject to severe erosion of the stator (Fig. 11) after relatively short operating times and initially required replacement of the stator after each test program. Up to 60 cm³ (3.7 in.³) of material was lost from each vane. However, even such severe erosion reduced its power-absorption capacity only slightly.

Investigation. The damage was clearly erosion by liquids, but it could not be firmly established—nor was there agreement in speculation—whether it was caused by cavitation or by liquid impact. It could be argued that cavitation is induced in the recirculating flow by the rotor vanes (acting as obstructions) or by the discontinuity of the water-discharge slot. It could also be argued that the acceleration forced on the water in the rotor, which makes the flow hug one side of each rotor pocket, causes a rotating discontinuous pattern of streams to emerge from the rotor, which then produces discrete liquid impacts on the stator vanes. A dynamic pressure transducer, installed in the working compartment, did show strong peaks at rotor vane frequency. Injection of air bubbles did not result in a reduction of fluctuation pressures in the working compartments, although this is a recognized method of controlling cavitation. Minor changes in geometry had little effect on erosion.

Recommendations. The remedy for this machine was a material substitution. The original stator casting material was changed to an Mo-13Cr-4Ni stainless steel (ACI designation CA-6NM). The original casting material has a normalized erosion resistance, N_e, of about 1; the CA-6NM, an N_e of 2 to 2.5. Consideration was given to Stellite cladding, but the CA-6NM

Fig. 11 Vanes of a dynamometer stator damaged by liquid erosion

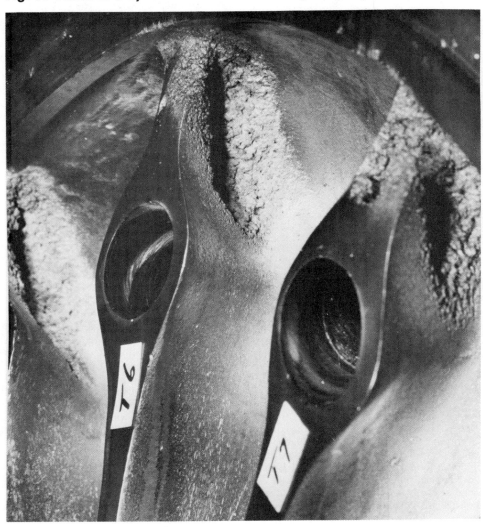

stators, while not erosion free, operated satisfactorily until the test facility was decommissioned. The dynamometer manufacturer has since radically redesigned its line of high-speed high-power dynamometers, with reduction of susceptibility to erosion (and to erosion-producing conditions) as one of the objectives.

Prevention of Erosion Damage

Damage from liquid erosion can be prevented or minimized by reducing the intensity of cavitation or liquid impingement, using erosion-resistant metals, or, under certain conditions, using elastomeric coatings.

Reduction of hydrodynamic intensity in devices subject to liquid impingement can be accomplished by reducing the quantity or size of liquid droplets in the gas stream, by reducing flow velocity, or both. In modern low-pressure steam turbines, for example, the problem of liquid erosion is attacked simultaneously by incorporating interstage moisture-removal devices, which reduce the amount of condensed water that can impinge on rotor blades in the following stage, by increasing the axial spacing between stator and rotor, resulting in smaller size and lower impact velocity of droplets, and by attaching shields of a Stellite alloy or hardened tool steel to the leading edges of rotor blades or locally flame hardening the leading edges, either of which provides the region subject to the greatest damage with a highly erosion-resistant surface layer. Stellite alloys have been used in the form of brazed-on strip, weld-deposited overlay, and laser cladding. Very hard wear-resistant coatings such as tungsten carbide often have not been successful for erosion protection.

In devices subject to cavitation, it may be possible to reduce the hydrodynamic intensity simply by increasing the radius of curvature of the flow path or by removing surface discontinuities. Both of these factors can significantly reduce the probability of cavitation. Increasing the cross section of flow passages will reduce flow velocity, thus reducing the intensity of cavitation. Also, entrained gas in a cavitating liquid reduces collapse pressures by a cushioning effect; consequently, damage can sometimes be reduced by the injection of controlled amounts of air into the liquid. Air injection is often used to reduce damage to certain hydraulic structures, such as dam spillways (additional information on cavitation in hydraulic equipment is available in Ref 1).

Use of Erosion-Resistant Metals. It may be impossible to reduce the hydrodynamic intensity significantly without seriously degrading performance. In such instances, the use of erosion-resistant metals may be the only practical solution to a problem of liquid erosion.

Many of the erosion-resistant metals can be applied as welded overlays; this makes salvage or repair of damaged surfaces easier or surface treatment of new components less costly than would be possible if the component had to be made entirely of erosion-resistant metal. Because liquid erosion is basically a surface phenomenon, the use of erosion-resistant overlays is effective in combating damage. The Stellite alloys and stainless steels are the alloys most widely used as overlays.

For example, Fig. 12 shows two portions of the leading edge of a blade from the last stage of a low-pressure steam turbine. One portion (at left, Fig. 12) was protected by an erosion shield of Stellite 6B; the other portion (at right, Fig. 12) was unprotected type 403 (modified) stainless steel. The shield made of 1-mm (0.04-in.) thick rolled strip and brazed onto the leading edge, resisted erosion quite effectively, but the unprotected base metal did not.

Both blade portions shown in Fig. 12 also illustrate the dependence of erosion on hydrodynamic intensity. Damage was most severe at the leading edge, where hydrodynamic intensity was greatest. Away from the leading edge, impacting droplets were smaller, impact velocity was lower, and impact occurred at oblique rather than right angles; therefore, damage was progressively less severe. The normal velocity component is important in erosion.

Many small parts are not amenable to protection by the use of erosion-resistant overlays. Therefore, the most effective means of combating erosion of small parts is to increase the hardness of the metal or to specify a more erosion-resistant metal.

Use of Elastomeric Coatings. Some devices, particularly those that operate in regions of low hydrodynamic intensity, have successfully resisted erosion when covered with a layer of a flexible material. Highly flexible materials, such as elastomers, are particularly resistant to cavitation erosion (and sometimes impact erosion), especially for low hydrodynamic intensities. In fact, under certain conditions, the erosion resistance of elastomers exceeds that of metal having considerably greater mechanical properties. For example, polyurethane coatings

Fig. 12 Two portions of a modified type 403 stainless steel steam-turbine blade damaged by liquid-impingement erosion

The portion at left was protected by a shield of 1-mm (0.04-in.) thick rolled Stellite 6B brazed onto the leading edge of the blade; the portion at right was unprotected. Compare amounts of metal lost from protected and unprotected portions. Both 2.5×

are widely used to protect radomes and some aluminum alloy surfaces of subsonic aircraft from rain erosion.

The resistance of elastomers to cavitation damage can be partly explained by the observed behavior of microjets during cavitation-bubble collapse near an elastomeric surface. In contrast to the tendency of microjets to be attracted toward rigid surfaces, microjets tend to be repelled from elastomeric or other highly flexible surfaces and dissipate their energy into the fluid rather than against the surface, due to effects on bubble-collapse dynamics near a flexible surface (Ref 1, 2). Theoretical analysis using a simple ideal fluid has verified that a flexible surface acts like a free surface, repelling collapsing bubbles, which are then attracted to any nearby rigid surface.

Elastomers can absorb energy by viscoelastic deformation. This allows them to resist liquid impacts at low hydrodynamic intensities. Impact stresses are attenuated because of the low acoustic impedance of the elastomer compared to that of metals, and the energy of individual impacts is dissipated within the elastomer. However, at high hydrodynamic intensities, the heat generated by the dissipation processes is excessive and causes decomposition and other forms of thermal failure characteristic of these materials.

Flexible coatings, especially rubber, have significant disadvantages; they are difficult to bond to some metals and to complex shapes and are susceptible to damage when short periods of high hydrodynamic intensity are encountered in an otherwise low-intensity environment. Intense cavitation sometimes destroys the bond between a relatively thin layer of rubber and the substrate, perhaps due to temperature buildup in the bond due to cold work.

REFERENCES

1. R.T. Knapp, J.W. Daily, and F.G. Hammitt, *Cavitation*, McGraw-Hill, 1970
2. F.G. Hammitt, *Cavitation an d Multiphase Flow Phenomena*, McGraw-Hill, 1980
3. F.J. Heymann, High-Speed Impact Between a Liquid Drop and a Solid Surface, *J. Appl. Phys.*, Vol 40, 1969, p 5113-5122
4. B. Vyas and C.M. Preece, Residual Stresses Produced in Nickel by Cavitation, in *Proceedings of the 4th International Conference on Rain Erosion and Allied Phenomena*, A.A. Fyall, Ed., Royal Aircraft Establishment, London, 1975
5. "Standard Method of Vibratory Cavitation Erosion Test," G 32, *Annual Book of ASTM Standards*, Vol 03.02, ASTM, Philadelphia
6. F.J. Heymann, *Toward Quantitative Prediction of Liquid Impact Erosion*, STP 474, ASTM, Philadelphia, 1970, p 212-248
7. "Standard Practice for Liquid Impingement Erosion Testing," G 73, *Annual Book of ASTM Standards*, Vol 03.02, ASTM, Philadelphia

SELECTED REFERENCES

- A Discussion on Deformation of Solids by the Impact of Liquids, *Philos. Trans. R. Soc. (London) A*, Part No. 1110, 1966
- *Characterization and Determination of Erosion Resistance*, STP 474, ASTM, Philadelphia, 1970
- P. Eisenberg, H.S. Preiser, and A.T. Thiruvengadam, How to Protect Materials Against Cavitation Damage, *Mater. Des. Eng.*, March 1967
- *Erosion by Cavitation or Impingement*, STP 408, ASTM, Philadelphia, 1967
- *Erosion, Wear, and Interfaces with Corrosion*, STP 567, ASTM, Philadelphia, 1974
- F.G. Hammitt, Cavitation, in *Handbook of Fluids and Fluid Machinery*, John Wiley & Sons, 1986
- F.G. Hammitt, Cavitation and Liquid Impact Erosion, in *Wear Control Handbook*, M.B. Peterson and W.O. Winer, Ed., American Society of Mechanical Engineers, New York, 1980, p 161-230
- C.M. Preece, Ed., *Treatise on Materials Science and Technology*, Vol 16, *Erosion*, Academic Press, 1979
- J.M. Robertson and G.F. Wislicenus, Ed., *Cavitation State of Knowledge*, American Society of Mechanical Engineers, New York, 1969
- T.R. Shives and W.A. Willard, Ed., *The Role of Cavitation in Mechanical Failures*, NBS 394, National Bureau of Standards, Washington, 1974
- *Symposium on Erosion and Cavitation*, STP 307, ASTM, Philadelphia, 1962

Corrosion Failures

Revised by W.G. Ashbaugh, Cortest Engineering Services, Inc.

CORROSION, which is the unintended destructive chemical or electrochemical reaction of a material with its environment, frequently leads to service failures of metal parts or renders them susceptible to failure by some other mechanism. This article deals with failures that proceed almost exclusively by corrosion at normal temperatures. Other articles in this Volume discuss analysis and prevention of several types of failure in which corrosion is a contributing factor. These include "Stress-Corrosion Cracking," "Liquid-Metal Embrittlement," "Hydrogen-Damage Failures," "Corrosion-Fatigue Failures," "Liquid-Erosion Failures," "Wear Failures," and "Elevated-Temperature Failures."

The rate, extent, and type of corrosive attack that can be tolerated in a part vary widely, depending on the specific application. Whether the observed corrosion behavior was normal for the metal used in the given service conditions helps determine what corrective action, if any, should be taken.

All corrosion reactions are electrochemical in nature and depend on the operation of electrochemical cells at the metal surface. This applies even to generalized uniform chemical attack, in which the anodes and cathodes of the cells are numerous, small, and close together. In addition to a basic understanding of electrochemistry, however, the analysis of corrosion failures and the development of suitable corrective measures requires the application of the principles of chemistry and metallurgy.

More detailed discussions of the fundamentals and mechanisms of corrosion are provided in the References listed at the end of this article. It should also be noted that Volume 13 of the 9th Edition of *Metals Handbook* will deal exclusively with the subject of corrosion.

Factors That Influence Corrosion Failures

Several factors, as well as the possibility of their interactions, must be considered by the failure analyst in determining whether corrosion was the cause of, or contributed in some way to, a failure and in devising effective and practical corrective measures. The type of corrosion, its rate, and the extent to which it progresses are influenced by the nature, composition, and uniformity (or nonuniformity) of the environment and the metal surface in contact with that environment. These factors usually do not remain constant as corrosion progresses, but are affected by externally imposed changes and by changes that occur as a direct consequence of the corrosion process itself.

Other factors that have major effects on corrosion processes include temperature gradients at the metal/environment interface, the presence of crevices in the metal part or assembly, relative motion between the environment and the metal part, and the presence of dissimilar metals in an electrically conductive environment. Processing and fabrication operations, such as surface grinding, heat treating, welding, cold working, forming, drilling, and shearing, produce local or general changes on metal parts that, to varying degrees, affect their susceptibility to corrosion.

The specific application determines the amount of metal that can be lost before a part is considered to have failed by corrosion. In some applications, especially where uniform corrosion occurs, a substantial reduction in thickness of a part can be tolerated. In applications where appearance is important or where discoloration or contamination of a food or other product in processing or storage is unacceptable, the dissolution of even a minute amount of metal constitutes failure (see Example 5 in this article).

Localized attack—for example, by pitting—can penetrate the walls of vessels, piping, valves, and related equipment to cause leakage that constitutes failure. Even relatively shallow localized attack can provide stress concentrations or can generate hydrogen on the metal surface, and the result may be failure by mechanisms other than corrosion.

Analysis of Corrosion Failures

In analysis of corrosion failures, the procedures described in the article "General Practice in Failure Analysis" in this Volume should be followed whenever they are applicable. An investigation should include a visit to the scene of the failure whenever possible; otherwise, the analysis of the failure is more difficult and more susceptible to error.

Not all corrosion failures require a comprehensive, detailed failure analysis. Often, the preliminary examination will provide enough information to show that a relatively simple procedure will be adequate. In general, the investigation should consider a broad range of possibilities without being so exhaustive as to be too costly or time-consuming.

Routine checks on whether the material used is actually that specified should not be ignored. For example, such checks have shown that seamless tubes that failed in service by developing longitudinal splits were actually welded tubes that corroded preferentially at the welds; forgings that failed in service were actually castings in which failure was initiated by corrosion at porous areas exposed at the surface of the metal; and Monel parts that corroded rapidly in an environment to which Monel is highly resistant were actually strongly magnetic and made of carbon steel.

History of Failed Part

Before taking samples or conducting tests that might destroy evidence relating to the failure, it is best to obtain and evaluate all the available information that time permits about the circumstances of the failure, the history of the failed part, and the seriousness or potential seriousness of the failure. Information about the type of environment to which the failed part was exposed is of primary concern. The corrosion behavior of the part is affected by both local and upstream chemical composition in the system, by whether exposure to the environment is continuous or intermittent, by temperature, and by whether these and other factors varied during the service life of the part.

If available, engineering drawings and material and manufacturing specifications for the part should be examined, with particular attention being paid to any part changes that may have been made. Missing information should be obtained from operating and inspection personnel, at the same time verifying the accuracy of any relevant documentary information, such as daily log sheets or inspection reports. The investigator should try to learn what, if any, tests or changes that could affect physical evidence relating to the failure may have already been made after the failure occurred. Often, only a small part of the desired information will be available to the failure analyst, information obtained may be of questionable accuracy, or it may be impossible to verify some of the information that is obtained.

On-Site Examination

On-site examination is in general the same for corrosion failures as for other types of failures. The region of failure itself should be visually examined using hand magnifiers and any other suitable viewing equipment that is available.

The areas immediately adjacent to and near the failure, as well as related components of the system, should be examined for possible causative effects on the failure. Remotely located but related equipment should also be examined, particularly in complex systems and where liquids or gases flow. Also, the possibility of the introduction of chemicals or other contaminants from upwind or upstream areas should be checked.

Photography. The failed component and related features of the system should be photographed before samples are removed. Color photographs are particularly useful when colored corrosion products are present; accurate color rendition is enhanced by use of a gray background, which is used as a guide in developing and printing. Detailed information on cameras, lighting arrangements, types of films, lens openings and exposure times, and processing of films will be provided in a subsequent Volume of *Metals Handbook* (see ''Photography of Fractured Parts and Fracture Surfaces'' to be published in Volume 12 of the 9th Edition of *Metals Handbook*).

On-Site Sampling

In conducting on-site sampling, the investigator should be guided by the information already obtained about the history of the failure. The bulk environment to which the failed part was exposed should be sampled, and suitable techniques should be used to obtain samples and make observations—for example, pH—on the local environment at the point of failure.

In addition to taking samples from the failed area, samples from adjacent areas or from apparently noncorroded regions should be obtained for comparison purposes. New or unused parts can provide evidence of the initial or unexposed condition of the part.

Precautions. Removal of specimens and of samples of corrosion product from the failed part requires careful consideration in the selection of locations and method of removal. Suitable precautions must be taken to avoid the destruction of evidence that could be of value in investigating the failure and to avoid further damage to the part or to other related components and structures.

Torch cutting is frequently used for removal of specimens because of its speed and conve-

nience. Cuts should be made at a sufficient distance from the failure site to prevent alteration of the microstructure, thermal degradation of residues that may be present, and the introduction of contaminants. If an abrasive cutoff wheel or saw is used, the same precautions to avoid overheating apply; also, coolants or lubricants that can contaminate or alter the part or any deposits present should not be used.

Protection of Samples. Small parts as well as samples taken from large parts can be protected during transportation to the laboratory by individual packaging. Glass vials and polyethylene film and bags are useful.

Preliminary Laboratory Examination

The procedures followed in the preliminary laboratory examination will vary, depending on whether an on-site examination has already been done by the failure analyst and the completeness of any such examination. An on-site examination by a well-equipped investigator will have included much of the work that would otherwise have to be done in the preliminary laboratory examination. When there has been no on-site examination by the analyst, records on the part and environment, the remainder of the failed part (or at least a good photographic record of it), along with undamaged or unused parts and related components, plus samples of the environment, will assist greatly in the performance of a complete and accurate failure analysis.

Preservation of Evidence. Whether or not an on-site investigation has been done, the course of action should be based on handling of all samples in such a way that maximum information can be gained before any sample is damaged, destroyed, or contaminated to the extent that other potentially useful tests cannot be performed on it. Also, a complete written and photographic record should be maintained through all stages of the investigation.

Visual Examination and Cleaning. Metal samples are first examined visually, preferably with the aid of a hand magnifier or other suitable viewing aid. At this stage, such features as the extent of damage, general appearance of the damage zone, and the color, texture, and quantity of surface residues are of primary interest. If substantial amounts of foreign matter are visible, cleaning is necessary before further examination. The residues can be removed in some areas, leaving portions of the failure region in the as-received condition to preserve evidence. When only small amounts of foreign matter are present, it is sometimes preferable to defer cleaning so that the surface can be examined microscopically before and after cleaning or to defer cleaning until necessary for surface examination at higher magnifications.

Washing with water or solvent, with or without the aid of an ultrasonic bath, is usually adequate to remove soft residues that obscure the view. Inhibited pickling solutions will re-

move adherent rust or scale. Usually, it is advisable to save the cleaning solutions for later analysis and identification of the substance removed. Alternatively, plastic replicas can be used for cleaning (as described in the article ''General Practice in Failure Analysis'' in this Volume); in addition, the replicas retain and preserve surface contaminants, thus making them available for analysis.

Nondestructive Examination. For parts in which internal damage may have resulted from corrosion or the combined effects of corrosion, stress, and imperfections in the metal, the application of nondestructive detection methods before cutting may be desirable. Radiography, ultrasonic flaw detection and measuring, liquid-penetrant inspection, magnetic-particle inspection, eddy-current testing, and holographic examination are the principal methods. The possible introduction of contamination on the test specimen when using some of these techniques should be recognized and considered before their use.

Microscopic Examination

Examination by both light microscopy and electron microscopy can be used to observe minute features on corroded surfaces, to evaluate microstructure of the metallic parts, and to observe the manner in which, and extent to which, the metal was attacked by the corrodent.

Corroded Surfaces. Viewing the cleaned surface with a stereomicroscope clearly shows gross topographic features, such as pitting, cracking, or surface patterns that can provide information about the failure mechanism—whether corrosion was the sole phenomenon involved, the type of corrosion, and whether other mechanisms, such as wear and fracture, were also operative.

If the features cannot be observed clearly using an optical (light) microscope or stereomicroscope, scanning electron microscopes, which produce images with a greater depth of field, can often resolve the features, especially on very rough surfaces. Transmission electron microscopy, using replicas, may be needed to resolve extremely fine features. These instruments, their capabilities, and their limitations are described in the articles ''Analytical Transmission Electron Microscopy'' and ''Scanning Electron Microscopy'' in Volume 10 of the 9th Edition of *Metals Handbook*.

Microstructure. Techniques of specimen preparation are described, and micrographs of both normal and abnormal structures for a wide variety of materials are presented in Volume 9 of the 9th Edition of *Metals Handbook*. Microscopic examination of polished or polished-and-etched sections can reveal not only microstructural features and additional damage, such as cracking, but also the manner in which the corrodent has attacked the metal, such as grain-boundary attack or selective leaching.

It is desirable to retain the corrosion products if they possess sufficient coherence and hardness to be polished. One method of keeping the surface material in place is to impregnate the sample with a castable resin, which is allowed to harden before cutting samples. Polishing on napless cloths with diamond abrasives is recommended to secure maximum quality of edge retention.

Identification and Analysis

To establish what role, if any, corrosion played in a failure, the following must be identified and analyzed: the metal or metals of which the failed part is made, the environment to which the failed part was exposed, inhomogeneities in the metal surface, and foreign matter and metal surface layers. Procedures and equipment usually used are described in the section "Chemical Analysis" in the article "General Practice in Failure Analysis" in this Volume. Chemical spot tests are often used as a first step and when equipment needed for other methods is not immediately available.

A number of techniques may be needed to identify accurately the composition and/or structure of corrosion products. Among these are classical wet analytical chemistry, emission spectroscopy, gas chromatography, Auger electron spectroscopy, x-ray photoelectron spectroscopy, secondary ion mass spectroscopy, x-ray diffraction methods, and electron diffraction methods, such as analytical transmission electron microscopy and electron probe x-ray microanalysis. Each of these techniques is described in separate articles in Volume 10 of the 9th Edition of *Metals Handbook*. An extensive compilation of standards on methods for chemical analysis and surface analysis of metals can be found in Ref 1 and 2, respectively.

Corrosion Testing

Various types of testing techniques are used in investigating corrosion failures and in evaluating the resistance to corrosion of metals and alloys for service in specific applications. These include accelerated tests, simulated-service tests, and electrochemical tests. Monitoring of performance in pilot-plant operations and in actual service is also done, usually in applications for which there is extensive knowledge and experience or where laboratory evaluation has given favorable results. Information on standard test methods for corrosion testing are available from ASTM (Ref 3) and the National Association of Corrosion Engineers (Ref 4).

Accelerated tests are commonly used in investigating failures and evaluating corrosion resistance. Some accelerated test methods have been accepted as standard both by the military and by industry. To shorten testing time, corrosion is accelerated in relation to naturally occurring corrosion, usually by increasing temperature or using a more aggressive environment. Because the various factors that influence natural corrosion processes differ widely in time dependence, the results of accelerated tests must be interpreted with extreme care and can be related to expected actual service behavior

only where close correlation with long-term service results has been established.

Simulated-service tests are frequently used in analyzing corrosion failures and evaluating the corrosion behavior of metals and alloys in specific applications. In these tests, either actual parts or test specimens are exposed to a synthetic or natural service environment.

A considerable number of laboratory corrosion tests that are not accelerated tests have been described in ASTM standards, in military specifications, and in standards issued by the National Association of Corrosion Engineers. These procedures provide guidance in reproducing service conditions as accurately as possible and in interpreting and evaluating the test results in relation to service failures.

Electrochemical tests provide data that can establish criteria for passivity or anodic protection against corrosion and determine critical breakdown or pitting potentials. Two general methods of electrochemical corrosion testing are used—controlled-current and controlled-potential methods. For either test method, ASTM G 3 provides useful guidance and standardization of the manner of recording and reporting electrochemical measurements (Ref 5).

In the controlled-current method, the current (a measure of the corrosion rate) is controlled, and the resulting corrosion potential is measured. A number of instruments are available for such tests, in which either logarithmic or linear polarization curves are developed. Both galvanostatic and galvanodynamic polarization measurements are used to plot anodic and cathodic polarization curves.

A limitation of controlled-current testing is that corrosion rates often change with time, and estimating continuing corrosion rates is not always accurate. Also, the applied current density in these tests depends on the absolute difference between the total oxidation and total reduction current densities. Consequently, if the total oxidation current density consists of two components, only one of which is corrosion, the apparent corrosion rate from the test will exceed the actual corrosion rate from weight loss.

In the controlled-potential method of electrochemical testing, the corrosion potential (oxidizing power) is controlled, and the resulting corrosion current is measured. Equipment is available for both constant-potential (potentiostatic) and varying-potential (potentiodynamic) testing to determine overall corrosion-rate profiles for metal-electrolyte systems over a range of potentials.

Corrosion Rates and Corrosion Types

To determine whether a corrosion failure was caused by the use of an unsuitable material, it is necessary to know whether the rate and type of corrosion were normal for the given metal-environment combination. An extensive com-

pilation of data on corrosion rates is given in Ref 6. This compilation is arranged by corrosive substances in alphabetical order (including about 1400 chemicals and chemical mixtures used in industry) and graphically shows corrosion rates for more than 25 metals and alloys (both ferrous and nonferrous). For each corrodent-metal combination, the corrosion rates at various concentrations and temperatures are given. Special notes are included on susceptibility to pitting, intergranular attack, stress-corrosion cracking (SCC), and crevice corrosion. Additional information on corrosion behavior and complications stemming from corrosion-rate data can be found in Ref 7 and 8.

Uniform Corrosion

Corrosion of metals by uniform chemical attack is the simplest and most common form of corrosion, and it occurs in the atmosphere, in liquids, and in soil, frequently under normal service conditions. The rate of attack can be rapid or slow, and the metal surface can either be clean or covered with corrosion products. Selection of a metal that has a suitable resistance to the environment in which the specific part is used and the application of paints and other types of protective coatings are two common methods used to control uniform corrosion.

Uniform corrosion commonly occurs on metal surfaces having a homogeneity of chemical composition and of microstructure. Access to the metal by the attacking environment is generally unrestricted and uniform. In uniform corrosion, electrochemical reaction between adjacent closely spaced microanode and microcathode areas is involved; consequently, uniform corrosion might be considered as localized electrolytic attack occurring consistently and evenly over the surface of a metal.

All metals are affected by this form of attack in some environments; the rusting of steel and the tarnishing of silver are typical examples of uniform corrosion. In some metals, such as steel, uniform corrosion produces a somewhat rough surface by removing a substantial amount of metal, which either dissolves in the environment or reacts with it to produce a loosely adherent, porous coating of corrosion products. In such reactions as the tarnishing of silver in air or the attack on lead in sulfate-containing environments, thin, tightly adherent protective films are produced, and the metal surface remains smooth.

Corrosion rate and expected service life can be calculated from measurements of the general thinning produced by uniform corrosion. However, because the rate of attack can change over a period of time, periodic inspection at suitable intervals is ordinarily carried out to avoid unexpected failures. The protection provided by paints and other resinous coatings and their limitations are discussed in the section "Use of Resinous and Inorganic-Base Coatings" in this article.

Fig. 1 Uniform corrosion of steel tubes in boiler feedwater containing oxygen (O₂) and a chelating water-treating chemical

Fig. 2 Effect of acid concentration on the corrosion rate of iron completely immersed in aqueous solutions of three inorganic acids at room temperature.

It should be noted that the scales for corrosion rate are not the same for all three charts. As discussed in text, the corrosion rate of iron (and steel) in nitric acid in concentrations of 70% or higher, although low compared to the maximum rate, is sufficient to make it unsafe to ship or store nitric acid in these metals. Source: Ref 9

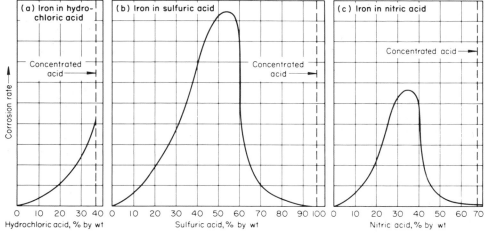

Modification of the environment by changing its composition, concentration, pH, and temperature or by adding an inhibitor are also effective and appropriate methods of controlling uniform corrosion in some situations.

Example 1. The carbon steel tubes shown in Fig. 1 have been corroded to paper thinness, revealing a lacelike pattern of total metal loss. These steel tubes are from a boiler feedwater heater feeding a deaerator. As part of the boiler-water treatment program, it was decided to inject a chelate to control scale formation in the boiler tubes. Unfortunately, the chelate was added ahead of the preheater, where the boiler water still contained oxygen (O₂). As the chelate removed iron oxide, the O₂ reformed more iron oxide, and this uniform dissolution of steel reduced the tubing to the totally corroded condition shown. Moving the chelate addition to a point after the deaerator stopped the corrosion and allowed reuse of steel in the preheater.

Effect of Concentration. The effect on corrosion rate of increasing or decreasing the concentration of corrodent in the environment to which a metal part is exposed does not follow a uniform pattern because of (1) ionization effects in aqueous solutions and the effects of even trace amounts of water in nonaqueous environments and (2) changes that occur in the characteristics of any film of corrosion products that may be present on the surface of the metal. Typical patterns of the types of corrodent-concentration effects on corrosion rate that may be encountered are illustrated in Fig. 2, which plots the variation of corrosion rate of iron as a function of the concentration of three common inorganic acids in aqueous solutions at room temperature.

The rate of corrosion of a given metal usually increases as the concentration of the corrodent increases, as shown in Fig. 2(a) for the corrosion of iron in hydrochloric acid. However, corrosion rate does not always increase with concentration of the corrodent; the effect often depends on the range of corrodent concentration, as shown in Fig. 2(b) and (c) for iron in sulfuric acid and in nitric acid, respectively.

Because corrosion is electrochemical and involves anodic and cathodic reactions, process variables influence corrosion rate if they influence one or both reactions. For example, the main cathodic reaction for iron corroding in dilute inorganic acids is $2H^+ + 2e \rightarrow H_2\uparrow$. The more hydrogen ions available, the faster the rate of the cathodic reaction. In turn, this permits a high rate of anodic dissolution: $(M \rightarrow M^{+n} + ne)$ where M stands for metal. This is what happens throughout the concentration range in hydrochloric acid solutions.

In both nitric acid and sulfuric acid solutions, the hydrogen-ion concentration increases with acid concentration, but at the higher levels of acid concentration, it decreases again. Iron may be used to handle concentrated sulfuric acid at ambient temperatures. Care must be taken to avoid any contamination with water, because this will dilute the acid and increase the rate of attack. Also, impurities in the acid or the iron can increase the rate of attack.

The corrosion rate of iron (and steel) at room temperature in nitric acid decreases with increasing concentration above about 35% because of the formation of a passive oxide film on the metal surface. However, this passive condition is not completely stable.

The rate of attack for concentrations of 70% or higher, although low compared to the maximum rate shown in Fig. 2, is still greater than 1.3 mm/year (50 mils/year), making iron and steel unsuitable for use in shipping and storing nitric acid at any concentration (see also the section "General Corrosion" in this article). Nitric acid in bulk is usually stored and shipped in type 304 stainless steel,

aluminum alloy 3003, or commercially pure titanium (grade 2).

Metals that have passivity effects, such as Monel 400 in hydrochloric acid solutions and lead in sulfuric acid solutions, corrode at an extremely low rate at low acid concentration at room temperature, but lose their passivity at a certain limiting acid concentration above which the corrosion rate increases rapidly with increasing acid concentration.

Effect of Temperature. When investigating the effect of temperature on the rate of corrosion of a metal in a liquid or gaseous environment, the temperature that must be considered is that existing at the metal/corrodent interface; this temperature often differs substantially from that of the main body of the corrodent. This difference is especially important for heat-transfer surfaces, where the hot-wall effect causes a much higher corrosion rate than is ordinarily encountered in the solution being heated. An increase of 10 °C (18 °F) in bulk temperature of the solution can increase the corrosion rate by a factor of two or more. Hot-wall failures are fairly common in heating coils and heat-exchanger tubes.

An example of the hot-wall effect is the corrosion rate of Hastelloy B in 65% sulfuric acid. At 120 °C (250 °F), the corrosion rate would be less than 0.5 mm/year (20 mils/year). However, a Hastelloy B heating coil in the same solution might have a surface-wall temperature of 145 °C (290 °F) and a corrosion rate greater than 5 mm/year (200 mils/year).

In some metal-corrodent systems, there is an approximately exponential rise in corrosion rate with an increase in temperature at the interface. In others, the corrosion rate is low for low corrodent concentrations at room temperature; increasing the temperature up to a certain point has practically no effect on the corrosion rate, but the corrosion rate increases very rapidly at temperatures above that point.

Temperature changes sometimes affect corrosive attack on metals indirectly. In systems where an adherent protective film on the metal may be stable in a given solution at room temperature, the protective film may be soluble in the solution at higher temperatures, with the result that corrosion can progress rapidly.

When boiling occurs in the solution, other factors also influence the rate of attack on a metal immersed in it. One factor may be simply an increase in velocity of movement of the liquid corrodent against the metal surface, which in turn increases the corrosion rate. In some situations, more radical changes occur at the metal surface, such as the substitution of steam and spray in place of liquid as the corrodent or the formation of a solid film on the metal surface—either of which may completely alter the metal/corrodent interface conditions.

There are some exceptions to the general rule that increasing temperature increases the corrosion rate. One is the reduction in rate of attack on steel in water as the temperature is increased, because the increase in temperature decreases the oxygen content of the water, especially as the boiling point of the water is approached. Other exceptions arise where a moderate increase in temperature results in the formation of a thin protective film on the surface of the metal or in passivation of the metal surface.

If thick deposits are formed on a heat-transfer surface, they have a twofold effect by changing the metal surface temperature and making crevice corrosion possible. Local differences in temperature on a heat-transfer surface exposed to steam can influence corrosion rates by causing differences in the duration of exposure to condensation. For example, the inner surface of a carbon steel tube that carried saturated steam at 234 °C (454 °F) corroded to a greater depth opposite exterior heat-transfer fins than elsewhere on the inner surface (Fig. 3). The heat loss through the fins lowered the temperature of the inner surface of the tube in the area beneath the fins, resulting in the presence of condensate on this area for a greater percentage of the time and in a correspondingly greater loss of metal than elsewhere on the inner surface. All corroded areas were fairly smooth.

Pitting Corrosion

Pitting of metals is extremely localized corrosion that generally produces sharply defined holes. The attack on the interior walls of the hole is usually reasonably uniform, but may be irregular where the specific conditions introduce a secondary intergranular attack. Every engineering metal or alloy is susceptible to pitting. Pitting occurs when one area of a metal surface becomes anodic with respect to the rest of the surface or when highly localized changes

Fig. 3 Carbon steel steam tube that corroded on the inner surface more rapidly opposite the exterior heat-transfer fin than elsewhere along the tube

Etched longitudinal section. 3 ×

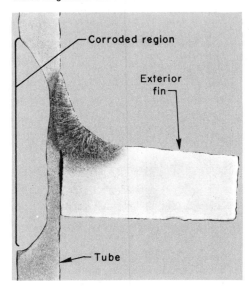

in the corrodent in contact with the metal, as in crevices, cause accelerated localized attack.

In general, when pitting occurs on a freely accessible clean metal surface, a slight increase in corrosivity of the environment will cause general or uniform corrosion. Pitting on clean surfaces ordinarily represents the start of breakdown of passivity or local breakdown of inhibitor-produced protection. When pits are few and widely separated and the metal surface undergoes little or no general corrosion, there is a high ratio of cathode-to-anode area, and penetration progresses more rapidly than when pits are numerous and close together.

Difficulty of Detection. Pitting is one of the most insidious forms of corrosion; it can cause failure by perforation while producing only a small weight loss on the metal. Also, pits are generally small and often remain undetected. A small number of isolated pits on a generally uncorroded surface are easily overlooked. A large number of very small pits on a generally uncorroded surface may not be detected by simple visual examination, or their potential for damage may be underestimated. When pits are accompanied by slight or moderate general corrosion, the corrosion products often mask them.

Pitting is sometimes difficult to detect in laboratory tests and in service, because there may be a period of months or years, depending on the metal and the corrodent, before the pits initiate and develop to a readily visible size. Delayed pitting sometimes occurs after an unpredictable period of time in service, when some change in the environment causes local

destruction of a passive film. When this occurs on stainless steels, for example, there is a substantial increase in solution potential of the active area, and pitting progresses rapidly.

Stages of Pitting. Immediately after a pit has initiated, the local environment and any surface films on the pit-initiation site are unstable, and the pit may become inactive after just a few minutes if convection currents sweep away the locally high concentration of hydrogen ions, chloride ions, or other ions that initiated the local attack. Accordingly, the continued development of pits is favored in a stagnant solution.

When a pit has reached a stable stage, barring drastic changes in the environment, it penetrates the metal at an ever-increasing rate by an autocatalytic process. In the pitting of a metal by an aerated sodium chloride solution, rapid dissolution occurs within the pit, while reduction of oxygen takes place on adjacent surfaces. This process is self-propagating. The rapid dissolution of metal within the pit produces an excess of positive charges in this area, causing migration of chloride ions into the pit.

Thus, in the pit there is a high concentration of MCl_n and, as a result of hydrolysis, a high concentration of hydrogen ions. Both hydrogen and chloride ions stimulate the dissolution of most metals and alloys, and the entire process accelerates with time. Because the solubility of oxygen is virtually zero in concentrated solutions, no reduction of oxygen occurs within a pit. Cathodic reduction of oxygen on the surface areas adjacent to pits tends to suppress corrosion on these surface areas. Thus, isolated pits cathodically protect the surrounding metal surface.

Because the dense, concentrated solution within a pit is necessary for its continuing development, pits are most stable when growing in the direction of gravity. Also, the active anions are more easily retained on the upper surfaces of a piece of metal immersed in or covered by a liquid.

Some causes of pitting are local inhomogeneity on the metal surface, local loss of passivity, mechanical or chemical rupture of a protective oxide coating, galvanic corrosion from a relatively distant cathode, and the formation of a metal ion or oxygen concentration cell under a solid deposit (crevice corrosion).

The rate of pitting is related to the aggressiveness of the corrodent at the site of pitting and the electrical conductivity of the solution containing the corrodent. For a given metal, certain specific ions increase the probability of attack from pitting and accelerate that attack once initiated. Pitting is usually associated with metal-environment combinations in which the general corrosion rate is relatively low; for a given combination, the rate of penetration into the metal by pitting can be 10 to 100 times that by general corrosion.

With carbon and low-alloy steels in relatively mild corrodents, pits are often generally distributed over the surface and change locations as they propagate. If they blend together, the individual pits become virtually indistinguishable, and the final effect is a roughened surface but a generally uniform reduction in cross section. If the initial pits on carbon steel do not combine in this way, the result is rapid penetration of the metal at the sites of the pits and little general corrosion.

The most common causes of pitting in steels are surface deposits that set up local concentration cells and dissolved halides that produce local anodes by rupture of the protective oxide film. Anodic corrosion inhibitors, such as chromates, can cause rapid pitting if present in concentrations below a minimum value that depends on the metal-environment combination, temperature, and other factors. Pitting also occurs at mechanical ruptures in protective organic coatings if the external environment is aggressive or if a galvanic cell is active.

Buried pipelines that fail because of corrosion originating on the outside surface usually fail by pitting corrosion. Another form of corrosion affecting pipelines involves a combination of pitting and erosion (erosion-corrosion). This form of corrosive attack will be discussed in the section "Velocity-Affected Corrosion in Water" in this article.

With corrosion-resistant alloys, such as stainless steels, the most common cause of pitting corrosion is highly localized destruction of passivity by contact with moisture that contains halide ions, particularly chlorides. Chloride-induced pitting of stainless steels usually results in undercutting, producing enlarged subsurface cavities or caverns (Example 2).

Example 2: Corrosion Failure by Pitting of Type 321 Stainless Steel Aircraft Fresh-Water Tanks Caused by Retained Metal-Cleaning Solution. Two fresh-water tanks of type 321 stainless steel were removed from aircraft service because of leakage. One tank had been in service for 321 h, the other for 10 h. The tanks were made of 0.81-mm (0.032-in.) thick type 321 stainless steel. Pitting and rusting had occurred only on the bottom of the tanks near a welded outlet, where liquid could be retained after draining. Most of the pits were about 6 to 13 mm (¼ to ½ in.) from the weld bead at the outlet.

Service History. Inquiry revealed that there had been departures from the specified procedure for chemical cleaning of the tanks in preparation for storage of potable water. First, the sodium hypochlorite sterilizing solution used was at three times the prescribed strength. Second, although the tanks were drained after the required 4-h sterilizing treatment, a small amount of solution remained in the bottom of the tanks and was not rinsed out immediately. Rinsing was delayed for 16 h for the tank that failed after 321 h of service and for 68 h for the tank that failed after 10 h of service, exposing the bottom surfaces of the tanks to hypochlorite solution that had collected near the outlet.

Chemical Tests. Samples of the tanks were subjected to a 5% salt-spray test for three days.

No corrosion was observed. Additional samples, which were subjected to one cycle of the Huey test (a test sample is boiled in 65% nitric acid for 48 h), showed no appreciable intergranular attack.

Samples of both tanks were tested for reaction with sodium hypochlorite at three strength levels:

- The prescribed concentration of 125 ppm available chlorine
- The triple-strength solution (375 ppm available chlorine) used in sterilizing both tanks
- The full-strength solution (5% available chlorine) from the stock bottle

The samples immersed in the first solution rusted and discolored the solution in approximately 2 h. Samples in the second solution showed no corrosion for one to two days but a slight reaction after four to five days. Samples in the third solution showed no corrosion after four to five days of immersion.

Samples from the tanks, including the heat-affected zones (HAZs) near the welds at the outlet, were pickled in an aqueous solution containing 15% by volume concentrated nitric acid and 3% by volume concentrated hydrofluoric acid to remove heat discoloration and scale, then immersed in the three sodium hypochlorite solutions for several days. None of the samples showed any sign of corrosion.

Metallographic inspection of specimens from the tanks disclosed a normal microstructure with no precipitated carbides. Cross sections through pits in the tanks showed subsurface enlargement of the pits (Fig. 4).

Conclusions. Oxidizing solutions, such as hypochlorites, and nitrates in certain concentration ranges cause rapid pitting of stainless steels. With greater concentration, passivation is complete; at lower concentrations, chloride does not penetrate the passive film. Apparently, the concentration of sterilizing solution used was in the range that promotes pitting.

Failure of the stainless steel tanks by chloride-induced pitting resulted from using an overly strong hypochlorite solution for sterilization and neglecting to rinse the tanks promptly afterward. Failure to remove the scale from the HAZ near the welds at the outlet and failure to passivate the interior of the tanks may have accelerated the pitting, although severe pitting also occurred outside the HAZ.

Corrective Measures. Directions for sterilization and rinsing were revised. The instructions for diluting the hypochlorite solution were clarified, and the importance of using the proper strength solution was emphasized. Immediate rinsing after draining the solution was specified, using six fill-and-drain cycles and leaving the tank full of potable water as the final step.

To improve resistance to chloride-induced pitting further, it was recommended that welding be carefully controlled and adequately protected with inert gas to restrict heat effects

Fig. 4 The bottom of a type 321 stainless steel aircraft fresh-water storage tank that failed in service as a result of pitting
This unetched section shows subsurface enlargement of one of the pits. 95 ×

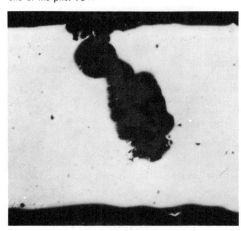

to formation of a golden-brown to deep-purple discoloration on the metal, avoiding the formation of black scale. It was also recommended that the scale and discoloration of the welds at the outlet and HAZs be removed by abrasive blast cleaning and that the interior of the completed tanks be passivated, using either 22% nitric acid at 60 °C (140 °F) or 50% nitric acid at 50 °C (125 °F). Three rinse cycles of water immersion and drainage were stipulated, to be followed by final rinsing inside and out with fresh water and then oven drying.

Pitting of Various Metals. Despite their good resistance to general corrosion, stainless steels are more susceptible to pitting than many other metals. High-alloy stainless steels containing chromium, nickel, and molybdenum are also more resistant to pitting, but are not immune under all service conditions.

Pitting failures of corrosion-resistant alloys, such as Hastelloy C, Hastelloy G, and Incoloy 825, are relatively uncommon in solutions that do not contain halides, although any mechanism that permits the establishment of an electrolytic cell in which a small anode is in contact with a large cathodic area offers the opportunity for pitting attack.

In active metals, such as aluminum and magnesium and their alloys, pitting also begins with the establishment of a local anode. However, a major effect on the rate of pitting penetration is that the corrosion products of these metals produce a high pH in the pit, thus accelerating the dissolution of the protective oxide film. Although the surrounding metal may be exposed to relatively pure water, the metal at the bottom of a pit corrodes at an appreciable rate, even in the absence of galvanic effects. Maintaining surface cleanness helps avoid pitting of these metals.

Table 1 Combinations of alloys and environments subject to selective leaching, and elements removed by leaching

Alloy	Environment	Element removed
Brasses	Many waters, especially under stagnant conditions	Zinc (dezincification)
Gray iron	Soils, many waters	Iron (graphitic corrosion)
Aluminum bronzes	Hydrofluoric acid, acids containing chloride ions	Aluminum
Silicon bronzes	Not reported	Silicon
Copper nickels	High heat flux and low water velocity (in refinery condenser tubes)	Nickel (denickelification)
Monels	Hydrofluoric and other acids	Copper in some acids, and nickel in others
Alloys of gold or platinum with nickel, copper, or silver	Nitric, chromic, and sulfuric acids	Nickel, copper, or silver (parting)
High-nickel alloys	Molten salts	Chromium, iron, molybdenum, and tungsten
Cobalt-tungsten-chromium alloys	Not reported	Cobalt
Medium-carbon and high-carbon steels	Oxidizing atmospheres, hydrogen at high temperatures	Carbon (decarburization)
Iron-chromium alloys	High-temperature oxidizing atmospheres	Chromium, which forms a protective film
Nickel-molybdenum alloys	Oxygen at high temperature	Molybdenum

Fig. 5 Difference in dezincification of inside and outside surfaces of a plated copper alloy C26000 (cartridge brass, 70% Cu) pipe for domestic water supply

Region A shows plug-type attack on the nickel-chromium-plated outside surface of the brass pipe that initiated below a break in the plating (arrow). Region B shows uniform attack on the bare inside surface of the pipe. Etched with $NH_4OH-H_2O_2$. 85×

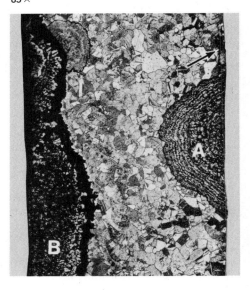

Pitting of aluminum and magnesium and their alloys in aqueous solutions is often accelerated by galvanic effects, which occur because these metals are anodic to most other metals, and by the effects of dissolved metallic ions and suspended particles in the solution. Pitting of these metals can be caused by even a few parts per billion of dissolved copper compounds as well as by particles of metallic iron (which produce both crevice and galvanic effects) that are embedded in or adhering to their surfaces.

General rules for avoiding pitting of any metal are as follows: (1) keep the surface clean, (2) avoid contact with stagnant solutions, (3) avoid galvanic couples, and (4) avoid rupture of any natural or applied protective coatings.

Selective Leaching

Selective leaching is the removal of an element from an alloy by corrosion. The most common example is dezincification, the selective removal of zinc in brasses. Many alloys are susceptible to selective leaching under certain conditions. The elements that are more resistant to the environment remain behind, provided they have a sufficiently continuous structure to prevent them from breaking away in small particles. Table 1 lists some of the alloys for which selective leaching has been reported, together with the corrosive environments and the elements removed by leaching.

Mechanisms. Two mechanisms have been described for selective leaching: (1) two metals in an alloy are dissolved, and one redeposits on the surface; and (2) one metal is selectively

dissolved, leaving the other metals behind. Dezincification of brasses occurs by the first mechanism; the loss of molybdenum from nickel alloys in the molten sodium hydroxide occurs by the second. In some alloys, selective leaching takes place by either mechanism, depending on temperature and on the type, concentration, and flow rate of the corrodent.

A special case of selective leaching is preferential attack on inclusions (Example 5). The metal in the affected area becomes porous and loses much of its strength, hardness, and ductility. Failure may be sudden and unexpected because dimensional changes are not always substantial and the corrosion sometimes appears to be superficial, although the selective attack may have left only a small fraction of the original thickness of the part unaffected.

Dezincification occurs in brasses containing less than 85% Cu. Zinc corrodes preferentially, leaving a porous residue of copper and corrosion products. Alpha brass containing 70% Cu and 30% Zn (copper alloy C26000) is particularly susceptible to dezincification when exposed in an aqueous electrolyte at elevated temperatures.

Dezincification proceeds as follows: the brass dissolves, the zinc ions stay in solution, and the copper plates back on. Dezincification can proceed in the absence of oxygen, as evidenced by the fact that zinc corrodes slowly in pure water. However, oxygen increases the rate of attack when it is present. Analyses of dezincified areas usually show 90 to 95% Cu, with some of it present as copper oxide.

Dezincification may be either uniform or of the plug type (Fig. 5). High zinc content in a brass favors uniform attack; relatively low zinc content favors plug-type attack. Composition

of the liquid in contact with the metal has a greater effect on the type of dezincification, but the pattern of behavior is neither completely consistent nor fully understood. Slightly acidic water, low in salt content and at room temperature, is likely to produce uniform attack, whereas neutral or alkaline water, high in salt content and above room temperature, often produces plug-type attack.

Access of corrodent, flow rate, and other factors are sometimes involved. As shown in Fig. 5, plug-type attack occurred on the exterior of a nickel-chromium-plated brass pipe for a domestic water supply; untreated municipal water leaking from the faucet packing gland ran down the pipe and attacked the brass through a break in the plating caused by mechanical damage. The water inside the pipe produced fairly uniform, layer-type dezincification on the pipe, which had an initial wall thickness of 0.8 mm (0.030 in.). Only about one-third of the original wall remained as sound metal in the corroded 0.6-mm (0.024-in.) thick region shown in Fig. 5.

Example 3: Failure of Copper Alloy C27000 Innercooler Tubes for Air Compressors Because of Dezincification. After about 17 years in service, copper alloy C27000 (yellow brass, 65% Cu) innercooler tubes in an air compressor began leaking cooling water, causing failure and requiring the tubes to be replaced. The tubes were 19 mm (¾ in.) in

Fig. 6 Copper alloy C27000 (yellow brass, 65% Cu) air-compressor innercooler tube that failed by dezincification

(a) Unetched longitudinal section through the tube. (b) Micrograph of an unetched specimen showing a thick uniform layer of porous, brittle copper on the inner surface of the tube and extending to a depth of about 0.25 mm (0.010 in.) into the metal, plug-type dezincification extending somewhat deeper into the metal, and the underlying sound metal. 75×. (c) Macrograph of an unetched specimen showing complete penetration to the outside wall of the tube and the damaged metal at the outside wall at a point near the area shown in the micrograph in (b). 9×

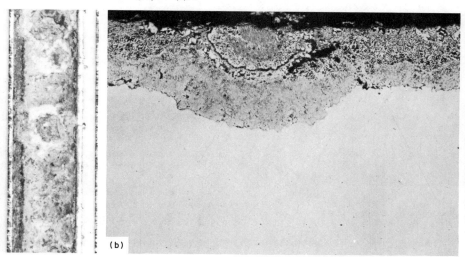

diameter and had a wall thickness of 1.3 mm (0.050 in.). The cooling water that flowed through the tubes was generally sanitary (chlorinated) well water; however, treated recirculating water was sometimes used. Sections of 12 cooler tubes were sent to the laboratory for a determination of the cause and mechanism of failure.

Investigation. The tubes were sectioned longitudinally (Fig. 6a) so that the inside surfaces could be examined. Visual examination of the inside surfaces disclosed a thick layer of porous, brittle copper. Figure 6(b) shows a uniform layer of the spongelike copper to a depth of about 0.25 mm (0.010 in.) on the inside surface and a plug-type deposit that penetrated somewhat deeper into the tube wall; the sound metal underlying this damaged surface layer is also shown in Fig. 6(b). Figure 6(c) shows a macrograph that illustrates the damaged metal at the outside wall of the tube at one of many points where the wall had been completely penetrated.

Spectrochemical analysis of the base material and the brittle layer disclosed that the tubes had been fabricated from copper alloy C27000 (yellow brass, 65% Cu) and that only a trace of the nominal 35% Zn remained in the brittle layer. It was concluded that failure of the tubes was the result of the use of an uninhibited brass that has a high zinc content and therefore is readily susceptible to dezincification.

Recommendations. The material for the tubes should be replaced with copper alloy C68700 (arsenical aluminum brass), which contains 0.02 to 0.06% As and is highly resistant to dezincification. Copper alloy C44300 (inhibited admiralty metal) could be an alternative

selection for this application; however, this alloy is not as resistant to impingement attack as copper alloy C68700.

Before the failure analysis was completed, the innercooler was retubed using tubes made from aluminum bronze (copper alloy C61400 or C62800) and was put back into service. However, copper alloy C68700 was the material used for all subsequent replacement tubes.

Copper alloys with high copper content, such as alloy C23000 (red brass, 85% Cu), are almost immune to dezincification in most water and salt solutions, as are inhibited brasses and other copper-zinc alloys containing 0.02 to 0.6% As, Sb, or P. Apparently, these inhibiting elements are redeposited on the alloy as a film and thereby hinder deposition of copper. For severely corrosive environments where dezincification occurs or for critical parts, copper nickels, such as copper alloys C70600 (10% Ni), C71000 (20% Ni), and C71500 (30% Ni) are used.

Graphitic Corrosion. Perhaps the second most frequently observed type of selective leaching is graphitic corrosion of gray iron, which occurs in relatively mild aqueous environments and on buried pipe. The graphite in gray iron is cathodic to iron and remains behind as a porous mass when iron is leached out. Graphitic corrosion usually occurs at a low rate. The graphite mass is porous and very weak, and graphitic corrosion produces little or no change in metal thickness. A corroded surface usually does not appear different from gray iron.

Graphitic corrosion does not occur in ductile iron or malleable iron, because no graphite network is present to hold together the residue.

White iron has essentially no free carbon and is not subject to graphitic corrosion.

Detection of Selective Leaching. On many alloys, selective leaching is not readily detected by visual examination. A copper flash is usually visible on dezincified copper alloys, but this is not positive evidence of dezincification because a copper flash can deposit from even small amounts of copper salts in aqueous solution without the occurrence of dezincification.

An area where selective leaching has occurred will sound dull when struck. However, severe intergranular corrosion may produce the same effect.

Metallographic examination will show the porous structure and different etching characteristics of the remaining metal. In some metals, selective leaching will not leave a porous surface layer, but will cause a change in microstructure, for example, decarburization of high-carbon steel. The low strength and low hardness of the selectively leached layer are usually detectable. This layer can often be scraped or chipped away.

Where other observations are inconclusive, an electron microprobe analysis can be made of the affected surface and original alloy; a loss in one or more elements will usually be very apparent. However, the electron microprobe is relatively insensitive to elements with an atomic number less than 11. (See the article ''Electron Probe X-Ray Microanalysis'' in Volume 10 of the 9th Edition of *Metals Handbook* for details on this analytical technique.)

A layer is sometimes formed on an alloy that appears, even under metallographic examination, to be selectively leached. Microprobe analysis can be used to determine if the layer is actually a new element introduced into the original alloy by diffusion—for example, carbon and iron diffusion into pure nickel in molten salts or carbon and oxygen diffusion into titanium at elevated temperatures—but only at fairly high concentrations (chemical analysis or metallographic examination are needed to detect trace amounts).

Intergranular Corrosion

Intergranular corrosion is preferential dissolution of the grain-boundary phases or the zones immediately adjacent to them, usually with slight or negligible attack on the main body of the grains. Grain boundaries are generally slightly more active chemically than the grains themselves, because they are areas of mismatch between the orderly and stable crystal lattice structure within the grains.

The preferential attack is enhanced by the segregation of specific elements or compounds, or enrichment of one of the alloying elements, in the grain boundaries, or by the depletion of an element necessary for corrosion resistance in the grain-boundary areas. Susceptibility to intergranular corrosion is usually related to thermal processing, such as welding or stress relieving, and can be corrected by a solution heat treatment, which redistributes alloying elements more uniformly; modification of the alloy; or the use of a completely different alloy.

When the attack is severe, entire grains may be dislodged because of the complete deterioration of their boundaries. In the presence of residual or applied stress, failure by SCC can occur before substantial intergranular attack has occurred.

Austenitic stainless steels become sensitized or susceptible to intergranular corrosion when heated in the temperature range of about 550 to 850 °C (1000 to 1550 °F). Heating at 650 °C (1200 °F) for 1 h is often used for intentionally sensitizing stainless steel specimens for testing purposes.

The extent of the sensitization effect is a function of both time and temperature. Exposure to temperatures near the middle of this range for a few minutes is equivalent to several hours near the upper and lower limits. The limits of the sensitizing temperature range cannot be defined exactly, because they are influenced by the composition, especially the percentage of carbon and of carbide-forming elements such as titanium, niobium, and niobium plus tantalum.

Also, whether sensitizing is damaging in a given instance depends on the requirements of the specific application and the environment and stresses to which the alloy is exposed as well as the prior thermal and mechanical working history of the alloy. Holding at temperatures as low as about 400 °C (750 °F) and as high as about 900 °C (1650 °F) for prolonged periods has been reported to cause sensitization of austenitic stainless steels.

Fig. 7 Sensitized 304 stainless steel exhibiting intergranular attack

100×

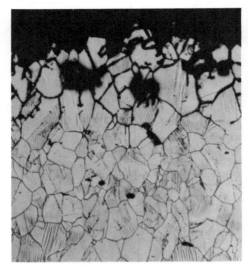

Depletion of chromium in the grain-boundary areas is usually the cause of intergranular corrosion in austenitic stainless steels. Generally, more than 10% Cr is needed to give stainless steels corrosion resistance substantially greater than that of carbon and low-alloy steels.

In the sensitizing range, chromium carbides and carbon precipitate out of solution if the carbon content is about 0.02% or higher. The result is metal with lowered chromium content in the area immediately adjacent to the grain boundaries. The chromium carbide in the grain boundary is not attacked, but in many corrosive environments, the chromium-depleted zone immediately adjacent to the grain boundary is attacked.

Type 304 (18-8) stainless steel usually contains from 0.06 to 0.08% C; thus, excess carbon is available for combining with the chromium and precipitating the carbide. Carbon diffuses toward the grain boundary readily at sensitizing temperatures, but chromium diffuses much more slowly.

The typical appearance of intergranular corrosion of stainless steels is as shown in Fig. 7 for an attack on sensitized type 304 stainless steel at 80 °C (180 °F) in water containing a low concentration of fluorides in solution. Intergranular corrosion of the type shown in Fig. 7 is more or less randomly oriented and does not have highly localized propagation, as does intergranular stress corrosion, in which cracking progresses in a direction normal to applied or residual stresses.

One method of reducing the susceptibility of austenitic stainless steels to intergranular corrosion is to use solution heat treatment, usually by heating to 1065 to 1120 °C (1950 to 2050 °F) and immediately water quenching. By this pro-

cedure, chromium carbide is dissolved and retained in solid solution, provided cooling of the steel from the solution heat-treating temperature is done without delay and proceeds rapidly. Solution heat treatment poses difficult problems on many welded assemblies and is generally impracticable on large equipment or in making repairs.

Using stainless steels that contain less than 0.03% C reduces susceptibility to intergranular corrosion sufficiently for serviceability in many applications. Somewhat better performance can be obtained from types 347 or 321 stainless steel, which contain sufficient titanium and niobium (or niobium plus tantalum), respectively, to combine with all of the carbon in the steel.

Example 4: Intergranular Corrosion of a Type 304 Stainless Steel Fused-Salt Pot Due to Sensitization. A fused-salt electrolytic-cell pot containing a molten eutectic mixture of sodium, potassium, and lithium chlorides and operating at melt temperatures ranging from 500 to 650 °C (930 to 1200 °F) exhibited excessive corrosion after two months of service. The pot, shown schematically in Fig. 8(a), was a welded cylinder with 3-mm (⅛-in.) thick type 304 stainless steel walls and was about 305 mm (12 in.) in height and diameter. There were severe localized corrosion and horizontal cracking below and immediately above the level of molten salts (melt level) adjacent to a vertical weld.

Investigation. Six specimens were removed from the wall of the pot for evaluation. Their location with respect to the melt level and the vertical weld is shown in Fig. 8(a). The structures of the specimens are shown in Fig. 8(b) to (f). The surface of each specimen examined corresponded to the inner peripheral surface of the pot wall.

All the specimens exhibited carbide precipitation in the grain boundaries, with continuous networks in most instances. Two specimens were taken from regions (b) in Fig. 8(a); both areas were 100 mm (4 in.) above the melt level and were exposed to temperatures substantially below the melt temperature. One of these specimens (not shown in Fig. 8) was 57 mm (2¼ in.) from the weld and outside one HAZ. The other (Fig. 8b), was in the HAZ. Both were similar in microstructure and showed no evidence of intergranular corrosion.

As shown in Fig. 8(c), no grain-boundary attack was observed in the specimen taken from region (c) in Fig. 8(a), which is located 45 mm (1¾ in.) above the melt level (at temperatures slightly below the melt temperature) and outside the HAZ. The specimen from region (d) in Fig. 8(a), which was the same distance above the melt level as specimen (c) but was within the HAZ, exhibited severe grain-boundary attack (Fig. 8d).

The specimens from regions (e) and (f) in Fig. 8(a), both of which had been sensitized in welding and were in direct contact with the fused salt and thus exposed to temperatures of

Fig. 8 Fused-salt electrolytic-cell pot of type 304 stainless steel that failed by intergranular corrosion as a result of metal sensitization

(a) Configuration and dimensions (given in inches). (b) to (f) Micrographs of corroded and uncorroded specimens taken from the correspondingly lettered areas on the pot shown in (a). Etched with CuCl$_2$. 500×

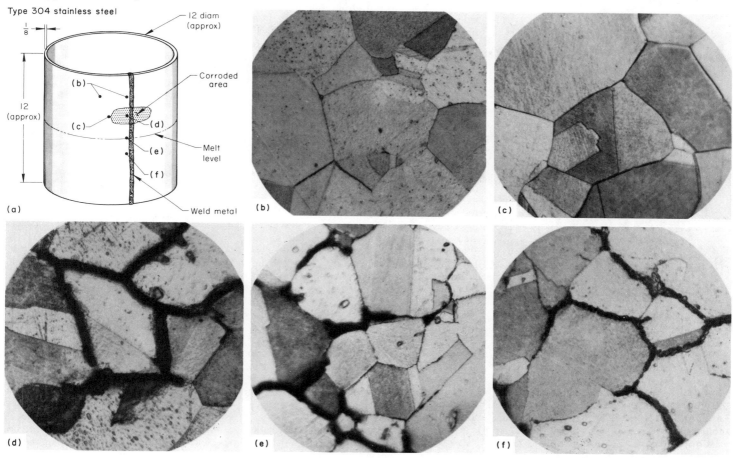

500 to 650 °C (930 to 1200 °F) in service, exhibited extensive grain-boundary penetration (Fig. 8e and f). Corrosion in these specimens was so severe that numerous regions were found where entire grains had fallen out, leaving behind microscopic cracks.

Conclusions. The pot failed by intergranular corrosion because an unstabilized austenitic stainless steel containing more than 0.03% C had been sensitized and placed in contact in service with a corrosive medium at temperatures in the sensitizing range.

Welding, followed by exposure in service to temperatures in the sensitizing range, caused severe intergranular corrosion in the HAZ below the melt level (region f), at the melt level (region e), and 45 mm (1¾ in.) above the melt level (region d). Although carbide precipitation was present in the remaining three areas, intergranular corrosion did not occur in any of them because none of these areas was exposed to both the welding heat and temperatures in the sensitizing range in service.

Recommendations. Material for the pot should be changed from type 304 stainless steel

to Hastelloy N (70Ni-17Mo-7Cr-5Fe). Maximum corrosion resistance and ductility are developed in Hastelloy N when the alloy is solution heat treated at 1120 °C (2050 °F) and is either quenched in water or rapidly cooled in air.

An alternative, but less suitable, material for the pot was type 347 (stabilized grade) stainless steel. After welding, this stainless steel should be stress relieved at 900 °C (1650 °F) for 2 h and rapidly cooled to minimize residual stresses while avoiding sensitization.

Corrective Measures. Type 347 stainless steel was selected to replace type 304, primarily because it cost less and was more available than Hastelloy N. Manufacturing techniques were similar to those used previously, but included the recommended stress-relief heat treatment. The new pot performed satisfactorily for six months, after which the operation was discontinued.

Subsequent visual examination of the interior wall revealed no localized corrosion or cracking like that observed on the original pot. In addition, the general surface corrosion extended

over wider areas, but was much more uniform and was shallower.

Although satisfactory service was obtained for six months with the pot made of type 347 stainless steel, this alloy has only marginal behavior for this application, in which the melt temperature can reach 650 °C (1200 °F) on occasions, and the use of alloy Hastelloy N should be considered if trouble-free performance for longer periods of time is required.

Nickel Alloys. Precipitation-hardenable nickel alloys corrode intergranularly in some environments. Inconel X-750 is susceptible to intergranular corrosion in hot caustic solutions, in boiling 75% nitric acid, and in high-temperature water containing low concentrations of chlorides or other salts. It can be made resistant to intergranular corrosion in these corrodents, but not necessarily in others, by heat treating the cold-worked alloy at 900 °C (1650 °F).

Solid-solution nickel-base alloys, such as Inconel 600, are subject to grain-boundary carbide precipitation if held at or slowly cooled through the temperature range of 540 to 760 °C

Fig. 9 Intergranular attack of Admiralty B brass in a hot water containing a small amount of sulfuric acid

150×

(1000 to 1400 °F). If thus sensitized, they are susceptible to intergranular corrosion in the same types of corrodents as cited above for Inconel X-750. Hastelloy B and C are susceptible after being heated at 500 to 705 °C (930 to 1300 °F), but are made immune to intergranular corrosion by heat treatment at 1150 to 1175 °C (2100 to 2150 °F) for Hastelloy B and at 1210 to 1240 °C (2210 to 2260 °F) for Hastelloy C, followed by rapid cooling in air or water.

Copper alloy C26000 (cartridge brass, 70% Cu) corrodes intergranularly in dilute aqueous solutions of sulfuric acid, iron sulfate, bismuth trichloride, and other electrolytes. Figure 9 shows the intergranular corrosion attack on an inhibited admiralty brass caused by hot water containing 0.1 to 0.2% sulfuric acid.

Light Metals and Alloys. The precipitated phases in high-strength aluminum alloys make them susceptible to intergranular corrosion. The effect is most pronounced for such alloys as 2014 containing precipitated $CuAl_2$ and somewhat less for those containing $FeAl_3$ (1100), Mg_2Si (2024), $MgZn_2$ (7075), and $MnAl_6$ (5xxx) along grain boundaries or slip lines. Solution heat treatment makes these alloys almost immune to intergranular corrosion, but substantially reduces their strength. Some magnesium alloys are similarly attacked unless solution heat treated, and die-cast zinc alloys are corroded intergranularly by exposure to steam and marine atmospheres.

Titanium. Intergranular corrosion and, in the presence of tensile stress, SCC of titanium and some titanium alloys occurs in fuming nitric acid at room temperature. A small addition (about 1%) of sodium bromide acts as an inhibitor. Commercially pure titanium is subject to similar corrosion and cracking in meth-anol solutions containing Br_2, Cl_2, or I_2; or Br^-, Cl^-, or I^-. A small addition of water acts as an inhibitor.

Various titanium alloys, including Ti-8Al-1Mo-1V, when heated in air while in contact with moist sodium chloride, for example, from fingerprints, at temperatures of 260 °C (500 °F) or higher, undergo intergranular corrosion or SCC, usually along grain boundaries. Pure titanium is resistant to both of these types of failure.

Selective Attack on Inclusions

Selective attack on inclusions by an environment to which the body of metal is resistant, in which only small amounts of material (as compared to the massive attack usually encountered in selective leaching) are preferentially corroded away, is a special case of selective leaching. The inclusions provide small anodic areas surrounded by large cathodic areas. Where the inclusions are in the form of elongated stringers and there is end-grain exposure to the environment, as in the following example, the attack is highly directional (unlike that shown in Fig. 7 and 8), and deep penetration of the metal is possible.

Example 5: Localized Corrosion of Inclusions in a Type 303 Stainless Steel Vending-Machine Valve. After about two years in service, a valve in contact with a carbonated soft drink in a vending machine occasionally dispensed a discolored drink with a sulfide odor, causing complaints from customers.

Manufacturing specifications called for the valve body to be made of type 303 stainless steel, a free-machining steel chosen because of the substantial amount of machining necessary to make the parts. Other machine parts in contact with the drink were made from type 304 stainless steel or inert plastics.

According to the laboratory at the bottling plant, the soft drink in question was one of the most strongly acidic of the commercial soft drinks, containing citric and phosphoric acids and having a pH of 2.4 to 2.5.

Investigation. The body of the valve, through which the premixed drink was discharged to the machine outlet, had an abnormal appearance on some portions of the end surface that were continuously in contact with the liquid, even during idle periods. These regions showed dark stains and severe localized corrosive attack in the stained areas. The remaining portions of the valve surface and other metal parts that had also been in contact with the liquid appeared bright and unaffected. Examination at low magnification confirmed the presence of severe highly localized attack in the stained areas.

The valve bodies were supplied by two different vendors. Chemical analysis of drillings from a corroded valve body supplied by vendor A and from an unused valve body supplied by vendor B showed that both parts met the composition requirements for type 303 stainless steel. This alloy had been specified for the valve bodies for at least nine years before complaints on its performance had come to the attention of the valve manufacturers.

Immersion Tests. Several of the valve bodies, one used and five unused, selected at random from parts on hand supplied by the two vendors, were cleaned in acetone and then continuously immersed in the highly acidic soft drink mixture for several days. The results were as follows:

Valve body	Vendor	Initial condition	Effect of test on valve
1	A	Stained and corroded	None
2	B	Unused	Black stain(a)
3, 4	A	Unused	None
5, 6	B	Unused	None

(a) The test also discolored the mixture and gave it a pronounced sulfide odor, making it completely unpalatable.

Metallographic Examination. The valve body that was stained and corroded (from vendor A, valve body 1 in the above table), the unused valve body that had stained in the immersion test (from vendor B, valve body 2 in the above table), and an unused valve body from vendor A that had not been subjected to immersion testing were all sectioned through the end that would be continuously exposed to the soft drink in the vending machine and were examined metallographically at magnifications of 50 to 400×. Numerous stringer-type inclusions of manganese sulfide were observed in each of the three metallographic specimens.

Figure 10 shows a micrograph of an unetched specimen from the corroded region on the end of the used valve body (valve body 1 in the immersion test). The corroded surface is an end-grain surface, and the corrosion began at the exposed ends of manganese sulfide stringer-type inclusions, which were anodic to the surrounding metal and readily attacked by the acidic soft drink.

The attack extended along the inclusion lines to a maximum depth of about 0.64 mm (0.025 in.), with the depth apparently depending on the length of the exposed sulfide stringers. A continuous line of attack that extended at least 0.58 mm (0.023 in.) below the surface of the metal is shown in Fig. 10, which also illustrates the distribution and dimensions of the sulfide stringers generally characteristic of the three specimens that were examined metallographically. No significant attack was found on the metallographic specimen taken from unused valve body 2 from vendor B, which had turned black during laboratory immersion in the soft drink mixture.

Conclusions. The failure of valve body 1 occurred because (1) manganese sulfide stringers were present in significant size and concentration in the type 303 stainless steel used, and large stringers were exposed at end-grain surfaces of the valve body, (2) the beverage was

Fig. 10 Unetched section through a type 303 stainless steel valve exposed to an acidic soft drink in a vending machine

Micrograph shows localized corrosion along manganese sulfide stringer inclusions at the end-grain surface. 100×

sufficiently acidic (having a pH of 2.4 to 2.5) to cause staining and continuing preferential attack on the exposed sulfide stringers, which were anodic to the surrounding metal, and (3) after the machine stood unused overnight or over a weekend, there was occasionally enough attack on exposed sulfide stringers on the end surface of the valve body to produce a hydrogen sulfide concentration in the immediately adjacent liquid, thus making at least the first cup of beverage dispensed discolored and unpalatable.

Discussion. Type 303 stainless steel has only marginal corrosion resistance for this application, because of the size and distribution of sulfide stringers found in some lots of standard grades of this alloy. The sulfide stringers are anodic to the surrounding stainless steel and are preferentially corroded away. This behavior can be considered a special case of selective leaching.

The inconsistent results obtained on immersing unused valve bodies 2 to 6 from the two vendors in the beverage for several days were not surprising in view of this marginal corrosion resistance, the locally varying distribution and dimensions of the sulfide stringers, and the use of only solvent cleaning before the immersion test. The complete resistance of the stained and corroded valve body (valve body 1 in the immersion tests) is not inconsistent with its past history. Any available sulfides on the corroded surface of the valve body could have already been consumed, and its passivity restored during exposure to the air after removal from service.

The extent of attack on the end-grain surface of the corroded valve body (Fig. 10), and the failure of an unused valve body in the laboratory immersion test make it highly probable that contamination of the soft drink by the corroded valve body occurred previously on one or more occasions during the year that it was in service without being reported to technically qualified personnel for investigation.

Recommendations. Specification of type 304 stainless steel (which contains a maximum of 0.030% S) for these and similarly exposed metal parts was recommended to avoid possible adverse effects on sales, even if the failures were infrequent. This alloy is generally satisfactory for processing and dispensing soft drinks and has been widely used for these purposes.

Concentration-Cell Corrosion

If a piece of metal is immersed in an electrolyte and there is a difference in concentration of one or more dissolved compounds or gases in the electrolyte, two areas of metal in contact with solution differing in concentration will ordinarily differ in solution potential, forming a concentration cell. Two electrically connected pieces of a given metal could also form a concentration cell in the same manner.

If the difference in potential is great enough, the more anodic area corrodes preferentially by concentration-cell corrosion. This type of corrosion can occur on a continuous, uniform, freely exposed surface at a rate depending on the difference in potential, the ratio of cathode area to anode area, the conductivity of the electrolyte, and the distance between the cathode and the anode. In a drop of an electrolyte on a metal surface, the concentration of dissolved gases on the metal surface at the center of the drop would differ from that at the edges of the drop in what is called a differential-aeration cell.

There are many other circumstances in which concentration cells are formed and cause corrosion. The rate of diffusion of air produces differential aeration in the layers of water or aqueous solutions just below the liquid level and causes concentration-cell corrosion in this region on partly immersed metal parts.

Differential concentration cells in which a metal part is partly immersed in a liquid electrolyte above which the gaseous phase consists of a gas or gases other than air are also common. Concentration-cell corrosion may be controlled by diffusion effects of dissolved gases or other substances, especially when pitting-type concentration-cell corrosion proceeds in stagnant solutions. Concentration-cell corrosion occurs on buried metals as a result of their being in contact with soils that have different chemical compositions, water contents, or degrees of aeration.

Fig. 11 Crevice corrosion pitting that has taken place where type 316 stainless steel bubble caps contact a type 316 stainless steel tray deck

This oxygen-concentration cell corrosion occurred in concentrated acetic acid with minimal oxidizing capacity. 1/8 actual size

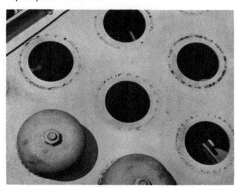

Crevice Corrosion

A crevice in a metal surface at a joint between two metallic surfaces or between a metallic and a nonmetallic surface or a crevice beneath a particle of solid matter on a metallic surface provides conditions that are conducive to the development of the type of concentration-cell corrosion called crevice corrosion. Crevice corrosion can progress very rapidly (tens to hundreds of times faster than the normal rate of general corrosion in the same given solution). For example, a sheet of stainless steel can be cut (corroded) into two pieces simply by wrapping a rubber band around it, then immersing the sheet in seawater or dilute ferric chloride solution. The open surfaces will pit slowly, but the metal under the rubber band will be attacked rapidly for as long as the crevice between the rubber band and the steel surface exists.

In a metal-ion concentration cell, the accelerated corrosion occurs at the edge of or slightly outside of a crevice. In an oxygen-concentration cell, the accelerated corrosion usually occurs within the crevice between the mating surfaces.

Figure 11 shows an example of crevice corrosion in which type 316 stainless steel bubble caps contacted the type 316 tray deck. At the contact points, the type 316 stainless steel was selectively corroded by the concentrated acetic acid solution. In this part of the distillation column, the oxidizing capacity of the acid stream is nearly exhausted, but is sufficient to prevent corrosion of the open surfaces. A change to a higher-alloy stainless was necessary to fully resist this corrosive service. Any layer of solid matter on the surface of a metal that offers the opportunity for exclusion of oxygen from the surface or for accumulation of metal ions beneath the deposit because of restricted diffusion is a probable site for crevice corrosion.

Differential aeration beneath solid deposits or at cracks in mill scale is a frequent cause of crevice corrosion in boilers and heat exchangers. Suitable water treatment to provide thin protective films, together with special attention to cleaning, rinsing, and drying when boilers are shut down, minimizes the occurrence of crevice corrosion in such equipment. Breaks (holidays) in protective organic coatings or linings on vessels containing corrosive chemicals are also likely sites for the development of crevice corrosion.

Liquid-Level Effects. Crevice corrosion often occurs underneath deposits of solid substances that sometimes collect just above the liquid level on a metal part that is partly immersed in an electrolyte. The deposits usually remain moist or are intermittently moist and dry.

Where the liquid level fluctuates or where the liquid is agitated, the area of the metal that is intermittently wetted is called the splash zone. Splash zones are encountered in all types of tanks and equipment containing liquids; corrosion occurring in these areas is called splash-zone corrosion.

A failure caused by this type of crevice corrosion occurred in a carbon steel tank that contained a saturated solution of sodium chloride (brine) at room temperature. When a pump, apparently in good condition, failed to deliver brine, inspection showed that a steel suction line, partly immersed in the liquid and used to transfer small amounts of brine as needed to other equipment, had corroded through in the intermittently wetted area. Further inspection showed that the tank had also been attacked in the intermittently wetted area. The corrosion problem was eliminated by replacing the steel suction line with a glass fiber reinforced epoxy pipe and by lining the tank with a coal-tar epoxy coating.

Rust films that have little opportunity to dry do not develop protective properties. Such unfavorable conditions exist in the splash zone above the high-tide level along the seashore. The corrosivity of the moist rust films is further aggravated by the high oxygen content of the splashing seawater. Observations of old steel piling along the seashore usually reveal holes just above the water line, where the rate of corrosion is several times greater than that for continuous immersion in seawater.

Corrosion Under Thermal Insulation. The application of thermal insulation to carbon steel and stainless steel equipment presents a special case of crevice corrosion. It is special because there are a number of misconceptions about the role of thermal insulation in this corrosion situation and the phenomenon is not, strictly speaking, crevice corrosion, but does not comfortably fit in any other category. Thermal insulation is used to maintain temperature of equipment, either below ambient or above ambient; therefore, the metal normally will operate at temperatures different from the surrounding temperature. Because of this situa-

tion, the thermal insulation system—which consists of the metal outer surface, the annular space between the metal and the insulation, the insulation itself, and its weather barrier—is a dynamic, breathing combination of materials. The weather barrier or weather proofing, coating, wrapping, or cladding, while intended to protect the insulation from the ravages of weather and other mechanical damage, does not and will not prevent the entry of humidity and moisture to the annular space between the metal surface and the inner side of the insulation.

Therefore, with moisture present on the metal surface, the insulation acts as a barrier in terms of the evaporation of the moisture and maintains it in a vapor phase in contact with the metal surface. There is also free access to oxygen as well as contaminants that might be in the insulation. This system can then become very aggressive when the metal temperature is operating in the range of 40 to 120 °C (100 to 250 °F). At these elevated temperatures, the humid annular space can corrode steel at rates as high as 0.8 mm/year ($\frac{1}{32}$ in./year) or, in the case of stainless steel, can initiate chloride SCC with relatively rapid failure times.

The approach to preventing corrosion under insulation has been examined by many investigators (Ref 10). The attempts to produce insulation materials of high purity and with silicate-type inhibitors added are being used. However, they offer limited protection against aggressive atmospheres that may be introduced beneath the insulation through cracks, crevices, and other openings in the insulation. The most widespread approach at this time is to use a protective coating on the metal to protect it from the water vapor existing in the annular space.

When selecting protective coatings, they should be considered as being in warm water immersion service because of the severity of the environment. The most severe environments develop when the vessels or piping operate with cycling temperatures, that is, operate below ambient for a period of time and then go through a heating cycle that tends to evaporate the water but goes through the hot metal corrosion temperature zone.

The problem of corrosion under thermal insulation has become widely recognized as one of significant concern. Because of the hidden nature of the problem, a well-insulated vessel may operate for many years without any apparent external signs of damage while significant corrosion is taking place under the insulation.

Effects of Solid Deposits. In power-generation equipment, crevice-corrosion failures have occurred in main-station condenser tubes cooled with seawater as a result of the formation of solid deposits and the attachment of marine organisms to the tube wall. These failures have occurred particularly in condensers with stainless steel tubing.

Riveted and bolted joints must be considered as possible sites for crevice corrosion; therefore, they require careful attention in design and assembly to avoid crevices, as well as

provisions to ensure uniform aeration and moderate but not excessive flow rates at the joints. Replacement with welded joints can eliminate crevice corrosion provided special care is taken in welding and subsequent finishing of the welds to provide smooth, defect-free joints.

Crater Corrosion. Type 347 stainless steel is sometimes subject to a type of crevice corrosion known as crater corrosion, which occurs at the stop point of a weld. The failure is related to microsegregation of certain constituents in the pool of molten metal that is the last to solidify on a weld. In a manner similar to zone-melting refinement, the moving weld pool continuously sweeps selected constituents ahead of it, and the concentration of these selected constituents in the pool increases continuously until the welding is stopped and the pool solidifies. The center of the stop point is attacked rapidly in oxidizing acids, such as nitric acid, in a form of self-accelerating crevice corrosion.

The final solidification pool need not extend through the thickness of the part for full perforation to occur, as illustrated by the occurrence described in the following report.

Beveled weld-joint V-sections were fabricated to connect inlet and outlet sections of tubes in a type 347 stainless steel heat exchanger for a nitric acid concentrator. Each V-section was permanently marked with the tube numbers by a small electric-arc pencil.

After one to two years of service, multiple leaks were observed in the heat-exchanger tubes. When the tubes were removed and examined, it was found that the general corrosion rate was normal for service of heat-exchanger tubes in a nitric acid concentrator, but that crater corrosion had perforated the tubes.

The crater corrosion occurred at two general locations. One location was at the stop point of the welds used to connect the inlet and outlet legs of the heat exchanger. The other location was at the stop points on the identifying numerals. The material was changed to type 304L stainless steel, in which the zone-melting concentration does not take place.

Differential-Temperature Cells

In electrolytic cells of the differential-temperature type, the anode and cathode consist of the same metal and differ only in temperature. If the anode and cathode are areas on a single piece of metal (or on two electrically connected pieces of the same metal) immersed in the same electrolyte, corrosion proceeds as in any short-circuited galvanic cell.

For copper in aqueous salt solutions, the area of the metal at the higher temperature is the cathode and the area at the lower temperature is the anode. In the preferential attack on the anode, copper dissolves from the cold area and deposits on the warmer area. Lead acts similarly, but for silver the polarity is reversed, with the warmer area being attacked preferentially.

For steel immersed in dilute aerated chloride solutions, the warmer area is anodic to the colder area, but as the reaction progresses, the polarity sometimes reverses, depending on aeration, solution velocity against the metal surface, and other factors.

Differential-temperature cell corrosion occurs most frequently in heat-transfer equipment and piping, where substantial temperature differences exist between the inlet and the outlet portions exposed to the same electrolyte.

Galvanic Corrosion

When dissimilar metals are in electrical contact in an electrolyte, the less noble metal (anode) is attacked to a greater degree than if it were exposed alone, and the more noble metal (cathode) is attacked to a lesser degree than if it were exposed alone. This behavior, known as galvanic corrosion, can often be recognized by the fact that the corrosion is more severe near the junction of the two metals than elsewhere on the metal surfaces. Galvanic corrosion is usually the result of poor design and selection of materials or the plating-out of a more noble metal from solution on a less noble metal.

The greater the difference in potential between the two metals, the more rapid the galvanic attack will be. The textbook electromotive-force series ranks the metals according to their chemical reactivity, but applies only to the laboratory conditions under which the reactivity was determined. In practice, the solution potential of metals is affected by the presence of passive or other protective films on some metals, polarization effects, degree of aeration, complexing agents, and temperature.

Galvanic Series in Seawater. A galvanic series based on immersion in seawater is more generally applicable than the electromotive-force series as an indication of the rate of corrosion between different metals or alloys when they are in contact in an electrolyte. In most electrolytes, the metal close to the active end of the galvanic-series chart will behave as an anode, and the metal closer to the noble end will act as a cathode. The amount of separation between two metals in the chart is a rough measure of the difference in potential that can be expected and is usually related to the rate of galvanic corrosion between the two metals in a given electrolyte.

This galvanic series, which includes most of the industrially important metals, is given in Table 2. In most cases, metals from one group can be coupled with other metals from the same group without causing a substantial increase in the corrosion rate of the more active metal.

Passivity Effects. Because some alloys (most notably, the stainless steels) can be either passive or active, they may occupy two places in the galvanic series: a relatively noble position for the passive state and a less noble position for the active state. Thus, such alloys can play complex roles in situations involving corrosion. Passivity is generally thought to be

due to a tightly bound film of oxide on the metal surface, rendering the surface much less susceptible to further reaction. Destruction of the film—for example, by halide ions—will restore the metal to the active state; thus, for a passive alloy, hydrochloric acid is often more dangerous than nitric acid.

Similar considerations apply to many instances in which passivity is destroyed locally—for example, on sensitized stainless steel where the grain boundaries have been depleted of chromium and are active, and the interior of the grains passive. Because any stainless steel in the active condition is widely separated in the galvanic series from the same alloy in the passive condition (Table 2), this is just a special case of galvanic attack. With a large cathode area and a small anode area (the depleted region at the grain boundary), the attack can be especially severe in certain environments.

Selection of Compatible Metals. In the design of a product, selecting metals that will be in electrical contact in an electrolyte, without giving adequate consideration to the possibility of galvanic action between the metals, is a frequent cause of failure.

The coupling of aluminum with brass in equipment that was immersed in ground water resulted in the failure described in the following report.

A sump pump failed to operate, and the resultant flooding of a basement in a manufacturing plant caused extensive damage when a float arm for a water-level control device failed as a result of galvanic corrosion. In this device, brass stops were attached to an aluminum vertical rod (float arm) to control the off-on switch. The aluminum rod corroded severely just below the lower brass stop and broke off at this point.

The selection of these two metals for this application was a gross error by the manufacturer of the equipment, because aluminum is known to be strongly anodic to brass in nearly all aqueous environments. In seawater, the difference in corrosion potential between the two metals is 0.6 V.

The aluminum flat arm was replaced with one made of type 316 stainless steel, which has a corrosion potential close to that of brass. No further problems were encountered during the ensuing five years of service.

Example 6: Galvanic-Corrosion Failure of a Malleable Iron Latch in a Valve for an Automatic Sprinkler System. One of three valves in a dry automatic sprinkler system tripped accidentally, thus activating the sprinklers. A check of the system indicated that the two other valves were about to trip. Maintenance records showed that the three valves had been in service approximately 21 months.

The valve consisted of a cast copper alloy clapper plate that was held closed by a pivoted malleable iron latch (Fig. 12a). The latch and top surface of the clapper plate were usually in

Table 2 Galvanic series in seawater

Corroded end (anodic, or least noble)
Magnesium
Magnesium alloys
Zinc
Galvanized steel or galvanized wrought iron
Aluminum alloys
5052, 3004, 3003, 1100, 6053, in this order
Cadmium
Aluminum alloys
2117, 2017, 2024, in this order
Low-carbon steel
Wrought iron
Cast iron
Ni-Resist (high-nickel cast iron)
Type 410 stainless steel (active)
50-50 lead-tin solder
Type 304 stainless steel (active)
Type 316 stainless steel (active)
Lead
Tin
Copper alloy C28000 (Muntz metal, 60% Cu)
Copper alloy C67500 (manganese bronze A)
Copper alloys C46400, C46500, C46600, C46700 (naval brass)
Nickel 200 (active)
Inconel alloy 600 (active)
Hastelloy B
Chlorimet 2
Copper alloy C27000 (yellow brass, 65% Cu)
Copper alloys C44300, C44400, C44500 (admiralty brass)
Copper alloys C60800, C61400 (aluminum bronze)
Copper alloy C23000 (red brass, 85% Cu)
Copper C11000 (ETP copper)
Copper alloys C65100, C65500 (silicon bronze)
Copper alloy C71500 (copper nickel, 30% Ni)
Copper alloy C92300, cast (leaded tin bronze G)
Copper alloy C92200, cast (leaded tin bronze M)
Nickel 200 (passive)
Inconel alloy 600 (passive)
Monel alloy 400
Type 410 stainless steel (passive)
Type 304 stainless steel (passive)
Type 316 stainless steel (passive)
Incoloy alloy 825
Inconel alloy 625
Hastelloy C
Chlorimet 3
Silver
Titanium
Graphite
Gold
Platinum
Protected end (cathodic, or most noble)

a sanitary-water environment (stabilized, chlorinated well water with a pH of 7.3) under stagnant conditions. Process make-up water that had been clarified, filtered, softened, and chlorinated and had a pH of 9.8 was occasionally used in the system.

Water pressure on the bottom side of the clapper plate varied from 240 to 1035 kPa (35 to 150 psi). During a fire emergency, heat detectors activated a switch, releasing a weight that pivoted the lip of the latch away from the

Fig. 12 Sprinkler system in which a malleable iron deluge clapper latch failed from galvanic attack caused by contact with a copper alloy clapper in stagnant water

(a) Schematic illustration of sprinkler system showing deluge clapper latch that failed. (b) Photograph of clapper latch showing effects of galvanic attack at area of contact (near top) with cast copper alloy clapper plate and crevice corrosion (lower left). (c) Micrograph of a cross section of the failure area on the clapper latch showing the pattern of the corrosion and elongated grains in the microstructure (indicative of a ductile type of failure). 250×

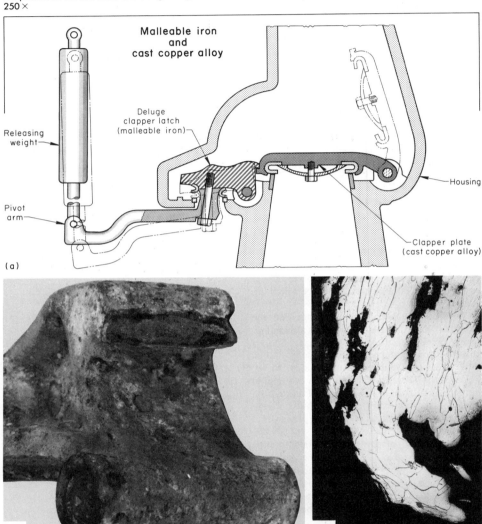

(a)

(b)

(c)

clapper. Water pressure forced the clapper plate upward, allowing water to flow through the system.

Investigation. Visual examination of the latch and clapper plate showed corrosion at the contact surfaces between the clapper plate and the latch. Figure 12(b) shows the large amount of metal dissolved from the clapper latch, leaving only about half of the original amount of metal to hold the clapper in a closed position. Also visible in Fig. 12(b) are indications of crevice corrosion where the latch was hinged to the main body of the valve; this attack, however, did not contribute to the failure.

A micrograph of a section through the contact area of the malleable iron clapper latch (Fig. 12c) showed elongated grains and substantial deformation, as well as the pattern and depth of the corrosion, which had drastically weakened the remaining metal in the contact area. Other areas of the failure surface showed transgranular cracks.

Discussion and Conclusions. Corrosion of the malleable iron latch was by galvanic action and greatly reduced the shear load the latch was capable of withstanding. If the latch were not properly adjusted or corrosion decreased the contact area considerably, the applied stress on

actuation could exceed the shear strength of the material. In this instance, the contact area was reduced to about half its normal size, and the shear strength of the remaining metal was reduced by the penetrating galvanic corrosion.

When the sprinkler valve accidentally tripped, plastic deformation, characteristic of ductile failure, occurred. Failure of the latch was caused by extensive loss of metal by galvanic corrosion and the sudden loading related to the tripping of the valve. The malleable iron latch had been corroded to such an extent that it failed mainly by plastic deformation. Failure in some regions of the contact area was by ductile (transgranular) fracture.

Corrective Measures. The latch material was changed from malleable iron to silicon bronze (C87300). The use of silicon bronze prevented corrosion or galvanic attack, and the latch was kept properly adjusted to maintain an adequate contact area. The replacement latches were still in satisfactory condition after more than 14 years.

Ratio of Cathode Area to Anode Area. Breaks (holidays) in organic protective coatings on an active metal component of equipment in which the remaining metal is made of a relatively noble or passive metal can result in extremely rapid galvanic attack, even in a weakly conductive electrolyte, because of a high ratio of cathode area to anode area.

In one instance of this type of behavior, a plant had replaced the leaking bottom of a type 316 stainless steel tank with a bottom of ASTM A285 steel and coated it with a vinyl-paint system. The tank contained pure condensate; therefore no problems were anticipated. However, after about six months several small leaks were noted through the painted-steel bottom. Inspection revealed breaks in the vinyl-paint coating. The unfavorable cathode-to-anode ratio, approximately 1000 to 1, resulted in vigorous attack on the carbon steel at the breaks and caused pinhole failures.

Metals Embedded in Concrete or Plaster. Galvanic corrosion of embedded metals has caused failures of aluminum conduit in steel-reinforced concrete and failures of gypsum plaster on steel lath coupled with copper tubing.

Aluminum conduit embedded in steel-reinforced concrete that contained calcium chloride corroded galvanically in the ceiling of a newly constructed stadium. Cracks in the concrete ceiling began to appear within a few months after completion of the stadium, causing spalling of the concrete. Repair of these defects involved chipping away the surrounding concrete, removing the corroded aluminum conduit, replacing it with steel conduit, and recasting fresh concrete. More than 4600 m (15 000 ft) of corroded aluminum conduit was replaced in this stadium.

The corrosion of steel tie wires in plaster that also contains copper tubing is another, although less frequent, example of damage to a cementing material caused by galvanic corrosion of

Fig. 13 Components for the mounting surface of a hydraulic actuator that failed in service because of galvanic attack on the aluminum alloy spacer

The galvanic attack occurred on the aluminum alloy spacer (a) when a vellum gasket (b) that separated the spacer from a nickel-plated steel housing (c) became saturated with moisture-containing molybdenum disulfide lubricant that acted as an electrolyte. (d) and (e) Micrographs of unetched specimens that were taken from a corroded area and an uncorroded area, respectively, of the aluminum alloy spacer. 500×

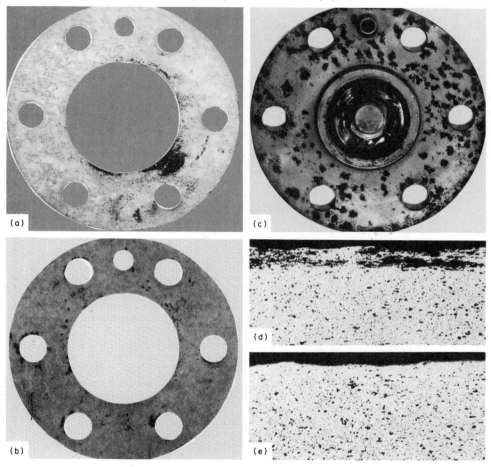

protected, while the steel becomes the anode and corrodes. When the steel tie wires in a plastered ceiling become sufficiently weakened by corrosion, the plaster (and metal lath) will fall under its own weight.

In the building-construction systems described, dissimilar metals will, for practical reasons, nearly always be in electrical contact, either externally to the cementing material or, more usually, both externally and internally. The galvanic-cell principle applies in both cases. The only difference is that when contact is internal, the corrosion will tend to be greater near the point of contact.

The degree to which the reaction of two dissimilar metals in a cementing material corrodes the anode depends on electrical conductivity (for cold-weather installation, up to 2% calcium chloride is added to concrete to speed up setting), the proximity of the dissimilar metals, and the ratio of anode area to cathode area.

If it is impossible to avoid the use of dissimilar metals in cementing materials, the following precautions should be observed:

- The electrical conductivity of the cementing material should be kept to a minimum. The use of calcium chloride or other ionic compounds should be forbidden, and contamination of the cementing material, such as by use of seawater for mixing water, should be avoided. Abnormally delayed drying and subsequent exposure to moisture should also be prevented as far as possible
- Dissimilar metals should be spaced as far apart as practical
- The ratio of anode area to cathode area should be at a maximum. Coating the anode should be avoided, unless the coating is continuous or unless the intense localized corrosion that may occur at a few uncoated points can be tolerated. A strict rule in corrosion control is if one of two dissimilar metals is to be coated, coat the cathode

Effect of Time-Dependent Factors. Some types of time-dependent changes that affect susceptibility to corrosion are readily overlooked, as in the galvanic-attack failure of a number of hydraulic actuators in the following example.

Example 7: Failure of Aluminum Alloy Spacers by Galvanic Attack. Immediately after installation, leakage was observed at the mounting surface (Fig. 13a) of several rebuilt hydraulic actuators that had been in storage for up to three years before installation. At each joint, there was an aluminum alloy spacer (Fig. 13a) and a vellum gasket (Fig. 13b). The mounting flanges of the steel actuators (Fig. 13c) had been nickel plated. During assembly of the actuators, a lubricant containing molybdenum disulfide had been applied to the gaskets to serve as a sealant. One actuator housing, several aluminum alloy spacers, and one used

embedded metals. In one case, a gypsum-plaster suspended ceiling in a military hospital fell a few months after installation. In addition to the steel tie wires in the system (used to tie the prepainted and therefore insulated metal lath to the upper supporting members), the ceiling contained a grid network of copper tubing for hot-water radiant heating. Failure was caused by galvanic corrosion of the steel tie wires. Analysis of the fallen plaster showed that it unaccountably contained 4400 ppm of chloride ions.

In both of the systems described above, the mechanism of corrosion is similar. Both contain the essential components of a galvanic cell: two electrically coupled dissimilar metals immersed in an electrolyte.

During the time that the concrete is damp, the cement matrix of the concrete is a fairly good electrolyte. The steel and aluminum serve as electrodes, and because of the different

positions of these metals in the galvanic series, an electrical potential develops between them. If an external connection is made between the steel and the aluminum, a current will flow in the circuit just as in any other short-circuited galvanic-cell battery. The aluminum becomes the anode, and it corrodes. The steel becomes the cathode, and it tends to be protected. The intensity and duration of the current flow depend on a number of variables. The concrete fails when the aluminum corrosion products occupy a larger volume than the metallic aluminum plus its corrodents. The increasing volume builds up an internal pressure around the conduit. If the conduit is thin walled and deeply buried in strong concrete, the pressure may collapse the conduit. Otherwise, the concrete will crack.

In the gypsum-plaster system containing steel tie wires and copper tubing, similar actions apply. Copper becomes the cathode and is and one new vellum gasket were sent to the laboratory for examination.

Investigation. Visual examination of the components disclosed corrosion deposits and staining (Fig. 13a, b, and c). The stained areas were dark in color and greasy to the touch. Samples of deposits scraped from the aluminum spacer and the vellum gasket were identified by x-ray diffraction as molybdenum disulfide. Deposits taken from the nickel-plated steel housing were not completely identified. Analysis indicated, however, that nickel formate dihydrate was one of the constituents. No molybdenum disulfide or nickel formate was found on the new gasket that was submitted for examination.

Metallographic examination of sections taken through a badly stained area on one of the aluminum alloy spacers showed that corrosion had penetrated to a depth of 64 μm (0.0025 in.) at some areas (Fig. 13d and e).

Test for Galvanic Action. A test was conducted to determine if the corrosion on the aluminum alloy spacers was the result of galvanic action. Sections of the actuator housing, the vellum gasket, and the aluminum alloy spacer were clamped with an insulated C-clamp. The vellum gasket was found to be electrically conductive, and a potential of 0.1 V at 60 000 Ω was measured between the aluminum alloy spacer and the actuator housing. This voltage could be increased or decreased by changing the force applied by the C-clamp.

The C-clamp was removed, and the testpieces were baked in an oven at 230 °C (450 °F) for seven days. After baking, no potential could be detected between the testpieces when they were reclamped by the insulated clamp.

Visual examination of the testpieces revealed staining similar to that observed on the failed spacers. Metallographic examination of cross sections through stained areas in the aluminum-spacer testpieces disclosed the presence of incipient corrosion.

Conclusions. Leakage was the result of galvanic corrosion of the aluminum alloy spacers while in storage. The molybdenum disulfide was apparently suspended in a volatile water-containing vehicle that acted as an electrolyte between the aluminum alloy spacer and the nickel-plated steel actuator housing.

Initially, there was no couple between the spacer and housing because the vellum gasket acted as an insulator, but the water-containing lubricant gradually impregnated the vellum gasket and established a galvanic couple between the spacer and the housing. The staining of the nickel-plated steel housing and the presence of hydrated nickel formate deposits did not contribute to the leakage, but indicated some attack on the housing and helped to confirm the nature of the corrosion mechanism.

Corrective Measures. Use of the molybdenum disulfide lubricant as a gasket sealer was discontinued, and the actuators were assembled using dry vellum gaskets. A satisfactory seal was obtained with the dry vellum gaskets, and

no delayed failures by leakage at the gaskets were observed on units assembled in this manner.

Velocity-Affected Corrosion in Water

The attack on metal immersed in water may vary greatly, depending on the relative velocity of movement of water against the metal surface. Where attack occurs, the effects of differences in water velocity are most pronounced for metals that show passivity behavior or form other protective films in water.

Effects in Slow-Moving and Stagnant Waters. Slow-moving and stagnant waters allow loosely adherent solid corrosion products to form on metal surfaces and aggravate corrosion. Also, in closed systems where a corrosion inhibitor is used, the effectiveness of the corrosion inhibitor is reduced where the water is stagnant or nearly so.

In designing for corrosion control, stagnant zones should be eliminated by the following methods:

- Allowing free drainage of water and suspended solids
- Installing baffling to eliminate stagnant liquid zones
- Increasing the frequency of cleaning
- Providing gas-vent lines
- Providing strainers or separators to remove foreign material (dirt)

Any of the common varieties of iron or steel, including low-carbon or high-carbon steel, low-alloy steel, wrought iron and cast iron, corrode in slow-moving fresh water or seawater at almost the same rate, which is about 0.13 mm/year (0.005 in./year). At high temperatures, the rate increases, but it remains relatively low on an absolute scale; therefore, steel can be used for boilers in contact with deaerated water. Commercially pure aluminum corrodes in aerated or deaerated fresh water at still lower rates than iron, making it a suitable material for handling distilled water.

Swift-moving water may carry dissolved metal ions away from corroding areas before the dissolved ions can be precipitated as protective layers. Gritty suspended solids in water scour metal surfaces and continually expose fresh metal to corrosive attack.

In fresh waters, as water velocity approaches very high values, it is expected that corrosion of steel first increases, then decreases, and then increases again. This occurs because erosive action serves to break down the passive state.

The corrosion of steel by seawater increases as the water velocity increases. The effect of water velocity at moderate levels is shown in Fig. 14, which illustrates that the rate of corrosive attack is a direct function of the velocity until some critical velocity is reached, beyond which there is little further increase in corro-

Fig. 14 Effect of velocity of seawater at atmospheric temperature on the corrosion rate of steel

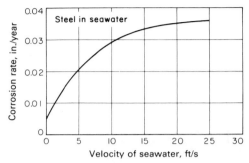

sion. At much higher velocities, corrosion rates may be substantially higher.

The effect of changes in water velocity on the corrosion resistance of stainless steels, copper alloys, and nickel alloys shows much variation from alloy to alloy at intermediate velocities. Type 316 stainless steel may pit severely in seawater at velocities of less than 1 or 1.5 m/s (4 or 5 ft/s), but is usually very corrosion resistant at higher velocities. Copper alloy C68700 (aluminum brass) has satisfactory corrosion resistance if the seawater velocity is less than 2.4 m/s (8 ft/s).

Stainless steels perform much better when in contact with seawater at high velocities (over 1.5 m/s, or 5 ft/s). This is probably because of the absence of adherent organisms or other deposits. The austenitic stainless steels containing molybdenum, such as type 316, are superior to other grades of stainless steel in slow-moving seawater.

Copper alloy C71500 (copper nickel, 30% Ni) has excellent resistance to swift-moving seawater and to many types of fresh water. Because copper alloy C71500 also is subject to bio-fouling, velocities should be kept above 1.5 m/s (5 ft/s).

In seawater at high velocity, metals fall into two distinctly different groups: those that are velocity limited (carbon steels and copper alloys) and those that are not velocity limited (stainless steels and many nickel alloys).

Metals that are not velocity limited are subject to virtually no metal loss from velocity effects or turbulence short of cavitation conditions. The barrier films that form on these metals seem to perform best at high velocities with the full surface exposed and clean. It is in crevices and under deposits that form from slow-moving or stagnant seawater that local breakdown of the film and pitting begin.

Erosion-Corrosion. When movement of a corrodent over a metal surface increases the rate of attack due to mechanical wear and corrosion, the attack is called erosion-corrosion. It is encountered when particles in a liquid impinge on a metal surface, causing the removal of

Fig. 15 Erosion pitting caused by turbulent river water flowing through copper pipe

The typical horseshoe shaped pits point up stream. ½×

protective surface films, such as air-formed protective oxide films or adherent corrosion products, and exposing new reactive surfaces that are anodic to uneroded neighboring areas on the surface. This results in rapid localized corrosion of the exposed areas in the form of smooth-bottomed shallow recesses.

Nearly all flowing or turbulent corrosive media can cause erosion-corrosion. The attack may exhibit a directional pattern related to the path taken by the corrodent as it moves over the surface of the metal.

Figure 15 shows the interior of a 50-mm (2-in.) copper river waterline that has suffered pitting and general erosion due to excessive velocity of the water. The brackish river water contained some suspended solids that caused the polishing of the copper pipe surface. The horseshoe-shaped pits (facing up stream) are typical of the damage caused by localized turbulence. The copper pipe was replaced with fiberglass-reinforced plastic piping.

Impingement corrosion is a severe form of erosion-corrosion. It occurs frequently in turns or elbows of tubes or pipes or on surfaces of impellers or turbines where impingement is encountered and erosion is more intense. It occurs as deep, clean, horseshoe-shaped pits with the deep, or undercut, end pointing in the direction of flow. Impingement-corrosion attack can also occur as the result of partial blockage of a tube. A stone, a piece of wood, or some other object can cause the main flow to deflect against the wall of the tube. The impinging stream can rapidly perforate tube walls. Water that carries sand, silt, or mud will have an additional severely erosive effect on tubes.

Steam erosion is another form of impingement corrosion. It occurs when high-velocity wet steam contacts a metal surface. The resulting attack usually produces a roughened surface showing a large number of small cones with the points facing in the direction of flow.

Example 8: Impingement-Corrosion Failure of a Ferritic Malleable Iron Elbow.

Leakage was detected in a malleable iron elbow after only three months in service. Life expectancy for the elbow was 12 to 24 months. The 21-mm (0.824-in.) inside-diameter 90° elbow connected segments of 19-mm (¾-in.) pipe with a wall thickness of 2.9 mm (0.113 in.) for a line through which steam and cooling water were alternately supplied to a tire-curing press. The supply line and elbow were subjected to 14 heating and cooling cycles per hour for at least 16 h per day, or a minimum of 224 cycles per day. Steam pressure was 1035 kPa (150 psi), and water pressure was 895 kPa (130 psi). Based on pump capacity, the water-flow rate was estimated at 1325 L/min (350 gal./min). Water-inlet temperature was 10 to 15 °C (50 to 60 °F); water-outlet temperature was 50 to 60 °C (120 to 140 °F). The water had a pH of 6.9.

The elbow was cast from ASTM A47, grade 35018, malleable iron and had a hardness of 76 to 78 HRB. Composition of the iron was:

Element	Composition, %
Carbon	1.95
Manganese	0.60
Silicon	1.00
Sulfur	0.15
Phosphorus	0.05
Copper	0.17
Chromium	0.03
Nickel	0.02
Molybdenum	0.001

Investigation. Specimens were cut from two areas on the elbow, one just below the point of leakage (region A, Fig. 16a) and another further downstream (region B, Fig. 16a). The deepest penetration in the first specimen was at the top, just below the point of leakage, where the wall thickness had been reduced to 1.6 mm (1/16 in.). Maximum wall thickness was 9.5 mm (3/8 in.).

Metallographic examination of the first specimen showed that moderate but irregular attack had occurred (Fig. 16b). A small area of ferrite remained at the top, but the surface ferritic zone (light areas, Fig. 16b) had been eroded and corroded away on the remainer of the surface, exposing the pearlitic zone (dark areas, Fig. 16b). The interior of the specimen showed a ferritic malleable microstructure.

The second specimen, which showed no signs of attack, had a typical ferrite zone at the surface, then a subsurface pearlitic zone with about twice the thickness of the ferrite zone and a ferritic malleable microstructure in the interior of the specimen (Fig. 16c).

Conclusions. Examination of the micrographs (Fig. 16b and c) indicated that the elbows had been given the usual annealing and normalizing treatment for ferritizing malleable iron. This resulted in lower resistance to erosion and corrosion than pearlitic malleable iron.

Recommendations. It was recommended that replacement elbows be heat treated to produce a

Fig. 16 Malleable iron elbow in which impingement corrosion caused leakage and failure at the bend

(a) Section through the elbow showing extent of corrosion and point of leakage. Regions A and B are locations of specimens shown in micrographs (b) and (c), respectively. (b) Micrograph of a nital-etched specimen from region A (just below the failure area) showing ferritic surface (light areas) corroded away, exposing the subsurface pearlitic zone (dark areas). 67×. (c) Micrograph of a nital-etched specimen from region B (an uncorroded area of the elbow) showing a typical ferrite zone at the surface, a subsurface pearlitic zone (with twice the thickness of the ferrite zone), and a ferritic malleable microstructure in the interior. 67×

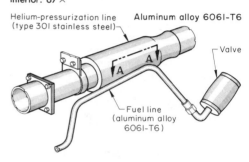

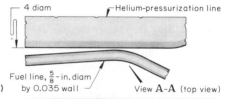

(a)

(b)

(c)

pearlitic malleable microstructure, which has longer life under the given conditions of service (additional information on the heat treatment and properties of ferritic and pearlitic malleable irons is provided in the article "Malleable Iron" in Volume 1 of the 9th Edition of *Metals Handbook*).

In piping systems, erosion-corrosion can be reduced by increasing the diameter of the pipe, thus decreasing velocity and turbulence. The streamlining of bends is useful in minimizing the effects of impingement. Inlet pipes should not be directed onto the vessel walls if this can be avoided. Flared tubing can be used to reduce problems at the inlet tubes in a tube bundle.

Cavitation erosion is the most severe form of erosion-corrosion. It occurs principally when relative motion between a metal surface and a liquid environment causes vapor bubbles to appear. When the bubbles collapse, they impose hammerlike blows simultaneously with the initiation of tearing action that appears to pull away portions of the surface. Although the tearing action can remove any protective oxide film that exists on the surface of a metal, exposing active metal to the corrosive influence of the liquid environment, corrosion is not essential to cavitation erosion, as discussed in the article ''Liquid-Erosion Failures'' in this Volume. Whenever high velocities give rise to extremely low-pressure areas, as in a jet or rotary pump, vapor bubbles collapse at high-pressure areas and destroy the protective film on the metal surface or disrupt the metal itself.

Cavitation erosion occurs typically on rotors or pumps, on the trailing faces of propellers and of water-turbine blades, and on the water-cooled side of diesel-engine cylinders. Damage can be reduced by operating rotary pumps at the highest possible head of pressure to avoid formation of bubbles. For turbine blades, aeration of water serves to cushion the damage caused by the collapse of bubbles. Neoprene or similar elastomer coatings on metals are reasonably resistant to damage from this cause. To reduce cavitation damage to diesel-engine cylinder liners, the addition of 2000 ppm sodium chromate to the cooling water has proved effective, as has the use of Ni-Resist (high-nickel cast iron) liners. Table 3 rates some metals frequently used in seawater in four groups on the basis of their resistance to cavitation erosion in seawater.

Bacterial and Bio-Fouling Corrosion

Biological organisms affect corrosion processes on metals by directly influencing anodic and cathodic reactions, by affecting protective surface films on metals, by producing corrosive substances and by producing solid deposits. These organisms include microscopic forms, such as bacteria, and macroscopic types, such as algae and barnacles. Microscopic and macroscopic organisms have been observed to live and reproduce in mediums with pH values between 0 and 11, at temperatures between 0 and 80 °C (30 and 180 °F), and under pressures up to 103 MPa (15 ksi). Thus, biological activity may influence corrosion in a variety of environments, including soil, inland water and seawater, crude oil and petroleum products, and in oil-emulsion cutting fluids.

Table 3   Ratings of some metals for resistance to cavitation erosion in seawater(a)

Group 1:   Most resistant. Subject to little or no damage. Useful under extremely severe conditions

Stellite hardfacing alloys
Titanium alloys
Austenitic and precipitation-hardening stainless steels
Nickel-chromium alloys such as Inconel alloys 625 and 718
Nickel-molybdenum-chromium alloys such as Hastelloy C

Group II: These metals are commonly used where a high order of resistance to cavitation damage is required. They are subject to some metal loss under the most severe conditions of cavitation.

Nickel-copper-aluminum alloy Monel K-500
Nickel-copper alloy Monel 400
Copper alloy C95500 (nickel-aluminum bronze, cast)
Copper alloy C95700 (nickel-aluminum-manganese bronze, cast)

Group III: These metals have some degree of cavitation resistance. They are generally limited to low-speed low-performance applications.

Copper alloy C71500 (copper-nickel, 30% Ni)
Copper alloys C92200 and C92300 (leaded tin bronzes M and G, cast)
Manganese bronze, cast
Austenitic nickel cast irons

Group IV: These metals normally are not used in applications where cavitation damage may occur, except in cathodically inhibited solutions or when protected by elastomeric coatings.

Carbon and low-alloy steels
Cast irons
Aluminum and aluminum alloys

(a) Applies to normal cavitation-erosion intensities, at which corrosion resistance has a substantial influence on the resistance to damage.
Source: Ref 11

Effects of Anaerobic Bacteria. Probably the most important anaerobic bacteria that affect the corrosion of buried steel are those of the sulfate-reducing type, which produce sulfides. They are often found in wet clay, boggy soils, and marshes. The sulfide they produce accelerates the dissolution of iron and also retards cathodic reactions, especially hydrogen evolution.

The buried structure is often coated with asphalt, enamel, plastic tape, or concrete to prevent contact between the steel structure and the environment. Concrete is less satisfactory than the other coating materials in the presence of sulfur-oxidizing bacteria, because it is also rapidly attacked by the sulfuric acid environment.

Because it is not possible to avoid pinholes or breaks from accidental damage to coatings, cathodic protection of coated buried structures is necessary to minimize or prevent microbiological corrosion. Substitute materials, such as asbestos and plastic pipe in place of steel pipe, have also been used as an effective means of preventing the detrimental effects of microbiological activity in certain undesirable soil locations.

Anaerobic bacteria also affect such operations as the secondary recovery of petroleum by waterflooding, which involves injection of very large amounts of water into the oil well. Even with closed systems using water from deep-source wells, localized pitting corrosion occurs and has resulted in complete wall perforation.

In one well, the corrosion product in the pits underneath a calcium carbonate scale contained 31% iron sulfide. The source water contained 225 ppm sulfate. The bacterial action formed iron sulfide by the reaction $4Fe + SO_4^{2-} + 4H_2O \rightarrow 3Fe(OH)_2 + FeS + 2OH^-$. Thus, localized pitting corrosion resulted from direct anaerobic bacterial reduction of the sulfates at the metal surface. Corrosion of water-injection lines can be controlled by the use of bactericides, but their effectiveness must be monitored with regularly scheduled microbiological tests and the use of corrosion-monitoring test equipment.

Effects of Aerobic Bacteria. Aerobic bacteria can oxidize elemental sulfur or sulfur-containing compounds to sulfuric acid. The reaction with elemental sulfur is $2S + 3O_2 + 2H_2O \rightarrow 2H_2SO_4$.

These organisms thrive at low pH and can produce localized sulfuric acid concentrations up to 5 wt%, creating extremely corrosive conditions. They are frequently found in sulfur fields, in oil fields, and in and about sewage-disposal piping that contains sulfur-bearing organic waste products. In sewage lines, sulfur-oxidizing bacteria cause rapid acid attack of cement piping. Aerobic bacterial corrosion can occur in the operation of the source wells in a secondary-recovery waterflood program unless steps are taken to prevent leaking in packing glands and valves to guard against aeration of the water supply.

A heavy hydrated iron oxide scale, together with a complete wall perforation in a 90 mm (3½-in.) diam waterflood flow line, is shown in Fig. 17. Microbiological examination of a water sample from the line showed the concentration of aerobic viable bacteria to be 22 000/mL after three days and 28 000/mL after seven days. No sulfate-reducing bacteria were found.

The oxygen content of the water in waterflood systems should be monitored, and the water should be deoxygenated on the basis of corrosion-monitoring data. The most effective method involves elimination of the sources of aeration.

Corrosion in Water-Containing Fuels. Bacterial contamination also causes corrosion in integral fuel cells of jet aircraft, where attack occurs on structural surfaces of lower sump areas made of aluminum alloy. Fungus or bacterial microorganisms originate in stagnant fuel in fuel dumps and are introduced into the aircraft during fuel transfer and transportation. The microorganisms (several hundred species have been identified) become attached to the structure and in the presence of water and fuel proliferate to form a tenacious mat.

Fig. 17 An oil-well waterflood flow line that failed in service from oxide scaling and perforation

The scaling and perforation were caused by aerobic bacterial corrosion that was initiated by aeration of the water supply. Longitudinal section

The microorganisms survive at the fuel/water interface, and their biological by-products develop an acidic aqueous solution in the small amounts of water that are always present. The matlike formation retains water locally along with other foreign materials, such as rust particles entrained with the fuel. The organic coating systems used are not completely impervious, and the corrosive solution eventually attacks the aluminum and causes perforation.

The solution to this problem in jet-aircraft fuel cells requires special fuel-handling and filtration equipment and techniques, together with frequent inspection and drainage of fuel cells. In addition, aluminum fuel cells are coated with an aromatic polyurethane paint to prevent corrosion.

Fouling by Marine Organisms. Marine organisms, such as barnacles and mussels, attach themselves to any surface; they grow and effectively seal off a small part of the surface from its environment. This is termed fouling, or bio-fouling. Concentration cells form underneath the barnacles and produce deep pits.

Fouling on ship hulls is a function of environmental conditions. The most severe problem occurs in relatively shallow water, because in deeper water there are no natural surfaces, such as rocks in the tidal zone, to which the organisms may adhere. Thus, harbor conditions are particularly conducive to the initiation of fouling on ship hulls. In general, warm water temperatures favor long breeding seasons and rapid multiplication of macroorganisms that cause fouling.

Relative motion between an object and water usually inhibits the attachment of organisms. Thus, rapidly moving vessels accumulate only small quantities of organisms, and the major portion of fouling occurs when the vessel is docked.

A similar velocity effect also occurs in heat exchangers that use seawater as a coolant. Rapid fluid flow tends to suppress fouling of heat exchangers, whereas rapid accumulation occurs at low fluid rates or during shutdown periods. Also, the nature of the surface strongly influences the attachment of macroorganisms. Smooth, hard surfaces offer an excellent point for adhesion, whereas rough, flaking surfaces

inhibit adhesion. For example, the fouling of stainless steel and iron occurs initially at about the same rate in seawater. However, after some exposure, the surface of the iron is covered by a loosely adhering iron oxide, and fouling is generally less on iron than on stainless steel after long exposure periods.

Copper and copper alloy C70600 (copper nickel, 10% Ni) are highly resistant to fouling in quiet seawater, whereas many of the more noble alloys foul, and pit deeply, in quiet seawater. However, as velocities reach the range of 0.9 to 1.8 m/s (3 to 6 ft/s), fouling diminishes, and pitting of the more noble alloys slows down and even ceases. As velocities continue to increase, the corrosion-barrier film is stripped away from copper and copper nickel, while the stainless steels and many nickel-base materials remain passive and inert. The complete reversal in the tolerance of metals for the marine environment as velocities change is the source of much seemingly conflicting information on actual experience with metals in marine service.

Control of Bio-Fouling. Fouling by organisms is most effectively inhibited by the use of antifouling paints. These paints contain toxic substances, usually copper compounds. They function by slowly releasing copper ions into the aqueous environment, which poisons the mussels, barnacles, and other creatures. A similar technique is used in closed systems, where various toxic agents and algicides, such as chlorine and chlorine-containing compounds, are added to the environment. These methods work more or less successfully, depending on their application. However, under conditions conducive to the growth of aqueous organisms, periodic cleaning is almost always necessary to ensure unimpeded fluid flow and to prevent crevice attack.

Corrosion of Buried Metals

Major influences on the corrosion of uncoated metallic objects buried in the earth include galvanic effects; chemical composition, oxygen content, and pH of the soil; alloy selection; and stray currents.

Galvanic Effects. Failures of the threads in steel pipe have been encountered when brass valves have been screwed directly into the pipe. An insulating coupling should always be used to prevent this type of attack.

Example 9: Failure of a Buried Type 304L Stainless Steel Drain Line by Galvanic Attack. One of five underground drain lines intended to carry a highly acidic effluent from a chemical-processing plant to distant holding tanks failed in just a few months, before all the lines were completed. Type 304L stainless steel was selected from a number of alloys after extensive laboratory testing of corrosion resistance to the highly corrosive liquid. Each line was made of pipe 73 mm (2⅞ in.) in diameter with a 5 mm (0.203 in.) wall thickness to provide additional assurance of long life in

contact with corrosive effluent. Five small drain lines instead of just one large line were installed to reduce the risk of depending on one outlet to discharge effluent (inability to discharge effluent would make it necessary to shut down the plant).

Pipelaying began at the processing building; the lengths of pipe were joined by shielded metal arc welding. Soundness of the welded joints was determined by water back-pressure testing after several lengths of pipe had been installed and joined. Backfilling of the pipe trench was begun after several series of welding and back-pressure testing cycles were complete.

Before completion of the pipeline, a pressure drop was observed during back-pressure testing. An extreme depression in the backfill near the beginning of the lines and near the building revealed the site of failure. A leak was found in one of the five pipes.

Investigation. The failed section of pipe was removed and found to have a hole approximately 19 mm (¾ in.) in diameter on the outside, narrowing down to a diameter of 6 mm (¼ in.) on the inside. Measurements showed that the voltage decrease between the soil and the pipe averaged about 1.5 V. The soil was not consistent in composition, but was a coarse intermixture of sandy loam and heavy clay; the voltage decreases between the pipe and these conglomerates varied greatly for the different types of soil.

The generally smooth contour of the hole in the pipe, together with the voltage-decrease measurements, showed that the failure had resulted from galvanic corrosion at a point where the corrosivity of the soil was substantially greater than the average, resulting in a voltage decrease near the point of failure of about 1.3 to 1.7 V.

Corrective Measures. Because of the necessity for very high reliability of continued operation of the line without failure, three corrective measurements were implemented:

- The pipelines were asphalt coated
- The pipelines were enclosed in a concrete trough with a concrete cover
- Magnesium anodes, connected electrically to each line, were installed at periodic intervals along their entire length to provide cathodic protection

Effect of Soil Composition. Soil containing organic acids derived from humus is relatively corrosive to steel, zinc, lead, and copper. The measured total acidity of such a soil appears to be a better index of its corrosivity than pH alone. High concentrations of sodium chloride and sodium sulfate in poorly drained soil make the soil very corrosive. A poorly conducting soil, whether low in content of moisture, dissolved salts, or both, is generally less corrosive than a highly conducting soil. However, conductivity alone is not a sufficient index of corrosivity; the anodic or cathodic polarization characteristics of a soil are also a factor.

Cinders constitute one of the most corrosive environments. It has been reported that in exposures of four or five years, corrosion rate in cinders for steel and zinc was five times, for copper eight times, and for lead 20 times as high as the average rates for the same metals in 13 different soils.

Pitting corrosion of steel and cast iron pipe is prevalent in areas where backfill has consisted of coal ashes, which normally contain sulfides. Corrosion may be a combination of galvanic (steel-to-carbon couple) corrosion and corrosion caused by the formation of weak sulfuric acid from the ashes. Backfill should always be well compacted to exclude air. The presence of moisture plus air results in oxidation of metallic objects, particularly those made from iron or steel. Chemical pollution of the soil, such as from leaks or spills, often results in very aggressive corrosion of metallic objects, such as pipe and underground structures.

The soils of the East coast of Florida have a pH of about 6.5 to 8.3. The variation in silica sand, oyster and clam shells, imported clay, and shale does not appear to have an appreciable effect on the corrosion rate of cast iron pipe and steel pilings. Cast iron pipes examined in one study were not significantly damaged by corrosion. Any pipe failures appear to have been caused by impact and over-pressurization. Steel pilings in service for ten years were examined. The condition of the pilings was good, with little or no corrosion. The deepest pits were 1.3 mm (0.050 in.) over a 6.45-cm^2 (1-in.^{2}) area. These pits were found on a spot of exposed bare metal on a mill-scale surface. The pilings were of 405-mm (16-in.) diam low-alloy steel pipe with a wall thickness of 9.5 mm (3/8 in.).

Alloy Selection. Minor changes in composition and microstructure are not important to corrosion resistance. Therefore, copper-bearing steel, low-alloy steel, low-carbon steel, and wrought iron corrode at approximately the same rate.

Increases in chromium content of low-alloy steel decrease observed weight loss in a variety of soils, but above 6% Cr, depth of pitting increases. In 14-year tests, 12% Cr and 18% Cr stainless steels were severely pitted. Type 304 stainless steel was not pitted or was only slightly pitted. Type 316 stainless steel did not pit in any of 15 soils to which the alloy was exposed for 14 years.

Zinc coatings are effective in reducing weight loss and pitting rates of steel exposed to soils. A major source of protection appears to result from the alloy layer formed between zinc and the steel surface in hot dip galvanizing.

Copper, on the average, corrodes at about one-sixth the rate of iron. However, in tidal marsh, for example, the rate is comparatively higher than in most other soils, being one-half that of iron.

Lead also corrodes less on the average than steel does. In poorly aerated soils or soils high in organic acids, the corrosion rate may be four to six times the average.

Atmospheric Corrosion

The metals ordinarily used in equipment and structures corrode at a negligible rate when exposed to the atmosphere in the absence of moisture to serve as an electrolyte. For example, metal parts exposed in the desert air remain free from corrosion for long periods of time. Also, metal parts exposed to the air at temperatures below the freezing point of water or of aqueous condensates on the metal do not corrode to a significant extent, because ice is a poor electrolytic conductor.

Corrosivity of Different Atmospheres. Table 4 compares the general corrosivity of rural, industrial and marine atmospheres in temperate climates and gives the average corrosion rates for a number of metals exposed for 10 and 20 years in the three types of atmospheres. Three generalizations can be drawn from the data for the nonferrous metals in Table 4:

- The industrial atmosphere is much more corrosive than the rural atmosphere (except on lead) and is more corrosive or at least as corrosive as the marine atmosphere
- The rural atmosphere is less corrosive (on some metals, much less) than the marine atmosphere, and the two atmospheres have low and approximately equal rates of attack on the more resistant metals
- The corrosion rate on a given metal shows only a slight increase or decrease between the 10- and 20-year periods, with no apparent pattern

For the most part, the corrosion rate depends on the chemical behavior of each metal in relation to the moisture content and the amount and nature of the particulate matter and of the gaseous impurities in a given type of atmosphere.

Corrosion-Product Films. One factor that accounts for many of the differences in corrosion rates in Table 4 is the formation of corrosion-product films that provide different amounts of protection against corrosion. Only lead, which readily forms tightly adherent, insoluble, highly protective films in the air, is equally resistant to rural, industrial, and marine atmospheres. These films usually consist mainly of oxycarbonates, but contain equally protective sulfates and oxychlorides in industrial and marine atmospheres.

The protection provided by corrosion-product films causes the corrosion rates given in Table 4 to level off to approximately steady-state values after an initial period of more rapid attack. The slopes of the curves shown in Fig. 18 illustrate this behavior for steels exposed to an industrial atmosphere. Figure 18 also shows the effects of the compositions of the steels (Table 5) on the protective characteristics of the rust film formed, with the greatest protection being provided by the films on the high-strength low-alloy steels. General experience with atmospheric-exposure tests on steel and zinc has demonstrated that the conditions of exposure during the early stages of a test also have a major effect on the long-term corrosion behavior of the metals, further confirming the generally protective nature of the corrosion-product films.

In one series of tests, weight losses of zinc specimens exposed for several weeks to an industrial atmosphere depended primarily on the humidity and presence of moisture during the first few days of exposure and were affected less by conditions during the remainder of the test. A similar primary dependence on the initial exposure conditions has been observed on steel when the specimens were exposed for a year, with the exposures beginning in different months of the year. In winter, the greater surface accumulation of combustion products, in particular sulfuric acid, produces a less protective initial corrosion product, which influences the subsequent corrosion rate. One-year exposure of aluminum and aluminum alloys in an industrial atmosphere has resulted in substantially deeper

Table 4 Average corrosion rates of various metals exposed for 10 and 20 years to three types of atmospheres(a)

Metal	Rural atmosphere(b) 10-year exposure	Rural atmosphere(b) 20-year exposure	Industrial atmosphere(c) 10-year exposure	Industrial atmosphere(c) 20-year exposure	Marine atmosphere 10-year exposure	Marine atmosphere 20-year exposure
Aluminum	0.001	0.003	0.032	0.029	0.028	0.025
Copper	0.023	0.017	0.047	0.054	0.052	0.050
Lead	0.019	0.013	0.017	0.015	0.016	0.021
Tin	0.018	...	0.047	0.052	0.091	0.112
Nickel	0.006	0.009	0.128	0.144	0.004	0.006
Monel 400(e)	0.005	0.007	0.053	0.062	0.007	0.006
Zinc, 99.9%	0.034	0.044	0.202	0.226	0.063	0.069
Zinc, 99.0%	0.042	0.043	0.193	0.218	0.069	0.068
Carbon steel(f)	...	...	0.48	...	...	...
Low-alloy steel(g)	...	...	0.09	...	...	...

(a) Source: Ref 12. Data on steels, Ref 13. (b) State College, PA. (c) New York City, for nonferrous metals; Kearny, NJ, for carbon and low-alloy steels. (d) La Jolla, CA. (e) 65Ni-32Cu-2Fe-1Mn. (f) 0.2C-0.02P-0.05S-0.05Cu-0.02Ni-0.02Cr. (g) 0.01C-0.2P-0.04S-0.03Ni-1.1Cr-0.4Cu

Fig. 18 Variation in weight loss of several steels with exposure time in a corrosive industrial atmosphere

See Table 5 for compositions of the steels. Source: Ref 12

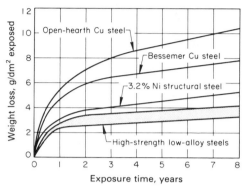

Table 5 Compositions of steels for which corrosion data are plotted in Fig. 18

Steel	C	Mn	P	S	Si	Cu	Ni	Cr	Mo
Open-hearth Cu steel 0.02		0.39	0.006	0.018	0.005	0.20	...	0.07	...
Bessemer Cu steel 0.10		0.40	0.112	0.059	0.018	0.21	0.003	0.03	...
3.2% Ni structural steel 0.19		0.53	0.016	0.022	0.009	0.07	3.23	0.10	...
High-strength low-alloy steels									
Ni-Cu steel. 0.05		0.36	0.054	0.016	0.008	1.14	1.99	0.01	...
Cu-Ni-Mn steel. 0.09		0.86	0.008	0.025	0.019	1.41	0.95	0.03	0.09
Cr-Si-Cu-P steel 0.09		0.24	0.154	0.024	0.80	0.43	0.05	1.07	...

pitting when begun in winter than in summer but, unlike the results for steel and zinc, only slightly greater weight loss.

Passivity. Among the metals for which atmospheric corrosion rates are given in Table 4, nickel and aluminum have a natural passivity in uncontaminated air (steels and zinc can be passivated by chemical-oxidizing treatments). The passivity of nickel is destroyed in the industrial exposure, and that of aluminum is destroyed in both the industrial and the marine atmospheres. The surface films generally considered to be the source of passivity are only about 3 nm or less in thickness, while ordinary corrosion-product films are several orders of magnitude thicker.

Rural atmospheres are generally free of the particulate matter and corrosive gases found in industrial atmospheres, and this condition accounts for the low corrosion rates in rural atmospheres (Table 4). The normal carbon dioxide content of air has little or no effect on metals exposed to the atmosphere.

Industrial atmospheres contain dust as a primary contaminant, in concentrations of about 2 mg/m^3 for average city air to 1000 mg/m^3 for heavily industrialized areas. It is estimated that more than 35 000 kg of dust/km^2 (100 tons of dust/square mile) settles in urban industrial areas in a month.

Dust deposited on metal surfaces in these areas generally contains particles of carbon and carbon compounds, metal oxides and metal salts (chiefly sulfates and chlorides), and sulfuric acid. The combination of moisture with dust particles bearing soluble contaminants produces crevice corrosion by forming differential aeration cells and other types of concentration cells.

Most of the soluble contaminants are hygroscopic and absorb moisture from the air when the relative humidity is substantially less than 100%. The critical level of relative humidity (that at which moisture absorption takes place on metals exposed to even relatively mild industrial

atmospheres) is usually about 50 to 70% for steel, copper, nickel, zinc, and most metals that are used in structures and objects intended for industrial atmospheric exposure. This takes into account the effect of the normal fluctuations in temperature between day and night.

Such gases as sulfur trioxide, sulfur dioxide (which readily oxidizes to sulfur trioxide), hydrogen chloride, oxides of nitrogen, hydrogen sulfide, and the halogens in the air accelerate the crevice-corrosion effects of moist dust deposits on these metals. Even at low concentrations, such gases also corrode these metals in the absence of dust deposits, if the humidity reaches or exceeds the critical level.

Metal surfaces located where they become wet but where rain cannot wash the surfaces may corrode more rapidly than if fully exposed. For example, the rusting of steel in partly sheltered locations in moist air containing oxides of sulfur, which form sulfuric acid, is apparently accelerated in a self-perpetuating sequence of reactions. The acid attacks the steel, producing iron sulfate that is retained on the moist rust and hydrolyzes to form more sulfuric acid and iron oxide, thus catalyzing the rusting process.

Marine Atmospheres. At and near the seacoast, the deposition of salt-water spray is the most corrosive aspect of marine atmospheric exposure. The rate of attack on exposed metals varies widely, depending on distance from the ocean, prevailing wind direction, relative humidity, and temperature fluctuations that can produce condensation.

The penetration of protective film by chloride ions and the high solubility and hygroscopic nature of metal chlorides cause rapid corrosion on carbon and low-alloy steels. Zinc and cadmium plating extend the life of steel hardware to a useful but limited extent; large steel structures must be protected by painting.

The high conductivity of moisture that contains dissolved salt accelerates crevice corrosion and galvanic corrosion, making the use of sealants mandatory at joints and the use of single-metal systems good practice.

Brasses undergo rapid dezincification unless alloyed with small amounts of arsenic, antimony or phosphorus in inhibited grades. Copper nickels, titanium, and such alloys as Incoloy 800 perform well in marine atmospheres, but most metals corrode severely unless protected by organic coatings.

In general structure applications of aluminum alloys and stainless steels in thick sections, pitting and crevice corrosion are not usually serious problems (however, stress corrosion can cause problems, as discussed in the article "Stress-Corrosion Cracking" in this Volume). Pitting and crevice corrosion can produce rapid penetration and failure by leakage in thin-wall vessels, tubes, and pipes made of these alloys.

Corrective and Preventive Measures

Where a corrosion failure has occurred, economical and practical measures for prevention of future failures of the same type are required. The major types of corrective and preventive measures are:

- Change in alloy, heat treatment, or product form
- Use of resinous and inorganic-base coatings
- Use of inert lubricants
- Use of electrolytic and chemical coatings and surface treatments
- Use of metallic coatings
- Use of galvanic protection
- Design changes for corrosion control
- Use of inhibitors
- Changes in pH and applied potential
- Continuous monitoring of variables

Change in Alloy, Heat Treatment, or Product Form

Where the environment cannot be changed to solve a corrosion problem, a change in alloy, heat treatment, or product form may be required. Under some circumstances, both alloy and product form may be changed.

Extreme temperatures and high operating stresses favor the use of special-purpose cast alloys. For example, the tubes in gas reformers are usually made of cast heat-resisting alloys, such as HK-40 or HP-modified, to obtain maximum life at operating temperatures of 870 to 1010 °C (1600 to 1850 °F). Similarly, the use of high-alloy castings enables turbine engines to operate at higher temperatures and thus at higher efficiencies than weldments would permit.

Solution annealing of austenitic stainless steels minimizes the risk of intergranular attack and SCC. For stainless steel weldments that cannot be annealed and that are to be used in applications where either intergranular corrosion or intergranular SCC is of concern, a quenching procedure can be used. In this procedure, the weld is rapidly quenched to prevent the precipitation of the chromium carbides. If the quenching procedure proves ineffective, either a low-carbon or a stabilized stainless steel may be substituted, depending on the tensile requirements.

Selection of the proper alloy and product form requires a complete knowledge of the operating conditions, including chemical environment and operating temperatures and stresses. Sometimes, intermediate stages of chemical reactions produce an environment that will cause failure of an alloy shown by laboratory testing to be resistant to the initial components and the end products. Therefore, it is advisable to field test where possible.

Use of Resinous and Inorganic-Base Coatings

Vinyls, acrylics, epoxies, phenolics, furanes, and urethanes are used extensively for corrosion protection in the form of paints, adhesives, coatings, and linings. Their chemical resistance makes them suitable for many applications.

However, especially for fairly corrosive immersion service, consideration must always be given to the possible presence of pinhole porosity in paint-type coatings, which may result in pitting and crevice corrosion, and to the possibility of accelerated galvanic attack where dissimilar metals are present (see the section "Ratio of Cathode Area to Anode Area" in this article). The versatility of these organic compounds for being formulated for different methods of application (as liquids, as solids applied by bonding, and as powders that can be fused) permits their extensive use in protection of metal structures and equipment against corrosion.

Applications for resinous and inorganic coatings include the following:

- Vinyl, acrylic, epoxy, phenolic, furane, and urethane primers and paints can be used to coat any metal for atmospheric protection
- Clear acrylics and urethanes are used to protect surfaces that must remain visible, such as nameplates and dials
- Thick, viscous epoxy coatings are used to embed delicate electrical connections in printed circuit boards.
- Sealant materials, if properly applied, are highly effective in preventing crevice corrosion; if not properly applied, they can make it more likely to occur
- Elastomeric or solid-form sheets of rubber, polypropylene, polyvinyl chloride, and other resins are used for corrosion-protective liners or containers of almost unlimited size or shape

Certain primers that form cross-linked polymeric films have superior adhesion and thus minimize local rusting and underfilm (filiform) corrosion.

Zinc-Rich Coatings. Organic and inorganic coatings containing zinc dust give excellent protection to steel structures. They provide sacrificial protection because the zinc particles are in intimate contact with one another; therefore, the coating film is electrically conductive. This is achieved by very high zinc loading with a relatively small amount of binder, such as an ethyl silicate. To protect the steel, the coating itself must be in electrical contact with the substrate and not insulated by any rust, scale, old paint, or pretreatment chemicals. Sand blasting to produce a near-white surface is the required surface preparation.

As the zinc is sacrificed, zinc corrosion products in the form of efflorescing salts—commonly called white rust—appear on the film, making the coating thicker and reducing its electrical conductivity. These products of corrosion then act as a barrier between the active zinc and the corrodents. However, if the coating is damaged, the fresh zinc metal that is exposed provides renewed anodic action. Essentially, then, a zinc-rich coating is a self-healing film, because any damage to the barrier reinitiates the anodic action.

The inorganic vehicles used in these coatings are usually ethyl silicates, which are available in both solvent-reducible and water-reducible types. The organic vehicles used include chlorinated rubber, styrene, epoxies, phenoxies, urethanes, and silicones.

In a seacoast environment, the performance of zinc-rich coatings in inorganic vehicles is far superior to that of zinc-rich coatings in organic vehicles. One disadvantage of the coatings that are based on inorganic vehicles is that they require more critical surface preparation than those based on organic vehicles.

When zinc-rich coatings are used as primers for a top coat of an essentially pore-free resinous coating, the difference between the two types of vehicle becomes much less dramatic, and neither zinc-rich undercoat has much effect when a large break in the film occurs. Both organic and inorganic vehicles are effective in healing small, narrow breaks in the film and in preventing underfilm corrosion.

Use of Inert Lubricants

Certain chemically inert resins, such as silicones, esters, and fluorocarbons, can serve both as effective lubricants and as corrosion-resistant coatings and linings. In seacoast environments, lubricants must frequently perform this dual role of lubrication and corrosion protection. The role of corrosion protection is often overlooked when selecting a lubricant for a specific function, as for wire rope on exposed sliding surfaces.

In a lubricant-testing program at an aerospace facility, 45 oils and greases of widely varying composition were evaluated for effec-

tive lubricity, corrosion protection, and conformance to specifications. Most of the lubricants performed satisfactorily except in providing corrosion protection. A total of four corrosion tests were performed in this program: (1) ASTM standard method D 1743 (Ref 14), which involves the determination of the corrosion-prevention properties of greases using grease-lubricated tapered-roller bearings stored under wet conditions, (2) specimens exposed for 2541 h in a humidity cabinet, (3) specimens exposed for 1003 h in a 5% salt-fog environment, and (4) specimens exposed for one year on a seaside test rack. The most severe of these tests was exposure on the seaside test rack. Some of the results of this test are described below.

Specialized Inert Lubricants. The highly specialized inert lubricants required by gaseous and liquid oxygen systems and by hypergolic systems (systems of two components that react explosively on contact) offered little corrosion protection to carbon steel test panels. This was probably caused by the failure to include effective inhibitors in these lubricants to prevent sensitizing the lubricants to oxygen or hypergolics. However, one such grease, when provided with an organic-base inhibitor (in an amount less than 0.5%) that did not compromise the inertness of the grease to oxygen, performed well in the seaside corrosion test by protecting the carbon steel panel for a full year and permitting only minor corrosion.

Conventional petroleum-base lubricants varied widely in the seaside-exposure test. However, it was found that earth-gel-thickened greases performed very poorly in this test, possibly because the earth gel may be capable of transporting moisture and contaminants to the surface of the metal through the oil film. This was found to be true for greases based on petroleum oils, polychlorotrifluoroethylene oils, or silicone oils.

Thickened Greases. Some of the more effective corrosion-preventing greases were those thickened with lithium soaps, alkyl ureas, or organic polymers. Greases containing graphite or molybdenum disulfide varied widely between affording protection and accelerating corrosion, depending almost entirely on the effectiveness of the inhibitor. Polychlorotrifluoroethylene lubricants must be used with extreme caution, because they can detonate when in contact with aluminum or magnesium subjected to shear stresses.

Use of Electrolytic and Chemical Coatings and Surface Treatments

The major types of electrolytic and chemical coatings and surface treatments include anodizing, chemical conversion coatings, and passivation treatments. These processes vary widely in their effectiveness in protecting treated metals against corrosion. Anodic and chemical conversion coatings also serve as excellent bases for organic coatings.

Anodizing of aluminum and aluminum alloys provides effective protection in natural environments but not in aggressive environments, especially acidic and alkaline environments.

Most aluminum components for aircraft applications are anodized for added corrosion protection; the anodic coating also serves as an excellent base for paint. The sulfuric acid method should not be used on assemblies that can entrap liquids; the chromic acid method must be used on such assemblies. Detailed information on anodizing of aluminum can be found in the section "Anodizing Processes" in the article "Cleaning and Finishing of Aluminum and Aluminum Alloys" in Volume 5 of the 9th Edition of *Metals Handbook*.

Anodizing is also done on magnesium alloys (see the section "Anodic Treatments" in the article "Cleaning and Finishing of Magnesium Alloys" in Volume 5 of the 9th Edition of *Metals Handbook*). It has little protective value against corrosion, but provides an excellent base for corrosion-resistant paints and other resinous coatings.

Chemical Conversion Coating. Chromate conversion coatings are extensively used on steel products that have been electroplated with cadmium or zinc, and they improve corrosion resistance substantially. Similar coatings are also produced by electrolytic processing, with the parts acting as the anode in the electrolytic chromating bath. Chromate coatings, however, are very thin and can be readily removed by abrasion or impact. Local bare areas may then corrode preferentially. Chromate, phosphate, and other conversion coatings are also used on aluminum, magnesium, steel, and other metals, but primarily as a base to improve the adhesion and protective value of organic coatings to be applied over them.

The chemical conversion coatings offer limited corrosion protection and should not be exposed to severely corrosive environments. More information on chromate conversion coatings can be found in the article "Chromate Conversion Coating" in Volume 5 of the 9th Edition of *Metals Handbook*.

Passivation is common practice in the manufacture of stainless steel components and assemblies. Stainless steels, whether martensitic, ferritic, or austenitic, often show rust spots on the surface after exposure to humid conditions. This results from the embedding of small particles of iron or steel in the surface from cutting, machining, fabrication, and handling. If the particles are not removed, the stainless steel is susceptible to local rusting and pitting in the presence of moisture or other electrolytes.

Passivation involves the removal of particles of iron or steel by chemical methods (pickling) or by mechanical methods and permits the formation of a very thin but highly effective passive film on the stainless steel surface upon exposure to a clean, dry atmosphere. Pickling must be followed by thorough neutralization, rinsing, and drying.

Special chemical-passivating treatments produce an effective passive film more rapidly than exposure to the atmosphere does. The complete absence of rusting after treated metal has been exposed to moisture or high humidity for 24 h is an indication that the passivating treatment was effective. Procedures and solutions for passivation of stainless steels are given in the section "Passivation" in the article "Cleaning and Finishing of Stainless Steels" in Volume 5 of the 9th Edition of *Metals Handbook*.

Use of Metallic Coatings

A wide variety of metallic coatings are applied, mainly by electroplating, hot dipping, electroless processes, and cladding. Selective brush plating, spray metallizing, vacuum deposition, gas plating, and cathode sputtering are also used. Selection of a coating process is based on the availability of equipment, the criticality of the particular part, and the overall cost of the type of protection needed.

Electroplated Coatings. Electrodeposition of zinc or cadmium is widely used to protect steel from corrosion. Zinc provides better performance in industrial areas; cadmium is preferred for marine environments. These coatings offer sacrificial protection to the steel substrate and will minimize dissimilar-metal effects when the coated part is joined to aluminum or magnesium. Tin, nickel, chromium, and copper are other metals that are readily applied by electroplating, but coatings of these metals are much less effective than sacrificial coatings in providing corrosion resistance.

The possibility of producing hydrogen damage (hydrogen embrittlement) must be considered in applying electroplated coatings, especially on high-strength steels. Additional information is provided in the article "Hydrogen-Damage Failures" in this Volume. Various plating materials and procedures, many of which are applicable in corrosive atmospheres, are discussed in the section "Plating and Electropolishing" in Volume 5 of the 9th Edition of *Metals Handbook*.

Sprayed Metal Coatings. Metal spraying can provide thick protective coatings. Multiple coating minimizes, but does not eliminate, the occurrence of voids and weak spots in the coatings. Aluminum spray coatings are discussed in the section "Spray Coatings" in the article "Aluminum Coating of Steel" in Volume 5 of the 9th Edition of *Metals Handbook*.

Cladding. For heavy-duty corrosive service, a metallurgically bonded or clad surface of alloy metal can be used to protect the steel substrate. When vessel design dictates heavy wall construction, that is, 25 mm (1 in.) or greater, it is economical to clad the heavy steel with a corrosion-resistant alloy. Most corrosion-resistant alloys can be bonded to steel by either hot rolling or explosion cladding. Particular care must be taken in fabricating the clad metals to ensure the corrosion-resistant integrity of the alloy cladding.

Use of Galvanic Protection

Galvanic protection can be either cathodic protection, in which the object to be protected is made cathodic, or anodic protection, in which the object to be protected is made anodic. Cathodic protection is most commonly used.

Cathodic protection may be of two different types: impressed direct current or sacrificial anode. In the impressed direct-current type, the structure to be protected is made the cathode in a direct-current (dc) electrical circuit. The anode in the circuit is an auxiliary electrode, usually of iron or graphite, that is located some distance away from the structure to be protected. The positive terminal of the source of direct current is connected to the auxiliary electrode, and the negative terminal is connected to the structure to be protected. Current then flows from the electrode through the electrolyte to the structure, and the structure does not corrode. The applied voltage need only be high enough to supply an adequate current density to all parts of the structure to be protected. Soils or waters of high resistivity will require higher voltages than those of lower resistivity, and higher voltages will be required when anode spacing is increased in soil or water with a given resistivity.

In the sacrificial-anode system, the structure to be protected is made the cathode in a galvanic-corrosion cell, and current is supplied by the corrosion of anodes that are commonly made of zinc or magnesium. Voltage and current are limited by the corrosion of the sacrificial anodes; the number and location of anodes are much more critical in this method than in the impressed direct-current method. Periodic anode replacement may be required where corrosion rates of the structure would be high without cathodic protection.

The choice of the system (impressed direct current versus sacrificial anode) depends on a variety of factors, including availability of direct current, corrosion rate of the unprotected structure, ease of anode replacement, and total lifetime system cost.

Cathodic protection is used to protect such metals as steel, copper, lead, and brass against corrosion in all soils and in almost all aqueous media. Cathodic protection is ordinarily used in conjunction with resinous coatings, which greatly reduce the area to be cathodically protected and the current required.

Cathodic protection cannot be used to avoid corrosion above the waterline, because the impressed current cannot reach metal areas that are out of contact with the electrolyte. Moreover, the protective current does not enter electrically screened areas, such as the interior of water-condenser tubes (unless the auxiliary anode enters the tubes), even though the water box may be adequately protected.

Fig. 19 Cathodic-protection system for a buried steel tank

(a) The original design that caused local failure of a nearby unprotected buried pipeline by stray-current corrosion. (b) Improved design. Installation of a second anode and an insulated buss connection provided protection for both tank and pipeline, preventing stray currents.

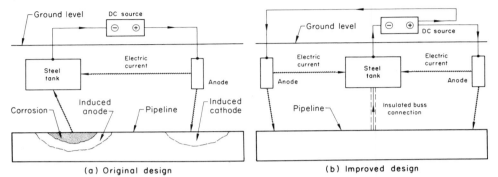

(a) Original design

(b) Improved design

For buried pipelines, cathodic protection costs far less than any other means offering equal assurance of protection. Assurance that no leaks will develop on the soil side of a cathodically protected buried pipeline has made it economically feasible, for example, to transport oil and high-pressure natural gas across half the continent of North America.

Stray currents are frequently encountered problems in cathodic-protection systems. Figure 19(a) shows an arrangement in which stray currents were produced when the owner of a buried tank installed cathodic protection, not knowing of the presence of a nearby pipeline. The pipeline rapidly failed by corrosion because of the stray currents. If the pipeline had been cathodically protected, stray-current attack could have caused the buried tank to fail. The stray-current problem shown in Fig. 19(a) was corrected by electrically connecting the tank and the pipeline by an insulated buss connection and by installing a second anode (Fig. 19b). Thus, both pipe and tank were protected without stray-current effects.

Anodic Protection. By imposing an external potential to make them anodic, some metals can be prevented from corroding in an electrolyte in which they would otherwise be attacked. This technique, termed anodic protection, is applicable only to metals and alloys that show active-passive behavior. It has been applied to iron, titanium, aluminum, and chromium, but mostly to steel and stainless steel. Anodic protection is not applicable to zinc, magnesium, cadmium, silver, copper, and copper-base alloys.

In anodic as in cathodic protection, the corrodent must be an electrolyte. The passive potential is automatically maintained, usually electronically, by a potentiostat.

The anodic technique has been used for protecting low-alloy steel against uniform corrosion in ammonium nitrate fertilizer mixtures, carbon steel in 86% spent sulfuric acid at temperatures up to 60 °C (140 °F), and carbon steel in 0.1 to 0.7 M oxalic acid at temperatures up to 50 °C (120 °F).

Because passivity of iron and the stainless steels is destroyed by halide ions, anodic protection of these metals is not possible in hydrochloric acid or in acidic chloride solutions. Also, if Cl^- should contaminate the electrolyte, the metal may corrode by pitting. In the latter case, however, it is necessary only to operate in the potential range below the critical pitting potential for the mixed electrolyte.

Titanium, which has a very noble critical pitting potential over a wide range of Cl^- concentration and temperature, is passive in the presence of Cl^- and can be anodically protected without danger of pitting, even in solutions of hydrochloric acid.

Current densities to initiate passivity are relatively high, but current densities for maintaining passivity are usually low. Corrosion rates are commonly in the range of 0.2 to 25 mg/dm^2 per day.

Ability to maintain the desired potential accurately over the entire structure is very critical in anodic control. Fortunately, an anodic-protection system has high throwing power, and it is possible to protect quite complex structures with proper cathode placement. It is essential that the whole structure remain within the passive range. This may be difficult to achieve in deep crevices, and active corrosion could occur at the bottom of a crevice.

The cathode material must be one that does not suffer gross corrosion in the environment. A platinum-clad metal or a corrosion-resistant alloy is commonly used.

Operating costs for anodic protection are high, but they can be reduced by proper planning of the system and its operation. Once passivity has been achieved, it is often unnecessary to apply the current continuously. In some cases, it may be possible to operate by applying the current for only about 1% of the time—5 s on, 500 s off. In addition to saving power, this can be useful when there is more than one anodic-protection system in the vicinity. By installing an automatic switching device, it is possible to protect two or more structures with one power unit.

The effect of agitation in anodic-protection systems is complex. In some systems, agitation will lower current requirements by enhancing passivity; in other systems, it has the opposite effect.

The current required to maintain passivity is monitored during protection. An increase in the current shows that the corrosion rate is increasing and that corrective measures may be needed.

Anodic Versus Cathodic Protection. Anodic and cathodic protection tend to complement one another; each method has its own specific areas of application. Anodic protection can be used in corrosives ranging from weak to very aggressive; cathodic protection is restricted to moderately corrosive conditions because of its high current requirement, which increases with the corrosivity of the environment. Therefore, it is not practical to protect metals cathodically in very aggressive media. Anodic protection, on the other hand, uses a very low applied current and can be used in strong corrosive media.

The installation of a cathodic-protection system is relatively inexpensive because the components are simple and easily installed. Anodic protection requires more complex instrumentation, including a potentiostat and reference electrode, and its installation cost is higher. The operating costs of the two systems differ because of the difference in current requirements noted above. The throwing power of cathodic protection is generally low, thus requiring numerous closely spaced electrodes to achieve uniform protection. Anodic-protection systems have high throwing power; consequently, a single auxiliary cathode can be used to provide protection to extensive areas, such as long lengths of pipe.

Anodic protection has two unique characteristics. First, the current required is directly related to the rate of corrosion for the protected system. Thus, anodic protection not only protects but also offers a means of monitoring instantaneous corrosion rate. Second, operating conditions for anodic protection can be precisely established by laboratory measurements of polarization.

Design Changes for Corrosion Control

A change in design is sometimes the most appropriate way to eliminate a corrosion problem. Welded construction can eliminate the crevices normally found in bolted or riveted assemblies, thus preventing crevice corrosion. Modifying equipment to permit agitation of a mixture may prevent pitting attack under deposits that might otherwise settle out in a stagnant system. The relocation of drain lines in such a way that complete drainage of a vessel is possible can eliminate pitting corrosion, which might occur in stagnant solutions.

When a solution containing dissolved ions that might cause corrosion at high concentrations is being heated, it is desirable to pass the solution through the tube side of the heat exchanger rather than the shell side. This procedure avoids possible concentration either in joints of tubes to tube sheets or in the vapor phase that usually exists in the shell side of heat exchangers even when special precautions are taken to eliminate it. Drip lips on tank inlet lines are desirable so that any concentration of solutions (as by simple evaporation), which might cause corrosion, occurs on the extension of the inlet line rather than on the tank wall.

Cavitation, erosion, and impingement-corrosion problems can be minimized by providing for smooth flow of liquids or gases without abrupt changes in direction or velocity. In some cases, it may be possible to transfer the corrosion to an acceptable location by a relatively simple design change, as in the following situation.

Frequent problems were encountered with rapid failure of the inlet line of a type 304L stainless steel continuous concentrator. The concentrator feed was 0.5 M nitric acid containing approximately 10 g/L of a polyvalent metal ion; the concentration factor through the unit was approximately 10. The original inlet line, which was a simple flanged connection, failed by what appeared to be a combination of severe general corrosion and pitting attack that was limited to the first 150 to 305 mm (6 to 12 in.) of the line just ahead of the concentrator.

Corrosion tests revealed that the inlet line was the anodic member of a massive concentration cell made up of the solution circulating in the concentrator and the relatively dilute solution entering through the feed line.

It was decided to replace the entire unit with one made of titanium, but while the new unit was being fabricated, a temporary correction was made. This consisted of simply extending the inlet line into the concentrator loop.

The severe concentration-cell corrosion then took place on the inlet-line extension rather than on the inlet line itself. The unit performed satisfactorily under these conditions until it was replaced with the new titanium unit.

Redesign may lead to corrosion problems that were not present in the old equipment. The introduction of a stainless steel tank or agitator may cause galvanic corrosion of associated carbon steel components. The application of an organic coating on the interior of a carbon steel vessel may result in serious pitting problems at pinholes in the coating, whereas only mild general corrosion had been encountered before the coating was applied.

Occasionally, a change in design or materials for purposes of corrosion control may have serious effects that are not intrinsically related to corrosion. The material of construction of a concentrator handling a nitric acid solution was changed from type 304L stainless steel to unalloyed titanium because of corrosion problems with the stainless steel. No further corrosion problems were encountered with the unit, but the frequency of rejection of product batches

because of excess silica content increased significantly, even though there was no significant increase in the silica content of the feed stream to the concentrator.

It was subsequently discovered that the silica in the feed stream deposited preferentially on the titanium surfaces and then broke loose, raising the silica content of a particular batch above the acceptable limit. No such deposition of silica had occurred on the stainless steel unit because the general corrosion rate before changing the material was such that the substrate metal was being continually but slowly dissolved.

Changes in design may necessitate a change in materials. For example, a decision to replace an existing shell-and-tube heat exchanger with a plate-type exchanger requires careful consideration. The plate-type exchanger has crevices, is highly stressed in the embossed areas, and is made of thinner material. Thus, materials that have been successful in the shell-and-tube exchanger may fail in the plate-type exchanger.

In other cases, a less expensive engineering material may give satisfactory life if design is revised to eliminate such features as crevices, pockets where scale or mud can accumulate, condensation points, areas of high velocity, phase separations within a vessel, and high stresses, such as those imposed by improperly designed supports. A material may be satisfactory for a tank but may fail as a heating coil because of the hot-wall effect. A pump may have an impeller made of an alloy different from that of the pump casing because of velocity effects. Thus, the engineer must consider the possibility of galvanic corrosion. When making design changes to solve a corrosion problem, consideration must be given to the mechanical properties of the material involved and to the chemical resistance of the new material in the existing chemical environment.

Use of Inhibitors

Inhibitors, which are chemical substances added to a liquid (usually water or an aqueous solution) to prevent corrosion or to control it at an acceptably low rate, are used mainly in closed or recirculating systems. They are selected for their effectiveness in protecting the specific metal or combination of metals in a given system. A list of inhibitors used to protect various metals in a variety of environments is provided in Ref 15. Inhibitors that function by stifling the anodic corrosion reaction are called anodic inhibitors; those that function by stifling the cathodic corrosion reaction are called cathodic inhibitors.

Anodic inhibitors may be divided into two types: oxidizing and nonoxidizing. Inhibition by oxidizing inhibitors is not a direct function of oxidizing power. Thus, on steel, chromates and nitrites act in the absence of oxygen; molybdates and tungstates are effective as inhibitors only in the presence of air; and pertechnetates are good inhibitors at concentra-

tions as low as 5 to 10 ppm, although permanganates have little inhibitive action.

Anodic inhibitors stifle the anodic reaction, usually forming sparingly soluble substances as adherent protective films. There is often no change in the appearance of the metal, although it carries a very thin film that may be isolated by the use of special techniques.

Such salts as hydroxides, silicates, borates, phosphates, carbonates, and benzoates are effective on steel only in the presence of dissolved oxygen. The maintenance of inhibition by these materials is indicated by the more noble potential attained by the metal, which approaches that of the cathode.

When present in insufficient amounts, anodic inhibitors, except benzoate, can be dangerous because they permit the formation of small anodic areas without appreciably decreasing the amount of metal dissolution. Thus, they produce intense local attack.

Cathodic inhibitors stifle the cathodic reaction, either by restricting the access of oxygen or by poisoning sites favorable for cathodic hydrogen evolution. Cathodic inhibitors that decrease the corrosive action of aqueous solutions on steel include salts of magnesium, manganese, zinc, and nickel. The increase in alkalinity near the vessel walls by reduction of oxygen to OH^- leads to the precipitation of the hydroxides of these metals as a reasonably adherent porous deposit that retards the diffusion of oxygen to the steel.

The presence of calcium bicarbonate in water gives a general precipitate of calcium carbonate if the water is supersaturated or gives a local deposit on or near cathodic areas where the pH is high. The addition of lime to water increases the pH and serves as a cathodic inhibitor.

Cathodic inhibitors form a visible film on the metal, are generally not as efficient as anodic inhibitors, and do not completely prevent attack. On the other hand, cathodic inhibitors are less likely than anodic inhibitors to intensify attack if added in insufficient amounts. Many waters contain both magnesium and calcium salts as natural constituents with inhibitive possibilities.

Factors Affecting Inhibitor Systems. The successful application of inhibitors requires knowledge and understanding of their chemical behavior and of the corrosion processes in the system under consideration. A given substance may inhibit corrosion in one environment and increase it in another. The choice and concentration of inhibitor depend on the type of system, the composition of the water, the temperature, the rate of movement, the presence of residual or applied stresses, the composition of the metal, and the presence of dissimilar metals. The presence of natural crevices, loose scale, and debris must also be taken into account.

Installations in which inhibitors are used vary from small, closed systems using recirculated water to large cooling systems using more than a million gallons of water per day. On economic grounds alone, the choice of inhibitor in large installations is relatively restricted. The concentration of inhibitor will, in general, be greater the higher the concentration of aggressive corrodents, such as chloride and sulfate, that interfere with the formation and maintenance of a passivating film on the metal.

Selection of Inhibitors. When choosing an inhibitor, the corrosion engineer must consider the difference in behavior of safe and dangerous inhibitors. Safe inhibitors reduce the total amount of corrosion without increasing the intensity on unprotected areas; dangerous inhibitors produce increased rates of attack on unprotected areas. Intensification of attack by dangerous inhibitors can be caused by a number of factors: lack of sufficient inhibitor, the presence of enough chlorides and sulfates to prevent complete protection, and the presence of crevices and dead-ends into which renewal of inhibitor by diffusion is not rapid enough.

Anodic inhibitors are, for the most part, dangerous inhibitors. Cathodic inhibitors are generally safe, but zinc sulfate (for use with steel), in high enough concentrations, results in intensified attack along the waterline.

Effect of Velocity. The amount of inhibitor required depends on the velocity of movement of the water or solution and the relative ratio of volume of liquid to area of metal surface. The higher the velocity of movement, the thinner the diffusion-boundary layer and the greater the amount of inhibitor reaching the surface. The overall effect of increasing the velocity is the same as increasing the concentration of inhibitor.

In addition, moving water is less likely to deposit debris and to screen the surface from the action of the inhibitor. Because inhibition is a dynamic process, the inhibitor, at least initially, is consumed during build up of the cathodic or anodic film. Therefore, the safe inhibitor concentration is influenced by the ratio of volume of liquid to the area of metal exposed.

Great care is needed in applying the results of laboratory tests to industrial equipment. Apart from the factors already mentioned, an increase in temperature and the presence of dissimilar metals both require an increase in inhibitor concentration.

The condition of the metal surface and its accessibility to the inhibitor are also of the greatest importance. In industrial equipment containing water or aqueous solutions, there usually are crevices or re-entrant corners at which the replenishment of inhibitors is slow. Good design can reduce these to a minimum, but some sites of this type may be unavoidable and should be considered when applying inhibition. A higher concentration of inhibitor is required if the surface is rough, has been sand blasted, or is covered with grooves and scratches. The same applies to surfaces under stress or on which impurities that could cause galvanic effects are present.

Compatibility of inhibitors with the liquid in the system and with nonmetallic materials with which the liquid may be in contact must also be considered. The presence of organic matter may lead to the rapid depletion of oxidizing inhibitors, such as chromate, and bacteria may flourish in nitrite and phosphate solutions.

Changes in pH and Applied Potential

Other changes in environment to reduce corrosion rates are sometimes more convenient than the addition of inhibitors. Also, the presence of inhibitors cannot be tolerated in potable and some process waters or in aqueous solutions. Changing the pH of the water or aqueous solution, in conjunction with the application of a suitable external potential (as discussed in the section "Use of Galvanic Protection" in this article), is often helpful in reducing corrosion rates.

Pourbaix diagrams are sometimes used to provide guidance in making appropriate adjustments. These diagrams, which are derived from electrochemical measurements and thermodynamic data, are potential-pH diagrams that relate to the electrochemical and corrosion behavior of metals in water and in aqueous solutions.

Pourbaix diagrams show the conditions of potential and pH under which the metal either does not react (immunity) or can react to form specific oxides or complex ions. They do not provide information on rates of reaction.

Pourbaix diagrams indicate the conditions for which diffusion-boundary films may form on an electrode surface, but they provide no measure of how effective such barrier films may be in the presence of specific anions, such as SO_4^{2-} or Cl^-. Similarly, they do not indicate the detailed conditions under which nonstoichiometric metal compound films influence corrosion rates.

However, Pourbaix diagrams outline the nature of the stoichiometric compounds into which any less stable compounds may transform. They relate the possibility of corrosion to the pH of the corrodent and the potential difference between the pure metal and its ions in solution. They show under what pH-potential conditions corrosion might occur and those under which it will not occur, thus providing guidance as to how corrosion can be minimized or avoided.

Figure 20 shows a Pourbaix diagram for iron in water and dilute aqueous solutions. Note that there are zones of corrosion, immunity from corrosion, and passivity.

Some metals besides iron for which Pourbaix diagrams have proved to be especially helpful include tantalum, titanium, aluminum, and magnesium. The diagrams correlate well in a general way with the well-established facts that tantalum is relatively inert, titanium is resistant to a wide range of conditions, aluminum is amphoteric and is attacked by acids and alkalis over a wide range

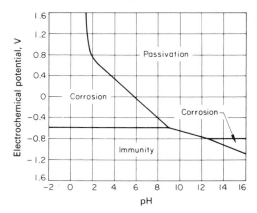

Fig. 20 Pourbaix diagram showing the theoretical conditions for corrosion, passivation, and immunity of iron in water and dilute aqueous solutions

of conditions, and magnesium is very active over a wide range of conditions. Additional information on Pourbaix diagrams can be found in Ref 16.

Continuous Monitoring of Variables

Continuous monitoring of critical variables in both batch operations and process streams is a valuable tool in controlling the environment to avoid the development of unacceptably high corrosion rates. Some of the variables monitored and their applications are noted and described below.

Electrical Resistance. Corrosion rate is indicated by the electrical resistance of wire made of a metal similar or identical to that present in the system. Electrical resistance has been used to indicate corrosion rates in liquid and vapor phases, but has a relatively slow response time. An example of the use of an electrical-resistance probe would be in anhydrous hydrochloric acid in steel vessels. The presence of small amounts of water would increase the corrosion rate considerably, and an electrical-resistance probe would indicate when a water leak has occurred.

Linear Polarization Resistance. Monitoring by linear polarization resistance is based on the linear relationship that exists between a small amount of electrochemical polarization and the corrosion rate; the probe must be immersed in a conductive liquid phase. The linear polarization resistance instruments have a relatively fast response time and have been used in cooling-water systems to control the rate of inhibitor additions.

Electrical Conductivity. The corrosion rate of many metals in organic systems increases significantly if a corrodent is present and the electrical conductivity increases above 10^{-7} S. Conductivity measurements have a very fast response time. They do not measure corrosion rate directly, only the changes in the properties of the environment. Conductivity gages have been used to monitor the presence of water in chlorinated hydrocarbons.

Continuous Chemical Analysis. Various types of instrumentation, such as infrared spectroscopy, ultraviolet/visible absorption spectroscopy, and gas chromatography, have been used to monitor or to control the presence of various chemical substances in process streams (these techniques are discussed in separate articles in Volume 10 of the 9th Edition of *Metals Handbook*). Measurements of pH would be included in this area. In some operations, the amount of inhibitor added to a process batch or stream can be controlled by such instrumentation.

Monitoring metal ion concentration through routine chemical or spectroscopic analysis gives direct correlation to corrosion rates. Actual metal-loss rates are difficult to estimate, but real trends can be easily established.

Temperature. Continuous monitoring and close control of temperature are readily done and are often needed to minimize corrosion.

Electronic Hydrogen Analysis. An electronic hydrogen probe can continuously monitor an environment that generates hydrogen by corrosion and can be equipped to plot corrosion rate against time. A thin-wall probe made of the metal being monitored is connected to a getter-ion pump, which maintains a vacuum in the probe, into which some fraction of the corrosion-generated hydrogen diffuses. The pump current, which is proportional to the amount of hydrogen entering the probe, is monitored. The response time can be low enough that a change in corrosive conditions can be detected within a period of a few minutes. This technique has been used in the laboratory for hydrogen sulfide and oxygen corrosion research and inhibitor testing. It has been used in the field for monitoring corrosion of carbon and low-alloy steels in sour-gas pipelines, drilling mud, and refinery streams.

Corrosion of Specific Metals and Alloys

A large variety of materials are used by the engineer to construct bridges, automobiles, chemical-processing equipment, pipelines, and powerplants (both fossil fuel and nuclear), as well as other applications for which corrosion is a concern. This section will review a number of important ferrous and nonferrous alloys used in corrosive environments. Additional information on the mechanical, physical, and corrosion properties of these materials can be found in Volumes 1, 2, and 3 of the 9th Edition of *Metals Handbook*.

Carbon and Low-Alloy Steels

Carbon and low-alloy steels are ordinarily expected to corrode in service, and they fre-

quently fail from that cause. Uniform generalized corrosion can be predicted with reasonable accuracy in most systems and can be checked or monitored to prevent unexpected failures.

Pitting. The most common form of premature failure encountered in the carbon and low-alloy steels is pitting corrosion. Any condition that removes the moderately protective oxide film normally present on these materials may allow pitting corrosion to occur. Pitting rates under the least damaging conditions are usually about ten times the normal corrosion rate of the metal, and they may be as high as 100 to 1000 times the uniform generalized corrosion rate.

On bare metal surfaces, there are three means of minimizing pitting attack: frequent cleaning, inhibition, and galvanic protection. A clean surface is much less likely than a contaminated surface to develop pitting attack because there is free access of oxygen to a clean metal surface for the repair of ruptures in the oxide film. Also, there is little opportunity for formation of a small anode beneath a deposit while the remaining surface acts as a large cathode that intensifies the corrosion by galvanic action. The use of frequent cleaning to combat pitting corrosion is limited to surfaces that are accessible at suitable intervals for cleaning in some manner.

The use of inhibitors is limited to those systems for which an inhibitor compatible with the product is available, to closed systems, and to the use of specific inhibitors that can be disposed of or recovered economically from once-through systems. Poor control of anodic inhibitors can cause corrosion, instead of preventing it, with chromates being a prime example.

Galvanic protection can be either anodic or cathodic. The use of galvanic protection as a means of minimizing or preventing attack of steels by pitting and other forms of corrosion is described in the section "Use of Galvanic Protection" in this article.

General Corrosion. Premature failure of carbon and low-alloy steels may occur from general corrosion in situations where the actual environment to which the material is exposed is significantly different from the environment in which it was designed to operate. While carbon steel exhibits a low corrosion rate in 100% sulfuric acid at ambient temperature, the rate increases very rapidly with dilution of the acid.

The rate of attack on iron and steel by nitric acid at concentrations of 70% by weight and higher is reported to be greater than 1.3 mm/year (50 mils/year) at room temperature, making it unsafe to ship or store the acid at any concentration in these metals. The rate of attack is increased by the presence of even low concentrations of nitrogen oxides in the acid.

In addition, differential aeration (described in the section "Concentration-Cell Corrosion" in this article) causes concentration-cell corrosion just below the liquid level. On steel partly immersed in the acid, a vigorous self-accel-

erating attack has been observed to develop gradually at the liquid level in the presence of air.

Temperature Effects. Carbon steel may also show abrupt changes in corrosion resistance with increasing temperature and is susceptible to differential-temperature cell attack in some applications, for example, in nonuniformly heated pipelines carrying 25 to 50% sodium hydroxide solutions.

Galvanic and stray-current corrosion accounts for a large number of failures of buried or submerged carbon and low-alloy steel components. Galvanic protection systems are usually used in conjunction with heavy bituminous coatings to provide protection against breaks in the coatings (see the section "Use of Galvanic Protection" in this article).

Importance of Oxide Film. Carbon and low-alloy steels are reactive metals on the thermodynamic scale and will corrode rapidly in the absence of the normally present thin film of iron oxide. Any conditions that prevent the formation and repair of this oxide film or that remove the film can cause rapid corrosion of these materials.

Cast Iron

Cast iron varies greatly in its resistance to corrosion, depending on the type and composition of the metal and on the service environment. It is widely used in buried pipelines and in chemical-processing equipment.

Gray iron, particularly when exposed to underground corrosion or some other relatively mild form of attack, is subject to a selective process called graphitic corrosion. The iron is selectively leached out, leaving a residue that has the shape of the original structure but the appearance of graphite and practically no mechanical strength. Superficial examination may not detect this situation, as in a gas or water pipe, and very hazardous circumstances may result. Details of graphitic corrosion are described in the section "Selected Leaching" in this article. The interlocking network of graphite flakes retains its shape and integrity, but only as long as the network interlocks.

Malleable and Ductile Irons. When malleable and ductile irons corrode, both the graphite and the matrix are swept away by erosion-corrosion attack. Pearlitic malleable iron is more resistant to erosion-corrosion than ferritic malleable iron.

Cast iron equipment is sometimes used in chemical processes because of its low cost, although the corrosion rate is very high. Large gray iron kettles (205 mm, or 8 in., in wall thickness) have been used in producing hydrochloric acid from the reaction of sulfuric acid with sodium chloride even though the corrosion rate exceeds 100 mm/year (4 in./year). Other materials of construction also have high corrosion rates in this application and are considerably more expensive.

High-silicon cast iron is used in a wide variety of highly corrosive chemicals, including sulfuric acid, hydrochloric acid, and hydrochloric acid plus oxidizing salts. One notable exception is hydrofluoric acid, in which it is attacked at a substantial rate.

High-silicon cast iron is one of the most universally resistant of the commercial (nonprecious) metals and alloys. However, it is very brittle, which limits its use in some equipment or requires special precautions to avoid severe impacts.

High-nickel cast irons, such as Ni-Resist austenitic iron, have been used to solve problems of severe corrosion and erosion-corrosion in pumps and related equipment exposed to seawater. Several compositions of high-nickel cast irons are described in Ref 17.

Stainless Steels

The most important aspect of the corrosion of stainless steels is that their corrosion resistance is due to a film of chromic oxide less than about 10 nm thick. In the absence of this oxide film, stainless steels corrode at rates comparable to those for carbon steels. Two of the major aspects of the corrosion-related behavior of the stainless steels are SCC, which is discussed in a separate article in this Volume, and the phenomenon of intergranular corrosion.

General Corrosion Behavior. The entire class of stainless steels is quite similar in general corrosion behavior. At chromium contents below about 12%, the corrosion resistance is generally like that of other alloy steels. However, chromium content above about 13% makes these alloys resistant to attack by most mild corrodents and prevents rust in moist air.

The corrosion rate in aggressive corrodents, such as nitric acid, is reduced rapidly as the 13% Cr level is exceeded, then declines more slowly until a content of about 18% Cr is reached. At this point, there is another large reduction in corrosion rate in aggressive corrodents.

In general, stainless steels pit in halide solutions, corrode in a uniform manner in hydrochloric and sulfuric acids, and, if sensitized, corrode intergranularly in oxidizing acids, such as nitric acid, but there are exceptions. The addition of 1 g/L of CrO_3 to a boiling solution of nitric acid will increase the corrosion rate of an 18Cr-8Ni stainless steel by a factor of about 100, while the addition of about 1% nitric acid to a boiling solution of sulfuric acid will reduce the corrosion rate by a factor of about 100.

Pitting attack of stainless steels is a major problem that results from localized breakdown of the oxide film. Pitting failures of stainless steels occur most frequently in heat exchangers operating in seawater, brackish water, or acid-polluted water. Usually, the unit appears to operate satisfactorily until one or more pumps are shut down and some area of the exchanger goes stagnant. Shortly after returning to operation, multiple tube leaks are a common occurrence, and pitting is almost invariably the mode of corrosive attack. For operation of stainless steels in salt water, it is important to keep surfaces clean and to keep the water moving.

Sensitization is a serious problem encountered with stainless steels. It makes stainless steels susceptible to rapid intergranular corrosion in certain environments in which they would be expected to be immune, and it occurs as a result of improper heat treating or welding (sensitization of austenitic stainless steels was discussed in the section "Intergranular Corrosion" in this article). In stainless steels in the as-welded condition, this attack occurs in a limited area within the HAZ because of loss of chromium level in the grain-boundary material by precipitation of chromium carbides.

All sensitization can be eliminated by solution heat treating at 1050 to 1100 °C (1920 to 2010 °F) and cooling rapidly. Sensitization associated with welding can be eliminated by using a low-carbon alloy, such as type 304L, or a stabilized alloy, such as type 347.

Types of Preferential Attack. Another form of sensitization-related corrosion occasionally encountered in as-welded type 347 is known as knifeline attack. In this form of attack, a very thin layer of metal immediately adjacent to the fusion line of a weld is preferentially corroded in an intergranular manner. In addition to knifeline attack, type 347 also shows preferential corrosion at the stopping point of weld beads (crater corrosion) and end-grain corrosion. Knifeline attack, crater corrosion, and end-grain corrosion usually occur only in strongly oxidizing environments.

Knifeline attack can be eliminated by reheating the weldment to a temperature of about 1060 °C (1940 °F), then quenching. No heat treatment is known to eliminate the end-grain attack on type 347 or on types 304 and 316, which also are susceptible to this form of corrosion. End-grain attack can be avoided by not exposing the ends of bar stocks or tubing or the parting lines of forgings to corrodents. When a design cannot accomplish this, the exposed ends should be fused with a gas tungsten arc welding torch or should be covered with weld-metal deposit.

Severe general corrosion, when encountered at ordinary temperatures in stainless steels, suggests the presence of either hydrochloric or dilute sulfuric acid. Very low concentrations of these acids cause severe attack in the absence of an oxidizer, such as nitric acid. Severe general intergranular attack can occur in noncorrosive environments if grain-boundary diffusion of readily soluble elements, such as sulfur, has occurred at high temperatures during the fabricating process.

Hot-Salt Corrosion. At elevated temperatures, the stainless steels are subject to hot-salt corrosion. This is a rapid form of corrosion attack caused by molten salts that act as a flux, continuously removing the protective oxide film. One aggressive mixture that has caused problems with jet-engine components consists primarily of sodium sulfate contaminated with sodium chloride. Solutions to a hot-salt corrosion problem are to eliminate the corrodent, to change the material to one resistant to such attack, or to apply a resistant coating.

Heat-Resisting Alloys

Heat-resisting alloys include several groups: solid-solution nickel-base alloys, precipitation-hardenable nickel-base alloys, dispersion-strengthened nickel-base alloys, iron-nickel-chromium and iron-chromium-nickel alloys, and cobalt-base alloys.

Hot Corrosion. One problem peculiar to heat-resisting alloys in many applications is hot corrosion; this term refers to any corrosion behavior in which the rate of attack is significantly accelerated over and above that expected from oxidation. Difficulties from hot corrosion have been experienced in gas turbines and jet engines, particularly those operating in marine atmospheres, and in petroleum-refining and petrochemical equipment.

The two most common types of hot corrosion are vanadium corrosion and sulfidation; both originate in impurities in the petroleum-base fuel or feed stock. In both types, the impurity results in molten compounds on the metal surface. If vanadium is present in the form of vanadium pentoxide (V_2O_5), complex sodium vanadates ($nNa_2O \cdot V_2O_5$) or sodium vanadyl-vanadates ($nNa_2O \cdot V_2O_4 \cdot mV_2O_5$) may form. If sulfur is present, the deposit depends on the source of sulfur and operating conditions. For turbines operating on sour (high-sulfur) fuel in salt air, the result is sodium sulfate (Na_2SO_4).

In both of these types of hot corrosion, the threshold temperature for damage corresponds fairly well to the melting point of the compound. The molten compound fluxes, destroys, or disrupts the normal protective oxide. In petroleum processing, such as cracking and reforming, sulfur may be present in the feed in many forms, including organic sulfur compounds and hydrogen sulfide. The damage mechanisms in such cases are often analogous to oxidation, but are much more rapid because mass transport is much more rapid through sulfide scales, which are not very protective.

Remedial measures for hot corrosion consist of the elimination, whenever possible, of the impurities in fuel or feed, limiting operating temperatures, selection of more-resistant alloys, and the use of protective coatings. High-chromium stainless steels and aluminum diffusion coatings have met with some success where hydrogen sulfide is the problem; in turbines and aircraft applications in general, aluminum coatings over a nickel-aluminum base layer are used to minimize or prevent sulfidation corrosion. The main method of coping with vanadium corrosion is to avoid using vanadium-bearing fuel in turbines operating above the threshold temperature.

Behavior at Room Temperature. The heat-resisting alloys, as a group, have excellent corrosion resistance at ordinary temperatures, and suitably resistant alloys for even the most aggressive environment at room temperature and moderately elevated temperatures can usually be found among them.

Aluminum and Aluminum Alloys

Aluminum and aluminum alloys develop a protective oxide film that, in most atmospheric exposures, prevents or retards corrosion. However, this film, like most surface coatings, is often not complete in its coverage. In seacoast areas where salt-water deposits occur, localized corrosion of aluminum will occur at small breaks or defects in the protective film, and this localized corrosion may result in the development of large pits. Pitting is a characteristic form of corrosion in aluminum alloys, as is its related form, crevice corrosion. Crevice corrosion occurs at laps, seams, and between faying surfaces, where the access of oxygen is restricted. Both pitting and crevice corrosion can result in severe localized metal removal and thus in structural damage.

The seacoast environment also contributes to the occurrence of SCC among the heat-treatable high-strength aluminum alloys (see the article "Stress-Corrosion Cracking" in this Volume). A form of corrosion sometimes encountered in these alloys is exfoliation, which causes separation of layers in sheet, plate, forgings, and extrusions. Exfoliation corrosion, which is relatively common in seacoast areas, begins at exposed edges, where residual stress promotes delamination. Accumulation of the voluminous corrosion products also serves to propagate exfoliation by a wedging action. Figure 21 shows the exfoliation of an aluminum alloy structural member from a tanker truck operating in the Gulf Coast. The 5154 aluminum alloy tank was unaffected, but the 2014 alloy extrusion was badly attacked. The trailer carried organic acids; small spills of the acid, plus the seacoast environment, resulted in the exfoliation attack.

Galvanic corrosion, from the contact of aluminum alloys with more noble metals, can be very damaging structurally if salt is available to contribute to the electrical conductivity of moisture films in contact with the metal couple.

Corrosion control for aluminum alloys is best begun with alloy selection. If strength considerations permit, the strain-hardenable alloys of the 1100, 3000 or 5000 series should be used, because they are generally resistant to corrosion and almost insusceptible to SCC. Among the heat-treatable grades, alloy 6061 performs well.

The methods used to minimize or prevent corrosion failures involve the application of surface treatments or coatings. These are of four general types: (1) cladding with a thin surface layer of almost pure aluminum, (2) anodizing to obtain a relatively thick surface layer of aluminum oxide, (3) treatment to obtain a surface layer of chemically combined corrosion-inhibiting compounds, and (4) paint systems (usually two or more coats) using a

Fig. 21 Exfoliation corrosion of a 6-mm (¼-in.) thick aluminum alloy (2014) plate

Cross-sectional view showing the separation of layers of the aluminum resulting from selective corrosion along the elongated grain boundaries developed during rolling or extrusion. 2×

zinc chromate primer. The selection of the most suitable protective system depends on the configuration of the part, the alloy, and the severity of the exposure.

Copper and Copper Alloys

Copper and copper alloys have excellent corrosion resistance in many environments. Copper forms many commercially useful alloys that differ widely in composition and therefore in corrosion resistance; the most suitable copper alloy for use in a specific application must be selected with care.

Pure copper has good resistance to corrosion by sulfide-free hydrocarbons, dry gases, and nonoxidizing acids and salts, but does not resist corrosion by oxidizing acids or salts, ammonia, and moist halogens or sulfides. Copper usually corrodes by general thinning (erosion-corrosion) and by pitting when exposed to wet steam traveling at high velocities or when exposed to fresh waters at velocities above 1.2 m/s (4 ft/s) and to seawater at velocities above 0.9 m/s (3 ft/s).

Copper-zinc alloys have better physical properties than copper alone, and they are also more resistant to impingement attack. Therefore, brasses are used in preference to copper for condenser tubes. Among the brasses, resistance to impingement attack increases with the zinc content of the alloy. Corrosion failures of brasses usually occur by selective leaching (dezincification), pitting, and SCC.

Cartridge brass, 70% Cu (copper alloy C26000), which is used where easy machining and casting are desirable, gradually dezincifies in seawater and in soft fresh waters. This tendency is retarded by the addition of 1% Sn, the resulting alloy is called admiralty metal. The addition of 0.02 to 0.10% As, Sb, or P (inhibited admiralty metal) further retards the rate of dezincification and permits the use of these alloys in condenser tubes. In this application, conditions favoring dezincification are in contact with slightly acid or alkaline water not

highly aerated, low rates of flow of the circulating liquid, relatively high tube-wall temperatures, and permeable deposits or coatings over the tube surface.

Inhibited naval brasses, which contain a nominal 0.8% Sn and 0.02 to 0.10% As, Sb, or P and have a higher zinc content than admiralty metal, are somewhat lower in resistance to impingement attack and dezincification. They are used in marine hardware, ship-propeller shafts, and condenser plates.

Aluminum bronzes have good resistance to erosion-corrosion and to SCC and can be used at elevated temperatures. The alloy and the sequence of mechanical working and thermal treatments must be selected carefully for good performance in exposure to steam. With inhibitors, aluminum bronzes have excellent resistance to brackish water, to clear and polluted seawater, and to many types of fresh water. However, because these alloys are subject to bio-fouling, the water velocity should be kept above 1.5 m/s (5 ft/s).

Silicon bronzes are resistant to many organic acids and compounds, to dry gas, and to a number of inorganic compounds. They have good resistance to both fresh and salt water, but should never be used with ammonia, which induces SCC.

Copper nickels contain iron to form protective films that improve the resistance to impingement corrosion. Copper nickel, 10% Ni (copper alloy C70600), has been used for years by the power and marine industries. The alloy has excellent resistance to clean flowing seawater at velocities up to 2.4 m/s (8 ft/s) and to brackish water at velocities up to 1.5 to 1.8 m/s (5 to 6 ft/s). It is used extensively in the food-processing industry and in the conversion of saline water. This alloy, like other copper alloys, is attacked by oxidizing acids and salts and by sulfur. It is subject to corrosive attack by pitting and general thinning.

Copper nickel, 30% Ni (copper alloy C71500), has excellent resistance to rapidly moving seawater and to many types of fresh water. The cooling-water velocities should be kept above 1.5 m/s (5 ft/s) to avoid biofouling. It is widely used in the power and marine industries and for condenser service in naval use. It is used in oil refineries and in chemical plants for its high corrosion resistance. Copper nickel (30% Ni) is the most resistant of all the copper-base alloys to SCC, but it can be corroded by general metal thinning and pitting.

Selective leaching of this alloy can occur under conditions of high heat flux in condensers when the water velocity is low (less than about 1.5 m/s, or 5 ft/s). In wet or dry steam at low velocity, it usually corrodes at rates less than 2.5 μm/year (0.1 mil/year). A modified copper nickel, 30% Ni, containing about 3% Cr, is used where higher yield strength is required.

Erosion-Corrosion. The movement or turbulence of water at the inlet ends of condenser tubes frequently leads to localized corrosion, commonly called inlet-end corrosion, which is a form of erosion-corrosion. It results in clusters of deep pits, which are usually undercut on the downstream side and often take the shape of a horseshoe around a raised point of relatively unattacked metal. The attacked areas are usually bright and free from scale, corrosion products, or other visible films.

The harmful effects of turbulence, aeration, and high velocity of the circulating water in promoting erosion-corrosion of the tube walls may be minimized by using copper nickels, limiting velocity to a relatively safe value, controlling turbulence, eliminating entrained or separated air, eliminating debris in the circulating water, and using plastic inserts at the inlet ends of tubes. The plastic inserts must be tightly fitted to the tubes in order to avoid the possible development of crevice corrosion.

Nickel and Nickel-Copper Alloys

Nickel has the best resistance to strongly alkaline solutions of all the common metals, but it is readily attacked by strong oxidizers. It has good corrosion resistance in dilute nonoxidizing acids if the solution is deaerated, but it has poor resistance to complexing agents, such as ammonia. Nickel is resistant to most waters, including seawater, but may pit deeply in stagnant waters or in the presence of biological fouling. Alkaline hypochlorites also pit nickel.

Nickel-copper alloys (Monels) have corrosion resistance similar to that of nickel in most environments. Monels are readily attacked by strong oxidizers and are corroded most rapidly in aerated solutions. Corrosion rates in aerated nonoxidizing acids may be more than 100 times those in unaerated solutions. Monels are not resistant to strong hot caustic solutions or to ammonium hydroxide in concentrations greater than 3%.

Monels are resistant to flowing seawater and brackish water, but they pit in stagnant waters. Both nickel and copper are selectively leached by various media. Oxidizing conditions and exposure to ammonia, mercury, and sulfur or sulfur-bearing reducing environments at temperatures of more than about 315 °C (600 °F) should be avoided.

Titanium

Commercially pure titanium is outstanding among structural materials in its resistance at ordinary temperatures to strongly oxidizing acids, aqueous chloride solutions, moist chlorine gas, sodium hypochlorite, seawater, and brine solutions. It resists many other corrosive substances well enough to be used in contact with them. Titanium alloys are generally less resistant to corrosion than commercially pure titanium.

Titanium is corroded by hydrofluoric, hydrochloric, sulfuric, oxalic, and formic acids, but attack may be inhibited by additions to the solution, except for hydrofluoric acid. When titanium is passivated (as is usual), it is the noble metal in a galvanic couple with all other structural alloys except the Monels and stainless steels.

Pure titanium is the preferred material of construction for much of the equipment built to handle industrial brines. It is used for pumps, piping, thermowells, heat exchangers, crystallizers, evaporators, condensers, and many other items that are subject to the corrosive action of these brines.

Crevice Corrosion and Pitting. In brine solutions, however, titanium may suffer crevice corrosion and pitting at sufficiently high temperatures. In 50% sodium chloride solutions, the temperature threshold for pitting is between 125 and 130 °C (260 and 265 °F). Special titanium alloys have been developed that extend this temperature threshold for brine solutions to higher levels. Crevice corrosion can occur where the pH of the solution in the crevice drops below 1.5 to 2.0, as can occur in cleaning titanium equipment with sulfamic acid solutions.

Cathodic attack of titanium may occur if it is coupled to a less noble metal and picks up hydrogen. When titanium is made the cathode in a reducing environment in which it has only marginal stability, hydrogen discharged on the surface destroys the thin protective film. The corrosion rates of both titanium and aluminum when coupled in dilute oxalic or sulfuric acids are accelerated. When made a cathode, titanium is also known to absorb hydrogen, forming a brittle hydride layer that spalls with time. Spalling of this hydride film on a corrosion coupon may appear at first glance to be due to direct dissolution, because the coupon loses weight.

Harmful Alloying Elements and Corrosive Substances. Some alloying elements affect the corrosion resistance of titanium alloys adversely. Iron is probably the most critical element. High iron content (more than 0.20%) lowers the corrosion resistance to nitric acid. Multivalent oxidizing metallic ions are powerful inhibitors against corrosion of these alloys. Most other soluble ions either improve the corrosion resistance or have little effect. However, concentrations of fluoride ions as low as 100 ppm significantly lower the corrosion resistance of titanium and its alloys, which are also attacked at significant rates by reducing acids and, at temperatures above 80 °C (175 °F), hot alkaline solutions.

REFERENCES

1. *Chemical Analysis of Metals and Metal-Bearing Ores*, Vol 03.05, *Annual Book of ASTM Standards*, ASTM, Philadelphia, 1984
2. *Emission Spectroscopy; Surface Analysis*, Vol 03.06, *Annual Book of ASTM Standards*, ASTM, Philadelphia, 1984
3. *Metal Corrosion, Erosion and Wear*, Vol 03.02, *Annual Book of ASTM Standards*, ASTM, Philadelphia, 1984
4. National Association of Corrosion Engineers, P.O. Box 218340, Houston, TX 77218. See standard test methods listed in the NACE Publications Catalog.
5. "Standard Practice for Conventions Applicable to Electrochemical Measurements in Corrosion Testing," ASTM G 3, *Annual Book of ASTM Standards*, Vol 03.02, ASTM, Philadelphia, 1984, p 100-106
6. D.L. Graver, Ed., *Corrosion Data Survey*, 6th ed., National Association of Corrosion Engineers, Houston, 1985, p 1-192
7. *Metal Corrosion in the Atmosphere*, STP 435, ASTM, Philadelphia, 1968
8. *Materials Performance and the Deep Sea*, STP 445, ASTM, Philadelphia, 1969
9. M. Henthorne, *Corrosion Causes and Control*, Carpenter Technology Corp., Reading, PA, 1972, p 30
10. *Corrosion of Metals Under Thermal Insulation*, STP 880, ASTM, Philadelphia, 1985
11. A.H. Tuthill and C.M. Schillmoler, Guidelines for Selection of Marine Materials, in *Ocean Science and Ocean Engineering Conference*, Marine Technology Society, Washington, June 1965
12. H.H. Uhlig, *Corrosion and Corrosion Control*, 2nd ed., John Wiley & Sons, 1971, p 166
13. C.P. Larrabee, *Corrosion*, Vol 9, Aug 1953, p 259
14. "Standard Test Method for Corrosion Preventive Properties of Lubricating Greases," D 1743, *Annual Book of ASTM Standards*, Vol 05.02, ASTM, Philadelphia, 1984, p 36-41
15. M.G. Fontana and N.D. Greene, *Corrosion Engineering*, 2nd ed., McGraw-Hill, 1978
16. M. Pourbaix, *Atlas of Electrochemical Equilibria in Aqueous Solutions*, Pergamon Press, 1966
17. "Standard Specification for Austenitic Gray Iron Castings," A 436, *Annual Book of ASTM Standards*, Vol 01.02, ASTM, Philadelphia, 1984, p 256-261

SELECTED REFERENCES

- W.H. Ailor, Jr., *Handbook on Corrosion Testing and Evaluation*, John Wiley & Sons, 1971
- H. Leidheiser, Jr., Corrosion Control by Organic Coatings, National Association of Corrosion Engineers, Houston, 1981
- J.B. Lumsden, Ed., *Corrosion*, National Association of Corrosion Engineers, Houston
- C.G. Munger, *Corrosion Prevention by Protective Coatings*, National Association of Corrosion Engineers, Houston, 1985
- L.C. Rowe, Ed., *Mater. Perform.*, National Association of Corrosion Engineers, Houston
- H.H. Uhlig, *The Corrosion Handbook*, John Wiley & Sons, 1948

Stress-Corrosion Cracking

Revised by B.E. Wilde, Fontana Corrosion Center, Ohio State University

STRESS-CORROSION CRACKING (SCC) is a failure process that occurs because of the simultaneous presence of tensile stress, an environment, and a susceptible material. Although manifest mostly in metals, it can also occur in other engineering solids, such as ceramics and polymers. Removal of or changes in any one of these three factors will often eliminate or reduce susceptibility to SCC and therefore are obvious ways of controlling SCC in practice, as will be discussed later.

Failure by SCC is frequently encountered in seemingly mild chemical environments at tensile stresses well below the yield strength of the metal. The failures often take the form of fine cracks that penetrate deeply into the metal with little or no evidence of corrosion on the nearby surface. Therefore, during casual inspection no macroscopic evidence of impending failure is seen.

Several other processes in addition to SCC can cause failure of metals exposed to an environment, such as corrosion fatigue, hydrogen embrittlement, and fretting corrosion, all of which are reviewed in other articles in this Volume.

Mechanisms of SCC

Several theories have been advanced to explain in detail the mechanisms of SCC. Two major theories are the electrochemical and stress-sorption theories.

Electrochemical Theory. According to the electrochemical theory, galvanic cells are set up between regions of exposed metal surfaces, for example, metal grains and heterogeneous phases; anodic paths are thus established. For example, the precipitation of $CuAl_2$ from an Al-4Cu alloy along grain boundaries produces copper-depleted paths in the edges of the grains. When the alloy, stressed in tension, is exposed to a corrosive environment, the ensuing localized electrochemical dissolution of metal, combined with localized plastic deformation, opens up a crack. With sustained tensile stress, protective films that form at the tip of the crack rupture, causing fresh anodic material to be exposed to the corrosive medium, and the SCC is propagated.

Supporting this theory is the existence of a measurable potential in the metal at grain boundaries, which is negative (or active) with respect to the potential of the grains. Furthermore, cathodic polarization will stop the cracking. This theory has been extended to include metals that do not form intermetallic precipitates, but for which phase changes or segregation of alloying elements or impurities can occur during plastic deformation of metal at the crack tip; the resulting composition gradient then sets up galvanic cells.

Stress-Sorption Theory. According to the stress-sorption theory, SCC generally proceeds by weakening of the cohesive bonds between surface-metal atoms through adsorption of damaging substances in the environment. Because chemisorption is specific, damaging components are also specific. The surface energy of the metal is said to be reduced, increasing the probability that the metal will form a crack under tensile stress. Adsorption of any kind that reduces surface energy should favor crack formation.

Only a monolayer of adsorbate is needed to decrease significantly the affinity of surface-metal atoms for each other or for atoms of substances in the environment. The only adsorbates presumed to be effective are those that reduce the attractive force of adjoining metal atoms for each other.

Inhibiting anions compete with particles of damaging substances for adsorption sites, thereby making it necessary to apply a more positive potential to the metal to reach a concentration of damaging substances that is adequate for adsorption and resulting cracking of the metal.

Crack Initiation

The site of initiation of a stress-corrosion crack may be submicroscopic and determined by local differences in metal composition, thickness of protective film, concentration of corrodent, and stress concentration. A pre-existing mechanical crack or other surface discontinuity, or a pit or trench produced by chemical attack on the metal surface, may act as a stress raiser and thus serve as a site for initiation of SCC. Laboratory investigation has shown that generally there is some localized plastic deformation before cracking occurs.

The tip of an advancing crack has a small radius, and the attendant stress concentration is great. Using audio-amplification methods, it has been shown in certain materials that a mechanical step or jump can occur during crack propagation. In fact, in one test (Ref 1), pings could be heard with the unaided ear. In this test, it was demonstrated that the action of both stress and corrosion is required for crack propagation. An advancing crack was stopped when cathodic protection was applied (corrosion was stopped, but the stress condition was left unchanged). When cathodic protection was removed, the crack again started to propagate. This cycle was repeated several times. In this test, the progress of the crack was photographed and projected at the actual speed of propagation.

The role of tensile stressing is important in the rupture of protective films during both initiation and propagation of cracks. These films may be tarnish films, such as those on brasses; thin oxide films; layers richer in the more noble element of a binary alloy, such as a copper-gold alloy; or other passive films. Breaks in the passive film or enriched layer at various points on the surface initiate plastic deformation or incipient cracking.

Breaking of a film ahead of an advancing crack permits crack propagation to continue in the metal. Rapid local breaking of the film without particles of film filling the exposed crack is required for rapid crack propagation. Intergranular cracking occurs when grain-boundary regions are anodic to the main body of the metal and therefore less resistant to corrosion because of precipitated phases, depletion, enrichment, or adsorption.

General Features of Stress-Corrosion Cracks

Stress-corrosion cracks ordinarily undergo extensive branching and proceed in a general direction perpendicular to the stresses contributing to their initiation and propagation. Figure 1 shows intergranular SCC in a specimen of sensitized AISI type 304 stainless steel, which occurred at 82 °C (180 °F) in water containing 70 ppm of dissolved chlorides.

There are exceptions to the general rule that stress-corrosion cracks are branched. For exam-

Fig. 1 Etched section through a sensitized specimen of AISI type 304 stainless steel

Micrograph shows intergranular SCC produced at 82 °C (180 °F) in water containing 70 ppm of dissolved chlorides. 100×

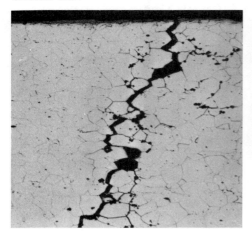

Fig. 2 Picral-etched specimen of structural steel that was exposed to contaminated agricultural ammonia showing nonbranched stress-corrosion cracks

75×

Fig. 3 Effect of alloy composition on threshold stress

Effect is shown by the relationship of applied stress to average time-to-fracture for two 18-8 stainless steels (AISI types 304 and 304L) and two high-alloy stainless steels (AISI types 310 and 314) in boiling 42% magnesium chloride solution.

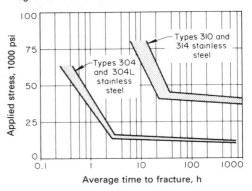

ple, some nonbranched cracks have been observed along with branched cracks in structural steel exposed to contaminated agricultural ammonia (Fig. 2).

The surfaces of some stress-corrosion cracks resemble those of brittle mechanical fractures, although they actually are the result of local corrosion in combination with tensile stress. In some metals, cracking propagates intergranularly; in others, transgranularly. In certain metals, such as high-nickel alloys, iron-chromium alloys, and brasses, either type of cracking can occur, depending on the metal-environment combination. Features of stress-corrosion cracked surfaces revealed by macroscopic and microscopic examination are discussed in the sections ''Macroscopic Examination'' and ''Microscopic Examination'' in this article.

In wrought high-strength heat-treatable aluminum alloys, paths of stress-corrosion cracks are always intergranular, because the thermal treatments required to achieve high strength never completely eliminate the electrochemical potential between grain boundaries and grain centers. Stress-corrosion cracking in high-strength aluminum alloys is probably the most widely known example of the effect of grain orientation on the path of cracking. Where the grains are substantially elongated in the rolling direction, stress-corrosion cracks cannot easily propagate perpendicular to this direction, but they propagate very rapidly parallel to the rolling direction. However, newer alloys and tempers have been developed to provide a high resistance to SCC when rolled, extruded, or forged products are stressed in the short-transverse direction.

In wrought austenitic stainless steels, crack paths are usually transgranular if proper heat treatment has been employed. However, if thermal processing has produced sensitization

because of carbide precipitation, SCC frequently progresses intergranularly because the chromium-depleted zones along the boundaries of a grain are anodic to the main body of the grain. In wrought martensitic stainless steels, such as AISI types 403, 410, and 431, intergranular cracking is the rule when heat-treatment procedures, although not necessarily improper, result in similar carbide precipitation.

The path of SCC in some metals is not governed completely by composition and structure of the metal but is also influenced by the environment. For example, the path of cracking in copper-zinc alloys can be made transgranular or intergranular by adjusting the pH of aqueous solutions in which these alloys are immersed.

In extensive empirical investigations, several special characteristics of SCC have been observed. Although the electrochemical and stress-sorption theories and their modified versions are generally in accord with most of the facts, no single theory that has been proposed to date completely explains all the special characteristics, which include the following:

- Only certain specific environments contribute to this type of failure for a given metal or alloy, with no apparent general pattern
- Pure metals are much less susceptible to this type of failure than impure metals, but pure binary alloys, such as copper-zinc, copper-gold, and magnesium-aluminum alloys, are generally susceptible
- Cathodic protection has been successful in preventing initiation of SCC and in stopping propagation of cracking that has already progressed to a limited extent
- Addition of soluble salts containing certain specific anions can inhibit the crack-producing effect of a given environment on a given alloy
- Certain aspects of the metallurgical structure of an alloy, such as grain size, crystal structure, and number of phases, influence the susceptibility of the alloy to SCC in a given environment

Sources of Stresses

Stresses, either applied or residual, may result from manufacturing or actual service applications. The length of time required to produce SCC is shorter for higher stresses. The relation between the direction of stressing and the grain direction of the metal influences SCC. Transverse stressing is considerably more detrimental than longitudinal stressing, and short-transverse stressing is more detrimental than long-transverse stressing. In aluminum alloy 7075-T6 extrusions, for example, the threshold stresses in the longitudinal, long-transverse, and short-transverse directions are 414, 221, and 48 MPa (60, 32, and 7 ksi), respectively. For forgings, stressing across a flash line is generally regarded as at least as severe as stressing in the short-transverse direction..

Threshold stress (the minimum stress at which the probability is extremely low that cracking will occur) depends on temperature, on the composition and metallurgical structure of the alloy, and on the composition of the environment. In some tests, cracking has occurred at an applied stress as low as about 10% of the yield strength; for other metal-environment combinations, threshold stress is about 70% of yield strength (Ref 2).

The effect of alloy composition on threshold stress is typified by the graph in Fig. 3, which illustrates the relationship between applied stress and average time-to-fracture in boiling 42% magnesium chloride solution for two 18-8 stainless steels (AISI types 304 and 304L) and two high-alloy stainless steels (AISI types 310 and 314). As indicated by the nearly level portions of the curves, the threshold stress in this environment is about 241 MPa (35 ksi) for high-alloy stainless steel and about 83 MPa (12 ksi) for 18-8 stainless steel.

Threshold Stress Intensity. For most metal-environment combinations that are susceptible to SCC, there appears to be a threshold stress intensity, K_{Iscc}, below which SCC does not occur. Average values of this parameter for maraging steels heat treated to various yield strengths and then exposed to aqueous environments are given later in this article, along with a general description of the relationship of the rate of cracking to stress intensity for maraging steels and for other high-strength steels. Additional information on stress intensity and the role of fracture mechanics in studying the kinetics of SCC is provided in the article ''Failure Analysis and Fracture Mechanics'' in this Volume and Ref 3 in this article.

Sources of Stresses in Manufacture

The principal sources of high local stresses in manufacture include thermal processing, stress raisers, surface finishing, fabrication, and assembly.

Thermal Processing. One of the most frequently encountered sources of thermal-processing stresses is welding. Shrinkage of weld metal during cooling and the restraint imposed by the adjacent metal and by rigid welding fixtures can produce residual tensile stresses as high as 207 to 276 MPa (30 to 40 ksi). Other thermal-processing effects that often produce stresses during manufacture include solidification of castings, especially those having large differences in section thickness and those made with cast-in inserts, and improper heat-treating practices, such as failure to preheat when required, overheating during austenitizing or solution treating, failure to provide required temperature uniformity in furnaces, use of quenching practices too severe for a specific alloy or part shape, and undue delay in transferring workpieces from the quenchant to the tempering furnace.

Stress raisers that result from various types of deficiencies in manufacture often contribute to mechanical-environmental failures. Some common types of such stress raisers are geometrical stress raisers or notches related to design; notches caused by accidental mechanical damage or electric-arc strikes; cracks produced by incorrect heat treatment, such as quench cracks, or by deficiencies in welding and in associated preweld and postweld treatments; inclusions and hydrogen blisters; interfaces in layer-type bonded materials (applied by cladding, rolling, electroplating, spraying, brazing, or soldering); case-core interfaces in case-hardened steels; and severe surface irregularities produced in grinding and rough machining.

Surface Finishing. Residual tensile stresses that are damaging or potentially damaging in conjunction with environmental attack are produced in many different types of surface-finishing treatments. These treatments include electroplating, electrical discharge machining, and, under some conditions, conventional grinding and machining. When hydrogen is produced in the finishing process and diffuses

into the metal, internal stresses can be produced.

Shot peening and surface rolling are often used to produce compressive stresses in metal surfaces, but the magnitude of stress imposed by peening is sometimes not great enough to overcome the effects of extremely high local tensile stresses.

Fabrication. High residual tensile stresses sometimes result from bending, stamping, deep drawing, and other cold-forming operations. Residual tensile stresses of 207 to 414 MPa (30 to 60 ksi) have been measured on the surfaces of cold-bent steel tubes. Expansion rolling of boiler and heat-exchanger tubes into holes in the tube sheets produces tensile stresses that may be damaging (see Example 3 in this article). Another source of residual tensile stresses is straightening to remove deformation caused by heat treatment.

Under some circumstances, severe uniform cold working improves the resistance of a metal to SCC. For example, cold-drawn steel wire is more resistant to SCC than oil-tempered wire having equal mechanical properties. Also, cold reduction of low-carbon steel to 50% or less of its original thickness makes it relatively immune to cracking in boiling nitrate solutions at 100 to 200 °C (212 to 390 °F) for thousands of hours.

Assembly. Fit-up and assembly operations are often sources of tensile stresses. Press fitting, shrink fitting, and assembly by welding are among the major operations in this category. Dimensioning and designing interference fits so as to keep the amount of interference small is desirable in order to avoid creating tensile stresses that could lead to SCC, but this procedure makes the assembly more likely to fail by such mechanisms as fretting, fatigue, and corrosion fatigue; accordingly, a compromise in dimensioning is often the safest procedure. An interference fit of a bushing in a hole was the source of stresses that resulted in service failure by SCC of a high-strength aluminum alloy part in a marine atmosphere (see Example 7 in this article).

Forming operations used in assembly to retain components can produce residual tensile stresses that can induce SCC, particularly when the parts are used or stored in a corrosive atmosphere. In the following example, copper alloy ferrules that had not been stress relieved after drawing and were assembled in fuses by crimping failed by SCC.

Example 1: SCC of Copper Alloy C27000 Ferrules in Storage and in Service in Chemical Plants. A substantial number of copper alloy C27000 (cartridge brass, 70%) ferrules for electrical fuses cracked while in storage and while in service in paper mills and other chemical-processing plants. The ferrules, made by three different manufacturers, were of several sizes. One commonly used ferrule was 3.5 cm (1⅜ in.) long × 7.5 cm (3 in.) in diameter and was drawn from 0.5-mm (0.020-in.) thick strip.

Investigation. Ferrules from fuses in service and storage in different types of plants, plus ferrules from newly manufactured fuses, were evaluated by visual examination, by determination of microstructure, by examination of any existing cracks, and by the mercurous nitrate test for copper and copper alloys (ASTM B 154) (Ref 4). This is an accelerated test used to detect residual stress in copper and copper alloy stock and in fabricated parts that might bring about failure of the material by SCC. Ferrules that had been stress relieved after forming, but not assembled to fuses, passed the mercurous nitrate test; similar unassembled ferrules that had not been stress relieved after forming cracked before the required 30-min immersion period was completed.

The mercurous nitrate test also showed that ferrules crimped to fuses were susceptible to cracking whether or not they had been stress relieved after forming. This last observation established that the crimping operation in assembly produced residual stresses, even on stress-relieved ferrules, that were high enough to make the ferrules susceptible to SCC. The cracks introduced during forming that were found in ferrules on fuses in service or storage were of the multiple-branched type characteristic of SCC. The higher incidence of cracking of ferrules in the paper mills was apparently related to a higher concentration of ammonia there, in conjunction with a humid atmosphere.

Conclusions. The ferrules failed by SCC resulting from residual stresses induced during forming and the ambient atmospheres in the chemical plants. The humid, ammonia-containing atmosphere in the paper mills was the most detrimental.

Corrective Measures. The fuses were specified to meet the requirements of ASTM B 154 (Ref 4). The three manufacturers used different methods to solve the problem. One changed to copper ferrules and fastened them to the fuse tube with epoxy cement (copper has insufficient strength for crimping). Another changed the ferrule material to a copper-iron alloy. The third manufacturer began using copper alloy C23000 (red brass, 85%), began subjecting the crimped ferrules to a stress-relief anneal using an induction coil, and made long-range plans to change to plated steel ferrules.

Other Manufacturing Examples. The use of pipe-threaded fittings in making connections to an aluminum alloy 2024-T431 valve body in a missile launch unit produced excessive stress in the threads and resulted in failure of the valve body by SCC in a service environment of hydraulic fluid that contained small amounts of moisture and chloride. Similarly, stresses on tapered threads on a cast aluminum-zinc alloy connector exposed to a semi-industrial atmosphere caused failure by SCC in less than one month. This connector had been overtorqued during tightening to seal the assembly for a preservice hydrostatic test. In both instances, the substitution of straight-threaded for pipe-threaded connections to reduce the stresses was one of the recommended corrective measures.

Sources of Stresses in Service

Stresses in addition to those that a metal part or assembly was designed to withstand are introduced in service in various ways. Of major concern in connection with failure by mechanical-environmental processes is the combined effect of stress raisers and environment.

Stress raisers are frequently introduced in service by damage from accidental mechanical impact or local electrical arcing. Stress raisers or notches also frequently result from local wear, fretting, erosion, cavitation, and spalling.

Stress raisers can also result from any form of localized corrosion, such as pitting, selective leaching, intergranular attack, and concentration-cell, crevice, or galvanic corrosion. In Example 9 in this article, dezincification of copper alloy fittings provided stress raisers that led to failure by SCC.

Environmental effects of various types introduce stresses in service. Exposure of metal parts to high and low temperatures, which is often accompanied by nonuniform heating rates and sharp thermal gradients, is a major source of stress in service.

In heat exchangers, for example, thermal gradients can create strains having equivalent elastic stresses ten to twenty times greater than residual or applied stresses. If the environment on the cold side of the heat exchanger is conducive to stress corrosion, these thermal stresses can produce cracking on this side. Because of the general distribution of the thermal stress, the result is extensive cracking and sometimes complete fragmentation.

In an AISI type 321 stainless steel ammonia converter in which the temperature produced by the chemical reaction was 910 °C (1670 °F), the temperature of boiler feedwater in an integral cooling jacket surrounding the converter was 205 °C (400 °F). The water jacket had deaerator tubes at the top to remove any steam formed in the jacket, but the tubes did not have sufficient capacity to carry away all the steam formed in the jacket. The resulting accumulation created a steam pocket and a vapor/liquid interface at a level opposite the 910-°C (1670-°F) interior of the converter, and leaks developed at and near the level of the interface.

Removal and examination of portions of the wall in the damaged area revealed the presence of extensive SCC that had completely penetrated the wall from the water side. The tensile stress on the water side of the wall at the vapor/liquid interface was calculated to be about 90 MPa (13 ksi); the boiler feedwater in the cooling jacket contained 1 ppm or less of chlorides.

To correct the problem, it was recommended either that tail gas, which also required heating, be fed through the jacket instead of boiler water or that the deaerator tubes be of a size to ensure flooding of the hot wall at all times. Diffusion

of carbon, gaseous carbon compounds, hydrogen, nitrogen, oxygen, and other gases from the environment into the interior of metal parts is also a major source of stress in service.

The formation of corrosion products by local attack in confined spaces produces high stress levels in metals because the corrosion products occupy a larger volume than the metal from which they are formed. The wedging action of corrosion products in cracks and between tight joints in assemblies has been known to generate stresses of 28 to 48 MPa (4 to 7 ksi) and to initiate and propagate failure by mechanical-environmental processes. Where lamellar structures are subject to chemical attack or where weak lamellar structures are subject to stress from entrapped solid or gaseous corrosion products, exfoliation occurs; blisters are produced by gaseous corrosion products entrapped locally just below the surface of a metal part.

Cyclic Stresses. The sources and effects of cyclic stress in service are described in the article "Fatigue Failures" in this Volume. The relation of cyclic stress to mechanical-environmental failure is discussed in the article "Corrosion-Fatigue Failures" in this Volume.

Cyclic stresses of mechanical origin generally arise from the rotary or reciprocating motion of mechanical devices, either from normal operation or from abnormal effects, such as vibration and resonance. Other major sources of cyclic stresses in service are fluid-flow effects, including the von Karman effect and cavitation, and thermal effects.

It has been shown recently that a particularly damaging situation is developed when a small-amplitude cyclic flutter is superimposed on a mean sustained tensile load, such as occurs in the vicinity of a pump station on a pressurized pipeline. Threshold stresses under such situations can markedly decrease. Low-frequency tensile loading, such as occurs during hydrotesting, has been found in some circumstances to enhance SCC crack growth due to the induced dynamic straining.

Metal Susceptibility

Stress-corrosion cracking can be produced in most metals under some conditions. Susceptibility of a given metal to SCC in a specific environment depends on its overall and local chemical composition and on its metallurgical structure. The sensitization behavior of the various classes and individual types of stainless steels is a good example of such a dependency.

Effects of Composition. In considering the effects of metal composition on susceptibility to SCC, the presence of alloying elements and impurities in low concentrations (and even in trace amounts) must also be considered. In general, binary alloys that contain only very small amounts of elements other than the two major constituents are quite susceptible to failure by SCC.

High-purity metals are generally much less susceptible to this type of failure than most commercial grades of metals and alloys. In fact, pure metals were long considered immune to SCC on the basis of experience and several theories of the mechanism of this failure process. However, it was later shown that SCC can be produced in almost any alloy or pure metal in certain environments and at a corrosion potential specific to the metal-environment combination. One example is the SCC of 99.999% pure copper in aqueous ammoniacal solutions containing $Cu(NH_3)_4^{2+}$ complex ions.

Commercial grades of low-carbon steels, high-strength steels, austenitic stainless steels and other austenitic alloys (especially in the sensitized condition), high-strength aluminum alloys, and brasses and certain other copper alloys are among the metals in which SCC frequently occurs (see Tables 1 and 2 and the sections of this article on various classes of metals).

The effect of interstitial elements on the susceptibility of metals to SCC is adequately shown by the behavior of austenitic stainless steels containing trace amounts of elemental nitrogen when exposed to a boiling aqueous solution of magnesium chloride. Below 500 ppm of nitrogen, no cracking occurs; above 500 ppm, cracking occurs with great rapidity. The cracking is transgranular; therefore, the effect of the nitrogen must be attributed either to an influence on the slip process or to the stability of the protective film.

Similar effects are observed in a broad range of alloys when cracking is intergranular. In this circumstance, it is reasonable to assume that impurities are concentrated in grain boundaries and that the chemical reactivity is associated with the segregation.

Depending on the chemical interaction between these enriched grain boundaries and the environment, certain chemical reactions occur preferentially at the boundaries. This reactivity influences both the susceptibility of an alloy to SCC and the paths followed by any cracks that develop.

The concentration of impurities in grain boundaries discussed in the above two paragraphs does not refer to grain-boundary precipitation effects but rather to concentration of impurities that remain in solid solution at the grain boundaries and make the local chemical composition substantially different from that of the bulk of the material in the grains.

Effects of Metal Structure. In general, any metal with a small grain size is more resistant to SCC than the same metal having a large grain size. This relationship, which applies whether crack propagation is intergranular or transgranular, is shown in Fig. 4 for copper alloy C26800 (yellow brass, 66%) in ammonia.

Elongated grain structures characteristic of high-strength wrought aluminum alloy mill products markedly affect the path of SCC in susceptible alloys and tempers. Alloy 7075-T6,

Table 1 Specific ions and substances that have been known to cause SCC in various alloys when present at low concentrations and as impurities

Damaging specific ions and substances	Alloys susceptible to SCC	Temperature
Halogen group		
Fluoride ions	Sensitized austenitic stainless steel	Room
Gaseous chlorine	High-strength low-alloy steel	Room
Fused chloride salt	Zirconium alloys and titanium alloys	Above melting point of fused salts
Gaseous iodine	Zirconium alloys	300 °C (570 °F)
Gaseous HCl and HBr	High-strength low-alloy steels (rapid crack growth)	Room
Halides in aqueous solutions	High-strength aluminum alloys	Room
	High-strength steels	Room
	Austenitic stainless steels	Hot
Oxygen group (H_2O-O_2-H_2 systems)		
O_2 dissolved in liquid H_2O	Sensitized stainless steels	300 °C (570 °F)
Gaseous hydrogen at ambient pressure	High-strength low-alloy steels	Room
Gaseous hydrogen at high temperature and pressure	Low-strength and medium-strength steels	>200 °C (>390 °F)
Gaseous H_2O	High-strength aluminum alloys	Room
Gaseous H_2O-O_2-H_2	High-strength uranium alloys	...
Hydroxides (LiOH, NaOH, KOH)	Carbon steels; Fe-Cr-Ni alloys (caustic cracking)	>100 °C (>210 °F)
Oxygen group (S, Se, Te systems)		
Polythionic acids ($H_2S_nO_6$)	Sensitized stainless steels, sensitized Inconel 600	Room
H_2S gas	High-strength low-alloy steels	Room
Sulfide impurities in aqueous solutions	Medium-strength to high-strength steels (accelerated hydrogen-induced cracking)	Room
MnS and MnSe inclusions	High-strength steels (initiation sites for cracking)	Room
SO_2 gas with moisture	Copper alloys	Room
Nitrogen group		
N_2O_4 liquid	High-strength titanium alloys	50 °C (120 °F)
Fuming nitric acid	Pure titanium; high-strength aluminum alloys	>100 °C (>210 °F)
Nitrates in aqueous solution	Carbon steels	>100 °C (>210 °F)
Nitrogen oxides with moisture	Copper alloys	Room
Aerated aqueous NH_3 and ammonium salts in aqueous solution	Copper alloys	Room
Nitrogen, phosphorus, arsenic, antinomy, and bismuth, as alloying species in metal	Stainless steels (in presence of Cl^-) and copper alloys (in presence of aerated aqueous NH_3); accelerated cracking	Room
Arsenic, antimony, and bismuth, as ions in aqueous solutions	High-strength steels, accelerated hydrogen entry and hydrogen-induced cracking	Room
Carbon group (C, Si, Ge, Sn, Pb)		
Carbonate ions in aqueous solutions	Carbon steel	100 °C (210 °F)
CO-CO_2-H_2O gas	Carbon steel	...
Lead ions in aqueous solutions	High-nickel alloys	...

for example, in the form of sheet and plate has a high resistance to SCC when stressed in the direction of rolling or in the long-transverse direction, but has a relatively low resistance when stressed in the short-transverse direction (normal to the plane of the plate). A high resistance in the short-transverse direction can be obtained, however, by an extended stabilization treatment to obtain a T7-type (lower-strength) temper.

In some metals, SCC follows special crystallographic planes. For example, SCC of titanium proceeds at a slight angle to the basal plane. If the basal plane is perpendicular to the potential plane of crack propagation, resistance to SCC is at a maximum.

Two-phase brasses (those containing more than about 40% Zn) may crack merely in water. Single-phase brasses are resistant to cracking in water, and the presence of a corrodent, such as

ammonia or an amine, is necessary to cause cracking in them.

Crystal structure also has an effect on SCC. For example, ferritic stainless steels (body-centered cubic, bcc) are much more resistant to SCC when exposed to chlorides in aqueous solutions than austenitic stainless steels (face-centered cubic, fcc). The resistance of ferrite-containing cast austenitic stainless steels to cracking when exposed to 205-°C (400-°F) condensate from water containing 800 ppm of chloride is roughly proportional to the volume percentage of ferrite, because pools of ferrite in the austenite matrix interfere with or block the propagation of stress-corrosion cracks.

Environmental Effects

The onset of SCC often depends on the subtleties of environment composition and on alloy composition and structure. Frequently, minor changes in the environment or the alloy can prevent cracking. Hence, the relationships between environments and alloy types presented in this section provide only general guidelines as to the probability of SCC.

Specific Ions and Substances

Stress-corrosion cracking is often caused by specific ions that are present as impurities in the environment. For example, even when in the solution heat-treated or the annealed condition, austenitic stainless steels crack in aqueous solutions that contain as little as 2 ppm of chloride at 200 °C (390 °F). Sensitized austenitic stainless steels crack at room temperature in water containing about 10 ppm of chloride or 2 ppm of fluoride. Some copper alloys crack in environments that contain ammonia in similarly low concentrations.

The dependence of the general pattern of SCC on the presence of one or more specific corrosive substances is sometimes called the specific-ion effect. The presence of ions known to induce cracking in a given type of alloy is widely used as a criterion for predicting whether or not cracking will occur. For example, the presence of chlorides in contact with austenitic stainless steels is always regarded as hazardous. Although this is a reasonable criterion for predicting SCC, it is known that specific ions are not always required and that environments as simple as pure water, dry hydrogen, and other pure substances can contribute to SCC. The combination of oxygen and chlorides in trace concentrations in alkaline phosphate-treated boiler water is sufficient to cause SCC of austenitic stainless steels in areas exposed to steam and intermittently wetted by boiler water.

Aircraft parts, both of steel and of aluminum alloys, have failed because of the mildly corrosive action of the atmosphere; water vapor and trace contaminants in air can initiate and propagate SCC and can cause premature failure of such parts. Marine and industrial environments containing small quantities of halide ions are

Table 2 Substances in atmospheric environments that contribute to SCC of various alloys

Damaging substance	Alloy	Special aspects
Oxygen	Copper alloys	Depends on oxygen concentration
Hydrogen sulfide	Many commercial alloys	Sources: anaerobic bacteria; breakdown of organic products
Oxides of nitrogen	Carbon steels; copper alloys	The oxides produce nitric acid and nitrates
CO or CO_2, plus moisture	Carbon steels	...
Ammonia	Copper alloys	...
Arsenic and antimony compounds	Many commercial alloys	Contained in insecticides and other sprays
Sulfur dioxide	Copper alloys	The oxide produces H_2SO_3, H_2SO_4
Chlorides plus moisture	Aluminum alloys, austenitic stainless steels, titanium alloys, high-strength steels	Marine exposure

Fig. 4 Effect of grain size on time-to-fracture in ammonia atmosphere

Data are for copper alloy C26800 (yellow brass, 66%) at various values of applied stress. Source: Ref 5

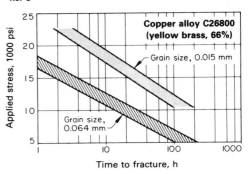

particularly aggressive in causing cracking of high-strength steels, corrosion-resistant steels, and aluminum alloys.

Specific ions and substances that are known to have caused SCC of various alloys when present in low concentrations and as impurities are conveniently classified on the basis of the periodic table. Table 1 lists these damaging specific ions and substances, which are classified as the halogen group, the oxygen group (H_2O-O_2-H_2 systems and S, Se, Te systems), the nitrogen group, and the carbon group. For each damaging substance, the alloys that are susceptible to SCC and the approximate temperature or temperature range at which cracking has been observed are given. Stress-corrosion cracking will not always occur for each listed combination of damaging species, alloy type, and temperature, but these combinations should be considered suspect when the possibility of this type of failure is being assessed.

In environments containing the damaging species of the halogen group, SCC can be caused by these species in the gaseous molecular form, in acids, in fused salts, and as ions in aqueous solutions.

The data presented in Table 1 for the oxygen group demonstrate that simple environments, not necessarily specific ions, are sufficient to cause SCC. The oxygen group is divided into two systems. The first system includes the species based on water, oxygen, hydrogen, and combinations of these and is treated separately because these species are so widely present in the environments to which alloys are exposed. Caustic cracking of low-carbon steels in boilers by hot solutions containing sodium hydroxide, which is listed in this system, is described later in this article. In the second system of the oxygen group, the substances that most frequently cause cracking are those that contain sulfur.

The nitrogen group includes as damaging species some substances that are present not in the environment but as alloying species in stainless steels and copper alloys. These alloying species are the elements nitrogen, phosphorus, arsenic, antimony, and bismuth. The other damaging species in the nitrogen group are present in the environment.

Service Environments

Commercial and consumer equipment is generally operated in two classes of service environments: controlled environments and noncontrolled environments. In controlled environments, the composition of the fluid is controlled; these environments include working (energy-transfer) fluids as well as process streams and batches in the chemical and food industries. In operations of this type, the composition of the environment is controlled either to meet and maintain product-quality requirements or to prevent corrosion of containers and other types of equipment.

Noncontrolled environments include those to which products of the transportation and manufacturing industries are exposed. Here, structures are exposed to a broad range of environments that can be anticipated by the design engineer but can rarely be controlled. As a prerequisite to investigating the possibility of SCC, the properties of a noncontrolled environment should be defined as accurately as possible.

Even in controlled environments, unexpected impurities are sometimes introduced by accidents or improper operations, and SCC can be caused by transient as well as prolonged periods of deviation from normal operating conditions and from normal composition of the environment.

Atmospheric environments contain a variety of damaging or potentially damaging chemical substances. Some of the substances in the atmosphere that contribute to SCC, the alloys they affect, and their special characteristics are given in Table 2.

Example 2: SCC of Aluminum Alloy Fittings in a Marine Atmosphere. During a routine inspection, cracks were discovered in several aluminum alloy coupling nuts (Fig. 5a) on the fuel lines of a missile. The fuel lines had been exposed to a marine atmosphere for six months while the missile stood on an outdoor test stand near the seacoast. A complete check

was then made, both visually and with the aid of a low-power magnifying glass, of all coupling nuts of this type on the missile.

One to three cracks were found in each of 13 nuts; two of these nuts had been supplied by one vendor, and the other 11 had been supplied by another vendor. There were no leaks at any of the cracked nuts, and further checking using liquid-penetrant techniques did not reveal any additional cracks. Subsequent examination of the remaining missiles on test stands at this site showed that some coupling nuts of this type on each of the missiles that had been exposed to the atmosphere for more than about a month had cracked in a similar manner.

Macroexamination. The thirteen cracked coupling nuts, which were on 6.4- to 19-mm (¼- to ¾-in.) outside-diameter fuel lines, were removed for further inspection. The nuts had been anodized, then dyed for identification.

Visual and low-magnification examination showed that the cracks had originated in the round section at the threaded end and had propagated parallel to the axis of the nut across one of the flat surfaces. An actual-size view of one of the cracked nuts and a view of the crack at 6× are shown in Fig. 5(a) and (b), respectively.

Analysis of Materials. Spectrographic analysis of the 13 cracked nuts identified the material as aluminum alloy 2014 or 2017 (the composition ranges of these two alloys overlap, which prevented exact identification by spectrographic analysis). Chemical analysis of a white deposit removed from one of the cracks showed the presence of a high concentration of chloride.

Metallographic examination of mounted specimens in the polished and polished-and-etched conditions revealed that the cracks were similar to stress-corrosion cracks observed previously in aluminum alloys of this type—primary and secondary intergranular cracks showing intergranular corrosion. A cross section taken through the failure region of one nut, near the origin of the crack, is shown in Fig.

Fig. 5 Aluminum alloy coupling nut that cracked by stress corrosion in a marine atmosphere.

(a) Overall view of coupling nut. (b) View of the crack. 6×. (c) and (d) Micrographs of a section through the crack near the origin, showing appearance before and after etching. Both 100×

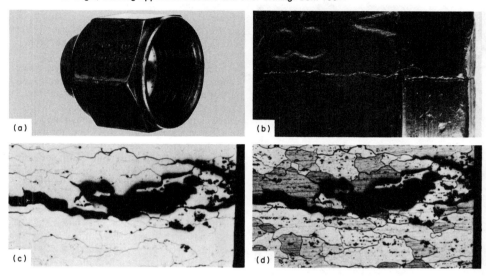

5(c) as it appeared before etching and in Fig. 5(d) after etching. The microstructure and longitudinal (axial) grain orientation identified the material of the nuts as bar stock.

Discussion. Tightening of the nuts resulted in sustained tensile stresses in the material. These were in the form of hoop stresses, which are higher than those encountered in most applications of this type of coupling nut because the parts had to be liquid oxygen clean (completely free of lubricants or organic materials); therefore, the nuts had to be tightened to high torque in order to obtain liquid-tight seals.

Conclusions. Cracking of the aluminum alloy coupling nuts was caused by stress corrosion. Contributing factors included use of a material (aluminum alloy 2014 or 2017) that is susceptible to this type of failure, sustained tensile stressing in the presence of a marine (chloride-bearing) atmosphere, and an elongated grain structure transverse to the direction of stress. The elongated grain structure transverse to the direction of stress was a consequence of following the generally used procedure of machining this type of nut from bar stock.

Corrective Measures. The material specification for new coupling nuts for this application was changed to permit use of only aluminum alloys 6061-T6 and -T651 and 2024-T6, -T62, and -T851 (for identification purposes, the anodic coating on coupling nuts of the new materials was dyed a color different from that used on the aluminum alloy 2014 or 2017 nuts). Alloy 6061 is not susceptible to SCC, and alloy 2024 in the tempers selected has been shown to withstand much higher transverse tensile stresses than alloy 2014-T6 in alternate-immersion tests in 3.5% sodium chloride solution. These tests have shown excellent correla-

tion with exposure to marine atmospheres. The 7xxx-series aluminum alloys, which are generally less resistant to corrosion and stress corrosion, are not acceptable materials for use in fluid-connection fittings.

It was recommended that specified procedures be followed closely in tightening both the alloy 2014 or 2017 nuts still in service and the new nuts. Because proper tightening alone would not eliminate failure of the 2014 or 2017 nuts, it was also recommended that any of these nuts that were exposed to marine atmospheres be inspected for cracks at least once every two weeks and that those exposed to inland atmospheres be inspected at least once a month. Visual inspection using a low-power magnifying glass and adequate lighting was to be supplemented by liquid-penetrant inspection of any nuts that were suspect.

Subsequently, shear failures occurred in the shoulders of 6061-T6 nuts during the tightening operation, and alloy 2024-T6 became the preferred material. Failure of the 6061-T6 nuts was attributed to the lower strength of this alloy, the high-torque tightening required for liquid oxygen clean nuts, and the thin shoulder webs permitted by wide tolerances on the shoulder dimensions.

Working Fluids. Some types of working fluids are the liquids and gases used in power-production systems. An important example is pure water used in production of electricity by steam-turbine systems. In such systems, water is converted to steam, and the expanded steam is passed through a turbine and subsequently condensed.

Such environments are usually treated with corrosion inhibitors to minimize attack of structural materials. A common method is to maintain pH within a range where iron-base alloys

are stable by adding such chemicals as phosphates, alkali or ammonium hydroxides, or amines. In addition to control of the pH of such working fluids, additions of such reducing substances as hydrazine and sodium sulfite are made to maintain a low concentration of oxygen, because oxygen accelerates corrosion.

Other heat-transfer fluids include liquid sodium, which is used in liquid-metal-cooled fast-breeder nuclear reactors, and high-temperature gases, such as helium, nitrogen, and carbon dioxide, which are used in gas-cooled reactors. Any of these heat-transfer fluids may contain impurities that are possible causes of SCC.

Another type of heat-transfer fluid is the water used in cooling towers. Failure by SCC of copper alloy fittings in a system of this type is described in Example 9 in this article.

In addition to the use of working fluids under essentially steady-state conditions in normal operation, transient changes in the composition of fluids can produce chemical effects that lead to SCC. Such transient changes can be caused by leakage in condensers, pumps, valves, seals, and fittings—for example, leakage of chlorides into a condenser cooled by seawater.

Temporary use of a substitute working fluid in a given application can also have unexpected harmful effects. A stress-corrosion failure related to occasional use of a fluid different from that normally employed in a heat exchanger is described in the example that follows.

Example 3: SCC of Copper Alloy Tube Sheet. Tube sheets of an air-compressor aftercooler (Fig. 6a) were found to be cracked and leaking about 12 to 14 months after they had been retubed. Most of the tube sheets had been retubed several times previously because of tube failures (the tube failures were not related to the tube-sheet failures). Sanitary (chlorinated) well water was generally used in the system, although filtered process make-up water (river water) containing ammonia was occasionally used.

The tube sheets were 5 cm (2 in.) thick. The tubes, which were 19 mm (¾ in.) in outside diameter and 1.3 mm (0.050 in.) in wall thickness, had been expansion rolled into the holes in the tube sheets.

Examination of Failed Part. One of the cracked tube sheets was removed and submitted for laboratory examination. Chemical analysis showed that the tube sheet was made of copper alloy C46400 (naval brass).

Metallographic examination of sections cut parallel and perpendicular to the tube-sheet surface revealed cracks that had penetrated through more than 90% of the tube-sheet thickness. The cracks had propagated intergranularly through the β phase, as shown in Fig. 6(b), a micrograph of a section taken parallel to the tube-sheet surface. Some dezincification of the β phase can also be seen in this micrograph. Figure 6(c), a macrograph of a section through a crack that was more than 2.5 cm (1 in.) deep, shows multiple branching.

Fig. 6 Tube sheet from an air-compressor after-cooler that failed by SCC

(a) Configuration of tube sheet. (b) Micrograph of a specimen etched in 10% ammonium persulfate solution showing intergranular crack propagation. 250×. (c) Macrograph of an unetched specimen showing multiple branching of cracks. 5×

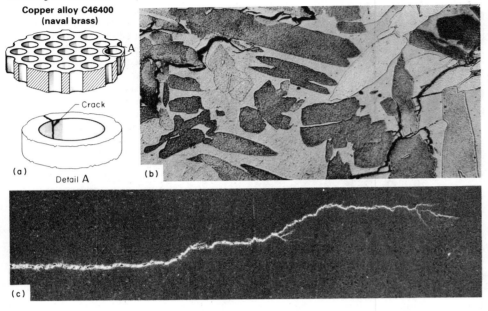

Copper alloy C46400 (naval brass)

(a)

Detail A

(b)

(c)

Test for Residual Stresses. Sample sections of the tube sheet were subjected to the mercurous nitrate test. Cracks induced in the sample sections during the test indicated the presence of high residual stresses in the tube sheet.

Discussion. The presence of ammonia in the river water used occasionally and the presence of residual internal stresses provided the conditions necessary for SCC of the tube sheets. The multiple branching of the cracks was characteristic of SCC, and the intergranular crack propagation corresponded to the usual course of SCC in brass in the presence of ammonia.

Conclusions. The tube sheets failed by SCC as a result of the combined action of internal stresses and a corrosive environment. The internal stresses had been induced by retubing operations, and the environment had become corrosive when ammonia was introduced into the system by the occasional use of process make-up water. The slight amount of dezincification observed did not appear to have contributed substantially to the failures.

Corrective Measures. It was made a standard procedure to stress relieve tube sheets before each retubing operation. The stress relieving was done by heating at 275 °C (525 °F) for 30 min and slowly cooling for 3 h to room temperature. No SCC failures have occurred in this equipment in the several years since the adoption of stress relief before retubing.

Process streams and batches differ from working fluids in that they usually are chemicals, foods, or other materials that are being processed into some final product form. The process streams and batches in themselves may not be aggressive with respect to SCC, but may become aggressive if damaging impurities are introduced.

Leached Substances. Stress-corrosion cracking of a metal is frequently caused by damaging substances that have been leached from nonmetallic materials in contact with the metal. Sometimes, the substances that cause cracking are products of thermal decomposition of a material that is innocuous unless heated to its decomposition temperature.

Damaging substances, such as chlorides or sulfur compounds, can be leached from concrete, gasket materials, insulating materials, and polyvinyl chlorides and similar plastic materials.

Moisture can leach chlorides from the calcium chloride frequently added to concrete to aid in curing. Sulfur compounds and chlorides can leach from gasket materials under conditions of pressure and elevated temperature. Gasket materials include those composed of asbestos, cork, cellulose, or other organic fibers in combination with various binders or impregnants, and the vulcanized elastomeric materials that are used for automotive applications.

The binders used in asbestos and fiberglass insulation frequently contain chloride compounds and are leached in the presence of moisture at room temperature or at elevated temperatures, depending on composition (Example 5 in this article describes SCC of stainless steel piping that resulted from leaching of chlorides from insulation in the presence of moisture). Polyvinyl chloride materials are usually harmless unless they are heated above the decomposition temperature or are exposed to ultraviolet radiation.

Concentration of Crack-Inducing Substances. In any environment (liquid, gas, or moist solid), substances capable of inducing SCC are frequently present at harmless levels, either as normal components or as impurities. However, several factors that are related to design or to service conditions can cause these substances to concentrate locally and produce SCC.

Intermittent wetting and drying is a major cause of concentration of damaging and potentially damaging soluble substances in films of moisture on metal surfaces. Where a dilute aqueous solution is transmitted to a metal surface by capillary action through an absorbent fibrous material, the process is called wicking. Tests have been done on the effect of temperature on the time required to produce SCC in 1.6-mm (0.062-in.) thick AISI type 304 stainless steel U-bends in a wicking experiment. Cracking was found to occur at much lower temperatures when alternate wetting and drying was used than when the specimens were kept wet continuously.

In service, intermittent wetting and drying is often caused by changes in weather, rise and fall of tides, and fluctuations in levels of liquids in storage. Also, intermittent condensation and drying occur on metal surfaces in many service environments because of fluctuations in temperature. If soluble substances are available, solutions sufficiently concentrated to cause SCC are readily produced on the metal surfaces.

This phenomenon was found to have caused SCC failure of carbon steel cables on a bridge over a river, where the cables were exposed to an industrial atmosphere that contained ammonium nitrate. The cables, which contained 0.7% C, failed after 12 years of service. Tests showed that when specimens from the cables were stressed in tension while immersed in 0.01 N ammonium nitrate or sodium nitrate at room temperature, they failed after 3½ to 9 months. No cracking was produced in similar tests in distilled water and in 0.01 N solutions of sodium chloride, ammonium sulfate, sodium nitrate, and sodium hydroxide.

Example 4: SCC of Carbon Steel Hoppers by Ammonium Nitrate Solution. After 10 to 20 months of service, the carbon steel hoppers on three trucks used to transport bulk ammonium nitrate prills developed extensive cracking in the upper walls. The prills were used in blasting operations and required transportation over rough roads to a mining site.

The prills were discharged from the steel hoppers using air superchargers that generated an unloading pressure of about 11 kPa (7 psi). A screw conveyor at the bottom of the hopper assisted in the unloading operation. Each hopper truck held from 9100 to 11 800 kg (10 to 13 tons) of prills when fully loaded and handled approximately 90 700 kg (100 tons) per month.

The interior surfaces of all three hoppers had been painted before placement in service, after

7 days of service, and after 60 days of service. All exterior surfaces were covered with a gray epoxy coating of the type used at the prill-manufacturing plant. There were no specifications for the interior coatings, but one appeared to be a fluorocarbon and another an aluminum-pigmented coating.

Investigation. The walls of the hoppers were made of 2.7-mm (0.105-in.) thick flat-rolled carbon steel sheet of structural quality, conforming to ASTM A 245.

Visual inspection of the internal surfaces of the hoppers revealed extensive rusting under the paint, especially along edges and at corners. This inspection also revealed that most of the cracking was confined to the top areas of the hoppers, around manholes and at circumferential welds in the sheet-steel walls.

The cracking occurred in areas where bare steel had been exposed by chipping and peeling of paint. The cracks were 5 to 20 cm (2 to 8 in.) long and were located in or adjacent to welds. Some cracks were perpendicular to the weld bead, but most followed the weld. In one hopper, cracks were found along the full length of approximately 20 intermittent welds that attached internal bracing to the walls. All cracks in the hopper walls could be seen from the outside, where they appeared as straight lines of rust breaking through the gray external coating.

A rectangular section that included a weld was removed from the upper front edge of the wall of one hopper for metallurgical examination. A transverse section approximately 3.2 mm (⅛ in.) from the weld was mounted to permit examination of both the interior and exterior edges of the hopper wall. Examination of metallographic specimens after polishing and etching with nital revealed an extensive pattern of intergranular cracking (Fig. 7) that had originated at the interior surface.

Ammonium nitrate solutions of certain concentrations are known to crack steel. A concentrated nitrate solution was produced on the upper walls of the carbon steel hoppers by dissolution of ammonium nitrate dust in atmospheric water vapor that had condensed when moist air had entered the hoppers during unloading and during standby periods in which the manhole covers were left open.

Conclusions. Failure of the hoppers was the result of intergranular SCC of the sheet-steel walls because of contact with a highly concentrated ammonium nitrate solution. The solution was produced by dissolution of ammonium nitrate dust in condensed atmospheric water vapor that formed intermittently on the upper walls of the tank.

Corrective Measures. A three-coat epoxy-type coating with a total dry thickness of 0.3 mm (0.013 in.) was applied to the interior surfaces of the hoppers. Although high stress levels in the steel resulting from welding, unloading, and traveling on uneven roads contributed to the cracking, it was more economical to coat the interior surfaces of the steel

Fig. 7 Nital-etched specimen of ASTM A245 carbon steel

Micrograph shows SCC that occurred in a concentrated solution of ammonium nitrate. 100×

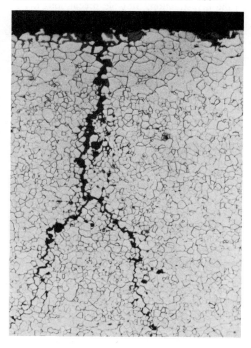

hoppers properly than to design for minimum stress levels or to make a change in material. For example, equivalent aluminum hoppers would cost approximately 2½ times as much as the carbon steel hoppers, and stainless steel hoppers would cost five times as much.

Gradual concentration of a damaging substance on a metal surface can result from diffusion effects caused by an imposed electric field or an electric field produced by corrosion processes or by some other local condition, and can lead to SCC in gases, liquids, and moist solid environments, such as soil, concrete, or plaster.

Concentration can also occur because of part shape, especially on assemblies in which close fit of parts creates crevices. Crevices provide both sites and proper conditions for concentration of impurities on metal specimens in aqueous media. Such crevices are common in assemblies using gaskets or sealants. Because of the tightness of such crevices, it is impossible for impurities to be removed from them. The impurities are thus concentrated by leaching, decomposition, or corrosion and frequently result in SCC. In heated crevices, such as in tube sheets of steam generators, impurities are concentrated when water flashes to steam, and SCC may occur if hydroxides are present.

Dried processing chemicals on metal surfaces, such as those that remain on poorly rinsed surfaces after pickling, cleaning, and similar operations, are available for corrosion attack at a later time when the surfaces are

exposed to their normal environments. One example is the effect of residual fluoride on zirconium in corrosion testing. When zirconium specimens exposed to fluoride solutions in corrosion tests are properly cleaned after removal from the solution, they gain weight at a low rate when subsequently exposed to high-temperature water; however, when there is sufficient residual fluoride, the specimen loses weight because of corrosion that can initiate SCC.

Preservice Environments

Many SCC failures are caused by exposure of metal equipment to corrosive environments before the equipment is put into service. These environments include those associated with fabrication, testing, shipment, storage, and installation.

Fabrication Environments. Substances to which metal parts and equipment are exposed during fabrication and processing often leave residues on the metal surfaces or introduce impurities into the metal. Stress-corrosion cracking caused by these residues or impurities frequently begins, and occasionally results in failure, during the production process. Sometimes, impending failure is discovered during fabrication; at other times, it is not discovered until the equipment is in service.

Examples of sources of such deleterious impurities are machining lubricants; protective coatings; fumes from adjacent processing lines; atmospheric impurities; welding flux; cleaning solutions; inspection environments, such as liquid penetrants; containers for processing; lubricants, such as copper and lead, used in tube drawing; and plating and chemical-surface treatment solutions.

Testing Environments. In product research and development, metal parts and assemblies are tested to determine load capacity and conformance to other requirements. Failures are sometimes initiated and/or propagated by test environments, such as in hydrostatic testing of pressure vessels, piping, and related equipment. For example, a titanium alloy Ti-6Al-4V vessel that was to contain N_2O_4 in service failed after about 13 h of hydrostatic testing at about 43 °C (110 °F) at 75% of yield stress. Here, the testing environment, which closely simulated the service environment, was the cause of failure. In this case, failure was fortuitous, because it established that the material-environment combination was not suitable for reliable performance of the intended application.

Other environmental impurities that may be harmful in hydrostatic testing include microbiological and bacterial contaminants, chlorides, and fluorides. Stress-corrosion cracking of titanium in methanol was first observed when methanol was used as a medium for hydrostatic testing.

Shipping and Storage Environments. During shipping or storage, parts and assemblies may or may not be protected from the atmosphere. There have been instances in which allegedly protective materials have in fact contained species that caused, rather than prevented, SCC. Where suitable protection is not provided, the components can be continually or intermittently in contact with impurities in the surroundings. For example, metal parts shipped by sea are in danger if not fully protected from chlorides. Thus, to prevent initiation and propagation of SCC during shipment or storage, expected exterior environments and protective systems must be thoroughly evaluated and effective protective measures taken.

Example 2 in this article describes the SCC of aluminum alloy couplings, which occurred in a missile that was on a test stand near the seacoast for several months. The combination of tensile stresses and marine atmosphere caused the failure.

Installation Environments. During installation of equipment, critical materials are often exposed to damaging or potentially damaging environments. Some of these environments are welding flux, marking materials, lubricants (applied to moving parts and to large bolts during installation), machining oils, test environments, and contaminants from fingerprints. Sometimes, installation procedures, such as bolting or crimping, inadvertently introduce residual stresses that can lead to SCC in service.

Failure Analysis

The procedures for analyzing failures suspected to have occurred by SCC are in general the same as those described in the article ''General Practice in Failure Analysis'' in this Volume. Ideally, the failed part should be brought to the laboratory immediately after failure, with no further disturbance and with all of the physical evidence intact. The part should be accompanied by a complete record of the manufacturing history and results of prior testing. Attention must also be given to chemical environments to which the part was exposed during manufacturing, shipment, storage, and service, particularly to such factors as concentration, temperature, pressure, acidity or alkalinity, normal or accidentally introduced impurities, cyclic conditions, and any deliberate or accidental changes in the control of these factors.

On-site examination provides an opportunity to observe and test the environmental conditions and mechanical processes involved. The relationship of the failed part to the overall operation as well as possible sources of stress can generally be established. Photographs and sketches are indispensable for recording the location of failure, the extent of fracture or cracking, the presence of corrosion products or foreign deposits, evidence of mechanical abuse, and other significant details.

Sampling. The samples required may be failed parts, related parts, specimens removed from parts, and the environment to which the failed parts were exposed. Selection of samples and removal of specimens should be carefully planned, keeping in mind the main objectives of the investigation. Suitable precautions must be taken to avoid (1) destruction of evidence, for example, corrosion products or other surface deposits, that could be of value in investigating the failure, (2) overheating, contaminating, and damaging the part and adjacent and remote related components and structures, and (3) contamination of samples of the environment.

In sampling the environment to which the failed part was exposed, the investigator should be guided by the information already obtained about the operating conditions and history of the failure. The general environment should be sampled. Also, appropriate techniques should be used to obtain samples of and to characterize (for example, by pH and the presence of specific contaminants) the environment in immediate contact with the failure region. General information on chemical substances and environments likely to cause SCC in specific metals is provided in Table 1.

Following visual examination, the fracture surface should be stripped with cellulose acetate tape to entrap any fracture-surface deposits. The tape can be used for later microanalysis, with the aid of an electron microprobe, a scanning electron microscope with an energy-dispersive system, or other suitable analytical equipment. After stripping, the fracture surface should be cleaned (ultrasonically, if possible) and degreased in reagent-grade acetone, which leaves minimal amounts of organic residues on the surface while removing almost all traces of cellulose acetate from stripping.

Observation of Fracture-Surface Characteristics

Crack patterns, initiation sites and topography of crack surfaces, and evidence of corrosion all provide clues as to the causes, origins, and mechanisms of fractures.

Crack Patterns. As discussed in this article and illustrated in Fig. 1, SCC is almost without exception characterized by multiple branching. The nature of the crack patterns is best observed by microscopic examination of sections in the failure region.

Initiation Sites and Topography of Crack Surfaces. These features are crucial in determining the causes and mechanisms of fractures. When a primary fracture has been damaged or corroded, secondary cracks and incomplete primary cracks are opened to expose their surfaces for examination.

Corrosion of Fracture Surfaces. The techniques for determining the presence of corrodents and corrosion products and for identifying them are noted in the article ''Corrosion Failures'' in this Volume. Besides establishing the presence and identity of these substances on the fracture surface, an investigation (often extensive) of the environment and service history of the part must be carried out to determine the exact role (if any) of these substances in the failure.

Where information on the environment and service history of the part are incomplete or where abnormal transient conditions may have existed, results of the investigation may be inconclusive. In some circumstances, tests that simulate service and investigations of the effects of variations in composition of the environment and in operating conditions are warranted.

Macroscopic Examination

Macroscopically, fractures produced by SCC always appear brittle, exhibiting little or no ductility even in very tough materials; in this way, they resemble corrosion-fatigue fractures. Many transgranular stress-corrosion cracks characteristically change fracture planes as they propagate, producing flat facets. Both these flat facets and the grain facets of intergranular stress-corrosion cracks are ordinarily observable at low magnification.

The fracture surfaces usually contain easily identifiable regions of crack initiation, slow crack propagation, and final rupture. Final rupture usually occurs by tensile overload. Thus, the area of final fracture often shows some evidence of ductility, such as a shear lip or a herringbone pattern emanating from the zone of slow cracking.

The area of slow crack growth often contains corrosion products or is stained or otherwise discolored with respect to the area of final fast fracture. However, the presence of staining or corrosion products on the fracture surface is by no means positive proof of SCC. Some SCC fractures are not stained or discolored, especially in materials with good corrosion resistance; in addition, many fractures become corroded before inspection can be accomplished.

The zones of slow propagation of stress-corrosion cracks are usually much rougher in appearance than those of corrosion-fatigue cracks and do not contain fatigue beach marks or macroscopic evidence of cold work. However, if either the stress component is removed or the environment becomes inactive, crack propagation will discontinue until these factors again become operative. This sometimes results in macroscopic markings similar in appearance to the beach marks characteristic of fatigue fractures.

Both types of fracture can, and often do, initiate at corrosion pits or other stress raisers on the surface of the part. However, if the environment is sufficiently corrosive and if tensile stresses are present, SCC can be initiated even on smooth surfaces, whereas corrosion fatigue is often initiated at a stress raiser. Fractures caused by either of these mechanisms always initiate at the surface of the part or at some location where the aggressive species can contact the metal. By corollary, if the crack is found to have a subsurface origin,

Fig. 8 Macrograph and TEM (transmission electron microscopy) fractographs of the fracture surface of an aluminum alloy 7075-T6 aircraft landing-gear component that failed by SCC

(a) Fracture surface. 1.35×. Arrows show locations of multiple crack origins. Regions A and B are representative of early and advanced stages of stress-corrosion crack growth, respectively. Region C is the area of final fast fracture. (b) TEM fractograph from area A showing mud-crack pattern. 1350×. (c) TEM fractograph from region B showing intergranular separation. 1700×. (d) TEM fractograph from region C showing dimples in region of ductile final fracture. 3400×. TEM fractographs are of plastic-carbon replicas.

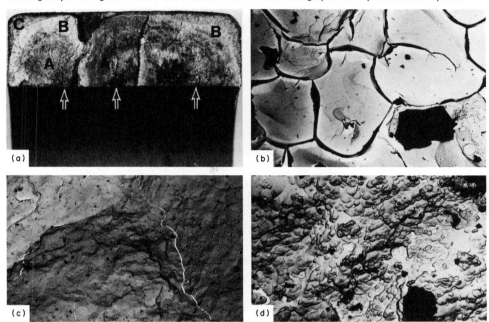

corrosion fatigue and SCC can usually be eliminated as possible causes of failure.

Cracks produced by stress corrosion and by some hydrogen-damage processes generally have macroscopically rough fracture features, which are usually discolored from reaction with the environment. These mechanisms are easily recognized as macroscopically different from corrosion fatigue, but identifying the exact mechanism by macroexamination alone is not always possible. These cracks do not always have distinct origins; there may be much pitting in the region of crack initiation, and the crack may exhibit branching in this region.

Microscopic Examination

Careful correlation of microscopic fracture-surface topography with macroscopic fracture-surface features is essential. The crack-initiation region and the directions of crack growth must be identified accurately so that information concerning the sequence of events and the micromechanisms of fracture, as observed with electron microscopy, can be correlated with the circumstances of crack initiation and the mechanism of crack propagation. Frequently, several different fracture micromechanisms are observed on a single fracture surface. Accordingly, correct identification of the initiating fracture mechanism and of any changes in micromechanism dur-

ing fracture propagation is of vital importance in arriving at a correct understanding of the failure.

In one case, an aluminum alloy 7075-T6 aircraft landing-gear component that was stressed only by loads from the dead weight of the aircraft and exposed to a corrosive marine environment failed by SCC. The fracture surface, showing the locations of multiple crack origins and areas representing three stages of crack propagation, is illustrated in Fig. 8(a). The areas of the early stages of crack growth were heavily coated with gray corrosion products, which were thickest in the crack-nucleation regions indicated by the arrows in Fig. 8(a).

Figure 8(b) shows a TEM fractograph of a two-stage plastic-carbon replica and reveals a mud-crack pattern from a region representing an early stage of crack growth (region A, Fig. 8a). Mud cracks are replicated cracks in a layer of corrosion products on a fracture surface, but are sometimes misinterpreted as intergranular. At the lower right corner of Fig. 8(b) is a particle of corrosion product (dark) extracted from the surface and retained in the replica.

As crack growth progressed, the concentration of corrosion products decreased, and curved macroscopic crack-progression lines were evident in areas representing the latter stages of crack growth (region B, Fig. 8a). These lines are a form of beach mark and, at

first glance, could lead to the erroneous conclusion that the component failed by fatigue. However, microfractographic examination established the actual mechanism at this stage of fracture propagation. Figure 8(c), a TEM fractograph of a portion of region B in Fig. 8(a), shows clear evidence of intergranular separation. The dimpled surface visible in Fig. 8(d), which is a TEM fractograph of region C in Fig. 8(a), was the region of final fast fracture by ductile rupture.

It is difficult to distinguish between service failures by SCC and by hydrogen damage in steel solely from microfractographic evidence. Fractures of both types mainly follow intergranular paths, although SCC is sometimes transgranular, and the surfaces of both types of fracture may be substantially corroded. Also, hydrogen evolution during SCC may be a factor in the cracking process so that the combination of hydrogen embrittlement and corrosion at the crack tip may be instrumental in producing grain-boundary separation.

Chemical Analysis

Chemical analysis is used in investigating cracking failures that may have been caused by SCC or other mechanical-environmental failure processes to help determine the cause and mechanism of failure. Chemical analysis of the environment, the metal, and surface deposits or scale on the metal is discussed in the article "General Practice in Failure Analysis" in this Volume.

Impurities and substances present at low concentrations or in trace amounts in the environment are of special interest where SCC is a possibility because they are the substances that usually induce this type of failure. Identification of the chemical compounds present in any surface deposits on the metal is important in determining the identity of corrodents and the nature of the damaging corrosion processes. However, great care should be taken to ensure that surface deposits are representative of the service environment and are not residues of extraneous materials introduced during such operations as liquid-penetrant or magnetic-particle inspections performed after the failure. In analyzing SCC failures, the nominal chemical composition of the metal is usually of less concern than the substances present at low concentrations or in trace amounts (usually as impurities) and the structure of the metal.

Metallographic Analysis

Aspects of metallographic analysis that require special attention include the following:

- Preferred grain orientation such that the direction of maximum stress is perpendicular to elongated grain faces
- The possible existence of highly hardened surfaces resulting from improper machining practices, wear or localized working, which

can initiate stress-corrosion cracks in high-strength materials

- The possibility that impurities have been dissolved in the surface layers as a result of high-carbon or high-sulfur activity in the environment
- The possibility that processing has introduced hydrogen into the metal, which can result in hydrogen-induced delayed fracture
- Detailed analysis of grain-boundary regions to determine whether composition or structure was not controlled or was changed such that the grain boundaries are preferential regions for stress-corrosion attack (sensitization of austenitic stainless steel has such an effect)
- The possibility that an impurity may be present that is not specified for the alloy and that would render the alloy particularly susceptible to SCC. This might include, for example, the presence of arsenic or antimony in stainless steel—elements for which chemical analysis is usually not conducted

It is not sufficient to consider only nominal chemical composition in assessing the potential for SCC. The possibility of surface contamination or changes in surface mechanical properties, as well as other structural changes resulting from heat treatment, must also be carefully considered:

Simulated-Service Tests

Once the cause and mechanism of a failure have been established, it is desirable to verify the conclusions experimentally so that appropriate corrective measures can be taken. These tests should simulate as closely as possible the environmental and mechanical conditions to which the failed part was subjected in service. Ideally, this involves testing a part in service; however, such testing is frequently not feasible because of such factors as time-to-failure or extraneous damage caused by failure of the part. Thus, simulated-service tests may have to be performed in the laboratory. The following factors should be carefully considered:

Environmental factors

- Temperature, which may be steady or fluctuating, and may also affect stress
- Single-phase or two-phase environment, which may involve alternate wetting and drying
- Composition, including major and minor environmental constituents, concentration and changes thereof, dissolved gases, and pH
- Electrochemical conditions, which may involve galvanic coupling or applied cathodic protection

Mechanical factors

- Loading, which may be static or cyclic. If cyclic, mean stress may be zero, tensile, or compressive; also, if cyclic, stress-wave shape and period must be defined

- Surface damage, which may occur by fretting or abrasion

Combined mechanical and environmental factors

- Surface damage, which may occur by cavitation or liquid-impingement erosion

The common methods for evaluating the susceptibility of metals to SCC are described in Ref 6 and 7.

Test Conditions. When conducting simulated-service tests, exact duplication of service conditions cannot always be achieved. Also, the size and shape of the failed part may not be conducive to laboratory testing. In this case, carefully selected test specimens should be machined from the original part, if possible, incorporating any design features that may have contributed to the failure, such as sharp bend radii or crevices.

Because many parts fail only after long periods of time, it is often not feasible to perform a simulation test of similar duration. In these cases, the test may be accelerated by increasing the stress, temperature, or concentration of the corrodent. However, changing these conditions may produce a type of failure completely different from that observed in service. Increasing the stress may produce failure by overload rather than SCC, and increasing the temperature or concentration of the corrodent may produce general corrosion or pitting corrosion rather than SCC. Meaningful accelerated testing is accomplished largely by trial and error. Once failure of the test part or specimen has occurred, the failure must be investigated, using appropriate techniques to ensure that the failure mechanisms and appearance are representative of that previously encountered.

Wrought Carbon and Low-Alloy Steels

The susceptibility to SCC of carbon steels, which usually contain at least 0.10% C, generally increases as carbon content decreases. However, decarburized steel and bulk pure iron are resistant to cracking; thus, susceptibility must be at a maximum at some concentration of carbon between 0 and about 0.10%. The exact carbon content at which cracking occurs most readily is not known, but the susceptibility to cracking is high in steels containing 0.05% C (Ref 8). The nature and concentration of other alloying elements in low-carbon steels have less effect on general susceptibility to SCC than microstructure does.

Surveys of the stress-corrosion behavior of high-strength low-alloy steels in a variety of environments have shown that the strength of the steel is the most important single indication of sensitivity to SCC. Steels with yield strengths of about 1379 MPa (200 ksi) or higher are especially susceptible to this type of failure.

Caustic Cracking in Boilers. Caustic cracking of steel is a serious stress-corrosion problem and causes many explosions and other types of failures in steam boilers. The general conditions that cause caustic cracking are reasonably well understood, but failure by such cracking is still a problem.

Caustic-cracking failures frequently originate in riveted and welded structures near faying surfaces where small leaks permit soluble salts to accumulate high local concentrations of caustic soda and silica. Crack propagation is ordinarily intergranular. Failures of this type have been produced at concentrations of sodium hydroxide (NaOH) as low as 5%, but a concentration of 15 to 30% NaOH, plus a small amount of oxygen, is usually required to induce caustic cracking. Failures typically take place when the operating temperature is in the range of 200 to 250 °C (390 to 480 °F). The concentration of NaOH necessary for producing cracking increases as temperature decreases.

Cracking occurs where corrosion potential is such that only part of the steel surface is covered with an oxide film. Cracking can be prevented by anodic protection, which forms a continuous and more stable oxide film, or by cathodic protection, which completely reduces the oxide film.

Reducing pH and adding strong oxidizing agents to passivate the steel surface are common modifications of boiler-water treatment that are used to prevent caustic cracking. Inhibitors, such as nitrates, sulfates, phosphates, and tannins, have been used. However, adjustment of boiler-water composition and addition of inhibitors should be done only if the chemical behavior of the materials involved is fully understood and only after suitable laboratory testing, because adjustment of boiler-water composition can destroy the effectiveness of many inhibitors. Common practice for most steam boilers is to control pH and to monitor the water supply or maintain an effective concentration of inhibitors.

Reducing grain size has been found to be helpful in decreasing the susceptibility to caustic cracking of the low-carbon steels generally used in boilers. Minor changes in steel composition are of little value because they do not alter sensitivity at grain boundaries—areas that are preferentially attacked in caustic cracking. Stress relieving after welding is advisable in order to avoid excessive residual stress around the welds.

Cracking in Nitrate Solutions. Stress-corrosion cracking of carbon and low-alloy steels in nitrate solutions has occurred in tubing and couplings used in high-pressure condensate wells and in storage tanks containing radioactive wastes.

Cracking in nitrate solutions follows an intergranular path. Boiling solutions of several nitrates, including NH_4NO_3, $Ca(NO_3)_2$, $LiNO_3$, KNO_3, and $NaNO_3$, have been found to produce cracking. In general, more acidic solutions have more potent effects. The threshold stress necessary to produce cracking decreases with increasing concentration of the nitrate in the solution. This threshold stress can be quite low. For example, exposure to boiling 4 N solutions of the above nitrates has produced cracking in some carbon and low-alloy steels at tensile stresses lower than 69 MPa (10 ksi). Decreasing temperature increased time-to-failure. By extrapolation of these test results, it can be estimated that failure would occur in about 1000 h at room temperature. Room-temperature tests of bridge-cable wire in 0.01 N nitrate solutions have produced failures after exposure for several months.

Decreasing pH enhances nitrate cracking, and resistance to cracking can be improved by raising pH. Sodium hydroxide, which causes caustic cracking by itself, can be added to nitrate solutions to retard cracking. The reverse is also true: nitrate additions retard caustic cracking. Cathodic protection can prevent SCC in many nitrate solutions. Anodic polarization is harmful. In addition to sodium hydroxide, several other inhibitors prevent cracking.

In low-carbon steels, carbon content has a strong effect on susceptibility to nitrate-induced cracking. Initiation of SCC is minimized when carbon content is lower than 0.001% (extremely low) or higher than about 0.18%. The threshold stress for cracking is low at carbon contents of about 0.05%, but increases as carbon contents above about 0.10%. Accordingly, decarburization of steel surfaces can lead to cracking in nitrate solutions. This effect of carbon content applies primarily to crack initiation. A stress-corrosion crack that starts in a decarburized surface layer will continue to propagate into the higher-carbon interior region of the metal. Tests performed on notched specimens have shown that pre-existing cracks will propagate in steels of high carbon content.

Crack initiation and propagation in nitrate solutions appear to be closely related to carbon and, possibly, nitrogen in grain-boundary regions. It has been suggested that cracking proceeds by a mechanism in which carbon-rich areas act as cathodes that sustain dissolution of the adjacent ferrite. As would be expected, changes in composition or processing history of the steel so as to alter carbon distribution can appreciably influence susceptibility to SCC. Strong carbide-forming elements, such as chromium, tantalum, niobium, and titanium, are generally helpful in improving resistance to cracking. Reducing grain size and cooling slowly after annealing are also helpful. Steels with pearlitic or coarse spheroidized microstructures are usually less susceptible to cracking than steels containing fine carbides. Moderate cold working may increase susceptibility to cracking, but severe cold working is beneficial. Just as in caustic cracking, full stress-relief annealing after welding can prevent cracking of welded assemblies, which may contain fairly high residual stresses in the as-welded condition.

Cracking in Ammonia. Carbon steel tanks containing ammonia have developed leaks because of SCC. In one case, cracking developed in the interior surface of the head regions of tanks having cold-formed heads, but not in tanks that had been stress relieved after fabrication. Thus, residual tensile stresses, in combination with the applied stresses, were evidently sufficient for cracking to occur. Crack growth in the tanks that failed was slow, and the average service time before detection of leakage was three years.

It has been shown that both plain carbon steels and quenched-and-tempered low-alloy steels are susceptible to SCC in ammonia. In this investigation, both intergranular and transgranular cracking occurred in the quenched-and-tempered steels, but only intergranular cracking in the carbon steels. The higher-strength steels were more susceptible to cracking.

These failures were not produced in pure ammonia, but occurred in ammonia mixed with air or air plus CO_2. It is now known that the critical ingredient for cracking is oxygen and that nitrogen further aggravates cracking. It was discovered that addition of water (0.2 wt%) to liquid ammonia completely eliminated susceptibility to cracking. Water additions have become mandatory by Department of Transportation ruling for interstate transportation of ammonia in high-strength steel vessels (>620 MPa, or 90 ksi, yield strength). Since the adoption of this ruling, the problem of SCC of steel in ammonia has disappeared in the United States.

Cracking in Other Environments. Stress-corrosion cracking of carbon and low-alloy steels has also occurred in other environments, including carbon dioxide and carbonate solutions; mixtures of water, carbon monoxide, and carbon dioxide; mixtures of ammonia, hydrogen, hydrogen sulfide, hydrogen cyanide, and carbon dioxide; hydrogen cyanide; organic liquids, including methanol, butyl alcohol, acetone, and carbon tetrafluoride; phosphorus trifluoride; sodium phosphate; nitric acid; sulfuric acid; and ferric chloride. In general, little is known about the conditions necessary for either initiation or prevention of SCC in these environments.

Austenitic Stainless Steels

Although the SCC of austenitic stainless steels has been studied extensively, it is not possible to predict the exact conditions under which failure will or will not occur. However, certain behavior patterns have been noted.

Austenitic stainless steels should be quenched rapidly from high temperatures to avoid sensitization. Slow cooling from about 850 °C (1560 °F) to about 550 °C (1020 °F) in heat treatment or welding increases susceptibility to intergranular SCC (or to intergranular corrosion), but rapid cooling through this range

prevents such damage. The extent of damage is a function of both time and temperature. In general, the damage is attributable to the precipitation of chromium carbides at grain boundaries, resulting in a depletion of chromium in the matrix. The limits of the so-called sensitizing-temperature range cannot be defined exactly, because they are influenced by the composition of the stainless steel, especially the percentages of carbon and of stabilizing carbide-forming elements or combinations of such elements as titanium, niobium, and niobium plus tantalum.

Also, whether sensitizing is damaging in a given instance depends on the requirements of the specific application, on the environment and stresses to which the steel is exposed, and on the prior thermal and mechanical working history of the steel. Holding at temperatures as low as about 400 °C (750 °F) and as high as about 900 °C (1650 °F) for prolonged periods has been reported to cause sensitization of unstabilized austenitic stainless steels.

In austenitic stainless steels, SCC proceeds transgranularly in solutions containing chlorides. It appears that cracking in chlorides usually occurs only at temperatures above about 70 °C (about 160 °F). Heat transfer intensifies stress-corrosion problems, probably by affecting the concentration of chlorides at the metal surface. Even without apparent increased concentration, very low chloride content is sufficient to cause cracking, especially at higher temperatures. As the nickel content of austenitic stainless steels is increased above about 10%, resistance to cracking in chloride solutions is improved.

Example 5: SCC of Type 316 Stainless Steel Piping. A 680 000-kg (750-ton) per day ammonia unit was shut down following fire near the outlet of the waste heat exchanger. The fire had resulted from leakage of ammonia from the AISI type 316 stainless steel outlet piping.

As Fig. 9(a) shows, the outlet piping immediately downstream from the waste heat exchanger consisted of a 46-cm (18-in.) diam flange, a 46- to 36-cm (18- to 14-in.) outside-diameter reducing cone, a short length of 36-cm (14-in.) outside-diameter pipe, and a 36-cm (14-in.) outside-diameter 90° elbow, all joined by circumferential welds. The flange was made from a casting; the reducing cone, pipe, and elbow were made of 13-mm (½-in.) thick plate. A 33-cm (13-in.) outside-diameter × 1.6-mm (¹⁄₁₆-in.) thick liner wrapped with insulation was welded to the smaller end of the reducing cone. All of the piping up to the flange was wrapped with insulation.

Investigation. Visual examination disclosed cracks through the weld joining the flange and cone and into the base metal on either side. The cracks extended only a short distance into the flange, but well into the reducing cone. Cracks also were found in a short section of the longitudinal weld in the 36-cm (14-in.) outside-diameter pipe between the cone and the elbow.

Fig. 9 AISI type 316 stainless steel piping that failed by SCC at welds

Cracking was caused by exposure to condensate containing chlorides leached from insulation. (a) View of piping assembly showing cracks on inner surface of cone. Dimensions given in inches. (b) Macrograph of failure region at flange-to-cone weld. Actual size. (c) Macrograph of a longitudinal section through flange-to-cone weld showing branched cracks. Etched in oxalic acid. 5×. (d) Macrograph of an unetched section through cone base metal near flange-to-cone weld showing branched cracks. 10×. (e) Macrograph of an unetched section through longitudinal weld showing shallow, nonbranched cracks. 10×. (f) and (g) Micrographs of sections through weld shown in (e) and through adjacent base metal, respectively, exhibiting intergranular scale-lined cracks. Both etched in oxalic acid. Both 100×

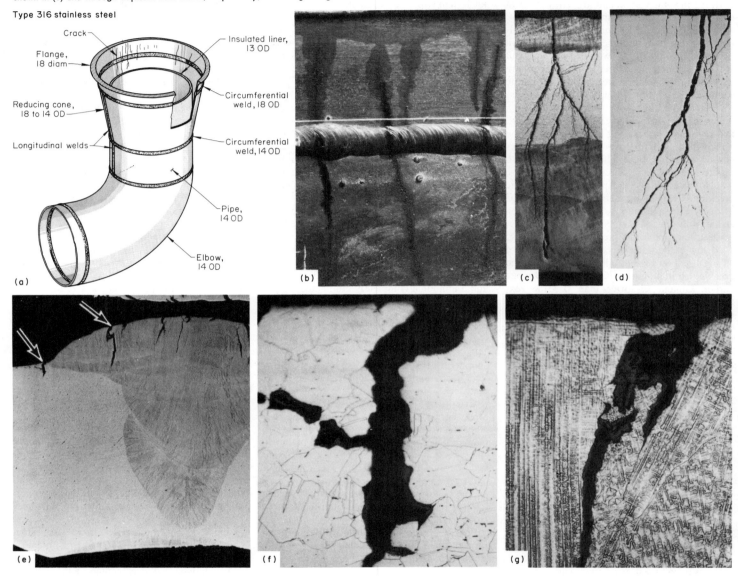

The piping assembly was cut into the individual components for ease of examination. The components were then liquid penetrant inspected.

At the flange end of the cone, there were numerous cracks across the inner top surface from the 8 o'clock position to the 4 o'clock position. The remaining area across the bottom of the cone was free of cracks. Only one of these cracks was visible on the outer surface. On the inner surface, this crack extended across the weld metal and into the flange, but on the outer surface it did not entirely penetrate the weld metal. Also, the crack was shorter and discontinuous on the outside, indicating that it had originated at the inner surface. Figure 9(a)

shows several of the cracks on the inner surface of the cone.

Visible cracks in the pipe section were confined to a 5- to 7.5-cm (2- to 3-in.) long area in the longitudinal weld. Most of these cracks were in a generally transverse direction in the weld metal, but did not appear to extend beyond the weld metal. This area also had a blue tint as a result of the fire, which was reported to have reached behind the insulation.

Analysis of the insulation wrapped around the liner revealed chlorides in the form of NaCl, at a concentration of 190 to 300 ppm, produced by selective leaching.

Metallographic examination of specimens from the cracked area of the cone shown in Fig.

9(b) revealed that both the weld joining the flange and the cone (Fig. 9c) and the base metal of the cone (Fig. 9d) contained several branched, transgranular cracks, all of which had originated at the inner surface. These cracks were characteristic of SCC in austenitic stainless steel.

A macrograph of a cross section through the longitudinal weld in the pipe section is shown in Fig. 9(e). Cracks in this weld were much different from those in the weld joining the flange and cone. The cracks were in the outside surface, relatively straight and shallow, and nonbranched. At higher magnification (Fig. 9f and g), it could be seen that the cracks were intergranular and lined with scale. These characteristics indicated that the cracks had been formed at a high temperature.

Discussion. The branched, transgranular cracks in the flange-to-cone weld and in the base metal of the cone were characteristic of stress-corrosion cracks. Stress-corrosion cracking in AISI type 316 stainless steel is most commonly caused by aqueous chlorides. Concentrated caustic can also cause this type of damage, but would not have been present in the waste heat exchanger. There were two possible sources of chlorides: boiler feedwater and insulation. The boiler feedwater was an unlikely source because it was monitored regularly to guard against chlorides. However, the insulation material between the cone and its inner liner was known to contain leachable chlorides.

Chlorides will not cause SCC unless an aqueous phase is present; therefore, the temperature of the outlet stream must have dropped below the dew point at some time. It is also possible that the absence of insulation at the flange caused a cool spot at which condensation could occur.

Conclusions. The outlet piping failed in the area of the flange-to-cone weld by SCC as the result of aqueous chlorides; the chlorides were leached from the insulation around the liner by condensate. The cracks in the longitudinal weld in the pipe section were caused by exposure of the weld metal to the high temperature of the fire behind the insulation.

Recommendations. To prevent future failures, it was recommended that the chlorides be eliminated from the system, that the temperature of the outlet stream be maintained above the dew point at all times, or that the AISI type 316 stainless steel be replaced with an alloy more resistant to chloride attack. Use of a more resistant alloy, such as Incoloy 800, would be a reliable approach.

Caustic Environments. Austenitic stainless steels are also susceptible to SCC in caustic environments. Both transgranular and intergranular failures have been observed. Fortunately, the conditions leading to caustic cracking are more restrictive than those leading to chloride cracking. Temperatures near or above the boiling point at ambient atmospheric pressure are required, and very concentrated caustic solutions are usually necessary.

Austenitic stainless steels have also suffered intergranular SCC in polythionic acids ($H_2S_nO_6$; n = 2 to 5). Sensitized steels are most susceptible, but transgranular cracking of nonsensitized steels is also possible. Proper care to prevent the entry of moisture during shutdown prevents formation of polythionic acid in refinery equipment and is the most effective measure used in the petroleum industry for prevention of this type of failure.

Intergranular SCC of austenitic stainless steels in boiling-water nuclear reactors has been reported. In most instances, the steel was sensitized, but cracking has also occurred in heavily cold-worked steels free of grain-boundary carbides.

The usual preventive measures (stress relief and cold reduction or peening to produce residual compressive stresses) are helpful in preventing SCC of austenitic steels. Measures aimed at preventing chloride concentration at the metal surface are also helpful. It is advisable to use insulation materials that do not yield aggressive solutions by leaching or to incorporate a suitable inhibitor in the insulation. Sensitization of the steel should be avoided even in environments that do not cause intergranular corrosion in the absence of stress. Welding practice should be adjusted to minimize residual stress and sensitization in the heat-affected zone (HAZ).

Ferritic Stainless Steels

Ferritic stainless steels are highly resistant to SCC in the chloride and caustic environments that crack the common austenitic stainless steels. However, laboratory investigations have shown that additions of small amounts of nickel or copper to ferritic steels may make them susceptible to cracking in severe environments. This is especially true for as-welded steels. Nevertheless, cracking in chloride or caustic environments in service occurs very rarely.

Ferritic stainless steels with chromium contents of 14% or more are subject to embrittlement at temperatures from about 370 to 540 °C (700 to 1000 °F). Embrittlement results in loss of room-temperature toughness.

Martensitic and Precipitation-Hardening Stainless Steels

Higher-strength martensitic and precipitation-hardening stainless steels are subject to SCC and, to a greater degree, hydrogen damage. Actually, failure of these alloys under conditions that appear conducive to SCC is so similar to failure caused by hydrogen damage that it is often questionable whether distinguishing between the two mechanisms is meaningful. Susceptibility to failure is determined primarily by metallurgical structure.

As expected, increasing the yield strength of martensitic and precipitation-hardening stainless steels increases the probability of cracking. The environments that cause failure are not specific; almost any corrosive environment capable of causing hydrogen evolution can also cause cracking. Even as mild an environment as fresh water at room temperature may cause failure in especially susceptible alloys.

Example 6: SCC Failure of a Sensitized Valve Stem. A 9-cm (3.5-in.) diam valve stem made of 17-4 PH (AISI type 630) stainless steel, which was used for operating a 61-cm (24-in.) gate valve in a steam power plant, failed after approximately four months of service, during which it had been exposed to high-purity water at about 175 °C (350 °F) and 11 MPa (1600 psi). The valve had been subjected to 25 hot cycles to 300 °C (575 °F) and

12.6 MPa (1825 psi) by the manufacturer and to occasional operating tests at 260 °C (500 °F) during service. Valve stems of 17-4 PH stainless steel had given satisfactory performance in similar service conditions. The bottom of the stem, from which the valve gate was suspended, was in the shape of a T.

Manufacturing History. The valve stem was forged from 15-cm (6-in.) square billet stock and is reported to have been solution heat treated at 1040 ± 14 °C (1900 ± 25 °F) for 30 min and either air quenched or oil quenched to room temperature. It was then rough machined and inspected by magnetic-particle, liquid-penetrant, and ultrasonic methods. At this stage, the stem was reportedly aged at a temperature from 550 to 595 °C (1025 to 1100 °F) for 4 h before final machining. The final hardness to be expected from this treatment is less than 35 HRC. Final inspection was done by the liquid-penetrant method.

Laboratory Examination. Each of the two surfaces of the fracture showed a semicircular stain (Fig. 10a) in a large area where fracture had propagated along prior-austenite grain boundaries without deformation. The stains remained even after ultrasonic cleaning in a solution of detergent in hot water.

Also shown in Fig. 10(a) is a narrow cup-and-cone shear-type area of failure that extends almost around the entire circumference of the fracture surface. This portion of the fracture was transgranular, with some deformation, but not enough to affect outside-diameter measurements 6.4 mm (¼ in.) away. In one narrow region of the periphery, intergranular fracture extended to the surface with no cup-and-cone shearing; intergranular SCC may have begun here. This point also coincided with the junction of the stem with a tapered fillet at the T-head, which could have been a site of stress concentration.

Microscopic examination of the structure adjacent to the fracture revealed a pattern of typical stress-corrosion cracks that radiated from the fracture surface along the grain boundaries (Fig. 10b). A macroetched cross section displayed a region of coarse-grain material (ASTM 2) surrounded by a structure of finer and more uniform grain size that extended to the surface. The coarse grain size was abnormal, suggesting that the solution heat treatment that should have followed forging may have been either omitted or carried out at too high a temperature. In either case, coarse grain size would be conducive to low ductility.

The matrix was found to contain small areas of ferrite, which in turn contained fine precipitate. This type of structure has been found in specimens that have been slowly (furnace) cooled through the range of 455 to 400 °C (850 to 750 °F) after aging. Brinell hardness readings along the length of the stem indicated a uniform hardness throughout. Hardness readings taken from a cross section of the stem 19 mm (¾ in.) from the fracture surface averaged 42 HRC.

Fig. 10 17-4 PH stainless power-plant gate-valve stem that failed by SCC in high-purity water
(a) A fracture surface of the valve stem showing stained area and cup-and-cone shearing at perimeter. 0.7×. (b) Micrograph showing secondary intergranular cracks branching from fracture surface. 50×

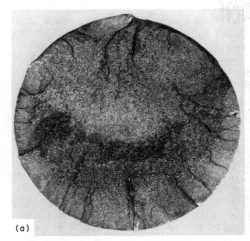

(a)

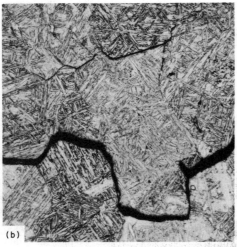

(b)

Because the hardness seemed too high for the reported aging temperature, a reheat treatment was undertaken as a double check. A portion of the failed stem was re-solution treated at 1040 °C (1900 °F) for 1 h, oil quenched, then sectioned to permit aging at a variety of times and temperatures. The hardness values resulting from these aging treatments were as follows:

Aging treatment	Hardness, HRC
1 h at 480 °C (900 °F), air cool	44
4 h at 495 °C (925 °F), air cool	41
4 h at 525 °C (975 °F), air cool	39
1 h at 550 °C (1025 °F), air cool	39
2 h at 550 °C (1025 °F), air cool	37
4 h at 550 °C (1025 °F), air cool	35

These results indicated that the stem had been aged at 495 °C (925 °F) or less, rather than at the reported 550 to 595 °C (1025 to 1100 °F).

Conclusions. Failure was by progressive SCC that originated at a stress concentration, followed by ultimate shear of the perimeter when the effective stress exceeded the yield strength of the structure. Heat treatment had been improper, creating a structure having excessive hardness, inadequate ductility, and unfavorable sensitization to SCC.

Corrective Measures. To prevent failures of this type in other 17-4 PH stainless steel valve stems, the following heat treatments were recommended: after forging, solution heat treat at 1040 °C (1900 °F) for 1 h, then oil quench. To avoid susceptibility to SCC, age at 595 °C (1100 °F) for 4 h, then air cool.

Maraging Steels

Stress-corrosion cracking of maraging steels has been studied extensively, using U-bends, unnotched tensile specimens, machined notched specimens, and fatigue-precracked fracture-toughness test specimens. As with high-strength steels, the more severe the notch, the greater the susceptibility to SCC.

Effect of Stress Intensity. As yield strength is increased, the threshold stress intensity needed for SCC of maraging steels in aqueous environments decreases, as shown below:

Yield strength		Threshold stress intensity	
MPa	ksi	MPa√m	ksi√in.
1379	200	110	100
1724	250	44	40
2069	300	11	10

This general trend of decreasing threshold values with increasing yield strength is similar to what is observed for other high-strength steels. At a given yield strength, maraging steels generally have somewhat higher threshold values than other high-strength steels.

Crack-growth rates for maraging steels usually show the three-stage curve characteristic of other high-strength steels. At lower stress intensities, the rate of cracking increases exponentially with stress intensity. At higher stress intensities, a plateau is reached at which crack-growth rate is constant with increasing stress intensity. At still higher stress intensities, near the values for fracture toughness in air, there is a further increase in the rate of SCC with stress intensity (Ref 3). Crack-growth rate at a given stress intensity increases as yield strength increases.

Path of cracking is usually intergranular along prior-austenite grain boundaries. Transgranular cracking has also occurred, and may preferentially follow the martensite-platelet boundaries. Whether or not the crack path depends on stress intensity, as in low-alloy steels, is not clear.

Microscopic branching of the main crack into two cracks inclined to the plane of the precrack

occurs frequently, but the conditions that cause branching are not known. Crack branching in specimens broken in air has been reported and is believed to be associated with banding of the microstructure. Thus, the appearance of crack branching by itself cannot be taken as proof of SCC.

Effects of Environment. For smooth specimens tested in water, adding chloride ions and increasing pH from 9 to 13 accelerates SCC, probably by accelerating pitting. With precracked specimens, these variables generally have much less effect. Cathodic protection has not been found to be a consistently effective way to prevent SCC of maraging steels.

Effects of Heat Treatment. High annealing temperatures that coarsen the microstructure are deleterious. Underaging heat treatments are also generally deleterious. Overaging heat treatments have produced mixed results. Generally, the best practice is probably the use of a low annealing temperature and the standard maraging heat treatment.

Effects of Cold Working and Prestressing. Cold working either before or after aging improves resistance to SCC. Prestressing so as to leave residual compressive stresses at the notch roots is also helpful.

Effects of Composition. Variations in major alloying elements have little or no effect on cracking behavior. Of the minor elements, carbon and particularly sulfur are very deleterious. Marked improvement in resistance to cracking can be obtained if the concentration of these elements is held to very low levels.

Prevention of SCC. As with other high-strength steels, there is no completely satisfactory way to prevent SCC in maraging steels other than by reducing the strength level. In some cases, cathodic protection, surface coatings, and prestressing can prevent SCC, but such techniques must be used with care.

Aluminum Alloys

Stress-corrosion cracking occurs in high-strength aluminum alloys in ordinary atmospheric and aqueous environments. Both initiation and propagation of cracking are accelerated by moisture, temperature, chlorides, and other industrial contaminants.

Although proper selection of both material and structural design and regular inspection in service normally prevent such failures from operating stresses, there are some factors that may be overlooked. The loading conditions responsible for corrosion-fatigue cracking are generally known, but the stress situations that generally cause initiation of SCC are not known to the designer because they usually involve sustained tensile stresses that are not anticipated and cannot be measured with precision. These situations frequently result from poor fit-up during assembly of structural components or from residual stresses introduced into individual components during such operations as cold

Fig. 11 Aluminum alloy 2014-T6 hinge bracket that failed by SCC in service

(a) Hinge bracket. Actual size. Arrow indicates crack. (b) Micrograph showing secondary cracking adjacent and parallel to the fracture surface. Etched with Keller's reagent. 250×

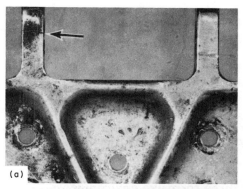

(a)

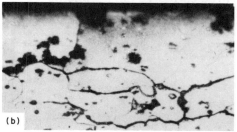

(b)

forming, welding, and quenching or straightening after heat treatment.

Example 7: SCC of Aircraft Hinge Brackets. Forged aluminum alloy 2014-T6 hinge brackets in naval-aircraft rudder and aileron linkages were found cracked in service. The cracks were in the hinge lugs, adjacent to a bushing made of cadmium-plated 4130 steel.

Investigation. Examination of the microstructure in a lug of a typical failed hinge bracket revealed intergranular corrosion and cracking in the region of the cracks discovered in service. The fracture was neither located along the flash line nor oriented in the short-transverse direction, both of which are characteristic of most stress-corrosion failures of forgings.

It was evident from the interference fit of the bushing in the hole of the lug that a hoop stress of sustained tension existed in the corroded area. The overall condition of the paint film on the hinge bracket was good, but the paint film on the lug was chipped. Figure 11(a) shows the crack in the lug (arrow), and Fig. 11(b) shows branched secondary cracking adjacent to and parallel to the fracture surface.

Conclusions. Failure of the hinge brackets occurred by SCC. The corrosion was caused by exposure to a marine environment in the absence of paint in the stressed area. The stress resulted from the interference fit of the bushing in the lug hole.

Corrective measures taken to prevent further failures consisted of the following:

• All hinge brackets in service were inspected for cracks and for proper maintenance of paint
• Aluminum alloy 7075-T6 was substituted for alloy 2014-T6 to provide greater strength and resistance to SCC. Surface treatment for the 7075-T6 brackets was sulfuric acid anodizing and dichromate sealing
• The interference fit of the bushing in the lug hole was discontinued. The bushings, with a sliding fit, were cemented in place in the lug holes

Example 8: SCC of a Forged Aircraft Lug. During a routine shear-pin check, the end lug on the barrel of the forward canopy actuator on a naval aircraft was found to have fractured. The lug was forged from aluminum alloy 2014-T6.

Investigation. As shown in Fig. 12(a), the lug had fractured in two places; the original crack occurred at the top, and the final fracture occurred at a crack on the left side of the lug. The surface of the original crack was flaky, with white deposits that appeared to be corrosion products. Apparently, the origin of the failure was a tiny region of pitting corrosion on one flat surface of the lug (back surface, Fig. 12a; arrows there show location of pitting-corrosion region). The opposite flat surface (front, Fig. 12a) had a shear lip at the fracture edge.

A metallographic section through the initial fracture displayed intergranular branched cracking that had originated at the fracture surface, which was also intergranular (Fig. 12b). Electron microscope fractography confirmed the intergranular nature of the fracture, revealing evidence of corrosion and typical characteristics of SCC (Fig. 12c). The bolt from the failed lug gave no indication of excessive loading and showed no evidence of deformation.

Conclusions. The cause of failure was SCC resulting from exposure to a marine environment. The fracture occurred in normal operation at a point where damage from pitting and intergranular corrosion acted as a stress raiser, not because of overload. The pitting and intergranular attack on the lug were evidence that the surface protection of the part had been inadequate as manufactured or had been damaged in service and not properly repaired in routine maintenance.

Recommendations. To prevent future failures of this type, the lug and barrel should be anodized in sulfuric acid and given a dichromate sealing treatment, followed by application of a coat of paint primer. During routine maintenance checks, a careful examination should be made for damage to the protective coating, and any necessary repairs should be made by cleaning, priming, and painting. Severely corroded parts should be removed from service.

Fig. 12 Forged aluminum alloy 2014-T6 actuator barrel lug that failed by SCC

(a) View of the lug. 2×. Fracture at top was the initial fracture; arrow indicates location of a tiny region of pitting corrosion (on back side of lug) at which failure originated. Final fracture is at left. (b) Micrograph of an etched (Keller's reagent) section through surface of initial fracture showing branched cracking. 140×. (c) TEM fractograph showing corroded areas and features typical of SCC. 2770×

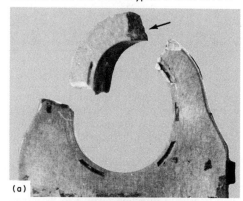

(a)

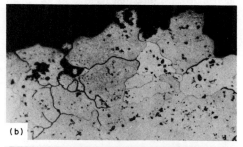

(b)

(c)

Effect of Alloy Selection. High-purity and commercially pure aluminum and the relatively low-strength aluminum alloys are not susceptible to SCC. Failures by this mechanism are chiefly associated with heat-treated wrought products of the higher-strength alloys used in load-carrying structures, such as aluminum-copper, aluminum-zinc-magnesium, aluminum-zinc-magnesium-copper, and aluminum-magnesium (3% or more Mg) alloys. With these alloys, processing and heat treatment must be controlled to ensure high resistance to SCC. The relative resistance to SCC of various wrought aluminum alloys in relation to strength is given in Table 3.

Effect of Direction of Stressing. Cast products are isotropic with regard to cracking,

Table 3 Ratings of resistance to SCC of wrought commercial aluminum alloys

Alloy series	Type of alloy	Strengthening method	Tensile strength MPa	ksi	SCC rating(a)
1xxx	Al	Cold working	69-172	10-25	A
2xxx	Al-Cu-Mg (1-2.5Cu)	Heat treatment	172-310	25-45	A
2xxx	Al-Cu-Mg-Si (3-6Cu)	Heat treatment	379-517	55-75	B
3xxx	Al-Mn-Mg	Cold working	138-276	20-40	A
5xxx	Al-Mg (1-2.5Mg)	Cold working	138-290	20-42	A
5xxx	Al-Mg-Mn (3-6Mg)	Cold working	290-379	42-55	B
6xxx	Al-Mg-Si	Heat treatment	152-379	22-55	A
7xxx	Al-Zn-Mg	Heat treatment	379-503	55-73	B
7xxx	Al-Zn-Mg-Cu	Heat treatment	517-620	75-90	B

(a) A, no instance of SCC in service or in laboratory tests. B, SCC has occurred in service with certain alloys and tempers; service failures can be avoided by careful design and assembly and proper selection of alloy and temper.

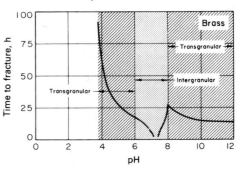

Fig. 13 Effect of pH on time-to-fractureby SCC

Data are for brass in ammoniacal copper sulfate solution at room temperature.

and directionality is generally not important in the performance of sheet (except in structural applications, such as the stressed skins of aircraft wings), but resistance to cracking of wrought products can vary markedly with direction of stressing. Service failures of extrusions, rolled plate, and forgings have been caused chiefly by tensile stresses acting in the short-transverse direction relative to the grain structure. It is important to consider the direction of stress relative to the grain structure when designing structural components and performing failure analyses. Exceptions may occur in thin forged sections that recrystallize during solution heat treatment and have low directionality.

Effect of Precipitation Heat Treatment. For thick wrought sections, special precipitation heat treatments, such as those that produce T7 tempers in aluminum-zinc-magnesium-copper alloys and T8 tempers in aluminum-copper alloys, have been developed to provide relatively high resistance to SCC in the more critical short-transverse stressing direction. Evaluation of the resistance of improved alloys and tempers in accelerated corrosion tests, however, can be markedly influenced by test procedures, and reliance should therefore be placed mainly on exposure to outdoor atmospheres or other service environments.

Effects of Temperature. The resistance to SCC of the nonheat-treatable aluminum-magnesium alloys and the naturally aged (T3 and T4 tempers) heat-treated aluminum-copper alloys can be adversely affected in structures subjected to elevated temperatures. The effects depend on the specific alloy and temper, on temperature, and on time at temperature.

Relative ratings of resistance to SCC for various alloys and tempers are published by the Aluminum Association in Aluminum Standards and Data, and more detailed guidelines for comparing the high-strength alloys, tempers, and product forms with respect to stressing direction are presented in MIL-Handbook-5.

Copper and Copper Alloys

Copper and copper alloys have excellent corrosion resistance in many industrial environ-

ments, in seawater, and in marine atmospheres, but are susceptible to SCC in some important industrial environments. These environments include those containing ammonia, citrate, tartrate, moist SO_2, and mercury.

Effect of Environment. Stress-corrosion cracking in copper alloys occurs most frequently in environments that contain ammonia or amines, in either aqueous solutions or moist atmospheres. Cracking occurs at room temperature and at stress levels as low as 1% of the tensile strength of the alloy.

In aqueous solutions, pH has a strong influence on susceptibility to cracking and on whether crack paths are intergranular or transgranular, as shown for brass in ammoniacal copper sulfate solution in Fig. 13. Cracking occurs most rapidly in nearly neutral solutions, where the crack path is intergranular. The crack path is transgranular in both alkaline and acidic solutions, and the alloy is highly resistant to cracking when pH is less than 4.

Failure of copper and copper alloys will occur at low stresses in aqueous solutions or moist atmospheres. Figure 14 plots the relation between time-to-fracture at room temperature and initial tensile stress for brass partly immersed in concentrated ammonium hydroxide (curve A), exposed to the vapor of concentrated ammonium hydroxide (curve B), and exposed to a gaseous mixture of ammonia, oxygen, carbon dioxide, and water vapor (curve C).

The presence of such oxidizing substances as dissolved oxygen and cupric, ferric, and nitrate ions accelerates SCC of copper alloys in ammoniacal aqueous solutions. In stress-corrosion tests in which brass was exposed for 20 days to atmospheres containing (by volume) 0.1 to 1% SO_2 the specimens cracked, but similar specimens that had been pretreated with the inhibitor benzotriazole were unaffected. Pretreatment of brass specimens with benzotriazole failed to prevent cracking in tests in which the stressed specimens were exposed for 72 h in an atmosphere containing (by volume) 2% ammonia and in an aged solution of ammonium hydroxide. Aqueous solutions of mercurous nitrate are widely used in testing copper alloys for the presence of residual stresses that make the alloys susceptible to SCC. Detailed information

Fig. 14 Effect of initial tensile stress on time-to-fracture by SCC at room temperature of brass in three corrosive environments

Curve A, partly immersed in concentrated ammonium hydroxide; B, exposed to the vapor of concentrated ammonium hydroxide; C, exposed to a gaseous mixture of ammonia, oxygen, carbon dioxide, and water vapor

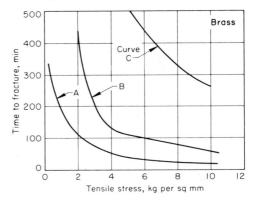

on stress-corrosion tests for copper and other metals and alloys is available in the article "Tests for Stress-Corrosion Cracking" in Volume 8 of the 9th Edition of *Metals Handbook*.

Effect of Alloying Elements. High-purity copper is almost immune to SCC in most environments and in the practical range of service stresses. However, intergranular cracking of high-purity copper has been observed under some conditions, apparently as a result of segregation of trace impurities at the grain boundaries.

The resistance of copper to SCC is greatly reduced by the presence of low concentrations of arsenic, phosphorus, antimony, and silicon as alloying elements. Time-to-fracture for copper containing various concentrations of these alloying elements when stressed at an applied tensile stress of 69 MPa (10 ksi) is plotted in Fig. 15. As the concentration of each alloying element is increased, time-to-failure at first decreases, reaching a minimum between about 0.1 and 1%, then increases.

Fig. 15 Effects of arsenic, phosphorus, antimony, and silicon on SCC of copper

Effect of presence of low concentration of arsenic, phosphorus, antimony, and silicon on time-to-fracture of copper by SCC under an applied tensile stress of 69 MPa (10 ksi) in a moist ammoniacal atmosphere. Composition of test atmosphere was 80% air, 16% ammonia, and 4% water vapor; temperature was 35 °C (95 °F), which was above the dew point.

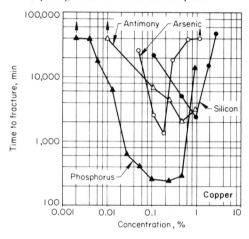

Fig. 16 Effects of aluminum, nickel, tin, and zinc on SCC of copper

Effect of presence of low-to-moderate concentrations of aluminum, nickel, tin, and zinc on time-to-fracture of copper by SCC under an applied tensile stress of 69 MPa (10 ksi) in a moist ammoniacal atmosphere. Composition of test atmosphere was 80% air, 16% ammonia, and 4% water vapor; temperature was 35 °C (95 °F), which was above the dew point.

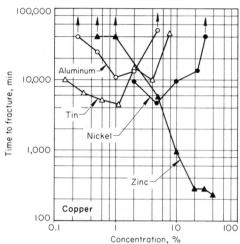

Results of similar tests on copper containing higher concentrations of the alloying elements aluminum, nickel, tin, and zinc are plotted in Fig. 16. Except for zinc, a minimum time-to-fracture is reached between about 1 and 5%; time-to-failure for copper containing zinc does not show a minimum for concentrations of 0.5 to 40%, but decreases with increasing zinc content for the concentration range shown in Fig. 16.

The effect of zinc content on susceptibility of copper-zinc alloys to SCC when exposed to cooling-tower water containing amines and dissolved oxygen is discussed in the next example. The alloys that failed in this application had nominal zinc contents of 25, 26, and 40%. Two replacement cast silicon bronzes that gave satisfactory service contained less than 5 and 1.5% Zn; a satisfactory replacement wrought silicon bronze contained less than 1.5% Zn.

Example 9: Failure of Copper-Zinc Alloy Cooling-Tower Hardware. After 14 months of service, cracks were discovered in castings and bolts used to fasten together braces, posts, and other structural members of a cooling tower, where they were subjected to externally applied stresses.

Selected specimens were removed and then examined in a laboratory to determine the nature and extent of the failures. A total of 35 samples of cracked hardware (21 castings and 14 bolts and nuts) from different zones of the cooling tower were removed for examination. The castings were made of copper alloys C86200 and C86300 (manganese bronze). The bolts and nuts were made of copper alloy C46400 (naval brass, uninhibited). The water

that was circulated through the tower had high concentrations of oxygen, carbon dioxide, and chloroamines.

Investigation. Twenty-one representative castings, 16 of which had failed in service, were submitted for laboratory examination. After a preliminary examination, the castings were divided into two categories: those that had completely separated into two or more parts in service, and those that had not completely fractured but that contained cracks.

Figure 17(a), (b), and (c) show some of the castings that broke into two or more parts in service. The casting pictured in Fig. 17(a) broke apart in four places. In Fig. 17(b), the separation second from the left and second from the right are complete breaks that occurred in service. The casting shown in Fig. 17(c) broke into three pieces in service.

Some of the castings had broken completely and were so badly corroded that it was impossible to gain information concerning the failure mechanism. Five of the castings appeared to be sound.

The unfailed fittings and unfailed areas of the other fittings were subjected to laboratory bend tests, which produced the two end and two central fractures shown in Fig. 17(b) and the fractures of which surfaces are shown in Fig. 17(d), (e), and (f).

The fractures resulting from the bend tests were of three types:

- Fractures of samples in which cracking had been initiated in service
- Brittle fractures through apparently sound areas containing large dross inclusions,

some as great as 90% of the cross-sectional area
- Ductile fractures through apparently sound areas containing small casting defects

Figure 17(d) shows a macrograph of a surface of a stress-corrosion crack (arrows) that had propagated most of the way through the casting. The lighter area along the lower edge and at the right was a ductile fracture produced in a bend test in the laboratory. Evidence of corrosion and dezincification was found in the upper parts of the fracture surface.

Gas pockets, or blowholes, such as the one shown in Fig. 17(e), in otherwise ductile metal were also the cause of fracture of laboratory bend-test specimens.

Figure 17(f) shows a fracture surface containing large dross inclusions and extreme shrinkage porosity. The dark-gray area at bottom has been completely dezincified.

Figure 17(g) shows a section through a fracture surface of a broken brace-connection casting, illustrating an intergranular fracture path.

A micrograph of a section through a local area of plug-type dezincification is shown in Fig. 17(h). The copper in the corrosion product was redeposited in a porous, friable, and weak mass on the metal surface. Cracks are clearly in evidence in the deposit.

A review of the fractures in 15 of the hardware castings showed that 26 fractures had occurred in service by SCC and dezincification and that four had been produced in laboratory bend tests. One service fracture and six laboratory fractures had occurred in areas that had been weakened by inclusions or shrinkage porosity. Two of the laboratory fractures had taken place near letter impressions that reduced the cross-sectional area. Ductile fracture had occurred in 14 instances during laboratory bend testing. One ductile fracture had occurred in an area containing blowholes.

Bolting Material. Only two of the 14 cracked bolts had broken during service. One had broken through the threaded portion, and one through the shank adjacent to the bolt head. Both fractures were brittle in nature. Laboratory bend tests failed to reveal other brittle areas in the bolts.

Corrosion attack in the form of dezincification was found at the innermost extremities of forging laps near the bolt heads and in the threaded sections. A micrograph of a section through one of the bolt threads that had been dezincified is shown in Fig. 17(j). Initial attack occurred in the β phase, followed by attack in the α (white) phase. The sequence of attack in the two-phase structure is more easily observed in Fig. 17(k), which shows at higher magnification the area enclosed in the rectangle in Fig. 17(j).

Conclusions. The castings and bolts failed by SCC caused by the combined effects of dezincification damage and applied stresses.

Although the fracture surfaces of the failed castings were badly damaged, laboratory exam-

Fig. 17 Copper-zinc alloy cooling-tower hardware that failed by SCC and dezincification

(a), (b), and (c) Photographs showing some of the castings that broke into two or more parts in service. 1/3×. In (b), separations other than those second from left and second from right were produced in laboratory bend tests. (d) Fractograph of a broken casting showing area of stress-corrosion crack that occurred in service (arrows) and area of ductile fracture produced in a laboratory bend test. 2½×. (e) Fractograph showing a gas pocket in a bend-test specimen. 2½×. (f) Fractograph showing dross inclusions and porosity. 2½×. (g) Micrograph of a section through a fracture surface showing intergranular fracture path. Etched with NH₄OH. 100×. (h) Micrograph of an unetched section through an area of plug-type dezincification. 100×. Note cracks in deposit. (j) Micrograph of an unetched section through a partly dezincified bolt thread showing initial attack in β phase and subsequent attack in α (white) phase. 50×. (k) Micrograph of area enclosed in the rectangle in view (j) showing more clearly the sequence of attack in this two-phase structure. 500×

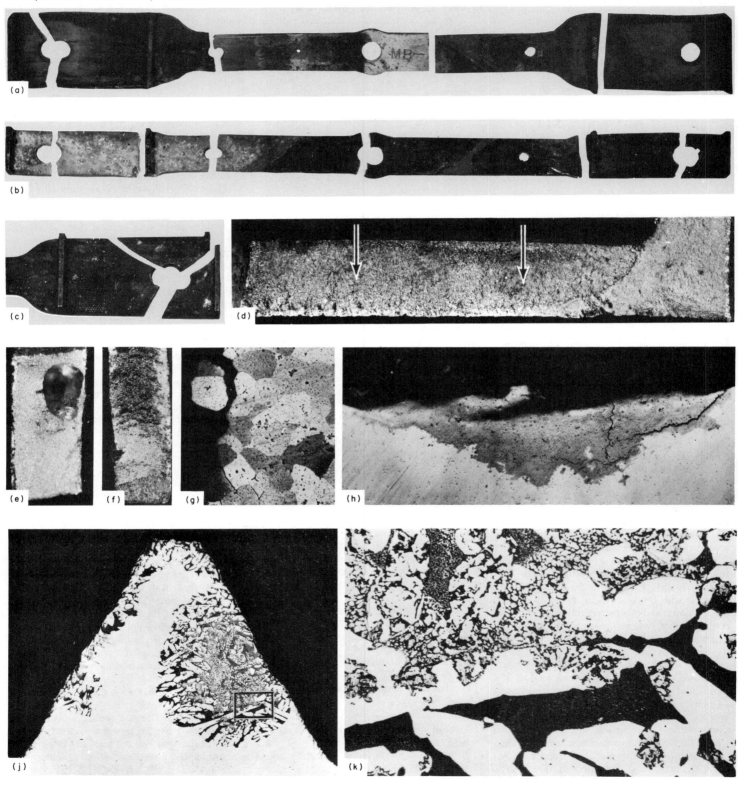

ination revealed the numerous areas containing SCC. Brittle fractures at dross inclusions showed evidence of dezincification, and ductile fractures were produced in the castings during bend tests. The bolts tested in the laboratory were generally of sound material, although evidence of dezincification was found in the threads and at the bases of small forging laps.

Corrective Measures. Because of the susceptibility of copper-zinc alloys that contain relatively high percentages of zinc (manganese bronze castings and naval brass fasteners) to SCC in the recirculating water in the cooling tower, the castings were replaced with copper alloy C87200 (cast silicon bronze) castings. Replacement bolts and nuts were made from copper alloy C65100 or C65500 (wrought silicon bronze).

The castings were inspected after 11 months of service, and no evidence of SCC was found. Visual examination after approximately 24 months of service revealed some red stain on the surfaces of the castings and bolts. Metallographic examination revealed the stains to be superficial. Bend tests indicated that there had been no loss of strength in the material and that no cracks had developed. The replacement hardware made of silicon bronze was in service for more than 16 years with no reported failures.

Magnesium Alloys

Commercial alloys of magnesium are generally resistant to failure by SCC in their major applications, in which they are exposed to the atmosphere.

Effects of Environment and Alloy. Exposure to the atmosphere has been shown to have little effect on SCC of magnesium alloys, whether in marine, industrial, or rural environments. High-purity magnesium as well as magnesium alloys containing manganese, rare earths, thorium, or zirconium can be made to fail by SCC only by exposure to such corrodents as dilute aqueous fluoride solutions at stresses higher than those encountered in normal service. Experience and testing of the commonly used commercial alloys have shown that in atmospheric exposure, only wrought ZK60A [5.5% Zn, 0.45% Zr (min)] and the alloys containing more than approximately 1.5% Al are significantly susceptible to SCC.

Effect of Product Form. When judged on the basis of safe stress as a fraction of tensile yield strength, castings are less susceptible to SCC than wrought products of the same composition. Although wrought magnesium alloy products usually show some anisotropy in their mechanical properties, their resistance to SCC is not appreciably influenced by the direction of applied stress in relation to the direction of working.

Test Methods. There is no standard accelerated test recommended for assessing the susceptibility of magnesium alloys to SCC; exposure of stress specimens to the atmosphere

has generally been used to determine stress-corrosion susceptibility of specific products. In most tests, spring-loaded constant-tension fixtures have been used, and results of these tests have been supplemented by atmospheric tests on structures designed to simulate fabrication stresses.

Path of cracking in magnesium alloys is generally transgranular, but in magnesium-aluminum alloys, intergranular cracking sometimes occurs. The crack path may change with changes in environment for a given alloy and processing and with changes in alloy processing for a given alloy composition and environment.

Causes and Preventive Measures. Stress-corrosion failures of magnesium alloy structures in service, which are infrequent, are usually caused by residual tensile stresses introduced during fabrication. Sources of such stresses are restrained weldments, interference fits, and casting inserts.

Effective preventive measures include stress-relief annealing of structures that contain residual stresses produced by welding or other methods of fabrication, cladding with a metal or alloy that is anodic to the base metal, and applying protective inorganic coatings, paint, or both to the surface.

Nickel and Nickel Alloys

Commercially pure wrought nickel is very resistant to SCC in various chloride salts and has excellent resistance to all the nonoxidizing halides. Oxidizing acid chlorides, such as ferric, cupric, and mercuric, are extremely corrosive and should be used only in low concentrations with wrought nickel. Stannic chloride is less strongly oxidizing, and dilute solutions at room temperature are resisted by commercially pure wrought nickel.

The susceptibility of austenitic nickel-chromium alloys to transgranular cracking in chloride solutions generally decreases as the nickel content of the alloy is increased. Inconel 600, which has a minimum nickel content of 72%, is virtually immune to chloride-ion SCC. This alloy is subject to SCC in high-temperature caustic alkalis in high concentrations. When the alloy is used in this type of service environment, it should be fully stress relieved before use, and operating stresses should be kept to a minimum. Stress-corrosion cracking may also occur in the presence of mercury at elevated temperatures. The recommendation given for caustic-alkali service should be followed if the alloy is used in an application that involves contact with mercury at elevated temperatures.

Inconel 601 has exhibited good resistance to SCC in various corrosive media, including 45% magnesium chloride and 10 to 98% concentrations of sodium hydroxide. Nickel-chromium-molybdenum alloys (e.g., alloy C-276, alloy 625) are often employed for their resistance to SCC in many industrial chemicals.

Table 4 Environments and temperatures that may be conducive to SCC of titanium alloys

Environment	Temperature
Nitric acid, red fuming	Ambient
Hot dry chloride salts	260-480 °C (500-900 °F)
Cadmium, solid and liquid	Ambient to 400 °C (750 °F)
Chlorine	Elevated
Hydrogen chloride	Elevated
Hydrochloric acid, 10%	Ambient to 40 °C (100 °F)
Nitrogen tetroxide	Ambient to 75 °C (165 °F)
Methyl and ethyl alcohols	Ambient
Seawater	Ambient
Trichloroethylene	Elevated
Trichlorofluorethane	Elevated
Chlorinated diphenyl	Elevated

Titanium and Titanium Alloys

Titanium is an inherently active metal that forms a thin protective oxide film. The apparent stability and high integrity of the film in environments that cause SCC of most structural alloys make titanium and its alloy resistant to SCC in boiling 42% magnesium chloride and boiling 10% sodium hydroxide solutions, which are commonly used to induce SCC in stainless steels.

Effects of Environment and Temperature. A number of environments in which some titanium alloys are susceptible to SCC and the temperatures at which cracking has been observed are listed in Table 4. Some of these environments are discussed in the following paragraphs.

Hot Dry Chloride Salts. Hot-salt SCC of titanium alloys is a function of temperature, stress, and time of exposure. In general, hot-salt cracking has not been encountered at temperatures below about 260 °C (500 °F); greatest susceptibility occurs at about 290 to 425 °C (about 550 to 800 °F), based on laboratory tests. Time-to-failure decreases as either temperature or stress level is increased. All commercial alloys, but not unalloyed titanium, have some degree of susceptibility to hot-salt cracking.

Cadmium. Titanium alloys crack when in contact with cadmium at a temperature of 320 °C (610 °F) or higher if the protective oxide film on the titanium alloy is ruptured. These alloys will crack at ambient temperatures when solid cadmium is pressed tightly against the titanium alloy with sufficient force to rupture the oxide film, thus permitting direct metal-to-metal contact, provided the titanium alloy is under relatively high tensile stress.

Chlorine, Hydrogen Chloride, and Hydrochloric Acid. In these environments, the mechanism of cracking is not completely understood, although it appears that both oxygen and water must also be present for cracking to occur.

Nitrogen tetroxide (N_2O_4) containing small amounts of dissolved oxygen causes cracking of titanium and some titanium alloys. No cracking occurs if the nitrogen tetroxide contains a small percentage of nitric oxide (NO). The cracking may be transgranular, intergranular, or both, depending on alloy composition.

Methyl and ethyl alcohols containing small amounts of water, chloride, bromide, and iodide promote cracking at ambient temperatures. Greater concentrations of water inhibit cracking. Higher alcohols may induce cracking, but to a lesser extent; the longer the chain, the less reactive the alcohol becomes.

Seawater. Using standard smooth-surface U-bend and four-point loaded specimens, no susceptibility of titanium and its alloys to cracking in seawater and other chloride-containing solutions is found. Under plane-strain conditions using prenotched fatigue-cracked specimens, susceptibility to crack propagation depends on alloy composition and heat treatment. For example, alloys containing more than 6% Al are especially susceptible to rapid crack propagation in seawater. Tin, manganese, cobalt, and oxygen have adverse effects, but isomorphous β stabilizers, such as molybdenum, niobium, or vanadium, reduce or eliminate susceptibility to cracking.

REFERENCES

1. W.M. Pardue, F.H. Beck, and M.G. Fontana, Propagation of Stress-Corrosion Cracking in a Magnesium-Base Alloy as Determined by Several Techniques, *Trans. ASM*, Vol 54, 1961, p 539-548
2. M.G. Fontana and N.D. Greene, *Corrosion Engineering*, McGraw-Hill, 1967
3. B.F. Brown, Ed., *Stress-Corrosion Cracking in High-Strength Steels and in Titanium and Aluminum Alloys*, Naval Research Laboratory, Washington, 1972
4. "Standard Method of Mercurous Nitrate Test for Copper and Copper Alloys," B 154, *Annual Book of ASTM Standards*, Vol 02.01, ASTM, Philadelphia, 1984
5. H.H. Uhlig, *Corrosion and Corrosion Control*, 2nd ed., John Wiley & Sons, 1971
6. W.H. Ailor, Jr., *Handbook on Corrosion Testing and Evaluation*, John Wiley & Sons, 1971
7. M. Henthorne, *Corrosion Causes and Control*, Carpenter Technology Corp., Reading, PA, 1972
8. K. Bohnenkamp, Caustic Cracking of Mild Steel, in *Fundamental Aspects of Stress Corrosion Cracking*, R.W. Staehle, A.J. Forty, and D. van Rooyen, Ed., National Association of Corrosion Engineers, Houston, 1969, p 374-383

SELECTED REFERENCES

* R.D. Barer and B.F. Peters, *Why Metals Fail*, Gordon and Breach Science Publishers, 1970
* C.D. Beachem, Microscopic Fracture Processes, in *Fracture*, Vol 1, *Microscopic and Macroscopic Fundamentals*, H. Liebowitz, Ed., Academic Press, 1968
* F.H. Cocks, Ed., *Manual of Industrial Corrosion Standards and Control*, STP 534, ASTM, Philadelphia, 1974
* H.L. Craig, Ed., *Stress Corrosion Cracking of Metals—A State of the Art*, STP 518, ASTM, Philadelphia, 1972
* M.G. Fontana and R.W. Staehle, Ed., *Advances in Corrosion Science and Technology*, Vol 3, Plenum Press, 1973
* H. Godard, W.B. Jepson, M.R. Bothwell, and R.L. Kane, *The Corrosion of Light Metals*, John Wiley & Sons, 1967
* N.E. Hamner, Ed., *Corrosion Data Survey*, 5th ed., National Association of Corrosion Engineers, Houston, 1974
* J. Hochmann, J. Slater, and R.W. Staehle, Ed., *Stress-Corrosion Cracking and Hydrogen Embrittlement of Iron Base Alloys*, National Association of Corrosion Engineers, Houston, 1975
* H. Leidheiser, Jr., *The Corrosion of Copper, Tin, and Their Alloys*, John Wiley & Sons, 1971
* H.L. Logan, *The Stress Corrosion of Metals*, John Wiley & Sons, 1966
* J.C. Scully, *The Theory of Stress Corrosion Cracking in Alloys*, North Atlantic Treaty Organization, Brussels, 1971
* R.W. Staehle, A.J. Forty, and D. van Rooyen, Ed., *Fundamental Aspects of Stress Corrosion Cracking*, National Association of Corrosion Engineers, Houston, 1969
* J.K. Stanley, "The Current Situation on the Stress-Corrosion and Hydrogen Embrittlement of High Strength Fasteners," Paper No. 72-385, American Institute of Aeronautics and Astronautics, New York, 1972
* *Stress Corrosion Testing*, STP 425, ASTM, Philadelphia, 1967
* A.S. Tetelman and A.J. McEvily, Jr., Fracture of Structural Materials, in *Fracture Under Static Loading*, John Wiley & Sons, 1967
* J.A. Whittaker, *A Survey on the Stress Corrosion of Copper Based Alloys*, International Copper Research Association, New York, 1965

Liquid-Metal Embrittlement

M.H. Kamdar, Benet Weapons Laboratory,
U.S. Army Armament Research, Development, and Engineering Center

LIQUID-METAL EMBRITTLEMENT (LME) is the catastrophic brittle failure of a normally ductile metal when coated with a thin film of a liquid metal and subsequently stressed in tension. The fracture mode changes from a ductile to a brittle intergranular or brittle transgranular (cleavage) mode; however, there is no change in the yield and flow behavior or in the stress-strain curve of the solid metal. As shown in Fig. 1, embrittlement manifests itself as a reduction in fracture stress, strain, or both. Fracture can occur well below the yield stress of the solid. The stress needed to propagate a sharp crack or a flaw in liquid is significantly lower than that necessary to initiate a crack in the liquid-metal environment. In most cases, the initiation or the propagation of cracks appears to occur instantaneously, with the fracture propagating through the entire test specimen. The velocity of crack or fracture propagation has been determined and, in some cases, estimated to be 10 to 100 cm/s (4 to 40 in./s).

Examination of LME fracture surfaces shows complete coverage by the liquid metal. Liquid metal is in intimate contact with the solid and is usually difficult to remove. The fracture mode becomes apparent when special techniques are used to remove the liquid. It should be emphasized that gross amounts of liquid are not necessary; even micrograms of liquid lead can cause LME in 75-mm (3-in.) thick steel tubes (Fig. 2). The fracture is usually brittle intergranular, with little indication of crack branching or striation to indicate slow crack propagation.

The presence of liquid at the moving crack tip appears necessary for fast fracture. However, in high-strength brittle metals, the possibility exists that the crack initiated in liquid may propagate in a brittle manner in the absence of liquid. In low-strength metals, a transition may occur from brittle-to-ductile failure under similar circumstances. Although far less frequent than brittle fracture, embrittlement can also occur by a ductile dimpled rupture mode in certain steels, copper alloys, and aluminum alloys. The embrittlement then manifests itself as the degradation of the mechanical properties in the solid metal (see the section ''Mechanisms of Embrittlement'' in this article for a discussion of ductile-fracture LME). Fracture is not limited to polycrystalline metals and the pres-

Fig. 1 Effects of environment on the yield stress and strain-hardening rate on various iron-aluminum alloys tested in air and mercury-indium solutions

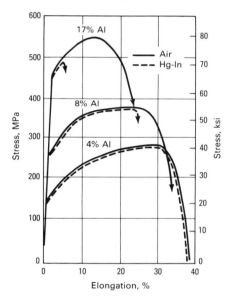

ence of grain boundaries. Single crystals of zinc and cadmium are also embrittled by liquid metal and result in fracture by a brittle cleavage mode (Fig. 3). These are but a few of the manifestations of LME.

Embrittlement is not a corrosion, dissolution, or diffusion-controlled intergranular penetration process, but is considered to be a special case of brittle fracture that occurs in the absence of an inert environment and at low temperatures. Time- and temperature-dependent processes are not considered to be responsible for the occurrence of LME. Also, in most cases of LME, little or no penetration of liquid metal into the solid metal is observed. The embrittlement of the solid metal coated with liquid metal or immersed in the liquid does not depend on the time of exposure to the liquid metal before testing (Fig. 4) or on whether the liquid is pure or presaturated with the solid. Usually, one of

Fig. 2 Liquid lead induced brittle fracture of a 75-mm (3-in.) section of a 4340 steel pressure vessel tube

Fig. 3 Cleavage fracture of cadmium monocrystals at 25 °C (75 °F) following a coating with Hg-60In (at.%) solution

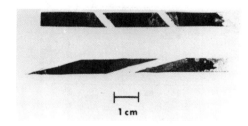

the prerequisites is that the solid has little or no solubility in the liquid and forms no intermetallic compound to constitute an embrittlement couple. However, exceptions to this empirical rule have been noted.

An increase in test temperature decreases embrittlement, which leads to a brittle-to-ductile transition. Severe embrittlement occurs near the freezing temperature of the liquid. In fact, in iron-lead, iron-indium (Fig. 5), and many other metallic couples, embrittlement

Fig. 4 Fracture stress of polycrystalline aluminum and Cu-30Zn brass as a function of exposure to liquid mercury before testing in this environment

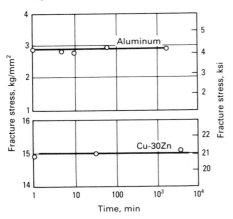

Fig. 5 Embrittlement of 4140 steel by various liquid metals below their melting point

Comparison of normalized true fracture strength (TFS) ratio and the reduction-of-area (RA) ratio as a function of homologous temperature T_H where T and T_M are the test and the melting temperature of the liquid metal, respectively. M.P., melting point

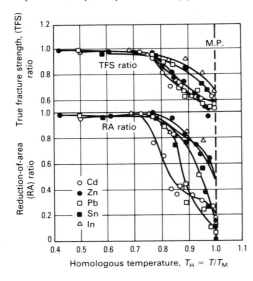

Fig. 6 Influence of lead on the fracture morphology of a 4340 steel

(a) Ductile failure after testing in argon at 370 °C (700 °F). (b) The same steel tested in liquid lead at 370 °C (700 °F) showing brittle intergranular fracture

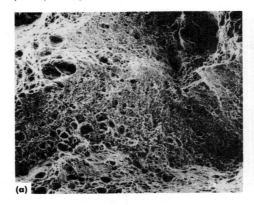
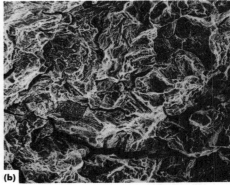

machinability may lead to embrittlement (Fig. 6 and 7). Internally leaded steels have cracked at lead inclusions. For example, leaded steel gears have cracked during induction-hardening heat treatments, and warm punching of leaded steel shafts has resulted in unexpected fracture during the forming operation

- Cadmium-plated titanium and steels are embrittled during high-temperature service by molten cadmium (Fig. 8 and 9)
- Indium, used as a high-vacuum seal in steel chambers, has caused cracking during bakeout operations
- Zircaloy tubes used in nuclear reactors have been cracked by both solid and liquid cadmium
- Although infrequent, LME also occurs in petrochemical plants and in the steel industry during heat treatment, hot rolling, brazing, soldering, and welding operations
- Embrittlement of steel occurs by electroplated or dipped cadmium, zinc, or tin—all of which provide corrosion resistance
- In liquid metal cooled reactors, liquid lithium can cause both corrosion and LME in metals and alloys

One of the more disastrous events associated with LME was an explosion at a chemical processing plant in Flixborough, England, in 1975, that killed 28 people. The explosion was caused by dripping zinc that came in contact with a 200-mm (8-in.) diam type 316L stainless steel pipe, which led to LME-induced cracking.

A nose landing gear socket assembly that failed catastrophically is shown in Fig. 10. As described in the figure caption, failure was attributed to tin/cadmium-induced LME.

The distinctive features of embrittlement and of the resultant fracture surfaces are a significant loss in mechanical properties and usually a brittle fracture mode, although ductile-fracture morphologies have been reported. The fracture surfaces and their appearances are easily distinguished from those due to stress-corrosion cracking (SCC), which are similar to fracture

due to hydrogen or temper embrittlement. The embrittlement is severe, and the propagation of fracture is very fast in the case of LME as compared to that in SCC. Thus, the predominant occurrence of brittle fracture surfaces, fast or catastrophic fracture, significant loss in ductility and strength, and the presence of liquid at the tip of the propagating crack are some of the characteristics that may be used to distinguish LME from other environmentally induced failures.

Detailed information on the mechanisms of LME and the susceptibility of specific metals and alloys to liquid metal induced failures can be found in the Selected References provided at the end of this article. The article "Failures of Locomotive Axles" in this Volume also discusses LME.

Mechanisms of Embrittlement

Several mechanisms have been proposed to explain LME, including stress-assisted dissolution of the solid at the crack tip and reduction in the surface energy of the solid by the liquid metal. It has been suggested that embrittlement is associated with liquid-metal adsorption induced localized reduction in the strength of the atomic bonds at the crack tip or at the surface of the solid metal at sites of stress concentrations.

With this possibility in mind, consider the crack shown in Fig. 11. Crack propagation will occur by the breaking of A-A_0 bonds at the crack tip and, subsequently, the breaking of similar bonds at the propagating crack tip by the chemisorbed liquid-metal atom B (a vapor phase from a solid in a solid-solid metal embrittlement couple or an elemental gas, such as hydrogen, may also provide the embrittling atom B). Next, assume that liquid-metal atom B at the crack tip reduces the cohesive strength of A-A_0 bonds. The chemisorption process presumably occurs spontaneously or only after the A-A_0 bonds have been strained to some critical value. In any event, electronic rearrangement occurs because of adsorption, which

occurs below the melting temperature of the liquid. This new phenomenon is called solid metal induced embrittlement (SMIE) of metals and is described in the article so named in this Volume.

The occurrence of LME is not a laboratory curiosity. Such failures have been observed in metals and alloys in the following industrial applications and processes:

- Small amounts of alloying elements, such as lead and tellurium, added to steel to improve

Fig. 7 Macrophotograph of AISI 12L14 + Te free-cutting steel sample rolled within the embrittlement range (1030 °C, or 1885 °F) showing cracks on the surface

(a) Top view. (b) Side view. 9×

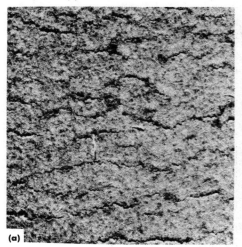

(a)

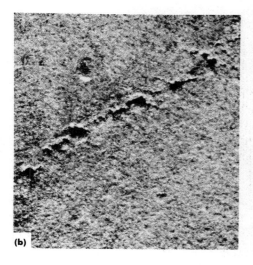

(b)

weakens the bonds at the crack tip. When the applied stress is increased so that it exceeds the reduced breaking strength of A-A$_0$ bonds, then the crack propagates. The liquid-metal atom becomes stably chemisorbed on the freshly created surfaces. The surface diffusion of the liquid-metal atoms over the chemisorbed liquid-metal atoms feeds the advancing crack tip, thereby propagating the crack at reduced stress and causing complete failure of the specimen.

In the above mechanism, it is assumed that reduction in the tensile cohesion is predominantly responsible for the occurrence of embrittlement, although both tensile as well as shear cohesion are reduced. Embrittlement can also occur by the adsorption-induced reductions in the shear strength of the atomic bonds at the

crack tip. The reduced shear strength facilitates nucleation of dislocations or slip at low stresses at or near the crack tip. This localized increase in plasticity produces a plastic zone with sufficiently large strains such that a void is nucleated ahead of the crack tip at precipitates, inclusions, or at subboundaries in a single crystal. The void will grow, and crack growth will occur. The fracture occurs by the ductile rupture mode, with the appearance of dimples on the fracture surfaces. The localized increased plasticity results in an overall reduction in the strain at failure, as compared to that in the absence of liquid metal, thus causing embrittlement. For embrittlement due to reduced shear cohesion, lower strain at failure or subcritical crack growth by linkage of voids is the measure of LME.

A schematic representation of the above process is given in Fig. 12, and supporting fractographic evidence is given for high-strength steel (Fig. 13a) and beryllium-copper alloys (Fig. 13b) broken in liquid-metal environments. Conversely, Fig. 13(c) to (e) provide strong support for the reduced tensile cohesion mechanism.

In general, most fractures in LME occur by a brittle intergranular or cleavage fracture mode at reduced tensile stresses and thus support the tensile decohesion mode. However, the dimpled ductile failure mechanism of LME should be considered as an embrittling process in a failure analysis in which the reduction in the mechanical parameters at fracture is the measure of embrittlement. Although LME can occur by either a brittle or a ductile fracture mode, both degrade the mechanical properties of the solid metal.

Role of Liquid in Crack Propagation

In the mechanism of embrittlement described above, it was implied that once a crack is nucleated subsequent crack propagation occurs mechanically with liquid absent at the crack tip or occurs by the continuous presence of the liquid-metal atoms at the propagating crack tip caused by the surface diffusion of liquid-metal atoms over chemisorbed liquid-metal atoms. The role of liquid in embrittlement should be investigated by measuring the crack-growth rate as a function of temperature and stress intensity. The velocity of crack propagation, or crack-growth rate, in SCC has been extensively investigated. However, such investigations in liquid metal have been reported only recently for brass (Fig. 14) and aluminum in liquid mercury. The velocity of crack growth in brass in liquid mercury was at least two orders of magnitude higher than that in an inert environment. The crack-propagation activation energy for brass in liquid mercury is only 3 to 5 kcal/mol (Fig. 15). This corresponds to diffusion of liquid mercury over mercury adsorbed on brass. The very high velocities of crack propagation and very low activation energies

are considered to be distinguishing features from similar investigations in SCC environments in the same metals. Such characteristics can be used to differentiate liquid metal induced crack propagation from stress-corrosion or hydrogen damage.

Occurrence of LME

Susceptibility to LME is unique to specific metals. For example, liquid gallium embrittles aluminum but not magnesium, and liquid mercury embrittles zinc but not cadmium. The equilibrium phase diagrams of most embrittlement couples show that they form simple binary systems with little or no solid solubility. Also, they are immiscible in the liquid state, and they usually do not form intermetallic compounds. This is an empirical observation. Although valid for many embrittlement couples, exceptions have been reported.

The severity of embrittlement is not necessarily related to the above criteria, but it is related to the chemical nature of the embrittling species. For example, zinc is more severely embrittled by liquid gallium than by mercury. In addition, the severity of embrittlement is related to the properties of the solid and depends upon such factors as strength, alloying elements, grain size, and strain rate. The severity or occurrence or nonoccurrence of embrittlement depends upon the type of test and whether the solid contains stress concentrators, such as pre-existing cracks or flaws, or is smooth and free of stress raisers. By increasing the possibilities of maintaining high stress concentrations, the susceptibility of the solid to LME is also increased. For example, smooth specimens of low-alloy low-strength steel are not embrittled by liquid lead, but the same steel containing a fatigue precrack is severely embrittled by liquid lead.

The most critical and mandatory condition for LME is that the liquid should be in intimate contact with the surface of the solid to initiate embrittlement and should subsequently be present at the tip of the propagating crack to cause brittle failure. Even a thin oxide film several angstroms thick may prevent interaction between the solid and the liquid so that embrittlement is not observed. Thus, whatever conditions promote intimate contact and enhanced wetting of the liquid by the solid, such as freshly created surfaces by plastic deformation in a liquid environment, breaking of oxide films, or other such factors, will lead to LME of the solid.

Certain other prerequisites must be fulfilled before fracture can initiate in a solid. For a ductile nonprecracked metal, these are an applied tensile stress, plastic deformation, and the presence of a stable obstacle to slip serving as a stress concentrator. This obstacle can be pre-existing, for example, a grain boundary, or can be created during deformation, for example, a twin or kink band.

Fig. 8 Service failure of a low-alloy steel nut by LME

Cadmium-plated, 4140 low-alloy steel (44 HRC) nuts were inadvertently used on bolts for clamps used to join ducts that carried hot (500 °C, or 930 °F) air from the compressor of a military jet engine. (a) The nuts were fragmented or severely cracked. (b) The fracture surfaces of the nuts were oxidized and tarnished; purple, blue, and golden-yellow colors were sometimes evident. (c) SEM revealed brittle intercrystalline fracture surfaces. (d) Energy dispersive x-ray analysis showed that the fracture surfaces were covered with a thin layer of cadmium. Failure was therefore attributed to LME by cadmium.

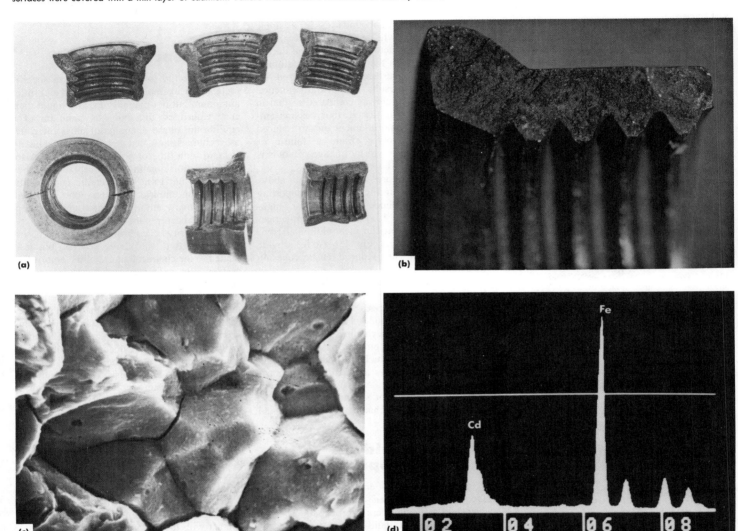

Fig. 9 Failed Ti-6Al-4V shear fastener

The fasteners were cadmium plated for galvanic compatibility with the aluminum structure. (a) Photograph showing failure at the head-to-shank fillet. (b) Intergranular fracture morphology. Failure was attributed to LME caused by excessive temperature exposure while under stress. SEM split screen. Left: 105×. Right: 1050×

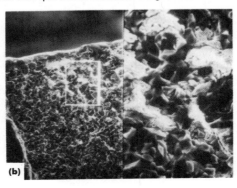

In addition, there should be an adequate supply of liquid metal to adsorb at the obstacle and subsequently at the propagating crack tip (only a few monolayers of liquid-metal atoms are necessary for LME). Plastic deformation or yielding may mean localized deformation in few grains rather than general yielding of the solid metal. If a specimen contains a pre-existing crack, then adsorption of the liquid at the crack tip and some tensile stress are necessary to propagate a crack. If the solid is notch brittle, LME will occur at reduced stress. However, it may not be necessary for the liquid to be present at the propagating crack tip, because cracks, once initiated, may propagate in a brittle manner in the absence of the liquid at the tip. Thus, the occurrence of LME in an embrittlement couple requires good wetting of the

Fig. 10 Failed nose landing gear socket assembly due to LME

(a) Overall view of the air-melted 4330 steel landing gear axle socket. Arrow A indicates the fractured lug; arrow B, the bent but unfailed lug. Arrow C indicates the annealed A-286 steel interference-fit plug containing the grease fitting that was removed from the fractured lug. (b) The segment of the fractured lug (A) that remained attached to the launch bar. (c) The forward section of the lug shown in (b) after removal from the launch bar. Arrow A indicates the primary fracture side; arrow B, the secondary overload side. (d) The fracture face looking at the forward end of the lug [arrow A in (c)]. Region A is the grease hole that accepts the interference-fit plug; the hole was inadvertently cadmium plated approximately three-fourths of its length during original part processing and before the required brush tin plate. Arrows B indicate approximate origins; arrows C, the extent of intergranular crack growth predominantly from liquid-metal embrittlement; Regions D, final catastrophic overload. (e) Shown are the interference-fit plug (arrow A) with a brush tin plate and a dry film (MoS$_2$) finish, the threaded grease fitting (arrow B), and the sealing washer (arrow C). Arrow D indicates an upset area that occurred during shrink-fit insertion of the plug. The upset area provided high contact pressure between the plug and the hole, resulting in a high hoop tension stress condition. (f) A typical section adjacent to the interference-fit hole (arrow B) showing intergranular crack growth and the liquid-metal diffusion of almost pure tin with a trace of cadmium at the grain boundaries (arrows C). SEM. 320×

(a)

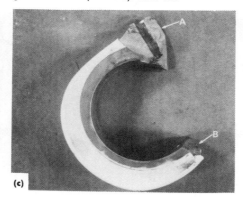

(c)

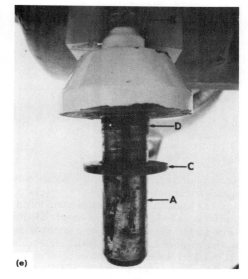

(e)

(b)

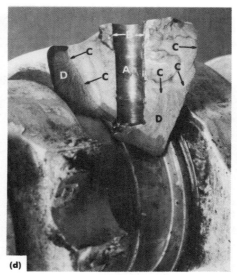

(d)

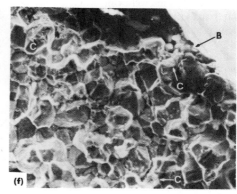

(f)

surface of the solid by the liquid, the presence of sufficient stress concentration to initiate or propagate a crack in liquid, and possible concurrence with the empirical observations mentioned above.

Effects of Metallurgical, Mechanical, and Physical Factors

It has been shown that the prerequisites for liquid metal induced brittle fracture are the same as those for brittle fracture in an inert environment at low temperatures, that is, the need for plastic flow. A barrier for crack nucleation, sufficient tensile stress, and adsorption of liquid metal are necessary to initiate and/or propagate a crack. From the investiga-

tions reported for many embrittlement couples, including classic zinc-mercury couples, it has been concluded that adsorption-induced embrittlement can be regarded as a special case of brittle fracture that normally occurs in brittle materials at low temperatures in an inert environment (ductile fracture can also result by the adsorption-induced reductions in the shear strength of the atomic bonds at the crack tip, as described in the section ''Mechanisms of Embrittlement'' in this article). Thus, the effects of yield stress, grain size, strain rate, temperature, and so on, in a liquid-metal environment follow trends similar to those noted for metals tested in an inert environment.

Effects of Grain Size. The grain size dependence of fracture stress has been investigated for zinc-mercury (Fig. 16), cadmium-gallium, brass and copper alloys in mercury,

and low-carbon steel in lithium. In these and other instances, the fracture stress varied linearly with the reciprocal of the square root of grain size and followed the well-known Cottrell-Petch relationship of grain size dependence on fracture stress. For the zinc-mercury couple shown in Fig. 16, fracture is nucleation controlled in region I, whereas in region II it is propagation controlled.

Nucleation-controlled embrittlement means that once a crack is initiated it propagates to failure in the presence or absence of the liquid metal at the crack tip. In propagation-controlled failure, microcracks are formed at some low stress in a liquid-metal environment, but the microcracks propagate to failure only when a higher stress is reached. Thus, in propagation-controlled failure, unpropagated microcracks may be found beside the main fracture, and

Fig. 11 Schematic illustrating displacement of atoms at the crack tip

The bond A-A₀ constitutes the crack tip, and B is the liquid-metal atom.

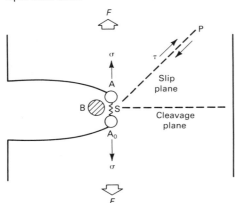

Fig. 12 Schematic illustrating the mechanisms of crack growth by microvoid coalescence

(a) Inert environment. (b) Embrittling liquid-metal environment

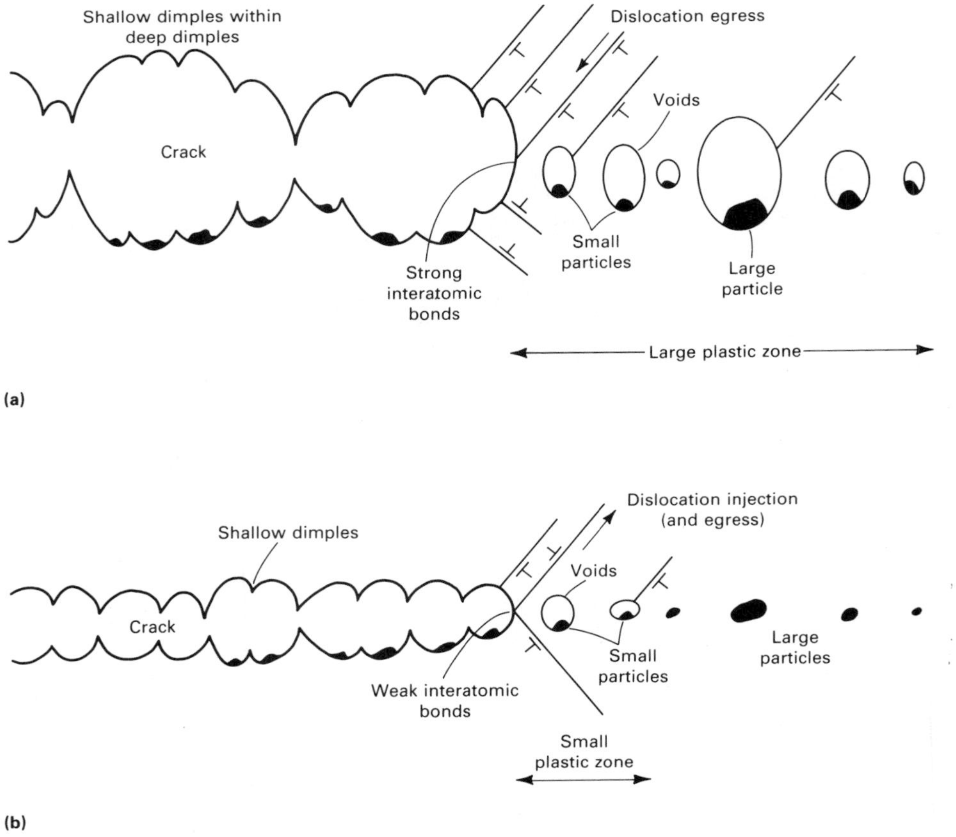

complete fracture of the specimen occurs only in the presence of liquid metal. The reverse is observed when fracture is initiation controlled.

Effects of Temperature and Strain Rate. Liquid-metal embrittlement occurs at the melting point of liquid metal (there are quite a few instances in which embrittlement occurs below the melting point of liquid by SMIE, which is described in the following article in this Volume). In most cases, the susceptibility to embrittlement may remain unchanged with temperature. At a sufficiently high test temperature, a brittle-to-ductile transition occurs when embrittlement ceases and ductility is restored to the solid. Such transitions do not occur at a sharply defined temperature and are not predictable on a theoretical basis or in terms of diffusion, dissolution, or other such embrittlement processes. It is generally accepted that a significant increase in ductility with temperature counteracts the inherent propensity for fracture in the embrittling liquid-metal environment.

In smooth steel specimens tested in liquid lead, brittle-to-ductile transition occurs at approximately 510 °C (950 °F), some 335 °C (600 °F) above the melting point of lead. In similar steel fracture mechanics type test specimens containing sharp notches tested in liquid lead, the transition occurs at approximately 650 °C (1200 °F), some 140 °C (250 °F) higher than that for smooth specimens. Transition temperature in this case depends upon the presence or absence of a stress raiser in the specimen and on the type of test method used. Such brittle-to-ductile transitions have been reported for zinc-mercury, aluminum-mercury, brass-mercury, titanium-cadmium, steel-lead, and steel-lead-antimony solution embrittlement couples.

Transition temperature also varies with strain rate and grain size in a manner observed for metals tested in an inert environment. Thus, an increase in strain rate and a decrease in grain size increase the transition temperature. The effects of grain size on transition temperature have been

reported for aluminum in Hg-3Sn solutions and brass-mercury embrittlement couples. The effects of strain rate on transition temperature have been reported for cadmium-gallium, zinc-mercury, aluminum-mercury, zinc-indium, brass-mercury, and titanium-cadmium embrittlement couples. Changes in the strain rate by orders of magnitude are usually required to change the transition temperature by 50 to 100 °C (90 to 180 °F). For example, in the titanium-cadmium couple, a change of approximately 60 °C (110 °F) occurs with a change of one to two orders of magnitude in the strain rate and corresponds to a comparable increase in yield stress with the strain rate. The effects of strain rate appear to be related to the increase in the yield stress. This corresponds to an increase in the embrittlement susceptibility and a decrease in the brittle-to-ductile transition temperature.

Inert Carriers and LME. In some cases, investigation of the embrittlement behavior of a potential solid-liquid metal couple is not possible. The reason is that the solid metal at temperatures just above the melting temperature of the liquid metal is either too ductile to initiate and propagate a brittle crack or the solid is excessively soluble in the liquid metal. It is possible, however, that a solid metal can be embrittled by a liquid metal of high melting

point by dissolving it in an inert-carrier liquid metal of a lower melting point. Thus, the embrittling species is effectively in a liquid state at some temperature far below its melting point. In addition, it provides a means of investigating the variation in the degree of embrittlement induced in a solid metal as a function of the concentration and as a function of the chemical nature of the several active liquid-metal species dissolved separately in a common inert-carrier liquid metal.

For example, mercury does not embrittle cadmium, but indium may embrittle cadmium (melting point: 325 °C, or 617 °F). However, cadmium is quite ductile at 156 °C (313 °F), the melting temperature of indium. Mercury dissolves up to 70% In at room temperature. The indium in the mercury-indium solution at room temperature embrittles cadmium (Fig. 17). In fact, cadmium single crystals, which are ductile even at liquid helium temperature, can be embrittled by indium and cleaved in indium-mercury solutions (Fig. 3). Another example of embrittlement by inert carriers is that of Zircaloy nuclear-fuel claddings used in liquid metal cooled nuclear reactors. In liquid cesium, Zircaloy claddings fail in a ductile manner; in cesium saturated with cadmium, Zircaloy is subject to brittle fracture.

Fig. 13 Ductile and brittle fracture morphologies resulting from LME

(a) Fracture surface produced by subcritical cracking in D-6ac steel (tempered at 650 °C, or 1200 °F) in liquid mercury showing predominantly dimpled intercrystalline fracture along prior-austenite grain boundaries. (b) Fracture surface produced by rapid subcritical crack growth (~1 mm/s, or 0.4 in./s) in a Cu-1.9Be alloy in liquid mercury at 20 °C (70 °F) showing predominantly dimpled intercrystalline fracture. (c) Macrograph of a cadmium-plated fastener made from 1040 steel. Aerodynamic heating during descent of the solid-fuel rocket engine resulted in brittle intergranular fracture. (d) Scanning electron micrograph of the failed machine screw shown in (c). (e) Fracture surface of a Monel specimen that failed in liquid mercury. The fracture is predominantly intergranular with some transgranular contribution.

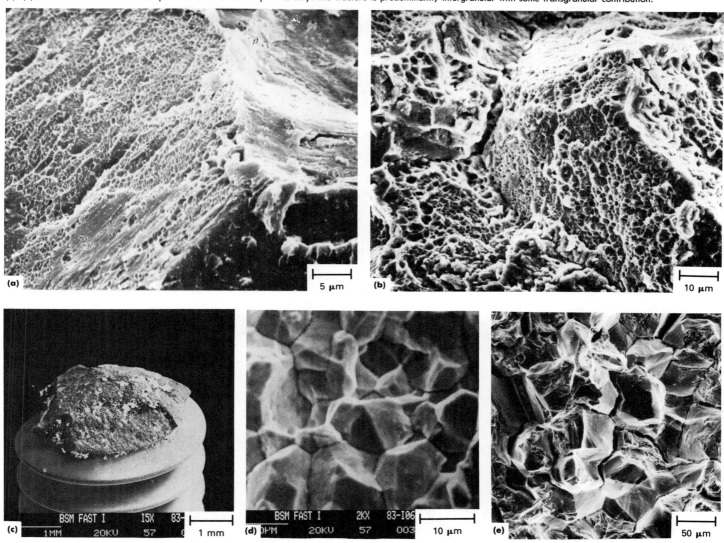

The concept of inert-carrier LME is useful in failure analysis where LME has occurred in a liquid solution, for example, solder rather than a pure liquid metal. The susceptibility of various species in the solder in causing LME can be investigated by incorporating each species in an inert liquid. This may suggest a means of preventing embrittlement by replacing the most potent embrittler with a nonembrittling species.

Fatigue in Liquid-Metal Environments

Most studies of LME have been concerned with the effects of tensile loading on fracture. Investigations of fatigue behavior are important because this test condition is more severe than other test conditions, including the tensile testing of metals. Thus, a solid tested in fatigue may become embrittled, whereas the same solid tested in tension may not exhibit embrittlement. This may be because at the high stress level corresponding to yield stress, which is a prerequisite for the occurrence of embrittlement in tough metals, the solid may be sufficiently ductile to prevent initiation of a crack in the liquid.

For example, smooth specimens of high-purity chromium-molybdenum low-alloy steel (yield stress: ~690 MPa, or 100 ksi) are not embrittled by liquid lead, but the same steel specimens containing a fatigue precrack are severely embrittled by liquid lead. The fatigue life in liquid lead is reduced to 25% of that in an inert argon environment. Fatigue testing of smooth specimens of copper-aluminum alloy in liquid mercury has revealed a significant loss in endurance limit relative to tests in air, with maximum fatigue stress decreasing with increasing grain size. Crack propagation in a liquid-metal environment under fatigue and tensile loading can be significantly different. The stress intensity at failure of 7.7 MPa$\sqrt{m}$ (7 ksi$\sqrt{in.}$) for a 4340 steel specimen containing a fatigue precrack tested in cyclic fatigue in liquid lead was five times lower than that for the same specimen tested in tension in static fatigue and was twenty times lower than that in an inert argon environment. Furthermore, the 7.7-MPa$\sqrt{m}$ (7-ksi$\sqrt{in.}$) stress intensity

Fig. 14 Crack-growth rate versus stress-intensity factor for brass in liquid mercury at various temperatures under load and displacement control conditions

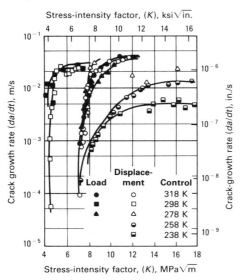

Fig. 15 Crack-growth rate versus reciprocal of temperature for brass in liquid mercury

The activation energy, Q, is 4 cal/mol.

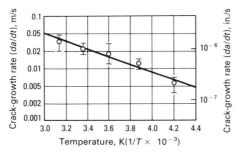

Fig. 16 Effect of grain size on LME

Variation of the flow stress of amalgamated zinc polycrystalline specimens, $\sigma_{f_{Zn}}$, and fracture stress of amalgamated zinc specimens, $\sigma_{F_{Zn-Hg}}$, as a function of grain size at 298 K

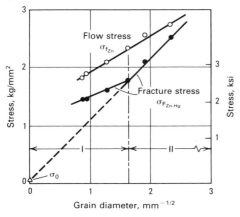

Fig. 17 Variation in ductility of polycrystalline cadmium as a function of indium content of mercury-indium surface coatings

Specimens tested at 25 °C (75 °F) in air and in mercury-indium solution

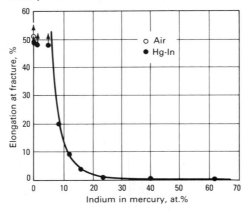

was the same whether the specimen had a machined notch (0.13-mm, or 0.005-in., root radius) or had a fatigue precrack at the root of the notch; that is, embrittlement was independent of root radius. Similar results have also been reported for the same steel tested in liquid mercury.

These results clearly indicate that fatigue testing of a notched or a fatigue precracked specimen provides the most severe test condition and therefore causes maximum susceptibility to embrittlement in a given liquid-metal environment. To determine whether a solid is susceptible to LME in a particular liquid, it is advisable to test the solid metal with a stress raiser or a notch in a tension-tension fatigue test.

Static fatigue or delayed-failure tests are conducted to study the effects of time-dependent processes, such as diffusion penetration of the grain boundary, as the cause of embrittlement. Also, such tests are performed because they simulate service conditions of constant stress in an embrittling environment as a function of time and assist in the failure analysis. Delayed failure has been reported for notch-insensitive aluminum-copper and beryllium-copper alloys in liquid mercury. Penetration of mercury did not occur, and embrittlement was independent of time-to-failure; but failure time increased with an increased stress level of the tests. In notch-sensitive zinc tested in mercury, the time-to-failure was very sensitive to the stress level, with complete failure occurring instantly at a stress slightly higher than a threshold stress. Penetration of mercury was not observed.

These tests indicate that time-dependent diffusion penetration or dissolution processes are not the cause of LME. In isolated instances, grain-boundary penetration by the liquid has been the cause of embrittlement. For example, unstressed aluminum or commercial aluminum alloys disintegrate into individual grains when

contacted with liquid gallium. Liquid antimony embrittles steel, and embrittlement increases with temperature; thus, diffusional processes cause antimony to embrittle steel. Liquid copper and liquid lithium embrittle steel in a similar manner (Fig. 18). However, in these cases, both diffusional and classic LME have been observed. Such behavior can be differentiated by the temperature dependence, by the susceptibility to embrittlement, or by embrittlement occurring in solid-metal environments (Fig. 19).

Failure Analysis of LME-Induced Fracture

Failure by LME is somewhat distinct and more easily recognized than that due to SCC, hydrogen embrittlement, or temper embrittlement of metals. The distinctive features of LME include very drastic degradation in the mechanical properties of ductile metals. This may mean no reduction of area at failure, failure occurring at very low stresses or strains, and significant reduction in the fracture toughness of the solid metal.

The fracture is fast, and the velocity of crack propagation is several orders of magnitude higher than that in the other embrittlement processes mentioned above or in an inert environment. Fracture usually occurs in a brittle intergranular mode, with less crack branching. Cleavage fracture occurs in hexagonal close-packed (hcp) and body-centered cubic (bcc) metals. The fracture surface is usually wetted with liquid metal, which is quite difficult to remove.

The fracture appearance or mode does not become evident until the liquid is removed by evaporation in a vacuum chamber, film stripping, chemical means, or other techniques. It is not necessary for the fracture surface to be coated with liquid. A thin film either a few atoms or a few microns thick can cause just as severe an embrittlement as when the surface is covered with large amounts of liquid. Energy-dispersive x-ray analysis in a scanning electron microscope or Auger electron spectroscopy (AES) should clearly identify the presence of thin layers of liquid on brittle fracture surfaces. The site or sites of initiation of fracture should be at the surface or at the liquid/solid metal interface, except when the source of the liquid is internal, such as inclusions or a second phase in the solid. In ductile metal, liquid will usually follow the propagating crack and should therefore be present on the entire fracture surface, including secondary cracks below the fracture surface. In brittle material or at high loading rates or high strain rates, fracture may initiate in the liquid-metal environment, propagate a short distance, and subsequently propagate mechanically in the absence of liquid metal. The subsequent fracture path can either be brittle or ductile, depending upon the nature of the material.

Decrease or Elimination of LME Susceptibility

The reduction in the cohesion mechanism of embrittlement indicates that the electronic interactions, resulting in possible covalent bonding due to electron redistribution between the solid- and the liquid-metal atoms, reduces cohesion and thereby induces embrittlement. Such interactions at the electronic level are the inherent properties of the interacting atoms and are therefore difficult or impossible to change in order to reduce the susceptibility to LME.

One possibility is to introduce impurity atoms in the grain boundary that have more affinity for sharing electrons with the liquid-metal atoms than for sharing electrons with the solid-metal atoms. For example, additions of phosphorus to Monel segregate to the grain boundaries and reduce the embrittlement of Monel by liquid mercury. Additions of lanthanides to internally leaded steels reduce lead embrittlement of steel. However, in general, the best alternative is to electroplate or clad the solid surface with a metal as a barrier between the embrittling solid-liquid metal couple, making sure that the barrier metal is not embrittled by the liquid metal.

Another possibility is that a ceramic or a covalent material coating on the solid surface will inhibit embrittlement. Apparently, only materials with metallic bonding are susceptible to LME. The severity of embrittlement could be reduced by decreasing the yield stress of the solid below the stress required to initiate a crack or by cladding with a high-purity metal of the alloy that is embrittled but that has a very low yield stress. Thus, Zircaloy, which is clad with a high-purity zirconium, becomes immune to embrittlement by cadmium. The obvious possibility is to replace embrittling liquid metal or solutions by nonembrittling metals or solutions.

Embrittlement of Nonferrous Metals and Alloys*

Zinc is embrittled by mercury, indium, gallium, and Pb-20Sn solder. Mercury decreases the fracture stress of pure zinc by 50% and dilute zinc alloys (0.2 at.% Cu or Ag) by five times that in an inert environment. The fracture-propagation energy for a crack in zinc single crystals in mercury is 60% and in gallium is 40% of that in air. Zinc is embrittled more severely by gallium than by indium or mercury.

Aluminum. Mercury embrittles both pure and alloyed aluminum. The tensile stress is decreased by some 20%. Fatigue life of 7075 aluminum alloy is reduced in mercury, and brittle-to-ductile transition occurs at 200 °C (390 °F). Additions of gallium and cadmium to

*Adapted from M.G. Nicholas, A Survey of Literature on Liquid Metal Embrittlement of Metals and Alloys, in *Embrittlement by Liquid and Solid Metals*, M.H. Kamdar, Ed., The Metallurgical Society of AIME, Warrendale, PA, 1984, p 27-50

Fig. 18 Copper-induced LME of 4340 steel

(a) and (b) Copper-plated specimen that was pulled at 1100 °C (2010 °F) in a Gleeble hot tensile machine showing liquid copper embrittlement of steel. (c) Scanning electron micrograph of the pulled specimen. 1600×. (d) Computer-processed x-ray map showing the presence of copper in prior-austenite grain boundaries

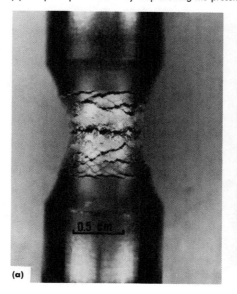
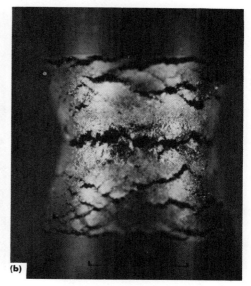
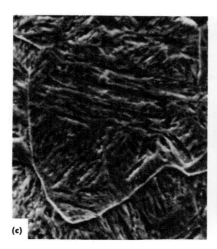
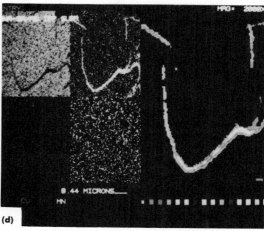

mercury increase embrittlement of aluminum. Delayed failure by LME occurs in mercury. Dewetting of aluminum by mercury has been found to inhibit embrittlement. The possible cause of dewetting is the dissolution of aluminum by mercury and oxidation of fine aluminum particles by air and formation of aluminum oxide white flowers at the aluminum/mercury interface.

Aluminum alloys are embrittled by tin-zinc and lead-tin alloys. The embrittlement susceptibility is related to heat treatment and the strength level of the alloy. Gallium in contact with aluminum severely disintegrates unstressed aluminum alloys into individual grains. Therefore, grain-boundary penetration of gallium is sometimes used to separate grains and to study topographical features and orientations of grains in aluminum. There is some uncertainty about whether zinc embrittles aluminum. However, indium severely embrittles aluminum. Alkali metals, sodium, and lithium are known to embrittle aluminum. Aluminum alloys containing either lead, cadmium, or bismuth inclusions embrittle aluminum when impact tested near the melting point of these inclusions. The severity of embrittlement increases from lead to cadmium to bismuth.

Copper. Mercury embrittles copper, and the severity of embrittlement increases when copper is alloyed with aluminum and zinc. Antimony, cadmium, lead, and thallium are also reported to embrittle copper. Apparent absence of embrittlement of copper, if observed, should be attributed to the test conditions and metallurgical factors discussed in this article. Mercury embrittlement of brass is a classic example of LME. Embrittlement occurs in both tension and fatigue and varies with grain size and strain rate. Mercury embrittlement of brass and cop-

Fig. 19 Solid-copper embrittlement of 4340 specimen tested at 1000 °C (1830 °F) in a Gleeble tensile-test machine

(a) Brittle failure due to SMIE. (b) Ductile failure of the same specimen in an inert environment. Both at 8×

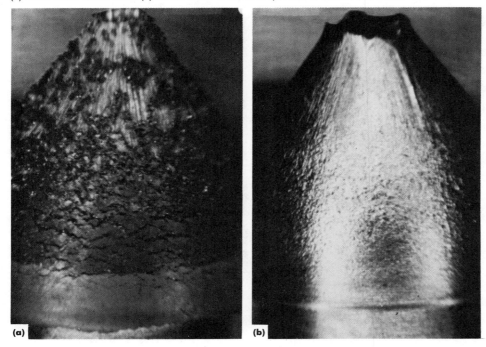

(a)   (b)

Fig. 20 Fracture face of an arc-cast W-25Re alloy with a fully recrystallized structure

Tensile specimen that was pulled in molten plutonium at 900 °C (1650 °F) at a strain rate of 2.5×10^{-4} s^{-1}. Intergranular attack contributes to individual grains becoming dislodged from the specimen. 250×

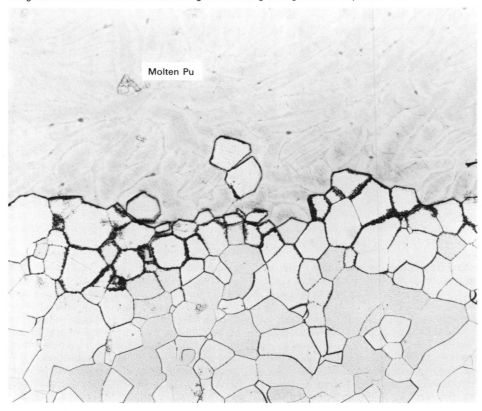

Molten Pu

per alloys has been extensively investigated. Lead and tin inclusions in brass cause severe embrittlement when tested near the melting point of these inclusions. Lithium reduces the rupture stress and elongation at failure of solid copper. Sodium is reported to embrittle copper; however, cesium does not. The embrittling effects of bismuth for copper and their alloys are well documented.

Gallium embrittles copper at temperatures ranging from 25 to 240 °C (75 to 465 °F). Gallium also embrittles single crystals of copper. Indium embrittles copper at 156 to 250 °C (313 to 480 °F).

Other Nonferrous Materials. Tantalum and titanium alloys are embrittled by mercury and by Hg-3Zn solution. Refractory metals and alloys, specifically W-25Re, molybdenum, and Ta-10W, are susceptible to LME when in contact with molten Pu-1Ga. Figure 20 shows a tungsten alloy specimen that failed due to intergranular LME. Cadmium and lead are not embrittled by mercury. However, indium dissolved in mercury embrittles both polycrystalline and single crystal cadium (Fig. 3). The embrittlement of titanium and its alloys by both liquid and solid cadmium is well recognized by the aircraft industry. Cadmium-plated fasteners of both titanium and steel are known to fail prematurely below the melting point of cadmium (Fig. 8 and 9). Cadmium-plated steel bolts have good stress-corrosion resistance, but cadmium is known to crack steel. Therefore, such bolts are not recommended for use. Zinc is reported to embrittle magnesium and titanium alloys. Silver, gold, and their alloys are embrittled by both mercury and gallium. Nickel is severely embrittled by cadmium dissolved in cesium. The fracture mode is brittle intergranular with bright grain-boundary fracture.

Failure of Zircaloy tubes used as cladding material for nuclear-fuel rods has been suspected to result from nuclear-interaction reaction products, such as iodine and cadmium carried by liquid cesium, which is used as a coolant in the reactor. Systematic investigation in the laboratory has shown that cadmium, both in the solid and the liquid state or as a carrier species dissolved in liquid cesium, causes severe liquid and solid metal induced embrittlement of zirconium (Fig. 21) and Zircaloy-2. Embrittlement of Zircaloy by calcium, strontium, zinc, cadmium, and iodine has also been reported.

Embrittlement of Ferrous Metals and Alloys

Embrittlement by Aluminum. Tensile and stress-rupture tests have been conducted on steels in molten aluminum at 690 °C (1275 °F). For short-term tensile tests, a reduced breaking stress and reduction of area were found as compared to the values in air. In the stress-rupture tests, the time-to-failure was dependent upon the applied stress.

Embrittlement by Antimony. AISI type 4340 steel tested in fatigue in liquid Pb-35Sb at

Fig. 21 Cleavage fracture of a zirconium crystal bar liner from a Zircaloy-2 fuel cladding due to cesium-cadmium LME

Failure occurred during a localized ductility test at a constant extension rate in uniaxial loading of 4.2 × 10⁻³ mm/s.

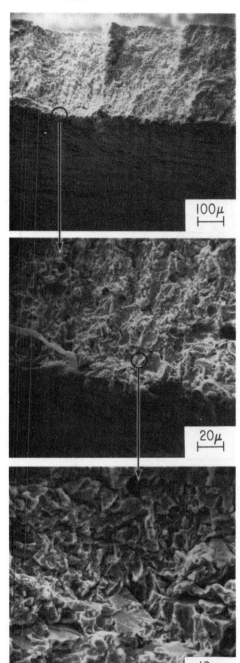

540 °C (1000 °F) and in antimony at 675 °C (1250 °F) was very severely embrittled. The embrittlement in lead-antimony occurred 165 to 220 °C (300 to 400 °F) higher than that observed for high-purity lead. Small additions of antimony (~5 wt%) to lead had no effect on embrittlement when tested in fatigue, although 0.002 to 0.2% Sb additions have caused a significant increase in the embrittlement of smooth 4140 steel specimens tested in tension. Embrittlement by antimony increases with temperature and is thought to occur by the grain-boundary diffusion of antimony in steel.

Embrittlement by Bismuth. Upon testing in liquid bismuth at 300 °C (570 °F), no embrittlement was noted in bend tests on a quenched-and-tempered steel. The stress-rupture data obtained on the low-carbon steel showed that the time-to-failure and reduction of area increased with decreasing load, but no intercrystalline attack was noted.

Embrittlement by Cadmium. In several studies of the embrittlement of low-alloy AISI type 4340 steel, embrittlement occurred at 260 to 322 °C (500 to 612 °F) but not at 204 °C (399 °F) for high-strength steel. Cracks were observed in samples loaded to 90% of their yield stress at 204 °C (399 °F). The threshold stress required for cracking decreases with an increase in temperature. Cracking at 204 °C (399 °F) was strongly dependent on the strength level of the steel, and embrittlement was not observed for strength levels less than 1241 MPa (180 ksi). Delayed failure occurs in cadmium-plated high-strength steels (AISI type 4340, 4140, 4130, and an 18% Ni maraging steel). Failures were observed at 232 °C (450 °F), which is about 90 °C (160 °F) below the melting point of cadmium. Static failure limits of 10% and 60% of the room-temperature notch strength have been reported for electroplated and vacuum-deposited cadmium, respectively, at 300 °C (570 °F). A discontinuous crack-propagation mode was observed, consisting of a series of crack-propagation steps separated by periods of no apparent growth. This slow crack-growth region was characterized by cracks along the prior-austenite grain boundaries. Once the cracks reached a critical size, a catastrophic failure occurred that was characterized by a transgranular ductile fracture.

Embrittlement of high-strength steels occurs when they are stress rupture tested in liquid and solid cadmium. Cadmium produces a progressive decrease in the reduction of area at fracture of AISI type 4140 steel over the temperature range of 170 to 321 °C (338 to 610 °F). Cadmium was identified as a more potent solid-metal embrittler than lead, tin, zinc, or indium.

Embrittlement by Copper. The embrittlement of low-carbon steel by copper plate occurred at 900 °C (1650 °F) during a slow-bend test. The embrittling effects of the copper plate exceeded those encountered with brazing alloys. Similar observations have been made for plain carbon steels, silicon steels, and chromium steels at 1000 to 1200 °C (1800 to 2190 °F).

The surface cracking produced during the hot working of some steels at 1100 to 1300 °C (2010 to 2370 °F) also has several characteristics of LME. It is promoted by surface enrichment of copper and other elements during oxidation and subsequent penetration along the prior-austenite grain boundaries. Elements such as nickel, molybdenum, tin, and arsenic that affect the melting point of copper or its solubility in austenite also influence embrittlement. A ductility trough has also been noted, with no cracking produced at temperatures above 1200 °C (2190 °F). Steel plated with copper and pulse heated to the melting point of copper in milliseconds was embrittled by both liquid and solid copper. Hot tensile testing in a Gleeble testing machine at high strain rates produced severe cracking in 4340 steel by both the liquid and solid copper (Fig. 18 and 19). Detailed information on copper-induced LME of steels can be found in the article "Failures of Locomotive Axles" in this Volume.

Embrittlement by Gallium. The alloy Fe-3Si is severely embrittled by gallium, as are the solid solutions of iron and 4340 steel.

Embrittlement by Indium. Indium embrittles pure iron and carbon steels. Embrittlement depends on both the strength level and the microstructure. Pure iron was embrittled only at temperatures above 310 °C (590 °F), appreciably above the melting point of indium. However, other steels, such as AISI type 4140 (ultimate tensile strength, UTS, of 1379 MPa, or 200 ksi), were embrittled by both solid and liquid indium. Surface cracks were detected at temperatures below the melting temperature of indium. This was interpreted as a local manifestation of the underlying embrittlement mechanism and it was assumed that the cracks must reach a critical size before the gross mechanical properties are affected.

Embrittlement by Lead. The influence of lead on the embrittlement of steel has been extensively investigated and has been found to be sensitive to both composition and metallurgical effects. The studies fall into two major classifications: (1) LME due to contact with an external source of liquid lead and (2) internal LME in which the lead is present internally as inclusions or a minor second phase, as in leaded steels. Both external and internal lead-induced embrittlement exhibit similar characteristics, but for the purpose of simplicity, each will be treated individually.

LME Due to External Lead. AISI type 4145 and 4140 steels exposed to pure lead exhibit classical LME, as shown by substantial decreases in both the reduction of area and the elongation at fracture. The fracture stress and the reduction of area decreased at temperatures considerably below the melting point of lead and varied continuously through the melting point, suggesting that the same embrittlement mechanism is operative for both solid- and liquid-metal environments.

Additions of zinc, antimony, tin, bismuth, and copper increase the embrittling potency of lead. Additions of up to 9% Sn, 2% Sb, or 0.5% Zn to lead increased the embrittlement of AISI type 4145 steel. In some cases, the embrittlement and failure occurred before the UTS was reached. The extent of embrittlement increases with increasing impurity content. No correlation was observed between the degree of embrittlement and wettability. The lead-tin alloys readily wetted steels, whereas the more embrittling lead-antimony alloys did not.

LME Due to Internal Lead. Leaded steels are economically attractive because lead increases the machining speed and the lifetime of the cutting tools. The first systematic investigation of embrittlement of leaded steels was reported in 1968 (Mostovoy and Breyer). Embrittlement characteristics similar to those promoted by external lead were observed. The degradation in the ductility began at approximately 120 °C (215 °F) below the melting point of lead, with the embrittlement trough present from 230 to 454 °C (446 to 849 °F). This was followed by a reversion to ductile behavior at about 480 °C (895 °F).

The severity of the embrittlement and the brittle-to-ductile transition temperature T_R have been shown to be dependent upon the strength level of the steel, with the degree of embrittlement and T_R increasing with strength level. In these studies, intergranular fracture was produced, which was propagation controlled at low temperatures and nucleation controlled at high temperatures. The degree of embrittlement was critically dependent upon the lead composition, and the influence of trace impurities completely masked any variations due to different carbon and alloy compositions of the steel. Lead embrittlement of a steel compressor disk was induced by bulk lead contents of 0.14 and 6.22 wt%. The lead is associated with the nonmetallic inclusions, and upon yielding, microcracks form at the weak inclusion/matrix interface, releasing a source of embrittling agent to the crack tip that aids subsequent propagation. An electron microprobe analysis of the nonmetallic inclusions identified the presence of zinc, antimony, tin, bismuth, and arsenic. With the exception of arsenic, all these trace impurities have been shown to have a significant effect on the external LME of steel.

The two most promising methods of suppressing LME are (1) control of sulfide composition and morphology and (2) cold working of the steel. The addition of rare-earth elements to the steel melt modifies the sulfide morphology and composition and can eliminate LME.

Embrittlement by Lithium. Exposure of AISI type 4340 steel to lithium at 200 °C (390 °F) resulted in static fatigue, with the time-to-failure depending on the applied stress. A decreasing fracture stress and elongation to fracture were noted with increasing UTS of variously treated steels, and catastrophic failure occurred for those steels with tensile strengths exceeding 1034 MPa (150 ksi). The tensile ductility of low-carbon steel at 200 °C (390 °F) was drastically reduced in lithium, with intergranular failure after 2 to 3% elongation, but there was no effect on the yield stress or the initial work-hardening behavior. The fracture

stress was shown to be a linear function of $d^{-1/2}$, where d is the average grain diameter, in accordance with the Petch relationship.

Embrittlement by Mercury. It has been shown that mercury embrittlement is crack nucleation controlled and can be induced in low-carbon steel samples by the introduction of local stress raisers. The fracture toughness of a notched 1Cr-0.2 Mo steel was significantly decreased upon testing in mercury. The effective surface energy required to propagate the crack was 12 to 16 times greater in air than in mercury. The fatigue life of 4340 steel in mercury is reduced by three orders of magnitude as compared to that in air.

The addition of solutes (cobalt, silicon, aluminum, and nickel) to iron, which reduced the propensity for cross-slip by decreasing the number of active slip systems and changed the slip mechanism from wavy to planar glide, increased the susceptiblity to embrittlement. Iron alloys containing 2% Si, 4% Al, or 8% Ni and iron containing 20% V or iron containing 49% Co and 2% V have been shown to be embrittled by mercury in unnotched tensile tests. The degree of embrittlement behavior was extended to lower alloy contents by using notched samples. No difference in the embrittlement potency of mercury or a saturated solution of indium in mercury was noted.

Effect of Selenium. Selenium had no embrittling effect on the mechanical properties of a quenched-and-tempered steel (UTS: ~1460 MPa, or 212 ksi) bend tested at 250 °C (480 °F).

Embrittlement by Silver. Silver had no significant effect on the mechanical properties of a range of plain carbon steels, silicon steels, and chromium steels tested by bending at 1000 to 1200 °C (1830 to 2190 °F). However, a silver-base filler metal containing 45% Ag, 25% Cd, and 15% Sn has been reported to embrittle A-286 heat-resistant steel in static-load tests above and below 580 °C (1076 °F), the melting point of the alloy.

Effect of Sodium. Unnotched tensile properties of low-carbon steel remained the same when tested in air and in sodium at 150 °C (300 °F) and 250 °C (480 °F). Similarly, Armco iron, low-carbon steel, and AISI type 316 steel were not embrittled by sodium at 150 to 1600 °C (300 to 2910 °F).

Embrittlement by Solders and Bearing Metals. A wide range of steels are susceptible to embrittlement and intercrystalline penetration by molten solders and bearing metals at temperatures under 450 °C (840 °F). Tensile tests have revealed embrittlement as a reduction in ductility. The embrittlement increased with grain size and strength level of the steel, except for temper-embrittled steels.

The tensile strength and ductility of carbon steel containing 0.13% C were decreased upon exposure to the molten solders and bearing alloys. The embrittlement was concomitant with a change to a brittle intergranular fracture mode and penetration along prior-austenite

grain boundaries. Ductile failure was observed with samples tested in air or in the liquid metal at temperatures exceeding 450 °C (840 °F). No intercrystalline penetration of solder was noted in carbon steels containing 0.77% and 0.14% C at 950 °C (1740 °F).

It has been reported that solder embrittles steel more than Woods metal (a bismuth-base fusible alloy containing lead and tin), particularly if it contains 4% Zn. The bearing metals produced embrittlement similar to the solder 4% Zn alloy.

Embrittlement by Tellurium. Tellurium-associated embrittlement has been reported for carbon and alloy steels (Fig. 7). Hot shortness occurs in AISI 12L14 + Te steel, with the most pronounced loss in ductility between 810 and 1150 °C (1490 and 2100 °F), embrittlement being most severe at 980 °C (1795 °F). The embrittlement has been shown to occur by the formation of a lead-telluride film at the grain boundary, which melts at 923 °C (1693 °F). The mechanical-test data and the examination of fracture surfaces by AES and SEM indicated LME of steel by the lead-telluride compound.

Effect of Thallium. Thallium had no embrittling effects on the mechanical properties of a quenched-and-tempered steel (UTS: ~1460 MPa, or 212 ksi) tested in bending at 325 °C (615 °F).

Embrittlement by Tin. The embrittling effect of tin has been observed in a range of austenitic and nickel-chromium steels, the degree of embrittlement increasing with their strength level. Embrittlement depends on the presence of a tensile stress and is associated with intercrystalline penetration.

The embrittlement by solid tin occurs at approximately 120 °C (215 °F) below its melting point. The fracture surfaces exhibited an initial brittle zone perpendicular to the tensile axis that followed the prior-austenite grain boundaries. Layers of intermetallic compound present at the steel/tin interface did not impede the embrittlement process. Embrittlement has been observed in delayed-failure tests down to 218 °C (424 °F), (14 °C, or 25 °F, below the melting point of tin); however, in tensile tests, embrittlement by solid tin was effective at temperatures as low as 132 °C (270 °F), which is 100 °C (180 °F) below melting point.

AISI type 3340 steel doped with 500 ppm of phosphorus, arsenic, and tin has been tested in the presence of tin while in the segregated (temper embrittled) and the unsegregated states. Temper-embrittled steels were found to be more susceptible to embrittlement than steel heat treated to a nontempered state.

A lower fatigue limit and lifetime at stresses below the fatigue limit for low-carbon steel and for 18-8 stainless steel have been noted when tested in tin at 300 °C (570 °F). The exposure time to the tin before testing had no influence on the fatigue life of the steel.

Embrittlement of Austenitic Steels by Zinc. Two main types of interaction of zinc and austenitic stainless steel have been observed.

Fig. 22 Brittle cracks induced in 4140 steel by liquid zinc at 431 °C (808 °F)

See also Fig. 20.

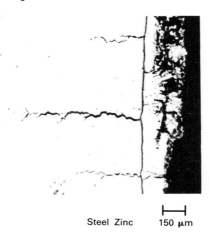

Steel Zinc   ⊢—⊣ 150 μm

Fig. 23 Liquid metal induced crack in 4140 steel at 431 °C (808 °F)

(a) Crack magnified at 140×. (b) Higher magnification (265×) of the circled area shown in (a). The mottled material in the crack is zinc. (c) SEM x-ray fluorescence of zinc from the crack at 1150×. The field represented in (c) is shown by the insert in (b). The concentrated white dots are zinc. The randomly scattered white dots are background noise.

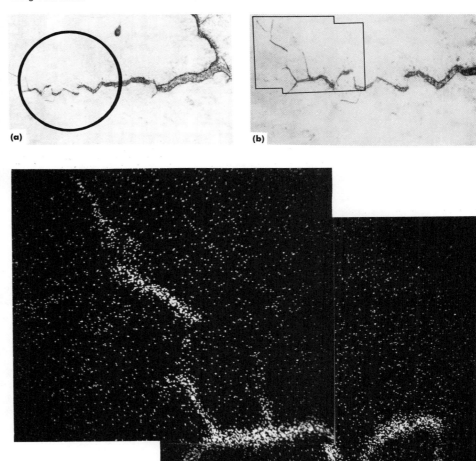

Type I relates to the effects on unstressed material in which liquid-metal penetration/erosion is the major controlling factor, and Type II relates to stressed materials in which classic LME is observed.

Type I Embrittlement. Zinc slowly erodes unstressed 18-8 austenitic stainless steel at 419 to 570 °C (786 to 1058 °F) and penetrates the steel, with the formation of an intermetallic β nickel-zinc compound at 570 to 750 °C (1060 to 1380 °F). At higher temperatures, penetration along the grain boundaries occurs, with a subsequent diffusion of nickel into the zinc-rich zone. This results in a nickel-exposed zone adjacent to the grain boundaries, reducing the stability of the γ phase and causing it to transform to an α-ferrite; the associated volume change of the γ → α transformation produces an internal stress that facilitates fracture along the grain boundaries. Similar behavior has been observed in an unstressed 316C stainless steel held 30 min at 750 °C (1380 °F), in which penetration occurred to a depth of 1 mm (0.4 in.), and in an unstressed AISI type 321 steel held 2 h at 515 °C (960 °F), in which a penetration of 0.127 mm (0.005 in.) was observed.

Type II embrittlement occurs in stainless steel above 750 °C (1380 °F) and is characterized by an extremely fast rate of crack propagation that is several orders of magnitude greater than that of Type I, with cracks propagating perpendicular to the applied stress. In laboratory tests, an incubation period was observed before the propagation of Type II cracks, suggesting that they may be nucleated by Type I cracks formed during the initial contact with zinc.

At 800 °C (1470 °F), a stressed 316C stainless steel failed catastrophically when coated with zinc. Cracking was produced at a stress of 57 MPa (8 ksi) at 830 °C (1525 °F), and 127 MPa (18 ksi) at 720 °C (1330 °F), but failure was not observed at a stress of 16 MPa (2 ksi) at 1050 °C (1920 °F).

Liquid-metal embrittlement may be produced by the welding of austenitic steels in the presence of zinc or zinc-base paints. Intercrystalline cracking has been observed in the heat-affected zone in areas heated from 800 to 1150 °C (1470 to 2100 °F), and electron microprobe analysis has been used to identify the grain-boundary enrichment of nickel and zinc, together with the formation of a low-melting nickel-zinc compound. The embrittlement of sheet samples of austenitic steel coated with zinc dust dye and zinchromate primer occurs at stresses of the order of 20 MPa (3 ksi).

Embrittlement of Ferritic Steels by Zinc. Embrittlement of certain ferritic steels and Armco iron by molten zinc has been reported in the temperature range of 400 to 620 °C (750 to 1150 °F). Long exposures and intercrystalline attack were needed to cause a reduction in the elongation to fracture; an iron-zinc intermetallic layer was formed that inhibited embrittlement until the layer was ruptured.

High-alloy ferritic steels exhibit embrittlement by zinc at temperatures above 750 °C (1380 °F).

Delayed failure occurs in steel in contact with solid zinc at 400 °C (750 °F), which is 19 °C (34 °F) below the melting point of zinc. The slow crack-growth region is characterized by intergranular mode of cracking.

Exposing AISI type 4140 steel to solid zinc results in a decrease in the reduction of area and in fracture stress at 265 °C (510 °F), with no significant changes in the other mechanical properties. The tensile fracture initially propagates intergranularly, with the final failure occurring by shear. Liquid zinc embrittles 4140 steel at 431 °C (808 °F), and it has been shown that zinc is present at the crack tip (Fig. 22 and 23). Figure 24 shows a compilation of liquid-metal embrittling and nonembrittling couples based on the theoretical calculations of the solubility parameter and the reduction in the fracture-surface energy. The embrittlement curve is separated by brittle and ductile frac-

Table 1 Summary of embrittlement couples

Solid		Hg		Cs	Ga		Na	In		Li	Sn		Bi		Tl	Cd	Pb		Zn	Te	Sb	Cu
		P	A	P	P	A	P	P	A	P	P	A	P	A	P	P	P	A	P	P	P	P
Sn	P	X	...	...	X	...	...	...	...	...	...	...	...	...	...	...	...	...	...	...	...	...
Bi	P	X	...	...	...	...	...	...	...	...	...	...	...	...	...	...	...	...	...	...	...	...
Cd	P	...	...	X	X	X	...	...	...	X	...	...	...	...	...	...	...	...	...	...	...	...
Zn	P	X	X	...	X	X	...	X	X	...	X	...	...	...	...	X	...	...	...	...	...	...
	LA	...	X	...	...	...	...	...	...	...	...	...	...	...	...	...	...	...	...	...	...	...
Mg	CA	...	...	...	...	...	X	...	...	...	...	...	...	...	...	...	...	...	X	...	...	...
Al	P	X	X	...	X	...	X	...	...	...	X	...	X	...	...	X	X	...	...	...	...	...
	CA	X	X	...	X	...	X	X	...	...	X	...	X	X	...	X	...	...	X	...	...	...
Ge	P	...	...	...	X	...	X	...	...	...	X	...	X	...	X	X	X	...	...	...	X	...
Ag	P	X	X	...	X	X	...	...	...	X	...	...	...	...	...	...	...	...	...	...	...	...
	LA	...	...	...	...	...	...	...	...	X	...	...	...	...	...	...	...	...	...	...	...	...
Cu	CP	X	X	...	...	...	X	...	...	X	...	...	X	X	...	X	...	...	...	...	...	...
	LA	X	...	...	X	...	...	...	...	...	...	...	...	...	...	...	...	...	...	...	...	...
	CA	X	X	X	...	X	...	X	X	X	X	(?)	X	...	...	...	...	...	...	...	...	...
Ni	P	X	...	...	...	...	...	...	...	X	...	...	...	...	...	X	...	...	...	...	...	...
	LA	...	...	...	...	...	...	...	...	X	...	...	...	...	...	...	...	...	...	...	...	...
	CA	...	...	...	...	...	...	...	X	...	...	...	...	...	...	X	...	...	...	...	...	...
Fe	P	...	...	...	...	...	...	...	...	...	...	...	...	...	...	...	...	...	...	...	...	X
	LA	X	X	...	X	...	...	X	X	X	...	...	...	...	...	...	...	...	...	...	...	...
	CA	...	...	...	...	...	...	X	...	X	X	...	...	...	...	...	X	X	X	X	X	X
Pd	P	...	...	...	...	...	...	...	...	X	...	...	...	...	...	...	...	...	...	...	...	...
	LA	...	...	...	...	...	...	...	...	X	...	...	...	...	...	...	...	...	...	...	...	...
Ti	CA	...	X	...	...	...	...	...	...	...	...	...	...	...	...	...	X	...	...	...	...	...

P, element (nominally pure); A, alloy; C, commercial; L, laboratory

Fig. 24 Calculated reduction in the fracture-surface energy relating to solubility parameter for many solid-liquid embrittlement couples

Note that the curve separates embrittlement couples from nonembrittled solid-liquid metal couples.

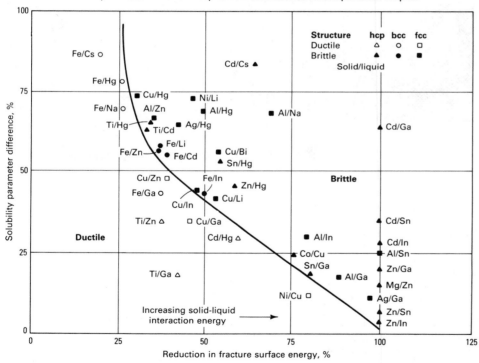

search and Development; and C.E. Price, Oklahoma State University, for providing various photomicrographs used in this article.

SELECTED REFERENCES

- M.H. Kamdar, Ed., *Embrittlement by Liquid and Solid Metals*, The Metallurgical Society of AIME, Warrendale, PA, 1984
- M.H. Kamdar, Liquid Metal Embrittlement, in *Treatise on Materials Science and Technology*, Vol 25, C.L. Briant and S.K. Banerji, Ed., Academic Press, 1983, p 361-459
- M.H. Kamdar, *Prog. Mater. Sci.*, Vol 15, 1973
- V.I. Kikhtman, P.A. Rebinder, and G.V. Karpeno, *Effects of Surface Active Medium on Deformation of Metals*, Her Majesty's Stationary Office, London, 1958
- V.I. Likhtman, E.D. Shchukin, and P.A. Rebinder, *Physico-Chemical Mechanics of Metals*, Academy of Sciences U.S.S.R., Moscow, 1962
- S. Mostovoy and N.N. Breyer, The Effect of Lead on the Mechanical Properties of 4145 Steel, *ASM Q.*, Vol 61 (No. 2), 1968
- W. Rostoker, J.M. McCaughney, and H. Markus, *Embrittlement by Liquid Metals*, Van Nostrand Rheinhold, 1960
- N.S. Stoloff, Metal Induced Embrittlement, in *Embrittlement by Liquid and Solid Metals*, M.H. Kamdar, Ed., The Metallurgical Society of AIME, Warrendale, PA, 1984
- N.S. Stoloff, Recent Developments in Liquid Metal Embrittlement in Environment Sensitive Fracture of Engineering Materials, in *Proceedings of the AIME Conference*, 1980
- A.R.C. Westwood, C.M. Preece, and M.H. Kamdar, *Fracture*, Vol 13, H. Leibowitz, Ed., Academic Press, 1971

ture. A concise summary of embrittlement couples is provided in Table 1. Both pure and alloyed solids are listed.

ACKNOWLEDGMENT

The author is grateful to Stanley Lynch, Defense Metallurgical Laboratories, Melbourne, Australia; D. Meyn, Naval Research Laboratory; R.M. Fisher, U.S. Steel Research Laboratories; R. Maller, Gruman Aerospace Company; D. Bhattarcharya, Inland Steel Research Laboratory; S. Buzolits, SPS Technologies; O.R. Lesuer and J.B. Bergin, Lawrence Livermore National Laboratory; R.J. Schwinghamer, Marshall Space Flight Center; L.F. Coffin and T. Grubb, General Electric Re-

Embrittlement by Solid-Metal Environments

M.H. Kamdar, Benet Weapons Laboratory,
U.S. Army Armament Research, Development, and Engineering Center

EMBRITTLEMENT occurs below the melting temperature of the solid in certain liquid-metal embrittlement (LME) couples. The severity of embrittlement increases with temperature, with a sharp and significant increase in severity at the melting point, T_m, of the embrittler (Fig. 1). Above T_m, embrittlement has all the characteristics of LME. The occurrence of embrittlement below the T_m of the embrittling species is known as solid metal induced embrittlement (SMIE) of metals.

Although SMIE of metals has not been mentioned or recognized as an embrittlement phenomenon in industrial processes, many instances of loss in ductility, strength, and brittle fracture of metals and alloys have been reported for electroplated metals and coatings or inclusions of low-melting metals below their T_m (Ref 1). Delayed failure of cadmium-plated high-strength steel has been observed below the T_m of cadmium (Ref 2, 3). Accordingly, cadmium-plated steel bolts, despite their excellent resistance to corrosion, are not recommended for use above 230 °C (450 °F). Notched tensile specimens of various steels are embrittled by solid cadmium. Solid cadmium, silver, and gold embrittle titanium. Leaded steels are embrittled by solid lead, with considerable loss in ductility below the T_m of lead; this phenomenon accounts for numerous elevated-temperature failures of leaded steels, such as radial cracking of gear teeth during induction-hardening heat treatment, fracture of steel shafts during straightening at elevated temperature, and heat-treatment failure of jet-engine compressor disks. Liquid and solid cadmium metal environments as well as cadmium dissolved in inert nonembrittling coolant liquid serve to embrittle the Zircaloy 2 nuclear-fuel cladding tubes used in nuclear reactors (Ref 4). Inconel vacuum seals are cracked by solid indium. These reports of brittle failure clearly indicate the importance of SMIE in industrial processes.

Solid metal induced embrittlement was first recognized and investigated in 1967 by studying the delayed failure of steels in a solid cadmium environment (Ref 5). In 1970, the embrittlement of various titanium alloys by solid cadmium was reported (Ref 6, 7) (Fig. 2 to 5). In 1974, a systematic investigation of SMIE of steel by a number of solid-metal embrittling species (Fig. 1) was conducted (Ref 8). Solid metal induced embrittlement was concluded to be a generalized phenomenon of embrittlement, at least in steel. Thus, solid metal as an external environment can cause embrittlement.

Mostovoy and Breyer (Ref 1) were the first researchers to recognize that SMIE can occur when the embrittling solid is an internal environment, that is, present in the solid as an inclusion. They clearly demonstrated that inter-

Fig. 1 Embrittlement of 4140 steel by various liquid metals below their melting point and their effects on normalized true fracture strength and reduction of area as a function of homologous temperature T_H

T and T_m are the test and melting temperatures of the liquid metal. Source: Ref 8

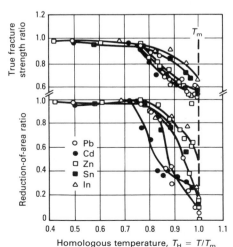

Fig. 2 Crack depth as a function of exposure time to solid cadmium environment for various titanium alloys at three stress levels

STA, solution treated and aged; ST, solution treated; VAC STA, vacuum solution treated and aged; βA, β-annealed; A, annealed; MA, mill annealed. Source: Ref 3

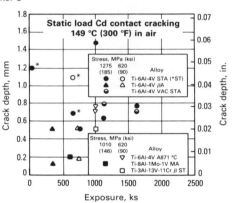

nally leaded steel is embrittled below the T_m of the lead inclusion of steel (Fig. 6). It was also suggested that the reduction-in-cohesion mechanism proposed for LME can also account for SMIE, but the kinetics of the cracking processes are determined by the thermally activated processes. However, the most recent and substantial work on the mechanisms of SMIE suggests that stress-induced diffusion penetration by the embrittling species in grain boundaries causes crack initiation, whereas multilayer self-diffusion of the embrittler controls crack propagation (Ref 9, 10).

The brittle fracture in LME and SMIE is of significant scientific interest because the embrittling species are in the vicinity of or at the tip of the crack and are not transported by dislocations or by slip due to plastic deformation into the solid, as is hydrogen in the hydrogen embrittlement of steels. Also, embrittling species are less likely to be influenced by

Fig. 3 Crack morphology for Ti-6Al-4V (solution treated and aged)

(a) Typical specimen with multiple cracks in the indented area. (b) Fracture surface of (a) showing the depth of cadmium-induced cracking. (c) Cross section showing mixed intergranular cracking and α cleavage in cadmium-induced crack. Etched with Kroll's reagent. (d) TEM fractograph (two-stage replica) of similar area showing intergranular cracking and cleavage. Courtesy of D.A. Meyn

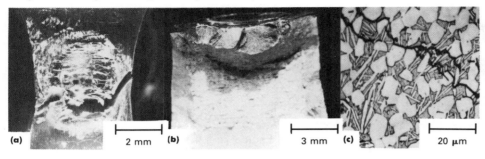

the effects of grain-boundary impurities, such as antimony, phosphorus, and tin, which cause significant effects on the severity of hydrogen and temper embrittlement of metals. Investigations of SMIE and LME, therefore, can be interpreted less ambiguously than similar effects in other environments, such as hydrogen and temper embrittlement of metals. Thus, solid-liquid environmental effects provide a unique opportunity to study embrittlement mechanisms in a simple and direct manner under controlled conditions.

It is believed that a common mechanism may underlie solid, liquid, and gas phase induced embrittlement. The interactions at the solid/ environment interface and the transport of the embrittling species to the crack tip may characterize a specific embrittlement phenomenon. A study of SMIE and LME may provide insights into the mechanisms of hydrogen and temper embrittlement. It is apparent that the phenomenon of SMIE is of both industrial and scientific importance. This article will review the investigations of the occurrence and mechanisms of SMIE.

Characteristics of SMIE

To date, SMIE has been observed only in those couples in which LME occurs, suggesting that LME is a prerequisite for the occurrence of SMIE. Solid metal induced embrittlement may occur in the absence of LME if a brittle crack cannot be initiated at the T_m of the embrittler. A recent compilation of SMIE couples is given in Tables 1 and 2, which show that all solid-metal embrittlers are also known to cause LME.

Solid metal induced embrittlement and liquid-metal embrittlement are strikingly similar phenomena. The prerequisites for SMIE are the same as those for LME: intimate contact between the solid and the embrittler, the presence of tensile stress, crack nucleation at the solid/embrittler interface from a barrier (such as a grain boundary), and the presence of embrittling species at the propagating crack tip. Also, metallurgical factors that increase brittleness in metals, such as grain size, strain rate, increases in yield strength, solute strengthening, and the presence of notches or stress raisers, all appear to increase embrittlement. The susceptibility to SMIE is stress and temperature sensitive and does not occur below a specific threshold value. Embrittlement by delayed failure is also observed for both LME and SMIE (Table 3).

Some differences also exist. Multiple cracks are formed in SMIE; in LME, a single crack usually propagates to failure. The fracture in SMIE is propagation controlled. However, crack-propagation rates are at least two to three orders of magnitude slower than in LME.

Brittle intergranular fracture changes to ductile shear because of the inability of the embrittler to keep up with the propagating crack tip. Incubation periods have been reported, indicating that the crack-nucleation process may not be the same as in LME (Ref 9-11). Nucleation and propagation are two separate stages of fracture in SMIE.

These differences may arise because of the rate of reaction or interactions at the metal/ embrittler interface and also because the transport properties of solid- and liquid-metal embrittlers are of significantly different magnitudes. It has been suggested that reductions in the cohesive strength of atomic bonds at the tip are responsible for both SMIE and LME (Ref 1, 9, 10, 12, 13). However, transport of the embrittler is definitely the rate-controlling factor in SMIE. Another possibility is that stress-assisted penetration of the embrittler in the grain boundaries initiates cracking, whereas surface self-diffusion of the embrittling species similar to that proposed for LME controls crack propagation.

Investigations of SMIE

The first investigations of delayed failure were reported in cadmium-, zinc-, and indium-plated tensile specimens of 4340, 4130, 4140, and 18% Ni maraging steel in the temperature range of 200 to 300 °C (390 to 570 °F) (Fig. 7 and 8) (Ref 2). The results indicated that 4340 was the most susceptible and that 18% Ni maraging steel was the least susceptible alloy to cadmium embrittlement. The activation energy for steel-cadmium embrittlement was 39 kcal/ mol (Fig. 9), which corresponded to diffusion of cadmium in the grain boundary. A thin plated layer of nickel or copper has been reported to act as a barrier to the embrittler and to prevent SMIE of steel (Ref 2). Solid indium is reported to embrittle steels (Ref 10, 11), and an incubation period exists for crack nucleation.

Embrittlement of different steels in lead was investigated as both an external and internal (leaded steel) environment; SMIE was conclusively demonstrated to be a reproducible effect (Ref 1, 8). It was also demonstrated that SMIE is an extension of LME (Ref 1, 10). Internally leaded high-strength steels are susceptible to severe embrittlement in the range of the T_m of lead, and this embrittlement is a manifestation of LME. However, it was found that the onset of embrittlement occurs some 95 °C (200 °F) below the T_m of lead (330 °C, or 625 °F) and is continuous up to the T_m, with no discontinuity or anomalies in the variation of the embrittlement with temperature (Ref 1). At the T_m of lead, a sharp increase occurs in the severity of embrittlement, and ductile-to-brittle transition occurs in the temperature range of 370 to 450 °C (700 to 840 °F).

The same behavior was noted for pure lead as an external environment soldered onto 4140 steel, and such embrittlement has also been

Fig. 4 Solid cadmium induced crack morphology for β-annealed Ti-6Al-4V

The structure consists of coarse Widmanstätten α phase in a β phase. (a) Cross section showing cracking along α plate boundaries and cleavage through α plates. Etched with Kroll's reagent. (b) TEM fractograph showing cleavage and planar α plate boundary facets. (c) Same as (a) but at high magnification. Courtesy of D.A. Meyn

Fig. 5 Solid cadmium induced crack morphology for β solution-treated Ti-3Al-13V-11Cr

The structure is all β phase. (a) Cross section showing mixed intergranular cracking and cleavage. Etched with Kroll's reagent. (b) TEM fractograph showing mostly large intergranular facets with some cleavage. (c) Same as (b) except more cleavage. Courtesy of D.A. Meyn

(a) 25 μm

(a) 100 μm

(b) 5 μm

(b) 5 μm

(c) 5 μm

(c) 5 μm

Fig. 6 Brittle fracture surfaces of steel in the region between 205 and 325 °C (400 and 620 °F)

(a) Leaded steel. (b) The same steel showing ductile failure tested in air at the same temperatures. Source: Ref 1

(a)

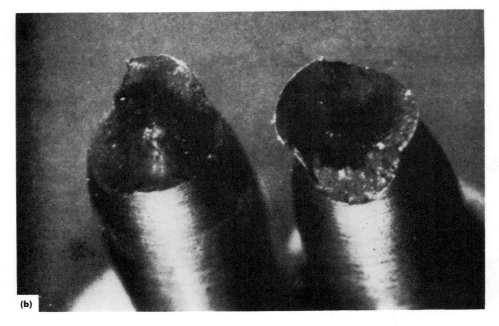

(b)

Fig. 7 Embrittlement behavior of cadmium-plated 4340 steel

Specimens were tested in delayed failure at 300 °C (570 °F) and unplated steel in air at 300 °C (570 °F). Source: Ref 2

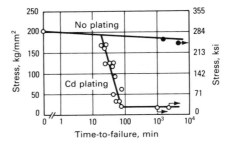

Fig. 8 Embrittlement behavior of cadmium-plated 4340 steel

Specimens were tested in delayed failure at temperatures ranging from 360 to 230 °C (680 to 445 °F). Source: Ref 2

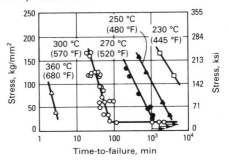

diffusion of embrittler over embrittler and suggest that diffusion, not vapor transport, is the rate-controlling process.

Delayed Failure and Mechanism of SMIE

If a metal in contact with the embrittling species is loaded to a stress that is lower than that for fracture and is tested at various temperatures, then either the environment-induced fracture initiates and propagates instantaneously, as in a zinc-mercury LME couple, or such a failure occurs after some time; that is, delayed failure or static fatigue is observed. Examples of these two types of fracture processes are given in Table 3. Investigations of delayed failure provide an opportunity to separate crack nucleation from crack propagation and, specifically, to evaluate the transport-related role of the embrittling species. Solid metal induced embrittlement is a propagation-controlled fracture process, and the time, temperature, and stress dependence of embrittlement presents an opportunity to study the kinetics of the cracking process.

The most critically investigated embrittlement system is 4140 steel embrittled by solid and liquid indium. Measurement of the decrease in electrical potential has been used to

observed for surface coatings of solid zinc, lead, cadmium, tin, and indium on steel (Ref 1). The embrittlement manifests itself as a reduction in tensile ductility over a range of temperatures extending from about three-quarters of the absolute T_m of the embrittler up to the T_m (Fig. 1). It has been shown that embrittlement is caused by the growth of stable subcritical intergranular cracks and that crack propagation is the controlling factor in embrittlement (Ref 1, 10). This indicated that trans-

port of embrittlers to the crack tip was either by vapor phase or by surface or volume diffusion. The vapor pressures of the embrittling species at T_m varied widely and ranged from 20 to 6×10^{-36} Pa (1.5×10^{-1} to 4.5×10^{-38} torr) (Table 4). However, the crack-propagation times for all embrittlers were similar. The estimated values of the diffusion coefficients ranged in the vicinity of 10^{-4} to 10^{-6} cm²/s (6.5×10^{-4} to 6.5×10^{-6} in.²/s). These values are comparable to surface or self-

Table 1 Occurrence of SMIE in steels

Base metal	Embrittler (melting point)	Onset of embrittlement °C	°F	Test type(b)	Specimen type(c)
1041	Pb (327 °C, or 621 °F)	288	550	ST	S
1041 leaded	Pb	204	399	ST	S
1095	In (156 °C, or 313 °F)	100	212	ST	S
3340	Sn (232 °C, or 450 °F)	204	399	ST	N
	Pb	316	601	ST	N
4130	Cd (321 °C, or 610 °F)	300	572	DF	N
4140	Cd	300	572	DF	N
	Pb	204	399	ST	S
	Pb-Bi (NA)(a)	Below solidus		ST	S
	Pb-Zn (NA)	Below solidus		ST	S
	Zn (419 °C, or 786 °F)	254	489	DF	N
	Sn	218	424	DF	N
	Cd	188	370	DF	N
	Pb	160	320	DF	N
	In	Room temperature		DF	N
	Pb-Sn-Bi (NA)	Below solidus		ST	S
	In	80	176	DF	S
	Sn	204	399	ST	S
	Sn-Bi (NA)	Below solidus		ST	S
	Sn-Sb (NA)	Below solidus		ST	S
	In	110	230	DF	S
	In	93	199	DF	S
	In-Sn (118 °C, or 244 °F)	93	199	DF	S
4145	Sn	204	399	ST	S
	In	121	250	ST	S
	Pb-4Sn (NA)	204	399	ST	S
	Pb-Sn (NA)	204	399	ST	S
	Pb-Sb (NA)	204	399	ST	S
	Pb	288	550	ST	S
4145 leaded	Pb	204	399	ST	S
4340	Cd	260	500	DF	N
	Cd	300	572	DF	N
	Cd	38	100	DF	S
	Zn	400	752	DF	N
4340M	Cd	38	100	DF	S
8620	Pb	288	550	ST	S
8620 leaded	Pb	204	399	ST	S
A-4	Pb	288	550	ST	S
A-4 leaded	Pb	204	399	ST	S
D6ac	Cd	149	300	DF	N

(a) NA, data not available. (b) ST, standard tensile test; DF, delayed-failure tensile test. (c) S, smooth specimen; N, notched specimen
Courtesy of Dr. A. Druschitz

Table 2 Occurrence of SMIE in nonferrous alloys(a)

Base metal	Embrittler (melting point)	Onset of embrittlement °C	°F	Test type(b)
Ti-6Al-4V	Cd (321 °C, or 610 °F)	38	100	DF
	Cd	149	300	BE
Ti-8Al-1Mo-1V	Cd	38	100	DF
	Cd	149	300	BE
Ti-3Al-14V-11Cr	Cd	149	300	BE
Ti-6Al-6V-2Sn	Cd	149	300	BE
	Ag (961 °C, or 1762 °F)	204-232	399-450	BE
	Au (1053 °C, or 1927 °F)	204-232	399-450	BE
Cu-Bi(c)	Hg (−39 °C, or −38 °F)	−84	−119	ST
Cu-Bi(d)	Hg	−87	−125	ST
Cu-3Sn(d)	Hg	−48	−54	ST
Cu-1Zn(d)	Hg	−46	−51	ST
Tin bronze	Pb (327 °C, or 621 °F)	200	392	IM
Zinc	Hg	−51	−60	ST
Inconel	In (156 °C, or 313 °F)	Room temperature		RE
Zircaloy 2	Cd	300	572	ST

(a) All test specimens were smooth type. (b) DF, delayed-failure tensile test; BE, bend test; ST, standard tensile test; IM, impact tensile test; RE, residual stress test. (c) Heat treated to uniformly distribute solutes. (d) Heat treated to segregate solutes to grain boundaries
Courtesy of Dr. A. Druschitz

Table 3 Delayed failure in SMIE and LME systems

Base metal	Liquid	Solid
Type A behavior: delayed failure observed		
4140 steel	Li	Cd
4340 steel	Cd	In
4140 steel	In	Cd
4140 steel	...	Pb
4140 steel	...	Sn
4140 steel	...	In
4140 steel	...	Zn
2024 Al	Hg	...
2424 Al	Hg-3Zn	...
7075 Al	Hg-3Zn	...
5083 Al	Hg-3Zn	...
Al-4Cu	Hg-3Zn	...
Cu-2Be	Hg	...
Cu-2Be	Hg	...
Type B behavior: delayed failure not observed		
Zn	Hg	...
Cd	Hg	...
Cd	Hg + In	...
Ag	Hg + In	...
Al	Hg	...

Courtesy of Dr. A. Druschitz

Table 4 Embrittler vapor pressures and calculated vapor-transport times at the embrittler-melting temperatures

Embrittler	Vapor pressure Pa	torr	Time, s
Zn 20		1.5×10^{-1}	5×10^{-3}
Cd 10		7.5×10^{-2}	1×10^{-2}
Hg(a) 0.3		2.25×10^{-3}	3×10^{-1}
Sb .... 4×10^{-4}		3.0×10^{-6}	3×10^{2}
K .... 1×10^{-4}		7.5×10^{-7}	1×10^{3}
Na .... 2×10^{-5}		1.5×10^{-7}	5×10^{3}
Ti .... 4×10^{-6}		3.0×10^{-8}	3×10^{4}
Pb .... 5×10^{-7}		3.75×10^{-9}	2×10^{5}
Bi .... 2×10^{-8}		1.5×10^{-10}	5×10^{6}
Li .... 2×10^{-8}		1.5×10^{-10}	5×10^{6}
In .... 3×10^{-19}		2.25×10^{-21}	3×10^{17}
Sn .... 8×10^{-21}		6.0×10^{-23}	1×10^{19}
Ga .... 6×10^{-36}		4.5×10^{-38}	2×10^{35}

(a) At room temperature
Courtesy of P. Gordon

monitor crack initiation and propagation and to investigate the effects of temperature and stress level on delayed failure in 4140 steel by liquid and solid indium (Ref 9). The effects of temperature and stress level on the initiation time (incubation period for crack initiation) for both SMIE and LME are given in Fig. 10. The activation energy of the crack-initiation process is approximately 37 kcal/mol and is essentially independent of the applied stress level. The activation energy is considered to represent the energy for stress-aided self-diffusion of both liquid and solid indium in the grain boundaries of the steel. Thus, crack initiation in both SMIE and LME occurs first by the adsorption of the embrittler at the site of crack initiation.

However, because of the presence of the incubation period, the rate-controlling process is stress-aided diffusion penetration of the base-metal grain boundaries. Such penetration re-

Fig. 9 Arrhenius plots of delayed failure in steel tested in cadmium and zinc

The activation energy Q is 39 kcal/mol for cadmium, 70 kcal/mol for zinc. Source: Ref 2

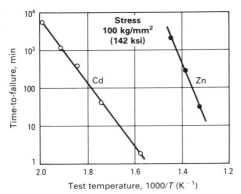

Fig. 10 Initiation time versus temperature in SMIE and LME of 4140 steel in indium at two stress levels

Source: Ref 10

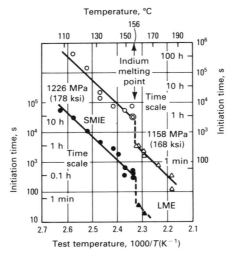

Fig. 11 Propagation time versus temperature for 4140 steel in indium at various temperatures and stress levels

Note the occurrence of SMIE and LME and little dependence on stress level of test. Source: Ref 10

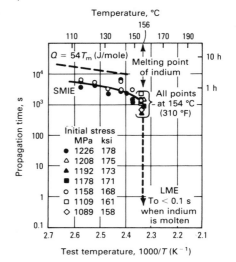

duces the stress necessary to initiate a crack, causing embrittlement of the base metal. However, on this basis, it will be difficult to explain embrittlement of single crystals, which do not contain grain boundaries. It is possible that, rather than stress-aided diffusion penetration, the embrittling species reduces cohesion of atoms at the surfaces of the grain boundaries and thus is in accord with the reduction-in-cohesion mechanism. The crack-propagation time as a function of temperature and stress level for both SMIE and LME of steel by indium is plotted in Fig. 11. The activation energy for crack propagation in SMIE was 5.6 kcal/mol, which represents the energy for self-diffusion of indium over indium. Thus, propagation is controlled by the diffusion of indium over several multilayers of indium adsorbed on the crack surface or by the so-called waterfall mechanism of embrittlement. This mechanism is a variation of that proposed and discussed in detail as the reduction-in-cohesion mechanism for LME.

Adsorption of the embrittler at the solid surface is followed by the rate-controlling step of stress-aided diffusion of the embrittler in the grain boundary of the base metal. The embrittler reduces cohesion of the grain-boundary base-metal atoms, thereby initiating a crack at a lower stress, with the result being embrittlement. Crack propagation occurs by multilayer self-diffusion of the embrittling species. These processes are valid for both LME and SMIE in the steel-indium embrittlement couple. It is clear that SMIE is similar to LME except that SMIE is a slower, embrittler transport-limited process. In this regard, the study of SMIE is important in eliminating the possibility in LME that a crack, once nucleated, may propagate in

a brittle manner in the absence of the embrittling species at the tip of the crack. Solid metal induced embrittlement is a recent phenomenon, and further research is needed concerning the effects of metallurgical, mechanical, and chemical parameters on embrittlement. However, it is clear that SMIE must be recognized as yet another phenomenon of environmentally induced embrittlement.

REFERENCES

1. S. Mostovoy and N.N. Breyer, Effect of Lead on the Mechanical Properties of 4145 Steel, *Trans. ASM*, Vol 61 (No. 2), 1968, p 219-232
2. Y. Asayama, Metal-Induced Embrittlement of Steels, in *Embrittlement by Liquid and Solid Metals*, M.H. Kamdar, Ed., American Institute of Mining, Metallurgical, and Petroleum Engineers, Warrendale, PA, 1984
3. D.A. Meyn, Solid Cadmium Induced Cracking of Titanium Alloys, *Corrosion*, Vol 29, 1973, p 192-196
4. W.T. Grubb, Cadmium Metal Embrittlement of Zircaloy, *Nature*, Vol 265, 1977, p 36-37
5. Y. Iwata, Y. Asayama, and A. Sakamoto, Delayed Failure of Cadmium Plated Steels at Elevated Temperature, *J. Jpn. Inst. Met.*, Vol 31, 1967, p 73 (in Japanese)
6. D.N. Fager and W.F. Spurr, Solid Cad-
mium Embrittlement in Steel Alloys, *Corrosion*, Vol 27, 1971, p 72
7. D.N. Fager and W.F. Spurr, Solid Cadmium Embrittlement of Titanium Alloys, *Corrosion*, Vol 26, 1970, p 409
8. J.C. Lynn, W.R. Warke, and P. Gordon, Solid Metal Induced Embrittlement of Steels, *Mater. Sci. Eng.*, Vol 18, 1975, p 51-62
9. P. Gordon, Metal Induced Embrittlement of Metals—An Evaluation of Embrittler Transport Mechanisms, *Metall. Trans. A*, Vol 9, 1978, p 267-272
10. P. Gordon and H.H. An, The Mechanisms of Crack Initiation and Crack Propagation in Metal-Induced Embrittlement of Metals, *Metall. Trans. A*, Vol 13, 1982, p 457-472
11. A. Druschitz and P. Gordon, Solid Metal Induced Embrittlement of Metals, in *Embrittlement by Liquid and Solid Metals*, M.H. Kamdar, Ed., American Institute of Mining, Metallurgical, and Petroleum Engineers, Warrendale, PA, 1984
12. A.R.C. Westwood and M.H. Kamdar, Concerning Liquid Metal Embrittlement, Particularly of Zinc Monocrystals by Mercury, *Philos. Mag.*, Vol 8, 1963, p 787-804
13. N.S. Stoloff and T.L. Johnston, Crack Propagation in Liquid Metal Environments, *Acta Metall.*, Vol 11, 1963, p 251-256

Hydrogen-Damage Failures

C.D. Kim, U.S. Steel Corporation

THE TERM HYDROGEN DAMAGE has been used to designate a number of processes in metals by which the load-carrying capacity of the metal is reduced due to the presence of hydrogen, often in combination with residual or applied tensile stresses. Although it occurs most frequently in carbon and low-alloy steels, many metals and alloys are susceptible to hydrogen damage. Hydrogen damage in one form or another can severely restrict the use of certain materials.

Because hydrogen is one of the most abundant elements and is readily available during the production, processing, and service of metals, hydrogen damage can develop in a wide variety of environments and circumstances. The interaction between hydrogen and metals can result in the formation of solid solutions of hydrogen in metals, molecular hydrogen, gaseous products that are formed by reactions between hydrogen and elements constituting the alloy, and hydrides. Depending on the type of hydrogen/metal interaction, hydrogen damage of metal manifests itself in one of several ways.

Types of Hydrogen Damage

Specific types of hydrogen damage, some of which occur only in specific alloys under specific conditions, are:

- Hydrogen embrittlement
- Hydrogen-induced blistering
- Cracking from precipitation of internal hydrogen
- Hydrogen attack
- Cracking from hydride formation

The first three types are usually observed at ambient temperatures and are closely related to one another. Hydrogen damage usually manifests itself as hydrogen embrittlement in high-strength steels and as hydrogen-induced blistering in low-strength steels. The solubility and diffusivity of hydrogen in steel sharply decrease with lowering temperatures; therefore, when a heavy section of steel containing hydrogen at elevated temperature is rapidly cooled to ambient temperature, the hydrogen remaining in the steel precipitates out in the gaseous state. The pressure of the hydrogen gas is often great enough to produce internal cracks. Hydrogen attack is an elevated-temperature phenomenon, in which hydrogen reacts with metal substrates or alloy additions. A number of transition and rare-earth metals form hydrides, and the formation of metal hydrides can result in cracking.

Hydrogen Embrittlement

When high-strength steel containing hydrogen is stressed in tension, even if the applied stress is less than the yield strength, it may fail prematurely in a brittle manner. This type of hydrogen damage occurs most often in high-strength steels, primarily quenched-and-tempered steels, and precipitation-hardened steels. The presence of hydrogen in steel reduces the tensile ductility and causes premature failure under static load that depends on the stress and time. This phenomenon is known as hydrogen embrittlement.

Steel can be embrittled by a very small amount of hydrogen, often a few parts per million, and hydrogen may come from various sources. Unlike stress-corrosion cracking, cracks caused by hydrogen embrittlement usually do not branch, and the crack path can be either transgranular or intergranular. Failure by the hydrogen-embrittlement mechanism is accompanied by very little plastic deformation, and the fracture mode is usually brittle cleavage or quasi-cleavage fracture. The susceptibility of a material to hydrogen embrittlement generally increases with increased strength level. For a given hydrogen content, the tendency to embrittlement increases with decreased strain rate, and the embrittlement is most prevalent at room temperature. The cracking tendency decreases with increasing temperature, and above 200 °C (390 °F), the hydrogen-embrittlement phenomenon disappears entirely in steels.

Although many mechanisms of hydrogen embrittlement have been proposed, all of them can be classified according to one of three theories: the pressure theory, the reduced surface energy theory, and the decohesion theory. The pressure theory (Ref 1-3) assumes that the embrittlement is caused by the pressure exerted by gaseous hydrogen in a Griffith crack. When steel saturated with hydrogen at elevated temperature is cooled, gaseous hydrogen will precipitate in microvoids, and an extremely high pressure of the gas can be developed. Flakes in heavy forgings and underbead cracks in weldments can be explained by the pressure theory. Hydrogen charged into steel during aqueous corrosion or cathodic charging can also produce a very high pressure inside internal voids.

The reduced surface energy theory states that the absorption of hydrogen decreases the surface free energy of the metal and enhances the propagation of the Griffith crack (Ref 4). This theory may explain the crack propagation of high-strength steels in low-pressure gaseous hydrogen.

The decohesion theory (Ref 5-7) holds that dissolved hydrogen migrates into a triaxially stressed region and embrittles the lattice by lowering the cohesive strength between metal atoms.

Hydrogen embrittlement has been referred to as hydrogen-induced delayed failure, hydrogen-stress cracking, flakes, fisheyes, underbead cracking, and hydrogen-assisted cracking, depending on where and how it is observed. Because failure by hydrogen embrittlement occurs under a wide range of conditions, it is useful to examine several circumstances that are frequently encountered with steels.

Cracking From Hydrogen Charging in an Aqueous Environment. When metal corrodes in a low-pH solution, the cathodic partial reaction is reduction of the hydrogen ion. Although most of the reduced hydrogen reacts to form H_2 and leaves the metal surface as gaseous hydrogen, a part of the reduced hydrogen enters into the metal as atomic hydrogen. The presence of certain chemical substances that prevent the recombination of hydrogen to form molecular hydrogen enhances the absorption of nascent (atomic) hydrogen into the metal. These substances are called cathodic poisons, and they include phosphorus, arsenic, antimony, sulfur, selenium, tellurium, and cyanide ion. Among the cathodic poisons, sulfide is the most common. Environments containing hydrogen sulfide can cause severe embrittlement of steels and some other high-strength alloys. On the other hand, corrosion inhibitors lower the corrosion rate and thus the amount of hydrogen charged into the metal.

Atmospheric and Aqueous Corrosion. Most high-strength steels are susceptible to hydrogen embrittlement when they are stressed and exposed to fresh or sea water and even during atmospheric exposure. The susceptibility of steels to hydrogen embrittlement generally increases with increasing tensile strength.

Steels having a tensile strength greater than about 1034 MPa (150 ksi) are susceptible to embrittlement. Above a tensile-strength level of 1241 MPa (180 ksi), most high-strength low-alloy steels, such as AISI 4130 and 4340, and precipitation-hardening stainless steels are susceptible to hydrogen-embrittlement cracking in marine atmospheres when the residual or applied tensile stresses are sufficiently high, and the cracking usually occurs in a form of delayed failure. Steels with tensile strengths less than 690 MPa (100 ksi) appear to be resistant to hydrogen-embrittlement cracking, and the structures made with such steels have been used in service without serious problems in various environments that do not contain hydrogen sulfide.

Hydrogen-embrittlement cracking of high-strength steel in aqueous environments is sometimes erroneously called stress-corrosion cracking. Cracking of high-strength steel in aqueous environments is caused by hydrogen absorbed into the steel during corrosion. It is not caused by an anodic dissolution mechanism, such as the chloride stress-corrosion cracking of austenitic stainless steels or caustic cracking of low-alloy steels.

The electrochemical conditions at the tip of a pit or an advancing crack are not the same as those of the bulk solution because the solution chemistry in this region may be quite different from that of the bulk solution. In a laboratory study, when wedge-opening loading specimens were exposed to sodium chloride solutions, the pH value of the solution at the crack tip was measured to be about 3.5 regardless of that of the bulk solution. At the crack tip, where diffusion is limited, the pH value of the solution is lowered by the acidic hydrolysis reaction, and in such a low-pH solution, the cathodic partial reaction is the reduction of the hydrogen ion. As a result, nascent hydrogen is generated at the tip of the pit or crack and absorbed into the metal. It has also been shown that hydrogen can be generated at the crack tip even when an anodic potential is applied to the bulk metal.

Environments Containing Hydrogen Sulfide. High-strength steel pipes used in drilling and completion of oil and gas wells may exhibit delayed failure in environments containing hydrogen sulfide. This type of failure is referred to as sulfide stress cracking. The basic cause of sulfide stress cracking is embrittlement resulting from hydrogen absorbed into steel during corrosion in sour environments. The presence of hydrogen sulfide in the environment promotes hydrogen absorption into steel, thereby making the environment more severe and thus more likely to cause hydrogen embrittlement. Although hydrogen sulfide gas, like gaseous hydrogen, can cause embrittle-

ment, water ordinarily must be present for sulfide stress cracking to occur.

The susceptibility to sulfide stress cracking increases with increasing hydrogen sulfide concentration or partial pressure and decreases with increasing pH. The ability of the environment to cause sulfide stress cracking decreases markedly above pH 8 and below 101 Pa (0.001 atm) partial pressure of hydrogen sulfide. The cracking tendency is most pronounced at ambient temperature and decreases with increasing temperature. For a given strength level, tempered martensitic steels have better sulfide stress cracking resistance than normalized-and-tempered steels, which in turn are more resistant than normalized steels. Untempered martensite demonstrates poor resistance to sulfide stress cracking. It is generally agreed that a uniform microstructure of fully tempered martensite is desirable for sulfide stress cracking resistance.

The effect of alloying elements on the sulfide stress cracking resistance of carbon and low-alloy steels is controversial, except for one element. Nickel is detrimental to sulfide stress cracking resistance. Steels containing more than 1% Ni are not recommended for service in sour environments.

The sulfide stress cracking susceptibility of weldments appears to be greater than that of the base metal, and the high hardness and residual stresses resulting from welding are believed to increase the susceptibility. In a laboratory study of the sulfide stress cracking resistance of submerged-arc weldments in a hydrogen sulfide saturated aqueous solution of 0.5% acetic acid and 5% sodium chloride, no failures were observed for the welds with hardness values below 191 HB; all of the welds with hardness values of 225 HB (20 HRC) or higher failed. It should be noted that sulfide stress cracking occurred in the weldments having hardnesses lower than 22 HRC. Steels with hardness values less than 22 HRC are considered acceptable for sour service.

The National Association of Corrosion Engineers (NACE) issued a standard (Ref 8) for metallic materials resistant to sulfide stress cracking for oil-field equipment. This standard covers metallic-material requirements for sulfide stress cracking resistance for petroleum production, drilling, gathering, and transmission equipment and for field-processing facilities to be used in hydrogen sulfide bearing hydrocarbon service.

Pickling and Electroplating. Pickling alone is not a serious problem (unless internal voids or other imperfections lead to local formation of molecular hydrogen), because the metal is not being stressed and much of the absorbed hydrogen diffuses out of the metal. Heating at 150 to 200 °C (300 to 390 °F) hastens the removal of hydrogen. Also, the addition of suitable inhibitors to the pickling solution eliminates or minimizes attack on the metal and the consequent generation of nascent hydrogen. Salt baths operated at about 210 °C

(410 °F) can be used for descaling titanium alloys, superalloys, and refractory metals to avoid the possibility of hydrogen charging associated with pickling.

Plating solutions and plating conditions selected to produce a high-cathode efficiency minimize the amount of hydrogen generated on the metal. Because the metallic coatings plated on metal often act as barriers to effusion of hydrogen, elevated-temperature baking after plating is generally required for removal of hydrogen. Baking at 190 °C (375 °F) usually suffices, unless the coating is cadmium, through which hydrogen diffuses less readily than through other electrodeposited metals. Raising the baking temperature to accelerate effusion is not possible without reducing the protective qualities of the cadmium plate. Thus, baking time must be lengthened.

In one hydrogen-damage failure caused by improper baking after electroplating, several cadmium-plated AISI 8740 steel nuts cracked in seven days after installation on an aircraft wing structure (Fig. 1). Examination of the fracture surface (Fig. 2) showed predominantly intergranular fracture typical of hydrogen embrittlement. The nuts had not been baked for a sufficient time; thus, hydrogen picked up during the plating process was not released. Procedures for prevention of hydrogen damage in electroplating and preparation for electroplating are described in Ref 9.

Cathodic Protection. Marine structures and underground pipelines are cathodically protected either with sacrificial anodes or with impressed current. Although the cathodic protection of such structures usually does not cause

Fig. 1 Cadmium-plated AISI 8740 steel nut that failed by hydrogen embrittlement

Failure occurred seven days after installation on an aircraft wing structure. See also Fig. 2. 5×. Courtesy of Lockheed-Georgia Company

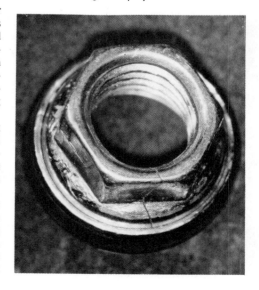

Fig. 2 Fracture surface of failed cadmium-plated nut in Fig. 1

Macrograph of fracture surface. 15×. (b) Scanning electron micrograph of the area in the box in (a) showing typical intergranular fracture; 3950×. Courtesy of Lockheed-Georgia Company

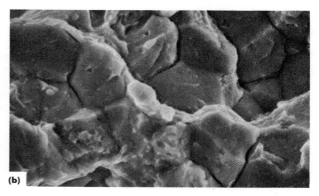

Fig. 3 Cross section through a hydrogen blister in a 1020 steel specimen

Molecular hydrogen precipitated in the matrix-particle boundaries surrounding a subsurface inclusion, causing the blister. Approximately 385×

hydrogen embrittlement, high-strength steel can be embrittled by cathodic protection.

Cracking From Gaseous Hydrogen. Steel vessels and other equipment that contains hydrogen gas at high pressures at ambient temperature are susceptible to failure by hydrogen embrittlement. Many high-strength steels are severely embrittled in tension tests conducted in high-pressure hydrogen gas, and the cracking tendency increases with increasing pressure of hydrogen. The cracking susceptibility of steels generally increases with increasing strength.

Generally, high-strength steels and high-strength nickel alloys display severe degradation of tensile properties in hydrogen gas. Austenitic stainless steels, aluminum alloys, and alloy A-286 show very little embrittlement in this environment; most other engineering metals and alloys are affected to a lesser degree. Seamless steel vessels have been used for the containment of hydrogen gas at moderately high pressure without any reported problems. When austenitic stainless steels, such as AISI 304L and 309S, are exposed to high-pressure hydrogen gas (69 MPa, or 10 ksi), the tensile ductility is markedly reduced, although the fracture mode is ductile.

Hydrogen gas at atmospheric pressure can cause embrittlement in martensitic high-strength steels. Subcritical crack growth is observed with high-strength steel at very low stresses in pure hydrogen gas, and oxygen in small quantities stops the crack growth. Oxygen is believed to form an oxide barrier at the crack tip by preferential adsorption, thereby protecting the crack tip from hydrogen.

Hydrogen-Induced Blistering

Hydrogen-induced blistering is most prevalent in low-strength alloys, and it is observed in metals that have been exposed to hydrogen-charging conditions, for example, acid pickling or corrosion in environments containing hydrogen sulfide. During pickling, atomic hydrogen generated at the metal surface is absorbed by the metal. When hydrogen is absorbed into metal and diffuses inward, it can precipitate as molecular hydrogen at internal voids, laminations, or inclusion/matrix interfaces, and it can build up pressure great enough to produce internal cracks. If these cracks are just below the surface, the hydrogen-gas pressure in the cracks can lift up and bulge out the exterior

layer of the metal so that it resembles a blister (Fig. 3). The equilibrium pressure of the molecular hydrogen in the void, which is in contact with the atomic hydrogen in the surrounding metal, is great enough to rupture any metal or alloy. The absorbed hydrogen may come from various sources.

Corrosion-generated hydrogen causes blistering of steel in oil-well equipment and in petroleum-storage and refinery equipment. In the refinery, hydrogen-induced blistering has been found most frequently in vessels handling sour (hydrogen sulfide containing) light hydrocarbons and in alkylation units where hydrofluoric acid is used as a catalyst. Storage vessels with sour gasoline and propane are highly prone to blistering, but sour crude storage tanks are less prone to blistering, apparently because the oil film of the heavier hydrocarbons acts as a corrosion inhibitor. In storage vessels, blistering is generally at the bottom or in the vapor space where water is present. Gas-plant vessels in catalytic hydrocarbon-cracking units are particularly prone to blistering because the cracking reaction generates cyanides. The presence of cyanides, hydrogen sulfide, and water enhances hydrogen absorption into steel. Hydrogen-induced blistering also occurs on steel plates used as cathodes in industrial electrolysis.

Line pipe transmitting wet sour gas can develop hydrogen-induced cracking in the pipe wall. When line-pipe steel is exposed to sour brine in the absence of applied stress, a number of cracks parallel to the longitudinal axis of the pipe may develop through the wall, and the tip of one crack may link up with another in stepwise fashion. This type of cracking is called stepwise cracking, and it can significantly reduce the effective wall thickness of the pipe. It is believed that a single, straight longitudinal crack is less harmful than stepwise cracks. The presence of hydrogen sulfide in the corrodent greatly promotes the absorption of hydrogen into steel. Thus, the fugacity of hydrogen generated during sulfide corrosion is extremely high—often of the order of 10^3 MPa (10 000 atm).

Hydrogen-induced cracks in line-pipe steels are always associated with certain metallurgical features, such as inclusions, large precipitate particles, or martensite bands. Elongated inclusions, such as type II manganese sulfides and glassy silicates, are particularly detrimental to hydrogen-induced cracking resistance. The oxygen content in molten steel markedly affects the shape of sulfide inclusions in rolled steel. Manganese sulfides in semikilled steel deform to a lesser extent during hot rolling than those in silicon-aluminum killed steels. Semikilled steels usually contain ellipsoidal manganese sulfides (type I MnS) and generally have better hydrogen-induced cracking resistance than fully-killed steels having elongated manganese sulfides (type II MnS). Although the hydrogen-induced cracking resistance of steel increases with decreased sulfur content, lowering the sulfur level alone does not provide immunity to hydrogen-induced cracking.

The presence of martensite banding in line-pipe steel increases the susceptibility to hydrogen-induced cracking. The formation of martensite can be controlled by selecting suitable steel compositions and the proper thermomechanical treatment. Addition of about 0.3% Cu is known to improve the hydrogen-induced cracking resistance of steel, but the beneficial effect of copper prevails only in environments with relatively high pH (about 5 or more) and diminishes in low-pH (3.5 or less) solutions.

Cracking From Precipitation of Internal Hydrogen

Flakes. Heavy steel forgings often contain a number of hairline cracks in the center part, and such cracks are called flakes. On fracture surfaces, flakes appear as small areas of elliptical bright cracks. Flakes are formed during cooling after the first forging or rolling and not during cooling after solidification. Flakes are caused by localized hydrogen embrittlement resulting from internal hydrogen.

The primary source of hydrogen in steelmaking is moisture in the atmosphere and in the additives, and the hydrogen content of molten steel after refining can be as high as 5 to 8 ppm. Hydrogen dissolves more in γ-iron (face-centered cubic, fcc) than in α-iron (body-centered cubic, bcc), and the solubility decreases exponentially with decreasing temperature. The lattice solubility of hydrogen in steel is much smaller than 0.1 ppm at room temperature. Therefore, upon cooling, hydrogen precipitates as molecular hydrogen at inclusions or micropores, and because such regions are already embrittled by hydrogen, flakes are readily formed by the gaseous hydrogen pressure. It is generally known that flakes form at temperatures below 200 °C (390 °F). Flakes are generally oriented within the forging grain or segregated bands. Flaking sensitivity increases with increasing hydrogen content.

Fisheyes are another example of localized hydrogen embrittlement. The term is used to describe small shiny spots occasionally observed on the fracture surface of tension specimens from steel forgings or plates having a high hydrogen content. When fisheyes appear on the fracture surface of a tensile specimen, the tensile ductility is also reduced. Fractographic examination usually reveals fracture-initiation sites, such as inclusions or pores, associated with fisheyes. Baking or prolonged room-temperature aging of tension specimens frequently eliminates fisheyes and restores tensile ductility.

When this type of hydrogen damage occurs in welding, it is called underbead cracking. This form of cracking develops in the heat-affected zone (HAZ) of the base metal and runs roughly parallel to the fusion line. Because cracking caused by hydrogen may occur hours or days after welding, it is also known as delayed cracking. The factors controlling this type of cracking are dissolved hydrogen, tensile stress, and low-ductility microstructure, such as martensite. Hydrogen can be supplied to the arc atmosphere by the shielding gas, flux, or surface contamination and is dissolved in the weld metal. As the weld metal cools, it becomes supersaturated with hydrogen, which diffuses into the HAZ. As the austenite transforms into martensite upon rapid cooling, the hydrogen is retained in this region, and as a result, the metal is embrittled by its presence. The stresses generated by external restraint and by volume changes due to the transformation can easily produce cracks in this region. More information on hydrogen damage in welds is available in the article "Failures of Weldments" in this Volume.

Hydrogen Attack

Steel exposed to high-temperature high-pressure hydrogen appears to be unaffected for days or months and then suddenly loses its strength and ductility. This type of damage is called hydrogen attack. It is important to note that hydrogen attack is different from hydrogen embrittlement. Hydrogen attack is irreversible damage, and it occurs at elevated temperatures; whereas hydrogen embrittlement is often reversible and occurs at temperatures below 200 °C (390 °F).

In hydrogen attack of steel, absorbed hydrogen reacts internally with carbides to produce methane bubbles along grain boundaries; these bubbles subsequently grow and merge to form fissures. Failure by hydrogen attack is characterized by decarburization and fissuring at grain boundaries or by bubbles in the metal matrix. This type of hydrogen damage is most commonly experienced in steels that are subjected to elevated temperatures in petrochemical-plant equipment that often handles hydrogen and hydrogen-hydrocarbon streams at pressures as high as 21 MPa (3 ksi) and temperatures up to 540 °C (1000 °F).

The severity of hydrogen attack depends on temperature, hydrogen partial pressure, stress level, exposure time, and steel composition. Additions of chromium and molybdenum to the steel composition improve the resistance to hydrogen attack. On the basis of industrial experience, the American Petroleum Institute (API) prepared a material-selection guide for hydrogen service at elevated temperatures and pressures (Ref 10). This is known as the Nelson Diagram, and it is very useful for proper selection of materials for high-temperature hydrogen service.

Moisture in hydrogen enhances decarburization of steel. Hydrogen gas with a dew point of −45 °C (−50 °F) does not decarburize hypoeutectoid steels. Also, the decarburization effect of hydrogen on these steels at temperatures below 700 °C (1290 °F) is negligible. Although hydrogen can severely decarburize steel under certain conditions of temperature and dew point, decarburization is not limited to hydrogen atmospheres only, and it may occur in a variety of hydrogen-free heat-treatment atmospheres, such as oxygen, air, and carbon dioxide, and in molten salt baths.

Failure by hydrogen embrittlement or blistering may also occur in steel when the steel is rapidly cooled after prolonged exposure to high-temperature high-pressure hydrogen. The excess hydrogen cannot then escape during cooling and causes cracking or blistering.

Hydrogen attack can also occur in copper. At elevated temperature, hydrogen absorbed in tough-pitch copper internally reduces Cu_2O particles present in the copper forming pockets of steam. Because water vapor cannot diffuse away, the steam pressure in the pockets can be very high, causing cavities or blisters. Similar behavior can occur in silver.

Cracking From Hydride Formation

A number of transition, rare-earth, alkaline-earth metals and the alloys of these metals are subject to embrittlement and cracking due to hydride formation. Among these metals, the commercially important ones are titanium, tantalum, zirconium, uranium, thorium, and their alloys. The presence of hydrides in these metals can cause significant increases in strength and large losses in ductility and toughness. As in other types of alloys, excess hydrogen is readily picked up during melting or welding, and hydride formation takes place during subsequent cooling. The use of vacuum melting and the modification of compositions can reduce susceptibility to hydride formation. Hydrogen can often be removed by annealing in vacuum. Welding generally requires the use of inert-gas shielding to minimize hydrogen pickup.

The hydride particles often have the form of platelets and show preferred orientation within the parent lattice, depending primarily on the metal or alloy composition. The large volume change associated with hydride formation leads to a strong interaction between the hydride-formation process and externally applied stresses. Applied stresses can cause preferential alignment of hydrides or realignment. In most cases, the hydride phase has a much lower ductility than the matrix.

Susceptibility of Various Metals

Most metals and alloys are susceptible to hydrogen damage, and many are susceptible to more than one type of hydrogen damage.

Carbon and low-alloy steels can fail by several types of hydrogen damage. Failure by hydrogen embrittlement is often encountered with high-strength steels, especially when the tensile strength is above 1034 MPa (150 ksi). Low-strength steels are considered resistant to hydrogen embrittlement, but they are susceptible to hydrogen-induced blistering. In high-temperature hydrogen environments, carbon and low-alloy steels are also subject to hydrogen attack. However, steels are not susceptible to cracking caused by hydride formation.

Stainless Steels. The susceptibility of the different types of stainless steels (austenitic, ferritic, martensitic, and precipitation hardening) to failure by hydrogen damage varies widely.

Austenitic Stainless Steels. Although austenitic stainless steels are very susceptible to chloride stress-corrosion cracking, they are very resistant to hydrogen damage. A major factor in the resistance of austenitic stainless steels to hydrogen damage may be the very low hydrogen diffusivities in these steels. Hydrogen diffusivity in austenite is several orders of magnitude lower than that in ferrite. When austenitic stainless steel is exposed to corrosive environments, a very small amount of hydrogen is generated on the metal surface during corrosion due to superior corrosion resistance, and as a result, the amount of hydrogen absorbed into the metal is also very small.

Ferritic Stainless Steels. In the annealed condition, ferritic stainless steels are very resistant to hydrogen damage because of their low hardness. However, in the cold-worked or as-welded condition, ferritic stainless steels are susceptible to hydrogen embrittlement.

Martensitic and precipitation-hardening stainless steels have high strength and are subject to hydrogen embrittlement. The susceptibility to cracking increases with increasing yield strength, and almost any corrosive environment can cause failure by hydrogen embrittlement in these types of stainless steels. Even a mild environment, such as fresh water at room temperature, may cause cracking in especially susceptible alloys. Figure 4 shows a precipitation-hardened stainless steel bolt that failed by hydrogen embrittlement after exposure to a warm seacoast atmosphere.

Heat-Resistant Alloys. In general, hydrogen embrittlement has not been a serious problem with heat-resistant alloys. However, laboratory tests have shown that certain nickel-base alloys are susceptible to embrittlement when exposed to highly oxidizing environments or to pure hydrogen at a pressure of 34 MPa (5 ksi) and a temperature of 680 °C (1250 °F). Figure 5 illustrates hydrogen-embrittlement failure of bolts fabricated from an iron-base heat-resistant alloy.

Aluminum and Aluminum Alloys. Hydrogen damage occurs occasionally in aluminum and aluminum alloys, but it is not a serious problem. When high-strength aluminum alloy is cathodically charged, its ductility is reduced. Hydrogen embrittlement of aluminum alloys can result in intergranular or transgranular cracking. Dry hydrogen gas does not cause significant hydrogen embrittlement in aluminum alloys, but stress-corrosion cracking of aluminum-zinc-magnesium alloys in moist gases probably involves hydrogen embrittlement. Problems with hydrogen in aluminum alloys arise mostly from formation of gas-filled voids during solidification. These voids can affect both cast and wrought products. In ingots for wrought products, the presence of hydrogen gas in voids inhibits healing on subsequent working and is responsible for such defects as bright flakes in thick sections and blisters on the surface of annealed or heat-treated material. The principal effect of bright flakes is a reduction in short-transverse ductility.

Titanium and Titanium Alloys. Hydrogen damage in titanium alloys results from embrittlement caused by absorbed hydrogen. Hydrogen may be supplied by a number of sources, including water vapor, pickling acids, and hydrocarbons. The amount of absorption depends primarily on the titanium oxide film on the metal surface, and an adherent unbroken film can significantly retard hydrogen absorption.

Fig. 5 Unitemp 212 bolts that failed in service from hydrogen-induced delayed cracking

(a) Type of bolts that failed. 0.75×. (b) Cross section through failed bolt that broke in the unthreaded shank (top) near the last full thread. 9×. (c) Transmission electron fractograph from a plastic-carbon replica of a fracture surface of a bolt showing quasi-cleavage. 7000×

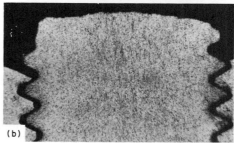

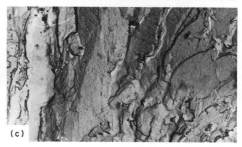

Fig. 4 17-4 PH (AISI type 630) stainless steel bolt that cracked from hydrogen embrittlement

The bolt was exposed for about nine months in a warm seacoast atmosphere and was in contact with aluminum alloy 7075-T6. (a) Overall view of failed bolt. (b) Micrograph of an etched section through a secondary crack showing the transition from unbranched to branched cracking (left). 75×

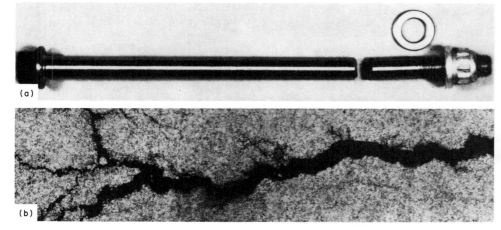

Titanium and its alloys will become embrittled by hydrogen at concentrations that produce a hydride phase in the matrix. The exact level of hydrogen at which a separate hydride phase is formed depends on the composition of the alloy and the previous metallurgical history. In commercial unalloyed material, this hydride phase is normally found at levels of 150 ppm of hydrogen; however, hydride formation has been observed at levels as low as 40 or 50 ppm of hydrogen.

At temperatures near the boiling point of water, the diffusion rate of hydrogen into the metal is relatively slow, and the thickness of the layer of titanium hydride formed on the surface rarely exceeds about 0.4 mm (0.015 in.) because spalling takes place when the hydride layer reaches thicknesses in this range.

Hydride particles form much more rapidly at temperatures above about 250 °C (480 °F) because of the decrease in hydrogen solubility within the titanium lattice. Under these conditions, surface spalling does not occur, and the formation of hydride particles through the entire thickness of the metal results in complete embrittlement and high susceptibility to failure. This type of embrittlement is often seen in material that has absorbed excess hydrogen at elevated temperatures, such as during heat treatment or welding, and subsequently has formed hydride particles during cooling.

There have been instances of localized formation of hydrides in environments where titanium has otherwise given good performance. Investigations of such instances suggest that the localized formation of hydrides is the result of impurities in the metal (particularly the iron content) and the amount of surface contamination introduced during fabrication.

There is a strong link between surface iron contamination and formation of hydrides of titanium. Severe hydride formation has been noted in high-pressure dry gaseous hydrogen around particles of iron present on the surface. Anodizing in a 10% ammonium sulfate solution removes surface contamination and leads to thickening of the normal oxide film.

In chemical-plant service, where temperatures are such that hydrogen can diffuse into the metal if the protective oxide film is destroyed, severe embrittlement may occur. For example, in highly reducing acids where the titanium oxide film is unstable, hydrides can form rapidly. Hydrogen pickup has also been noted under high-velocity conditions where the protective film erodes away as rapidly as it forms.

Hydrogen contents of 100 to 200 ppm may cause severe losses in tensile ductility and notched tensile strength in titanium alloys and may cause brittle delayed failure under sustained loading conditions. The sensitivity to hydrogen embrittlement from formation of hydrides varies with alloy composition and is reduced substantially by alloying with aluminum.

Care should be taken to minimize hydrogen pickup during fabrication. Welding operations generally require inert-gas shielding to minimize hydrogen pickup. Hydrogen can be removed from titanium by annealing in vacuum.

Transition and Refractory Metals and Alloys. Tantalum, zirconium, uranium, thorium, and alloys of these metals, when exposed to hydrogen, can sustain severe damage due to hydride formation.

Tantalum absorbs hydrogen at temperatures above 250 °C (480 °F), resulting in the formation of tantalum hydride. Hydrogen absorption also can occur if tantalum is coupled to a more active metal in a galvanic cell, particularly in hydrochloric acid. As little as 100 ppm of hydrogen in tantalum will cause severe embrittlement by hydride formation.

Zirconium and its alloys are highly susceptible to embrittlement by hydride formation. Absorption of gaseous hydrogen into zirconium and its reaction with zirconium to produce the hydride ZrH_2 occur at an extremely fast rate at 800 °C (1470 °F). This hydride is very brittle, and it may be crushed into a powder. The hydrogen may then be pumped off, leaving a powder of metallic zirconium.

Zirconium alloys can pick up substantial amounts of hydrogen during exposure to high-pressure steam. Zirconium with 2.5% Nb will corrode in water at 300 °C (570 °F) at a rate of approximately 25 μm per year and pick up hydrogen at the rate of 2 to 4 mg/cm^2 (13 to 26 mg/in.^{2}) per day. Hydrogen can be removed from zirconium and its alloys by vacuum annealing.

Uranium, uranium alloys, and thorium are susceptible to hydrogen embrittlement. Hydrides in thorium have been identified as ThH_2 and Th_4H_{15}. The hydride in uranium is UH_3. Uranium absorbs hydrogen from three sources: fused salt-bath annealing, etching and electroplating operations, and corrosion. A hydrogen pickup of 1 to 2 ppm as a result of salt-bath annealing noticeably reduces elongation. Hydrogen absorbed during etching and electroplating is concentrated at the surface and does not degrade tensile properties. Corrosion of uranium in water or water vapor reduces tensile ductility. Uranium is embrittled in water over the pH range of 5 to 10.

Failure Analysis

Before proceeding to a failure analysis, it is prudent to review the history of the failed part and to determine whether the metal or part has a known susceptibility to hydrogen-damage failure, whether the failed part had been exposed to a source of hydrogen during manufacturing and service, and the type and magnitude of stress in the part during service.

Crack Origin. When a hydrogen-damage failure is caused by hydrogen embrittlement or hydrogen-induced blistering, the crack always originates in the interior of the part or metal. Hydrogen-embrittlement cracking often originates very close to the metal surface. In metals of relatively low strength and hardness and in

high-strength metals that are not subjected to significant levels of applied or residual tensile stresses, cracking almost always originates in the interior of the metal. In high-strength metals subjected to applied or residual tensile stresses, especially if a severe stress raiser, such as a sharp fatigue or stress-corrosion crack, is present at the surface, cracking is likely to originate at the surface.

Crack Morphology. A hydrogen-assisted crack is usually a single crack that shows no significant branching. The crack path can be either intergranular or transgranular, and the crack path sometimes changes from one to the other as it propagates. Fracture mode is usually of the brittle cleavage or quasi-cleavage type, especially for high-strength materials, but it can be also of the dimple type for low-strength materials. Therefore, the results of metallographic and electron fractographic examinations should be carefully interpreted for failure analysis. The crack morphology alone does not provide sufficient evidence for hydrogen damage.

Deposits on the Fracture Surface. Corrosion products are ordinarily not present on the surface of hydrogen-damaged fractures unless the surfaces have been exposed to corrosive environments. Foreign matter, if present, is in trace amounts only unless the fracture surfaces had been exposed to contaminants. Energy-dispersive x-ray analysis allows the determination of the elements on the fracture surface.

Laboratory cracking tests are used to verify that the material is susceptible to hydrogen damage under a given environmental condition or to distinguish between hydrogen damage and stress-corrosion cracking. One method of distinguishing between hydrogen embrittlement and stress-corrosion cracking is to note the effect of small impressed electric currents on time to failure in constant-load test. If cracking has occurred by a stress-corrosion mechanism, application of a small anodic current shortens the time to failure. When hydrogen embrittlement is the cracking mechanism, a cathodic current will accelerate cracking (Ref 11).

An electrochemical method of hydrogen-permeation measurement is also used to determine whether hydrogen can be charged into metal in given solutions (Ref 12).

Prevention of Failure

The factors controlling hydrogen damage are material, stress, and environment. Hydrogen damage can be often prevented by using more resistant materials, changing the manufacturing processes, modifying the design to lower stresses, or changing the environment.

Selection of materials resistant to hydrogen damage is often possible. In many applications, a lower-strength material will function just as well mechanically as a higher-strength material, and the use of such a material may eliminate a hydrogen-embrittlement problem.

Fig. 6 Hydrogen-induced toe cracking in the HAZ of a shielded metal-arc weld in low-carbon steel

Etched with 2% nital. 18×. Courtesy of The Welding Institute

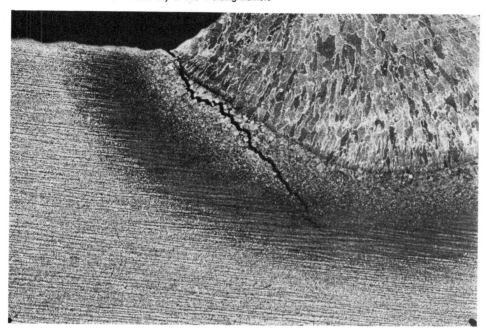

Coatings and linings can be applied to the surface of the metal to shield it from the environment. Such coating materials should be stable in the environment.

Tempering. In high-strength steels, tempering at a higher temperature for a longer time lowers the strength and improves the toughness, and such a heat treatment would also help to reduce hydrogen content. Although such tempering treatments can improve resistance to hydrogen embrittlement, care should be exercised to avoid temper embrittlement. Temper-embrittled materials are more prone to cracking. Baking of electroplated parts enhances the resistance to hydrogen embrittlement.

Surface preparation techniques that impart residual compressive stresses to the surface are used to improve resistance to cracking when susceptible metals are subjected to the combination of hydrogen embrittlement and residual or applied tensile stress. These techniques include shot peening, grit blasting, and face milling.

Low-hydrogen welding rods should be specified for welding if hydrogen embrittlement is a potential problem (Fig. 6). Also, it is important to maintain dry conditions in storing the rods before use and during welding, because water and water vapor are major sources of hydrogen. Preheating and postweld heat treatment are also important when welding high-strength steels to prevent high residual stresses or microcracks.

Lowering the stress below the threshold value for hydrogen embrittlement is often possible. This can be achieved by increasing the cross section of parts, avoiding stress raisers in design, reducing residual stresses through heat treatment, or reducing the load. Corrosion inhibitors reduce corrosion and therefore the amount of hydrogen absorbed into the metal, and they often eliminate the cracking problem. Proper use of inhibitors in the service environment can prevent hydrogen-embrittlement failure.

REFERENCES

1. C. Zappfe and C. Sims, Hydrogen Embrittlement, Internal Stress and Defects in Steel, *Trans. Met. Soc. AIME*, Vol 145, 1941, p 225
2. F. de Kazinczy, A Theory of Hydrogen Embrittlement, *J. Iron Steel Inst.*, Vol 177, 1954, p 85
3. A.S. Tetelman and W.D. Robertson, Direct Observation and Analysis of Crack Propagation in Iron-3% Silicon Single Crystals, *Acta Metall.*, Vol 11, 1963, p 415
4. N.J. Petch and P. Stables, Delayed Fracture of Metals Under Static Load, *Nature*, Vol 169, 1952, p 842
5. A.R. Troiano, The Role of Hydrogen and Other Interstitials in the Mechanical Behavior of Metals, *Trans. ASM*, Vol 52, 1960
6. R.P. Frohmberg, W.J. Barnett, and A.R. Troiano, Delayed Failure and Hydrogen Embrittlement in Steel, *Trans. ASM*, Vol 47, 1955, p 892
7. R.A. Oriani, A Mechanistic Theory of Hydrogen Embrittlement of Steels, *Ber. Bunsenges. Phys. Chem.*, Vol 76, 1972, p 848
8. "Sulfide Stress Cracking Resistant Metallic Material for Oil-Field Equipment," MR-01-75, National Association of Corrosion Engineers, Houston, 1980 (revision)
9. "Standard Recommended Practice for Safeguarding Against Embrittlement of Hot-Dip Galvanized Structural Steel Products and Procedure for Detecting Embrittlement," A 143, *Annual Book of ASTM Standards*, Vol 01.06, ASTM, Philadelphia, 1984, p 39-42
10. "Steel for Hydrogen Service at Elevated Temperatures and Pressures in Petroleum Refineries and Petrochemical Practice," API 941, American Petroleum Institute, Dallas, July 1970
11. H.J. Bhatt and E.H. Phelps, *Corrosion*, Vol 17 (No. 9), 1961, p 430
12. M.A.V. Devanathan and Z. Stachurski, *Proc. Roy. Soc. A*, Vol 270, 1962, p 90

SELECTED REFERENCES

- C.D. Beacham, Ed., *Hydrogen Damage*, American Society for Metals, 1977
- I.M. Bernstein and A.W. Thompson, Ed., *Effect of Hydrogen on Behavior of Materials*, American Institute of Mining, Metallurgical, and Petroleum Engineers, Warrendale, PA, 1976
- I.M. Bernstein and A.W. Thompson, Ed., *Hydrogen in Metals*, American Society for Metals, 1974
- J. Hochmann, J. Slater, and R.W. Staehle, Ed., *Stress-Corrosion Cracking and Hydrogen Embrittlement of Iron Base Alloys*, National Association of Corrosion Engineers, Houston, 1975
- C.G. Interrante and G.M. Pressouyre, Ed., *Current Solutions to Hydrogen Problems in Steel*, American Society for Metals, 1982
- R.D. Kane, Role of H₂S Behavior of Engineering Alloys, *Int. Met. Rev.*, Vol 30, 1985, p 291-301
- F.K. Naumann, *Failure Analysis: Case Histories and Methodology*, C.G. Goetzel and L.K. Goetzel, Trans., H. Wachob, Ed., American Society for Metals, 1983
- R.W. Staehle, A.F. Forty, and D. van Rooyen, Ed., *Fundamental Aspects of Stress-Corrosion Cracking*, National Association of Corrosion Engineers, Houston, 1969
- D.E.J. Talbot, Effects of Hydrogen in Aluminum, Magnesium, Copper and Their Alloys, *Int. Met. Rev.*, Vol 20, 1975, p 166-174

Corrosion-Fatigue Failures

Revised by Peter S. Pao, McDonnell Douglas Corporation, and
Robert P. Wei, Department of Mechanical Engineering and
Mechanics, Lehigh University

CORROSION FATIGUE is a term that is used to describe the phenomenon of cracking, including both initiation and propagation, in materials under the combined actions of a fluctuating, or cyclic, stress and a corrosive environment. It is recognized that corrosion fatigue depends strongly on the interactions among loading, metallurgical, and environmental parameters. An aggressive environment usually has a deleterious effect of fatigue life, producing failure in fewer stress cycles than would be required in a more inert environment. In certain instances, however, such as exposure of nickel-base superalloys to high-temperature oxidizing environments, an aggressive environment can slow the fatigue-fracture process, increasing the number of stress cycles to failure.

The basic principles of corrosion and fatigue in metals are discussed in the articles "Corrosion Failures" and "Fatigue Failures," respectively, in this Volume. Detailed information on corrosion-fatigue crack propagation in specific environments, as well as the fracture mechanics approach to corrosion fatigue, can be found in the article "Environmental Effects on Fatigue Crack Propagation" in Volume 8 of the 9th Edition of *Metals Handbook*. This article will describe the interaction of these two processes, especially the influence of environment on crack initiation and propagation under cyclic or repeated loading. It is important to recognize that corrosion fatigue involves the joint action of cyclic stress and short-time reactions of the environment with the bare metal surfaces exposed during fatigue. Corrosion fatigue is not necessarily involved if cyclic stress and exposure to a corrosive medium occur successively or alternately. Furthermore, corrosion does not always reduce fatigue life, and cyclic stressing does not always increase rates of corrosion.

The microscopic processes by which a fatigue crack is initiated in the presence of an aggressive chemical substance and by which the environment affects growth of the crack are not completely understood. Many different models of these processes have been proposed to explain the macroscopic effects observed. No single model proposed to date explains all the observed effects adequately. In fact, it appears likely that different processes are responsible for the behavior observed with various combinations of material and environment.

In corrosion fatigue, the magnitude of cyclic stress and the number of times it is applied are not the only critical loading parameters. Time-dependent environmental effects are also of prime importance. When failure occurs by corrosion fatigue, the stress-cycle frequency, stress-wave shape, and stress ratio must be known to define the loading conditions satisfactorily, (for a discussion of the concepts and terms related to fatigue, see the article "Fatigue Failures" in this Volume).

Effects of Environment on Fatigue Strength

For a given material, the fatigue strength, or fatigue life at a given value of maximum stress, generally decreases in the presence of an aggressive environment. The effect varies widely, depending primarily on the characteristics of the material-environment combination. The environment affects crack-growth rate, the probability of fatigue-crack initiation, or both. For many materials, the influence of environment is more pronounced at lower stress ranges.

Corrosion-fatigue tests on smooth specimens of high-strength steel indicate that very large reductions in fatigue strength or fatigue life can occur in salt water. For instance, the fatigue strength at 10^6 cycles could be reduced to as little as 10% of that in dry air. In these tests, the main role of the environment was corrosive attack of the polished surface, creating local stress raisers that initiated fatigue cracks. Salt water also increases the crack-growth rate in steels.

Stainless steels have a lower fatigue strength in seawater than in fresh water, probably because of the presence of chloride ions in seawater. Chlorides are known to attack the protective oxide surface films on stainless steels and therefore expose the underlying material to the environment and affect fatigue strength.

Aluminum alloys are highly susceptible to corrosion fatigue. For example, the small

Fig. 1 Fatigue strength (at 10^6 cycles) of magnesium alloy AZ31B-F plotted as a function of relative humidity

Tests were conducted in tension-tension fatigue at about 3 Hz with a stress ratio, *R*, of 0.25.

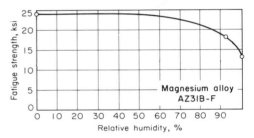

amount of water vapor normally present in the atmosphere severely reduces the fatigue life of several aluminum alloys. The fatigue life in ambient air has been observed to be only about one-tenth the fatigue life in vacuum or in dry air (relative humidity of less than 1%).

The fatigue strengths of alloys of many other systems are reduced by aggressive environments. This appears to be true particularly for those systems that are susceptible to stress-corrosion cracking (SCC), although susceptibility to SCC is not a prerequisite for corrosion fatigue. Some typical effects are shown in Fig. 1 for magnesium alloy AZ31B-F (Mg-3Al-1Zn-0.2Mn) extruded and machined rod that was tested in air with varying relative humidity and in Table 1 for a series of lead alloys tested in air and in a 38% solution of sulfuric acid in water. In another series of tests, pure lead achieved two to four times longer fatigue life in oil than in air, indicating that air, or perhaps the moisture in air, accelerates fatigue failure substantially for lead alloys.

Corrosion-Fatigue Crack Initiation

The influence of an aggressive environment on the fatigue-crack initiation of a material can be illustrated in Fig. 2(a) by comparing the

Table 1 Corrosion-fatigue limits for lead alloys in air and in a 38% H_2SO_4 solution

Alloy	Corrosion-fatigue limit, kPa (psi) In air	In 38% H_2SO_4
Pure lead	2965	0
	(430)	(0)
Pb-0.05Te-0.06Cu	4135	2760
	(600)	(400)
Pb-1Sb	5930	5100
	(860)	(740)
Pb-9Sb	14 135	12 755
	(2050)	(1850)

Fig. 2 Typical fatigue behavior in an aggressive environment compared with fatigue behavior in an inert environment or at high frequency

(a) Data plotted as an *S-N* curve. (b) Data plotted as the crack-growth rate versus stress-intensity range

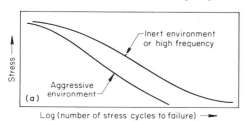
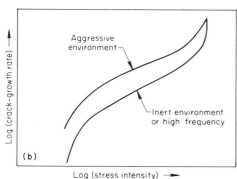

Fig. 3 Corrosion-fatigue cracks in carbon steel

A nital-etched section through corrosion-fatigue cracks that originated at hemispherical corrosion pits in a carbon steel boiler tube. Corrosion products are present along the entire length of the cracks. 250×

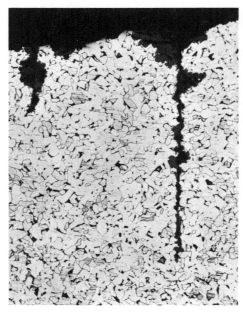

Fig. 4 Fatigue-fracture surfaces of aluminum alloy 7075-T6

SEM fractographs show that river patterns produced in air (a) are supplanted by more brittle morphology in an aerated 3% sodium chloride solution (b).

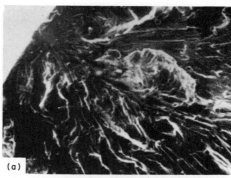

(a)

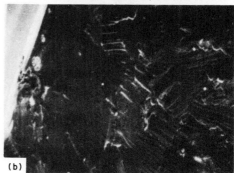

(b)

smooth-specimen stress-life (*S-N*) curves obtained from inert and aggressive environments. Because as much as 95% of the structure life is spent on fatigue-crack initiation, *S-N* curve comparison will provide a good indication of the effect of environment on the crack initiation. As shown in Fig. 2, an aggressive environment can promote crack initiation and can shorten the fatigue life of the structure.

Corrosion-fatigue cracks are always initiated at the surface unless there are near-surface defects that act as stress-concentration sites and facilitate subsurface-crack initiation. Surface features at origins of corrosion-fatigue cracks vary with the alloy and with specific environmental conditions. In carbon steels, cracks often originate at hemispherical corrosion pits and often contain significant amounts of corrosion products (Fig. 3). The cracks are often

transgranular and may exhibit a slight amount of branching. Surface pitting is not a prerequisite for corrosion-fatigue cracking of carbon steels, nor is the transgranular fracture path; corrosion-fatigue cracks sometimes occur in the absence of pits and follow grain boundaries or prior-austenite grain boundaries.

In aluminum alloys exposed to aqueous chloride solutions, corrosion-fatigue cracks frequently originate at sites of pitting or intergranular corrosion. Initial crack propagation is normal to the axis of principal stress. This is contrary to the behavior of fatigue cracks initiated in dry air, where initial growth follows crystallographic planes. As shown in the fractographs in Fig. 4, the river patterns normally observed in stage I propagation in dry air (Fig. 4a) are supplanted by more brittle morphology in aerated chloride solutions (Fig. 4b). Initial corrosion-fatigue cracking normal to the axis of principal stress also occurs in aluminum alloys exposed to humid air, but pitting is not a requisite for crack initiation.

Corrosion-fatigue cracks in copper and various copper alloys initiate and propagate intergranularly (Fig. 5). Corrosive environments have little additional effect on the fatigue life of pure copper over that observed in air, although they change the fatigue-crack path from transgranular to intergranular. Alternatively, copper-zinc and copper-aluminum alloys exhibit a marked reduction in fatigue resistance, particularly in aqueous chloride solutions. This type of failure is difficult to distinguish from

SCC except that it may occur in environments that normally do not cause failures under static stress, such as sodium chloride or sodium sulfate solutions.

Environmental effects can usually be identified by the presence of corrosion damage or corrosion products on fracture surfaces or within growing cracks. Corrosion products, however, may not always be present. For example, corrosion-fatigue cracking of high-strength steel exposed to a hydrogen-producing gas, such as water vapor, may be difficult to differentiate from some other forms of hydrogen damage, such as hydrogen-induced slow-strain-rate embrittlement and sustained-load cracking. At sufficiently high frequencies, the fracture-surface features produced by corrosion-fatigue crack initiation and propagation do not differ significantly from those produced by fatigue in nonaggressive environments.

Corrosion-Fatigue Crack Propagation

The current emphasis in corrosion-fatigue testing is measurement of rates of fatigue-crack propagation in different environments. One of the significant developments in the understanding of fatigue-crack growth and the use of crack-growth data in design is associated with

Fig. 5 Intergranular fatigue cracking in a Cu-8Al alloy tested in 3% sodium chloride solution

500×

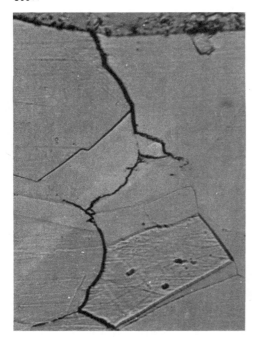

the introduction of fracture mechanics technology (see the articles "Fracture Mechanics" in Volume 8 of the 9th Edition of *Metals Handbook* and "Failure Analysis and Fracture Mechanics" in this Volume). Through linear fracture mechanics, an appropriate crack driving force has been defined as a conjugate to the rate of fatigue-crack growth. The more commonly used parameter to characterize crack driving force is the crack-tip stress-intensity factor, K. However, valid linear-elastic analysis would require the imposition of a condition of limited plasticity at the crack tip (in reference to specimen thickness and crack size). Thus, if a significant plastic zone develops at the crack tip, other appropriate parameters, such as those from R-curve and J-integral analyses, should be used to represent crack driving force.

The environmentally assisted fatigue-crack growth in metals is a complex phenomenon and is influenced by loading, metallurgical, and environmental variables. In a given alloy, in addition to the environmental effects, crack-growth rates can vary widely with variations in loading parameters, such as stress state, stress-intensity range, and stress-wave shape and frequency. Figure 2(b) illustrates schematically the effect of an aggressive environment on the fatigue-crack growth behavior. As shown in Fig. 2(b), the environmental influence is most pronounced at low and intermediate stress-intensity levels. At high stress intensity, the mechanical crack-growth driving force becomes dominant, and the influence of environ-

ment is less significant. As the stress intensity approaches the fracture toughness of the material where instability occurs, the effect of the aggressive environment becomes significant.

Crack-propagation paths in steels may be intergranular or transgranular (including quasi-cleavage) and depend on the frequency, temperature, and environment. Extensive corrosion often damages fracture surfaces and can make positive identification of the fracture mechanism difficult. At low frequencies and low temperatures, fatigue-crack growth in hydrogen, for example, follows primarily prior-austenite grain boundaries, although some quasi-cleavage fracture is also observed. At high frequencies and temperatures, crack growth tends to follow a transgranular path.

Corrosion-fatigue cracking susceptibility in the aluminum alloys does not appear to be sensitive to grain directionality and grain shape. This is in contrast to the propagation of stress-corrosion cracks in susceptible wrought aluminum alloys, which shows greater susceptibility in the short-transverse direction.

Short-Crack Corrosion-Fatigue Behavior

As discussed above, a unique relationship between the fatigue-crack growth rate and the stress-intensity range can be obtained in deleterious and inert environments by using a fracture mechanics approach. Although this relationship can be used to compute the remaining life or to determine the inspection interval of structures, the use of conventional methodology, however, is questioned when the crack length is small.

Experiments on the growth of small cracks in air have clearly shown that the cracks whose dimensions are comparable to the material microstructure grow at rates that are by and large higher than those predicted from the fracture mechanics relationship established for large cracks. The upper limit of the size of cracks that show this anomalous growth behavior ranges from 10 to 500 μm, depending on the material microstructure. The anomalous growth of small cracks arises from the interaction of the crack with the microstructure and from the reduced amount of crack closure. Experiments on corrosion fatigue indicate that the environmental effect on small cracks can be substantially larger than that on large cracks even when the crack is large in the mechanical and metallurgical sense. The peculiar electrochemical conditions near the tips of small cracks may be responsible for this observed size effect. Although crack-size effect is expected in most cases of corrosion fatigue, basic kinetic data of crack growth and chemical analyses of chemical conditions and reactions at the crack tip are not yet available and are under intense investigation.

Most previous studies of the growth of small fatigue cracks have been conducted in air at room temperature. Direct application of a con-

ventional fracture mechanics approach indicates some discrepancy in growth behavior between small and large cracks in most cases. For low-alloy steel and 7075-T6 aluminum alloy, the growth rate of small cracks was higher than that of large cracks, and the amount of increase became larger with decreasing crack length. A similar increase was found in a cast nickel-aluminum bronze. On the other hand, small cracks grew slower than large cracks in quenched-and-tempered steel and in IN100 nickel-base superalloy.

Crack size and geometry may alter the electrochemical environment near the crack tip and consequently change the crack-growth rate and may, in extreme cases, change the controlling process and growth mechanism. For example, the fatigue-crack growth rate for a small crack in a high-strength (4130) steel in a 3% sodium chloride solution is about ten times faster than that of long cracks at similar stress-intensity levels in the same bulk environment, although the short-crack fatigue data obtained in air do coincide with those for long cracks. The difference, therefore, is purely chemical in nature. The magnitude of the environmental enhancement of crack growth varies, depending on crack size and on applied stress. At a low stress range, crack-growth rates exceed those observed in air by two orders of magnitude and are faster than the values reported for long cracks. While at a high stress range, the rate of cracking is nearly equivalent to that reported for long cracks. For low-strength steels, the crack-size effect does not appear to be as large. The electrochemical effect will be discussed in the section "Electrochemical Effects" in this article.

Effect of Loading Parameters

Effect of Cyclic Stress. The service life of a component or structure is often determined by the number of stress cycles required for a crack to initiate plus the number of stress cycles required for the crack to grow to critical size before final fracture occurs. Nominal section stress is not of great value in defining the conditions under which a crack will grow, although it may prove adequate for defining the conditions of crack initiation. Local stress is important for both crack initiation and crack growth. During crack growth, however, local stress is further affected by the crack itself—a significant stress concentrator. The local stress at the tip of a crack can be defined most effectively by the use of linear-elastic fracture mechanics, as discussed in the article "Failure Analysis and Fracture Mechanics" in this Volume. By describing local stress in terms of applied stress intensity, it is possible to predict when a growing fatigue crack of subcritical size will reach critical size. Thus, sudden final fracture of a cracked component can be avoided by removing the part from service before the crack reaches critical size.

Effect of Frequency. In nonaggressive environments, the cyclic frequency generally has little effect on fatigue behavior. On the other hand, in aggressive environments, fatigue strength is strongly dependent on frequency. An observed dependence of fatigue strength or fatigue life on frequency is often considered definitive in establishing corrosion fatigue as the mechanism of a failure.

For most materials, the frequency dependence of corrosion fatigue is thought to result from the fact that the interaction of a material and its environment is essentially a rate-controlled process. Low frequencies, especially at low stress amplitudes or when there is substantial elapsed time between changes in stress levels, allow time for interaction between material and environment and cause greater fatigue damage. High frequencies do not, particularly when a high stress amplitude is also involved. However, high frequencies or high strain rates induce adiabatic heating, which may alter material properties and change the fatigue-crack growth response. Frequency also affects the crack-tip electrochemistry by mechanical-pumping effects.

Changing the cyclic frequency will invariably affect the strain rate at the crack tip and the effective time at load. In some materials, such as titanium alloys, the higher strain rate associated with higher cyclic frequency may promote hydride formation and may deleteriously affect the corrosion-fatigue resistance. For example, the corrosion-fatigue crack-growth rate for β-annealed Ti-6Al-4V alloy in 3.5% sodium chloride solution at a given stress-intensity level increases with increasing frequency (strain-rate effect), reaches a maximum, then decreases with further increase in frequency (time effect). The range of frequency in which the environmental effect is most pronounced depends on the stress-intensity factor level, material, and other environmental parameters. For other alloys, such as steels, the frequency effect reflects only the reaction-time effect and has no apparent effect on strain rate. Additional information on the effect of high frequency on corrosion-fatigue behavior can be found in the article "Ultrasonic Fatigue Testing" in Volume 8 of the 9th Edition of *Metals Handbook*.

Effect of Stress Amplitude/Stress-Intensity Factor Range. In general, a low amplitude of cyclic stress/stress intensity favors relatively long fatigue life, permitting greater opportunity for involvement of the environment and larger environmental effect in the failure process. Where stresses/stress intensities are sufficiently high, environmental interaction may be reduced, because mechanical crack driving force becomes dominant.

Stress amplitude must be considered together with mean stress and frequency. For the stress-intensity factor range, stress must be considered in conjunction with the current crack size in determining the stress-intensity factor range. Low stress/stress-intensity levels may allow

adequate time for environmental interaction, but if the frequency is high, the crack tip may be exposed to the environment in a very short period. Thus, only limited environmental reaction may occur, and the full environmental effect cannot be realized. Figure 2 shows typical behavior, which illustrates the effect of frequency and/or stress as discussed above. Additional information on the effect of stress amplitude/stress-intensity factor range can be found in the article "Environmental Effects on Fatigue Crack Propagation" in Volume 8 of the 9th Edition of *Metals Handbook*.

Effect of Mean Stress. High tensile mean stress will raise the general level of stress intensity at the crack tip and will increase the duration that the crack will open, both of which could enhance corrosion-fatigue damage during each cycle.

Effect of Stress-Wave Shape. The shape of the cyclic stress wave influences the effect of a surrounding aggressive environment. Prolonged exposure at peak stress contributes to accelerated failure. In corrosion fatigue, a primary concern is whether SCC may occur. In a system that is susceptible to stress corrosion, cracking can occur at any load above the critical stress intensity for SCC, K_{Iscc}; in other words, holding times during the high-load portion of the stress cycle are detrimental.

A second consideration in evaluating the influence of stress-wave shape is the strain rate associated with various stress waves. Such strain-rate effects on the corrosion-fatigue cracking in titanium alloy have been briefly discussed in the section "Effect of Frequency" in this article. For corrosion fatigue of stainless steel in aqueous solution, strain rate may control the rate of rupture of the crack-tip protective oxide film. In conjunction with the anodic dissolution rate, these two processes are believed to control corrosion-fatigue crack growth. More research is needed on the joint effect of stress-wave shape and aggressive environment on the corrosion-fatigue characteristics.

Effect of Load Ratio and Overloading. The effect of load ratio on corrosion fatigue is more complex. On one hand, lower load ratio effects higher load amplitude, which may provide a high mechanical crack driving force. On the other hand, a higher load ratio will result in a higher mean load. This would affect environmental species transport to the crack tip through its influence on the effective crack opening. The influence of load ratio on corrosion fatigue should be evaluated based on the interaction of environment with the particular material system.

Applying tensile overloading will cause delay in fatigue-crack growth in either inert or aggressive environments. The number of delayed cycles will depend on a number of variables, such as overload ratio, maximum stress intensity, environment, temperature, and material. Generally, the number of delayed cycles is fewer if an aggressive environment is present.

For titanium alloys in salt solution, however, the number of delayed cycles could be substantially more.

Effect of Mixed-Mode Loading. Mixed-mode loading can significantly change the corrosion-fatigue response of a material. Generally, mode I (or uniaxial stress in the smooth specimen) loading represents a worst-case condition, and the corrosion-fatigue crack-growth rate is higher than that of mixed-mode loading in an aggressive environment.

Effect of Environment

Influence of Chemical Activity of the Environment. Increasing the chemical activity of the environment—for example, by lowering the pH of a solution, by increasing corrodent concentration, or by increasing the pressure of vapor/gas—generally decreases the resistance of a material to corrosion fatigue, whereas decreasing the chemical activity improves resistance to corrosion fatigue.

The corrosion-fatigue behavior of high-strength steels and aluminum alloys are related to the relative humidity or partial pressure of water vapor present in the air. Corrosion-fatigue crack-growth rates in these materials generally increase with increasing water-vapor pressure until a saturation condition (at high vapor pressure) has been reached. Saturation conditions result from the completion of reaction with the surface. The fatigue-crack growth rate normally remains constant past the saturation condition. Highly reactive gases, such as hydrogen sulfide, tend to produce severe losses in corrosion-fatigue resistance. The loss of resistance is evident at high cyclic frequencies and relatively low pressures, for example, at 100 Hz and 10 Pa, or 0.075 torr, for corrosion fatigue of steels in hydrogen sulfide. For less reactive gases, the loss of resistance may become evident only at low frequencies.

Effect of Temperature. Temperature can have a significant effect on corrosion fatigue. It can change the rate of surface chemical reaction, the rate of transport of environment to the crack tip, or even the embrittlement process itself. However, the exact effect of temperature on the corrosion-fatigue crack-growth rate depends on a number of parameters, including temperature range, material/environment combination, and test frequency. Corrosion-fatigue crack-growth rates tend to increase with increasing temperature. For example, fatigue-crack growth rates of HY130 steel in distilled water, 3.5% sodium chloride solutions, and acetate buffer solutions increase with increasing temperature. There are also indications that the rate-controlling process for crack growth may be thermally activated.

The effect of corrosive environments on corrosion fatigue caused by thermal cycling has not received much attention. For some materials, a corrosive environment may improve corrosion-fatigue resistance, especially when a coherent bulk oxide bridges a crack or blunts the crack tip. Conversely, when thermal cycling cracks an oxide scale, exposing the underlying metal to the environment, corrosion-fatigue resistance is usually reduced.

Electrochemical Effects. Electrochemical variables can affect corrosion-fatigue crack initiation and crack propagation. These variables include electrode potential, solution pH, and ionic concentration. The specific effects depend on the material and the electrolyte, and corrosion fatigue may be caused by the presence of specific ions. Therefore, no general guidelines can be given at this time.

For corrosion-fatigue crack growth of steels in chloride and sulfate solutions, the cracking rates tend to be increased by cathodic polarization and by decreasing pH (more acidic solutions). In acetate buffer solutions, on the other hand, there appears to be no effect of either pH or applied potential. The effectiveness of these variables in influencing crack propagation appears to be determined by the ability for altering the crack-tip environment through environmental transport and reactions (see the section "Effect of Crack-Tip Chemistry" in this article).

Recent results suggest that the specificity of ionic species may reflect their influences on reaction rates and therefore the frequency range over which environmental effects are observed. Specificity, therefore, does not necessarily mean inherent susceptibility or immunity in a given environment, but rather the presence or absence of a significant influence at the particular cyclic-load frequency. Generalizations should be avoided, and verification over a broad range of frequencies of interest should be made.

Anodic polarization of steels, aluminum alloys, and copper alloys tends to decrease resistance to corrosion-fatigue crack initiation, whereas cathodic polarization tends to increase initiation resistance. Accordingly, bimetallic couples that make the cyclically stressed material the anode would decrease its resistance to corrosion-fatigue crack initiation. Conversely, the resistance is increased when the stressed material is the cathode. Corrosion-fatigue initiation lives under these conditions are ordinarily inversely related to the magnitude of the galvanic effect. The influence of polarization on crack propagation is less clear, with cathodic polarization tending to increase propagation rates and thereby reduce fatigue life.

Effect of Crack-Tip Chemistry. The chemical composition of solutions and/or gases inside growing corrosion-fatigue cracks is usually different from that in the surrounding environment. Also, the crack-tip chemistry in large cracks can be expected to be different from that in small cracks. Environmental enhancement in crack growth, both in gaseous and aqueous environments, is proportional to the amount of hydrogen produced. In small-crack cases, as in large-crack cases, the concentration of hydrogen is an important factor.

For the case of transport-controlled crack growth in a gaseous environment, the gas pressure at the crack tip may be higher in small cracks because of the shorter transport distance. Consequently, a more pronounced acceleration might be expected. In aqueous environments, the hydrogen production processes are more complex and depend on hydrolysis and hydrogen ion reduction and on crack solution pumping. Electrochemical analysis of these processes indicates a higher concentration of hydrogen ions in small cracks. For crack pumping, an idealized model based on perfect mixing shows less convective dispersal of the products of hydrolysis. On the other hand, in corrosion fatigue of a 12Ni-5Cr-3Mo steel in 3.5% sodium chloride solution, the pH at the tip near the center of the specimen was about 3, but rose to 7 near the specimen surface. These results cannot be applied to small cracks directly, because of the difference in pumping efficiency. The difference in chemistry between the crack tip and bulk environment and the effect of crack size should be properly recognized in evaluating corrosion-fatigue phenomena. Otherwise, a nonconservative design and possible structure failure may result.

Effect of Metallurgical Parameters

Effect of Chemical Composition. Alloying adjustment that improves general corrosion and/or SCC resistance can be effective in reducing corrosion-fatigue susceptibility. For example, chromium steels are generally more resistant to corrosion fatigue than carbon steels. The 7000-series aluminum alloys with high zinc-to-magnesium ratios and/or lower magnesium concentrations have shown indications of lower corrosion-fatigue susceptibility.

Effects of Microstructure and Heat Treatment. Microstructural features, such as grain size and second-phase particles, can affect corrosion fatigue in various ways. Small-grain materials are generally more resistant to crack initiation because stress concentration at the surface is less as compared to the coarse-grain material.

For fatigue-crack propagation, coarse-grain materials generally have a more tortuous fracture path that may restrict the transport of environmental damaging species to the crack tip. Rougher fracture surfaces may also cause roughness-induced crack closure, which reduces the effective crack driving force; both of which would reduce the corrosion-fatigue crack-growth rate.

Heat treatment that alters the microstructure and the distribution of alloying and impurity elements can significantly influence the corrosion-fatigue behavior. For example, artificially aged 7075-T73 and 2024-T6 aluminum alloys exhibit corrosion-fatigue crack-growth rates in a water-vapor environment that are a factor of two or three times lower than those of peak-aged 7075-T6 and naturally aged 2024-T3, respectively.

Analysis of Corrosion-Fatigue Failures

General procedures and techniques for analysis of failures, regardless of mechanism, are discussed in the article "General Practice in Failure Analysis" in this Volume. Additional information, particularly regarding collection, identification and analysis of corrosion products, and examination of corroded surfaces, is provided in the article "Corrosion Failures" in this Volume. In this section, only those aspects of failure analysis that bear directly on the identification of corrosion fatigue as the mechanism of failure are discussed.

Macroscopic Examination of Fractures

After historical information about the failure has been obtained, on-site examinations performed, and samples for laboratory investigation collected, detailed examination usually begins with visual and low-magnification examination of the failed part, with special attention to features of exposed fracture surfaces. Because analysis of fracture-surface deposits can be vital to the identification of corrosion fatigue, the first step in the macroscopic investigation should be to strip the fracture surface with cellulose acetate tape or plastic replicating material to entrap any deposit for later microanalysis. Then, whether complex electron optics or simple photographic documentation is to be used, the piece should be cleaned and degreased in an ultrasonic cleaner containing reagent-grade acetone, or similar organic solvent, as the tank medium.

The main advantage of reagent-grade acetone is that it leaves a minimal amount of organic residue on the fracture surface yet removes almost all traces of the cellulose acetate left from stripping. Also, acetone is an excellent solvent for wet magnetic-particle suspensions and for most liquid penetrants, which are often used in the field for nondestructive detection of cracks. If the part is too large or otherwise not amenable to ultrasonic cleaning, a solvent may be sprayed onto the surface of the part using a squeeze bottle or an aerosol can; the solvent should not be applied with a brush.

Instrumentation. By far the most useful instrument for macroscopic examination is a good binocular microscope, preferably one with a zoom-type objective lens. Many corrosion-fatigue cracks characteristically change fracture planes, or they may be intergranular; therefore, a laboratory microscope must be capable of providing a three-dimensional field of view that is not distorted by lens aberrations.

A scanning electron microscope is also valuable for macroscopic examination. Corrosion-

Fig. 6 A single-origin corrosion-fatigue crack in type 403 stainless steel exposed to steam showing rubbed origin (arrow) and beach marks

SEM fractograph (secondary electron image). 60×

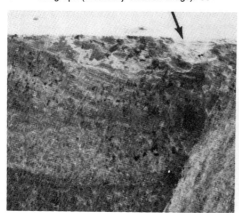

Fig. 7 Two views of the fracture surface of a forged 17-4 PH stainless steel steam-turbine blade that failed by corrosion fatigue originating at severe corrosion pitting

(a) Light fractograph showing primary origin (arrow) and three secondary origins (along right edge below primary origin). 7½×. (b) SEM fractograph (secondary electron image) of area surrounding primary origin (in circle). 50×. White areas are corrosion pits. Black areas, one of which is indicated by arrow, are remnants of corrosion product left after the fracture surface was electrolytically cleaned.

(a)

(b)

fatigue fractures are often sufficiently rough that a standard optical binocular microscope cannot focus the entire area of view, but a scanning electron microscope has a greater depth of field. In addition, most scanning electron microscopes have controls for continuous zoom from magnifications of 25 to 50 000×.

Most environmentally affected fracture surfaces exhibit a tenacious oxide formed on the metal substrate, which interferes with direct observation. Excellent macroscopic examination of these surfaces can often be conducted using the backscattered electron image of a scanning electron microscope. The only possible disadvantage of the backscattered electron image is that re-entrant angles ordinarily resolved by the secondary electron image generally cannot be seen using backscattered electrons. The secondary electron image on unprepared fracture surfaces will often charge up and obscure portions of an oxidized fracture, but the surface often can be sputtered with a film of a conductive material, such as carbon, to minimize this effect. Information on the instrumentation and principles of SEM, including a discussion of backscattered and secondary electron images, can be found in the article "Scanning Electron Microscopy" in Volume 10 of the 9th Edition of *Metals Handbook*.

Fracture Origins. Origins of corrosion-fatigue fractures are often surrounded by crack-arrest lines, or beach marks (Fig. 6). Often, there are several origins, particularly when fracture has been initiated by pitting (Fig. 7). Occasionally, no well-defined origin can be resolved.

Features at the origin of a corrosion-fatigue fracture are often indistinct because the compression portion of each stress cycle has forced mating fracture surfaces together and has formed an extremely rubbed, discolored origin (arrow, Fig. 6). Also, the area of the origin is

exposed to the environment for the longest time and thus may exhibit more extensive residues of a corrosion product than the rest of the fracture surface. A corrosion product at fatigue origins can be misleading when viewed macroscopically. The oxide may be flat and tenacious, globular, or nodular. At magnifications of less than 60×, globular or nodular oxide particles are easily mistaken for intergranular facets; therefore, no conclusions should be drawn until other metallographic or fractographic techniques confirm the mechanism of crack initiation. Occasionally, oxidation is so severe that no information other than location of origin can be obtained.

Other features that can be observed macroscopically are secondary cracks, pits, and fissures, all of which are often adjacent to the main origin of a particular fracture. In corrosion-fatigue failures, cracks and fissures adjacent to the primary origin are indicative of a uniform state of stress at that location. Therefore, the primary fracture simply propagated from the flaw that either induced the most severe stress concentration or was exposed to the most aggressive local environment. Sometimes, the primary crack origin is related more to heterogeneous microstructure than to the stress distribution.

Crack propagation by corrosion fatigue sometimes proceeds from cracks or pits that were started by another mechanism. For example, the fracture shown in Fig. 8 was initiated by SCC, but crack propagation occurred by corrosion fatigue. In instances where SCC and corrosion fatigue are competing mechanisms, as might be expected in a component exposed to both cyclic loading and static tensile loading in a corrosive environment, the mechanism of propagation is determined by the stress-intensity factor and the time of exposure. When the maximum stress-intensity factor exceeds K_{Iscc}, both SCC and corrosion fatigue will take place. Corrosion fatigue can still take place when the maximum stress intensity is less than K_{Iscc} even though SCC will not occur under such conditions.

Both corrosion fatigue and mechanical fatigue may exhibit multiple origins that ultimately join to form a single crack front on a single plane. However, it is sometimes possible to distinguish corrosion fatigue from purely mechanical fatigue by the number of cracks propagating through the part. Frequently, several corrosion-fatigue cracks form and propagate simultaneously along parallel paths (Fig. 9). On the other hand, mechanical-fatigue cracks may initiate at several points in the same general region on a part, but one crack usually becomes dominant, or several cracks join to form a single front, before cracking has progressed very far into the part.

Final Fracture. The region of fast fracture is usually macroscopically distinct from the region of slow crack propagation. The region of slow crack propagation almost always exhibits

Fig. 8 Fracture surface of a type 403 stainless steel steam-turbine blade

Fracture originated in a weld that had not been stress relieved. (a) Light fractograph of corrosion-fatigue crack. 7½×. Arrow indicates origin; white area along right edge is overload fracture produced during a laboratory examination. (b) Light fractograph of intergranular stress-corrosion crack (outlined by arrows) that initiated the corrosion-fatigue crack. 60×

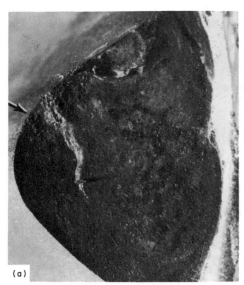

(a)

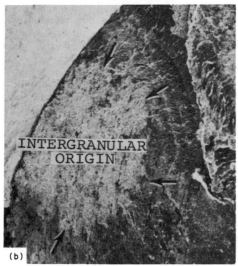

INTERGRANULAR ORIGIN

(b)

signs of corrosion, whereas the fast-fracture region is corroded only when sufficiently exposed to the environment following rupture. For example, the white area along the right edge of the fracture surface shown in Fig. 8(a) is a clean overload fracture produced by breaking a cracked turbine blade in the laboratory during failure analysis. More often, complete separation occurs in a service environment, and there is at least a slight amount of corrosion on the overload portion of the fracture surface.

Fig. 9 Corrosion-fatigue cracking in oval wire of hardened 1065 steel

(a) Macrograph showing multiple corrosion-fatigue cracks that propagated along parallel paths. Approximately 8×. (b) Light fractograph of a fracture surface adjacent to the region shown in (a) Approximately 8×. Arrow indicates origin.

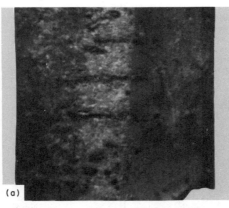

(a)

(b)

Microscopic Examination of Fractures

Studies of fracture surfaces are actually studies of fracture paths. Many of the features found at high magnification bear little resemblance to those found at low magnifications, primarily because local factors that influence the fracture process, such as environment, stress state and magnitude of stress at the crack tip, and microstructure, are different from their macroscopic counterparts.

Most corrosion-fatigue fractures start as small flaws and grow as stress intensity increases, as the environment at the crack tip alters, and as the plastic zone ahead of the crack front increases in size. In many instances, crack initiation occurs at low stress under dominant crystallographic influences, producing clear-cut cleavage-type or intergranular fracture with a minimum of observed plasticity. As the crack grows, more plasticity is associated with the cracking process, more grains become involved in producing each fracture-surface feature, and the fine-scale features become less distinctly crystallographic. Thus, a typical crack begins at a local stress raiser, grows as a crystallographic crack, and subsequently changes to a noncrystallographic crack as it propagates through the component.

Scanning electron microscopes and, to a lesser degree, transmission electron microscopes are widely used in failure analysis. Detailed information on these instruments, as well as their applicability in solving materials problems, is presented in Volume 10 of the 9th Edition of *Metals Handbook*. It is often important to recognize features that can be produced during overload fracture or fatigue without concurrent environmental effects, because it is sometimes the absence of these features that identifies an environmentally assisted fracture.

Microfractographic features that are clear indicators of fatigue include striations (Fig. 10a), tire tracks (Fig. 10b), and plateaus separated by shear steps, tear ridges, or both (Fig. 10c). Any or all of these features may be absent on a fatigue-fracture surface. When fracture occurs by corrosion fatigue, these features may be indistinct or may be obscured by corrosion products. For example, fatigue in a mildly corrosive service environment produced the features shown in Fig. 10(a); the striations are somewhat indistinct, and residues of corrosion product, possibly oxide, produced by fretting appear as black spots in the fractograph. On the other hand, the features shown in Fig. 10(c), which were produced in a less aggressive laboratory environment, are more clearly delineated. Striations are difficult to observe on the surfaces of fatigue fractures produced in high vacuum.

Determination that a fracture has been initiated by corrosion fatigue does not necessarily involve identification of fine-scale fatigue features on the fracture surface. Figure 11(a) shows a light fractograph of a broken Ti-6Al-4V aircraft component. Fracture was initiated by corrosion fatigue at the thumbnail-shaped area indicated by the arrow. At least three other, smaller thumbnail-shaped areas were found on the periphery in the general region of the primary origin. As shown in Fig. 11(b), clearly defined striations were not observed in the fatigue areas. The remainder of the fracture was intergranular, with the grain facets exhibiting dimpled rupture (Fig. 11c).

In the fractograph shown in Fig. 12(a), which illustrates the fracture surface of a sand-cast magnesium alloy AZ91C-T6 aircraft-generator gearbox, the beach marks appear to represent the deposit of corrosion products at successive positions of the crack front. Extensive corrosion in the region of crack initiation at a series of filing or grinding grooves in a fillet (arrow, Fig. 12a) had removed all of the fracture features (Fig. 12b). Although corrosion obscured much of the fine fracture detail over most of the fracture surface, fatigue striations were seen in regions corresponding to later stages of crack growth (Fig. 12c).

Although fatigue striations are usually well formed in most metals, they are often not very well formed in steels. In addition, steels are particularly susceptible to atmospheric corrosion. Thus, even if exposed for only a short time to humid air, minute features on the fracture surfaces of a steel component may be

Fig. 10 TEM fractographs of plastic-carbon replicas showing fracture-surface features that are clear indicators of fatigue

(a) Fatigue striations and corrosion products (black spots) on a fracture surface of an aluminum alloy 2014-T6 aircraft wheel that cracked in service. 3500×. (b) Tire cracks caused by surfaces rubbing together as fatigue crack opened and closed. 18 000×. Tire tracks are associated with high-stress low-cycle fatigue. (c) Plateaus (patches of fatigue striations) separated by shear strips and tear ridges on a fracture surface of 12.5-cm (5-in.) thick aluminum alloy 2024-T851 plate that was fatigue tested in air. 2000×. Arrow indicates direction of crack propagation.

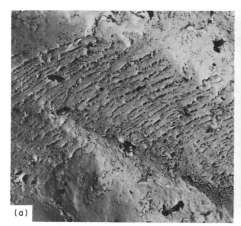
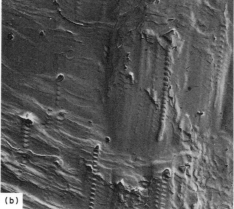
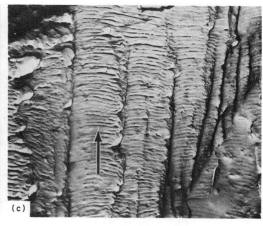

masked. Longer exposure to humid air or exposure to a more oxidizing atmosphere, such as hot exhaust gases, may destroy much of the detail or may cover the fracture surface with a tenacious crystalline oxide. Fatigue in steels is sometimes intergranular; Fig. 13 shows an intergranular corrosion-fatigue fracture in 4340 steel that occurred during testing in humid air.

Corrective Measures

The results of failure analysis, including duplication of the original failure by either in-service or laboratory testing of a similar part, should establish and confirm the mode of failure. Further analysis and testing should reveal the basic reasons for the failure and should establish whether initial material selection, original design, or possibly both were inappropriate and whether incorrect fabrication or unusual service conditions contributed to failure.

Once the cause of failure has been determined, a temporary or permanent solution can be considered. Both temporary and permanent solutions for corrosion fatigue involve one or more of the following: (1) reducing or eliminating cyclic stress, (2) increasing corrosion-fatigue strength of the material, and (3) reducing or eliminating corrosion. These objectives are accomplished by changes in material, design, or environment.

For permanent solutions, if a new material is selected (the same alloy with a different heat treatment or fabrication method, another alloy for the same system, or an alloy from a completely different system), basic compatibility with design requirements and with the environment to be encountered in service usually must be proved by testing. If changes in design or in mode of fabrication are selected, test specimens should reflect these changes. Any environmentally affected properties must be checked in a simulated service environment. Any alteration of the environment should also be tested at this stage. Finally, the ultimate test of a redesigned component is its service life in comparison with the life of the component that is replaced.

It may be possible to reduce operating stress or increase corrosion-fatigue strength so that a replacement part will have a longer service life. It may also be possible to coat the part to prevent access of corrosive environment (at least temporarily) or to modify the environmental or electrochemical conditions.

Operating stress may be lowered by reducing either the mean stress or the amplitude of the cyclic stress. This almost always involves a change in component design. Sometimes, only a minor change is required, such as increasing a fillet radius to reduce the amount of stress concentration at a critical location. In other instances, more extensive changes are required, such as significantly increasing the cross-sectional area or adding a strengthening rib. If failure has occurred because of stress raisers introduced during manufacture, changes in manufacturing specifications and quality requirements may be necessary.

Shot peening is usually effective in prolonging fatigue life in air by introducing residual compressive stresses in a metal surface, thus reducing the mean tensile stress at potential crack-initiation sites. In more aggressive environments, however, shot peening may have only limited value, because general corrosion can eventually remove the surface layer and thus the beneficial compressive residual stresses.

Nitriding, which also introduces compressive residual surface stresses, can improve the corrosion-fatigue resistance of steels, particularly when a relatively short life is required. For applications requiring extended life, other measures are more effective.

Material strength is generally increased by alloying, changes in heat treatment, or selection of a material from a different alloy system. Because corrosion may reduce the fatigue strength of a material to only a small fraction of its strength in air, alloying additions or heat treatments that only increase strength, without altering the corrosion resistance of the material, may be of marginal value. Conversely, alloying that improves corrosion resistance can be effective in combating corrosion fatigue. For instance, chromium steels are generally more resistant than carbon steels. Surface alloying, such as laser surface melting and ion-implantation techniques, not only alters the surface composition and microstructure but may also introduce compressive residual stress. These techniques have shown promise in reducing general corrosion and improving crack-initiation resistance. The effectiveness of these surface modification techniques is limited, however, in reducing the corrosion-fatigue crack-propagation susceptibility.

Corrosion effects can be lessened by alloying, by providing galvanic protection, or by altering or removing a corrosive environment. Galvanic protection by sacrificial anodes or applied cathodic currents has been successful in reducing the influence of corrosion on fatigue of metals and alloys exposed to aqueous environments, except in such alloys as high-strength steels that are subject to hydrogen-induced delayed cracking. In similar instances, anodic polarization has been used for protection of stainless steels. With passive alloys or alloys that can be polarized to produce passive behavior, crevices must be avoided because corrosion within crevices may actually proceed more rapidly due to the anode-cathode relationship. For example, in a part made of low-carbon steel, an area under an O-ring can fail by corrosion fatigue in a caustic solution

Fig. 11 Features on the fracture surface of a broken Ti-6Al-4V aircraft component

(a) Light fractograph showing predominantly intergranular fracture. 2×. Thumbnail-shaped area indicated by arrow is the primary origin of corrosion fatigue. (b) TEM fractograph of a plastic-carbon replica showing absence of striations in fatigue area. 5000×. (c) TEM fractograph of a plastic-carbon replica showing dimples on a grain facet. 4500×.

Fig. 13 Intergranular corrosion-fatigue fracture in 4340 steel tested in humid air

SEM fractograph. 425×

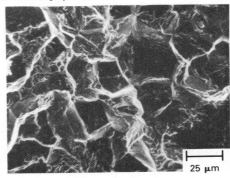

25 μm

with a pH of 12 at a location where only low cyclic stresses exist. Where pitting corrosion has occurred, corrosion pits may act as stress raisers and thus accelerate fatigue failure.

Inhibitors are sometimes added to the environment or included in organic coatings to eliminate corrosion fatigue. The effect of inhibitors is believed to depend solely on their ability to reduce corrosion rates to acceptable values.

Corrosion fatigue cannot occur if there is no contact between the surface of a susceptible material and a corrosive environment; however, ordinary fatigue can occur. In most instances, the environment cannot be prevented from contacting the component; therefore, coatings are necessary. Continuity of the coating is important; organic coatings, such as paint or plastic, are only physical barriers (unless they contain inhibitors) and therefore must be absolutely continuous to be effective. The density and the thickness of coatings are also significant factors in the prevention of corrosion fatigue. Noble-metal coatings can be effective, but only if they remain unbroken and are of sufficient density and thickness. The relatively low corrosion-

fatigue strength of carbon steel is reduced still further when local breaks in a coating such as nickel occur. Presumably, noble-metal coatings that contain residual compressive stresses are more effective than coatings that contain residual tensile stresses.

Corrosion-Fatigue Failure Analyses

Example 1: Corrosion-Fatigue Cracking in an AMS 6415 Steel Aircraft Shaft. The hollow, splined alloy steel aircraft shaft shown in Fig. 14 cracked in service after more than 10 000 h of flight time. The crack was detected during a routine overhaul that was conducted about 1000 h after the previous overhaul.

In operation, the shaft was subjected to complex loading that consisted mainly of steady radial and cyclic torsional loading and reversed cantilever bending. The inner surface of the hollow shaft was continuously exposed to hydraulic oil at temperatures of 0 to 80 °C (30 to 180 °F).

The shaft was specified to be machined from an AMS 6415 steel forging (approximately the same composition as 4340 steel), then quenched and tempered to a hardness of 44.5 to 49 HRC. After tempering, the tapered interior of the 150-mm (6-in.) diam shank (Fig. 14a) was machined to close tolerances and had a smooth finish.

Investigation. Preliminary visual and low-magnification examination of the shaft revealed a crack about 45 mm (1¾ in.) long on the tapered interior of the shank (Fig. 14a). The crack extended deep into the part and penetrated completely through the shank in one region. Scattered pitting corrosion had occurred on the interior surface of this shank, and corrosion pits were visible along the cracks. The pitting corrosion indicated that a corrosive substance, probably water in the hydraulic oil, had

Fig. 12 Fracture-surface features of a sand-cast magnesium alloy AZ91C-T6 aircraft-generator gearbox that failed by corrosion fatigue

(a) Light fractograph showing origin (arrow) and beach marks. 1.2×. (b) TEM fractograph (plastic-carbon replica) showing lack of fracture detail at origin. 2700×. (c) TEM fractograph (plastic-carbon replica) showing fatigue striations formed during later crack-growth stages. 5000×

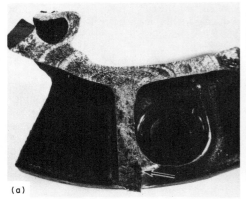

Fig. 14 Hollow, splined alloy steel shaft that failed by corrosion fatigue in aircraft service because of exposure to hydraulic oil that was contaminated with water

(a) View of a portion of the shaft showing the location of the corrosion-fatigue crack on the tapered interior of the shank. Dimensions given in inches. (b) Light fractograph showing fracture origins at corrosion pits (arrows O) and direction of fast fracture (arrows R). 4½×. Region where crack was opened in the laboratory, by sawing and then breaking by hand, is shown at upper right corner.

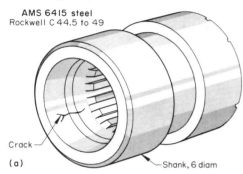

AMS 6415 steel
Rockwell C 44.5 to 49

Crack
(a)
Shank, 6 diam

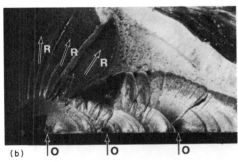

(b)   O   O   O

been present in service. No material or fabrication defects were detected, nor was there any other service damage that could be related to the failure.

Chemical analysis showed that the steel conformed in composition to the requirements of the specified material, AMS 6415 steel. Hardness measurements were within the range required for this part. The microstructure was free from any abnormalities or defects.

Visual and low-magnification (15 to 30×) examination of the fracture surfaces showed that the failure had originated at several corrosion pits. Several small corrosion-fatigue cracks has propagated from these nuclei and had joined to form a single corrosion-fatigue crack in the tapered shank (Fig. 14b). Figure 14(b) shows specific features and successive stages in the development of the crack, together with a region of fast, brittle fracture through the remaining 3 mm (⅛ in.) of shank thickness.

Conclusions. The shaft cracked in a region subjected to severe static radial, cyclic torsional, and cyclic bending loads. Cracking originated at corrosion pits on the smoothly finished surface and propagated as multiple small corrosion-fatigue cracks from separate

nuclei. After progressing separately to a substantial depth, the small cracks combined into a single larger crack that enlarged until the normal service stresses produced brittle cracking through to the opposite surface.

The originally noncorrosive environment (hydraulic oil) became corrosive in service because of the introduction of water into the oil. Corrosion pits acted in combination with cyclic torsional and reversed-bending stresses to cause corrosion-fatigue cracking.

Recommendations. It was recommended that additional precautions be taken in operation and maintenance to prevent the use of oil containing any water through filling spouts or air vents. Also, polishing to remove pitting corrosion (but staying within specified dimensional tolerances) was recommended as a standard maintenance procedure for shafts with long service lives. Inspection and polishing should be incorporated into normal overhaul procedures.

Example 2: Corrosion-Fatigue Fracture of an H21 Tool Steel Safety-Valve Spring in Moist Air. The safety valve on a steam turbogenerator was set to open when the steam pressure reaches 2400 kPa (348 psi). The pressure had not exceeded 1790 kPa (260 psi) when the safety-valve spring shattered into 12 pieces, two of which are shown in Fig. 15(a). The steam temperature in the line varied from about 330 to 400 °C (625 to 750 °F). Because the spring was enclosed and mounted above the valve, its temperature was probably slightly lower. The 195-mm (7¾-in.) outside diameter × 305-mm (12-in.) long spring was made from a 35-mm (1⅜-in.) diam rod of H21 hot-work tool steel. It has been in service for about four years and had been subjected to mildly fluctuating stresses.

Investigation. Visual examination of the broken spring revealed that all the individual fractures had thumbnail-shaped origins along the top surface of the wire (Fig. 15b). The origins were typical of fatigue, but the remaining portions of the fractures were brittle. The shape of the fracture surfaces and the fact that the spring had shattered into 12 pieces suggested that the spring had been uniformly loaded, predominantly in torsion, but that the fatigue cracks had introduced stress raisers at multiple points. Chemical analysis of the metal in the spring disclosed that it was within the composition range for H21 tool steel, except that the tungsten content was 7.8% instead of the usual 9 to 10%.

Sections were taken approximately 90° to the fracture surface and through the origins of three of the fractures. Examination of these sections revealed corrosion pits adjacent to the fracture surfaces, with cracks emanating from the bases of some of them. In one section, a corrosion pit approximately 1 mm (0.045 in.) in diameter had initiated two cracks 0.7 and 1.5 mm (0.028 and 0.060 in.) deep. The cracks contained an iron oxide corrosion product that enclosed spheroidal carbide particles.

Fig. 15 H21 tool steel safety-valve spring that fractured from corrosion fatigue in moist air

(a) Photograph of two of the 12 pieces into which the spring shattered. 0.3×. (b) Light fractograph showing typical corrosion-fatigue origin (arrow) and brittle final fracture. 0.7×

(a)

(b)

The microstructure of the spring consisted of spheroidal carbide particles in a matrix of tempered martensite. The surface of the spring had been decarburized, and the surface grains were larger than those in the interior. This condition probably resulted during heat treatment and may or may not have accelerated the rate of corrosion.

Conclusions. The spring failed by corrosion fatigue that resulted from application of a fluctuating load in the presence of a moisture-laden atmosphere.

Corrective Measures. All safety valves in the system were replaced with new open-top valves that had shot-peened and galvanized steel springs. No further failures occurred for more than four years following this action. Both the zinc coating, which provided galvanic protection for the spring, and the open-top valve construction, which allowed free circulation of air with a reduced concentration of moisture, were considered instrumental in eliminating corrosion fatigue of these valve springs. Alternatively, the valve springs could have been made from a corrosion-resistant metal—for example, a 300 series austenitic stainless steel or a nickel-base alloy, such as Hastelloy B or C.

SELECTED REFERENCES

- C.J. Beevers, Ed., *Fatigue 84: Conference Proceedings of the 2nd International Conference on Fatigue and Fatigue Thresholds*, The Chameleon Press Ltd., 1984
- J.J. Burke and V. Weiss, Ed., Fatigue: Environment and Temperature Effects, *Sagamore Army Mater. Res. Conf. Proc.*, Vol 27, 1983
- T.W. Crooker and B.N. Leis, Ed., *Corrosion Fatigue*, STP 801, ASTM, Philadelphia, 1983
- O.F. Devereux, A.J. McEvily, and R.W. Staehle, Ed., *Corrosion Fatigue: Chemistry, Mechanics and Microstructures*, National Association of Corrosion Engineers, Houston, TX, 1971
- J.T. Fong, Ed., *Fatigue Mechanisms*, STP 675, ASTM, Philadelphia, 1979
- Z.A. Foroulis, Ed., *Environment-Sensitive Fracture of Engineering Materials*, Conference proceedings, American Institute of Mining, Metallurgical, and Petroleum Engineers, Warrendale, PA, 1979
- R.P. Gangloff, Ed., *Embrittlement by the Localized Crack Environment*, Conference proceedings, American Institute of Mining, Metallurgical, and Petroleum Engineers, Warrendale, PA, 1984
- K.L. Maurer and F.E. Matzer, Ed., *Fracture and the Role of Microstructure: Proceedings of the 45th European Conference on Fracture*, The Chameleon Press Ltd., 1982
- P.R. Swann, F.P. Ford, and A.R.C. Westwood, Ed., *Mechanisms of Environment Sensitive Cracking of Materials*, The Metals Society, London, 1977
- J.M. Wells and J.D. Landes, Ed., *Fracture: Interactions of Microstructure, Mechanisms, and Mechanics*, Conference proceedings, American Institute of Mining, Metallurgical, and Petroleum Engineers, Warrendale, PA, 1985

Elevated-Temperature Failures

AT ELEVATED TEMPERATURE, the service life of a metal component subjected to either vibratory or nonvibratory loading is predictably limited. In contrast, at lower temperatures and in the absence of a corrosive environment, the life of a component in nonvibratory service is unlimited, provided the operational loads do not exceed the yield strength of the metal. Stress imposed at elevated temperature produces a continuous strain in the component and results in a phenomenon known as creep. Creep, by definition, is time-dependent strain occurring under stress. After a period of time, creep may terminate in fracture by stress rupture (also called creep rupture).

The conditions of temperature, stress, and time under which creep and stress-rupture failures occur depend on the metal or alloy, its microstructure, and on the service environment. Consequently, elevated-temperature failures may occur over a wide range of temperatures. In general, however, creep occurs in any metal or alloy at a temperature slightly above the recrystallization temperature of that metal or alloy; at such a temperature, atoms become sufficiently mobile to allow time-dependent rearrangement of structure.

It has been suggested that for a given metal elevated temperature begins at about one-half the absolute melting temperature, T_m, of that metal, but this is an oversimplification. Actually, the temperature at which the mechanical strength of a metal becomes limited by creep rather than limited merely by yield strength is not directly related to melting temperature; consequently, elevated temperature must be determined individually for each metal or alloy on the basis of behavior. Elevated-temperature behavior begins approximately at $0.54T_m$ (205 °C, or 400 °F) for aluminum alloys, $0.3T_m$ (315 °C, or 600 °F) for titanium alloys, $0.36T_m$ (370 °C, or 700 °F) for low-alloy steels, $0.49T_m$ (540 °C, or 1000 °F) for austenitic iron-base high-temperature alloys, $0.56T_m$ (650 °C, or 1200 °F) for nickel-base and cobalt-base high-temperature alloys, and 0.4 to $0.45T_m$ (980 to 1540 °C, or 1800 to 2800 °F) for refractory metals and alloys.

The principal types of elevated-temperature mechanical failure are creep and stress rupture, low-cycle or high-cycle fatigue, thermal fatigue, tension overload, and combinations of these, as modified by environment. Generally, the type of a failure is established by examina-

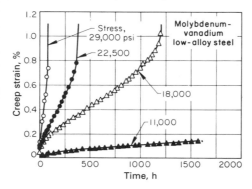

Fig. 1 Creep curves for a molybdenum-vanadium low-alloy steel under tension at four stress levels at 600 C (1110 F)

tion of fracture surfaces and comparison of component-operating conditions with available data on creep, stress-rupture, tension, elevated-temperature fatigue, and thermal fatigue properties. Such an analysis is usually sufficient for most failure investigations, but a more thorough analysis may be required when stress, time, temperature, and environment have acted to change the metallurgical structure of the component (see the section "Metallurgical Instabilities" in this article).

Creep

Some typical creep curves are plotted in Fig. 1. Most creep curves consist of three distinct stages (Fig. 2). Following initial elastic strain resulting from the immediate effects of the applied load, there is a region of increasing plastic strain at a decreasing strain rate (first-stage, or primary, creep). Following first-stage creep is a region of nominally constant rate of plastic strain (second-stage, or secondary, creep). Finally, there is a region of drastically increased strain rate with rapid extension to fracture (third-stage, or tertiary, creep).

The region of initial elastic strain under load has not been given a numerical stage designation, because it is common practice to ignore this contribution to total strain when plotting creep curves. Consequently, creep curves generally show only the time-dependent plastic strain that follows the initial elastic (and possi-

bly plastic) strain. Although this procedure is acceptable for research or investigation, the initial strain, which may amount to a substantial fraction of the total strain, should not be omitted from design studies and failure analyses. Thermal expansion may impose additional strain and must also be considered.

Primary creep, also known as transient creep, represents a stage of adjustment within the metal during which rapid thermally activated plastic strain, which occurs in the first few moments after initial strain, decreases in rate as crystallographic imperfections within the metal undergo realignment. This realignment leads to secondary creep.

Secondary creep is an equilibrium condition between the mechanisms of work hardening and recovery; it is also known as steady-state creep. A constant creep rate that is the lowest for the metal under given service conditions of stress, temperature, and environment is assumed to be characteristic of this stage. This constant creep rate is generally known as the minimum creep rate (MCR) and is widely used in research and engineering studies. Minimum creep rate does not, in fact, continue for a significant period of time.

Figure 3, which illustrates the variation of strain rate with time in a constant-load creep test carried to fracture, shows that strain rate begins at a high value, then decreases rapidly until a minimum is reached. This minimum is within the range of steady-state creep, although it amounts to no more than a bottoming-out of the strain-rate-time curve, and is promptly followed by an increase in strain rate that continually increases until the metal fractures. Secondary creep, therefore, is essentially a transition between primary and tertiary creep even though it often occupies the major portion of the duration of the creep test. On the strain-rate-time curve, this transition zone in creep-resistant materials is sufficiently flat so that the minimum creep rate is applicable to virtually all of secondary creep. For these materials, the minimum creep rate can be empirically related to rupture life.

Tertiary Creep. Primary creep has no distinct end point, and tertiary creep has no distinct beginning. Tertiary creep refers to the region of increasing rate of extension that is followed by fracture. Principally, it may result from metallurgical changes, such as recrystallization under load, that promote rapid in-

Fig. 2 Schematic creep curves showing the three stages of creep

Curve A, constant-load test; curve B, constant-stress test

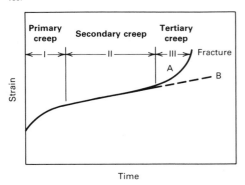

Fig. 3 Relationship between strain rate, or creep rate, and time during a constant-load creep test

The MCR is attained during secondary creep.

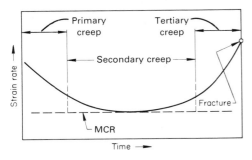

Fig. 4 Creep curves showing no primary creep and no tertiary creep

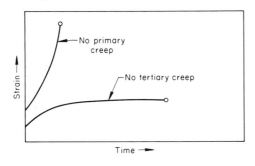

Fig. 5 Intergranular cracks in stress rupture

(a) Schematic view of cracking due to grain-boundary sliding. Arrows along a grain boundary indicate that this boundary underwent sliding. (b) Cracks and voids in Al-5.10Mg that was stress rupture tested at 260 °C (500 °F). Electrolytically polished. 60×

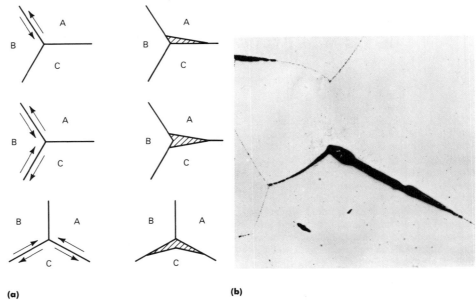

(a)   (b)

creases in deformation, accompanied by work hardening that is insufficient to retard the increased flow of metal. In service or in creep testing, tertiary creep may be accelerated by a reduction in cross-sectional area resulting from cracking or localized necking. Environmental effects, such as oxidation, that reduce cross section may also initiate tertiary creep or increase the tertiary-creep rate. In many commercial creep-resistant alloys, tertiary creep is apparently caused by inherent deformation

processes and will occur at creep strains of 0.5% or less.

In designing components for service at elevated temperatures, data pertaining to the elapsed time and extension that precede tertiary creep are of the utmost importance; design for creep resistance is based on such data. However, the duration of tertiary creep is also important because it constitutes a safety factor that may allow detection of a failing component before catastrophic fracture.

Modified Creep. Under certain conditions, some metals may not exhibit all three stages of plastic extension. For example, at high stresses or temperatures, the absence of primary creep is not uncommon, with secondary creep or, in extreme cases, tertiary creep following immediately upon loading. At the other extreme, notably in cast alloys, no tertiary creep can be observed, and fracture may occur with only minimum extension. The creep curves in Fig. 4 illustrate both of these phenomena.

Stress Rupture

A component under creep loading will eventually fracture (rupture) if the strain occurring during creep does not relieve the stress. Although time-versus-extension data for the plotting of a creep curve can be measured during a stress-rupture test, often only the stress, temperature, time-to-fracture, and total elongation (stress-rupture ductility) are recorded.

Stress-Rupture Fracture. Depending on the alloy, the appearance of the stress-rupture fracture (creep fracture) may be macroscopically brittle or ductile. Brittle fracture is inter-

granular and occurs with little or no elongation or necking. Ductile fracture is transgranular and is typically accompanied by discernible elongation and necking. Intergranular cracking may not be readily discernible on the surface of a part; however, if the oxide scale developed during elevated-temperature stressing in air is removed, such cracking will usually be visible. It can be readily seen in a polished longitudinal section and may extend beyond the fracture zone. Fractographs of some stress-rupture fractures will exhibit both transgranular and intergranular fracture paths. In such cases, it is usually found that the transgranular fractures were initiated by prior intergranular fissures that decreased the cross-sectional area and increased the stress.

Intergranular fractures typically start at grain-boundary triple points—intersections of single boundaries with free surfaces, intersections of single boundaries with second phases, or intersections of two or more boundaries. Wedge-shaped cracks (Fig. 5) form at triple points. Voids or creep cavities can form along grain boundaries, particularly those having high densities of precipitated second-phase particles, without the need for intersecting boundaries. This process is called cavitation creep. The presence of grain-boundary voids in the microstructure on either side of an elevated-temperature fracture is evidence that it is a stress-rupture fracture.

The type of fracture depends on temperature and on strain rate. At constant temperature, the occurrence of either transgranular or intergranular fracture depends on strain rate. Conversely, at constant strain rate, the type of fracture depends on temperature. In general,

Fig. 6 Logarithmic plot of stress-rupture stress versus rupture life for alloy S-590

See text for details.

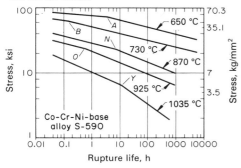

Table 1 Typical elevated-temperature ductility of a 1.25Cr-0.5Mo low-alloy steel and type 316 stainless steel

Time-to-rupture, h	Total elongation in 38 mm ($1\frac{1}{2}$ in.)	Reduction of area, %
1.25Cr-0.5Mo low-alloy steel at 540 °C (1000 °F)		
8.7	19.6	41
47.0	12.1	29
259.4	14.0	20
660.6	10.5	13
2162.0	17.6	32
Type 316 stainless steel at 705 °C (1300 °F)		
3.1	26.6	26
3.7	24.4	27
37.5	17.8	28
522.6	56.2	41
881.6	39.4	33

Fig. 7 Relation of elongation and rupture life for alloy S-590 tested at two temperatures

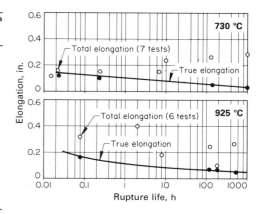

Fig. 8 Schematic creep curves for alloys having low and high stress-rupture ductility

The curves show the increased safety margin provided by the alloy with high stress-rupture ductility.

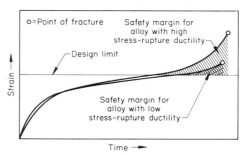

lower creep rate and higher temperatures promote intergranular fractures.

Other significant microstructural features are also associated with stress-rupture fracture. When intergranular cracking occurs, the crack paths follow grain boundaries at and beneath the fracture surfaces, and the grains appear equiaxed even after considerable plastic deformation and total elongation. In contrast, transgranular fracture results in severely elongated grains near the fracture. The degree of elongation varies with certain metallurgical factors, particularly the prior condition of the metal and its susceptibility to recrystallization or grain-boundary migration under given service or test conditions.

Stress-Rupture Curves. The creep curve describes the behavior of material under one set of conditions. It is desirable to illustrate graphically the stress-rupture behavior over a wide range of conditions both for design purposes and for improved metallurgical knowledge of the failure process. Stress-rupture is a variable-rate situation, but logarithmic plots of stress (σ) versus rupture life give straight-line approximations and provide an indication of metallurgical instabilities. These instabilities are delineated by the occurrence of straight-line segments of changed slope on the log-log plot, as shown by points A, B, N, O, and Y in Fig. 6.

Stress-Rupture Ductility. The two common measures of ductility are elongation and reduction of area. In the standard tension-creep curve (Fig. 2), two measures of elongation are of interest: true elongation, which is defined as the elongation at the end of the second stage of creep, and total elongation, which is the elongation at fracture (and which may be measured as total strain, or the last creep reading recorded before fracture). In some cases, elongation at fracture is mainly extension caused by crack separation. In others, it is mainly necking extension or extension resulting from other tertiary-creep processes. True elongation, on the other hand, nominally consists of extension resulting from nonlocalized creep processes only, although it frequently includes some elongation due to intergranular void formation.

Ductility data from stress-rupture tests are generally erratic, even in replicate tests, and they are more erratic for castings than for wrought products. In some metals and alloys, the values of total elongation follow a smooth curve that increases or decreases with increasing rupture time and temperature. However, many metals exhibit an apparently unpredictable series of maximums and minimums that are essentially without significance.

Table 1 lists data on typical total elongation and reduction of area for a low-alloy steel at 540 °C (1000 °F) and a stainless steel at 705 °C (1300 °F). These ductility data do not reflect the microstructural changes occurring in these alloys at the test temperatures. Despite obvious difficulties in interpretation of ductility data, it is common practice to plot total elongation versus rupture life. The open circles shown in Fig. 7 represent such data for copper-chromium-nickel-base alloy S-590 at 730 and 925 °C (1350 and 1700 °F). At both temperatures, considerable scatter of data points is observed.

True elongation is claimed to give a more accurate representation of the ductility behavior of metals than total elongation. As shown by the solid circles in Fig. 7, the variation of true elongation with rupture life for alloy S-590 (an iron-nickel-chromium-cobalt heat-resistant alloy) follows a smooth curve, with true elongation decreasing with decreasing strain rate (longer rupture life). On the other hand, total elongation for alloy S-590 exhibits no clear-cut correlation with rupture life.

The large differences between total and true elongation, such as those shown in Fig. 7, are a function of crack volume and distribution in tertiary creep. The differences in crack initiation and growth can make only relatively small differences in rupture time, but they account for large differences in total elongation. Replicate tests may thus exhibit large differences in total elongation but very small differences in true elongation.

Stress-rupture ductility is an important factor in alloy selection. In conventionally cast nickel-base superalloys, for example, a stress-rupture ductility of about 1% is common at 760 °C (1400 °F), compared to values above 5% for the strongest wrought superalloys. Because component designs are frequently based on 1% creep, a low stress-rupture ductility may preclude use of an alloy to its full strength potential. As shown by the schematic creep curves in Fig. 8, a higher rupture ductility for the same load and temperature conditions means a higher safety margin. Premature failures have resulted from lack of ductility during tertiary creep. Thermal fatigue resistance may also be related to rupture ductility. Generally, superalloys with the highest ductilities for a given strength level show the greatest resistance to thermal fatigue.

Stress Concentration. Strength and stress-rupture ductility must be considered in the design of components for service at elevated temperatures. Normally, these properties are determined quantitatively by smooth-bar stress-rupture tests. In most service applications, components are subjected to complex stresses, including stress concentrations from inherent flaws in the metal as well as those from design.

Results of smooth-bar tests do not account for such stress conditions. Fortunately, these conditions can be partly reproduced in the laboratory with the notched-bar test. Notched-bar stress-rupture tests are of principal interest because of the relationship between stress-rupture ductility and the ratio of notched-bar to smooth-bar stress-rupture strength.

It was discovered that alloys with smooth-bar stress-rupture ductilities of less than 5% (as measured by reduction of area) were usually notch sensitive, whereas alloys with smooth-bar stress-rupture ductilities above 5% were rarely notch sensitive (Ref 2). A better correlation was later obtained by direct use of notched-bar stress-rupture ductility (also measured by reduction of area); on the basis of this correlation, notch sensitivity was determined to occur at notched-bar stress-rupture ductilities below 3% (Ref 3).

In components that contain notches, stress-rupture behavior is determined by notch configuration and by the ductility available to relieve the stress concentration at the roots of notches. For materials ordinarily used in high-temperature components, a ratio of notched-bar to smooth-bar stress-rupture strength of less than 1 indicates marginal ductility. Obviously, notch configuration can affect the notch sensitivity of real structures (notch sensitivity increases with notch sharpness). However, it is doubtful that a simple relationship is entirely satisfactory for applying data on stress-rupture ductility to all types of high-temperature structures.

Elevated-Temperature Fatigue

In service, the steady loads, or strains, to which components are subjected are often accompanied by mechanically induced cyclic loads that are responsible for failure by fatigue. The fatigue properties of metals are normally presented on an *S-N* curve, which plots maximum stress against the number of cycles before fracture. The effect of temperature on fatigue strength is marked: fatigue strength decreases with increasing temperature. However, the precise relationship between temperature and fatigue strength varies widely, depending on the alloy and the temperature to which it is subjected.

The fatigue behavior of structures that are loaded in combined steady-state and vibratory loading is complex. Combined creep-fatigue loads result in substantially decreased life at elevated temperatures as compared with that anticipated in simple creep loading, and this effect must be considered in failure analysis.

Thermal Fatigue

Mechanical vibration is not the only source of cyclic loads. Transient thermal gradients within a component can induce plastic strains, and if these gradients are repeatedly applied, the resulting cyclic strain can induce compo-

Fig. 9 Schematic of the distortions in shape that occur in the airfoils of turbine blades as a result of temperature cycling or uneven cooling

Failure by fracture in thermal fatigue is caused by these cyclic thermal stresses.

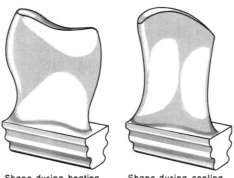

Shape during heating   Shape during cooling

nent failure. This process is known as thermal fatigue. Figure 9 shows the effects of the strains induced by thermal transients on the airfoils of blades for gas-turbine engines. The thermal strains are generated in the airfoils because the outer surfaces change temperature more rapidly than the metal within. The effects of thermal fatigue may develop in airfoils after a relatively short service life (10 000 to 100 000 cycles). In turbine airfoils, creep strains are superimposed on thermal strains and thus account for a further reduction in life expectancy.

Cracking can occur in many structures during high-temperature operation. An example is the cracking that occurs in heat-treating fixtures, as described in the section "Heat-Treating Fixtures" in this article.

Thermal Fatigue Fracture. Thermal fatigue cracks initiate along the surface and progress inward. They are oriented normal to the surface and may occur singly or in multiples. Because the crack initiates externally, the amount of corrosion or oxidation along the surface of a thermal fatigue crack is inversely proportional to the depth of the crack. In stress-rupture cracks, on the other hand, oxidation is more uniform. Thermal fatigue cracks may progress intergranularly, but they usually progress transgranularly.

Excessive creep produces numerous subsurface cracks, whereas thermal fatigue usually produces relatively few surface-oriented cracks. Fracture surfaces of components cracked in stress rupture are typically irregular and discontinuous, in contrast to the planar continuous surfaces of thermal fatigue fractures.

Creep-Fatigue Interaction

Creep-fatigue interaction can have a detrimental effect on the performance of metal parts or components operating at elevated temperatures. At high temperatures, creep strains and

cyclic (fatigue) strains can be present; interpretation of the effect that one has on the other becomes extremely important. For example, it has been found that creep strains can seriously reduce fatigue life and/or that fatigue strains can seriously reduce creep life. For detailed information, see the article "Creep-Fatigue Interaction" in Volume 8 of the 9th Edition of *Metals Handbook*.

Metallurgical Instabilities

Stress, time, temperature, and environment may change the metallurgical structure during testing or service. They may thus contribute to failure by reducing strength, although some changes may enhance strength. These structural changes are also referred to as metallurgical instabilities and, although they influence all types of failure, are most conveniently described in terms of their influence on stress-rupture properties. A sharp change in the slope of a log σ versus log RL (rupture life) curve, that is, a break in the curve, can be ascribed to metallurgical instability. Sources of instabilities include transgranular-intergranular fracture transition, recrystallization, aging or overaging, phase precipitation or decomposition of carbides, borides, or nitrides, intermetallic-phase precipitation, delay transformation to equilibrium phase, order-disorder transition, general oxidation, intergranular corrosion, stress-corrosion cracking (SCC), slag-enhanced corrosion, and contamination by trace elements.

Transgranular-intergranular fracture transition is the primary metallurgical factor in stress-rupture behavior. The temperature at which the transition (frequently called the equicohesive transition) occurs is called the equicohesive temperature (ECT). The transition occurs because the properties of grain-boundary regions differ from those of grains. At low temperatures, grain-boundary regions are stronger than grains; therefore, deformation and fracture are transgranular. At high temperatures, grain boundaries are weaker than grains, and deformation and fracture are largely intergranular.

The ECT varies with exposure time and stress. For each combination of stress and rupture life, there is a temperature above which all stress-rupture fractures will be intergranular. For lower stresses, this transgranular-to-intergranular transition will occur at higher temperatures, and for higher stresses, the transition will occur at lower temperatures. This effect is illustrated by points *A* and *B* in Fig. 6 (points *N*, *O*, and *Y* represent other types of metallurgical instability). Under certain conditions, features of both transgranular fracture and intergranular fracture will be found; consequently, an analysis of rupture-life data or component failure is not complete without a thorough metallographic examination to establish the initial failure mechanism.

Fig. 10 Logarithmic plot of stress-rupture stress versus rupture life for nickel-base alloy B-1900

The increasing slope of the curves to the right of the γ' break is believed to be caused by γ' coarsening due to overaging.

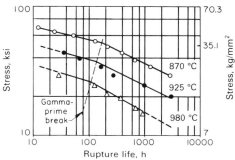

Fig. 11 Logarithmic plot of stress-rupture stress versus rupture life for nickel-base alloy U-700 at 815 °C (1500 °F)

The increasing slope of the curve to right of the σ break is caused by σ-phase formation.

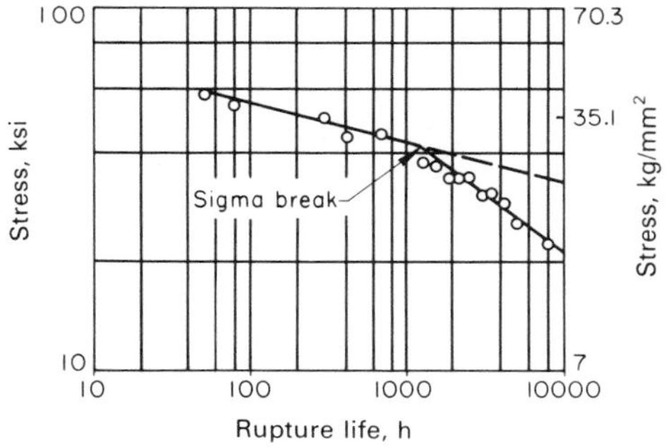

Fig. 12 Grain-boundary carbide films in a Waspaloy forging

The films substantially reduced stress-rupture life. The specimen was electropolished before replication in a solution containing (by volume) 100 parts hydrochloric acid, 50 parts sulfuric acid, and 600 parts methanol. Transmission electron micrograph. 4000×

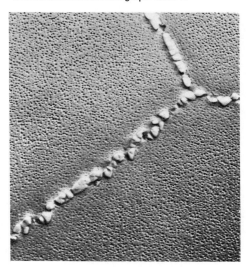

Aging and Overaging. Age-hardening alloys are characteristically unstable; structurally, they are in a state of transition to a stable (equilibrium) condition. Consequently, under creep conditions, it is likely that additional temperature- and stress-induced atomic migration will cause aging to continue, resulting in reduced strength. The extent and nature of this change will depend on several factors, including the condition of the alloy before creep, temperature, stress, and time of exposure.

Some of the more common high-temperature structural alloys that harden as a result of decomposition of highly supersaturated solid solutions include the Nimonic alloys (nickel-chromium-aluminum-titanium alloys), austenitic steels that do not contain strong carbide formers, and secondary-hardening ferritic steels. These alloys are widely used for their creep resistance, but are not immune to reduced rupture life due to overaging.

Chief among the age-hardening phases precipitated in nickel-base systems is γ' $Ni_3(Al,Ti)$ phase. When a high-strength nickel-base alloy is exposed for 1000 to 2000 h to stress at elevated temperatures, a considerable amount of γ' coarsening may occur. The rupture life of a typical cast high-strength nickel-base alloy at three temperatures is plotted in Fig. 10. The dashed line in Fig. 10 designated γ' break indicates the beginnings of slight increases in slope of the curves, which are believed to be caused by γ'-coarsening due to overaging. Actually, the coarsening was not apparent in the microstructure at the points where the breaks occurred, but was observed metallographically after longer tests were completed.

Intermetallic-Phase Precipitation. Topologically close-packed (tcp) phases, such as σ, μ, and Laves phases, form at elevated temperatures in austenitic high-temperature alloys. Not all of the effects of such phases on rupture life are well known. Figure 11 shows the effect of σ-phase formation on the stress-rupture life of nickel-base alloy U-700 at 815 °C (1500 °F). Starting at about 1000 h, a pronounced break was found in the slope of the rupture curve. The difference between the extrapolated life and the actual life at 207 MPa (30 ksi) was about 5500 h, representing a decrease of about 50% in expected life. Sigma phase was identified in this alloy system and was clearly associated with the failure because the voids formed by creep occurred along the periphery of σ-phase particles. However, it was found that σ did not have a similar effect on alloys 713C and U-520. Sigma phase, therefore, does not seem to have a universally deleterious effect on stress-rupture behavior. The amount, location, and shape of σ-phase precipitation determine whether σ strengthens or weakens an alloy or has no effect.

The inconsistency of the effect of σ-phase formation on creep and stress-rupture properties may arise from the simultaneous presence of other phases, such as carbides. The shape and distribution of carbide particles can influence crack initiation and propagation (and hence the resultant stress-rupture ductility and rupture life) in a pronounced manner. It is improbable that σ phase affects ductility to a significant degree at low strain rates. Consequently, σ need not always result in deterioration of creep and stress-rupture properties unless it is present in relatively large amounts. Because the presence of σ does not automatically result in decreased rupture life, it is imperative that careful metallographic work be carried out on failures to ensure discrimination between σ-promoted failures and other types of failures in which σ or other tcp phases are merely present.

Carbide Reactions. A variety of carbide types are found in steels and superalloys. Although temperature and stress affect carbides found within grains and grain-boundary carbides, their effects on grain-boundary carbides are usually a much more significant factor in altering creep behavior. The presence of carbides is considered necessary for optimum creep and stress-rupture behavior in polycrystalline materials; however, the subsequent alteration in their shape or the breakdown and transition to other carbide forms may be sources of property degradation.

In acicular form, grain-boundary carbides do not appear to act as brittle notch formers that might directly affect rupture life at elevated temperatures, but they may reduce impact strength. Indirectly, compositional changes in the vicinity of carbides can alter rupture strength. In general, acicular M_6C carbides are not believed to affect the properties of nickel-base high-temperature alloys greatly unless the alloying elements involved in the carbide reaction alter the matrix composition noticeably.

Carbide films formed at grain boundaries can decrease rupture life. Figure 12 shows an electron micrograph of grain-boundary carbide films in a Waspaloy forging; these films substantially reduced the stress-rupture life of the alloy.

Interaction of Precipitation Processes. Frequently, carbide and tcp phases do not act independently; in fact, they may interact with each other as well as with other precipitating phases. The relationship between σ, carbide, nitride, and ferrite precipitation from austenite has been studied in 18Cr-8Ni stainless steels with high and low chromium contents and high and low carbon-plus-nitrogen contents. Stress-rupture curves for these alloys are shown in Fig. 13 and 14. The curves in Fig. 13 reflect the effects of varying chromium content when the carbon-plus-nitrogen content is low (0.05%) and constant. The curves in Fig. 14 reflect the effects of varying carbon-plus-nitrogen content when the chromium content is high (20%) and constant.

The curves in Fig. 13 indicate that ferrite forms in alloys with low interstitial content

Fig. 13 Effect of varying chromium content on stress-rupture life of an 18Cr-8Ni stainless steel with low carbon-plus-nitrogen content

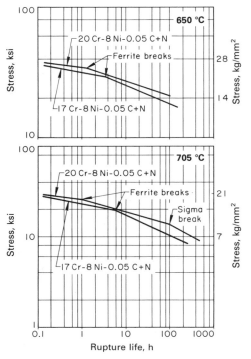

Fig. 14 Effect of varying carbon-plus-nitrogen content of stress-rupture life of an 18Cr-8Ni stainless steel with high chromium content

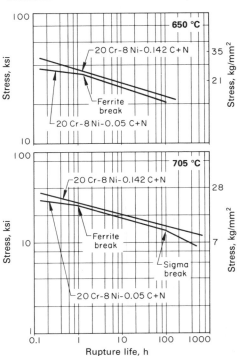

when the chromium content is high (20%) or low (17%). However, high chromium content promotes σ-phase formation. Ferrite and σ are interrelated; precipitation of ferrite ultimately promotes formation of σ. Modifying the composition of the steel inhibits ferrite and σ-phase formation.

A significant feature in Fig. 13 is the evidence that small amounts of ferrite or σ may enhance rupture life, although increased amounts formed by long exposure result in deterioration. The strengthening action of σ can be seen by comparing the curves after the ferrite breaks in Fig. 13. The slope of the σ-prone 20% Cr alloy is less than that of the 17% Cr alloy, indicating that the rate of strength decrease with time is less when small amounts of σ are acting to dispersion harden the alloy. Eventually, the small initial grain-boundary σ-phase precipitates agglomerate, and larger acicular σ also forms, reducing strength.

Environmentally Induced Failure

A critical factor in the performance of metals in elevated-temperature service is the environment and resulting surface-environment interactions. In fact, the most important source of elevated-temperature failure requiring premature replacement of a component is environmental degradation of material. Control of en-

vironment or protection of materials (by coating or self-protective oxides) is essential to most elevated-temperature applications.

General oxidation can lead to premature failure; grain-boundary oxidation may produce a notch effect that can also limit life. Some environments may be more harmful than others; attack of fire-side surfaces of steam-boiler tubes by ash from vanadium-bearing fuel oils can be quite severe. Vanadium-ash attack and hot corrosion are equally harmful in gas turbines. For example, Inconel 700 is used for gas-turbine blades under such conditions, and significant loss in rupture life is reported to occur because of hot corrosion at 700 to 750 °C (1290 to 1380 °F).

Salt-containing atmospheres exert a deleterious effect on materials other than steels and high-temperature alloys. For example, hot-salt (stress) corrosion can be induced in titanium alloys kept in contact with salt during high-stress exposure at elevated temperatures. The resistance of titanium alloys to hot-salt corrosion is frequently determined by exposing specimens to salt while creep testing them for a fixed time, for example, 100 h, at the temperature of interest. Subsequent tensile testing reveals susceptibility to hot-salt corrosion cracking. Hot-salt corrosion effects normally are not observed in titanium at temperatures below about 290 °C (550 °F), and in some

alloys, the threshold (crack/no-crack) line is essentially equivalent to the line for 0.1% creep in that alloy for the same exposure time. Thus, in creep tests, no hot-salt instability may be observed. Furthermore, actual components are not likely to be exposed to heavy salt-containing deposits. If they are exposed to such an atmosphere, cyclic-temperature rupture testing indicates that an incubation time exists for the occurrence of hot-salt corrosion instability in titanium.

Corrosion and Erosion-Corrosion

The alloys used in the hot sections of gas-turbine engines are normally exposed to environments that promote corrosion and erosion-corrosion. Consequently, these alloys are selected on the basis of high strength and resistance to surface degradation due to operating environment.

With improvements in the design of gas-turbine engines, gas and metal temperatures were necessarily increased to improve efficiency. The maximum temperatures, however, were primarily limited by the high-temperature strength of the alloys used for turbine applications. In nickel-base high-temperature alloys, increased strength at elevated temperature was attained by lowering the chromium content and by increasing the aluminum and titanium contents to form high volume percentages of the strengthening precipitate γ'. However, the decrease in chromium content also resulted in a reduction of the hot-corrosion resistance of these alloys. Cobalt-base alloys, which are also used for gas-turbine rotors, are susceptible to hot corrosion in a manner similar to the nickel-base high-temperature alloys, although to a lesser degree.

Hot corrosion of nickel-base and cobalt-base alloys is accelerated oxidation caused by the presence of sodium chloride and sodium sulfate. The main sources of corrosive elements in these compounds are the fuel used in the engine and sea salt in the air from operation near seawater. Numerous concepts of the mechanism of hot corrosion have been developed. According to one, a molten sulfur-bearing slag forms on the surface of the engine components, fluxing the normally protective oxide scales on the alloys and resulting in accelerated oxidation.

With the introduction of the lower-chromium, higher-strength nickel-base alloys, hot corrosion became the major cause for the premature removal of first-stage and second-stage turbine blades and vanes in gas-turbine engines. Higher-chromium lower-strength alloys had previously been used with very few reports of sulfidation, probably because the problem was not recognized and because of lower engine-operating temperatures.

Figure 15 shows both sides of two IN-713 turbine blades that were removed from service because of severe corrosion. These blades show evidence of hot corrosion—the characteristic swelling, together with splitting and flaking

Fig. 15 Both sides of two IN-713 turbine blades removed from service because of severe hot corrosion

(a) and (c) Swelling at the trailing edge on the concave sides (at right). (b) and (d) Splitting and flaking along leading and trailing edges of the convex sides. See also Fig. 16.

(a)   (b)   (c)   (d)

along the leading and trailing edges of the airfoil. Metallographic examination of these blades revealed four zones: an oxide layer on the outside surface, then a layer of oxide intermingled with an alloy-depleted nickel matrix, then a zone of chromium sulfide particles in an alloy-depleted matrix, and finally a zone of normal γ' precipitation from the γ matrix. These zones, except for the outer oxide layer, are shown in Fig. 16 for one of the blades in Fig. 15. Electron probe x-ray microanalysis of the oxidized layer showed that the sulfur was associated with chromium in an alloy-depleted zone and that the depleted matrix metal was intermixed with the oxides.

Accelerated hot corrosion has been produced during engine testing by injecting simulated sea salt through the fuel nozzles to provide a salt concentration of 0.75 ppm and by operating the engine at takeoff temperatures. The appearance and surface condition of uncoated IN-713 turbine blades after 45 cycles of this hot-corrosion test were very similar to those of blades returned from service.

In another laboratory hot-corrosion test, a regression equation relating loss in base-metal volume to alloy composition was derived for nickel-base alloys. Chromium was shown to be beneficial to hot-corrosion resistance, whereas molybdenum was detrimental. The severity of hot corrosion increased with temperature throughout the entire testing range (925 to 1040 °C, or 1700 to 1900 °F) in an approximately logarithmic manner (Fig. 17). Limited testing at 1095 °C (2000 °F) indicated that the severity of hot corrosion continued to increase up to this temperature.

Fig. 16 Microstructure of one of the IN-713 turbine blades in Fig. 15 showing corrosion products

Top: Mixed oxides and an alloy-depleted nickel matrix. Center: Sulfide particles in an alloy-depleted matrix. Bottom: γ matrix with γ' precipitate. Etched with ferric chloride. 1000×.

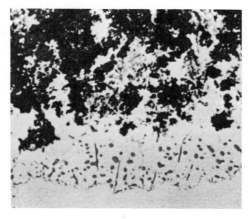

Protective coatings provide a partial solution to the hot-corrosion problem, supplementing the inherent corrosion resistance of most high-temperature alloys. Aluminide-type coatings successfully extend the life expectancies of gas-turbine engine blades and vanes (Fig. 18). Principal processes for applying aluminum-rich coatings are vacuum coating, thermal spray coating, pack cementation, hot dipping, slurry, and electrophoresis; coatings applied by these processes are usually diffused at elevated temperatures. Regardless of the methods used for

Fig. 17 Average volume loss of ten nickel-base high-temperature alloys as a function of hot-corrosion test temperature

The alloys were tested in a laboratory test rig with sodium sulfate as the corrodent.

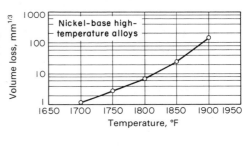

application, basic similarities exist among all of the aluminum coatings. They are composed predominantly of nickel-aluminum on nickel-base alloys or cobalt-aluminum on cobalt-base alloys, with minor additions of an alloying element, usually chromium. Another class of high-temperature coatings is obtained by applying corrosion-resistant alloys, such as cobalt-chromium-aluminum-yttrium and iron-chromium-aluminum-yttrium, to the surfaces of turbine components by means of electron beam vapor deposition.

A nickel-aluminum protective coating on a nickel-base alloy will eventually degrade by hot corrosion and will spall during subsequent operation. Figure 18(c) shows the surface appearance of an aluminide-coated IN-713 turbine blade after accelerated hot-corrosion testing in an engine. The concave and convex surfaces

Fig. 18 Effect of an aluminide-type (nickel-aluminum) coating on the hot-corrosion resistance of an IN-713 turbine blade compared with that of an uncoated blade

(a) Uncoated turbine blade after 118 programmed cycles. 1.8×. (b) Severe degradation of the uncoated blade by hot corrosion. Etched with ferric chloride. 450×. (c) Aluminide-coated blade after accelerated hot-corrosion testing in an engine. 1.8×. (d) Slight degradation of the aluminide coating. Etched with ferric chloride. 450×.

Fig. 19 Localized degradation of the aluminide coating on an Inconel 738 turbine blade after accelerated hot-corrosion testing in an engine

Etched with ferric chloride, hydrochloric acid, and water. 500×

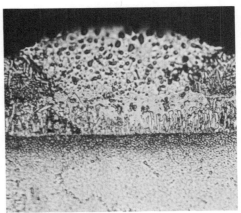

Fig. 20 Erosion-corrosion on leading edge of an aluminide-coated MAR-M 246 alloy turbine blade after simulated flight testing

4×

have bulged and flaked as a result of hot corrosion. Metallographic examination shows that the coating was slightly degraded by hot corrosion (Fig. 18d). Nevertheless, the life of a coated blade represents about a four-to-one improvement over that of an uncoated blade.

Hot corrosion can also attack a coating in a localized area (Fig. 19). Localized attack may result from variations in the chemical composition of the coating or from localized concentration of the corrodents. In addition, erosion-corrosion degradation of the coating can occur at one specific location, such as the leading edge of a turbine blade, because of exposure to the high-temperature high-velocity gases in the turbine. Subsequent attack of the base metal occurs, requiring premature removal of the component from the engine. Figure 20 shows an example of erosion-corrosion.

Coating cracks that result when a coating is strained beyond its ductility limit can also cause premature failure of a turbine blade because hot corrosion, oxidation, or both can penetrate to the base metal and eventually cause spalling of the coating. Figure 21 shows hot-corrosion and oxidation attack of the base metal at the bottom of a crack in the aluminide coating of a turbine blade.

In an oxidizing atmosphere, aluminide coatings form a protective aluminum oxide layer on

Fig. 21 Hot-corrosion and oxidation attack at the bottom of a crack in the aluminide coating of a turbine blade

Etched with ferric chloride, hydrochloric acid, and methanol. 500×

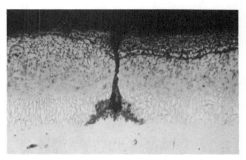

Table 2 Suggested maximum temperatures for operation of wrought stainless steel components in oxidizing atmospheres without excessive scaling

AISI type of steel	Temperature, max	
	°C	°F
(7Cr-Mo)	650	1200
(9Cr-Mo)	675	1250
302, 304	900	1650
302B	980	1800
309	1095	2000
310	1150	2100
316, 321, 347	900	1650
403, 410	705	1300
430	845	1550
442	955	1750
446	1095	2000
501	620	1150

Fig. 22 Effect of carburization on the microstructure of stainless steel

(a) Typical microstructure after carburization showing massive carbides in the austenite matrix. (b) Microstructure of the same steel before carburization. In (a), massive carbides are formed by the reaction of carbon with chromium, which depletes the matrix of chromium in regions adjacent to the carbides. Both etched with a mixture of 10 mL HNO₃, 20 mL HCl, 20 mL glycerol, 10 mL H₂O₂, and 40 mL water. 50×

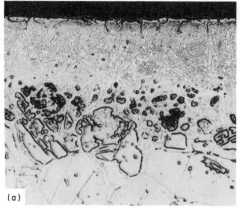 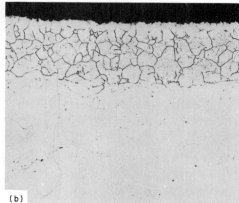

(a) (b)

the blade surface. The cyclic operation of a gas-turbine engine causes the oxide to spall off; subsequently, the coating forms another protective layer of aluminum oxide. The rate of coating degradation depends on the operating environment and will vary for each coating-alloy combination.

General Oxidation

Oxidation, as well as all other reactions of metals with gaseous environments, has long been recognized as a severe limitation to the use of metals at high temperatures. The development of alloys, therefore, is usually aimed at improving oxidation resistance and mechanical properties.

Table 2 provides suggested maximum temperatures for operation of wrought stainless steel components in oxidizing atmospheres without excessive scaling. These temperatures may require modification, depending on the severity of the oxidation rate as influenced by the type of atmosphere and by other factors that can increase the loss of metal. The scaling-temperature or operating-temperature limits given are based on arbitrary standards and indicate the approximate temperatures at

which, after prolonged heating in a strongly oxidizing atmosphere, a scale of sufficient thickness to cause flaking or spalling will form. These stainless steels oxidize superficially at lower temperatures, forming a series of temper colors, until a thin, more or less adherent layer of scale is formed. This action proceeds very slowly until the scaling temperature is reached; at this point, progressive thickening of the scale layer, with spalling, causes much more rapid deterioration of the metal, thus reducing the service life or necessitating the selection of a more highly alloyed material for economical and continued operation. Because of their lower coefficient of thermal expansion, the ferritic stainless steels may be damaged less by scale formation and subsequent flaking during repeated heating and cooling than austenitic steels.

Carburization

The problem of carburization of steel, particularly the carburization of stainless steels used in elevated-temperature furnace environments, is common to many industrial applications. A typical example is the occurrence of carburization, with or without oxidation, in stainless steel resistance-heating elements and various components and fixtures of heat-treating furnaces. The simultaneous carburization and oxidation of stainless steel heating elements results in a form of attack sometimes referred to as green rot. Basically, this results from precipitation of chromium as chromium carbide, followed by oxidation of the carbide particles, and is common in nickel-chromium and nickel-chromium-iron alloys.

Another typical industrial example involves the carburization of certain petrochemical-plant components, such as heater tubes. These tubes may be subjected to carburization, carburization-oxidation, or both. These two processes generate uneven volume changes, which result in

very high internal stresses, together with metal loss due to erosion-corrosion or with embrittlement due to carbon pickup and consequent carbide formation. Ultimately, the corrosion or erosion-corrosion results in a loss of supporting thickness in the structural members, or the carbide formation results in a loss of ductility that renders the component susceptible to brittle fracture. Heavy carburization also eliminates the possibility of repair welding.

Effect of Carburization on Stainless Steels. Diffusion of carbon in stainless steel results in the formation of additional carbides; these carbides may take the form of M_7C_3, $M_{23}C_7$, or M_3C_2. Figure 22(a) shows a typical example of a microstructure developed in stainless steel as a result of carburization; islands of massive carbides have formed in the austenite matrix. In the same steel before carburization, carbides are small and are restricted to the grain boundaries (Fig. 22b). In the carburized steel, chromium has migrated to the carbides, depleting the matrix. As chromium depletion progresses, the relative percentages of nickel and iron increase. These changes can be readily detected by electron microprobe scanning.

The primary significance of carburization is its effect on properties. There is a slight increase in the creep strength of the carburized alloy, together with a change in volume that results from an increased amount of carbide, which has a lower density than that of the original alloy. A density gradient, predicated on carbon content, develops across the carburized zones and into those zones that have not been carburized. In stainless steel heater tubes used in petroleum refineries, carburization occurs on the interior walls of the tubes. This leads to the development of an inner layer of carburized metal and an outer layer of uncarburized metal, each with a different density and a different coefficient of thermal expansion. During thermal cycling, these differences

Fig. 23 Section through a stainless steel weld (bottom) showing typical creep fissuring in carburized weld metal and extending to uncarburized base metal

Etched with a mixture of 10 mL HNO₃, 20 mL HCl, 20 mL glycerol, 10 mL H₂O₂, and 40 mL water. Approximately 1.3×

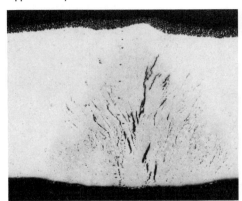

Fig. 24 Wall of a stainless steel tube after exposure to carburizing and oxidizing conditions

(a) After one year of exposure. (b) After several years of exposure. The wall thickness in (b) has been reduced by about one-third by grain detachment and subsequent erosion. Etched with a mixture of 10 mL HNO₃, 20 mL HCl, 20 mL glycerol, 10 mL H₂O₂, and 40 mL water. 4×

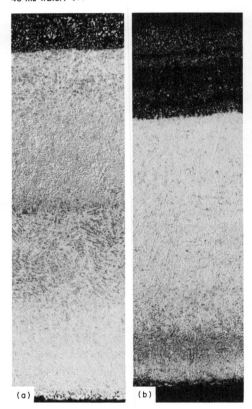

(a)   (b)

Fig. 25 Metal-dusting attack on the inner wall of a stainless steel tube

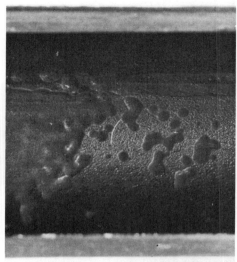

Fig. 26 Magnetic permeability versus chromium content of the matrix of a cast HK-40 alloy tube after carburization

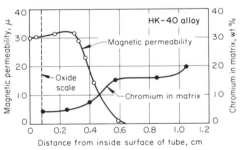

promote the generation of high thermal stresses that can result in tube failure at elevated temperatures. Thermal stresses due to carburization can also be generated in weld metals, as is evidenced by the creep fissuring in the carburized weld metal extending to uncarburized base metal in Fig. 23.

In addition to its adverse effects on density and thermal expansion, carburization of stainless steels contributes to embrittlement by producing a high volume percentage of carbide and an increased susceptibility to attack by oxidation. Oxidation attack is promoted by the depletion of chromium from the matrix by preferential formation of chromium carbides. The attack occurs when metal surfaces are simultaneously or intermittently exposed to heavily carburizing and oxidizing environments. Under these conditions, it is thermodynamically possible for the dual action of carburization and oxidation to weaken the metal. Specifically, in carburized metal, oxidation attack occurs at grain boundaries. Loss of grain-boundary strength is followed by detachment of grains and subsequent erosion. This form of attack slowly eats away the metal grain by grain; an example is shown in Fig. 24, in which the original wall thickness of the tube has been reduced by approximately one-third.

Carburization is also a problem in various nuclear applications in which stainless steel tubing is used in contact with carburizing environments. These environments may be gaseous (CO-CO₂) or liquid, such as liquid sodium, which is capable of transporting carbon. Both types of environment can promote carbide formation and severe embrittlement.

Metal dusting, a type of attack somewhat similar to oxidation attack of carburized metal, has occurred in petrochemical-plant equipment.

Metal dusting can be distinguished from conventional oxidation attack of carburized metal as a result of the fact that it occurs randomly in localized areas and progresses more rapidly. An example of metal-dusting attack is shown in Fig. 25; random pitting is evident on the interior wall of the tube. The cause of this highly accelerated attack is believed to be related to the high localized stresses surrounding the pits—the stresses resulting from volumetric changes.

Metal dusting generally takes place at temperatures from 480 to 815 °C (900 to 1500 °F), although it has occurred at temperatures as high as 1095 °C (2000 °F) in strongly reducing atmospheres, such as those containing large amounts of hydrocarbon gases. Atmospheres that are alternately oxidizing and reducing appear to promote metal dusting, whereas those containing sulfur-bearing compounds, such as hydrogen sulfide, deter it.

Detection in Austenitic Alloys. Several methods have been explored for detecting carburization in austenitic alloys. The most successful of these is a method for detecting magnetic changes; specifically, carburization is measured and defined by measuring changes in magnetic permeability, with the aid of a magnetic eddy-current device. When an austenitic alloy is carburized, metallurgical changes result in the formation of a new matrix composition—a magnetic iron-nickel-chromium alloy. The matrix alterations that occur are proportional to the magnetic change; therefore, carburization and magnetic measurements can be directly related. As shown in Fig. 26 for a tube cast from HK-40 alloy, the actual percentage of chromium depleted from the matrix is proportional to the magnetic permeability. The effectiveness and accuracy of the magnetic method depend on the development and use of standard reference specimens—carburized specimens made from the same alloy and having the same wall dimensions as the tubes or other components being tested.

Heat-treating fixtures, such as trays and baskets, that are used in carburizing and carbonitriding atmospheres usually fail as a

Fig. 27 Effect of section size on resistance to thermal fatigue in alloy RA 330 bars

9.5- and 13-mm (⅜- and ½-in.) diam bars that were taken from the same furnace basket show evidence of carburization, but only the 13-mm diam bar exhibits severe cracking due to thermal fatigue. Etched with mixed acids. Approximately 5×

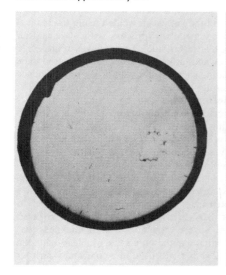 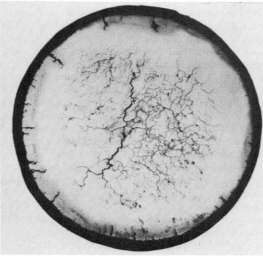

result of carburization where the cross section is relatively thin. When the trays or baskets, which are used to contain or support parts, are subjected to cyclic quenching in a liquid medium, there is also the possibility of failure by thermal fatigue. In fact, fixtures with heavier cross sections are more likely to fail as a result of thermal fatigue than as a result of carburization when they are subjected to both cyclic quenching and a carburizing atmosphere. Failure caused by carburization generally occurs at room temperature when the component embrittled by chromium carbide breaks as a result of impact or jarring. Failure from thermal fatigue is generally characterized by gradual crack growth that proceeds with each thermal cycle until final fracture occurs. Such fracture usually occurs at the central portion of a fixture.

Those furnace trays and containers made of relatively thin sheet (about 3 mm, or 0.120 in. thick) are not subjected to very high thermal stresses during quenching from austenitizing temperatures (about 815 to 925 °C, or 1500 to 1700 °F) and are more susceptible to failure by embrittlement caused by carburization. Failures of pans used in shaker-hearth furnaces are exceptional in that they generally fail in thermal fatigue. The high thermal stresses to which they are subjected arise from the loading of cold workpieces onto the pans. However, at the time of failure, the pans are usually severely carburized.

Also subject to failure by thermal fatigue in carburizing and carbonitriding applications are heat-treating furnace baskets made of round bar. The likelihood of failure by thermal fatigue increases with increasing diameter of the bar used to construct the basket. Under identical service conditions, however, alloys with a lower mean thermal coefficient of expansion,

such as Inconel 600 and 601, Hastelloy X, and RA 333, will fare better than those alloys with a slightly higher mean thermal coefficient of expansion, such as Incoloy 800 and RA 330. Figure 27 shows the effect of section size on resistance to thermal fatigue. These micrographs show two RA 330 alloy bars—one 9.5 mm (⅜ in.) in diameter and the other 12.5 mm (½ in.) in diameter—taken from the same furnace basket after many service cycles of heating and quenching. Obviously, 9.5-mm (⅜-in.) diam bar exhibited the better resistance to thermal fatigue.

The effect of carbonitriding atmospheres on heat-treating fixtures made of heat-resisting alloys is usually more severe than that of carburizing atmospheres. Consequently, to obtain longer service life, high-temperature alloys with higher chromium and silicon contents, such as RA 333 (25Cr-1.25Si), are preferred for carbonitriding service. The alloys with higher chromium and silicon contents also resist carburization more effectively. For example, AISI type 446 stainless steel (25% Cr) is more resistant to carburization than AISI type 430 stainless steel (17% Cr) under identical conditions of environmental exposure. At a given nickel content, alloys with a high chromium content, such as RA 333 (25% Cr) and the cast HL alloy (30% Cr), can be expected to resist carburization to a greater extent than can alloys with lower chromium contents.

Carbon-Nitrogen Interaction

Centrifugally cast furnace tubes of alloys HK-40 (26Cr-20Ni) and HL-40 (30Cr-20Ni) have been used extensively in ethylene-pyrolysis furnaces for many years. In service, thick deposits of carbon or coke form on the intertube walls, and it is necessary to burn these

deposits away periodically by means of steam-air decoking. Examination of failed furnace tubes has established an interaction between carbon diffused from the surrounding furnace atmosphere and nitrogen already present in the tubes. Metallographic examinations and chemical analyses have shown that a high nitrogen content can cause microscopic voids along grain boundaries, which in turn lead to premature failure by stress-rupture cracking. Void formation occurs only when carbon and nitrogen exceed a certain critical level.

It has been determined that, in general, the inward diffusion of carbon from the carburizing atmosphere forces the migration of nitrogen present in the tube toward the outside surface. However, there may be exceptions. In one hydrogen-reformer furnace, for example, there was a joint inward diffusion of carbon and nitrogen because the methane-gas feed contained 18% nitrogen. Analyses showed that the tubes contained 0.83% C and 0.156% N at the inside surface and only 0.42% C and 0.094% N at the outside surface.

Contact With Molten Metal

Molten metals used in coating and in other industrial processes, such as aluminum, copper, zinc, and their alloys, cannot be contained in vessels made of high-temperature alloys. Molten lead is the exception, provided the lead is covered with a protective layer of powdered charcoal. When molten lead is not protected from air, a lead oxide forms; this oxide is highly corrosive to most high-temperature alloys. Among the alloys that have been used successfully to contain molten lead, however, are Inconel 600 and RA 330.

Molten zinc, which is used in hot-dip galvanizing of fabricated articles, is commonly contained in tanks or vats made from carbon steel plate of boiler-plate quality. Aside from strength, the principal requirement of a galvanizing-tank material is the ability to resist the corrosive attack of molten zinc. When a layer of flux is maintained on the surface of the bath, a collar of firebrick or other suitable ceramic material normally surrounds and abuts the top 15 to 18 cm (6 to 7 in.) of the tank to retard heat transfer in this area and thus reduce attack by the flux on the steel tank wall. The remainder of the interior of the tank, however, is directly exposed to the molten zinc.

Example 1: Failure of a Carbon Steel Galvanizing Vat. A steel galvanizing vat at a shipbuilding and ship-repair facility failed after only three months of service. The source of the failure was a leak that spilled 6800 kg (15 000 lb) of molten zinc into the firebox and onto the floor. The vat, which measured 3 × 1.2 × 1.2 m (10 × 4 × 4 ft), was made of 19-mm (¾-in.) thick carbon steel plate (ASTM A285, grade B). Welded joints were of T construction; each fillet weld was made in a single pass with semiautomatic submerged arc welding. Full fillet-to-fillet penetration was not obtained.

Fig. 28 Microstructures of weld metals in T joints of ASTM A285, grade B, steel

(a) Submerged arc weld in a galvanizing vat that failed by molten-zinc corrosion along elongated ferrite bands such as those shown. Etched with 2% nital. 100×. (b) Multiple-pass manual shielded metal arc weld in a test specimen, which resisted molten-zinc corrosion and which shows a refined microstructure. Etchant and magnification same as (a)

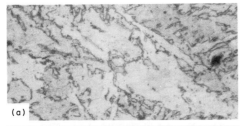

(a)

(b)

Investigation. When the failed vat was visually examined, severe channeling and pitting attack were observed on the inside fillet welds. Chemical analysis of the weld metal and the plate indicated that the welds contained considerably more silicon than the plate. The micrograph shown in Fig. 28(a) is of a specimen taken from an area adjacent to an area of channeling and exhibits elongated ferrite bands. The damage was most severe along the ferrite bands, possibly because silicon was dissolved in the ferrite and thus made it more susceptible to attack by the molten zinc. To verify this assumption, two T joints were made in 12.5-mm (½-in.) thick ASTM A285, grade B, steel plate. One joint was welded using the semiautomatic submerged arc process with one pass on each side. A second joint was welded manually by the shielded metal arc process using E6010 welding rod and four passes on each side; the refined microstructure of this weld is shown in Fig. 28(b). Chemical analyses of the welds indicated that the silicon content of the manual (shielded metal arc) weld was 0.54%, whereas that of the semiautomatic (submerged arc) weld was 0.86%.

After being weighed, the specimens were submerged in molten zinc for 850 h, after which they were again weighed, polished, and examined under the microscope. The specimen welded by the submerged arc process exhibited a weight loss 12 times that of the shielded metal arc welded specimen; the bulk of corrosive attack was directed toward the silicon-rich ferrite bands.

In comparing the two welding processes, it was determined that the flux used in submerged arc welding was high in silicon and that the large amount of weld metal deposited in a single pass remained molten long enough to dissolve much larger amounts of silicon than could be dissolved in manual shielded metal arc welding. Furthermore, the coating on the welding rods used in shielded metal arc welding was relatively low in silicon content, and because completion of the weld required four manual passes, the lesser amount of weld metal deposited during each pass solidified quickly, thereby limiting dissolution of silicon. Finally, each pass refined the ferrite grain developed in a previous pass, thus avoiding ferrite banding.

Corrective Measures. Based on the test results, the original welds were removed from the vat, and the vat was rewelded using the manual shielded metal arc process with at least four passes on each side. The vat was returned to service and, after nine years, was reported to be operating successfully with little or no evidence of pitting.

Molten lead can be contained in tanks fabricated from high-temperature alloys, including austenitic stainless steels, provided the lead is covered by a layer of charcoal to prevent a severe oxidizing reaction. In the following example, premature failure of a stainless steel lead-containing tank was traced to the high moisture content of the charcoal cover.

Example 2: Premature Failure of a Type 309 Stainless Steel Pan for a Lead Bath. Severe reduction of wall thickness was encountered at the liquid line of a lead-bath pan (Fig. 29) that was used in a continuous strip or wire oil-tempering unit. Replacement of the pan was necessary after six months of service. The pan, 6.9 m (22.5 ft) long, 0.6 m (2 ft) wide, and 38 cm (15 in.) deep with a 2.5-cm (1-in.) wall thickness, was a type 309 stainless steel weldment. The pans were previously cast from the same alloy. Operating temperatures of the lead bath in the pan ranged from 805 °C (1480 °F) at the entry end to 845 °C (1550 °F) at the exit end.

Investigation. Visual examination of the pan disclosed that thinning of the walls (shaded area in perspective view, Fig. 29) was most severe at the surface of the molten lead along the

Fig. 29 Stainless steel pan (top) for a molten-lead bath that failed as a result of wall thinning (shaded area) caused by oxidation and decarburization

Overall view of pan and section taken through a second pan (Section A-A) showing the relation of the thinned region to coke level. Dimensions given in inches. Micrograph (at right) shows a section taken through the failure area.

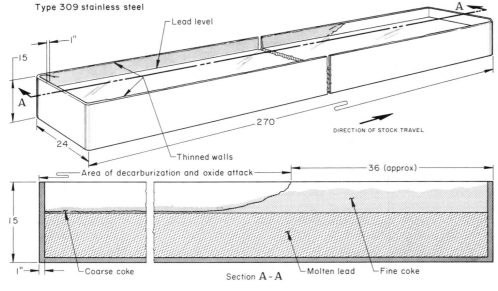

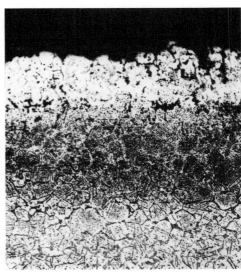

length of the pan, except for the last 0.9 m (36 in.) In this area, very little reduction in wall thickness had occurred.

Because the failed pan had been taken out of service, operating conditions were reviewed by examining a second pan of welded construction that had been in service for about three months. This pan was found to have a thin, uneven layer of coarse coke over its entire length, except for the last 0.9 m (36 in.) at the exit end, which was covered with a layer of fine coke approximately 15 to 20 cm (6 to 8 in.) deep (Section A-A, Fig. 29). The thin layer of coke at the entrance end required frequent replenishing from a pile of damp, coarse coke. In contrast, the exit end of the pan required less frequent replenishing because of the greater depth of coke and the drag-out of the coarse coke. Like those of the failed pan, the sidewalls of the second pans also exhibited a pattern of thinning.

Moisture content of the coarse coke in the supply pile was 12%; that of the fine coke, 7%. Coke ready for shipping normally contains 2 to 3% water.

Metallographic examination of specimens taken from a greatly thinned area of the failed pan, when etched with Murakami's reagent [10 g $K_3Fe(CN)_6$, 10 g NaOH, 100 mL H_2O], disclosed decarburization to a depth of approximately 0.13 mm (0.005 in.) and heavy intergranular oxide attack. Beneath the decarburized layer was a carburized zone approximately 0.25 mm (0.010 in.) deep. The core area of all specimens had uniformly distributed acicular carbide. When specimens were etched with modified Murakami's reagent [15 g K_3Fe $(CN)_6$, 2 g NaOH, 100 mL H_2O], σ phase was found to occur uniformly throughout the core material, appearing as a gray constituent that in general followed the grain boundaries (micrograph, Fig. 29). This phase generally causes embrittlement; however, its effect was negligible in this application.

Only traces of decarburization and oxide attack were observed at the exit end of the pan, where thinning at the liquid line was slight. Attack of the metal was confined to the liquid line and did not extend up the sidewall as in the more severely attacked areas. The carburized layer was narrower and closer to the surface.

Discussion. Type 309 stainless steel is susceptible to carburization and embrittlement in the temperature range of 760 to 815°C (1400 to 1500 °F). The carbon diffuses inward from the surface and forms chromium carbide. Carbide formation results in depletion of chromium from the austenite matrix and buildup of carbides at the grain boundaries. The lower chromium content at the grain boundaries makes the steel susceptible to intergranular oxide attack. This type of attack was observed in all the specimens that were taken from areas in which the sidewalls of the pan were greatly reduced in thickness.

The high water content in the coke ensured the presence of a moist, strongly oxidizing and decarburizing environment at the metal surface—a result mainly of reactions between water and carbon, which produce hydrogen and carbon monoxide, and between carbon, carbon monoxide, and oxygen, which produce carbon dioxide. The strongly decarburizing effect of moist hydrogen explains the observed carbon depletion at the surface, whereas the other environmental constituents—mainly carbon and oxygen and their reaction products—account for the observed carburization and oxidation.

Conclusions. Thinning of the pan walls at the surface of the molten lead resulted from using coke of high moisture content and from the low fluctuating coke level. In comparison, the exit end of the pan, where the coke level was high, encountered much less reduction in wall thickness.

Recommendations. The use of dry (2 to 3% moisture content) coke would reduce the supply of oxygen attacking the grain boundaries and the hydrogen that readily promoted decarburization. Maintaining a thick layer of coke over the entire surface of molten lead in the pan would exclude atmospheric oxygen from the grain boundaries.

Molten Iron. Over limited time intervals, permanent molds also serve as containers for molten metals. In the following example, it was determined that failure of a permanent mold after limited service life was caused by excessive pouring temperature of the metal in combination with inadequate thickness of mold-wash coating.

Example 3: Failure of a Mold for Centrifugal Casting of Gray and Ductile Iron Pipe. The forged 4130 steel cylindrical permanent mold shown in Fig. 30 was used for centrifugal casting of gray and ductile iron pipe. In operation, the mold rotated at a predetermined speed in a centrifugal casting machine while the molten metal, flowing through a trough approximately 5.8 m (19 ft) long, was poured into the mold, beginning at the bell end. The mold continued to spin while moving at a steady rate away from the end of the trough so that the spigot end of the mold was poured last. After the pipe had cooled to 760 to 870 °C (1400 to 1600 °F), it was pulled out from the bell end of the mold, and the procedure was repeated. When pulling of the pipe became increasingly difficult, examination revealed that the mold contained three spall pits about 6.5 × 11 × 2.4 mm (¼ × 7/16 × 3/32 in.) deep located 43 to 56 cm (17 to 22 in.) from the spigot end.

The pits were found after 277 pipes had been cast. The average performance for molds of this size was about 2100 castings. Normal pouring temperatures were 1290 °C (2350 °F) for gray iron and 1370 °C (2500 °F) for ductile iron.

Investigation. Visual inspection of the bore of the mold revealed three large score marks originating at the spalls. The spalls were not all located at the same distance from the end face of the spigot, but were helically distributed on the inner surface at the 1, 6:30, and 11 o'clock positions. The spall at the 6:30 o'clock position was the most inward. All score marks terminated within 81 cm (32 in.) of the spigot end. A large number of small high spots, which were quite shiny in appearance, were observed around the entire bore of the mold 25 to 50 cm (10 to 20 in.) from the spigot end face. The remainder of the bore appeared to be in fairly good condition.

Chemical analysis showed that the metal in the mold was within specification for 4130 steel, except that the manganese content was 0.04% above maximum. Rockwell C hardness values ranged from 23 to 30, with a mean of 27.5.

The mold was sectioned transversely at 81 and 106 cm (32 and 42 in.) from the spigot end. These two sections were then split longitudinally at the 3 and 9 o'clock positions to provide access to the bore. Detail A in Fig. 30 shows the spall at the 6:30 o'clock position and the typical shiny spots. The two other spalled areas were similar in appearance.

The mold sections were then cut to obtain specimens as follows:

- One 25-cm (10-in.) long transverse specimen of good metal (no bore defects) from the mold section located 81 to 106 cm (32 to 42 in.) from the spigot end
- Longitudinal specimens adjacent to areas of score marks at both the 11 and 1 o'clock positions
- One longitudinal specimen 20 cm (8 in.) long that included the spigot end of mold
- One transverse specimen across the score marks located at the 6:30 o'clock position
- Transverse specimens through shiny spots

Several specimens were polished and etched with ammonium persulfate [$(NH_4)_2S_2O_8$] for macroscopic examination. All specimens exhibited a heat-affected zone (HAZ) on the bore approximately 3.2 mm (⅛ in.) deep, beginning 13.6 cm (5⅜ in.) from the spigot end and extending about 60 cm (24 in.) toward the bell end. At this point, the HAZ was shallower but still present. The 25-cm (10-in.) long specimen of good metal (no bore defects) had an HAZ that was 0.2 to 0.3 mm (0.010 to 0.015 in.) deep. Several specimens were etched in 3% nital for 20 s for metallographic examination at magnifications of 3, 10, and 100×.

The specimen of good metal exhibited a bainitic structure, which is typical for 4130 steel and indicates correct and effective heat treatment. Metal at the bore surface showed indications that the temperature there had exceeded the tempering temperature of the mold and may have been higher than 650 °C (1200 °F).

A section taken through an area of the mold containing shiny spots is shown as Section B-B in Fig. 30. Here, shiny spots appear as two dark plateaus separated by depressions. A 2.5- to 3-mm (0.10- to 0.12-in.) deep HAZ is visible.

Fig. 30 4130 steel permanent mold for centrifugal casting of gray and ductile iron pipe that failed because of localized overheating

The failure was caused by splashing of molten metal at the spigot end. Subsequent overheating resulted in mold-wall spalling and scoring, details of which are shown in the photographs. Dimensions given in inches.

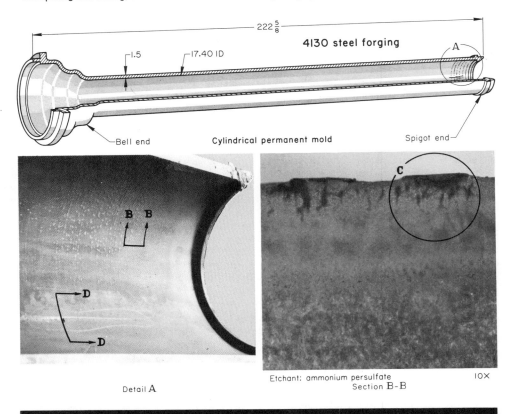

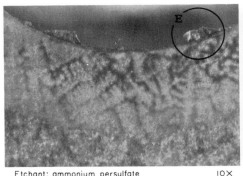

Detail A

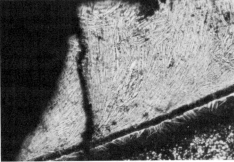

Etchant: ammonium persulfate 10×
Section B-B

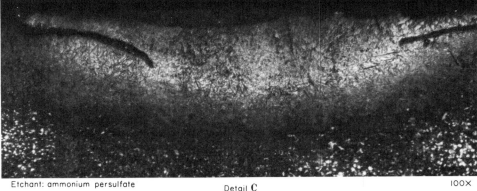

Etchant: ammonium persulfate Detail C 100×

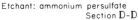

Etchant: ammonium persulfate 10×
Section D-D

Etchant: ammonium persulfate 100×
Detail E

Detail C in Fig. 30, an enlarged view of the area at upper right in Section B-B, shows a crescent-shaped area that consists of (from top) tempered martensite, a near-white arc of untempered martensite, and a banded gray and black mixture of tempered and overtempered material in the form of globular pearlite. The cracks in the martensite indicate the presence of very high stresses. Propagation of similar cracks in other martensitic areas resulted in spalling, with some of the untempered structure being exposed on the surface of the bore.

A micrograph of a section through a scored, spalled area is shown as Section D-D in Fig. 30. Overheated material was in evidence at the left edge of the depressed spall mark, the bottom of which was about 2.3 mm (0.090 in.) below the original bore surface. On the surface of the depression are several protuberances. Examination at a magnification of 100× (Detail E, Fig. 30) revealed the protuberances to be gray iron welded to the bore surface. The first layer (approximately 0.08 mm, or 0.003 in., thick) beneath the gray iron had an untempered martensitic structure. Between this layer and the original quenched-and-tempered bainite structure there was a high-carbon diffusion area of undetermined structure, which was about 1 mm (0.040 in.) thick, followed by a band approximately 1.5 mm (0.060 in.) thick that exhibited an overtempered structure of ferrite and pearlite. Thus, at this point, overheating had penetrated approximately 4.8 mm (0.190 in.) beneath the bore surface.

The defects in the bore (spalled and scored areas, shiny areas, and overheated surfaces) were found near the spigot end, where hot metal impinged on the bore surface. Temperature of the molten metal was probably kept rather high, resulting in relatively deep overheating of metal in the mold at the bore surface. The overheating may have been accentuated by an inadequate thickness of insulating mold-wash coating.

Conclusions. Failure of the mold surface was the result of localized overheating caused by splashing of molten metal on the bore surface near the spigot end. The remainder of the mold wall being cold caused very rapid heat extraction and thus created localized spots of martensitic structure. Because of very high stresses in the hard martensitic structure, spalls and score marks resulted. Apparently, the mold-wash compound (a bentonite mixture) near the spigot end was too thin to provide the proper degree of insulation and to prevent molten metal from sticking to the bore surface.

Recommendations. To reduce failure of centrifugal casting molds, it was recommended that the pouring temperatures of the molten metal be reduced and that a thicker insulating coating (bentonite mixture) be sprayed onto the mold surface.

Contact With Molten Salts

The life of high-temperature alloys in contact with molten salts is usually erratic. For example, thermocouple-protection tubes under simi-

Fig. 31 Specimen from an RA 330 alloy salt pot that failed because of intergranular corrosion and chromium depletion

Specimen was taken from an area where chromium content was only 11.9% instead of the normal 19%. Note the chromium-depleted grain boundaries and the depth of penetration of intergranular oxidation attack. Etched with 10% oxalic acid. 60×

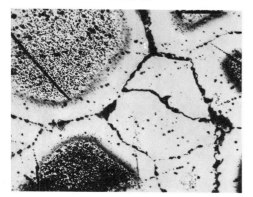

lar service conditions may survive a hundred immersions before replacement is necessary, or they may fail after only five or ten immersions. Failure of components exposed to molten salts usually occurs by perforation or by structural disintegration. Intergranular attack and chromium depletion often accompany failure in molten salts. Failure is sometimes related to operating variables, including hot spots, lax control of temperature, accumulation of sludge, and method of salt replenishment. Type 446 stainless steel protection tubes that provided acceptable life and those that failed prematurely have been subjected to exhaustive investigation, but no correlation could be found between any of the variables investigated that would account for the marked differences in service life. Salt pots used in heat-treating applications also exhibit erratic service behavior, as described in the following example.

Example 4: Failure of Three Wrought Heat-Resisting Alloy Salt Pots. Over a period of about 1½ years, three RA 330 alloy salt pots from a single heat-treating plant were submitted to failure analysis. All of the pots, which had 9.5-mm (3/8-in.) thick walls, were used primarily to contain neutral salts at temperatures from about 815 to 900 °C (1500 to 1650 °F). However, some cyaniding was also performed in these pots, which, when not in use, were idled at 760 °C (1400 °F). It was reported that sludge was removed from the bottom of the pots once a day. Normal pot life varied from about 6 to 20 months.

The pots were removed from the furnace, visually inspected, and rotated 120° every three weeks to ensure that no single location was overheated for a prolonged period of time.

The pots were fired by four natural-gas burners, which were mounted tangentially, with two burners located approximately one-

third of the way up the pot and the remaining burners located two-thirds of the way up from the cylindrical side weld at the bottom head of the pot. The two sets of burners were located at right angles to each other. The distance from the gas flames to the outside surface of the pot varied from 10 to 15 cm (4 to 6 in.), and there was supposedly no flame impingement.

Pot 1 failed at the knuckle radius of the flanged and dished head to the pot—an area well away from the locations of the burners. The grain size of the RA 330 alloy was determined to be ASTM 0, and at the outset, it was thought that the coarse grain size may have contributed to failure by providing less resistance to intergranular attack than would be provided by a finer-grain material. Chemical analyses of material were made at several locations near the failure. Although analyses of most elements corresponded closely with the nominal RA 330 composition, chromium contents varied markedly.

The nominal chromium content of RA 330 alloy is 19%. Analysis of metal taken from an inside surface halfway up the pot indicated a chromium content of only 11.9%. Figure 31 shows a micrograph of the chromium-depleted structure observed in this area. X-ray analysis of the outer surface of the pot indicated a chromium content of 18.6%. Further analyses of drillings taken from an inside surface near the location of failure indicated a chromium content of only 5.7%. It was concluded that most of the pot was inherently sound and free of mill-related defects, but that contamination of the molten salts may have been responsible for the severe corrosive attack and chromium depletion, particularly in the grain boundaries, in the general area where failure occurred.

Pot 2 was returned about a year after the failure of pot 1. The second pot failed at two locations: one near the liquid/air interface and the second in the head area at the bottom of the pot. A micrograph revealed that attack near the liquid/air interface was intergranular and extensive. Attack was so severe that the alloy would no longer conduct electricity, thus making spectrographic analysis impossible. Chemical analyses of drillings obtained at the liquid/air interface indicated chromium contents of 9.7% at the inner surface, 13.4% at the center section, and 11.1% on the outer surface; the chromium depletion at the outer surface was presumed to have been caused by high-temperature oxidation resulting from the initial heating of the salt pot by the combustion of natural gas. The chromium content of a specimen taken from the bottom of the pot was 10.7% at the inner surface, 11.4% at center, and 16% on the outer surface. It was concluded that the cause of failure was severe intergranular corrosion accompanied by substantial chromium depletion.

Pot 3 was returned a few months after the second pot. This pot had been used exclusively for cyaniding at a temperature of 870 °C (1600 °F). Severe scaling was noted on the sides of

the pot, and wall thickness had decreased from 9.5 to 1.6 mm (3/8 to 1/16 in.) at some locations. A micrograph revealed that severe attack had occurred at both the inner and outer walls of the pot. The inner wall showed evidence of severe intergranular corrosion accompanied by chromium depletion. There was also evidence of pronounced carbon buildup on the inner wall.

Chemical analysis indicated that the chromium content of the alloy at a central portion of the pot was only 9.7%. Depletion of chromium and, to a lesser extent, depletion of silicon were comparable to those observed in the other failed pots. However, grain size in the third pot was relatively fine (ASTM 5 to 6), thereby discounting the theory that coarse grain size was a principal cause of erratic service life or premature failure.

Conclusions. The cause of failure of each of the three salt pots was severe intergranular corrosion accompanied by substantial chromium depletion.

Salt baths composed of molten barium chloride are commonly used for austenitizing high-speed tool steels and, for this application, may be operated at temperatures as high as 1290 °C (2350 °F). The containers used at these temperatures are usually made of, or lined with, a suitable refractory material. However, the electrodes used for heating the baths and the thermocouple-protection tubes used in the baths are made of high-temperature alloys. The alloys generally selected for protection tubes are type 446 stainless steel and a nickel-base alloy containing 75% Ni, 15% Cr, and about 8% Fe. Metallurgically, these alloys are notably dissimilar. Type 446 stainless steel contains 25% Cr, which is a relatively low alloy content. However, users of this steel report that it is well suited to service in molten barium chloride at temperatures above 1095 °C (2000 °F) because of its low residual nickel content. Others prefer the nickel-base alloy.

Techniques for Analyzing Elevated-Temperature Failures

Although components that fail at elevated temperatures may, and often do, exhibit certain metallurgical, physical, or fractographic characteristics that are distinctive, the procedures and techniques used to analyze elevated-temperature failures are similar to those used to analyze failures that are not related to temperature. These procedures and techniques are described in detail in the article "General Practice in Failure Analysis" in this Volume.

At the outset, it is essential to recognize that not all causes of failure at elevated temperatures are related to temperature. Obviously, damage caused by careless or improper handling, errors in design or fabrication, faulty material, or tensile overload can promote premature failure. All these factors should be considered as part of the service history of the component. Also, a complete history of the base metal should be developed, including chemical composition and all related material-specification requirements. Service history, wherever possible, should include data on elapsed time in service, operating temperatures, stress or loading conditions, environmental factors, and pertinent design factors, such as notches and other mechanical stress raisers.

The laboratory failure analysis techniques that have been applied to elevated-temperature failures include:

- Visual examination by eye, low-power magnifying glass, or binocular microscope at 1 to 30×
- Photomacrography of portions of the failed part and failed surfaces
- Electron fractography with a scanning electron microscope or a transmission electron microscope
- Nondestructive inspection with ultrasonic, liquid-penetrant, magnetic-particle, eddy-current, radiographic, or other techniques
- Measurement of residual stress by x-ray diffraction
- Bulk chemical analysis
- Determination of mechanical properties and physical properties, including hardness
- Microexamination using optical microscopy at 100 to 1500× or SEM or TEM at 10 to ~450 000
- X-ray diffraction analysis or chemical analysis of surface compounds or contaminants
- Analysis of selected phases or regions by electron probe x-ray microanalysis, ion probe microanalysis, or Auger electron spectroscopy
- Phase extraction by electrochemical methods and analysis by x-ray and chemical methods

Only rarely would a complete evaluation of an elevated-temperature failure require the use of more than five or six of the above techniques. Nevertheless, all available procedures should be considered for a definitive analysis.

Equipment and Tests for Analysis of Failures at Elevated Temperatures

The equipment and tests used in analysis of elevated-temperature failures (or to supplement failure analysis) range from the simple to the highly complex and from the general to the highly specialized. The number of special tests applied to specific products and components alone, although incalculable, can be assumed to be very large. Therefore, the coverage given such equipment and tests in this section constitutes only a selected sampling.

Some Basic Tests. Magnetic tests are widely used to analyze failures involving normally nonmagnetic nickel-containing heat-resisting alloys and austenitic stainless steels.

Many of these alloys become magnetic as a result of chromium depletion caused by oxidation, carburization, or both. Consequently, when these alloys are tested with a magnet and show evidence of magnetism, some insight is gained into the possible causes of failure, and additional tests can be selected more judiciously on the basis of this initial evidence.

Macroetching of cross sections constitutes another simple test; it is useful in detecting the voids and cracks that are characteristic of thermal fatigue failures. When thermal fatigue is severe, sectioning alone will reveal the characteristic signs. Metallographic examination is helpful in detecting microstructural changes and corrosion products associated with common modes of failure. Microconstituents and corrosion products can then be identified more positively by chemical analysis, x-ray diffraction, or electron probe microanalysis.

Various types of tests that simulate service conditions or that exceed them in severity are helpful in reconstructing failures under carefully controlled conditions. For example, failures of engine poppet valves can be reconstructed in a special fatigue-testing machine that simulates loading, operating temperatures, and corrosive atmospheres. All known conditions of bending, overloading, and overheating can be readily reproduced. This machine is also used to predict valve life under extreme conditions and to evaluate experimental valve alloys and coatings. Thermal fatigue failures can be simulated in a test rig that advances and retracts a standard engine valve alternately from the flame of a torch to a water-cooled copper chill block under controlled conditions of heating, cooling, and cycling. The first visual evidence of cracking constitutes failure.

Tests for Gas-Turbine Components

Several specialized rigs have been developed for testing gas-turbine blades, vanes, and disks under simulated operating conditions. Several of these test rigs are described below.

Thermomechanical Fatigue Tests. Thermal fatigue is a major contributor to the failure of gas-turbine engine components, particularly the airfoils of turbine blades and vanes, which are subjected to high mechanical and thermal strains. These strains, together with other operating conditions, have been successfully reproduced in thermomechanical strain-cycling tests performed in a closed-loop servohydraulic fatigue-testing machine (Fig. 32). Strain and temperature can be programmed for running a variety of thermomechanical strain cycles. Load, strain, and temperature are recorded on a digital data-acquisition system and punched on paper tape for plotting of mechanical strain versus temperature and stress versus strain in various cycles. Initiation and propagation of cracks in the components tested are monitored in periodic inspections by taking plastic replicas of the affected surfaces.

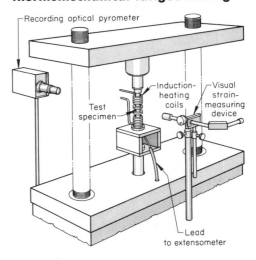

Fig. 32 Closed-loop servohydraulic fatigue-testing machine used in thermomechanical fatigue testing

In the testing machine, the specimen is surrounded by an induction-heating coil that facilitates rapid heating of the specimen to temperatures as high as 1205 °C (2200 °F). Air jets force cool the specimen at a rate that permits completion of the heating-cooling cycle in about 2 min. Strain cycling is normally controlled to equal the heating cycle (about ½ cycle per minute, cpm). Dwells (extended heating cycles) can be readily introduced when extended creep-interaction studies are to be made. Test conditions are such that failures always occur in low-cycle fatigue, usually at less than 5000 cycles.

Fluidized-Bed Thermal Fatigue Tests. Fluidized beds, consisting of high-temperature air-fluidized zircon sand, have been used to evaluate the thermal fatigue capability of alloys for gas-turbine blades and vanes. The beds, which can be used to test specimens or small components (small enough not to disturb the heat-transfer characteristics of the beds), can operate at temperatures as high as 1095 °C (2000 °F) or as low as −75 °C (−100 °F). Specimens or parts are inserted alternately and automatically into the hot and cold beds in accordance with a desired cycle frequency. The strain cycle developed depends on the temperature of the test bed, the alloy, and the design of the specimen or component. The strain cycle cannot be programmed independently. Test conditions result in failure in low-cycle fatigue—usually at less than 10 000 cycles.

The carousel-type thermal fatigue test rig shown in Fig. 33 is also used for testing turbine blades and vanes. This rig can test eight specimens at one time. During each half-cycle, four specimens are heated to the desired temperature as the other four are cooled. During the next half-cycle, the rig turntable rotates one-eighth of the way around, and the heating and cooling conditions imposed on the test speci-

Fig. 33 Carousel-type rig for thermal fatigue testing

See text for description of operation

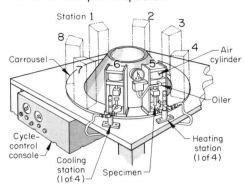

Fig. 34 Crack produced on the trailing edge of a turbine blade by thermal shock

As-polished. 500 ×

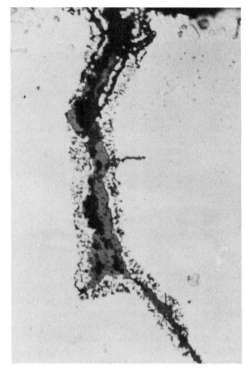

Fig. 35 Ferris-wheel rig for testing gas-turbine engine disks under simulated conditions of low-cycle fatigue

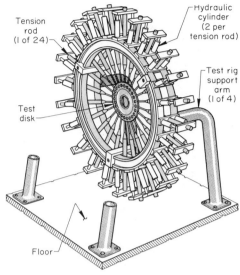

mens are reversed. Because the rig rotates 360°, all eight specimens are heated by each of four burners; consequently, any variations in burner patterns are averaged on all test specimens in like manner.

Test temperatures are measured by optical pyrometer, and the test temperature is the maximum recorded on a specimen at the end of each heating cycle. Each test cycle consists of rapid heating, usually in less than 1 min, to the desired test temperature, followed by cooling in an air blast for the same period of time. Heating is concentrated at the center of the test specimen or turbine component to develop the high localized thermal stresses that cause cracking. With experience, cracking can be observed during cycling, at which time the number of cycles is read on an automatic counter and recorded. Alternatively, test specimens are subjected to fluorescent liquid-penetrant inspection every 100 cycles to detect indications of thermal cracking.

Metallographic examination of the thermal cracks produced in the test rig has shown that they are very similar to those produced during engine operation. Figure 34 shows a typical thermal-shock crack produced on the trailing edge of a turbine blade.

The carousel-type thermal fatigue test rig is also used to evaluate various turbine-blade alloys and coatings. Standard and hollow tensile-type test bars can be tested in the stressed condition, simulating the centrifugal forces imposed on rotating parts during engine operation. These stresses are imposed by air cylinders that maintain a constant stress during thermal expansion or contraction.

Simulated-Bolthole Test. The purpose of bolthole testing is to evaluate resistance to cracking at or around boltholes in tension-tension low-cycle fatigue at various elevated temperatures. The bolthole specimen is a plate with a simulated bolthole at its center. Half-bolthole specimens are also tested. In these, the half bolthole is located at one edge of the test plate and is readily accessible for producing replicas following crack initiation and propagation.

The test rig is hydraulically loaded and is programmed for rates ranging from ¼ to 15 cpm. This test was developed for evaluating boltholes in gas-turbine engine disks, which operate at lower temperatures than turbine blades and vanes. The rig can induce vibratory and steady-state loads over a range of temperatures, thus simulating actual engine conditions.

The ferris-wheel disk test makes it possible to subject full-scale gas-turbine engine disks to simulated low-cycle fatigue conditions. The test rig, with a disk in position, is shown in Fig. 35. The disk is loaded hydraulically by extensions connected to each attachment. A common manifold is used to ensure equal distribution of loads around the circumference of the disk during loading or unloading cycles. The larger test rigs have a total radial-load capacity of 1800 Mg (2000 ton), a cyclic rate of 8 to 20 cpm, and a temperature range from room temperature to 480 °C (900 °F). A special type of nondestructive inspection is used to ensure detection of the first indications of a crack. The procedure consists of interrupting the test periodically, applying fluorescent liquid penetrant to suspect areas, then cycling slowly several times at half the normal test load, wiping the area to which the penetrant was applied, resuming the slow half-load cycling, and inspecting under ultraviolet light. Cracks appear as pinpoints or lines

that dim under load as penetrant is drawn into the crevice and that brighten as the load is released and the closing crevice squeezes out the penetrant.

Spin Test. Gas-turbine engine disks and blades can also be tested by applying centrifugal loads in a spin test (Fig. 36). Disks may be spun at speeds and temperatures sufficient to cause creep or even stress rupture. Blades can be tested for creep under high centrifugal loads. The spin-test rig and pit permit rotation of a bladed disk at any speed or temperature up to those at which the disk or blade will burst. The disk is suspended in an evacuated pit and is driven by a 20- or 35-cm (8- or 14-in.) steam turbine. Heat is provided to the disk by either pan-type resistance heaters, pan-type induction heaters, or resistance-type oven heaters.

Oxidation-Corrosion Tests. Oxidation-corrosion burner rigs have been developed to test turbine-blade alloys and coatings under environmental conditions and temperatures encountered in gas-turbine engines. These test rigs (Fig. 37) can test eight to twelve specimens under engine conditions at temperatures from 790 to 1260 °C (1450 to 2300 °F).

The basic steady-state oxidation test simulates cruising conditions, but higher-than-cruising temperatures are obtainable to accelerate degradation. A burner, using jet-engine fuel, is controlled to maintain specimen temperatures by adjusting fuel pressure. Specimen temperatures are monitored and maintained by an optical pyrometer. The specimens are of a modified airfoil configuration; that is, they have simulated leading and trailing edges. The specimens are weighed before and after each test and are evaluated on the basis of weight

Fig. 36 Spin-test rig and pit used for creep testing of gas-turbine engine disks and blades

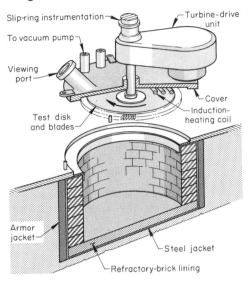

Fig. 37 Oxidation-corrosion test rig used for evaluation of turbine-blade alloys and coatings

loss. Often, oxygen penetration is also evaluated as a means of determining overall metal attack in oxidation-corrosion tests.

A wider range of corrosion studies of alloys and coatings can be made by modifying the burner nozzle to permit the introduction of contaminants through the throat of the nozzle into the burner flame. As is common practice in oxidation testing, uncoated specimens are weighed before and after testing, and weight loss is the criterion of failure. When tested, coated specimens are inspected visually, and corrosive attack of the base metal is the indicator of failure.

Fig. 38 Fixture and test blades used in a rig for hot-corrosion testing to evaluate sulfidation-oxidation resistance of turbine blades or airfoil test specimens

Oxidation-corrosion rigs may have the capability of thermal cycling and may therefore be able to duplicate more accurately the thermal conditions in an engine. For example, a typical blade-corrosion test cycle consists of 3 min at 845 °C (1550 °F), idling; 2 min at 1010 °C (1850 °F), takeoff; then removal from the flame for 2 min, engine shutdown. Typical steady-state oxidation testing is conducted at a constant 1095 °C (2000 °F).

Test for Resistance to Hot Corrosion by Sulfidation-Oxidation. A hot-corrosion test rig used in testing the sulfidation-oxidation resistance of turbine blades consists principally of two chambers—one for heating and one for cooling. However, the rig is modified to spray a solution of deionized water contaminated with sulfur on the rotating test specimens during the cooling portion of a cyclic hot-corrosion test. The test fixture (Fig. 38) has been adapted for evaluation of either eight airfoil or paddle-type test specimens or 16 turbine blades.

The heating chamber is lined with firebrick and is heated by two city-gas burners on opposite sides of the chamber. Air-fuel mixtures are adjusted to maintain oxidizing conditions in the heating chamber. In a typical test, 16 turbine blades mounted on a special fixture and rotated by a motor at a speed of 1800 rpm are heated to a predetermined temperature in the range of 925 to 1095 °C (1700 to 2000 °F). To ensure accurate and reproducible temperature control, the maximum temperature is always measured at the tips of the rotating specimens by an optical pyrometer. After heating, the specimens and fixture are retracted into the cooling chamber, where the specimens are sprayed with an aspirated solution of deionized water containing 1.4% sodium sulfate, equivalent to a sulfate-ion concentration of 1.0%. Each test cycle consists of a 90-s heating cycle followed by a 30-s cooling and spraying cycle.

When test blades are to be evaluated by the weight-loss method, a test consists of 500 cycles, and the blades are weighed before and

Fig. 39 Hot-corrosion attack produced on an uncoated IN-713 specimen in a laboratory test rig (a) and an IN-713 blade that was removed from a service engine (b)

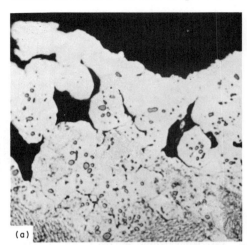

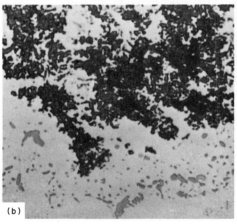

after testing. The surface oxide is removed by immersing the blades in molten potassium hydroxide before final weighing. Weight loss can be plotted as a function of time and temperature. When the blades are to be evaluated metallographically, they are examined with a binocular magnifier after each 100 cycles, and when a total corrosion area of approximately 0.03 cm^2 (0.005 in.^{2}) is observed, the blade is removed from the rig, sectioned, and examined metallographically. The hot-corrosion attack produced on turbine blades in the test rig is basically comparable to that produced in service in commercial or military engines or in hot-corrosion test engines (Fig. 39).

Cascade Tests. Because oxidation-corrosion burner test rigs usually operate at 101 kPa (1 atm) pressure, they cannot provide the close simulation of engine conditions desired for some tests. To overcome this limitation and to provide for simulated testing of actual engine components, thermal fatigue test rigs of the cascade type have been developed. In these rigs, as many as six turbine airfoils can be subjected to normal engine gases under conditions that are more severe than those imposed by the oxidation-corrosion test rig. In addition to providing variations in pressure, the cascade rigs can impose thermal shocks that are more severe than those encountered in normal operation.

A typical rig consists of standard engine fuel nozzles, ignition equipment, and a rotary-vane air supply that blows air to a burner using jet-type fuel. The engine burner can supply the hot gases that are applied to the test vanes at elevated-temperature differentials. The exhaust from the burner is directed through water-cooled ductwork over the test vane and out to an atmosphere discharge. Automatic cycling equipment turns the fuel to the burner on and off, making it possible to evaluate, on a cyclic basis, the resistance of vanes to oxidation-corrosion, cracking, melting, or fracturing.

One type of high-pressure cascade-type thermal fatigue test rig can test at an absolute pressure of about 2 MPa (300 psi) and at a range of gas-stream temperatures from 400 to 1760 °C (750 to 3200 °F). The major components of the annular six-vane cascade rig are a preburner, a main burner with a water-cooled tubular transition duct, coupled with a lower air-cooled section, a vane-cascade pack with individual vane-cooling air supply, and an exhaust duct. Choked-flow venturis have been installed in the air-supply lines of the burner to provide constant air-flow conditions during cyclic operation. The burner supplies hot gases at pressures up to 4.7 MPa (685 psi) at a flow rate of 11 kg (25 lb) per second. Airfoil metal temperatures are monitored with fixed and traversing probes. Standard test instrumentation is provided throughout the rig.

Thermal-Shock Test. A thermal-shock testing machine (Fig. 40) uses cast or machined triangular test specimens with 2.5-mm (½-in.) sides to evaluate the resistance of alloys and coatings to thermal cracking. The specimen is shifted between a heating source and a forced-air cooling vent in accordance with a predetermined cycle. Heating is accomplished with a gas-air flame that can be adjusted to produce either oxidizing or reducing conditions. Both heating and cooling temperatures are automatically recorded. The test is discontinued when the first crack appears on the knife-edge surface of the test specimen.

Tests for Cannon Tubes

Most failures of cannon tubes occur at ambient temperature and are caused by overpressures, in-bore detonations, fatigue, or erosion. The high-temperature failures that do occur are in thin-wall tubes where rapid firing rates cause the tube to overheat and to fail by bulging due to yield or creep acting independently or in combination. Erosion is caused when the hot gases heat the metal at the bore surface of the tube to its melting point and the high pressures from the propellant force the molten metal from

Fig. 40 Diagram of a thermal-shock testing machine used to evaluate the resistance of alloys and coatings to thermal cracking

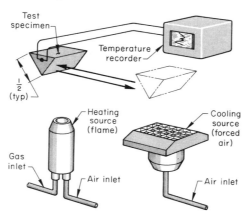

the tube. Most weapons are test fired at their maximum firing rates to determine the maximum operating temperatures for testing purposes. Then, the various cannon-tube alloys are tested at these temperatures to reproduce actual operating conditions. The standard short-term tensile tests, usually 15 min for gun tubes, are generally conducted in the transverse direction to simulate the actual mode of failure.

Three other special high-temperature tests for cannon tubes are the high-temperature hydrodynamic tension test, the high-temperature high-pressure fatigue test, and the erosion test. These tests and the equipment used are described below.

High-Temperature Hydrodynamic Tension Test. The load in a cannon tube is applied at a rapid rate by the exploding propellant. Thus it became necessary to determine the transverse yield strength of several prospective high-strength cannon-tube alloys as a function of strain rate over the range of strain rates from 10^{-4} to 10 s^{-1}. The effects of strain rates on yield strength as a function of composition, microstructure, and yielding mechanism were investigated.

A hydrodynamic loading unit was built that can load a tensile specimen to a force of 27 000 kg (60 000 lb) in 0.5 ms. It is capable of elastic strain rates to 10 s^{-1} and plastic strain rates of 50 s^{-1}. The various high-strength and ultrahigh-strength alloy steels investigated exhibited increases in 0.2% offset yield strength varying from 4 to 11% over the 10^{-4} to 10 s^{-1} range of strain rates. This increase is a function of both the composition and microstructure of the alloy. Coupled with the variable-strain-rate loading equipment is a high-rate heating system to reproduce the temperatures encountered in the thin-wall tubes. This system can heat a 6.4-mm (0.250-in.) diam tensile specimen to about 870 to 1095 °C (1600

to 2000 °F), depending on the resistivity of the specimen. Heating can be at any desired constant rate up to 17 °C (30 °F) per second, with higher heating rates obtainable on smaller specimens. The specimens can be held at peak temperature for any desired time before testing, loading, or both at a specified strain rate.

High-Temperature High-Pressure Fatigue Test. Because of the experimental difficulties involved, there are insufficient experimental data on the fatigue behavior of cannon tubes under conditions of repeated high-pressure cycling at elevated temperatures. To reproduce high-temperature high-pressure fatigue-failure conditions and to generate the fatigue data necessary for design of thin-wall pressure vessels for high-temperature service, a system was developed for testing pressurized cylinders at a maximum temperature of 815 °C (1500 °F) and a maximum pressure of 69 MPa (10 ksi).

The system consists of a hydraulically driven displacement ram that compresses the gas entrapped in the test cylinder. The cylinder is heated by a zoned electric furnace, and a controlled region is maintained at the test temperature. A control system provides remote operation, automatic fail-safe shutdown at cylinder failure or system malfunction, and monitoring of cylinder temperature and pressure. A start-up time of approximately 2 h is used to make necessary adjustments and to bring the cylinder to test temperature. Under initial test conditions of 705 °C (1300 °F) and 34.5 MPa (5 ksi), a continuous cyclic rate of 12 cpm has been achieved and maintained for 24 h. The major components of the system are the test cylinder, the hydraulic system, the pneumatic system, the temperature-control system, and the operating controls.

Erosion Test. Lengthening of the vent in a cannon tube permitted the development of a test for the high-temperature erosion characteristics of low-alloy steels differing slightly in chemical composition and mechanical properties. The test assembly (Fig. 41) consists of a cannon breechblock and a small part of the cannon tube, which contains the nozzle and nozzle insert of the alloy to be tested. Calculated amounts of powder are placed into the breech chamber and fired to produce the required pressure and temperature in the nozzle insert of the material to be tested. After each firing, the nozzle insert is inspected and weighed to determine the erosion loss.

Causes of Premature Failure

Although components used in elevated-temperature applications can fail prematurely because of inherent defects, many premature failures result from misuse. Misuse need not be intentional; it is often inadvertent or accidental. For example, a premature failure caused by excessive operating temperature may be traced to failure of a temperature controller or to

Fig. 41 Breech assembly developed for testing the high-temperature erosion characteristics of low-alloy steels used in cannons

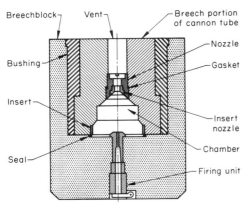

Fig. 42 Effect of increasing temperature from 980 to 1205 °C (1800 to 2200 °F) on the microstructure of nickel-base alloy MAR-M 246

γ′ precipitate, clearly visible as tiny particles in (a), goes into solution at a progressive rate with increasing temperatures, (b) through (f). Heating conditions and times were: (a) 980 °C (1800 °F), 15 min; (b) 1065 °C (1950 °F), 15 min; (c) 1095 °C (2000 °F), 1 min; (d) 1150 °C (2100 °F), 1 min; (e) 1150 °C (2100 °F), 15 min; (f) 1205 °C (2200 °F), 1 min. All specimens were air cooled from the testing temperatures. TEM micrographs of single-stage positive replicas. 2500×

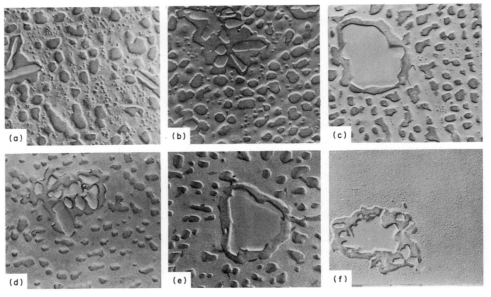

breakdown of a cooling system. The introduction of unanticipated contaminants to the elevated-temperature environment, which results in accelerated corrosion, may also be inadvertent or unavoidable. On the other hand, some premature failures at elevated temperature can be traced to mechanical abuse at room temperature.

Exceeding normal operating temperature is one of the most common causes of premature failure in virtually all elevated-temperature applications. For example, when nickel-base high-temperature alloys are used in turbine-rotor applications, the normal operating temperatures range from 815 to 870 °C (1500 to 1600 °F), although short-term operation at 925 °C (1700 °F) to obtain maximum power is permissible. These temperature limits are imposed because the strength and stability of these alloys at elevated temperatures depend on preservation of particles of γ′ precipitate in the solid-solution γ matrix, as well as the preservation of carbide particles in the grain boundaries.

Exceeding prescribed temperature limits reduces the number of particles of both γ′ and carbide, resulting in a marked deterioration in elevated-temperature properties, particularly creep and stress-rupture properties. The effects of an increase in temperature from 980 °C (1800 °F) to 1205 °C (2200 °F) on the microstructure of MAR-M 246, a nickel-base casting alloy, are shown in the series of micrographs presented in Fig. 42. Excessive temperature can also develop high thermal stresses that result in thermal fatigue cracking. Other harmful effects that can result from excessive operating temperatures include severe grain coarsening, excessive oxidation or erosion, increased hot-corrosion rate, melting, and abnormal dimensional growth.

Example 5: Failure of a Hastelloy X Reactor-Vessel Wall. A portion of the wall of a reactor vessel used in burning impurities from carbon particles failed by localized melting. The vessel was made of Hastelloy X (Ni-22Cr-9Mo-18Fe). Considering the service environment, melting could have been caused either by excessive carburization (which would have lowered the melting point of the alloy markedly) or by overheating.

Investigation. A small specimen containing melted and unmelted metal was removed from the vessel wall and examined metallographically. It was observed that the interface between the melted zone and the unaffected base metal was composed of large grains and enlarged grain boundaries. An area a short distance away from the melted zone was fine grained and relatively free of massive carbides.

Conclusions. The vessel failed by melting that resulted from heating to about 1230 to 1260 °C (2250 to 2300 °F), which exceeded normal operating temperatures, and carburization was not the principal cause of failure.

Example 6: Failure of a Muffle for a Brazing Furnace. A brazing-furnace muffle 34 cm (13¼ in.) wide, 26 cm (10⅜ in.) high, and 198 cm (78 in.) long, was fabricated from nickel-base high-temperature alloy sheet and installed in a gas-fired furnace used for copper brazing of various assemblies. The operating temperature of the muffle was reported to have been closely controlled at the normal temperature of 1175 °C (2150 °F); a hydrogen atmosphere was used during brazing. After about five months of continuous operation, four or five holes developed on the floor of the muffle, and the muffle was removed from service.

Investigation. Copper was found in areas near the holes on the floor; a specimen taken from one of these areas was analyzed with x-ray spectrometry and was found to have an abnormally high copper content. Metallographic examination of a cross-sectional specimen, also from the failed area, revealed the presence of a microconstituent foreign to the alloy; this constituent, probably a nickel-copper intermetallic phase, was observed in the grain boundaries throughout the entire thickness of the muffle floor. The concentration of this phase was heaviest at the inner surface and decreased progressively toward the outer surface. The presence of a dendritic structure in some areas indicated that some melting had occurred. Melting was also observed in areas that were remote from the locations of failure.

Conclusion. The muffle failed by localized overheating in some areas to temperatures exceeding 1260 °C (2300 °F). The copper found near the holes had dripped to the floor from assemblies during brazing. The copper diffused into the nickel-base alloy and formed a grain-boundary phase that was molten at the operating temperature. The presence of this phase caused localized liquefaction and weakened the alloy sufficiently to allow formation of the holes.

Gas-Turbine Components

The principal sections of a gas-turbine engine and their approximate operating temperatures are shown in Fig. 43. Because operating tem-

Fig. 43 Principal sections of a gas-turbine engine and their approximate operating temperatures as related to their position in the engine

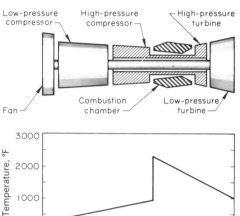

Fig. 45 Turbine shroud showing tip wear (arrow A) and notch wear (arrow B) that resulted when blades were used past their allowable creep limits

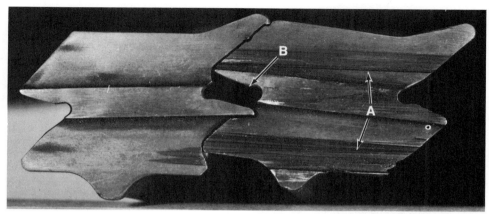

Fig. 44 Creep damage (bowing) resulting from overheating in a cobalt-base alloy turbine vane

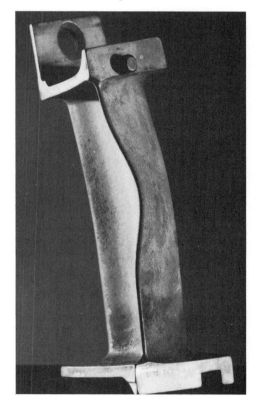

peratures range from ambient to above 1205 °C (2200 °F), engine components are made of various metals. Steels and titanium alloys are used for the relatively cool components, such as those in the fan and low-pressure compressor sections. Nickel-base, iron-nickel, and iron-base high-temperature alloys are used for warm parts, such as shafts, turbine disks, high-pressure compressor disks, and cases. Nickel-base and cobalt-base high-temperature alloys are used for the hot parts, such as burners, turbine blades, and vanes.

Hot Components. Turbine blades and vanes are designed for high load-carrying capacity at elevated temperatures. First-stage turbine vanes are exposed to hot gases from the burner having temperatures in excess of 1205 °C (2200 °F); consequently, airfoil metals temperatures may exceed 980 °C (1800 °F). These vanes and first-stage blades must be cooled. Metal temperatures in later stages may not exceed 650 to 760 °C (1200 to 1400 °F); therefore, these blades and vanes can operate without being cooled. Nevertheless, creep is a problem in that blades and vanes will stretch. Stretch is acceptable provided it remains within the limits established by engine design. Occasionally, however, components will be exposed to excessive temperatures and will stretch beyond limits.

Figure 44 shows creep damage of a cobalt-base alloy vane that resulted from overheating; the reduction in creep strength at the high temperatures resulted in bowing. Excessive creep can also result in damage to sections other than the airfoil. For example shroud wear can result when blades are used past their allowable creep limits (Fig. 45). Figure 45 also shows notch wear, which results from rubbing of the turbine shrouds. Notch wear is undesirable because it can lead to ''untwisting'' of turbine blades.

Wrought turbine blades may also stretch excessively as a result of improper heat treatment. Figure 46 shows extensive creep damage to a turbine blade that was improperly heat treated and then tested to failure in an experimental engine. The stretching and necking due

Fig. 46 Stretching and necking due to creep in a nickel-base alloy turbine blade that was improperly heat treated and then tested to failure in an experimental engine

Note also the wear at the blade tip. Approximately actual size

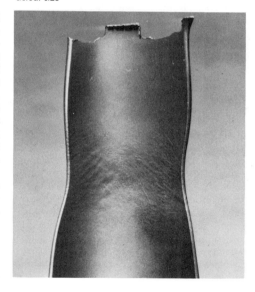

to creep are readily apparent, and the shroud at the blade tip has been worn off.

Intergranular cracking is an indicator of creep damage even when very little change has occurred in the external dimensions of the component. In fact, when a component is improperly heat treated, the creep stretch obtained is above average. In cast alloys, cracking is more likely to result before much stretch is apparent. Figure 47 shows a cast nickel-base alloy turbine blade that was run in an experimental engine to produce cracking. The intergranular cracking produced was extensive, and

Fig. 47 Cracking due to creep in a cast nickel-base alloy turbine blade

1.5×

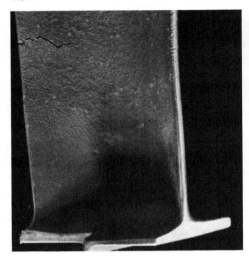

Fig. 48 Fatigue fracture in a cast cobalt-base alloy turbine blade

The fracture surface indicates that fatigue progressed from the trailing edge of the blade, terminating in overload fracture at the leading edge (left). Approximately 3.3×

Fig. 49 Thermal fatigue cracks on the internal surface of a nickel-base alloy forward liner of a gas-turbine burner

One crack extends from a keyhole slot, and another can be seen in the area adjacent to an airhole. 1.5×

as is typical of creep and stress-rupture failure, the surfaces along the cracks were oxidized and alloy depleted.

Creep and stress rupture are not the only causes of cracking in turbine blades and vanes; fatigue is a fairly common cause. Figure 48 shows a fatigue fracture that developed in a cast cobalt-base alloy turbine blade. The fatigue fracture progressed from the trailing edge, resulting in a decrease in effective cross-sectional area and subsequent fracture due to ductile tension overload.

Components of the burner section are susceptible to thermal fatigue cracking, which, like fatigue cracking, also produces cracks that may be difficult to distinguish from cracking produced by creep and stress rupture. Cracks induced by thermal fatigue in a forward liner, which produced typical coarse oxidized fracture surfaces, are shown in Fig. 49; one extends from a keyhole slot, and another is evident adjacent to an airhole. Among the other types of burner-component failures encountered are distortion, extreme oxidation, and melting.

Warm Components. At temperatures from about 425 to 705 °C (800 to 1300 °F), steel and nickel-base alloy components are subjected to high loads, and wherever stress concentrations are high, fatigue is the most common mechanism of failure. Fatigue is induced in several ways. For example, Fig. 50 shows cracking in the rivet slot of a nickel-base alloy disk that was induced in low-cycle fatigue by thermal straining. In disks, low-cycle fatigue stems from stresses imposed by the combined effects of centrifugal loading by blades, the body load of the disk, and the thermally induced load between disk rim and bore. Discontinuities, such as rim slots and boltholes, are typically the sites for initiation of low-cycle fatigue.

Among the cracks that are not initiated by fatigue are those that originate during forging. Generally, these cracks are detected by inspection techniques. Occasionally, however, forging cracks may be so small or so masked as to escape detection, and they may not become visible until after several cycles of service operation. Cracks can interact with fatigue loads to propagate after relatively few cycles. Interactions can sometimes be produced between types of fatigue processes, particularly low-cycle fatigue and high-cycle fatigue. Under laboratory conditions, a crack can be initiated in low-cycle fatigue and then, after a given number of cycles, can be propagated in high-cycle fatigue. The crack can then be further propagated by alternating periods of low-cycle and high-cycle fatigue. It is also possible for a crack to be initiated and propagated by low-cycle fatigue for some finite number of cycles and then to be further propagated by high-cycle fatigue.

Cool Components. At still lower temperatures, up to 425 °C (800 °F), there may be failure of other components, principally those fabricated from titanium alloys. In the operation of certain components, such as compressor-blade attachments, galling can occur on the root-attachment surface as a result of severe loading and vibration (Fig. 51). The blade shown in Fig. 51 had multiple fatigue origins on the fracture surface; these resulted from galling. Galling is not restricted to blades; it can sometimes occur also on disk-rim slots. Figure 52 shows a close-up of a fracture through a disk and illustrates fatigue progressing from an origin in galled material produced by locking-tab contact with the disk.

Fig. 50 Low-cycle fatigue cracking induced by thermal straining in the rivet slot of a nickel-base alloy turbine disk

3×

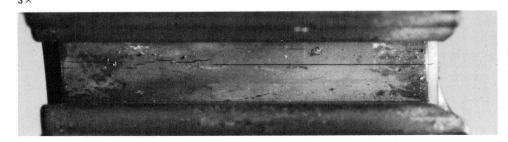

Fig. 51 Galling and cracking (arrow) on the root-attachment surface of a titanium alloy compressor blade

The damage was the result of severe loading and vibration. 1.5×

Fig. 52 Surface of a fatigue fracture in the galled rim-slot area of a titanium alloy disk

Fatigue progressed from an origin in the galled material that was produced by locking-tab contact with the disk. 1.5×

Fig. 53 Incoloy 901 turbine spacer that was removed from service because of cracking of the radial rim

(a) Fracture surface of the turbine spacer showing the fatigue crack, which progressed aft from the forward side of the rim. 9×. (b) and (c) TEM fractographs showing the irregular striations indicative of fatigue. The striations are indistinct because of rubbing between the mating surfaces. 5000 and 10 000×, respectively

(a)

(b)

(c)

Example 7: Failure of a Turbine Spacer. A turbine spacer made of AMS 5661 alloy (Incoloy 901; composition: Fe-43Ni-13Cr-6Mo-2.5Ti) was removed from service because of a crack in the forward side of the radial rim. The crack extended axially for a distance of 16 mm (⅝ in.) across the spacer rim; radially, it extended to a depth of 6.4 mm (¼ in.) into the web section.

Investigation. The spacer was sectioned to expose the fractured surfaces of the crack; these surfaces showed features of fatigue starting from the forward rim side adjacent to, and at, the inside surface of the rim. The fatigue features progressed aft for a distance of about 8.9 mm (0.350 in.) before changing to features of a tension fracture. Figure 53(a) shows a light fractograph of the crack that illustrates macroscopic features near the origin (at left).

Replica studies of the fracture, using a transmission electron microscope, revealed striations that indicated a fatigue-type failure progressing from the forward side of the rim (Fig. 53b and c). A striation count conducted at a magnification of 10 000×, showed that the fatigue had progressed at a rate of 180 cycles per 0.025 mm (0.001 in.) at a location 0.25 mm (0.010 in.) from the forward face of the rim and at a rate of 100 cycles per 0.025 mm (0.001 in.) at a location 2.0 mm (0.080 in.) from this face.

Metallographic examination of a cross section of the fracture on the forward side of the rim in the fatigue-origin area indicated that separation was transgranular, which is typical of fatigue cracking. The microstructure was normal for AMS 5661. Further testing revealed that the chemical composition, hardness, tensile strength, and stress-rupture properties of the material conformed with the requirements of the AMS 5661 specification.

Conclusions. Cracking on the forward rim of the spacer occurred in fatigue that initiated on the forward rim face and that progressed into the rim and web areas. Because there was no apparent metallurgical cause for the cracking, the problem was assigned to engineering.

Example 8: Failure of a Turbine Vane. A turbine vane made of cast cobalt-base alloy AMS 5382 (Stellite 31; composition: Co-25.5Cr-10.5Ni-7.5W) was returned from service after an undetermined number of service hours because of crack indications on the airfoil sections. This alloy is cast by the precision investment method.

Investigation. Metallographic examination revealed that the cracks were thermal fatigue cracks and had emanated from the leading edges of the airfoil and progressed along grain boundaries (Fig. 54). The microstructure also showed evidence of age hardening by intragranular precipitation of carbide particles that was induced by engine-operating temperatures. Bend tests conducted on specimens removed from the airfoil section of a vane indicated extreme brittleness; the specimens fractured with no measurable bend radius—a condition attributed to age hardening.

In addition, the airfoil sections exhibited an unusual type of selective subsurface oxidation. This condition appeared to be related to the intragranular carbides. It was also determined that extensive residual tensile macrostresses could have contributed to cracking at the leading edges.

Conclusions. Cracking of the airfoil sections was caused by thermal fatigue and was contributed to by low ductility due to age hardening, subsurface oxidation related to intragranular carbides, and high residual tensile macrostresses. No further conclusions could be drawn because of the lack of detailed service history.

Example 9: Premature Failure of Turbine Blades by Corrosion. Aluminide-coated and uncoated IN-713 turbine blades were returned for evaluation after service in a marine environment because of severe corrosion. Based on service time, failure of these blades by corrosive deterioration was considered to be premature.

Investigation. The airfoils of the uncoated blades exhibited splitting along the leading edges and swelling on the surfaces, indicating severe hot corrosion (Fig. 55a). Metallographic examination of an uncoated blade confirmed hot-corrosion damage by revealing the charac-

Fig. 54 Airfoil segment from a cast Stellite 31 turbine vane that failed by thermal fatigue

(a) and (b) Thermal fatigue cracks emanating from a leading edge and progressing along grain boundaries. The microstructure shows evidence of age hardening by intragranular precipitation of carbide particles. Etched with a mixture of ferric chloride, hydrochloric acid, and methanol. (a) 100×. (b) 500×

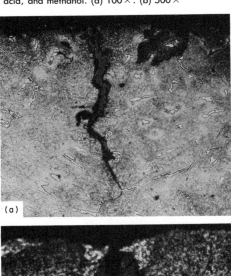

(a)

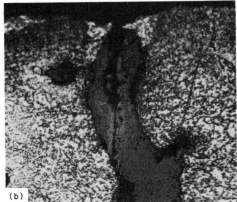

(b)

Fig. 55 Uncoated and aluminide-coated IN-713 turbine blades that failed by hot corrosion in a marine environment

(a) An uncoated blade showing splitting along the leading edge and swelling on the surface of the airfoil. 2.7×. (b) Section taken through the leading edge of an uncoated blade showing a discontinuous surface layer of solid oxide at the blade tip, a layer of oxides in the alloy-depleted nickel-base material, and gray globules of chromium sulfide. (c) and (d) Sections taken through the leading edges of aluminide-coated blades. Corrosion in (c) has penetrated up to the base metal, and in (d) has penetrated into the base metal. (b), (c), and (d) Etched with ferric chloride and hydrochloric acid in methanol

(a)

(b)

(c)

(d)

teristic sulfidation-oxidation attack of the base metal (Fig. 55b). The outer surface at the leading and trailing edges showed a discontinuous layer of solid oxide; the next layer, varying considerably in depth, was composed of oxides intermingled with alloy-depleted nickel-base material. A third layer was an alloy-depleted zone containing gray globules of chromium sulfide adjacent to the normal IN-713 structure. These corrosion products did not readily flake from the surface of the blade, because of the composition of the mixed oxides and that of the alloy-depleted base metal.

The response of the aluminide-coated blades to the hot-corrosion environment of the turbine rotor varied considerably. Metallographic examination of two blades from a high-temperature region of the rotor revealed that, although both blades were attacked by corrosion, attack of one blade was limited to the coating, whereas attack of the other blade penetrated both coating and base metal (Fig. 55c and d). In contrast, an uncoated IN-713 blade operating at

a lower temperature in the same rotor showed only slight hot-corrosion attack on the airfoil surface, and an aluminide-coated blade in the same location was relatively unaffected—indicating that the severity of hot-corrosion attack increases with temperature. Because none of the blades showed evidence of a structural change, such as partial solution of precipitated γ', it was not possible to estimate the temperatures to which the blades had been subjected.

Conclusions. The blades failed by hot-corrosion attack. Variation in rate of attack on coated blades was attributed to variation in integrity of the aluminide coating, which had been applied in 1966, when these coatings were relatively new. It is evident that maintaining the integrity of a protective coating could significantly increase the life of a nickel-base alloy blade operating in a hot and corrosive environment.

Example 10: Premature Failure of a Turbine Blade by Thermal Fatigue Fracture. During disassembly of an engine that was

to be modified, a fractured turbine blade was found. When the fracture was examined at low magnification, it was observed that a fatigue fracture had originated on the concave side of the leading edge and had progressed slightly more than halfway from the leading edge to the trailing edge on the concave surface before ultimate failure occurred in dynamic tension.

Investigation. The fracture was examined by SEM to determine the fracture characteristics at the origin and at the transition from fatigue to tensile fracture at the concave surface. This study revealed that the fatigue fracture was initiated at an intergranular crack in the concave side of the leading edge. Oxidation of the fracture surface at the origin prevented replication for fractography by TEM.

Metallographic examination of a longitudinal specimen cut through the fracture-origin area revealed several oxide-filled cracks (characteristic thermal fatigue cracks) below the fracture surface (Fig. 56a). In high-temperature alloys, surface oxidation at elevated temperatures can deplete alloying elements (notably chromium)

Fig. 56 Micrographs of two turbine blades that failed by thermal fatigue

(a) Longitudinal section taken through origin of failure (upper left corner) of fractured blade showing the fracture surface in profile (top), oxidation on blade surface (left), and oxide-filled crack (arrow). 500×. (b) Section taken through the leading edge of another blade showing oxide-filled crack following a grain boundary. 250×. Both etched with ferric chloride and hydrochloric acid in methanol.

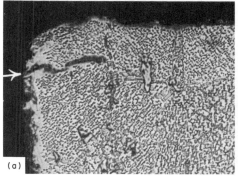

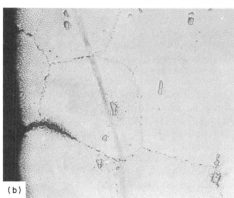

Fig. 57 ASTM A356, grade 6, steel turbine casing that failed by cracking

(a) Segment removed from the casing showing the fracture surface at right. A large porosity defect can be seen at the upper right corner near the broken-open tapped hole. (b) to (e) TEM fractographs of four locations on the fracture surface. 7500×. Intergranular modes of fracture are shown in (b) and (c); a transgranular mode, typical of corrosion fatigue, is shown in (d). The fractograph in (e) was taken in the region of crack arrest and indicates an intergranular mode of fracture.

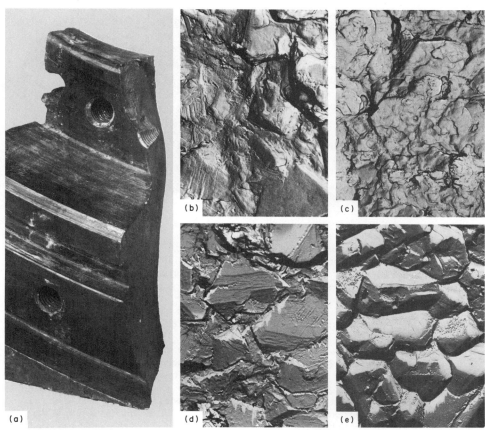

in the base metal, thereby reducing the strength of the metal and making it susceptible to thermal fatigue cracking.

Some thermal fatigue cracks are intergranular and originate at points where grain boundaries intersect the surfaces. Figure 56(b) shows a thermal fatigue crack at the edge of another turbine blade from the same rotor; it is an oxide-filled intergranular crack. The oxide prevents the crack from closing, and additional stresses are imposed on the crack tip, which may accelerate intergranular cracking or act as a stress raiser for nucleation of a transgranular fatigue crack.

None of the blades examined showed evidence of sustained metal temperatures exceeding 1040 °C (1900 °F). The microstructures of the blades in the airfoil sections were essentially identical to those in the base sections.

Conclusions. Application of a protective coating to the blades, provided the coating was sufficiently ductile to avoid cracking during operation, would be beneficial by preventing surface oxidation. Such a coating would also alleviate thermal differentials, provided the

thermal conductivity of the coating exceeded that of the base metal. It was also concluded that directionally solidified blades could minimize thermal fatigue cracking by eliminating intersection of grain boundaries with the surface. However, this improvement would be more costly than applying a protective coating.

Steam-Turbine Components

Large numbers of thick-wall low-alloy steel castings are used in steam-turbine power-generating equipment. Among the common cast components are casings, cylinders, valve chests, and throttle valves.

Example 11: Failure of a Thick-Wall Casing for a Steam Turbine by Cracking. When a crack developed in a cast steel steam-turbine casing, a small section of the casing was removed by torch cutting to examine the crack completely and to determine its origin. The crack was discovered during normal overhaul of the turbine. The chemical composition and processing of the casting were in accordance with ASTM A356, grade 6. The

mechanical properties of the casting were yield strength, 352 MPa (51 ksi); tensile strength, 490 MPa (71 ksi); elongation, 29.5%; and reduction of area, 70.5%, all of which exceeded the requirements of the specification. Charpy V-notch impact-energy values for specimens taken from the casting were 14.6 J (10.75 ft · lb) at 25 °C (75 °F), 88 J (6.5 ft · lb) at 0 °C (32 °F), and 5.4 J (4.0 ft · lb) at −18 °C (0 °F).

When the fracture surface was examined visually, an internal-porosity defect was observed adjoining the tapped hole. A second, much larger cavity was also detected (upper right portion, Fig. 57a).

Metallographic Examination. An examination of the microstructure revealed a concentration of precipitates at slip planes within grain boundaries. Fatigue increases the concentration of points and line defects in the active slip planes, and these defects assist diffusion. Therefore, precipitation, which is a diffusion-controlled process, will be more advanced in the slip bands of a material subjected to fatigue. In this case, the fatigue was thermal fatigue caused by repeated temperature changes in the turbine over a period of about ten years. These temperature changes were more frequent than usual because the turbine was on standby service for several years.

In regions near the main fracture surface, small subsidiary cracks—both transgranular and intergranular—were noted. At surfaces contacted by steam, there was evidence of SCC and intergranular attack.

Electron Fractography. Replica impressions were taken at eight locations on the main fracture surface for examination in an electron microscope. Depending on location, there was evidence of intergranular or transgranular cracking. For example, Fig. 57(b) and (c) indicate an intergranular mode of cracking; however, Fig. 57(d), taken farther down the fracture surface, indicates a transgranular mode typical of corrosion fatigue. Figure 57(e), which was taken from the region of crack arrest, indicates a completely intergranular mode.

Conclusions. Failure occurred through a zone of structural weakness that was caused by internal casting defects and a tapped hole. The combination of cyclic loading (thermal fatigue), an aggressive service environment (steam), and internal defects resulted in gradual crack propagation, which was at times intergranular—with or without corrosive attack—and at other times was transgranular.

Valves in Internal-Combustion Engines

Valves in internal-combustion engines are subjected to elevated temperatures and high loads. Among the types of failures encountered in valves are those caused by burning (rapid oxidation by hot gases), fatigue, thermal fatigue, corrosion, overstressing, and combinations of these. Burning is the most common mode of valve failure. The next most significant mode of valve failure is fatigue at the junction of the stem and head, the point at which the hot gases from the combustion process impinge on the valve stem during the exhaust cycle. The temperatures in this area normally range from about 730 to 760 °C (1350 to 1400 °F), and loads, which are due to firing pressure and valve-spring pressure, are low.

Figure 58 shows the fracture surface of a typical failed valve stem, illustrating multiple crack-initiation points. The crack propagated by fatigue through three-quarters of the stem before final fast fracture occurred in a single cycle. The source of stress concentration was a metallurgical notch resulting from a crack caused by oxidation.

If, in addition to normal loading, a bending load is imposed on a valve, the combined stresses may lead to failure at the point of maximum temperature and stress, as in the following example.

Example 12: Failure of a Truck-Engine Valve. The exhaust valve of a truck engine

Fig. 58 Fracture surface of a typical valve-stem failure with multiple crack-initiation points
4.5×

failed after 488 h of a 1000-h laboratory endurance test; unleaded gasoline was used during the entire test period. The valve was made of 21-2 valve steel (Fe-21Cr-2Ni-8Mn-0.5C-0.3N) in the solution-treated and aged condition and was faced with Stellite 12 alloy (Co-30Cr-8W-1.35C). The failure occurred by fracture of the underhead portion of the valve.

X-ray analysis of the stem portion of the valve confirmed that the composition of the 21-2 valve steel was acceptable (19.48Cr-2.04Ni-8.7Mn). Hardness measurements and examination of the microstructure in the stem region indicated that the valve had been correctly heat treated.

Macroscopic Analysis. When the valve was examined, there was no evidence of severe wear of the valve-seat face, and seat recession appeared uniform. The fractograph shown in Fig. 59 illustrates the valve-stem fracture surface, which clearly exhibits beach marks typical of fatigue fracture. The presence of several fatigue-crack origins concentrated on one side of the fracture surface (arrows, Fig. 59) suggested that loading was not applied uniformly. This hypothesis is supported by the change in curvature of the beach marks, which became concave, indicating unidirectional bending with low nominal stress and a mild stress concentration. The position and appearance of the zone of final fast fracture are characteristic of this stress state and mode of loading.

Metallographic Examination. A longitudinal section of the stem taken through the fracture surface was metallographically examined. The area of the stem in which the fatigue cracks nucleated exhibited corrosion pits. Such pits act as notches and are likely sites for nucleation of fatigue cracks. The corrosion pits were the result of intergranular attack by the corrosive environment and the depletion of chromium from the grain-boundary areas. Depletion of

Fig. 59 Fracture surface of a 21-2 steel valve stem
Arrows indicate several crack origins on one side; also visible are fatigue beach marks and the final-fracture zone (bottom). 4.5×

chromium ultimately resulted in the formation of microcracks.

A second type of corrosive action also occurred and resulted in the formation of nodule-type oxides of chromium. The nodules formed at regular intervals in the stem and the underhead region. The material adjacent to the oxides was not depleted of chromium, indicating a different type of corrosive attack. Corrosion pitting and chromium depletion produce a more severe notch than oxide formation does because nodular oxide formation is intragranular. Although the nodules extended over a large area, their notch intensity was insufficient to nucleate cracks.

When the microstructure near the stem surface was examined, it was apparent that carbide spheroidization had occurred. Also, there was a coarsening of the carbide network within the austenite grains. The microstructure indicated that the underhead region of the valve was heated to about 930 °C (1700 °F) during operation. The cause of fatigue fracture, therefore, was a combination of nonuniform bending loads and overheating.

Electron probe x-ray microanalysis was conducted on the valve stem to determine the nature of the corrosion products and to verify the absence of lead. The results showed the nodules to be chromium-rich oxides. The fact that no chromium depletion was observed adjacent to the nodules suggested that other elements, notably manganese and iron, were oxidized and subsequently scaled off, leaving the remaining material rich in chromium. The microanalysis also confirmed that lead-free gasoline was used.

Electron probe x-ray microanalysis was also conducted at the sites of microcrack formation. The results indicated increases in sulfur and

Fig. 60 Poppet valve that failed by a mode of thermal fatigue known as valve guttering

(a) Head of the valve. 0.5×. Severe oxidation-erosion of the valve seat was caused by the hot gases of the combustion process. (b) Portion of the overhead side of the failed valve showing several fine cracks that developed at sharp corners in the valve-seat facing material, Stellite F (SAE VF5) alloy. Actual size. (c) Etched longitudinal section through a portion of the head of the failed valve showing cracks that propagated through, and parallel to, dendrites in the welded structure of the valve-seat facing alloy. 40×

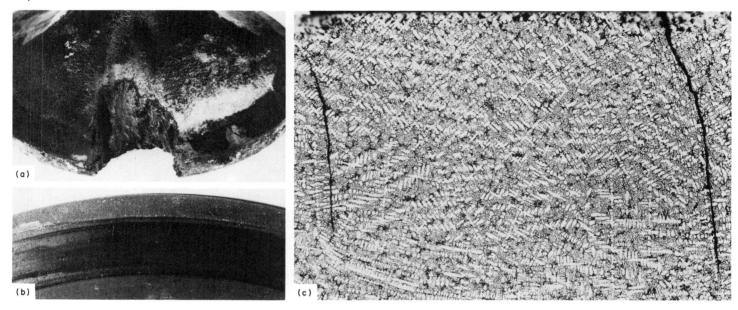

oxygen contents and depletion of chromium and manganese. In several areas, segregates of manganese and sulfur were detected, indicating the presence of manganese sulfide inclusions in the material. These were attributed to contamination of the engine oil.

Conclusions. Failure of the valve stem occurred by fatigue as a result of a combination of a nonuniform bending load, which caused a mild stress-concentration condition, and a high operating temperature (about 930 °C, or 1700 °F) in a corrosive environment.

Thermal fatigue, often referred to as valve guttering, is another common mode of failure for valves used in gasoline or diesel-fuel engines. Thermal fatigue in poppet valves begins with crack initiation, usually at sharp corners in the valve facing material, which is applied to increase corrosion resistance and to combat wear. Thermal stress arises from a cooling action that results from the seating of the valve face on the cylinder head. The operating temperature of the valve head may range from about 705 to 760 °C (1300 to 1400 °F), but the valve face is cooled to about 650 °C (1200 °F) by contact with the cooler cylinder. The temperature gradient produces tangential stresses in the valve face. As the valve alternately opens and closes, cyclic heating and cooling leads to crack initiation and propagation.

With continued cycling, crack propagation leads eventually to gas leakage. Ultimately, the escaping gas raises the temperature in the crack zone to about 1925 °C (3500 °F), the combustion temperature. Oxidation and erosion progress rapidly, and failure is detected as a result of loss of compression. Valve life can be extended either by reducing thermal stresses, which improves fatigue life, or by selecting valve materials with greater dynamic ductility, which would increase the number of service cycles to failure.

The poppet valve of which the head is shown in Fig. 60(a) failed after more than 5000 h of engine operation, or about 600×10^6 cycles. Hot gases from the combustion processes oxidized and eroded a path through the valve-seat facing alloy, Stellite F (SAE VF5), and into the valve base material, 21-12N (SAE EV4) steel. Some of the cracks that developed on the valve seat are shown in Fig. 60(b). At higher magnification (Fig. 60c), two cracks are visible that propagated through, and parallel to, dendrites in the welded structure of the VF5 facing alloy.

Molten-Solids Failures. The addition of tetraethyl lead, an antiknock agent, to gasoline contributes to the failure of valves by a mechanism known as molten-solids failure. The function of the additive as an antiknock agent depends on the oxidation of tetraethyl lead to lead oxide and lead. However, in addition to these products, intermediate compounds of lead are also formed, and these react with scavenging agents (halogens and sulfates) and metallic additives (barium and calcium) in the engine lubricating oil.

At elevated temperature, these intermediate compounds of lead act as liquid-phase corrodents of the protective oxide scale that forms on valves. Constant fluxing of the oxide scale reduces the oxidation resistance of the valve alloy so that oxidation proceeds at a markedly accelerated rate. In addition, lead oxide also reacts directly with the valve alloys to produce low-melting eutectics. Typical low-melting eutectics include $PbO \cdot SiO_2$, $PbO \cdot MoO_3$, and compounds of tungsten, boron, and aluminum. The resulting oxidation attack is both intergranular and intragranular.

Petroleum-Refinery Components

Some petroleum-refining operations involve the use of, or production of, hydrogen at high pressures and at temperatures of 230 °C (450 °F) or above. These service conditions lead to the deterioration of steel and result in the failure of components, notably pressure vessels. Under certain conditions of temperature and hydrogen partial pressure, atomic hydrogen has been observed to permeate and decarburize steel by reducing iron carbide (Fe_3C) to form methane (CH_4). The methane does not diffuse from the steel, and its pressure may exceed the cohesive strength of the steel and cause fissuring between grains. When fissuring occurs, the ductility of the steel is significantly and permanently lowered. The severity of hydrogen attack increases with increasing temperature and hydrogen partial pressure.

Hydrogen attack usually occurs in three stages. First, atomic hydrogen diffuses into the steel; second, decarburization occurs; and finally, intergranular fissuring occurs. A steel that has undergone only the first stage of hydrogen attack suffers a loss in ductility that is considered to be temporary, because ductility can be restored by heat treating at a relatively low temperature. Permanent, irreversible embrittlement results when the attack has pro-

Fig. 61 Comparison of the normal microstructure of a low-carbon steel (a) with that of the same low-carbon steel following hydrogen attack (b)

Arrow in (b) indicates a fissure generated by entrapped methane.

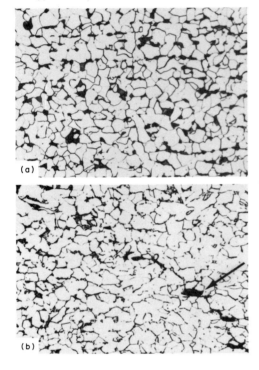

(a)

(b)

gressed to the second or third stage. Consequently, the term hydrogen attack is intended to denote the permanent embrittlement that occurs in the second and third stages.

In Fig. 61, the typical structure of a low-carbon steel is compared to that having undergone the microstructural changes associated with hydrogen attack. Hydrogen attack has transformed the pearlite colonies to ferrite by reducing iron carbide to form methane. In turn, the internal pressure generated by the entrapped methane has exceeded the strength of the steel and resulted in formation of the fissure shown in Fig. 61(b). Because a loss in ductility actually precedes a change in microstructure, bend tests and tensile tests are the most sensitive indicators of hydrogen attack.

Components of Steam Reformers

Reformers are used to produce a hydrogen-rich synthesis gas from a mixture of steam and natural gas or steam and another hydrocarbon gas at high pressures and elevated temperatures. A typical reformer consists of a brick-lined furnace or combustion chamber containing a series of oil-fired or gas-fired burners that heat cast heat-resistant alloy tubes containing the steam-gas mixture and a suitable catalyst, such as nickel or nickel oxide. The steam-gas mixture, preheated at 425 to 650 °C (800 to 1200 °F), is introduced into the tubes, which are in turn heated to 705 to 1040 °C (1300 to 1900 °F). Thus, the mixture is catalytically and endothermically reformed to synthesis gas, which is then fed into waste-heat-recovery equipment for further processing.

Furnace Tubes. Surveys of operating plants have indicated that stress rupture is the principal mode of failure in cast heat-resistant alloy steam-reformer furnace tubes. The stress pattern developed in the wall of the reformer tube during service ultimately results in fissuring in the region extending from the inside surface to midwall; there is no fissuring at the outside surface. The sequence of steam-reformer tube failure has been studied systematically by examining specimens taken from tubes that have been subjected to varying amounts of service.

The first signs of creep damage can be observed metallographically, as shown in the

Fig. 63 Effect of exceeding prescribed (base) service temperature on the expected life of cast HK-40 alloy steam-reformer furnace tubes

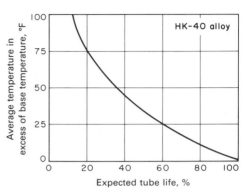

micrograph of an HK-40 alloy tube in Fig. 62(a). These early signs are observed after a tube has completed about 25% of its total service life. Creep damage in the form of microvoids (Fig. 62b) becomes readily visible to the unaided eye after completion of about 50% of total service life. At this stage, the creep voids are becoming aligned and are beginning to form fissures. Finally, after completion of about 75% of total service life, cracks develop through most of the tube wall (Fig. 62c). Nondestructive testing techniques have been developed to detect the various stages of tube damage.

Overheating of tubes will result in a reduced service life. As shown in Fig. 63, when a tube is subjected to a service temperature 55 °C (100 °F) above the prescribed (base) temperature, its service life will be reduced by approximately 90%. Control of furnace temperature, therefore, is of prime importance in ensuring normal

Fig. 62 Three stages of creep damage in alloy HK-40 steam-reformer furnace tubes

Unetched specimen showing early signs of creep damage in the form of microvoids (black dots) at two grain boundaries. 50×. (b) Unetched specimen showing the extensive network of microvoids present after about 50% of total service life. 7×. (c) Unetched specimen showing the advanced stage of cracking present after about 75% of total service life. 2×

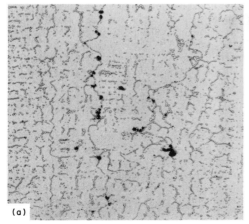

(a)

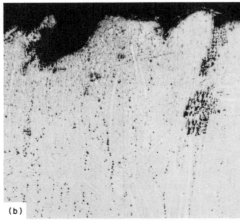

(b)

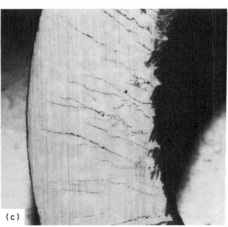

(c)

Fig. 64 Comparison of microstructure of two HK-40 alloy steam-reformer furnace tubes that failed in stress rupture

The structure in (a) shows fewer particles of primary carbides than that in (b), indicating that it was subjected to a lower service temperature. Arrows indicate primary-carbide particles. Etched using 10 mL HNO_3, 20 mL HCl, 20 mL glycerol, 10 mL H_2O_2, 40 mL water. 250×

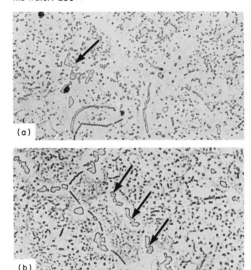

Fig. 65 Schematic of Incoloy 800 outlet-piping system for a steam-reformer unit showing the four welds that failed by cracking

Dimensions given in inches

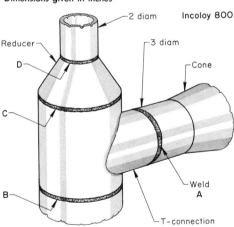

Fig. 66 Three radiant tubes from a carburizing furnace that failed by corrosion

(a) Corroded portions of tubes made of Hastelloy X (top), experimental alloy 634 (center), and RA 333 alloy (bottom). 0.3×. (b) to (d) Unetched sections from corroded portions of the tubes: (b) alloy 634 tube showing heavy scale, (c) Hastelloy X tube showing heavy carburization and severe subsurface oxidation, and (d) RA 333 alloy tube showing heavy carburization and slight subsurface oxidation. All 100×

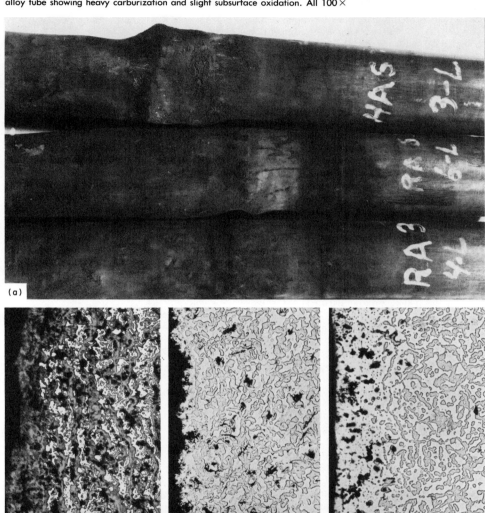

tube life. Evidence of overheating can be detected in the microstructure of the affected tube, thus providing a useful standard in failure analysis.

For example, Fig. 64 shows the microstructure of two HK-40 alloy tubes that failed by stress rupture. The first tube (Fig. 64a) failed after 70 months of service, whereas the second (Fig. 64b) failed after only 57 months. The microstructure of the second tube shows a greater amount of primary carbide than that of the first tube. The amount of primary carbide in the microstructure increases with an increase in service time or service temperature. Consequently, because the second tube failed after a short service life (57 versus 70 months), its failure was temperature related. It was estimated that this tube had been operating at a service temperature 28 °C (50 °F) above the prescribed temperature.

Outlet Piping. A steam-reformer outlet-piping system, operating at 815 °C (1500 °F), failed after a service life of about one year. As described in the following example, the princi-pal causes of failure were sensitization and the development of cracks at welds.

Example 13: Failure of an Incoloy 800 Piping System. The outlet-piping system of a steam-reformer unit failed by extensive crack-ing at four weld locations, identified as A, B, C, and D in Fig. 65. The welded system consisted of Incoloy 800 (Fe-32Ni-21Cr-0.05C) pipe and fittings. The exterior surfaces of the system were insulated with rock wool that did not contain weatherproofing. On-site visual examination and magnetic test-ing indicated severe external corrosion of most of the piping. The insulation formed a bond with corroded surfaces and was difficult to remove. The system showed extensive cracking in weld HAZ.

Investigation. Representative specimens of the piping, typical in terms of both external corrosion and cracking, were examined. A specimen taken from area A in Fig. 65 indicated that corrosion extended to a depth of 3.2 mm (⅛ in.); cracks were seen at the edge of the cover bead and in the HAZ of the weld. Metallographic examination of the cracked region showed that cracking was intergranular and that adjacent grain boundaries had undergone deep intergranular attack. Examination at higher magnification of areas in the vicinity of attack revealed heavy carbide precipitation, primarily at grain boundaries, indicating that the alloy had been sensitized. Sensitization resulted from heating during welding.

Electron probe x-ray microanalysis showed that the outside surface of the tube did not have the protective chromium oxide scale normally found on Incoloy 800. The inside surface of the tube had a thin chromium oxide protective scale that attached to the base metal at the grain boundaries.

Conclusions. The deep oxidation greatly decreased the strength of the weld HAZ and cracking followed. A detailed stress analysis showed that failures had occurred in the areas of highest stress in the reducer and the T connection, namely areas A, B, C, and D in Fig. 65.

Components of Heat-Treating Furnaces

Of the several types of heat-treating furnace components exposed to gaseous environments at elevated temperatures, radiant tubes, which are in effect the furnace heating elements, are sometimes subjected to the most severe service. The radiant tubes used in carburizing and hardening furnaces are representative of components subjected to this degree of service. These tubes may operate normally at temperatures as high as 1040 °C (1900 °F) and may be exposed to atmospheres that are alternately carburizing and oxidizing, as in the following example.

Example 14: Failure of Radiant Tubes in a Batch-Carburizing Furnace. Three radiant tubes, made of three different high-temperature alloys, were removed from a carburizing furnace after approximately 8½ months of service when they showed evidence of failure by collapsing (telescoping) in a region 30 cm (12 in.) from the tube bottoms in the vicinity of the burners. The affected areas of the three tubes, which vary in degree of corrosion damage sustained, are shown in Fig. 66(a).

The tubes had an original wall thickness of 3.0 mm (0.120 in.) and were made of three different alloys. The first was Hastelloy X:

Element	Composition, %
Nickel	48
Chromium	22
Iron	18.5
Molybdenum	9

Fig. 67 Section of an RA 333 radiant tube that corroded in a heating furnace

Top: hole corroded in tube wall and locations where three metallographic specimens were removed. Bottom: specimens etched with mixed acids showing microstructure of specimen 1 at various depths in corroded region and normal microstructure of specimen 3. 250×

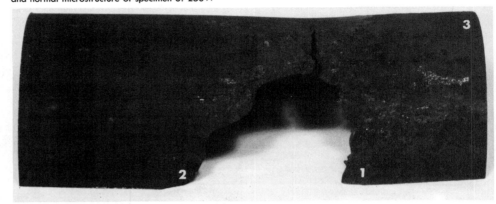

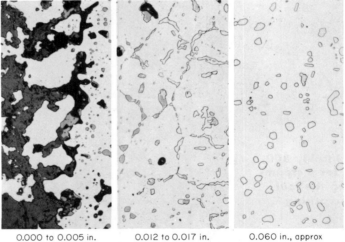 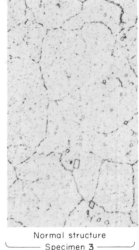

0.000 to 0.005 in.	0.012 to 0.017 in.	0.060 in., approx	Normal structure

Specimen 1 at various distances from outer surface — Specimen 3

The second alloy was RA 333, a wrought nickel-base heat-resistant alloy:

Element	Composition, %
Nickel	45
Chromium	25
Iron	18
Molybdenum	3
Tungsten	3
Copper	3
Manganese	1.5
Silicon	1.25
Carbon	0.05

The third was experimental alloy 634, which contained 72% Ni, 4% Cr, and 3.5% Si. The three radiant tubes had been operated at a temperature of about 1040 °C (1900 °F) to maintain furnace temperatures of 900 to 925 °C (1650 to 1700 °F). The normal life of radiant tubes in this furnace ranged from 6 to 20 months.

Investigation. Specimens for metallographic examination were removed from the heavily corroded area of each tube. Micrographs taken from areas near the surfaces of the three tubes are shown in Fig. 66(b) to (d). As shown in Fig. 66(b), the tube made of alloy 634, which contained the highest nickel and lowest chromium contents of the three alloys, suffered severe oxidation in the form of heavy surface scale. A micrograph of the Hastelloy X tube is shown in Fig. 66(c). This alloy contained the second-highest nickel content (48%) and 22% Cr; the micrograph shows heavy carburization and considerable subsurface oxidation. A micrograph of the RA 333 alloy tube is shown in Fig. 66(d) and exhibits heavy carburization but only slight evidence of subsurface oxidation. This alloy contained 25% Cr and only 45% Ni. In radiant tubes, selective oxidation is a far more serious problem than carburization. Measurement of the sound wall thickness (metal carburized but not oxidized) remaining in each of the three tubes confirmed the evidence shown in the micrographs: in the alloy 634 tube it was 1.7 to 2.2 mm (0.067 to 0.087 in.); in the Hastelloy X tube, 2.2 to 2.9 mm (0.087 to

0.114 in); and in the RA 333 alloy tube, 2.8 mm (0.110 in.).

Conclusions. From the information obtained in the investigation described above, it was decided that tube life might be extended by selecting an alloy, such as RA 333, with a higher chromium content and with an additional element, like silicon, resistant to carburization-oxidation.

Example 15: Corrosion Failure of a Radiant Tube in a Furnace. One of 14 vertical radiant tubes in a heat-treating furnace failed prematurely when a hole about 5 × 12.5 cm (2 × 5 in.) corroded completely through the tube wall (Fig. 67). The radiant tubes were arranged in rows of seven along each side of the furnace and were spaced on 30-cm (12-in.) centers. They were made of RA 333 alloy.

The tube measured 183 cm (72 in.) in length and 8.9 cm (3½ in.) in outside diameter and had a wall thickness of about 3 mm (0.120 in.). Failure occurred where the tube passed through the refractory hearth (floor) of the furnace.

When the furnace was examined, large amounts of soot were observed at various locations, especially on the hearth near the vertical tubes and covering the refractory-cement packing that sealed the tubes to the hearth. Preliminary examination *in situ* of the 14 tubes showed some pits on the outer surfaces of several tubes just below the roof of the furnace.

Service History. The furnace had been rebricked when new radiant tubes were installed. The tubes had been in service for only five months when the failure occurred, although the service lives of similar radiant tubes used under the same operating conditions in an identical furnace at the same plant ranged from 24 to 30 months, with a maximum life of about five years.

The two furnaces were used exclusively for clean hardening of H13 tool steels and similar alloy steels in an endothermic atmosphere at about 1010 °C (1850 °F). Although the atmosphere was neutral with respect to the work, it had a carburizing potential relative to the radiant tubes.

Sampling. An 18-cm (7-in.) long sample piece that contained the failure region (Fig. 67) was cut from the failed tube for examination. Specimens for spectroscopic analysis were taken from near the hole in this sample (a heavily scaled and corroded region) and from an apparently unaffected portion of the sample. In addition, a quantity of loose scale was scraped from the heavily scaled region to be analyzed chemically for sulfur. Specimens for metallographic examination (specimens 1, 2, and 3) were also cut from the sample in the locations shown in Fig. 67.

Visual Examination. Scaling and corrosion on the outside surface of the failed tube were observed around the hole in the tube and were most severe immediately adjacent to the hole. In this region, the severe attack extended about 15 cm (6 in.) around the circumference of the tube.

Less severe scaling and pitting extended slightly above the hearth. The upper end of the sample was covered with a glossy adherent oxide layer.

The lower end of the sample (the portion below the hearth) did not bear any visible signs of attack or deterioration. The outside surface at this end of the sample was coated with a light layer of soot in some areas. Wiping away the soot revealed undisturbed surface scratches and other fabrication marks, but there was no evidence of scaling in this region. The remainder of the failed tube (above the sample) showed no evidence of damage or deterioration.

Because of the pits observed on several tubes at the roof line in the preliminary examination and because of the severity of the attack on the failed tube at the hearth, the 13 remaining tubes were removed from the furnace. Ten of these tubes were found to be pitted to varying degrees in the regions that had been enclosed within the bricked roof of the furnace, and pinhole penetration was detected at a few isolated points on two or three of these tubes in the pitted regions. Further examination and testing were restricted to the failed tube and the refractory cement.

Spectroscopic and Chemical Analysis. The composition of the failed tube was found by spectrographic analysis to be normal for RA 333 alloy. Chemical analysis of scale taken from the heavily corroded outside surface of the tube near specimen 1 (Fig. 67) showed that the scale contained 0.25% S. The refractory cement used around the tubes when they were installed was not analyzed, but was described by the supplier to be a type having a relatively high sulfur content.

Metallographic examination was performed on specimen 1, which had been taken from the region just above the hole in the tube (Fig. 67). Examination of a cross section of the tube wall taken from specimen 1 showed that several distinct zones existed at different depths inward from the outside surface, as shown in the three photomicrographs at the bottom left in Fig. 67 (specimen 2 showed a generally similar pattern of microstructure and was not photographed).

The micrograph at the extreme left, extending to a depth of 0.1 mm (0.005 in.) inward from the surface (from which the loose scale had first been removed), showed the outermost portion to be a carbon-depleted corrosion zone. The outer 0.1 mm (0.005 in.) consisted mostly of sulfide (gray constituent), remnants of the austenitic matrix (off-white), and small amounts of a silicate phase (black constituent).

The second micrograph of specimen 1 in Fig. 67 shows a region 0.3 to 0.4 mm (0.012 to 0.017 in.) in from the outer surface of the tube. This region had a fine grain size (ASTM 4 to 7) and was heavily carburized, showing extensive carbide separation along the grain boundaries (light-etching constituent) and a few scattered small gray particles of sulfide and black particles of silicate.

The third micrograph of specimen 1 in Fig. 67 shows a carburized coarse-grain (ASTM 000 to 3) structure at a depth of approximately 1.5 mm (0.060 in.), the midpoint of the wall thickness, in which the carbide was present as scattered small globular and elongated particles. The carbon content at this level was substantially lower than at the level shown in the second micrograph of specimen 1.

The micrograph of specimen 3 (an apparently unaffected region away from the failure region, as shown in Fig. 67) corresponded to the normal fine-grain microstructure of the alloy (ASTM grain size 4 to 7, but predominantly 5 to 7).

Discussion. Extreme grain coarsening was observed in the corroded region of the failed tube at the approximate midpoint of the 3-mm (0.120-in.) wall thickness (third micrograph of specimen 1, as shown in Fig. 67) and extending inward to the interior or fireside wall of the tube. The microstructure in this region of the tube was evidence that local overheating to about 1175 to 1205 °C (2150 to 2200 °F) had occurred in service.

The exceptional severity of the local overheating on this tube within the hearth (as compared to the 13 other tubes) apparently resulted from failure to position the burner so as to heat the tube uniformly around its circumference and high enough to avoid concentrating heat on the portion of the tube within the refractory hearth. As on all the tubes, the insulating packing and brick of the hearth interfered with loss of heat by radiation, which was unhindered on the portions of the tubes that were freely exposed to the furnace atmosphere.

Within the hearth, the metal of the tube wall near its outer surface and in the corroded area, as illustrated by the second micrograph of specimen 1 in Fig. 67 (labeled 0.012 to 0.017 in.), showed no grain coarsening. The metal near the outer surface (which was exposed to soot and to the endothermic atmosphere), before being overheated, had apparently undergone sufficient grain-boundary carburization to prevent subsequent grain growth in the outer half of the wall. In the inner half of the wall, grain growth from overheating occurred before sufficient grain-boundary carburization took place to impede grain growth. This early grain growth resulted in the type of coarse-grain carburized structure shown in the third micrograph of specimen 1 in Fig. 67 (labeled 0.060 in., approximately).

The corrosion observed on ten of the tubes where they passed through the roof of the furnace resulted from attack by the sulfur in the refractory cement, in combination with a local tube temperature estimated to be about 1120 °C (2050 °F) in this region. The attack was less severe than that observed at the hearth region on the failed tube, where the tube temperature reached about 1175 to 1205 °C (2150 to 2200 °F) because of improper burner placement and adjustment.

The formation of a glossy adherent oxide, as observed on intermediate-temperature areas of the 18-cm (7-in.) specimen removed from the failed tube, is a phenomenon that is frequently encountered on nickel-base heat-resisting alloys when they are exposed to temperatures at which their oxidation resistance is marginal.

Conclusions. The premature failure of the tube by perforation at the hearth level resulted from (1) corrosion caused by sulfur contamination from the refractory cement in contact with the tube and (2) severe local overheating at the same location.

Corrective Measures. All of the tubes were replaced. To lengthen the service life of the replacement tubes, a low-sulfur refractory cement was used in installation, and burner positioning and regulation were controlled more closely to avoid excessive heat input at the hearth level. The replacement tubes had a life well in excess of two years, which was normal for RA 333 alloy tubes in this application.

Cement-Mill Equipment

An important part of the cement-making process involves roasting or firing the raw materials in a horizontal rotating kiln. In this operation, equipment failures are frequent and usually result from abrasion at elevated temperatures coupled with some corrosion (corrosion is rarely a major factor). In time, abrasion thins out guides, crosses, and other components to the extent that they must be replaced. Cast HH alloy (25Cr-12Ni) is widely used in kiln components because it provides good high-temperature strength and abrasion resistance at a reasonable cost.

Incinerator Equipment

Municipal and industrial incineration plants have become more numerous in an increasing effort to control pollution and waste disposal and to develop additional sources of combustible gases for heating. In many of these plants, waste materials are charged into the top zone of refractory-lined circular furnaces. Rotating, bladed rabble arms force the refuse through the various stages of burning, and the ash is discharged at the bottom of the furnace. The rabble arms and blades rotate around a vertical axis and are forced air cooled to maintain a temperature of 427 to 760 °C (800 to 1400 °F).

Cast HH alloy is the material most commonly used in incinerator furnace components. Because of the abrasiveness and corrosiveness of the refuse, components usually fail as a result of corrosive wear. Some of the refuse and the products formed during incineration remove the protective oxide skin that normally forms on stainless alloys and thus accelerate wear and corrosion. Among the corrosive materials common to incinerators are chlorides, which are released when chlorinated plastics—such as polyvinyl chloride—are burned, lead and zinc compounds that arise from volatilization of metal scrap, and sulfur compounds and salts.

Some corrosive gases, low-melting chlorides, and sulfur-containing salts exert a fluxing action on the protective films on metal surfaces.

Ordnance Hardware

During test firing, 81-mm mortar tubes bulged because of the pressures and temperatures that were developed during firing. When bulging occurred, the mortar tubes could no longer be used. The cause of these failures was apparent: the alloy from which the tubes were made did not have sufficient yield strength at the maximum tube operating temperature of 620 °C (1150 °F). The following example is not concerned with failure analysis of the bulged tubes, but rather with an evaluation that was conducted to select a suitable tube material.

Example 16: Evaluation of Alloys for Use in 81-mm Mortar Tubes. When bulging occurred in mortar tubes made of British I steel (see Table 3 for chemical composition) during elevated-temperature test firing, a test program was formulated to evaluate the high-temperature properties (at 540 to 650 °C, or 1000 to 1200 °F) of the British I steel and of several alternative alloys. These alloys included a maraging steel (18% Ni, grade 250), a vanadium-modified 4337 gun steel (4337V), H19 tool steel, and high-temperature alloys René 41, Inconel 718, and Udimet 630. The compositions of these alloys are given in Table 3, and the heat treatments that the alloys were given before being tested for evaluation are listed in Table 4. All the alloys evaluated had been used in mortar tubes previously or were known to meet the estimated requirement of a minimum yield strength of 552 MPa (80 ksi) at 620 °C (1150 °F).

Test Program. Tests were made to determine (1) transverse yield strength (at 0.1% offset) and ductility at room temperature and at 540,

595, 620, and 650 °C (1000, 1100, 1150, and 1200 °F), (2) transverse yield strength and ductility at 595 and 620 °C (1100 and 1150 °F) on specimens cycled 1, 5, 10, and 25 times for 15 min at temperature, and (3) impact and fracture toughness at −40 °C (−40 °F) and at room temperature. Fatigue-strength data were also obtained from manufacturers.

Except the vanadium-modified 4337 gun steel (4337V) and the British I steel, which were air melted and vacuum degassed, all alloys came as vacuum-melted bar stock (induction or arc remelted or both) in sizes equal to that of the 81-mm mortar tube. Transverse disks cut from each bar were machined into three tensile and two V-notch Charpy specimens for each test condition.

For the elevated-temperature tests, tensile-test bars were pulled after being held at temperature for 15 min in an electric furnace mounted on the tensile-testing machine. Specimens tested at 595 and 620 °C (1100 and 1150 °F) were cycled 1, 5, 10, and 25 times; each cycle consisted of heating in an electric furnace for 4 min to equalize the temperature of the bar, holding at temperature for 15 min, and cooling in still air. In each group, the last cycle was performed during hot tensile testing.

Test Results. Table 5 shows how transverse yield strengths of the alloys vary with cycling from room temperature to 595 and 620 °C (1100 and 1150 °F). Ductility of alloys cycled between room temperature and 595 °C (1100 °F), as measured by elongation and reduction of area, is shown in Table 6. The alloys fall in this order of decreasing strengths: Udimet 630, Inconel 718, René 41, H19 tool steel, British I steel, 4337V gun steel, and maraging steel. When cycled between room temperature and 540 to 650 °C (1000 to 1200 °F), only Udimet 630, Inconel 718, and René 41 retained yield strengths higher than the minimum of 552 MPa

Table 3 Compositions of alloys evaluated for use in mortar tubes

Alloy	C	Mn	Si	Cr	Ni	Mo	Fe	Other
Maraging steel 0.02	0.08	0.06	...	18.5	4.7	70.0	0.3 Ti, 7.7 Co, 0.04 Al
4337V gun steel 0.32	0.59	0.22	0.94	2.3	0.58	94.0	0.12 V
British I steel 0.45	0.62	0.30	2.8	0.43	0.90	94.0	0.20 V
Inconel 718 . . . . . . .0.05	0.01	0.10	18.2	53.0	3.1	18.0	1.1 Ti, 5.4 Nb + Ta, 0.5 Al
H19 tool steel 0.40	0.29	0.23	4.1	...	0.44	86.0	4.1 Co, 4.0 W, 2.1 V
Udimet 630 . . . . . . .0.03	0.15	0.10	17.3	57.0	2.9	17.5	1.0 Ti, 0.10 Co, 0.6 Al
René 41 . . . . .0.09	0.04	0.10	18.8	55.0	9.8	1.4	3.2 Ti, 11.3 Co, 1.6 Al

Table 4 Heat treatments of mortar-tube alloys before evaluation

Alloy	Heat treatment(a)	Hardness, HRC
Maraging steel	ST at 815 °C (1500 °F) 1 h, AC, A at 495 °C (925 °F) 3 h, AC	50
4337V gun steel	A at 855 °C (1575 °F) 4 h, OQ, T at 595 °C (1100 °F) 5 h, WQ	40
British I steel	A at 915 °C (1680 °F) 5 h, OQ, T at 610 °C (1130 °F) 5 h, AC	40
Inconel 718	ST at 980 °C (1800 °F) 4 h, AC, A at 720 °C (1325 °F) 16 h, FC at 55 °C (100 °F)/h to 620 °C (1150 °F), hold 8 h, AC	40
H19 tool steel	A at 1165 °C (2125 °F) 1 h, TT at 650 °C (1200 °F) 1 h, AC	44
Udimet 630	ST at 1025 °C (1875 °F) 4 h, AC, A at 760 °C (1400 °F) 8 h, AC, A at 650 °C (1200 °F) 10 h, AC	44
René 41	ST at 1080 °C (1975 °F) 4 h, AC, A at 760 °C (1400 °F) 16 h, AC	37

(a) ST, solution treat; AC, air cool; A, age; OQ, oil quench; T, temper; WQ, water quench; TT, triple temper

Table 5 Transverse yield strength of mortar-tube alloys after cycling between room and test temperatures

Number of 15-min cycles	At 595 °C (1100 °F) MPa	Transverse yield strength(a) ksi	At 620 °C (1150 °F) MPa	ksi
Maraging steel				
1	420	61	276	40
5	393	57	228	33
10	434	63	276	40
25	400	58	296	43
4337V gun steel				
1	531	77	407	59
5	496	72	345	50
10	455	66	338	49
25	448	65	345	50
British I steel				
1	531	77	407	59
5	490	71	379	55
10	455	66	338	49
25	448	65	345	50
Inconel 718				
1	903	131	841	122
5	869	126	841	122
10	848	123	779	113
25	827	120	765	111
H19 tool steel				
1	586	85	552	80
5	627	91	558	81
10	607	88	558	81
25	586	85	531	77
Udimet 630				
1	993	144	1000	145
5	965	140	965	140
10	979	142	972	141
25	1013	147	979	142
René 41				
1	683	99	696	101
5	724	105	690	100
10	690	100	655	95
25	703	102	703	102

(a) Values are averages for two or three specimens tested at each number of cycles.

Table 6 Ductility of mortar-tube alloys at 595 °C (1100 °F) after cycling between room temperature and 595 °C (1100 °F)

Number of 15-min cycles	Ductility, as measured by: Elongation, %(a)	Reduction of area, %(a)
Maraging steel		
1	33.6, 34.8, 47.9	87.5, 90.3, 96.0
5	42.1, 45.7, 55.0	91.6, 94.8, 97.2
10	35.0, 43.3, 50.7	90.0, 93.1, 97.7
25	45.0, 52.9, 58.2	92.9, 94.7, 96.6
4337V gun steel		
1	16.8, 19.0, 25.0	65.0, 71.4, 76.8
5	19.3, 21.4, 24.7	76.5, 77.5, 79.6
10	18.6, 21.0, 21.8	71.6, 71.9, 75.4
25	21.8, 23.2, 34.7	70.7, 78.3, 84.4
British I steel		
1	15.4, 16.1, 16.9	34.6, 35.8, 40.9
5	20.0, 21.4, 22.2	43.6, 58.4, 67.0
10	25.6, 29.2	68.8, 70.4
25	20.4, 33.2	66.0, 79.9
Inconel 718		
1	9.3	21.7
5	12.8, 18.3	23.2, 27.5
10	12.5, 15.4, 18.3	20.7, 23.2, 25.6
25	10.7, 17.1, 21.8	26.1, 26.6, 29.5
H19 tool steel		
1	3.6, 5.4, 8.3	7.1, 8.1, 12.5
5	6.4, 7.6, 8.3	9.7, 11.4, 12.0
10	6.1, 6.8, 9.3	10.9, 12.5, 13.5
25	6.8, 7.6	11.2, 12.0
Udimet 630		
1	5.4, 5.7, 6.4	10.3, 12.0, 12.0
5	4.3, 6.8, 6.8	7.1, 12.0, 13.5
10	2.9, 5.7	3.8, 9.2
25	2.9, 5.0	7.1, 9.2
René 41		
1	3.5, 4.3	4.9, 6.5
5	3.6	3.8
10	1.8, 4.0	4.3, 5.4
25	3.3	3.5

(a) Values are for one, two, or three specimens tested at each number of cycles.

(80 ksi) at 620 °C (1150 °F). Also, these three alloys maintained high strengths over the tested range, whereas the others decreased in yield strength as cycling progressed.

In general, the decrease in yield strength with time was greater at higher temperatures. However, short-term exposure at a given temperature lowered yield strength appreciably, while longer exposures had less effect. In fact, some alloys actually increased in yield strength with time at temperature. This phenomenon, more noticeable in the highly alloyed grades, may be due to the development of a critical size of precipitates either during aging of the precipitation-hardening alloys or by tempering at elevated temperature. The large decrease in yield strength at elevated temperatures is due to overaging in the maraging steel and to overtempering in H19, British I steel, and 4337V.

Tensile tests revealed the following order of decreasing ductility: maraging steel, 4337V, British I steel, Inconel 718, H19 tool steel, Udimet 630, and René 41. The ductility of the maraging steel, 4337V, and British I steel rose with decreasing yield strength or increasing temperature, whereas the ductility of H19,

René 41, Inconel 718, and Udimet 630 remained constant from room temperature to 650 °C (1200 °F).

Data revealed that the impact-toughness values of 4337V and Inconel 718 were, respectively, 17 and 21 J/cm^2 (950 and 1210 in. · lb/in.^{2}) at −40 °C (−40 °F) and 27 and 19 J/cm^2 (1550 and 1110 in. · lb/in.^{2}) at room temperature, or about triple the values for maraging steel, Udimet 630, and René 41. Note that impact-toughness values for the three nickel-base alloys are about the same at −40 °C (−40 °F) as at room temperature. From limited fatigue data, it appears that Inconel 718 is superior to René 41 in fatigue strength at 650 °C (1200 °F).

Conclusions. Because the mortar tubes made with maraging, 4337V, and British I steels bulged when test fired in earlier field tests, a suitable tube material should have a minimum yield strength of 552 to 586 MPa (80 to 85 ksi) at 540 to 620 °C (1000 to 1150 °F). The evaluation indicated that H19 tool steel, René 41, Inconel 718, and Udimet 630 alloys met this requirement. However, the properties of H19 tool steel were marginal, and this alloy

was not considered suitable for use in mortar tubes.

Because the remaining three alloys had uniform high yield strength at elevated temperatures, other criteria were used to determine their suitability. Udimet 630 had the highest yield strength (965 MPa, or 140 ksi) with low ductility (6% elongation) and low toughness. Also, it was not readily available and was second most expensive of the three alloys. René 41 was the most expensive of the alloys, was also limited in availability, and had the lowest yield strength (690 MPa, or 100 ksi), lowest ductility (4%), lowest toughness, and lowest fatigue strength.

Based on the test data, therefore, Inconel 718 was considered best suited for 81-mm mortar tubes. It had an optimum combination of

Fig. 68 Service-temperature ranges and comparative emf values, in millivolts, for various thermocouple materials

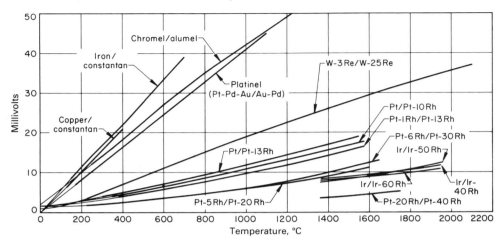

Fig. 70 Unetched specimen from a platinum thermocouple that was exposed to silicon contamination

A network of platinum-silicide second phase (gray) and voids (black) where the eutectic melted are visible. 100×

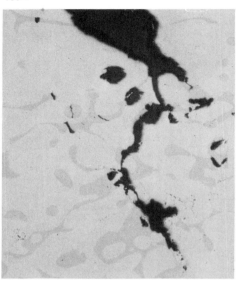

elevated-temperature yield strength, ductility, impact toughness, and fatigue strength. Widespread industrial use ensured its availability. Although a high-cost material, it was less expensive than the two alternative alloys.

Platinum and Platinum-Rhodium Components

Industrially, platinum and platinum-rhodium crucibles are used as containers for melting glass, for growing single crystals, and for making determinations of ash content in the cement and flour industries. Platinum and platinum-rhodium wires are used in thermocouples for sensing temperatures of 400 to 2400 °C (752 to 4352 °F) in furnaces and other types of heating equipment. Because of the very high cost of platinum and its alloys, crucibles and thermocouples made of these materials should be properly used and maintained to ensure maximum service life. Failure analysis is helpful in providing guidance for proper selection of alloys for, and proper care of, crucibles and thermocouples.

Crucibles. Based on analysis of several hundred failures, it is possible to summarize the causes of failure in crucibles. About 85% of all crucible failures result from the formation of low-melting eutectics and brittle intermetallic compounds along grain boundaries, which arise from interaction with metallic vapors, such as those of tin, arsenic, antimony, lead, and silicon. The remaining 15% are the result of miscellaneous causes, such as hammering the crucible to restore its shape after it has distorted as a result of thermal cycling. Hammering often results in cracking. Another cause of failure can be traced to the differences in thermal-expansion coefficients that exist between the crucible material and the substance being heated and cooled in it. This problem arises when glass is melted in Pt-10Rh and Pt-20Rh

Fig. 69 Bamboo-type surface that developed on the platinum leg of a thermocouple as a result of excessive loading at elevated temperatures

40×

crucibles, the latter of which are selected for their rigidity and resistance to deformation in thermal cycling. This problem can be solved by using platinum crucibles of a thicker gage; unalloyed platinum has better ductility than the Pt-Rh alloys and can better accommodate differences in coefficients of thermal expansion.

Thermocouples. By definition, a thermocouple is a device for measuring temperatures, and it consists of two dissimilar metals that produce an electromotive force (emf) roughly proportional to the temperature difference between their hot and cold junctions. Thermocouples made of platinum and a platinum-rhodium alloy usually fail as a result of mechanical breakage or deterioration of thermoelectric properties, both of which involve contamination. Grain growth can also contribute to mechanical failure of thermocouple wires, especially unalloyed platinum wires, after prolonged service at elevated temperatures. Small strains in the wire contribute to grain growth.

Platinum/platinum-rhodium thermocouples are generally used at temperatures to 1800 °C

Fig. 71 Sheathed thermocouple consisting of a platinum or platinum-rhodium sheath surrounding Pt/Pt-Rh wires embedded in compacted magnesium oxide or aluminum oxide

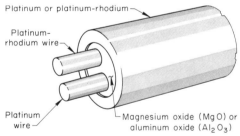

(3270 °F) in oxidizing atmospheres and require several precautions to avoid premature failure. The user should always observe the service-temperature ranges and comparative emf values of the various noble-metal thermocouples given in Fig. 68. Even at room temperature, platinum is a very soft metal and is commonly fractured by rough usage. At elevated temperatures, the strength of platinum is very low; therefore, platinum wire should not be subjected to high-temperature loading. Figure 69 shows the bamboo-type surface that develops on platinum wire and results in grain-boundary fracture because of excessive loading at elevated temperature. If loading is unavoidable, a mechanically stronger thermocouple, such as Pt-6Rh/

Pt-30Rh, should be used. The addition of rhodium increases the stress-rupture strength of the wire.

Thermocouples containing platinum should not be used in reducing atmospheres in refractory-lined furnaces, because they can be contaminated by pickup of silicon from the refractory lining. The solubility of silicon in platinum is low, and a low-melting eutectic is formed with a very small silicon content. Contamination will occur even when the platinum is not in contact with the silicon-bearing material, because, in a reducing atmosphere, the silicon can be transferred through the vapor phase. Figure 70 shows the typical structure that results from silicon contamination. The presence of the platinum-silicide network eventually leads to localized melting and catastrophic failure.

When temperatures must be checked in a refractory-lined furnace containing a reducing atmosphere, the risk of premature failure can be minimized by protecting the thermocouple with an impervious sheath, ensuring free access of air within the sheath to maintain oxidizing conditions in the event of leakage. Sheathing will also protect the thermocouple from other volatile impurities, including volatile metals, that adversely affect mechanical strength and thermoelectric properties.

One type of sheathed thermocouple is shown in Fig. 71. It consists of a platinum or platinum-rhodium sheath surrounding Pt/Pt-Rh wires embedded in compacted magnesium oxide (MgO) or aluminum oxide (Al_2O_3). This type of sheathed thermocouple is widely used in various protective furnace atmospheres and as an immersion thermocouple in the glass industry.

In general, observing the following precautions will ensure the accuracy and extend the life of thermocouples. A single length of double-bore insulating tubing should be used along the entire sheathed length of the thermoelements. Ample air space should be provided between the inner sheath and the outer tube of the double-bore tubing. If this is not practical, a type of sheathed thermocouple, such as that shown in Fig. 71, should be used. Once placed in service, thermocouples should be disturbed as little as possible, and their mountings should provide for uniform movement of the thermocouple wires when they expand or contract. Information on properties and selection of thermocouple materials can be found in the article "Thermocouples for Industrial Applications" in Volume 3 of the 9th Edition of *Metals Handbook*.

REFERENCES

1. A.W. Mullendore and N.J. Grant, Grain-Boundary Behavior in High-Temperature Deformation, in *Deformation and Fracture at Elevated Temperatures*, N.J. Grant and A.W. Mullendore, Ed., MIT Press, 1965, p 165-211
2. W.F. Brown, Jr., and G. Sachs, "A Critical Review of Notch Sensitivity in Stress Rupture Tests", TN 2433, National Advisory Committee for Aeronautics, Aug 1951
3. W.F. Brown, Jr., M.H. Jones, and D.P. Newman, Influence of Sharp Notches on the Stress-Rupture Characteristics of Several Heat-Resisting Alloys, in *Symposium on Strength and Ductility of Metals at Elevated Temperatures*, STP 128, ASTM, Philadelphia, 1952

SELECTED REFERENCES

- M.K. Booker, Analysis of Creep and Creep-Rupture Data, in *Mechanical Testing*, Vol 8, 9th ed., *Metals Handbook*, American Society for Metals, 1985, p 308-310
- N.L. Carroll, A. Fox, and R. McDemus, Creep Stress-Rupture, and Stress-Relaxation Testing, in *Mechanical Testing*, Vol 8, 9th ed., *Metals Handbook*, American Society for Metals, 1985, p 311-328
- J.B. Conway, Creep-Fatigue Interaction, in *Mechanical Testing*, Vol 8, 9th ed., *Metals Handbook*, American Society for Metals, 1985, p 685-694
- H.R. Voorhees, Assessment and Use of Creep-Rupture Properties, in *Mechanical Testing*, Vol 8, 9th ed., *Metals Handbook*, American Society for Metals, 1985, p 329-342
- J.D. Whittenberger, An Introduction to Creep, Stress-Rupture, and Stress-Relaxation Testing, in *Mechanical Testing*, Vol 8, 9th ed., *Metals Handbook*, American Society for Metals, 1985, p 301-310

Failures in Sour Gas Environments

Jeffery A. Colwell, Battelle Columbus Laboratories

THE PRESENCE of hydrogen sulfide and carbon dioxide in hydrocarbon reservoirs can create significant materials problems. Materials degradation can take many forms, including sulfide-stress cracking, hydrogen-induced cracking, stress-corrosion cracking, and weight-loss corrosion. Each of these must be understood if appropriate measures are to be taken to minimize its effect in field operations.

Sulfide-Stress Cracking (SSC)

Sulfide-stress cracking can occur when H_2S is present in the reservoir and is in contact with high-strength steels commonly used in drilling, completing, and producing wells. Sulfide-stress cracking is an embrittlement phenomenon in which failures can occur at stresses well below the yield strength of the material. For SSC to occur, three conditions must be met. The first is that a surface tensile stress must be present. It is important to remember that tensile stresses can be both applied and residual. The second requirement is that the particular material must be susceptible. In oil- and gas-production environments, this includes some of the standard casing and tubing alloys, such as API 5AX, grade P-110. The third requirement is that an embrittling agent—in the case of SSC, hydrogen sulfide—must be present in the environment.

Sulfide-stress cracking is basically a hydrogen-embrittlement phenomenon. Atomic hydrogen enters the steel to cause cracking. Many mechanisms have been proposed to explain this occurrence (Ref 1, 2). The hydrogen is generated on the surface of the steel because of a corrosion reaction. Iron reacts with H_2S to form iron sulfide and hydrogen (Ref 2). This hydrogen is generated in atomic form on the steel (or sulfide) surface, where either it can combine to form molecular hydrogen and leave the surface as bubbles or the hydrogen atoms can diffuse into the steel. This latter process may result in hydrogen embrittlement or SSC. Hydrogen sulfide prevents hydrogen recombination and thus promotes entry of atomic hydrogen into the steel. It is important to note that water must be present for this mechanism to occur; without it, SSC will not be observed, because the ionization of the hydrogen sulfide

Fig. 1 SSC of API grade L-80 oil well production tubing

SSC initiated the tool marks on the tube OD. The well contained about 15% H_2S and 4% CO_2 at 48 MPa (7000 psi). $0.3\times$. Courtesy of S.W. Ciaraldi, Amoco Corporation

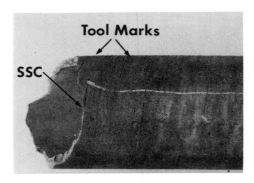

is required. Figure 1 shows an example of SSC in API grade L-80 downhole production tubing.

Factors That Influence SSC

A number of factors influence the SSC resistance of steels. Some of these include H_2S concentration, pH, temperature, strength level, and cold work.

Hydrogen Sulfide Concentration. In general, lower H_2S concentrations take longer to promote cracking than higher concentrations; lower concentrations also require high-strength materials before SSC is observed. The concentration of hydrogen sulfide in a produced fluid, for example, brine, is a function of the hydrogen sulfide partial pressure in the gas phase, which is a function of the total gas pressure. It should be noted that in reservoirs containing small amounts of H_2S there may be some difficulty in obtaining an accurate analysis; the reaction of H_2S with tubular goods and other components could result in a reading that is too low. Readings over several weeks, or even months, are sometimes required to obtain accurate results.

Effect of pH. The tendency toward SSC is a function of the pH of the system. With decreasing pH, the corrosion rate of the steel tends to increase, which causes more hydrogen to be produced. This causes more hydrogen to enter

the steel and increases the susceptibility to cracking. It is generally agreed that increasing the pH above 8 is beneficial in reducing the tendency toward SSC. Although fluid pH control cannot be easily accomplished with produced fluids, control of drilling environments is common. During drilling operations in sour reservoirs, the pH is usually maintained in the 10 to 11 range, thus providing the opportunity to use high-strength steels.

Temperature has been found to have a substantial effect on SSC resistance. As temperature increases, the resistance to SSC also increases. This is due to a reduction in the hydrogen-permeation rate at elevated temperatures (Ref 3). This effect allows materials that are susceptible to SSC at room temperature to be used at elevated temperatures. It has been found that SSC is most severe at room temperature; below room temperature, resistance to SSC again begins to increase.

The beneficial effect of temperature on SSC resistance is of practical significance in completing wells. For example, API, grade P-110, steel is susceptible to SSC at room temperature and therefore is not used in such applications. However, when the temperature is kept above about 80 °C (175 °F), P-110 can be successfully used. Therefore, this grade of steel is often used in the deeper portions of a well, where the temperatures are higher. However, caution must be exercised because SSC would be expected to occur if the temperature should ever decrease.

Strength Level. Various metallurgical variables can be controlled to maximize the resistance of a material to SSC. The strength level (commonly measured nondestructively by hardness) is probably the most widely used criterion in ensuring that steels and stainless steels do not fail by SSC. In most cases, carbon and low-alloy steels are used at hardnesses of 22 HRC or below. The exception is the quenched-and-tempered AISI 41xx series, which has been used at hardnesses to 26 HRC. Materials should be pretested in the anticipated service environment if there is any doubt about their suitability.

The microstructure of the steel is very important and essentially controls the observed SSC properties. It has been found that the microstructure that provides the best SSC resistance is that of tempered martensite; other transformation products reduce SSC resistance. Consequently, it is very important to ensure through hardening in the component. It is common practice when quenching tubular products for this application to implement both inside-diameter and outside-diameter quenches to ensure that the entire cross section is transformed to martensite.

Cold-Work. It is widely known that cold work can adversely affect the SSC resistance of materials. The hardness is locally increased, and residual stresses can also be generated. Hardnesses that greatly exceed 22 HRC (some as high as 40 HRC have been measured) can be produced by improper straightening and handling. Even identification stamping has been reported to cause enough cold work to initiate sulfide-stress cracks.

Stress. Because SSC is a stress-dependent phenomenon, the actual stress to which the component is subjected will affect the SSC resistance of the material. It must be remembered that the total stress, which includes both applied and residual stresses, must be considered. It is generally agreed that there is a threshold stress below which SSC is not expected. The threshold stress is a function of the material as well as environmental parameters. However, it is difficult to design to this threshold stress because of the uncertainty of controlling the service conditions at all times. Another factor that contributes to high stresses and the initiation of sulfide-stress cracks is the presence of stress concentrations, such as those found in threaded connections.

Hydrogen-Induced Cracking (HIC)

Hydrogen-induced cracking, also called stepwise cracking or blister cracking, is primarily found in lower-strength steels, typically with tensile strengths less than about 550 MPa (80 ksi). It is primarily found in line-pipe steels.

This type of degradation also begins with a reaction between steel and hydrogen sulfide in the presence of water. Again, hydrogen atoms enter the steel, but with HIC, as opposed to SSC, these hydrogen atoms combine to form hydrogen gas at internal defects. These internal discontinuities can be hard spots of low-temperature transformation products or laminations. However, manganese sulfide inclusions are the primary sites for this to occur. These inclusions tend to become elongated during pipe manufacture and give rise to high stresses at the tip of the inclusion when hydrogen gas forms there. As cracks initiate and propagate, they begin to link up with others, and a series of stepwise cracks can propagate through the material (Fig. 2). An applied stress is not required for this mechanism to occur. More information on the various forms of hydrogen attack is available in the article "Hydrogen-Damage Failures" in this Volume.

Factors That Affect HIC. Aside from reducing the amount of hydrogen being generated

Fig. 2 HIC of API, grade X-60, line-pipe steel after 28 days of exposure to ASTM seasalt brine, 20% H₂S, 20% CO₂, 1 atm, 24 °C

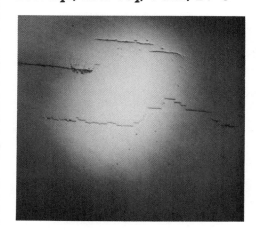

by reducing the corrosion reaction, another way of controlling HIC is through material processing. Shape control of sulfide inclusions is perhaps the best way to minimize the tendency toward HIC in line-pipe steels. Elongated manganese sulfide inclusions promote crack initiation and propagation due to the high stresses at the tips of the inclusions. However, the addition of calcium or rare earths to the steel makes the sulfides spherical, and because of their hardness, they remain spherical after processing. In addition, reduction of the sulfur content is also beneficial in reducing the susceptibility of steels to HIC. Other alloying additions that reduce hydrogen permeation, such as copper up to about 0.25%, are also beneficial (Ref 4).

Stress-Corrosion Cracking (SCC)

With this mechanism of degradation, a tensile stress is again required, together with a susceptible material and an environment that promotes cracking. In the oil and gas industries, the materials that are most generally found to be susceptible to SCC are the austenitic stainless steels and nickel-base alloys. Many different mechanisms have been proposed for SCC. The differences among these mechanisms depend upon the material and the environment, which would suggest that a single unified theory probably does not exist.

Austenitic stainless steels and nickel-base alloys are used in oil and gas production because they form protective films and therefore have very low corrosion rates. However, chloride ions (hence the term chloride stress-corrosion cracking), either in combination with hydrogen sulfide or alone, can attack this film, causing small pits to form. These small pits act as anodes, while the remainder of the oxide film acts as a cathode; the unfavorable area ratio causes the pit to grow. Also, the solution

Fig. 3 SCC of AISI type 304 austenitic stainless steel
As-polished. 100×

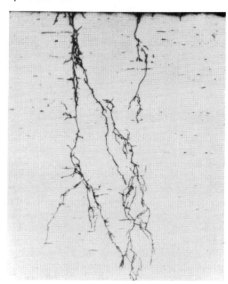

inside the pit is acidified because of the corrosion reaction, which also tends to increase the corrosion rate. Finally, a crack is initiated at the base of the pits because of the stress concentration, and propagation occurs because of the tensile stress. The crack often grows along grain boundaries because grain boundaries are electrochemically more active than the bulk grains. This is called active-path corrosion and is one of many possible mechanisms. Chloride SCC is usually observed at temperatures exceeding 65 to 95 °C (150 to 200 °F). Figure 3 shows an example of chloride SCC in an austenitic stainless steel. More information on SCC is available in the article "Stress-Corrosion Cracking" in this Volume.

Factors That Influence SCC. There has been considerable interest in minimizing weight-loss corrosion in sour environments by using corrosion-resistant alloys. Corrosion-resistant alloys include nickel-base alloys, austenitic stainless steels, and duplex stainless steels. Use of these alloys is extremely attractive for offshore applications, where the corrosion inhibition of carbon and low-alloy steels may be difficult and expensive. Corrosion-resistant alloys possess very low corrosion rates because of the presence of a passivating oxide layer that protects the base metal from further corrosion. In general, it has been found that the risk of SCC of these alloys in production environments increases as temperature and chloride and hydrogen sulfide concentrations increase. In many cases, a synergism exists between H₂S and chloride. Separately, each will not cause SCC, but together they promote it.

The presence of elemental sulfur also increases susceptibility to SCC. The reason for this is complicated, but it is probably due in part to a shift in potential into a range in which cracking is observed. If the pH is decreased, the material becomes more susceptible to SCC. The tendency of a material to exhibit SCC is also very temperature dependent. As temperature increases, susceptibility to SCC in production environments also increases. Unfortunately, this is where corrosion-resistant alloys are of most use, because corrosion inhibitors often become ineffective at elevated temperatures.

Weight-Loss Corrosion

Weight-loss corrosion is not related to the other forms of embrittlement discussed above, but it can cause severe problems during oil and gas production. Weight-loss corrosion is simply the loss of metal thickness due to an electrochemical reaction with the environment. The two basic types of weight-loss corrosion are uniform and localized. Uniform corrosion (Fig. 4) is characterized by metal loss at an essentially constant rate across the entire exposed surface. This is in contrast to localized corrosion, which is characterized by pitting and crevice corrosion. Both H_2S and chloride generally increase weight-loss corrosion rates. This can limit the potential usefulness of materials by reducing the load-carrying capability of many different components, or localized corrosion sites could serve as initiation points for SCC. Generally, increasing H_2S, CO_2, chloride, or temperature causes an increase in the corrosion rate.

The most widely used method for controlling weight-loss corrosion is the application of inhibitors. Many different formulations are available from manufacturers that are designed to be effective under different environmental conditions. There are also a number of ways that the corrosion inhibitor can be delivered to the area where it is needed. The two primary methods are batch treatment and continuous injection. In the batch method, the corrosion inhibitor is added on a regular basis, such as weekly. The treatment frequency depends upon how fast the concentration decreases with time. With continuous injection, the inhibitor is continuously added at the desired concentration throughout production.

Comparison of SSC and SCC

Stress-corrosion cracking is usually considered to be an anodic process, while SSC is considered to be a cathodic process. This can be very important in analyzing failures because the success of the method chosen to eliminate future failures may depend to a great extent on the determination of the proper failure mechanism. For example, it may have been decided that a particular failure was due to chloride SCC, and cathodic protection is being considered to increase the resistance of the material. However, if the actual failure mechanism is SSC, this remedy may actually worsen the

Fig. 4 Carbon steel pipe showing uniform corrosion

The 89-mm (3½-in.) OD pipe contained crude oil with a small, unknown quantity of H_2S at 370 °C (700 °F). Internal corrosion was very uniform, with the initial wall thickness of 5.7 mm (0.226 in.) reduced to 0.9 mm (0.035 in.) in the pipe body and 2.3 mm (0.09 in.) at the weld seam. Actual size. Courtesy of L. Wolfe, Conoco Inc.

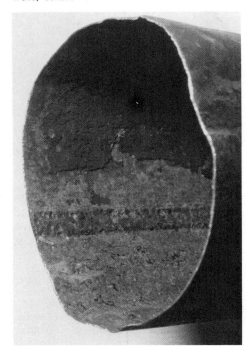

situation because SSC is a cathodic process. The application of a cathodic-protection current will tend to add even more hydrogen to the lattice, which will increase the likelihood of another failure. It should be remembered that failures of low-alloy steels will most often be the result of SSC, while the high-temperature embrittlement of high-alloy stainless materials will most likely be due to SCC. Failure classification of intermediate alloys is sometimes more difficult.

Another important difference between SCC and SSC is the effect of temperature. Decreasing temperature usually causes a decrease in SCC susceptibility, while decreasing temperature causes an increase in SSC susceptibility. This can be useful when trying to determine the actual failure mechanism.

Materials Selection to Minimize Failures in Sour Service

It is clear that safety issues are of primary importance in selecting materials for sour service. Even a small leakage of H_2S into the environment can cause extreme danger to the surrounding area. The National Association of Corrosion Engineers (NACE) has issued standard MR-01-75 (Ref 5) that is intended to

guide the selection of metallic materials for sour service, but it does not address the problems of HIC or SCC.

Reference 5 provides curves that can be used to determine when materials need to be selected for resistance to SSC (Fig. 5). The environments above the line in Fig. 5(a) and (b) are where SSC is expected and where materials should be selected to be resistant to SSC. For example, in sour gas systems, when the total pressure exceeds 0.45 MPa (65 psi) and the partial pressure of H_2S exceeds 0.3 kPa (0.05 psi), materials resistant to SSC should be used. Reference 5 lists acceptable materials in several alloy classes. The areas below the curves do not necessarily imply that SSC is not a problem there, because Ref 5 clearly states that environments in this area are beyond the scope of the standard. Alternatively, instead of using materials that are resistant to SSC, the environment can be controlled, but this is often not very practical, except in the case of drilling environments.

Reference 5 gives acceptable materials and conditions for a number of cases, some of which are highlighted in this article. This document should be reviewed for further information and clarification.

Carbon and low-alloy steels are acceptable at a maximum hardness of 22 HRC. However, steels in the AISI 41xx series are acceptable as tubes and tubular components at a maximum hardness of 26 HRC if they are in the quenched-and-tempered condition. Careful attention to the composition and the heat treatment is required to ensure SSC resistance of these alloys at hardness levels above 22 HRC.

Austenitic stainless steels, such as AISI type 304 and type 316, can be used to a maximum hardness of 22 HRC, provided they are in the annealed condition and are free of cold work. Ferritic stainless steels are also acceptable under the same conditions.

Martensitic stainless steels —AISI type 410, for example—are acceptable at hardness levels of 22 HRC maximum, provided they are normalized or austenitized and quenched, followed by double tempering. Both tempering temperatures should be 620 °C (1150 °F) minimum, with the second temperature lower than the first.

Nonferrous alloys are also considered and include, for example, many different nickel-, chromium-, and molybdenum-containing alloys. Reference 5 should be reviewed for processing requirements and specific alloys, but in a number of cases these alloys can be used up to 35 HRC.

When welding materials for sour service, the weldment must comply with the requirements that are specified for the base material. Many carbon steels meet this requirement in the as-welded condition. Weld overlays are allowed provided the temperature of the substrate does not exceed the lower critical temperature during application. If the lower

Fig. 5 Definition of sour environments

(a) Sour gas systems. (b) Sour multiphase systems. Source: Ref 5

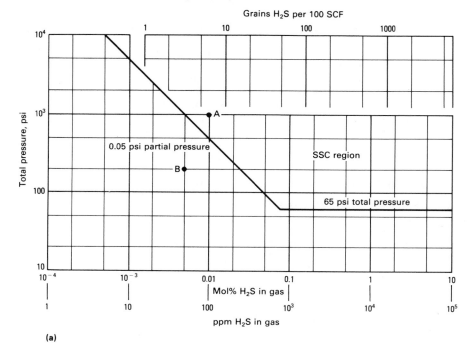

(a)

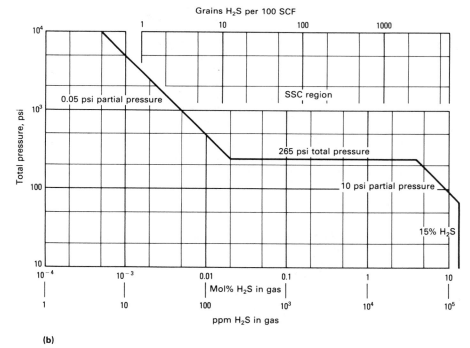

(b)

critical temperature is exceeded, the component must be heat treated to restore the base material to a maximum hardness of 22 HRC.

Because Ref 5 covers many more topics than those discussed in this article, the latest revision should be reviewed to become familiar with the various aspects. Tables are presented in the document that list materials of many different classes that are acceptable under certain conditions. Equipment design is often very important in obtaining resistance to SSC because threshold stresses can be low for certain materials.

Test Methods

As part of the materials-selection process, laboratory testing (or field testing with labora-tory specimens) often plays a major role. Because of the risk associated with using new and as yet unproven materials in the field, most companies conduct a test program to compare the results with those of alloys that are already known to perform satisfactorily in the field. Many different test methods are available, each having advantages and disadvantages. These are reviewed in Ref 6.

Environmental Cracking Tests. Both SSC and chloride SCC fall under this general category. However, the results of the tests do not distinguish between the mechanisms. The load applied to the specimen can be either static or dynamic, depending upon the test method selected. Statically loaded specimens include U-bend specimens, tensile bars, C-rings, bent beams, and double-cantilever beams. The most widely used dynamic test is the slow strain-rate test.

U-bends are made from sheet and bent into a "U" shape to stress the material plastically. The specimen is usually held in this position by a bolt while taking care to ensure electrical isolation between the two if they are not of the same material. The technique is discussed in ASTM G 30 (Ref 7). Specimens are exposed to the environment for the desired time and then examined with metallographic techniques. This test provides information only on whether or not the material cracked during the exposure. The advantage of this test is that a large number of specimens can be run at the same time, which lends itself to the screening of various alloys.

Tensile bars, C-rings, and bent beams are all stressed to known values, which are usually reported as a percentage of yield strength. Use of any of the three of these specimen types results in a threshold stress below which cracking is not expected. This is a useful way of ranking materials; a material with a high threshold is assumed to perform better in field environments.

Of the three specimen types listed above, only the tensile bar is associated with a standard test method for sour environments. However, a NACE committee (T-1F-9) is currently developing suitable standards for all three specimen types.

A great deal of testing has been conducted using tensile bars in accordance with the NACE TM-01-77 test method (Ref 8). This method uses a test solution (called NACE solution) consisting of 5% NaCl, 0.5% acetic acid, and 1 atm H_2S being bubbled through the solution at room temperature. The test is run for 720 h with the time-to-failure being plotted at various stress levels. From this plot, a threshold stress can be determined. The test specimens can be cut from almost any product form.

C-rings are made from tubular products and are bolt loaded to the desired stress level. ASTM G 38 (Ref 9) gives a procedure for making and stressing these samples. These specimens are usually placed in environments that more closely simulate field conditions than does the NACE solution. This requires the use of high-temperature and high-pressure autoclaves. The specimens are examined metallographically at the conclusion of the test for evidence of cracking. Again, a threshold stress can be determined for the exposure period in order to rank alloys.

Bent beams can be stressed in three-point or four-point bending. The stress level is again reported as a percentage of yield strength, and samples are tested in autoclaves to determine the susceptibility to cracking of the material over some particular time period. Beams can be machined from almost any product form.

Double-cantilever beam specimens are fracture mechanics specimens that allow development of crack-propagation data after the samples are exposed to the test environment (Ref 10). Other types of fracture mechanics specimens could be used, but the double-cantilever beam is the most common and probably the most convenient. These specimens are precracked by fatigue loading, then wedge-loaded to develop an initial stress intensity at the crack tip. If the crack propagates during the test, the stress intensity decreases, which reduces the driving force for further crack propagation. After a certain period of time, the crack is arrested. By measuring the length of the crack at arrest, a threshold value for the stress intensity can be calculated. This value is rarely used in minimum flaw tolerance design of drilling equipment. Instead, it provides a numerical value to be used for ranking alloys.

The slow strain-rate test, or constant extension rate test (CERT), is a technique that is used to determine the susceptibility of a material to cracking under dynamic loading conditions. The specimen is a tensile bar with a gauge diameter of about 2.5 mm (0.1 in.). A specimen is tested to failure in the environment of interest, and another is tested in an inert environment, such as air. Strain rates between 10^{-4} and 10^{-8} s^{-1} can be used, with rates between 10^{-6} and 4×10^{-6} s^{-1} being common.

After the experiment, the fracture surface can be examined for evidence of embrittlement. Also, the ductility of the sample tested in the environment (measured by reduction of area or percent elongation) is compared to that of the test performed in air. A decrease in ductility is an indication that the material may be susceptible to embrittlement.

The slow strain-rate test is often considered to be a severe test because it will usually indicate embrittlement when other tests do not. This may be related to the fact that any passive films on the specimen surface are continually broken during the test, which exposes fresh surface to the environment. Detailed information on environmental cracking test specimens, procedures, and results can be found in the article "Tests for Stress-Corrosion Cracking" in Volume 8 of the 9th Edition of *Metals Handbook*.

HIC Tests. The National Association of Corrosion Engineers has published a procedure for testing steels to determine their HIC resistance. Standard TM-02-84 (Ref 11) calls for exposing unstressed coupons to a synthetic seawater solution saturated with hydrogen sulfide at ambient temperature at a pH between 4.8 and 5.4. The test is run for 96 h and is designed to accelerate the formation of cracks.

At the conclusion of the test, the samples are polished, and any cracks are measured at $100\times$. Crack thickness, length, and sensitivity ratios are calculated and are used to rank the various materials.

Failures in Field Environments

Duplex Stainless Steel Wireline (Ref 12). It is well known that cold-worked duplex stainless steel wirelines are susceptible to cracking in sour environments. In one case in which alternative materials were not available and the wireline needed to be run, it was decided that failure would probably be unlikely because the exposure time would be minimal. However, a failure did occur that resulted in lost production and a costly fishing job. The well conditions were:

H_2S	3.5-4.0%
CO_2	12-20%
Pressure	19 MPa (2750 psi)
Depth	4210 m (13 800 ft)
Liquid level	3660-3965 m (12 000-13 000 ft)
Chlorides	73 000-85 000 ppm
pH	~7
Temperature	91 °C (195 °F)

The wireline diameter was 2.3 mm (0.092 in.) and was stressed to about 360 MPa (52 ksi) by the hanging weight. The mechanical properties are unknown. The line parted at a depth of 3800 m (12 500 ft), which was very near the liquid level. The failure occurred after less than 9 h of exposure to the environment.

Metallographic examination revealed longitudinal and transverse cracks. The longitudinal cracks were primarily intergranular, while the transverse cracks were mostly transgranular. Microhardness measurements were made, with hardnesses of 380 to 480 HK being reported. This implied a tensile strength of approximately 1380 MPa (200 ksi).

The failure was caused by SCC due to the presence of chlorides and hydrogen sulfide. This is in contrast to laboratory experiments that showed that failures would not be expected at 95 °C (200 °F). It was speculated that wetting and drying cycles may have played a role in causing the failure because the failure occurred so close to the liquid level.

API 5AC, grade L-80, Casing Failure (Ref 12). Standard MR-01-75 (Ref 5) states that API 5AC, grade L-80, steel is suitable for use in sour environments at all temperatures. The document further states that this grade can be used at hardnesses up to 22 HRC. However,

API allows L-80 to be manufactured at hardnesses up to 23 HRC; one should be aware of this when ordering this material.

This failure occurred with L-80 at a hardness level near this limit. Because of a packer-seal leak, excessive pressure developed in the casing annulus. Together with the stresses from the weight of the string, it was estimated that the axial tension in the casing at the failure location was close to the specified minimum yield strength. The total pressure was approximately 31 MPa (4500 psi), with a hydrogen sulfide content of 2.3% and carbon dioxide content of 3.0%. The diameter of the casing was 178 mm (7 in.), with a wall thickness of 9.5 mm (0.375 in.). The failure occurred at a depth of 31.4 m (103 ft) in the top joint on this string. Heavier wall casing was used above this joint of pipe.

It was found that the casing failed due to SSC. The metallographic analysis showed that the hardness of the casing at the inside diameter was probably about 23 HRC. In addition, banding was observed that was associated with hardnesses near 25 HRC.

The failure apparently occurred in two stages. The cracks initiated at the inside diameter and were the result of the presence of microcracks that resembled stepwise cracks observed from HIC. These created stress raisers and led to the formation of sulfide-stress cracks.

Laboratory tests were carried out on this particular steel using the NACE TM-01-77 method (Ref 8). The resulting threshold was determined to be 85% of the specified minimum yield strength (69% of the actual yield strength). Consequently, the NACE tensile test predicted that failure would have occurred because this casing was stressed near its yield strength.

Rupture of a Sour Gas Pipeline. The analysis of a sour gas pipeline that failed in Germany in 1982 has been reported (Ref 13). The 219 mm (8⅝ in.) line transported dehydrated gas (23% H_2S and 8.5% CO_2) at 8.8 MPa (1275 psi). The fracture occurred at the 12 o'clock position and was about 1.3 m (51 in.) in length.

By examining the failed area, it was found that small HIC-like cracks were present in an array that was vertical to the pipe surface. Because the gas was supposed to have been dehydrated with triethylene glycol (TEG), this was unexpected. Near the fracture area, elemental sulfur was also found. After further investigation, it was discovered that a gas containing up to 5000 ppm O_2 was being used in the glycol stripper. The oxygen can react with hydrogen sulfide to form water and sulfur. Consequently, there was water in the lines because of the TEG.

Laboratory experiments showed that the water content of the TEG proved to be the important factor for corrosion and hydrogen uptake. The presence of sulfur and chloride increased the tendency for HIC by about one order of magnitude. With 1% NaCl and sulfur, HIC was observed in TEG with only about 2.7% water.

It was concluded that restricting the use of the oxygen-containing stripper gas and frequent use of an inhibitor-triethanolamine-TEG mixture would minimize future problems. In addition, new lines will be constructed using low-sulfur inclusion shape controlled steel.

REFERENCES

1. I.M. Bernstein and A.W. Thompson, Ed., *Hydrogen Effects in Metals*, The Metallurgical Society of AIME, Warrendale, PA, 1981
2. S.A. Golovanenko, V.N. Zikeev, E.B. Serebryana, and L.V. Popova, Effect of Alloying Elements and Structure on the Resistance of Structural Steels to Hydrogen Embrittlement, in *H₂S Corrosion in Oil and Gas Production—A Compilation of Classic Papers*, R.N. Tuttle and R.D. Kane, Ed., National Association of Corrosion Engineers, Houston, 1982, p 198-204
3. C.M. Hudgins and R.L. McGlasson, The Effect of Temperature (75-400 °F) on the Aqueous Sulfide Stress Cracking Behavior of an N-80 Type Steel, in *H₂S Corrosion in Oil and Gas Production—A Compilation of Classic Papers*, R.N. Tuttle and R.D. Kane, Ed., National Association of Corrosion Engineers, Houston, 1982, p 90-94
4. G.J. Biefner, The Stepwise Cracking of Line-Pipe Steels in Sour Environments, *Mater. Perform.*, Vol 21, 1982, p 19
5. Sulfide Stress Cracking Resistant Metallic Materials for Oil Field Equipment, NACE MR-01-75, National Association of Corrosion Engineers, Houston
6. R.D. Mack, S.M. Wilhelm, and B.G. Steinberg, Laboratory Corrosion Testing of Metals and Alloys in Environments Containing Hydrogen Sulfide, STP 866, G.S. Haynes and R. Baboian, Ed., ASTM, Philadelphia, 1985, p 246-259
7. "Recommended Practice for Making and Using U-Bend Stress Corrosion Test Specimens," G 30, *Annual Book of ASTM Standards*, Vol 03.02, ASTM, Philadelphia, 1984, p 166-176
8. Testing of Metals for Resistance to Sulfide Stress Cracking at Ambient Temperatures, NACE TM-01-77, National Association of Corrosion Engineers, Houston
9. "Recommended Practices for Making and Using C-Ring Stress Corrosion Cracking Test Specimen," G 38, *Annual Book of ASTM Standards*, Vol 03.02, ASTM, Philadelphia, 1984, p 219-227
10. R.B. Heady, Evaluation of Sulfide Corrosion Cracking Resistance in Low-Alloy Steels, *Corrosion*, Vol 33, 1977, p 98
11. Evaluation of Pipeline Steels for Resistance to Stepwise Cracking, NACE TM-02-84, National Association of Corrosion Engineers, Houston
12. S.W. Ciaraldi, Materials Failures in Sour Gas Service, in *Corrosion/85*, Paper 217, March 1985
13. W. Bruckhoff, O. Geier, K. Hofbauer, G. Schmitt, and D. Steinmetz, Rupture of a Sour Gas Line Due to Stress Oriented Hydrogen Induced Cracking; Failure Analyses, Experimental Results and Corrosion Prevention, in *Corrosion/85*, Paper 389, March 1985

Products of Principal Metalworking Processes

Failures of Cold Formed Parts

THE TERM COLD FORMED can have broad meaning, depending largely on the product forms to which it is applied. Cold forming is frequently defined as forming that occurs below the recrystallization temperature of the specific metal being deformed, and it usually begins at or near room temperature. Depending mainly on the severity of the forming, the temperature of the work metal may increase as much as several hundred degrees Farenheit during the forming operation.

In some cases, limited heat is applied to the work metal before forming begins in order to increase its formability. Some metals—for example, certain magnesium alloys—can be difficult to form at room temperature. Preheating to temperatures as low as 120 °C (250 °F) will make these alloys much more formable. The practice of applying some heat to the material being formed without surpassing or closely approaching the recrystallization temperature is known as warm forming.

Regardless of the initial temperature, forming must be performed within the boundaries set by the material and tooling used. It is obviously possible to exceed the strength or elastic limits of a given material during a drastic deformation process. Care must therefore be taken to ensure that material specifications are accurately and realistically prepared in order to provide a product that will fulfill the desired expectations. Designing a process that operates at the extreme upper limit of the capability of any material will lead to an unacceptably high level of part failure because of the inevitable variations that occur from heat to heat or lot to lot.

Conversely, a number of material defects can also produce an unacceptably high percentage of failure even though the process and specification have been properly prepared and administered. Some of the areas that are important in cold-forming operations are described below.

Materials Defects That May Cause Failure

The grain size of a material to be formed is critical to both its ductility and strength. Because many carbon and low-alloy steels are used in the hot-rolled or subcritically annealed condition, grain size is often determined in the final hot-rolling operation. In general, grains larger than ASTM 3 are not desirable for most cold-forming operations. A duplexed grain size—that is, significant differences in grain size depending on location—should also be avoided.

Center Defects. Large inclusions, remnants of secondary pipe, or internal bursts can also be troublesome with cold-formed materials. Because the cold-forming operation typically involves severe stresses and deformation, such internal defects frequently lead to cracks and breaks that propagate to the surface of the parts being made, leading to their rejection.

Surface Defects. Laps, seams, and other surface defects formed during manufacture of the blanks can also lead to major problems during cold forming of the actual parts. Once any surface defect is exposed to the atmosphere, its surfaces become oxidized, providing an initiation point for failure during future operations. When such cracks form early in the steel manufacturing process, the surfaces of such breaks can become completely decarburized, forming what is known as a ferrite finger. Even if the original crack about which they formed is then removed by conditioning, the remaining ferrite is extremely low in tensile strength and will often split during the metal movement necessary during cold forming.

Segregation. A certain degree of alloy segregation occurs in all steel products. However, excessive segregation to the point of significant variations in hardness from one area of a bar to another can lead to premature failure and extreme difficulties during cold forming.

One of the simplest and most effective tests for incoming material defects such as those described above is a simple standard upset test. The details of such a test can be worked out between the supplier and the cold forger.

It is best to develop a well-understood specification between the steel supplier and cold forger. Such a specification should stipulate the forming and strength requirements of the final cold-forged part. With mutual effort, the strength and ductility requirements of the incoming stock and the quality control effort required can be accurately determined and the above difficulties avoided.

Design or Process Problems That May Cause Failures

In general, parts produced by cold forming are susceptible to the same causes of failure as parts of the same metal composition and function produced by other forming processes. There are, however, some unique characteristics of cold-formed parts that may influence their susceptibility to failure.

Grain Deformation. Cold forming of any metal part deforms the grains; the amount of deformation depends on the severity of forming. Severe deformation of the grains in locations of sharp bends or where the metal has become stretched, as in deep-drawn parts, results in a condition that is more vulnerable to failure from such mechanisms as corrosion and fatigue. Susceptibility to corrosion is easily demonstrated by observing simple parts formed from low-carbon steel and exposed to the weather. It will often be observed that the severely deformed areas will rust before areas on the same part that have been subjected to little or no deformation. This condition prevails even when there are no signs of cracks in severely deformed areas.

Frequently, there are fine cracks in certain areas of formed parts where deformation was severe. Although such cracks are harmless in many applications, they help to accelerate any type of corrosion, and they will also initiate fatigue cracks if the part is subjected to repeated stress.

Sharp radii or lack of generous fillets in drawn parts sometimes leads to failure. This occurred for the drawn container illustrated in Fig. 1. Although the engineering drawing for this container showed a large radius at the fillet at the change in diameters and a relatively long transition (Proposed design, Fig. 1), the dimensions were nót specific. During routine maintenance, the die radius was reduced to about 3.2 mm (0.125 in.), and the transition was shortened to improve metal flow over the die ring (Actual design, Fig. 1). This resulted in a sharp fillet along the most highly stressed location on the step. This sharp fillet had no effect on short-time hydrostatic strength; however, it constituted a stress raiser at which a fatigue crack initiated and propagated as a result of in-service pressure fluctuations (Fig. 1). After several such containers had accumulated a large number of pressure cycles, field failures were encountered because of fatigue cracks through the wall of the container. Corrective action included enlarging the die radius and instructing the incoming-inspection department that the radius in question must be checked on all new parts.

Fig. 1 Drawn 1008 steel container that failed by fatigue cracking

The cracking, which originated at a sharp-radius fillet, occurred because the generous fillet radius in the proposed design was reduced during die maintenance to improve metal flow over the die ring. The sharp radius fillet in the actual design acted as a stress raiser. Dimensions given in inches

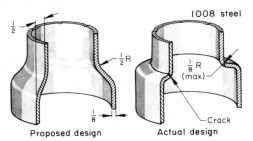

Proposed design   Actual design

Other characteristics that are generally unique to cold-formed parts include certain types of tool marks, particularly indentations at inside bend radii, scoring that may occur during deep drawing or forming, and scoring on outer surfaces such as often occurs when forming in a press brake using V-dies.

Orange-peel results when a very coarse-grain metal is stressed beyond its yield strength, increasing vulnerability to corrosion and helping to initiate fatigue cracks if the part is subjected to repeated stress.

Corner thinning causes concentration of stress flow lines at corners, which may help to cause failure by one or more mechanisms.

Work Hardening. Another unique characteristic of cold-formed parts involves local areas that have become severely work hardened, such as those areas that have been thinned by stretching or localized compression. Such areas are vulnerable to stress cracking.

Plating. Many formed parts are plated to protect them from corrosion. This often causes difficulties because the open areas of the part where there was little or no deformation usually receive the thickest plating, whereas severely formed areas may be plating-starved—for example, the inside corners of a box-shaped part. Thus, some formed parts are more susceptible to various forms of corrosion simply because the more vulnerable areas were not sufficiently protected.

Prevention of Failures

To a great extent, measures for preventing failures in formed parts are not different from the generally well-known preventive or corrective measures employed for parts that were shaped by other means; such measures include alloy-composition changes, design changes, process changes (tooling, lubricant, etc.), elimination of surface discontinuities, and use of (or change of) heat-treating procedures.

However, because of the unique characteristics of formed parts discussed above, there are

some preventive procedures that apply for forming parts from flat-rolled metal. Regardless of the metal being formed, there are several interrelated factors concerned with selecting the metal for formability. For manufacturing reasons and for prevention of failure in service, it is essential that the grade of metal be selected for maximum formability.

As mentioned above, grain size is of special importance and generally increases in degree of importance as the metal being formed becomes thinner. As a rule, the finer the grain, the better the formability. In some instances of coarse-grain metal, the locations where forming was the most severe may have only a very few highly deformed grains in a given cross section. It is important to select a fine-grain metal to accommodate the tendency of grains of varying orientation to deform in preferential directions but this is not always sufficient; it may be necessary to follow forming with full annealing (recrystallization) or at least with a stress-relieving heat treatment.

Certain precautions must be taken in specifying a recrystallization-annealing treatment because of the possibility of producing very large grains, which may result in undesirable properties for some applications. Subcritical strain for recrystallization can be achieved (1) in the transition region from a bend radius to a straight section in parts formed from sheet, strip, or wire, (2) near the neutral axis in some bent structural shapes, such as I-beams or channels, and (3) in some deep-drawn parts.

Various metals have their own specific peculiarities. For example, in the forming of certain stainless steels or iron-base heat-resisting alloys, severe forming encourages grain-boundary carbide precipitation if the metal is reheated to a temperature within the sensitizing-temperature range (400 to 870 °C, or 750 to 1600 °F). In the sensitized condition, the metal is extremely vulnerable to stress-corrosion cracking (SCC). Such a condition can be corrected only by the use of a recrystallizing anneal after forming, or it can be avoided by using a grade of stainless steel that is less susceptible to sensitization.

The following examples describe the difficulties encountered and the corrective measures taken or recommended for parts formed from carbon steel, a heat-resisting alloy, silicon bronze, and aluminum alloys.

Example 1: Fatigue Failure of a Steel Channel-Shaped Retainer Because of Vibration.
The governor on an aircraft engine failed and was returned to the manufacturer to determine the cause of failure. Upon disassembly of the unit, it was discovered that the retainer for the flyweight pivot pins was broken (Fig. 2).

The channel-shaped retainer (Fig. 2a) was made of 0.8-mm (0.030-in.) thick 1018 or 1020 steel. The part was plated with copper, which acted as a stop-off during carburizing of the offset, circular thrust-bearing surface surrounding the 16-mm (0.637-in.) diam hole (Fig. 2a).

Fig. 2 Fatigue-fractured low-carbon steel retainer (a) for the pivot pins of a flyweight assembly (b) used in an aircraft-engine governor

Dimensions given in inches

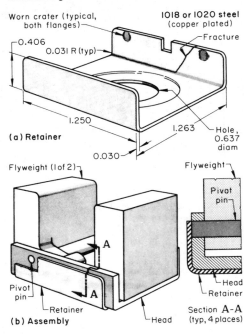

The bearing surface was case hardened to a depth of 0.05 to 0.1 mm (0.002 to 0.005 in.), then austempered to obtain a minimum hardness of 600 Knoop (1-kg, or 2.2-lb, load).

Considerable vibration was created in the installation because of the design of the mechanical device used to transmit power to the governor. Figure 2(b) illustrates how the pivot pins for the flyweight were retained. The pins were permitted to slide axially a small distance. The sources of stress in the retainer were vibration and centrifugal force, which hammered the pins against the retainer.

Investigation. Part of one flange broke completely off the retainer (Fig. 2a). The features of the fracture surface on the loose piece were obliterated by severe postfailure damage. Most of the fracture surfaces on the retainer were intact and suffered negligible damage subsequent to failure.

The loose piece from the retainer became lodged in other mechanical members of the governor and resulted in failure. The feed pivot pin extended out against the wall of the cast aluminum housing that encased the governor and eroded away part of the wall. If the loose piece had not caused the damage that resulted in failure of the mechanism, failure could have been caused by the accumulation of aluminum chips.

Microscopic examination of the fracture surface indicated that progressive cracking was initiated at the edge of the retainer at the 0.8-mm (0.031-in.) bend radius and propagated

Fig. 3 Heat-resistant alloy clamp for securing the hot air ducting system on fighter aircraft that failed by stress corrosion

(a) Configuration and dimensions (given in inches). (b) Section through the fracture area showing an intergranular crack. 540×. (c) Specimen of the work metal showing carbide precipitation at grain boundaries and within grains. 2700×. (b) and (c) Electrolytically etched with oxalic acid

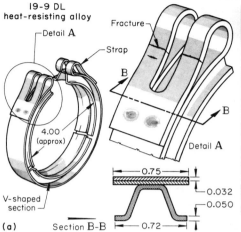

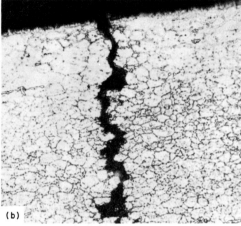

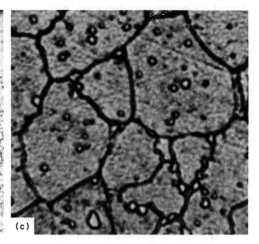

along the bend radius approximately half the length of the part, then extended across the flange. The 0.8-mm (0.031-in.) bend radius (approximately 1t) acted as a stress concentrator at the point of fracture initiation.

Ductility of the metal in the retainer was found to be excellent when the remaining part of the flange was intentionally bent. Craters 2.1 to 2.4 mm (0.082 to 0.096 in.) in diameter (Fig. 2a) were formed at four areas corresponding to the locations of the pivot pins. Surrounding each crater was a slight rim of jagged metal that had been extruded out of the crater, and the copper plating had been worn through at the bottom of the crater. Copper plating on the underface of the retainer had also been worn through by the motion between the retainer and the mating part.

On the edge of the hole in the bottom of the retainer, there was evidence of motion between the shaft and retainer. Spline teeth on the shaft had formed impressions 0.1 mm (0.005 in.) deep on the edge of the hole. The wear resistance of the thrust-bearing surfaces around this hole was judged to be satisfactory, as indicated by the absence of excessive wear.

Conclusion. Failure of the retainer was the result of fatigue caused by vibration in the flyweight assembly. Impact of the pivot pins on the retainer also contributed to failure.

Corrective Measures. The flyweight assembly was redesigned, and the channel-shaped retainer was replaced with a spring-clip type of pin retainer. The spring clip extended around the ends of the head and was held in place by notches in each end of the head.

Example 2: Stress-Corrosion Failure of a Strap-Type Clamp Made of 19-9 DL Heat-Resisting Alloy. The clamp shown in Fig. 3(a) was used for securing the hot air ducting system on fighter aircraft. The strap was 0.8 mm (0.032 in.) thick, and the V-

section was 1.3 mm (0.050 in.) thick; both were made of 19-9 DL heat-resisting alloy with the following composition:

Element	Composition, %
Carbon	0.3
Manganese	1.1
Silicon.	0.6
Chromium	19
Nickel	9
Molybdenum	1.25
Tungsten	1.2
Niobium	0.4
Titanium	0.3
Iron.	rem

The operating temperature of the duct surrounded by the clamp was 425 to 540 °C (800 to 1000 °F). The life of the clamp was expected to equal that of the aircraft. After 2 to 3 years of service, the clamp fractured in the area adjacent to the slot near the end of the strap (Fig. 3a) and was returned to the manufacturer to determine the cause of fracture.

Investigation. Micrographic examination was made of a section through the strap near the fracture area and including an intergranular secondary crack. This section (Fig. 3b) revealed a region of large grains at the surface that was about 0.05 mm (0.002 in.) deep. The large grains could have been the result of a sizing pass on the hot-rolled surface. Knoop microhardness values (50-g load) were 345 (25.6 HRC) for the metal in the core and 370 (28.6 HRC) for the metal near the surface.

Electrolytic etching in oxalic acid revealed carbide precipitation at the grain boundaries (Fig. 3c) throughout the material but more prevalent in the area of large grains.

Annealing a portion of the clamp by heating at 980 °C (1800 °F) and water quenching removed the intergranular network of carbides.

Conclusions. The clamp fractured by SCC because the work metal was sensitized. Sensitization occurred during long-term exposure to the service temperature; the effects of sensitization were intensified as a result of cold forming.

Recommendations. Failure of the clamp could have been prevented by using a work metal that was less susceptible to intergranular carbide precipitation.

Example 3: Effect of Alloy Selection and Part Design on SCC of Formed Silicon Bronze in Marine-Air Atmosphere. Electrical contact-finger retainers blanked and formed from annealed copper alloy C65500 (high-silicon bronze A) failed prematurely by cracking while in service in switchgear aboard seagoing vessels. In this service, they were sheltered from the weather, but were subject to indirect exposure to the sea air.

About 50% of the contact-finger retainers failed after 5 to 8 months of service aboard ship. Figure 4(a) shows one of the contact retainers that failed and the location of the crack.

Nine contact fingers were mounted in each of the two slots of each retainer, and a bank of four to eight retainers held the electrical contacts for a typical circuit-breaker system. When a retainer cracked, the usual consequences were open circuits across some contacts, short circuits, and arcing, with resultant overloading and damage to adjacent circuits as well.

Investigation. Emission spectrographic analysis of failed retainers removed from the vessels showed that the composition was within the prescribed range for copper alloy C65500. On each defective retainer examined, a single crack was visible to the unaided eye (Fig. 4a). Each crack was on an end of the retainer, extending between an outer corner of one of the rectangular slotted openings and the periphery of the

Fig. 4 Silicon bronze contact-finger retainer that failed from SCC in shipboard service

(a) Overall view of retainer showing cracking in corner (arrow). (b) Specimen taken from failure region showing secondary cracks (arrows). Etched with equal parts NH₄OH and H₂O₂. 250 ×

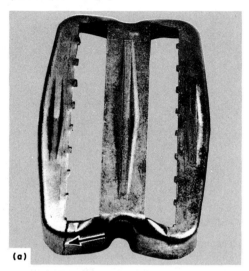

(a)

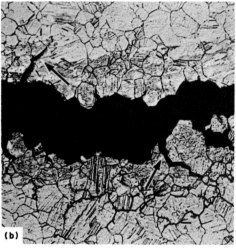

(b)

retainer, and nearly in line with the lengthwise direction of the slot.

Viewing with a low-power stereomicroscope showed that each crack had originated at a slot corner and progressed outward and through the side walls, extending almost or completely to the outermost edge of the retainer. Examination of crack regions at higher magnifications with an optical microscope showed faceted, intergranular fracture patterns and extensive side cracking. No evidence of corrosive attack or corrosion products was visible, and no signs of incipient cracking could be detected at the three other corners of cracked retainers.

The results of the preceding examination of the cracks were confirmed by microscopic examination of polished-and-etched specimens taken from the cracked regions of several of the retainers. When observed at a magnification of 250 ×, each of the specimens examined showed an intergranular crack pattern and a microstructure closely resembling that shown in Fig. 4(b).

Conclusions. The cracking was produced by stress corrosion as the combined result of residual forming and service stresses, the concentration of tensile stress at outer square corners of the pierced slots, and preferential corrosive attack along the grain boundaries as a result of high humidity and occasional condensation of moisture containing a fairly high concentration of chlorides (seawater typically contains about 19 000 ppm of dissolved chlorides) and traces of ammonia. Exposure to significant concentrations of ammonia and other corrodents could also have occurred while the vessels were in harbor in industrial locations.

The absence of general surface attack and the formation of faceted, intergranular fracture patterns and extensive side or branch cracking were consistent with the characteristic pattern for failure by SCC of this alloy.

The occurrence of cracking on only one corner of a retainer apparently reflected a characteristically uneven distribution of stress on the four corners of the workpiece during forming and in service, by which one corner was more highly stressed than the three others. Cracking on one corner appeared to relieve the stress on the other corners.

Corrective Measures. Three changes were made to improve the resistance of the contact-finger retainers to SCC. The slots were redesigned to have inside corner radii of 3.2 mm (0.125 in.) to reduce the stress concentration associated with the square corners in the original design, and the formed retainers were shot blasted to remove uneven surface tensile stresses and to put the surface layer in compression.

In addition, the retainer metal was changed to a different type of silicon bronze—copper alloy C64700. This alloy has a nominal alloying content of 2% Ni and 0.6% Si instead of the nominal 3% Si content in the alloy C65500 originally used for the contact-finger retainers. Selection of copper alloy C64700 was made on the basis of tests for formability and resistance to SCC (moist ammonia atmosphere) done by the manufacturer after the service failures were encountered.

All of the alloy C65500 retainers in shipboard service were replaced with retainers made to the improved design using alloy C64700. The parts were blanked and formed in the solution, heat-treated and 37% cold-reduced temper, aged after forming, then shot blasted.

Over a period of several years, only a few isolated SCC failures of the replacement retainers were observed, and these occurred on other types of ships where there was exposure to high concentrations of ammonia and high humidity.

Discussion. Of the various copper alloys considered for the retainers, all except alloy C64700 had to be formed in the annealed temper, with consequent lower strength and excessively unbalanced stress distribution, whereas alloy C64700 could be formed without difficulty in the half-hard (37% cold-reduced) condition. Retainers made from alloy C64700, although not completely immune to SCC in the tests performed by the manufacturer, were much less susceptible to this type of failure than retainers made from alloy C65500.

The excellent formability of 37% cold-reduced alloy C64700 was a key factor in the successful use of this alloy for the retainers. Cold reduction, either alone or in combination with artificial aging, has been found to minimize the susceptibility of alloy C64700 and other silicon bronzes to SCC. Better as-annealed resistance to SCC and greater improvement by cold reduction were found for alloy C64700 than for alloy C65500.

Plastic deformation produced slip planes, lattice rotation, and curved bands in the grains, thus providing many new sites for attack that were "blind alley" paths for preferential corrosion. These alternate paths allowed only limited penetration of the new sites and decreased corrosion attack at grain boundaries.

Example 4: Fatigue Failure of an Aluminum Alloy Assembly at Spot Welds Because of Improper Heat Treatment.

Postflight inspection of a gas-turbine aircraft engine that had experienced compressor stall revealed that the engine air-intake bullet assembly shown in Fig. 5(a) had dislodged and was seated against the engine-inlet guide vanes at the 3 o'clock position. The engine was restarted, operated at 95% rotational speed, and immediately shut off when the port fire-warning light came on.

The bullet assembly consisted of an outer aerodynamic shell and an inner stiffener shell, both of 1.3-mm (0.050-in.) thick aluminum alloy 6061-T6, and four attachment clips of 1-mm (0.040-in.) thick alclad aluminum alloy 2024-T42. Each clip was joined to the outer shell by 12 spot welds (Fig. 5a) and was also joined to the stiffener (by spot welds not shown in Fig. 5). Service life of the assembly was about 680 h.

Investigation. The outer shell of the assembly had separated from the stiffener because of fracture of the four attachment clips through the shell-to-clip spot welds (View A-A, Fig. 5a). Visual examinations of the fracture surfaces with the unaided eye and with a low-power binocular microscope were not definitive, because the legs of the clip that were attached to the outer shell were severely mutilated. This damage indicated a large degree of relative movement between the clip and the shell either during or after complete separation.

Dye-penetrant inspection of the assembly revealed cracks in the clip spot welds on the shell, adjacent to the fractures, and in the spot welds joining the clip to the stiffener. Microscopic examination of sections through the cracked spot welds revealed that the cracks had originated at the edge of the weld nugget and at

Fig. 5 Aluminum alloy engine air-intake assembly that failed because of fatigue fracture of spot-welded attachment clips

(a) Configuration and dimensions (given in inches). Note typical location of cracks in View A-A. (b) Crack (arrow A) initiating at a notch of spot weld and faying surfaces of outer shell (top) and clip (bottom). Also shown are excessive penetration in the clip, excessive indentation of the shell, and the large HAZ (heat-affected zone) (arrow B) in the shell. 10×. (c) Region of crack origin. 150×. (b) and (c) Etched with Keller's reagent

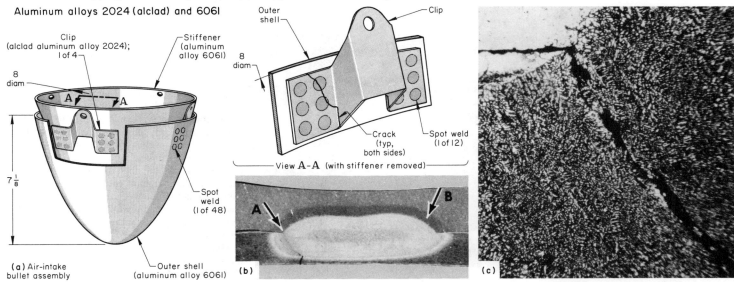

the faying surfaces of the shell and the clip (Fig. 5b and c). The shape of the weld nugget at this point created a region of peak stress. The cracks propagated transversely through the spot welds and progressed in a transgranular manner through the clips.

The microstructure of the aluminum alloy 6061 in the shell and stiffener, combined with a hardness of 50 to 52 HR15T (required: 76 to 84 HR15T) indicated that the material was in the annealed (O) temper, rather than the T6 temper as specified by the engineering drawing. The metallurgical properties of the clad aluminum alloy 2024-T42 clip were satisfactory with respect to microstructure and to hardness (required: 78 to 84 HR15T; observed: 81 to 84 HR15T). The chemical compositions of the alloy 2024 clip and the alloy 6061 shell and stiffener were satisfactory.

As Fig. 5(b) shows, the spot welds joining the clip to the outer shell exhibited excessive penetration (0.8 mm, or 0.03 in.) of the 1-mm (0.040-in.) thick clip and excessive indentation of the shell; also, a large HAZ (0.25 mm, or 0.01 in., deep) was present at the portion of the weld in the shell. These conditions were the result of spot welding the shell in the annealed condition. During welding, the weld zones distorted under the welding pressure and increased the contact area, which changed the current distribution and resulted in low or inconsistent weld strength.

Conclusion. The outer shell of the bullet assembly separated from the stiffener because the four attachment clips fractured through the shell-to-clip spot welds. Fracture occurred by fatigue that initiated at the notch created by the intersection of the faying surfaces of the clip

and shell with the spot weld nuggets. The 6061 aluminum alloy shell and stiffener were in the annealed (O) temper rather than T6, as specified. This resulted in excessive penetration and indentation during spot welding and was a contributing factor in the bullet-assembly failure.

Recommendations. The shell and stiffener should have been heat treated to the T6 temper after forming. A review of the spot welding operation also may have been desirable; in this review, the size and shape of the electrode face that made contact with the shell and stiffener should have been considered, as well as the magnitudes of welding current and of current-on time.

Example 5: Aluminum Alloy 7178-T6 Aircraft Deck Plate That Failed in Service by Fatigue Cracking. Two cracks were discovered in a deck plate of an aircraft during overhaul and repair after 659 h of service. As shown in Fig. 6, the cracks were on opposite sides of the deck plate in the flange joggles. The plate had been formed from 7178-T6 aluminum alloy sheet.

Investigation. Visual inspection of the deck plate disclosed that the fastener holes adjacent to each crack (five holes transverse to and three holes parallel to the length of the plate, as shown in Fig. 6) had been slightly elongated. The deck plate was cut to expose the crack surfaces, and several fracture origins were found, all adjacent to the undersurface of the deck plate. The evidence indicated that both cracks had initiated in the inner curve of the flange joggles.

The crack surfaces were replicated by a two-stage plastic-carbon technique, with chro-

mium shadowing at 45°. The predominant features of electron microscope fractographs made of the replicas were fatigue striations at the fracture origins.

Microscopic examination of sections through the fractures showed that fracture had been transgranular and that several secondary transgranular cracks existed in the curved inner surface of the flange. These were parallel with the main cracks and originated at surface defects. These defects could have been shallow corrosion pits, perhaps produced during cleaning, or could have been notches originating in the forming operation—the examination did not determine which. In other respects, the structure was normal in both areas.

The condition of the plate was checked by hardness tests and by measurements of electrical conductivity. Hardness values ranged from 92 to 93 HRB. The electrical-conductivity values extended from 30.8 to 33.5% IACS (Fig. 6a). The values for both hardness and electrical conductivity were satisfactory for this alloy and temper, indicating that the deck plate had been properly heat treated.

Conclusions. The failure was caused by fatigue cracks originating on the inside curved surface of the flanges. The cracks had initiated in surface defects caused by either corrosion pitting or forming notches, acting in combination with lateral forces evidenced by the moderate distortion of the fastener holes.

Recommendations. The surface defects should be eliminated by revised cleaning and/or forming procedures. The lateral forces should be alleviated by reviewing the design and installation and by making appropriate revisions.

Fig. 6 Aluminum alloy 7178-T6 aircraft deck plate that failed in service by fatigue cracking

(a) Deck plate showing location of cracks at opposing flange joggles. Percentages are IACS values of electrical conductivity as measured at three locations. Approximately 0.2×. (b) Detail of crack 1 in the plate. 2×. (c) Detail of crack 2. 3×

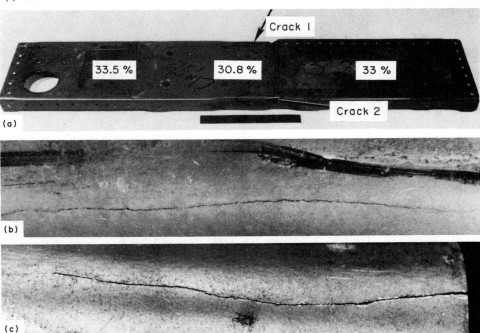

(a)
Crack 1
33.5 %   30.8 %   33 %
Crack 2

(b)

(c)

Fig. 7 Failed aluminum alloy 6061-T6 connector tube from a water-cooling system

(a) Macrograph of the tube showing blow out that appears to involve the loss of a single large grain. (b) Structure near the neutral axis of the bend. The huge grain at lower left (light area) occupied almost the entire wall thickness of the tube and was 5 to 7.6 mm (0.2 to 0.3 in.) long. 65×

(a)

(b)

Example 6: Failure by Blowout of Aluminum Alloy 6061-T6 Connector Tubes From a Water-Cooling System. Several of the aluminum alloy 6061-T6 drawn seamless tubes connecting an array of headers to a system of water-cooling pipes failed from causes described as blowouts, circumferential tears, and longitudinal cracks. The aluminum alloy was supplied to conform with the requirements of ASTM B 234.

The tubes were 2.5-cm (1.0-in.) outside-diameter tubes with a wall thickness of 1.7 mm (0.065 in.) and were supplied in the (O) temper. They were bent to the desired curvature (approximately 6.4-cm, or 2.5-in., centerline

bend radius), preheated at 510 °C (950 °F) for 30 min, then solution treated for 30 min at 520 °C (970 °F) and water quenched. They were then aged at 177 ± 3 °C (350 ± 5 °F) for 8 to 10 h. Before shipment, the tubes were pressure tested at 7 MPa (1000 psi) and dimensionally inspected.

Investigation. Visual inspection showed that some of the tubes failed by blow out of a section of the wall; some blow outs produced edges with sharp cleavage planes typical of a shear failure in a single crystal. Figure 7(a) shows an example believed to involve the loss of a single large grain. The edges of other blow outs were rough and jagged, as though caused

by rapid fatigue. All tube blow outs were found in or adjacent to the bends, and in most instances, cracks extended from the blow outs through the full wall thickness. The size of the blow outs varied from about 3.2 × 9.5 mm (0.125 × 0.375 in.) to a maximum of 1.9 × 5 cm (0.75 × 2 in.).

Other failures involved circumferential tears in which the fractures were extremely ragged and appeared to follow the boundaries between very large grains and adjacent very small ones; these fractures were also associated with tube bends. Incipient failures were also discovered in the form of both longitudinal and circumferential cracks that varied in size from a small fraction of an inch to several inches along the tube length, in or close to the bends.

The tubes exhibited a number of defects on the internal surface, including striations, cold shuts, hairline cracks emanating from cold shuts, and both circumferential and longitudinal cracks in the bends. It appeared that the as-received tubes possessed numerous stress raisers in the inner surface.

Macroetching revealed that the structure of the tubes had recrystallized in heat treatment. The macrostructure of the straight sections was uniformly fine grained. In the bends, there were exceedingly large grains, in some cases adjacent to very small grains. The size of the very large grains was consistent with the theory that some of the blow outs involved single grains that occupied the full wall thickness of the tube.

Spectrographic analysis of six specimens revealed that the tubes conformed to the ASTM specification except for the magnesium content, which ranged from 1.1 to 1.3%. This degree of excess beyond the specified range of 0.8 to 1.2% Mg did not appear to be serious or to explain the failures.

Rockwell hardness tests, microhardness tests, and tension tests of material from the failed tubes gave results that were typical of aluminum alloy 6061-T6. However, flattening tests on the tubing (using strip specimens cut from the tubing), slow-bend test, notched bend test, and a bend test in which the specimens were bent by hand around a relatively small radius, indicated that the tubing material was notch sensitive and had a type of low ductility that had not been detected in the tension tests. Specimens of aluminum alloy 6061-T6 sheet were notch sensitive and had a much greater ductility in all the bend tests.

Microscopic examination confirmed the presence of extremely coarse grains in the bends of the tube connectors, in contrast to a uniformly fine grain size in the straight sections. In some instances, coarse grains were measured as large as 5 to 13 mm (0.2 to 0.5 in.) in diameter. The evidence indicated that the bending operation achieved a critical degree of strain near the neutral axis of the tube. With less than this amount of strain, no grain growth would occur upon heating to the solution-treatment temperature, as was the case with some of the neutral-axis grains. Regions receiving the critical strain or slightly more would experience maximum grain growth upon solution treatment. The portions of the tube at 90° to the neutral axis received the maximum amount of cold work and therefore recrystallized, but did not undergo any appreciable change in grain size. The wide variation in grain sizes observed near the neutral axis is illustrated in Fig. 7 (b).

Conclusions. The bending of the connector tubes in the annealed condition induced critical strain near the neutral axis of a tube, which resulted in excessive growth of individual grains during the subsequent solution treatment. The wide variation in grain size within the bends was considered unfavorable for resistance to flexural fatigue. The procedures used in fabricating the connectors created many defects on the interior surfaces, which were potential stress raisers and possible origins of fracture. The specific cause of the notch sensitivity and brittle behavior was not evident except to the extent that discontinuities on the inside surface and the heterogeneous grain size may have been contributing factors. In any event, the slow-bend tests offered a means of detecting notch-sensitive material.

Corrective Measures. The connector tubes were bent in the T4 temper, as early as possible after being quenched from the solution temperature. The tubes were stored in dry ice after the quench until bending could be done. The tubes were aged immediately after being formed. Flattening and slow-bend tests were specified to ensure that the connector tubes had satisfactory ductility. Macroetching of random samples was instituted, and a limit on the maximum acceptable grain size was imposed. Limits were placed on the number and size of permissible surface defects.

Representative samples of connector tubes supplied under the revised specifications were sectioned and examined metallographically. It was determined that the new tubes had a uniformly fine grain size throughout the entire length. The grain size in the bends of the new tubes was two to three times smaller than the grain size in the straight sections of the original tubes. The improved quality of the new connector tubes was effective in securing trouble-free operation.

Failures of Forgings

Vito J. Colangelo and Peter A. Thornton, Benet Weapons Laboratory, Watervliet Arsenal

THE TERM FORGING usually connotes an orange-colored, hot metal being pounded into some usable shape in a noisy, dimly lit forge shop that is covered with a layer of dark gray, powdery grime. In a technical sense, however, a forging can be generically described as a metal mass that has been worked or wrought to a desired configuration by controlled plastic deformation through hammering, pressing, upsetting, rolling, and extruding or some combination of these processes. Historically, forging is traceable back to about 8000 B.C. in the Middle East, where gold and copper were simply hammered into useful shapes. In fact, until the development of the water-powered tilt hammer in the 13th century and the advent of the steam hammer during the Industrial Revolution, metalsmiths forged objects almost exclusively by hammering the metal on an anvil.

Until the 1800s, most structural forgings were produced from iron. Today, however, forgings are made from a wide variety of metals and alloys, including aluminum, copper, steel, titanium, magnesium, high-temperature alloys, refractory metals, and metal-matrix composites. These materials are forged from ingot stock and from preforms, such as a powder metal (P/M) compact, in a number of ways, including open- and closed-die pressing, rotary forging, ring rolling, swaging, drop and counter-blow hammering, and extrusion methods. Present-day forgings range in size and complexity from minute close-tolerance computer parts weighing less than an ounce to large structural components, such as bridge girders and papermill rolls weighing several tons. A summary of forging processes, materials, and applications is provided in Ref 1 and 2.

Forgings can be produced by hot working, which is carried out at temperatures in excess of 60% of the absolute melting point of the metal; by warm working, which is conducted between 25 and 60% of the absolute melting temperature; and by cold working, in which temperatures are less than 25% of the absolute melting point of the metal. Although most wrought metals can be worked over a wide range of temperatures, hot forging takes a minimum of force or forging load and results in maximum workability. However, dimensional control of the workpiece is often difficult in hot-working operations because the metal contracts nonuniformly during cooling. This particular problem is not encountered in the cold-working temperature range; thus, many forgings are produced by cold working even though workability is less than that of hot forging. Paradoxically, the workability of metals in the warm-forging regime is considerably lower than that of cold forging (Ref 3). Because the article "Failures of Cold Formed Parts" in this Volume addresses cold-formed parts and because the bulk of thermally worked products are produced at temperatures in the hot-work regime, this article will primarily deal with forgings manufactured by hot working.

When a metal is hot worked, the individual grains are deformed; the deformed grains then immediately begin to recrystallize, that is, nucleate new stress-free grains that are roughly equiaxed in morphology. The hotter the metal, the more plastic it behaves and the more easily it deforms. At excessively high temperatures, however, grain growth, incipient melting, phase transformation, and changes in composition can occur, which may cause degradation of the properties in the forging.

At lower hot-forging temperatures, the metal is more difficult to work; yet the resultant grain size may be finer, and the forging may have better mechanical properties. If the forging temperature is reduced still further, well below the recrystallization temperature, the deformed grains will not break up and form new grains but will remain deformed and highly stressed. When the metal is in this condition, it can behave in a brittle manner, and cracking may occur.

Occasional failures are sometimes unavoidable in commercial products. The more complex a part design is, or the more complex its processing history, the greater its potential for failure. These statements are not intended as an indictment of forging products, because forgings generally perform well in terms of critical applications and demanding service requirements.

The point is that, due to the large number of factors typically involved in the manufacture of an engineering component and the synergism that inevitably exists between such factors during a component failure, the information to be gained from a comprehensive failure analysis can be very valuable and must be given the utmost consideration. The proper application of information gained in a failure analysis can provide a valuable adjunct to the design and manufacture of a product such as a forging. These studies help identify design inadequacies, material flaws, fabrication or processing defects, and certain limitations of the product. Because they are evidently the last piece of information in the product chain, these data should be routed back to the appropriate state of production. Such a procedure will certainly improve the overall integrity and reliability of any product.

Imperfections From the Ingot

Except for forged powder metal components, most forgings originate from cast ingots or continuous-cast stock. Many large open-die forgings are forged directly from ingots, while most closed-die and upset forgings are produced from billets, bar stock, or a preform that has received some previous mechanical working.

Regardless of the number or variations of hot-working operations that take place, many of the imperfections found in forgings, as well as problems that occur in forgings during service, can be traced to conditions that existed in the original ingot product. Some of the more important imperfections that persist from the ingot stage and can seriously impair the performance and reliability of the finished forging are reviewed below.

Chemical Segregation. The elements in a cast alloy are seldom distributed in a uniform manner. Even commercially pure metals contain various amounts of impurities in the form of tramp elements or dissolved gases; these impurities are likewise seldom distributed uniformly. Thus, the composition of the metal or alloy will vary from location to location in the cast product. Unfortunately, such variation in chemical composition can often be significant and produce deleterious material conditions.

Deviation from the mean composition at a particular location in a cast or wrought product

Fig. 1 Microstructural banding due to chemical segregation and mechanical working

Fig. 2 Schematic showing piping in top-poured ingots

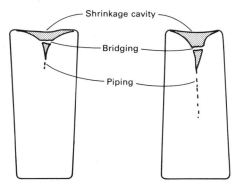

Fig. 3 Longitudinal section through an ingot showing extensive centerline shrinkage

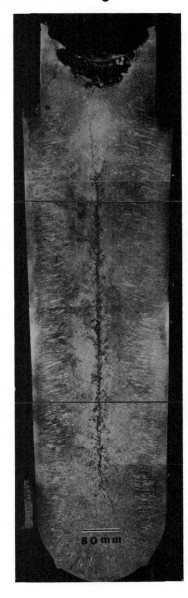

80 mm

is an imperfection termed segregation. In general, segregation is the result of solute rejection at the solid/liquid interface during solidification. For example, a compositional gradient typically exists from the cores of dendrites to the interdendritic regions, with the latter enriched in alloying elements (solute) and low-melting contaminants. Furthermore, due to convection and gravity effects in large ingots, variations in chemical composition can reach much larger proportions. Thus, segregation produces a material having a range of compositions and consequently a material that exhibits variation in mechanical and physical properties.

Forging can partially alleviate chemical segregation by recrystallizing or breaking up the grain structure and promoting diffusion, thereby developing a more homogeneous material. However, the effects of segregation cannot be totally eliminated by forging; rather, the segregated regions tend to be altered by the working operation, forming bands as shown in Fig. 1.

The presence of localized regions that deviate from the nominal composition in metals and alloys can affect corrosion resistance; working operations, such as forging; joining processes, such as welding; and mechanical properties, especially fracture toughness and fatigue behavior. In heat-treatable alloys, variations in the compositions can produce unexpected responses to heat treatment that result in hard or soft spots, quench cracks, or other defects. The degree of degradation that results from segregation depends on the type of alloy and the severity of the segregation, plus a number of interrelated processing variables.

Most metallurgical processes operate on the basis that composition is nominal and reasonably uniform. However, the specified chemical composition of the materials, as reported for

qualification purposes, is generally measured in the ladle before solidification. Therefore, the premise of reasonably uniform composition in the solidified or the converted product may be seriously in error and can lead to problems in processing or to unexpected premature failures in service.

Ingot Pipe and Centerline Shrinkage. A common imperfection in ingots is the shrinkage cavity, commonly referred to as pipe, often found in the upper central portion of the ingot. During freezing, contraction of the metal occurs, and there may eventually be insufficient liquid metal to feed the last remaining portions as they contract. As a result, a central cavity forms, usually approximating the shape of a cylinder or cone; hence the term pipe. Piping is illustrated in Fig. 2. In addition to the primary pipe near the top of the ingot, secondary regions of piping and centerline shrinkage may extend deeper into an ingot (Fig. 3).

Primary piping is generally an economic concern, but if it extends sufficiently deep into the ingot body and goes undetected, it can eventually result in a defective forging. Detection of the pipe can be obscured in some cases if bridging has occurred.

Piping can be minimized by pouring ingots with the big end up, providing risers in the ingot top, and applying sufficient hot-top material (insulating refractories or exothermic materials) immediately after pouring. These techniques extend the time that the metal in the top regions of the ingot remains liquid, thereby minimizing the shrinkage cavity produced in this portion of the ingot.

On the other hand, secondary piping and centerline shrinkage can be very detrimental because they are harder to detect in the mill and may subsequently produce centerline defects in bar and wrought products. Such a material condition may indeed provide the flaw or stress concentrator for a forging burst in some later processing operation or for a future product failure.

High Hydrogen Content. A major source of hydrogen in certain metals and alloys is the reaction of water vapor with the liquid metal at

high temperatures. The water vapor may originate from the charge materials, slag ingredients and alloy additions, refractory linings, ingot molds, or even the atmosphere itself if steps are not taken to prevent such contamination. The resulting hydrogen goes into solution at elevated temperatures; but as the metal solidifies after pouring, the solubility of hydrogen decreases, and it becomes entrapped in the metal lattice.

Hydrogen concentration in excess of about 5 ppm has been associated with flaking, especially in heavy sections and high-carbon steels (Ref 4). Hydrogen flakes (Fig. 4) are small cracks produced by hydrogen that has diffused to grain boundaries and other preferred sites,

Fig. 4 Hydrogen flaking in an alloy steel bar

(a) Polished cross section showing cracks due to flaking. (b) Fracture surface containing hydrogen flakes. Note the reflective, faceted nature of the fracture. (c) SEM micrograph showing the intergranular appearance of the flakes in this material

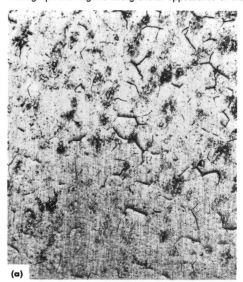

(a)

(b)

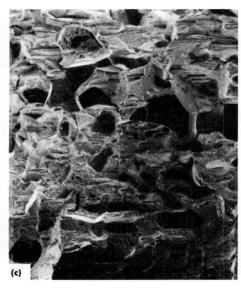

(c)

for example, inclusion/matrix interfaces. However, hydrogen concentrations in excess of only 1 ppm have been related to the degradation of mechanical properties in high-strength steels, especially ductility, impact behavior, and fracture toughness.

Metals can also possess a high hydrogen content without the presence of flakes or voids. In this case, the hydrogen may cause embrittlement of the material along selective paths, which can drastically reduce the resistance of a forged part to crack propagation resulting from impact loading, fatigue, or stress corrosion.

In cases where hydrogen-related defects can serve as the initiation site for cracking and thus increase the likelihood of future failures, it is advisable to use a thermal treatment that can alleviate this condition. For example, slow cooling immediately following a hot-working operation or a separate annealing cycle will relieve residual stresses in addition to allowing hydrogen to diffuse to a more uniform distribution throughout the lattice and, more importantly, to diffuse out of the material. Additional information on hydrogen-induced failures is available in the article "Hydrogen-Damage Failures" in this Volume.

Nonmetallic inclusions are an inevitable consequence of commercial alloys and their respective melting or casting practices. Inclusions that originate in the ingot are carried onto forged products even though their shapes may be appreciably altered. Furthermore, additional nonmetallic matter, such as oxides, may develop during intermediate hot-working stages and also end up in the finished forging.

Two categories of nonmetallics are generally distinguished in metals: (1) those that are entrapped inadvertently and originate almost exclusively from foreign matter, such as refrac-

tory linings, that is occluded in the metal while it is molten or being cast and (2) those inclusions that form in the metal because of a change in temperature or composition. The second category of nonmetallics is produced by separation or precipitation of certain compounds when the solubility of these elements in the metal is exceeded. Although the nonmetallics that typically occur in steels, such as oxides, sulfides, nitrides, and so on, have been thoroughly documented (Ref 5), other alloy systems are also susceptible to significant inclusion formation. These include the alloys of aluminum, copper, titanium, and high-temperature alloys based on nickel and cobalt; the latter systems suffer from their own particular potentially harmful second-phase formations, that is, σ-phase and Laves phase. Typical inclusions found in forged products are illustrated in Fig. 5.

Because these compounds are products of reactions occurring within the metal, they are normal constituents of it, and conventional melting practices cannot completely eliminate such inclusions. However, it is desirable to keep inclusions to a minimum so that their harmful effects are reduced.

Nonmetallic inclusions are unquestionably one of the most common imperfections involved in problems or failures of forgings. In many applications, the presence of these flaws adversely affects the properties of a material to withstand applied loads, impact forces, fatigue, and, occasionally, corrosion or stress corrosion. Ductility, as measured by reduction of area (% RA) in the transverse direction, can also be seriously impaired by inclusions that have been elongated in the direction in working.

Inclusions can easily act in the manner of a stress concentrator because of their discontinu-

ous nature and incompatibility with the surrounding matrix. From the fracture mechanics standpoint, such a combination of features may very well yield a flaw of critical size or one that is capable of growing to critical size under the appropriate loading conditions. This can result in a system breakdown or a catastrophic failure.

Anisotropy Caused by Forging

Preferred orientations can develop when a metal is plastically deformed by mechanical working. Preferred orientations consist of (1) crystallographic texturing, which is produced by crystallographic reorientation of the grains during severe deformation, and (2) mechanical fibering, which is brought about by the alignment of nonmetallic inclusions, voids, chemical segregation, and second-phase constituents parallel to the main direction of mechanical working (Ref 3). Both types of preferred orientations cause the mechanical and physical properties of a metal to exhibit different properties in different directions; this condition is termed anisotropy.

Direct metallographic evidence of mechanical fibering is often present in the form of slag stringers in wrought iron or in the form of ferrite-pearlite banding, elongated grain structures, and flow lines in steels. These flow lines appear as fiberlike lines on polished-and-etched sections of forgings. They are caused by orientation of the constituents and inclusions in the metal during forging. Whenever any of these obvious displays of structural anisotropy is observed, there are usually measurable differences in strength and ductility in test specimens that are oriented parallel and transverse to the direction of working. Improper grain-flow ori-

Fig. 5 Nonmetallic inclusions in wrought steel

(a) Manganese sulfide stringers. (b) Silicate stringer. Both 200×

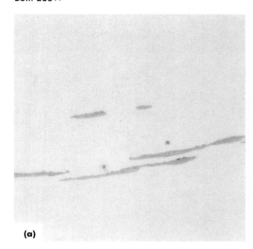

(a)

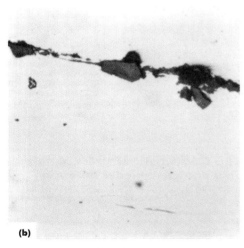

(b)

Fig. 6 Micrograph of a forging lap

Note the included oxide material in the lap. 20×

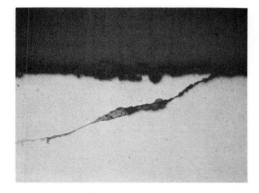

entation relative to the stress direction can be a serious problem in forgings; however, proper die design and the use of upsetting and cross-forging techniques can usually improve mechanical properties and reduce anisotropy.

Probably the most dramatic evidence for the effects of mechanical fibering are those observed during fracture, either in service or during laboratory testing or process deformation. There is usually a substantial decrease in the impact strength, a reduction of area, and reduced fracture toughness when the fracture is parallel to the plane of forging.

Anisotropy of fracturing in wrought products is extremely widespread, occurring in both ductile and brittle materials and including some that are essentially pure metals. The literature contains numerous reports of investigations concerning anisotropy and mechanical properties. The results of some of the more specific investigations are reported in Ref 6 to 9.

Defects Caused by Forging

Laps are surface irregularities that appear as linear defects and are caused by the folding over of hot metal at the surface. These folds are forged into the surface, but are not metallurgically bonded (welded) because of the oxide present between the surfaces. Thus, a discontinuity with a sharp notch is created. Figure 6 shows a forging lap.

A seam is a surface defect that also appears as a longitudinal indication and is a result of a crack, a heavy cluster of nonmetallic inclusions, or a deep lap (a lap that intersects the surface at a large angle). A seam may also result from a defect in the ingot surface, such as a hole, that becomes oxidized and is prevented from healing during working. In this case, the hole simply stretches out during forging or rolling, producing a linear cracklike seam in the workpiece surface.

Hot tears in forgings are surface cracks that are often ragged in appearance. They result from rupture of the material during forging and are often caused by the presence of low-melting or brittle phases.

Bursts. During forging operations, substantial tensile stresses are produced in addition to the applied compressive stress. Where the material is weak, for example, from pipe, porosity, segregation, or inclusions, the tensile stresses can be sufficiently high to tear the material apart internally, particularly if the forging temperature is too high (Ref 10). Such imperfections are known as forging bursts. Similarly, if the metal contains low-melting phases resulting from segregation, these phases may rupture during forging.

Figure 7 shows a large forging burst that occurred during the forging of an electroslag-remelted (ESR) ingot. The cause was traced to a weak solidification plane near the bottom of the ingot combined with higher forging temperatures than normal.

Thermal cracks occur as a result of nonuniform temperatures in the forging. Quench cracks are one example of such thermal cracks (see the section "Quench Cracks" in

this article). Internal cracks, another type of thermal crack, may occur when forgings are heated too rapidly. These occur as a result of unequal temperatures of the surface relative to the center of the mass, and the resulting differences in the degree of thermal expansion produce tensile stresses near the center. The formation of such cracks depends on both the section size and the thermal conductivity of the material. Large section sizes and poor thermal conductivity promote thermal gradients and favor crack formation.

Causes of Failure in Forgings

A failure in a forging may be caused by such factors as material defects, deficiencies in design, improper processing and fabrication, or deterioration resulting from service conditions. Moreover, the actual failure may be attributable to one of several fundamental mechanisms, such as brittle fracture, fatigue, wear, or corrosion. All of these are discussed in detail in the articles "Ductile and Brittle Fractures," "Fatigue Failures," "Wear Failures," and "Corrosion Failures" in this Volume.

Although a failure can result from the independent action of any one of these factors, the final failure is typically a consequence of the combined action, either simultaneously or sequentially, of more than one factor. This concept—interaction between separate factors or failure mechanisms—is very important and should not be neglected during a postfailure investigation.

Finally, a comprehensive analysis of the failure, whatever its cause, must be conducted to determine the reasons and factors for the failure. Although one may not be inclined to spend much time and effort on a postmortem, an accurate assessment of the causative factors can provide valuable insight into design problems or material limitations so that corrective measures can be taken and similar failures avoided in the future. In fact, such a procedure—that is, routing failure analysis information back to the design or engineering stage of product development—inevitably improves the safety, reliability, and usefulness of a product.

Relationship Between Design and Failure

Few industrial or commercial products are intentionally designed to last indefinitely. Indeed, some are purposely designed to serve for relatively short periods of time (Ref 11). Nevertheless, one of the first deliberations conducted when any new design is being considered for production is its expected service life. Later, when that item has passed through the various stages of development and manufacturing and has entered service, one of the primary considerations that will determine the success of the design is whether it actually achieves this intended life.

Fig. 7 Cross section of a forged bar showing a forging burst

The burst is located approximately at the centerline of the workpiece. Arrow indicates the direction of working.

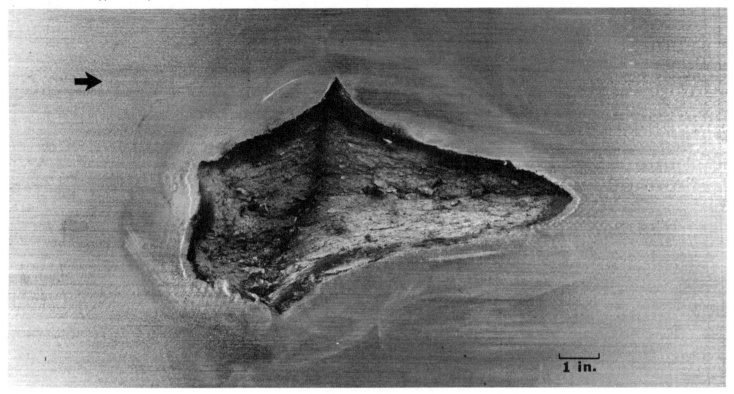

1 in.

Failures of a product are generally distributed over a period of time and usually follow a pattern. Initially, the rate of failure is relatively high, followed by, in successful designs, a longer period during which failures occur at a low and fairly regular level. Eventually, the rate of failure rises steadily when the product has been in service for a relatively long time. Therefore, with an item that is produced in quantity, the occasional failure of different parts may not be cause for serious concern, but repeated failures of a particular component usually indicate a design inadequacy or a faulty manufacturing process.

Failures associated with design may be broadly related to the following factors:

- Basic faults in design concept
- Incorrect selection of materials
- Improper specification of welding design and procedure
- Disregard of anisotropy in mechanical properties
- Fatigue due to stress concentrators
- Neglect of the factors associated with brittle fracture and fatigue failure
- Lack of consideration of operating conditions and service environment
- Inadequate provision for suitable protection and safety devices in the detailed design

Stress Concentration. Although stress concentrators may influence the failure of forgings in general, fatigue fractures due to the stress-concentrating effects of sharp corners, small radii, and abrupt changes in section size constitute a large percentage of design-induced failures. Parts and assemblies frequently contain threads, slots, fillets, holes, notches, and other geometric irregularities that concentrate or increase stresses in their vicinity.

Machined components may have severe stress concentrators formed by tearing or gouging of the workpiece when the cutting tool is not performing properly. Even ordinary tool marks can result in serious damage to a component if they occur in a region of high stress concentration, such as a fillet (Ref 12).

Figure 8 shows the various stresses produced in a notched component undergoing tensile loading. This schematic illustrates that the longitudinal and tangential normal stresses σ_L and σ_T and the maximum shear stress τ_{max} peak at the root of the notch, while the radial normal stress σ_R peaks below the root of the notch. More severe notches or stress concentrators simply intensify these values. Thus, the presence of any notch or discontinuity, whether intentional or accidental, must be carefully considered with respect to the operation of a component because stresses encountered in service tend to be highest at the surface.

Immediately below the base of the notch, the maximum shear stress drops to a low level. Therefore, when plastic flow begins at the root

Fig. 8 Schematic of stress patterns in a notched cylinder in tension

The height above the notch root of any point on the curves indicates the relative magnitude of stress at that radial position. σ_L, longitudinal normal stress; σ_T, tangential normal stress; σ_R, radial normal stress; τ_{max}, maximum shear stress. Source: Ref 12

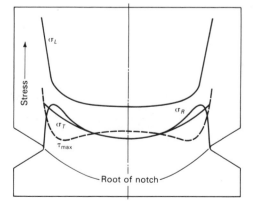

surface of a notch, an even greater degree of triaxiality occurs. This is due to additional tangential and radial restraint, causing an increased σ/τ ratio just below the notch surface. Under a static load, highly stressed metal yields plastically at a notch root or hole edge, thereby conveying the high stresses on to other sections until fracture occurs. However, under fatigue or repeated loads, most of the material is stressed below its elastic limit, yielding locally on a much smaller scale. Highly localized plastic deformation may then initiate a crack, before the stress pattern changes, to relieve concentrated stresses.

Example 1: Fatigue Fracture of a Plunger Shaft That Initiated at a Sharp Fillet. Plunger shafts machined from 4150 steel bar stock were involved in a series of fatigue failures. The fractures consistently occurred at two different locations on the shafts—the shaft fillet (arrow A) and either side of a machined notch (arrow B) (Fig. 9a). The material specification for the shafts required 41*xx* series steel with a carbon content of 0.38 to 0.53%, a hardness of 35 to 40 HRC for the shaft, and a hardness of 50 to 55 HRC for the notch (which was case hardened).

The plunger was subjected to sliding and torque during service. This motion probably resulted in some bending loads. Also, the loads were cyclically imposed when the plunger was operated.

Investigation. Chemical analysis of the shaft material showed it to be of 4150 steel within the required range of chemical composition. The hardness was measured as 30 to 34 HRC on the shaft and 50 to 55 HRC on the notch.

Inspection of the fracture surface at arrow A (Fig. 9b) revealed the beach marks indicative of fatigue loading. The curvature of these marks implied that some torsional loading was also involved. A large number of ratchet marks are visible at the outer edge of the fracture surface, which indicates multiple fatigue origins with a high stress concentration due to the sharp radius. The final-fracture area is near the surface (outside diameter) opposite the initiation site; therefore, the stress that caused failure was probably not much above the fatigue strength of the material. This near-to-normal stress allowed for slow crack growth until fracture occurred.

Several shafts had broken at the notch. The surfaces of these fractures also had a fatigue-fracture appearance (Fig. 9c). The fractures originated in the sharp fillets along the edges of the notch; these fillets are indicated by arrow B in Fig. 9(a). Machining marks, also evident of the surface, added to the stress concentration at the fillets.

Several shafts that had broken at the fillet (arrow A) were inspected by the magnetic-particle method. Cracks were found in the fillets of the notch (arrow B) in some of these shafts.

The ends of the plunger shafts (arrow C, Fig. 9a) were significantly deformed by peening. The peening had been caused by the impact of the cam against the shaft end during service. Shaft hardness was found to be slightly lower than the minimum required by the specification, indicating improper heat treatment. However, the peening did not halt the operation of the components and was considered to be of secondary importance.

Fig. 9 4150 steel plunger shaft that failed in service from fatigue fractures in two locations

(a) Fracture sites on the plunger shaft: fillet (arrow A) and machined notch (arrow B). Shaft end (arrow C) was deformed by peening, but did not fracture. (b) Fracture surface at arrow A showing beach marks characteristic of a fatigue failure. (c) Fracture surface at arrow B showing fatigue marks indicative of initiation at the notch root (bottom) and progressive advancement to the outside diameter (top)

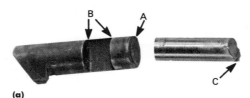

(a)

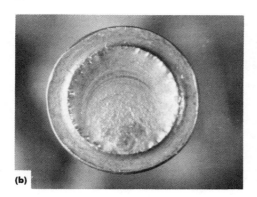

(b)

(c)

Conclusions. All the fractures were fatigue-induced failures due to sharp radii in the fillets. The stress-concentrating effects of the fillets caused fatigue cracks to initiate and grow under cyclic loading until the crack depth was critical, causing the shaft to fail and rendering the assembly inoperative.

Corrective Measures. The cause of the failures was eliminated by increasing the radii of the notch and shaft fillets, thereby reducing the effects of stress concentration to an acceptable level. If fatigue cracking had continued to be a problem with this component, shot peening of the subject radii would be appropriate. This process produces residual compressive stresses in the surface of the part, thereby retarding initiation of fatigue cracks.

Anisotropy results from mechanical fibering or banding and from crystallographic texturing when metals and alloys are mechanically worked. These material conditions can in turn play a very important role in the final properties of forgings and their performance in service.

Texturing refers to the orientation of crystal unit cells in a preferred direction and is intentionally produced in such materials as silicon steels to enchance magnetic properties. Although texturing can influence anisotropy in mechanical properties, this condition is usually considered in the design and selection stage and therefore should not be a serious problem with respect to forging failures.

On the other hand, mechanical fibering, or banding, as it is referred to in severe cases, is produced by the alignment of microstructural features, such as nonmetallic inclusions, second phases, and chemical segregation, in the direction of working when wrought products are produced. The occurrence and severity of fibering varies with such factors as composition, extent of chemical segregation, and the amount of work or reduction the workpiece receives. Furthermore, this condition is a consequence of processing and is often not considered during the design of a particular product. Fibering can have very deleterious consequences on the performance and reliability of wrought parts in service. Differences in the etching reaction of the aligned microstructural constituents will reveal patterns called flow lines associated with the direction of mechanical fibering. Ideally, the flow lines in forgings should be aligned in the direction of maximum applied stresses, because forgings typically exhibit the greatest strength in a direction parallel to the fibering. If fibering is properly considered during the design stage, the greatest stresses placed on a component act along, rather than across, the fibers in the same manner as a fiber-reinforced composite material.

Figure 10 shows a comparison of desirable and undesirable flow patterns in a forging. In contrast to the pattern shown in Fig. 10(a) in which the flow lines are mainly parallel to the free surface, the pattern in Fig. 10(b) contains a large number of flow lines that intersect the surface at the bottom of the recess. This condition creates many potential crack-initiation sites that, under appropriate service conditions, could result in brittle or ductile fracture, fatigue fracture, or stress-corrosion cracking (SCC).

Efficient design of structural components generally requires the use of the lightest, least expensive materials that will safely carry the applied loads. Anisotropy, therefore, is an important design consideration for safety reasons and as a means of reducing the weight of a given component. For example, longitudinally oriented forged material is more efficient from the standpoint of strength-to-weight ratio versus stress concentration than transversely oriented material from the same forging. If this efficiency can be used in the initial design, the best

Fig. 10 Comparison of flow patterns in forging

(a) Desirable flow pattern. (b) Undesirable flow pattern

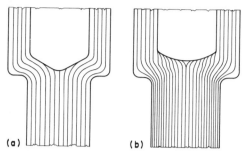

(a)   (b)

combination of material and processing is achieved.

Example 2: Brittle Failure of Locking Collar Produced From Rolled 4140 Grade Steel Plate. The locking collar on a machine failed suddenly when the shaft it restrained was inadvertently subjected to an axial load slightly higher than the allowable working load (Fig. 11a). The locking collar fractured abruptly, producing four large fragments (Fig. 11b). This allowed the shaft to be propelled forcefully in the direction of the load, causing substantial damage to other machinery components in the vicinity. The failed component, which was 43 cm (17 in.) in diameter, was machined from 4140 plate and heat treated to 34 to 36 HRC.

Investigation. The fracture surface of the locking collar fragments exhibited a brittle appearance with no evidence of ductile or fibrous fracture. The cracks initiated in the fillets and propagated to the outside corners, as shown schematically in Fig. 11(a) and (b). No pre-existing cracks or flaws were observed on the fractures or at the initiation sites. Scanning electron microscopy (SEM) examination corroborated the macroexamination observations in that the failure mode consisted entirely of cleavage through the grains (Fig. 11c).

Mechanical-property analysis revealed that, although the yield strengths were comparable (793 to 834 MPa, or 115 to 121 ksi) in the S-T and L-T orientations shown in Fig. 11(b), there were significant differences in ductility and fracture toughness between these orientations. For example, reduction of area averaged 4% in the S-T orientation as opposed to 57% in the L-T, while K_{Ic} (plane-strain fracture toughness) averaged 66 MPa$\sqrt{m}$ (60 ksi$\sqrt{in.}$) in S-T and 104 MPa$\sqrt{m}$ (95 ksi$\sqrt{in.}$) in L-T. A composite photomicrograph (Fig. 11d) taken from the locking collar shows that the material contains microstructural bands that are oriented in the rolling direction (RD).

Conclusions. The alloy steel plate used in this application contained significant microstructural fibering or banding. This condition produced considerable anisotropy in ductility and toughness as revealed by mechanical test-

Fig. 11 Failure of a 4140 steel locking collar machined from rolled plate

(a) Collar assembly showing fracture location. (b) Illustration of the failure with respect to the plate orientation. (c) SEM micrograph showing the cleavage-failure mode. (d) Composite micrograph showing microstructural orientation produced by rolling. RD, rolling direction

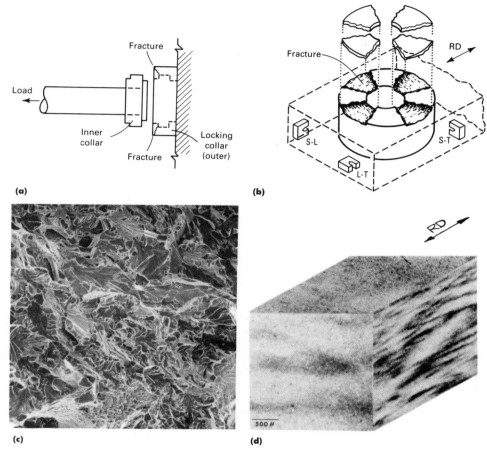

(a)   (b)

(c)   (d)

ing. Unfortunately, the potential effects of anisotropy were apparently neglected when this component was designed and manufactured from the plate stock, because the loading was applied in a direction that stressed the weakest planes in the material, that is, a direction normal to the fibering.

Failures Related to Material Selection

Although material selection is sometimes given less-than-adequate attention, the choice of an appropriate starting material and the design/manufacture of a forging are inseparable factors. Material selection is generally based on such criteria as applied stresses, operating temperatures, service environment, availability, processing parameters, and economics. To make the best material selection for a particular forged product, it is necessary to consider the processing requirements in addition to functional and economic factors. The material that displays the greatest number of positive attributes should then be selected.

Fig. 12 Stainless steel poppet-valve stem that fractured in service

17-4 PH stainless steel
Rockwell 15N 80 to 85

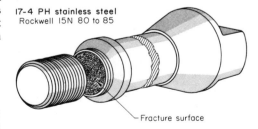

Fracture surface

Ideally, the net result of such an analysis should be a forging that will satisfy design requirements, operate safely and reliably, fulfill its intended function, and be produced at a reasonable cost.

Example 3: Fracture of Poppet-Valve Stems Due to Incorrect Material Selection. A series of poppet-valve stems fabricated from 17-4 PH (AISI type 630) stainless steel (Fig. 12) failed prematurely in service during the development of a large combustion assem-

Fig. 13 Axle shaft that fractured at the journal from reversed-bending fatigue

(a) The fracture at the journal. (b) Fracture surface of the axle shaft showing fatigue beach marks with a final fast-fracture zone near the center. Source: Ref 13

(a)

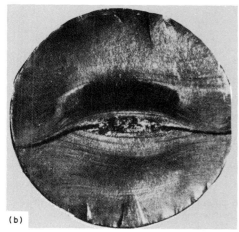

(b)

Fig. 14 Surface of a typical rotating-bending fatigue fracture

Final fracture is located just left of the center. Source: Ref 13

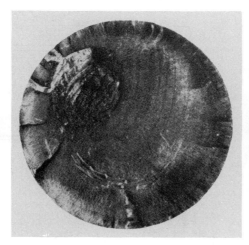

Fig. 15 Surface of a typical fatigue fracture

Arrow indicates origin of fatigue crack at a sharp keyway

bly. The poppet valves were components of the scavenging system that evacuated the assembly after each combustion cycle. The function of the valve is to open and close a port; thus, the valve is subjected to both impact and tensile loading.

Investigation. The stems had fractured in the upper threaded portion (Fig. 12). The threads of the stems exhibited slight corrosion. Also, a few threaded crests were flattened, and machining scratches were detected on the sides of some threads. Damaging as they were, these defects did not appear to have initiated the failure.

Because the valve stem was subjected to heating from the combustion cycles, the microstructure of the failed stem was compared with that of a new part. There was no significant difference between microstructures in the new and in the used stems. The microstructures were found to be typical of a martensitic precipitation-hardening stainless steel in the H900 condition.

Because of the small size of the stems, hardness was the only mechanical property that could be measured. Requirements called for a hardness of 80 to 85 HR15N. Actual values were 84.0, 81.8, and 84.5 HR15N.

Finally, a stress analysis was conducted on the root of the first thread in the valve stem (region of fracture). Results of the analysis showed an approximate tensile stress of 1207 MPa (175 ksi). Therefore, the existence of an overstressed condition is likely because the yield strength of 17-4 PH stainless steel is approximately 1172 MPa (170 ksi).

Conclusions and Recommendations. In service, the valve stems were impact loaded to stresses in excess of their yield strength. That they failed in the threaded portion also suggests a stress-concentration effect. However, major changes in design can be avoided by selecting a higher-strength material with greater impact strength. In this case, it was recommended that the stems, despite any possible design changes, be manufactured from an alloy such as PH 13-8Mo, which can be processed to a yield strength of 1379 MPa (200 ksi), with impact energies of the order of 81 J (60 ft · lb) at room temperature. Thus, a better combination of tensile and impact properties is possible without serious loss of corrosion protection.

Fatigue Failures

Many forged components, including shafts, gears, bolts, springs, pins, bars, and rods, are subjected to cyclic or fluctuating loads during service. Such alternating forces may be an intentional element of the design, and these loads should then result in stresses lower than the fatigue strength or the endurance limit of the material. However, in some cases, engineering components may be inadvertently subjected to cyclic loading conditions that result in fatigue failure.

The nature and progress of a fatigue crack can vary considerably, depending on geometric, metallurgical, mechanical, and environmental factors. For example, with alternating tension and compression, or bending stresses,

the crack tends to grow in a plane perpendicular to the principal stresses. Figure 13 shows an axle shaft that fractured at the journal by reversed-bending fatigue. The fracture surface of this shaft (Fig. 13b) displays the characteristic appearance of reversed-bending fatigue, that is, beach marks that indicate the cyclic advancement of the crack from the outside diameter toward the center and a region of fibrous-appearing fracture at the location that acted as a hinge point for the reversed bending. The latter feature actually represents the final ligament of material that held the shaft together just before it failed.

Similarly, a fracture surface typical of rotating-bending fatigue is shown in Fig. 14. This fracture surface also displays evidence of a progressive crack growth and a final-fracture zone, or terminus, when the fatigue crack has advanced to the critical stage. Rotating-bending fatigue fractures usually initiate at several points around the circumference; the individual fatigue cracks subsequently link together to form a single crack front.

The effects of stress concentrators in forging failures were discussed in the section "Stress Concentration" in this article. Therefore, it is understandable that stress concentrators also play a key role in the initiation of fatigue cracks in forged components. For example, notches in the surface of a cyclically loaded part can substantially reduce its resistance to fatigue-crack initiation (Ref 14-17).

Abrupt changes in cross section, sharp edges, threads, grooves, machining marks, and so on, can all act as notches. Correspondingly, the resistance to fatigue-crack initiation decreases as the acuity of the notch increases. Stress concentration amounting to two or three times the nominal applied stress can easily be developed at sharp notches. Figure 15 shows a fatigue failure that originated at the radius of a sharp keyway.

Unfortunately, the higher the strength of an alloy, the greater its notch sensitivity. Therefore, the inherent design advantages of high-strength alloys can be nullified or can even become potentially deleterious when the material is used in an inappropriate design or application and contains acute stress concentrators.

Example 4: Failure of a Structural Bolt Due to Reversed-Bending Fatigue. A portion of a large (19-mm, or 0.75-in., diam) structural steel bolt was found on the floor of a manufacturing shop. This shop contained an overhead crane system that ran on rails supported by girders and columns (Fig. 16b). Fortunately, an astute employee reported the broken bolt to a supervisor.

Investigation. Inspection of the crane system revealed that the bolt had come from a joint in the supporting girders and could be considered one of the principal fasteners in the track system. An analysis of the remaining portion of the failed bolt showed fracture-surface markings clearly indicative of fatigue, including beach marks and a final fast-fracture zone (hinge) in the center (Fig. 16c) characteristic of reversed bending. The cracks initiated on opposite sides of the shank at the radius between the head and shank. Mechanical damage, or wear, was present on the shank portion, and these areas of wear were located on opposite sides of the shank (180° apart).

A metallographic examination of the bolt disclosed a microstructure consisting of ferrite and fine pearlite. No unusual or gross metallurgical imperfections were present. The hardness of the material was 75 HRB, which places this bolt in grade 1 or 2 (Ref 18).

The mechanical damage, or wear marks, on the shank of the bolt and the fracture-surface features indicated that the bolt was subjected to cyclic loading in reversed bending during the operation of the crane; that is, the crane passed back and forth over the bolted joint. The fillet between the bolt head and shank facilitated the fatigue failure by concentrating the stresses at this location. Furthermore, a bolt placed in such a structural assembly should have higher strength than that indicated by a hardness of 75 HRB.

Conclusions. Fatigue induced by the overhead movement of the crane produced failure of the bolt. The bolt was deficient in strength for the cyclic applied loads in this case and probably was not tightened sufficiently.

Recommendations. Remaining bolts in the crane support assembly should be removed and replaced with a higher-strength, more fatigue-resistant bolt, for example, SAE grade F, (104 to 108 HRB). The bolts should be tightened according to the specifications of the manufacturer, and the system should be periodically inspected for correct tightness.

Failures Related to Nonmetallic Inclusions

Nonmetallic inclusions, both indigenous and exogenous, are frequently present in ingot and

Fig. 16 Failure of a structural steel bolt in the rail assembly of an overhead crane

(a) Illustration of the crane rails and attendant support beams. (b) Shank portion of the failed bolt. (c) Fracture surface of the bolt showing evidence of reversed-bending fatigue

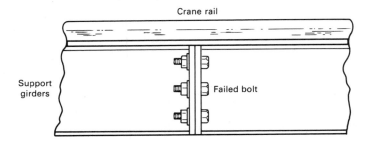
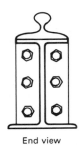

cast products. Such inclusions are a frequent consequence of alloying and melting/casting practices. The metal-producing industry attempts to control nonmetallics because they have the potential to impair seriously the performance and reliability of finished products in service.

The deleterious nature of nonmetallic inclusions depends on several factors, including chemical composition of the inclusion, volume percentage, shape, orientation, and the mechanical/physical properties of the inclusion as compared to its surrounding matrix (Ref 19). One study has demonstrated that the local constraint of the matrix by hard inclusions during tensile loading can produce severe stress concentrations that depend on the elastic moduli, size, shape, and orientation of the inclusions (Ref 20). For example, inclusions such as manganese sulfide, which are very deformable at hot-working temperatures, elongate in the direction of working. This results in an anisotropic material condition, which, depending on its severity, can produce substantial decreases in transverse mechanical properties, such as ductility, fatigue life, and fracture toughness.

Sulfide inclusions that have been shape controlled by the addition of calcium or rare-earth metal treatments are much less detrimental to mechanical properties. These inclusions are not plastic or deformable at typical hot-working temperatures and retain their globular shape. Therefore, they are less injurious with respect

to ductility, toughness, and fatigue life in finished wrought products.

However, inclusions that are very brittle and tend to fracture and fragment during working operations can be very detrimental to the above-mentioned properties. In fact, with certain refractory and slag inclusions, fragmentation frequently occurs when the amount of cross-sectional reduction in the workpiece is large.

Example 5: Stub Axles That Were Rejected Because of Slag Inclusions (Ref 21). An automobile manufacturer rejected several 1035 steel stub axles because of what appeared to be short longitudinal cracks in the surfaces of the pins. (Fig. 17a). The cracks were found when six axles were examined for defects by magnetic-particle inspection.

Investigation. Further magnetic-particle inspections conducted on the six axles showed large quantities of short straight lines in the surfaces of the pins that were at first interpreted as cracks. Metallographic examination showed, however, that these lines were not cracks but slag inclusions at and immediately below the surface. The inclusions at the surface had been cut during machining. The inclusions consisted of multiphase spinels within a transparent glassy matrix, which resembled slaglike material such as that used in channel bricks. Coarse slag streaks like these are often called sand spots because they are friable and trickle out when the metal is worked. Figures 17(b) and (c)

Fig. 17 1035 steel automobile stub axle that was rejected because of slag inclusions at and below the surface

(a) View of axle showing inclusions at the surface (circled). (b) and (c) Longitudinal sections (unetched) showing inclusions at 100 and 500×, respectively. Source: Ref 20

(a)

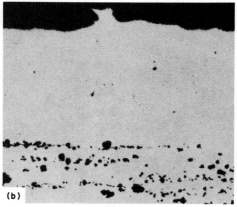

(b)

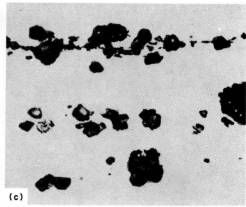

(c)

show slag streaks at two magnifications in the longitudinal (working) direction. The inclusions were crumbled (fragmented) during forging, but were not deformed. The particles consisted primarily of aluminum oxide or silicates rich in aluminum oxide. Inclusions of this type could not be found in the interior of the cross section; only oxide and sulfide inclusions of normal size were found in the interior.

Conclusions. The inclusions consisted of pieces of fireclay from channel brick that were flushed into the ingot mold. Although no true cracks were present, rejection of the stub axles was nevertheless justified. Slag streaks could reduce the strength of the stub axles and lead to the formation of fatigue fractures during operation.

Example 6: Diesel-Engine Crankshaft That Fractured in Fatigue Because of Subsurface Inclusions. A 1050 steel crankshaft with 6.4-cm (2.5-in.) diam journals that measured 87 cm (34.25 in.) in length and weighed 31 kg (69 lb) fractured in service. The shaft had been quenched and tempered to a hardness of 19 to 26 HRC, then selectively hardened on the journals to a surface hardness of 40 to 46 HRC.

Investigation. The fractured shaft was examined for chemical composition and hardness, both of which were found to be within prescribed limits. The fracture surface (Fig. 18a) shows a complex type of fatigue failure initiated from subsurface inclusions in the transition zone between the induction-hardened surface and the softer core. A concentration of inclusions near the fracture origin is shown in Fig. 18(b).

Conclusions. The failure was caused by fatigue cracks that initiated in an area having an excessive amount of inclusions. The inclusions were located in a transition zone, which is a region of high stress.

Fatigue cracks propagate at right angles to maximum tensile stress; thus, a change in the direction of cracking or in the crack-plane angle indicates a change in the direction in which

stresses were applied. For example, when a landing wheel is removed from an aircraft to change a tire, it may not be reinstalled on the same side of the aircraft, and changes in direction of wheel rotation cause changes in the crack-plane angle.

Example 7: Fatigue Cracking That Originated at a Material Defect in a Forged Aircraft Wheel Half. A commercial aircraft wheel half was removed from service because a crack was discovered in the area of the grease-dam radius during a routine inspection. The wheel half (Fig. 19) was machined from an aluminum alloy 2014 forging that had been heat treated to the T6 temper. Neither the total number of landings nor the roll mileage was reported, but about 300 days had elapsed between the date of manufacture and the date the wheel was removed from service.

Investigation. The portion of the hub that contained the crack was broken open to reveal the fracture surface. The area in which the crack occurred and a view of the fracture surface are shown in Detail A in Fig. 19.

Visual examination of the fracture surface revealed five distinct elliptical beach marks that are characteristic of fatigue failure. The outer boundary of each ellipse (points 1, 2, 3, 4, and 5, View B-B, Fig. 19) was a point of crack arrestment, indicating a change in crack-plane angle, which resulted from a change in the direction of wheel rotation. The total length of the crack lies between points 5 (View B-B, Fig. 19). Point 0 indicates the origin of the crack, which was at a prior material imperfection. The type of imperfection was not initially identified, but was possibly an undissolved grain refiner or an area of unhealed porosity, both of which are common in aluminum alloys.

The crack was apparently exposed at the inner surface of the wheel half early in the propagation cycle but at the outer surface later in the cycle. Detection was delayed because the inner surface was not readily accessible for inspection. Propagation of the fatigue crack to the outer surface of the wheel half may have

been retarded because the surface had been shot peened during manufacture.

A flat, shiny area (arrow 6, View B-B, Fig. 19) was observed in the fracture surface that was intentionally produced when the hub was broken open. A micrograph of a section through this flat, shiny area showed that the area was stringerlike in appearance. Prior experience indicated that such shiny areas were likely to be scattered throughout the forging.

The stringerlike area was analyzed by electron microprobe. The results of the analysis showed that the area contained relatively large amounts of manganese oxide, iron oxide, and titanium oxide; medium amounts of copper oxide and aluminum oxide; and traces of chromium oxide, silicon oxide, and magnesium oxide. The electron microprobe analysis (mainly, the presence of the titanium oxide) indicated that the shiny stringerlike area in the fracture surface was an undissolved grain refiner.

Conclusions. The wheel half failed by fatigue. The fatigue crack originated at a material imperfection and progressed in more than one plane because changes in the direction of wheel rotation altered the direction of the applied stresses.

Recommendations. Because the standard inspection techniques used in the machining plant were unable to detect subsurface imperfections in the as-received forging, it was recommended that the specifications be rewritten to require sound forgings.

Failures Related to Chemical Segregation

Chemical segregation originates in alloys during the solidification stage. Such deviations from the nominal composition are due to convection currents in the liquid, gravity effects, and redistribution of the solute during the formation of dendrites. The eventual result of dendritic solidification is a concentration gradient from the dendrite core to the interdendritic region, with the last metal to solidify being

Fig. 18 Fatigue fracture in a 1050 steel crankshaft

(a) Mating surfaces of the fracture, which initiated at subsurface inclusions (arrows). (b) Micrograph showing concentration of inclusions near the fracture origin. 100×

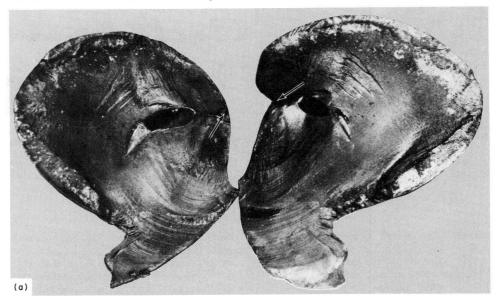

(a)

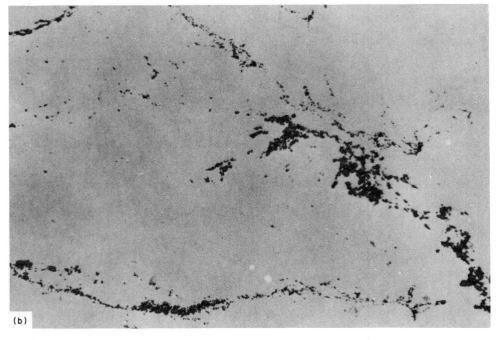

(b)

enriched with solute and other phases, for example, low-melting compounds and nonmetallic inclusions.

One function of forging is to break up the cast (dendritic) structure and promote chemical homogeneity. Therefore, a minimum amount of cross-sectional reduction is usually required from the cast ingot to the billet. However, although this working can alleviate some of the inhomogeneity, it cannot eliminate it entirely if the ingot is badly segregated.

Figure 20(a) shows schematically the transverse distribution of a solute that might typically exist in an ingot immediately after solidification. Figure 20(b) shows the transverse distribution of the solute in an ingot after some degree of working or forging. The differential in solute concentration (amplitude) remains essentially undiminished; however, the peaks are brought closer together. The reason for this is that the dendrite arms are generally lower in impurities, such as sulfur and

phosphorus in steel, than the interdendritic regions. Consequently, the dendrite arms are stronger and, upon forging, do not deform and flow as readily as the matrix in which they are incorporated. Initial working causes matrix flow that tends to reorient the dendrites in the direction of working. With increased mechanical working, the dendrites deform and fracture, thus becoming increasingly elongated. The elongated dendrite arms, together with the interdendritic regions, form a quasi-composite structure. Also included in this composite are both ductile and brittle inclusions as well as oxides and gases.

The overall effect is that the forging contains regions that exhibit varying compositions on a microscale and therefore varying physical and mechanical properties. These microstructural features are illustrated in Fig. 21. Segregation, with its attendant variation in mechanical properties, can manifest itself in forging failures, as shown in the following example (Ref 22).

Example 8: Fracture of a Forging Die Caused by Segregation. A cross-recessed die of D5 tool steel fractured in service. The die face was subjected to shear and tensile stresses as a result of the forging pressures from the material being worked. Figure 22(a) illustrates the fractured die.

Investigation. A longitudinal section was taken through the die to include one arm of the cross on the recessed die face. The specimen was polished and examined in the unetched condition. Examination revealed the presence of numerous slag stringers.

The polished specimen was then etched with 5% nital. A marked banded structure was evident even macroscopically (Fig. 22b). Microscopic examination revealed that the pattern was due to severe chemical segregation, or banding (Fig. 22c).

Hardness measurements were then conducted across the face of the specimen in locations corresponding to the banded and nonbanded regions. These results (Fig. 22a) showed that the segregated region is considerably harder than the neighboring material. The reason is that the increased carbon content of the segregated region, together with its higher alloy content, makes the region more responsive to what would have been normal heat treatment for this grade of tool steel.

The high-hardness material is also subject to microcracking upon quenching; microcracks can act as nuclei for subsequent fatigue cracks. Examination of the fracture surface revealed that the fracture originated near the high-stress region of the die face; however, no indications of fatigue marks were found either on a macroscale or a microscale.

Conclusions. Failure of the die was the result of fracture that originated in an area of abnormally high hardness. Although fatigue marks were not observed, the fact that the fracture did not occur in a single cycle but required several

Fig. 19 Aluminum alloy 2014-T6 aircraft wheel half that was removed from service because it developed a fatigue crack at a material defect

Detail A shows the area where the crack occurred and a view of the fracture surface revealed when the hub was broken open to examine the crack. View B-B, a macrograph of the fracture surface, shows the origin of the fatigue crack (0), major points of crack arrestment (1 to 5), and total crack length (between 5s). Arrow 6 indicates an undissolved grain refiner. Dimensions given in inches

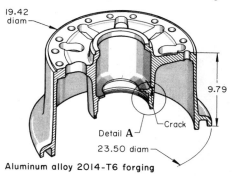

Aluminum alloy 2014-T6 forging

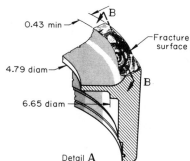

Detail A

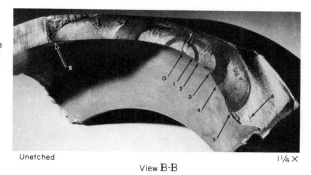

Unetched

View B-B

1¼×

Fig. 20 Schematic illustrating the transverse distribution of a solute in an ingot

(a) Immediately after solidification. (b) After some degree of working during forging. Note the difference in dendrite arm spacing.

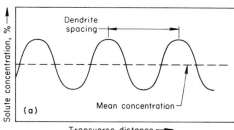

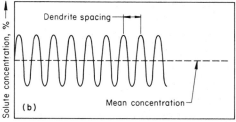

cycles to cause failure defines the failure as low-cycle fatigue.

Failures Related to Microstructure

The principal factors affecting the properties of metals are composition (alloy content) and microstructure. The unique ability of many engineering alloys to have their structure and thus their properties controlled or changed almost at will by heat treatment can also introduce problems for the engineer and manufacturer. Small deviations from optimum processing procedures in alloying and heat treating frequently result in large discrepancies in the final structure and properties of forgings. Such

Fig. 21 Composite micrograph showing microstructural orientation (fibering) in a steel forging

The arrow indicates the direction of working. 15×

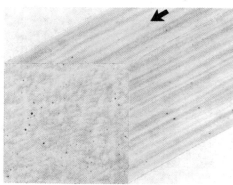

discrepancies may be due to insufficient or excessive alloy content and to improper heating or cooling, both of which produce an incorrect microstructure for a particular application.

Incorrect Microstructure. Steel that is heat treated to form a tempered martensite microstructure is generally considered to possess the best combination of mechanical properties (Ref 23), including tensile strength, ductility, and toughness. Typically, the higher the tempering temperature, the greater the ductility and toughness of the tempered martensite product. Correspondingly, the hardness, tensile strength, and the ductile-to-brittle transition temperature decrease as the tempering temperature increases.

In such industries as consumer electronics, transportation, aerospace, biomedical, and pressure vessels, the current trend is toward achieving higher strength and toughness with little or no increase in weight. Such goals can in some cases be achieved by modifying the structures and properties of existing alloys. In extending

the usable property range of available materials, heat treatment and the resultant microstructure are considered to be of primary importance. However, uniform quenching is a major difficulty encountered with large forgings and with smaller, more intricately shaped forgings. Whenever the cooling rates vary significantly within a part due to size and shape, the resultant overall microstructure can consist of a combination of structures that may eventually lead to premature failure of the component.

Example 9: Fracture of a Lifting-Fork Arm Due to Microstructural Deficiency (Ref 24). A forged alloy steel arm of a lifting fork with an approximate cross section of 150 × 240 mm (5.92 × 9.45 in.) fractured after only a short service life on a lift truck.

Investigation. The fracture surface had the appearance of a fracture originating from a surface crack (Fig. 23a). A large number of these surface cracks became visible after paint was removed from the surface of the arm. The cracks penetrated the metal to a maximum depth of about 3 mm (0.2 in.) and appeared to have originated during forging of the fork. Paint that had penetrated into the cracks confirmed this conclusion. Forging at too low a temperature could have caused these cracks.

Results of a chemical analysis showed the steel to be EN-25 (nominal composition: 0.30C-0.65Cr-2.55Ni-0.55Mo). The steel is suitable for this type of application and can be satisfactorily heat treated even in relatively large components.

Metallographic examination revealed a rather coarse bainitic structure (Fig. 23b), which indicated low resistance of the steel to shock loading. Shock loads are predictable in the daily operation of a fork-lift vehicle. Charpy V-notch impact tests on specimens at room temperature confirmed the suspected low shock resistance.

An improvement in the impact toughness of this material was then attempted by a suitable heat treatment consisting of quenching in oil from 860 °C (1580 °F), followed by tempering

Fig. 22 D5 tool steel forging die that failed in service because of segregation

(a) Hardness traverse correlated with the microstructure of the die. (b) Section through one arm of the cross on the recessed die face showing a severely segregated (banded) structure. Etched with 5% nital. (c) Micrograph of the segregated area. Etched with 5% nital. 200×. Source: Ref 22

Fig. 23 Failure of an alloy steel lifting-fork arm

(a) Fracture surface of the arm. Arrow indicates the fracture origin. (b) Micrograph showing the coarse bainitic structure of the arm. 200×. (c) Micrograph showing the tempered martensite structure that resulted from revised heat treatment. 200×. Source: Ref 24

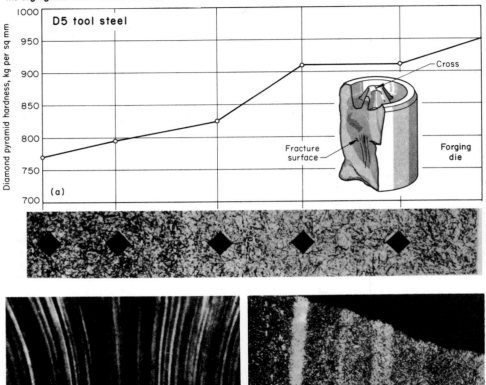

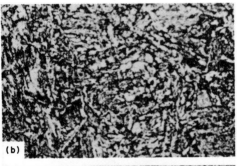

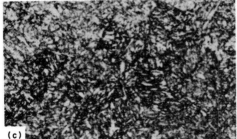

at 650 °C (1200 °F). This resulted in a considerably finer microstructure of tempered martensite (Fig. 23c) and a seven-fold improvement of Charpy V-notch energy absorption as confirmed in impact tests.

Conclusions. The primary cause of the failure was the brittleness (lack of impact toughness) of the steel. The coarse bainitic microstructure was inadequate for the service application. The microstructure resulted from either improper heat treatment or no heat treatment after the forging operation. The surface cracks in the lifting-fork arm acted as starter notches (stress raisers), assisting in the initiation of fracture.

Retained Austenite. Certain alloying elements, called austenite stabilizers, tend to depress the transformation of austenite to martensite in steel. These elements include manganese, nickel, and cobalt. In some cases, this behavior is desirable, for example, Hadfield (austenitic manganese) steels, 9Ni steel, and the 18-8-type stainless steels. However, in other steels, such as the low- and medium-alloy classes, undesirable microstructures and properties may result when the transformation of austenite is incomplete. In hardened steels containing more than 0.55% C, some austenite is usually retained after quenching, particularly when nickel and manganese are present in relatively large concentrations. In fact, in presence of such austenite stabilizers, austenite may be retained even when the carbon content is low (Ref 25).

Some alloys are refrigerated to transform austenite to martensite when the transformation temperature is below room temperature. These materials include certain grades of tool steels, precipitation-hardening stainless steels, high-carbon and high-nickel martensitic stainless, and 25Ni maraging steels.

Retained austenite can also occur in parts that have been carburized to increase their surface hardness (Ref 26). The increase in surface-

carbon content is generally accompanied by an increase in the amount of retained austenite unless precautions are taken to ensure its transformation. For example, Fig. 24 shows the microstructure of a carburized gear containing about 45% retained austenite in the case-hardened region.

In general, when the amount of retained austenite is small, no appreciable effect on properties is noticed, although the elastic limit and yield strength may be lowered slightly. Large percentages of retained austenite, on the other hand, can transform to untempered mar-

Fig. 24 Microstructure in the surface of a carburized steel

White areas are retained austenite; dark needles are martensite. 500×

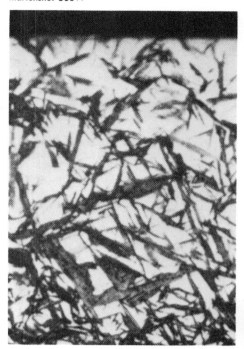

Fig. 25 Roll assembly on which the stainless steel sleeve cracked because of improper microstructure

(a) Roll assembly consisting of a type 440A stainless steel sleeve shrink-fitted over a 4340 steel shaft and secured by keyways. The sleeve cracked from keyway to keyway. Dimensions given in inches. (b) Micrograph of a transverse section through the sleeve showing a continuous network of massive carbide particles at grain boundaries. Etched with a mixture of 2 parts HNO_3, 2 parts acetic acid, and 3 parts HCl. 150×

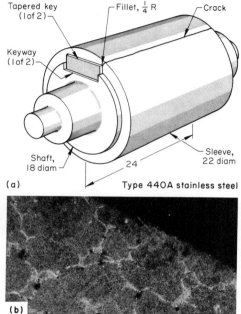

tensite when subjected to cold working and low service temperatures. Such a transformation product is inherently brittle and can promote cracking or failure. Furthermore, if significant amounts of austenite are retained after quenching, transformation to upper bainite during subsequent tempering can result in poor mechanical properties.

Continuous second-phase networks in a steel matrix are also considered to be microstructural deficiencies that can contribute to failures. The following example illustrates such a condition.

Example 10: Brittle Fracture of a Roll-Assembly Sleeve Due to Improper Microstructure. A roll manufacturer had successfully used the following procedure for many years to make roll assemblies: a sleeve, or hoop, forged from AISI type 440A stainless steel was shrink fitted over a 4340 steel shaft (Fig. 25a). Tapered keys were inserted into keyways on the ends of the sleeve/shaft interface to secure the sleeve further (Fig. 25a). A roll assembly of this design was crated and shipped by air. Upon arrival, the sleeve was found to have cracked longitudinally from keyway to keyway.

Examination. The sleeve was torch cut on the opposite side of the crack to expose the fracture surface. Examination revealed that the fracture originated in a fillet of the notch formed by the keyway. The radius of the fillet was generous (6.4 mm, or ¼ in.) and should not have posed a problem, particularly in view of its long

history of successful use. In addition, no defects, such as tool marks, were found on the root of the fillet that might account for crack initiation at that location. Examination of the fracture surface by electron microscopy indicated that the fracture was brittle in nature with considerable evidence of intergranular fracture.

Estimates of the residual stress were made on the basis of the interferences resulting from the shrink-fitting operation. Similar calculations were made considering the stress resulting from differential expansion and contraction due to thermal differences occurring during transit by air. These stresses alone could not cause cracking.

A transverse specimen was polished and examined metallographically. The results revealed that, although the surface of the forged sleeve exhibited a microstructure with carbide particles dispersed in a martensite matrix, the interior of the forging, beginning at a depth of 25 mm (1 in.) below the surface, exhibited a continuous network of massive carbide particles (Fig. 25b).

Conclusion. Superficial working of the metal, probably insufficient hot working, produced a microstructure in which the carbide particles were not broken up and evenly distrib-

uted throughout the structure. Instead, the grains were totally surrounded with brittle carbide particles. This facilitated the formation of a crack at a fillet in the keyway. Crack growth was rapid once the crack had initiated.

Failures Related to Forging Defects

A successfully forged part is the result of a series of metalworking operations designed to develop the final shape required and simultaneously achieve grain refinement, internal soundness, and improved mechanical properties. Initial working of any ingot product, usually referred to as cogging, removes flutes, ripples, or corrugations that were formed on the ingot by the mold contour to prevent cracking of the ingot surface during solidification and cooling (Ref 27). Light drafts (small reductions) are taken all over the ingot until the surface irregularities are smoothed. Heavier drafts are then taken, and working continues, ultimately converting the cross section from that of the original ingot into the desired final shape.

During forging operations, several types of defects may be created or worsened (from the ingot stage). Among these are pipe, laps, folds, seams, cracks, tears, bursts, and forged-in scale. Pipe is a solidification-shrinkage defect often carried through from the ingot stage and is detected as a small round cavity located near the center of an end surface (see the section "Ingot Pipe and Centerline Shrinkage" in this article). Laps, folds, seams, cracks, bursts, and tears are produced during forging.

Forged-in scale is one of the most prevalent surface defects. The scale is formed in a previous heating operation and has not been eliminated before or during forging. The following example describes a pressure vessel that failed from forged-in scale.

Example 11: Cracking of a Pressure Vessel During Autofrettage Because of Scale Worked Into Forging Laps and Seams. Certain forged pressure vessels are subjected to autofrettage during their manufacture to induce residual compressive stresses at locations where fatigue cracks may initiate. The results of the autofrettage process, which creates a state of plastic strain in the material, is an increase in the fatigue life of the component.

During autofrettage of a thick-wall steel pressure vessel, a crack developed through the wall of the component. Figure 26(a) shows the cracked region (arrow), which extends about 37 m (12 ft).

Investigation. The mechanical properties, composition, and microstructure of the component were evaluated to determine if material deficiencies contributed to this processing failure. Correspondingly, the strength, ductility, and toughness displayed no significant deviations from the reported values of the manufacturer and met the specified requirements. Chemical composition was also within specifications.

Fig. 26 Alloy steel pressure vessels that developed cracks in the wall during autofrettage

Failure was due to scale worked into the surface of the vessel. (a) Overall view of the vessels; the cracked component is the second from left (arrow indicates crack). (b) Section cut for examination. Arrow indicates suspected area of fracture initiation. (c) Micrograph of an unetched specimen showing nonmetallic inclusions (light gray) along the fracture surface. 50×. (d) Portion of area shown in (c); note agglomerated appearance of nonmetallics, which were identified as iron oxides. 500×

Metallographic specimens were prepared from the area of the pressure vessel containing a suspected crack origin. This region is indicated by an arrow in Fig. 26(b). The specimens were taken from sections perpendicular and parallel to the long axis of the forging. Metallography revealed the tempered martensite structure normally obtained in this steel after proper quenching and tempering.

Figure 26(c) shows the fracture profile parallel to the longitudinal axis of the component. This edge, which is adjacent to the outer surface, displays a somewhat continuous light-gray network of nonmetallic included matter. Figure 26(d) shows at a higher magnification a portion of the area shown in Fig. 26(c) and illustrates the agglomerated appearance of this feature. Subsequent electron microprobe analysis identified the included material as iron oxide.

Conclusion. Iron oxide of this magnitude is highly abnormal in vacuum-degassed steels. Included matter of this nature (exogenous) most likely resulted from scale worked into the surface during forging. Therefore, it is understandable that failure occurred during autofrettage when the section containing these defects was subjected to plastic strains. Because the inclusions were sizable, hard, and extremely irregular, this region would effect substantial stress concentration. Consequently, the fracture

toughness of the steel was exceeded, and failure through the wall resulted.

Laps and folds are forging defects that can initiate failures, as illustrated in the following three examples.

Example 12: Fatigue Fracture That Initiated at a Forging Lap in a Connecting Rod for a Truck Engine. A connecting rod from a truck engine failed after 73 000 km (45 300 miles) of service. The rod (Fig. 27a) was forged from 15B41 steel and heat treated to a hardness of 29 to 35 HRC. The connecting rod was sent to a laboratory for examination. A piece of the I-beam sidewall about 6.4 cm (2½ in.) long was missing when the connecting rod arrived at the laboratory.

Investigation. The sidewall and web areas of the I-beam section of the connecting rod were inspected for surface defects by fluorescent magnetic-particle testing. An indication of a defect was found extending along the edge of a sidewall for approximately 3.8 cm (1½ in.) and was 6.4 mm (¼ in.) deep, starting at, and perpendicular to, the fracture surface. The defect was identified as a forging lap (Fig. 27a).

Visual examination of the fracture surface (Fig. 27b) revealed beach marks typical of fatigue failure. The origin was along the forging lap in the sidewall of the I-beam section and was approximately 4.7 mm (³⁄₁₆ in.) below the

forged surface. The fatigue indications extended slightly beyond a lubrication hole drilled longitudinally through the web. The remaining portion of the fracture surface was severely mutilated, but an area in the web contained indications of brittle fracture. The missing portion of the sidewall appeared to be the result of an extension of the forging lap.

Metallographic examination of the metal in the rod disclosed an acceptable tempered martensitic microstructure with a hardness of 30 to 31 HRC. A micrograph of a section taken perpendicular to the lap and parallel to the web (Fig. 27c) showed that the metal was severely decarburized for a depth of 0.25 mm (0.010 in.) from the lap surface and partially decarburized for a depth of 0.5 mm (0.020 in.). Oxides were scattered for a depth of 0.5 mm (0.020 in.) from the lap surface.

Figure 27(d) shows a micrograph of a section taken near the intersection of the two branches of the forging lap. Oxides and decarburization are clearly visible along the lap surface. One branch of the lap was on a 45° angle; the other was vertical.

Conclusions. The rod failed in fatigue with the origin along the lap and located approximately 4.7 mm (³⁄₁₆ in.) below the forged surface. The presence of oxides may have been a partial cause for the defect. To correct the problem, the forgings should be more carefully inspected by fluorescent magnetic-particle testing before machining.

Example 13: Connecting Rod That Fractured Because of a Forging Fold. A motorboat-engine connecting rod forged from carbon steel fractured in two places and cracked at the small end during service. The locations of the fracture surfaces and of the crack are shown as arrows A, B, and C in Fig. 28(a). The fracture at arrow A was caused by fatigue resulting from operating stresses; the fracture at arrow B is a secondary tensile fracture. At arrow C, the transition on the other side of the rod is cracked symmetrically to the fatigue fracture. Figure 28(b) shows the fracture surfaces, with the fatigue fracture at top and the tensile fracture at bottom.

Investigation. Magnetic-particle inspection revealed indications of cracking at the transition area, between the rod and the small end. Six other connecting rods inspected from the same batch also had these indications.

Metallographic examination of one of the connecting rods selected by the magnetic-particle tests revealed deep folds in the flash zone. As shown in Fig. 28(c), these folds are filled with scale and flanked by a decarburized zone. The microstructure of the remaining material reflected correct annealing practice. Also, because the flash zone was ground after forging, no decarburization could be detected outside the fold zone.

Conclusions. The connecting rods were rendered susceptible to fatigue-crack initiation and propagation by the notch effect of coarse folds formed during the forging operation (Ref 28).

Fig. 27 15B41 steel forged truck connecting rod that failed in service from fatigue initiated at a forging lap

(a) Connecting rod and a detail of the I-beam portion showing the forging lap in one wall. Dimensions given in inches. (b) Fracture surface showing origin of the fracture (arrow). 1½×. (c) Micrograph of a nital-etched section taken perpendicular to the forging lap and parallel to the I-beam web. Oxides (black) and decarburization (light) are present along the forging lap; surface of the lap is at right. 100×. (d) Nital-etched section taken near the intersection of the two branches of the forging lap. One branch is on a 45° angle; the other is vertical (edge at right). Both oxides and decarburization are evident. 37½×

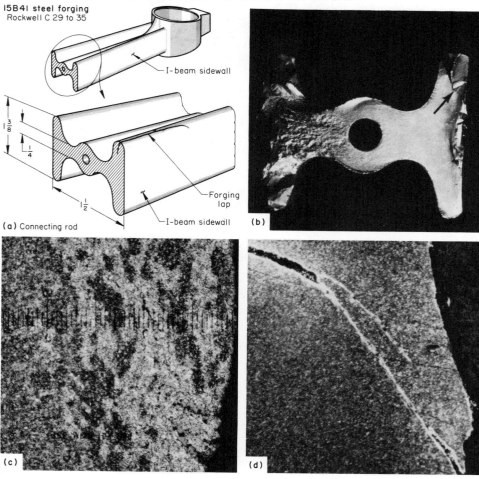

15B41 steel forging
Rockwell C 29 to 35

I-beam sidewall

Forging lap

I-beam sidewall

(a) Connecting rod

(b)

(c)

(d)

Example 14: Cracking of a Steel Socket Spanner Head Because of Forging Folds. A head of a socket spanner made of heat-treated 0.40C-0.34Cr steel cracked in service. One half of the head, longitudinally sectioned, is shown in Fig. 29(a) in the as-received condition.

Investigation. The pronounced fibrous structure of the component became evident as soon as it was etched with 2% nital. Folds in the material originating from the shaping process were visible at the point marked with an arrow in Fig. 29(a). The micrograph in Fig 29(b) shows that cracks run along these folds oriented according to the fiber. One exception is the longest crack, of which only the first third follows the direction of the fiber; the remainder runs transversely and shows the typical features of a hardening crack.

The fissures, with the exception of the hardening crack, were partly filled with oxide and showed signs of decarburization at the edges (Fig. 29c). From this it could be assumed that parts of the external skin had been forced into the folds during forging.

Conclusions. Even though there was some indication of chemical segregation, the folds made during forging initiated the main crack. Furthermore, even if the steel had been more homogeneous, hardening cracks would probably have been promoted by the coarse fissures at the fold zones (Ref 29).

Seams can be difficult to detect because they may appear as scratches on the forging or because a machining process may obliterate them. When the constraint exerted by the bulk of material is removed from the neighborhood of a seam, there is a strong likelihood that the seam will open, rendering the forged part inoperative.

Example 15: Prongs on Forceps That Split and Fractured Due to Forging

Seams. The pointed ends of several stainless steel forceps split or completely fractured where split portions broke off (Fig. 30). All the forceps were delivered in the same lot. The pointed ends of the forceps are used for probing and gripping very small objects and must be true, sound, and sharp.

Investigation. Analysis showed the failures to be the result of seams in the steel that were not joined during hot working.

Conclusion. Closer inspection of the product at all stages is necessary. Inspection at the mill will minimize discrepancies at the source, and the inspection of the finished product will help detect obscure seams.

Example 16: Surface Indications in Hot-Rolled 4130 Steel Bars. Routine magnetic-particle inspection revealed crack indications in a number of shafts produced from hot-rolled 4130 steel bar (Fig. 31a and b). A pronounced indication of this size is cause for rejection if the defect is not eliminated during subsequent machining.

Investigation. A microstructural analysis of the shaft cross section revealed that the crack was approximately 0.5 mm (0.020 in.) deep and oriented in a radial direction (Fig. 31c). Furthermore, no stringer-type nonmetallic inclusions were observed in the vicinity of the flaw, which did not display the intergranular characteristics of a quench crack. The defect did, however, contain substantial amounts of oxide (Fig. 31d), which evidently resulted from the hot-working operation.

Conclusion. Overall, the appearance of this discontinuity, with the long axis parallel to the working direction and radial orientation with regard to depth, strongly suggests a seam produced during rolling. Flaws of this nature are not uncommon in rolled bars and generally do not present a serious problem if allowance has been made for adequate stock removal before finishing. Unfortunately, the shaft in this case was to be used without further machining and was therefore rejected.

Use of components with surface-defect indications as small as 0.5 mm (0.02 in.) can be risky in certain circumstances. Depending on the orientation of the flaw with respect to applied loads, the nature of the applied forces (for example, cyclic), and the operating environment, such a surface flaw can become the initiating site for a fatigue crack or a corrosion-related failure.

Abnormal grain flow is often associated with subsurface defects in forgings that can lead to fracture. The two examples that follow describe forged aluminum alloy aircraft wheels that failed in service because of subsurface defects. Both wheels exhibited abnormal grain flow.

Example 17: Cracks in Forged Aircraft Wheel Halves That Originated From Forging Defects. Two outboard main-wheel halves from a commercial aircraft were removed from service because of failure. One wheel half was in service for 54 days and had

Fig. 28 Steel connecting rod end that fractured and cracked in service because of a forging fold

(a) Reassembled rod end showing locations of fractures and crack. Arrow A shows a fatigue fracture; arrow B, a secondary tensile fracture; arrow C, the crack. (b) Fracture surfaces of the broken-off shell. Fatigue fracture is at top; tensile fracture (fibrous) is at bottom. (c) Micrograph of the scale-filled forging fold. Note decarburization flanking the fold. Etched with 2% nital. 50×. Source: Ref 28

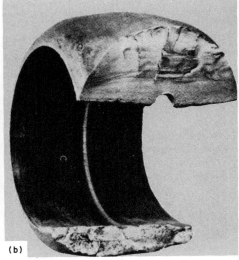
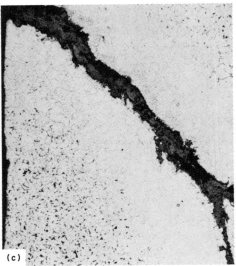

made 130 landings (about 1046 roll km, or 650 roll miles) when crack indications were discovered during eddy-current testing. The flange on the second wheel half failed after only 31 landings, when about 46 cm (18 in.) of the flange broke off as the aircraft was taxiing. A sectioned wheel half is shown in Fig. 32(a). Both wheel halves were made from aluminum alloy 2014-T6 forgings. Stains on the fracture surfaces were used to determine when cracking was initiated.

Investigation of Cracked Wheel Half. After a cursory examination of the wheel half removed from service because of cracking, it was concluded that the reported crack was superficial, and the following rework was authorized:

* Liquid penetrant inspect the entire wheel half
* Polish surface at known crack indication. Etch and liquid penetrant inspect as needed to ensure that the crack indication is removed
* Chemically treat repaired area in preparation for painting, then paint

In initial liquid-penetrant inspection, the area containing the crack indication did not bleed out. However, the crack became apparent after light surface polishing.

The wheel half was then forwarded to the laboratory for failure examination. The portion containing the, crack was saw cut from the wheel half and broken open. No evidence of crack propagation by fatigue was visible, but as shown in Fig. 32(b), the fracture surface was marked by a slight gray discoloration that was judged to be the result of aluminum oxidation. Examination of the fracture surface revealed that the crack was associated with disrupted and

transverse grain flow. Figure 32(c) shows an enlarged section through the crack that illustrates the abnormal grain flow.

Investigation of Broken Wheel Half. Examination of the fracture surface on the wheel half with the broken flange revealed that fracture was initiated by a crack about 19 cm (7½ in.) long. Surface detail indicated disrupted and transverse grain flow, which was verified by deep etching in aqueous 20% sodium hydroxide. Grain flow at the crack plane changed abruptly from longitudinal to transverse, similar to the pattern shown in Fig. 32(c). The surface showed definite chromic acid stains and evidence of rapid fatigue-crack propagation before final failure.

Conclusion. Failure on both wheel halves was by fatigue caused by a forging defect resulting from abnormal transverse grain flow. The crack in the first wheel half occurred during service, and the surfaces became oxidized. Because the fracture surface of the second wheel half had chromic acid stains, it was obvious that the forging defect was open to the surface during anodizing.

The forging defects may have been the result of any of the following factors, singly or in combination:

* The forging blank contained more than the optimum volume of metal in the flange area
* The forging blank contained less than the optimum volume of metal in the flange area, resulting in an underfill condition
* Forging-die design was inefficient
* The forging operation was unduly fast for the specific design
* Some metal was removed from the surface during the forging process to eliminate a

specific surface defect, resulting in an underfill condition in the flange area

After the forging supplier was notified of the defects, the frequency of failure decreased markedly. However, forgings with this defect were still occasionally found at incoming inspection.

Example 18: Fatigue Fracture of an Aircraft Wheel Half That Was Initiated at a Subsurface Defect. The flange on an outboard main-wheel half on a commercial aircraft fractured during takeoff. The wheel half (Fig. 33) was made from an aluminum alloy 2014-T6 forging. The failure was discovered later during a routine enroute check. The flange section that broke away was recovered at the airfield from which the plane took off and was thus available for examination.

Failure occurred after 37 landings (about 298 roll km, or 185 roll miles). Routine inspection of the wheel half before takeoff did not disclose evidence of incipient failure. However, a thorough inspection of the surface after failure revealed an unusual grain-flow pattern at numerous points along the crack.

Investigation. Examination of the fracture surfaces revealed that a forging defect was present in the wall of the wheel half. The paint was stripped from the fracture area, and the wheel surfaces were examined. The anodized coating showed distinct twin-parallel and end-grain patterns between which the fracture occurred (section A-A, Fig. 33). This type of grain pattern has been observed in other wheels that had forging defects.

Macroscopic examination of the fracture surfaces indicated that the forging defect was not open to either side of the wall; therefore, the crack was not detected during routine inspec-

Fig. 29 Steel socket spanner head that cracked in service because of forging folds

(a) Longitudinal section through the head showing region of crack (arrow). (b) Cracked region. 20×. Arrow indicates forging folds that initiated the crack. Etched with 2% nital. (c) Crack showing oxides (black) and some decarburization (light). Etched with 2% nital. 500×. Source: Ref 29

(a)

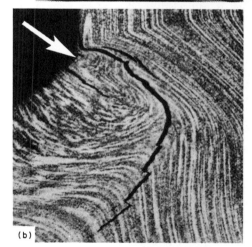

(b)

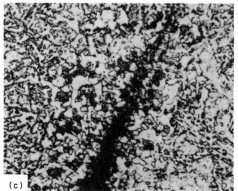

(c)

Fig. 30 Forged stainless steel forceps that fractured or split at prong tips

(a) Overall view of forceps. (b) Prongs showing a split (arrow). (c) Prong on which a split top broke off (arrow)

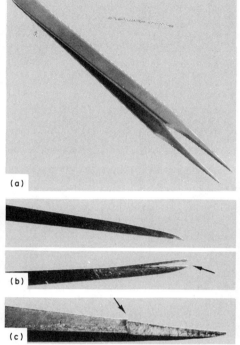

(a)

(b)

(c)

tion. As shown in View B-B in Fig. 33, the periphery of the defect was the site of several small fatigue cracks that eventually progressed through the remaining wall. Rapid fatigue then progressed circumferentially to a length of about 15 cm (6 in.) before final fracture occurred. The original forging defect was about 5 cm (2 in.) long.

Sections through the fracture plane were prepared for metallographic examination using Keller's reagent. The microstructure was judged normal for aluminum alloy 2014-T6. The work metal had a hardness of 82 to 83 HRB, which surpassed the minimum hardness of 78 HRB required for aluminum alloy 2014-T6. An abrupt change in the direction of grain flow across the fracture plane indicated that the wall had buckled during forging.

Conclusions. The results of the investigation showed that the wheel half failed in the flange by fatigue as the result of a rather large subsurface forging defect.

Failures Related to Postforging Processes

A forging may be properly designed and fabricated from sound material and yet fail from a defect created by improper processing or handling after forging. The range of the defects

Fig. 31 Seam in rolled 4130 steel bar

(a) Overall view of bar showing location of seam (arrow). (b) Closeup of seam. Note the linear characteristics of this flaw. (c) Micrograph showing cross section of the bar. Seam is normal to the surface and filled with oxide. 30×

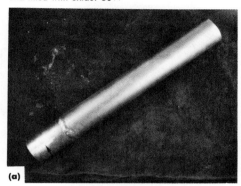

(a)

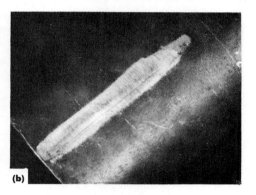

(b)

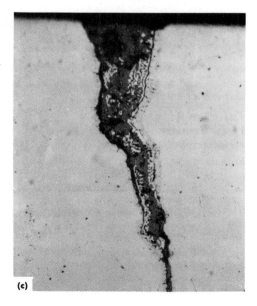

(c)

can vary from engraved identification marks, which can nucleate fatigue cracks, to undesirable microstructure resulting from variations in the time-and-temperature schedules for heat treatment to hydrogen embrittlement resulting from surface-finishing operations, such as elec-

Fig. 32 One of two aluminum alloy aircraft wheel halves that cracked in service from defects during forging

(a) Sectioned wheel half showing location of fracture. Dimensions given in inches. (b) Fracture surface showing what appeared to be the result of aluminum oxidation (dark gray patches on light gray fracture area). (c) Section through the crack and the disrupted grain flow. Etched with sodium hydroxide. Approximately 8½×

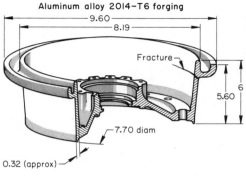

(a) Aircraft outboard main-wheel half

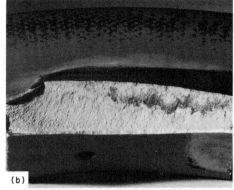

(b)

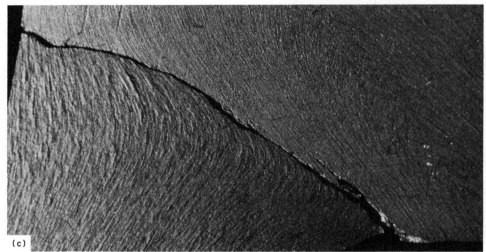

(c)

troplating. Table 1 lists some of the defects that may result from various postforging processes.

Failures Related to Heat Treatment

Improper heat treatment can result in failure to attain the desired microstructure in the metal and therefore the desired levels of mechanical or physical properties. Such deficiencies can cause failures in service.

Overheating. As alloys are heated above their recrystallization temperatures, grain growth occurs. As the temperature increases, so does grain growth, becoming quite rapid and resulting in large grains, often accompanied by many undesirable characteristics. The impairment that usually accompanies large grains is caused not only by the size of the grains but also by the more continuous films that are formed on larger grains by grain-boundary impurities, such as preferential precipitates and evolved gases. Finer grains, on the other hand, present a greater amount of total grain-boundary area over which the impurities may be distributed.

The detrimental effects of overheating depend on the temperature and the time of exposure as well as on the chemical composition of the alloy. For example, a short exposure of a high-speed tool steel to temperatures around 1250 °C (2280 °F) is required to dissolve the carbides, but prolonged heating at that temperature will cause grain growth and loss of mechanical properties. The damage caused by overheating is particularly significant in the high-carbon and medium-carbon steels, in which both strength and ductility are affected.

One of the most conspicuous indications that a metal has been overheated is the coarse-grain fracture surface that results. Usually, examination with a stereoscopic microscope will show the characteristic faceted surface of an intergranular fracture. This is demonstrated by the carbon steel hook (Fig. 34) used as a fixturing device in a process tank. During forming, this hook was heated far in excess of the recrystallization temperature, resulting in large austenitic grains that promoted brittle fracture through the grain boundaries. The rock-candy fracture

appearance (Fig. 34b) exemplifies this intergranular fracture path.

Microscopically, the large grain size resulting from overheating is quite evident and can be measured and compared to a similar metal with a normal grain size. In addition to large grain size, fine oxide particles are often dispersed throughout the grains, particularly near the surface. These oxides result from internal oxidation and are particularly evident in overheated copper forgings (Fig. 35a).

A Widmanstätten structure is often associated with coarse grains in an overheated steel forging. Two conditions are required for the presence of Widmanstätten structure: a controlled rate of cooling (neither extremely fast nor extremely slow) and a large grain size (Ref 30).

Burning is a term applied when a metal is grossly overheated and permanent irreversible damage to the structure occurs as a result of intergranular penetration of oxidizing gas or incipient melting. The micrograph in Fig. 35(b) shows burning at the grain boundaries in a specimen of copper C11000 (electrolytic tough-pitch copper) heated to 1065 °C (1950 °F). Copper oxide migrated to grain boundaries, forming a continuous network that severely reduced strength and ductility.

In steels, burning may manifest itself with the formation of extremely large grains and incipient melting at the grain boundaries. Melting is particularly evident where segregation has occurred with the marked formation of low-melting phases in the interdendritic regions and at the grain boundaries.

Burning cannot be readily detected by visual examination. However, a metallographic examination of the structure shows the enlarged grains and the pronounced grain-boundary network.

The typical causes of burning include excessive furnace temperatures; welding and flame-cutting operations that are not properly controlled; and operations involving the direct heating of a workpiece surface, such as flame hardening, induction hardening, shrink fitting, and so on, that are inadequately controlled. Occasionally, burning will occur in adequately controlled furnaces simply because the flame is allowed to impinge upon the metal surface, causing localized overheating. Another source of overheating, though less common, can be the conversion of mechanical energy into heat. If there are segregated areas near the center of the billet and if the initial forging temperature is close to the melting point of the segregated regions, the additional heat supplied by the transformation of work into heat can cause localized burning during the forging operation. Burned material cannot be salvaged and should be scrapped, because the metallurgical changes that have occurred are irreversible.

Example 19: Brittle Fracture of a Clamp Because of Burning During Forg-

Fig. 33 Failed aluminum alloy aircraft wheel half

Illustration shows an overall view of the wheel and the flange portion that broke off during takeoff because of a large subsurface forging defect in the flange. Dimensions given in inches. Section A-A: A view of the crack that occurred between the twin-parallel and end-grain patterns. View B-B: The fracture surface with arrows indicating sites of several small fatigue cracks

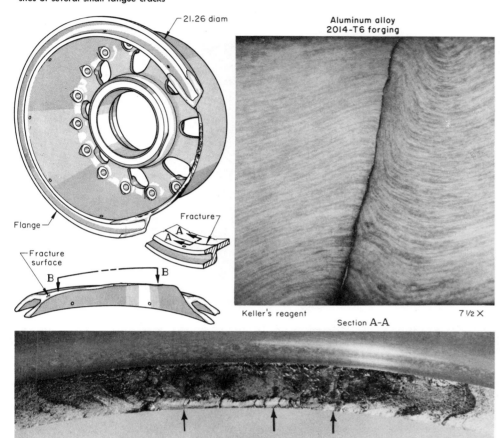

21.26 diam

Flange

Fracture surface

Fracture

Aluminum alloy
2014-T6 forging

Keller's reagent   Section A-A   7 1/2 ×

View B-B

Table 1 Defects that may result from various postforging processes

Process	Possible defects
Electroplating	Hydrogen embrittlement, galvanic corrosion
Heat treatment	Excessive grain growth, burning of grain boundaries, brittle structure, carburization, decarburization, quench cracks
Electrolytic cleaning	Pitting
Surface hardening, nitriding, carburizing, anodic hard coating	Excessive case thickness, microcracks, embrittled material at stress raisers
Machining	Tool marks, grinding cracks
Welding	Weld-metal defects, hydrogen-induced cracking, inclusions, improper structure

Fig. 34 Carbon steel hook that fractured in a brittle manner from overheating

(a) Overall view of the hook. Arrow indicates the fracture. (b) Fracture surface showing the faceted, intergranular appearance. 3 ×

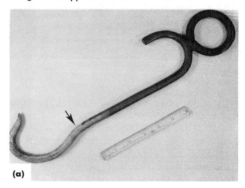

(a)

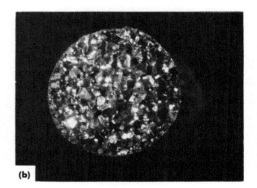

(b)

ing. The ring clamp shown in Fig. 36(a) was used for attaching ducts to an aircraft engine. After 3 h of service, the clamp became loose. When the clamp was removed from the engine, the hinge tabs on one clamp half were found to be broken. The clamp half had been machined from an 8740 (AMS 6322) steel forging and was cadmium plated.

Investigation. Visual examination of the clamp half revealed that it had fractured through the hinge tabs (Fig. 36a). The appearance of the fracture surface indicated a brittle intergranular fracture (Fig. 36b). Microscopic examination of a section through the fracture surface revealed burning (incipient melting) and decarburization at the grain boundaries (Fig. 36c), which indicates gross overheating of the metal.

Macroscopic inspection of the area adjoining the fractures, after it had been stripped of cadmium plate and swab etched with 5% nital, confirmed that the burning had been confined to the tab end of the clamp half (Fig. 36d). This end of the clamp had been hot upset to provide material for the hinge tabs.

The metallurgical quality of the clamp, remote from the burned area, was satisfactory with respect to hardness, microstructure, and chemical composition for this grade steel. The thickness of the hinge tabs was within blueprint requirements.

Conclusions. Both hinge tabs on the clamp half fractured in a brittle manner as the result of gross overheating, or burning, during forging. The mechanical properties of the metal, especially toughness and ductility, were greatly reduced by burning. Evidence that burning was confined to the hinge end of the clamp indicated that the metal was overheated before or during the upset forging operation.

Recommendations. The forging supplier should be notified of the burned condition on the end of the clamp. The clamps should be macroetched before cadmium plating to detect overheating. The clamps in stock should be inspected to ensure that the metal has not been weakened by overheating during the upset forging operation.

Quench cracks are often the cause of failure in steel forgings. These cracks may be

obvious and may prompt the removal of a component during production, or they may be obscured during manufacture and may be present in the shipped component. Quench cracks in steel result from stresses produced by the volume increase accompanying the austenite-to-martensite transformation. When a steel forging is quenched, untempered (hard

Fig. 35 Micrographs showing the effects of overheating and burning on microstructures of copper forgings

(a) Overheated copper C10200 forging showing oxides (black particles). The forging was heated to 1025 °C (1875 °F). (b) Burning (black outlines) at grain boundaries of a copper C11000 forging heated to 1065 °C (1950 °F)

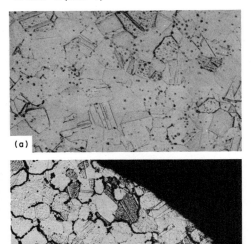

(a)

(b)

Fig. 36 Cadmium-plated 8740 steel aircraft wing clamp that failed because of burning during forging

(a) View of assembled clamp and detail showing locations of fractures. Dimensions given in inches. (b) Fracture surfaces showing brittle, intergranular nature of fracture. Approximately 2×. (c) Section through fracture surface showing incipient melting (large arrow) and decarburization at grain boundaries (small arrow). Etched with 2% nital. 100×. (d) Area adjoining the fractures after being stripped of cadmium plate showing coarse grains (between arrows). Etched with 5% nital. 1½×

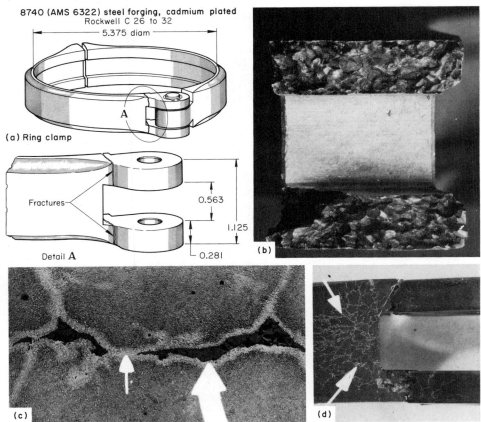

8740 (AMS 6322) steel forging, cadmium plated
Rockwell C 26 to 32

5.375 diam

A

(a) Ring clamp

Fractures

0.563

1.125

Detail A

0.281

(b)

(c)

(d)

and brittle) martensite is formed at the outer surfaces first.

As cooling continues, the austenite-transformation reaction progresses toward the center of the workpiece, accompanied by a volumetric expansion that can place the center portions of the part in tension. When the workpiece cannot adjust to the strains produced by this process, cracking generally occurs. Such cracking may take place during or after the quench because the transformation product is hard and brittle before tempering.

This susceptibility to cracking is increased by the presence of stress raisers, such as sharp fillets, tool marks, or other notches, massive inclusions, or voids. Quench cracks in forgings may also occur near the trim line of a closed-die forging where some localized burning may have occurred. Other factors that affect quench cracking are the hardenability of the steel, the rate of cooling, and elapsed time between quenching and tempering. In general, an increase in any of these factors increases the likelihood of quench cracking.

Quench cracks have several recognizable features. Macroscopically, they are usually straight and extend from the surface (often initiating from a fillet or tool mark) toward the center of the component. The margins of the crack may have scale present if the part was tempered after quenching. The cracks are open to the surface and may be detected by magnetic-

particle, ultrasonic, or eddy-current inspection. Microscopically, quench cracks are invariably intergranular and may be free of decarburization.

Example 20: Failure of Seamless Tubing Due to a Quench Crack. Hydraulic cylinder housings were being fabricated from 4140 grade seamless steel tubing. During production, magnetic-particle inspection indicated the presence of circumferential and longitudinal cracks in a large number of cylinders (Fig. 37a).

Investigation. Visual examination of the cracks in several pieces revealed that the fractures exhibited a dull gray appearance over most of their surfaces. Such a feature is indicative of an elevated-temperature oxide in this material. A subsequent SEM analysis of the fracture surfaces confirmed that a fairly heavy oxide had formed on the crack surfaces and that the fracture mode was entirely brittle, consisting of intergranular fracture and cleavage (Fig. 37b and c).

Microstructural analysis showed a tempered martensite structure typical of a heat-treated

4140 grade steel. Furthermore, large concentrations of sulfide stringers were also present (Fig. 37d). It is plausible to suspect that these second-phase particles could act as stress raisers and contribute to quench-crack initiation. Finally, a cross section of one of the cracks is shown in Fig. 37(e). This view shows the oxide layer associated with the crack surface and the intergranular (branching) nature of the fracture.

Conclusions. The cracking problem in these components was identified as quench cracks due to their brittle, intergranular nature and the characteristic temper oxide on the fracture surfaces. Although the steel met the compositional requirements of SAE 4140, the sulfur level was 0.022% and would account for the formation of the sulfide stringers observed. Apparently, the combination of the clustered, stringer-type inclusions and the quenching conditions were too severe for this component geometry. The result was a high incidence of quench cracks that rendered the parts useless.

Corrective Measures. The steel should have a specification that will require lower sulfur concentrations. Magnetic-particle cleanliness

Fig. 37 Alloy steel seamless tubing that failed because of quench cracks

(a) Cross section of tube showing extensive cracking revealed by dye-penetrant inspection. (b) SEM micrograph showing intergranular fracture at a crack origin. 90×. (c) SEM micrograph illustrating the brittle mode of failure associated with the fracture. 50×. (d) Micrograph showing the typical concentrations of nonmetallic stringers in the tube material. (e) Micrograph showing a quench crack. Note the intergranular branching and heavy oxide. 400×

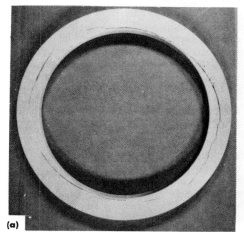

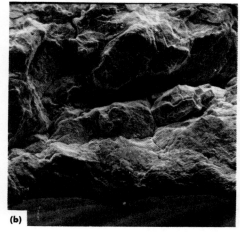

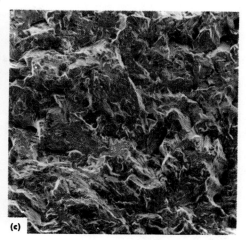

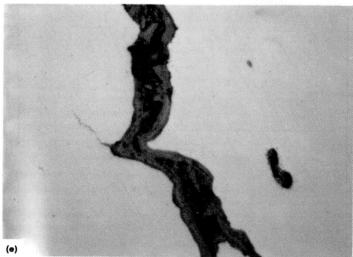

standards should be imposed that will exclude material with harmful clusters of sulfide stringers, for example, modified AMS 2301.

Temper Embrittlement. Certain steels, particularly the chromium-nickel grades, show a severe decrease in impact strength when tempered in the range of approximately 350 to 575 °C (660 to 1065 °F). The exact range depends largely on the composition of the steel. The degree of segregation and certain variables related to making the steel also influence this range. This embrittlement, which is called temper embrittlement, is believed to result from the precipitation of a complex phase at prior-austenite grain boundaries; however, this is not always observable when examining a specimen known to be embrittled.

Temper embrittlement increases the ductile-to-brittle transition temperature and reduces the energy required for brittle fracture. Brittle characteristics are not easily identified by conventional tensile testing and are more pronounced at high strain rates and low test temperatures. It has also been shown that temper embrittlement significantly increases the growth rates of both fatigue cracks and stress-corrosion cracks (Ref 31). When the fracture surfaces of temper-embrittled metal are examined under a microscope, the appearance is always that of an intergranular fracture with the crack path along the prior-austenite grain boundaries.

The effect of alloying elements and impurities on temper embrittlement has been investigated thoroughly. The presence of increased amounts of phosphorus, antimony, arsenic, and tin generally increase susceptibility to temper embrittlement. It has been demonstrated that, although temper embrittlement depends on the interaction of the major alloying elements, the effect of minor elements was more severe in a chromium-nickel steel than in either a chromium steel or a nickel steel (Ref 32). Embrittlement depends on the composition of the metal and on the thermal treatment, including the rate of cooling. Information on several types of embrittlement, including temper embrittlement, that commonly occur in structural alloys is provided in the article "Ductile and Brittle Fractures" in this Volume.

Carburization and Decarburization. Although surface treatment by carburization is commonly used to improve the wear resistance and overall strength characteristics of many forged steel components, particularly axles, the accidental alteration of the surface structure during heat treatment can produce disastrous results. Heating a metal surface that is contaminated with oil or carbon or heating in a carbon-rich atmosphere can produce a surface with a high carbon content. The mechanical properties of a carburized surface layer are different from those of the core, and cracking problems can arise because a carburized surface was not considered in the original design. For example, components subject to severe impact

Fig. 38 Forged 1.60Mn-5Cr steel gear and pinion that failed in service because of improper carburization

(a) Gear showing irregular wear on teeth. (b) Microstructure on the periphery of a gear tooth on the unworn side showing coarse martensite and large amounts of retained austenite (white). (c) Spalled teeth of a pinion. (d) Intergranular cracking at the tip of a pinion tooth. Source: Ref 33

are rarely carburized, because the hardened surface layers generally possess low toughness.

Even when carburizing is intentional, care must be taken to control the processing variables, because the depth of case (hardened zone) is often vital to good performance. If the case is excessive, the component may fracture in a brittle manner. If the case is inadequate, it may crush under load. Fatigue failures may initiate from microcracks in the carburized case or from stresses at the case/core interface, then propagate during cyclic loading.

The opposite condition (decarburization) can occur by heating in an oxidizing atmosphere such as air. The result in this case is a surface layer with a lower carbon content than that of the core. Either carburization or decarburization can be the cause of service problems, as described in the following example (Ref 33).

Example 21: Gear and Pinion Failures Due to Improper Carburization. A gear manufacturer experienced service problems with various gears and pinions that had worn prematurely or had fractured. All gears and pinions were forged from 1.60Mn-5Cr steel and were case hardened by pack carburizing.

Gear Failure. One of the gears showed severe wear on the side of the teeth that came into contact with the opposing gear during engagement (Fig. 38). This contact point is where high shock loads had to be accommodated at large differences in peripheral speed. The microstructure at the periphery of a worn tooth at its unworn side (Fig. 38b) consisted of coarse acicular martensite with a large percentage of retained austenite. Parts with such a structure usually have a relatively low wear resistance because of low hardness. They are also very sensitive to shock because of their large grain size.

It was concluded that the high wear rate was caused by spalling of the coarse-grain surface layer. The underlying cause of the wear was overheating during the carburization.

Pinion Failures. The teeth of the pinion in Fig. 38(c) exhibited severe spalling; the fracture surfaces at spalled regions were coarse grained and lustrous. The microstructure at the surface consisted of coarse acicular martensite with retained austenite, similar to that shown in Fig. 38(b). Also, a coarse network of precipitated carbide particles showed that the carburization of the case had appreciably exceeded the most favorable carbon content, resulting in increased sensitivity to shock. This structure indicates that the pinions were overheated during carburizing and that the carbon content of the case was excessive.

Another pinion showed wear on one side of the teeth, similar to the wear on the gear teeth shown in Fig. 38(a). Cracks followed coarse prior-austenite grain boundaries in regions along tips of teeth (Fig. 38d).

Examination of the fracture surfaces of the pinions, together with examination of the microstructures, showed that the failures resulted from overheating combined with excessive case carbon content. These factors resulted in a large grain size with a brittle grain-boundary network. The overall effect is that the steel had low resistance to shock loading, wear, and surface fatigue.

Failures Related to Hydrogen Damage

Many of the metals used in forgings are susceptible to embrittlement by hydrogen. The embrittling hydrogen originates from various sources. A major source of hydrogen in steels is the reaction of water vapor at high temperatures with the liquid metal during melting and pouring. The water vapor may come from the scrap used to charge the furnace, the slag ingredients in the charge, or the refractory materials lining the furnace, transfer vessels, and molds. A similar problem can occur with ESR steels. In this case, the hydrogen problem has been traced to such causes as condensation in the copper crucibles or moisture pickup by the slags. The hydrogen that results can then be trapped during solidification. Hydrogen may also develop during acid pickling or plating operations. Exposure in service to process fluids bearing hydrogen, as in catalytic cracking, can also cause embrittlement. Similarly, hydrogen may be generated as a corrosion product in certain environments and thus become available to cause embrittlement.

The metals most susceptible to hydrogen embrittlement are those with body-centered cubic (bcc) crystal structures, such as steel and the refractory metals, and those with hexagonal close-packed (hcp) crystal structures, such as titanium and zirconium.

Hydrogen Damage in Steel. Hydrogen entrapment is the most common cause of embrittlement in bcc metals, such as steel. Hydrogen damage in steels may be evident as flakes or as general embrittlement (Ref 34). Flakes are small internal fissures that occur in the interior of a forging parallel to the forging direction. They may be detected by ultrasonic inspection or destructively by etching of transverse sections (Fig. 39).

On a fracture surface, these flakes appear as bright, highly reflective areas (Fig. 40a). Fractographic examination shows that the flakes contain some flat areas combined with other areas that appear to be contoured grain boundaries (Fig. 40b). The striations visible in Fig. 41 are characteristic of hydrogen flakes (Ref 35).

Hydrogen embrittlement in steel also manifests itself as a decrease in tensile strength and ductility when the steel is tested under static loads at low strain rates (Ref 36). This is demonstrated by the data shown in Fig. 42. Fractographic examinations indicate that the

Fig. 39 Hydrogen flakes on a ground cross section of a steel bar

Dye-penetrant inspection was used to reveal the hydrogen damage. Approximately 2×

Fig. 40 Macroscopic appearance of hydrogen flakes on a fracture surface

(a) Flaking in a 4340 steel. Arrows indicate individual flakes (fisheyes). Actual size. (b) Closeup of fracture surface containing the mating halves of a flake. Note the distinctive shiny appearance of the flake.

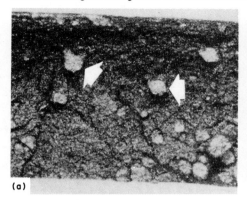

(a)

(b)

Fig. 41 TEM fractograph of contoured boundary and striations in a hydrogen flake

The specimen is 4340 steel. 6000×. Source: Ref 35

Fig. 42 Effect of dissolved hydrogen on the tensile ductility of nickel-molybdenum-vanadium rotor steel

Source: Ref 34

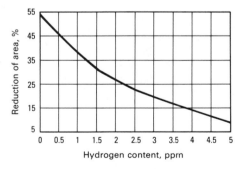

Fig. 43 Intergranular cracks originating from an electroplated coating

Source: Ref 37

visible features vary. Microcracks begin internally, often near inclusions or other interfaces, and propagate integranularly for an indeterminate distance, or they may originate from electroplated surfaces (Fig. 43). The regions between the adjacent microcracks fracture in a ductile manner, presenting evidence of microvoid coalescence (dimpling). When the effective cross section has been sufficiently reduced, final fracture occurs as the result of overstress.

It is now believed that much of what has been reported in the literature as the stress-corrosion failure of high-strength steel has often been due to hydrogen embrittlement resulting from hydrogen generated as a corrosion product. Based on fractographic examination, the differences between the two mechanisms are very subtle, which often makes it difficult to distinguish between them. Detailed discussions of these mechanisms are provided in the articles "Hydrogen-Damage Failures" and "Stress-Corrosion Cracking" in this Volume.

Example 22: Fracture of an Accumulator Ring Due to Hydrogen Embrittlement. Fracture of a cadmium-plated accumulator ring forged from 4140 steel was discovered during inspection and disassembly of a hydraulic-accumulator system stored at a depot. The ring had broken into five small and two large segments.

The small segments of the broken ring displayed very flat fracture surfaces with no apparent yielding, but the two large segments did show evidence of bending (yielding) near the fractures. In addition, some segments contained fine radial cracks (arrow, Fig. 44a).

Investigation. Optical microscopy on polished-and-etched specimens cut from the ring revealed the microstructure to be very fine-grain tempered martensite, typical of quenched-and-tempered low-alloy steel.

Hardness testing of the broken ring yielded a range of 43 to 48 HRC. This hardness ranged above the required 40 to 45 HRC, but this small variation from requirements was not regarded as sufficient to have caused the fracture.

Inspection of fracture-surface replicas by electron microscopy disclosed the fracture mode to be predominantly intergranular (Fig. 44b). Also observed were many hairline indications on the fracture facets and some partly formed dimples. All these features indicate the cracking associated with hydrogen embrittlement.

Chemical analysis showed that all elements were within the specified limits for 4140 steel. Hydrogen-gas analysis of four specimens yielded the following information:

Specimen	Hydrogen content, ppm
1	0.15 ± 0.03
2	0.23 ± 0.03
3	2.29 ± 0.07
4	0.29 ± 0.04

Although three values are low, specimen 3 exhibited a hydrogen content in the range of those generally associated with hydrogen embrittlement (>1 ppm). Also, these data indicate a localized hydrogen concentration rather than a uniform distribution throughout the ring.

Conclusions. Results of the examination point to a brittle fracture mechanism, as evidenced by the intergranular nature of the fracture path. Also, the hydrogen content of one tested specimen indicates that localized embrittlement of grain boundaries by hydrogen is a very strong possibility. The fine cracks in some of the intact segments most likely existed where the fractures occurred. Because these fractures were clean—that is, they showed no corrosion products, discoloration, or obvious contamination—the hydrogen most likely resulted from a processing operation. The processing history

Fig. 44 Hydrogen-embrittlement fracture in a cadmium-plated 4140 steel forged accumulator ring

(a) Ring segment showing a fine crack (arrow).
(b) Electron fractograph showing the predominantly intergranular fracture. 7800×

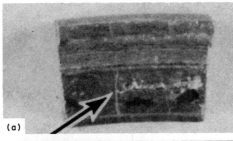

for the fractured component was reviewed for potentially detrimental operations.

The review of processing revealed that the components required cadmium plating, after

which they were scheduled for stress relieving at 260 °C (500 °F) for 3 h. An investigation showed that there was a strong possibility that one batch of rings did not receive the required stress-relieving treatment. The conclusion, therefore, is that the hydrogen penetration occurred during the plating operation and was not relieved subsequently as required. The slightly higher-than-specified hardness may have helped to promote embrittlement, but was not considered significant.

Hydrogen Damage in Nonferrous Alloys. Alloys other than the iron-base alloys can also be adversely affected by hydrogen. Titanium and zirconium alloys have an hcp crystal structure, and both exhibit a marked propensity toward hydrogen embrittlement through the formation of stable hydrides.

Hydrogen is also known to exert a damaging effect on the bcc refractory metals, such as tungsten, tantalum, and niobium. The evidence indicates that hydrogen exists in a supersaturated solution and that stable hydrides are not present. Figure 45 shows the effect of hydrogen content on the ductility of tantalum and niobium as measured by reduction of area. These curves show that even low hydrogen content can be detrimental. Hydrogen in the refractory metals promotes brittle cleavage fractures.

Failures Related to Service

Many forging failures occur that are not necessarily a result of a deficiency in material or design but because consideration was not given to the service environment to which the forgings would be subjected. For example,

degradation by corrosion or wear may occur in service and seriously impair the function of a forging either by altering its performance or by initiating defects that can cause failure.

Corrosion is the degradation of metal by chemical or electrochemical dissolution that occurs as a result of the interaction of the metal with its environment. Occasionally, a component with a history of satisfactory performance in one environment will fail when subjected to another environment. The failure may be due to general corrosion, although this is not likely because the general effects of a specific environment are reasonably well known and can be anticipated. The damage that does occur is usually due to localized attack resulting from crevice corrosion, pitting, corrosion of highly worked or segregated regions, or SCC. More information on corrosion failures is available in the articles "Corrosion Failures" and "Corrosion-Fatigue Failures" in this Volume.

Intergranular Attack. In certain cases, the grain boundaries in a forging are more susceptible to corrosive attack than the grain interior. The preferential dissolution suffered by these areas may be related to several factors, depending on the particular circumstances.

The primary cause of intergranular attack is an inhomogeneous condition at the grain boundary, which may be the result of a segregation mechanism or of intergranular precipitation. These conditions may also be modified by enhanced diffusion effects operating within the grain boundary or by the selective absorption of certain solutes, such as hydrogen.

The overall effect of this preferential dissolution is that great damage to the structure can occur, with only slight corrosive damage occurring to the grains themselves. Because dissolution is confined to such small regions, the actual weight losses are small, penetration rates are high, and destruction can be quite rapid.

Exfoliation is a special form of intergranular attack that primarily affects aluminum and magnesium alloys. It is markedly directional and is characterized by attack of the elongated grains on a plane parallel to the rolled, extruded, or forged surface. This results in a characteristic delamination or stratification of the surface structure.

Example 23: Failure of Rifle Receivers Caused by Exfoliation (Ref 38). The lower receiver of the M16 rifle is an anodized forging of aluminum alloy 7075-T6. Degradation of these receivers was observed after 3 years of service in a hot, humid atmosphere. The affected areas were those in frequent contact with the hands. One of the failed receivers is shown in Fig. 46(a).

Investigation. Because there was no question that the material failed as a result of exfoliation corrosion, the investigation centered around the study of thermal treatments that would increase the exfoliation resistance and still develop the required 448-MPa (65-ksi) yield strength.

Several factors relating to the processing were investigated using forgings made from

Fig. 45 Effect of hydrogen content on the ductility of wrought and recrystallized tantalum and niobium as measured by the reduction of area at the temperatures shown

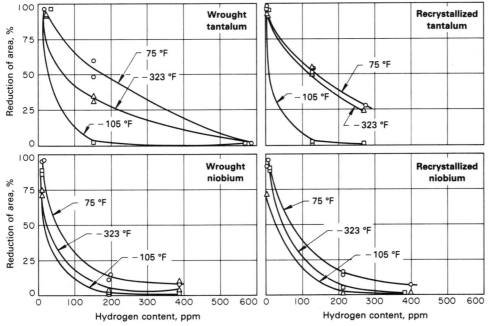

Fig. 46 Forged aluminum alloy 7075-T6 receiver from an M16 rifle that failed by exfoliation corrosion

(a) Rifle receiver. 0.7×. Similar receivers were forged from three different materials to investigate the effects of processing on exfoliation resistance (Table 2). Section A-A: (b), (c), and (d) are sections through as-received forgings of the three materials. Etched with concentrated Keller's reagent. 2×. Letters A through G indicate locations of specimens shown in correspondingly lettered micrographs at bottom. Etched with Keller's reagent. 80×. Compare the grain structures of the three materials. Source: Ref 38

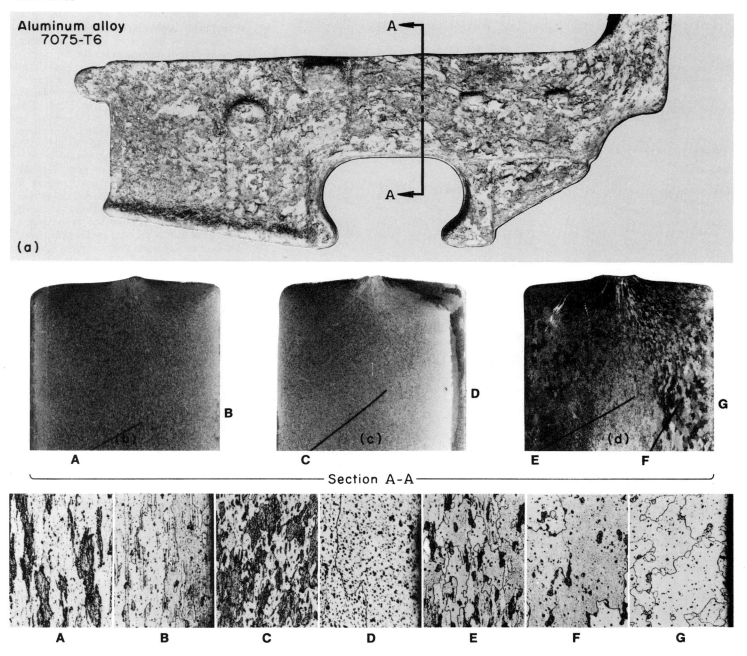

three different starting materials (referred to here as materials 1, 2, and 3). Specifically, the factors investigated were the grain structures developed by thermomechanical treatments, thermal treatment, and the quenching rate subsequent to the solution heat treatments.

Effect of Grain Structure. A section through a forging fabricated from material 1 (extruded stock) in the as-received (T6) temper is shown in Fig. 46(b). The metal had an interior struc-

ture composed of highly deformed elongated grains (micrograph A, Fig. 46) and a thin, partly recrystallized grain structure at the surface (micrograph B, Fig. 46). The yield strength of the as-received (T6) forging was about 552 MPa (80 ksi); exfoliation was found to be severe.

Material 2 was also extruded stock, forged in the as-received (T6) temper. However, the observed microstructure of material 2 (Fig.

46c) is somewhat different from that of material 1. On one side (right side, Fig. 46c), there is a completely recrystallized surface layer about 3.2 mm (0.125 in.) wide; on the other side, the layer is only a few grain diameters in width. The surface grains are elongated vertically (micrograph C, Fig. 46), and the structure in the interior of this forging (micrograph D, Fig. 46) is similar to that observed in material 1. The yield strength of material 2 was also similar to

Table 2   Effect of thermal treatments on properties of aluminum alloy 7075-T6 forgings made from three different materials

Thermal treatment	Longitudinal strength				Electrical conductivity, %IACS	Exfoliation rating
	Yield		Tensile			
	MPa	ksi	MPa	ksi		
Material 1(a)						
As-received (T6)	547	79.4	609	88.3	32.9	Severe
Re-solution treat, age to T6	561	81.3	614	89.1	34.0	Medium
Convert as-received T6 to T7x1	543	78.7	587	85.2	36.0	Severe
Re-solution treat, age to T6, convert to T7x1	541	78.4	581	84.3	37.8	Severe
Convert as-received T6 to T76	515	74.7	560	81.3	36.7	Mild
Re-solution treat, age to T76	510	73.9	554	80.3	39.0	Mild
Re-solution treat, age to T6, convert to T76	509	73.8	553	80.2	38.5	Slight
Re-solution treat, age to T6, convert to T7x2	513	74.4	558	80.9	38.3	Slight
Convert as-received T6 to T7x2	476	69.0	534	77.5	40.0	Immune
Convert as-received T6 to T73	447	64.8	505	73.2	42.9	Immune
Re-solution treat, age to T73	485	70.3	533	77.3	40.5	Immune
Re-solution treat, age to T6, convert to T73	454	65.8	510	74.0	40.5	Immune
Material 2(b)						
As-received (T6)	555	80.5	606	87.9	33.0	Medium
Re-solution treat, age to T6	544	79.0	607	88.0	32.5	Medium
Convert as-received T6 to T76	492	71.3	552	80.1	38.3	Medium
Re-solution treat, age to T76	500	72.5	556	80.7	38.9	Medium
Re-solution treat, age to T73	504	73.1	563	81.7	38.3	Slight
Convert as-received T6 to T73	463	67.1	522	75.8	40.4	Immune
Material 3(c)						
As-received (T6)	521	75.6	588	85.3	32.5	Medium
As-forged, solution treat and age to T6	499	72.4	574	83.2	31.4	Slight
Convert as-received T6 to T76	501	72.7	554	80.4	35.5	Medium
As-forged, solution treat and age to T6, convert to T76	492	71.4	544	78.9	36.7	Medium
As-forged, solution treat and age to T76	479	69.5	535	77.6	37.8	Medium
As-forged, solution treat and age to T73	487	70.6	544	78.9	38.0	Slight
As-forged, solution treat and age to T6, convert to T73	460	66.7	521	75.6	39.2	Immune
As-received (T6) converted to T73	459	66.5	512	74.2	40.2	Immune

(a) Extruded stock; composition Al-5.67Zn-2.40Mg-1.58Cu-0.18/0.30Cr-0.05Mn-0.1/0.2Si-0.1/0.3Fe-0.05Ni. (b) Extruded stock; composition Al-5.67Zn-2.47Mg-1.47Cu-0.18/0.30Cr-0.05Mn-0.1/0.2Si-0.1/0.3Fe-0.05Ni. (c) Rolled stock; composition Al-5.67Zn-2.44Mg-1.56Cu-0.18/0.30Cr-0.05Mn-0.1/0.2Si-0.1/0.3Fe-0.05Ni. Source: Ref 38

that of material 1 (approximately 552 MPa, or 80 ksi), but exfoliation of material 2 was rated as intermediate.

A third forging, fabricated from rolled stock (material 3), had a microstructure (Fig. 46d) quite different from that of either material 1 or 2. For the material 3 forging, the structure on both surfaces was composed of recrystallized grains with no directionality. The microstructure changed from highly deformed grains at the center (micrograph E, Fig. 46) to large equiaxed grains near and at the surface (micrographs F and G, Fig. 46). The yield strength of this material in the T6 temper was 524 MPa (76 ksi), and the exfoliation resistance was intermediate. Laboratory reheat treatment further enhanced the improvement in exfoliation resistance of the material 3 forging when all materials were retreated to the T6 temper.

Thermal Treatment. Table 2 shows the compositions and mechanical properties of materials 1, 2, and 3 and the exfoliation resistance of forgings from the three materials that was obtained with various thermal treatments. In general, forgings in the T6 temper had the highest strength, but also had the lowest resistance to exfoliation. Thermal softening of the as-received T6 forgings as induced by the

intermediate tempers was accompanied by a slight increase in exfoliation resistance. With two exceptions, all forgings heat treated to the T73 temper were immune to exfoliation and had a minimum longitudinal yield strength of 448 MPa (65 ksi). All forgings with an electrical conductivity of at least 40.0% IACS were immune to exfoliation.

Quench Rate. Data obtained by quenching in water at four different temperatures showed that susceptibility to exfoliation increased markedly as the temperature of the quenching medium increased.

Conclusions. The results of the study show that differences in grain structure of the forgings, as induced by differences in thermal-mechanical history of the forged material, can have a significant effect on susceptibility to exfoliation corrosion. This factor, which was previously considered to be relatively minor, appears to be very important for material in the T6 temper. It was observed that forgings fabricated from rolled stock, identically treated to T6, had a higher resistance to exfoliation than forgings from extruded stock. The increased resistance to exfoliation observed on forgings from rolled stock is attributable to a more favorable randomized grain structure induced

by the rolling process. Thus, for optimizing exfoliation corrosion resistance of high-strength forgings, rolled bar stock should be preferred to extruded bar stock.

Regarding thermal treatment, the results show conclusively that large changes in strength and exfoliation characteristics of 7075 forgings can be induced by changes in temperature or time of thermal treatment.

With regard to the effect of quenching rate on exfoliation characteristics, a cold-water quench (<25 °C, or 75 °F) would appear to be far superior to an elevated-temperature quench to minimize exfoliation for 7075 forgings in the T6 temper. Under conditions where the temperature of water approaches 80 °C (180 °F), exfoliation could become very severe.

Wear is a major mechanism in the degradation of equipment with moving components. If the wear occurs at a predictable and steady rate, the useful life of the component can be readily established and provisions made for its replacement. However, if the wear mechanism is such that the wear rate is not steady or predictable, or results in catastrophic failure, the consequences obviously are more serious. Additional information is available in the article "Wear Failures" in this Volume.

Pitting, in a wear situation, is the result of surface (rolling-contact) fatigue under high contact stresses. One major difference between the more conventional fatigue and surface fatigue is that no apparent endurance limit exists in surface fatigue; that is, there seems to be no stress level below which the material remains unaffected by surface-fatigue damage. Another difference is that test data are subject to wider scatter than that obtained in conventional fatigue tests. Consequently, on the basis of test results, it is exceedingly difficult to design a highly stressed bearing component with the positive knowledge that surface fatigue has been eliminated. Pitting caused by surface fatigue is discussed in detail in the articles "Wear Failures" and "Failures of Rolling-Element Bearings" in this Volume.

Cracks may originate at a subsurface location when the maximum contact stresses are coincident with an inclusion site. Maximum contact stresses occur slightly below the surface (Fig. 47). Vacuum arc remelting, vacuum induction melting, electroslag melting, and electron beam refining have resulted in materials with significantly reduced inclusion counts. In these materials, the importance of inclusions relative to crack initiation has been minimized so that studies have been directed more toward the effects of segregation, retained austenite, and the role of banded or fibered structures on crack initiation.

The type of surface movement, that is, rolling versus sliding, also helps to determine whether the crack will originate at the surface or subsurface. With pure rolling, the maximum shear stress occurs below the surface. When a sliding compound is added, however, the maximum shear stresses may occur at the surface.

Fig. 47 Location of maximum stress for elastic contact of a sphere and a flat object relative to the surface of the flat object
Source: Ref 38

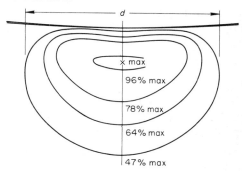

Fig. 48 Schematic of shape and orientation of a surface pit in a rotating component with respect to the direction of rotation

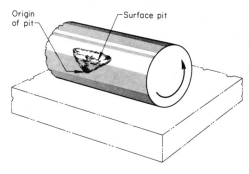

Fig. 49 Cross section of a pitted area at the surface of a steel forging
Note the gradual incline of the pit to the surface, near the top of the micrograph. 150×. Source: Ref 40

Fig. 50 Fretting damage on rollers and inner race of a bearing assembly
Source: Ref 41

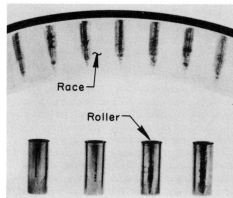

Fig. 51 Effect of cyclic displacement on total weight loss resulting from fretting
Data are for a low-carbon steel. Source: Ref 42

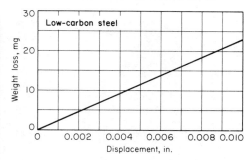

Pits are generally triangular or fan-shaped in appearance, with the apex of the triangle pointing in the direction of rotation of the surface in which they appear (Fig. 48). A micrograph of an actual surface pit in the initial stages of development is shown in Fig. 49; the gradual incline to the surface near the origin is visible.

Electron microscopy has been used in fractography to study pitting (and spalling) under conditions of rolling-contact fatigue. These studies show that the fracture surface of the leading edge of a pit consists primarily of equiaxed and elongated cavities typical of ductile fracture by the growth and coalescence of microvoids. This area extends approximately one-third the distance to the trailing edge. The remainder of the surface exhibits parabolic tongues. It has been suggested that the initial crack results from fatigue at the leading edge, and upon some extension, the crack becomes fast-growing and completes the fracture (Ref 40). This change from fatigue extension to fast growth is believed to result in the change in fracture appearance from the leading-edge area to the trailing-edge area.

Spalling results when several pits join or when a crack runs parallel to the surface for some distance rather than running into it (subsurface). Consequently, spalling defects are typically large. Spalling from cracks is often associated with surface-hardened parts and occurs near the core/case interface. Spalling occurs because the shear strength of the subsurface material is inadequate to withstand the shear stress to which it is subjected. Spalling can often be prevented by increasing the case depth or the core hardness.

Fretting, often called fretting corrosion, occurs as the result of a slight oscillatory motion between two mating surfaces under load and manifests itself as pits in the surface surrounded by oxidation debris. An example of fretting damage is shown in Fig. 50.

The relative motion required to produce fretting damage may be quite small. Displacements of 0.03 μm are sufficient to cause damage, but the amplitudes of displacement usually seen in service are of the order of a few thousandths of an inch. Figure 51 shows the effect of cyclic displacement on total weight loss for a low-carbon steel.

The total amount of fretting damage increases with increasing number of cycles of oscillation. The increase is essentially linear except for the initial stage of fretting, when very little abrasive material is available to cause damage. Frequency effects are observed, but are usually small. The wear rate is higher at low frequencies, but decreases to a constant value as frequency increases (Ref 42).

Although fretting damage alone is sufficient to cause malfunction in many close-tolerance components, the problem is made even more serious because fatigue fractures are frequently initiated from fretting pits.

Erosion may result from cavitation or from the impingement of liquid or solid particles on a surface. Cavitation erosion and liquid impingement are discussed in the article "Liquid-Erosion Failures" in this Volume. Erosion by solid particles (abrasive erosion) is discussed in the article "Wear Failures" in this Volume.

Erosion can be prevented or minimized by changing the material to one that is more erosion resistant or by modifying the design to decrease erosion severity. However, it is impossible to make absolute statements regarding the selection of materials for erosive application without testing, except to state that materials that possess good corrosion resistance to the environment in question and have high hardness will, in general, perform satisfactorily. Tests should always be conducted to verify the selection.

Design changes made to reduce turbulence or to reduce velocities, such as removing protuberances or increasing tube diameters, are beneficial. Also beneficial are such modifications as increased radii on elbows and streamlined baffles, which reduce impingement severity.

Erosion-corrosion is exemplified by an increase in corrosion rate caused by relative motion between a metal surface and a corroding environment. The good corrosion resistance of many metals is due to an insoluble film that is created and maintained on the metal surface. Although increases in velocity of relative motion between the metal and its environment may

Fig. 52 1030 steel tube that failed by erosion-corrosion

(a) Progressive thinning (left to right) in the tube wall that resulted in a rupture due to internal pressure.
(b) Portion of the tube wall showing longitudinal rupture

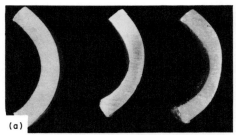

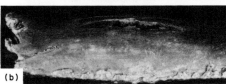

occasionally decrease the rate of corrosion because of kinetic factors, the general effect of increasing velocity is that film maintenance is impaired; corrosion rates increase as the surface film is destroyed. Erosion-corrosion presents a distinct appearance. The metal surface usually exhibits severe weight loss, with many hollowed-out regions; the overall surface has a carved appearance.

Figure 52 shows progressive thinning caused by erosion-corrosion in the wall of a 1030 steel tube. Progressive thinning of the wall occurred until the hoop stress generated by the internal pressure exceeded the yield strength of the material and the wall ruptured.

Failures Related to Mechanical Damage

Frequently, failure of a component may occur that is obviously the result of a specific failure mechanism, such as fatigue, but no abnormality in materials or design may become apparent during the investigations. A search of the records and history of the component may show that the component suffered mechanical damage during the course of service or during handling.

Marks produced by electric engraving tools may produce transformation of the base metal, embrittlement, and subsequent fatigue-crack initiation and fracture. Surface abrasion of cold-drawn wire during the handling of the coils may produce fatigue fracture in springs fabricated from the wire. Impact to an axle during service can result in distortion that can alter the performance of the axle, result in bearing damage, and, sometimes, cause fatigue fracture.

Example 24: Fatigue Fracture of a Steering Knuckle Caused by Deformation. A steering knuckle used on an earthmover failed in service. The component fractured

Fig. 53 Fracture surface of a 4340 steel steering knuckle flange that failed in service by fatigue

Arrows indicate fracture origins.

into a flange portion and a shaft portion. The flange was approximately 27.9 cm (11 in.) in diameter and had 12 evenly spaced 16-mm (⅝-in.) diam boltholes around its diameter. The shaft was hollow, approximately 10.5 cm (4⅛ in.) in outside diameter, with a wall thickness of approximately 17 mm (¹¹⁄₁₆ in.). The steering knuckle was made of 4340 steel and heat treated to a hardness of about 415 HB (yield strength of about 1069 MPa, or 155 ksi).

Investigation. The fracture was examined through a stereoscopic microscope. Figure 53 shows the fracture surface on the flange side of the fracture. Examination of the fracture surface revealed the presence of numerous beach marks emanating from several fracture origins.

The fracture originated in the fillet at the junction between the flange and the shaft portions. This is an unexpected location because of the large radius of the fillet, which minimizes stress concentration. Had the fillet been sharp, the stress concentrations would have been high and the fatigue fracture more likely. In addition, the fillet was shot peened, which further reduced tensile stresses. Both of these factors (the large radius and the shot-peened surface) indicated that the manufacturer had used good design practice.

Examination of the fracture surface also verified that the component was subjected to a low stress concentration at the origin and to moderate loading. These factors were established by the location of the zone of final fracture. Microstructural examination revealed no unusual discontinuities or inclusions that might act as stress raisers.

The service history of the earthmover indicated that the vehicle had been involved in a field accident 6 months before the steering knuckle failed. Several components, including portions of the frame, had been damaged and replaced as a result of this accident, but there was no observed damage to the steering.

Conclusions. The failure was due to a fatigue fracture. No evidence of a defective design was

observed; on the contrary, the generous radius of the fillet and the shot-peened surface indicate good design practice. In addition, there was no evidence of an improper microstructure, high inclusion count, or other stress-raising condition that might account for fatigue-crack initiation. It was concluded from the service history that the fracture was the result of the prior accident, the most likely explanation being that the shaft was bent in the accident and that continued use caused a crack to initiate and propagate to fracture.

Failure Types and Frequencies

The more important aspects of forging failures and their analysis have been discussed in this article. To include every type of forging failure would be impossible; however, the following tabulation illustrates the variety and frequency of failures that commonly occur in industrial plants:

Service or operational failures	36%
Failures caused by faulty design	34%
Failures resulting from shop and mill practice	17%
Failures due to metallurgical aspects	13%

This information is a result of a review of some 470 failures in a single plant over several years (Ref 43). In the most frequent category, service or operational failures, overload or overstress was predominant for various reasons, with 28% of the total number of failures reviewed; wear and corrosion were next within this category. In the design-related failure category, sharp corners and sharp fillets alone accounted for 12% of the total number of failures reviewed; misapplication of materials or use of the wrong type of steel for the design was responsible for approximately 8%. Shop and mill practices resulting in failure included such items as improper grinding, welding stress, defective welds, faulty lubrication, and misalignment. The final category, metallurgical aspects, comprised improper heat treatment, quench cracks, forging defects, and residual stress. Although these data apply strictly to a particular plant, they show a good correspondence with the preponderance of forging failures.

It is evident that failures of forgings can usually be attributed to more than one mechanism or factor, even though a particular mechanism may control the failure process. Furthermore, as the technological complexity of designs, materials, and processing methods increases, the concept of interactive failures will surely become more prevalent. In this case, the failure analyst must isolate and define all the contributory factors or problems. Only when this type of definitive analysis has been accomplished can complete corrective measures be instituted and the recurrence of failure prevented.

REFERENCES

1. T.G. Byrer, S.L. Semiatin, and D.C. Vollmer, Ed., *Forging Handbook*, Forging Industry Association, Cleveland, and American Society for Metals, 1985
2. T. Altan, S. Oh, and H. Gegel, *Metal Forming—Fundamentals and Applications*, American Society for Metals, 1983
3. S.L. Semiatin and G.D. Lahoti, *Sci. Am.*, p 9-106
4. *Making, Shaping and Treating of Steel*, United States Steel Corp., Pittsburgh, 1964, p 1079
5. R. Kiessling and N. Lange, *Nonmetallic Inclusions in Steel*, Parts 1 to 4, The Metals Society, London, 1978
6. V.J. Colangelo, *Trans. AIME*, Vol 233, 1965, p 319
7. J.T. Ransom and R.F. Mehl, *Proceedings of the American Society for Testing and Materials*, Vol 52, 1952, p 779
8. B.M. Kapadia, A.T. English, and W.A. Backofen, *Trans. ASM*, Vol 55, 1962, p 389
9. F.A. Heiser and R.W. Hertzberg, *J. Basic Eng.*, Vol 93 (No. 2), 1971, p 211
10. V.N. Whittacker, *Nondestr. Test.*, Oct 1971, p 320
11. V. Packard, *The Wastemakers*, David McKay Co., 1960, p 53
12. D.J. Wulpi, Effects of Variables, *Met. Prog.*, Vol 88 (No. 6), 1965, p 66
13. F.K. Naumann and F. Spies, *Prakt. Metallogr.*, Vol 7, 1969, p 447-456
14. R. Hertzberg, *Deformation and Fracture Mechanics of Engineering Materials*, John Wiley & Sons, 1976, p 459-462

15. A.S. Tetelman and A.J. McEvily, *Fracture of Structure Materials*, John Wiley & Sons, 1967, p 356
16. H.O. Fuchs and R.I. Stephens, *Metal Fatigue in Engineering*, John Wiley & Sons, 1980, p 105-123
17. S.T. Rolfe and J.M. Barsom, *Fracture and Fatigue Control in Structures—Applications of Fracture Mechanics*, Prentice-Hall, 1977, p 208-229
18. SAE J429, *SAE Handbook*, Vol 11, *Materials*, Society for Automotive Engineers, Warrendale, PA, 1983
19. P.A. Thornton, *J. Mater. Sci.*, Vol 6, 1971, p 347
20. D.V. Edmonds and C.J. Beevers, *J. Mater. Sci.*, Vol 3, 1968, p 457
21. F.K. Naumann and F. Spies, *Prakt. Metallogr.*, Vol 10, 1973, p 532
22. E. Kauczor, *Prakt. Metallogr.*, Vol 8, July 1971, p 443-446
23. D.J. Wulpi, Effects of Variables, *Met. Prog.*, Vol 88 (No. 6), 1965, p 70
24. G. Paul, *Prakt. Metallog.*, Vol 8, 1971, p 254
25. E.C. Bain and H.W. Paxton, *Alloying Elements in Steel*, American Society for Metals, 1961, p 124
26. J.A. Parrish, Factors Affecting the Properties of Carburized and Hardened Gears, *Met. Prog.*, Vol 124 (No. 4), 1983, p 45-52
27. H.E. McGannon, *The Making, Shaping and Treating of Steel*, United States Steel, Pittsburgh, 1964, p 999
28. E. Kauczor, *Prakt. Metallogr.*, Vol 9, 1972, p 298
29. E. Kauczor, *Prakt. Metallogr.*, Vol 11, 1974, p 36
30. R.F. Mehl and C.S. Barrett, *Studies Upon the Widmanstätten Structure*, Part 1, AIME 353, American Institute of Mining, Metallurgical, and Petroleum Engineers,

Warrendale, PA, 1930
31. V.J. Colangelo, "Effect of Temper Embrittlement on Fatigue and SCC Growth Rates," Unpublished paper, Westec Conference, American Society for Metals, 1973
32. J.R. Low, D.F. Stein, A.M. Turkalo, and R.D. LaForce, *Trans. AIME*, Vol 242, 1968, p 14-24
33. F.K. Naumann and F. Spies, *Prakt. Metallogr.*, Vol 11, 1974, p 40-44
34. J.E. Steiner, Control of Flaking and Other Hydrogen Problems in Heavy Forgings, in *Current Solution to Hydrogen Problems in Steels*, American Society for Metals, 1982, p 55-62
35. A. Phillips and V. Kerlins, Analyzing Fracture Characteristics by Electron Microscopy, *Met. Prog.*, Vol 95 (No. 5), 1969, p 81-85
36. G.C. Interrante, Basic Aspects of the Problems of Hydrogen in Steels, in *Current Solutions to Hydrogen Problems in Steels*, American Society for Metals, 1982, p 3-17
37. A.S. Tetelman, Ph.D. thesis, Yale University
38. J.V. Rinnovatore, K.F. Lukens, and J.D. Corrie, *Corrosion*, Vol 29 (No. 9), 1973, p 364-372
39. E. Robinowitz, *Friction and Wear of Materials*, John Wiley & Sons, 1965, p 193
40. W.D. Syniuta and C.J. Corrow, *Wear*, Vol 15, 1970, p 187-199
41. R.L. Widner and J.O. Wolfe, Valuable Results From Bearing Damage Analysis, *Met. Prog.*, Vol 93 (No. 4), 1968, p 79-86
42. H.H. Uhlig, I.M. Feng, W.D. Tiernay, and A. McClellan, *Fundamental Investigation of Fretting Corrosion*, NACA Technical Note No. 3029, National Advisory Committee for Aeronautics, Washington, DC, Dec 1953
43. *Republic Alloy Steels*, Republic Steel Corp., Cleveland, 1968, p 352

Failures of Iron Castings

James J. Snyder, Gould Inc., Ocean Systems Division

A PART is considered failed when it becomes inoperable, cannot satisfactorily perform its intended function, or becomes unreliable or unsafe. Although unfavorable foundry practice can result in various casting imperfections that are detrimental in service and that may contribute to failure, many of the common causes of failures of iron castings are not foundry related. Failure in cast iron, as in other materials, can occur from one or more aspects of design, materials selection, casting imperfections, faulty processing, improper assembly, or service conditions not initially anticipated.

The types of failures that can occur in iron castings include fracture (ductile, brittle, or fatigue), distortion, wear, corrosion, erosion, chemical attack, or combinations of these. Analysis of a failure usually requires identification of the type of failure. A detailed description of types of failures is provided in the article "Identification of Types of Failures" in this Volume.

Failure Analysis Procedures

This article will describe the methods of finding the cause of failure by following failure analysis procedures. In general, the procedures, techniques, and precautions in failure analysis described in the article "General Practice in Failure Analysis" in this Volume can be applied to castings without modification. Other useful information on failure analysis procedures is available throughout this Volume.

One step in specimen preparation that may require special consideration for iron castings is macroetching. The macroetching of cast iron is used to reveal such phenomena as chill formation and segregation. Eutectic cells in gray irons can be revealed by etching to exhibit cell boundaries. The technique is described in the article "Metallographic Technique for Cast Irons" in Volume 8 of the 8th Edition of *Metals Handbook*.

In failure analysis, it is fundamental to establish the objectives of the designer by reviewing drawings, specifications, and purchase orders. It is also important to determine what was manufactured, delivered, and installed and to find if it agrees with the drawings. It is then necessary to ascertain what happened to the component before and at the time of failure (history, performance, environment, and so on). The evidence and investigation should lead to determination of the cause of the failure, which is usually classified into one of the following categories:

- Faulty design or materials selection
- Faulty material or processing
- Service conditions

To determine the true cause of failure, the investigator must fully consider the interplay among design, fabrication, material properties, environment, and service conditions. Once the failure is accurately classified, corrective action and recommendations can be made. Appropriate solutions may involve redesign, change in material selection or processing, quality control, changes in maintenance or inspection schedules, protection against environment, or restrictions on service loads or service life.

Failures Related to Faulty Design or Material Selection

Materials Selection. Proper selection of a material for a given application requires a number of steps. These include a review of the process/operating conditions (temperatures, pressures, flow rates, corrosive environments, and so on), a review of the type/design of the component, an evaluation of the material through lab tests or in-field testing, and a review of applicable codes, standards, and quality assurance requirements.

Example 1: Fracture of Teeth in an Oil-Pump Gear Because Ductility Was Inadequate for Shock Loading in Service. The two oil-pump gears shown in Fig. 1 broke after 4 months of service in a gas compressor that operated at 1000 rpm and provided a discharge pressure of 7240 kPa (1050 psi). The compressor ran intermittently, with sudden starts and stops. The large gear was sand cast from class 40 gray iron with a tensile strength of 290 MPa (42 ksi) at 207 HB. The smaller gear was sand cast from ASTM A536, grade 100-70-03, ductile iron with a tensile strength of 696 MPa (101 ksi) at 241 HB.

Investigation. Visual examination of the larger gear revealed that several teeth had cracked and fractured at the root and that many other teeth were crushed and pitted at the top. The smaller gear displayed one area in which a segment of the rim had broken away. The remainder of the teeth on this gear showed relatively little wear in light of the battered condition of the larger gear with which it meshed.

Metallographic examination disclosed that both gears had normal microstructures. The ductile iron consisted of a pearlitic matrix containing well-shaped nodules of graphite generally surrounded by ferrite envelopes. The gray iron exhibited a pearlitic matrix containing type A graphite flakes. There were no metallurgical causes of failure other than the inherent brittleness of gray iron, and there was no evidence of machining imperfections that could have contributed to the fractures.

Conclusions. Brittle fracture of the teeth of the gray iron gear resulted from high-impact loading that arose from the sudden starts and stops of the compressor. During subsequent rotation, fragments broken from the gray iron gear damaged the mating ductile iron gear.

Fig. 1 Sand-cast oil-pump gears

(a) ASTM A536, grade 100-70-03, ductile iron. (b) Class 40 gray iron that fractured because of improper material selection. 0.25 ×

Ductile iron, grade 100-70-03
Diametral pitch, 8; pitch diameter, 6.5 in.; face width, 0.75 in.; bore diameter, 0.875 in.

(a)

Gray iron, class 40
Diametral pitch, 8; pitch diameter, 8.0 in.; face width, 1.25 in.; bore diameter, 4.43 in.

(b)

Fig. 2 Gray iron cylinder block that cracked due to casting stresses

(a) External view. (b) Internal view showing crack site (arrow). 0.25×. Source: Ref 1

Excessive beam loading and the lack of ductility in the gray iron gear teeth were the primary causes of fracture.

Corrective Measures. Both gears were subsequently cast from ASTM A536, grade 100-70-03, ductile iron and were normalized at 925 °C (1700 °F), air cooled, reheated to 870 °C (1600 °F), and oil quenched. The larger gear was tempered to 200 to 240 HB, and the smaller gear to 240 to 280 HB. These hardness levels provided the desired resistance to shock and wear yet retained high strength and good ductility. No further gear failures occurred.

Casting Design. The shape of a casting can contribute to residual stresses if it entails sections of markedly different thicknesses, which can undergo different cooling rates in the mold. This effect can be minimized if the thinner sections are well insulated so that they cool at a lower rate. The use of chills to accelerate cooling of heavy sections can also be helpful. Stress relief can diminish internal stresses if the cast alloy is amenable to such treatment.

Example 2: Cracking in Gray Iron Cylinder Blocks Caused by Casting Stresses (Ref 1). During the operation of tractors with cantilevered bodies, the lateral wall of the cylinder blocks cracked repeatedly. Three of the blocks were examined.

Investigation. Figures 2(a) and (b) exhibit one of the failure cases as seen from the outside and inside of the cylinder block. The crack was always located at the approximate center of the sidewall, and it ran in a horizontal direction. The sectional view (Fig. 3a) shows more clearly that the thin case wall is reinforced at this location by a thicker rib. The crack was located in the thin-wall portion next to the transition to the rib, as shown in the larger scale

Fig. 3 Sections through gray iron cylinder block

(a) Original design. 0.25×. (b) Improved design. 0.25×. (c) Enlarged section from (a). Actual size. Source: Ref 1

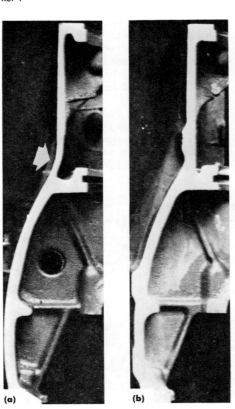

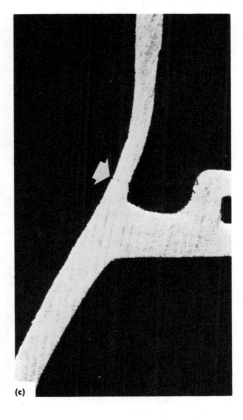

view in Fig. 3(c). At this location, high residual stresses may be expected due to the earlier solidification in the thin wall compared with the

heavier secondary sections during cooling of the casting. The stresses reduced wall thickness at this point by stretching the iron while at a high temperature. The arrow in Fig. 3(a) indicates this reduced section.

Chemical analysis of the three failed blocks gave the following compositions:

	Composition, %					
Block	C	Mn	P	S	Si	Fe
1.....	3.30	0.66	0.104	0.054	2.48	rem
2.....	3.29	...	0.100	...	2.44	rem
3.....	3.30	...	0.168	...	2.28	rem

Note: Manganese and sulfur were not determined for blocks 2 and 3.

From this data, the carbon-saturation values were determined (methods of determining carbon saturation are discussed later in this article in the section "Effects of Foundry Practice and Processing"). The three values were less than 1 (0.96, 0.96, and 0.95), demonstrating that the cast iron was hypoeutectic. The composition was acceptable for this gray iron.

Metallographic examination was performed on sections from these castings to determine if cracking was promoted by material defects. The microstructure of the thick-wall portion consisted of uniformly distributed, medium flake size, type A graphite in a matrix of pearlite with little ferrite (Fig. 4a). The thin-wall part had a structure of type D graphite, with dendrites of austenite forming and later transforming to a pearlitic-ferritic structure; the interdendritic residual liquid formed the type D graphite with a ferritic matrix (Fig. 4b). The structure is formed by supercooling of the molten iron before solidification. In this case, it is promoted by fast cooling of the thin wall.

Conclusions. The fracture formation in the cylinder blocks can be attributed primarily to casting stresses, which could be alleviated by better filleting of the transition cross section.

Recommendations. The case wall at the thin-wall section should be reinforced. Shrinkage stresses and supercooling could be alleviated or avoided by prolonged cooling of the casting inside the mold. Based on these recommendations, the design was changed with slower cooling (Fig. 3b). Figure 4(c) shows the microstructure of the reinforced wall with the revised design. This is a clear improvement over the previous design (Fig. 4b).

Stress Raisers Related to Design. Other aspects of design that affect the load-bearing ability of a casting, or its capacity to resist externally applied stress, are fillet radii (whether the fillets serve as stress raisers) and the adequacy of thickness of sections.

Example 3: Fatigue Fracture of a Stuffing Box That Originated at the Inner End of a Lubrication Hole. The stuffing box shown in Fig. 5(a) was sand cast from ASTM A536, grade 60-45-10, ductile iron and began leaking water after 2 weeks of service. The machine was operating at 326 rpm, with discharge water pressure of 21.4 MPa (3100 psi). The stuffing box was removed from the machine after a crack was discovered in the sidewall.

Investigation. The vertical portion of the crack passed through a tapped 8-mm (⁵⁄₁₆-in.) diam lubrication-fitting hole that was centered on the parting line of the casting (Fig. 5a). The direction of the crack changed abruptly to nearly horizontal at an internal shoulder about 100 mm (3⁷⁄₈ in.) below the top of the externally threaded end (View A-A, Fig. 5a). A second nearly horizontal crack (faintly visible in View A-A) at the bottom of the vertical crack extended to the right and slanted slightly downward from the parting line for at least 19 mm (¾ in.). This second crack was very tightly closed.

The surface of the fracture was closely examined, and indications of fatigue beach marks were found. As is normal for fatigue of cast iron, these indications were faint. The beach marks may originally have been slightly more distinct, but if so, they were partly eroded by leakage of water through the crack at high velocity.

A metallographic specimen cut from the casting displayed a structure consisting of well-rounded graphite nodules in a ferrite matrix that also contained remnants of a network of pearlite (Fig. 5b). Evidently, the heat treatment of the stuffing box had not been fully effective in eliminating the pearlite. The mechanical properties, however, were acceptable: 527 MPa (76.5 ksi) tensile strength, 393 MPa (57 ksi) yield strength, and 12% elongation in 50 mm (2 in.) at 179 HB.

Examination of the fracture surface suggested that the crack had been initiated at the inner edge of the lubrication hole and had propagated toward both the threaded and flange ends of the casting. An appreciable residual-stress concentration must have been present and caused propagation of the crack, because the operating pressure of 21.4 MPa (3100 psi), combined with the inside radius of 41 mm (1⁵⁄₈ in.) and the section thickness at the roots of the external threads of 16 mm (⁵⁄₈ in.), indicated a hoop stress of only about 69 MPa (10 ksi). The residual stress may have been caused when a fitting was tightly screwed into the lubrication hole, and it may have been concentrated by notches at the inner end of the hole that were created when the drill broke through the sidewall to the stuffing box.

Conclusion. The stuffing box failed by fatigue cracking that originated at the intersection of the casting bore and the lubrication hole.

Corrective Measure. The material for the stuffing box, ASTM A536 ductile iron, was changed from grade 60-45-10 to grade 86-60-03. This eliminated subsequent failures except those that occurred in extremely corrosive applications.

Example 4: Premature Failure of a Deburring Drum Initiated by Fatigue at a Stress Concentration Caused by a

Fig. 4 Micrographs of the gray iron cylinder block

(a) Section from the thick-wall part. (b) Section from the thin-wall part. (c) Thin-wall part of reinforced case. All etched with picral. 100×. Source: Ref 1

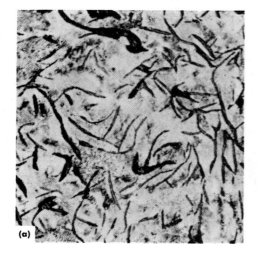
(a)

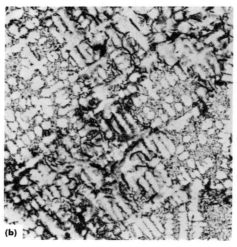
(b)

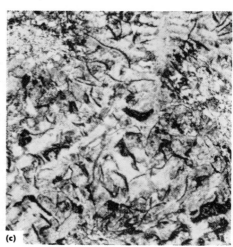
(c)

Fig. 5 Stuffing box sand cast from ASTM A536, grade 60-45-10, ductile iron

(a) Configuration and dimensions (given in inches). (b) Micrograph showing the structure consisting of graphite nodules in a ferritic matrix with remnants of a pearlite network. Etched with nital. 100×

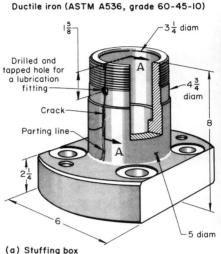

Ductile iron (ASTM A536, grade 60-45-10)

(a) Stuffing box

View A-A

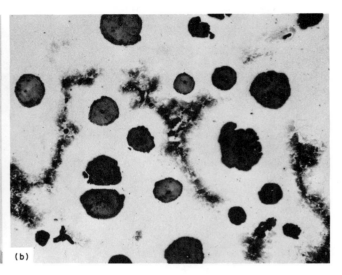

(b)

Sharp Corner at a Bolt Hole. Deburring drums are filled with abrasive, water, and small parts, such as roller bearing rollers, and they rotate on their axis at 36 rpm. Cracks were discovered very early in the service lives of these castings manufactured of high-chromium white iron.

Investigation. All of the fractures were through the bolt holes of the mounting flange. Figure 6(a) shows the probable fracture origin at a sharp corner of the bolt hole. This sharp corner is a stress-concentration site. Figure 6(a) also shows an uneven wear pattern on the inside diameter of the hole. In operation, the mounting bolts were frequently found to be loose and in at least one case broken off. Figure 6(b) shows a scanning electron microscopy (SEM) fractograph from near this fracture-initiation area. This secondary electron image illustrates fatigue striations. No casting or metallurgical structural defects were found that could explain the failures. The chemical composition of this sample was:

Element	Composition, %
Carbon	2.79
Manganese	0.77
Phosphorus	0.019
Sulfur	0.019
Silicon	0.90
Nickel	0.18
Chromium	25.2
Molybdenum	0.24
Iron	rem

Conclusions. Cracking was a result of the stress-concentration site at the bolt holes of the mounting flange where a fatigue-initiated fracture occurred.

Recommendations. It was recommended that the radii be increased at the sharp corners of the

bolt holes. Also suggested was the use of lock-wiring to secure against bolt loosening.

Example 5: Failed Bearing Caps Caused by the Combination of a Stress Raiser and a Low-Strength Microstructure (Ref 1). Cast iron bearing caps in tractor engines fractured repeatedly after brief operating periods. Fractured bearing caps, as well as others that gave satisfactory service over longer operating periods, were examined.

Investigation. Initially, two fractured bearing caps were examined. Both exhibited the same phenomena, externally as well as in metallographic section. The fracture originated in a cast-in groove and ran approximately radially to the shaft axis (Fig. 7a). The smallest cross section was at the point of fracture. Furthermore, a shape-conditioned stress peak may have been formed at the bottom of the groove that had a rounded-off radius of only 2 mm (0.08 in.).

For metallographic examination, sections were made at the notch base parallel to the fracture. Figure 7(b) shows one of the sections at 3×. The larger ferrite clusters are especially noticeable. They are permeated by fine type D graphite, which could be seen at higher magnification (Fig. 8a and b). In one piece, a narrow streak with a ferritic matrix (Fig. 9) extended under the entire surface. This indicated that the casting had been annealed at a comparatively high temperature for stress relief. The core structure of the caps consisted of graphite in a pearlitic-ferritic matrix (Fig. 10a and b). The average hardness was 160 HB, from which a low strength could be inferred.

Chemical analysis was subsequently performed on three other failed bearing caps and five others that performed satisfactorily over longer operating periods of service. The mean compositions were:

| | Element, % | | | | | Graphitic | |
Cap	C	Mn	P	S	Si	carbon	Fe
Failed	3.55	0.65	0.109	0.064	2.74	3.05	rem
Not fractured	3.25	0.66	0.312	0.124	2.62	2.73	rem

From this data, the carbon-saturation values, S_c, were calculated as 1.06 for the fractured caps and 0.97 for the unfractured caps.

In addition to the stated determinations, spectrographic analyses were conducted for additions of such elements as chromium, nickel, copper, vanadium, and titanium, but in no case could additions be found that exceeded the customary quantities. The carbon-saturation value S_c signifies the position of the alloy relative to the eutectic composition where $S_c = 1$. This value is important because it is a measure of the tendency toward carbide decomposition and of the strength of the cast iron.

Carbon, graphite, and silicon contents were significantly higher for the fractured caps than for the satisfactory caps, while phosphorus and sulfur contents were lower. The carbon-saturation value was also higher for the fractured caps than for the satisfactory ones despite a lower phosphorus content. All unsatisfactory caps showed a hypereutectic composition with $S_c = 1.03$ to 1.09, while all satisfactory ones had a hypoeutectic composition with $S_c = 0.96$ to 0.98.

According to one relationship, the tensile strength in megapascals equals $100 - 80 S_c$ (Ref 2). Relative to this relationship, the fractured caps should show an average strength of 15 MPa (2.2 ksi), while the good caps should have an average strength of 22 MPa (3.2 ksi). Expressed in terms of hardness, this indicated

Fig. 6 Failed cast iron deburring drum

(a) Light fractograph showing the probable fracture origin at a sharp corner of a bolt hole. 0.25×. (b) SEM fractograph showing fatigue striations at the area of suspected fracture origin. 1000×

(a)

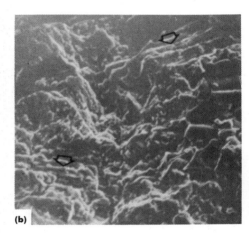

(b)

Fig. 7 Failed gray iron bearing cap

(a) Side view showing crack. Actual size. (b) Section through groove parallel to fracture. 3×. See also Fig. 8 to 10. Source: Ref 1

Fig. 8 Microstructure of the failed bearing cap shown in Fig. 7

(a) Structure at edge zone consists of type D graphite mixed with ferrite. The dark zones are pearlitic. 100×. (b) Same as (a) except at 500×. Source: Ref 1

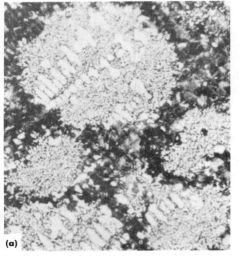

(a)

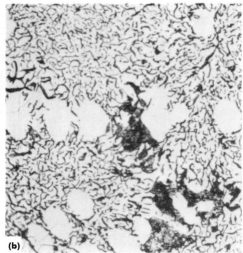

(b)

an average of 164 HB for fractured caps and 224 HB for satisfactory ones.

A comparison of metallographic sections of the groove bottoms showed that the good caps have less ferrite at the edge and core (Fig. 11a and b) compared with the broken ones (Fig. 12a and b). This indicates that the chemical composition of the cast iron should be held in the hypoeutectoid range.

The effects of stress-relief annealing was explored by annealing a number of untreated caps for 2 h at 550, 575, 600, 650, and 750 °C (1020, 1065, 1110, 1200, and 1380 °F), respectively, after their structures were confirmed in the as-cast condition. Even the nonannealed specimens showed ferrite clusters with type D graphite under the surface as well as in the core in minor proportion correspond-

ing to the favorable composition of the casting. This indicates that ferrite did not result from carbide decomposition during annealing. No change was noticeable in the structure of the annealed specimens at temperatures of up to

625 °C (1160 °F) (Fig. 13a), but most of the pearlitic matrix had decomposed after annealing for 2 h at 650 °C (1200 °F) (Fig. 13b). For 625 °C (1160 °F), the hardness had already decreased from 189 to 171 HB, and for 650 °C

Fig. 9 The surface of the failed bearing cap shown in Fig. 7

Note the ferrite skin at the top. 100×. Source: Ref 1

Fig. 10 Microstructure at the core of the failed bearing cap shown in Fig. 7

(a) Flake graphite in a pearlite/ferrite matrix. 100×. (b) Same as (a) except at 500×. Source: Ref 1

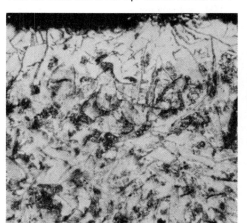
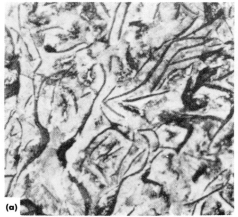
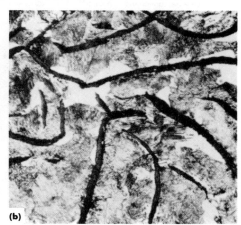

Fig. 11 Structure of bearing cap cast from hypoeutectic cast iron

(a) Section through cap. 3×. (b) Microstructure of cap before annealing. 200×. Both specimens etched with picral. Compare with Fig. 12 and 13. Source: Ref 1

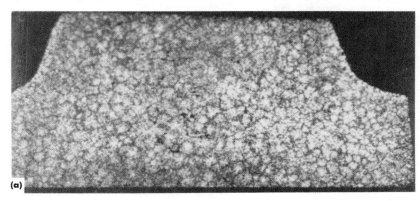
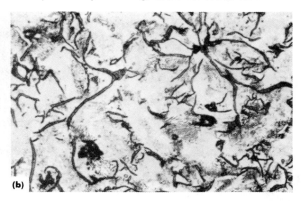

Fig. 12 Structure of bearing cap cast from hypereutectic gray iron

(a) Section through cap. 3×. (b) Core microstructure of (a). 200×. Compare with Fig. 11. Source: Ref 1

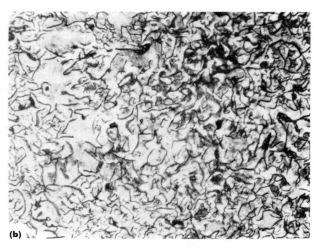

Fig. 13 Structures of hypoeutectic bearing caps after annealing

(a) After annealing 2 h at 625 °C (1160 °F) in air. 100×. (b) After annealing 2 h at 650 °C (1200 °F) in air. 200×. Compare with Fig. 11. Source: Ref 1

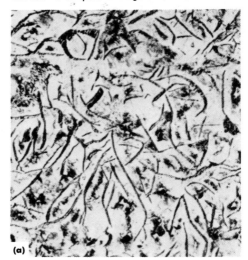

(a)

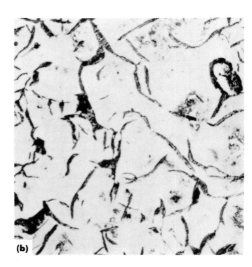

(b)

Fig. 14 Failed malleable iron rocker lever

(a) Illustration showing failure location (arrow). (b) Typical appearance of fracture face with dark overload fracture at top, light-gray fatigue fracture, and black casting defect (arrow). 1.5×. See also Fig. 15.

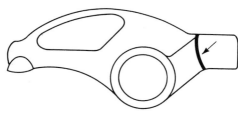

(a)

(b)

Fig. 15 SEM fractographs of the failure surface of the rocker lever shown in Fig. 14

(a) Dark overload fracture surface. (b) Light fatigue fracture surface. Both 250×

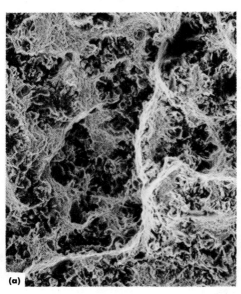

(a)

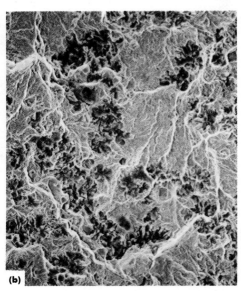

(b)

(1200 °F), it had dropped even further to 132 and 121 HB, respectively. This indicates that stress-relief annealing is not objectionable if the temperature does not exceed 600 °C (1110 °F).

Conclusions. Two factors exerted an unfavorable effect in addition to the comparatively low strength of the material. First, the operating stress was raised locally by the sharp-edged groove, and, second, the fracture resistance of the cast iron was lowered at this critical point by the existence of a ferritic bright border. The fine type D graphite with ferrite microstructure is a consequence of delayed solidification due to undercooling. The ferrite formation may have been further accelerated by the annealing. This ferritic rim structure showed low strength that had a particularly damaging effect on the bottom of the groove at high operating stress.

Casting stresses did not play a decisive role because of the simple shape of the pieces that were without substantial cross-sectional variations.

Recommendations. To avoid the occurrence of this failure mode, the recommendations consisted of elimination of the grooves, which would simultaneously strengthen the cross section and relieve the notch effect; removal of the ferritic bright border; slow cooling in the mold to avoid supercooling; inoculation of finely powdered ferrosilicon into the melt for the same purpose; and annealing at lower temperature or elimination of subsequent treatment in consideration of the uncomplicated shape of the castings.

Example 6: Fatigue Failure of Malleable Iron Diesel-Engine Rocker Levers Originating at Spiking Defects. Several diesel-engine rocker levers failed at low hours in over-speed, over-fuel, highly loaded developmental engine tests. The sand-cast rocker levers were malleable iron similar to ASTM

A602, grade M7002. Identical rocker levers had performed acceptably in normal engine tests.

Investigation. The rocker levers were failing through the radius of an adjusting screw arm (Fig. 14a). The typical fracture face exhibited two distinct modes of crack propagation (Fig. 14b). The upper portion, extending approximately 6.4 mm (0.25 in.) from the top surface, appeared very dark and rough, which would indicate overload at final fracture (Fig. 15a). The majority of the fracture face was very light gray and smooth in appearance. The fanlike fracture surface would be characteristic of the

Fig. 16 SEM fractographs of spiking defects at the fracture origin of the rocker lever shown in Fig. 14

(a) 50×. (b) Heavy oxidation (scaly appearance) and an interdendritic void are visible within the defect. 300×

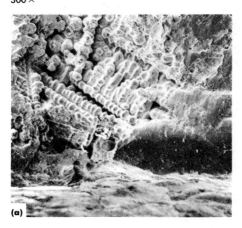

(a)

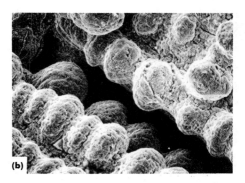

(b)

Fig. 17 Section through a spiking defect in rocker lever similar to that shown in Fig. 14

(a) 30×. (b) 60×. See also Fig. 18 and 19.

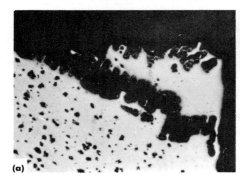

(a)

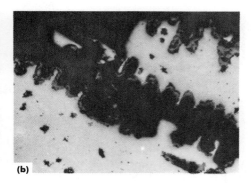

(b)

Fig. 18 Microstructure at and near the spiking defect shown in Fig. 17

(a) Heavy oxidation within an interdendritic void. As-polished. (b) Partial decarburization of the tempered martensitic matrix adjacent to a spiking defect. Etched with nital. Both 300×

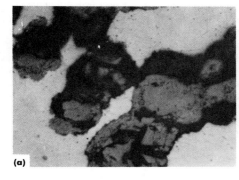

(a)

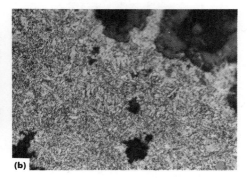

(b)

Fig. 19 SEM fractographs showing a spiking defect in a rocker lever without oxidation

(a) The surrounding fracture was induced by intentional mechanical overload. 25×. (b) Higher magnification view of (a). 150×

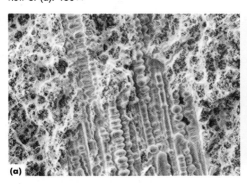

(a)

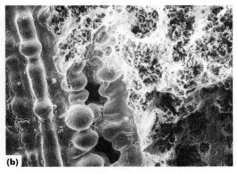

(b)

brittle, transgranular propagation of a fatigue fracture (Fig. 15b).

The fatigue fracture originated at a small black-appearing casting defect, approximately 1 mm (0.04 in.) maximum depth × 1 mm (0.04 in.) maximum width, adjacent to the radius on the underside of the adjusting screw arm, near the casting parting line (Fig. 16a and b). The black appearance was a result of heavy oxidation along interdendritic voids connected to the surface by dendrite ribs. A cross section through a similar defect in another rocker lever revealed interdendritic-appearing voids surrounded by heavy oxidation and partial decarburization of the adjacent tempered martensitic matrix structure (Fig. 17 and 18).

Efforts to sort defective rocker levers by means of magnetic-particle inspection and radiography proved ineffective. Because fatigue testing is costly and time-consuming, a method was devised to evaluate the effectiveness of foundry efforts to correct casting defects. The test method involved sectioning the rocker levers to reduce the cross section in the region of the defects and mechanically

overloading to initiate fracture in the general location of the defects. A large percentage of old and new castings were found to contain the dendritic-appearing defects. The defects located, both with and without oxidation (Fig. 19), ranged in size to a maximum of 2.5 mm (0.10 in.) deep × 1 mm (0.04 in.) wide on the fractured surfaces.

Conclusions. The rocker levers failed in fatigue, with casting defects acting as stress raisers to initiate failures in highly loaded engine tests. The casting defects, known as spiking, were attributable to a solidification condition by which large austenite dendrites develop that can create an impossible feeding condition and thus the interdendritic voids.

Fig. 20 Failed gray iron drier head

(a) Overall view. (b) Inside surface. Both approximately 0.05×. See also Fig. 21 to 24.

(a)

(b)

Limiting the size of austenite dendrites by increased nucleation should eliminate the spiking defects.

Corrective Measures. Because initial foundry efforts to eliminate the spiking condition were not totally effective, shot peening of the levers was introduced as an interim measure to reduce the possibility of failure. Redesign to increase the cross-sectional area of the levers has since eliminated failures through the adjusting screw arm, as verified by fatigue testing.

Failures Related to Surface Discontinuities

Unless a surface discontinuity is of critical size or is located at a portion of the casting surface that is subjected to major stress, it is unlikely to decrease the service life of the casting. This section will discuss several types of surface discontinuities, some of which usually contribute only to the poor appearance of a casting. Definitions of casting imperfections can be found in the "Glossary of Terms" at the front of this Volume.

The most damaging types of surface discontinuities are those that are linear in nature, those that generate notch-type stress raisers, and combinations of these. Among these are hot tears, cold cracks, and cold shuts (also surface porosity, inclusions, or shrinkage porosity extending to the surface). Such discontinuities as

misruns, cold shots, scabs, and buckles are among those that contribute only to poor casting appearance and that rarely initiate failure.

Cold shuts can form in all types of castings. In general, they are caused by pouring metal at too low a temperature or running it over too long a distance. Cold shuts occur when two streams of molten metal meet but do not completely merge, and they may serve as a site of stress concentration (notches) that can originate fatigue cracks.

Example 7: Gray Iron Paper-Drier Head Removed From Service. A paper-drier head manufactured from gray cast iron was removed from service as a result of non-destructive evaluation (NDE) detection of crack-like surface discontinuities. These imperfections were not removed by grinding. This component is subjected to internal steam pressure to provide heat energy for drying. Because the vessel is under pressure, a crack could present a safety hazard; therefore, it was removed from service for evaluation. The material specification of this cast iron is unknown.

Investigation. Visual-optical examination was used to identify areas that exhibited the undesired NDE magnetic-particle indications. Figure 20 shows overall outside and inside views of the paper-drier head. Figure 21 shows close-up views of the surface casting defects. Sections of this casting at the area designated as 12 o'clock were removed and subjected to magnetic-particle inspection. These sections have cracklike indications (Fig. 22). These sections were further studied by chemical analysis, mechanical testing, and metallography.

The chemical composition of a section near the 12 o'clock position was:

Element	Composition, %
Carbon	3.45(a)
Manganese	0.49
Sulfur	0.072
Phosphorus	0.713
Silicon	2.61
Nickel	<0.05
Chromium	<0.05
Molybdenum	<0.03
Copper	0.39
Tin	<0.02
Vanadium	0.01

(a) 0.12% combined carbon

The carbon equivalent was 4.56. The analysis indicates a satisfactory composition for gray cast iron. However, the phosphorus content is significantly higher than the normal 0.02 to 0.07% range common for gray iron. The high phosphorus content can cause the formation of steadite, a binary eutectic consisting of ferrite (containing some phosphorus in solution) and iron phosphide (Fe_3P). This undesirable, hard, brittle constituent degrades the strength of the material.

Tensile specimens prepared from this same region have a tensile strength of 101 MPa (14.7 ksi); the material had a hardness of 107 HB. These properties indicate a strength level lower

than a low-strength gray iron (ASTM A48, grade 20). This specification has a minimum expected tensile strength of 138 MPa (20 ksi). Another specification, ASTM A159, grade G1800, has an expected hardness value of 187 HB maximum. Both of these cast iron grades are expected to have tensile-strength values exceeding 124 MPa (18 ksi). The low strength measured indicates that the casting may have substandard properties. However, the specifications for the material were not known and thus cannot be verified.

Metallographic sections were made to include material at and beneath the crack. The graphite structure shown in Fig. 23(a) is essentially ASTM type A, size 4; ASTM type B graphite was also observed in the vicinity of the surface near the crack indications. The matrix microstructure was observed to be principally ferritic. The matrix also contained approximately 10% pearlite and 10 to 15% steadite (Fig. 23b and c). Ferritic grades of gray cast iron are not common, but they do exist, for example, ASTM A159 G1800. The matrix explains the low strength and hardness. The presence of steadite reflects the higher phosphorus content.

At the crack-indication site revealed by NDE, a demarcation line was observed progressing from the surface of the casting to the interior. Figure 24 shows this demarcation line for the condition observed in Fig. 22(a). The condition shown in Fig. 22(a) was found to be much deeper and also reflected this demarcation line separating one portion of the casting from the other.

This examination further revealed that the structure of the metal on one side of the demarcation line was different than that observed on the other side (Fig. 25). This condition generally results when a cold shut forms during the casting process. A cold shut reflects two different solidification fronts in the mold, which explains the reason for the structure variances observed on either side of the demarcation line. Although not a crack, the demarcation can serve as a site for crack initiation. In this instance, however, the demarcation line had not propagated beyond the point at which the structure was different. The demarcation line terminates at approximately the same area where the difference in structure terminates (Fig. 25). This condition further indicates that the demarcation had not propagated beyond the original condition that occurred during the casting operation.

Conclusions. The NDE indications were the consequence of a cold-shut condition in the casting. This conclusion is based on the presence of two significantly different microstructures representing two separate solidification fronts that met but did not merge. The cold shut serves as a stress-concentration site and is therefore a potential source of crack initiation. The combination of low material strength and a casting defect is a potential source of unexpected fracture during service

Fig. 21 Close-up views of surface casting defects on the drier head shown in Fig. 20

(a) At the 12 o'clock position. See also Fig. 22. (b) At the 9 o'clock position, with arrow indicating a surface defect. (c) At the 6 o'clock position. All approximately 0.2×

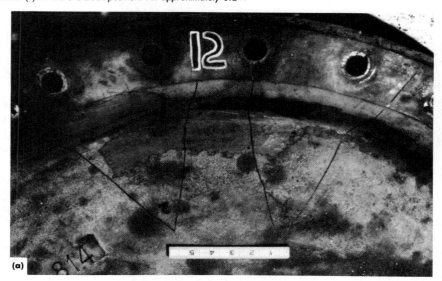

(a)

(b)

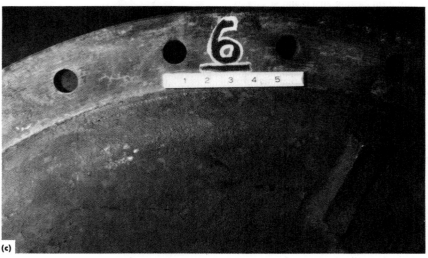

(c)

because the component is under pressure from steam.

Recommendations. Other dryer heads exhibiting similar discontinuities and/or material quality should be removed from service.

Misruns and Cold Shots. Misruns usually do not result in rejection of castings unless good surface appearance is of primary importance. If a misrun is not severe, it should offer little difficulty in service provided it is not in a highly stressed region of the casting. Cold shots, which are the result of turbulence, occur when metal splatters during pouring, adheres to the mold wall, and fails to fuse with the metal that engulfs it as the mold is filled. Thus, cold shots are globules embedded in the casting and normally produce no ill effect except for appearance. Under certain conditions, however, the interface between the cold shot and the casting can serve as a corrosion cell that leads to pitting and the initiation of fatigue cracks.

Hot tears and hot cracks are sometimes produced in external surfaces, such as in the fillet of a flange or a rib, when overall contraction of the casting is hindered by the mold. Hot tears and hot cracks should normally be discovered during inspection. Hot tears and hot cracks that escape detection and are present in castings put in service may propagate into larger cracks.

Mechanical (cold) cracks may be generated by rough treatment or by thermal shock in cooling, either at shakeout or during heat treatment. These cracks may be very difficult to detect without the aid of nondestructive testing. Magnetic-particle or liquid-penetrant inspection should reveal them clearly. Depending on the value and weldability of the casting, mechanical cracks should be removed and repair welded unless they are large enough to warrant scrapping of the casting. Removal by grinding, chipping, or other methods must be complete, or the remaining portion of the crack will be a stress raiser beneath the weld. Magnetic-particle or liquid-penetrant inspection after grinding is effective in determining if a crack has been completely removed.

Pits and Sand Holes. Dirt or sand inclusions in the surface of a casting are generally removed in cleaning. The resulting cavities affect appearance, but generally do not affect performance in service unless they are larger and are in a critical area.

Scabs, buckles, rattails, and pulldowns are surface discontinuities that result from local expansion of a sand mold. A scab is a thin, flat layer of metal formed behind a thin layer of sand and held in place by a thin ligament of metal. When the scab is removed (readily done by chipping), an indentation called a buckle is exposed. Rattails are minor buckles in the form of small irregular lines on the casting surface. A pulldown is a buckle in the cope portion of the casting. All these discontinuities affect casting appearance, but usually have no influence on service life.

Fig. 22 Higher magnification view of the casting defects shown at the 12 o'clock position in Fig. 21(a)

(c) Cracks at section 1 in Fig. 21(a) indicated by magnetic-particle inspection. 0.45×. (b) Magnetically inspected area at section 2 in Fig. 21(a). 1.1×

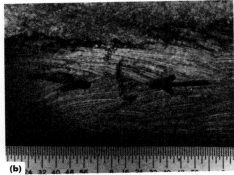

Fig. 23 Microstructure of the gray iron drier head shown in Fig. 20

(a) Typical graphite distribution in the casting—type A, size 4 graphite. As-polished. 54×. (b) The matrix microstructure consists of ferrite with approximately 10% pearlite and 15% steadite. Etched with nital. 54×. (c) Higher magnification view of (b). Etched with nital. 215×

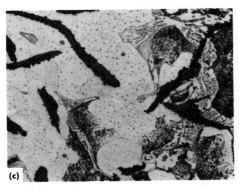

Failures Related to Internal Discontinuities

The general types of internal discontinuities are shrinkage porosity, gas porosity, steam bubbles, air bubbles, inclusions, and carbon flotation. Because the major stresses in castings are normally at or near the surface, internal discontinuities, unless severe, are less likely than surface discontinuities to be direct causes of fracture, although they may contribute to it. The extension of a fracture through a discontinuity is not sufficient evidence that the discontinuity caused failure. Overload will cause fracture along easy crack paths, which often include internal discontinuities. Overload failures are discussed in the section "Failures Related to Service Conditions" in this article.

Shrinkage porosity can result from use of an insufficient number of pouring gates, inadequate gate cross section, excessive pouring temperature, or inadequate feeding. Unless the casting is of simple design, the use of only one pouring gate will often result in shrinkage near the gate. Shrinkage may be caused by too much metal passing over one area in the mold, which results in localized overheating of the mold wall and a slowdown in the solidification rate of the metal in that area of the casting. This type of shrinkage can usually be eliminated by use of two or more gates. Use of only one gate may be satisfactory when a lower pouring temperature can be used. A lower pouring temperature results in an improvement in the surface appearance of the casting.

Shrinkage is often aggravated by high pouring temperatures. If the casting will not run with a moderate pouring temperature, it may be because of back pressure due to inadequate venting, use of low-permeability molding material, or inadequate gating. Correction of one or more of these factors will allow the pouring temperature to be lowered and will alleviate the shrinkage difficulty.

In some instances, surface shrinkage cavities may be caused by inadequate feeding at the junction of a thin section and a thick section where risering is insufficient to feed the shrinkage occurring in the thick section. The shape, location, and size of the risers govern the degree to which they will feed the casting. Small errors in designing the risers often cause great differences in the results.

Gating can often affect the efficiency of a riser. A riser must contain hot metal, and the metal must remain liquid until after the casting has been adequately fed and has solidified. Where possible, it is generally good practice to gate through a riser into the casting, causing the riser to the be the last portion of the mold to be filled. Risers that are located on the side of a casting opposite the gates are usually ineffective (or less effective), because they will contain the coldest metal in the mold and will be unable to feed the casting adequately.

Chills are an effective tool in controlling shrinkage. One function of a chill is to equalize the solidification rates of a thick section and an adjacent thin section, minimizing the likelihood of shrinkage. Chills also are useful in promoting directional solidification so that a riser can act more effectively.

Depending on casting shape and size, variations in section thickness, and types of junctions between component members, dendritic growth during solidification may generate dendrite arms that isolate local internal regions, preventing complete feeding by the risers. Generally, the microscopic voids that may form between the dendrite arms do not significantly reduce density or adversely affect mechanical properties.

Even a small amount of shrinkage porosity formed near the surface can sometimes lead to failure. Consider, for example, a Y-shaped casting with a small included angle. If a small fillet is used between the arms of the casting, there will be a thin, sharp edge of sand at the junction, which will be heated to a high temperature by the contacting molten metal. This wedge of hot sand will then act as a thermal insulator, retarding solidification of the metal around it. As a result, a region will be isolated from the risers by solid metal, and a shrinkage cavity will form just under the skin of the Y-junction. This cavity may remain undetected and become a site for crack initiation or corrosion penetration in service. Use of a larger-radius fillet could eliminate this problem.

Figure 26 illustrates shrinkage porosity at bolt-boss sites of an alloy ductile iron cylinder head. This defect was detected by a leak test after completion of machining operations. If this porosity goes undetected, contamination of oil in the engine can result.

Fig. 24 Demarcation line at the crack shown in Fig. 22(a)

Note the differences in graphite type and distribution on either side of the crack. Etched with nital. 33×. See also Fig. 25.

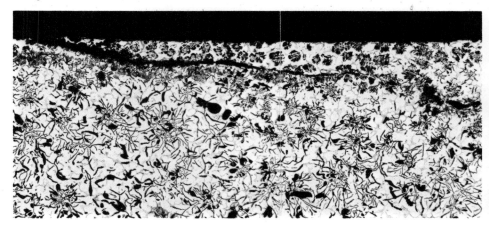

Fig. 25 Higher magnification view of Fig. 24 showing the two different solidification structures

Etched with nital. 230×

Example 8: Damaged Austenitic Cast Iron Impellers in a Rotary Pump Caused by Cracks Initiated at Zones of Localized Shrinkage Porosity (Ref 1).

Two damaged impellers manufactured from austenitic cast iron came from a rotary pump used for pumping brine mixed with drifting sand. On impeller 1, pieces were broken out of the back wall in four places at the junction of the blades. Impeller 2 had cracks at the same location, but did not yet break apart.

Investigation. Figure 27(a) shows the fractured edges following the shape of the blade. Numerous cavitation pits on the inner surface of the front wall are visible. Impeller 2, which did not exhibit such deep cavitation pits, was cracked in places along the line of the blades (Fig. 27b).

The microstructure shown in Fig. 28, illustrating lamellar graphite and carbides in an austenitic matrix, can be considered normal for the specified material (GGL NiCuCr 15 6 2). A specimen for metallographic examination was taken from one of the fracture edges. The material in this region was found to be in the final stages of disintegration (Fig. 29). A further section was taken through an as yet unbroken region, the essential portion of which is shown at low magnification in Fig. 30(a). Cavitation pits can be seen on the internal surface of the front wall (arrow at top left), and casting pores (arrow at bottom right) are visible in the region corresponding to Fig. 29. It should be mentioned that even when the specimens were cut carefully with a water-cooled cutting wheel, the stem in the brittle porous zone crumbled away, and only on the third attempt was it possible to obtain a suitable whole specimen for examination. A section through a cracked region of impeller 2 revealed a porous zone in the corresponding position. Figure 30(b) shows a section of a cavitation region. The absence of corrosion products confirms that the surface has been damaged by cavitation.

Conclusions. The cause of impeller damage from cracking is the shrinkage porosity at the junction between the back wall and blades. The shrinkage porosity is attributed to the casting process. Cavitation could have led to long-term damage even in a sound casting, although it did not contribute to the fracture.

Recommendations. Casting practices should be implemented to avoid porosity, and an alloy austenitic cast iron (GGL NiCuCr 15 6 3) that has a higher chromium content and is more resistant to cavitation should be used.

Example 9: Coolant Leakage Through a Cylinder-Head Exhaust Port Caused by Shrinkage Porosity.

An engine cylinder head failed after operating just 3.2 km (2 miles) because of coolant leakage through the exhaust port.

Investigation. The left-bank cylinder head was submitted for examination because of coolant leakage into the exhaust port. Visual examination of the exhaust ports revealed a casting defect on the No. 7 exhaust-port wall. A longitudinal section through the defect indicated shrinkage porosity. This defect was found to interconnect the water jacket and the exhaust gas flow chamber (Fig. 31).

No cracks were found by magnetic-particle inspection. The gray iron cylinder head had a hardness of 229 HB on the surface of the bottom deck. The microstructure consisted of type A size 4 flake graphite in a matrix of pearlite with small amounts of ferrite.

Conclusion. The cylinder-head failure resulted from the presence of a casting defect (shrinkage) on the No. 7 cylinder exhaust-port wall interconnecting the water jacket with the exhaust-gas flow chamber.

Example 10: Cracking in a Gray Iron Cylinder Head Caused by Microporosity.

A cracked cylinder head was removed from an engine after approximately 16 000 km (10 000 miles) of service. This cylinder head was sub-

Fig. 26 Shrinkage porosity at bolt-hole bosses in a ductile iron cylinder head

mitted for failure analysis to determine the cause of cracking.

Investigation. The cylinder head was magnetic particle inspected by both the head and coil methods. No indications were present on the cylinder head except those shown in Fig. 32 and 33. The head was cracked on the rocker-arm pan rail next to the No. 3 intake port. The crack extends into the water jacket on the rocker-arm side of the head. Visual inspection did not reveal any indication of material deformation or mishandling. The crack was opened for examination, and no indication of internal discontinuities could be seen during optical examination. Wall thickness at the thinnest section of failure measured 5.3 mm (0.210 in.). The specification minimum is 4.6 mm (0.180 in.).

The hardness at the bottom deck of the cylinder head was 217 HB. The hardness at the pan rail adjacent to the crack was 217 HB. This hardness is within the specified range of 192 to 241 HB. The chemical composition of this cylinder head as determined by spectrographic analysis was:

Fig. 27 Photographs of a damaged cast iron pump impeller

(a) Breaks in the back wall of impeller 1 and cavitation pits on the front wall. (b) Cracks following the shape of the blades on the back wall of impeller 2. Both 0.7×.
Source: Ref 1

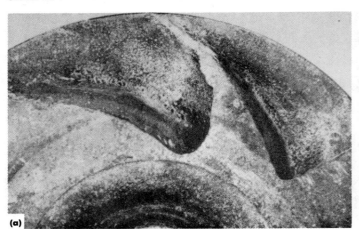

(a)

(b)

Element	Composition, %
Carbon	3.34
Manganese	0.55
Phosphorus	0.019
Sulfur	0.081
Silicon	2.22
Chromium	0.16
Nickel	0.038
Molybdenum	0.024
Copper	0.06
Antimony	0.03
Iron	rem

Fig. 28 Micrograph of impeller specimen

This structure, which consists of lamellar carbides in an austenitic matrix, is considered normal for this material. Etched with V2A reagent. 500×. Source: Ref 1

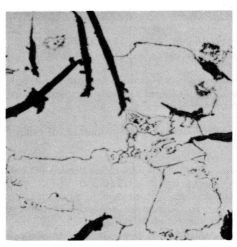

Fig. 29 Porous zone under a fracture edge of impeller 1

See also Fig. 30(a). 20×. Source: Ref 1

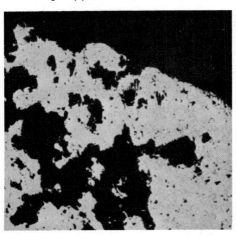

Three specimens for metallographic examination were taken from the crack: one from the rocker-arm pan rail, one from the area next to the plug in the water jacket, and one from the area next to the No. 3 intake port. All sections were made perpendicular to the fracture plane. Microporosity was found intersecting the crack in the sections taken from the water jacket next to the plug and the area next to the No. 3 intake port. Figures 33(b) and (c) show the microstructure of the specimen taken from near the plug. A wave of microporosity travels midway between the inner and outer surfaces of the casting. It varies in width from 3.8 mm (0.150 in.) at the fracture to 7.6 mm (0.30 in.) at the specimen section edge on the sample taken next to the plug. There was no indication of microporosity in the specimen taken from the rocker-arm rail.

The microstructures consist of predominantly types A and B graphite, flake size 4 to 6, in a matrix of medium-to-fine lamellar pearlite. The specimen taken near the No. 3 intake port has a higher-than-usual amount of free ferrite, approximately 5 to 10%.

Conclusion. The cracking is attributed to microporosity in the casting material. The high operational stresses of the engine acting on a material discontinuity, such as microporosity in a web member, can produce cracking.

Gas Porosity. Formation of cavities resulting from the presence of gas is influenced by many factors, including melting practice, pouring practice, and type of mold to be used. Three common causes of gas holes during solidification of castings are exceeding the solubility necessary to keep the gas in solution, a reaction-type pinhole formed by segregation during solidification, and a reaction product between the metal and mold material. In the cast iron family, the susceptibility to form pinholes is related to the rate of solidification. Ductile iron having the slowest solidification rate is most susceptible, while white iron having the most rapid rate is least susceptible. By analyzing the pinhole shape and character-

istics, it can be learned which gases are present. For example, a hydrogen-reaction pinhole will have a graphite-free layer and a graphite film. A carbon monoxide (CO) reaction type will have no graphite film and no graphite-free layer. It is necessary to establish the source from appearance to eliminate the problem (Ref 3-5).

Pinholes resulting from insufficiently dried sand molds may lower the endurance limit of steel castings, or they may cause only poor surface appearance; however, if the number of pinholes is high, the casting may corrode in sevice. Internal gas cavities may reduce the effective cross section appreciably and may therefore impair load-carrying ability. In general, however, internal gas cavities have smooth surfaces and do not act as stress concentrators. All but the very smallest cavities can

Fig. 30 Impeller specimen showing cavitation damage and porous zones

(a) Cavitation pits (top-left arrow) and porous zones (bottom-right arrow) in an unetched transverse section of impeller 1. 2.5×. (b) Microsection from a cavitation zone. Etched with V2A reagent. 100×. Source: Ref 1

(a)

(b)

Fig. 33 Additional views of the failed cylinder head shown in Fig. 32

(a) Fracture surface observed when the crack shown in Fig. 32 was opened. 0.5×. (b) Specimen taken from area adjacent to the vent plug showing microporosity. Etched with 2% nital. 100×. (c) Same as (b), but at a higher magnification. 400×

(a)

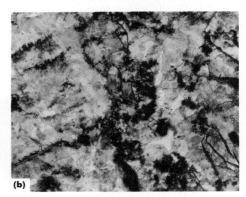

(b)

(c)

Fig. 31 Section from a failed cylinder-head exhaust port

The shrinkage porosity allowed engine coolant to leak into the exhaust port. Not polished, not etched. 0.9×

Fig. 32 Crack in a gray iron cylinder head

(a) Crack on side of head next to intake manifold No. 3. (b) Another view of the same crack, which ends at the water jacket vent plug. Both 0.5×

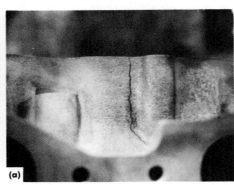

(a)

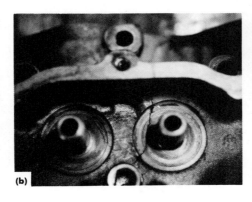

(b)

be readily revealed by radiography, and quality can be assessed in terms of the intended service. A measurement of casting density will serve a similar purpose.

Steam bubbles result from exposure of any liquid metal to moisture. Moisture in the air, in gases of combustion, in furnace and ladle linings, and in mold materials provides sources of steam bubbles. Whatever the source, the entrained water is released as steam bubbles upon freezing of the molten metal, resulting in porosity in the castings that varies with the manner in which the water was entrained. If the water content in a molding sand is too high or if water exists as condensate on the mold surface, porosity will form at the surface of the casting. If a slight amount of moisture is entrained in the melting furnace, it may be dissipated during cooling in the ladle, and the castings may be free of porosity. If a ladle lining is inadequately dried, the entire heat can be ruined by gross gas evolution during freezing in the molds, which causes all the risers to puff and overrun, creating large cavities throughout the casting.

Air bubbles as a cause of casting porosity are less detrimental than steam bubbles. Air bubbles usually arise when the conditions of pouring and the nature of the ingate passages combine to produce a turbulent flow that sucks air into the mold cavity. The amount of air in the cavity is sometimes low enough for all of it to be carried to the topmost riser, causing no porosity in the casting proper. In most instances, smoother pouring and a more streamlined ingate system will minimize air bubbles.

Inclusions of refractory, slag, sand, oxides, and other compounds in the casting metal can affect the service life of the casting. Some of these, such as refractory and sand, are inadvertently introduced and can be minimized by proper maintenance of linings and by good housekeeping. The presence of slag can be controlled by careful pouring and the use of more effective skimming techniques. Oxides of the casting metal are formed largely during melting and can be avoided by thorough deoxidation. Some elements—for example, chromium in some steels—will oxidize preferentially on the surface of the pouring stream. Skillful skimming is the only means of preventing this oxidized metal from entering the mold. Other compounds, such a phosphides and sulfides, are detrimental to mechanical properties unless kept to a minimum. Globular manganese sulfides are beneficial as replacements of iron sulfides, which can form deleterious films at grain-boundary junctions.

Although it is desirable to minimize inclusions in cast structures, for most applications there is an added expense in preparing castings that are essentially inclusion free. Methods of reducing inclusions to very low levels include the use of special ceramic filters.

Inclusions played a major role in thread defects that occurred in ductile iron tire-mold castings. In this case, the defect was found to be caused by stringers of magnesium silicate inclusions, which are known to be hard and detrimental to tool life. This resulted in poor machinability and improper finish in a localized area of the tire-mold castings. Figure 34 shows an SEM secondary image of a representative area of the surface exhibiting an uneven finish and an oxidized appearance.

Figure 35 illustrates optical photomicrographs of subsurface areas immediately beneath the defective casting surface. These micrographs show the presence of inclusion stringers adjacent to the surface (Fig. 35a) and the deterioration of the graphite nodule shape in the metal below (Fig. 35b and c). The long stringers of inclusions are very similar to film-type magnesium silicate inclusions reported in the literature (Ref 6).

Carbon flotation, sometimes called kish graphite, is another type of internal discontinuity. This form of segregation appears at and near the upper surfaces of a casting, always in the cope portion. This is caused by the carbon equivalent being too high. Carbon flotation is free graphite separated from molten iron during solidification. This gross segregation occurs in spheroidal graphite metal (ductile iron). Carbon flotation is remedied by increasing the pouring temperature, by lowering the carbon equivalent value by closer chemistry control, and by increasing the cooling rate (Ref 7, 8).

Example 11: Premature Engine Failure Due to Improper Surface Finish Caused by Carbon Flotation. Premature failures were occurring in a certain engine within the first 1600 km (1000 miles) of operation. The failures involved the nodular cast iron crankshaft and its main-bearing inserts. These failures were indicated by internal noise, operation at low pressure, and total seizing. Concurrent with the incidence of engine field failures was a manufacturing problem: the inability to maintain a similar microfinish on the cope and drag sides of a cast main-bearing journal. The failure analysis and investigation of the manufacturing problem showed that the cause was related to this manufacturing problem.

Investigation. During the in-plant investigation, it was found that the surface finish of the cope side was rougher than that of the drag side. The surface roughnesses of the 50 sample crankshaft main bearings checked ranged from 0.4 to 0.8 μm (15 to 30 μin.) on the cope side and from 0.25 to 0.5 μm (10 to 20 μin.) on the drag side. The requirement was 0.5 μm (20 μin.) maximum.

Surface photomicrographs were taken for comparison purposes at the cope and drag sides

Fig. 34 SEM secondary electron image of a surface defect in a tire-mold casting

1000×. See also Fig. 35.

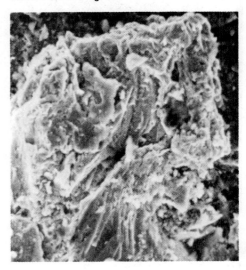

of the main-bearing journals to identify the cause of the nonuniform surface finish. Figure 36 illustrates the finishes on the cope and drag sides. The cope side had areas where the surface was broken open, and it showed some raised areas. The drag-side indicated fewer raised areas and almost none of the broken-open surface areas.

Longitudinal metallographic sections were made of the cope and drag portions of the main-bearing journal. Figure 37 shows that the cope portion had larger and more numerous graphite nodules and ferrite envelopes than the drag portion.

Fig. 35 Internal discontinuities in a tire-mold casting

(a) Stringer-type inclusions adjacent to the surface. As-polished. (b) Structure below the surface. Note the change in graphite shape. Unetched. (c) Ferrite matrix with degenerate vermicular graphite nodules. Etched with 2% nital. All 135×

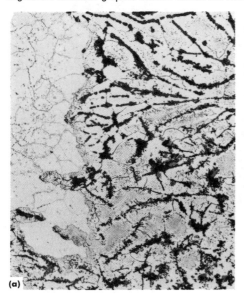

(a)

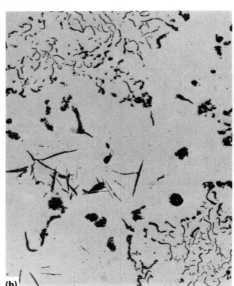

(b)

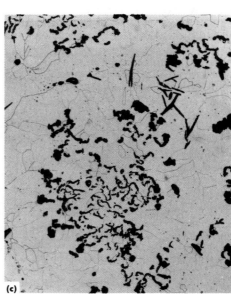

(c)

Fig. 36 Micrographs of a failed crankshaft main-bearing journal

(a) Cope side showing the surface broken open with raised areas. (b) Drag side, with fewer raised areas and almost no broken open surface. Both 50×

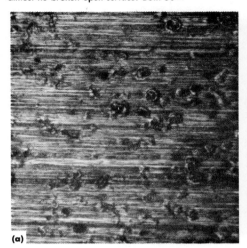 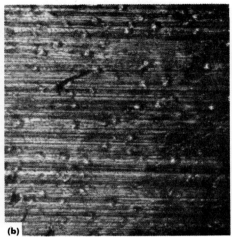

(a) (b)

Fig. 38 Micrograph from the cope side of the main-bearing journal

Ferrite caps, which partially cover the graphite nodules and cause the broken-open surface condition, are shown. The burrs rise above the surface from 4 to 13 μm. Etched with nital. Both at 220×

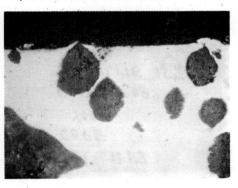

Fig. 37 Longitudinal sections from the main-bearing journal

(a) From the cope side. (b) From the drag side. Note the more numerous graphite nodules in (a). Both etched with nital. 60×

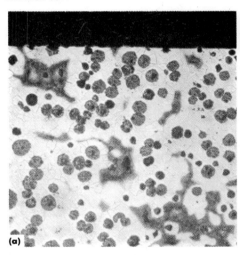 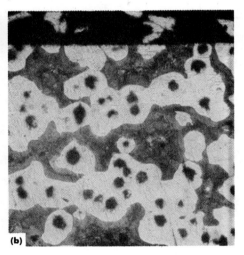

(a) (b)

ally, some nonmetallurgical practices in journal-finishing techniques were modified to ensure optimum surface finish.

Effect of Microstructure

Aspects of microstructure that may be detrimental to the performance of iron castings in service include embrittling intergranular networks (carbides); graphite of unfavorable shape, size, or distribution (in gray iron); surface-structure gradients, such as accidental carburization or decarburization; and generally unsatisfactory microstructures resulting from improper materials selection, improper compositions for section size (in gray and ductile irons), or incorrect heat treatment.

Intergranular Networks. Grain-boundary networks of either proeutectoid ferrite (in hypoeutectic alloys) or iron carbide (in hypereutectic alloys) are sometimes encountered in as-cast iron castings that have been cooled through the austenite temperature range at a slow rate. Such netwoks impart low ductility and low toughness to castings and constitute paths of weakness because of the brittleness of the carbides. The presence of such networks is indicated by low impact energy in a Charpy V-notch test and can be detected by metallographic examination. A normalizing treatment can eliminate networks, provided austenitizing has been done at the proper temperature and time. The casting can then be tempered to the desired hardness or, if higher strength is desired, can be quenched from just above the critical temperature and tempered appropriately.

Graphite. The size, shape, orientation, and distribution of graphite in castings of gray iron, ductile iron, and malleable iron determine the mechanical properties of the castings. For any given gray iron composition, the rate of cooling from the solidification temperature to below about 650 °C (1200 °F) determines the ratio of combined carbon to graphitic carbon contents; this ratio controls the hardness and strength of

At the higher magnification shown in Fig. 38, the areas of the cope side that had the appearance of a broken-open surface in Fig. 36(a) were confirmed. The broken-open areas are called ferrite caps. Due to the pressure and heat of polishing during manufacture, the larger graphite nodules break open, and the ferrite envelopes (caps) become actual burrs. These burrs rise 4 to 13 μm above the normal surface (Fig. 38).

The height of these protrusions above the normal surfaces explains the variance in surface finish of cope versus drag areas. This evidence confirms the cause of crankshaft and crankshaft-bearing failures. The ferrite caps (burrs) can shear the lubricating oil film. Also, the burrs can actually penetrate the bearing-insert material. In either case, the net result is failure of the crankshaft and crankshaft-bearing insert interface due to heat generation or physical damage.

Conclusions. The root cause of the failure is carbon flotation. Chemical analysis of the crankshafts involved in the failures showed a higher-than-normal carbon content and/or carbon equivalent. Therefore, during the casting process, the carbon or graphite would tend to rise in the mold to the cope side, resulting in larger and more numerous graphite nodules in that area and in the presence of more ferrite and less pearlite on the cope side. This layer could be 2.5 mm (0.1 in.) below the surface. The larger nodules break open, causing the ferrite caps or burrs. They then become the mechanism of failure by breaking down the oil film and eroding the bearing material. A by-product is heat, which assists the failure.

Recommendations. Steps were taken to establish closer control of chemical composition and foundry casting practices to alleviate the carbon-flotation form of segregation. Addition-

Fig. 39 Piston for a gun recoil mechanism, sand cast from ductile iron conforming to MIL-I-11466, grade D7003, that fractured in fatigue because of vermicularity of graphite

(a) and (b) Two different views of the piston showing fractures; A and B indicate orifices (see text). Approximately 0.35×. (c) Flat-plan view showing composite pattern of fractures in several pistons. (d) Fracture surface showing mottled structure caused by vermicular graphite; arrow points to a fatigue zone. Approximately 0.25×. (e) to (h) Micrographs showing graphite with nodularity of rank 2, at 50× (e); rank 8, at approximately 60× (f); rank 13, at 50× (g); rank 16, at approximately 65× (h). Specimens in (e) and (g) were not etched; specimens in (f) and (h) were etched with 2% nital.

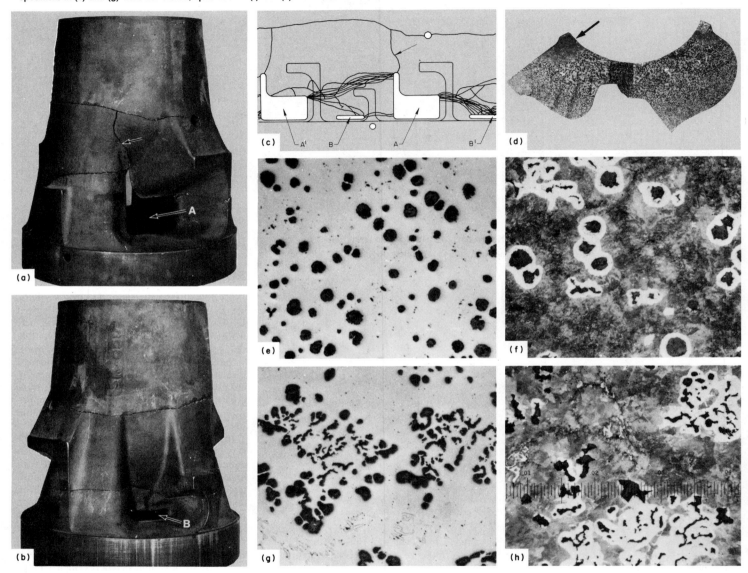

the iron. Therefore, the effect of section size is considerably greater for gray iron castings than it is for castings made of more homogeneous ferrous metals for which the rate of cooling does not affect the form of the carbon content on a macroscopic scale.

The typical microstructure of gray iron is a matrix of pearlite with graphite flakes dispersed throughout. Too-rapid cooling (relative to carbon and silicon contents) produces mottled iron, which consists of a pearlite matrix with both primary cementite (iron carbide) and graphite. Very slow cooling of irons that contain larger percentages of silicon and carbon is likely to produce considerable ferrite as well as

pearlite throughout the matrix, together with coarse graphite flakes. Under some conditions, very fine graphite (type D) may be formed near casting surfaces or in thin sections. Frequently, such graphite is surrounded by a ferrite matrix; therefore, soft areas result.

Type D graphite is usually associated with a ferrite matrix. This structure occurs due to a lack of nucleation sites, a large undercooling, and usually at thin casting sections. This formation is also called granular graphite or coral-like graphite. Its three-dimensional configuration resembles the shape of sea coral (Ref 9-11).

Example 12: Fracture of Ductile Iron Pistons for a Gun Recoil Mechanism as

Affected by Type of Graphite. The gun mount used for two types of self-propelled artillery (175-mm guns and 205-mm howitzers) involved a recoil cylinder and a sand-cast ductile iron piston containing orifices through which oil was forced in order to absorb the recoil energy of the gun. A rod attached to the gun tube engaged a thread inside the small end of the piston. The piston was stressed in tension by oil pressure on the flange at the large end, opposing the direction of motion of the rod. Several pistons that had cracked or fractured in service were submitted for analysis.

Investigation. Figures 39(a) and (b) show two views of a piston, illustrating the orifices specially designed to control oil flow and the pattern of cracks that developed. In Fig. 39(c), which shows a composite of the fractures found in several pistons, the piston surface has been developed as a flat-plan view. Orifices A and B correspond to those shown in Fig. 39(a) and (b), respectively; orifices A′ and B′ are 180° from orifices A and B. The vertical fracture (arrow) shown in Fig. 39(a) is unusual in origin and direction of propagation: most of the fractures originated at the upper right-hand corners of the large orifices (A and A′, Fig. 39c) and propagated in an approximately horizontal direction. Some of these nearly horizontal cracks intersected the small orifices (B and B′, Fig. 39c); others propagated around the piston to the opposite large orifice, joining it at the left side of the narrow opening at top left.

Figure 39(d) shows the surface of one of the horizontal fractures, which has a blotchy, mottled appearance. The pattern was created by the graphite present; there was relatively little of the massive, hypereutectic carbide that is by definition present in iron with a mottled structure. The arrow in Fig. 39(d) indicates a location that exhibits the characteristics of a fatigue zone.

The uppermost fracture shown in Fig. 39(a) is met by the more or less vertical break at virtually a right angle. This signifies that the upper fracture took place earlier (additional information is provided in the article "General Practice in Failure Analysis" in this Volume).

Because some of the pistons exhibited regions with an appearance suggesting fatigue (Fig. 39d), SEM fractographs were prepared to attempt to locate fatigue striations and to estimate the number of cycles to failure. These fractographs showed configurations similar in appearance to fatigue striations, but it was concluded that they were pearlite lamellae because the spacings agreed with pearlite dimensions observed in optical micrographs. Scanning electron microscope views of the area of final fracture revealed dimples in some regions and features resembling cleavage facets in others. However, what appeared to be facets may have been interfaces of pearlite lamellae oriented nearly parallel to the direction of crack propagation.

The castings were made according to the specifications of MIL-I-11466, grade D7003, which requires minimum properties of 690 MPa (100 ksi) tensile strength, 483 MPa (70 ksi) yield strength, and 3% elongation in 50 mm (2 in.); these properties are usually acquired by heat treatment (normalizing for these castings). The specification stipulated that the graphite be substantially nodular and required a metallographic test for each lot of castings. There was no requirement concerning the matrix. Of 17 pistons that were sectioned for testing (2 or more test specimens per piston), 11 averaged 814 MPa (118 ksi) tensile strength but the remaining 6 averaged only 535 MPa (77.6 ksi); 13 met the yield-strength requirement, averaging 552 MPa (80 ksi), while 4 did not, averag-

ing 399 MPa (58 ksi); in elongation, 9 pistons were satisfactory, averaging 4.7%, but 8 averaged only 2.1%.

Examination of the microstructures of these 17 pistons, along with those of additional pistons, revealed that 77% of those that failed did not contain the required nodular graphite. The range in nodularity was sufficient for establishment of a microstructure-rating system in which quality of graphite was ranked from 1 (nodular) to 16 (vermicular with essentially no nodules). Structures containing graphite ranked 2, 8, 13, and 16 are shown in Fig. 39(e) through (h), respectively. Figure 39(h), which shows the structure of the piston with the mottled surface in Fig. 39(d), exhibits massive carbides, but not enough to form a mottled structure in the usual sense of the term. For this application, graphite ranked from 1 to 7 was considered acceptable, graphite ranked from 8 to 12 borderline, and graphite ranked from 13 to 16 unacceptable.

The degree of nodularity of the graphite was correlated with the results obtained in ultrasonic testing in order to appraise the quality of pistons that had already been installed. It was found that vermicular graphite damped the back-reflection signal to a significantly greater degree than nodular graphite did. Comparison of the results obtained in the ultrasonic tests and in the metallographic examinations yielded a correlation coefficient of 0.84—an indication of a high degree of correlation.

Smooth-bar rotating-beam tests (R.R. Moore tests, as described in the article "Fatigue Crack Initiation" in Volume 8 of the 9th Edition of *Metals Handbook*) were performed on specimens prepared from pistons specifically selected for a comparison of fatigue-test behavior and graphite structure. For specimens with rank 1 graphite tested at 207 MPa (30 ksi), the number of cycles to failure was about 10 times that for specimens with rank 12 graphite and 18 times that for specimens with rank 16 graphite. When tests were performed at 310 MPa (45 ksi), the disparity was not as great, but the order of decreasing fatigue life was the same as at 207 MPa (30 ksi). These findings indicated that low fatigue strength due to vermicular graphite had caused the service fractures.

These tests did not simulate service, because they involved reversed bending; gun recoil loadings were primarily in tension. Fatigue tests in simple tensile loading were conducted on new, complete pistons containing nodular graphite. Four tests at a stress corresponding to the gun recoil load produced fatigue fractures at 20 000 to 35 000 cycles. All these fractures, unlike most of those observed in service, occurred at the base of the threads in the top of the piston, which is the region of the upper crack shown in Fig. 39(a) and (b). The section through the threads had been calculated during design as the location of maximum stress under simple tension. Because nearly all service fatigue cracks did not occur at the threads but initiated from the oil orifices, cracking may be attributed to slight misalignment that caused

bending as well as tension and to the force of the oil on the surface immediately above and below the large orifices, increasing the stresses at the corners of these orifices. These tests showed that the life of the pistons was about 20 000 recoil loadings instead of the indefinite service life that had been anticipated.

Conclusions. Fracture of the pistons occurred under high strain rate axial tension modified by superimposed bending loads from oil pressure and/or normal alignment tolerance. Most of the service fractures occurred in pistons containing vermicular graphite instead of the specified nodular graphite. Brittleness caused by massive carbides may have shortened the service lives of a few pistons.

Corrective Measures. Ultrasonic testing was adopted for inspection of pistons already in the field to identify and reject those containing vermicular graphite. Metallographic control standards were applied to a production of new pistons to ensure structures containing substantially nodular graphite and no massive carbides.

Excess Ferrite. A generally acceptable microstructure occasionally may not perform satisfactorily because of some particular aspect. Figure 40 shows a micrograph of a section through the tip of a gray iron cam lobe that sustained rapid lubricated wear because of an excessive amount of ferrite. The microstructure contained a dendritic network of pearlite and islands of free ferrite in a field consisting of type D graphite in a ferrite matrix. The large amount of free ferrite present was detrimental to wear resistance, although the cam surface was well lubricated. A lower silicon content and a slightly higher chromium content would reduce the amount of free ferrite and suppress the formation of type D graphite in the gray iron.

Carburization and Decarburization. Meticulous control of the furnace atmosphere is needed to prevent unintentional carburization and decarburization. Such control is not diffi-

Fig. 40 Section through the worn tip of a gray iron cam lobe
The white islands are free ferrite in a field of type D graphite. The matrix is ferritic. Etched with 2% nital. 235×

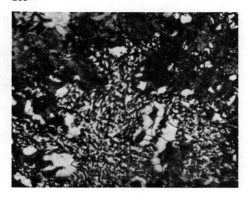

Fig. 41 Grinding cracks at 0.6× in a section of a briquet roll

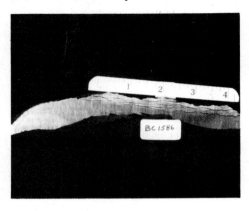

cult with properly instrumented heat-treating equipment.

In shell-mold castings of gray iron, the high thermal conductivity of the mold produces an initial chill of the iron surface. Subsequent annealing of the surface by the heat of the casting results in a skin that is almost 100% ferrite; such a skin has poor wear properties and should be removed by machining for applications requiring wear resistance.

Inverse chill is a condition in gray iron, ductile iron, or malleable iron in which the interior is chilled or white, while the outer sections are mottled or gray. This condition is observable on fracture casting specimens and generally occurs in lighter casting sections. Some of the causes of inverse chill are (1) sulfur content not balanced by manganese, (2) too low of a degree of nucleation for the section arising from high melting and pouring temperatures, and (3) the presence of tellurium, chromium, and other severe carbide stabilizers (Ref 7, 8).

Effects of Foundry Practice and Processing

This section will discuss processing problems that can lead to casting failures. Some common sources of casting failures are foundry practice, machining effects, residual stresses, grinding cracks, improper heat treatment, improper chemical composition, and careless assembly.

Foundry Practice. The procedures used in the foundry can influence the creation of residual stresses in castings. Very high pouring temperatures will retard cooling rates of thick sections, but may not significantly affect the cooling of thin sections. Shakeout at too high a temperature will accelerate cooling of thin sections. Stress relief, if performed properly, can alleviate residual stresses.

Machining of castings can introduce stress raisers; tool marks, notches, and sharp edges such as those around drill holes are among the most common stress raisers produced during machining. The effects of machining, however, also apply to wrought products and are not peculiar to castings.

One problem unique to cast iron is pull-out of graphite particles, which impart a pock-marked finish. Pull-out is a function of the graphite flake size and force applied on the surface by the tool. The presence of carbides in microstructure can also have a deleterious effect on surface finish.

Grinding Cracks. Some cast iron components have smooth hard surfaces with close tolerances for resistance to wear in service. These parts include shafts, journals, rolls, piston pins, and bearing surfaces. After machining, these cast parts are hardened by heat treatment, then ground to close tolerances. Cracks can result from grinding burns caused by localized frictional heat during grinding.

A high-chromium white iron cast iron briquet roll suffering from thermally induced grinding cracks is shown in Fig. 41. The microstructure was comprised of primary eutectic (Cr_7C_3) carbides in a matrix of martensite, retained austenite, and secondary carbides. This 2.8% C, 27% Cr alloy is extremely susceptible to cracking from thermal shock to the brittle eutectic-carbide network in the microstructure. This problem was alleviated by ensuring proper grinding practice.

Stress Concentration and Residual Stresses. The presence of significant residual internal stresses in a casting may be detrimental. Residual stresses generally result from local plastic deformation imposed by thermal gradients created during processing. These gradients may be caused by local cooling when the casting is very hot (in the mold or during shakeout, or in quenching from heat treatment) or by local heating, such as in repair or assembly welding. The effect is plastic stretching (or compression) of the weakest (hottest) portion of the casting. By the time the temperature has equalized, the stretched or compressed region is too long or too short to match the size of the other portion of the casting. The result is a sizable residual internal stress that may be compressive or tensile. The behavior in service then depends on whether these residual internal stresses are added to or subtracted from the applied service loads.

Example 13: Cracked Gray Iron Crankcases Caused by Casting Stresses (Ref 1). The front wall of a gray cast iron crankcase cracked at the transition from the comparatively minor wall thickness to the thick bosses for the drilling of the bolt holes. This transition was abruptly formed (Fig. 42). The dark coloration of a part of the fracture plane indicated that the crack originated in the thin-wall part. Four failed crankcases were examined to determine the cause of failure.

Investigation. Figure 42(b) shows the abrupt change in section thickness. These adjoining

Fig. 42 Failed gray iron crankcase
(a) Fragments of crankcase. (b) Fracture surface of crankcase. 1×. Source: Ref 1

(a)

(b)

Fig. 43 Microstructure of failed crankcase shown in Fig. 42

(a) Normal flake graphite from a thick-wall section. (b) Type B rosette graphite from a thin-wall section. Both etched with picral. 200×. Source: Ref 1

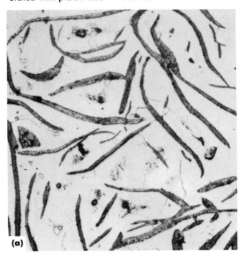 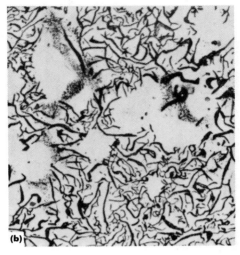

Fig. 44 Microstructure of crankcase 2

(a) Section through the thick-wall part. (b) Section through the thin-wall part with crack. Both etched with picral. 100×

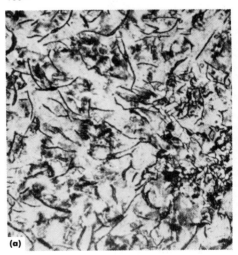 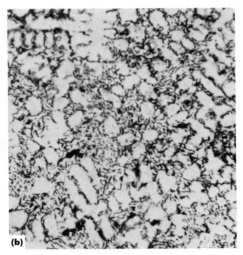

Fig. 45 Crankcase 3 with cracks

Source: Ref 1

thick- and thin-wall sections solidify and shrink at different rates. Metallographic examination indicates a differential structure at the thick- and thin-wall sections (Fig. 43). The thick-wall section of the first sample (Fig. 43a) has a more favorable graphite size and shape, while the thin-wall section (Fig. 43b) has a rosette-type graphite. Crankcase 2 also had the differential graphite condition at thick- and thin-wall sections, which can be observed by comparing Fig. 44(a) (thick-wall) with Fig. 44(b) (thin-wall).

Crankcase 3 (Fig. 45) had fractures at the frontal end wall (arrow a) and at the thinnest point between two bore holes (arrow b). Figures 46(a) and (b) show in larger scale the sectors of the front wall with the fracture, which is stained with alumina for easier recognition. In the cross-sectional cut of Fig. 46(c), it can be seen more easily that the fracture occurred at the point at which the crankcase is only 5 mm (0.2 in.) thick; the crankcase becomes thicker toward both sides. The structure of this crankcase as well as that of the thin-wall part of crankcase 2 consisted of pearlitic dendrites and a connecting enveloping network of ferrite and fine type D graphite (Fig. 47a). The same structure could be found at the other point of fracture (arrow a, Fig. 45), where the wall between the borings in the raw casting was only 6 mm (0.24 in.) thick and was later worked down to 2.3 mm (0.09 in.) (Fig. 47b). However, the thick-wall places had a pearlitic structure comparatively deficient in ferrite (Fig. 47c).

Crankcase 4 had cracked in the borehole of the frontal face. One crack was so narrow that it was discernible only by magnetic-particle

inspection. The structure of this casting was permeated with clusters of types B and D graphite and ferrite near the surface as well as in the interior (Fig. 48).

Conclusions. In the first three cases, casting stresses caused by unfavorable construction and rapid cooling were responsible for the crack formation. In the fourth case, the cause of the fracture is considered to be casting stresses at the bore hole. Crack propagation led to the conclusions that shrinkage stresses were the cause of the cracks, because the adjoining thick- and thin-wall cast parts solidify and shrink at different rates. The faster a piece cools, the more damage is caused by differential shrinkage, that is, the higher the stresses at the cross-section transition. The amount of stress built up at the transition is also proportional to the abruptness of the change in section thickness. These stresses may already lead to crack formation during cooling, or they may remain in the piece as residual stresses, which are later superimposed on operating stresses and thus favor the fracture at this particular point.

Effects of Chemical Composition. Improper melting or foundry practice can cause variations in composition that adversely affect the microstructure and resulting mechanical properties. Improper microstructure and deviations from specified hardness or mechanical-property requirements may result in inadequate load-bearing capacity, brittle behavior, or poor wear resistance—deficiencies that may contribute to premature failure.

Example 14: Fracturing of Gray Iron Door-Closer Cylinder Castings Caused by Lack of Foundry Control Over Chemistry. Door-closer cylinder castings manufactured of class 30 gray iron were breaking during machining. The manufacturing source reported that a random sampling of castings from this lot had hardnesses from 180 to 210 HB. Based on the color of the components, heat treatment of these castings was suspected. Metallurgical examination on two

Fig. 46 Additional views of crankcase 3

(a) External view. 0.25×. (b) Internal view. 0.25×. (c) Section through crack in Fig. 45. Etched with picral. 1×. Source: Ref 1

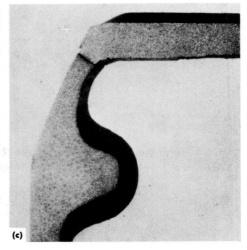

representative castings revealed the cause of failure.

Investigation. The first casting (arbitrarily identified as fractured) had separated into three pieces (Fig. 49a). The outer-edge portion of the fractured surfaces exhibited a dark gray color, while other regions of the fractured surfaces had a light gray color. No defects or discontinuities were evident on this fractured casting.

The second casting (identified as cracked) was cracked along the parting line and through a region exhibiting the presence of a void. Further examination revealed that the bore was not centered, indicating a core-shift condition during casting. The void had characteristics of a gas defect. Figure 49(b) illustrates the conditions observed in this casting.

The chemical composition of the so-called fractured casting was:

Element	Composition, %
Carbon	3.21(a)
Manganese	0.62
Phosphorus	0.022
Sulfur	0.064
Silicon	2.05
Nickel	0.05
Chromium	0.09
Molybdenum	0.03
Copper	0.55
Tin	<0.02
Vanadium	0.05
Iron	rem

(a) 0.39% combined carbon

This casting had a carbon equivalent of 3.90. The composition of the cracked casting was:

Element	Composition, %
Carbon	3.25(a)
Manganese	0.61
Phosphorus	0.024
Sulfur	0.060
Silicon	2.21
Nickel	<0.05
Chromium	0.09
Molybdenum	0.04
Copper	0.55
Tin	<0.02
Vanadium	0.05
Iron	rem

(a) 0.43% combined carbon

The carbon equivalent of the cracked casting was 3.99. These data indicate that both castings have a composition closer to class 40 than to class 30 gray iron. Although class 40 gray iron is normally stronger than class 30, the low combined carbon values (<0.8%) indicate a probable high-ferrite matrix. Because ferrite tends to lower the strength of the material, heat treatment may have been attempted to achieve the desired hardness. Hardness values were: 193 HB for the fractured casting and 184 HB for the cracked part. Both were within the expected hardness range for class 30 gray iron.

Metallographic examination of the fractured casting revealed an ASTM type B and D graphite structure, with type B predominating (Fig. 50a). Type A graphite is the proper type for classes 30 and 40 gray iron. Types B and D graphite indicate a rapid cooling rate during solidification. The matrix microstructure was a mixture of pearlite and ferrite (Fig. 50b). The approximate 70% ferrite content is excessive for this type of cast iron.

The outer and fractured surfaces of the fractured casting had a predominantly ferritic condition indicative of decarburization. Although a ferrite skin can be expected on the outside cast surface, a ferrite layer along a fractured surface indicates the presence of the fracture during heat treatment with an oxidizing atmosphere. This suggests that the fracture was present before machining. Figure 50(c) shows the decarburized surface.

The cracked casting had an essentially type D graphite structure with very little type B and no type A graphite. Figure 51(a) illustrates the graphite formations. Near the gas defect, temper carbon (Fig. 51b) was detected. Temper carbon gives evidence that the primary carbides had dissociated due to a heat treatment. In the area slightly removed from the gas defect, type D graphite predominated (Fig. 51c). The matrix structure consisted of pearlite, with approximately 70% ferrite (Fig. 52a). At the region of the observed gas defect, carbides were observed in the microstructure (Fig. 52b).

Conclusions. The cracks in these gray iron door closers that were present either before or

Fig. 47 Microstructures of crankcase 3

(a) Structure from the thin-wall part at crack region a in Fig. 45. 100×. (b) Structure in the thin-wall part near crack region b in Fig. 45. 500×. (c) Structure in the thick-wall part near crack region a in Fig. 45. 500×. All etched with picral. Source: Ref 1

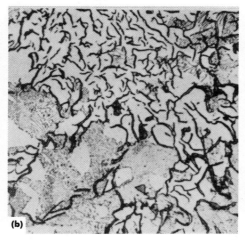
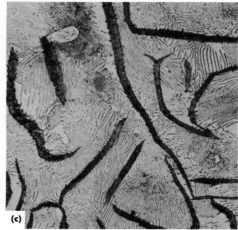

Fig. 48 Microstructures of crankcase 4

(a) Specimen taken from near the surface. (b) The internal structure. Both etched with picral. 100×. Source: Ref 1

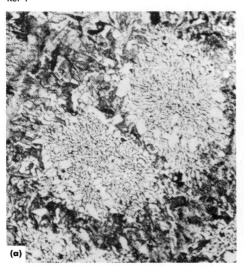
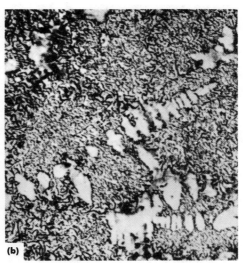

Fig. 49 Failed gray iron door-closer cylinder

(a) Overall view showing positions of fractures. 0.3×. (b) Parting line that was cracked on another casting. 1.5×

during the heat treatment are attributed to a substandard microstructure of the wrong type of graphite combined with excessive ferrite. This anomalous structure is caused by shortcomings in the foundry practice of chemical composition, solidification, and inoculation control. Judging from the microstructure, the strength of the material was lower than desired for class 30 gray iron, and the suspected heat treatment further reduced the strength.

Recommendations. The chemistry and inoculation should be controlled to produce type A graphite structure. The chemistry control should aim for a carbon equivalent close to 4.3% to achieve adequate fluidity for thin sections and to alleviate gas defects.

Example 15: Wear of Cast Iron Pump Parts Due to Improper Composition and Resulting Microstructure (Ref 1). A slide and the two guideways of a pump had to be prematurely disassembled after 20 h of operation. They had galled completely before the rated speed of 800 rpm had been reached.

Investigation. Visual examination indicated worn surfaces (Fig. 53). For metallographic examination, transverse sections through the galled regions A-A, D-D, and F-F and horizontal sections parallel to the galled surfaces of all three parts were made.

Figure 54(a) shows the structure under the gliding surface of the slide (section A-A, Fig. 53), which contains considerable amounts of ferrite. At the edges that form the gliding surface with the unworked windows (points C, Fig. 53), the structure consists predominantly of type D graphite in a ferritic matrix (Fig. 54b). The ferrite content is still higher in the steps between the windows (sections D-D and E-E, Fig. 53), in the center (Fig. 54c), and especially in the vicinity of the edge (Fig. 54d). In contrast, the structure of the guideways (section F-F, Fig. 53) that had a considerably thicker cross section is almost entirely pearlitic (Fig. 55).

The chemical composition of the slide was determined to be Fe-3.60C-0.51Mn-0.485P-0.112S-2.49Si. The graphitic carbon content was 3.22%. This corresponds to a carbon-saturation value of 1.08, which indicates that the cast iron was distinctly hypereutectic.

The ferritic structure results in parts with poor wear resistance and a tendency toward galling. This, of course, also encompasses the opposite surface. The type D graphite in a ferritic matrix is formed by supercooling of the melt during solidification. It appears, therefore, in thin-wall places during fast cooling or in the vicinity of the edges. Cast iron in which the carbon content and carbon-saturation value are high has a strong tendency toward this type of supercooling, especially if the iron contains few crystallization nuclei.

Conclusions. The galling of the pump parts was favored by an unsuitable structure caused by improper composition and fast cooling. Distortion by casting stresses may have contributed or may have played the principal part.

Recommendations. To prevent a repetition, the use of hypoeutectic or eutectic iron, slower cooling of the casting, inoculation of the melt with finely powdered ferrosilicon, and possibly rounding off the edges or machining of the surfaces was recommended.

Carbon Equivalent and Carbon Saturation. Carbon equivalent (CE) is a simplified method of evaluating the effect of composition. Several equations are used; Eq 1 is the most common:

$$CE = TC + \frac{\%Si + \%P}{3} \qquad (Eq\ 1)$$

where TC is total carbon content. This value is important because it can be compared with the eutectic composition (4.3% C) to indicate whether the cast iron will behave as a hypoeutectic or hypereutectic alloy during solidification.

Carbon saturation, S_C, can be calculated as:

$$S_c = \frac{TC}{4.3 - \left(\frac{\%Si + \%P}{3}\right)} \qquad (Eq\ 2)$$

This value is important because it is a measure of the tendency toward carbide decomposition and it can be used to estimate the tensile strength for a given composition and section size of cast iron. Hypereutectic irons tend to form graphite, while hypoeutectic irons tend to form carbides during solidification (Ref 12, 13).

Improper Heat Treatment. The methods and types of heat treatment applied to castings (or any other type of part) constitute sources

Fig. 50 Microstructures of the fractured door-closer cylinder casting

(a) Casting showing types B and D graphite. Fracture is at top, and outside surface is at left. As-polished. 100×. (b) Pearlite/ferrite microstructure of casting containing about 70% ferrite. Etched with nital. 400×. (c) Same as (a), but etched with nital to reveal the ferrite skin indicative of decarburization. 53×

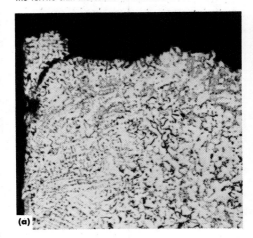

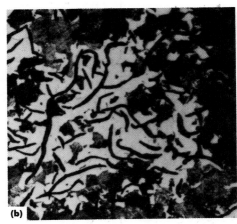

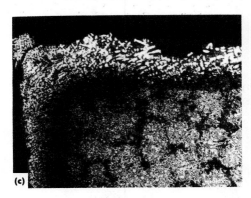

of residual internal stresses. For example, carburized-and-hardened surface layers are sources of compressive stresses that are generally limited to the carburized zone. Over-heating by flame or induction hardening can cause localized, mottled white iron micro-

Fig. 51 Microstructures of the cracked door-closer cylinder casting

(a) Fracture region showing predominantly type D graphite. This region was along the partling line of the thin section. 100×. (b) A primary type D graphite structure adjacent to the gas defect (bottom). The outside of the casting is at the top. 63×. (c) A zone remote from the gas defect; type D graphite predominates. 100×. All as-polished

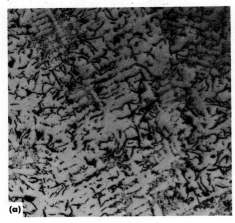

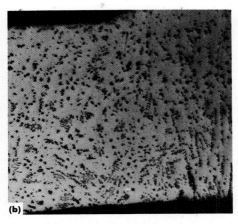

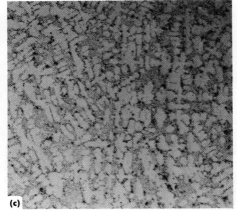

structures and can create a propensity for cracking. Improperly austenitized cast irons, which subsequently develop retained austenite, are subject to low hardness and rapid wear.

Fig. 52 Microstructures of the cracked casting

(a) Ferrite/pearlite matrix is representative of the cracked casting. 100×. (b) Structure of cracked casting adjacent to the gas defect. The white constituents in the dark pearlite zone are carbides. White regions in the fine type D graphite are ferrite. 400×. Both etched with nital

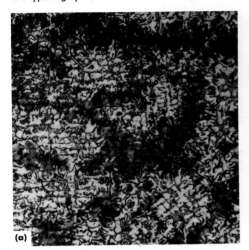 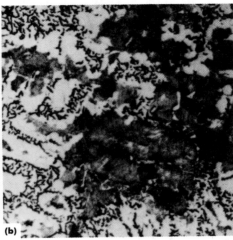

Fig. 53 Worn surfaces of cast iron pump parts

Galled surfaces are at left. See also Fig. 54. Source: Ref 1

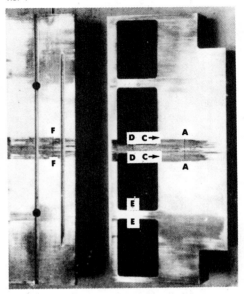

Example 16: Rapid Wear of an Impact Breaker Bar Due to Excessive Retained Austenite. The nominal composition of this chromium alloy cast iron was Fe-2.75C-0.75Mn-0.5Si-0.5Ni-19.5Cr-1.1Mo. The measured hardness of this bar was 450 to 500 HB. The desired hardness for this material after air hardening is 600 to 650 HB. The microstructure shown in Fig. 56 consists of eutectic chromium carbides (Cr$_7$C$_3$) in a matrix of retained austenite and martensite intermingled with secondary carbides.

Conclusions. The low hardness resulted from an excessive amount of retained austenite. This caused reduced wear resistance and thus rapid wear in service.

Recommendations. To reduce retained austenite, recommendations consisted of avoiding an excessive austenitizing temperature and excessive cooling rates from the austenitizing temperature and controlling the chemical composition to avoid excessive hardenability for the section size involved.

Failures Related to Service Conditions

Among the common causes of failures in cast iron components are overload, thermal-stress overload, fatigue, thermal fatigue, wear, and environmental effects (stress state, temperature, corrosive environment). This section will present case histories of service-related failures.

Overload fractures occur by ductile fracture, brittle (cleavage) fracture, and mixed ductile/brittle modes. Methods of recognizing these failure modes are discussed in the articles "Identification of Types of Failures" and "Ductile and Brittle Fractures" in this Volume. The following two examples describe overload-type failures. Example 17 discusses a ductile overload fracture and Example 18 a brittle-type overload fracture.

Example 17: Ductile Overload Failure of a T-Hook That Fractured in Service. Two ductile iron T-hooks were submitted for evaluation. One of the hooks (Fig. 57) was reported to have fractured in service. It was

Fig. 54 Microstructures of worn pump parts

(a) Structure of slide under the gliding surface (Section A-A, Fig. 53). (b) Structure at edge with casting skin (horizontal section taken from points C). (c) Structure at center of gliding surface (Transverse sections D-D and E-E). (d) Structure at edges of gliding surface (Transverse sections D-D and E-E). All etched with picral. 200×. Source: Ref 1

 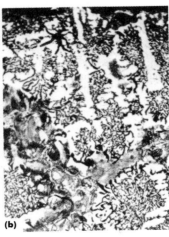  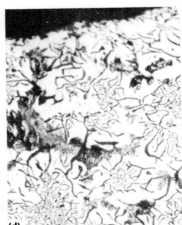

Etched with picral. 215×. Source: Ref 1

Fig. 56 Microstructure of a cast iron breaker bar that suffered premature wear

The structure is eutectic chromium carbide (Cr_7C_3) in a matrix of retained austenite and martensite. Some patches of martensite are intermingled with secondary carbides. Etched with Marble's reagent. 500×

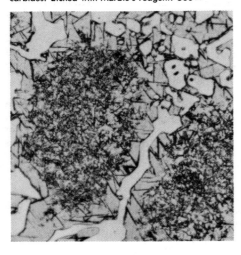

further reported that the hook had been subjected to a load that did not exceed 5900 kg (13 000 lb) at the time of fracture. No information was provided regarding the type of metal used to manufacture the hook. A failure analysis was requested to determine the cause of fracture.

Investigation. Examination by light fractography (Fig. 58) revealed the presence of graphite nodules, indicating the probability that the component was manufactured from a ductile cast iron. The fractured surface also exhibited the presence of a dark region toward the center, suggesting the possibility of a shrinkage defect. The estimated area of this defect was approxi-

mately 3% of the cross-sectional area. No other defects or discontinuities were observed.

The circular section at the fracture had a diameter of 19 mm (0.75 in.) and a cross section of 2.85 cm² (0.44 in.²). The chemical composition of a portion of the T-hook was determined to be:

Element	Composition, %
Carbon	3.57
Manganese	0.32
Phosphorus	0.024
Sulfur	0.013
Silicon	2.46
Nickel	0.28
Chromium	0.05
Molybdenum	<0.03
Copper	0.18
Tin	<0.03
Magnesium	0.043
Iron	rem

This composition is typical for ferritic ductile cast iron, grade 60-40-18 or 65-45-12.

Scanning electron microscope fractography indicated a dimpled fracture morphology. The central portion of the fracture displayed interdendritic shrinkage porosity, confirming that observed visually. Figure 59 shows the dimpled fracture and the shrinkage porosity. An identifiable fracture-origin site was not found. No fatigue striations or evidence of embrittlement was observed. The T-hook hardness was 143 HB, the minimum acceptable level for a 60-40-18 grade of ductile iron.

Metallographic examination included a section through the fractured surface, including the dark region at the center of the fracture. The structure in the unetched condition shown in Fig. 60 has a spheroidal graphite distribution of an estimated 90% ASTM type I and type II nodules with 150 nodules/mm². The matrix microstructure was predominantly ferritic with 3 to 5% pearlite (Fig. 61a to c). The darkened region toward the center exhibited a curved profile, as shown in these photomicrographs. This provided additional evidence of shrinkage at the dark region observed on the fractured surface. This region also exhibited an oxide coating (Fig. 61b) not visible in other regions of the fracture.

Close examination of the pearlite shown in Fig. 61(d) revealed the initial stage of decomposition to spheroidization. This suggested that the component had been subjected to a subcritical anneal treatment to promote a ferrite matrix structure. This is an acceptable procedure to achieve the 60-40-18 grade of ductile iron.

The profile of the fractured surface exhibited a rough texture that is typically associated with the ductile tensile overload observed using SEM. No evidence was found to indicate that this texture had an orientation other than perpendicular to the fracture surface. Such a condition suggested that shear was not a factor in the fracture. Except the relatively minor amount of shrinkage that was detected, the

Fig. 57 Overall view of fractured T-hook

See also Fig. 58 to 61. 0.35×

structure was not found to contain any other undesired defects or discontinuities.

Conclusions. This component fractured in service as a consequence of ductile tensile overload. Evidence indicates that the fractured region was subjected to a load exceeding the capacity of the material. The dimpled fracture morphology observed by SEM fractography and the rough fracture-surface texture confirm a ductile tensile overload fracture.

The hardness and microstructure indicated that the T-hook was manufactured from a ductile iron heat treated to the 60-40-18 grade. This grade has minimum expected properties of 414 MPa (60 ksi) tensile strength, 275 MPa (40 ksi) yield strength, and 18% elongation in 50 mm (2 in.). With this minimum strength value and a 2.85-cm² (0.44-in.²) cross-sectional area, a minimum load of 12 000 kg (26 500 lb) would be necessary to cause the fracture.

Because the information available from the service application indicated that the component had not been subjected to a stress that exceeded 5900 kg (13 000 lb), the observations made in this investigation suggested that either the load was underestimated or that the indicated load was applied at a more rapid rate (perhaps with a jerk), which would tend to increase the effective force of the load. Although some shrinkage was found on the fractured surface, the amount of area occupied by this shrinkage was estimated to be not more than 5%. This defect would detract from the strength available to the cross-sectional area and would reduce the ability of the component to support the load. However, the amount of shrinkage observed was not sufficient to lower the strength of this cross section to the 5900-kg (13 000-lb) load reported from the field.

Other grades of ductile iron with higher strength capacities are available. However, if this component was originally specified to be manufactured from a 60-40-18 grade of ductile iron, the material for this component satisfies those requirements. No other metallurgical rea-

Fig. 58 Additional views of the fractured T-hook shown in Fig. 57

(a) Fractograph at 1.6×. (b) Close-up view of fracture showing dark region (the result of a shrinkage defect) near the center. 2.7×

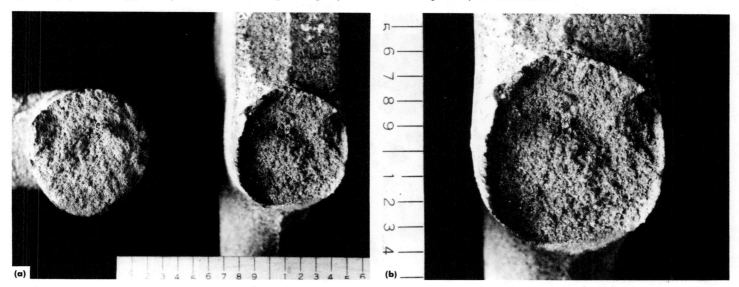

Fig. 59 SEM fractographs of the T-hook

(a) Dimpled morphology is evident. 25×. (b) Central portion of fracture showing shrinkage and porosity. 110×

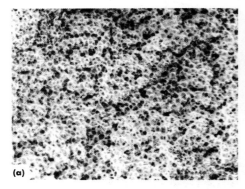
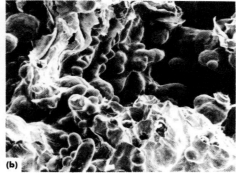

Fig. 60 Graphite structure in the areas of the shrinkage condition

An estimated 90% is ASTM types I and II, with 150 nodules/mm². As-polished. 54×

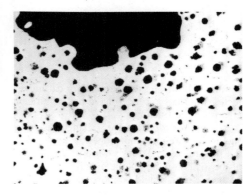

son was observed that could account for the fractured condition of the component. It was concluded that the T-hook was overloaded.

Example 18: Brittle Fracture of a Gray Iron Nut Due to Overload Caused by Misalignment in Assembly. The sand-cast gray iron flanged nut shown in Fig. 62(a) was used to adjust the upper roll on a 3.05-m (10-ft) pyramid-type plate-bending machine. The flange broke away from the body of the nut during service.

Investigation. Visual examination of the flange disclosed that the fracture was circumferential, extending diagonally from the threaded interior to the fillet where the flange joined the body of the nut (Section B-B, Fig. 62a); the fracture surface intersected the threads about 25 mm (1 in.) from the flange face. There were also four radial cracks in the flange, one of which (crack 4, View A-A, Fig. 62a) passed through the flange completely. None of the cracks entered the body of the nut.

Metallographic examination of a nital-etched specimen taken from the flange (Fig. 62b) revealed a matrix of coarse pearlite. Graphite flakes were distributed and oriented randomly, but were undesirably large. Several types of inclusions were observed, including sulfides, oxides, phosphides, and silicates. A large inclusion, believed to be manganese sulfide, was found along a prior-austenite grain boundary located on the surface of crack 2 at an inside corner. Sizable areas of iron sulfide were identified in the surface of crack 4. No evidence of ductility was found in any of the fracture surfaces, a condition to be expected in this material.

Discussion. The coarseness of the graphite flakes and the presence of sizable inclusions constituted features of the microstructure that were not entirely satisfactory yet did not appear severe enough to have caused failure of such a large flanged nut. Because gray iron has little ductility and is susceptible to fracture by shock loads in bending, the roughly symmetrical array

of radial cracks leading from the main fracture face suggested that the flange had been subjected to bending.

The evidence indicated that the nut had been seated improperly against the roll holder (a result of careless installation, not of improper contours of the nut), which created a bending load at a point between cracks 1 and 4. Because of the low ductility of the material, redistribution of the load could occur only through crack initiation. This behavior may have been encouraged by inclusions near the surface, which served as stress raisers. After cracks 1 and 4 had started, the load shifted, and stresses were built up on the opposite side of the flange such that cracks 2 and 3 were initiated.

Conclusions. Brittle fracture of the flange from the body was the result of overload caused by misalignment between the flange and the roll holder. The microstructure contained graphite flakes of excessive size and inclusions in criti-

Fig. 61 Microstructures from the T-hook

(a) At the fracture surface, nodular graphite and a ferrite matrix. 27×. (b) At the shrinkage site, a ferrite matrix with a surface oxide film. 110×. (c) and (d) Higher magnification views of the fracture surface and of the interior of the part, respectively. Both 215×. All etched with nital

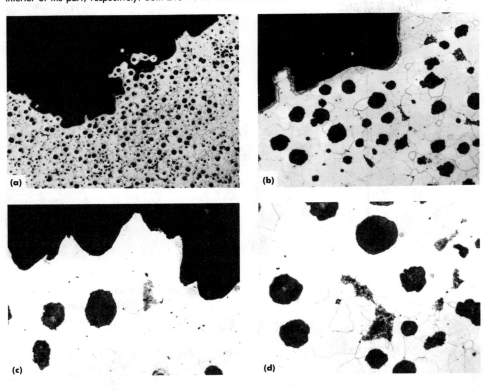

Fig. 62 Sand-cast gray iron flanged nut that failed by brittle fracture

(a) Flanged nut, which was used to adjust a plate-bending roll, and the flange that fractured from the body. Dimensions given in inches. (b) Micrograph of a specimen from the flange showing coarse pearlite matrix, large graphite flakes (arrow A), manganese sulfide inclusions (arrow B), and silicate inclusions (arrow C). Etched with nital. 150×

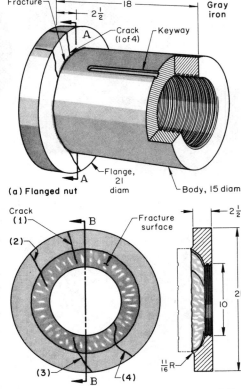

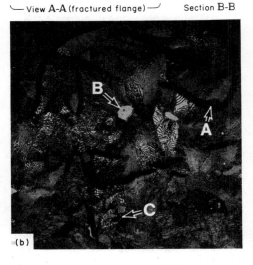

cal areas; however, these metallurgical imperfections did not appear to have had significant effects on the fracture.

Recommendations. The flange surface should be carefully and properly aligned with the roll holder to achieve uniform distribution of the load. A more ductile metal, such as steel or ductile iron, would be more suitable for this application and would require less exact alignment.

Thermal-Stress Overload. The most common interpretation of the term overstress involves mechanically applied stress. There is, however, another source of overload that is perhaps even more destructive and that is provided by hindered thermal expansion or contraction. If a stress from a thermal change demands a specific change of dimension, the yield strength of the material may be exceeded. This will cause distortion, yielding, or fracture.

Example 19: Brittle Fracture of a Ductile Iron Brake Drum by Thermal-Contraction Overload. A 58.4-cm (23-in.) diam heavy-duty brake drum that was a component of a cable-wound winch broke into two pieces during a shutdown period. Other drums had failed in a similar fashion. Average service life of these drums was 2 weeks; none had failed by wear. The drums were sand cast from ductile iron because this material had

been successful in a similar application. This brake drum, however, was a relatively new design.

The winch was used to move large loads, which were carried by gears and clutch plates during haul-in and by the brake drum during haul-out. Figure 63(a) shows the arrangement of the parts. The brake drum was aligned by and bolted to both the clutch mount and the disk, which was bolted to the cable drum. The disk and the clutch mount fitted radially against the inside edge of the web of the brake drum with a clearance of 0.064 to 0.19 mm (0.0025 to 0.0075 in.).

During haul-in, the brake drum idled freely. During haul-out, the cable on the cable drum drove the brake drum, and resistance was provided by brake bands (not shown in Fig. 63a) applied to the outside surface of the brake drum. According to the operator, the friction during heavy service was sufficient to heat the brake drum, clutch mount, and disk to a red color.

Investigation. The severity of the friction on the surface of the brake drum is illustrated by the heat checks shown in Fig. 63(b). The heat checks completely penetrated the rim of the drum, as shown by regions A in Fig. 63(c). The final-fracture area (region B, Fig. 63c) comprised the entire web and part of the rim. The regions of the fracture surface produced by heat

checks were oxidized. Final fracture was the result of a single tensile overload.

Measurements of the two fragments of the drum revealed that the diameter of the drum had expanded along the plane of the fracture, indi-

Fig. 63 Sand-cast ductile iron brake drum from a cable-wound winch that fractured from overload caused by thermal contraction

(a) Schematic of the clutch/brake drum assembly. Dimensions given in inches. (b) Heat checks on the surface of the drum. (c) A fracture surface of the drum showing regions affected by heat checking (A) and final-fracture region (B)

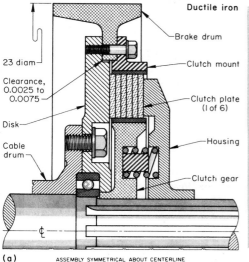

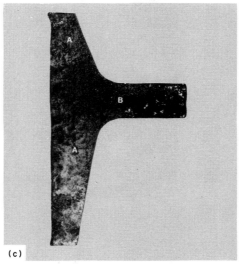

(a) ASSEMBLY SYMMETRICAL ABOUT CENTERLINE

(c)

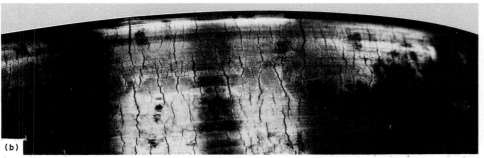

(b)

Fig. 64 Cast iron brake drum at 2.5 × that suffered thermal fatigue cracking

cating the release of internal stress. Cracks found between the inside surface and the bolt holes also indicated the existence of internal stress, but these cracks were considered of secondary importance. A ridge of heavy oxidation was observed along the inside surface of the web, located between the areas of contact with the clutch mount and the disk. This oxide confirmed the observation of the red color during haul-out, indicating a service temperature of at least 650 °C (1200 °F).

Examination of the assembly indicated that the brake drum would cool faster than its mounts and would therefore contract onto them. The contraction of the inside surface of the web as the drum cooled from about 650 °C (1200 °F) would be 0.33 mm (0.013 in.) per hundred degrees Fahrenheit, and the maximum clearance of 0.19 mm (0.0075 in.) would be closed tightly within less than the first hundred degrees Fahrenheit of temperature decrease. Therefore, cooling of the brake drum would generate thermal tensile stresses of such a magnitude that the thickness of the rim remaining after heat checking would be insufficient to withstand the stresses, and the result would be fracture.

Conclusion. Brittle fracture of the brake drum occurred as a result of thermal contraction of the drum web against the clutch mount and the disk.

Corrective Measures. The inside diameter of the drum web was enlarged sufficiently to allow for clearance between the web and the clutch mount and disk at a temperature differential of up to 555 °C (1000 °F). Aluminum spacers were used for alignment during assembly, but melted when heated in service and thus caused no interference. With the adoption of this procedure, brake drums failed by wear only.

Fatigue. The most common cause of failure of machine parts is fatigue under cyclic-stress loading, either alone or in combination with corrosion. If the fatigue strength of the part has been reduced by a surface discontinuity, such as a crack, a tool mark, or a decarburized layer, the failure is attributable to the condition of the casting, not to the service conditions. Method of recognizing fatigue fractures are reviewed in the article "Fatigue Failures" in this Volume.

Thermal Fatigue. This situation is produced by service conditions that repeatedly heat and cool a portion of a casting, causing it to expand and contract. When the thermal gradi-

ents produced in the casting are steep, the large stresses generated by the expansion or contraction can produce failure by fatigue. Figure 64 shows a gray cast iron brake drum that suffered thermal fatigue cracking. This brake drum was from a group with a high rate of premature thermal fatigue failures that occurred when a new type of high-friction brake-lining material was introduced. The brake-drum material was an SAE J431c, grade 3500b, pearlitic gray iron with type A graphite flakes. The microstructure immediately below the degraded surface contained some spheroidized pearlite, confirming exposure to elevated temperatures.

Corrosion Fatigue. Cracking initiated by corrosion fatigue depends upon the synergistic action of cyclical stress and a corrosive medium. Corrosion fatigue is characterized by a plane crack containing corrosion products. The mechanism and identification are more fully explained in the article "Corrosion-Fatigue Failures" in this Volume.

Example 20: Corrosion-Fatigue Cracking of Gray Iron Cylinder Inserts From a Water-Cooled Locomotive Diesel Engine (Ref 14). On cylinder inserts from a water-cooled locomotive diesel engine, cracks formed on the water side in the neck between the cylindrical part and the collar.

Investigation. Cracks in the region between the cylindrical part and the collar were revealed by magnetic-particle inspection (Fig. 65a). As a rule, several parallel cracks had appeared, some of which were very fine. These cracks are shown in Fig. 65(b) on a section of the collar subjected to magnetic-particle inspection.

Fig. 65 Cracked cylinder insert collar

(a) Collar with cracks. 0.75×. (b) Section of the collar shown in (a); a magnetic-particle test on a white-painted section. 0.6×. Source: Ref 14

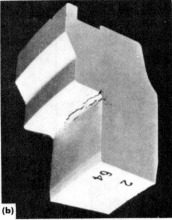

Fig. 66 Structure of the insert collar shown in Fig. 65

(a) Corrosion-fatigue crack; as-polished section showing the flake graphite form. 20×. (b) Section etched with 3% nital; at the corrosion-fatigue crack, the structure is pearlite and ferrite. 75×. Source: Ref 14

Figure 66 shows a representative metallographic section of a crack in the unetched and etched conditions. The surface regions of the cracks widened into funnel form, as shown. The widening at the surface is a result of the corrosive influence of the cooling water. Actual corrosion pits, which often occur in corrosion-fatigue cases, could not be found. This observation indicated that in the present case the vibrational stresses had a greater share in the damage than the corrosive influence. Although all of the inserts investigated had the same mechanical properties, cracks appeared initially only in those engines in which no corrosion inhibitor had been added to the cooling water. It must therefore be assumed that the fatigue strength of the cast iron was considerably reduced by the corrosiveness of the cooling medium. In this case also, the part played by corrosion in the formation of the cracks could be demonstrated only with the help of metallographic techniques.

Conclusion. The cracking was caused by corrosion fatigue. The combined presence of a corrosive medium and cyclical operating stress was needed to cause cracks. When corrosion inhibitor was added to the cooling water, no cracks appeared.

Graphitic Corrosion. Gray iron is susceptible to a form of selective leaching known as graphitic corrosion when immersed in soft waters, salt waters, mine waters, or very dilute acids or when buried underground in some soils, particularly those containing sulfates. Corrosion eats away the matrix of the iron, lowering mechanical strength and leaving behind a soft, black graphitic residue on the surface. If the residue is permitted to remain on the corroded surface, it serves as a protective coating and effects an appreciable reduction in corrosion rate (Ref 15).

In the first of three examples that follow, erosion washes away the graphitic residue, and the process of corrosion is permitted to continue. In the second example, graphitic corrosion develops a hole through a cast iron watermain pipe. The third example describes graphitic corrosion enhanced by stray galvanic currents.

Example 21: Failure of a Gray Iron Pump Bowl Because of Graphitic Corrosion from Exposure to Well Water. Deterioration of the vanes and a wearing away of the area surrounding the mainshaft-bearing housing of the pump bowl for a submersible water pump used in a well field (Fig. 67a) were noticed during a maintenance inspection. The bowl was sand cast from gray iron and had been in service approximately 45 months. An inspection of the pump after 24 months of service did not reveal a serious condition, although some wear was noticed. Several pumps of the same design and material but of different sizes were operating in the well field.

Investigation. Visual examination of the vanes and the area surrounding the mainshaft-bearing housing revealed a dark corrosion product that was soft, porous, and of low mechanical strength. Also, there were areas in which severe erosion had occurred (Fig. 67a). Macrographs of sections through the pump shell and a vane are shown in Fig. 67(b) and(c). The darker areas on both photographs represent graphitic residue and corrosion products that were not removed by erosion. Chemical analysis of the pump material confirmed that it was gray iron.

Conclusion. Exposure of the pump bowl to the well water resulted in graphitic corrosion, which generated a soft, porous graphitic residue impregnated with insoluble corrosion products. Failure of the pump bowl, however, resulted from the continuous erosion of the residue by action of the water within the pump.

Example 22: Graphitic Corrosion of a Gray Iron Water-Main Pipe Resulting in a Corroded-Through Hole. A section of cast iron water-main pipe containing a hole 6.4 cm long × 3.8 cm wide (2.5 × 1.5 in.) was submitted to determine the cause of failure. This pipe was laid in a clay-type soil.

Investigation. Visual-optical examination revealed severe pitting on the outside diameter of the pipe around the hole and at the opposite side of the outside diameter (Fig. 68a to d).

A macroscopic examination of a pipe section at the hole area showed that the porosity ex-

Fig. 67 Sand-cast gray iron pump bowl that failed due to graphitic corrosion and erosion

(a) Section through the pump bowl. (b) and (c) Macrographs of sections through the corroded areas in the pump shell and vane, respectively, showing graphitic residue not eroded by the action of water with in the pump. Both 7×

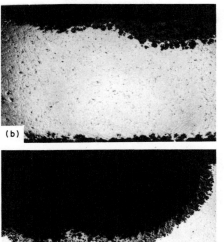

(a)

(b)

(c)

the pipe. Toward the center of the wall thickness, types A, B, and D graphite were observed. This graphite structure distribution is expected in a typical centrifugally cast iron with a hypoeutectic carbon equivalent. Figure 69(a) exhibits a lacy-network microstructure coinciding with a type E dendritic graphite distribution. Figure 69(b) shows the porous condition and its association with the graphite in the structure.

Chemical analyses of the pipe were made in two regions near the hole. One sample exhibited a pronounced porous condition, and a second sample exhibited no porosity. The chemical analyses of these two regions were Fe-11.34C-4.30Si-0.322P-0.44S for the porous sample and Fe-3.75C-1.36Si-0.175P-0.024S for the nonporous sample. The nonporous sample had a composition typical of cast iron pipe. The porous region, however, had a substantial increase in the four elements analyzed. The carbon and silicon ratios were of the order of 3:1. The sulfur increase was extremely high (18:1).

Conclusions. The porous appearance, the soft porous residue, and the composition of the residue confirmed graphitic corrosion. The environmental reaction started from the outside. The presence of a sulfate-reducing bacteria is frequently associated with clay-type soils. The selective leaching of iron by this region leaves a residue rich in carbon, silicon, and phosphorus. The high sulfur

tended a considerable distance into the pipe wall from the hole surface (Fig. 68e and f). This indicated that the porous condition at the inside diameter of the pipe was merely an extension of the more severe porosity at the outside diameter.

Metallographic examination revealed a type E graphite structure at the outside diameter of

Fig. 68 Views of a failed gray iron water pipe

(a) Outside surface of pipe at the region of the side-wall hole. 0.2×. (b) Close-up of hole from the outside. 0.4×. (c) Outside surface of the pipe 180° from the hole. 0.2×. (d) Close-up of hole from inside. 0.5×. (e) Section of pipe at hole area showing porosity and graphitic residue. 5×. (f) Section of pipe at hole area showing porosity progressing from surface toward interior of pipe. 5×

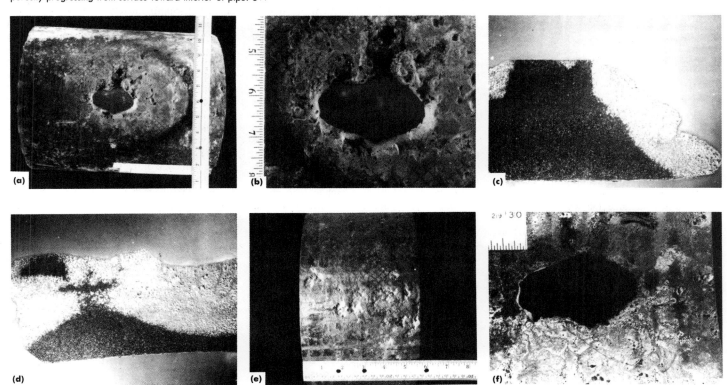

(a)

(b)

(c)

(d)

(e)

(f)

Fig. 69 Microstructure of the failed water pipe shown in Fig. 68
(a) A lacy network of porous residue. 27×. (b) Porosity associated with the graphitic structure. 135×

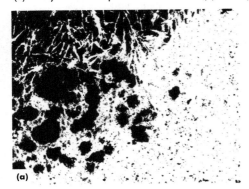

(a)

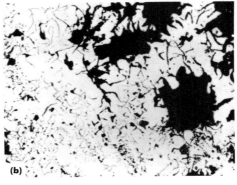

(b)

Fig. 70 Section through pump impeller showing graphitic corrosion (dark edges)
Source: Ref 1

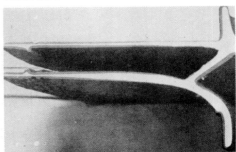

content is attributed to the accumulation of ferrous sulfide from the reaction of the bacteria (Ref 15, 16).

Recommendations. Reinforced coal tar protective coating was recommended to prevent this corrosion associated with the soil condition.

Example 23: Graphitic Corrosion of a Cast Iron Pump Impeller (Ref 1). A cast iron pump impeller suffered significant corrosion after an operating period of only 6 months. The pump had been used to move scrubbing water from a gas generator.

Investigation. According to the reported analysis, the water had the following properties:

pH value	7.2
m value	3.13
Total hardness	32.4 °dH(a)
Carbonate hardness	8.8 °dH(a)
Noncarbonate hardness	23.6 °dH(a)
Total carbonic acid	74.3 mg/L
Combined carbonic acid	68.8 mg/L
Free carbonic acid	5.5 mg/L
Chlorides	99.4 mg/L
Sulfates	78.5 mg/L
Iron	2.7 mg/L
Sulfides	trace

(a) Degrees of hardness (°dH) for water are given in grains per gallon. For example, 1° (French unit) = 0.583 grains per U.S. gallon; 1° (German unit as given above) = 1.044 grains per U.S. gallon, and 1° (U.S. unit) = 0.83 grains per gallon.

A section through the pump impeller illustrated in Fig. 70 indicates that corrosion had substantially penetrated the wall thickness of the thin sections without loss of material. The darker edges characterize graphitic corrosion. The impeller and pump housing were manufactured of an alloy cast iron containing nickel. A chemical analysis of the noncorroded portion of the impeller revealed a composition of Fe-3.14C (2.55 graphitic carbon)-3.12Ni-0.15S.

The microstructure at thicker wall sections consisted of flake graphite in a pearlitic matrix (Fig. 71a). Some ternary phosphide eutectic (steadite) is also present. In thinner sections where a solidification with large undercooling occurred, the graphite precipitated into a type D

form in a predominantly ferrite matrix. This microstructure at thin, undercooled sections (Fig. 71b) should not affect the corrosion resistance.

The corrosion zone is shown in Fig. 71(c). The corrosion attack took place at the graphite/matrix interface both in the normal structure and the fine graphite, supercooled structure. The corrosion selectively leached the ferrite and pearlite matrix constituents while not affecting the graphite, steadite, or sulfide inclusions. Figure 71(d) shows the structure of the corroded layer. This selective form of cast iron corrosion is known as graphitic corrosion. This selective leaching form of corrosion in cast iron does not cause material removal, but markedly lowers the strength such that a knife can readily cut or scrape off a black powder. This is the dark edge shown in Fig. 70. The black powder thus produced had a composition of 10.85C-1.8S-1.45P.

Discussion and Conclusion. It is known that graphitic corrosion can occur through the effect of chloride solutions or weak acids, such as those occurring in acidic soil or hydrogen sulfide-containing water. It appears possible that in the scrubbing water transported from the generator at least a part of the sulfate found was actually sulfide or hydrogen sulfide and was primarily responsible for the corrosion. Graphitic corrosion is normally a slow process. Comparable occurrences were found, however, in a relatively short time during simulated tests to isolate electrolytically graphite, phosphide, carbides, and nonmetallic inclusions in cast iron. Therefore, it is reasonable to assume that in this failure corrosion was accelerated by galvanic currents.

Temperature Effects. Prolonged exposure to excessively high temperatures can cause expansion and distortion in cast iron components (Example 24). Additional information on high-temperature failures can be found in the articles "Distortion Failures" and "Elevated-Temperature Failures" in this Volume.

Example 24: Failure of High-Temperature Rotary Valve Due to Expansion and Distortion Caused by the

Effects of Excessive Operating Temperature. An experimental high-temperature rotary valve was found stuck in the housing due to growth and distortion after approximately 100 h. Gas temperatures were suspected to have been high due to overfueled conditions. Both the rotor and housing were to have been an annealed ferritic ductile iron similar to ASTM A395.

Investigation. Visual examination of the rotor revealed unusually heavy oxidation and thermal fatigue cracking along the edge of the gas passage (Fig. 72). Material properties, including microstructure, composition, and hardness, of both the rotor and housing were evaluated to determine the cause of failure.

The microstructure of the rotor was examined in three regions. The shaft material consisted of spheroidal graphite in a matrix of ferrite with approximately 50% partially spheroidized pearlite (Fig. 73). Because the shaft material would not be directly exposed to the high-temperature gas, it would be most representative of the rotor material before use, which should exhibit a ferritic matrix. The heavy section next to the gas passage exhibited a ferritic matrix, with only a small amount of badly decomposed pearlite remaining (Fig. 74). The oxide layer on the outside diameter of the rotor was approximately 0.3 mm (0.012 in.) deep (Fig. 75).

The thin edge of the rotor adjacent to the gas passage exhibited a ferritic matrix with transformation products that were lacy in appearance (Fig. 76a and b). The fine distribution of spheroidal graphite interspersed with more massive graphite nodules would be a feature of the original casting rather than a result of operating temperatures. The thermal fatigue cracks in this section were surrounded by regions of oxidation that appeared to facilitate crack propagation (Fig. 76c).

The general microstructure of the housing was satisfactory. It consisted of spheroidal graphite in a ferritic matrix with a small amount of spheroidized pearlite (Fig. 77).

The chemical composition of the rotor was:

Element	Composition, %
Carbon	3.47
Manganese	0.24
Phosphorus	0.014
Sulfur	0.008
Silicon	2.7
Nickel	0.27
Chromium	0.06
Molybdenum	<0.01
Copper	0.32
Aluminum	0.02
Magnesium	0.052
Iron	rem

The composition of the housing was:

Element	Composition, %
Carbon	3.59
Manganese	0.2
Phosphorus	0.014
Sulfur	0.005
Silicon	2.5
Nickel	0.30
Chromium	0.07
Molybdenum	<0.01
Copper	0.36
Aluminum	0.02
Magnesium	0.05
Iron	rem

The compositions conform to the material requirements. Average hardness values of the rotor and housing were 145 HB and 156 HB, respectively.

Conclusions. The excessive gas temperatures were responsible for the expansion and distortion that prevented rotation of the rotor. The hardness and composition conformed to the material requirements and were acceptable for the intended use. However, the actual operating temperatures exceeded those intended for this application. Both the decomposition of pearlite and the heavy surface oxidation would have contributed to growth of the rotor. The presence of transformation products in the brake-rotor edge indicated that the lower critical temperature had been exceeded during operation. The excess pearlite in the microstructure was not compatible with the experienced temperature regime.

Wear Failures. The various modes of wear failures are discussed in the article "Wear Failures" in this Volume. The wear of cast iron pump parts due to improper chemical composition and microstructure is discussed in Example 15 in this article. The following example concerns abrasive wear in an ore-crushing operation.

Example 25: Rapid Wear of Shell Liner Due to Severe Abrasion. This high-chromium white cast iron shell liner was used in an ore-crushing operation to process taconite rocks. Impact from the rocks fractured particles from the surface of the liner.

Investigation. Visual-optical examination revealed that particles were fractured off the surface, with cracks aligned horizontally in the shell liner. Metallographic examination indicated a heavily deformed surface layer with chip formation at the wear surface. Deforma-

Fig. 71 Microstructure of the pump-impeller section shown in Fig. 70

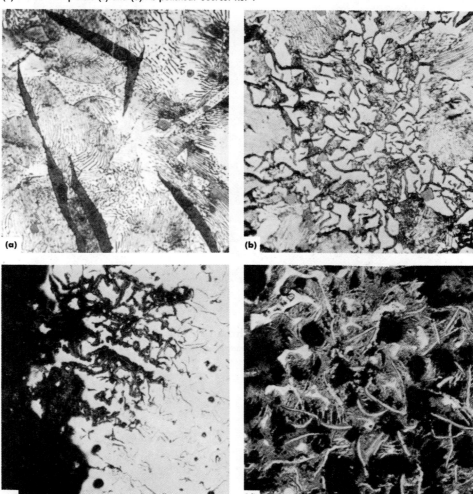

(a) Specimen from a thick-wall section showing flake graphite in a matrix of pearlite with steadite and some sulfide inclusions. 500×. (b) Specimen from a thin-wall section. 500×. (c) Structure in the area between the attacked layer and the unattacked core material. 100×. (d) Structure of the attacked layer. 200×. (a) and (b) Etched with picral. (c) and (d) As-polished. Source: Ref 1

tion and chip formation at the wear surface is shown in Fig. 78. The chemical composition of this shell liner was Fe-2.74C-0.75Mn-0.55Si-0.51Ni-19.4Cr-1.15Mo. This alloy has low ductility and toughness and is excellent for resistance to abrasive wear. However, little plastic displacement will occur with this material, resulting in brittle cracking (chipping) of the surface material. Also, impacting by a hard substance will result in chipping.

Conclusions. Rapid wear was a result of severe abrasion caused by the impact of taconite rock. This was a material-selection problem in that the wrong alloy was used for a condition not anticipated in the original choice.

Recommendations. Because the exact service condition was not known at the time, a similar alloy with improved toughness should be applied. Also, the proper installation of the shell liners should ensure sufficient backing to reduce chipping damage from impact.

Mechanical Abuse. Localized overstress conditions can also lead to failure of cast iron components. Example 26 discusses the failure of a white iron casting due to severe localized impact.

Example 26: Failure of High-Chromium White Iron Shell Liners Due to Mechanical Abuse. Two broken ball-mill liners from a copper-mine ore operation were submitted for failure analysis. These liners failed prematurely, having reached less than 20% of their expected life.

Investigation. The chemical composition of the liners was reported to be within specifications for high-chromium white cast iron. The two broken liners were sand blasted for visual inspection and subsequent metallography and hardness testing. Many cracks, several of which are illustrated in Fig. 79, were found externally and on the undersides. These views also show signs of mechanical damage that

Fig. 72 Oxidation and thermal fatigue cracking of a cast ductile iron rotor

See also Fig. 73 to 78.

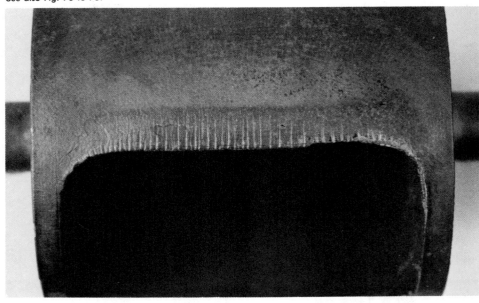

Fig. 73 Microstructure of the rotor shown in Fig. 72

(a) General structure. 60×. (b) Graphite nodules in a matrix of ferrite and pearlite. 300×. Both etched with nital

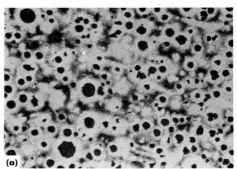 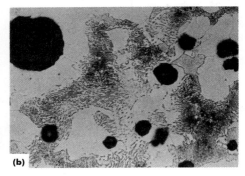

Fig. 74 Microstructure of the heavy section adjacent to the gas passage in the rotor

(a) General structure. 60×. (b) Nodular graphite in a ferrite matrix with some decomposed pearlite. 300×. Both etched with nital

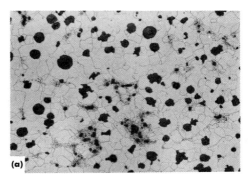 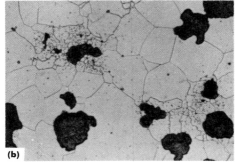

Fig. 75 Oxide layer on the outside diameter of the rotor

Etched with nital. 100×

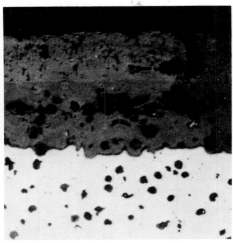

occurred inside the mill before detection of the failures. The presence of considerable cracking at the underside is significant because the user advised that the liners were not backed in the installation.

Radiographic examination revealed a complete absence of shrinkage or other major defects. Hardness surveys on liner sections ranged from 550 to 600 HB for liner 1 and from 600 to 650 HB for liner 2. These values are within the specified limits for this part. Retained austenite content as determined by x-ray diffraction with corrections applied for 30% estimated volume fraction of carbide was 24.6% for liner 1 and 22.0% for liner 2. These values are within specifications.

The microstructures of both liners near the crack location consisted of massive primary chromium carbide in a matrix of martensite, austenite, and secondary carbide particles. The primary carbide is the light-etched constituent in Fig. 80(a). Cracking was present in the microstructures of both liners (Fig. 80a and b). The cracks occurred mostly in the massive carbides, with their propagation direction essentially independent of carbide-phase orientation. The shattered appearance of these carbides indicated the presence of several localized stresses. These cracks tend to fracture the brittle carbide phase first; once nucleated, the sharp cracks can propagate and grow to critical dimensions, which eventually induces complete failure to the load-bearing section. Small amounts of ferrite (the dark-etching constituent shown in Fig. 80c) were observed in both liners. The presence of small amounts of pearlite was attributed to slow cooling rates after the solution treatment. Although pearlite may slightly degrade abrasion resistance, its minor presence did not contribute to the failures.

Conclusion. The premature failure of these liners was caused by severe localized overstress conditions due to localized impact in service.

Fig. 76 Microstructure near the thin edge of the rotor adjacent to the gas passage
(a) General structure. 60×. (b) Ferritic matrix with transformation products. 300×. (c) Oxidation surrounding a thermal fatigue crack. 60×. All etched with nital

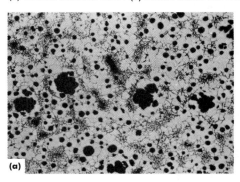 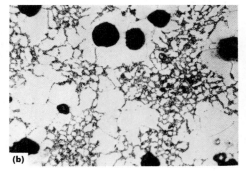 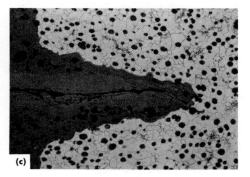

Fig. 77 Microstructure of the rotor housing
(a) General microstructure. 100×. (b) Ferritic matrix with spheroidized pearlite. 500×. Both etched with nital

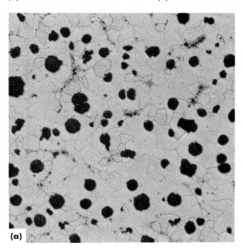 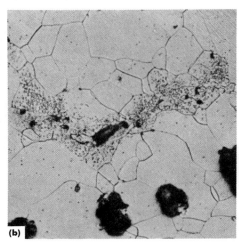

Fig. 79 Fractured liners from a ball mill
(a) Fractured surface. (b) Cracks at underside. Both approximately 0.2×

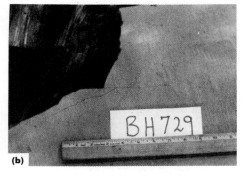

Fig. 78 Wear surface of a shell liner cast from high-chromium white iron
The chip formation initiates at points of the brittle carbide (Cr_7C_3) fracture. 125×

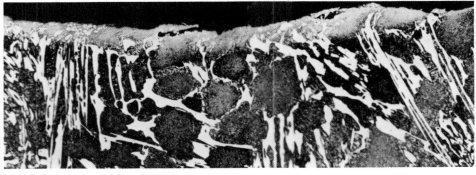

Recommendations. Proper backing of shell liners should be ensured to reduce the effect of impact forces in the ball mill.

Cavitation. The mechanism of cavitation and its characterization are covered in detail in the article "Liquid-Erosion Failures" in this Volume. Two common ways that cavitation damage originates are (1) the introduction of air or a gaseous substance into a flowing liquid system and (2) the bubble formation and collapse at a propeller site. In either case, the bubble collapse results in a shock impulse to a material surface similar to a localized miniature-hammer blow. However, although cavitation damage appears to result basically from a mechanical cause, the exact mechanism is not entirely clear. Two schools of thought have developed, one supporting an essentially erosive and the other an essentially corrosive nature (Ref 15, 17).

Example 27: Damage to a Cylinder Lining from a Diesel Motor Caused by Cavitation (Ref 1). The cylinder lining of a diesel motor suffered localized damage on the cooling-water (outer) side.

Investigation. Localized damage occurred on the outer cooling-water side of the cylinder lining, leading to edge serration and heavy pitting (Fig. 81a). Beyond the heavily damaged zone, the external wall was coated with deposits. These deposits consisted principally of magnetite (Fe_3O_4) with small amounts of SiO_2, graphite, and CaO, as determined by

Fig. 80 Microstructures of the fractured ball-mill liners shown in Fig. 79

(a) Primary carbides are the light-etching constituent; cracks are also present. 275×. (b) Crack morphology and shattered appearance of many primary carbides attributed to severe, localized stresses in service. 275×. (c) Pearlite present in the matrix. 500×. (a) and (c) Etched with Marble's reagent. (b) As-polished

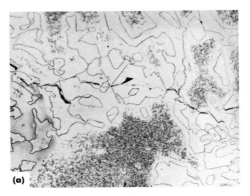

(a)

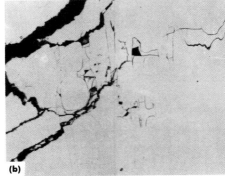

(b)

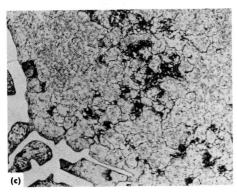

(c)

Fig. 81 Damaged diesel engine cylinder lining

(a) General view of the liner. 0.3×. (b) Microstructure composed of flake graphite in a pearlitic matrix. Etched with nital. 500×. Source: Ref 1

(a)

(b)

attain such high values that the sluggish liquid can no longer keep pace. As the body returns, vacuum bubbles form that implode on the next outward movement of the medium. This produces such a high pressure on a microscopically small area that the material disintegrates and the particles of material are knocked out of the surface.

The conditions for cavitation are set up at the cylinder linings in a running diesel motor because at the moment of reversal the cylinder is excited to bending vibrations by the lateral pressure of the piston. If the cooling water can no longer follow the vibration of the cylinder, vacuum bubbles form at the most strongly vibrating point (impact side). These bubbles implode upon reversal of the vibration direction and produce cavities on the cylinder surface as a result of this violent cyclic stressing.

Recommendations. The measures recommended in the literature consist of reducing piston play, reducing the amplitude of vibration with thicker wall linings, hard chromizing of the cooling-water side, and adding a protective oil to the cooling water. The effect of the protective oil is presumably based on a film of oil that forms on the cylinder surface and that is not so easily scoured off during vibration. The effect of the imploding vacuum bubbles is reduced by the oil film, which can renew itself from the emulsion.

ACKNOWLEDGMENT

The following contributors provided case histories or other useful information: Charles O. Cambridge, Navistar International, Inc.; George M. Goodrich, Taussig Associates, Inc.; Renita G. King, Cummins Engine Company; Dilip K. Subramanyam, Abex Corporation; John F. Wallace, Case Western Reserve University; Charles F. Walton, Professional Engineer; Peter F. Wieser, Case Western Reserve University; and Frank J. Worzala, University of Wisconsin.

chemical analysis. The material beneath the deposit had not been attacked, suggesting that foreign particles were responsible. These were possibly carried by the circulating cooling water, but could also have originated in part from the damaged zone, as indicated by the presence of graphite and SiO_2. The microstructure consisted essentially of type A

graphite in a pearlitic matrix (Fig. 81b), indicating that the base material was gray iron.

Conclusions. This heavy localized damage is the cavitation damage frequently observed in diesel motor cylinders. Cavitation must always be dealt with on a body vibrating in a liquid when the amplitude and frequency of vibration

REFERENCES

1. P.M. Unterweiser, Ed., *Case Histories in Failure Analysis*, American Society for Metals, 1979
2. P.A. Heller and H. Jungbluth, *Giesserei*, Vol 42, 1955
3. B. Chalmers, *Principles in Solidification*, John Wiley & Sons, 1982
4. M.C. Flemings, *Solidification Processing*, McGraw-Hill, 1974
5. B. Hernandez and J.F. Wallace, Mechanisms of Pinhole Formation in Gray Iron, *Trans. AFS*, Vol 87, 1979, p 335
6. S.I. Karsay, *Ductile Iron-Production Practices*, American Foundrymen's Society, Des Plaines, IL, 1975, p 167
7. F. Hudson *et al.*, *Atlas of Defects in Castings*, Institute of British Foundrymen, London, 1961
8. G.W.S. Anselman *et al.*, *Analysis of Casting Defects*, 2nd ed., American Foundrymen's Society, Des Plaines, IL, 1966
9. C.F. Walton, *The Properties of Gray Iron*, Gray and Ductile Iron Founder's Society, Des Plaines, IL, 1965
10. E. Campomanes and R. Goller, *Cast Iron Containing Intermediate Forms of Graphite, Properties and Potential Applications*, Technical Reports System No. 76-01, American Society for Metals, 1976
11. V.H. Patterson, Inoculants Can Improve Gray Iron Properties, *Foundry*, Vol 100, June 1972, p 68
12. H.T. Angus, *Cast Iron: Physical and Engineering Properties*, 2nd ed., Butterworths, 1976
13. C.F. Walton, *Gray and Ductile Iron Castings Handbook*, Iron Founder's Society, Des Plaines, IL, 1971
14. L.E. Alban *et al.*, *Source Book in Failure Analysis*, American Society for Metals, 1974
15. M.G. Fontana and N.D. Greene, *Corrosion Engineering*, 2nd ed., McGraw-Hill, 1978
16. H.H. Uhlig, *The Corrosion Handbook*, 9th ed., John Wiley & Sons, 1966
17. F.R. Hutchings and P.M. Unterweiser, Ed., *Failure Analysis: The British Engine Technical Reports*, American Society for Metals, 1981

SELECTED REFERENCES

- L. Eldoky and R.C. Voigt, Fracture of Ferritic Ductile Iron, *Trans. AFS*, Vol 93, 1985, p 365
- N. Lazardis, N.K. Nanstad, F.J. Worzala, and C.R. Loper, Jr., Determination of the Fracture Behavior of Ductile Cast Irons for Failure Analysis Purposes, *Trans. AFS*, Vol 85, 1977, p 277
- *Metallography and Microstructures*, Vol 9, 9th ed., *Metals Handbook*, American Society for Metals, 1985
- V.H. Patterson, *Foote Foundry Facts*, Foote Mineral Company, Exton, PA, 1970
- *Properties and Selection: Irons and Steels*, Vol 1, 9th ed., *Metals Handbook*, American Society for Metals, 1978
- D.F. Socie, J.W. Fash, and S.D. Downing, *Fatigue of Gray Iron*, University of Illinois, Urbana, IL, Oct 1982
- R.C. Voigt, Microstructural Analysis of Austempered Ductile Cast Iron Using the Scanning Electron Microscope, *Trans. AFS*, Vol 91, 1983, p 253

Failures of Steel Castings

Robin K. Churchill, John R. Dillon, David Eyolfson, William E. Fuller, Christopher Z. Oldfather, William D. Scott, and Laurence J. Venne, ESCO Corporation

FAILURES OF CASTINGS, like the failures of wrought materials, can be divided into two broad categories. Some fail because of design deficiencies; others fail because of defects in the castings that the designer did not (could not) anticipate. A design-related failure is one in which even a perfect casting (free from unexpected deficiencies, that is, one meeting the specified requirements) will not perform up to the requirements of the application. Correspondingly, a defect-related failure is one resulting from a defect that falls outside of the established acceptance criteria when castings meeting the specified requirements have performed up to standard in the past. In many instances, both design and manufacture of the casting contribute to premature failure.

Clearly, both types of failure are preventable. Upon first inspection, it seems clear that the designer must accept responsibility for failures of the first type, and the founder the second. No such distinction should be drawn. The key to serviceable castings is communication between the designer and the foundry engineers. Foundrymen are an invaluable source of information on just what can be produced in cast form, both in terms of geometry and material selection. This input should be sought early in the design process to incorporate the suggestions where practical. This can also have a dramatic impact on the manufacturability of the final part, influencing the ultimate quality of the casting and thus reducing the likelihood of failure due to defective parts.

Included in the design process is the determination of specifications to which the foundry must work. Here, too, the foundry can be invaluable in providing the designer with input as to what materials should be used, what specifications are called out, what nondestructive testing techniques are applicable, what level of metallurgical testing should be performed, and, finally, what quality-assurance documentation is needed. A teamwork approach between the designer and foundry (and, where possible, the ultimate user) is the best (and lowest cost) method of avoiding casting failures.

This article has been broken down into a number of subsections to help organize the material (casting defects and failures due to heat treatment, composition, weld repair, hydrogen, high temperatures, overload, and corrosion). Many of the failures overlap into several sections; that is, they could be categorized under several titles. This is typical of failure analysis work in that several factors may have contributed to the final failure. Attempts have been made to provide examples that clearly represent the specified mode of failure to be illustrated. This has not always been possible. Because specific examples of failures were sometimes not available, a general description of the probable mechanisms is included to assist in future investigations.

Casting Defects

Although foundrymen favor referring to the deviations in less-than-perfect castings as discontinuities, these imperfections are more commonly referred to as casting defects. Some casting defects may have no effect on the function or the service life of cast components, but will give an unsatisfactory appearance or will make further processing, such as machining, more costly. Many such defects can be easily corrected by shot blast cleaning or grinding. Other defects that may be more difficult to remove can be acceptable in some locations. It is most critical that the casting designer understand the differences and that he write specifications that meet the true design needs.

Classification of Casting Defects. Foundrymen have traditionally used rather unique names, such as rat tail, scab, buckle, snotter, and shut, to describe various casting imperfections. Unfortunately, several foundrymen may use different nomenclature to describe the same defect. The International Committee of Foundry Technical Associations (CFTA) has standardized the nomenclature, starting with the identification of seven basic categories of casting defects (Ref 1):

- Metallic projections
- Cavities
- Discontinuities
- Defects
- Incomplete casting
- Incorrect dimension
- Inclusions or structural anomalies

In this scheme, the term discontinuity has the specific meaning of a planar separation of the metal, that is, a crack.

Table 1 presents some of the common defects in each category. Those that can be associated with the failure of steel castings are noted. In general, defects that can serve as stress raisers or crack promoters are the most serious. These would include pre-existing cracks, internal voids, and nonmetallic inclusions. It is useful to review this classification system when performing defect or failure analysis.

Control of Casting Defects. Defect-free castings can be produced—at a price. The multitude of process variables, such as molding mediums, binder, gating and risering, melting and ladle practice, pouring technique, and heat treatment, that must be controlled to produce such sound castings requires a thorough understanding of the processes used and strict engineering supervision. The price of a casting often reflects the cost of such efforts. The attention to detail necessary to make good castings can reduce total cost of a manufactured part by significantly reducing machining, repair, and fit problems later in the assembly. Proper specification—as opposed to overspecification or underspecification—is the key to the successful application, that is, freedom from failure, of castings.

Finally, steel castings generally lend themselves to being repaired. With proper attention to the details of these procedures, imperfections can be removed and a suitable weld deposit substituted to give a finished product fit for the intended service. Unfortunately, many failures of steel castings result from improper repairs of casting defects (see the section "Failures Related to Welding" in this article).

Many defects are simply a problem from an appearance standpoint and would not normally lead to premature failure of a casting in service.

Table 1 International classification of common casting defects

No.	Description	Common name	Sketch	No.	Description	Common name	Sketch

Metallic Projections

A 100: Metallic projections in the form of fins or flash

A 110: Metallic projections in the form of fins (or flash) without change in principal casting dimensions

No.	Description	Common name
A 111	Thin fins (or flash) at the parting line or at core prints	Joint flash or fins
A 112	Projections in the form of veins on the casting surface	Veining or finning
A 113	Network of projections on the surface of die castings	Heat-checked die
A 114(a) . . .	Thin projection parallel to a casting surface, in re-entrant angles	Fillet scab
A 115	Thin metallic projection located at a re-entrant angle and dividing the angle in two parts	Fillet vein

A 120: Metallic projections in the form of fins with changes in principal casting dimensions

No.	Description	Common name
A 123(a) . . .	Formation of fins in planes related to direction of mold assembly (precision casting with waste pattern); principal casting dimensions change	Cracked or broken mold

A 200: Massive projections

A 210: Swells

No.	Description	Common name
A 212(a) . . .	Excess metal in the vicinity of the gate or beneath the sprue	Erosion, cut, or wash
A 213(a) . . .	Metal projections in the form of elongated areas in the direction of mold assembly	Crush

A 220: Projections with rough surfaces

No.	Description	Common name
A 221(a) . . .	Projections with rough surfaces on the cope surface of the casting	Mold drop or sticker
A 222(a) . . .	Projections with rough surfaces on the drag surface of the casting (massive projections)	Raised core or mold element cutoff
A 223(a) . . .	Projections with rough surfaces on the drag surface of the casting (in dispersed areas)	Raised sand
A 224(a) . . .	Projections with rough surfaces on other parts of the casting	Mold drop
A 225(a) . . .	Projections with rough surfaces over extensive areas of the casting	Corner scab
A 226(a) . . .	Projections with rough surfaces in an area formed by a core	Broken or crushed core

(continued)

(a) Defects that under some circumstances could contribute, either directly or indirectly, to casting failures. Adapted from *International Atlas of Casting Defects*, American Foundrymen's Society, Des Plaines, IL

Table 1 (continued)

No.	Description	Common name	Sketch	No.	Description	Common name	Sketch

Cavities

B 100: Cavities with generally rounded, smooth walls perceptible to the naked eye (blowholes, pinholes)

B 110: Class B 100 cavities internal to the casting, not extending to the surface, discernible only by special methods, machining, or fracture of the casting

B 111(a) . . . Internal, rounded cavities, usually smooth-walled, of varied size, isolated or grouped irregularly in all areas of the casting — Blowholes, pinholes

B 112(a) . . . As above, but limited to the vicinity of metallic pieces placed in the mold (chills, inserts, chaplets, etc.) — Blowholes, adjacent to inserts, chills, chaplets, etc.

B 113(a) . . . Like B 111, but accompanied by slag inclusions (G 122) — Slag blowholes

B 120: Class B 100 cavities located at or near the casting surface, largely exposed or at least connected with the exterior

B 121(a) . . . Exposed cavities of various sizes, isolated or grouped, usually at or near the surface, with shiny walls — Surface or subsurface blowholes

B 122(a) . . . Exposed cavities, in re-entrant angles of the casting, often extending deeply within — Corner blowholes, draws

B 123 . . . . . Fine porosity (cavities) at the casting surface, appearing over more or less extended areas — Surface pinholes

B 124(a) . . . Small, narrow cavities in the form of cracks, appearing on the faces or along edges, generally only after machining — Dispersed shrinkage

B 200: Cavities with generally rough walls, shrinkage

B 210: Open cavity of Class B 200, sometimes penetrating deeply into the casting

B 211(a) . . . Open, funnel-shaped cavity; wall usually covered with dendrites — Open or external shrinkage

B 212(a) . . . Open, sharp-edged cavity in fillets of thick castings or at gate locations — Corner or fillet shrinkage

B 213(a) . . . Open cavity extending from a core — Core shrinkage

B 220: Class B 200 cavity located completely internal to the casting

B 221(a) . . . Internal, irregularly shaped cavity; wall often dendritic — Internal or blind shrinkage

B 222(a) . . . Internal cavity or porous area along central axis — Centerline or axial shrinkage

B 300: Porous structures caused by numerous small cavities

B 310: Cavities according to B 300, scarcely perceptible to the naked eye

B 311(a) . . . Dispersed, spongy dendritic shrinkage within walls of casting; barely perceptible to the naked eye — Macro- or micro-shrinkage, shrinkage porosity, leakers

(continued)

(a) Defects that under some circumstances could contribute, either directly or indirectly, to casting failures. Adapted from *International Atlas of Casting Defects*, American Foundrymen's Society, Des Plaines, IL

Table 1 (continued)

No.	Description	Common name	Sketch

Discontinuities

C 100: Discontinuities, generally at intersections, caused by mechanical effects (rupture)

C 110: Normal cracking

C 111(a) . . . Normal fracture appearance, sometimes with adjacent indentation marks	Breakage (cold)		

C 120: Cracking with oxidation

C 121(a) . . . Fracture surface oxidized completely around edges	Hot cracking		

C 200: Discontinuities caused by internal tension and restraints to contraction (cracks and tears)

C 210: Cold cracking or tearing

C 211(a) . . . Discontinuities with squared edges in areas susceptible to tensile stresses during cooling; surface not oxidized	Cold tearing		

C 220: Hot cracking and tearing

C 221(a) . . . Irregularly shaped discontinuities in areas susceptible to tension; oxidized fracture surface showing dendritic pattern	Hot tearing		
C 222(a) . . . Rupture after complete solidification, either during cooling or heat treatment	Quench cracking		

C 300: Discontinuities caused by lack of fusion (cold shuts); edges generally rounded, indicating poor contact between various metal streams during filling of the mold

C 310: Lack of complete fusion in the last portion of the casting to fill

C 311(a) . . . Complete or partial separation of casting wall, often in a vertical plane	Cold shut or cold lap		

C 320: Lack of fusion between two parts of casting

C 321(a) . . . Separation of the casting in a horizontal plane	Interrupted pour		

C 330: Lack of fusion around chaplets, internal chills, and inserts

C 331(a) . . . Local discontinuity in vicinity of metallic insert	Chaplet or insert cold shut, unfused chaplet		

C 400: Discontinuities caused by metallurgical defects

C 410: Separation along grain boundaries

C 411(a) . . . Separation along grain boundaries of primary crystallization	Conchoidal or "rock candy" fracture		
C 412(a) . . . Network of cracks over entire cross section	Intergranular corrosion		

Defective Surface

D 100: Casting surface irregularities

D 110: Fold markings on the skin of the casting

D 111 . . . . . Fold markings over rather large areas of the casting	Surface folds, gas runs		

(continued)

(a) Defects that under some circumstances could contribute, either directly or indirectly, to casting failures. Adapted from *International Atlas of Casting Defects*, American Foundrymen's Society, Des Plaines, IL

Table 1 (continued)

No.	Description	Common name	Sketch	No.	Description	Common name	Sketch
D 112	Surface shows a network of jagged folds or wrinkles (ductile iron)	Cope defect, elephant skin, laps		D 134	Casting surface entirely pitted or pock-marked	Orange peel, metal mold reaction, alligator skin	
D 113	Wavy fold markings without discontinuities; edges of folds at same level, casting surface is smooth	Seams or scars		D 135	Grooves and roughness in the vicinity of re-entrant angles on die castings	Soldering, die erosion	
D 114	Casting surface markings showing direction of liquid metal flow (light alloys)	Flow marks		**D 140:**	**Depressions in the casting surface**		
D 120:	**Surface roughness**			D 141	Casting surface depressions in the vicinity of a hot spot	Sink marks, draw or suck-in	
D 121	Depth of surface roughness is approximately that of the dimensions of the sand grains	Rough casting surface		D 142	Small, superficial cavities in the form of droplets of shallow spots, generally gray-green in color	Slag inclusions	
D 122	Depth of surface roughness is greater than that of the sand grain dimensions	Severe roughness, high pressure molding defect		**D 200:**	**Serious surface defects**		
				D 210:	**Deep indentation of the casting surface**		
D 130:	**Grooves on the casting surface**			D 211	Deep indentation, often over large area of drag half of casting	Push-up, clamp-off	
D 131	Grooves of various lengths, often branched, with smooth bottoms and edges	Buckle		**D 220:**	**Adherence of sand, more or less vitrified**		
D 132	Grooves up to 5.1 mm (0.2 in.) in depth, one edge forming a fold which more or less completely covers the groove	Rat tail		D 221	Sand layer strongly adhering to the casting surface	Burn on	
D 133	Irregularly distributed depressions of various dimensions extending over the casting surface, usually along the path of metal flow (cast steel)	Flow marks, crow's feet		D 222	Very adherent layer of partially fused sand	Burn in	

(continued)

(a) Defects that under some circumstances could contribute, either directly or indirectly, to casting failures. Adapted from *International Atlas of Casting Defects*, American Foundrymen's Society, Des Plaines, IL

Table 1 **(continued)**

No.	Description	Common name	Sketch	No.	Description	Common name	Sketch
D 223	Conglomeration of strongly adhering sand and metal at the hottest points of the casting (re-entrant angles and cores)	Metal penetration		D 243	Scaling after anneal	Scaling	
D 224	Fragment of mold material embedded in casting surface	Dip coat spall, scab					

D 230: **Plate-like metallic projections with rough surfaces, usually parallel to casting surface**

Incomplete Casting

E 100: **Missing portion of casting (no fracture)**

E 110: **Superficial variations from pattern shape**

No.	Description	Common name	Sketch
D 231(a) . . .	Plate-like metallic projections with rough surfaces parallel to casting surface; removable by burr or chisel	Scabs, expansion scabs	
D 232(a) . . .	As above, but impossible to eliminate except by machining or grinding	Cope spall, boil scab, erosion scab	
D 233(a) . . .	Flat, metallic projections on the casting where mold or core washes or dressings are used	Blacking scab, wash scab	

No.	Description	Common name	Sketch
E 111	Casting is essentially complete except for more or less rounded edges and corners	Misrun	
E 112	Deformed edges or contours due to poor mold repair or careless application of wash coatings	Defective coating (tear-dropping) or poor mold repair	

E 120: **Serious variations from pattern shape**

No.	Description	Common name	Sketch
E 121	Casting incomplete due to premature solidification	Misrun	
E 122	Casting incomplete due to insufficient metal poured	Poured short	
E 123	Casting incomplete due to loss of metal from mold after pouring	Runout	
E 124	Significant lack of material due to excessive shot-blasting	Excessive cleaning	

D 240: **Oxides adhering after heat treatment (annealing, tempering, malleablizing) by decarburization**

No.	Description	Common name	Sketch
D 241	Adherence of oxide after annealing	Oxide scale	
D 242	Adherence of ore after malleablizing (white heart malleable)	Adherent packing material	

(continued)

(a) Defects that under some circumstances could contribute, either directly or indirectly, to casting failures. Adapted from *International Atlas of Casting Defects*, American Foundrymen's Society, Des Plaines, IL

Table 1   (continued)

No.	Description	Common name	Sketch
E 125 . . . . . Casting partially melted or seriously deformed during annealing	Fusion or melting during heat treatment	

E 200:   Missing portion of casting (with fracture)

E 210:   Fractured casting

No.	Description	Common name	Sketch
E 211 . . . . . Casting broken, large piece missing; fractured surface not oxidized	Fractured casting	

E 220:   Piece broken from casting

| E 221 . . . . . Fracture dimensions correspond to those of gates, vents, etc. | Broken casting (at gate, riser, or vent) | |

E 230:   Fractured casting with oxidized fracture

| E 231 . . . . . Fracture appearance indicates exposure to oxidation while hot | Early shakeout | |

Incorrect Dimensions or Shape

F 100:   Incorrect dimensions; correct shape

F 110:   All casting dimensions incorrect

| F 111 . . . . . All casting dimensions incorrect in the same proportions | Improper shrinkage allowance | |

F 120:   Certain casting dimensions incorrect

| F 121 . . . . . Distance too great between extended projections | Hindered contraction | |

No.	Description	Common name	Sketch
F 122 . . . . . Certain dimensions inexact	Irregular contraction	
F 123 . . . . . Dimensions too great in the direction of rapping of pattern	Excess rapping of pattern	
F 124 . . . . . Dimensions too great in direction perpendicular to parting line	Mold expansion during baking	
F 125 . . . . . Excessive metal thickness at irregular locations on casting exterior	Soft or insufficient ramming, mold-wall movement	
F 126 . . . . . Thin casting walls over general area, especially on horizontal surfaces	Distorted casting	

F 200:   Casting shape incorrect overall or in certain locations

F 210:   Pattern incorrect

| F 211 . . . . . Casting does not conform to the drawing shape in some or many respects; same is true of pattern | Pattern error | |
| F 212 . . . . . Casting shape is different from drawing in a particular area; pattern is correct | Pattern mounting error | |

(continued)

(a) Defects that under some circumstances could contribute, either directly or indirectly, to casting failures. Adapted from *International Atlas of Casting Defects*, American Foundrymen's Society, Des Plaines, IL

Table 1 (continued)

No.	Description	Common name	Sketch
F 220:	**Shift or Mismatch**		
F 221	Casting appears to have been subjected to a shearling action in the plane of the parting line	Shift	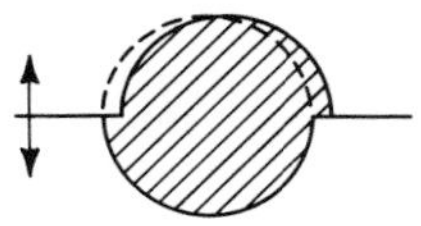
F 222	Variation in shape of an internal casting cavity along the parting line of the core	Shifted core	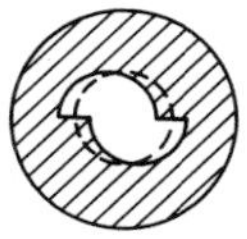
F 223	Irregular projections on vertical surfaces, generally on one side only in the vicinity of the parting line	Ramoff, ramaway	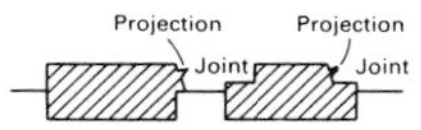
F 230:	**Deformations from correct shape**		
F 231	Deformation with respect to drawing proportional for casting, mold, and pattern	Deformed pattern	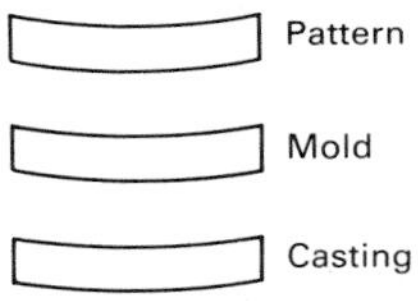
F 232	Deformation with respect to drawing proportional for casting and mold; pattern conforms to drawing	Deformed mold, mold creep, springback	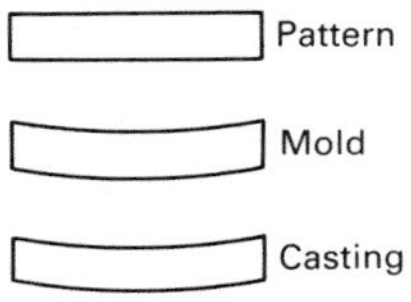
F 233	Casting deformed with respect to drawing; pattern and mold conform to drawing	Casting distortion	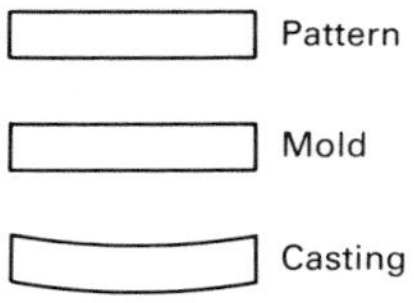
F 234	Casting deformed with respect to drawing after storage, annealing, machining	Warped casting	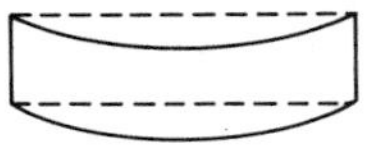

Inclusions or Structural Anomalies

G 100: Inclusions

G 110: Metallic inclusions

No.	Description	Common name	Sketch
G 111(a) . . .	Metallic inclusions whose appearance, chemical analysis or structural examination show to be caused by an element foreign to the alloy	Metallic inclusions	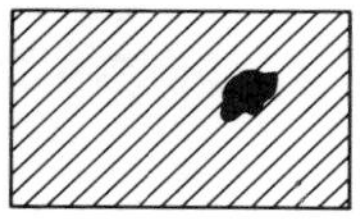
G 112(a) . . .	Metallic inclusions of the same chemical composition as the base metal; generally spherical and often coated with oxide	Cold shot	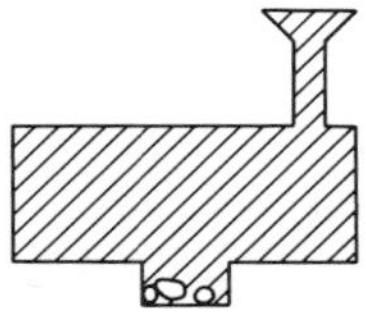
G 113	Spherical metallic inclusions inside blowholes or other cavities or in surface depressions (see A 311). Composition approximates that of the alloy cast but nearer to that of a eutectic	Internal sweating, phosphide sweat	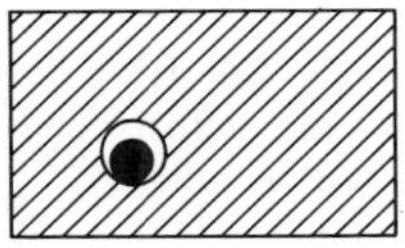

G 120: Nonmetallic inclusions; slag, dross, flux

| G 121(a) . . . | Nonmetallic inclusions whose appearance or analysis shows they arise from melting slags, products of metal treatment or fluxes | Slag, dross or flux inclusions, ceroxides |  |
| G 122(a) . . . | Nonmetallic inclusions generally impregnated with gas and accompanied by blowholes (B 113) | Slag blowhole defect | 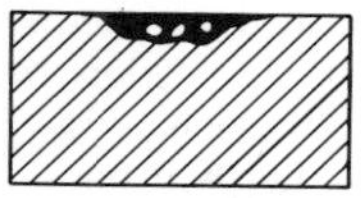 |

G 130: Nonmetallic inclusions; mold or core materials

| G 131(a) . . . | Sand inclusions, generally very close to the surface of the casting | Sand inclusions | 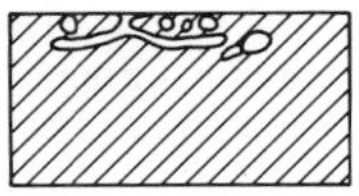 |
| G 132(a) . . . | Inclusions of mold blacking or dressing, generally very close to the casting surface | Blacking or refractory coating inclusions | 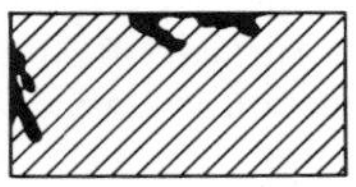 |

(continued)

(a) Defects that under some circumstances could contribute, either directly or indirectly, to casting failures. Adapted from *International Atlas of Casting Defects*, American Foundrymen's Society, Des Plaines, IL

Table 1   (continued)

No.	Description	Common name	Sketch	No.	Description	Common name	Sketch
G 140:	**Nonmetallic inclusions; oxides and reaction products**						
G 141	Clearly defined, irregular black spots on the fractured surface of ductile cast iron	Black spots		G 143(a) . . .	Folded films of graphitic luster in the wall of the casting	Lustrous carbon films, or kish tracks	
G 142(a) . . .	Inclusions in the form of oxide skins, most often causing a localized seam	Oxide inclusion or skins, seams		G 144	Hard inclusions in permanent molded and die cast aluminum alloys	Hard spots	

(a) Defects that under some circumstances could contribute, either directly or indirectly, to casting failures. Adapted from *International Atlas of Casting Defects*, American Foundrymen's Society, Des Plaines, IL

As the following examples illustrate, however, casting defects can sometimes be the direct cause of failures.

Example 1: Fracture of a Cast Stainless Steel Lever Because of a Cold Shut. The lever shown in Fig. 1 was a component of the main fuel-control linkage of an aircraft engine. After a service life of less than 50 h, the lever fractured in flight. The lever had been machined from a casting made of AISI type 410 stainless steel, then surface hardened by nitriding.

Investigation. Examination revealed that the lever had broken at a cold shut (Fig. 1) that extended through about 95% of the cross section of the lever arm. The remaining 5% of the cross section appeared to be sound metal.

Specifications for both the casting and the finished lever required radiographic inspection. This inspection method will reveal internal voids, such as gas porosity or shrinkage porosity, but will not detect a crack or a cold shut unless the plane of the imperfection is nearly parallel to the x-ray beam and unless there is a finite opening; a cold shut that has its faces in contact will rarely be revealed by radiography.

Conclusions. The lever broke at a cold shut extending through approximately 95% of the cross section. The normally applied load constituted an overload of the remainder of the lever.

Corrective Measures. Magnetic-particle inspection was added to the inspection procedures for this cast lever. Because cold shuts extend to the surface of a casting, magnetic-particle inspection is a relatively reliable method of detecting them in ferromagnetic materials.

Example 2: Fatigue Fracture of a Sand-Cast Steel Axle Housing That Originated at a Hot Tear. A sand-cast medium-carbon steel (Fe-0.25C-1Mn-0.40Cr-0.35Mo) heavy-duty axle housing, which had been

quenched and tempered to about 30 HRC, fractured after almost 5000 h of service.

Investigation. Figure 2 shows the fracture surface of the axle housing. Study of the casting revealed that the fracture had been initiated by a hot tear (region A, Fig. 2) about 10 cm (4 in.) long that had completely penetrated the 2.5-cm (1-in.) thick wall; this tear formed during solidification of the casting. The mass of a feeder-riser system located near the tear retarded cooling in this region, creating a hot spot.

Because this tear formed at a high temperature and was open to the surface, it was subject to oxidation, as indicated by the dark scale in region A in Fig. 2. Exposure to air also resulted in decarburization below the oxide scale and penetration of oxide along secondary cracks.

Both the size of the hot tear and the long service life of the housing indicated that the service stresses in the region of the tear had been relatively low. Stress concentrations at the tip of the tear, however, were sufficient to initiate the formation of two fatigue cracks (note the beach marks visible in regions B in Fig. 2). Ultimately, the effective thickness of the housing wall was reduced sufficiently for the applied load to cause final fracture. In this instance, the hot tear was located at an area of low stress, and the casting served a satisfactory life. However, had the local stress been high, even a very small discontinuity could have generated an early fracture.

Conclusions. Fracture of the axle housing originated at a hot tear that formed during solidification by hindered contraction and was enlarged in service by fatigue.

Corrective Measures. Changing the feeder location eliminated the hot spot and thus the occurrence of hot tearing.

Example 3: Fracture of Cast Steel Equalizer Beams. Two sand-cast low-alloy

Fig. 1 Cast type 410 stainless steel fuel-control lever that fractured at a cold shut

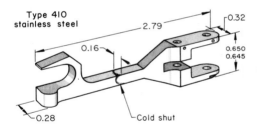

steel equalizer beams designed to distribute the load to the axles of a highway truck broke after an unreported length of service. Normal service life would have been about 805 000 km (500 000 miles) of truck operation. The cast equalizer beams were made according to ASTM A 148, grade 105-85, which specifies that all castings be heat treated (either fully annealed, normalized, normalized and tempered, or quenched and tempered, usually at the option of the foundry).

Investigation. Figure 3(a) shows a fracture surface of one of the beams. Internal shrinkage porosity and evidence of a cold shut were found near the tip of one flange (Detail A, Fig. 3). Decarburization of the fracture surface indicates that the initial crack (region B, Detail A, Fig. 3) was formed before heat treatment—most likely as a mechanical crack resulting from rough handling during shakeout. The crack originated in a region of internal shrinkage porosity and entrapped oxides.

The appearance of the fracture surface suggests that propagation of the crack occurred in a sequence of four large increments (regions

Fig. 2 Fracture surface of a sand-cast medium-carbon steel heavy-duty axle housing

Failure originated at a hot tear (region A), which propagated in fatigue (region B) until final fracture occurred by overload. 0.4×

strength, 712 MPa (103.25 ksi); yield strength (0.2% offset), 381 MPa (55.25 ksi); elongation in 5 cm (2 in.), 20%; reduction of area, 48%; and room-temperature impact strength, 32.5 J (24 ft · lb). These properties indicated that the steel was too soft for the application—probably due to improper heat treatment.

Conclusions. Fracture of the equalizer beams resulted from growth of mechanical cracks in one flange that were formed before the castings were heat treated. The pattern of crack growth suggested a service condition involving many relatively low applied stresses and occasional high loads.

Recommendations. Three changes in processing were recommended: (1) better gating and risering in the foundry to achieve sounder castings, (2) better shakeout practice to avoid mechanical damage and better inspection to detect imperfections, and (3) normalizing and tempering to achieve better mechanical properties.

Example 4: Tensile Fracture That Originated at Shrinkage Porosity in a Cast Low-Alloy Steel Connector. A sand-cast steel eye connector (Fig. 4a) used to link together two 54 430-kg (120 000-lb) capacity floating-bridge pontoons broke prematurely in service. The pontoons were coupled by upper and lower eye and clevis connectors that were pinned together.

The eye connector was 8.9 × 12.7 cm (3½ × 5 in.) in cross section and was cast from low-alloy steel conforming to ASTM A 148, grade 150-125. Expected life of such connectors was 5000 or more full-load cycles.

Investigation. Examination of the fracture surface established that the crack had originated along the lower surface (arrow O, Fig. 4b), initially penetrating a region of shrinkage porosity (region A, Fig. 4b) that extended over the lower right quarter of the fracture surface. Cracking then propagated in tension through sound metal (region B, Fig. 4b) and terminated in a shear lip at the top of the eye (region C, Fig. 4b). The stress configuration that produced the fracture approximated simple static tension.

Conclusions. Fracture of the eye connector occurred by tensile overload because of shrinkage porosity in the lower surface of the casting. The design of the connector was adequate because normal service life was more than 5000 load cycles.

Corrective Measures. Subsequent castings were inspected radiographically to ensure sound metal.

Example 5: Brittle Fracture of Cast Low-Alloy Steel Jaws Because of Shrinkage Porosity and Low Ductility of Case and Core. Eight pairs of specially designed sand-cast low-alloy steel jaws that had fractured were submitted for laboratory examination. The jaws were implemented to stretch the wire used in prestressed concrete beams. The wire was gripped at one end of the stretching machine between stationary semicylindrical jaws and at the other end between movable wedge-shaped

B, C, D, and E, Detail A, Fig. 3). The detectable difference in darkening of these fracture segments indicates that they existed in turn for sufficient lengths of times to undergo different cumulative amounts of surface oxidation. This behavior indicates infrequent but high loads. The fracture segment indicated as region E in Detail A in Fig. 3 shows a band of fatigue beach marks that nearly penetrated the fillet between the upper portion of the left flange and the web of the beam. Final fracture occurred at this stage because the remaining section of the beam was unable to sustain the applied load or as the result of a high load, such as those that caused earlier crack advances.

Examination of the fracture surface of the second beam showed that only two increments of crack propagation occurred before final fracture and that there was no fatigue cracking. Failure in the second beam occurred after far less crack penetration than in the companion beam, indicating that a high load, rather than a uniformly applied load, caused final fracture.

Both beams exhibited evidence of internal shrinkage porosity and entrapped oxides; a typical condition is shown in Fig. 3(b). Areas

of decarburization were found on the fracture surfaces of both beams. Internal cracking and cold shuts, such as those shown in Fig. 3(c), were evident adjacent to the fracture surfaces.

The microstructure of the beams consisted of transformation products and ferrite (Fig. 3d). The microconstituents exhibited some evidence of dendritic segregation.

Chemical analysis of the two beams, which were cast from separate heats of steel, showed close agreement in composition. The average composition was:

Element	Composition, %
Carbon	0.32
Manganese	0.93
Phosphorus	0.014
Sulfur	0.023
Silicon	0.46
Nickel	<0.05
Chromium	0.95
Molybdenum	0.25

A tensile specimen and a Charpy V-notch test bar were machined from one flange of each beam. The properties of the two beams were in close agreement. Average values were tensile

Fig. 3 Highway-truck equalizer beam, sand cast from low-alloy steel, that fractured because of mechanical cracking

(a) Fracture surface; Detail A shows increments (regions B, C, D, and E) in which crack propagation occurred sequentially. Dimensions given in inches. (b) Micrograph of an unetched specimen showing internal shrinkage porosity and entrapped oxides. 65×. (c) Micrograph of a specimen etched in 1% nital showing internal tensile crack and cold shut adjacent to fracture surface; note faint evidence of dendritic segregation at lower left. 65×. (d) Micrograph showing nonhomogenized microstructure consisting of transformation products and ferrite. 65×

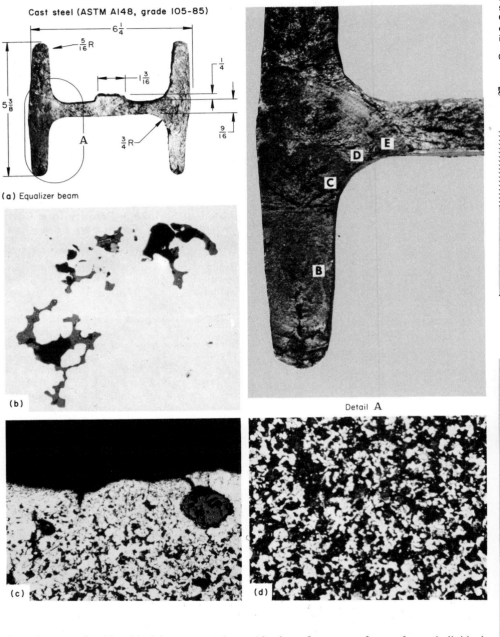

Cast steel (ASTM A148, grade 105-85)

(a) Equalizer beam

(b)

(c)   (d)

Detail A

Fig. 4 Sand-cast low-alloy steel eye connector from a floating-bridge pontoon that broke under static tensile loading

(a) Schematic illustration of pontoon bridge and enlarged view of eye and clevis connectors showing location of fracture in eye connector. (b) A fracture surface of the eye connector. Fracture origin is arrow O, area containing shrinkage porosity is at region A, area of sound metal is at region B, and the shear lip is at region C.

Cast steel (ASTM A148, grade 150-125)

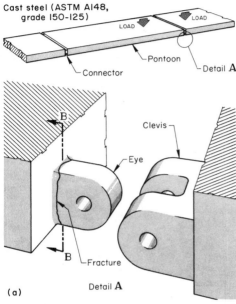

(a)   Detail A

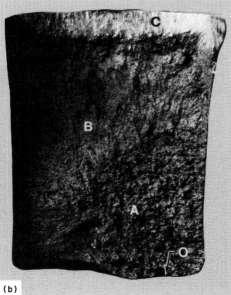

(b)

View B-B

jaws that were fitted in a block in a manner that forced them together as tension was applied to the wire. Figure 5(a) shows two pairs of movable jaws; Fig. 5(b), two pairs of stationary jaws.

Investigation. It was found that all the fractures were brittle and exhibited very little evidence of deformation. Two of the jaws contained surface cracks (not distinguishable in Fig. 5), but these cracks did not appear to be related to the main fractures. Figure 5(c) and

(d) show fracture surfaces of two individual movable jaws (from different pairs).

Little is known about the possible variations of stress that may have existed in this application. For example, the fit of the jaws in the block may have been imperfect, or there may have been too large an opening in the block, allowing the tips of the jaws to emerge without support. In such circumstances, the section of wire within the jaws might not have been in axial alignment with the wire under tension,

which would have resulted in one jaw being subjected to a side thrust at the tip.

All surfaces of the jaws had been case carburized and heat treated to a hardness of 62 to 64 HRC in the case and 36 to 39 HRC in the

Fig. 5 Wire-stretching jaws that broke because of shrinkage porosity and low ductility of case and core

The jaws, sand cast from low-alloy steel, were used to stretch wire for prestressed concrete beams. (a) Two pairs of movable jaws. 0.7×. (b) Two pairs of stationary jaws. 0.7×. (c) and (d) Fracture surfaces of two movable jaws. Approximately 2×

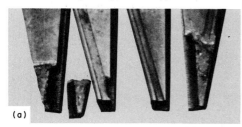

(a)

(b)

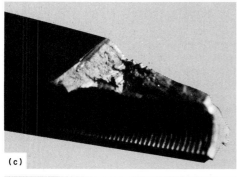

(c)

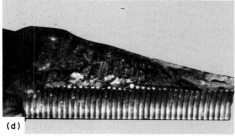

(d)

core. Metallographic examination of a specimen prepared from a jaw established that (1) the surface had been case carburized to a depth of 0.8 to 1.6 mm (¹⁄₃₂ to ¹⁄₁₆ in.), (2) the case structure was martensite containing small spheroidal carbides but no carbide network, (3) the structure of the core consisted of martensite plus some ferrite, and (4) the fracture was related to shrinkage porosity, indicating that the jaws were made from defective castings.

Chemical analysis of the core showed that the composition corresponded to 3310 steel except for low manganese, slightly high chro-

mium, and lower silicon than would normally be expected in a cast steel.

The hardness range of the core conforms to data in end-quench hardenability curves for 3310 steel. It represents, in fact, essentially the maximum hardness values attainable and as such indicates that tempering was probably limited to about 150 °C (300 °F). In view of the hardenability of the core and assuming that the tapered slot in the block had a hardened surface, the low tempering temperature used may have contributed to the brittleness.

Conclusions. Failure of the jaws was primarily attributed to the presence of internal microshrinkage, although the brittleness of case and core probably contributed to failure.

Recommendations. The procedures used for casting the jaws should be revised to eliminate the internal shrinkage porosity. Tempering at a slightly higher temperature to reduce surface and core hardness would be desirable. Finally, the fit of the jaws in the block should be checked to ensure proper alignment and loading.

Failures Due to Composition

While improper chemical composition would not appear to be a defect, so many problems can be traced back to composition that this subject should be discussed at some length. Errors in chemical composition can occur in three ways. The first involves selecting an inappropriate composition for a part and the service conditions it will encounter. The second way involves making an error in manufacturing, leading to the part or casting being produced to improper chemistry. Finally, trace elements, such as arsenic, tin, antimony, aluminum, nitrogen, and hydrogen, which are not a part of most specifications, can lead to unexpected failures.

Failure Due to Poor Material Selection. Consider the first case, in which a carbon or low-alloy casting is ordered to an inappropriate composition. If the part is for a low-strength low-toughness application, little damage is done except for the possible higher cost of the part. Premature failure is unlikely in this case.

In a high-strength casting, however, the improper selection of composition can lead to a myriad of problems. The composition may be such that it is not possible to obtain the strength and/or toughness required to prevent failure in service. In high-strength low-alloy steel castings, improper compositions could also lead to cracking during riser removal, quench cracking, or weldability problems.

Example 6: Poor Alloy Selection as a Cause of Failure. Induction-hardened teeth on a sprocket cast of low-alloy steel wore at an unacceptably high rate. A surface hardness of 50 to 51 HRC was determined; 55 HRC minimum had been specified. Analysis revealed that the alloy content of the steel was adequate for

the desired hardenability but that the specified carbon content (0.29%) was too low.

Conclusions. The low specified carbon content resulted in unacceptably low hardness. Because hardness largely controls wear rate, an early failure occurred.

Recommendations. The specification for this part was changed so that a higher carbon content (0.45% C) was required.

In stainless steels, of course, the selection of an improper composition can have even more severe consequences. Because of the relatively high cost of stainless, it is not unusual for a purchaser to request the minimum chemistry that will do the job under ideal conditions. These ideal conditions seldom occur, and the result is often a casting failure.

For example, it is not unusual for a customer to order an ACI type CF-8 casting that must be welded in service. A more expensive ACI type CF-3 is often a better choice when the part is to be welded, and postweld heat treatment is impractical. With the type of CF-8 casting, corrosion in the heat-affected zone (HAZ) of the weld is likely (see the section "Corrosion Failures of Cast Materials" in this article); with type CF-3 castings, such corrosion is very rare. This is merely one example of the imperfections that can be avoided if care is used in selection of composition. Most foundries have adequate technical staffs to advise users of stainless steel castings; consultation with them on details of type of service could in many cases lead to a better selection of composition.

Off Analysis as a Cause of Failure. Manufacturing-related errors in composition are usually due to an equipment malfunction or to an operator error. If not detected, they can lead to processing or service problems. Probably the most severe problems arise with deviations in carbon content (low or high), but difficulties can occur with other elements, such as nickel, chromium, or manganese. These are generally less severe. Fortunately, with foundries reasonably equipped to produce castings, off-analysis heats are rarely a major problem. Still, as the following example illustrates, they do occur.

Example 7: Brittle Fracture of a Roadarm Weldment of Two Steel Castings Because of Excessive Carbon-Equivalent Content. A roadarm for a tracked vehicle failed during preproduction vehicle testing. The arm was a weldment of two cored low-alloy steel sand castings specified to ASTM A 148, grade 120-95. A maximum carbon content of 0.32% was specified. The welding procedure called for degreasing and gas metal arc welding; neither preheating nor postheating was specified. The filler metal was E70S-6 continuous consumable wire that was bare except for a copper coating to protect it from atmospheric oxidation while on the reel.

Investigation. Analysis of the two castings revealed that the carbon content was higher than specified, ranging from 0.40 to 0.44%. Except for the carbon content, composition of

Fig. 6 Section through weld in a roadarm (a weldment of low-alloy steel castings)

The roadarm fractured in the HAZ because of high carbon-equivalent content. Fracture surface is at arrow. 0.8×

Fig. 7 A low-alloy steel casting exhibiting conchoidal or rock-candy fracture

Failure was caused by aluminum nitride embrittlement. In this case, high aluminum content (0.23%) and low nitrogen (36 ppm) produced the characteristic fracture appearance. 0.3×

Fig. 8 Scanning electron micrograph illustrating the characteristic rodlike artifacts from the failed low-alloy casting shown in Fig. 7

This characteristic appearance is confirmation of aluminum nitride embrittlement as opposed to ferrite films or temper embrittlement, which also lead to intergranular failure.

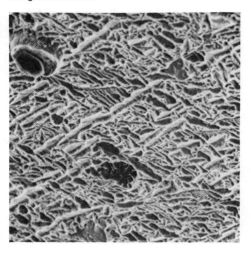

the castings corresponded to that of a high-manganese 8640 steel. The fracture occurred in the HAZ (Fig. 6), where quenching by the surrounding metal had produced a hardness of 55 HRC.

Some roadarms of similar carbon content and welded by the same procedure had not failed; it was determined that this was because they had been tempered during a hot-straightening operation. Rather than attempt to identify and segregate tempered and untempered roadarms, all production arms were tempered, including those already installed. A split induction coil was used to encompass the region of the weld and the adjacent hardened HAZ, thus avoiding the need to dismantle the arm, which would have been required for furnace heat treatment. As a result of tempering, there were no additional failures.

Analysis of these castings revealed a composition unfavorable for welding, particularly without preheating or postheating. The carbon equivalent for an average carbon content of 0.42% was 0.87% using the formula for low-alloy steel and 1.10% using the formula for carbon steel. These values, which are so high that they lie off the graph frequently used to relate weldability to carbon content, indicate that preheating and postheating are mandatory.

Conclusions. Brittle fracture of the roadarm was caused by a combination of too high a carbon equivalent in the castings and the lack of preheating and postheating during the welding procedure, which resulted in HAZs that were extremely hard and brittle.

Corrective Measures. Steps were taken to ensure that future roadarm castings would meet the specified carbon-equivalent content; a preheat was added to the welding procedure, and tempering after welding was specified for all roadarms.

Unspecified Elements and Trace Elements. Another area of composition that is probably less understood but is of increasing importance concerns the gases nitrogen, oxygen, and hydrogen and the residuals, such as sulfur, phosphorus, aluminum, antimony, arsenic, and tin. It is well documented that high nitrogen contents in low-alloy steels can be very detrimental in combination with aluminum (which is used for deoxidation) through the formation of aluminum nitride along grain boundaries. The intergranular fracture that results is the so-called rock-candy structure that can cause catastrophic failures under severe conditions of embrittlement if not detected. Figure 7 shows a typical rock-candy fracture. A high-resolution scanning electron microscope image shows the relics of the rodlike precipitate responsible for the embrittlement (Fig. 8). Scanning electron microscopy (SEM) analysis for this mode of embrittlement is one way to differentiate this mode of failure from temper embrittlement or grain-boundary ferrite. In the case shown, a nickel-chromium-molybdenum low-alloy steel was embrittled by very high levels of aluminum (0.23%) combined with good low-nitrogen (36 ppm) content.

Because much has been written on the effect of high sulfur contents on mechanical properties, this subject needs no elaboration except to state that this is recognized and should be of decreasing importance as a defect cause.

Relatively little is known of the effect of arsenic, tin, and antimony on castings except that it is generally agreed that low levels, that is, 0.005% or lower, are important to minimize failures due to mechanical properties in high-strength castings.

Much has been written on the effects of hydrogen. As in wrought products, hydrogen can be a cause of failure in low-alloy steel castings. Hydrogen-related defects can vary from open flakes to fisheyes that appear in fractured samples. This can be extremely detrimental in heavy section castings and particularly in those heat treated to high strength (yield strengths above about 690 MPa, or 100 ksi). Additional information is provided in the section "Failures Related to Hydrogen-Assisted

Cracking" in this article. Extreme care in manufacturing is the only way to minimize the potential for hydrogen problems.

Oxygen, although not such a potentially dangerous residual, can be cause for dirty steel and low mechanical properties when care is not taken in the manufacturing process to keep oxygen under control.

Failures Due to Improper Heat Treatment

Heat treating often contributes to casting failures when it is done improperly or incompletely. All cast alloys can be detrimentally affected by incorrect heat treatment.

Carbon steel parts are normally supplied in the normalized or normalized-and-tempered condition, but can be quenched and tempered or annealed. Some are also used in the as-cast condition. Heat treating of carbon steel should be selected to ensure adequate toughness and weldability. Carbon steel castings, if they fail due to improper heat treatment, most often fail in a service that requires high impact toughness, particularly at low temperatures.

Low-alloy steel castings are susceptible to heat-treating problems. In addition to toughness and ductility, strength is also determined by the heat-treat cycles, and these can be varied over a wide range. The best combinations of mechanical properties are developed in the quenched-and-tempered condition, with the tempering conditions (temperature, time at temperature, and quenching temperature) controlling the hardness and ductility.

Fig. 9 The bainitic microstructure of a 0.27% C low-alloy steel casting that was pulled from the water quench too early

Etched with Vilella's reagent. 500×. See also Fig. 10.

Fig. 10 The same steel shown in Fig. 9 after a good water quench

The microstructure is predominantly martensitic. Etched with Vilella's reagent. 500×

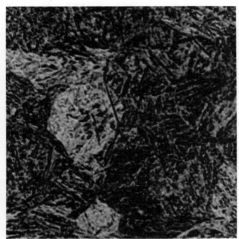

Many failures in low-alloy steel castings are due to lack of strength, leading to fatigue or overload fractures. Low hardness (strength) can also lead to premature wear of wear-resistant low-alloy castings. Low strength can be caused by several heat-treating errors. A slack quench caused by lack of good tank circulation, overloaded heat-treat baskets, or early (hot) removal from the quench will lead to low strength. This is due to the lower-hardness bainitic microstructures that are formed instead of the harder martensitic ones. Figure 9 shows a bainitic microstructure in a digger tooth that was pulled early from the water quench; Fig. 10, the normal fully martensitic microstructure. Low strength can occur with a good quench when it is followed by tempering at too high a temperature. An extended hold time in the heat-treat furnace at the tempering temperature may produce the same result.

Low-alloy steel castings can suffer from temper embrittlement when they are tempered in the region of 260 to 540 °C (500 to 1000 °F). Slow cooling through this region from a higher temperature can also lead to the problem. It is advisable to specify a water quench from the tempering temperature whenever practical to avoid this. The low toughness caused by temper embrittlement can lead to brittle fractures. More highly alloyed steels, such as the 4300 series and the HY steels, tend to be more susceptible to the problem (see the article "Embrittlement of Steels" in Volume 1 of the 9th Edition of *Metals Handbook* for a more detailed explanation of temper embrittlement).

Temper embrittlement is one example of a heat-treating problem that is not detected by the common acceptance criteria of hardness testing. Annealed steels also differ from normal-

ized-and-tempered steels in toughness but not in hardness. Data from Volume 1 of the 8th Edition of *Metals Handbook* (Table 2, page 124) illustrate this. Two heats of 0.22C-0.12Mo steel were treated and tested. One was annealed at 925 °C (1700 °F) and furnace cooled. The other was normalized at 925 °C (1700 °F), air cooled, then tempered at 670 °C (1240 °F). While the tensile strengths of about 483 MPa (70 ksi) and the hardnesses of about 140 HB were the same, the impact results were quite different. The annealed material exhibited only 41- to 46-J (30- to 34-ft · lb) Charpy V-notch impact toughness, while the normalized-and-tempered heat had 83- to 87-J (61- to 64-ft · lb) impact toughness. Toughness testing is a good method of ensuring correct heat treatment.

It is hoped that the effect of section size on mechanical properties is well understood by those responsible for the design and specification of castings. This is particularly important when considering low-alloy materials. The mechanical properties of a part are determined by the quench rate during the hardening cycle and the hardenability of the specified alloy. It is standard practice in the foundry industry to use separately cast keel blocks, for example, ASTM A 370 dimensions, to determine the mechanical properties of a heat of steel. Where a historical relationship exists, tying test bar properties to field service, this approach is a sound one.

There is always a danger, however, that the assumption will be made that the mechanical properties obtained from a 3.8-cm (1.5-in.) thick test bar are equivalent to those found in a casting with a much greater thickness. In the vast majority of the cases, this will not be true. When no experience exists in correlating heat treatment/alloy/section size/service/test bar properties, it is prudent to request mechan-

ical properties from test bars roughly equivalent in section thickness to the critical sections of the castings. Alternatively, this relationship should be established through field trials under controlled conditions or by destructively testing actual castings. In many cases, the foundry may have established this information.

Austenitic Steels. Care is also essential in heat treating more highly alloyed castings. An improperly heat treated stainless steel, such as ACI CF-8, will corrode faster in service. This occurs because carbides precipitate in grain boundaries, leaving other areas depleted of the chromium that gives the steel its corrosion resistance. This is called sensitization, and failure to heat treat properly is a frequent cause of the problem (see Example 18 in this article). Too low a temperature or slow cooling from the solution-annealing temperature can produce the same detrimental effect.

Austenitic manganese (Hadfield) steels can suffer from a similar problem microstructurally, but in this case, poor wear resistance and decreased ductility are the results. Austenitic manganese steel exhibits its best properties when it gets a good water quench from 1095 °C (2000 °F). Otherwise, carbides precipitate heavily at the grain boundaries or, in severe cases, within the austenite, reducing the work-hardening capability of the matrix and lowering overall toughness and ductility. Reheating manganese steel can give a similar effect, as shown in the following example.

Example 8: Brittle Fracture of a Cast Austenitic Manganese Steel Chain Link. The chain link shown in Fig. 11(a), a part of a mechanism for transferring hot or cold steel blooms into and out of a reheating furnace, broke after approximately 4 months of service. The link was cast from 2% Cr austenitic manganese steel and was subjected to repeated heating to temperatures of 455 to 595 °C (850 to 1100 °F).

Investigation. Chemical analysis revealed that the link had been cast from 13% Mn austenitic manganese steel to which 1.5 to 2.0% chromium was added to improve wear resistance by forming chromium carbides.

Examination with a hand magnet indicated that the entire link was magnetically responsive. When properly heat treated, austenitic manganese steel is a tough nonmagnetic alloy; however, composition changes that occur at the surface during heating may produce a magnetic skin.

Further examination of the magnetic properties of the link was performed with the aid of a Tinsley thickness gage, which is normally used for measuring the thickness of nonmagnetic coatings on ferritic steel surfaces. With this instrument, a completely nonmagnetic substance is given a rating of 10, and a magnetic material, such as plain carbon steel, is assigned a rating of 0. Specimens cut from the link were tested for magnetism after various heat treatments with the following results:

Fig. 11 Reheating-furnace chain link, sand cast from austenitic manganese steel, that failed by brittle fracture because material was not stable at operating temperatures

(a) Chain link showing location of fracture. Dimensions given in inches. (b) Macrograph of a nital-etched specimen from an as-received chain link. 1.85×. (c) and (d) Micrographs of a nital-etched specimen from an as-received chain link. 100 and 600×, respectively. (e) Normal microstructure of as-cast standard austenitic manganese steel. 100×. (f) Micrograph of a nital-etched specimen that had been austenitized 20 min at 1095 °C (2000 °F) and air cooled. 315×. (g) Micrograph of the specimen in (f) after annealing 68 h at 480 °C (900 °F). 1000×. (h) Chart showing time at temperature needed to embrittle austenitic manganese steel after heat treating for 2 h at 1095 °C (2000 °F) and water quenching

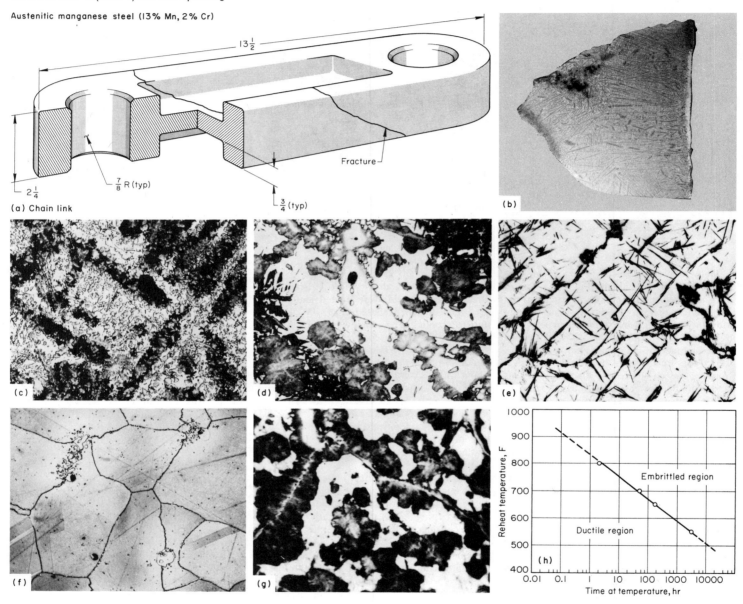

Condition(a)	Tinsley reading	Degree of magnetism
Web (as-received)	4.5	Partial
Flange (as-received)	4.65	Partial
1095 °C (2000 °F), WQ	10.0	Nonmagnetic
1095 °C (2000 °F), AC	10.0	Nonmagnetic
1095 °C (2000 °F), WQ; 4 h at 455 °C (850 °F)	9.0-9.6	Trace at edges
1095 °C (2000 °F), FC	8.0	Partial
1095 °C (2000 °F), FC; 68 h at 480 °C (900 °F)	4.8	Partial
1095 °C (2000 °F), AC; 68 h at 480 °C (900 °F)	4.75	Partial

(a) WQ, water quenched; AC, air cooled; FC, furnace cooled

Figure 11(b), a macrograph of a polished and nital-etched specimen from the link as received at the laboratory, shows dendritic segregation and microshrinkage pores (dark spots at upper left). Micrographs of specimens from the as-received link are shown in Fig. 11(c) and (d). These micrographs display a segregated structure of complex carbides (black needlelike phase), $(FeMn)_3C$ carbides (white-on-white area), and pearlite (gray areas) in an austenite matrix, which accounts for the macrostructure shown in Fig. 11(b). The segregation follows the grain boundaries, but the appearance is modified by dark oxide patches left by the nital etch.

The effects of the repeated heating of the link on the characteristics of the precipitated carbides can be observed by comparing the micrographs shown in Fig. 11(c) and (d) with that shown in Fig. 11(e), which illustrates the normal microstructure of as-cast standard austenitic manganese steel (nominal composition, 1.15C-12.8Mn-0.50Si), displaying pearlite and/or martensite with carbide precipitation at grain boundaries and along crystallographic planes.

Figure 11(f) shows the microstructure of a specimen from the chain link that was austenitized for 20 min at 1095 °C (2000 °F) and air cooled. In this small specimen, the cooling rate from the austenitizing temperature was sufficient to retain the austenitic structure. The globular carbides did not dissolve at this temperature. The effect of reheating this structure to 480 °C (900 °F) for 68 h is illustrated in the micrograph shown in Fig. 11(g), which depicts carbides and austenite decomposed to pearlite in an austenite matrix. Many patches of fine pearlite formed both along the grain boundaries and within the grains. Note the similarity of the microstructure in Fig. 11(g) to that in Fig. 11(d).

Brinell hardness tests performed on the link adjacent to the fracture surface yielded values of 415 HB for the web section and 363 HB for the flange; many fine cracks were found in the hardness indentations that attested to a brittle condition. These hardness values were in distinct contrast to the normal level of 180 to 200 HB expected in as-quenched austenitic manganese steel.

One question arising from the data on magnetism and from the results of metallographic analysis was whether the broken link had been properly heat treated before installation in service and magnetized as the result of service temperatures or whether it had inadvertently been shipped from the foundry as-cast and thus partly magnetic. This question was resolved by the pronounced dendritic segregation observed in the microstructure shown in Fig. 11(c). The presence of the dendritic pattern indicates that the link was not heat treated after casting, because austenitizing followed by rapid cooling would have homogenized the microstructure and largely eliminated the dendritic pattern.

Although quantitative data are lacking concerning the mechanical-property changes that result from tempering of austenitic manganese steel, metallographic data do exist that predict when embrittlement may be expected during reheating at various temperatures. Figure 11(h) shows a graph of the time at temperature needed to embrittle austenitic manganese steel after an initial heat treatment of 2 h at 1095 °C (2000 °F), followed by water quenching. The data used to plot this curve are based on the first evidences of transformation products visible under the microscope.

Conclusions. The chain link failed in a brittle manner because the austenitic manganese steel from which it was cast became embrittled after being reheated in the temperature range of 455 to 595 °C (850 to 1100 °F) for prolonged periods of time. The alloy was not suitable for this application, because of its metallurgical instability under service conditions.

The chain links had not been solution annealed after casting. This was not as significant a factor in the failure as reheating above 425 °C (800 °F).

Unanticipated Heating After Final Heat Treatment. Thermal processing of cast-ings after normal heat treatment can lead to other problems. One example is preheating for straightening. Castings that require quenching and tempering for maximum properties often require straightening afterwards due to warpage in the quench operation. A common example is a bulldozer blade. To restore flatness, it is common practice to reheat the parts with gas-fired torches so that the high-hardness part does not break during pressing. The problem is that if the tempering temperature is exceeded in the local region heated by the torch, the material will be softened, and premature wear will occur. In extreme cases, the tempered martensite will be transformed to a softer ferritic, pearlitic mixed microstructure that is subject to breakage in use. This same problem can occur in field welding and hardfacing, in which preheating with torches is even more common.

Failures Related to Surface Hardening. Many castings receive selected surface hardening in areas of high wear. This can be done with a variety of processes, including flame, induction, laser, and electron beam. Process controls must be tight, as the following failures illustrate.

Cast steel gears are frequently surface hardened, which can in some cases lead to casting failures. Induction hardening of the internal teeth in a geared winch drum produced transverse cracking (Fig. 12). Because of their orientation, the cracks did not propagate during service and thus were judged not detrimental to service life. The drum was cast from low-alloy steel to meet the mechanical-property requirements of ASTM A148, grade 105-85. The teeth were machined and then induction hardened to 45 HRC minimum with a minimum case depth of 1.6 mm (1/16 in.). The gear shown in Fig. 12 and others exhibiting similar cracks were tested. Required service life was 20 h at full load. The tests were halted after 42 h (2 × 10⁵ load cycles); no changes in the cracks had occurred. These data permitted establishment of an acceptance criterion for cracks that would normally have been cause for rejection of the castings.

The teeth were induction hardened one tooth at a time. The expansion stresses created by heating an individual tooth may have been sufficient to crack the adjacent hardened tooth if there were underlying defects. Hardening of the tooth contours by case carburizing with subsequent quenching and tempering might avoid this problem.

Example 9: Fatigue Fracture of a Cast Chromium-Molybdenum Steel Pinion. A cast countershaft pinion on a continuous reversible car puller for a blast furnace broke after 1 month of service; normal expected life was 12 months. Although operating conditions during the month in question were reported as normal, overloading did occur. The pinion was specified to be made of 1045 steel and to be heat treated to a hardness of 245 HB.

Fig. 12 Internal teeth in a cast ASTM A148, grade 105-85, winch drum that cracked transversely during induction hardening
The cracks were judged not detrimental to service life.

Investigation. The pinion steel was analyzed and found to have the following composition:

Element	Composition, %
Carbon	0.32
Manganese	0.70
Phosphorus	0.10
Sulfur	0.023
Silicon	0.54
Chromium	1.57
Molybdenum	0.37

This composition was considered to be a satisfactory alternative to 1045 steel. The pinion was annealed before flame or induction hardening of the teeth, which resulted in a surface hardness of 363 HB and a core hardness of 197 HB.

Figure 13(a) shows a portion of the broken pinion with one tooth missing; the missing tooth failed by fatigue fracture through the tooth root. The fracture surface contained beach marks typical of fatigue-crack propagation. Magnetic-particle inspection revealed similar cracks at the bases of the adjacent teeth. Tool marks were found along the roots of these teeth.

A heavy wear pattern was visible on one end of the flank of the pressure side of the teeth; at the other end, the original paint was undisturbed. This indicated misalignment of the pinion and therefore abnormally heavy loading at the end from which the fatigue cracks originated.

Figure 13(b) shows a section through a broken tooth that was etched to reveal the macrostructure. The surface hardening did not extend to the tooth root, leaving this highly stressed area with a low-strength annealed structure. Microscopic examination of the hardened surface area confirmed that it had a quenched-and-tempered structure.

Conclusions. The pinion teeth fractured by fatigue because of the low strength of the tooth

Fig. 13 Countershaft pinion that fractured in fatigue at roots of teeth because of incomplete hardening at tooth surfaces

(a) Schematic illustration of the pinion, which was sand cast from a chromium-molybdenum steel. Dimensions given in inches. (b) Macrograph of a nital-etched section through a broken tooth showing surface hardening on sides and top of tooth. 2½ ×

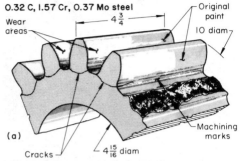

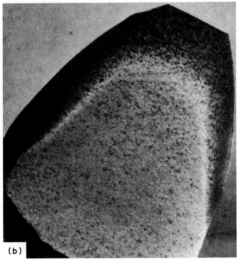

(b)

root regions that resulted from incomplete surface hardening of the tooth surfaces. Contributing factors to the fracture included uneven loading because of misalignment and stress concentrations in the tooth roots caused by tool marks.

Corrective Measures. Greater strength was provided by oil quenching and tempering the replacement pinions to a hardness of 255 to 302 HB. Machining of the tooth roots was revised to eliminate all tool marks. Surface hardening was applied to all tooth surfaces, including the root. Proper alignment of the pinion was ensured by carefully checking the meshing of the teeth at startup.

Carburizing. Another method of surface hardening is carburizing in a controlled-atmosphere furnace. This type of operation is usually well handled, and no problems occur. However, if the fuel-fired burners of a normal-atmosphere oven are adjusted too far on the reducing side, a similar but uncontrolled carburization can occur. This can give the part

surface hard spots that may be brittle. Similarly, if the burners are adjusted too much on the oxidizing side, surface decarburization will result. This leads to soft spots on the surface that will be low in fatigue and/or wear resistance. The depth of either a carburized or a decarburized surface layer is best determined by metallography and careful hardness testing. Both can be minimized by carefully controlling the furnace atmospheres with properly instrumented heat-treating equipment.

Overload Failures of Steel Castings

The majority of steel-casting failures are the result of overload. The solution to these problems is primarily the responsibility of the design engineer and the user of the equipment. The failure analyst must determine if failure is due to a defective part, inadequate design, or a combination of these. He must also assist the engineer in the selection of improved materials, if available. To accomplish this, the analyst must know and understand most of the failure mechanisms associated with steel, specifically those related to castings.

Fracture-Surface Analysis. Low-alloy steel castings that require moderate-to-high strength and good shock resistance should be quenched to a martensitic or lower bainitic microstructure. The steel is then tempered to the required strength, taking care to avoid the temper-embrittlement zone.

In tensile or bending overloads, these castings will exhibit a shear zone around the outer edge of the fracture. This zone, often referred to as a shear lip, is usually quite large at lower strength levels, becoming less prominent at higher strengths (Fig. 14).

The existence of a reasonable shear lip is a positive clue that the steel behaved in a ductile manner under the conditions of failure. Also, if the hardness indicates a proper strength level, these two factors are strong indicators of proper heat treatment.

The initiation area of the fracture must be carefully observed for discontinuities or defects. The fracture surface will have a chevron pattern that will radiate outward from the origin.

Figure 15 shows the chevron pattern radiating from a hot tear. A typical overload fracture will show a chevron pattern starting just under the tension surface, with a shear lip at the surface. The initiation zone should be free of excessive shrinkage, hot tears, inclusions, laps, hydrogen flakes, and prior cracks. Figure 16 shows an example of such a fracture.

Fatigue Failures. A common type of overload failure is fatigue. This occurs during cyclic loading at stress levels below the yield strength of the steel. The chances of fatigue occurring are increased by design (stress concentrations), moist or corrosive environments, and rough surface conditions.

Example 10: Fatigue Failure Due to Improper Design. Figure 17 shows a fracture

Fig. 14 Shear lips (arrows) indicate the toughness of a material and the mode of failure

As strength increases, the size of the shear lip generally decreases. (a) Good shear in a casting with hardness of 500 HB. (b) and (c) Increasing shear lip size as hardness drops to 400 HB

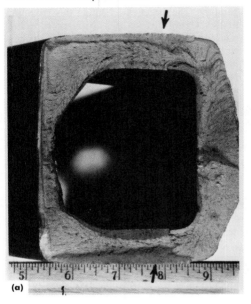

(a)

(b)

(c)

Fig. 15 A fracture that initiated at a hot tear (arrows)

The chevron pattern radiating from the fracture origin is evident. See also Fig. 16.

Fig. 18 Fracture surface of a cast steel rear-axle housing for an off-road truck that fractured by fatigue due to overload

Fracture origin is at region O, fatigue is at region A, and the final-fracture is at region B.

Fig. 16 A second illustration of the chevron pattern radiating from origin (arrow)

In this case, no defect is observed at the origin.

Fig. 17 Fatigue failure of an anchor link

The arrow indicates the sharp edge that acted as the stress concentrator that led to fatigue-crack growth.

of an anchor link that rides on a post with a spherical surface. No bevel or radius was called for, leaving an extremely sharp edge. A fatigue crack grew from the top sharp edge (note beach marks in Fig. 17). The casting had good toughness and strength (1448 MPa, or 210 ksi, tensile strength).

Conclusions and Recommendations. This fatigue failure was the result of poor design, which incorporated a notch that acted as a stress raiser, leading to fatigue-crack growth. A generous radius was recommended to avoid similar failures.

Example 11: Fatigue Fracture of a Cast Steel Axle Housing. A fractured steel sand casting that was part of a rear-axle housing from an off-road oil-rig truck was submitted for laboratory analysis of the failure. It was reported that the axle had been subjected to overload before failure. There had been a considerable history of successful service of these castings without failure.

Investigation. Figure 18 shows a view of the fracture surface. The crack initiated at region O and propagated for approximately two-thirds of the total fracture surface as a fatigue crack (region A, Fig. 18). The remainder of the fracture (region B, Fig. 18) was produced by a single overload. No evidence of surface defects was detected at the site of the crack origin.

Chemical analysis of the housing established a composition of 0.23C-1.20Mn-0.11V, which was within the specification. Mechanical properties of the casting were as follows: yield strength, 345 MPa (50 ksi); tensile strength, 524 MPa (76 ksi); elongation in 5 cm (2 in.), 31%; reduction of area, 51.9%; hardness, 163 to 167 HB. The hardness met the specification satisfactorily (163 HB minimum), but the strength was below specification (minimum yield strength, 365 MPa, or 53 ksi; minimum tensile strength, 586 MPa, or 85 ksi). The ductility was acceptable (22% minimum elongation; 35% minimum reduction of area). The specification was not consistent in that 163 HB corresponds to approximately 545 MPa (79 ksi) tensile strength, and 586 MPa (85 ksi) tensile

strength corresponds to a hardness of about 174 HB. Because Brinell hardness was generally used as a quality-control check, it is probable that many castings meeting the hardness specification were accepted for this service and were considered to be of acceptable strength.

Evidence of corrosion on the fracture surface indicated that crack growth had been very slow and that the casting had withstood the usual service stresses for a long time despite the presence of the fatigue crack. No material defect was found that could have caused the original crack or hastened its subsequent growth; therefore, operation of the truck over rough terrain must have brought about a condition of shock loading, resulting in excessive stress that initiated the crack.

Conclusions. Fatigue fracture of the rear-axle housing was the result of overstress and cyclic loading when the truck was traveling over rough terrain. This was considered to have been a unique failure, and because axles of this design had performed satisfactorily, no corrective action was taken.

Other Examples of Fatigue Failures. Design considerations are extremely important in fatigue-overload problems. Figure 19 shows an example of a cast machined pin with a drilled oil hole. Fatigue started at the sharp outer edges of the hole and grew through 90% of the section before final failure. A generous radius at the edges of the hole would have greatly reduced the stress concentration.

Fatigue failure was studied in a stainless steel rotor (centrifuge) bowl. After several years of use in a high-speed spinning operation, cracks were discovered at the roots of sharply machined corners (Fig. 20).

Fatigue was suspected, but proof was required. The crack was opened for study of the surface. Because of the sharpness of the corners, steps at initiation were not apparent, and only a suggestion of beach marks could be seen at the top of the crack (Fig. 21). The crack surface was examined using a scanning electron microscope. At a magnification of 20×, steps were apparent at the initiation zone, but beach marks could not be resolved. Due to the slow

Fig. 19 Fatigue failure of a cast pin

Cracking initiated (arrows) at both ends of the oil hole and grew through 90% of the section. A generous radius to the oil hole eliminated failure.

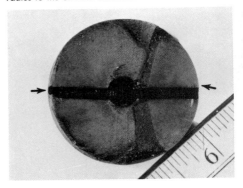

Fig. 20 Fatigue cracking in a cast stainless steel rotor bowl

Initiation occurred at the sharp machined corner; crack was made visible using dye-penetrant inspection. See also Fig. 21.

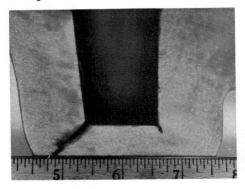

Fig. 21 Cracks visible in Fig. 20 that were opened to reveal the fracture

There is little evidence of beach marks except at the initiation zone (arrows).

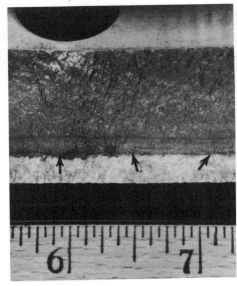

Fig. 22 Fatigue failure showing multiple steps on the fracture surface

The steps were the result of cracks that begin at different locations on the fracture surface.

Fig. 23 Steel chain links that failed by fatigue

In both cases, cracking initiated on the inside surface. (a) and (b) Fracture surfaces showing beach marks and zone of final fast fracture.

Fig. 24 Tooth adapter that failed by overload

These types of failures are often seen as bent or deformed parts.

crack growth, individual crack fronts could be resolved only at high magnification.

In another fatigue failure, an adapter was attached to a lip of an earthmoving bucket by an insertable shank held in place with a wedge. Due to lip wear, the adapter was not seated properly, allowing repeated stress to be transmitted to the wedge slot. Fatigue cracking started at the interior surfaces of the slot, growing through a quarter of the section before final overload failure occurred. Note the multiple steps at the fatigue-initiation surface shown in Fig. 22. Wear after fracture or corrosion will often erase the beach marks. The stepped initiation is a good indication of fatigue. Chain links often fail by fatigue due to heavy cyclic loading. Figure 23 shows two examples, both starting on the inside surfaces. The links are 1034-MPa (150-ksi) tensile steel.

Overload failures can also be determined by deformation. Due to the excellent toughness of quenched-and-tempered cast steels, bending instead of breaking often occurs. Figure 24 shows an example of a bucket point adapter with bending (arrow) due to significant overload. The tensile strength is 1103 MPa (160 ksi).

In nearly all of the above cases, the solution to fatigue-related failures involves lowering the local stress (or stress intensity) by eliminating or reducing stress concentrations. In some cases, advantage can be gained by increasing the strength of the part through alloy modifications. Similarly, material cleanliness on a microscopic scale can influence performance, but to a much lower extent—for example, argon-oxygen decarburized (AOD) produced alloys as compared to electric arc melted steels.

Overload Failures Characterized by Shear. On occasion, dramatic sudden overloads result in fracture surfaces that are 100% shear. Figure 25 shows an anchor link that failed due to a twisting action, resulting in a 100% shear fracture. Such fractures have no

Fig. 25 Overload fracture that failed in torsion

These failures often exhibit a 100% shear fracture mode. Arrows indicate the twisting direction.

shear lips with only a flat shear plane. Overload (design) failures are characterized by initiation at defects that are allowable according to the governing specifications. Mechanical-property measurements may be needed to confirm a hardness measurement. Hardness, combined with a careful inspection of the fracture surface for signs of good ductility (shear lips and ductile dimple fracture mode), will often suffice.

Failures Related to Welding

Although it is often not appreciated by those outside the foundry industry, welding is an integral part of casting production. Welding is used for repairing casting defects and for fabricating larger assemblies from component castings. Virtually all of the fusion-welding processes are used for welding steel castings. Because welding is so widely used, it should not be surprising that a significant proportion of steel-casting failures are welding related.

Failures related to welding are covered in the article "Failures of Weldments" in this Volume. Although many of the examples given in that article involve wrought materials, much of the information can be applied directly to the corresponding cast versions of those materials. It may be helpful, however, to discuss and describe several of the more common types of welding-related failures that occur in steel castings.

Weld failures related to hydrogen-assisted cracking are most frequently observed in hardenable steels having moderate-to-high degrees of hardenability. These materials include carbon steels containing more than about 0.35% C, most low-alloy and medium-alloy steels, most tool steels, and the martensitic stainless steels. The manner in which these types of failures occur is basically as follows.

First, hydrogen is introduced into the weld metal while it is in the molten state. In arc-welding processes, this usually comes about when hydrogen-bearing contaminants are decomposed in the arc. The most common of these contaminants is water, which may be present in electrode coatings, flux, shielding gas, wire lubricants, or rust on electrodes or base metal. Grease, oil, and paint are other potential sources of hydrogen. Second, some of the hydrogen remains in the weld metal as it solidifies and cools. Third, a portion of the hydrogen in the weld deposit diffuses into the HAZ during cooling. Finally, a crack initiates as a result of the dissolved hydrogen in combination with stress (residual welding stress, service stress, or both). Cracking is most common in the HAZ. When high-strength filler materials are used, however, cracking may also occur in the weld deposit.

There are three commonly used ways for reducing the risk of failures related to hydrogen-assisted cracking of welds. The first involves obtaining low hydrogen contents. The second consists of improving the resistance of the weld to hydrogen-assisted cracking. Finally, stress levels can be reduced.

The hydrogen contents of welds can be minimized in a number of ways. One method is the use of welding processes that inherently tend to produce lower weld-metal hydrogen contents, for example, gas tungsten or gas metal arc welding using solid wire. Another is to use low-hydrogen filler materials, such as EXX16 or EXX18 coated electrodes, for flux-cored arc welding. Another important step is to employ the appropriate storage, handling, and bake-out procedures for welding consumables, such as low-hydrogen electrodes and shielded arc welding flux. To ensure low hydrogen contents, it is also necessary to remove any hydrogen-containing contaminants, such as rust, grease, paint, and oil, before welding. A final method of obtaining low hydrogen contents is to bake the weldment to allow dissolved hydrogen to escape. However, this is rarely practical for thick-section welds due to the extensive times required.

Probably the single most effective way to decrease the susceptibility of a weld to hydrogen-assisted cracking is to lower its strength level. This is the basis for the common use of preheating. Preheating lowers the cooling rate in the HAZ and the weld deposit, thereby allowing transformation to softer microstructures. A postweld heat treatment can be even more effective in lowering susceptibility to hydrogen-assisted cracking.

Without stress, hydrogen-assisted cracking cannot occur. Relief of welding stresses, therefore, is an important way of reducing the risk of weld-related hydrogen failures. In most medium-strength carbon steels, low-alloy steels, and martensitic stainless steels, stress relief is usually accomplished by postweld heat treating at a temperature slightly below the tempering temperature of the base metal. When these materials are to be used in the high-strength conditions (necessitating low temper-

ing temperatures), complete re-heat treatment is preferred.

A proper stress-relief heat treatment will not only relieve welding stresses but will also improve the resistance of the material to cracking. A common error in postweld heat treatment of some alloy steels is to air cool from the postweld heat-treatment temperature. When this is done, stresses will be lowered, but the material may also suffer temper embrittlement. Because temper embrittlement greatly increases susceptibility to hydrogen-assisted cracking, the net result of air cooling from the postweld heat-treatment temperature may be increased risk of failure.

Example 12: Failure of a Large Cast Dragline Bucket Shackle. A large shackle used in operating a dragline bucket failed in service. The shackle was made of a cast low-alloy steel (similar to AISI 4320) and had been heat treated to a hardness of 415 BN.

Investigation. The shackle failed by fracturing through the load-bearing region. Examination of the fracture surface revealed that at the time of final failure a fatigue crack had grown through about one-third of the cross section (Fig. 26). A secondary fatigue crack, perpendicular to the main fracture, was also observed.

A saw cut was made about 13 mm (½ in.) behind the fracture surface. The cut surface was ground, then macroetched with an ammonium persulfate solution. This revealed a repair weld (Fig. 27). No HAZ was observed, indicating that the weld had been rehardened and tempered after welding. It was obvious, however, that the previously mentioned fatigue cracks had initiated in the former HAZs of this weld. A crack was also observed in the weld deposit itself.

The composition of the weld deposit was analyzed, and the results were as follows:

Element	Composition, %
Carbon	0.18
Manganese	1.00
Silicon	0.43
Chromium	0.42
Nickel	1.37
Molybdenum	0.31
Sulfur	0.015
Phosphorus	0.007

This corresponded to a heat-treatable flux-cored arc welding filler material that was known to have been used for repair welding of these products.

Conclusion. This shackle failed because of fatigue initiating at hydrogen cracks that had occurred in the HAZ of a repair weld. According to good practice for high-strength steels, the weld had been made with a heat-treatable filler material, and a full postweld heat treatment had been performed. However, a low-hydrogen filler material had not been used to make the weld.

Fig. 26 Stereo pair showing fracture surface of cast dragline shackle

See also Fig. 27.

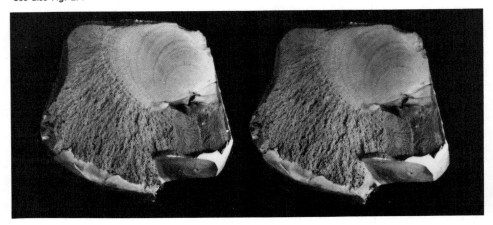

Fig. 27 Macroetched cross section taken 13 mm (½ in.) behind fracture shown in Fig. 26

Note repair weld and weld-deposit crack. Etched with ammonium persulfate

Recommendations. Repair welds in high-strength steel castings should always be made with low-hydrogen filler materials. The weld associated with this failure should have been made with a low-hydrogen wire.

Failures Arising From Stress Concentration Due to Welding Defects. Casting failures resulting from welding defects are not uncommon. Such welding defects as porosity, trapped slag, undercut, cracks, and lack of fusion often act as fracture-initiation sites because of their stress-concentration effects. Methods of combating these types of failure include welding with qualified procedures and welders as well as various forms of nondestructive examination.

Example 13: Failure of a Dragline Bucket Tooth. A cast dragline bucket tooth failed by fracturing after a short time in service. The tooth was made of medium-carbon low-alloy steel that had been heat treated to a hardness of 555 HB.

Investigation. The fracture surface was covered with chevron marks. These converged at several sites on the surface of the tooth. A hardfacing deposit was located at each of these sites. Visual inspection of the hardfacing deposits revealed numerous transverse cracks, which are characteristic of many types of hardfacing.

Conclusions. This failure was caused by cracks present in hardfacing deposits that had been applied to the ultrahigh-strength steel tooth.

Corrective Measures. Given the small critical crack sizes characteristic of ultrahigh-strength materials, it is generally unwise to weld them unless absolutely necessary. It is particularly inadvisable to hardface ultrahigh-strength steel parts with hard, brittle, crack-prone materials when high service stresses will be encountered. The operators of the dragline bucket were warned against further hardfacing of these teeth.

Example 14: Failure of a Cast Conveyor Chain Link. The conveyor chain link

shown in Fig. 28 failed by fracturing through a fabrication weld after some time in service. The link was made of cast low-alloy steel heat treated to a hardness of 285 HB. The fabrication weld was made using E7018 electrodes.

Investigation. The fracture surfaces were almost entirely covered with beach marks, indicating fatigue cracking (Fig. 29). The beach marks emanated from a 6.4-mm (¼-in.) wide band of unusual appearance running across the center of the fracture. This band was covered with parallel marks, such as those produced by grinding.

Saw cuts were made perpendicular to both of the fracture surfaces and across the unusual-looking bands. The cut surfaces were ground and etched with ammonium persulfate solution. This procedure revealed the fabrication weld (Fig. 30). It became obvious at this point that complete penetration had not been achieved at the weld root. There was no evidence that any attempt had been made to back-gouge the reverse side of the groove before welding. The 6.4-mm (¼-in.) wide band observed in the fracture was apparently part of the unfused weld-preparation surface at the root of the weld.

Conclusion. This failure was caused by fatigue that initiated at an incomplete penetration defect at the root of the fabrication weld. The defect served as a site of stress concentration, allowing fatigue cracks to initiate and grow through the weld deposit.

Corrective Measures. As a result of this failure, steps were taken to ensure that good welding practice (back-gouging) would be followed in the future fabrication of these links.

Failures associated with corrosion of welds are occasionally observed in cast corrosion-resistant alloys. Probably the most commonly encountered failures of this type are those involving sensitization of austenitic stainless steels.

The term sensitization refers to the formation of chromium-rich carbides at grain boundaries as a result of heating to temperatures in the

Fig. 28 Cast conveyor chain link that fractured through fabrication weld (bottom)

Secondary fracture at upper left. See also Fig. 29.

Fig. 29 Surface of fracture through weld in link shown in Fig. 28

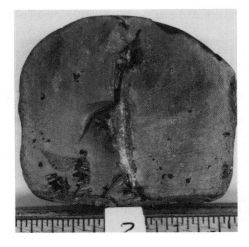

Fig. 30 Macroetched cross section through fracture of conveyor chain link

Note incomplete penetration at weld root. Etched with ammonium persulfate

Fig. 31 Photomicrograph showing typical appearance of HAZ hot-crack in cast IN-625 heat-resistant alloy

Weld metal is at bottom. Electrolytically etched in 10% chromic acid

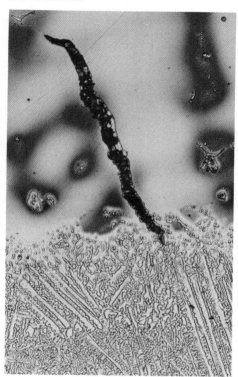

range of about 480 to 815 °C (900 to 1500 °F). When a weld is being made, a portion of the HAZ will reach these temperatures. As a result of the formation of these carbides, the regions adjacent to the grain boundaries are depleted in chromium, making them more susceptible to corrosion. If the weld is then exposed to an adequately corrosive environment, corrosion may occur in the sensitized region.

Welding of any corrosion-resistant material will usually result in some degree of sensitization. This problem is of most concern, however, in such materials as austenitic stainless steels, which contain more than about 0.03% C. In these materials, the corrosion resistance in the HAZ can be seriously degraded. It should be pointed out, however, that many cast stainless steels are less susceptible to this problem than corresponding wrought stainless steels. The reason for this is that most cast austenitic stainless steels actually contain some ferrite in their microstructures, while most wrought austenitic stainless steels do not. With a duplex microstructure, carbide precipitation occurs preferentially along ferrite-austenite boundaries (as opposed to austenite-austenite grain boundaries). As a result, there is no continuous path for corrosion to follow.

There are a number of ways to avoid corrosion problems associated with sensitization. The best method is to perform an appropriate heat treatment that will dissolve the carbides and restore the chromium to the chromium-depleted regions. In most cast austenitic stainless steels, for example, such a heat treatment consists of heating to about 1095 °C (2000 °F), then water quenching. Another method for

avoiding sensitization-corrosion problems is to use stabilized or low-carbon grades of material. The stabilized grades contain additions of strong carbide-forming elements (titanium or niobium) that combine with the carbon and thus prevent formation of chromium carbides. The low-carbon grades contain so little carbon (less than 0.03%) that the amount of chromium-carbide precipitation that occurs usually has a negligible effect on corrosion resistance.

Failures Related to HAZ Hot-Cracking. A welding defect that occasionally leads to failure of high-alloy steel castings is HAZ hot-cracking. This type of defect occurs primarily in alloys having a fully austenitic microstructure, such as the corrosion-resistant ACI alloy CN-7M. The most common form of HAZ hot-cracking begins with melting of base-metal grain boundaries next to the fusion line. Because of the thermal strains associated with welding, separation occurs along the melted grain boundary, and a crack is formed. Figure 31 shows a typical HAZ hot-crack.

In many cases, HAZ hot-cracks are completely innocuous. They are often only a few thousandths of an inch long, and the materials in which they occur often have exceptional ductility and toughness. Under some circumstances, however, hot-cracks can provide fa-

vorable sites for crevice corrosion or stress-corrosion cracking (SCC).

Fully austenitic alloys are particularly susceptible to HAZ hot-cracking for two primary reasons. First, their compositions promote grain-boundary segregation of carbon, sulfur, phosphorus, silicon, lead, tin, antimony, arsenic, and so on. This can significantly suppress the melting temperature of the grain-boundary regions. The second reason for the high HAZ hot-crack susceptibility of cast fully austenitic materials is that they have inherently large grain sizes. This minimizes the grain-boundary area to which the contaminant elements can segregate, thus increasing their concentration at grain boundaries.

At this point, it is appropriate to emphasize that HAZ hot-cracking is a serious problem only in fully austenitic materials. Most grades of cast stainless steel are intentionally balanced in composition in order to obtain at least some ferrite in their microstructures. When ferrite is present, the risk of HAZ hot-cracking is greatly reduced (in this respect most cast stainless steels have a weldability advantage over their ferrite-free wrought conterparts). When service conditions allow, the best method for minimizing the risk of HAZ hot-cracking is to select an alloy that is at least partially ferritic.

When cast fully austenitic materials must be used, two measures for minimizing the risk of hot-cracking are most effective. These are careful selection of charge materials and appropriate processing of the heat to ensure the highest practical degree of purity. Another step that can be quite beneficial is to pour at low temperatures in order to achieve as fine a grain size as possible in the casting. In some cases, the risk of HAZ hot-cracking can be slightly reduced by welding with low heat inputs, by keeping the interpass temperature low, and by minimizing the degree of restraint on the weld.

Corrosion Failures of Cast Materials

In the third quarter of 1985, high-alloy castings accounted for approximately 10%, by weight, of the castings sold by members of the Steel Founders' Society of America (Ref 1). The percentage in monetary terms greatly exceeds that estimate. Similarly, given the critical role of high-alloy castings as valves, pumps, and so on, in critical and often severe service, the overall importance of these parts is also underestimated.

Examination and analysis of corrosion-related failure of cast components should be approached in the same manner applied to wrought components. These techniques are covered in the articles "General Practice in Failure Analysis" and "Corrosion Failures" in this Volume. The fundamental corrosion problems are the same for both cast and wrought components with a few exceptions. Casting defects can play a major role in corrosion-related failures. Porosity can easily act as an initiation site for localized corrosion or as a stress concentrator leading to SCC. The same is true for other defects, such as hot tears, cold laps, oxide laps, and inclusions. Although wrought materials have their own weaknesses, they are free of many of the defects that may be associated with a casting.

It is significant to note that castings also have some striking advantages over comparable alloys in wrought form. Cast 18-8 type stainless steels—for example, ACI CF-8 and CFR-8M—contain small amounts of ferrite in cast form (typically 5 to 15%), while their nearest wrought equivalents (AISI type 304 and 316, respectively) do not. This yields a material with enhanced resistance to chloride SCC due to the crack-arrest (blunting) properties of this duplex microstructure. Inclusions in cast alloys tend to be spherical; these are much less harmful in terms of stress concentration than the planar inclusions characteristic of wrought materials. Finally, foundries offer relatively small heat sizes, which allows considerable freedom to modify the composition of an alloy to meet a specific application.

One group vitally interested in the performance of cast corrosion-resistant components is the Materials Technology Institute of the Chemical Process Industries, Inc. (Ref 2). They surveyed foundries, pump and valve manufacturers, and their members on the causes of premature failure in castings. Their conclusions are listed below:

- Intergranular corrosion accounts for about 50% of the premature service failures of chromium-bearing corrosion-resistant castings
- Solidification-type discontinuities (hot tears, porosity, and shrinkage) account for about 30 to 40% of the premature service failures of corrosion-resistant castings
- Some premature service failures have been associated with repair welds
- The data obtained from the surveys were not sufficient to identify precisely the causes of the intergranular corrosion reported. Intergranular corrosion results from sensitization, which in turn could be caused by inadequate control of the carbon content of the castings, inadequate heat treatment, or no heat treatment. The technology is available to eliminate this type of discontinuity
- Hot tears and cracks in CN-7M, CD-4MCu, and some of the nickel-base alloys are particularly troublesome to the producing foundries and may be responsible for some premature service failures. Although information is available on the cause of hot tearing, improvements in the technology are probably required to produce acceptable castings more consistently
- Crevice corrosion and SCC were seldom cited as the cause of premature service failures

- Much better communication is needed among casting users, valve and pump manufacturers, and casting producers
- Present-day specifications and nondestructive inspection techniques are not adequate in themselves to ensure consistent performance of corrosion-resistant castings in severe service. Consideration should be given to an acceptance-type corrosion test, such as ASTM A 262, Practice B

The article "Corrosion Failures" in this Volume offers many detailed accounts of failure investigations and should be used as reference for cast-component failures. The same mechanisms listed in that article are at work on cast components.

Uniform corrosion is the simplest form of chemical attack. It is also the most avoidable (through proper alloy selection) or at least the most controllable, because the rate of material loss is generally predictable. The chemical attack occurs over the entire exposed surface, avoiding localized attack that can lead to unexpected and sometimes catastrophic failure. In some cases, simply making the part thicker to reduce replacement frequency is an adequate solution to general corrosion problems.

Erosion-Corrosion. The effects of uniform corrosion can be dramatically increased when the corroding fluid is moving across the component. In most corrosion-resistant alloys, resistance to corrosion is related to formation of a layer of corrosion products, which then protects the material from further corrosion.

In erosion-corrosion, the velocity of the fluid acts to remove the protective oxide or other coating from the surface under attack, thus increasing the rate of penetration. Cavitation is a special case of this general form of corrosion. In this case, the rapid flow of a liquid over a changing surface, as in an impeller or propeller, creates such drastic pressure changes that gas bubbles form in the liquid on the surface of the part. When these bubbles collapse, great pressures (estimated to be as high as 414 MPa, or 60 ksi) (Ref 3) are exerted locally on the component, destroying the protective surface film of the alloy. Detailed information on this form of attack is available in the articles "Liquid-Erosion Failures" and "Corrosion Failures" in this Volume.

An example of erosion-corrosion is shown in Fig. 32. The combined effects of high velocities and corrosion are clearly visible on the cast CF-8M pump impeller. This particular part is a standard CF-8M, but unfortunately, nothing is known about the service conditions. A material change is required to stand up to this service. Clearly, flow conditions must be considered when selecting alloys for corrosive service.

Example 15: Failure of a Fan Support Casting. A fan support casting failed unexpectedly while running at 1800 rpm in pulp at 65 °C (150 °F). The leading edge of the blade

Fig. 32 The classic appearance of erosion-corrosion in a CF-8M pump impeller

exhibited deep spongy holes leading to reduced section and finally to fracture of the part when the remaining section size was insufficient to support the load. Analysis of the support casting showed it to be a standard 8620 type composition with a hardness of 311 HB. The design of the casting was not streamlined; that is, there were several square corners present where great pressure differences could be generated.

Conclusions. This was a case of erosion-corrosion with the classic spongy appearance of cavitation.

Recommendations. Two changes were proposed. A significant improvement could be realized by streamlining the part to avoid abrupt changes in fluid flow. In addition, a change in alloy to a more corrosion-resistant material (304 or preferably 316) would help by increasing the tenacity of protective films.

Stress-corrosion cracking mechanisms are described in the article "Stress-Corrosion Cracking" in this Volume. In simplest terms, SCC is cracking caused by the joint action of stress and environmental attack. In general, the environment alone might not cause failure, and the stress levels may be well below yield. Yet together, combined with a susceptible material, disastrous cracking may occur. A part exhibiting these cracks will often show little corrosion over most of its surface. The stress levels at which this cracking may occur are generally well within normal design ranges. Therefore, it is most important to analyze specific environment-alloy combinations for potential problems.

Stresses leading to failure may arise from sources other than the imposed loads of service.

Fig. 33 Illustration of neck liner removed from a pulp digester vessel

Note the abrupt change in cross section that led to a caked-on buildup. See also Fig. 34 to 36. Dimensions given in millimeters (inches)

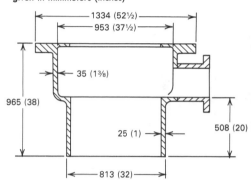

Fig. 34 Four sections taken from the neck liner shown in Fig. 33

(a) and (b) Extensive repair welds (in service) on the original casting. (c) A location where a piece of AISI type 317 plate was used to repair the casting. (d) Extensive SCC into the 317 plate on the inner surface. See also Fig. 33, 35, and 36.

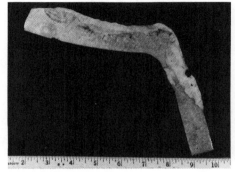

(a)

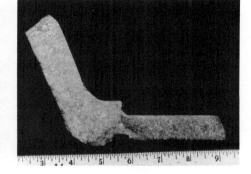

(b)

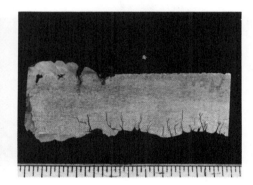

(c)

(d)

In some cases, residual stresses due to welding or heat treating are the cause of premature failure. Casting defects may play a role as stress concentrators that act as initiation sites for failure and so may play a role in SCC.

Example 16: SCC in a Neck Liner. A neck liner or manway was removed from the top of a digester vessel. Repeated attempts to repair the part in the field during its life cycle of many years had failed to keep the unit from leaking. Figure 33 shows the general appearance of the part. Visual inspection showed extensive weld repair and evidence of a heavy buildup of material at the abrupt change in diameter to the lower neck of the body. Figure 34 shows some cross sections through the part that illustrate the weld metal, wrought plate, and the original casting. Chemistries of three constituents are shown in Table 2. The casting is a CF-8M modified with the molybdenum level at the top end of the range. The plate is standard 317L material.

The filler metal is type 316, although marginal in molybdenum content. Figures 35 and 36 show closeup views of the cracks. Figure 35 shows evidence that cracks in the cast body were not completely removed but rather welded over (see crack extending into weld metal), and Fig. 36 shows SCC of the weld metal. The cracking and repair work were concentrated above the change in diameter where the heavy corrosion products and caked-on pulp had collected.

Conclusions. It appears the original casting suffered SCC in the area where there was a buildup of debris and where a low-oxygen area high in chlorides from repeated wet/dry cycles caused the cracking. Attempts at repair had worsened the situation by increasing the stress levels by the addition of residual stresses from the repairs and by leaving stress concentrators under the repairs by not arcing all the way to clean uncracked metal. In addition, if redesign was impossible, then an alloy more resistant to chloride SCC should be used, such as the new high-strength duplex stainless steels or high-molybdenum (4 to 6%) austenitic stainless steels.

Example 17: SCC in Mixer Paddle Shafts. Small paddles used to mix pulp had experienced a high incidence of breakage through the shafts. In some of the shanks, shrinkage was found relatively close to the surface where threads had been cut all the length of the shaft. Chemistries were run on all the parts, and all were found to fall within normal CF-8M ranges. Metallography showed the parts to be correctly heat treated. Cross sections of several of the parts showed pitting corrosion, and beneath the pits, stress-corrosion cracks were found (Fig. 37). These occurred in areas where the shafts had been bent during use.

Conclusions. All the samples showed deep SCC in the areas where bending had occurred. In several cases, centerline shrinkage from inadequate risering had decreased life by reducing the cross-sectional area. Type CF-8M is not resistant to chloride SCC where the chloride concentration is considerable.

Recommendations. The biggest problem was the tendency for bending of these parts. This deformed material with high residual stresses would always be susceptible to SCC. Redesign

Table 2 Chemistries of failed neck liner

Element	Casting	Plate	Weld
Carbon	0.02	0.02	0.043
Manganese	1.26	1.69	1.50
Silicon	0.93	0.54	0.36
Chromium	19.03	18.47	17.89
Nickel	10.32	13.92	11.54
Molybdenum	2.81	3.10	1.91
Copper	0.09	0.16	0.13
Sulfur	0.009	0.018	0.21
Phosphorus	0.030	0.026	0.031

to lower stresses was essential. In addition, it was recommended that a change to a high-strength duplex stainless steel with its higher strength and greater resistance to chlorides would be helpful. Finally, the part must be adequately risered to produce solid shanks free from shrinkage.

Intergranular corrosion is one of the most common causes of failure in cast corrosion-resistant components. In most cases, its cause is some step in its thermal history that leaves it susceptible. This could include missing the essential solution-annealing heat treatment or the heat associated with an unheat-treated weld. Both welding and improper heat treatment can

Fig. 35 Stress-corrosion cracks extending into the casting, with weld metal deposited over the cracks

It would appear that the repairs were needed because of the SCC, but the cracks were not completely removed before repair. See also Fig. 33, 34, and 36.

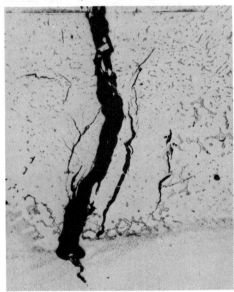

Fig. 36 Close-up view of the cracks shown in Fig. 34(d)

Note the classic branched appearance of the SCC in the AISI type 316 weld metal. See also Fig. 33 to 35.

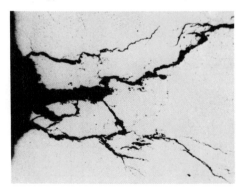

give rise to sensitization. This involves chromium-rich carbide precipitation at grain boundaries, accompanied by depletion of chromium in the adjacent region. The low-chromium area is then preferentially attacked by the corrosive solution. Intermetallic second phases, such as σ phase, can also give rise to preferential grain-boundary corrosion. This is covered in the article "Corrosion Failures" in this Volume.

Example 18: Corrosion of a Neck Fitting. A neck fitting exhibiting extreme corrosion with large deeply pitted areas was examined. It had been in service in a sulfite digester

Fig. 37 Stress-corrosion cracks in cast CF-8M mixer paddle shafts

Cracking was most extensive in deformed area of the shafts

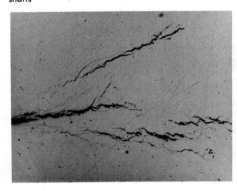

Fig. 38 Sample from a cast stainless steel neck fitting that failed by intergranular corrosion

See also Fig. 39 and 40.

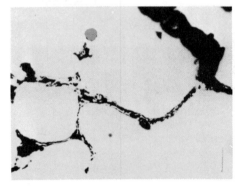

at 140 °C (205 °F) and 689 kPa (100 psi). The liquor was calcium bisulfite with 9% total weight of SO_2 at a pH of 1.5 to 1.8. Chloride content was reported to be low. The composition was a cast equivalent of AISI type 317 (18Cr-10Ni-3.2Mo). A sample was taken from the deeply attacked area (Fig. 38). The corrosion was occurring along grain boundaries. Micrographs from these regions showed σ phase present in the ferrite (Fig. 39) and also carbides (Fig. 40).

A 10 N KOH electrolytic etch was used to bring out the σ phase, while Murakami's reagent showed the carbides at the grain boundaries. It appears that this casting never received a proper solution anneal.

Recommendations. Solution treating is essential for CF-8M castings to avoid both carbide and intermetallic precipitates. Corrosion-screening tests per ASTM A 262 (Ref 4) may be needed to ensure adequate corrosion resistance.

Example 19: Intergranular Corrosion in an Unknown Part. The origins of the casting shown in Fig. 41 are unknown. It was returned from chemical service with no information about service conditions or part identi-

Fig. 39 Region adjacent to the intergranular corrosion shown in Fig. 38

Electrolytic etching using 10 N KOH revealed extensive σ-phase precipitation at grain boundaries. See also Fig. 38 and 40.

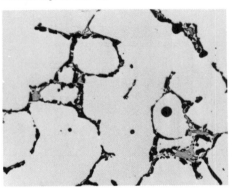

Fig. 40 Same area as Fig. 39 after repolishing and etching with Murakami's reagent

This etchant revealed substantial grain-boundary carbide precipitation. See also Fig. 38 and 39.

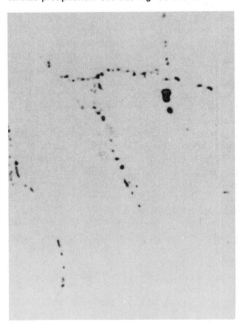

fication. It is included here as a classic case of intergranular corrosion. The part (apparently a pump outlet) was named the "rubber casting" because of the severity of the intergranular attack. Every grain boundary has been attacked to the extent that the casting could be twisted and stretched as though made of rubber. A close-up view of the neck area (Fig. 42) shows how extensive the corrosion is, with individual grains actually falling out of the part.

The chemistry of the casting was acceptable for CN-7M. The carbon content was at the top

Fig. 41 Severe intergranular corrosion in a CN-7M casting

The exact function and service conditions of the part (probably a pump outlet) are unknown. Intergranular corrosion was so severe that the part could be twisted and stretched; thus, it was termed the rubber casting. See also Fig. 42 and 43.

Fig. 42 Close-up view of the casting shown in Fig. 41

Individual grains stand out, and many fall free when the part is moved. See also Fig. 43.

Fig. 43 Grain-boundary carbide film in the part shown in Fig. 41 and 42

A sharp crack runs parallel to the grain-boundary precipitate. The crack apparently follows the sensitized (chromium-depleted) region adjacent to the carbide.

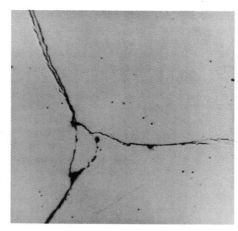

Fig. 44 Preferential oxidation of the grain boundaries in a cast high-temperature alloy

See also Fig. 45.

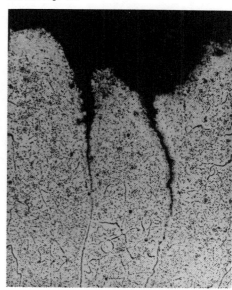

end of the CN-7M range (0.06), but within specification.

Figure 43 shows the reason the part failed so dramatically. A continuous film of carbide is shown with a continuous crack running parallel to the carbides. This sensitized structure produces an area depleted in protective chromium, making it susceptible to corrosion.

Two solutions to this problem are available. The simplest approach is to ensure correct heat treatment to dissolve grain-boundary carbide film and return the protective chromium to the depleted zone. Alternatively, a low-carbon (0.03% maximum C, for example, CF-3) grade can be specified, which is further protection against sensitization. Finally, procedures are given in Ref 4 for screening castings that may be susceptible to intergranular corrosion due to processing errors.

Failures During Exposure to Elevated Temperatures

Many castings are produced for high-temperature service. The control of composition inherent in castings, coupled with the ability to increase the carbon contents of standard high-temperature alloys beyond the wrought steel levels, has led to a significant number of specialty cast grades. Nonetheless, the techniques used for analysis and the general failure modes are shared with the wrought alloys.

These include general oxidation, sulfidation, metallic oxide attack, carburization, creep, precipitation embrittlement, and thermal fatigue. These are covered in the article "Elevated-Temperature Failures" in this Volume. Most oxidation-resistant materials depend on formation of a layer of oxidation products in order to shield the underlying metal from further attack.

General Oxidation. Low-alloy steels can be used effectively to temperatures as high as 540 to 650 °C (1000 to 1200 °F) for extended periods. As these temperatures are exceeded, general oxidation increases to rates that become unacceptable. A fluffy oxide or scale occurs and sloughs off, leading to decreased service life. A whole series of alloys has been developed to cover a wide range of elevated-temperature applications. Most of these alloys contain chromium or chromium and nickel plus other elements, depending on the application requirements. Also, there are nickel- and cobalt-base alloys used in critical high-temperature applications. It is not the purpose of this article to develop the reasoning for each alloy and its appropriate applications, but to examine the types of failures caused by high-temperature environments. As mentioned in the beginning of this article, general oxidation rate is one of the primary concerns of extended high-temperature service. The first endeavor in examining a high-temperature failure should be to determine what portion of the original section is left unaffected by the general oxidation. If the oxidation rate is found to be excessive, the alloy should be replaced with an alloy of greater oxidation resistance.

Sulfidation and Preferential Oxidation. However, not all oxidation is general in nature. There are several phenomena leading to premature failure that should be considered. The first is sulfidation attack at the grain boundaries. Sulfidation attack is a preferential deterioration at the grain boundaries due to a combination of sulfur with nickel in high-nickel alloys. This type of attack can be found in the stack areas of systems burning fuel oils or coal high in sulfur.

This type of preferential attack may be confirmed by electron microprobe analysis of the products found in the attacked grain boundary. One must be careful to differentiate between this type of attack enhanced by sulfur in the environment and perferential oxidation of the grain boundaries. Because most high-temperature alloys contain higher carbon than can be dissolved in the structure, these alloys tend to form eutectic carbides in grain boundaries and can show preferential oxidation of this area even in the absence of sulfur (Fig. 44).

Preferential oxidation can also be interdendritic in nature and is primarily due to the

Fig. 45 Interdendritic preferential oxidation

See also Fig. 44.

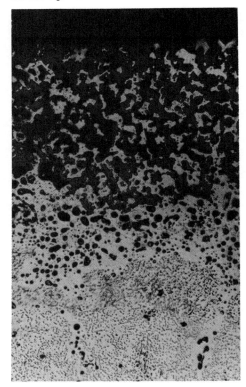

chemistry gradients across the dendrites (Fig. 45). While sulfidation attack can be intergranular and very rapid in high-nickel alloys, sulfidation also takes on rapid scaling or exfoliation characteristics in low alloys and some other alloys. This is attributed to the modification of the protective film produced on the alloy. In most cases, sulfidation resistance is improved by increasing chromium content, and aluminum and silicon have also been shown to be beneficial. Where hydrogen sulfide is the primary environmental factor, it has been found that intermediate amounts of chromium (5 to 12%) are actually deleterious and increase the rate of sulfidation. Chromium contents greater than 17% have been found to be adequate in H_2S atmospheres, and protective layers of aluminum have been found to protect chromium steels totally from H_2S attack.

Metal Oxide Attack. Vanadium pentoxide, compounds containing sodium, welding fluxes, and some other reactive compounds can create a fluxing action of the normal protective oxide on the metal. These compounds tend to drop the melting temperature of the protective film and to allow the transport of oxygen or sulfur to the active metal surface. In this fashion, the contaminating compound acts like a catalyst, promoting the accelerated attack in the localized area involved. It should be noted that these fluxes can continue to cause damage even after the source has been removed

because they will remain as a part of the oxide system. This type of problem is not easily handled, because there are no known metallurgical solutions. Most efforts to solve this problem require complete removal of all oxides from the affected surfaces, keeping these metal surfaces below the fusion temperature of the combined oxides, and require addition of additives to raise the fusion temperature in the case of vanadium oxide in residual fuels.

A similar phenomenon, known as catastrophic oxidation, is encountered in those high-temperature alloys containing significant quantities of molybdenum and tungsten. These elements are widely used for enhancing creep-resistance properties in high-temperature alloys; however, they form oxides, such as molybdenum trioxide, that are gaseous at relatively low temperatures. The vaporization temperature of molybdenum appears to be 800 °C (1475 °F). Again, the molybdenum trioxide tends to change the character of what would otherwise be protective oxide films. Under the proper conditions of stagnation and temperature, alloys containing these metals will produce thick, fluffy oxide coatings at a very rapid rate, resulting in both general and pitting-type metal loss. In general, this can be prevented by total protection from oxide production or by the removal of gaseous molybdenum trioxide by entrainment in a moving gas stream, including air (Ref 5).

Carburization is another mode of surface deterioration. At elevated temperatures, carbon in contact with the surface metal can react to form chromium carbides and, to a lesser degree, iron carbides, resulting in carbide enrichment of the surface layer of metal. This enrichment can go quite deep into the metal, with resulting increases in creep strength and decreases in metal ductility and toughness. Surface oxidation (Fig. 46) is not a primary concern under carburizing-type conditions.

A reducing atmosphere that will produce carburization can remove significant amounts of chromium, allowing such phenomena as sulfidation to become significant. Also, the depletion of chromium can allow normal corrosion to occur during shutdown periods; during these cooling down periods, liquids can wet and therefore corrode these surfaces. While the corrosion properties of these carburized surfaces are important, a loss of ductility in heavily carburized parts has been the cause of many failed parts. Increasing chromium content has been shown to slow the diffusion rate of carbon into the metal. Increasing nickel content has also been found to be beneficial. There are several alloys designed for carburizing service.

Creep Failures. Tensile and compressive loads tend to cause failures at high temperature just as they do at room temperature and below. However, at elevated temperatures, a new phenomenon tends to come into play and become quite significant. This phenomenon is known as creep.

Fig. 46 Surface carburization

While this condition may not lead directly to rapid oxidation, the surface may be left susceptible to sulfidation and other forms of attack.

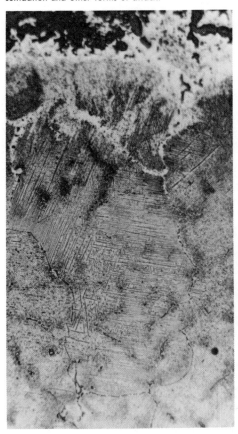

Metals at high temperature, while under load, even very light loads, undergo continual and slow plastic deformation or flow. The load required to cause a creep rate or deformation rate of 0.0001% per hour can vary from less than 4% to more than 28% of the short-time yield strength of that material at temperatures greater than 650 °C (1200 °F). For example, HK-40, a cast heat-resistant alloy, has a yield strength of 60 MPa (8.7 ksi) at 980 °C (1800 °F); the stress necessary to cause a creep rate of 0.0001% per hour is 17 MPa (2.5 ksi). Obviously, in the design of high-temperature systems, loads and life expectancy are determined by creep rate and ductility and not the normal engineering units usually used for room-temperature design work.

Another measure used for high-temperature alloys is stress-to-rupture. This measurement tends to combine both creep rate and ductility into one measurement, giving the design engineer a load-to-cause failure in a specified number of hours, such as 10 000 or 100 000 h. High-temperature fractures tend to be complicated by the oxidation of the metal surface or by contamination from environmental sources after the fracture has occurred. These fractures can also be quite complex; however, in many

Fig. 47 Casting failure believed to have been caused by incipient melting (left) compared to a tensile-test specimen tested at 1290 °C (2350 °F)
Note the similarities between the fracture surfaces.

Fig. 48 Cast heat-resistant alloy HH, type II, showing the effects of long-term exposure to temperatures between 705 and 925 °C (1300 and 1700 °F)

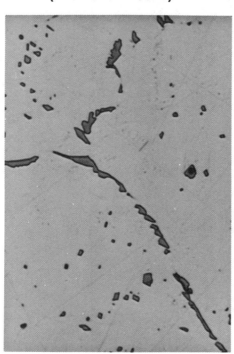

Fig. 49 Same alloy as Fig. 48 showing cracking through intergranular carbides

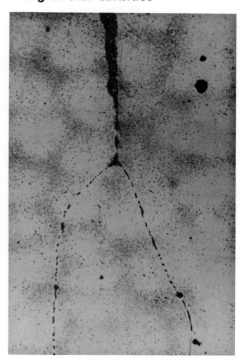

failures of high-temperature materials, the fracture mode is intergranular, and in castings, this usually means easily observable intergranular patterns. In tensile overload situations where the strain rate is approaching that of a normal tensile test, ductilities are high, and fractures are mostly ductile. Under normal creep conditions, elongation and reduction of area are much lower, and the fractures have the appearance of brittleness. Also, it must be noted that when the nil-ductility temperature (the onset of incipient melting) is reached, no ductility will be shown, and the metal will fail at the grain boundaries due to their lower-melting composition (Fig. 47).

Precipitation Embrittlement at Elevated Temperatures. Low-carbon austenitic stainless steels have some usefulness at elevated temperatures due to their oxidation resistance. Most of the alloys designed for elevated-temperature service, however, have much higher carbon contents than the typical 304 type of austenitic stainless steel. They often contain such carbide-stabilizing elements as titanium or niobium, and they intentionally contain precipitated carbides to help increase their creep resistance. Other elements, such as molybdenum and tungsten, are often added to increase the creep resistance, and the matrix microstructure is usually austenitic for greater strength and stability.

The amount of carbides, the size, and the distribution are dependent on the operating temperature and time. However, there is a general tendency for a semicontinuous carbide film to be present in the grain boundaries (Fig. 48). This carbide film also tends to influence the overall toughness of these materials and the

general fracture mode both at elevated temperatures and all the way down to ambient temperature (Fig. 49). Weld repair of these alloys can become difficult, especially if the surface material has become contaminated, and they often require a high-temperature heat treatment (1095 °C, or 2000 °F, or higher), with relatively fast cooling to dissolve the carbides and improve the ductility to allow for acceptable welding characteristics.

Intermetallics, such as σ phase, are important sources of failures in high-temperature materials. This phase is a particular combination of iron and chromium that produces a hard, brittle second phase in the temperature range of 480 to 955 °C (900 to 1750 °F). Chromium contents above 20%, especially with the addition of significant amounts of silicon, molybdenum, or tungsten in an iron-base material, tend to allow the formation of this phase. It tends to precipitate after long exposures at the indicated temperature range, and most of the time, it can be eliminated by reasonable cooling rates from a heat treatment at or above 995 °C (1850 °F).

Sigma phase tends to precipitate in those regions of high chromium content, such as the chromium carbides in the grain boundaries and (where present) within ferrite pools. A relatively small quantitiy of σ phase, when it is nearly continuous at a grain boundary, can lead

to very early failure of high-temperature parts. It is important to understand that σ phase cannot undergo any significant plastic deformation; instead, it fractures even at relatively low strain levels. This is true at elevated temperatures, but even more so at ambient temperatures. Where toughness and ductility are an important part of the system design, σ phase cannot be tolerated (Fig. 50). However, there are applications where σ phase has been present and has not played a significant role regarding failure.

Thermal Fatigue. Within a high-temperature system, there is often a temperature gradient either within a part or between parts. Because of the nature of the alloys used at high temperatures and because of creep, relatively low stresses can create failure. Thermal gradients can be one of the most significant causes of failure. It is especially important to realize that austenitic steels have almost twice the thermal-expansion coefficient of ferritic and martensitic steels. This is important when combining two different materials, but is always significant when considering the potential stress development within a single part cast in austenitic steel. Consistent section size and even heating rates tend to become very important considerations.

Thermal fatigue is a problem specific to parts that go through heating and cooling cycles on a regular basis. Obviously, the stresses can be created by several sources, but these stresses are usually created by a thermal gradient caused

Fig. 50 σ phase in cast heat-resistant alloy HH, type II

Intermetallic phases, such as σ, can greatly reduce the ductility of many high-temperature alloys in service at temperatures from 480 to 955 °C (900 to 1750 °F).

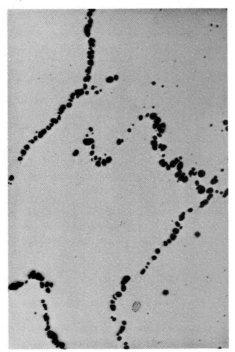

by unequal heating or cooling of the subject part. Fracture can be transgranular; however, it usually tends to be intergranular (Fig. 51).

This mode of failure is often termed craze cracking. Almost invariably, the crack is filled with oxide, indicating a somewhat slow growth pattern. Thermal fatigue is usually considered a low-cycle phenomenon, requiring less than 50 000 cycles to failure. Most of the normal fractographic indications of fatigue are destroyed by continued environmental exposure. In the case of thermal fatigue, it can be beneficial to move toward greater ductility instead of higher strength, especially if that greater ductility comes from a cleaner microstructure (lack of carbides in the grain boundary).

Failures Related to Hydrogen-Assisted Cracking

Hydrogen-assisted cracking is an important category of failure for products made of medium-, high-, and ultrahigh-strength steels. Castings made of these materials are no exception.

Failures related to hydrogen can be particularly insidious in that they usually occur quite unexpectedly. One reason for this is that the

Fig. 51 Thermal fatigue in a cast low-alloy steel ingot tub

Etched in 2% nital. 150×

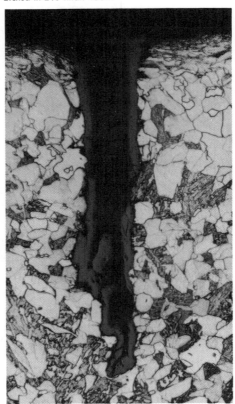

cracking process involves a time delay. Frequently, the cracking will initiate some time after a part has been inspected, certified to be crack free, and placed in service. Another reason is that in the higher-strength steels, hydrogen-assisted cracking can occur at surprisingly low stress intensities and at very low bulk concentrations of hydrogen. A third reason is that hydrogen-assisted cracking is difficult to address from an engineering standpoint. There is very little available quantitative data relating to hydrogen concentrations, strength levels, and threshold stress intensities for hydrogen-assisted cracking (particularly for cast materials). Even if such data were available, there would still be the problems associated with determining accurate hydrogen concentration and actual service stresses.

The basic characteristics of hydrogen-assisted cracking are well known and are described in detail in the article "Hydrogen-Damage Failures" in this Volume. Briefly, however, these characteristics include the following.

The Effect of Hydrogen Concentration. For hydrogen-assisted cracking to occur, hydrogen must be contained in the material. The higher the hydrogen concentration, the lower the cracking resistance of the material.

The Effect of Strength Level. The higher the strength of a material, the lower its resistance to hydrogen-assisted cracking. The specific microstructure of the material does have some influence on cracking resistance; however, the strength of the material as a result of its microstructure is thought to be a much more significant factor.

The Effect of Stress Level. Stress is required for hydrogen-assisted cracking to occur. Cracking occurs only above a certain stress intensity level K_{th}. This threshold stress intensity decreases rapidly as the strength level of a material increases.

The Effect of Temperature. Hydrogen-assisted cracking tends to occur only in the temperature range of approximately −128 to 205 °C (−200 to 400 °F). Hydrogen has the greatest influence near room temperature.

The Effect of Strain Rate. Hydrogen exerts the greatest influence on cracking resistance under static loading or slow strain-rate conditions. Hydrogen does not affect cracking resistance under impact-loading conditions.

The Effect of Time. Time is required for hydrogen-assisted cracking to occur. The amount of time required for crack initiation varies with many of the factors listed above.

In combating hydrogen-related failures, it is possible to make use of any of the characteristics described above. It is usually most practical, however, to prevent this type of failure by changing the hydrogen concentration, the stress level, or the strength level of the material.

Most hydrogen-related failures that occur in steel castings can be separated into three categories:

- *Internal hydrogen failures*: These are due to hydrogen contained in the metal as a result of melting, tapping, pouring, solidification in the mold, and so on
- *Environmental hydrogen failures*: These are due to hydrogen that is charged into the metal as a result of contact with a corrosive environment
- *Welding-related hydrogen failures*: These are due to hydrogen introduced during welding

Internal-hydrogen failures are usually confined to steel castings having section sizes greater than 7.5 cm (3 in.) and yield strengths greater than 690 MPa (100 ksi). Lower-strength steels generally have K_{th} levels adequate to resist this type of failure given reasonable service stresses and the normal hydrogen contents arising from standard casting production. Thinner-section castings rarely suffer this type of failure because most of their contained hydrogen rapidly escapes by effusion.

Internal-hydrogen failures have a number of fracture features that help to distinguish them from other types of failures. One of these is that the initiation sites of internal-hydrogen failures are usually deep inside a part, often near the center of the section (this is significant in that the highest service stresses and most serious defects usually occur near the surface of the part). Often, no significant defect is visible at the initiation site. Internal-hydrogen cracks often assume a relatively flat, nearly circular configuration with the initiation site at the center. This configuration is maintained as the crack grows under the effect of hydrogen and stress. Usually, final failure of the part occurs by a fracture mechanism other than hydrogen-assisted cracking (simple fast fracture is probably the most common). Because of the different fracture mechanisms, the circular hydrogen crack is often clearly visible on the fracture surface. These circular fracture features are known by a variety of names, but hydrogen flake is probably the most common.

Hydrogen flakes have a characteristic fracture appearance. They are usually covered with fine chevron marks that radiate from the initiation site. The fineness of these chevron marks varies with the distance from the initiation site, reflecting variations in stress intensity at different states of crack growth. In higher-strength steels, the region near the initiation site may be nearly featureless. In cases where the fracture resulted from residual stress or static-loading conditions, the chevron marks become progressively coarser with increasing distance from the initiation site. In cases where the failure has involved loads of varying magnitude, it is not uncommon to observe faint concentric rings reminiscent of the beach marks of a fatigue fracture. It is also not uncommon to observe hydrogen flakes where some of the crack growth has actually occurred by fatigue. In these cases, rings of smooth, beach-marked fracture may be visible within the flake.

Example 20: Failure of a Cast Swage Press Arm. The arm of a 4.5×10^5-kg (500-ton) capacity hydraulic swage press failed by fracturing after several years in service. Figure 52 illustrates the configuration of the press arm and the location of the fracture. The press arm was cast from low-alloy steel (0.27C-0.90Mn-0.50Si-0.90Cr-0.20Mo) that had been normalized, hardened, and tempered to give a yield strength of approximately 690 MPa (100 ksi).

Investigation. Examination of the fracture surfaces revealed the presence of a large hydrogen flake that had initiated near the center of the section (Fig. 53). The flake had grown through nearly the entire cross section before final failure took place. Large shear lips surrounded most of the fracture, indicating good toughness.

Conclusions. This failure was caused by an excessive hydrogen concentration in the metal given the loading conditions and the threshold stress intensity of the material for hydrogen-assisted cracking.

Corrective Measures. Subsequently produced press arms were cast from heats of steel that had been processed to give low hydrogen contents. Steps taken included controlling the percentage of carbon removed during the boil, AOD processing, and short tap-to-pour times.

Fig. 52 Internal hydrogen-assisted cracking in a cast medium-strength low-alloy steel press arm
See also Fig. 53.

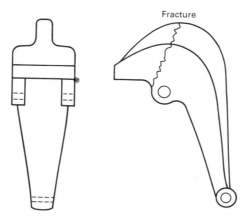

Fig. 53 Fracture surface of press arm that failed because of internal hydrogen-assisted cracking
Note characteristic hydrogen flake. 0.2×

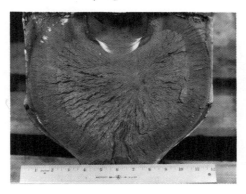

Example 21: Failure of a Cast Chain Link. A 10-cm (4-in.) chain link used in operating a large dragline bucket failed after several weeks in service. The link was made of cast low-alloy steel (similar to ASTM A487, class 10Q) that had been normalized, hardened, and tempered to give a yield strength of approximately 1034 MPa (150 ksi).

Investigation. A hydrogen flake approximately 5 cm (2 in.) in diameter was observed at the center of the fracture surface. Beach marks indicative of fatigue encircled the hydrogen flake and covered nearly all of the remaining fracture surface (Fig. 54).

Conclusions. The failure of this link was caused by an excessive hydrogen content given the service stresses and the threshold stress intensity of the material for hydrogen-assisted cracking.

Corrective Measures. Two steps were taken to combat this type of failure. First, when service conditions did not require high hardness to combat wear, the links were produced of a

Fig. 54 Fracture surface of cast 10-cm (4-in.) high-strength low-alloy steel chain link that failed because of internal hydrogen-assisted cracking
Note hydrogen flake. 0.25×

steel having a yield strength of about 690 MPa (100 ksi) rather than 1034 MPa (150 ksi). The higher K_{th} value of the lower-strength material was adequate to prevent cracking. Second, a design change was made that lowered the peak stresses in the link.

Environmental hydrogen failures of steel castings are most often observed in high- and ultrahigh-strength steels subjected to high, sustained service stresses. Cracking usually initiates at a site of stress concentration located on or near the surface. The basic mechanism of fracture is the same as for any other type of hydrogen-assisted cracking, except that the hydrogen is supplied by a corrosion reaction.

Hydrogen evolution is a common cathodic process in many corrosion reactions. Most of the hydrogen escapes as gas bubbles, but a portion is charged (absorbed) into the metal. Some environments, such as those containing H_2S, greatly increase the amount of hydrogen charged into the metal during corrosion. Large concentrations of hydrogen can also be charged into the metal under conditions of cathodic protection.

Several fracture morphologies are observed in environmental hydrogen-assisted cracks. In cast high-strength and ultrahigh-strength steels, however, the most common fracture morphology appears to be intergranular with respect to prior-austenite grains. When sections through environmental hydrogen-assisted cracks are examined metallographically, they are often found to be branched and discontinuous (Fig. 55).

Fig. 55 Environmental hydrogen cracks in a cast ultrahigh-strength steel part

Etched in Vilella's reagent. 150×

Fig. 56 Ultrahigh-strength steel dragline bucket tooth that failed due to environmental hydrogen-assisted cracking

See also Fig. 57.

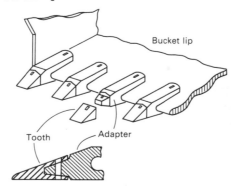

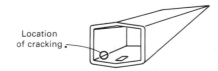

Fig. 57 SEM micrograph showing intergranular fracture along prior-austenite grain boundaries at fracture-initiation site in ultrahigh-strength steel dragline bucket tooth

200×

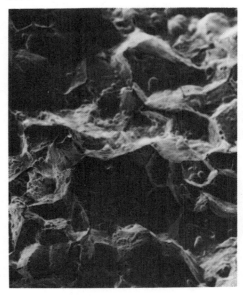

Example 22: Failure of a Cast Dragline Bucket Tooth. A tooth used on the digging edge of a large dragline bucket (Fig. 56) failed after several weeks in service. The tooth was cast in ultrahigh-strength low-alloy steel that had been heat treated to a hardness of 555 HB.

Investigation. Chevron marks on the fracture surface led to an initiation site on one of the interior corners of the socket portion of the tooth. Examination of the initiation site by SEM revealed that this portion of the fracture had occurred by intergranular fracture along prior-austenite grain boundaries (Fig. 57).

Examination of a region of the fracture away from the initiation site revealed a ductile-dimple fracture mode. Metallographic examination of a cross section taken through the initiation site revealed several secondary cracks similar to those shown in Fig. 55. The microstructure was found to consist of tempered martensite, which is the desired microstructure in this product. The hardness of the tooth was checked and found to be 555 HB, meeting the specification. A Charpy V-notch impact specimen taken near the initiation site gave a toughness value of 19 J (14 ft · lb) when tested at −40 °C (−40 °F). This is quite acceptable for material of this strength level. The impact specimen fracture was examined by SEM. It was found to have a ductile-dimple fracture morphology similar to that of the tooth fracture in regions away from the initiation site.

Conclusions. The failure of this tooth was caused by environmental hydrogen-assisted cracking. The stress necessary to cause fracture was due to the tooth becoming wedged onto the adapter in service.

Corrective Measures. The problem was corrected by changing the design of the tooth to prevent the wedging action. This eliminated the high stress that had led to failure.

REFERENCES

1. R. Monroe, private communcations
2. D.B. Roach and F.H. Beck, *Performance and Reliability of Corrosion-Resistant Alloy Castings*, MTI Manual No. 5, The Materials Technology Institute at The Chemical Process Industries, Inc., Columbus, OH, 1981
3. M.G. Fontana and N.D. Green, *Corrosion Engineering*, McGraw Hill, 1967, p 84
4. "Standard Practices for Detecting Susceptibility to Intergranular Attack in Austenitic Stainless Steels," A 262, *Annual Book of ASTM Standards*, Vol 01.05, ASTM, Philadelphia, 1984, p 82
5. W.C. Leslie and M.G. Fontana, Mechanism of the Rapid Oxidation of High Temperature, High Strength Alloys Containing Molybdenum, *Trans. ASM*, Vol 41, 1949, p 1213

Failures of Weldments

Revised by Alan G. Glover, Welding Institute of Canada;
Daniel Hauser, Edison Welding Institute; and Edward A. Metzbower, Naval Research Laboratory

WELDMENT FAILURES may be divided into two classes: those rejected after inspection and mechanical testing and those discovered in service. Failures in service may arise from fracture, wear, corrosion, or deformation. In this article, major attention is directed toward analysis of service failures.

Causes for rejection during inspection may be either features visible on the weldment surface or subsurface indications that are found by nondestructive-testing methods.

Surface features that are causes for rejection include:

- Excessive mismatch at the weld joint
- Excessive bead convexity and bead reinforcement
- Excessive bead concavity and undersized welds
- Sharp undercut and overlap at the weld toe
- Cracks—hot or cold, longitudinal or transverse, crater and at weld toe
- Gas porosity
- Incomplete fusion
- Arc strike
- Spatter

Subsurface features that are causes for rejection include:

- Underbead cracks
- Gas porosity
- Inclusions—slag, oxides, or tungsten metal
- Incomplete fusion
- Inadequate penetration

Failure to meet strength requirements is another cause for rejection of weldments. Details of test methods for welds are provided in the *Standards of the American Welding Society*, particularly ANSI/AWS B4.0-85, "Standard Methods for Mechanical Testing of Welds." A general treatment of evaluation methods for weldments is given in Ref 1.

Poor workmanship and improper selection of welding procedures and filler-metal composition account for numerous arc-weld failures. Other reasons for weldment failure include:

- Inappropriate joint design
- Improper weld size
- Unfavorable heat input

- Improper preweld and postweld heating
- Improper fit-up
- Incorrect parent (base) material composition
- Alloy segregation and embrittlement
- Unfavorable cooling rate in the weld metal or heat-affected zone (HAZ)
- High residual stresses
- Environmental conditions not contemplated in the design of the weld. These comprise accidental overload, continual loads higher than intended, fatigue, abnormal temperatures, and marine or other corrosive atmospheres

Resistance-weld failures will be discussed in the section "Failures in Resistance Welds" in this article.

Analysis Procedures

The initial task of the failure analyst is to seek out and compile as complete a history as possible of the failed weldment and its preparation. The success in arriving at a correct determination of the cause of failure will be greatly influenced by the amount of information that can be obtained as early as possible. There are no set rules as to the sequence in which the details are sought, but it is important to secure promptly all oral reports of the failure while the event is still fresh in the minds of the observers. The following is a suggested check list of information that will be useful later in the analysis:

- Determine when, where, and how the failure occurred. Interview all operators involved. How was the part treated after failure? Was it protected? How was the fracture handled? Did the failure involve any fire, which could have altered the microstructure of the weld or of the base metal?
- Establish the service history—loads, atmospheric exposure, service temperature, and length of service. Was an accident involved? Have there been other similar failures? Is the service history consistent with design criteria? Were operating parameters exceeded?
- Obtain drawings of the weldment design, calculations of service stresses, estimation of service life. What were the specified and actual base metal and filler metal? Obtain, if

possible, the actual chemical composition, heat treatment and mechanical properties of the base metal, and the actual chemical composition of the filler metal
- Ascertain the cleaning and fit-up procedures specified and those actually used. Obtain details of the welding procedures, specified and used, and any repairs made. Was preweld or postweld heating applied?
- Establish how the weldment was finished and what tests were performed. How long was the weldment stored, and under what conditions? When and how was it shipped for installation? Was it promptly and properly installed? Request copies of inspection procedures and inspection reports

Additional information on failure analysis procedures is provided in the article "General Practice in Failure Analysis" in this Volume.

Examination Procedures. After the background information relating to the failed weldment has been secured, the analyst is ready to undertake the study of the failed part itself. The study should begin with visual examination of the weldment and fracture-surface features, accompanied by preparation of sketches and photographs (both general and closeups) to provide a complete record. This should be followed by careful examination under a low-power (5 to 30×) stereomicroscope. These procedures should be followed for all failures, regardless of whether the failure is caused by fracture, wear, corrosion, or deformation.

The analyst should determine whether the failure was located in the weld, the HAZ, or the base metal; whether the weld met specifications; and whether those specifications were appropriate for the service conditions encountered. The subsequent steps used for identifying the exact cause of the failure will be greatly influenced by the outcome of these early observations.

If the failure is not a fracture, it may be necessary to follow low-power microscopic examination with examination at higher magnification. If so, this should be undertaken before chemical analysis of the weldment surface alters the surface characteristics. Examination of the weldment surface at high magnification should reveal whether there has been selective attack of the part by corrosion, erosion, or other failure mechanisms and may provide clues as to its nature.

The macroscopic and microscopic features that can aid in identification of a failure mechanism are described in the article "General Practice in Failure Analysis" in this Volume and in the articles on failures from various mechanisms and related environmental factors in this Volume.

If the failure is a fracture, much can be learned from the location of fracture origins, the mechanisms and directions of crack propagation, and the types of loading that were involved, by replica or direct examination in a transmission or scanning electron microscope.

Once the macroscopic and microscopic examinations are completed, analysis of the material and of the weldment surface may be undertaken by wet chemistry, by electron microprobe, or by x-ray diffraction, Auger electron, or ion-scattering spectrometers. More information on these and other analytical techniques is available in Volume 10 of the 9th Edition of *Metals Handbook*.

Metallographic Sectioning. When study of the fracture surface is complete, sectioning may be performed. The microstructure of each zone of the weldment can be revealed and compared to that expected for the welding procedure reported and the composition of base and filler metals. Sectioning will also permit appraisals of inclusion distributions and contours, disclosure of whether the fracture and any secondary cracks were transgranular or intergranular, and observation as to whether the weld fusion was incomplete or the penetration was inadequate. Such a microscopic examination of a distorted but unfractured weldment would also be appropriate as a means of discovering if the distortion was caused by loading at an excessively high temperature. Microhardness traverses of the weld zones could yield valuable corroborative data as to whether or not the microstructure had acceptable properties. Additional information on sectioning of welded joints can be found in the article "Weldments" in Volume 9 of the 9th Edition of *Metals Handbook*.

Nondestructive Testing. Certain nondestructive tests must be used with caution to avoid altering evidence or creating confusing evidence. For example, liquid-penetrant tests are very effective in revealing fine cracks, but if the failure resulted from stress-corrosion cracking (SCC), the use of a penetrant might provide false clues, for example, by introducing the penetrant or a corrodent into the crack. Application of the penetrant should be delayed until after completion of the pertinent chemical analyses and metallographic examination.

Magnetic-particle testing can also create problems. In one instance, welds were being made successfully on a steel containing more than 0.35% C, but because the normal borderline weldability of this material made it susceptible to cracking, a magnetic-particle test was done to check for cracks. No cracks were

Fig. 1 Two types of poor contours in arc welds

(a) Fillet weld showing two forms of undercut plus weld spatter and uneven leg length. (b) Butt weld showing a high-sharp crown

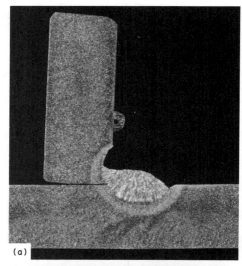

(a)

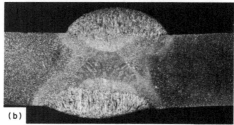

(b)

found, but a few weeks later, the weldments were returned because cracks had initiated at minute arc strikes that were caused by the prods used to make electrical contact for the magnetic-particle test. Due to their small size, the arc strikes cooled rapidly and created a hard, untempered martensitic structure that was susceptible to cracking. Magnetic-particle tests conducted with a magnetic field induced by a yoke that does not pass current through the part will avoid this difficulty.

Analysis of Information. The final stage in analysis of a weldment failure consists of comparing and integrating all of the information that has been gathered. It is particularly important to avoid arriving at a decision concerning the cause of failure before this final stage. This is because nearly all welds contain discontinuities of one type or another that at initial investigation might be considered to be a contributing factor to the failure. Therefore, it is important for the analyst to keep in mind that many welds containing discontinuities may have given satisfactory performance over long service lives.

Fractography can be used to identify the origin or origins of fracture and the paths that the propagating crack followed. The crack origin should reveal whether a stress concentrator might have contributed to failure. The metal-

lographic specimens should disclose the existence of any improper microstructures, such as decarburized layers, carburized skin, alloy segregation, martensite in the HAZ, and inadequate weld penetration. Any one of these items might prove to be the critical contributing factor in a particular instance, but generally the two most important items of information are the location of the fracture origin and the characteristics of the fracture surface. From these two items it may be deduced how the fracture began and how it propagated; whether it was generated by a single overload or by repetitive stresses; whether the loading was in tension, torsion, bending, shear, or some combination; whether fracture was ductile by microvoid coalescence or by tearing or was brittle cleavage or by integranular separation. However, all of the other data are necessary to ensure that a more obscure cause of failure is not missed.

The outcome of the failure analysis should usually be a recommendation that will avoid recurrence. Such recommendations generally consist of an eliminate-the-cause procedure, which will necessarily differ as the cause differs. The procedure may consist of using a better weld contour, a more favorable filler metal, a more appropriate heat input (a different cooling rate for the weld metal and the HAZ), a more accurate fit-up, a higher-quality base metal, or different service or maintenance conditions.

Failure Origins in Arc Welds

Some discontinuities that can serve as failure origins of arc-welded parts are found only in welds made by a particular process, but most discontinuities may be produced by any of the welding processes. Those of general origin are discussed first, and the types originating from specific processes are described later in this article. Information on process variables for control of weld quality is provided in articles on specific welding processes in Volume 6 of the 9th Edition of *Metals Handbook*. Additional information on the classification and causes of discontinuities in welded joints can be found in the article "Weld Discontinuities" in Volume 6 of the 9th Edition of *Metals Handbook*.

The discontinuities found in arc welds vary considerably in their importance as failure origins. Their locations (surface or subsurface) and geometries are the factors that must be considered in evaluating the significance of weld discontinuities. In moderate amounts, slag inclusions, porosity, groove overfill or underfill, or similar discontinuities will not reduce the static fatigue strength of the weld sufficiently to warrant their classification as potential causes of fracture. In fact, fatigue cracks that initiate at the toe of a weld will propagate around areas of gas porosity or slag inclusions as often as through them.

However, such features as hot (liquation) cracks, cold (hydrogen-induced) cracks, and lamellar tears are much more serious, because they are severe stress raisers and will affect fatigue strength to a degree that cannot be compensated for in design (Ref 2, 3). Hot cracks and lamellar tears normally occur during or immediately after welding and can be detected by nondestructive testing and eliminated from finished weldments. However, the formation of cold cracks resulting from the presence of hydrogen may be significantly delayed. For this reason, inspection is usually delayed at least 48 h after welding.

Porosity produced in arc welds can be grouped into three types: isolated, linear, and cluster. Isolated porosity is caused by a phenomenon similar to boiling when the arc power is too far above the ideal level. Linear or cluster porosity can result from interaction of components of the shielding gas, such as oxygen, hydrogen, or carbon dioxide, with the weld puddle to evolve a gas, such as hydrogen sulfide. Cluster porosity will also be formed when the cover of shielding gas is inadequate or when the welding is done on wet base metal.

Rust, if present on the base-metal surface, is a source of moisture. This moisture will either be dissolved as steam vapor by the molten weld pool or will be partly dissociated by the arc and dissolved as hydrogen. Where there is a possibility of porosity or embrittlement from rust, it can be avoided simply by removing the rust layer before attempting to weld or by using a special electrode.

Oxidation of shielded arc welds occurs only when there has been improper or inadequate shielding. Titanium, for example, is sufficiently reactive to require shielding of both the face and the root sides of the weld, as well as of the tungsten electrode (if gas tungsten arc welded) to prevent oxygen and nitrogen contamination and embrittlement of both the weld and the HAZ. Such additional shielding is not mandatory for the root side of steel or aluminum welds, but it can improve the quality of the weld.

Formation of Compounds. Another effect of the shielding material may be the formation of compounds, such as oxides or nitrides, in the molten puddle from reaction with minute impurities, such as oxygen or nitrogen, in the shielding gas. Such compounds have been shown to lower the toughness of the weld compared to that of welds made with pure inert gas. Considerable advances have been made with shielding-gas mixtures for gas metal arc welding. These new mixtures have overcome some of the problems of lack of toughness.

Hot cracks, or solidification cracks, in welds may be caused by joint design and the restraint imposed on the weld. However, hot cracks are commonly caused by the presence of low-melting constituents that extend the temperature range of low hot strength and low ductility to temperatures below that of the alloy. In steels, these constituents are com-

pounds, such as phosphides and sulfides, and elements, such as copper, which can segregate at grain boundaries and cause grain-boundary tearing under thermal-contraction stresses. Hot cracks can be minimized or eliminated through residual-element control and control of the weld pool shape. Weld beads with a high depth-to-width ratio can promote the buildup of low-melting phases at the pool centerline and thus cause hot cracking. The same type of hot crack is a crater crack. The material in the crater region is richer in solutes than the remainder of the weld, because of the directional nature of the weld-metal solidification, and is very prone to cracking. The best method of control is to use welding materials with minimum contents of residual elements, such as phosphorus, sulfur, and copper in steel, thus reducing the possibility of segregation. If this is not feasible, the next best procedure is to avoid the slow cooling rates that favor segregation between the liquidus and solidus temperatures and accelerate cooling until the crater has solidified. In some hot-cracking situations, a reduction in travel speed can alter the solidification directions in the weld deposit and thus reduce susceptibility to hot cracking.

In wrought heat-treatable aluminum alloys, the relation between working and weld direction will influence cracking. Welds made in the short-transverse direction often develop grain-boundary microcracks in the HAZ.

Cold cracks, or hydrogen cracks, form in welds after solidification is complete. In steel, cold cracks depend on the presence of a tensile stress, a susceptible microstructure, and dissolved hydrogen (Ref 4, 5). The stress may arise from restraint by other components of a weldment or from the simple thermal stresses created by welding in a butt, groove, or T-joint. The susceptibility of the microstructure to cold cracking relates to the solubility of hydrogen and the possibility of supersaturation. Austenite, in which hydrogen is highly soluble, is least susceptible to cold cracking, and martensite, in which the solubility of hydrogen is lower, is most susceptible, because the rapid cooling necessary for the austenite-to-martensite transformation traps the hydrogen in a state of supersaturation in the martensite.

The presence of hydrogen in a weld is generally due to moisture that is introduced in the shielding gas, dissociated by the arc to form elemental hydrogen, and dissolved by the molten weld puddle and by the adjacent region in the HAZ. In the supersaturated state, the hydrogen diffuses to regions of high stress where it can initiate a crack. Continued diffusion of the hydrogen to the region of stress concentration at the crack tip extends the crack. This behavior means that hydrogen-induced cold cracking is time dependent, that is, time is needed for hydrogen diffusion, and the appearance of detectable cracks can be delayed until long after the weld has passed inspection. The remedy is to eliminate as far as possible the sources of hydrogen—water, oils, greases,

waxes, and rust that contains hydrogen or hydrates. Preweld and postweld heating will promote the escape of hydrogen by diffusion and will reduce the level of residual tensile stress that may be present.

Poor weld contours include overlaps, which result when the weld puddle is too cold for fusion or when the puddle is too hot and solidifies after protruding beyond the toe or root of the weld; undercutting, from too high a welding current or too slow a welding speed (Fig. 1a); and a bead with a very high, sharp crown from too low a current or too fast a travel speed (Fig. 1b). In addition to proper control of welding current, the use of a weave type of bead is helpful in contour control. The weave technique must be used with caution, however, because low weld-metal toughness can result.

Embrittlement. Preweld and postweld heating are used for relief of internal stress and elimination of hydrogen. However, temperatures to which parts are heated should be selected with care; carbon steels and many alloy steels undergo a decrease in notch toughness when heated in certain critical temperature ranges. A discussion of this and other embrittlement phenomena is provided in the article "Ductile and Brittle Fractures" in this Volume.

Inadequate joint penetration and incomplete fusion in arc welds usually result from improper fit-up or groove design or from improper welding procedures. The presence of inadequate penetration and incomplete fusion can be critical because, in addition to reducing the effective cross-sectional area, it can produce discontinuities that can be almost as sharp as cracks and can even display crack-extended tips. In welding some alloys, inadequate penetration and incomplete fusion can result from an oxide layer on the joint sidewall.

Arc strikes and flame gouges can cause considerable damage to many metals, and care must be taken to avoid them. Arc strikes can result from accidental contact between the workpiece and the electrode or electrode holder. Also, loosely connected ground cables can cause arc burns. Careless handling of a cutting torch can result in flame gouging of a part. In hardenable alloys, arc strikes, arc burns, and flame gouges cause a variety of thermal responses that degrade the properties of base metal.

These responses can include the generation of defects (for example, hot cracks and porosity) that can result in stress concentrations that may initiate cracks. Figure 2(a) shows a micrograph of a section through an arc strike that displays porosity and shrinkage cracks. Figure 2(b) shows a fatigue fracture that had its origin in an arc strike. In another instance, a welder was making a flame cut on a steel section and allowed the flame to come in contact with an adjacent part. The result was a 13-mm (1/2-in.) deep gouge in the part, which eventually led to its fracture.

Shielded Metal Arc Welding. The condition of the electrode coating is important to the

Fig. 2 Typical damage to type 410 stainless steel caused by accidental arc strikes

(a) Nital-etched metallographic section through the crater of an arc strike showing porosity and shrinkage cracks below the surface. 35×. (b) Fatigue fracture (in profile at top) that originated in the center of arc strike. 5×

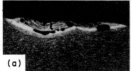

(a)

(b)

Fig. 3 Solubility of hydrogen in steel as a function of temperature

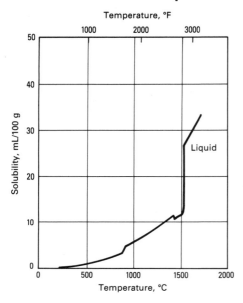

quality of shielded metal arc welds. There are two general types of coatings: one is a cellulose-type, which will break down in the arc to produce hydrogen; the other is a cellulose-free coating that produces very low hydrogen levels. The first type of electrode coating is used for welds on noncritical structures in which hydrogen in moderate amounts is acceptable or in situations where its fast-freeze and good penetration characteristics are desirable and hydrogen is controlled by preheat requirements. The second type of electrode coating is used for welds on structures for which even small amounts of hydrogen are not tolerable.

The low-hydrogen electrode coatings are hygroscopic and pick up moisture from the atmosphere, particularly if the relative humidity is high. This occurs especially if the electrodes are exposed to the air for more than 2 to 4 h. Moisture or other hydrogen compounds in the coating are dissociated by the arc temperature, and the resulting atomic hydrogen is readily dissolved in the weld puddle. Water vapor may also be directly dissolved by the molten steel with very little dissociation. The solubility for hydrogen in

the steel decreases abruptly and drastically (as much as 4:1 for hydrogen) when the molten phase freezes (Ref 6). A further, significant decrease in solubility for hydrogen occurs when the austenite transforms to either ferrite or martensite, depending on the alloy content and the cooling rate. During freezing, the hydrogen may escape by diffusion or formation of bubbles, but a portion will remain in the austenite in a supersaturated state. Any dissolved water will be largely ejected as bubbles during the freezing process, which may form a "picture frame" of wormholes if the quantity of dissolved water and the rate of freezing are both moderate. If a great deal of vapor has been dissolved, the metal will puff seriously, resulting in an extremely porous structure.

Although hydrogen solubility in molten iron is high compared to solidified iron (Fig. 3), not all of the potential hydrogen is transferred to the weld deposit. Much of the hydrogen in the weld bead diffuses to the surrounding material during cooling. The redistribution of hydrogen, however, does not depend only on the thermal history of the weld and the diffusion coefficient of hydrogen. The stresses in the weld area also affect the redistribution behavior and thus the cold-cracking tendency.

Hydrogen cracking can occur in either the HAZ or the weld metal and can be either transverse or longitudinal to the weld axis. The level of preheat or other precautions necessary to avoid cracking will depend on which region, weld metal or HAZ, is the more sensitive. In the older carbon-manganese medium-strength structural steels, the HAZ was usually the more critical region and weld metal rarely caused a problem. Modern steels with low carbon are now very resistant to hydrogen cracking; therefore, the weld metal may be the controlling factor.

Hydrogen cracking in the weld metal depends on the same fundamental factors as in the HAZ, that is, hydrogen content, microstructure, and stress. The controlling variables in practice are usually strength level, hydrogen content, restraint, and tensile stress level. In single-pass welds and the root run of multiple-pass welds, the root gap provides a stress concentration to the transverse stress; this leads to longitudinal cracks in the weld metal.

The higher dilution of the root run may result in a harder weld bead that is more likely to crack. This is the predominant form of cracking in such applications as pipelines. In cases where transverse notches exist—for example, an unspliced backing bar—the longitudinal stress becomes the controlling factor, and transverse cracks may occur. In heavy multiple-pass welds, cracking will generally be in the transverse direction either normal to the surface or at an angle of about 45°. The latter type has been called chevron cracking. The solution to these cracking problems rests with proper control of hydrogen through such practices as handling of electrodes, drying of fluxes and electrodes, and adequate preheat.

The characteristics of impact fractures of welds are related to the compositions of the electrodes used. A quantitative relationship is given in the table that accompanies Fig. 4, which summarizes data from a series of weld-cracking test specimens prepared from 25-mm (1-in.) thick 3.5% Ni steel plates cut with circular grooves, as shown in Fig. 4. In these weld-cracking tests, shielded metal arc welding was performed semiautomatically on a turntable, producing a 270° weld bead at 175 A, 26 V, and a travel speed of 150 mm/min (6 in./min) with a 5-mm (³/₁₆-in.) diam electrode.

The weld in specimen A in Fig. 4 was made with a low-hydrogen covered electrode (E6015) from a sealed container that was opened immediately before the test weld was made. Two rods were taken from the package at the same time, the covering of the two rods was stripped, and the moisture content of the coverings was determined by heating them for 1 h at 1400 °C (2550 °F). The temperature of the welding room was 25 °C (75 °F), and the relative humidity during the several tests was 35 to 40%. Other rods, used for the test welds in specimens B and C in Fig. 4, were exposed to the welding-room air for 168 h (7 days) and for 744 h (1 month), respectively. The crack lengths in the three welds were correlated with determinations of moisture content, as shown in the table with Fig. 4. Although the electrode coverings were of the low-hydrogen type, the moisture content increased with additional time of exposure to room air, and the welds exhibited increases in crack lengths.

Hydrogen damage to weld deposits (either as porosity or as cracking) can be avoided simply by proper control of hydrogen through handling of electrodes, drying of fluxes and electrodes, and adequate preheat. For low-hydrogen shielded metal arc electrodes, this means storing in holding ovens to prevent moisture from re-entering the electrode coating from the air. Normally, such ovens should be controlled between 120 and 175 °C (250 and 350 °F). Electrodes may require baking to remove moisture at temperatures as high as 425 °C (800 °F) before storing, but excessive baking can oxidize ferroadditions in the electrode coating and affect deoxidation in the weld pool.

Most carbon steels can be welded using coated electrodes and appropriate welding procedures, including preheat when required. Electrodes are classified on the basis of chemical composition, mechanical properties, and type of coating.

In flux-cored arc welding, the operating characteristics of the electrodes vary with the core ingredients and the shielding gas (if used). Care should be taken to remove the slag in some cases and to maintain the low-hydrogen characteristics of the tubular electrodes. Adequate shielding must be maintained through either an external source or a self-shielding type of electrode.

Gas Metal Arc Welding. Typical problems that can occur in this type of welding are

Fig. 4 Circular-groove specimens used in weld-cracking tests

(a) Configuration and dimensions (given in inches). Specimens A to C are 3.5% Ni steel test specimens that were shielded metal arc welded with low-hydrogen coated electrodes (E6015) that were exposed to welding-room air for different lengths of time before welding. Test results are given in the accompanying table.

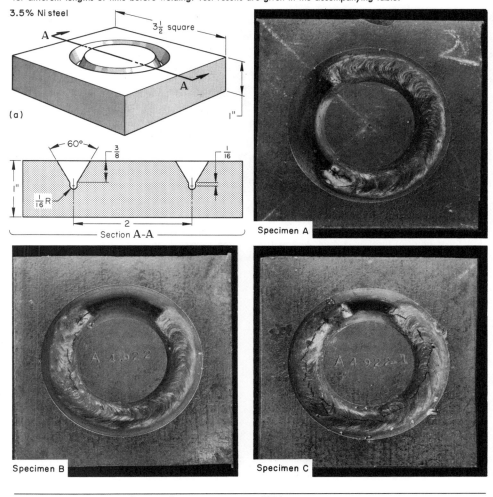

Specimen A

Specimen B

Specimen C

Test specimen	Exposure of electrode to air, h	Moisture content of covering, wt%	Length of crack in test weld mm	in.
A	0	0.25	12.7	0.5
B	168 (7 days)	0.56	38	1.5
C	744 (1 month)	0.70	114	4.5

Fig. 5 Macrograph of a defective plasma arc weld

The weld was made with improperly programmed reduction of welding current and orifice-gas flow, which resulted in surface irregularity and subsurface porosity. 6×

lack-of-fusion defects caused during short-circuit transfer and porosity from shielding-gas problems. Adequate deoxidizers are required in the electrode to prevent porosity in the weld. The low-hydrogen nature of the process will be lost if the filler metal or shielding gas are contaminated. Electrodes must be properly stored and cleaned.

Submerged Arc Welding. Factors that contribute to the formation of discontinuities in other arc-welding processes generally also prevail for the submerged arc process. However, in submerged arc welding, flux is separate from the electrode, and the quantity of the flux layer can affect results. For example, if the flux layer is too deep, it will produce a ropey reinforcement, whereas too shallow a layer contributes to the formation of porosity and gas pockets. Also, a flux layer that is too narrow causes a narrow reinforcement. Flux must be kept dry to maintain low-hydrogen characteristics.

Gas Tungsten Arc Welding. A potential but infrequent problem in gas tungsten arc welding is the pickup of tungsten particles from the electrode caused by excessive welding current or physical contact of the electrode with the weld puddle. Tungsten inclusions are extremely brittle and fracture readily under stress and can thus act as failure origins. Tungsten inclusions can be caused by excessively high welding currents, incorrectly sized or shaped electrodes, and oxygen contamination of the shielding gas. Tungsten particles are easily detected in radiographs.

Plasma arc welding is closely related to gas tungsten arc welding. A principal difference between these two welding processes is the keyhole effect produced by the constricted arc column of a plasma arc weld. This effect is created by displacement of molten metal by the forceful plasma jet, permitting the arc column to pass completely through the workpiece. If the welding current is turned off abruptly at the end of a welding pass, the keyhole will not close, and a hole or a large region of subsurface porosity will be produced. This is not a problem, however, if the weld can be terminated on a runoff tab.

Plasma arc welding has been used, for example, for the sealing of the circumferential joints of a high-performance pressure vessel made of modified type 410 stainless steel. The plasma arc completely penetrated through the 8-mm (0.3-in.) thick joint, which did not require the special edge preparation that would have been necessary if gas tungsten arc welding had been used, and permitted completion of the joints with fewer weld passes. To avoid introducing porosity in the weld-joint overlap region, it was necessary to program a careful welding current slope-down to reduce the welding current and the orifice-gas flow gradually at the end of the weld. In this way, a smooth transition was achieved from the deep, keyhole mode of penetration to shallow penetration. Figure 5 shows a macrograph of a longitudinal section through a plasma arc weld, illustrating the surface irregularity and subsurface porosity that can result when the welding current and orifice-gas flow

Fig. 6 Lamellar tear beneath a T-joint weld that joined two low-carbon steel plates

(a) Fractograph of lamellar tear showing separation that has followed flattened inclusions. Approximately 0.3×. (b) Section through fracture (top), which occurred in the coarse-grain reaustenitized region (A) of the HAZ. Region B is a fine-grain reaustenitized area. Region C is a partly reaustenitized region of ferrite and refined pearlite. Region D is an unaffected region of ferrite and pearlite. Approximately 8×. (c) SEM fractograph showing inclusion platelets (arrow E) and dimples (arrow F). 1300×

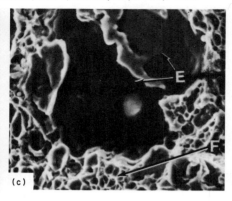

are not properly reduced through the overlap region.

Failures in Arc-Welded Low-Carbon Steel

Low-carbon steels are generally easy to weld and are quite ductile; therefore, the existence of porosity and slag inclusions in shielded metal arc welds is usually not critical. These steels can be welded using any of the common welding processes if adequate precautions are taken.

At low temperatures, the tolerance of low-carbon steels for weld discontinuities is critically affected by the microstructure. For instance, fine-grain pearlite formed by normalizing is much more tolerant of weld discontinuities than coarse pearlite formed by furnace cooling. The existence of a coarse pearlitic structure was the basis for the brittle failures experienced by the welded Liberty ships of World War II, which split at welds while still at dockside. The fracture origins in the Liberty ships were at weld discontinuities, and crack propagation was through furnace-cooled plate. The cracking was minimized by specifying normalized plate. Example 1 in this article describes another instance of brittle low-temperature fracture of a weld.

Embrittlement of low-carbon steel is not common, but proper care should be taken in applying peening to low-carbon steel weld deposits. High-strength low-alloy steels are readily welded, usually using low-hydrogen processes and preheat (see the article "Arc Welding of Hardenable Carbon and Alloy Steels" in Volume 6 of the 9th Edition of *Metals Handbook*).

Inclusions. The presence of flattened inclusions, such as silicates or manganese sulfides, in rolled carbon-steel plate can cause difficulties, particularly in welding T-joints. The cooling stresses may cause rupture by decohesion between the inclusions and the steel matrix, producing a fracture with a laminated appearance (Fig. 6a). The nature of fracture through the coarse-grain reaustenitized structure of the HAZ is shown in profile in Fig. 6(b). The fine

details of the fracture can be seen in the SEM fractograph in Fig. 6(c), which displays the dimpled rupture of the low-carbon steel matrix and the exposed surface of silicate and sulfide platelets.

Hot cracking in gas tungsten arc welds of low-carbon steel depends primarily on the steel composition, the degree of joint restraint, the cooling rate, the energy input, and the solidification structure. Susceptibility to cracking rises with increases in carbon, sulfur, and phosphorus content and with decreases in the manganese-to-sulfur ratio. Steels with carbon contents up to 0.25% are normally weldable by the gas tungsten arc process without hot cracking.

A very rigid joint restraint concentrates the thermal contraction in the weld deposit and thereby promotes hot cracking. The formation of a columnar structure during solidification of the weld is more likely to promote hot cracking than if an equiaxed structure is formed.

Example 1: Brittle Fracture of a Soybean-Oil Storage Tank Caused by High Service Stresses. A riveted 0.25% C steel oil-storage tank located in Oklahoma was dismantled, the edges of the plates were beveled for welding, and the plates were reassembled in Minnesota by shielded metal arc welding to form a tank for storage of soybean oil. The reconstructed tank, 18 m (60 ft) in diameter and 12 m (40 ft) high, was set on a rather yielding subsoil base, with minimum density requirements. There was gravel fill over the subsoil with no compaction. After storage of soybean oil for a year, the tank was cleaned to remove the heavy semisolid residues. For ease of access, a rectangular opening was cut in the side of the tank large enough to admit a front-end loader. After the tank was cleaned, a frame of heavy angle iron was prepared for the opening. One leg of the angle iron was fillet welded to the tank at a 90° angle to the tank surface along each edge of the opening, projecting outward; the other leg was drilled to permit bolting a 19-mm (¾-in.) thick steel plate to it as an external closure that could be removed for periodic cleaning.

During the following year, the tank was filled to a record height. In mid-January the temperature dropped to −31 °C (−23 °F), accompanied by winds of 47 km/h (29 mph). During this cold period, the tank split on both sides of the opening. The primary origin of the failure was at the upper-left corner of the opening, from which the crack rapidly propagated in a brittle-type fracture vertically to the roof of the tank and then turned to follow either side of the roof in a ductile manner, allowing the roof to fall in and the sides to spread out.

Investigation. In addition to the primary origin, the fractures were found to have two secondary origins—at the upper right and lower right corners of the opening. All three origins were located at the toes of the fillet welds attaching the angle-iron frame to the tank wall. The direction of crack propagation was clearly indicated by chevron marks in each instance.

The welding of the frame was done in the fall and early winter, using the shielded metal arc process with E6010 electrodes. Use of this type of electrode would suggest that weld porosity, hydrogen embrittlement, or both were probably present. At the subzero temperatures existing at the time of failure, the steel was certainly below its ductile-to-brittle transition temperature. These combined circumstances thus suggested a possible brittle condition.

The record level to which the tank was filled imposed a high stress on the tank walls, which was accentuated at the corners by the rectangular shape of the opening. Calculated values of the stress at the welds were found to approach the yield strength of the steel. It was therefore not surprising that the fracture, once initiated, propagated at a speed that simulated an explosion.

Conclusions. The tank failed by brittle fracture because very high service stresses were applied to a steel at well below its transition temperature. Fracture was initiated by stress concentrations at the sharp corners of the opening and at the toes of the fillet welds that attached the angle-iron frame to the wall of the tank. The choice of E6010 electrodes also contributed to a brittle condition in the fillet weld.

Recommendations. Several steps essential to avoid a repetition of such a failure are as follows:

- For greater suitability to the climate, the steel plate should have a low carbon content and a high manganese-to-sulfur ratio and should be in a normalized condition
- Low-hydrogen electrodes and welding practices should be used
- All corners of the opening should be generously radiused to avoid stress concentrations at the toe of the welds
- The welds should be inspected and ground or dressed (if necessary) to minimize stress concentrations
- Postweld heating would be advisable
- Radiographic and penetrant inspection tests should be performed

Example 2: Observatory Column That Cracked Because of High Residual Stresses and Stress Raisers in the Welds.

During construction of a revolving sky-tower observatory, a 2.4-m (8-ft) diam cylindrical column developed serious circumferential cracks overnight at the 14-m (46-ft) level where two 12-m (40-ft) sections were joined by a girth weld. The winds were reported as varying from 7 to 19 km/h (4½ to 12 mph) during that night, with directions that began from the south but veered to the north between 1 and 4 a.m. The temperatures recorded at a nearby weather station ranged from 12 °C (53 °F) to 7 °C (45 °F) between 10 p.m. and 4 a.m. that night.

The column was shop fabricated in 12-m (40-ft) long sections of 19-mm (¾-in.) thick steel plate conforming to specifications for ASTM A36 steel. At the time of the fracture, the foundation had been poured, a 1.8-m (6-ft) high base section installed, and assembly of five 12-m (40-ft) sections had been partly completed, reaching an overall height of approximately 61 m (200 ft). The design of the section-to-section weld joint is shown in Fig. 7(a). This called for a bevel on the edge of the upper cylinder and an external backing band to facilitate full penetration of the girth weld. The band was to be circumferentially fillet welded at each edge to the outer surface of the cylinders.

Investigation. Figure 7(b) shows the outside surface of the north side of the column that contains the backing band, a vertical pair of guide rails, assorted welded braces, and steadying guy wires. At A and B in Fig. 7(b) are two openings where samples were cut out for examination. The portion of the section-to-section joint beneath the guide rails had not yet been welded, and the backing band does not extend over this section. A similar unwelded portion was found beneath a pair of guide rails diametrically opposite to those shown in Fig. 7(b). The inside of the north side of the column is

Fig. 7 Cylindrical column of arc-welded ASTM A36 steel for a sky-tower observatory that cracked during construction because of high residual stresses and stress raisers in the welds

(a) Design of section-to-section weld joint showing the circumferential girth welds and backing band joined by two continuous, full-penetration 4.7-mm (³⁄₁₆-in.) fillet welds. Dimensions given in inches. (b) Outside surface of the north side of the column showing sites of cutout samples A and B and unwelded portion at N. (c) Inner surface of the north side of the column showing sites of cutout samples A, C, and D; unwelded portion N; and crack F

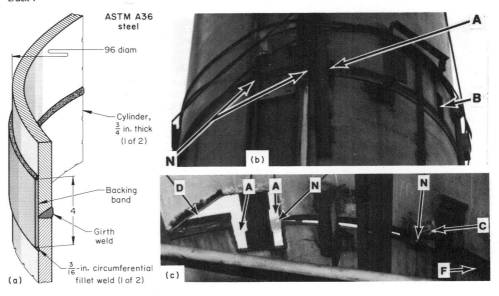

shown in Fig. 7(c). The unwelded portion is clearly visible at the 13-mm (½-in.) gap between the upper and lower sections of the column wall. The I-beam at near right is a vertical internal stiffener attached to the column wall by intermittent fillet welds. Two short lengths of angle beam that span the joint were welded to the walls of the column. Three butts, two triangular and one rectangular, were made in this region to remove samples for investigation.

Study of the samples established that there were two crack origins, one in or adjacent to the weld at each end of the unwelded portion shown in Fig. 7(b) and (c). Cutout sample C (Fig. 7c) contained the fracture origin to the right of the unwelded portion; the continuation of the crack to the right is shown in Fig. 7(c). The crack origin to the left of the unwelded portion was contained in cutout sample D; an I-beam hides the continuation of that crack to the left. The complete extent of cracking in the column wall was as follows. The crack on the west side continued to nearly the southwest side of the column, propagating at about 100 mm (4 in.) above the girth weld. The crack on the east side of the column propagated in the weld to about the northeast side of the column, then emerged to grow in the upper wall at a slight angle to the weld to reach again a distance of about 100 mm (4 in.) above the girth weld. This crack halted at a welded angle on the east side of the column. A 100-mm (4-in.) long crack was found on the south side of the column that had initiated in the girth weld at the east end of the unwelded section beneath

the guide rails. This crack appeared to be about to emerge from the weld into the column wall, but was not opened for study.

Much of the length of the long cracks displayed chevron marks, which suggested that most of the crack propagation had been fairly rapid. The fact that the cracks halted instead of continuing for the full circumference of the column suggests that there was longitudinal tension across the girth weld on the north side of the column that did not exist at anywhere near the same magnitude on the south side of the column.

Hardness measurements on the column wall outside the weld zone gave an average of 85 HRB. This is approximately equivalent to the tensile strength of 552 MPa (80 ksi), or the upper limit required by ASTM A36. Four Charpy V-notch impact test specimens were prepared from one of the cutout samples and tested, with the following results:

Test specimen	Testing temperature		Impact strength	
	°C	°F	J	ft · lb
1	24	76	54	40
2	16	60	46	34
3	7	45	19	14
4	0	32	16	12

These data show that by 4 a.m. on the night of failure the column steel had cooled below its ductile-to-brittle transition temperature.

Examination of the girth welds at the 14-m (46-ft) level revealed that many did not completely fill the groove. Many contained porosity, and areas of weld-bead cracking were seen at the fractures on the north side of the column near the two origins. The cracking band had not yet been fillet welded along its upper edge.

Discussion. The exact details of securing fit-up of the two column sections are not recorded. However, the fact that the backing band was fillet welded along its bottom edge but not its top suggests that the backing band may have been applied before the upper section was hoisted in place and thus used as a positioning ring. There were at least ten short angle beams spanning the joint, which were fillet welded on both sides to the column walls. These were apparently welded in place after the upper section had been set precisely vertical and served as anchors until the girth weld could be applied. How much of the joint had a gap as large as that visible at the unwelded portion in Fig. 7(c) cannot be determined, but the two sections must have been in contact in some areas. However, there was a large gap on the north side of the column, which was maintained by the rigidity of the angle-beam anchors. This combination of rigid restraint and a large amount of weld metal contracting from its solidification temperature showed that there were high residual stresses across the weld. The additional unfavorable factor was the termination of the girth welds, which in effect provided notches that were stress raisers.

Conclusions. Crack initiation was caused by high residual stress from undue restraint during girth welding, combined with the presence of notches formed by the termination of the incomplete welds. Continuation of the cracks in the portion of the base metal that was well removed from the girth weld was attributed to the brittle condition of the steel when cooled by the night air and acted on by the critical notch created by the crack emerging from the weld bead. The wind was not thought to have significantly contributed to the stresses causing failure.

Recommendations. For this type of structure, in which the safety of a human life would be at stake, a steel with a much lower ductile-to-brittle transition temperature is essential. The other necessary steps include better control of the girth-welding conditions, such as more accurate fit-up to provide a uniform gap, pre-weld and postweld heating to minimize thermal stresses, welding at a slower rate (with perhaps less metal laid down per pass) combined with careful peening to reduce internal stress, perhaps the choice of a more favorable electrode to avoid porosity, careful termination of all welds interrupted by a shift change to avoid formation of notches, and completion of all welds before other sections of the column are erected.

Example 3: Fatigue Fracture of a Shaft for an Amusement Ride Because of Undercuts in Welds. An amusement ride in a shopping center failed when a component in the ride parted, permitting it to fly apart. The ride consisted of a central shaft supporting a spider

of three arms, each of which was equipped with an AISI 1040 steel secondary shaft about which a circular platform rotated. Each platform carried six seats, each set being capable of holding up to three persons. The whole assembly could be tilted to an angle of 30° from horizontal. The main shaft rotated at about 12 rpm and the platforms at a speed of 20 rpm.

The accident occurred when one of the secondary shafts on the amusement ride broke, allowing the platform it supported to spin off the spider. The point of fracture was adjacent to a weld that attached the shaft to a 16-mm (⅝-in.) thick plate, which in turn bore the platform-support arms.

Investigation. The fracture surface of the shaft was found to possess the beach marks characteristic of fatigue fracture (Fig. 8). It was apparent that the fillet weld had undercut the shaft slightly, creating points of stress concentration that served as nuclei for the fatigue cracks. Crack propagation occurred from two directions, penetrating all of the shaft section except the roughly triangular region (Fig. 8) that was the zone of final fast fracture.

The shaft at the location of the fracture was 108 mm (4¼ in.) in diameter; at the bearing raceway, 64 mm (2½ in.) below the weldment, it was 98 mm (3⅞ in.) in diameter. This indicated that cross-sectional area was not the determining factor in the stress level, but rather that the weld contained appreciable residual stress mainly from the restraint and chilling effect exerted by the heavy plate and shaft on the weld as it underwent thermal contraction.

A section through the fracture region revealed a small secondary crack (which did not contribute to fracture but which arose from the same causes) and clearly defined HAZs. The structure confirmed a report from the manufacturer that there had been no postweld heat treatment.

Conclusions. The shaft fractured in fatigue from the combination of residual stresses generated in welding and centrifugal stresses from operation that were accentuated by areas of stress concentration at the undercuts. Without the excessive residual stress, the shaft dimensions appeared ample for the service load.

Recommendations. The fillet weld should be applied with more care to avoid undercutting. The residual stresses should be minimized by preweld and postweld heat application.

Example 4: Fracture of Supplementary Axle-Support Channels for a Highway Trailer Caused by Restricting Welds. A supplementary axle, which was used as an extension to a highway-trailer tractor to increase its load-bearing capacity, failed in service. The rolled steel channel extensions that secured the axle assembly to the tractor main-frame I-beams fractured transversely, with the crack in each instance initiating at a weld that joined the edge of the lower flange to the support-bracket casting. The cracks propagated through the flange on each side until the effective cross-sectional area had been reduced suf-

Fig. 8 AISI 1040 steel shaft for an amusement ride

Fatigue fracture originated at weld undercuts. Two sets of beach marks and a triangular final-fracture zone are visible. Approximately 0.4×

ficiently to bring about sudden and complete fracture of the remaining web and upper flange.

Investigation. Figure 9(a) and (b) show the mating fracture surfaces of the left channel. Figure 9(b) shows the fracture origin in the weld along the toe of the lower flange of the channel extension that attached it rigidly to the support-bracket casting. At left in Fig. 9(a) is the cross I-beam of the supplementary axle assembly; nesting in the channel extension is a support plate, which was bolted through the channel to the support-bracket casting and welded to the I-beam at its web and flanges. The fracture surface in general appears smooth with little evidence of deformation and little visible detail, suggesting that the fracture was brittle. The fracture of the right channel exhibited the same characteristics.

A vertical transverse section through the flange and the weld attaching the channel to the support-bracket casting was prepared for metallographic examination. Figure 9(c) shows the weld, the HAZ, and the flange cross section at low magnification. Of particular significance is the light-shaded layer about midway between the top and bottom surfaces of the flange. This was a region of carbon segregation where chemical analysis showed the steel to contain 0.34% C and 1.16% Mn. Near the surfaces, the composition was 0.23% C and 1.06% Mn. As a result of this variance, the microstructure in the light-shaded layer showed much more pearlite than the lower-carbon regions near the surfaces. The residual-element contents of the steel were low: 0.05% Si, 0.05% Ni, 0.05% Cr, and 0.05% Mo.

The higher carbon content of the segregated central layer had its most marked effect in the HAZ of the weld at the toe of the flange, producing a band of very hard martensite in contrast to the lower-carbon pearlitic micro-

Fig. 9 Steel supplementary axle assembly for a highway trailer that broke in service because of restricting welds

(a) Forward left side of the assembly showing fracture surface of the channel extension. Also visible are the cross I-beam, supporting plate, and support-bracket casting. (b) Photograph of section toward the rear showing mating fracture surface and the fracture origin. (c) Macrograph of the channel flange showing weld metal, HAZ, and non-heat-affected zone. Polished, and etched with 3% nital. 8×

(a)   (b)   (c)

Fig. 10 Flange-to-pipe assembly of a carbon steel header, used for handling superheated water, that cracked by fatigue because of notches at welds

(a) Section through butt-welded joint showing crack (arrow A) that originated at toe of weld on inner surface, incomplete weld penetration (arrow B), and difference in thickness of flange (left) and pipe. Etched with 2% nital. Approximately 1.7×. (b) Section through same weld as in (a), but showing no crack indications. Etched with 2% nital. Approximately 1.7×

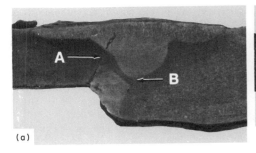 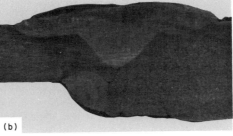

(a)   (b)

structure of unaffected areas of the flange. The hardness in the HAZ measured 60 HRC in the carbon-segregated layer and about 53 HRC elsewhere. In regions not influenced by the weld and outside the central segregation, the hardness was between 10 and 20 HRC.

The rigid attachment of the channel to the support-bracket casting provided a stress concentration at the weld by creating a steep gradient in section modulus at that point. The natural swaying of the load carried by the trailer imposed high bending-moment stresses on the channels of the supplementary axle assembly, raising the stress concentration at the weld to above the yield strength of the material. Had

the HAZ been ductile, fracture might have been delayed by yielding, but the hard and brittle midsection layer was susceptible to cracking under these conditions. Once the crack was formed, its growth became rapid.

Conclusions. Fatigue fracture was caused by a combination of high bending stresses in the bottom flanges of the channels due to the heavy load being carried, concentration of stresses due to the rapid change in section modulus of the channel at its point of attachment to the support-bracket casting, and brittleness of the high-hardness HAZ of the weld associated with the abnormally high carbon content in the central part of the channel.

Recommendations. Welding of channel edges is generally viewed as contributing to harmful gradients in section moduli and should be avoided in future assemblies. The segregation of carbon appears to be such an unlikely occurrence that use of inspection steps to avoid it seems unwarranted.

Example 5: Fatigue Cracking of Headers for Superheated Water Because of Notches at Welds. A system of carbon steel headers, handling superheated water of 188 °C (370 °F) at 2 MPa (300 psi) for automobile-tire curing presses, developed a number of leaks within about 4 months after 2 to 3 years of leak-free service. All the leaks were in shielded metal arc butt welds joining 200-mm (8-in.) diam 90° elbows and pipe to 200 mm (8-in.) diam welding-neck flanges. A flange-elbow-flange assembly and a flange-pipe assembly that had leaked were removed for examination.

Investigation. Magnetic-particle tests confirmed the existence of the cracks in the welds through which leaking had occurred and revealed the presence of other cracks that had not yet penetrated to the weld surface. X-ray inspection was conducted with the source at the centerline of each flange and the film wrapped around the weld bead on the outside. The film showed the cracks indicated by the magnetic-particle inspection as well as additional, incipient cracks and regions of incomplete weld penetration that could not be correlated with the major crack locations.

The butt welds were sectioned, with some cuts being made through regions of pronounced magnetic-particle indication and others at regions where no indications were found. The polished-and-etched surfaces of two such sections from the flange-pipe weld are shown in Fig. 10(a) and (b).

It was immediately apparent upon sectioning that, although the outside diameter of the flanges matched those of the elbows and pipe (nominal 220 mm, or 85⁄8 in.), there was a considerable difference in the inside diameters. Actually, only the elbows had been of schedule 80 wall (194-mm, or 75⁄8-in., nominal inside diameter), as called for in the specification for flanges, elbows, and pipe. The welding flanges that were used had been bored to 202.5-mm (7.975-in.) inside diameter (for use with schedule 40 pipe) and the pipe had been of schedule 100, with a 189-mm (7.437-in.) nominal inside diameter. The result was a disparity in inside diameter of about 8.9 mm (0.350 in.) between flange and elbow and about 13.7 mm (0.538 in.) between flange and pipe. This was the reason for depositing an internal fillet weld, which created a stress-concentrating notch at the toe of the weld against the flange inner surface.

The welding procedure had been as follows. The edges were prepared by machining a 6.4-mm (¼-in.) deep groove with a 75° included angle at the outside surface. The initial weld deposit was a root pass applied from the outside. Second and third passes were laid on top of this, the latter depositing a wide weave-type bead. Finally, an attempt was made to reconcile the difference in wall thickness by depositing a fillet-weld bead on the inside. In some instances, penetration was deep enough so that the fillet weld merged with the root pass of the exterior weld. In others, as in Fig. 10(a) and (b), this did not occur; as a result, a short span existed where there was incomplete penetration, and the original edges of the flange and elbow were still visible in contact. Such a region of incomplete penetration is normally considered a site for stress concentration, but it can be seen in Fig. 10(a) and (b) that there was no crack growth at this point. All the major cracks found in either elbow or pipe welds originated at the toe of the fillet weld in the inner surface of the flange (Fig. 10a). As illustrated in Fig. 10(b), some fillet-weld toes did not produce such cracks.

A hardness survey was taken of each weld joint. The results showed that the weld beads and the HAZs were similar in hardness to the base metal and that there was no evidence of hard spots. Tensile and bend-test properties of the welds were not determined, because the stress-concentrating notches present at the toes of the fillet welds would have rendered any test result meaningless. However, the welds were sound, fine grained, and free from slag inclusions and porosity.

Chemical analysis of the five components of the two assemblies confirmed that these headers conformed to the requirements of the applicable specifications (flanges and elbow of ASTM A105, grade II, and pipe of ASTM A106, grade B; similarly numbered ASME SA specifications also apply).

A powdery, green deposit found in the interior of an iron-body gate valve of the system was determined by chemical analysis to consist of a trivalent chromium oxide or salt plus iron oxide or salt and 22% combined water. No organic matter was detected and no copper was found.

Conclusions. The failures of the butt welds were the result of fatigue cracks caused by cyclic thermal stresses that initiated at stress-concentrating notches at the toes of the interior fillet welds on the surfaces of the flanges. Several of these penetrated to the exterior surface and caused leaks, but others existed that were still in an incipient growth stage. Incomplete penetration existed in some of the welds; but no evidence was found of cracks originating from these sites, and they were not a factor in the failures.

The powdery, green deposit was attributed to reduction of some of the chromate used as a corrosion inhibitor in the superheated water. It was not regarded as a contributing cause of the failures.

Recommendations. It was concluded that all other joints of the system possessing the same mismatch in wall thickness, and thus in all probability the same stress raiser at the toe of the interior fillet weld, could be expected to fail

Fig. 11 Weld attaching the head to the shell of a steam preheater that cracked because of poor root penetration in original and first replacement joint designs

(a) and (b) Sections taken through the head-to-shell joint. Etched in hot 50% hydrochloric acid. Actual size. The section in (a) is the original design; (b) shows the first replacement design. (c) The final design, which achieved full root penetration

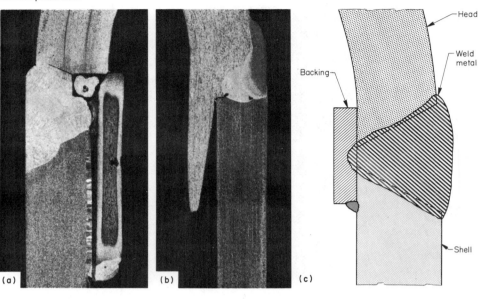

by the same mechanisms. Repair of such joints was not considered feasible, and it was recommended that they be replaced. Ultrasonic testing should be used to identify the joints by detecting the differences between the wall thicknesses of the flange and the pipe and of the flange and the elbow. Special attention to accuracy of fit-up in the replacement joints was also recommended to achieve smooth notch-free contours on the interior surfaces.

Example 6: Cracking of a Weld That Joined the Head to the Shell of a Steam Preheater Because of Poor Root Penetration. A weld that attached the head to the shell of a preheater containing steam at 1.4 MPa (200 psi) and used in the manufacture of paper cracked in service. The length of service was not reported. A section of the failed preheater, plus a section of proposed new design, were selected for examination.

Investigation. The preheater was approximately 0.9 m (3 ft) in diameter and about 1.2 m (4 ft) long. The original joint (Fig. 11a) contained a 6.4- × 50-mm (¼- × 2-in.) backing ring that had been tack welded to the inside surface of the shell in a position to project about 16 mm (⅝ in.) beyond the fully beveled top edge of the shell. The ring served the dual purpose of backing up the weld and positioning the head; the projecting edge of the ring fitted against a 9.5-mm (⅜-in.) wide, 3.2-mm (⅛-in.) deep undercut on the inner corner of the rim of the head. The internal 90° angle in this undercut was sharp, with almost no fillet. A bevel from the lower edge of the undercut to the outside of the head completed

the groove for the circumferential attachment weld.

The head was joined to the shell by arc welding, in three passes. In all sections that were taken through the welded joint and examined, the fusion was found to range from poor to none at the root of the weld and at the backing ring. The weld did not penetrate to the undercut rim of the head, and in some locations, the weld bead was standing free (not fused with the backing ring). In all sections of the joint examined, there was a crack beginning at the sharp, internal 90° angle of the undercut and extending toward the outer surface of the head. This crack had to penetrate only 50% of the full thickness of the head before final fracture occurred.

The first proposed replacement design (Fig. 11b) was more or less the reverse of the failed joint. A deep offset had been machined in the exterior of the rim of the head, reducing the wall thickness by 65%. The outer portion of the offset had been beveled to receive the weld deposit; below the offset, a horizontal step, about 2.5 mm (0.10 in.) deep, had been cut to align the head against the square-cut shell edge. Below the step was a long taper almost to the inner surface of the head. A number of sections through the proposed replacement joint showed incomplete root penetration of the weld, and in one section, a crack was discovered extending from the outer edge of the positioning step through the root-pass weld bead.

Conclusions. Cracking occurred in the HAZ in the head of the original design, originating in the sharp corner of the undercut, which was an inherent stress raiser. In the first replacement design, beveling of the head and shell was insufficient to permit full root-weld penetration, thus creating an inherent stress raiser. In both designs, the stress raisers would not have been present had full root-weld penetration been obtained.

Corrective Measures. To ensure full root penetration, the joint design shown in Fig. 11(c) was adopted. This joint retained the full thickness of both the head and the shell and, with the gap between the root edges, made it possible to obtain full root penetration, thereby reducing any notch effect to a minimum. A suggested alternative that would avoid the possibility of crevice corrosion was the use of a single-V-groove joint without a backing ring, with welding being done from the outside in two or more passes, ensuring that the root pass attained full penetration.

Example 7: Failure During Fabrication of an Armature Because of Lamellar Tearing. During the final shop welding of a large armature for a direct-current motor (4475 kW, or 6000 HP), a large bang was heard and the welding operation stopped. When the weld was cold, nondestructive evaluation revealed a large crack adjacent to the root weld.

Investigation. Figure 12(a) shows the design of the weld and the region where cracking occurred. The 100-mm (4-in.) thick plate had a chemical composition of 0.21C-0.77Mn; composition of the 125-mm (5-in.) plate was 0.19C-0.76Mn, giving carbon equivalents (CE) of 0.34 and 0.32, respectively (CE = C + Mn/6 + (Cr + Mo + V)/5 + (Ni + Cu)/15). The welds were made with a preheat of 65 °C (150 °F), with the heat applied only on the outside wall at point B. The weld was deposited using a single-head submerged arc process, and the failure occurred when 32 mm (1.25 in.) of weld depth had been laid. Sections through the failure revealed that the cracking had initiated at the root of the weld at the intersection of the plate and backing bar. The crack had then propagated at right angles to the fusion zone through the HAZ in a series of steps. The main crack had propagated parallel to the fusion boundary along the subcritical HAZ and was associated with long stringers of type II manganese sulfide (MnS) inclusions (Fig. 12b).

The morphology of the cracking is typical of lamellar tearing. This can occur in fillet or groove welds where welding stresses are effectively applied across the thickness of at least one of the plates being joined. If the through-thickness ductility is very low, as in this case because of the presence of large stringers of type II manganese sulfides, then cracking results. The presence of hydrogen in the weldment can also enhance the likelihood of cracking. In this case, the deposition of an unbalanced groove weld where rotation can occur around the root, together with the superimposition of thermal strains from the one-sided application of preheat, produces a high

Fig. 12 Armature for a large dc motor that failed during fabrication

The cause of failure was lamellar tearing.
(a) Schematic diagram of the weld preparation and the position of the failed weld. (b) Schematic of crack initiation and propagation from the root of the weld

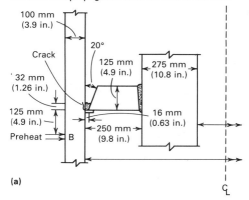

(a)

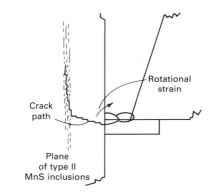

(b)

enough strain to initiate and propagate the lamellar tear.

Some of the major solutions available for preventing lamellar tearing—for example, changing the design or using prior destructive testing, such as the tensile test—cannot be used in this particular case. The high level of restraint in this particular welding procedure may also have given problems with low risk plate, for example, vacuum degassed. However, some benefit can be obtained from the use of welding procedures that reduce the effect of restraint. In this case, the use of a balanced welding technique, together with a higher preheat of 150 °C (300 °F) uniformly applied to both plates, will help reduce the restraint.

Conclusion. The weld failed by lamellar tearing as a result of the high rotational strain induced at the root of the weld caused by the weld design, weld sequence, and thermal effects.

Corrective Measure. The immediate solution to the problem was to use a ''butter'' repair technique. This involved removing the old weldment to a depth beyond the crack and replacing this with a softer weld-metal layer before making the main weld. In this manner,

the peak stresses are generated in a ductile region, and the high contractional strains are withstood. A successful repair was effected.

Example 8: Cracking of a Field Girth Weld. During the construction of a large-diameter pipeline, several girth welds had to be cut out as a result of radiographic interpretation. The pipeline was constructed of 910-mm (36-in.) diam × 13-mm (0.5-in.) wall thickness grade X448 (X65) line pipe. The girth welds were fabricated using standard vertical down stove pipe-welding procedures with E7010 cellulosic electrodes.

Investigation. Examination of the radiographs indicated that the defects were located near the 6 o'clock region of the weld. This region of the weld was removed and sectioned at the location where defects were indicated. Visual examination revealed a crack on the internal root bead side of the weld. Macrosections of the weld showed that the crack had initiated from the root and propagated to 50% of the wall thickness (Fig. 13a). The crack started partially as a result of incomplete fusion on the pipe side wall (Fig. 13b), which in turn was a result of misalignment of the two pipes. The crack was essentially transgranular and consisted of short straight segments, typical of hydrogen cracking.

Conclusions. In this failure, the hydrogen content will be relatively high because of the nature of cellulosic electrodes. The susceptibility of the materials is low; however, the stress state is high, and this led to cracking. It is significant that the cracks were located in the 6 o'clock region. During the pipelining operation, significant strain is put on this region when the next pipe length is being lined into position. Movement of the pipe up and down during alignment locally increases the strain in the root-pass region. This in itself may not be sufficient to cause cracking, but the presence of misalignment considerably increases the local strain and leads to cracking.

A large number of factors appear to influence the probability of hydrogen cracking, including type of joint, type of electrode, welding process, and preheat level. However, research has shown that these factors influence cracking through their effect on the following fundamental factors:

- The susceptibility of the weld metal or HAZ to hydrogen embrittlement, which is primarily related to the chemical composition and microstructure
- The hydrogen content
- The stress or strain level at the point of crack initiation
- The temperature

Girth welds can be made using cellulosic electrodes; however, the risk of cracking increases with increasing carbon equivalent, increasing strain (misalignment, lift height, wall thickness, and so on), and changes in welding

Fig. 13 Failed girth weld in a large-diameter pipeline

(a) Crack that initiated at the root and propagated through 50% of wall thickness. 4×. (b) Detail showing crack-initiation site due in part to incomplete fusion on pipe side wall. 12.5×

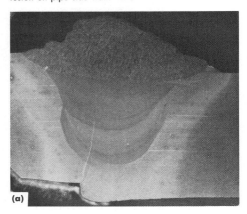

(a)

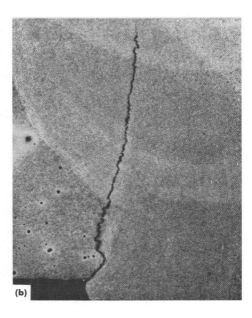

(b)

procedure. For high-risk girth welds, an increase in preheat and/or a reduction in the local stress by controlling lift height or depositing the hot pass locally before lifting may be required.

Failures in Arc-Welded Hardenable Carbon Steel

Welding of medium-carbon and high-carbon steels is limited mainly by the hardenability that these steels possess. Cooling rates must be more carefully adjusted, preheating is usually required, postweld heating is essential in some instances, and the presence of stress raisers must be minimized.

Welds in these steels are subject to severe underbead cracking as a result of hydrogen embrittlement; an example is illustrated in Fig. 14. The base metal was AISI 1045 steel, and the weld deposit was made by the shielded metal arc process using 3.2-mm (⅛-in.) E6010 electrodes without preheat. The cellulose covering of the E6010 electrodes emitted hydrocarbons, which were first volatilized and then dissociated by the welding arc. The hydrogen was dissolved in the weld puddle and diffused into the HAZ. The heat of welding raised the temperature of this adjacent metal to the austenitic temperature range, and when this metal was quickly cooled by the underlying mass of metal, a martensitic region was produced in the coarse-grain HAZ. Local stresses generated by the transformation, along with the presence of hydrogen, are sufficient to cause the underbead cracking shown in Fig. 14. The normal procedure for arc welding a hardenable steel is to use preheat and a low-hydrogen type of electrode, for example, E7018.

Example 9: Failure of a Throttle-Arm Assembly Because of Thread-Root Cracks in the HAZ. A throttle arm of an aircraft engine fractured and caused loss of engine control. The broken part consisted of a 6.4-mm (¼-in.) diam medium-carbon steel rod with a ¼-20 thread to fit a knurled brass nut that was inserted into the throttle knob.

Investigation. The threaded rod had been welded to the throttle-linkage bar by an assembly-weld deposit that had been made on the rod adjacent to the threaded portion. Examination of the threads revealed the presence of thick oxide scale at and near the fracture and extending three threads back from the fracture. The fracture surface exhibited a coarse-grain brittle texture with an initiating crack at a thread root. The crack was coated with black iron oxide typical of scale formed at high tempera-

ture. The brass nut, which had been in contact with the three threads adjacent to the fracture, showed no sign of corrosion or oxidation on its mating threads.

The thread roots in the rod contained many intergranular microcracks (Fig. 15a). A grain-boundary network of ferrite is also visible in Fig. 15(a) and at the tip of the thread next to the fracture, shown in Fig. 15(b). Also shown in Fig. 15(b) is a Widmanstätten structure of ferrite, which is often found in the HAZ of welded medium-carbon steel. The microhardness of the cross-section of the rod, measured by Knoop and diamond pyramid indenters and converted to Rockwell C, was 27 to 31 HRC. In the tooth-tip area, the hardness values as converted were 95 to 100 HRB (about 16 to 23 HRC).

Conclusions. The throttle-arm failed by brittle fracture because of the presence of cracks at the thread roots that were within the HAZ of the adjacent weld deposit. The heat of welding had generated a coarse-grain structure with a weak grain-boundary network of ferrite that had not been corrected by postweld heat treatment. The combination of the cracks and this unfavorable microstructure provided a weakened condition that resulted in catastrophic, brittle fracture under normal applied loads.

Corrective Measure. The design was altered to eliminate the weld adjacent to the threaded portion of the rod.

Example 10: Failure of a Repair Weld on a Crankshaft Because of Inclusions and Porosity. The AISI 1080 steel crankshaft of a large-capacity double-action stamping press broke in service and was repair welded. Shortly after the crankshaft was returned to service, the repair weld fractured. To determine what corrective action could be taken, the repair-weld fracture was examined ultrasonically.

Investigation. Ultrasonic testing revealed many internal reflectors, thus indicating the presence of slag inclusions and porosity. The reflectors were so numerous that the ultrasonic waves could not penetrate to the far side of the 455-mm (18-in.) diam journal. The slag inclusions and porosity were present because the repair weld had to be made in a very cramped position and the slag was difficult to remove.

Fig. 14 Underbead crack, the result of hydrogen embrittlement, in the HAZ of a shielded metal arc weld in AISI 1045 steel

The weld was made with 3.2-mm (⅛-in.) E6010 electrodes at 100 A, 26 V, and a 40-mm/min. (10-in./min) rate of travel. Etched with 2% nital. 30×

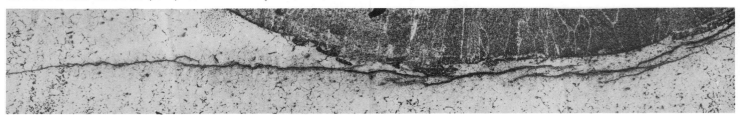

Fig. 15 Coarse-grain microstructure in the HAZ formed by assembly-weld deposit in the threaded portion of a broken throttle-arm rod of medium-carbon steel

(a) Micrograph showing intergranular cracks that originated in the thread root. 200×. (b) Micrograph showing Widmanstätten structure at the thread tip. 100×. Both etched with 2% nital

(a)

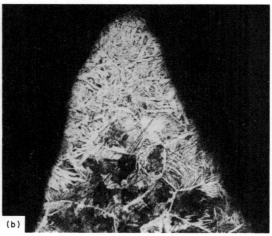
(b)

A low-carbon steel flux-cored filler metal was used in repair welding the crankshaft, without any preweld or postweld heating. This resulted in the formation of martensite in the HAZ.

Conclusions. The repair weld failed by brittle fracture, which was attributed to the combination of weld porosity with many slag inclusions and to the formation of brittle martensite in the HAZ. It was decided that a second repair weld was worth trying if all possible precautions were taken to avoid causes of cracking.

Corrective Measure. A repair weld was made using an E312 stainless steel electrode, which provides a weld deposit that contains considerable ferrite to prevent hot cracking. Before welding, the crankshaft was preheated to a temperature above which martensite would form. After completion, the weld was covered with an asbestos blanket, and heating was continued for 24 h; during the next 24 h, the temperature was slowly lowered. The result was a crack-free weld.

Failures in Arc-Welded Alloy Steel

The high hardenability that characterizes most alloy steels often presents difficulties in welding. The formation of untempered martensite in the HAZ is compounded by the large grain size created in the weld-bead side of the HAZ, with the overall result being a region with a severe loss in toughness. This condition may contribute to immediate cracking and lead to fatigue fracture, corrosion-fatigue cracking, or brittle fracture at low temperatures. Welding may also produce alloy segregation, which can lead to hot cracking, or the presence of hard and soft spots, which can be failure-initiation sites. Alloy segregation also can cause formation of galvanic cells in segregated areas; this can

generate pitting corrosion, causing service-induced stress raisers that sometimes lead to final fracture by fatigue or overloading.

Welds in alloy steels must be stress relieved, and during the stress-relieving treatment, the steels are susceptible to temper embrittlement. The fracture of an embrittled weld is usually intergranular and is recognizable as such at high SEM or transmission electron microscopy (TEM) magnification. Whether the fracture is brittle or ductile, the origin in alloy steels may be very small, and considerable examination may be required to locate it. Temper embrittlement can be avoided by proper attention to the cooling rate of the thermal cycle.

The formation of martensite in the HAZ can be avoided by use of preheating, interpass temperature control, and postheating, which will maintain the zone at a temperature that will ensure transformation to lower bainite rather than to martensite. Use of these procedures minimizes cracking and provides greatly increased toughness and ductility (Ref 7).

Repair welding sometimes produces side effects that limit its success. In one instance, a number of improperly machined parts of D-6AC steel, heat treated to a tensile strength of 1517 to 1655 MPa (220 to 240 ksi), were to be repair welded. Tests showed that the ultimate tensile strength of gas tungsten arc welded joints was only slightly less than that of the base metal. The fatigue strength of the welds, however, proved to be considerably lower than the fatigue strength of the base metal, with sizable scatter of data points; fatigue cracking was initiated in general at microporosity sites in otherwise acceptable welds. Because the fatigue strength of the parts was a critical requirement, repair welding had to be abandoned.

Weld cracking may occur not only during the welding process but also in subsequent heat

treatment. In one instance, cracking occurred in welded AISI 4340 steel box assemblies for aircraft seats during quenching from the austenitizing temperature. The intended minimum tensile strength was 1241 MPa (180 ksi). Cracks attributed to nonuniform cooling were found at interfaces between weld metal and base metal in thick-section areas. Nonuniform cooling had occurred in the enclosed box assembly, which was a 6.4-mm (¼-in.) thick trapezoidal plate welded around its edges to a hollowed support fitting. The fact that there were only three 6.4-mm (¼-in.) diam holes in the box did not permit adequate flow of the quenching medium to the box interior, which was necessary for uniform cooling. The resultant stresses caused cracks that initiated at the notches of the fillet-weld toes.

Example 11: Failure of Welds in an Aqueduct Caused by Poor Welding Techniques (Ref 8). A 208-cm (82-in.) ID steel aqueduct fractured circumferentially at two points 152 m (500 ft) apart in a section above ground. A year later, another fracture occurred in a buried section 6.4 km (4 miles) away. Both pipes fractured during January at similar temperatures and pressures. The pipe had a 24-mm (¹⁵/₁₆-in.) wall thickness, and the hydrostatic head was 331 m (1085 ft). The air temperature was about −13 °C (9 °F), the water temperature about 0.6 °C (33 °F), and the steel temperature about −4 °C (25 °F). The fractures occurred at bell-and-spigot slip joints (Fig. 16a) that, at the two initial fractures, had been fillet welded both inside and out; at the third break, the joints had been fillet welded only on the inside. The pipe had been shop fabricated of ASTM A572, grade 42, type 2, steel in 12-m (40-ft) lengths, then shop welded into 24-m (80-ft) lengths. Field assembly was with the bell-and-spigot joints.

Investigation. All three fractures occurred at a bell section and initiated at the toe of the inside fillet weld (Fig. 16a). The welds in the first two fractures had been made in the field; the third weld that failed had been made in the shop. Also, both of the original inside welds at the early fractures had been gouged out to permit relocation of the pipe to maintain the supporting ring girder legs in a vertical position. The joints had then been rewelded on the inside. This had not been done in the third, later weld that failed.

Charpy V-notch impact tests on the steel used for this pipeline revealed a ductile-to-brittle transition temperature that was unfavorably high—approximately 10 °C (50 °F). At the temperatures existing at the time of the failures, the impact strength of the steel was about 13.5 J (10 ft · lb), which is dangerously brittle. Inspection of the interior field welds showed that instead of using continuous circumferential stringer beads, the welder had used a wash-pass technique, depositing a heavy bead from toe to crown. This permitted faster welding, but gave inconsistent penetration and a considerable number of slag inclusions and imperfections,

Fig. 16 Bell-and-spigot joint used in an aqueduct of steel pipe

(a) The original design cracked because of poor welding technique and poor choice of metal. (b) Improved design showing modification of weld beads. Dimensions given in inches

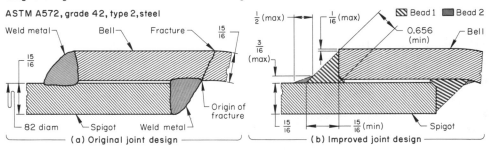

ASTM A572, grade 42, type 2, steel

(a) Original joint design

(b) Improved joint design

such as deep valleys, grooves, and undercuts, at the toe of the weld, which served as notches and stress raisers.

Examination of the third fracture at the shop weld disclosed that areas in the HAZ and at the toe of the weld had hardness levels as high as 34 HRC. When test welds were made on metal preheated to 95 °C (200 °F) before welding, the hardness of the HAZ was 22 HRC.

Conclusions. Brittle fracture of the aqueduct pipe was attributed to a combination of stress concentrations at the toes of the fillet welds due to poor welding technique, including shop welds made without preheat, and a brittle condition of the steel at winter temperatures caused by a high ductile-to-brittle transition temperature.

Corrective Measures. The following changes were made to eliminate welding and material problems:

- Any area to be welded was preheated to at least 65 °C (150 °F)
- In any weld having an undercut or abrupt transition to the base metal at the toe, an additional contour weld bead was deposited at the toe (Fig. 16b)
- Tensile loadings of above-ground sections were reduced by the installation of expansion joints
- Where the pipe was buried and pipe temperatures varied considerably, butt welds were used
- All ASTM A572 steel plate used was rolled from fully killed ingots made with a grain-refining practice and normalized after rolling, which provided a ductile-to-brittle transition temperature of about −32 °C (−25 °F)

Example 12: Fatigue Cracking of Welded Tubular Posts in a Carrier Vehicle Because of the Presence of Inclusions That Acted as Stress Raisers.
Two tubular steel posts in a carrier vehicle failed by cracking at the radius of the flange (Fig. 17) after 5 weeks of service. The posts were two of four that supported the chassis of the vehicle high above the wheels so that the vehicle could straddle a stack of steel or lumber for pickup and transport. Over a period of about 4 years,

Fig. 17 AISI 1025 steel tube post for a carrier vehicle

The post failed in fatigue because of improper design and choice of flange metal. Dimensions given in inches

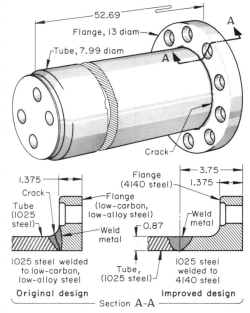

Section A-A

47 other posts had failed in nine vehicles. The latest failures occurred in posts of an improved design (Fig. 17) that had been used in an attempt to halt the pattern of cracking.

The original design involved a flat flange of low-carbon low-alloy steel that was fillet welded to an AISI 1025 steel tube with a wall thickness of 22 mm (⅞ in.). The fillet weld was machined to a smooth radius; but the region was highly stressed, and fatigue accounted for the numerous failures of posts.

In an effort to improve the strength of the joint, the improved design was adopted. The improvement was in the flange, which was machined from a 100-mm (4-in.) thick plate of normalized low-carbon low-alloy steel so that the welded joint would be about 50 mm (2 in.) away from the flange fillet. However, after the brief service of 5 weeks, the two posts showed

the same fatigue cracks in the fillet as the original design.

Investigation. Chemical analysis of the flange metal showed that it met specification requirements. Examination of a metallographic specimen from the fracture area revealed a banded structure in the direction of rolling and aluminum oxide stringers at the fillet surface, which undoubtedly served as stress raisers, in the direction of crack extension.

Conclusions. The failures in the flanges of improved design were attributed to fatigue cracks initiating at the aluminum oxide inclusions in the flange fillet. It was concluded that a higher-strength flange was necessary.

Corrective Measures. The design of the flange with the weld about 50 mm (2 in.) from the fillet was retained, but the metal was changed to a forging of AISI 4140 steel, oil quenched and tempered to a hardness of 241 to 285 HB. Preheating to 370 °C (700 °F) before and during welding with AISI 4130 steel wire was specified. The weld was subjected to magnetic-particle inspection and was then stress relieved at 595 °C (1100 °F), followed by final machining. No further failures of posts were reported in 4 years of service after the changes.

Example 13: Fatigue Fracture of a Rebuilt Exciter Shaft That Was Accelerated by Weld-Deposit Cracks.
The shaft of an exciter that was used with a diesel-driven electric generator broke at a fillet after 10 h of service following resurfacing of the shaft by welding. The exciter, which turned at 1750 rpm, was top mounted (instead of floor mounted) on the same base as the main generator, which operated at 450 rpm. Therefore, the vibrations from the main generator also caused vibrations in the exciter unit.

The shaft had been previously installed in an exciter for another diesel-generator unit, where it had experienced turning of a pulley on the shaft. The shaft was polished, the pulley was bored out, and a bushing was inserted, but after indeterminate service, the pulley turning recurred. At this time, the shaft was removed for resurfacing. After belt grinding, the keyway was filled in and the surface of the shaft was built up by gas metal arc welding using a low-carbon steel filler wire (0.04C-1.15Mn-0.50Si) and a shielding gas containing 75% argon and 25% carbon dioxide. Preheating and postheating were not used, nor was the rebuilt surface nondestructively tested for cracks. The shaft was remachined to size, and a new keyway was cut in a different location. The shaft was placed in stock (length of time not reported) until used to replace another exciter shaft, which had been removed for maintenance. At the rated operational speed, the rebuilt shaft had been subjected to just over 10^6 revolutions before failure.

Investigation. Visual inspection established that there were no nicks or toolmarks and no evidence of corrosion in the general area of the fracture, which was located at a fillet where the diameter of the shaft was reduced from 53 to

Fig. 18 Resurfaced AISI 4130 steel exciter shaft that failed by fatigue accelerated by weld-deposit cracks

(a) Section through shaft at fracture site. Original keyway is at left, recut keyway at top. Arrow A indicates resurfacing weld deposit; arrow B, HAZ. Etched with nital. 2×. (b) Section near the filled-in keyway showing incomplete fusion (arrow C) at the bottom of the keyway and subsurface cracks at arrow D. 8×. (c) Stereo pair of SEM fractographs of the weld-deposit surface near the fracture showing a very granular appearance. A deep, wide hot crack progresses from upper to lower left. 1750×

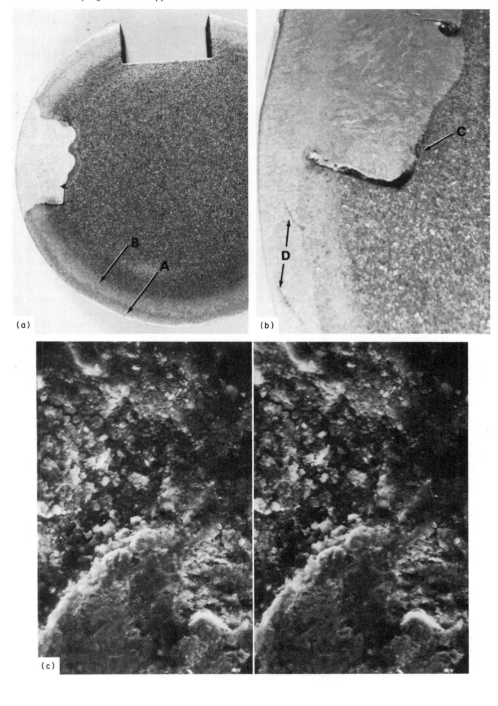

(a)

(b)

(c)

48 mm (2.1 to 1.9 in.). The fracture surface contained a dull off-center region of final ductile fracture surrounded by regions of fatigue that had been subjected to appreciable rubbing. The fracture appeared to be typical of rotary-bending fatigue under conditions of a low nominal stress with a severe stress concentration. It appeared that the fatigue cracks initiated in the surface-weld layer. Liquid-penetrant inspection of the entire shaft revealed an abundance of weld porosity and one shallow surface crack about 13 mm (½ in.) from the fracture, but no detectable deep cracks.

A polished-and-etched cross section of the shaft taken near the fracture surface is shown in Fig. 18(a), in which the surface-weld layer and the 9.5-mm (⅜-in.) thick HAZ (arrows A and B, respectively, Fig. 18a) are quite apparent. The thickness of the weld layer varies from 1.6 mm (1/16 in.) to none at all. The weld deposit in the original keyway (at left, Fig. 18a) displays a lack of fusion at the bottom corner (arrow C, Fig. 18b). Immediately below the filled-in original keyway are two small cracks (arrow D, Fig. 18b) that were beneath the surface and thus were not reached by the liquid penetrant. The top crack was entirely in the resurfacing weld; the lower crack appeared to be mostly in the resurfacing weld but with a small tail in the HAZ.

The microstructure of the base metal was a normal mixture of ferrite and pearlite, except for having a coarse grain size (ASTM 3). Analysis showed the metal to be AISI 4130 steel. The weld metal was finer grained, largely pearlite with only a small quantity of ferrite. This could have resulted only from appreciable diffusion of carbon from the HAZ into the weld, because the filler metal should produce a structure of ferrite. The average hardness of the base metal taken from 44 readings was 88.5 HRB; two readings in the filled-in keyway were 89 and 94 HRB, perhaps reflecting either the higher proportion of pearlite or an insufficient number of measurements. There was, however, no hardness value in the base metal higher than 90.5 HRB.

An SEM examination of the weld surface near the fracture surface showed a granular appearance (Fig. 18c). The deep hot crack that progresses from upper to lower left in Fig. 18(c) is significant because, although this crack did not initiate the fracture, it is believed that a similar one did. In SEM fractographs of the fracture surface, pearlite lamellae were visible in the weld area and rub marks in the shaft interior.

Discussion. In considering the bending stresses that might have caused the failure, it should be realized that the subbase of the diesel-generator set was mounted on springs, which would reduce vertical vibrations much more than if mounted on a firm base. Vibrations were present, however, because the bolts connecting the exciter baseplate to its support plate often fractured. The pulley of the exciter was driven by a matched set of seven V-belts, but the amount of tension used could not have been a significant source of vibrations because the area of fast fracture was small compared to the area of fatigue, indicating a condition of low nominal stress. Mutual misalignment of the two pulleys would not appear to have been a source of vibration unless the misalignment was quite large. Because both pulleys were supported in cantilever fashion at the end of the respective shafts, the possibility of vibration causing some degree of shaft whip could have been a significant factor in the failure.

Conclusions. Fatigue fracture of the shaft resulted from stresses that were created by vibration acting on a crack or cracks formed in the weld deposit because of the lack of preheating and postheating. The cracks in the weld deposit provided severe stress concentrations so that the fatigue cracks almost completely penetrated the shaft before fast fracture occurred.

Recommendations. Rebuilding of exciter shafts should be discontinued, and the support plate of the exciter should be braced to reduce the amount of transmitted vibration. Also, the fillet in the exciter shaft should be carefully machined to provide an adequate radius.

Example 14: Hydrogen Sulfide SCC of Welds in Pressure Vessels.
After operating satisfactorily for about a year, the outer insulation on a series of large pressure vessels was discovered to be leaking. The vessels were used for hydrogenation and desulfurization of petroleum crude oils at a pressure of about 21 MPa (3 ksi) at 315 °C (600 °F) or higher. The incoming hydrogen was conducted along the interior surface of the pressure vessel to maintain the wall temperature of the vessel below about 230 °C (450 °F). The vessels were approximately 3 m (10 ft) in diameter, and the walls ranged from 100 to 230 mm (4 to 9 in.) in thickness.

The vessels were fabricated of quenched-and-tempered plates of 2.25Cr-0.5Mo steel, rolled and welded vertically with electroslag welding. The welds in these sections were quenched and tempered before assembly girth welding with the submerged arc process. The girth welds were then stress relieved to a final hardness of 28 to 32 HRC.

Investigation. The outer insulation was stripped, which exposed perforations through the vessel walls. Examination of the vessel interiors revealed numerous cracks in the welds, some of which were transverse to the weld but did not penetrate the HAZ on either side.

Initially, it was concluded either that delayed hydrogen embrittlement resulted from the welding process or that failure resulted from hydrogen sulfide attack. Tests showed that sulfur-bearing hydrogen was capable of cracking 2.25Cr-0.5Mo steel at hardnesses of 28 to 32 HRC in an atmosphere that is essentially moisture free.

Before these pressure vessels failed, valves, valve fittings, and welded nipples in the system had experienced hydrogen sulfide SCC. An effort was made to halt the SCC by stress relieving the pressure vessels in place and decreasing their hardness to 22 HRC, because it was believed that no SCC induced by hydrogen sulfide would occur below this hardness. The stress relief was not effective in preventing further cracking.

Conclusions. The vessels failed by SCC resulting from attack by hydrogen sulfide. Reducing the hardness of 2.25Cr-0.5Mo steel to less than 22 HRC does not prevent SCC in a hydrogen sulfide atmosphere when the service stress is high—as it was with these vessels. Also, the pre-existing cracks would produce stress concentrations that in turn would cause further crack growth under the high service stress, regardless of the hardness of the steel. Thus, the second stress-relief treatment would not be expected to halt the SCC, and a different corrective action must be devised.

Example 15: Failure in Stainless Steel Welds Joining Low-Carbon Steel Handles to Type 502 Stainless Steel Covers Because of Martensite Zone in the Welds.
Handles welded to the top cover plate of a chemical-plant downcomer broke at the welds when the handles were used to lift the cover. The handles were fabricated of 19-mm (¾-in.) diam low-carbon steel rod; the cover was of type 502 stainless steel [0.10% C (max), 5% Cr, 0.5% Mo] plate. The attachment welds were made with type 347 stainless steel filler metal to form a fillet between the handle and the cover.

Investigation. A metallographic specimen was prepared to show a cross section through the handle, the fillet weld, and the cover. The structure was found to contain a zone of brittle martensite in the portion of the weld adjacent to the low-carbon steel handle; fracture had occurred in this zone.

Sufficient heat had been generated during welding to fuse an excessive amount of the carbon steel and thereby dilute the weld deposit locally to create a layer of low-alloy steel between the handle and the weld metal. This low-alloy zone had sufficient hardenability to form martensite when mass quenched by heat transfer to the cover.

Conclusions. Failure was due to brittle fracture. A brittle martensite layer in the weld was the result of using too large a welding rod and too much heat input. The high heat input resulted in excessive melting of the low-carbon steel handle, which diluted the austenitic stainless steel filler metal and formed martensitic steel in a local zone in the weld.

Corrective Measures. Because it was impractical to preheat and postheat the type 502 stainless steel cover plate, the 19-mm (¾-in.) diam low-carbon steel handle was welded to a 13- × 100- × 150-mm (½- × 4- × 6-in.) low-carbon steel plate, using low-carbon steel electrodes. This plate was then welded to the type 502 stainless steel plate with type 310 stainless steel electrodes. This design produced a large weld section over which the load was distributed.

Example 16: Fracture of a Chip-Conveyor Pipe at a Flange Weld as a Result of Poor Fit-Up.
A pipe in a chip conveyor cracked at the toe of an exterior fillet weld connecting a flange to the pipe.

The chip conveyor consisted of several spool sections. Each section was made up of a length of 560-mm (22¹⁄₁₆-in.) OD × 546-mm (21½-in.) ID low-alloy steel pipe and two 713-mm (28¹⁄₁₆-in.) OD × 562-mm (22⁹⁄₆₄-in.) ID flanges of 13-mm (½-in.) thick low-carbon steel, which were welded to the two ends of the pipe. The wall thickness of the pipe was 7 mm (0.275 in.). The composition specified for the pipe steel was 0.25C-0.98Mn-3.52Ni-1.34Cr-0.24Mo, which approximates a 9300 steel with high molybdenum. The hardness of the pipe was reported as 495 HB, and of the flange as 170 HB. The flanges were joined to the pipe by shielded metal arc welding, using E7018 electrodes.

Investigation. It was found that one end of the pipe had been cut before assembly at an angle of 76° 39′ to its axis, which presented an ellipse (Fig. 19) over which the flange was forced to fit by hydraulic pressure. A 90° angle would have given a radial clearance between the pipe and flange of about 1.1 mm (³⁄₆₄ in.), but the long outside diameter of the elliptical pipe was 576 mm (22.662 in.) compared to the inside diameter of the flange of 562 mm (22.141 in.), which placed both components in a condition of severe stress.

A schematic illustration of a section cut through the weld in the area of maximum interference is shown in detail A in Fig. 19. Underbead cracking began at the toe on the pipe wall of the first fillet-weld pass at a point where there was incomplete fusion and inadequate penetration of the third and final pass. Several cracks initiated at the toe of the third pass on the pipe wall, where there was a slight undercut.

Conclusions. The conveyor pipe failed by brittle fracture, which was attributed to the stresses induced in forcing the circular flange over the elliptical section of the pipe. The toe of the weld and the adjacent undercut were stress raisers that determined the point of major crack origin. Under residual stress, the internal point of incomplete fusion also initiated additional cracks.

Recommendation. A proper fit between an elliptical flange and pipe end would eliminate the cracking.

Failures in Arc-Welded Stainless Steel

The difficulties encountered in welding stainless steels arise from reactions of the microconstituents of the steels to welding temperatures or from interactions, such as alloy dilution, between the steels and dissimilar alloys.

There are four basic groups of stainless steels: austenitic, ferritic, martensitic, and precipitation hardening. The metallurgical features of each group generally determine the weldability characteristics of steels in that group. The weldability of martensitic stainless steels is greatly affected by hardenability that can result in cold cracking. Welded joints in ferritic stainless steels have low ductility as a result of grain coarsening. The weldability of austenitic stainless steels is governed by their susceptibility to hot cracking. With precipitation-hardening

Fig. 19 Low-alloy steel conveyor pipe that cracked at fillet welds securing a carbon steel flange because of poor fit-up

Dimensions given in inches

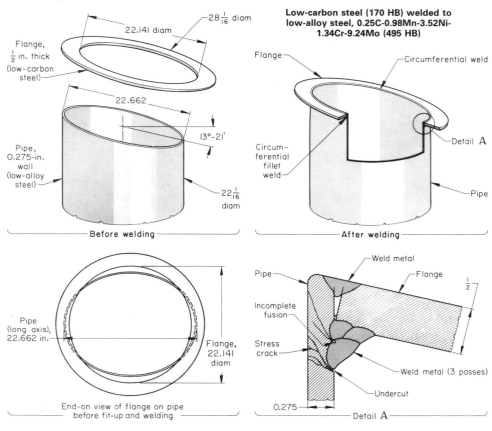

Fig. 20 Stress-rupture crack at the interface of a weld bead (region A) and the HAZ (region B) in a 2.25Cr-1Mo low-alloy steel pipe

The weld was made with ER308 stainless steel filler metal. Arrows indicate the propagation of stress-rupture cracks into the weld metal.

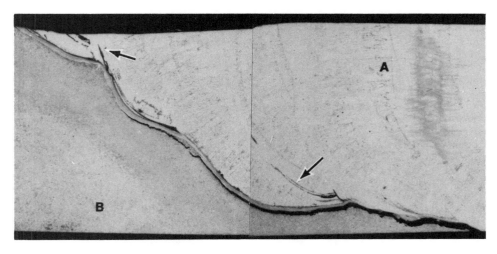

stainless steels, weldability is related to the mechanisms associated with the transformation (hardening) reactions.

It is sometimes necessary to weld a stainless steel part to a low-alloy steel part using a stainless steel filler metal (Example 15). If excessive heat is introduced by the arc, the weld metal can be diluted by excessive melting of the low-alloy steel. At the same time, the heat of welding causes marked grain growth in the HAZ of the low-alloy steel. The result can be regions of intermediate alloy content, too low to be austenitic but high enough to transform to martensite unless the temperature is lowered very carefully. The carbon that diffuses into the stainless steel weld can precipitate as grain-boundary carbide particles. These changes can result in the creation of a metallurgical stress raiser that is susceptible to stress-rupture cracking if sufficient stress is applied at elevated temperature. Figure 20 shows a stress-rupture crack at the interface of a stainless steel weld bead and a HAZ in a low-alloy steel pipe welded with ER308 stainless steel filler metal.

To minimize dilution, it is necessary to use a low heat input during the first weld pass. This can be done with a small diameter electrode and a lower welding current. Also, instead of type 308 stainless steel filler metal, types 309 and 312 can be used—both of which provide a higher alloy level to compensate for possible alloy dilution.

Martensitic Steels. Welding of type 410 stainless steel can cause SCC at a later date if it is not feasible to give the HAZ of the weld a satisfactory postweld heat treatment. If type 410 stainless steel is tempered above 500 °C (930 °F), there should be almost no susceptibility to SCC, but if this steel is tempered at a temperature lower than 500 °C (930 °F), it will be susceptible to SCC when exposed to chlorides and other environments. Figure 21(a) shows a micrograph that illustrates intergranular stress-corrosion cracks in the HAZ of an ER308 stainless steel weld deposit on type 410 stainless steel that was not tempered following welding. The hardness was 40 to 45 HRC in the HAZ adjacent to this weld, whereas it should have been approximately 20 HRC for good resistance to SCC.

Underbead cracks in martensitic stainless steels may be caused by quenching stresses or, with some welding processes, stresses generated by supersaturation of hydrogen. Once they have been formed, such cracks frequently serve as nuclei for fatigue-cracks (Fig. 21b). The tip of a type 410 stainless steel tool was welded with an ER308 electrode. A crack that formed in the HAZ was shown by examination in a scanning electron microscope to be entirely intergranular.

Austenitic Steels. In austenitic grades of stainless steel, such as type 304, the corrosion resistance of the HAZ of a weld may be seriously reduced by the precipitation of carbides along the grain boundaries, which results in sensitization. The depletion in chromium of the region adjacent to the grain boundaries (because of the quantity precipitated as carbide) reduces the corrosion resistance of the matrix. The existence of the sensitized condition can be demonstrated by subjecting a specimen of the material to the acidified copper sulfate (Strauss) test. If the material is sensitized, the attack will be completely along the grain boundaries, frequently leaving individual grains entirely detached. The nature of this attack is shown in

Fig. 21 Cracks in HAZs of type 410 stainless steel beneath weld deposits of ER308 stainless steel

(a) Section through the HAZ that was not tempered after welding showing an intergranular stress-corrosion crack. The weld deposit is at upper right. Electrolytically etched with HCl-methanol. 140×. (b) Fatigue crack initiated by an intergranular underbead crack (arrow). Note beach marks in fatigue region. 3×

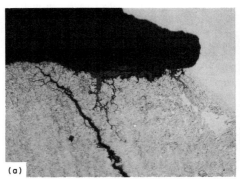

(a)

(b)

Fig. 22 Corrosive attack in the sensitized HAZ of a weld in type 304 stainless steel that occurred during an artificial copper sulfate (Strauss) test

(a) Micrograph showing the intergranular nature of the corrosion, which follows the chromium-depleted regions adjacent to grains. 85×. (b) SEM micrograph showing individual grains that have been isolated by dissolution of grain-boundary material. 725×

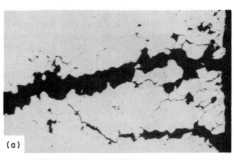

(a)

(b)

Fig. 22. If size of the part permits, complete solution heat treatment and quenching after welding will redissolve the chromium carbides and restore corrosion resistance. This cannot be

accomplished by localized heating. Sensitization can be controlled by carbon content, alloying additions, or postweld heat treatment.

Ferritic steels are nonhardenable by heat treatment. The early ferritic stainless steels—for example, types 430 and 422—contain chromium as a ferrite stabilizer and have high carbon contents. Weldments are therefore subject to intergranular corrosion unless a postweld heat treatment is applied. Some ferritic stainless steels, such as types 405 and 409, have lower chromium and carbon contents, but contain powerful ferrite formers. These types can form some martensite when welded and are generally low in toughness. Ferritic types such as 444 are not susceptible to intergranular corrosion and have improved toughness after welding.

Precipitation-hardening steels need an aging heat treatment after welding. This may not always be practical. Precipitation-hardening stainless steels do not require preheat, and the martensitic and semiaustenitic types are not susceptible to cracking. Austenitic types, however, suffer from hot cracking in the HAZ. This group of steels is very notch sensitive, and sharp stress concentrators in design and during welding should be avoided.

Example 17: Intergranular Fracture of Martensitic Welds. Several fractures occurred in flange studs used for remote handling of radioactive equipment. The studs, of quenched-and-tempered type 414 stainless steel, fractured in the HAZs produced in the studs during the circumferential welding that joined the studs to the flanges (Fig. 23). The weld deposits were of type 347 stainless steel, and the flanges were type 304 stainless steel.

Investigation. Metallographic examination of the failed studs revealed that the HAZs contained regions of martensite and that intergranular cracks, which initiated at the stud surfaces during welding, propagated in the manner indicated in Fig. 23 to complete separation under subsequent loading. Microscopic examination showed that these cracks extended into the stud about 4.8 mm (³⁄₁₆ in.) and about 100° around

Fig. 23 Type 414 stainless steel stud, welded to type 304 stainless flange, that fractured in service

Fracture occurred because of intergranular postweld cracking in a martensitic region of the HAZ.

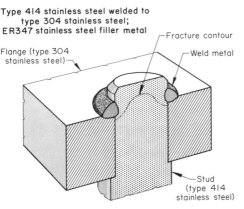

Type 414 stainless steel welded to type 304 stainless steel; ER347 stainless steel filler metal

Flange (type 304 stainless steel)
Fracture contour
Weld metal
Stud (type 414 stainless steel)

the circumference. The welding heat had reaustenitized a portion of the HAZ of the stud. After welding this portion was quenched by the heat-sink effect of the surrounding mass of cooler metal in the remainder of the stud and in the flange to produce martensite. The stresses imposed by the thermal gradients and the phase changes were sufficient to generate the cracks. High-temperature preheat and postheat treatments would have retarded the cooling rate and prevented the formation of martensite, but preheating and postheating were not considered practical for this application. The studs fractured under service loads as a result of intergranular crack propagation in the HAZ that reduced the effective cross section.

Conclusions. Rapid heating and cooling during attachment welding produced a martensitic structure in the HAZ of the stud, which cracked circumferentially from the combination of thermal-gradient and phase-change stresses.

Recommendations. Joining the studs to the flanges by welding should be discontinued. They should be attached by screw threads, using a key and keyway to prevent turning in service.

Example 18: Failure of a Weld Due to Cracking in the HAZ. An aircraft fuel-nozzle-support assembly exhibited cracks along the periphery of a fusion weld that attached a support arm to a fairing in a joint that approximated a T-shape in cross section. The base metal was type 321 stainless steel. The exterior appearance of the fusion weld and of some of the fairing surface is shown in Fig. 24(a).

Investigation. Examination of a section removed perpendicular to the cracks in the weld visible in Fig. 24(a) showed a good-quality weld penetrating to the support arm beneath, but it revealed notch configurations at the inner mating surfaces at each edge of the fairing (arrows 2 and 3, Fig. 24b). These configurations were the result of welding a poor fit-up of

Fig. 24 Cracks that occurred at the margin of a weld joining a fairing to a support arm of a type 321 stainless steel fuel-nozzle-support assembly

(a) Photograph of the weld showing fatigue cracks along both edges. The support arm, welded to the fairing at a T-shape joint, is directly beneath the weld contours. Approximately 5×. (b) Section through the fairing, weld, and support arm showing a crack at the junction of the weld and the fairing (arrow 1) and interior notches at arrows 2 and 3. Electrolytically etched in 10% oxalic acid. Approximately 7×

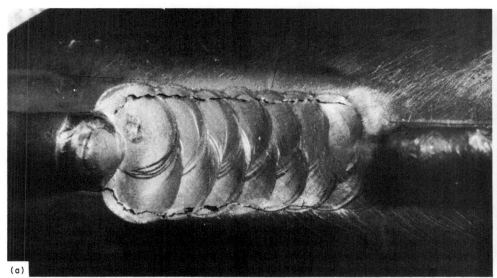

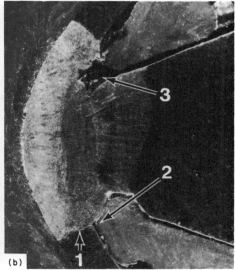

the support arm to the fairing. The crack adjacent to the weld in the lower section of the fairing is shown at arrow 1 in Fig. 24(b).

Fractures that originated at the cracks were examined by stereomicroscope and were found to contain fatigue marks that indicated crack propagation from multiple origins at the inner surface of the weld edge. Microscopic enlargement of the weld-edge cracks, such as those at arrow 1 in Fig. 24(b), disclosed that the rupture progressed along the fusion-zone boundary of the weld with some incursions into the HAZ and others, to a lesser degree, into the weld bead.

Conclusions. Fatigue cracking was initiated at stress concentrations created by the notches at the inner surfaces between the support arm and the fairing, enhanced by poor fit-up in preparation for welding.

Example 19: Hot Cracks in a Repair Weld. The fuel-nozzle-support assembly discussed in Example 18 also showed transverse indications after fluorescent liquid-penetrant inspection of a repair-welded area at a fillet on the front side of the support neck adjacent to the mounting flange. Visual examination disclosed that the fluorescent-penetrant indications were due to an irregular crack (Fig. 25a).

Investigation. The crack through the neck was sectioned; examination showed that the crack had extended through the repair weld (Fig. 25b) as previously assumed on the basis of visual examination. It was evident that the crack had followed an intergranular path. The crack was opened, and binocular-microscope examination of the fracture surface showed that, to a depth of 3.2 mm (⅛ in.), the surface contained dendrites with discolored oxide films that were typical of exposure to air when very

hot. Several additional subsurface cracks, typical of hot tears, were observed in and near the weld. In making the repair weld, there had been too much local heat input, perhaps with too large a welding electrode, and probably an excessive deposition rate. The result was localized thermal contraction that was not shared with abutting colder sections, with the outcome being hot tearing at a temperature at which the weld deposit yielded to the contraction stresses by cracking rather than by plastic flow.

Conclusions. The cracking of the repair weld was attributed to unfavorable welding practice that accentuated thermal contraction stresses and caused hot tearing.

Recommendations involved use of a small-diameter welding electrode, a lower heat input, and deposition in shallow layers that could be effectively peened between passes to minimize internal stress.

Example 20: Fatigue Fracture of a Fuel Line at a Butt Weld. A weld in a fuel-line tube broke after 159 h of engine testing. The 6.4-mm (0.25-in.) OD × 0.7-mm (0.028-in.) wall thickness tube and the end adapters were all of type 347 stainless steel. The butt joints between tube and end adapters were made by automated gas tungsten arc (orbital arc) welding, using a very thin outer surface lip on the adapters to ensure concentric alignment with the tube. This outer lip was fused during welding, eliminating any need for filler metal. This procedure has been successful in preparing weld beads with good penetration and smooth root and weld-face contours, particularly when more than one weld pass has been used.

Investigation. It was found that the tube had failed in the HAZ, the crack propagating cir-

cumferentially about 180° parallel to the weld before turning approximately 90° to penetrate the weld transversely, parallel to the tube axis. It then turned 90° in the other direction to resume circumferential growth parallel to the weld in the HAZ of the adapter.

Examination of a plastic replica of the fracture surface in a transmission electron microscope established that the crack origin was at the outer surface of the tube. The crack growth was by fatigue; closely spaced fatigue striations were found near the origin, and more widely spaced striations near the inner surface.

A section through the tube revealed a transgranular fracture of a normal microstructure. The quality of the weld and the chemical composition of the tube both conformed to the specifications. However, the fuel-line assembly, which was rigidly supported at both end adapters, had vibrated excessively in service. The service condition thus confirmed the findings of fatigue fracture obtained from study of the fracture-surface replica.

Whether the weld determined the site of the fracture is not clear, because the mode of vibration of the tube is not accurately known. Although the microstructure was found normal in an examination at 500×, it would seem probable that the HAZ would suffer some slight loss of properties either by formation of a few very fine precipitate particles or by a minute amount of phase transformation or both.

Conclusion. The fuel-line fracture was caused by fatigue induced by severe vibration in service.

Corrective Measure. Additional tube clamps were provided to damp the critical vibrational

Fig. 25 Crack in the repair-welded area of a type 321 stainless steel fuel-nozzle-support assembly because of incorrect welding procedure

(a) Photograph showing the crack in the fillet on the front side of the support neck. Approximately 6×. (b) Section through the support neck. Arrow indicates the crack, which was found to be intergranular. Electrolytically etched with 10% oxalic acid. 7×

(a)

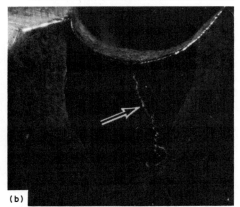

(b)

Fig. 26 Type 347 stainless steel inlet header for fuel-to-air heat exchanger that cracked due to poor welding technique and unfavorable joint design

Dimensions given in inches

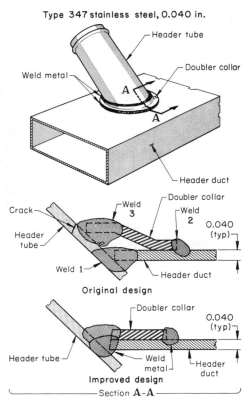

The header was sectioned to open the crack and expose the fracture surface for examination. The tube and duct were initially welded together (weld 1, Original design, Section A-A, Fig. 26). A doubler collar, 16 mm (⅝ in.) wide and of the same material and thickness as the duct, was then welded to the duct along the outer edge of the collar (weld 2). A third weld joined the inner edge of the collar to the header tube, thus creating an area of twice the normal header thickness adjacent to the joint. This third weld, throughout much of the collar circumference, penetrated the original tube-to-duct weld (weld 1); but in the area of the fracture there was a gap between the two sheets of stainless steel, and the third weld was separate (weld 3).

Examination of the underside of the joint showed incomplete fusion of the weld between the tube and duct. Distortion of the base metal adjacent to the weld was also visible from this side.

Examination of the surface of the opened crack revealed the origin of the crack at a weld undercut at the toe of weld 3. The crack then progressed through the HAZ of the base metal.

Conclusions. The crack in the header tube was the result of a stress concentration at the toe

stresses. No further fuel-line fractures were encountered.

Example 21: Fatigue Failure of an Inlet Header Because of Poor Welding Technique and Unfavorable Weld-Joint Design. While undergoing vibration testing, a type 347 stainless steel inlet header for a fuel-to-air heat exchanger cracked in the header tube adjacent to the weld bead between the tube and header duct (Fig. 26).

Investigation. The weld was relatively massive, which was apparently necessary to fill large clearances between the mating parts. Collapse of the base metal in several areas adjacent to the weld indicated that there was damage from excessive heat.

Liquid-penetrant inspection of the weld between the header tube and the header duct revealed a 29-mm (1⅛-in.) long crack in the weld on the tube side. Several weld-bead undercuts were visible, and the crack appeared to propagate through each of these small defects.

of the weld joining the double collar to the tube. The stress concentration was caused by undercutting from poor welding technique and an unfavorable joint design that did not permit a good fit-up.

Corrective Measures. The doubler collar was made so that it could be placed in intimate contact with the header duct (Improved design, Section A-A, Fig. 26). The two sheets were beveled, where necessary, to form a V-groove joint with the tube. Two weld passes in the V-groove were used to join the two sheets to the tube. This procedure resulted in a smaller, controlled, homogeneous weld joint with less distortion.

Example 22: Corrosion Failure of Stainless Steel in Sensitized HAZ of Assembly Weld. Two aircraft-engine tailpipes of 19-9 DL stainless steel (AISI type 651) developed cracks along longitudinal gas tungsten arc butt welds after being in service for more than 1000 h. A crack 43 cm (17 in.) long occurred in one tailpipe. The second tailpipe cracked longitudinally in a similar manner to the first, then fractured circumferentially, which resulted in the loss of the aft section of the tailpipe. The gas temperatures during the major portion of the service life were approximately 400 to 500 °C (750 to 1020 °F), but the metal temperatures were probably lower.

Investigation. Binocular-microscope examination of the cracks in both tailpipes revealed granular, completely brittle-appearing surfaces that were confined to the HAZs of the welds. The crack surfaces were severely discolored, apparently from gaseous blow-by following failure. There also were additional longitudinal cracks in the HAZs, where partial weld separation had occurred, and transverse cracks formed by bending during failure.

Microscopic examination of sections transverse to the weld cracks showed severe intergranular corrosion, primarily in the HAZ. The corrosion had been intense enough to cause grain detachment and, in some regions, complete grain dissolution. There was no evidence suggesting that manufacturing procedures before welding could have contributed to the failure. A resistance seam weld remote from the gas tungsten arc weld showed only superficial (0.02 to 0.025 mm, or 0.0008 to 0.001 in. deep) and sporadic surface attack. The 19-9 DL stainless steel skin remote from the gas tungsten arc weld had a satisfactory microstructure and chemical composition and was of the specified thickness.

The fractures gave every appearance of having been caused by loss of corrosion resistance due to sensitization. This condition could have been induced by the temperatures attained during gas tungsten arc welding.

To determine the degree of sensitization present in the HAZs of the tailpipes, specimens were subjected to 72-h immersion in boiling, acidified copper sulfate (Strauss test). These specimens showed separation of the gas tungsten arc weld from the base metal with com-

Fig. 27 HAZ of a weld in 21Cr-6Ni-9Mn stainless steel part showing evidence of cracking caused by liquid-copper penetration

(a) SEM micrograph of section cut parallel to weld and perpendicular to crack showing the intergranular nature of the crack. 165×. (b) Cu Kα x-ray fluorescence scanning micrograph of crack showing copper concentration at the surface (left) and along crack contours. 165×

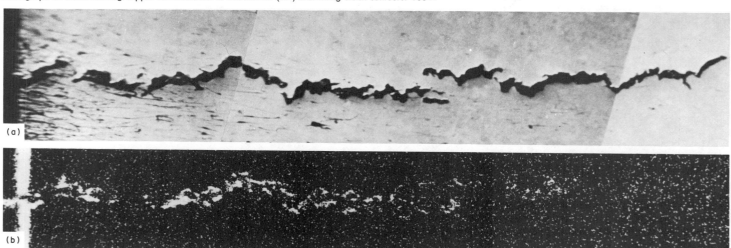

plete dissolution of the HAZ. A specimen from the resistance seam weld remote from the gas tungsten arc weld showed only scattered superficial surface attack (as deep as 0.06 mm, or 0.0023 in.). Specimens from the base metal completely remote from any weld showed slight attack that varied from 0.018 to 0.05 mm (0.0007 to 0.0019 in.) deep. These results decisively demonstrated the presence of sensitization in the HAZ of the gas tungsten arc weld.

Conclusions. The aircraft-engine tailpipe failures were due to intergranular corrosion in service of the sensitized structure of the HAZs produced during gas tungsten arc welding.

Recommendations. All gas tungsten arc welded tailpipes should be postweld annealed by re-solution treatment to redissolve all particles of carbide in the HAZ. Also, it was suggested that resistance seam welding be used, because there would be no corrosion problem with the faster cooling rate characteristic of this technique.

Example 23: Embrittlement of Stainless Steel by Liquid Copper From a Welding Fixture. Parts of 21Cr-6Ni-9Mn stainless steel that had been forged at about 815 °C (1500 °F) were gas tungsten arc welded. During postweld inspection, cracks were found in the HAZs of the welds. These cracks were small and somewhat perpendicular to the weld. Welding had been done using a copper fixture that contacted the steel in the area of the HAZ on each side of the weld but did not extend under the tungsten arc.

Investigation. Sections of the parts were taken parallel to the weld and perpendicular to the cracks. In SEM examination, the cracks appeared to be intergranular (Fig. 27a) and extended to a depth of approximately 1.3 mm (0.05 in.). The crack appearance suggested that

the surface temperature of the HAZ could have melted a film of copper on the fixture surface and that this could have penetrated the stainless steel in the presence of tensile thermal-contraction stresses. A Cu Kα x-ray fluorescent scan of the crack revealed a heavy copper concentration within the crack as well as on the external surface of the HAZ (Fig. 27b). Residual stresses in the weld away from the cracks were measured at 345 MPa (50 ksi) in tension, which were considered adequate to cause cracking in the presence of liquid copper.

Conclusions. The cracks in the 21Cr-6Ni-9Mn weldments were a form of liquid-metal embrittlement caused by contact with superficially melted copper from the fixture and subsequent grain-boundary attack of the stainless steel in an area under residual tensile stress.

Corrective Measure. The copper for the fixtures was replaced by aluminum. No further cracking was encountered.

Example 24: Corrosive Attack of Stainless Steel Welds in Hot Brine. Type 316L stainless steel pipes carrying brine at 120 °C (250 °F) and at a pH of about 7, failed by perforation at or near circumferential butt-weld seams (Fig. 28a).

Investigation. The failure was examined optically and radiographically in the field. Specimens were removed and examined metallographically and with a scanning electron microscope in the laboratory. The examinations revealed a combination of failure mechanisms. The weld itself had been perforated by pitting corrosion. Preferential attack of the cast structure of the weld is shown in Fig. 28(b). The high chromium interdendritic region was the last to dissolve. The adjacent base metal had stress-corrosion cracks that extended through the pipe wall. The transgranular nature of the cracking is shown in Fig. 28(c).

Discussion. The pitting corrosion reaction between type 316L stainless steel and the environment in which it was used is typical of the factors that determine whether the surface of a chromium-nickel alloy will be activated or passivated by a solution. The amount of oxygen that is locally present in the brine is particularly critical. A neutral hot chloride solution may locally activate the alloy, causing continuous pitting. Once the reaction has started, ferric chloride can be produced, which has an autoacceleration action. This apparently was the main cause of pitting attack of the seam weld in the pipes. Once the surface has been activated, the metal around the pit becomes cathodic, and penetration within the pit is rapid because the area of effective cathode is very large compared to that of the anodic pit.

In addition to the pitting corrosion of the weld itself, the base metal adjacent to the weld suffered SCC. The stress that produced these cracks was the result of residual stresses from weld shrinkage. The source of corrodent that caused the cracking was not positively identified. Increased oxygen content of the brine could result in cracking. Another possibility is the action of the pitting corrosion product as the stress-corroding medium.

Conclusions. The pitting failure of the welds was attributed to localized attack of an activated surface, in which anodic pits corroded rapidly. Additionally, SCC driven by residual welding stresses occurred in the base metal adjacent to the welds.

Recommendations. Use of highly stressed austenitic stainless steels in high-chloride environments having a temperature above 65 °C (150 °F) should be discouraged. Solution annealing or shot peening to reduce residual stresses may be advisable. If heat treatment is not feasible after welding, the substitution of a

Fig. 28 Type 316L stainless steel pipe that fractured by localized attack in welds during exposure to hot brine

(a) Interior surface of pipe showing perforation at the weld seam caused by pits and stress-corrosion cracks. Actual size. (b) SEM micrograph of corrosion pits in the weld showing penetration between high-chromium interdendritic boundaries. 2000×. (c) SEM micrograph showing transgranular facets of a stress-corrosion crack that formed in the base metal after exposure to a chloride-containing environment at 120 °C (250 °F). 500×

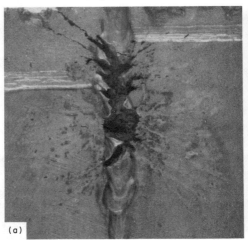 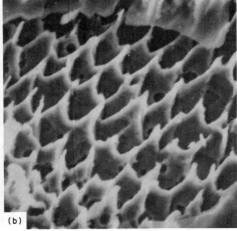 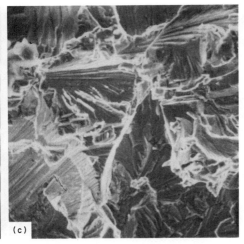

(a)   (b)   (c)

Fig. 29 Section through a type 309S (Nb) stainless steel evaporator tube containing a defective seam weld

The cavity shown resulted from crevice corrosion and extended longitudinally below the inner surface of the weld. 60×

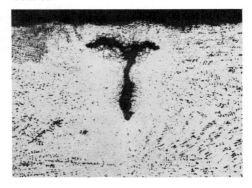

Fig. 30 Section of stainless steel tube/fin assembly removed from economizer

Note the weld geometry and fin weld locations.

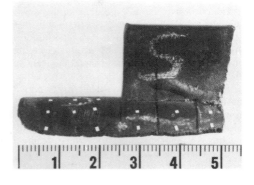

more corrosion-resistant alloy, such as Incoloy 800 or 825, may be necessary.

Example 25: Crevice-Corrosion Failure of Evaporator Tubes Because of Defective Seam Welds. Several tubes in a tube bundle in an evaporator used to concentrate an acid nitrate solution failed by leakage. The feed to the evaporator contained about 6% nitrate, and the discharge about 60% nitrate. The tube bundle was comprised of 751 type 309S (Nb) stainless steel drawn-and-welded tubes 25 mm (1 in.) in outside diameter, 2.8 mm (0.109 in.) in wall thickness, and about 4.3 m (14 ft) long. The tubes were expanded and welded into two type 304L stainless steel tube sheets 64 mm (2½ in.) thick.

Investigation. The leaks were located at the seam welds. Sectioning of a defective tube revealed shallow, longitudinal cracks in the seam weld. Metallographic examination showed that crevice corrosion had attacked the weld cracks on the inner surface of the tube, ultimately penetrating the full thickness of the weld. In some areas, the penetration was also longitudinal, extending the cavities axially within the weld and below the inner surface to regions where no trace of surface cracks existed (Fig. 29).

Conclusions. The tubes failed by crevice corrosion. The failed tubes were defective as-received, and the establishment of concentration cells within the longitudinal cracks in the seam welds led to ultimate corrosive penetration of the wall. There was no evidence of crevice corrosion or any localized penetration of tubes that had sound welds.

Recommendations. The leaking type 309S (Nb) welded tubes should be replaced with seamless tubes of type 304L stainless steel to minimize the areas requiring welding and to provide maximum weldability for the tube-sheet joints.

Example 26: Corrosion-Fatigue Failure of Finned Stainless Steel Economizer Tubes Due to Differential Thermal Expansion. After 10 years of satisfactory operation, several economizer-tube failures occurred in a large black liquor recovery boiler for a paper mill. Several other areas of the boiler had experienced problems, but this was the first unscheduled outage due to economizer problems.

The economizer contained 1320 finned tubes approximately 13.7 m (45 ft) long. These tubes were specified to be 50 mm (2 in.) OD × 4.2 mm (0.165 in.) minimum wall thickness. Two fins ran longitudinally 180° apart for most of the tube length and were attached by fillet welding on one side. The fins were about 6.4 mm (¼ in.) thick and 50 mm (2 in.) high. The weld was started and stopped by running up on the fin for the last 13 mm (0.5 in.). This technique left the ends of fins unwelded (Fig. 30). This fin/tube configuration is commonly used in black liquor recovery boilery.

The economizer began experiencing tube leaks that resulted in unscheduled outages. These leaks occurred at the end of the fin near the bottom of the economizer. After the initial failure, the frequency of failures increased, the time between failures decreased, and a failure analysis was eventually initiated.

Investigation. During an unscheduled outage, a sample was removed from a tube that had not failed. The inside of the tube showed heavy pitting attack, probably due to excess oxygen in the feedwater. After cathodic cleaning, the tube was penetrant tested, and numerous longitudinal cracks were noted on the inside in the area of the fin tip (Fig. 31). Cracking at the end of the fin-to-tube fillet weld was noted during visual examination of the outside surface.

Fig. 31 Interior surface of economizer tube after penetrant testing

Note indications of cracking (arrows).

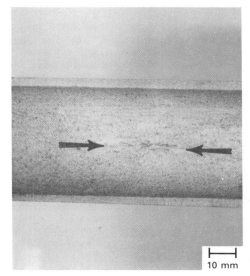

10 mm

Metallographic examination of the section near the end of the fin showed cracking on the inside under the fin and on the outside at the root and toe of the weld (Fig. 32a). A cross section under the fin tip showed one large straight crack on the inside-diameter surface that penetrated approximately one-third of the wall; heavy pitting was also visible (Fig. 32b). Examination of the inside-diameter crack at higher magnification revealed numerous small cracks adjacent to it that initiated at the inside-diameter surface (Fig. 32c). All of these cracks were transgranular, straight, unbranched, and contained oxide typical of corrosion fatigue.

Discussion. A general review of the stresses in the tube fin area indicated that failure was most probably due to thermal stress from differential expansion of the fin and tube. Specifically, the outside of the fin approaches the temperature of the flue gas, while the tube is nearer to the temperature of the cooler inlet water.

This temperature difference results in more expansion of the fin than the tube. Furthermore, the outside of the fin expands more than the portion welded to the tube due to the temperature gradient in the fin. These actions cause the fin to bow such that the outside surface becomes convex. This distortion forces the unrestrained fin end to press down on the outside surface of the tube. At the fin tip, this pressing subjects the tube to a localized bending stress in which the outside is in compression and the inside in tension.

Conclusions. The results indicate the failures were due to corrosion fatigue whose stresses were primarily thermally induced. Corrosion, probably due to a high oxygen content in the feedwater, caused pitting, which resulted in localized stress concentration. Corrosion also lowered the threshold for fatigue. The fatigue cycles were associated with start-ups and shutdowns. Problems in other areas of the boiler resulted in numerous shutdowns over the 10 years of operation. Increased cycling of the boiler for repairs caused accelerated crack propagation in the tubes. This increased the frequency of failure and shortened the time interval between failures. Accelerated shutdown procedures associated with the unscheduled outages caused severe thermal gradients, which further raised the stress level.

Recommendations. A temporary solution included inspecting all 1320 tubes with shear-wave ultrasonics to determine the extent of damage (many tubes contained cracks). Tubes with the most severe cracking were ground and repair welded. The square corners of the fins were trimmed back with a gradual taper so that expansion strains would be more gradually transferred to the tube surface. Water chemistry was closely evaluated and monitored, especially with regard to oxygen content. After the inspection, repair, and fin rework was completed, the boiler ran for approximately 1½ years without any additional failures. The economizer was then replaced due to efficiency and other considerations.

Failures in Arc-Welded Heat-Resisting Alloys

The arc welding of heat-resisting alloys has many aspects in common with the arc welding of stainless steels. The nickel-base and cobalt-base superalloys, as well as highly alloyed iron-base metals, can be welded by arc-welding processes; specific procedures vary with composition and strengthening mechanism. Joint design, edge preparation, fit-up, cleanness of base metal and filler metal, shielding, and welding technique all affect weld quality and must be carefully controlled to prevent porosity, cracks, fissures, undercuts, incomplete fusion, and other weld imperfections.

Example 27: Fatigue Fracture of a Gas-Turbine Inner-Combustion-Chamber Case Assembly Because of Unfused Weld Metal and Undercuts. The case and stiffener of an inner-combustion-chamber case assembly failed by completely fracturing circumferentially around the edge of a groove arc weld joining the case and stiffener to the flange (Fig. 33a). The assembly consisted of a cylindrical stiffener inserted into a cylindrical case that were both welded to a flange. The case, stiffener, flange, and weld deposit were all of nickel-base Alloy 718. It was observed that a manual arc weld repair had been made along almost the entire circumference of the original weld.

Investigation. Microscopic examination of the fracture site revealed unfused weld-metal surfaces and severe reductions (undercuts) in thickness of the stiffener in many areas. The thickness of the case had also been undercut in several areas. Fatigue cracks had originated at multiple sites along the weld interfaces of the case and stiffener. Also, the groove weld contained areas of mismatch that were greater than those allowed by the specifications.

Metallographic specimens from sections of the fracture site showed the double weld beads that were caused by a repair weld and also indicated that many of the interfaces were unfused, as evidenced by films of oxide dross (Fig. 33b).

Fig. 32 Inside-diameter cracking in stainless steel economizer tube

(a) Cross section of tube near the end of fin fillet weld. Note cracking at the root and toe of the weld. (b) Cross section of tube under unwelded fin end showing crack below fin and pitting on inside-diameter surface. (c) Detail of inside-diameter cracking showing one large oxide-filled crack and two smaller cracks. 45×. All as-polished

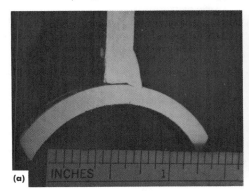

(a)

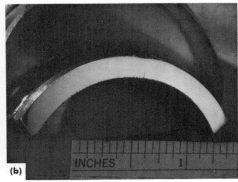

(b)

(c)

Fig. 33 Alloy 718 inner-combustion-chamber case assembly that fractured by fatigue in the weld joining the flange to the case and stiffener

(a) Exterior surface of the assembly showing the circumferential fracture of the case (arrow). 0.5×. (b) Section through the fracture showing the weld (region A) that originally joined the flange to the case (region B) and the stiffener (region C). The external repair weld (region D) had only partial fusion with the earlier bead. The arrow shows a film of oxide slag at the interface. Etched with 2% chromic acid plus HCl. 10×

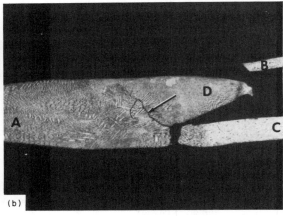

Conclusions. Failure was by fatigue from multiple origins caused by welding defects. The combined effects of undercutting the case wall, weld mismatch, and unfused weld interfaces contributed to high stress concentrations that generated the fatigue cracks. Ultimate failure was by tensile overload of the sections partly separated by the fatigue cracks.

Recommendation. Correct fit-up of the case, stiffener, and flange is essential, and more skillful welding techniques should be used to avoid undercutting and unfused interfaces.

Failures in Arc-Welded Aluminum Alloys

Aluminum alloys can be joined by most fusion welding processes. Submerged arc welding is not used, and shielded metal arc welding is used only where high strength and quality are not required for the intended service. Recommended procedures for arc welding of aluminum alloys are described in Ref 9 and 10 and in the article ''Arc Welding of Aluminum Alloys'' in Volume 6 of the 9th Edition of *Metals Handbook*.

Aluminum alloys of the 1*xxx*, 3*xxx*, 5*xxx*, and 6*xxx* series are easily welded. Those of the 2*xxx* and 4*xxx* series require more care, which may include the use of special techniques to reduce ductility. In a few aluminum alloys, such as 7075, 7079, and 7178, welding results in HAZs that are sufficiently brittle to warrant the recommendation that these grades usually not be welded. Aluminum alloys 7005 and 7039, however, were developed specifically for welding and are not subject to this problem.

The response of an aluminum alloy to the heat input of welding varies, depending on whether or not the alloy is heat treatable and thus can be strengthened by precipitation hard-ening. In non-heat-treatable alloys, such as the 1*xxx*, 3*xxx*, 4*xxx*, and 5*xxx* series, as well as several alloys in the 7*xxx* and 8*xxx* series, the strength and hardness in the fusion zones and HAZs will approximate the values typical of the annealed condition; that is, a weld in an annealed non-heat-treatable alloy will not alter the base-metal properties. However, if the alloy has been strain hardened, the weld will soften an area within approximately 25 mm (1 in.) from the weld centerline, returning this area to the annealed condition. The original hardness and strength may be restored by re-strain hardening, if feasible.

If welding conditions are kept constant or stable, results of the welding operation are also consistent and reproducible. For instance, in junctions of plates of aluminum alloys 5456-H321 and 6061-T6 that were repair welded six times, the tensile properties across the welds were not significantly affected by the number of times the plates were rewelded, nor was the width of the HAZ significantly increased, and the effect of the heat of welding consistently extended no more than 38 mm (1½ in.) from the weld centerline (Ref 11).

Defects that may occur in gas metal arc and gas tungsten arc welds of aluminum alloys include gas porosity, inadequate joint penetration, cracks, undercuts, and inclusions. Distortion may also occur, especially if the weldment is a subassembly of a large structure.

Gas Porosity. The occurrence of gas porosity in aluminum alloy welds is almost always caused by entrapment of hydrogen gas. Solidification shrinkage is not a significant contributor to weld porosity. Hydrogen has high solubility in molten aluminum, but its solubility in solid aluminum is only a small fraction of its solubility in solid steel or solid titanium. Hydrogen dissolved in the weld puddle during welding is released during solidification. The high freezing rate associated with gas metal arc welding, for example, can prevent the evolved hydrogen from rising to the surface of the weld puddle, with the result that porosity occurs. Hydrogen from an extremely limited source, for example, a fingerprint, can cause a significant amount of porosity.

Hydrogen sources are the water, grease, oil, or other hydrocarbons that may contaminate the base-metal surfaces, the ambient atmosphere, the shielding gas, or the electrode or filler wire. The causes of hydrogen in the weld include poor quality of electrode or filler wire; dirty filler material because of careless storage, unsatisfactory cleaning of base-metal surfaces, or contamination of welding room by paint spray or oil drips; water condensate on base metal; water leakage into the gas shield through poor torch seals; leakage of moist air into the shielding gas through defective hoses; and use of a shielding gas with a high dew point.

Gas porosity is usually well distributed in overhead-position welds, is frequently found along the top edge of welds made in the horizontal position, and is least likely to be encountered in welds made in the vertical-up position. Linear porosity is most often found in welds where the root-pass penetration is inadequate. Aligned or layered porosity exerts a much greater effect on mechanical properties than does uniformly distributed porosity. Microporosity, too fine to be detected by radiographic procedures, typically occurs in layers along the weld fusion line and at weld-pass interfaces. This type of porosity lowers mechanical properties appreciably.

A study of the effect of porosity on the tensile properties of aluminum alloy welded panels 9.5 and 25 mm (⅜ and 1 in.) thick had made the following deductions possible, based on welds in which gas bubbles were formed by the deliberate introduction of hydrogen (Ref 12):

- Yield strength was little affected, even by as much as 30% porosity
- Tensile strength was mainly determined by the choice of alloy. However, porosity generally reduced tensile strength because of the loss in effective cross-sectional area
- Elongation was also affected by the choice of alloy, but in general was markedly diminished by porosity in a roughly linear fashion. At 20% total porosity in the tensile-test fracture, the median strain averaged about 60% of that for sound material
- Most macroporosity was in the cover passes, and removal of the reinforcement frequently removed some of the severest porosity. Up to 5% microporosity could be present without any macroporosity; however, at 5% macroporosity, the microporosity in the fracture could be as much as 40%. Radiography was successful in detecting macroporosity, but gave no indication of microporosity

Incomple fusion and inadequate penetration are usually the result of too low a welding current, poor welding techniques, or poor preparation of the base-metal surface. Specific causes may include inadequate wire brushing between weld passes, poor fit-up in preparing the joint, unfavorable torch angle for the travel speed and current being used, joint design that does not afford proper torch accessibility, and unsuitable angle of impingement between the base metal and the welding arc.

Incomplete fusion occurs if the refractory aluminum oxide film on the base metal or the surface of the previous pass is not completely removed by the scouring action of the arc. If the edges being welded fit too tightly together, the arc may be unable to remove the film. Incomplete weld-to-base-metal fusion and interpass fusion in multiple-pass welds in thick plates are of special concern because this may be difficult for the welder to detect while welding and is always difficult to detect in finished welds. Such incomplete fusion occurs most often in horizontal- and overhead-position welding because of the need to use a smaller puddle and therefore less current for these positions. Occluded aluminum oxide is so thin (0.005 μm) that it is difficult to detect with nondestructive-testing procedures. Inadequate root-pass penetration, however, can be discovered by visual, liquid-penetrant, or radiographic inspection.

Weld Cracks. Cold cracking is usually not encountered in aluminum alloy welds except when a bead has been laid down that is too small to resist the cooling stresses. However, both crater cracks and longitudinal hot cracks can occur in aluminum alloy welds if the welding technique is not favorable.

Crater cracks, formed in the residual puddle after the arc has been broken, are potential sources of failure because, although frequently small, they usually exist at the end of the weld where stress concentration is greatest. Their occurrence may be minimized by breaking and restarting the arc several times to feed the shrinkage pipe or by terminating welding on runoff tabs that are cut away later. If crater cracks are found, they should be chipped out and the area rewelded because it is very difficult to melt out a crater crack.

Longitudinal hot cracking during welding is often called hot-short cracking, because sensitivity to weld cracking is strongly related to the composition of the weld metal. In most aluminum alloy systems, there is a range of compositions in which maximum crack sensitivity exists. To avoid longitudinal hot cracking, it is necessary to select a filler metal that will provide a satisfactory final weld composition. Figure 34 schematically illustrates the effects of various combinations of magnesium and silicon and of copper and silicon in the weld metal on the relative cracking resistances of several 6xxx and 2xxx series aluminum alloys. Cracking resistance is affected by the heat input during welding. The heat input determines what fraction of the weld is derived from the base metal,

Fig. 34 Schematic of the different degrees of resistance to weld-crack formation provided by various combinations of alloy content in aluminum-magnesium-silicon (6xxx) and aluminum-copper-silicon (2xxx) alloys

Source: Ref 13. See also Ref 14.

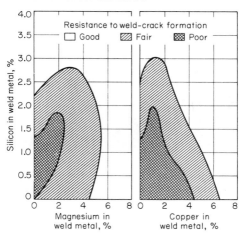

and this approaches 50% only at very high heat input. The type of groove (square versus V) also influences the amount of melting of the base metal.

Studies of welds in 5xxx and 6xxx series alloys have shown that susceptibilities to cracking were lowered when aluminum-magnesium filler alloys containing up to 5.5% Mg were used (Ref 15). Minor alloying elements, particularly titanium and zirconium, are employed as grain-refining agents in filler metal to reduce further the crack sensitivity of welds. Alternatively, minor alloying constituents may accentuate hot shortness by forming low-melting phases at the grain boundaries. Increased copper in 7xxx series alloys, for example, produces this effect, contributing to hot shortness under the thermal stresses present during weld solidification. In some circumstances when the stress on a weld bead is low, cracks may not form immediately, but may open under the heat and stress from a subsequent weld overlay.

Of the various grades of aluminum alloys, the 5xxx series has a very low crack sensitivity. However, under particularly unfavorable conditions and with abusive welding practices, three particular types of weld cracks (aggravated hot-short cracking, cracking of previously deposited welds, and weld-shrinkage cracks) have occurred in aluminum alloy 5083 plate welded with aluminum alloy 5183 filler metal.

Aggravated hot-short cracking can occur as a result of applied tensile forces across the solidifying weld metal, producing the type of defects seen in Fig. 35. These cracks may be caused by restraining fixtures, by the mass of the connected aluminum alloy members, or by release

Fig. 35 Micrograph of hot-short cracks in a gas metal arc weld of aluminum alloy 5083 plate

Etched. 100×

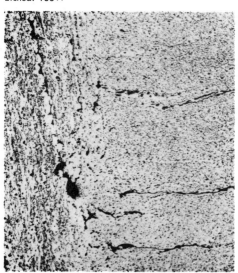

of stored energy in formed parts. A revision of the fixturing, of the welding conditions, or of the welding sequence on the assembly usually eliminates this problem.

Cracking of previously deposited welds is generally traced to the buildup of residual tensile stresses that finally causes fracture. This type of failure has been observed in large fillet welds and in multiple-pass welds in thick aluminum alloy 5083 plate. This can be avoided by changing the welding conditions or welding sequence.

Weld-shrinkage cracks are actually weld-shrinkage pores that are similar in shape to cracks and are usually associated with oxide inclusions. This type of defect is uncommon, but has occurred in welds made by inexperienced operators who used excessive welding current, which formed extremely large, deep weld puddles.

Undercuts (grooves in the base metal along the edge of a weld) are serious defects in aluminum alloys because they reduce the cross-sectional area of the welded zone, introduce a stress concentration, and consequently lower the load-bearing capacity of the weldment. Undercuts result from use of unfavorable welding conditions, such as excessive welding current, insufficient arc-travel speed, or improper electrode-holder angle, or from dirt or oxidation on the work-metal surface.

Inclusions. Aluminum alloys contain particles of the intermetallic compounds and second phases that make desirable increases in strength and hardness possible by precipitation (in heat-treatable alloys) or by solid-solution strengthening and strain hardening (in non-heat-treatable alloys). These microconstituent particles, however, are integral parts of the

Fig. 36 Aluminum alloy air bottle that failed by intergranular cracking caused by overheating before assembly welding

(a) Machined air bottle showing hot-spun region, circumferential butt weld, and location of crack indications. Dimensions given in inches. (b) Longitudinal intergranular crack and triangular void (arrows) typical of those at the failed areas of the boss and tank. Etched with Keller's reagent. 250×. (c) Fracture surface of one of the tank-wall cracks showing typical glossy surfaces. 10×

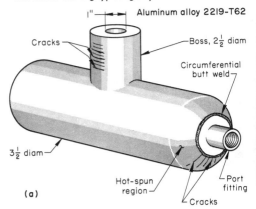
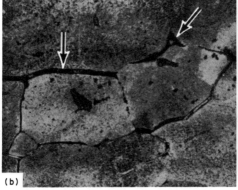

alloys and are therefore not considered as extraneous inclusions.

There are two types of extraneous inclusions in aluminum alloys: metallic and nonmetallic. Metallic inclusions consist principally of particles of tungsten and copper. In gas tungsten arc welding, the use of excessive current for a specific electrode size will cause melting and deposition of tungsten in the weld. When present as fine, uniformly scattered particles, tungsten has little effect on the mechanical properties of the weld. Many codes, however, require that the welds be essentially free of such inclusions.

Copper inclusions may be found in gas metal arc welding if the electrode suffers burn-back to the contact tube. Such inclusions produce a brittle weld and can be a serious corrosion hazard; therefore, they should be removed from the weld deposit. Both tungsten and copper inclusions can be identified visually, and x-ray inspection will detect tungsten particles because of their high density.

A third type of metallic inclusion can be caused by wire-brush bristles that become caught in the weld groove in preweld cleaning or in cleaning between weld passes.

Nonmetallic inclusions in welds of aluminum alloys are seldom derived from the base metal. They are almost always caused by deficiencies in the welding procedures, principally by inadequate protection from the ambient atmosphere because of insufficient quantity or purity of the shielding gas.

Distortion of aluminum alloy welds stems from elastic and plastic (permanent) deformation of the base metal by thermal-contraction stresses in both the longitudinal and transverse directions. It can be controlled somewhat by favorable adjustment of joint design, joint fit-up, welding conditions, and welding sequence, but it cannot be eliminated without resort to postweld cold flow, as in stretcher straightening or weld planishing.

Elastic distortion is caused by residual stresses in the vicinity of the weld that are less than the minimum yield strength of the alloy in the HAZ. These stresses do not affect the static or impact strength of aluminum alloys, but may have an adverse effect on fatigue strength, may cause secondary distortion in assemblies machined after welding, and may reduce SCC resistance of certain alloys. In non-heat-treatable alloys, the residual welding stresses can be reduced to a low level by a thermal treatment with little sacrifice of the original work-hardened strength.

In the assembly of large structures, distortion in welded subassemblies can cause serious mismatch in final assembly, and a force fit of mismatched sections introduces new problems in local damage and new residual stresses. These problems can be minimized by adjusting dimensions, welding procedures, and welding sequences to anticipate the direction and magnitude of possible distortion.

Example 28: Intergranular Cracking in an Air Bottle Because of Torch Overheating Before Welding. An air bottle, machined from a solid block of aluminum alloy 2219-T852, displayed liquid-penetrant crack indications after assembly welding (Fig. 36a). The air bottle was machined to rough shape, which consisted of a 3.8-mm (0.15-in.) wall thickness cylindrical cup with a 19-mm (3/4-in.) wall thickness integral boss on one side, and was annealed at 415 °C (775 °F). The lip of the cup was then hot spun to a dome shape, the metal being preheated and maintained at 205 °C (400 °F) with a torch, and was subsequently liquid penetrant inspected. After a second anneal, a port fitting for the dome opening was attached by four tack welds. The assembly was torch preheated to 120 to 150 °C (250 to 300 °F) using a contact pyrometer to check the temperature. Presumably, this included heating of the boss, which constituted a massive heat sink. The port fitting was then welded in place. Final

full heat treatment to the T62 temper was followed by machining, pressure testing, and radiographic and liquid-penetrant inspection.

Investigation. The crack indications were found only on one side of the boss and on the lower portion of the hot-spun dome region (Fig. 36a). Sections were cut from both the boss and the tank wall for metallographic examination of the crack indications. Other tank-wall cracks were broken open and studied fractographically.

The metallographic specimens revealed triangular voids and severe intergranular cracks that progressed longitudinally (Fig. 36b). The cracks that were opened displayed very few fracture features, but did display the glossy surfaces (Fig. 36c) that are typical of melted and resolidified material.

At regions away from the crack indications, the material was found to be metallurgically satisfactory with a hardness of 69 to 70 HRB, electrical conductivity of 32 to 33% IACS, and a chemical composition conforming to aluminum alloy 2219.

There were indications of localized overheating. A review of the production procedures indicated that preheating for welding was the only operation in which local overheating of both the boss and dome could have occurred. Also, a substantial undercut was found on the inside of the weld, suggesting a higher-than-normal adjacent base-metal temperature at the time of welding. The conclusion was that the torch heating caused local grain-boundary melting.

Conclusions. The localized cracks in the air bottle were the outcome of grain-boundary eutectic melting caused by local torch overheating used in preparation for assembly welding of a port fitting.

Corrective Measure. A change in design was scheduled to semiautomatic welding without the use of preheating for the joining of the port fitting for the dome opening.

Example 29: Failure of an Oil-Line Subassembly Because of Poor Welding.
Several elbow subassemblies comprising segments of oil-line assemblies that recycled aircraft-engine oil from pump to filter broke in service. The components of the subassemblies were made of aluminum alloy 6061-T6. The expected service life of each subassembly was the same as that of the aircraft, but actual life was between 6 months and 1 year.

Two subassemblies were returned to the laboratory for determination of cause of failure. In one (Fig. 37a), the threaded boss had separated from the elbow at the weld made with the gas tungsten arc process using aluminum alloy 4043 filler metal. In the second, the failure was by fracture of the elbow near the flange (Fig. 37b).

Investigation. The fracture through the weld (Fig. 37a) reveals clearly that penetration of the elbow wall section was entirely lacking. The result was a continuous notch beneath the weld, which provided a stress concentration that overloaded the inadequate weld during normal service stresses. The load on the assembly also caused cracking at the apex on each side of the V-notch in the tube that had been cut to receive the threaded boss. The specifications required that the subassembly be heat treated to the T6 temper after welding; hardness readings of 102 HB (normal for T6) taken near the weld confirmed that this had been done. Examination of the weld fracture surface showed evidence of fatigue and overstressing, but no sign of corrosion.

Study of the second elbow (Fig. 37b) revealed that the welded joint attaching the threaded boss had also failed but had been rewelded by the user and returned to service. The second failure (Fig. 37b) then occurred in the tube near the flange. The tube hardness near the fracture was less than 53 HB, representing a condition harder than fully annealed (30 HB) but softer than a T4 temper (65 HB). The flange had been attached to the elbow by torch brazing, using BAlSi-3 filler metal (aluminum alloy 4145) and AWS type 1 brazing flux. According to the manufacturing procedure, following the brazing of the flange the V-notch should be cut, the threaded boss should be welded in place, and the subassembly should be solution treated and aged to the T6 temper. However, the rewelding of the threaded boss joint undoubtedly overaged the elbow, reducing its strength to below that of the weld. The fracture showed evidence of fatigue and overstress, suggesting that the service loading had been excessive for the softened tube with failure occurring in the weakest region. No evidence of corrosion was found.

Conclusions. The separation of the threaded boss from the elbow was due to a poor welding procedure that failed to achieve penetration of the elbow. The crack propagation was accelerated by fatigue caused by cyclic service stresses.

The fracture of the second elbow near the flange was caused by overaging during repair welding of the boss weld, with consequent loss of strength in the tube. Had the elbow been re-heat treated following the repair welding, it is likely that the tube would not have fractured.

Corrective Measures. Satisfactory weld penetration at the threaded boss joint was achieved by improved training of the welders plus more careful inspection. Repair welding was prohibited, to avoid any recurrence of overaging from the welding heat. An additional support for the oil line was installed to reduce vibration and minimize fatigue of the elbow.

Failures in Arc-Welded Titanium and Titanium Alloys

Commercially pure titanium and most titanium alloys can be arc welded satisfactorily by a wide range of welding processes. Unalloyed titanium and all the α-titanium alloys, being substantially single-phase materials, may be welded with little effect on the microstructure. For this reason, the mechanical properties of a correctly welded joint of these alloys are equal to those of the base metal and have good ductility.

The two-phase microstructures of the α-β titanium alloys respond to thermal treatment and consequently can be altered by welding. The result is extreme brittleness in some alloys that renders them nonweldable. The most commonly used α-β titanium alloys that respond well to welding include Ti-6Al-4V, Ti-3Al-2.5Sn, and Ti-6Al-6V-2Sn.

Most β-titanium alloys can be successfully welded. These, however, require particular care if heat treatment is to be employed to strengthen the welds. Aging of certain β-titanium alloy welds, such as Ti-3Al-13V-11Cr alloy, renders the welds susceptible to embrittlement.

Titanium is a highly reactive metal, combining readily at elevated temperatures with carbon, hydrogen, nitrogen, and oxygen to form interstitial solid solutions. If these contaminate the fusion zones and HAZs of a weld, embrittlement, cracking, and weld failure can result. To guard against this, the parts to be welded must first be carefully cleaned to remove scale, oxide, oil, grease, dust, fingermarks, and grinding-wheel grit and binders before welding begins. A solution of 25 to 30% nitric acid plus 2 to 3% hydrofluoric acid in water will remove scale. Acetone or alcohol will effectively degrease the surface; however, chlorinated solvents should never be used, because chloride residues cause SCC in and around weld zones. During welding, a protective gas shield must be used—for example, an inert atmosphere to protect the arc, the weld puddle, and the HAZ during weld deposition. This is readily accomplished during automatic open-air welding with secondary and backup shielding-gas inlets that ensure a complete inert-gas blanket for the operation. Manual welding is done either in an atmosphere-controlled chamber or, when this is

Fig. 37 Fractures in aluminum alloy aircraft-engine oil-line elbows caused by poor welding practice

(a) Fracture of gas tungsten arc weld joining threaded boss to oil-line elbow showing lack of penetration through the surface of the V-notch and cracks at both sides of the apex of the notch. (b) Fracture near the brazed flange (left) after repair welding of the boss V-joint

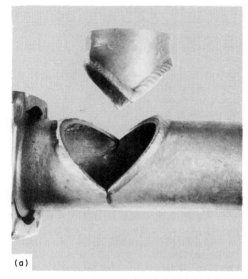

(a)

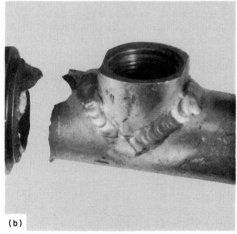

(b)

not feasible, inside a plastic bag; auxiliary gas shielding is awkward and unreliable for open-air manual welding unless containment of some type is employed.

The factors that cause failure of titanium or titanium alloy welds may be metallurgical, mechanical, or a combination of both. Basically, the metallurgical origin is the presence of a phase or combination of phases that possess a very limited capacity to tolerate strain within a critical temperature range. The critical phase or phase mixture will be influenced by the composition of the alloy, the thermal and mechanical processing the alloy has received, the welding conditions used, and the weld shape, which affects the temperature distribution in the adjacent metal. The extent to which either the metallurgical structure or the imposed stresses contribute to a failure can vary considerably.

Mechanical origins include the occurrence of thermal and restraint stresses that, as affected by the presence of weld defects and within this critical temperature range, produce strain in excess of the tolerance of the microstructure. The majority of weld failures in titanium alloys, however, arise from residual stresses in the welded joint from normal weld shrinkage or from weldment restraint. If these residual stresses are multiaxial and exist in areas of limited ductility because of contamination by carbon, oxygen, hydrogen, or nitrogen, the result may be cracking immediately after welding or later under applied service loads. Abuse in service is another common source of mechanical origins for failures of weldments made of titanium alloys.

An example of a failure caused by a combination of contamination and stress is the fracture of a heat-treated Ti-6Al-4V alloy pressure vessel containing helium (Ref 16); this vessel parted at the weld fusion line of hemispheres with a wall thickness of 10 mm (0.4 in.). The joining was done by multiple-pass automatic gas tungsten arc welding in an atmosphere-controlled chamber with inadvertent use of unalloyed titanium filler metal. Although the filler metal and the Ti-6Al-4V alloy base metal each contained less than 100 ppm of hydrogen, a band of concentrated titanium hydride, representing a segregated level of 600 to 1600 ppm H_2 was found at the interface between the weld and the HAZ (Fig. 38). The vessel had previously survived proof testing to pressures well above the service pressure; therefore, the use of low-strength unalloyed titanium in the weld did not appear to be a direct cause of failure. Rather, the fracture, which was of the brittle-cleavage type, appeared to have resulted from hydride precipitation during service. The unusual aspect of the weld was the almost complete lack of dilution; at the interface between the weld and the HAZ, the composition changed from essentially pure titanium to the Ti-6Al-4V alloy.

The use of unalloyed filler metal in welding Ti-6Al-4V alloy is common practice, being employed quite satisfactorily in fabricating many thin-wall pressure vessels. However, there is usually a significant amount of weld dilution resulting in an appreciable content of aluminum and vanadium in the weld deposit. There have been no failures attributed to hydride formation in these welds.

Example 30: Fracture of Welds in a Pressure Vessel Because of Atmospheric Contamination. A Ti-6Al-4V alloy pressure vessel failed during a proof-pressure test, fracturing along the center girth weld. The pressure vessel, about 1.2 m (4 ft) in diameter and 3 m (10 ft) long, consisted of a dome section, two cylindrical segments, and an aft-adapter section. The girth-weld joint was prepared as a 90°

Fig. 38 Gas tungsten arc weld of a pressure vessel wall that had been stress relieved 1 h at 540 °C (1000 °F)

A band of titanium hydride needles is evident at the interface between unalloyed titanium filler metal (left) and the HAZ of titanium alloy Ti-6Al-4V forging (right). Etched with Kroll's reagent. 100 ×

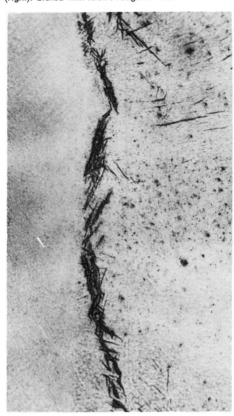

single-V-groove joint with a 1-mm (0.040-in.) root face; the wall thickness of the vessel was 4.5 mm (0.175 in.). The parts were solution treated and aged before welding. The girth joints were welded with the automatic gas tungsten arc process utilizing an auxiliary trailing shield attached to the welding torch to provide inert-gas shielding for the exterior surface of the weld. A segmented backup ring with a gas channel was used inside the vessel to shield the weld root during weld deposition. The vessel was stress relieved 4 h at 540 °C (1000 °F) after welding. This welding procedure had been satisfactory for the open-air welding of large titanium vessels.

Investigation. Preliminary examination of the failed weld revealed that a distinct discoloration existed on the HAZ, suggesting the possibility of atmospheric contamination during welding. The discoloration seemed concentrated at the interface line of the segmented backup ring. A fluorescent liquid penetrant applied to the interior weld surface revealed many small transverse cracks in the fusion zone and HAZ (Fig. 39a).

Fig. 39 Center girth weld of a Ti-6Al-4V pressure vessel that failed during proof testing because of weld embrittlement resulting from oxygen contamination

(a) Interior surface of the weld illuminated with ultraviolet light, which reveals fluorescent liquid-penetrant indications of transverse cracks in the fusion zone and HAZ. 3 ×. (b) Fracture origin (top) formed by welding stresses on the contaminated weld metal. The light region below the origin is the rupture that occurred during the proof test. Approximately 5 ×

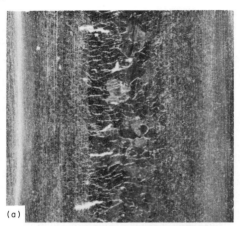
(a)

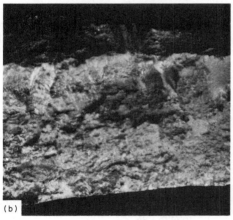
(b)

The fracture origin was at a surface crack, similar to those in Fig. 39(a), which penetrated through approximately one-quarter of the weld, as shown by the dark, discolored region at the top of the fracture surface in Fig. 39(b). Many of the small transverse cracks in other regions were also considered to be the result of crack initiation that occurred in contaminated areas of the girth weld during proof testing.

Conclusions. The pressure vessel failed due to contamination of the fusion zone by oxygen, which resulted when the gas shielding the root face of the weld was diluted by air that leaked into the gas channel. Thermal stresses, which were inherent in the temperature gradients created by welding, cracked the embrittled weld in various locations, exposing the crack surfaces to oxidation before cooling. One of these cracks

Fig. 40 Ti-5Al-2.5Sn gas-turbine fan duct that failed because of contamination of a repair weld in an arc weld in the front flange-duct segment

(a) The circumferential 75-mm (3-in.) crack in the repair weld in the arc weld. 1.5×. (b) Longitudinal section through the repair weld showing an oxide-rich α-case surface layer that contributed to cracking. 500×

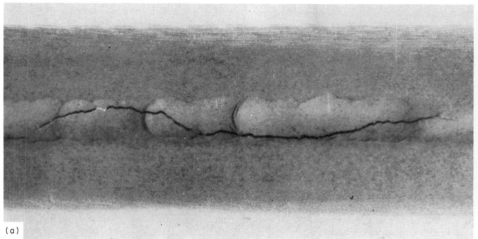

(a)

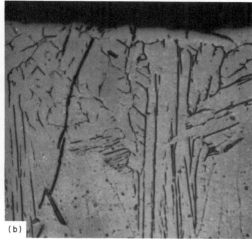

(b)

(Fig. 39b) provided a critical notch that caused a stress concentration so severe that failure of the vessel wall during the proof test was inevitable.

Corrective Measures. A sealing system at the split-line region of the segmented backup ring was provided, and a fine-mesh stainless steel screen diffuser was incorporated in the channel section of the backup ring to prevent air from leaking in. A titanium alloy color chart was furnished to permit correlation of weld-zone discoloration with the degree of atomspheric contamination that may have occurred during a particular welding procedure.

Example 31: Cracking in a Gas-Turbine Fan-Duct Assembly Because of Contamination of a Repair Weld. An outer fan-duct assembly of titanium alloy Ti-5Al-2.5Sn (AMS 4910) for a gas-turbine fan section cracked 75 mm (3 in.) circumferentially through a repair weld in an arc weld in the front flange-duct segment (Fig. 40a). The flange was also made of titanium alloy Ti-5Al-2.5Sn but with the designation AMS 4926.

Investigation. Examination of the crack with a binocular microscope revealed no evidence of fatigue. A blue etch-anodize inspection (Ref 17) showed the presence of an α case along the edges of the repair weld. The α case, a brittle oxide-enriched layer, forms when welds are inadequately shielded from the atmosphere during deposition. The presence of the layer was confirmed by metallographic examination of longitudinal sections transverse to the weld and the crack (Fig. 40b). The brittleness of this layer (48 to 52 versus 29 to 33 HRC for the base metal) caused transgranular cracks to form and propagate in tension under the thermal stresses created by the repair-weld heat input.

Conclusions. The crack resulted from contamination and embrittlement of a repair weld

that had received inadequate gas shielding. Thermal stresses cracked the oxide-rich layer that formed.

Corrective Measures. The gas-shielding accessories of the welding torch were overhauled to ensure that leak-in or entrainment of air was eliminated. Also, the purity of the shielding-gas supplies was rechecked to make certain that these had not become contaminated.

Thermal Stresses. Cracks in titanium alloy parts can also be caused by the thermal stresses produced by welds that are not contaminated. For example, a bracket-brace assembly of Ti-5Al-2.5Sn (AMS 4966) welded with unalloyed titanium filler metal (AMS 4951) cracked in a corner formed between a weld (Fig. 41) and the bracket. The crack propagated diagonally across the assembly to a second weld and then to a nearby slot. Microscopic examination of the fracture surface revealed that the crack propagation was by fatigue under cyclic loading. The origin of the crack, however, was attributed to thermal stresses induced by welding and concentrated by the sharpness of the radius at the corner.

Failure Origins in Electroslag Welds

Electroslag welding is most commonly used for joining relatively thick sections of low-carbon steels, but it can also be used to weld medium-carbon steels, such as AISI 1045 and 1050, and to a lesser extent for welding high-strength structural steels, high-strength alloy steels (such as D-6AC), stainless steels, and nickel alloys.

Because an arc is used to start the welding in the electroslag process and is used until a pool of molten flux (slag) has been formed that is deep enough to quench the arc, the discontinui-

Fig. 41 Ti-6Al-2.5Sn bracket-brace assembly that failed by fatigue

The fracture originated at a weld (arrow) from concentrated thermal stresses.

ties typical of arc welds are usually present in the initial deposit. These possible defects, however, can be isolated from the welded joint by the use of a starting lug that can be cut away later, leaving only metal deposited by electrical resistance heating of the molten flux. This heat serves to melt both the filler metal and the surface of the base metal that are in contact. Once the metal has been deposited by resistance heating of the flux, welding must be continuous until the joint has been completely formed if welding defects are to be avoided. A halt at any intermediate stage in completing the joint weld will probably result in a region containing porosity, nonmetallic inclusions, cracks, and undercuts when welding is restarted. In welds completed without interruption, discontinuities can be detected visually or by nondestructive testing. (Detailed information on the electroslag welding process is provided in the article "Electroslag Welding" in Volume 6 of the 9th Edition of *Metals Handbook*.)

Inclusions. Slag inclusions derived from the flux pool may be found at the weld interface. These inclusions can be best avoided by making certain that the guide tube for the electrode wire is centered between the base-metal surfaces and between the dams.

Other inclusions may result from melting of nonmetallic laminations in the steel being welded. These are detectable by radiography. They can be avoided by magnetic-particle or liquid-penetrant inspection of the faying surfaces of the steel to be welded. When located, the laminations should be gouged to a depth greater than the expected weld penetration, and the cavities filled by repair welds using low-hydrogen electrodes.

Porosity. Because electroslag welding is basically a low-hydrogen process, gas porosity is usually not encountered. However, if gas porosity should be detected and if the flux used is of the composite or bonded type, the flux should be prebaked. Electroslag welding establishes directional solidification in the welds, which is favorable for the production of welds that are free of shrinkage porosity.

Incomplete fusion or inadequate penetration can lead to a major failure of the weld by cracking from thermal contraction stresses on cooling. This is usually overcome by using a predetermined butt-joint gap or by increasing the depth of the flux pool.

Poor Weld Contour. The weld-face contours are formed by the copper dams or shoes that bridge the gap between the plates being welded. If the clamping pressure on the shoes is too great, the resulting flexure will cause an improper weld contour. In some circumstances, the weld face may contain some undercutting. Undercutting is easily detected by visual, magnetic-particle, or liquid-penetrant inspection and can be repaired by grinding or by arc welding plus grinding.

Cold Cracks in Base Metals. In thick plates, microfissures may be produced by forming operations close to or below the nil-ductility temperature. If such microfissures lie close to or in the HAZ of an electroslag weld, they will grow to macrofissures that can be detected by visual inspection or by examination using magnetic-particle, liquid-penetrant, or radiographic methods.

Hydrogen Cracks in Weld Metal. Hydrogen cracks have been observed in electroslag welds. These have resulted from moisture in the flux or from moisture picked up from the cooling shoes. For this reason, dried fluxes should be used at all times.

Excessive Grain Growth. Because of their slow cooling rates, electroslag welds usually have a large grain size in the fusion zone. This may lower toughness, particularly at low temperatures, and may make electroslag welds more sensitive than other types of weldments to contamination.

Failure Origins in Electrogas Welds

Many aspects of electrogas welding resemble electroslag welding, but there are certain distinct differences (the electrogas technique is described in detail in the article "Electrogas Welding" in Volume 6 of the 9th Edition of *Metals Handbook*. The discontinuities and defects that may be encountered in electrogas welds are described below.

Inclusions. Because a flux is not employed in electrogas welding (except in flux-cored electrodes), this possible source of inclusions is frequently absent in electrogas welds. Nonmetallic inclusions introduced by base-metal laminations are sometimes found by radiographic examination. The detection and removal of such laminations are covered in the section "Inclusions" in the preceding discussion on electroslag welding. The depth of gouging necessary to remove base-metal laminations in preparation for electrogas welding is greater than for electroslag welding because of the deeper penetration achieved by the electrogas process.

Porosity. Gas porosity is almost always the result of an inadequate volume of shielding gas or because air was aspirated in the flow of shielding gas. A large gas flow and an increased height to gas coverage should correct this condition.

Highly directional solidification exists in electrogas welding to the same degree as in the electroslag process. Thus, the welds should be free of shrinkage porosity.

Incomplete fusion or inadequate penetration will sometimes cause gross cracking during cooling as the result of thermal-contraction stresses. The usual remedial procedure is to reduce the gap between the plates being welded, increase the welding current, or decrease the oscillation speed, thereby deepening the molten-metal pool.

Poor Weld Contour. The causes and correction of poor weld contour in electrogas welding are the same as for electroslag welding (see the section "Poor Weld Contour" in the discussion of electroslag welding).

Hot Cracks in Weld Metal. If hot cracks are formed in the weld, they will be observed almost immediately as the weld cools and the weld dams move up. Such defects usually mean rejection of the weldment. Their occurrence can be related to the alloy content of the electrode filler metal and can be eliminated by selection of the proper combination of filler metal and shielding gas.

A filler metal should be similar to the base metal in composition, tensile strength, and elongation. Selection of the shielding gas depends somewhat on the composition of the base metal.

Cold Cracks in Base Metal. The occurrence of cold cracks in the base metal is the same as for electroslag welding (see the section "Cold Cracks in Base Metal" in the discussion of electroslag welding).

Failures in Resistance Welds

The general term resistance welding includes resistance spot welding, resistance seam welding, and projection welding, which are all closely related. Flash welding (a resistance butt welding process) is dealt with separately in this article, predominantly because the extensive arcing and plastic upsetting that characterize this process make it appreciably different from conventional resistance-welding processes. The techniques and equipment that are appropriate for resistance welding are described in the Section "Resistance Welding" in Volume 6 of the 9th Edition of *Metals Handbook*.

The qualities of resistance welds are affected by many variables, including the properties of the material to be welded, the surface smoothness and cleanness, the electrode size and shape, and the welding-machine settings that determine welding time, pressure, and current. Successful welding depends on consistent weld properties, which in turn require uniform welding conditions. Experience has shown that a change in any single variable of more than 10% is sufficient to make the weld unacceptable. Unacceptability may represent failure to meet a specified property limit, such as a minimum tensile or impact strength, or it may indicate actual fracture of the weld. Some possible causes of resistance weld failures are described below.

Inclusions. The sources of inclusions in a resistance weld include the surfaces of the parts being joined and their internal structures. Surface contaminants that may cause inclusions are dirt, rust, scale, certain types of coatings, and sometimes oil and grease. These may be between the surfaces to be joined or may be between a part and the electrode contacting the part. Whether or not the inclusions cause failure depends on their quantity, size, location within the weld microstructure, and such properties as melting point or softening range. An inclusion that is likely to become fluid enough at the nugget-fusion temperature to penetrate a grain boundary would be particularly damaging.

Porosity. Some inclusions may cause weld porosity, which is generally considered undesirable but is totally unacceptable if certain size limitations are exceeded. Other inclusions may generate blowholes, usually a cause for immediate rejection. Improper machine settings, particularly those causing excessive current and insufficient pressure, can cause porosity or blowholes. These defects at the weld surface can cause part rejection, but moderate porosity near the center of the weld nugget is usually acceptable.

Inadequate Penetration. Except in aircraft-quality spot welding, limits of nugget penetration in welded parts are commonly not specified. Even for aircraft quality, the range may be from 20 to 80%. Inadequate penetration results from too low a current density in the weld, perhaps from a malfunction or improper setting of the machine or other cause, such as mushroomed electrodes, partial shunt of welding current through adjacent welds, excess pressure, or improper material.

The effect of inadequate current density was observed when a bench-type resistance spot welder was being used to join two crossed wires by direct-current welding for an electronic assembly. The measured weld strength was 4 kg (9 lb). After maintenance on the weld head, the cables to the electrodes were accidentally reversed, and the subsequent welds showed a strength of 1.4 to 2 kg (3 to 4½ lb). This was due to the effect of polarity in welding dissimilar metals with direct current. This effect can also be observed in very short-time welds (1 to 2 cycles) made with alternating current.

In another instance in which inadequate current density prevented satisfactory welding of a crossed-wire junction, a tinned-copper wire was to be welded to a nickel wire. The welding operation, in which power input was 25 W · s, was unsuccessful in producing a weld. Inspection of the wires for cleanness with emery paper revealed a copper color beneath the surfaces of both wires, and a hand magnet showed that both wires were nonmagnetic. Another tinned-copper wire had been accidentally substituted for the nickel wire; a power input of 65 W · s would have been necessary to weld together the two copper wires.

Insufficient current density resulted in inadequate joint penetration when three spot welds were made in attaching a small angle to a sheet; both the angle and the sheet were of cold-rolled steel 1.6 mm (1/16 in.) thick. The spot welds were made 13 mm (½ in.) apart. However, welds 2 and 3 were often weak. Because shunting was suspected as the cause of the weak welds, test were performed, which showed that weld 1 was always good while welds 2 and 3, when made in that order, were both weak, because part of their current passed through the earlier made welds. Conversely, it was shown that if weld 3 was made after weld 1, they were both strong, whereas if weld 2 was made last, it was extremely weak. It was decided to eliminate weld 2 and use two welds 25 mm (1 in.) apart, which would minimize shunting, ensure adequate current density, and provide two strong welds consistently.

Poor weld shape encompasses a variety of defects that may result from part configurations causing undersized nuggets that can lead to failure. Examples of these conditions include poor fits, inadequate flange width or contacting

Fig. 42 Schematic of the failure of spot welds to join a bracket to a baseplate assembly

The failure was caused by inadequate welding current density resulting from extension of the electrode beyond the edge of the bracket flange.

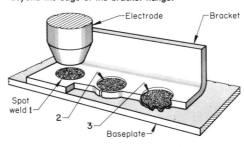

overlap, edge bulges, cracks, burn-throughs, and distortion. In many such instances, the defects are often the result of carelessness on the part of the machine operator. Less frequently, the fault may be the result of an improper design, such as the insufficient flange width of the bracket shown in Fig. 42.

The bracket shown in Fig. 42 had been difficult to spot weld to the baseplate. The size of the welding electrode was determined by the bracket thickness, which required that the electrode be of a specific size that definitely related to the electrode-face diameter and to the electrode-shank diameter. The narrowing flange caused the electrode used to extend over the edge of the bracket, producing a bulge and crack at spot weld 2 and an unsound weld at position 3 because of metal expulsion. The operator could have selected a smaller-diameter electrode, but then the spot welds, although sound, would have been smaller and perhaps unable to carry the load on the bracket. The correct solution is to redesign the bracket with a wider flange or to emboss the flange with projections to concentrate the current flow at proper distances from the flange edge.

Poor weld surfaces that can cause weld failure, usually because of unsatisfactory nugget formation, are often the result of poor machine operations or adjustments that lead to weld spatter, blowholes, electrode pickup, or excessive indentation.

Cracks. Welds containing cracks usually result from overheating, improper loading during the welding cycle, or the use of welding programs that are unfavorable for crack-sensitive metals. Hot cracks are uncommon in resistance welding because the time at high temperature is so brief. Cold cracks may occur because the weld metal froze under insufficient pressure and was therefore forced to undergo thermal contraction relative to the surrounding matrix that demanded more deformation than the metal could tolerate. Such cracks weaken the weld considerably if they are at the weld rim, but are innocuous if at the weld center. These cracks may be avoided by proper pressure control, especially by application of forging pressure at the end of the weld cycle.

Unfavorable fabrication procedures or inadequate bend radii sometimes cause cracks that propagate in parts to induce failures that can be confused with weld-crack failures. There is a clear need to determine the true crack origin before assigning the cause of failure in such circumstances. Water quenching during resistance seam welding can lead to quench cracks from martensite transformation and contraction stresses.

Inadequate Weld Properties. Conditions that affect nugget formation include surface coatings, preweld cleaning, electrode overlap, spot-weld-to-edge distance, sheet thickness, and wall thickness of embossed projections. Any variation in properties that influences the electrical resistance between the parts will influence the weld quality.

Intermittent weld failures can result from variations in surface conditions that are not machine controlled. Examples include the presence of rust or of die-casting flash or mold-release compound in layers of varying thickness. Careful surface cleaning and inspection before welding are the only preventatives.

Parts made of hardenable alloys may be hardened by the high heat and rapid quench of a typical resistance welding cycle. Such parts will be brittle, and the welds can easily be broken unless they are tempered, either as an added portion of the weld cycle or in a separate operation outside the welding machine.

Example 32: Failure of Resistance Spot Welds in an Aircraft Drop Tank Because of Poor Fit-Up. A series of resistance spot welds joining Z-shape and C-shape members of an aircraft drop-tank structure failed during ejection testing. The members were fabricated of alclad aluminum alloy 2024-T62. The back surface of the C-shape members showed severe electrode-indentation marks off to one side of the spot weld, suggesting improper electrode contact.

Investigation. Visual examination of the weld fractures showed that the weld nuggets varied considerably in size, some being very small and three exhibiting an HAZ but no weld. Metallographic examination of sections through the spot welds indicated normal metallurgical structure with no sign of embrittlement. Three welds showed lack of penetration (the cladding being intact at the joint interface), and most of the welds were irregular in size and shape. Of 28 welds, only nine had acceptable nugget diameters and fusion-zone widths.

The weld deficiencies were traced to problems in forming and fit-up of the C-shape members and to difficulties in alignment and positioning of the weld tooling. In particular, the marks on the backs of the C-shape members were due to curvature of the faying surface that interfered with electrode contact at the correct point. It was evident that the lack of precise electrode alignment created an unfavorably long current path, reducing the heat input at the weld and the size of the nugget.

Conclusions. The failure of the resistance spot welds was attributed to poor weld quality caused by unfavorable fit-up and lack of proper weld-tool positioning.

Recommendations. The problem could be solved by better forming procedures to provide an accurate fit-up that would not interfere with electrode alignment.

Example 33: Failure of Resistance Spot Welds Joining Stiffeners to the Skin of an Aircraft Drop Tank Because of Poor Fit-Up.
Resistance spot welds joining aluminum alloy 2024-T8511 stiffeners to the aluminum alloy 6061-T62 skin of an aircraft drop tank failed during slosh and vibration testing.

Investigation. Visual examination of the fracture surfaces showed that the failure was by tensile or bending overload, which was subsequently confirmed by TEM. Measurements of the diameter of fractured spot welds established that all welds were below specification size, which indicated that the joints had lower-than-required strength. Review of the assembly procedures revealed that there had been poor fit-up between the stiffeners and the tank skin, which probably caused shunting of the welding current and resulted in weak, undersize weld nuggets.

Conclusions. The spot welds failed because of undersize nuggets that were the result of shunting caused by poor fit-up.

Corrective Measures. The forming procedures were revised to achieve a precise fit between the stiffener and the tank wall. Also, an increase in welding current was suggested.

Example 34: Crack in a Resistance Seam Weld in a Titanium Alloy Stator Vane Because of Metal Expulsion That Caused Fatigue.
A fluorescent liquid-penetrant inspection of an experimental stator vane of a first-stage axial compressor revealed the presence of a longitudinal crack over 50 mm (2 in.) long at the edge of a resistance seam weld. The vane was made of titanium alloy Ti-6Al-4V (AMS 4911).

Investigation. The crack was opened by fracturing the vane. The crack surface displayed fatigue beach marks emanating from multiple origins at the inside surface of the vane at the seam-weld interface. Both the leading-edge and trailing-edge seam welds exhibited weld-metal expulsions up to 3.6 mm (0.14 in.) in length. Metallographic examination of transverse sections taken through the main fatigue origin and at other random points confirmed that metal expulsion from the resistance welds was generally present. The surface imperfections created by the metal expulsion provided points of stress concentration that lowered the fatigue strength and led to failure in experimental testing.

Conclusions. The stator vane failed by a fatigue crack that initiated at internal surface discontinuities caused by metal expulsion from the resistance seam weld used in fabricating the vane.

Recommendations. Expulsion of metal from seam welds should be eliminated by a slight reduction in welding current to reduce the temperature, by an increase in the electrode force, or both.

Failure Origins in Flash Welds

Flash welding is a variation of resistance welding, in which arcing (flashing) is used to supply the major portion of the initial heat for the formation of a butt weld. The techniques are described in the article ''Flash Welding'' in Volume 6 of the 9th Edition of *Metals Handbook*. Some characteristics of flash welds that may lead to failure are described below.

Poor Surface Conditions. Preweld cleaning of the workpieces at the die-contact area is important to ensure proper current flow and prevent local overheating of the workpiece surface. No surface coating (plated metal, conversion coating, or anodized coating) should be present before welding, because it will result in weld-area contamination or die burns. Oxides and foreign particles can cause small pits in the surface as they overheat and burn into the workpiece.

Inclusions. During the flashing action, craters are formed that contain molten metal and possibly oxides. If the energy input that produces the flashing is properly controlled, these oxides should be expelled with the flashing molten-metal particles that give the process its name. When the upsetting force is then applied, most of the impurities not expelled by the flashing will be expelled with plastic upset metal. Any nonmetallics that are not expelled usually remain at the fusion line with little apparent depth back into the base metal on either side. Most static and fatigue failures that occur in flash welds originate at such discontinuities. These discontinuities normally have little effect on static strength, but can measurably reduce fatigue life. Excessive energy input and insufficient upsetting force or travel speed are the most common causes of such inclusions.

An example of failure of a flash-welded joint attributed to the presence of nonmetallic inclusions is the fracture surface of an aircraft arresting-hook stinger (Fig. 43). This stinger, which was fabricated of 300M steel with a tensile strength of 1931 to 2103 MPa (280 to 305 ksi), was one of several that failed at flash-welded joints during proof loading. The light-colored radial lines visible in the fracture surface in Fig. 43 are nonmetallic impurities that penetrated the fusion line and that were not completely expelled during the upsetting action following flashing. Further occurrence of these inclusions, which were identified as manganese silicate stringers, was eliminated by changing the protective atmosphere from natural gas to propane to provide a greater deoxidizing potential and by increasing the upset pressure and reducing the rate of the energy input.

Fig. 43 A failed flash-welded joint in a 300M steel arresting-hook stinger
Light-colored radial manganese sulfide inclusions are evident. 0.5×

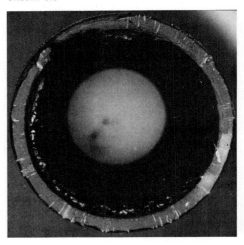

Porosity. Insufficient upsetting force or travel may leave porous areas of cast metal in the weld. Excessive electrical energy input can also lead to porosity by the formation of large craters by the expulsion of molten metal. Where large-diameter pieces are to be joined, one end face should be slightly chamfered so as to start the flashing at the center of the cross section. This helps to avoid trapping of particles in the weld.

Incomplete Fusion. Inadequate heating will result in incomplete fusion, but it may also result from an upsetting travel that is so slow that the metal does not become sufficiently plastic to forge properly during upsetting. Another cause of inadequate heating can be the shunting of electrical current around the closed side of a ring to be flash welded; sufficient current may be by-passed to make flashing difficult, even resulting in a melt-through of the closed side of the ring.

Poor Weld Contours. Whenever heating is not uniform over the intended joint, an unfavorable temperature gradient is established that contributes to misalignment after welding. If parts are not properly aligned by the dies, they may slip past each other during upsetting, creating a lap instead of a proper weld. Workpieces of different cross sections that have not been adjusted in design to achieve a good heat balance, such as tubes of different diameters or wall thicknesses, may slide (telescope) over each other and create a poor-quality weld.

The internal flash and upset metal formed in flash welding of tubes or pipe are normally not removed (often, they cannot be), which leaves a reinforcement with two sharp notches adjacent to the fusion line. These notches act as stress raisers that, under cyclic loading, will reduce the fatigue strength. They also restrict fluid flow and can serve as concentration-cell sites for corrosion.

Hot Cracks. Alloys that possess low ductility over a temperature range below the melting point may be susceptible to hot cracking. Such alloys are more difficult to flash weld, but can usually be welded if the most favorable welding conditions are selected. Tension obviously should be avoided, and because the rim of the upset metal is in circumferential tension as upsetting progresses, it may be essential to keep the upsetting force to the minimum acceptable. In such an operation, some cracks are likely to form that are shallow enough to allow complete removal when the upset metal is machined away. Precise coordination of current cessation and upsetting travel are important, and the joint should be under moderate compression during cooling.

Cold Cracks. Insufficient heat during upsetting or excessive upsetting travel causes colder metal to be forced into the weld zone. The metal will crack transversely to the weld line in the upset zone. This effect is also encountered in the flash welding of hardenable steels, because of the mass quench provided by the remainder of the workpiece. The use of a slow cooling rate, as well as heat treatment following welding, will alleviate transformation stresses that could cause cracking under service loads.

Failure Origins in Upset Butt Welds

Upset resistance butt welding is similar to flash welding, but is less widely used. The advantages of resistance butt welding are:

- There is little weld spatter because there is no flashing
- Less metal is used in the weld upset
- The upset is usually smooth and symmetrical, although it may be pronounced

Upset butt welding is used in joining small ferrous and nonferrous strips, wires, tubes, and pipes end to end.

Because there is no flashing of abutting workpieces, joint preparation for upset butt welding is more critical than for flash welding. Shearing to provide clean, flat, parallel ends is the usual procedure for small parts and wire. Beveled or other special shapes, such as spherical, are used to restrict the area of initial contact in butt welding larger parts, especially if they have oxidized ends. Wires are sometimes pinch cut in wedge shapes and the opposing ends rotated 90°.

Unlike a flash weld, for which the proper upset slope is 45° to 80°, the slope of the burr of an upset butt weld is normally 80° to 90°, quite thin, and often split into two, three or even four petal-shaped segments (Fig. 44); the burr diameter should be at least twice the wire diameter. A split burr with a diameter three to four times the wire diameter is a better burr. The burr can be easily removed by hand or by cutting pliers.

Fig. 44 Upset butt welded steel wire showing typical acceptable burrs on the welds

Dimensions given in inches

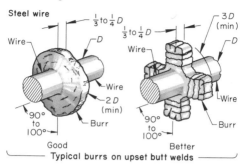

Typical burrs on upset butt welds

Example 35: Failures of Upset Butt Welds in Hardenable High-Carbon Steel Wire Because of Martensite Formation and Poor Wire-End Preparation. Extra high strength zinc-coated 1080 steel welded wire, 2.0 ± 0.08 mm (0.080 ± 0.003 in.) in diameter, was wound into seven-wire cable strands 6.1 ± 0.15 mm (0.240 ± 0.006 in.) in diameter for use in self-supporting aerial cables and guy wires. The wires and cable strands failed to meet tensile, elongation, and wrap tests, with wires fracturing near welds at 2.5 to 3.5% elongation and through the welded joints in wrap tests. The property requirements, a modification of ASTM A 475, included: (1) 3016-kg (6650-lb) minimum tensile strength for the strand, with or without a weld in one wire, one joint allowed per 45.7 m (150 ft), (2) 454-kg (1000-lb) minimum tensile strength for the wire, (3) 4% minimum elongation in 61 cm (24 in.) for a strand without welds, 3.5% for a strand with one weld, and (4) no fracture of an individual wire with or without a weld when wrapped in a close helix of at least five turns at a rate not exceeding 15 turns per minute around an 8-mm (⁵⁄₁₆-in.) diam cylindrical mandrel.

The welded wire was annealed by resistance heating with a 3-kV·A transformer and control separate from the welding transformer. The wire was clamped in annealing jaws that were 0.9 m (36 in.) apart (outside the upset jaws), and the automatic cycle included six pulses of 3 s on and 5 s off. The wire ends had a chisel shape, produced by the use of sidecutters (Fig. 45a).

Investigation. Tests of wire and weld properties gave the following values:

Wire condition	Tensile strength		Elongation in 61 cm (24 in.), %
	kg	lb	
Unwelded			
As-received	506-510	1115-1125	4.5
Welded			
As-welded	113	250	0
Heat treated	206-225	454-496	4.0-9.5
Heated at 80% voltage	224	494	3.0
Heated at 88% voltage	231	510	0.5-1.5

Fig. 45 Wire-end preparation for upset butt welding

The preparation was changed from chisel end (a) to square end (b) to eliminate test failures in welded zinc-coated AISI 1080 or 1055 steel wire.

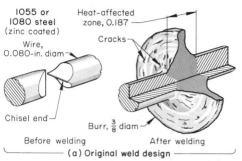

(a) Original weld design

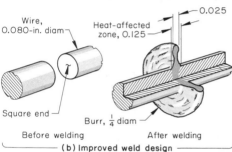

(b) Improved weld design

Tests of the heat-treatment temperatures showed that the wire exceeded 775 °C (1425 °F) at a point 305 cm (12 in.) from the weld area. Metallographic examination revealed that there was martensite present in the weld area even after the heat treatment. The wrap test was being poorly applied, with the wire and mandrel gripped together in a vise and wrapping done with pliers; this subjected the wire to variations in magnitude and type of load as well as of speed of wrap. The weld burrs were excessive, cracked, and poorly formed (Fig. 45a).

At this stage, the manufacturer changed the specified steel from AISI 1080 to 1055 and increased the minimum tensile strength for individual wires (unwelded) from 454 to 522 kg (1000 to 1150 lb). Although this single-wire strength was more stringent, it meant that the 3016-kg (6650-lb) strand-strength requirement could be met even if one wire broke prematurely. Tests were conducted to determine the appropriate postweld heat treatment, using a resistance-heating setup employing a 7.5-kV·A transformer equipped with digital time controls from 1.0 to 9.9 s and digital heat control from 10 to 99%. Heat-sensitive crayons and a hot-wire pyrometer were used to measure wire temperature. The effects of heat-treatment time and temperature on room-temperature tensile strength and elongation are listed in Table 1.

The outcome of the tests was that the welds were brittle unless they were heated at high enough temperatures to undergo transformation to austenite. Although the heat treatment at

Table 1 Effects of heat-treatment conditions on room-temperature tensile strength and elongation of 1055 steel wire strands

Wire condition	Tensile strength		Elongation in 250 mm (10 in.), %
	kg	lb	
Unwelded			
As-received	526	1160	5.0
Welded			
As-welded(a)	113	250	0
Resistance heated(a)			
3-5 s at 595 °C (1100 °F)	356-445	785-980	1.0-2.0
6.2-6.5 s at 595 °C (1100 °F)	290-302	640-665	1.8-2.0
7-8 s at 650 °C (1200 °F)	238-252	525-555	3.0-8.0
7.7-8.2 s at 705 °C (1300 °F)	233-245	514-540	9.0-12.5
7.9-8.4 s at 705 °C (1300 °F)	249-252	548-555	5.5-11.5
8.5-9.3 s at 760 °C (1400 °F)	245-248	540-546	9.0-12.5
9.3-9.5 s at 760 °C (1400 °F)	236	520	9.2-9.8
9.7-9.8 s at 785 °C (1450 °F)	236-240	520-530	8.5-12.5

(a) 1055 steel heated for an adequate time at 605 to 670 °C (1120 to 1240 °F) should undergo stress relief; at 670 to 725 °C (1240 to 1340 °F), spheroidization; at 725 to 765 °C (1340 to 1410 °F), transformation to austenite; and at 765 to 800 °C (1410 to 1475 °F), full annealing

760 °C (1400 °F) did not eliminate all martensite, it did produce enough ductility so that the weld could satisfactorily pass the wrap test.

Conclusions. The test failures of the AISI 1080 steel wire butt-welded joints were due to martensite produced in cooling from the welding operation that was not tempered adequately in postweld heat treatment and to poor wire-end preparation for welding that produced poorly formed weld burrs. The heat treatment of the 1055 steel welds that produced transformation to austenite was successful in minimizing the martensite content sufficiently for the welds to pass the wrap test.

Corrective Measures. The postweld heat treatment was standardized on the 760 °C (1400 °F) transformation treatment to obtain consistent tensile strengths of about 245 kg (540 lb) and elongations above 8%. In addition, the shape of the wire ends, which was a chisel shape produced by the use of sidecutters, was abandoned in favor of flat, filed ends (Fig. 45b) to make the weld heating more uniform and to improve the alignment. The jaw spacing was reduced to half of the initial opening, which further improved the alignment, reduced the amount of upset (thereby improving the shape of the burr), and decreased the time at weld heat. The wrap test was improved by adopting a hand-cranked device that made it possible to conduct the wrapping about the mandrel at a uniform speed and load. Under these conditions the welded joints withstood the tensile and wrap tests.

Failure Origins in Friction Welds

Welding by the direct conversion of mechanical energy to thermal energy at the interface of the workpieces without heat from other sources is called friction welding. A detailed description is given in the article "Friction Welding" in Volume 6 of the 9th Edition of *Metals Handbook*. The types of defects that may lead to failure of friction welds are described below.

Center Defects. A small unwelded area that can occur in friction (inertia) welding at the center of the interface of the workpieces is called a center defect. It can range from 0.25 to 6.4 mm (0.010 to 0.250 in.) in diameter, depending on the size of the workpiece. Center defects are more common in low-carbon and medium-carbon steels than in high-carbon or high-strength alloy steels. Their cause is insufficient friction at the center of the interface, giving rise to heat generation that is inadequate to achieve a complete weld. An increase in peripheral velocity or a decrease in axial pressure will avoid creation of a center defect. Another remedy is to provide either a center projection or a slight chamfer to one of the workpieces.

Restraint Cracks. A type of crack that is most commonly encountered in friction welding a bar to a larger bar or plate is called a restraint crack. This type of crack, which is usually observed in medium-carbon, high-carbon, and alloy steels, occurs during cooling and is caused by thermal stresses, particularly if the HAZ is thicker at the center than at the periphery. Decreasing the peripheral velocity or increasing the axial pressure to gain a thinner HAZ at the center will eliminate this type of crack, but care must be exercised to avoid overcompensation, which would create a center defect.

Weld-Interface Carbides. Carbide particles may be present in a friction weld interface in tool steels or heat-resisting alloys. In both metals, the particles impair weld strength. If a friction weld of tool steel is fractured, circumferential shiny spots will be seen in the weld interface. The formation of these spots is attributed to the lack of adequate time and temperature of the process to take the carbide particles into solution. In welding of heat-resisting alloys, precipitation of carbide particles at the interface at welding temperature can be prevented by increasing the amount of upset.

Poor Weld Cleanup. If the exterior surface of a friction weld does not clean up completely in machining to final diameter, it is the result of hot tearing at or near the surface. In ferrous metals, this usually happens when welding two alloys of different forgeability. The best procedure to avoid this is to increase the section size of the workpiece that has the better forgeability. If the workpieces must be of the same size, an increase in surface velocity and in axial pressure will be helpful. For instance, in welding of a low-alloy steel engine-valve stem to a heat-resisting alloy head (using the flywheel method), a stepped pressure cycle has proved useful. The second pressure is applied to the end of the weld cycle to stop the flywheel rapidly, thereby eliminating the hot tearing.

In friction welding of nonferrous alloys, poor weld cleanup is usually caused by hot tearing at the back edge of the HAZ, especially in copper alloys. An increase in speed will help avoid this condition. For aluminum alloys, an increase in speed plus an increase in axial pressure will normally eliminate such hot tearing.

Hot-Shortness Cracks. Brittleness in the hot-forging temperature range requires special precautions if friction welds free of hot-shortness cracks are to be made in a hot-short metal. The welding conditions must be adjusted to keep the weld temperature to a minimum, using high axial pressures and very low surface velocities.

Trapped oxide occurs occasionally. Oxide entrapment is usually caused by a concave initial weld interface of a remnant of a hole for a machine center in one or both workpieces. As welding starts, the pieces will behave like tubes, with flash flowing inward toward the axis. In normal circumstances, the trapped air will not be able to escape and will remain either under pressure as gas pores or combined with the metal as oxide and nitride. If pieces with machine-center holes must be used, the cavities should be large enough so that the air is not compressed more than 90%. Concave surfaces should be avoided.

Porosity. If cast workpieces that contain shrinkage or gas porosity at or near the weld interface are friction welded, the porosity will normally be evident in the final interface. These defects will usually be gray or black, but will differ in appearance from center defects, which have a metallic appearance. Porosity cannot be eliminated, but careful inspection of the workpieces will reveal the defect.

Failure Origins in Electron Beam Welds

Electron beam welding (vacuum and nonvacuum) is used to weld essentially any metal that can be arc welded, and the weld quality in most metals is either equivalent to or better than that achieved in the best gas tungsten arc or plasma arc welding. The advantages of vacuum electron beam welding include: (1) the capability of making deeper, narrower, and less tapered welds with less total heat input than is possible with arc welding and (2) a superior

Fig. 46 Gas porosity in electron beam welds of low-carbon steel and titanium alloy

(a) Gas porosity in a weld in rimmed AISI 1010 steel. Etched with 5% nital. 30×. (b) Massive voids in weld centerline of 50-mm (2-in.) thick titanium alloy Ti-6Al-4V. 1.2×

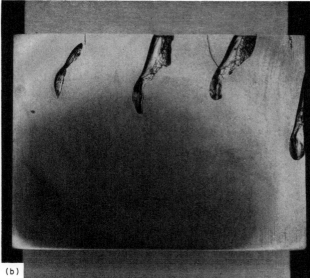

control of depth of penetration and other weld dimensions and properties. Vacuum and nonvacuum processes are discussed in detail in the article "Electron Beam Welding" in Volume 6 of the 9th Edition of *Metals Handbook*. Defects that may cause difficulties in electron beam welds are described below.

Inclusions are minimal in vacuum electron beam welds because of the protection from oxidation and because many impurities are vaporized at high temperature and very low pressure. If foreign metal is inadvertently positioned in the path of the electron beam, it will be added to the melt in a dispersed form. If this occurs, that portion of the weld should be gouged out and rewelded.

Porosity. Electron beam welds are particularly susceptible to the formation of porosity resulting from the release of gases dissolved or trapped in the base metal or from impurities remaining on the metal surface because of inadequate cleaning. To a limited extent, welding conditions, such as the power input, can influence the degree of porosity created. In the welding of carbon steels, for example, aluminum-killed or silicon-killed grades should be selected to minimize internal porosity. The porosity that can occur in an electron beam weld in a rimmed low-carbon steel is illustrated in Fig. 46(a). The material was AISI 1010 steel welded at a speed of 25 mm/s (1 in./s) at 150 kV·A with a current of 4.3 mA.

An unusual variety of porosity called massive voids is sometimes found in electron beam welded joints that are thicker than 25 mm (1 in.). These cavities are similar to wormhole porosity in steel and may result from gas evolution from the base metal during welding

or from the presence of an inverted V-shape gap in the joint before final welding. This inverted V-shape (joint edges in contact at weld-face surface, separated by perhaps 0.8 mm, or 0.03 in., at root surface) may be caused by poor fit-up or may be produced when partial-penetration tack welds or locking passes are used before full-penetration passes in 50-mm (2-in.) thick material. The appearance of massive voids in a centerline section of an electron beam weld in 50-mm (2-in.) thick titanium alloy Ti-6Al-4V is shown in Fig. 46(b). These defects are readily detected by x-ray inspection. Corrective measures to avoid their occurrence include more careful check of joint fit-up and the procedures cited above for general porosity.

Incomplete fusion and inadequate penetration are caused either by insufficient power input or by misalignment between the electron beam and the seam. Such a defect may be eliminated by rewelding the area with increased power or a more accurate alignment. Two types of missed-seam defects are shown in Fig. 47, one being an angular misalignment (Fig. 47a) and the other a beam parallel to the seam but with its position displaced sufficiently to one side to miss the seam (Fig. 47b).

A different type of incomplete-fusion defect, called spiking, is often associated with the irregularly shaped melting pattern at the root of a partial-penetration electron beam weld. Labels such as cold shuts, root porosity, and necklace porosity have been used for defects of this type. They are typically conical in shape and appear similar to linear porosity when examined radiographically in a direction parallel to the electron beam orientation. Therefore, they have the characteristics of solidification

and shrinkage defects in the weld root at the tips of a portion of the spikes. An illustration of spiking that occurred in a 75-mm (3-in.) thick titanium alloy Ti-6Al-4V plate is shown in Fig. 48. A sound weld existed to a depth of 55 mm (2.15 in.) below the weld face.

The peak-to-valley depths can vary over a wide range, depending on the welding conditions employed; these depths can sometimes amount to 25% of the depth of penetration.

The primary procedure for preventing spiking is to defocus the beam so that beam crossover is near the surface instead of being buried 38 mm (1½ in.) deep in the metal. Also helpful is a reduction in travel speed to aid fusion of the root spike. Usually, the depth of a partial-penetration pass in titanium alloys cannot exceed 35 mm (1⅜ in.) without the occurrence of some spiking, although the exact depth is a function of the capability of the equipment.

Example 36: Fatigue Fracture of an Electron Beam Weld in the Web of a Steel Gear Because of Incomplete Fusion. A 9310 steel gear was found to be defective after an unstated period of engine service. A linear crack approximately 25 mm (1 in.) long was discovered by routine magnetic-particle inspection of an electron beam welded joint that attached a hollow stub shaft to the web of the gear. The shaft had a step 2.5 mm (0.10 in.) wide and 50 mm (2 in.) in diameter, which was about 19 mm (¾ in.) larger than the outside diameter of the shaft. Thus, the joint in the web was circumferential, about 50 mm (2 in.) in diameter and with a butt-joint thickness of 2.5 mm (0.10 in.). The assembly procedure permitted a variation in diameter between an interference of 0.025 mm (0.001 in.) and a clearance of 0.05 mm (0.002 in.). In setting up the welding procedure, it had been necessary to apply a cosmetic weld pass on top of the initial full-penetration weld. Gears were welded by this method over a period of several years with no other known service failure. Magnetic-particle and radiographic inspection were used on a routine basis.

Investigation. The crack in the gear web was opened, and the fracture surface (Fig. 49) was examined. One zone of the welded joint showed incomplete fusion; this zone is shown at region A in Fig. 49. Surrounding the incomplete-fusion zone were two zones containing fatigue beach marks (region B, Fig. 49), which indicated that the incomplete-fusion zone was the site at which primary fracture originated. Secondary fatigue patterns that originated on both surfaces of the gear web were evident in several small areas; one of the areas is at region C in Fig. 49. These were interpreted as indicating bending fatigue.

The composition, microstructure, and hardness of the gear were as specified. The design was adequate because it was known that the service stresses in the weld joint region were low.

Review of the conditions of failure suggested that this particular gear assembly contained an

Fig. 47 Sections through electron beam welds in Ti-6Al-4V titanium alloy showing two types of missed-seam defects

(a) Angular misalignment. (b) Position misalignment. Tops of welds at right. Both approximately 13×

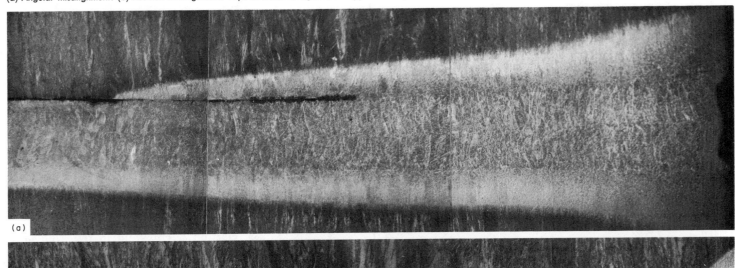

(a)

(b)

Fig. 48 A spiking defect in a centerline section at the root of an electron beam weld

The partial-penetration weld was made in 75-mm (3-in.) thick Ti-6Al-4V titanium alloy plate. 1.6×

interference fit before welding. The possible causes of the incomplete-fusion defect include localized magnetic deflection of the electron beam, a momentary arc-out of the electron beam, and eccentricity in the small weld diameter. Failure to detect the defect in final inspection in the shop probably occurred because

interference fits may not be revealed by radiography and because the cosmetic weld pass may have covered the weld face sufficiently to have prevented detection by magnetic-particle inspection.

Conclusions. The failure was attributed to fatigue originating at the local unfused interface of the electron beam weld, which had been the result of a deviation in the welding procedure.

Careful examination of the possible causes of failure gave no evidence that a recurrence of the defect had ever occurred. Thus, there appeared to be no basis on which to recommend a change in design, material, or welding procedure.

Poor Weld Contours. Poor contours of the face of the weld in electron beam welds may be improved by using cosmetic weld passes. Relatively poor root-surface contours are inherent in the process at the beam exit, especially as base-metal thickness is increased. Thin sections (less than 3.8 mm, or 0.15 in., thick) can be welded by partial penetration of the thickness, which entirely avoids the rough root side at the beam exit of a full-penetration weld.

Hot Cracks. Craters and crater cracks are typically produced in weld-stop areas on full-penetration passes. Use of runoff tabs or pro-

cedures for slope-down and pull-out of the beam are helpful in solving this problem.

Hot cracks in the weld may mean that the base-metal alloy is not weldable (because of its metallurgical structure) or that a foreign metal has contaminated the melt. The use of filler metal (skin stock) to provide weld metal that is less brittle than the base metal (or metals) can prevent cracking; this is especially helpful in joining dissimilar metals if a filler metal is chosen that has a composition that is compatible with both members of the joint.

In welding some materials, such as the heat-treatable aluminum alloys, the grain direction of the base metal relative to the weld can affect the susceptibility to cracking and should be critically reviewed when planning welding. Welds made or stressed in the short-transverse direction, for instance, often show a tendency to develop grain-boundary microcracks in the HAZ.

Bursts. A disk-shaped solidification defect that occurs at the electron beam weld centerline is called a burst. It is usually found near a change in mass, for example, near corners or near adjacent lugs or stiffeners. Figure 50(a) shows the shape, location, and appearance of a

Fig. 49 Fracture surface of an electron beam weld in the web of an AISI 9310 steel gear

The weld fractured because of incomplete fusion. The incomplete-fusion zone where fracture originated is shown at region A; region B shows fatigue beach marks, and region C exhibits bending-fatigue patterns. Root of weld is at bottom surface. 2×

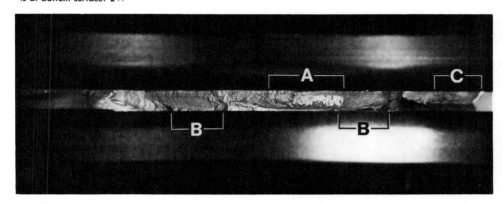

Fig. 50 A burst in an electron beam butt weld

(a) Shape, location, and appearance of burst.
(b) Cross section of burst in a weld in a Ti-6Al-4V plate. 20×

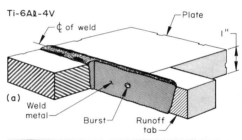

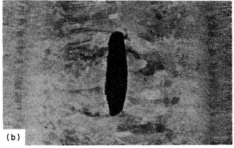

burst and Fig. 50(b) shows the defect in the last part of the weld to freeze, that is, the portion at centerline and near the end.

A typical burst in a 25-mm (1-in.) thick joint has a length (in weld direction) of 1.3 to 3.2 mm (0.050 to 0.125 in.), a height (in the direction of the joint thickness) that is comparable to the length, and a thickness (maximum dimension transverse to weld direction) of approximately 0.25 to 0.4 mm (0.010 to 0.015 in.). Occasionally, burst lengths to 25 mm (1 in.) have occurred. Bursts can be prevented by reducing the travel speed as the beam nears the change in mass or by welding on the runoff tabs to reduce the change in mass.

Failure Origins in Laser Beam Welds

Laser beam welding is used in the automotive, electronics, consumer products, and aerospace industries to join wire to wire, wire to sheet, tube to sheet, and for stud welding. Principles, applications, equipment, cutting, and safety practices are discussed in Ref 18 and in the article "Laser Beam Welding" in Volume 6 of the 9th Edition of *Metals Handbook*. Problems experienced in continuous-wave high-power laser beam welding are very similar to those in electron beam welding processes. Several defects that may cause difficulties in pulsed laser beam welds are discussed below.

Inclusions. Because the laser equipment itself does not contact the workpieces, the presence of any inclusions that could affect the quality of a laser beam weld would be due to foreign materials either on the surface or within the structure of the metal being joined. In most instances, any foreign substances on the workpiece surfaces are vaporized by the high heat of welding. Careful selection, cleaning, and preparation of materials are the best means of minimizing failures from inclusions of any type.

Porosity. Weld pulse-time schedules must be very carefully developed to avoid the formation of voids and porosity. Cleanness of workpiece surfaces and avoidance of low-boiling alloy constituents, such as zinc, are also important in minimizing voids.

Inadequate Penetration. Generally, the penetration that can be achieved with a laser beam increases with the square root of the available pulse time and of the thermal diffusivity of the metal being welded. However, at the same time, high surface reflectivity may be a cause of reduced energy input and inadequate penetration. Simply raising the energy level to compensate for high reflectivity may exceed the intensity limits and produce vaporization and holes. One technique to overcome this diffi-

culty is to apply a suitable thin absorptive coating, such as an anodize or black oxide, that will decrease the surface reflection without affecting the metallurgy of the weld. Metal surfaces can sometimes be roughened to obtain lower reflectivity. An ideal technique when the workpiece shape permits, however, is to arrange the surfaces to be joined so that the weld is in a crevice or hole that can act as a black body, which is a perfect absorber. In this case, the high reflectivity is an advantage, guaranteeing that melting will occur only at the intended area.

Example 37: Failure of a Laser Beam Attachment Weld Because of Inadequate Penetration in Joint Between Cooling Components for a Jet Turbine Blade. Airfoil-shape impingement cooling tubes were fabricated of 0.25-mm (0.010-in.) thick Hastelloy X sheet stock, then pulsed laser beam butt welded to cast Hastelloy X base plugs. Each weldment was then inserted through the base of a hollow cast turbine blade for a jet engine. The weldments were finally secured to the bases of the turbine blades by a brazing operation. At operating speed, the centrifugal force exerted on the welded joint was 207 MPa (30 ksi) at a temperature of approximately 595 °C (1100 °F). One of the laser beam attachment welds broke after a 28-h engine test run.

Investigation. Exposure of the fracture surface for study under the electron microscope revealed that the joint had broken in stress rupture; the fracture surface indicated a ductile tensile fracture. A metallographic section taken through an unfailed area of the weld disclosed that penetration at the weld root had been clearly inadequate, creating a sharp notch that was a severe stress raiser (Fig. 51a). A second section (Fig. 51b), through the fracture, shows that the fracture initiated at the sharp notch and propagated outward through the cooling-tube wall, following the edge of the weld fusion zone.

Although the turbine blades were 100% radiographically inspected prior to engine assembly, the sharp root notch (region D, Fig. 51a) was not detected, possibly because of less-than-optimum orientation of the x-ray beam. After the failure occurred, several other tubes were radiographed with improved orientation of the x-ray beam, which revealed the presence of discontinuities at similar locations.

Conclusions. Failure was caused by tensile overload from stress concentration at the root of the laser beam weld, which was caused by the sharp notch created by the lack of full weld penetration.

Corrective Measures. Careful radiographic inspection of all cooling-tube weldments was made mandatory, with rejection stipulated for any joints containing subsurface weld-root notches. In addition, all turbine blades containing cooling-tube weldments were reprocessed by back-brazing in vacuum at 985 °C (1800 °F) using a small hairpin preform of an 82Au-18Ni

Fig. 51 Sections through a fractured laser beam weld joining a Hastelloy X cooling tube to a base plug of the same material

The weld fractured by stress rupture from tensile overload, which resulted from stress concentrations at a notch left by inadequate joint penetration. (a) Unfailed portion of the weld (region C) joining cooling tube (region A) to base plug (region B); notch caused by inadequate penetration is at region D. 100×. (b) Profile of stress-rupture fracture following upper margin of fusion zone of weld (region C). The sharp notch between the tube wall and the base plug (region B), where the crack started, is shown at region D. 100×. (c) Laser beam weldment that was back brazed. The tube wall is at region A, the base plug region B, the laser weld region C, the initial notch at the weld root region D, and the brazing alloy region E. 170×

Fig. 52 Schematic of a hook crack in a pipe caused by pipe-wall delamination after high-frequency welding

The "hook" has turned outward to follow the direction of metal flow in the outer portion of the upset weld zone.

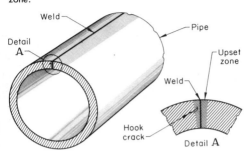

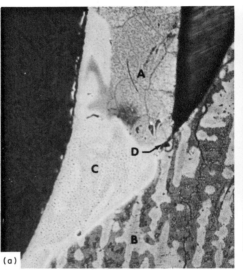

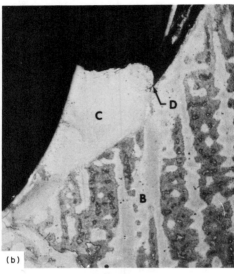

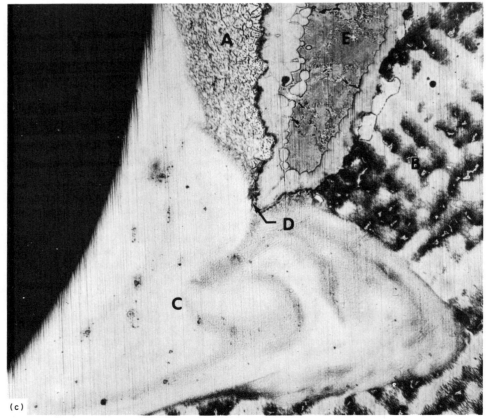

weld, and the base plug is shown in Fig. 51(c) in the region of the weld root. The turbine blades that had been back brazed were reinstalled in the engine and withstood the full 150-h model test run without incident. The back-brazing operation was incorporated into the procedure to fill any notches that might be undetected, as well as to provide a small internal fillet, thus reducing the stress concentration.

Poor Weld Shape. The most favorable joint configuration for pulsed laser beam welding is probably the lap joint, although sound welds can also be made with other types of joints. In any configuration, the important factors are process variations, because these have a definite influence on weld quality. At the desired laser energy output, for example, a 10% change in output produces a 4% change in weld strength. Focusing distance is likewise important to weld strength. For a 50-mm (2-in.) focal length, variations in position of up to ±1.3 mm (±0.05 in.) will still produce acceptable welds. Even though no separation between wires in a lap joint is recommended, some degree of separation can be tolerated.

Poor welds can also be caused by variations in alignment of the axis of the joint with respect to the axis of the laser beam; weld failures may thus be attributed to the use of fixtures that were inadequate to position and hold the workpieces with the necessary precision.

A further factor influencing weld shape and weld strength is the environment in which the laser functions. The laser beam will be modified by any condition that will absorb radiant energy, such as smoke, weld spatter, water vapor, or condensates on the ruby, lens, or mirror surfaces of the equipment. Clean deionized cooling water through the optical cavity, dry nitrogen gas flow through the laser chamber, an exhaust fan for the welding chamber, and regular replacement of the glass slide over the objective lens all represent desirable procedures to maintain constant laser beam input energy to the weld joint.

alloy. This preform could easily be preplaced through the small cooling holes in the base of the turbine blade and was chosen for its excel-

lent oxidation resistance and strength at the 595 °C (1100 °F) service temperature. The bond between the back braze, the cooling tube, the

Base-Metal Cracking. When laser beam energy is put into the weld at too high a rate, the surface may not only suffer from vaporization but base-metal cracking may also occur from thermal shock. This relates to the limitations of thickness that can be welded, which in turn depend on the thermal properties of the metal, its vaporization characteristics, and reflectivity. Generally, longer pulse times are needed to weld the lower thermal conductivity metals and higher pulse-energy densities are necessary in welding high-reflectivity and high-melting-point alloys. The need for precisely developed weld schedules and controls is thus apparent.

Failure Origins in High-Frequency Induction Welds

Pipes for high-pressure service, pipes for well casings, and structural beams have been welded by the high-frequency induction process. This process is described in the article "High Frequency Welding" in Volume 6 of the 9th Edition of *Metals Handbook*.

Hook cracks are a type of discontinuity associated with these welds. They are difficult to detect and require extremely careful inspection for assurance that they are not present.

Delamination of the pipe wall just outside of the narrow weld zone can be the cause of hook cracks. The slight upset of the metal at the weld acts to turn the ends of these usually very small subsurface cracks in the direction of the metal flow (Fig. 52). Because the metal flow of the upset is toward both surfaces the "hook" of the hook crack may turn toward either the outside or the inside surface, depending on which is closer to the delamination and in which direction the greatest amount of upset occurs. Generally, upset is greater toward the outer surface, thus causing hooking to turn outward. Use of the cleanest steel available and the lowest possible heat input will greatly reduce the possibility of hook cracks.

Depending on the materials and welding parameters used, high-frequency weldments are susceptible to other forms of cracking. In one case, weld-centerline cracking was encountered during flaring and flattening tests of high-frequency induction-welded type 409 stainless steel tubing for automobile exhaust systems. Simulation of the welding process and metallographic examinations of simulated and actual welds from several different heat compositions demonstrated that the heat composition, which controlled the type of phase transformations occurring during cooling, determined whether weld cracking occurred. Compositions having a high nickel equivalent (Ref 19) or low chromium equivalent contained martensite at and adjacent to the weld, which strengthened the welds and forced plastic straining to occur in the ductile sheet outside the weld during testing. On the other hand, compositions having a lower nickel equivalent or higher chromium equivalent transformed to weaker ferrite at the weld centerline, which permitted bondline defects to initiate fracture.

REFERENCES

1. J.J. Vagi, R.P. Meister, and M.D. Randall, *Weldment Evaluation Methods*, DMIC 244, Defense Metals Information Center, Battelle Memorial Institute, Columbus, OH, Aug 1968
2. K.G. Richards, *Weldability of Steel*, The Welding Institute, Abington, Cambridge, UK, 1972
3. *Lamellar Tearing in Welded Steel Fabrication*, The Welding Institute, Abington, Cambridge, UK, 1972
4. B.A. Graville, *The Principles of Cold Cracking Control in Welds*, Dominion Bridge, Hanover, 1975
5. F. Coe, *Welding Steels Without Hydrogen Cracking*, The Welding Institute, Abington, Cambridge, UK, 1973
6. D.J. Carney, Gases in Liquid Iron and Steel, in *Gases in Metals*, American Society for Metals, 1953, p 69-118
7. M.A. Pugacz, G.J. Siegel, and J.O. Mack, The Effect of Postheat in Welding Medium-Alloy Steels, *Weld. J. Res. Supp.*, Oct 1944
8. R.V. Phillips and S.M. Marynick, Brittle Fracture of Steel Pipeline Analyzed, *Civ. Eng.*, July 1972, p 70-74
9. *Aluminum Welder's Training Manual and Exercises*, The Aluminum Association, New York, 1972
10. *Welding of Aluminium*, 6th ed., Alcan Ltd., Canada, 1984
11. F.G. Nelson, Effects of Repeated Repair Welding of Two Aluminum Alloys, *Weld. J.*, April 1961
12. F.V. Lawrence, Jr. and W.H. Munse, Effects of Porosity on the Tensile Properties of 5083 and 6061 Aluminum Alloy Weldments, *Weld. Res. Counc. Bull.*, No. 181, Feb 1973
13. P.H. Jennings, A.E.R. Singer, and W.I. Pumphery, Hot-Shortness of Some High-Purity Alloys in the Systems Al-Cu-Si and Al-Mg-Si, *J. Inst. Met.*, Vol 74, 1948, p 235-238
14. H.E. Adkins, Selection of Aluminum Alloys and Filler Alloys for Welding, in *Aluminum Welding Seminar Technical Papers*, Paper No. 3, The Aluminum Association, New York, 1973
15. J.D. Dowd, Weld Cracking in Aluminum Alloys, *Weld. Res. Supp.*, Oct. 1952
16. D.N. Williams, B.G. Koehl, and R.A. Mueller, Hydrogen Segregation in Ti-6Al-4V Weldments Made With Unalloyed Titanium Filler Metal, *Weld. J. Res. Supp.*, May 1970, p 207s-212s
17. D.J. Baron and W.A. Snell, Detecting Segregation in Titanium Alloys, *Met. Prog.*, Aug 1971, p 92-95
18. *AWS Welding Handbook*, 7th ed., Vol 3, American Welding Society, Miami, p 217-238
19. J. LeFevre, R. Tricot, and R. Castro, Nouveaux Aciers Inoxydables a 12% Chrome (New Stainless Steels With 12% Chromium), *Rev. Métall.*, Vol 70, April 1973, p 259-268

Failures of Brazed Joints

Revised by S.J. Whalen, Aerobraze Corporation, J.J. Kozelski, Vacuum Furnace Systems Corporation, and T. Hikido, Pyromet Industries

BRAZING comprises a group of joining processes that produces coalescence of materials by heating them to a suitable temperature and by using a filler metal having a liquidus above 450 °C (840 °F) but below the solidus of the base metal. The filler metal is distributed between closely fitted surfaces of the joint by capillary action. Brazing is a frequently implemented fabrication process that is used not only to join a wide variety of ferrous and nonferrous alloys but also to join metals to ceramics. Procedures for brazing cast irons, steels, stainless steels, heat-resistant alloys, aluminum alloys, copper alloys, reactive and refractory metals, and carbon and graphite are described in Volume 6 of the 9th Edition of *Metals Handbook*.

The various brazing methods (furnace, torch, induction, resistance, dip, and laser) are used to produce a wide range of highly reliable brazed assemblies. However, defects that can lead to braze failure can result if proper attention is not paid to materials properties, joint design, brazing procedures, and quality control. Factors that must be considered include brazeability of the base metals; joint design and fit-up; filler-metal selection; brazing temperature, time, atmosphere, or flux; conditions of the faying surfaces; and service conditions.

Brazeability. Low-carbon steels, copper, and copper alloys (except the high-lead brasses) are generally considered to be the metals most easily brazed. In contrast, stainless steels and heat-resistant alloys, particularly those containing titanium and/or aluminum, are among the metals most difficult to braze. As the brazeability decreases, the likelihood of joint defects increases. With lower brazeability, additional care must be given to surface preparation and brazing-atmosphere control or flux selection.

The properties of brazed joints in copper alloys can be degraded if low-melting elements, such as lead, tellurium, and sulfur, which are added to improve machinability, make the alloys susceptible to hot cracking. The susceptibility of leaded brasses to hot cracking varies directly with lead content. Brazing results are poor at a lead content of 3 wt%; alloys containing more than 5% Pb should not be brazed.

Joint Design and Fit-up. As stated above, brazed joints are formed by drawing filler metal into closely fitted surfaces by capillary action. Because most brazed joints are designed as lap joints, they are subjected mainly to shear stresses in service. In general, the overlap should be at least three times the thickness of the thinnest member, and the joint gap should be 0.025 to 0.075 mm (0.001 to 0.003 in.). For critical applications, the actual overlap should be determined by stress analysis or by testing representative joint sections. The joint gap will be influenced by filler-metal selection and manufacturing economics. For example, tight joints are preferred when brazing with copper filler metal. It may not be feasible to maintain optimum clearances in large sheet metal assemblies. In the past 20 years, there has been a tendency to incorporate T joints and butt joints, provided there is sufficient metal-to-metal contact and due consideration is given to the stresses and strains that these joint types will be subjected to during operation.

Filler-Metal Selection. To prevent brazed-joint failures, the filler metal must be compatible with the base metals, the brazing process, and the service environment. Fortunately, there are numerous commercially available alloys with a wide range of compositions and properties. The demands of the aerospace industry for improved joining processes have resulted in the development of new heat- and corrosion-resistant braze filler metals, such as nickel- and cobalt-base alloys and precious metals (gold, platinum, and palladium). These new filler materials, when used with improved atmospheres (such as ultradry hydrogen and vacuum) and equipment (such as vacuum furnaces with gas-quenching capabilities), have raised the upper temperature brazing limit to above 1650 °C (3000 °F). Before these developments (circa 1950), the upper limit of the brazing process did not exceed 1150 °C (2100 °F).

Some examples of problems to consider in filler-metal selection are the liquid-metal stress-cracking tendency of nickel-base alloys brazed with silver alloy filler metals, the possibility of joint porosity when filler metals with volatile constituents are used in vacuum brazing, and the susceptibility of silver-brazed stainless steel to interface corrosion if the filler metal does not contain nickel or tin. Significant amounts of arsenic and phosphorus in the filler metal should be avoided when brazing carbon and low-alloy steels, because these alloying elements form brittle compounds in the brazed joint. Interface corrosion and phosphorus embrittlement will be discussed in sections so named in this article.

Brazing Conditions. Brazed-joint failures can occur unless proper brazing practices are followed. If the brazing temperature is too high, excessive interaction between the filler metal and base metal can result because of either general alloying or intergranular penetration. On the other hand, temperatures that are too low can cause incomplete flow of the filler metal into the joint and inadequate wetting of the base metal. The heating rate is also important, especially if the filler metal melts over a wide temperature range. In this case, if the heating rate is too slow, liquation of lower-melting constituents can occur, with the higher-melting phases remaining as a rough porous mass. Incomplete joints can also result from use of the wrong flux or from atmosphere quality that is less than adequate.

Surface Conditions. Capillary flow of the brazing filler metal into the joint depends on base-metal surfaces that are clean at the brazing temperature. Generally, fluxes, reducing-gas atmospheres, and vacuum are intended to maintain surface cleanliness. Contaminants, such as oil, graphite, molybdenum disulfide, or lead lubricants used during machining or forming, must be removed before assembly. Oil and grease are usually removed by vapor degreasing. Heavy oxide or mill scale should be removed by either chemical pickling or mechanical cleaning, such as grinding, grit blasting, or wire brushing. When grit blasting, such materials as sand or aluminum oxide should not be used, because they impart a refractory barrier that will prevent wetting and capillary flow of the molten braze filler metal. Such defects as porosity, voids, inclusions, and lack of braze filler metal (incomplete brazing) can be eliminated or minimized by implementing compatible cleaning operations before assembly and brazing.

The use of fluxes or salts (aluminum dip brazing) is also a contributer to producing defects similar to those due to improper cleaning. Fluxes have been the mainstay of low-temperature torch and induction brazing and are responsible for additional defects, such as flux inclusions, flux entrapment, and oxidation of the faying surfaces, which are attributed to improper flux, improper application of flux, and nonuniform heat control resulting in localized underheating or overheating.

Testing and Inspection. Some of the defects that have been discussed above are easily detected with visual inspection. Radiography and/or ultrasonic testing have proven beneficial for detecting internal defects. Many brazements are subjected to pressure, vacuum, or other practical methods of testing that are more appropriate to the service application of the assembly, but periodic destructive testing of brazed assemblies is also suggested to avoid failures. Such defects as oxide or flux inclusions would not be responsible for braze failures unless heavy concentrations had passed undetected. Many engineering specifications will allow a limited number of these defects to remain; however, not a single defect is permitted for any brazement to be used in handling food and/or dairy products, because each defect is considered a potential bacteria trap.

Defects Resulting from Chemical, Physical, or Metallurgical Reactions

Major defects that are the result of a chemical, physical, or metallurgical reaction may not be detectable at braze inspection, because some are caused by environments during operation. Many of these so-called defects or conditions evolved during the years when stainless steels were being brazed with the same methods and techniques used to braze carbon and low-alloy steels. These defects are described below, along with some recommended solutions involving atmosphere/vacuum furnace brazing with nickel-base, cobalt-base, or precious-metal braze filler metals.

Carbide precipitation is a metallurgical reaction that is common to stainless steels and other alloys that contain both carbon and chromium. The precipitate that is formed at the grain boundaries is chromium carbide, which by itself is a very brittle constituent. The great affinity that carbon has for chromium causes the adjacent grains to lose their corrosion-resistant properties and to render the brazement susceptible to corrosion in certain corrosive environments. The chromium carbide precipitate is formed by slow cooling through the temperature range of 870 °C (1600 °F) down through 425 °C (800 °F) and by prolonged heating in this temperature range.

Although this condition was first reported in heat-treating operations, such as annealing of

stainless steels, it became more pronounced in the brazing of stainless steels because of the limited selection of braze filler metals and the lack of equipment that could provide rapid cooling through the carbide-precipitation range. Brazing temperatures of the silver braze filler metal were well within the carbide-precipitation range, and prolonged heating within that range would be the prime reason for a chromium carbide precipitate to form.

Manufacture of the stabilized grades (AISI 321 and 347) of stainless steel helped to curtail carbide precipitation because the stabilizing elements titanium and niobium exhibited a greater attraction for the carbon than the chromium did. The chromium, therefore, was retained in the matrix and was left to perform its intended function of providing corrosion resistance for the stainless steel. Rate of cooling was not an important factor for the stabilized grades. The introduction of the low-carbon stainless steels has also helped curtail carbide precipitation by limiting the carbon content so that a minimum of the precipitate could form.

The introduction of the nickel braze process in ultradry hydrogen atmospheres with elevated brazing temperatures was a major step in producing brazements with corrosion- and heat-resistant braze filler metals. In recent years, inert-gas quenching through the carbide-precipitation range has made engineers and fabricators disregard the fear of carbide-precipitation formation on regular grades of stainless steels.

A typical and economical cycle implemented for concurrently brazing and heat treating AISI-type 410 stainless steel using a nickel-base braze filler metal, AWS-BNi-2 (AMS 4777), would be:

- Hard vacuum pump down
- Heat at 17 °C/min (30 °F/min) to 1040 to 1065 °C (1900 to 1950 °F)
- Hold at 1040 to 1065 °C (1900 to 1950 °F) for 30 min
- Furnace vacuum cool to 925 °C (1700 °F)
- Gas (nitrogen, argon) quench to room temperature

Depending on the configuration and the amount of exposed molten braze filler metal, it is sometimes possible to gas quench directly from the austenitizing temperature. Tempering to the desired hardness would follow after inspection of the brazed joint.

Interface corrosion, also known as crevice corrosion and cavitation erosion, occurs at the faying surfaces of the brazed joint. Interface corrosion is a metallurgical and chemical reaction that develops in brazements of stainless steels, especially those of the AISI 400 series, when these materials are brazed with some of the low-temperature silver braze filler metals (AWS BAg-1, AWS BAg-2) and with flux. This reaction takes place when the stainless steels are heated under molten braze flux with a selective depletion of chromium from the fay-

ing surfaces, thereby exposing a thin layer of chromium-free iron. This surface layer is readily attacked, particularly by the presence of the silver brazed alloy to which iron would be anodic. The attack on this thin layer produces a crevice. Once this crevice is established, the electrochemical corrosion is accelerated, especially if any fluoride flux residues or dissolved salts containing chlorides remain after cleaning. The brazed joints thus made are subject to very rapid corrosion at the brazing alloy/steel interface in rather mild corrosion media.

This defect surfaced during World War II on proximity fuses that had been torch silver brazed with flux, inspected, and stored for shipment. Upon withdrawal a few weeks later, it was noted that the faying surfaces were not bonded and were completely free of any traces of the silver braze filler metals.

Fabricators were again faced with a limited choice of braze filler metals to join the stainless steels. Manufacturers of low-temperature silver alloys quickly resolved this problem by adding 2 to 3% Ni (AWS BAg-3) to the problematic braze filler metals. A more desirable solution came about with the introduction of the ultradry hydrogen atmosphere (AWS-7), which was a prime ingredient in the success of the nickel braze process in the late 1940s. Fluxless brazing in the dry hydrogen atmosphere with the silver braze alloys was another solution in preventing the electrochemical corrosion that was primarily due to the absence of flux in the furnace brazing operation. With the wide selection of braze filler metals that are more compatible for corrosion and heat resistance with stainless steel, it seems proper to choose a nickel-base alloy, such as AWS BNi-2, and to conform to furnace cycles as described as a solution for elimination of carbide precipitation.

Stress-corrosion cracking (SCC) is a defect that was first noticed during process-annealing operations on sheet metal details of stainless steels, nickel alloys, and copper-nickel alloys that had undergone severe cold-working operations, such as forming, drawing, and spinning; SCC placed these details in a highly stressed condition. After annealing, it was noted that some of the details, especially those that were spun, showed evidence of craze cracking, which became known as SCC. This form of cracking was precipitated by residues of brass that had been burnished into the surfaces of the details during spinning operations when forming heads of brass were used. Brass is composed of copper and zinc, and it has a much lower melting point than the annealing temperatures of the base materials in question. The solution to preventing SCC was to eliminate forming heads and dies that were made of brass or to submit details so formed to a pickling operation to remove all traces of the brass before heat processing.

These same materials are also susceptible to SCC if they are in a highly stressed condition and if brazing is attempted with a low-

Fig. 1 Segments of a type 321 stainless steel radar coolant-system assembly that broke at a brazed joint between a bellows and a cup because of inadequate bonding between the brazing alloy and the stainless steel

(a) Portions of the broken coolant-system assembly; bellows is at A, cup at B. (b) Fractograph showing residual brazing alloy at the bottom of the cup. 24×. No brazing alloy was found on the sides of the cup. The surface of the braze appeared oxidized and contaminated. Location of section shown in (c) is indicated by arrows. (c) Micrograph of a section through residual brazing alloy and cup, from location between arrows in (b). 350×. Braze metal (region C) appears to be mechanically bonded to oxide layer (region D) on surface of cup (region E).

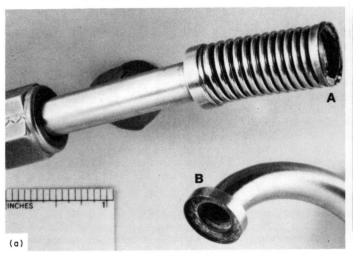

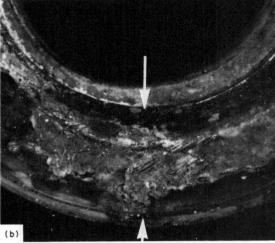

temperature silver braze filler metal containing ≥16% Zn. Again, with the limited selection of braze filler metals, the solution to the elimination of SCC was to remove all stresses, internally and externally (by fixturing), before assembly and brazing.

With the introduction of high-temperature braze filler metals, such as nickel-base and precious metals, brazing could take place at high temperatures that were compatible with the annealing temperatures. These suggested braze filler metals do not contain zinc, which is the prime candidate in the silver braze filler metal for producing SCC.

Alloying or excessive alloying is a general term that is intended to be an overall description of almost all aspects of interactions between the base metals and the braze filler metals. Not all of the interactions are detrimental to the brazement, but the same interaction can become detrimental if amounts of braze filler metal, braze temperatures, and time at braze temperature are excessive.

Mutual Solubility. The molten braze filler metal can diffuse and dissolve the base metals, and the base metal can diffuse into the braze filler metal. Diffusion and dissolution can increase or decrease the liquidus or solidus temperature of the braze filler metal, depending on its composition and thermal cycle.

This mutual solubility was noted in the brazing of a high-nickel (Inconel) base material in an inert-atmosphere furnace using copper braze filler metal. The resulting brazement was not of the distinctive copper color, but was of a color resembling Monel, which is composed of copper and nickel. In the copper brazing of such base materials as nickel, Monel, or the copper-nickels, the solidus of the molten braze filler metal is increased, and the flow through the joint can be terminated. The remelt temperature of the brazed joint thus becomes higher than the original solidus temperature.

If these same base metals are thin in section size, the diffusion and dissolution of mutually soluble elements can lead to undercutting and erosion and can sometimes culminate in the complete destruction of the base metal adjacent to the braze application. The undercutting and erosion effects may also occur in the brazing of aluminum and magnesium as well as the high-temperature alloys.

Mutual solubility of copper and silver is very pronounced during the silver brazing of copper when the brazing temperature is well above the liquidus of the silver braze filler metal. Diffusion of the gold-nickel braze filler metals (AWS-BAu 4) into stainless steel or nickel-base alloys can lead to erosion of the base materials, as experienced with other braze alloys and base materials that exhibit mutual solubility. To control excessive base-metal dissolution, temperatures, time at temperatures, and amount of braze filler metals should be kept at a minimum, but should be sufficient to produce a complete brazement with sufficient diffusion.

Intermetallic Compounds. The molten braze filler metals and base materials can interact and form intermetallic compounds that are usually brittle. The formation of these intermetallic compounds depends on base- and filler-metal compositions, time, and temperature. Intermetallic compounds are more likely to form when using nickel-base braze filler metals containing boron, which is the low-melting constituent responsible for grain-boundary penetration.

Controlled diffusion cycles at temperatures below the original brazing temperature can create a homogeneous structure at the brazed joint.

Phosphorus Embrittlement. Braze filler metals containing phosphorus, such as AWS BCuP, BNi-6, and BNi-7, should not be used to braze any iron or nickel-base materials, because they form phosphides, which are brittle compounds. However, AWS BNi-6 and BNi-7 nickel-base braze filler metals are sometimes used to braze heat-resistant metals, provided proper diffusion time has been afforded.

Oxide Stability. Despite the normal precleaning operations, residual oxides remain on the surface of metals and alloys. Each group or classification of base materials contains elements that form oxides; therefore, each group of these materials must be processed with more active fluxes and/or reducing atmospheres as these oxide formations become more stable. Carbon and low-alloy steels present no problems, and there are numerous fluxes and gaseous atmospheres that will reduce and remove these oxides and protect the surfaces from further oxidation so that brazing can take place.

Stainless steels contain chromium, and the chromium oxides formed during brazing are of a refractory nature. The gaseous atmosphere best suited to the high-temperature brazing of stainless steels with nickel-base or precious-metal filler metals is limited to either ultradry hydrogen (AWS-7) or vacuum (AWS-10C).

Superalloys or high-temperature alloys that contain reactive elements (titanium, aluminum, beryllium, zirconium, and so on) of the order of ≥1% form very stable oxides, and once formed, these oxides can be removed only by mechanical or chemical means. There has been some success in brazing these materials in vacuum if fit-up is intimate and if the furnace equipment can attain very low pressures (1.3×10^{-3} Pa, or 10^{-5} torr). Techniques to aid in preventing oxide formation include bake-out cycles with partial pressures of hydrogen, brazing within an inner retort, and inclusion of titanium alloy turnings in close proximity to the brazements. These turnings serve as getters and absorb or chemically bind the elements or compounds from the brazing atmosphere that are damaging to the final product.

Many specifications will require these materials to be nickel plated and then brazed in vacuum with argon as a backfill or cooling gas. This appears to be the best procedure to follow, especially if the vacuum furnace is old and is used to process various materials in a job-shop heat-treating operation. Nitrides on the surfaces must be removed in precleaning operations because they are a definite barrier to the flow and wetting action of the braze filler metal.

Examples

Very few actual examples of braze failures are available. If the joint is properly designed, if the proper filler metal is selected, and if the brazing procedure is correct, then failures rarely occur in the brazed joint proper. Example 1 deals with a failure in the brazed joint of stainless steel because of poor bonding due to improper brazing procedure. The second example is not a braze failure as such, but was caused by the corrosive flux that was not removed after brazing. Again, this failure must be attributed to improper brazing procedure. The third example was due to lack of braze coverage, which subsequently led to braze failure. In Example 4, failure was due to excessive solution or penetration of the braze alloy into the tube wall. Failure occurred in the tube but not in the brazed joint proper.

Example 1: Failure of a Brazed Joint in a Type 321 Stainless Steel Assembly Due to Inadequate Cleaning. A radar coolant-system assembly, made of type 321 stainless steel and fabricated by torch brazing with AWS type 3A flux, broke at the brazed joint when subjected to mild handling before installation, after being in storage for about 2½ years. Before being put in storage, the assembly had withstood a 1035-kPa (150-psi) pressure test using tap water. All surfaces were to have been passivated according to MIL-S-5002 before assembly, but no other cleaning techniques were stipulated.

Investigation. The failed braze had joined a convolute bellows to a cup on the end of a tube elbow (Fig. 1a). Visual examination revealed that the filler metal had not covered all the mating surfaces and that the surface of the braze metal was not bright and shiny; no fresh fractures were visible. The surfaces of the joint were covered with what appeared to be residues of oxidation products from the filler metal and of flux. Spectrographic analysis of the oxide

residue indicated that it contained major amounts (more than 10%) of silver and copper and minor amounts (0.5 to 10%) of cadmium and zinc, which confirmed that it had been derived from the filler metal. However, the residue conjectured to be flux contained a major amount of zinc and a minor amount of cadmium, which are unlikely as constituents of flux except in trace amounts. Also, the boron content (0.01 to 0.5%) was too low for a material composed of boric acid, borates, fluoborates, and fluorides. The presence of boron in the residue, however, indicated that a flux had been used on the faying surfaces.

Examination of the broken joint at the cup (Fig. 1b and c) revealed that much of the filler metal had flowed into the cup but had not formed a metallurgical bond. Instead, the filler metal had been mechanically bonded to an oxide layer on the stainless steel surface. It was evident that the flux had not protected the steel surface from oxidation during torch heating, because the oxide layer was about 0.008 mm (0.0003 in.) thick and thus could not have been only the normal passivated film, which is too thin to measure.

During examination of this joint, another such assembly broke at the brazed joint when the bellows was subjected to slight deflection. Visual examination also disclosed a lack of bonding in the second joint, which was attributed to the tough, tenacious oxide film that is formed by the heat of brazing. This film is attributed to preferential oxidation of the titanium present in this steel.

The lack of metallurgical bonding explained the joint failure, but appeared to contradict the success of the 1035-kPa (150-psi) water-pressure test before storage. However, the end thrust, which was only about 151 N (34 lb), may have been absorbed by a fixture; no information is available on the way in which the assembly was supported for testing.

Conclusions. The joint failed because of a lack of metallurgical bonding between the brazing alloy and the stainless steel due to an oxide film on the steel, produced during brazing, that the flux did not clean away.

Recommendations. Type 347 stainless steel was recommended for the components of the coolant-system assembly instead of type 321, because of the better brazeability of type 347 using silver brazing alloys. Also recommended was deflection testing at several orientations during final inspection of the brazed joints. Other recommendations were use of type 3B instead of 3A flux to achieve better filler-metal flow and use of a larger torch tip to allow wider distribution of heat in the joint area.

Example 2: SCC of a Brazed Joint in a Type 321 Stainless Steel Pressure-Tube Assembly. A pressure-tube assembly of type 321 stainless steel (AMS 5570), which contained a reinforcing liner attached by brazing, leaked air during a pressure test. Fluorescent liquid-penetrant inspection revealed a circumferential crack extending approximately 180°

Fig. 2 Section through the cracked wall of a type 321 stainless steel pressure tube (region A) showing the branched transgranular nature of the crack

The crack origin (arrow B) was at the inner surface of the tube, next to the braze (region C) joining the tube to the reinforcing liner (region D), also of type 321 stainless steel. 80×

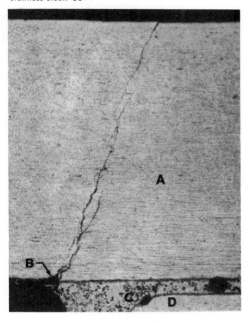

around the tube parallel to, and at a distance of 2.5 mm (0.10 in.) from, the fillet of the brazed joint.

Investigation. The tube was deliberately fractured to expose a portion of the crack surface, which was found to be relatively smooth, with discolorations and multiple crack origins on the inside surface. A longitudinal metallographic section taken through another portion of the crack (Fig. 2) showed that the crack was branched and transgranular, that it had originated adjacent to the braze joining the tube and the reinforcing liner, and that it had propagated through the wall of the tube to its outer surface.

Residues on the inner surface of the tube were identified by chemical analysis as fluorides from the brazing flux. Because it was known that the tube end had been swaged before brazing without an intervening stress-relief treatment, because the residual fluorides represented a potential for corrosion, and because of the nature of the crack, it was deduced that the failure had resulted from SCC.

Conclusions. Brazing-flux residues provided fluoride attack in a region of the pressure-tube assembly where residual swaging stresses existed, producing failure by SCC.

Recommendations. The tube should be stress relieved after swaging but before brazing, and the joint should be thoroughly cleaned of all residual fluorides immediately after brazing.

Fig. 3 Type 347 stainless steel pressure-probe housing that failed by fatigue fracture because of voids in a brazed joint

(a) Photograph of opening produced in housing by detachment of a segment at time of fracture. 5¼×. Large region indicated by arrows is devoid of braze metal. (b) Fractograph showing marks of fatigue-crack propagation from multiple origins (two are indicated by arrows) on inside surface of housing. 10½×. (c) Micrograph of a specimen taken from an area remote from the fracture and etched in aqueous FeCl₃ showing large voids (arrow) and joint separation (lower left) resulting from poor bonding between housing and braze metal. 50×

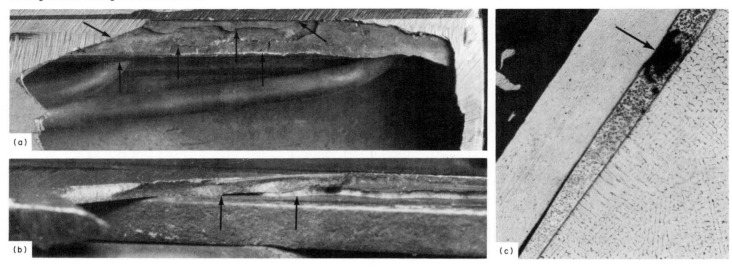

Fig. 4 Waspaloy (AMS 5586) spray-manifold tube that failed by fatigue fracture because of embrittlement by penetration of molten braze metal

(a) Macrograph showing fracture at the edge of the brazed joint between the tube and the sleeve. 4×. (b) Fractograph showing granular, discolored region of fracture origin (between arrows). 12×. (c) Micrograph of a longitudinal section through granular region of fracture showing penetration of molten braze metal (arrows) along grain boundaries. Etched with Kalling's reagent. 100×

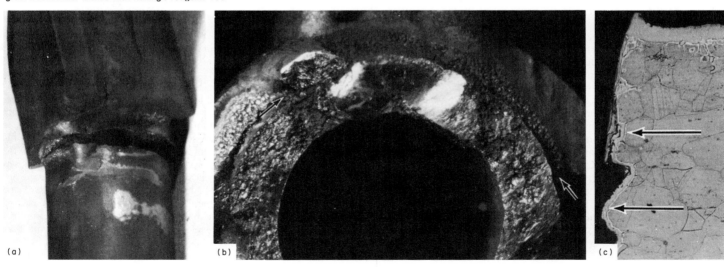

Example 3: Fatigue Fracture of a Type 347 Stainless Steel Pressure-Probe Housing Originating at Voids in a Brazed Joint. A rectangular segment broke from the type 347 stainless steel housing of a pressure-probe assembly used in experimental instrumentation. The pressure probe and the housing had been joined by brazing with AMS 4772D filler metal.

Investigation. Visual examination of the upper portion of the opening produced by detachment of the segment (Fig. 3a) revealed that a large region of the joint between the pressure probe and housing was devoid of braze metal at the fracture. Further study of the fracture surfaces revealed fatigue marks emanating from multiple crack origins on the inside surface of the housing at the brazed joint (Fig. 3b). Sections taken at the fracture and well away from it (Fig. 3c) showed large irregular voids and some trapped flux in the braze metal, plus separation between the housing and the braze, which indicated a poor metallurgical bond.

Conclusions. The failure of the pressure-probe housing was attributed to fatigue cracking that originated at multiple origins at the interface between the housing and the braze metal in areas where there were large irregular voids in the braze metal and separation of the joint because of poor bonding.

Example 4: Fatigue Fracture of Waspaloy Spray-Manifold Assembly Because of Embrittlement by Penetration of Molten Braze Metal. The inner ring of a spray-manifold assembly fabricated from

Waspaloy (AMS 5586) fractured transversely through the manifold tubing at the edge of the brazed joint between the tube and support sleeve (Fig. 4a). The assembly was brazed with AWS BAu-4 filler metal (AMS 4787).

Investigation. Microscopic examination of the fracture surface revealed fatigue beach marks that had propagated from the extremities of a granular gold-tinted surface region (between arrows, Fig. 4b) adjacent to the tube-to-sleeve brazed joint and extending circumferentially in the tube wall for approximately 110°.

Figure 4(c) shows a micrograph of a section through the granular portion of the fracture, displaying evidence that, during the brazing operation, molten braze metal (arrows) penetrated along the grain boundaries, embrittling the tube. This penetration provided a continuous intergranular layer that reduced the strength of the tube locally to that of the braze metal.

From the profile of the initial fracture (at left, Fig. 4c), it is evident that the fatigue crack began as an intergranular separation and subsequently became transgranular. The original inner-ring sleeve halves had been removed by melting the braze metal with a torch, and then new, longer sleeves were brazed to the inner ring.

Conclusions. Failure of the tube was caused by excessive alloying between the braze metal and the Waspaloy, producing a grain-boundary network of braze metal in the tube and the resultant embrittlement that rendered the penetrated region vulnerable to fatigue cracking. Fatigue stresses acting in combination with additional heating and possible overtempering of the tube during torch debrazing and subsequent rebrazing may have contributed to failure.

Recommendation. Reduction of the temperature to which the components of the joint were heated, either in torch debrazing or in rebrazing, would make the tube less susceptible to penetration by molten braze metal (liquid-metal embrittlement).

Manufactured Components and Assemblies

Failures of Shafts

Revised by Donald J. Wulpi, Metallurgical Consultant

A SHAFT is a metal bar—usually cylindrical in shape and solid, but sometimes hollow—that is used to support rotating components or to transmit power or motion by rotary or axial movement. Even fasteners, such as bolts or studs, can be considered to be stationary shafts, usually with tensile forces, but sometimes combined with bending and/or torsional forces. In addition to failures in shafts, this article will discuss failures in connecting rods, which translate rotary motion to linear motion (and conversely) and in piston rods, which translate the action of fluid power to linear motion.

Shafts operate under a broad range of service conditions, including dust-laden or corrosive atmospheres and temperatures that vary from extremely low, as in arctic or cryogenic environments, to extremely high, as in gas turbines. In addition, shafts may be subjected to a variety of loads—in general, tension, torsion, compression, bending, or combinations of these. Shafts are also sometimes subjected to vibratory stresses.

Apart from wear by bearings, which can be a major contributor to shaft failure (see the section "Wear" in this article), the most common cause of shaft failure is metal fatigue. Fatigue is a weakest link phenomenon; hence, failures start at the most vulnerable point in a dynamically stressed area—typically a stress raiser, which may be mechanical, metallurgical, or sometimes a combination of the two. Mechanical stress raisers include such features as small fillets, sharp corners, grooves, splines, keyways, nicks, and press or shrink fits. Shafts often break at edges of press-fitted or shrink-fitted members, where high degrees of stress concentration exist. Such stress concentration effectively reduces fatigue resistance, especially when coupled with fretting. Metallurgical stress raisers may be quench cracks, corrosion pits, gross nonmetallic inclusions, brittle second-phase particles, weld defects, or arc strikes.

Occasionally, brittle fractures are encountered, particularly in low-temperature environments or as a result of impact or a rapidly applied overload. Brittle fracture may thus be attributable to inappropriate choice of material because of incomplete knowledge of operating conditions and environment or failure to recognize their significance, but it may also be the result of abuse or misuse of the product under service conditions for which it was not intended.

Surface treatments can cause hydrogen to be dissolved in high-strength steels and may cause shafts to become embrittled even at room temperature. Electroplating, for instance, has caused failures of high-strength steel shafts. Baking treatments applied immediately after plating are used to ensure removal of hydrogen.

Ductile fracture of shafts is usually caused by accidental overload and is relatively rare in normal operation. Creep, a form of distortion at elevated temperatures, can lead to stress rupture and can also cause shafts having close tolerances to fail because of excessive changes in critical dimensions.

Fracture Origins

Fractures of shafts originate at points of stress concentration either inherent in design or introduced during fabrication or operation. Design features that concentrate stress include ends of keyways, edges of press-fitted members, fillets at shoulders, and edges of oil holes. Stress concentrators produced during fabrication include grinding damage, machining marks or nicks, and quench cracks resulting from heat treating operations.

Frequently, stress concentrators are introduced during hot or cold forming of shafts; these include surface discontinuities, such as laps, seams, pits and forging laps, and internal imperfections, such as bursts. Internal stress concentrators can also be introduced during solidification of ingots from which forged shafts are made. Generally, these stress concentrators are internal discontinuities, such as pipe, segregation, porosity, shrinkage, and nonmetallic inclusions.

Fractures also result from bearing misalignment that is either introduced at assembly or caused by deflection of supporting members in service, from mismatch of mating parts, and from careless handling in which the shaft is nicked, gouged, or scratched.

To a lesser degree, shafts can fracture from misapplication of material. Such fractures result from use of materials having high ductile-to-brittle transition temperatures, low resistance to hydrogen embrittlement, temper embrittle-ment, or caustic embrittlement, or chemical compositions or mechanical properties other than those specified. In some instances, fractures may originate in regions of partial or total decarburization or excessive carburization, where mechanical properties are different because of variations in chemical composition.

Examination of Failed Shafts

As with any failure, examination of a failed shaft should include gathering as much background information as possible about the shaft. This information should include design parameters, operating environment, manufacturing procedures, and service history. Detailed knowledge of these factors can often be helpful in guiding the direction of failure investigation and corrective action.

Design Parameters. The failure analyst should have copies of the detail and assembly drawings as well as the material and testing specifications that involve the shaft. Potential stress raisers or points of stress concentration, such as splines, keyways, cross holes, and changes in shaft diameter, should be noted. The type of material, mechanical properties, heat treatment, test locations, nondestructive examination used, and other processing requirements should also be noted. Special processing or finishing treatments, such as shot peening, fillet rolling, burnishing, plating, metal spraying, and painting, can influence performance, and the analyst should be aware of such treatments.

Mechanical Conditions. How a shaft is supported or assembled in its working mechanism and the relationship between the failed part and its associated members can be valuable information. The number and location of bearings or supports, as well as how their alignment may be affected by deflections or distortions that can occur as a result of mechanical loads, shock, vibrations, or thermal gradients, should be considered.

The method of connecting the driving or driven member to the shaft, such as press fitting, welding, or use of a threaded connection, a set screw, or a keyway, can influence failure. It is also important if power is transmitted to or taken from the shaft by gears, splines, belts, chains, or torque converters.

Manufacturing records may indicate when the part was made and the material supplier, heat, or lot number. Inspection records may provide the inspection history of the part and indicate any questionable areas.

Service History. Checking the service records of an assembly should reveal when the parts were installed, serviced, overhauled, and inspected. These records should also show whether service or maintenance operations were conducted in accordance with the manufacturer's recommendations. Often, talking with operators or maintenance personnel may reveal pertinent unrecorded information.

Initial Examination. Ideally, the entire assembly that was involved in the failure should be made available for examination. However, in the real world this is not always possible. Samples of oil, grease, and loose debris should be carefully removed from all components, identified, and stored for future reference. The components can then be cleaned with a solvent that will not remove or obliterate any rust, oxidation, burnishing marks, or other pertinent evidence.

Surfaces of all areas that may have been involved in or may have contributed to the failure should be examined, noting scuff marks, burnished areas, abnormal surface blemishes, and wear. These marks should be associated with some abnormal service condition, if possible. In addition, the failed part should be examined to establish the general area or location of failure, noting the proximity to any possible stress concentrations found when examining the part drawings.

Fracture surfaces should be examined visually to determine if there are indications of one or more fracture mechanisms and if there is an apparent crack origin. Surfaces of the component adjacent to the fracture surface should be examined for secondary cracks, pits, or imperfections. Photographs should be taken to record the condition of the pertinent parts before physical evidence is destroyed by subsequent examinations.

Nondestructive methods of inspection, such as ultrasonic inspection, can sometimes provide useful information. Such methods may reveal other cracks that have not progressed to rupture; these cracks may have other cracks or fracture surfaces that are not as badly damaged as the primary fracture and that can be diagnosed more readily.

Also, some machines and structures may have other shafts similar to the one that failed; these shafts may have the same service history as the failed shaft, and examination may reveal cracks that can provide useful information. Whenever possible, the analyst should also compare shafts that are from machines in service but have not failed.

Macroscopic Examination (from less than actual size to approximately 50×). Many characteristic marks on fracture surfaces, though visually identifiable, can be better distinguished with macroscopic exami-

nation. A magnifying glass or binocular microscope can be used to study the unique vestigial marks left on tensile-fracture surfaces in the form of fibrous, radial, and shear-lip zones from which the relative amounts of ductility and toughness possessed by the metal can be appraised. In the study of fatigue fractures, macroscopic examination of such features as beach marks, ratchet marks, fast-fracture zones, and crack-initiation sites, if present, may yield information relative to the kinds and magnitude of the stresses that caused failure.

Microscopic Examination (50 to 2000×.) Metallographic sections taken through fractures are used to classify fracture paths (transgranular or intergranular), to establish the mode of fracture (shear or cleavage), and to locate and identify crack-initiation sites. Plating before mounting can be used to preserve the edge or edges of the fracture surface. In addition, metallographic examination can reveal microstructure near the fracture surface, the grain size of the material, and the presence of undue segregation, inclusions, alloy concentrations, brittle grain-boundary phases, decarburization, and fabricating imperfections.

Scanning Electron Microscopy (SEM) and Transmission Electron Microscopy (TEM) Examination (10 to More Than 20 000×). The optical metallograph is limited for fracture studies by its restricted depth of field (the scanning electron microscope has a depth of field about 300 times that of the optical microscope). Thus, the electron microscope is better suited for fractographic work. With it, fractures may be classified by fracture path, fracture mechanism, and fracture features. There are two fracture paths: transgranular (or transcrystalline) and intergranular (or intercrystalline).

The fracture mechanisms and related microscopic features of transgranular fracture are microvoid coalescence (dimples, associated with ductile fracture), tearing (tear ridges), cleavage (river patterns, feather marks, Wallner lines, and cleavage tongues, all associated with brittle fracture), and fatigue (striations and tire tracks).

The fracture mechanism of intergranular fracture is grain-boundary separation. The causes may be the presence of grain-boundary phases, alloy-depleted boundaries, and environmental or mechanical factors such as stress-corrosion cracking, hydrogen damage, heat damage, and triaxial stress states.

Mechanical testing and chemical analysis occasionally pinpoint the cause of a failure as wrong material, improper heat treatment, or in-service changes in properties. Hardness testing and spectroscopic analysis should be performed as a matter of course. Impact tests, tensile tests, and other special mechanical tests may be performed if manufacturing procedures are in doubt or if other paths of investigation are not fruitful.

Stress Systems Acting on Shafts

The stress systems acting on a shaft must be clearly understood before the cause of a fracture in that shaft can be determined. Also, both ductile and brittle behavior under static loading or single overload, as well as the characteristic fracture surfaces produced by these types of behavior, must be clearly understood for proper analysis of shaft fractures.

Figure 1 shows simplified, two-dimensional, free-body diagrams illustrating the orientations of the normal-stress and shear-stress systems at any internal point in a shaft loaded in pure tension, torsion, and compression. Also, the single-overload fracture behavior of both ductile and brittle materials is illustrated for each type of load.

A free-body stress system may be considered to be a square of infinitely small dimensions. Tensile and compressive stresses act perpendicular to each other and to the sides of the square to stretch and squeeze the sides, respectively. The shear, or sliding, stresses act on the diagonals of the square, 45° to the normal stresses. The third-dimension radial stresses are ignored in this description. The effects of the shear and normal stresses on ductile and brittle materials under the three types of loads illustrated in Fig. 1 and under bending load are discussed below.

Tension. Under tension loading, the tensile stresses, σ_1, are longitudinal, whereas the compressive-stress components, σ_3, are transverse to the shaft axis. The maximum-shear-stress components, τ_{max}, are at 45° to the shaft axis (Fig. 1a).

In a ductile material, shear stresses developed by tensile loading cause considerable deformation (elongation and necking) before fracture, which originates near the center of the shaft and propagates toward the surface, ending with a conical shear lip usually about 45° to the shaft axis. However, in a brittle material, a fracture from a single tensile overload is roughly perpendicular to the direction of tensile stress, but involves little or no permanent deformation. The fracture surface is usually rough and crystalline in appearance.

The elastic-stress distribution in pure tension loading, in the absence of a stress concentration, is uniform across the section. Thus, fracture can originate at any point within the highly stressed volume.

Torsion. The stress system rotates 45° when a shaft is loaded in torsion (Fig. 1b). Both the tensile and compressive stresses are 45° to the shaft axis and remain mutually perpendicular. One shear-stress component is parallel with the shaft axis; the other is perpendicular to the shaft axis.

In a ductile material loaded to failure in torsion, shear stresses cause considerable deformation before fracture. However, this deformation is usually not obvious, because the shape of the shaft has not been changed. The distortion will be obvious if there were axial grooves

Fig. 1 Free-body diagrams showing orientation of normal stresses and shear stresses in a shaft and the single-overload fracture behavior of ductile and brittle materials

(a) Under simple tension. (b) Under torsion. (c) Under compression loading. See text for discussion.

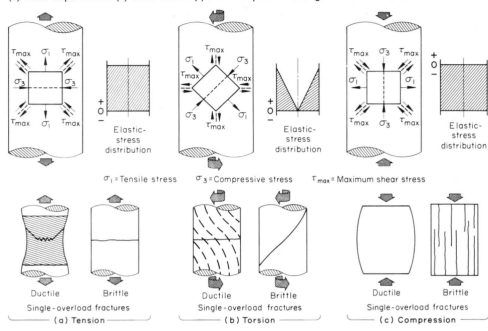

Fig. 2 Fatigue marks produced from single origins at low and high nominal stresses and from multiple origins at high nominal stresses

Fatigue marks are typical for a uniformly loaded shaft subjected to unidirectional bending. Arrows indicate crack origins; final-fracture zones are shaded.

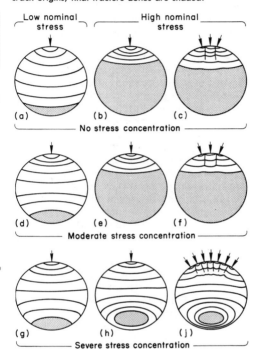

or lines on the shaft before twisting, or if the metal is hot etched to reveal grain-flow twisting. If a shaft loaded in torsion is assumed to consist of an infinite number of infinitely thin disks that slip slightly with respect to each other under the torsional stress, visualization of deformation is simplified. Torsional single-overload fracture of a ductile material usually occurs on the transverse plane, perpendicular to the axis of the shaft. In pure torsion, the final-fracture region is at the center of the shaft; the presence of slight bending will cause it to be off-center.

A brittle material in pure torsion will again fracture perpendicular to the tensile-stress component, which is now 45° to the shaft axis. The resulting fracture surfaces usually have the shape of a spiral.

The elastic-stress distribution in pure torsion is maximum at the surface and zero at the center of the shaft. Thus, in pure torsion, fracture normally originates at the surface, which is the region of highest stress.

Compression. When a shaft is loaded in axial compression (Fig. 1c), the stress system rotates so that the compressive stress, σ_3, is axial and the tensile stress, σ_1, is transverse. The shear stresses, τ_{max}, are 45° to the shaft axis, as they are during axial tension loading.

In a ductile material overloaded in compression, shear stresses cause considerable deformation but usually do not result in fracture. The shaft is shortened and bulges laterally under the influence of shear stress. A brittle material loaded in pure compression, if it does not

buckle, again will fracture perpendicular to the maximum tensile-stress component. Because the tensile stress is transverse, the direction of brittle fracture is parallel to the shaft axis.

The elastic-stress distribution in pure compression loading, in the absence of a stress concentration, is uniform across the section. If fracture occurs, it will likely be in the longitudinal direction, because compression loading increases the shaft diameter and stretches the metal at the circumference.

Bending. When a shaft is stressed in bending, the convex surface is stressed in tension and has an elastic-stress distribution similar to that shown in Fig. 1(a). The concave surface is stressed in compression and has an elastic-stress distribution similar to that shown in Fig. 1(c). Approximately midway between the convex and concave surfaces is a neutral axis, where all stresses are zero.

Fatigue Failures

Fatigue in shafts can generally be classified into three basic subdivisions: bending fatigue, torsional fatigue, and axial fatigue. Bending fatigue can result from these types of bending loads: unidirectional (one-way), reversed (two-way), and rotating. In unidirectional bending, the stress at any point fluctuates. Fluctuating stress refers to a change in magnitude without changing algebraic sign. In reversed bending and rotating bending, the stress at any point alternates. Alternating stress refers to cycling between two stresses of opposite algebraic sign,

that is, tension ($+$) to compression ($-$) or compression to tension. Torsional fatigue can result from application of a fluctuating or an alternating twisting moment (torque). Axial fatigue can result from application of alternating (tension-and-compression) loading or fluctuating (tension-tension) loading. More complete information is available in the article "Fatigue Failures" in this Volume.

Unidirectional-Bending Fatigue. The axial location of the origin of a fatigue crack in a stationary cylindrical bar or shaft subjected to a fluctuating unidirectional-bending moment evenly distributed along the length will be determined by some minor stress raiser, such as a surface discontinuity. Beach marks (also called clamshell, conchoidal, and crack-arrest marks) of the form shown in Fig. 2(a) and (b) are indicative of a fatigue crack having a single origin at the point indicated by the arrow. The crack front, which formed the beach marks, is symmetrical relative to the origin and retains a concave form throughout. Both the single origin and the smallness of the final-fracture zone in Fig. 2(a) suggest that the nominal stress was low. The larger final-fracture zone in Fig. 2(b) suggests a higher nominal stress.

Figure 2(c) shows a typical fatigue crack originating as several individual cracks that ultimately merged to form a single crack front. Such multiple origins are usually indicative of high nominal stress. Radial steps (ratchet marks) are present between crack origins.

Figures 2(d), (e), and (f) show typical fatigue beach marks that result when a change in section in a uniformly loaded shaft provides a moderate stress concentration. With a low nominal stress, the crack front changes from concave to convex before rupture (Fig. 2d). At higher nominal stresses, the crack front flattens and may not become convex before final fracture (Fig. 2e and f).

A change in section in a uniformly loaded shaft that produces a severe stress concentration will lead to a pattern of beach marks such as that shown in Fig. 2(g), (h), or (j). An example of a severe stress concentration is a small-radius fillet at the junction of a shoulder and a smaller-diameter portion of a shaft or at the bottom of a keyway. Such a fillet usually results in the contour of the fracture surface being convex with respect to the smaller-section side. The crack-front pattern shown in Fig. 2(g) was produced by a low nominal stress. The crack front in Fig. 2(h) developed more rapidly because of a higher stress in the peripheral zone. Multiple crack origins, high nominal stress, and unidirectional bending usually produce the beach-mark pattern shown in Fig. 2(j).

Example 1: Unidirectional-Bending Fatigue Failure of an A6 Tool Steel Shaft. The shaft shown in Fig. 3 was part of a clamping device on a tooling assembly used for bending 5.7-cm (2.25-in.) outer-diameter tubing on an 8.6-cm (3.375-in.) radius. The assembly contained two of these shafts, both of which failed simultaneously and were sent to the laboratory for examination. The maximum clamping force on the assembly was 54 430 kg (120 000 lb). The material specified for the shafts was a free-machining grade of A6 tool steel.

The shafts were subjected to a tensile stress imposed by the clamping force and a bending stress resulting from the nature of the operation. Unidirectional-bending stresses were imposed on one shaft when a right-hand bend was made in the tubing and on the other shaft when a left-hand bend was made. Approximately 45 right-hand and 45 left-hand bends were made per hour on the machine; the total number of bends made before the shafts failed was not known. The tensile stress on the shafts was also cyclic, because the clamping force was removed after each bend was made.

Investigation. Analysis of the steel, using wet chemical and spectroscopic techniques, showed that the composition was within specifications. The average hardness of the steel was 48 HRC. A 1.3-cm (0.505-in.) diam tensile specimen removed from the center of one of the shafts failed in a brittle manner at a tensile stress of 1572 MPa (228 ksi).

The microstructure of the steel was fine, dispersed, tempered martensite with elongated stringers of manganese sulfide. Also present

were spheroidized white particles that were identified as high-alloy complex carbides (M_6C) corresponding to the double carbides Fe_4Mo_2C and Fe_4Cr_2C. Microscopic examination of the edge of the fracture surface at 100 and 1000× revealed some nonmetallic oxide-sulfide segregation.

Visual examination of the fracture surface revealed both a smooth area and a coarse, granular area (View B, Fig. 3). The dull, smooth area is typical of some fatigue fractures and resulted as the crack was opened and closed by the bending stress. Beach marks on the smooth area of the fracture surface also indicate fatigue fracture. The coarse, bright, crystalline-appearing area is the final-fracture zone. The smooth-textured fatigue zone is relatively large compared with the crystalline-textured final-fracture zone, which indicates that the shaft was subjected to a low overstress. The final-fracture surface at bottom shows that a one-way bending load was involved.

The fatigue crack was initiated in a 0.25-mm (0.010-in.) radius fillet at a change in section ("Original design," Section A-A, Fig. 3). Cracking was nucleated by a nonmetallic inclusion that intersected the surface at a critical location in the fillet.

Conclusions. The shafts fractured in fatigue as the result of a low-overstress, high-cycle unidirectional-bending load. The small radius of the fillet at the change in section resulted in a stress concentration that, in conjunction with the oxide-sulfide inclusion that intersected the surface of the fillet, initiated a crack.

Corrective Measures. New shafts were made with a 2.4-mm (0.09-in.) radius fillet at the critical change in section ("Improved design," Section A-A, Fig. 3). The larger-radius fillet minimized stress concentration in this region and prevented recurrence of failure.

Reversed-Bending Fatigue. When the applied bending moment is reversing (alternating), all points in the shaft are subjected alternately to tension stress and compression stress; while the points on one side of the plane of bending are in tension, the points on the opposite side are in compression. If the bending moment is of the same magnitude in either direction, two cracks of approximately equal length usually develop from origins diametrically opposite each other and often in the same transverse plane. If the bending moment is greater in one direction than in the other, the two cracks will differ in length.

Figure 4 shows typical fatigue marks on the fracture surface of a stationary (nonrotating) shaft subjected to a reversing bending moment evenly distributed along its length. The crack origins (arrows) are shown diametrically opposite each other, but sometimes they are slightly displaced by minor stress raisers. The pattern shown in Fig. 4(a) is typical of that for a single-diameter shaft with no stress concentration. The bending moment is equal in both directions.

Fig. 3 A6 tool steel tube-bending-machine shaft that failed by fatigue fracture

Section A-A: Original and improved designs for fillet in failure region. Dimensions are in inches. View B: Fracture surface showing regions of fatigue-crack propagation and final fracture

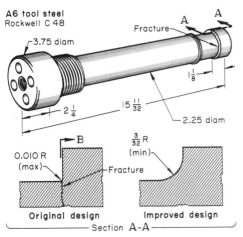

View B

Fig. 4 Typical fatigue marks on the fracture surface of a uniformly loaded nonrotating shaft subjected to reversed-bending stresses

(a) No stress concentration. (b) Moderate stress concentration. (c) Severe stress concentration. Arrows indicate crack origins; shaded areas are final-fracture zones.

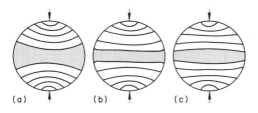

A large-radius fillet at a change in shaft diameter imposes a moderate stress concentration. Figure 4(b) shows the pattern on the surface of a fracture through such a fillet. A small-radius fillet at a change in diameter results in a severe stress concentration. Figure 4(c) shows the typical pattern on the surface of a fracture through a small-radius fillet. The reason for the fatigue-pattern changes is that the fatigue crack propagates faster in more severe stress concentrations at each end than in the interior.

Under the above loading conditions, each crack is subjected alternately to tensile and compressive stresses, with the result that the surfaces of the crack are forced into contact with one another during the compression cycle, and rubbing occurs. Rubbing may sometimes be sufficient to obliterate many of the characteristic marks, and the crack surfaces may become dull or polished.

Rotating-Bending Fatigue. The essential difference between a stationary shaft and a rotating shaft subjected to the same bending moment is that in a stationary shaft the tensile stress is confined to a portion of the periphery only. In a rotating shaft, every point on the periphery sustains a tensile stress, then a compressive stress, once every revolution. The relative magnitude of the stresses at different locations is determined by conditions of balance or imbalance imposed on the shaft.

Another important difference introduced by rotation is asymmetrical development of the crack front from a single origin. There is a marked tendency of the crack front to extend preferentially in a direction opposite to that of rotation. The crack front usually swings around about 15° or more (Fig. 5a and c). A third difference arising from rotation is in the distribution of the initiation sites of a multiple-origin crack.

Fig. 5 Typical fatigue marks on the fracture surface of a uniformly loaded rotating shaft

Marks are produced from single and multiple origins (arrows) having moderate and severe stress concentration; shaded areas are final-fracture zones. Shaft rotation is clockwise.

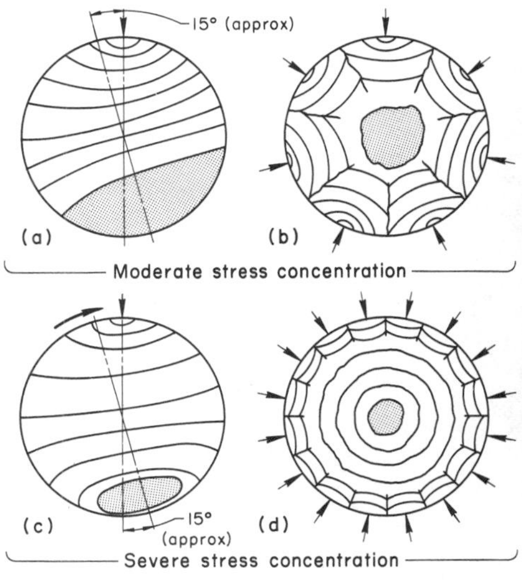

In a nonrotating shaft subjected to unidirectional bending, the origins are located in the region of the maximum-tension zone (Fig. 2). In a nonrotating shaft subjected to reversed bending, the origins are diametrically opposite each other (Fig. 4). In rotary bending, however, every point on the shaft periphery is subjected to a tensile stress at each revolution; therefore, a crack may be initiated at any point on the periphery (Fig. 5b and d).

The crack surfaces are pressed together during the compressive component of the stress cycle, and mutual rubbing occurs. A common result of final fracture is that slight movement of one side of the crack relative to the other side frequently causes severe damage to the fracture surfaces and tends to obliterate many marks. However, although the high spots on one surface may rub the high spots on the other, the marks in the depressions are retained. Because the depressions are negative images of the damaged high spots on the opposing surface, they provide useful evidence; therefore, it is desirable to examine both parts of a cracked or fractured shaft.

The similarity in macroscopic appearance of fractures in shafts resulting from rotating-bending fatigue and from single overload torsional shear of a relatively ductile metal frequently results in misinterpretation. The fracture surface shown in Fig. 6(a) was the result of fatigue, as evidenced by the ratchet marks around the periphery and the pronounced beach marks. Under the low magnification of the fractograph shown in Fig. 6(b), beach marks are not visible, because they were obliterated by rubbing. The presence of ratchet marks around the periphery is also an indication of rotating-bending fatigue. However, the metal smearing apparent on the fracture surface and the twisting deformation of the shaft shown in Fig. 6(c) indicate torsional shear and would preclude mistaking this fracture for a fatigue fracture.

The fracture shown in Fig. 6(d) also exhibits a superficial similarity to a fatigue fracture. However, it is evident that this fracture was the result of torsional shear, because the entire fracture surface has a smooth texture and no well-defined final-fracture area.

Fig. 6 Fracture surfaces of failed shafts

(a) and (b) Failure by fatigue. (c) and (d) Failure by torsional shear. See text for discussion.

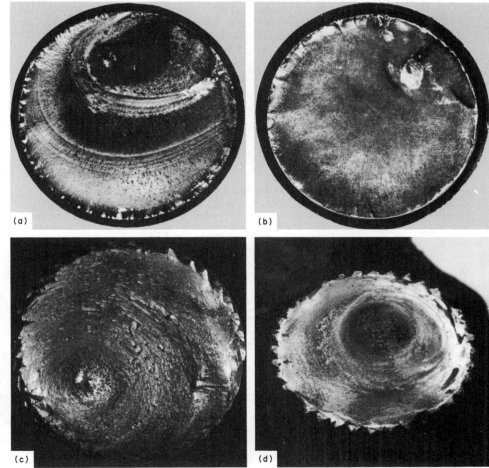

(a)

(b)

(c)

(d)

In splined shafts, fracture resulting from torsional shear is frequently accompanied by deformation of the splines not engaged by the mating part. However, the portion of the shaft not engaged by the mating part is sometimes unavailable for examination. When macroscopic examination affords only inconclusive evidence, use of an electron microscope may reveal fatigue striations or elongated shear dimples. Also, metallographic examination of a section through the fracture surface may reveal microdeformation from torsional shear in the rotary direction that would not be present in a fatigue fracture.

Torsional Fatigue. Fatigue cracks arising from torsional stresses also show beach marks and ridges. Longitudinal stress raisers are comparatively harmless under bending stresses, but are as important as circumferential stress raisers under torsional loading. This sensitivity of shafts loaded in torsion to longitudinal stress raisers is of considerable practical importance because inclusions in the shaft material are almost always parallel to the axis of rotation. It is not unusual for a torsional-fatigue crack to originate at a longitudinal inclusion, a surface mark, or a spline or keyway corner and then to branch at about 45°.

When a stress raiser such as a circumferential groove is present, different states of stresses exist around the stress raiser, and the tensile stress is increased to as much as four times the shear stress. Therefore, the tensile stress on the 45° plane will exceed the tensile strength of the steel before the shear stress reaches the shear strength of the steel. Fracture occurs normal to the 45° tensile plane, producing a conical or star-shaped fracture surface.

Example 2: Torsional-Fatigue Fracture of a Large 4340 Steel Shaft That Was Subject to Cyclic Loading and Frequent Overloads. The 4340 steel shaft shown in Fig. 7(a), which was the driving member of a large rotor subject to cyclic loading and frequent overloads, broke after three weeks of operation. The shaft was also part of a gear train that reduced the rotational speed of the driven member. The driving shaft contained a shear groove at which the shaft should break if a sudden high overload occurred, thus preventing damage to an expensive gear mechanism. The rotor was subjected to severe chatter, which was an abnormal condition resulting from a series of continuous small overloads at a frequency of about three per second.

Investigation. Examination disclosed that the shaft had broken at the shear groove and that the fracture surface contained a star-shaped pattern (Fig. 7b). Figure 7(c) shows the fracture surface with the pieces fitted back in place. The pieces were all nearly the same size and shape, and there were indications of fatigue, cleavage, and shear failure in approximately the same location on each piece. The cracks were oriented at approximately 45° to the axis of the shaft, which indicated that final fracture was caused by a tensile stress normal to the 45° plane and not by the longitudinal or transverse shear stress that had been expected to cause an overload failure.

Examination of the surfaces of one of the pieces of the broken shaft revealed small longitudinal and transverse shear cracks at the smallest diameter of the shear groove. Also, slight plastic flow had occurred in the metal adjacent to these cracks. Cracking occurred at many points in the groove in the shaft before several of the cracks grew to a critical size.

No surface irregularities were present in the shear groove at any of the shear cracks. The structure of the metal was normal, with a uniform hardness of 30 to 30.5 HRC across the section, indicating a strength in the expected range for quenched-and-tempered 4340 steel

shafts. A hot-acid etch showed the steel to be free from pipe, segregation, or other irregularities.

Conclusion. The basic failure mechanism was fracture by torsional fatigue, which started at numerous surface shear cracks, both longitudinal and transverse, that developed in the periphery of the root of the shear groove. These shear cracks resulted from high peak loads caused by chatter. Stress concentrations developed in the regions of maximum shear, and fatigue cracks propagated in a direction perpendicular to the maximum tensile stress, thus forming the star pattern at 45° to the longitudinal axis of the shaft. The shear groove in the shaft, designed to prevent damage to the gear train, had performed its function, but at a lower overload level than intended.

Corrective Measures. The fatigue strength of the shaft was increased by shot peening the shear groove, and chatter in the machine was minimized.

The relative extent of development of two torsional-fatigue cracks mutually at right angles can indicate the magnitude of the torque reversals that have been applied. If the cracks are of approximately the same length, the indications are that the torque reversals have been of equal magnitude, but only if the cracks are in a comparatively early stage of development. Beyond this stage, one crack usually takes the lead, and such inferences are no longer justified. If the shaft transmits a unidirectional torque but two cracks develop mutually at right angles, it can be presumed that the torque was of a reversing character. If a bending stress is applied to a shaft that is transmitting torque, the angle at which any fatigue crack develops will be modified. Therefore, if the angle differs significantly from 45° to the shaft axis, the presence of a bending stress is indicated.

Fig. 7 4340 steel rotor shaft that failed by torsional fatigue

(a) Shear groove designed to protect gear mechanism from sudden overload. Dimensions are in inches. (b) Star-shaped pattern on a fracture surface of the shaft. (c) Longitudinal and transverse shear cracks on the surface of the shear groove, which resulted from high peak loads caused by chatter

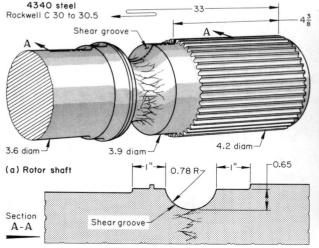

Contact Fatigue

Contact fatigue occurs when components roll, or roll and slide, against each other under high contact pressure and cyclic loading. Pitting occurs after many repetitions of loading and is the result of metal fatigue from the imposed cyclic contact stresses. Factors that govern contact fatigue are contact stress, relative rolling/sliding, material properties, and metallurgical, physical, and chemical characteristics of the contacting surfaces, including the oil film that lubricates the surfaces.

The significant stress in rolling-contact fatigue is the maximum alternating shear stress that undergoes a reversal in direction during rolling. In pure rolling, as in antifriction bearings, this stress occurs slightly below the surface and can lead to the initiation of subsurface fatigue cracks. As these cracks propagate under the repeated loads, they reach the surface and produce cavities, or pits.

When sliding is superimposed on rolling, as in gear teeth, the tangential forces and thermal gradient caused by friction alter the magnitude and distribution of stresses in and below the contact area. The alternating shear stress is increased in magnitude and is moved nearer to the surface by friction resulting from the sliding action. Additional information is provided in the articles ''Fatigue Failures'' and ''Failures of Rolling-Element Bearings'' in this Volume.

Wear

Wear of metal parts is commonly classified into either of two categories: abrasive wear or adhesive wear.

Abrasive wear, the undesired removal of material by a cutting mechanism, can reduce the size and destroy the proper shape of a shaft. The shaft may then fail by another means, such as by fracture, or may cease to perform its designed function. Foreign particles, such as sand, dirt and other debris, in the lubricant can cause wear of a shaft.

Example 3: Wear Failure of a Fuel-Pump Drive Shaft Caused by the Presence of Sand, Metallic Particles, and Vibration. The fuel pump in a turbine-powered aircraft failed, resulting in damage to the aircraft. The pump is shown in Fig. 8(a) and (b).

This particular model of fuel pump had a history of wear failures and had been redesigned to incorporate a shaft of case-hardened steel (composition not reported). Vibration was common during operation, but generally was not excessive for the aircraft.

Investigation. The pump and the filter chamber were found to be dry and free of any debris or contamination, except for some accumulated deposit on the filter cartridge, when the pump was disassembled in the laboratory. The drive-shaft splines that engaged the impeller were almost completely destroyed down to the roots (Detail A, Fig. 8). Extensive damage to the splines was apparent because on subsequent reassembly the shaft could be rotated without rotating the impeller.

The pressure side of each spline tooth in the impeller also exhibited some damage. Relatively smooth cavities and undercutting of the flank on the pressure side of the spline teeth indicated that the damage had not been caused by wear from metallic contact between the

splines but by an erosion or abrasion mechanism.

Hardness readings taken at several axial locations on the drive shaft showed a reasonably uniform hardness of approximately 570 HV. The impeller and the retaining ring each had a hardness of approximately 780 HV, and the parts surrounding the impeller (including the vanes) exhibited a hardness of approximately 630 HV. The microstructure of the metal in the impeller exhibited scattered porosity and carbide particles and appeared to be a sintered powder metallurgy compact.

Metallographic examination of a section through the damaged splines and of a section through the adjacent undamaged part of the same splines disclosed no material defects. The microstructure indicated that the shaft had been satisfactorily heat treated by quenching and tempering, but there was no evidence of case hardening on the spline surfaces, a treatment commonly given to shafts of this type.

The worn surfaces of the splines showed evidence of cold work at the edges. Also, there was a relatively smooth worn area at the center of each tooth that appeared to be free of cold work and that appeared to have been caused by an abrasive action. The damaged side of each spline appeared as an undulating outline with some undercutting rather than a jagged or deformed shape.

The residue on the filter cartridge was brown, and when viewed under a low-power microscope, particles of sand, paint, or plastic, fibers from the cartridge, brass, and steel could be identified. Application of a magnet to the sample showed that it contained a large amount of iron.

Fig. 8 Fuel pump that failed by vibration and abrasion

(a) Configuration and dimensions (given in inches). (b) Splines on the drive shaft and in the impeller were worn away by vibration in the presence of sand and metallic particles. Detail A: Enlarged view of failure area showing worn splines

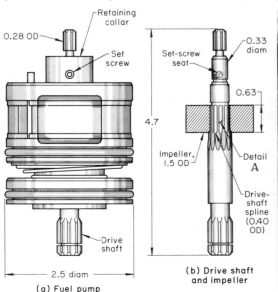

(a) Fuel pump

(b) Drive shaft and impeller

Detail A

Chemical analysis indicated that the deposit contained about 20% sand, 30% iron, and 30% organic material (paint, plastic, and filter fibers). The reddish-brown color of the deposit suggested that some of the iron present was an oxide (rust), but this was not confirmed.

Discussion. Vibration in the fuel pump could be expected to initiate damage, particularly when combined with an abrasive action. Under these conditions, fretting or abrasive wear can be expected on sliding-contact surfaces that are not sufficiently abrasion resistant. The residue from the filter contained significant quantities of sand and iron; the iron probably originated from the damaged shaft.

The examination indicated that the splines on the drive shaft had been damaged by abrasion, which could have been caused by the combined effect of vibration and abrasives, such as sand and the metal particles removed from the splines. The same action would also damage the internal splines of the impeller, but to a lesser extent because of its greater hardness and thus greater abrasion resistance. However, the internal splines exhibited significant damage, which appeared to have been produced by erosion. Thus, the internal splines suffered slight abrasive damage and somewhat more damage by erosion.

Conclusions. Failure of the shaft was the result of excessive wear on the splines caused by vibration and the abrasive action of sand and metal particles.

Recommendations. To increase resistance to wear and abrasion, the surfaces of the spline teeth should be case hardened. Although the drive shaft exhibited reasonably high strength, its resistance to wear and abrasion was inadequate for the conditions to which it was exposed.

Wire Wooling. Abrasive-wear failure of shafts by wire wooling has been observed under certain circumstances where contact occurs between the shaft and a stationary part, resulting in removal, by machining, of fine wire shavings that resemble steel wool. This type of failure has been found on turbine and turbine-generator shafts made of 3Cr-0.5Mo steels, on 12% Cr stainless steels, on 18Cr-8Ni stainless steels, and on nonchromium steels in the presence of certain chloride-containing oils. Wire wooling has also been observed on thrust bearings and on centrifugal compressor shafts.

Figure 9 shows the end of a 13.3-cm (5.25-in.) diam shaft that was worn by wire wooling. The worn surface was in contact with a labyrinth seal and was not a bearing surface. The shaft had operated for about 3½ years before the damage was discovered. Although records were not available, the machine was known to have been opened at least once during that time, but damage under the seal had not been noticed. Figure 9(b) shows a micrograph of a section through the material found in the circumferential grooves in the shaft. The material was loose in the grooves; some was almost identical in

Fig. 9 Shaft that was severely damaged by wire wooling

(a) End of shaft. (b) Micrograph of an unetched section through loose material found in grooves worn in shaft by wire wooling. 50×

appearance with steel wool, whereas other pieces were coarser and more like slivers.

Although the mechanisms of wire wooling are not clearly understood, it is known that the process requires contact between a shaft (or shaft sleeve) and a labyrinth (or bearing), either directly or through buildup of deposits. If the deposit or the stationary part contains hard particles, fine slivers can be cut or spun off the shaft surface. As fine slivers or pieces come off, additional hard particles may be formed by reaction between the steel and the oil or gas present if the resultant friction and heat are sufficient. Scabs or solid chunks of laminated or compacted slivers and other deposits are sometimes formed (Fig. 9b). Bearings or labyrinths of babbitt or copper alloys have been associated with this type of failure. Contact-area atmospheres, in addition to bearing oils, have contained air and methyl chloride.

Methods for prevention of wire wooling include changing the shaft material, using a softer bearing or labyrinth material, changing to a different oil, eliminating the deposits, and providing greater clearance.

Adhesive Wear of Shafts. Adhesive wear, sometimes called scoring, scuffing, galling, or seizing, is the result of microscopic welding at the interface between two mutually soluble metals, such as steel on steel. It frequently occurs on shafts where there is movement between the shaft and a mating part, such as a gear, wheel, or pulley.

Fretting on a shaft can be a source of serious damage. Production of a reddish-brown powder is characteristic of fretting on steel. Typical

locations for fretting are at splined or keyed hubs, components that are press fitted or shrink fitted to a shaft, and clamped joints. Shot peening, glass-bead peening, and surface rolling are methods of reducing the possibility of fatigue fracture of shafts because of fretting of joints. Fretting is discussed in the article "Wear Failures" in this Volume.

Adhesive wear has a characteristic torn appearance because the surfaces actually weld together, then are torn apart by continued motion, creating a series of fractures on both surfaces. This indicates that metal-to-metal contact took place between clean, uncontaminated mating surfaces.

Because excessive frictional heat is generated, adhesive wear often can be identified by a change in the microstructure of the metal. For example, steel may be tempered or rehardened locally by the frictional heat generated. Additional information is provided in the article "Wear Failures" in this Volume.

Influence of Bearings and Seals. If acceptable wear resistance is not obtained by optimizing selection of shaft material and heat treatment, consideration should be given to the sliding (sleeve) bearing and seal material and its compatibility with the shaft. Very often, the high-wear area of a shaft may be chromium plated or may be covered with a sleeve that can be discarded and replaced when worn (rather than throwing away an expensive shaft or replacing the bearings). These methods involve risks, and steps should be taken to avoid or offset their effects.

Brittle Fracture of Shafts

Brittle fractures are associated with the inability of certain materials to deform plastically in the presence of stress at the root of a sharp notch, particularly at low temperatures. Brittle fractures are characterized by sudden fracturing at extremely high rates of crack propagation, perhaps 1830 m/s (6000 ft/s) or more, with little evidence of distortion in the region of fracture initiation. This type of fracture is frequently characterized by marks known as herringbone or chevron patterns on the fracture surface. The chevrons point toward the origin of the fracture. Additional information on brittle fracture can be found in the article "Ductile and Brittle Fractures" in this Volume.

Ductile Fracture of Shafts

Ductile fractures, which result from microvoid coalescence, exhibit evidence of distortion (plastic flow) at the fracture surface similar to that observed in ordinary tensile-test or torsion-test specimens. When a shaft is fractured by a single application of a load greater than the strength of the shaft, there is usually considerable plastic deformation before fracture. This deformation is often readily apparent upon visual inspection of a shaft that fractured in tension, but is often not obvious when the shaft fractured in torsion. This ability of a material to deform plastically (permanently) is a property known as ductility. The appearance of the fracture surface of a shaft that failed in a ductile manner is also a function of shaft shape, the type of stress to which the shaft was subjected, rate of loading, and, for many alloys, temperature. In general, ductility is decreased by increasing the strength of the metal by cold work or heat treatment, by the presence of notches, fillets, holes, scratches, inclusions, and porosity in a notch-sensitive material, by increasing the rate of loading, and for many alloys, by decreasing the temperature.

Ductile fracture of shafts occurs infrequently in normal service. However, ductile fractures may occur if service requirements are underestimated, if the materials used are not as strong as had been assumed, or if the shaft is subjected to a massive single overload, such as in an accident. Fabricating errors, such as using the wrong material or using material in the wrong heat treated condition (for example, annealed instead of quenched and tempered), can result in ductile fractures.

Distortion of Shafts

Distortion of a shaft can render the shaft incapable of serving its intended function.

Permanent distortion simply means that the applied stress has exceeded the yield strength (but not the tensile strength) of the material. If it is not feasible to modify the design of the shaft, the yield strength of the shaft material must be increased to withstand the applied stress. Yield strength may be increased either by using a stronger material or by heat treating the original material to a higher strength.

Creep, by definition, is time-dependent strain (distortion) occurring under stress imposed at elevated temperature, provided the operational load does not exceed the yield strength of the metal. If creep continues until fracture occurs, the part is said to have failed by stress rupture. Creep can result from any type of loading (tensile, torsion, compression, bending, and so on).

Some high-temperature applications, such as gas turbines and jet aircraft engines, require materials to operate under extreme conditions of temperature and stress with only a limited amount of deformation by creep. In other high-temperature applications, the permissible deformation is high and may not even be limited as long as rupture does not occur during the intended life of the part. For this type of service, stress-rupture data, rather than long-term creep data, are used for design.

Buckling, a third type of distortion failure, results from compressive instability. It can occur if a long slender rod or shaft collapses from compressive axial forces. The load required to cause buckling can be changed only by design changes, not by metallurgical changes, such as heat treatment, in a given type of metal.

Corrosion of Shafts

Most shafts are not subjected to severe reduction in life from general corrosion or chemical attack. Corrosion may occur as general surface pitting, may uniformly remove metal from the surface, or may uniformly cover the surface with scale or other corrosion products. Corrosion pits have a relatively minor effect on the load-carrying capacity of a shaft, but they do act as points of stress concentration at which fatigue cracks can originate.

A corrosive environment will greatly accelerate metal fatigue; even exposure of a metal to air results in a shorter fatigue life than that obtained under vacuum. Steel shafts exposed to salt water may fail prematurely by fatigue despite periodic, thorough cleaning. Aerated salt solutions usually attack metal surfaces at the weakest points, such as scratches, cut edges, and points of high strain. To minimize corrosion fatigue, it is necessary to select a material that is resistant to corrosion in the service environment or to provide the shaft with a protective coating.

Most large shafts and piston rods are not subject to corrosion attack. However, because ship-propeller shafts are exposed to salt water, they are pressure rolled, which produces residual surface-compressive stresses and inhibits origination of fatigue cracks at corrosion pits. Also, rotating parts, such as centrifugal compressor impellers and gas-turbine disks and blades, often corrode. Centrifugal compressors frequently handle gases that contain moisture and small amounts of a corrosive gas or liquid. If corrosion attack occurs, a scale is often formed that may be left intact and increased by more corrosion, eroded off by entrained liquids (or solids), or thrown off from the rotating shaft.

Stress-corrosion cracking occurs as a result of corrosion and stress at the tip of a growing crack. Stress-corrosion cracking is often accompanied or preceded by surface pitting; however, general corrosion is often absent, and rapid, overall corrosion does not accompany stress-corrosion cracking.

The tensile-stress level necessary for stress-corrosion cracking is below the stress level required for fracture without corrosion. The critical stress may be well below the yield strength of the material, depending on the material and the corrosive conditions. Evidence of corrosion, although not always easy to find, should be present on the surface of a stress-corrosion-cracking fracture up to the start of final rupture.

All of the common materials used in shafts may undergo stress-corrosion cracking under certain specific conditions. Factors that influence stress-corrosion cracking, either directly or indirectly, include microstructure, yield strength, hardness, corrodent(s), concentration of corrodent(s), amounts and nature of water, pH, and applied and residual stresses, degree of cold working, and chemical composition of the base metal. Additional information is available in the article "Stress-Corrosion Cracking" in this Volume.

Corrosion fatigue results when corrosion and an alternating stress—neither of which is severe enough to cause failure by itself—occur simultaneously; this can cause failure. Once such a condition exists, shaft life will probably be greatly reduced. Corrosion-fatigue cracking is usually transgranular; branching of the main cracks occurs, although usually not as much as in stress-corrosion cracking. Corrosion products are generally present in the cracks, both at the tips and in regions nearer the origins. The article "Corrosion-Fatigue Failures" in this Volume contains more detailed information on the effect of combined corrosion and fluctuating stress.

Common Stress Raisers in Shafts

Most service failures in shafts are largely attributable to some condition that intensifies stress. In local regions, the stress value is raised above a value at which the material is capable of withstanding the number of loading cycles that corresponds to a satisfactory service life. Only one small area needs to be repeatedly stressed above the fatigue strength of the material for a crack to be initiated. An apparently insignificant imperfection, such as a small surface irregularity, may severely reduce the fatigue strength of a shaft if the stress level at the imperfection is high. The most vulnerable zone in torsional and bending fatigue is the shaft surface; an abrupt change in surface configuration may have a damaging effect, depending on the orientation of the discontinuity to the direction of stress.

All but the simplest shafts contain oil holes, keyways, or changes in shaft diameter (threads, fillets, annular grooves, and so on). The transition from one diameter to another, the location and finish of an oil hole, and the type and shape of a keyway exert a marked influence on the magnitude of the resulting stress-concentration and fatigue-notch factors, which often range in numerical value from 1 to 5 and sometimes attain values of 10 or higher.

Types of Stress Raisers. The majority of stress raisers can be placed into one of the following general groups:

- *Group 1*: Nonuniformities in the shape of the shaft, such as steps at changes in diameter, broad integral collars, holes, abrupt corners, keyways, grooves, threads, splines, and press-fitted or shrink-fitted attachments
- *Group 2*: Surface discontinuities arising from fabrication practices or service damage, such as seams, nicks, notches, machining marks, identification marks, forging laps and seams, pitting, and corrosion
- *Group 3*: Internal discontinuities, such as porosity, shrinkage, gross nonmetallic inclusions, cracks, and voids

Most shaft failures are initiated at primary (group 1) stress raisers, but secondary (group 2 or 3) stress raisers may contribute to a failure. For example, a change in shaft diameter can result in stress intensification at the transition zone; if there is a surface irregularity or other discontinuity in this zone, the stress is sharply increased around the discontinuity.

Influence of Changes in Shaft Diameter

A change in shaft diameter concentrates the stresses at the change in diameter and in the smaller-diameter portion. The effects of an abrupt change and three gradual changes in section on stress concentration are shown schematically in Fig. 10. The sharp corner at the intersection of the shoulder and shaft in Fig. 10(a) concentrates the stresses at the corner as they pass from the large to the small diameter. The large-radius fillet shown in Fig. 10(d) permits the stresses to flow with a minimum of restriction. However, the fillet must be tangent with the smaller-diameter section, or a sharp intersection will result, overcoming the beneficial effect of the large-radius fillet.

Example 4: Fatigue Fracture of a 6150 Steel Main Shaft in a Coal Pulverizer. Unusual noises were noted by the operator of a ball-and-race coal pulverizer (shown schematically in Fig. 11a), and the unit was taken out of service to investigate the cause. The pulverizer had been in service for ten years.

The lower grinding ring of the pulverizer was attached to the outer main shaft (Fig. 11, left side). The upper grinding ring was suspended by springs from a spider that was attached to the main shaft. The weight of the upper grinding ring and the load imposed by the springs resulted in a total force of 8165 kg (18 000 lb) on the pulverizer balls. The shaft was made of 6150 steel normalized to a hardness of about 285 HB.

Investigation. Visual examination of the shaft revealed a circumferential crack in the main shaft just below the upper radial bearing at an abrupt change in shaft diameter. The shaft was set up in a lathe to machine out the crack for repair welding, and the smaller end (17.8 cm, or 7 in., in diameter) was found to be slightly eccentric with the remainder of the shaft. The crack did not disappear after a 6.3-mm (0.25-in.) deep cut, and the crack was opened by striking the small end of the shaft. The shaft broke about 1.3 cm (0.5 in.) from crack. Examination of the fracture surface (Fig. 11, right side) revealed a previous fracture, almost perpendicular to the axis of the

Fig. 10 Effect of size of fillet radius on stress concentration at a change in shaft diameter

See text for discussion.

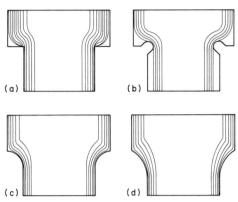

shaft, that resulted from torsional loading acting along a plane of maximum shear.

Although the shaft operated at the relatively low speed of 82 rpm, the weight of the upper portion of the shaft, the spider, and the upper ring caused the shaft to repair itself by friction welding in a band about 2.5 cm (1 in.) wide beginning at the periphery of the shaft. Thus,

Fig. 11 6150 steel coal pulverizer shaft that failed by fatigue

(Left) Section through pulverizer showing the inner main shaft that fractured, repaired itself by friction welding, and fractured a second time. (Right) Photograph of the friction welded surface

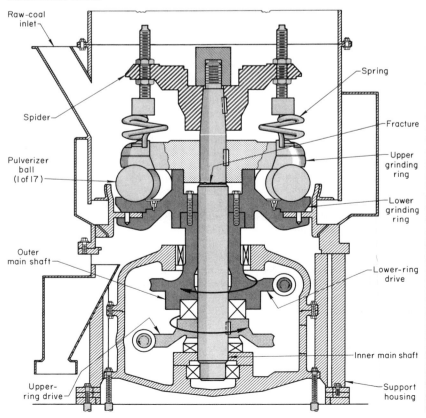

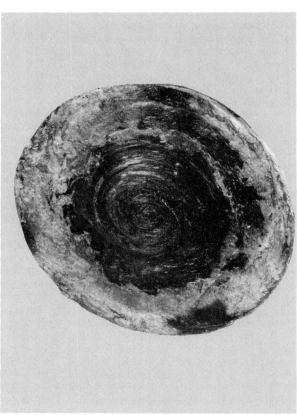

friction welding was confined to a small part of the 17.8-cm (7-in.) diam cross section. The center region of the fracture surface of the shaft (Fig. 11, right side) contained swirls of metal softened by frictional heat as a result of rotation of the lower part of the shaft. Welding of the plasticized outer ring of metal on the shaft and solidification of the swirls probably occurred while the pulverizer was shut down.

Examination of the machined surface revealed faint lines parallel to the visible crack. These faint lines generally followed bands containing large grains. The lines were thought to be fatigue cracks. The hardness of the shaft in the friction welded area was 38 to 49 HRC, and inward from the weld area, the hardness was 31 to 33 HRC.

Conclusion. The second fracture of the shaft was by fatigue, which resulted from an eccentric condition after the shaft was friction welded and from the inherent vibrations within the machine. Shafts in other coal pulverizers at the same plant fractured in the same region after similar service lives.

Corrective Measures. The shaft was repaired by welding a new 20.3-cm (8-in.) diam section to the older, longer section and machining to the required diameters. The fatigue cracks in the longer section were machined away before repair welding. Liquid-penetrant inspection was performed to ensure soundness of the metal in the older section. The repaired shaft was machined to provide a taper between the two diameters, rather than reproduce the sharp change in diameter in the original design. The taper section was carefully blended with the smaller diameter so that there were no sharp corners. The repaired coal-pulverizer shaft operated satisfactorily for more than five years. As shafts in other units failed, they were repaired by the same technique.

Example 5: Bending-Fatigue Fracture of an Inertia-Welded Alloy Steel Pushrod Originating at a Sharp Internal Corner and a Decarburized Surface. The pushrod shown in Fig. 12 fractured two weeks after it was installed in a mud pump operating at 160 rpm with a discharge pressure of 14.4 MPa (2.1 ksi). The pushrod was made by joining two pieces of bar stock by inertia welding. Each piece was rough bored or drilled to produce a wall thickness of about 1.3 cm (0.5 in.) at the surfaces to be inertia welded. The flange portion was made of 94B17 steel, and the shaft was made of 8620 steel. After welding, the rod was machined, carburized, and hardened to 60 HRC; the shaft was then chromium plated to within 1.3 cm (0.5 in.) of the flange.

Investigation. Visual examination of the pushrod disclosed that fracture had occurred in the shaft portion at the intersection of the 1.3-cm (0.5-in.) thick wall and the tapered surface at the bottom of the hole. This intersection was about 1.3 cm (0.5 in.) from the weld.

The fracture surface contained indications that a fatigue crack had initiated at the inner surface and progressed around the entire inner circumference, forming a narrow plateau, before propagating outward to final fracture. Operation of the pushrod after it had fractured obliterated some detail of the fatigue pattern on the fracture surface, but it still was evident that one-way bending stresses had influenced cracking. The amount of coarse texture on the fracture surface indicated that fatigue cracking had occurred for a short time.

Detail A in Fig. 12 shows a macrograph of a section through the weld and fracture region. A heavily decarburized layer was found on the inner surface of the flange portion, but was not detectable in the shaft portion. The inner surface had a poor finish, and there was a sharp corner at the intersection of the sidewall and bottom of the hole. The weld flash protruded from the inner wall, producing a small-radius fillet at the wall surface. Except for the metal extrusion, the weld zone was almost undetectable.

A metallographic specimen taken from near the fracture surface showed that the microstructure of the metal at the inner surface contained acicular ferrite and bainite, indicative of decarburization (Fig. 12b) The microstructure of the metal approximately 1.6 mm (0.0625 in.) from the inner surface contained low-carbon martensite and bainite (Fig. 12c).

Hardness of the flange was equivalent to 27 HRC at the bottom of the hole and 37 HRC on the inner-wall surface. The shaft had a hardness equivalent to 22 HRC at the bottom of the hole and 27 HRC at the inner-wall surface.

Fig. 12 Pushrod that fractured in bending fatigue after being fabricated by inertia welding

(a) Configuration and dimensions (given in inches). (b) and (c) Micrographs showing structure of decarburized inner surface and sound metal below the decarburized layer

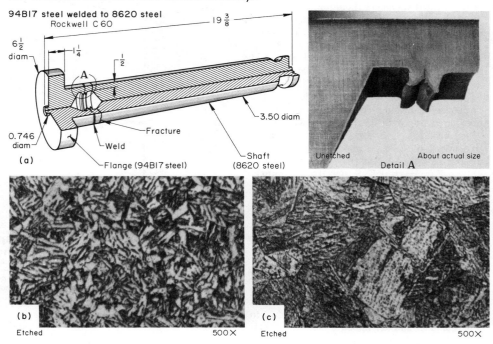

Sample parts having shapes and dimensions similar to those of the fractured pushrod were joined by inertia welding, then sectioned so that the weld area and inner surfaces could be examined. The sample weldments were not heat treated or machined. Sharp corners and rough surfaces were found that were similar to those on the fractured part (Detail A, Fig. 12). The cut surface was ground smooth and etched, which revealed the heat-affected zone of the weld but no indications of abnormal oxidation or decarburization of the inner surfaces.

Conclusion. The pushrod failed in fatigue primarily from one-way bending stresses. Cracking was initiated in a sharp, rough corner at the intersection of the sidewall and the bottom of a drilled hole that acted as a stress raiser. The stress raiser created by the abrupt section change was accentuated by the roughness of the drilled surface and by the decarburized layer. Decarburization occurred during heat treatment and was caused by the atmosphere trapped in the cavity formed by friction welding.

Corrective Measures. The design of the pushrod was changed to a one-piece forging. To produce a high-quality welded pushrod, it would be necessary to provide radii at the corners of the drilled holes, finish machine the inner surface, and drill a hole in the flange end of the rod to permit circulation of atmosphere during heat treatment.

Press-Fitted Members. Gears, pulleys, wheels, impellers, and similar components are often assembled on shafts by means of press fitting or shrink fitting, which can result in stress raisers under bending stress. Figure 13(a) shows typical stress flow lines in a plain shaft at a press-fitted member. Enlarging the shaft at the press-fitted component and using a large-radius fillet would produce a stress distribution such as that shown schematically in Fig. 13(b). A small-radius fillet at the shoulder would result in a stress pattern similar to that shown in Fig. 10(a).

An investigation was conducted to study the influence of absolute specimen dimensions on the fatigue strength of specimens with press-fitted bushings. The tests were carried out on specimens of St 35 carbon steel [0.18% C (max), 0.40% Mn (min), 0.35% Si (max), 0.05% P (max), 0.05% S (max)] with ultimate tensile strength of 559 MPa (81 ksi) and high-quality chromium-nickel-molybdenum alloy steel with ultimate tensile strength of 726 MPa (105 ksi) (similar to AISI 9840). Specimen blanks were cut from the surfaces of large forgings that were to be used for propeller shafts. Two groups of specimens were tested: (1) smooth cylindrical specimens 5, 12, 27, and 50 mm (0.2, 0.5, 1.1, and 2 in.) in diameter, with gage portions equal in length to four specimen diameters and shoulder radii equal to one diameter and (2) smooth cylindrical specimens 5, 12, 27, and 50 mm (0.2, 0.5, 1.1, and 2 in.) in diameter with press-fitted bushings. The outside diameter and length of each bushing were equal to two specimen diameters; thus, the bushings were similar in size to actual machine parts of this kind. The bushings were made of normalized (195 HB) St 34 steel [0.15% C (max), 0.08% P (max), and 0.06% S (max)] and a brass alloy of unreported composition. Press fitting was done to class 2 tolerances.

The specimens were tested in rotating bending at a constant frequency of 3000 rpm. The test base (in air) was 10^7 cycles for specimens without bushings and 5×10^7 cycles for specimens with bushings. The results (Table 1) showed that the fatigue limit of steel specimens was sharply reduced by the presence of press-fitted bushings and that the magnitude of this reduction increased with steel strength.

The endurance of specimens with press-fitted bushings depends on specimen diameter to a greater extent than does the endurance of specimens without bushings. The effect of specimen diameter on the fatigue strength of both plain carbon steel and alloy steel specimens with press-fitted bushings was more pronounced when the bushings were made of steel than when they were made of brass.

The main cause of the reduction in the fatigue strength of the specimens with bushings was friction between the bushing and the shaft, which resulted from the cyclic loads in rotating bending, and the resulting wear of the shaft. The wear, which depends on the frictional force, that is, on the distribution of specific pressure, should vary (as did the pressure)

Fig. 13 Schematic illustration of stress distribution in two types of rotating shafts with press-fitted elements under a bending load

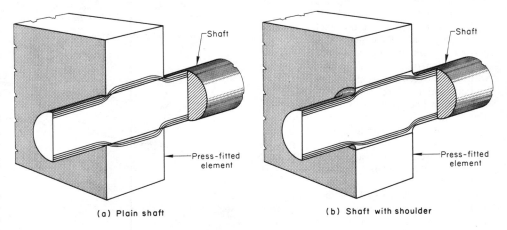

(a) Plain shaft   (b) Shaft with shoulder

Table 1 Fatigue strength of steel specimens without bushings, with steel bushings, and with brass bushings

| Specimen diameter | | St 35 carbon steel | | | | | | Chromium-nickel-molybdenum alloy steel | | | | | |
| | | Without bushings | | Steel bushings | | Brass bushings | | Without bushings | | Steel bushings | | Brass bushings | |
mm	in.	MPa	ksi	MPa	ksi	MPa	ksi	MPa	ksi	MPa	ksi	MPa	ksi
5	0.2	265	38	157	23	147	21	319	46	196	28	147	21
12	0.5	255	37	157	23	142	20.6	314	45.5	167	24	152	22
27	1.1	221	32	123	18(a)	123	18	294	43	123	18	147	21
50	2.0	201	29	103	15	108	16	294	43	103	15	123	18

(a) 20-mm (0.8-in.) diam specimen

along the shaft surface. When a shaft with a press-fitted bushing is bent, the maximum pressure should be near the bushing edge, at a point whose location will vary depending on the fit.

Friction produces wear of the shaft, leads to wear-induced surface roughness, and causes a local temperature increase, all of which promote nucleation and growth of cracks. Friction also leads to the destruction of oxide films, which passivate and strengthen metal parts, and activates metal surfaces as a result of plastic deformation. These factors facilitate chemical reaction between the metal surfaces and the working environment; the resulting damage is known as fretting.

The fatigue strength of specimens up to 20 mm (0.8 in.) in diameter was reduced more by the brass bushings than by the steel bushings. Conversely, the fatigue strength of specimens greater than 20 mm (0.8 in.) in diameter was reduced less by the brass bushings than by the steel bushings. The steel bushings, which are harder and more rigid than brass bushings, exhibited a higher friction coefficient when in contact with steel and produced more intense wear than the brass bushings. Other factors being equal, the environment had a more damaging effect on small-diameter specimens; the fatigue strength of large-diameter specimens was influenced predominantly by stress raisers produced as a result of friction-induced wear.

Longitudinal grooves in shafts, such as keyways and splines, are the origins of many service failures of shafts subjected to torsional stress. Most of these failures result from fatigue fracture where a small crack has initiated at a sharp corner because of stress concentrations. The crack gradually enlarges as cycles of service stress are repeated until the remaining section breaks. A sharp corner in a keyway can cause the local stress to be as much as ten times the average nominal stress.

Failures of this kind can be avoided by using a half-round keyway, which permits the use of a round key, or by using a generous fillet radius in the keyway. Good results are obtained by the use of fillets having radii equal to approximately one-half the depth of the keyway. A half-round keyway produces a local stress of only twice the average stress, thus providing greater load-carrying ability than that permitted by a square keyway. Many shafts with square keyways do not fracture in service because stresses are low or because fillets with generous radii are used.

Shafts with tapered, or conical, ends may also have keyways for alignment and/or redundancy. Integrity of the tapered joint relies on the friction fit between the wedgelike taper and the tapered hole in the hub of the part fitting to the shaft. It is essential to keep the fit very tight by using a threaded end and a locknut or other

Fig. 14 Peeling-type cracks in shafts

(a) and (b) Cracks originated at keyways.

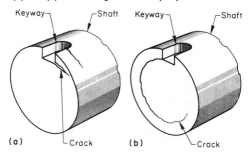

(a)   (b)

Fig. 15 Stress fields and corresponding torsional-fatigue cracks

(a) and (b) Shaft with keyway. (c) Shaft with splines

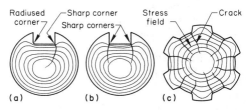

(a)   (b)   (c)

Fig. 16 4340 steel compressor shaft that fractured because of peeling-type fatigue cracking

(a) Configuration and dimensions (given in inches). (b) and (c) Photographs showing crack path, failure origin, and beach marks on fracture surfaces of shaft and key

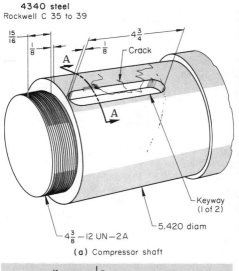

(a) Compressor shaft

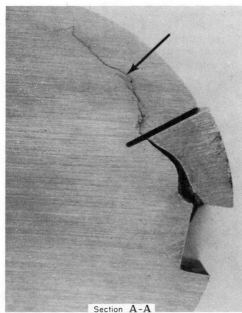

Section A-A

(b)

(c)

mechanism to avoid relying on the key and keyway to transmit the torque.

In an assembly in which a keyed member was loosely fitted to a shaft, nearly all the alternating torque was transmitted through the key, resulting in initiation of cracks at the bottom edge of the keyway of the shaft and producing a peeling type of fracture (Fig. 14a). Occasionally, peeling progresses entirely around the shaft and results in a sharp-edged shell like that shown in Fig. 14(b).

Figure 15 shows stress fields and corresponding torsional-fatigue cracks in shafts with keyways or splines. In Fig. 15(a), the fillet in one corner of the keyway was radiused and the fillet in the other corner was sharp, which resulted in a single crack; this crack progressed approximately normal to the original stress field. In Fig. 15(b), both fillets in the keyway corners were sharp, which resulted in two cracks that did not follow the original stress field, a condition that results from the cross effect of cracks on the stress field.

A splined shaft subjected to alternating torsion can crack along the bottom edges of the splines (Fig. 15c). This is another instance of a highly localized stress field strongly influencing crack development.

Example 6: Fatigue Cracking of 4340 Steel Compressor Shafts Because of Cyclic Stresses Induced by Gear-Type Couplings. A plant utilized high-horsepower electric motors to drive large compressors required for a manufacturing process. Eight compressors had shafts made of 4340 steel

quenched and tempered to a hardness of 35 to 39 HRC and gear-type couplings. The compressors operated for short periods at power levels up to 3730 MW (5000 hp) and for longer periods at slightly above 2980 MW (4000 hp). Figure 16(a) shows the design of the shaft and the keyways. Two keys, spaced 180° apart, were expected to transmit the load so that an interference fit between shaft and coupling would not be required, thus simplifying maintenance. After only a few months of operation at above 2980 MW (4000 hp), six of the eight compressor shafts were found to be cracked in a keyway, and one of the six had fractured.

Investigation. Visual examination of the fractured shaft disclosed cracks propagating from one of the keyways. Figure 16(a) shows the cracks that originated in the keyway corner and propagated to the surface. Examination of a cross section of a shaft revealed that the crack propagated from the keyway corner circumferentially around the shaft but remained near the surface (arrow, Section A-A, Fig. 16). This type of cracking occurs when there is slippage between a shaft and a coupling.

Fretting was found on the shaft surface both in and near the keyways. Upset keys were found, indicating that slippage had occurred. Fretting was noted on the nonloaded side of the keys, which suggested that appreciable chattering had occurred. The fretting greatly reduced the fatigue limit of the shaft metal and resulted in initiation of fatigue cracks.

Examination of the fractured shaft revealed beach marks around the 3.2-mm (0.125-in.) radius keyway corner (arrow, Fig. 16b). These beach marks indicated that fatigue cracking had initiated at the corner radius and propagated across the shaft. As shown in Fig. 16(c), similar fatigue marks were observed on the key that fractured, radiating from one corner (arrow). Propagation of fatigue cracks was the result of repetitive impact loading.

Chemical analysis confirmed that the shafts were made of 4340 steel. Hardness traverses indicated that the shaft metal was within an acceptable hardness range.

High cyclic bending stresses were caused by misalignment between the electric motor and compressor and were transmitted to the shaft through the geared coupling. Measurements on shafts using a flexible disk-type coupling indicated cyclic bending stresses of less than 7 MPa (1 ksi).

Conclusions. The shafts failed by fatigue cracking that initiated at the corner radii of the keyways. The cyclic stresses that caused fretting and fatigue cracking in the keyways were the result of bending stresses produced by motor-to-compressor misalignment and were transmitted to the shaft through the gear-type coupling.

Corrective Measures. Flexible-disk couplings capable of transmitting the required horsepower were installed on the shafts. After three years of operation at 2235 to 2980 MW (3000 to 4000 hp) with the new couplings, no failures of shafts or keys had been reported.

Size and mass of shafts have considerable influence on residual stresses, load distribution, and mechanical properties. Mechanical properties and residual stresses in quenched-and-tempered shafts are greatly influenced by both mass and diameter, which affect the cooling rate during heat treatment. The difference between surface and internal properties becomes greater where the metal does not harden throughout the section.

Influence of Fabricating Practices

Surface discontinuities produced during manufacture, repair, or assembly (into a machine) of a shaft can become points of stress concentration and thus contribute to shaft failure. Operations or conditions that produce this type of stress raiser include:

- Manufacturing operations that introduce stress raisers, such as tool marks and scratches
- Manufacturing operations that introduce high tensile stresses in the surface, such as

improper grinding, repair welding, electromachining, and arc burns
- Processes that introduce metal weakening, such as forging flow lines that are not parallel with the surface, hydrogen embrittlement from plating, or decarburization from heat treatment

Fatigue strength may be increased by imparting high compressive residual stresses to the surface of the shaft. This can be accomplished with such processes as surface rolling or burnishing, shot peening, tumbling, coining, or induction hardening.

Improper Machining. There are many ways in which improper machining can lead to shaft failures, and unless they are recognized, correction of service-failure problems can be difficult. The metal on the surface of a machined part can be cold worked and highly stressed to an appreciable depth (approximately 0.5 to 0.8 mm, or 0.020 to 0.030 in.). Occasionally, the heat generated in machining, particularly in grinding, is sufficient to heat a thin layer of the steel above the transformation temperature and thus cause martensitic hardening at the surface upon cooling. Stresses resulting from thermal expansion and contraction of the locally heated metal may even be great enough to cause cracking of the hardened surface layer (grinding cracks). Rough-machining operations can produce surface cracks and sharp corners, which concentrate stresses.

Example 7: Fatigue Fracture of a 4140 Steel Forged Crankshaft Resulting From Stress Raisers Created During Hot Trimming. Textile-machine crankshafts like that shown in Fig. 17 were usually forged from 1035 steel, but because of service conditions, the material was changed to 4140 steel. The forgings were made from 5.5-cm (2.15625-in.) diam bar stock by cutting to length, hot bending, upsetting, hot-trimming flash, hot pressing, and visually inspecting before shipping.

The crankshafts were failing by transverse fracture of one cheek after one to three years of service. The expected life was 20 years of continuous service. One complete forging that had fractured (No. 1 in this Example) and a section containing the fractured cheek on the shorter shaft of another forging (No. 2) were sent to the laboratory so that the cause of failure could be determined.

Investigation. Visual examination of the fracture surfaces of both crankshafts revealed indications of fatigue failure; however, the origins were not readily visible (Fig. 17b). The surfaces had a clean fine-grain structure, but the edges were peened—evidently the result of damage after fracture.

The surfaces of the cheeks at the parting line contained rough grooves from hot trimming of the flash and from snag grinding (Fig. 17c and d). Forging 2 contained the most severe of such markings.

Longitudinal and transverse sections were prepared and etched with hydrochloric acid at a

temperature of 71 to 77 °C (160 to 170 °F). The steel was of good quality and contained the normal amount of nonmetallic inclusions but no segregation or pipe.

Examination of an as-polished specimen revealed no intergranular oxidation. Etching the specimen with a 10% sulfuric acid and 10% nitric acid solution revealed no evidence of burning or overheating of the steel.

An area containing shallow surface folds was found on the outer face of one cheek of the throw on forging 1 (Fig. 17e). As shown in Fig. 17(f), the metal around one of the folds contained some ferrite, and the forged surface was slightly decarburized. Also, a fatigue crack had initiated in the fold and was propagating across the cheek. Examination of a section through the fracture surface disclosed cold working of the surface, which could have been the result of a rather extended period of fatigue cracking.

Chemical analysis of the two forgings found the metal to be 4140 steel, as specified. The hardness of the two forgings at the subsurface, midradius, and core ranged from 19 to 22 HRC. The specified hardness of the machined forging was unknown.

Tensile strength of the forgings ranged from 790 to 817 MPa (114.5 to 118.5 ksi), and yield strength at 0.2% offset was 580 to 620 MPa (84 to 90 ksi). Elongation in 3.8 cm (1.5 in.) was 20.6 to 22%, and reduction of area was 56.2 to 59.4%. These properties were representative of 4140 steel quenched and tempered to a hardness of 20 to 22 HRC. The general microstructure of the forgings was tempered bainite; the grain size was ASTM 6 to 8.

Conclusions. Fatigue cracking resulted in the transverse fracture of one cheek in each of the two crankshafts submitted for examination. In crankshaft 1, fatigue cracks were initiated at a shallow hot-work defect. A rough surface resulting from hot trimming or snag grinding of the forging flash was the point of initiation of fatigue cracks in crankshaft 2.

Corrective Measures. Before being machined, the forgings were normalized, hardened and tempered to 28 to 32 HRC to increase fatigue strength. The quenching procedure was changed to produce a more complete martensite transformation and to increase the ratio of yield strength to tensile strength. The surfaces were inspected by the magnetic-particle method, and shallow folds, notches, or extremely rough surfaces were removed by careful grinding.

Identification Marks. Excessive stresses may be introduced in shafts by stamped identification marks that indicate manufacturing date or lot number, steel heat number, size, or part number. The location of such a mark and the method by which it is made can be important. Identification marks should not be placed in areas of high bending or torsional stresses. For shafts, this often requires that they be located either on the end face or on an adjoining collar, but surface-finish requirements on the end of the shaft cannot be ignored if thrust loads are

Fig. 17 Forged 4140 steel textile-machine crankshaft that fractured in fatigue originating at machining marks and forging defects

(a) Configuration and dimensions (given in inches). (b) Fracture surface. (c) Hot trim marks. (d) Snag grinding marks. (e) Hot folds. (f) Section through a hot fold. Etched with 2% nital. 100×

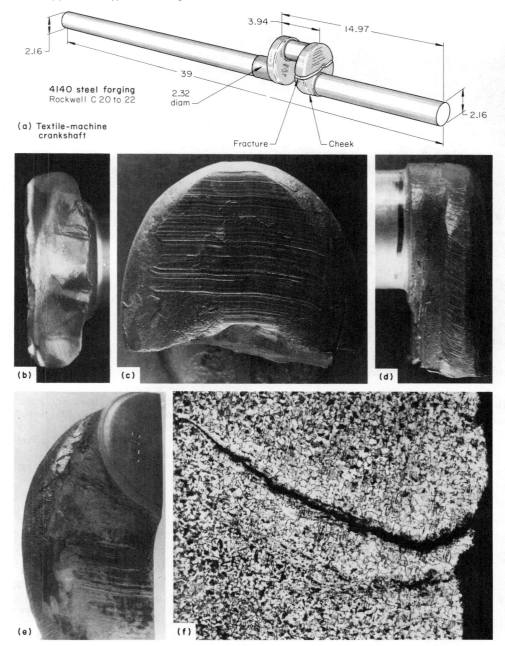

Fig. 18 Cracks in a shaft radiating from deep identification marks made with steel stamps

Cracking occurred during heat treatment of the shaft.

cleated at an electroetched numeral 5 on one of the flange surfaces (arrow B, Fig. 19a). Around the nucleus was a series of concentric beach marks (arrows A and C, Fig. 19a) that extended almost the full width of the flange and about one-half the width of the web.

Metallographic examination of a polished-and-etched section through the fracture origin revealed a notch (arrow E, Fig. 19b) that was caused by arc erosion during electroetching. The microstructure of the metal at the origin consisted of a remelted zone (arrow D, Fig. 19b) and a layer of untempered martensite 0.4 mm (0.015 in.) deep (arrow F, Fig. 19b). Small cracks were observed in the remelt area and were the result of thermal stresses caused by the high temperatures developed during electro-etching.

The hardness of the untempered martensite was 56 HRC, and the hardness of the tempered-martensite core was 40 HRC. The hardness of the rod (40 HRC) and the microstructure of the tempered martensite remote from the electroetched area indicated that the material had been properly heat treated.

Conclusions. Fatigue fracture of the rod was caused by a metallurgical notch that resulted from electroetching of an identification number on the flange.

Corrective Measures. Marking of the articulated rods by electroetching was discontinued because of the detrimental effect of electroetching such highly stressed parts.

Residual Surface Stresses. Most machining operations produce notches and may also cause residual tensile stresses on the surface of the workpiece. Grinding can cause local high-temperature heating followed by very rapid cooling as the surface of the workpiece leaves the grinding-wheel-contact area. Grinding cracks can result, and these cracks can provide points of fatigue-crack initiation. However, grinding under properly controlled conditions provides a surface with a minimal amount of residual stress.

Grinding cracks are sometimes visible in oblique light, but are often so tight that they are impossible to see. They can be detected readily by magnetic-particle inspection, by fluorescent

taken at that location. Stamping of marks with straight-line portions is the most likely to cause cracks, although characters with rounded contours also can cause cracking (Fig. 18). Stamping of metal shaft surfaces should be avoided because it is impossible to predict on which surface stamp marks will cause cracking in service (more information is provided in the article "Fatigue Failures" in this Volume.)

Example 8: Fatigue Fracture of a 4337 Steel Articulated Rod Originating at an

Electroetched Numeral. The articulated rod in an aircraft engine fractured after being in operation for 138 h since the engine had been overhauled; total operating time was unknown. The rod was made from a 4337 steel (AMS 6412) forging and was quenched and tempered to 36 to 40 HRC.

Investigation. Visual examination disclosed that the rod had broken into two pieces, approximately 6.4 cm (2.5 in.) from the center of the piston-pin-bushing bore. The fracture had nu-

Fig. 19 Fracture surface of an articulated rod made from a 4337 steel forging

(a) Fracture origin (arrow B) at an electroetched numeral, and beach marks (arrows A and C). (b) Micrograph through fracture origin showing remelted zone (arrow D), notch produced by electroetching (arrow E), and untempered martensite (arrow F). Etched with 2% nital. 100×

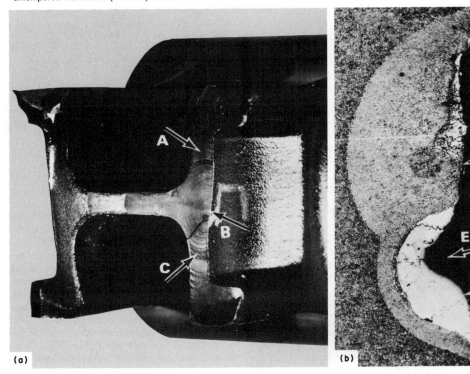

(a)

(b)

Fig. 20 A2 tool steel mandrel for a tube-expanding tool

Fracture originated at a 6.3-mm (0.25-in.) diam hole in the square end that was drilled by EDM. The fractograph shows a crack pattern on the fracture surface that originated at the hole.

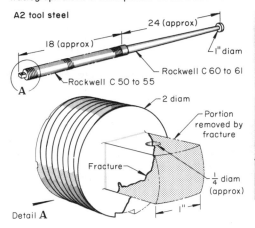

Light fractograph   About 1¼×
Fracture surface of portion removed

liquid-penetrant inspection, or by etching in cold dilute nitric acid. If grinding cracks are not detected before the shaft is put in service, the cracks may enlarge such that the shaft can fail by fatigue or brittle fracture. Grinding cracks often show a characteristic pattern; light grinding cracks occur as parallel cracks at 90° to the direction of grinding, whereas heavy grinding cracks have a rectangular network pattern.

Residual surface stresses also are produced by electrical discharge machining (EDM), which is metallurgically analogous to torch (flame) cutting but on a much smaller scale. The cutting effect is produced by a succession of sparks between the electrode and the workpiece. Each spark heats a small volume of metal to a temperature well above its melting point. Most of the molten metal is removed from the part by the action of the spark and the surround-ing liquid dielectric, but a thin white surface layer remains on the workpiece and resolidifies as it is quenched by the surrounding mass of metal. The thickness of the white (untempered martensite) layer varies from 0.005 mm (0.0002 in.) on finishing cuts to 0.08 mm (0.003 in.) on roughing cuts. This untempered martensite will be very hard and brittle, possibly cracking and leading to fracture.

Beneath the quenched surface layer is another layer of steel that is heated during EDM; thus, if the workpiece was originally in the hardened condition, the heated layer is tempered. The lowest tempered hardness is adjacent to the quenched layer, and the hardness gradually rises in subsequent layers until the original base hardness is reached. The tempered layer is similar metallurgically to the heat-affected zone in welding and is impossible to avoid. Electrical discharge machining often produces an as-quenched martensitic layer of base metal that was not molten, but only heated above the critical temperature to a depth about the same as the white layer. Frequently this layer cannot be distinguished as separate from the remelted white layer. The tempered zone, when hardened steel is cut by EDM, varies in depth from about 0.05 mm (0.002 in.) for finishing cuts up to 0.8 mm (0.030 in.) for roughing cuts.

Example 9: Fatigue Fracture of a Rolling-Tool Mandrel Initiated at Cracks Formed by Machining of a Hole. The mandrel shown in Fig. 20 was part of a rolling tool used for mechanically joining two tubes before they were installed in a nuclear reactor. The operation consisted of expanding the end of a zirconium tube into a stainless steel cylinder having an inside diameter slightly larger than the outside diameter of the zirconium tube.

Difficulty was experienced in withdrawing the tool, and a 6.4-mm (0.25-in.) diam hole was drilled by EDM through the square end of the hardened mandrel. The mandrel, which fractured after making five rolled joints, was made of A2 tool steel. The tapered end was hardened to 60 to 61 HRC, and the remainder of the mandrel to 50 to 55 HRC.

Investigation. Fracture had occurred at approximately 45° through the 6.4-mm (0.25-in.) diam hole in the square end and progressed into the threaded section to form a pyramid-shape fragment (Detail A, Fig. 20). The fracture surfaces exhibited brittle fracture characteristics but with clearly defined beach marks. The fracture pattern was characteristic of a torsional fatigue fracture.

The fracture originated on both sides and near the top of the hole in the square end, as shown in the fractograph in Fig. 20 (the poor surface finish of the hole is visible). Examination at approximately 10× revealed that the rough surface was the result of the metal having been melted.

Metallographic examination (at 250×) of specimens taken through the fracture origin revealed that melting had occurred around the

Fig. 21 Diesel-engine crankshaft that broke because of misalignment

(a) Fatigue marks on the fracture surface. (b) Micrograph of a section through the fracture origin showing a small crack (arrow) and some inclusions. Etched with nital. 500×

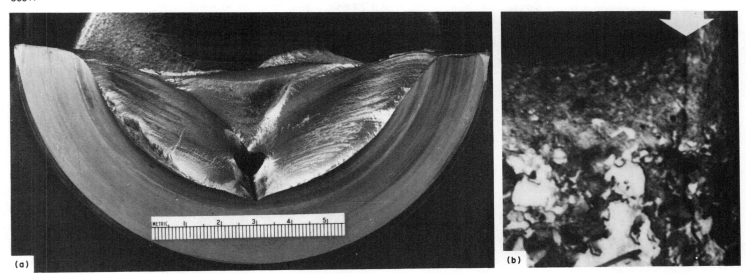

(a)　　(b)

hole, resulting in an irregular zone of untempered martensite with cracks radiating from the surface of the hole. The core material away from the hole exhibited a microstructure of fine tempered martensite containing some carbide particles. The martensitic zones around the hole had a hardness of 68 to 70 HRC. The core structure had a hardness of 60 to 61 HRC.

Conclusion. Failure of the mandrel was the result of torsional fatigue initiated by cracks formed by the EDM process used to drill the hole in the square end. Propagation of the fatigue crack was accelerated by the hardness of the material, which was considered exceptionally high for this application.

Corrective Measures. The hole through the square end of the mandrel was incorporated into the design of the tool and was drilled and reamed before heat treatment. The specified hardness of the threaded portion and square end of the mandrel was changed to 45 to 50 HRC, from the original hardness of 50 to 55 HRC. The specified hardness of the tapered end remained at 60 to 61 HRC.

Damage During Assembly. Operations performed during assembly can sometimes introduce misalignment of components, which can be detrimental to the performance of shafts. Proper alignment of bearings, seals, couplings, pedestals, and foundations has an influence on vibration and bending loads and thus on the fatigue performance of shafts.

Misalignment or even a change in alignment within tolerance limits in conjunction with a stress raiser, such as a deep scratch or a large nonmetallic inclusion, may initiate a fatigue crack. Figure 21(a) shows the fracture surface of a crankshaft for a diesel engine that had been operating for several years. The engine was overhauled, and the bearing journals were

Fig. 22 Shaft assembly in which the height of the pillow-block bearing caused misalignment of the extension shaft with the drive shaft, resulting in bending-fatigue fracture

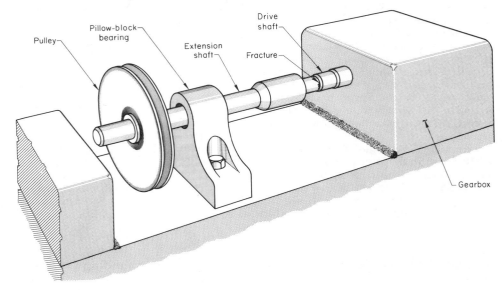

chromium plated. The crankshaft failed after the engine had operated only a few hundred hours.

No scratches or tool marks were found at the fracture origin. Metallographic examination of a section through the fracture surface revealed a very fine crack parallel to the fracture surface and a line of inclusions (Fig. 21b).

Misalignment of the extension shaft with the drive shaft of the assembly shown in Fig. 22 resulted in bending-fatigue fracture of the shaft at a change in shaft diameter. The drive shaft protruded from a gearbox that was

mounted rigidly on the base of the machine, but the extension shaft was supported in a pillow-block bearing. Tightening the pillow block deflected the extension shaft downward, imparting significant bending stress to the drive shaft.

Heat Treatment. The mechanical properties of the material in most shafts are developed through heat treatment or cold drawing. For quenched-and-tempered shafts less than 7.6 cm (3 in.) in diameter, a predominantly martensitic microstructure tempered to a surface hardness of 235 HB or higher is typical.

Fig. 23 8640 steel shaft for a fuel-injection-pump governor that fractured by fatigue through a lubrication hole

Fatigue life of the shaft was increased by nitriding the critical surface. Dimensions given in inches

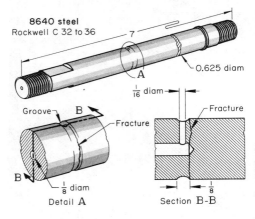

Where the properties of the shaft material throughout its entire cross section are important, use of the lowest austenitizing temperature consistent with the strength required is usually good practice. The higher the fracture toughness of a shaft (high K_{Ic}), the larger the crack needed to cause fracture.

For most shafts, where surface stresses are high and crack initiation is of concern, surface decarburization can be a major problem. For very large shafts, rotational speeds are often high enough that internal centrifugal stresses are of major concern. For such shafts, heat treatments are designed to give optimum strength and toughness uniformly across the full section.

Example 10: Fatigue Fracture of an 8640 Steel Shaft From a Fuel-Injection-Pump Governor Because of Insufficient Fatigue Strength.

The shaft shown in Fig. 23 was from a fuel-injection-pump governor that controlled the speed of a diesel engine used in trucks and tractors. Shafts in newly installed governors began breaking after only a few days of operation. Specifications required the shaft to be made of 8640 steel heat treated to a hardness of 32 to 36 HRC.

The shaft had a cross hole and a groove that were part of a force-feed lubricating system for a sleeve that moved longitudinally on the shaft to control the amount of fuel delivered to the engine. The shaft rotated at relatively high speed and was subjected to shock loading; therefore, the mechanism that drove the shaft included a slip clutch designed to eliminate transmission of shock to the governor shaft.

Investigation. Visual examination of the broken shaft showed that fracture had initiated in the sharp corner at the bottom of the 3.2-mm (0.125-in.) diam longitudinal hole (Section B-B, Fig. 23). Beach marks were observed on the fracture surfaces. The shafts had been made of the prescribed metal and heat treated as specified.

Further investigation disclosed that, in an effort to reduce costs, the slip clutch had been eliminated from the drive mechanism, thus removing the cushioning effect that had been provided for the governor shaft. Restoration of the original design in a short time was not feasible.

Shafts were taken from stock and fatigue tested in a rotating-beam machine in a manner that would concentrate the stress at the circumferential groove. By this procedure, the fatigue limit for the shaft (survival for 10^7 cycles) was found to be about 480 MPa (70 ksi). This fatigue limit was insufficient for the application because shafts were fracturing after operating a few hours to a few days.

To improve the fatigue life of the shafts, finished shafts were taken from stock and nitrided for 10 h at 515 °C (960 °F). The nitrided shafts were then fatigue tested. Results of the tests indicated that the nitriding treatment had increased the fatigue limit from a minimum of 480 MPa (70 ksi) to a minimum of 760 MPa (110 ksi). Tests showed this strength to be satisfactory for the application.

Conclusions. Fatigue fracture of the shaft was the result of increased vibration and shock loading after the slip clutch had been eliminated from the drive mechanism. The increased vibration produced stresses that exceeded the fatigue strength of the metal in the shaft.

Corrective Measures. The shafts were nitrided, and the portion of the surface covered by the sleeve was lightly buffed after nitriding. No further changes were made in the manufacture of these parts. No failures occurred during several months of operation, and the use of nitrided shafts became standard practice. Nitriding the shafts was more economical than restoring the slip clutch.

Example 11: Fracture of a 1040 Steel Fan Shaft Resulting From Use of an Improper Material.

The fan drive support shaft shown in Fig. 24 fractured after 3600 km (2240 miles) of service (minimum expected life, 6440 km, or 4000 miles). Specifications required that the shaft be made of cold-drawn 1040 to 1045 steel with a minimum yield strength of 586 MPa (85 ksi). The fractured shaft was sent to the laboratory to determine if failure had resulted from inherent stress raisers in the design or from nonconformance to specifications.

Investigation. Visual examination of the shaft revealed that the fracture had been initiated near the fillet at an abrupt change in shaft diameter. Examination of the fracture surface disclosed that cracks had originated at two locations approximately 180° apart on the outer surface of the shaft (arrows, Fig. 24b) and propagated toward the center. Final fracture occurred near the center of the shaft. The fracture surface exhibited features typical of reversed-bending fatigue (Fig. 24b). Analysis of the material in the shaft showed that it conformed with the composition of 1040 steel.

The mechanical properties of a tensile specimen machined from the center of the shaft were: tensile strength, 631 MPa (91.5 ksi); yield strength, 369 MPa (53.5 ksi); and elongation, 27%. The yield strength was much

Fig. 24 1040 steel fan shaft that fractured in reversed-bending fatigue

(a) Overall view of shaft. Dimensions given in inches. (b) Fracture surface showing diametrically opposed origins (arrows)

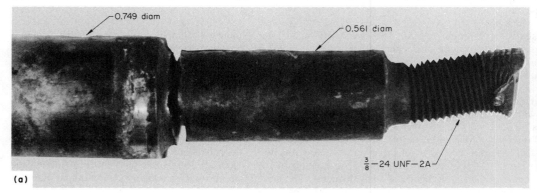

lower than the specified minimum of 586 MPa (85 ksi). Hardness of the shaft was 179 HB.

Metallographic examination disclosed that the microstructure was predominantly equiaxed ferrite and pearlite, indicating that the material was in either the hot-worked or normalized condition. The grain size was ASTM 6 to 7, which is a fine-grain structure. No nonmetallic stringers or segregation were visible. Severe corrosion, which probably occurred after fracture, was found on the fracture surfaces of the shaft.

Conclusions. The shaft failed in reversed-bending fatigue. Fracture was initiated near a fillet at an abrupt change in shaft diameter, which is considered a region of maximum stress. The metal did not conform to specifications; it was either hot rolled or normalized and had a yield strength of only 369 MPa (53.5 ksi).

Recommendations. The development of a quenched-and-tempered microstructure would result in an improvement of the fatigue strength of the shaft. A 40% increase in the fatigue limit could be effected by quenching and tempering the steel to a hardness of 30 to 37 HRC (286 to 344 HB) after machining. Shot peening the fillet after machining would provide residual compressive stresses at the surface, which would inhibit the formation of fatigue cracks and increase the fatigue limit.

Influence of Metallurgical Factors

The fatigue properties of a material depend primarily on microstructure, inclusion content, hardness, tensile strength, distribution of residual stresses, and severity of the stress concentrators that are present.

Internal discontinuities, such as porosity, large inclusions, laminations, forging bursts, flakes, and centerline pipe, will act as stress concentrators under certain conditions and may originate fatigue fracture. To understand the effect of discontinuities, it is necessary to realize that fracture can originate at any location—surface or interior—where the stress first exceeds material strength. The stress gradient must be considered in torsion and bending because the stress is maximum at the surface, but is zero at the center or neutral axis. In tension, however, the stress is essentially uniform across the section.

If discontinuities such as those noted above occur in a region highly stressed in tension by bending or torsional loading, fatigue cracking may be initiated. However, if the discontinuities are in a low-stress region, such as near a neutral axis, they will be less harmful. Similarly, a shaft stressed by repeated high tensile loading must be free from serious imperfections, for there is no neutral axis; any imperfection can be a stress concentrator and can be the origin of fatigue cracking if the stress is high with respect to the strength.

Example 12: Fatigue Cracking of a Forged 4337 Steel Master Connecting

Fig. 25 Forged 4337 steel master connecting rod for a reciprocating aircraft engine that failed by fatigue cracking in the bore section between the flanges

(a) Configuration and dimensions (given in inches). (b) Fractograph showing inclusions (arrows) and fatigue beach marks

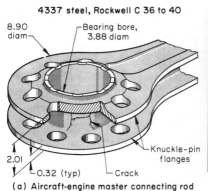

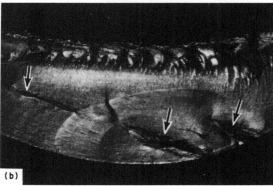

Rod Because of Nonmetallic Inclusions. Routine inspection of a reciprocating aircraft engine revealed cracks in the master connecting rod. Cracks were observed in the channel-shaped section consisting of the knuckle-pin flanges and the bearing-bore wall. The rods were forged from 4337 (AMS 6412) steel and heat treated to a specified hardness of 36 to 40 HRC.

Investigation. Visual examination revealed H-shaped cracks in the wall between the knuckle-pin flanges (Fig. 25a). The cracks originated as circumferential cracks, then propagated transversely into the bearing-bore wall. Magnetic-particle and x-ray inspection before sectioning did not detect any inclusions in the master rod.

Macroscopic examination of one of the fracture surfaces revealed three large inclusions lying approximately parallel to the grain direction and fatigue beach marks around two of the inclusions. The inclusions and beach marks are shown in Fig. 25(b).

Microscopic examination of a section through the fracture origin showed large nonmetallic inclusions that consisted of heavy concentrations of aluminum oxide (Al_2O_3). These inclusions were of the type generally associated with ingot segregation patterns.

The hardness of the rods, 36 to 40 HRC, and the microstructure of the heat-treated alloy steel were satisfactory for the application.

A preliminary stress analysis indicated that the stresses in the area of cracking, under normal operating conditions, were relatively low compared with other areas of the rod, such as in the shank and the knuckle-pin straps.

Conclusions. The rod failed in fatigue in the bore wall between the knuckle-pin flanges. Fatigue was initiated by the stress-raising effect of large nonmetallic inclusions. The nonmetallic inclusions were not detected by routine magnetic-particle or x-ray inspection because of their orientation.

Recommendations. The forging vendors were notified that nonmetallic inclusions of a size in excess of that expected in aircraft-quality steel were found in the master connecting rods. Forging techniques that provided increased working of the material between the knuckle-pin flanges to break up the large nonmetallic inclusions were not successful. A nondestructive-testing procedure for detection of large nonmetallic inclusions was established.

Example 13: Fatigue Cracking of a 1040 Steel Crankshaft Because of Excessive Segregation of Nonmetallic Inclusions. The crankshaft in a reciprocating engine had been in operation for less than one year when the engine was shut down for repairs. Examination of the engine components disclosed that the crankshaft had suffered fairly severe cracking. The crankshaft was sent to the laboratory for a complete examination.

Investigation. The journals of the main and crankpin bearings were inspected by the magnetic-particle method. At least four of the main-bearing journals had three to six indications of discontinuities 1.5 to 9.5 mm (0.0625 to 0.375 in.) long. Another main-bearing journal had approximately 20 similar but generally shorter indications. One main-bearing journal had a crack along the fillet and almost entirely through the web (Fig. 26). Another main-bearing journal had a 1.3-cm (0.5-in.) long crack in the fillet. Several of the crankpin journals had cracks 10 to 12.7 cm (4 to 5 in.) long, primarily in the fillets.

A metallographic section was taken through the No. 4 main-bearing journal at the primary crack. The surface was macroetched, which disclosed numerous large, coarse segregates identified as sulfide inclusions (Fig. 26).

Macroscopic examination of the crack surface disclosed indications of fatigue cracking with low-stress high-cycle characteristics. Sul-

Fig. 26 Forged 1040 steel main-bearing journal that failed in fatigue

Top: Section showing cracks originating at coarse sulfide inclusions. Dimensions given in inches. Bottom: Macrograph of a 5%-nital-etched section showing the segregated inclusions (dark areas). 4×

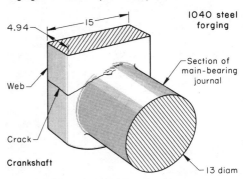

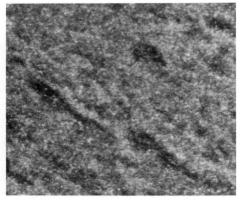

fide inclusions were present in the region where cracking originated.

The crankshaft was made from a 1040 steel forging. The mechanical properties of the metal met the specifications.

Conclusions. The crankshaft failed in fatigue at the main-bearing and crankpin-bearing journals because of excessive segregation of sulfide inclusions, which acted as stress raisers at which fatigue cracks initiated.

Corrective Measures. Ultrasonic inspection was used in addition to magnetic-particle inspection to detect discontinuities.

Surface Discontinuities. During primary and secondary mill operations, a variety of surface imperfections often result from hot plastic working of material when lapping, folding, or turbulent metal flow is experienced. The resultant surface discontinuities are called laps, seams, and cold shuts. Similar discontinuities also are produced in cold-working operations, such as fillet and thread rolling. Other surface imperfections develop from embedding of foreign material under high pressures during the working process; for example, oxides, slivers, or chips of the base material are occasionally rolled or forged into the surface.

Most of these discontinuities are present in the metal before final processing and are open to the surface. Standard nondestructive testing procedures, such as liquid-penetrant and magnetic-particle inspection, will readily reveal the presence of most surface discontinuities. If not detected, discontinuities may serve as sites for corrosion or crack initiation during fabrication, in addition to their deleterious effect on fatigue strength.

Because fatigue-crack initiation is the controlling factor in the life of most small shafts, freedom from surface imperfections becomes progressively more important in more severe applications. Similarly, internal imperfections, especially those that are near the surface, will cause fatigue cracks to grow under cyclic loading and result in failure when critical size is attained. Service life can be significantly shortened when such imperfections cause premature crack initiation in shafts designed on the basis of conventional fatigue-life considerations. Surface or subsurface imperfections can cause brittle fracture of a shaft after a very short service life when the shaft is operating below the ductile-to-brittle transition temperature. When the operating temperature is above the transition temperature or when the imperfection is small relative to the critical flaw size, especially when the cyclic-loading stress range is not large, service life may not be affected.

Example 14: Brittle Fracture of Splines on Induction-Hardened 1151 Steel Rotor Shafts Caused by a Seam in the Material.

Splined rotor shafts like that shown in Fig. 27 were used on small electric motors. It was found that one spline was missing from each of several shafts before the motors were put into service. The shafts were made of 1151 steel, and the surfaces of the splines were induction hardened to 58 to 62 HRC. Several shafts were sent to the laboratory for examination to determine the cause of failure.

Visual examination. The shafts were examined both visually and with the aid of a stereoscopic microscope. This examination revealed apparent peeling of splines on the induction-hardened end of each rotor shaft. Only one tooth on each shaft had broken off. Examination of the fracture surfaces revealed that they were dark in color, which indicated that cracking had occurred before or during oil quenching after induction heating.

The shafts were inspected by the fluorescent liquid-penetrant method for surface cracks or imperfections. This inspection revealed a longitudinal discontinuity extending from the fracture surface of the tooth through the heat-affected zone at the end of the spline and into the unheated portion of the shaft (Fig. 27). This defect exhibited the characteristic directionality of a seam.

Metallographic Examination. Specimens were removed at three locations along each of the fractured shafts and prepared for metallographic examination.

A micrograph (65×) of an unetched section through a fractured tooth (Section A-A, left, Fig. 27) showed that the fracture surface (arrow A, Fig. 27) was concave and had an appearance characteristic of a seam. The tooth separated from the shaft when a crack originated at the root of the seam (arrow B, Fig. 27) and propagated at 90° to the seam toward the root fillet. Etching the specimen in 1% nital and examining it at a magnification of 260× disclosed partial decarburization of the surface (Section A-A, right, Fig. 27). The microstructure of the metal in this region of the shaft was characteristic of a free-machining medium-carbon steel.

Examination (at 65×) of an unetched section through the region at the end of the splines that was affected by induction heating (Section B-B, Fig. 27) showed that the seam had been opened as the result of thermal stresses produced during induction hardening. The curvature of the surface at the seam was similar to that on the fractured tooth.

Examination (at 65×) of an unetched section through the shaft in an area unaffected by induction heating (Section C-C, left, Fig. 27) disclosed the presence of a crack extending to a depth of approximately 0.6 mm (0.025 in.). The root of the crack contained oxides typical of those found in seams. Etching the specimen in 1% nital and examining it at 65× revealed partial decarburization along the crack surfaces similar to that found on the fracture surface of the tooth (Section C-C, right, Fig. 27). The curvature of the crack surface was similar to that observed in the heat-affected zone and on the fracture surface of the spline tooth.

Decarburization of the crack surfaces and oxides in the root of the crack indicated that the seam had been present before the shaft was heat treated and are characteristic of seams produced during the manufacture of steel billets, bars, rods, and wires.

The average hardness across the tooth surface was 60 HRC, which was within the specified range.

Conclusions. The shafts failed by brittle fracture because of deep seams produced during processing of the steel. The seams acted as stress raisers during induction hardening.

Recommendations. Specifications should require that the shaft material be free of seams and other surface imperfections.

Grain size, composition, and microstructure of a shaft material largely define the strength and toughness of that material and thus the performance of a shaft of that material. Generally, a fine-grain low-temperature transformation product (martensite or bainite) is desirable. However, where elevated temperatures are involved, a coarse grain size may be preferred. Grain size, composition, and microstructure affect material toughness at any given strength level; only when these factors combine to produce low toughness do they contribute to brittle fracture of a shaft.

Toughness and transition temperature are generally more important in large-diameter shafts than in small-diameter shafts. The life of

Fig. 27 Induction-hardened 1151 steel rotor shaft in which a spline fractured because of a seam

Top left: Configuration and dimensions (given in inches). Section A-A: Micrographs of section through broken spline, showing shape of fracture (arrow A), root of seam (arrow B), and decarburized surface. Sections B-B and C-C: Micrographs of section through seam showing regions where metal was affected and not affected by induction heating

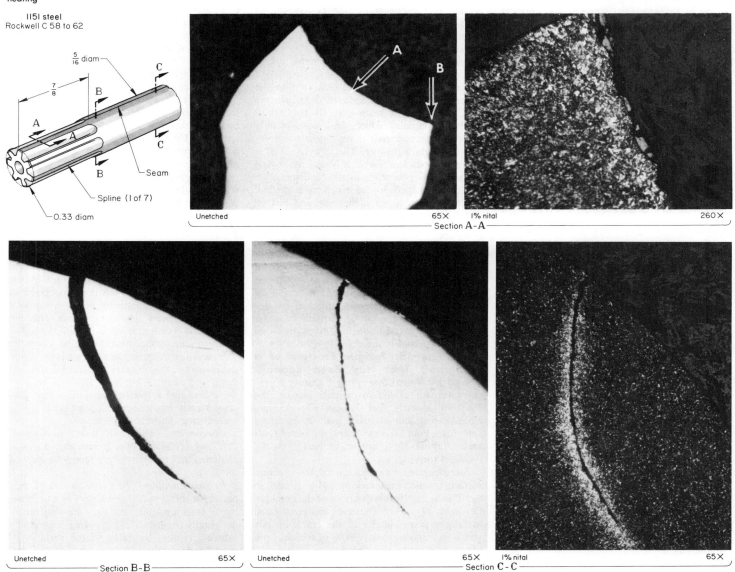

a small-diameter shaft is usually determined both by the time required for a fatigue crack to initiate and by the rate of crack propagation. Large-diameter shafts are more likely to contain flaws of critical size. Thus, the life of a large-diameter shaft is often determined solely by the rate of propagation of cracks originating at pre-existing flaws. For large-diameter shafts with high internal stresses, the minimum level of toughness that can be expected in the shaft at minimum operating temperature must be known in order to determine the maximum size of flaw that can be tolerated.

Temper embrittlement of certain shaft steels, especially the nickel-chromium, nickel-chromium-molybdenum, and nickel-chromium-molybdenum-vanadium alloy steels used

in larger rotor shafts, results from prolonged exposure at temperatures of 350 to 575 °C (660 to 1070 °F) or from slow cooling after exposure above this temperature range. In most instances, a short exposure time within, or rapid cooling through, this temperature range will minimize temper embrittlement. However, when heat treating thick sections, this procedure may not be possible, and temper embrittlement may occur. Fortunately, temper embrittlement can be reversed by retempering above the temperature range of 350 to 575 °C (660 to 1070 °F) followed by rapid cooling through this range to restore toughness (details of various types of embrittlement are provided in the article "Ductile and Brittle Fractures" in this Volume).

Temper embrittlement decreases impact toughness by increasing the ductile-to-brittle transition temperature and decreases the amount of ductile fracture; thus, it is of considerable concern for larger shafts. For steels that exhibit a sensitivity to temper embrittlement, the degree of embrittlement that develops is largely dependent on composition, heat treatment, and service conditions.

With respect to composition, molybdenum in amounts from 0.20 to 0.30% significantly retards temper embrittlement. Greater amounts of molybdenum yield no additional improvement. The effect of molybdenum on suppressing temper embrittlement decreases as the purity of the steel increases. The total content of impurities (sulfur plus phosphorus plus nonferrous elements plus gases), expressed in atomic parts per million, ranges in value from about 1500 ppm in steels of conventional purity (corresponding to the usual amounts found in air-arc-melted steel) to 1000 ppm in very clean steel (corresponding to vacuum-melted steel) and to about 500 ppm in the extraclean steels (corresponding to vacuum-melted steel using a very pure charge). Hence, when relatively impure steel is used (with over 1500 ppm impurities and over 0.01% P), the unique role of molybdenum in minimizing temper embrittlement is very important. Molybdenum does not appear to be a necessary alloying element in the production of high-purity steels (under 500 ppm impurities and under 0.001% P), which are not susceptible to temper embrittlement.

Effect of Surface Coatings

Metallic coatings of several types and compositions are often used to protect shaft surfaces from wear and corrosion and to repair worn shafts. The types of coatings include, but are not limited to, electroplating, metal spraying (with or without subsequent fusing), catalytic deposition, and metal overlaying. Typical coating materials are chromium, nickel, iron, alloy steels, aluminum alloys, copper alloys, and chromium-nickel-iron alloys. Care must be taken in applying these coatings, because harmful residual stresses and/or stress concentrators may be inadvertently left in or produced in the shaft.

Many electroplating processes leave harmful residual stresses in the plating and therefore in the adjacent base metal. Chromium, iron, and nickel platings generally contain high residual tensile stresses, which reduce the fatigue strength of the base metal of a shaft. Residual tensile stresses reduce the corrosion resistance of both the plating and the base metal. Piston rods and other parts made of high-strength steel and plated with metallic coatings can crack and fracture at unexpectedly low stress levels in simple fatigue.

During acid cleaning and electroplating, hydrogen is usually produced and may be absorbed into the base metal. If the hydrogen is not removed by a subsequent heat treatment, severe embrittlement of the base metal may occur, especially in base metals with hardnesses above approximately 35 HRC. High-strength highly stressed parts can crack and fracture as a result of hydrogen embrittlement. Failure by hydrogen embrittlement is even more likely to occur if the shaft contains high residual tensile stresses before being plated. More information on failures caused by hydrogen embrittlement can be found in the article "Hydrogen-Damage Failures" in this Volume.

Metal spraying to obtain specific surface properties is sometimes used on shafts and, if proper bonding is obtained, is usually satisfactory. However, spraying heavy thicknesses of metals that have fairly high coefficients of thermal expansion on large-diameter shafts results in high residual tensile stresses; in the as-sprayed condition, the metal sometimes cracks. If subsequent heating is required to bond the coating to the base metal, high residual stresses and/or thermal cracking of the base metal or coating may occur.

Sprayed metal is mechanically weak and, if overloaded in compression, tends to spread laterally, thus disturbing its adhesion to the underlying base metal. Sprayed metal adheres well to crankshaft main journals but not to crackpin journals, which are more heavily loaded than crankshaft main journals and are subject to shock loading. The sprayed metal is entirely dependent on a mechanical keying action for its adherence to the base metal. A shallow V-shaped groove, such as a shallow small-pitch thread, is inadequate and has an undesirable stress-raising effect. The most satisfactory method of surface preparation is shot or grit blasting, which must be done in a manner so that the surface is roughened and not simply peened. The stress-raising effects of blasting are negligible because of the small size of the indentations. However, adhesion of sprayed metal to a surface prepared by grit blasting may not always be satisfactory under severe loading conditions, such as those to which a crankpin is subjected in service.

Example 15: Fatigue Fracture of a Crankshaft That Had Been Reconditioned by Metal Spraying.* During a general overhaul of a four-cylinder engine, the crankpin journals and 7.6-cm (3-in.) diam main-bearing journals were built up by metal spraying. Four weeks later, the crankshaft broke through the crankpin farthest from the flywheel (driving) end.

Investigation. Examination of the fracture surface revealed that cracking had originated in the fillet at the "inside" surface of the crankpin (arrow A, Fig. 28a). The crack had propagated in fatigue perpendicular to the crankpin axis and across approximately 80% of the crankpin before final rupture occurred.

It was found that adhesion of the sprayed-metal coating to the fractured crankpin was very poor and that the coating could be detached easily by tapping a chisel inserted between the coating and the pin. Portions of the sprayed coating were removed from all the pins and journals so that the surface preparation carried out before metal spraying could be examined. In each of two pins, a helical groove approximately 0.8 mm (0.03 in.) wide and deep, with a pitch of 10 per inch, had been cut in the surface (arrow B, Fig. 28b). The grooving extended over one-half of the pin periphery only; the remaining portion of the surface had been indiscriminately indented with a tooth punch (Fig. 28b).

*Adapted from "Fatigue Failure of a Metal-Sprayed Crankshaft," *Failure Analysis: The British Engine Technical Reports,* American Society for Metals, 1981, p 131-133

From one pin, a large portion of the sprayed metal, extending the whole width of the crankpin and covering one-third of its circumference, came away in one piece. Examination indicated that the tongues of metal that filled the grooves had broken away from the main layer at the level of the original pin surface. The average thickness of the sprayed layer was 0.8 mm (0.03 in.). The hardness of the crankpin was only 163 HB, indicating that it was readily machinable.

However, the surfaces of the main journals had not been grooved, but appeared to have been roughened by shot or grit blasting before spraying. The sprayed coating was more firmly adherent to these surfaces than to those of the pins.

With regard to the sequence of failure, there was a possibility that the crack might have been developing at the time restoration of the shaft was carried out. Careful examination showed that this was unlikely, because the origin of the crack was situated at the bottom of one of the turned grooves and the crack followed the groove about halfway around the pin periphery. Figure 28(c) is a view of a portion of the crankpin, showing that the crack occurred in the groove. The continuation of the groove beyond the point where the crack changed direction is shown at arrow C in Fig. 28(c). The crack was not located near the region where the fillet was tangent with the shaft surface, as are the majority of such cracks in crankshafts, but developed about 3.2 mm (0.125 in.) away in the cylindrical region of the pin, which also suggests that its location was determined by the pre-existing machined groove.

Conclusions. The shaft failed in fatigue. Cracking originated in a groove that was machined in the crankpin to help retain the sprayed metal.

Recommendations. Surfaces should be prepared for metal spraying by shot or grit blasting in such a manner that they are roughened, not simply peened. Metal spraying is generally satisfactory for rebuilding the surfaces of main-bearing journals, but not for the surfaces of crankpins that are subject to shock loading.

Shaft Repair by Welding. Shafts made of annealed materials can usually be repaired without difficulty using well-established welding practices. Generally, these practices include matching of base-metal and filler-metal compositions, proper preparation of the region to be welded, and thermal procedures that include preheating, postheating, and stress relieving. The repaired shaft is then heat treated in the conventional manner.

Welding of hardened materials involves compromises and should be avoided whenever possible. For instance, the heat-affected zone usually contains areas with hardness 10 to 20 points (HRC) lower than that of the base metal, thereby causing stresses in the shaft due to a so-called metallurgical notch. To minimize these stresses, the material should be slowly

Fig. 28 Crankpin of a crankshaft that fractured in fatigue after the crankpin journals and main-bearing journals were rebuilt by metal spraying

(a) Fracture surface showing fracture origin (arrow A). (b) Machined grooves and punch marks (arrow B) on surface of crankpin. (c) Groove in crankpin, where cracking was initiated

(a)

(b)

(c)

A new stub shaft was made by welding a type 316 stainless steel cap to a forged steel shaft that had a chemical content of 0.25% C, 0.25% Mo, and 1% Cr and a tensile strength of 600 MPa (87 ksi). The grooves machined in the replacement shaft for the weld overlay had radii instead of sharp corners (Detail A, "Original design" and "First revised design," Fig. 29).

Investigation. Examination revealed that the forged steel shaft had fractured at approximately 90° to the shaft axis (Detail A, "Original design," Fig. 29). The face of the outboard section of the shaft had a groove approximately 6.4 mm (0.25 in.) deep by 7.9 mm (0.3 in.) wide around its circumference. This step followed the shape of the weld metal on the mating piece. The fracture surface along the weld metal contained no forged steel, indicating that the failure was in the weld metal and not in the heat-affected zone of the forged steel shaft. Worn and smeared metal on both mating surfaces indicated that the stub shaft had continued to rotate after it had fractured.

Inspection of the PVC reactor revealed that the drive coupling between the shaft and motor was purposely misaligned 1.3 mm (0.05 in.) in the vertical direction with the shaft. This misalignment was done to counter the misalignment caused by bow of the concrete slab base, which occurred when the reactor was loaded.

Cracks in the heat-affected zone or cracks in the martensite transition zone at the fusion line were possible, but were not evident from microscopic inspection of the weld area. Presence of the martensite transition zone offered a path of easy crack propagation.

Chemical analyses of the steel shaft and the stainless steel end cap at various locations are given in the table that accompanies Fig. 29. Analyses of metal at the fusion line were made from scrapings and drill shavings that included metal on either side of the fusion line; therefore, the values are average values for the general area rather than actual values for the metal at the fusion line.

The chromium, nickel, and molybdenum contents of the weld metal were lower than those of the filler metal, whereas the contents of the same elements in the heat-affected zone of the shaft were higher than those of the base metal, indicating that the weld metal had been diluted by the carbon steel shaft.

The stainless steel used for the end cap was similar to type 316 stainless steel except for a higher chromium content. The carbon content of the forged steel used for the shaft was greater than the maximum specified for ASTM A105, grade 2.

Microstructure. The base metal that was not affected by the heat of welding had a typical pearlite structure, with ferrite surrounding each grain. The microstructure changed as the fusion line was approached: the amount of ferrite decreased, and the microstructure became martensitic.

The hardness of the metal in the steel shaft near the fusion line was 358 HV. Because a

cooled after welding, then tempered immediately using a temperature slightly lower than the original tempering temperature.

Weld overlays resulting in a diameter larger than the adjacent shaft diameter should have a tapered transition. This would minimize the stress concentration at an otherwise abrupt change of section that may coincide with the heat-affected zone of a weld. Any subsequent grinding of the weld area to produce the desired dimensions should be done carefully, using a soft grinding wheel and numerous light passes to avoid overheating that would cause grinding cracks.

Example 16: Ductile Fracture of a Forged Steel Shaft at a Change in Section and at a Stainless Steel Weld. The stub-shaft assembly shown in Fig. 29, which was part of the agitator shaft in a PVC (polyvinyl chloride) reactor, fractured in service after a nut that retained a loose sleeve around the smaller-diameter section of the

shaft had been tightened several times to reduce leakage. Removal of the stainless steel sleeve revealed that tightening of the retaining nut had forced the end of the sleeve against the machined stainless steel shoulder on the shaft and had permanently flattened the O-ring seal.

The shaft was made of ASTM A105, grade 2 steel, and the larger-diameter section was covered with a type 316 stainless steel end cap. The cap was secured at each end by welding, using type ER316 stainless steel filler metal (Detail A, "Original design," Fig. 29). The radial surface at the change in diameter was a weld overlay, a continuation of the weld metal securing the cap to the shaft. Transverse fracture of the shaft occurred near this abrupt change in shaft diameter.

The stub shaft was removed from the agitator so that a replacement stub shaft could be installed. Both sections of the fractured shaft were sent to the laboratory for analysis.

Fig. 29 Stub-shaft assembly, for agitator in a polyvinyl chloride reactor, that failed by ductile fracture

Top left: Configuration and dimensions (given in inches). Detail A: Sections through failure area showing original design, first revised design, and final design

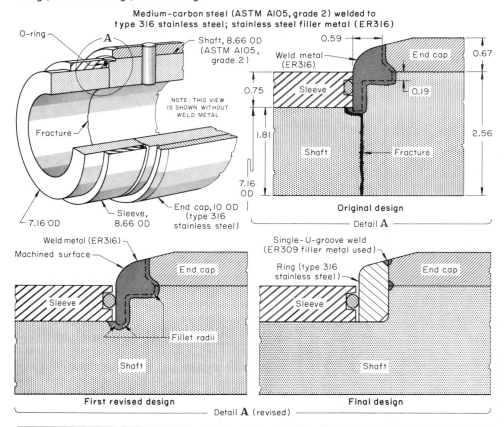

Chemical analysis of steels in shaft assembly						
	ASTM A105, grade 2, steel		Type 316 stainless steel			Nominal composition of type 316 stainless steel(a)
	Shaft			End cap		
Element	Typical	At fusion line	At fusion line	At center of weld	In end cap	
Carbon	0.456	...	0.55	0.054	0.037	0.08(b)
Manganese	0.25	0.18	0.20	0.73	0.53	2.0
Phosphorus	...	...	...	0.010	0.016	0.045
Sulfur	0.017	...	...	0.016	0.022	0.03
Silicon	0.35	0.35	0.25	0.29	0.607	1.00
Nickel	0.10	0.75	10.00	10.16	11.06	10-14
Chromium	0.068	2.00	16.00	19.43	19.46	16-18(c)
Molybdenum	0.023	0.65	2.00	2.86	2.88	2-3
Iron	rem	rem	rem	rem	rem	rem

(a) Maximum content unless range is given. (b) Carbon content is 0.03% max for type 316L stainless steel. (c) Chromium content is 18 to 20% for type 316 stainless steel filler metal.

maximum hardness of approximately 520 HV can be obtained with a carbon steel of 0.45 to 0.50% C content, the martensite may have been tempered by the welding heat during subsequent weld passes.

Discussion. The sharp corners at the root of the weld, which were shown on the original shaft drawing, were not evident on the shaft,

and there was no evidence of a sharp corner on either fracture surface. Therefore, it can be assumed that the fracture was initiated at the surface (the welding practices employed and penetration of the weld metal may have removed evidence of the sharp corners).

Some features on the fracture surface were removed by rotation after fracture, which was

unfortunate because the surface may have contained indications of whether fracture originated on the surface in the heat-affected zone or at the fusion line of the weld. An initial crack on the surface of the shaft could have been caused by inherent defects in the weld or by tensile stresses resulting from rotation of the misaligned shaft.

Conclusions. The forged steel shaft failed by ductile fracture. The fracture surface was approximately 90° to the axis of the shaft. The fracture was the result of:

- Reduction of the effective cross-sectional area of the shaft by a crack that followed the fusion line of a circumferential weld that served as an overlay and as a means of securing one end of a stainless steel cap
- A stress concentration from the notch effect of the crack created at the root of the weld
- Tensile stresses induced by misalignment of the shaft
- In addition to the normal torsional and tensile stresses of the shaft, a tensile stress that was applied to this area by the sleeve being forced against the machined surface of the weld when the retaining nut was tightened excessively.

Corrective Measures. The replacement shaft for the immediate failure followed the first revised design shown in Detail A in Fig. 29. Shafts for new agitators were made using the final design in detail A in Fig. 29. Both the end cap and the ring were made of type 316 stainless steel. Welding was done with type ER309 stainless steel filler metal, and the end cap and end ring were preheated to 150 °C (300 °F). The cap was joined to the shaft using a single U-groove weld, which was machined flush before the stainless steel ring was installed. Another single U-groove weld was used to join the ring to the end cap. This design eliminated welding at an inside corner, reducing the stress-concentration factor at that area.

Manganese and zinc phosphate are conversion coatings capable of producing stress concentrations on hardened ferrous material. This condition results from failure to follow proper processing procedures, which include control of bath composition, bath temperature, and processing time. Camshaft lobes and bearings are sometimes coated with manganese phosphate to assist when lubrication is marginal and during break-in. If one or more control parameters are not observed, the surface becomes grossly etched and pitted. This intergranular surface attack leads to flaking in service and ultimately to complete deterioration of the highly stressed lobe surface.

Failures of Sliding Bearings

Kenneth C. Ludema, Department of Mechanical Engineering, University of Michigan

MECHANICAL DEVICES usually contain some moving parts. Bearings are usually provided where a specific spatial relationship (alignment) must be maintained between the parts or where a force is to be transmitted from one part to the other. Mechanical bearings are usually made of selected material and in special geometries in order to effect easy and reliable motion for a specific task. The major types are rolling-element (ball or roller) bearings and sliding bearings. This article will discuss sliding bearings.

Classification of Sliding Bearings

Sliding bearings are usually conformal; that is, the components of a sliding pair fit together fairly closely. The surfaces may be flat, spherical, or cylindrical. Flat surfaces include the slideways of machine tools and the thrust bearings that limit the axial movement of rotating shafts. Spherical surfaces allow rotary motion around more than one axis, such as that provided by the hip joint of the human skeleton. The cylindrical surface pair (the shaft and journal pair) is the most common type in machines. These are also called plain bearings or simple bearings.

Sliding surfaces may be in motion for long periods of time, as in clocks. Some start and stop frequently, but always slide along one axis, as in engines; others oscillate, as in door hinges. They may operate dry (photo film-making machinery) or lubricated with available moisture (some water pumps) or a specifically applied lubricant. It is usually the case that the more severe the duty of bearings the more precisely they will be made and the more likely there will be a support (lubrication) system to ensure long life and to prevent catastrophic failure.

Bearing Materials

The lowest cost bearings are bushings of molded polymers (plastics) in which steel shafts rotate. The lowest cost polymers are the nylons and acetals. At low sliding speeds, they usually wear by abrasion due to dirt or the roughness of the surface of the shaft.

Polymers have two disadvantages—their low thermal conductivity and their tendency to cold flow and become distorted if loaded very heavily. If polymers are used at high speeds, the surfaces may melt, possibly even with some of the molten polymer being squeezed out. There are three solutions to the thermal problems. One is to operate the bearing with a low contact pressure, p, and at a low speed, V. A recommended pV limit is 0.2 MPa · m/s (Ref 1). Another is to lay a thin film of polymer on a metal backing to increase the heat-flow rate. A third is to fill the polymer with a material that has high thermal conductivity. The third solution to the thermal problem also offers resistance to cold flow. For these purposes, the polymer can be filled and reinforced with particles or fibers of glass, metal, and minerals. However, these fillers usually cause more wear of the shaft than the polymer does.

For low friction, the more expensive polytetrafluoroethylene (PTFE) is available. Polytetrafluoroethylene is sometimes added to acetals to reduce friction. Graphite and molybdenum disulfide (MoS_2) are also added; although they decrease friction, these solid lubricants usually do not prevent wear.

Another low-cost bushing material is porous bronze, which can be impregnated with lubricant. For a little higher cost, cast iron bearings can be used, particularly for large and flat surfaces.

Polymers, gray cast iron (with large amounts of graphite), and bronzes (see the section "Bronze Bearing Materials" in this article) are selected as bearing materials because of their ability to survive even though they may be made to poor tolerances or may be operated with occasional lubricant starvation or without maintenance. Poor treatment of these bearings will often appear as adhesive grooving and galling. Where these materials are run against a harder counter-surface, there may be metal transfer to that surface.

Almost all materials, such as carbon, ceramics, white cast iron, and metal alloys of all kinds, can be used as bearings. These materials may be chosen primarily for corrosion resistance, for prevention of sparking, for resistance to high volumes of abrasives, or simply because they are available. High-performance bearings are usually made with considerable care and are comparatively more expensive.

High-Performance Bearings. The soft metals, in various combinations, are used in high-performance bearings, and they are virtually always lubricated. The soft metals are sometimes used in thicknesses up to 6 mm (0.24 in.), but to increase fatigue resistance, the soft metals are coated or plated on a steel (or other stronger metal) backing in layers as thin as 25 μm.

The major groups of soft metal bearing materials are:

- *Tin-base or lead-base babbitts or white metals*: These materials are good for embedding hard contaminant particles and for resistance to galling, but fatigue quickly. Lead alloys also corrode readily
- *Copper-lead alloys*: These are superior to the white metals for corrosion resistance in many applications
- *Aluminum alloys that contain tin and lead*: The alloying elements do not mix, thus forming small solid droplets in the aluminum that smear over the aluminum surface at times of inadequate lubrication. The aluminum alloys are superior in corrosion resistance

Babbitts are of two types: tin-base with 4 to 8% Cu and 4 to 8% Sb (SAE 11 and 12; ASTM B23, grades 1, 2, 3, and 11) and lead-base with a maximum of 10.7% Sn and 9.5 to 17.5% Sb (SAE 13-16; ASTM B23, grades 7, 8, 13, and 15). These babbitts generally cause the least damage to steel shafts when operated with inadequate lubrication or with contaminants. Tin-base babbitt is more corrosion resistant than lead-base babbitt and is easier to bond to a steel backing. Babbitts are inherently weak and strongly affected by high temperatures. For this reason, thin layers of babbitt are often applied to backing materials, such as bronze, steel, or cast iron.

Copper-lead bearing alloys usually contain between 20 and 30% Pb, up to 3% Sn, and may contain up to 5.5% Ag and less than 0.1% Zn. Typical alloys are SAE 48, 49, 480, and 481 (CDA alloys C98200, C98400, C98600, and C98800, respectively). These alloys have all been used as main bearings and connecting-rod bearings of engines. However, in recent years, corrosion has been found to be a problem with these alloys. They are, therefore, usually covered with a thin layer of lead-tin, lead-tin-copper, or lead-indium alloy to reduce the corrosion of the substrate. In some instances, a nickel barrier about 1 μm thick may be placed between the substrate and the overlay to prevent diffusion of tin from the overlay to the copper in the substrate.

Bronze bearing materials are either leaded bronze, tin bronze, or aluminum bronze. Strength and high-temperature resistance increase in the order lead to tin to aluminum, but embeddability and scuff resistance decrease in that order.

The leaded bronzes contain from 5 to 25% Pb and up to 10% Sn. The high-lead alloy is soft and has a temperature limit of about 230 to 260 °C (445 to 500 °F). The addition of tin increases the fatigue resistance and hardness. Zinc in place of tin improves ductility but decreases strength and hardness. Some wrought leaded bearing bronzes are SAE alloys 791 to 794, 797, and 799. The alloy SAE 795 has especially good fatigue resistance. Some castable leaded bearing bronzes are SAE 40 (CDA C83600), SAE 660 (CDA C93200), SAE 66 (CDA C93500), SAE 64 (CDA C93700), CDA C94100, and CDA C94300.

Tin bronzes contain more than 20% Sn, with less than 10% Pb added to aid machinability. Small amounts of zinc, nickel, or both are sometimes added to increase strength.

Aluminum bronzes have iron, nickel, silicon, and manganese as the principal alloying elements. The hardness of these alloys may range to 240 HB. They have excellent shock resistance and will operate at comparatively high temperatures. They tend to scuff more readily than the leaded bronzes. Typical alloys are CDA C86200, C86300, C95200, C95300, C95400, and C95500.

Aluminum alloys other than bronzes contain about 7% Sn and 1% Cu, with 1 or 2% Si or Mg, plus some lead and cadmium. The copper is retained in solid solution in the aluminum, but the other alloys form soft-solid particles in the matrix. These alloys are usually relatively weak and require strong backing. They do not require an overlay for protection from corrosion; however, an overlay is often provided for start-up protection. Typical alloys are 850.0-T5, A850.0-T5, and B850.0-T5 (formerly 750-T5, A750-T5, and B750-T5).

Fluid-Film Lubrication

Fluid films can be provided in a bearing by surface-tension retention of a fluid in a gap, by pumping fluid into a contact region (hydrostatic lubrication), or by hydrodynamic action.

Surface Tension. If a drop of liquid is placed on a flat surface and then another flat surface is laid upon the wetted surface, some, but not all, of the liquid will be squeezed out. Surface tension, the same force that makes

Fig. 1 Schematic of the transition from inadequate (boundary) lubrication at start-up to hydrodynamic lubrication at full speed

(a) Steady load at start-up. (b) Steady load at full speed

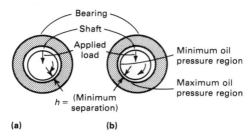

liquid rise in a very small diameter glass tube, makes complete exclusion of the liquid impossible. The amount that will be retained in the gap between two surfaces is related to the wettability of the liquid (lubricant) on the surface of interest. If a drop of the lubricant spreads out completely and spontaneously on the surface, that lubricant will also run out of the bearing completely. If the drop of lubricant stands up as water does on a waxed surface, retention will be low. If the drop has a base diameter of about twice the height of the drop, retention will be satisfactory. The drop-spreading test is convenient for selecting materials and lubricants for applications in which small quantities of lubricant are applied throughout the life of the product. This method is frequently used for flat surfaces, rolling-element bearings, low-speed bushings, and the myriad of surfaces that are lubricated by the infrequent drop of oil.

Hydrostatic Lubrication. Two sliding surfaces can be separated by pumping a fluid into the contact region at a sufficient pressure. The contact region must be properly designed so that there is a symmetric or balanced contact-pressure distribution. A large volume of fluid will separate the sliding surfaces a great distance, thereby producing a low resistance to movement. However, the energy required to pump the fluid must be considered in the overall economy of the bearing system. Another factor to consider is that hydrostatic lubrication requires a pump, which may of course fail and ruin a bearing. The principles of hydrostatic and hydrodynamic bearings are detailed in Ref 2 to 4.

Hydrodynamic Lubrication. If one surface slides along another at a moderately high speed and if the shape of the leading edge of the moving surface is such that fluid can be gathered under the sliding surface, the two surfaces will be separated and will slide easily. This is hydrodynamic lubrication.

Figure 1(a) shows the start-up of a shaft in a bearing. A shaft at rest, with a downward load on it, will contact the bearing at the bottom. As

the shaft begins rotating, it climbs one side of the bearing, but it eventually begins to slide. If the shaft and bearing are immersed in oil, the sliding shaft will drag oil underneath itself to begin forming the hydrodynamic wedge. It is not a visible wedge because the entire system is immersed. Rather, it is a pressurized region in which the pressure lifts the shaft. When the wedge of fluid is fully developed, the shaft takes the position shown in Fig. 1(b), with a minimum separation h. The fluid-film pressure builds behind the location of minimum separation between the shaft and bearing, taking the shaft surface as the reference.

The mathematics of the fluid wedge are embodied in equations of hydrodynamics. Some of the variables of interest for a shaft of diameter D rotating in a journal bearing of inner diameter $D + 2c$ are:

$$\frac{\eta N}{p}\left(\frac{D}{c}\right)^2\left(\frac{L}{D}\right)^2 = \frac{(1-\epsilon^2)^2}{\epsilon\pi\sqrt{(1+0.62\epsilon^2)}} \qquad \text{(Eq 1)}$$

where η is the lubricant viscosity (which will change with temperature and pressure), N is the shaft speed (in rpm), $p = W/DL$ or the average pressure between the shaft and bearing due to the applied load W, L is the length of contact between the shaft and bearing, c is the clearance, and $\epsilon = 1 - h/c$ (h is the minimum separation). The term on the left is one form of the Sommerfeld number and is sometimes referred to as the bearing characteristic. It is clear from Eq 1, at least, that bearings with the same Sommerfeld number will operate with the same ϵ.

Many calculations have followed upon the work of Sommerfeld in attempts to optimize bearing design. The result is a number of guidelines, the most important of which relate to efficiency and stability. For efficiency, Ref 2 recommends operating with the ratio $h/c \approx 0.3$. The consequence of this recommendation would be a particular set of values for the adjustable variables $\eta N/p$ for a given bearing. These values can be readily calculated, and this is done in design. The procedure is rather detailed and requires some expertise in dimensions and units (see Ref 2-4). It is sufficient for this discussion to suggest how to recognize the likely regimes of operation of bearings from the damage done to them.

Miscellaneous Methods of Lubrication. Low-speed bearings do not need large quantities of lubricant flowing through to remove heat. Therefore, some are lubricated by a drip-feed device, some by a wick that carries lubricant from a reservoir to the shaft surface, and some use grease. One design for a horizontal shaft uses a large ring that rests on top of the shaft, dips into a supply of lubricant, and carries lubricant to the shaft as the shaft rotates.

Shaft Whirl

Equation 1 was described in the context of a shaft that was carrying a constant downward load. It could also apply to a weightless shaft that is out of balance. In this case, the contact is restricted to one location on the shaft, but rotates on the inner surface of the bearing. The fluid wedge develops as in the former case except that it does so ahead of the location of minimum separation between the shaft and bearing.

An interesting situation develops when a shaft has a small load on it and when an intermittent load is also imposed such that the position of the shaft in the bearing changes. The shaft will move around in the bearing as if it were carrying an unbalanced load, which makes the location of minimum separation move in the direction of shaft motion. This amounts to a receding of the wedge-forming surface, which decreases h momentarily until the wedge ahead of minimum separation develops. However, because this film cannot be sustained unless there is a constant out-of-balance load, the film soon diminishes as well. The result is a whirl of the shaft in the direction of shaft rotation but at half the shaft speed. The value of h is much smaller than for either a constant load or a constant out-of-balance load. Whirl can be prevented by reducing $\eta N/p$ or by reducing the load-carrying capacity of the bearing.

Load-Carrying Capacity, Surface Roughness, Contaminants, and Grooves

The maximum load-carrying capacity of a bearing, according to Eq 1, would appear to be when $h = 0$, but this is not practical. First of all, the above equations are based on perfectly smooth sliding surfaces. Second, in practice some account should be taken of the inevitable dirt, wear debris, and contaminants in lubricated systems.

One practical limit on the minimum value of h is the point at which the asperities or small bumps on all sliding surfaces begin to interact. This is usually expressed in terms of the lambda ratio, $\Lambda = h/\sqrt{(\sigma_1^2 + \sigma_2^2)}$. In this equation, the root mean square (rms) surface roughness of the shaft is σ_1, and that of the bearing surface is σ_2. Where $\Lambda > 1$, there should be no contact and thus no wear, but such bearings are not economical for uses other than in large capital machinery, such as electric generators. Most other bearings usually operate with Λ near 1 and even less. The consequence of operating with $\Lambda < 1$, as in the case of overload or where the viscosity of the lubricant is low because of overheating, is that the surface topographical features (bumps) begin to interact through a very thin local fluid film. This interaction is popularly referred to as metal-metal contact, which is not accurate. In any case, this interaction causes deformation and fracture of the bearing surface on both micro- and macroscopic scales.

Fig. 2 Approximate relationship of coefficient of friction to Sommerfeld number, $\eta N/p$

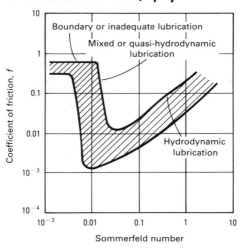

If a bearing is operating in the thick fluid-film regime and if $\eta N/p$ decreases, h will increase. Eventually, as Λ decreases, the friction increases. A widely used method of representing this transition is the plot of the coefficient of friction, f, versus $\eta N/p$ (Fig. 2). This type of curve is often referred to (although not strictly accurately) as the Stribeck or McKee-Petroff curve.

The curve implies that the minimum f and the transition to inadequate hydrodynamic lubrication occur at one critical Sommerfeld number, which is not the case. Any single curve is for a single value of Λ as well as for particular values of D, c, and L. However, the curve is illustrative and is widely used. The type of damage seen in bearings is related to these factors. A whirling shaft may be operating to the left of the transition in Fig. 2, although the design parameters would indicate adequate fluid-film lubrication.

Grooves and depressions are often provided in bearing surfaces to supply or feed lubricant to the load-carrying regions. It is often stated that such grooves are reservoirs for lubricant, but they can also be pathways for lubricant to escape. They function best when lubricant is pumped into them at the pressure prevailing at their location in the bearing. In a hydrodynamic bearing carrying a steady load, the grooves should be located in the half of the bearing behind the location of the minimum fluid-film thickness. Grooves may be circumferential, axial, or spiral in orientation. If they are supplied with lubricant from a pump, they should not be open ended because pressure cannot be established in such grooves. The circumferential groove in the center of the bearing is best for the highest performance bearing supplied with large quantities of lubricant. Spiral grooves are often preferred where the supply volume is low, but axial grooves are easier to make.

Squeeze-Film Lubrication

The above discussion relates to a near steady state of bearing operation. In some instances, a shock load may be imposed on a bearing in which an equilibrium fluid film was established. The value of h cannot decrease instantaneously: Time is required to decrease the fluid-film thickness. Thus, a bearing will operate for a few microseconds in a state that is sometimes referred to as the squeeze-film regime. The fluid is squeezed out at a rate that is proportional to $1/h^2$. There are few guidelines for calculating the effects of such shock loads.

Elasto-hydrodynamics

There is another class of surfaces for which adequacy of a fluid film is calculated by equations of elasto-hydrodynamics. These are cases in which the contacting surfaces are of counterformal shape, as, for example, with a ball on a bearing race, and the contact surfaces are deformed elastically an amount that is a significant fraction of the fluid-film thickness. The principles of lubrication are the same as for the hydrodynamic case, but the governing equations are much more complicated

Lubricants

In hydrodynamic theory, only the viscosity of the lubricant is considered. Actually, most lubricants are chemically complex. They often react with the sliding surfaces to some extent, and they change in use. For example, the pH of a lubricant usually changes during use, which changes both the chemical reactivity and the viscosity. Some lubricants have fatty acids added in order to achieve desirable friction properties. Others have additives that are chemically reactive under extreme pressure to prevent scuffing. These chemical reactions are usually carefully controlled to minimize detrimental effects.

Each type of additive in oil produces selected effects that are too numerous to tabulate here. One effect is to alter friction. Figure 3 shows schematically the effect of three general classes of oils: without additive, with fatty acid added, and with additives of the general types known as antiwear and extreme-pressure (EP) additives. Each is responsive to temperature in a manner different from the others. The temperature increase may be effected by general heating or by localized friction effects.

There are some graphic representations of the effects of additives to be found in the literature, but there is no method of determining how any one additive will work except to test it. It is thought that the fatty acids attach to a sliding surface without reacting chemically with the surface. The antiwear and EP additives are thought to form a thin film of reaction products that has low friction.

Fig. 3 The friction of sliding bearing pairs for three different lubricants

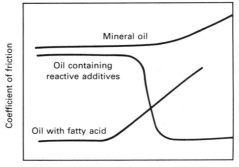

Fig. 3 The friction of sliding bearing pairs for three different lubricants

Debris and Contaminant Particles

Virtually all bearings contain particles that were not deliberately placed there. Some of the particles are left from the manufacturing process; some are dirt; some are the products of chemical reaction, such as oxidation; and some are wear-debris particles. Hydrostatic bearings are flushed continuously by the flow required to carry a load; hydrodynamic bearings are often flushed by a flow of lubricant to remove heat, which develops mostly in the fluid film itself due to viscous shear. It is convenient, then, to supply filtered lubricant to keep the contact areas reasonably clean.

The effect of contaminants in bearings depends on the hardness of the contaminant, the number of such particles, and their size. In general, most dirt, usually SiO_2 and related minerals, is harder than most shafts and bearings and will wear them away. One solution is to make the bearing very soft so that particularly offensive contaminant particles will become embedded in the metal and not cause damage. The alternative is to make the surfaces so hard that they resist abrasion. In general, a particle will abrade another surface if it is 1.3 times (or one Mohs number) harder than the other material.

Contaminant particles have been found to be of a size range from less than one micron to tens of microns. For perspective, hydrodynamic bearings are usually designed with a clearance that is between 0.1 and 0.2% of the shaft diameter. Thus, for a 50-mm (2-in.) diam shaft, the clearance will be between 50 and 100 μm. The design value of minimum film, h, will be between about 15 and 30 μm at normal load and down to 5 μm when under very heavy load. By using automated particle count devices, it has been found that used engine oil may contain, for example, 31 particles in the size range of 30 to 40 μm, 850 particles in the size range of 15 to 20 μm, and 58 469 particles in the size range of 5 to 10 μm. Under normal conditions, the small particles flow through the bearing with no effect. The larger ones can be filtered out without difficulty, but the midrange particles may not be adequately removed. As a rule, those particles of about the dimension of h will cause the most wear.

A surface can be examined to determine whether contaminants may be a major source of wear. This is done by comparing the surface roughness of the worn surface with that of a new surface. For example, a crankpin surface may begin with a roughness of 0.25 μm rms. If the crankpin is worn but has the same roughness as before, it has probably been worn by large contaminant particles of the order of 30 μm. If the crankpin is worn but has a roughness of 0.05 μm rms, it has been worn with contaminants of about 6 μm in diameter.

Failure Analysis Procedure

A general procedure for failure analysis of sliding bearings involves the following steps:

- Do not clean the machine or bearing before the first examination
- Obtain a history of operation of the bearing, lubricants used, other failures in the system, and histories of bearings in similar machines. It is often helpful to get comments on machine behavior from technicians and mechanics instead of from engineers and managers
- Examine the bearing in its normal operating position, and look for evidence of abnormal environment and surrounding heating and damage. Photograph the machine, both close up and overall
- Examine all parts, wear debris, decomposed lubricant, and so on, during disassembly. Retain samples of debris and lubricant, and retain both halves of a sliding pair
- Use a low-power (binocular, if possible) optical microscope for observing the uncleaned parts, and try to establish the progression of damage
- Use a higher-power optical microscope if something interesting appears and it is not yet appropriate to clean the parts thoroughly
- If the parts can be cleaned, it is usually most convenient to use the scanning electron microscope, even for magnifications down to 20×, because the greater depth of field of the scanning electron microscope is far less likely to obscure details on a rough surface than an optical microscope at the same magnification
- X-ray microanalysis is a convenient tool for detecting material transfer, the composition of embedded particles, and loss of layers of multilayer bearings
- Look for evidence of fretting, poor bonding between layers in the bearing, oxidation, plastic flow, fatigue, high temperature, scuffing, and embedded particles (often found in regions of low pressure in the bearing)

- Analyze the lubricant for debris particles and by spectrographic methods to measure the concentration of dissolved elements, and compare the results with a normal lubricant. The lubricant supplier may have access to the characteristics of such normal lubricant. Determine the possible sources of high amounts of dissolved elements

Bearing Failures

Wear Mechanisms in General. There is a limited number of modes or mechanisms by which small amounts of material can be removed from a larger body. These are ductile fracture, brittle fracture, high-cycle (ordinary) fatigue, low-cycle (plastic) fatigue, melting, and chemical dissolution. Hydrogen embrittlement is an example of a hybrid mechanism in which ordinarily ductile material may fracture in a brittle manner. It is often helpful to try to classify the type of observed wear in terms of the modes of material loss, because it is in these terms that materials are selected to resist wear. However, the emphasis in this article is on the analysis of bearings that have failed or are failing so that solutions may be found for the problem.

Fatigue. Segments of a long-lived bearing may fail by fatigue. A bearing may have many cycles of varying load on it, as in engines, or may have an unbalanced load, as in other machinery. Fatigue failure will be hastened by high stresses and by a corrosive environment.

Fatigue cracking of bimetal bearings occurs in stages, beginning with a hen-track pattern on the bearing surface. As fatigue progresses, the soft metal separates from the hard backing in a pattern shown in Fig. 4. This may occur even though lubrication is adequate, but operation with very thin fluid films could cause heating and weakening of the bearing material.

The fatigue limit of engine bearings is a major topic. The performance of such bearings in service is very difficult to quantify, but there are practical guidelines for selecting material types based on applied stress. Table 1 lists such information.

Electrical Erosion. Some bearings, particularly those in steam turbines, may have an electrical potential applied across them, usually inadvertently. When the gap spacing in the bearing is small and the applied potential is high enough, sparks will pass through the oil film and cause erosion. The bearing surfaces will not be reflective, but will appear frosted (Fig. 5).

Fretting. When oscillatory sliding occurs with an amplitude of the order of 0.05 μm, fretting occurs. This may happen on the outside of the bearing backing, against the housing of the machine. The bearings may not have fit tightly, or the housing of the machine may have been too flexible, or the bearing may have been assembled with some dirt in it. A fretted surface is shown in Fig. 6.

Corrosion. Several products of degradation of oil (alcohols, aldehydes, ketones, and acids),

Fig. 4 Sliding bearing surfaces that failed by fatigue

(a) Wormhole appearance in a lead-base babbitt bearing. (b) Flaking (between arrows) of overlay in a trimetal bearing

(a)

(b)

Table 1 Peak stresses for normal fatigue life of various bearing materials used in automotive applications(a)

Bearing metal	Peak stress (max)	
	MPa	psi
Solid-metal bearings		
Aluminum alloy SAE 770-T101 4.1 mm (0.162 in.) thick	34.5	5000
Bimetal bearings		
Tin babbitt 0.6 mm (0.022 in.) thick on steel...........................	8.3	1200
Lead babbitt 0.6 mm (0.022 in.) thick on steel..........................	9.7	1400
Tin babbitt 0.1 mm (0.004 in.) thick on steel...........................	15.2	2200
Lead babbitt 0.1 mm (0.004 in.) thick on steel..........................	16.5	2400
Copper-lead 0.6 mm (0.022 in.) thick on steel	34.5	5000
Aluminum alloy SAE 780 0.8 mm (0.03 in.) thick on steel.................	41.4	6000
Trimetal bearings		
Lead-tin 0.025 mm (0.001 in.) thick on copper-lead 0.6 mm (0.022 in.) thick on steel	51.7	7500
Lead-tin 0.025 mm (0.001 in.) thick on silver 0.3 mm (0.013 in.) thick on steel............................	51.7	7500
Lead-tin 0.025 mm (0.001 in.) thick on aluminum alloy SAE 780 0.8 mm (0.03 in.) thick on steel	55.2	8000
Lead-tin-copper 0.4 mm (0.015 in.) thick on aluminum alloy SAE 781 0.4 mm (0.015 in.) thick on steel	69	10 000

(a) Fatigue strength of bearing materials was determined under laboratory conditions of excellent alignment, excellent oil flow, a high degree of cleanness, ideal clearance, and normal operating temperature of 80 to 105 °C (180 to 220 °F). Normal fatigue life of bearings is considered to be 161 000 km (100 000 miles) in gasoline automobile engines and 402 000 km (250 000 miles) in diesel truck engines.

Fig. 5 Electrical wear on the surface of a thrust bearing from a steam turbine

The microscopic pitting is caused by sparking from electric current. Actual size

Fig. 6 Fretting on the outside surface of steel backing of a sliding bearing

contaminants (such as water), and some antiwear additives in oil cause corrosion. A corrosive environment may selectively attack an alloy, removing only one element. In some environments, lead and zinc may be more readily dissolved than copper or aluminum, for example, leaving a surface such as that shown in Fig. 7. The remaining copper matrix is unsupported and can readily fatigue and fracture away.

Example 1: Corrosion of Copper-Lead Alloy Sliding Bearings by Sulfur Compounds. A connecting-rod shell bearing from a six-cylinder gasoline engine was returned to the factory for examination. The bearing was made of copper-lead alloy SAE 485 bonded to a low-carbon steel backing. The recommended lubrication oil was used in the engine. The length and type of service was not reported.

Investigation. Material loss was measurable over most of the bearing halves, but the steel backing of the upper bearing half in particular was exposed in a wide region at the centerline. No dirt or other foreign particles were found. Shallow pits and a darkened region were visible, and both were partly covered with a brittle waxlike substance that was identified to be a mixture of copper and lead sulfides.

Conclusion. The sulfides were probably formed by the action of sulfuric acids in the oil on the copper and lead in the bearing. The sulfides are soft and easily removed from a surface, and this removal constitutes a wearing away of the bearing.

Recommendations. The best solution to the problem is to change oil with greater frequency to prevent the buildup of sulfur compounds. If the problem persists, bearing halves that have corrosion-resistant overlay materials should be used.

Corrosion Fatigue. Under some conditions of corrosion, a material that would not fail without the corrosion can fail by fatigue.

Example 2: Corrosion-Fatigue Failure of a Bearing in a Marine Environment. A support bearing of a hydrofoil vessel failed after only 220 h of operation. The bearing (Fig. 8) consisted of an outer ring made of chromium-plated AISI type 416 stainless steel and an inner ring with a spherical outer surface made of AISI type 440C stainless steel, with a plastic material between the two. The plastic material was bonded to the outer ring.

Fig. 7 Micrograph of a section through a copper-lead alloy bearing that failed by deleading

Light area at the upper surface is the copper matrix that remained after the alloy was depleted of lead. As-polished. 100×

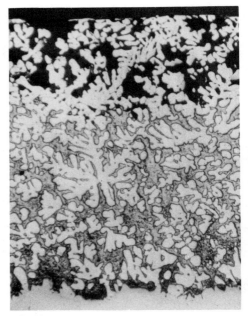

Fig. 8 Plastic-lined stainless steel spherical bearing for a hydrofoil that failed by corrosion fatigue

(a) Construction of bearing and location of fractures. Dimensions given in inches. (b) Fracture surface showing multiple fatigue origins (arrows) at edge of bore and on the spherical surface. 2.5×. (c) Picral-etched section through one of the fatigue origins at the edge of the bore showing intergranular corrosion (arrow A) and intergranular attack (arrow B) on the fracture surface. 250×

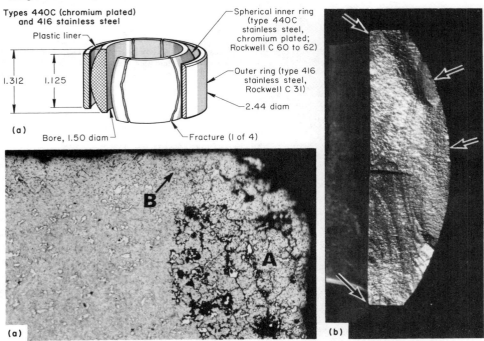

Investigation. The inner ring had failed in four places, as shown, and several less serious cracks had begun as well. The plastic material had disappeared, allowing the two metallic rings to come in contact with each other. Examination of the fracture surfaces of the inner ring showed that it had failed by fatigue and that the fatigue had initiated in corrosion pits. The fracture was mostly transgranular, and it was corroded by seawater.

Conclusions. Corrosion fatigue was the cause of failure. Corrosion pitting of the spherical inner ring produced local stress concentrations and formed cracks under repeated stressing. The cyclic stressing and corrosion in the cracks led to the failure on the inner ring.

Recommendation. The inner and outer rings should both be made from 17-4 PH (AISI type 630) stainless steel, which is more resistant to corrosive marine environments than either type 440C or type 416 stainless steel.

Cavitation. In certain areas of fluid flow along a solid boundary, high pressures and low pressures can develop. If the negative pressure is high, cavities can form in the fluid. If the pressure then increases as the cavity travels to an area of higher pressure, the cavities collapse, and the surrounding fluid impinges on the solid surface. This may produce large stresses in the solid, and if the stress is repeated, the solid may fail by fatigue. Cavitation often occurs in the unloaded region of a bearing or on the surface of high-speed gear teeth. Its appearance is shown in Fig. 9.

Fig. 9 Typical cavitation damage on the surface of a sliding bearing

Cavitation may be reduced by using fatigue-resistant metals, by increasing the lubricant viscosity, and by increasing the pressure in the system. Entrained air or water in oil can increase the likelihood of cavitation. More information of this phenomenon is available in the article ''Liquid-Erosion Failures'' in this Volume.

Fig. 10 Sliding bearings damaged by abrasion by foreign particles

(a) Unetched specimen showing a large foreign particle (arrow) at the interface of a bearing and shaft. 110×. (b) Bearing surface scored by metal chips

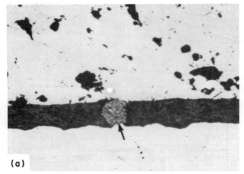

(a)

(b)

Fig. 11 Burrs around a hole drilled through the wall of a spherical bearing at a lubrication groove

The burrs contacted the mating surface, resulting in rough operation of the bearing.

Foreign Particles. As discussed earlier, dirt, chips, and abrasive grit cause wear by abrasion. This is a problem in all applications of sliding bearings. The foreign matter in office machines is often eraser residue, which is composed of 50% glass beads by weight and clay from high-quality paper. Some of this material can be kept out with felt wipers and

Fig. 12 Barrel-shape and hourglass-shape journals

These types of journals can penetrate the oil film and result in bearing failure because of increased loading in localized areas.

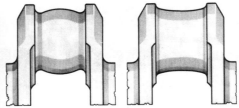

Barrel-shape journal   Hourglass-shape journal

covers, but it is sometimes better to make the bearings of hardened, sintered iron rather than bronze. One insidious effect of foreign particles is that they may scratch or score the shaft surface rather severely, which adds more foreign particles to the system (Fig. 10).

Misalignment. Parts are sometimes improperly manufactured, as in the case of a burr left from drilling (Fig. 11). Another example of improper manufacture is when the shaft is not cylindrical (Fig. 12). Some bearings are improperly assembled and fracture very early in use. Some bearing-holder systems deflect excessively, as shown in Fig. 13. The result of

misalignment is a distorted wear pattern, as shown in Fig. 14.

Long bearings are more likely to be misaligned than short bearings. Sharp edge bearings are likely to fail on the edge, particularly when there is a heavy overhanging load on the shaft.

REFERENCES

1. G.C. Pratt, Materials for Plain Bearings, *Int. Met. Rev.*, Vol 18 (No. 17), 1973
2. F.T. Barwell, *Bearing Systems*, Oxford Press, 1979
3. A. Cameron, *The Principles of Lubrication*, John Wiley & Sons, 1966
4. D.F. Wilcock and E.R. Booser, *Bearing Design and Application*, McGraw-Hill, 1957

Fig. 14 Bearing halves (a) that failed by fatigue resulting from localized overloading after bearing cap shifted position (b)

Note the damaged areas (arrows A and B) on diametrically opposite sides of the bearing halves where they contacted the shaft after the cap shifted.

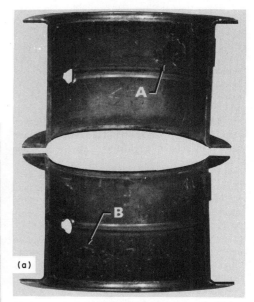

(a)

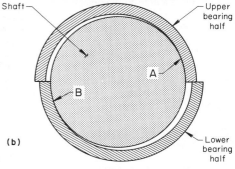

(b)

Fig. 13 Lower (left) and upper (right) halves of trimetal bearings for a diesel engine that failed because of distortion of the crankcase

The center three bearing halves sustained greater damage than the end halves.

Failures of Rolling-Element Bearings

Revised by Ronald L. Widner, The Timken Company

ROLLING-ELEMENT BEARINGS use rolling elements (either balls or rollers) interposed between two raceways, and relative motion is permitted by the rotation of these elements. Bearing raceways that conform closely to the shape of the rolling elements are normally used to house the rolling elements. The rolling elements are usually positioned within the bearing by a retainer, cage, or separator; in ball bearings of the filling-slot type and in needle bearings, they occupy the available space, locating themselves by contact with each other.

Ball bearings can be divided into three categories: radial contact, angular contact, and thrust. Radial-contact ball bearings are designed for applications in which loading is primarily radial with only low axial (thrust) loads. Angular-contact bearings are used in applications that involve combinations of radial loads and high axial loads and require precise axial positioning of shafts. Thrust bearings are used primarily in applications involving axial loads.

Roller bearings have higher load capacities than ball bearings for a given envelope size and are usually used in moderate-speed heavy-duty applications. However, in recent years, improved materials and special designs have allowed use of cylindrical and tapered-roller bearings in high-speed applications. The principal types of roller bearings are cylindrical, needle, tapered, and spherical.

Cylindrical-roller bearings have rollers with approximate length-to-diameter ratios of 1:1 to 3:1. Needle-roller bearings have cylindrical rollers (needles) with greater length-to-diameter ratios (approximately 4:1 to 8:1). The rolling elements of tapered-roller bearings are truncated cones. Spherical-roller bearings are available with both barrel- and hourglass-shape rollers. Figure 1 shows the principal components of ball and roller bearings.

Bearing Materials

A significant improvement in air-melted bearing steels has occurred in recent years. Bearing steels are currently made by electric-furnace melting, ladle refining, and casting by either bottom or top pouring of ingots or continuous casting. The air-melted steel quality is approaching that of electroslag remelting (ESR) or vacuum arc remelting (VAR). However, steel from these secondary refining processes is still required for bearings in high-reliability applications, such as aircraft engines.

The bearing industry has used through-hardened material, such as 52100, for ball bearings since early introduction in Europe and carburized materials, such as 8620, for roller bearings since the turn of the century in the United States. A commonly accepted minimum surface hardness for most bearing components is 58 HRC. The carburizing grades have a core hardness range of 25 to 45 HRC.

At surface-hardness values below the minimum 58 HRC, resistance to pitting fatigue is reduced, and the possibility of brinelling (denting) of bearing raceways is increased. Because hardness decreases with increasing operating temperature, the conventional materials for ball and roller bearings can be used only to temperatures of approximately 150 °C (300 °F). Although ball bearings made of high-temperature materials, such as M50 (Fe-0.80C-4Cr-1V-4.25Mo), or roller bearings made of CBS1000M (Fe-0.13C-0.5Mn-0.5Si-1.05Cr-3Ni-4.5Mo-0.4V) are usable to approximately 315 °C (600 °F), the practical limit is actually determined by the breakdown temperature of the lubricant, which is 205 to 230 °C (400 to 450 °F) for the synthetic lubricants that are widely used at elevated temperatures.

Molybdenum high-speed tool steels, such as M1, M2, and M10, are suitable for use to about 425 °C (800 °F) in oxidizing environments. Grades M1 and M2 maintain satisfactory hardnesses to about 480 °C (900 °F), but the oxidation resistance of these steels becomes marginal after a long exposure at this temperature. An important weakness of these highly alloyed materials is their fracture toughness. Also, regardless of operating temperature, bearings require adequate lubrication for satisfactory operation.

For bearings that operate in moderately corrosive environments, AISI type 440C stainless steel should be considered. Its maximum obtainable hardness is about 62 HRC, and it is recommended for use at temperatures below 175 °C (350 °F). However, the dynamic-load capacity of bearings made from type 440C stainless steel is not expected to be comparable to that of bearings made from 52100 steel. The carbide structure of 440C is coarser, the hardness generally lower, and the fracture toughness about one-half that of 52100 steel.

Bearing-Load Ratings

Bearing-load ratings are based on results of laboratory rolling-contact fatigue tests that have been conducted under conditions as near ideal

Fig. 1 Principal components of rolling-element bearings

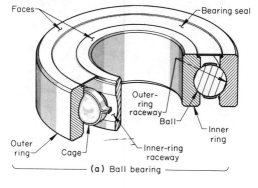

(a) Ball bearing

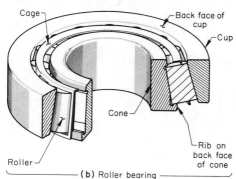

(b) Roller bearing

as possible. Any departure from these reasonably ideal conditions, such as misalignment, vibration, shock loading, insufficient or inefficient lubrication, extremes of temperature, or contamination, will reduce life expectancy and may cause a type of failure other than by rolling-contact fatigue.

The life of a bearing is expressed as the number of revolutions, or the number of hours at a given speed, a bearing will complete before developing fatigue spalling. Life may vary from bearing to bearing, but conforms to a statistically predictable pattern for large numbers of bearings of the same size and type operating under the same conditions. The L_{10} rating life of a group of such bearings is defined as the number of revolutions (or hours at a given constant speed) at which at least 90% of the tested bearings will survive. Similarly, the life reached or exceeded by 50% of the tested bearings is called the L_{50} life, or median life. The L_{50} life is about 3.5 to 5 times the L_{10} life.

The basic dynamic-load rating for a radial rolling-element bearing is the calculated constant radial load that 90% of a group of apparently identical bearings with stationary outer rings can theoretically endure for one million revolutions of the inner ring. The basic load rating is a reference value only because, as a load, it usually exceeds the yield limit of the bearing material. To avoid harmful plastic deformation, static-load capacities are published along with dynamic-load ratings in most catalogs.

Formulas used by most manufacturers to calculate basic load ratings of rolling-element bearings are given in Anti-Friction Bearing Manufacturers Association (AFBMA) standards. The basic load-rating life may be modified by life-adjustment factors, such as reliability, material, lubrication, alignment, temperature, and other environmental factors. Additional information regarding use of these formulas is provided in the AFBMA-ANSI Standards 9 and 11 and in literature from bearing manufacturers.

Experimental data have provided a simple relationship between load and bearing life. The bearing life, L_{10}, in millions of revolutions equals the ratio of bearing rating, C, to applications load, P, raised to an exponent, y, expressed as $L_{10} = (C/P)^y$, where $y = 3$ for ball bearings and $y = {}^{10}/_3$ for roller bearings. The exponential character of this relationship between basic load and bearing life indicates that, for any given speed, a change in load may have a substantial effect on life in hours.

Examination of Failed Bearings

If a bearing fails to meet its predicted life requirement, the analyst must discover the cause of damage that led to failure and recommend measures that will eliminate or control this damage. The influence of uncontrolled or unknown factors that can overshadow the effect

of the controlled variable in tests must be determined.

The analyst should first obtain complete information about the application and operating environment and, if possible, get samples showing the earliest detectable stages of damage. The complete bearing including the separator should be obtained rather than the failed part only. Second, the bearings and, if possible, the machine should be visually examined to determine the general condition of all related parts and the specific condition of the bearing components so that the type of damage can be identified. The bearing and the installation should be photographed to document the appearance of the damaged area. Such pictures are valuable for preserving the bearing condition prior to cleaning for future reference. Parts of this step may be carried out by an experienced field engineer. Third, samples of lubricant and debris should be collected for analysis. Fourth, the bearings should be thoroughly examined at low magnification (3 to 50×) to classify the type of damage. Significant areas may be photographed. Fifth, representative areas associated with the damage should be selected for metallographic study at high magnification (100× or higher) to verify the results of visual examination. A scanning electron microscope, if available, is an important tool that can provide important analysis and record photographs over a wide range of magnification. An entire small bearing can be placed on some microscope stages. Sixth, available information on load, speed, lubrication, operating temperature, and other environmental factors should be considered in comparing actual and expected life.

The analyst needs information about the application for comparison with previous experience to expedite analysis. The following information on the application and environment should be available:

- Description of the application, including the method of mounting the bearing
- Speed of rotation and whether rotation is constant or intermittent or of variable speed
- Lubrication system, lubricant, and filtering, if used
- Temperature of environment, lubricant, and bearing components
- Potential sources of debris
- Potential sources of electrical current passing through the bearing
- Potential sources of water and other corrosive fluids
- Load spectrum and sources of potential-unknown loads
- Misalignment of deflections that influence load distribution within the bearing system

Not all of this information may be needed for every analysis, but it is desirable because the necessary data cannot be established until considerable analytical work has been done. Later

acquisition of pertinent data is frequently difficult or impossible. Generally, the data come from the field representative, the bearing user, or both. The analyst should note comments by the field representative or bearing user regarding overheating, excessive noise, frequency of replacement, vibration, looseness, and resistance to shaft rotation.

Before a failed bearing is removed from its housing, the orientation of the outer race with respect to the housing and that of the inner race with respect to the shaft should be marked so that any distress pattern noted in subsequent inspection can be related to the conditions that might have produced it. Frequently, such orientation markings can be compared with similar markings applied at the time of original assembly.

Ball or Roller-Path Patterns. Damage to bearings usually results from subjection to loads or conditions other than those for which the bearings were designed. For example, misalignment or improper fit can result in loading that differs considerably, both in magnitude and direction, from that anticipated by the designer. Determination of such abnormal conditions by inspection of the location and distribution of damage on bearing components is often helpful. Most types of damage to ball bearings will be located on the path of ball travel. Because each type of loading produces its own characteristic ball path, the conditions of loading under which the damage occurs can be determined from an inspection of this ball path. Similar observations can be made on roller bearings.

Satisfactory operation of a bearing that has functioned under radial loads can be easily recognized. In radial ball bearings that have been properly mounted, operated under good load conditions, and kept clean and properly lubricated, the paths of the balls on the highly polished raceways appear as dulled surfaces (similar to lapped surfaces) on which microscopic grinding scratches have been smoothed out. No appreciable amount of material has been removed from the surface of the raceways or balls, as indicated by the fact that there is no measurable decrease in the diameter of the balls, although the entire surface has been dulled. Other indications of satisfactory operation are uniformity; exact parallelism with the side of the raceway, which indicates correct alignment; and centering of the ball path in the raceway, which indicates that loading of the bearing has been purely radial.

With unidirectional radial loading and a fixed outer ring (Fig. 2a), the load is normally carried on slightly less than half of the outer-ring raceway. The inner ring rotates through the loaded region; thus, the ball path on the inner-ring raceway extends around its entire circumference. Ball paths produced by unidirectional radial loading of a fixed inner-ring bearing are shown in Fig. 2(b). The load is carried by slightly less than half of the inner-ring raceway; the outer ring rotates through the load region

Fig. 2 Ball paths produced on raceways of ball bearings by axial and unidirectional radial loads

Small arrows indicate rotating elements. See text for discussion. Source: Ref 1

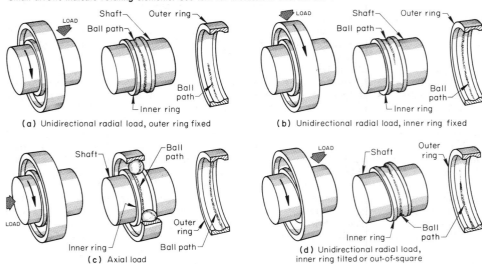

(a) Unidirectional radial load, outer ring fixed

(b) Unidirectional radial load, inner ring fixed

(c) Axial load

(d) Unidirectional radial load, inner ring tilted or out-of-square

and develops a ball path that extends around the entire outer-ring raceway.

The ball paths produced by unidirectional radial loading of a bearing whose fixed outer ring has crept rotationally in its housing are centered in the raceways, but the path on the outer raceway is longer than that shown in Fig. 2(a) by an amount equal to the length of the arc through which rotational creep has occurred. Any rotational creep that occurs is usually sufficient to make the ball path around the outer raceway of the bearing continuous. Evidence of such creep rotation may be visible on the surfaces at which creep occurs.

A radial-type load can be imposed on a bearing by the manner in which the bearing is mounted. Too tight a press fit of the outer ring in its housing or of the inner ring on its shaft will result in a more uniform ball path around the circumference of each raceway than would be seen with adequate internal clearance.

The ball paths produced by pure axial (thrust) loading are shown in Fig. 2(c). In an angular-contact bearing, the distance that the paths are displaced from the center of the raceways is greater than for a radial-contact bearing carrying an axial load, because of the greater angle of contact in the radial type. If the axial load is applied continuously, the balls do not have an opportunity to change their axis of rotation, and circumferential banding of the balls results (Fig. 2c).

If both radial and axial loads act together on a bearing, the path produced will depend on the relative magnitude of each load. If the axial load is large compared with the radial load, the paths will be similar to those produced by pure axial loading. However, if the radial load is greater than the axial load, the path on the inner raceway will be wider than the path on the outer raceway, and the outer-raceway path will be

tilted by an amount depending on the relative magnitudes of the radial and axial loads.

Another type of ball path is caused by misalignments, such as bearings being out of line, outer rings being tilted or out of square, or inner rings being tilted or out of square. Tilting of the outer ring in its housing causes the ring to become slightly oval, which results in a squeeze across one diameter and a relief across the diameter at right angles to the first. This causes narrowing of the ball path at the opposite sides of the diameter across which the relief occurs. If the tilt is not great enough to take up all the clearance, there will be no ball path on the upper (nonloaded) half of the outer raceway.

Figure 2(d) shows a bearing whose inner ring was tilted or out of square with respect to the axis of rotation. Again, the amount of tilt and the amount of bearing clearance affect ball-path characteristics. The paths shown are produced when the load is taken by the bottom portion of the bearing, that is, when some clearance still exists at the top of the bearing after tilting. The inner-ring ball path is at an angle in the raceway as shown and becomes narrower at two opposite points on the raceway. This is due to the ovality caused by pressing the ring on the shaft in a tilted manner. If the tilting were great enough to take up all the clearance at the top of the bearing, the path on the outer raceway would extend around the entire circumference.

Failure Types

Failure of rolling-element bearings can occur for a variety of reasons. Accurate determination of the cause of a bearing failure depends to a large extent on the ability of the analyst to recognize and distinguish among the various types of failures. In most instances, this recog-

nition will enable the analyst to determine the primary cause of failure and to make suitable recommendations for eliminating the cause.

The major factors that, singly or in combination, may lead to premature failure in service include incorrect fitting, excessive preloading during installation, insufficient or unsuitable lubrication, overloading, impact loading, vibration, excessive operating or environmental temperature, contamination by abrasive matter, entry of harmful liquids, and stray electric currents. The deleterious effects resulting from the above factors are as follows, though not necessarily in the same related order: flaking or pitting (fatigue), cracks or fractures, rotational creep, smearing, wear, softening, indentation, fluting, and corrosion (Ref 2).

Two or more failure mechanisms can be active simultaneously and can thus be in competition with one another to terminate the life of the bearing. Also, a mechanism that is active for one period in the life of a bearing can lead to or can even be supplemented by another mechanism, which then produces failure. Thus, in some instances, a single mechanism will be obvious; in others, indications of several mechanisms will be evident, making exact determination of the primary cause difficult. When more than one mechanism has been active, proper determination depends not only on careful examination of the failed components but also on analysis of the material and on the manufacturing, installation, and operating history of the bearing.

Before discussing the characteristics of the various types of failure in detail, it is pertinent to mention that a rolling-element bearing does not have an unlimited life—a fundamental fact that does not appear to be generally appreciated. Even if a bearing is run under the recommended conditions of load, speed, and lubrication and is protected against adverse external influences that would otherwise tend to reduce its life, failure will ultimately result by some process, such as fatigue, wear, or corrosion.

When a bearing is loaded, elastic deformation occurs at the zones of contact between the rolling elements (balls or rollers) and the raceways. Although the stresses involved appear to be compressive only, just below the surface there are high shear stresses and low tensile stresses. Subsurface stresses are described in more detail later in this article. Repeated stressing in shear resulting from rotation of the bearing ultimately leads to the initiation of fatigue cracks, and fragments of metal become detached to produce the effect described as pitting or flaking. This type of failure is similar to that found on the flanks of gear teeth. In addition to load, the pitting or flaking of material from the surface of a bearing race depends on the lubricant viscosity and/or film thickness. Various theories have been proposed that attribute the spalling to hydrostatic pressure of the lubricant being forced into surface cracks to propagate them. Another theory states that the propagation is caused by the tractive forces in the contact causing the subsurface shear stresses to be at the surface.

Experimental data indicate that a rolling-element bearing, considered as a composite whole, does not appear to possess a fatigue limit, such as is found with the materials from which it is constructed; that is, a rolling-element bearing has no lowest value of loading below which failure will not occur no matter how many stress cycles are applied. Typical characteristics and causes of several types of failures are given in Table 1.

Failure by Wear

Rolling-element bearings are designed on the principle of rolling contact rather than sliding contact; frictional effects, although low, are not negligible, and lubrication is essential. If a bearing is adequately and correctly lubricated, wear should not occur. In practice, however, the entry of dirt, hard particles, or corrosive fluids will initiate wear, causing increases in the running clearances of the various parts of the bearing, which may lead to noisy operation and early failure.

Significant amounts of material can be removed from contact surfaces under relatively mild operating conditions if (1) abrasive contaminants are present in the contact area, (2) there is highly inadequate lubrication, or (3) the contact surfaces are subjected to a high rate of sliding. This failure mode is characterized by the presence of wear debris in the lubricant and by evidence of abrasion on the contact surfaces of the raceways and rolling elements. Generally, the surface character is impaired only gradually, and some rolling-element bearings can sustain considerable wear before they become unsuitable for operation. Wear failure can be evaluated by dimensional or weight measurement of the components, by observation of loose wear particles near the contact area, or by spectrographic analysis of the lubricant for wear particles.

Dirt. Ball bearings are particularly sensitive to dirt or foreign matter, which is always more or less abrasive, because of the very high unit pressure between the balls and raceways and because of the rolling action of the balls, which entraps foreign particles, especially if they are small. Foreign matter may get into the bearing during initial assembly of the machine, during repairs, or by seepage from the atmosphere into the bearing housing during operation of the machine. Dirt can also enter bearings as adulterants in the lubricant.

The character of the damage caused in bearings by different types of foreign matter varies considerably. Foreign matter that is fine, or soft enough to be ground fine by the rolling action of the balls or rollers, will have the same effect as that resulting from the presence of a fine abrasive or lapping material. Both raceways and rolling elements become worn, and the bearings become loose and noisy. The lapping action increases rapidly as the fine steel debris

Table 1 Characteristics and causes of some common failures of rolling-element bearings

Characteristic and/or location	Cause and remarks
Wear	
Dull raceway surfaces	Coarse abrasive matter in bearing
Shiny raceway surfaces	Fine abrasive matter in bearing
Dull indentations in raceway at same spacing as rolling elements	See *Fretting* and *Indentation*
Wear, together with discoloration	Running without lubricant or with lubricant that has hardened
Wear at points of contact with cage	Inertia forces acting on cage and insufficient lubricant
Wear of ball cage	Inner or outer rings out of square; abrasive matter in bearing
Internal looseness from wear of raceway surfaces and rolling elements	Abrasive matter in bearing because of insufficient or inefficient lubrication or ineffective filtration of lubricant
Uneven wear in ball path	Vibration in presence of abrasive matter
Shaft wear or wear in bearing or housing bore	Rotational creep because of loose fit
Electrical Pitting	
Small craters uniformly distributed over ball paths	Continuous passage of electric current
Small craters randomly oriented or in bandlike order along ball paths	Intermittent passage of electric current
Fluting	
Ridges with burnt craters	Vibration together with electrical current passing through rotating bearings
Ridges without burnt craters	Vibration together with wear or excessive overload on rotating bearings
Smearing	
Cage material smeared onto the rolling elements	Jamming of rolling elements in cage pockets; too high speed; inefficient lubrication; inertia forces
Smeared peripheral streaks in bore or on periphery of outer ring	Rotational creep from inner ring being loose on shaft or outer ring being loose in housing bore
Smeared peripheral streaks on end face of ring	Rotational creep as above, or sliding under pressure against shoulder or other surface
Axial smear marks on raceway and/or rolling elements of cylindrical and spherical roller bearings	Mounting under load; inner and outer rings forcibly assembled out of square with each other; axial displacement between inner and outer rings while under load
Axial smear marks on spherical raceway of self-aligning bearings	Angular movements of shaft while bearings are stationary under load
Axial smear marks on spherical raceway of self-aligning thrust bearings	Radial movement in unloaded bearings; improper mounting; inefficient lubrication; faulty manufacture of parts adjacent to bearing
Peripheral streaks on raceways, cage pockets, or rolling elements	Inefficient lubrication or light loads and high speeds
Spiral streaks on ends of rollers or locating flange	Inefficient lubrication under axial load; running under no load because of large radial clearance
Fretting	
Red or black oxide spots, usually with shiny borders, in bore or on periphery of outer ring	Incomplete contact and slight movement or vibration between ring and its seat
Red oxide on surface where ring contacts shoulder	Slight movement due to shaft flexure
Dull indentations on raceway at same spacing as rolling elements	Vibration in stationary bearing, particularly in presence of abrasive particles; referred to as false brinelling
Corrosion	
Local spots or pits on raceways, same spacing as rolling elements	Moisture or acid in bearing while stationary for extended periods
Spots or pits on surfaces	Corrosive lubricant or free water in lubricant; moisture on unprotected surfaces; corrosive atmosphere
Indentation	
Shiny indentations on raceway at same spacing as rolling elements	Improper mounting procedure, hammer blows during mounting, or excessive load on stationary bearings; referred to as true brinelling
Dull indentations on raceway at same spacing as rolling elements	Vibration in stationary bearing, particularly in presence of abrasive particles; referred to as false brinelling
Indentations near filling slot or other isolated area on raceway	Excessive external force when assembling rolling elements
Irregular indentations all over rolling elements	Abrasive products or other foreign matter

(continued)

Table 1   (continued)

Characteristic and/or location	Cause and remarks
Flaking	
At isolated areas on raceway	Early stage of fatigue originating at indentations from foreign matter, bruise on raceway, or corrosion
All around one raceway	Progressive fatigue from overload or insufficient clearance because of inner-ring expansion or outer-ring contraction
At diametrically opposite points on radial bearings	Ring oval through distortion by mounting on out-of-round shaft or in out-of-round housing bore
On only one side of raceway surface	Improper mounting or excessive axial load
At one end of roller raceway	Ring misaligned with cylindrical roller bearings
Equally spaced apart as the spacing of rolling elements	Either true brinelling or false brinelling, depending on appearance of surface; see *Indentation*
Oblique flaking on raceway of rotating shaft	Misalignment or deflection of shaft; ring out of square
Oblique flaking on raceway of stationary shaft	Shaft deflection or ring out of square
On rolling element	Forced assembly; overloaded or insufficient lubrication
Eccentric pitting on raceway of thrust bearing	Eccentric mounting or loading
Cracking and Fracturing	
Cracks through ring	Fit too tight; nonuniform seating surface; deformation or ovality of housing; rotational creep or fretting
Axial cracks in inner-ring bore or on periphery of outer ring	Rotational creep or fretting
Circumferential cracks in rings	Deformation of housing; nonuniform seating surface; excessive overload
Radial cracks on end face of rings	Smearing resulting from rotational creep
Radial cracks on end face of rotating ring	Fouling or rubbing on housing or shoulder during operation
Fractured flange on ring of roller bearing	Mounting pressure unevenly distributed around flange; hammer blows during mounting
Crack or fracture at root of cage tongues or across cage pocket	In bearings generally: insufficient lubricant; too high speed or inertia forces; smearing; fracture of rolling elements; misalignment. In thrust bearings: ring mounted eccentrically or out of square; one row of balls not under load
Miscellaneous	
Enlargement of inner and outer rings	Operation for prolonged periods at high or low temperature; aging of ring material
Superficial discoloration of material without reduction in hardness	Lubricant affected by temperature rise; film deposited by extreme-pressure (EP) lubricants
Abnormal temperature increase	Low lubricant viscosity; too much lubricant, resulting in churning; insufficient internal clearance; speed too high; excessive load; incipient failure
Pronounced high noise level	Raceways indented by blows during mounting or by vibration when bearings are stationary
Piping or metallic sound	Insufficient internal clearance; inadequate lubrication
Irregular noise	Foreign matter in bearing; incipient flaking or other surface discontinuities on the raceways; acoustical properties of adjacent parts

Source: Ref 13

from the bearing surfaces adds more lapping material.

A full-complement (uncaged) drawn-cup needle-roller bearing that failed because of wear by fine abrasive material in the bearing is illustrated in Fig. 3(a). Wear decreased each roller diameter by more than 0.05 mm (0.002 in.), and a block of rollers eventually turned in the cup. The flow of lubricant (and abrasive) through the bearing was sufficient to forestall development of heat, although polished flats are much in evidence on the rollers.

Hard particles, usually nonmetallic, of any size that will scratch, cut, or lap the surfaces of a bearing must be regarded as abrasive. Figure 3(b) shows the effects of natural diamond particles (≤ 5 μm) that were added to the lubricant in a laboratory test. The wear, known as third-particle abrasive wear, was very extensive on all load-carrying surfaces. Some sub-stances that might not normally be considered as hard or even abrasive, such as iron oxide powder, are excellent lapping compounds.

Hard, coarse foreign particles or iron, steel, or other metals introduced during assembly of the machine produce small depressions considerably different from those produced by overload failure, acid etching, or corrosion. Jamming of such hard particles between rolling elements and raceways may cause the inner ring to turn on the shaft or the outer ring to turn in the housing.

Corrosive Fluids. Water, acid, or other corrosive fluids, including corrosives formed by deterioration of the lubricant, produce a type of failure that is characterized by a reddish-brown coating and very small etch pits over the entire exposed surfaces of the raceways. Frequently, such etching is not evident on the ball path because the rolling action of the balls pushes the lubricant, loaded with corrosive, away from the ball path. The corrosive oxides formed act as lapping agents that cause wear and produce a dull gray color on the balls and the ball paths, in contrast to the reddish-brown color of the remainder of the surface.

Figure 4(a) shows a failure traceable to vibration in the presence of dirt and moisture. General rusting and deep pits can be seen in this drawn cup, as well as polished indentations resulting from abrasive action under vibratory conditions. The roller spacing is indicated by the spacing of the polished indentations.

Some lubricants decompose after extended service without replacement or replenishment, forming acidic by-products. In the presence of moisture, bearing surfaces may be badly attacked by what is referred to as black-acid etching. Such attack produces numerous stress raisers that, under subsequent heavy service loads, can lead to spalling. Figure 4(b) shows an example of water-etched raceways of tapered-roller bearings.

Electrical pitting is produced by passage of electrical current between two surfaces. In such applications as electrical equipment and railway cars and locomotives, there is a possibility of electric current passing through the bearings. When the current is broken at the contact surfaces between raceways and rolling elements, arcing or sparking occurs, producing high temperatures and localized damage. The overall damage is proportional to the number and size of localized points.

A peculiar characteristic of the passage of electric current for prolonged periods of time is the fluting that sometimes occurs. Fluting is pitting in which cavities occur in a regular pattern, forming grooves (flutes). It can apparently occur in any type of bearing. Flutes sometimes develop considerable depth, producing noise, vibration, and eventual fatigue from local overstressing. The cause of fluting is not definitely known. However, it is believed that arcing takes place on the trailing edges of the rolling elements, resulting in local vaporization of metal and the production of craters and pits. This spark-erosion effect frequently gives rise to rapid vibrations that assist in producing the characteristic fluting effect. In some instances, under high magnification, it is possible to observe the burned or fused metal associated with individual craters. Figure 5 shows an example of fluting, including a micrograph of the hundreds of individual arc spots that help create the flutes. The effects of the passage of electric current are usually shown on raceways, but they may occasionally be found on rolling elements, particularly if sparking occurred while the bearing was stationary.

If fluting is found in bearings in rotating electrical machinery, it is reasonable to suspect that it resulted from passage of electric current associated with defective insulation or induction effects. Remedial measures to eliminate such stray currents include incorporation of insulation within bearing housings or at bearing

Fig. 3 Bearings that failed because of wear by abrasive material in the bearing

(a) Needle-roller bearing. Note that flats have been worn on the rollers. (b) Abrasive wear caused by natural diamond dust (≤5 μm) that was deliberately introduced into the lubricant in the laboratory. Deep grooving of the raceways and excessive wear of the rib and roller ends is evident. Similar wear can occur from hard third particles, such as bearing or gear materials, sand, or machining chips. Source: Ref 4

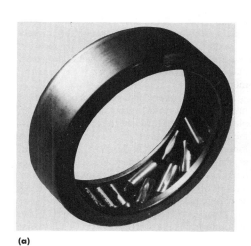

(a)

(b)

pedestals and use of suitable short-circuiting connections between the rings or cones.

Pitting of bearing elements by electrical action causes noisy bearing operation. In the following example, use of an electrically nonconductive grease prevented proper grounding.

Example 1: Pitting Failure of Ball Bearings in an Electric Motor by Static Electrical Discharges. The electric motor in an office machine was producing intermittent noise. Ball bearings were suspected as the source of the noise and were removed from the motor and sent to the laboratory for examination.

The bearings were made of type 440C stainless steel hardened to 60 HRC and measured 41 mm OD × 19 mm ID × 12.7 mm wide (1.625 × 0.750 × 0.500 in.). Specifications required the bearings to be lubricated with a grease that could conduct electricity. Bearing life was approximately 200 h.

Investigation. The ball-bearing rings were sectioned for examination in an optical microscope. The surfaces of the balls were observed in a scanning electron microscope. A number of spots were found on the inner-ring raceway (Fig. 6a). An enlarged view of one spot (Fig. 6b) shows more clearly an area from which metal was removed and evidence of a heat-affected zone (HAZ) around the area.

Figure 6(c) shows a spot on a ball where metal was removed. The metal in the area around the spot showed evidence of having been melted. Another spot on a ball is shown in Fig. 6(d). This ball was welded to the raceway in the region shown at the top on Fig. 6(d) and was separated from the raceway as it rotated. Melting of the metal and welding were indicated by small round beads found in deep

indentations in both the ball and the inner-ring raceway. The metal was melted and joined together for only a brief period of time and was then pulled apart by the force of rotation of the ball and the inner-ring raceway.

Because welded areas were found on the raceways of both the inner and outer rings and were randomly spaced and not continuous, it was thought that welding was the result of short electrical discharges between the bearing raceways and balls. The electrical discharges were suspected to have resulted from use of an electrically nonconductive lubricant in the bearings, which permitted a static charge to be accumulated and discharged. To confirm this, a motor made by the same manufacturer was obtained from stock for testing. The electrical resistance between the rotor and the motor frame was measured and was found to be 2 to 3 MΩ while the motor was running. A set of bearings lubricated with a grease known to conduct electricity was installed in the motor. The resistance between rotor and motor frame with this arrangement was only 0.01 to 0.02 MΩ. When not running, the motor had a resistance of less than 0.5Ω (5 × 10⁻⁷ MΩ).

As a further check on the electrical conductivity of the lubricant in the failed motor, the resistance across two copper plates was measured using each type of grease as a separator. The result indicated that the grease in the failed bearings was a poor conductor of electricity.

Conclusions. Intermittent electrical discharges across the nonconductive lubricant in the ball bearings resulted in melting and welding of metal at several locations. Because welding was momentary, pits were produced in the ball and ring surfaces.

Corrective Measures. The lubricant was replaced by an electrically conductive grease. This permitted proper grounding of the rotor and eliminated welding between balls and raceways.

Example 2: Failure of a Low-Alloy Steel Bearing in an Electric Motor Because of Stray Electric Currents. The roller bearing shown in Fig. 7 was one of two that were mounted in an electric motor/gearbox assembly. Rough operation of the bearing was observed, and the unit was shut down. The bearing was removed and submitted for laboratory analysis.

The bearing components were made of a low-alloy steel, such as 4620 or 8620. The cup, cone, and rollers were carburized and hardened and tempered to 59 to 64 HRC.

Investigation. Visual examination of the cup, cone, and rollers revealed that their contact surfaces were uniformly electrolytically etched. This could have occurred only as a result of an electric current and the presence of an electrolyte, such as moisture, between the cup and roller surfaces of the bearing. The action was similar to anodic etching, as in electroetching of metallographic specimens, only more severe.

Conclusion. Failure of the bearing was the result of electrolytic etching of the contact surfaces by stray currents in the electric motor, which was not properly grounded. The etching was caused by uncontrolled input of current for a long period of time and the presence of ample electrolyte (moisture).

Corrective Measures. The motor was electrically grounded, and the bearing was insulated for protection from stray currents. Both bearings in the assembly were then sealed to keep out moisture.

Fig. 4 Bearings that were damaged by corrosive fluids

(a) Drawn cup for a needle-roller bearing damaged by vibration in the presence of dirt and moisture. Roller spacing is indicated by polished indentations. (b) and (c) Corrosion or water etch on bearing components resulting from a defective seal in a grease-lubricated bearing application. (b) 1.5×. (c) 75×. Source: Ref 4

(a)

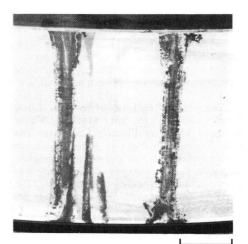

(b)

(c)

0.1 mm

100 μm

Fig. 5 Tapered-roller bearing damaged by electrical pitting

(a) Fluting damage caused by continuous passage of electrical current. (b) A roller from (a) polished on the outside diameter and etched with nital to show the many individual arc marks that led to the destruction of the raceway surfaces. Source: Ref 4

(a)

(b)

Cumulative material transfer, a type of failure sometimes called smearing or galling, results from considerable sliding between bearing contact surfaces. It is characterized by the formation of welded junctions between the contact surfaces. This welding usually occurs on a localized microscopic scale and results in removal of material from one or both bearing surfaces by a tearing action. Wear particles may form, which are sometimes work hardened and cause abrasive wear. However, metal buildup usually occurs on one or more of the bearing surfaces.

Smearing failures appear to be confined to contact surfaces that have undergone considerable sliding. Smearing is generally a self-aggravating condition and, if allowed to progress, often is followed by a temperature-imbalance failure resulting from the high friction between the smeared surfaces. If the operating conditions that lead to smearing are relaxed, the failure may be arrested, and there will be some plastic smoothing and possibly wearing away of the transferred material, which may permit continued operation. The transferred metal appears as a bright fracturelike patch.

One cause of sliding is hardening of grease lubricant due to deterioration or contamination by dirt or other foreign matter. As a result of hardening of grease, rotation of the balls or rollers within the cage may be effectively prevented, and in some instances, the bearing can freeze, for example, to such an extent that it will prevent rotation of an electric motor.

Sliding of rollers and subsequent smearing may also occur in a bearing if wear has taken place to a degree sufficient to permit skewing of the rollers so that their axes of rotation are no longer parallel to that of the shaft. Smearing, in a characteristic cycloidal pattern, may be frequently observed on the end faces of rollers as a result of sliding contact between the end faces and their retaining flanges. The possibility of smearing, not the fatigue properties of the material, limits the amount of axial loading that a roller bearing fitted with flanges on both cones can withstand. Inadequate lubrication or a lubrication failure can be the cause of smearing of cage material on rolling elements or adhesive wear between the roller end and cone rib (Fig. 8).

Rotational creep, when the term is applied to bearing-failure analysis, refers to an effect in which relative motion takes place between a press-fitted bearing ring and its shaft or housing. Rotational creep occurs principally on the inner ring and may result from the ring being initially loose on the shaft as a result of fitting errors or from plastic or elastic deformation due to either abnormal loads or severe imbalance in service. Under rotational-creep conditions, the assembly behaves like a friction gear, with rotation resulting from the very slight difference in the circumferences of the shaft and the inner ring.

Rotational creep may also occur in thrust bearings if the shaft shoulder against which the ring abuts is not perpendicular to the shaft axis. The relative motion that occurs produces heavy score marks and polished regions, both in the bore of the inner ring and on the shaft. The end face of the ring, which is in frictional contact with the shoulder on the shaft, shows a characteristic scuffing effect typified by heavy circumferential score marks and, in severe cases, by short radial cracks. This effect is similar to local seizure, and it is believed that the cracking arises from thermal stresses induced by intermittent local heating. In some instances, cracks can extend completely through the ring.

Scuffing effects, in addition to those resulting from rotational creep, may occur if a bearing locks solid. Scuff marks may be present on an outer ring, particularly if the bearing rotates as a whole, and a high polish is generally produced on the outer cylindrical surface where it rubs against the housing.

Fig. 6 Weld spots on contact surfaces of a type 440C stainless steel ball bearing

The spots are caused by static electrical discharges resulting from use of an electrically nonconductive grease. (a) and (b) Photographs of inner-raceway surface. 10 and 100×, respectively. (c) and (d) SEM micrographs of ball surfaces. Both 1000×

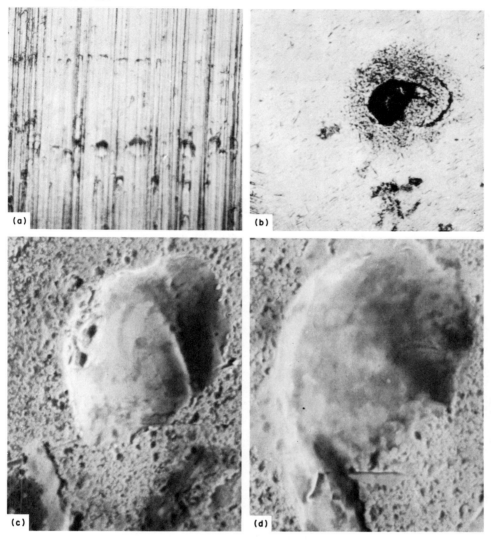

Fig. 7 Low-alloy steel roller bearing from an improperly grounded electric motor that was pitted and etched by electrolytic action of stray electric currents in the presence of moisture

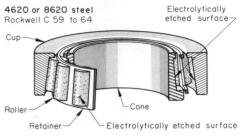

4620 or 8620 steel
Rockwell C 59 to 64

machines subjected to vibration while at rest.

There are several features that permit false brinelling to be distinguished from true brinelling (see the section "Brinelling" in this article). False brinelling is found on those regions of a bearing that are subjected to the heaviest loads, for example, at the 6 o'clock position in a bearing supporting a horizontal shaft; the effect diminishes in a fairly regular manner on either side of this position.

False brinelling, although most often found in machines on standby duty subjected to vibration from nearby running machinery, has also been found to occur during transit. Transport by rail or truck is conducive to this form of damage. Even transport by sea does not guarantee immunity, particularly where pronounced vibrations from diesel engines may be present. False brinelling has also been found in the roller-bearing axle boxes of railway cars that have been standing for some time in sidings and that are subjected to vibrations from passing trains.

In many instances, false brinelling remains undetected until the machine is started, at which time noisy operation is immediately evident. The phenomenon is thought to be a manifestation of fretting. The minute movements initiated by vibration cause wear of the contact surfaces, and the fine particles produced rapidly oxidize and result ultimately in production of characteristic grooves, with the oxide acting as an abrasive. To prevent false brinelling in bearings of standby equipment, it may be necessary to arrange for continuous slow rotation of shafts while nearby machines are running.

For a given angular rotation of a shaft resulting from vibration, slightly greater movements of rolling elements relative to raceways will occur in a bearing with a greater number of elements. These movements may be greater than those at which false brinelling develops. Therefore, where false brinelling is not severe, changing to a bearing with a larger number of

Failure by Fretting

Fretting is variously referred to as fretting corrosion, vibrational false brinelling, friction oxidation, and even chafing or wear corrosion. Such damage is common where there is vibration or low-radial angle oscillation between the bearing elements such that relative slip of the involved bearing-element surface takes place. Corrosion, which is part of two of the terms for fretting given above, describes the resultant wear product, red iron oxide (Fe_2O_3)—the usual evidence of fretting of opposing steel contact surfaces. However, ordinary red rust ($Fe_2O_3 \cdot H_2O$) looks the same. The by-product of fretting may also be black iron oxide (FeO or Fe_3O_4).

Types of Fretting. Fretting may be of two types. Type 1 is the contact corrosion (fit rust) that takes place between the bore of the bearing and the shaft or between the outside surface of the bearing and the bore of the housing. Type 2 is fretting damage within the bearing-contact area. This type of fretting is frequently referred to as false brinelling.

False brinelling is caused by vibrations or oscillations over a few degrees of arc between rolling elements and raceways in a nonrotating bearing. At the contact areas between the rolling elements and raceways, lubricant is squeezed out, resulting in metal-to-metal contact and localized wear. False brinelling does not occur during normal running, but is found in the bearings of

Fig. 8 Examples of adhesive-type wear caused by inadequate lubrication

(a) Metal pickup on a roller outside diameter from sliding contact on the cage shown in (b). Approximately actual size. (c) The end of the same roller showing the scoring damage from the rolling-sliding contact of the thrust rib shown in (d). 3×. Source: Ref 4

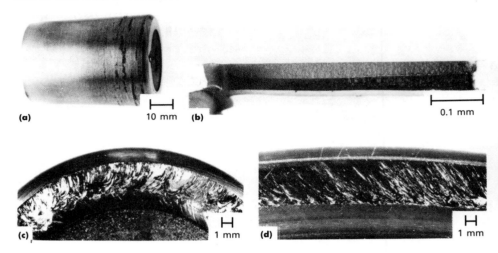

Fig. 9 Severe damage from fretting (false brinelling) on the surface of a shaft that served as the inner raceway for a needle-roller bearing

rolling elements or to one of the needle-roller type may be sufficient.

When equipment is to be transported, one of the following preventive measures should be adopted: (1) dismantle the machine, (2) remove the rolling-element bearings and fit temporary wooden packings in their place, or (3) use clamps to lock rotors of electric motors rigidly to the frames and, at the same time, relieve the bearing of the dead-weight load. In other instances, the machinery could be packed so that the axis of the rotating portion is in the vertical plane, thereby preventing damage by stationary indentation.

Recommendations for reducing fretting in rolling-element bearings include:

• Keep radial play in the bearings at the lowest practical value
• Increase the angle of oscillation (if possible) to secure roller or ball overlap in order to drag fresh lubricant into the area. If the surfaces can be separated by lubricant, fretting of the metal cannot occur
• Relubricate frequently to purge the red iron oxide debris and reinstate the lubricating film
• Use a larger bearing of higher capacity to reduce contact loads
• Increase the hardness of the elements as much as possible. The commercial antifriction bearing already has fully hardened components. For best results, the shaft, if used as an inner raceway, should be hardened to 58 HRC minimum
• Use a grease that has been specially formulated to provide maximum feeding of lubricant to areas susceptible to fretting damage. If the bearing is oil lubricated, flood it if possible

The corrugated surfaces produced by fretting or false brinelling form stress raisers that, under

subsequent conditions of rotation, may produce excessive noise and may cause premature spalling by rolling-contact fatigue.

Figure 9 shows a portion of a shaft that served as the inner raceway for a drawn-cup needle-roller bearing. The rollers left deep and clearly defined impressions on the shaft. The damage was identified as fretting, or false brinelling, because of the dull surface with little or no trace of the original surface finish remaining at the bottoms of the indentations, although the shaft was also brinelled by a very heavy overload.

Example 3: Fretting Failure of Raceways on 52100 Steel Rings of an Automotive Front-Wheel Bearing. The front-wheel outer angular-contact ball bearing shown in Fig. 10 generated considerable noise shortly after delivery of the vehicle. The entire bearing assembly was removed and submitted to the laboratory for failure analysis.

The inner and outer rings were made of seamless cold-drawn 52100 steel tubing, the balls were forged from 52100 steel, and the retainer was stamped from 1008 steel strip. The inner ring, outer ring, and balls were austenitized at 845 °C (about 1550 °F), oil quenched, and tempered to a hardness of 60 to 64 HRC.

Investigation. Visual examination of the outer raceway revealed severe fretting and pitting in the ball-contact areas at spacings equivalent to those of the balls in the retainer. The areas were elongated, indicating that only a slight oscillation of the ring or wheel had occurred. Similarly spaced but less severely damaged areas were observed on the inner raceway.

Conclusions. Failure was caused by fretting due to vibration of the stationary vehicle position without bearing rotation. These vibrations were incurred during transportation

Fig. 10 Automotive front-wheel bearing that failed by fretting of raceways on inner and outer 52100 steel rings

Dimensions given in inches

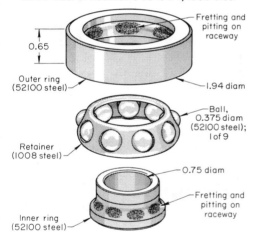

of the vehicle. This bearing had experienced little if any service, but showed conclusive evidence of fretting between the balls and raceway.

Recommendations. Methods of securing the vehicle during transportation should be improved to eliminate vibrations.

Failure by Corrosion

Corrosion of any or all components of a bearing can result from the entry of water or other liquids or from deterioration of the lubricant. The formation of deleterious decomposition products in lubricants is discussed in the section "Corrosive Fluids" in this article.

General corrosion of the hardened bearing surfaces often takes the form of minute pitting, which leads to noisy operation and provides surface discontinuities at which cracks can originate. Corrosion of bearings can take place during nonoperating periods or even when the bearings are being stored waiting installation, although corrosion in storage should not occur if the bearings are properly packaged.

Corrosion damage may not be externally visible and may show itself only by subsequent noisy operation. In this respect, it is generally recommended that the protective lubricant in which the bearings have been packed by the manufacturer not be removed unless, in the case of very small bearings, it is desirable to replace the lubricant by one having lower viscosity. The presence of a small amount of moisture in a lubricant, although perhaps not sufficient to cause general corrosion, may result in a lowering of the fatigue life of a bearing ring.

If a cold bearing is degreased or washed in a cold, volatile solvent, moisture will condense on the cold bearing surfaces. The moisture may not be noticeable, but if the bearing is then dipped in cold oil or is covered with a cold slushing compound, the moisture cannot evaporate and will expend itself in corroding the steel. If the bearing is immersed in hot oil or slushing compound of suitable temperature and assumes the temperature of the bath, the moisture will be driven off, and no corrosion should occur. The presence of water or acids in the lubricant of a running bearing will sometimes cause lines of corrosion to develop in the direction the lubricant is squeezed out from between the rolling elements and raceways.

Failure by Plastic Flow

Bearing components, such as balls, rollers, and rings, are generally made of high-strength steels with a hardness of 58 to 64 HRC. However, even at this hardness, overloading of metal components in rolling contact will cause plastic flow. The geometric relationship of the contact surfaces can be altered by cold flow or by unstable overheating.

Cold Plastic Flow. The primary effect of plastic flow on the operation of a bearing is geometric distortion. The contact surfaces can be damaged by formation of indentations if the bearing is stationary or by distortion if the bearing is rotating. Plastic flow on a small scale occurs under normal contact loads and operating temperatures. This has been recognized in defining a static capacity for rolling-element bearings. The absence of all plastic flow is not required, but flow must be limited to a tolerable level. Plastic flow in a rolling-element bearing may be evaluated by measuring the distortion of the contact area.

Rolling-element bearings can be damaged by indentations of several forms, each from a different cause: (1) shallow, random marks from rolled-in particles (Fig. 11), (2) shallow indentations in the raceways corresponding to regions in contact with rolling elements when

Fig. 11 Denting damage to cone, cup, and rollers from debris entering the bearing from spalling in another component, through defective seals, or from improperly cleaned housings

10 mm

the bearing is stationary, and (3) well-defined, uniformly spaced indentations.

The first form of indentation results in rolling-contact fatigue, and the second causes bulk damage (damage outside the contact zone). The third form of indentation is a type of cold plastic flow. The well-defined indentations spaced at a uniform pitch generally result from balls or rollers having been inadvertently forced into the raceways and often arise from carelessness or the implementation of incorrect procedures in withdrawing or installing rings. In the most common situation, the force required to assemble the bearing is applied indirectly through the balls or rollers instead of directly to the ring itself. Careless handling of the machine in transit can also damage bearings, particularly where impact loading has been sustained. Similar, regularly spaced marks may be produced on the raceways of a bearing that has been forcibly assembled with the axis of one ring at a slight angle to the other. These effects (to which the term brinelling is sometimes applied) usually manifest themselves as noisy operation.

Brinelling, sometimes referred to as true brinelling or denting, is permanent deformation

produced by excessive pressure or impact loading of a stationary bearing. Hollows or dents are produced by plastic flow of metal. Such indentations in raceways will be spaced in correspondence to ball or roller spacing, thus differentiating true brinelling from dents made by debris, such as dirt or chips, which usually are spaced randomly. The final grinding marks made during the manufacture of raceways can be seen, virtually undisturbed, within the true brinelled areas of Fig. 12. Improper mounting procedures, such as forcing of a tight-fitting or cocked outer ring into a housing, are common causes of true brinelling. Improper handling, such as dropping or pounding of bearings, will cause true brinelling. True brinelling can also be caused by vibrations in ultrasonic cleaning.

Example 4: True Brinelling of Ball-Bearing Raceways During Ultrasonic Cleaning. During the early stages of production, randomly selected dictating-machine drive mechanisms, which contained small ball bearings, were found to exhibit unacceptable fluctuations in drive output. This seemed to indicate that the bearing raceways were being true brinelled before or during installation of the bearings.

Fig. 12 Comparison of true brinelling with false brinelling

(a) In true brinelling, the surface texture is essentially undisturbed. Note grinding marks in the impression. (b) In false brinelling (fretting), the surface is worn away or material is transferred. No grinding marks are visible in the impression. Both 15×. Source: Ref 4

(a)

(b)

Fig. 13 Surface of a shaft, which served as an inner bearing raceway, that failed by spalling initiated at true-brinelling indentations

Investigation. The preinstallation practices and the procedures for installing the bearings were carefully studied to determine the cause of failure. New control practices were instigated with no net gain in bearing performance.

One preinstallation procedure involved removing the bearing lubricant applied by the bearing manufacturer and relubricating the bearings with another type of lubricant. The bearings were ultrasonically cleaned in trichlorethylene to ensure extreme cleanliness. Careful examination of the bearing raceways at a moderate magnification revealed equally spaced indentations resembling true brinelling. Further examination showed that the ultrasonic cleaning technique was improper in that the ultrasonic energy transmitted to the balls brinelled the raceways enough to cause fluctuations in machine output.

Conclusion. The bearing raceways were true brinelled during ultrasonic cleaning.

Corrective Measures. The lubricant was removed from the bearings by solvent-vapor cleaning instead of by ultrasonic cleaning.

Spalling that originated at true-brinelling indentations in the surface of a shaft is shown in Fig. 13. The shaft surface in the failed area served as the inner raceway of a bearing.

Softening is a plastic-flow phenomenon observed in rolling-element bearings with unstable thermal balance resulting from the generation of more heat in the bearing than is being removed. The maximum permitted temperature in bearings depends on the material, but is generally quoted as approximately 120 °C (250 °F). An increase in operating temperature, if undetected, may result in lubricant failure and seizure of the bearing or in softening of the bearing steel with a reduction in L_{10}. Gross overheating of bearings above the temperature

at which the rolling elements and rings were tempered during manufacture will result in rapid softening of these parts with subsequent plastic deformation. Occasionally, if the applied torque is sufficient to overcome the inherent resistance of a seized bearing or to cause the shaft to rotate within the inner ring, the heat generated may result in temperatures approaching a dull red heat, leading to loss of strength in the shaft followed by rapid failure from shear.

Gross destruction of heat-softened components can sometimes be recognized by discoloration and a predominantly plastically deformed appearance (Fig. 14). Therefore, it may be different in appearance from fracture that results from cracking. However, temperature-imbalance failures are often so devastating that little is left of the bearing from which to identify a failure source.

Failure by Rolling-Contact Fatigue

Failures caused by rolling-contact fatigue are the result of cyclic shear stresses developed at or near bearing-contact surfaces during operation. The effects on the shear-stress field below the surface of a bearing raceway as a cylindrical roller travels across the surface are shown schematically, from right to left, in Fig. 15. The roller is shown at position 1 in Fig. 15(a). In the material directly below the roller, the shear stresses along the x, y, and z axes are equal to zero and are not shown in the stress diagram below the roller, but the maximum shear stress, τ_{max}, is indicated on planes at 45° to the coordinate axes. The material just to the left of the roller (position 2, Fig. 15a) is

subjected to shear stresses, τ_{zy} and τ_{yz}, in the directions of the coordinate axes. In position 3, Fig. 15(a), the shear stresses are zero. In Fig. 15(b), the roller is shown in position 2. The 45° shear stress in the material below the roller is at a maximum, and τ_{zy} and τ_{yz} are zero. The material just to the right and left of position 2 is subjected to stresses τ_{zy} and τ_{yz}, which have finite values.

Figure 15(c) depicts the roller in position 3, and τ_{max} is shown as being in the material below the roller. The shear stresses along the major axes at position 2 once again have finite values, while the stress at position 1 has dropped to zero.

It will be noted after studying Fig. 15(b) that the shear-stress field reverses itself at position

Fig. 14 An example of burnup with plastic flow in a tapered-roller bearing

This type of failure may result from loss of lubrication or gross overload. The damage begins as heat generation followed by scoring, and if the lubricant is not replenished or the load reduced, the excessive heat buildup results in hot plastic flow. Source: Ref 4

Fig. 15 Shear stresses produced by a cylindrical roller below the surface of a bearing raceway

Source: Ref 5

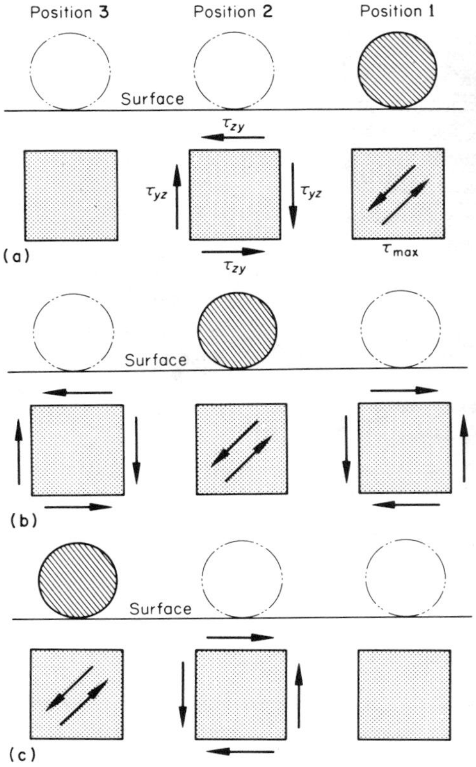

Fig. 16 Damage from surface deterioration and spalling in the drawn-cup outer raceway of a needle-roller bearing because the rollers were overloaded at one end

2. This reversal is believed to be the primary contributor to bearing failure and is the phenomenon on which the Lundberg-Palmgren theory (Ref 5) is based. A maximum value for τ_{zy} is reached at approximately the same distance below the surface as is τ_{max}. The absolute value of this maximum, because it extends from τ_{zy} to $-\tau_{zy}$, becomes $|2\tau_{zy}|$ and thus is greater than the maximum shear stress, τ_{max}.

The probability of survival of a bearing is assumed to be a function of the stressed volume, alternating shear stress, number of stress repetitions, and depth of alternating shear stress. A discussion of the theoretical relationships involved is provided in Ref 6. The products of the above assumptions are mathematical relationships that allow life predictions based on geometric properties of the bearing.

Alternating shear stresses reach maximum values at a distance below the surface determined by the load on the bearing as well as by other design factors. Subsurface cracks are usually initiated in regions of surface contact where these alternating shear stresses acting in the material are at a maximum. Pure rolling loads generate maximum shear stresses slightly below the surfaces of the contacting components. With the addition of frictional or sliding forces, maximum shear stresses move closer to the surface. Sliding forces of sufficient magnitude can cause failure to originate at the surface and can significantly reduce bearing life.

Subsurface-initiated fatigue cracks usually propagate parallel to the surface and then abruptly to the surface. They are easily distinguishable provided subsequent operation does not destroy the identifying features. As metal spalls out, steep-sided flat-bottomed pits remain. These result in progressive deterioration of the material by one or more cumulative damage mechanisms that eventually cause initiation and propagation of fatigue cracks. In some failures, the initial stages are characterized by polished contact surfaces in which small pits are often observed; this damage may

be serious enough to preclude further satisfactory bearing operation. Such surface distress, if allowed to continue, can also lead to spalling, in which metal fragments break free from the components, leaving cavities in the contact surfaces. In other instances, subsurface cracks are initiated with little or no observable surface deterioration until the final stages of the process, when cracking and spalling become evident. The mechanisms of rolling-contact fatigue are discussed in more detail in the articles "Fatigue Failures" and "Wear Failures" in this Volume.

Rolling-contact fatigue failures can be separated into two general categories: surface initiated and subsurface initiated. These two categories may be further subdivided according to the appearance and location of the fatigue spalling, as well as the factors that led to crack initiation. These factors may be related to material, manufacture, mounting, operation, lubrication, and care of the bearing. The recognition of the interplay of such factors is achieved through fractography.

Surface-Initiated Fatigue. Fatigue cracks may be initiated at or very near the contact surfaces of bearing components as the result of several causes. Geometric stress concentrations, for example, can produce abnormally high stresses in localized regions of

bearing raceways because of improper contact configuration, misalignment, or distortion. Eventually, such situations can lead to surface deterioration, crack initiation, and spalling. Figure 16 shows the damage on the outer raceway of a roller bearing due to surface deterioration and spalling that resulted from overloading of the rollers at one end. Inclusions and other defects are seldom responsible for surface damage, although they sometimes act as sites for crack initiation. In some failures, cracks are formed at the surface; in others, slightly below it.

Corrosion pits, handling scratches, surface inclusions, and surface dents can also cause stress concentrations and lead to surface-initiated rolling-contact fatigue cracks. Surfaces of fractures (spalls) resulting from such origins usually have an arrowhead type of appearance and point in the direction of load approach. The cracks are open to the surface at the earliest stages of development and thus can be reservoirs for the lubricant. Crack propagation may then be accelerated by hydraulic pressure developed in the cracks during each stress cycle. Corrosive contaminants such as water and organic acids resulting from lubricant deterioration, may also expedite crack propagation if they infiltrate the cracks.

Minor surface discontinuities, such as peeling, are usually associated with ineffective lubrication. Shallow flakes of material are removed from the contact zone, whereas relatively deep cavities are produced by spalling. Micropits may form on the surface and can initiate fatigue cracks. Crack propagation is usually at an acute angle to the contact surface, but results in a spall cavity that is parallel to the contact surface with only shallow penetration (Fig. 17).

Fig. 17 Microspalling (peeling) on a tapered-roller bearing caused by a thin lubricant film compared to the composite surface roughness

(a) Cup showing fatigue on the peaks of surface texture. (b) Cone showing fatigue on the peaks of surface texture. (c) Roller with a general spalled area. All 65×. (d) Transverse section through a slightly spalled area on a roller showing typical shallow cracking. Nital etch. 350×. Source: Ref 4

Fig. 18 Fine flaking damage on the surface of a shaft that served as a roller-bearing inner raceway

The flaking originated along the ridges of the surface finish of the shaft.

Damage involving large regions of a bearing surface indicates marginal oil-film thickness. Damage limited to areas surrounding surface irregularities, such as scratches and other depressions, results from localized metal-to-metal contact; if surface deterioration is allowed to continue, deep spalling will eventually occur.

In an application in which a portion of the surface of a shaft served as a roller-bearing inner raceway, flaking on the raceway (Fig. 18) contributed to failure of the shaft. The lubricating oil was very light, and the film was of insufficient thickness to separate the raceway and the rollers. Fine flaking propagated along the ridges of the surface finish; as the ridges flaked away, the valleys began to flake, and surface breakup became general.

Example 5: Contact-Fatigue Failure of a Raceway for a Thrust Bearing. The service life of a production gearbox had decreased drastically from that normally encoun-

tered. The axial load on a bevel gear was taken by a thrust-type roller bearing in which a ground surface on the back of the bevel gear served as a raceway. The gear was made of 8620 steel and was carburized and case hardened to 60 HRC; case depth was 0.9 mm (0.036 in.). The outside diameter of the bevel gear was 9.2 cm (3⅝ in.). The thrust bearing had a 3.8-cm (1½-in.) inside diameter and a 5.6-cm (2³⁄₁₆-in.) outside diameter.

Investigation. The gearbox was dismantled, and inspection of the bevel gear disclosed spalling damage on the ground bearing raceway at five equally spaced zones.

The gear was mounted on an arbor, and the bearing raceway was checked for runout. The raceway was found to undulate to the extent of 0.008 mm (0.0003 in.) total indicator reading (TIR). A spalled area was observed at each of the high points.

The outside surface of the gear hub was a 3.8-cm (1½-in.) diam raceway for a radial-

contact roller bearing. Examination of this surface disclosed a five-lobe contour having a 0.007-mm (0.00028-in.) (TIR) radial deviation. A chart trace on a rotary profile recorder was made of the two surfaces. The high surfaces on the radial-bearing raceway coincided exactly with the beginning of the spalled area on the thrust-bearing raceway and therefore the high points on that raceway.

Magnetic-particle inspection of the thrust-bearing raceway revealed the presence of numerous cracks that resembled grinding cracks. Microscopic examination of a section through the thrust-bearing raceway also disclosed numerous fissures that resembled grinding cracks.

The thrust bearing operated under a heavy load, which was intensified by the local support at the five high zones. This high nonuniform loading, in conjunction with grinding cracks, was sufficient to produce spalling.

Corrective Measures. The spindle of the grinding machine was reconditioned to eliminate the five equally spaced undulations. Heat treatment of this gear was more carefully controlled to minimize retained austenite. Fatigue life of the bearing returned to normal with these changes.

Flaking is preceded by formation of pits, which appear initially on one of the raceways rather than on the balls or rollers. In the early stage of flaking, isolated pits are formed, but these join rapidly, forming lines or bands of confluent pits and progressively destroying the raceways. The process of destruction is self-accelerating because the pits themselves, as well as the particles of metal that become embedded in the raceways, cause vibration and noisy running. Impact loading at the edges of the pits produces further deterioration. In addition, metallic particles in the lubricant increase the overall wear rate.

Fig. 19 SEM views of the surface of the outer-ring raceway of a type 440C stainless steel radial-contact ball bearing that failed by rolling-contact fatigue and of particles found in the lubricant

(a) Lubricant-residue particles that flaked off the raceway. 100×. (b) Flaked surface showing nearness of flaking to edge of raceway (arrow). 200×. (c) Flaked surface. 1000×

 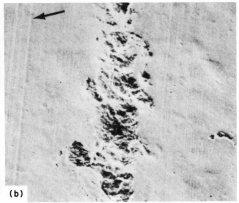 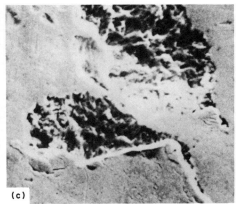

As a general rule, fatigue flaking first becomes evident in the stationary ring of a bearing. For a horizontal shaft supported by bearings at each end, with the inner rings rotating and the outer rings stationary (Fig. 2a), loading is radial and the highest stress occurs opposite the point of loading. On each outer (stationary) raceway, there is a specific zone that is subjected to a cycle of maximum stress every time a rolling element passes over it, but because of the rotation of the inner ring, the maximum stress is exerted at a different region of the inner raceway with every cycle. In circumstances in which the outer ring revolves and the inner ring is stationary (Fig. 2b), the converse is true. For the same speed of rotation, a bearing having a fixed inner ring has a lower life expectancy than a bearing in which the inner ring rotates.

The cause of flaking can frequently be determined by its location on a raceway. Ideally, the load in a roller bearing is uniformly distributed across the rollers and raceways. Flaking in a narrow ring in a plane corresponding to one end of the rollers would indicate grossly uneven loading and would suggest that the axes of the two raceways are not parallel. Flaking restricted to one portion of a stationary raceway is usually caused by overloading from dead weight, excessive belt tension, gear-tooth reaction, or misalignment of driving couplings. Service life can be extended by partially rotating the stationary ring at appropriate intervals, shorter than those at which flaking was found to occur.

Shaft-bearing assemblies required to withstand a significant amount of axial loading usually incorporate angular-contact bearings arranged in opposition and initially adjusted to a certain degree of preload. In this type of application, flaking can be caused by the stresses induced by an increase in preload resulting from thermal expansion as the operating temperature is increased.

Flaking damage may also be distributed around a raceway at intervals corresponding approximately to the distance between the balls or rollers. In these instances, flaking may be initiated at indentations (true-brinelling marks) produced by impact when the bearing is mounted. Another cause for flaking located at roller spacing is corrosion pitting. Both of these modes of damage were described earlier.

Example 6: Rolling-Contact Fatigue Failure of Type 440C Stainless Steel Radial-Contact Ball Bearings Because of Excessive Axial Load. The radial-contact ball bearings supporting a computer microdrum became noisy and were removed for examination. A sample of the lubricant used with the bearings was sent to the laboratory with four sets of bearings.

Two sizes of bearings were used for the microdrum: 20 mm ID × 42 mm OD × 12 mm wide (0.8 × 1.7 × 0.5 in.), and 35 mm ID × 62 mm OD × 14 mm wide (1.4 × 2.4 × 0.6 in.). The bearings were made of type 440C stainless steel and were hardened to 60 to 62 HRC. For accurate positioning of the microdrum, which rotated at 3600 rpm, a spring washer that applied a 22.6-kg (50-lb) axial load on the smaller bearing was installed in contact with the inner ring.

Investigation. The seals of one set of bearings (one small and one large) were removed so that the inner surfaces of the bearings and seals could be examined. Also, samples of the grease in the bearings were taken for analysis and comparison with the sample submitted with the bearing. Comparison of infrared spectrophotometer patterns indicated that the two greases were similar, but not of the same composition.

The bearings were soaked and washed in a petroleum solvent until they were absolutely clean of grease. The solution was evaporated by heating. The residue contained particles that were attracted to a magnet. These magnetic particles were recleaned and mounted for viewing in a scanning electron microscope. Observation at a magnification of 100× indicated that the particles had flaked off the outer-raceway surface and were not foreign contaminants (Fig. 19a).

The surfaces of the inner and outer raceways were examined in a scanning electron microscope. Smearing, true-brinelling marks, and evidence of flaking were found off-center on one side of the outer-ring raceway. This pattern is commonly found in a radial ball bearing subjected to an axial load that moves the contact area to one side of the raceway. Views of the flaked surface in the outer-ring raceway are shown in Fig. 19(b) and (c). Surface damage to the inner raceway was less severe.

The true-brinelling marks on the raceways were caused by excessive loading when the bearing was not rotating or during installation. These marks were not prominent and were smeared when the raceway surfaces were flaked.

A new bearing of the smaller size was installed in a test stand with a preload of 22.6 kg (50 lb). After 4000 h of operation, the temperature of the bearing was 44 °C (111 °F), and the bearing was noisy. Normal operating temperature for these bearings did not exceed 31 °C (88 °F).

Conclusions. The bearings failed in rolling-contact fatigue, as indicated by flaking in the raceway surface. The 22.6-kg (50-lb) preload moved the zone of contact between the balls and raceway to one side of the raceway, reducing the normal contact area. Metallic particles found in the lubricant in the bearing were from the outer-ring raceway. Some flaking occurred at the true-brinelling marks in the raceways.

Corrective Measures. The preload was reduced to 13 kg (28 lb) using a different spring washer. This eliminated the noise, reduced the operating temperature to normal, and extended bearing life to an acceptable level.

Spalling failure is the result of fatigue and is recognizable by craterlike cavities in the surfaces of raceways or rolling elements. The cavities have sharp edges, generally steep walls, and more or less flat bottoms. Spalling

Fig. 20 Spalling damage on the end of a shaft that served as roller-bearing raceway

The spalling was initiated at subsurface inclusions.

may be obliterated by further destruction of the part or by rolling of surrounding metal into the cavities. Spalling can originate either at or below the contact surface. Figure 20 shows severe spalling damage on the end of a shaft that served as a raceway for a roller bearing. The spalling did not extend all the way around the shaft, only about 225°. The general pattern of spalling damage indicated good alignment and load distribution. However, failure was premature. Linear marks in the damaged surface indicated that the shaft was of poor-quality material. The raceway surface was underlaid with inclusions, which served as stress concentrators for fracture initiation.

Subsurface-Initiated Fatigue. Subsurface fatigue results from the shear stresses described above acting in the region below the contact surface of the rollers or balls on the bearing races. Nonmetallic inclusions act as stress concentrators and are the cause for the most common mode of subsurface-initiated fatigue.

Microstructural alterations that occur at high-stress levels or very high lives at more moderate stress levels have been judged to cause planes of weakness resulting in fatigue spalling. Subcase fatigue in case-carburized components is also related to highly stressed applications.

Effect of Inclusions. Nonmetallic inclusions in the load-carrying areas of bearing elements can reduce resistance to contact-fatigue fracture. An inclusion is foreign material with properties different from those of the matrix. Nonmetallic inclusions are compounds, sometimes complex compounds, that are present in all steels as a result of steelmaking practices, additions for machinability, or additions for strengthening or grain-size control. The primary inclusion types are oxides, nitrides, sulfides, and carbides.

The most detrimental inclusions are the hard (compared to the steel matrix) nondeformable types with an angular shape (oxides, nitrides, and some carbides) that are sometimes loosely bonded to the steel matrix. Oxide inclusions have the greatest influence on rolling-contact fatigue because of their size and frequency and their poor bond with the steel matrix. Oxide inclusions result from the steelmaking process, which requires the use of aluminum or silicon for deoxidation of the molten steel during the refining of steel. Aluminum is also required to ensure that a steel is fine grained and remains that way during subsequent thermal processing. Modern bearing steelmaking processing is aimed at elimination of large oxide inclusions and minimizing the frequency and severity of the smaller ones.

Sulfur is added to steel to improve its machinability. The sulfur forms inclusions by combining with manganese, an alloying element added for hardenability. Fortunately, manganese sulfide inclusions are deformable and form an intimate bond with the steel matrix. Historically, they have not been found to influence rolling-contact fatigue life. In wrought products, the inclusion formations are usually discontinuous or semicontinuous stringers oriented parallel to the direction of working. Thus, the most adverse effects of nonmetallic inclusions occur when the stress direction is perpendicular to the lay of the stringer. Under cyclic or fluctuating load, inclusions locally intensify stresses to a degree depending on the shape, size, hardness, and distribution of the inclusions. Initiation and propagation of fatigue cracks from a nonmetallic inclusion origin does not automatically mean the bearing has been prematurely damaged. The true bearing load and the material specification for the particular application must be known.

Example 7: Fracture of Ball-Bearing Components by Rolling-Contact Fatigue Because of Subsurface Nonmetallic Inclusions. The pilot of an aircraft reported illumination of the transmission oil-pressure light and an accompanying drop in pressure on the oil-pressure gage. The aircraft was grounded so that the transmission could be checked.

Investigation. Teardown analysis of the transmission revealed no discrepancies in assembly of the bearings and related components. The oil strainer contained numerous fragments of bronze similar to that used in the bearing cage.

The center bearing of the transmission input-shaft ball-bearing stack had a broken cage and one ball that had been split into several pieces along paths resembling the seams of a baseball. Also observed were several scored balls and flaking damage in the raceways of the inner and outer rings. Total operating time for this bearing stack was 770 h.

The origin of flaking in the raceway of the rotating inner ring was identified by the following distinguishing features (Fig. 21a). The origin (area in rectangle), oriented axially in the raceway, was dark and discolored, and was flanked by areas of markedly different-textured flaking damage, that is, shallow fine-texture flaking damage extending from the origin in the

Fig. 21 Inner-ring raceway of an aircraft-transmission ball bearing that failed by rolling-contact fatigue because of subsurface nonmetallic inclusions

(a) Macrograph of inner-ring raceway showing fine-texture flaking damage (arrow A), coarse-texture flaking damage (arrow B), and origin of flaking (rectangle). 3×. (b) Micrograph of section through inner ring at origin of flaking showing inclusions (arrow C). 440×

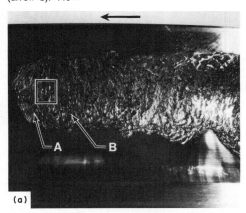

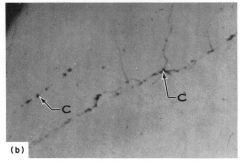

direction of inner-ring rotation (arrow A) and deep coarse-texture flaking damage extending from the origin in the direction opposite to the rotation of the inner ring (arrow B). Both areas had a convex texture in relation to the axial origin.

A section parallel to the axially oriented origin was cut at the origin area. The specimen was ground and polished for metallographic examination. As shown in Fig. 21(b), stringers of nonmetallic inclusions (arrow C) were revealed at the origin. The chemical composition, microstructure, and hardness of the bearing elements were checked and found to be acceptable.

Conclusions. Failure of the bearing occurred through a contact-fatigue mechanism (flaking) activated by the presence of subsurface nonmetallic inclusions (stringers).

Microstructural alterations are another type of inhomogeneity sometimes found in bearing materials. Although inclusions and other kinds of discontinuities are initially present in bearing materials, microstructural alterations are formed during actual operation of the bearing. These alterations result from localized plastic deformation caused by the action of subsurface cyclic stresses. When observed in a microscope, microstructural alterations appear as a white-etching constituent, either in a characteristic patchy form known as butterflies (Fig. 22) or in the form of narrow parallel bands at an acute angle to a contact surface (Fig. 23a) beneath a spall. At present, it is uncertain to what extent such alterations affect the fatigue life of bearings. Although fatigue cracks have been observed in or very near alterations and sometimes propagate along them, it is not certain that alterations actually cause fatigue-crack initiation. Nevertheless, microstructural alterations are associated with rolling-contact fatigue and may yet prove to be responsible for the initiation of cracks in certain instances.

Microstructural alterations create planes of weakness along which cracks can propagate. Figure 23 illustrates a spalled area that was produced by rolling-contact fatigue in which cracks propagated alternately along and between bandlike microstructural alterations inclined at 23° to the surface.

Subcase fatigue is also the result of subsurface-initiated cracking in case-hardened components. Cracks are initiated in or below the lower-carbon portion of the case near the junction of case and core. Once formed, such cracks usually propagate parallel to the case-hardened region until branching cracks propagate to the surface and form spalling cavities. This type of failure is rarely encountered in normal service; rather, it appears under gross-overload conditions, such as may be encountered in accelerated tests. Increasing case depth usually minimizes subcase fatigue.

Failure by Damage

Bulk-damage failures are the result of damage to bearing components outside the contact zone. Although evidence of material deterioration may be observed on the rolling-contact surfaces, the most severe damage is found at other locations on bearings. Deterioration of the bearing results from one or more causes, including overloading, overheating, bulk fatigue, fretting, and permanent dimensional changes.

Fractured rings, rollers, balls, and cages are examples of bulk-damage failures. Such failures may be the result of overloading the bearing beyond its design limits, but they can also occur because of misalignment, improper mounting, and improper fitting of a ring on a shaft or in a housing. Severe overloading is usually accompanied by permanent dimensional changes in the bearing as evidenced by distortion of the components. Figure 24 shows a drawn-cup needle-roller bearing that was mounted side by side with another in a sleeve,

Fig. 22 Stress butterflies (microstructural alterations) in a steel bearing ring

4% nital etch. Approximately 425×

Fig. 24 Drawn-cup needle-roller bearing that failed by gross overload

As the cup increased in width under overload, the oil hole became elongated, and circumferential cracks developed in the outer surface.

or tire. The bearing failed because of gross overload. The core material, being relatively soft, suffered bulk plastic flow. As the cup increased in width, circumferential stretch cracks developed on the outside surface, and the oil hole became elongated. The cup wall thinned out sufficiently that radial play became excessive and, coupled with noisy operation, resulted in termination of the service life of the bearing. A similar but perhaps more dramatic example of bulk damage due to high impact loads on a stationary tapered-roller bearing cone is shown in Fig. 25.

Cracks and Fractures. Significant cracking in rings and rolling elements takes several forms and results from several causes. Although fracture can result from fatigue, fatigue that leads to fracture differs from the very localized rolling-contact fatigue cracking that causes pitting, flaking, and spalling.

Circumferential cracks that develop in the outer ring of a bearing may indicate of a lack of uniform support from a housing not truly cylindrical but shaped like a barrel or an hourglass. Under these conditions, flexing can take place

Fig. 23 Microstructural alterations in a 52100 steel bearing ring

(a) Micrograph of picral-etched section through a spalled area showing elongated bandlike microstructural alterations. 200×. At the top is a profile of the bottom of the spalled area. (b) Macrograph of part of the spalled area. 7¼×. Note parallel ridges that were formed because subsurface cracking partly followed microstructural alterations.

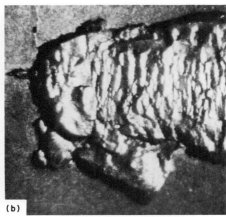

(a)

(b)

under load, leading to the development of fatigue cracks that, in a ball bearing, are usually found at the bottom of the ball raceway. This form of cracking ultimately leads to separation of the ring into two or more pieces. Crowned races can be used to compensate for bending in the shaft.

Cracking of inner rings frequently occurs in service. Such cracking generally takes place in an axial direction and may be caused by an abnormally high hoop stress resulting from an excessive interference fit on the shaft. On the other hand, a loose fit can lead to movement between the ring and the shaft, giving rise to radial cracking on the end faces—an effect discussed in the section ''Rotational Creep'' in this article.

Cracking of any component of a bearing can result from gross overloading, but may also be associated with other effects, such as overheating, wear, and flaking. Fatigue cracks are sometimes initiated at surface locations where pitting, flaking, and spalling have occurred. In other instances, cracks may start from regions where there is evidence of fretting, which has an adverse effect on fatigue endurance.

Cracks that lead to detachment of small, roughly semicircular pieces from the retaining flanges (back-face ribs) of roller bearings may be caused by excessive loading due to (1) incorrect assembly and dismantling procedures, (2) general abuse in service, or (3) abnormally severe axial impact, which brings the ends of the rollers into heavy contact with the cup or cone back-face rib.

Fracture and cracking of balls and rollers occur occasionally. Balls generally fail by splitting into portions of approximately equal size. Beach marks on the fracture surfaces indicate normal fatigue cracking, that is, not cracking due to rolling-contact fatigue; the cracks often originate in regions of incipient flaking on the outer surface. Splitting of balls is initiated primarily in the area of fiber flow essentially perpendicular to the rolling surface. Fiber orientation arises because balls are manufactured by upsetting slugs of round stock between hemispherical dies.

A similar form of failure is shown by rollers. Most rollers fracture axially into two equal portions, but others, although extensively cracked, remain intact. The region of final fracture is approximately circular and is situated on the central axis of the roller adjacent to the one end face. The basic cause of this particular form of cracking is not obvious, but it most likely arises from gross overloading in both the radial and axial directions.

Rollers subjected to considerable end thrust, with one face being forced into heavy contact with one of the flanges of the inner raceway, frequently fail by smearing and wear. Final fracture is usually located adjacent to the smeared end, but may be near the opposite end face.

Another form of failure involves fatigue cracking in a diametral plane and in a helical

Fig. 25 Bulk damage to a stationary tapered-roller bearing cone resulting from gross impact loading that yielded the cone material and cracked the case-carburized surface
Source: Ref 7

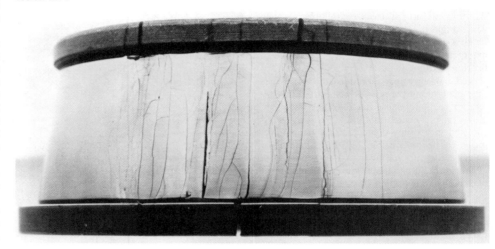

direction at 45° to the axis of the roller. The pattern of cracking is symmetrical, with a second, helical crack on the remote side of the roller. Usually, cracking in the diametral plane develops first, and the cracks turn when they reach the opposite edge of the roller and proceed in a helical direction. In general, such cracking develops as a result of excessive compressive loads, which produce tensile stresses at right angles to the direction of loading. In addition, torsional stresses are involved, as indicated by the helical cracks. In theory, a roller should not be subjected to torsional stresses, but such stresses can result from misalignment or inherent dimensional discrepancies, particularly if the load on the inner and outer rings is carried by opposite ends of the rollers.

Effects of Fabrication Practices

Correct assembly of a rolling-element bearing into a machine, as well as correct design and fabrication of the machine elements contiguous with the bearing, are essential for satisfactory operation. Satisfactory life and operation of a bearing depend on correct mounting design, accuracy of the machine elements that support the bearing, cleanness of the bearing assembly, and proper mounting on the shaft and in the housing.

Mounting Design. The rotating ring of a rolling-element bearing is generally press fitted onto a shaft or into a housing, while a sliding fit is maintained on the other ring. Tightness of fit between a bearing ring and a shaft or housing varies with the use to which the bearing will be subjected. The type of fit depends on the magnitude and direction of the load (radial or axial) and on whether the load will be applied to

the inner or the outer ring or whether application of load to both members may occur. Shaft and housing tolerances have been standardized by ANSI and AFBMA.

The standard internal looseness in a bearing is sufficient to allow for recommended tightness of fit between the inner ring and the shaft or between the outer ring and the housing. If tight fits are required in mounting both rings or if the inner ring will be substantially warmer than the outer ring during operation, the internal clearance in the unmounted bearing must be greater than normal to compensate for press fitting or thermal expansion. A smaller internal clearance than normal is used when radial and axial displacements must be minimized.

Selection of a bearing with inadequate internal clearance for conditions where external heat is conducted through the shaft can result in overheating or noisy operation due to expansion of the inner ring. An oversize shaft to undersize bore results in an excessive press fit and reduces the internal bearing clearance sufficiently to make the bearing tight, causing the bearing to overheat or become noisy.

Rotational creep will occur when the inner ring is loose on the shaft or the outer ring is loose in the housing. This will result in overheating, excessive wear, and contact erosion (or fretting) between the ring and the shaft or housing.

Shaft Shoulders and Fillets. The height of the shaft shoulder should be about half the thickness of the inner ring at the face. If the shaft shoulder is too low, the corner radius of the inner ring will push up against the shaft shoulder (Fig. 26a). Inadequate support of the bearing at the shaft shoulder can result in bending of the shaft or cocking of the bearing, which can cause overheating of the bearing, vibration, hard turning of the shaft, and general unsatisfactory operation of the equipment.

Fig. 26 Incorrect and correct shaft-shoulder heights and fillet radii for mounting rolling-element bearings on shafts

(a), (b), and (c) Incorrect designs. (d) Correct design. See text for discussion.

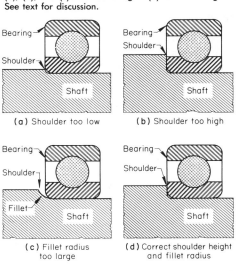

(a) Shoulder too low

(b) Shoulder too high

(c) Fillet radius too large

(d) Correct shoulder height and fillet radius

Fig. 27 Four types of misalignment of rolling-element bearings

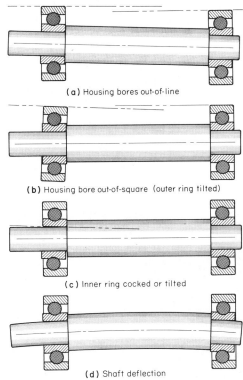

(a) Housing bores out-of-line

(b) Housing bore out-of-square (outer ring tilted)

(c) Inner ring cocked or tilted

(d) Shaft deflection

A shaft shoulder that is too high (Fig. 26b) results in rubbing or distortion of the bearing seals. Also, it is sometimes necessary to remove the bearing from the shaft. This operation requires that a puller be placed against the inner ring, and a surface must be left free on which the puller can operate. A slight corner break should be used on the shaft shoulder and on the bearing seat to avoid raising burrs that would interfere with assembly of the bearing.

Either an undercut or a fillet is used between the bearing-seat surface and the shaft shoulder. An undercut is usually the simplest design at the corner between the shaft seat and the shoulder provided the weakening effect of the undercut can be tolerated. Frequently, however, a fillet must be used; the shaft fillet radius should be less than the bearing bore corner radius indicated in the bearing catalog for the particular bearing being considered. With too large a fillet radius, the face of the bearing will not seat against the shaft shoulder (Fig. 26c), and misalignment of the bearing, bending of the shaft, inadequate support, and improper shaft and bearing location may result.

The proper relationship between the diameter of the shaft and the outside diameter of the inner ring and between the shaft fillet radius and the corner radius on the inner ring are shown in Fig. 26(d). Generally, these same conditions apply to the housing shoulder and fillet radius.

Shaft Alignment and Deflection. Alignment of shaft, bearings, bearing seats, and the machine itself should be within certain tolerances to provide efficient mechanical operation. Misalignment can originate from any of the four general situations shown in Fig. 27. Carefully calibrated micrometers and dial indi-

cators, with the appropriate mounting and clamping accessories, are used to evaluate misalignment.

In actual engineering applications, misalignment does not always cause failure. A single-row deep-groove ball bearing can withstand ¼° angular misalignment without cramping or binding. This is less than 1.6 mm (¹⁄₁₆ in.) per foot, and a 30-cm (1-ft) long shaft out of line by as much as 1.6 mm (¹⁄₁₆ in.) is unusual. Figure 27(a) illustrates tilting of inner rings when housing bores are out of line. Much more prevalent is an out-of-square bearing-housing bore (Fig. 27b) or housing shoulder; these can cause tilting of the outer ring. A cocked or tilted inner bearing ring (Fig. 27c) is also common. The same effect is created by out-of-square lockwashers and locknuts holding the bearing inner ring. On a 2.5-cm (1-in.) diam shaft, a shoulder being out of square by 0.1 mm (0.005 in.) corresponds to ¼° misalignment; thus, difficulties with tilted inner and outer bearing rings are not uncommon. Shaft deflection (Fig. 27d) is rarely critical, because many shafts would break before their angularity at the bearing reached ¼°. All these conditions should be checked if the ball path on a raceway indicates misalignment; if any of these conditions exist, changes in bearing clearances may be required, or inspection procedures with improved measurement techniques might be necessary.

Misalignment is indicated by a ball path not parallel to the edge of the raceway (Fig. 2d). When the misaligned path is on the outer raceway, such as in applications where the inner ring rotates, the bore of the housing is not parallel to the shaft axis (Fig. 27b). If the path on the inner raceway is not parallel with the edge of the raceway, it is generally because the inner ring is cocked on the shaft (Fig. 27c), the shaft shoulder is not square with the bearing seat, or the shaft is bent (Fig. 27d).

Misalignment of either ring of a deep-groove bearing will impose an additional load on the bearing. This additional load, together with the normal load, causes overstressing and overheating, resulting in early failure, as in the following example.

Example 8: Failure of a Bearing for a Jet Engine Because of Misalignment Between the Bearing and a Shaft. An engine on a jet aircraft produced excessive vibration in flight and was immediately shut down.

Dismantling the engine revealed complete failure (fracture) of the cage in one of the two main-shaft ball bearings. The ball bearings were both of the single-row deep-groove type with split inner rings, manufactured for use in jet-aircraft engines. The bearings were made of vacuum-melted 52100 steel and were designed to operate at a maximum temperature of 175 °C (350 °F). The bearings were side by side in the engine.

Investigation. Hardness surveys were conducted on the inner and outer rings of both bearings. Hardness of the inner and outer rings of the bearing having the unfailed cage averaged 58 and 59 HRC, respectively, which is consistent with the specified hardness of 58 to 64 HRC. The inner ring of the bearing with the failed cage had a hardness of 55 HRC, and the outer ring 57 HRC. At 175 °C (350 °F), these parts would have a hot hardness of 50 HRC, perhaps indicative of overtempering.

All components of both bearings were visually inspected for evidence of excessive heating, structural damage, and misalignment. Figure 28(a) shows a photograph of the bearing with the fractured cage. Damage caused by severe scoring, scuffing and plastic deformation of the material surrounding the ball pockets of the cage is shown in Fig. 28(b). The cage was also cracked between pockets at several locations around the periphery; one of these cracks (arrow) is visible between the two pockets at top in Fig. 28(b).

Examination of the outer raceway of the bearing with the failed cage revealed severe damage to approximately 20% of the load-bearing surface. As shown in Fig. 28(c), more damage occurred on one shoulder of the groove than on the other, indicating misalignment of the bearing during at least part of its operation. Further evidence of misalignment was found in the ball pockets of the cage, which were unevenly worn. This misalignment caused severe stressing of the cage material, resulting in fracture of the cage. Discoloration

Fig. 28 52100 steel jet-engine ball bearing that failed because of overheating resulting from misalignment

(a) Photograph of bearing components showing fractured cage. (b) Enlarged view of cage showing damage caused by scoring, scuffing, and plastic deformation around ball pockets and a circumferential crack (arrow). (c) Segment of ball groove in outer raceway, showing off-center damage from misalignment. (d) Micrograph of section through spalling cavity in inner raceway of unfailed bearing showing inclusion (arrow) about 0.013 cm (0.005 in.) below surface

(a)

(c)

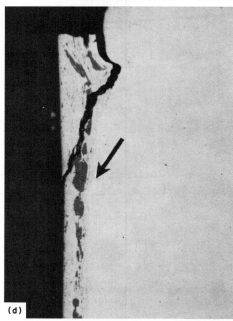

(b)

(d)

of the bearing components indicated that the bearing had been heated to temperatures in excess of the design limit. The cause of misalignment is uncertain, but may have been improper mounting of the bearing on the main shaft.

The second bearing exhibited no damage other than several small spalling cavities in the inner-ring raceway. Metallographic examination of a circumferential section through the largest cavity revealed an elongated subsurface inclusion (arrow, Fig. 28d). This inclusion was in the region of maximum shear stress in the bearing (approximately 0.1 mm, or 0.005 in., below the surface), which indicated that it had initiated the spalling cavity.

Conclusions. Failure of the cage resulted from overheating and excessive stressing of the cage caused by misalignment of the bearing on the main shaft. The bearing that failed was slightly tempered by overheating during operation.

The small spalling cavities found on the inner ring of the backup bearing were caused by an elongated subsurface inclusion located several thousandths of an inch below the surface and in the region of maximum shear stress. These cavities were not large enough to contribute to vibration in the engine at the time it was shut down.

Abuse before or during mounting can cause bearing raceways to become ball dented. Bearings operated in this condition are very noisy. Ball denting, or true brinelling, is caused by excessive pressure exerted on the raceways through the balls when the bearing is not rotating (see the section "Brinelling" in this article). Denting commonly occurs in small bearings that are pressed into the housing while already mounted on the shaft. If the outer ring is tight or becomes cocked in the housing and pressure to force it into position is applied through the inner ring, indentations are produced.

Impact in an axial direction can force the balls against the edge of the groove, causing dents and nicks. Brinelling can be easily distinguished from dents produced by chips or dirt because ball dents are usually spaced the same distance apart as the balls. Ball dents can be produced in the bottom of a groove if sufficient impact loads are transmitted through the balls in a radial direction.

Mounting bearings on a shaft by applying blows or force to the outer raceway will also cause denting. If the force is sufficient, the raceway may crack. With less force, dents are produced near the center of the raceway, causing rough and noisy operation. As operation continues, pitting followed by flaking will develop at the individual dents. A nick on a ball or a raceway resulting from impact by some sharp object causes noise and vibration, and pitting is likely to follow.

Metal chips and dirt can adhere to and enter bearings that are placed on dirty benches or floors. Such contaminants can cause noise, denting, and locking of one or more of the rolling elements, which ultimately produces failure. Heating bearings with a torch to facilitate mounting or removal is hazardous. A severe temperature gradient is produced that may cause cracking or warping, and excessive heating will lower the hardness, change dimensions, and induce early failure.

Punctured and bent bearing shields interfere with bearing performance and eventually cause bearing damage. Shield damage usually results from use of an improper tool in mounting or removing a bearing from a shaft or spindle.

Broken ball retainers in ball bearings may be the result of misalignment during installation. Excessive shaft deflection can also be responsible for broken retainers. A ball-path pattern alternating from one side of a raceway to the other will be in evidence when a retainer has failed.

When bearings are pressed onto a shaft, the bore must be in perfect alignment with the shaft surface before force is applied. If a misaligned bearing is forced onto a shaft, the inner ring may be distorted or cracked. Forcing a misaligned hardened ring onto a shaft can burr or score the shaft seat. An arbor press is the most satisfactory means of applying the mounting force; use of impact tools should be avoided.

Heat Treatment and Hardness of Bearing Components

The most common through-hardened ball bearing materials are 52100 low-alloy steel and

M50 high-speed tool steel. Table 2 lists average hardness, amount of retained austenite, and typical austenite grain size for eight ball bearing materials (three heat-treatment lots of each material).

Hardness is an important variable in rolling-contact fatigue and can significantly affect bearing life (see the section ''Bearing Materials'' in this article). In general, bearing hardness should be at least 58 HRC for adequate bearing life. Investigations have shown that, within limits, rolling-contact fatigue life increases as hardness of the bearing components is increased (Ref 9).

Significant differences in hardness between rings and balls can affect bearing life. Experimental results (Ref 10) indicate that balls should be approximately 1 to 2 Rockwell C points harder than rings.

Because bearings operate at elevated temperatures and because bearing life depends on hardness, the hardness of the bearing material at operating temperature is significant. Data on short-term hot hardness (Ref 11) indicate that there is a significant difference between bearing components of low-alloy steels, such as 52100, and those of high-speed tool steels, such as M50, in ability to maintain hardness as temperature increases.

Most bearing alloys fall within the low-alloy or high-speed tool steel categories; however, a heat treatment that is optimum for a material in a tool is not necessarily optimum for the same material when used in a bearing component. Unfortunately, most heat treatments for these bearing materials are based on the heat treatments that have been developed for the materials as tools.

Example 9: Bearing Failure Caused by Improper Heat Treatment of Outer-Ring Raceway. A large bearing from a radar antenna was replaced because of deformation, surface cracking, and spalling on the raceway of the outer ring. Figure 29(a) shows a sectional view of the bearing.

Specifications required that the rings be made of 4140 steel. The raceway surfaces were to be flame hardened to 55 HRC minimum and 50 HRC 3.2 mm ($\frac{1}{8}$ in.) below the surface. Other surfaces of the rings were to have a hardness of 24 to 28 HRC. The bearing capacity at 1.67 rpm was a radial load of 713 000 kg (1 572 000 lb) and an axial (thrust) load of 459 000 kg (1 012 800 lb). Samples of the inner and outer rings 15 to 20 cm (6 to 8 in.) long were sent to the laboratory for examination.

Investigation. Examination revealed that the raceway of the outer ring was damaged by deformation, surface cracking, and spalling (Fig. 29b). The raceway of the inner ring exhibited little or no damage.

Wet chemical analysis of the material in the inner and outer rings showed it to be 4140 steel. Molybdenum contents were 0.26% in the inner ring and 0.31% in the outer ring, both slightly above the specified range of 0.15 to 0.25%. However, this deviation is not significant and

Table 2 Variations among heat-treatment lots in average Rockwell C hardness, amount of retained austenite, and austenite grain size, for eight bearing materials

Bearing material	Average hardness, Rockwell C	Retained austenite, vol %	Austenite grain size (ASTM E 112)	Bearing material	Average hardness, Rockwell C	Retained austenite, vol %	Austenite grain size (ASTM E 112)
52100 steel				**M50 tool steel**			
Lot A	62.5	4.90	13	Lot A	62.6	1.90	10.3
B	62.0	4.10	13	B	62.2	2.90	9
C	62.5	0.80	13	C	62.3	1.50	10
Halmo				**M10 tool steel**			
Lot A	60.8	0.60	8	Lot A	62.2	1.10	9
B	60.8	1.00	8	B	62.0	2.40	6
C	61.1	1.70	8	C	61.8	1.60	6
T1 tool steel				**M1 tool steel**			
Lot A	61.4	7.30	11	Lot A	63.3	2.90	10
B	61.4	5.20	9	B	63.4	3.30	9
C	61.0	9.50	10	C	63.5	1.00	8
M42 tool steel				**M2 tool steel**			
Lot A	61.8	1.00	9	Lot A	63.4	1.70	6
B	61.3	4.40	10	B	63.4	2.40	10
C	61.3	4.90	8	C	63.4	2.30	9

Source: Ref 8

would not affect strength properties or hardenability.

A cross-sectional specimen about 9.5 mm ($\frac{3}{8}$ in.) thick was taken through the center of each sample; after cutting, both surfaces of each specimen were ground smooth. Hardness of the raceway of the outer ring was 29.8 to 11.7 HRC. A horizontal traverse through the center of the ring showed a hardness of 26.1 to 18.7 HRC. A vertical traverse 4.1 cm ($1\frac{5}{8}$ in.) from the outer surface showed a hardness of 25.2 to 18.9 HRC. The low hardness values for the outer-ring raceway indicated that it had not been properly flame hardened. The hardness of the top and outer surfaces of the rings was within specifications. The hardness traverse of the inner ring raceway showed a hardness of 46.8 to 54.8 HRC, which was below the specified hardness of 55 HRC minimum.

The specimens were etched in 3% nital to differentiate between hardened and unhardened areas. The inner ring showed a well-developed hardened case at the raceway; the outer ring did not.

Metallographic examination of an etched specimen showed that the structure of the inner ring adjacent to the raceway was a mixture of tempered martensite and some ferrite. This suggested that heat treatment was insufficient to make the structure completely austenitic before quenching; therefore, the full hardening capability of the steel was not attained. The microstructure of the material away from the hardened surface was a mixture of finely divided pearlite and ferrite, which resulted in the metal being relatively soft.

The microstructure of the material in the outer ring adjacent to the raceway (Fig. 29c) was a mixture of white ferrite, scattered patches of pearlite, and martensite, which showed that the steel had been improperly austenitized, producing very low hardness.

Displacement of metal on the outer raceway is shown in Fig. 29(d). The grain structure had

been elongated, indicating metal movement. Rolled-out and embedded metal particles were also found in the surface.

Hardenability of the metal in the outer ring was checked by heating a specimen 25 × 25 × 6.4 mm (1 × 1 × $\frac{1}{4}$ in.) along one edge with an oxyacetylene torch to 870 to 900 °C (1600 to 1650 °F), then quenching in water. Hardness in the heated area was 57 to 60 HRC; in the unheated area, 23 HRC.

Conclusions. Failure of the raceway surface of the outer ring of the bearing was the result of incomplete austenitization. The steel in the inner-ring raceway had been hardened to slightly below the specified hardness of 55 HRC minimum, but the outer-ring raceway had a maximum hardness of only 29.8 HRC. The raceway surface in the outer ring was not properly heated by the flame-hardening process; therefore, subsequent quenching and tempering operations, if any, would have had virtually no effect on hardness.

Lubrication of Rolling-Element Bearings*

The functions of lubricants for rolling-element bearings are to provide a thin film of oil between bearing contact surfaces to reduce metal contact, to prevent excessive heating by acting as a coolant, to remove wear debris and other foreign substances from the contact surfaces, and to help seal out dust, dirt, and other environmental contaminants. Experience has shown that reductions in bearing life will occur if a lubricant is ineffective in performing any of these functions. Lubricants must have the proper physical and chemical characteristics under the conditions of operation. A program of

*This section is a condensation of an unpublished article by C.J. Polk, Mobil Research and Development Corp.

Fig. 29 Large-diameter 4140 steel radar-antenna bearing that failed because of improper heat treatment of outer-ring raceway

(a) Configuration and dimensions (given in inches). (b) Fractograph showing typical damage on outer-ring raceway. (c) Micrograph of section through metal in outer ring adjacent to raceway showing ferrite, scattered patches of pearlite, and tempered martensite. (d) Micrograph of section through outer ring at raceway showing grains elongated by metal movement (raceway surface is at top)

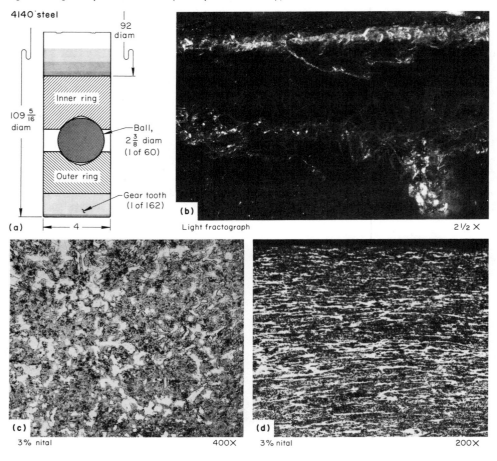

Fig. 30 Schematic representation of areas on raceways deformed by a ball (a), a cylindrical roller (b), and a tapered roller (c)

Source: Ref 12

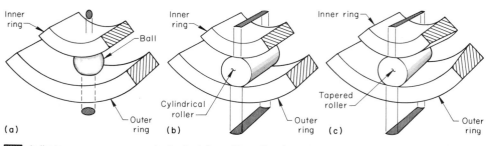

Indicates areas on raceways elastically deformed by rolling elements

regular inspection and maintenance is necessary to ensure that both the lubricant and its distribution system, as well as other mechanical components, remain in satisfactory condition.

Thin-Film Lubrication. When the proper amount of lubricant is supplied to a bearing, a thin film is formed on the bearing surfaces. During operation, extremely small and randomly distributed surface irregularities penetrate the thin film, causing metal-to-metal contact. The operational life of a bearing is inversely related to the number of such contacts

per revolution. Other factors being equal, the fewer the contacts made per cycle, the longer the life of the bearing.

Under load, the area of contact between a ball and its raceway is elliptical in shape (Fig. 30a). The zone of contact between a cylindrical roller and its raceway is rectangular (Fig. 30b), whereas a tapered roller forms an area that is wider at one end (Fig. 30c). As load is applied, the elements deform elastically, and contact is spread over an increasingly larger area. With a suitable lubricant between the surfaces, such an increase in contact area and the increase in fluid viscosity caused by the load result in the formation of a pressurized (elastohydrodynamic) film that separates the surfaces during operation (Ref 13).

Elastohydrodynamic films are only a few micro-inches in thickness. Film thickness can be increased by increasing the rotational velocity of the bearing, increasing the viscosity of the oil, or improving the conformity of rolling elements to their raceways. Increasing either temperature or load decreases film thickness. Consequently, it is detrimental to bearing life to operate a bearing at too low a speed, too high a load or temperature, or with an oil of insufficient viscosity.

The surface finish of bearing elements is also an important consideration in elastohydrodynamic lubrication. In addition to the periodic dimensional deviations resulting from machining, bearing surfaces contain nonperiodic irregularities, including scratches, debris, and indentations with raised edges. Ordinarily, these features are quite small and have a statistical size distribution; they range in height from a few micro-inches to a few hundredths of a micro-inch or less. However, elastohydrodynamic films are also only a few micro-inches in thickness, and for a given film thickness, some of the surface features protrude through the film and contact the opposing surface. Under conditions of insufficient film thickness, the number of contacts per revolution is excessive, and bearing life suffers accordingly.

Temperature. Sources of heat in bearings include frictional contact between component surfaces, hysteresis losses because of cyclic stressing of the metal, and fluid friction caused by displacement and churning of the lubricant. Sources external to the bearing may also contribute to heating, as in hot-rolling mills, paper-mill dryers, and jet-aircraft engines.

Lubricating oils and greases are available for applications over a wide range of temperatures. As shown in Fig. 31, oils of higher viscosity are better suited for bearing applications in which rotational velocites are low and operating temperatures are high. Oils of lower viscosity are more suitable for bearings operated at higher speeds and lower operating temperatures. Petroleum-base oils and greases are available for both subzero and high-temperature operations. Many lubricants are formulated to function over a wide temperature range. For example, certain greases can be used at temperatures

Fig. 31 Viscosities of several lubricants for ball and roller bearings, as related to operating speed and temperature

Source: Ref 14

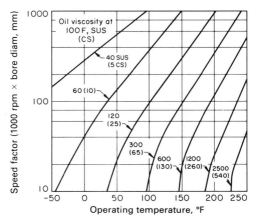

Table 3 Guides for choosing between grease and oil lubrication for bearings

Operating condition	Grease	Oil
Temperature	Below 93-121 °C (200-250 °F)	Above 93-121 °C (200-250 °F)
Speed factor (bore, mm, × rpm)	Below 200 000-300 000	Above 200 000-300 000
Load	Low to moderate	High, where oil is necessary for cooling
Long periods without attention.	Yes	No, rate of leakage and other losses higher
Central oil supply for other machine elements	No	Yes
Dirty conditions	Yes, if proper design seals out contaminants	Yes, if circulated and filtered
Lowest torque	No, soap structure offers resistance	Yes, except excessive amount of oil and churning should be avoided. Use oil mist.

Source: Ref 14

from −55 to 175 °C (−65 to 350 °F) (Ref 15). Under certain conditions, greases made with nonsoap thickeners can operate at temperatures up to 260 °C (500 °F). If oil-type lubricants are required for continuous high-temperature operation, as in jet-engine applications, specially formulated ester-base fluids may be required.

Contamination. Contaminants that may enter a bearing include dust, dirt, fine metallic particles, water, and acid fumes. Lubricants help to keep such contaminants out of the bearing and restrict the action of those that reach the raceway and rolling-element surfaces. Hydrophobic oil films and antirust additives inhibit the effects of water contamination. Lubricating oils may also be used to flush foreign particles from the contact zone; these particles are removed from the oil by filtration.

Abrasive dust or dirt in a bearing causes wear of bearing components. The resulting increase of internal clearance causes radial and axial play, reduced accuracy and rigidity of shaft positioning, and increased vibration and noise. Furthermore, metallic particles and other hard debris can become embedded in contact surfaces. Repeated impact of rolling elements on these particles causes dents and scratches that deteriorate the surface finish. The resulting raised or sharp edges produce highly localized stress concentrations that can lead to early fatigue failure. Adequate lubrication minimizes the effects of such abrasive particles by establishing films of sufficient thickness to allow easy passage of the contaminants between bearing surfaces.

Corrosion can result if condensation occurs in an idle bearing, or if water enters the housing during operation. Rust particles can flake off the components, forming surface pits; in addition, an abrasive lapping compound may be formed by mixing of the particles with the lubricant. Water contamination can also cause

corrosion fatigue, stress-corrosion cracking (SCC), galvanic corrosion, and hydrogen embrittlement of high-strength bearing steels (Ref 16). Corrosion from other sources, such as acid fumes, can produce effects similar to those caused by water. Although corrosion inhibitors are included in most lubricant formulations, moisture and other corrosive elements must be excluded from the bearing and the lubricating system.

Housings of open-type bearings are provided with shaft seals to control the flow of lubricant. Outward flow of a lubricant aids in preventing dirt, water, and other contaminants from entering the bearing. Shaft seals may be of the clearance or rubbing-contact type. Seals must be examined periodically to inspect for excessive wear and to evaluate effectiveness in preventing contamination. Slingers can sometimes be installed on shafts for centrifugal removal of water or other liquids.

Example 10: Failure of Bearings Because of Wear of Labyrinth Seals. A large number of drive-shaft hanger bearings failed after 300 to 400 h in service. The bearings normally lasted 600 h or more. The shaft was made of aluminum 2024-T3 tubing 2.5 cm (1 in.) in diameter and 1.2 mm (0.049 in.) in wall thickness. The shafts were supported by labyrinth-sealed single-row radial ball bearings of ABEC-1 tolerances. The bearings had spot-welded two-piece retainers and were lubricated with 1.5 ± 0.2 g of a paste-type mineral-oil lubricant or a grease conforming to MIL-G-81322. The lubricant contained molybdenum disulfide and polytetrafluoroethylene particles having a size of about 5 μm. The grease contained a thickening agent and synthetic hydrocarbons.

Investigation. Examination of the failed bearings revealed that the rings were blackened, deformed, and smeared. The balls were embedded in the inner-ring raceway, which had been softened by the elevated temperatures reached during the failure. The retainers were broken, and the seals were worn and bent out of shape.

Examination indicated that the following factors contributed to the failures:

- Insufficient lubrication
- Contamination of the bearings by gritty particles (dirt)
- Intrusion of a corrosive agent (water)
- Corrosion pitting of rings and balls
- Contact-fatigue mechanisms

Bearings that had been in service about 300 h were examined. These bearings exhibited light corrosion pits throughout the rings and balls. However, pitting occurred primarily in the areas of contact between balls and raceways. Areas surrounding these pits contained evidence of contact fatigue. The bearings were blackened by the high temperatures, but the amount of coke residue or burned lubricant was minimal. The seals exhibited wear, which had permitted penetration of gritty particles; water and other corrosive agents; and leakage of lubricant out of the bearing.

Maintenance reports contained notes regarding grease slinging by these bearings. This was substantiated by the dry conditions of the failed bearings.

New bearings were tested under simulated environmental conditions. Half of these bearings had labyrinth seals and were lubricated with the mineral-oil paste; the remainder had a positive rubbing seal and the MIL-G-81322 grease. The bearings with the positive rubbing seals and MIL-G-81322 grease lubricant had 30 to 100% longer life than those with the labyrinth seals and mineral-oil-paste lubricant.

Conclusions. Wear of the labyrinth seals permitted the lubricant to flow out of the bearing and dirt and corrosive agents to enter, resulting in overheating.

Corrective Measures. Bearings containing a positive rubbing seal and a MIL-G-81322 grease lubricant were installed. These bearings had satisfactory life.

In certain applications, contamination problems require a change in the method of lubrication. For example, where bath oiling is used and considerable trouble is encountered with dirt, it may be advisable to change to circulation oiling. The lubricant can then be cleaned by passing it through a full-flow filter. Where dust, moist air, or acid fumes are a problem, changing to an oil-mist system may be beneficial. This type of system can maintain clean, dry air under positive pressure in the housing and thus prevent contaminants from entering. In all instances, passages for supply and drainage of oil must be free of dirt and other clogging debris.

Effective Oil and Grease Lubrication. General guides for determining if oil or grease should be used as a lubricant are listed in Table 3. Lubricating oils must be of suitable viscosity for the particular application to minimize the frequency of contact between the bearing surfaces, to reduce frictional heating from all sources, and to protect against wear. Adequate film strength is necessary to resist the wiping action between components, particularly in tapered-roller bearings carrying heavy axial and radial loads. Lubricating oils should have the highest possible resistance to oxidation to prevent the formation and accumulation of sludge during long periods of operation. Corrosion inhibitors are needed to provide maximum protection when corrosive agents are present.

Greases suitable for bearing applications should be of high quality to last for long periods of service. In addition to the requirements listed for oils, greases should have proper structural stability to resist both softening and stiffening in service and should have the appropriate consistency for the method of application. Greases must have the proper slump characteristics at operating temperature, or controlled oil-bleeding properties, for adequate penetration into the small clearances between separators and raceways. Greases used in permanently packed bearings must be exceptionally stable.

Although the housings of some bearings are packed by hand, most have grease fittings. Except when it is necessary to exclude water by complete filling with grease, no housing should be packed so full that the constant churning by the rolling elements causes unnecessary lubricant friction and high operating temperatures. Excessive lubrication, like lubricant starvation of bearings, is detrimental to optimum performance and long bearing life.

REFERENCES

1. H.N. Kaufman, Classification of Bearing Damage, in *Interpreting Service Damage in Rolling Type Bearings—A Manual on Ball and Roller Bearing Damage*, American Society for Lubrication Engineers, Chicago, April 1958
2. F.R. Hutchings, ''A Survey of the Causes of Failure of Rolling Bearings,'' Technical Report Vol VI, British Engine Boiler & Electrical Insurance Co., Ltd., Manchester, 1965, p 54-75
3. R.K. Allan, *Rolling Bearings*, Sir Isaac Pitman & Sons, Ltd., 1945
4. R.L. Widner and W.E. Littmann, *Bearing Damage Analysis*, NBS 423, National Bureau of Standards, Washington, April 1976
5. A.G. Lundberg and A. Palmgren, Dynamic Capacity of Roller Bearings, *Acta Polytech.*, No. 210, 1952
6. A. Mendelsen, *Plasticity: Theory and Applications*, Macmillan, 1968
7. M.R. Hoeprich and R.L. Widner, ''Environmental Factors and Bearing Damages,'' SAE Paper 800678, Society of Automotive Engineers, Warrendale, PA, April 1980
8. J.L. Chevalier and E.V. Zaretsky, *Effect of Carbide Size, Area, and Density on Rolling-Element Fatigue*, NASA TN D-6835, National Aeronautics and Space Administration, Washington, June 1972
9. T.L. Carter, E.V. Zaretsky, and W.J. Anderson, *Effect of Hardness and Other Mechanical Properties on Rolling-Contact Fatigue Life of Four High-Temperature Bearing Steels*, NASA TN D-270, National Aeronautics and Space Administration, Washington, March 1966
10. E.V. Zaretsky, R.J. Parker, and W.J. Anderson, Component Hardness Difference and Their Effect on Bearing Fatigue, *J. Lubr. Technol. (Trans. ASME)*, Jan 1967, p 47
11. J. Chevalier, M. Dietrich, and E. Zaretsky, *Short-Term Hot Hardness Characteristic of Rolling-Element Steels*, NASA TN D-6632, National Aeronautics and Space Administration, Washington
12. *Anti-Friction Bearings and Their Lubrication*, Technical Bulletin 3-92-002, Mobil Oil Corp., New York, 1963
13. T.E. Tallian, Rolling Contact Failure Control Through Lubrication, *Proc. Inst. Mech. Eng. (London)*, Vol 182 (Pt 3A), 1967-68, p 205
14. D.F. Wilcock and E.R. Booser, *Bearing Design and Application*, 1st ed., McGraw-Hill, 1957
15. Rolling Element Bearings, *Mach. Des.*, Vol 44, June 22, 1972, p 59-75
16. I.M. Felsen, R.W. McQuaid, and J.A. Marzani, Effect of Seawater on the Fatigue Life and Failure Distribution of Flood-Lubricated Angular Contact Ball Bearings, *Am. Soc. Lubr. Eng. Trans.*, Vol 15, 1972, p 8-17

SELECTED REFERENCES

- E.N. Bamberger, Effect of Materials—Metallurgy Viewpoint, in *Interdisciplinary Approach to the Lubrication of Concentrated Contacts*, NASA SP-237, National Aeronautics and Space Administration, Washington, July 1969, p 409
- H.R. Bear and R.H. Butler, *Preliminary Metallographic Studies of Ball Fatigue Under Rolling-Contact Conditions*, NASA TN3925, National Aeronautics and Space Administration, Washington, 1957
- J.T. Burwell, Jr., Survey of Possible Wear Mechanisms, *Wear*, Vol 1, 1957-58, p 119-141
- R.E. Cantley, ''The Effect of Water in Lubricating Oil on Bearing Fatigue Life,'' Preprints 76-AM-7B-1, American Society of Lubrication Engineers, Chicago, May 1976
- T.L. Carter, R.H. Butler, H.R. Bear, and W.J. Anderson, Investigation of Factors Governing Fatigue Life with the Rolling-Contact Fatigue Spin Rig, *Am. Soc. Lubr. Eng. Trans.*, Vol 1 (No. 1), 1958, p 23-32
- D.V. Culp and R.L. Widner, ''The Effect of Fire Resistant Hydraulic Fluids on Tapered Roller Bearing Fatigue Life,'' SAE Paper 770748, Society of Automotive Engineers, Warrendale, PA, Sept 1977
- D.V. Culp and J.D. Stover, ''Bearing Fatigue Life Tests in a Synthetic Traction Lubricant,'' Preprint 75-AM-1B-1, American Society of Lubrication Engineers, Chicago, May 1975
- C.H. Danner, Fatigue Life of Tapered Roller Bearings Under Minimal Lubricant Films, *Trans. ASLE*, Vol 13, 1970
- W.J. Davies and K.L. Day, Surface Fatigue in Ball Bearings, Roller Bearings and Gears in Aircraft Engines, in *Fatigue in Rolling Contact*, Institute of Mechanical Engineers, London, 1963, p 23-40
- T.A. Harris, Predicting Bearing Performance Under Nonstandard Operating Conditions, *Mach. Des.*, Aug 17, 1967, p 158-162
- J.J.C. Hoo, *Rolling Contact Fatigue Testing of Bearing Steels*, STP 771, ASTM, Philadelphia, 1982
- P.L. Hurricks, The Mechanism of Fretting—A Review, *Wear*, Vol 15, 1970, p 389-409
- R.F. Jacobs, Sleeve-Bearing Fatigue, *Mach. Des.*, Dec 19, 1963, p 134-139
- C.F. Jatczak, Specialty Carburizing Steels for Elevated Temperature Service, *Met. Prog.*, Vol 113 (No. 4), April 1978
- N.H. Kaufman and H.O. Walp, *Interpreting Service Damage in Rolling Type Bearings—A Manual on Ball and Roller Bearing Damage*, American Society of Lubrication Engineers, Chicago, 1953
- B.W. Kelly, Lubrication of Concentrated Contacts—The Practical Problem, Interdisciplinary Approach to the Lubrication of Concentrated Contacts, NASA symposium held at Rensselaer Polytechnic Institute, Troy, NY, July 15-17, 1969
- R.L. Leibensperger and T.M. Brittain, Shear Stresses Below Asperities in Hertzian Contact as Measured by Photoelasticity, *J. Lubr. Technol.*, July 1973, p 277-286
- W.E. Littmann, R.L. Widner, J. Stover, and J.O. Wolfe, The Role of Lubrication in the Propagation of Contact Fatigue, *J. Lubr. Technol.*, Vol 90 (No. 1), Jan 1968, p 89-100

- W.E. Littmann and C.A. Moyer, "Competitive Modes of Failure in Rolling Contact Fatigue," Preprint 620A, Society of Automotive Engineers, Warrendale, PA, Jan 1963
- W.E. Littmann and R.L. Widner, Propagation of Contact Fatigue from Surface and Subsurface Origins, *J. Basic Eng.*, Vol 88 (No. 3), Sept 1966, p 624-636
- W.E. Littmann, The Mechanism of Contact Fatigue, in *Interdisciplinary Approach to the Lubrication of Concentrated Contacts*, NASA SP-237, National Aeronautics and Space Administration, Washington, July 1969
- T.W. Morrison, T. Tallian, H.O. Walp, and G.W. Baile, The Effect of Material Variables on the Fatigue Life of AISI 52100 Steel Ball Bearings, *Trans. ASLE*, Vol 5 (No. 2), Nov 1962, p 347-364
- C.A. Moyer and H.R. Neifert, A First Order Solution for the Stress Concentration Present at the End of Roller Contact, *Trans. ASLE*, Vol 6, 1963, p 324
- C.A. Moyer, "The Use of Elastohydrodynamic Lubrication in Understanding Bearing Performance," Preprint 710733, Society of Automotive Engineers, Warrendale, PA, Sept 1971
- R. Pederson and S.L. Rice, Case Crushing of Case Carburized and Hardened Gears, *Trans. SAE*, Vol 68, 1960, p 187
- J.C. Skurka, Elastohydrodynamic Lubrication of Roller Bearings, *J. Lubr. Technol.*, April 1970, p 281-291
- F.F. Simpson and W.J.J. Crump, Effect of Electrical Currents on the Life of Rolling-Contact Bearings, in *Lubrication and Wear Convention*, Institution of Mechanical Engineers, 1963, p 299-304
- F.F. Simpson, Failure of Rolling Contact Bearings, in *Lubrication and Wear, Second Convention*, Institution of Mechanical Engineers, Vol 179 (Part 3D), 1965, p 248
- J.O. Smith and C.K. Liu, Stresses Due to Tangential and Normal Loads on an Elastic Solid with Application to Some Contact Stress Problems, *J. Appl. Mech.*, June 1953, p 157-165
- K. Sugino, K. Miyamoto, M. Nagumo, and K. Aoki, Structural Alterations of Bearing Steels under Rolling Contact Fatigue, *Trans. Iron Steel Inst. Jpn.*, Vol 10 (No. 2), 1970, p 98-111
- T.E. Tallian, On Competing Failure Modes in Rolling Contact, *Trans. ASLE*, Vol 10, 1967, p 418-439
- T.E. Tallian and J.I. McCool, An Engineering Model of Spalling Fatigue Failure in Rolling Contact, II. The Surface Model, *Wear*, Vol 17, 1971, p 447-461
- N.M. Wickstrand, Depth of Permanent Indentations in Flat Plates Due to Loaded Cylindrical Rollers, *J. Lubr. Technol.*, Jan 1970
- R.L. Widner and J.O. Wolfe, "Analysis of Tapered Roller Bearing Damage," Technical Report C7.11.1, American Society for Metals, Oct 1967
- R.L. Widner, W.K. Dominik, and A.J. Jenkins, "High Performance Bearings," SAE Paper 780784, Society of Automotive Engineers, Warrendale, PA, Sept 1978
- D.C. Witte, Operating Torque of Tapered Roller Bearings, *Trans. ASLE*, Vol 16 (No. 1), Jan 1973, p 61-67
- H. Zantopulos, The Effect of Misalignment on the Fatigue Life of Tapered Roller Bearings, *J. Lubr. Technol.*, Vol 94 (No. 2), April 1972, p 181-187
- C.V. Zaretsky and J.W. Anderson, Effect of Materials—General Background, in *Interdisciplinary Approach to the Lubrication of Concentrated Contacts*, NASA SP-237, National Aeronautics and Space Administration, Washington, July 1969, p 379
- E.V. Zaretsky, W.J. Anderson, and R.J. Parker, *Effect of Nine Lubricants on Rolling-Contact Fatigue Life*, NASA TN D-1404, National Aeronautics and Space Administration, Washington, Oct 1962

Failures of Lifting Equipment

Frank L. Jamieson, Stelco Inc. (retired)

LIFTING EQUIPMENT is used for raising, lowering, and transporting materials, parts, and equipment, generally within a limited area. The types of metal components used in lifting equipment include gears, shafts, drums and sheaves, brakes and brake wheels, couplings, bearings, wheels, electrical switchgear, chains, steel wire rope, and hooks. This article will primarily deal with many of these metal components of lifting equipment in the following three categories:

- Cranes and bridges, particularly those for outdoor and other low-temperature service
- Attachments used for direct lifting, such as hooks, chains, wire rope, slings, beams, bales and trunnions, and the members to which they are attached, such as lifting lugs or eyes
- Built-in members that are the items necessary for the operation of lifting equipment, such as shafts, gears, and drums

Most of the failures discussed are related to the more common and critical components of lifting equipment used in steel mills and similar industrial applications, but the problems encountered and the methods of analysis are the same as for lifting equipment used in other industries.

Failure Mechanisms. Failures of lifting equipment commonly result from fatigue, ductile fracture, brittle fracture, and wear. A part may fail from any one of these mechanisms or from a combination of two or more of them. A member may initially fracture by fatigue, with final fracture being of a ductile or brittle nature. Total brittle fracture of a part is most undesirable, because it is unpredictable and often has catastrophic results. Periodic inspection of a part that is susceptible to brittle fracture is often of little purpose. It is preferable for fracture of a component to begin by fatigue rather than suddenly in a brittle manner. Properly timed inspections will usually reveal a slowly propagating crack, and the equipment can be removed from service before failure occurs.

Wear is easily recognized; excessive wear can usually be corrected by changing the material or its processing. However, the complete elimination of wear of lifting equipment may require selection of a material that is subject to brittle fracture—an alternative that may well be unacceptable.

Failure Origins. Failures occur for a number of reasons, which may be related to operation, design, material selection, material quality, and manufacturing practices. The largest portion of lifting-equipment failures are of operational origin. Overloading of a lifting mechanism is a common practice and often leads to either ductile or brittle fracture or to fracture by fatigue from repeated overstressing. Figure 1 shows a steel wire rope that fractured in tension due to overloading. With a fatigue type of failure, the member will usually fail at a load well below its specified load limit; thus, a false sense of security exists when dealing with a member that is fracturing by fatigue. Periodic inspection of such parts is therefore very important. Items that show signs of excessive wear or abuse should be removed from service for replacement or repair. Figure 1 shows the end of a steel wire rope that failed in tension because of overloading. Excessive wear on a 16-mm (⅝-in.) diam 6 × 37 (6 strands of 37 wires each) fiber-core improved plow steel wire rope is described in Example 3 in this article. Corrosion failures are also relatively frequent because of the environments in which lifting equipment is operated.

Section changes, keyways, lubrication holes, rough-machining marks, and threaded sections are often points of initiation of cracks. Small fillets, sharp corners, grooves, and threads act as stress raisers from which fatigue cracks initiate and fractures propagate. Correct design of a threaded shank on a hook, for example, and the stress-relief groove at the base of a threaded shank are very important to failure prevention. Failures have also occurred where names, numbers, and other identification marks are die stamped or imprinted on highly stressed surfaces. Information on the effects of identification marks is provided in the articles "Fatigue Failures" and "Failures of Shafts" in this Volume.

Another common cause of lifting-equipment failures is improper material selection. The choice of a material is related to such aspects as its basic quality, composition, overall mechanical properties, notch toughness, corrosion re-

Fig. 1 End of a steel wire rope that failed in tension because of overloading

Necking at the ends of the wires indicates ductile fracture; no worn or abraded areas were found at the break.

sistance, weldability, and machinability. In most applications, many materials may be satisfactory for a particular part, but only a few materials will be optimum.

In hardenable steels, tempered martensite has greater fatigue resistance than mixed structures. From a practical and theoretical aspect, alloy steels are a better material selection than carbon steels because of the deeper-hardening characteristics of alloy steels.

The most encompassing cause of failures of lifting equipment is poor manufacturing, assembly, and maintenance practices. Inferior machining, defective welds, residual welding stresses, misalignment, and improper and insufficient lubrication are all common and critical causes of lifting-equipment failures. Metallurgically, the most common cause of failures is improper heat treatment. Quench cracks and residual heat-treat stresses contribute to such failures. The examples in this article illustrate many of these causes.

Investigation of Failures. The investigation of a failure can provide valuable information that can be used for design improvement, material selection, and other factors related to the efficient operation of lifting equipment. Much information can be obtained by visual examination. Close, careful visual inspection should always precede any technique requiring destruction or cutting of a failed part.

A frequent error in investigating a failure is to interpret it automatically on the basis of the most commonly known causes indicated from past experience. Although this cannot be avoided entirely, efforts should be made to be alert for other or unique details that may be contributory or related factors and that may significantly alter the interpretation.

In the investigation of a fatigue failure, the following should be considered: material, design of the entire machine as well as that of the failed part, fabrication practice (machining, welding, forging, or casting), heat treatment, types of loading to which the equipment is subjected, and operating environment, such as corrosive or high-temperature atmospheres.

Many failures may be the result of a combination of several faults or contributing factors. Correcting one fault may not be sufficient or may even have a strongly adverse effect in the presence of other uncorrected faults. It is important that the investigator sort out the contributory causes of failure from the associated but immaterial aspects of the failed part to determine the proper corrective measures. The techniques for conducting failure analyses are provided in the article ''General Practice in Failure Analysis'' in this Volume.

Materials for Lifting Equipment

Recommendations given here are for materials commonly used in the manufacture of lifting equipment for the steel industry; other industries may use different materials as required by the application. The recommendations are general; thus, each situation must be considered individually. Transition temperatures of the steels used in the equipment should be considered in all cases, particularly for applications that involve temperatures below 20 °C (68 °F). For a discussion of ductile-to-brittle transition, see the article so titled in this Volume.

Chains, Chain Hooks, and Fittings. Low-carbon steel chains are in wide use and are generally acceptable. However, to ensure a greater margin of safety, alloy steel chains with a hardness of 302 to 352 HB are recommended.

Chain hooks are usually made of a grade of steel similar to that of the chain assembly, except in special cases. The design of chain hooks is often characteristic of the manufacturer rather than the application. Materials for chains for use in acid or other corrosive environments must be specified for the particular application.

Shafts. Misalignment of shafts resulting in failure is a common occurrence and is difficult to eliminate when dealing with large equipment. Frequently, 1040 or 1045 steel is specified for shafts. Failures in shafts made of these

steels can often be minimized by heat treatment. In other cases, it may be necessary to make shafts from quenched-and-tempered alloy steels, such as 4140 or 8640, which have excellent fatigue properties. For severe applications and large section sizes, 4340 or a comparable high-alloy steel may be used with excellent results. More information is available in the article ''Failures of Shafts'' in this Volume.

Bridges and Cranes. The use of a steel having a high impact resistance at low temperatures is required for the prevention of brittle fracture of structures, especially welded ones, that are subjected to cold-temperature conditions in service. Use of a fine-grain, normalized, relatively low-carbon (0.20% C) steel will generally ensure adequate impact resistance and freedom from high hardness in the heat-affected zone (HAZ) of welds. Fully quenched-and-tempered low-carbon steels and similarly treated alloy steels are also used for such applications.

Miscellaneous Equipment. Lifting booms, coil hooks, lugs and rings, boxes and containers, trunnions, and miscellaneous attachments should be made of either normalized or quenched-and-tempered fully killed fine-grain steels. For equipment operated at temperatures below 20 °C (68 °F), the transition temperature of the material must also be considered; the material selected should have a transition temperature below the operating temperature. When the weight of these components is a limiting factor, plain carbon steels may not be suitable. A high-strength low-alloy steel can be used instead, the grade and fabrication depending on the requirements of the specific application.

Flame-cut parts and welded components, insofar as possible, should be stress relieved at 600 to 650 °C (1100 to 1200 °F); flame cutting may produce hard edges that are deleterious to machining and may act as stress raisers in a weldment. Nondestructive inspection of welded assemblies is highly recommended. In the fabrication of a member, the material should be oriented so that the rolling direction will be parallel to the direction of principal service stress.

Bolts, nuts, and washers should be made in accordance with ASTM A 325 (Ref 1) or other recognized specifications for applications in which they are used only as fasteners. Pins and bolts used as pivots should be of fully killed fine-grain steel of compositions that will satisfy processing and application requirements. Rivets should be made in accordance with ASTM A 502 (Ref 2). Particular designs or service conditions may require the use of heat-treated alloy steels.

Steel Wire Rope

Many factors must be considered when investigating a failure of steel wire rope used on lifting equipment. Environment is of obvious concern because wire rope may be subjected to

Fig. 2 Components of a steel wire rope
Source: Ref 3

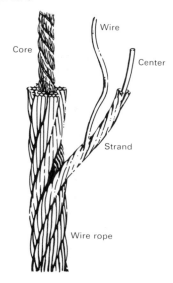

corrosive atmospheres, such as water, acid, and various chemicals, or to elevated temperatures. Another important factor that must be considered is the flexibility of wire rope. When sheaves or drums are involved, and they almost always are, the diameter of the sheave or drum has a direct relationship to the flexibility of the wire rope. The degree of flexibility of a wire rope is determined by its construction. Figure 2 illustrates the component parts of a wire rope.

Strength and Stretch. The ultimate breaking strength of a wire rope is by design less than the aggregate strength of all the wires and will vary, depending on the construction of rope and grade of wire used. All manufacturers of wire rope publish catalogs that list minimum breaking strengths for the various sizes, constructions, and grades of ropes. The proper design factor for a wire rope demands consideration of all loads. These loads should include (when applicable) acceleration, deceleration, rope speed, rope attachments, number and arrangement of sheaves and drums, conditions producing corrosion and abrasion, and length of rope.

Cold-drawn high-carbon steel wire, the type generally used in wire rope, has a modulus of elasticity of approximately 1.93 to 1.99 × 10⁵ MPa (28 000 to 29 000 ksi). Modulus of elasticity is the measure of the degree to which the wire will stretch under increasing load. The amount of rope stretch caused by the relative movement of all the wires in their attempt to adjust their positions to a stable condition corresponding to the load imposed varies with the construction of the rope. Generally, the more flexible ropes, which contain the greatest number of wires and have fiber cores, will stretch more than all-metal ropes with fewer wires and therefore less flexibility.

Stretch resulting from the movement of the wires is of two types: constructional and elastic. When a rope is first placed under load, the resulting slight rearrangement of the wires will cause a permanent lengthening, known as constructional stretch, and a recoverable lengthening, which is elastic stretch.

The constructional stretch, depending to some degree on the magnitude of the load imposed, amounts to approximately 0.5 to 0.75% of the length of the rope for six-strand ropes with fiber cores. For six-strand ropes with steel cores, the corresponding value is 0.25 to 0.5%; for eight-strand ropes with fiber cores, 0.75 to 1%.

The preceding remarks apply to bright ropes of a type intended to operate over sheaves and drums, and it is assumed that the ends of the rope are restricted against rotation, are free from corrosion, and have not deteriorated appreciably either internally or externally. Corrosion and other kinds of rope deterioration have a marked effect on stretch properties.

Sheaves. The size of a sheave is usually expressed in terms of its diameter measured at the base of the groove. This is its tread diameter. It can also be expressed as a ratio determined by dividing the tread diameter by the nominal rope diameter.

As sheave size is decreased, the stresses resulting from bending and the contact pressure between rope and sheave are increased. Higher bending stresses cause more rapid fatigue of the wires in the rope. Increased pressure also accelerates rope deterioration and, at the same time, increases sheave wear.

As sheave size is increased, rope-to-sheave pressure is decreased, and bending becomes easier. If bending alone were involved, an increase in rope life could obtained by increasing sheave size up to a limit of about 90 to 100 times the rope diameter for a 6 × 19 (6 strands of 19 wires each) classification rope. However, except on some shaft hoist installations, sheaves this large are seldom in actual use for two reasons. First, bending is seldom, if ever, the only factor involved. Large, expensive sheaves would be unnecessary if the abrasion or scrubbing encountered by the rope on some other part of the installation were the determining factor in its ultimate service life. Second, most machines would not be practical with such large sheaves, and many would never have been designed if sheaves of this size were mandatory.

In practice, many factors other than bending influence rope life, such as repeated stressing, abrasion, pounding, impact, vibration, twisting, speed, drum-winding abuse, corrosion, and lack of maintenance. For many applications, one or more of these factors affects rope life more than sheave size alone.

The optimum sheave size is determined by evaluating the factors affecting safe, economical operation. On high-speed mobile-type equipment and on equipment requiring low initial cost, it is customary to use smaller sheaves. The need for more frequent rope replacements is outweighed by the speed and flexibility of movement provided by the more compact equipment.

Although no definite minimum sheave size can be established for all types of installations, one important factor must not be overlooked. The heavy pressures between the wires at the contact points in the rope, combined with the high bending stresses resulting from operation over small sheaves, have a definite effect on rope deterioration. Under such adverse conditions, wire breakage often occurs at and between the points where the strands contact each other and contact the core of the rope. Broken wires in these sections are difficult, and sometimes impossible, to detect.

Therefore, for applications in which a high degree of safety is essential, sheaves should be liberal in size to provide better assurance that the wires will deteriorate progressively on the surface of the rope where they can be readily seen and evaluated. Extremely small sheaves should be limited to those types of equipment on which such usage is in line with acceptable practice.

Sheave size is such an important factor in wire-rope life that many laboratory bending or fatigue tests have been conducted to evaluate this relationship. Laboratory bending life can be determined; however, field service life can seldom be judged by these results. There are too many factors existing under actual field conditions to be simulated under controlled laboratory tests.

Laboratory tests have their value for research and development, particularly to the wire-rope engineer. It may also be important at times to show how bending alone, as performed in such tests, affects wire rope.

One series of tests was made on 16-mm (⅝-in.) diam wire-rope specimens in the 6 × 7, 6 × 19, 6 × 37, and 8 × 19 classifications. The sheaves ranged from 20 to 50 cm (8 to 20 in.) in diameter. The load approximated a design factor of five, and specimens were operated until they broke. The number of cycles of operation was the measure of bending life.

The laboratory tests showed that as the ratio of sheave diameter to rope diameter was reduced, there was a reduction in bending life, with the reduction in bending life showing the same trend for all specimens tested, and that the reduction in bending life was faster than the reduction of the sheave-to-rope diameter ratio.

Sheave Grooves. To allow wire rope to perform the maximum amount of useful work, it is essential that sheave grooves be of sufficient diameter to provide and maintain proper rope clearance. All wire rope, when new, is slightly oversize to allow for some pulling down while becoming adjusted to its normal working tensions.

Table 1 lists the recommended groove clearances for the usual sheave applications. In some cases, departures from these recommendations

Table 1 Groove clearance in sheaves based on nominal wire-rope diameter

| Nominal rope diameter | | Groove clearance(a) | |
mm	in.	mm	in.
6.4-7.9	¼-⁵/₁₆	0.4	¹/₆₄
9.5-19	⅜-¾	0.8	¹/₃₂
20.6-28.6	¹³/₁₆-1⅛	1.2	³/₆₄
30-38	1³/₁₆-1½	1.6	¹/₁₆
40-57	1⁹/₁₆-2¼	2.4	³/₃₂
59-76	2⁵/₁₆-3	3.2	⅛
<76	<3	4.0	⁵/₃₂

(a) Recommended groove diameter equals nominal rope diameter plus groove clearance.

are advisable. For example, with some deflectors or equalizers, smaller clearances can be used, or with conditions involving large fleet angles, larger clearances may be necessary.

The diameter of a sheave groove does not remain constant. Its change is influenced by the rope and, to a great extent, by the sheave material and the pressures involved. Because a groove of the recommended size becomes worn to a smaller diameter by the time a rope is taken out of service, the question arises as to whether or not a sheave should be regrooved each time a new rope is installed. In theory, best service will be obtained when a new rope starts with all sheave grooves exactly proper for the conditions involved. In practice, it would be false economy in most installations to change sheaves with every rope replacement. In most cases, the economical procedure is to allow the new rope to take additional wear at first, even at the expense of some rope life. For most installations, the limit below which a sheave should be regrooved is a clearance value in excess of the nominal rope diameter equal to half the clearance listed in the second column of Table 1.

When a sheave is regrooved, it is generally necessary to dress the flanges and to cut the base of the groove to the proper diameter. If only the base is machined, the throat angle near the base may be left too small to provide proper clearance where the rope makes a fleet angle to one side or the other.

Rope Pressure. In addition to the bending stresses introduced by operation over sheaves, wire rope is subjected to radial pressure by its contact with the sheave. This pressure sets up shearing stresses in the wires, distorts the rope structure, and affects the rate of wear of the sheave grooves. Therefore, the magnitude of the pressure and the wear resistance of the sheave material should be considered when selecting the most suitable rope construction.

The radial pressure, in pounds per square inch of projected area of the rope, can be determined using:

$$P = \frac{2T}{Dd} \qquad \text{(Eq 1)}$$

where P is radial pressure (in pounds per square inch), T is tension in the rope (in pounds), D is tread diameter of the sheave (in

Table 2 Allowable pressures on sheaves of three materials for various classifications of wire rope(a)

| Rope classification | Allowable pressures on sheaves made of: | | | | | |
| | Cast iron | | Cast steel | | 11-13Mn steel | |
	MPa	psi	MPa	psi	MPa	psi
6 × 7	2.1	300	3.8	550	10.3	1500
6 × 19	3.4	500	6.2	900	17.2	2500
6 × 37	4.1	600	7.4	1075	20.7	3000
8 × 19	4.1	600	7.4	1075	20.7	3000
Flattened strand	5.5	800	10	1450	27.6	4000

(a) Values are for regular-lay rope; for lang-lay rope, these values may be increased 15%, except for flattened-strand rope, which is normally lang lay.

Fig. 3 Three reeving diagrams for double-drum hoists

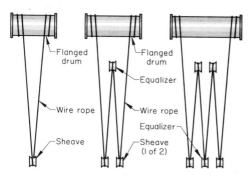

inches), and *d* is diameter of the rope (in inches).

Just as is true for bending stresses, the magnitude of the stresses resulting from the radial pressure increases as the size of the sheave decreases. High bending stresses generally indicate the need for flexible rope constructions. However, the resulting relatively small-diameter wires have less ability to withstand heavy pressures than the larger wires in the less flexible rope constructions. Both factors should be considered in selecting the most suitable type of wire rope.

Furthermore, the pressure of the rope against the sheave tends to flatten the rope structure. This type of rope distortion can be controlled to a large extent by proper sheave-groove contour.

Contact of the rope with the sheave results in wear in the rope and the sheave groove. The rate of wear is influenced by the magnitude of the rope pressure against the sheave groove.

Sheave Material. If the sheave material has insufficient wear resistance, the grooves will wear to the diameter of the rope operating in it. When a new full-size rope is installed, the groove worn small by the previous rope will subject the new rope to unnecessary abrasion in its attempt to grind the groove to its own diameter. To avoid this condition, the sheave-groove material should be selected to resist the wear corresponding to the rope pressure that will be present.

Table 2 lists allowable radial pressures calculated from Eq 1 for three cast, annealed sheave materials. Heat treatment to improve wear resistance will permit greater pressures. The values listed in Table 2 are of necessity approximate, but do represent desirable limits for avoiding excessive groove wear. These values are for regular-lay wire rope, in which the wires are laid in the direction opposite to the twist of the strands in the rope. The values can be increased for lang-lay wire rope (wires and strands laid in the same direction).

Drums. A common method of driving a wire rope is by a drum. One end of the rope is fastened to the drum, and the rope is wound and stored as the drum revolves. In some applications, both ends of the rope are attached to the drum.

Most drums are cylindrical with flanges at the ends. Flanges may not be necessary with grooved drums when the rope will never wind in more than one layer and when the rope cannot get out of the grooves. Flanges are necessary for multiple-layer winding and for cases where the rope might slip or wind off the face.

Drum size, groove contour, pressure, and drum material relate closely to the matters previously discussed on sheaves. The groove contour in grooved drums should be the same as in sheaves. The radial pressure between rope and drum on the first layer of a properly grooved drum is calculated by Eq 1. For these conditions, the same limitations apply to the materials listed in Table 2.

High pressures are exerted on the face of the drum with two or more layer winding. This has little influence on the choice of drum material because the additional pressure is applied after the first-layer wraps are fixed in position on the drum.

Higher pressures resulting from multiple-layer winding should be considered in the structural design of the drum. A drum for two-layer winding does not have to be twice as strong as that for single-layer winding, nor does one for three layers have to be three times as strong. When a second layer is wound on a drum, some tension is lost in the rope on the first layer. With the application of a third layer, additional tension is lost in the rope on the second layer. Studies have indicated that the combined rope tension with two layers is about 1¾ times the tension with one layer, with three layers about 2⅛ times the tension with one and with four layers about 2¼ times the tension with one.

With plain-faced drums, the actual radial pressures are higher because the rope does not have support around part of its circumference. On drums of this type, wear to both rope and drum would be more severe than that on grooved drums, and rope deterioration, resulting from pressure, would be more pronounced.

The diameter of the drum should be influenced by considerations of safety and economy. For general conditions, the ratios for sheaves should be considered, but for some applications, departures are made from these values.

Where bending is the primary factor affecting the service life of the rope, the drum may be slightly smaller in diameter than the sheaves in the system, because in traveling in each direction the rope bends only once at the drum, whereas it bends twice at each sheave. In traveling over each sheave, the rope is bent the first time when it conforms to the curvature of the sheave and the second time when it straightens out and leaves the sheave. The number of bends is decreased on those machines in which the rope travel is not very great and the section of rope operating on and off the drum does not reach the sheaves. The reeving diagrams shown in Fig. 3 can be used to determine the number of bends in a wire rope when multiple-pulley blocks are used.

When wire rope is wound on a drum, it exhibits a slight rotation because of the spiral lay of the strands. Standing behind the drum and looking toward an oncoming overwind rope, rotation of a right-lay rope is toward the left, whereas rotation of a left-lay rope is toward the right. This rotation is extremely small and seldom of any significance.

However, with a plain-faced drum, in which the only other influence to the rope is winding on the first layer is the fleet angle, this slight rotation can sometimes be used to advantage in keeping the windings close and uniform. Right-lay rope is standard and the one most readily available from stock. Therefore, all machines or installations having plain-faced drums should be designed for use with right-lay rope.

With a plain-faced drum, a right-lay rope, and overwind reeving, the attachment should be made at the left flange. With a right-lay rope and underwind reeving, the attachment should be made at the right flange.

When grooved drums are used, there generally is sufficient control by the grooving to wind the rope properly whether it is right or left lay. With either an overwind or an underwind installation or a left or right flange attachment, the standard right-lay rope construction can be used. Only in special applications should a change to left-lay rope be considered. One instance would be when opposite rotation of the rope might help prevent open winding, or a piling up at the flange under adverse fleet-angle conditions.

Grooved drums offer greater control and uniformity of rope winding compared with plain-faced drums. Moreover, when several layers are involved, grooved drums also influence the winding of the second and subsequent layers, provided the change from each layer to the next is properly accomplished. Grooving also provides some degree of circumferential support for the rope, which is advantageous to drum and rope life.

Maximum support is provided by grooving that has a depth equal to half the diameter of the rope, if the contour is proper. However, it is seldom possible or advisable to make the grooves this deep. For some applications, deep grooves have another advantage: they help keep the rope winding properly where a swinging load or some other condition of abnormal displacement causes the rope to lead improperly to the drum.

Deep grooves can be a disadvantage by adding to rope abrasion and restricting freedom of movement as the rope enters the groove. In leading to the drum, there is only one point at which the lead is absolutely straight into the groove. On either side of this point, some fleet angle is encountered. This causes rope and groove wear if the rope is confined around too much of its circumference.

There is seldom any advantage in designing a groove to have more than 150° circumferential support at the base, and in most cases, it would be disadvantageous to do so. If single-layer winding is involved and if deep grooves are necessary, this can be best accomplished by spreading the wraps apart and providing higher ridges between grooves.

It is important that grooved drums be designed with the proper pitch or distance from center to center of grooves to allow ample but not excessive clearance between successive wraps of rope. This is essential to prevent crowding and scrubbing of the oncoming rope against the rope already on the drum. To provide proper conditions for multiple-layer winding, the pitch between grooves should only be enough to prevent rope contact when winding under the maximum angle of fleet.

Example 1: Bending-Fatigue Failure of a Steel Wire Hoisting Rope for a Stacker Crane. A 13-mm (½-in.) diam 18 × 7 fiber-core improved plow steel nonrotating wire rope broke after 14 months of service on a stacker crane. Previously, 13-mm (½-in.) diam 6 × 37 improved plow steel ropes with independent wire-rope cores had been used. These ropes were in service for 12 months. The change to an 18 × 7 rope was made because of difficulties caused by twisting of the 6 × 37 rope.

Investigation. Chemical analysis of the steel wire indicated a carbon content of 0.51 to 0.55%, as specified. The tensile strength of the individual wires averaged 1234 MPa (179 ksi), which was above average. The quality of the steel was satisfactory, with the crown wires showing very little wear.

The hoist arrangement for this crane consisted of one rope with each end attached to a separate drum; the rope wound around two 30-cm (12-in.) diam sheaves in the block and

Fig. 4 13-mm (½-in.) diam 18 × 7 fiber-core improved plow steel nonrotating wire rope that failed in bending fatigue

The rope was operated over a sheave that was too small in diameter.

back up and around an equalizer sheave (Fig. 3b). The section of the rope that had been in contact with the sheaves was found by measurement checks. Reverse bending of the section of the rope normally subjected to this flexing revealed the presence of broken wire ends (Fig. 4), which indicated that the rope failed by fatigue. The minimum sheave diameter for a 13-mm (½-in.) diam 18 × 7 wire rope should be 43 cm (17 in.). Thus, the 30-cm (12-in.) diam sheaves were too small.

Conclusions. Failure of the wire rope occurred by bending fatigue. Continually running the same section of the rope over a sheave too small in diameter resulted in excessive bending stresses.

Corrective Measures. The sheave diameter could not be increased; therefore, the flexibility of a 6 × 37 rope was required. The 13-mm (½-in.) diam 18 × 7 rope was replaced by two 13-mm (½-in.) diam 6 × 37 steel-core ropes stranded side by side, one with left lay and the other with right lay. The twisting problem was eliminated by the use of the two counter-stranded cables.

Corrosion is another common cause of wire-rope failure. The corrosive atmospheres in which wire ropes operate are created by blast furnaces, cleaning tanks, plating tanks, and exposure to outdoor elements. Figure 5 shows corrosion failure of a steel wire rope that operated partially underwater.

Shock loading of a wire rope can cause it to vibrate, producing high-frequency cyclic bending stresses in the rope. Vibration is most severe at the connection end of the rope, and can result in fatigue failure.

Example 2: Fatigue Failure of a Steel Wire Rope Resulting From Shock Loading. The wire rope on a cleaning-line crane broke while lifting a normal load of coils. This rope, which was specified for the application, was 11-mm (⁷⁄₁₆-in.) diam 8 × 19 fiber-core rope of improved plow steel wire. Service life of the rope was 5 weeks; average expected life was 6 weeks. The rope was inspected weekly.

Fig. 5 Steel wire rope with heavy corrosion and broken individual wires resulting from intermittent underwater service

Fig. 6 Steel wire rope, used on a cleaning-line crane, that failed from fatigue resulting from vibration caused by shock loading

(a) Section of the wire rope adjacent to the fracture. Approximately 1½×. (b) Unetched longitudinal section of a wire from the rope showing fatigue cracks originating from both sides. 75×

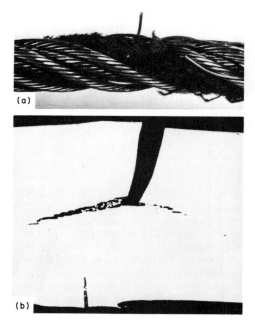

Investigation. Visual examination of a section of the wire rope adjacent to the fracture revealed several broken wires and fraying of the fiber core (Fig. 6a). The construction and mechanical properties of the rope were as specified.

Metallurgical examination of several wires revealed a uniform cold-drawn microstructure with no evidence of severe abrasion or mar-

Fig. 7 Transverse section through 2.6-mm (0.102-in.) diam steel wire

Light-etching surface layer (top) is untempered martensite; adjacent dark-etching zone is self-tempered martensite. The matrix was composed of deformed pearlite. Etched with 5% nital. 265×

Fig. 8 Wire rope, made of improved plow steel with a fiber core, that failed because of heavy abrasion and crushing under normal loading

(a) Crushed rope showing abraded wires and crown wear. 1.8×. (b) Nital-etched specimen showing martensite layer (top) and uniform, heavily drawn microstructure. 500×

(a)   (b)

tensite. Microscopic examination of a longitudinal section of a wire revealed fatigue cracks originating from both sides of the wire (Fig. 6b). One crack changed direction and propagated parallel to the centerline of the wire. The diameter of the sheave on the bale, 27 cm (10⅝ in.), was slightly below that specified for the 11-mm (7/16-in.) diam rope.

Observation of the crane in operation revealed that in rolling the coils over the edge of the rinse tank after pickling the hook received a sudden shock load, which was transmitted to the rope, causing vibrations. The vibrations were most severe at the clamped end of the rope.

Conclusions. Failure of the rope was attributed to fatigue, resulting from vibration caused by shock loading.

Corrective Measures. Pitched roll plates were installed between the tanks where rolling of coils was required. The plates reduced the free fall of the coils and aided in rolling. Also, the diameter of the sheave was increased to 33 cm (13 in.).

Scrubbing of a wire rope against a foreign object can cause excessive heating of the crowns of the outside wires. Because these regions are small in comparison to the total area of the wire, they may be rapidly quenched to a martensitic structure by the adjacent and underlying cooler metal. Martensite has very little ductility and will readily crack under the slightest bend.

Martensite is more easily formed in rope made of the higher-carbon grades of steel wire. Two ropes were analyzed that had operated under similar conditions as main lines in a logging machine. The ropes were 3.2 cm (1¼ in.) in diameter and of 6 × 19 independent wire rope core (IWRC) construction. One of the ropes had a carbon content of 0.685% and a manganese content of 0.607%; the other rope had a carbon content of 0.751% and a manganese content of 0.607%.

Both ropes developed martensite on the crowns of the outside wires. The martensite was caused by the friction and heat generated by the action of the rope sliding against the sides of the sheave. The outside wires were

rapidly heated, then rapidly quenched by the adjacent metal. The difference in carbon contents manifested itself in the layer of martensite formed being greater in the wires that contained the greater amount of carbon. Figure 7 shows a micrograph that illustrates the martensite layer developed by the wires having 0.685% carbon content. The ropes containing the 0.751% C wires displayed a deeper layer of martensite. The rope made of the lower-carbon wires had a life of about three times the other; however, this life was not considered satisfactory. The formation of martensite must be avoided, which means that the cause of failure was the abuse the ropes received and not necessarily the carbon content or the martensite formation.

Example 3: Fatigue Fracture of Individual Steel Wires in a Hoisting Rope. The wire rope on a crane in a scrapyard broke after 2 weeks of service under normal loading conditions. Expected life of the wire rope was 2 months. The rope was 16-mm (5/8-in.) diam 6 × 37 fiber-core improved plow steel. This type of rope is made of 0.71 to 0.75% C steel wires, with a tensile strength of 1696 to 1917 MPa (246 to 278 ksi).

The rope became damaged while it was attached to a chain for pulling jammed scrap from the baler. The chain broke during this operation, skipping the cable up onto the edge of the sheave. The rope then apparently seated itself in a kink on the flange of the sheave.

Investigation. As shown in Fig. 8(a), the rope was heavily abraded (scrubbed) and crushed. Also, several of the individual wires were broken.

Metallographic examination of several of the broken wires revealed a uniform cold-drawn microstructure, with patches of untempered martensite in regions of severe abrasion and crown wear.

Chemical analysis of the wire showed a carbon content of only 0.46%, whereas 0.71 to 0.75% C was specified. The tensile strength (1620 MPa, or 235 ksi) was below specification.

A hard layer of martensite (Fig. 8b) was formed on the wires as a result of abrasion. This surface layer, being very brittle, was susceptible to fatigue cracking while bending around the sheave.

Conclusions. The wire rope failed in fatigue because of the brittle layer of martensite on the wires. Both the carbon content and the tensile strength of the wires were below specifications.

Corrective Measures. Because of a record of poor service life on this wire rope, 6 × 19 rope was substituted because it withstands abrasion and drum crushing better than 6 × 37. Service life was doubled after this change in rope construction.

Example 4: Failure of Steel Wire Rope Because of Overheating. A 3.8-cm (1½-in.) diam 6 × 37 rope of improved plow steel wire broke in service during dumping of a ladle of hot slag.

Investigation. Examination of the rope showed a heavy blue oxide extending 0.6 to 0.9 m (2 to 3 ft) back from each side of the break. The broken ends of the rope showed tensile fractures. Microscopic examination of the wires adjacent to the break revealed that the steel had been recrystallized. The rope had thus been heated in excess of 700 °C (1300 °F). The tensile strength of the wires in the rope that broke was 896 MPa (130 ksi), whereas the specification required 1724 MPa (250 ksi).

Conclusions. Failure of the wire rope was attributed to overheating, which resulted in a 50% loss of tensile strength.

Recommendations. Wire ropes must not be subjected to extremes of heat, such as exposure to hot slag, except for brief, intermittent peri-

Fig. 9 Fatigue fracture of a steel 8 × 19 elevator cable

The fracture resulted from cyclic torsional and tensile stresses. (a) Conical shape at end of cable, and end of broken cable. (b) As-received 1.2-mm-diam wire. 25×. (c) Same wire after cleaning with a cold aqueous solution of 10% HCl. 25×. A indicates a nick in side of wire; B, a bright, smooth area containing fatigue marks. (d) Longitudinal section of 0.6-mm-diam wire etched in 2% nital showing necked region. 55×. (e) 2% nital-etched longitudinal section through 1.6-mm-diam wire showing cold working at A, flat-type fracture surface at B, and longitudinal cracks. 100×. (f) Fractured end (left) and longitudinal view (right) of 1-mm-diam wire. Both 34×. Arrows A indicate nick in a region showing cold working and wear in service; arrow at far right, a secondary crack origin.

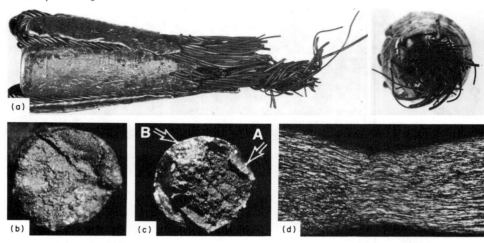

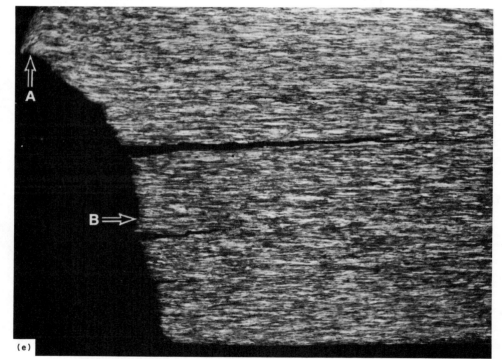

ods of time. Prolonged or continuous exposure will cause a rapid deterioration of service life.

Example 5: Fatigue Fracture of a Steel 8 × 19 Elevator Cable. Fracture occurred in one of six cables on a passenger elevator. The elevator continued to operate on the remaining five cables until routine inspection discovered the broken cable. The cable was made of 16-mm (⅝-in.) diam steel wire rope designated as 8 × 19 G Preformed Extra High Strength Special Traction Elevator Cable with fiber core. The cables had been in service for 1½ years. The end of the wire rope was sealed into a conical shape (Fig. 9a) in a low-melting alloy. Fracture occurred at the shackle where the end of the cable was socketed.

The fractured end, a length of wire rope away from the fracture, and several mounted specimens of various wires from the rope were examined. Samples of wire received were 1.2-mm (0.0465-in.) diam heart wire, 1-mm (0.0405-in.) diam outer wire, and 0.6-mm (0.023-in.) diam inner wire.

Investigation. Close examination of the wire rope under a stereomicroscope revealed two general types of fracture:

- A flat-type fracture in the samples of larger-diameter 1.2- and 1-mm (0.0465- and 0.0405-in.) wires
- A cup-and-cone type of fracture in the samples of smaller-diameter 0.6-mm (0.023-in.) wire

Generally, the larger wires, which failed with a flat-type fracture, were rusted; the smaller wires were oxidized but were not as heavily rusted. Figure 9(b) illustrates the fractured ends typical of the larger wires as received; Fig. 9(c), after removal of the rust. Arrow A in Fig. 9(c) indicates what appeared to be a nick in the side of the wire; arrow B indicates a bright, smooth area containing beach marks radiating inward, which indicated fatigue cracking. Flat-type fractures were believed to result from cyclic stresses, with the major stress component being torsional.

The nature of ductile behavior in the 0.6-mm (0.023-in.) diam wires is illustrated in Fig. 9(d), which shows a reduction of area in a necked-down region under excessive tensile stress without fracture. The fracture surfaces of the smaller-diameter wires were slightly oxidized; however, some bright regions could be seen, which indicated that the smaller wires broke later than the larger-diameter wires.

Numerous specimens were examined microscopically on a longitudinal plane at the failed ends. Figure 9(e) shows the typical microscopic characteristics of the larger wires. The important features are the cleanness of the material, microstructure, and the nature of the failure. No evidence of material or processing defects could be associated with the failure. It appears that as transverse cracking occurs under cyclic torsional stresses, longitudinal cracking follows as the transverse cracking progresses, placing

Fig. 10 Composite micrograph of a transverse section through a type 303(Se) stainless steel eye terminal for a wire rope showing corroded crack surface and final-fracture region

75×

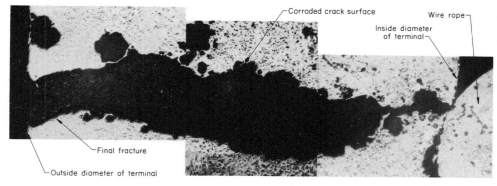

the maximum torsional fiber stress below the surface. Because of the fibrous nature of cold-drawn materials, longitudinal crack propagation would be expected under conditions of cyclic torsional loading with transverse crack propagation. Restriction of free movement of the socket-end in the shackle would be expected to promote fracture by increasing the magnitude of the torsional stress under cyclic loading.

Evidence of torsional stress was also observed in the 1-mm (0.0405-in.) diam wires. Figure 9(f) shows nicks and cold working on the surfaces of the wire in the region of fracture. This particular wire probably failed under greater tensile stress after the 1.2-mm (0.0465-in.) diam heart wires broke. Ductility of this wire is indicated by some necking and by cup-and-cone characteristics in the fracture surface. Most of the wires examined showed essentially flat fracture surfaces. The cleanness of the wire in the region of fracture is also illustrated in Fig. 9(f).

Conclusions. Mechanical damage to the surfaces of the wires was sufficient to cause fatigue cracking under the stresses encountered in service. The wire rope was composed of several sizes of wire; rust on the fracture surfaces of the larger-diameter wires indicated that they had failed first. The smaller-diameter wires behaved in a ductile manner under excessive loads before ultimate failure. The microstructure and cleanness of the material appeared normal for the application and could not be associated with the broken wires examined.

Stress-corrosion cracking (SCC) frequently occurs in terminals that have been roll swaged on ends of wire ropes for marine use. In the following example, rolling lines from the swaging operation acted as stress raisers. Cracks developed along these stress raisers, and corrosion by seawater occurred in the cracks.

Example 6: SCC of a Stainless Steel Wire-Rope Terminal. An eye terminal made of AISI type 303(Se) stainless steel that was roll swaged on the end of a 9.5-mm (⅜-in.) diam wire rope was found to have cracked exten-

sively after 1 year of service. The terminal was sectioned, and specimens were mounted for metallographic examination.

Investigation. Examination of one specimen revealed that a hairline crack had initiated at the inner surface of the fitting. Holes in the region adjoining the crack and the rough texture of the crack surface (Fig. 10) indicated that a corrosive medium (presumably seawater) had entered the crack from the inner surface of the fitting and, coupled with the hairline crack, developed crevice corrosion. High residual stresses in the swaged metal caused the crack to propagate toward the outer surface, followed closely by corrosion. The last 0.3 mm (0.012 in.) of metal thickness failed in pure tension, as evidenced by the smooth appearance of the crack suface in this region.

Swaging creates a few small cracks in the surface of the hole in a terminal, particularly when swaging onto stranded wire rope. Residual stresses plus load stress and corrosion frequently result in stress corrosion.

Conclusions. The terminal failed by SCC as a result of residual stresses from swaging, load stresses, and corrosion by seawater.

Recommendations. Rotary swaging or swaging in a punch press is recommended instead of roll swaging. Roll swaging is accomplished by nonsymmetrical radial metal deformation. The size of internal cracks is likely to be greater under these conditions than when swaging is accomplished in a more uniform symmetrical reduction, such as by rotary swaging or press swaging. Corrosion then combines with the internal stress to cause SCC.

Chains

Chains made of resistance-welded plain low-carbon steel are in wide use and are generally acceptable. However, to ensure safety and to minimize chain failures, heat-treated alloy steel chains with a hardness range of 302 to 352 HB should be used. This hardness range is recommended to reduce the frequency of brittle frac-

tures and to avoid premature wear and excessive stretching. However, it is much more desirable for a chain to stretch or to exhibit wear than to be so hard that it can fail by brittle fracture from a small fatigue crack, as often happens to chains with hardnesses of 375 HB and higher.

Fatigue cracks in chains are usually initiated at one of the following origins:

* A heavily cold-worked zone resulting from misuse of the chain
* A weld zone with entrapped inclusions or with inadequate weld penetration
* The portion of a weld at the inside surface of a link, which, because of its position, cannot be planed as smoothly as the outside surface

In the investigation of chain failures, the following possibilities must be considered in determining the cause of failure:

* Subjection of the chain to sudden impact loads or unbalanced loads
* Kinks, twists, or knots in the chain, which create severe stresses that can bend, weaken, and break chain links
* Deterioration of the chain by strain, usage, corrosion, or operation at elevated temperatures. The specified load limit should be reduced when alloy steel chains are used above 260 °C (500 °F). Chains should be replaced when wear has exceeded the specified limit of the manufacturer
* Stretching of the chain, which usually indicates overloading and which can be avoided by the use of proper chain size. Stretch is often expressed in terms of percentage of overall length. This is not recommended, because individual links may often be severely elongated; thus, only a small portion of the entire chain is stretched. To determine the degree of stretching, the chain should be inspected link by link
* Improper repair of the chain. Alloy steel chains should always be identified as such to ensure that any repairs—for example, replacement of damaged section or joiner links—are made properly. Alloy steel chains are usually returned to suppliers for repair and proof testing

Regular inspection of chains is imperative. Where many chains are in service, inspection is usually done visually. There are critical applications, however, in which magnetic-particle or liquid-penetrant inspection should be implemented. Inspection of individual links by these methods is a slow and tedious process and is carried out only when chain failure would cause injury to personnel and damage to equipment.

Example 7: Fracture of a 4615 Steel Chain Link Because of a Weld Defect. A resistance-welded chain link made from 16-mm (⅝-in.) diam 4615 steel broke while lowering a 9070-kg (10-ton) load of billets into a rail car. The acceptable load limit of the chain was approximately 13 600 kg (15 ton). The chain had been in service for 13 months. Because of the large number of chains in service and the fact that chains were left around billets in storage, no routine inspection was carried out.

Observations. The link broke at the weld. Beach marks, typical of fatigue, originated at the inside of the link (arrow, Fig. 11a). Metallographic examination of a section through the fracture surface revealed cracks (arrows A, Fig. 11b) in the weld zone that were up to 1.2 mm (0.0465 in.) deep. Examination of this region at a magnification of 65× showed that the cracks were filled with scale, indicating that they had formed during resistance welding of the link. The chain was made of fine-grain quenched-and-tempered 4615 steel with a hardness of 285 HB, as specified.

Conclusions. Failure of the chain link was attributed to weld defects, which acted as stress raisers from which fatigue cracks originated.

Corrective Measures. Inspection revealed welding laps in all chains in service. All such chains were removed from service and replaced with defect-free chains. The chain manufacturer changed the welding method to ensure defect-free chain links.

Example 8: Brittle Fracture of Alloy Steel Chain Links Because of Excessive Hardness. Over a 1-year period, chain-link fractures occurred in many of several thousand 16-mm (⅝-in.) diam alloy steel sling chains used for handling billets. Several shipments of new chains had been acquired because of expansion of production facilities. No failures had occurred before delivery of the new chains.

Observations. The links broke at the weld, with the breaks originating at the inside of the link. The breaks were often undetected because the broken links opened only slightly at the fracture line. There had been no fracture-related damage to the links. Metallographic examination revealed no weld irregularities. All failures occurred in links having hardness values in the range of 375 to 444 HB.

Upon contacting the supplier, it was learned that the hardness level of the new chain links was 375 to 444 HB, a change from the previous hardness level of 302 to 375 HB. Hardness was increased to minimize wear, but it made the links notch sensitive, resulting in fractures that initiated at the butt-weld flash on the inside surfaces of the links. It is believed that most of the fractures occurred during the winter months; the low temperatures undoubtedly caused a further reduction in ductility.

Conclusions. The chain links failed in a brittle manner because the high hardness of the material made it notch sensitive, particularly at low temperatures.

Corrective Measures. All chains were visually inspected, and those with broken links were returned to the supplier for repair. All chains were retempered to a hardness of 302 to 375 HB.

All new alloy steel sling chains were subsequently ordered to a hardness of 302 to 352 HB.

Fig. 11 Resistance-welded 4615 steel chain link that broke because of a weld defect

(a) A fracture surface of the chain link showing fatigue beach marks (arrow) progressing across the surface from the inside of the link. (b) Nital-etched longitudinal section through the link showing fracture origin (arrow O) and weld cracks (arrows A). Weld cracks at top were approximately 1.2 mm (0.0465 in.) deep.

(a)

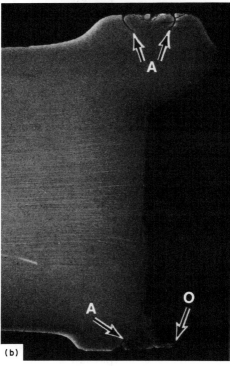

(b)

No failures of this type were reported since the chains were retempered and the hardness requirement of 302 to 352 HB was introduced.

Hooks

Hooks are designed for individual applications to suit the type of load, the size and weight of the load, the intended service environment, and the degree of versatility required. Failure of hooks occurs frequently, and hooks are continually being altered to avoid future failures. Among these alterations are changes in method of fabrication, grade of steel, heat treatment, and design.

In general, the most important features required of a hook material are excellent fatigue properties and resistance to brittle fracture. It is desirable for a hook to exhibit wear rather than to fracture in a brittle manner because of the possible serious injury and damage that can result. Such a hook should be evaluated visually while in service, and if signs of excessive wear are found, the hook should be removed from service for repair or replacement. If worn in the saddle region, a hook may be ground smooth within the specified reduction limit (usually, a maximum of 20% of the cross-sectional depth), renormalized if originally in a normalized condition (or otherwise thermally treated, depending on prior treatment), and returned to service. If fatigue cracks are found by magnetic-particle inspection, it may be possible to eliminate these cracks by grinding within the specified limits; if not, the hook should be scrapped.

Welding on crane hooks is to be discouraged. When welding must be done to attach safety devices or to join laminated sections, it should be followed by a heat treatment that will eliminate any hard spots or residual stresses in the HAZ.

Frequently, hooks that have been in service for a considerable length of time are given a stress relief to eliminate the effects of cold working from repeated lifting contacts by chains or rings.

Materials for Crane Hooks. C-hooks, pin hooks, coil hooks, and sister hooks are usually forged from fully killed fine-grain carbon or alloy steels. Extremely large ladle hooks and coil hooks are made of laminated plate sections. Specifications governing forgings for industrial use are given in ASTM A 668 (Ref 4); these specifications may be augmented as required.

Low-carbon (1018, 1020, or 1025) and medium-carbon (1045) steels are widely used in the normalized or the quenched-and-tempered condition. Alloy steels used include the 8620, 4320, 4130, and 4140 grades, generally in the quenched-and-tempered condition. The low-carbon steels usually have greater resistance to crack propagation. Other alloy steels, or other materials, may be required in certain corrosive atmospheres or because of weight limitations placed on the hook.

Some hooks have been made of cast steel. Derrick hooks, for example, may be of quenched-and-tempered 4140 cast steel. This heat-treated steel has excellent fatigue properties. L-shape hooks, such as coil hooks, are

Fig. 12 Typical design of a 45 360-kg (50-ton) capacity 1020 steel C-hook with a stress-relief groove at end of threads and well-proportioned radii in body

Dimensions given in inches

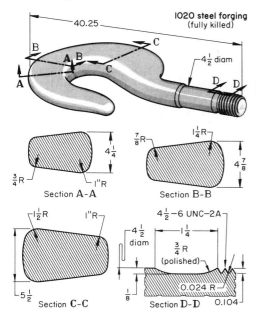

Fig. 13 13 600-kg (15-ton) 1020 steel crane hook that failed in fatigue

View of a fracture surface of the hook showing beach marks. Original and improved designs for the nut and the threaded end of the hook are also shown. Dimensions given in inches

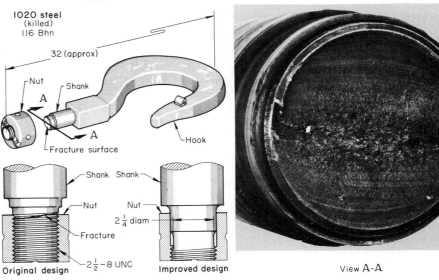

View A-A

usually constructed of low-alloy steel. Hooks that are torch cut from plate often fail because of cracks initiated in the torch-hardened zone. Such items should be thermally treated to eliminate this condition. A riveted, laminated L-shape hook is much more desirable. Depending on their ultimate service requirements, hooks are frequently magnetic particle inspected or ultrasonically tested during processing and at final inspection.

Design of C-Hook. The design of the hook shown in Fig. 12 has been developed from years of experience, and since its employment in a large steel complex, no failures of hooks of this design have occurred. The hook design may be adjusted to accommodate any load limit. The hook is forged from fully killed fine-grain 1020 steel and is normalized after forging. This steel exhibits excellent fatigue properties and will wear rather than fail in a brittle manner.

Many hooks break at large changes in section or at poorly blended radiused section changes in the shank. The design shown in Fig. 12 avoids severe section changes. An important feature of this hook is the stress-relief groove at the end of the threads. If the nut is turned past the last thread—as it should be so as not to concentrate the stress at the bottom of the exposed thread groove—the stresses will be distributed over the length of the stress-relief groove, minimizing stress concentrations. The depth of a properly designed groove exceeds that of the thread roots. The thread design illustrated has also

been found to be optimum. It is important, as with all hooks, to avoid rough-machining marks in the threads. Also, the nuts should be individually fitted to hooks.

Control and Testing. All hooks must be clearly identified, and accurate records must be kept of their characteristic features and location in the plant. Hooks should be inspected at least once a year, if not more often, depending on the predicted service life and the severity of the application. Magnetic-particle and liquid-penetrant inspection are commonly used for hooks. If feasible, new hooks should be checked by magnetic-particle inspection and should be spark tested to ensure that the correct grade of steel has been used.

Example 9: Fatigue Fracture of a 1020 Steel Crane Hook. The crane hook shown in Fig. 13 broke in the threaded shank while lifting a load of 9072 kg (10 ton). The crane was nominally rated at 13 000 kg (15 ton). Service life of the hook was not known. The nut, still threaded on the broken shank, and the hook were sent to the laboratory for failure analysis.

Investigation. Chemical analysis of the metal in the hook showed that it was killed 1020 steel. The steel had a hardness of 116 HB and was judged to be satisfactory for the application.

Visual examination disclosed that fracture had occurred at the last thread on the shank. Rough-machining marks and chatter marks were evident on threads. The fourth thread from the fracture had a small crack at the root. Beach marks emanating from the thread-root locations on opposite sides of the fracture surface (View A-A, Fig. 13) identified these locations as the origins of the fracture.

Metallographic examination revealed a medium-coarse slightly acicular structure, indi-

cating that the material was in the as-forged condition. Light segregation and nonmetallic inclusions were present, but the material was considered sound.

Conclusions. Fracture of the hook resulted from fatigue cracking that originated at stress concentrations in the root of the last thread. Life of the hook was shortened considerably because the material was in the as-forged condition, which resulted in low fatigue strength.

Corrective Measures. The crane hook was normalized after forging to produce the most desirable structure. A stress-relief groove with a diameter slightly smaller than the root diameter was placed at the end of the thread, and a large-radius fillet was machined at the change in diameters of the shank. Also, the nut was redesigned to extend beyond the threaded portion to relieve the stresses (Improved design, Fig. 13). Undercutting the last thread, using large-radius fillets, and ensuring that all crane hooks were properly normalized prevented additional failures for about 16 years.

Example 10: Fatigue Fracture of a 1040 Steel Coil Hook. A 10 890-kg (12-ton) coil hook failed after 8 years of service while lifting a load of 13 600 kg (15 ton). The hook had been torch cut from 1040 steel plate.

Investigation. The inner surface of the hook exhibited the normal ironing (wear) marks. Visual examination of the fracture surface indicated that cracking had originated at the inside radius of the hook (Fig. 14a). Beach marks, typical of fatigue fracture, extended over approximately 20% of the fracture surface. The surface containing the beach marks was stained; the surface of the final-fracture region was bright.

Macroscopic examination of the torch-cut surfaces revealed numerous cracks. Figure

Fig. 14 Torch-cut 1040 steel coil hook that failed by fatigue

(a) Fracture region of the 10 890-kg (12-ton) hook. (b) Macrograph of a nital-etched section showing cracks propagating from the surface (top), which was hardened and embrittled during torch cutting. 7½×. (c) Replacement hook of a laminated design made of ASTM A242 fine-grain steel plate. Dimensions given in inches

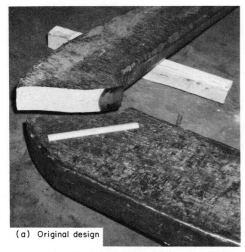

(a) Original design

(b)

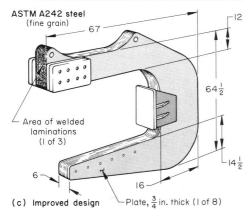

ASTM A242 steel (fine grain)

Area of welded laminations (1 of 3)

(c) Improved design — Plate, ¾ in. thick (1 of 8)

14(b) shows a macrograph of an etched section through the hook, illustrating cracks that were initiated in a hardened martensitic zone at the torch-cut surface. One crack is shown extending into the coarse pearlite structure beneath the martensitic zone.

Conclusions. The hook fractured by fatigue that originated at the brittle martensitic surface produced in torch cutting. The 25% overload contributed to the failure. The hook should have been normalized after torch cutting.

Corrective Measures. The coil hooks were remade to a laminated design (Fig. 14c); these hooks were flame cut from ASTM A242 fine-grain steel plate, then ground to remove the material damaged by flame cutting. The hooks were stress relieved at 620 °C (1150 °F) after welding but before riveting the wear pads. No failures were reported since adoption of the laminated design.

Example 11: Failure of a Forged Semikilled 1015 Steel Hook on a 13-mm (½-in.) Diam Chain Sling. A hook on a two-leg chain broke while lifting a 4990-kg (11 000-lb) load. The included angle between the two 13-mm (½-in.) diam chains was 60°. The service life of the sling was 2½ years.

Investigation. The safe load limit for each leg of the sling was 2630 kg (5800 lb). The hook broke at the junction of the eye and shank. The diameter of the hook at this junction was approximately 22 mm (⅞ in.).

Visual examination of the fracture region revealed light intergranular oxidation at the surface on the side of the hook where cracking started. Approximately 50% of the fracture surface contained beach marks; the remainder contained cleavage facets.

Metallographic examination showed a medium-coarse acicular as-forged structure. The internal cleanness and soundness of the metal were satisfactory. Chemical analysis showed the metal to be semikilled 1015 steel.

Conclusions. Fatigue fracture was initiated at a region of intergranular oxidation that developed during forging. The acicular as-forged structure provided poor fatigue and impact properties, which contributed to failure of the hook.

Corrective Measures. The chain-sling hook was replaced with one made of normalized, fully killed, fine-grain 1020 steel.

Shafts

The most common or recurring initiation sites for fatigue fractures in shafts are section changes with insufficient fillet radii, roughly machined keyways and section changes, intersections of keyways with section changes, improperly located grease holes, and fretting corrosion from connecting attachments.

Shafts are usually subjected to bending or torsional loading or both. Careful visual examination of the fracture surfaces can provide the information necessary to identify the actual manner of loading. Bending stresses are largely the result of misalignment of bearings, fittings, or couplings and are usually difficult to eliminate completely. Although a particular steel may be satisfactory for a certain application under normal conditions, misalignment will

Fig. 15 Fatigue-fracture surface and keyway of a broken 1030 steel pinion shaft

impose further stresses that result in the shaft being unsuitable for the application.

To provide a greater margin of safety, which is justified for lifting equipment, it is recommended that the shaft material be upgraded from the normally used hot-rolled medium-carbon steel to quenched-and-tempered carbon or alloy steel. Such materials, particularly alloy steels when properly heat treated, display superior fatigue properties and are much less susceptible to service failures. The danger and downtime resulting from a shaft failure make upgrading by heat treatment or a change in material economically feasible. It is presumed, of course, that any required improvement in design, such as increasing fillet radii or removing keyways from shoulders, would also be considered before or incorporated with any change in processing or material. For a more complete discussion of shaft failures, see the article "Failures of Shafts" in this Volume.

Example 12: Fatigue Fracture of a 1030 Steel Crane Shaft. A drum pinion shaft, part of the hoisting gear of an 18 140-kg (20-ton) capacity crane operating in a blooming mill, broke while the crane was lifting a 9070-kg (10-ton) load. Specifications indicated that the shaft was made of 1030 steel in the as-rolled condition.

Investigation. The end of the broken shaft containing the keyway was visually examined (Fig. 15). The keyway extended into a shoulder at a change in diameter; chatter marks, rough-machining marks, and sharp corner radii were visible in the keyway. Also, at each end of the keyway was a circular recess below the normal keyway surface—an outdated method of machining a keyway that is infrequently used. Both the bottoms and the sides of the recesses contained tool marks.

Examination of the fracture surface (Fig. 15) revealed that the origin of fracture was a sharp corner at the end of the keyway. Beach marks radiated from the origin over a large portion of

Fig. 16 Change in section in a 1040 steel main hoist shaft where a fatigue crack (arrow) was initiated at rough-machining marks and a break in a fillet

Fig. 17 4140 steel cross-travel shaft that failed in service

(a) Broken end of the shaft from a derrick showing the star-type fracture that results from reversed torsional loading. (b) Transverse section through the spline showing cracks initiated at sharp corners at the roots of the spline teeth

(a)

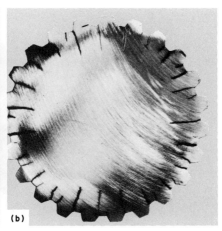

(b)

the fracture surface. The final-fracture zone, a small ductile shear lip, was approximately 30° off-center, indicating low stress and rotational bending. The shaft material was 1030 steel, as specified, and had a slightly acicular fine-grain structure.

Conclusion. The shaft failed by fatigue fracture, which originated at a sharp corner at the end of a keyway at a change in section.

Corrective Measures. A replacement shaft was made of 4140 steel, quenched and tempered to a hardness of 286 to 319 HB. The keyway was moved away from the change in section and was machined with a 1.6-mm (1/16-in.) radius in the bottom corners. A larger-radius fillet was machined at the change in section.

Example 13: Fatigue Cracking of a 1040 Steel Main Hoist Shaft. The 14-cm (5½-in.) diam main hoist shaft of a mobile shovel was found to have multiple crack indications when ultrasonically inspected in the field. A crack indication located approximately 15 cm (6 in.) from the gear end extended around the shaft at a change in section adjacent to the bearing surface. The shaft was removed

for further examination. It had been in service for 3 years. A previous shaft had failed after a short service life.

Investigation. Magnetic-particle inspection of the shaft showed a crack around the entire circumference at the change in section (Fig. 16). The fillet at the change of section where the crack was located was well polished and generally free of rough-machining marks. However, the crack coincided with the junction of the fillet and the smaller diameter (10.8 cm, or 4¼ in.) at this change in section. A slight step, or break, in the continuity of the 8-mm (5/16-in.) radius fillet and some machining marks were noted at this junction. Microscopic examination of a section through the crack indication revealed a fine crack extending 2.5 mm (0.10 in.) from the surface, originating at the machining marks.

Chemical analysis of the shaft indicated that it was made of 1040 steel. Hardness was 170 HB. Metallographic examination revealed a fine, uniform, normalized structure.

Conclusions. The shaft failed by fatigue; the step at the base of the fillet was the point of initiation of the fatigue crack. The shaft was underdesigned from a material standpoint; 1040 steel with a hardness of 170 HB was too low in fatigue strength.

Corrective Measures. Shaft material was changed to 4140 steel oil quenched and tempered to a hardness of 302 to 352 HB. All discontinuities related to fillet machining were also removed. With the above changes, no further failures were encountered.

Reversed Torsional Loading. Shafts used in lifting equipment are frequently subjected to reversed torsional loading. Insufficient fillets at the roots of spline teeth act as stress raisers, and fracture occurs frequently when the shaft is subjected to torsional loading.

Example 14: Fatigue Fracture of a 4140 Steel Cross-Travel Shaft. The horizontal cross-travel shaft on a derrick broke after

2 years of service. Specifications required the shaft to be made of 4140 steel quenched and to a hardness of 302 to 352 HB.

Investigation. Examination revealed that the shaft had fractured approximately 13 mm (½ in.) from the change in section between the splined end and the shaft proper. At this point, the cracks propagated in the longitudinal and transverse directions until failures occurred (Fig. 17a). A transverse section through the spline (Fig. 17b) showed that the longitudinal cracks had been initiated at the sharp corners at the roots of the spline teeth. The cracks visible in Fig. 17(b) were highlighted by techniques used in magnetic-particle inspection. Operation of the derrick had subjected the shaft to reversed torsional loading.

Conclusions. The shaft fractured in fatigue from reversed torsional loading; the sharp corners at the roots of the spline teeth acted as initiation sites for fatigue cracks.

Corrective Measures. The fillets at the roots of the spline teeth were increased in size and polished to minimize stress concentrations in these areas, and no further shaft failures occurred.

Cranes and Related Members

Failures resulting in the immobilization of cranes are most often related to shafts (which have been discussed previously), structural members (as related primarily to grade of steel and fabrication practice), crane wheels, and rail runways. Figure 18 shows the fracture surface of a 1055 steel crane wheel that failed by fatigue. The most common cause of failure is poor welding practice. Poor welding practice, such as a lack of preheat or postheat or use of the wrong filler metal, often results in hardening from the heat of welding, weld porosity, incomplete fusion, inadequate joint penetra-

Fig. 18 1055 steel wheel from a stripper crane that failed by fatigue

The wheel failed after about 1 year of service in the 544 320-kg (600-ton) crane. In the center of the fracture is a fatigue zone showing beach marks that are concentric around the crack origin, which evidently was an internal flaw about 3.8 cm (1½ in.) beneath the wheel-tread surface. Chevron patterns are visible in the area of final fast fracture that surrounds the fatigue zone. 0.3 ×

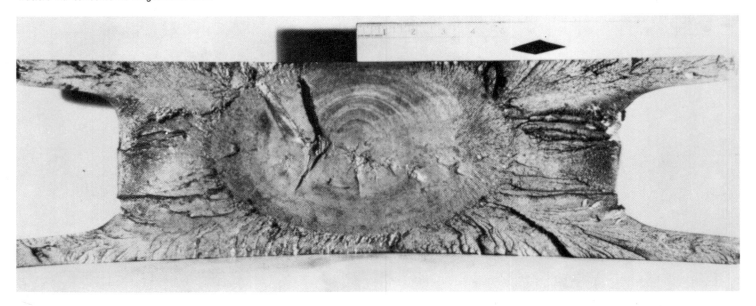

tion, and weld cracks. Poor welding practice may occur not only during repair but also in initial fabrication or assembly.

The most common method of repairing structural members is welding. The part must be welded properly, or it may fail soon after being returned to service. Such faults as microcracks and macrocracks, mechanical gouges, and excessive wear may be corrected by welding. If a crack is present, it should be removed by grinding. Magnetic-particle or liquid-penetrant inspection should be done to ensure complete removal of the crack before repair welding. Generally, all parts should be preheated before welding to prevent hardening from the heat of welding. This is most critical in medium-carbon and high-carbon steels because of their sensitivity to hardening by welding heat. It is essential that the correct welding filler metal be used. After welding, low-carbon steel members may be air cooled if they have been preheated correctly. To prevent hardening from the heat of welding, the cooling rate of medium-carbon and high-carbon steel members should be controlled—for example, by wrapping them in asbestos.

The susceptibility of a crane to failure is also heightened by a poor choice of material. In general, structural members of a crane should be designed so as not to fail in a brittle manner. The members should be made from impact-resistant fine-grain steels to resist brittle fracture and to exhibit excellent fatigue properties. This becomes more critical in cold-temperature applications.

Crane members should be inspected periodically for fatigue cracks to ensure that a failure is not imminent. A standard specification covering all items related to cranes is difficult to establish because of the wide variety of associated items. Those specifications that do exist must be viewed as a minimal guideline.

Example 15: Fracture of a Steel Tram-Rail Assembly. Figure 19(a) shows a hoist-carriage tram-rail assembly fabricated by shielded metal arc welding the leg of a large T-section 1020 steel beam to the leg of a smaller T-section 1050 steel rail. Several of the welds failed in one portion of the assembly. Cracks in weld regions of other portions of the assembly were found by magnetic-particle inspection.

Investigation. The weldment was inspected using the magnetic-particle method. This inspection revealed four weld cracks and several indefinite indications. All cracks were located at the toes of welds that joined the rail to the beam, such as the crack shown at the arrow in Fig. 19(b).

Metallographic examination of longitudinal sections through the welds revealed, in every instance, that cracks originated in HAZs in the rail section. A crack (arrow, Fig. 19c) was initiated in a localized HAZ produced by feathering the weld, which left a thin deposit of weld metal on the upper edge of the rail. Cracks were also found in HAZs resulting from weld spatter (Fig. 19d). Additional macrographs revealed cracks in welds and in HAZs resulting from arcing the electrode adjacent to the weld.

A relatively broad HAZ containing tempered and untempered martensite was present at each of the weld beads because of multiple-pass welding. The rail and beam both exhibited normal as-rolled steel structures with no irregularities. Chemical analysis revealed the rail to be made of 1050 steel and the beam of 1020 steel.

Hardness surveys revealed the following values: beam, 132 HB; rail, 255 HB; weld metal, 285 HB; tempered martensite, 321 HB; and untempered martensite, 578 HB.

Conclusions. The tram-rail assembly failed by fatigue cracking in HAZs. The size and shape of HAZs adjacent to welds indicated that the assembly had been neither preheated nor postheated during the welding operation. Welding workmanship was generally very poor. There was excessive feathering of weld metal and careless arcing during welding, which resulted in HAZs containing hard untempered martensite. Thus, vibration of the tram rail by movement of the hoist carriage on the rail could easily initiate and propagate fatigue cracking in the HAZs.

Corrective Measures. Welding procedures were improved, and the replacement rail assemblies were preheated and postheated.

Example 16: Brittle Fracture of a 1020 Steel Stop-Block Guide on a Crane Runway. A section broke from a stop-block guide (Fig. 20) on a crane runway and fell to the floor. A system consisting of wire ropes, pulleys, a counterweight, and a lever was used to raise the stop block from the crane runway to permit the crane to pass. When the crane was to be isolated within an area, the stop block was returned to the down position. Because the weight of the stop block and guide was about 30% greater than that of the counterweight, there was some impact against the rail each time the block was returned to the down position. One section of the fractured guide was sent to the laboratory for determination of the cause of failure.

Investigation. Examination of the fracture surface disclosed a brittle crystalline-type

Fig. 19 Tram-rail assembly that fractured because of poor welding practices

(a) Section of tram rail as fabricated. T-section beam (1020 steel) is at top, T-section rail (1050 steel) is at bottom. (b) Enlarged view of welded area showing crack at toe of weld (arrow). (c) Crack in rail initiated in the HAZ produced by feathering of the weld deposit. (d) Crack in rail initiated in the HAZ caused by weld spatter

(a)

(b)

(c)

(d)

break. The point of initiation was in a hardened heat-affected layer that developed during flame cutting and welding. Hardness of the heat-affected layer was 248 HB.

Chemical analysis showed the metal to be 1020 steel. The base metal (hardness, 156 HB) had a coarse as-rolled structure and a grain size of ASTM 00 to 4. The coarse grain size indicated that the weldment (stop block and

Fig. 20 Welded stop-block assembly for a crane runway showing stop-block guide that failed by brittle fracture

Dimensions given in inches

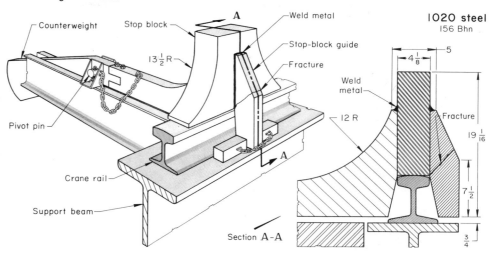

guide) had not been normalized.

Conclusions. The stop-block guide failed in a brittle manner. Failure was initiated at a metallurgical and mechanical notch that was produced by flame cutting and welding.

Corrective Measures. New stop-block weldments were made from fully silicon-killed 1020 steel with a maximum grain size of ASTM 5. The weldments were normalized at 900 °C (1650 °F) after flame cutting and welding to improve microstructure and impact strength. All flame-cut surfaces were ground to remove notches. Stop-block weldments in service were checked for grain size and heat treatment and were normalized where required.

Example 17: Brittle Fracture of an Aluminum Alloy Lifting-Sling Member.
The T-section cross member of the lifting sling shown in Fig. 21(a) broke in service while lifting a 966-kg (2130-lb) load. The L-section sling body and the cross member were made of aluminum alloy 5083 or 5086 and were joined by welding using aluminum alloy 4043 filler metal. Design load for the sling was 2722 kg (6000 lb).

Investigation. Visual examination of the sling showed that fracture occurred at the weld joining the sling body and the cross member. Examination of the ends of the failed cross member showed that a rotational force had been applied on the cross member, causing it to fracture near the sling body. The fracture surfaces of the sling body and the cross member are shown in Fig. 21(b). Macrographic examination of the weld disclosed inadequate joint penetration and porosity (Fig. 21c).

Chemical analysis showed that the metal was within specifications for aluminum alloy 5083 or 5086. Analysis of the weld metal indicated that aluminum alloy 4043 filler metal had been used. The silicon content was lower and the magnesium and manganese contents were

higher than normal for alloy 4043 filler metal. These differences were attributed to dilution of the weld metal by the base metal.

A similar sling was fabricated in the laboratory, set up in a test fixture, and subjected to three-point loading duplicating that intended for the sling that failed. The yield strength of the test sling (stress causing permanent deflection in the cross member) was 3538 kg (7800 lb), and the ultimate load was 7031 kg (15 500 lb).

The initial break in the test sling started at the welded joint between the sling body and the edge of the flange of the cross member and propagated through the body to failure. The cross member of the sling that failed in service sustained permanent deformation; however, the same member of the test sling exhibited no evidence of deformation.

Conclusions. The sling failed in a brittle fracture at the weld when the cross member was overloaded. Overloading was attributed to misalignment of the sling during loading, which concentrated the entire load on the cross member.

Recommendations. In welding of aluminum alloy 5083 or 5086, aluminum alloy 5183 or 5356 filler metal should be used to avoid brittle welds. Aluminum-silicon filler metals, such as alloy 4043, should not be used to weld aluminum alloys in which the magnesium content exceeds 3%, because an excess of Mg_2Si produces brittle welds.

Example 18: Fracture of a 1055 Steel Crane-Bridge Wheel.
A bridge wheel on a crane fractured after 1 year of service. The wheel was forged from 1055 steel. The wheel fractured in the web between the hub and the rim.

Investigation. Visual examination of the fracture surface revealed a small area containing beach marks that originated in a heavily burned area on the web surface. Metallographic examination of a section taken through the

Fig. 21 Aluminum alloy lifting sling that fractured because of improper welding of the cross member to the sling body

(a) Configuration and dimensions (given in inches). (b) Fracture surfaces of the sling body (top) and the cross member (bottom). (c) Enlarged view of a fracture surface of the sling body showing weld defects

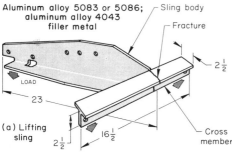

(a) Lifting sling

Aluminum alloy 5083 or 5086; aluminum alloy 4043 filler metal — Sling body — Fracture — LOAD — 23 — 2½ — 2½ — 16½ — Cross member

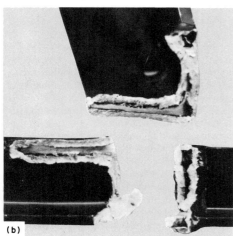

(b)

(c)

region that contained the beach marks showed surface burning to a depth of approximately 0.8 mm (0.030 in.) (Fig. 22). The degree of decarburization and oxide dispersion that were visible indicated a forging defect.

Conclusions. Failure occurred because of surface burning during the forging operation. To prevent this type of failure, heating practice during forging should be more closely controlled to eliminate surface burning. If burning does occur, the burned region, if not too se-

verely damaged, should be removed by machining.

Example 19: Fatigue Fracture of a Forged 1055 Steel Crane-Bridge Wheel. A bridge wheel from a 272 160-kg (300-ton) stripper crane failed after 1½ years of service. The wheel was forged from 1055 steel, and the tread, hub faces, and hub bore were machined. The wheel fractured in the web near the rim.

Investigation. Macroscopic examination of the fracture surfaces revealed beach marks indicative of fatigue at ten locations. Fatigue cracking extended across about 15% of the surface before final fracture occurred. The dark areas on the fracture surface shown in Fig. 23(a) are the portions that failed by fatigue. The surface of the web was heavily scaled and decarburized.

Chemical analysis showed that the material was equivalent to 1055 steel. Metallographic examination revealed that the wheel had been normalized and that the tread had been hardened to a depth of 5 mm (0.2 in.). The tread was tempered martensite; the core, fine pearlite.

Examination of a micrograph of a section through one of the fatigue origins showed a gross forging defect extending about 1.8 mm (0.072 in.) along the fracture surface (at right, Fig. 23b). Shallower forging defects and the decarburized surface were visible along the web surface (bottom, Fig. 23b).

Conclusions. Fatigue cracking of the wheel initiated at forging defects in the web.

Corrective Measures. Replacement wheels were machined all over and were magnetic particle inspected to detect any cracks that could act as stress raisers.

Fig. 22 Surface burning that initiated fracture in the web of a crane-bridge wheel forged from 1055 steel

Etched with 2% nital. Approximately 35×

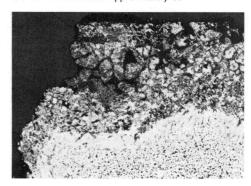

REFERENCES

1. "Standard Specification for High-Strength Bolts for Structural Steel Joints," A 325, *Annual Book of ASTM Standards*, ASTM, Philadelphia
2. "Standard Specification for Steel Structural Rivets," A 502, *Annual Book of ASTM Standards*, ASTM, Philadelphia
3. A.B. Dove, Ed., *Steel Wire Handbook*, Vol 3, The Wire Association, Inc., Branford, CT, 1972
4. "Standard Specification for Steel Forgings, Carbon and Alloy, for General Industrial Use," A 668, *Annual Book of ASTM Standards*, ASTM, Philadelphia

Fig. 23 1055 steel crane-bridge wheel that failed by fatigue

(a) Fracture surface of the crane-bridge wheel. Fatigue originated at forging defects. Dark areas are fatigue beach marks. (b) Micrograph of a nital-etched section through the fatigue origin showing a gross forging defect along the fracture surface (at right) and decarburized web surface (bottom). 35×

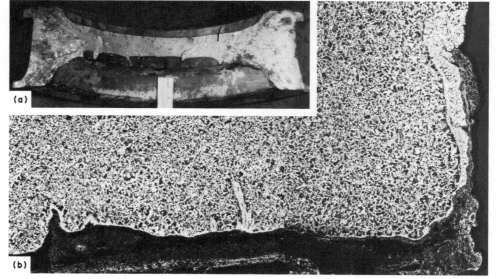

(a)

(b)

Failures of Mechanical Fasteners

Walter J. Jensen, Lockheed-Georgia Company

THE PRIMARY FUNCTION of a fastener system is to transfer load. Many types of fasteners and fastening systems have been developed for specific requirements, such as higher strength, easier maintenance, better corrosion resistance, greater reliability at high or low temperatures, or lower material and manufacturing costs.

The selection and satisfactory use of a particular fastener are dictated by the design requirements and conditions under which the fastener will be used. Consideration must be given to the purpose of the fastener, the type and thickness of materials to be joined, the configuration and total thickness of the joint to be fastened, the operating environment of the installed fastener, and the type of loading to which the fastener will be subjected in service. A careful analysis of these requirements is necessary before a satisfactory fastener can be selected. The selection of the correct fastener or fastener system may simply involve satisfying a requirement for strength (static or fatigue) or for corrosion resistance. On the other hand, selection may be dictated by a complex system of specification and qualification controls. The extent and complexity of the system needed is usually dictated by the probable cost of a fastener failure.

Adequate testing is the most practical method of guarding against failure of a new fastener system for a critical application. The designer must not extrapolate existing data to a different size of the same fastener, because larger-diameter fasteners have significantly lower fatigue endurance limits than smaller-diameter fasteners made from the same material and using the same manufacturing techniques and joint system.

Types of Mechanical Fasteners. For descriptive purposes, mechanical fasteners are grouped into threaded fasteners, rivets, blind fasteners, pin fasteners, special-purpose fasteners, and fasteners for composites. Rivets, pin fasteners, and special-purpose fasteners are usually designed for permanent or semipermanent installation.

Threaded fasteners are considered to be any threaded part that, after assembly of the joint, may be removed without damage to the fastener or to the members being joined.

Rivets are permanent one-piece fasteners that are installed by mechanically upsetting one end.

Blind fasteners are usually multiple-piece devices that can be installed in a joint that is accessible from only one side. When a blind fastener is being installed, a self-contained mechanism, an explosive, or other device forms an upset on the inaccessible side.

Pin fasteners are one-piece fasteners, either solid or tubular, that are used in assemblies in which the load is primarily shear. A malleable collar is sometimes swaged or formed on the pin to secure the joint.

Special-purpose fasteners, many of which are proprietary, such as retaining rings, latches, slotted springs, and studs, are designed to allow easy, quick removal and replacement and show little or no deterioration with repeated use.

Mechanical fasteners for composites are frequently used in combination with adhesive bonding to improve the integrity of highly stressed joints. The usual bolts, pins, rivets, and blind fasteners are used for composites; however, the many problems encountered have stimulated the development and testing of numerous special-purpose fasteners and systems. Some of these problems are drilling of and installation damage to the composite, delamination of the composite material around the hole and pullout of the fastener under load, differences in expansion coefficients of the composite and the fastener, galvanic corrosion between the composite hole wall and the fastener, moisture and fuel leaks around the fastener, and fretting.

Load-carrying capacity has been significantly improved by increasing the head diameter of fasteners in order to distribute loads over a greater surface area. The bases of nuts and collars were similarly increased.

Fasteners for composites should fit the holes with no clearance in order to avoid fretting, but an interference fit may cause delamination. This problem has been alleviated to some extent by the development of several special sleeve/fastener combinations.

The extreme temperature changes experienced by aircraft can cause differential thermal expansion and contraction of composites and their fasteners, thus altering clamp-up loads.

Therefore, all new joint designs should be thoroughly tested to ensure satisfactory service. Although adhesives are generally used to join composites, mechanical fasteners are added to increase reliability when severe service conditions are involved. Table 1 compares fastener usage for several composite materials.

Specifications are used to outline fastener requirements, to control the manufacturing process, and to establish functional or performance standards. Their goal is to ensure that fasteners will be interchangeable, dimensionally and functionally. The most common fastener specifications are product specifications that are set up to govern and define the quality and reliability of fasteners before they leave the manufacturer. These specifications determine what material to use; state the objectives for tensile strength, shear strength, and response to heat treatment; set requirements for environmental temperature or atmospheric exposure; and define methods for testing and evaluation. To ensure proper use of fasteners, additional specifications are necessary as a guide to proper design application and installation.

All ASTM and SAE specifications covering threaded fasteners require that the heads be marked for grade identification. Grade markings are a safety device that provides a positive check on selection, use, and inspection. The markings reduce the possibility of selecting and using a bolt of insufficient strength, which might lead to a failure and cause damage to equipment or injury of personnel. The failure analyst should be familiar with the different grade markings required by the specifications.

In an effort to standardize the reporting of fastener test failures, a failure-identification code has been established for reporting the failure of each fastener resulting from tests under MIL-STD-1312. Failures of threaded and blind fasteners and of fastened sheet covered by this identification code are illustrated in Fig. 1.

Failure Origins

The most common locations for fastener failure are in the head-to-shank fillet or, on threaded fasteners, through the first thread inside the nut or at the transition from the thread to the shank. Failure origins are often imperfections in the metal caused by segregation in

Table 1 Fasteners for advanced composites

Fastener type	Fastener material	Surface coating	Suggested application			
			Epoxy/graphite composite	Kevlar	Fiberglass	Honeycomb
Blind rivets(a)	5056 aluminum	None	Not recommended(h)	Excellent(h)	Excellent(h)	(e)
	Monel	None	Good(h)	Excellent(h)	Excellent(h)	(e)
	A-286	Passivated	Good(h)	Excellent(h)	Excellent(h)	(e)
Blind bolts(b)	A-286	Passivated	Excellent(h)	Excellent(h)	Excellent	(e)
	Alloy steel	Cadmium	Not recommended(h)	Excellent(h)	Excellent	(e)
Pull-type lockbolts	Titanium	None	Excellent(d)	Excellent(c)	Excellent(c)	Good or not recommended(f)
Stump-type lockbolts	Titanium	None	Excellent(d)	Excellent(c)	Excellent(c)	Good or not recommended(f)
Asp fasteners	Alloy steel	Cadmium/ nickel	Good(g)	Excellent	Excellent	Excellent
Pull-type lockbolts	7075 Aluminum	Anodized	Not recommended	Excellent	Excellent	Not recommended

(a) Blind rivets with controlled shank expansion. (b) Blind bolts are not shank expanding. (c) Fasteners can be used with flanged titanium collars or standard aluminum collars. (d) Use flanged titanium collar. (e) Performance in honeycomb should be substantiated by installation testing. (f) Depending on fastener design; check with manufacturer. (g) Nickel-plated Asp only. (h) Metallic structure on backside. Source: Ref 1

the form of inclusions, such as sulfides and oxides in the ingot, or by folds, laps, or seams that have formed because of faulty working either in the semifinishing or finishing mills.

Seams are crevices in the surface of the metal that have been closed, but not welded, by working the metal. Seams seldom penetrate to the core of bar stock. Seams may cause additional cracking in hot forging and quench cracking during heat treatment. Oxides in a seam can be the result of exposure to elevated temperatures during processing. Seams are sometimes difficult to detect in an unused fastener, but they are readily apparent after a fastener has been subjected to installation and service stresses, as discussed below.

During periodic inspection of an airplane, a longitudinal crack was found in one of the wing-attachment bolts. The bolt was 22 mm (⅞ in.) in diameter and 89 mm (3½ in.) in length and conformed to National Aerospace Standard 634. Chemical analysis verified that the bolt material was 4340 steel, as called for in specifications, and hardness measurements indicated that it had been heat treated to a tensile strength of 1379 MPa (200 ksi).

As indicated by the arrows on the photograph of the bolt in Fig. 2(a), the crack extended almost the full length of the bolt shank, but it branched circumferentially at the head-to-shank fillet (arrows, Fig. 2b) and extended about halfway through the bolt-head diameter (arrows, Fig. 2c). The bolt was sectioned to expose a portion of the crack surface, which was found to be heavily oxidized (Fig. 2d). Figure 2(e) shows a micrograph of a cross section transverse to the crack that illustrates the thickness and uniformity of the oxide coating. This heavy oxide coating and the longitudinal direction of the crack indicated that a seam was present in the bolt bar stock. Hot heading had disrupted the linear direction of the seam in the bolt-head areas so that the crack branched in a circumferential direction. Installation and service loads apparently opened the seam, which made it visually detectable.

Corrective action was not deemed necessary, because the periodic scheduled inspection of the aircraft was considered adequate to identify any faulty part for replacement and thus prevent failure of any safety-of-flight components.

Causes of Fastener Failures

In a well-designed mechanically fastened joint, the fastener may be subjected to either static loading (overload) or dynamic fatigue loading. Static loading may be tension, shear, bending, or torsion—either singly or in combination. Dynamic forces may result from impact or from cyclic fatigue loading, including vibration. Pure shear failures are usually obtained only when the shear load is transmitted over a very short length of the member, as with rivets, screws, and bolts. When the ultimate shear stress is relatively low, a pure shear failure may result, but in general, a member subjected to a shear load fails under the action of the resulting combined stresses. In addition to fatigue and overload, other common causes of fastener failures include environmental effects, manufacturing discrepancies, and improper use or incorrect installation. The following illustrates a common manufacturing discrepancy.

During an assembly operation involving HL 22-8 fasteners, the fastener heads would sometimes crack (Fig. 3a and b). Aluminum alloy 7075-T73 nuts were being installed on type 7075-T6 fasteners with an installation torque range of 7.3 to 8.5 N · m (65 to 75 in. · lb). Dimpled fracture features indicated that the head cracks were caused by overload; however, chemical analyses, hardness tests, and microstructural examination showed that the fasteners were within specifications. Unused fasteners and nuts from the same batch were installed in a simulated joint in an effort to duplicate the head cracking. Overtorquing of the fasteners caused nut-thread stripping at torques of 10.2 to 11.3 N · m (90 to 100 in. · lb). Additional torque tests using stronger steel nuts stripped

the fastener threads but did not produce any head cracks. However, when the fastener holes were drilled 2 to 4° off normal to the bearing surface, the head cracks shown in Fig. 3(c) were produced with torques of 7.9 to 10.2 N · m (70 to 90 in. · lb). The mandatory use of a drill guide to prevent off-line holes eliminated the problem.

A threaded fastener may be either a male or a female threaded cylinder—both of which are necessary as a two-element device for applying force, primarily to clamp two members together. Bolts, screws, and studs are male threaded cylinders; nuts or tapped holes are female threaded cylinders. For this discussion, bolts and nuts are considered representative of the two.

The primary force in the bolt is tension, which is set up by stretching the bolt during tightening, whereas the most important stress in the nut is the shear stress in the threads. If both the bolt and the nut had perfectly matched threads and both elements were inelastic, the torque load would be evenly distributed over all of the engaged threads. However, because the economy of mass production of fasteners requires that all fasteners be produced with practical dimensional tolerances and because all fastener materials are elastic, all threaded fastener devices experience nonuniform distribution of load over the threads. The bolt tension causes elongation, which results in more load on the threads near the bearing face of the nut, and the compression load of the nut is likely to have the same load-concentrating effect. To obtain a more uniform distribution of load along the threaded cylinder and thus minimize the chance of failure, a more ductile material must be provided for one of the elements. The usual combination is to provide material with high tensile strength for the bolt and material with good ductility for the nut (Ref 3).

In the theory of joint design, the ideal application can be represented as a situation in which there is a rigid structure and a flexible or elastic fastener. The bolt then behaves like a spring

Fig. 1 Types of failures in threaded and blind fasteners and in fastened sheet
Source: Ref 2

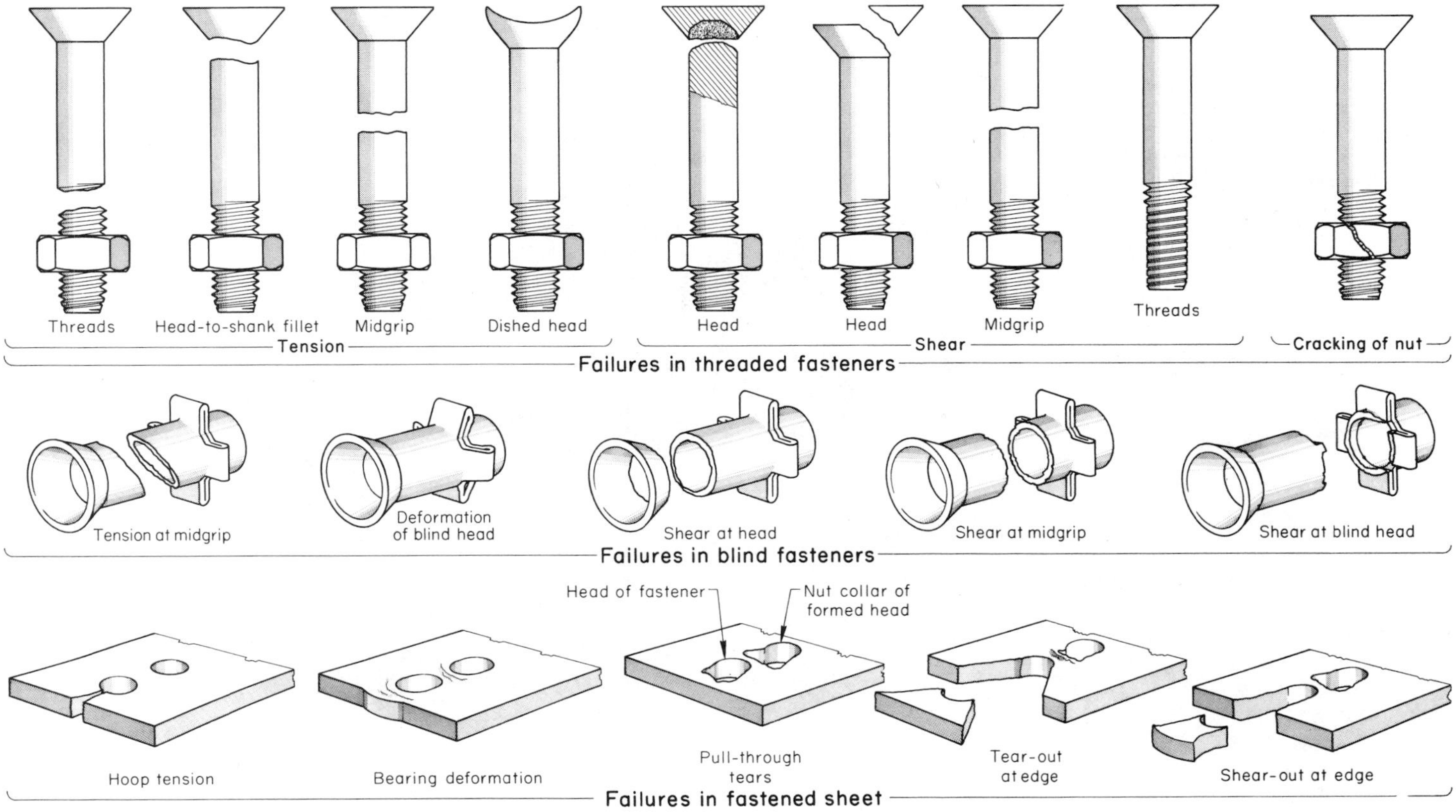

Threads | Head-to-shank fillet | Midgrip | Dished head | Head | Head | Midgrip | Threads | Cracking of nut
——Tension—— | | | | ——Shear—— | | | |
Failures in threaded fasteners

Tension at midgrip | Deformation of blind head | Shear at head | Shear at midgrip | Shear at blind head
Failures in blind fasteners

Head of fastener | Nut collar of formed head

Hoop tension | Bearing deformation | Pull-through tears | Tear-out at edge | Shear-out at edge
Failures in fastened sheet

(Fig. 4a). When the bolt is preloaded or when the spring is stretched, a stress is induced in the bolt (and a clamping force in the structure) before any working load is encountered. As the working load is applied, the preloaded bolt does not encounter an additional load until the working load equals the preload in the bolt (Fig. 4b). At this point, the force between members is zero. As more load is applied, the bolt must stretch. Only beyond this point will any cyclic working load be transmitted to the bolt.

For schematic representation of an actual condition, the structural material would also be represented by springs. The bolt would also be preloaded, but before application of the working load, the bolt encounters a slight increase in load because of the elasticity of the structural members. Actually, some of the additional load is encountered before the working load equals the preload. As the dynamic working load cycles, a fatigue condition is established.

When the bolt encounters an increase in load, being elastic, it stretches. If this stretch (strain) exceeds the elastic limit, the bolt yields plastically, taking a permanent set. The result is a loss in preload or clamping force (dashed curve, Fig. 4c). With a fluctuating load, this situation can cycle progressively, with continued loss of preload and possibly rapid fatigue

failure (the effect of preload on steel automotive wheel studs and the fracture that resulted from the preload are described in Example 1 in this article).

To eliminate fatigue problems that occur at room temperature, the designer should specify as high an initial preload as practical. The optimum fastener-torque values for applying specific loads to the joint have been determined for many high-strength fasteners. However, these values should be used with caution, because the tension produced by a selected torque value depends directly on the friction between the contacting threads. The use of an effective lubricant on the threads may result in overloading of the fastener, whereas the use of a less effective lubricant may result in a loose joint. With proper selection of materials, proper design of bolt-and-nut bearing surfaces, and the use of locking devices, the assumption is that the initial clamping force will be sustained during the life of the fastened joint. This assumption cannot be made in elevated-temperature design.

At elevated temperature, the induced bolt load will decrease with time as a result of creep, even if the elastic limits of the materials are not exceeded, and this can adversely affect fastener performance. Therefore, it is necessary to compensate for high-temperature conditions in ad-

vance when assembling the joint at room temperature. Failures that occur at elevated temperatures necessitate evaluation of the properties of the fastener material relative to the applied loads, operating temperature, and time under load at temperature. Detailed information on the types of failures that occur at high temperatures is available in the article "Elevated-Temperature Failures" in this Volume.

Fatigue in Threaded Fasteners

Fatigue is one of the most common failure mechanisms of threaded fasteners. Insufficient tightening of fasteners can result in flexing, with subsequent fatigue fracture. Higher clamping forces make more rigid joints and thus increase fastener fatigue life. The fatigue origin is usually at some point of stress concentration, such as an abrupt change of section, a deep scratch, a notch, a nick, a fold, a large inclusion, or a marked change in grain size; however, fatigue failures are most frequently located at the washer face of the nut, at the threaded runover, or at the head-to-shank fillet.

Example 1: Fatigue Fracture of 4520 Steel Wheel Studs. Each of the ten studs on one wheel of a semitrailer that was used to haul coal broke in half while the trailer was in

Fig. 2 4340 steel wing-attachment bolt that cracked along a seam

(a) Bolt showing crack (arrows) along entire length. (b) Branching cracks (arrows) at head-to-shank radius. (c) Head of bolt showing cracking (arrows) about halfway through bolt-head diameter. (d) Section through bolt showing heavy oxidation on crack surface (between arrows). (e) Section through crack showing thickness and uniformity of oxide coating. Etched with 2% nital. 1000×

operation. Both halves of each of three studs were sent to the laboratory to determine the cause of failure.

Investigation. Visual examination of the fracture surfaces of the studs disclosed beach marks, indicative of fatigue cracking, starting at opposite sides of each stud, with final fracture occurring across the stud (Fig. 5a). Each failure occurred in the first thread of the stud. Figure 5(b) shows a micrograph of a cross section of the stud at the thread next to the one that failed; it illustrates that cracking was also present at the root of this thread.

The studs were 19 mm (¾ in.) in diameter. Spectrographic analysis revealed that the studs were made from a steel containing molybdenum, such as 4520 steel, and heat treated to a hardness of 30 HRC and that the wheel nuts were made of carbon steel.

Because all ten of the wheel studs fractured, it was throught that the wheel nuts were not tightened adequately and uniformly on the wheel. If the wheel were not secured snugly by the studs, a slight movement of the wheel relative to the studs could initiate fatigue cracking. Each time the wheel made one revolution, reversed bending would occur on the studs. After fatigue started, the loosening of any stud would increase the stress on the remaining studs until they all failed.

Conclusion. The wheel studs fractured by reversed-bending fatigue.

Corrective Measures. To minimize the possibility of a recurrence of this type of failure, the wheel nuts were tightened with an air-impact wrench to a torque of 610 to 678 J (450 to 500 ft · lb) dry. All wheel studs were checked at normal maintenance periods to ensure uniform and proper loading, and no further failures occurred.

The bolt-thread root immediately adjacent to the edge of the nut on the washer side is a common fatigue-initiation site in threaded fasteners. This stress-concentration site occurs because the bolt elongates as the nut is tightened, thereby producing increased loads on the threads nearest the bearing face of the nut, which add to normal service stresses. This condition is alleviated to some extent by using nuts of a softer material that will yield and distribute the load more uniformly over the engaged threads. Significant additional improvement in fatigue life is also obtained by rolling (cold working) the threads rather than cutting them, as suggested in Example 5 in this article. It is important that the rolling operation follow, rather than precede, heat treatment.

The head-to-shank fillet is another site for fatigue fracture of tension-loaded fasteners. Several techniques can be used to avoid failure

at this location. The heads of most fasteners are formed by hot or cold forging, depending on the type of material and size of the bolt. In addition to being a relatively low-cost manufacturing method, forging provides smooth, unbroken metal flow through the head-to-shank fillet, which closely follows the external contour of the bolt (Fig. 6) and thus minimizes stress raisers, which promote fatigue cracking. In hot forging of fastener heads, temperatures must be carefully controlled to avoid overheating, which may cause grain growth. Several failures of 25-mm (1-in.) diam type H-11 airplane-wing bolts quenched and tempered to a tensile strength of 1793 to 1931 MPa (260 to 280 ksi) have been attributed to stress concentration that resulted from a large grain size in the shank. Other failures in these 25-mm (1-in.) diam bolts, as well as in other similarly quenched-and-tempered steel bolts, were the result of cracks in untempered martensite that formed as a result of overheating during finish grinding.

The shape and size of the head-to-shank fillet also have an important influence on fatigue performance. In general, the radius of this fillet should be as large as possible while at the same time permitting adequate head-bearing area. This requires a design tradeoff between the head-to-shank radius and the head-bearing area

Fig. 3 Cracked HL22-8 aluminum alloy 7075-T6 fasteners

(a) and (b) Typical cracked fastener head. (c) Typical head cracks produced by installing fasteners in misaligned holes during testing

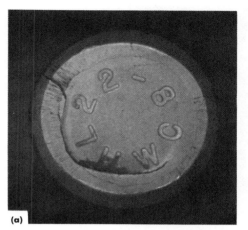

(a)

(b)

(c)

Fig. 4 Schematic showing the springlike effect of loading conditions on bolted joints

(a) Theoretical load condition for an elastic fastener and a rigid structure. (b) Ideal relationship of bolt load to working load with an elastic fastener and a rigid structure. (c) Actual relationship (both fastener and structure elastic) of bolt load to working load

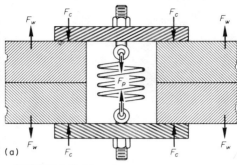

(a)

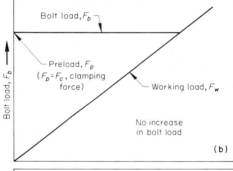

(b)

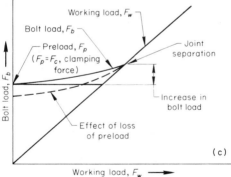

(c)

showed a single origin of cracking, a concave crack front symmetrical with the origin, and a small zone of final rupture—all of which suggested a low nominal stress. This screw was apparently the first to fracture.

The fracture surface of the other cap screw (Fig. 7c) contained two diametrically opposite origins of cracking from which two crack fronts of similar size and shape extended that exhibited beach marks characteristic of high-stress reversed-bending fatigue failure. The width of the final-rupture area was slightly less than half the screw diameter. Fracture of this screw followed that of the first screw. Fracture of the two cap screws allowed the adjacent bearing cap (the unfailed cap) to become free from the spider. As a result, the bearing cap on the opposite side of the spider was overloaded, and it fractured. With the two screws and the bearing cap fractured, the drive shaft was free at one end. The effect of the free drive shaft was to overload the universal joint at the opposite end of the shaft, resulting in failure of the spider and the two remaining bearing caps.

Chemical analysis and hardness determinations were made of the bearing-cap, spider, and screw materials. The bearing caps were made of 1144 steel and had a hardness of 55 HRA. The spider was made of 8620 steel and had been carburized and hardened to 62 to 63 HRC. The screws were made of a steel containing 0.36% C, 1.09% Mn, 0.01% P, 0.17% S, 0.21% Si, 0.01% Ni, and 0.16% Cr, which is a modified 1035 steel. The screw material had a tensile strength of 855 MPa (124 ksi), a yield strength of 634 MPa (92 ksi), and an approximate endurance limit (smooth specimen) of 427 MPa (62 ksi) at a hardness of 36 HRC.

Microscopic examination of a longitudinal section through one of the screws revealed the structure to be tempered martensite. Micrographs at magnifications of 200 and 400× revealed cracks extending from the thread root.

Conclusions. The two cap screws failed in fatigue under service stresses. The three bearing caps and the universal-joint spider broke in a brittle manner.

The properties of the material in the cap screws did not fulfill the specifications. The modified 1035 steel was of insufficient alloy content. Also, the tensile strength and endurance limit were lower than specified and were inadequate for the application.

Corrective Measure. The material for the cap screw was changed from modified 1035 steel to 5140 steel.

Example 3: Fatigue Fracture of 1045 Steel U-Bolts on a Tractor Fitting. When a farm tractor is to be used on sticky or wet soils, it is common practice to attach dual driving wheels to the rear axles of the tractor by using fittings such as that shown in Fig. 8. SAE, grade 5, U-bolts are used to fasten the dual wheels to the axles and under typical farm usage are expected to have infinite life. However, several U-bolts made of 29-mm

to achieve optimum results. Cold working of the head fillet is another common method of preventing fatigue failure because it induces a residual compressive stress and increases the material strength. Finally, fatigue resistance can be improved by changing to a material with higher strength and endurance limit, as shown in the following example.

Example 2: Fatigue Fracture of Modified 1035 Steel Cap Screws. The drive-line assembly shown in Fig. 7(a) failed during vehicle testing. The vehicle had traveled 9022 km (5606 miles) before the failure occurred. Both the intact and fractured parts of the assembly were sent to the laboratory to determine the cause and sequence of failure. Specifications for the threaded fasteners in the assembly called

for ⅜-24 UNF, SAE, grade 8, hexagon-head cap screws made of medium-carbon alloy steel, quenched and tempered to a hardness of 32 to 38 HRC and a minimum tensile strength of 1034 MPa (150 ksi).

Investigation. Visual examination of the assembly showed that three of four bearing caps, two cap screws, and one universal-joint spider had fractured. Examination of the three fractured bearing caps and the spider showed no evidence of fatigue, but showed that fracture occurred in a brittle manner. The bearing cap that was not destroyed still contained portions of the two fractured cap screws.

Macrographs of the fracture surfaces of the screws revealed beach marks typical of fatigue-type failures. One of the screws (Fig. 7b)

Fig. 5 4520 steel semitrailer wheel studs that failed by fatigue

The wheel nuts were improperly tightened, resulting in reversed-bending stresses each time the wheel turned. This eventually caused all ten wheel studs to fracture. (a) Fracture surfaces of three of the studs. (b) Cross section of a stud showing cracking at a thread root away from the failure area. 100×

(a)

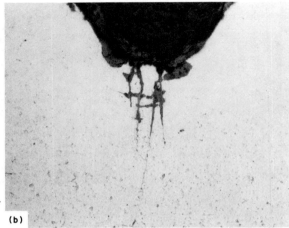

(b)

Fig. 6 Section through the forged head of a threaded fastener

The uniform grain flow minimizes stress raisers and unfavorable shear planes. Source: Ref 4

(1⅛-in.) diam rod broke after less than 100 h of service.

The material for the U-bolts was commercial quality cold-finished 1045 steel bar 29 mm (1⅛ in.) in diameter. The bar stock was cut to length, threads were cut on both ends, and the parts were bent to the specified shape. The U-bolts were then austenitized, oil quenched, and tempered to a hardness of 30 to 35 HRC at the surface to obtain a uniform tempered-martensite microstructure. Oil quenching, rather than water quenching, was used to avoid cracking of the bolts through the cut threads.

Investigation. The bolt legs in which the failures occurred were all in the same position relative to the direction of wheel rotation. This leg was designated the leading leg of the U-bolt. Visual examination showed that the break was a fairly flat transverse fracture in the threaded section between the washer and the nut. The appearance of the fracture surfaces was characteristic of failure by low-cycle fatigue, with a smooth matte fatigue-failure re-

gion showing beach marks and generally extending over about 40 to 60% of the fracture surface, which indicated severe overload. The point of initiation of fatigue was at the root of the last thread at the edge of the nut on the side toward this washer.

Metallographic examination (at 100 to 1000×) of cross sections of broken U-bolts (polished, and etched in 1% nital) showed that the structure of the outer surface of the bolts, to a depth of about 6.4 mm (¼ in.), was tempered martensite. The structure of the interior of the bolts was pearlitic. Microhardness measurements on the polished surfaces of the metallographic specimens showed the hardness of the surface layer to correspond to the specified range of 30 to 35 HRC and that of the interior to be about 25 HRC.

The observed microstructure was not that specified for this application, namely uniformly tempered martensite with a hardness of 30 to 35 HRC. The design and processing for the fractured bolts had been based on a similar arrangement in which 19-mm (¾-in.) diam and 22-mm (⅞-in.) diam U-bolts made of 1045 steel and heat treated by the same procedure had not broken after being in service on smaller tractors for several years. When the 29-mm (1⅛-in.) diam U-bolts were selected for service on the larger tractors, it had not been considered that oil quenching instead of water quenching would require them to be made of a higher-hardenability steel.

The sequence of failure was based on the assumption that the loading of the U-bolt in service included three main components: tension caused by the tightening of the nuts during assembly (by the farmer) of the bolt and hub to the axle, cyclic tension and release from supporting the weight of the tractor as the wheels turned, and a somewhat random tension in the leading leg of the U-bolt caused by driving the axle and wheels. Superposition of the stresses

from the three major components of loading resulted in a net cycle stress (which was always tensile) on the U-bolts during normal operation of the tractor. The net stress was greater on the leading leg of the bolt.

Conclusions. The U-bolts fractured in fatigue because the bolt material had poor hardenability relative to the diameter of the bolts. The oil-quench hardenability of 1045 steel was adequate for 19-mm (¾-in.) diam and 22-mm (⅞-in.) diam bolts, but not for 29-mm (1⅛-in.) diam stock.

Corrective Measures. The bolt material was changed from 1045 steel to 1527 steel, a warm-finished low-alloy steel. The diameter of the bolts was reduced to 27.2-mm (1.070 in.), and the threads were rolled rather than cut. The 1527 steel was sufficiently ductile as-received to permit bending into the U shape, yet strong enough not to need heat treatment.

No failures of the improved bolts had been reported 4 years after the indicated changes were made. Manufacturing cost of the bolts was reduced by about 30%.

Fretting

Fretting results from a very small movement of one highly loaded surface over another. The slight movement prevents the formation of, or destroys, any protective oxide film. Thus, continually clean surfaces become available for oxidation. Also, because there is no oxide film to prevent true metal-to-metal contact, local bonding of microscopic areas can occur, ultimately resulting in dislodgment of extremely small metal particles from the surfaces. Because of their relatively small size, these metal particles are then rapidly oxidized by the ambient atmosphere.

Bolted machine parts that are subjected to vibration are the most susceptible to fretting. Fretting can be prevented by minimizing clear-

Fig. 7 Drive-line assembly that failed because of fatigue fracture of two cap screws

The screws were made of modified 1035 steel instead of the specified medium-carbon alloy steel. (a) Drive-line assembly showing fractured components. (b) Fracture surface of one of the two cap screws showing beach marks typical of a fatigue-type fracture. 4×. (c) Fracture surface of the other failed cap screw showing beach marks typical of high-stress reversed-bending fatigue failure. 4×

(a)

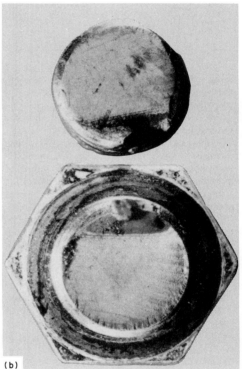

(b)

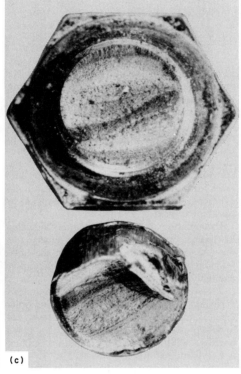

(c)

Fig. 8 Fitting for attaching dual wheels to a tractor axle that failed when 1045 steel U-bolts in the assembly were overstressed

The bolts failed because of poor hardenability relative to their diameter.

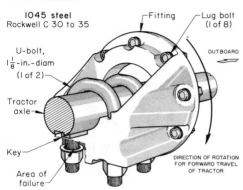

Corrosion in Threaded Fasteners

Corrosion is electrochemical in nature; therefore, an electrolyte must be present. Water is an excellent electrolyte, especially if it contains dissolved minerals. Temperature is also important because little reaction occurs below 5 °C (40 °F). Oxygen is required and is present in the atmosphere and, to a lesser extent, underwater and underground.

Types of corrosion include atmospheric corrosion, liquid-immersion corrosion, crevice corrosion, galvanic corrosion, stress-corrosion cracking, and hydrogen damage. Detailed information on these types of failures is available in the articles "Corrosion Failures," "Stress-Corrosion Cracking," and "Hydrogen-Damage Failures" in this Volume.

Atmospheric Corrosion. Atmospheric contaminants vary widely with location; therefore, rates of atmospheric corrosion vary widely. For ferrous materials, because of sulfur and chloride compounds, most severe corrosion generally occurs in highly industrialized and severe marine locations, with rural exposure being much less corrosive.

Liquid-Immersion Corrosion. Both fresh water and saltwater are corrosive. Saltwater is usually more severe than fresh water because saltwater is a better electrolyte and because the electrochemical current generated by inhomogeneities in the same metal or by different metals is greater. Also, the chloride ions present break down the protective oxide film and corrosion products. Completely submerged locations are not as corrosive as those at the waterline or splash zone, where more oxygen is present and where the surface is alternately wet and dry.

Crevice Corrosion. Every bolted connection offers the potential for crevice corrosion. Corrosion may be of two types, depending on the tightness of the joint or crevice opening. If there is a gap or opening, dirt can collect and retain moisture and thus enhance corrosion. On the other hand, in a correctly made tight joint, corrosion can result because of an oxygen deficiency or oxygen-starvation cell. The area of low oxygen becomes an anode, and the high-oxygen area becomes a cathode; thus, corrosion occurs. Fortunately, the ferrous metals most used for fasteners are not highly susceptible to the second type of crevice corrosion. Stainless steels and other metals that, in the presence of oxygen, form protective passive surface films are much more susceptible as the film breaks down and creates an active metal area.

Galvanic Corrosion. When dissimilar metals are in contact or are electrically connected together, galvanic corrosion results because of the electrical potential existing between the two metals. Current thus produced has a great effect on metal corrosion and must be considered whenever different metals are to be fastened together.

Galvanic corrosion particularly should be kept in mind at all times when designing structures using fasteners. Too often, errors are made—for example, using ferrous fasteners to join more noble metals such as copper alloys. In this case, galvanic corrosion and the size effect are involved, as described below.

For a structure to be well designed, the anode or corroding-metal area should be large relative

ance in boltholes and by using high-strength fasteners properly tensioned to prevent relative motion of contacting surfaces.

to the cathode or protected metal in order to ensure reasonable anode life. With fasteners, this is sometimes difficult to accomplish because fasteners are always smaller than the structure they join. In one example of size effect, iron nails were used to fasten copper sheathing to the bottom of wooden ships to prevent marine growth. At the end of each voyage, most of this sheathing had fallen off because of the accelerated corrosion of the iron nails.

Galvanic corrosion can be avoided by using the same metal for both fastener and structure. Several alternatives exist when this is impractical or impossible: the structure may be painted, the fastener may be isolated by non-conducting materials or plated with a more noble metal, moisture may be prevented from contacting the couple, or fasteners may be made of a metal that will be protected by sacrifical corrosion of the much larger mass of the metal being fastened.

Stress-corrosion cracking (SCC) is an intergranular fracture mechanism that sometimes occurs in highly stressed fasteners after a period of time, and it is caused by a corrosive environment in conjunction with a sustained tensile stress above a threshold value. An adverse grain orientation increases susceptibility of some materials to stress corrosion. Consequently, SCC can be prevented by excluding the corrodent, by keeping the static tensile stress of the fastener below the critical level for the material and grain orientation involved, or by changing to a less susceptible material or material condition (for example, aluminum alloy 7075 is less susceptible to SCC in the T73 or T76 temper than in the T6 temper). In some cases, the corrodent can be the seemingly innocuous ambient atmosphere; in others, it can be unwittingly added, as in the following example.

Example 4: SCC of Stainless Steel Bolts. Stainless steel bolts broke after short-term exposure in boiler feed-pump applications. Specifications required that the bolts be made of a 12% Cr high-strength steel with a composition conforming to that of AISI type 410 stainless steel.

Several bolts from three different installations were sent to the laboratory for examination. One lot consisted of ten 100-mm (4-in.) long 1-8 UNC bolts, the second lot nine 100-mm (4-in.) long 1-8 UNC bolts, and the third lot five 70-mm (2.75-in.) long ⅝-11 UNC bolts. An antiseizure compound consisting of metallic copper flakes in a silicone vehicle had been applied to the bolts during assembly.

All of the bolts had fractured in the threaded portion, but not all at the same thread. None of the 25-mm (1-in.) diam bolts had fractured under the head, but some of the 16-mm (⅝-in.) diam bolts had broken under the head and in the threads.

Visual Examination. The 25-mm (1-in.) diam bolts exhibited external damage, which varied in degree, depending on the time that

Table 2 Results of stress-corrosion tests of 12% Cr steel bolts in a 3.5% NaCl solution to determine effect of antiseizure coating

Bolt	Hardness, HRC	Surface	Appearance(a)	Tensile strength MPa	ksi	Remarks
1	41-42	Uncoated	Localized pitting, no cracks	1524	221	No corrosion attack on fracture surface
2	18-20	Uncoated	Localized pitting, no cracks	862	125	No corrosion attack on fracture surface
3	20-22	Coated(b)	Heavy pitting, no cracks	896	130	No corrosion attack on fracture surface
4	41-42	Coated(b)	Bolt completely failed in test fixture	Not applicable		Heavy corrosion on fracture surface

Bolt 1   2   3   4

(a) After 3 months of intermittent dipping in a 3.5% NaCl solution. (b) Coated with an antiseizure compound consisting of metallic copper flakes in a silicone vehicle

had been required to shut down the unit. No external damage was observed on the 16-mm (⅝-in.) diam bolts.

The first lot of 25-mm (1-in.) diam bolts exhibited flat fracture surfaces with chevron patterns. The second lot was similar in fracture-surface appearance; however, the fracture surfaces were badly stained by the corrosive boiler feedwater. The fracture surfaces of the 16-mm (⅝-in.) diam bolts exhibited stained brittle-fracture features in starlike patterns.

Examination of bolts in all three lots also disclosed a copper-colored residue adhering to the threads. This residue was from the antiseizure compound that had been applied during assembly.

Accelerated Corrosion Testing. Laboratory corrosion tests were conducted on four bolts taken from stock. The purpose of the tests was to determine the effects of the antiseizure compound in acidified 3.5% NaCl solution on bolts at two hardness levels.

Two bolts, one with a hardness of 41 to 42 HRC and the other of 20 to 22 HRC, were coated with the antiseizure compound. Another bolt that had a hardness of 41 to 42 HRC and a bolt having a hardness of 18 to 20 HRC were left uncoated for the test. All four bolts were assembled in a static test fixture and tightened to a torque value equal to 90% of their yield strength and then intermittently dipped in an acidified solution of 3.5% NaCl. Intermittent dipping consisted of a 10-min total immersion and a 50-min air dry for 8 h per day, 5 days per week for 3 months.

After 3 months of testing, the coated bolt having a hardness of 41 to 42 HRC fractured completely across at the third thread from the head. The three other bolts had not fractured, but were tensile tested to destruction. The results of the test and appearance of the fracture surfaces are presented in Table 2. This test demonstrated the damaging effects of the me-

tallic copper antiseizure compound to high-hardness 12% Cr steels in the presence of a corrosive medium.

Microscopic examination showed that the microstructure of the metal in the bolts of lot 1 (Fig. 9a) contained tempered martensite. The metal in the two other lots (Fig. 9b) showed a two-phase structure consisting of tempered martensite, ferrite, and nonmetallic sulfide stringers. Ferrite content of the matrix was about 38%.

The examination also showed that the threads on some of the bolts had been produced by cutting, and on others by rolling. Both types of threads had fractured on the bolts in service.

Longitudinal sections were taken through the fractured bolts and examined microscopically. Branched intergranular cracking at the thread root next to the fracture, similar in appearance to the crack shown in Fig. 9(a), was found in several of the bolts. Intergranular cracking and oxide entrapment were found under the bolt heads and at the fracture surfaces.

Fractographic examination by electron microscopy disclosed evidence of corrosion products on the fracture surfaces with localized pitting and second-phase carbide particles. The carbide particles (probably chromium carbide, $Cr_{23}C_6$) and evidence of intergranular fracture are shown in Fig. 9(c). The fractograph shown in Fig. 9(d) illustrates brittle intergranular fracture with corrosive residue on the surface. Large clusters of carbide particles and evidence of intergranular fracture are shown in Fig. 9(e). No fatigue indications were in evidence. The branching intergranular cracks shown in Fig. 9(a) and the intergranular fracture surface with corrosion products shown by the electron fractographs in Fig. 9(d) and (e) are evidence of stress corrosion.

Chemical analysis of the metal in a fractured bolt from each lot revealed the following compositions:

Fig. 9 Microstructures of stainless steel bolts that failed from SCC

(a) Branched intergranular cracking in a type 410 stainless steel bolt from lot 1 (see Example 4). Etched with picral plus HCl. 250×. (b) Microstructure of a type 416 stainless steel bolt typical of those in lots 2 and 3 (see Example 4). Same etchant as (a). 100×. (c) to (e) Typical fracture surfaces of the bolts showing carbide particles (c), intergranular fracture (d), and intergranular fracture with corrosive residue (e). Electron fractographs of unetched specimens. 6500×

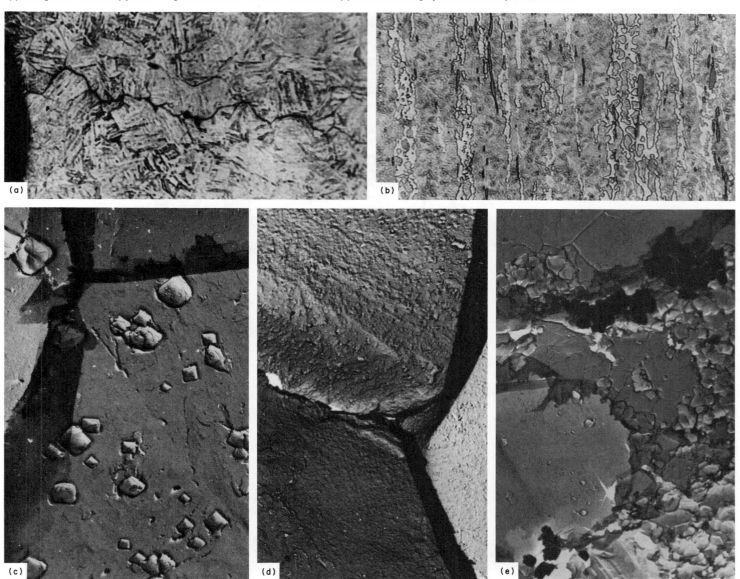

Element	Composition, %			
	Lot 1	Lot 2	Lot 3	Nominal(a)
C	0.140	0.058	0.061	0.15
Mn	0.560	0.370	0.410	1.00
P.	0.010	0.017	0.018	0.04
S.	0.016	0.251	0.241	0.03
Si	0.392	0.437	0.395	1.00
Cr	12.09	12.54	12.14	11.5-13.5

(a) Nominal composition for type 410 stainless steel (to which bolts were specified to conform). Percentages, except for chromium, are maximums.

Bolts from lot 1 conformed to specifications, but a deviation from prescribed chemical composition was found in lots 2 and 3, which also contained a sulfur addition to enhance machinability.

Conclusions. Fracture of the bolts was by intergranular stress corrosion. A metallic copper-containing antiseizure compound on the bolts in a corrosive medium set up an electrochemical cell that produced trenchlike fissures or pits for fracture initiation. Because the bolts were not subjected to cyclic loading, fatigue or corrosion fatigue was not possible.

Corrective Measures. Bolts were required to conform to the specified chemical composition. The hardness range for the bolts was changed from 35 to 45 HRC to 18 to 24 HRC. Petroleum jelly was used as an antiseizure lubricant in place of the copper-containing compound. As a result of these changes, bolt life was increased to more than 3 years.

High static tensile stresses above the threshold for SCC can be minimized by introducing compressive stresses on the fastener surfaces where SCC is most likely to originate. Cold rolling and shot peening are two means of keeping static tensile stresses at the operating surface below the safe maximum. However, the beneficial effects of either of these will be negated if the fastener is subsequently heat treated, if the normal service temperature is above the tempering or aging temperature of the material, if the fastener is loaded above its yield point, or if general corrosion removes the compressively stressed layer.

Fig. 10 AISI type 431 stainless steel T-bolt that failed by SCC

(a) T-bolt showing location of fracture. Dimensions given in inches. (b) Fracture surface of the bolt showing shear lip (arrow A), fine-grain region (arrow B), and oxidized regions (arrows C). (c) Longitudinal section through the bolt showing severe intergranular carbide precipitation. Etched electrolytically with 10% NaCN. 540×

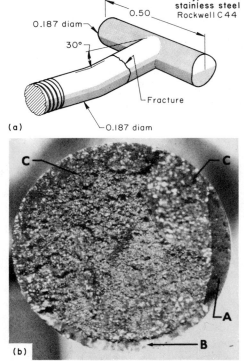

(a)

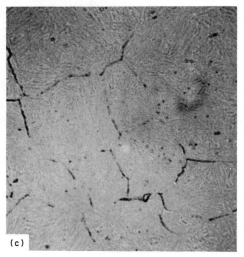

(b)

(c)

Fig. 11 Die-cast zinc alloy nuts from a water tap

(a) Nut for the cold-water tap that failed by SCC. (b) Mating nut for the hot-water tap that shows only isolated areas of corrosion. (c) Unetched section showing metal in the cold-water tap after corrosion testing. 600×

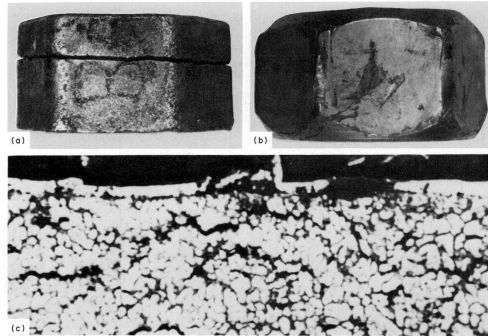

(a)

(b)

(c)

Example 5: SCC of a T-Bolt. The T-bolt shown in Fig. 10(a) was part of the coupling for a bleed air duct of a jet engine on a transport plane. Specifications required that the 4.8-mm (³/₁₆-in.) diam T-bolt be made of AISI type 431 stainless steel and heat treated to 44 HRC. The operating temperature of the duct was 425 to 540 °C (800 to 1000 °F), but that of the bolt was

lower. The T-bolt broke after 3 years of service. The expected service life was equal to that of the aircraft.

The bolt was made by cold upsetting bar stock to a T-shape, chasing the threads, bending to the specified curvature, then heat treating. Heat treatment consisted of heating to 1010 to 1040 °C (1850 to 1900 °F) and oil quenching, tempering at 275 to 415 °C (525 to 575 °F) for 2 h, and retempering for the same time and temperature. In the manufacture of similar bolts, the threads were rolled after heat treatment, and the bolt shank was shot peened.

Investigation. Visual examination of the fracture surface revealed intergranular cracking and evidence of brittle fracture. Macrographs were taken of the outer surface of the bolt and of the fracture surface. The outer surface contained circumferential cracks on the lower (tension) side of the T-bolt. The fracture surface (Fig. 10b) exhibited three distinct fracture areas—a shallow 45° shear-lip section along one side (arrow A, Fig. 10b), a thin region containing fine grains along part of the surface that could have been the result of stress corrosion or fatigue (arrow B, Fig. 10b), and regions that appeared to be oxidized as the result of a prior crack (arrows C, Fig. 10b). Hardness of the T-bolt was found to be 44 HRC, as specified.

Examination of micrographs of longitudinal sections through the T-bolt revealed the basic microstructure to be acicular martensite with no δ-ferrite present. However, there was evidence of severe intergranular carbide precipitation (Fig. 10c), which indicated that the material

had been subjected to the sensitizing-temperature range after the bolt had been manufactured.

The micrographs also showed some intergranular surface and subsurface cracking near, but not at, the primary fracture. Although not confirmed, chlorine or chlorides in some form could have been present around the duct.

Conclusions. The bolt broke as a result of SCC. Thermal stresses were induced into the bolt by intermittent operation of the jet engine. Mechanical stresses were induced by tightening of the clamp around the duct, which in effect acted to straighten the bolt. The action of these stresses on the carbides that precipitated in the grain boundaries resulted in fracture of the bolt.

Corrective Measures. Because of the operating temperatures of the duct near the bolt, the material was changed to A-286, which is a heat-resisting alloy and is less susceptible to carbide precipitation. The bolt was strengthened by shot peening the shank and rolling the threads after heat treatment. Avoidance of heating in the sensitizing range would have been desirable, but difficult to ensure because of the application.

Example 6: SCC of a Die-Cast Zinc Alloy Nut. The two nuts shown in Fig. 11(a) and (b) were used to secure the water-supply pipes to the threaded connections on hot-water and cold-water taps. The nut used on the cold-water tap (Fig. 11a) fractured about 1 week after installation.

The cold-water tap and fractured nut were submitted to the laboratory for examination to determine the cause of failure. The mating hot-water tap was forwarded for comparison.

Investigation. Visual examination of the two taps indicated that they were of similar design. As shown in Fig. 11(a) and (b), the two nuts on the threaded connections were of a different style, although they had the same size of thread. The two nuts did not exhibit any evidence of installation abuse, such as stripping of the threads. Various other marks and indentations were found on the flats of the nuts, particularly on the hot-water nut, such as would be made during normal installation.

Examination of the fracture surfaces of the cold-water nut did not reveal any obvious defects to account for the fracture, but there were indications of excessive porosity in the nut. The fracture had occurred through the root of the first thread that was adjacent to the flange of the tap.

Metallographic examination of the cold-water nut revealed evidence of considerable corrosive attack, mainly intergranular, extending into the material from the surface. This was present over the entire surface, including the threads, and extended into the material as deep as 0.25 mm (0.010 in.). Chemical spot tests indicated that the nut had been zinc plated.

Microscopic examination of the fracture surfaces of the cold-water nut indicated that considerable intergranular corrosion had also occurred at both sides of the fracture surface to a substantial depth, somewhat more than at the zinc-plated outer surface. Because this was a recent fracture, it would appear that corrosion on the fracture surface occurred before the failure rather than after.

Visual and microscopic examination of the hot-water nut also revealed some evidence of corrosion, but only in isolated areas mainly associated with the threads. This nut was plated with a copper-nickel coating. Corrosion was present only at breaks in the plating and then only to a limited extent.

Radiographic examination indicated that both nuts had excessive porosity in similar amounts and distribution. Metallographic examination revealed that both nuts had a cast structure and were of a similar basic composition of zinc-aluminum die-casting alloy.

Corrosion Testing. It appeared that the cold-water nut was highly susceptible to corrosive attack; the hot-water nut did not exhibit this characteristic to the same degree. Samples of nuts of both styles were subjected to a corrosion test designed to indicate whether a particular zinc-alloy casting was susceptible to intergranular attack in atmospheric service. The test consisted of exposing the nuts in air saturated with water vapor at a temperature of 90 °C (195 °F) for 10 days.

If the metal was attacked during this test, it indicated that there were deficiencies in magnesium content or that the content of the impurities (lead, cadmium, and tin) were above 0.004%, 0.003%, and 0.002%, respectively. In

either case, fracture during the test would indicate that the parts are likely to fail in service.

After 4 days of testing, the cold-water nut was noted to be severely cracked and was therefore removed for further examination. Slight pressure between the fingers caused the nut to crumble into several fragments. Metallographic examination of the material in the cold-water nut indicated that it had completely disintegrated, as shown by the cracks and corrosion network in the micrograph in Fig. 11(c). The zinc plating on the outer surface appeared to be relatively unattacked.

The corrosion test for the hot-water nut was run for the full 10 days without any outward signs of attack. Metallographic examination revealed that there had been some additional corrosion, but not to as great an extent as in the cold-water nut.

Conclusions. The nut from the cold-water tap failed by SCC. Apparently, sufficient stress was developed in the nut to promote this type of failure by normal installation because there was no evidence of excessive tightening of the nut.

Corrosion testing of the nuts indicated that the fractured nut was highly susceptible to intergranular corrosion because of either a deficiency in magnesium content or excessive impurities, such as lead, tin, or cadmium. The primary cause of failure of the fractured nut was, therefore, an incorrect alloy composition.

Corrective Measure. This composition problem with zinc alloys was recognized many years ago, and particular attention has been directed toward ensuring that high-purity zinc is used. This corrective measure was reported to have resulted in virtual elimination of this type of defect.

Hydrogen Damage. Many fasteners require some kind of protective coating against corrosion. This need is sometimes satisfied by an electrodeposited coating, such as cadmium or zinc. An acid pickling bath is usually used to clean ferrous fasteners before plating. This pickling bath and the subsequent plating operation both supply nascent (atomic) hydrogen to the fastener surface, and some of the hydrogen is absorbed by the steel. In high-strength steel components that have been electroplated but have not been baked for expulsion of hydrogen, hydrogen diffuses to the area of the highest triaxial stress, such as at a notch. Figure 12 shows a cadmium-plated 8740 steel nut that failed by hydrogen embrittlement because the nuts were not baked after electroplating. After an incubation period, a crack initiates and, with time, propagates through the section. A characteristic of this phenomenon is that there is a stress value below which hydrogen-induced delayed failure will not occur. This is sometimes referred to as the static-fatigue limit.

This difficulty, resulting from hydrogen charging during the pickling or electroplating process, can sometimes be eliminated without appreciable change in hardness by baking after plating at about 205 °C (400 °F) in air for a

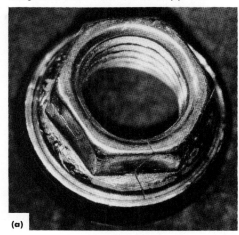

Fig. 12 Cadmium-plated 8740 steel aircraft-wing assembly nut that failed by hydrogen embrittlement
The nut was not baked after electroplating to release hydrogen. (a) Overall view. 5×. (b) Fracture surface. 9×. (c) Scanning electron micrograph of typical intergranular fracture shown in box in (b). 3600×

(a)

(b)

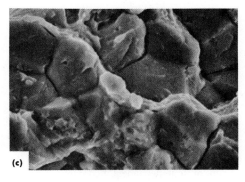

(c)

suitable time at temperature, depending on the size and finish of the fastener. In the following examples, cadmium-plated high-strength steel bolts that had not been properly baked failed when normally stressed in service, and a case is described in which baking did not alleviate the hydrogen-embrittlement problem.

Example 7: Delayed Failure of an Electroplated Steel Bolt Because of Hydrogen Embrittlement. Cadmium-plated high-strength steel bolts were used to facilitate quick disassembly of a vehicle. One bolt was found fractured across the root of a thread after being torqued in place for 1 week. The bolts were made of 8735 steel heat treated to a tensile strength of 1241 to 1379 MPa (180 to 200 ksi) with a hardness of 39 to 43 HRC, followed by cadmium plating.

The bolt that failed and several that did not were sent to the laboratory for determination of the cause of failure. It was reported that the bolts that did not fail were stressed less than the specified torque level when attached to the assembly.

Investigation. It was assumed that the bolts had not been given a baking treatment for removal of hydrogen. Because there is no nondestructive test that will confirm such an assumption, two of the bolts were baked at 205 °C (400 °F) for 24 h, then a series of tests was performed on both the baked and the as-plated bolts.

First, short-term tensile tests were performed on one of the as-plated bolts and on one that had been baked. The two bolts exhibited similar notched tensile strengths (1448 MPa, or 210 ksi, for the as-plated bolt and 1510 MPa, or 219 ksi, for the baked bolt); therefore, this test did not disclose the presence of hydrogen. Consequently, the remaining bolts were tested under static loading.

The bolts were static loaded in tension on a creep frame. Delayed failure occurred in the as-plated bolts, but no failure occurred in the bolt that was baked to remove the hydrogen. Test results were as follows:

Condition	Applied stress MPa	ksi	Time-to-failure, h
As-plated	690	100	5.1
As-plated	517	75	5.6
As-plated	517	75	1.1
As-plated	345	50	67(a)
Baked(b)	517	75	598(a)

(a) No failure occurred; tests were concluded after time shown.
(b) Baked 24 h at 205 °C (400 °F)

These results indicated that the static-fatigue limit was between 345 and 517 MPa (50 and 75 ksi) for the as-plated (hydrogen-charged) bolts. The properly baked bolt did not fail when it was subjected to a torque stress of 517 MPa (75 ksi).

Conclusions. Failure of the bolts was the result of time-dependent hydrogen embrittlement. Had the remaining bolts been torqued to the normal stress levels, all would have failed within 2 weeks.

Corrective Measures. The bolts were baked, as specified by ASTM B 242, at 205 °C (400 °F) for 30 min. No further failures occurred. Baking for 30 min is the minimum baking time; however, baking times up to 24 h are recommended for greater safety.

Example 8: Hydrogen Embrittlement of Cadmium-Plated Alloy Steel Self-Retaining Bolts in a Throttle-Control Linkage. Two clevis-head self-retaining bolts used in the throttle-control linkage of a naval aircraft failed on the aircraft assembly line (a failed bolt is compared with an unused bolt in Fig. 13a). Specifications required the bolts to be heat treated to a hardness of 39 to 45 HRC,

Fig. 13 Cadmium-plated alloy steel self-retaining bolts that fractured because of hydrogen damage

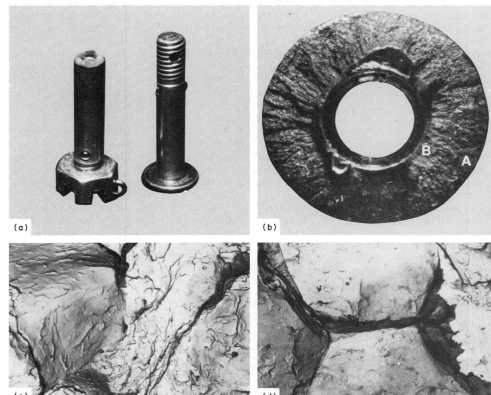

(a) Fractured and unused intact bolt. (b) Fractured bolt; brittle fracture surface is indicated by A and B. (c) and (d) Electron fractographs of surfaces A and B, respectively, showing brittle intergranular fractures. 5000×

followed by cleaning, cadmium electroplating, and baking to minimize hydrogen embrittlement. The bolts broke at the junction of the head and shank. The nuts were, theoretically, installed fingertight.

Investigation. Inspection of the bolts and nuts showed excessive burrs in the threads at the cotter-pin holes and the nut slots, which made it necessary to use a wrench in assembly. This obviously had set up a tensile stress, which caused the bolts to fail. The fillet at the junction of the head and shank was undesirably sharp, but met the manufacturer's specification of 0.13 to 0.25 mm (0.005 to 0.010 in.) for the fillet radius.

Macroscopic examination of the fracture surface (Fig. 13b) gave no evidence of corrosion, fatigue marks, or rapid ductile failure. Electron fractographs of the fracture surface (Fig. 13c and d) showed a typical brittle intergranular fracture, suggestive of hydrogen embrittlement. The microstructure was found to be sound and normal, and the hardness was within specifications.

Considerations of the event of failure before actual service and the possibility of hydrogen embrittlement that is inherent in cleaning and electroplating procedures led to the question of whether the bolts had been adequately baked after plating and whether postplating treatment had been performed as specified. To explore this possibility, bolts from the same lot as the failed bolts were tested in prolonged tension in three conditions: as-received, baked at 205 °C (400 °F) for 4 h, and baked at 205 °C (400 °F) for 24 h. The bolts in the first two conditions failed under a sustained tensile load, but the bolts baked for 24 h did not, indicating that baking by the manufacturer had either been omitted or had been insufficient.

Conclusions. The failure was attributed to hydrogen embrittlement (resulting from cleaning, cadmium plating, or both) that had not been satisfactorily alleviated by subsequent baking. The presence of the burrs on the threads prevented assembly to fingertightness, and the consequent wrench torquing caused the actual fractures. The very small radius of the fillet between the bolt head and the shank undoubtedly accentuated the embrittling effect of the hydrogen.

Corrective Measures. The cleaning and cadmium-plating procedures were stipulated to be low-hydrogen in nature, and an adequate postplating baking treatment at 205 °C (400 °F), in conformity with ASTM B 242, was spec-

Fig. 14 Cadmium-plated AISI 8740 alloy steel fasteners that failed by hydrogen embrittlement
See also Fig. 15.

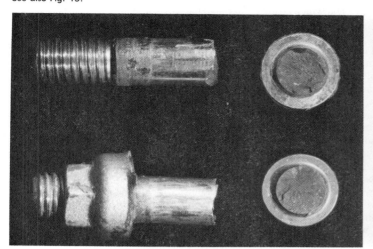

Fig. 15 Scanning electron micrograph of fracture surface of fasteners shown in Fig. 14
880×

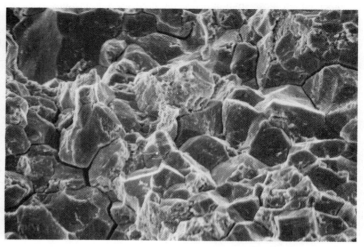

ified. A minimum radius for the head-to-shank fillet was specified at 0.25 mm (0.010 in.). All threads were required to be free of burrs. A 10-day sustained-load test was specified for a sample quantity of bolts from each lot.

Example 9: Hydrogen Embrittlement of Alloy Steel Fasteners (Ref 5). During an inspection of a structure 2 weeks after assembly, the heads of several cadmium-plated AISI 8740 steel fasteners were found to be completely separated from their respective shanks (Fig. 14). Scanning electron microscopy (SEM) examination of the fracture surfaces revealed a brittle, intergranular fracture mode (Fig. 15), indicating hydrogen embrittlement. An investigation was conducted to determine the extent of hydrogen embrittlement in the various lots of cadmium-plated 8740 steel fasteners.

Mechanical Testing. To simulate installation conditions, fasteners from several lots were torqued to 80% of rated ultimate tensile strength into a predrilled aluminum alloy plate. This high torque (greater than normal installation) was used to speed the failure rate of the fasteners.

After 335 h, a head was lifted with fingernail pressure from the shank of a fastener. Another was removed by inserting the blade of a pocket knife under the head. After 41 days, the test rig was disassembled. Nine more fasteners were found to be cracked in the head-to-shank radius. All of the failures were brittle in nature.

Stress-rupture tests. When torque tests were abandoned, all subsequently tested fasteners were installed in fixtures in which a constant load equivalent to 90% of the rated ultimate tensile strength could be maintained on the specimens. Stress-rupture testing was performed on lever-arm-type creep machines. Testing of any lot of pins was suspended when one pin failed (indicating hydrogen embrittle-

ment) before the end of the 72-h test period. Four pins per lot were used as a representative sample. If none of the four pins failed during the 72 h, the lot was considered to be free of hydrogen embrittlement and was released for production. If a lot was embrittled, one of the fasteners being tested generally failed within the first 4 h.

Metallurgical examination showed that the microstructure of the steel base metal was tempered martensite, supposedly heat treated to an ultimate tensile strength range of 1240 to 1380 MPa (180 to 200 ksi). Testing revealed that the actual ultimate tensile strength of the fasteners varied from 1310 to 1480 MPa (190 to 215 ksi). The grain-flow pattern in the head-to-shank radius was satisfactory.

Condition of cadmium plating. The thickness of the cadmium plating on the fasteners varied from 7.9 to 11 μm (0.31 to 0.43 mils). The coating was applied from a plating bath to which brighteners were added. Bright, impervious coatings increase the possibility of hydrogen embrittlement in steels by forming a barrier to outward hydrogen diffusion from plated parts.

An accepted practice for removing hydrogen picked up during plating operations is to bake the cadmium-plated parts for 23 h at 190 ± 14 °C (375 ± 25 °F). This treatment was implemented, as well as baking fasteners at 220 °C (425 °F) for periods of 4, 12, and 24 h. No baking cycle eliminated embrittlement in pins plated with a bright coating that did not allow the escape of mobile hydrogen.

Conclusions. Hydrogen embrittlement is caused by the use of a bright, impervious cadmium electroplate that hinders the diffusion of mobile hydrogen outward from the surface of the pin. After the cadmium layer was removed, the mobile hydrogen contained on the surface of the steel and in the electroplated deposit was

released, and the embrittlement problem was alleviated.

Recommendations. The bright cadmium layer was stripped from the pins, which were then baked and replated with a dull, porous cadmium layer that allowed outward diffusion of hydrogen. The pins were baked again after deposition of the porous cadmium layer. This eliminated the problem of hydrogen embrittlement.

Corrosion Protection

The most commonly used protective metal coatings for ferrous metal fasteners are zinc, cadmium, and aluminum. Tin, lead, copper, nickel, and chromium are also used, but only to a minor extent and for very special applications.

In many cases, however, fasteners are protected by some means other than metallic coatings. They are sometimes sheltered from moisture or covered with a material that prevents moisture from making contact, thus drastically reducing or eliminating corrosion. For fasteners exposed to the elements, painting is universally used. The low-alloy high-strength steel conforming to ASTM specification A 242 forms its own protective oxide surface film. This type of steel, although it initially corrodes at the same rate as plain carbon steel, soon exhibits a decreasing corrosion rate, and after a few years, continuation of corrosion is practically nonexistent. The oxide coating formed is fine textured, tightly adherent, and a barrier to moisture and oxygen, effectively preventing further corrosion. Plain carbon steel, on the other hand, forms a coarse-textured flaky oxide that does not prevent moisture or oxygen from reaching the underlying noncorroded steel base.

Steels conforming to ASTM A 242 are not recommended for exposure to highly concentrated industrial fumes or severe marine conditions, nor are they recommended for applications in which they will be buried or submerged. In these environments, the highly protective oxide does not form properly, and corrosion is similar to that for plain carbon steel.

Zinc Coating. Zinc is the coating material most widely used for protection of fasteners from corrosion. Hot dipping is the most often used method of application, followed by electroplating and, to a minor extent, mechanical plating.

Hot dipping, as the name implies, involves immersing parts in a molten bath of zinc. Hot-dip zinc coatings are sacrificial by electrochemical means. These coatings for fasteners meet ASTM specification A 394, which specifies an average coating weight of 1.25 oz/ft^2 (2.2-mil thickness). Zinc electroplating of fasteners is done primarily for appearance, where thread fit is critical, where corrosion is not expected to be severe, or where life expectancy is not great.

Specification ASTM B 633 for electrodeposited zinc coatings on steel specifies three coating thicknesses: GS, 25 μm (0.0010 in.); LS, 13 μm (0.0005 in.); and RS, 4 μm (0.00015 in.)—corresponding to coating weights of 0.59, 0.29, and 0.09 oz/ft^2, respectively. These electrodeposited coatings are often given supplemental phosphate or chromate coatings to develop a specific color and, to a minor extent, enhance corrosion resistance. Corrosion life of a zinc coating is proportional to the amount of zinc present; thus, the heaviest electrodeposited coating (GS) would have only about half the life of a hot-dip galvanized coating.

Mechanical (nonelectrolytic) barrel plating is another method of coating fasteners with zinc. Coating weight can be changed by varying the amount of zinc used and the duration of barrel rotation. Such coatings are quite uniform and have a satisfactory appearance.

Cadmium coatings are also applied to fasteners by an electroplating process similar to that used for zinc. These coatings meet ASTM specification A 165. As is true for zinc, cadmium corrosion life is proportional to the coating thickness. The main advantage of cadmium over zinc is in marine environments, where the corrosion life of cadmium is longer. Cadmium-plated steel fasteners also are used in aircraft in contact with aluminum because the galvanic characteristics of cadmium are more favorable than those of zinc. Chromate coatings are also used over cadmium coatings for the reasons given for zinc-plated fasteners.

Aluminum coating on fasteners offers the best protection of all coatings against atmospheric corrosion. Aluminum coating also gives excellent corrosion protection in seawater immersion and in high-temperature applications.

Aluminum coatings are applied by hot-dip methods at about 675 to 705 °C (1250 to 1300 °F). Aluminum alloy 1100 is usually used because of its general all-around corrosion resistance. As with any hot-dip coating, a metallurgical bond is formed that consists of an intermetallic alloy layer overlaid with a coating of pure bath material.

Aluminum coatings do not corrode uniformly, as do zinc and cadmium coatings, but rather by pitting. In some cases, these pits may extend entirely through the coating to the base metal; in others, only through the overlay to the intermetallic layer. Pits, which may occur in a part soon after exposure, sometimes discolor the coated surface, but cause little damage. The complex aluminum and iron oxide corrosion product seals the pits, and because the corrosion product is tightly adherent and impervious to attack, corrosion is usually limited. There is little tendency for corrosion to continue into the ferrous base, and there is none for undercutting and spalling of the coating.

Aluminum coatings will protect steel from scaling at temperatures up to about 540 °C (1000 °F); the aluminum coating remains substantially the same as when applied, and its life is exceptionally long. Above 650 °C (1200 °F), the aluminum coating diffuses into the steel to form a highly protective aluminum-iron alloy. This diffusing or alloying is time-temperature dependent; the higher the temperature, the faster the diffusion. However, scaling will not take place until all the aluminum is used up, which may take a thousand or more hours even at temperatures as high as 760 °C (1400 °F).

Prevention of galling at elevated temperatures is another characteristic of aluminum coatings. Stainless steel fasteners for use at 650 °C (1200 °F) have been aluminum coated just to prevent galling. Coated nuts can be removed with an ordinary wrench after many hours at these temperatures, which is impossible with uncoated nuts.

Fastener Performance at Elevated Temperatures

The temperature, the environment, and the materials of the structure are normally fixed; therefore, a design objective is to select a bolt material that will give the desired clamping force at all critical points in the joint. To do this, it is necessary to balance the three time- and temperature-related factors (relaxation, thermal expansion, and modulus) with a fourth factor—the amount of initial tightening or clamping force.

Modulus of Elasticity. As temperature increases, the modulus of elasticity decreases; therefore, less load (or stress) is needed to impact a given amount of elongation (or strain) to a material than at lower temperatures. This means that a fastener stretched a certain amount at room temperature to develop preload will exert a lower clamping force at higher temperature. The effect of reducing the modulus is to reduce clamping force, whether or not the bolt

and structure are of the same material, and is strictly a function of the bolt metal.

Coefficient of Expansion. With most materials, the size of the part increases as the temperature increases. In a joint, both the structure and the fastener increase in size with an increase in temperature. If the coefficient of expansion of the fastener exceeds that of the joined material, a predictable amount of clamping force will be lost as temperature increases. Conversely, if the coefficient of expansion of the joined material is greater, the bolt may be stressed beyond its yield or even fracture strength, or cyclic thermal stressing may lead to thermal fatigue failure. Thus, matching of materials in joint design can ensure sufficient clamping force at both room and elevated temperatures without overstressing the fastener.

Relaxation. At elevated temperatures, a material subjected to constant stress below its yield strength will flow plastically and permanently change dimensions in the direction of the stress. This phenomenon is called creep. In a joint at elevated temperature, a fastener with a fixed distance between the bearing surfaces of the male and female members will produce less and less clamping force with time. This characteristic is called relaxation. It differs from creep in that stress changes because the elongation or strain remains constant. Such elements as material, temperature, initial stress, manufacturing method, and design affect the rate of relaxation.

Relaxation is the most important of the three time- and temperature-related factors. It is also the most critical consideration in design of fasteners for service at elevated temperature. A bolted joint at 650 °C (1200 °F) can lose as much as 50% of preload. Failure to compensate for this could lead to fatigue failure or a loose joint, even though the bolt was properly tightened initially.

Other Considerations. Bolt preloading is not the only aspect of joint design that must be considered for elevated-temperature service. Other considerations include time, environment, serviceability, coatings, manufacturing methods, fastener design, and material stability. The element of time that is introduced with exposure to heat requires an assessment of the service life of the structure (or at least of the fastener) and of the amount of this life that will be spent at elevated temperature.

Galling and seizing are serious considerations if there is to be subsequent disassembly, which is usually necessary for threaded connections. Many of the materials with desirable elevated-temperature properties, such as the stainless steels, are inherently prone to galling during sliding contact of one member over another, even at room temperature. These materials require protective lubricating coatings. Additional problems of seizing of mating materials develop after exposure to temperature. Ease of disassembly requires that a suitable coating be used and that clearance sometimes be provided at the pitch diameter.

Fig. 16 Cadmium-plated 4140 steel nuts from a military jet engine that failed by LME

(a) Fragmented and cracked nuts. (b) Typical fracture surface. (c) Electron fractograph showing brittle intergranular fracture

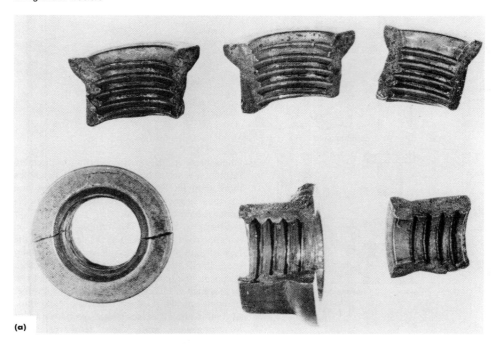

(a)

(b)

(c)

Standard high-strength bolts of medium-alloy steel have a usable temperature range up to about 230 °C (450 °F). The medium-alloy chromium-molybdenum-vanadium steel conforming to ASTM A 193, grade B16, is a commonly used bolt material in industrial turbine and engine applications to 480 °C (900 °F). An aircraft version of this steel, AMS 6304, is widely used in fasteners for jet engines.

For many industrial applications, corrosion- and heat-resistant stainless steels are usable to 425 °C (800 °F). Stabilized grades, such as types 321 and 347, can be used at slightly higher temperatures, at lower strength levels.

The 5% Cr tool steels, most notably H11, are used for fasteners having a tensile strength of 1517 to 1793 MPa (220 to 260 ksi). They retain excellent strength through 480 °C (900 °F).

From 480 to 650 °C (900 to 1200 °F), corrosion-resistant alloy A-286 is used. Alloy 718, with a room-temperature tensile strength of 1241 MPa (180 ksi), has some applications in this temperature range.

The nickel-base alloys René 41, Waspaloy, and alloy 718 can be used for most applications in the temperature range of 650 to 870 °C (1200 to 1600 °F). Between 870 and 1095 °C (1600 to 2000 °F) is a transition range in which the nickel- and cobalt-base alloys do not retain sufficient strength to give satisfactory service.

Mechanical fasteners are available for short-term exposure at ultrahigh temperature. Some refractory-metal fasteners are in service in carefully controlled applications at temperatures of 1095 to 1650 °C (2000 to 3000 °F).

Liquid-Metal Embrittlement (LME)

Liquid-metal embrittlement is the brittle failure of a normally ductile metal when it is coated with a thin film of a liquid metal and subsequently stressed in tension. Embrittlement manifests itself as a reduction in fracture stress, strain, or both; failure occurs well below the yield stress of the solid. Even small amounts of liquid metal can cause catastrophic failure by LME. Fracture is usually intergranular, with little crack branching or striations to indicate slow crack propagation. More information on this subject is available in the article "Liquid-Metal Embrittlement" in this Volume, and the article "Failures of Locomotive Axles" in this Volume gives examples of steel locomotive axles that failed by liquid-metal contamination by copper alloy bearing materials.

In one case of LME of mechanical fasteners, cadmium-plated 4140 steel nuts (hardness: 44 HRC) were inadvertently used on bolts for clamps used to join ducts that carried hot (500 °C, or 930 °F) air from the compressor of a military jet engine. The nuts were fragmented or severely cracked (Fig. 16a). Fracture surfaces were oxidized and tarnished (Fig. 16b), and fracture was brittle and intergranular (Fig. 16c). Failure occurred when the hot air melted the electroplated cadmium (melting point:

At moderate temperatures, where cadmium and zinc anticorrosion platings might normally be used, the phenomenon of stress alloying becomes an important consideration. Conventional cadmium plating, for example, is usable only to 230 °C (450 °F). At somewhat above that temperature, the cadmium is likely to melt and diffuse into the base material along the grain boundaries, causing cracking by liquid-metal embrittlement, which can lead to rapid failure. For corrosion protection of high-strength alloy steel fasteners used at temperatures between 230 and 480 °C (450 and 900 °F), special nickel-cadmium coatings such as that described in AMS 2416 are often used.

At extremely high temperatures, coatings must be applied to prevent oxidation of the base material, particularly with refractory metals. These must be selected on the basis of the environments. In static air, suitable coatings are available. However, at partial pressure or in dynamic air, there will be a significant reduction in service life.

Fastener-manufacturing methods can also influence bolt performance at elevated temperature. The actual design and shape of the threaded fastener are also important, particularly the root of the thread. A radiused thread root is a major consideration in room-temperature design, being a requisite for good fatigue performance. However, at elevated temperature, a generously radiused thread root is also beneficial in relaxation performance. Starting at an initial preload of 483 MPa (70 ksi), a Waspaloy stud with square thread roots lost a full 50% of its clamping force after 20 h, with the curve continuing downward, indicating a further loss. A similar stud made with a large-radiused root lost only 36% of preload after 35 h.

Selection of fastener material is perhaps the single most important consideration in elevated-temperature design. Most materials are effective over a limited temperature range. No fastener material is suitable for service at temperatures ranging from cryogenic to ultrahigh.

Fig. 17 Five basic types of rivets used to fasten assembled products

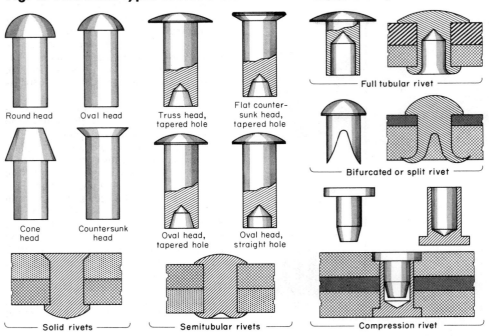

Round head   Oval head   Truss head, tapered hole   Flat countersunk head, tapered hole

Cone head   Countersunk head   Oval head, tapered hole   Oval head, straight hole

Full tubular rivet

Bifurcated or split rivet

Solid rivets   Semitubular rivets   Compression rivet

Fig. 18 Schematic of buckling failure of a thin sheet in a riveted joint

Countersinking the top sheet formed a sharp edge at the faying surface.

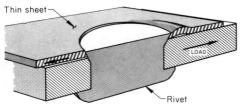

Thin sheet

LOAD

Rivet

---

321 °C, or 610 °F), causing LME of the steel base metal.

Rivets

Solid, tubular, and split rivets are among the most commonly used fasteners for assembled products. Rivets are used as mechanical fasteners because they have a favorable relative cost and weight, because they are available in a wide variety of materials, head styles, and sizes, or because they have excellent hole-filling ability. To achieve these benefits, however, a product must be designed for riveting from the beginning, with consideration given to production as well as product design. The five basic types of rivets used are shown in Fig. 17.

Rivet Positioning. The location of the rivet in the assembled product influences both joint strength and clinching requirements. The important dimensions are edge distance and pitch distance.

Edge distance is the interval between the edge of the part and the centerline of the rivet and is controlled by the bearing strength of the material being joined. Maximum bearing strength is obtained with an edge distance equal to twice the diameter of the rivet shank. At less than this distance, the bearing strength of the joint decreases approximately directly with edge distance under some shear loadings. Even in the absence of shear load, edge distance of twice rivet-shank diameter should be specified to avoid buckling of the material while clinching. The recommended edge distance for plastic materials, either solid or laminates, is between

two and three diameters, depending on the thickness and inherent strength of the material.

Pitch distance, which is the interval between centerlines of adjacent rivets, should be determined by the load to be carried across the joint and by the size of the fastener. Unnecessarily high stress concentrations in the riveted material and buckling at adjacent empty holes can result if the pitch distance is less than three times the diameter of the largest rivet in the assembly. The maximum pitch distance is limited by the tendency of the sheet to buckle between rivets. Multiple rows of rivets, sometimes staggered, are frequently necessary to provide the desired joint strength.

Types of Rivet Failures. Standard design analyses of riveted joints consider two primary types of failures: (1) shear in the shank of the rivet and (2) bearing or crushing failure of the metal at the point where the rivet bears against the joined members. The load per rivet at which each of these two types of failures may occur is separately calculated, and the lower value of the two calculations governs the design. However, the practical goal should be to design all riveted joints as bearing critical. This can be accomplished by selecting the correct material composition, condition, and thickness. For optimum performance, the rivet and the members being joined should have the same properties, but from the driving standpoint, it is often necessary to have a softer rivet.

In addition to the design considerations, rivets may also be susceptible to failure by many of the same failure mechanisms experienced by threaded fasteners. For example, rivets are especially susceptible to fatigue failure

when used in the same joint with threaded fasteners. This is because rivets fill the holes more completely than threaded fasteners and are thus subjected to more than their proportionate share of the load.

Shear in Shank. The number of planes on which a rivet may fail in shear depends on the type of joint that is riveted. Rivets are normally used in single or double shear. For design purposes, the load required to shear a rivet can be assumed to increase directly with the number of planes on which shear loading occurs; thus, in double shear, a rivet is approximately twice as strong as in single shear. However, if the members being joined are relatively thin and have significantly greater hardness than the rivets, the rivets will fail at a lower value of shear stress than that determined in tests with thicker material.

Machine-countersunk flush-head rivets are subject to failure of the sheet or to shear and tension failure of rivet heads. Riveted joints with thin outer sheets are critical because of the instability of the outer sheets, which can buckle around the rivet before full bearing stress is developed (Fig. 18).

Bearing-Surface Failure. In riveted joints involving sheet of moderate thickness, the controlling factor is the strength of the rivet head, because induced shear and tension in the rivet head are more critical than shear through the shank. Determining factors are the magnitude of the induced loads in the rivet head, the deflection of the rivet head and sheet under load, and the bearing of the knife-edge of the countersunk hole in the top sheet of the rivet. Typical loading of a riveted joint is shown in Fig. 19(a); failure generally occurs as shown in Fig. 19(b). Because of these inherent limitations of flush-head rivets, protruding-head rivets are preferred. Flush-head rivets should be used only when a clean surface is required for appearance or for aerodynamic efficiency or when clearance for an adjacent structure is inadequate to accommodate protruding-head fasteners.

Stress-Corrosion Cracking. In aluminum alloys containing more than 4% Mg, the introduction of plastic deformation followed by aging at a slightly elevated temperature, such as might occur in the tropics, can lead to SCC. A

Fig. 19 Schematic of typical behavior of flush-head rivets

(a) Loading of rivet. A, bearing area of the upper sheet; B, bearing area of the lower sheet; L, load; P, shear component; P_t, tension component; P_r, resultant of shear and tension components. (b) Shear and tension failure mode

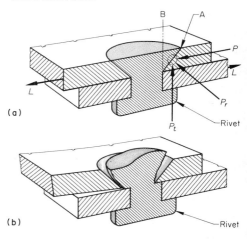

Fig. 20 Alcan aluminum alloy B54S-O rivet that failed by stress corrosion after being heated for 7 days at 100 °C (212 °F)

(a) Section through rivet showing shape of fracture surface. $5\times$. (b) Micrograph showing an intergranular fracture path. $450\times$

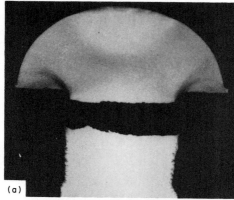

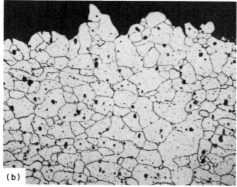

rivet of Alcan aluminum B54S-O, shown in cross section in Fig. 20(a), failed during a stress-corrosion test in the laboratory and illustrates this effect. The rivet was heated for 7 days at 100 °C (212 °F) before testing. The residual deformation from the heading operation, followed by the treatment at elevated temperature, resulted in failure from SCC. The path of the fracture along the grain boundaries is illustrated in Fig. 20(b). The specimen was etched in dilute phosphoric acid, which, on alloys of this type, outlines the grains in a material that is sensitive to stress corrosion. Electron microscope investigation has shown that this type of grain-boundary attack prevails when a fine, closely spaced precipitate is present at the grain boundary (Ref 6). If the precipitate is coarser and more widely spaced, as when treated at 230 °C (445 °F) for 6 h, small etch pits that resemble a string of pearls at the grain boundaries can be seen with the optical microscope. Such a structure is not susceptible to SCC.

Blind Fasteners

Blind fasteners are mechanical fasteners that can be installed in a joint that is accessible from one side only. When a blind fastener is set, a self-contained mechanical, chemical, or other device forms an upset on the inaccessible, or blind, end of the fastener and also expands the fastener shank, thereby securing the parts being joined.

Blind fasteners are classified according to the methods by which they are set, either pull-mandrel, threaded, drivepin, or chemically expanded. Rivets that conform to these classifications are illustrated in Fig. 21.

A blind-fastened joint is usually in compression or shear, both of which the fasteners can

support somewhat better than tensile loading. The amount of loading that blind fasteners subjected to vibration can sustain is influenced by minimum hole clearance and such installation techniques as applying compression to the assembly with clamps before the fasteners are set.

Some blind fasteners can be used in material as thin as 0.5 mm (0.020 in.) if the heads are properly formed and if shank expansion is carefully controlled during setting. However, in joining sheets that are dissimilar in thickness, the best practice is to form the blind head against the thicker sheet. Also, if one of the components being joined is of a compressible material, rivets with extralarge-diameter heads should be used.

Tolerances applied to blind-fastener holes vary according to the type of rivet used. Ordinarily, it is more economical to select a fastener that clamps against the surfaces of the components being joined than to select one that depends only on hole filling.

Pin Fasteners

Pins are effective fasteners where loading is primarily in shear. They can be divided into

two groups: semipermanent pins and quick-release pins.

Semipermanent Pins

Semipermanent pin fasteners require the application of pressure for installation or removal. The two basic types are machine pins and radial-locking pins.

General rules that apply to all types of semipermanent pins are:

- Avoid conditions in which the direction of vibration is parallel to the axis of the pin
- Keep the shear plane of the pin a minimum distance of one pin diameter from the end of the pin
- Where the engaged length is at a minimum and appearance is not critical, allow pins to protrude at each end for maximum locking effect

Machine Pins. The four important types of machine pins are dowel pins, taper pins, clevis pins, and cotter pins. Machine pins are used either to retain parts in a fixed position or to preserve alignment. Under normal conditions, a properly fitted machine pin is subjected to shear only, and this occurs only at the interface of the surfaces of the two parts being fastened. Taper pins are commonly used to fasten parts that are taken apart frequently, where the constant driving-out of the pin would weaken the holes if a straight pin were used.

Example 10: Fatigue Fracture of a Taper Pin in a Clutch-Drive Assembly. The drive wheel on the clutch-drive support assembly shown in Fig. 22(a) was slightly loose and caused clutch failures in service after 6.8×10^6 cycles. After failure, removal of the taper pin holding the drive wheel on the shaft was difficult, indicating that the pin was tight in the assembly. The taper pin was made of 1141 steel, the shaft 1117 steel, and the drive wheel 52100 steel.

Investigation. Upon removal from the assembly, the taper pin was found to be broken at the small end (Section A-A, Fig. 22). In assembly, the fracture occurred near the interface of the shaft and the drive wheel. As shown in Fig. 22(b), the surface of the taper pin contained axial grinding marks, some of which were partly obliterated by movement of the components during operation, before and after the pin broke. Evidence of fretting was also found on the pin. Surface finish on the pin was 1 µm (45 µin.); specifications required 0.8 µm (32 µin.). The fracture surface of the pin contained ductile tears, fatigue striations, and shear lips.

Examination of the clutch shaft revealed that fretting had occurred on the portion of the shaft covered by the drive wheel (Fig. 22c). It did not appear that fretting alone had impaired machine operation.

During the investigation, a second clutch-drive support assembly was received at the laboratory. This second assembly exhibited a loose fit similar to the first. The assembly was

Fig. 21 Types of blind fasteners used in assembled components

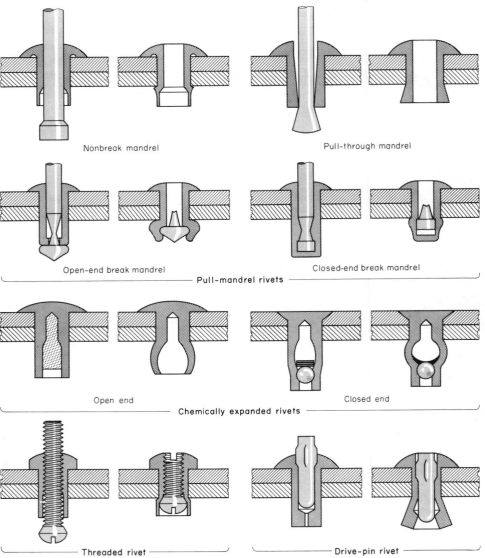

Nonbreak mandrel

Pull-through mandrel

Open-end break mandrel

Closed-end break mandrel

—— Pull-mandrel rivets ——

Open end

Closed end

—— Chemically expanded rivets ——

—— Threaded rivet ——

—— Drive-pin rivet ——

Fig. 22 Steel clutch-drive support assembly that failed in service after a taper pin fractured

The pin fractured because of a loose fit between components. (a) Drive support assembly. Section A-A shows the break in the small end of the taper pin (arrow). (b) Fractured pin; note axial grinding marks on the surface. (c) Fretting on the portion of the shaft covered by the drive wheel

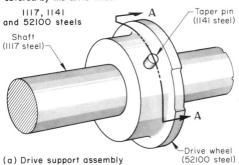

1117, 1141 and 52100 steels

Shaft (1117 steel)

Taper pin (1141 steel)

Drive wheel (52100 steel)

(a) Drive support assembly

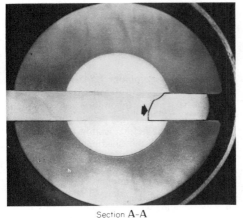

Section A-A

(b)

(c)

sectioned transversely at the taper pin, and the pin was found to be fractured in the same area as the first. There was a tight line-to-line fit between the shaft and the taper pin. However, there was a maximum clearance of 0.08 mm (0.003 in.) between the drive wheel and the large end of the pin, and there was a lesser amount at the small end. Also, the clearance between the shaft and the drive wheel was 0.04 mm (0.0015 in.) (specifications permitted a maximum clearance of 0.025 mm, or 0.001 in.). Because of the excessive clearance between the shaft and the drive wheel, alternating stresses were transmitted to the pin.

Conclusions. Failure of the clutch-drive support assembly occurred as a result of fatigue fracture of a taper pin. A loose fit between the drive wheel and the shaft and between the drive wheel and the pin permitted movement that resulted in fatigue failure. Fretting of the pin and drive shaft was observed but did not appear to have contributed to the failure.

Recommendations. The assembly should be redesigned to include an interference fit between the shaft and the drive wheel. The drive wheel and the shaft should be taper reamed at assembly to ensure proper fit. In addition, receiving inspection should be more critical of the components and accept only those that meet specifications.

Example 11: SCC of 300M Steel Jack-screw Drive Pins. Both jackscrew drive pins on a landing-gear bogie failed suddenly when the other bogie on the same side of the airplane was kneeled for a tire change. The pins were smooth cylindrical tubes that fastened the top tubular ends of the jackscrew to the plane. The primary service stress on the pins was shear.

The pins had nominal dimensions of 41-mm (1.6-in.) outside diameter, 25-mm (1-in.) inside diameter, and 178-mm (7-in.) length, and they had a smooth push fit with a hole tolerance of +0.04 to +0.1 mm (+0.0015 to +0.0040 in.). They were shot peened and chromium plated on the outside surface and were cadmium plated and painted with polyurethane on the

Fig. 23 300M steel jackscrew drive pins that failed by SCC

(a) Four views of aft-pin locations of individual origins (numbers), directions of fracture (arrows), and final-fracture regions (wavy lines). (b) Same as (a) except for forward pin. (c) Top surface of forward pin showing slight bend at left end. (d) Micrograph showing pits in base metal under cracks in chromium plating. Etched with nital. 100×. (e) Electron fractograph showing corroded intergranular fracture surface at origin 1 on forward pin (see b). 4000×. (f) Same as (e) except at 6500× showing origin 1A on aft pin. (g) Micrograph showing intergranular crack in base metal propagating from corrosion pit adjacent to crack. 400×

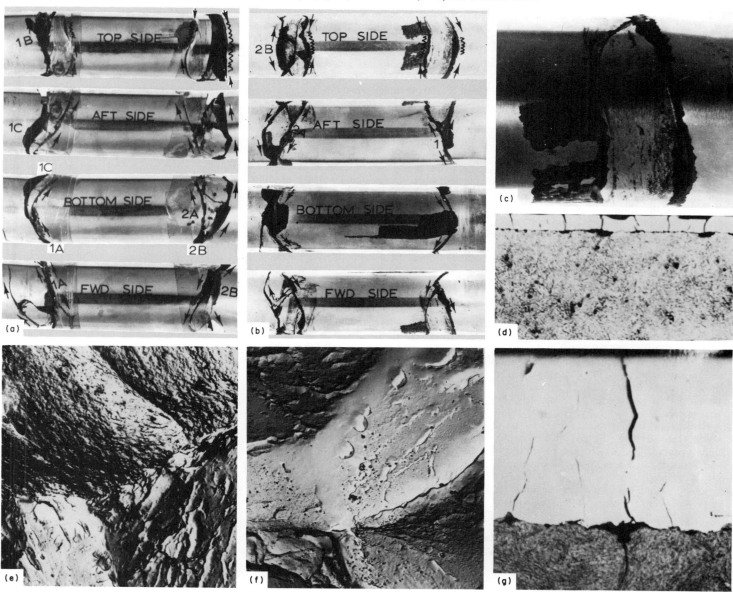

inside surface. The pins were made of 300M steel heat treated to a tensile strength of 1931 to 2069 MPa (280 to 300 ksi). The top of the jackscrew was of 6150 steel heat treated to a tensile strength of 1241 to 1379 MPa (180 to 200 ksi).

Investigation. After the retrieved portions of the fractured pins had been reassembled (Fig. 23a and b), it was evident that they had sustained prior mechanical damage. Both ends of the pins were dented where the jackscrew had pressed into them, probably as a result of overdriving the jackscrew at the end of

an unkneeling cycle. One end of the forward pin was noticeably bent (Fig. 23c). The chromium plating was missing from the dented areas, and these areas were heavily corroded.

Chromium-carbon replicas of the areas of fracture origin showed a heavily corroded intergranular fracture mode (Fig. 23e and f). Subsequent metallographic examination revealed deep corrosion pits adjacent to the fracture origins and directly beneath cracks in the chromium plate (Fig. 23d). Intergranular cracks that had propagated from several of the pits

under the cracks in the chromium plate are shown in Fig. 23(g).

The dimensions, surface finish, and composition of the pins were within the specified limits. Microhardness traverses and metallographic examination of sections through the pins confirmed that neither carburization nor decarburization had occurred on the surface of either pin. Alkaline chromate etching confirmed that no intergranular oxidation products were present. The microstructure was typical of tempered martensite and was satisfactory.

Fig. 24 Zinc-electroplated 1060 steel fastener that failed by hydrogen embrittlement

The part was used to secure fabric to lawn-furniture framework. Dimensions given in inches

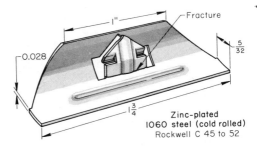

Conclusions. Overdriving the jackscrew dented the drive pins and caused cracking and chipping of the chromium plate. The unprotected metal rusted, and stress-corrosion cracks grew out of some of the rust pits. The terminal portion of the fracture resulted from overstress when the remaining uncracked cross section was not adequate to carry the load.

Corrective Measures. The jacking-control system was modified to prevent overdriving and subsequent damage to the drive pins. The pin material was changed from 300M steel to PH 13-8 Mo stainless steel, which is highly resistant to rusting and SCC. The inside diameter of the pins was reduced to compensate for the lower strength of the PH 13-8 Mo steel. The protective plating and paint were eliminated because of the good corrosion resistance of the new material.

Radial-locking pins are of two basic styles: (1) solid, with grooved surfaces, and (2) hollow spring pins, which may be slotted or spiral wound. In assembly, radial forces produced by elastic action at the pin surface develop a friction-locking grip against the hole wall. Spring action at the pin surface

also prevents loosening under shock and vibration.

Locking of the solid grooved pins is provided by parallel, longitudinal grooves uniformly spaced around the pin surface. When the pin is driven into a drilled hole corresponding in size to the nominal pin diameter, elastic deformation of the raised groove edges produces a force fit with the hole wall. Optimum results under average conditions are obtained with holes drilled the same size as the nominal pin diameter. Undersize holes should be avoided, because they can lead to deformation of the pin in assembly, damage to hole walls, and shearing-off of the raised groove edges.

Resilience of hollow cylinder walls under radial compression forces is the principle of spiral-wound and slotted tubular pins. Compressed when driven into the hole, the pins exert spring pressure against the hole wall along their entire engaged length to develop locking action. For maximum shear strength, slotted pins should be placed in assemblies so that the slots are in line with the direction of loading and 180° away from the point of application.

Quick-Release Pins

Quick-release pins are of two types: push-pull and positive locking. Both use some form of detent mechanism to provide a locking action for rapid manual assembly and disassembly.

Push-pull pins are made with a solid or hollow shank containing a detent assembly, such as a locking lug, button, or ball, which is backed by a resilient core, plug, or spring. These pins fasten parts under shear loading, ideally at right angles to the shank of the pin.

Positive-locking pins use a locking action that is independent of insertion and removal forces. These pins are primarily suited for shear-load applications. However, some tension loading can usually be tolerated without adversely affecting the function of these pins.

Special-Purpose Fasteners

Special-purpose fasteners, many of which are proprietary, are frequently used in nonstructural applications to provide some unique feature, such as quick release, snap action, or cam action. These fasteners are subject to the same failure mechanisms experienced by threaded fasteners and rivets.

Spring clips are a type of special-purpose fastener that perform multiple functions, are generally self-retaining, and, for attachment, require only a flange, panel edge, or mounting hole. The spring-tension principle of fastening eliminates loosening by vibration, allows for design flexibility, compensates for tolerance buildup and misalignment, and minimizes assembly damage.

The basic material for spring-clip fasteners is steel with 0.50 to 0.80% carbon content. Fasteners are generally formed in the annealed stage, then hardened and tempered to 45 to 50 HRC. Zinc or cadmium electroplating is specified for certain applications. Other applications may use a zinc mechanical plating to eliminate hydrogen embrittlement.

Example 12: Failure of a Zinc-Electroplated 1060 Steel Fastener. The fastener shown in Fig. 24 was used to secure plastic fabric or webbing to the aluminum framework of outdoor furniture. Several clips were required on each piece of furniture, and these had to hold the fabric for several seasons of varying weather conditions.

The fasteners were made in high-production progressive dies from 0.7-mm (0.028-in.) thick cold-rolled 1060 steel. They were barrel finished to remove burrs and hardened and tempered to 45 to 52 HRC. The fasteners were electroplated with zinc and coated with clear zinc dichromate.

During attachment of the fabric to the furniture, approximately 30% of the fasteners

Fig. 25 Static tensile failures in carbon-graphite composite samples

Failure was by fastener pullout. (a) Single-lap shear specimen. (b) Double-lap shear specimen. Both 1⅓ ×

(a)

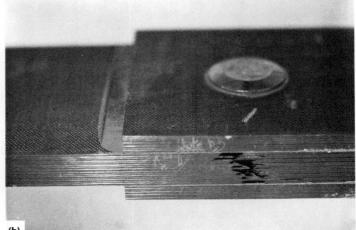

(b)

Fig. 26 Fatigue failure of fasteners in single-lap shear carbon-graphite composite joints

(a) Fastener pullout resulting from a static tensile load. (b) Fatigue failure of fasteners initiated by cocking of the fasteners. Both 1⅓×

(a) (b)

cracked and fractured as they were compressed to clamp onto the framework prior to springback, which held the fabric in place. Fasteners on some of the furniture that was successfully assembled had fractured in warehouses, in retail stores, and in service, allowing the plastic fabric or webbing to come off the chair.

Discussion. The heat treatment of the fasteners consisted of austenitizing at 790 °C (1450 °F) for 10 min, quenching in oil at 80 °C (180 °F), rinsing, then tempering at 345 °C (650 °F) to obtain a microstructure of tempered martensite. After being acid cleaned, zinc electroplated to a thickness of 0.005 mm (0.0002 in.), and coated with a clear dichromate, the fasteners were baked 8 h at 165 °C (330 °F) to remove the nascent hydrogen. Fasteners treated in this manner were very brittle and had a limited capacity of being compressed without fracture. Low ductility at room temperature and delayed failure are characteristic of steel that has been damaged by hydrogen embrittlement.

Conclusions. The fasteners failed in a brittle manner as the result of hydrogen embrittlement. Hydrogen was induced into the surface of the steel when the hardened parts were acid cleaned before electroplating and during the electroplating operation. Baking at 165 °C (330 °F) for 8 h failed to remove all of the nascent hydrogen that had diffused interstitially into the steel.

Corrective Measures. The heat treatment was changed from austenitizing, quenching,

and tempering to austenitizing at 790 °C (1450 °F) for 15 min, quenching in a potassium chloride salt bath at 345 °C (650 °F) for 20 min, air cooling, and rinsing. This heat treatment produced a more ductile and bainitic structure than the original heat treatment, as well as a hardness of 48 to 52 HRC. In addition, the method of plating the fastener with zinc was changed from electroplating to a mechanical deposition process.

These two changes eliminated the possibility of hydrogen embrittlement, because acid cleaning was not needed after austempering and before plating. Also, mechanical deposition of zinc coating does not produce hydrogen embrittlement in hardened steel.

Fastener Failures in Composites

The typical failure mechanism for statically loaded graphite composite joints is fastener pullout (Fig. 25a and b). The oversize fastener head is evident in Fig. 25(b); the collar flange, Fig. 25(a). Collars are used, rather than nuts, to avoid galling, which is frequently a problem when titanium nuts and bolts are installed. Figure 26(a) shows the typical pullout-type failure of a flush-head fastener in a composite joint. However, when this type of joint is fatigue tested, failure is usually first initiated by cocking of the fasteners, followed by fatigue failure of one or more of the fasteners (Fig. 26b). The fasteners in Fig. 26(b) were installed

with close-fitting sleeves in the holes to improve fatigue life. After the sleeve is placed in the hole, the interference-fit fastener is pulled into the sleeve; this expands the sleeve to provide an interference fit between it and the composite and reduces the possibility of delamination which can occur if interference-fit fasteners are forced into unsleeved holes. Nondestructive inspection of some type is needed to detect delamination or other internal damage without removal of the fasteners.

REFERENCES

1. Mach. Des., 1985 Fastener and Joining Reference Issue, Vol 54 (No. 27), 1985, p 141
2. *Metallic Materials and Elements for Aerospace Vehicle Structures*, Vol II, Military Handbook, Figure 9.4.1.7(a)
3. J.S. Davey, Why Nuts Should Have Their Own Specifications, *Fasteners*, Vol 13 (No. 3), Fall 1958
4. *SPS Bolts for the Aerospace Industry*, 1964-1965 ed., Standard Pressed Steel Company, Fort Washington, PA
5. E.L. Williams and D.M. Anderson, Hydrogen Embrittlement of AISI 8740 Alloy Steel Fasteners, *Mater. Perform.*, Vol 24 (No. 12), 1985, p 9-12
6. A.T. Thomas, Etching Characteristics and Grain Boundary Structures of Aged Aluminum-Magnesium Alloys, *J. Inst. Met.*, Vol 94, 1966

Failures of Springs

James H. Maker, Consultant, Associated Spring

SPRINGS are made in many types, shapes, and sizes, ranging from delicate hairsprings for watches to massive buffer springs for railroad equipment. The vast majority of springs are made from one of several grades of carbon or alloy steel. However, to accommodate the broad spectrum of applications, springs have been made from a wide variety of metals and alloys, including stainless steels, heat-resisting alloys, copper alloys and titanium alloys.

The broad range of compositions, sizes, and types of springs is equaled by the required range of quality and characteristics. Quality can range from that required for low-cost ballpoint pens to the extremely high quality necessary for aerospace applications.

Ordinary springs, such as those used in noncorrosive environments and at temperatures not much higher than room temperature, are produced to high strength levels that often can be obtained entirely by severe coldworking (drawing or rolling). This approach to producing springs generally dictates that no metal be removed from the wire or strip during working, which allows any existing defects in the raw material to remain. Springs produced in this manner are adequate for many applications because minor surface discontinuities are of little or no consequence.

More rigorous applications are likely to require springs with defect-free surfaces. Such surfaces can be obtained by conditioning of billets or even by grinding of the wire or strip stock from which the springs are formed. These practices, however, significantly increase cost and decrease availability. For more detailed information on compositions, properties, methods of manufacture, and major applications for springs, see the article "Steel Springs" in Volume 1 of the 9th Edition of *Metals Handbook*.

Operating Conditions

Although most spring applications involve operating temperatures that are not far above or below room temperature, many applications require springs to operate over an appreciable temperature range. For instance, valve springs in internal-combustion engines must operate in frigid weather for start-up, then, within minutes, must function at the normal operating temperature of the spring chamber, which may approach 90 °C (195 °F). This temperature range, and even a range with a maximum temperature somewhat higher, can be accommodated by springs made of carbon or low-alloy steels (for examples of springs that failed because of excessive operating temperatures, see the article "Distortion Failures" in this Volume). However, some springs are required to operate at much higher temperatures and thus must be made of heat-resisting materials. Such applications must be evaluated individually before a suitable selection of material can be made.

Corrosive Environments. Most springs are not subjected to environments more corrosive than humid air; there are, however many exceptions. As for elevated-temperature service, applications that involve corrosive environments must be evaluated individually before selection of materials for springs or of suitable protective coatings.

Common Failure Mechanisms

Fatigue is the most common mechanism of failure in springs, as demonstrated by the examples in this article. Any of the causes noted in the section "Common Causes of Failures" in this article may result in failure by fatigue; more than one cause is sometimes involved.

Relaxation is another common failure mechanism. This condition is caused by overstressing and often results from use of a grade of material that is marginal for the application. Relaxation from operation at excessively high temperatures is also a common failure mechanism.

Although some spring materials are quite notch sensitive and have limited ductility, brittle fracture is not a major problem in springs in the usual sense. Apparently, this is true because the section sizes are usually small, an extreme degree of working has been used (which creates a homogeneous structure), and care in minimizing stress concentrations has been applied.

One form of brittle fracture occasionally encountered results from hydrogen absorption in martensitic structures, which occurs in electroplating or other cleaning or finishing processes. Corrosion contributes to spring failure in much the same way as it contributes to failure of other highly stressed structures, although in springs it is most often connected with fatigue. Nevertheless, static stress-corrosion failures can occur at high stresses in the presence of electrolytes.

Examination of Spring Failures

For the most part, procedures that apply to analysis of failures in other parts also apply to analysis of spring failures (see the article "General Practice in Failure Analysis" in this Volume). Loading and other service conditions as well as all details of the material should be determined, and a preliminary visual examination should be made to identify fracture-surface features, wear patterns, contact marks, surface deposits, and temperature colors. Fracture features often are damaged by clashing of the portions of the broken spring; therefore, it may be necessary to analyze several similar failures to arrive at the correct conclusion as to the cause of failure. For small springs, a stereoscopic bench microscope is essential because the entire cross section may be only as large as the origin of failure in many large sections.

Another important factor is lighting. Spot lighting is sometimes required, but because of reflection into and away from the objective lens, such lighting can be misleading. What is best for fractography is often not best for stereo viewing. A small ring-type fluorescent lamp around the objective is suitable for general use if supplemented by a more concentrated source of light. By exploration at all possible viewing angles with this lighting, features may be revealed that otherwise could escape detection.

Common Causes of Failures

Overstressing is a common cause of failure of springs. Settling and other types of distortion are generally caused by operation of springs at stresses that are higher than expected. It must be kept in mind, however, that the stresses a given spring can withstand are greatly affected by the operating environment. For example, helical springs made of 6150 steel provided failure-free service in fuel-injection pumps when the fuel oil being pumped was a normal low-sulfur grade, but several of these springs failed under identical stress conditions when the fuel oil contained substantial amounts of hydrogen sulfide. In addition, operating temperatures that are higher than those anticipated often result in failures of springs without changes in stress.

Design deficiencies, material defects, processing errors or deficiencies, and unusual operating conditions are common causes of spring failures. In most cases, these causes result in failure by fatigue.

In the remainder of this article, 19 examples of specific failures of springs are presented. Although most of the springs discussed in these examples failed by fatigue, the fatigue cracks were initiated by a wide variety of causes. In some of the examples, more than one cause contributed to failure.

Failures Caused by Errors in Design

Rules governing design of the more common types of springs are generally well understood and closely followed. Proper distribution of stresses and avoidance of sharp bends are among the principal requirements of a well-designed spring. Because the rules of design are generally understood and followed, the number of failures of springs that occur solely as the result of faulty design is comparatively small. However, there are occasional exceptions.

For example, one of a test set of titanium alloy Ti-13V-11Cr-3Al torsionally stressed compression springs failed by fatigue after 12×10^6 cycles at a maximum stress of 640 MPa (93 ksi). The surface of the fracture, which occurred one turn from the end, is shown in Fig. 1. Radial marks emanate from the fracture origin (arrow), which was in an area of contact with the end turn of the spring. In this area, small elongated gall marks were visible, indicating that there was sliding motion over a region about 0.1 mm (0.004 in.) in length. It was concluded that fretting in this area had lowered the fatigue strength. In this instance, and in many others, the proximity of one coil to another does not permit some areas to be shot peened properly, and this can contribute to a reduction in fatigue life.

Failures Caused by Material Defects

Seams frequently contribute to failures of springs, usually by initiating fatigue cracks. Pipe, inclusions, and pits may also contribute to premature failures. The following three examples describe fatigue failures that resulted from material defects of these types. Additional examples of failures due to defects can be found in the section "Failures That Occur During High Stress Fatigue" in this article (see Examples 5 to 8).

Example 1: Fatigue Failure of a Locomotive Suspension Spring that Initiated

Fig. 1 Fractograph of the surface of a fracture that occurred one turn from the end of a titanium alloy spring

The arrow indicates the point of contact with the end of the next coil. 10 ×

at a Seam. Hot-rolled hardened-and-tempered 5160 bars for suspension springs are purchased to a restricted seam-depth requirement. Outdoor exposure can result in rusting in the seam and on the fracture surface (Fig. 2a). This macrograph shows a smooth portion of the fracture that exhibits a step due to a seam, which can be seen on the surface.

Investigation. Figure 2(b) shows a detailed view of the area around the step. The thumb nail looks off-center from the step, but there is a smaller thumb-nail shape that is concentric with the step and a second stage of growth, very slightly rougher, that spreads principally to the right of the step. The demarcation line is nearly invisible because of rust on the fracture. The rapid stage of failure, which begins at the edge of the thumb nail, is much rougher and exhibits rays that diverge approximately radially from it.

Figure 2(c) shows the seam wall at 15 × from a section next to that shown in Fig. 2(b). This was accomplished by cutting the metal away beneath the seam with an abrasive cutoff disk oriented such that the cut was aligned with the seam. Cutting was continued until the wheel had penetrated to about 0.8 mm (1/32 in.) from the bottom of the seam. The section of the spring was then placed in a vise so that the abrasively cut walls could be pushed toward each other and the seam walls pulled apart by using the sound metal in between as a fulcrum. Under these conditions, the sound metal just under the seam is placed in tension, and it fractures in a ductile manner. The seam wall has two zones, the lower zone being mottled.

At the base of the seam wall are dozens of spear-head-shaped areas pointing away from the seam. These facets are all parallel to the final thumb-nail surface and are fatigue cracks. Although they extend laterally into the matrix metal, they are deeper than they are wide, showing that the fracture fronts did not develop at a uniform rate in all directions in their early

stages. The orientation of these origins is normal to the direction of resultant tensile stress from torsional stressing of the spring material.

Figure 2(d) shows a metallographic view of a cross section of the seam. The light areas near the surface are predominantly ferrite resulting from decarburization. The dark areas in the seam are nonmetallics that were not identified.

Conclusion. The failure in this spring initiated at the base of a seam. Many fracture origins developed, which oriented normally to the direction of tensile stress at 45° to the wire axis, resulting from the torsional stressing of the spring material. As failure progressed, one or more of these origins developed into a semicircular fracture plane at the same 45° angle. This eventually caused complete fracture.

Example 2: Fatigue Fracture of Alloy Steel Valve Springs Because of Pipe. Two outer valve springs broke during production engine testing and were submitted for laboratory analysis. The springs were from a current production lot and had been made from air-melted 6150 pretempered steel wire. The springs were 50 mm (2 in.) in outside diameter and 64 mm (2½ in.) in free length, had five coils and squared-and-ground ends, and were made of 5.5-mm (7/32-in.) diam wire. Because both failures were similar, the analysis of only one will be discussed.

Investigation. The spring (Fig. 3a) broke approximately 1½ turns from the end. Fracture was nucleated by an apparent longitudinal subsurface defect. Magnetic-particle inspection did not reveal any additional cracks or defects.

Microscopic examination of transverse sections of the spring adjacent to the fracture surfaces revealed that the defect was a large pocket of nonmetallic inclusions at the origin of the fracture. The inclusions were alumina and silicate particles. Partial decarburization of the steel was evident at the periphery of the pocket of inclusions. The composition and appearance of the inclusions and the presence of the partial decarburization indicated that the inclusions could have been associated with ingot defects. The defect was 0.8 mm (0.03 in.) in diameter (max), 25 mm (1 in.) long, and 1.3 mm (0.05 in.) below the surface (Fig. 3b).

The steel had a hardness of 45 to 46 HRC, a microstructure of tempered martensite, and a grain size of ASTM 6 or 7, all of which were satisfactory for the application. The fracture surface contained beach marks typical of fatigue and was at a 45° angle to the wire axis, which indicated torsional fracture.

Conclusions. The spring fractured by fatigue; fatigue cracking was nucleated at a subsurface defect that was longitudinal to the wire axis. The stress-raising effect of the defect was responsible for the fracture.

Example 3: Fatigue Fracture of a Carbon Steel Pawl Spring That Originated at a Delamination at a Rivet Hole. The pawl spring shown in Fig. 4(a) was part of a selector switch used in telephone equipment.

Fig. 2 Fatigue fracture of a locomotive spring with a bar diameter of 36 mm (1⁷⁄₁₆ in.)

(a) A step (shown by arrow) is visible at the fracture surface near the inner diameter of the spring. A seam is visible, extending from the step. 0.6×. (b) Higher magnification (1.7×) of thumb-nail position of the fracture. The step is in the center of a smaller semicircle indicated by arrows. Growth occurred principally to the right of the step until the area was approximately doubled, at which time rapid fracture (rough texture) began. (c) The seam wall (exposed by a specimen-fracture technique) is very dark in places because of scale. Below this is a mottled area exhibiting rust and scale. At the base of this area are dozens of fatigue origins with faces all parallel to the final fracture. At the bottom is the area of rapid, rough fracture. 14×. (d) Metallographic cross section of the seam. The light flecks near the surface are ferrite; the gray areas in the seam are unidentified nonmetallics, probably oxides. 100×

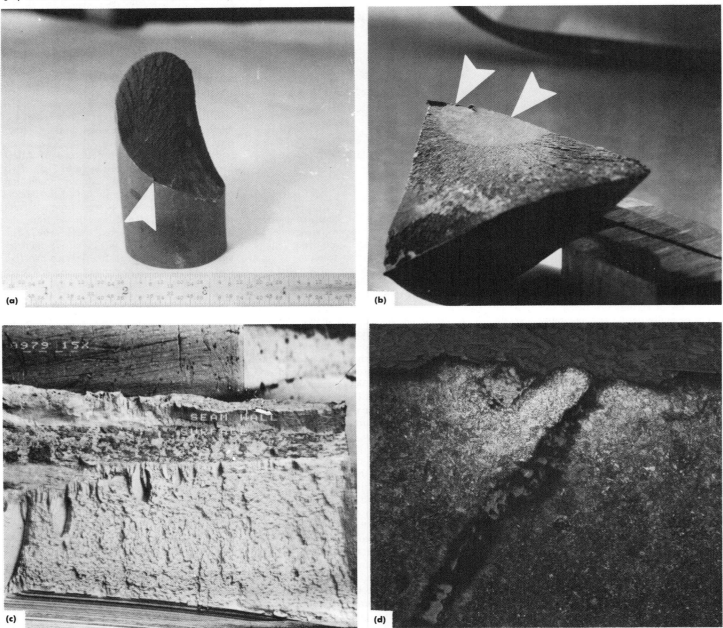

Three of these springs that broke and strip specimens of the raw material used to fabricate similar springs were examined to determine the cause of failure. The springs had been blanked from 0.4-mm (0.014-in.) thick tempered 1095 steel, then nickel plated.

Investigation. Longitudinal specimens from the three broken springs were mounted and processed, and microscopic examination of these specimens revealed numerous pits around the rivet holes. These pits are readily visible in Fig. 4(b), a close-up view of one rivet hole, because their surfaces were covered with nickel plating that was not removed during grinding and etching of the specimen. The pits as well as other defects found at the rivet holes and at the bottom end of the spring were produced by a dull blanking-and-punching tool.

A longitudinal section through a rivet hole is shown in Fig. 4(c). This micrograph reveals delaminations that were formed at inclusion sites during punching of the rivet holes and that were filled with nickel during the plating operation. Each of the delaminations acted as a stress raiser, and one of them initiated the fracture (Fig. 4d).

Longitudinal specimens were sectioned from the raw material and bent slightly for mounting purposes. Examination of the stock revealed numerous long, narrow sulfide stringers, which

Fig. 3 Valve-spring failure due to residual shrinkage pipe

(a) Macrograph showing fracture as indicated by arrow. (b) Fracture surface; pipe is indicated by arrow.

(a)

(b)

Fig. 4 Nickel plated 1095 steel pawl spring that fractured by fatigue

(a) Configuration and dimensions (given in inches) of the failed component. (b) Micrograph showing pits at edge of rivet hole. 45×. (c) Micrograph of area adjacent to rivet hole, showing delaminations (arrows) filled with nickel plating. 250×. (d) Micrograph showing delamination, at A, that initiated fracture; edge of rivet hole is at B. 250×

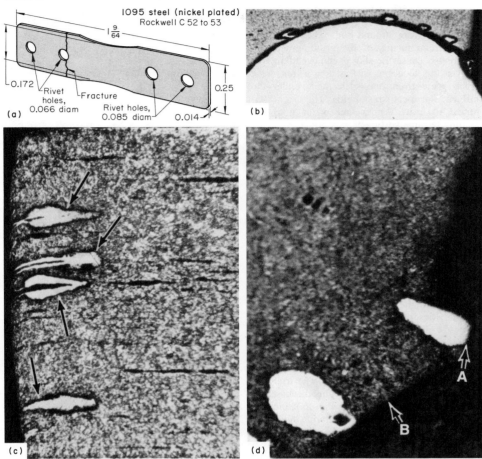

(c)

(d)

probably were the main cause of delamination in this spring material.

The broken springs and the specimens of raw material had a hardness of 52 to 53 HRC, which was acceptable because specifications required a minimum hardness of 45 HRC. However, the hardness was higher than normal and may have contributed to delamination.

Conclusions. Fracture of the springs occurred by fatigue that originated at delaminations around the rivet holes. The fatigue was caused by stresses from blanking and rivet compression and by use of poor-quality spring material. Occasional failures have resulted from the stress pattern alone even though the material is satisfactory.

Failures That Occur During High-Stress Fatigue

Figure 5 shows a modified 5160 (0.5% Cr and 0.1 to 0.2% V) hardened-and-tempered valve spring that failed due to fatigue after running 10 to 15 × 10^6 cycles at deflections considerably larger than those encountered in production engines. Depending on the size of the engine, typical loads and deflections are 70 to 110 kgf (150 to 250 lbf) and 13 mm (0.5 in.), respectively. The failure origin is at the inner

diameter, and the fracture plane at that point is at an angle to the wire axis. It is not at 45° to this axis, as would be predicted for failure from torsional loading, but is much nearer the normal of the wire axis. Examples 4 to 8 in this article discuss valve-spring failures due to various causes that occurred during high-stress fatigue. Because all the springs are similar in appearance and composition to the valve spring shown in Fig. 5, no photographs of the individual springs will be shown; only scanning electron microscopy (SEM) fractographs of the failure origins will be given in this section. No specific data on stresses or life to failure will be given in the following examples. The failures are from a variety of automobile valve springs made on four continents and from modified valve springs run in laboratory tests.

The methods for listing stresses differ. The dynamic stresses and range may be considerably greater than the calculated stresses due to harmonic vibrations excited by high engine speeds. If the material in a valve spring has a tensile strength of 1655 MPa (240 ksi), if there

Fig. 5 Valve spring that failed due to fatigue

Fractographs of similar valve springs are shown in Fig. 6 to 11. 0.8×

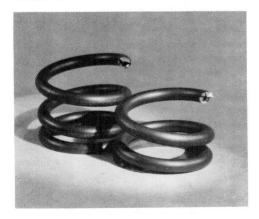

are no flaws in the surface deeper than a few tenths of a thousandth of an inch, and if the spring is made with a careful manufacturing process, there may still be a few field failures if the dynamic operation (torsional) stress range is of the order of 275 to 1035 MPa (40 to 150 ksi).

Example 4: Failure of Valve Spring Because of Grinding and Shot Peening Operations. The valve spring shown in Fig. 6 was made from ground rod. The transverse marks are remnants of the grinding and are apparent because the shot-peening pattern was light at this location. In this area a transverse crack grew from one such mark under the influence of local stress fields. The crack continued to grow deeper and wider until the loading became sufficient to reorient the crack front to the plane normal to the major tensile axis. The transverse crack is somewhat galled because of the sliding motion of the crack surfaces on each other as the crack opened and closed with successive cycles.

Recommendations involved altering the shot-peening procedure to create adequate surface compression at all stressed points on the springs. The base of surface flaws should ideally be in the zone of high compressive stress from the peening.

Example 5: Influence of Inclusions on Failure. Figure 7(a) shows two areas that are dark due to the electrostatic charging of the nonmetallics present. In Fig. 7(b), the inclusions are visible immediately below the surface (marked BB). The fracture in this zone consists of a number of triangular facets with the apexes pointing inward toward the second dark area (marked AA) with its inclusion. The triangles are much like the multiple origins discussed in Example 1 in this article (see Fig. 2c), indicating multiple origins associated with the inclusion cluster. The second, deeper inclusion appears to be a secondary origin surrounded by a faint partial ring formation like a beach mark, with a ray pattern faintly visible. The inclusion at AA would probably not have caused a failure. Two of the small inclusions at the surface were less than 0.025 mm (0.001 in.) in length on the fracture, and because they are in the zone of maximum compression, either one might not have caused failure; however, the group created a stressed zone that fostered eventual fatigue failure.

Figure 8 shows a failure origin formed at a nonmetallic inclusion located 0.2 mm (0.008 in.) below the wire surface. This is well below the zone of peak compression produced by shot peening; consequently, the protective effect is considerably lessened. The beach marks form nearly complete ellipses with a centroid near but not exactly at the inclusion. The first crack opening probably began at the boundary of the inclusion nearest the wire surface, progressing toward the centroid. A larger inclusion at the bottom of the fractograph took no part in developing the ellipse pattern. In this spring, the applied stress at this point would be about 20% lower than at the surface, and the metal would be essentially free of anything but a modest residual tension from peening. It was,

Fig. 6 Transverse failure origin in a valve spring made from ground rod

The transverse marks (arrow) are remnants of the grinding operation. 8×

Fig. 7 Spring failure originating at a cluster of inclusions

(a) Two adjacent dark areas (boxed zone) indicate presence of nonmetallics. 9×. (b) Two failure origins are located at BB, and one at AA. 43×

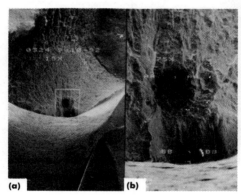

(a) (b)

therefore, not an appreciable factor in this fatigue failure.

Note also the smaller inclusion just below the surface at the left of the fractograph. The metal at this point shows a trace of postfailure damage, which did not break the hard inclusion. This nonmetallic had nothing to do with the failure, and may not have caused any difficulty, because it is in the zone of maximum compression from peening.

Example 6: Fracture at a Transverse Wire Defect. A failure originating at a transverse wire defect is shown as a band extending across the central part of Fig. 9. Above it is the shot-peened surface, and below is a zone of marks, each delineating the progress of the crack front after an increased number of cycles. The defect is about 0.13 mm (0.005 in.) deep and 0.64 mm (0.025 in.) long. This defect could have resulted from a scratch or gash incurred in handling the rod. Drawing causes the sides of such defects to be pushed into close contact.

Fig. 8 Failure origin at an inclusion (arrow) located 0.2 mm (0.008 in.) below the spring surface

Note beach marks surrounding inclusion. 55×

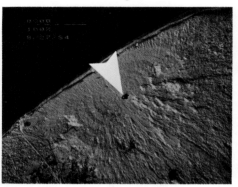

Fig. 9 Fracture at a wire defect

Beach marks are prominent beginning at the base of the flaw, which is indicated by the arrow. 39×

The striations below the defect indicate progressive crack development and are rarely seen in steel as hard as valve-spring wire. At the base of this area, several faceted areas developed, pointing away from the surface. A tensile-mode fracture plane then formed around all these fronts. This fills the bottom area of the fractograph.

Example 7: Spring Failure Due to a Surface Defect. Figure 10 shows light streaks arranged in a diagonal direction on the wire surface. The streaks are parallel to the wire axis. A darker depressed area is visible between the streaks and below the center of the fractograph. The light areas were raised and rubbed smooth by the other half of the spring. This occurred because the engine ran after failure of the spring. Careful examination of the depressed areas revealed distinct outlines that represent sharp corners in the depressions. The origin at one of the sharp corners is shown immediately above the center of the fractograph. The depressed area associated with this was in the other half of the spring that sustained more extensive postfailure damage. Apparently, a hard material (possibly mill scale) was impressed during drawing of the wire, but was

Fig. 10 Spring failure originating at a sharp-edged pitted area

Arrows indicate the location of the sharp-edged areas. 28×

broken out during peening, leaving the depressions with sharp-bottomed corners.

Example 8: Spring Failures Originating at a Seam. The failure shown in Fig. 11(a) began at the seam that extended more than 0.05 mm (0.002 in.) below the wire surface. The fatigue-fracture front progressed downward from several origins. Each one of these fronts produces a crack that is triangular in outline and is without fine detail due to sliding of the opposing surfaces during the later stages of fracture. This occurs when the fracture plane changes to an angle with the wire axis in response to the torsional strain. These surfaces are seen in the lower part of Fig. 11(a).

The failure shown in Fig. 11(b) has many of the characteristics of that shown in Fig. 11(a), except that the seam is scarcely deeper than the folding of the surface that results from shot peening. Observation of it requires close exam-

ination of the central portion of the fractograph. This spring operated at a very high net stress and failed at less than a million cycles.

Failures Caused by Fabrication

Fabrication errors or deficiencies often result in spring failures. The most common of these are:

- Split wire, usually caused by overdrawing of the wire
- Tool marks
- Arc burns, sometimes occurring during welding and sometimes during plating; either can cause hard, brittle spots
- Hydrogen damage, usually resulting from improper pickling or plating practice
- Improper winding, which may allow formation of flats or scratches
- Improper heat treating
- Improper relation of spring index and hardness. A small index and high hardness may result in incipient cracks
- Unfavorable residual-stress pattern

The examples that follow demonstrate how improper fabrication can result in failure.

Example 9: Failure of Carbon Steel Springs During Testing Because of Split Wire. The springs shown in Fig. 12 failed to comply with load-test requirements. These springs were formed from 3.8-mm (0.148-in.) diam cold-drawn carbon steel wire.

During load testing, most of the springs from a production lot supported a load of 90 kg (200 lb) without going solid. However, the springs shown in Fig. 12, which were from the same lot, supported only 60 to 70 kg (130 to 150 lb).

Investigation revealed split wire, as shown in the spring at top in Fig. 12. The spring at

bottom in Fig. 12 is intact except for what appears to be a seam (arrow) that runs the entire length of the spring. The upper spring (originally similar to the lower) has been deliberately distorted to expose the split interfaces. The fracture shows a smooth heat-tinted longitudinal zone, which was the original split, and a rougher bright-appearing longitudinal zone, which resulted from tearing during testing and examination. The course of the defect can be followed for a full turn in Fig. 12.

Conclusion. The springs failed the load test because of split wire. This condition can vary from the extreme demonstrated here to occasional short lengths of one or two turns that have no visible or magnetically detectable opening to the surface and that cannot be detected until the spring is stretched out of shape. The cause of this condition is overdrawing, which results in intense internal strains, high circumferential surface tension, and decreased ductility.

Example 10: Fatigue Fracture of a Phosphor Bronze Spring Because of Tool Marks. Premature failure of a copper alloy C51000 (phosphor bronze, 5% A) spring occurred during life testing of several such springs. As shown in Fig. 13(a), the fracture occurred in bend 2, which had an inside radius of 0.38 mm (0.015 in.). The wire used for the springs was 0.46 mm (0.018 in.) in diameter and was in the spring-temper condition (tensile strength, 1000 MPa, or 145 ksi).

Test Procedure. These springs were formed, then assembled and tested as shown in Fig. 13(a) for up to 50×10^6 cycles. During testing, the springs were subjected to cyclic loading, mainly in the horizontal and vertical planes, with displacements of approximately 3.2 mm (0.127 in.) and 0.5 mm (0.020 in.), respectively. Some torsional loading also re-

Fig. 11 Failures in wire springs

(a) Longitudinal failure originating at a seam. 45×. (b) Origin of failure at a very shallow seam. The arrow indicates the base of the seam. 115×

(a)

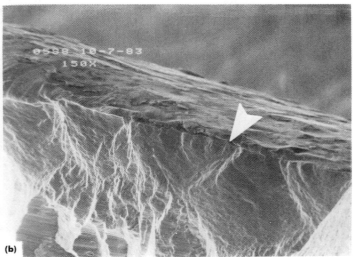

(b)

Fig. 12 Split wire in a 3.8-mm (0.148-in.) diam carbon steel spring (top)

The spring at bottom appears to have a seam along its entire length, as indicated by the arrow.

Fig. 13 Phosphor bronze (C51000) spring that failed prematurely during fatigue testing

Failure was due to the presence of a tool mark (indentation) at a bend. (a) Setup for fatigue testing, and detail of the spring showing location of crack at bend 2. (b) A broken end of the spring, 40×; the tool mark (indentation) is just to the right of fracture surface. (c) Spiral marks on spring surface. 450×. (d) and (e) Two areas of the fracture surface, 225×; flat along edge of surface in (d) is tool mark. (f) Micrograph of a longitudinal section through spring at bend 2, 145×; arrow points to crack that originated at surface at inside of bend

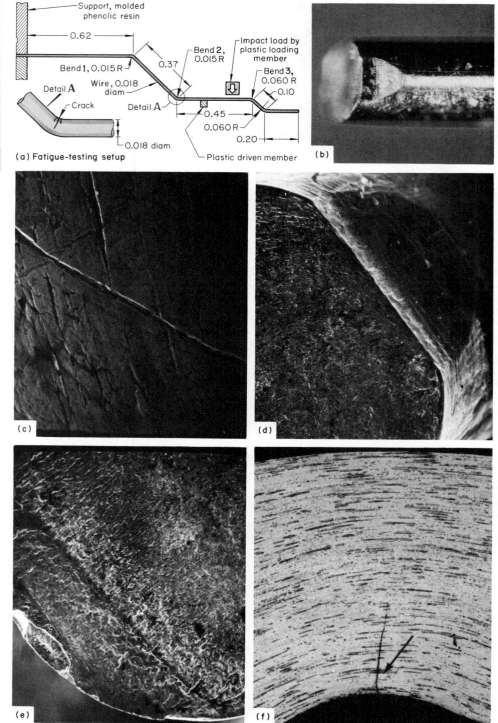

sulted. The vertical loading was applied by means of a typical cantilever arrangement. Near bend 3 the springs were subjected to impact loading by a plastic loading member, which moved a driven member approximately 0.5 mm (0.020 in.). This complex loading system developed very complicated stress patterns along the length of the springs.

The springs were examined by high-power microscopy at various intervals during testing to note the time at which cracks were initiated. After approximately 150 000 cycles, a crack was observed at the inside of bend 2 in one of the springs, as shown in detail A in Fig. 13(a).

Investigation. Microscopic examination of the surface of the fractured spring revealed an indentation at the inner surface of the bend, where fracture occurred (Fig. 13b). This indentation was presumed to have been made by the bending tool during forming. Spiral marks and other surface defects, such as those shown in Fig. 13(c), were observed on the surfaces of all the springs being tested. These spiral marks are similar to those that can be produced on springs during rotary straightening. The indentation and the other marks were, in effect, notches and would have some detrimental effect on fatigue life because of the reduced cross-sectional area in those zones.

Microscopic examination of the fracture surface (Fig. 13d) revealed two areas of prominent characteristics. One area was smooth, discolored, and rippled, indicating the gradual propagation of a crack from one or more origins. The indentation developed by the bending tool is also shown in Fig. 13(d). The remainder of the surface (Fig. 13e) had either a crystalline or a fibrous appearance, which indicated final fracture.

A longitudinal section was taken through bend 2. A micrograph of this section (Fig. 13f) revealed a crack that had originated at the surface at the inside bend and had propagated toward the outside of the bend. The crack had

been initiated at a spiral mark where the fatigue limit had been decreased by a weak skin.

The small bend radius could create a condition that would result in straining at the bend zone and therefore render the section weak. It is difficult to determine to what extent the small bend radius contributed to spring fatigue, but this condition is recognized as an important factor in fatigue life.

The microstructure of the metal appeared normal for drawn wire and contained no nonmetallic inclusions that would act as stress raisers.

Conclusions. The spring failed by fatigue that was initiated at a bending-tool indentation and at spiral rotary-straightener marks. The small bend radius at the fracture zone contributed to the failure.

Recommendations. The springs should be made of wire free from straightener marks, and the bending tool should be redesigned so as not to indent the wire. The small bend radius should be increased, particularly at bend 2, the area of maximum stress.

Example 11: Fatigue Fractures of Toggle-Switch Springs That Originated at Tool Marks. Several electrical toggle switches failed by fracture of the conical helical spring sealed within each switch enclosure. The springs were fabricated from 0.43-mm (0.017-in.) diam AISI type 302 stainless steel wire to the configuration shown in Fig. 14(a). In qualification testing of the springs at room temperature, fractures occurred after 11 000 to 30 000 switching cycles. Two broken springs and two unbroken springs were submitted for laboratory examination.

Investigation. Inspection of a fracture surface of one of the broken springs revealed a typical fatigue fracture that originated at a tool mark on the wire surface. This fracture surface is shown in Fig. 14(b); the fracture origin is indicated by the arrow at the right edge of the fracture surface. Scanning electron microscope fractographs showed regions displaying beach marks around the fracture origin and parallel striations within the beach-mark regions. Outside the area containing beach marks the fracture surface was dimpled in a fashion characteristic of overload or rapid ductile fracture. Figure 14(c) is a fractograph of the fracture origin in Fig. 14(b). An appreciable amount of scale was observed on the fracture surfaces of both broken springs.

The inner surfaces of one unbroken spring and both broken springs had coarse textures and contained tool marks (Fig. 14d); the tool marks had been formed during the spring-winding operation. The other unbroken spring had a relatively smooth-textured surface and no tool marks. The microstructure of the wire in all four springs was normal. Chemical analysis of the material established that its composition was in conformity with specifications for type 302 stainless steel.

Conclusions. Fracture of the springs was caused by fatigue and was initiated at tool marks.

Corrective Measures. The spring-winding operation was altered to eliminate the tool marks. No further fatigue failures of the toggle-switch springs occurred.

Fig. 14 Stainless steel toggle-switch spring that fractured by fatigue originating at a tool mark

(a) Configuration and dimensions (given in inches) of the spring. (b) Fracture surface, 85×; fracture origin (arrow) is at lower edge of tool mark. (c) SEM fractograph of fracture origin, 1000×. (d) SEM micrograph of surface of an unbroken spring, showing area around a tool mark, 300×

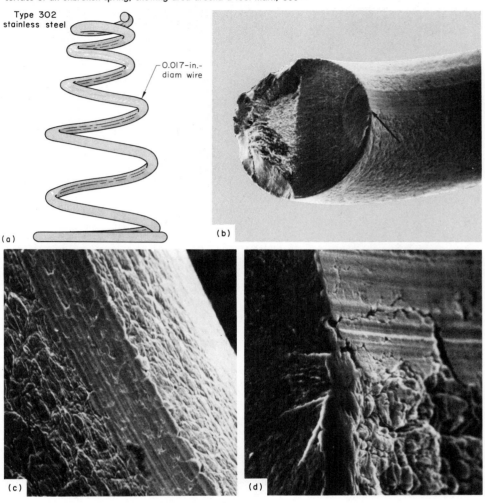

Example 12: Fatigue Fracture of a Music-Wire Spring Caused by Poor Electroplating Practice. A cadmium-plated music-wire return spring that operated in a pneumatic cylinder was designed for infinite life at a maximum stress level of 620 MPa (90 ksi). Several such springs broke after only 240 000 cycles. No flaws were revealed by visual inspection.

Investigation. Figure 15(a) shows the surface of one spring at the origin of failure. The origin was at the outside surface of the coiled spring despite higher stress at the inside surface (due to the Wahl effect).

Figure 15(b) shows the fracture surface, which has a small kernel (arrow) at its center. This kernel appears as a circle on the surface and looks like a weld deposit. Stroking this kernel with a file showed it to be extremely hard, indicating that the kernel may have resulted from extreme localized overheating.

It was learned that these springs had been barrel electroplated after fabrication. It is well known that barrel loads that are too small or that tangle can result in intermittent contact with the dangler (suspended cathode contact) as the barrel rotates. This allows high local currents when the last contact is broken, resulting in an arc that causes local melting of the metal being plated. Because the motion of the barrel rapidly moves the spring away from the dangler, the spot of molten metal caused by arcing is quenched instantly by the plating solution and by the mass of the cold metal of the spring.

Conclusion. Fatigue fracture of the spring resulted from a hard spot that was caused by arcing during plating.

Recommendations. Hard spots can be avoided either by resorting to rack plating, which is more expensive than barrel plating, or by using barrels that have fixed button contacts at many points instead of dangler-type contacts.

Fig. 15 Fractured music-wire spring

(a) Fracture surface at 9.2×; arrow indicates fracture origin. (b) Fracture surface, 9.2×; arrow indicates hard kernel that caused fracture

(a)

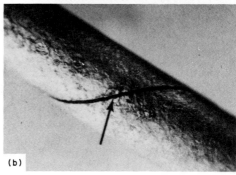

(b)

Example 13: Fatigue Fracture of a Carbon Steel Counterbalance Spring Caused by Hydrogen Damage.

During fatigue testing, the power-type counterbalance spring shown in Fig. 16(a) fractured at the two locations indicated by the arrows. The spring had been formed from hardened-and-tempered carbon steel strip and subsequently subjected to a phosphating treatment. In the normal manufacture of these springs, winding to solid is performed to take out set (provide stable torque output) before phosphating. Although this operation creates residual tensile stresses in the interior surfaces of the coils, phosphating of this spring had not produced sufficient hydrogen absorption to cause embrittlement.

Investigation disclosed a dark band at the inside edge of the fracture surface, as indicated by the arrow in Fig. 16(b). This dark band was rust colored and extended through about 20% of the spring thickness. Examination of the cleaned surface revealed etch pits 0.05 to 0.1 mm (0.002 to 0.004 in.) deep (Fig. 16c). Such pits were never observed on properly phosphated springs; therefore, the spring that failed must have been subjected to an abnormal acid attack in pickling or phosphating. This attack resulted in considerable absorption of hydrogen by the metal.

Conclusion. The dark band in Fig. 16(b) could have been rust colored only by the

Fig. 16 Carbon steel counterbalance spring that failed during fatigue testing

(a) Macrograph showing fracture locations (arrows). ⅓×. (b) Fracture surface showing dark band (arrow) that nucleated fracture. 6×. (c) Etch pits in surface. 100×

(a)

(b)

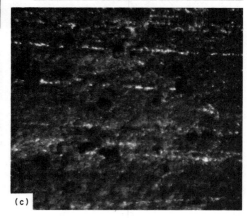

(c)

chemical action of some aqueous medium, which means that the part cracked in a water-base solution—either the phosphating solution or the pickling solution.

Cracks can result from vigorous pickling. The sum of the tensile stress from setout (which deforms the metal beyond the yield point) and the internal hydrogen pressure exceeds the strength of the metal. When the crack reaches the edge of the area of residual tensile stress, it stops. In normal practice, when hydrogen pickup is low, no fracture occurs. It was therefore suspected that the spring had been subjected to excessive acid pickling before phosphating and that this excessive pickling was responsible for hydrogen absorption by the spring. Normal phosphating treatments do not cause hydrogen embrittlement.

Example 14: Fatigue Fracture of a Carbon Steel Wiper Spring Because of Stress Concentration at a Sharp Corner.

Parts such as the one shown in Fig. 17(a) were used as grease-wiper springs for cams. The springs were formed from stampings of 0.25-mm (0.010-in.) thick carbon spring steel (0.65 to 0.80% C) and were hardened to HR15N 84 to 86. The springs were fracturing at the 0.025-mm (0.001-in.) radius on the stamped 135° corner at a 90° bend after 5×10^6 cycles; normal life was 2.5×10^9 cycles.

Four broken springs were sent to the metallurgical laboratory to determine the cause of fracture. Three new springs from stock and six used springs from reconditioned machines were also sent to the laboratory for inspection and comparison. The grease-wiper springs that fractured were of an old design, but they had been made in a new die.

Investigation. Visual examination of all the springs in the laboratory disclosed tool marks 2 to 2.3 mm (0.080 to 0.090 in.) from the center of the stamped bend. The fracture surfaces were nearly parallel to these marks, but the fractures had not been initiated by them.

Examination of the fracture surface by SEM revealed fatigue striations originating from cracks at the 0.025-mm (0.001-in.) radius inside corner at the bend. Examination of a new part from stock and one from a reconditioned machine also showed cracks originating at the small radius inside corner (Fig. 17b). This micrograph also shows the condition of the cut edge of the stamping.

On the new parts, the cracks extended approximately 0.5 mm (0.020 in.) along the width. On the used parts, the cracks extended across approximately 75% of the width and completely through the stock thickness (Fig. 17c).

Stress calculations indicated the maximum stress at the bend, in stock of maximum thickness (0.3 mm, or 0.012 in.), to be 283 MPa (41 ksi) without considering the stress-concentration factor in the 135° corner. The minimum stress under the same conditions would be 83 MPa (12 ksi). Based on 0.3-mm (0.012-in.) thick stock and a maximum stress of 283 MPa (41 ksi), calculated stress as a function of the radius of the 135° corner was:

Radius		Stress concentration factor	Maximum stress	
mm	in.		MPa	ksi
0.025	0.001 .... 3.0		841	122
0.38	0.015 .... 1.82		510	74
0.64	0.025 .... 1.56		434	63
0.76	0.030 .... 1.52		427	62
0.89	0.035 .... 1.46		407	59

The maximum allowable fluctuating stress for this material was 945 MPa (137 ksi). Therefore, the 841 MPa (122 ksi) maximum stress provides a very small factor of safety.

Fig. 17 Wiper spring that fractured at a small-radius corner of a stamped bend

(a) Configuration and dimensions (given in inches) of the spring. (b) SEM micrograph showing a forming crack (arrow) in the 135° corner on a new spring. 200×. (c) SEM micrograph showing a crack (arrow) that originated at the 135° corner in a used spring. 20×

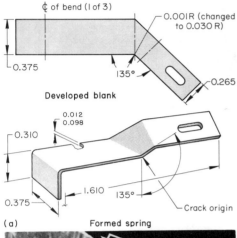

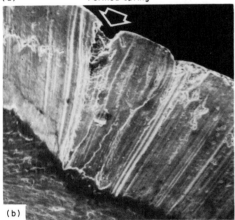

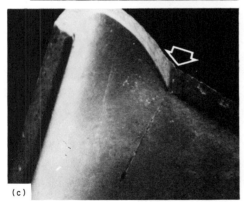

Fig. 18 Helical spring that failed by fatigue

Weld spatter (arrows) was believed to have caused the failure (see text). Arrow FO indicates fracture origin. 29×

Fig. 19 Failure origin (arrow) on the edge of a flat spring

58×

Conclusions. The wiper springs fractured in fatigue that originated at 0.025-mm (0.001-in.) inside radius at the corner of a stamped bend. The very small radius resulted in a high stress-concentration factor in the spring, which caused the maximum applied stress to approach the allowable limit. Tool marks at the bend were not considered major causes of fracture. Cyclic loading resulted from cam rotation.

Corrective Measures. The corner radius was increased to 0.76 mm (0.030 in.), and the tools were repolished to avoid tool marks. These changes increased fatigue life of the springs to an acceptable level.

Example 15: Fatigue Failure Caused by Weld Spatter. A medium-carbon helical spring was installed in a machine assembly that was welded into its final location. During welding, which was conducted several inches from the spring, no shield was used to prevent spatter from landing on the wire surface. Figure 18 shows an elongated drop of metal a short distance from the fracture and two tiny droplets between it and the fracture. Although no spatter was observed at the origin, visual observation revealed an oxide color difference on the surface at the origin. The other half of the fracture was not available, but it was assumed that a drop of molten metal landed at the origin. The degree of heating involved would lessen or eliminate the compressive stress produced by shot peening, and contraction of the surface in contact with the tightly adherent weld metal during cooling could generate other stresses. Adherence of the spatter drop could be affected by the opening and closing of the fatigue crack.

Conclusion. No evidence of other reasons for failure could be found. It was therefore concluded that a weld spatter bead had resulted in the fatigue failure.

Example 16: Flat-Spring Failure Due to an Edge Defect. The face of a 6150 flat spring was under tensile stress. Although small flaws were visible on the surface (Fig. 19), the failure began at the dark spot on the edge, where roughness resulted from shearing during the blanking operation. On the fracture surface at the left, rays approximately 0.18 mm (0.007

in.) below the surface can be seen diverging from the origin.

Failures Caused by Corrosion

Corrosive environments very often are principal contributors to failures of springs. Any metal is more likely to be attacked when under stress. Consequently, springs are extremely susceptible to stress-corrosion cracking.

Example 17: Stress-Corrosion Cracking of Inconel X-750 Springs. Springs such as that shown in Fig. 20(a) were used for tightening the interstage packing ring in a high-pressure turbine. After approximately seven years of operation, the turbine was opened for inspection, and several of the springs were found to be broken. Operation of the turbine was not impaired by the broken springs. The springs were made of 1.1-mm (0.045-in.) diam Inconel X-750 wire. The interstage packing rings were machined from centrifugally cast leaded nickel brass (German silver). The broken springs were submitted for laboratory analysis.

There were chemical deposits in the turbine starting at about the fifth stage. These deposits were judged to have come from the water or boiler chemicals. There were no broken springs in the fifth or sixth stages. The temperatures attained in the turbine stages were:

- *Stage 1*: 462 °C (864 °F)
- *Stage 2*: 405 °C (761 °F)
- *Stage 3*: 372 °C (702 °F)
- *Stage 4*: 351 °C (664 °F)
- *Stage 5*: 330 °C (626 °F)
- *Stage 6*: 313 °C (595 °F)

Investigation. Many of the springs were broken into small pieces that had a white deposit on their rustlike surface scale. The white deposit was 100% water soluble and had a pH of 9.6 (slightly alkaline). Spectrographic examination showed the surface scale to contain a high amount of sodium, a small amount of tin,

Fig. 20 Inconel X-750 spring that failed by stress-corrosion cracking

(a) Configuration and dimensions (given in inches) of the spring. (b) and (c) Unetched longitudinal sections showing intergranular cracking. 100× and 500×, respectively

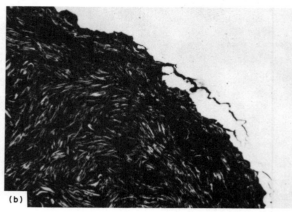
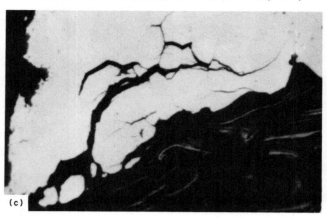

Inconel X-750
Rockwell C 43 to 48
←— 0.500 diam —→

0.045-in.-diam wire

(a)

(b)

(c)

a high trace of zinc, and traces of lead and calcium.

Chemical analysis of the spring material showed that it was within specifications for Inconel X-750. The hardness values were within the specified range of 43 to 48 HRC.

Metallographic examination of a longitudinal section through the wire revealed intergranular cracks (Fig. 20b and c) about 1.3 mm (0.05 in.) in depth and oriented at an angle of 45° to the axis of the wire. In some places, these cracks were 90° apart on the wire circumference. A light-gray phase had penetrated the grains on the fracture surfaces. A fine lacey attack was also observed along the surface of the wire to a depth of about 0.013 mm (0.0005 in.). This general attack had the appearance of liquid-metal corrosion. When bent in the laboratory, the spring wires fractured in a brittle manner along a plane 45° to the wire axis, which is characteristic of brittle fracture from torsional loading.

High-nickel alloys are susceptible to embrittlement by some low-melting metals and alloys. Tin, lead, and zinc melt below the maximum service temperature of the turbine, which was 462 °C (864 °F). One Sn-Zn system had a eutectic melting point of less than 204 °C (400 °F). Such metals can cause liquid-metal embrittlement at temperatures above their melting points.

Sodium, tin, and zinc could have been present in a pigmented oil or grease used as a lubricant during spring winding and could have been left on the Inconel wire by the manufacturer. The white deposit on the surface of some of the springs was slightly alkaline and could have initiated stress-corrosion cracking by caustic penetration.

Conclusions. The springs fractured by intergranular stress-corrosion cracking during service. The cracks were promoted by the action of liquid zinc and tin in combination with static and torsional stresses on the spring wire. Stress-corrosion cracking by caustic action was an alternative but less likely mechanism of fracture.

Corrective Measures. The springs were cleaned thoroughly by the spring manufacturer before shipment to remove all contaminants and were cleaned again before installation in the turbine. Failure of the springs was thus reduced to a minimum.

Protective Coatings. Electroplated coatings of any of several metals or organic coatings, such as paint, are commonly used to protect springs that must operate in corrosive environments. These means of protection can be very effective, but either or both of two deficiencies are likely to prevail, particularly for helical springs that have closely spaced coils. The portions of springs that have the greatest need for protection (often about 1½ to 2 coils from the ends) usually get the least protection because of the proximity of the coils to each other and are the areas from which protective coatings are most likely to be rubbed or broken away in operation.

Failures Caused by Operating Conditions

In some instances of spring failure, the cause is not evident, even after an exhaustive examination of material, processing, and design factors. When a thorough investigation in these areas does not reveal the cause, it is necessary to investigate operating conditions. Changes in operating conditions that are thought to be only minor deviations from normal sometimes prove to be the causes of spring failure.

Example 18: Fracture of a Landing-Gear Flat Spring. A flat spring for the main landing gear of a light aircraft broke after safe execution of a hard landing. The spring (Fig. 21a) was submitted for laboratory examination. No information was furnished concerning design, material, heat treatment, or length of service.

Investigation. Fracture occurred near the end of the spring that was inserted through a support member about 25 mm (1 in.) thick and attached to the fuselage by a single bolt. The required fit between the spring and the support member was

obtained by inserting wedges beneath the spring at each side of the support. These wedges were bolted to the support member, and the degree of tightness of the assembly was controlled by the

Fig. 21 Landing-gear spring, 6150 steel, that broke during a hard landing

(a) Configuration and dimensions (given in inches) of the spring. (b) Fractograph showing fatigue crack that initiated the brittle fracture. 7×

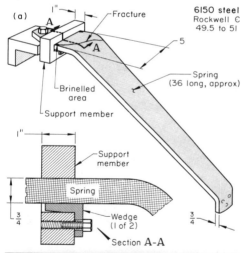

(a)

Fracture

A

A

6150 steel
Rockwell C 49.5 to 51

5

Spring (36 long, approx)

Brinelled area

Support member

Support member

Spring

Wedge (1 of 2)

Section A-A

¾

¾

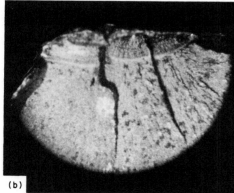

(b)

tension on the bolts. Details of the assembly are shown in section A-A in Fig. 21(a).

The spring had broken laterally at an angle of about 60° to its longitudinal axis (Fig. 21a). One end of the fracture corresponded to the point at which the forward edge of the spring contacted the support member. The other end of the fracture was in an unsupported area.

Brinelling (plastic flow and indentation due to excessive localized contact pressure) had occurred on the upper surface of the spring where the forward and rear edges of the spring contacted the support member. There was no evidence of brinelling on the bottom surface of the spring where it contacted the support member.

Visual examination of the fracture surface revealed that the spring had failed by mainly brittle fracture. Chevron marks indicated that fracture had started beneath the brinelled area at the forward edge of the upper surface of the spring. Figure 21(b), a light microscope fractograph of this corner, shows that the origin of the brittle fracture was in fact a small fatigue crack that had been present for a considerable period of time before final fracture occurred.

Microscopic examination of longitudinal and transverse sections near the fracture confirmed that the material had suffered extensive grain flow from cold working in the brinelled areas. The general microstructure was fine-grained tempered martensite of good quality. There was no surface decarburization.

Chemical analysis identified the spring material as 6150 steel. The hardness of the material was 49.5 to 51 HRC.

Discussion. Normal recurring vertical landing loads cause the top surface of the spring at the edge of the support to be stressed in compression. These probably were the major loads applied to the spring. On the other hand, side and drag loads produce tensile stresses at the top surface of the spring. When vertical, side, and drag loads are applied simultaneously during a landing, the resultant stress at the support is compressive. However, after the aircraft has touched down and the vertical load has decreased to the static load of the aircraft, the continuing application of side and drag loads during taxiing causes a tensile stress at the top surface of the spring.

Compressive stresses in the top surface of the spring at the support are generally localized at the edges of the support by the wedges and by any lack of straightness across the section that may have been introduced during fabrication or that may have occurred when the complicated loading pattern caused a torsional deflection in the region of failure. Also, it was considered probable that the assembly could have been more tightly compressed than it actually was. This would explain the brinelling observed near the forward and the rear edges of the top surface. The localized plastic deformation under compressive loading (brinelling) created high residual tensile stresses in the deformed region, reducing the capacity of the material to withstand fluctuating tensile stresses due to applied loads.

Ultimately, a crack started at the corner where the fluctuating tensile stress was highest under combined side and drag loads. This crack progressed by fatigue over a relatively long period of time (many landings). Final fracture occurred when the applied stress (primarily tensile stress due to side and drag loads) exceeded the critical fracture stress for the partially cracked spring. Under normal conditions, such a cracked part might last a long time, although complete fracture would occur eventually. An abnormal condition, such as a sharp change of direction on a runway or taxiway, rolling over the edge of a runway, or encountering a rut, could impose a sufficient amount of additional loading to cause final, fast fracture.

Conclusions. Fracture of the landing-gear spring was caused by a fatigue crack that resulted from excessive brinelling at the support point. The fatigue crack significantly lowered the resistance of the material to brittle fracture under critical side and drag loads.

Although the loading at the time of final fracture may have been unusual, a spring that did not contain a fatigue crack similar to the one found in the failed spring probably would not have broken. The fatigue crack had been present in the part for some time, and suitable inspection would have revealed it.

Recommendations included regular visual examinations to detect evidence of brinelling and wear at the support in aircraft with this configuration of landing-gear spring and means of attachment. If visual examination showed evidence of brinelling or wear, the spring should be magnetic-particle inspected for evidence of cracks. Liquid-penetrant inspection might be adequate if the amount of brinelling were minor, although care should be used in this type of inspection because brinelling can hide small cracks.

Cracked parts should be replaced. Brinelling itself need not be cause for retiring a part. The presence of a crack was considered necessary for failure to occur in the manner described in this example.

Example 19: Failure of a Retainer Spring Because of Cyclic Loading and Torsional Vibration. The part shown in the "original design" portion of Fig. 22 was a valve-seat retainer spring from a fuel control on an aircraft engine. This spring was found to be broken during disassembly of the fuel control for a routine overhaul after 3980 h of service. The two inner tabs had broken off but were not recovered. There had been no known malfunction of the fuel control.

The spring, made of 0.23-mm (0.009-in.) thick 17-7 PH stainless steel, was thrust loaded in assembly, but there was no thrust load on the tabs. The thrust load was the primary function of the spring. The radial, or torsional, load was

Fig. 22 Original and improved designs of a 17-7 PH stainless steel valve-seat retainer spring
As originally designed, the inner tabs on the spring broke off as a result of fatigue, and the outer tab exhibited wear.

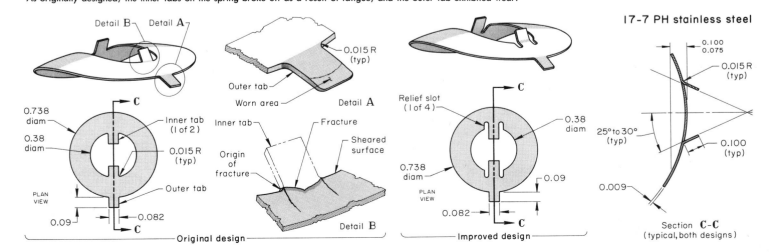

rotational vibration of the assembly in which the spring imparted a relatively constant thrust load. The three tabs were antirotational devices and were subjected to unintentional torsional vibrations. These torsional vibrations created the cyclic loading that resulted in fracture.

Investigation. As received at the laboratory, the spring was clean and free from corrosion products, with no evidence of chemical attack. Radial wear marks on the convex surface indicated that the part had been in relative rotation against its contacting member. The outer tab was worn in the area indicated in detail A in Fig. 22, which was evidence of slippage and apparent rotation against the contacting member. There was a wear mark on the concave surface 180° from the outer tab.

The fracture zone of an inner tab that broke off is shown in detail B in Fig. 22. Examination of the fracture surfaces of the washer in the area of the tabs revealed beach marks indicating that fatigue fracture had been initiated at the convex surface of the washer and had propagated across to the concave surface. The cracks originated in the 0.38-mm (0.015-in.) radius fillet between the tab and the body of the washer. The fracture started at three places at radial cracks emanating from the radii at the intersections of the tab and washer, and at one place on the sheared edge of the stamping.

Microscopic examination of the radial cracks showed a slight amount of plastic deformation, with the short surface between the fractures pushed toward the concave side. The two cracks on the sides of one tab were opened by sectioning. The fractures were smooth with arc-shape crack fronts.

Analysis of this compound fracture indicated that it was composed of fatigue fractures caused by the formed tab being loaded so as to compress the spring along the axis of its centerline and produce torsional vibrations. After the radial cracks had started as a result of torsional vibration, a straight bending mode of stress resulted in initiation of fatigue cracking in the concave side of the tab at the bend.

This was not an isolated failure; shortly after this part had been examined, three additional springs that had broken were returned from service. Although there were some minor variations in the type of fracture, the basic cause appeared to be the same for all the springs. Examination of a mating part revealed worn impressions that had been generated by the torsional motion of the tabs against the walls of the mating slot. Twelve springs from stock were inspected and found to be in satisfactory condition except for some malformed areas that were unrelated to this problem.

Conclusion. The two inner tabs broke in fatigue as the result of cyclic loading that compressed and torsionally vibrated the spring. The fracture resulted in relative rotation between the spring and the mating parts, both the one in contact with the convex side and the one in contact with the outer tab, as evidenced by the wear on the appropriate surface.

Corrective Measures. The springs were redesigned as shown in the "improved design" view in Fig. 22. The fillets were replaced with slots to minimize stress concentration at the corners. Fatigue testing indicated that springs of the improved design would have a service life 2½ times as long as that of springs of the original design.

Failures of Tools and Dies

George F. Vander Voort, Carpenter Technology Corporation

FAILURE MECHANISMS in tool and die materials that are very important to nearly all manufacturing processes are discussed in this article. A wide variety of tool steel compositions are used. Properties and selection of tool steels are described in the Section "Tool Materials" in Volume 3 of the 9th Edition of *Metals Handbook*; microstructures and metallographic techniques for tool steels are detailed in the article "Tool Steels" in Volume 9 of the 9th Edition of *Metals Handbook*. This article is primarily devoted to failures of tool steels used in cold-working and hot-working applications.

Tool and Die Characteristics

Steels used for tools and dies differ from most other steels in several aspects. First, they are used in the manufacture of other products by a variety of forming processes (stamping, shearing, punching, rolling, bending, and so on) and machining processes (drilling, turning, milling, and so on). Second, tools and dies are generally used at higher hardnesses than most other steel products; 58 to 68 HRC is a typical range. Dies for plastic molding or hot working are usually used at lower hardnesses, typically from 30 to 55 HRC.

These high hardness requirements are needed to resist anticipated service stresses and to provide wear resistance. However, the steels must also be tough enough to accommodate service stresses and strains without cracking. Premature failure due to cracking must be avoided, or at least minimized, to maintain minimum manufacturing costs. Unexpected tool and die failure can shut down a manufacturing line and disrupt production scheduling. This leads to higher costs.

Tools and dies must also be produced with the proper size and shape after hardening so that excessive finishing work is not required. Heat-treatment distortion must be controlled, and surface chemistries must not be altered. Because of the careful balance that must be maintained in heat treatment, control of the heat-treatment process is one of the most critical steps in producing successful tools and dies. Heat treating of tool and die steels is discussed in the Section "Heat Treating of Tool Steels" in Volume 4 of the 9th Edition of *Metals Handbook*. In addition to controlling the heat-treatment process, tool and die design and steel selection are integral factors in achieving tool and die integrity.

Analytical Approach

Analysis of tool and die failures is substantially aided by knowledge of the manufacturing and service history of the failed part. In many instances, however, such information is sketchy, and the analyst must rely on experience and engineering judgment. The examples discussed in this article are typical of tool and die failures and illustrate many of the features specific to such failures. Additional information can be found in the Selected References that follow this article.

A basic approach can be followed for most tool and die failures that will maximize the likelihood of obtaining reasons for the failure. At times, however, no reason for failure can be discovered. In such cases, the failures are normally attributed to unknown service conditions.

Before beginning the investigation, a complete history of the manufacture and service life should be compiled. This is often difficult to accomplish. Next, the part should be carefully examined, measured, and photographed to document the extent and location of the damage. Relevant design features, as well as some machining problems, are generally apparent after this study. The origin of the failure, when due to fracture, will also usually be determined. Only after careful visual examination has been completed should any destructive work be considered. In certain cases, various nondestructive examination techniques—for example, ultrasonics, x-ray, or magnetic-particle inspection—should be implemented before cutting to obtain a more complete picture of the damage, either internal or external.

When this work is completed, several other phases of the study can begin. First, the composition of the component should be verified by a reliable method. Tool and die failures occasionally result from accidental use of the wrong grade of steel. While this work is underway, the analyst continues macroscopic examination of the fracture features by opening tight cracks (when present). Because quench cracking is a very common cause of failures, the fracture surfaces should always be checked for temper color. Scale on a crack wall would indicate that it was exposed to temperature higher than those used in tempering.

Visual inspection of the damage done to a tool and die is usually adequate to classify the type of tool and die failure, as illustrated by the examples in this article. High-magnification fractographic examination is required in only a small percentage of the cases. A simple stereomicroscope generally suffices.

Microstructural examination at and away from the damage and at the origin is imperative. This generally requires good edge-retention preparation, which is relatively easy for tool steels. A large percentage of tool and die failures are due to heat-treatment problems, as illustrated by the examples in this article; the value of proper metallographic procedures cannot be stressed enough.

Other techniques are also very important. Hardness testing is used to confirm the quality of the heat treatment and often reveals problems. Macrostructural examination, either by cold etching (more common) or hot etching, is very useful for detecting gross problems. Prior-austenite grain size is frequently evaluated by the Shepherd fracture grain size technique, a simple but accurate approach. X-ray diffraction can be used to determine the amount of retained austenite present.

In some cases, it is necessary to perform simulations or to conduct experimental heat treatments to determine if a part was tempered. Chemical analysis of millings or turnings from the surface will define variations in surface-carbon content. Electron metallographic devices using either wavelength- or energy-dispersive x-ray analysis can be used to identify inclusions or segregates. It is also helpful in some studies to compare the characteristics of good parts to those of failed parts.

When these data are compiled and analyzed, the analyst is ready to prepare a report that details the cause of failure with the supporting facts. In many such cases, it is also necessary to make recommendations regarding corrective action for future parts or for existing parts.

Causes of Tool and Die Failures

A number of factors are responsible for tool and die failures including:

- *Mechanical design*: Must be compatible with the steel grade selected, the procedures required to manufacture the tool or die, and the use of the tool or die
- *Grade selection*: Must be compatible with the design chosen, the manufacturing processes used to produce the tool or die, and the intended service conditions and desired life
- *Steel quality*: Must be macrostructurally sound, free of harmful inclusions to the degree required for the application, and free of harmful surface defects
- *Machining processes*: Must not alter the surface microstructure or surface finish and must not produce excessive residual stresses that will promote heat-treatment problems or service failures
- *Heat-treatment operation*: Must produce the desired microstructure, hardness, toughness, and hardenability at the surface and the interior
- *Grinding and finishing operations*: Must not impair the surface integrity of the component
- *Tool and die setup*: Alignment must be precise to avoid irregular, excessive stresses that will accelerate wear or cause cracking
- *Tool and die operation*: Overloading must be avoided to ensure achievement of the desired component life

The above classification of factors, while helpful in categorizing problems, will not necessarily deduce the cause of a particular failure. For example, a failure due to one of the above factors may have been caused by other problems earlier in the processing sequence. To illustrate, the majority of cracking problems attributed to abusive grinding practices are caused by failure to temper the part or are due to overaustenitization. Both of these problems will make a tool steel virtually impossible to grind without producing surface burning and cracking regardless of the care taken during grinding. Thus, the above factors are interdependent, much like the links of a chain. If one step is poorly executed, the tool or die will exhibit limited service life regardless of how carefully all of the other steps are conducted. This sensitivity to processing stems from the use of tools and dies at these very high hardness levels, at which minor deficiencies in processing exert major influences on performance.

Mechanical Testing of Tool Steels. For the majority of tool steels, tensile tests at room temperature are not conducted because of the technical difficulties of obtaining valid results at these very high hardnesses. Also, because most tools and dies are subjected to high compressive stresses, it would be more useful

Fig. 1 AISI W1 (0.85% C) tool steel concrete roughers that failed after short service (2 min for S, 7 min for S11)

Failures of these and other concrete roughers all occurred at the change in section (arrows indicate cracks).

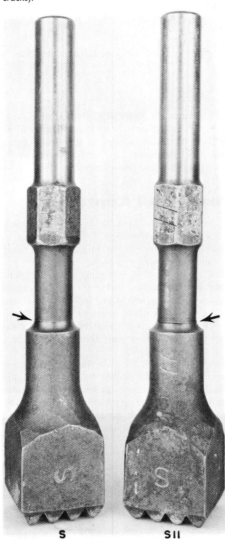

S   SII

to obtain compressive yield strength data to aid the designer. However, such tests are quite difficult to perform, and few useful data are available. Consequently, the simple indentation hardness tests, generally the Rockwell C test or the Vickers test, are used to define strength differences. Such data are simple to produce and are extremely useful. The lower carbon, high-strength hot-work tool steels can be tested in tension, and such data are available for both room temperature and elevated temperatures. Tension, compression, and hardness testing are discussed in separate articles in Volume 8 of the 9th Edition of *Metals Handbook*.

As a class, tool steels are not noted for high toughness. Impact-type tests are commonly

Fig. 2(a) Front view of an AISI O1 tool steel die that cracked during oil quenching

The die face contains holes that are too close to the edge for safe quenching. See also Fig. 2(b). 0.6×

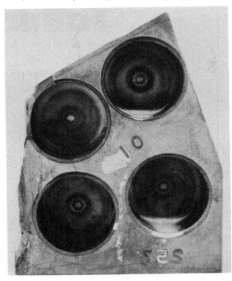

Fig. 2(b) Side view of broken die halves showing the mating fracture surfaces and temper color (arrow) on the crack surfaces

A front view of this component is shown in Fig. 2(a). 0.5×

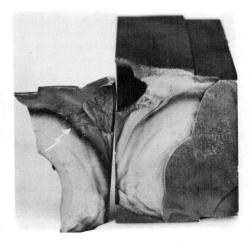

used, but most test specimens either are not notched or have a C-type notch rather than the more common V-notch. Use of the latter type of specimen is reserved for the higher-toughness grades or is used with low-blow fracturing devices. Fracture-toughness data are available for many tool steel grades, particularly the higher-toughness grades for which such data are useful. For most tool steels, plane-strain fracture toughness predicts very small critical flaw sizes, which emphasizes the need for minimizing stress concentrators and maintaining smoothly machined surfaces.

Fig. 3 AISI O1 tool steel die that cracked during oil quenching

Note the cracks emanating from the sharp corners. The four holes, which are close to the edge, also contributed to cracking. Temper color was observed on the crack walls.

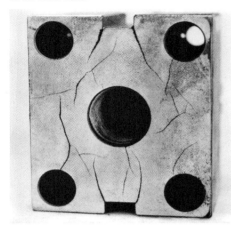

Types of Failures. Tool and die failures due to breakage, which are generally catastrophic, are the most spectacular and draw the most attention from the failure analyst. Such failures are usually the easiest to diagnose, but the analyst should be cautioned not to halt the investigation when one obvious problem is observed. It is not uncommon to observe several factors that contribute to the failure in varying degrees. The goal of any failure analysis should be to provide a total picture of all the problems present so that complete corrective action, rather than partial corrective action, is taken on future parts.

Aside from cracking, a wide variety of problems can be encountered that cause limited tool or die life. These problems include, but are not limited to, distortion (during heat treatment, machining, or service), excessive wear, galling, pick-up, erosion, pitting, cosmetic problems, and corrosion problems.

Influence of Design

Any examination of tool or die failures must begin with a careful reexamination of the design to determine if shortcomings are present and if improvements can be made. Tools and dies that perform satisfactorily over the desired service life may not produce the same performance if a new manufacturing process is adopted, the grade is changed, or the service conditions are altered. Consequently, in the study of the existing design, it is imperative to obtain complete details on the history of the component relative to its past performance, manufacturing line-up, service conditions, and so on. All too often, however, such data are sketchy.

The importance of a good design cannot be overemphasized. Poor design can cause or promote heat-treatment failures before any service life is obtained or may reduce service life, sometimes dramatically. In designing a tool or die, a host of factors must be considered. In practice, it is difficult to separate the design stage from grade selection because the two steps are interdependent. The choice of a certain grade of steel, such as one that must be brine or water quenched, will have a very substantial bearing on all aspects of design and manufacture. In general, any steel grade that requires liquid quenching demands very conservative, careful design. Air-hardening grades tolerate some design and manufacturing aspects that could never be tolerated with a liquid-quenching grade. The design must also be compatible with the equipment available—for example, heat-treatment furnaces and surface-finishing devices.

Designing tools and dies is more difficult than designing components made from struc-

Fig. 4 Fixture made from AISI O1 tool steel that cracked during oil quenching

This design is poor for liquid quenching. A nick in the fillet region helped to initiate cracking. 0.75×

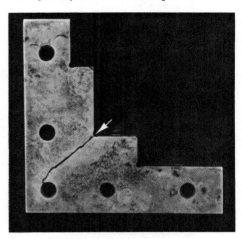

tural steels because of the difficulty in predicting the service stresses. Despite advances made in design procedures, much of the design work is still empirically based. Such experience is primarily based on past failures; therefore, it is important that the findings of the failure analyst be incorporated into future work. Despite the shortcomings of the empirical approach, there is a vast body of common sense engineering knowledge available for guidance (see the Selected References).

Effect of Sharp Corners and Section-Mass Changes. Analysis of many tool and die failures shows that two relatively simple design problems cause the most failures. These design shortcomings are the presence of sharp corners and the presence of extreme changes in section mass. A sharp corner concentrates and

Fig. 5 Threaded part made from AISI W2 tool steel that cracked during quenching at an undercut at the base of the threads

(a) The two pieces that separated during fracture. (b) Cold-etched (10% aqueous nitric acid) disk cut through the threaded portion showing the hardened surface zone, which is also visible on the fracture faces shown in (a)

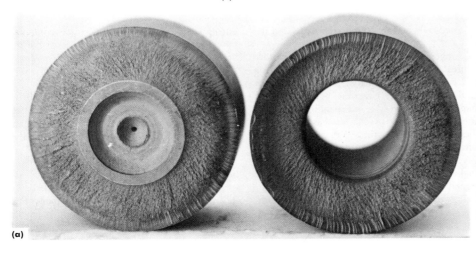

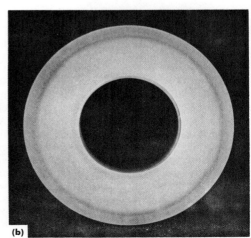

Fig. 6 Punch made of AISI S7 tool steel that cracked during quenching

Temper color was observed on the crack walls. Cracking was promoted by and located by the very coarse machining marks. Magnetic particles have been used to emphasize the cracks. 0.5×

Fig. 7 The surface of an AISI A4 primer cup plate showing spalling at one of the 3.2-mm (⅛-in.) diam holes made by EDM

The surface was etched with 10% aqueous nitric acid to bring out the influence of the EDM operation at the spall. 2.5×

Fig. 8 Microstructures associated with the spalled hole (Fig. 7) caused by improper EDM technique

Etched wtih 3% nital

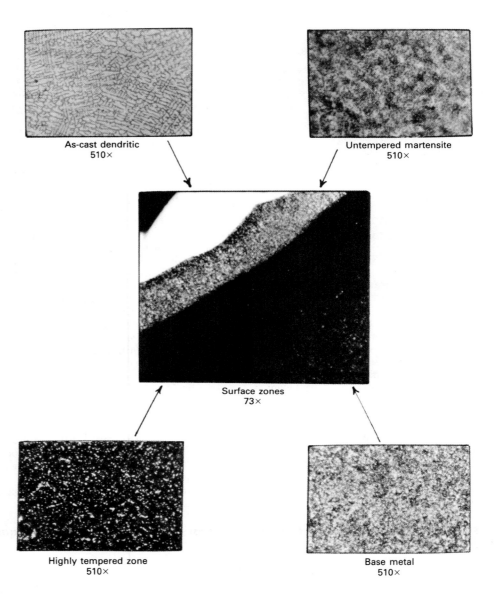

As-cast dendritic
510×

Untempered martensite
510×

Surface zones
73×

Highly tempered zone
510×

Base metal
510×

magnifies applied stresses, stresses that arise in tool and die manufacturing (such as during quenching), or stresses that occur during service. In addition to promoting cracking during liquid quenching, sharp corners promote buildup of residual stresses that may not be fully relieved by tempering and can therefore reduce service life. The largest possible fillet should be used at all sharp corners. Air-quenching grades are more tolerant of sharp corners than liquid-quenching grades and are

preferred when only minimal fillets can be used.

Changes in section size can locate premature failures. Figure 1 shows two AISI W1 carbon steel concrete roughers that failed after a few minutes of service. Cracking occurred at the change in section due to bending stresses. Although the section change has a smooth, filleted surface, it is still a very effective stress concentrator. Subsequent design changes involved a tapered change in section at the cracked location and later at the start of the wrench flats above the cracked region.

Holes placed too close to the edges of components are a common source of failure during heat treatment or in service. Figures 2(a) and (b) show an AISI O1 tool steel die that cracked

during oil quenching. The die face contained numerous fine cracks. The left side of the die broke off during quenching. Figure 2(b) shows both sides of the fracture. Temper color (arrow), typical of the 205 °C (400 °F) temper used, is apparent. This indicates the depth of the crack produced during quenching that was open during tempering. Coarse machining marks and deep stamp marks were also present.

Sharp, unfilleted corners may also promote quench cracking. Figure 3 shows a 76- × 87- × 64-mm (3- × 3⁷⁄₁₆- × 2½-in.) AISI O1 tool steel die that cracked during oil quenching. The crack pattern (emphasized using magnetic particles) that emanates from the sharp corners is visible. A few cracks are also associated with

Fig. 9 Plastic mold die made from AISI S7 tool steel that was found to be cracked before use

A crack followed the lower recessed contour of the large gear teeth and had an average depth of 1.6 mm (1/16 in.). Smaller cracks were also observed on the flat surfaces. (a) Actual size. (b) Etching the surface with 10% aqueous nitric acid revealed a white-etching appearance at the teeth. 2×. (c) Micrograph showing a surface layer of as-quenched martensite (from the EDM operation) and an overaustenitized matrix structure (unstable retained austenite and coarse plate martensite). Etched with 3% nital. 420×

(a)

(b)

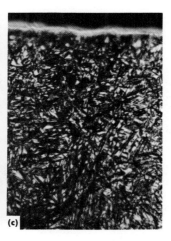
(c)

Fig. 10 Grinding cracks caused by failure to temper a part

(a) Two dies made from AISI D2 tool steel that cracked after finish grinding (cracks accentuated with magnetic particles). (b) Macroetching (10% aqueous nitric acid) of the end faces revealed grinding scorch. These dies were not tempered after hardening.

(a)

(b)

the holes that are rather close to the edges. Temper color was observed on the crack surfaces, indicating that the cracks were present before tempering.

Figure 4 shows another example of a quench crack initiated by a sharp corner. This fixture was also made of AISI O1 tool steel that was oil quenched. In this case, the corner was filleted, but there was a nick in the corner where cracking began. The shape of this fixture is also poor for a steel that must be oil quenched. The thinner outer regions cool more rapidly, forming martensite first, while the more massive central region cools at a slower rate. An air-hardenable steel would be a better choice for this part.

Another example of a poor design for liquid quenching is shown in Fig. 5. This 76-mm (3-in.) diam × 76-mm (3-in.) long threaded part made of AISI W2 carbon tool steel cracked in half at an undercut at the base of the threads. Figure 5 shows the two broken halves, along with a cold-etched disk taken from the hollow portion of the part. The hardened outer case can be seen in the fracture detail and in the cold-etched disk. Similar parts, without the undercut, were successfully hardened.

Influence of Steel Grade

Selection of the optimum grade for a given application is generally a compromise between toughness and wear resistance, although other factors may be more important in certain situations. Because most tools and dies operate under highly stressed conditions, toughness must be adequate to prevent brittle fracture. It is usually better for a tool or die to wear out than to break in service prematurely. Thus, in a new application, it is best to select a grade that will definitely have adequate toughness. When some experience is gained that shows freedom from breakage but perhaps excessive wear, a

Fig. 11 Two views of an AISI S1 tool steel cutter die that cracked and spalled after regrinding

(a) The as-received condition. (b) The cracks have been accentuated by use of magnetic particles. Also note the grinding scorch pattern (the dark parallel lines perpendicular to the cracks).

Fig. 12(a) Two AISI A6 tool steel parts that shattered during finish (abusive) grinding

See also Fig 12(b)

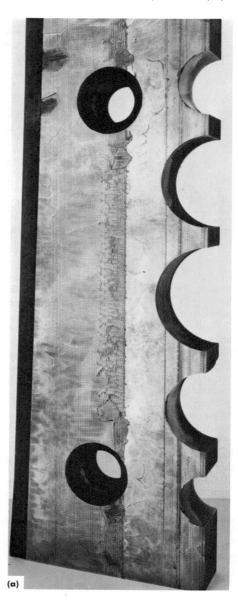

(a)

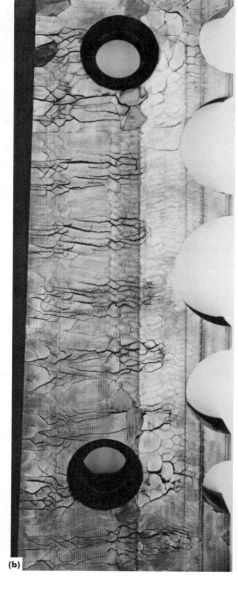

(b)

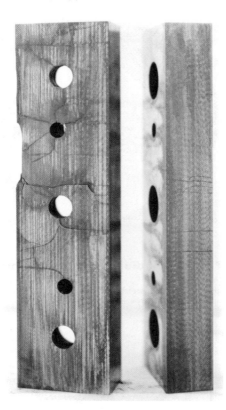

Fig. 12(b) Photomicrograph of the ground parts shown in Fig. 12(a)

A reaustenitized region (white) and a back-tempered zone (dark) at the ground surface are shown. Etched with 3% nital. 70×

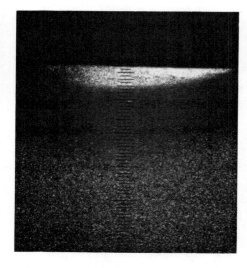

different grade can be chosen that would provide better wear resistance but somewhat less toughness. As experience is gained, the best grade choice can be made by balancing the required toughness and wear resistance.

Influence of Machining

Machining problems are a common cause of tool and die failures. It is generally best to avoid machining directly to the finish size unless a prehardened die steel is used. It is difficult to obtain perfect control of surface chemistry and size during heat treatment. Thus, some final grinding is usually needed

after heat treatment. The presence of decarburization is generally quite detrimental. Also, because stresses are high in heat treatment and in service, rough machining marks must be avoided. Identification stamp marks are another common source of failures in heat treatment and in service; they should be avoided.

Quench-Crack Failures. Two quench-crack failures promoted by machining problems have been illustrated. Quench cracking of the fixture shown in Fig. 4 was located by a nick in the fillet, while failure of the threaded part shown in Fig. 5 was due to an undercut at the base of the threaded region.

Rough machining marks are another common cause of quench cracking. Figure 6 shows a punch made of AISI S7 tool steel that cracked during quenching. Because of the section size,

Fig. 13(a) AISI S5 tool steel hammer head that cracked during heat treatment

The fracture was caused by quench cracking that was promoted by the decarburized surface (Fig. 13b) and deep stamp mark (arrows). Actual size

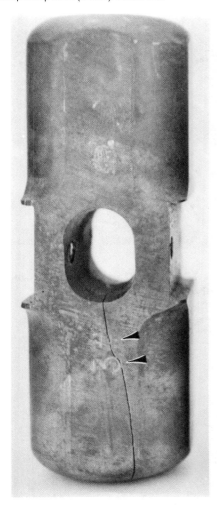

Fig. 14 A quench crack promoted by the presence of a deep, sharp stamp mark in a die made of AISI S7 tool steel

This die had not been tempered, or was ineffectively tempered, after hardening. 2×

Fig. 15 AISI W2 carbon tool steel (1.05% C) component that cracked during quenching due to the presence of soft spots (arrows) on the surface

These soft spots were revealed by cold etching the surface with 10% aqueous nitric acid. 0.4×

the punch was oil quenched to 540 °C (1000 °F), then air cooled. The crack pattern has been emphasized with magnetic particles. Temper color was observed on the crack walls.

Failures Due to Electrical Discharge Machining (EDM). Die cavities are often machined by EDM. The technique has many advantages, but failures have been frequently observed due to failure to remove the as-cast surface region and associated as-quenched martensitic layer. Cavity surfaces must be stoned or ground, then tempered to prevent such failures.

A classic example of a failure due to improper EDM technique is shown in Fig. 7 and 8. Figure 7 shows four 3.2-mm (⅛-in.) diam EDM holes in an AISI A4 tool steel primer cup plate. The holes were finished by jig-bore

grinding, during which spalling was observed at many of the holes (see upper-right hole). The surface was swabbed with 10% aqueous nitric acid (HNO_3) to reveal regions affected by EDM. Figure 8 shows the microstructure of these regions. An as-cast region was present at the extreme edge (about 35.5 HRC). Beneath this layer was a region of as-quenched martensite (about 63.5 HRC). Next was a back-tempered region (about 56 HRC) and then the

Fig. 13(b) Macroetched disk cut from the head of the sledge hammer shown in Fig. 13(a)

The heavily decarburized surface is revealed by macroetching. Actual size

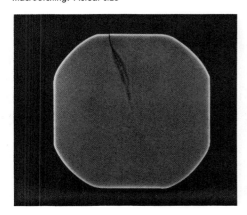

Fig. 16(a) AISI O1 tool steel ring forging that cracked during quenching

The forging was overaustenitized (unstable retained austenite was present) and was decarburized to a depth of about 0.5 mm (0.020 in.). Temper color was present on the crack walls. See also Fig. 16(b).

Fig. 16(b) Interior microstructure of the cracked ring forging shown in Fig. 16(a)

Unstable retained austenite (white) and coarse plate martensite (dark) can be seen. The amount of residual carbide was negligible compared to what should have been present. Etched with 3% nital. 700×

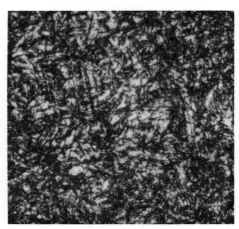

Fig. 17 AISI M2 roughing tool that cracked just after heat treatment

(a) Cracks accentuated with magnetic particles. (b) Microstructural examination revealed a badly overaustenitized condition with a heavy grain-boundary carbide film, coarse plate martensite, and unstable retained austenite. Etched with 3% nital. 490×

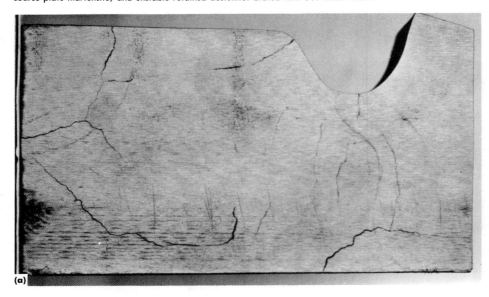

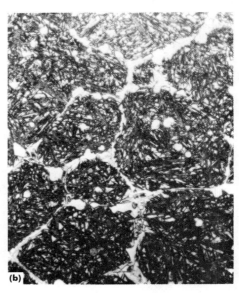

Fig. 18 AISI O6 graphitic tool steel punch machined from centerless-ground bar stock that cracked after limited service

(a) Cracks (arrows) accentuated with magnetic particles. (b) Microstructural examination revealed an overaustenitized structure consisting of appreciable retained austenite and coarse plate martensite. Etched with 3% nital. 700×

Fig. 19 AISI P20 mold made from prehardened stock that was carburized and rehardened

After heat treatment, it was found to be cracked (arrow). See also Fig. 20.

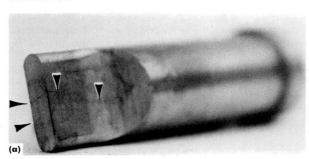

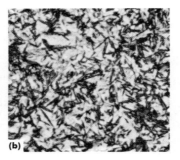

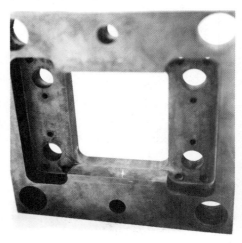

base, unaffected interior (59 to 61 HRC). The brittle nature of the outer layers and the associated residual-stress pattern caused the spalling.

In many EDM-related failures, the as-cast layer is not observed, because of the technique used or because of subsequent machining. In these failures, however, an outer layer of brittle as-quenched (white-etching) martensite is present. Such a failure is shown in Fig. 9. This failure occurred in a plastic-mold die made from AISI S7 tool steel. The crack followed the lower, recessed contour of the larger-diameter gear teeth and extended to a depth of about 1.6 mm (¹⁄₁₆ in.). Etching of the surface revealed a light-etching rim around the teeth. Microstructural examination revealed an as-quenched martensite surface layer (thin, white layer), while the internal structure was grossly overaustenitized (note the retained austenite, white, and coarse plate martensite). Both factors lead to cracking. If the EDM

surface layer was not present, poor service life would have resulted anyway due to the poor microstructural condition.

Failures Due to Finish Grinding. Cracking due to the stresses and microstructural alterations caused by grinding is a relatively common problem. In many cases, the grinding technique is not at fault, because the microstructure of the part rendered it sensitive to grinding damage due to failure to temper the part or because the part was overaustenitized and contained substantial unstable retained austenite.

Figure 10 shows an example of grinding cracks due to failure to temper the part. Two AISI D2 tool steel dies, which measured 57 × 60 × 29 mm or 51 mm thick (2¼ × 2⅜ × 1⅛ in. or 2 in. thick), were observed to be cracked after finish grinding. The cracks are emphasized with magnetic particles. Macroetching of the surfaces revealed the classic scorch pattern indicative of abusive

grinding. The failure, however, was not due to poor grinding practice, but was caused by failure to temper the dies. The interior hardness was 63 to 64 HRC, typical for as-quenched D2. The scorched surface was back tempered to 55 to 58 HRC. It is difficult to grind as-quenched high-hardness tool steels without damaging the surface.

Grinding damage can also occur in parts that have been properly heat treated, but such cases are less common. Figure 11 shows two views of a 32- × 152- × 381-mm (1¼- × 6- × 15-in.) AISI S1 tool steel cutter die that cracked extensively during regrinding after an initial service period. The as-received surface is

Fig. 20 Metallographic section from the AISI P20 mold shown in Fig. 19

(a) Top part of a macroetched (10% aqueous nitric acid) disk cut from the mold revealing a heavily carburized case. Actual size. (b) Micrograph showing gross carbide buildup at the surface with an underlying region having a continuous grain-boundary carbide network. Etched with 3% nital. 100×

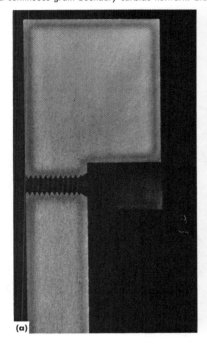
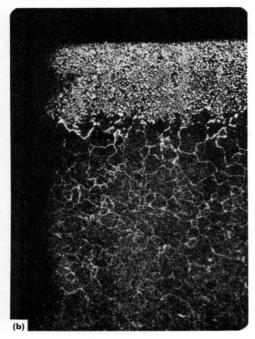

shown in Fig. 11(a); the crack pattern is vividly revealed by magnetic particles (Fig. 11b). The grinding scorch pattern is also visible. This die had been carburized and the surface hardness was 62 to 64 HRC.

Figure 12(a) shows two AISI A6 tool steel parts that shattered during finish grinding. These parts were properly hardened (57 HRC), and cracking was due to the grinding practice. The parts have been etched to show the scorch

pattern (Fig. 12b). The surface hardness was erratic, varying between 48 and 56 HRC. The photomicrograph shows a typical surface region affected by the grinding heat. Spots, such as the white-etching martensitic region shown, were about 0.08 mm (0.003 in.) deep. The back-tempered region beneath the as-quenched martensite extended to a depth of about 0.38 mm (0.015 in.). Cracking is often present in these rehardened zones.

Influence of Heat Treatment

Improper heat-treatment procedures are the single largest source of failures during heat treatment, in subsequent processing steps, or in service. Each tool steel grade has a recommended austenitizing temperature range (generally rather narrow), a recommended quench medium, and recommended tempering temperatures and times for optimum properties. Some grades are more forgiving than others regarding these parameters.

Handling of Samples. One of the most common sources of problems arises in the handling of samples between the quench and the temper. As soon as the part reaches a temperature of about 65 °C (150 °F), it should be quickly transferred to the tempering furnace. The heat treater will sometimes check the hardness of the part after quenching and avoid the tempering treatment if the as-quenched hardness equals the desired hardness. This is a very poor practice for two reasons. First, tool steels must always be tempered to reduce the quenching stresses to an acceptable level and to improve the toughness of the steel. Untempered tools and dies nearly always fail prematurely in service. Second, the as-quenched hardness may be low due to testing on a decarburized surface. If a spot is ground on an as-quenched surface, cracking may result. Double and triple tempering is required for the more highly alloyed grades to stabilize the microstructure.

Quench Cracking. Numerous heat-treatment problems can promote quench cracking. Figure 13(a) shows a 41-mm (1⅝-in.) square, 1.4-kg (3-lb) AISI S5 tool steel sledge-hammer head that cracked during quenching. A disk cut from the head was macroetched, revealing a heavily decarburized surface (Fig. 13b). Such a condition promotes quench cracking, particu-

Fig. 21 Failed AISI S7 jewelry striking die

(a) Crack (arrows) that formed shortly after the die was placed in service. (b) and (c) Microstructural examination revealed that the surface was slightly carburized and that the die had been overaustenitized. Note the coarse plate martensite and unstable retained austenite in the carburized surface region and the coarse martensitic matrix structure beneath this layer. (b) 75×. (c) 530×. Both etched with 3% nital

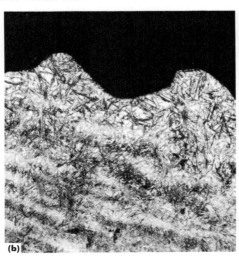
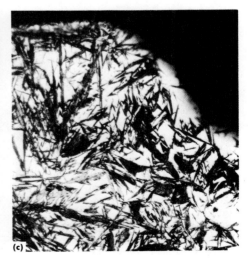

Fig. 22 AISI S7 punch that had a low surface hardness after heat treatment and was given a second carburizing treatment, then rehardened

Cracking was observed after this retreatment (the cracks have been accentuated with magnetic particles). Coarse circumferential machining marks were present on the lower portion of the punch. See also Fig. 23 and 24.

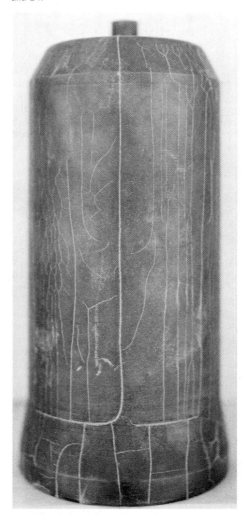

Fig. 23 Crack pattern on the bottom of the punch shown in Fig. 22

Many of the cracks are located by the deep stamp marks (the cracks have been accentuated with magnetic particles). Actual size

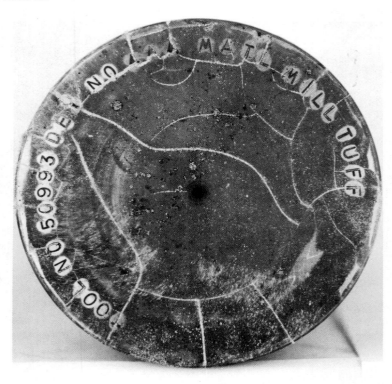

larly in liquid-quenching grades such as S5 (oil quenched), due to differential surface stresses. A deep stamp mark also helped promote cracking.

Stamp marks, such as that shown in Fig. 14, commonly promote quench cracks. This was present on an air-quenched die made from AISI S7 tool steel. In this case, the die was not tempered, another prime cause of quench cracking. The hardness of the die was 61.5 to 62 HRC. The surface was slightly decarburized; the hardness was 59 to 60 HRC at the surface.

Quench cracking of water- or brine-quenched tool steels can be promoted by soft spots on the surface of the part. These soft regions may be due to the fixture used to hold the part during quenching, or from tongs, or may be caused by vapor pockets during quenching due to inadequate agitation or contamination of the quench. Figure 15 shows a component made from AISI W2 tool steel that quench cracked due to soft spots (unhardened regions). The surface of the part was macroetched, revealing these zones. The hardness numbers reveal the difference in hardness between the case and the soft spots.

Failures Due to Overaustenitizing. Another very common heat-treatment problem is the use of an excessively high austenitizing temperature. This may be due to improper furnace-temperature control or the combining of several parts made of different grades in one furnace batch using an austenitizing temperature chosen as a compromise between the recommended temperatures for the grades. Excessively high austenitizing temperatures promote grain growth and excessive retained austenite. Most tool steel grades have high carbon contents and rely on the undissolved portion of the carbides to control grain growth. An excessive austenitizing temperature puts more carbon in solution, thus permitting grain growth as well as excess retained austenite due to suppression of the martensite start, M_s, and finish, M_f, temperatures.

Tool steel parts that have been overaustenitized, producing coarse plate martensite and unstable retained austenite, often fail by quench cracking. Figure 16(a) shows a 44-cm (17½-in.) OD × 33-cm (13-in.) ID × 5-cm (2-in.) thick ring forging made of AISI O1 tool steel that cracked during quenching. Temper color was present on the crack surface. The interior microstructure (Fig. 16b) revealed an overaustenitized condition, coarse plate martensite, and retained austenite (white). The hardness was 61 to 62 HRC. A section was cooled in liquid nitrogen, which transformed much of the retained austenite to martensite and increased the hardness to 64 to 65 HRC. The surface was decarburized with a hardness of 55 to 57 HRC, which did not change after refrigeration.

Figure 17(a) shows an AISI M2 roughing tool that cracked during hardening (the cracks were accentuated with magnetic particles). Microstructural examination revealed an overaustenitized condition with a heavy grain-boundary film, coarse plate martensite, and unstable retained austenite (Fig. 17b). The Snyder-Graff intercept grain size was 4.5 (ASTM 7), which is quite coarse for this grade.

Improper control of the austenitizing temperature is a common problem. Figure 18(a) shows an AISI O6 graphitic tool steel punch that cracked after limited service. Examination of the microstructure revealed an overaustenitized condition (Fig. 18b). All of the carbide has been dissolved; only martensite and retained

Fig. 24 Microstructure of the heavily carburized cracked punch shown in Fig. 22 and 23

(a) Massive carbide enrichment at the surface. (b) Excess carbides at the base of the crack, about 0.7 mm (0.0275 in.) deep. (c) Structure at about 1.08-mm (0.0425-in.) depth. (d) Coarse overaustenitized structure about 19 mm (0.75 in.) below the surface. All etched with 3% nital. All 700×

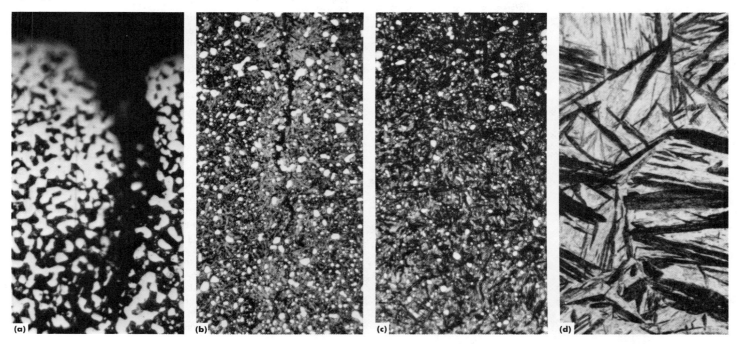

austenite are observed. The hardness of the punch was 59 to 60 HRC at the surface and 60 to 62 HRC in the interior. Cooling in liquid nitrogen raised the hardnesses to 63 to 64 and 63 to 65 HRC, respectively.

Excessive Carburization. While many tool steels have high carbon contents, a few low-carbon P-type mold steels are carburized before heat treatment. Control of the carburizing cycle, as well as subsequent heat treatment, is required to obtain good results. Figure 19 shows a mold made of AISI P20 prehardened tool steel measuring 20.3 × 20.3 × 3.5 cm (8 × 8 × 1⅜ in.) that cracked sometime after a carburizing/heat-treatment procedure. Figure 20(a) shows a disk cut from the mold that was cold etched to reveal a dark-etching surface case about 3.2 mm (⅛-in.) thick. Microstructural examination revealed that the cracks (arrow, Fig. 19) on the outside diameter and the inside diameter of the mold were 1.8 and 2.9 mm (0.070 and 0.113 in.) deep. The external surfaces were carburized to a depth of at least 2.5 mm (0.10 in.). Figure 20(b) reveals the heavy concentration of carbide at the surface and the extensive grain-boundary carbide networks. Analysis of millings taken from the surface revealed that the carbon content of the outer 0.25 mm (0.010 in.) was 1.96 wt%. This is far too high, thus rendering the surface layer extremely brittle. Coarse machining marks were also present on the mold surface in the cracked region.

Furnace atmospheres are sometimes not properly controlled during heat treatment, and either decarburization or carburization results. The latter is less common, but when it occurs, failure may result. Some tool steel grades can tolerate minor, unintentional carburization better than others. Figure 21(a) shows a 64- × 56- × 81-mm (2½- × 2³⁄₁₆- × 3³⁄₁₆-in.) jewelry striking die made from AISI S7 tool steel that cracked shortly after being placed in service. Cracking occurred in the die cavity and extended along a recessed groove and down the side. The hardness of the die surface varied between 50 and 55 HRC, while the interior was 56 to 57 HRC. Cooling of a section in liquid nitrogen increased the surface hardness to 62 to 65 HRC, indicating the presence of substantial retained austenite at the surface. The interior hardness did not change. Examination of the microstructure (Fig. 21b and c) revealed extensive retained austenite (white) at the surface, while the interior microstructure was coarse, indicating that the austenitizing temperature was excessive. The excess retained austenite at the surface indicated that carburization had occurred during hardening. Chemical analysis of milling from the outer 0.13 mm (0.005 in.) revealed a carbon content of 0.79 wt% as opposed to an interior carbon content of 0.53 wt%. The carburized case was about 0.51 mm (0.020 in.) deep.

Another example of the detrimental influence of excessive carburization is shown in Fig. 22 to 24. This was a punch made from AISI S7 tool steel that was used in the first stage of cold forming of 105-mm shells that cracked after a second heat treatment. The desired hardness was not achieved with the initial treatment. The punch was supposed to be lightly carburized to produce a surface hardness of 62 HRC (higher than the usual hardness of 58 HRC for noncarburized S7).

Figure 22 shows the extensive crack pattern (accentuated with magnetic particles) on the outer surface of the punch. Coarse circumferential machining marks were also present. Figure 23 shows the bottom of the punch, which is also extensively cracked. The deep stamp marks promoted cracking. The hardness of the punch varied from the surface inward. At the extreme surface it was 63 to 64 HRC. The hardness decreased gradually to 55 HRC 25 mm (1 in.) below the surface and was 53 HRC at the center. Cooling in liquid nitrogen raised the subsurface hardness substantially, indicating the presence of unstable retained austenite. The fracture grain size was very coarse—equivalent to ASTM 4.5—which indicates overaustenitization.

Microstructural examination revealed a deeply carburized condition with a very high surface-carbon content and a coarse martensitic structure below the region of massive carbide content. Figure 24 shows the microstructure at the surface (Fig. 24a) and at three different levels below the surface (Fig. 24 b to d). The structure shown in Fig. 24(d) is very coarse and is undesirable. Chemical analysis of incremental turnings revealed that the carbon content between the surface and a depth of 0.13 mm (0.005 in.) was 2.94 wt%. The total depth of decarburization exceeded 2.5 mm (0.10 in.).

Fig. 25 Failed chromium-plated blanking die made from AISI A2 tool steel

(a) Cracking (arrows) that occurred shortly after the die was placed in service. (b) Cold-etched (10% aqueous nitric acid) disk cut from the blanking die (outlined area) revealing a light-etching layer. Actual size. (c) Micrograph showing the decarburized layer that was unable to support the more brittle, hard chromium plating. Etched with 3% nital. 60×

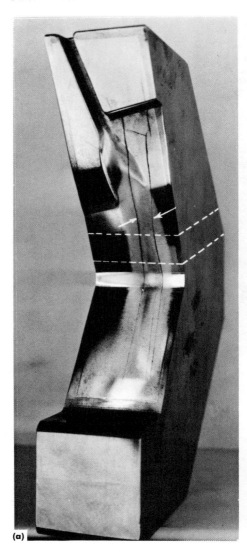

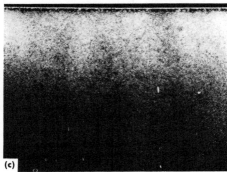

Thus, AISI S7 was not a good choice for this application. It is difficult to carburize this grade lightly unless a liquid-carburizing treatment is used.

Decarburization may also occur at the surface of components during heat treatment, and it will affect service life. Figure 25(a) shows a chromium-plated blanking die made from AISI A2 tool steel that cracked after limited service. Cold etching of a disk cut from the blanking die revealed a light-etching layer that is particularly prominent at the working face and along the adjacent sides (Fig. 25b). Microscopic examination (Fig. 25c) revealed that the surface at the working face was decarburized to a depth of about 0.05 mm (0.002 in.). The soft zone beneath the hard chromium

plating permitted the plating to flex under the influence of the blanking stresses, thus cracking the plating and surface region.

As another example of the influence of decarburization on service life, Fig. 26 shows a fractured pin (16 mm or ⅝ in., in diameter) and gripping cam (19 mm, or ¾ in., thick), both made of AISI S5 tool steel, that were used in plate-lifting clamps that broke after limited service. The rough chamfered edges of the hole of cam No. 2 are visible. The pin surface was heavily decarburized, as shown by the cold-etched longitudinal section through the pin (Fig. 26b). The flat surfaces of the cam were also heavily decarburized. Measurement of the parts indicated that the mill bark present on the surfaces of the stock used to make these parts

was not removed. The standard machining allowances for hot-rolled bars were not taken. Decarb-free stock would have been a better choice for the material for these parts.

Failures of Nitrided Parts. Although not used as commonly as carburizing, nitriding is occasionally used to improve the surface properties of tool steel parts. As with any process, the conditions must be properly controlled to produce good results. Failure of nitrided components can arise if the nitriding operation produces a heavy, white-etching nitride surface layer. Figure 27(a) shows an AISI 4150 alloy steel chuck jaw that broke due to the presence of such a detrimental surface layer. A section from a broken tooth of the chuck jaw is shown in Fig. 27(b). The nitride layer is present as a grain-boundary film beneath the surface layer. This is a very brittle condition.

Failures Due to Improper Quenching. Carbon tool steels are widely used in many tool and die applications. For best results, the hardened case must be properly developed. Figure 28 shows the striking face of a 38-mm (1.5-in.) diam × 64-mm (2.5-in.) long header die made from AISI W1 (1.0% C) tool steel that failed early in service due to chipping at the striking surface. Macroetching of a disk cut longitudinally through the die at the broken region revealed that the case around the striking face region was quite shallow and unable to support the working stresses. The flush-quenching procedure used to harden the striking face and bore did not produce adequate case depth at the striking face. The sharp corners of the striking-face cavity act as a stress concentrator; even a minor fillet at the corner would be helpful.

Figure 29 shows another example of a failure of a carbon tool steel (AISI W1) component due to improper quenching. This part, which measured 38 mm (1.5 in.) at the base, 21 mm (¹³⁄₁₆ in.) at the top, 22 mm (⅞ in.) thick, and 92 mm (3⅝ in.) long, was a wire-forming die that broke prematurely in service. Macroetching of a disk cut behind the fracture (Fig. 29b) revealed that the bore of the die was not hardened. Only a region along the corner of the die (the dull gray region on the macroetch disk) was hardened, and this only superficially (35 to 38 HRC in this region; the balance of the disk was at 29 to 30 HRC). A hardened chill zone should have been present at the bore surface (working surface of the die) if it had been properly quenched.

Failures Due to Improper Furnace-Atmosphere Control. Other surface-related problems may occasionally be encountered in tool and die failures. Figure 30(a) shows the mottled surface of a moil point, made from AISI W1 (0.80% C) tool steel, that was detected after the surface was sand blasted following heat treatment. This surface condition was due to excessive scaling (oxidation) resulting from improper furnace-atmosphere control. Figure 30(b) shows that the surface case was poorly developed, probably due to the influence of the scaled surface.

Fig. 26 Fractured pin and gripping cam made from AISI S5 tool steel

(a) These fractures occurred when the plate-lifting clamps containing these parts failed early in service. (b) As shown by this macroetched (10% aqueous nitric acid) pin, both the cam and the pin were heavily decarburized. (c) The chamfered cam holes were quite rough. 6.5×. The as-rolled surface decarburization had not been removed before heat treatment, and the surfaces were not cleaned-up after heat treatment.

Fig. 27(a) AISI 4150 alloy steel chuck jaw that broke because of the presence of a brittle white-etching nitride surface layer

The part was hardened and tempered before nitriding. A micrograph of a broken tooth (arrows) of this chuck jaw is shown in Fig. 27(b).

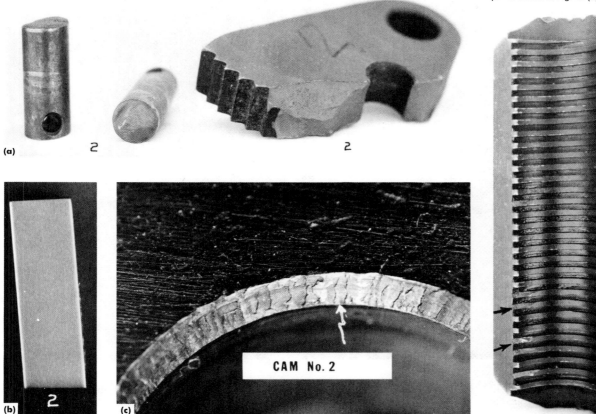

Failures Due to Burning and Melting. Temperature-control problems can cause difficulties other than overaustenitization: burning or incipient melting can also occur in extreme cases. Figure 31(a) shows an end view of a powder metallurgy die component made from AISI D2 tool steel that melted and deformed due to highly excessive localized temperatures during heat treatment. The microstructure in this region (Fig. 31b) shows the melted region, heat-affected zone, and base microstructure.

High-speed steel components are particularly susceptible to burning/localized-melting problems due to heat-treatment irregularities because the austenitizing temperatures required are very high, much higher than for other tool steel grades. Figure 32(a) shows a part made from AISI M2 tool steel whose surface exhibited a rippled appearance after hardening. This condition was usually observed on the top of these parts, sometimes on the sides, but not on the bottoms. Examination of the top surface of the die on a transverse section (Fig. 32b) revealed a remelted surface layer. Beneath this zone was a layer containing as-quenched mar-

Fig. 27(b) Surface of a broken tooth from the chuck jaw shown in Fig. 27(a)

The white layer at the surface is brittle iron nitride, which is also present as a grain-boundary film surrounding many of the grains near the surface. Note that the crack (arrows) follows the brittle white-etching nitride that formed in the prior-austenite grain boundaries. Etched with 2% nital. 290×

Fig. 28 Header die made from AISI W1 tool steel that failed prematurely in service

(a) The striking face of the carbon tool steel die chipped. The die had been flush quenched through its center hole to harden the working surfaces. (b) Cold etching (10% aqueous nitric acid) of a longitudinal disk through the cracked region revealed sharp corners at the striking face and an insufficient chill zone to support the edge under service conditions.

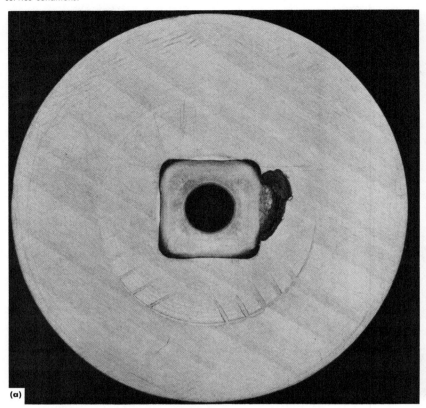

tensite and retained austenite. The outer 0.13 mm (0.005 in.) of the top surface was found to be enriched in carbon (1.28%). Immediately below this zone, the carbon content was normal for the grade. The higher carbon content lowered the melting point of the surface. The as-quenched martensite beneath the remelted zone probably formed due to conversion of some of the retained austenite at the surface during the tempering operations. The interior microstructure was tempered martensite.

Brittle Fracture of Rehardened High-Speed Steels. If high-speed steels must be rehardened, they must be annealed first. If they are not, fracture generally occurs. Such fractures have a fish-scale appearance due to the shiny, coarse-grain facets on their fracture surfaces. This is due to production of a coarse grain structure, which leads to a very brittle condition. Figure 33 shows the duplex, coarse-grain structure that results when a high-speed steel is rehardened without an intermediate anneal. Resistance to grain growth is provided by the presence of many fine carbides in the structure that are produced by the anneal but are not present after hardening. Thus, a second hardening treatment, without annealing, produces rapid grain growth at the very high austenitizing temperatures required for hardening high-speed steels.

Influence of Steel Quality

Despite the care taken in the manufacture and inspection of tool steels, faulty materials occasionally cause tool and die failures. However, the incidence of such problems is low. The most common of these defects are voids from secondary pipe, hydrogen flakes, surface cracks (such as seams or laps), porosity or microvoids, cooling cracks, segregation, and poor carbide distributions. Improper control of annealing may also produce nonuniform carbide distributions or carbide networks that may influence heat-treatment uniformity, lower ductility, or impair machinability.

Failures Due to Seams or Laps. Figure 34 shows a coil spring made from AISI H12 tool steel bar stock. Although the bar was centerless ground before coiling and heat treatment, a tight seam (arrows) that was still present opened during quenching. Figure 35 shows part of a ring forging made of AISI A8 tool steel that measured 46 cm (18⅜ in.) OD × 32 cm (12⅝ in.) ID × 3.8 cm (1½ in.) thick. As shown in Fig. 35, cracks were found after forging and annealing. Scale was present on the crack wall, showing that the crack was open during forging. Figure 36(a) shows a macroetched disk cut transverse to the ring, illustrating the crack depth and

decarburization around the cracks. Alkaline chromate etching (Fig. 36b) shows oxygen enrichment around the crack, verifying that it was open during high-temperature exposure. Figure 37(a) shows another example of a forging lap. This 3.6-kg (8-lb) sledge-hammer head, which measured 16.8 × 6 × 5.7 cm (6⅝ × 2⅜ × 2¼ in.), cracked shortly after being placed in service. The crack surface was oxidized, and the alkaline chromate etch (Fig. 37b) revealed oxygen enrichment. These two forging laps were caused by the forging operation and were not present in the original forging billet.

Hydrogen flaking is a potential problem for carbon tool steels and for some of the medium-carbon low-alloy steels, such as those used as prehardened plastic-mold alloys. Vacuum degassing is commonly used for such grades to reduce the hydrogen content to a safe level. Figure 38(a) shows a die made from AISI O1 tool steel that was found to be cracked after heat treatment. When opened (Fig. 38b), these cracks exhibited a coarse, shiny, faceted appearance. The cracks in this part were longitudinally oriented and were confined to the center of the section. Flakes are not observed in the outer region of sections because there is apparently more time for the hydrogen in the outer region to diffuse to a safe level. Flakes that

Fig. 29 Failure caused by improper quenching

(a) AISI W1 tool steel wire-forming die that broke prematurely during service. (b) Cold etching (10% aqueous nitric acid) of a disk cut behind the fracture revealed that the bore-working surface was not hardened; only the dull gray region was partially hardened.

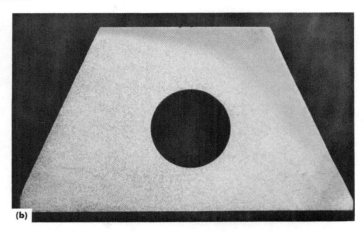

Fig. 30(a) Moil point made of AISI W1 tool steel that exhibited a rough, scaled surface after heat treatment

The actual size of the moil point is shown at left. An enlarged view (3×) at the surface condition, which resulted in erratic surface hardness, is shown at right. See also Fig. 30(b).

Fig. 30(b) Cold-etched (10% aqueous nitric acid) disk cut from the moil point shown in Fig. 30(a)

A nonuniform chill is evident; the dark areas are hardened. 2×

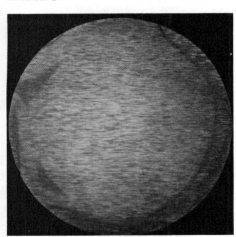

were close to the surface of the die exhibited temper color, but some of the deeper flakes did not. The temper-colored flakes apparently opened during quenching.

Failures Due to Overheating During Hot Working. Heating for hot working must be carefully controlled to prevent overheating, burning, or incipient melting problems. Overheating involves higher-than-tolerable hot-working temperatures such that most or all of the sulfides are put into solution and grain growth is excessive. Heavy hot reduction can repair most of the damage due to overheating, but if hot reduction is limited, ductility suffers. Figure 39 shows some of the typical characteristics of overheating. Tensile testing of an AISI H12 tool steel forged liner revealed poor tensile ductility. The coarse nature of the tensile fracture is apparent in Fig. 39(b). A section cut from the liner was notched and fractured, re-

Fig. 31 AISI D2 powder metallurgy die component that melted and deformed because of flame impingement during heat treatment

(a) End view. 4.5×. (b) Microstructure in the affected region. Etched with Marble's reagent. 150×

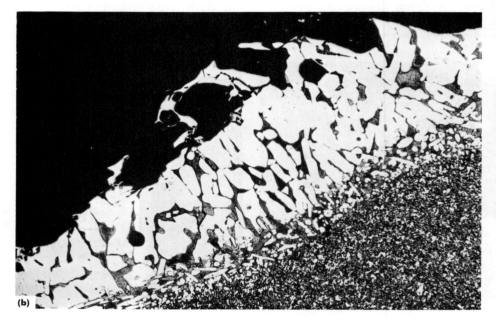

Fig. 32 Localized melting at the surface of a part made from AISI M2 tool steel

(a) Rippled surface appearance after hardening. 0.75×. The surface was slightly carburized, which lowered the melting point. (b) Microstructure shows the melted surface region and a zone beneath it containing retained austenite and as-quenched martensite. The matrix was tempered martensite. Etched with Marble's reagent. 70×

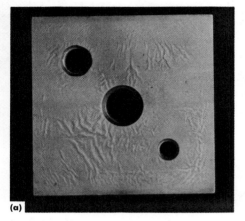

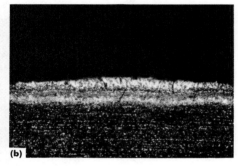

vealing similar, coarse fracture features (Fig. 39a). The microstructure was also coarse, with a grain size coarser than ASTM 1.

Failures Due to Unconsolidated Interiors. It is possible, particularly with large section sizes that receive limited hot reduction, to experience failures due to material with an unconsolidated interior. A study of failures conducted over a 20-year period revealed three such failures, one of which is illustrated in Fig. 40(a) and (b). Figure 40(a) shows a fractured section from part of an AISI W2 die insert that cracked during rehardening. This die insert was made from a 15- × 25-cm (6- × 10-in.) billet. The horizontal arrow in Fig. 40(a) indicates the origin of the fracture, which was near the center of the section. A disk was cut through the part shown in Fig. 40(a) and hot acid etched. This revealed an unconsolidated region along the centerline (Fig. 40b).

Failures Due to Carbide Segregation and Poor Carbide Morphology. Large sections of high-alloy tool steels may exhibit very fine porosity or microvoids, which are often associated with carbide segregation and poor carbide distribution and morphology. These conditions reduce ductility and may lead to failures, depending on the application. Figure 41(a) shows a roll made from a 23-cm (9-in.) diam bar of AISI D2 tool steel. The flange edge of the roll that chipped off during initial use was due to poor carbide morphology (Fig. 41b).

Figure 42(a) shows a 7-cm (2¾-in.) diam scoring die made from AISI A2 tool steel. The scoring edge near the center of the die spalled during either the stoning or grinding step after heat treatment. Microscopic examination at the spalled region revealed a band of carbides running down the center of the profile of the scoring die (Fig. 42b). Such carbide bands are not unusual for AISI A2; a different grade of steel should be used for such an application.

Influence of Service Conditions

Tools and dies, although made from the correct grade, well designed, and properly machined and heat treated, can fail after limited service due to improper operation or mechanical problems. In such cases, the failure analyst may spend considerable effort evaluating the design, grade selection, machining, and heat treatment without uncovering any abnormalities. In many cases, the analyst has little information concerning the use of the tool and die and can only conclude that service problems were responsible for the failure. In some cases, there may be evidence of the mechanical or service problem.

Mechanical factors that may cause premature failures include overloading, overstressing, or alignment/clearance problems. Excessive temperature may be a factor in hot-working die failures, perhaps due to inadequate cooling between operations. Failures have also occurred during assembly—for example, during shrink fitting of one part onto another. Stamp

Fig. 33 Examples of the microstructure of AISI M2 high-speed steel

(a) Desired quenched-and-tempered condition: 1200 °C (2200 °F) for 5 min in salt, oil quench, double temper at 595 °C (1100 °F). Etched with 3% nital. 500×. (b) Grain growth caused by reaustenitizing without annealing: 1220 °C (2225 °F) for 5 min in salt, oil quench, 1175 °C (2150 °F) for 5 min in salt, oil quench (not tempered). Etched with 10% nital. 400×. (c) Overaustenitization and onset of grain-boundary melting: 1260 °C (2300 °F) for 5 min in salt, oil quench, double temper at 540 °C (1000 °F). Arrows indicate regions of grain-boundary melting. Etched with 3% nital/Villella's reagent. 1000×

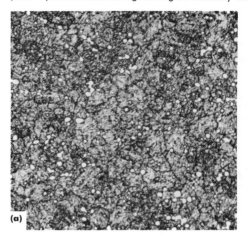

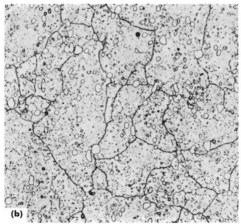

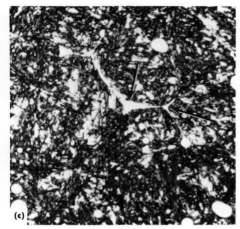

Fig. 34 Coil spring made from AISI H12 tool steel that cracked after heat treatment

A tight seam that was not removed by centerless grinding before heat treatment opened during hardening (arrows). 0.3×

Fig. 35 Portion of an AISI A8 ring forging that cracked (arrows) after forging and annealing.

Scale was present on the crack walls to a depth of 13 mm (½ in.) beneath the surface. Note the area that was ground to probe the crack depth.

Fig. 36 Metallographic samples from the ring forging shown in Fig. 35

(a) Macroetching (10% aqueous nitric acid) of a disk cut through the ring forging revealed decarburization around the cracks. 0.75×. (b) Etching of a polished section with a hot alkaline chromate solution revealed oxygen enrichment indicative of a defect open during forging. 80×

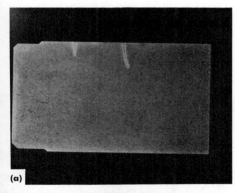

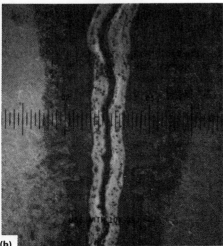

Fig. 37 Failure caused by a forging lap in a sledge-hammer head

(a) Cracks on the striking face soon after the hammer was first used. (b) A hot alkaline chromate etch revealed oxygen enrichment (white region) adjacent to the crack. 65×

(a)

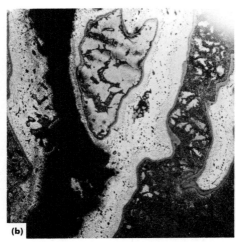

(b)

marks, in addition to causing heat-treatment failures, can cause service failures due to stress concentration. Alignment problems are a common cause of failure of tools used in shearing operations. While most tools and dies fail in a brittle manner, fatigue failures are sometimes encountered. In most cases, the fatigue failure is located at a change in section size, at a sharp corner, or at stamp marks.

Tools used in hot-working applications generally fail due to development of a craze-crack network referred to as heat checks. These result due to thermal stresses from alternate heating and cooling. Certain hot-working grades are particularly susceptible to heat checking if rapid cooling—for example, water cooling—is used.

As an example of a severe-wear condition that caused premature failure of a proprietary air-hardening die steel, Fig. 43 shows a die measuring 92 mm (3⅝ in.) max OD × 90 mm (3½ in.) high whose surface exhibits a crazed and eroded condition. The vertical scratch marks indicate inadequate clearance. The severe rubbing action reaustenitized a portion of the surface, resulting in a layer of brittle as-quenched martensite, a commonly observed condition in such failures. This die was properly machined and heat treated with no apparent problems other than service abuse.

Shear knives operate under particularly difficult service conditions. Figure 44 shows two AISI H13 shear knives used to grip bars after hot rolling so that they can be separated. Bar temperatures were typically from 815 to 980 °C (1500 to 1800 °F). Heat from contact and friction was excessive enough to reaustenitize the gripping edge, and this promoted spalling. Figure 45 shows the as-quenched martensite produced in this manner at the tip of one of the knives.

The heat pattern produced by service conditions can generally be revealed by macroetching. Figure 46 shows the face of a 12.7-cm (5-in.) wide AISI S7 cutter blade that was cold etched (10% aqueous nitric acid) after re-sharpening. The heat pattern is visible along the working edge and at the clamping locations.

Surface-chemistry changes can also occur during service. For example, Fig. 47 shows a cold-etched disk (10% aqueous nitric acid) from a 13.5-cm (5⁵⁄₁₆-in.) diam AISI H13 mandrel that was used to pierce and extrude brass. The mandrel cracked after about 30% of its anticipated service life. Cracking was due to heavy decarburization that occurred during service.

Fig. 38 Die made from AISI O1 tool steel that was found to be cracked after heat treatment

(a) Longitudinal cracks after the surface was swabbed with 5% nital. (b) One of the cracks opened, revealing features typical of hydrogen flakes. 6.5×

(a)

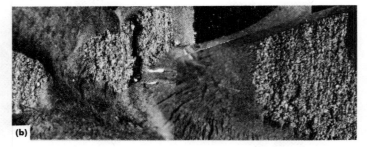

(b)

Fig. 39 Microstructural characteristics of overheating

(a) Test fracture and (b) tensile-bar fracture from an overheated forged liner made from AISI H12 tool steel. Both 2×. (c) Micrograph illustrating the very coarse martensitic grain structure due to overheating during forging. Etched with 3% nital. 75×

(a)

(b)

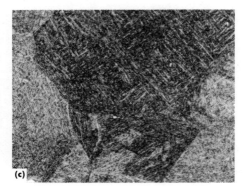

(c)

Fig. 40(a) Fractured section of an AISI W2 die insert that cracked during rehardening

The horizontal arrow shows the origin of the failure, which corresponds to the center of the billet used to make the insert. 0.3×

Fig. 40(b) Disk that was cut through the origin of the insert section shown in Fig. 40(a)

Hot-acid etching (50% aqueous hydrochloric acid at 70 °C, or 160 °F) revealed an unsound center condition that promoted the fracture.

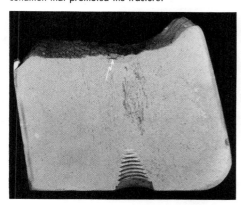

Aluminum die casting places severe demands on hot-work die steels. The type of damage that results during such service is illustrated by the hot-work die steel runner block shown in Fig. 48. This part was used to produce over 100 000 aluminum transmission case covers. Areas 1 and 2, illustrating heat checking and wash-out, respectively, are shown at higher magnifications in Fig. 49.

Die-casting dies usually fail by erosion of the die cavity, runners, and gates or by heat check-

Fig. 41(a) The flange edge of a roll made from AISI D2 tool steel that chipped off during its initial use

Failure was due to poor carbide distribution and morphology, which embrittled the material. See also Fig. 41(b).

Fig. 41(b) Micrograph showing the poor carbide distribution and morphology in the roll shown in Fig. 41(a)

The grain size, ASTM 6.75, was coarser than desired. Etched with 3% nital. 700×

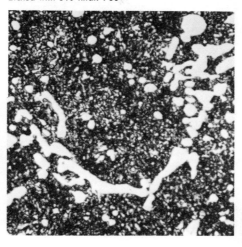

ing of the die cavity. Zinc die-casting dies are particularly prone to erosion-related problems. Heat checking is less of a problem because service temperatures are lower than for magnesium or aluminum die casting. Figure 50(a) shows an AISI H13 nozzle used in zinc die casting that failed due to erosion. A furrow about 7.6 cm (3 in.) long in the longitudinal direction by about 1.6 mm (¹⁄₁₆ in.) wide was present from the base of the nozzle to the outside diameter surface. Intrusion of the zinc into the crack was apparent. Figure 50(b) shows a longitudinal section through the nozzle, revealing rough machining marks and a misalignment of the bore.

Fig. 42 Failure due to a band of carbides

(a) AISI A2 scoring die spalled at the cutting edge during either the stoning or final grinding step after heat treatment. (b) Sectioning through the spalled region revealed a band of carbides intersecting the edge profile that promoted cracking. Etched with 3% nital. 180×

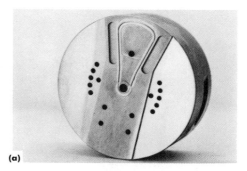

(a)

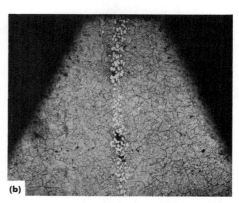

(b)

Figure 51 shows an AISI H26 exhaust-valve punch that split longitudinally due to fatigue after producing 1007 parts. The progressive nature of the crack growth is evident on the fracture faces. Heat checking on the working face initiated the fatigue failure. The punch surfaces were nitrided after heat treatment. No material or manufacturing abnormalities were detected. Heat checking was promoted by the service conditions.

SELECTED REFERENCES
General

- C.S. Azzalina, Analysis of Tool Steel Failures, *Am. Mach.*, Sept 8, 1969, p 133-134; Sept 22, 1969, p 121-122
- Errors in Heat Treatment Can Cause Tool Failures, *Tool Steel Trends*, Winter 1972, p 2-7
- *Improving Production From Tools and Dies*, Handbook 538, Bethlehem Steel Corp., Bethlehem, PA, 1960
- S. Kalpakjian, Ed., *Tool and Die Failures: Source Book*, American Society for Metals, 1982
- M.K. Nath, Failure Analysis, *Tool and Alloy Steels*, Vol 14, 1980, p 231-232
- J.Y. Riedel, Causes of Tool Failures. I— Mechanical Factors, *Met. Prog.*, Vol 58,

Fig. 43 Die failure caused by severe wear

(a) Die made from air-hardening tool steel that exhibited a crazed and eroded condition. Areas A and B are shown in (b) and (c), respectively. Both 10×. (d) Microstructural examination of area B revealing a layer of as-quenched martensite at the surface and a back-tempered region beneath it caused by frictional heat. Etched with 3% nital. 295×

(a)

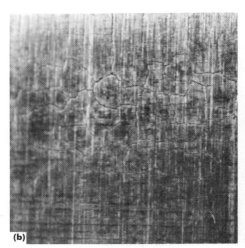

(b)

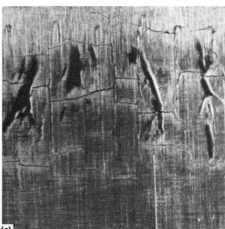

(c)

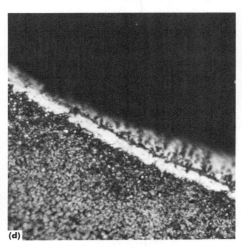

(d)

Fig. 44 Two shear knives made from AISI H13 tool steel

The knives were used to grip hot-rolled bars after rolling so that they could be separated. The knives failed by spalling of the gripping edge after normal service life.

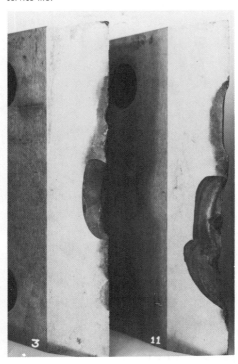

Fig. 46 Macroetched (10% aqueous nitric acid) face of a cutter blade made from AISI S7 steel

Macroetching reveals the influence of frictional heat from service (dark-etching areas) that produce localized back-tempering (softening).

Fig. 45 A typical example of freshly formed martensite at the tip of a failed shear blade

The hardness was 59 to 60 HRC. Etched with 3% nital. 50×

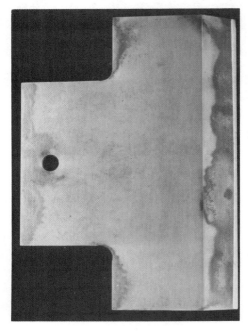

Fig. 47 AISI H13 mandrel used to pierce and extrude brass that failed after 298 pushes, about 30% of its expected life

The disk shown below, cut from the mandrel, was macroetched (10% aqueous nitric acid), revealing a heavily decarburized surface. The decarburization occurred during service.

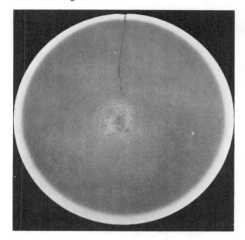

Fig. 48 Runner block made from a proprietary hot-work tool steel that was used to die cast aluminum transmission case covers

Macrograph shows the worn out surface of the die. 0.25×. Close-up views of areas 1 and 2 are shown in Fig. 49.

Fig. 49 Close-up views of areas 1 and 2 shown in Fig. 48

(a) Area 1 reveals heat checking on the working surface. 6×. (b) Area 2 illustrates a washed-out area. 3×

Aug 1950, p 171-175; II—Improper Heat Treatment, *Met. Prog.*, Vol 58, Sept 1950, p 340-344
- J.Y. Riedel, Why Tools and Dies Fail, *Met. Prog.*, Vol 97, April 1970, p 101-104
- A.M. Schindler *et al.*, Metallurgical Factors Affecting the Service Life of Tool Steels and High Speed Steels, *Tool and Alloy Steels*, Vol 8, 1974, p 23-32
- J.W. Sullivan, Preventing Failures in Cold Forming Tools—Heat Treatment and Design, *ASM Met. Eng. Q.*, Vol 13, Feb 1973, p 31-41
- *The Tool Steel Troubleshooter*, Handbook 2828, Bethlehem Steel Corp., Bethlehem, PA
- J. Wallbank and V.B. Phadke, Some Metallurgical Aspects of Die Failure, in *Towards Improved Performance of Tool Materials*, Book 278, The Metals Society, London, 1982, p 56-64
- C.B. Wendell, Metallurgical Factors Affecting the Service Life of Tool Steels, *Met. Treat.*, Dec-Jan 1968-1969, p 3-12
- L.H. Williamson, Die Failures, Their Diagnosis and Cure, *Met. Prog.*, Vol 28, July 1935, p 32-37
- W.H. Wills, Causes and Avoidance of Tool Steel Failures, *Met. Alloys*, Vol 2, Sept 1931, p 112-116
- W. Young, "Hints for Best Tool Performance," Paper MR68-192, American Society of Tool and Manufacturing Engineers, Dearborn, MI, 1968
- W. Young, How to Increase Steel Tool Performance, *Cut. Tool Eng.*, Vol 21, May 1969, p 8-12

Heat Treatment

- E. Ameen, Dimension Changes of Tool Steels During Quenching and Tempering, *Trans. ASM*, Vol 28, 1940, p 472-512

Fig. 50(a) Erosion damage from the bore to just below the outside-diameter surface of an AISI H13 nozzle from a zinc die-casting die

Actual size

Fig. 50(b) Erosion damage and misaligned bore of the AISI H13 tool steel zinc die-casting nozzle shown in Fig. 50(a) after longitudinal splitting

Actual size

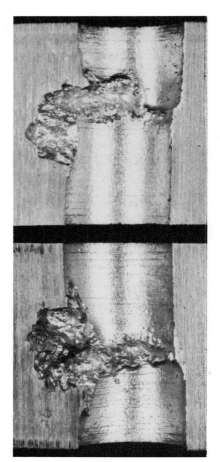

Fig. 51 Failed AISI H26 exhaust-valve punch

(a) and (b) Longitudinal splitting of the punch caused by fatigue. Note the fracture progression starting from the top center at the punch. The punch surfaces were nitrided. (c) Top surface. 100×. (d) Extreme top surface. Note secondary crack (arrows). 700×. (e) Case/core interface. 700×. All etched with 3% nital

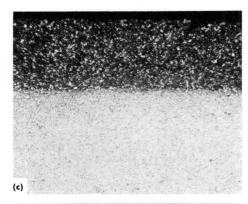

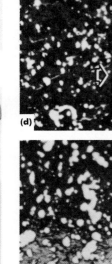

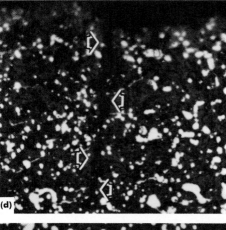

- M. Cohen, Retained Austenite, *Trans. ASM*, Vol 41, 1949, p 35-94
- R.F. Harvey, Overcoming Distortion and Cracking in Heat Treating Tool Steels, *Met. Prog.*, Vol 79, June 1961, p 73-75
- L.D. Jaffee and J.R. Hollomon, Hardenability and Quench Cracking, *Trans. AIME*, Vol 167, 1946, p 617-626
- T. Kunitake and S. Sugisawa, The Quench-Cracking Susceptibility of Steel, *Sumitomo Search*, No. 5, May 1971, p 16-25
- L.A. Norström *et al.*, On the Role of Retained Austenite in Martensitic Cold Work Tool Steels, *Scand. J. Metall.*, Vol 9 (No. 2) 1980, p 79-82
- J.Y. Riedel, Retained Austenite and Dimensional Stability, *Met. Prog.*, Vol 88, Sept 1965, p 78-82
- K. Sachs, Effect of Carbide Stringers on the Distortion of Die Steels During Heat Treatment, *Met. Treat. Drop Forg.*, Vol 27, Oct 1960, p 395-408; Nov 1960, p 455-460; Dec 1960, p 487-492; Vol 28, Jan 1961, p 31-36; Feb 1961, p 59-62; March 1961, p 115-119; April 1961, p 157-164
- J.W. Spretnak and C. Wells, An Engineer-

ing Analysis of the Problem of Quench Cracking in Steel, *Trans. ASM*, Vol 42, 1950, p 233-269
- G. Steven, Nonuniform Size Changes of High-Speed Steels During Heat Treatment, *Trans. ASM*, Vol 62, 1969, p 130-139

Grinding

- W.R. Harwick, Grinding Hazards, *Metallurgia*, Vol 49, Jan 1954, p 21-25
- W.E. Littmann and J. Wulff, The Influence of the Grinding Process on the Structure of Hardened Steel, *Trans. ASM*, Vol.47, 1955, p 692-714
- L.P. Tarasov and C.O. Lundberg, Nature and Detection of Grinding Burn in Steel, *Trans. ASM*, Vol 41, 1949, p 893-939

Hot-Working Dies

- J.L. Aston and A.R. Muir, Factors Affecting the Life of Drop Forging Dies, *J. Iron Steel Inst.*, Vol 207, Feb 1969, p 167-176
- C.L. Gibbons and J.E. Dunn, Investigations of Reduced Service Life of Hot Work (Cr-Mo) Die Steel Pieces, *Ind. Heat.*, Aug 1980, p 6-9
- E. Glenny, Thermal Fatigue, *Met. Rev.*, Vol 6 (No. 24), 1961, p 387-465
- A. Kasak and G. Steven, Microstructural Considerations in Heat Checking of Die Steels, in *Proceedings of the 6th SDCE International Die Casting Congress*, Paper 112, Society of Die Casting Engineers, Detroit, MI, 1970
- L. Northcott, The Craze-Cracking of Metals, *J. Iron Steel Inst.*, Vol 184, Dec 1956, p 385-408
- R.E. Okell and F. Wolstencroft, A Suggested Mechanism of Hot Forging Die Failure, *Metal Form.*, Feb 1968, p 41-49
- B.W. Rooks *et al.*, Temperature Effects in Hot Forging Dies, *Met. Technol.*, Vol 1, Oct 1974, p 449-455
- H. Thielsch, Thermal Fatigue and Thermal Shock, *WRC Bull.*, No. 10, April 1952
- T.C. Yen, Thermal Fatigue—A Critical Review, *WRC Bull.*, No. 72, Oct 1961

Die-Casting Dies

- J.C. Benedyk *et al.*, Thermal Fatigue Behavior of Die Materials for Aluminum Die Casting, in *Proceedings of the 6th SDCE International Die Casting Congress*, Paper 111, Society of Die Casting Engineers, Detroit, MI, 1970
- J.A. Dankovich and W.H. Oberndorfer, Effects of Heat Treatment on Aluminum Die Cast Die Life, in *Proceedings of the Third National Die Casting Exposition and Congress*, Paper 62, Society of Die Casting Engineers, Detroit, MI, 1964
- M.A.H. Howes, Heat Checking in Die Casting Dies, *Die Cast. Eng.*, March-April 1969, p 12-16
- A.I. Kemppinen, Die Pickup Can Be Explained, *Met. Prog.*, Vol 93, June 1968, p 147, 148, 150
- S. Malm and J. Tidlund, Increased Life for Die Casting Dies, in *Proceedings of the 10th International Die Casting Congress*, Paper G-T79-051, Society of Die Casting Engineers, Detroit, MI, 1979
- H. Nichols and C. Dorsch, Die Maintenance to Improve Die Performance, in *Die Tech '84*, American Die Casting Institute and Die Casting Research Foundations, Des Plaines, IL, 1984, p 315-350
- S.J. Noesen and H.A. Williams, The Thermal Fatigue of Die Casting Dies, Paper 801, *Trans. SDCE*, Vol 4, 1966
- Preventing Failure in Die Casting Dies, *Precis. Met.*, Vol 39, May 1981, p 102-103, 107-108, 110
- G.A. Roberts and A.H. Grobe, Service Failures of Aluminum Die-Casting Dies, *Met. Prog.*, Vol 69, Feb 1956, p 58-61
- A. Schindler and R. Breitler, Die Care Improves Die Life, in *Die Tech '84*, American Die Casting Institute and Die Casting Research Foundation, Des Plaines, IL, 1984, p 293-314
- W.R. Wollering and L.C. Oertle, Thermal Fatigue as a Cause of Die Failure, Paper 804, *Trans. SDCE*, Vol 4, 1966
- W. Young, "Are You Getting Maximum Performance From Your Die Casting Dies?," Technical Paper CM68-587, American Society of Tool and Manufacturing Engineers, Dearborn, MI, 1968
- W. Young, Better Performance from Die Casting Dies, *Precis. Met. Mold*, Vol 23, May 1965, p 58-59, 75-79
- W. Young, Die Casting Die Failure and its Prevention, *Precis. Met.*, March 1979, p 28-31
- R.M. Young *et al.*, Thermal Fatigue on Die Steels: A Brief Summary and Perspective, *Ind. Heat.*, July 1981, p 24-27
- W. Young, How To Obtain Better Performance from Die Casting Dies, *Die Cast. Eng.*, Vol 13, July-Aug 1969, p 32-34
- W. Young, Why Die Casting Dies Fail, in *Proceedings of the 10th International Die Casting Congress*, Paper G-T79-092, Society of Die Casting Engineers, Detroit, MI, 1979

Plastic-Molding Dies

- W. Young, Hints for Better Mold Performance, *SPE J.*, Vol 21, Dec 1965, p 1-3

High-Speed Steels

- R. Brownsword *et al.*, Studies of Wear in High-Speed Steel Tools, *ISI P126*, Iron and Steel Institute, London, 1971, p 38-42
- E.D. Doyle, Effect of Different Heat Treatments on the Wear of High Speed Steel Cutting Tools, *Wear*, Vol 27, 1974, p 295-301
- E.D. Doyle, Influence of High Speed Steel Microstructure on Tool Failure, in *International Conference on Production Technology*, National Conference Publication 74/3, Institute of Engineers, Melbourne, 1974, p 302-305
- W.E. Henderer, Strengthening Mechanisms in High-Speed Steel as Related to Tool-Life, *J. Eng. Ind., (Trans. ASME)*, Vol 101, May 1979, p 217-222
- F. Jandos, Fractography of Hardened High-Speed Steel, *Kovové Mater.*, No. 6 (BISI 8237), 1969, p 510-524
- K. Lassota, The Structure of 18-4-1 Type High Speed and its Influence on the Life of Cutting Tools, *Rev. Met.*, Vol 63, (BISI 4883), Feb 1966, p 105-116
- E. Niesielski, Metallurgical Defects in High Speed Steels, *Iron Steel*, Special issue, 1968, p 63-75
- H. Optiz and W. Konig, Basic Research on the Wear of High-Speed Steel Cutting Tools, *ISI P126*, Iron and Steel Institute, London, 1971, p 6-14
- C. Oxford, Causes of Twist Drill Breakage, *Tool. Prod.*, Vol 43, 1978, p 88-90
- C.O. Smith, Failure of a Twistdrill, *J. Eng. Mater. Technol.*, April 1974, p 88-90
- S. Söderberg, A Study of Twist Drill Wear, *Tool. Prod.*, Vol 45, 1979, p 99-102
- R.W. Thompson *et al.*, Wear and Failure of High-Speed Steel Cut-Off Tools, in *Proceedings of the Third North American Metalworking Research Conference*, Carnegie Press, 1975, p 385-400

Failures of Gears

Revised by Lester E. Alban, Metallurgical Consultant

GEARS can fail in many different ways, and except for an increase in noise level and vibration, there is often no indication of difficulty until total failure occurs. In general, each type of failure leaves characteristic clues on gear teeth, and detailed examination often yields enough information to establish the cause of failure. Despite the variety of ways in which gears fail, service failures of gears are relatively rare.

This article will deal primarily with the common types and causes of gear failures and the procedures employed in analyzing them. First, however, the major types of gears and the basic principles of gear-tooth contact must be reviewed briefly.

A gear is a machined component that transmits motion and force from one element in a working unit to another element in the same unit or to another working unit either in the same plane and direction or in a completely different plane or direction. The force due to this transmission may either increase or decrease in magnitude from one element to the next. Design and function are closely associated because a gear is designed with a specific function in mind. The question is, Will this gear perform the function that was intended by the designer?

Types of Gears

Spur gears (Fig. 1a) are used to transmit motion between parallel shafts or between a shaft and a rack. The teeth of a spur gear are radial, uniformly spaced around the outer periphery, and parallel to the shaft on which the gear is mounted. Contact between the mating teeth of a spur gear is in a straight line parallel to the rotational axes, lying in a plane tangent to the pitch cylinders of the gears (a pitch cylinder is the imaginary cylinder in a gear that rolls without slipping on a pitch cylinder or pitch plane of another gear).

Helical gears (Fig. 2a) are used to transmit motion between parallel or crossed shafts or between a shaft and a rack by meshing teeth that lie along a helix at an angle to the axis of the shaft. Because of this angle, mating of the teeth occurs such that two or more teeth of each gear are always in contact. This condition permits smoother action than that of spur gears.

Fig. 1 Sections of a spur gear (a) and a spur rack (b)

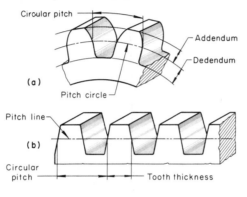

Fig. 2 Sections of a helical gear (a) and a helical rack (b)

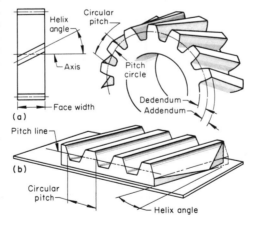

However, unlike spur gears, helical gears generate axial thrust, which causes slight loss of power and requires thrust bearings.

Herringbone gears (Fig. 3a), sometimes called double helical gears, are used to transmit motion between parallel shafts. In herringbone gears, tooth engagement is progressive, and two or more teeth share the load at all times. Because they have right-hand and left-hand helixes, herringbone gears are usually not subject to end thrust.

Fig. 3 Illustration of herringbone and helical gears

(a) One-piece herringbone gear; the opposed helixes allow multiple-tooth engagement and also eliminate end thrust. (b) Mating crossed-axes helical gears

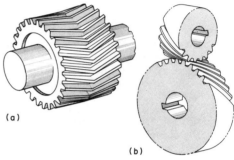

Crossed-axes helical gears transmit motion between shafts that are nonparallel and nonintersecting (Fig. 3b). The action between the mating teeth has a wedging effect, which results in sliding on tooth flanks. These gears have low load-carrying capacity, but are useful where shafts must rotate at an angle to each other.

Worm-gear sets are usually right-angle drives consisting of a worm gear (or worm wheel) and a worm. Figure 4 shows a double-enveloping worm-gear set. Worm-gear sets are used where the ratio of the speed of the driving member to the speed of the driven member is large and where compact right-angle drive is required. If a worm gear such as that shown in Fig. 4 engages a straight worm, the combination is known as single-enveloping worm gearing.

Internal gears are used to transmit motion between parallel shafts. The teeth of internal gears are similar in form to those of spur gears and helical gears, but point inward toward the center of the gear (Fig. 5a). Common applications for internal gears include rear drives for heavy vehicles, planetary gear systems, and speed-reducing devices. Internal gears are sometimes used in compact designs because the center distance between the internal gear and its mating pinion is smaller than that required for two external gears. Figure 5(b) shows the

Fig. 4 Mating of worm gear (worm wheel) and worm in a double-enveloping worm-gear set

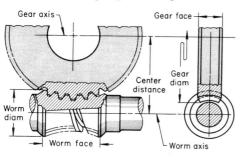

Fig. 5 Illustration of internal gears

(a) Section of a spur-type internal gear. (b) Relation of internal gear and mating pinion

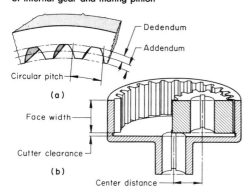

relationship between an internal gear and a mating pinion.

A rack is a gear whose teeth lie in a straight line (pitch circle of infinite radius). The teeth may be at right angles to the edge of the rack and mesh with a spur gear (Fig. 1b) or may be at some other angle and engage a helical gear (Fig. 2b).

Bevel gears transmit rotary motion between nonparallel shafts that are usually at 90° to each other.

Straight bevel gears (Fig. 6a) have straight teeth that, if extended inward, would intersect at the intersection of gear and pinion axes. Thus, the action between mating teeth resembles that of two cones rolling on each other (see Fig. 7 for angles and terminology). The use of straight bevel gears is generally limited to drives that operate at low speeds and where noise is not important. If the speeds are to remain the same with only a 90° change of direction, the set is called a miter gear set. Any change in the number of teeth will change speed as well as direction.

Spiral bevel gears (Fig. 6b) have teeth that are curved and oblique and that lie along a spiral at an angle to the shaft. The inclination of the teeth results in gradual engagement and continuous line contact or overlapping action; that is, two or more teeth are in contact at all

Fig. 6 Four types of bevel gears

See text for discussion.

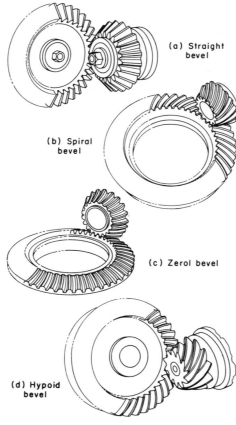

(a) Straight bevel

(b) Spiral bevel

(c) Zerol bevel

(d) Hypoid bevel

Fig. 7 Angles and terminology for straight bevel gears

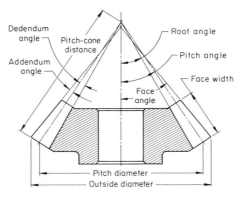

times. Because of this continuous engagement, the load is transmitted more smoothly from the driving gear to the driven gear than with straight bevel gears. Spiral bevel gears also have greater load-carrying capacity than straight bevel gears. Spiral bevel gears are usually preferred over straight bevel gears for operation at speeds greater than 5 m/s (1000 surface feet per minute, sfm) and particularly for very small gears.

Zerol bevel gears (Fig. 6c) are curved-tooth bevel gears with zero spiral angle. They differ from spiral bevel gears in that their teeth are not oblique. They are used in the same way as spiral bevel gears and have somewhat greater tooth strength than straight bevel gears.

Hypoid bevel gears (Fig. 6d) are similar to spiral bevel gears in general appearance, the important difference being that in a hypoid-gear set the axis of the pinion is offset somewhat from the axis of the gear. This feature provides many design advantages. In operation, hypoid gears run even more smoothly and quietly than spiral bevel gears and are stronger. However, they undergo more sliding action along the tooth-profile axes than spiral bevel gears and, for many applications, may require extreme-pressure lubricants.

Gear-Tooth Contact

The way in which tooth surfaces of properly aligned gears make contact with each other is responsible for the heavy loads that gears are able to carry. In theory, gear teeth make contact along lines or at points; in service, however, because of elastic deformation of the surfaces of loaded gear teeth, contact occurs along narrow bands or in small areas. The radius of curvature of the tooth profile has an effect on the amount of deformation and on the width of the resulting contact bands. Depending on gear size and loading, the width of the contact bands varies from about 0.38 mm (0.015 in.) for small, lightly loaded gears to about 5 mm (0.2 in.) for large, heavily loaded gears.

Gear-tooth surfaces are not continuously active. Each part of the tooth surface is in action for only short periods of time. This continual shifting of the load to new areas of cool metal and cool oil makes it possible to load gear surfaces to stresses approaching the critical limit of the gear metal without failure of the lubricating film.

The maximum load that can be carried by gear teeth also depends on the velocity of sliding between the surfaces, because the heat generated varies with rate of sliding as well as with pressure. If both pressure and sliding speeds are excessive, the frictional heat developed can cause destruction of tooth surfaces. This pressure-velocity factor, therefore, has a critical influence on the probability of galling and scoring of gear teeth. The permissible value of this critical factor is influenced by gear metal, gear design, character of lubricant, and method of lubricant application.

Lubrication is accomplished on gear teeth by the formation of two types of oil films: (1) the reaction film, also known as the boundary lubricant, is produced by physical adsorption and/or chemical reaction to form a desired film that is soft and easily sheared but difficult to penetrate or remove from the surface and (2) the elastohydrodynamic film that forms dynamically on the gear-tooth surface as a function of the surface speed. This secondary film is very

Fig. 8 Tooth contact lines on a spur gear (a), a bevel gear (b), and a low-angle helical gear (c)

Fig. 9 Lines of contact on a stepped spur gear

The heavy line on a tooth face of each gear section represents the instantaneous line of contact for that section. This offset-contact pattern is typical for helical, spiral bevel, and hypoid gears.

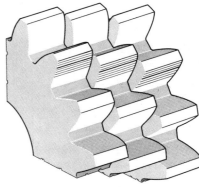

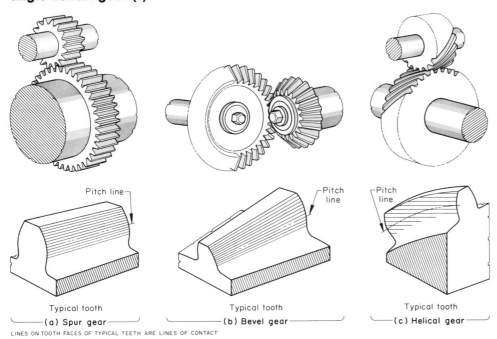

Pitch line

Pitch line

Pitch line

Typical tooth

Typical tooth

Typical tooth

(a) Spur gear

(b) Bevel gear

(c) Helical gear

LINES ON TOOTH FACES OF TYPICAL TEETH ARE LINES OF CONTACT

LINES ON TOOTH FACES ARE LINES OF CONTACT

thin, has a very high shear strength, and is only slightly affected by compressive loads as long as constant temperature is maintained.

Certain rules about a lubricant should be remembered in designing gearing and analyzing failures of gears. First, load is transferred from a gear tooth to its mating tooth through a pressurized oil film. If not, metal-to-metal contact may be detrimental. Second, increasing oil viscosity results in a thicker oil film (keeping load, speed, and temperature constant). Third, heat generation cannot be controlled above a certain maximum viscosity (for a given oil). Fourth, breakdown of the oil film will occur when the gear-tooth surface-equilibrium temperature has reached a specific value. Fifth, the scuffing load limit of mating tooth surfaces is speed dependent. With increasing speed, the load required to be supported by the reaction film decreases while the load that can be supported by the increasing elastohydrodynamic film increases. The result is a decreasing scuffing load limit to a certain speed as the reaction film decreases; then, as the speed picks up to where the elastohydrodynamic film increases, the scuffing load limit increases. This allows an increase in overall load-carrying capacity (assuming no change in temperature that would change viscosity). Finally, at constant speed, surface-equilibrium temperature increases as load increases, which lowers the scuffing load limit of the reaction film (surface-equilibrium temperature is attained when the heat dissipated from the oil is equal to the heat extracted by the oil). Damage to and failures of

gears can and do occur as a direct or indirect result of lubrication problems.

Spur and Bevel Gears. Spur-gear teeth are cut straight across the face of the gear blank, and the mating teeth theoretically meet at a line of contact (Fig. 8a) parallel to the shaft. Straight teeth of bevel gears also make contact along a line (Fig. 8b) that, if extended, would pass through the point of intersection of the two shaft axes. As teeth on either spur or bevel gears pass through mesh, the line of contact sweeps across the face of each tooth. On the driving tooth, it starts at the bottom and finishes at the tip. On the driven tooth, the line of contact starts at the tip and finishes at the bottom.

Helical, Spiral Bevel, and Hypoid Gears. Gear-tooth contact on these gears is similar to that developed on a stepped spur gear (Fig. 9). Each section, or lamination, of the spur gear makes contact with its mating gear along a straight line; each line, because of the offset between sections, is slightly in advance of its adjacent predecessor. When innumerable laminations are combined into a smoothly twisted tooth, the short individual lines of contact blend into a smoothly slanted line (Fig. 8c), that extends from one side of the tooth face to the other and sweeps either upward or downward as the tooth passes through mesh. This slanted-line contact occurs between the teeth of helical gears on parallel shafts, spiral bevel gears, and hypoid gears.

The load pattern is a line contact extending at a bias across the tooth profile, moving from one end of the contact area to the other end. Under

load, this line assumes an elliptical shape and thus distributes the stress over a larger area. Also, the purpose of spiral bevel gearing is to relieve stress concentrations by having more than one tooth enmeshed at all times.

The greater the angle of the helix or spiral, the greater the number of teeth that mesh simultaneously and share the load. With increased angularity, the length of the slanted contact line on each tooth is shortened, and shorter but more steeply slanted lines of contact sweep across the faces of several teeth simultaneously. The total length of these lines of contact is greater than the length of the single line of contact between straight spur-gear teeth of the same width. Consequently, the load on these gears is distributed not only over more than one tooth but also over a greater total length of line of contact. On the other hand, the increased angularity of the teeth increases the axial thrust load and thus increases the loading on each tooth. These two factors counterbalance each other; therefore, if the power transmitted is the same, the average unit loading remains about the same.

Helical gears on crossed shafts make tooth contact only at a point. As the teeth pass through mesh, this point of contact advances from below the pitchline of the driving tooth diagonally across the face of the tooth to its top and from the top of the driven tooth diagonally across its face to a point below the pitchline. Even with several teeth in mesh simultaneously, this point contact does not provide sufficient area to carry an appreciable load. For this reason, helical gears at angles are usually used to transmit motion where very little power is involved.

Worm Gears. In a single-enveloping worm-gear set, in which the worm is cylindrical in shape, several teeth may be in mesh at the same time, but only one tooth at a time is fully engaged. The point (or points) of contact in this type of gear set constitutes too small an area to carry an appreciable load without destruction of the metal surface. As a result, single-enveloping worm-gear sets are used in applications similar to those for helical gears on crossed shafts: to transmit motion where little power is involved.

Considerable power must be transmitted by commercial worm-gear sets; therefore, the gears of these sets are throated to provide a greatly increased area of contact surface. The gear tooth theoretically makes contact with the worm thread along a line curved diagonally across the gear tooth. The exact curve and slant depend on tooth design and on the number of threads on the worm relative to the number of teeth on the gear. Usually, two or more threads of the worm are in mesh at the same time, and there is a separate line of contact on each meshing tooth. As meshing proceeds, these lines of contact move inward on the gear teeth and outward on the worm threads. To secure smooth operation from a gear of this type, the teeth of the gear and sometimes the threads of the worm are usually altered from theoretically correct standard tooth forms. These alterations result in slightly wider bands of contact, thus increasing the load-carrying capacity of the unit. Load-carrying capacity also depends on the number of teeth in simultaneous contact. Exact tooth design varies from one manufacturer to another, and for this reason, the patterns of contact also vary.

In a double-enveloping worm-gear set, the worm is constructed so that it resembles an hourglass in profile. Such a worm partly envelops the gear, and its threads engage the teeth of the gear throughout the entire length of the worm. The teeth of both the worm and the gear have straight-sided profiles like those of rack teeth, and in the central plane of the gear, they mesh fully along the entire length of the worm.

The exact pattern of contact in double-enveloping (or double-throated) worm-gear sets is somewhat controversial and seems to vary with gear design and with method of gear manufacture. It is generally agreed, however, that contact is entirely by sliding with no rolling and that radial contact occurs simultaneously over the full depth of all the worm teeth.

Operating Loads

Gears and gear drives cover the range of power transmission from fractional-horsepower applications, such as hand tools and kitchen utensils, to applications involving thousands of horsepower, such as heavy machinery and marine drives. However, neither the horsepower rating nor the size of a gear is necessarily indicative of the severity of the loading it can withstand. For example, the severity of tooth loading in the gear train of a 186-W (¼-hp) hand drill may exceed that of the loading in a 15-MW (20 000-hp) marine drive. Factors other than horsepower rating and severity of loading can affect gear strength and durability, particularly duration of loading, operating speed, transient loading, and such environmen-

Fig. 10 Photoelastic study of two mating pinion teeth receiving full load

Note the high concentration of compressive stress at the point of contact, the tensile stress at the root radius, and the zero-stress point at the tooth centerline below the root circle.

tal factors as lubrication, temperature, contaminants, and mechanical stability.

Gear Stresses. Figure 10, a photoelastic study of a loaded gear in action, emphasizes the importance of root fillets in gear loading. The areas where the stress patterns are close together and concentric indicate very high stress gradients. High stress gradients are usually indicative of high stress levels. Figure 10 shows two matching teeth in contact at or near the pitchline. Note the high concentration of compressive stress at the pitchline or point of contact, the concentration of tensile stress at the root radius of the loaded side, the concentration of compressive stress at the root radius of the opposite or nonloaded side, and the zero-stress point below the root circle at or near the tooth centerline. The progression of any fatigue crack initiated at the root radius will head directly for this zero-stress point, which is the path of least resistance.

Associated Parameters

The importance of the function of the matching part has been emphasized. There are also components in the design and structure of each gear and/or gear train that must be considered in conjunction with the teeth.

A round bore, closely toleranced and ground, may rotate freely around a ground shaft diameter, rotate tightly against a ground diameter by having a press fit, or be the outer race of a needle bearing that gives freedom of rotation. Each application has its own unique problems. A ground bore is always subject to tempering, burning, and checking during the grinding operation. A freely rotating bore requires a good lubrication film, or seizing and galling may result. The bore used as a bearing outer race must be as hard as any standard bearing surface and is subject to all conditions of rolling contact, such as fatigue, pitting, spalling, and galling. The bore that is press fit onto a shaft is subject to a definite amount of initial tensile

stress. Also, any tendency for the bore to slip under the applied rotational forces will set up a unique type of wear between the two surfaces that leads to a recognizable condition of fretting damage.

A spur of a helical round bore gear acting as an idler, or reversal, gear between the input member and the output member of a gear train has an extremely complicated pattern of stresses. The most common application is the planet pinion group in wheel-reduction assemblies or in planetary-type speed reducers. The photoelastic study shown in Fig. 11 reveals certain facts:

- The tensile and compressive stresses in the bore are due to bending stresses of the gear being loaded as a ring
- The maximum tensile and compressive stresses in the bore increase as the load on the teeth increases
- The maximum tensile and compressive stresses in the bore increase as the clearance between the bore and the shaft is increased
- The maximum tensile and compressive stresses in the bore increase as the ratio of the size of the bore to the root diameter of the teeth is increased
- During one revolution of the gear under load, the teeth go through one cycle of complete reversal of stresses, whereas each element of the bore experiences two cycles of reversals

There are three modifications of the round bore that alter stress patterns considerably:

- *Oil holes* that extend into the bore are intended to lubricate the rotating surfaces. Each hole may be a stress raiser that could be the source of a fatigue crack
- *Tapered bores* are usually expected to be shrink fitted onto a shaft. This sets up a very high concentration of stresses not only along the ends of the bore but also at the juncture on the shaft
- *A keyway* in the bore also creates a stress-concentration area. A keyway is also required to withstand a very high load that is continuous. In fact, the applied load to the side of the keyway is directly proportional to the ratio of tooth pitchline radius to the radius of the keyway position. Fatigue failure at this point is common

Splined Bores. The loads applied to the splined-bore area are also directly proportional to the ratio of pitchline radius of the teeth to the pitchline radius of the splines. However, the load is distributed equally onto each spline; therefore, the stress per spline is usually not excessive. It is possible for an out-of-round condition to exist that would concentrate the load on slightly more than two splines. Also, a tapered condition would place all loadings at one end of the splines, which would be detrimental both to the gear splines and to the shaft.

Fig. 11 Photoelastic pattern of idler gear between load input pinion and load output gear

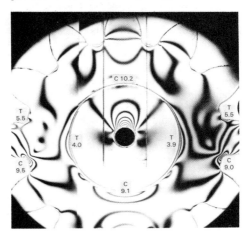

Heat treating of the splined area must also be monitored closely for quench cracks. Grinding of the face against the end of the splines can also cause grinding checks to radiate from the corners of the root fillets.

Shafts. The shafts within a gear train, as well as the shank of a pinion that constitutes a shaft in function, are very important to load-carrying capacity and load distribution. They are continually exposed to torsional loads, both unidirectional and reversing. The less obvious stressed condition that is equally important is bending. These bending stresses can be identified as unidirectional, bidirectional, or rotational. When the type of stress is identified, one can explore the causes for such a stress.

There can be a number of stresses applied to a shaft that are imposed by parts riding on it. For example, helical gears will transmit a bending stress, as will straight and spiral bevel gears; round bores may be tight enough to cause scoring and galling; a splined bore may cause high stress concentration at the end face; runout in gears may cause repeated deflections in bending of the shaft; and loose bearings may cause excessive end play and more bending.

Gear Materials

A variety of cast irons, powder-metallurgy materials, nonferrous alloys, and nonmetallic materials are used in gears, but steels, because of their high strength-to-weight ratio and relatively low cost, are the most widely used gear materials. Consequently, steel gears receive primary consideration in this article.

Among the through-hardening steels in wide use are 1040, 1060, 4140, and 4340. These steels can also be effectively case hardened by induction heating. Among the carburizing steels used in gears are 1018, 1524, 4026, 4118, 4320, 4620, 4820, 8620, and 9310

(AMS 6260). Many high-performance gears are carburized. Some special-purpose steel gears are case hardened by either carbonitriding or nitriding. Other special-purpose gears, such as those used in chemical or food-processing equipment, are made of stainless steels or nickel-base alloys because of their corrosion resistance, their ability to satisfy sanitary standards, or both. Gears intended for operation at elevated temperatures may be made of tool steels or elevated-temperature alloys.

Most gears are made of carbon and low-alloy steels, including carburizing steels and the limited number of low-alloy steels that respond favorably to nitriding. In general, the steels selected for gear applications must satisfy two basic sets of requirements that are not always compatible—those involving fabrication and processing and those involving service. Fabrication and processing requirements include machinability, forgeability, and response to heat treatment as it affects fabrication and processing. Service requirements are related to the ability of the gear to perform satisfactorily under the conditions of loading for which it was designed and thus encompass all mechanical-property requirements, including fatigue strength and response to heat treatment.

Because resistance to fatigue failure is partly dependent upon the cleanness of the steel and upon the nature of allowable inclusions, melting practice may also be a factor in steel selection and may warrant selection of a steel produced by vacuum melting or electroslag refining. The mill form from which a steel gear is machined is another factor that may affect its performance. Many heavy-duty steel gears are machined from forged blanks that have been processed to provide favorable grain flow consistent with load pattern rather than being machined from blanks cut from mill-rolled bar.

Classification of Gear Failures

Systematic analysis of a gear failure begins with classification of the failure by type, or mode. The mode of failure is determined from the appearance of the failed gear and from the process or mechanism of the failure. A mode of gear failure is a particular type of failure that has its own descriptive identification. One mode may or may not be unique to a specific failure, because the origin may be of one mode, progression may be of a second mode, and final fracture may be of a third mode. After the mechanism of a failure has been established, the cause of the failure remains to be determined. In general, an understanding of the failure mechanism is of considerable assistance in isolating the cause or causes of a failure.

Several failure analysts in various fields of expertise have compiled their findings as to the frequency of failure modes. Table 1 lists these findings in order of decreasing frequency (the primary cause of failure is fatigue, the second-

Table 1   Failure modes of gears

Failure mode	Type of failure
Fatigue	Tooth bending, surface contact (pitting or spalling), rolling contact, thermal fatigue
Impact	Tooth bending, tooth shear, tooth chipping, case crushing, torsional shear
Wear	Abrasive, adhesive
Stress rupture	Internal, external

ary cause is impact, and so on). In a composite analysis of more than 1500 studies, the three most common failure modes, which together account for more than half the failures studied, are tooth-bending fatigue, tooth-bending impact, and abrasive tooth wear.

Fatigue

Fatigue failure results from cracking under repeated stresses much lower than the ultimate tensile strength. This type of failure:

- Ordinarily depends upon the number of repetitions of a given stress range rather than the total time under load
- Does not occur below some stress amplitude called the fatigue limit
- Is greatly encouraged by notches, grooves, surface discontinuities, and subsurface imperfections that will decrease the stress amplitude that can be withstood for a fixed number of stress cycles
- Is increased significantly by increasing the average tensile stress of the loading cycle

There are three stages within a fatigue failure that must be studied closely: the origin of the fracture, the progression under successive cycles of loading, and final rupture of the part when the spreading crack has sufficiently weakened the section. Most attention is devoted to the first stage to answer the question, Why did it start at this point? The second stage is observed to determine the direction of the progression. It appears that a fatigue crack will follow the path of least resistance through the metal. The final fracturing may be by shear or tension, but in either case, the size of the final-fracture area can be used to calculate the apparent magnitude of stress that had been applied to the part.

Tooth-bending fatigue is the most common mode of fatigue failure in gearing. There are many variations of this mode of failure, but the variations can be better understood by first presenting the classical failure.

The classical tooth-bending fatigue failure is that which occurs and progresses in the area designed to receive the maximum bending stress. Figure 12 shows such an example. There are five conditions that make Fig. 12 a perfect tooth-bending fatigue failure:

Fig. 12 Spiral bevel pinion showing classic tooth-bending fatigue

The origin is at midlength of the root radius on the concave (loaded) side. 0.4×

- The origin is at the surface of the root radius of the loaded (concave) side of the tooth
- The origin is at the midpoint between the ends of the tooth where the normal load is expected to be
- One tooth failed first, and the fracture progressed slowly toward the zero-stress point at the root, which shifted during the progression to a point under the opposite root radius and then proceeded outward to that radius
- As the fracture progressed, the tooth deflected at each cycle until the load was picked up simultaneously by the top corner of the next tooth, which (because it was now overloaded) soon started a tooth-bending fatigue failure in the same area. The fracture of the second tooth appears to be of more recent origin than the first fracture
- The material and metallurgical characteristics are within specifications

If any of the above five conditions is changed, there will be a variation of tooth-bending fatigue away from the classical example.

The rudiments of the classical tooth-bending fatigue can be studied using photoelastic models. Once the crack at the root radius is initiated and is progressing toward the zero-stress point (Fig. 13), the point moves away from the leading edge of the crack and shifts laterally until it reaches a position under the opposite root. At that time, the shortest remaining distance is outward to the root, and the point terminates there. In the meantime, as the failing tooth is being deflected, the top of the adjacent tooth has picked up the load. The load on the first tooth has now been relieved, allowing a slower progression due to lower cyclic stress. It is now only a matter of time until the next tooth initiates a crack in the same position. Variations from the classical tooth-bending fatigue appear for many reasons.

Tooth-bending fatigue of a spur-gear tooth with the origin along the surface of the root radius of the loaded side but at a point one-third

Fig. 13 Photoelastic study of mating teeth indicates the shift of the zero-stress point during crack propagation until final fracture reaches the opposite root radius

At the same time, deflection of the tooth allowed the adjacent tooth to pick up the load.

Fig. 14 Spur tooth pinion at 0.5× (top) and 1.5× (bottom)

Tooth-bending fatigue originating at the root radius (arrows), loaded side, one-third the distance from the open end. Progression was to the bore.

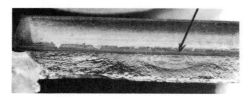

the distance from the open end (Fig. 14) can be considered a classical failure in that the maximum crown of the teeth had been placed intentionally at the area where the failure occurred; therefore, the maximum applied loads were at the same area. This, then, is the area that was designed for the point of initiation of normal tooth-bending fatigue. The progression was directly to the bore, which was the shortest direction of least resistance.

Tooth-bending fatigue of a spur-gear tooth meeting all but one of the conditions for a classical failure is shown in Fig. 15. The origin was at one end of the tooth and not at the midlength. There could have been several reasons for this point of initiation; in this instance, a severe shock load had twisted the parts until a

Fig. 15 Spur pinion

Tooth-bending fatigue with origin at root radius of loaded side at one end of the tooth. 0.6×

Fig. 16 Spur pinion

Tooth-bending fatigue is at midlength of the tooth at the root radius, but the origin is at an inclusion located in the case/core transition. 55×

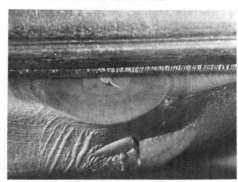

momentary overload was applied at the end of the tooth, causing a crack to form. The crack became a high stress-concentration point from which fatigue progression continued.

An example of tooth-bending fatigue that did not conform to classical conditions is shown in Fig. 16. Crack initiation was at a nonmetallic inclusion at the case/core interface; it did not originate at the surface. Subsequent progression was through the case to the surface as well as through the core toward the zero-stress point.

Another example of tooth-bending fatigue that did not conform is shown in Fig. 17. The failure was at the root radius and at the midlength, but the origin was not at the surface. The origin was at the apex of a tapped bolt hole that had been drilled from the back face of the gear and had terminated less than 6.4 mm (0.25 in.) from the root radius. Subsequent progres-

Fig. 17 Spiral bevel gear tooth

Tooth-bending fatigue with origin at the apex of the drilled bolt hole, which terminated just below the root radius. 0.5×

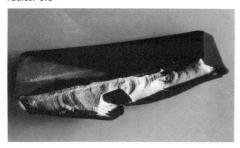

sion of the fatigue crack was to the surface at the root radius as well as into the core toward the zero-stress point.

Surface-Contact Fatigue (Pitting). As a surface or near-surface failure, pitting is recognized as a fatigue mode of failure. As elements of the structure at and near the surface are subjected to alternating compressive stresses, there is plastic deformation of some regions of the microstructure and elastic deformation in others. The differences of plasticity between grains will encourage fatigue cracks to initiate under pulsating and alternating stresses. Also, each grain is randomly oriented as to the direction of its shear plane. Under a specific compressive load, some grains will tend to break apart in tension, others will tend to shear, and yet another group will be unaffected. It is these internal stress raisers that will allow crack initiation at a specific point when the resultant strain exceeds the elastic limit of that point. The initiated cracks flow together and accumulate to form a plane of progression; then, the path of least resistance is followed, and a particle falls away from the surface. Pitting has begun.

All too often, there may be present at or near the surface an embedded inclusion that will act as a nucleus for a crack progression. These inclusion sites are usually much weaker and will allow crack initiation first at that point (Fig. 18).

The initiation of pitting is mostly confined to three areas along the profile of a gear tooth. First, the pitchline is the only area of line contact that receives pure rolling pressure. The pits forming directly at the pitchline are usually very small and may not progress beyond the point of original initiation. Some analysts claim that this type of pitting repairs itself and is not detrimental. That may be true when the pitting is formed by the breakdown of a rough surface. Note, too, that lubrication will not deter the origin of pitting along the pitchline. Oil is noncompressible and will not cushion the pressures exerted in the area of pure rolling.

Second, the area immediately above or below the pitchline is very susceptible to pitting. Not only is the rolling pressure great at this point but sliding is now a real factor. The

Fig. 18 Spur-gear tooth

An internal crack originating at an oxide-type inclusion below the surface at the pitchline. A pit is being formed. Nital etch. 90×

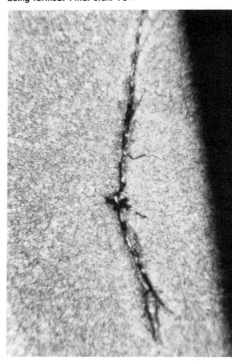

mechanics of surface-subsurface pitting can be best understood when looking at the resultant applied stresses illustrated in Fig. 19. The extra shear stress of the sliding component when added to that of the rolling component often results in near-surface fatigue at the point of maximum shear below the surface. In many instances, it is difficult to determine whether some pitting cracks actually initiate at or below the surface. Figure 20 shows initiation of pitting fatigue both at the pitchline of a helical-gear tooth and directly above the pitchline. Progression up the addendum in some areas makes it difficult to differentiate the two. A surface-pitted area near the pitchline is illustrated in Fig. 21. Although the surface was in a rolling-sliding area, the immediate surface shows no catastrophic movement. It is then most probable that cracking initiated by shearing fatigue below the surface. The freedom from surface movement also speaks well for the lubrication of the gear teeth.

Third, the lowest point of single-tooth contact is that point which receives the tip of the mating tooth as it makes first contact low on the active profile. Tip contact will produce high pressures, even if a small fraction of the actual load is transferred by the tip. Also, the maximum sliding speed, both in approach and recess, occurs at the tip contact.

There are two types of pits that prevail at this area. One is the type shown in Fig. 22, which appears to be a swift-shear and lift-out type.

Fig. 19 Stress distribution in contacting surfaces due to rolling, sliding, and combined effect

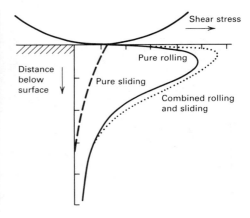

Fig. 20 Helical-gear tooth

Pitting initiated along and immediately above the pitchline. In some areas, the progression has been continuous. Actual size

Perhaps the question whether this is fatigue or impact can be answered, but if it is shearing fatigue, it is certainly rapid and would call for a study of tip interference and lubrication. The second type of pit is the same as that illustrated in Fig. 21 and shows failure only by compressive loads. This is not uncommon in spur-gear teeth or helical-gear teeth but, when found, is puzzling because one does not expect to find tip interference pits other than those previously discussed. However, it must be remembered that dynamic effects are to be taken into account. Gear vibration is a fact, but is seldom ever recognized as such. Studies indicate that gear vibration at high speeds, excited by static transmission error, may cause corner contact even though the tip relief is sufficient to prevent corner contact at low speed. Also, with sufficient lubrication, the pressures of tip interference will result in compressive-type pitting (Fig. 21) rather than the adhesive type (Fig. 22).

Other than the three specific and most common areas of surface-fatigue pitting just discussed, there are always the random areas that crop up to demand recognition, such as pitting at only one end of the face, at opposite ends of

Fig. 21 Spiral pinion tooth

Near-pitchline pitting fatigue. Origin is subsurface at plane of maximum shear. 180×

Fig. 22 Spiral bevel tooth

Pitting at the lowest point of single-tooth contact illustrating contact path of the tip of the mating tooth. Nital etch. 90×

Fig. 23 Gear-tooth section

Rolling-contact fatigue. Crack origin subsurface. Progression was parallel to surface and inward away from surface. Not etched. 60×

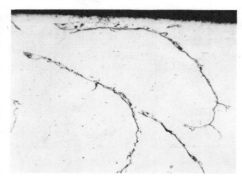

Fig. 24 Gear-tooth section

Rolling-contact fatigue. Crack origin subsurface. Progression was parallel with surface, inward, and finally to the surface to form a large pit or spall. Not etched. 60×

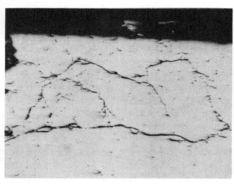

Fig. 25 Gear-tooth section

Rolling-contact fatigue distinguished by subsurface shear parallel to surface. Note the undisturbed black oxides at the surface, indicating no surface-material movement. Not etched. 125×

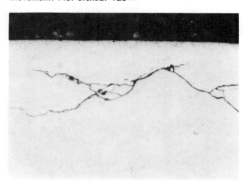

Fig. 26 Same sample as in Fig. 25, showing details of submicrostructure called butterfly wings

3% nital etch. (a) 125×. (b) 310×

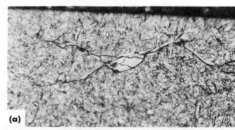

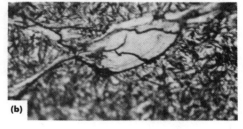

opposite faces, along the top edge of the addendum, or at one area of one tooth only. The random incidences are without number and are most baffling; but each one is present because of a reason, and that reason must be found if at all possible.

Rolling-contact fatigue can only be described by using Fig. 19. Under any conditions of rolling, the maximum stress being applied at or very near the contact area is the shear stress parallel to the rolling surface at some point below the surface. For normally loaded gear teeth, this distance is from 0.18 to 0.30 mm (0.007 to 0.012 in.) below the surface just ahead of the rolling point of contact. If sliding is occurring in the same direction, the shear stress increases at the same point. If the shear plane is close to the surface, then light pitting will probably occur. If the shear plane is deep due to a heavy rolling load contact, then the tendency is for the crack propagation to turn inward (Fig. 23). The cracks will continue to propagate under repeated stress until heavy pitting or spalling takes place (Fig. 24).

There is always one characteristic (and often two) of rolling-contact fatigue that will distinguish it from other modes of surface-contact fatigue. Both characteristics can be observed only by an examination of the microstructure. In rolling contact, the surface will not show a catastrophic movement; it will remain as the original structure. For example, an unetched, polished sample taken near the origin of a subsurface fatigue crack (Fig. 25) very clearly shows undisturbed black oxides at the surface. The subsurface cracking could not have been caused by either abrasive or adhesive contact; it had to be by rolling. This is always the first evidence to look for when determining the type of applied stress.

The second characteristic is common only in a martensitic case that contains very little or no austenite and is found only at, along, or in line with the shear plane. This is a microstructural feature that has been termed butterfly wings. If the sample in Fig. 25 is etched properly with 3% nital, the result is the microstructure shown in Fig. 26(a). Increasing the magnification to 310× shows more detail (Fig. 26b).

The original discussion presented by rolling-element bearing manufacturers had no explanation as to the origin of this microstructural alteration except that it was always in conjunction with and radiated from an inclusion. It is

Fig. 27 Spiral bevel gear teeth

Original pitting low on the active profile gives initiation to a fast and extensive progression of spalling over the top face and down the back profile. This is often called the cyclone effect. 0.25×

Fig. 28 Applied stress versus case depth (net strength)

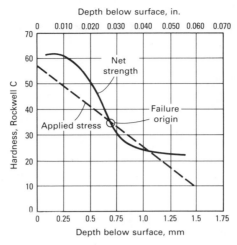

Fig. 29 Subsurface cracking that subsequently resulted in spalling at a gear-tooth edge

Unetched section of a carburized AMS 6260 steel gear tooth. Cracking initiated in the transition zone between the carburized case and the core. 500×

Fig. 30 Spalling on a tooth of a steel spur sun gear shaft

(a) Overall view of spalled tooth. (b) Micrograph of an unetched section taken through the spalled area showing progressive subsurface cracking. 100×

(a)

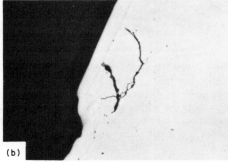

(b)

ting and spalling. For example, Fig. 27 shows a spiral-gear tooth with pitting low on the profile that subsequently progressed until spalling occurred over the top face and back side profile. This apparent rapid and extensive progression can be referred to as the cyclone effect.

Spalling, in the true sense of the word, is a mode of fatigue failure distinct in itself and unique in its origin. It originates below the surface, usually at or near the case/core transition zone. As illustrated in Fig. 28, the origin is at that point where the sum of all applied and misapplied stresses intersect the net strength of the part. The applied stresses are substantially in shear, and the intercept point is most likely under a carburized case. Fatigue fractures will generally progress under the case (Fig. 29 and 30) and will eventually spall away from the gear tooth.

Thermal fatigue in gearing, although not necessarily correct, is most often synonymous with frictional heat. This is perhaps the only type of alternating heating and cooling that is applied in the field. Then, too, it is not commonly found on the active profile of the gear teeth, but on a rotating face that is exerting a high amount of thrust. Figure 31(a) shows an

are caused when, under an extreme shearing stress, movement is called for but is restrained and contained to such an extent that the energy absorbed institutes a change in the microstructure ahead of a progressing crack. They are never observed when significant amounts of austenite are present; austenite will quickly absorb the energy and will be converted to untempered martensite. They are also in an area that has not been deformed but has definitely been transformed. Each area has distinct boundaries, and the oncoming cracks appear to follow these boundaries. It has been noted that some academic studies refer to this same structure as being a transformed shear band product formed by adiabatic shear.

Contact Fatigue (Spalling). Spalling, in general, is not considered an initial mode of failure but rather a continuation or propagation of pitting and rolling-contact fatigue. It is very common to reference this failure mode as pit-

Fig. 31 Thermal fatigue cracking of a spur gear

(a) Radial cracking due to frictional heat against the thrust face. 0.4×. (b) Progression of thermal fatigue produced by the frictional heat. 1.5×

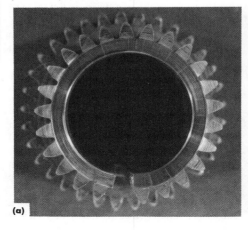

(a)

(b)

true that many times there is an inclusion present, but not always. One could argue that the fracture is a plane and not a line, such as could be seen on a polished sample; therefore, an inclusion could be present somewhere else along the plane being observed.

Two other extensive and very detailed studies of rolling-contact fatigue refer to the gray substructure as white bands of altered martensite. It is believed that these substructures example of high end thrust that resulted in frictional heat. The thermal expansion and contraction initiated a large number of radial cracks that progressed as a fatigue fracture (Fig. 31b) into the bore of the part and along the roots of the teeth.

Impact

Tooth-Bending Impact. When a tooth is removed from a gear within a very few cycles (usually one or two), the fracture is uniform in structure and does not show the striations common to the fatigue mode of failure. They are usually random, due to a sudden shock load, either in a forward or reverse direction, and do not necessarily originate at the root radius. In fact, if the fracture did originate at the root radius, it would follow a rather flat path across to the opposite root radius rather than travel downward toward the zero-stress point.

Tooth Shear. When the impact is very high and the time of contact is very short, and if the ductility of the material will allow it, the resultant tooth failure mode will be shear. The fractured area appears to be highly glazed, and the direction of the fracture will be from straight across the tooth to a convex shape. For example, a loaded gear and pinion set were operating at a high rate of speed when the pinion stopped instantaneously (Fig. 32). The momentum of the gear was strong enough to shear the contacting pinion teeth from the reverse direction, leaving the remaining teeth in excellent condition. The gear teeth were partially sheared and were all scrubbed over the top face from the reverse direction.

Most commonly, bending impact and shear are found in combination (Fig. 33). Tooth-bending impact fractures supersede the shearing fractures; thus, the direction of impact can be determined.

Tooth chipping is certainly a type of impact failure, but is not generally considered to be tooth-to-tooth impact. It is usually accomplished by an external force, such as a foreign object within the unit, a loose bolt backing out into the tooth area, or even a failed tooth from another gear. Figure 34 shows a chipping condition that existed during the operation of the part. Most chipping, however, is caused by mishandling before, during, or after the finished parts are shipped from the manufacturer and are not field failures. This is one of the modes that must be identified by a cause.

Case crushing will occur when an extreme overload is applied to a carburized case. For it to occur, there are four factors to be considered because case crushing is a combination of events. Case crushing is dependent upon the stress applied at the point of contact, the radius of curvature of the contacting surfaces, the thickness of the case, and the core hardness of the material. The resultant failure is due to excessive compressive loads per unit area for the existing conditions. The fracture will start at the case/core interface and will continue to

Fig. 32 Spiral bevel gear and pinion set that sheared in the reverse direction

The pinion came to a sudden and complete stop at the instant of primary failure of the unit, allowing the gear to shear the contacting teeth and to continue rotating over the failed area. 0.4×

Fig. 33 Spur-gear tooth showing combination failure modes

(a) Tooth-bending impact. (b) Tooth shear. Arrows indicate direction of applied force.

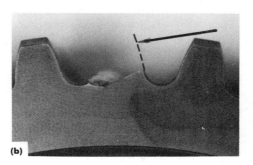

(a)

(b)

Fig. 34 Tooth chipping as a field failure generally shows a pattern

In this case, some object of impact was within the assembled unit.

Fig. 35 Case crushing at midprofile of a spiral bevel gear tooth

Progression is from the subcase area into the core and outward to the surface.

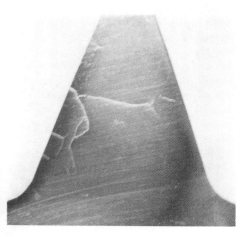

shear into the core and outward to the surface. Figure 35 shows the subsurface propagation of the crushing effect.

Wear

Surface deterioration of the active profile of the gear teeth is called wear. There are two distinct modes of wear—abrasive and adhesive.

Abrasive wear occurs as the surface is being freely cut away by hard abrasive particles. It can happen only as two surfaces are in sliding contact. The dissociated material must be continually washed away and not allowed to accumulate on the sliding surfaces, or adhesion may take place. The first evidence of abrasive wear is the appearance of light scratches on the surface, followed by scuffing. As the scuffing continues to deepen, the result is scoring. Figure 36 illustrates an excellent example of pure abrasive wear that shows the entire tooth

Fig. 36 Spiral bevel gear teeth showing contact wear

Insert A shows a tooth area exhibiting no wear. Insert B shows abrasive wear clearly cutting away 3.2 mm (⅛ in.) of the surface without damage to underlying material.

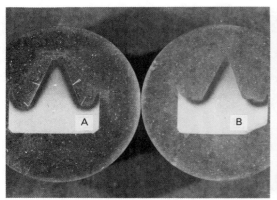

Fig. 39 Carburized AMS 6260 steel gear damaged by adhesive wear

(a) Overall view of damaged teeth. (b) Etched end face of the gear showing excessive stock removal from drive faces of teeth

(a)

(b)

Fig. 37 Pinion tooth profile

Glazed surface showing the start of catastrophic movement of surface material. Frictional heat has already started to temper the surface. 75×

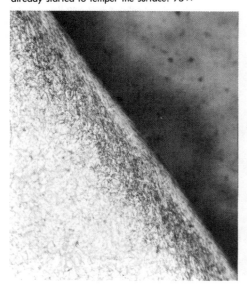

Fig. 38 Pinion tooth profile

The pinion was plastically deformed by frictional heat and sliding pressures. The surface layer has locally rehardened, and galling is evident. 80×

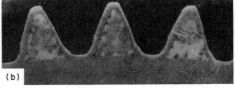

come sharp points that cut into the matching parts. As the adhesive wear continues with the parts still in service, there is the distinct possibility that the part will ultimately fail (Fig. 40).

Adhesive wear on gear teeth may not always point to that set as the culprit. There have been instances when a severe case of adhesion is of a secondary nature. In other words, particles of material from a primary failure of another component in the assembly may have impinged upon the gear-tooth surfaces and thus started the sequence of events leading to seizure and destruction.

Stress Rupture

When the internal residual stresses build to a magnitude beyond the strength of the material, the part will rupture. The rupture will occur at the point where this critical value is exceeded, either internally or externally.

Internal Rupture. The most likely point within a gear that would attract a buildup of residual stresses is the case/core interface near the top face or the corner of a tooth. Figure 41 shows an excellent example of case/core separation due to internal residual stresses exceeding the strength of material at the case/core transition zone. In severe cases, the entire top of the tooth might pop off (Fig. 42).

External rupture is much easier to understand because it almost always originates from a stress raiser. For example, a part not yet in service ruptured starting at the end face (Fig.

profile cleanly cut away with no microstructural damage to the underlying material.

Abrasive wear cannot be deterred by lubrication, because the lubricant is often the vehicle that contains and continually supplies the abrasive material as a contaminant. When this is the case, all moving parts within the assembly are affected, such as seals, spacers, bearings, pumps, and matching gears.

When abrasive wear is isolated to only one part, that is, either gear or pinion, it is imperative to examine closely the surface of the matching part. For example, a very large amount of massive carbides embedded in the surface may easily cut into a softer matching surface.

Adhesive wear occurs on sliding surfaces when the pressure between the contacting asperities is sufficient to cause local plastic de-

formation and adhesion. Whenever plastic deformation occurs, the energy expended to produce the deformation is converted to heat.

The first indication of trouble is a glazed surface, followed by galling, then by seizure. A glazed surface may not have any dimensional changes, but the microstructure shows a catastrophic movement of surface material (Fig. 37). As frictional heat increases, the surface becomes softer, the adhesive tendency becomes greater, there is more plastic deformation, the local temperature becomes high enough to change completely the microstructure at the surface (Fig. 38), and galling occurs. Galling continues, the sliding surfaces begin welding together, and adhesion takes place. Large areas may be pulled away from the surface (Fig. 39). Some of these welded particles are so hard due to change of microstructure that they be-

Fig. 40 Extreme wear of steel pinions

(a) Hypoid pinion with teeth worn to a knife-edge from both sides. (b) A similar pinion with teeth worn completely away by adhesive wear

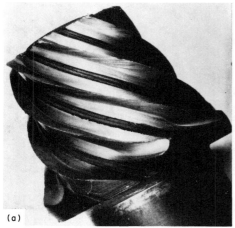

(a)

(b)

Fig. 41 An internal rupture in a gear tooth at the case/core transition zone

The rupture does not reach the surface. This condition can be discovered by ultrasonic testing.

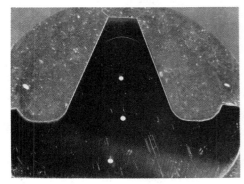

43) from a grinding crack; this was confirmed by examination of the fractured face (Fig. 44). The grinding cracks were not the entire cause of the failure, but they supplied the notch in the highly stressed area.

Fig. 42 Spiral bevel gear tooth

Internal rupture is lifting the entire top of a tooth.

Fig. 43 Spur gear

External rupture (assembled, but not in service) with origin at the end face. See also Fig. 44.

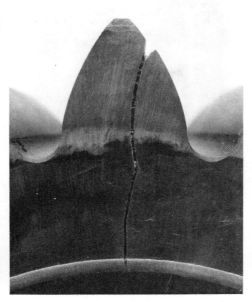

Causes of Gear Failures

Analysis of the actual part that failed will show the mode of failure, but the cause of failure is often found in the mating or matching part. One should always inspect both parts very closely for the correct answer to the problem. In fact, if there are several components to an assembly, each component must be suspect until eliminated.

The cause of a specific gear failure may be obvious enough to be recognized by anyone or subtle enough to defy recognition by the most experienced analyst. At times, one almost believes that the number of causes may equal the number of failures. This is close enough to being correct that each failure must be treated individually as a complete case in itself.

It is very difficult in many instances to separate cause from mode. In fact, a single cause can initiate several different modes, depending upon the forces being applied. Conversely, a specific mode of failure could have been initiated by one of several causes. It must be emphasized that one should not match a fracture pattern of a failed gear to a picture in

Fig. 44 Spur gear

External rupture originating from a grinding check. See also Fig. 43.

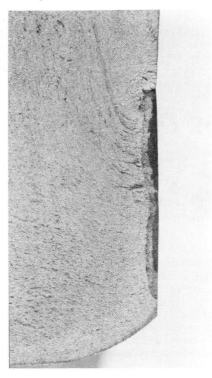

Table 2 Summary of a statistical report on causes of gear failures (%) over a 35-year period

Basic material	0.8
Steel	
Forgings	
Castings	
Engineering	6.9
Design	
Material selection	
Heat treat specifications	
Grinding tolerances	
Manufacturing	1.6
Tool undercutting, sharp notches	
Tooth characteristics	
Grinding checks, burns	
Heat treat changes	
Heat treatment	16.0
Case properties	
Core properties	
Case/core combinations	
Hardening	
Tempering	
Miscellaneous operations	
Service application	74.7
Set-matching	
Assembly, alignment, deflection, vibration	
Mechanical damage, mishandling	
Lubrication	
Foreign material	
Corrosion	
Continuous overloading	
Impact overloading	
Bearing failure	
Maintenance	
Operator error	
Field application	

Fig. 45 Destructive wear of an 1113 steel worm used in a silo hoist

Fig. 46 Pitting and wear pattern on a carburized AMS 6263 steel impeller drive gear
Approximately 2.3×

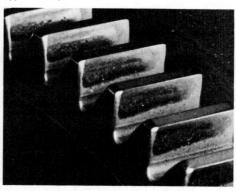

this Handbook and conclude that the cause and effect are the same. They may be, but not necessarily. The pictures of a reference book should be used only as a guide. Causes, as listed in Table 2, are under five major headings and several subheadings. Undoubtedly, there may be more. The headings are not listed in order of importance (they are all important), but in sequence of probable occurrence.

The Final Analysis

No failure examination has been successfully completed until an evaluation of the results is made and it is only when the failure mechanism is understood that effective corrective measures can be devised. The purpose now is to take the known facts of a specific failure, to place them in a systematic context to arrive at a logical conclusion, and to indicate a corrective measure. Obviously, one cannot detail the hundreds of examinations available, but a chosen few will be presented to illustrate a systematic approach to the examination, regardless of its apparent simplicity or complexity.

Example 1: Catastrophic Failure of a Hoist Worm Caused by Destructive Wear Resulting From Abusive Operation. A farm-silo hoist was used as the power source for a homemade barn elevator. The hoist mechanism consisted of a pulley attached by a shaft to a worm that, in turn, engaged and drove a worm gear mounted directly on the hoist drum shaft. The worm was made of leaded cold-drawn 1113 steel with a hardness of 80 to 90 HRB; the worm gear was made of class 35-40 gray iron that had been nitrided in an aerated salt bath. Driving power was applied to the pulley, which actuated the worm and worm gear and rotated the drum to which the elevator cable was attached. After less than 1 year of service, the hoist failed catastrophically from destructive wear of the worm (Fig 45). The hoist was rated at 905 kg (2000 lb) at a 200-rpm

input. It was determined that at the time of failure the load on the hoist was only 325 kg (720 lb).

Investigation. When examined, the gearbox was found to contain fragments of the worm teeth and shavings that resembled steel wool. It was determined that both fragments and shavings were steel of the same composition as the worm, ruling out the possibility that these materials may have derived from a source other than the gearbox. The teeth of the worm had worn to a thickness of approximately 1.6 mm (1/16 in.) from an original thickness of 3.2 mm (1/8 in.). More than half of the worm teeth had been sheared off.

Further investigation revealed that the drive pulley had been replaced with a pulley of different diameter and that the pulley was rotating at 975 rpm instead of at the rated maximum of 200 rpm. By calculation, with a load of 325 kg (720 lb), the gearbox was forced to develop 0.24 kW (0.32 hp)—almost twice the specified rating of 0.13 kW (0.18 hp) (2000 lb at 200 rpm).

Conclusions. The failure mode was adhesive wear of the steel worm, which resulted from operation of the gearbox at a power level that far exceeded the maximum specified by the gearbox manufacturer. The cause of failure was misuse in the field.

Example 2: Failure of Carburized Steel Impeller Drive Gears Due to Pitting and a Wear Pattern. Two intermediate impeller drive gears were submitted for metallurgical examination when they exhibited evidence of pitting and abnormal wear after production tests in test-stand engines. Both gears were made of AMS 6263 steel and were gas carburized, hardened, and tempered. Figure 46 shows the pitting and the wear pattern observed on the teeth of gear 1.

According to the heat-treating specification, the gears were required to satisfy the following requirements: (1) a carburized case depth of 0.4

to 0.6 mm (0.015 to 0.025 in.), (2) a case hardness of 77 to 80 HR30N, and (3) a core hardness of 36 to 44 HRC.

Investigation. Sections of both gears were removed with a cutoff wheel and examined for hardness, case depth, and microstructure of case and core. Results were as follows:

	Gear 1	Gear 2
Hardness		
Case, HR30N	70–73	77–78
Core, HRC	40	40
Case depth, mm (in.)	0.5–0.6 (0.02–0.024)	0.6–0.7 (0.024–0.028)
Microstructure		
Case	Lean carbon (<0.85% C)	Normal
Core	Normal	Normal

Both the pitting and the wear pattern were more severe on gear 1 than on gear 2.

Conclusions. Gear 1 failed by surface-contact fatigue (pitting) because the carbon content of the carburized case and consequently the case hardness were below specification and inadequate for the loads to which the gear was subjected. The pitting and wear pattern observed on gear 2 were relatively mild because case-carbon content and case hardness were within specified requirements. Nevertheless, it was evident that case depth, as specified, was not adequate for this application—even with case-carbon content and hardness being acceptable.

Recommendations. It was recommended that the depth of the carburized case on impeller drive gears be increased from 0.4 to 0.6 mm (0.015 to 0.025 in.) to 0.6 to 0.9 mm (0.025 to 0.035 in.), an increase that would still allow a minimum of 30% core material across the tooth section at the pitchline. This increase would ensure improved load-carrying potential and wear resistance. It was also recommended that hardness readings on gears be made using a

Fig. 47 Metallurgical causes of destructive pitting that occurred in a carburized AMS 6263 steel gear

(a) Specimen etched 15 to 20 s in 2% nital showing a surface layer of decarburized material. 500×.
(b) Same specimen repolished and etched 3 s in 2% nital showing a heavy subsurface layer of oxide scale. 500×. The white band immediately above the decarburized and oxidized layers is electrodeposited nickel that was applied to prevent edge-rounding during polishing.

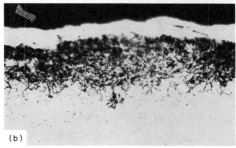

(a)

(b)

60-kg (132-lb) load (Rockwell A scale) and that the minimum case-hardness requirement be set at 81 HRA, an additional safeguard to ensure adequate load-carrying capacity.

Example 3: Failure of a Carburized Steel Generator-Drive Idler Gear by Pitting Due to Decarburization and Subsurface Oxidation. Following test-stand engine testing, an idler gear for the generator drive of an aircraft engine exhibited evidence of destructive pitting on the gear teeth in the area of the pitchline; as a consequence, the gear was submitted for metallurgical examination. The gear was made of AMS 6263 steel and was gas carburized to produce a case 0.4 to 0.6 mm (0.015 to 0.025 in.) deep.

Investigation. Sections of the gear were removed and examined for hardness, case depth, and microstructure of case and core. The results of the examination and the specified requirements were:

	Required	Actual
Case depth, mm (in.)	0.4–0.6 (0.015–0.025)	0.6–0.7 (0.024–0.028)
Hardness		
Case, HR30N	77–80	74–76 (HR15N88–89)
Core, HRC	36–44	40–41

When it was determined that the case hardness of the gear was below specification, additional hardness measurements were made using a 15-kg (33-lb) load. These measurements confirmed the low hardness and suggested that it had resulted from one or more surface defects.

A specimen from the pitted area of one gear tooth was prepared for metallographic examination. The results of this examination are shown in Fig. 47. When the specimen was etched in 2% nital for 15 to 20 s, a decarburized surface layer was observed in the vicinity of pitting (Fig. 47a). When the specimen was repolished and then etched in 2% nital for only 3 s, a heavy subsurface layer of oxide scale was observed in the vicinity of pitting (Fig. 47b).

Conclusions. Based on low case-hardness readings and the results of metallographic examination, it was concluded that pitting had resulted due to a combination of surface decarburization and subsurface oxidation. Because of previous failure analyses and thorough investigation of heat-treating facilities, the source of these defects was well known. The oxide layer had been developed during the carburizing cycle because the furnace retort had not been adequately purged of air before carburization. Decarburization had occurred during the austenitizing cycle in a hardening furnace containing an exothermic protective atmosphere because of defects in the copper plating applied to the gear for its protection during austenitizing. These defects had permitted leakage of exothermic atmosphere to portions of the carburized surface, and, because of the low carbon potential of the exothermic atmosphere, the leakage had resulted in decarburization.

Recommendations. It was recommended that steps be taken during heat treatment to ensure that furnaces are thoroughly purged before carburizing and that positive atmospheric pressure be maintained throughout the carburizing cycle. Further, it was recommended that more effective control be exercised over all aspects of copper plating of carburized gears, including final inspection, before releasing the gears for hardening.

Example 4: Fatigue Failure of a Carburized 4817 Steel Spiral Bevel Gear at Acute-Angle Intersections of Mounting Holes and Tooth-Root Fillets. A spiral bevel gear set was returned to the manufacturer because the gear broke into three pieces after about 2 years of service. Damage to the pinion was minor. The gear broke along the root of a tooth intersected by three of the six 22-mm (0.875-in.) diam holes (Fig. 48) used to mount the gear to a hub. The intersection of the hole and root fillet created an acute-angle condition (Section A-A, Fig. 48).

The ring gear, machined from a forged blank made of 4817 steel, had been gas carburized at 925 °C (1700 °F), cooled to 815 °C (1500 °F), press quenched in oil at 60 °C (140 °F), then

Fig. 48 Carburized 4817 steel spiral bevel gear

The gear broke from fatigue at acute-angle intersections of mounting holes and tooth-root fillets as a result of through hardening. Dimensions given in inches

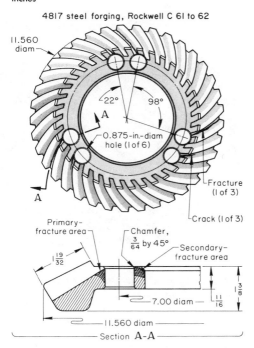

tempered at 175 °C (350 °F) for 1½ h to produce a case hardness of 61 to 62 HRC, as measured at a tooth land.

Investigation. Examination of the gear revealed fatigue progression for about 6.4 mm (¼ in.) at acute-angle intersections of three mounting holes with the root fillets of three teeth. Subsequent fracture progressed rapidly until the entire section failed. Cracking between the holes and the bore of the gear started before the tooth section failed. Magnetic-particle inspection revealed cracks at the intersections of the remaining three mounting holes and the adjacent tooth-root fillets.

Microstructure of the carburized and hardened case was fine acicular martensite with about 5% retained austenite. The core was composed principally of low-carbon martensite.

Metallographic examination revealed that the acute-angle intersections of the mounting holes and tooth-root fillets were carburized to a depth that comprised the entire cross section and consequently were through hardened.

Conclusions. The gear failed in fatigue, slowly to a depth of about 6.4 mm (¼ in.), then rapidly until final fracture. Design of the gear and placement of the mounting holes were contributing factors in the failure because they resulted in through hardening at the acute-angle intersections of the mounting holes and tooth-root fillets.

Fig. 49 Spiral bevel pinion of 4820H steel

(a) Rippled surface for three-fourths of the length of the tooth starting from the toe end. Pitting originated low on the active profile 50 mm (2 in.) from the toe end. (b) The pitting area extended in both directions and broadened. The central profile has an area of spalling that appears to be contiguous with the pitting. The microstructure indicates that the two modes occurred independently. 0.5×. See also Fig. 50 and 51.

(a)   (b)

Example 5: A Spiral Bevel Pinion That Had Two Failure Modes Occurring Independently of Each Other. This spiral bevel set was the primary drive unit for the differential and axle shafts of an exceptionally large front-end loader in the experimental stages of development.

Visual Examination. The tooth contact area of the pinion appeared to have been heavy at the toe end and low on the profile of the concave (drive) side. The entire active profile showed surface rippling for more than half the length of the tooth from the toe end (Fig. 49). Pitting originated at the lower edge of the profile 50 mm (2 in.) from the toe end of the concave side (Fig. 49a) and subsequently progressed in both directions along the lower edge of the active profile and upward toward the top of the tooth, changing the mode from pitting to spalling (Fig. 49b). Both photographs show evidence of a double contact: the original setting, and a secondary pattern with the pinion moved out and away from the center of the gear. The convex (reverse) sides were slightly worn and scuffed high over the profile.

All of the gear teeth were intact and highly polished along the top of the profile toward the toe end of the convex (forward) side. Some scuffing was evident low on the profile of the concave sides.

Physical Examination. Magnetic-particle inspection showed no evidence of tooth-bending fatigue on either part. The several cracks indicated are associated with the spalling surfaces on the concave sides of the pinion teeth. The gear teeth showed no crack indications.

Tooth Characteristics. The original setting test charts taken on this set in December 1979 are on file and document a very good and normal tooth contact pattern. The secondary

Fig. 50 Section normal to surface of tooth profile taken near the spalled area shown in Fig. 49(b)

The surface shows no catastrophic movement; the butterfly wings are generally parallel to the surface, but extend 0.7 mm (0.027 in.) below the surface. Microstructure is very fine acicular martensite retaining less than 5% austenite. 200×

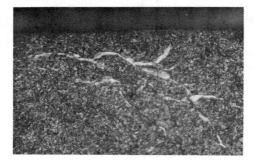

tooth pattern that prevails at this time could not be duplicated on the gear tester. It appears that the pinion had moved away from the center of the gear, thus activating a very heavy low-profile contact toward the toe end.

Surface-Hardness Tests. The surface hardness of both parts was 58 to 59 HRC, which is within the specification of 58 to 63 HRC.

Metallurgical Examination. The procedures to implement are determined as follows. The rippling (Fig. 49) was extensive and covered the entire area of the initial contact pattern. Rippling is a reflection of apparent movement or of a tendency to move. It can be a superficial adjustment of a surface being heavily rolled while sliding, or it can be an adjustment of the surface during absorption of

Fig. 51 Case-hardness traverse of section used for Fig. 50 taken from tooth shown in Fig. 49(b)

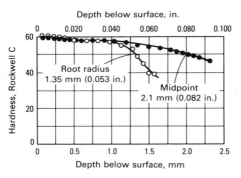

energy by retained austenite. Rippling is not always associated with a failure, nor is it detrimental in itself. It does usually indicate, however, a rolling-sliding condition that is highly compressive.

The pitting originated in a localized area, and the spalling appeared to be a contiguous event, although this type of spalling could easily have been found as an independent mode without the association of surface pitting. It was therefore necessary to section the pinion tooth normal to the active profile near the spalled area shown in Fig. 49(b) and to prepare this section for a microhardness traverse of the case and for a microscopic examination.

Microscopic Examination. The core structure of the pinion showed an equal admixture of low-carbon martensite and lamellar pearlite. The case structure consisted of very fine acicular martensite retaining less than 5% austenite.

Fig. 52 2317 steel bevel pinion that failed by fatigue breakage of teeth
(a) Configuration and dimensions (given in inches). (b) View of area where two teeth broke off at the root.
(c) Fracture surface of a broken tooth showing fatigue marks

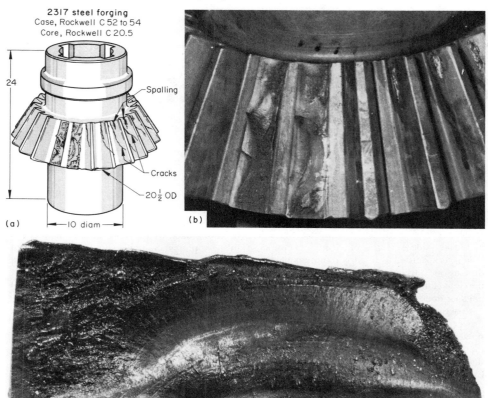

(a) 2317 steel forging
Case, Rockwell C 52 to 54
Core, Rockwell C 20.5

24

Spalling

Cracks

20½ OD

10 diam

(b)

(c)

The pitted area low on the active profile showed evidence of a heavy metal-to-metal sliding action which tended to produce adhesion. The central profile area, being subjected to spalling, had a surface that was relatively unaltered (Fig. 50). In the same area, there was a subsurface display of butterfly wings reaching to a depth of 0.7 mm (0.027 in.) from the surface.

Case-Hardness Traverse. Figure 51 shows the case-hardness traverse of the section used for Fig. 50. The results are self-explanatory.

Conclusions. The material and process specifications were met satisfactorily. The primary mode of failure was rolling-contact fatigue of the concave (drive) active tooth profile. The spalled area was a consequence of this action. The pitting low on the profile appeared to have originated after the shift of the pinion tooth away from the gear center. The shift of the pinion is most often due to a bearing displacement or malfunction. The cause of this failure appeared to be continuous high overload, which may also have contributed to the bearing displacement.

Example 6: Fatigue Failure of a Carburized Steel Bevel Pinion Because of Misalignment. The bevel pinion shown in Fig. 52(a) was part of a drive unit in an edging mill. The pinion had been in service about 3 months when several teeth failed. Specifications required that the pinion be made from a 2317 steel forging and that the teeth be carburized and hardened to a case hardness of 56 HRC and a core hardness of 250 HB (24.5 HRC).

Investigation. Chemical analysis of the metal in the pinion showed that it was 2317 steel as specified. Case hardness was 52 to 54 HRC, and core hardness was 229 HB (20.5 HRC). Both hardness values were slightly lower than specified, but acceptable.

Visual inspection of the pinion showed that two teeth had broken at the root (Fig. 52b). The surfaces of these fractures exhibited fatigue marks extending across almost the entire tooth (Fig. 52c). Magnetic-particle inspection of the pinion showed that all of the teeth were cracked. Each crack had initiated along the tooth root, at the toe (small) end of the tooth, extending to the center of the crown. Spalling was also noted on the pressure (drive) side of each tooth at the toe end. A metallographic specimen taken transversely through a broken tooth showed it to be case carburized to a depth of 4.8 mm (3/16 in.), hardened, and tempered.

Conclusions. The pinion failed by tooth-bending fatigue. Some mechanical misalignment of the pinion with the mating gear caused a cyclic shock load to be applied to the toe ends of the teeth, as exhibited by the spalling. This continuous pounding caused the teeth to crack at the roots and to break off.

SELECTED REFERENCE

L.E. Alban, *Systematic Analysis of Gear Failures,* American Society for Metals, 1985

Failures of Boilers and Related Equipment

Revised by David N. French, David N. French, Inc., Metallurgists

FAILURES IN BOILERS and other equipment in stationary and marine power plants that use steam as the working fluid are discussed in this article. The discussion is mainly concerned with failures in Rankine-cycle systems that use fossil fuels or a nuclear reactor as the primary heat source, although many of the principles that apply to Rankine-cycle systems also apply to systems using other steam cycles or to systems using working fluids other than steam.

It is important to learn as much as possible from each failure. In any metallurgical evaluation, the general aim is to understand the root cause of the failure in terms of both the material and the boiler operation. To that end, estimates can sometimes be made of metal temperatures and implied boiler conditions at the time of failure. This information may then be useful in the prevention of future failures.

Procedures for Failure Analysis

Procedures for analysis of failures in steam power plants do not differ significantly from procedures for failure analysis in general or for analysis of specific types of failure. These procedures are presented in other articles in this Volume, particularly those dealing with basic mechanisms of failure; failures of specific product forms, such as castings, forgings, weldments, and pressure vessels; and elevated-temperature failures. Consequently, this article will discuss the main types of failure that occur in steam-power-plant equipment, with major emphasis on the distinctive features of each type that enable the failure analyst to determine cause and to suggest corrective action.

Service Records. Most power-plant operators maintain relatively complete records. Records of operating conditions and preventive maintenance for a component that has failed, and for the system as a whole, are relatively good sources of background information. These records can provide valuable information, such as operating temperature and pressure, normal power output, fluctuations in steam demand, composition of fuel, amount of excess combustion air, type and amount of water-conditioning chemicals added, type and amount of contaminants in condensate and make-up feedwater, frequency and methods of cleaning fire-side and water-side surfaces of steam generators, materials specified as to alloy requirements and dimensions, frequency and location of any previous failures, length of service, and any unusual operating history.

Precautions in On-Site Examination. Power plants are vital to most industries, and power-plant downtime has an adverse effect on the entire operation. Thus, the individual conducting the on-site examination should have experience in making preliminary determinations of cause and recommending corrective action on the spot.

Certain deductions, such as the exact cause or mechanism of failure, frequently cannot be made without laboratory examination. However, some determinations—for example, which of several damaged components failed first—can be made on the basis of careful on-site examination. The relationship of the location of the failure to the locations of other system components—for example, the location of a tube rupture in relation to those of burners or soot blowers—is an important phase of on-site examination.

It is usually helpful for the individual making the on-site examination to have an assortment of plastic bags or paper envelopes in which samples can be placed. Identification of each sample, with its location and orientation marked on photographs or sketches, is recommended because the failed component usually must be repaired or replaced with as little delay as possible so that normal system operation can be resumed. This need for prompt return to service may also preclude a second on-site examination; therefore, the first examination should be thorough and complete.

Precautions in Sampling. Because of the massive size and the fixed, sometimes remote location of power plants, detailed examination usually cannot be carried out at the scene and must be performed on selected samples taken from the failed equipment. Therefore, the methods used to obtain samples for laboratory examination are of utmost importance. For example, if the failure involves rupture of one or more boiler or superheater tubes, there is a high probability that adjacent tubes may have been degraded by the same conditions that led to the ruptures. Samples should be taken from adjacent, apparently undamaged tubes so that the total extent of damage can be assessed.

The effect of the method of sample removal should be considered when choosing the size and location of samples. Mechanical methods of sample removal, such as cutting with a tube cutter, sawing, or drilling, are less likely to alter either the microstructure or the characteristics of steam-side or fire-side scale, deposits, or corrosion products than is flame cutting. Mechanical methods, although generally preferred, are slow and sometimes cannot be used, because samples are inaccessible to the necessary tools; flame cutting may be the only reasonable alternative. When flame cutting is used, larger samples are necessary than when mechanical methods are used, and cuts must be made farther from the location of interest.

Corrosion is frequently associated with failure in steam-power-plant equipment, but the active corrosive agent is not always obvious. Consequently, collection and preservation of corrosion products, particularly loose, flaky, or powdery deposits, may be vital because chemical analysis of corrosion products can be a key factor in determination of cause.

Characteristic Causes of Failure in Steam Equipment

Boilers and other types of steam-power-plant equipment are subject to a wide variety of failures involving one or more of several mechanisms. Most prominent among these mechanisms are corrosion, including pitting and erosion; mechanical-environmental processes, including stress-corrosion cracking (SCC) and hydrogen damage; fracture, including fatigue fracture, thermal fatigue fracture, and stress rupture; and distortion, especially distortion involving thermal-expansion effects or creep.

The causes of failure can generally be classified as design defects; fabrication defects, including wrong material; improper operation, including improper maintenance and inadequate water treatment; routine normal operations, such as too frequent soot blowing; and miscellaneous causes. Of these general types of causes, improper operation, which includes most incidents of overheating, corrosion, and fouling, and fabrication defects, which include most incidents of poor workmanship, improper material, and defective material, together account for more than 75% of all failures of steam-power-plant equipment.

Most steam-generator failures occur in pressurized components, that is, the tubing, piping, and pressure vessels that constitute the steam-generating portion of the system. With very few exceptions, failure of pressurized components is confined to the relatively small-diameter tubing making up the heat-transfer surfaces within the boiler enclosure.

Overheating is the main cause of failure in steam generators. For example, a survey compiled by one laboratory over a period of 12 years, encompassing 413 investigations, listed overheating as the cause in 201 failures, or 48.7% of those investigated. Fatigue and corrosion fatigue were listed as the next most common causes of failure, accounting for a total of 89 failures, or 21.5%. Corrosion, stress corrosion, and hydrogen embrittlement caused a total of 68 failures, or 16.5%. Defective or improper material was cited as the cause of most of the remaining failures (13.3%). Although defective material is often blamed for a failure, this survey indicates that, statistically, it is one of the least likely causes of failure in power-plant equipment.

Defective material does not always cause a component to fail soon after being put into service. Figure 1 shows cracking at the root of a longitudinal mill defect in a stainless steel superheater tube. This tube ruptured after 18 years of service because the normal operating pressure caused stress-rupture cracking to initiate at the mill defect.

Boiler design is inherently conservative; thus, even massive defects may be present in some areas without causing fracture to occur until after a considerable period of operation. Some imperfections may be present without ever causing failure. Figure 2 shows a poorly made weld with incomplete fusion amounting to more than half the tube wall thickness, yet this weld gave more than 27 years of service without failure. Nevertheless, fabrication and repair procedures should be aimed at producing defect-free systems.

Failures Involving Sudden Tube Rupture

In the basic design of a boiler, the heat input from the combustion of fuel is balanced by the formation of steam in the furnace and the heating of steam in a superheater or reheater. The heat-flow path through a clean boiler tube has three components. First, fire-side heat

Fig. 1 Micrograph showing stress-rupture cracking at the root of a longitudinal mill defect in a stainless steel superheater tube

The tube ruptured after 18 years of service. Approximately 25×

Fig. 2 Weld defect (lack of fusion)

This defect did not cause a failure even after 27 years in a reheater.

transfer from the flame or hot flue gases is by both radiation and convection. Radiation predominates in the furnace, where the gas temperatures may be close to 1650 °C (3000 °F). By the time the flue gas has left the furnace, it

has been cooled to 925 to 1095 °C (1700 to 2000 °F), and convection is the predominant mode of heat transfer. Second, conduction through the steel boiler tubes transfers heat to the internal fluid. Although conduction is important, boiler tubes are not chosen for thermal conductivity but for strength, especially creep or stress-rupture strength. Therefore, the temperature gradient through the steel is not controlled by design but accepted as a consequence of material selection. Third, at the fluid interface with the inside-diameter surface is a second convective heat-transfer mode. The steam-side heat-transfer coefficient is a function of fluid velocity, viscosity, density, and tube bore diameter.

Boilers in service for some time have a fourth component to the heat-flow path: internal scale or deposits. Steam reacts with steel to form iron oxide:

$$4H_2O + 3Fe \rightarrow 3Fe_3O_4 + 4H_2 \qquad \text{(Eq 1)}$$

Furnace walls may also have other deposits from impurities in the boiler feedwater. Because these deposits and scale have a lower thermal conductivity than the steel tube, the net effect is an increase in tube metal temperatures. In a superheater or a reheater, such temperature increases can lead to premature creep failures, dissimilar-metal weld failures, and accelerated ash corrosion or oxidation. In furnace walls, deposits may also lead to hydrogen damage (additional information is provided in the article "Hydrogen-Damage Failures" in this Volume).

An upset in any stage along the heat-flow path can upset the balance and cause a sudden tube rupture. Sudden rupture of a tube in a steam generator is a serious failure, because the steam generator must be shut down immediately to minimize or avoid erosion of adjacent tubes and furnace sidewalls by escaping steam, overheating of other tube banks due to loss of boiler circulation, and damage to other components in the system resulting from loss of working fluid. The downtime resulting from boiler failure and subsequent repair may require other operations to be curtailed or shut down, with an attendant economic loss.

Tube ruptures (excluding cracks caused by stress corrosion or fatigue, which usually result in leakage rather than sudden fracture) may be classified as ruptures caused by overheating and ruptures caused by embrittlement. Each type has characteristic features.

Ruptures Caused by Overheating

When water is boiled in a tube having uniform heat flux (rate of heat transfer) along its length under conditions that produce a state of dynamic equilibrium, various points along the tube will be in contact with subcooled water, boiling water, low-quality steam, high-quality steam, and superheated steam. A temperature gradient between the tube wall and the fluid within the tube provides the driving force for heat transfer at any point.

The design of a steam-generating unit balances the heat input from the combustion of a fossil fuel with the formation and superheating of steam. Within the furnace, flame temperatures may approach 1650 °C (3000 °F). Heat absorption by the furnace walls reduces the temperature of the flue gases to 925 to 1095 °C (1700 to 2000 °F). The heat absorbed is converted into steam at its saturation temperature, a function of the operating boiler pressure. Within the convection passes, the flue-gas temperature is further reduced by the superheating or reheating of steam in superheaters and reheaters. To extract more heat and to improve overall thermal efficiency, an economizer preheats the boiler feedwater to a temperature close to its boiling point. The flue gas travels through an air preheater, which heats the combustion air, then makes its way up the stack. The steady-state heat transfer can be given by (Ref 1, 2):

$$\frac{Q}{A_0} = U_0 \Delta T \qquad \text{(Eq 2)}$$

where Q/A_0 (in Btu/h $\cdot$ ft^2) is the heat flux per unit area, U_0 (in Btu/h $\cdot$ ft^2 $\cdot$ °F) is the overall heat-transfer coefficient, and ΔT (in degrees Fahrenheit) is the temperature difference that drives the heat flow. The reciprocal of U_0 is R_0, the combined resistance to the heat flow. Figure 3 shows a schematic representation of the temperature gradient in the heat transfer from hot flue gas or flame to steam or water. The individual thermal resistances from flue gas to steam are:

Gas side: $R_1 = \dfrac{1}{h_0}$ (Eq 3)

Tube wall: $R_2 = \dfrac{r_0 \ln (r_0/r_i)}{k_m}$ (Eq 4)

Inside scale or deposit:

$$R_3 = \frac{r_0 \ln (r_i/r_s)}{k_s} \qquad \text{(Eq 5)}$$

Fig. 3 Schematic representation of the temperature profile from flue gas temperature (T_0) to bulk steam temperature (T_s) for the clean tube and internally scaled conditions

Note that the effect of inside-diameter scale is to raise tube metal temperatures.

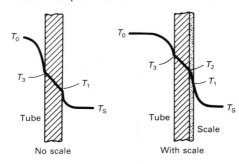

Steam side:

$$R_4 = \frac{r_0}{r_s h_i} \qquad \text{(Eq 6)}$$

Because these resistances are in series, they are additive, and the overall heat-transfer coefficient can be expressed as:

$$U_0 = \cfrac{1}{\cfrac{1}{h_0} + \cfrac{r_0 \ln (r_0/r_i)}{k_m} + \cfrac{r_0 \ln (r_i/r_s)}{k_s} + \cfrac{r_0}{r_s h_i}} \qquad \text{(Eq 7)}$$

Individual temperature gradients through each step can be given by:

$$\Delta T_G = T_1 - T_2 = \frac{Q}{A_0}\left(\frac{1}{h_0}\right) \text{ Gas side} \qquad \text{(Eq 8)}$$

$$\Delta T_M = T_2 - T_3 = \frac{Q}{A_0}\left(\frac{r_0 \ln (r_0/r_i)}{k_m}\right) \begin{matrix}\text{Tube}\\\text{wall}\end{matrix} \qquad \text{(Eq 9)}$$

$$\Delta T_S = T_3 - T_4 = \frac{Q}{A_0}\left(\frac{r_0 \ln (r_i/r_s)}{k_s}\right) \begin{matrix}\text{ID}\\\text{scale}\end{matrix} \qquad \text{(Eq 10)}$$

$$\Delta T_f = T_4 - T_5 = \frac{Q}{A_0}\left(\frac{r_0}{r_i h_i}\right) \begin{matrix}\text{Steam}\\\text{side}\end{matrix} \qquad \text{(Eq 11)}$$

For convenience, the gas-side heat-transfer coefficient, h_0, combines both convection and radiation effects and any contribution by any ash or soot deposit on the external tube surface. A clean boiler tube will have little or no internal scale, and that term will then be absent from the value of U_0. Because this resistance is on the tube inside diameter, the net effect will be an increase in tube metal temperature. Equation 10 can be used to estimate the temperature increase due to the inside-diameter scale, because $r_i - r_s$ is the inside-diameter scale thickness. Inspection of Eq 8 to 11 indicates that changes in heat flux or h_0 (for example, flame impingement), increases in inside-diameter scale thickness, or decreases in steam-side heat-transfer coefficient caused by reduced flow, will lead to tube metal temperature increases. Table 1 lists the range of values for heat-transfer coefficients and thermal-conductivity values for internal deposits.

In steam-cooled tubes, the internal magnetite scale thickness may be used to estimate the metal-temperature increase due to the internal scale using Eq 10 and to estimate the average operating metal temperature from the scale thickness. Figure 4 plots scale thickness as a function of the Larson-Miller Parameter (LMP), T °R(20 + log t) where T °R = T(°F) + 460). The equation of this scale-thickness curve is:

$$\log X = 0.00022 (T + 460) \\ (20 + \log t) - 7.25 \qquad \text{(Eq 12)}$$

where X is the scale thickness (in mils), T is temperature (in degrees Rankine), and t is operating time (in hours). Thus, by measuring X from metallographic specimens and knowing t from plant operating records, T may be estimated.

Caution should be exercised in the calculation of average operating temperatures. Equation 12 gives an average value, but scale thickness is a function of the cumulative thermal history of the particular tube. Excessive metal temperatures during a start-up for a short period of time on each start-up will increase scale thickness. One recent example indicated tube metal temperatures above 815 °C (1500 °F) on initial boiler operation. No failure occurred,

Table 1 Heat-transfer factors and thermal conductivities of internal scale deposits

Heat-transfer factors(a)	h_i		U_0		Q/A_0	
	W/m² · K	Btu/h · ft² · °F	W/m² · K	Btu/h · ft² · °F	W/m²	Btu/h · ft²
Economizers	5680	1000	28-40	5-7	15 770-31 540	5000-10 000
Water walls	22 700-45 400	4000-8000	115-125	20-22	126 160-394 250	40 000-125 000
Reheaters	1700-2270	300-400	60-95	11-17	19 000-38 000	6000-12 000
Superheaters	1135-2840	200-500	60-100	11-18	31 540-63 000	10 000-20 000

Thermal conductivities(b)	k					
	W/m · K	Btu · in./h · ft² · °F				
Calcium phosphate · · · · · · · · · · ·	3.6	25				
Calcium sulfate · · · · · · · · · · · ·	2.3	16				
Magnetic iron oxide · · · · · · · · ·	2.9	20				
Silicate scale (porous) · · · · · · · ·	0.09	0.6				
Steel ·	24.5-44.7	170-310				

(a) Source: Ref 1. (b) Source: Ref 3

Fig. 4 Plot of scale thickness and oxide penetration versus LMP

X = 595 °C (1100 °F)/2500 h, Y = 620 °C (1150 °F)/2500 h, Z = 650 °C (1200 °F)/2500 h. Source: Ref 4

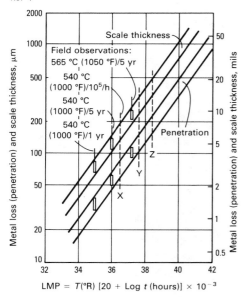

$$LMP = T(°R) [20 + Log\ t\ (hours)] \times 10^{-3}$$

Fig. 5 Variation of fluid temperature and tube-wall temperature as water is heated through the boiling point with low, moderate, high, and very high heat fluxes (rates of heat transfer)

See text for discussion. Source: Ref 5

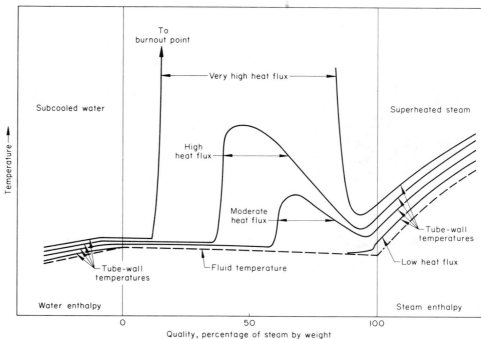

because pressure was substantially lower than normal. At above 815 °C (1500 °F), the tube was renormalized. No chemical cleaning was performed, and the boiler operated for 16 years. After this service history, the microstructure was mildly spheroidized carbides, with pearlite colonies still well defined. Inside-diameter scale thickness indicated an average metal temperature of nearly 570 °C (1060 °F). For the material in question, the two values were not in agreement; the microstructure, after 16 years of service at 570 °C (1060 °F), should have been fully spheroidized. It was not. Therefore, these calculations of tube metal temperature are estimates—useful to be sure, but estimates only.

Figure 5 indicates the effects that different heat fluxes have on tube-wall temperature. In the region where subcooled water contacts the tube (at left, Fig. 5), the resistance of the fluid film is relatively low; therefore, a small temperature difference sustains heat transfer at all heat-flux levels. However, the resistance of a vapor film in steam of low quality is relatively high; therefore, at the onset of film boiling, a large temperature difference between the tube wall and the bulk fluid is required to sustain a high heat flux across the film. The effect of the onset of film boiling on tube-wall temperature appears as sharp breaks in the curves for moderate, high, and very high heat fluxes in Fig. 5. With increasing heat flux, the onset of unstable film boiling, also known as departure from nucleate boiling (DNB), occurs at lower steam qualities, and tube-wall temperatures reach higher peak values before stable film boiling, which requires a lower temperature difference to sustain a given heat flux, is established.

At very high heat fluxes, DNB can occur at low steam quality, and the temperature difference between tube wall and bulk fluid at a point slightly downstream from DNB is very high. Under these conditions, tube failure can theoretically occur by melting of the tube wall, although in reality the tube will rupture because metal loses its strength, and therefore its ability to contain pressure, before it melts. Departure from nucleate boiling is an important consideration in the design of fossil-fuel boilers and nuclear reactors because heat flux can quickly exceed the failure point (burnout point) at local regions in a tube if the tube does not receive an adequate supply of incoming feedwater.

In superheaters and reheaters, which normally operate at temperatures 30 to 85 °C (50 to 150 °F) higher than the temperature of the steam inside the tubes, heat transfer is primarily controlled by the conductance of fluid films at the inner and outer surfaces. Although higher heat fluxes cause higher tube-wall temperatures, deposits have a greater effect on tube-wall temperatures and therefore on overheating.

A tube rupture caused by overheating can occur within a few minutes, or can take several years to develop. A rupture caused by overheating generally involves fracture along a longitudinal path, typically with some detectable plastic deformation before fracture. Longitudinal fracture may or may not be accompanied by secondary, circumferential fracture. The main fracture usually has a fishmouth appearance

(Fig. 6a) and is either a thick-lip or a thin-lip rupture.

Thick-lip ruptures in steam-generator tubes occur mainly by stress rupture as a result of prolonged overheating at a temperature slightly above the maximum safe working temperature for the tube material.

The failures are characterized by thick-edged fracture lips, little ductility or tube swelling, excessive internal scale, and other evidence of oxidation or corrosion. The fracture is normal to the tube surface and parallel with the tube axis (Fig. 6a). The microstructure (Fig. 6b and c) exhibits evidence of creep damage, creep voids or cavitation, grain-boundary separation, and outside-diameter and/or inside-diameter intergranular cracking or oxide penetration of the grain boundaries. The carbide phase of ferritic steels is fully spheroidized.

Temperatures slightly above the design condition are caused by six factors. The first is increases in heat flux. In a superheater or a reheater, partial blockage of the convection pass by fly ash will increase the flue-gas flow to certain regions. Higher velocity will increase the steam-side heat-transfer coefficient, h_0, and will increase metal temperature. Flame impingement on a furnace wall tube will do the same thing.

The second factor is internal scale buildup. The thermal conductivity of steam or water-side scale or deposit can be about 5% that of steel. An insulating layer on the inside-diameter surface acts as a barrier to heat transfer. The net

Fig. 6 Type 321 stainless steel (ASME SA-213, grade TP321H) superheater tube that failed by thick-lip stress rupture

(a) Overall view showing a typical fishmouth rupture. Approximately ½×. (b) Unetched section from location between arrows in (a) showing extensive transverse cracking adjacent to the main fracture (at right). Approximately 4½×. (c) Specimen etched electrolytically in 60% HNO_3 (nitric acid) showing intergranular nature of cracking. 100×

Fig. 7 Thin-lip rupture in a boiler tube that was caused by rapid overheating

This rupture exhibits a "cobra" appearance as a result of lateral bending under the reaction force imposed by escaping steam. The tube was a 64-mm (2½-in.) outside-diameter × 6.4-mm (0.250-in.) wall thickness boiler tube made of 1.25Cr-0.5Mo steel (ASME SA-213, grade T-11).

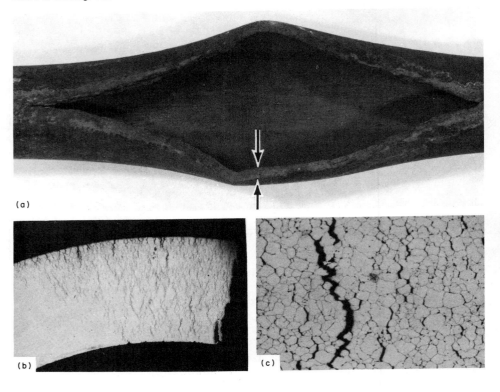

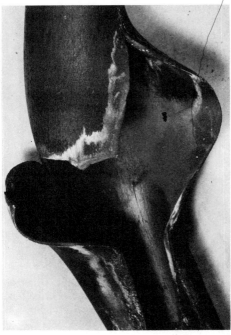

effect is an increase in tube metal temperature. An estimate of this temperature increase can be calculated from Eq 10.

The third factor is reduced steam flow. Partial pluggage of steam-cooled tubes by exfoliated internal oxide scale will reduce steam flow in certain tubes. Lower steam flow means a smaller steam-side heat-transfer coefficient and high metal temperatures.

The fourth factor is uneven steam attemperation. One side of a high-temperature superheater or reheater may have steam temperature up to 55 °C (100 °F) higher than the other due to uneven spray flow in the attemperation stage. Because the steam-temperature increase from inlet to outlet of a high-temperature superheater should be uniform across the superheater, differences in inlet temperature will lead to differences in tube metal temperature.

The fifth factor is uneven burner adjustments. Uneven flue-gas or oxygen distribution due to maladjusted or worn fuel burners will have the same effect as an increase in heat flux.

Sixth is nonuniform steam flow. In the usual design of a superheater or a reheater, a tube-to-tube imbalance of steam flow is factored into the metal-temperature calculations. On occasion, reality and prediction are quite different, and steam temperatures of ±40 °C (±75 °F)—

as opposed to the desired ±20 °C (±40 °F)—from the average occur.

Thin-lip ruptures (Fig. 7 and 8) are usually transgranular tensile fractures occurring at metal temperatures from 650 to 870 °C (1200 to 1600 °F). These elevated-temperature tensile fractures exhibit macroscopic and microscopic features that are characteristic of the tube alloy and the temperature at which rupture occurred. A tensile fracture results from rapid overheating to a temperature considerably above the safe working temperature for the tube material and is accompanied by considerable swelling of the tube in the regions adjacent to the rupture that have been exposed to the highest temperatures. As shown in Fig. 7, steam escaping at high velocity through the rupture will sometimes impose a reaction force on the tube that is sufficient to bend it laterally. The higher and more uniform the degree of overheating, the greater the likelihood of lateral bending.

Ruptures caused by rapid overheating exhibit an obvious reduction in wall thickness caused by yielding adjacent to the rupture, often to a knife-edge at the fracture surface, as shown in the inset in Fig. 8. Thinning also occurs in areas of swelling adjacent to ruptures; near the rupture shown in Fig. 8, the tube wall was 67% as thick as it was opposite the rupture, where the tube was shielded from exposure to hot gases.

There are three causes of these rapid failures. The first is tube blockage. In a superheater or a reheater, condensate collects at low points in the steam circuit. A rapid start-up will lead to high metal temperature because no steam flows before the condensate has time to evaporate. The second cause of rapid failure is tube leaks. In a water-wall tube, an undetected tube leak low in the furnace will starve that tube. Higher in the furnace, reduced fluid flow allows DNB to occur at normal firing conditions. Last is flame impingement. Very high heat fluxes will cause DNB and rapid high-temperature failure.

Even though tube-wall thinning characterizes all rapid-overheating failures, rapid overheating is not necessarily the cause of all ruptures that exhibit tube-wall thinning. Erosion and corrosion are mechanisms that can also cause thinning and subsequent rupture. Overheating may or may not occur in tubes thinned by erosion or corrosion.

Microstructural Features. Prolonged overheating, usually at temperatures below Ac_1 (the temperature at which austenite begins to form) in carbon and low-alloy steels, causes decomposition of pearlite into ferrite and spheroidal carbides, which weakens tube materials. If continued overheating persists, it can cause formation of voids along grain boundaries (Fig.

Fig. 8 Thin-lip rupture in a 64-mm (2½-in.) outside-diameter × 2.7-mm (0.105-in.) wall thickness carbon steel furnace-wall tube that was caused by rapid overheating

Knife-edge wall thinning at longitudinal main rupture is shown in cross section in the inset. Note secondary, circumferential fracture at left end of the longitudinal main rupture.

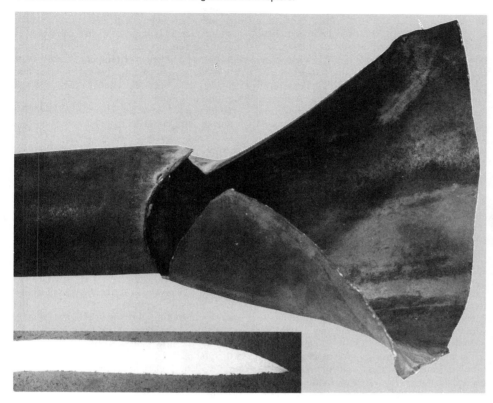

Fig. 9 Microstructures of specimens from carbon steel boiler tubes subjected to prolonged overheating below Ac₁

(a) Voids (black) in grain boundaries and spheroidization (light, globular), both of which are characteristic of tertiary creep. 250×.
(b) Intergranular separation adjacent to fracture surface (top). 50×. Mottled areas in both specimens are regions where pearlite has decomposed into ferrite and spheroidal carbides. Both etched with 2% nital

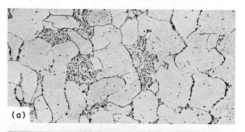

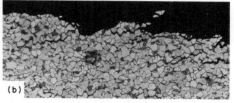

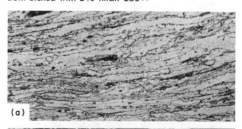

Fig. 10 Typical microstructures of 0.18% C steel boiler tubes that ruptured as a result of rapid overheating

(a) Elongated grains near tensile rupture resulting from rapid overheating below the recrystallization temperature. (b) Mixed structure near rupture resulting from rapid overheating between Ac₁ and Ac₃ and subsequent quenching by escaping water or steam. Both etched with 2% nital. 250×

9a) and eventual grain separation (Fig. 9b), resulting in stress-rupture failure of the tube.

Austenitic stainless steels can generally exhibit three types of metallurgical instability at elevated temperatures: carbide formation or modification, ferrite precipitation, and σ-phase or χ-phase formation. The effects of σ phase are discussed in detail in the article "Elevated-Temperature Failures" in this Volume. All three types of metallurgical instability may shorten the stress-rupture life of stainless steels and can enhance the probability of tube failure with prolonged overheating. Regardless of whether metallurgical instabilities occur, tube materials can fail by stress rupture if subjected to excessive temperatures over sufficiently long periods of time.

Rapid overheating of boiler tubes made of carbon and low-alloy steels usually results in failure because of a decrease in yield strength. If rupture occurs at a temperature below the recrystallization temperature, the microstructure near the fracture will exhibit severely elongated grains (Fig. 10a). Rupture that occurs at a temperature between Ac₁ and Ac₃ may exhibit a mixed microstructure of ferrite and upper transformation products (pearlite and bainite) resulting from the quenching effect

of escaping water or steam on the partly austenitic structure existing at the instant of rupture (Fig. 10b).

Microstructural evidence of overheating in the area of rupture is not conclusive proof that overheating occurred in service. The microstructural features discussed above may have been present at the time the tube was installed in the boiler. Thus, microstructural examination should be performed on the ruptured tube at a point some distance from the rupture, on an adjacent tube, or, better still, on a sample of unused tubing from the same mill lot.

Example 1: Rupture of Low-Carbon Steel Boiler Tubes Because of Severe Overheating. After 7 months of service, two low-carbon steel boiler tubes ruptured during a start-up period, causing extensive secondary damage to a two-drum marine reheat boiler. The tubes were 50-mm (2 in.) in diameter and 5.6 mm (0.220 in.) in nominal wall thickness and were made to ASME SA-192 specifications.

Reports indicated that there had been sufficient water in the boiler 2 h before start-up. Six hours after start-up, the center portions of two adjacent tubes in the boiling surface ruptured. The pressure and temperature of the steam at the time of failure were reported to be 8.6 MPa

(1.25 ksi) and 399 °C or 750 °F (boiling temperature: 300 °C, or 572 °F).

Examination of the boiler revealed no blocked tubes, and as determined with an audio gage, tube-wall thickness ranged from 4.7 to 5.5 mm (0.185 to 0.215 in.). The ruptured section of one tube and sections of two appar-

Fig. 11 Microstructures of three specimens taken from severely overheated carbon steel boiler tubes

The specimens were taken adjacent to a rupture in a tube (a), about 250 mm (10 in.) from the rupture in the same tube (b), and from a nearby unruptured tube (c). All three structures contain martensite or other lower transformation products. All etched with nital. 500×

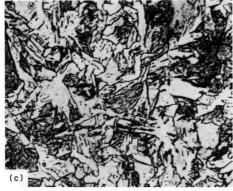

ently unaffected tubes were submitted for laboratory examination.

Investigation. The outer surfaces of all of the tubing exhibited a general oxidized condition. The inner surfaces were covered with an evenly distributed, thin, tightly adherent scale. Chemical analysis of the metal in the tube that ruptured indicated that it was in compliance with ASME SA-192 specifications.

Polished and nital-etched specimens of the metal in both the ruptured and the unruptured tubes were examined metallographically at a magnification of 500×. Near the rupture edge in the failed tube, the microstructure consisted of ferrite and acicular martensite or bainite (Fig. 11a). There were no indications of a cold-worked structure in this area; hardness of the metal was 33 to 34 HRC. The microstructure and the lack of cold-worked metal indicated that a temperature above the transformation temperature of 727 °C (1340 °F) had been reached. Examination of the microstructure opposite the rupture revealed a lesser amount of lower transformation products; this indicated that the maximum temperature was lower in this area of the tube than at the rupture, but was still higher than 727 °C (1340 °F). The hardness of the metal opposite the rupture was 89 HRB.

The microstructure of the metal about 250 mm (10 in.) from the rupture (Fig. 11b) contained some small, randomly located areas of acicular martensite or bainite in a ferrite matrix. Hardness of the metal was 80 to 82 HRB. The microstructure of one of the unruptured tubes (Fig. 11c) revealed a considerable amount of martensite or bainite in a ferrite matrix (hardness was 82 to 85 HRB), but the other unruptured tube had a structure of spheroidized carbide in a ferrite matrix and a hardness of 72 to 73 HRB.

In order for lower transformation products to form in low-carbon steel, the metal must be heated above 727 °C (1340 °F) and rapidly quenched. The microstructure of the ruptured tube indicated that this temperature was exceeded in the general area of failure and was greatly exceeded at the rupture, where a temperature of at least 870 °C (1600 °F) had been reached. Apparently, only one of the unruptured tubes had been heated above 727 °C (1340 °F).

Measurements of the wall thickness and outside diameter of the tubing showed that both the ruptured and the unruptured tubes were swollen. The diameter of the failed tube near the rupture ranged from 52.4 to 52.9 mm (2.062 to 2.083 in.), which was about 3½% greater than the diameter of the tubing as shipped from the mill (50.6 to 51.0 mm, or 1.992 to 2.010 in.). Records indicated that the original wall thickness of the tubing was 5.7 to 7.3 mm (0.226 to 0.286 in.), whereas the wall thickness in the swollen regions was only 4.6 to 5.6 mm (0.182 to 0.221 in.).

The outside diameters of the two unruptured tubes could not be accurately measured because they had been split before being shipped to the laboratory; however, the diameter was judged to be 51.8 mm (2.038 in.) minimum. The wall thickness of the unruptured tubes was 5.1 to 5.7 mm (0.201 to 0.225 in.), which was further evidence that swelling due to overheating had occurred in portions of the tube bank where no ruptures occurred.

Conclusions. The tubes failed as the result of rapid overheating. The microstructure, hardness, and scale thickness indicated that temperatures above 727 °C (1340 °F) were reached in a large section of the boiler. The tubing at the rupture area had reached a temperature of about 870 °C (1600 °F).

The boiler was heated too rapidly and to an excessively high temperature, at which time two tubes burst. This released a flood of water that quenched the overheated tubes, which accounts for the martensite and other lower transformation products in the microstructure. Because no evidence of tempered transformation products was found, it was concluded that the tubes had been overheated only once—at the time of failure.

Evidence of severe overheating was also found in unruptured tubes. This indicated that widespread damage had occurred. Consequently, the entire tube bank was replaced.

Effects of Deposits. During operation, a wide variety of deposits can form on both sides of heat-transfer surfaces. The chemical properties of these deposits and the means by which they are accumulated on heat-transfer surfaces are discussed in the section "Failures Caused by Corrosion or Scaling" in this article. Deposits can cause overheating failures by changing the heat-transfer characteristics of a tube bank or of an individual tube.

Fuel-ash deposits on surfaces exposed to flue gases (fire-side deposits) can cause local hot spots by insulating portions of the heat-transfer surface or, if the deposits are thick enough, by changing the flow pattern of flue gases through a tube bank. In either instance, metal temperatures will be higher in regions devoid of fire-side deposits than in regions where deposits are present.

When water-side deposits are present, tube-wall temperatures increase in the region of the deposits. A local region of scale on the inside surface of a boiler tube interposes resistance to heat flow between the tube wall and the working fluid, thus causing a temperature gradient across the scale and increasing tube-wall temperature in direct proportion to the thickness of the scale (Fig. 12). It is important to keep water-side surfaces free of adherent scale, particularly in regions of high heat flux, because of the effect on tube-wall temperatures. For example, as can be seen in Fig. 12(d), adherent scale 0.4 mm (0.016 in.) thick or thicker can cause the maximum metal temperature at the outside surface of the tube to exceed the maximum allowable temperature for a 50-mm (2.0-in.) diam carbon steel boiler tube at a heat flux of 315 000 W/m² · h; 1W = 1 J/s (100 000 Btu/ft² · h). The effect of water-side scale on metal temperatures is more pronounced for smaller-diameter tubing and higher heat fluxes.

Many boilers are designed so that heat transfer at the boiling surface is controlled by fluid-film conductance. The presence of scale on internal surfaces of tubes inhibits the vaporization process and can make it difficult to maintain proper operating conditions in the boiler.

Causes of Overheating. Overheating can result from restriction of flow within a heated

Fig. 12 Plots of scale thickness versus temperature for two sizes of boiler tubes and two values of heat flux

(a) and (b) The effect of scale thickness on the temperature gradient across the scale. (c) and (d) The effect of scale thickness on the temperature of the metal at the outer surface of the tube. The graphs are based on a continuous, uniform scale, with a thermal conductivity of 14.4 W/m · K (10 Btu · in./h · ft² · °F) on tubes in the film-boiling region of a 690-kPa (100-psi) system (saturation temperature 285 °C, or 545 °F).

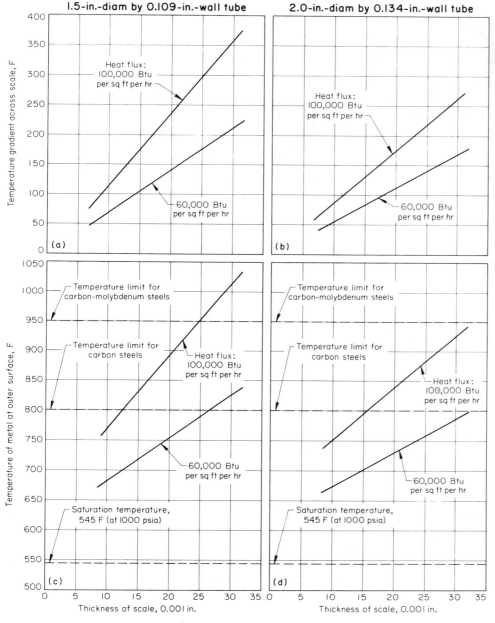

Fig. 13 Superheater tubes made of chromium-molybdenum steel (ASME SA-213, grade T-11) that ruptured because of overheating

(a) Tube that failed by stress rupture. (b) Resultant loss of circulation and tensile failure

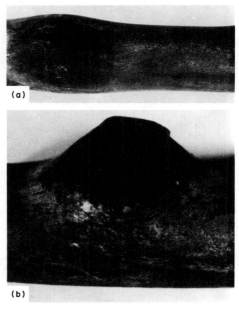

superheater rupture occurred about 10 days apart in a stationary industrial boiler. Three tubes failed the first time, and four tubes the second. The tubes, 64 mm (2.5 in.) in diameter and made of 1.25Cr-0.5Mo steel (ASME SA-213, grade T-11), failed by two different types of rupture, and both types occurred in each instance of failure.

Investigation. Visual examination of each type of rupture revealed noticeable swelling of the tubes in the area of rupture (Fig. 13). Relatively slight scaling, probably caused by high-temperature oxidation, was visible on fire-side surfaces, and black deposits and white deposits were found on steam-side surfaces. Subsequent analyses by scanning electron microscopy (SEM) and x-ray diffraction identified magnetic iron oxide (Fe_3O_4) as the major constituent in the black deposit, and a complex sodium aluminum silicate as the major constituent in the white deposit. The nature and extent of the scale did not suggest that it was a contributing factor in the failures.

Metallographic examination indicated that the tubes with slight longitudinal splits (Fig. 13a) had failed by stress rupture resulting from prolonged overheating at 540 to 650 °C (1000 to 1200 °F); the microstructure exhibited extensive spheroidization and coalescence of carbides. In contrast, the larger ruptures (Fig. 13b) were tensile failures resulting from rapid overheating to 815 to 870 °C (1500 to 1600 °F); a completely martensitic structure existed at the edges of the ruptures in these tubes because of rapid quenching by escaping fluid.

tube or from localized hot spots in a tube wall. Water or steam within a tube extracts heat from the metal, thus cooling it. Mild restriction of flow favors a small degree of overheating and failure by stress rupture; a sudden or severe restriction favors rapid overheating and tensile failure. Mild restrictions can result from such causes as local imbalance of flow among tubes joined to a common header, localized deposits near a tube inlet, or local variations in inside diameter because of variations in fabrication

techniques. Rapid overheating, which is most likely to occur at start-up, during periods of rapid fluctuation in steam demand, or during periods of full-power operation, may be caused by sudden loss of circulation resulting either from general loss of circulation, for example, feed-pump failure, or from local loss of circulation affecting only a few tubes.

Example 2: Rupture of Chromium-Molybdenum Steel Superheater Tubes Because of Overheating. Two instances of

Fig. 14 Ruptured tubes from a pendant-style reheater

(a) As-received sections from the toe of the reheater. (b) Creep-type failure typical of all the failed tubes. See also Fig. 15.

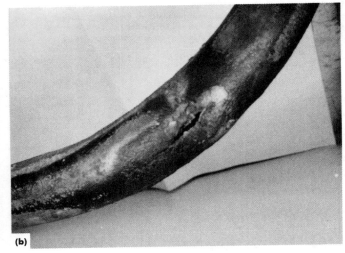

Conclusions. Based on the evidence—especially that both types of rupture had occurred in each instance of failure—it was concluded that the prolonged-overheating failures had been the primary ruptures and that local loss of circulation had caused rapid overheating in adjacent tubes. The prolonged overheating was believed to have resulted from poor boiler circulation and high furnace temperatures.

Rapid overheating frequently involves improper operation—specifically, delivering heat to the system more rapidly than the boiler or superheater heat-transfer surface can transmit heat to the working fluid. At the beginning of this section, DNB and its effect on tube-wall temperature was discussed. If a boiler is operated in a manner that causes DNB to occur in a region of high heat flux where the system cannot accommodate this condition, a rapid-overheating failure is inevitable. It is also possible that heat transfer at the boiling surface will be significantly lower than anticipated in design of the boiler, with the result that the flue gases in the superheater or reheater section will be too hot and an overheating failure in the superheater or reheater may occur. Uneven firing of a multiburner furnace can cause a similar effect, especially when furnace conditions do not favor high turbulence of the flue gases.

Example 3: Rupture of a 1.25Cr-0.5Mo Steel Reheater Tube Because of Localized Overheating. This particular pendant-style reheater, constructed of ASME SA-213, grade T-11, steel, had been plagued by a rash of failures. All failures were similar in appearance and locations. The tubes measured 64 mm (2.5 in.) outside diameter × 3.4 mm (0.134 in.) specified minimum wall thickness. A set of four tubes from several pendants was submitted for metallurgical evaluation. Figure 14(a) shows the four-tube set from the bottom of the pendant.

Investigation. Visual examination showed that all of the ruptures were similar. Figure 14(b) illustrates one of these narrow fissure-type failures. Dimensional measurements showed no swelling except at the failure. Wall thickness away from the rupture was 3.8 to 3.9 mm (0.148 to 0.154 in.), depending on precise location and on the tube—well above the specified 3.4 mm (0.134 in.). No corrosion or other excessive wastage was observed at any location, nor was any inside-diameter deposit noted. Inside-diameter scale was quite uniform away from the fissure and measured 0.13 mm (0.005 in., or 5 mils). The inside-diameter scale at the fracture region measured up to 1.3 mm (0.050 in., or 50 mils). A small quantity of loose debris was removed from the inside of one of the tubes; the particles were simply poured out as chunky bits.

Samples for microstructural analysis were taken through the failure lip, in line with the failure on the bottom of the pendant 205 mm (8 in.) from the failure, and in the same plane as the failure but 180° around the perimeter of the tube. The structure at the failure was completely spheroidized carbides in a ferrite matrix (Fig. 15a). The samples from locations 2 and 3 are nearly normal ferrite and pearlite (Fig. 15b and c). The carbide phase was spheroidized, but the pearlite colonies were still clearly defined.

Energy-dispersive x-ray spectrometry (EDS) analysis of the debris showed the major constituent to be iron with traces of phosphorus, manganese, sodium, calcium, copper, zinc, potassium, silicon, chromium, and molybdenum. Thus, the debris is probably scale from the inside diameter of the tube (iron, manganese, silicon, chromium, and molybdenum), with boiler feedwater chemicals (phosphorus, sodium, and calcium) from the attemperation spray. The other elements detected (copper and zinc) also come from the spray, but indicate a feedwater heater or condenser-tube leak.

Using Eq 12 with X either 5 or 50 mils and a time, t, of 37 000 h (80% availability and 5½ years of service on the reheater), the average metal temperature is estimated to be about 650 °C (1200 °F) at the failure and about 540 °C (1000 °F) elsewhere.

Conclusion. The likely cause of failure is exfoliation of the scale from the inside-diameter surface of the tube. These small particles collect at the lowest point of the pendant, effectively insulating the tube. Thus, a small localized area has a metal temperature about 110 °C (200 °F) hotter than the rest of the circuit. At 650 °C (1200 °F), ASME SA-213, grade T-11, is above its usual oxidation limit or maximum service temperature of 550 °C (1025 °F). Such service temperatures resulted in the creep failures and the low reliability reported on this unit.

Recommendations. Because reheaters are more apt to experience exfoliation than superheaters and because prevention is difficult, the most effective solution is to replace the affected tubes. Recommendations involved inspecting the tubes by radiography to find the circuits with the greatest accumulation of debris and replacing them as necessary on an annual basis.

Example 4: Creep Failure of a 2.25Cr-1Mo Steel Superheater Tube. This example concerns a creep failure taken from a radiant superheater. There have been no reported creep failures in nearly 17 years of service. The boiler is coal fired. Operating drum pressure is 18 MPa (2.6 ksi), and the outlet steam temperature is 540 °C (1005 °F). The failed tube is specified as 50 mm (2 in.) outside diameter × 7.6 mm (0.300 in.) minimum wall thickness ASME SA-213, grade T-22, material. Figure 16(a) shows the failure.

Investigation. Chemical analysis of the tube sample for chromium and molybdenum revealed:

Fig. 15 Microstructures of the failed reheater tube in Fig. 14(b)

(a) Section through the failure lip showing a complete spheroidization of the carbide phase in ferrite. (b) Section in the same plane as the failure, but 180° around the circumference of the tube. Structure is nearly normal pearlite and ferrite. (c) Section taken 205 mm (8 in.) away from the failure is also normal ferrite and pearlite. These three structures indicate the elevated temperature is confined to a small area of the failure. All etched with nital. 500×

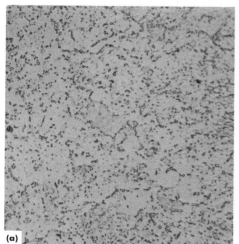  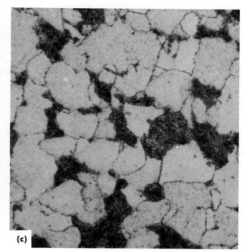

Element	Sample, %	SA-213, grade T-22 specification, %
Cr	2.35	1.90–2.60
Mo	0.91	0.87–1.13

Fig. 16 2.25Cr-1Mo steel superheater tube that failed by creep

(a) As-received failure. (b) Microstructure of the whole tube section is spheroidized carbides in ferrite. Etched with nital. 500×

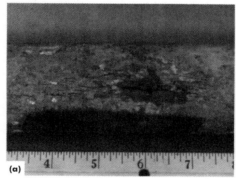 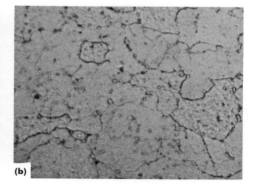

This confirmed the material as T-22. Dimensional measurements in the area of the failure were 52.2/53.0 mm (2.055/2.085 in.) outside diameter × 5.8/8.0 mm (0.230/0.315 in.) wall thickness. There was limited but measurable swelling of the tube diameter of about 2.4 mm (0.094 in.). The measured wall thickness of 5.8 mm (0.230 in.) in line with the failure and 8.0 mm (0.315 in.) 180° around the tube perimeter from the failure suggested that the tube wastage was caused by corrosion or erosion.

The microstructure through the tube some 50 mm (2 in.) from the failure showed a complete spheroidization of the carbides in a ferrite matrix (Figure 16b). Inside-diameter scale was measured as 0.34 mm (0.0134 in., or 13.4 mils).

At the time of the first creep failure, the question of the remaining life of the rest of the superheater must be addressed. The log stress versus LMP curves given in Fig. 17 are useful in answering this query. There are three plots: minimum, mean, and maximum failure lines. These curves were drawn from published stress-rupture data. The stress-rupture data have a range of values, and the three curves in Fig. 17 reflect the uncertainty in these data. These curves show that for a given stress, when a combination of time and temperature satisfies the LMP, a creep failure will occur:

$$LMP = (T + 460)(20 + \log t) \qquad \text{(Eq 13)}$$

where LMP is the Larson-Miller Parameter, T

is temperature (in degrees Fahrenheit), t is time (in hours), and 20 is an empirical constant.

To use Fig. 17, the tube metal temperature and stress must be estimated. Service or operational time—113 000 h in this case—can be obtained from plant records. If chemical cleaning had been performed on this superheater, then t would be the operational hours since the cleaning. Equation 12 is used to estimate the tube metal temperature with $X = 13.4$ mils (0.34 mm) and $t = 113\,000$ h, and it gives $T = 1060$ °F (570 °C).

The stress for a cylinder with internal pressure is:

$$S = \frac{PD_M}{2W} \qquad \text{(Eq 14)}$$

where S is stress (in pounds per square inch), P is internal pressure (in pounds per square inch), W is wall thickness (in inches), D_M is mean tube diameter (in inches) and is equal to

$D - W$, and D is tube outside diameter (in inches). An estimate of the original stress when the tube was new is:

$$S = \frac{(2600)(2.0 - 0.315)}{(2)(0.315)} = 6950 \text{ psi (48 MPa)}$$

The present stress just before failure is:

$$S = \frac{(2600)(2.085) - 0.230}{(2)(0.230)}$$
$$= 10\,500 \text{ psi (72 MPa)}$$

The LMP at the time of failure is:

$$LMP = (1060 + 460)(20 + \log 113\,000)$$
$$= 38\,000$$

The estimated design temperature, based on the tube geometry, a design pressure of 2.85 ksi (20 MPa), and the ASME Boiler and Pressure Vessel Code rules for allowable stress given in

Fig. 17 Log stress versus LMP = (T + 460)(20 + log t) for SA-213, grade T-22

Source: Ref 6. The three data points are explained in Example 4. T °F + 460 = T °R

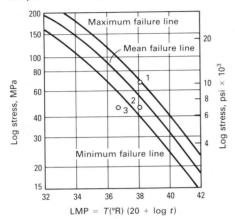

Fig. 18 Micrograph of an etched specimen from a carbon steel boiler tube

Decarburization and discontinuous intergranular cracking resulted from hydrogen damage. 250×

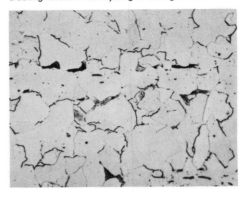

Fig. 19 Window fracture

Typically results from hydrogen damage in carbon or low-alloy steel boiler tubes

Carbon or low-alloy steel

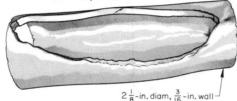

$2\frac{1}{8}$-in. diam, $\frac{3}{16}$-in. wall

Fig. 20 Hydrogen damage (dark area) in a carbon steel boiler tube

The tube cross section was macroetched with hot 50% hydrochloric acid.

Table PG 23.1 of the Code, is 1000 °F (540 °C) at this location of the superheater. It is evident that the estimated operating temperature of 1060 °F (570 °C) is higher than the estimated design temperature of 1000 °F (540 °C) and that the tube wastage has increased the actual operating stress from 6950 psi (48 MPa) to 10 500 psi (72 MPa).

There are three points plotted on Fig. 17:

1. The conditions at failure, 10 500 psi (72 MPa) and 38 000 LMP
2. The operating conditions if wastage had not occurred, 6950 psi (48 MPa) and 38 000 LMP
3. The operating conditions if the estimated design temperature of 1000 °F (540 °C) had been the operating temperature, 6950 psi (48 MPa) and 36 600 LMP

Conclusion. Tube wastage hastened the failure. However, even in the absence of metal loss, the operating temperature was too high, and failure was not surprising.

Recommendations. If this tube is representative, then all of the hottest tubes are suspect. To verify this, other samples should be removed for a better understanding of the material condition of this superheater.

Ruptures Caused by Embrittlement

Tube ruptures caused by embrittlement of the metal result from metallurgical changes within the tube metal that affect its ability to sustain service loads. The main mechanisms are hydrogen damage and graphitization, which make normally satisfactory alloys susceptible to brittle fracture.

Hydrogen damage in steam systems occurs primarily in steel components. At low to moderate temperatures, cracking caused by hydrogen damage often resembles SCC, except

that hydrogen-damage failures may exhibit little or no crack branching. At the high temperatures common in steam generators and most high-pressure piping, hydrogen damage is more likely to manifest itself as discontinuous intergranular cracking, often accompanied by decarburization (Fig. 18).

In high-temperature hydrogen damage, discontinuous cracking occurs because of the precipitation of molecular hydrogen (or methane resulting from hydrogen decarburization of the steel) along grain boundaries. Tubes that have undergone this type of hydrogen damage often rupture as shown in Fig. 19, in which a portion of the tube wall is detached; this type of rupture is sometimes termed a window fracture. High-temperature hydrogen damage can be confirmed by macroetching with a hot 50% solution of hydrochloric acid; regions of hydrogen damage appear black and porous (Fig. 20).

Hydrogen is one of the normal products of the basic corrosion reaction between iron and water. Atomic hydrogen formed at a corrosion site can combine with other hydrogen atoms to form molecular hydrogen, which becomes dispersed in the working fluid; can become dissolved as ions in surrounding liquid; or can pass into the metal as highly mobile atoms. Hydrogen damage results only from the last phenomenon.

Hydrogen damage is usually associated with thick internal deposits. Between the tube metal

and the deposit, the steel is actively being corroded either in acidic or basic conditions:

$$2H^+ + Fe = Fe^{2+} + 2H \quad \text{(acidic conditions)}$$

$$2OH^- + Fe = FeO_2^{2-} + 2H \quad \text{(basic conditions)}$$

Under these deposits, the tube metal temperature is higher than that for a clean deposit-free tube (note preceding discussion). Hydroxide ion is concentrated by local boiling at the metal/deposit interface. The normal boiler water pH is increased by the localized boiling. Corrosion rates increase with pH (Fig. 21). If all internal deposits are not removed during chemical cleaning, the porous deposits will retain the cleaning acid. During start-up, the low pH will also lead to rapid corrosion and will produce atomic hydrogen.

Most of the dissolved atoms migrate through the tube wall and pass through the opposite surface, where they are carried away by another fluid, usually flue gas. However, some of the dissolved hydrogen, the amount of which is determined primarily by the concentration of the hydrogen and by the temperature of the metal, participates in the damaging reactions.

Nascent or atomic hydrogen diffuses into the steel and reacts with iron carbide to form ferrite and methane:

$$4H + Fe_3C = 3Fe + CH_4$$

Methane is a large molecule and cannot diffuse out of the steel. It collects at ferrite grain boundaries and, when the pressure is high enough, leaves cracks behind (Fig. 18).

Example 5: Rupture of a Carbon Steel Tube Because of Hydrogen-Induced Cracking and Decarburization. A 75-mm (3-in.) outside-diameter × 7.4-mm (0.290-in.) wall thickness carbon steel boiler tube ruptured

Fig. 21 Effect of pH on the corrosion rate of steel by water at 310 °C (590 °F)

Source: Ref 7

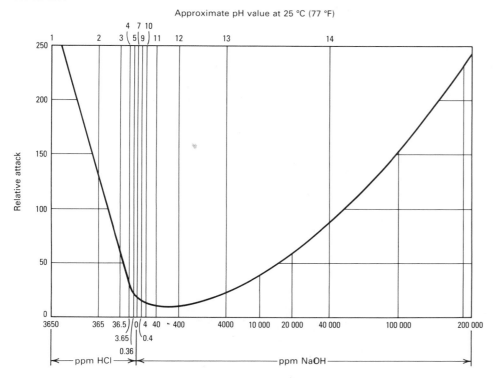

Fig. 23 Graphitized microstructure of SA-210-A-1 plain carbon steel

The structure is ferrite and graphite with only a trace of spheroidized carbon remaining. Etched with nital. 500×

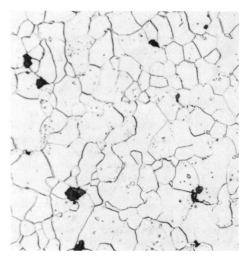

Fig. 22 Carbon steel boiler tube that ruptured due to hydrogen damage

as shown in Fig. 22. The fracture was of the brittle window type, without the fishmouth appearance characteristic of overheating failures.

Investigation. Visual examination disclosed a substantial degree of corrosion on the water-side surface, leaving a rough area in the immediate vicinity of the rupture. Microscopic examination of a cross section through the tube wall at the fracture revealed decarburization and extensive discontinuous intergranular cracking (similar to the structure shown in Fig. 18).

Conclusion. The combination of water-side corrosion, cracking, and decarburization observed in this instance led to the conclusion that rupture had occurred because of hydrogen dam-

age involving the formation of methane by the reaction of dissolved hydrogen with carbon in the steel. Hydrogen was produced by the chemical reaction that corroded the internal tube surface.

Recommendation. Steel embrittled by hydrogen can be restored to its original ductility and notched tensile strength only if grain-boundary cracking or decarburization has not occurred. A low-temperature bake (3 h or more at 175 to 205 °C, or 350 to 400 °F) is usually sufficient to drive dissolved hydrogen out of a steel part, thus restoring ductility. However, hydrogen damage involving internal cracking is irreversible; material embrittled in this manner is permanently degraded and should be replaced.

Graphitization is a microstructural change that sometimes occurs in carbon or low-alloy steels that are subjected to moderate temperatures for long periods of time. Graphitization results from decomposition of pearlite into ferrite and carbon (graphite) and can embrittle steel parts, especially when the graphite particles form along a continuous zone through a load-carrying member. Graphite particles that are randomly distributed throughout the microstructure cause only moderate loss of strength.

Figure 23 presents graphitized microstructures in plain carbon steels, ASME SA-210, grade A-1. The addition of more than 0.5% Cr stabilizes the iron carbide and prevents decom-

Fig. 24 Temperature-time plot of pearlite decomposition by the competing mechanisms of spheroidization and graphitization in carbon and low-alloy steels

The curve for spheroidization is for conversion of one-half of the carbon in 0.15% C steel to spheroidal carbides (Ref 8, 9). The curve for graphitization is for conversion of one-half of the carbon in aluminum-deoxidized 0.5% Mo cast steel to nodular graphite (Ref 10).

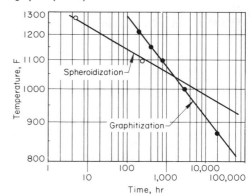

position to graphite. Regardless of operating temperatures, ASME SA-213, grades T-11 and T-22, do not show graphitized structures.

Graphitization and the formation of spheroidal carbides are competing mechanisms of pearlite decomposition. The rate of decomposition is temperature dependent for both mechanisms, and the mechanisms have different activation energies. As shown in Fig. 24, graphitization is the usual mode of pearlite decomposition at temperatures below about 550 °C (1025 °F), and formation of spheroidal carbides can be expected to predominate at higher temperatures. Because graphitization involves prolonged exposure at moderate temperatures, it seldom occurs in boiling-surface tubing. Economizer tubing, steam piping, and other components that are exposed to temperatures from about 425 to 550 °C (800 to 1025 °F) for several thousand hours are more likely than boiling-surface tubing to be embrittled by graphitization.

The heat-affected zones (HAZ) adjacent to welds are among the more likely locations for graphitization to occur. Figure 25(a) shows a carbon-molybdenum steel tube that ruptured in a brittle manner along fillet welds after 13 years of service. Investigation of this failure revealed that the rupture was caused by the presence of chainlike arrays of embrittling graphite nodules (Fig. 25b and c) along the edges of HAZs associated with each of the four welds on the tube. Arrays of graphite nodules were also found in the same locations on welds in several adjacent tubes, necessitating replacement of the entire tube bank.

Failures Caused by Corrosion or Scaling

The mechanisms of corrosion, specific techniques for analyzing corrosion failures, and degradation of various alloys in some of the more common corrosive mediums are discussed in detail in the article "Corrosion Failures" in this Volume. In this section, only some of the more common types of corrosion and corrosion-related failures that occur in steam-power-plant components and equipment will be discussed.

Water-Side Corrosion

By far the most common corrodent encountered on water-side surfaces is oxygen. The three means by which oxygen may be admitted to the water side of a steam system are:

- During operation, air can leak into a closed system in regions where the internal pressure is less than atmospheric pressure, that is, in regions between the outlet of the low-pressure turbine and the boiler feed pump
- Usually, air is admitted to a system each time the system is opened for repair or cleaning
- Free oxygen is released as a product of the dissociation of water molecules

Raw water used for initial filling and for make-up feed may contain one or more of the following: dissolved or emulsified organic substances; suspended organic and inorganic substances; dissolved inorganic solids (silica and a wide variety of inorganic compounds); and dissolved oxygen, nitrogen, and carbon dioxide. In addition, many industrial processes involve organic and inorganic chemicals that can leak into a steam system and contaminate

Fig. 25 Carbon-molybdenum steel tube that ruptured in a brittle manner after 13 years of service, because of graphitization at weld HAZs

(a) View of tube showing dimensions, locations of welds, and rupture. (b) Macrograph showing graphitization along edges of a weld HAZ (A); this was typical of all four welds. 2×. (c) Micrograph of a specimen etched in 2% nital, showing chainlike array of embrittling graphite nodules (black) at the edge of an HAZ. 100×

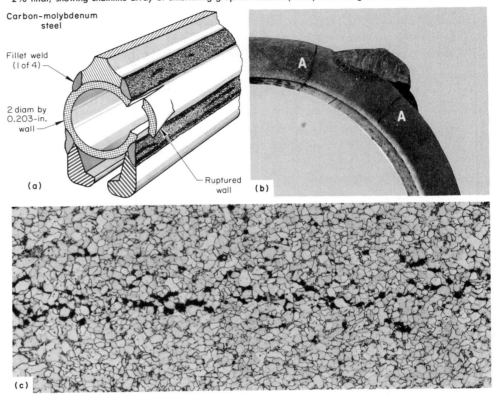

the condensate. Many of the organic and inorganic contaminants in raw water or condensate can form scale or varnish on heated boiler tubes. Contaminants also cause foaming in vapor-generating regions. Foaming causes gross mechanical carryover, or the entrainment of water droplets in the steam. Carryover reduces turbine efficiency and may result in heavy deposition of solids from the feedwater on superheater and reheater heat-transfer surfaces, in piping, and in turbines. Certain dissolved substances, especially oxygen, carbon dioxide, chloride ions, and hydroxide ions, can cause corrosion or stress-corrosion failures.

Because boiling-surface and economizer components are made mainly of carbon and low-alloy steel, they are corroded by water or by oxygen dissolved in the water. Make-up water is usually purified, and chemicals are added to boiler feedwater to control water-side corrosion; the type of purification and the type and concentration of the chemical additives vary with operating pressure, type of boiler, and type and concentration of contaminants in the raw water and the recirculating condensate. However, even with the most carefully controlled program of make-up purification and feedwater conditioning, water-side corrosion cannot be eliminated entirely.

Iron reacts with oxygen and water to form one of the iron oxides or hydroxides. The corrosion product formed depends primarily on the temperature of the metal at the corrosion site and on the concentration of oxygen in the surrounding environment. In steam systems, general corrosion is rarely a mode of failure, because water treatment is employed to prevent it. Despite all precautions, however, certain types of corrosion can still cause problems.

Types of Water-Side Corrosion in Boilers. In economizers and in boiling-surface components, pitting corrosion and crevice corrosion are the main types of severe corrosion that occur.

In pitting corrosion, a small area is attacked because it becomes anodic to the rest of the surface or because of highly localized concentration of a corrosive contaminant in the water. Pitting is often the direct result of local breakdown of normally passive surface films. Pitting that occurs at relatively few and scattered sites can result in rapid perforation because of the large ratio of cathode-to-anode area.

In crevice corrosion, oxygen is excluded from between two close-fitting surfaces or from beneath a deposit or particle on a surface. The area within the crevice, or beneath the deposit, is anodic to the surrounding area because of the

Fig. 26 Carbon steel superheater tube

Pitting corrosion and perforation were caused by the presence of oxygenated water during idle periods.

difference in oxygen concentration. The anodic areas are subject to relatively rapid attack, mainly because they are small in relation to the surrounding cathodic area.

Corrosion of Components Exposed to Steam. Surfaces exposed to steam in superheaters, reheaters, steam piping, and turbines usually do not corrode. However, certain conditions that exist during start-up, shutdown, and idle periods may cause corrosion of these components.

The main cause of corrosion of components that are normally exposed to dry steam is moisture. When a high concentration of oxygen is also present, as often occurs when a system has been opened to the atmosphere for maintenance or repair, the potential for corrosion is enhanced. Moisture can be present in normally dry portions of the system if strict procedures for exclusion of moisture are not followed during start-up, shutdown, and idle periods. It is important to balance the flow of working fluid that is required to prevent overheating of superheater and reheater tubes with the minimum flow required to maintain proper boiler circulation at low firing rates during start-up and shutdown. This balance of flow is usually obtained by use of a steam-bypass system. It is important to avoid admitting steam of excessively low quality to superheaters, reheaters, or turbines, especially during shutdown, because the high moisture content can leave a film of water on normally dry surfaces. If metal surfaces are allowed to cool more rapidly than the steam inside a component, condensation will occur, leaving small pools of water at low points in the component.

Example 6: Pitting Corrosion of a Carbon Steel Superheater Tube Caused by Oxygenated Water Trapped in a Bend. A resistance-welded carbon steel superheater tube made to ASME SA-276 specifications failed by pitting corrosion and subsequent perforation (Fig. 26), which caused the tube to leak. The perforation occurred at a low point in a bend near the superheater outlet header after about 2 years of service in a new plant.

During these 2 years, the plant had experienced a number of emergency shutdowns, which resulted in the boiler being idle for periods of several days to several weeks. The superheater tubes could not be completely drained because of low points.

Investigation. When the perforated bend was cut open, water-level marks were noticed on the inside surface above the area of pitting. In addition, above the water-level marks, some slight pitting had occurred in an area where it appeared that water had condensed and run down the surface. Microscopic examination disclosed that localized pitting had resulted from oxidation.

Conclusions and Recommendations. Water containing dissolved oxygen had accumulated in the bends during shutdown. Because the bends at low points could not be drained, cumulative damage due to oxygen pitting resulted in perforation of one of the tubes.

This type of corrosion can be best avoided by completely filling the system with condensate or with treated boiler water. Alkalinity should be maintained at a pH of 9.0, and approximately 200 ppm of sodium sulfite should be added to the water.

Alternatively, dry nitrogen should be admitted to the system after steam pressure has dropped below 517 kPa (75 psi). The dry nitrogen will purge the system, keeping internal surfaces dry while the temperature drops to ambient. The boiler then should be sealed under 35-kPa (5-psi) nitrogen pressure, with the nitrogen feed line left connected to compensate for any slight loss in pressure due to leakage.

Corrosion of Condensers and Feedwater Heaters. Corrosion of components that are exposed to condensate, primarily condensers, feedwater heaters, and boiler-feed piping, is caused mainly by the presence of carbon dioxide in the condensate. Carbon dioxide is produced when carbonates in the feedwater dissociate in the boiler. The carbon dioxide, being gaseous, is carried along in the steam through the turbine and into the condenser, where it dissolves in the liquid condensate. The mildly acidic solution (dilute carbonic acid) that results from contamination of condensate by carbon dioxide is corrosive to the copper alloy or low-carbon steel components in the preboiler portions of the system.

The main problem caused by corrosion of copper alloys is that soluble copper in the feedwater may be plated out on steel surfaces in the boiler or economizer, where galvanic corrosion can result in pitting of steel tubing. If

soluble copper salts are present during chemical cleaning, metallic copper can be readily deposited on steel surfaces. The resultant layer of flash plating can be harmful if it covers a sufficiently large area and is continuous. Although extensive deposition of copper in mud drums, attemperators, and similar components has occurred in some instances without subsequent galvanic corrosion of adjacent steel surfaces, removal of copper and copper salts from the components before the acid-treatment stage of chemical cleaning avoids this problem.

Condensers are subject to corrosion not only by the working fluid but also by the cooling medium. Lakes, rivers, and other sources of fresh water are highly oxygenated and often contain significant concentrations of various industrial wastes; these contaminants are occasionally sources of significant corrosion. When external corrosion of condenser elements occurs, leakage of the contaminated water into the working fluid can cause severe disruption of the chemical balance in high-temperature regions of the system, leading to corrosion or fouling of boiler tubes. Marine systems and stationary plants at seaside locations are subject to special problems because of the salinity of the cooling water.

One instance of condenser corrosion involved a heat exchanger that was used during shutdowns for rapid cooling of a steam generator. During long periods between shutdowns, the heat exchanger was allowed to remain idle with stagnant, highly oxygenated lake water on the shell side. Pitting corrosion caused premature failure of the carbon steel tubes. The entire tube bundle was replaced using copper alloy C70600 (copper nickel, 10%), which resists this type of attack.

Protection From Corrosion During Operation. Control of corrosion and scaling generally involves three processes: removal of dissolved solids, oxygen scavenging, and control of pH. Solids are removed primarily by chemical treatment or demineralization of make-up feedwater. Oxygen scavenging and control of pH are mainly accomplished by chemical treatment of boiler feedwater. Feedwater treatments commonly used for control of pH include conventional phosphate, coordinated phosphate, chelant, and volatile treatments. The first three also control scaleforming hardness constituents by reacting with the dissolved solids to form substances that can be removed by blowdown. In volatile treatment, which is used mainly when boiling pressure exceeds 12.5 MPa (1.8 ksi), pH is controlled by the addition of a volatile amine, such as ammonia; total solids are maintained below 15 ppm for drum boilers, or below 50 ppb for once-through boilers, by demineralization. Conventional phosphate treatment is the least sensitive to upsets resulting from leakage of contaminants into the system. Volatile treatment cannot remove hardness constituents and is therefore extremely sensitive to upsets.

Oxygen and other dissolved gases are removed primarily through the use of deaerating heaters. Typical dissolved-oxygen concentrations at the deaerator outlet seldom exceed 7 ppb. Chemical scavenging of the remaining dissolved oxygen is accomplished by addition of sodium sulfite or hydrazine in drum boilers that operate below 12.5 MPa (1.8 ksi) or by addition of hydrazine in once-through boilers or in drum boilers that operate at pressures above 12.5 MPa (1.8 ksi).

Corrosion Control in Nuclear Systems. Coolant for primary loops of pressurized-water reactors consists of high-purity water to which a rather large quantity of boric acid is added. Because the primary loop is constructed largely of stainless steel, halogen ions must be eliminated to avoid halide cracking. Oxygen is scavenged chiefly by pressurizing with hydrogen. Other contaminants are limited by using high-purity make-up feedwater and purified additives.

Secondary-loop treatment generally follows the methods used for supercritical boilers, that is, control of solids by demineralization, oxygen scavenging with hydrazine, and control of pH with ammonia.

Boiling-water reactors make use of a side-stream loop, in which a portion of the circulating coolant in the reactor is continually purified; oxygen is removed by mechanical deaeration, and no chemical scavengers are used.

Protection From Corrosion During Idle Periods. The oxygen scavengers that are normally added to feedwater during operation usually provide sufficient protection from corrosion for short idle periods, probably not exceeding 24 h. For longer idle periods, components should be drained and filled with dry inert gas (nitrogen, helium, or argon) or should be protected by wet lay-up. Depending on the characteristics of the system, wet lay-up may consist of filling the system with deoxygenated water treated with caustic to a high pH, deoxygenated water treated with about 200 ppm of an oxygen scavenger, or treated deoxygenated water plus a cover of nitrogen gas called a nitrogen cap. Continuous recirculation of demineralized deoxygenated water is often used for protection against corrosion during lay-up of large central-station boilers.

To a certain extent, the volatile amines used to control corrosion of surfaces contacted by steam or condensate will also help to protect these surfaces during idle periods, but use of dry inert gas is preferable for this purpose.

Scaling

In most systems, the pH of the feedwater is maintained between 8.0 and 11.0. The optimum pH level, which is established by experience for each installation, is the level that keeps the amounts of iron and copper corrosion products in the boiler to a minimum. When water treatment is inadequate or control of pH is lax, there is a potential for excessive water-side corrosion in economizer and boiling-surface

tubes. If thick corrosion scale forms, overheating failures can result.

However, the major source of water-side scale in boilers is not corrosion but dissolved solids in the feedwater. Compounds containing iron, calcium, magnesium, and sodium cations normally exist in most raw water; the usual anions are bicarbonate, carbonate, sulfate, and chloride radicals. Calcium and magnesium compounds (water hardness compounds) become less soluble in water as temperature is increased. These hardness compounds separate from solution at temperatures common in economizers and boilers and usually precipitate as tube-wall deposits, or hardness scale. As is true for corrosion scale, hardness scale is undesirable in regions of high heat transfer because it can cause rapid overheating of tube material. In addition, hardness scale may be somewhat porous. Dissolved solids in the boiler water, as well as strong alkalis sometimes used to treat feedwater, can be concentrated in the pores, which may lead to severe pitting corrosion of the tube surface beneath the hardness scale.

Scaling in Superheaters. Deposition of solids as scale in superheater tubing occurs primarily as the result of carryover of water droplets in wet steam. When carryover occurs, water droplets vaporize in the superheater, and any nonvolatile solids present in these droplets are deposited on heat-transfer surfaces. Deposits may not be as likely to cause rapid-overheating failures in superheater tubing as in boiling-surface tubing because of the lower heat flux in superheaters. However, because superheaters operate at higher metal temperatures than boilers, scaling may lead to stress-rupture failure or to loss of ductility because of such metallurgical changes as σ-phase formation or spheroidization of carbides.

Deposition in Turbines. Silica is the substance most commonly deposited in turbines, although other feedwater solids may also be found. At the higher boiling temperatures associated with high-pressure subcritical systems and with supercritical systems, a portion of the silica in the feedwater will volatilize and will be carried into the turbine along with the steam. As heat is extracted from the working fluid, steam temperature drops; volatile silica becomes nonvolatile and deposits out of the vapor stream onto turbine blades and surfaces of interstage passages. These deposits interfere with fluid flow in the turbine, resulting in a loss of power.

Silica, which is present in most raw water, is not removed by the same treatments used for control of water hardness; consequently, when silica deposition is excessive, special treatments must be used.

Removal of Water-Side Scale. Periodic removal of scale from internal surfaces in boilers and economizers is part of the normal maintenance of this equipment. Removal of scale can minimize unscheduled outages caused by tube rupture due to overheating. In addition, internal cleanness is an important factor in

maintaining high thermal efficiency and low operating cost.

Although mechanical methods of scale removal can be used, chemical cleaning with a suitable inhibited acid offers the advantages of less downtime; lower cost; ability to clean otherwise inaccessible areas, such as sharp bends, irregular surfaces, crevices, and small-diameter tubes; and ability to clean internal surfaces without dismantling the unit. The particular acid and inhibitor that are appropriate for a specific situation depend on the composition and adherence of the deposits, the type of cleaning method (circulation or soaking), and the type of materials being cleaned—especially in applications that involve alloys with high chromium contents.

Chemical cleaning should be supervised by a competent chemist to ensure maximum effectiveness and safety and to ensure that cleaning solutions do not excessively corrode base metal under deposits. In chemical cleaning of nuclear steam generators, additional precautions must be taken in disposing of the solutions and in ensuring that personnel are not exposed to radioactive wastes from the cleaning process.

Chemical cleaning should always be followed by thorough flushing to remove loosened deposits and residual acid, after which the boiler should be filled with a neutralizing and passivating solution to prevent the clean metal surfaces from rusting. Chemical cleaning is not a substitute for adequate water treatment. Corrosion and scale deposition can lead to severe deterioration of tube materials, and the chemical-cleaning process itself removes a thin layer of base metal along with the deposits.

Fire-Side Corrosion

Except for most gaseous fuels, combustion of fossil fuels produces solid, liquid, and gaseous compounds that can be corrosive to structural components and heat-transfer surfaces. In addition, deposits of solid and liquid residues in gas passages can alter the heat-transfer characteristics of the system, with potentially severe effects on system efficiency and tube-wall temperatures.

Residues from the combustion process, called ash, normally constitute 6 to 20% of bituminous coals, but may run as high as 30%. The composition of coal ash varies widely, but is composed chiefly of silicon, aluminum, iron, and calcium compounds, with smaller amounts of magnesium, titanium, sodium, and potassium compounds.

Wood, bagasse (crushed juiceless remains of sugar cane), and other vegetable wastes used as fuel in some industrial plants contain lower amounts of ash than coal does. In many respects, however, the compositions of these vegetable ashes resemble that of coal ash.

Fuel oils have ash contents that seldom exceed 0.2%. Even so, corrosion and fouling of oil-fired boilers can be particularly troublesome because of the nature of oil-ash deposits. The main contaminants in fuel oil are vanadium,

Fig. 27 Schematic illustration of the three layers formed during coal-ash corrosion of a superheater tube

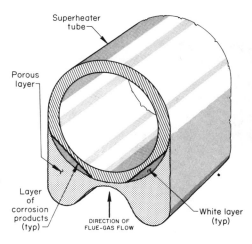

Fig. 28 2.25Cr-1Mo steel superheater tube that ruptured because of thinning by coal-ash corrosion

sodium, and sulfur—elements that form a wide variety of compounds, many of which are extremely corrosive.

Regardless of the fuel, there are similarities in all rapid fire-side corrosion circumstances. A liquid phase forms in the ash deposit adjacent to the tube surface. Once a liquid phase forms, the protective oxide scale on the tube surface is dissolved, and rapid wastage follows. In coal-fired boilers, the liquid phase is a mixture of sodium and potassium iron trisulfate—$(Na_3Fe(SO_4)_3$ and $K_3Fe(SO_4)_3$, respectively. Mixtures of these melt at as low as 555 °C (1030 °F). In oil-fueled boilers the liquid phase that forms is a mixture of vanadium pentoxide (V_2O_5), with either sodium oxide (Na_2O) or sodium sulfate (Na_2SO_4). Mixtures of these compounds have melting points below 540 °C (1000 °F).

Coal-Ash Corrosion. During combustion of coal, the minerals in the burning coal are exposed to high temperatures and to the strongly reducing effects of generated gases, such as carbon monoxide and hydrogen. Aluminum, iron, potassium, sodium, and sulfur compounds are partly decomposed, releasing volatile alkali compounds and sulfur oxides (predominantly SO_2, plus small amounts of SO_3). The remaining portion of the mineral matter reacts to form glassy particles known as fly ash.

Coal-ash corrosion starts with the deposition of fly ash on surfaces that operate predominantly at temperatures from 540 to 705 °C (1000 to 1300 °F), primarily surfaces of superheater and reheater tubes. These deposits may be loose and powdery, or may be sintered or slag-type masses that are more adherent. Over an extended period of time, volatile alkali and sulfur compounds condense on the fly ash and react with it to form complex alkali sulfates, such as $KAl(SO_4)_2$ and $Na_3Fe(SO_4)_3$, at

the boundary between the metal and the deposit. The reactions that produce alkali sulfates are believed to depend in part on the catalytic oxidation of SO_2 to SO_3 in the outer layers of the fly-ash deposit. The exact chemical reactions between the tube metal and the ash deposit including the alkali iron trisulfates are not well defined. However, in every example studied, the ash and corrosion products contain both carbon and sulfides. The sulfides cannot exist in a strongly oxidizing environment. Thus, at least on the tube surface, the conditions are reducing. The following coal-ash corrosion characteristics are common:

- Rapid attack occurs at temperatures between the melting temperature of the sulfate mixture and the limit of thermal stability for the mixture
- Corrosion rate is a nonlinear function of metal temperature, being highest at temperatures from 675 to 730 °C (1250 to 1350 °F)
- Corrosion is almost always associated with sintered or glassy slag-type deposits
- The deposit consists of three distinct layers (Fig. 27). The porous, outermost layer comprises the bulk of the deposit and is composed essentially of the same compounds as those found in fly ash. The innermost layer (heavy black lines, Fig. 27) is a thin, glassy substance composed primarily of corrosion products of iron. The middle layer, called the white layer, is whitish or yellowish in color, is often fused, and is largely water-soluble, producing an acid solution. The fused layers contain sulfides as part of the corrosion product and can be easily detected by means of a sulfur print
- Coal-ash corrosion can occur with any bituminous coal, but is more likely when the coal contains more than 3.5% sulfur and 0.25% chlorine
- None of the common tube materials is immune to attack, although the 18-8 austenitic stainless steels corrode at slower rates than lower-alloy grades. Although they vary widely depending on environment and operating conditions, corrosion rates of 2.25Cr-

1Mo steel are in the range of 0.4 to 1 mm (15 to 40 mils, or 0.015 to 0.040 in.) per year, while the 18-8 stainless steels similar to 304 corrode in the range of 0.08 to 0.1 mm (3 to 5 mils, or 0.003 to 0.005 in.) per year
- The overall appearance of the wasted tube surface is rough, with what is termed an alligator hide appearance

Because the layer next to the tube contains sulfides, the likely net corrosion reaction is:

$$2C + SO_2 + Fe = FeS + 2CO$$

It also seems probable that coal-ash corrosion of superheaters and reheaters can be reduced by better, more complete combustion of the coal. Because the corrosion occurs in the presence of carbon, removal of the carbon would seem to reduce the wastage.

Particles of fly-ash deposit on superheater and reheater tubes in a characteristic pattern (Fig. 27) in relation to the direction of flue-gas flow. The tube surfaces are corroded most heavily beneath the thickest portions of the deposit. When deposits are removed, shallow macropitting can be seen. Eventually, the tube wall becomes thinned to the point at which the material can no longer withstand the pressure within the tube, and the tube ruptures.

Example 7: Coal-Ash Corrosion of a Chromium-Molybdenum Steel Superheater Tube. The top tube of a horizontal superheater bank in the reheat furnace of a steam generator ruptured as shown in Fig. 28 after about 7½ years of service (normal service life would have been 30 years). The tube consisted of 64-mm (2½-in.) diam × 5.6-mm (0.220-in.) wall thickness ferrite steel tubing welded to 64-mm (2½-in.) diam × 3.4-mm (0.134-in.) wall thickness austenitic stainless steel tubing. The rupture occurred in the ferritic steel tubing near the welded joint.

The ferritic steel tubing, used in the low operating temperature portion of the tube, was made of 2.25Cr-1Mo steel (ASME SA-213, grade T-22); the austenitic steel tubing was made of type 321 stainless steel (ASME SA-213, grade TP321H).

The surface temperature of the tube was 620 to 695 °C (1150 to 1280 °F), which was higher than the operating temperature in use up to about 9 months earlier. At that time, the exit-steam temperature had been raised from about 520 °C (970 °F) to about 540 °C (1000 °F).

Investigation. The ferritic steel portion of the tube had split longitudinally for about 38 mm (1½ in.) along the top surface. Heavy corrosion had occurred in the region of the rupture (portion at left of rupture, Fig. 28).

The top surface of the tube in the rupture area was free of deposits, but the sides and bottom were covered with a heavy accumulation of red and white deposits that had a barklike appearance (Fig. 28). The nearby stainless steel portion was free of deposits, but did show some pitting corrosion.

Spectrographic analysis showed that the ferritic tube steel contained 2.10% Cr and 1.10% Mo, both of which were within specified limits.

The white deposit was identified as sodium-potassium sulfate [$(Na,K)SO_4$], containing about 3% alkali acid sulfate [$(Na,K)HSO_4$]. The red deposit was found to be mainly Fe_2O_3 plus about 5% SiO_2, 2% Al_2O_3, and 0.5% Cr_2O_3. Attack of the steel by the alkali acid sulfate, which has a low melting point, thinned the tube wall and produced the red deposit. This type of corrosion is known to be severe on chromium-molybdenum steel at temperatures of about 595 to 705 °C (1100 to 1300 °F). Where the tube surface had been exposed to intermittent blasts of air from soot blowers, the corrosion products had been carried away. Although stainless steel is also vulnerable to this type of attack, the reaction is slower, and only a slight amount of tube-wall thinning was detected on the stainless steel portion of this tube.

The operating records showed two probable causes of the corrosion: the increase in operating temperature and the use of coal with an ash-fusion temperature 28 °C (50 °F) lower than normal during the week before the failure. Use of this coal could have caused severe slagging conditions in the boiler.

Sectioning of the tube through the region of failure showed that the attack by the alkali acid sulfate had significantly reduced tube-wall thickness. Metallographic examination revealed a structure that was largely normal except for marked spheroidization of the carbides, which indicated appreciable service time at temperatures close to the Ac_1 temperature.

Conclusions. Rupture of the tube was attributed to thinning of the wall by coal-ash corrosion to the point at which the steam pressure in the tube constituted an overload of the remaining wall. The attack was accelerated by the increase in operating temperature and may have been influenced by temporary use of an inferior grade of coal.

Corrective Measures. Adjacent tubes, which also showed evidence of wall thinning, were reinforced by pad welding. Type 304 stainless steel shields were welded to the stainless steel portions of the top reheater tubes and were held in place about the chromium-molybdenum steel portions of the tubes by steel bands. These shields were intended as a temporary expedient against coal-ash corrosion, not as a permanent solution. Because the shields would not be cooled by steam flow, their temperatures were expected to be about 705 °C (1300 °F), the maximum at which coal-ash corrosion would occur, and little corrosion was anticipated. The replacement for the ruptured tube was left bare to permit assessment of the degree of corrosion that would ensue after the improvements in coal quality and operating temperature were adopted.

The superheater operated for about 10 years with no further ruptures or additional repairs. At that time, the tubes were inspected, and those that were thinned the most were replaced; this amounted to about 10% of the superheater bank.

Furnace-wall corrosion is also a problem on occasion in coal-fired boilers. The first principal cause of furnace-wall corrosion is the reducing conditions. Just above the stoker or around wall burners, low-oxygen conditions may exist. Radiation from the flame can lead to high skin temperatures. Reduction of the protective oxide film on the steel tube exposes the metal and leads to quite rapid metal loss. Corrosion rates up to 6.4 mm (0.250 in., or 250 mils) per year are reported. Worst-case examples are steam leaks after 2000 h of operation in a stoker-fired unit. Corrosion morphology is smooth, fairly uniform wastage.

The second principal cause is low-melting ash constituents. Rapid wastage by components of the coal ash has also been frequently observed. Differential-scanning calorimeter measurements have measured ash melting points between 335 and 410 °C (635 and 770 °F), well within the operating temperatures of water walls. Chordal thermocouples have measured temperature spikes of 30 to 55 °C (50 to 100 °F) during soot-blower operation and during slag falls. Such rapid temperature excursions coupled with liquid-ash constituents lead to a thermal fatigue component in these failures. Circumferential grooves spaced along the length of the tube are characteristic of this corrosion attack (Fig. 29).

Oil-Ash Corrosion. During combustion of fuel oils, organic compounds (including those containing vanadium or sulfur) decompose and react with oxygen. The resulting volatile oxides are carried along in the flue gases. Sodium, which is usually present in the oil as a chloride, reacts with the sulfur oxides to form sulfates. Initially, vanadium pentoxide (V_2O_5) condenses as a semifluid slag on furnace walls, boiler tubes, and superheater tubes—in fact, virtually anywhere in the high-temperature region of the boiler. Sodium oxide reacts with the vanadium pentoxide to form complex compounds, especially vanadates ($nNa_2O\cdot V_2O_5$)

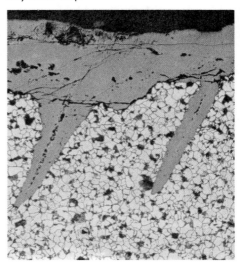

Fig. 29 Thermal fatigue plus liquid-ash corrosion on water walls leads to circumferential grooving

The cross section in an axial plane nearly parallel to the tube axis shows the deep fingerlike penetrations into the wall. Etched with nital. 210×. Courtesy of Riley Stoker Corp.

and vanadylvanadates ($nNa_2O\cdot V_2O_4\cdot m\,V_2O_5$). These complex compounds, some of which have melting temperatures as low as 249 °C (480 °F), foul and actively corrode tube surfaces.

Slag of equilibrium thickness (3.0 to 6.4 mm, or 0.12 to 0.25 in.) has developed in experimental furnaces within periods of time as short as 100 h. Slag insulates the tubes, resulting in an increase in the temperature of the slag, which in turn increases the rate of corrosion and promotes further deposition of ash. Thicker slag deposits generally lead to greater corrosion, because slag temperatures are higher and more of the corrodent is present to react with tube materials. However, higher slag temperatures also make the slag more fluid so that it will flow more readily on vertical surfaces. Consequently, slag generally builds up in corners and on horizontal surfaces, such as at the bases of water walls and around tube supports in the superheater.

Oil-ash corrosion affects all the common structural alloys. Even highly corrosion-resistant materials, such as 60Cr-40Ni and 50Cr-50Ni cast alloys, which are sometimes used for superheater-tube supports, are not immune. In addition, nonmetallic refractory materials used for furnace linings are attacked by vanadium slag; the mechanism of this attack appears to be a dissolution or slagging type of attack rather than the direct chemical attack that characteristically occurs with metals.

Low-Temperature Corrosion. In low-temperature zones of flue-gas passages, corrosion is caused chiefly by condensed water vapor containing dissolved SO_3 and CO_2. The dew point of sulfuric acid, which is the most active

Fig. 30 Suggested minimum metal temperatures to avoid flue-gas corrosion in economizers and air heaters plotted as a function of as-fired sulfur contents of various fuels

Source: Ref 1

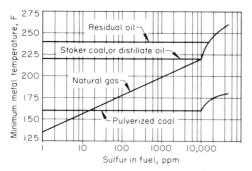

corrodent, ordinarily ranges from 120 to 150 °C (250 to 300 °F) for SO_3 concentrations of 15 to 30 ppm, which are common for coal-fired boilers. The dew point of the acidic vapors depends on the amount of moisture in the fuel and in the combustion air, the quantity of excess air, the amount of hydrogen in the fuel, and the amount of steam used for soot blowing—all of which influence the amount of water vapor in the flue gas.

Condensation of acidic vapors is most prevalent in air heaters, precipitators, stack coolers, flues and stacks, and near the inlets of economizers in units without feedwater heaters. Factors that increase the likelihood of acid condensation include low flue-gas flow, such as occurs during start-up and during periods of low-load operation, excessively low flue-gas exit temperatures during normal operation, too great an amount of excess air of high humidity, and very low atmospheric temperatures. Figure 30 shows a graph of suggested minimum metal temperatures for avoiding external corrosion in economizers and air heaters; the temperatures are plotted as functions of the type of fuel and the amount of sulfur in the fuel.

Low-alloy steels, particularly those containing copper, have been used successfully in economizers and air heaters that are prone to low-temperature corrosion. An additional advantage of these alloys is their good resistance to pitting caused by excess oxygen in boiler water.

The deposits that form as a result of low-temperature corrosion may contain corrosion products, fly ash, and the products of chemical reaction of the condensed acid with fly-ash constituents. Many of the compounds in these deposits are water soluble and can therefore be removed conveniently by washing the affected areas with water. However, the deposits sometimes become difficult to remove, especially when they are allowed to accumulate until they completely plug passages

or when they contain insoluble compounds, such as calcium sulfate.

Low-temperature corrosion is encountered more often in oil-fired units than in coal-fired units, because the vanadium in oil-ash deposits is a catalyst for oxidation of SO_2 to SO_3 and because oil firing produces less ash than coal firing. In coal-fired units, a substantial portion of the sulfur oxides is absorbed by fly-ash deposits in the high-temperature regions, where the sulfur oxides participate in coal-ash corrosion. Furthermore, coal ash, which is composed chiefly of basic compounds, partly neutralizes the condensed acids when it deposits on moist surfaces in low-temperature regions. Oil ash, which is mainly acidic, is incapable of neutralizing condensed acids.

Control of Corrosion. Fire-side corrosion can rarely be eliminated altogether. Obviously, if alkali and sulfur could be removed from coal, or vanadium and sulfur from oil, prior to combustion, fire-side corrosion would largely disappear. However, although there are some processes for removing contaminants from these fuels, they are expensive and partly ineffective. Reduction of fire-side corrosion can be accomplished by one or more of the following methods, which are listed in approximately the descending order of their effectiveness:

- Fuel selection
- Combustion control
- Boiler design and construction
- Periodic ash removal
- Use of fuel additives

Frequently, several of these measures are required rather than just one.

Fuel Selection. Use of ashless or sulfur-free fuel is perhaps the most direct means of minimizing fire-side corrosion. Most natural and manufactured gaseous fuels are ashless, whereas most oils, coals, cokes, vegetation, residues, and waste gases have some ash content. Any type of fuel may contain sulfur; few coals, oils, tars, or burnable wastes are sulfur free, whereas gases are more likely to be sulfur free or to have low sulfur contents. Some fuels, such as the black liquor burned in kraft-mill recovery furnaces, have very substantial sulfur contents.

Some boilers are fitted with burners that can use more than one type of fuel; others are fitted with auxiliary burners to permit firing of secondary fuels (generally, main and auxiliary burners are not fired together). Most boilers are built to burn only one type of fuel; thus, selection of low-ash or low-sulfur fuel involves selection of a specific type of boiler as well. The boiler operator should always be aware of the composition of the fuel being burned so that operating conditions can be adjusted to minimize ash deposition.

Use of a high-sulfur fuel sometimes cannot be avoided. In such instances, close control of operating conditions is mandatory. For example, in a fire-tube boiler that burned sulfur-

contaminated refinery gas, leaks in the carbon steel tubes in the fifth pass caused the boiler to be taken out of service. Sulfuric acid corrosion on the internal surfaces of the tube was determined to have caused the leaks. Apparently, the operator had failed to purge the boiler by burning fuel gas with a low sulfur content for a sufficient length of time prior to several previous shutdowns, and acid had condensed in the tubes as the boiler cooled through the dew point of the acid in the flue gas. Corrosion of this type normally is not a problem, provided that clean fuel gas is burned long enough to expel all sulfur compounds from the tubes before the boiler is allowed to cool down.

Combustion control—that is, adjustment of firing rate, amount of excess air, air temperature, and amount of recirculated flue gas (applicable only to boilers equipped for gas tempering)—can be very effective in controlling the amount and composition of ash deposits. For example, in an experimental boiler burning high-vanadium high-sulfur residual oil, it was found that reducing the amount of excess air from 7% down to 1 to 2% reduced the rate of corrosion of 18-8 stainless steel superheater tubes by as much as 75%. This reduction in corrosion rate apparently occurred because the low amount of excess air did not permit oxidation of vanadium and sulfur to their highest states of oxidation and thus prevented formation of the most corrosive oil-ash compounds. Low-temperature corrosion was virtually eliminated by maintaining excess air at 1 to 2%. This effect of low excess air on low-temperature corrosion has been widely substantiated, but the results are not as conclusive for high-temperature corrosion.

Boiler design and construction, including tube size and spacing, furnace configuration, size and direction of flue-gas passages, and size and location of baffles or shields, can help control ash deposition. The ability to make changes in boiler design specifically for the purpose of corrosion control is somewhat limited by the need to maintain proper gas-temperature distribution in order to facilitate efficient generation of steam.

Periodic ash removal, although less effective in controlling fire-side corrosion than procedures that reduce ash deposition, is nevertheless an important aspect of boiler operation. Ash removal prevents development of conditions that alter the distribution of metal temperatures on heat-transfer surfaces, which can lead to such problems as failure by overheating. Almost all water-tube boilers are provided with soot blowers or air lances. Soot blowers and lances direct jets of air or steam against heat-transfer surfaces to remove fire-side deposits. The effectiveness of soot blowers in performing this task depends on the design and location of the soot blowers; the blowing medium and its temperature; the properties of the deposit, especially its friability and adherence; and the frequency and duration of soot-blower operation. It is important to coordinate combustion

Table 2 Coefficients of thermal expansion for ferritic and austenitic steels

Temperature range		Ferritic steels		Austenitic steels	
°C	°F	× 10⁻⁶ m/m · K	× 10⁻⁶ in./in. · °F	× 10⁻⁶ m/m · K	× 10⁻⁶ in./in. · °F
−18 to 315	0-600 . . . . 11.5		6.4	15.5	8.6
−18 to 425	0-800 . . . . 12.6		7.0	16.6	9.2
−18 to 540	0-1000 . . . . 13.3		7.4	17.3	9.6
−18 to 650	0-1200 . . . . 13.8		7.7	17.8	9.9

control with the soot-blowing schedule to ensure that deposits are initially friable, or loosely adherent and powdery, and that they are removed before they are transformed into sintered masses.

Additives are used to control both coal-ash and oil-ash corrosion. Finely powdered magnesia, dolomite, and alumina are injected into furnace gases or blown onto heat-transfer surfaces through soot blowers. Additives are sometimes mixed with fuel oil before combustion. Additives promote the formation of ash deposits that are easily removed. Additives also reduce low-temperature corrosion by reacting with SO_3 to form innocuous sulfates. This reduces the SO_3 content of the flue gas, thus reducing the likelihood that acid will condense in low-temperature regions.

Dissimilar-Metal Welds

Weld failures are more completely covered in the article "Failures of Weldments" in this Volume. However, there are topics unique to the long-term elevated-temperature operation of the modern steam generator. One such topic is dissimilar-metal welds (DMW) between austenitic stainless steel and low-alloy ferritic steels, usually SA-213, grade T-22, material.

In the 1950s, when steam temperatures reached 540 °C (1000 °F), the need for stainless steels in the final stages of a superheater or reheater became apparent. Because stainless is required in only a portion of a superheater or reheater, the principal pressure-part tubing remained low-alloy ferritic steels. Problems of premature failure of welds between austenitic stainless steels and ferritic steels have plagued the boiler industry ever since. There are three root causes of these creep failures: carbon migration from the HAZ of the T-22 into the weld metal, expansion differences between the two varieties of steel, and the differences in corrosion resistance to flue gas leading to the formation of an oxide wedge on the outside diameter of the T-22 next to the weld.

Carbon Diffusion. When these DMWs are made with iron-base alloys of stainless steel, carbon will migrate from the T-22 into the weld metal. Particles of chromium-iron carbides will form along the weld interface in the stainless steel. The strength of T-22, especially the creep strength, is a function of carbon content. As the chromium-iron carbides form, the metal immediately adjacent to the carbides is weakened by the loss of carbon. The loss of carbon reduces the creep strength just where the temperature-induced strain is the largest. These two effects

will lead to the formation of creep voids next to these carbide particles and ultimately to premature creep failure. To reduce the carbon migration and the formation of carbide particles, it is desirable to use a welding alloy that does not form a carbide as readily as stainless steel does. Such alloys are the nickel-base materials. Nickel does not form a carbide, and while these weldments do contain chromium, the tendency to form carbides is substantially reduced. Carbon remains in the T-22, the creep properties of the HAZ remain high, the temperature-induced stresses are lower due to the compatibility of thermal coefficients, and so the formation of creep voids and failure is retarded.

Expansion Differences. The coefficients of thermal expansion for ferritic and austenitic steels differ by about 30% (Table 2). Because of these expansion differences, a 50-mm (2-in.) diam tube of stainless steel will be 0.1 mm (0.004 in.) larger in diameter at 540 °C (1000 °F) than a similar tube of T-22. This difference in expansion, the temperature-induced strain, must be accommodated at the weld.

There are three likely combinations of expansion: (1) the weld metal may have the same coefficient of thermal expansion as the austenitic stainless, (2) in between the expansion of the two, and (3) the same as the ferritic steels. When the weldment has the same expansion as the stainless steel, the temperature strain must be absorbed at the interface between the weld deposit and the T-22. As noted above, the long-term creep strength of the T-22 is adversely affected by the loss of carbon. When the expansion coefficients are split (case 2, above), the temperature strains are reduced but not to zero. Potential premature creep failures may still occur, but at a reduced frequency and over longer service times. However, if the welding alloy has expansion similar to the ferritic steel, there is only limited temperature-induced strain on the T-22. The maximum temperature strain is transferred to the stainless interface. Both the weld and stainless have much stronger creep strengths, suffer no carbon loss, and are much better able to last. No DMW has failed on the stainless steel side of the weld.

In the past, the most popular welding electrode was an E-309, an austenitic stainless steel. The use of ferritic alloy welding rods, for example, E-9018, did not solve the problem. The creep voids and failures would still originate in the ferritic alloy—in this case, the weld metal next to the stainless steel. The current choice is a nickel-base welding electrode. These nickel-base alloys have thermal coeffi-

cients very close to ferritic steels of T-11 and T-22. Thus, the temperature-induced strains are nearly zero at the ferritic interface.

Corrosion. At the juncture of the weld and the T-22 on the outside-diameter surface, an oxide wedge forms whose volume is larger than the volume of metal from which it formed. Thus, the wedge acts as a stress raiser and creates its own stress as well. With time, this wedge penetrates the T-22 and may lead to failure. The cause of the formation of this oxide wedge is related to the difference in corrosion resistance between T-22 and the weld metal. The weld metal is more noble than the ferritic alloy. In effect, a galvanic cell is created that leads to rapid oxidation at this juncture. The use of nickel-base welding alloys will not prevent the formation of these external defects.

Stresses, regardless of origin, are additive. The life expectancy of DMWs may be improved by careful attention to stresses higher than the pressure stress. To minimize the system stresses on the DMWs:

- Place the welds close to a support; the support will often carry the load, not the weld
- Make the welds in a vertical run of the tube rather than in the horizontal run to reduce the nonuniform bending stresses
- Locate the DMWs in the penthouse, out of the convection pass. The temperature gradient through the tube adds yet another stress that may be removed entirely
- Fabricate the welds with as few welding defects as possible: no undercut; a smooth radius fillet; without backing rings, suck back, or lack of fusion at the root, and so on
- Use nickel-base welding alloys

All of the above considerations will not guarantee three decades of trouble-free operation, but will, on average, yield 2½ times the life of stainless steel alloy weldments.

Once trouble begins in existing DMWs, the method of repair depends on whether the superheater or reheater is expected to last a few or many more years. If the condition is such that an overall general replacement is scheduled in the next few years, the oxide wedge can be ground out to a depth of about three-quarters of the tube-wall thickness and rewelded in place with a nickel-base alloy. If carefully done, this procedure will last several years and will give satisfactory results until the replacement is made. However, if more than 10 years of service are expected, the defective DMWs should be removed.

Although the previous discussion has focused on the pressure parts, principally tube-to-tube butt welds, no less important are the attachment welds. Spacers, supports, and alignment clips of stainless steel are routinely welded to T-11 and T-22. The preferred welding metals are also the nickel-base alloys, and for exactly the same reasons. However, extra care must be taken with the shape of the fillet;

a smooth blending to the tube with no undercut is especially important. Full-penetration welds are usually not necessary or required, but it is better to have a continuous bead all around the attachment wherever possible. If an open gap is left, that is, welded only from the sides and not around the ends, flue gas and elevated temperature will fill the space with an oxide and fly ash. In time, this oxide will wedge open the attachment weld and cause failure. Because the load on a spacer is largest at the end, it is necessary to begin and finish the weld pass toward the center of the weld rather than at an end so that any crater cracks are away from the highest stress zones.

Failure by Fatigue

Fatigue is often considered the most common mechanism of service failure of mechanical devices. Basically, fatigue involves nucleation and propagation of brittle-appearing cracks under fluctuating (cyclic) loading at stresses that have peak values less than the tensile strength of the material. Macroscopic and microscopic features of fatigue fractures are discussed in the articles "Fatigue Failures" and "Identification of Types of Failures" in this Volume. Additional information, especially on high-temperature fatigue and thermal fatigue, can be found in the article "Elevated-Temperature Failures" in this Volume.

Steam equipment may fail by fatigue if mechanical service stresses are fluctuating or vibratory or if thermal cycles or thermal gradients impose sufficiently high peak stresses. Vibratory loads in steam equipment are fairly common—for example, as a result of a slight imbalance in rotating equipment, rapidly fluctuating fluid pressures (water hammer), vibration transmitted through settings or mountings, or turbulent fluid flow, especially at high flow rates. All but a few of the various components in a steam plant are subjected to pressure and thermal cycling during start-up and shutdown. Transient thermal gradients of a lesser degree may occur during operation, especially during changes in power output.

In contrast to most mechanical stresses, thermal stresses are often difficult to avoid and, if severe, can lead to fracture in relatively few cycles. Cyclic strain is more important than cyclic stress in thermal fatigue, because nonuniform or differential thermal expansion is the usual source of the cyclic loading. Frequently, extended periods of high-temperature operation ensue between thermal cycles, and thermally activated processes, such as stress relaxation, stress oxidation, and second-phase precipitation, may alter material properties between successive cycles, thus changing resistance to thermal fatigue and to other failure mechanisms.

Identification of Fatigue Fracture. On a fracture surface, features that indicate fatigue as the fracture mechanism include beach marks, multiple origins, ratchet marks, fatigue stria-

tions, and smooth, rubbed regions. Of these five types of features, striations are the most reliable indicators of fatigue.

Example 8: Fatigue Failure of Carbon-Molybdenum Steel Boiler Tubes Caused by Vibration. Tubes in a marine boiler on a new ship failed after brief service lives. Circumferential brittle cracking occurred in the carbon-molybdenum steel tubes near the points where the tubes were attached to the steam drum.

Investigation. Visual, macroscopic, and metallographic examination did not disclose the reason for the failures. There was no evidence of overheating, excessive corrosion, or metallurgical damage caused by improper fabrication. However, when the fracture surfaces were examined by electron microscopy at high magnification, fatigue striations were found.

Conclusion. It was concluded that the fatigue failures were caused by vibrations resulting from normal steam flow at high steam demand. The tubes were supported too rigidly near the steam drum, resulting in concentration of vibratory strain in the regions of failure.

Recommendations. The method of supporting the tubes was changed to reduce the amount of restraint and thus the strain concentration, and no further failures occurred. In similar circumstances, it might be necessary to lengthen or shorten tubes (or pipes) or to reduce operating load in order to eliminate fatigue caused by vibrations resulting from normal fluid flow.

Mechanism of Fatigue Fracture. Fatigue results from reversed plastic strain in metallic crystals. If plastic straining is confined to microscopic or submicroscopic regions in an otherwise elastically stressed component, it is likely that a single crack will occur, originating at the point of maximum local stress and minimum local strength in the entire structure. Microstructural discontinuities, such as inclusions, grain boundaries, and intersections of slip planes with free surfaces, are the usual microscopic sites of crack initiation. Cracking occurs only after many cycles of loading (high-cycle fatigue), and usually there is no macroscopic evidence of plastic flow. On the other hand, when plastic straining is more extensive, there is a greater likelihood that cracks will initiate at many discontinuities after fewer cycles of loading (low-cycle fatigue) and that there will be macroscopic evidence of plastic flow.

Fatigue-crack initiation most often occurs at structural features that concentrate macroscopic stress, such as notches, fillets, holes, joints, changes in section thickness, and material discontinuities. In addition, service-induced damage, such as dents, gouges, and corrosion pits, frequently raise local stress sufficiently to initiate fatigue cracks.

Example 9: Fatigue Failure of a Carbon Steel Water-Wall Tube Because of an Undercut at a Welded Joint. In a new stationary boiler, furnace water-wall tubes were

(a) Illustration of a portion of the boiler showing location of failure. Dimensions given in inches. (b) Photograph of fractured tube; fatigue crack is at arrow A, and ductile fracture is at arrow B.

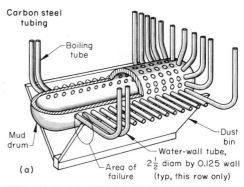

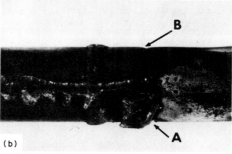

welded to the top of a dust bin for rigid support (Fig. 31a). The tubes measured 64 mm (2½ in.) outside diameter × 3.2 mm (0.125 in.) wall thickness and were made of carbon steel to ASME SA-226 specifications. Shortly after start-up, one of the tubes broke at the welded joint.

Investigation. The portion of the fracture surface adjacent to the attachment weld around the bottom half of the tube had a flat, brittle appearance. Over the remainder of the tube wall, the fracture was ductile, with a noticeable shear lip and necking. Beach marks were evident on the brittle portion of the fracture surface, indicating that fatigue was the fracture mechanism. The origin was identified as a small crevicelike undercut at the edge of the weld bead. In Fig. 31(b), arrow A indicates the origin; arrow B, local necking opposite the origin.

Conclusion. Cracking began by fatigue from a severe stress raiser. As the fracture progressed out of the region of high restraint where the tube was welded to the dust bin, the fracture changed from fatigue to ductile overload. It was concluded that the crevicelike undercut was the primary cause of the fracture and that the source of the necessary fluctuating stress was tube vibration inherent in boiler operation.

Recommendations. The remaining water-wall tubes in the row were inspected by the

Fig. 32 Type 304 stainless steel tee fitting that failed by low-cycle thermal fatigue

Top: original design. Inset shows the locations of thermocouples used in analyzing thermal gradients and the typical temperatures at each thermocouple location. Bottom: the analysis resulted in an improved design. Dimensions given in inches

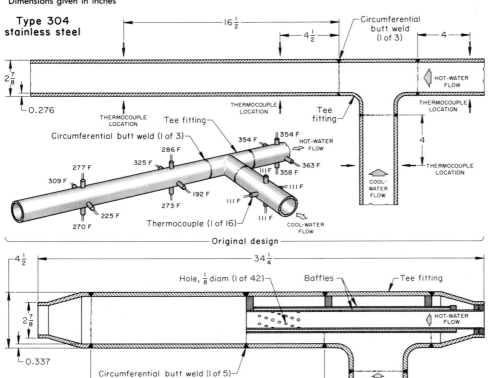

magnetic-particle method; several other tubes were found to have small fatigue cracks adjacent to the welds attaching the tubes to the dust bin. The broken tube was replaced, cracks in other tubes were repaired by welding, and the attachment welds were blended smoothly into the tube surfaces for the entire length of the dust bin. No further failures occurred after the severe undercut was eliminated.

Low-Cycle Fatigue. Pressure and thermal cycling during start-up and shutdown are the most common causes of low-cycle fatigue failures. Design factors that concentrate strain influence low-cycle fatigue as well as high-cycle fatigue; however, the effect is not quite the same. When service stresses are high enough to be conducive to low-cycle fatigue, the presence of a severe stress raiser will often cause complete fracture to occur on the first load cycle. Mild stress raisers usually do not lead to fracture on the first cycle, but only induce localized yielding. The resultant plastic flow may redistribute the stress, but usually not enough to avoid further yielding on successive cycles.

In piping exposed to high temperatures, for example, restraint produces bending moments

(and strain) in the piping system when it expands at start-up; strain reversal occurs at shutdown. Elbows and U-bends frequently undergo larger bending moments than straight sections. The mild stress concentration associated with these bends can, under certain conditions, induce plastic strain in the bends, resulting in cracking after a sufficient number of thermal cycles. The thermal expansion of the entire system is important in determining the magnitude of strain concentration and consequently the number of cycles to failure. Lack of flexibility is a frequent cause of low-cycle fatigue in piping systems. Major redesign of the system is sometimes required to remove or reduce restraint, thus allowing the system to accommodate thermal strains by deflecting elastically over a greater portion of the system than was possible in the original design.

Example 10: Low-Cycle Thermal Fatigue Failure of a Type 304 Stainless Steel Tee Fitting. Several failures occurred in 64-mm (2½-in.) schedule 80 type 304 stainless steel (ASME SA-312, grade TP304) piping in a steam-plant heat-exchanger system near tee fittings at which cool water returning from the

heat exchanger was combined with hot water from a bypass (Original design, Fig. 32). During operation, various portions of the piping were subjected to temperatures ranging from 29 to 288 °C (85 to 550 °F).

Investigation. Visual and metallographic examination showed that each of the failures consisted of cracking in and/or close to the circumferential butt weld joining the tee fitting to the downstream pipe leg, where the hot bypass water mixed with the cool return water. The cracking was transgranular and was located within the weld, adjacent to the weld, or in the base metal of either the pipe or the tee. Several cracks resulted in leakage. In one instance, the welds were radiographed before any leak was observed, and a long crack was found in the weld between the tee fitting and the downstream pipe leg. It was established that this crack had developed in service and had not been an original welding defect. In another instance, a crack occurred in a pipe segment about 100 mm (4 in.) from the nearest weld. This crack was outside the HAZ of the weld; however, additional cracks were found in the tee fitting and in the welds.

Because propagation of the cracks was entirely transgranular, thermal fatigue was a more likely cause of failure than SCC, in which some intergranular cracking would be expected.

To investigate thermal stresses, an array of thermocouples was installed in two tee fittings in the arrangement shown in Fig. 32. The temperatures at different points along the length and around the circumference of the piping are also shown for one of the two instrumented fittings; temperatures were similar for the other fitting. It was concluded that circumferential temperature gradients, in combination with inadequate flexibility in the piping system as a whole, had caused all of the failures.

Corrective Measures. To alleviate the thermal-stress pattern, the tee fitting was redesigned (Improved design, Fig. 32). A larger tee fitting, made of 100-mm (4-in.) schedule 80S type 316H stainless steel pipe (ASME SA-312, grade TP316H), was used, and a 455-mm (18-in.) long extension of 100-mm (4-in.) schedule 80 type 316H pipe was added to the downstream leg. Two concentric tubular baffles, both made of type 316H stainless steel, were placed in the run direction of the interior of the tee fitting. The inner baffle (455 mm, or 18 in., long × 38 mm, or 1½ in., outside diameter × 2.1 mm, or 0.085 in., wall thickness, and perforated at the downstream end) carried the full flow of the hot water from the bypass. The outer baffle (400 mm, or 16 in., long × 50 mm, or 2 in., outside diameter × 2.1 mm, or 0.085 in., wall thickness), which was open at each end, served to insulate the inner baffle from the cool water entering the fitting from the heat exchanger and provided a region in which the cool water could form an annulus to receive the core of hot bypass water entering the downstream extension beyond the baffles. Mixing in that region was possible without generation of damaging circumferential thermal gradients.

Fig. 33 Crazed pattern of thermal fatigue cracking on the outer surface of a stainless steel tube

See also Fig. 37. Approximately 4 ×

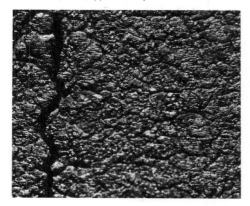

Thermal fatigue may result in either low-cycle or high-cycle failure. Low-cycle thermal fatigue is usually associated with large plastic strains and is most often caused by large changes in temperature or large differences in thermal expansion between two structural members. If a change in temperature is especially severe and if it occurs rapidly, it is called thermal shock; cracks due to thermal shock generally initiate in ten thermal cycles or less—sometimes in only one thermal cycle when the material is sufficiently brittle.

High-cycle thermal fatigue frequently results from intermittent wetting of a hot surface by liquid having a considerably lower temperature. In such instances, the wetted surface contracts rapidly, whereas the metal below the surface does not; this produces large biaxial tensile stresses in the wetted surface. As the water absorbs heat and evaporates, the temperature of the surface returns to its previous value, and surface stresses are relaxed. After a sufficient number of thermal cycles, a crazed pattern of cracks appears on the surface (Fig. 33). Eventually, one or more of the surface thermal fatigue cracks may propagate completely through the section by fatigue or by another mechanism, as discussed in the section "Multiple-Mode Failure" in this article.

Cracking that is caused by thermal fatigue sometimes resembles cracking caused by corrosion fatigue or stress corrosion. However, in contrast to corrosion fatigue and stress corrosion, which can be retarded by use of inhibitors, coatings, or galvanic protection, thermal fatigue will proceed regardless of what is done to alter the environment or to coat the metal to prevent direct contact with the environment. Thermal fatigue can be prevented only by eliminating excessive strain caused by thermal cycling.

Ferritic steels are generally considered to be more resistant to thermal fatigue than austenitic steels. This has been confirmed in several instances in which parts made of ferritic steels have withstood service conditions under which austenitic steel parts have failed. One such instance involved two flanged 18Cr-12Ni austenitic stainless steel valves that were bolted to 5Cr-0.5Mo ferritic steel piping in a petroleum plant. The valves were alternately exposed to atmospheric temperature and to an elevated temperature between 595 and 620 °C (1100 and 1150 °F). Each valve was subjected to 1½ thermal cycles per day, that is, three cycles every 2 days; about 8 h elapsed between the start of the heating portion of the cycle and the start of the cooling portion, and conversely. After 10 days of operation, cracks were found radiating from the bore into the flange of one valve; the other valve failed shortly after. The mating 5Cr-0.5Mo ferritic steel piping gave excellent service, as did replacement valves made of 9Cr-1Mo ferritic steel.

The superiority of ferritic alloy steel over several wrought austenitic stainless steels in applications involving thermal cycling is a result of differences in thermal conductivity and coefficient of thermal expansion between the two types of steel. The higher thermal conductivity of ferritic alloy steel results in a lower temperature gradient between the surface and the interior when the surface of a cool part is heated or when the surface of a hot part is cooled. This lower temperature gradient results in less thermal stress in the surface layer. In addition, because of the lower coefficient of expansion of ferritic steel, a given temperature gradient produces less expansion, and thus less thermal stress, in ferritic alloy steel than in wrought austenitic stainless steel. The combination of these two factors can result in appreciably less thermal stress and longer life for ferritic steel under a given set of cyclic thermal conditions.

Corrosion fatigue is discussed in detail in the article "Corrosion-Fatigue Failures" in this Volume. Strictly speaking, few service failures occur by pure fatigue. Most fatigue failures actually occur by corrosion fatigue, because most parts are exposed simultaneously to cyclic stress and to a medium that is somewhat aggressive toward the material; in this strict sense, even air can be considered an aggressive medium. In a more practical sense, however, because fatigue properties are normally determined in air, corrosion fatigue can be thought of as fatigue that takes place in an environment in which the combined action of corrosion and cyclic loading results in more rapid crack initiation and/or propagation than would occur in air. The microscopic mechanism of corrosion fatigue appears to be related to the galvanic or preferential dissolution processes that are believed to be responsible for SCC. In fact, all alloys that are susceptible to stress corrosion in a given medium appear to be susceptible to rapid corrosion fatigue in the same medium. However, corrosion fatigue has been identified as a failure mechanism in alloys that are considered immune to stress corrosion (but not immune to ordinary corrosion in the same environment).

The length of time necessary for a crack to be initiated and propagated by corrosion fatigue can be increased—sometimes substantially—by reducing or eliminating either the fluctuating stress or the corrosiveness of the environment. In addition to reduction of stress or stress concentration, the following measures have been used successfully to combat corrosion fatigue: (1) use of corrosion inhibitors, (2) cathodic protection, (3) shielding of the surface of a part from contact with the environment by seals, by paint, plastic or other nonmetallic coatings, or by plating with corrosion-resistant metals, and (4) use of a more corrosion-resistant material. The last method should be used with care, however, because some of the ordinarily corrosion-resistant materials can be more susceptible to failure by corrosion fatigue in certain environments than the materials they replace. Coatings should be used to combat corrosion fatigue only when local strains are uniform and low. The protection afforded by coatings, either metallic or nonmetallic, is only skin deep, and local breaches of this protective skin expose the metal underlying the coating to the same conditions as those that ordinarily lead to failure of uncoated parts.

Failures Caused by Erosion

Erosion involves impact of large numbers of small solid or liquid particles against a surface, or it is caused by the collapse of gas-filled bubbles in a cavitating liquid. Erosion by solid particles is a form of abrasive wear, whereas liquid-impingement erosion—for example, as caused by the water droplets in wet steam—is more like cavitation erosion.

The microscopic mechanisms of material removal by erosion are discussed in the articles "Wear Failures" and "Liquid-Erosion Failures" in this Volume. Failures caused by these mechanisms can be readily recognized by visual examination.

Abrasive erosion of screen tubes or superheater tubes results from impact by particles of fly ash entrained in the flue gases. Erosion is enhanced by high flow velocities; thus, partial fouling of gas passages in tube bands by deposition of fly ash can lead to erosion by forcing the flue gases to flow through smaller passages at higher velocity. This effect, sometimes called laning, exposes tube surfaces to a greater probability of impact by particles having higher kinetic energy, thus increasing the rate of damage. Erosion by fly ash causes polishing, flat spots, wall thinning, and eventual tube rupture.

Fly-ash erosion can be controlled by coating tube surfaces with refractory cements or other hard wear-resistant materials, although this reduces the heat-transfer capability of the surface. An alternative method is the channeling of gas flow away from critical areas with baffles.

Liquid-impingement erosion occurs chiefly in components that are subjected to

Fig. 34 Cast iron suction bell, 455 mm (18 in.) in diameter, from a low-pressure general service water pump that failed by cavitation erosion after about 5 years of service
Note the deeply pitted surface and the irregular shape of the erosion pattern, both of which are typical characteristics of caviation damage.

Fig. 35 Inside surface of an austenitic stainless steel superheater tube showing a tight crack caused by stress corrosion
Arrows indicate ends of crack.

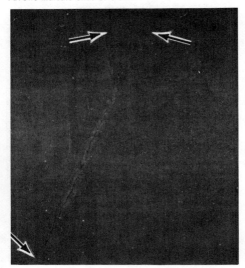

high-velocity flow of wet steam. Among the components most susceptible to liquid-impingement erosion are low-pressure turbine blades, low-temperature steam piping, and condenser or other heat-exchanger tubes that are subjected to direct impingement by wet steam. Corrosion may or may not occur simultaneously with liquid-impingement erosion.

Liquid-impingement erosion in tubing or piping is most likely when fluid velocities exceed 2.1 m/s (7 ft/s). Damage occurs first at locations where direction of flow changes, such as elbows or U-bends. Large-radius bends are less susceptible to such damage; however, use of erosion-resistant materials, such as austenitic stainless steel, is often more effective. Erosion of heat-exchanger tubes caused by impingement of wet steam can sometimes be eliminated by redirecting flow with baffles.

Cavitation erosion occurs in regions of a system where a combination of temperature and flow velocity causes growth and subsequent collapse of large quantities of vapor-filled bubbles in a flowing stream of liquid. Damage occurs at the points of bubble collapse. In steam plants, the components most often damaged by cavitation erosion are feed lines, pump casings, and pump impellers. Figure 34 shows the typical appearance of a part extensively damaged by cavitation erosion; the surface is deeply and irregularly pitted.

Cavitation erosion is most effectively avoided by reduction of either flow velocity or

temperature (reduction of flow velocity is preferable). Other measures, including selection of material and the use of erosion-resistant welded overlays, are discussed in the article "Liquid-Erosion Failures" in this Volume.

Failure by SCC

Stress-corrosion cracking occurs in specific alloys exposed to specific environments, as discussed in detail in the article "Stress-Corrosion Cracking" in this Volume. Stress-corrosion cracking is a type of failure defined as fracture of a structural member under the combined effects of static tensile stress and a corrosive environment in circumstances where, if either the tensile stress or the corrosive environment were absent, fracture would not occur.

Stress-corrosion cracking can occur in boiler and superheater tubes, piping, valves, turbine casings, and other parts, especially where feedwater or condensate can collect. Concentration of caustic in portions of systems that use conventional or coordinated phosphate methods of feedwater treatment is probably the most common cause of stress-corrosion (caustic) cracking in steam equipment. Aqueous solutions containing chloride ions can cause SCC of austenitic stainless steel (Example 12 in this article), and solutions containing ammonia can cause SCC of copper alloys. Stress oxidation in high-temperature regions can also occur.

Many cracks that result from stress corrosion are tight and difficult to detect visually (Fig. 35); they are particularly difficult to detect on scaled surfaces. Liquid-penetrant examination is useful in locating tight stress-corrosion cracks, but may not reveal all such cracks. Because the substances found in liquid penetrants are similar to some of the substances that cause SCC, use of a liquid penetrant to detect a crack can interfere with the identification of fracture-surface deposits. Consequently, liquid penetrants should not be used without due consideration of their undesirable side effects on the overall analysis.

Microscopic examination is required to confirm SCC as the mechanism of failure. Stress-corrosion cracks may be transgranular or intergranular, depending on the alloy and the environment, but are usually extensively branched, becoming more branched as cracking progresses.

Although service conditions may produce static stress that exceeds the threshold for SCC, residual stress is more often responsible. Residual stress may result from metallurgical transformations or from uneven heating or cooling, especially in weldments. Hot- or cold-forming processes can introduce residual stress of high magnitude, as can assembly processes, such as press fitting, bolting, or riveting. Incidental damage, such as bending or hammering during assembly, is less frequently a source of residual stress.

Example 11: SCC of a Type 304 Stainless Steel Pipe Caused by Residual Welding Stresses. A 150-mm (6-in.) schedule 80S type 304 stainless steel pipe (170 mm, or 6⅝ in., outside diameter × 11 mm, or 0.432 in., wall thickness), which had served as an equalizer line in the primary loop of a pressurized-water reactor, was found to contain several circumferential cracks 50 to 100 mm (2 to 4 in.) long. Two of these cracks, which had penetrated the pipe wall, were responsible for leaks detected in a hydrostatic test performed during a general inspection after 7 years of service.

Investigation. The general on-site inspection had included careful scrutiny of all pipe welds by both visual and ultrasonic examination. Following discovery of the two leaks, the entire line of 150-mm (6-in.) pipe, which contained 16 welds, was carefully scanned. Five additional defects were discovered, all circumferential and all in HAZs adjacent to welds. In contrast, scans of larger-diameter pipes in the system (up to 560 mm, or 22 in., in diameter) disclosed no such defects.

Samples of the 150-mm (6-in.) pipe were submitted to three laboratories for independent examination. Inspection in all three laboratories disclosed that all the defects were circumferential intergranular cracks that had originated at the inside surface of the pipe and that were typical of stress-corrosion attack. A majority of the cracks had occurred in HAZs adjacent to circumferential welds. Some cracks had penetrated the entire pipe wall; others had reached a depth of two-thirds of the wall thickness. In general, the cracks were 50 to 100 mm (2 to 4 in.) long. Branches of the cracks that had approached weld deposits had halted without invading the structure (austenite plus δ-ferrite) of the type 308 stainless steel welds. All HAZs examined contained networks of precipitated carbides at the grain boundaries, which revealed that welding had sensitized the 304 stainless steel in those local areas.

Additional cracks, also intergranular and circumferential and originating at the inner surface of the pipe, were discovered at locations remote from any welds. These cracks penetrated to a depth of only one-sixth to one-fourth of the wall thickness through a solution-annealed structure showing evidence of some cold work at the inner surface.

The water in this heat exchanger was of sufficient purity that corrosion had not been anticipated. The water conditions were as follows:

- *Temperature*: 285 °C (545 °F)
- *Pressure*: 7 MPa (1000 psi)
- *pH*: 6.5 to 7.5
- *Chloride content*: <0.1 ppm
- *Oxygen content*: 0.2 to 0.3 ppm
- *Electrical conductivity*: <0.4 μmho/cm

Analysis of the pipe showed the chemical composition to be entirely normal for type 304 stainless steel. The material was certified by the producer as conforming to ASME SA-376, grade TP304. All available data indicated that the as-supplied pipe had been of acceptable quality.

Several types of stress-corrosion tests were conducted. In one group of tests, samples of

150-mm (6-in.) type 304 stainless steel pipe were welded together using type 308 stainless steel filler metal. Two different welding procedures were used—one consisting of three passes with a high heat input and the other consisting of ten passes with a low heat input. The welded samples were stressed and exposed to water containing 100 ppm dissolved oxygen at 285 °C (545 °F) and 7 MPa (1000 psi). Samples that were welded with the high heat input procedure, which promotes carbide precipitation and residual stress, cracked after 168 h of exposure, whereas samples that were welded with the low heat input procedure remained crack free. These results were considered significant, because service reports indicated that the cracked pipe had been welded using a high heat input and a small number of passes and that the larger-diameter piping in the primary loop, which did not crack in service, had been welded with the use of a low heat input, multiple-pass procedure.

Conclusions. The intergranular SCC of the 150-mm (6-in.) schedule 80S type 304 stainless steel pipe exposed to high-temperature high-pressure high-purity water in 7 years of heat-exchanger service was believed to have been caused by the following factors:

- *Stress*: The welding procedure, which employed few passes and high heat input, undoubtedly generated unacceptably high levels of residual stress in the HAZs. The cold working of the internal surface in a sizing operation also set up residual stresses in areas apart from the welds. Both conditions were conducive to SCC
- *Sensitization*: The high heat input welding caused precipitation of chromium carbides at the grain boundaries in the HAZs. This rendered the steel sensitive to intergranular attack, which, combined with the residual stress, afforded a completely normal setting for stress-corrosion cracking
- *Environment*: Although the level of dissolved oxygen in the water was quite low, it was considered possible that, on the basis of prolonged exposure, oxygen in the range of 0.2 to 1.0 ppm could have provided the necessary ion concentration

Corrective Measures. All replacement pipe sections were installed using low heat input multiple-pass welding procedures. When stress corrosion is identified as the mechanism of a failure, the proper corrective action can be either a reduction in stress (or stress concentration) or an alteration of the environment. When the cause of failure is inadvertent concentration of a corrodent or inadvertent exposure of the part to a corrosive foreign substance, such as use of a steam line to feed concentrated caustic during chemical cleaning, the preventive measure may be no more complicated than exclusion of the corrodent from that region of the system. In many instances, however, reduction of the level of residual stress is the most effective means of minimizing or preventing SCC.

Fig. 36 Section through type 316 stainless steel tubing that failed by SCC because of exposure to chloride-contaminated steam condensate

Micrograph shows a small transgranular crack that originated at a corrosion pit on the inside surface of the tubing and only partly penetrated the tubing wall. Etched with aqua regia. 250×

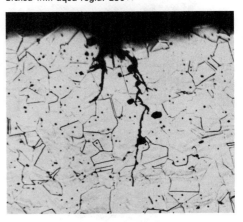

Example 12: SCC of Type 316 Stainless Steel Tubing. A steam-condensate line made of type 316 stainless steel tubing that measured 19 mm (0.75 in.) outside diameter × 1.7 mm (0.065 in.) wall thickness had been in service for 5 to 6 years when leakage occurred. The line was buried, with no wrapping, in a sandy, caustic soil. The line carried steam condensate at 120 °C (250 °F) with a 2-h heat-up/cool-down cycle. No chemical treatment had been given to either the condensate or the boiler water. Two sections of the tubing, with a combined length of approximately 330 mm (13 in.), were submitted for laboratory analysis of the cause of the leakage.

Investigation. The outside surface of the tubing exhibited no signs of corrosion; however, on one of the sections there was a transverse crack along the edge of a 6.4-mm (¼-in.) diam pitted area. Two adjacent areas were raised, resembling blisters, and also had transverse cracks along their edges. The cracks were irregular and had penetrated completely through the tube wall. The pitted area was believed to be a spall, and the blisters incipient spalls, resulting from SCC that originated at the inner surface of the tube. The cracks, the spall, and the blisters did not appear to have been related to any effect of the external environment.

The inside surface of the tubing was covered with a brown, powdery scale, was slightly rough, and contained small pits. Some cracks were found that had originated at pits on the inner surface but had not completely penetrated the tubing wall (Fig. 36). The cracks were transgranular and branching, which is typical of chloride-induced SCC of austenitic stainless steel.

Fig. 37 Stainless steel superheater tube that failed by thermal fatigue and stress rupture

(a) Photograph of the tube showing thick-lip rupture. (b) Macrograph of a section taken transverse to a fracture surface of the tube showing that thermal fatigue cracking started at the outside surface (top) and was followed by stress-rupture cracking in the inner half of the tube wall. 10×. Thermal fatigue also produced cracking on the outside surface of the tube in the crazed pattern shown in Fig. 33

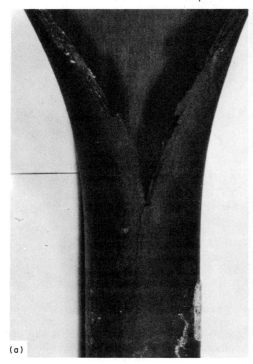

To check for chlorides, the inside of the tubing was rinsed with distilled water, and the rinse water was collected in a clean beaker. A few drops of silver nitrate solution were added to the rinse water, which clouded slightly because of the formation of insoluble silver chloride. This test indicated that there was a substantial concentration of chlorides inside the tubing.

Conclusion. The tubing failed by chloride-induced SCC. Chlorides in the steam condensate also caused corrosion and pitting of the inner surface of the tubing. Stress was produced when the tubing was bent during installation.

Recommendations. Water treatment should be provided to remove chlorides from the system. Continuous flow should be maintained throughout the entire tubing system to prevent concentration of chlorides. No chloride-containing water should be permitted to remain in the system during shutdown periods. Bending of tubing during installation should be avoided to reduce residual stress.

Multiple-Mode Failure

Many failures in steam systems involve more than one failure process—one that initiates the failure and another (or others) that eventually destroys the part. For example, fatigue failures that cause tubes or pipes to leak may start at corrosion pits. A failure that involved a com-

bination of thermal fatigue and stress-rupture cracking is described below.

The tube rupture shown in Fig. 37(a) has the macroscopic characteristics of a thick-lip rupture caused by overheating. However, as shown in Fig. 37(b), stress-rupture cracking was confined to the inner half of the tube wall. There was no microstructural evidence of overheating.

The outside surface of this 54-mm (2⅛-in.) diam stainless steel superheater tube exhibited extensive crazing by thermal fatigue, as shown in Fig. 33. Apparently, cyclic thermal stress initiated cracking at the fire-side surface. Pressure stresses assisted in defining the preferential direction of fatigue-crack propagation (parallel to the tube axis and transverse through the tube wall), and stress-rupture cracking became the ultimate fracture process in the weakened tube.

The effects of internal and external scaling on tube ruptures caused by overheating are discussed in the section "Scaling" in this article. In tube and other high-temperature structural components, thinning caused by corrosion or erosion can also lead to overheating-type failures. Corrosion or erosion, particularly on the fire-side surfaces of screen tubes and superheater tubes, may reduce the tube-wall thickness to the extent that the tube can no longer sustain the pressure stress without creep. Continued thinning will accelerate the onset of

tertiary creep, ultimately leading to stress-rupture fracture. Because localized loss of material is responsible for increasing the local tube-wall stress, overheating-type ruptures often occur with little or no swelling. Internal grain separation will be evident in the microstructure only in the thinned region, and macroscopic examination of the fire-side surface is usually sufficient to reveal the extent of external attack. Localized overheating is not necessary for tube rupture to occur. Consequently, a lack of microstructural evidence of overheating would be a positive indicator that corrosion or erosion was a factor. Conversely, the presence of such evidence does not rule out corrosion or erosion as a cause of failure.

REFERENCES

1. D.N. French, *Metallurgical Failures in Fossil Fired Boilers*, John Wiley & Sons, 1983, p 155-165
2. W.H. McAdams, *Heat Transmission*, McGraw Hill, 1952, p 184-189
3. *Principles of Industrial Water Treatment*, Drew Chemical Co., Boonton, NJ, 1979, p 194
4. I.M. Rehn, W.R. Apblett, Jr., and J. Stringer, Controlling Steamside Oxide Exfoliation in Utility Boiler Superheaters and Reheaters, *Mater. Perform.*, June 1981, p 27-31
5. *Steam*, 38th ed., Babcock & Wilcox Co., New York, 1972
6. *High Temp. Bull.*, No. 5, Babcock & Wilcox Co., Tubular Products Divison, Beaver Falls, PA
7. H.A. Grabowski and H.A. Klein, Corrosion and Hydrogen Damage in High Pressure Boilers, in *Proceedings of the 2nd Annual Educational Forum on Corrosion*, National Association of Corrosion Engineers, Houston, Sept 1964
8. R.W. Bailey, *Engineering*, Vol 129, 1930, p 265
9. R.W. Bailey, *Engineering*, Vol 133, 1932, p 261
10. W.L. Hemingway, *The Study of Graphitization*, Edwards Valve Co., 1952

SELECTED REFERENCES

- *Combustion-Fossil Power Systems*, Combustion Engineering, Inc., Windsor, CT, 1981
- O. Devereux, A. McEvily, and R. Staehle, Ed., *Corrosion Fatigue: Chemistry, Mechanics and Microstructure*, National Association of Corrosion Engineers, Houston, 1972
- B.E. Gatewood, *Thermal Stresses*, McGraw-Hill, 1957
- S.S. Manson, *Thermal Stress and Low-Cycle Fatigue*, McGraw-Hill, 1966
- *Manual for Investigation and Correction of Boiler Tube Failures*, EPRI CS-3945, Electric Power Research Institute, Palo Alto, CA, April 1985

- C.C. Osgood, *Fatigue Design*, John Wiley & Sons, 1970
- *Proceedings of the International Conference on Fatigue of Metals*, Institution of Mechanical Engineers, London, 1956
- *Proceedings of the Joint International Conference on Creep*, Institution of Mechanical Engineers, London, 1963
- P.D. Stevens-Guille, *Steam Generator Tube Failures: World Experience in Water-Cooled Nuclear Power Reactors During 1972*, Chalk River Nuclear Laboratories, Ontario, March 1974
- H. Thielsch, *Defects and Failures in Pressure Vessels and Piping*, Reinhold, 1965
- H. Thielsch, Why High Temperature Piping Fails, *Met. Eng. Q.*, Vol 10 (No. 1), 1970, p 7-11
- H.H. Uhlig, *Corrosion and Corrosion Control*, 2nd ed., John Wiley & Sons, 1971

Failures of Heat Exchangers

Revised by Robert J. Franco, Exxon Company, U.S.A.

HEAT EXCHANGERS are generally used to transfer heat from combustion gases, steam, or water to gases, vapors, or liquids of various types. Heat exchangers are of tube, plate, or sheet construction. Tubular heat exchangers are generally used for large fluid systems, whereas heat exchangers of plate or sheet construction are often preferred for smaller fluid streams. Heat exchangers are usually separate units, but can sometimes be incorporated as components in larger vessels.

In selecting materials for heat exchangers, corrosion resistance, strength, heat conduction, and cost must be considered. The demand for corrosion resistance is particularly difficult to meet because the material may be exposed to corrosive attack by various mediums. Therefore, damage to heat exchangers is often difficult to avoid.

Characteristics of Tubing

The primary function of tubes in a heat exchanger is to transfer heat from the shell side of the unit to the fluid on the inside of the tube, or conversely. Therefore, thermal conductivity, wall thickness, and resistance to scaling are extremely important. In most cases, tensile strength and yield strength of the tubes are not significant factors; loading, because it is opposed by the internal pressure, is so low that tubes with a high allowable stress are not required.

Stiffness and resistance to denting are properties of tubing that are important in the manufacture of heat exchangers. Stiffness of a tube is proportional to its wall thickness, diameter, and modulus of elasticity. Resistance to denting is related to the same factors plus the yield strength of the material.

Extended-surface (finned) tubing is sometimes used to provide the proper thermal conductance for effective transfer of heat from gas to liquid or from liquid to liquid. Fins on much of the tubing used in air coolers and hydrogen coolers are applied by soldering, brazing, or resistance welding; thus, the ability of the tube material to accept these processes is important. Oil coolers often use integral-finned (extruded) tubing. This type of tubing requires a material with a significant amount of ductility to ensure that the extended surface can be satisfactorily extruded.

Corrosion Resistance. To meet corrosion requirements, tubing must be resistant to general corrosion, stress-corrosion cracking (SCC), selective leaching (for example, dezincification of brass), and oxygen-cell attack in whatever environments are encountered before service, in service, and during periods in which the equipment is not operating.

The environment to which tubes will be exposed is often difficult to predict. Many heat exchangers operate with river water, seawater, brackish water, or recirculated cooling-tower water in the tubes. The compositions of these waters used for cooling can change over a period of time, through seasonal variations, or from an unpredictable accident. Long lead time in power-plant erection can have a considerable effect on the tubes. Heat exchangers may be completed and sent to the field years before the remainder of the plant is completed and put into operation. Under such circumstances, the heat exchangers should be adequately protected from the corrosion that might occur before start-up. Wet lay-up using a suitably buffered or inhibited solution can be used to retard corrosion. Another method is to keep the heat exchanger dry and, if necessary, fill it with an inert gas. Desiccants or vapor-phase corrosion inhibitors can also be added, and the heat exchanger can then be sealed until start-up.

After start-up, consideration must be given to the many unscheduled shutdowns that generally affect the early life of any new plant. Suitable precautions must be taken to protect equipment temporarily out of service if damage is to be avoided during downtime.

Examination of Failed Parts

Full background information on a part is important for proper failure analysis. The information should describe the circumstances of failure, operation, or other pertinent details. Inadequate information on the circumstances surrounding the failure often results when the investigation is not conducted at the failure site and samples are sent to the laboratory for examination. A list of the type of background information useful for a proper analysis

Table 1   Data needed for analysis of failures in heat exchangers

Process name and function
Unit name and function
Material size, grade, specification
Sketches of unit, failures or other particulars
Environment (inside and out):
Temperatures, max and min
Substances contacting tube:
Common name
Chemical composition
Concentration
Impurities
pH
Velocity
Aeration
Contact with other metals
Particulars of failure:
Type
Location
Circumstances at time of failure
Service life
History:
Service behavior of other materials
Unusual conditions before failure
Fabricating operations, sequence
Maintenance operations, sequence
Heat-treating operations, sequence

of failures in heat exchangers is given in Table 1.

Sample Selection. The selection of samples that are representative of the failure is important. Samples should be taken not only from the failure area but also from areas remote from the site of the failure so that comparisons can be made. Care should be taken in selecting and removing samples from heat exchangers. Samples that have been burned or battered in removal or have been modified during the preliminary examination do not adequately illustrate the failure and cannot be considered representative. Removal of rolled tubes from tube sheets requires particular care.

Visual Examination. Much information concerning a failure can be obtained from a careful visual examination. Dimensional changes reflect swelling or thinning. The pattern, extent, or nature of corrosion or cracking is often an important clue to the cause of failure (consequently, the examination may be hindered by a sample that is too small). Discoloration or rusting on stainless steel may be a sign of iron contamination. Heavy deposits on a tube, such as scale or corrosion product, could be caused by overheating.

Visual examination will indicate the direction of further work; the detailed investigation should be planned at this stage. Care must be taken so that evidence is not inadvertently altered or destroyed. Provision should be made for preserving some of the original samples to provide material for a fresh start if an answer to the failure is not found. Sketches and photographs prepared at this time are helpful for later reference in interpreting results.

Microscopic Examination. For metallurgical investigation, microstructural details provide much information about thermal history, operating temperatures, chemical environment, manner of attack, and cracking.

Microstructural examination of a section through the fracture surface is an aid in determining if the fracture is transgranular or intergranular. Also, variations in material between the failure area and a remote region can reveal specific local microstructural effects, such as decarburization, alloy depletion, graphitization, or precipitation of embrittling constituents at the failure area. Quality of soldered or brazed joints, depth of penetration of welds, or occurrence of gas porosity can be revealed by proper sectioning. A critical combination of strain and subsequent heating can occasionally cause excessive grain growth. This has occurred in copper tubing, aluminum alloy tubing, or copper-clad steel tubing used for medium-pressure duty.

Materials may be identified by microstructural examination. For example, various grades of carbon steel pipe or tubing may be distinguished according to the deoxidation practice and resulting microstructure. Similarly, seamless and welded pipe or tubing may be differentiated. A gross error in material usage, such as using low-carbon instead of medium-carbon steel, can be determined quickly by microstructural inspection. In low-carbon steels, rimmed steels can be distinguished from killed steels by their larger-grain rimmed zone at the surface and concentration of carbon and impurities at the center of the section. The presence or absence of inclusions can also be checked.

In failures involving a weld, careful evaluation of the heat-affected zone (HAZ) is necessary. In carbon and alloy steels, HAZs can be abnormally hard. In austenitic stainless steels, HAZs can be sites for precipitated intergranular carbides; the resulting alloy-depleted zones surrounding the precipitates are potential origins for intergranular corrosion or cracking of the weldment.

Chemical Analysis. Information on corrosive agents is usually obtained by analysis of corrosion products or metal surfaces. For example, silver nitrate tests may be performed on residues clinging to a pitted stainless steel surface to confirm the presence of chloride ions—a common cause of pitting. Wet chemical methods are often supplemented by the use of x-ray diffraction or electron probe x-ray microanalysis, using wavelength- or energy-dispersive x-ray spectrometers for identification of compounds, which in turn may identify corrosives. These techniques are often used to analyze for the presence of contaminants on crack surfaces or at the roots of pits. More information on these analytical techniques is available in Volume 10 of the 9th Edition of *Metals Handbook*.

Mechanical Tests. Tensile or stress-rupture tests can be used to determine if the metal had mechanical properties suitable for the intended service. Hardness tests can be used to check whether heat treatment was correct or uniform or if hardening occurred from cold working, overheating, carburization, or phase changes in service. Impact tests indicate brittle tendencies.

Causes of Failures in Heat Exchangers

Failures in heat exchangers are commonly associated with methods of manufacturing pipe and tubing, handling methods during fabrication, testing methods in the shop and in the field, and the total environment to which the unit is exposed after fabrication—including conditions during shipment, storage, start-up, normal operation, and shutdown.

Fabricators and suppliers of pipe or tubing strive to minimize or eliminate latent surface or subsurface imperfections produced during manufacture. Nondestructive inspection, such as by eddy-current, ultrasonic, air-under-water, and hydrostatic tests, is used to detect these imperfections. Each test has its limitations. Hydrostatic tests and air tests are useful for detecting discontinuities extending completely through the tube wall, but they are of little use in detecting discontinuities that extend only partly through the wall. In these cases, ultrasonic and eddy-current methods are more useful. These test methods allow the sensitivity of the test instrument to be set at a level consistent with the type and size of imperfection being sought. Where an imperfection extends through the wall but is very small and tight, the tube may not leak in a hydrostatic test, but it may leak in subsequent service. Air tests under water can be used to reveal imperfections of this type.

Fortunately, relatively few imperfections produced during manufacture escape detection and lead to failures in service. The combination of eddy-current and hydrostatic testing, whether conducted by the fabricator or tube supplier, effectively minimizes service failures resulting from imperfections produced during manufacture.

Secondary manufacturing techniques result in more service failures than do surface and subsurface imperfections produced during tubing manufacture. The methods used to draw, clean, heat treat, and straighten the tubes may be quite significant to service performance, but are often overlooked in specifications and inspection procedures. Drawing, heat treating, and straightening operations determine the level of residual hoop and bending stresses induced in the tube. By varying the severity of these operations, it is possible to produce tubes with very low residual stresses or with very high residual stresses that are near the yield strength of the metal. Low residual stresses are particularly important when tube materials must have maximum resistance to SCC in service. The residual stresses on the inner and outer surfaces of the tube and through the tube wall must be low; residual stresses at the surfaces should be compressive.

Cleaning methods, control of heat-treating atmospheres, and pickling operations affect the condition of the oxides on the inner surface of the tube. The oxide should be very thin and flexible and should completely cover the inner surface of the tubing because fractures in the oxide layer create anodic and cathodic sites that can lead to pitting corrosion. If the surface oxide scale is to be removed, it should be completely removed in order not to establish anodic and cathodic sites, which are starting points for corrosion during service. The inner surfaces of some tubes are abrasive blasted to remove the oxides produced during heat treatment and thus eliminate anodic and cathodic sites. The strip stock from which welded tubing is made can be inspected immediately before forming and welding and can be rejected if it does not meet quality requirements.

Alignment and drilling of tube sheets and tube-support plates to reasonable tolerances will ease tube insertion and will minimize stresses induced during the fabrication of heat exchangers. Overrolling the tube past the tube sheet may introduce residual stresses. Tubes rarely fail within the tube sheet, but failure behind the tube sheet because of overrolling is relatively common. The tube-expander cage must be properly located and the rolling process must be controlled to minimize overrolling.

Effects of Inspection Procedures. All heat-exchanger tubes purchased to ASTM specifications are hydrostatically tested at the mill; the purchaser sometimes specifies that a nondestructive test also be used. Hydrostatic testing is also performed after the heat exchanger has been fabricated. One advantage of not hydrostatic testing at the tube mill is that tubes can be kept dry during shipment and storage. The presence of corrosives in the water used in hydrostatic testing, inadequate drying after hydrostatic testing, and improper boxing of the tubes at the tube mill can lead to corrosion. Most tube mills are aware of the problems associated with hydrostatic testing and drying and take steps to ensure that the test water is free of corrosives and that tubes are dry before boxing.

Some copper alloys are very sensitive to stagnant-water conditions, especially if the water contains biological material or chemicals that can generate or decompose into ammonia. Hydrostatic-testing procedures that permit wa-

Fig. 1 Pitted inside-diameter surface of AISI type 410 stainless steel tube
(a) Typical example of pitting. Approximately 2½×. (b) Enlargement of pit shown in (a). Approximately 50×

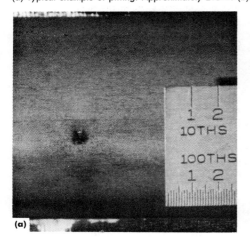

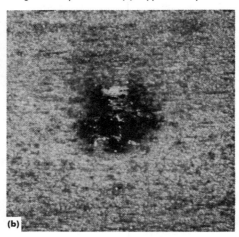

ter to become stagnant in the tubes can result in serious pitting and SCC.

Failure to remove shop-hydrostatic-test water that contained contaminants has caused tubing failures to occur before start-up. The water used for hydrostatic testing of units in the field is also important, as is the method used to drain and dry the units after field hydrostatic testing. The following example describes pitting in ferritic stainless steel heat-exchanger tubes caused by the presence of chlorides in the hydrostatic test water.

Example 1: Pitting of Stainless Steel Heat-Exchanger Tubes Due to Chloride Ions in Flush Water. Many tube bundles in a heat exchanger fabricated from AISI type 410 stainless steel experienced leakage during hydrostatic testing. The tubes had not yet been in service, and no metallurgical deficiencies that could have caused the failure were found.

Investigation. Figure 1 shows a typical example of the pitting observed on the inside surfaces of the tubes. A cross section of a pit (Fig. 2) revealed undercutting characteristic of chloride-ion pitting. X-ray spectrometry of two of the pits revealed evidence of chlorides.

Chemical analysis revealed that the tube material was within specifications for type 410 stainless steel. The microstructure consisted of annealed ferrite free from any inclusions or carbides. Pitting was random and could not be correlated with any sensitization or metallurgical deficiencies in the tubes.

Conclusions. Pitting failures of both 300- and 400-series stainless steels can occur after brief exposures to chloride environments. In this case, failure was caused by chlorides in the water used to flush the tubes before service.

Recommendations. Pitting could be reduced or eliminated by avoiding the use of brackish water to flush or test stainless steel equipment.

Boxing of tubing for shipping should be done with care. All tubes should be wrapped in waterproof paper or plastic for resistance to corrodents that may be present in the air during

Fig. 2 Cross section of a pit as shown in Fig. 1
6×

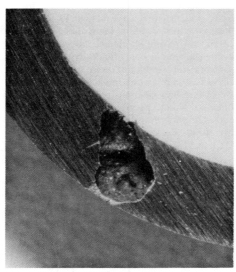

shipping or that may be present in outdoor storage and may be washed through the tubes by rainwater. It is also possible for preservatives in wooden boxes to leak out and wash over the tubes.

A problem with an airborne corrodent occurred in one case in which packing boxes became contaminated with calcium chloride from road-deicing and dust-control operations. The corrodent washed over stainless steel condenser tubes, causing ferric chloride pitting, which perforated the tubes before they were removed from the boxes. Wrapping with waterproof paper probably would have prevented corrosion of these tubes. In another case, a zinc chloride wood preservative leached out of wooden boxes containing tubes and caused pitting of the tubes before their installation in a heat exchanger. Exposure to ammonia or other

atmospheric contaminants in industrial areas could result in damage to stored components.

Corrosion

General corrosion resulting in uniform wall thinning causes considerable damage to heat-exchanger tubes. However, the principal corrosion damage in heat exchangers is usually from localized attack. Stray electrical currents may also contribute to significant corrosion of heat exchangers. Localized corrosive attack can occur as pitting, impingement, thinning, or selective leaching. More information on corrosion is provided in the article "Corrosion Failures" in this Volume.

General corrosion can be the result of deliberately designing a heat exchanger with a limited life. Otherwise, it can be the result of a mistaken choice of materials or of miscalculation of the corrosive effect of the medium circulating through the heat exchanger. For example, an acid-containing vapor stream may not be corrosive above its dew point, but if the stream is cooled below its dew point, severe general attack can result from acid condensation on internal surfaces of tube walls. On the outside surfaces of tubes, corrosion may be concentrated in the bottom row of tubes or in other areas where condensates can accumulate. Water vapors containing acids, hydrogen sulfide, carbon dioxide, and ammonia can be very corrosive environments.

Another cause of inadvertent general corrosion is improper chemical cleaning using uninhibited acids, excessive temperatures, or prolonged contact time. Water used to wash fire-side deposits where high-sulfur fuel oil or gas was fired will also create very acidic conditions and lead to general corrosion in areas that are not easily drained or flushed, such as under refractory linings.

Example 2: Corrosion Failure of a Tee Fitting. Wet natural gas was dried by being passed through a carbon steel vessel that contained a molecular-sieve drying agent. The gas, which contained 15% H_2S, was passed through the vessel at a pressure of 4.5 MPa (650 psi) and at a temperature of 40 to 45 °C (100 to 110 °F). After several hours, the drying agent became saturated and was taken off the line and regenerated by gas that was heated to 290 to 345 °C (550 to 650 °F) in a salt-bath heat exchanger. Figure 3(a) shows the arrangement of the 75-mm (3-in.) diam schedule 40 piping for the dehydrator system. After 12 months of service, the tee joint failed and a fire resulted.

Investigation. Inspection of the piping between the heat exchanger in the salt bath and the molecular-sieve bed revealed a hole in the tee fitting (Fig 3a and b) and a corrosion product (scale) on the inner surface of the pitting. This scale occurred in four layers (Fig. 3c).

Chemical analysis of the scale revealed it to be iron sulfide of various compositions in four distinct layers. Table 2 lists the probable composition and thickness of each layer. The

Table 2 Compositions and thicknesses of layers of corrosion product found on inner surface of low-carbon steel gas-dryer piping

Layer (see Fig. 4c)	Carbon, %(a)	Sulfur, %(b)	Stoichiometric ratio	Probable composition	Approximate thickness mm	in.
1	10	36.9	FeS$_{1.08}$	FeS$_{1.1}$	0.5	0.02
1 and 2 (composite)	...	41.7	FeS$_{1.15}$	FeS$_{1.1}$ + FeS$_{1.2}$	...	...
2	...	...	...	FeS$_{1.2}$(c)	2.0	0.08
2 and 3 (composite)	6	45.0	FeS$_{1.46}$	FeS$_{1.2}$ + FeS$_2$	...	...
3	...	...	...	FeS$_{1.2}$ + FeS$_2$(c)	2.0	0.08
4	...	...	...	FeS$_2$(c)	(d)	(d)

(a) Determined by combustion methods. (b) Determined by gravimetric methods. (c) Determined by x-ray diffraction. (d) Layer too thin and fragile to measure its thickness

layered structure of the scale indicated hydrogen sulfide attack on the carbon steel. The sulfur content was lowest in the layer in contact with the inner surface of the carbon steel pipe.

For several months before the failure, the temperature of the gas heated by the salt-bath heat exchanger was below 290 °C (550 °F) and was about 230 °C (450 °F) at the molecular sieves.

Corrosion of carbon steel by hydrogen sulfide occurs only at temperatures above 260 °C (500 °F). At 315 °C (600 °F), 15% H$_2$S would cause rapid attack on carbon steel. However, the carbon and sulfur found in the scale on the piping and in the molecular sieves indicated that oxygen was present in the system. At pressures of 2070 kPa (300 psi) or more, the presence of small amounts of oxygen promotes hydrogen sulfide corrosion of carbon steel, even at room temperature. Thus, oxygen combined with moisture produces conditions for attack of hydrogen sulfide on carbon steel at temperatures below 260 °C (500 °F).

Failure in the wall of the tee fitting was probably the result of turbulence, with some effect from the coarse grain size usually found in pipe fittings. Grain size usually affects corrosion rate.

Conclusions. The piping failed by corrosion in the tee fitting because of the presence of hydrogen sulfide, moisture, and oxygen in the natural gas that was dried in the system. Turbulence in the tee fitting and coarse grain size both contributed to the corrosion.

Corrective Measures. The piping material was changed from carbon steel to AISI type 316 stainless steel, which is readily weldable and resistant to corrosion by hydrogen sulfide at temperatures to 400 °C (750 °F). Chloride concentration in the plant was very low; however, postweld stress relief was used to minimize residual stresses and to avoid the possibility of SCC.

A less expensive alternative would have been to use a 5% Cr steel or a 9% Cr steel, neither of which is susceptible to SCC in the presence of chlorides. These steels are susceptible to corrosion by hydrogen sulfide, although much less so than carbon steel.

Example 3: Pitting of a Condenser Tube in a Saltwater Heat Exchanger Due to Hydrogen Sulfide Contamination. A tube sample from an aluminum brass seawater surface condenser was received for analysis. This condenser had failed due to pitting after less than 1 year of service.

Investigation. Metallographic analysis, energy-dispersive x-ray spectrometry, and x-ray diffraction were used to analyze the tube and deposit samples. A thick, nonuniform black scale was present over the entire inside surface of the tube. When this scale was removed, large pits filled with a green deposit were evident. Figure 4 shows the pitting, the thickness of the scale, and the intergranular nature of the attack. Energy-dispersive x-ray analysis of the base metal indicated that the material was aluminum brass, which is commonly used for seawater condensers. X-ray analysis of the scale, which constituted the bulk of the corrosion deposit, yielded the following results:

Element	Composition, %
Copper	69
Zinc	17
Sulfur	9
Aluminum	2.5
Chlorine	1.5
Calcium	1
Iron, silicon, magnesium	trace

The black deposit was identified as primarily copper sulfide, with zinc and aluminum sulfides also present. The green deposit was also analyzed and identified as copper chloride. Copper, chlorine, zinc, and aluminum were the major constituents.

Discussion. Most copper alloys depend on the formation of a protective oxide scale for corrosion protection. When hydrogen sulfide is present, it interferes with the formation of such a layer, forming a sulfide layer instead. Unlike the oxide layer, the sulfide scale is not protective, and it allows corrosion, as well as diffusion and concentration of such corrosive species as chlorides, to continue beneath the deposit.

Conclusions. The combination of sulfide and chloride attack on the tubes resulted in accelerated failure. The major cause of the shortened tube life was hydrogen sulfide, which was present in seawater at this particular site at about 5 to 10 ppm.

Recommendations. Injection of ferrous sulfate just upstream of the condenser can aid the formation of protective oxide films. From the

Fig. 3 Low-carbon steel tee fitting in a line leading to a natural-gas dryer that failed from hydrogen sulfide corrosion

(a) Arrangement of piping showing point of leakage in the tee fitting. (b) Inner surface of the tee fitting showing corrosion deposit and area of complete penetration through the tube wall. (c) Positions of layers of corrosion product on inner surface of piping (see Table 2 for compositions of layers)

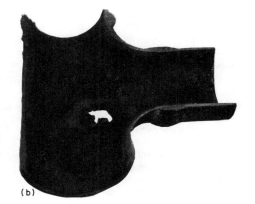

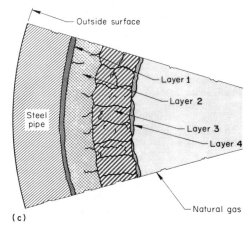

preceding discussion, the most obvious corrective measure would be to eliminate sulfide contamination. The prime source of such contamination appears to be the accumulation of sulfide-containing sludge in coral formations near the condenser inlets. Consideration might also be given to relocation of the pump suction to an area where there is continuous entry of fresh seawater, which would reduce sulfide concentrations.

Crevice Corrosion in Water. A common cause of deterioration in tubing containing water is corrosion pitting caused by differential

Fig. 4 Failed aluminum brass condenser tube from a saltwater heat exchanger

The tube failed from pitting caused by hydrogen sulfide and chlorides in the feedwater. (a) Cross section of tube showing deep pits and excessive metal wastage. 2¾×. (b) Higher magnification view of a pit showing the depth of attack and thickness of the corrosion deposit. 28×. (c) Micrograph showing the intergranular nature of the attack and undercutting of metal particles. Etched with ammonium persulfate. 110×

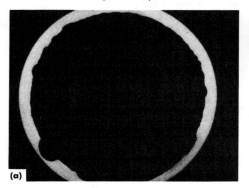

aeration or concentration-cell corrosion (crevice corrosion). This type of corrosion occurs frequently, but is often not identified as crevice corrosion.

Corrosion by differential aeration arises from regional differences in the concentration of dissolved gases (principally oxygen) or of soluble contaminants, such as chlorine, fluorine, or sulfur, in water that contacts a part or component. The result is the establishment of an electrochemical cell, with the surface in the region of lowest concentration becoming anodic relative to the surrounding surface. This situation can develop under conditions of stagnant flow between two overlapping surfaces or beneath corrosion products or sludge on the surface. Anything that restricts free access of oxygen is likely to cause crevice corrosion.

Other factors, such as gas bubbles on the surfaces, inhomogeneity of the material, cracks in mill scale, and local breakdown of protective films, may also initiate crevice-corrosion attack. Corrosion of this type is often associated with water that contains dissolved solids in comparatively small concentrations, the amounts present being insufficient or unsuitable to cause formation of protective films or deposits but sufficient to raise the electrical conductivity of the water so that it becomes an electrolyte. Invariably, the cathodic zone is larger than the anodic zone; consequently, the rate of attack at the small anodic zone is high. Corrosion extends preferentially in depth, forming pits and cavities. The affected part sometimes becomes completely penetrated even though the total area affected by corrosion is relatively small. The corrosion products commonly form a nodule over the cavity, increasing the crevice effect. When the corrosion is associated with elevated temperature—for example, at the hottest parts of heat-exchanger surfaces—the nodule that is formed is hard and is difficult to remove until it has attained considerable size. The temperature may be sufficient to evaporate some of the liquid within the nodule; the

corrosive action then becomes intermittent, and partial drying of the corrosion products occurs.

In the corrosion of iron-base alloys, dissolved oxygen in the water diffuses into the precipitated ferrous hydroxide and, in combination with the ferrous hydroxide and water, forms hydrated ferric hydroxide (rust). If the supply of oxygen is limited, the corrosion product may be green hydrated magnetite or black anhydrous magnetite. The nodule that forms from the corrosion product is porous; therefore, electrochemical attack proceeds as iron enters into solution in the aqueous medium beneath the nodule. The oxygen dissolved in the water is mainly consumed over the large cathodic surface in oxidizing the hydrogen evolved there; that is, it acts as a depolarizer, with the rate of oxygen consumption determining the rate at which ions leave the metal beneath the nodule.

Because of the intense corrosive attack, the corrosion nodule frequently has a stratified structure, with the composition changing from rust-colored ferric hydroxide on the outside surface of the nodule to layers of ferrous hydroxide, ferric oxide, and magnetite on the inside. The ferric oxides permit the passage of hydroxide ions and the diffusion of oxygen, with the result that electrochemical action is able to proceed, and the rate of attack is increased with an increase in temperature because of the resultant alteration in the nature of the corrosion products, the improved electrical conductivity of the water, and the enhanced rate of chemical reaction that follows an increase in temperature.

The following features are characteristic of crevice-corrosion attack:

- Nodules will generally form on the corroded surfaces. The nodules will be hard if formed on a hot surface, but may be softer if formed on a cold surface
- The composition of the nodules varies from the outside surface to the inside. Externally, they often present a rusty appearance. Inter-

nally, they are mostly black. In some cases, the appearance may be modified by the presence of salts deposited from the water, giving various colors
- The nodules are generally isolated from one another, although they may sometimes be found in confluent groups
- When a nodule is removed, the cavity in the plate beneath it is often sharply defined, and the walls may be almost perpendicular to the metal surface
- The metal surface between the cavities generally shows little or no corrosion, and in some cases the original mill scale may be present

New installations are particularly vulnerable to crevice corrosion because of insufficient time for a protective film or scale to form on the surface before the onset of corrosion.

Copper alloys are also susceptible to crevice corrosion, forming similar corrosion products and taking part in similar chemical reactions. For example, in the corrosion of a copper alloy in an aerated-water medium, copper carbonate hydroxide $[CuCO_3 \cdot Cu(OH)_2]$ would be the expected corrosion product. Bicarbonate (HCO_3) and carbonate (CO_3) radicals, which are needed for the formation of copper carbonate hydroxide, are present as contaminants in most naturally occurring waters.

Example 4: Crevice Corrosion of Tubing in a Hydraulic-Oil Cooler. Leakage from the horizontal heat-exchanger tubes in one of two hydraulic-oil coolers in an electric-power plant occurred after 18 months of service. Under normal operating conditions, service life in similar units is generally about 10 years. In this plant, river water was used as the coolant in the heat-exchanger tubes. The tubes were 9.5 mm (⅜ in.) in diameter with a 0.6-mm (0.025-in.) wall thickness and were made of copper alloy C70600 (copper nickel, 10%). Five of the tubes that were leaking were removed and sectioned lengthwise. One section of each tube was sent to the manufacturer of the cooler, and the other was sent to the power company laboratory for determination of the cause of failure.

Investigation. Visual examination of the sectioned tubes revealed several nodules on the inner surface and holes through the tube wall, which appeared to have formed by pitting under the nodules. In Fig. 5(a), arrow A shows a pit that penetrated the wall of the tube, and arrow B shows a typical nodule.

Metallographic examination was performed on three distinct areas of the tube: areas that exhibited no effects of corrosion, areas that were pitted, and areas containing nodules. The tubing contained several nodules and pits, but all were isolated from each other. A cross section through an unaffected area of the tube showed no corrosion on the inner surface and an undamaged microstructure. Most of the inner surface of the tube was in good condition.

Figure 5(b) shows a micrograph of a section through a pit that had penetrated the tube wall. The pit had steep sidewalls, which indicated a high rate of attack. A micrograph of a section through the tube beneath one of the nodules is shown in Fig. 5(c). Corrosion beneath the nodule had penetrated approximately 65% of the tube wall. The pit had a low slope angle, which indicated a low or intermittent rate of corrosion.

Spectrographic analysis of the tube metal indicated that it was copper alloy C70600. A reddish deposit was removed from the inner surfaces of the tubes and analyzed. The major constituent was iron oxide, and less than 1% was manganese dioxide. The presence of these and other constituents indicated that effluent from steel mills upstream was the source of most of the solids found in the tubes.

A greenish nodule was analyzed by x-ray diffraction. The major compound was copper carbonate hydroxide [$CuCO_3 \cdot Cu(OH)_2$]. This compound would be expected in corrosion of a copper alloy exposed to aerated natural water.

Flow of cooling water from the river through the tubes was not uniform, but ebbed and flowed in accordance with cooling demand. Under these circumstances, deposits would form on, or particles would drop on, the inner surfaces of the horizontal tubes.

Conclusions. The tubing failed by crevice corrosion. Particles of dirt in the river water were deposited inside the tubes during periods of low flow. The difference in oxygen content under the deposit and on the exposed surface of the tube established an electrolytic cell under the deposit. The end results were nodules of a corrosion product and pits that in some cases eventually became holes in the tube wall.

Corrective Measures. The tubing in the cooler was replaced, and the cooling-water supply was changed from river to city water, which contained no dirt to deposit on the tube surfaces. After the first cooler was back in service, the second cooler was shut down, and the tubes removed for investigation. Nodules, pits, and deposits similar to those found on tubes from the first cooler were found on the inner surfaces of the tubes. Although the tubes were not leaking, they would have been within a short time. For this cooler also, the tubes were replaced, and city water was substituted for river water as the coolant. At the time of this report, both coolers had been in service for about 3 years with no evidence of leakage.

An alternate solution to the problem would have been to install replacement tubes in the vertical position and to continue to use river water. With the tubes in the vertical position, solids would be less likely to be deposited on the tube walls during low flow of the river water.

Selective leaching, or dealloying, is localized corrosion of copper alloys that leaves a spongy, structurally weak mass of the more noble alloying element in place at the site of corrosion attack. Selective leaching can occur in brasses (dezincification) and in copper-nickel alloys (denickelification). The net result of selective leaching is that sound metal is gradually changed to a brittle, porous mass of copper that has little mechanical strength. Selective leaching can sometimes be detected visually by the reddish appearance of the deposited copper; it is favored by waters having a high content of oxygen, carbon dioxide, or chloride and is accelerated by elevated temperatures and low water velocities. Excessive chlorination of cooling water may also lead to dezincification of brass.

Dezincification occurs in two forms, plug type and layer type. Plug-type dezincification is less spread out on the surface than layer type and can cause pitlike wall perforation more rapidly than layer type. An acidic environment favors layer-type attack. High-zinc brasses of the α variety, such as copper alloy C26000 (cartridge brass, 70%), are more susceptible to dezincification than copper alloys having a lower zinc content, for example, copper alloy C24000 (low brass, 80%). In copper alloys that show an α-plus-β structure, such as copper alloy C28000 (Muntz metal, 60%), the β constituent, which has the higher zinc content, may suffer dezincification.

Figure 6 shows a section through a copper alloy C26000 steam-turbine condenser tube.

Fig. 5 Copper alloy C70600 tube from a hydraulic-oil cooler

The cooler failed from crevice corrosion caused by dirt particles in river water that was used as a coolant. (a) Inner surface of hydraulic-oil cooler tube containing a hole (arrow A) and nodules (one of which is indicated by arrow B) formed from a corrosion product. (b) Etched section through the pit at arrow A, which penetrated the tube wall. 100×. (c) Etched section through a pit beneath the nodule at arrow B, with about 35% of the wall thickness remaining. 100×

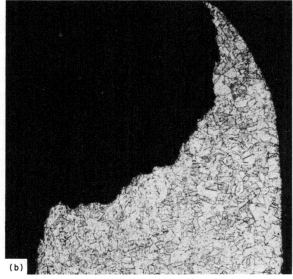
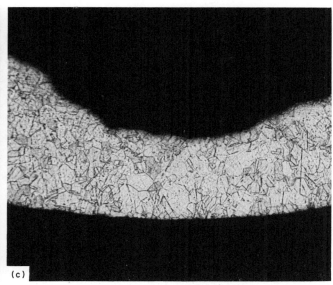

(a) (b) (c)

Fig. 6 Copper alloy C26000 steam-turbine condenser tube that failed by dezincification

(a) Section through condenser tube showing dezincification of inner surface. 3½×. (b) Etched specimen from the tube showing corroded porous region at the top and unaffected region below. 100×

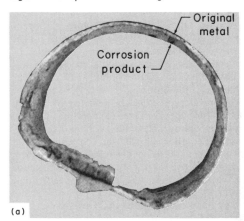

(a)

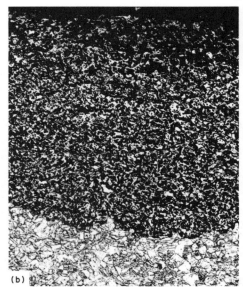

(b)

Uniform dezincification of the bore extended from one-half to two-thirds through the wall of the tube. The porous nature of the affected portion of the tube is shown in Fig. 6(b).

Figure 7(a) and (b) show a section through a heat-exchanger tube made of copper alloy C71000 (copper nickel, 20%). Water passing through the tube and steam in contact with the outer surface caused denickelification of the alloy at both inner and outer surfaces. The dealloying occurred along the grain boundaries, which have higher energy than the grains and are more susceptible to attack. Another form of denickelification is shown in Fig. 7(c) for a copper alloy C71500 (copper nickel, 30%) heat-exchanger tube.

Impingement attack, or erosion-corrosion, occurs where gases, vapors, or liquids

Fig. 7 Copper-nickel alloy heat-exchanger tubes that failed from denickelification due to attack by water and steam

(a) Etched section through a copper alloy C71000 tube showing dealloying (light areas) around the tube surfaces. Etched with NH₄OH plus H₂O. 3.7×. (b) Unetched section through the outer surface of the tube shown in (a); dark pattern shows dealloyed region. 85×. (c) Etched section through a copper alloy C71500 tube showing the surface that was denickelified (top) and unaffected base metal (bottom). Etched with equal parts nitric and acetic acid. 375×

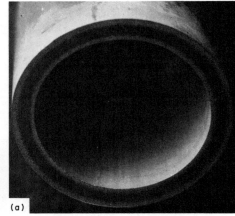

(a)

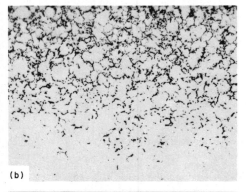

(b)

(c)

impinge on metal surfaces at high velocities. The erosive action removes protective films from localized areas at the metal surface, thereby contributing to the formation of

differential cells and localized pitting of anodic areas.

Impingement attack often produces a horseshoe-shaped pit. This form of attack frequently occurs in condenser systems handling seawater or brackish water that contains entrained air or solid particles and circulates through the system at relatively high velocities and with turbulent flow. The pits are elongated in the direction of flow and are undercut on the downstream side. When the condition becomes serious, a series of horseshoe-shaped grooves with the open ends on the downstream side may be formed. As the attack progresses, the pits may join one another, forming fairly large patches of undercut pits.

When impingement attack occurs in heat-exchanger or condenser tubing, it is usually confined to a short distance on the inlet end of the tube where the fluid flow is turbulent. Impingement attack can be controlled by use of corrosion-resistant alloys, such as copper alloy C71500 (copper nickel, 30%).

Impingement attack has been somewhat relieved by inserting relatively short tapered plastic or alloy sleeves inside the tube on the inlet end of the tube bundle. Other corrective measures include decreasing the velocity, streamlining the fluid flow, and decreasing the amount of entrained air or solid particles. These may be attained by a streamlined design of water boxes, injector nozzles, and piping. Abrupt angular changes in direction of fluid flow, low-pressure pockets, obstruction to smooth flow, and any other feature that can cause localized high velocities or turbulence of the circulating water should be minimized as much as possible. Sacrificial impingement baffles can minimize damage to the shell-side surfaces of tubes. More information on liquid impingement is provided in the article "Liquid-Erosion Failures" in this Volume.

Example 5: Failure of Copper Alloy 443 Heat-Exchanger Tubes. Tubes in heat exchangers used for cooling air failed in an increasing number after 5 to 6 years of service. Air passed over the shell-side surface of the tubes and was cooled by water flowing through the tubes. Although sanitary well water was generally used, treated recirculating water was also available and was sometimes used in the heat exchanger. Water vapor in the air was condensed on the tube surfaces during the cooling process. The pH of this condensate was approximately 4.5. Air flow over the tubes reversed direction every 585 mm (23 in.) as a result of baffling placed in the heat exchangers.

The tubes were 19 mm (¾ in.) in diameter with a wall thickness of 1.3 mm (0.050 in.) and were made of copper alloy C44300 (arsenical admiralty metal). Samples of the tubes were sent to the laboratory to determine the mechanism of failure.

Investigation. Visual examination of the damaged area revealed an uneven ridgelike thinning and perforation of the tube wall (Fig. 8a) on the leeward side of the tube. An enlarged

Fig. 8 Copper alloy C44300 heat-exchanger tube that failed by impingement corrosion from turbulent flow of air and condensate along the shell-side surface

(a) Shell-side surface of tube showing damaged area. (b) Damaged surface showing ridges in affected area. 4×. (c) Unetched section through damaged area showing undercut pitting typical of impingement attack. 100×

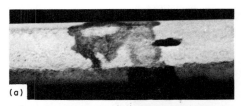

(a)

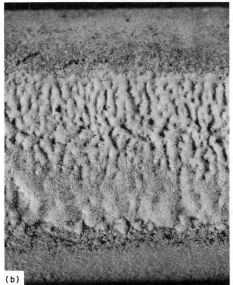

(b)

(c)

view of the ridges on the surface of the tube is shown in Fig. 8(b). Metallographic examination of a cross section of the failed area (Fig. 8c) disclosed undercut pits on the outer surface of the tube. Both the ridgelike appearance of the damaged area and the undercut pitting indicate of impingement attack. Spectrochemical analysis of the tube confirmed that the material was copper alloy C44300 (arsenical admiralty metal).

Conclusion. The tubes failed by impingement corrosion that resulted in perforation. A combination of air and water (the condensate) was rapidly agitated by changes in direction at the metal surface and accelerated the impingement corrosion. This action and the low pH (4.5) of the condensate resulted in removal of

the protective film that normally formed on the tube surface during initial corrosion and ultimately produced a hole in the tube.

Corrective Measures. The heat exchanger was retubed with tubes made of aluminum bronze, for example, copper alloy C61400. At the time of this report, these tubes had been in service for more than 12 years without failure.

Stress-Corrosion Cracking

Stress-corrosion cracking (SCC) occurs as a result of corrosion and stress at the tip of a growing crack. The stress level necessary for SCC is below that needed for fracture without corrosion. Additional information is provided in the article "Stress-Corrosion Cracking" in this Volume.

Evidence of corrosion, although not always easy to find, should be present on the surface of an SCC fracture up to the start of final rupture. Generally, one of several specific corrosive substances must be present for SCC to occur in a given alloy. For example, ammonia compounds have been identified as agents that produce SCC in copper-zinc alloys.

The arsenic, antimony, or phosphorus additions for inhibiting dezincification do not prevent SCC of inhibited admiralty metals (copper alloys C44300, C44400, and C44500). Cracking can be transgranular, intergranular, or mixed on the same metal part, and it may be accompanied by deposition of copper within the cracks. Arsenic does not prevent SCC in copper alloy C68700 (arsenical aluminum brass) subjected to ammonia- and copper-dosed seawater, although arsenic can have beneficial effects at stresses below the elastic limit. Low-temperature SCC has been found in brasses. The copper nickels (copper alloys C70100 to C72900), despite their proven susceptibility to ammonia, are much superior in service with respect to SCC in condenser-tube service than brasses are.

Carbon steels can experience SCC in several environments (see the article "Stress-Corrosion Cracking" in this Volume). In heat exchangers, these environments may be present at such low concentrations that they appear to be harmless. However, concentration of cracking species by evaporation can lead to equipment failure. For example, sodium hydroxide may be present in boiler water in part per million concentrations. Waste-heat boilers can allow concentration of hydroxide ions by steam blanketing, by incomplete water wetting of a vertical boiler, or by flange leaks that allow water to become steam. Failure usually occurs in welds or in highly stressed areas. Carbon steel heat-exchanger equipment can also fail by SCC in wet hydrogen sulfide environments if the hardness of the component exceeds approximately 22 HRC. Flange bolts or studs that are improperly heat treated for service in such an environment are particularly susceptible. Additional information is available in the articles "Hydrogen-Damage

Failures" and "Failures in Sour Gas Environments" in this Volume.

The austenitic stainless steels are subject to SCC that is induced by acidic chlorides and fluorides or by polythionic ($H_2S_xO_6$; $x = 2$ to 6) acids if the microstructure is sensitized. In hot systems in which there is a possibility of chloride contamination, austenitic stainless steels should not be considered. If these steels must be used, they should be in the annealed condition or should be clad over a base metal that resists cracking. In very corrosive environments, higher-nickel austenitic alloys may be required to resist both general attack and SCC. In less corrosive environments, ferritic or duplex stainless steels can be considered.

Example 6: SCC of a Stainless Steel Integral-Finned Tube. A tubular heat exchanger in a refinery reformer unit was found to be leaking after 1 month of service. The exchanger contained 167 type 304 stainless steel U-bent integral-finned tubes, 19 mm (¾ in.) in outside diameter.

Investigation. Examination revealed cracks in the tube wall about 75 mm (3 in.) from the tube sheet and at the first full-depth fin on the tube (Fig. 9a). Hardness of the tube was 30 HRC at the inside surface and up to 40 HRC at the base of the fin midway between the roots. These hardness values indicated that the fins were cold formed and not subsequently annealed. Cold working without subsequent annealing made the tubes susceptible to SCC because of a high residual-stress level.

Metallographic examination of a section through the tube wall established that fracture was predominantly by transgranular branched cracking (Fig. 9b) and had originated from the inside surface. Transgranular branched cracking of austenitic stainless steels is frequently caused by chlorides in the presence of high residual stresses.

The presence of chlorides is not unusual in a reformer unit. Although steam and steam condensate injected into reformer systems are normally free of chlorides, there can be accidental carry-over of dissolved solids with the steam under periods of high demand.

Conclusion. The tubes failed in SCC caused by chlorides in the presence of high residual stresses. The fins were cold formed on the tubes and not subsequently annealed.

Corrective Measure. The finned tubes were ordered in the annealed condition. This minimized the possibility of residual stresses in the tubes.

Improper Materials Specification and Design. Heat exchangers sometimes fail by SCC because of improper specification of materials and design. In one case, a heat exchanger was part of a separator, which was designed as a single combination vessel for construction economy. The exchanger was of a vertical-tube design and was located directly beneath the separator. The separator had a 6.4-mm (¼-in.) thick AISI type 316 stainless steel shell that extended below the partition

Fig. 9 Type 304 stainless steel integral-finned tube that cracked from chlorides and high residual stresses
(a) Section of integral-finned tube showing major crack (circumferential crack between fins). Dimension given in inches. (b) Branched transgranular cracking propagating from major crack. 45×

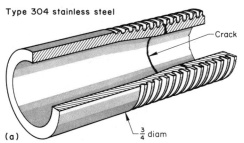

Type 304 stainless steel

(a) ── Crack ── ¾ diam

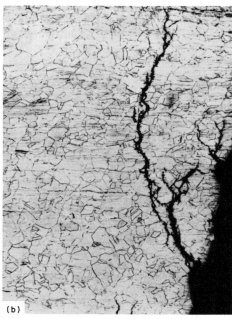

(b)

Fig. 9 Type 304 stainless steel integral-finned tube that cracked from chlorides and high residual stresses
(a) Section of integral-finned tube showing major crack (circumferential crack between fins). Dimension given in inches. (b) Branched transgranular cracking propagating from major crack. 45×

head of the vessel, forming the shell of the heat exchanger. The designer had not provided for complete filling of the water space between the upper tube sheet and the partition head. As a result, the type 316 stainless steel shell of the exchanger was constantly wetted with water in the vapor space below the partition head. Heat from the separator caused continuous flashing of water into steam inside the exchanger, building salt deposits on the hot surfaces of the shell. The deposits contained chlorides from the city water used in the exchanger. Acting in combination with the normal fabricating stresses inherent in the vessel, chloride salts caused the type 316 stainless steel shell at the vapor space to fail because of the presence of numerous small stress-corrosion cracks. Some of the cracks penetrated completely through the wall of the shell.

This problem could have been avoided by providing for complete filling of the water

space above the tube sheet or by providing sufficient distance between the separator and exchanger so that heat from the separator would not cause steam flashing inside the exchanger. Finally, it is not considered good practice to use austenitic stainless steels in hot-water service under conditions of alternate wetting and drying; the drying favors the accumulation of corrosive salt deposits.

As a corrective measure, carbon steel doubler plates were welded over the cracked region of the exchanger shell. The exchanger was eventually replaced with a vessel designed to eliminate the vapor space between the tube sheet and the partition head.

Corrosion Fatigue

Corrosion fatigue is the combined action of an aggressive environment and a cyclic stress, which results in premature failure of metals by cracking. Neither cyclic stress in air nor environmental attack applied separately produces the same damaging results as combined action. Thus, a precorroded specimen does not necessarily show appreciable reduction in fatigue life, nor does prefatiguing in air increase the corrosion rate of metals.

In general, corrosion fatigue should not be confused with stress corrosion. All metals that are susceptible to corrosion are susceptible to corrosion fatigue in any corrosive environment, whereas SCC occurs only in alloys subjected to very specific environments and normally occurs under conditions of static tensile stress rather than dynamic stress. However, the application of cyclic stresses under SCC conditions may also influence corrosion-fatigue behavior.

In general, more materials are susceptible to corrosion fatigue than to stress corrosion. Stress-corrosion cracking mainly affects the brasses and the austenitic stainless steels, but corrosion fatigue can occur in most metallic materials used in the chemical industry.

Two general sources of stress and environment are necessary for corrosion fatigue: (1) where the cyclic stress and corrosive environment result from the same sequence of events and (2) where the cyclic stress and corrosive environment arise independently. The first source is more common in high-temperature applications and relies on changes in temperature that produce thermal stresses and corrodent concentration. In the second source, cyclic stresses are usually mechanical in origin, such as rotating bending, cyclic pressurization, and start-stop centrifugal loading. Because corrosion fatigue is by definition restricted to the simultaneous action of a cyclically varying stress and a corrosion process, the damage mechanisms not covered by this term are purely mechanical stress, followed by a corrosive action in the unstressed state, and corrosion in the nonloaded state, followed by purely mechanical cyclic stressing without interaction of corrosion.

Fracture Appearance. For carbon and low-alloy steels, the corrosion-fatigue fracture surface is typically jagged, whereas the fracture surface of rust- and acid-resisting steels—for example, austenitic stainless steels—is usually flat. The number of secondary cracks influences whether the fracture surface is flat or jagged; with a great number of cracks, the fracture is jagged. The jagged surface results when several cracks penetrate, very close to one another, into the interior of the material. Because the cracks are not in the same plane, the fracture jumps from one crack plane to another, causing the jagged appearance. When the distance between each crack is large, which is usually the case with a small number of cracks, the fracture cannot jump from one plane to another. Consequently, the crack that penetrates to the greatest depth into the material continues to grow until the component fails. The flat form of fracture is, therefore, determined by the number of cracks per unit length and is not a specific property of austenitic steels.

On the other hand, the form of a fracture in an austenitic corrosion-resistant steel will be similar to that of a carbon steel if the rate of corrosion is increased. Tests carried out with an austenitic stainless steel that was exposed to cyclic loading and a spray of saturated sulfurous acid showed severe corrosion conditions, producing numerous surface cracks and a jagged fracture surface.

The number of cracks, which considerably influences the form of fracture, depends not only on the rate of corrosion but also on the state of the surface from which the crack begins to propagate. On rough surfaces, many cracks are formed, resulting in a jagged fracture surface. With alloys that are not corrosion resistant, the jagged form of fracture becomes more characteristic when the mechanical stress decreases. With a small stress, the life of a part is long; consequently, there is enough time for the corrosion to roughen the surface. Thus, the conditions for the formation of many cracks and a jagged fracture are created.

Thermal Stresses. Thermally generated cyclic stresses are as important in corrosion fatigue as mechanical stresses. Transient thermal conditions can arise when a fluid is brought into contact with a surface having a markedly different temperature or, more commonly, due to imposed system operations. Such thermal stresses result from constrained thermal expansion or contraction.

Mechanical Stresses. Mechanical cyclic stresses arise from rotating machinery and cyclic pressurization. Vessels subject to cyclic pressurization are usually penetrated by pipes welded to the vessel shell. In addition to being regions of stress concentration, these welded connections can act as sources of residual stresses, and any lack of penetration or fusion at the weld may be a site for environmentally assisted cracking. Mechanical cyclic stresses superimposed on any residual stresses and any additional thermal stresses resulting from operational transients are similarly additive. Other significant sources of cyclic stresses are fans or pumps, which generate high-frequency vibrations in coolant circuits; turbulent fluid flow through or over tubes, such as steam flow over condenser tubes; and such upsets as water hammer in steam-generating equipment (see the article "Failures of Boilers and Related Equipment" in this Volume).

Example 7: Corrosion-Fatigue Failure of U-Bend Heat-Exchanger Tubes. Two admiralty brass heat-exchanger tubes from a cooler in a refinery catalytic reforming unit were sent to the laboratory for analysis. The tubes had been in service approximately 10 years and had cracked circumferentially in the area of U-bends in the tubes.

Investigation. Both tubes showed cracks extending circumferentially about 180° on the tension (outer) side of the U-bends. Tube 1 exhibited a relatively smooth, uniform black deposit on its surfaces; tube 2 had an additional buildup of corrosion deposits on both the inside- and outside-diameter surfaces. Inspection data indicated that the samples received were typical of many cracked U-bends through the tube bundle.

Metallography of the fracture area of tube 1 showed blunt transgranular cracking with minimal branching propagating from the inside surface of the tube (Fig. 10). In addition to the large crack, several smaller cracks were found near the fracture. The appearance of these cracks was typical of a corrosion-fatigue mechanism.

Energy-dispersive x-ray analysis of the corrosion deposits from tube 2 indicated the presence of copper, zinc, iron, as well as small amounts of chloride, sulfur, silicon, tin, and manganese. The dark brown and black deposits covering the surfaces of the tubes were identified as copper (II) oxide. Also present were reddish deposits identified as copper (I) oxide, as well as iron oxides. The green color of some deposits indicated that the chlorine and sulfur present were most likely in the form of copper chloride and copper sulfate.

Hardness measurements on the tube specimens ranged from 80 HRF near the crack to 99 HRF on the tension side of the tube. Typical hardness for annealed copper alloy tubing is 75 HRF. This indicated that the tubes may not have been annealed after the U-bends were formed. The appearance of cracks on the tension side of the U-bends showed that residual stresses played a role in crack formation.

Conclusions. The tubes failed by corrosion fatigue initiated by pitting at the inside-diameter surface. The presence of chlorides and sulfates in the cooling water contributed to corrosion by the formation of basic copper chlorides and sulfates. The corrosion pits acted as stress raisers that, in combination with residual stresses from tube bending and mild cyclic stresses in service, promoted crack growth.

Recommendations. Admiralty brass was recommended for continued use in the tubes due to its generally good corrosion resistance in the catalytic reforming environment. The tubes should be annealed after bending to reduce residual stresses from the bending operation to an acceptable level.

Effects of Design

The shape of the components in heat exchangers is important in producing regions of stress concentrations. The ASME Boiler and Pressure Vessel Codes are generally used for the design of pressure vessels. These codes contain relatively simple formulas that can be used for design. If the formulas are properly used, the vessel will withstand the operating pressures and temperatures anticipated. If used improperly or if used for the wrong application, these formulas can result in designs that can cause failures.

One member of the brine heater described in the following example had a high discontinuity stress that was not taken into account in the design. This discontinuity stress exceeded the tensile strength of the carbon steel shell material. Although initial cracking probably occurred during hydrostatic testing, the vessel operated at such low pressure that it was in service for about 2 years, during which time a crack propagated nearly 2.1 m (7 ft) before failure. Investigation revealed that a single welded joint, although properly made, provided a notch that acted as a site of crack initiation.

Example 8: Fracture of a Brine-Heater Shell at Welds. The brine-heater shell (Fig. 11) in a seawater-conversion plant failed by bursting along a welded joint connecting the hot well to the heater shell. Failure occurred approximately 2 years after the plant had been brought on stream.

Specifications required the heater shell to be made of ASTM A285, grade C, steel, which has a tensile strength of 379 to 448 MPa (55 to 65 ksi), a yield strength of 207 MPa (30 ksi), and elongation in 50 mm (2 in.) of 27% minimum. The tube side of the heater was designed for a pressure of 1205 kPa (175 psi) and a temperature of 150 °C (300 °F); the shell side, 1035 kPa (150 psi) and 185 °C (365 °F). Both sides were hydrostatically tested at 150% of design pressure. The heater tubes were made of copper alloy C70600 (copper nickel, 10%) per ASTM B 466.

On-site and laboratory investigations were conducted to determine the cause of failure and to develop corrective measures to minimize future failures.

Service History. The steam supplied to the plant was normally controlled at 517-kPa (75-psi) line pressure. For the first 18 months of service, the brine heater operated at a maximum temperature of 120 °C (250 °F) and a steam temperature of 130 °C (265 °F) with a pressure of 170 kPa (25 psi). One exception was during the scaling of the brine-heater tubes

Fig. 10 Failed admiralty brass heat-exchanger tubes from a refinery reformer unit

The tubes failed by corrosion fatigue. (a) Circumferential cracks on the tension (outer) surface of the U-bends. Approximately 1¼ ×. (b) Blunt transgranular cracking from the water side of tube 1. 40 ×

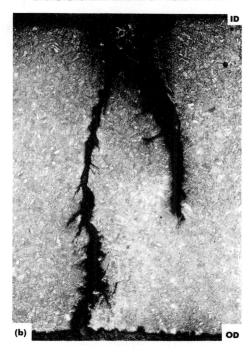

about 1 month after start-up, when the brine temperature exceeded 145 °C (290 °F) and the steam pressure was greater than 386 kPa (56 psi).

During the 6 months before failure, the maximum brine temperature was 140 °C (280 °F) and the steam pressure was 331 MPa (48 psi). About 5 months before failure, a second scaling of the brine-heater tubes occurred, and the steam pressure approached that of the steam line (517 kPa, or 75 psi).

Fig. 11 Failed brine-heater shell of ASTM A285, grade C, carbon steel
The shell fractured at welded joints because of overstress during normal operation. Dimensions given in inches

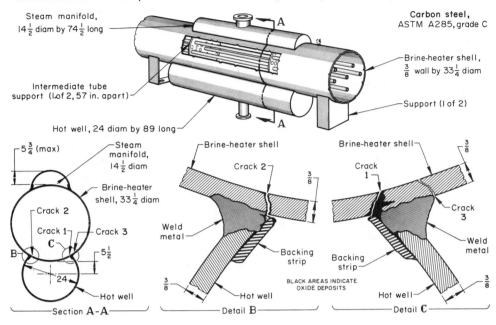

There were 44 recorded pressure surges in the steam line. Most of these were minor (plus 69 kPa, or 10 psi, maximum). However, pressures of 621 kPa (90 psi) were recorded three times; more than 690 kPa (100 psi), four times. The last 690-kPa (100-psi) pressure surge occurred about 6 weeks before the failure.

When the brine heater failed, the operators immediately went to the control room and found that the steam-temperature valve that admits steam to the brine heater was secured. However, the chart indicated that the brine temperature was 60 °C (140 °F) and that the pressure in the steam supply had increased. The steam valve was immediately closed, and within 3 min, the valve in the steam line at the power plant was closed.

The relief valve on the steam-supply line at the power plant, which was set at 862 to 896 kPa (125 to 130 psi), was inspected; there were no indications that the valve had relieved. Also, in the power-plant control room was an alarm that sounded if the pressure upstream of the final reducing station increased to approximately 2070 kPa (300 psi). According to the power-plant operators, this alarm did not sound. During the day the failure occurred, the steam supply pressure at the conversion plant did not exceed 655 kPa (95 psi).

Visual inspection of the heater shell disclosed three cracks or failure areas in the welded joints between the heater shell and the hot well. The locations of these cracks, numbered in the probable order of occurrence, are shown in Section A-A in Fig. 11.

Crack 1 may have initiated at the built-in notch (Detail C, Fig. 11), which is characteristic of welds made with a backing strip. The

fracture surfaces were covered with high-temperature oxidation, indicating that the surfaces may have been separated for some time. The crack, which extended the length of the hot well, was completely internal and could not be detected from outside the vessel. The intermediate tube supports were tack welded to the heater shell. The welds nearest the crack were broken and were covered with high-temperature oxidation, indicating that they had been broken for some time.

The heater shell burst along crack 3 (Detail C, Fig. 11), which was in the HAZ of the longitudinal weld. All the fracture surfaces of this crack exhibited bright metal, and there did not appear to be any one point or defect at which this failure started.

Crack 2 (Detail B, Fig. 11) was on the side of the hot well that did not rip open. This crack was approximately 760 mm (30 in.) long and was propagating toward the ends of the hot well. These fracture surfaces were also covered with high-temperature oxidation. The temperature-control valve was removed and inspected, but no damage was noted.

Metallurgical Tests. A section was taken through the longitudinal weld on the same side of the hot well as crack 2. A crack was observed that extended from the notch formed at the root of the weld and the backing strip to about the center of the heater-shell plate. This crack terminated at what appeared to be an inclusion in the metal. The hardness of the metal in the heater-shell plate at this section ranged from 81.5 HRB in the base-metal zone to 85 HRB in the HAZ adjacent to the weld. Hardness near the inclusion, which was in the HAZ, was 72 HRB. The hardness value of

81.5 HRB indicated that the steel should have a tensile strength of about 517 MPa (75 ksi).

Tensile Tests. The tensile strength of three specimens from the brine-heater shell was 531 to 545 MPa (77 to 79 ksi), yield strength was 372 to 414 MPa (54 to 60 ksi), and elongation in 50 mm (2 in.) was 22 to 27%. Compared to the specified requirements for ASTM A285, grade C, steel, these strengths exceeded the maximum, and ductility was less than or equal to the minimum. Chemical analysis showed a silicon content of 0.48%, which indicated a killed steel. Carbon content was 0.20%.

Stress analysis was conducted on the original design of the brine heater. A very high discontinuity stress existed at the longitudinal welds between the hot well and the heater shell. A lesser discontinuity stress existed at the longitudinal weld between the brine-heater shell and the steam manifold. The stress values were calculated for the heater-shell wall between the welds and the hot well outside the welds. These values exceeded the actual yield strength of the shell material at operating pressures of 331 kPa (48 psi) and greater. The material was no longer in the elastic region, and it deformed plastically to relieve or transfer the stress.

It was assumed that the section of the heater shell between the welds attempted to straighten. When this occurred, the root of the weld would have been under tension, which would create an ideal site for a crack to initiate.

Discussion. Cracks 1 and 2 probably originated during hydrostatic tests of the heater or shortly after the heater was put into operation. These cracks were internal and undetectable from the outside. During the first 18 months, when the brine heater was operated at 172 kPa (25 psi), these cracks would propagate at a very slow rate, if at all. During the last 6 months, the brine heater was operated at 331 kPa (48 psi), where the discontinuity stresses were greater than the yield strength of the material, and the cracks propagated at a more rapid rate. The brine heater was subjected to 44 recorded steam-pressure surges, which increased the operating pressure above the nominal 172 and 331 kPa (25 and 48 psi) needed to maintain the required brine temperature. During plant shutdown, there were probably other pressure surges similar to the one that caused the final failure that were not recorded. In addition to the high discontinuity stresses, the fact that the cracks did exist increased the stress concentrations in the heater. The loss of the support by the intermediate tube sheets at the failed portion of the shell increased the stress in the remaining shell.

Conclusions. The brine heater cracked and fractured because it was overstressed in normal operation. The two internal cracks propagated by continuing along the root of the longitudinal weld. The crack on the side that fractured had extended nearly 2 m (7 ft) before failure.

Corrective Measures. The heater design was modified to make the heater shell and the hot well two separate units. Also, the heater-shell expansion section was redesigned to minimize stress on the tubes and to minimize the resulting distortion. A tight shutoff valve was added to the brine-heater steam line to prevent steam from leaking into the heater. Holes in the support plates were enlarged to allow the heater to move in any horizontal direction.

Recommendations. A relief valve should be installed in the heater or in the steam line near the heater. At the time of failure, the relief valve was some distance upstream of the brine heater. Additional hydrostatic tests of the new brine heater should be conducted at ambient temperature and 1035 kPa (150 psi) (design pressure) at approximately 6-month intervals.

Mechanical Joints. The coefficient of expansion and yield strength at high temperature of metal components used in an assembly are important design considerations. Bolted assemblies require fasteners with a yield strength that will allow the fasteners to be tightened to such a prestress value that the assembly would remain leakproof even at high temperatures. In one case, leaks developed in an assembly that consisted of a flanged bonnet bolted to the shell of a heat exchanger and that was operated at a temperature of 510 °C (950 °F). The bonnet was made of a ferritic steel; the tube sheet, heat-exchanger shell, and studs were made of an austenitic stainless steel. When leakage was not stopped by repeated tightening of the nuts on the studs, the joint was disassembled.

Initial inspection was done to check the surface finish on the flanges and the condition of the gasket. Both were satisfactory. The joint dimensions and metals used were then checked for compliance to specifications; these were also satisfactory. During the examinations and review of the design calculations, it was noted that the materials in the assembly had greatly differing coefficients of expansion and that the studs had a coefficient of expansion that matched only the components made of austenitic stainless steel. Studs made of conventional stainless steel have a low yield strength and high ductility, which leads to stud elongation because of the operating temperature and load. These differing coefficients of expansion caused the tension on the studs on the assembly to relax, permitting leakage. The joint was made tight by replacing the conventional austenitic stainless steel studs with those conforming to ASME specification SA-453, which calls for a high-temperature bolting material with a coefficient of expansion comparable to austenitic stainless steel but with a much higher yield strength. Use of material conforming to SA-453 permitted tightening the nuts on the studs to a torque load that would maintain tightness at the operating temperature.

Effects of Welding Practices

Welding, which is commonly used to join components of heat exchangers, produces metal in the joints that is not homogeneous. The weld consists of a dendritic structure (cast metal) and intermediate zones between fused metal and unaffected base metal. In some cases, the entire heat exchanger can be normalized or annealed, and the metallurgical structure of the weld metal and base metal may become nearly identical. However, because heat treating of the heat exchanger is not always feasible, there generally is a gradation in metallurgical condition from unaffected base metal through an HAZ to weld metal.

In addition to this intrinsic change in homogeneity of material, other factors may provide metallurgical or mechanical stress raisers. Weld defects may include voids, porosity, slag or oxide inclusions, poor penetration, and shrinkage cracks. Many welds have rough surfaces and may join the base metal with an undercut or reinforcement. If metallurgical discontinuities are minimal, if defects are eliminated, and if the external contour does not contain mechanical notches, the properties of a welded joint can be expected to approach those of the base metal.

Preparation of joint surfaces before welding can be a source of weld defects in heat exchangers. For example, a pinhole leak developed in a heat-exchanger shell adjacent to a weld deposit that joined a nozzle, made of 19-mm (¾-in.) diam schedule 80 pipe, to the shell. When the nozzle was removed, it was found to have been installed by torch cutting a 38-mm (1½-in.) diam hole in the shell, inserting the nozzle and welding it in place. There was no evidence of cleanup after torch cutting the hole or of beveling of the wall edges around the hole to permit good weld penetration. Examination of the weld deposit revealed that the weld metal was generally oxidized and contained several flux inclusions. There was little or no bonding of the weld bead to the surfaces that had been cut with the torch. Examination revealed evidence that the weld was made under conditions of poor fit-up, lack of cleanness, and poor workmanship. A good joint was made by removing old weld metal and properly beveling the edges of the hole before rewelding.

Joint design can be responsible for mechanical notches at welds in heat-exchanger components. Designs that incorporate sharp corners and heavy longitudinal welds are sources of stress raisers.

Example 9: Fracture of a Carbon Steel Pipe in a Cooling Tower. A 455-mm (18-in.) diam, 7.9-mm (⁵⁄₁₆-in.) wall carbon steel discharge line for a circulating-water system at a cooling tower fractured in service; a manifold section cracked where a Y-shaped connection had been welded. The steel pipe was made to ASTM A 53 specifications.

Investigation. Examination of the pipe revealed that cracking occurred in the HAZ parallel to the weld and across the weld as shown in Fig. 12(a). The end of the branch pipe was prepared for welding by making a transverse cut; a second cut was then made at about 15° to

Fig. 12 Carbon steel discharge line at a cooling tower that failed because of poor fit-up at Y-joint and poor-quality welds

(a) Original joint design of pipe connection and location of cracks. Photograph is an oblique view of a section through the weldment showing the abrupt intersection of pipe walls and the voids and crevices in the weld metal. (b) Improved joint design. Dimensions given in inches

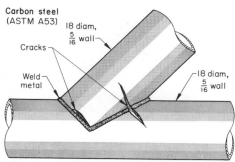

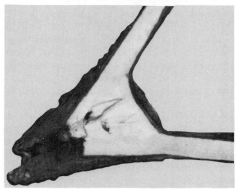

(a) Original design

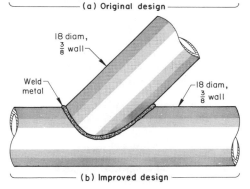

(b) Improved design

the longitudinal axis of the pipe, resulting in a sharp corner (Fig. 12a). The main pipe was then notched to fit the end of the branch pipe. Fitting of the saddle connection was poor, and no backing strips were used even though the pipe wall was thin; as a result, the condition of the root-weld bead was below standard, and numerous deep crevices and pits were in the weld area exposed to the inner surfaces of the manifold (Sectional view, Fig. 12a).

Conclusions. The pipe failed by fatigue. Cracks originated at crevices and pits in the weld area that acted as stress raisers, producing high localized stresses because of the sharp-

Fig. 13 Weld in AISI type 316 heat-exchanger shell that failed due to hot shortness

(a) Longitudinal section of weld; the dotted line indicated how the sample was sectioned for microexamination. Approximately 2½×. (b) Micrograph of section from weld. Hot shortness resulted in intergranular cracking along the large columnar grain boundaries of the root-pass fusion zone. Etched with 10% oxalic acid. 55×

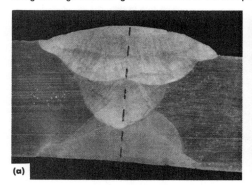

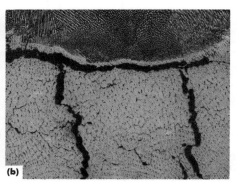

Fig. 14 Type 321 stainless steel heat-exchanger bellows that failed by fatigue originating at heavy weld reinforcement of a longitudinal seam weld

(a) A section of the bellows showing locations of the longitudinal seam weld, the circumferential welds, and the fatigue crack. Dimensions given in inches. (b) Mating fracture surfaces of the bellows. Crack initiated at the longitudinal seam weld (region A) and propagated circumferentially along the HAZ of the circumferential weld (region B). Fatigue beach marks are present along edges (at center). The fracture surfaces are oriented to show the inner surface of the bellows (top) and the outer surface (bottom). 2×

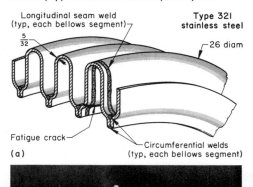

radius corner design. Abnormally high structural stresses and alternating stresses resulting from the pump vibrations contributed to the failure.

Corrective Measures. The joint design was changed to incorporate a large-radius corner (Fig. 12b), and fitting of the components was improved to permit full weld penetration. Backing strips were used to increase weld quality. Also, the pipe wall thickness was increased from 7.9 to 9.5 mm (⁵⁄₁₆ to ⅜ in.).

Example 10: Intergranular Cracking in Heat-Exchanger Welds Due to Hot Shortness. Samples from an AISI type 316 stainless steel heat-exchanger shell were submitted for examination after radiographic testing revealed the presence of subsurface cracks in a longitudinal weld seam of the shell. The heat exchanger had been in service for 12 years.

Investigation. The as-received sample was dye penetrant tested before sectioning, and no cracks were indicated. Longitudinal sections were then taken along the centerline of the weld (Fig. 13).

Metallographic examination revealed numerous intergranular cracks associated with the root pass of the weld. The cracks had propagated both parallel to and normal to the weld seam (Fig. 13). This type of cracking is characteristic of a phenomenon termed hot shortness (see the article "Failures of Weldments" in this Volume).

Energy-dispersive x-ray spectrometry of the areas surrounding the cracks showed that a type 316 electrode was not used for the root pass. Analysis indicated that the root pass was instead made using a nickel-copper alloy electrode. Dilution effects and subsequent welding with type 316 electrodes accounted for the observed variations in elemental concentration.

Conclusions. Cracking was due to the use of an incorrect electrode for the root pass of the weld. Fully austenitic electrode materials, such as the nickel-copper alloy used for the root pass, are quite crack sensitive, especially if overheated. The subsurface defects in the fu-

sion zone of the weld were probably too small to be detected by nondestructive inspection at the time of fabrication, but long-term service and thermal stresses propagated the fissures to the extent that they were detected by radiographic inspection 12 years later.

Recommendations. This weld seam and others were completely ground out and were replaced with the correct electrode material.

Stiffening, which results when the weld metal is thicker than the adjacent base metal, can also lead to failure, as discussed in the following case. The design of expansion bellows in a heat exchanger incorporated a longitudinal weld across each corrugation and a circumferential weld joining each corrugation (Fig. 14a). The bellows had an outside diameter of 660 mm (26 in.) and were made of 3.9-mm (⁵⁄₃₂-in.) thick type 321 stainless steel. The heat exchanger was shut down when leakage occurred in the bellows. Circumferential cracks were found along the HAZ of the weld and through the corrugation wall. These are areas of highest stress.

Examination of the fracture surface revealed fatigue beach marks (Fig. 14b) that initiated at the heavy weld reinforcement of the longitudinal seam weld. The unusually heavy reinforcement had a stiffening effect, which caused overstressing and transverse cracking of the longitudinal seam. Grinding the reinforcement of the longitudinal weld flush reduced the stiffening. As a corrective measure, the bellows were formed from a single sheet large enough to provide the proper number of convolutions without a circumferential weld in the working section. The only weld required was one longitudinal seam the full length of the bellows. This change in design eliminated the points of high stress that had led to failure.

Effects of Elevated Temperature

Failures associated with elevated temperature generally involve a material that is too weak for

the service temperature or a material that develops unexpected embrittlement at or above the service temperature. Thermal cycling can cause failures where adjacent materials differ significantly in composition and coefficient of thermal expansion. Embrittlement may also be the result of metallurgical changes that leave the material relatively ductile and tough at elevated temperature but brittle at or slightly above room temperature.

Another temperature-related problem is scaling and oxidation due to excessive temperatures. Creep and stress rupture also occur because of excessive temperatures, as discussed in the article "Elevated-Temperature Failures" in this Volume. These temperature-related failure mechanisms must be accounted for in selection of material, in fabrication, and in operating procedures.

Example 11: Embrittlement of a Titanium Heater Tube. A flow stoppage in a leach heater caused brief overheating of the

Fig. 15 Titanium heat-exchanger tube (ASTM B337, grade 2) that became embrittled and failed because of absorption of hydrogen and oxygen at elevated temperatures

(a) Section of the titanium tube that flattened as a result of test per ASTM B 337; the first crack was longitudinal along the top inside surface. ¾×. (b) Side of flattened tube section showing cracks. 1¼×. (c) High-temperature (370 °C, or 700 °F) tension-test specimens: top, specimen from normally heated tube; bottom, specimen from overheated tube. (d) Normally heated tube specimen etched in Kroll's reagent showing fine grain size and hydride particles (faint black needles). 200×. (e) Overheated tube specimen etched in Kroll's reagent showing large grain size and hydride particles (small needles). 200×. (f) Transverse section of overheated tube specimen at outside surface showing 0.05-mm (0.002-in.) thick oxygen-embrittled layer (light band near top)

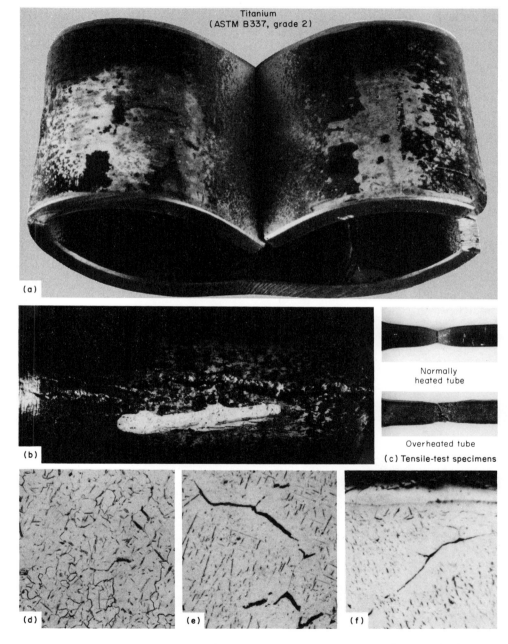

Titanium
(ASTM B337, grade 2)

(a)

(b)

Normally
heated tube

Overheated tube

(c) Tensile-test specimens

(d)   (e)   (f)

89-mm (3½-in.) outside-diameter, 6.4-mm (0.25-in.) wall thickness titanium heater tubes (ASTM B337, grade 2). Many of the tubes showed colorful surface oxides, and some sagged slightly.

A short section of an oxidized and sagged tube was sent to the metallurgical laboratory for evaluation. The purpose of the examination was to determine whether the mechanical properties of the tube had been altered by overheating, particularly whether the tube had been embrittled, making it susceptible to fracture. A section from a similar tube removed previously from a leach heater that had not been heated above normal operating temperatures was sent to the laboratory for comparison.

Visual examination of the section of the overheated tube disclosed blue-tinted areas and patches of flaky white, yellow, and brown oxide scale. There was a dark-gray scale on the inside surface with no evidence of corrosion. The presence of the colored scale on the outer and inner surfaces was an indication of heating to approximately 815 °C (1500 °F).

Mechanical Tests. A 75-mm (3-in.) long section of the overheated tube was subjected to a flattening test. Specification ASTM B 337 requires that a welded or seamless grade 2 titanium tube be capable of withstanding, without cracking, flattening under a load applied gradually at room temperature until the distance between the platens is seven times the nominal wall thickness—in this case, 45 mm (1.75 in.). In the test, when the platens were 66 mm (2.6 in.) apart, cracking initiated at both inner and outer surfaces of the tube; when the platens were 61 mm (2.4 in.) apart, the tube fractured, with fracture initiating at the inner-surface cracks. Thus, the tube no longer met ASTM B 337 specifications. Figure 15(a) shows an end view of the flattened tube. The cracks on the outer surface are shown in Fig. 15(b).

Longitudinal tensile tests were conducted at 370 °C (700 °F) on specimens taken from the overheated and normally heated tubes. The results are listed in Table 3. As shown in Fig. 15(c), the specimen from the normally heated tube necked down more than the overheated tube specimen, although elongation was approximately the same.

Metallographic examinations were made of specimens from both the normally heated and the overheated tubes. The normally heated tube had a fine grain size and only a few faint hydride particles in the microstructure (Fig. 15d). By comparison, the microstructure of the overheated tube (Fig. 15e) exhibited a much larger grain size and more numerous needlelike hydride particles. Also, the overheated tube had a very hard brittle surface layer approximately 0.05 mm (0.002 in.) thick (Fig. 15f), which was present on both the inner and outer surfaces. The layer was extremely hard (51 HRC) at the surface and decreased to almost normal hardness (33 HRC) at 0.076 mm (0.003 in.) from the surface. At the approximate center of the tube wall, the hardness was 83 HRB.

The flaky white, yellow, and brown oxide scale and the increased grain size indicated that the tube had been heated to approximately 815 °C (1500 °F). At this temperature, the metal readily absorbs hydrogen and oxygen. The hydride formation and the shallow surface embrittlement were indications of hydrogen and oxygen absorption. The oxygen absorbed at the surfaces created very hard and therefore brittle layers that made the tubes susceptible to cracking at ambient temperatures.

Although the high-temperature (370 °C, or 700 °F) tensile strength of the overheated tube was significantly lower than that of the nor-

Table 3   Results of longitudinal tensile test

Specimen	Yield strength, 0.2% offset		Tensile strength		Elongation in 50 mm (2 in.), %	Reduction of area, %
	MPa	ksi	MPa	ksi		
Overheated tube.............	97	14	146	21.2	36	69
Normally heated tube	100	14.5	197	28.5	38	82

mally heated tube, the yield strengths were about the same, and the ductility of the overheated tube was slightly lower. At 370 °C (700 °F), there was no evidence of the severe embrittlement shown in the flattening test at room temperature. However, at temperatures below 205 °C (400 °F), embrittlement from hydrogen and oxygen absorption occurred, as evidenced by the poor performance of the flattening test.

Conclusions. The titanium tubes had been embrittled by being heated to a temperature of about 815 °C (1500 °F), which is excessive for ASTM B337, grade 2, titanium. The most serious effect of overheating the tubes was the generally embrittled condition at low temperature, and particularly the severe surface embrittlement resulting from oxygen absorption. The brittle surface layers made the overheated tubes susceptible to cracking under start-up and shutdown. At operating temperatures of 370 °C (700 °F), the overheated tubes would not have been brittle, but the overall mechanical properties of the tubes would have been somewhat degraded.

Recommendations. If the overheated tubes remain in service, thermal stresses should be avoided during start-up and shutdown, particularly at temperatures below 205 °C (400 °F).

Although at the operating temperature of 370 °C (700 °F) the overheated tubes are not in an embrittled condition, the tubes should be replaced because there is a possibility that cracks formed during start-up and shutdown would propagate through the tube wall under operating conditions. Replacement tubes should be made of a heat-resistant alloy (e.g., Hastelloy C-276), rather than of titanium.

ACKNOWLEDGMENT

R.J. Franco wishes to acknowledge G.M. Buchheim, Exxon Research & Engineering Company, for his assistance in preparing several of the examples in this article.

SELECTED REFERENCE

• S.D. Reynolds and F.W. Pement, Corrosion Failures of Tubing in Power Plant Auxiliary Heat Exchangers, *Mater. Perform.*, Vol 13 (No. 9), Sept 1974

Failures of Pressure Vessels

Roy T. King, Cabot Corporation

PRESSURE VESSELS, piping, and associated pressure boundary items of the types used in nuclear and conventional power plants, refineries, and chemical-processing plants are discussed in this article. These types of equipment operate over a wide range of pressures and temperatures and are exposed to a wide variety of operating environments.

The National Board of Pressure Vessel Inspectors compiles domestic statistics on reported accidents involving pressure vessel material. These data indicate the prevalent causes and types of failures and are summarized in Table 1 for a 4-year period beginning in 1981. Faulty design, fabrication, and inspection contribute to significant numbers of failures, particularly for shell components. Damage during shipment and storage must be considered because air and its contaminants may be corrosive. Field fabrication and erection practices should also be considered. Simple matters such as specifying and verifying the proper material and material condition are critical. One company has used portable spectroscopy equipment to examine a number of bulk items supposedly made of alloy steels and found 1 to 3% of the total items to be carbon steel. A survey of equipment that required 320 000 individual instrument readings found that 1.8% of the material was supplied incorrectly, and the most frequent error was substitution of carbon steel for alloy or stainless steel, or vice versa (Ref 1).

Normal operating conditions contribute to changes in materials and structures that should be accounted for in design. Operating temperatures are important because they determine whether the material is stressed above or below any ductile-to-brittle transition temperature below which relatively low energy fractures can occur. Other degradation mechanisms, such as corrosion and scaling; stress corrosion; erosion from flowing gases, liquids, and solids; hydrogen damage; creep and stress rupture; fatigue and creep fatigue; and aging, are generally accelerated with increasing temperature. Operating pressure is another important design consideration because pressure combined with structural constraints determines the stresses in pressure boundaries both at steady-state operating conditions and during transients. Verifying whether failures occurred as a result of, or during, off-normal conditions is an important part of the analysis.

Cyclic operation greatly affects the life of pressure boundaries. Intermittent operation can result from normal demand schedules or from operation during peak demand periods only. The cyclic thermally induced stresses associated with frequent start-ups and shutdowns are frequently more severe than stresses of steady-state operations, and designers cannot always predict the use cycles of equipment that owner-operators impose. Economics frequently dictate that use of capital equipment be extended beyond the original design life in many industries, with the result that repairs and renovations are attempted. Accounting for all the relevant practices of several decades that can affect life-extension programs is difficult, and failures can arise from obscure combinations of circumstances.

Intermittent operation can also affect the life of pressure vessels. Long periods of inactivity greatly enhance the probability of internal or external corrosion. During long shutdown periods, the recommended practice is to fill the system with inert or nonreactive gases, such as helium or nitrogen, in order to avoid unacceptable corrosion. Wet lay-up using deoxygenated water or the continuous circulation of demineralized deoxygenated water can also be used to minimize corrosion. Recordkeeping during such periods is as important as record-keeping during operating periods.

Unanticipated effects resulting from the design of other components of the system, such as piping and valves, can be important. When all components are incorporated into a system, the service conditions may be unacceptable for one of the components. Water pulses and vibration can damage piping systems, including hangers and supports. Heavy valves, instruments, and associated components should be properly supported so that the connecting piping system is not overstressed under fatigue loading.

Failures in other parts of a system can cause pressure boundaries to fail. Burner failures, pressure control valve failures, low water cut-offs that lead to improper coolant levels, and simple operator errors or improper maintenance of equipment account for a sizable fraction of recorded pressure boundary failures.

A large number of failures result from causes designated as "other" in the failure statistics. Unique experiences and failures due to multiple causes are not uncommon, and the analyst must be aware of multiple, interconnected mechanisms leading to failures.

Procedures for Failure Analysis

The basic procedures for analyzing failures of pressure vessels and pressure piping are not much different from those for investigating failures of other types of equipment (see the article "General Practice in Failure Analysis" in this Volume). At the outset, a preliminary examination should be made to provide a clear overall picture of what happened. If the failure involved an explosion, a careful search should be made for all material components and fragments. The location and mass of each component or fragment should be recorded, and all burn discolorations or impact marks on surrounding objects or surfaces should be located. This ballistic information can be used to estimate launch angles and trajectories and to calculate the approximate pressure in the vessel at the instant of failure.

Because an explosive failure of a pressure vessel may result in extensive damage, system downtime, and expense, it is natural that the preliminary examination will generate considerable expression of opinions as to the possible cause of failure. Casual opinions should be avoided, and certainly not solicited. Subsequent recollections of witnesses could be colored by the offhand comments or speculation of investigators.

As in any failure analysis, it is important that the investigator gain a clear understanding of the entire system involved, including specifications, intended functions, normal methods of operation, thermomechanical history, maintenance history, manufacturing history, and operator experience. This information, plus information from subsequent examinations and from laboratory analyses of specimens, should permit accurate determination of how and why the failure occurred.

Techniques used during a preliminary examination should not affect any subsequent exam-

Table 1   Summary of statistics on pressure vessel failures compiled from 1981 through 1984 by the National Board of Boiler and Pressure Vessel Inspectors

Initial part failure	Causes							Type of failures							Numbers		
	Low water cut-off	Faulty design fabrication or installation	Corrosion or erosion	Operator error or poor maintenance	Burner failure	Pressure control failure	Other	Burned or overheated	Collapsed inward	Combination explosion	Cracked	Torn assunders (rupture)	Leakage	Other	Accidents	Injuries	Deaths
1981																	
Shell	5	18	25	36	...	12	21	...	6	21	57	26	23	3	94	19	3
Head	1	7	5	9	...	3	1	...	6	2	12	7	1	...	23	...	...
Attachments	7	5	...	4	...	7	2	...	...	2	7	5	1	...	31	1	1
Piping	10	2	1	3	1	5	18	1	...	2	10	7	15	4	41	1	...
Safety valves	...	...	...	1	...	...	2	...	...	...	...	3	...	...	3	1	1
Miscellaneous	...	2	4	7	...	4	166	1	2	5	4	9	12	...	232	5	4
1982																	
Shell	4	17	29	13	1	3	34	4	10	3	62	45	35	0	129	7	4
Head	1	4	3	11	0	4	5	0	2	2	13	8	7	1	19	3	2
Attachments	2	4	3	4	0	2	2	1	0	0	10	9	11	0	21	1	0
Piping	0	3	8	18	0	5	5	0	0	1	14	30	24	0	33	17	0
Safety valves	0	18	25	12	0	3	4	0	0	1	1	2	26	0	5	5	1
Miscellaneous	1	6	14	24	0	86	43	0	1	0	11	2	23	3	89	3	2
Total															296	36	9
1983																	
Shell	1	26	64	44	1	12	1143	15	13	1	61	31	41	22	167	75	13
Head	1	3	9	9	0	3	12	2	2	3	14	6	9	3	22	18	4
Attachments	0	2	3	2	0	2	2	0	0	0	15	3	16	4	19	2	0
Piping	1	7	14	54	2	6	24	1	0	1	16	18	72	4	62	16	0
Safety valves	0	0	28	1	0	2	23	0	0	0	1	0	39	14	3	11	0
Miscellaneous	2	25	12	47	0	6	56	1	1	0	17	20	33	15	149	18	0
1984																	
Shell	3	77	75	53	16	10	151	4	9	17	265	98	92	96	625	209	43
Head	...	17	15	22	7	2	17	...	3	...	57	10	5	11	95	124	13
Attachments	1	4	5	6	1	2	12	1	...	1	27	2	9	12	57	20	12
Piping	...	15	22	21	...	4	47	9	...	1	10	17	34	25	110	46	4
Safety valves	9	6	...	11	7	12	4	...	1	2	...	9	1	7	51	11	1
Miscellaneous	1	25	15	36	2	11	189	14	2	1	44	22	22	147	372	27	...
Total															1310	437	73

inations. For instance, liquid-penetrant inspection can make it difficult to analyze for chlorides in deposits on the part surfaces or on fracture surfaces.

Effects of Using Unsuitable Alloys

Specification of an alloy that is not suited to the application is frequently the cause of failure of a pressure vessel. The composition of the alloy may be incorrect, or the metallurgical properties may not meet requirements. This often results from a misunderstanding of the operating conditions by the designer.

A common cause of failure of pressure vessels is use of an alloy other than the one specified. Sometimes, bars or plates are not properly marked or are accidentally stored with another alloy. Mistakes can be made when care is not taken to ascertain that the alloy used complies with the alloy specified.

In one case, several low-carbon steel nipples were inadvertently used in place of 5Cr-0.5Mo steel nipples in a pipeline containing hydrogen at 400 °C (750 °F). Failure occurred after approximately 4 years of service. The micrograph shown in Fig. 1, which illustrates a specimen from one of the low-carbon steel nipples, reveals fissuring at the grain boundaries. This phenomenon is characteristic of hydrogen attack when low-carbon steel is exposed to hydrogen at high pressures and temperatures. The piping system was completely surveyed to ensure that only the specified metal was used. Replacements were made when the wrong metal was found.

A common experience for major oil refiners is failures or near-failures that can be attributed to insufficient alloy content in steels for hydrogen service. In one example, a carbon steel pipe section failed that was inadvertently substituted for the specified 1.25Cr-0.5Mo steel. This 305-mm (12-in.) diam line ruptured after 16 years of service at conditions slightly above the Nelson curve limit for carbon steel, resulting in an expensive fire. The failed end of the pipe was jagged but undistorted, indicating a brittle failure (Fig. 2a). Metallographic examination of a fragment of the pipe revealed extensive cavities through about three-quarters of the wall thickness, with the outer surface of the pipe wall relatively unaffected (Fig. 2b). At higher magnification, decarburization and fissuring are visible in the damaged microstructure (Fig. 2c). Such incidents of hydrogen damage occur in other branches of the petrochemical- and chemical-processing industries. A severe explosion resulted from the erroneous installation of a 305-mm (12-in.) carbon steel pipe at a German ammonia plant in 1974. The pipe section, which had been installed in a 17-MPa (2500-psig) alloy steel loop, fragmented and released a jet that toppled a 254-Mg (280-ton) ammonia converter.

Welding Problems. In the fabrication of pressure vessels, welding problems, such as brittle cracking in the heat-affected zone (HAZ), often result from the use of steels containing excessive amounts of residual elements that increase hardenability and susceptibility to cracking. Such steels are sometimes produced inadvertently when low-carbon steel is made in a furnace normally used to make high-alloy steels. The refractory lining of the furnace may impart sufficient residual chromium and other alloying elements to a heat of low-carbon steel that problems occur when components made from that heat are joined using standard welding procedures. Therefore, complete control of composition is of the utmost importance when welding is involved.

As an example of improper material specification or use, galvanized steel clips were

Fig. 1 A specimen from a low-carbon steel nipple showing fissuring at grain boundaries (top) caused by hydrogen attack
80×

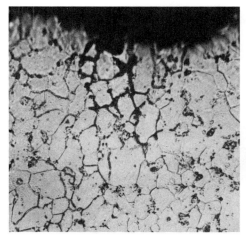

Fig. 2(a) Ruptured 305-mm (12-in.) carbon steel pipe, inadvertently installed in a 1.25Cr-0.5Mo circuit, that was severely damaged by hydrogen embrittlement
On-stream failure caused extensive fire damage.

Fig. 2(b) Micrograph of outside-diameter surface of failed pipe in Fig. 2(a)
Hydrogen attack had progressed about three-fourths through the wall thickness. 18×

Fig. 2(c) Micrograph of inside-diameter surface of failed pipe in Fig. 2(b) showing severe decarburization and fissuring
180×

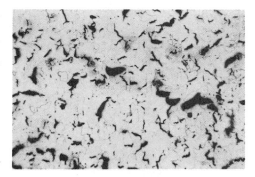

welded to stainless steel piping. The galvanized (zinc) coating became molten during the welding process, and the liquid zinc attacked the grain boundaries in the stainless steel, causing intergranular cracks. The piping was dimpled and leaked at some of the welds. The locations and depths of the cracks were related to the welding procedure and the amount of welding heat used. Excessive welding heat had caused dimpling (distortion) of the stainless steel piping and further crack penetration into the piping wall. The cracks continued to propagate during service from stresses induced in the piping.

In another case, a transfer line carrying 73% sodium hydroxide developed a leak at a weld joining a cast Monel coupling to wrought Monel pipe. Metallographic examination disclosed many small microfissures in the weld metal, but no microfissures were found in the HAZs of the wrought Monel. This section of line contained high thermal stresses that resulted from injection of steam a short distance upstream from the coupling. Chemical analysis revealed that the cast Monel coupling had a silicon content of 1.97%, whereas the wrought Monel piping had a silicon content of 0.10%. Castings of high-silicon Monel generally have poor weldability and are often hot short, or sensitive to cracking in the weld HAZ. Fissures in the cast coupling were formed during welding and subsequently penetrated the section when the coupling was subjected to the thermal shock of steam injection. Where welding is required, a weldable-grade casting should be specified to ensure freedom from hot fissuring.

A welded-on blank cap began leaking during a hydrostatic test because it had cracked in the weld. Analysis showed that the cap had a carbon content of 0.56%, whereas a maximum

carbon content of 0.30% was specified. The excessive carbon content had resulted in a brittle HAZ. It was found that construction personnel had substituted ASTM A53 type F butt-welded piping, which does not have a specification for maximum carbon content. Because the higher-carbon steel had been used on a random basis throughout the piping system, all piping was spark tested in the field to expedite repair. When a spark-tested section generated a questionable spark pattern, a specimen was removed for chemical analysis. All sections having a high carbon content were replaced by piping with a maximum carbon content of 0.30%.

Failure of several high-pressure nipples was attributed to the occasional inadvertent use of 4140 steel instead of the specified 1035 steel. The presence of approximately 1% Cr and 0.15 to 0.25% Mo in 4140 steel enhanced the hardenability and contributed to hardness and brittleness in HAZs adjacent to welds. All failures in the nipples occurred in these hardened zones. A spot test for chromium was made, and those nipples showing chromium contents exceeding 0.3% (estimated) were removed and replaced with nipples fabricated from 1035 steel. Although 4140 steel is a weldable alloy steel, welding procedures suitable for 1035 steel were not appropriate for the higher-alloy material.

Specification of the filler metal used for welding components of pressure-piping systems is important, as demonstrated in the following example.

Example 1: Brittle Fracture of a Clapper Weldment for a Disk Valve Due to Improper Filler Metal. After a service life of only a few months, the clapper in a 250-mm (10-in.) diam valve fractured. During operation, the valve contained a stream of gas consisting of 55% H_2S, 39% CO_2, 5% H_2, and 1% hydrocarbons at 40 °C (100 °F) and 55 kPa (8 psi). Specifications required the clapper to be

made of 13-mm (½-in.) thick ASTM A36 steel, stress relieved and cadmium plated.

Investigation. Fracture occurred at the welded joint between the clapper and a 20-mm (¾-in.) diam support rod, which was also made of ASTM A36 structural steel. Examination revealed voids on the fracture surface and evidence of incomplete weld penetration. The overall appearance of the fracture surface indicated brittle fracture, but there was some evidence of fatigue beach marks. Fracture originated at a slag inclusion in the weld metal. In some areas, particularly on the weld metal, the plating material had flaked off the clapper.

Vickers microhardness surveys across the weld area revealed very narrow bands of high hardness (400 to 650 HV) at the edges of the weld metal. The normal hardness of the base metal was 150 to 160 HV, hardness in the HAZ was 150 to 180 HV, and hardness of the weld metal was 230 to 250 HV.

Chemical analysis of the material in the high-hardness bands revealed a composition of 70Fe-17Cr-9Ni-1.2Mn. This composition indi-

Fig. 3 Return bend made of ASTM A213, grade T11, ferritic steel that ruptured because it contained a large number of inclusions

(a) Overall view of the return bend showing rupture. (b) Micrograph of an unetched specimen showing high concentration of inclusions. 400×. (c) Micrograph of an unetched specimen showing inclusions and cracks containing scale. 435×

(a)

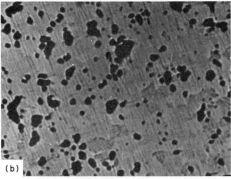

(b)

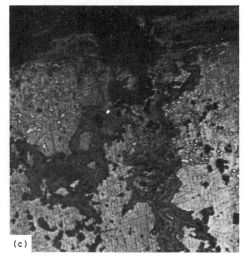

(c)

cated that a stainless steel filler metal had been used, which produced a mixed composition at the weld boundaries. Although the weldment may have been stress relieved as specified, stress-relief annealing was apparently incapable of reducing the high hardness of the localized bands caused by alloy mixing. Chemical analysis of the plating material showed it to be

nickel, probably electroless nickel because it had a high hardness.

Conclusions. The clapper failed by fatigue and brittle fracture because it was welded with an incorrect filler metal. Fatigue cracking was initiated at voids and high-hardness bands in the weld deposit. Also, the clapper was plated with nickel instead of cadmium.

Corrective Measure. A clapper assembly was welded with a low-carbon steel filler metal, then cadmium plated. The new clapper was installed in the valve and subsequently provided satisfactory service.

Effect of Metallurgical Discontinuities

Nonmetallic inclusions, seams, laps, bursts, and pipes are common discontinuities found in wrought products that may cause premature failure. Shrinkage, gas porosity, and cold shuts are likely to occur in castings and can lead to failure, usually leakage in cast components.

Discontinuities can and sometimes do cause failures and should never be overlooked during failure analyses. However, except in fatigue, the number of failures caused by microstructural defects is relatively small compared with the number resulting from macroscopic defects.

Example 2: Rupture of a Ferritic Steel Return Bend Because of Inclusions. After 2 years of service, a return bend from a triolefin-unit heater ruptured (Fig. 3a). The bend was made from 115-mm (4½-in.) schedule 40 (6.0 mm, or 0.237-in., wall thickness) pipe of ASTM A213, grade T11, ferritic steel. The unit operated at 2410 kPa (350 psi), with a hydrocarbon feed stream (85% propylene) entering at 260 to 290 °C (500 to 550 °F) and leaving at 425 to 480 °C (800 to 900 °F). The temperature of the combustion gas was 900 °C (1650 °F).

This rupture occurred on the inner surface of the bend, rather than on the outer surface where erosion-induced failures usually occur. The fracture was limited to the return bend and terminated at the welds that joined the bend to the pipeline.

Investigation. Examination of the fracture surfaces with a low-power stereomicroscope revealed that very little thinning of the metal had taken place, indicating low ductility and little corrosive attack.

Metallographic specimens were prepared from regions adjacent to the fracture and at the outer surface of the bend. The steel near the fracture exhibited a high concentration of both small and large inclusions (Fig. 3b). The steel in the outer surface contained relatively few inclusions, most of which were very small.

A micrograph of a section through the fireside edge of the fracture surface is shown in Fig. 3(c). Branched cracks similar to those produced by stress corrosion of steel were

observed. These cracks had apparently propagated between the inclusions in the steel.

Scale was observed over most of the crack path, even down to the fine tip of the crack. The scale provided stress raisers in two ways. First, the volume of the oxide exceeded that of the steel it replaced, placing the metal at the crack tip in tension and promoting a stress-corrosion reaction. Second, the effect of the oxide was magnified during thermal cycles because of differential thermal expansion, with the steel having a greater expansion coefficient than the scale. The high density of inclusions was assumed to be the result of improper cropping of the ingot from which the tube for the return bend was made.

Conclusions. The return bend failed as a result of stress-corrosion cracking (SCC), which propagated because of the presence of numerous inclusions in the metal. Scale formed in the cracks, abetted by thermal cycling, contributed to stresses that caused cracking.

Corrective Measures. There is no general remedy for this relatively rare type of failure, which is difficult to control because of its localized nature. It is not cost-effective to nondestructively examine all fittings for such defects, but material that is intended for critical applications where failure cannot be tolerated should be examined.

Effects of Fabrication Practices

Mechanical or metallurgical imperfections in a workpiece can be introduced, and detected, at any stage of construction, including mill processing, fabrication, shipping, and erection. Imperfections do not always result in failure and are in fact a relatively infrequent cause of failures of pressure vessels during operation. The ASME Boiler and Pressure Vessel Codes recognize that perfection is seldom obtained in commerical materials and define acceptability standards for the producer.

Many imperfections are detected before actual operation. For example, a crack indication was found in a forged nozzle by magnetic-particle inspection before welding of the nozzle into a pressure vessel. The nozzle manufacturer had found the crack and, instead of rejecting the part, had attempted to repair it by welding, which was not permitted in the specification.

In another case, a piece of 50-mm (2-in.) diam schedule 80 (5.5 mm, or 0.218-in., wall thickness) ASTM A106 seamless carbon steel pipe that had been hydrostatically tested was about to be welded into a pipeline when the operator found a lap in the outer surface. The lap was about 75 mm (3 in.) long and extended through almost the entire wall. A slight ridge on the inner surface had prevented leakage through the lap during hydrostatic testing. A section across the lap showed light mill scale adhering to the sides with some decarburization. There was no evidence of branched cracking nor any

Fig. 4 Crack path in a failed utility boiler drum

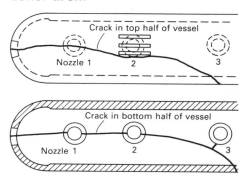

Fig. 5 Section through arc-gouged drain groove showing thin layer of dendritic structure containing cracks

Etched with nital. 55×

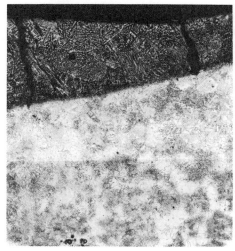

indication of SCC. The condition of the lap surfaces indicated that the lap had been formed at a temperature above 870 °C (1600 °F), which would have occurred early in the tube-forming process. It was concluded that a foreign object had fallen on the tube as it was being formed and that the object had passed through the rollers, producing a dent in the tube wall. Subsequent forming operations produced a lap.

Example 3: Failure of a Utility Boiler Drum During Hydrotesting. A 150-cm (60-in.) inside-diameter boiler drum fabricated of 18-cm (7-in.) thick ASTM A515, grade 70, steel failed during final hydrotesting at a pressure of approximately 26 MPa (3.8 ksi). The shock wave from the vessel rupture also damaged several connected headers and piping runs. A schematic of the crack path, including the location of three SA-106C nozzles, is shown in Fig. 4. The nozzles are numbered 1 to 3, moving away from the head.

Investigation. Field examination occurred 2 days after the fracture. The fracture surfaces between nozzles 1 and 2 and through the manway (access port) in the head were classic brittle fractures. The remainder of the fracture involved tearing, and the crack led off from parallel to the axis of the vessel.

Samples were taken to allow examination of the brittle-fracture area and the fracture-initiation sites; they were flame cut and then sawed into manageable sizes for laboratory work. Chemical analyses of the materials of construction demonstrated that the original materials met the specified chemistries. Drop-weight tests, dynamic tear tests, and Charpy V-notch impact tests of the shell plate material, the head material, and the nozzle material indicated that these materials had ductile-to-brittle transition temperatures above room temperature.

Macroscopic observations of the fracture surfaces at nozzles 1 and 2 showed chevron patterns that allowed tracing the fracture to its initiation points. The fracture apparently initiated at the nozzle 1 to shell junction on the head side and on the shell side, propagated around the head to the back side of the vessel, and

continued down the back side of the shell. On the nozzle side, the fracture reinitiated on the far side of each of nozzles 1 and 2 and continued down the shell.

Drain grooves had been arc gouged at the nozzle sites, and examination of the fracture surface at the grooves revealed short, flat segments of fracture area indicative of pre-existing cracks. These features were consistent with crack indications found during magnetic-particle and scanning electron microscopy (SEM) examination of the groove lips. The remainder of the edge along the groove surface exhibited small segmented shear lips.

Metallographic sections were taken through the drain groove and fracture surface at the forward sides of nozzles 1 and 2. In these sections, a thin layer of material with a dendritic structure was observed at the groove surface, and the HAZ resulting from the arc gouging was obvious (Fig. 5). Cracks were visible in the groove surface; most were confined to the outer dendritic layer, but at least one crack was observed to extend about 6 mm (0.25 in.) into the base metal. The microhardness of the dendritic layer was 489 to 591 HK and 47 to 52 HRC, while the HAZ was 179 to 258 HK and 85 HRB to 21 HRC. A qualitative microprobe analysis of the dendritic layer revealed that it contained over 1% C, higher than the carbon content of the base metal. This high hardness is consistent with the carbon content and indicates a high susceptibility to cracking upon cooling or deformation.

The microstructure of the main fracture surface in unaffected base metal tended to be straight and transgranular, characteristic of cleavage fracture. In regions in which the fracture intersected a weld HAZ, the fracture

path became irregular and exhibited features suggesting ductile fracture within the zone. Ductile dimples were also observed in the HAZ at the arc-gouged drain grooves.

Conclusions. The cracks in the drain groove surface could have occurred (1) immediately after arc gouging as a result of differential thermal contraction between the resolidified layer and the base metal, (2) during the subsequent stress-relieving operation, or (3) during the hydrostatic test due to localized stresses exceeding the yield strength of the base metal. The ragged surface of the drain groove probably contributed to even higher local stresses than would a smooth groove.

Recommendations. Flame cutting of surfaces in these steels should be avoided because it can lead to local embrittlement and stress raisers that initiate major failures. In addition, the ductile-to-brittle transition temperature of materials of construction should be below the hydrostatic-test temperature.

Effect of Stress Raisers. Designers generally avoid configurations that can provide local stress raisers, but macroscopic stress raisers can occur as a result of fabrication errors. The most common are those that result from lack of complete weld penetration. Inspection practices are aimed at preventing pressure vessels containing inadvertently incorporated stress raisers from reaching service, but they occasionally do.

Example 4: Failure of a Steam Accumulator Due to Lack of Complete Weld Penetration. A 1.2-m (4-ft) OD × 4.3-m (14-ft) long steam accumulator, constructed with 10.3-mm (¹³⁄₃₂-in.) thick SA515-70 steel heads and an 8-mm (⁵⁄₁₆-in.) thick SA455A steel shell, ruptured after about 3 years of service, launching the body of the vessel into a neighboring house. The accumulator was used in a plastic molding operation and operated at a pressure of 827 kPa (120 psi), dropping to 780 kPa (113 psi) during the injection cycle. It was designed for 1035-kPa (150-psi) operation and had presumably been pressure tested at 1550 kPa (225 psi). Because the seals on both safety relief valves were intact, the valves were tested and found to operate properly at the required 1035-kPa (150-psi) set point.

Investigation. A field examination showed that the manhole end of the vessel had ruptured around approximately 270° of the circumference of the head-to-shell girth weld before continuing to complete separation in the head material. Visual examination indicated an extensive lack of weld penetration in this girth weld. Both sides of the weld were removed for laboratory examination (Fig. 6).

The initial laboratory examination after cleaning of the samples disclosed that the weld penetration ranged from less than 50 to 75% of the head thickness at the joint, rather than the 100% penetration required by the ASME Boiler and Pressure Vessel Code. At no location was there 100% weld penetration. Typical longitu-

Fig. 6 Both sides of the fracture surface from a failed steam accumulator
Section of the vessel is the upper piece, and the mating head is the lower piece.

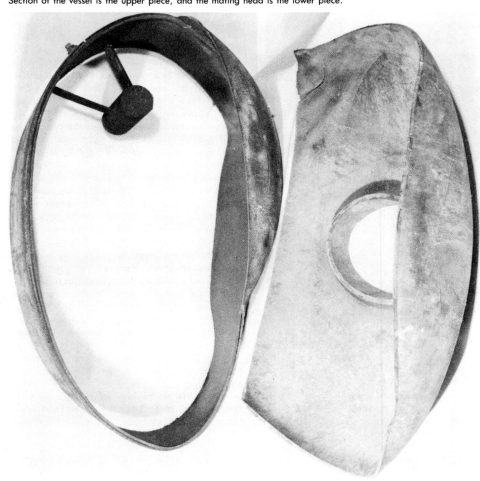

Fig. 7 Several sections through a girth weld in a failed steam accumulator showing lack of weld penetration

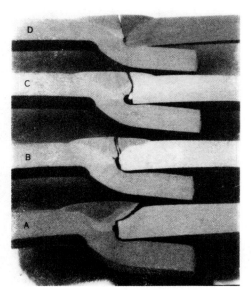

Fig. 8 Rimmed steel tube that failed by brittle fracture after being strain aged by cold swaging

dinal cross sections through the weld are shown in Fig. 7. The joint had apparently been made in two passes, but the root pass was not always centered over the joint. There was also some misalignment of the head to the shell because of radial displacement of the shell and head centerlines. This resulted in excessive clearances between the two parts and a slight difference in the thicknesses of the parts.

A hardness traverse on the shell yielded values from 84.1 to 86.5 HRB, corresponding to an approximate tensile strength of 550 MPa (80 ksi), which is within the 515- to 620-MPa (75- to 90-ksi) range required for SA455 material. The hardness of the head material ranged from 94.2 to 95.7 HRB, which indicated a tensile strength of approximately 695 MPa (101 ksi), which exceeds the 480- to 585-MPa (70- to 85-ksi) range for SA515-70 material. This higher hardness was probably due to cold forming of the head flange. The weld-metal hardness ranged from 84.8 HRB at the root pass to 99.3 HRB at the finish-pass crown—acceptable values for this material.

There were no anomalies in the microstructure of either the weld metal or the base metal.

The fracture path was clearly transgranular, with occasional secondary branches. Some small, secondary cracks were also observed. There were relatively few beach marks on the fracture surface, indicating the occurrence of only a few loading cycles. In all cases, the fracture path extended from the root of the weld through the weld metal or HAZ (up to the point at which the final-fracture path was in the head).

A stress analysis of the joint was performed, taking into account the various degrees of weld penetration and misalignment, as well as the differences in wall thickness of the shell and head. For proper alignment and complete weld penetration, the stresses in the failure region are quite low—about 25 MPa (3.57 ksi). Lack of penetration can cause the longitudinal stress to approach the yield strength of the material, and a small amount of misalignment added to lack of penetration increases the stresses to near the tensile strength of the material.

Conclusions. It was concluded that the failure could be attributed to two fabrication errors: lack of weld penetration and misalignment of mated parts. The failure was judged

to be a short-cycle high-stress notch-fatigue failure.

Heat treatment may be required subsequent to a fabricating operation. Deviations from specifications are potential causes of failure. Figure 8 shows a section of tubing made of rimmed steel that had been cold swaged from 65 to 50 mm (2½ to 2 in.) in diameter. Stress relief at 620 °C (1150 °F) after swaging was specified but was inadvertently omitted, and when the tube was expanded in a hole in a tube sheet, it broke in a brittle manner. The fracture was the result of strain aging of the steel, which can occur at room temperature and markedly decreases formability.

Repeated heat treatments, although performed at the proper temperature, may produce undesirable microstructures that can result in premature failure. For example, a stainless steel tube was solution heat treated, straightened, then reheat treated. The light surface deformation between heat treatments caused severe localized grain growth, which resulted in a coarse and less ductile grain structure. In service, grain-boundary cracking developed with relatively little swelling. Generally, this condition is harmful only when the tubing material has marginal strength or is stressed to an extent that results in measurable deformation.

An instance of improper heat treatment involved a pressure vessel made from forged HY-100 high-strength alloy steel plates that were subsequently heat treated. Heat-treating specifications called for water quenching, with the vessel being lowered into the bath in the vertical position to develop uniform properties. After heat treating, six slots were machined through the vessel shell to serve as seats for wedges designed to retain a removable closure. The plugs cut from the slots in the vessel wall were subjected to mechanical tests, which showed large variations in tensile strength and impact properties depending on circumferential location in the shell. In some locations, the values met specification; in other locations, they were considerably below specification. Investigation revealed that the vessel had been quenched in the horizontal position instead of the vertical position because there was not enough headroom over the available quench tank to permit lowering the vessel into the tank in the vertical position. The vessel was reheated and quenched in the vertical position to develop the specified properties.

An instance in which heat treatment was required after forming involved a gas-plant debutanizer tower made of 30-mm (1¼-in.) thick high-strength steel plate with properties as specified by ASTM A516, which permits, as an option, a refined normalized microstructure. Radiographic inspection of the entire vessel disclosed no imperfections. The vessel was designed to operate at 1550 kPa (225 psi) and 345 °C (650 °F). The vessel was hydrostatically tested using water at a temperature of not less than 15 °C (60 °F). When the test pressure reached 2760 kPa (400 psi), the vessel ruptured, and about one-third of the head was torn out. Chemical analysis and tensile testing showed that the composition and tensile properties of the steel complied with ASTM A516. Charpy impact tests made from 22 to 80 °C (72 to 175 °F) indicated that the ductile-to-brittle transition temperature was above 40 °C (100 °F). Metallographic examination showed a microstructure with a grain size coarser than ASTM 1, which indicated that the head had not been normalized after forming. The coarse grain size contributed to a reduction in fracture toughness of the steel. The vessel was completely checked ultrasonically for other plates with large grains, and the head was replaced. The vessel was hydrostatically tested again, then returned to service. No further difficulties were encountered.

Overheating during fabrication can result in cracking of susceptible materials. Leakage was detected in the U-bend of a 50-mm (2-in.) diam stainless steel tube during preoperational hydrostatic testing. The tube had been heated

Fig. 9 Fire-extinguisher case that failed because of ferrite streaks resulting from overheating during spinning

(a) Top of the case. Dimensions given in inches. (b) Micrograph showing ferrite streaks. 150×

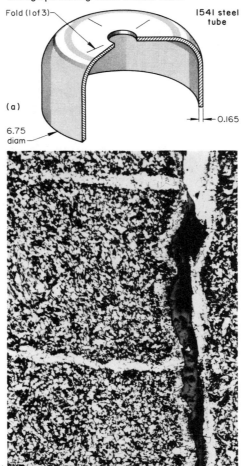

for bending, with a specified maximum temperature of 980 °C (1800 °F). Metallographic examination of a section through the leakage area showed evidence of partial melting, which indicated that the specified forming temperature had been considerably exceeded.

Example 5: Cracking of a Fire-Extinguisher Case Because of Overheating. The top of a fire-extinguisher case (Fig. 9a) was closed by spinning. The case was made from 1541 steel tubing 170 mm (6.75 in.) in diameter and 4.2 mm (0.165 in.) in wall thickness. Leakage from the top of the case was observed during testing, and a 75-mm (3-in.) long specimen was sent to the laboratory for analysis to determine the cause.

Investigation. Visual examination of the specimen did not reveal any surface cracks. However, three small folds were observed on the surface, and one was sectioned for microscopic examination. A very fine transverse fissure through the section was visible. Examination of a micrograph (Fig. 9b) showed streaks of ferrite, which can be formed when steel is heated approximately to the melting point. Examination of specimens from other areas of the case exhibited a similar structure. The normal microstructure of 1541 steel for the specified heat treatment was annealed ferrite and pearlite.

The chemical composition was normal for 1541 steel. No steelmaking or tubemaking defects were found in the specimen.

Conclusions. Cracking of the top of the fire-extinguisher case was the result of ferrite streaks that were formed when the metal was overheated. The initial forming temperature of the case plus frictional heat generated during spinning resulted in extreme overheating and some localized melting.

Recommendations. The temperature of the metal must be more closely controlled so that the spinning operation is done at a lower temperature, thus avoiding incipient melting and formation of ferrite streaks.

Example 6: Failure of Piping System Cross by Intergranular Cracking Traceable to Improper Heat Treatment (Ref 2). During a routine maintenance procedure at a power station, dye-penetrant examination revealed linear indications on the outer surface of a cross in a piping system. The cross was specified as SA403 type WP 304 stainless steel that had originally been extruded as a 65-cm (26-in.) diam, 75-mm (3-in.) thick pipe section. The material had been bright dipped, sand blasted, dye penetrant inspected, and passivated by the manufacturer. A spare cross from a sister power station was examined and found to have similar indications. Both crosses had been subjected to an induction-heating stress improvement—induction heating the outer wall of the pipe to 550 to 600 °C (1020 to 1110 °F) maximum for a maximum of 12 min while the inside is cooled with water. This treatment leaves a residual tensile stress on the outer surface and a residual compressive stress on the inner surface of the pipe.

Investigation. The linear indications on the cross in service were located in 150- to 180-mm (6- to 7-in.) wide bands running circumferentially below the cross-to-cap weld and above the cap-to-discharge-pipe weld. The indications were random in orientation, measured up to 20 mm (0.75 in.) in length (excluding one very large indication that measured 90 mm, or 3.5 in., in length), and averaged about 6 mm (0.25 in.) in depth. The indications were predominantly located in the thicker wall sections of the cross.

The available specimen of the unused cross was visually inspected. It consisted of about a 180-mm (7-in.) square that had a maximum thickness of 60 mm (2.37 in.). It included a weld-preparation area that was about 16 mm (0.625 in.) thick. One side of the specimen had been flame cut, while the other had been sawed. The outside surface appeared to be coarse ground and had cracks visible on the surface. Record photographs were taken.

A dye-penetrant examination was performed using a two-step penetrant method. There were numerous indications on the flame-cut side; none were on the saw-cut side of the specimen. The longest single indication was approximately 19 mm (0.75 in.), while the average length was about one-half that. The cracks appeared to be intergranular.

The pipe was chemically analyzed and found to conform to SA182 type F 304 stainless steel specifications. It contained 0.2% C, 1.56% Mn, <0.005% O, 0.016% S, 0.66% Si, 18.0% Cr, 9.6% Ni, and 0.12% Mo.

Knoop hardness measurements were made with a 300-g load on a cross section of the material. The minimum hardness for 28 readings was 215, the maximum hardness was 298, and the average was 267, corresponding to 23.1 HRC. This hardness corresponds to a tensile strength of 807 MPa (117 ksi), and the material should therefore meet the 517-MPa (75-ksi) tensile-strength requirement.

A cross section was ground, polished, lightly etched in 10% oxalic acid, and examined optically. Intergranular cracks filled with oxide were observed to have depths up to about 8 mm (0.3 in.); one crack appeared to be transgranular when viewed with an optical (light) microscope. The grain size of the material exceeded the 00 ASTM grain size defined by ASTM Standard E 112 (Ref 3).

To determine whether the material had been sensitized by the induction-heating stress-improvement treatment, the oxalic acid etch test, Practice A of ASTM A 262 (Ref 4), was implemented. The polished specimen was etched at a current density of 1 A/cm² (6.5 A/in.²) for 1.5 min in 10% oxalic acid-demineralized water, using a piece of type 304 stainless steel as the anode. There was no evidence of continuous grain-boundary precipitates typical of sensitized material. The cross section was repolished and etched electrolytically with NaOH to search for other second phases, such as σ and χ; no evidence of these phases was found.

Three mechanical test specimens with cracks were cut from the specimen and notched. The specimens were soaked in liquid nitrogen and broken. The large grain size of the material was apparent on the intergranular fracture surface (Fig. 10). A rust-colored oxide film was visible on the part of the fracture surface that corresponded to the cracked volume of material. At higher magnification, evidence of some plastic deformation before fracture was found. Energy-dispersive spectrometry (EDS) scans of an oxidized area of the fracture revealed traces of sulfur and chlorine, probably from the dye penetrant, and the iron, chromium, and nickel

Fig. 10 Fracture surface of mechanical test specimen from piping cross

Fracture is intergranular. The coarse grain size of the material is evident. Note 0.75-in. scale.

normally present. At another location, traces of potassium and titanium were found. Analyses of other areas did not disclose definitive evidence of contaminants to support SCC as the failure mechanism.

The possibility of intergranular and hot cracking during welding is high for materials containing high δ-ferrite. Ferritescope measurements on this material yielded results in the range from 0.3 to 1.6% δ-ferrite, too low for this mechanism to have been operative.

Conclusions. The results of the analysis eliminated intergranular SCC; the material was not in a sensitized condition, and no definitive evidence of contaminants (other than the dye-check fluid) was found. No detrimental second phases were found, nor was there significant ferrite to cause hot cracking during welding. The remaining possibility was overheating or burning of the forging, which classically results in large grain size, intergranular fractures, and fine oxide particles dispersed throughout the grains. The induction-heating stress-improvement process was suspected to have induced tensile stresses into the outside of the cross that opened pre-existing cracks so that they were found by dye penetrant.

Multiple Causes. Extensive examination of a failed vessel sometimes reveals that failure was caused not by one but by several interrelated inadequacies in fabrication, as in the following example.

Example 7: Failure of a Thick-Wall Alloy Steel Pressure Vessel Caused by Cracks in Weld HAZ (Ref 5). A large pressure vessel (Fig. 11a) designed for use in an ammonia plant at a pressure of 35 MPa (5.1 ksi)

at 120 °C (250 °F) failed at 34 MPa (5.0 ksi) during hydrostatic testing. The intended maximum test pressure was 48 MPa (6.95 ksi). The vessel weighed 166 Mg (183½ ton), measured 18.2 m (59 ft, 8⁹⁄₁₆ in.) in overall length, and had an outside diameter of 2.0 m (6 ft, 6¾ in.). It was fabricated from ten manganese-chromium-nickel-molybdenum-vanadium steel plates 150 mm (5⅞ in.) thick, which were rolled and welded to form ten cylindrical shell sections and three forgings of similar composition. The forgings formed two end closures and a flange for attaching one of the forged end closures to the vessel.

The failure did extensive damage to one end forging and three adjacent shell sections. Four large pieces were blown from the vessel; the largest, weighing about 2 Mg (2¼ ton), penetrated the shop wall and traveled a total distance of 46 m (152 ft). Pieces of the failed vessel are shown in Fig. 11(b).

Fabrication History. The cylindrical shell sections were hot formed so that the rolling direction of the plates was perpendicular to the axis of the vessel. All plates were supplied in the normalized-and-tempered condition. The forgings were annealed, normalized, and tempered at 645 °C (1195 °F) to obtain the desired mechanical properties.

The longitudinal seams in the cylindrical shells were electroslag welded, and the welds were ground to match the curvature of the shell. The cylindrical shell sections were heated at 900 to 950 °C (1650 to 1740 °F) for 4 h, rounded in the rolls within that temperature range, and then cooled in still air for examination of the seams.

Circumferential welding was done by the submerged arc process, using preheating at 200 °C (390 °F). Each subassembly was stress relieved by being heated to 620 to 660 °C (1150 to 1220 °F) for 6 h.

Final joining of the three subassemblies followed the same welding procedures except that localized heating for stress relief was employed. During various stages of manufacture, all seams were examined by gamma radiography, automatic and manual ultrasonic testing, and magnetic-particle inspection.

Hydraulic Testing. The vessel was closed at the top, and water was admitted to the vessel through a fitting in the cover plate until the vent that was within 3 mm (⅛ in.) of the inside surface at the top showed a full-bore discharge of water, at which time it was blanked off. The specified hydrostatic proof test called for a pressure of 48 MPa (6.95 ksi) at an ambient temperature of not less than 7 °C (45 °F).

At 34 MPa (5.0 ksi), the customary halt was made, and about 30 s later, the flange end of the vessel exploded without warning. The flange forging was found to be cracked completely through in two locations, and the first two cylindrical sections were totally shattered.

Metallurgical Examination. The fracture surfaces were typical for brittle steel fracture.

Fig. 11 Large thick-wall pressure vessel that failed because of cracking in weld HAZ

(a) Configuration and dimensions (given in inches). (b) Shattered vessel. (c) General appearance of one fracture surface; arrow points to facet at fracture origin. (d) Enlarged view of region at arrow in (c). 5.5×. (e) Charpy V-notch impact-strength values of specimens from weld area, both as-fractured and after retempering at 650 °C (1200 °F) for 6 h. Source: Ref 5

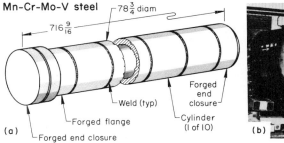

Mn-Cr-Mo-V steel

(a)

(b)

(c)

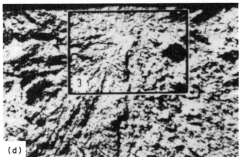

(d)

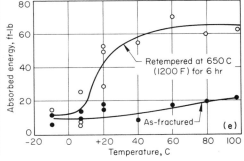

(e)

There were two points of origin, both associated with the circumferential weld that joined the flange forging to the first shell section.

The general appearance of one fracture surface of the flange forging is shown in Fig. 11(c). The arrow near the center indicates a flat facet (with major dimensions of roughly 9 mm, or 3/8 in.) approximately 14 mm (9/16 in.) below the outer surface of the vessel, which was partly in the HAZ on the forging side of the circumferential seam weld. Figure 11(d) shows an enlarged view of the region containing this facet. A slightly larger facet was found on another fracture surface of the flange, near a fracture origin 11 mm (7/16 in.) below the outer surface. This facet was also situated partly on the forging side of the HAZ. These flat, featureless facets in the HAZ were the fracture-initiation sites.

Metallographic examination revealed that the structure immediately below each facet in the HAZ was a mixture of bainite and austenite having a hardness (1-kg load) of 426 to 460 HV. Elsewhere, the structure of the HAZ was coarse and jagged bainite with a hardness of 313 to 363 HV.

Examination of a section transverse to the weld disclosed that the structure of the flange forging contained pronounced banding, whereas the plate did not. The structure between the bands (and away from the weld) consisted of ferrite and pearlite with a hardness of 180 to 200 HV (10-kg load). In the bands, the structure was upper bainite with a hardness of 251 to 265 HV. Specimens cut from this area were austenitized at 950 °C (1740 °F) and quenched in a 10% solution of NaOH in water to produce pure martensite. A Vickers hardness

traverse (1-kg load) across a band showed average values of 507 on one side of the band, 549 within the band, and 488 on the other side. A microprobe scan across the banded area showed the following differences in chemical composition: 1.56% Mn outside the band, 1.94% within it; 0.70% Cr outside, 0.81% within; and 0.23% Mo outside, 0.35% within. These differences indicated higher hardenability within the bands, which suggested a greater susceptibility to cracking, especially where the bands met the HAZ of the circumferential weld.

The hard spots within the forging suggested that during stress relief the vessel had not actually attained the specified temperature. To check this, specimens from the HAZ were accordingly heated to a variety of tempering temperatures to observe at what point softening would occur. Vickers hardness tests (1-kg load) on hard-spot specimens showed no deviation from the as-failed hardness scatter band for temperatures up to 550 °C (1020 °F), but gave definitely lower values after retempering at or above 600 °C (1110 °F). A similar behavior occurred in specimens taken outside the hard spots. It was therefore concluded that stress relief had not been performed at the specified temperature level.

Other studies of microstructure established that the mode of crack propagation was transgranular cleavage, which occurred not only in prior-austenite grains but also in ferrite grains and pearlite colonies. Branching and subsidiary cracks were also found, all very close to the main fracture surface. The distance from the fracture to the crack farthest away from it was 0.5 mm (0.020 in.).

To further investigate the effect of the stress relief on the properties of the vessel, standard Charpy V-notch specimens were prepared from the flange forging, the plate, and the weld. All specimens were oriented with their length parallel to the circumference of the vessel so that the induced fractures would be parallel to the main fracture. Tests were performed on as-received vessel material over the temperature range of −10 to 100 °C (15 to 212 °F), yielding impact values that met the specifications for the forging and plate, but the weld metal exhibited inferior impact strength (lower curve, Fig. 11e). None of the materials displayed a sufficient capacity for energy absorption (41 J, or 30 ft · lb, for the forging; 69 J, or 51 ft · lb, for the plate; and 16 J, or 12 ft · lb, for the weld) to have arrested the crack propagation that led to failure of the pressure vessel. Other specimens of the weld metal, which were retempered at 650 °C (1200 °F) for 6 h and tested over the same temperature range as those discussed above, showed a marked improvement in impact strength at or above 20 °C (70 °F) (upper curve, Fig. 11e). This confirmed the previous deduction that the stress relief of the vessel was at a lower temperature than the specified temperature.

Conclusions. Failure of the pressure vessel stemmed from the formation of transverse fabrication cracks in the HAZ of the circumferential weld joining the flange forging to the first shell section. The occurrence of the cracks was fostered by the presence of bands of alloying-element segregation in the forging, which had created hard spots, particularly where they met the HAZ. Stress relief of the vessel had been inadequate, leaving residual stresses and hard spots and providing low notch ductility, especially in the weld.

Recommendations. The final normalizing temperature should be $Ac_3 + 50$ °C ($Ac_3 + 90$ °F), and the tempering temperature should be 650 °C (1200 °F), where Ac_3 represents the temperature at which transformation of ferrite to austenite is complete during heating. After completion of each seam weld, the weld preheat should be continued for a short distance along the seam. The critical flaw size for this alloy and section thickness should be determined by fracture mechanics, and the capacity of available nondestructive testing methods for detecting flaws of this size should be assessed. The possible benefits of a more effective solution treatment or a revised forging practice should be explored to eliminate the banding in the flange forging.

Example 8: Failure of a Reheat Steam Piping Line at a Power-Generating Station. A 75-cm (30-in.) OD by 33-mm (1.31-in.) thick pipe in a horizontal section of a hot steam reheat line ruptured explosively with serious damage and loss of life after approximately 15 years of service. The line nominally operated at 4135 kPa (600 psi) and 540 °C (1000 °F). The section that failed was manufactured from rolled plate of material specification SA387, grade C. The longitudinal seam weld was a double butt weld that was V welded from both sides. The failure propagated along the longitudinal seam and its HAZ for a distance of 2.7 m (9 ft.) in both directions from the point of origin, creating a clam-shell opening 2.1 m (7 ft) wide. On one end, the failure traveled through a circumferential seam that deflected its path and stopped about 150 mm (6 in.) past the weld. On the other end, the fracture stopped several feet short of a circumferential seam while traveling in the longitudinal seam HAZ. The fracture in the pipe was pointed directly toward occupied regions of the plant.

Investigation. A systematic review of the circumstances of the failure revealed a number of factors that may have played a role in the failure. In the design of the system, the original material specified was ASTM A155 fusion-welded pipe that had been substituted with SA387, grade C, plate, rolled and welded with filler metal added. At the time of the design of the system, the allowable design stress for SA387, grade C, material was 48 MPa (6.9 ksi) at 540 °C (1000 °F). Approximately 3 years after the plant was put into service, this was reduced to 46 MPa (6.6 ksi) at 540 °C (1000 °F).

The hydrostatic test pressure specified for the system was sufficiently high that, if it was

actually applied, it is possible that the pipe cylinder and its weldment were overstressed, perhaps to the point of being damaged. The actual test pressure could not be determined in the original investigation. Two other unconfirmed possibilities are operation of the pipe at higher-than-anticipated service temperatures for short times during its service life and earthquake loadings acting on the pipeline.

Several metallurgical factors were found during a laboratory analysis of samples from the failed pipe. However, two key pieces, the fracture area containing the fracture-initiation point and the end of the crack in the longitudinal seam weld, were not available for this analysis. Three samples containing parts of the fracture surface were received: a 165- × 285-mm (6.5- × 11.25-in.) section near the end of the fracture in the longitudinal seam weld as well as 205- × 405-mm (8- × 16-in.) and 180- × 330-mm (7- × 13-in.) sections from either side of the point of origin of the fracture. One 150- × 305-mm (6- × 12-in.) section, including the undamaged longitudinal weld and a circumferential weld to a fitting, was also received. These samples were removed by flame cutting, and care was taken to avoid the heat-affected cut zone in all testing.

The fracture surface near the inner wall of the pipe had a bluish gray appearance, while the fracture surface near the outer wall was rust colored. Two of the samples had a distinct gray color halfway through the thickness of the reheat pipe. Before sectioning, all fracture surfaces were coated with clear acrylic, which was later removed with acetone for optical microscope examination. The rust colored oxide could not be cleaned off, while the bluish gray coating appeared thick at magnifications of 10 to 70×. It was very brittle with a mud-crack appearance, and appeared to consist of a mixture of oxides.

The base metal, which was analyzed in three locations, contained:

Element	Composition, %
Carbon	0.14
Manganese	0.51-0.54
Phosphorus	0.016-0.018
Sulfur	0.011-0.015
Silicon	0.54-0.57
Copper	0.023-0.028
Tin	0.003-0.006
Nickel	0.022-0.024
Chromium	1.39-1.43
Molybdenum	0.50-0.54
Aluminum	0.000
Vanadium	0.005-0.006
Niobium	0.006-0.011
Zirconium	0.001-0.002
Titanium	0.000
Boron	0.0001-0.0003
Cobalt	0.004-0.006

The weld filler metal, including both inside-diameter and outside-diameter samples, contained:

Element	Composition, %
Carbon	0.096-0.10
Manganese	0.81-0.91
Phosphorus	0.015-0.018
Sulfur	0.013-0.015
Silicon	0.47-0.54
Copper	0.13-0.40
Tin	0.005-0.013
Nickel	0.10-0.12
Chromium	1.28-1.43
Molybdenum	0.50-0.54
Aluminum	0.003-0.007
Vanadium	0.007-0.008
Niobium	0.009-0.013
Zirconium	0.001-0.002
Titanium	0.000
Boron	0.0003-0.0005
Cobalt	0.007-0.009

These contents conform to the requirements of SA387-67, grade B and C, as well as SA387-82, grade 11. The weld-metal analyses were from the centers of the welds and may differ from other areas of the weld; the weld metal and base metals were thought to be compatible.

Each of the samples was polished and etched for metallographic examination in the planes normal to the longitudinal and/or circumferential welds. Cracks were found in and around the root-pass areas of the three samples containing the longitudinal seam weld. There were several secondary cracks on the fracture side and a short internal crack in one specimen (Fig. 12a).

The sample containing the circumferential weld was examined using radiography and was found to be acceptable. Radiography of the samples containing the longitudinal seam weld detected only one sample with a crack, which ran parallel to the weld seam.

Longitudinal and transverse tensile specimens were cut from the pipe, and transverse-to-the-weld specimens were cut from the longitudinal and circumferential welds from the undamaged specimen. All room-temperature tensile properties were within specification: tensile strengths ranged from 476 to 517 MPa (69 to 75 ksi), yield strengths ranged from 265 to 305 MPa (38.5 to 44.3 ksi), and elongations ranged from 24 to 27%, except for the transverse-to-the-weld specimen from the longitudinal seam weld, which had a value of 19% (compared with 22% from the specification). That sample has 34% reduction of area and a shear-type failure rather than the cup-cone failures in the 63 to 70% reduction-of-area range of the other samples. It was concluded that the welded longitudinal seam exhibited embrittlement.

A Rockwell hardness traverse across the longitudinal weld from the undamaged region did not disclose a systematic variation in hardness across the weldment; hardnesses from 72 to 75 HRB were reported. A similar traverse across the circumferential weld revealed higher hardness in the weld metal and HAZ, reaching a value of 87 HRB. Hardnesses along the

Fig. 12(a) Fracture surface of reheat steam pipe showing corrosion products covering early-fracture region and freshly exposed fracture surface of weld metal

Fig. 12(b) Close-up of weld metal showing intergranular cracks and appearance of brittle failure

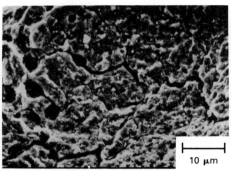

Fig. 12(c) Fracture surface near root pass showing corrosion products on part of fracture surface

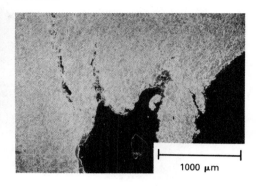

Fig. 12(d) Microstructure of longitudinal weld metal near fracture-initiation point

Note the white phase along grain boundaries.

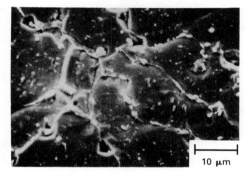

fracture edge of a longitudinal weld sample ranged from 70 to 80 HRB.

Several sections were cut from the fracture-surface samples and prepared for both optical microscopy and SEM. The rust spots were removed with phosphoric acid, which also removed or attacked some of the other corrosion products, changing their color from bluish gray to black. The microstructure of the material is predominantly an equiaxed ferritic matrix, with corrosion pits and grain-boundary precipitates. The fracture surface in the vicinity of the origin of fracture, near the outside diameter, is mainly ductile dimples, but corrosion pits and cracks are also present; by contrast, the tensile sample from the undamaged region had only ductile dimples. The pictures taken of the fracture using a scanning electron microscope (Fig. 12a to d) were consistent with a low-ductility intergranular fracture that progressed through the weld metal and had been in existence long enough to allow the formation of corrosion products on the fracture surface. Energy-dispersive x-ray analyses (EDAX) of the corrosion products showed them to be high in iron, with silicon, phosphorus, calcium, and manganese in small amounts, together with light elements, such as oxygen or carbon. Examination of the structure of the longitudinal weld near the fracture-initiation point revealed the presence of a grain-boundary phase (Fig. 12d), whose EDAX spectrum was high in silicon, sulfur, chromium, manganese, and iron.

Conclusions. The metallurgical analyst concluded that low ductility was responsible for the original initiation of cracks in the pipe. The low ductility may have been due to the presence of carbides and sulfides in grain boundaries, and corrosion may have played a major role in the deterioration of the steel in the weld-joint area. The extent and amount of corrosion products indicated that the cracks were in existence for some time before the final failure. Further work was deemed desirable to confirm these find-

ings. It was not clear which of the design and construction, operating, and metallurgical factors played the dominant role in this failure, and further investigation of these factors may be required.

Example 9: Shell and Head Cracking in Gray Cast Iron Paper Machine Dryer Rolls. Unlike the individual examples of failures discussed throughout this article, several failures of dryer rolls will be addressed in this example. Most of these gray iron structures were cast in the 1890-1920 time frame when they were not covered by pressure vessel codes. The rolls are typically 0.3 to 0.9 m (1 to 3 ft) in diameter and range to about 3.7 m (12 ft) in length, with ground outer cylindrical surfaces on which the paper web is dried. They are unique pressure vessels in that they rotate about their longitudinal axes at speeds from 50 to 250 rpm while containing saturated steam from 35 to 380 kPa (5 to 55 psi). During its service life, a roll may be subjected to a variety of unrecorded operating conditions. Repair, regrinding, and wear in service all change the dimensions and stresses applied under given loads. Several constructions are found. The heads may be cast integrally with the cylindrical bodies or cast separately and bolted to the shells. The wall thicknesses of the shells, and especially those of the heads, may vary; artistically cast company names, logos, and raised reinforcements are frequently found.

Service failures typically occur in the shell body, in a head near a hand hole or a manhole opening, or in a head near the journal-to-head interface. Examples of these failures are shown in Fig. 13 and 14. Detailed fractographic analyses using SEM techniques invariably show a cleavage fracture, regardless of the driving stress for failure. Fracture surfaces may exhibit the chevron marks typical of fatigue or may have raised points or tears pointing in the direction of the probable origin of failure. Heavy rust, patches of oil, or other contaminants can indicate the existence of cracks older than the final failure.

When analyzing one of these failures, the analyst should be aware of the following characteristics of thin-wall cast iron structures:

- Castings are frequently poured from multiple ladles; therefore, the chemical composition and composition-dependent properties can vary widely within a given roll
- The cooling rates and solidification directions can vary dramatically with locations, producing anisotropic structures with variations in tensile strength up to 20% in different principle directions. The shapes and orientations of inclusions and other defects also vary widely. A single orientation of samples for metallographic analysis is not generally sufficient to characterize a microstructure
- The combinations of these factors can lead to tensile-strength variations up to 20% within a shell (excluding anisotropy effects)

For any specific failure, an immediate cause may be determinable, either in the material or in the recent service history. However, the analyst should be wary of the long histories of these vessels. Owners and operators of these

Fig. 13 Failures of gray cast iron paper-roll driers
(a) Axial-shell failure. (b) Circumferential-shell failure

(a)

(b)

vessels are concerned with developing end-of-life criteria and inspection methods for these and similar older pressure vessels.

Inspection Practices. The lower limits of detectability of various nondestructive inspection tools are of major importance in failure prevention. The effectiveness of radiographic, ultrasonic, and eddy current techniques depends largely on size and orientation of the discontinuity in the metal.

Touching the surface of a pressure vessel with an energized coil or probe used to magnetize a component for magnetic-particle inspection may produce an arc burn, which is a potential stress raiser.

The water used in hydrostatic proof testing and the liquid vehicles used in liquid-penetrant inspection can leave chloride residues that can later cause cracking unless the parts are carefully cleaned and dried after testing. For exam-

ple, a type 304L stainless steel pipeline used for transferring caustic materials was hydrostatically tested after fabrication and was found acceptable. However, after about 2 months of service, leaks were detected at several circumferential weld joints. The pipe had been installed in a horizontal position, and the leaks occurred predominantly at the bottoms of the joints (6 o'clock position), with a higher incidence of leakage at low points in the pipeline. Leakage occurred in both shop and field circumferential welds; none was observed in the longitudinal seam in the welded pipe.

Investigation revealed that the leaks had resulted from chloride pitting. The source of the chlorides was determined to be the water used in hydrostatic testing of the pipeline. After the hydrostatic test had been completed, the water, which originally contained less than 100 ppm chlorides, was drained from the system. It was determined that some water had collected at low points in the pipeline and at discontinuities resulting from the weld reinforcement on the pipe inner surface. The piping was allowed to sit in this condition for about 4 to 5 months before being used to transfer caustic. Evaporation and chloride concentration, followed by corrosion and leakage, could have been eliminated if the pipeline had been completely drained after hydrostatic testing.

Pressure Vessels Made of Composite Materials

Composite materials are used in pressure vessels to reduce cost and to conserve strategic material. Also, they provide vessels strong enough to resist deformation under pressure with surfaces that resist attack by the contained substance. Rigidity is usually provided by a carbon steel shell. A stainless steel or corrosion-resistant alloy lining is common.

Linings used for pressure vessels made of composite materials are applied by surface welding, roll cladding, plug welding, or explosive bonding. Surface-welded linings are produced by facing the interior carbon steel surface with an overlay of a corrosion-resistant alloy, using one of several processes for surface welding. These processes include manual shielded metal arc welding, automatic and semiautomatic submerged arc welding, and plasma arc welding.

Roll cladding is usually accomplished by one of two methods. In one method, stainless steel is placed on a carbon steel ingot and bonded to it by hot rolling. The other method is essentially the same except that a flux is placed between the two materials before hot rolling. Plug-welded linings are produced by securing roll-formed sections of a corrosion-resistant alloy to the interior surface of a carbon steel shell by plug welding.

Explosive bonding produces an excellent bond and an excellent product. The layer of cladding material is bonded to the base metal by

Fig. 14 Both sides of a journal-to-head failure in a cast iron paper-roll dryer

(a) Bottom of failed dryer. (b) The failed bolted-on head

(a)

(b)

a shock wave generated in a suitable medium by a controlled explosion.

Welding. When composite materials are joined by welding, special techniques are required, and metallurgical problems can arise. With roll-clad material, the overlay is machined away from the weld site, and the carbon steel shell is welded by the usual methods from the outside. The inside welds are made with stainless steel filler metals, such as ER308, ER309, ER310, or ER312. In plug weldings, similar techniques are implemented. Either 300 or 400 series stainless steel sheets may be used, and these sheets are welded together and to the carbon steel pressure vessel with stainless steel filler metal.

Metallurgical problems arise from the effects of dilution of the alloying constituents in the corrosion-resistant alloy coating by the carbon steel base metal. Fully austenitic stainless steel filler metals are susceptible to microfissuring during welding. A duplex structure of austenite and a small amount of ferrite in the weld deposit prevents microfissuring. Therefore, the ferrite content is usually kept between 5 and 9%. This duplex structure may be obtained by suitably balancing the chromium and nickel contents in the weld metal. Alloy dilution can be controlled by varying the welding technique and the electrode composition. Small dilutions can produce fully austenitic structures, but too much produces structures containing martensite.

Sigma phase is sometimes found in austenitic stainless steel weld metal. Although σ phase can occur in the as-welded condition, it is more commonly encountered after long exposures at temperatures of 595 to 870 °C (1100 to 1600 °F). Sigma phase may embrittle austenitic stainless steel at room temperature; at elevated temperatures, it can cause low ductility failures during creep deformation. Sigma phase is usually associated with ferrite; alloying elements, such as chromium, niobium, and molybdenum, which stabilize ferrite, also promote σ-phase formation.

Care must be used when welding Monel to carbon steel to avoid excessive dilution of the Monel with iron, such as in weld overlays. Excessive iron dilution embrittles the Monel and causes cracking. Iron dilution can be minimized by using a barrier layer of pure nickel weld deposit or by carefully controlling the welding procedure with the use of small-diameter electrodes and a stringer-bead technique.

Example 10: Cracking in Plug Welds That Joined a Stainless Steel Liner to a Carbon Steel Shell (Ref 6). Repeated cracking occurred in the welds joining the liner and shell of a fluid catalytic cracking unit that operated at a pressure of 140 kPa (20 psi) and a temperature of 480 °C (895 °F). The shell was made of ASTM A515 carbon steel welded with E7018 filler metal. The liner was made of type 405 stainless steel and was plug welded to the shell using ER309 and ER310 stainless steel filler metal.

Investigation. Examination of the microstructure in the plug-weld zone revealed a good duplex structure with what appeared to be fine cracks starting inside the weld zone and spreading outward through the weld and toward the surface (Fig. 15a). At higher magnifications, some of the fine cracks appeared to be precipitates (Fig. 15b). When cracking had completely penetrated the plug weld, corrosion of the carbon steel occurred, spreading radially. Also, there was decarburization and graphitization of the carbon steel at the interface.

The body of the weld had a hardness of 45 HRC. This, combined with the decarburization of the carbon steel, was assumed to be indicative of carbon migration out of the HAZ and into the weld metal, thus preventing the forma-

Fig. 15 Cracks in pressure vessel made of ASTM A515 carbon steel lined with type 405 stainless steel
Failure occurred at plug welds because of dilution of weld metal. (a) Micrograph of specimen through weld area etched in acid cupric chloride showing ASTM A515 carbon steel (top), interface region (center), and duplex weld structure containing fine cracks (bottom). 100×. (b) Micrograph of an area of the specimen in (a) showing cracks and precipitates. 250×. Source: Ref 6

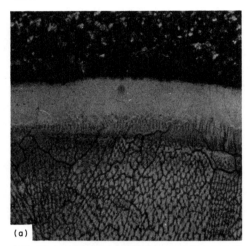
(a)

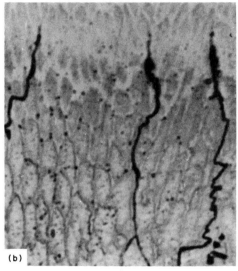
(b)

tion of martensite. Further into the weld metal, the high carbon level allowed martensite to form. In the area where the grain-boundary precipitates appeared heaviest, between the martensite band and the duplex structure, the structure was probably austenitic. These phase changes are predicted from the Schaeffler diagram (page 322 in Volume 6 of the 9th Edition of *Metals Handbook*).

The precipitates followed the austenitic regions into the duplex structure and eventually disappeared. They undoubtedly contributed to the cracking that occurred.

The composition of the precipitates was analyzed using an electron microprobe. Silicon was not present in significant amounts, nor was oxygen, except near a filled branch crack that also showed small amounts of carbon. Phosphorus was absent, and iron, chromium, and nickel were present in amounts commensurate with the weld material. Approximately 4% S was present in the filled branch crack, and the precipitates contained about 12% S.

Thus, several mechanisms were operative in cracking and damaging of the vessel, many of which were the result of microstructural changes in the weld alloy at the interface caused by dilution of the alloy. Austenite was formed at a low dilution level, and the presence of sulfur produced grain-boundary precipitation of sulfides, thus causing hot shortness. Further dilution produced a brittle martensitic phase, which is more susceptible to failure and is a likely source of cracking initiation. This phase was not present at the interface, probably because of carbon migration into the weld, as evidenced by the decarburization and graphitization of the base material. Finally, the differential thermal expansion of the carbon and stainless steels produced the necessary internal stress to produce extensive cracking. Once the carbon steel was exposed to the contents of the pressure vessel, corrosion occurred.

Conclusions. Cracking in the plug welds was the result of excessive dilution of the weld metal. Dilution of the weld metal permitted formation of a martensitic phase in which cracks initiated.

Corrective Measures. Periodic careful gouging of the affected areas followed by repair welding proved sufficient to keep the vessel in service. However, in vessels operated at higher pressures or containing more aggressive corrosive materials, such measures would not suffice, and careful selection of the filler metal and welding procedure would be necessary to prevent formation of the martensitic phase. A lower sulfur content would prevent the grain-boundary precipitate. Carbon migration would still occur, but in itself might not cause a major problem.

Stress-Corrosion Cracking (SCC)

Stress-corrosion cracking occurs when stress and a specific corrodent for the material are present in amounts large enough to cause failure by the combined action of both but too small to cause failure by either one acting alone. Highly branched transgranular cracks are characteristic of stress corrosion; under certain conditions, intergranular cracking occurs. Equipment that fails by SCC usually must be replaced because the extent of cracking is difficult to determine and because the cracks are difficult to repair by welding.

The source of the stress may be internal or external. Internal stress can occur on a microscopic scale—for example, the stresses induced by a martensitic transformation. Also, residual stresses produced during thermal cycling, weld-ing, and straightening or bending may be very important. Externally applied stresses arise from applied mechanical loads, vibrations, or pressure.

Effect of Dissolved Solids. Corrosion may arise from many sources, some of which appear to be innocuous or nonaggressive. Even relatively pure water can be corrosive or can accelerate corrosion. If any concentration mechanism is at work, very small amounts of dissolved solids can be disastrous.

Example 11: SCC of a Nuclear Steam-Generator Vessel at Low Concentrations of Chloride Ion (Ref 7, 8). A small leak was found in a nuclear steam-generator vessel constructed of 100-mm (4-in.) thick SA302, grade B, steel [0.25% C (max), 1.15-1.5% Mn, 0.035% O (max), 0.04% S (max), 0.15-0.4% Si, 0.45-0.6% Mo] during a scheduled shutdown. The leak originated in the circumferential closure weld joining the transition cone to the upper shell. The unit had provided approximately 3 years of effective full-power operation in the 6 years following its installation.

Investigation. Examination of the leak area disclosed a hole approximately 16 mm (0.6 in.) long and 5 mm (0.2 in.) wide. Further visual inspection, ultrasonic examination, and magnetic-particle examination of that closure weld, as well as others on similar steam generators at the plant, disclosed about 100 cracks associated with each closure weld, although the only through-crack was the leak. Figure 16 shows a macrophotograph of a polished-and-etched plug sample of the failure. The details are obscured by erosion damage from the escaping coolant, but the failure clearly passes through both weld and base metal.

The welds had been fabricated from the outside by the submerged arc process with a backing strip. The backing was back gouged off, and the weld was completed from the inside with E8018-C3 electrodes by the shielded metal arc process. The weld was continuously stress relieved for 12 h at 540 °C (1000 °F). Hardness traverses on the weldments suggested that the actual heat-treatment temperature may have been less than 540 °C (1000 °F), leaving relatively high residual stresses in the weldments. Metallographic analyses demonstrated that the base metal had a tempered martensitic structure consistent with A302, grade B, steel and that the weld metal had dendritic structures normal for both the shielded metal arc and submerged arc portions of the weld. The HAZ had a more acicular structure normally associated with untempered martensite.

Scanning electron microscopy of the failure surface of the plug sample revealed striations of the type normally associated with progressive or fatigue-type failures (Fig. 17), including beach marks that allowed tracing the origin of the fracture to the pits on the inner surface of the vessel. Examination of discolorations with EDS disclosed that copper deposits with zinc were also present.

Fig. 16 Cross section through leak in steam-generator wall
Crack extends across weld metal, base metal, and HAZ. Erosion obliterated much of the original crack detail.

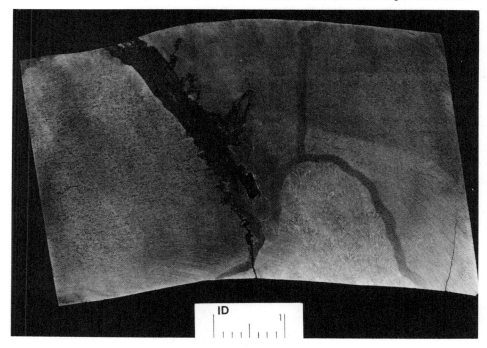

Fig. 17 Fracture morphology of steam-generator wall in a region where regularly spaced striations typical of fatigue are evident

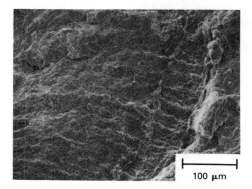

100 μm

Three samples were also taken from cracked regions of another steam generator. They exhibited similar features, including varying degrees of pitting at the origin of the cracks (Fig. 18). Examination of the leading edges of the cracks by EDS revealed traces of copper and silicon, as well as a few scans with zinc or nickel present. The area of initiation of one crack had iron, chromium, nickel, silicon, sulfur, copper, and zinc.

Several historical factors were found to influence the purity level of the coolant water in the steam generators. Brackish river water had corroded copper alloys in the condenser tubes, and some leakage into the system was indi-

cated. Both cuprous oxide (Cu_2O) and α-hematite were found in the sludge, indicating that oxygen control in the steam generators had been poor for some time and that hydrazine levels had been kept low due to environmental concerns. These factors were thought to contribute to pitting. In addition, a turbine-blade failure occurred and damaged about 50 condenser tubes, releasing sufficient chloride into

the steam generators that values up to 350 ppm were measured. This may have also played a role in pitting.

At this point, the analyst was uncertain whether crack propagation was due to corrosion fatigue or stress corrosion; therefore, a series of tests was performed on welded plate of the same specification grade that failed. A series of controlled crack-propagation-rate stress-corrosion tests was performed at temperatures ranging from 30 °C to 268 °C (85 to 514 °F) for flat tensile specimens in various heat-treatment conditions. These tests led to the conclusion that A302, grade B, steel is susceptible to transgranular stress-corrosion attack in constant extension rate testing with as low as 1 ppm chloride present (as cupric chloride) at 268 °C (514 °F). Welded SA302, grade B, in the stress-relieved condition is susceptible to stress-corrosion failure at stress levels of about 70% of the yield stress in 268-°C (514-°F) water containing 325 ppm of chloride (as cupric chloride). At lower concentrations of cupric chloride (1 and 5 ppm of chloride), there is a beneficial effect of stress relief on weldments in minimizing this kind of attack; however, the base metal is susceptible to SCC in this environment. A base-metal specimen that failed in water containing 1 ppm of chloride had a fracture morphology nearly identical to that of the failed steam generator.

Recommendations. To alleviate the possibility of additional cracking in the field, the coolant environment should be maintained low in oxygen and chloride, and the copper ions in solution should be eliminated or minimized.

Deaerating Feedwater Heater Failures. Recently, there have been several catastrophic failures and some through-wall leaks in deaerating feedwater heaters used to degas

Fig. 18 Fracture surface of a failed steam-generator sample
The fracture morphology is characteristic of fatigue cracking, but 1 ppm of Ce⁻ under constant extension rate testing produced this same fracture morphology in laboratory tests. The initiation site is on the inside surface of the steam generator and apparently emanates from surface pits.

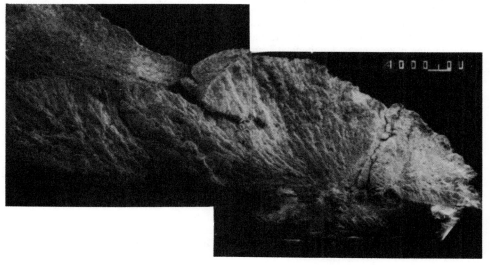

Fig. 19 Typical cracking found by fluorescent magnetic-particle inspection of the internal surface of a feedwater heater

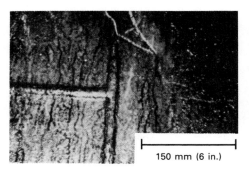

150 mm (6 in.)

Fig. 20 Typical micrographs of cracks in feedwater heater steels

(a) Cracks identified as corrosion fatigue mixed with SCC. 50×. (b) Corrosion-fatigue crack morphology alternating with corrosion pits and transgranular cracking. 100×

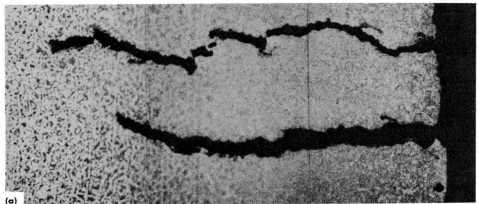

(a)

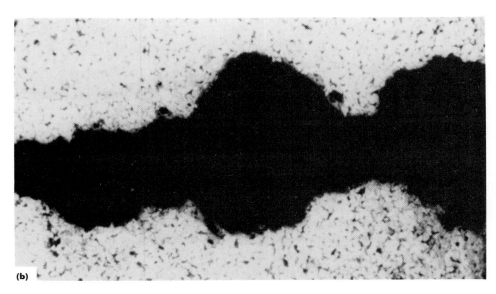

(b)

water for tube boiler installations operating at pressures exceeding 1035 kPa (150 psi). It has been estimated that up to 60% of the vessels that have been properly inspected in service have shell cracks. The deaerators operate at pressures ranging from subatmospheric to about 1380 kPa (200 psi). No general correlation has yet been established with such factors as vessel age, size, capacity, manufacturing techniques, service pressure, materials, and mode of operation.

The typical deaerating feedwater heater has both the deaerating zone vessel and the storage vessel made of low-carbon steel. For older vessels, SA212, grade A, B, and C, was commonly used, and for recent vessels, SA285, grade C, as well as all grades of SA515 and SA516 are frequently specified. The vessels, which are usually designed and fabricated in accordance with Section VIII, Division 1, of the ASME Boiler and Pressure Vessel Code, are usually of welded construction and are frequently thin enough that postweld heat treatment or stress relieving is not required.

Cracks in deaerating feedwater heaters have been found on internal surfaces at welds and in the HAZs of the welds; the cracks run both parallel and transverse to the welds. Head-to-shell circumferential welds, circumferential welds joining plate courses, longitudinal seam welds, and nozzle welds have all been found cracked. Failures have been examined *in situ* and in laboratories using both metallographically prepared samples (boat samples) and whole sections of plates. Figure 19 shows the types of cracks that are typically disclosed by field inspection of the insides of vessels. Wet fluorescent magnetic-particle examination is effective on a surface that has been brushed, ground, or blasted free of oxide and dirt without closing the surface indications or cracks. The typical cross section of a crack is shown in Fig. 20. The crack morphologies include both transgranular and intergranular modes of separation; both types of cracks are usually filled with corrosion products.

The mechanism of crack initiation and propagation has been classified by various authors as stress corrosion and corrosion fatigue. The exact mechanism of cracking has not been defined, nor have reliable methods been developed for preventing these failures. Current practice is to inspect the tanks periodically and repair or replace them as needed.

Caustic Embrittlement. The embrittling effects of mercury on brass and aluminum alloys are well known. Ammonia produces season cracking in cold-worked brass. Alkaline compounds attack high-strength aluminum alloys. Stainless steels are readily attacked by hot chlorides in aqueous solution, which even in very small amounts can cause severe damage to stressed austenitic stainless steels (see the section "Stress-Corrosion Cracking From Hot Chlorides" in this article). Such steels can also be attacked by concentrated sodium hydroxide (NaOH), as can low-carbon steels. Stress-corrosion cracking of Hastelloy C-276

has occurred in environments containing organic compounds and aluminum chloride ($AlCl_3$) at 100 °C (212 °F). No cracking occurred after the $AlCl_3$, present as an impurity, was removed.

Example 12: Failure of a Low-Carbon Steel Pressure Vessel From Caustic Embrittlement by Potassium Hydroxide. A large pressure vessel (Fig. 21a) that had been in service for about 10 years as a hydrogen sulfide (H_2S) absorber developed cracks and began leaking at a nozzle. The vessel contained a 20% aqueous solution of potassium hydroxide (KOH), potassium carbonate (K_2CO_3), and arsenic at 6550 kPa (950 psi) maximum and 33 °C (91 °F). The cylindrical portion of the vessel was 168 cm (66 in.) in inside diameter, 1020 cm (402 in.) in length, and 64 mm (2½ in.) in wall thickness. A few months before failure occurred, portions of the exterior surface of the vessel had been exposed to a fire.

Fig. 21 Large enclosed cylindrical pressure vessel that failed by SCC because of caustic embrittlement by potassium hydroxide

(a) View of vessel before failure and details of nozzle and tray support. Dimensions given in inches. (b) Micrograph showing corrosion pits at edge of fracture surface and inclusions. 600×. (c) Micrograph showing auxiliary cracks in tensile-overload region of main crack. (d) Micrograph showing auxiliary cracks in stress-corrosion zone of main crack. 100×

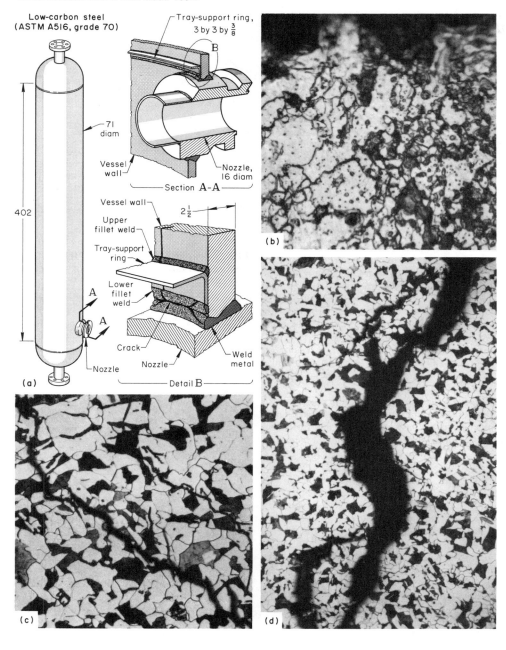

The vessel was examined at the job site and was found to be empty, but the inside surface was covered with a layer of sulfur. A sample was removed from near the end of the crack at the nozzle and was sent to the laboratory. The nozzle itself was not cracked. The vessel was returned to the fabricator and was stress relieved before repair work was begun.

The vessel wall, in which the fracture occurred, consisted of 64-mm (2½-in.) thick ASTM A516, grade 70, low-carbon steel plate. To provide a tray support, a steel angle, 75 × 75 × 10 mm (3 × 3 × ⅜ in.), had been formed into a ring and continuously welded to the inside wall of the vessel at the heel and the toe of one leg (Section A-A, Fig. 21a).

Investigation. In the groove formed by the junction of the lower tray-support weld and the top part of the weld around the nozzle was a crack about 55 mm (2¼ in.) long. Each end of the crack branched into a Y (Detail B, Fig. 21a). The crack portions in the weld metal around the nozzle were arrested, and the nozzle was not damaged.

The field sample submitted for laboratory analysis contained the remote end of one of several cracks that propagated into the upper tray-support weld. The surfaces of this crack were covered by a tightly adhering dark-brown and black scale. The surfaces contained two triangular regions that were smooth and that showed no evidence of ductility, one at the top of the weld and the other at the outer surface of the wall. The remainder of the fracture surface showed cleavage planes with some evidence of ductility, indicating that failure had resulted from tensile overload.

A dirty yellow scale was found under a piece of steel angle that was part of the field sample, but no scale was noticeable elsewhere after the angle had been removed by air carbon-arc gouging. No additional cracks were found in the welds for the tray support.

Tensile and impact tests performed on specimens from the sample indicated that the physical properties, after stress relief, were adequate for the service conditions.

Because of the proximity of the crack to the nozzle, which was saved for reuse, only fragments from the tray-support side of the crack were available. How representative these were of the original microstructure was a moot question because they had been annealed during removal. Microscopic examination revealed pits and scale near the crack origin; the pits contained corrosion products. Although difficult to characterize because of the small grain size, cracking was principally transgranular.

The surface of a section through a crack that started in a groove formed by the junction of the lower tray-support weld and the weld around the nozzle was examined under a microscope. The first 6 mm (0.25 in.) of the crack surface sloped at an angle of about 45° to the cut surface, was relatively smooth, had a thin orange-brown scale, and, at higher magnification, exhibited plastic flow at some points and corrosion pits at others. Figure 21(b) shows corrosion pits in the edge of the fracture surface as well as inclusions consisting of large dark orange-brown conglomerates and smaller gray globules that appear to have nucleated in the grain boundaries. These features suggested that periods of corrosion alternated with sudden instances of cleavage, under a tensile load, along preferred slip planes. Cracking in such a manner is typical of the stress-corrosion process, according to one theory.

After the first 6 mm (0.25 in.), the crack progressed abruptly straight downward into the weld and was filled with a thick gray scale. The crack surface was wavy and contained branching cracks filled with corrosion products. In this crack, corrosion had obviously predominated, with cleavage playing a lesser role or none at all. This condition is in accordance with an alternative, strictly electrochemical, theory on the mechanism of stress corrosion.

Micrographs of sections across the remote end of the crack in the field sample that propagated into the material showed both a zone that failed by tension overload (Fig. 21c) and a stress-corrosion zone (Fig. 21d). These two distinguishable fracture zones were confirmed by a gray corrosion product similar to that found near the crack origin, which was present in auxiliary cracks in the stress-corrosion zone but not in the tension-overload zone. Cracking in both zones was clearly transgranular.

Discussion. Because the surface of the steel angle adjacent to the vessel wall had been sealed and because only one point of seepage under it had been found, the crack at the grooved area between the lower weld joining the angle to the wall and the weld around the nozzle must have formed first. Although the vessel had been postweld heat treated before being put into service, residual stress was apparently sufficient in this grooved area for it to act as a notch or a stress raiser. With this configuration and enough time, the combination of the residual plus operating stresses and the amount of KOH present would have caused stress corrosion—in this case, as a result of caustic embrittlement.

The fire to which the vessel had been exposed before failure, the only evidence of which was observable bulging of the vessel skirt, could have been instrumental in initiating cracks by any or all of the following conditions affecting the critical grooved area: (1) an increase in temperature to above 120 °C (250 °F), (2) an increase in concentration of the caustic solution to greater than 50%, and (3) addition of thermal stresses to existing residual stresses. On the other hand, cracking may have been initiated simply by the combination of existing residual stresses and normal operating conditions over a long period of time.

Caustic embrittlement appeared most plausible as the cause of failure because of the presence of KOH and the known role of KOH in stress-corrosion failures in plain carbon steel vessels. However, the presence of arsenic can induce hydrogen embrittlement under alkaline conditions. The probability of sulfur, a lesser by-product of the removal of H_2S, inducing hydrogen embrittlement under alkaline conditions is slight. Hydrogen embrittlement from arsenic may therefore have been the added influence that helped produce transgranular corrosion rather than the simpler mechanism of intergranular corrosion.

Conclusion. Failure was caused by stress corrosion, which probably resulted from caustic embrittlement by KOH.

Recommendations. The tray support should be installed higher on the vessel wall to prevent coincidence of the lower tray-support weld with the nozzle weld, thus avoiding the formation of a notch.

Stress-Corrosion Cracking From Hot Chlorides. Equipment made of austenitic stainless steel may sustain SCC when contacted by hot chlorides in aqueous solution—even in concentrations of only a few parts per million. Therefore, chloride attack is generally insidious, and operating personnel may not be aware of such a problem until failure occurs.

Chlorides are commonly introduced into pressure vessels as dissolved salts in water or wet (saturated) steam. At temperatures below about 50 °C (120 °F), small quantities of chloride salts create few problems. At higher temperatures in systems containing austenitic stainless steels, chlorides can be disastrous. Water is most safely introduced to pressure vessels made of austenitic stainless steel in the form of either clean steam condensate or dry superheated steam. Generally, there is no problem as long as ample quantities of condensate or dry steam are available. Unfortunately, during refinery turnarounds, superheated steam can be in short supply because the heaters are shut down. As the various units are put back on stream, the boiler plant may lack sufficient capacity to meet all immediate needs. A related problem can occur in steam methane reformers during start-up. Unless special start-up procedures are carefully followed, steam introduced to the reformer may contain water droplets with entrained chloride salts (carryover). When the wet steam subsequently flashes against a hot surface farther down the system, a residue of chloride salts remains. If the surface is hot, SCC will occur quickly in some grades of austenitic stainless steel.

In one case, SCC caused failure of a type 347 stainless steel shaft in a hydrogen-bypass valve in a steam methane reformer plant. Chloride-salt residues were found on the fracture surfaces, together with highly branched cracks typical of stress-corrosion failure (Fig. 22). Wet steam introduced to the system during start-up operations was the source of the chlorides. The shaft had a hardness of 245 HB, indicating that it had been made from cold-finished bar stock. Undoubtedly, the higher surface residual stresses inherent in cold finishing made the shaft more sensitive to SCC. The fractured shaft was replaced with an annealed type 316 stainless steel shaft having a hardness of 169 HB. Also, changes were made in the reformer start-up procedure, minimizing the probability of introducing wet steam.

Example 13: SCC of an Inconel 600 Safe-End on a Reactor Nozzle. Cracking occurred in an ASME SB166 Inconel 600 safe-end forging on a nuclear-reactor coolant-water recirculation nozzle while it was in service. The safe-end was welded to a stainless steel-clad carbon steel nozzle and a type 316 stainless steel transition metal pipe segment (Fig. 23). An Inconel 600 thermal sleeve was welded to the safe-end, and a repair weld had obviously been made on the outside surface of the safe-end to correct a machining error. Initial visual examination of the safe-end disclosed

Fig. 22 Chloride SCC in a type 347 stainless steel shaft in a hydrogen-bypass valve

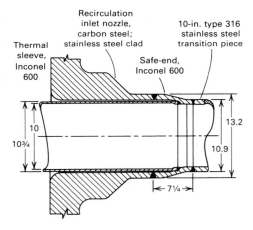

Fig. 23 Cross section through recirculation inlet nozzle of reactor vessel

Shown are the nozzle, the safe-end that failed, and the thermal sleeve that created susceptibility to crevice corrosion. Dimensions given in inches

that the cracking extended over about 85° of the circular circumference of the piece.

Investigation. After the safe-end had been cut from the reactor, it was radiographically inspected on-site to confirm the extent of cracking. This inspection indicated major cracking over about 280° of the circumference. Limited ultrasonic inspection performed in the laboratory with a 1.5-MHz transducer, directing the incident sound beam from the large end of the safe-end toward the small end, demonstrated that the cracking could be detected nondestructively over about the same region that was indicated radiographically.

Eleven longitudinal cross sections were prepared for macroscopic examination, including locations near each end of the through-wall

Fig. 24 Polished-and-etched cross section through the failed safe-end of the recirculation inlet nozzle

Note the intergranular nature of the failure. Etched in 8:1 phosphoric acid. 53×

portion of the crack and sections selected on the basis of radiographic crack indications. Major cracking was found in all sections examined. At all locations, the crack originated near the tip of the crevice between the thermal sleeve and the safe-end, adjacent to the attachment weld for the safe-end, and extended outward through the safe-end wall. Over about 35° of the circumference, cracking propagated through the repair weld, while over a different 50° of the circumference, the propagation path was entirely within the safe-end base metal, penetrating the outside surface adjacent to the fusion line of the repair weld.

Metallographic examination of selected cross sections showed the effects of repeated thermomechanical cycling during welding on the precipitate structure in the HAZ of the Inconel 600. There were very few matrix precipitates immediately adjacent to the fusion line of the heat shield attachment weld; this condition was true of the entire HAZ near the root pass. The cracking generally initiated in the HAZ some distance away from the tip of the tight crevice and, in its early stages, generally followed a path parallel to the fusion line (Fig. 24). Depending on the location, the crack could be found in regions of heavy or light carbide precipitation. Metallographic cross sections and SEM studies of the fracture surface were in agreement: the cracking was all intergranular in the base metal and intercolumnar where it penetrated the repair weld metal. No secondary cracking was found, nor was any discontinuity found that might have initiated the cracking.

Microanalyses disclosed the presence of sulfur at locations corresponding to the later stages of cracking, but not on the crevice surface between the safe-end and the heat shield. Chlorine was present at one point on the crevice surface. Other deposits on the crevice surface and the crack surface near the crevice were rich in iron.

Conclusions. These observations led to the conclusion that the cracking mechanism was intergranular SCC. The literature indicates that Inconel 600 is susceptible to intergranular stress corrosion in high-purity water, but it is only likely at stresses above the yield stress of the material. The analysis by the reactor vendor showed the primary bending plus membrane stresses due to service loading to be 78% of the yield strength. Further, butt welds in austenitic piping materials can leave residual stresses of the order of the yield strength; therefore, the mechanism is reasonable. Whether or not the material is in a sensitized condition does not greatly influence susceptibility to cracking, and cracks propagated in the lightly sensitized region adjacent to the fusion line and in the heavily sensitized regions some distance from the fusion line. Therefore, the furnace-sensitized condition of the material was not considered to be a significant contributing factor in the cracking problem.

The role, if any, of the observed contaminants could not be clearly identified from this study, although there is a suspicion that they contributed to the failure. Further surface analysis would be required to clarify this point.

The role of the tight crevice between the safe-end and the thermal sleeve thus comes into question. It was considered unlikely that a classical differential aeration cell could develop in the crevice in the uniform high-purity water environment, but such regions could entrap air during outages, causing locally high oxygen levels upon start-up. The presence of anionic contaminants could cause acidification of the crevice region, which might also enhance stress-corrosion mechanisms. The crevice is considered to be the principal contributing factor in this case.

Protection During Off-Stream Periods. When subjected to stress in aqueous chloride solutions, austenitic stainless steels fail by highly branched transgranular cracking. When the same steels are exposed in refinery processing units to weak sulfur acids, more commonly known as polythionic acids, intergranular cracking can occur. The austenitic stainless steel must be in a sensitized condition—that is, with heavy intergranular carbides—for polythionic acids to cause cracking. Polythionic acids form directly from sulfide corrosion products inside desulfurizing, hydrotreating, and hydrocracking units at ambient temperatures during refinery shutdowns, but are not a problem at operating temperatures. These types of units are frequently clad with austenitic stainless steel for on-stream corrosion protection, but the steel must be protected during off-

stream periods. Procedures used for this protection are discussed below.

First, when a stainless-clad desulfurizer, hydrotreater, or similar reactor is shut down but not opened, it should be kept under positive nitrogen pressure. If the reactor is opened but not dumped, it should be turned over with ammonia three or four times before opening. Also, the work should be completed before the walls of the reactor reach the ambient dew point. If this is not practical, the reactor should be kept under positive nitrogen pressure during the entire downtime. Second, during dumping of the reactor catalyst, 5000 ppm of ammonia should be added to the nitrogen purge. Once the catalyst is removed, a 2% solution of soda ash with 0.5% sodium nitrate should be used to fill the reactor and soak all stainless steel. Third, any austenitic stainless steel in such components as recycle-furnace tubes, feed-heater furnace tubes, feed or effluent piping, and tubes or cladding in feed-effluent exchangers is also susceptible to intergranular cracking. Provisions should be made to keep this equipment under positive nitrogen pressure if unopened or to fill and soak with soda ash solution before opening.

Because polythionic acids are most likely to cause cracking in austenitic stainless steels that are in a sensitized condition, cracking can be minimized in pressure vessels of welded construction by using stabilized stainless steels, such as type 321 or type 347, or by using low-carbon grades, such as type 304L or type 316L.

Hydrogen Embrittlement

When corrosive reactions occur at a metal surface, hydrogen may be produced, particularly if acids or strong alkali solutions are involved. The hydrogen evolved is preferentially absorbed by steel and possibly by other metals. This absorption can cause hydrogen embrittlement. Consequently, SCC and hydrogen embrittlement are not only associated but also sometimes indistinguishable, particularly in sour gas service in which H_2S is the active corrodent.

Example 14: Hydrogen-Embrittlement Cracking in a Large Alloy Steel Vessel. Figure 25(a) shows a vessel that had been in service about 3 years as a hydrogen reformer when cracking occurred in the weld between the shell and the lower head. As shown in Detail A in Fig. 25, one crack, which was 150 mm (6 in.) long, developed at the bottom of the weld at the inner surface of the head (crack A), and another, which was 0.9 m (3 ft) long, developed at the top of the weld at the outer surface of the shell (crack B). Crack B later extended to a length of 2.7 m (9 ft). Also, some small blisters were observed on the inner surface near the weld after attempts had been made to repair the crack.

The vessel contained ceramic balls in the lower head and a catalyst in the shell section.

Fig. 25 Vessel made of ASTM A204, grade C, steel that failed as the result of hydrogen embrittlement

(a) Portion of tank; Detail A shows locations of cracks, at welds joining shell to lower head. Dimensions given in inches. (b) Schematic illustration of the weld area showing locations of cracks and of samples removed for examination. (c) Macrograph of a nital-etched specimen from the inner surface of the head showing a crack in the HAZ. 6.5×. (d) Micrograph of a nital-etched specimen showing normal microstructure and hydrogen embrittlement. 100×

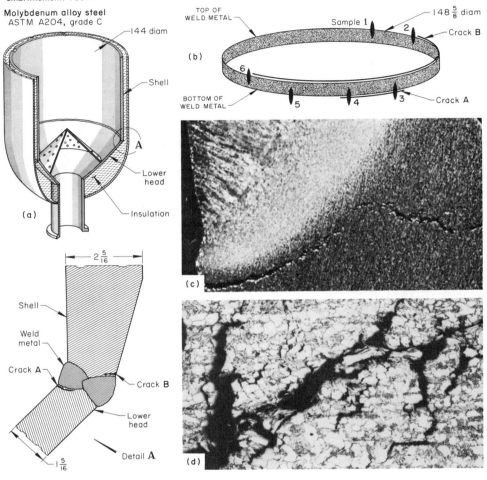

Temperature of the incoming gas was 690 °C (1275 °F); the outgoing gas, 620 °C (1150 °F). Design temperature of the vessel, which was insulated inside, was 120 °C (250 °F), and design pressure was 4480 kPa (650 psi). No liquid phase was present. A temperature of 340 °C (640 °F) had occurred at two hot spots 100 mm (4 in.) in diameter on the outside surface just above the area of the failure. The vessel was made of ASTM A204, grade C, molybdenum alloy steel. The head was 33 mm (1⁵⁄₁₆ in.) thick, and the shell was 59 mm (2⁵⁄₁₆ in.) thick.

Metallurgical Investigation. Six samples were removed from the outside surface of the vessel at right angles to the weld, using an abrasive cutoff wheel; as shown in Fig. 25(b), three were cut across the weld line on the shell side and three on the head side. The samples were triangular in cross section, about 13 mm (½ in.) wide on each surface, and 75 mm

(3 in.) long. A sample at the end of the crack on the inner surface of head was removed parallel to the weld. This sample was also triangular in shape, 22 mm (⅞ in.) wide at the head surface, 16 mm (⅝ in.) deep, and 240 mm (9½ in.) long. Also, three 21-mm (¹³⁄₁₆-in.) diam plugs were trepanned through the complete wall thickness, one each from the head, weld, and shell.

Of the six samples taken from the outer surface, only sample 3 was visibly cracked. This crack, about 5 mm (³⁄₁₆ in.) deep, initiated at the edge of the weld in the coarse-grain portion of the HAZ. Microscopic examination revealed a small crack in sample 1, which was taken from the shell side. This crack was also located in the coarse-grain portion of the HAZ. No cracks were discovered in the other samples taken from the outer surface. No microfissures, decarburization, or characteristics of hydrogen embrittlement were ob-

served in any of the six samples from that surface.

Examination of a section through the sample from the inner surface of the head showed that the crack was at the edge of the weld, completely in the HAZ of the head (Fig. 25c). Fusion of the weld metal on both the head and shell sides was good on both the inner and outer surfaces. A micrograph of the start of the crack, which was at the inner surface, revealed elongated grains, indicating that the metal had undergone plastic flow before failure. Figure 25(d) shows a micrograph of a portion of the same section farther along the crack. The microstructure was normal in appearance, but was severely embrittled, as shown by the region near the fracture surfaces. Additional micrographs near the crack showed that the metal had become severely gassed in an area extending up to 3 mm (⅛ in.) on each side of the crack.

Microscopic examination of the plug trepanned from the head showed decarburization, which was present in nearly equal amounts on both inner and outer surfaces and therefore was not the result of service conditions. The inner surface was coated with a dark-brown scale. The microstructure of the metal in the head was banded, except near the surfaces, and contained spheroidal carbides, indicating that the metal had been held at temperatures above 540 °C (1000 °F) for a long period of time.

The plug from the shell also had a dark-brown scale on the inner surface and showed a microstructure that was uniform throughout and typical for the metal. The carbides in the pearlite areas had lost some of their lamellar form, which indicated that the material had been held at a high temperature for some time, but the breakdown was not as severe as that observed in the head.

Hardness tests were performed on one of the samples taken from the outer surface, the sample from the inner surface, and the three trepanned plugs. Rockwell B hardness values and equivalent tensile-strength values are given in Table 2.

Discussion. Hardness data and microscopic examination indicated that the lower head had been overheated for a sufficiently long time to reduce the tensile strength below the 517-MPa (75-ksi) minimum required for ASTM A204, grade C, steel. The wide difference in tensile strength between head and weld metal (including HAZ) formed a metallurgical notch that enhanced the diffusion of hydrogen into the metal in the cracked region. The resultant embrittlement and associated fissuring led to further weakening and eventually to failure.

Because the fracture was brittle, internal cracking must have taken place before plastic deformation on the inner surface. Also, the major crack must have propagated internally because it did not intersect the inside surface near its end. Cracking on the outside surface was simply the result of overloading the embrittled material after internal failure had already occurred.

Table 2 Hardness and tensile properties of samples taken from a failed pressure vessel

Location of hardness test	Hardness, HRB	Equivalent tensile strength MPa	ksi
Outer-Surface Sample			
Shell plate	86	586	85
Head plate	86	586	85
HAZ(a)	92-98	724	105
Weld metal	92-96	690	100
Inner-Surface Sample			
Head plate	76	465	67.5
HAZ(b)	89	565	85
Weld metal	90	638	92.5
Trepanned Plugs			
Shell:			
Inner wall	84	569	82.5
Center	81	517	75
Outer wall	85	579	84
Head:			
Inner wall	76	465	67.5
Center	78	490	71
Outer wall	79	500	72.5
Weld metal:			
Inner wall	96	769	111.5
Center	94	748	108.5
Outer wall	95	758	110

(a) Both shell and head sides. (b) Head side

The total absence of decarburization on the trepanned plugs at the inside surface of the shell and the minor decarburization of the head suggested that hydrogen diffusion was limited to the head regions adjacent to the HAZ or that hydrogen diffused into the metal at temperatures below 205 °C (400 °F). The tensile tests of the head material would have detected any embrittlement. Had hydrogen diffusion occurred at higher temperatures, more fissuring (gassing) or considerably more decarburization would have been in evidence. Because no serious damage was detectable in the trepanned plugs, heat treatment would relieve any embrittlement that may have occurred in the shell and head. Grains need not be separated for embrittlement to occur, and provided no fissures have formed, the pressure can be relieved by allowing the gases to diffuse out of the metal during careful heat treatment; otherwise, blisters will result. The damaged region adjacent to the weld on the head side should be removed, because these fissures cannot be repaired by heat treatment.

Conclusion. Failure of the vessel was chiefly the result of hydrogen embrittlement with overheating as a contributory factor.

Corrective Measures. The vessel was wrapped in asbestos and heated to 455 °C (850 °F) at a rate of 55 °C (100 °F) per hour, held at 455 to 480 °C (850 to 900 °F) for 2¼ h, heated to 595 °C (1100 °F) at a rate of 55 °C (100 °F) per hour, and held for 2¼ h. The vessel was then cooled at a rate prescribed by the ASME code. This treatment caused hydrogen to be diffused out of the metal, thus removing the source of embrittlement.

The weld metal on both sides of the crack and 5 mm (³⁄₁₆ in.) of base metal in the shell and

the head were removed. The area thus prepared was repair welded using E7018 filler metal, then stress relieved.

Example 15: Cracking in Carbon-Molybdenum Desulfurizer Welds. Welds in two carbon-molybdenum (0.5% Mo) steel catalytic gas-oil desulfurizer reactors cracked under hydrogen pressure-temperature conditions for which hydrogen damage would not have been predicted by the June 1977 revision of the Nelson Curve for that material. As a result of this experience, a major refiner instituted regular ultrasonic inspections of all welds in its Fe-C-0.5Mo steel desulfurizers.

Investigation. During a routine examination of a naphtha desulfurizer by ultrasonic shear wave techniques, evidence of severe cracking was found in five weld-joint areas (Ref 9). The inspection involved stripping all the insulation from the vessel and testing all seams with a beam propagated perpendicular to the welds. Defect indications were found in longitudinal and circumferential seam welds of the 113-mm (4⁷⁄₁₆-in.) thick ASTM A204, grade A, Fe-C-0.5Mo steel sheet. Indications were also found in the 57-mm (2¼-in.) thick top and bottom heads that were formed by welding pie sections of plate. No records were available on the welding method used in the mid 1950s. The vessel had a 2.8-mm (0.109-in.) thick type 405 stainless steel liner for corrosion protection that was spot welded to the base metal at about 38-mm (1½-in.) centers, and all vessel welds had been hand overlaid with type 309 stainless steel.

The vessel was emptied and visually inspected, but no surface damage was evident. To inspect the weld metal, the type 309 stainless steel weld overlay was ground off along a 965-mm (38-in.) strip of a longitudinal seam weld in an area that had ultrasonic indications. This exposed a 330-mm (13-in.) long longitudinal crack in the weld metal, as well as transverse cracks. When about 3 mm (⅛ in.) of the base-metal weld was ground off and the surface was dye checked, hundreds of transverse cracks were observed. An external shear wave test was then performed on the outside of the vessel, but perpendicular to the normal scan direction, including weld areas that had not shown indications in the original testing; indications of transverse cracks were found. The testing was extended to include all welds, and all were found to have indications of transverse cracks.

Conclusions. Samples were taken from a girth weld for examination. It was found that weld metal from the cracked region contained 0.09% C, whereas the base plate contained 0.18% C as compared with a maximum 0.25% C for ASTM A204 material. This was taken as an indication of severe decarburization in the crack area. Subsequent metallurgical examination of a cross section of a longitudinal crack after polishing and etching with 2% nital disclosed a decarburized region on either side of the crack. The cracking was clearly intergranu-

lar in nature (Fig. 26). It was concluded that the damage was caused by a form of hydrogen attack.

Corrective Measures. The nominal service conditions for the desulfurizer were approximately 345 °C (650 °F) and 2690 kPa (390 psig) of hydrogen partial pressure, conditions that would not be expected to produce hydrogen damage based on the limits of the 1977 revision of the Nelson Curve for Fe-C-0.5Mo steel. The solution to this material problem was installation of a used 2.25 Cr-1Mo steel vessel with a type 347 stainless steel weld overlay. The use of an available vessel satisfied certain time constraints on the allowable down time for the plant.

Brittle Fractures

Materials that are ordinarily ductile can undergo brittle fracture suddenly and without warning, depending on certain conditions and factors, which are discussed in the article "Ductile and Brittle Fractures" in this Volume. Brittle fractures are most likely to occur in body-centered cubic (bcc) structures, such as ferritic steels, and in relatively thick parts that contain adequately large flaws. Strains rates required for brittle fracture depend somewhat on the temperature of the part; at low temperatures, strain rates are generally low, whereas at high temperatures strain rates are higher.

The origin of a brittle fracture will almost always be a pre-existing crack or similar type of notch or stress raiser because the concentrated triaxial stresses developed at such points are the single most important factor in determining whether a crack can propagate under the applied load. Brittle fracture generally occurs under an applied stress that is less than the yield strength, and the energy expended in propagating the crack is low.

Most sudden brittle fractures occur by cleavage, which produces a fracture surface with a bright granular or crystalline appearance. A characteristic chevron pattern, which points to the origin of fracture, is often present. Brittle fractures can be differentiated from ductile fractures by the short time required for the crack to achieve critical size and by the high rate of crack propagation. Although most brittle fractures are transgranular, intergranular brittle fracture can occur if the grain boundaries are embrittled by chemical segregation or the presence of a brittle second phase.

Brittle fractures in pressure vessels can be correlated with the notch toughness of carbon or low-alloy steels, and they occur at temperatures below the transition temperature measured by such testing. The transition temperature can increase in service due to microstructural or metallurgical changes. In carbon and low-alloy steels, the ductile-to-brittle transition temperature can be near or above the ambient temperature during fabrication and during hydrostatic testing. Factors controlling the transition temperature include chemical composition and the

Fig. 26 Cracks in naphtha desulfurization reactor attributed to hydrogen damage

(a) Note decarburized region adjacent to crack. 25×. (b) Higher-magnification view. 200×

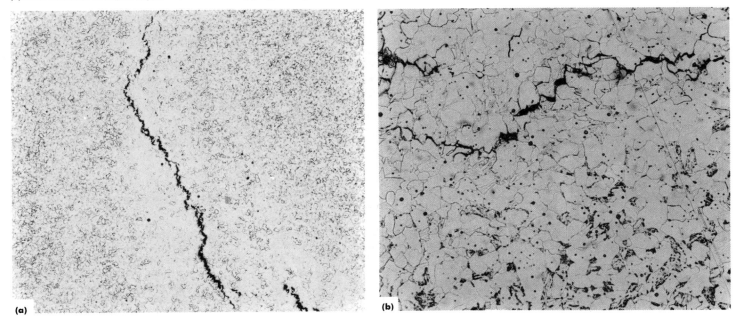

(a)　　　(b)

effects of deoxidation, forming, and heat treatment. Poor-quality steel or improper welding practice are frequently the material factors that cause brittle fracture, but externally applied stresses are required to cause failure.

Corrective action that can be used to guard against the possibility of brittle fracture in pressure vessels includes attention to all the factors noted above plus stress relief after fabrication. Metallurgical changes in the material during service should be avoided. Such changes might result if, for example, carbon steel were used instead of alloy steel in an application in which temperatures occasionally exceed the limit of operating temperature for carbon steel.

Embrittlement can result from graphitization in carbon steels and 0.5% Mo steels, welding or fabrication steps, exposure to atmospheres favoring hydrogen embrittlement, or formation of σ phase in stainless steels. Thermal fatigue caused by different amounts of expansion in adjacent dissimilar metals can result in crack propagation and eventual brittle fracture. Severe and repeated temperature changes in such components as chemical-system quench tanks can result in thermal shock, the severity of which might be sufficient to cause a brittle failure.

Brittle fractures are not restricted to steels, but have also occurred in components made of other materials, such as welded aluminum rocket-fuel tanks, because of residual shrinkage stresses. These stresses were caused by weld design that was improper for the relative stiffness of the elements being joined. Microcracks that formed during cooling of the welds pro-

vided the brittle fracture initiation sites that occurred during hydrostatic proof testing.

Example 16: Brittle Failure of a Pressure Vessel After Welded Modifications Without Stress-Relief Heat Treatment (Ref 10). A spherical carbon steel fixed-catalyst bed reactor failed after 20 years of service while in a standby condition (Fig. 27). At the time of failure, the unit contained primarily hydrogen at about 13 °C (55 °F) and 2400 kPa (350 psig) and was operating at an ambient temperature of 5 °C (41 °F) with no gas flow. The reactor fractured into six pieces in a single event. There was no evidence of an upset in operating conditions or introduction of oxygen to form an explosive mixture.

The reactor had been fabricated in 1958 from French steel A42C-3S, approximately equivalent to ASTM A201, grade B, with a 434-MPa (63-ksi) tensile strength and a 241-MPa (35-ksi) yield strength. The steel mill tests showed a Charpy V-notch energy absorption of 30 J (22 ft · lb) at 20 °C (68 °F). The plates had been hot formed into segments during construction of the reactor. The inside of the vessel was refractory lined to keep its temperature below 275 °C (527 °F) during operation. The vessel had a type 304 stainless steel shroud around the catalyst bed as protection against the overheating that was possible if the gas bypassed the bed through the refractory material.

Investigation and Conclusions. A visual examination of the fracture showed that the failure propagated through all of the plates of the shell (Fig. 28). The fracture followed the line of the shroud-support ring in four plates (plates A to D, Fig. 28), then branched into the rest of the

Fig. 27 Configuration and dimensions of spherical fixed-catalyst bed reactor that failed after 20 years of service

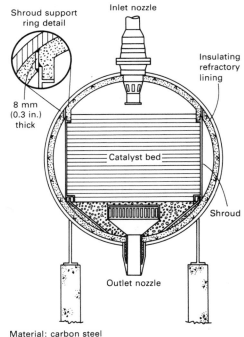

Material: carbon steel
Design pressure: 3.7 MPa (540 psig)
Design temperature: 275 °C (527 °F)
Diameter: 4.3 m (14 ft)
Shell thickness: 41 mm (1.6 in.)

Fig. 28 Fracture path of failed pressure vessel
The arrows indicate the direction of crack propagation as determined from the chevron markings on the fracture faces. The letters identify the individual plates.

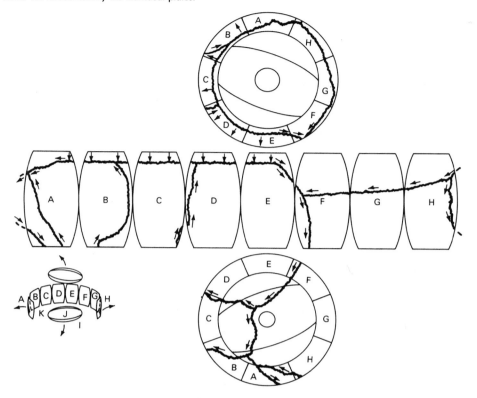

Fig. 29 Lamellar tearing at root of weld for shroud-support ring

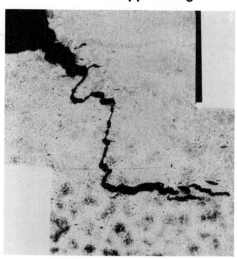

vessel. All fracture surfaces were classical flat fractures, with no shear lip and chevron markings on all fractures except the fracture immediately at the shroud-support ring. At the ring, the chevron patterns suggested that the crack had propagated from the inside of the reactor to the outside wall. Weld seams did not provide a path for fracture. The failure clearly began at the toe of the shroud-support ring weld, but no evidence of a large pre-existing crack that could have triggered the failure could be found.

The shroud-support ring weld was ultrasonically inspected where it had not failed and was found to contain a number of 3-mm (0.12-in.) or smaller cracks at the root of the weld. Subsequent metallographic investigation revealed these indications to be lamellar tears that originated at nonmetallic inclusions on planes parallel to the plate surface, caused by weld shrinkage stresses. Figure 29 shows a typical lamellar tear.

Scanning electron microscopy confirmed the cleavage mode of fracture and the origin of the failure. Cleavage failure clearly initiated at the root of the weld, and it continued through the wall thickness. Along the root of the weld were indications of multiple sites of crack initiation at the pre-existing shallow lamellar tears and minor weld discontinuities resulting from arc breaks and excessive or incomplete penetration.

Metallographic cross sections of the fracture surfaces revealed the presence of extensive secondary cracking and twinning where the fracture followed the line of the shroud-support ring. The twinning was identified as Neumann bands, which in ferritic steels are caused by the highly localized deformation associated with fast fracture or dynamic stresses. This led to the conclusion that the fracture initiated at multiple sites within one plate, starting at the root of the weld, then initiated in neighboring plates in rapid succession.

The possibility of hydrogen attack was eliminated by the fact that the subsurface cracks were transgranular and were only present immediately below the fracture surface. This conclusion was substantiated by bend tests and toughness tests performed on specimens of the vessel material.

Chemical analyses were performed on the plates, which were found to be within specifications for this variety of carbon-manganese steel. Analyses for arsenic, molybdenum, nickel, copper, and chromium indicated no abnormality. The steel was not required to be fine grained; in addition, the soluble aluminum content responsible for controlling grain size and reducing susceptibility to strain aging varied among the plates, but was very low for three of the plates (<0.005%). The hydrogen contents of the steels were normal.

Toughness tests were performed on numerous plate samples from locations near the bottom of the reactor, at the middle, and at the top adjacent to where the fracture followed the shroud-support weld. The lowest values measured at 5 °C (41 °F) were 6.8 and 9.5 J (5 and 7 ft · lb), respectively. The plate in which the fracture initiated had a 27-J (20-ft · lb) transition temperature of 30 °C (86 °F), while the plates in which the fracture developed shear lips and ran off into the vessel had 27-J (20-ft · lb) transition temperatures below −5 °C (23 °F). The coarsest grained steels exhibited the lowest toughness values. In addition, the Pellini drop-weight test procedure (Ref 11) was used to study the nil-ductility temperature (the highest temperature at which small flaws can propagate as brittle fractures) of three plates. Two plates were at or below their nil-ductility temperature.

Refinery maintenance and inspection records showed that the carbon steel shroud-support ring had been removed for a hydrotest, the existing weld metal ground off, and the shell surface inspected using dye penetrant. The new ring was welded from one side. The weld was made with six passes by using low-hydrogen electrodes equivalent to E60-18 without preheat. The root pass was made with 2.4-mm (³/₃₂-in.) electrodes, subsequent passes with 3.6-mm (%₄-in.) electrodes. The completed weld was visually inspected and not postweld heat treated, because postweld heat treatment would not have been required by applicable pressure vessel codes for new construction.

A numerical stress analysis of the weld was made for the design geometry and geometries that included sharp notches at the shroud-support ring-to-shell junction. These analyses included a condition meant to simulate a hot spot from the missing refractory lining and a combination of hot spot plus notch. The operating conditions investigated included thermal loadings from start-up and shutdown, internal pressure, axial friction from the catalyst bed, and a distributed load on the shroud due to the catalyst weight. The finite element analyses indicated a peak stress of 110 MPa (16 ksi) under an applied pressure of 2400 kPa (348 psig). The maximum stress occurred at the weld-to-ring junction, rather than at the weld-to-vessel junction, where failure occurred; the lower stresses on the reactor shell eliminated fatigue as a failure mechanism, in agreement with the metallographic observations.

The residual-stress pattern in the shroud-support ring attachment weld was measured at about 180° around the vessel from the crack-initiation site, where no cracking of the weld had occurred. Surface stresses were measured using a center-hole rosette strain gage technique, and through-thickness stresses were measured by a block-removal and layering technique. The maximum transverse tensile stress that would be applied across a radial defect in the shell near the root of the attachment weld was on the inner surface of the vessel wall near the toe of the weld. It was at least 276 MPa (40 ksi), which is equivalent to the yield strength of the base metal, and decreased linearly over the first 40% of the wall thickness.

Fracture mechanics tests were used to predict the mode of failure of the various plates. Measurements of crack-tip opening displacement indicated that ductile failure would occur in at least one plate that sustained brittle failure. No pre-existing defect could be found in the size range of $\geq$11 mm (0.43 in.), which would have been required to cause a brittle failure.

All of the evidence pointed to a change in the susceptibility of the plates to brittle failure near the point of failure initiation. Charpy V-notch specimens that were aged under either tensile or compressive stress at 180 °C (360 °F) for 168 h were found to lose about half their impact energy absorption, from about 11 J (8 ft · lb) to about 5.5 J (4 ft · lb), compared with unaged controls. A series of subsize and full-size crack-tip opening displacement specimens employing a 0.5-mm (0.020-in.) deep by 0.15-mm (0.006-in.) wide welded-over notch was prepared. Large decreases in ductility could be ascribed to the welded-over notch. Complete cleavage failure could be induced in these samples by testing at high strain rates. A small crack that initiated in the embrittled zone could thus propagate to critical size and precipitate an unstable fracture.

The strain-aging tendencies of these steels were demonstrated by making microhardness traverses adjacent to the tips of welded-over notches. The microhardness near a welded-over notch was much higher than that in the adjacent base metal. Thin-foil transmission electron microscopy (TEM) was also used to demonstrate that high dislocation densities occurred in the vicinity of welded-over notches. The dislocations were bowed and pinned, an observation that is not exclusively associated with strain-aging embrittlement but that supports the proposed failure mechanism.

The final cause of failure was felt to be cracking that developed during the installation of the new shroud ring. The cracks caused by the root weld pass were not by themselves large enough to initiate fracture, but subsequent thermomechanical cycling during the remainder of the welding process caused dynamic strain aging and localized embrittlement at the crack tips. Three plates had relatively low toughness. Under the combination of high residual stresses, pressurization stresses, and a low metal-shell temperature, the cracks extended and caused total destruction of the vessel.

The recommendation from this analysis was to perform stress-relief heat treatments to reduce residual-stress levels after welding non-pressure retaining components to certain pressure shells. This treatment should reduce residual stresses and lessen the embrittlement effects of strain aging.

Ductile Fractures

Ductile fracture occurs with a significant amount of macroscopic plastic deformation. It is considered to take place by a shear mechanism in which the metal is torn with considerable expenditure of energy. Even though final rupture may occur suddenly, most ductile failures of pressure vessels or piping develop gradually and give warning during propagation by means of a bulge or a leak.

The surface of a ductile fracture generally exhibits a dull, fibrous appearance, and there is evidence of plastic deformation, such as local bulging or distortion. A good indicator of deformation is a significant reduction of area at the fracture. When such an indicator is present, the failure may have resulted from some long-term change or deterioration of the material or from a gradual increase in the load. However, the mere classification of a fracture as ductile does not explain why it occurred, but simply suggests the factors that may have caused it. For example, gradual reduction in wall or section thickness by corrosion, erosion, or creep can render the material unable to sustain the operating load. To this extent, the classification of a failure as ductile depends on the time span over which the failure takes place.

Creep and Stress Rupture

Creep, or time-dependent deformation, is normally anticipated by designers of equipment that operates at elevated temperatures. The useful life of a pressure vessel or pressure boundary material can occur as a result of reaching or exceeding some design limit on strain. This is known as a creep failure, and it may involve only excessive deformation, not a fracture. Stress ruptures, complete failures under creep loadings, can breach a pressure boundary. The main variables determining the rate of deformation under creep conditions are temperature, time, and stress. It is not unusual to find other contributing factors, such as corrosion, fatigue, or material defects involved in creep and stress-rupture failures.

Creep usually occurs in time-temperature regimes for which diffusion-controlled processes are important. Within grains, the interactions of dislocations with solute atoms, second phases, and other defects can determine deformation rates; microstructures can change dramatically during the creep process. Deformation at grain boundaries, together with diffusion processes, can lead to the nucleation and coalescence of voids, causing grain boundaries to separate, sometimes at very low strains. The interpretation of the remaining useful life of a material that has undergone creep is a difficult, but important, concept.

Equipment that operates in the creep regime is frequently too expensive to discard without consideration of repair or partial replacement of material that has suffered creep damage. Life-extension programs are common in many industries and have recently become important in the electric power industry. While many other failure mechanisms must be considered in such programs, examination for creep damage is critical.

Example 17: Preventive Analyses of Croloy 1 1/4 Pressure Parts. It is generally not possible to predict the maximum life of base-loaded headers and piping until they have already developed microcracking. When cracking does occur, it can be monitored over a period of several years by a periodic inspection program (Ref 12). Typical elements of such a program include the following:

- An engineering stress analysis of the header, tube legs, and the associated piping system is conducted. Numerical analysis, a modern tool, can be used to determine which areas of the component should be tested and inspected during downtimes and to determine static and cyclic stresses accurately
- Dimensional examinations can be performed to determine the extent of time-dependent swelling, bowing, or other deformation that may have occurred
- Visual examination is performed to disclose the extent of other defects. Borescopes and fiberscopes can be used to examine otherwise inaccessible areas of a component. These examinations may be supplemented with dye-penetrant inspections in areas suspected of cracking
- Field replication is recommended to examine metallurgical microstructure *in situ*. Surfaces to be examined must be polished and etched by metallographic methods, usually under difficult field conditions and without regard to specimen flatness. The surface is then recorded by coating the surface with a resin that hardens and is stripped off, bearing the image of the specimen topography. The image is then coated to enhance detail and studied by optical or electron microscopy. Field replication is useful for detecting the early stages of creep, fatigue, grain-boundary separations, and other problems.

It can also be used to detect spheroidization or graphitization of steels
- Boat samples of headers and piping are generally removed only after cracking has been detected by other means. Tube stubs may be removed to examine the more highly stressed regions of headers

In one example, this type of routine disclosed cracks caused by creep swelling in the stub-to-header welds in the secondary superheater outlet headers of a major boiler. The header was constructed of SA335-P11 material (1.25Cr-0.25Mo silicon steel). A review of the design of this unit showed that it had been designed according to allowable stresses that had been established in 1951 from a relatively small data base. As more information became available, the ASME Boiler Code allowable stresses for the material were reduced in mid 1965 for temperatures above 525 °C (975 °F); therefore, the unit was underdesigned. After about 30 years of service, repairs had been necessary on 61 of the 222 header welds, and a permanent solution to the problem was sought.

To document the amount of creep swelling that had occurred, the outside diameter of the header was measured and found to have increased 1.6% since its installation. It was remeasured after another 6 months and found to have swelled another 0.14%. Surface inspection revealed ligament cracks extending from tube seat to tube seat (Fig. 30). Cracks originated from inside the header, extending axially in the tube penetrations and radially from those holes into the ligaments. The ligament cracking of the header was a separate event from the stub-to-header weld cracking, although both were caused by high stresses in the header. Dye-penetrant inspection of the latter welds revealed cracks in 94 locations, ranging from small radial cracks to full 360° cracks. An example of this cracking is shown in Fig. 31. The unit was operated under reduced-temperature conditions and with less load cycling than previously until a redesigned SA335-P22 (2.25Cr-1Mo-0.25-Si) header was installed.

Example 18: Failure of a Main Steam Line of a Power-Generating Station. A main steam pipe was found to be leaking during a start-up after a 1-day shutdown. The leak was a large circumferential crack in a pipe-to-fitting weld in one of two steam leads between the superheater outlet nozzles and the turbine stop valves, a line made of SA335-P22 material (2.25Cr-1Mo steel) with an outside diameter of 475 mm (18.75 in.) and a wall thickness of 95 mm (3.75 in.). The design operating conditions were 25 MPa (3.6 ksi) at 540 °C (1000 °F).

Investigation. A boat sample was cut from one side of the main crack and subjected to an initial examination. Visual and magnetic-particle examinations revealed secondary cracks roughly parallel to the main failure in the weld metal and adjacent base metal. The main

Fig. 30 Interligament cracking in a failed secondary superheater outlet header from a boiler

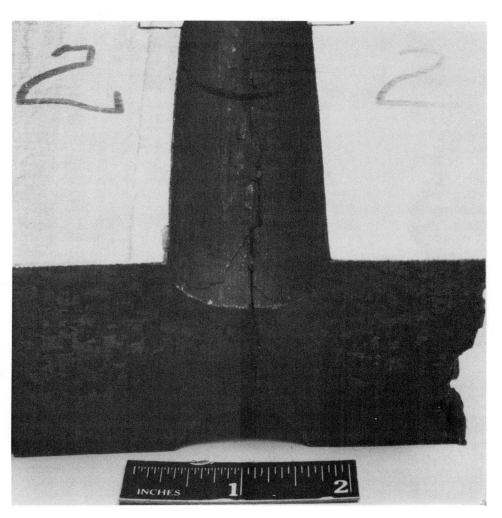

crack surface was rough, oriented about normal to the outside surface, and had a dark oxidized appearance. The topography of the crack surface reflected the weld-bead patterns. A metallographic section of this sample shows the main crack surface and the secondary cracks (Fig. 32). The cracking is predominantly intergranular, with some grain boundaries exhibiting unlinked intergranular voids that should eventually link to form complete grain-boundary separations. This microstructure was thought to be typical of elevated-temperature stress rupture in the material.

Microscopic examination of the cross section revealed distinct shiny bands that etched slower than the remainder of the sample at the top of each individual weld bead. These bands were located within the interbead HAZ and consisted of localized areas with a large-grain ferrite matrix and relatively large carbide particles. The bands frequently contained small cracks and microvoids that did not extend into the surrounding fine-grain material. These features

Fig. 31 Microstructure, linked voids, and split grain boundaries in the failed outlet header shown in Fig. 30

400×

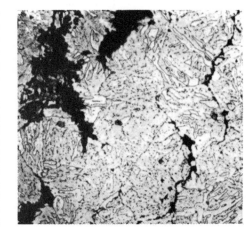

Fig. 32 Metallographic cross section through failure in 2.25Cr-1Mo weld main steam line of power plant

Secondary cracking in base metal indicates that failure is not uniquely the result of weld-metal properties.

suggest that the ferrite bands served as initiation sites for the failure and that they lowered the rupture strength of the weld metal. The Knoop hardnesses of the bands tended to be in the range 138 to 144 HK, while the adjacent areas of the interbead HAZ were 161 to 168 HK. Qualitative analyses of the ferrite bands by EDS revealed no significant differences in composition between the bands and surrounding material.

While these observations identified the mechanism of cracking at the main failure, they did not provide insight into the cause of the premature service failure. The work was therefore expanded to include other weld metal and base metal of the steam line. The samples included a full ring section containing the failed weld and a quadrant of a weld that had not cracked. Smaller boat samples and pyramid samples were removed from other welds. Each was characterized, at a minimum, by chemical analyses and microstructural examinations. Stress-rupture tests and tensile tests were performed on some samples to determine the strength properties of the materials of construction. The results of these examinations are summarized as follows:

- The chemical analyses and mechanical properties were within specifications or reasonable for the materials of construction after service
- The failed weld and three others had an unusually high manganese content that identified them as having been fabricated by the

submerged arc process. Three other welds were low in manganese, identifying them as having been fabricated by the manual shielded metal arc process. The submerged arc welds were cracked in the weld metal, but the submerged arc welds were cracked in the adjacent base metal or HAZ. The difference in the failure loci was apparently a result of the relative strengths and ductilities of the specific base metal/weld metal combinations that were present

- All observations are consistent with a mechanism of intergranular creep rupture at elevated temperature
- The crack shape indicated that cracking initiated on the pipe exterior, then propagated inward and in the circumferential direction in response to a bending moment load. There was no indication of a pre-existing flaw or defect at the initiation sites other than the microscopic hardness heterogeneity noted above

Conclusions. The primary cause of failure was the occurrence of bending stresses that exceeded the stress levels predicted by design calculations and that were higher than the maximum allowable primary membrane stress specified in the ASME Boiler and Pressure Vessel Code, Section I, and the ANSI Power Piping Guide.

Fatigue

Fatigue fractures result from cyclic stressing, which progressively propagates a crack or

cracks until the remaining section can no longer support the applied load. Pressure vessels and pressure piping are subject to high static stresses arising from the pressure of contained liquids or gases, to stresses resulting from misalignment of components, and to residual stresses induced during welding. The cyclical component may be added mechanically—by vibration of associated equipment, pulsation from a compressor, or thermal cycling. Fatigue fractures generally originate at stress raisers, such as a discontinuity or a notch, which produces triaxial stressing in the material. Thermal fatigue occurs when cyclic stresses are produced by uneven heating or cooling, or when part of a vessel is restrained and cannot expand or contract uniformly.

Over a period of years, several low-carbon steel thick-wall vessels equipped with cooling jackets cracked when operated as batch reaction vessels at 400 °C (750 °F). These vessels operated with sharp thermal gradients across the walls. A small amount of sodium hydroxide could have been present in the vessels, and it was thought that cracking might have been the result of caustic embrittlement.

To increase the life of the vessels, the material was changed from low-carbon steel to a carbon-molybdenum steel. Occasional cracking still occurred. The material was then changed to 1Cr-0.5Mo steel. However, cracking occurred in a shorter time than with vessels made of carbon-molybdenum steel.

A stress analysis of these jacketed vessels indicated that cyclical stresses produced during operation might have been high enough to have caused cracking in the vessel walls. Furthermore, such stresses would be high enough to produce failure in any available grade of weldable structural steel. Thus, life of the vessels could be increased by reducing the stress but not by upgrading the material. Reducing the wall thickness would lower thermal stresses by decreasing the temperature differential in the wall.

Example 19: Failure of a Main Steam Line by Thermal Fatigue (Ref 13). Visual inspection of a type 316 stainless steel main steam line of a major utility boiler system revealed cracks on its outer surface near a hanger lug. At that time, the system had accumulated 130 520 h of high-temperature operation and had been subjected to 326 start-up cycles, many of which were from forced shutdowns. The steam line was a 230-mm (9.05-in.) outside-diameter pipe with a 64-mm (2.525-in.) minimum wall thickness.

Investigation. One third of the pipe was immediately subjected to demineralized water penetrant visual inspection, including all sections of pipe from the heat that had cracked. No further cracking was found.

A 3.4-m (11-ft) section including the cracked area was removed from the line. From this, boat samples and several ring sections were removed for examination, including the cracked region. Conventional metallographic

Fig. 33 Oxide-filled intergranular cracks oriented normally to the hoop stress direction in the main steam line

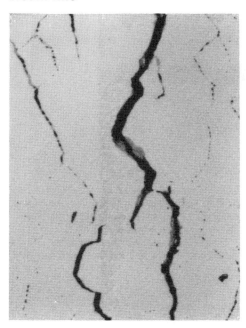

techniques were used to assess the extent of cracking and the nature of the cracks. The damage in all specimens was essentially the same: cracking initiated at the outside of the pipe wall or immediately beneath the surface. The cracks were all intergranular, oriented normally or nearly so to the main hoop stress direction. The cracks were usually filled with oxide (Fig. 33).

Examination of the metallographic samples at higher magnification revealed that the microstructure of the failed pipe consisted of a matrix precipitate array, probably $M_{23}C_6$, and large σ-phase particles in the grain boundaries. Although void formation and propagation along the σ/austenite interface is cited as a potential failure mode for this σ-phase morphology, the cavitation here occurs primarily in the ligament region between σ-phase particles. Scanning electron microscopy of the polished-and-etched cross sections of areas away from the cracks clearly showed this. The cavities were found to be decorated with small particles having a high sulfur content and were found to be faceted.

Charpy impact specimens and compact-tension specimens were cut from a section of the steam line that contained several cracks. The Charpy impact tests showed the inside pipe wall to be tougher than the outside, where the cracking and grain-boundary separations had occurred. There was no indication of a fracture-mode transition. Only slow, stable crack growth occurred in the compact-tension tests. One compact-tension specimen contained a natural crack from the steam line, but it behaved the same as specimens that had fatigue-sharpened notches on their crack fronts.

The insulation was examined to determine whether the impurities contained might have contributed to the failure. Its pH, leachable sodium, potassium, and lithium were analyzed. Colorimetric and the x-ray fluorescence spectroscopy were used to determine sulfur and chloride contents. The material was basic and had sufficiently low leachable impurities that intergranular stress-assisted corrosion was eliminated as a cause of failure.

Field metallographic techniques were developed, using a repeated polish-and-etch technique, to examine the remainder of the piping. A portable grinding tool was used to prepare the surface, followed by swab etching. A total of 72 locations in the piping system were examined, with the result that all material upstream of the boiler stop valve had oriented cracking of the type previously mentioned, while all locations downstream of the valve were free of damage. The surface metallography was validated by removal of boat samples that demonstrated the replication technique to provide microstructures representative of bulk material. A chemical analysis of the piping showed compliance with the material specification.

Significantly, the pipe showed no evidence of swelling due to creep. This fact, coupled with the lack of damage downstream from the stop valve and the high ductility of the inner-wall material, cast doubt on classical creep due to overstressing or overtemperature as the cause of failure.

Residual-stress measurements were made using a hole-drilling technique and strain gage rosettes. The holes were drilled with 50-μm alumina abrasive to minimize stresses introduced during drilling. Relatively large tensile axial residual stresses were measured at nearly every location investigated, but a large residual hoop stress was found for locations before the stop valve.

Conclusions. It was theorized that one or more severe thermal downshocks might cause the damage pattern that was found. A transient heat transfer and thermal-stress analysis was performed with numerical methods and software identified as CREPLACYL and with material data taken from the *Liquid Metal Fast Breeder Reactor Materials Handbook* (U.S. Department of Energy). The resultant elastic/plastic/creep deformation analysis indicated that a severe thermal downshock could explain the pattern and that one or more downshocks of less than 260 °C (500 °F) but more than 65 °C (150 °F) could have been involved. Thus, the root cause of the failure was thermal fatigue, with associated creep relaxation.

REFERENCES

1. G. Sorell and M.J. Humphries, High Temperature Hydrogen Damage in Petroleum Refinery Equipment, in *Corrosion/78*, National Association of Corrosion Engineers, Houston, 1978
2. C.J. Czajkowski, "Evaluation of Intergranular Cracks on the Ring Header Cross at Grand Gulf Unit #1," Technical Report A-3500, Prepared for the U.S. Nuclear Regulatory Commission, Region II, under contract, Brookhaven National Laboratory, Upton, NY, March 1986
3. "Standard Methods for Determining Average Grain Size," E 112, *Annual Book of ASTM Standards*, Vol 03.03, ASTM, Philadelphia, 1984, p 120-152
4. "Standard Practices for Detecting Susceptibility to Intergranular Attack in Austenitic Stainless Steels," A 262, *Annual Book of ASTM Standards*, Vol 03.02, ASTM, Philadelphia, 1984, p 1-27
5. Brittle Fracture of a Thick Walled Pressure Vessel, *BWRA Bull.*, Vol 7 (No. 6), June 1966
6. R.D. Wylie, J. McDonald, Jr., and A.L. Lowenberg, "Weld Deposited Cladding of Pressure Vessels," paper presented at the 19th Petroleum Mechanical Engineering Conference, Los Angeles, Sept 1963
7. C.J. Czajkowski, "Investigation of Shell Cracking on the Steam Generators at Indian Point Unit No. 3," NUREG/CR-3281, also BNL-NUREG-51670, U.S. Nuclear Regulatory Commission, Washington, June 1983
8. C.J. Czajkowski, "Constant Extension Rate Testing of SA302 Grade B Material in Neutral and Chloride Solutions," NUREG/CR-3614, also BNL-NUREG-51736, U.S. Nuclear Regulatory Commission, Washington, Feb 1984
9. D.B. Bird, "Cracking in Carbon 1/2 Moly Desulfurizer Reactor Welds," paper presented at the 1985 Fall Meeting of the American Petroleum Institute Operating Practices Committee, Subcommittee on Process Operations
10. A.R. Cuiffreda and R.D. Merrick, Brittle Fracture of a Pressure Vessel: Study Results and Recommendations, in *Proceedings of the American Petroleum Institute*, Vol 62, 1983
11. "Standard Method for Conducting Drop-Weight Test to Determine Nil-Ductility Transition Temperature of Ferritic Steels," E 208, *Annual Book of ASTM Standards*, Vol 03.01, ASTM, Philadelphia, 1984, p 346-365
12. G. Harth, Header and Piping Inspection Guides Boiler Life Extension, *Power*, Vol 109, March 1965, p 99
13. J.E. Bynum, F.V. Ellis, M.H. Rafiee, W.F. Siddall, T. Daikoku, H. Haneda, and J.F. DeLong, "Failure Investigation of Eddystone Main Steam Piping," paper presented at the 65th AWS Convention, Dallas, 1984, and publication TIS-7540, CE Power Systems, Windsor, CT

Failures of Metallic Orthopedic Implants

Ortrun E.M. Pohler, Institut Straumann AG

ORTHOPEDIC IMPLANTS are unique in that they are exposed to the biochemical and dynamic environment of the body and their design is dictated by anatomy and restricted by physiological conditions. Orthopedic implants are artificial devices (allentheses) that are mounted to the skeletal system of the human body or an animal for various purposes, such as supporting bone, replacing bone or joints, and reattaching tendons or ligaments.

Depending on the duration of their function, the two major categories of such devices are prostheses and internal fixation implants. Prosthetic devices (Fig. 1) are implants designed to remain in the body for a lifetime. They serve as joint replacements (replacing the total joint or only one component) or as bone replacements (replacing larger sections of bone, for example, after tumor resection). Internal fixation devices (Fig. 2) are implants designed to provide temporary stabilization. They are used for maintaining the shape of the reconstructed bone for treatment of fractures or for corrective orthopedic operative procedures. After healing, internal fixation devices, having served their purpose, are commonly removed.

Prosthetic Devices

Artificial joints are mainly used to replace painful joints that are degenerate and inflicted with diseases. Severe joint injuries are occasionally treated with prosthetic components.

When prostheses replace the entire joint, two artificial joint surfaces are anchored at the bone ends that remain after resection, and the prosthesis simulates the joint motion. Occasionally, only one joint surface, such as the femoral head, is replaced by an artificial component, which works against a natural articulation surface. Most joint replacements are performed on the hip. Because the normal femoral head is almost spherical, its motion is relatively easy to simulate. More complicated is the articulation of the knee and the various types of knee prostheses that compromise the original motion range. Prostheses are also available for the humeral head (shoulder), elbow and finger joints, the ankle, and the jaw (Fig. 1).

Most total joint systems have articulation components made from dissimilar materials so that, for example, metal or ceramics work against a plastic material, such as polyethylene (Ref 1). Finger joints, as well as some other prostheses, are made completely from plastics or metal-reinforced plastics (Ref 2).

Many prostheses are anchored by using a bone cement in a bed that has been mechani-

Fig. 1 Typical examples for joint prostheses (schematic)

(a) Classic Moore hip endoprosthesis. (b) Müller total hip prosthesis (metal against polyethylene acetabular cap). (c) Weber total hip prosthesis with movable head and metal, ceramic, and polyethylene components. (d) Müller total hip prosthesis with straight stem. (e) Hinge-like knee joint prosthesis; metal against metal (G.E.U.P.A.R.). (f) Sliding knee joint prosthesis; metal against plastic (Geomedic). (g) Total shoulder joint prosthesis; metal against plastic (St. Georg). (h) Total finger joint prosthesis with metal and plastic components (St. Georg). (i) Total elbow joint prosthesis; metal-to-metal (McKee)

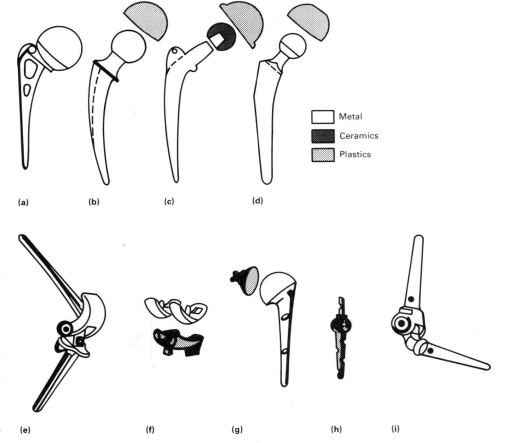

Metal
Ceramics
Plastics

(a)   (b)   (c)   (d)

(e)   (f)   (g)   (h)   (i)

cally prepared in the remaining bone. Recently, various designs have been developed for the cementless anchoring of prostheses by a tight fit. Certain prosthesis systems have porous surface coatings on their shafts to increase the surface area and to allow microanchorage of the bone. Developments in the field of joint prostheses are progressing rapidly. In this article, only metallic components will be considered.

In tumor resections, large defects are created by extensive bone and joint removals, which are often bridged with custom-made prostheses, modified standard joint prostheses, or fracture-treatment implants. Bone cement and/or other filling materials are sometimes added to the reconstruction for more stability. In the treatment of tumor resections, the surgeon is often forced to be inventive and to choose unorthodox techniques of reconstruction. Frequently, the remaining bone has been irradiated and lacks its normal regenerative vitality.

Implants for Internal Fixation

Implants for internal fixation are characterized by their versatility (Fig. 2). They are used to maintain the shape of the reconstructed bone in fracture treatment of the injured patient or, in corrective orthopedic surgery, to stabilize osteotomies (sectioning and realignment of bone) or to perform arthrodeses (stiffening of joints by fusion of bones). Various types and sizes of bone plates, bone screws, intramedullary rods (nails), pins, and orthopedic wire are available (intramedullary refers to use of the marrow space of the bone for support). Some implants have very specific uses; others have a wide range of applications with different functions.

Under the impact of a trauma, bones can fracture in any configuration and combination. In each patient, the fractures are unique and require individual treatment. This explains the necessity of versatile implants. Nevertheless, certain fracture classifications and methods for the treatment of typical fracture configurations have been established (Ref 3). In the selection of the appropriate implants and operation technique, not only the type of fracture and its location but also the condition of the soft tissue, the structure and consistency of the bone, the age and condition of the patient, and any present complication must be considered. Even when the best possible reconstruction has been achieved, the weight-bearing stability of an internal fixation device can vary widely, and the weight-bearing schedule of the patient must be correspondingly adjusted.

Depending on the situation, the patient may be allowed one of the following postoperative modes of motion: exercise only, partial weight bearing, or full weight bearing. The rate of bone healing depends on different factors and

Fig. 2 Typical examples for orthopedic internal fixation devices (schematic)

(a) and (b) Round hole bone plates (can be used with compression devices). (c) Classical Sherman bone plate. (d) to (g) Dynamic compression plates of various sizes. (h) Compression bone plate with glide holes. (i) Classical Bagby compression bone plate. (j) Cortical bone screw. (k) Cancellous bone screw (with shaft to produce compression). (l) Condylar angle blade plate. (m) Hip plate for osteotomies. (n) Jewett nail plate with three-flanged nail. (o) Two-component dynamic hip screw plate. (p) Miniature L-plate—for example, for hand surgery. (q) T-plate for the humeral and tibial head. (r) Straight Küntscher intramedullary femoral nail. (s) Intramedullary tibia nail. Note cloverleaf profile on both (r) and (s).

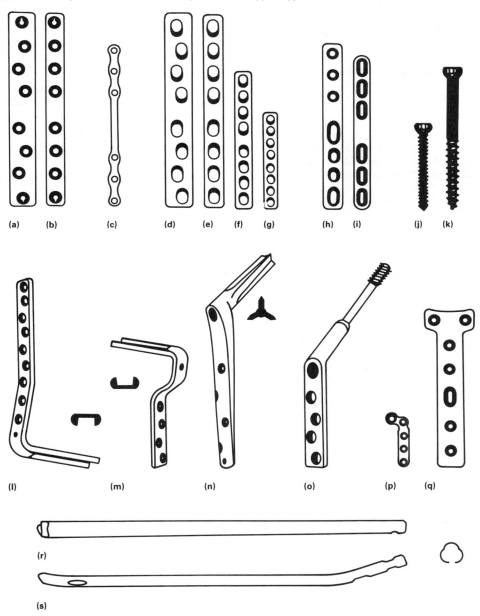

determines the further course of postoperative weight bearing. Because of the above-mentioned complexity, the possibility of an implant-related complication is comparably higher in fracture treatment than in corrective surgery or joint replacement, which are both usually carried out under standard conditions. Unless a pathological situation is present, the surgeon can achieve primarily stable biomechanical conditions when performing a corrective osteotomy or a joint replacement. When these conditions remain, the probability of an implant failure is minimal. The stabilization difficulties can be listed in ascending order, as follows:

- Joint replacements (endoprostheses)
- Corrective orthopedic surgery
- Fracture treatment
- Tumor resection (most difficult)

Complications Related to Implants

Internal fixation or joint replacement procedures can fail for such reasons as:

- Failure of the bone to heal
- Bone resorption
- Breakage of bone
- Loosening of implants
- Bending of implants
- Breakage or disintegration of implants

These events can be interrelated and usually require an intervention with reoperation to effect clinically successful results.

The term implant failure can be misleading. Therefore, in this article, the term implant failure is used to indicate that an implant has broken or disintegrated in a manner such that it cannot fulfill its intended function. Implant failure is not identical with a faulty implant, because there can be purely biological and biomechanical reasons for implant breakage or malfunction. In most cases, only a thorough analysis of the clinical conditions under which an implant failed can explain the reason for failure.

Implants can undergo surface attack by fretting, fretting corrosion, or wear. These types of attack can be relatively mild; they often occur only on the microscopic level, do not interfere with the functioning of the implant or the healing of the bone, and do not require reoperation. On prosthetic devices, however, wear of the articulation surfaces has occasionally been found to be so intense that the components have had to be replaced. Corrosion of implants involving dissolution that requires intervention is found only with materials that are excluded by the official standards for materials for orthopedic implants. In this article, the term degradation will be associated with the surface attack of implants.

Metallic Implant Materials

A number of metals and alloys have proven to be satisfactory as implant materials during years of surgical application. They are specified as implant materials by standards of the American Society for Testing and Materials (ASTM) and the International Organization for Standardization (ISO), as listed below, and by other national standards. These materials are corrosion resistant and well accepted by body tissues (biocompatible) and therefore satisfy two of the basic requirements for implants. These two properties are generally related because the less substance the metal surface releases, the better the material is accepted by the tissue. Experimentation has proven that the different pure metals of the periodic table exhibit cell toxicity at different concentrations in tissue and organ-culture tests (Ref 4-6). Some of the metals play important roles in the

body metabolism, although they can become toxic beyond certain concentrations (Ref 7). For other metals that behave indifferently in the body, such as titanium, no metabolic function is yet known. Information on the toxicity of various metals is available in the article "Toxicity of Metal Powders" in Volume 7 of the 9th Edition of *Metals Handbook*.

Fatigue resistance is another general requirement for implants, but the critical loading is different for the various types of implants and applications. The required mechanical strength of orthopedic implants also varies and depends on the shape of the implant and the application. Good ductility (discussed below) is often desired.

The following is a list of major standards for orthopedic implant materials:

- *Stainless steel*: ASTM F 55-82, ASTM 56-82, ASTM F 138-82, ASTM F 139-82 (contains remelted Special Quality), ISO/DIS 5832/1 (1986)
- *Unalloyed titanium*: ASTM F 67-83, ISO 5832/II (1984)

Fig. 3 Ionic compositions of blood plasma, interstitial fluid, and intracellular fluid

Source: Ref 9

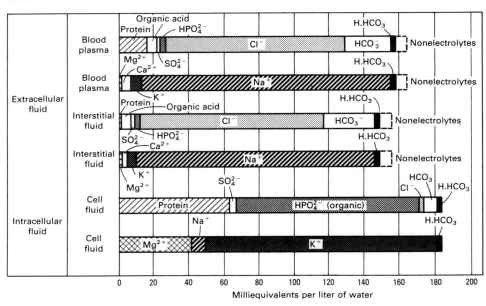

Table 1 Estimation of load cycles for one leg during walking

For these calculations, it is assumed that one load cycle per leg takes 2 s, which means that every second one step is made with alternating legs (other authors estimate 1 to 2.5 × 10⁶ load cycles per year per leg).

Motion per day, h	Cycles per day	Cycles per 1 month	Cycles per 4 months	Cycles per year
1	1 800	54 000	216 000	657 000
3	5 400	162 000	648 000	1 971 000
6	10 800	324 000	1 296 000	3 942 000

Fig. 4 Microradiograph from thin section of cortical bone

The varying x-ray density of the Haversian systems and the enlargement of some blood vessel cavities indicate that the bone is in a stage of remodeling. 77×

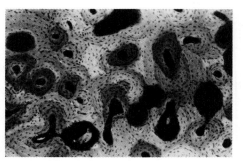

- *Titanium alloy Ti-6Al-4V ELI*: ASTM F 136-79, ISO 5832/III (1978)
- *Cast cobalt-chromium-molybdenum alloy*: ASTM F 75-82, ISO 5832/IV (1978)
- *Wrought cobalt-base alloys*: ASTM F 90-82, ASTM F 562-84, ISO/DIS 5832/6, ASTM F 563-83, ISO/DIS 5832/8

Fig. 5 Schematic of different degrees of stability in internal fixations

(a) Best possible stabilization with tension band fixation. The plate is on the tension side of the bone. The fracture site is compressed by means of the plate, which was applied under tension. If the bone surface is straight, the plate is prebent at the center to exert compression and on the opposite cortex when the plate is put under tension. (b) Good stability. The two screws that stabilize the small third fragment are inserted in lag screw fashion and compress the fracture gaps. If no bone defects are present, the reduced bone and the plate support each other. See Ref 3 for techniques. (c) Unstable situation where plate has only adapting and bridging functions because the fragments are too small to be stabilized individually. Primary bone grafting may be indicated. This situation does not differ significantly from that shown in (d). (d) Plate has bridging function in leg-lengthening procedure. The gap between bone ends has been created intentionally for lengthening. Except for very young patients in whom a bone regenerate appears spontaneously, the gap is filled with bone grafts. Implant breakage has been rarely reported for this technique.

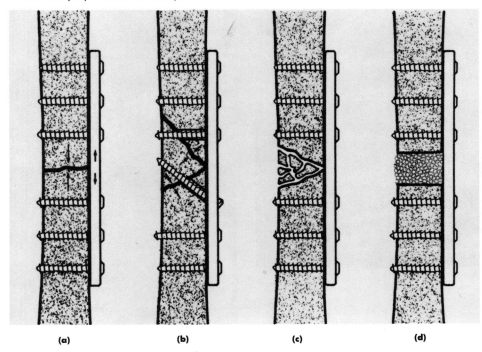

(a)   (b)   (c)   (d)

Certain applications—for example, in the skull—involve use of tantalum and niobium, both of which have high corrosion resistance and are biologically well accepted. Currently, clinical trials are being performed with modified austenitic stainless steels and titanium alloys, which may become accepted implant materials in the future. Additional ASTM standards specify cast and forged conditions for stainless steels and other materials; they are found among other implant-related standards in Ref 8.

The Body Environment and Its Interactions With Implants

The Biochemical Environment. Orthopedic implants are exposed to the biomechanical and biochemical forces of the body, and certain interactions take place between the implants and the biological environment. Figure 3 indicates the ionic composition of blood plasma and interstitial and intracellular fluids. Of the components shown, the chlorine ions are the most critical for metal implants. The normal pH of the body liquids is about neutral—in the range of 7.2 to 7.4 pH (Ref 10). At injured sites, the pH shifts to acidic values of about 5.2; in hematoma, it can drop to 4.0. In cases of infection, the pH shifts to alkaline values (Ref 11). At sites of local fretting corrosion, the environment can also become acidic through corrosion products and thus enhance further corrosion.

Chromium steels, low-grade austenitic stainless steels, and sensitized stainless steels are not sufficiently corrosion resistant in the body environment. This has been observed on retrieved implants that were manufactured 20 to 40 years ago. High-grade austenitic stainless steels as specified in the standards (ASTM F 138/139, ISO 5832/1; 1986) exhibit overall corrosion resistance and are presently the materials most frequently used for internal fixation devices.

The Dynamic Environment. Various types of forces act on the implants and the bone. In the intact musculoskeletal system, the acting forces are balanced. The macroscopic shape and the microscopic structure of the bone are adjusted to the forces of normal motion and loading. Muscles and tendons counteract bending forces that would overload the bone. When loaded physiologically under weight bearing, the bone undergoes small elastic deformations, which act as a fine stimulus that keeps the bone in a healthy condition. For example, a trained athlete develops heavier and stronger bones than an untrained person, or if a patient is bedridden for a long period of time, his bone structure becomes porous because it is not exposed to the normal loading forces.

When a bone is fractured, the balance of forces is destroyed. This can be well studied on the dislocations of injured bones; in these cases, the muscle forces pull the bone fragments in various directions. After operative reconstruction of a fractured bone, the fragments are stabilized with implants. If the bone is perfectly reduced, the entire implant is supported by bone, and the acting forces are again in equilibrium. Relatively small and uncritical loads are then exerted on the implant, and the risk of implant-related complications is minimal.

However, if the bone is not completely reconstructed—that is, fracture gaps are present or fragments of bone are missing—the weight-bearing forces are not completely balanced and evenly distributed. As a result, bending and torsional stresses can concentrate in areas of the implant where bone support is missing. Under weight bearing, the implant undergoes cyclic loading in these zones, and a potential risk of fatigue damage may arise. For fatigue cracks to develop, it is not necessary for the implants to be loaded in the plastic deformation range. Local stresses that occur under loading in the elastic deformation range of the material are sufficient to initiate fatigue cracks in the surface of the implant.

The development of fatigue damage depends on the number of load cycles and the intensity of loading. In clinical terms, this means that implant fatigue depends on the width of the bone gaps, on the length of the lever arms, and on the intensity and duration of the weight bearing, if a critical fatigue condition develops.

Table 1 gives an estimation of the number of cycles an implant may undergo during a given time period, during which a certain amount of motion per day is anticipated. To estimate the loading, static and dynamic forces must be considered. In one study, a maximum force of 2600 N (585 lbf) was determined on the femoral head in the loading phase during a test of a 60-kg (135-lb) patient (Ref 12). This is more than four times the body weight.

As long as bone healing progresses normally, an implant will not undergo fatigue fracture, because the loading decreases successively as the bone becomes more supportive. Normal bone has a remarkable inherent regeneration potential. However, certain conditions must be fulfilled so that the bone heals and regenerates. The course of bone healing is not fully predictable and depends on various factors.

Bone Healing. Normal bone regenerates continuously at a relatively slow rate. About 3% of the Haversian systems of the bone structure are usually in a stage of remodeling (Ref 13, 14). In the injured bone, the remodel-

Fig. 6 Failed historic Lane plate

(a) Heavily corroded Lane plate from chromium steel. Implant was retrieved after 26 years. (b) Longitudinal section parallel to plate with large corrosion holes. 190×. (c) Microprobe analysis of tissue surrounding the plate. Chromium and iron corrosion products have impregnated the tissue. At the location shown, the question is whether the corrosion products penetrated bone structures or the impregnated tissue mineralized.

Fig. 7 Retrieved screw of cast cobalt-chromium-molybdenum alloy (type ASTM F75)

(a) Defective screw threads from casting deficiencies. (b) Longitudinal section through threads showing porosity. 15×. (c) Enlarged thread of section shown in (b) with gas holes, segregation of primary phases, and dissolved oxides. 155×

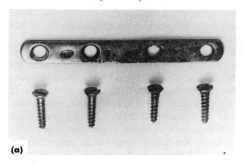

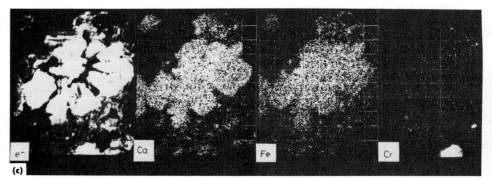

(a) (b) (c)

(a)

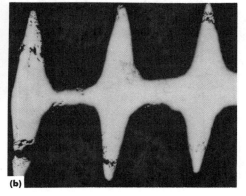

(b)

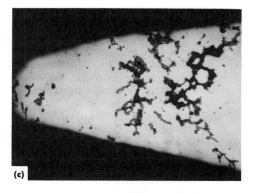

(c)

ing and regenerative processes are highly activated. Figure 4 shows a microradiograph of the cortical bone structure in the remodeling stage. One important precondition for bone healing is immobilization. Bone tissue tolerates only minimal amounts of elastic motion, particularly during repair. This explains the high success of operative fracture treatment by internal fixation. Even in severe joint injuries, functional healing is achieved.

Depending on the local conditions, different bone repair mechanisms can occur (Ref 13, 14). Generally, one can distinguish secondary and primary bone healing. Secondary bone healing takes place if relatively wide fracture gaps exist and a certain amount of mobility is present. The bone heals through callus formation; that is, fibrous tissue and cartilage form first, followed by bone. Under too much motion, complications can arise because bone does not form, and nonunion, or pseudarthrosis, develops (pseudarthrosis is the formation of a false joint between bone fragments). If this happens and an implant is present, enough load cycles can accumulate at the fracture zone to cause the implant to fail through fatigue. In most such cases, a reoperation is inevitable to treat the nonunion—for example, by application of compression—regardless of whether a broken implant is present or not.

Example 7 in this article discusses a case in which the implant was removed to treat a nonunion. Microscopic analysis revealed that a fatigue crack had already developed at the hole of the plate, which was situated near the unstable fracture site. Another plate was inserted under compression, and the bone healed without further complications.

Primary bone healing develops without callus formation, if the bone fragments are in contact with each other and are rigidly stabilized with implants. The bone fragments unite by means of transmigrating blood vessels and subsequent remodeling of the bone structure. These healing processes can be disturbed if the patient has a pathological bone condition, if an infection is present, or if necrotic (dead) bone fragments are resorbed.

Necrosis (the pathological death of living tissue) occurs when the blood supply of the bone fragments is damaged by the injury. Such fragments loose their nourishment and vitality. Complications can arise when necrotic fragments are resorbed: the fracture gap widens, relative motion increases, and bone healing is prevented or significantly delayed. Again, an implant can be subjected to a long period of increasing cyclic loading with a chance of fatigue failure. To treat such a situation and to enhance bone healing, implantation of cancellous (spongy) bone grafts and compression of the fracture site may be indicated. Although a second operation involves discomfort for the patient, it is sometimes necessary in order to

ensure bone healing and satisfactory functional results.

Bone Resorption Through Motion. Clinical and experimental observations indicate that persistent motion beyond a certain degree causes bone resorption when mechanical instability is present (Ref 15, 16). Thus, fracture gaps can widen and implants, or portions of implants, can become loose. The widening of fracture gaps means increased cyclic loading of the implant unless

Fig. 8 Two broken Moore pins from cobalt-chromium alloy

(a) Longitudinal section through fracture surface showing grain-boundary precipitates and a partially intercrystalline fracture. 63×. (b) SEM fractograph indicating grain-boundary separation. Compare with (e). (c) Longitudinal section through screw and nut. The nut shows as-cast structures of cobalt-chromium-molybdenum alloy (type ASTM F75). 160×. (d) Longitudinal section of the other broken pin in the cold-worked condition with fewer grain-boundary precipitates, lines of primary inclusions, and a small surface crack (there were more cracks along the surface). 100×. (e) Corresponding fracture surface with dimples and fatigue striations. Compare with (b).

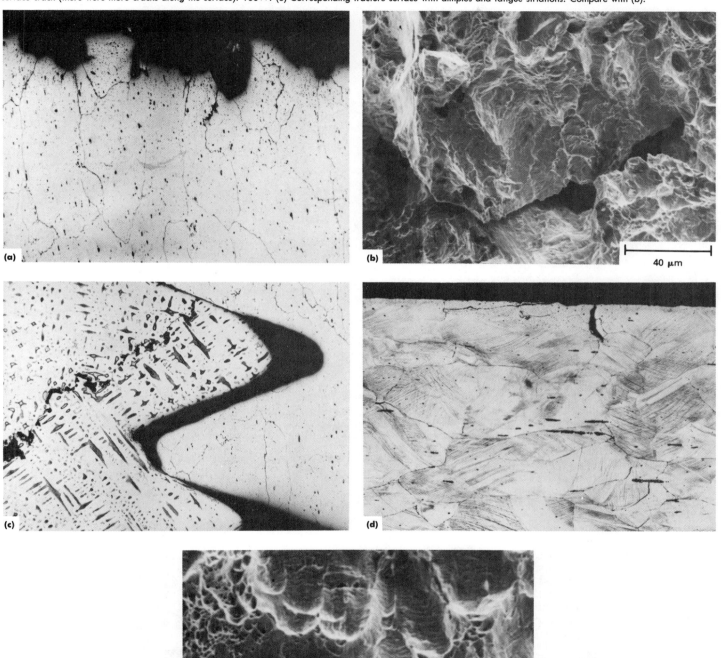

Fig. 9 Radiograph showing a tension band fixation containing a cerclage wire, two screws, and washers beneath the screw heads

See also Fig. 10.

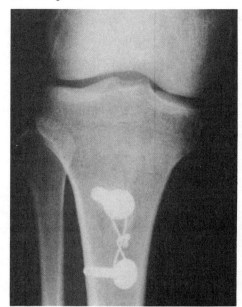

weight bearing is reduced. Implant loosening can cause either local stress concentration with increased risk of fatigue or relaxation of the stresses in the implant. Under the latter conditions, the chance of implant failure is reduced. Local bone resorption followed by mechanical degradation of bone cement and partial loosening of prosthesis stems is a typical sequence of events that is responsible for fatigue failures of prostheses.

Mechanical Properties of Bone. As a construction material, bone has properties that differ from those of metals. Bone is viscoelastic, can resorb or regenerate, and thus may change its mechanical properties and volume with time. Comparative mechanical measurements show that the ultimate tensile strength, the fatigue strength, and the elastic modulus of cortical bone are about ten times less than that of cold-worked stainless steel. Cancellous bone is spongy and much softer than cortical bone and exhibits mechanical behavior that is different from that of cortical bone, but has a higher healing rate (Ref 17). Bones derive their overall strength from their geometric configuration. The typical compressive strength of bone is the result of the composite structure of organic collagen and apatite crystals.

Combined Dynamic and Biochemical Attack. Dynamic loading in the presence of body liquids can cause wear and fretting corrosion at implant junctions, such as screw heads and plate holes, or at artificial articulation surfaces. Extensive research has been devoted to the study of wear on artificial joints and its degradation products. Corrosion fatigue

Fig. 10 Intercrystalline corrosion of a type 304 stainless steel cerclage wire

(a) Broken cerclage wire. (b) Energy-dispersive x-ray analysis under the scanning electron microscope. Lower spectrum: reference material type 316L screw material. Upper spectrum: type 304 steel with less nickel content and no molybdenum from investigated implant wire. (c) Cross section of sensitized wire, with grain boundaries and deformation lines heavily attacked by etching because of chromium carbide precipitates. 180×. (d) Fracture surface under scanning electron microscope indicating intercrystalline corrosion with pits on grain surfaces

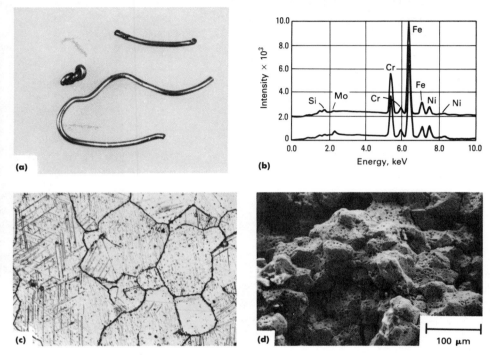

Fig. 11 Mechanically forced shearing fracture of type 316LR stainless steel screw

(a) Fracture surface with typical spiral deformation texture. SEM. (b) Close-up of fracture surface with shear dimples oriented in twisting direction. (c) Fracture edge with flow lines. (d) Longitudinal metallographic section through fracture surface. Deformation zone from shearing is adjacent to the fracture edge. Original deformation structure of the screw is visible at the bottom of the micrograph. 62×

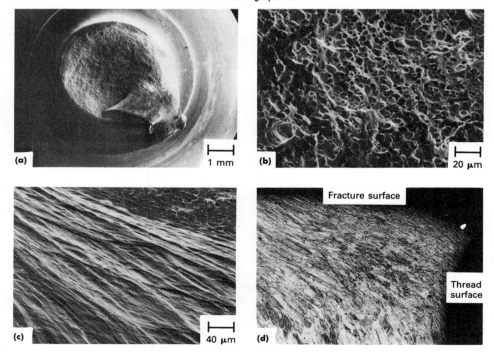

Fig. 12 Shear fracture of a commercially pure titanium screw

(a) SEM fractograph showing spiral textured fracture surface of sheared-off screw. Typical deformation lines are fanning out on the thread. (b) Uniformly distributed shearing tongues and dimples

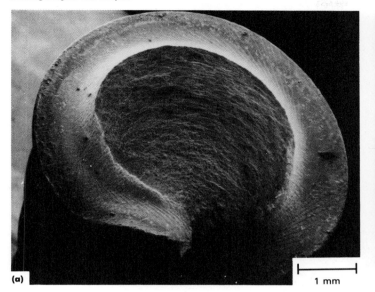 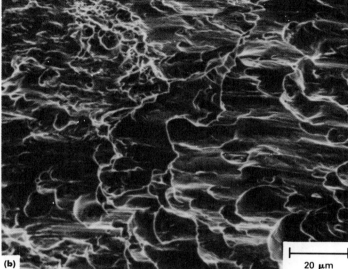

is always a subject of argument and will be discussed in the section "Fatigue Properties of Implant Materials" in this article.

It is generally agreed that stress-corrosion cracking is not a concern in high-quality implant materials. It can be observed only in combination with other failure modes, such as fatigue.

Design of Internal Fixation Devices

Leading internal fixation systems contain different types of devices that are available in various configurations and sizes. Some major implant types are specified by standard-bearing organizations, such as ASTM and ISO, but many established implants are not standardized. Figure 2 shows a selection of characteristic internal fixation devices.

Straight bone plates (a to i, Fig. 2) often must be contoured to fit the curvature of the bone. This requires ductility and moderate rigidity of the plates. In many fracture situations, it is necessary to compress the fracture site with the aid of the bone plate to achieve stability and to promote bone healing. Therefore, certain plates have specially formed screw holes (compare d to g, Fig. 2) that allow compression of the fracture gap while the screws are inserted. Certain types of plates can be applied with compression devices to achieve the same effect. One-piece angled plates, two-component angled plate systems, and plates with specially formed heads (l to q, Fig. 2) are designed for the fixation of short fragments at the bone ends, specifically the femur.

Bone screws have two characteristic types of threads (j and k, Fig. 2), depending on intended use in cortical or cancellous bone. Intramedullary nails (r and s, Fig. 2) are inserted in the bone cavity. They are increasingly used in an interlocking fashion for comminuted fractures, in which the bone is splintered or crushed into numerous pieces. Pins and wires of various dimensions belong to the standard equipment of the orthopedic surgeon. A variety of implants exist, and some are for specific fracture types and corrective operations.

When implants are designed, the following factors must be considered:

- General anatomy and its individual deviation
- Operative approach
- Operative technique
- Available space at the fracture site
- Variability of the fracture situation
- Local bone consistency
- Healing tendency of the particular bone regions
- Probable complications
- Physiology and biomechanics
- Dynamic loading and weight-bearing conditions
- Response of the bone to the implant
- Properties of the implant material

Due to all the physiological restrictions, the ideal implant design, from an engineering point of view, is sometimes not feasible. In such cases, only the best possible compromise can be reached.

In addition to biological compatibility, the endurance of the implant is one of the basic requirements. Because of the above listed factors, implants cannot be designed with high mechanical safety factors for all possible loading conditions, because the volume and rigidity would exceed the biological limits. For example, a bone plate displaces soft tissue; in areas where the soft tissue coverage is thin, a plate with too large a cross section could cause soft tissue damage and complications. An implant designed only for high mechanical and fatigue resistance could become very rigid and would shield the bone from the physiological loading stresses. Consequently, rarefaction of the bone structure could occur. If the elasticity of the plate and the screws are not in the right proportion to each other or to the bone, screws could be pulled out of the bone or could break. On the other hand, if a plate is too flexible, nonunion could occur. If too large an area of the bone surface is covered with a plate, the blood supply may be impaired.

Usually, implants are so designed that they maintain their shape unless a dramatic event takes place, such as a second trauma. As discussed above, the fatigue strength of implants is not unlimited. Stresses in the elastic loading range are sufficient to initiate fatigue. As long as the well-reduced bone fragments are in contact, the implants are supported. Critical cyclic loading occurs only when the internal fixation is unstable in the presence of fracture gaps and bone defects. Specific surgical techniques were developed to achieve stable fixation (Ref 3). Figure 5 shows the different degrees of stability in internal fixations.

There are clinical situations in which comminution (reduction to particles) or loss of fragments does not allow stable fixation.

Fig. 13 Sheared-off cast cobalt-chromium-molybdenum screw

SEM fractography. (a) Overview of portion of rough fracture surface. (b) Area with fracture planes of three differently oriented grains (single arrow, Fig. 13a). (c) Shearing structures and dimples in grain identified by the numeral 1 in Fig. 13(b). (d) Slip traces in grain identified by the numeral 2 in Fig. 13(b). (e) Gas holes with dendritic freezing surfaces (double arrows, Fig. 13a). (f) Longitudinal section through identical screw with starting shearing damage. 10×. (g) Shearing crack (arrow, Fig. 13f). 130×

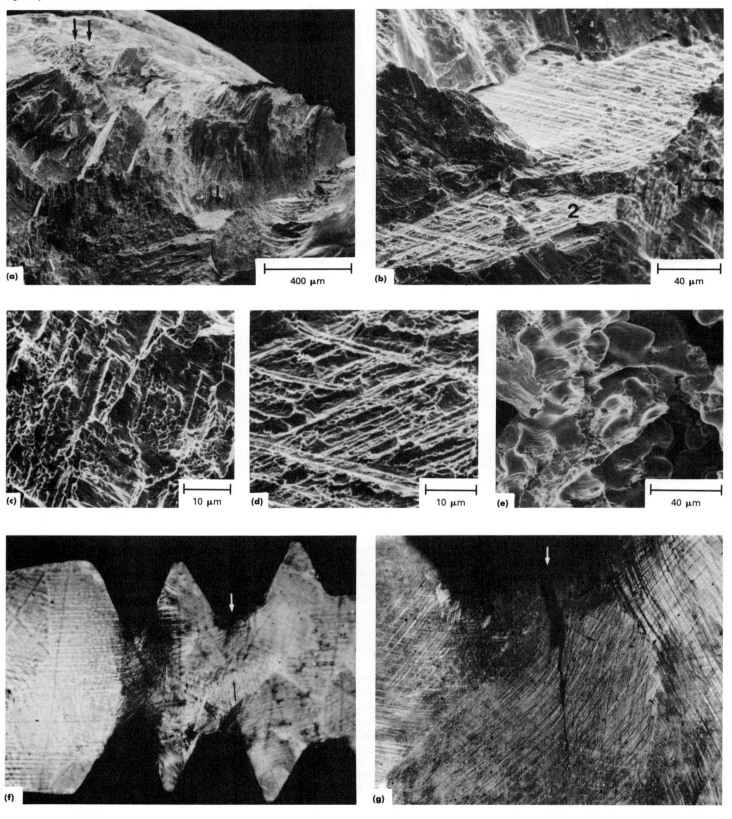

Fig. 14 Type 316LR stainless steel screws that failed by fatigue

(a) Fatigue fractures at different thread levels. (b) Longitudinal section perpendicular to fracture surface without deformation zone at fracture site. A small secondary crack is shown at thread site (arrow). 55×. (c) Fatigue striations on fracture surface

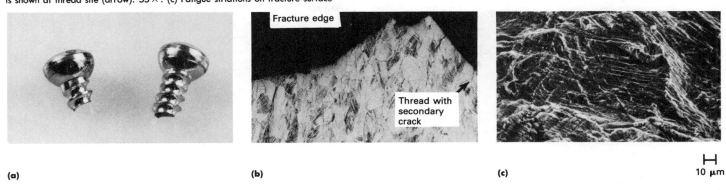

(a)

Fracture edge

Thread with secondary crack

(b)

(c)

⊢⊣
10 μm

Fig. 15 Crack initiation on type 316LR stainless steel dynamic compression plate

(a) Anterior-posterior radiograph. The plate was used to treat the nonunion of a fracture between the fourth and seventh screws. The plate was bent intraoperatively to fit the contour of the bone. (b) Radiograph with lateral view showing wide bone gap, indicating instability. The boxed area indicates the location of initiating fatigue cracks.

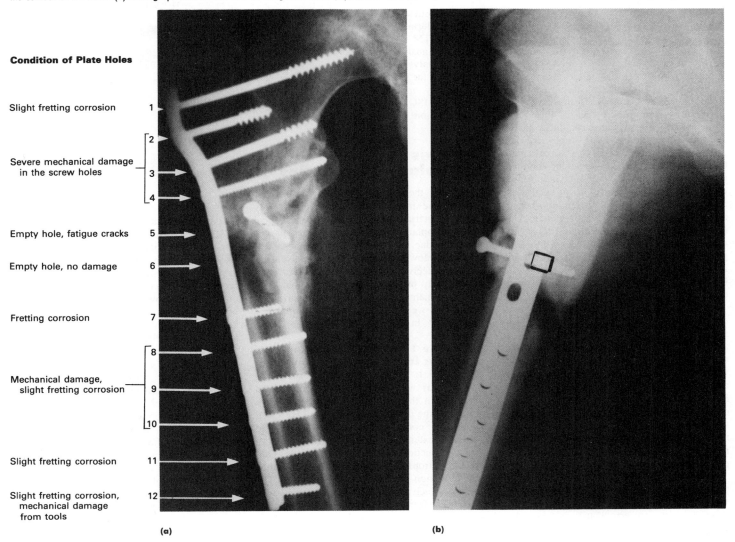

Condition of Plate Holes

Slight fretting corrosion — 1

Severe mechanical damage in the screw holes — 2, 3, 4

Empty hole, fatigue cracks — 5

Empty hole, no damage — 6

Fretting corrosion — 7

Mechanical damage, slight fretting corrosion — 8, 9, 10

Slight fretting corrosion — 11

Slight fretting corrosion, mechanical damage from tools — 12

(a)

(b)

Fig. 16 Two views of the plate surface with fatigue cracks at the fifth hole
White arrow in the left-hand figure indicates tool mark from bending that does not interfere with fatigue damage structure. See text for further description. Both 110 ×

 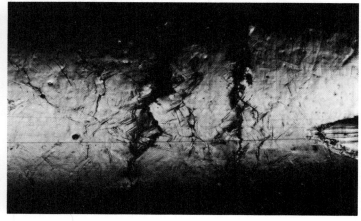

Implantation of bone grafts is then indicated to promote bone healing and to develop stability.

Analysis of Failed Internal Fixation Devices

The metallurgical investigation of failed implants can reveal the failure mechanism. Because of the complexity of the biological conditions discussed earlier in this article, the cause of implant failure can in most cases be assessed only when the clinical history and the corresponding radiographs are analyzed by a clinical expert. This is particularly true in cases of fatigue failure. One should be cautious not to mistake small deviations from material standard specifications or small flaws as the reason for an implant failure when the real cause is heavy cyclic overloading. As experiments demonstrate, fatigue cracks can initiate on the implant surface without the presence of flaws. Judging the implant design as a possible cause of failure necessitates biological, biomechanical, and surgical knowledge.

Implants should be disinfected before examination. Although implants are usually rinsed in the hospital, biological residues can adhere to the surface and obscure the details of the fracture surface. Ultrasonic cleaning with 10% oxalic acid, a clinical disinfectant, or an enzymatic digestion is suitable for the highly corrosion-resistant implants. Probable corrosion products must be detected before such a procedure. However iron-rich blood residues should not be mistaken for rust if local microanalysis of the chemical composition is performed.

Under the microscope, electropolished implant surfaces usually exhibit small craters where inclusions were dissolved. Occasionally, these structures are erroneously interpreted as corrosion pits.

During the internal fixation and later at the retrieval, implants are handled with metallic tools, and this handling results in scratches and mechanical marks on the implant. It is often not possible to determine whether the marks were produced during retrieval or fixation.

The ASTM standard F 561 on implant retrieval and analysis may be taken as a guideline for the analysis of failed implants. Not all portions of the suggested investigation protocol are equally applicable for the different implants. An ISO standard on the same subject is in preparation.

Failures Related to Implant Deficiencies

Most of the failed implants submitted for analysis are manufactured from high-quality implant materials and are free of metallurgical defects. Implant breakage of current materials is usually due to mechanical or biomechanical reasons. However, there are occasionally implants received that are manufactured from inadequately processed materials or that were produced many years ago when metallurgical techniques were less advanced. Examples 1 through 4 in this article document such cases.

Example 1: Severe General Corrosion on Historic Lane Plate Made From Chromium Steel. The four-hole Lane plate shown in Fig. 6(a) was inserted 46 years ago and remained in the body for 26 years. A large portion of the plate disintegrated and consisted mainly of corrosion products. Figure 6(b) shows the extensive corrosion holes on a metallographic section of the plate. The plate, manufactured from a chromium steel, exhibited transformation structures and carbides. The screws, which are formed like wood screws, were made from a soft austenitic 304 stainless steel and exhibited minimal surface corrosion. The corrosion products of the plate impregnate

Fig. 17 Top surface of broken plate of type 316LR stainless steel
Fatigue cracks parallel to the fracture edge and a wide area exhibiting primary fatigue deformation are visible. 65 ×

the surrounding tissues, as seen on the microprobe analysis shown in Fig. 6(c).

This is a case of general corrosion due to improper material selection. The corrosion may have been enhanced because of the austenitic stainless steel screws. In contrast, a recently retrieved intramedullary tibia nail of the type shown in Fig. 2(s), which had been in the body for 24 years, showed no signs of corrosion. This nail was produced from a low-carbon remelted type 316L stainless steel.

Example 2: Retrieved Bone Screw Made From Cobalt-Chromium-Molybdenum Alloy with Casting Defects. Portions of the threads of the screw shown in Fig. 7(a) had broken off and other threads had cracked. The screw was made from a cast cobalt-chromium-molybdenum alloy. A longitudinal section through the screw revealed gas porosity, segregation of primary inclusions, and oxides (Fig. 7b and c). The screw threads broke because of mechanical weakness from gas holes, brittleness, and dissolution of oxides. With today's advanced casting techniques and testing methods, such implant defects can be avoided and eliminated.

Fig. 18 Stainless steel bone plate with fatigue crack and broken screw

(a) Radiograph taken 13 weeks after operation. Anterior-posterior view. Arrows indicate crack in plate and open fracture gap. (b) Corresponding lateral view. Arrow indicates broken screw. (c) Bend in plate in the horizontal plane at the unfractured side of a screw hole (arrow). Deformation was due to dorso-lateral acting forces. Screws are not placed at original position. See also Fig. 19.

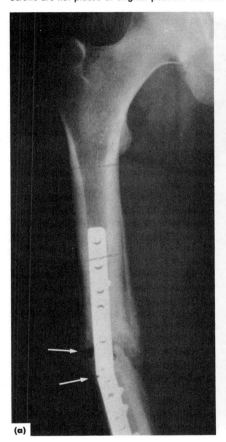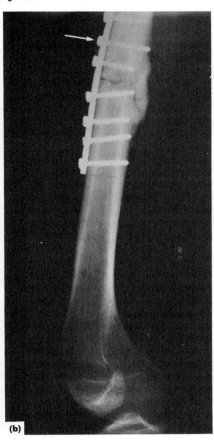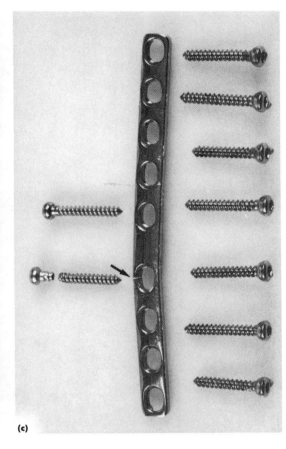

Example 3: Broken Adjustable Moore Pins Made From Cobalt-Chromium Alloy. Two of four adjustable Moore pins, which had been used to stabilize a proximal femur fracture, were broken and deformed at their threads. The pins were made from a cobalt-chromium alloy and were not in the same condition. One pin contained brittle precipitates in the grains and grain boundaries. Correspondingly, the fracture occurred partially along the grain boundaries (Fig. 8a). Scanning electron microscopy (SEM) showed grain-boundary separation (Fig. 8b).

The nut used with the pin was made from a cast cobalt-chromium-molybdenum alloy and had segregations in the microstructure and gas holes and precipitates in the grain boundaries (Fig. 8c). The nut was not specifically loaded and did not fail.

The other broken pin was made from a cold-worked cobalt-chromium alloy. The grain boundaries contained few precipitates, but lines of primary inclusions were randomly distributed (Fig. 8d). The fracture surface exhibited a mixed fracture mode with intermingled dimples and fatigue striations (Fig. 8e). The pins did not exhibit any corrosion.

This example demonstrates how the different conditions of cobalt-chromium alloys affected failure behavior. In one of the Moore pins, brittle fracture was aided by grain-boundary separation.

Example 4: Intercrystalline Corrosion on Cerclage Wire of Sensitized 304 Type Stainless Steel. The wire was used with two screws and washers for a tension band fixation (Ref 3) in a corrective internal fixation (Fig. 9). The cerclage wire was found broken at several points and corroded when it was removed after 9 months (Fig. 10a). Metallographic analysis revealed that the wire that was made of type 304 stainless steel without molybdenum, as seen in the energy-dispersive x-ray analysis (Fig. 10b); this material does not comply with the standards. The screws and washers were intact and were made of remelted implant-quality type 316L stainless steel corresponding to the standards. The microstructure of the wire, which was in the soft condition as required, showed signs of sensitization, with chromium carbide precipitates at the grain boundaries (Fig. 10c). Typical intercrystalline corrosion with pitted grains was evident through SEM fractography (Fig. 10d).

For this internal fixation, orthopedic wire of an insufficient stainless steel type was used. Improper heat treatment of the steel lead to intercrystalline corrosion and implant separation. The wire came from a source different from that of the screws and washers and did not meet current standards for orthopedic wire.

Implant Failures Related to Mechanical or Biomechanical Conditions

Bone screws fracture in two modes—by mechanical forces during insertion or removal or by fatigue failure. Examples for both are given below.

Mechanically forced shearing fractures are a typical failure in screws. Although stainless steel bone screws are normally manufactured from cold-worked material, their shearing ductility is usually considerable. Ductile shearing fractures are characterized by a spiral texture of the fracture surface, flow lines at the fracture edges, overload dimple structure on the fracture surface, and heavily deformed microstructure at the fracture edge in longitudinal sections. Occasionally, screw heads can

Fig. 19 Fracture surfaces of the failed screw and bone plate shown in Fig. 18

(a) Longitudinal section through fractured screw showing edge of fracture surface and high inclusion content. A large slag inclusion was present at the void under the fracture edge. 55×. (b) Fracture surface of screw showing fatigue striations (containing secondary cracks) and dimples. (c) Fracture surface of screw and zone showing overload fracture containing large dimples partly caused by inclusions. (d) Plate fracture surface. Arrow points to crack origin, which is indicated by beach marks. (e) Fatigue striations from center of plate fracture surface (running originally transverse to loading direction)

be sheared off by repeated tightening, particularly when the cortex (dense outer bone wall) is thick, because the thicker the cortex, the higher the retention of the screw. Excessive oblique positioning of the screw in a plate hole can also cause fracture if the screw head is bent by the plate hole during insertion.

Example 5: Shearing Fracture of a Type 316LR Stainless Steel Screw. The cortical bone screw shown in Fig. 11 broke during the internal fixation procedure. It may have been inserted very obliquely in the screw hole and therefore sheared off. The fracture surface (Fig. 11a) exhibited extensive spiral deformation. Figure 11(b) shows the dimple structure characteristic of a ductile failure mode. The dimples are oriented uniformly in the deformation direction. Figure 11(c) shows a magnification of the flow lines at the fracture edge. The longitudinal metallographic section perpendicular through the fracture surface (Fig. 11d) exhibited a zone of heavily deformed grains at the fracture edge. The microstructure and the hardness of the screw corresponded to the standards.

For comparison, the shearing fractures of a commercially pure titanium screw and a cast

cobalt-chromium-molybdenum alloy (ASTM F75) screw are shown in Fig. 12 and 13. The spiral-textured fracture surface of the titanium screw was smooth (Fig. 12a) and had uniform, shallow shearing tongues and dimples indicative of a ductile fracture mode. (Fig. 12b).

In contrast, the fracture surface of the cast cobalt-chromium-molybdenum alloy screw had an erratic structure (Fig. 13a), which can be explained by the large grain size and the tendency of the grains to fracture along crystallographic planes. Figure 13(b) shows such fracture planes of differently oriented grains. These planes exhibited crystallographically oriented mixed shearing structures and dimples (Fig. 13c), as well as traces of slip (Fig. 13d). Gas holes with dendritic freezing surfaces are shown in Fig. 13(e). In the longitudinal section of a similar screw (Fig. 13f), large grains can be observed, along with deformation between the threads, which leads to shear rupture. Figure 13(g) shows a starting shear crack propagating along slip planes.

Fatigue Failures of Screws. As discussed in the following example, fatigue failure at varying thread levels is also a problem some-

times encountered. One of the causes of such failures is unstable fixation.

Example 6: Characteristic Observations on Type 316LR Stainless Steel Screws That Failed by Fatigue. Fatigue fracture can occur on different thread levels, depending on the loading situation (Fig. 14a). No deformation zone was created at the fracture site shown in the longitudinal metallographic section in Fig. 14(b). Occasionally, the initiation of secondary fatigue cracks is found parallel to the fracture plane or in a nearby thread groove. Figure 14(c) shows one of the fracture surfaces with fatigue striations.

Corresponding to the energy-dispersive x-ray analyses, the screws (Fig. 14a) were manufactured from cold-worked remelted (R) type 316L stainless steel and showed no structural or manufacturing defects. The material meets ISO and ASTM F 318 Special Quality specifications. The screws were used with a relatively rigid plate to treat a fracture complication in the upper end of the femur. Radiographs indicated various signs of unstable fixation, which explains the fatigue failures. The two screws were situated at the distal end of a compression screw plate, where the proximal

Fig. 20 Angled blade plate that failed by fatigue

(a) Fatigue fracture at plate hole. (b) Top plate surface at fracture edge showing intense fatigue-surface damage (slip bands), indicating that the plate was heavily loaded. (c) SEM of fracture surface. Fatigue striations can be seen in the grain at the lower right-hand corner. Other fracture features are crystallographically oriented. (d) Microstructure of cross section reveals that plate was made from cold-worked stainless steel of high microcleanliness. 80 ×

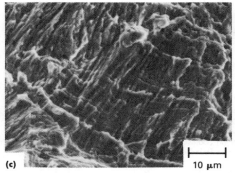

(a)

(c)   10 μm

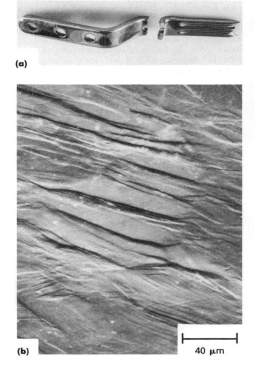

(b)   40 μm

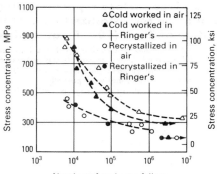

(d)

Fig. 21 Fatigue curves of type 316LR stainless steel implant material tested in bending mode

(a) *S-N* curves for stainless steel in cold-worked and soft condition that was tested in air and aerated lactated Ringer's solution. (b) Fatigue curve for number of cycles to failure as shown in Fig. 21(a) for cold-worked stainless steel that was fatigued in air. The two additional curves show the number of cycles to initiation of visible slip systems and the number of cycles to crack initiation. Source: Ref 20

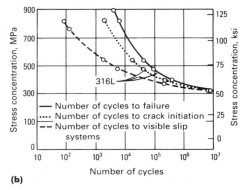

(a)

(b)

lever arm was the longest compared to the other screws.

Fatigue Damage of Bone Plates. Most broken plates have been loaded under unilateral bending, because bone defects in the opposite cortex are a typical cause of instability and breakage. Thus, tensile and shearing stresses are created in the top surface of the plate. Correspondingly, this is the surface where crack initiation is usually found. Depending on

Fig. 22 Secondary corrosion attack on fatigue-fracture surface

(a) Fracture surface of 5-mm (0.2-in.) long crack in an intramedullary tibia nail made of cold-worked type 316LR stainless steel. The crack developed during the early postoperative stage when the fixation was unstable and bending and rotational forces acted on the nail. The crack did not impair fracture healing or nail removal. The fracture surface was covered with fatigue structures except for a small zone where shallow corrosion pits superimposed the fatigue structures. Locally, critical crevice conditions must have developed in the crack. Under weight-bearing, small relative motion can take place in the gap, damaging the passive film. In addition, stagnation and acidification of the body liquid is possible. Arrows indicate the transition between fatigue and corrosion structures. The microstructure of the nail showed no irregularities and was not connected with the corrosion. (b) Shallow secondary corrosion pits superimposed on fatigue striations.

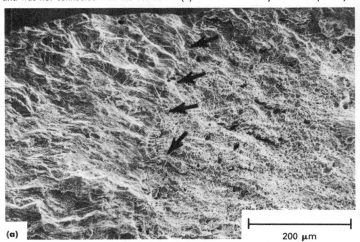

(a)   200 μm

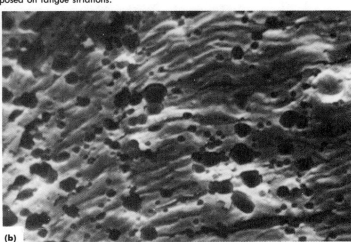

(b)

Table 2   Relationship between the ultimate tensile stress, yield stress, and fatigue limit for cold-worked and recrystallized type 316LR stainless steel

Material condition	UTS(a) MPa	UTS(a) ksi	Yield stress(b) MPa	Yield stress(b) ksi	Fatigue limit(b) MPa	Fatigue limit(b) ksi	Ratio of fatigue limit to UTS	Ratio of fatigue limit to yield stress
Cold worked	1050	152	994	144	300	43.5	0.28	0.30
Recrystallized	590	85.5	266	38.5	180	26	0.30	0.67

(a) UTS, ultimate tensile stress. (b) Yield stress = $\sigma_{0.2\%}$ yield stress

Fig. 23 Free surface replica showing the development of fatigue-surface damage on recrystallized type 316LR stainless steel in aerated Ringer's solution at 38 °C (100 °F), at applied stress of 250 MPa (35.5 ksi)

(a) The first visible slip systems developed at a triple point (decorated persistent slip steps) after 500 load cycles. 225×. (b) Increased density of multiple slip systems after 29 500 cycles. 275×. (c) Short cracks that developed in a zig-zag pattern between glide systems after 40 700 cycles. Loading axis is horizontal to the micrograph. 110×. (d) Major cracks that were partially developed through coalescence of short cracks after 60 000 cycles. Loading axis is vertical to the micrograph. 90×. The specimen failed after 69 113 cycles.

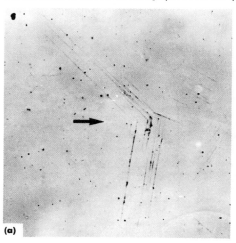

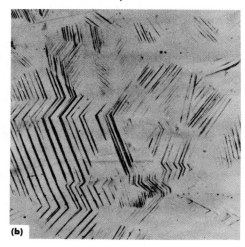

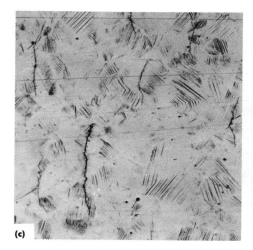

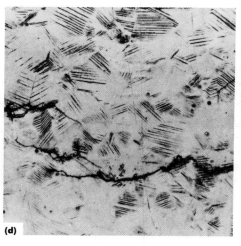

Fig. 24 Cold-worked type 316LR stainless steel that was fatigued in air at different stress levels

Surfaces of broken specimens at fracture edge are shown. (a) Failure at an applied stress of 330 MPa (47.8 ksi) after 7 682 434 load cycles. Only a few glide systems adjacent to the fracture edge are visible. 70×. (b) Failure at an applied stress of 600 MPa (87 ksi) after 105 308 cycles. Multiple glide systems and additional cracks near the fracture edge are shown. 55×. (c) Failure at an applied stress of 800 MPa (116 ksi) after 5308 cycles. Heavy deformation structure and multiple additional cracks appear. 55×

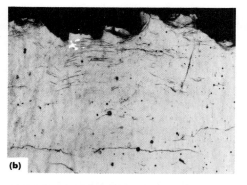

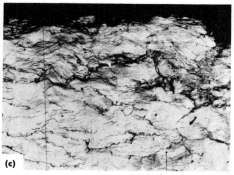

the fracture situation and the location, torsional stresses can also be created. This is especially true in forearm fractures because of the rotational motion that takes place. Examples 7 to 9 show three stages of fatigue damage on bone plates.

Example 7: Fatigue Initiation on Type 316LR Straight Bone Plate. The plate shown in Fig. 15(a) and (b) was used to treat a

pseudarthrosis in the proximal femur. Because healing did not progress, the plate was removed and submitted for investigation. The bone was replated with an angled blade plate, and compression was exerted on the pseudarthrosis to promote bone healing.

Investigation. The plate seemed to be macroscopically intact. At higher magnification, fatigue cracks were apparent on the top surface

of the small section of the plate (Fig. 16a and b) at the fifth screw hole indicated on Fig. 15(a). In the background of the surface, slip systems are visible, partially in conjunction with initial cracks.

From crack-initiation studies, it is known that persistant slip systems appear first on the surface that is loaded under critical cyclic stresses. Precracks and initial cracks develop from these slip systems. Some of the initiation cracks propagate further and become major

Fig. 25 Fatigue-fracture structures on wrought type ASTM F563 cobalt-alloy test specimens that fatigued in air

(a) Very fine fatigue striations are superimposed on crystallographically oriented fracture structures. 2480×. (b) Crystallographically oriented fracture morphology showing twin structures. 620×. (c) Close-up view of the twin structure indicated by arrow in Fig. 25(b). The fine parallel lines within the twin structure are slip (glide) bands, which are not easily distinguished from fatigue striations.

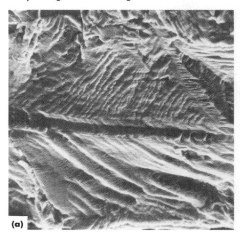

(a)

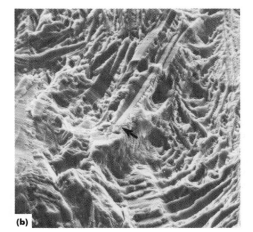

(b)

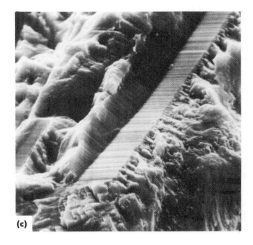

(c)

Fig. 26 Specimen surface of recrystallized titanium at fracture edge

Specimen was fatigued at a stress of 600 MPa (87 ksi) in air. Twinning, wavy glide deformation, and grain-boundary distortion are visible on this relatively heavily loaded specimen. Polarized light. 330×

cracks that penetrate the cross section. The extent of this fatigue surface damage is representative for the stress intensity.

Figure 16 shows structures as were described above for the fatigue initiation process. The intensity of these structures, compared to results of systematic crack-initiation experiments, suggested that the plate was rather heavily loaded. Morphological comparison and fatigue initiation curves indicated that a local stress concentration of 500 to 600 MPa (72 to 87 ksi) was present. This load would still be below the elastic limit of the cold-worked stainless steel. No fatigue cracks were visible on the broad section of the asymmetrically placed hole. Thus, fatigue damage occurred earlier in the portion with the smaller cross section. In addition, rotational forces are to be expected in this area.

Fatigue-bending tests on such broad plates showed that the fatigue life of plates with asymmetrically arranged holes is at least as long as for plates with holes situated in the center. At a centered screw hole, fatigue initiates at about the same time on both sides of the hole. At the asymmetrically placed holes, fatigue begins at the large section only after a fatigue crack begins to propagate into the small plate section.

Discussion. Although fatigue cracks are observed more frequently at the edges of the plate holes, crack initiation often takes place on the total surface area adjacent to the middle section of the plate hole. This is demonstrated on

Fig. 27 Fracture surface of commercially pure titanium test specimens that failed at an applied stress level of 600 MPa (87 ksi) in air

(a) Very fine fatigue striations. (b) Coarse fatigue striations probably in transition to glide bands. (c) Overload tearing structures

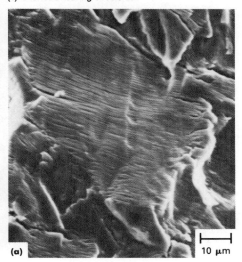

(a)   10 μm

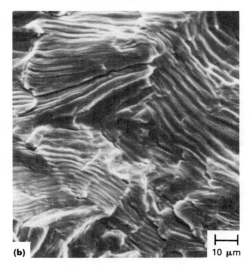

(b)   10 μm

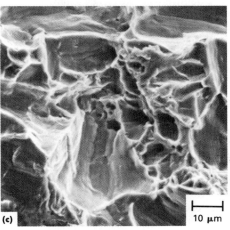

(c)   10 μm

Fig. 28 Fatigue-fracture surface of broken commercially pure titanium bone plate with mixed fracture morphology

(a) Fracture surface shows fatigue striations, terraces, and tearing ridges, depending on the local crystallographic orientation. 250×. (b) Higher magnification view of the area indicated by arrow in (a) showing large and small ledges with partial tearing configurations. 1250×

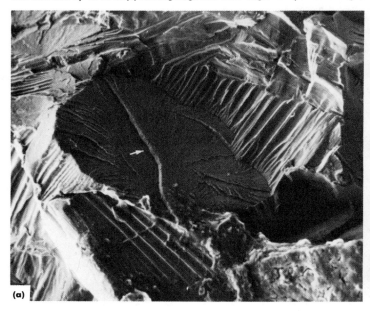
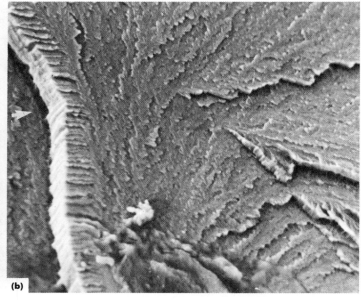

(a)   (b)

another broken plate (Fig. 17), in which a large secondary crack had developed parallel to the main crack in the center of the surface. Another small crack was visible on the edge of the plate hole. A large portion of the plate surface at the fracture edge showed deformation from fatigue damage. Crack initiation does not take place in the plate holes and is not related to the fretting that may occur between the screw head and plate. An exception may be some special plate hole designs for which corrosion fatigue was reported (Ref 18).

In Fig. 16, no crack initiation is visible at the pits caused by electropolishing. In addition, mechanical marks from the plate bending did not act as stress raisers. The stainless steel has no particular notch sensitivity, and the fatigue behavior is forgiving with respect to mechanical damage.

In the radiographs shown in Fig. 15(a) and (b), extensive bone deficiency is present in the area where two plate holes were not filled with screws. In Fig. 15(b), a wide gap is visible. The empty screw hole (No. 6 from the top) situated directly over the defect does not show fatigue damage, but hole No. 5, which sits at the edge of the proximal bone fragment, has the described fatigue cracks. Because screw hole No. 5 was situated at the transition between the supporting bone and the defect, stress concentration was higher here than at the empty hole No. 6, where the elastic deformation could occur more uniformly. In the defect zone of the original fracture, between screws No. 4 and 7, the plate is not supported by the bone, and the load is transmitted only through the plate, which caused cyclic loading of the implant and fatigue-crack initiation.

It is also characteristic that fretting corrosion occurred at the screw/plate interface at screw hole No. 7. This screw was closest to the bone defect and was secured only in one cortex. Compared to all the other screws, this one was the most likely to undergo relative motion. As outlined in the legend in Fig. 15(a), at all other contact areas between screw head and plate hole either minimal or no fretting, or fretting corrosion and mechanical damage, was found. For this analysis, only the two radiographs shown in Fig. 15 were available. They are not adequate to estimate the clinical postoperative history, but they are sufficient to explain the fatigue-crack initiation due to instability.

Example 8: Fatigue Crack on a Type 316 Stainless Steel Bone Plate and Corresponding Broken Screw. A narrow bone plate was used to stabilize an open midshaft femur fracture in an 18 year old patient.

Investigation. As seen in Fig. 18(a), a radiograph taken 13 weeks after the operation, the plate had a crack at a plate hole next to the fracture site. The plate was slightly bent in the horizontal plane, and the fracture gap was considerably open. In the lateral view, a broken screw is visible at the second plate hole above the fracture gap (Fig. 18b).

The screws and plate shown in Fig 18(c) were supplied by three different manufacturers. The compression holes of the plate are a modification of those of other implant systems. Energy-dispersive x-ray analyses indicated that all implants were made of type 316 stainless steel. However, the microcleanliness of the materials was different. The minimal primary inclusion content found in the plate and in one of the screws was typical of high-quality re-

melted steel that meets the requirements in ASTM standard F 138. The remaining screws from a different source, including the broken screw, had a high primary inclusion content and did not comply with this standard.

Figure 19(a) shows a longitudinal section perpendicular to the fracture surface of the broken screw. The local crack formation appeared to have been influenced by the presence of larger inclusions. From the striations found on the fracture surface under the scanning electron microscope, it was concluded that the screw failed through a fatigue mechanism. In addition to the fatigue striations, many dimples were present (Fig. 19b), which were created primarily through inclusions. In areas with final overload structures, the dimple formation was also influenced by inclusions (Fig. 19c).

The crack in the plate originated at the upper, outer corner of the plate, as seen from the beach marks in Fig. 19(d). This indicates the action of asymmetric bending and rotational forces. The beach marks were oxidized, and the fracture surface was covered with fatigue striations (Fig. 19e).

Discussion. Not all clinical details of this case are known, but from a biomechanical viewpoint based on the radiographs (Fig. 18a and b) the fatigue failures of the plate and screw can be attributed to the unusual placement of the narrow bone plate. Because of the open fracture situation with severe soft tissue damage, the narrow plate was placed on the frontal aspect of the femur, which is not a typical configuration. A broad plate is usually applied at the lateral side of the femur if plate fixation is indicated. Physiologically, the lateral side of the femur is under tension, and if

Fig. 29 Heavy pitting corrosion on type 304 stainless steel bone screw

(a) Longitudinal section through head of bone screw showing corrosion tunnels. (b) Etched longitudinal section showing the many primary inclusion lines and corrosion tunnels that follow the inclusions. (c) SEM overview of corrosion attack on screw head

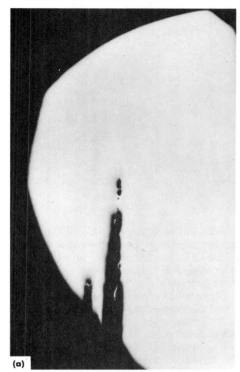

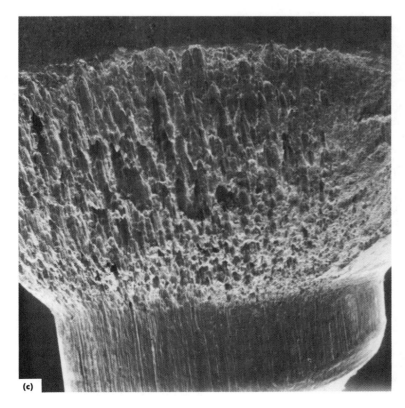

the plate is mounted laterally, exerting compression on the fracture site creates a tension band effect, which provides excellent stability (see Fig. 5a and Ref 3).

In this case, the plate was not on the tension side, and obviously, no compression (or not enough compression) had been applied. Because of the anatomic conditions and the fracture configuration, the plate was loaded under asymmetric bending and torsional forces, which created maximum stresses at the lateral edge of the plate hole near the fracture site. This is confirmed by the location of the fatigue-crack origin shown on the fracture surface in Fig. 19(d).

The instability of the internal fixation was indicated by the callus formation, by the rounded edges of the fracture ends, and by some screw loosening. The first screw over the fracture gap was definitely loose, as deduced from the resorption hole in the cortex beneath the plate. This caused additional loading on the next screw, which failed by fatigue. Because of the different sources of the implants, the head design of some of the screws was not compatible with the screw hole geometry of the plate. This did not cause the instability, but may have hastened its progress. Similarly, the inclusion content of the screw did not cause the fatigue failure, but may have reduced the fatigue strength to some extent.

This example shows that a smaller plate and an unusual positioning had to be chosen because of the critical soft tissue condition, which leads to decreasing stability and eventual fatigue failure. It is important to realize that at the time of injury the soft tissues have priority because of the risk of serious complications. After plate removal, the fracture was stabilized with an intramedullary nail, and bone healing proceeded.

Example 9: Fatigue Fracture of a Type 316L Stainless Steel Angled Plate. Figure 20(a) shows a broken angled plate that was applied to the proximal femur. The fatigue fracture occurred at a plate hole. The fatigue damage at the fracture edge at the top surface of the plate indicated that symmetric cyclic bending forces had acted. Figure 20(b) shows slip bands produced on the surface during cyclic loading. Figure 20(c) reveals fatigue striations.

The microstructure of the plate had a high degree of cleanliness (Fig. 20d). The deformation structure showed that the material was in the cold-worked condition, which was confirmed by Vickers hardness tests. Energy-dispersive x-ray analysis indicated a chemical composition corresponding to a type 316 stainless steel (the carbon content was not determined).

The plate material corresponded to the current standards for surgical implants and was not the cause of the implant failure. Because no radiographs or other clinical details were available, it could not be evaluated if the plate size was adequate for the clinical conditions.

Fig. 30 Fretting and fretting corrosion at the contact area between the screw hole of a type 316LR stainless steel bone plate and the corresponding screw head

(a) Overview of wear on plate hole showing mechanical and pitting corrosion attack. 15×.
(b) Higher-magnification view of shallow pitting corrosion attack from periphery of contact area. 355×.
(c) Contact area showing the effects of mechanical material transport (material tongue in right upper corner), fretting and wear (burnished areas), and corrosion (shallow pitting). 355×. (d) Higher-magnification view of Fig. 30(c). 650×

Fig. 31 Connective tissue near stainless steel bone plate with impregnation of corrosion products

These products are found extracellularly and in the connective tissue cells. 230×

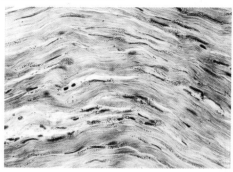

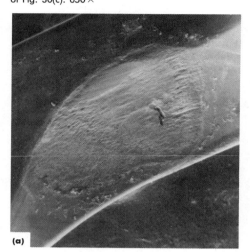
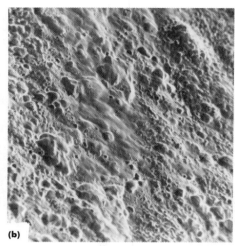
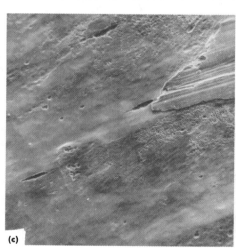
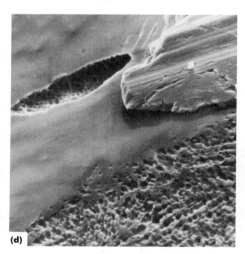

(a)
(b)
(c)
(d)

Fatigue Properties of Implant Materials

Remelted Type 316LR Stainless Steels.
Because metal fatigue accounts for most implant fractures, some fatigue properties of type 316LR stainless steel, based on experimental studies (Ref 19, 20), will be addressed.

The S-N fatigue curves for the steel in recrystallized and cold-worked conditions are shown in Fig. 21. As shown in Fig. 21(a), the fully reversed bending tests were performed on flat specimens in air and in Ringer's solution (a lactated 0.9% sodium chloride solution that contains about the same concentration of Cl⁻ ions as body fluids). The improved fatigue strength in the cold-worked condition is of particular interest for implants. Plates, screws, pins, and certain nails are produced mainly

from cold-worked material. The fatigue limits determined in rotating bending beam tests on the same materials are comparably higher than those found in the bending tests (Ref 19).

In the recrystallized condition, the fatigue strength of the steel is not significantly affected by Ringer's solution. In the cold-worked condition, fatigue life in the high-cycle range is reduced in Ringer's solution. Corrosion fatigue in the usual sense, however, does not occur, because, morphologically, no signs of corrosion are detectable by optical microscopy and SEM. Ringer's solution appears to accelerate the individual fatigue stages. This complies with observations of fracture surfaces of broken plates that do not show a corrosive component combined with their fatigue structures and which do not differ from structures produced experimentally in air.

Additions of 0.25% clottable fibrinogene to the Ringer's solution have been shown to neutralize its fatigue-life reducing effect. The corresponding S-N fatigue curves were similar to those generated in air (Ref 21).

On retrieved implants, however, it has been found that fatigue-fracture surfaces can secondarily corrode when crevice or fretting conditions develop in the fracture gap. Such corrosion structures can be locally observed. In the transition areas, fatigue striations with superimposed corrosion pits can be seen (Fig. 22a and b).

Implants rarely fail under conditions where fatigue loading exceeds the yield stress and plastic deformation takes place. Table 2 lists the relationships between the ultimate tensile stress, yield stress, and fatigue limits for the recrystallized and cold-worked stainless steel used to generate the fatigue curves shown in Fig. 21(a).

The fatigue endurance limit of the cold-worked stainless steel corresponds to about 30% of the yield stress, whereas that of the recrystallized steel condition corresponds to about 67% of the yield stress. If at stresses below the yield stress deformation structures develop on the specimen surface before fatigue-cracks form, one must conclude that these deformation structures are created by local repeated elastic deformations. Examination of test specimen surfaces during the fatigue tests described above shows how fatigue cracks develop from cumulative surface damage (Ref 20). This is illustrated on replicas of a recrystallized stainless steel specimen that was fatigued in Ringer's solution (Fig. 23a to d). The following stages of fatigue surface damage were observed: formation of persistent slip steps, slip bands, increasing quantities of multiple slip systems, precracks, short secondary cracks, and one or several major cracks. A main crack will then propagate through the cross section of the specimen. The same processes are morphologically observed on specimens fatigued in air; however, they proceed more slowly in air. This

Fig. 32 Wear on head of titanium screw

(a) Material transport and fretting zone. (b) Close-up view of wear structures showing fine wear products. 120×. (c) Wear structures showing generation of small wear particles. 1200×. (d) Wear structures with additional fretting structures. 305×

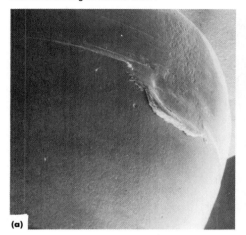

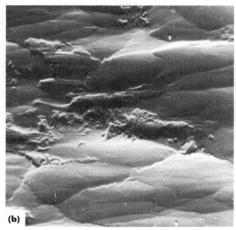

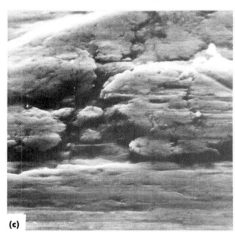

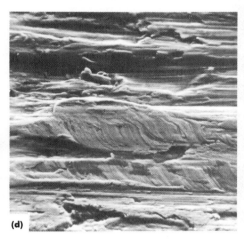

mechanism explains how fatigue cracks initiate without flaws or particular stress raisers.

The slip bands created by fatigue often have the character of extrusions that are accompanied by intrusions. These intrusions can act as micronotches and initiate cracks. On retrieved broken stainless steel implants, the fatigue damage structure described above is preserved on the surface in the vicinity of the fracture edge.

From the systematic evaluation of test specimens during the fatigue process, *S-N* curves of early fatigue damage were generated (Ref 20). As shown in Fig. 21(b) for the cold-worked stainless steel, curves for the number of cycles to visible slip system formation and crack initiation are similar to the *S-N* curve for the number of cycles to failure. The curves indicate that the lower the load, the longer the time spent in the initiation stage.

Micrographs of failed fatigue specimens of the cold-worked stainless steel show that the intensity of fatigue-surface damage adjacent to the fracture edge depends on the load (Fig. 24a to c). At low stress levels, few glide systems

develop. This surface morphology can be used to estimate if an implant was loaded in the high or low cyclic fatigue range.

On stainless steel implants, fatigue striations are abundant on the fracture surfaces, unless the surface has been subjected to secondary mechanical damage. In addition to fatigue striations, crystallographic fracture structures are found. The latter appears before a stable crack front forms. Secondary cracks, which run parallel to fatigue striations, do not indicate stress corrosion or corrosion fatigue (Ref 19).

Wrought Type ASTM F563 Cobalt Alloy. The fatigue-crack initiation mechanism in wrought cobalt alloys was assessed with the same type of tests described for stainless steel. Morphologically, the process is the same as that for stainless steel, but all initiation steps are shifted to higher stress values because of the higher strength and fatigue resistance of alloy ASTM F563.

Compared to stainless steels, the morphology of the fatigue-fracture surfaces is characterized by very dense patterns of fatigue striations. These patterns superimpose the normally dom-

inating crystallographic structures. It is sometimes difficult to distinguish fatigue striations from glide or tearing structures. This is particularly true when the glide systems have an orientation that accommodates the formation of striations (Fig. 25a to c). The toughness of these types of wrought cobalt alloys is assumed to account for this behavior.

Commercially Pure Titanium. The morphology of fatigue initiation and propagation of titanium is more complex than that of stainless steel. The limited number of glide systems of the hexagonal close-packed crystal structure of the titanium is assisted by twinning processes to allow plastic deformation (Ref 22). Correspondingly, the range of ductile behavior of titanium is more limited than that of stainless steel.

The corrosion resistance of titanium in the body and its tissue compatibility are excellent. There have been no reported cases of allergies to titanium. Unfortunately, the wear properties of titanium and its alloys are poor; therefore, they are not usually chosen as material for the articulation surfaces of prostheses.

The same fatigue experiments described for type 316LR stainless steels were performed on commercially pure titanium specimens. For comparison, Fig. 26 shows the fatigue-surface damage on a broken titanium test specimen. The polarized light reveals that twinning, wavy glide, and grain-boundary distortion contribute to the fatigue damage on the specimen surface.

The fatigue-fracture surfaces of titanium can show very different features, depending on overall and local stress conditions, crystallographic orientation, and material conditions. Figures 27(a) to (c) show fractographs of the surface of a cold-worked titanium specimen fatigued at a stress level of 600 MPa (87 ksi). Depending on local stress distribution and grain orientation at the crack front, extremely finely spaced fatigue striations (Fig. 27a) can form in the vicinity of very coarse striations (Fig. 27b), which appear to be in transition to glide formation. Figure 27(c) shows an area of overload fracture with prismatic and dimple tearing structures. These fatigue structures indicate ductile behavior.

Figures 28(a) to (c) illustrate an area of the fracture surface of a titanium bone plate that suffered fatigue failure in an unstable internal fixation situation. The bone plate showed mixed fracture structures, such as fatigue striations, tearing structures, and terraces. Figure 28(a) shows an overview, and Fig. 28(b) shows ledges and tearing structures of the terrace.

Degradation of Orthopedic Implants

Metallic substance can be released from implants by such mechanisms as:

- General corrosion
- Galvanic corrosion
- Intercrystalline corrosion

Fig. 33 Broken hip prosthesis of cast type ASTM F75 cobalt-chromium-molybdenum alloy

(a) Radiograph of total hip prosthesis. Circular wire marks acetabulum component made from plastics. Arrows (from top to bottom) indicate the area where the prosthesis stem is loosening at the collar, a stem fracture, and a fracture of bone cement at the end of the stem, respectively. (b) Fracture of prosthesis stem. Wear at end of stem (arrow) indicates stem movement due to loosening. (c) Wear at end of stem. (d) Close-up view of stem end showing material transfer and layering from wear. Corrosion signs are not observed. See also Fig. 34 to 36.

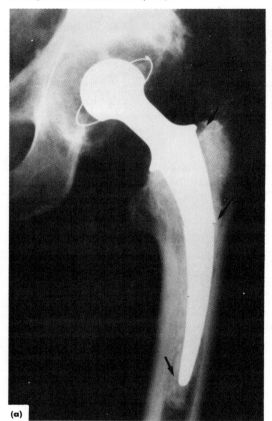

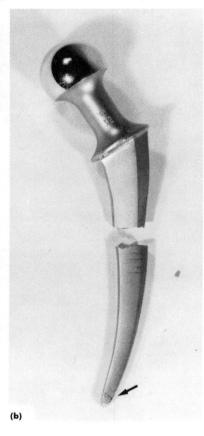

(a)

(b)

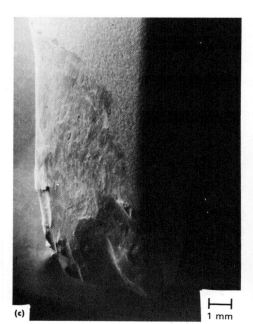

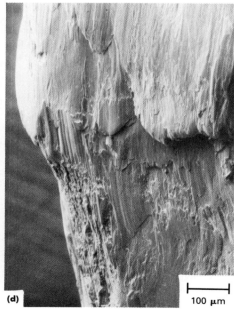

(c) 1 mm

(d) 100 μm

Fig. 34 Metallographic sections of failed hip prosthesis shown in Fig. 33

(a) Longitudinal section through fracture surface showing secondary fatigue crack parallel to fracture surface. 35×. (b) Cross section through prosthesis stem showing gas pores and second phase at grain boundaries and in grains. 105×

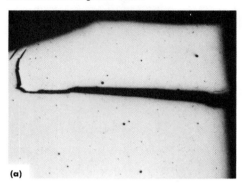

(a)

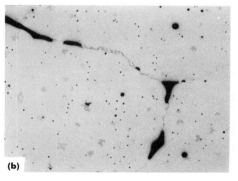

(b)

- Maintenance of passive film of highly corrosion-resistant implant materials
- Stress corrosion combined with fatigue
- Corrosion fatigue
- Crevice corrosion
- Fretting corrosion
- Fretting
- Wear

Reports on *in vivo* and *in vitro* studies and discussions of the degradation and corrosion of implants are found in Ref 23 to 25.

Implant degradation can lead to mechanical failure of the implant when severe disintegration takes place, as illustrated in Example 4. However, with the use of improved implant materials and metallurgical processing, visible general corrosion, intercrystalline corrosion, and galvanic corrosion are seldom observed today.

Combined attack of fatigue with stress corrosion, fretting, or fretting corrosion can also cause implant breakdown, but these failure modes are rare. Slight branching of cracks found on broken implants may not indicate stress corrosion, but can be explained by fatigue-propagation processes. In experiments with magnesium chloride solution, stress-corrosion cracks in type 316L stainless steel are characterized by extensive branching. The fracture surfaces of these cracks exhibit facets and

Fig. 35 Optical fracture analysis of the failed hip prosthesis shown in Fig. 33

(a) Fracture surface exhibiting large grains. The upper right grain is identical to grain A shown in Fig. 36(a). Crack originated at the lateral side of the stem. (b) Longitudinal section through fracture surface with glide systems activated by the fatigue process. Two cracks, one of which propagated through the grain, initiated along these glide systems. 195×. (c) Longitudinal section through fracture surface at the upper left area in Fig. 35(a) containing secondary cracks. Arrow indicates a gas pore. 270×

(a)

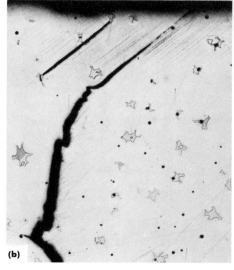

(b)

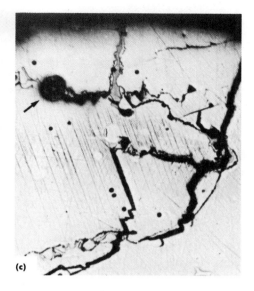

(c)

multiple fine cracks that in most cases are crystallographically oriented (Ref 19, 23). The combination of fatigue with stress corrosion has been observed in few cases on type 316L stainless steel. Very little metal is released into the tissue by this form of attack.

The generation of wear, fretting, or corrosion products can evoke local physiological reactions that may necessitate reoperation. This is, however, seldom experienced. Local reactions in relation to fracture-treatment implants are occasionally reported, but usually cease after resting of the limb and after routine removal of the implant upon healing of the fracture. The systemic effects of metals that are released from implants are the subject of much research and speculation. Unambiguously diagnosed general allergies to implant metals are rarely reported. However, mere contact with the metal is sufficient to create such reactions in patients susceptible to allergies. If implants remain in the passive condition, the release of metal ions is very low, corresponding to the current density related to the corrosion potential of the passive stage. These metal traces are usually below the concentrations that are detectable with modern analytical techniques.

Crevice corrosion, fretting corrosion, fretting, and wear typically occur at the contact areas of multiple-component implants. Wear is found on the articulation surfaces of prostheses. In a long-term study on total hip prostheses that were in the body for about 8.3 years, up to 2.8 mm (0.11 in.) of material loss was found, amounting to a wear rate of 0.34 mm/year (0.013 in./year). These wear rates were measured on radiographs and involve the abrasion of the metal head and the polymer acetabular cup, and the polymer creep.

The wear behavior of a prosthesis depends on the surface finish, the contact area, the loading, the loading directions, and the wear properties of the material. Many improvements have been made to minimize the wear on artificial joints, and researchers and producers have designed various wear simulators (Ref 26, 27). If prosthesis stems become loose, wear and fretting or, sometimes, shallow fretting corrosion can take place at various areas of the stem (compare Fig. 33c and d in Example 13).

The contact areas between screw heads and plate holes are usually susceptible to fretting and fretting corrosion when relative motion occurs between the screw and the plate. Compared to artificial joint surfaces, the contact areas between screws and plates are smaller and therefore are subject to higher surface pressures. Thus, it is difficult to prevent fretting at screw-head/plate-hole junctions by using special surface treatments. Mechanical attack is found on all implant materials at such junctions, but the degree of additional corrosion varies with the local condition and the type of material.

On commercially pure titanium, mechanical abrasion is found with oxidation of the wear particles (Example 12). On Ti-6Al-4V alloys, wear and wear fatigue are found; in these cases, wear fatigue is indicated by cracking of the worn material surface layers. Because fretting easily destroys the passive film, stainless steel can develop pitting corrosion in connection with fretting. In addition, the geometry between the plate hole and the screw head provides crevices. The intensity of pitting corrosion in connection with fretting depends on the quality of the steel. For example, type 304 stainless steel or steel with a

high inclusion content is prone to more severe pitting than a type 316LR steel (compare Examples 10 and 11). On high-quality stainless steels, mechanical attack and shallow pitting corrosion are observed. On screw-plate combinations in wrought cobalt alloys, pitting is observable only with SEM. Also, on cast cobalt-chromium-molybdenum alloys, fretting and slight corrosive attack have been observed.

On two-component hip-screw plates, fretting and fretting corrosion can also be observed. Reference 23 provides the results of analyses of fretting and fretting corrosion conditions on retrieved implants made from various materials. The following examples illustrate fretting and fretting corrosion conditions on screw-plate junctions for stainless steel and titanium.

Example 10: Heavy Pitting Corrosion on a Type 304 Stainless Steel Screw. Figure 29 shows a screw head that exhibits heavy pitting corrosion attack. Deep tunnels penetrated the screw head (Fig. 29a) and followed the inclusion lines (Fig. 29b). This screw was inserted in a plate made of type 316LR stainless steel that showed some mechanical fretting and very few corrosion pits.

This is another example where type 304 stainless steel is not satisfactory as an implant material and where inclusion lines contribute to corrosion, which is also observed on steels of higher quality but with larger amounts of primary inclusions.

Example 11: Screw Hole With Fretting and Fretting Corrosion of a Type 316LR Stainless Steel Plate. Figure 30 shows a plate hole with the area that was in contact with the screw head. In contrast to Example 10, the attack on this high-quality type

Fig. 36 SEM fracture-surface analysis of the failed hip prosthesis shown in Fig. 33

(a) Fracture surface showing three distinct grains labeled A, B, and C. (b) Grain A has a shallow crystallographically oriented fracture structure. (c) Grain B has a crystallographically oriented fracture structure with slip traces. Note vertical cracks running in the same direction as some of the fracture planes. (d) Grain C has a crystallographically oriented fracture structure and small vertical cracks similar to those shown in (c). A ruptured gas hole is visible at the center of the fractograph. 300×

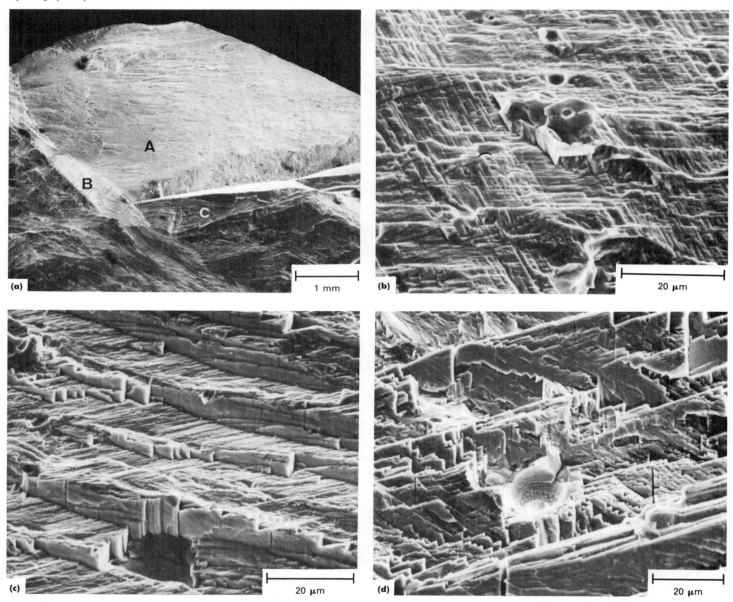

316LR stainless steel was only shallow. Figure 30(a) shows an overview indicating that a large portion of the contact area exhibited only mechanical grinding and polishing structures. The fine corrosion pits in the periphery are shown at a higher magnification in Fig. 30(b). The intense mechanical material transfer that can take place during fretting is visible in Fig. 30(c) and (d). In front of this structure is a corrosion pit that is surrounded by a burnished surface texture. The burnished surface may have broken open at this point. Observation of other implant specimens confirms that material layers are smeared over each other during wear and can be attacked by pitting corrosion. These pits are then covered by a new burnished

material film that can break open again. Thus, the material is worn and degradated.

Figure 31 illustrates how the corrosion and wear products are transported in the tissue. These products are found extracellularly and phagocytized in the fibrocytes of the connective tissue in this example.

Example 12: Titanium Screw Head With Fretting Structure at Contact Area With Plate Hole. Figure 32(a) shows a portion of a titanium screw head with a lip of material that was transported by fretting at a plate-hole edge. A flat fretting zone is visible on the screw surface over the material lip which is magnified in Fig. 32(b). A cellular wear structure containing wear debris is

present. Figure 32(c) shows material destruction and the formation of thin flakes of wear debris. Mechanical deformation is superimposed on wear structures in Fig. 32(d). No morphological signs of corrosion have been observed in connection with fretting structures. This has been typically found on all contact areas of plate holes and screw heads of titanium which were investigated.

Fractures of Total Hip Joint Prostheses

The hip joint prosthesis is the most common joint replacement. The function of the normal hip joint, the design of the hip joint prosthesis, and the related complications have been extensively studied compared to other joints.

A retrospective multicenter study on the performance of 39 000 total hip replacements over a period of 10 years was published in 1982 (Ref 28). A biomechanical concept of the total hip prosthesis based on clinical experience that gives a background for the understanding of complications can be found in Ref 29.

One of the complications encountered in total hip replacements is the fracture of prosthesis stems. Lateral bending and torsional forces act on the prosthesis stem when it is not supported. All fractures of the hip prosthesis stems are reportedly caused by a fatigue-failure mechanism (Ref 27). The fracture of prosthesis stems is a secondary event that follows loosening of the stem.

Different factors, which can also act in combination, promote stem loosening. These factors include bone resorption, degradation of bone cement, and unfavorable positioning of the prosthesis. Blood clot inclusions, holes in the cement, and overheating of the bone can also contribute to stem loosening.

Information on the performance and testing of hip stem material can be found in Ref 27, 30 to 32. Standard techniques for the cyclic bending test of hip prosthesis stems have also been developed (Ref 30) and an ISO standard is in preparation. Figures 33 to 36 show an example of a failed prosthesis stem that involves fatigue and stem loosening due to bone resorption and bone cement degradation.

Example 13: Broken Stem of Femoral Head Component of Total Hip Prosthesis Made From Cast Cobalt-Base Alloy. In a 65 year old male patient, radiotranslucency was visible around the collar of the femoral head prosthesis on radiographs taken 5 months after implantation. One month later, the bone cement was broken at the distal end of the prosthesis stem, and a small indentation on the lateral contour of the stem was visible where the stem had broken 2 weeks later.

Investigation. Figure 33(a) shows a radiograph illustrating the broken prosthesis. The dislocation of the fragment of the prosthesis indicated the degree of loosening and implant loading.

Under weight bearing, the proximal prosthesis fragment was pushed toward lateral, and the loosening in the medio-lateral plane can be seen as a gap between the lateral stem contour and the bone or cement. The upper arrow in Fig. 33(a) indicates the loosening at the collar. The bottom arrow marks the crack in the bone cement plug at the end of the stem. Figure 33(b) shows an overview of the broken prosthesis component. Because of the loosening, the end of the stem was heavily worn in contact with the bone cement (Fig. 33c to d).

In analyses of broken prosthesis stems, such wear or fretting traces are good indicators of the presence of motion and loosening. In other investigated prostheses, flat fretting zones have

also been found on other areas of the stem. On stainless steel prostheses, this wear appears occasionally as slight fretting corrosion attack. One may, however, distinguish wear that took place after fracture of the stem.

A metallographic specimen was taken parallel to the stem surface and perpendicular through the fracture surface of the distal fragment. This section revealed a secondary crack that had originated at the lateral aspect of the stem. The top contour in Fig. 34(a) represents the fracture edge, where two crack openings along slip planes can be observed. On a transverse metallographic section (Fig. 34b), gas pores are apparent in the grain and at the grain boundaries.

Discussion. Figure 35(a) shows an overview of the fracture surface. The grain size is considerable. The plateau in the upper right-hand corner represents a single grain, which is identical to grain A in Fig. 36(a). It is the same grain through which the secondary crack shown in Fig. 34(a) runs, and it is where the fracture originated. Figure 35(b) shows a higher-magnification view of the section through the fracture surface. The stresses created through the fatigue process activated glide systems that serve the formation of secondary cracks along glide planes. On the section through the fracture surface, another secondary crack formation is found (Fig. 35c). At these cracks, multiple slip bands also formed in connection with the fracture surface, and the cracks follow along these structures. One of the cracks propagated through a larger gas pore.

Figures 36(a) to (d) show details of the fracture morphology by SEM. In Fig. 36(a), three distinct grains are marked. These grains are oriented differently with respect to the propagating crack and therefore exhibit different crystallographic fracture patterns. Grain A (Fig. 36b) has a relatively flat texture, but also shows crystallographically oriented structures. Grain B (Fig. 36c) is characterized by a staircase pattern. Material separation occurred on preferred slip planes. The fine parallel line structures that run diagonally through the fractograph may be slip traces. Of special interest are the parallel microcracks that appear black and are superimposed on the other structures. On grain C (Fig. 36d), identical microcracks are found, although the grain is obviously oriented differently. A ruptured gas pore is visible in the center of Fig. 36(d). Typical fatigue striations were not found on the fracture surface. Glide systems appear to be activated at relatively low energies so that cracks can easily propagate along glide systems.

REFERENCES

1. D.C. Mears, *Materials and Orthopaedic Surgery*, The Williams & Wilkins Co., Baltimore, MD, 1979
2. L.R. Rubin, Ed., Chapters 23-27, in *Biomaterials in Reconstructive Surgery*, The C.V. Mosby Co., St. Louis, MO, 1983
3. M.E. Müller, M. Allgöwer, R. Schneider, and H. Willenegger, *Manual of Internal Fixation*, 2nd ed., Springer Verlag, 1979
4. H. Gerber et al., Quantitative Determination of Tissue Tolerance of Corrosion Products by Organ Culture, *Proceedings of the European Society for Artificial Organs*, Vol 1, 1974, p 29-34
5. H. Gerber and S.M. Perren, Evaluation of Tissue Compatibility of In Vitro Cultures of Embryonic Bone, in *Evaluation of Biomaterials*, G.D. Winter et al., Ed., John Wiley & Sons, 1980
6. V. Geret, B.A. Rahn, R. Mathys, F. Straumann, and S.M. Perren, A Method for Testing Tissue Tolerance for Improved Quantitative Evaluation, in *Evaluation of Biomaterials*, G.D. Winter et al., Ed., John Wiley & Sons, 1980
7. G.K. Smith, Systemic Aspects of Metallic Implant Degradation, in *Biomaterials in Reconstructive Surgery*, L.R. Rubin, Ed., The C.V. Mosby Co., St. Louis, MO, 1983
8. Medical Devices, *Annual Book of ASTM Standards*, Vol 13.01, ASTM, Philadelphia, 1985
9. A.C. Guyten, *Textbook of Medical Physiology*, W.B. Saunders Co., Philadelphia, PA
10. J.B. Park, *Biomaterials: An Introduction*, Plenum Press, 1979
11. P.G. Laing, Biocompatibility of Biomaterials, in *Orthopaedic Clinics of North America*, Vol 4, C.M. Evarts, Ed., 1973
12. F. Pauwels, *Der Schenkelhasbruch*, Enke, Stuttgart, 1935
13. R. Schenk, Fracture Repair—Overview, in *9th European Symposium on Calcified Tissues*, 1973
14. R. Schenk and H. Willenegger, Zur Histologie der primären Knochenheilung, *Unfallheilkunde*, Vol 80, 1977, p 155
15. S.M. Perren and M. Allgöwer, *Nova Acta Leopoldina*, Vol 44 (No. 223), 1976, p 61
16. S. Perren, R. Ganz, and A. Rüter, Mechanical Induction of Bone Resorption, in *4th International Osteol. Symposium*, Prague, 1972
17. H. Yamada, *Strength of Biological Materials*, The Williams & Wilkins Co., Baltimore, MD, 1970
18. L.E. Sloter and H.R. Piehler, Corrosion-Fatigue Performance of Stainless Steel Hip Nails—Jewett Type, in *STP 684*, ASTM, Philadelphia, 1979, p 173
19. O.E.M. Pohler and F. Straumann, Fatigue and Corrosion Fatigue Studies on Stainless Steel Implant Material, in *Evaluation of Biomaterials*, C.D. Winter et al., Ed., John Wiley & Sons, 1980
20. O.E.M. Pohler, "Study of the Initiation and Propagation Stages of Fatigue and Corrosion Fatigue of Orthopaedic Implant Materials," dissertation, Ohio State University, 1983

21. E. Powell, M.S. thesis, Ohio State University, 1983
22. R.W. Hertzberg, Deformation and Fracture Mechanics of Engineering Materials, John Wiley & Sons, 1976
23. O.E.M. Pohler, Degradation of Metallic Orthopaedic Implants, in *Biomaterials in Reconstructive Surgery*, L.R. Rubin, Ed., The C.V. Mosby Co., St. Louis, 1983
24. B.C. Syrett and A. Acharya, Ed., *Corrosion and Degradation of Implant Materials*, STP 684, ASTM, Philadelphia, 1979
25. A. Fraker and C.D. Griffin, Ed., *Corrosion and Degradation of Implant Materials*, STP 859, ASTM, Philadelphia, 1985
26. J.O. Galante and W. Rostoker, Wear in *Total Hip Prostheses, Acta Orthop. Scand. Suppl.*, Vol 145, 1973
27. M. Ungethüm, Technologische und biomechanische Aspekte der Hüft- und Kniealloarthroplastik, in *Aktuelle Probleme in Chirurgie und Orthopaedie*, Band 9, Verlag Hans Huber, Bern Stuttgart Wien, 1978
28. P. Griss *et al.*, Ed., Findings on Total Hip Replacement for Ten Years, in *Aktuelle Probleme in Chirurgie und Orthopaedie*, Band 21, Verlag Hans Huber, Bern Stuttgart Wien, 1982
29. R. Schneider, Die Totalprothese der Hüfte (Ein biomechanisches Konzept und seine Konsequenzen), in *Aktuelle Probleme in Chirurgie und Orthopaedie*, Band 24, Verlag Hans Huber, Bern Stuttgart Wien, 1982
30. M. Semlitsch and B. Panic, Corrosion Fatigue Testing of Femoral Head Prostheses Made of Implant Alloys of Different Fatigue Resistance, in *Evaluation of Biomaterials*, G.D. Winter *et al.*, Ed., John Wiley & Sons, 1980
31. M.F. Semlitsch, B. Panic, H. Weber, and R. Schoen, *Comparison of the Fatigue Strength of Femoral Prothesis Stems Made of Forged Ti-Al-V and Cobalt-Base Alloys*, STP 796, ASTM, Philadelphia, 1983
32. M. Semlitsch and B. Panic, Ten Years of Experience with Test Criteria for Fracture-Proof Anchorage Stems of Artificial Hip Joints, *Eng. Med.*, Vol 12, 1983, p 185

Failures of Pipelines

R.J. Eiber and J.F. Kiefner, Battelle Columbus Laboratories

PIPELINES have established the enviable record of having the fewest fatalities of any of the various modes of transportation. Failures do occur, however, for a variety of reasons. In this article, the known failure causes and failure characteristics of high-pressure long-distance pipelines will be summarized. In addition, the procedures used to investigate pipeline failures will be described. Finally, uses of fracture mechanics in failure investigations and in developing remedial measures will be described.

Characteristics of High-Pressure Long-Distance Pipelines

Three general types of pipelines exist: gathering lines, transmission lines, and distribution lines. Gas or crude oil gathering lines exist between a well and a treatment plant or collection point. These are usually relatively small-diameter lines (51 to 203 mm, or 2 to 8 in., in diameter) and operate at a variety of pressures, usually from 3490 to 8376 kPa (500 to 1200 psi). Cross-country transmission pipelines of over 1600 km (1000 miles) in length transport natural gas, natural gas liquids, liquid petroleum products, anhydrous ammonia, or crude oil. These are continuous lengths of pipe interrupted only by valves at roughly 16-km(10-mile) spacings and compression stations (gas lines) or pumping stations (liquid lines) at 80- to 160-km (50- to 100-mile) spacings. High-pressure lines generally operate at pressures up to 6895 kPa (1000 psi) and are made of steel pipes welded or mechanically coupled together. Since the 1940s, all of the lines have been assembled by welding. The third type of pipeline is a gas-distribution line that mainly transports natural gas within cities at pressures that vary from several tens of pounds per square inch to a few inches of water where a line enters a home. This article will not discuss failures associated with these latter lines, but will address only the high-pressure gathering and transmission lines.

The steels that have been used in the United States for line pipe range in yield strength from American Petroleum Institute (API) grade A (207 MPa, or 30 ksi) to API 5L-X70 (483 MPa, or 70 ksi). Generally, API 5L Specification steels with yield strengths above 359 MPa (52 ksi) are microalloyed and controlled rolled to achieve the desired strength and fracture properties. Pipeline wall thicknesses are based on the pressure in the line and on the allowable hoop stress levels. The allowable stress levels for gas pipelines vary from 40 to 72% of the specified minimum yield strength based on the population density in the area of the pipeline and are regulated by the U.S. Department of Transportation (DOT) in Title 49 of the Code of Federal Regulations (CFR) Part 192. Liquid petroleum pipelines are regulated by DOT in Title 49 CFR Part 195. Their maximum allowable hoop stress level is 72% of the specified minimum yield strength.

One of the reasons for the excellent safety record of the pipeline industry is that pipelines are pressure tested before being put into service. It is common to test pipelines hydrostatically to hoop stress levels of 90 to 105% of their specified minimum yield strength. In addition, this pressure is maintained for 8 to 24 h. This ensures that the pipeline, at the time it is put into service, contains no defects that might induce failure at its operating stress level. The ability to preservice test a pipeline to a high stress level is a unique characteristic of a pipeline. The lines are designed such that they can withstand this type of test without undergoing plastic deformation that would induce residual stresses or damage the pipe.

One of the failure problems unique to gas pipelines was the potential length of a failure. Because the pressurized gas contains an enormous amount of stored energy and because the energy-release rate upon rupture is often slow relative to the speed of a propagating fracture, a fracture may propagate along the pipe axis for a considerable distance. The longest fracture on record was brittle in nature and propagated for a distance of 13.3 km (8.3 miles). Thus, pipelines are among the few structures for which the extent of fracture propagation is of concern.

Pipeline Failure Investigation Procedures

Location of the Fracture Origin. The general procedures used to investigate pipeline failures are no different from those for other types of structures. The first step is to locate the fracture origin. This may be difficult when fractured pipe is scattered along the line for several hundred feet or more, but it is not impossible.

A study of the fracture surface is necessary to locate the origin. Features of the fracture surface make it possible to trace the fracture back to the origin. There are two basic modes of fracture propagation: brittle and ductile. Figure 1 shows a brittle, or cleavage, fracture. As shown in Fig. 1(a), brittle fractures in pipelines exhibit herringbone or chevron markings on the fracture surface. These markings are continuous and relatively uniform. This fracture pattern is produced because the fracture initiates first at the midwall of the pipe, then extends out to the surface, causing the V-shaped markings. As the result of stress-wave patterns established by the rupture event, brittle fractures in pipelines commonly exhibit a sinusoidal fracture path along the pipe axis. More than one fracture can often be observed due to the excessive energy available in the gas. If a brittle fracture propagates in a stress field that contains significant bending, the center of the chevron may move off-center (Fig. 1b). Brittle fractures that result from bending moments exhibit the appearance shown in Fig. 1(c).

A propagating ductile-fracture surface is shown in Fig. 2(a). Under gaseous pressurization due to the excessive stored energy, these fractures can also propagate long distances. Under liquid pressurization, both brittle and ductile fractures are usually short—of the order of a few feet. The fracture surface shown in Fig. 2(a) is typical of those found in API, grades B to X60, pipe manufactured by conventional rolling processes. The fracture surface is produced in a manner similar to a brittle fracture; that is, the fracture initiates first at midwall thickness and creates a deformation pattern on the fracture surface that can be likened to the chevron pattern of a brittle fracture but is not as distinct. This pseudo chevron pattern points toward the fracture origin and serves as a road map to locate the origin. Ductile fractures characteristically propagate axially along the pipeline, and in general, there is only one fracture.

Other types of ductile-fracture surfaces may be encountered in a failure investigation. One

Fig. 1 Modes of propagation for brittle fractures

(a) Fracture resulting from primary tensile loading. (b) Fracture resulting from combined tensile and through-thickness bending load. The center of the chevron is offset from the pipe midwall. (c) Fracture from bending load with schematic of resulting chevron pattern. Arrows indicate fracture-propagation direction.

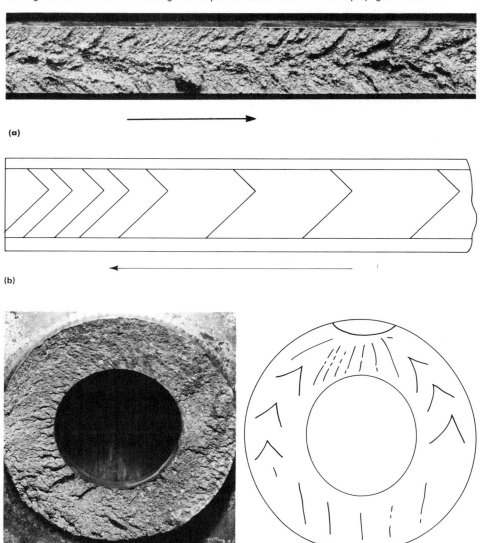

Fig. 2 Modes of propagation for ductile fractures

(a) Propagating ductile fracture. (b) Ductile fracture in separated material. (c) Ductile fracture with arrowheads, and illustration of chevron pattern. Arrows indicate fracture-propagation direction.

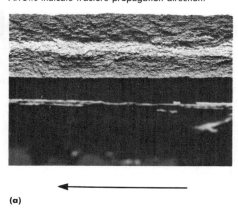

(a)

(b)

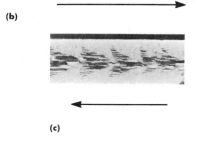

(c)

type (Fig. 2b) exhibits separations. Separations have been observed in the newer controlled-rolled X65 to X80 pipe steels due to crystallographic texturing of the steel during processing. This causes the steel to exhibit secondary fractures in the plane of the plate in a brittle mode under the influence of triaxial stress created as the primary ductile fracture propagates. One characteristic of a separated fracture surface is that it is more difficult to determine the direction of crack propagation based on the fracture-surface appearance. Another propagating ductile-fracture surface is the so-called arrowhead fracture surface shown in Fig. 2(c). The arrowheads on the fracture surface point away from the origin or in the direction of fracture propagation.

A third fracture appearance, called a tearing shear fracture, is found in conjunction with the arrest of ductile or brittle propagating fractures. Tearing shear fractures are pure shear fractures (Fig. 3), as opposed to the tensile fractures shown in Fig. 1 and 2. Tearing shear fractures occur at the arrest point, where an axially propagating fracture turns and spirals in a helical direction. Figure 4 shows an illustration of the location and configuration of tearing shear fractures. Tearing shear fractures occur because of the kinetic energy of the fracturing pieces of pipe and are therefore secondary to the main fracture.

Characteristics of an Origin. The origin of a fracture is a point at which there is generally only one fracture. The origin usually contains the defect that, in combination with the stress applied to the pipe steel, is responsible for the failure.

If the fracture propagation is brittle, it is possible that the fracture appearance will be brittle from the star-burst pattern at the origin to the tearing shear fracture at the arrest. Occasionally, the fracture at the origin will be ductile and then change to brittle as the fracture accelerates. The type of fracture that occurs is a function of the temperature of the pipe relative to the transition temperature of the steel.

Sample Selection. In failure investigations, a reasonable guess as to the cause of the accident can frequently be established in the field inspection. Based on the anticipated cause, it is possible to select samples logically for detailed examination. Normally, there are two objectives in selecting samples. The first is to select samples that help to define the cause of the failure, and the second is to select samples to investigate the mechanical and/or fracture-propagation properties of the pipe steel.

Fig. 3 Fracture surface of a tearing shear ductile fracture

Fig. 4 Tearing shear fractures

(a) In brittle material. (b) In ductile material

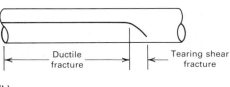

(a)

(b)

In selecting samples of the origin, large samples should be selected such that in flame cutting the heat of burning will not affect the fracture. Also, care should be taken to ensure that the fracture surface at the origin is not damaged during removal or shipment. Wire brushing of the origin fracture surface should not be permitted, because this removes the fine features of the fracture surfaces and eliminates the possibility of a fractographic examination in a scanning electron microscope. If the origin is associated with a weld, samples of the intact weld remote from the origin in the same length of pipe should be obtained to examine an unfractured section. If defects are associated with the weld, they may also be noted in the remote location. If an environmentally caused failure is suspected, consideration should be given to selecting and obtaining soil samples that are representative of the soil at pipe depth. Also, supplemental information regarding the cathodic-protection levels imposed on the pipe may be useful.

Samples selected for fracture-propagation property determination should be representative of the material that failed and should not be deformed or heated as a result of the failure. Samples measuring approximately 0.2 m² (2 ft²) are usually removed for material-property determination.

Causes of Pipeline Failures

One of the significant observations that have been made as a result of numerous failure investigations is that failures generally initiate at a defect, with the only exceptions being the few failures that have occurred from a gross overload, such as secondary stresses, internal combustion, or sabotage.

Table 1 lists the defects that have caused pipeline failures. The description of defects has been divided into two categories: those that cause failures in the preservice test and those that cause failures in service. The reason for the division is that the defects present in the pipe

Table 1 Types of defects that can cause pipeline failures

Causes of preservice test failures	Causes of service failures
Defects in the pipe body	**Defects in the pipe body**
Mechanical damage	Mechanical damage
Fatigue cracks	Environmental causes
Material defects	Corrosion (external or internal)
Longitudinal weld defects	Hydrogen-stress cracking
Submerged arc welds	External stress-corrosion cracking
Weld-area cracks	Internal sulfide-stress cracking
Incomplete fusion	Hydrogen blistering
Porosity	Fatigue
Slag inclusions	Miscellaneous causes
Inclusions at skelp edge	Secondary loads
Off seam	Weldments to pipe surface
Repair welds	Wrinkle bends
Incomplete penetration	Internal combustion
Electric welds(a)	Sabotage
Upturned fiber cracks	**Longitudinal weld defects**
Weld-line inclusions	High-hardness region(b)
Cold weld	
Excessive trim	
Contact burns	
Field weld defects	

(a) Electric flash welding, electric-resistance welding, or electric-induction welding without the addition of extraneous metal.
(b) These are a problem in gathering lines carrying sour gas (gas containing H_2S).

before the preservice test can generally be removed or detected by the preservice test if the test pressure is high enough.

Causes of Preservice Test Failures

As shown in Table 1, the defects that cause failures in line pipe in the preservice test can be broken down into three types, according to location: defects in the body of the pipe, longitudinal weld defects, or field girth weld defects.

Causes Associated With the Pipe Body. The three general causes of failure associated with the body of the pipe are mechanical damage, fatigue cracks, and material defects.

Mechanical damage consists of gouges or gouges and dents that have generally been

Fig. 5 Failure from lamination inclined with respect to the pipe wall

produced by excavation or handling equipment during construction. The failure characteristics will be presented in the section "Causes of Service Failures" in this article because this is the leading cause of service failures.

Material defects may be laminations, laps, seams, holes, hot tears, and so on. A few failures have resulted from foreign material being rolled into the pipe skelp. This results in a local region where the wall thickness is reduced or completely penetrated, which is generally detected as a leak during the preservice test.

Investigation of failures caused by material defects normally consists of one or two metallographic sections through the defective area. These metallographic sections are examined in an optical (light) microscope to determine the nature of the defect and the general features of the microstructure. When inclusions or foreign particles are present, examination in the scanning electron microscope using an energy-dispersive x-ray analysis will usually identify the composition of the microstructural constituents. Frequently, a deep etch will show the flow pattern of the steel grains during hot rolling, which can help to identify the cause. Figure 5 shows a lamination that was inclined to the pipe wall, which effectively reduced the wall thickness and resulted in a slant fracture in approximately one-half of the wall thickness.

These types of failures are generally related to (1) insufficient cropping of an ingot in the case of inclusions or a lamination in the steel, (2) foreign objects, such as parts of rolls or rolling equipment, being pushed into the steel during hot rolling, or (3) hot tears that form,

Fig. 6 Typical example of fatigue cracking adjacent to a longitudinal weld

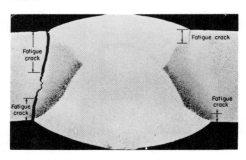

Fig. 7 Toe cracks in the HAZ of a double-submerged arc weld

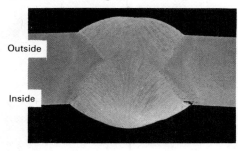

Fig. 8 Appearance of a hook crack in an electric-resistance weld

The thin white layer is a nickel coating that was applied to the fracture surface.

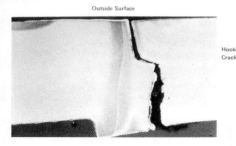

particularly during piercing and rolling of relatively thin-wall large-diameter seamless pipe. All of these problems can be controlled by careful mill practice and the application of adequate nondestructive inspection.

Fatigue cracks have occurred in the body of the pipe and adjacent to longitudinal welds. Figure 6 shows an example of a fatigue crack that occurred alongside the longitudinal weld. In this case, the cracks initiated at both the inside and outside surfaces of the pipe at the edge of the longitudinal weld. These cracks were detected in the preservice test of the line. Fatigue cracks have also been observed in the base metal. In these cases, the cracks have initiated from the outside surface in a region of mechanical damage on the pipe surface, which is usually produced during rail shipment.

Because these cracks were detected in pipe that had not been in service, the shipping of the pipe was subjected to investigation. The investigations have usually revealed that the fatigue cracks were associated with rail shipment over distances of 1125 to 1600 km (700 to 1000 miles). In pipe with diameter-to-thickness ratios of 80 or greater, the cracks were generally observed to be within 3 m (10 ft) of the end of a pipe length.

The fatigue cracks had been produced during rail shipment of the pipe as a result of the combination of static stresses produced in the lower layers of stacked pipe by the static weight of the pipe above and cyclic stresses produced by the vertical acceleration of the rail car. When mechanical-damage regions were evident, the pipe had usually contacted the rail car, frequently on a rivet head, which served as a stress concentrator.

A recommended practice for the loading of pipe in rail cars was developed by the API (Ref 1). This publication defines loading practices for rail car shipment to protect pipe from this type of failure. Since the introduction of this recommended practice, failures of this type have been significantly reduced.

Causes of Failures in Longitudinal Welds. Five types of longitudinal welds have been used to produce pipe. These are single-

and double-submerged arc welds, electric-resistance welds, furnace butt welds, flash welds, and lap welds. Currently, only double-submerged arc welds and electric-resistance welds are used. The following are brief descriptions of the welding processes and common causes of failures.

Double-Submerged Arc Welds. These welds are made with two passes (hence the term double): one on the inside and one on the outside. Either the inside or outside pass can be made first, depending on the producer. The weld passes can be made with one to four electrodes. The most common causes of failures in double-submerged arc welds are weld cracks, toe cracks, and lack of penetration.

Cracks may appear in either the weld metal or the heat-affected zone (HAZ). In the weld metal, the crack is typically in the center of the weld metal parallel to the columnar structure of the weld metal, and they generally occur when the weld metal is subjected to a high stress or restraint during solidification. Figure 7 shows cracks in the HAZ; these cracks are usually referred to as toe cracks at the edge of the weld reinforcement. Generally, these result when the edges of the plate are not adequately curved when the plate is formed into pipe. As the pipe is pressurized or cold expanded during manufacture, high strains occur at the edge of the weld reinforcement, causing cracks similar to those shown in Fig. 7. An excessive number of inclusions in the steel at the edge of the weld reinforcement have been observed to assist the formation of toe cracks.

Another type of defect in double-submerged arc welds results from lack of penetration, in which the inside and outside beads are too shallow and do not overlap. This leaves a nonfused portion, thereby reducing the effective wall thickness.

Electric-Resistance Welds and Flash Welds. These welding processes locally heat the edges of the plate to a suitable forging temperature; the edges are then pushed together, upsetting the wall thickness and forming a bond. In an electric-resistance welded seam, the weld is formed as the pipe passes through the welding station, with the edges making a vee before welding. In the flash-welding process, a complete length of pipe is welded at one time as the

edges of the plate are pushed together (flash welds are no longer used to produce pipe). In either type, after welding, the upset material is trimmed off.

Causes of failures common to both electric-resistance and flash welds are upturned fiber cracks, cold welds, and insufficient upset. Figure 8 shows the appearance of a metallographic section through an electric-resistance weld that contains a hook crack, or upturned fiber crack. Although it is somewhat difficult to see in Fig. 8, the steel has been upset such that the original plane of the pipe wall thickness extends through the thickness at the bond line. If inclusions (usually manganese sulfide inclusions) exist in the wall thickness before the upsetting, they may crack during the upsetting process, creating the upturned fiber defect. In this weld, it can also be observed that the upset removal on the inside of the weld was not uniform and that a notch was thereby created.

Figure 9 shows the origin fracture surface of one cause of failure in electric-resistance welded pipe. The smooth region is the result of inadequate upsetting and/or heating of the pipe edges (a cold weld). A stitched surface, or alternately bonded region, may also be observed at fracture origins associated with inadequate upsetting or too little heat.

Lap Welds. These welds are made by forging a weld between overlapped plate edges. The weld is made using forging rolls that form a low-angle weld through the thickness. Causes of failure in this type of weld are oxides on the bond line and lack of fusion due to too low a temperature or too little upsetting pressure. These defects reduce the effective thickness of the weld.

Regardless of the welding process used, the investigation of weld failures usually involves use of low-power (2 to 30×) magnification in either the optical microscope or the scanning electron microscope to locate the origin. Advantages of the optical microscope are that the sample orientation is not lost and that a relatively long length of weld fracture can be examined. Once an origin is identified, scanning electron microscopy (SEM) examination may be helpful in identifying fracture fea-

Fig. 9 Appearance of an unwelded section in electric-resistance welded line pipe

Fig. 11 Mechanical damage in a 406-mm (16-in.) diam API, grade X52, pipe

S, shear- or ductile-fracture surface; C, cleavage- or brittle-fracture surface. See also Fig. 12 to 14.

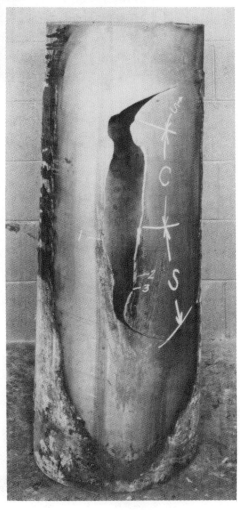

tures. Once the origin is identified, metallographic sections across the weld are prepared. In addition, sections across the intact weld from other locations in the same pipe length can be helpful. Intact weld sections can be used to:

Fig. 10 Underbead cracks (hydrogen cracking) in a girth weld

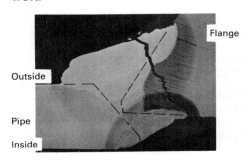

- Identify whether repairs have been made to the weld
- Determine whether a heat treatment has been adequate at the origin
- Determine the general configuration of the weld and bond lines before the failure occurred

The cause of failure can generally be determined if the manufacturing process is understood and the origin is properly identified and examined.

Causes of Failure in Field Girth Welds. Relatively few girth welds have failed either in service or in the preservice hydrostatic test. One of the main reasons for this is that girth welds are not subjected to high stress due to internal pressure. Girth welds may, on occasion, experience high stresses from such sources as soil movement, subsidence, or severe thermal fluctuations. Except for such unusual sources, the longitudinal stress in a pipeline is about 30% of the hoop stress or about 22% of the specified minimum yield stress (SMYS) for a line operating at 72% SMYS.

Typical manual metal arc weld defects, such as lack of penetration, hollow bead, porosity, alignment problems (high-low), weld-metal cracks, and underbead cracks, have caused leaks or ruptures in welds. Because the failure modes and investigation procedures are similar for all cases, only underbead cracks will be discussed.

Underbead Cracks. One situation that can cause a failure is an underbead crack. Underbead cracks occur when regions of high hardness form because of rapid cooling of the HAZ. High-hardness regions may result from welding pipes of two different wall thicknesses or a pipe and a heavy wall thickness fitting without adequate preheating. On occasion, due to fit-up or alignment problems, it is necessary to add a weld bead to the internal surface to provide a complete weld. Because this requires

a welder inside the pipe, it is difficult to preheat the weld area. Weld beads deposited under these conditions with cellulosic electrodes may result in rapid cooling of the HAZ with a tendency to produce untempered martensite. The untempered martensite is highly susceptible to cracks from hydrogen-stress cracking, producing underbead cracks. The same condition can result if welds are made with cellulosic electrodes between pipes that are at low temperatures, providing high-hardness HAZs due to the rapid cooling.

Figure 10 shows a metallographic section through a girth weld between a pipe and a heavy flange. The backup bead on the inside surface produced an HAZ with a maximum Knoop hardness of 405. A semicircular crack (underbead crack) occurred roughly parallel to the weld bond line. This crack caused the weld to fail during pressure testing.

Causes of Service Failures

Table 1 also lists the causes of service failures. Each category will be discussed to identify its characteristics, and investigative steps will be presented for selected causes.

Defects in the Pipe Body. Mechanical damage is the leading cause of failure in pipelines, accounting for more than one-half of all service failures. Generally, mechanical damage occurs from excavation equipment that accidentally contacts the pipe. The damage usually consists of a gouge and dent in the pipe surface.

After a failure, the dent is usually difficult to detect because the pressure in the pipe tends to remove it. Mechanical damage has resulted in penetration of the pipe wall thickness, producing an immediate leak, or in damage to the pipe, creating a potential for future failure as the dented area continues to deform due to creep from increased pressure. Frequently, the gouge appears to be relatively shallow. However, because of the strain imposed by the outward deflection of the dent, cracks may form in the cold-worked layer that are not visible on the gouged surface because of the smeared metal.

Figure 11 shows a photograph of a 406-mm (16-in.) diam API, grade X46, pipe that has been mechanically damaged. The damage is located in regions 1, 2, and 3. The chevron markings on the fracture surface locate the

Fig. 12 Macrograph of region 3 from Fig. 11

5×. See also Fig. 13 and 14.

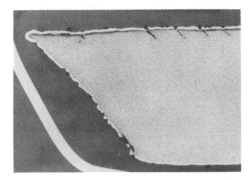

fracture origin between regions 2 and 3. The pipe surface was indented approximately 6 mm (0.25 in.) after the failure. In all probability, a much deeper dent existed at the time the damage was created.

The presence of a gouged surface will usually immediately identify the primary cause. Metallographic sections through the mechanical damage are necessary to identify the extent of the cold-worked steel, the depth of the gouges, the extent of cracking (if any), the hardness of the cold-worked steel, and whether or not foreign particles are attached. The identification of foreign particles, if they exist on the gouged surface, may be useful in identifying the source of the mechanical damage through chemical analysis of the foreign material (the

authors, however, have not been successful in identifying the specific damaging equipment). Examination of the fracture surface in a scanning electron microscope may be helpful in identifying the nature of the fracture, that is, transgranular versus intergranular and whether or not any evidence of fatigue can be observed. Most of the mechanical-damage failures have had transgranular fractures, but intergranular fracture surfaces have been observed and are believed to have resulted from hydrogen-stress cracking, with the hydrogen resulting from the cathodic potential applied to the pipeline for corrosion control. A problem in attempting to use the scanning electron microscope is that the fracture surfaces are often too corroded to permit examination.

Figure 12 shows a macrograph of one side of the fracture at region 3. It shows the fracture edge and the outside pipe surface (cracked surface). It should be noted that the section has been nickel coated, which is the white coating. The overall fracture in this region is a slant or ductile fracture, as evidenced by the deformation along the fracture edge. The slant cracks evident on the outside surface are believed to have formed as a result of the cold-worked region on the surface having been subjected to high strains by the outward deformation of the dent from the pressure in the pipe. These cracks are very significant; the depth of the gouge measured by the reduced wall thickness is approximately 0.2 mm (0.008 in.), while the maximum crack depth is 0.85 mm (0.033 in.). The crack that initiated the fracture was estimated to be 1.47 mm (0.058 in.) deep. Figure

13 shows a micrograph of the cold-worked outside pipe surface that illustrates the normal ferrite-pearlite microstructure of the carbon steel. Also shown is the cold-worked surface with a highly elongated grain structure, which has resulted from the mechanical damage.

Figure 14 illustrates another area of the surface in region 3 that shows a crack and the cold-worked structure. Near the surface in Fig. 14, there is evidence that, in addition to the cold work, the steel was heated to a temperature above the Ac_1 temperature (707 °C, or 1305 °F) and cooled at a fast rate. This layer of material has a metallurgical structure different from that of the severely cold-worked layer and different from that in the pipe. This structure is probably a bainitic structure, which is intermediate in hardness between ferrite and martensite. The heating was produced by friction between the pipe wall and the object that made the gouge. The rapid cooling is produced by the surrounding pipe steel. In some instances of mechanical damage, untempered martensite has been observed on the gouged surface.

An additional feature that is sometimes observed can be seen in Fig. 14. The light gray area just beneath the white nickel plating to the left of the crack is an area of extremely hard steel that is foreign to the pipe wall. The hardness is 870 HK and probably came from the hard surfacing materials used for construction equipment bucket and blade edges. This apparently became friction welded to the gouged surface during the damage process.

Failure of mechanically damaged areas can occur from a number of causes and may or may

Fig. 13 Micrograph of cold-worked pipe surface at region 3 in Fig. 11

Etched with nital. 500×. See also Fig. 12 and 14.

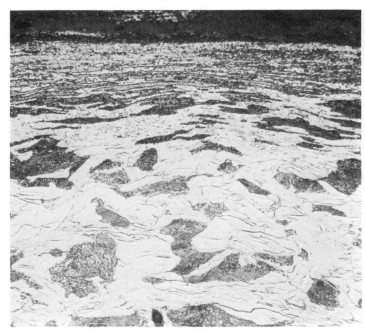

Fig. 14 Micrograph showing foreign material and secondary surface crack at region 3 in Fig. 11

Etched with nital. 100×. See also Fig. 12 and 13.

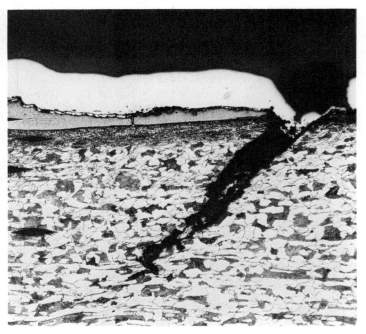

Fig. 15 Appearance of hydrogen-stress cracks in line pipe

(a) Fracture surface. 0.5×. (b) Results of Rockwell C hardness traverse of pipe

(a)

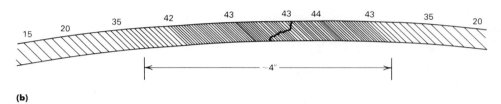

(b)

not occur at the time damage is done. The possible causes of failure are:

- The damaged area may crack due to the high strains imposed by the dent
- The pressure fluctuations in the pipeline may cause a low-cycle high-stress fatigue failure
- The normal cathodic protection applied to the pipe to prevent corrosion may charge the highly stressed steel with sufficient hydrogen to cause a hydrogen-stress crack

When a region of mechanical damage is detected, action should be taken to remove it or to reinforce it through the application of a sleeve. If a sleeve is applied, the gap between the sleeve and pipe should be filled with a compound, such as epoxy or polyester, that will prevent further outward deformation of the dent.

Environmental Causes of Pipeline Failures. There are a number of environmental causes of pipeline failures. The major causes are general corrosion, hydrogen-stress cracking in hard areas, external stress-corrosion cracking (SCC), internal sulfide-stress cracking, and hydrogen blistering. The failures break down into two general classes: those occurring from the inside surface of the pipe and those occurring from the outside surface of the pipe. Generally, the failures initiating on the inside pipe surface are associated with gathering lines containing moisture, hydrogen sulfide, carbon dioxide, or other corrosive impurities. The causes associated with the inside pipe surface are internal corrosion, sulfide-stress cracking, and hydrogen blistering. The remaining causes are associated with the external pipe surface. Investigation of this general category of failures is similar; therefore, only selected examples of these failure causes will be described.

Hydrogen-Stress Cracking. Figure 15 shows the surface of a fracture that was initiated from a hydrogen-stress crack in line pipe. The features of the fracture surface are:

- Extremely brittle fracture surface with very fine grains
- Fracture origin generally at the outside pipe surface or subsurface
- No out-of-plane deformation of the pipe wall thickness at the origin

The investigation consists of examining the fracture at low-power magnification to locate the origin. If hydrogen-stress cracking is suspected, hardness impressions are made on the surface to determine if a region of high hardness is associated with the origin area. The surface-hardness impressions generally identify the extent of the high-hardness region, which assists in the selection of metallographic specimens. In one failure, no region of high hardness was found on the surface, but alloy segregation at midwall thickness created a region of high hardness. Usually, two metallographic sections are prepared: one section through the fracture origin and one outside of the high-hardness region. The section outside of the suspected region of high hardness is taken for comparison purposes to determine the normal microstructure of the pipe steel.

Examination of the origin metallographic section usually reveals that this is a region of the plate that contains transformation products that may range from bainite to untempered martensite. These local regions of high hardness are believed to have been quenched during hot rolling. They generally exhibit a hardness gradient from the ferritic-pearlitic structure to the region of high hardness. Microhardness readings should be taken to determine hardness

level. Failures have occurred with hardnesses from 30 to 50 HRC in API, grade X52, steels. Generally, no secondary cracks have been observed on the fracture surfaces of these failures. Also, the fracture is usually partly intergranular and partly transgranular.

If hydrogen-stress cracking is suspected, information on the cathodic potential applied to the pipe for corrosion protection should be obtained as well as information on the coating condition. The atomic hydrogen necessary to cause the fracture is believed to result from the cathodic potential applied for corrosion control, which reaches the pipe surface through pores (commonly termed holidays in the pipeline industry) in the coating.

Hydrogen-stress cracks are caused by the combined factors of a very high-strength high-hardness steel, an applied hoop tensile stress from the pressure in the pipe, and atomic hydrogen in the steel from the cathodic-protection current applied to the pipe. Failures of this type may be prevented in several ways if the hard spots can be located. Success has been achieved by using internal nondestructive inspection devices based on the magnetic flux-leakage principle to locate hard spots. Once the hard spots are located, they can be removed, shielded to prevent the cathodic current from reaching them, or tempered to reduce the hardness. To date, failures have been observed only in areas with hardnesses exceeding 30 HRC. Also, hard spots are now prohibited by the API (Ref 2) and must be inspected for during the production of pipe. One of the best inspection techniques is to examine the pipe surface visually for flat spots.

Stress-Corrosion Cracking. Figure 16 shows the fracture surface of a stress-corrosion crack in line pipe. Failures of this type have primarily occurred on gas-transmission pipelines immediately downstream of compressor stations, where the pipe temperatures are the highest. The characteristic feature is a series of elliptical crack shapes, which initiated from the outside surface. Normally, a black oxide layer can be observed on the elliptical crack surfaces. Also, the cracks frequently overlap in the axial direction and link together. Figure 17 shows secondary cracks on the pipe surface. Information on the pipe-to-soil cathodic-potential level, pipe temperature, pressure, and pressure fluctuations in the pipeline should be obtained.

The investigation should begin with a low-power examination of the fracture surface to select locations for metallographic sections. The nature of the cracks can be determined by preparing several matched metallographic sections across the fracture surface. Also, analysis of any corrosion product on the elliptical crack surface using an energy-dispersive x-ray spectrometer may be helpful in defining the corrosive medium. In a high percentage of the failures, iron carbonate or bicarbonate has been detected on the crack surface (this environment is produced at a pore in the pipe coating as the hydroxide created by the cathodic potential is

Fig. 16 Appearance of fracture surface of line pipe that failed by SCC

Actual size. See also Fig. 17 to 19.

Fig. 17 Secondary cracking in pipe surface adjacent to stress-corrosion cracks shown in Fig. 16

Actual size. See also Fig. 18 and 19.

Fig. 19 Polished section through stress-corrosion cracks showing branching and intergranular cracking

See also Fig. 16 to 18.

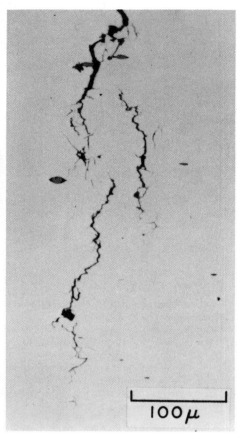

Fig. 18 Macrograph of unetched section through stress-corrosion cracks in line pipe

See also Fig. 16, 17, and 19.

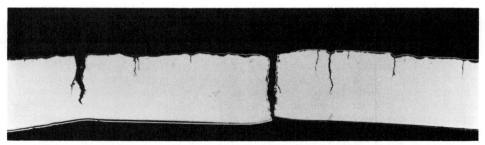

converted by the CO_2 in the soil to a carbonate-bicarbonate environment). Scanning electron microscopy examination of the fracture surface can also be helpful in identifying whether the fracture is transgranular or intergranular.

Usually, the fracture surfaces are primarily intergranular, and a number of branched secondary fractures are often evident in the metal-lographic sections (Fig. 18). Figure 19 shows an unetched section of a stress-corrosion crack that illustrates the branching and intergranular nature of the fracture.

The conditions that have been responsible for a high percentage of the stress-corrosion cracks in line pipe are the carbonate-bicarbonate environment, the pipe surface in the critical poten-tial range for cracking (−0.6 to −0.7 V versus the saturated calomel electrode, SCE, for a carbonate-bicarbonate environment), and the applied tensile stress.

The carbonate-bicarbonate environment is believed to be formed by the cathodic potential at pores in the coating where the current pene-trates to the pipe surface. The cathodic potential also creates the critical potential for SCC. The hoop stress in the circumferential direction of the pipe is created by the internal pressure in the pipe.

There is no simple control for external SCC, because one of the three conditions necessary for cracking must be prevented. This means keeping the cathodic potential outside of the critical range for cracking, which is difficult, if not impossible, at all points on a long pipeline.

Keeping the temperature as low as possible slows the reaction rate, which helps to prevent cracking. Many of the pipeline failures have been located in the first 16 km (10 miles) downstream of a gas compressor station, which is the high-temperature region on a gas pipe-line. Applying a good coating to the pipe surface is also a deterrent, as is shot peening the pipe surface to remove mill scale and to provide a good surface for coating bonding. Small pressure fluctuations have been found to promote cracking. Thus, pressure control is helpful, but is difficult to achieve in most situations.

Corrosion failures can initiate from either the inside or the outside surface of the pipe. Corrosion on the inside is generally associated with water in the pipe, which most commonly occurs in gathering pipelines where H_2S and CO_2 may also be present. Corrosion frequently takes place in the form of nonuniform metal loss whether it is on the internal or the external surface of the pipe. Both external and internal corrosion tend to be concentrated on the pipe bottom. General corrosion results in loss of metal over large areas, and pitting corrosion results in pits, which may be completely isolated or in overlapping arrays.

At the origin of a corrosion failure, the goal is to determine the nature of the corrosion and to rule out other environmental causes. In this discussion, rather than focusing on the investigation of the corrosion (which is covered in other discussions), the emphasis will be on the use of fracture mechanics to estimate the failure pressure of the corrosion flaw to verify that the failure resulted from the observed corrosion defect. This is also helpful in evaluating corrosion-defect severity when corrosion is detected on an operating pipeline.

An example of a corrosion flaw in 610- × 9.5-mm (24- × 0.375-in.) API, grade B, pipe will be described. The measured yield strength of the pipe was 336 MPa (49 ksi).

The corrosion defect that caused the failure is shown in Fig. 20. This defect occurred on the outside bottom surface of the pipe. The hoop stress in the pipe at the time the corroded area failed was 173 MPa (25 ksi). The failure was a rupture, with the fracture propagating a total axial distance of 5 m (17 ft).

The predicted failure pressure for the corroded area is based on treating the loss of metal as a part-through or surface flaw in the pipe (Ref 3). A grid is drawn on the corroded area, and remaining wall-thickness measurements are made from which a contour map of remaining wall thickness can be developed (Fig. 21). The next step is to develop a profile of the remaining wall thickness along the fracture edge (if this were a prediction being made for a nonfailed corroded area, the minimum remaining thickness path through the corroded region would establish the flaw profile). Figure 22 shows the measured profile in an exaggerated view.

Once the profile is available, the failure pressure can be predicted from Eq 1 (Ref 3):

$$\sigma_{hp} = \bar{\sigma}\,\frac{1 - A/A_0}{1 - (A/A_0)M_T^{-1}} = \bar{\sigma}M_p^{-1} \qquad \text{(Eq 1)}$$

where σ_{hp} is the failure stress of the part-through flaw; $\bar{\sigma}$ is the flow stress of the steel,

Fig. 20 Corroded area in 610-mm (24-in.) outside-diameter × 9.5-mm (0.375-in.) wall-thickness API, grade B, line pipe
See also Fig. 21 and 22.

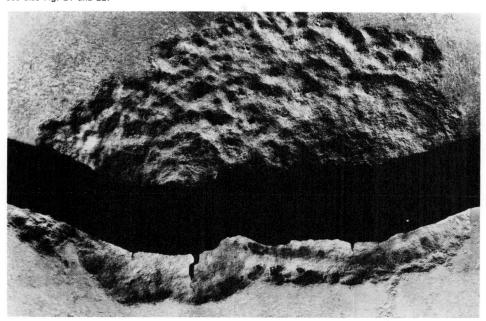

Fig. 21 Contour map of remaining wall thickness in corroded areas shown in Fig. 20
Dimensions are remaining wall thickness in mils. Fracture path is indicated by the heavy line. See also Fig. 22.

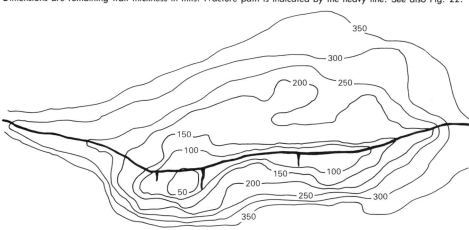

Fig. 22 Exaggerated depth profile of corroded area shown in Fig. 20 and 21

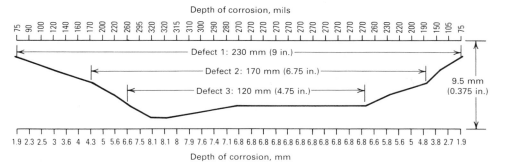

Table 2 Controlling parameters and predicted failure modes for three assumed defect lengths

| Defect 2c length | | | A | | | σ_{hp} | | Predicted failure mode |
mm	in.	M_T^{-1}(a)	cm²	in.²	M_p^{-1}(b)	MPa	ksi	
228.6	9.0 . . . . . . .0.390		13.5	2.09	0.503	204	29.6	Rupture
171.5	6.75 . . . . . .0.485		11.2	1.74	0.456	185	26.8	Leak
120.65	4.75 . . . . . .0.610		8.6	1.34	0.455	184	26.7	Leak

(a) M_T, bulging correction for an axial through-wall flaw. (b) M_p, bulging factor for part-through flaw

which has been taken as the yield strength plus 70 MPa (10 ksi); A is the area of the longitudinal profile of missing wall thickness; A_o is the area of full wall thickness for the length of the corrosion flaw; M_T is the bulging correction for an axial through-wall flaw in pressurized pipe; and M_p is the bulging factor for the part-through flaw.

Because corroded areas generally exhibit gradual transformation to full wall thickness along the critical axial profile, it is often difficult to preselect the axial length of the corroded region, which is termed the effective $2c$ length. By making several calculations of strength based on various trial values of crack length, one can usually obtain a minimum predicted failure stress level representing the effective crack length. This was done for the corroded region shown in Fig. 22. Using the trial lengths of 228.6, 171.5, and 120.65 mm (9, 6.75, and 4.75 in.), the values shown in Table 2 predict failure stresses for the three assumed flaw lengths.

The minimum of these three values is a predicted failure stress level, σ_{hp}, of 184 MPa (26.7 ksi) corresponding to the assumed 120.65-mm (4.75-in.) defect length. This value is 6% higher than the failure hoop stress of 174 MPa (25.2 ksi).

The mode of failure predicted for the 120.65-mm (4.75-in.) flaw in the 9.5-mm (0.375-in.) wall-thickness pipe is a leak. A leak is predicted because $M_p^{-1} < M_T^{-1}$ (0.455 versus 0.610). These M factors can be thought of as stress-concentration factors. When a part-through flaw fails, it becomes a through-wall flaw. Thus, if the failure stress, which is proportional to M_p^{-1}, for a part-through flaw is higher than for that same length through-wall flaw, the part-through flaw would be predicted to rupture. The reverse would predict a leak. Thus, if $M_p^{-1} > M_T^{-1}$, a rupture is predicted, and if the reverse, a leak. In the present case, the prediction is a leak, but a rupture occurred. In this case, however, the effective 120.65-mm (4.75-in.) flaw is in a region of much less than full wall thickness, as can be seen in Fig. 22. Computing the failure conditions for the 120.65-mm (4.75-in.) flaw length in a 5.1-mm (0.2-in.) average wall thickness results in an $M_T^{-1} = 0.8$ as contrasted with an $M_p^{-1} = 0.503$, which predicts a rupture.

The severity of a corrosion defect can be predicted using a fracture mechanics approach for line-pipe steels. This can be used to make

sure the origin identified is the origin responsible for the failure, because the predicted failure pressure should be within ±15% of service pressure at the time of failure, and to help in assessing the remaining strength of corroded pipe before failure.

The chances of this type of failure can be minimized by monitoring loss of wall thickness through nondestructive inspections from the inside of the pipe, using magnetic flux leakage devices as a means of detecting corroded areas. Monitoring the pipe-to-soil potential level can also identify areas of probable corrosion as areas of low potential levels.

Fatigue cracks resulting from service loadings are unusual in gas-transmission pipelines remote from compressor stations. This is because there are few pressure cycles and no mechanical vibration that would produce them. Pressure cycles on gaseous pipelines are usually a maximum of approximately 10% of the line pressure and may occur once or twice a day. Thus, in 40 years, this would represent only 14 400 to 28 800 cycles.

At compressor stations on gas pipelines, fatigue cracks have developed due to mechanical vibrations of the piping system. These are typical of the problems associated with rotating equipment.

On liquid pipelines, problems have been experienced from the relatively large and frequent pressure cycles often associated with liquid pipelines. Large pressure cycles can occur a number of times each day, and smaller cycles can occur more frequently, depending on how the line is operated and where deliveries are being made. The most common fatigue-crack problem occurs at the longitudinal weld—especially if there is a weld reinforcement, such as at a double-submerged arc weld. The appearance of these fatigue cracks is similar to the fatigue cracks due to shipping shown in Fig. 6. On liquid lines, the pumping stations also experience fatigue cracks due to mechanical vibrations and high-frequency pressure cycles similar to the potential problems at gas compressor stations.

Miscellaneous Failure Causes. A number of unusual failures have occurred. Many of these have been observed only once or twice, but can cause problems. Other causes observed are described below.

Secondary Loads. Soil instabilities, land slides, and mine subsidence have caused failures from excessive overloads in bending or tension.

Weldments to the Pipe. A few failures have been attributed to attachments that were improperly welded to the pipe. Before 1940, weldments to the pipe were common. Currently, no attachments are made to the pipe without careful design to minimize stress concentrations.

Wrinkles, Bends, or Buckles. At one time, it was not possible to obtain high-bend angles without wrinkling the pipe wall on the compressive side of the bend. Many of these are in service, and a number have cracked after years of service due to a low-cycle fatigue mechanism. This has been largely solved by the use of bending machines with internal mandrels. Also, because problems have developed, bends are carefully inspected for wrinkles and are rejected if any are present.

Internal Combustion. Several failures have occurred in new gas pipelines being tested with gas that were not adequately purged of air. The ignition source is unknown but may be static electricity. The characteristic of the fracture is that a detonation wave propagates along the inside of the pipe. The wave builds up a pressure front that causes pipe ruptures intermittently along the pipe and at major changes in direction. In between the points of rupture, the pipe is frequently expanded.

Sabotage. Several failures have occurred where the integrity of the pipeline was purposely breached. The suspected method is through the use of an explosive charge. The characteristic feature of the cause is that the pipe is usually severely fragmented at the origin area. Also, severe inward denting or deformation of the pipe is common, and multiple randomly oriented origins have been observed.

Longitudinal Weld Defects. The causes in this category are similar to those described previously, with the exception of the following two causes. First, selective corrosion of the weld bond line has been observed in electric-resistance welds. This forms an axial V-shaped groove centered on the bond line. This generally occurs because the bond line is ferritic. The ferritic region becomes anodic to the surrounding pearlite, creating a local corrosion cell.

Second, in the late 1940s and early 1950s, a limited quantity of electric-resistance weld pipe was produced with a martensitic weld region. It was believed that this high-strength region would be less prone to failure because it was higher in strength. Unfortunately, this region is also much more susceptible to failure from atomic hydrogen created by the cathodic potential applied to pipelines for corrosion control. The fracture surface near the origin exhibits an extremely brittle appearance.

Arrest of Pipeline Fractures

A propagating fracture in a gas pipeline can be extensive because of the relatively slow pressure decay associated with the compressed

Fig. 23 Scatter bands for Charpy V-notch impact specimens from 19 random pipe lengths through which a fracture propagated in a 762-mm (30-in.) outside-diameter × 9.5-mm (0.375-in.) wall-thickness API, grade X56, pipe

Data are for two-thirds thickness transverse specimens.

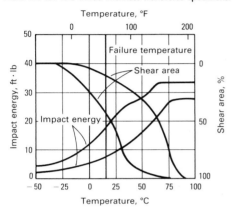

Fig. 24 Charpy V-notch curve for 762-mm (30-in.) outside-diameter × 9.5-mm (0.375-in.) wall-thickness API, grade X56, pipe in which a ductile fracture propagated and arrested

Data are for two-thirds thickness transverse specimens.

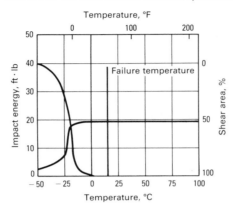

Fig. 25 Charpy V-notch impact energy and DWTT data for a failure in 915-mm (36-in.) outside-diameter × 10-mm (0.406-in.) wall-thickness API, grade X52, pipe

A length, ductile-fracture arrest; B and C lengths, cleavage-fracture propagation

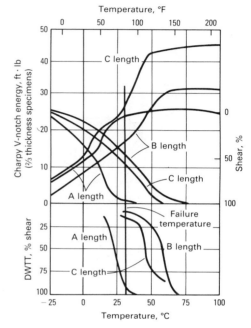

natural gas. Historically, in the early 1950s in the gas testing of pipelines, there were several cleavage fractures that propagated to distances of 914 m (3000 ft). Other than the length of the fracture, the unusual aspect of the failure was that there was no reduction in thickness at the fracture edge, and the fractures propagated along the pipe in a sinusoidal pattern.

A failure occurred in the early 1960s in which a cleavage fracture initiated from a fatigue crack alongside a longitudinal weld. The fatigue crack formed during rail shipment of the pipe. The pipe material was 762-mm (30-in.) diam × 9.5-mm (0.375-in.) wall-thickness API, grade X56, pipe. The fracture initiated when this section of line was being used to transfer gas to the next section to be tested. The pressure at the time of failure was equivalent to a hoop stress of 63% SMYS. This was reported to be the highest pressure that this section of pipe had experienced. The fracture extended from the origin as a cleavage fracture and propagated for a total distance of 13.3 km (8.3 miles). The origin was 5.1 km (3.2 miles) from one end. The fracture propagated in a sinusoidal pattern, with the number of simultaneous fractures varying from one to six, depending on the transition temperatures of the pipe steel. On one end, the fracture arrested as a ductile fracture in a pipe length with a low transition temperature, and on the other end, the fracture arrested at the transition from the 9.5-mm (0.375-in.) line pipe to a heavy-wall forged tee. The fracture did not penetrate the forged tee, but stopped at the girth weld.

Figure 23 shows Charpy V-notch impact curves from 19 random samples taken from pipe lengths through which the fracture propagated. The fracture appearance of the Charpy

specimens at the failure temperature ranged from approximately 10 to 40% shear. Figure 24 shows the Charpy V-notch impact curve for the end in which the fracture turned to a shear fracture, spiraled around the pipe, and arrested. At the failure temperature, the Charpy V-notch shear appearance was 100% shear. Thus, in this failure, the fracture arrested on one end because it encountered a length of pipe with a low transition temperature, and it arrested on the other end because it encountered a reduced stress level.

Another failure that occurred in 915-mm (36-in.) outside-diameter × 10-mm (0.406-in) wall-thickness API, grade X52, pipe was initiated by sabotage. The fracture propagated through 2.5 pipe lengths in one direction as a cleavage fracture and, upon crossing a girth weld, turned to shear and arrested. In the other direction, the fracture propagated through portions of three lengths as a cleavage fracture. It suddenly turned to tearing shear in the middle of a pipe length and arrested by spiraling around the circumference. Figure 25 shows the Charpy V-notch and drop-weight tear test (DWTT) curves for two pipe lengths through which cleavage fractures propagated and for the length in which a ductile fracture arrested. The DWTT results clearly show the arrest length as having an 85% shear-area transition temperature (SATT) below the failure temperature; thus, the shear fracture would be expected (SATT is used to identify the ductile-to-brittle transition temperature for pipeline steels). It should be noted that the transition temperature predicted by the DWTT (85% shear) is slightly above that predicted by the Charpy (85% shear) due to the 10-mm (0.406-in.) thickness of the DWTT. The two lengths through which the cleavage fracture propagated exhibited DWTT shear areas of 10 and 17%. This is consistent

with the fracture appearance of 12 and 25% shear, respectively, for the two pipe lengths.

In the opposite direction, the fracture propagated through parts of three lengths as a cleavage fracture and arrested in the middle of a pipe length as a tearing shear fracture. The Charpy and DWTT curves for the three pipe lengths are shown in Fig. 26. The DWTT predicts that the fracture appearance in all three lengths would be cleavage, as observed. The Charpy curves are in agreement, with the exception of length B, which exhibits an 85% SATT below the failure temperature. This discrepancy is probably due to thickness effects.

Comparing the energy available to drive the cleavage fracture with the energy absorbed in the fracture process, it is apparent that the cleavage-fracture arrest is to be expected. The energy U available to drive the fracture in length A (the cleavage-fracture arrest length) is determined by (Ref 4):

$$U = \frac{\sigma^2 \pi R}{E} = \frac{(33\ 250)^2 \pi 1.5}{30 \times 10^6} = 174 \text{ ft} \cdot \text{lb/in.}^2$$

(Eq 2)

where $\sigma = 750(18)/0.406 = 33.25$ ksi hoop stress (229.3 MPa), $E = 3 \times 10^4$ ksi (2.07×10^5 MPa), and R is the pipe radius (457 mm, or 18 in.). The energy absorbed in the fracturing process can be determined from the Charpy energy at the shear area equivalent

Fig. 26 Charpy V-notch and DWTT test data for failure in a 915-mm (36-in.) outside-diameter × 10-mm (0.406-in.) wall-thickness API, grade X52, pipe

A length, cleavage arrest; B and C lengths, cleavage-fracture propagation

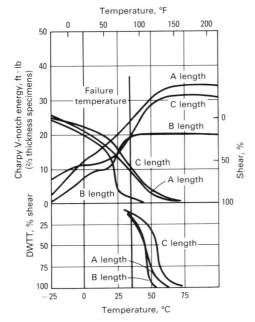

Fig. 27 Charpy V-notch impact data for pipe lengths in which ductile fracture propagated and arrested

End A, fracture arrest; End B, fracture arrest

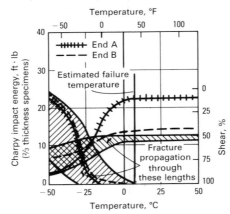

to the fracture appearance of the failed pipe (Ref 4). The average fracture appearance of the failed pipe was 22% shear. The Charpy energy at 22% shear is 19 J (14 ft · lb); converting this to Joules per square centimeter by dividing by the Charpy specimen area, the energy per unit area is 36 J/cm^2 (169 ft · lb/in.^{2}). Thus, the energy available to drive the cleavage fracture is very close to the energy absorbed in the fracture process, and the fracture arrest is highly probable.

Cleavage fractures have been observed to arrest in several ways. In pipe lengths with low transition temperatures, cleavage fractures turn to ductile fractures and arrest. Cleavage frac-

tures have also arrested at an abrupt change in cross section, such as the transition to a forged tee or flange or at a road crossing casing, and by encountering a pipe length with sufficient energy-absorption capacity so that there is insufficient energy to drive the fracture.

A second mode of fracture propagation and arrest is ductile-fracture propagation and arrest. There have been several service failures in which ductile fractures have propagated up to 305-m (1000-ft) total length and then have arrested. In these failures, a single fracture propagated axially along the pipe. The fractures that have propagated unstably have generally propagated on the top of the pipe. The scatterband of the Charpy V-notch curves for one failure involving 914-mm (36-in.) diam pipe is shown in Fig. 27. The Charpy curves from the two lengths in which the fracture arrested are also shown. This fracture occurred at a hoop stress level of 307.6 MPa (44.6 ksi), and the total fracture length was 259 m (850 ft). All of the transition temperatures were below the operating temperature; therefore, the fractures were all ductile. Because the Charpy plateau

energy was relatively low in the propagate lengths, unstable ductile-fracture propagation occurred. The fracture arrested in the end lengths. The two-thirds thickness Charpy data indicate that ductile fractures propagated through lengths having 15- to 17-J (11- to 12.5-ft · lb) upper plateaus and arrested in pipe lengths with 19- and 30-J (14- and 22-ft · lb) plateaus. This type of failure indicates that material with a low transition temperature may not provide fracture control unless consideration is also given to the toughness level of the material. Equation 3 (Ref 5) provides an approximate relationship for estimating the Charpy V-notch plateau energy required for ductile-fracture arrest in various diameters, wall thicknesses, and stress levels between 60 and 80% of the specified minimum yield stress in gas-transmission pipelines:

$$CVN = 0.0108 \, \sigma^2 \, (Rt)^{1/3} \qquad \text{(Eq 3)}$$

where CVN is the Charpy V-notch plateau energy (in foot-pounds) for a 10- × 10-mm (0.4- × 0.4-in.) specimen, σ is the hoop stress (in kips per square inch), R is the pipe radius (in inches), and t is the pipe wall thickness (in inches).

REFERENCES

1. "Recommended Practice for Railroad Transportation of Line Pipe," API RP5L1, American Petroleum Institute, Dallas
2. "Specification for Line Pipe," API 5L, 35th ed., American Petroleum Institute, Dallas
3. J.F. Kiefner and A.R. Duffy, "Summary of Research to Determine the Strength of Corroded Areas in Line Pipe," American Gas Association, Arlington, VA, July 20, 1971
4. W.A. Maxey, J.F. Kiefner, and R.J. Eiber, "Brittle Fracture Arrest in Gas Pipelines," Catalog L51436, American Gas Association, Arlington, VA, April 4, 1983
5. W.A. Maxey, J.F. Kiefner, and R.J. Eiber, "Ductile Fracture Arrest in Gas Pipelines," Catalog L32176, American Gas Association, Arlington, VA, Dec 12, 1975

Failures of Bridge Components

John W. Fisher, Fritz Engineering Laboratory, Lehigh University

FROM 1960 TO 1986, a number of localized failures developed in steel bridge components due to fatigue-crack propagation, which resulted in some instances in brittle fracture from crack instability. More than 200 bridge sites have developed one or more types of cracking. Several types of cracking have often developed at a single site because of the existence of different details on several types of structures (Ref 1, 2).

The largest category of cracking is a result of out-of-plane distortion in a small gap, which is usually a segment of a girder web. When distortion-induced cracking develops in a bridge component, large numbers of cracks usually form nearly simultaneously in the structural system because the cyclic stress is high and the number of cycles needed to produce cracking is relatively small. Displacement-induced fatigue cracking has developed in a wide variety of bridge structures, including suspension, two-girder floor-beam, multiple-beam, tied-arch, and box-girder bridges. In general, the cracks formed in planes parallel to the stresses from loading and were not detrimental to the performance of the structure provided they were discovered and retrofitted before turning perpendicular to the applied stresses.

The next largest category of cracked members and components comprises large initial defects and cracks. In several cases, the defects in this category resulted from poor-quality welds that were produced before non-destructive test methods were well developed. However, the largest number of this type of crack resulted because the groove-welded component was considered a secondary member or attachment; consequently, weld-quality criteria were not established, and nondestructive test requirements were not imposed.

Most of the remaining cracks resulted from the use of low fatigue strength details that were not anticipated to have such low fatigue resistance at the time of the original design.

Details and Defects

Low Fatigue Strength Details. The possibility of fatigue cracks forming at the ends of welded cover plates was demonstrated at the American Association of State Highway Offi-

Fig. 1 Cracked girder at the end of the cover plate

Fig. 2 Crack at the end of the lateral connection plate

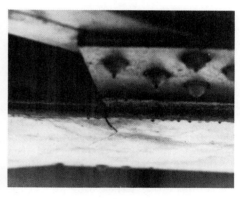

Fig. 3 Typical large crack at the weld toe at the end of the cover plate

cials (AASHO) Road Test in the 1960s (Ref 3). Multiple-beam bridges subjected to relatively high stress-range cycles (≈83 MPa, or 12 ksi) under controlled truck traffic experienced cracking after 500 000 vehicle crossings. In general, few cracked details were known to exist until a cracked beam was discovered in Span 11 of the Yellow Mill Pond Bridge in 1970 (Fig. 1). Between 1970 and 1981, the Yellow Mill Pond multibeam structures located at Bridgeport, CT, developed extensive numbers of fatigue cracks at the ends of cover plates (Ref 1). These cracks resulted from the large volume of truck traffic and the unanticipated low fatigue resistance of the large cover-plated beam members (Category E') (Ref 4).

Several other Category E or E' details have experienced fatigue cracking at weld termination (Ref 1). The crack shown in Fig. 2 typifies these structures and this type of cracking. It originated at the end of a longitudinal fillet weld that was used to attach a 0.9-m (3-ft) long lateral connection plate to the edge of a flange.

The Yellow Mill Pond Bridge was constructed in 1956-1957 and was opened to traffic

in January 1958. In October-November 1970, the steel superstructure of the Yellow Mill Pond Bridge was inspected after cleaning and repainting. On November 2, 1970, during a routine inspection of the repainting, it was discovered that a crack had developed in the eastbound roadway in Span 11. The crack started at the end of the cover-plated beam; it propagated through the flange and up to 400 mm (16 in.) into the web of one of the main girders (Fig. 1).

Additional cracks, such as that shown in Fig. 3, were detected between 1970 and 1981. The entire structural complex was retrofitted in 1981 by peening the weld toe at small cracks or uncracked details and by installing bolted splices at large cracks.

Fig. 4 Stress-range histograms

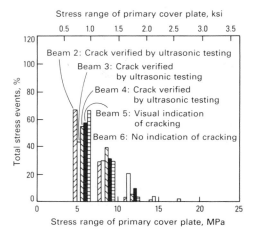

Fig. 5 Polished-and-etched core showing a fatigue crack in the beam flange

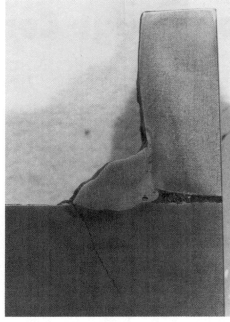

Fig. 7 Crack at a vertical butt weld

(a) Schematic showing insert and crack location. (b) Typical crack at a vertical groove weld. Source: Ref 9

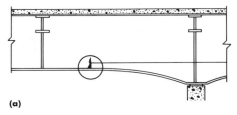

(a)

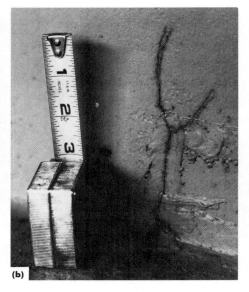

(b)

Fig. 6 TEM fractograph showing fatigue-crack growth striations
39 000×

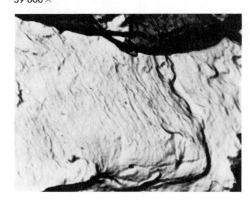

The eastbound and westbound bridges of Span 10 were selected for two major stress-history studies. The state of Connecticut and the Federal Highway Administration (FHWA) conducted both studies, the first in July 1971 (Ref 5) and the second between April 1973 and April 1974 (Ref 6). Figure 4 shows a typical stress histogram for the westbound lanes. The truck distribution and its position in the lanes were also concurrently recorded, and their weights were sampled. Some additional measurements were made in 1976 (Ref 7). The composition of the average daily truck traffic (ADTT) at the Yellow Mill Pond Bridge was about the same as the gross vehicle weight distribution developed from the 1970 FHWA nationwide Loadometer survey. From January 1958 to June 1976, approximately 259 million vehicles crossed the structure. Approximately 13.5% of this total flow was truck traffic. Therefore, approximately 35 million trucks crossed the eastbound and westbound bridges between 1958 and 1976.

The measurements acquired in June 1971, from April 1973 to April 1974, and in June 1976 were used to construct composite stress-range response spectra. Miner effective stress-range values S_{rMiner} were determined for the gage locations (Ref 7):

$$S_{rMiner} = [\Sigma \alpha_i S_{ri}^3]^{1/3} = 13.6 \text{ MPa (1.98 ksi)}$$

where α_i is the frequency of occurrence of stress-range level s_{ri}. The measurements indicated that the effective stress range in other girders varied from 7.6 to 13.6 MPa (1.1 to 1.98 ksi). Those beams located under the outside lane tended to provide the higher stress-range values.

An examination of the stress response to the passage of a truck also indicated that more than one stress cycle occurred. The variable stress spectrum appeared to correspond to about 1.8 events per truck. This corresponded to 62.8×10^6 cycles if it were applied to each vehicle passage.

During 1981, cores were removed from several of the beams to examine the crack surfaces and the retrofit conditions conducted in 1976. Figure 5 shows the center of a polished-and-etched core, and the crack at the weld toe can be seen to extend more than halfway through the beam flange. The crack surface of the core was exposed and examined with the electron microscope. Because the surface was relatively clean and not extensively corroded, the crack-surface features were not destroyed. Figure 6 shows a transmission electron microscopy (TEM) fractograph at a magnification of 56 000× at the crack tip, which was 16 mm (0.63 in.) deep. Striationlike features are visible on the crack surface at various locations. The random variable loading produced striations of variable height and spacing.

The cracks forming at the cover-plate weld toes were modeled as semielliptical surface cracks in the flange (Ref 1). The stress intensity K was defined as (Ref 7):

$$K = F_e F_s F_w F_g \sigma \sqrt{\pi a}$$

where

$$F_e = \frac{1}{E(k)}$$

$$E(k) = \int_0^{\pi/2} \left[1 - \left(\frac{c^2 - a^2}{c^2} \right) \sin^2 \theta \right]^{1/2} d\theta;$$

$$c = 5.46a^{1.133} \text{(in.)}$$

$$F_s = 1.211 - 0.186 \sqrt{\frac{a}{c}}$$

$$F_g = \frac{K_T}{1 + 6.79(a/t_f)^{0.44}}$$

$$K_T = -3.539 \ln \left(\frac{z}{t_f} \right) + 1.981 \ln$$

$$\left(\frac{t_{cp}}{t_f} \right) + 5.798$$

$$F_w = \sqrt{\sec \frac{\pi a}{2t_f}}$$

Fig. 8 Cracks in the Quinnipiac River Bridge after 9 years of service
(a) Crack in the girder web at the longitudinal stiffener groove weld. (b) Crack in the longitudinal stiffener weld

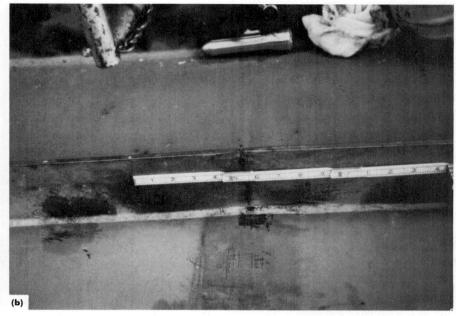

With $z = 16$ mm (0.63 in.), $t_f = 32$ mm (1.26 in.), $t_{cp} = 32$ mm (1.26 in.), $a_i = 0.75$ mm (0.03 in.), and $\Delta\sigma_{Miner} = 13$ MPa (1.9 ksi), the number of cycles N required to grow a fatigue crack 25 mm (1 in.) deep was estimated at:

$$N = \int_{0.03}^{1.0} \frac{da}{3.6 \times 10^{-10}\ \Delta K^3} = 36 \times 10^6 \text{ cycles}$$

This is in reasonable agreement with the larger fatigue cracks that formed between 1958 and 1976 when an estimated 35 million trucks crossed the bridges. It is also compatible with the experimental data on full-scale cover-plate beams (Ref 4, 8).

Large initial defects and cracks constitute the second largest category of cracked members and components. In several cases, the defects resulted because of poor-quality welds that were produced before nondestructive test methods were well developed. A larger number of these cracks resulted because the groove-welded component was considered a secondary member or attachment; therefore, weld-quality criteria were not established, and nondestructive test requirements were not imposed on the affected weldment. Splices in continuous lon-

gitudinal stiffeners represent a common condition that falls into this category. A related condition has occurred when backing bars were used to make a groove weld between transverse stiffeners and a lateral gusset plate. Lack of fusion often exists adjacent to the girder web in the transverse groove welds. If the transverse welds intersect with the longitudinal welds, they provide a path for a crack to enter the girder web.

Cracks were discovered at the vertical butt-weld detail at haunch inserts (Fig. 7a). One of these cracks extended 1.1 m (44 in.) into the girder web along a diagonal line starting from the vertical butt-weld detail.

The fatigue cracks that developed propagated from large initial weld imperfections or inclusions in the short transverse groove welds at the ends of the parabolic haunch inserts in the main girders (Fig. 7b). Of 24 welded details, 7 had cracked by 1973.

In November 1973, a large crack was discovered in the south facia girder of the suspended span in the center portion of the Quinnipiac River Bridge (Ref 10). The bridge had experienced approximately 9 years of service life before discovery of the crack. Figure 8(a) shows the crack that developed in the south facia girder web. The crack propagated approx-

imately to mid-depth of the girder and penetrated into the bottom flange of the girder before it was discovered. A second crack was detected toward the midspan about 9 m (29 ft) from the cracked section. This crack severed the stiffener but did not propagate through the web. Figure 8(b) shows this crack at the time it was detected.

A detailed study of the fracture surface was made on pieces that were removed from the section (Ref 10). This indicated that the fracture initiated at the unfused butt weld in the longitudinal stiffener.

Cracks that have developed in the web at lateral connection plates have generally occurred as a result of intersecting welds. The lateral connection plate is often framed around a transverse stiffener. It was used to connect diaphragms and lateral bracing members to the longitudinal girders of bridges.

One of the first bridge structures to exhibit this cracking was the Lafayette Street Bridge over the Mississippi River at St. Paul, MN (Ref 11). Figure 9 shows the cracked girder at the detail attached to the web. The primary problem was the large defect in the weld attaching the lateral connection plate to the transverse stiffener. Because this weld was perpendicular to the cyclic stresses and because the weld intersected with

Fig. 9 Cracked girder at the Lafayette Street Bridge

Fig. 10 Schematic of crack-growth stages

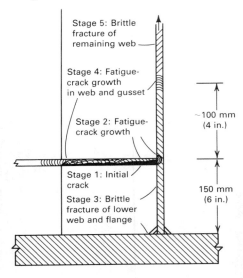

Stage 5: Brittle fracture of remaining web

Stage 4: Fatigue-crack growth in web and gusset

Stage 2: Fatigue-crack growth

Stage 1: Initial crack

Stage 3: Brittle fracture of lower web and flange

~100 mm (4 in.)

150 mm (6 in.)

Fig. 11 Cracked groove weld in the cover plate

the vertical welds attaching the stiffener to the web and the longitudinal welds of the connection plate, a path was provided into the girder web. A detailed study indicated that the crack originated in the weld between the gusset plate and the transverse stiffener as a consequence of a large lack-of-fusion discontinuity in the welded connection. Figure 10 shows a schematic of the crack surface. The intersecting welds at the corner permitted the transverse crack to penetrate into the girder web.

Typical of the cracks that have developed in flange groove welds is the splice in the cover plate at thickness transitions (Fig. 11). Numerous cracks caused by lack of fusion have been detected in groove welds (Fig. 12). In general, these cracks formed because of lack of fusion at the root (land) of the groove weld. Regarding Fig. 11, it is probable that the single-U groove weld was made using the beam flange as a

Fig. 12 Lack-of-fusion defect in the flange groove

backup. Figures 12 and 13 show a double-V groove weld made in the same era with lack of fusion at the root.

The design stress range at the cracked groove welds like that shown in Fig. 11 varied between 58.6 and 78.6 MPa (8.5 and 11.4 ksi). Total truck traffic crossing the structure was estimated at 15.7 million vehicles.

The actual service stresses were not available. However, the effective stress range can be estimated by assuming that the daily truck traffic corresponds to the nationwide distribution (Ref 12). The effective stress range becomes:

$$S_{rMiner} = \alpha S_r^D \left[\Sigma \gamma_i \left(\frac{S_{ri}}{S_r^D} \right)^3 \right]^{1/3}$$

$$= \alpha(0.35)^{1/3} S_r^D$$

$$= 0.7 \alpha S_r^D$$

An effective stress range of 14 to 21 MPa (2 to 3 ksi) is probable, because the factor γ_i is between 0.25 and 0.4. Assuming one stress cycle for each vehicle passage and no significant changes in total truck traffic between 1958 and 1980, the result would be 15.7×10^6 random variable stress cycles corresponding to the Miner effective stress range.

The probable size of the initial crack a_i is likely to be about 6 mm (0.25 in.). Because the cover plate is connected to the flange only along the edges, the applicable stress-intensity factor for an edge crack is:

$$K = 1.12\sigma\sqrt{\pi a}\sqrt{\sec \frac{\pi a}{2t_f}}$$

where t_f is the thickness of the thinner plate at the groove-weld splice. The predicted cycles of stress are given by:

$$N = \int_{a_i}^{a_f} \frac{da}{3.6 \times 10^{-10} \Delta K^3}$$

With $a_i = 6$ mm (0.25 in.), $S_{rMiner} = 14$ MPa (2 ksi), and $a_f = 12.5$ mm (0.5 in.), a predicted life of 16×10^6 cycles results.

Penetrations Through Box-Girder Webs. A related lack-of-fusion type of defect and crack among the most severe that has been encountered has occurred at details where a plate component was inserted through an opening cut into girder webs. The resulting detail was usually welded into place with either fillet or groove welds. In either case, large cracks resulted at the edge of the flange plate where

Fig. 13 Polished-and-etched section showing lack of fusion

Fig. 14 Section of box-girder bent showing the location of the crack

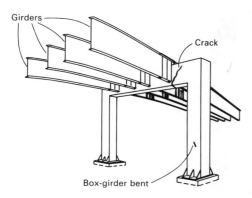

Fig. 15 Cracked box girder

Fig. 16 Fracture surface of the web plate

short vertical weld lengths resulted with large unfused areas.

A large crack in one of the steel box bents (rigid frames) supporting the elevated track of Chicago's Mass Transit Dan Ryan Line was discovered on January 4, 1978 (Ref 13). Subsequent inspection showed that two adjacent bents were also cracked.

The bents have a box-shaped cross section consisting of two column legs and a horizontal box member (Fig. 14). The top flanges of the plate girders pass over the top flanges of the boxes. The bottom flanges pierce the boxes through flame-cut slots near the bottom of the box side plates. In the cracked bents, the bottom flanges were welded to the box-girder web.

The initial field examination of the fracture indicated that all of the cracks started at the welded junction of the plate girder flange tip to the box side plate (Fig. 15). All three cracks completely severed the bottom flange of the box girders and the webs.

Fig. 17 Steel pier-cap section showing reaction brackets passing through the web

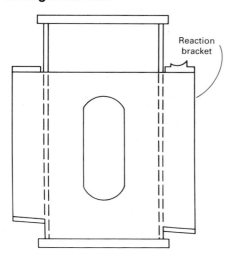

The crack surfaces of the side plates of the box girders adjacent to the flange tip were examined. Chevron markings found on the fracture surface indicated that the crack originated near the tip of the girder flange. Figure 16 shows a photograph of one fracture surface. The overall crack surface shows that the weldment connecting the flange tip to the box had a large lack-of-fusion area. Paint that had penetrated from the inside of the box was found in some of these areas. The largest lack-of-fusion area was about 18×83 mm (0.70×3.27 in.) and was approximately elliptical. Fatigue-crack propagation was detected at the inside and outside web surfaces (Ref 14).

Reaction diaphragm plates were also passed through box-girder webs (Fig. 17). Backup bars were used to make the groove welds between the girder webs and the insert plate (Fig. 18). Figure 19 shows a core removed from the box-girder web and reveals the crack that propagated from the initial flaw condition provided by the backup bar and slag at the root. Several bridges with these types of details have experienced fatigue-crack growth that was visible after 1 or 2 years of service.

Fig. 18 Schematic of the lack-of-fusion crack from the backup bar

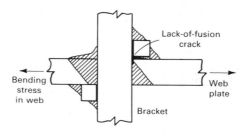

Out-Of-Plane Distortion

At least 100 bridge sites have developed fatigue cracks as a result of out-of-plane distortion in a small gap (Ref 1). Most often, this was a segment of a girder web. When distortion-induced cracking develops in a bridge, large numbers of cracks usually form before corrective action is taken because the cyclic stresses are often very high. As a result, many cracks form simultaneously in the structural system. Other types of cracking, such as a low fatigue strength detail or a large built-in defect, often result only in a single significant crack at a given bridge. Other potential crack locations can be identified and retrofitted before significant damage develops elsewhere.

Displacement-induced fatigue cracking has developed in such diverse structures as suspension bridges, two-girder floor-beam bridges, multiple-beam bridges, tied-arch bridges, and box-girder bridges. Cracks have formed mainly in planes parallel to the stresses from loading; therefore, these cracks have not been detrimental to the performance of the structure provided they were discovered and retrofitted before turning perpendicular to the applied stresses from loads. In some structures, the cracks have

Fig. 19 Investigation of cracking of a box-girder web

(a) Hole resulting from removal of the core. (b) Polished-and-etched surface showing crack, slag, and lack of fusion

(a)

(b)

Fig. 20 Crack along the web-flange weld toe at the end reaction

arrested when located in low-stress areas, and the crack has served to relieve the restraint condition.

Floor-Beam-Girder Connection Plates. One of the earliest and most common sources of fatigue cracks from distortion is the cracking in the web gaps at the ends of floor-beam connection plates (Ref 15, 16). These cracks have occurred in the web gap near the end reactions (Fig. 20) when the floor-beam connection plate was not welded to the bottom tension flange. However, the most extensive cracking is in the negative-moment regions of continuous bridges where the connection plate was not welded to the top flange of the girder.

The cracking illustrated in Fig. 20 and 21 has occurred in skewed, curved, and right bridges. Generally, it is more severe and occurs earlier in skewed and curved structures.

Strain measurements have verified that the primary cause of the cracking is the out-of-plane movement at the end of the connection plate (Fig. 22). Figure 23 shows the stress gradient that was observed in one bridge structure (Ref 15, 16). The end rotation of the floor beam develops from bending of the floor beam and differential vertical movement between its two ends. The end rotation forces the web out-of-plane in the gap

between the end of the connection plate and the web-flange weld.

Retrofit procedures for existing structures (Ref 1) have included removing a segment of the connection plate to increase the length of the web gap and welding or bolting the connection plate to the tension flange. Drilling holes to arrest the crack has provided only a temporary solution.

Multiple-Girder Diaphragm Connection Plates. Diaphragms and cross-frames in multiple-beam bridges often provide conditions at the transverse connection plates on the longitudinal girders to which they are connected that are similar to those of the floor-beam-girder connection plate. Generally, the diaphragms and cross-frames are connected to transverse stiffeners that are welded to the girder web. No connection has been provided between the stiffener and the girder tension flange. Sometimes, these stiffeners are not attached to either flange. Because adjacent beams deflect differing amounts, the differential vertical movement produces an out-of-plane deformation in the web gap at the stiffener ends that are not attached to the beam flange. The magnitude of this out-of-plane movement depends on the girder spacing, skew, and type of diaphragm or cross-frame. In general, skewed

bridges undergo larger relative vertical movement of adjacent girders, which results in larger out-of-plane movement in the web gap.

Various types of diaphragms and girder spacings have resulted in this type of cracking. The diaphragms have ranged from simple X-bracing using angles to more rigid rolled sections.

Figure 24 shows a five-girder bridge with relatively stiff diaphragms composed of a blocked rolled beam with haunched ends. Cracks formed in the web gap along the web-flange weld toe (Fig. 25). In all but one highway bridge, these cracks have been confined to the negative-moment region where the top flange is in tension. The slab has provided greater fixity to the top flange, and the adjacent web has experienced more deformation and has cracked first.

Fig. 22 Distortion in the web gap

M, bending moment; L, gap between surface of flange and weld to web

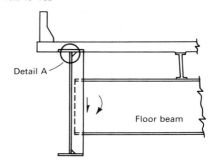

Floor beam

Detail A

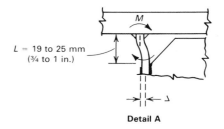

L = 19 to 25 mm
(¾ to 1 in.)

M

Detail A

Fig. 21 Cracking in floor-beam-girder connection plates

(a) Cracks in the outside web surface. (b) Cracks from the inside web surface

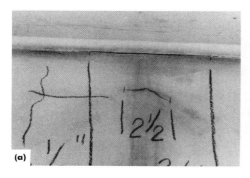

(a)

(b)

Fig. 23 Measured street gradient in the web gap

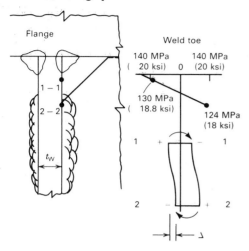

Fig. 24 Diaphragms and girders in a continuous-span structure

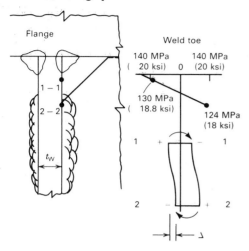

Fig. 25 Crack along the web-flange weld

Fig. 26 View of bracing and girders

Fig. 27 Fatigue cracks along the web-flange weld and at the end of the connection plate

Fig. 28 Crack in the girder web at the bottom of the transverse connection plate

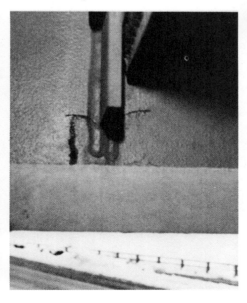

Fig. 29 Cracking in a floor-beam web above the end of the riveted angle end connection

(a) Floor beam/tie girder connection. (b) Crack along the web-flange weld above the end connection

(a)

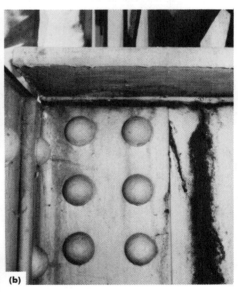

(b)

A more flexible diaphragm has been the X-type cross-frame shown in Fig. 26. Cracking has developed in the negative-moment region (top flange in tension) of continuous-span structures in the web gap (Fig. 27). The transverse stiffener to which the X-bracing was attached was not welded to the tension flange. This permitted the occurrence of out-of-plane displacement in the web gap and the formation of cracks along the web-flange weld and at the end of the connection plate.

Several simple-span railroad and mass-transit bridges with relatively rigid diaphragms have developed cracks in the web gap adjacent to the bottom flange (Fig. 28). These structures were usually skewed, with the diaphragms placed perpendicular to the longitudinal girders.

Cracking was initially observed at the ends of the transverse connection plate, which was cut short of the tension flange by 25 to 50 mm (1 to 2 in.). Cracking has since developed along the web-flange weld.

Tied-Arch Floor Beams. A number of tied-arch structures have been built with the floor beams framed into the tie girder with web shear connection alone. The tie girder is a bending tie that is deeper than the floor beams. No direct connection was provided between the floor-beam flange and the tie girder. In several older structures, the floor-beam end connections were riveted double angles attached to the floor-beam web, with the outstanding legs riveted to the tie girder. More recently, constructed arches have utilized welded transverse connection plates on the tie-girder webs. The floor beams have been bolted to the welded transverse connection plate.

Cracks have formed in the floor beams of structures with either riveted end connections or welded transverse connection plates. The cracks have formed in the floor-beam web along the web-flange connection at the floor-beam web gap between the end of the connection angles or at the end of the welded connection plates. These cracks extended parallel to the floor-beam flange along the length of the web gap, then began to turn and propagate toward the bottom flange.

Figure 29 shows a crack that formed in the floor-beam web above the end of the riveted angle end connection. Visual observation showed that a horizontal displacement in the direction of the bridge developed between the tie girder and the top flange of the floor beam at the end connection. This relative movement produces out-of-plane deformation in the small web gap between the end of the floor-beam

angle end connection and the top flange of the floor beam.

At least eight tied-arch structures have experienced cracking in the floor beams along the web-flange connection of the top flange. In some of these structures, the stringers rested on the top of the floor-beam flange; in others, the stringers framed into the floor beam.

REFERENCES

1. J.W. Fisher, *Fatigue and Fracture in Steel Bridges*, Wiley-Interscience, 1984
2. J.W. Fisher and D.R. Mertz, Hundreds of Bridges—Thousands of Cracks, *Civil Eng.*, April 1985
3. J.W. Fisher and I.M. Viest, *Fatigue Life of Bridge Beams Subjected to Controlled Truck Traffic*, Preliminary publications, 7th Congress, International Association for Bridge and Structural Engineering, Zurich, 1964
4. J.W. Fisher, H. Hausammann, M.D. Sullivan, and A.W. Pense, *Detection and Repair of Fatigue Damage in Welded Highway Bridges*, NCHRP report 206, Transportation Research Board, Washington, June 1979
5. D.G. Bowers, *Loading History Span 10 Yellow Mill Pond Bridge I95 Bridgeport, Connecticut*, Research project HPR 175-332, Connecticut Department of Transportation, Whethersfield, CT, 1972
6. R.L. Dickey and T.P. Severga, *Mechanical Strain Recorder on a Connecticut Bridge*, Report FCP 45G1-222, Federal Highway Administration Department of Transportation, Washington, 1974
7. J.W. Fisher, R.E. Slockbower, H. Hausammann, and A.W. Pense, Long Time Observation of a Fatigue Damaged Bridge, *J. Tech. Counc.*, Vol 107, (No. TC1), April 1981
8. J.W. Fisher, K.H. Frank, M.A. Hirt, and B.M. McNamee, *Effect of Weldments on the Fatigue Strength of Steel Beams*, NCHRP report 102, Transportation Research Board, Washington, 1970
9. J.P.C. King, P.F. Csagoly, and J.W. Fisher, Field Testing of Aquasabon River Bridge in Ontario, *Trans. Res. Rec.*, No. 579, 1976
10. J.W. Fisher, A.W. Pense, H. Hausammann, and G.R. Irwin, Analysis of Cracking in Quinnipiac River Bridge, *J. Struct. Div.*, Vol 106 (No. ST4), April 1980
11. J.W. Fisher, A.W. Pense, and R. Roberts, Evaluation of Fracture of Lafayette Street Bridge, *J. Struct. Div.*, Vol 103 (No. ST7), July 1977
12. J.W. Fisher, *Bridge Fatigue Guide—Design and Details*, Publication T112-11/77, American Institute of Steel Construction, Chicago, 1977
13. Engineers Investigate Cracked El, *Eng. News Rec.*, Vol 200 (No. 3), Jan 19, 1979
14. J.W. Fisher, J.M. Hanson, H. Hausammann, and A.E.N. Osborn, *Fracture and Retrofit of Dan Ryan Rapid Transit Structure*, Final report, 11th Congress, International Association for Bridge and Structural Engineering, Zurich, 1980
15. J.W. Fisher, Fatigue Cracking in Bridges From Out-Of-Plane Displacements, *Can. J. Civil Eng.*, Vol 5 (No. 4), 1978
16. W. Hsiong, Repair of Poplar Street Complex Bridges in East St. Louis, *Trans. Res. Rec.*, No. 664, 1978

Failures of Locomotive Axles

George F. Vander Voort, Carpenter Technology Corporation

FAILURES OF LOCOMOTIVE AXLES caused by overheated traction-motor support bearings are discussed in this article. These failures are of interest because the analysis shows an example of what can be done when the fracture face and origin are destroyed during the failure incident. In most failure analyses of broken components, it is generally assumed that conclusive results cannot be obtained if the fracture and the fracture origin cannot be identified and examined. In many failures this is true. However, the failures described in this article possess some unique characteristics that permit successful analysis despite the lack of a preserved fracture face and origin.

Failures of locomotive axles due to overheated friction bearings are rather common in the railroad industry and have been observed for more than 100 years. Because of this, such failures are usually diagnosed merely by visual inspection of the damage. Comprehensive metallographic studies, therefore, are not done and have not been accurately documented in the open literature for many years. However, detailed analysis of such failures reveals a number of significant features.

Background

Friction bearings have been used for many years and are perfectly adequate if they are lubricated. The bearing is essentially a bronze cylinder lined with babbitt metal. Typically, the bronze alloy composition is close to Cu-16Pb-6Sn-3.5Zn, while the babbitt composition is usually Pb-3.5Sn-8-11.5Sb. A window is cut into the bearing and packed with cotton waste, which trails down to an oil reservoir. Oil is drawn up the wick during service to lubricate the contacting surfaces.

Friction-bearing failures due to overheating have been examined by the metallurgist for many years. Perhaps the earliest example, documented in 1914, involved a failed Krupp railroad axle (Ref 1). The study revealed a rather complex crack pattern, evidence of exposure to very high temperatures, and ruptures in the overheated region. Bronze bearing metal was observed in the cracked surface region located beneath the support bearing. The bearing metal was molten when it penetrated the axle. In addition, the railroad that submitted the

failed axle had tried to remove surface evidence of overheating. Also mentioned was earlier work that had been done on broken axles that were overheated. In each of these cases, small particles of bearing metal had penetrated the axle in the overheated region.

In 1944, a review was published of railroad-car axle failures due to the absorption of molten copper (Ref 2). Most axle-journal failures occurred near the wheel hub, an area of high stress and temperature. A broken axle was shown in which the fracture was not destroyed after breakage had occurred. From the surface inward, the fracture surface was rough, indicating the depth of copper penetration. The central portion was smooth, indicative of a fatigue fracture that ultimately led to failure. Color photomicrographs revealed a yellow grain-boundary copper phase. It was pointed out that adequate lubrication is required to keep the operating temperature of the contacting surface below the melting point of the bearing materials.

Two reports were issued—one in 1947, the other in 1954—concerning copper-penetration axle failures and the use of nondestructive testing to detect surface cracks in such failures (Ref 3, 4). The 1947 paper (Ref 3) discusses twist-off failures due to overheated bearings. This failure mode is referred to as a hot-box in railroad terminology. Almost all axle-journal failures were claimed to be attributed to intergranular embrittlement of the steel by molten brass or copper. The steel in contact with the bearing must be heated to a temperature above the melting point of the brass journal bearing and must be under a stressed condition. Experiments were performed with 13-mm (½-in.) diam medium-carbon steel, loaded as cantilever beams. Samples heated to 925 °C (1700 °F) ran for hours without failure at 1750 rpm. However, the instant they were wetted with molten brass, catastrophic failure occurred. These samples were loaded above the yield point of the steel at 925 °C (1700 °F). Samples were also treated in the same manner, but the load was removed before complete rupture occurred. Microscopic analysis showed that molten brass entered the steel in a narrow canyon at the surface, then spread out in a delta pattern.

The 1954 report (Ref 4) studied two failed railroad axles (from the Macon, Dublin &

Savannah and the Chicago & Northwestern Railroads) with evidence of copper penetration. The axle from the Macon, Dublin & Savannah Railroad had gross cracking with visible copper-colored material in the cracks. Spectrographic analysis of samples from both axles containing copper penetration revealed that the major constituents were copper and lead with a minor amount of tin. These elements are the major constituents of the bronze friction bearing. The copper-penetration failure occurred in the following sequence:

- The bearing surface was heated by friction because of loss of lubrication
- The babbitt metal lining melts between about 240 and 315 °C (465 and 600 °F) and wets the surface, but penetration does not occur
- The babbitt metal is displaced, possibly by mechanical action or volatization
- The bronze backing is heated to its melting point (900 to 925 °C, or 1650 to 1700 °F) and penetrates the axle, causing failure

In 1959, the New York Central Railroad Company studied copper penetration in overheated journals (Ref 5). Two types of failures were observed. The first type, referred to as a burn-off, is indicative of a single continuous heating to failure due to penetration of bearing metals into journals at elevated temperatures. The second type, referred to as a cold break, also results from overheating, but is a two-stage failure process. These fractures exhibit an outer circumferential zone of irregular detail with evidence of thermal checks and intergranular separation and an inner fracture zone typical of a progressive-type fatigue fracture. In both cases, copper from the bronze-backed journal bearings was absorbed intergranularly into the hot steel journal. A surface analysis for copper indicated that the copper content exceeded the residual copper content of the steel to a depth of 1 mm (0.040 in.). A metallographic study confirmed the presence of grain-boundary copper penetration in the affected surface layer. This work was subsequently published (Ref 6).

An important source of information on copper-penetration failures is the reports of the Committee on Axle and Crank Pin Research, formed at the 1949 annual meeting of the

Fig. 1 Fracture surface at the drive-wheel side of axle 1611

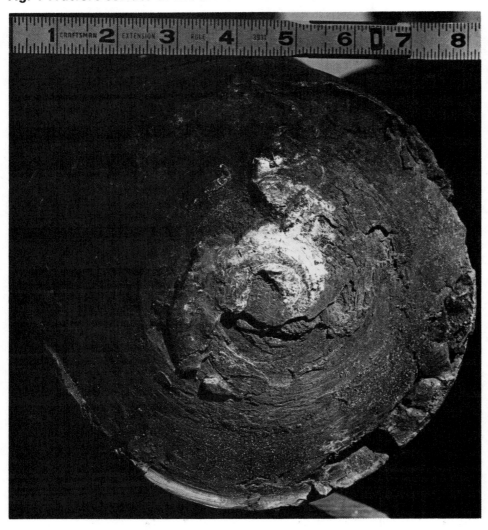

Association of American Railroads (AAR). A major problem faced by this group concerned the decision to scrap or to recondition overheated axles. Overheating can result from loss of lubrication of either friction or roller bearings and possibly from other mechanical problems. Because the number of overheated axles per year was considerable, outright scrapping of all overheated axles represented a significant expense.

In the 1951 proceedings of the AAR, two types of overheating failures were described (Ref 7). One type of burn-off failure exhibited a necked-down, elongated fracture that generally displayed large surface ruptures. The diameter may be necked down to half its original size before the axle twists off. Copper penetration is not observed in these failures. In the second type of overheating failure, the fracture is flat across the section with no reduction in diameter. Copper penetration and thermal cracks are usually found to be present in this type of failure. The Union Pacific Railroad reported that when hot steel and

copper are in contact, the steel is affected by the copper and breaks sharply without a reduction in diameter. The Atcheson, Topeka & Santa Fe Railroad reported that when an axle that is necked down and elongated is inspected, the brass is usually broken, and most of the journal brass is usually intact in the box.

In 1951 and in 1953, the Delaware, Lackawana & Western Railroad reported a revealing test (Ref 7, 8). A non-copper-containing (cast iron) bearing that was not lubricated was placed on one side of the axle, while a lubricated copper-base bearing was put on the other side. The car (a hopper car with a 63 500-kg, or 140 000-lb, load) was run for 108 km (67 miles) without burning off the journal, although it was red hot. The lubrication was then removed from the bronze bearing at the other end of the axle. The car traveled 55 km (34 miles) before the journal broke under the bronze bearing due to copper penetration (Ref 7). The journal under the nonlubricated cast iron bearing was still red hot, but had not broken.

The 1952 AAR proceedings discussed stress measurements made on a new 140- × 250-mm (5½- × 10-in.) standard black-collar freight-car axle fabricated from AAR M-126-49, grade F, steel (Ref 9). Strain gages were placed at five locations, and the journal loads were 69 to 138 MPa (10 to 20 ksi), with speeds from 65 to 135 km/h (40 to 84 mph). Results showed that the dynamic stresses on the journals were very low, in most cases less than one-half the level of stress at the wheel seat.

In this meeting, there was considerable discussion on the reuse of overheated axles. It was thought that if the bearing lining were melted out but surface cracks were not observed, the axle could be safely returned to service. However, when cracks were detected, it was thought that the axle surface could not be turned down to remove the crack without going below the minimum allowable diameter.

The Pennsylvania Railroad favored the scrapping of overheated axles when the babbitt metal was melted out. Experience indicated that about 90% of the overheated axles contained cracks that could not be turned out within the minimum diameter tolerance permitted. The expense of turning down all overheated axles when only 10% could be salvaged was concluded to be unjustifiable. The Southern Railway reported that 85% of its broken axles had been turned down previously, indicating that they had been overheated in earlier service. Removal of the cracking was concluded to be insufficient to guarantee that the axle was safe for additional service. The Atcheson, Topeka & Santa Fe Railroad reported on tests conducted using 19- to 25-mm (¾- to 1-in.) diam miniature journals that were heated, loaded, and subjected to melted bearing metal. The tests resulted in the type of break that 99 out of 100 burn-off journals show, and 99% of the journals exhibited copper penetration.

The 1952 proceedings (Ref 10) contained additional information on the Laudig iron-backed journal bearing that was developed to overcome the copper-penetration problem encountered with bronze-backed bearings. The Laudig journal bearing was recommended as an alternative to the bronze-backed journal (AAR M-501-48). In 1950, six railroads applied 600 test bearings successfully.

In the 1956 AAR proceedings (Ref 11), the Baltimore & Ohio Railroad stated that at least three factors must exist to cause copper-penetration fractures:

- A bearing that contains a metal or metals that will diffuse into the axle material in the grain boundaries
- Heat of the proper range
- Stress

Wetting of the axle surface was considered to be a possible fourth prerequisite. A comprehensive, cooperative study was suggested to solve the problem of copper-penetration fractures. The Chesapeake & Ohio Railroad reported on an investigation that began in 1952 to determine what could be done to reduce or eliminate wrecks and hot-boxes due to defective journals. The railroad developed a journal-inspection car that used ultrasonic inspection. During preliminary testing of this device, a cracked axle was found under a loaded car that was ready to be dispatched. The axle was removed and broken open. The crack had penetrated about an inch from the surface of the journal around approximately one-half of the axle circumference. Subsequent use of this inspection car showed that about one defective axle was found for every 580 cars inspected, based on inspection of about 70 000 cars. Testing of cars owned by other railroads revealed that about 40% of the axles tested were defective.

As a result of axle failures due to copper penetration, the AAR conducted a study to determine the effects of refined copper or alloyed copper on axles under hot-box conditions (Ref 12). The study attempted to answer the often raised question of whether the copper is a precursor of axle failure or is a post-fracture phenomenon, that is, does the molten copper flow into the grain boundaries and cause failure or does the molten copper flow into stress cracks after failure. In these tests, 19-mm (¾-in.) diam centerless-ground AISI 1045 carbon steel samples were left bare or wrapped with 0.13-mm (0.005-in.) thick copper foil and held in place with fine copper wire. In some tests, wrapping was done with 70Cu-30Zn. Specimens that fractured in the presence of brass failed suddenly and had sharp fracture faces; that is, no neckdown resulted. Copper penetration was found in these samples. The temperature above which failures occurred decreased as the applied bending stress increased. As the temperature increased, the probability of fracture increased.

Two distinct fracture modes were observed in these tests. When the failed samples exhibited reduced diameters (necking), evidence of twisting and the presence of cavernous cracks and fissures were due to high-temperature fatigue in an oxidizing atmosphere. When a brittle failure occurred, that is, a sharp, well-defined fracture, microscopic examination showed copper or brass in the grain boundaries. This type of fracture occurred at temperatures above the melting point of the penetrating metal. The penetration followed grain-boundary paths, as in stress-corrosion cracking (SCC). Branching of the penetration was found in the grain structure of several specimens. Copper penetration was as deep as 10 mm (0.4 in.) below the surface.

There is a considerable body of published information that shows that steels will fail by grain-boundary penetration of molten copper. The following conditions are required:

- Presence of applied tensile stress
- Wetting of the steel surface by molten copper
- Heating of the steel substrate to temperatures high enough to austenitize the steel

Fig. 2 Fracture surface at the commutator side of axle 1611

The failure of railroad axles due to the overheating of bronze-backed bearings has been well documented. It is well recognized that the steel used in these axles is sensitive to liquid-metal embrittlement (LME) by molten copper when the steel surface is heated above the austenitizing temperature by frictional heat due to loss of lubrication. Stress is present in the axle at this location due to the weight of the traction motor, but the stress level is less than one-half that at the gear seat.

Overheating of friction bearings results in LME and failure of the axle. Liquid-metal embrittlement is a complex phenomenon; the example described in this article is only one form of LME. Because of the different forms of LME that can occur, a brief review of the nature of LME is presented.

Liquid-metal embrittlement is a phenomenon in which the ductility or fracture stress of a solid metal is reduced by exposure of the surface to a liquid metal (Ref 13). The subject has been extensively reviewed (Ref 13-18), and specific aspects have been reported in numerous scientific papers. The first recorded recognition of the problem was made in 1914 by Huntington, who observed the

embrittlement of α-brass by mercury (Ref 19). Since then, LME has been identified in numerous failure analyses that include some notable cases: damage to aircraft by liquid-gallium embrittlement of aluminum and the embrittlement of stainless steel piping by molten zinc.

References 13 to 19 have helped to eliminate some of the confusion in the literature regarding LME. This confusion is partly due to the existence of several different forms of LME. Four distinct forms have been identified (Ref 13):

- Instantaneous fracture of a particular metal under an applied or residual tensile stress when in contact with particular liquid metals. This is the most common type of LME
- Delayed failure of a particular metal in contact with a specific liquid metal after a certain time interval at a static load below the ultimate tensile stress of the metal. This form of LME is related to grain-boundary penetration by the liquid metal and is less common than the first form
- Grain-boundary penetration of a particular solid metal by a particular liquid metal such

Fig. 3 Fracture surface at the commutator side of axle 2028

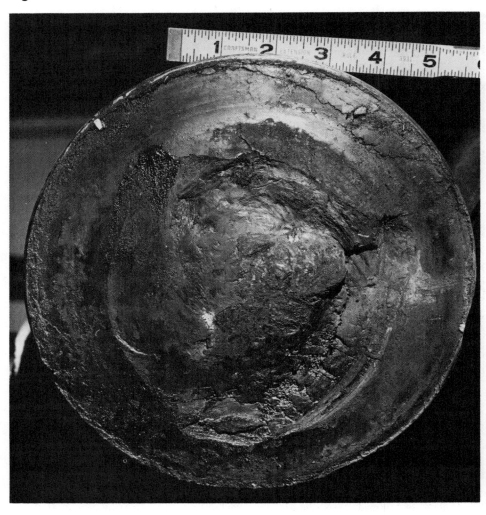

Fig. 4 Hot acid etched longitudinal centerline section from the drive-wheel side of axle 1611

Fig. 5 Hot acid etched longitudinal centerline section from the commutator side of axle 1611

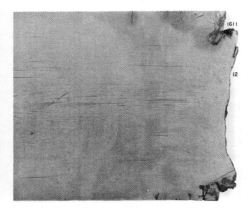

that the solid metal eventually disintegrates. Stress is not a prerequisite of this form of LME in all observed cases
* High-temperature corrosion of a solid metal by a liquid metal, causing embrittlement. This problem is entirely different from the others

In some respects, LME is similar to SCC. However, SCC always requires a measurable incubation period before fracture occurs. Hence, SCC is more similar to the second form of LME than to any of the other types. Like SCC, LME is a rather perplexing subject since embrittlement is restricted to only certain combinations of solid and liquid metals.

The following observations have been made regarding liquid metal/solid metal interactions (Ref 15):

* No apparent interaction
* Simple dissolution of the solid metal into the liquid metal
* Simple diffusion of the liquid metal into the solid metal

* Formation of intermetallic compounds at the liquid/solid interface
* Intergranular penetration of the liquid metal into the solid metal without the presence of applied or residual stress
* Brittle, premature failure of the solid metal due to intergranular penetration of the liquid metal into the solid metal under the influence of applied or residual tensile stress. Failure occurs after a finite incubation period
* Brittle, instantaneous failure of a stressed metal when a particular liquid metal is applied. Grain-boundary penetration is not necessarily observed

The last three types of interactions are the prime forms of LME. Most studies of LME discuss the last type, the instantaneous form of embrittlement, which is the most common form. Several authors refer to this form as adsorption-induced embrittlement. Other authors prefer to consider LME as merely a special type of brittle fracture and view the subject from a fracture mechanics approach.

A large number of metals and alloys can fail in this manner. Only specific liquid metals are known to embrittle each specific metal or alloy. Indeed, this is one of the more perplexing aspects of LME, and some investigators refer to this as specificity. Many of the studies of LME couples have involved tests consisting of only one temperature and fixed conditions regarding stress, grain size, and so on (Ref 19). Until all possible couples have been thoroughly studied using a wider range of test conditions, the specificity feature is of questionable validity (Ref 20). In addition, the LME tests have generally been "go, no-go" type tests rather than quantitative measurements of the degree of embrittlement.

The following conditions are required but are not necessarily sufficient to cause embrittlement (Ref 13):

* Wetting of the solid metal substrate by a liquid metal must be adequate
* An applied or residual tensile stress must be present in the solid metal
* A barrier to plastic flow must exist in the

Fig. 6 Hot acid etched longitudinal centerline section from the drive-wheel side of axle 2028

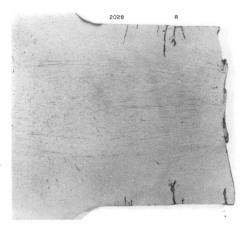

Fig. 7 Hot acid etched longitudinal centerline section from the commutator side of axle 2028

Fig. 8 Macrographs of two polished sections from the failed axles
(a) Axle 1611. (b) Axle 2028. Each was positioned along the fracture at the outside-diameter surface.

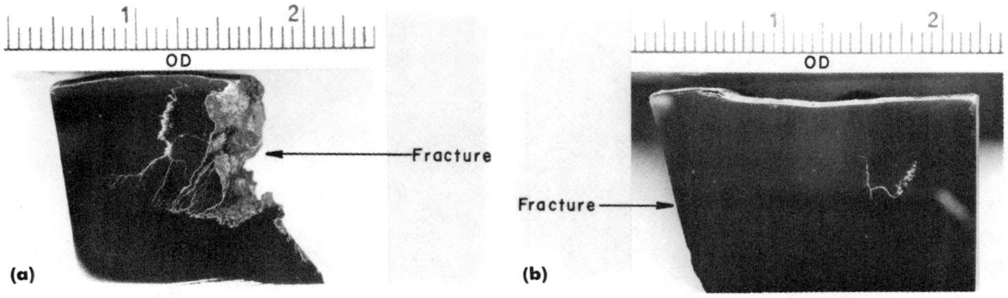

solid metal at some point in contact with the liquid metal

Obviously, for failure to occur the liquid metal must be of the correct composition to cause LME of the particular solid metal.

If the solid metal being embrittled is not notch sensitive, as in the case of face-centered cubic (fcc) metals, the crack will propagate only when the liquid metal feeds the crack. In a notch-sensitive metal—for example, in a body-centered cubic (bcc) metal, such as iron—the nucleated crack may become unstable and propagate ahead of the liquid metal.

In cases of LME, crack-propagation rates can be extremely rapid. Indeed, crack-propagation rates between 50 and 500 cm/s² (20 and 200 in./s²) have been measured. In most cases, the crack paths are intergranular; however, instances of transgranular crack paths have been observed.

Factors influencing the fracture stress and ductility of metals subject to LME include:

- Composition of the solid metal
- Composition of the liquid metal
- Temperature
- Strain rate
- Grain size

- Thermal-mechanical history of the solid metal

Studies of specific LME couples have empirically identified certain trends that are usually, but not always, obeyed. For example, most embrittlement couples have very little mutual solid solubility. Solid metals that are highly soluble in the liquid metal and solid/liquid metal combinations that form intermetallic compounds are usually immune to LME. These studies have also demonstrated that good wetting of the substrate is required in LME. Interestingly, the factors that promote the lowest interfacial energy and therefore the best wetting are high mutual solid solubility or intermetallic-compound formation, both of which are generally related to a low degree of LME.

The presence of notches or stress concentrators increases the severity of LME, and many current investigators have used a fracture mechanics approach to understand LME. Since stress raisers are known to influence brittle failure detrimentally under ordinary (dry) situations, the effect on LME is understandable.

In the previously discussed adsorption-induced mode of LME, grain-boundary penetration is not a prerequisite. There are a few well-known cases in which LME occurs by the

rapid penetration of a particular liquid metal along grain boundaries. Examples of this form of LME are much less common and not as well studied as the previously described form. Known examples include the wetting of aluminum by gallium (Ref 15), carbon steel by copper (Ref 15), and copper by bismuth (Ref 21).

Failures Caused by LME of Steel. The first reported case of LME of steel, in this instance by molten brass, was documented in 1927 (Ref 22). In 1931, tests were conducted on the influence of certain liquid metals on plain carbon steel, silicon steel, and chromium steel (Ref 23). These steels were embrittled at 1000 to 1200 °C (1830 to 2190 °F) by liquid tin, zinc, antimony, copper, 5% tin-bronze, and 10% zinc-brass. Liquid bismuth, cadmium, lead, and silver caused little or no embrittlement. In 1935, the bend strength of steel coated with a lead-tin solder at 250 °C (480 °F) and with a bearing metal at 350 °C (660 °F) was determined (Ref 24). Penetration of the liquid metals was observed on the tension side only of the bend specimens. In a similar study (Ref 25), the influence of liquid zinc at 425 to 500 °C (795 to 930 °F) on various steels was tested using bend, tensile, and creep-rupture specimens. Grain-boundary penetration of the zinc into the steels was observed. Another study investigated intercrystalline failures of a nickel-chromium steel airplane axle (Ref 26). The fracture occurred in the middle of the tubular axle where a brass number plate was attached by soft solder. When 50Pb-50Sn solder was applied to stressed rings cut from the axle, the specimen ruptured rapidly. The intergranular crack paths exhibited some solder penetration. The same results were observed when liquid tin, lead, zinc, cadmium, or Lipowitz alloy (a tin-lead-bismuth-cadmium alloy sometimes containing indium) was used.

In 1968, studies were conducted on the influence of cold work on the LME of pure iron and Fe-2Si (Ref 27). Sheet samples were tested in the annealed state and after 10, 25, 50, and 75% reduction in thickness. The degree of embrittlement increased with a small amount of cold work, but further working reduced the degree of embrittlement. For pure iron, maximum embrittlement occurred with the sample reduced 10% in thickness, while LME did not occur with the sample reduced 75%. For the Fe-2Si alloy, maximum embrittlement occurred with 25% reduction in thickness, and LME did not occur for either 50 or 75% reduction. The lack of LME at high levels of cold work was attributed to the fragmentation of the grain boundaries.

From 1970 to 1974, a series of studies was performed on the LME of iron (Ref 28-31). These tests were made under nonoxidizing conditions using notched tensile-creep test specimens. A thin copper coating (8.5 μm thick) was plated onto the notched surface using a cyanide bath. The amount of copper was increased by wrapping pure copper wire (0.18 mm, or 0.007 in., in diameter) around the

Fig. 9 X-ray elemental dot maps for copper and tin taken at three typical areas exhibiting penetration of the bearing elements

The three regions shown in the specimen current images (left) were quantitatively analyzed for copper and tin; results are given under the concentration maps. All 320×

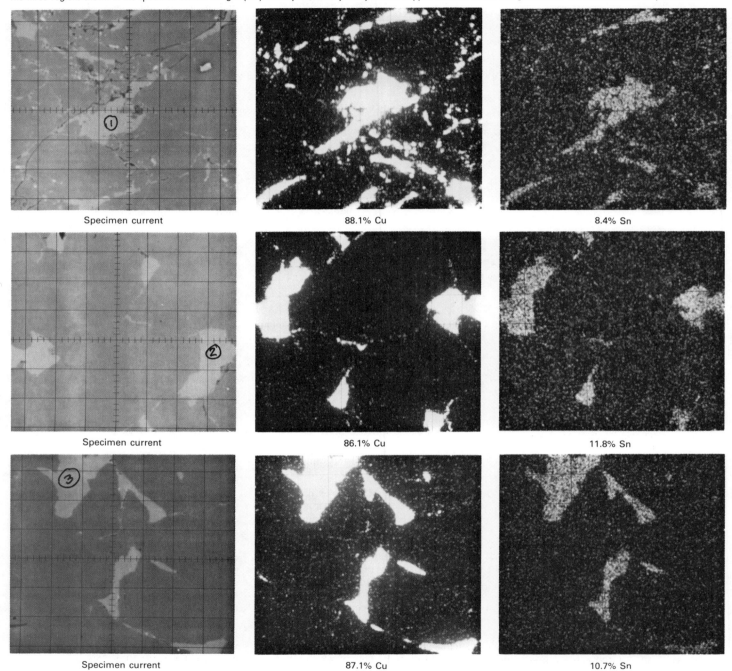

Specimen current	88.1% Cu	8.4% Sn
Specimen current	86.1% Cu	11.8% Sn
Specimen current	87.1% Cu	10.7% Sn

notch. Testing was done at 1100 and 1130 °C (2010 and 2065 °F) under stresses of 8.3 and 11 MPa (1.2 and 1.6 ksi). Copper significantly altered the creep behavior of pure iron and caused premature failure. An increase in applied stress or temperature reduced the time for LME failure. Embrittlement was of the delayed type and occurred by diffusion-controlled grain-boundary penetration of copper. The liquid copper appeared to be ahead of the growing

surface cracks. The depth of surface cracking was found to be controlled by the depth of copper penetration. Dihedral-angle measurements (Ref 30) for liquid copper in steel at 1100 and 1130 °C (2010 and 2065 °F) revealed that the most frequent angle was 34° for both temperatures. The liquid copper/austenite interfacial energy was calculated as 444 ergs/cm^2 at 1100 and 1130 °C (2010 and 2065 °F). Sessile drop measurements of the contact

angle between liquid copper and steel revealed similar results: 35° at 1100 °C (2010 °F) and 28° at 1130 °C (2065 °F).

Five earlier studies regarding the dihedral angle are significant. In the first, the dihedral angle of copper in steel was measured at 30° (Ref 32). In the second, a minimum value was observed for the dihedral angle in the region of the melting point of copper (Ref 33). The third study claimed that stress had little effect on the

Fig. 10 Structure of a specimen cut from the area near the center of the fracture of axle 1611

(a) Macrograph of specimen. Actual size. (b) Montage showing the microstructure from the fracture surface inward to just above region 3. 9 ×

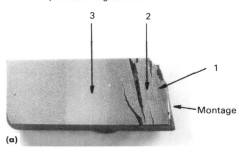

(a)

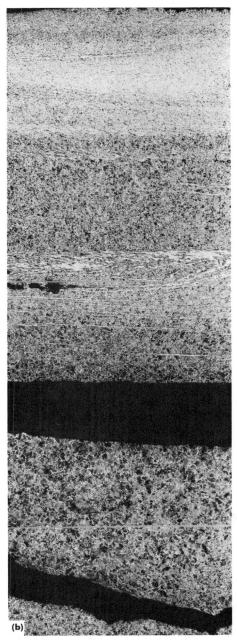

(b)

dihedral angle; however, stress was observed to promote spreading of the liquid metal along the grain boundaries (Ref 34). The fourth study maintained that the dihedral angle decreased with increasing applied stress (Ref 35). In the last study, it was found that in the presence of molten copper, the surface energy that must be applied to drive a crack along austenite grain boundaries is reduced by about a factor of 100, and the corresponding stress required is about one-tenth that required without the presence of copper (Ref 36).

There have been numerous studies of LME by copper occurring during welding and joining processes applied to steel. One such study involved the influence of copper deposition by welding of austenitic and ferritic stainless steels (Ref 37). In stainless steels with mixed ferritic-austenitic structures, the presence of the stable ferrite phase in austenite reduced the penetration of copper. When the stable ferrite content was greater than 30%, penetration of copper was not observed. When a wholly ferritic stainless steel was deposited with copper, cracking did not occur. The wettability of molten copper on austenitic and ferritic stainless steel was measured. The contact angle was 92 to 100° for the ferritic stainless (no wetting) and 22 to 28° for the austenitic stainless steel (wetting) at 1100 °C (2010 °F).

A series of reports has been published regarding the embrittlement of iron-base alloys by copper during welding (Ref 38-41). In a study of HY-80 steel welded with or without copper-nickel filler metal (Ref 38), infiltration of the grain boundaries by the copper-nickel deposit under an applied strain field was observed. The attack occurred in the heat-affected zone (HAZ) while the steel in this area was austenitic. In heated areas that remained bcc, penetration did not occur.

In another study involving copper deposition using a gas metal welding arc (Ref 39), alloy steels, such as AISI 4340 and 4140, and type 304 stainless steel were found to be extremely sensitive to copper penetration. In comparison, the penetration depths of AISI 1340, 1050, Armco iron, and carburized Armco iron were only about one-third those of the heat-treated alloy steels. A ferritic stainless steel, AISI 430, was almost completely immune to copper penetration. When stresses are absent, the ease of penetration by molten copper was concluded to be a function of the alloy content of the steel. When grain-boundary copper penetration occurs under an applied stress, partially filled or open cracks are formed in the steel. Notch-sensitive heat-treated steels lose much of their strength and ductility when copper penetration occurs. Steels that remain ferritic at the melting point of copper do not lose strength as a result of exposure to molten copper.

In a subsequent study (Ref 41), the Gleeble high strain rate hot-tensile test machine was used to determine the influence of temperature, atmosphere, stress, grain size, strain rate, and amount of copper on LME of iron- and cobalt-

Fig. 11 Structure of a specimen cut from the area near the center of the fracture of axle 2028

(a) Macrograph of specimen. Actual size. (b) Montage showing the microstructure in regions 1 and 2. 14 ×

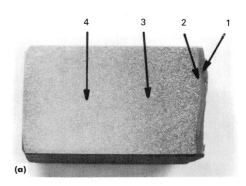

(a)

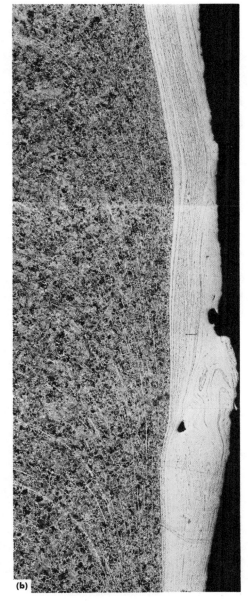

(b)

Fig. 12 Three views of the microstructure of axle 1611

The specimens were taken from the area near the center at different distances from the fracture face. (a) Near the fracture surface. (b) 6.4 mm (0.25 in.) from the fracture face. (c) 38 mm (1.5 in.) from the fracture face. The longitudinal axle direction is vertically oriented in each micrograph. All 500×

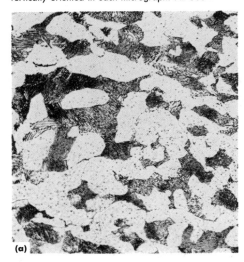

(a)

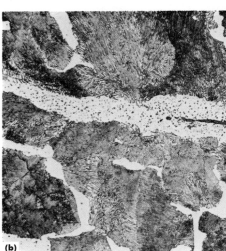

(b)

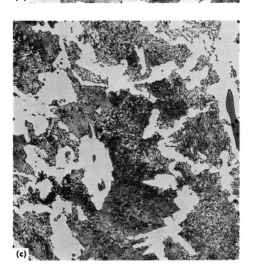

(c)

Fig. 13 Four views of the microstructure of axle 2028

The specimens were taken from the area near the center at different distances from the fracture face. (a) At the fracture surface. (b) At the boundary between the edge structure and the heat-affected structure. (c) At the heat-affected area. Arrows in (b) and (c) show sulfide inclusions. (d) 25 mm (1 in.) from the fracture surface

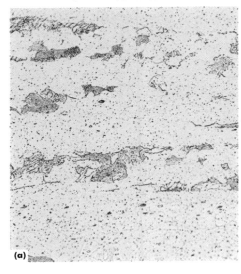

(a)

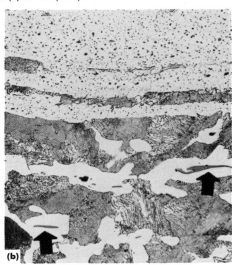

(b)

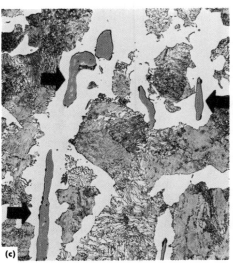

(c)

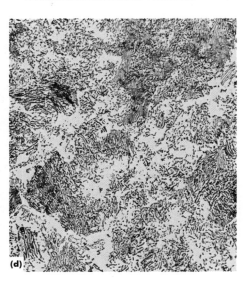

(d)

base superalloys. A copper contamination of only 0.08-mm (0.003-in.) thickness was found to be sufficient to cause hot cracking. Susceptibility to LME decreased with increasing temperature above some transition temperature in the HAZ. The intergranular LME cracks were oriented perpendicular to the direction of the principal stress. The hot cracks grew at high velocities after a critical amount of plastic strain was introduced in the HAZ.

It is well recognized that copper, as well as certain other elements in steel, can detrimentally influence hot workability (Ref 42-46). These studies have generally employed hot-torsion, bend, or cupping tests to assess the influence of the concentration of certain elements on hot workability. Much of the early work on the influence of copper on hot workability can be summarized (Ref 42):

- Because the solubility of copper in iron is

low, the concentration of copper appears as a layer of nearly pure copper between the metal and the scale
- If the rolling or forging temperature is above the melting point of copper, the molten copper on the surface of the steel will penetrate the steel along the grain boundaries

Elements such as tin, arsenic, and antimony decrease the melting point of copper and increase the sensitivity of steel to surface cracking. Tin reduces the solubility of copper in austenite by a factor as high as three (Ref 43). Hence, tin additions can cause the precipitation of a molten copper phase at a much lower level of copper enrichment than in the absence of tin.

It is quite clear that molten copper, or copper alloys, will embrittle steels under the proper conditions. The steel must be austenitic with tensile stresses present. Carbon steels contact-

Fig. 14 Average angle of sulfide inclusions relative to the longitudinal axle axis as a function of distance from the fracture face

Specimen was taken from the area near the centerline at the fracture face. Compare to the qualitative examples in Fig. 12 and 13.

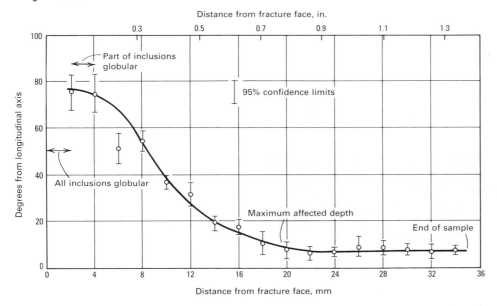

Fig. 15 Examples of the normal microstructure of the axles

Specimens were taken well away from the fracture surface. (a) Axle 2028. Etched with 2% nital. 100×. (b) Axle 2028. Etched with 4% picral. 500×. (c) Axle 1611. Same etchant and magnification as (a). (d) Axle 1611. Same etchant and magnification as (b)

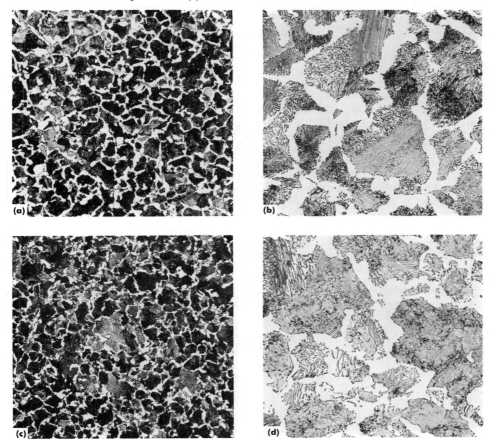

ing liquid copper will be heated into the fully austenitic range. Liquid copper will penetrate the austenitic grain boundaries. The speed of penetration depends on the magnitude of the applied tensile stresses. The depth of cracking depends on the depth of copper penetration. Thus, if the elements present in the friction-bearing sleeve are observed in the prior-austenite grain boundaries in failed axles, it can be safely concluded that the axle was intact and under normal loading when the bearing over-heated and that frictional heating due to loss of lubrication heated the axle and the bearing to a temperature at which the axle surface was austenitic and the bearing surface was molten. If the axle was broken before the bearing overheated, it would not be stressed in tension at the failure location, and penetration of bearing elements would not occur, irrespective of the crystallographic structure of the axle.

Results of Axle Studies

Three axles were examined. Two, coded 1611 and 2028, were provided by the Seaboard Coast Line Railroad. Visual examination of the failures indicated that overheating of the traction-motor support bearings, caused penetration of the bronze bearing material and failure. The third axle was the cause of a train derailment (Ref 47); a limited study was performed on this specimen.

Views of the fracture faces of axles 1611 and 2028 are shown in Fig. 1 to 3. Both axles exhibit little, if any, necking in agreement with the general fractures of such failures. Axle 1611 was used in a yard engine, and little damage was done to the fracture after rupture. Axle 2028 experienced considerable deformation after fracture. Figures 1 and 2 of axle 1611 show that the outer portion of the fracture face experienced considerable tearing and rupture, while the central portion exhibits hot torsional rupture (note the swirl pattern of the fracture). This suggests that LME occurred to a depth of nearly 25 mm (1 in.). The axle, which was heated to at least 925 °C (1700 °F), then ruptured to failure by twisting under the applied loads. The fracture features of axle 2028 are less distinct because of the rubbing action after fracture (Fig. 3). The outside-diameter surface of each of these axles in the area in contact with the traction-motor support bearing exhibited scale, and some of the friction-bearing material was present, as identified by x-ray fluorescence and diffraction.

Figures 4 to 7 show longitudinal sections cut from each side of the axle fractures after hot-acid etching (1:1 HCl in water at 70 °C, or 160 °F). This etching revealed a number of secondary cracks propagating from the outside-diameter surface inward along the portion of the axle that was in contact with the overheated bearing. The regions near the centerline of each axle behind the fracture face reveal evidence of the high temperatures and the metal flow from twisting to rupture.

Table 1 Chemical analysis of axles 1611 and 2028(a)

Axle	C(b)	Mn(c)	P(d)	S(b)	Si(e)	Ni(f)	Cr(f)	Mo(f)	Al(a)	Sn(g)	Cu(h)	V(f)	Nb(g)	Ti(g)	N(j)	O(k)
2028 . . . . .0.57	0.70	0.012	0.042	0.16	0.01	0.08	0.004	0.004	0.002	0.057	0.003	0.01	0.007	0.0045	0.0057
1611 . . . . .0.54	0.79	0.012	0.026	0.19	0.01	0.09	0.005	0.004	0.002	0.023	0.003	0.01	0.007	0.0058	0.0056
M-126-F . . .0.45-0.59	0.60-0.90	0.045	0.050	>0.15	...	...	...	...	...	...	...	...	...	...	...

(a) Weight percent. (b) Analysis by high-temperature combustion; infrared detection. (c) Peroxydisulfate-arsenite titrimetric analysis. (d) Alkalimetric analysis. (e) Gravimetric analysis. (f) Atomic absorption spectrometric analysis. (g) Plate spectrographic analysis. (h) Neocuproine photometric analysis. (j) Analysis by inert gas fusion; thermal conductivity detection. (k) Vacuum fusion analysis

Examination of the microstructure along the surface of the axle from the fracture face outward reveals extensive penetration of the bearing elements. Macrographs of two of the microsamples are shown in Fig. 8, which illustrates some of the gross crack patterns. Figure 9 shows electron microprobe views of three typical regions from axle 1611. Specimen current images and scans for copper and tin are illustrated. The three indicated areas were analyzed quantitatively; results appear under the appropriate dot maps. Minor amounts of lead, zinc, and antimony were also detected, but were not analyzed quantitatively.

Figure 10 shows an etched section from the central region of axle 1611 and a montage of the microstructure. The temperature in regions 1 and 2 was high enough to reaustenitize the steel. Region 3 was heated into the two-phase region, while the lower portion of the sample, below region 3, was not heated above the lower critical temperature. The montage shows that the original longitudinal fiber of the forged axis was removed by the high temperature. The twisting action produced a metal-flow pattern perpendicular to the original hot-working axis and subsequent rupture. Around the larger crack, recrystallization of the grain structure occurred.

Figure 11 shows a similar microsample cut from along the centerline of axle 2028 at the fracture face. The microstructure demonstrates the influence of the very high temperature and torsional rupture metal flow.

The influence of the high temperatures near the fracture and the rotational forces on axle 1611 are further illustrated in Fig. 12, which shows views of the microstructure taken at three locations with respect to the fracture along the centerline. All are oriented in the same manner; that is, the longitudinal axis is

Fig. 16 Optical micrograph of copper penetration (arrows) near the outside-diameter surface of the third axle

Etched with 2% nital. 55×

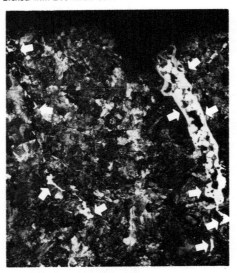

vertical in each illustration. At a distance of 38 mm (1½ in.) from the fracture face, the sulfide inclusions are parallel to the hot-working axis, as normally encountered. However, at 6.4 mm (0.25 in.) from the fracture, the sulfides are perpendicular to the longitudinal axis. In addition, they are much smaller, and many fine globular sulfides are observed. This indicates that the temperature was high enough at this location to dissolve most of the sulfides, which precipitated upon cooling as small globules. Near the edge of the fracture, only very fine spherical sulfides are observed, and considerable decarburization is evident.

Figure 13 shows similar micrographs from axle 2028 at regions 1 to 4 shown in Fig. 11. The microstructure in region D shows that the temperature was in the correct range to spheroidize the pearlitic structure to some extent (near the lower critical temperature). This same type of behavior was observed in samples from the third axle studied. Figure 14 shows measurements of the average angle of the sulfides with respect to the longitudinal axis as a function of the distance from the fracture face.

Table 1 shows the results of chemical analysis of axles 1611 and 2028 taken in unaffected regions compared with the specified range for M-126, grade F, axle steel. Figure 15 shows the normal microstructure of axles 1611 and 2028, taken well away from the affected regions. These axles are double normalized and tempered. The microstructure is fine lamellar pearlite with grain-boundary ferrite.

Penetration of bearing elements in the third axle is shown in Fig. 16. Color metallography reveals the characteristic copper color of the penetrating material. Figure 17 shows scanning electron microscope views of the area shown in Fig. 16, further confirming the presence of copper. More extensive elemental mapping of penetrating bearing material is shown in Fig. 18. Again, all of the typical bearing elements are observed.

Simulation of LME Mechanism

To provide additional information on the embrittlement mechanism, laboratory tests were conducted using material from one of the axles. Tensile specimens measuring 12.8 mm (0.505 in.) in diameter × 250 mm (10 in.) long were prepared. A 60° V-notch was machined

Fig. 17 SEM micrographs of the area from the third axle shown in Fig. 16

(a) Secondary electron image. (b) Backscattered electron image. (c) Copper x-ray dot map. All 55×

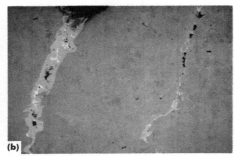

Fig. 18 X-ray elemental composition maps made with the electron microprobe using a region of copper penetration shown in Fig. 16 and 17

All 270×

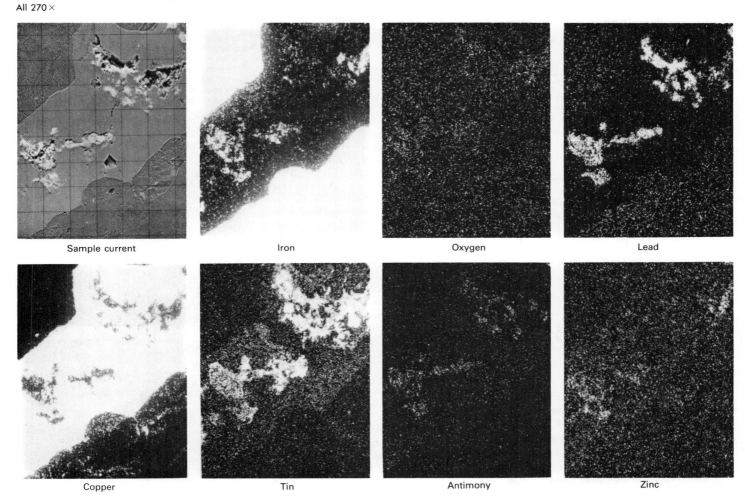

| Sample current | Iron | Oxygen | Lead |
| Copper | Tin | Antimony | Zinc |

into the periphery at midlength of each sample. The diameter at the root of the notch was 9.1 mm (0.357 in.). The notch surface, as well as some of the bar surface on each side, was electroplated with copper. The V-notch region was then wound with fine copper wire. One tensile specimen was left bare to determine the tensile strength of the steel at 1100 °C (2010 °F) without the influence of copper. All tests were performed using a Gleeble. The uncoated, notched sample was heated to 1100 °C (2010 °F) and pulled in tension with a crosshead speed of 2.5 mm/s (6 in./min). A maximum load of 259 kg (570 lb) was observed; that is, the ultimate tensile strength was 39 MPa (5.7 ksi).

Samples with copper in the notch were then heated to 1100 °C (2010 °F), high enough to melt pure copper, and held at constant loads of 32, 65, 78, 97, and 129 kg (71, 143, 171, 214, and 285 lb), that is, 12.5, 25, 30, 37, and 50% of the 1100 °C (2010 °F) tensile strength without copper. If the liquid copper had no influence on the steel, the tensile samples should have stayed intact for a long time.

The tests showed, however, that fracture occurred after 18 s when a 129-kg (285-lb) load was applied; this corresponds to a crack-growth rate of 907 mm/h (35.7 in./h). At a load of 97 kg (214 lb, or 37% of the normal tensile strength), complete rupture occurred in 24 s; this is a crack-growth rate of 680 mm/h (26.8 in./h). At 30% of the normal tensile strength (78 kg, or 171 lb), rupture occurred in 79 s. A number of tests were conducted at 32 kg (71 lb) and 65 kg (143 lb). These samples were held for different lengths of time, cooled to room temperature, sectioned, and examined to measure the crack depths. All of the test data are plotted in Fig. 19, which reveals a consistent relationship between the test load and crack-growth rate.

The microstructure of the sample that was not coated with copper when tested at 1100 °C (2010 °F) is shown in Fig. 20. The spherical cavities found at the grain boundaries are indicative of ductile overload rupture. Figure 21 shows examples of the fracture surface of a specimen tested to rupture when copper was present at 1100 °C (2010 °F). The intergranular

fracture pattern is indicative of LME. Two views of the microstructure of a partially broken specimen tested at 65-kg (143-lb) load (25% of normal tensile strength) are also shown. The white grain-boundary films are copper, showing the intergranular nature of the penetration, also shown by the crack path.

For comparison, a similar copper-filled V-notch specimen was heated to 1100 °C (2010 °F) in a laboratory furnace without any applied load. Figure 22 shows that no cracking occurred at the root of the V-notch. The high-magnification micrograph shows some dark-etching grain boundaries at the extreme surface where copper diffused into the specimen. As predicted by the literature, no evidence of LME was detected because there was no applied stress.

Conclusions

The carbon steel railroad axles examined failed because of LME due to friction-bearing overheating. After the babbitt metal had melted away, the bronze sleeve rubbed against the steel

Fig. 19 Graph showing the influence of load on the LME crack-growth rate for the experimental conditions described in text

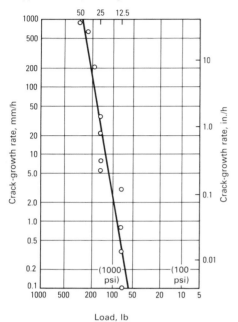

Copper-free tensile strength at 1100 °C (2010 °F), %

Fig. 20 Micrograph of the fractured end of the specimen tested with the Gleeble at 1100 °C (2010 °F) without copper present

The fracture is transgranular. Etched with 4% picral. 55×

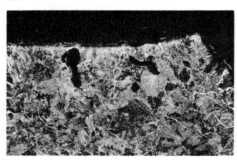

surface. Both overheated because of the frictional stresses, to a point where the bearing melted and the contacting steel surface became austenitic. In this region the axle is under stress because of the weight of the traction motor; hence, the liquid bearing material penetrated the axle. Further heating of the axle in the contact region continued while the copper penetrated the axle. Cracking around the periphery reduced the effective cross-sectional area of the axle. Because the axle was quite hot, its strength decreased substantially, and twisting began under the action of the rotational forces. Thus, fracture was initiated by LME and prop-

agated to final rupture by torsion. This is consistent with the observed depth of copper penetration and the observed metal flow in the central region of the axle.

Although the fracture face and origins were destroyed, there was evidence of LME along the entire surface in contact with the bearing surface. Fracture occurred in the axle under the approximate center of the bearing, where temperatures and stresses were highest.

REFERENCES

1. B. Straub, Mikroskopische Stahluntersuchung, *Stahl Eisen*, Vol 34 (No. 50), Dec 1914, p 1814-1820
2. F.H. Williams, Car Axle Failures Traced to Absorption of Bearing Metal, *Prod. Eng.*, Vol 15, Aug 1944, p 505-508
3. "The Mechanism of Copper Penetration Failure," Magnaflux Corp., Chicago, 1947
4. "Copper Penetration—New and Revealing Information Concerning the Analysis of the Penetrating Material Itself," Research Laboratory Technical Memo No. 220, Project 125, Magnaflux Corp., Chicago, Feb 1954
5. E.A. Unstich and E.J. Palinkas, "A Study of Copper Penetration in Overheated Journals," Report 69222, New York Central Railroad Company, New York, July 1959
6. New York Central Research Shows . . . Turning Removes Copper Penetration, *Rail. Locom. Cars*, Feb 1961, p 34-35
7. Report of the Committee on Axle and Crank Pin Research, in *Proceedings of the Association of American Railroads, Mechanical Division*, 1951, p 158-166
8. J.J. Laudig, Hot-Box Research Results . . . in the Iron-Back Journal Bearing, *Rail. Age*, Vol 134 (No. 12), March 1953, p 65-69
9. Report of the Committee on Axle and Crank Pin Research, in *Proceedings of the Association of American Railroads, Mechanical Division*, 1952, p 116-129
10. Laudig Iron Back Journal Bearing, in *Proceedings of the Association of American Railroads, Mechanical Division*, 1952, p 358-359
11. Report of Committee on Axles, in *Proceedings of the Association of American Railroads, Mechanical Division*, 1956, p 94-104
12. "The Failure of Railroad Axle-Journal Steel Including the Effects of Copper-Bearing Metals in Contact with Stressed Steel Rods at Elevated Temperatures," Report MR-434, Association of American Railroads, Research Department, Washington, Dec 1963
13. N.S. Stoloff, Liquid Metal Embrittlement, in *Surfaces and Interfaces, II*, Proceedings of 14th Sagamore Army Materials Research Conference, Syracuse University Press, Syracuse, NY, 1968, p 157-182

Fig. 21 Examples of axle steel tensile specimens tested at 1100 °C (2010 °F) with the Gleeble, with copper present

The liquid copper has penetrated prior-austenite grain boundaries, producing intergranular fracture. (a) Bottom of notch of a specimen held for 10 min at temperature at 65 kg, or 143 lb, load (25% of the normal tensile strength). 70×. (b) Below crack in the same specimen as (a). 70×. (c) Fracture surface of a specimen tested to rupture. 35×

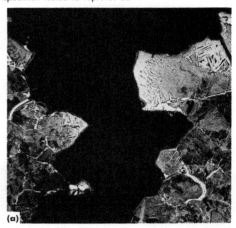

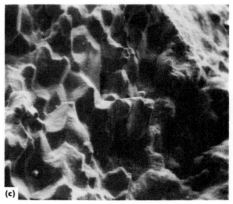

Fig. 22 Two views of a notched specimen that was electroplated with copper, wrapped with copper wire, and tested for 3 h at 1100 °C (2010 °F) without application of a load

No grain-boundary penetration of copper or intergranular cracking is evident, but there is evidence of minor bulk and grain-boundary diffusion [dark-etching lines in (b)]. (a) Root of notch. 55×. (b) Edge of specimen etched with 2% nital. 275×

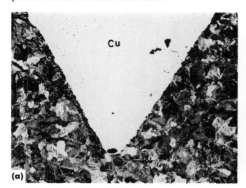

14. A.R.C. Westwood *et al.*, Adsorption-Induced Brittle Fracture in Liquid-Metal Environments, in *Fracture,* Vol III, Academic Press, 1971, p 589-644
15. W. Rostoker *et al.*, *Embrittlement by Liquid Metals*, Reinhold, 1960
16. M.H. Kamdar, Liquid Metal Embrittlement, in *Embrittlement of Engineering Alloys*, Vol 25, *Treatise on Materials Science and Technology*, Academic Press, 1983, p 361-459
17. M.G. Nicholas and C.F. Old, Review: Liquid Metal Embrittlement, *J. Mater. Sci.*, Vol 14, Jan 1979, p 1-18
18. M.H. Kamdar, Ed., *Embrittlement by Liquid and Solid Metals*, American Institute of Mining, Metallurgical, and Petroleum Engineers, Warrendale, PA, 1984
19. A.R. Huntington, discussion of paper by C.H. Desch, The Solidification of Metals From the Liquid State, *J. Inst. Met.*, Vol 11, 1914, p 108-109
20. F.A. Shunk and W.R. Warke, Specificity as an Aspect of Liquid Metal Embrittlement, *Scr. Metall.*, Vol 8, 1974, p 519-526
21. E. Scheil and K.E. Schiessl, The Diffusion of Liquid Bismuth Along Grain Boundaries in Copper, *Z. Naturforsch*, Vol 4a, 1949, p 524-526
22. R. Genders, The Penetration of Mild Steel by Brazing Solder and Other Metals, *J. Inst. Met.*, Vol 37, 1927, p 215-221
23. H. Schottky *et al.*, The Red-Shortening of Steels by Metals, *Arch. Eisenhüttenwes.*, Vol 4, 1931, p 541-547

24. W.E. Goodrich, The Penetration of Molten White Metals into Stressed Steels, *J. Iron Steel Inst.*, Vol 132, 1935, p 43-66
25. W. Radeker, The Production of Stress Cracks in Steel by Molten Zinc, *Stahl Eisen*, Vol 73 (No. 10), May 1953, p 654-658
26. L.J.G. van Ewijk, The Penetration of Steel by Soft Solder and Other Molten Metals at Temperatures up to 400 °C, *J. Inst. Met.*, Vol 56 (No. 1), 1935, p 241-256
27. J.V. Rinnovatore and J.D. Corrie, The Effect of Cold Work on the Liquid Metal Embrittlement of Iron and Iron-Silicon, *Scr. Metall.*, Vol 2 (No. 8), Aug 1968, p 467-470
28. R.R. Hough and R. Rolls, The High-Temperature Tensile Creep Behavior of Notched, Pure Iron Embrittled by Liquid Copper, *Scr. Metall.*, Vol 4 (No. 1), 1970, p 17-24
29. R.R. Hough and R. Rolls, Creep Fracture Phenomena in Iron Embrittled by Liquid Copper, *J. Mater. Sci.*, Vol 6, 1971, p 1493-1498
30. R.R. Hough and R. Rolls, Copper Diffusion in Iron During High-Temperature Tensile Creep, *Met. Trans.*, Vol 2, Sept 1971, p 2471-2475
31. R.R. Hough and R. Rolls, Some Factors Influencing the Effects of Liquid Copper on the Creep-Rupture Properties of Iron, *Scr. Metall.*, Vol 8 (No. 1), Jan 1974, p 39-44
32. L.H. VanVlack, Intergranular Energy of Iron and Some Iron Alloys, *Trans. AIME*,

Vol 191, March 1951, p 251-259
33. W.J.M. Salter, Surface Hot Shortness in Mild Steel, *J. Iron Steel Inst.*, Vol 200, Sept 1962, p 750-751
34. R.B. Waterhouse and D. Grubb, The Embrittlement of Alpha-Brasses by Liquid Metals, *J. Inst. Met.*, Vol 91, 1962-1963, p 216-219
35. C.A. Stickles and E.E. Hucke, The Effect of Stress on the Dihedral Angle in Leaded Nickel, *J. Inst. Met.*, Vol 92, 1963-1964, p 234-237
36. R. Eborall and P. Gregory, The Mechanism of Embrittlement by Liquid Phase, *J. Inst. Met.*, Vol 84, 1955, p 88-90
37. E.A. Asnis and V.M. Prokhorenko, Mechanism of Cracking During the Welding or Depositing of Copper on to Steel, *Weld. Prod.*, Vol 12 (No. 11), 1965, p 15-17
38. S.J. Matthews and W.F. Savage, Heat-Affected Zone Infiltration by Dissimilar Liquid Weld Metal, *Weld. J.*, Vol 50 (No. 4), 1971, p 174s-182s
39. W.F. Savage *et al.*, Intergranular Attack of Steel by Molten Copper, *Weld. J.*, Vol 57 (No. 1), 1978, p 9s-16s
40. W.F. Savage *et al.*, Copper-Contamination Cracking in the Weld Heat-Affected Zone, *Weld. J.*, Vol 57 (No. 5), May 1978, p 145s-152s
41. W.F. Savage *et al.*, Liquid-Metal Embrittlement of the Heat-Affected Zone by Copper Contamination, *Weld. J.*, Vol 57 (No. 8), Aug 1978, p 237s-245s
42. J.W. Halley, Residual Elements in Steel, in *Effect of Residual Elements on the Properties of Metals*, American Society for Metals, 1957, p 71-87
43. D.A. Melford, Surface Hot Shortness in Mild Steel, *J. Iron Steel Inst.*, Vol 200, April 1962, p 290-299
44. K. Born, Surface Defects in the Hot Working of Steel Resulting from Residual Copper and Tin, *Stahl Eisen*, Vol 73, HB 3255, 1953, p 1268-1280
45. G.G. Foster and J.K. Gilchrist, The Influence of Copper, Nickel and Tin on the Hot Working Properties of Mild Steel, *Metall.*, May 1952, p 225, 228, 230
46. W.J.M. Salter, Effects of Alloying Elements on Solubility and Surface Energy of Copper in Mild Steel, *J. Iron Steel Inst.*, Vol 204, May 1966, p 478, 487
47. "Railroad Accident Report—Derailment of Auto-Train No. 4 on Seaboard Coast Line Railroad, Florence, South Carolina, February 24, 1978," Report NTSB/RAR-84/1, National Transportation and Safety Board, Washington, June 1984

Engineered and Electronic Materials

Failure Analysis of Continuous Fiber Reinforced Composites

Brian W. Smith and Ray A. Grove, Boeing Commercial Airplane Company

CONTINUOUS FIBER REINFORCED COMPOSITES are seeing significantly expanded levels of use in hardware where reduced weight is critical. This is primarily a result of their tailorability as well as their high strength- and modulus-to-density ratios. In recent years, these materials have seen applications ranging from mass-produced tennis rackets to relatively complex structures, such as the wings of the AV-8B Harrier aircraft. As the use of these materials expands, so also does the likelihood of eventual fracture. As with their metal counterparts, the occurrence of fracture is likely to represent a relatively rare event that is not encountered with most hardware usage. However, when such fractures occur, the ability to determine their origin and cause constitutes a critical step that is necessary in providing valuable engineering feedback and ensuring the continued integrity of the components during service.

This article will review the basic methods required to carry out a rudimentary analysis into the cause of fracture for continuous fiber reinforced materials. Because these materials are significantly different from their metal counterparts, this article will deal with several considerations unique to composite materials. Specific topics that will be examined include potential fracture causes, typical fracture modes, and analysis methods for the identification of the origin and direction of fracture. However, because composite materials failure analysis studies have only recently been initiated, it is important to recognize that this article contains many procedures and data that are under development or are not wholly understood for all the possible composite structures and material systems that may be encountered.

Types of Composites

Composites, by definition, are those materials made up of an amalgamation of separate parts or microstructural elements. This amalgamation combines the attributes of each of the separate materials involved, typically resulting in improved properties for the system as a whole. As with metallic alloys, a wide range of discrete material phases and microstructural arrangements is available for composite materials (a wide variety of composite material microstructures are shown in the article "Fiber Composite Materials" in Volume 9 of the 9th Edition of *Metals Handbook*). The complexity presented is further complicated because composite materials can be easily tailored during fabrication, with the resulting structure and properties dependent upon the process and material forms used. Composite materials can be divided into two basic categories, depending upon their type of reinforcement: continuous fiber reinforced and particulate/short fiber reinforced composites. The first of these two materials typically uses a continuous array of oriented fibers, whereas the second category of material uses randomly dispersed particulates or chopped fibers. In this article, only continuous fiber reinforced composites will be considered.

Continuous Fiber Composites. Materials in this group are typically made up of 3- to 30-μm diam fibers that are oriented and surrounded in a supportive matrix material. Generally, the fibers used in these material systems are several orders of magnitude stiffer and stronger than the surrounding matrix. Fibers typically used for such applications include graphite, Kevlar, and fiberglass (Fig. 1). Encompassed within each of these generic fiber types are a number of specialized fibers with specifically tailored properties, for example, high-modulus or high-strength graphite. The high stiffness and strength of these fibers control the characteristic engineering properties, such as tensile, compressive, and shear moduli and strength.

The continuous matrix phase functions as a supportive element to the fibers. In this capacity, fiber orientation and alignment are maintained, load is transferred between fibers, and strength is provided in nonreinforced directions. A wide variety of matrix materials are available for use with each fiber type. By far the most extensive matrices in current use are those employing either elevated-temperature curing epoxies or room-temperature curing vinyl esters. Both of these matrix materials represent a broad class of polymers known as thermosets. These polymers are reacted during processing to form a stable matrix system. Generally, the choice of one thermosetting system over another depends upon a number of variables, including the environment under which usage is likely to occur. Typically, systems cured at elevated temperature exhibit better strengths under elevated temperature, absorbed moisture, and solvent attack than those cured at room temperature. More recently, thermally moldable matrices—thermoplastics—have begun to be used for continuous fiber composite applications. These matrices differ from thermosets in that their polymer network can be repeatedly molded and reformed with the application of heat and pressure.

While the principal engineering properties of a composite tend to reflect the constituent fiber used, it is important to recognize that the matrix material contributes significantly to these properties. In distributing fiber-to-fiber loads and supporting the fiber under axial compression, matrix properties directly control compression strength, damage resistance, and residual strength. As a result, both the matrix and the fiber must be considered when carrying out a failure analysis investigation.

One of the unique attributes of continuous fiber reinforced composites is their tailorability. With continuous fibers, control can be exercised over both the orientation and amount of load-carrying fiber reinforcement. This control is generally achieved during manufacture by using fibers arranged in either a fabric cloth or unidirectional tape. By stacking thin layers, a wide variety of thicknesses and fiber orientations can be produced.

Design. Because of their continuity and oriented structure, fiber-reinforced composites have highly anisotropic properties. For example, unidirectional tapes commonly exhibit moduli in the fiber direction 20 times greater than that of the transverse direction. In most designs, this anisotropy is tailored by arranging plies of material (either tape or fabric) at a

Fig. 1 Microstructures of common composite materials

(a) Graphite-epoxy tape. (b) Kevlar epoxy fabric.
(c) Fiberglass epoxy fabric. All at 200×

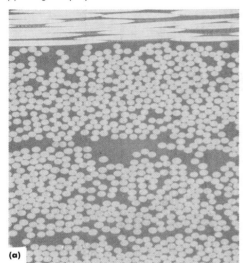

(a)

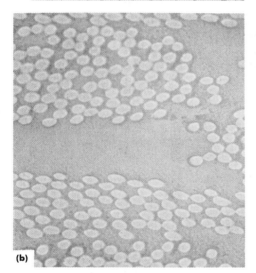

(b)

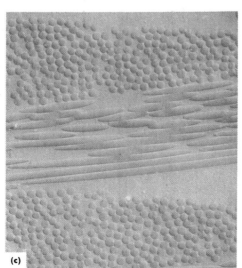

(c)

Fig. 2 Cross-sectional optical micrograph illustrating laminated construction typical of continuous fiber composite materials

50×

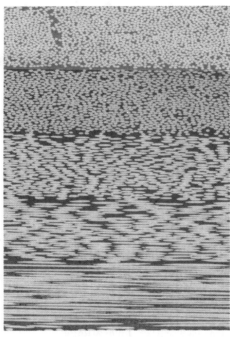

variety of angles. This structuring produces a stacked, laminated construction (Fig. 2). In the aerospace industry, the angle of plies used in the laminated stack are oriented at fixed angles to the direction of major load. The most common fixed angles are 0, 45, −45, and 90° to the major load axis. By selecting the proper number of plies at each of these angles, composite properties and their anisotropies can be designed for strength, modulus, or even the degree of thermal expansion along each principal direction.

Causes of Failure

Because of their relatively recent usage, new causes of failure in composite materials are still being uncovered as service experience is gained. Current knowledge, however, indicates that many of the basic sources of failure that occur in metals are likely to be observed in composites. These sources include three basic categories of causes: errors in design, fabrication and processing deficiencies, and anomalous service conditions. Given the construction, properties, and sensitivities of composite materials, the specific causes that may occur and should be considered during an analysis are worth reviewing.

Design Errors. Composite materials are somewhat unique in that both the fundamental properties of the material and the configuration of the component to be fabricated are subject to

design. Correspondingly, design errors can be made at both the material and structural levels of design. Engineering errors of the material may include a variety of problems. The more common of these include errors in analyzing the effect of individual ply anisotropies or the inadequate assessment of material damage and environmental sensitivities.

Because the level of stress carried by each ply in a uniformly strained laminate depends upon its modulus, large stress gradients and internal shear stresses can exist between plies oriented at significant angles to one another. Such stress gradients can lead to premature fracture, particularly where the magnitude of these gradients is large. Such large gradients are particularly common where groups of adjacent plies with the same orientation are oriented at 90° to another group of adjacent plies.

The highly anisotropic coefficient of thermal expansion of composite materials represents another area where design errors can be made at a material level. Many composite materials exhibit significantly large differences in thermal-expansion coefficients, depending upon their fiber orientation. As a result, changes in temperature, that is, temperatures that differ significantly from the curing temperature, can induce internal stress gradients where plies are oriented at significant angles to one another. These internal stress gradients are analogous to those generated under applied mechanical loads. Because many high-performance composites are cured or formed at elevated temperatures, the cooling of these parts during processing to ambient conditions can induce these internal stresses in the as-fabricated condition. For many designs, the magnitude of these stresses and those generated by additional temperature variations may be relatively inconsequential. However, high internal stress levels may develop in laminates with groups of adjacent plies oriented at large angles to one another or in such structures as space vehicles, in which extreme variations in temperature occur.

On a more general level, errors in material design can include many of the same problems encountered in metals. Fractures can be caused by inadequate understanding of environmental sensitivities, the effect of damage, or the fatigue sensitivity of the material used. Because the properties of composites depend upon their ply, or fiber, orientation and stacking sequences, the sensitivity of each design may vary significantly, posing a potential problem for design. In addition, synergistic effects may exist between these factors, giving rise to further sensitivities not considered during normal design practices.

Design errors relating to the component itself are likely to include unconsidered load sources, stress concentrations, and unanticipated buckling instabilities or modes. As with their metal counterparts, such failures may occur as a result of oversight. In most cases, thorough testing during design uncovers most of these errors.

Fig. 3 Effect of fiber orientation on flexural fracture modes in various graphite-epoxy lay-ups

(a) Quasi-isotropic. (b) 0°. (c) ±45°. (d) 0/90°

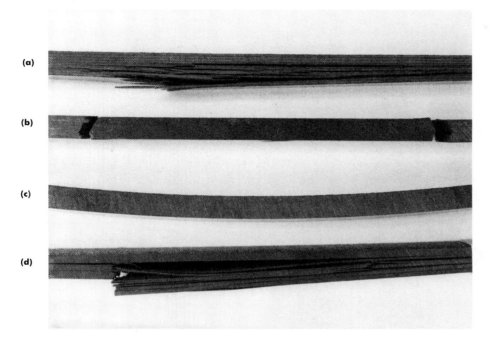

However, those related to fatigue or rarely attained load conditions may not become apparent until well into the life of the part.

Fabrication and Processing Deficiencies. Typically, the occurrence of defective or anomalous conditions is controlled and prevented by manufacturing controls and material-inspection testing imposed during the fabrication process. However, because absolute control and inspection are generally economically infeasible and because human errors do occur, these control and inspection methods sometimes allow occasional errors.

Continuous fiber reinforced composites are usually fabricated by laminating together and curing multiple plies impregnated with unreacted matrix resin. Within this fabrication operation, a number of errors can occur. Because each of the individual plies involved in a laminated composite has highly anisotropic properties, their placement and orientation can be critical in achieving the desired engineering properties. This is particularly true for composites in which each individual ply constitutes a significant percentage of the total laminate, that is, thin gage structures. For example, in a unidirectional laminate, a variation in overall fiber orientation of 15° can generate up to a 50% reduction in ultimate strength.

For thermosetting matrices, reacting the matrix resin represents one of the more critical steps. Either improper amounts of the two resin components or the inadequate application of heat during curing can produce conditions of undercure. Such conditions, when extensive,

can significantly degrade the properties of the matrix and its resistance to chemical or environmental exposure. Similarly, inadequate compaction during the lamination process can result in extensive porosity and reductions in material strength and durability.

Anomalous Service Conditions. Particular service anomalies include improper operation or use, faulty maintenance and repair, overloads due to failure of a related part, and environmental- or service-incurred damage beyond that reasonably anticipated. Many of these causes are not unique to composite materials. However, because of their construction, composites are particularly affected by some conditions more than other materials.

The engineering properties of composites can be significantly reduced by variations in temperature, foreign object impact damage, and, with some resin systems, chemical attack. With thermosetting matrices, the effect of temperature can become quite detrimental, particularly if moisture has been absorbed into the resin system. Property reductions due to foreign object damage can also be equally extreme. Studies have shown that moderate levels of impact energy can reduce material strength up to 60% (Ref 1).

Fracture Modes in Composites

Because of their laminated anisotropic construction, fractures in composites can occur in a number of complex ways. The types and modes

of fracture that can be encountered depend upon both the direction of applied load and the orientation of fibers (plies) making up the composite material. Figures 3 and 4 show that variations in either of these can produce strikingly different fracture appearances on a macroscopic scale. This range of diversity precludes the ability to assign well-defined macroscopic fracture types for most applications. The definition of fracture modes on a microscopic scale, however, provides a relatively useful means of classifying failure modes and fracture types in much the same way as with metals.

Fractures in continuous fiber reinforced composites can be divided into three basic fracture types: interlaminar, intralaminar, and translaminar (Fig. 5). As with the intergranular and transgranular terminology commonly used with metals, each of these classifications describes the plane of fracture with respect to the microstructural constituents of the material. Translaminar fractures are those oriented transverse to the laminated plane in which conditions of fiber fracture are generated. Interlaminar fracture, on the other hand, describes fractures oriented between plies, whereas intralaminar fractures are those located internally within a ply. Translaminar fractures involve significant fiber fractures, while interlaminar or intralaminar fractures occur in the laminate plane and therefore break few if any fibers.

Using this convention, failures in composites can be described in terms of the failure mechanism exhibited on these differing fracture surfaces. These failure mechanisms reflect the type of load under which microscopic separation occurs—tension, shear, or compression. The following sections discuss these fracture modes and their microscopic fracture features.

Interlaminar and Intralaminar Fractures

When considered on a microscale, interlaminar and intralaminar fracture types can be similarly described. In both cases, fracture occurs on a plane parallel to that of the fiber reinforcement. In a similar manner to that described for metals, fracture of either type can occur under mode I tension, mode II in-plane shear, mode III anti-plane shear, or any combination of these load conditions. Figure 6 illustrates load states I and II. All three fracture modes are still being investigated. As a result, such conditions as mode III anti-plane shear have not been thoroughly studied. However, for mode I tension and mode II in-plane shear, enough data exists to model their mechanisms of separation and to describe their fracture characteristics.

Because interlaminar and intralaminar fractures occur in the same plane as their fiber reinforcement, their fracture mechanism and appearance tend to be dominated by matrix fracture and fiber-to-matrix separation. In general, separation of the fiber from the matrix

Fig. 4 Differences in tensile fracture modes for various graphite-epoxy lay-ups

(a) 0/90° fabric. 0.5×. (b) 0° tape. 0.5×. (c) Open hole quasi-isotropic tape. 0.6×. (d) Open hole 0/90° tape. 0.6×

Fig. 5 Different planes of separation in continuous fiber reinforced composites

(a) Intralaminar fracture. (b) Interlaminar fracture. (c) Translaminar fracture

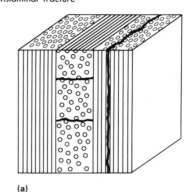

(a)

(a)

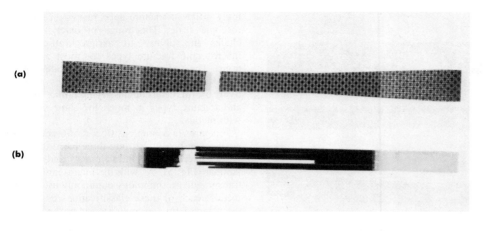

(b)

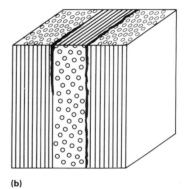

(b)

(c)

(c)

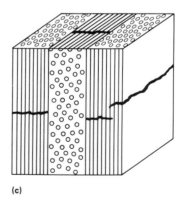

(d)

occurs at the interface for either mode I tension or mode II in-plane shear-loading conditions. As a result, when cohesive resins are involved, very little fracture occurs along the fiber. Fracture of the matrix resin between fibers exhibits pronounced cohesive fracture characteristics under both mode I tension and mode II in-plane shear loading.

For the majority of thermosetting matrices currently in use, cohesive-matrix failure occurs in a brittle manner. Common to brittle fracture in metals and unreinforced polymers, cohesive-resin fracture characteristically exhibits relatively flat fracture planes with very little evidence of material deformation. The plane of such brittle fractures is nearly always oriented normal to the direction of locally resolved tension. As shown in Fig. 6, separation under both mode I tension and mode II in-plane shear occurs by the same microscopic mechanism, that is, brittle tension. The only difference

between these two modes is the orientation of principal tension stress under which microscopic failure occurs.

In the case of mode I tension, the maximum principal tensile stress lies perpendicular to the plane of fracture. As a result, brittle cleavage results in distinctly flat areas of cohesive-resin fracture. Fractures produced under conditions of mode II in-plane shear, while also occurring by brittle-tensile separation, exhibit an appearance that is distinctly different from mode I tensile fractures. Under this load condition, the laminate planes formed on either side of the crack are laterally displaced with respect to one another. As described by Mohr's circle, the principal tensile stress for applied shear is oriented at 45° to the plane of fracture. Because brittle-matrix fracture occurs normal to this resolved tensile stress, a series of distinct inclined microcracks (Fig. 7) is formed. During the fracture process, these microcracks

coalesce, resulting in the formation of a series of upright curved platelets. Concave areas are found on the mating fracture surface, opposite to where platelet separation has occurred. Several terms have been used to describe each of these features, including lacerations or hackles for the upright platelets and scallops for concave areas. The more common use of hackles to describe platelets and scallops to describe depressed concave areas will be used in this article.

Fig. 6 Failure modes in fiber-reinforced composites

(a) Schematic of Mode I interlaminar-tension failure and the resulting fracture-surface appearance. (b) Schematic of Mode II interlaminar-shear failure and two possible fracture-surface morphologies

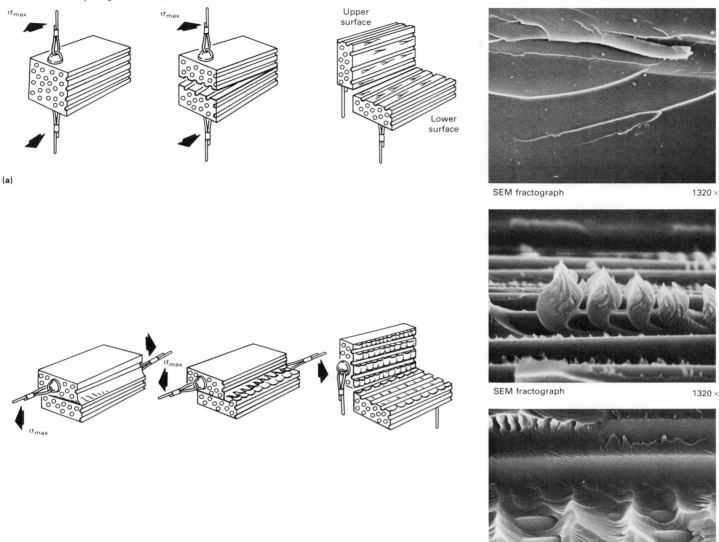

SEM fractograph   1320×

SEM fractograph   1320×

SEM fractograph   1320×

Crack Directions and Fiber Orientations. The basic fracture modes illustrated in Fig. 5 represent a simplified case of interlaminar or intralaminar fracture in which cracking progressed parallel with and between plies and in the same direction as the fiber reinforcement. In most actual applications, however, the direction of reinforcement is likely to be oriented at a variety of angles in order to achieve the specific properties desired. For such structures, the direction of crack propagation is likely to occur at a variety of angles to the fiber orientation, depending upon the direction of imposed stress and material anisotropy. As a result, it is important to understand how these factors alter the basic features illustrated in Fig. 6 and how they can be used to determine the direction of crack propagation.

Mode I Tension Fractures. The easiest mode to consider first is that generated under interlaminar tension. Fracture that is produced parallel to the direction of fiber reinforcement results in flat areas of brittle-matrix failure (Fig. 6a). Such areas typically exhibit distinct river marks. As with metals, these river marks correspond to fracture ridges formed by minutely displaced failure planes. As crack progression occurs, these planes link up, resulting in coalescence of this ridge structure to form a riverlike pattern. As with metals, the direction of this coalescence can be used to define the direction of crack propagation with growth during the direction of coalescence.

Areas of fractured matrix also exhibit a distinctly textured morphology (Fig. 8). This appearance is discernable only at higher magnifications and in most cases requires tilting of the scanning electron microscope specimen or transmission electron microscope replica for it to become visible. In many ways, this textured appearance is identical to the cleavage feathers characteristic of metallic brittle fractures. These feathers exhibit a distinctive chevron-type appearance, with the pointed end of the chevron oriented toward the origin of propagation.

Mode I tension fractures produced at various angles to the direction of fiber reinforcement typically exhibit both the river markings and feathering noted above. Significant alterations in the general fracture topography can occur, depending upon the number of fibers exposed by fracture and their orientation (Fig. 9). In some cases, the variations can produce relatively large areas of flat-resin fracture with distinct river marks oriented in a consistent direction. Alternatively, extensive amounts of fiber exposure can lead to the existence of extremely localized microscopic areas of fracture. This latter condition often results in river marks and cleavage feathers oriented in a variety of angles across the fracture surface. Variations in the direction of microscopic crack propagation can in most cases be averaged together to obtain an estimate of the overall direction of crack growth (Fig. 10).

Variations in the direction of microscopic crack growth depend upon several factors. The two most notable factors that must be considered are the formation of localized zones due to fiber intrusion and the magnitude of stress concentration involved in fracture. In the first case, fiber intrusion divides the crack tip into numerous microscopic zones. In general, these zones will exhibit differing growth rates and slightly displaced planes of fracture. Consequently, the resultant crack front formed by these zones can be highly irregular, with fingers of advanced growth extending beyond the main crack tip. This crack-front profile can produce locally divergent crack directions as zones of unfractured material located between the advancing fingers or behind the main crack tip grow toward immediately adjacent zones of existing fracture.

The second major condition that can lead to crack-direction divergence takes place when fracture occurs without the formation of any appreciable stress concentration. This condition is often associated with the initiation of ultimate material fracture in unnotched test specimens or components (Fig. 11). Material separation occurs in this case by the initiation of multiple origins without any preferred crack direction. As with ductile rupture in metals, failure occurs when these multiple fracture planes intersect. Such fractures are characterized by the formation of an extensive number of fracture planes throughout the laminate with extreme variations in the direction of crack propagation such that no overall direction of crack propagation often exists.

Mode II In-Plane Shear Fractures. The direction of crack growth for interlaminar-shear fractures cannot be established with the same confidence possible for interlaminar-tensile fractures. Separation under shear loading oc-

Fig. 7 Inclined microcracks found in short-beam shear specimen tested at 132 °C (270 °F)
500×. Source: Ref 2, 3

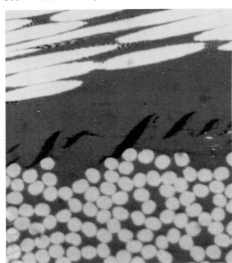

Fig. 8 Matrix feathering produced under interlaminar mode I tension
5000×. Source: Ref 2

Fig. 9 Mode I tension interlaminar fractures that propagated at various angles to the direction of fiber reinforcement
(a) Fracture between adjacent 0 and 90° plies. (b) Fracture between +45 and −45° plies. Both 2000×. Source: Ref 2

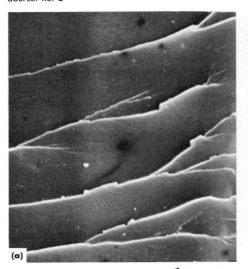
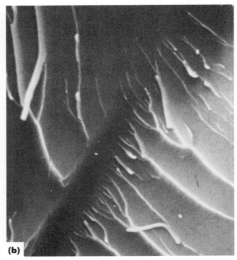

← Overall crack growth direction

curs by the coalescence of numerous tensile microcracks under continued shear displacement. Growth of each microcrack and subsequent coalescence can occur in one of two principal directions (Fig. 12). Microcrack growth can occur either coincident with or opposite to the direction of overall separation (toward or away from the crack tip). As a result, platelets formed under shear loading and their microstructural details, such as river marks and feathering, can be oriented either toward or away from the direction of overall

growth. Consequently, a single direction of crack growth cannot be absolutely defined for shear fractures. However, these features can be used to estimate the direction parallel to which actual crack growth occurred.

The orientation of fiber reinforcement at or adjacent to the delamination plane can have a significant effect on the morphology of both hackles (platelets) and scallops (concave resin fracture areas) that must be considered when determining the direction of crack propagation. Fibers intersecting the fracture surface parallel

Fig. 10 Variations in the crack growth of a graphite-epoxy (±45°) composite

The arrows shown in the fractograph on the right-hand side map the crack growth, which enables the overall crack-growth direction (shown below fractographs) to be determined. 400×. Source: Ref 2

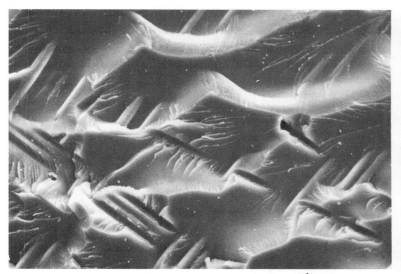

← Overall crack growth direction

Fig. 11 Schematic of interlaminar-tensile fracture without a stress concentration

Variations are produced in microscopic crack direction and multiple fracture planes.

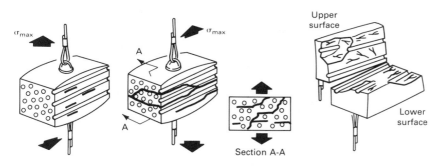

Fig. 12 Schematic of possible hackle-separation mechanisms

Mechanism A illustrates hackle formation coincident with the direction of crack propagation, whereas mechanism B illustrates the formation of hackles opposite to the direction of crack propagation. Source: Ref 2

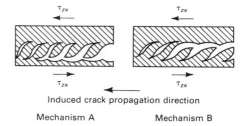

to the direction of crack growth tend to form roughly orthogonal hackles and scallops, whereas fibers intersecting the fracture surface at an angle to the direction of growth tend to form roughly triangular asymmetric hackles and scallops (Fig. 13). In the first case, a distinct branched structure exists on both sides of the hackles and scallops where they intersect adjoining areas of fiber-to-matrix separation. Because this symmetry and the roughly orthogonal shape of these features correspond to propagation parallel to the direction of exposed reinforcement, these features provide a relatively rapid and easy means of identifying the direction parallel to which crack growth occurred.

When crack propagation occurs at an angle to the direction of exposed reinforcement, the asymmetry of the features produced and the orientation of hackle tilting can be used to define the direction parallel to which fracture occurred. Figure 13 shows that hackles and

Fig. 13 Interlaminar mode II shear fractures that propagated at an angle to the direction of fiber reinforcement

(a) Delamination between 0 and 90° plies. 5000×. (b) Fracture between +45 and −45° plies. 2000×. Source: Ref 2

← Overall crack growth direction

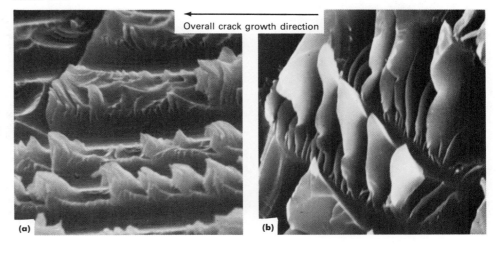

scallops produced under this condition exhibit a roughly triangular appearance, with distinct branched markings located at the apex of this triangle. For this condition, the crack-propagation direction is parallel to the direction of inclined hackle tilting.

Translaminar Fractures

While interlaminar fractures tend to be relatively planar, translaminar fractures are generally identified by their rough fiber-dominated morphologies. Due to the high translaminar fracture toughness of composite laminates, a considerable amount of gross damage typically occurs. As with fractures in metallic structures, macroscopic features generally reflect the con-dition of imposed load at failure. In contrast to interlaminar fractures, load states can be identified by visually examining the displacement between mating fracture surfaces and/or the general morphology of the fracture surfaces. Two distinct translaminar failure mechanisms that occur in laminates are tensile and compression microbuckling. One of these two separation mechanisms, or a combination, is operative in all translaminar fracture conditions. The following sections discuss the relationships between load states, fracture mechanisms, and characteristic fractographic features.

Translaminar-Tension Fractures. Macroscopically, translaminar-tension fractures exhibit an extremely rough fracture surface, with large amounts of fibers protruding out of the major fracture plane (Fig. 14a). Little or no delamination is evident at or near the fracture surface. Brittle-tensile fracture of individual fibers is the primary operative failure mechanism, with fracture of the surrounding matrix considered as secondary. Fibers fracture in groups (bundles) where the fibers within each bundle have a relatively flat, common fracture plane (Fig. 14b). Fiber pull-out, fiber-end fracture, and matrix fracture are the characteristic fractographic features of translaminar-tension failures. Figure 14(c) shows the typical radial topography found on broken fiber ends. This radial morphology is analogous to the chevron patterns in tension failures of metals, particu-

Fig. 14 Examples of translaminar-tension fractures

(a) Translaminar-tension fracture in graphite-epoxy composite. Note fiber bundles and individual fiber pull-out. 400×. Source: Ref 3. (b) Translaminar-tension failure with localized area of flat fracture. 2000×. Source: Ref 3. (c) Radial fracture topography of an individual graphite-fiber failure under translaminar tension. 10 000×. Source: Ref 3. (d) Variations in fiber fracture mapped to determine overall crack-growth direction. 2000×. Source: Ref 4

larly for rod or bar forms. The faint lines radiate from the point of fiber fracture initiation and thus indicate the direction of crack propagation for each individual fiber. Consistent with brittle fractures, the fiber origins are primarily located at flaws or notches in the crenelated fiber surfaces, although some initiate at internal flaws, such as voids.

Tension failure does not progress by a well-defined crack front. Due to flaw sensitivity, the fracture process involves multiple initiation zones in which all the fiber breaks originate at a single source. Thus, the crack front actually consists of several isolated fracture zones—at different axial planes—that coalesce and propagate in the overall growth direction. Fiber ends within each zone tend to fracture in a variety of directions, although they are often noticeably biased in a single direction. Through extensive mapping of the fiber ends (Fig. 14d), the macroscopic crack-growth direction can be determined. Extreme caution must be exercised in the evaluation of isolated microstructural details; therefore, a coherent approach is required to ensure accurate and unbiased crack mapping.

Translaminar-Compression Fractures. Macroscopically, fractures produced under uniaxial compression exhibit gross buckling, extensive delamination, and interlocking of the delamination planes (Fig. 15). An end-on view of the broken fiber ends reveals a distinct, flat fracture surface with extensive postfracture damage (Fig. 16a). This obliteration of fracture-surface details is due to relative postfailure motion between the fractured surfaces in contact. The surface is much flatter than that in translaminar-tension fractures and is virtually devoid of pulled-out fibers. Fiber buckling, fiber-end fracture, resin-matrix fracture, and postfracture damage are the characteristic fractographic features of translaminar-compression fractures.

Compression microbuckling is the primary operative failure mechanism. This mechanism involves local buckling of individual fibers at a point where a maximum localized lateral instability exists. Under compressive flexure, kinking of each fiber causes fracture at two locations (Fig. 16b), with each fracture separated by 5 to 10 fiber diameters. Short sections of fibers with this length can often be seen on the fracture surface. Figure 16(c) illustrates the typical flexural fracture morphology found on the fiber ends. The radial topography represents the tensile portion of fracture, while the smooth topography represents the compressive fracture. The distinct line shown intersecting the fiber end in Fig. 16(c) is the neutral axis line. For each individual fiber, the direction of flexure and failure occurs normal to the neutral axis line. For a given fiber, the compression and tension morphologies are reversed when comparing the two breaks. Therefore, a singular crack direction cannot be deduced, although the individual fiber fracture propagates perpendicularly to the neutral axis line.

The neutral axis lines are commonly found parallel to one another in a given region, indicating that microbuckling occurs on a local scale in a concerted manner and in a unified direction (Fig. 16d). Preliminary controlled crack-growth studies have shown that the neutral axis lines are often biased at an angle parallel to the direction of induced crack propagation. This indicates that flexural collapse and fracture propagation on a local scale often occur transverse to the gross crack direction; therefore, neutral axis lines cannot be used to determine the direction of crack propagation.

Procedures for Failure Analysis

In general, the procedures used in the failure analysis of composites are similar to those used for other material systems. As with any thorough analysis, the prudent investigation should follow a basic sequence that at the very least includes a review of available in-service records, a preliminary nondestructive examination of the part, verification of use of the intended material, and the fractographic determination of the origin and mode of fracture. In many cases, evaluation of the stress and fracture mechanics involved in failure will also be necessary to determine the cause adequately. Because this sequence of steps is basically the same for metals, the discussions in the article "General Practice in Failure Analysis" in this Volume are a good source of more detailed information. Most of the methods described in the aforementioned article can be directly applied to composites. However, because of their laminated construction and chemistry, some specific differences exist in the areas of nondestructive examination, materials verification, and fractography.

Nondestructive Examination. Because of their laminated construction, conditions of interlaminar fracture (delaminations) represent a relatively common mode of failure. Because these planes of fracture occur within the laminate, they are not generally visible during routine visual examinations. As a result, nondestructive examinations represent a particularly critical step required to define both the extent and type of damage involved in component failure.

The most successful methods of nondestructive inspection include through-transmission ultrasonics (TTU), pulse-echo ultrasonics, and dye-penetrant enhanced x-ray radiography. In the first two methods, internal damage is detected by the attenuation or reflection of 1- to 25-MHz acoustic waves transmitted into the panel through a transducer. The detection of damage by x-ray radiography is carried out in the same basic way as for metals, except a radiopaque penetrant is wetted into the crack surface to enhance damage visibility. In general, x-ray equipment operable in the 10- to 50-kV, 5-mA range with diiodobutane (DIB) penetrant yields the best results.

Fig. 15 Side view of a compression fracture with extensive delamination and interlocking
0.4×

Use of any of these nondestructive evaluation techniques requires that some forethought be given to the type of failure mode likely to be encountered. With both ultrasonic methods, liquids are generally used to carry acoustic waves into the panel. Similarly, the radiopaque penetrant used in radiography is also a liquid. The penetration of these liquids into the fracture surface can constitute a form of chemical contamination. In those cases where chemical surface-analysis techniques, such as x-ray photoelectron spectroscopy, are to be used, the presence of such contaminants should be eliminated or prevented from entering critical areas of failure by sealing areas of exposed fracture.

Materials Characterization. As indicated earlier in this article, errors in material lay-up, ply orientation, and degree of matrix curing can lead to or, at the very least, contribute to premature component failures. As a result, the analysis of these basic material and processing errors should be considered a standard operation in most failure analysis cases.

Because the matrices employed in most composites are organic in nature, the conventional test methods used for the evaluation of metals, such as hardness testing, do not translate well to composites. For these materials, assurance of material cure is generally best obtained by measuring both the matrix resin glass transition temperature, T_g, and residual heat of reaction, ΔH. Both of these measurements use standard thermodynamic instrumentation.

The temperature at which the matrix undergoes a transition from a glassy to a rubbery state is measured by monitoring the rate of change of one of several material properties (such as thermal expansion or stiffness) versus temperature. By comparing against known standards, a judgment as to the degree of material cure can be made. In those cases where this value appears lower than normal, an additional assessment of the degree of cure can be obtained by measuring the residual heat of reaction of the material. This measurement is made by using a calorimeter in which heat evolution as a func-

Fig. 16 Examples of translaminar-compression fractures

(a) Translaminar-compression fracture with extensive postfailure damage to fiber ends. 750×. (b) Translaminar-compression generated fiber kink in graphite-epoxy fabric. 100×. (c) Flexural fracture characteristics on fiber ends of compression specimen. 10 000×. (d) Translaminar-compression fracture illustrating parallel neutral axis lines representative of unified crack growth. 2000×

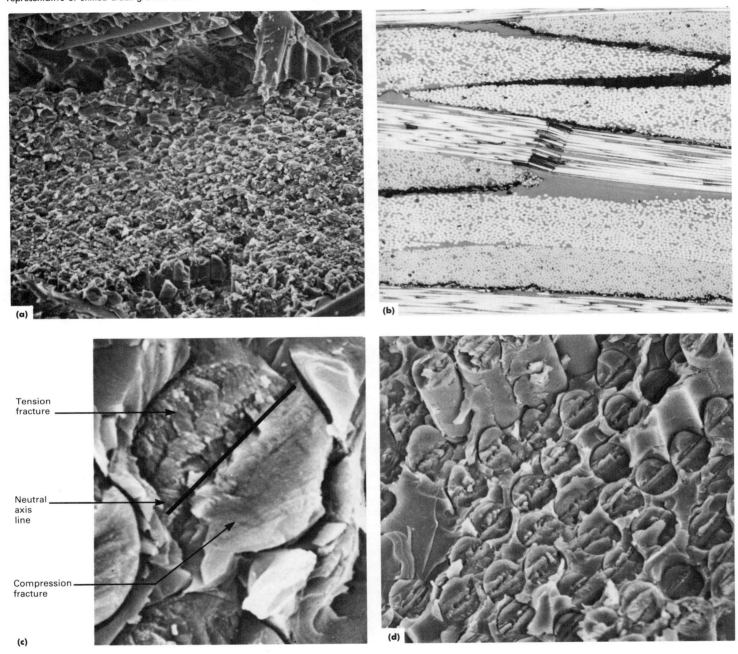

tion of temperature is measured. The magnitude of this measured value characteristically corresponds with the degree of material undercure.

In much the same manner as with metals, the number and orientation of plies within the laminate can be examined through simple metallographic cross-sectioning procedures. Polishing is best accomplished with wheels covered with cloths without appreciable nap, such as nylon or silk. Etching is generally not required, due to the distinct optical differences that exist between most fibers and matrix systems. An exception to this is fiberglass, for which a dilute hydrofluoric acid etch may be necessary to enhance fiber visibility. Because most plies are separated by a resin-rich region and because each fiber exhibits a different profile (round, elliptical, and so on) depending upon the cross-section angle, the identification of the number of plies and their orientation is relatively simple. Detailed information on metallographic procedures germane to composites can be found in the article "Fiber Composite Materials" in Volume 9 of the 9th Edition of *Metals Handbook*.

Fractography. As pointed out in this article, composites can produce complex fractures exhibiting interlaminar, intra-laminar, and translaminar planes of separation. Because the occurrence of such conditions as interlaminar fracture can often be quite extensive and can occur on multiple planes, the decision as to which surface to examine and how to examine it often represents one of the most difficult tasks involved in failure analysis.

Compression failures generally exhibit extensive interlaminar fracture, along with through-thickness translaminar-compression fracture. Because translaminar-compression fracture surfaces generally exhibit significant postcompression damage, examination tends to be the least informative. On the other hand, interlaminar fracture surfaces for this load state are often undamaged. As a result, the direction and origin of these fractures can often be established. Such interlaminar fractures may occur secondarily to translaminar fracture. However, their direction of propagation can often be used to infer the origin of primary fracture because their occurrence is related to the progression of compressive fracture.

In contrast to compression fractures, tensile fractures exhibit very little surface damage and lesser amounts of interlaminar fracture. As a result, interlaminar, intralaminar, and translaminar separation surfaces can all be examined. For these fractures, the selection of which surface to be examined should depend upon the predominant fracture mode exhibited and the amount of time available. Where possible, examinations should be carried out on each mode of separation exhibited.

Because the features used to map the direction and mode of fracture in composite materials are relatively fine, the examination of large areas of fracture surface can represent a sizable undertaking. As a result, it is important to recognize that many of the features described for interlaminar fractures, for example, river patterns and hackles, can be identified using medium-power optical microscopy (200 to $600 \times$). For such large-area failures, the use of optical microscopy before electron microscopy allows several orders of magnitude more area to be examined (details on the principles and applicability of optical microscopy can be found in the article ''Optical Microscopy'' in Volume 9 of the 9th Edition of *Metals Handbook*). The ability to survey a large area of fracture is particularly important because extreme variations in the direction of crack growth and fracture mode can occur on a local scale.

Scanning electron microscopy (SEM) of composites usually requires gold or goldpalladium coating of the fracture surface to prevent charging and to improve secondary electron yields (the use of conductive coatings in SEM analysis is described in the article ''Scanning Electron Microscopy'' in Volume 9 of the 9th Edition of *Metals Handbook*). Alternatively, many of the newer scanning electron microscopes can be operated at low electron voltages, eliminating the need for such coatings. Once within the microscope, the analysis of these materials can be carried out in the manner described in the section ''Fracture Modes in Composites'' in this article.

Transmission electron microscopy (TEM) of composite fracture surfaces has seen relatively little application to date. While this technique should be translatable to these materials, replication of the fracture surface may prove damaging (TEM replication procedures are described in the article ''Transmission Electron Microscopy'' in Volume 9 of the 9th Edition of *Metals Handbook*). This may be the case for delicate structures, such as hackles, or for those matrix systems susceptible to attack by replica solvents. For this reason, adequate optical and scanning electron microscopy should be carried out where possible before fracture-surface replication to prevent premature destruction of fracture evidence.

Stress Analysis. In most cases, an accurate understanding of the loads and stress levels in component operation is one of the critical ingredients involved in defining the source of failure. Although other methods of analysis may identify the origin and mode of crack propagation, stress analysis most often provides a quantitative explanation for the cause of failure. Through this analysis step, engineers involved in future or corrective redesigns are provided with direct feedback regarding the actual loads experienced by the part, poor design practices and configurations, and the effectiveness of the analysis methods used in design.

Stress analysis procedures for composite materials can be relatively complex due to several factors. Because composite materials are fabricated by the lamination of highly anisotropic plies, a nearly infinite variety of directional moduli and strengths can be achieved. Because of this tailorability, a different set of material properties must be considered for each failure case being examined. As a further complication, because of their laminated anisotropic construction, significant variations in stress can exist within the laminate itself. As a result, consideration must be given to failure at both the microstructural (individual ply) and the overall laminate levels. Reflective of this latter problem, the analysis of failures in composite materials has taken two basic approaches. Failures are generally analyzed at the individual ply level (lamination theories) or at the gross laminate level.

Typically, analyses at the individual ply level provide the greatest level of detailed information. This method predicts basic material properties by modeling the composite as a plate built up from individual plies, each with its own discrete lamina properties, that is, lamination theory (Ref 4). These lamina properties may involve considerable detail—for example, moduli, hygrothermal expansion coefficients, moisture diffusivity, and density—but can in general be estimated based upon individual fiber and resin properties using micromechanics equations. Through this method, the stiffness characteristics of the overall laminate can be determined for the general case of in-plane and/or flexural loads. Based upon this stiffness, the gross stress generated for any given load can be calculated.

However, as noted above, consideration must also be given to the stresses carried by each individual ply. These stresses depend on the modulus and orientation of each ply and can be determined using basic lamination theory. In most cases, individual ply stresses are generally different from the stress state of the laminate as a whole. This is due to their orientation, the constraint of neighboring plies, and the residual stresses from temperature or absorbed moisture. For example, uniaxial loading of a composite laminate that consists of plies with different orientations will result in a biaxial state of stress that varies from ply to ply. One of the chief advantages of lamination theory is its ability to quantify these stresses and to examine them on an individual ply level.

Although the lamination theory provides a powerful tool with which to examine the stress state operating within a composite, there are several disadvantages. The most notable of these is in its application in predicting the onset of failure. Such predictions are achieved by evaluating the stress within each individual ply against semiempirical models that define the maximum stress envelope to which a ply can be exposed without failure. However, predictions of the onset of interlaminar and intralaminar failure using such criteria have not been entirely successful. One of the causes of this lack of success centers around the fact that standard lamination theory is unable to predict the threedimensional state of stress that exists in laminates near free edges and other discontinuities, for example, holes, cut-outs, and ply drop-offs.

Various approaches can be used, however, to predict the three-dimensional state of stress within a laminate. Typically, these methods tend to be rather specialized, including finite element analysis, for example, generalized plane strain or three-dimensional elements, and approximate strength-of-materials methods (Ref 5). Using these methods, failure criteria based on principles of fracture mechanics (Ref 6) have been somewhat successful in predicting the onset and growth of interlaminar and intralaminar matrix damage, including fatigue effects.

The prediction of catastrophic failure at the laminate level has followed two distinctly different approaches. The first, which may be referred to as a statistical approach, uses coupon test data to determine the maximum allowable strains for each laminate design. Using a statistical approach, traditional methods of stress analysis can be applied (load versus area) to determine the occurrence of failure, without regard to individual ply or interlaminar stresses. In this case, failure occurs when the statistically determined maximum allowable strain is exceeded. The second approach to predicting translaminar fracture involves a more mechanistic approach in which the load redistribution due to intralaminar and interlaminar matrix damage is taken into account. In its simplest form, this approach can be applied directly with lamination theory and has been referred to as

Fig. 17 Schematic of fractured component showing orientation and direction of applied load and approximate fracture location
Source: Ref 10

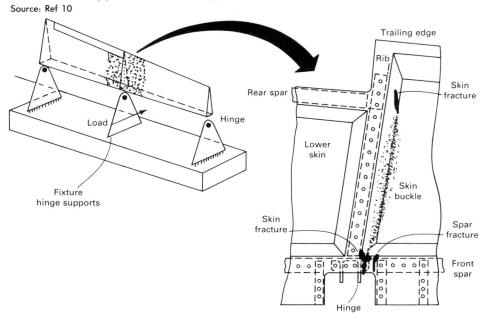

the global ply discount method (Ref 7). Applications of this method rely on a predetermined definition of characteristic intralaminar and interlaminar damage. Using the lamination theory, the state of stress surrounding this damage is defined, and the magnitude of the applied load necessary to create failure is determined.

Translaminar fracture near holes, bolts, and through-thickness cracks has been successfully modeled using modifications to fracture criterion that were used with metals. These modifi-

cations involve the determination of semiempirical correction factors that are referred to as characteristic dimensions. These characteristic dimensions describe the distance from a stress concentration for which complete laminate failure will occur if the ultimate strength of the material is reached at this point. A variety of characteristic dimension fracture criteria have been found to be functionally equivalent for engineering applications (Ref 8 and 9). Although these correction factors have been pro-

posed for both tension and compression loads, most of the work discussed in the literature has addressed the former.

Analysis of composite panel or column stability is similar to that used for metals. A finite element approach is used for postbuckling analysis of stiffened panels or box structures. Laminate stacking sequence and ply orientation are important to composite plate stability. Residual strength after impact damage has been found to be a controlling design parameter for composite laminates. This is particularly true for compression loads in which localized ply buckling and interlaminar damage growth can lead to overall panel failure at load levels that are as low as 40% of the undamaged strength. Methods to predict residual strength after impact have ranged from empirical to an approach involving three-dimensional finite element models of impact damage and fracture mechanics (Ref 9).

Example 1: Compression Fracture of a Graphite-Epoxy Test Structure Due to a Buckling Instability. Figure 17 illustrates a portion of a graphite-epoxy tapered-box structure that fractured during testing. This graphite-epoxy box consisted of two honeycomb skin panels fastened to a spanwise spar with intermediate chordwise ribs.

Investigation. A review of the test history revealed that premature fracture occurred during hingeline deflection of the front spar (Fig. 17). Initial nondestructive visual inspection of the fractured box revealed through-thickness cracks in the forward and trailing edges of the compression-loading skin panel. Upon further examination, some localized buckling of the skin panel was evident between each of these through-thickness fractures. To define areas of nonvisible damage, that is, delamination, a

Fig. 18 Crack-propagation direction and origin identified on fracture test component
(a) Optical micrograph. 0.4×. Crack origin is indicated by circled area. (b) Scanning electron micrograph of the circled area shown in (a). 5000×. Note river marks that coalesce in the direction of overall crack growth. Source: Ref 10

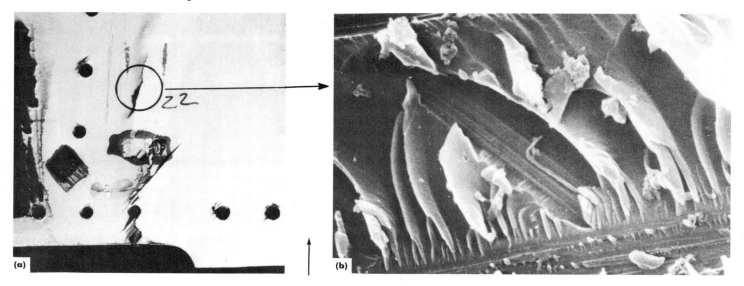

Crack direction

nondestructive evaluation was performed with TTU. This analysis revealed a roughly 10-cm (4-in.) wide band of delamination between the areas of through-thickness skin fracture at the front and rear spar.

Following the definition of the type and extent of fracture, tests were performed to determine if any major material discrepancies existed in either fabrication or processing. Accordingly, sections of the skin, spar, and rib panels were examined to verify the lay-up and to determine the overall panel quality. In addition, thermomechanical analyses (TMA) were performed to verify the extent of cure. Dimensions of panel, spar, and rib details were also measured and checked against required dimensions and tolerances. For each of these analyses, the spar, ribs, and skin panels were found to be in compliance with the drawing requirements.

Because no discrepancies were identified, fractographic examinations were selected as the next investigative operation. Primary emphasis was placed on identifying the direction of crack propagation, origin, and any anomalous conditions that could be associated with fractures. To facilitate examination, fractured areas of the panel were sectioned into roughly 15- × 15-cm (6- × 6-in.) squares and examined optically. These optical examinations were performed at 400×, which provided a rapid and effective means of identifying characteristic fracture features. Scanning electron microscopy was performed on selected areas of interest to examine and document specific fracture-surface features. The orientation of river patterns and resin microflow (Fig. 18) observed on the fracture surface generated a map of the local directions of crack propagation over the fracture surface.

Conclusions and Recommendations. By reconstructing the fracture process, it was discovered that crack initiation occurred at the periphery of a fastener hole located at the front spar. Subsequently, propagation occurred chordwise across the compression-loaded skin panel. Inasmuch as no anomalies were identified at the origin area that might explain premature fracture, detailed stress analyses of this area were initiated. Both the basic in-plane strains and the

buckling stability of the origin area were evaluated. These analyses revealed that premature skin buckling occurred because of a relatively large fastener spacing. As a result, further attention was paid to this design detail, and fastener spacing was reduced to prevent the buckling mode that precipitated fracture.

REFERENCES

1. M.B. Rhodes, Damage Tolerance Research on Composite Panels, in *Selected NASA Research in Composite Material and Structures*, NASA CP-2142, National Aeronautics and Space Administration, Washington, Aug 1980
2. B.W. Smith *et al.*, Fractographic Analysis of Interlaminar Fractures in Graphite-Epoxy Material Structures, in *Proceedings; International Conference: Post Failure Analysis Techniques for Fiber Reinforced Composites*, Air Force Wright Aeronautical Laboratories, MLSE, Wright Patterson Air Force Base, Dayton, OH, July 1985
3. A.G. Miller *et al.*, Fracture Surface Characterization of Commercial Graphite/Epoxy Systems, in *Nondestructive Evaluation and Flaw Criticality for Composite Materials*, STP 696, ASTM, Philadelphia, p 223-273
4. S.W. Tsai and H.T. Hahn, *Introduction to Composite Materials*, Technomic Publishing, 1980
5. W.S. Johnson, Ed., *Delamination and Debonding of Materials*, STP 876, ASTM, Philadelphia, 1985
6. A.S.D. Wang, Fracture Mechanics of Sublaminate Cracks in Composite Laminates in *AGARD Conference on Characterization, Analysis and Significance of Defects in Composite Materials*, AGARD-CP-355, Advisory Group for Aerospace Research and Development, Neuilly-Sur-Seine, France, 1983
7. K.L. Reifsnider, K. Schultz, and J.C. Duke, Long Term Fatigue Behavior of Composite Materials, in *STP 813*, ASTM, Philadelphia, 1983, p 136-159
8. J. Awerbuch and M.S. Madhukar, Notched Strength of Composite Laminates: Predictions and Experiments—A Review, *J. Reinf. Plast. Compos.*, Vol 4 (No. 1), 1985
9. A.A. Baker, R. Jones, and R.J. Callinan, Damage Tolerance of Graphite/Epoxy Composites, *Compos. Struct.*, Vol 4, 1985, p 15-44
10. B.W. Smith *et al.*, Composite Post-Fracture Analysis Experience as Related to a Developing and Evolving CPFA Methodology, in *Proceedings; International Conference: Post-Failure Analysis Techniques for Fiber Reinforced Composites*, Air Force Wright Aeronautical Laboratories, MLSE, Wright Patterson Air Force Base, Dayton, OH, July 1985

SELECTED REFERENCES

- S.L. Donaldson, Fracture Toughness Testing of Graphite/Epoxy and Graphite/PEEK Composites, *Composites*, Vol 6 (No. 2), April 1985
- T. Johannesson, P. Sjoblom, and R. Seldon, The Detailed Structure of Delamination Fracture Surfaces in Graphite/Epoxy Laminates, *J. Mater. Sci.*, Vol 19, 1984, p 1171-1177
- R.A. Kline and F.H. Chang, Composite Failure Surface Analysis, *J. Compos. Mater.*, Vol 14, Oct 1980, p 315-324
- K.M. Liechti *et al.*, *SEM/TEM Fractography of Composite Materials*, AFWAL-TR-82-4085, Air Force Wright Aeronautical Laboratories, MLSE, Wright Patterson Air Force Base, Dayton, OH, Sept 1982
- D. Purslow, Some Fundamental Aspects of Composites Fractography, *Composites*, Vol 12 (No. 4), Oct 1981
- J.H. Sinclair, Fracture Modes in High Modulus Graphite/Epoxy Angleplied Laminates Subjected to Off-Axis Tensile Loads, in *Rising to the Challenge of the 80's; 35th Annual Conference and Exhibit*, Society of the Plastics Industry, Inc., New York, 1980, p 12-C1 to 12-C8

Failure Analysis of Ceramics*

David W. Richerson, Ceramatec, Inc.

FAILURE ANALYSIS is extremely important in engineering, especially with ceramics, because it is the only means of isolating the failure-causing problem. In particular, failure analysis helps determine whether failure or damage occurred due to a design deficiency or a material deficiency. Until this has been determined, efforts cannot be efficiently directed toward finding a solution. The result is usually a shotgun approach that includes a little design analysis, a little empirical testing, and a little material evaluation and often ends up only in a repeat of the test or operating conditions that initially caused failure.

Much of the shotgun approach can often be avoided by fracture analysis. Fracture analysis or fractography is the examination of the fractured or damaged hardware in an effort to reconstruct the sequence and cause of fracture. The path a crack follows as it propagates through a component provides substantial information about the stress distribution at the time of failure. Features on the fracture surfaces provide further information, especially the position at which the fracture initiated (fracture origin), the cause of fracture initiation (impact, tensile overload, thermal shock, material flaw, etc.) and even the approximate local stress that caused fracture. The primary objective of this article is to acquaint the reader with these fracture surface features and the techniques used to interpret the cause of fracture in ceramic components.

Location of the Fracture Origin

The first step is to determine where the fracture initiated. Often, simply reconstructing the pieces will pinpoint the fracture origin and may even give useful information about the cause of fracture. After assembling the pieces, look for places where a group of cracks come together or where a single crack branches. Preston (Ref 1) has shown that the angle of forking is an indicator of the stress distribution causing fracture. Examples are shown schematically in Fig. 1.

*Adapted from Chapter 12 of *Modern Ceramic Engineering* by David W. Richerson, published by Marcel Dekker, Inc., 1982. With permission.

Fig. 1 Information available by examining crack direction and crack branching
Source: Ref 2

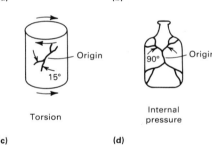

(a) Impact or point-loading
(b) Bending
(c) Torsion
(d) Internal pressure

The frequency of crack branching provides qualitative information about the amount of energy available during fracture. To branch, a crack must reach a critical speed. For glass, the critical speed is typically a little greater than half the speed of sound in the specific glass. At this instant of crack initiation, the crack velocity is zero, but quickly accelerates. The rate of acceleration is a function of the energy available either due to the stress applied or to energy stored in the part (such as residual stresses or prestresses, as in tempered glass). The more energy, the more rapidly the crack will reach its critical branching velocity and the more branching that will occur. A baseball striking a window will cause much more branching than a BB, due to the larger applied energy. Tempered glass will break into many fragments due to release of the high stored energy. On the other hand, a thermal shock fracture may not branch at all, especially if it initiates from a localized heat source and propagates into a relatively unstressed or compressively stressed region of the component. In this case, the fracture will tend to follow a temperature or stress contour

Fig. 2 Thermal shock fracture showing lack of branching

Fig. 3 Schematic showing the typical fracture features that surround the fracture origin
(a) Internal initiation. (b) Surface initiation

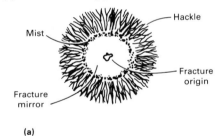

(a)

(b)

and will have a characteristic wavy or curved appearance, as shown in Fig. 2 for a thermally fractured ceramic setter plate for a furnace.

The pattern of branching will often lead the engineer to the vicinity of the fracture origin. The engineer will then have to examine the fracture surfaces in this region, often under a low-power optical binocular microscope, to

Fig. 4 Example of typical fracture mirrors for high-strength polycrystalline ceramics

(a) Initiation at a surface flaw in hot-pressed silicon nitride. (b) Initiation at an internal flaw in reaction-sintered silicon nitride

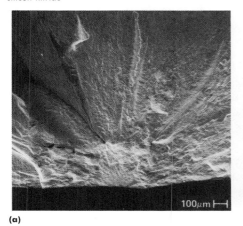

(a)   (b)

Fig. 5 Examples of fracture surfaces with indistinct fracture features

(a) Sintered silicon carbide. (b) Silicon carbide-carbon-silicon composite. (c) Porous lithium aluminum silicate. (d) Bimodal grain distribution reaction-sintered silicon carbide

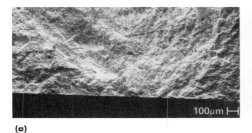
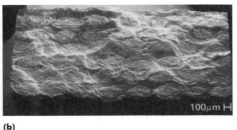

(a)   (b)

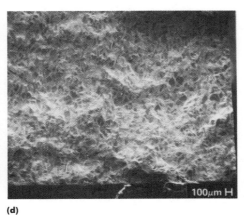

(c)   (d)

locate the precise point at which fracture initiated. This point of origin can be a flaw (such as a pore or inclusion in the material), a cone-shaped Hertzian surface crack resulting from impact, a crack in a surface glaze, an oxidation pit, intergranular corrosion, a position of localized high stress, or a combination. Location and examination of the fracture origin will help determine which of these factors is dominant and will provide specific guidance in solving the fracture problem.

As a fracture accelerates, it interacts with the microstructure, the stress field, and even acoustic vibrations and leaves distinct features on the fracture surface that can be used for locating the fracture origin (Ref 3). The most important features include hackle, the fracture mirror, and Wallner lines.

The Fracture Mirror and Hackle. When a crack initiates at an internal flaw, the crack front travels radially in a single plane as it accelerates. The surface formed is flat and smooth and is called the fracture mirror. When the crack reaches a critical speed, intersects an inclusion, or encounters a shift in the direction of principal tensile stess, it begins to deviate slightly from the original plane, forming small radial ridges on the fracture surface. The first of these are very faint and are referred to as mist. Mist is usually visible on the fracture surfaces of glass, but may not be on crystalline ceramics. The mist transitions into larger ridges called hackle. Hackle is also referred to as river patterns because the appearance is similar to the branching of a river into tributaries and the formation of deltas. The hackle region transitions into macroscopic crack branching such that the remaining portion of the fracture surface is often on a plane that is perceptibly different from the mirror and hackle. Sometimes, this gives the appearance that the fracture origin is either on a step or a pedestal.

Figure 3(a) shows schematically the fracture mirror, mist, and hackle for a fracture that initiated in the interior of a part. The mirror is roughly circular, and the fracture origin is at its center. Note that lines drawn parallel to the hackle will intersect at or very near the fracture origin. Similarly, Fig. 3(b) shows the fracture features for a crack that started at the surface of a part. The mirror is roughly semicircular for surface-initiated fractures, usually with slight elongation toward the interior of the material. The elongation results from a difference in stress intensity factor between the surface and the interior. The stress intensity factor is higher at the surface, causing the crack branching and the resulting appearance of the hackle to occur more quickly along the surface than inward.

The hackle lines surrounding the mirror result from velocity and stress intensity effects and are sometimes called velocity hackle. Another form of hackle, called twist hackle, usually forms away from the mirror and results from an abrupt change in the tensile stress field, such as going from tension to compression. Twist hackle points in the new direction of crack movement, appears more as parallel cracks than ridges, and does not necessarily point to the fracture origin. Twist hackle is an important feature for deducing the stress distribution in the ceramic at the time of fracture.

The size of the fracture mirror is dependent on the material characteristics and the localized stress at the fracture origin at the time of fracture. Studies by Terao (Ref 4), Levengood (Ref 5), and Shand (Ref 6) on glass suggest that the fracture stress σ_f times the square root of the mirror radius r_m equals a constant A for a given material:

$$\sigma_f \, r_m^{1/2} = A \qquad \text{(Eq 1)}$$

Kirchener and Gruver (Ref 7) determined that this relationship also provides a good approximation for polycrystalline ceramics, as long as the mirror is clearly visible and can be measured accurately. They obtained values of A

Fig. 6 Relationship of Wallner lines on a fracture surface to the stress distribution at the time of fracture

(a) Uniform tension. (b) Nonuniform tension. (c) Tension-compression. Source: Ref 2

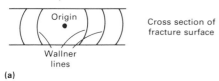

Cross section of fracture surface

(a)

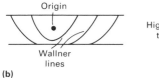

High tension on top surface

(b)

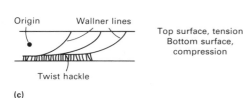

Top surface, tension
Bottom surface, compression

(c)

ranging from 2.3 for a glass to 9.1 for a sintered Al_2O_3 to 14.3 for a hot-pressed Si_3N_4.

It is therefore possible to estimate the stress causing failure of a ceramic component by comparing the mirror size with a graph of $r_m^{1/2}$ versus σ_f for the material. The graph can be compiled from bend strength or tensile strength data using scanning electron microscopy (SEM) of the fracture surfaces of the test bars to determine the mirror radius.

Griffith (Ref 8) proposed an equation for relating the fracture stress to material properties and flaw size:

$$\sigma_f = A \left(\frac{E\gamma}{c} \right)^{1/2} \qquad \text{(Eq 2)}$$

where σ_f is the fracture stress, E is the elastic modulus, γ is the fracture energy, c is the flaw size, and A is a constant that depends on the specimen and flaw geometries.

Evans and Tappin (Ref 9) have presented a more general relationship:

$$\sigma_f = \frac{Z}{Y} \left(\frac{2E\gamma}{c} \right)^{1/2} \qquad \text{(Eq 3)}$$

where Y is a dimensionless term that depends on the flaw depth and the test geometry, Z is another dimensionless term that depends on the flaw configuration, and c is the depth of a surface flaw (or half the flaw size for an internal flaw), and E and γ are defined as above. For an internal flaw that is less than $\frac{1}{10}$ of the size of the cross section under tensile loading, $Y = 1.77$. For a surface flaw that is much less than $\frac{1}{10}$ of the thickness of a cross section under bend loading, Y appproaches 2.0; Z varies according to the flaw shape, but is usually between 1.0 and 2.0.

Fig. 7 Difference in the fracture contour through the specimen thickness for bend loading versus pure tensile loading

(a) Bend (flexure) loading. (b) Pure tensile loading

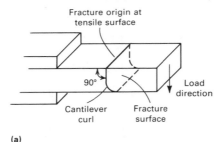

(a)

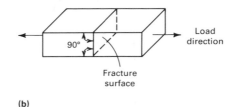

(b)

Fig. 8 Examples showing cantilever curl in four-point bend specimens

Specimens were 0.32 cm (0.125 in.) thick.

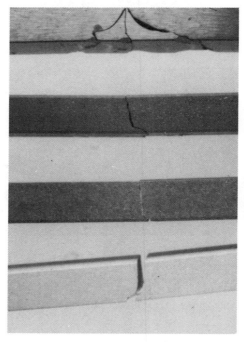

Because inclusions and pores and other flaws are not symmetrical and their boundaries are often not well defined, stress estimates based on flaw size are only approximate. If a knowledge of the local stress at the fracture origin is

Fig. 9 Major sequential steps in conducting fracture analysis

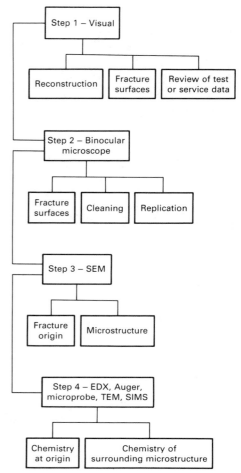

needed, perhaps it should be calculated both from the mirror size and the flaw size and then the most appropriate value selected by good engineering judgment.

Figure 4 shows examples of fracture mirrors and flaws on the fracture surfaces of strength-test specimens. The photomicrographs were taken using a scanning electron microscope. Since the ceramic specimens were not electrical conductors, a thin layer of gold was applied to the surface by sputtering to avoid charge buildup, which would result in poor resolution.

The mirror size cannot always be measured. If the fracture-causing stress is low and the specimen size is small, the mirror may cover the whole fracture surface. If the material has very coarse grain structure or a bimodal grain structure, the mirror and other fracture features may not be visible or distinct enough for measurement. Figure 5 shows examples of fracture surfaces with indistinct fracture features.

Wallner Lines. Sonic waves are produced in a material during fracture. As each succeeding wave front overtakes the primary fracture

Fig. 10 Failure analysis interaction to determine if the problem is design or materials oriented and to define a plan of action to solve the problem

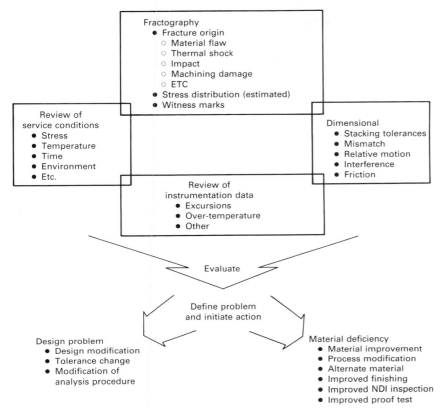

Fig. 10 Failure analysis interaction to determine if the problem is design or materials oriented and to define a plan of action to solve the problem

crack, the principal stress is momentarily disturbed. This results in a series of faint arc-shaped surface lines that are termed Wallner lines. The curvature of each line shows the approximate shape of the crack front at the time it was intersected by the sonic wave and provides information about the direction of crack propagation and the stress distribution. The direction is from the concave to the convex side of the Wallner lines. The stress distribution is inferred from the distance of each portion of a single line from the origin. If the stress distribution were of uniform tension, each portion of a line would be about equidistant from the origin. If a stress gradient were present, the distance of various portions of the Wallner line from the origin would vary, being farthest where the tensile stress was highest. These effects are shown schematically in Fig. 6.

Wallner lines are not always present. For high-energy fractures, where the fracture velocity is high and the surface is rough, Wallner lines often cannot be distinguished. In very slow crack velocities, such as occur in subcritical crack growth, Wallner lines are not present, because the sonic waves are damped and gone before the crack has propagated appreciably.

Other Features. Other fracture features besides the mirror, hackle, and Wallner lines

are useful in interpretation of a fracture. These include arrest lines, gull wings, and cantilever curl.

An arrest line occurs when the crack front temporarily stops. The reason for crack arrest is usually a momentary decrease in stress or a change in stress distribution. When the crack starts moving again, its direction invariably has changed slightly, leaving a discontinuity. This line of discontinuity looks a little like a Wallner line, but is usually more out of plane and more distinct. It is also called a rib mark. Arrest lines or rib marks provide essentially the same information as Wallner lines, that is, the direction of crack movement and the stress distribution. Twist hackle frequently is present after an arrest line.

The gull wing is a feature that occurs due to the crack intersecting a pore or inclusion. As the crack travels around the inclusion, two crack fronts result. These do not always meet on the same plane on the opposite side of the inclusion or pore, resulting in a ridge where the two link up and again become a single crack front. In some cases, the ridge is immediately in the wake of the inclusion or pore and looks like a tadpole. In other cases, two ridges resembling a gull wing form in the wake.

A cantilever curl, or compression lip, occurs when the material is loaded in bending. The fracture initiates on the tensile side perpendic-

ular to the surface and exits on the compression side, no longer perpendicular to the surface. This is illustrated in Fig. 7. If the part were fractured under pure tension, the crack would be straight through the thickness and would thus exit at 90°. This information can be valuable in diagnosing the cause of fracture. For instance, thermal fractures of plate-shaped parts typically approach a stress state of pure tension near the point of origin and will not result in a compression lip. These same parts fractured mechanically will normally have some bend loading and will thus have a compression lip. Another example is a part containing prestressing or residual internal stresses. The crack will not pass straight through the thickness, but instead will follow contours consistent with the stress fields it encounters. A third example is strength testing. A problem with testing a ceramic in uniaxial tension is avoiding parasitic bend stresses. Examination of the fracture surface for signs of cantilever curl after tensile testing will help determine if pure tension was achieved or not.

Figure 8 shows the cross sections of typical specimens tested in four-point bending. Note the variations in the shape of the compression lip.

Techniques of Fractography

The techniques of fractography are relatively simple and the amount of sophisticated equipment minimal. Often, the information required to explain the cause of fracture of a component can be obtained with only a microscope and a light source. In fact, an experienced individual can sometimes explain the fracture just by examining the fracture surfaces visually. At other times, a variety of techniques, including sophisticated approaches such as SEM, electron microprobe, and Auger analysis, are required (Ref 10-12). When extensive fractography is necessary, the steps and procedures are as shown schematically in Fig. 9.

Step 1 involves visual examination of the fractured pieces and review of data regarding the test or service conditions under which the hardware failed. These data usually provide some hypotheses to guide the evaluation. Reconstructing the broken pieces and sketching probable fracture origins and paths is also helpful.

A primary objective of visual examination is to locate the point of fracture origin. This can be done using Wallner lines, hackle, and the fracture mirror, as described previously.

Visual examination determines the extent of additional evaluation that will likely be required. It also determines if cleaning procedures are necessary prior to microscopy. Generally, one must be careful not to handle or damage the fracture surface. The origin and features can be fragile, and key information explaining the fracture can be lost by improper handling. For instance, fingerprints can be

Fig. 11 SEM photomicrographs of fracture-initiating material flaws in reaction-bonded Si₃N₄ (RBSN)

The flaws shown are typical and consistent with the normal microstructure and strength. Arrows point to the fracture origins.

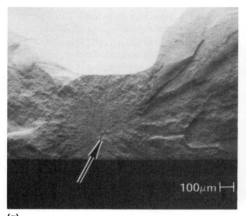

(a)

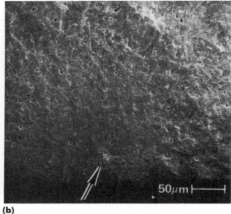

(b)

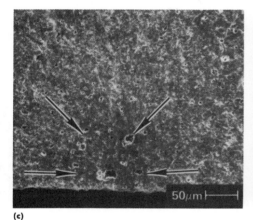

(c)

(d)

mistaken for Wallner lines. Debris on the surface can obscure the true fracture surface.

Cleaning can sometimes be accomplished with compressed air, but the source must be considered. Some compressors mix small amounts of oil with the air, which could produce a thin surface film that would later result in interpretation difficulties during SEM analysis. Compressed air should not be used unless it is known to be clean.

Ultrasonic cleaning in a clean solvent, such as acetone or methyl alcohol, is frequently used. Caution and judgment must be exercised, however, because ultrasonic cleaning is quite vigorous due to the cavitation action at the fluid/specimen interface and can damage the fracture surface. For instance, if fracture initiated at a low-density region or a soft inclusion, this material might be removed during ultrasonic cleaning and either prevent interpretation or lead to an erroneous interpretation.

It is apparent that cleaning and handling should be avoided unless absolutely necessary. Before trying other cleaning approaches, try a soft camel's hair brush.

Step 2 involves examination of the fracture surfaces under a low-power microscope. Usually, a binocular microscope with magnification up to 40× is adequate. Sometimes, higher power and special lighting or contrast features are required. Stereo photography may also be useful, since it accentuates the fracture surface features.

Preparation of replicas can also be useful; a replica often provides better resolution of the fracture features than does examination of the original part. Several methods are available for preparing replicas. A room-temperature method makes use of cellulose acetate, acetone, and polyvinyl chloride (PVC). A thin sheet of cellulose acetate is placed on a piece of PVC and submerged in acetone for about 15 s. It is then pressed against the fracture surface while being flooded with acetone and held for about 5 min. The acetone is then allowed to dry and the cellulose acetate replica peeled off.

Replicas can also be prepared with PVC alone. The specimen is heated and PVC pressed on with a Teflon rod. After cooling, the PVC replica is peeled off. This technique is quick,

but should not be used if there is a chance that heating of the specimen will alter the fracture surface.

The fracture origin can usually be located by low-power optical microscopy using either the original specimen or a replica, and an assessment can be made as to whether the fracture resulted from a material flaw or some other factor. However, details of the fracture origin, such as the nature of the flaw and interaction between the flaw and the microstructure, require higher magnification. Optical microscopes do not have adequate depth of focus at high magnification, so the scanning electron microscope must be used.

Examination of the fracture surface, in particular the fracture origin, by SEM is the third step in fractography. The scanning electron microscope provides extremely large depth of focus (compared to the optical microscope) and a range in magnification from around 10× to well over 10 000×. Most fractography is conducted between 25 and 5000×.

Scanning electron microscopy shows the difference between the fracture origin and the surrounding material and helps the engineer to develop a hypothesis of the cause of failure. One can frequently detect if the fracture initiated at a machining groove or at a material pore or inclusion. Scanning electron microscopy shows if the surface region is different from the interior and provides visual evidence of the nature or cause of the differences.

As with other techniques, SEM requires interpretation and must be used with caution. Because of the large depth of focus, it is sometimes difficult to differentiate between a ridge and a depression or to determine the angle of intersection between two surfaces. This can be better appreciated by comparing a single SEM photomicrograph with a stereo pair of the same surface. Only after doing this does one understand how easy it is to misinterpret a feature on an SEM photomicrograph.

Difficulty in interpretation is inevitable, but can be minimized if the engineer is present during the SEM analysis. This is especially true with respect to artifacts. An artifact is defined as extraneous material on the surface of the specimen. It can be a particle of dust or lint, a chunk of debris resulting from the fracture or handling, or a smeared coating resulting from oil contamination. If the engineer suspects that a feature is an artifact, he or she can instruct the SEM operator to look at it from different view angles and to examine surrounding areas in an effort to be sure.

As noted, SEM can usually locate the fracture origin and provide a photograph with a calibrated scale that allows accurate measurement of the size and shape of the flaw and the size of the fracture mirror. The engineer can then use Eq 1 and 3 to estimate the magnitude of the tensile stress that caused failure.

Once the fracture origin has been located, evaluation of the localized chemistry is often desirable. This leads to step 4 of fractography,

Fig. 12 SEM photomicrographs of abnormal fracture-initiating material flaws in RBSN traceable to improper processing prior to nitriding

(a) Large pore in slip-cast RBSN resulting from inadequate deairing. (b) Crack in greenware prior to nitriding. (c) and (d) Low-density regions in slip-cast RBSN resulting from agglomerates in the slip

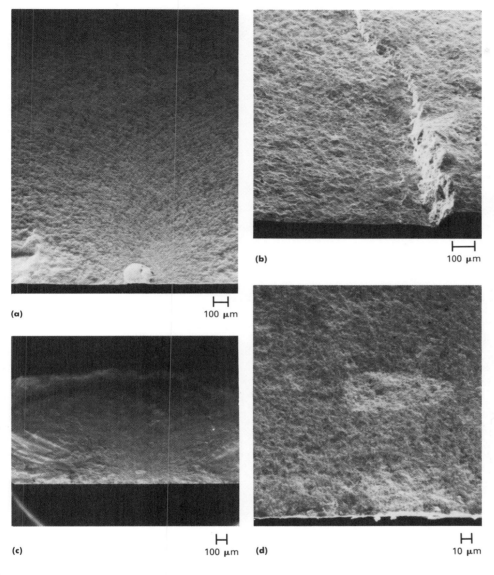

the use of instrumental techniques to conduct microchemical analysis. Most scanning electron microscopes have an energy-dispersive x-ray (EDX) attachment that permits chemical analysis of the x-rays that are emitted when the SEM electron beam excites the electrons within the material being examined. Each chemical element gives off x-rays under this stimulation that are characteristic of that element alone. These are detected by the EDX equipment and displayed as peaks by peripheral equipment. Comparison of the peak height of each element provides a semiquantitative chemical analysis of the microstructural feature being viewed.

Electron probe microanalysis works on the same principle as EDX, except an alternate x-ray detection mechanism is used that pro-vides better resolution and detects a wider range of elements. Energy-dispersive x-ray analysis cannot detect the lower-atomic-number elements.

Auger electron spectroscopy provides additional features for chemical analysis of microstructural features. It can remove surface material by sputtering while it is conducting chemical analysis and can thus determine changes in chemical composition as a function of depth. Auger electron spectroscopy is especially useful in cases where oxidation, corrosion, slow crack growth, or other intergranular effects are suspected. In some cases, fracture of specimens can be conducted in the Auger apparatus under high vacuum prior to conducting the chemical analysis. This allows examination of a fresh fracture surface

that has not had a chance to pick up contamination from the atmosphere and handling.

In addition to analysis of the fracture surface, other material tests can help determine the cause of failure. These include surface and bulk x-ray diffraction analysis, reflected-light microscopy of polished specimens and, in occasional cases, transmission electron microscopy. Detailed information on all of these techniques is available in Volume 10 of the 9th Edition of *Metals Handbook*.

There is no guarantee that the four steps of fractography will explain the cause of a failure and suggest a solution. However, it is still the most effective technology available and should be used routinely.

Determining Failure Cause

As mentioned above, determining the cause of failure is critical. It is obviously important in liability suits, where responsibilities for failure must be established, but it is also important for other reasons:

- To determine if failure is resulting from design or material limitations
- To aid in material selection or modification
- To guide design modifications
- To identify unanticipated service problems, such as oxidation or corrosion
- To identify material or material processing limitations and suggest direction for improvement
- To define specification requirements for materials and operating conditions

Figure 10 shows schematically how fractography interacts with other sources of information to determine the cause of failure and lead to action in the right direction to achieve a solution. It is imperative that all sources of data be considered and that the source of fracture be isolated to determine as quickly as possible whether it is design or materials oriented.

The following paragraphs review some of the common causes of fracture and describe the typical fracture surfaces that result.

Material Flaws. Flaws in a ceramic material concentrate stress. Fracture occurs when the concentrated stress at an individual flaw reaches a critical value that is high enough to initiate and extend a crack. Therefore, the first thing to look for on the fracture surface is whether there is a material flaw at the fracture origin. If a flaw, such as a pore or inclusion, is present, the engineer can do the following:

1. Compare the nature of the flaw with prior certification and service experience to determine if the flaw is intrinsic to the normal baseline material or is abnormal. In the latter case, a problem in the material fabrication process probably exists, and the help of the component manufacturer should be solicited

Fig. 13 SEM photomicrographs of abnormal fracture-initiating material flaws in RBSN traceable to nitriding process

(a) and (b) Porous aggregate rich in chromium and iron resulting from reaction of the silicon during the nitriding cycle with stainless steel contamination picked up during powder processing. (c) Energy dispersive x-ray analysis of (a) and (b). (d) Large aggregate of unreacted silicon resulting from localized melting due to local exothermic overheating. Arrows identify the fracture origins.

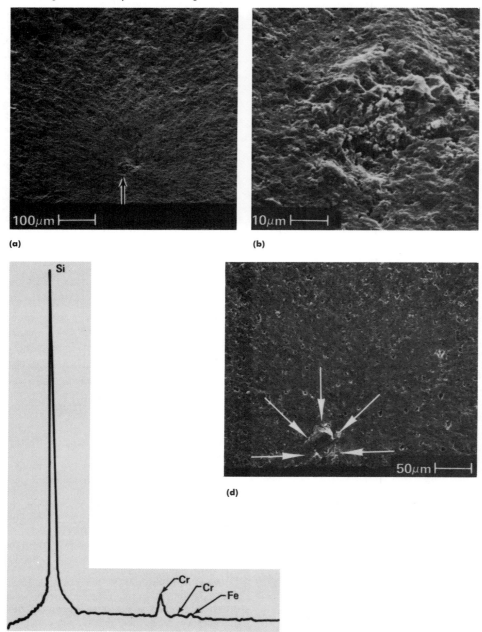

sure. Comparison of the chemical and physical nature of the flaw with the surrounding baseline microstructure will help in making this assessment

Figures 11 to 14 illustrate the types of intrinsic material flaws that can cause fracture and compare whether they are normal or abnormal for the material. Although the types of flaws vary depending on the material, the fabrication process and the specific step in the process in which they formed, it is relatively easy to distinguish between normal and abnormal flaws and thus to determine if errors in processing contributed to the failure (Ref 13).

Machining Damage. Surface flaws resulting from machining are a second common source of failure in ceramic components, especially in applications where high bend loads or thermal loads are applied in service. The most important flaws resulting from machining are median cracks and radial cracks.

The median crack is elongated in the direction of grinding and is like a notch in that fracture usually initiates over a broad front. A broad mirror usually results, but no flaw is readily visible because of the shallow initial flaw depth and because the median crack is perpendicular to the surface. The principal tensile stress is usually distributed such that the crack will extend in a plane perpendicular to the surface. If the flaw is perpendicular to the surface to start with, it will be in the same plane as the fracture and will be difficult to differentiate from the rest of the fracture mirror.

Figure 15 shows examples of specimens that fractured at transverse grind marks such that the median crack was likely the strength-determining flaw. Note the length of the surface involved in the fracture origin and the lack of a distinguishable flaw. Note also that the fracture origin is at a grinding groove and is elongated parallel to the direction of grinding.

Radial cracks produced by machining are roughly perpendicular to the direction of machining, are usually shallower than the median cracks, and are usually semicircular rather than elongated. The resulting fracture mirror is similar to one produced by a small surface pore or inclusion, but a well-defined flaw is not visible. However, by examining the intersection of the fracture surface with the original specimen surface at the center of the semicircular mirror under high magnification with the scanning electron microscope, the source can usually be seen and interpreted as machining damage. There are two features to look for. First, check to see if a grinding groove, especially an unusually deep one, intersects the fracture surface at the origin. Second, look for a small slightly out-of-plane region at the origin. This could indicate machining damage, but could also have other interpretations, such as contact damage or simply a tensile overload.

Machining damage often limits the strength of fine-grained ceramics and determines the

2. Measure the flaw size and/or mirror size and estimate the fracture stress σ_f using Eq 1 and 3. Compare this with the stresses estimated by design analysis and with fracture stress distribution projected for the material from prior strength certification testing. If the calculated fracture stress is within the normal limits specified for the material, a design problem should be suspected. If the fracture stress is below the normal limits specified for the material, a material problem is indicated

3. If the flaw is at or near the surface, assess whether the flaw is intrinsic to the material or resulted from an outside source, such as machining, impact, or environmental expo-

Fig. 14 SEM photomicrographs comparing normal and abnormal material flaws in sintered SiC

(a) and (b) Typical microstructure of high-strength material. (c) Large pore resulting from power agglomeration during powder preparation and shape forming. (d) Large grains resulting from improper control of temperature during sintering

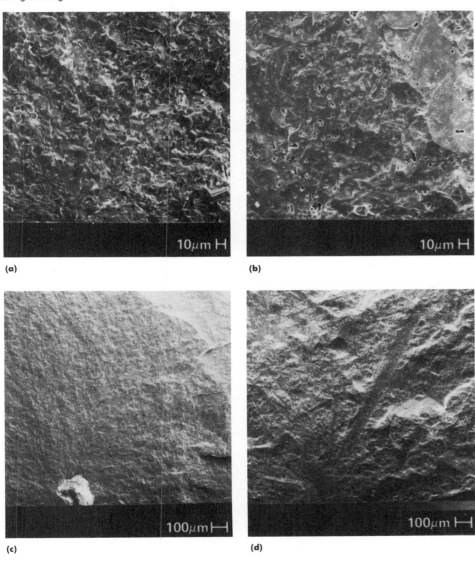

(a)

(b)

(c)

(d)

through the bottom surface. The fracture path and fracture surface features (Wallner lines and hackle) indicated that the surfaces were in compression and the interior was in tension and suggested that the problem was not linked to machining practice or material defects. Review of the processing history showed that the boron carbide was allowed to cool freely from 1950 to 1800 °C (3540 to 3270 °F) after hot pressing. This was followed by slow, controlled cooling thereafter. The free cooling was within the creep temperature range of the material, resulting in an effect comparable to tempering of glass, that is, formation of surface compression and internal tension. Based on this hypothesis, derived from fracture surface analysis alone, the hot-pressing operation was modified to permit slow cooling from 1950 to 1750 °C (3540 to 3180 °F). This eliminated the residual stress condition and permitted machining without cracking.

Thermal Shock. Little has been reported on identification of thermal shock fractures, and much additional work is needed. Therefore, the comments here will be fairly general, and the reader should be aware that the distinguishing features described may not always occur and may not be the best ones in a specific case. As in all other interpretive studies, the best approach is to apply known principles, thoroughly analyze all available data and options, and then make a decision, rather than depending on a cookbook procedure.

Thermal shock fractures tend to follow a wavy path with minimal branching and produce rather featureless fracture surfaces. This appears to be especially true for weak or moderate-strength materials, including both glass and polycrystalline ceramics. It is common for thermal shock cracks not to propagate all the way through the part. These characteristics suggest that a nonuniform stress field is present during thermal fracture; that is, the fracture initiates where the tensile stress is highest, follows stress or temperature contours (thus the wavy path), and stops when the stress drops below the level required for further extension or when a compressive zone is intersected. These conditions would also suggest a low crack velocity, which would account for lack of branching and lack of fracture surface features.

In line with the foregoing general considerations, several actual cases of fracture of material damage due to severe thermal transients are now analyzed.

The first case is the water quench thermal shock test for comparing the relative temperature gradient required to cause damage in simple rectangular test bars of different ceramic materials. The test bars are heated in a furnace to a predetermined temperature and then dropped into a controlled-temperature water bath. The bars generally do not break due to the quench. Instead, a large number of microcracks are produced. These become critical flaws that lead to fracture during subsequent bend testing;

measured strength distribution of certification specimens. Rice *et al.* (Ref 14) reported that hot-pressed Si_3N_4 specimens machined in the transverse direction fractured at machining flaws 98% of the time and that bars machined in the longitudinal direction failed at machining flaws greater than 50% of the time. This is similar to results for both Si_3N_4 and SiC (Ref 15-17).

Residual Stresses. Many ceramics have residual stresses, which usually result from the surface cooling faster than the interior after sintering or are due to chemical differences between the surface and interior. Often, the interior is under residual tension and the exterior is under compression. This can provide a strengthening effect if the material is used in

service in the as-fired, prestressed condition. However, most components require some finish machining. Once the compressive surface zone has been penetrated, the component is substantially weakened, often to the point of spontaneous crack initiation during machining.

Frechette (Ref 18) recounts a case where simple blanks of boron carbide cracked during machining. At the time, the cause of cracking was unknown, but improper machining was the primary suspect. Fracture analysis showed that the crack initiated at the point of machining and entered the material roughly perpendicular to the surface. However, the crack then quickly changed direction and propagated through the interior of the material parallel to the surface and finally changed direction again and exited

Fig. 15 SEM photomicrographs showing fractures initiating at transverse machining damage

(a) The fracture surface of a tensile specimen of hot-pressed silicon nitride that had been machined circumferentially. (b) The intersection of the fracture surface with the machined surface illustrating that the fracture origin is parallel to the grinding grooves. (c) and (d) The same situation for reaction-sintered silicon nitride

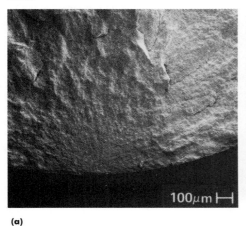

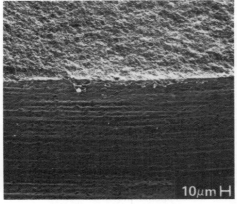

(a) | (b)

(c) | (d)

Fig. 16 SEM photomicrograph showing a typical featureless thermal shock fracture surface

(a) Overall surface at low magnification. (b) Fracture origin at higher magnification. Courtesy of Garrett Turbine Engine Company, Division of The Garrett Corporation

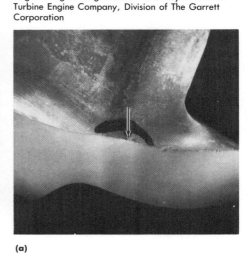

(a)

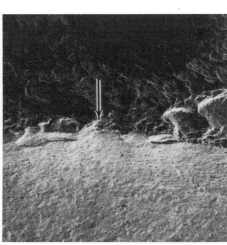

(b)

the size of the flaws is dependent on the material properties and the temperature change ΔT, and the residual strength is dependent on the size of these cracks. Further discussion of this test procedure and results on specific materials is available in Ref 19 and 20.

Ammann *et al.* (Ref 21) conducted cyclic fluidized bed thermal shock tests on wedge-shaped specimens of hot-pressed Si_3N_4. They reported that multiple surface cracking occurred and that the cracks grew in depth as a function of the total number of cycles.

Oxyacetylene torch thermal shock testing and gas turbine rig testing of reaction-bonded Si_3N_4 stator vanes has shown that thermal shock fractures of actual components under service conditions can be quite varied (Ref 22). Figure 16(a) shows a typical featureless fracture surface. The origin is shown by the arrow and the crescent-shaped ink mark and occurred in a region predicted by three-dimensional finite element analysis to have the peak thermal stress during rapid heat-up. No material flaw or machining damage is visible, even at high magni-

fication, as shown in Fig. 16(b). The crack apparently initiated at the material surface where the stress was maximum and propagated at moderate velocity without branching or making any abrupt changes in direction.

Figure 17(a) shows a thermal shock failure that initiated at the trailing edge of a stator vane airfoil. Finite element analysis also determined this position to be under high thermal stress during heat-up and steady-state service conditions. However, in this case the thermal shock fracture initiated at a preexisting material flaw, as shown more clearly in Fig. 17(b) at high magnification. The flaw was a penny-shaped crack nearly normal to the surface of the airfoil, but out of plane to the principal tensile stress. The fracture initiated at this crack due to the thermal stress and then quickly changed direction to follow the plane of maximum tensile stress. Hackle marks can be seen in the upper part of Fig. 17(a), which point to the vicinity of the origin.

The fracture surface shown in Fig. 17 is not what one would expect for thermal shock con-

ditions. It has relatively well-defined fracture features and could just as easily have been interpreted as a mechanical overload. Similarly, the fracture surface in Fig. 16 looks very much like fractures resulting from contact loading (discussed later). How does the engineer make the distinction? At our current state of knowledge, he or she does not, at least not based solely on fracture surface examination. The engineer needs other inputs, such as stress analysis, controlled testing (such as the calibrated oxyacetylene torch thermal shock tests), and a thorough knowledge of the service conditions. The engineer then needs to evaluate all the data concurrently and use his or her best judgment.

Impact can cause damage or fracture in two ways: (1) localized damage at the point of impact and (2) fracture away from the point of contact due to cantilevered loading. The former

Fig. 17 SEM photomicrograph of a thermal shock fracture initiating at a material flaw

(a) Overall surface at low magnification.
(b) Preexisting crack at fracture origin. Courtesy of Garrett Turbine Engine Company, Division of The Garrett Corporation

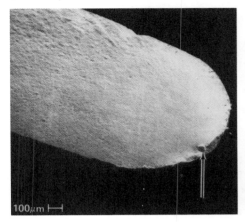

(a)

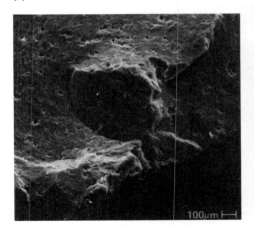

(b)

Fig. 18 Typical Hertzian cone crack resulting from impact and acting as the flaw that resulted in fracture under subsequent bend load

Shown at increasing magnification from (a) to (c). Arrow indicates fracture origin. Courtesy of Garrett Turbine Engine Company, Division of The Garrett Corporation

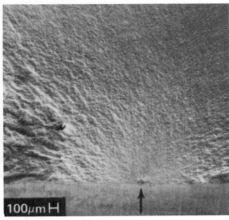

(a)

(b)

(c)

Fig. 19 Impact fracture of a ceramic rotor blade showing Hertzian cone crack

Courtesy of Garrett Turbine Engine Company, Division of The Garrett Corporation

(a)

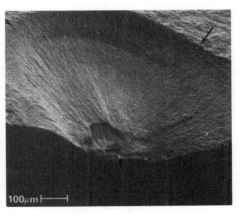

(b)

will have distinctive features and can usually be identified as caused by impact. The latter will appear like a typical bend overload (with a compressive lip on the exit end of the fracture) and can only be linked to impact by supporting data, such as location of the local damage at the point of impact.

The damage at the point of impact may be so little as to resemble a scuff mark on the ceramic or a smear of the impacting material (both cases referred to as a witness mark) or it may be as severe as complete shattering (Ref 23). The degree of damage depends on the relative velocity, mass, strength, and hardness of the impacting bodies. A baseball striking a plate glass window causes shattering; a BB striking a window only causes a series of concentric cone-shaped (Hertzian) cracks intersected by radial cracks. However, if the window shattered by the baseball could be reconstructed, it would also have conical and radial cracks.

The conical cracks are typical of impact and are thus a strong diagnostic fracture feature (Ref 24). Figure 18(a) shows a fracture surface of a polycrystalline ceramic where impact damage occurred first and was followed by fracture due to a bend load. The origin is indicated by an arrow and is easily located by observing the hackle lines and fracture mirror. Figure 18(b) and (c) show the origin at higher magnification. A distinct cone shape is present; the apex is at the surface where the impact occurred, and the fracture flared out as the crack penetrated the material.

Figure 19 shows another example of fracture due to impact. In this case, a ceramic rotor blade rotating at 41 000 rpm struck a foreign object and was fractured by the impact. Note the Hertzian crack extending from the fracture origin. Also note that additional damage is present at the origin, possibly a series of concentric cracks, providing evidence that the degree of impact was severe.

Fig. 20 Contact cracking in a ceramic

(a) and (b) Surface cracks resulting from relative movement between two contact surfaces under a high normal load and with a high coefficient of friction. (c) Typical multiple chipping resulting from contact loading and visible on a fracture surface

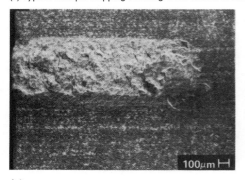

(a)

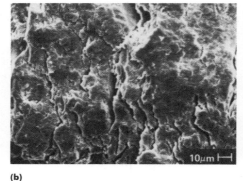

(b)

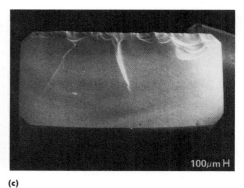

(c)

Fig. 21 Evidence of contact-initiated failure

(a) Witness mark on the surface of the ceramic adjacent to the fracture origin suggesting fracture due to contact loading. (b) and (c) Multiple cone features resulting from a contact fracture

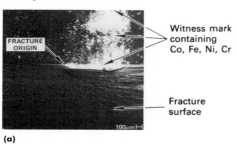

(a)

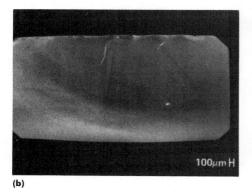

(b)

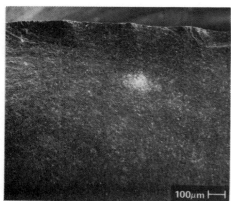

(c)

Fig. 22 SEM photomicrographs of the fracture surface of hot-pressed Si_3N_4 exposed to static oxidation for 24 h at 1100 °C (2010 °F)

(a) Overall fracture surface showing hackle marks and fracture mirror (the irregular dark spots on the fracture surface are artifacts). (b) Higher magnification showing the fracture mirror with an oxidation-corrosion pit at the origin. (c) Higher magnification showing the nature of the pit and the surface oxidation layer. Specimen size: 0.64 × 0.32 cm (0.25 × 0.125 in.). Source: Ref 16

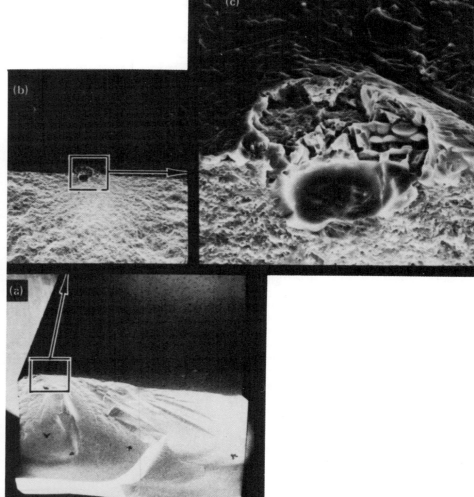

Fig. 23 SEM photomicrograph of the fracture-initiating oxidation-corrosion pit on the surface of reaction-bonded Si₃N₄

The EDX graph shows the relative concentration of chemical elements in the glassy region at the base of the pit. Courtesy of the Garrett Turbine Engine Company, Division of The Garrett Corporation

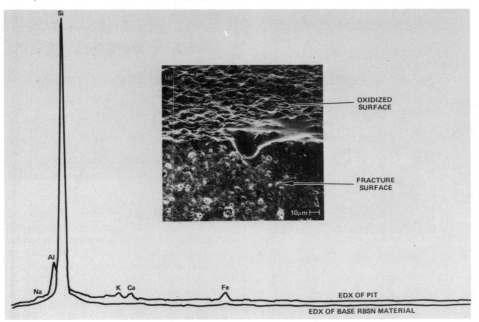

Fig. 24 SEM photomicrograph of hot-pressed Si₃N₄ that was exposed to combustion gases with sea salt additions showing that fracture initiated at the base of the glassy surface buildup

EDX analysis shows the chemical elements detected in the glassy material adjacent to the Si₃N₄. Courtesy of the Garrett Turbine Engine Company, Division of The Garrett Corporation

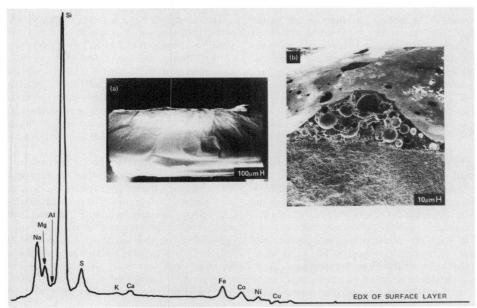

Biaxial contact refers to a situation where normal and tangential forces are being applied simultaneously at a ceramic surface. Examples where this might occur are numerous and include the following:

* Surface grinding

* Sliding contact (such as in bearings, seals, and many other applications)
* Applications involving a shrink fit
* Interfaces where the materials have different thermal expansion coefficients and operate under varying temperatures
* Any high-friction interface

Tensile stress concentration at a biaxially loaded interface is discussed in more detail in Ref 25 and 26. The fracture surface can be quite varied. If the contact is concentrated at a point, a Hertzian cone crack can form, and the resulting fracture surface will be similar to the one in Fig. 18. However, if the contact is more spread out, as in high-friction sliding contact, the damage will be spread over a larger surface area and will result in a relatively featureless fracture surface similar to the one shown in Fig. 16 for thermal shock fracture.

Distinguishing features that differentiate a contact failure from a thermal shock failure are just beginning to be defined (Ref 25) and appear to include the following:

* Contact cracks tend to enter the material surface at an angle other than 90°. An extreme case is shown in Fig. 20(a) for a ceramic specimen with a high-interference shrink fit axially loaded in tension
* Multiple parallel cracks often occur during a contact failure and can either be seen on the surface adjacent to the fracture origin or can show up as chips obscuring the fracture origin (Fig. 20b)
* Surface witness marks are often present at the fracture origin in cases of contact-initiated failure (Fig. 21a)
* One or more clamshell shapes, faint Hertzian cones, or pinch marks occur at the origin of some contact-initiated fracture surfaces (Fig. 21b)

Oxidation-Corrosion. Some types of oxidation or corrosion are easy to detect because they leave substantial surface damage that is clearly visible to the naked eye. In this case, the objective is to identify the mechanism of attack and find a solution. In other cases, especially where the oxidation or corrosion is isolated along grain boundaries, the presence and source of degradation may be more difficult to detect. In this case, the degree of attack may only be determined by strength testing, and the cause may be ascertained by controlled environment exposures and/or sophisticated instruments, such as Auger electron spectroscopy, which can detect slight chemical variations on a microstructural level.

Let us first examine some examples of oxidation and corrosion where visible surface changes have occurred. Figure 22 shows the surface and fracture surface of NC-132 hot-pressed Si₃N₄ after exposure in an SiC resistance-heated, oxide-refractory-lined furnace for 24 h at 1100 °C (2010 °F) (Ref 27). Figure 22(a) shows the complete cross section of the test bar. The fracture origin is at the surface on the left side of the photograph and is easily located by the hackle marks and the fracture mirror (the dark spots on the fracture surface are artifacts that accidentally contaminated the surface in preparing the sample for SEM). The specimen surface appears at low magnification to have many small spots that were not present prior to the oxidation exposure. At higher mag-

Fig. 25 SEM photomicrograph of reaction-bonded Si₃N₄ that was exposed to combustion gases with sea salt additions showing that fracture initiated at the base of the glassy surface buildup

EDX analysis shows the chemical elements detected in the glassy material adjacent to the Si₃N₄. Courtesy of the Garrett Turbine Engine Company, Division of The Garrett Corporation

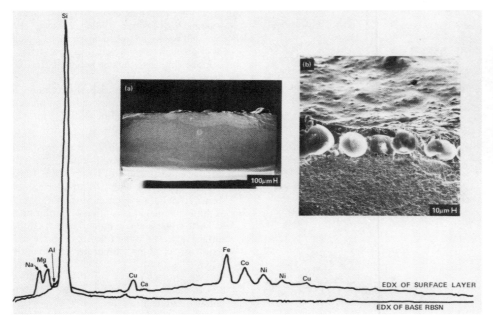

Fig. 26 SEM photomicrograph of the fracture surface of a low-purity Si₃N₄ material sintered with MgO and showing crack growth

Region of slow crack growth identified by arrows

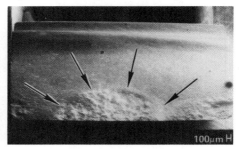

nification (Fig. 22b), these spots appear to be blisters or popped bubbles and one is precisely at the fracture origin. Still higher magnification (Fig. 22c) reveals that a glass-filled pit is at the base on the center of the blister. It also reveals that a surface layer less than 5 μm thick covers the specimen and that this layer appears to be partially crystallized.

By simply examining the specimen surface, especially the intersection of the oxidized surface and the fracture surface, we have obtained much insight into both the nature and sequence of oxidation. What else can we do to obtain further information? We can compare the strength of the oxidized specimen with that of unoxidized material. In this specific case, the oxidation exposure resulted in a reduction in strength from 669 MPa (97 ksi) to 497 MPa (72 ksi). We can also compare x-ray diffraction and chemical analyses for the original surface, the oxidized surface, and the bulk material. In this case, the oxidized surface contained much more magnesium and calcium than the original surface or the bulk material. Energy-dispersive x-ray analysis verified that the glassy material in the pit also had high concentrations of calcium and magnesium. X-ray diffraction revealed crystallized cristobalite (SiO₂) plus magnesium silicate and calcium magnesium silicate phases in the oxide layer. No sign of magnesium or calcium contamination was detected in the furnace.

Simultaneous evaluation of all the data led to a plausible hypothesis of the mechanism of oxidation degradation. Magnesium and calcium, present as oxide or silicate impurities in

the Si₃N₄, were diffusing to the surface, where they reacted with SiO₂ that was forming simultaneously at the surface from reaction of the Si₃N₄ with oxygen from the air. The resulting silicate compositions apparently locally increased the solubility or oxidation rate of the Si₃N₄. The reason for the formation of isolated pits was not determined, but could have resulted from impurity segregation or other factors and would have required additional studies to determine.

A similar example of static oxidation for reaction-bonded Si₃N₄ is illustrated in Fig. 23. In this case, the exposure was for 2 h at 1350 °C (2460 °F) plus 50 h at 900 °C (1650 °F) (Ref 13). Only isolated pits were present on the surface, and these appeared to occur where small particles of the furnace lining had contacted the specimen during exposure. The EDX analysis included in Fig. 23 was taken in the glassy region at the base of the pit, showing that aluminum, silicon, potassium, calcium, and iron were the primary elements present and again indicating a propensity for Si₃N₄ to be corroded by alkali silicate compositions. However, it should be noted that the size of the pit is much smaller than in the prior example and resulted in only a small strength decrease.

Figures 24 and 25 show examples of more dramatic corrosion of hot-pressed and reaction-bonded Si₃N₄ (Ref 27) resulting from exposure to the exhaust gases of a combustor burning jet fuel and containing a 5-ppm addition of sea salt. Exposure consisted of 25 cycles of 900 °C (1650 °F) for 1.5 h, 1120 °C (2050 °F) for 0.5 h, and a 5-min air quench. At 900 °C (1650

°F), Na₂SO₄ is present in liquid form and deposits along with other impurities on the ceramic surface. The EDX analyses taken in the glassy surface layer near its intersection with the Si₃N₄ document the presence of impurities, such as sodium, magnesium, and potassium from the sea salt; sulfur from the fuel; and iron, cobalt, and nickel from the nozzle and combustor liner of the test rig. An EDX analysis for the Si₃N₄ on the fracture surface about 20 μm beneath the surface layer is also shown in Fig. 25. Only silicon is detected (nitrogen and oxygen are outside the range of detection by EDX), indicating that the corrosion in this case resulted from the impurities in the gas stream plus the surface oxidation.

The strength of the hot-pressed Si₃N₄ exposed to the dynamic oxidation with sea salt additions decreased to an average of 490 MPa (71 ksi) from a baseline of 669 MPa (97 ksi). The reaction-bonded material decreased to 117 MPa (17 ksi) from a baseline of 248 MPa (36 ksi). Repeating the cycle with fresh specimens and no sea salt resulted in an increase to 690 MPa (100 ksi) for the hot-pressed Si₃N₄ and only a decrease to 207 MPa (30 ksi) for the reaction-bonded Si₃N₄.

The examples presented so far for oxidation and corrosion have had distinct features that help distinguish the cause of fracture from other mechanisms, such as impact or machining damage. Some corrosion-initiated fractures are more subtle. The corrosion or oxidation may only follow the grain boundaries and may be so thin that it is not visible on the fracture surface. Its effect may not even show up in room-temperature strength testing since its degradation mechanism may only be active at high temperature. How do we recognize this type of corrosion? The following suggestions may be helpful:

- Prepare a polished section of the cross section and try various etchants; this may enhance the regions near the surface where intergranular corrosion is present
- Conduct EDX, microprobe, or Auger anal-

ysis scans from the surface inward to determine if a composition gradient is present
- Use high-magnification SEM of the fracture surface to look for differences between the microstructure near the surface and in the interior; if the fracture surface near the specimen surface is intergranular and near the interior is transgranular, grain-boundary corrosion is a possibility
- Conduct controlled exposures under exaggerated conditions in an effort to verify if the material is sensitive to attack

Slow crack growth can occur under sustained loading and also under relatively fast loading, depending on the nature of the material, the temperature, the atmosphere, and the load. Figure 26 shows the fracture surface of a sintered Si_3N_4 material (which was developed for low- to moderate-temperature applications) after four-point bend testing at 980 °C (1800 °F) at a load rate of 0.05 cm/min (0.02 in./min). In spite of the rapid loading, substantial slow crack growth occurred. Examination of the fracture surface quickly tells the engineer that this material is not suitable for high-temperature application under a tensile load.

Sometimes, examination of the fracture surface by EDX or another surface chemical analysis technique can help identify the cause of slow crack growth. Specifically, the roughened region is analyzed separately from the rest of the fracture surface and bulk material. Chemical elements present in greater concentration in the slow-crack-growth region are probably associated with the cause. The tensile stress at fracture can also be approximated by assuming that the flaw size is equivalent to the slow-crack-growth region and by using Eq 2. However, it should be realized that this is only an approximation and that the reported elastic modulus and fracture energy values for the material, when used in Eq 2, may not be good approximations for the material under slow-crack-growth conditions.

There are other limitations to the information available from the fracture surface. The size of the slow-crack-growth region provides no information about the time to failure, the rate of loading, or the mode of loading (cyclic versus static).

REFERENCES

1. F.W. Preston, Angle of Forking of Glass Cracks as an Indicator of the Stress System, *J. Am. Ceram. Soc.*, Vol 18, 1935, p 175
2. V.D. Frechette, "Fracture of Heliostat Facets," The ERDA Solar Thermal Projects Semiannual Review, Energy Research and Development Agency, Washington, Aug 23-24, 1977
3. J.J. Mecholsky, S.W. Freiman, and R.W. Rice, in *Fractography and Failure Analysis*, STP 645, D.M. Strauss and W.H. Cullen, Jr., Ed., ASTM, Philadelphia, 1978, p 363-379
4. N. Terao, *J. Phys. Soc. Jpn.*, Vol 8, 1953, p 545
5. W.C. Levengood, *J. Appl. Phys.*, Vol 29, 1958, p 820
6. E.B. Shand, Breaking Stress of Glass Determined From Dimensions of Fracture Mirrors, *J. Am. Ceram. Soc.*, Vol 42, 1959, p 474
7. H.P. Kirchner and R.M. Gruver, Fracture Mirrors in Polycrystalline Ceramics and Glass, in *Fracture Mechanics of Ceramics*, Vol 1, R.C. Bradt, D.P.H. Hasselman, and F.F. Lange, Ed., Plenum Press, 1974, p 309-321
8. A.A. Griffith, The Phenomenon of Rupture and Flow in Solids, *Philos. Trans. R. Soc. (London) A*, Vol 221 (No. 4), 1920, p 163
9. A.G. Evans and G. Tappin, *Proc. Br. Ceram. Soc.*, Vol 20, 1972, p 275-297
10. R.W. Rice, Fractographic Identification of Strength-Controlling Flaws and Microstructure, in *Fracture Mechanics of Ceramics*, Vol 1, R.C. Bradt, D.P.H. Hasselman, and F.F. Lange, Ed., Plenum Press, 1974, p 323-345
11. H.L. Marcus, J.M. Harris, and F.J. Szalkowsky, Auger Spectroscopy of Fracture Surfaces of Ceramics, in *Fracture Mechanics of Ceramics*, Vol 1, R.C. Bradt, D.P.H. Hasselman, and F.F. Lange, Ed., Plenum Press, 1974, p 387-398
12. O. Johari and N.M. Parikh, in *Fracture Mechanics of Ceramics*, Vol 1, R.C. Bradt, D.P.H. Hasselman, and F.F. Lange, Ed., Plenum Press, 1974, p 399-420
13. K.M. Johansen, D.W. Richerson, and J.J. Schuldies, "Ceramic Components for Turbine Engines," Report 21-2974(08), Phase II Final Report, Prepared under Air Force Contract No. F33615-77-C-5171, AiResearch Manufacturing Company of Arizona, Phoenix, Feb 29, 1980
14. R.W. Rice, S.W. Freiman, J.J. Mecholsky, and R. Ruh, Fracture Sources in Si_3N_4 and SiC, in *Ceramic Gas Turbine Demonstration Engine Program Review*, J.W. Fairbanks and R.W. Rice, Ed., MCIC-78-36, Metals and Ceramics Information Center, Columbus, OH, 1978, p 665-688
15. D.W. Richerson, T.M. Yonushonis, and G.Q. Weaver, Properties of Silicon Nitride Rotor Materials, in *Ceramic Gas Turbine Demonstration Engine Program Review*, J.W. Fairbanks and R.W. Rice, Ed., MCIC-78-36, Metals and Ceramics Information Center, Columbus, OH, 1978, p 193-217
16. T.M. Yonushonis and D.W. Richerson, Strength of Reaction-Bonded Silicon Nitride, in *Ceramic Gas Turbine Demonstration Engine Program Review*, J.W. Fairbanks and R.W. Rice, Ed., MCIC-78-36, Metals and Ceramics Information Center, Columbus, OH, 1978, p 219-234
17. W.D. Carruthers, D.W. Richerson, and K. Benn, *3500 Hour Durability Testing of Commercial Ceramic Materials*, NASA CR-159785, Interim Report, National Aeronautics and Space Administration, Washington, July 1980
18. V.D. Frechette, Fractography and Quality Assurance of Glass and Ceramics, in *Quality Assurance in Ceramic Industries*, V.D. Frechette, L.D. Pye, and D.E. Rase, Ed., Plenum Press, 1979, p 227-236
19. R.W. Davidge and G. Tappin, *Trans. Br. Ceram. Soc.*, Vol 66, 1967, p 8
20. G.Q. Weaver, H.R. Baumgartner, and M.L. Torti, Thermal Shock Behavior of Sintered Silicon Carbide and Reaction-Bonded Silicon Nitride, in *Special Ceramics*, Vol 6, P. Popper, Ed., British Ceramic Research Association, England, 1975, p 261-281
21. C.L. Ammann, J.E. Doherty, and C.G. Nessler, *Mater. Sci. Eng.*, Vol 22, 1976, p 15-22
22. K.M. Johansen, L.J. Lindberg, and P.M. Ardans, "Ceramic Components for Turbine Engines," Vol 10, Report 21-2794, 8th Interim Report, prepared under Air Force Contract No. F33615-77-C-5171, AiResearch Manufacturing Company of Arizona, Phoenix, June 5, 1980
23. "Ceramic Gas Turbine Engine Demonstration Program Interim Report No. 11," Report 76-212188(11), prepared under Contract N00024-76-C-5352, AiResearch Manufacturing Company of Arizona, Phoenix, Nov 1978, p 3-47 to 3-64
24. J.M. Wimmer and I. Bransky, Impact Resistance of Structural Ceramics, *Ceram. Bull.*, Vol 56 (No. 6), 1977, p 552-555
25. D.G. Finger, "Contact Stress Analysis of Ceramic-to-Metal Interface," Final Report, Contract N00014-78-C-0547, Garrett Turbine Engine Co., Phoenix, Sept 1979
26. D.W. Richerson, W.D. Carruthers, and L.J. Lindberg, Contact Stress and Coefficient of Friction Effects on Ceramic Interfaces, in *Surfaces and Interfaces in Ceramic and Ceramic-Metal Systems*, J.A. Pask and A.G. Evans, Ed., Plenum Press, 1981
27. D.W. Richerson and T.M. Yonushonis, Environmental Effects on the Strength of Silicon Nitride Materials, in *Ceramic Gas Turbine Demonstration Engine Program Review*, J.W. Fairbanks and R.W. Rice, Ed., MCIC-78-36, Metals and Ceramics Information Center, Columbus, OH, 1978, p 247-271

Failure Analysis of Polymers

Abdelsamie Moet, Department of Macromolecular Science, Case Western Reserve University

POLYMERS are increasingly used in load-bearing structural applications. Frequently, they replace metals and other conventional engineering materials, and in many other cases, their properties create unique applications. Before discussing failure analysis of polymers, it is important to consider their basic molecular and morphological nature. In this respect, this article will highlight structural and mechanical characteristics that discriminate polymers as strong materials from metals. Subsequently, essentials of failure mechanisms will be examined, followed by examples of field failure. The fundamentals of fracture, such as toughness and fracture mechanics, ductile-to-brittle transition, fatigue, wear, fretting, and stress-corrosion cracking generally apply to strong materials, including polymers. More information on these topics can be found in the articles "Failure Analysis and Fracture Mechanics," "Ductile and Brittle Fractures," "Fatigue Failures," "Wear Failures," and "Stress-Corrosion Cracking" in this Volume.

Polymer Structure and Morphology

The special properties of polymers stem from the fact that they are composed of monomers that link together and form long chains. These long-chain molecules are sometimes branched or cross linked. A typical linear polymer compromises a backbone along which side groups are arranged at regular intervals. From a manufacturing point of view, there are two broad classes of polymers: thermoplastics, which can be melted and resolidified (such as polyethylene and polycarbonate, and thermosetting polymers, in which irreversible chemical changes occur during processing (such as vulcanized rubbers, epoxies, and phenol formaldehyde). The physical state of a polymer under a given condition is primarily governed by chain flexibility, which in turn is determined by its backbone and the size and shape of the side groups and their spatial organization. Details of these phenomena are simplified in Ref 1 to 3.

One of the most important differences between polymers and other materials is the statistical nature of their molecular chains. A

Fig. 1 Crazing in a thin polystyrene film

(a) Optical micrograph showing crazes developed under monotonic tensile loading applied perpendicular to the craze-propagation direction (arrow). 100×.
(b) TEM micrograph of a single craze. Oriented fibrillar material spans the craze boundaries. 5000×

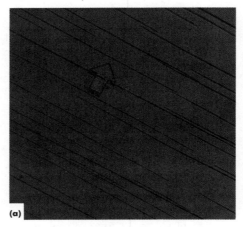

(a)

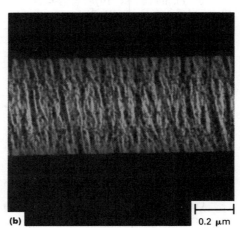

(b)   0.2 μm

distribution of chain lengths is a consequence of the random nature of most polymerization reactions. Thus, polymers are generally characterized by a molecular-weight distribution and the associated averages. A number-average molecular weight, which is the first moment of the number distribution, and a weight-average mo-

lecular weight, which is the second moment, are commonly employed to characterize a polymer.

Polymers exist in either of two physical states: amorphous, such as polystyrene and poly(methylmethacrylate), or semicrystalline, such as polyethylene and nylon. Fracture behavior is thus influenced by chain packing in amorphous polymers and by the average crystallite size and degree of crystallization in semicrystalline polymers. Relatively high mobility of chain molecules, even at room temperature, causes structural changes with time, a phenomenon known as physical aging. Some dimensional changes of an engineering component and functional property changes may be anticipated as a result of physical aging of certain polymers. Diffusion of certain liquids and vapors, particularly organics, may cause specific polymers to swell or shrink, which leads to dimensional changes and deterioration of their load-bearing capability.

Viscoelasticity of Polymers

Under circumstances in which metals may be regarded as purely elastic, polymers may deform nonelastically, that is, in a viscoelastic fashion. Their stress-strain response is nonlinearly time dependent; consequently, their fracture behavior is strongly influenced by the rate of load application. In addition, the considerable stress relaxation displayed by many polymers must be considered when assessing service loads in failure analysis. Another consequence of the viscoelasticity of polymers becomes obvious when fatigue loading is involved, particularly at high frequency. Hysteretic viscoelastic effects cause energy dissipation as heat, which can accumulate due to the low thermal conductivity of polymers, thus influencing failure mechanisms.

Deformation and Fracture Mechanisms

In most cases, failure of load-bearing structural components involves fracture, that is,

Fig. 2 TEM micrographs showing the deformation behavior of thin polystyrene film cast from a binary molecular-weight mixture

(a) Pure low molecular weight material deforms by microcracking. 130 000×. (b) The addition of 0.05% high molecular weight material causes sparse weak fibrils to span the microcrack. 44 000×.
(c) Numerous stronger fibrils develop a craze upon addition of 1.0% high molecular weight material. 44 000×

(a)

0.1 μm

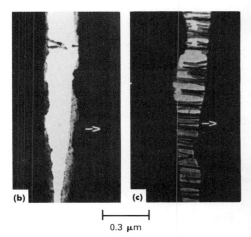

(b)   (c)

0.3 μm

Fig. 3 Section from a polystyrene sample that was deformed past its compressive yield

The section is viewed between cross polars, showing shear bands. 50×. Source: Ref 10

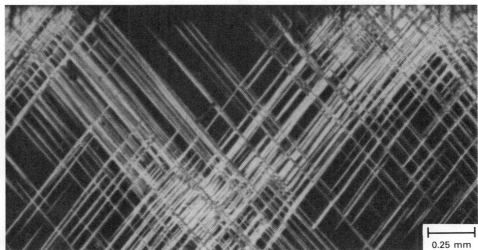

0.25 mm

Fig. 4 SEM micrograph showing fracture mechanisms in short glass fiber reinforced nylon

Fiber debonding, fiber fracture, and matrix deformation are evident. Courtesy of N. Sato

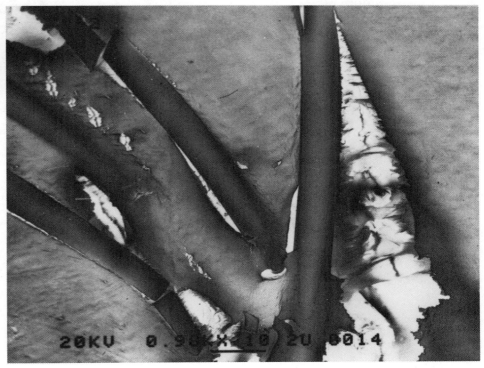

complete or partial separation of a critical member of the component under service loading that renders the component nonfunctional. The role of failure analysis, therefore, is to reconstruct the sequence of processes leading to failure, aiming at determining the primary cause. This effort requires knowledge of the various mechanisms by which the material responds to a loading environment similar to that under which a failure occurred. Frequently, laboratory experiments must be conducted to establish such cause and effect relationships. Again, failure mechanisms observed in a particular field failure should bear specific similarity to that observed in the laboratory to render comparison valid.

Fracture of load-bearing engineering components is generally a macroscopic phenomenon resulting from a series of microscopic and submicroscopic processes. Although molecular theories of polymer fracture are being advanced (Ref 4), less ambiguous fracture analysis is achievable from examining deformation events within the resolution of optical and scanning electron microscopy (SEM). Irreversible deformation mechanisms in polymers may fall into two basic categories: dilatational, such as crazes, voids, and

microcracks, or nondilatational, such as shear bands. Commonly, a mixture of both mechanisms may be encountered in particular polymeric materials (Ref 5, 6) or during certain loading conditions (Ref 6).

Crazing. Traditionally, crazes are believed to occur in type I (normally brittle in tension)

amorphous polymers, such as polystyrene and poly(methylmethacrylate), under monotonic tensile loading. Crazes, however, have been observed in semicrystalline polymers, such as polypropylene (Ref 6), and in type II (normally ductile in tension) amorphous polymers, such as polycarbonate (Ref 6), under tensile-fatigue

Fig. 5 Surface-microcracking network developed on a polyoxymethylene surface due to ultraviolet exposure

200×. Source: Ref 15

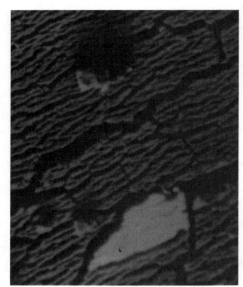

Fig. 6 Time-to-failure of high-density polyethylene pipes at different stresses and temperatures

Source: Ref 4

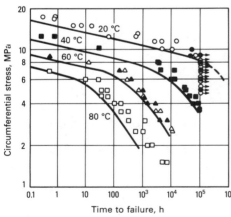

Fig. 7 A thinned section of fatigue-cracked polypropylene specimen

Crazes are visible surrounding and preceding the crack. 8×. Source: Ref 18

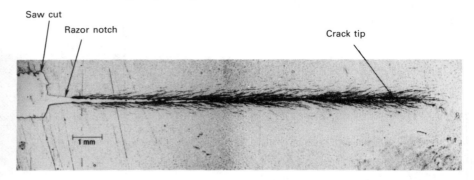

Saw cut

Razor notch

Crack tip

1 mm

Fig. 8 Fatigue-crack initiation in polystyrene from a V-notch

Note crazes surrounding and preceding the crack. 75×. Source: Ref 19

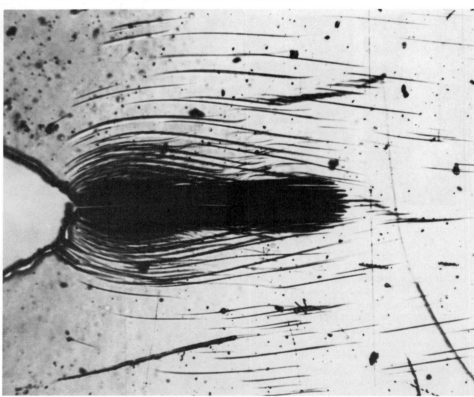

loading. Under monotonic tensile loading, polycarbonate is reported to deform by shear banding.

A craze is a slitlike microcrack (Fig. 1a) spanned by highly oriented fibrillar materials (Fig. 1b). Its width is of the order of 1 to 2 μm, and it may grow to several millimeters in length, depending upon its interaction with other heterogeneities. Being dilatational, crazes grow normal to the applied tensile component of the stress field. Craze fibrils are load-bearing elements whose strength and density depend to some extent on the molecular weight of the polymer (Fig. 2). Higher molecular weight polymers develop fewer, longer, and stronger, that is, stable, crazes. Although crazing is commonly associated with brittleness, higher molecular weight polymers have higher resistance to fracture. Implications of this phenomenon have been exploited commercially in the development of toughened polymers (Ref 7).

Shear-Deformation Bands. Like crazes, shear-deformation bands, or slip lines, are traditionally thought to be the mechanism of irreversible tensile deformation in type II amorphous polymers (Ref 5). Almost invariably, however, a compressive-stress state will cause shear deformation in polymers (Ref 8, 9). Shear bands (Fig. 3) are also microscopic localized deformation zones that propagate ideally along shear planes. They can be sharp or diffuse (Ref 5, 8). The nature and degree of orientation of the sheared material within a band remains controversial (Ref 8, 11, 12), although a strain magnitude reaching two to three is commonly accepted.

Other Deformation Processes. Yielding in a solid polymer under tensile force involves necking caused by extensive shear deformation, for example, polycarbonate, which may also be coupled with increased crystallization and chain

Fig. 9 Transmitted-light micrograph showing a yielded zone surrounding and preceding a fatigue crack in 0.25-mm (0.01-in.) thick polycarbonate sheet

20×

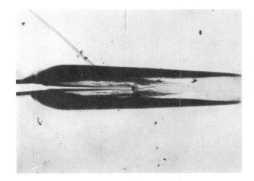

orientation, as in poly(ethyleneterephthalate) and polyethylene. Alternatively, yielding may occur by profuse crazing, as in high-impact polystyrene. Voids formed within small profuse crazes associated with the irreversible deformation and fracture of polymers scatter incident light, giving rise to the phenomenon known as stress whitening.

Deformation of Fiber Composites. The incorporation of high-modulus fibers within a polymer matrix causes a complex mixture of fracture-resistance mechanisms. These include matrix crazing, cracking, and shear deformation, in addition to fiber pull-out, debonding, and fracture (Fig. 4). The relative contribution of each mechanism depends on the specific nature of the fibers, their orientation, the polymer matrix, and the coupling agent.

Environmental Stress Cracking

Liquid Environments. A load-bearing structure is likely to serve in contact with environmental fluids. As organic materials, polymers are naturally soluble in organic fluids. A practical measure of the interaction between liquids and polymers is their solubility parameters. Expressed as the square root of Joules per cubic meter ($\sqrt{J/m^3}$), the solubility parameter is defined as the square root of the cohesive energy density (Ref 11, 12). The closer the solubility parameter of a liquid to that of a specific polymer, the stronger their interaction will be. This interaction is generally employed to assess stress crazing and cracking effects in polymers (Ref 13). Nonsolvents that possess specific physiochemical affinity leading to wetting of the polymer are also known to cause premature failure (Ref 9). However, stress-cracking effects on polymers are more complex and are currently subject to investigation. Environmentally enhanced failure generally involves crazing or cracking as the underlying mechanism.

Degradation by Heat and Light. A serious consideration in failure analysis of polymers is the fact that their mechanical properties may severely deteriorate by exposure to heat and/or light. Thus, structural components serving at elevated temperatures or in outdoor applications may undergo degradation if the polymers employed are not properly stabilized. Mechanical stress is known to enhance degradative effects. Degradation usually involves molecular-weight reduction by one of several mechanisms (Ref 14). Surface embrittlement and consequent microcracking occur. This promotes crack initiation and possibly assists crack propagation. Figure 5 illustrates surface microcracking induced in a polyoxymethylene specimen exposed to ultraviolet (UV) light in the laboratory for 1000 h. It should be noted, however, that some polymers are intrinsically more resistant to degradation than others. The following common polymers are ranked with respect to their relative resistance to the deterioration of mechanical properties due to photodegradative effects:

Poly(methylmethacrylate)	n
Poly(acrylonitrile)	n
Polyoxymethylene	m
Polyethylene	m
Poly(vinylchloride)	n
Polystyrene	w
Nylon 6	m
Polypropylene	vs
Polycarbonate	s
Polyurethane	m
Polysulphone	s
Poly(phenyleneoxide)	s

n, not significant deterioration; m, moderate; w, weak; s, severe; vs, very severe

Brittlelike Fracture

In terms of their fracture behavior, polymers are generally classified as brittle or ductile. Brittle polymers are those that are known to fracture at relatively low elongations in tension (2 to 4%). These include polystyrene, poly(methylmethacrylate), and rigid (unplasticized) poly(vinylchloride) (PVC). Crazing is the dominant mechanism of failure in such polymers. Highly cross linked polymers, such as epoxies and unsaturated polyesters, are also brittle, yet their fracture involves a microcracking mechanism.

On the other hand, semicrystalline polymers, such as polyethylene and nylons, and some amorphous polymers, such as polycarbonate and poly(ethyleneterephthalate), exhibit considerable post-yield plastic deformation and are thus classified as ductile. In spite of this classification, however, relatively low loads applied over long periods of time are reported to cause failure to occur even in the most ductile polymers. Indeed, early investigations on polyethylene, a ductile polymer (Ref 4), have shown that the failure mechanism switches from ductile to brittle at longer times and lower stress levels (Fig. 6).

In view of the fact that service loads in most engineering components are commonly taken as a small fraction of the yield stress, brittlelike fracture of polymers should be a major concern in their failure analysis. This notion perhaps underlies recent interest in applying fracture mechanics theory to study polymer failure (Ref 6, 16, 17).

Fig. 10 Reflected-light optical micrograph of the fracture surface of medium-density polyethylene pipe

Arrows indicate the direction of crack propagation. Source: Ref 22

4.5 mm

Fig. 11 Fracture band width as a function of crack length for the polyethylene pipe shown in Fig. 10

Source: Ref 22

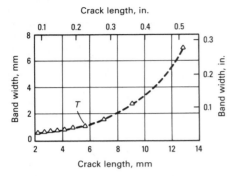

Fig. 12 Normalized fracture band width from Fig. 10 compared to that of a laboratory fatigue failure

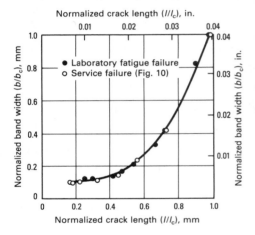

Crack Propagation in Polymers

On a macroscopic level, the crack-propagation behavior of polymers resembles that in metals. Microscopically, however, a propagating crack in a polymer is usually preceded by a zone of transformed material. This zone, commonly referred to as the plastic zone, involves any of the known irreversible deformation mechanisms. In most cases, a zone of crazes precedes crack propagation, as in polypropylene (Fig. 7) or in polystyrene (Fig. 8). It is important to note that polypropylene is a semicrystalline polymer that usually deforms by yielding and recrystallization under monotonic loading. The transformation at the crack tip may appear as a yielded zone, as in thin polycarbonate sheet material (Fig. 9). A pair of shear bands are also reported to evolve at the tip of propagating cracks in polycarbonate (Ref 20). Recently, it was observed that when the thickness of the polycarbonate sheet is 6 mm (0.25 in.) or more, brittle microcracking ahead of the propagating crack becomes the dominant

Fig. 13 Fracture in a PVC water filter

The fracture surface of the fatigue crack started from a fissure (arrow F). The lower dark zone is an artifact due to sectioning of the filter wall. 75×

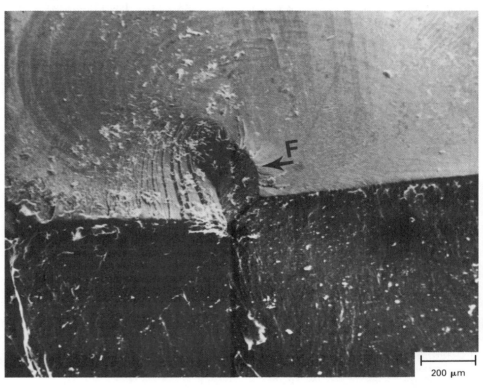

200 μm

mechanism of fracture. Thus, one polymer displays more than one fracture mechanism. In all cases, the crack appears to propagate through a craze at its very tip.

The magnitude of transformation preceding the crack in a particular polymer is influenced by the loading history and consequently determines the resistance to crack propagation (Ref 6, 17-19). Recent investigations show that damage evolution ahead of the crack tip accounts for the discrepancies in fracture toughness of polymers (Ref 21). These results suggest that examination of the material above and below the crack-propagation plane should be considered in failure analysis in addition to the commonly accepted fracture-surface studies. As was briefly shown above, vital information about the resistance of the material to crack propagation and its loading history can easily be decoded from a thinned section normal to the crack-propagation plane.

Case Studies

Several cases of field failure in various polymers will be considered to illustrate the applicability of available analytical tools in conjunction with an understanding of failure mechanisms.

Example 1: Failure of an Irrigation Pipe. Figure 10 shows an optical micrograph of the fracture surface of an irrigation pipe made of medium-density polyethylene that failed in

service (Ref 22). This pipe was subjected to severe cyclic-bending strain of the order of 6% while under combined tensile stress of about 6.9 MPa (1000 psi) and a hoop stress of the order 6.2 MPa (900 psi). These conditions of operation were far more stringent than those encountered in most applications of polyethylene pipes. A subsurface imperfection in the pipe wall (dark diamond-shaped spot, Fig. 10) acted as a crack starter. Contrary to the dominant belief that pipe failure initiates from surface defects, this example indicates that a critical-size flaw within the pipe wall can also initiate failure. This (flaw) crack starter was located closer to the outside wall, where compressive residual stresses may be dominant.

Concentric circular striations originate from the crack starter and grow simultaneously in radial and circumferential directions. It is well established that such striations represent crack-arrest lines, where the distance between two striations (a band) is due to a crack excursion. Thus, evolution of the band width reflects the nature of crack propagation. The band width measured from larger micrographs is plotted as a function of crack length in Fig. 11. A smooth transition is observed at a point T as catastrophic failure (pipe separation) is approached. This transition is indicative of considerable increase in crack speed and coincides with a transition in band geometry from circular to elliptical. Plausibly, this occurred when the crack-tip stress field interacted with the inside

Fig. 14 Optical micrograph showing a fractured automobile stick-shift

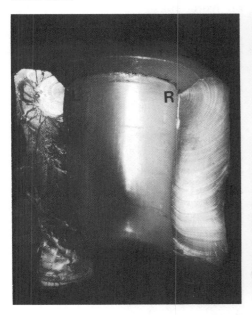

Fig. 15 Failed rubber office-chair roller

(a) Fracture surface; the arrow indicates site of crack initiation. (b) Wear microcracks (arrows) on the internal surface of the roller. Both approximately 3.5×

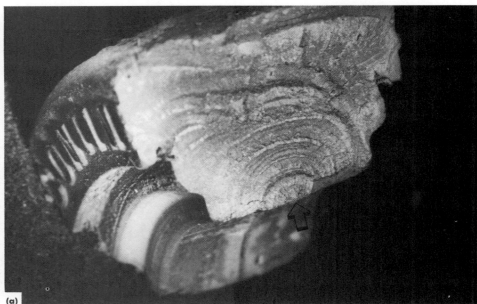

wall of the pipe. It should also be noted that maximum residual tensile stress dominates close to the inside wall.

A similar transition has been noted to occur when the sufficient thermodynamic condition for crack instability is fulfilled (Ref 23). The major axis of the ellipse increases faster than the minor axis until no more striations are observed and ultimate failure results in large-scale yielding (about 50%) of the pipe wall (not shown in Fig. 11).

Whenever possible, similarity criteria should be established between the fracture behavior of a component in service and that observed in the laboratory. In this case, the band width appears to be a suitable candidate. In Fig. 12, the band width b normalized with the critical (last) band width b_c is plotted as a function of the crack length l normalized with the critical crack length l_c for the field-failure specimen (Fig. 10) and for a polyethylene laboratory specimen fractured under fatigue loading (Ref 24). The agreement is surprisingly good, indicating that discontinuous crack-growth band width, when available, can be employed as a similarity criterion to establish correspondence in loading history.

Example 2: Failure of a PVC Water-Filter Housing. Figure 13 shows an injection-molded PVC water-filter housing that fractured in service (Ref 25). An initial fissure (arrow F) is believed to have started first due to residual stresses developed during injection molding. Failure seems to have occurred due to fatigue-crack propagation, as indicated by the presence of discontinuous crack-growth bands and their evolution. Although a tensile-stress component normal to the fracture surface was the dominant

cause of failure, considerable triaxial stress seems evident in the early stages of fracture, as indicated by the successively smaller fissures to the left of the crack starter. As would be expected in PVC, catastrophic failure occurred by brittle failure as opposed to large-scale yielding, as in the polyethylene pipe discussed in Example 1. This is evident from the relatively smooth appearance of the fracture surface beyond the last fatigue band in Fig. 13.

Example 3: Failure of an Automobile Transmission Stick-Shift. Figure 14 shows a stick-shift from an automobile transmission that failed in service (Ref 26). Obviously, fatigue loading is involved in this application. Failure

occurred by stable fatigue-crack propagation at locations R and L (Fig. 14). Plausibly, the crack initiated at location R due to maximum tensile stress, which was perhaps further concentrated by a surface imperfection. The crack appears to have started along the location R section to a certain extent at which service loads became mostly supported by the location L section. This overload on location L appears to have caused another fatigue crack that propagated first over the smooth-banded region. Catastrophic failure occurred when the component became unstable under the applied load.

Example 4: Failure of a Rubber Office-Chair Roller. A brittlelike crack propagation

Fig. 16 Wear marks on the surface of a nylon/polyethylene antifriction bearing

The bearing was in contact with a rotating steel shaft. 850×. Source: Ref 15

Fig. 18 Failed polyoxymethylene gear wheel that had been in operation in a boiler-room environment

305×. Source: Ref 15

Fig. 17 Pitting and surface microcracks on the tooth flank of an oil-lubricated nylon driving gear

75×. Source: Ref 15

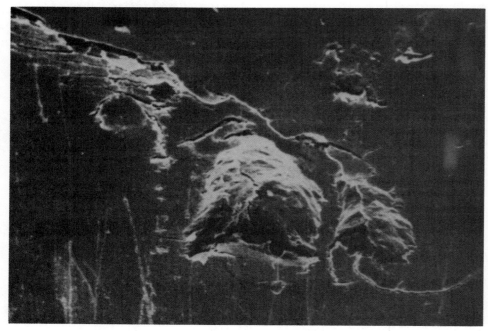

crack (Fig. 15a) might have initiated from similar surface microcracks.

In this case, compressive fatigue was a dominant mode of loading. Nevertheless, the fracture surface of the failure-causing crack (Fig. 15a) suggested that a tensile-stress component was involved in driving failure. Fatigue-crack propagation was evident from the striations visible on the fracture surface. The striated region, almost half the thickness of the roller, reflected stable (subcritical) crack propagation beyond which uncontrolled crack propagation occurred, causing separation of the component. The magnitude of subcritical crack propagation relative to the entire crack path is known to reflect the level of the applied load and its frequency, that is, the rate of load application. In the present case, subcritical crack propagation prevailed over 50% of the crack-propagation path, which indicated a relatively small stress level applied at a low frequency.

Wear of polymers due to friction constitutes an important aspect of their failure analysis and lifetime prediction in view of the rising number of industrial applications of polymers involving frictional forces. A summary of the laboratory analysis of wear of reinforced polymer composites is provided in Ref 27. Information on mechanistic analysis of wear field failure in the open literature remains sketchy. A few cases are discussed in Ref 15.

Example 5: Wear Failure of an Antifriction Bearing. Shown in Fig. 16 is the worn surface of an antifriction bearing made from a nylon/polyethylene blend. The bearing was worn in contact with a steel shaft (Ref 15).

caused failure in a rubber office-chair roller. Figure 15(a) shows a low-magnification optical micrograph of a segment cut from the failed roller. A crack initiated from the inside of the roller (arrow) and propagated in a discontinuous brittlelike fashion, as indicated from the evolution of concentric fracture striations. Examination of the internal surface of the roller (Fig. 15b) revealed other radial and circumferential surface microcracks (arrows). These result from frictional and wear effects. It is therefore reasonable to assume that the main

Movement of the shaft against the bearing caused abrasive marks (Fig. 16). Fine iron oxide particles acted as an abrasive, producing the failure mechanism observed.

Example 6: Failure of a Nylon Driving Gear. Figure 17 shows pitting on the tooth flank of a nylon oil-lubricated driving gear. The pitting produced numerous surface microcracks in association with large-scale fragmentation (frictional wear) (Ref 15). The stress-cracking effect of the lubricating oil is believed to have played a role in initiating the observed microcracks.

Example 7: Failure of a Polyoxymethylene Gear Wheel. A polyoxymethylene gear wheel (Fig. 18) exhibits a failure mechanism different from that discussed in Example 6. This component had been in operation in a boiler room and is believed to have failed because of considerable shrinkage. The oriented crystalline superstructures and the microporosity are reported to be due to post-crystallization. The porosity is attributed to the difference in densities between the amorphous (1.05 g/cm^3, or 0.04 lb/in.^{3}) and the semicrystalline (1.45 g/cm^3, or 0.05 lb/in.^{3}) states (Ref 15). Breakdown along the crystalline superstructure started mainly at the mechanically stressed tooth flanks. In addition, oil vapors, humidity, and other degradative agents could also have contributed to the observed failure.

REFERENCES

1. J.M.G. Cowie, *Polymers: Chemistry and Physics of Modern Materials*, International Text Book Company, Ltd., 1973
2. F.W. Billmeyer, Jr., *Text Book of Polymer Science*, John Wiley & Sons, 1984
3. H. Ulrich, *Introduction to Industrial Polymers*, Hanser Publishers, 1982
4. H.H. Kauch, *Polymer Fracture*, Springer-Verlag, 1978
5. A. Moet, E. Baer, and S. Wellinghoff, in *Structure and Properties of Amorphous Polymers*, A.G. Walton, Ed., Elsevier, 1980
6. A. Moet, Fatigue Failure, in *Failure of Plastics*, W. Brostow and R.D. Coneliussen, Ed., Hanser Publishers, 1986
7. C.B. Bucknall, *Toughened Plastics*, Applied Science Publishers Ltd., 1977
8. J.C.M. Li, *Polym. Eng. Sci.*, Vol 24, 1984, p 750
9. K. Matsushige, S.V. Radcliff, and E. Baer, *J. Mater. Sci.*, Vol 10, 1975, p 833
10. R.N. Haward, Ed., *The Physics of Glassy Polymers*, Applied Science Publishers Ltd., 1973
11. B. Escaig, *Polym. Eng. Sci.*, Vol 24, 1984, p 7373
12. D. Hull, Nucleation and Propagation Processes in Fracture, in *Polymeric Materials: Relationships Between Structure and Mechanical Behavior*, American Society for Metals, 1975
13. G.A. Bernier, and R.P. Kambour, *Macromolecules*, Vol 1, 1968, p 393
14. W. Schnabel, *Polymer Degradation*, Hanser International, 1981
15. L. Engel, H. Klingele, G.W. Ehrenstein, and H. Schaper, *An Atlas of Polymer Damage*, Prentice-Hall, 1981
16. R.W. Hertzberg and J.A. Manson, *Fatigue in Engineering Plastics*, Academic Press, 1980
17. J.G. Williams, *Fracture Mechanics of Polymers*, John Wiley & Sons, 1984
18. A. Chudnovsky, A. Moet, R.J. Bankert, and M.T. Takemori, *J. Appl. Phys.*, Vol 54, 1983, p 5562
19. J. Botsis, Ph.D. thesis, Case Western Reserve University, 1984
20. M.T. Takemori and R.P. Kambour, *J. Mater. Sci.*, Vol 16, 1981, p 1108
21. N. Haddaoui, A. Chudnovsky, and A. Moet, *Polym. Mater. Sci. Eng.*, Vol 49, 1983, p 117
22. K. Sehanobish, A. Moet, A. Chudnovsky, and P.P. Petro, *J. Mater. Sci. Lett.*, Vol 4, 1985, p 890
23. K. Sehanobish, J. Botsis, A. Moet, and A. Chudnovsky, *Int. J. Fract.*, in press
24. J.R. White and J.W. Teh, *Polymer*, Vol 20, 1979, p 764
25. W. Döll and L. Könzöl, Fraunhofer Institute für Werkstoffmechanik, Freiburg, F.R. Germany, private communications
26. K. Friedrich, Some Cases of Fatigue Damage of Polymeric Structural Components, in *Practical Metallography*, to be published
27. K. Friedrich, Ed., *Friction and Wear of Polymer Composites*, Elsevier, 1986

Failure Analysis of Integrated Circuits

C.M. Bailey, Jr. (retired) and D.H. Hensler, AT&T Bell Laboratories

THE BASIC FAILURE MECHANISMS that may be inherent in modern semiconductor devices are discussed in this article. Particular emphasis is placed on silicon integrated circuits, because they provide the tools for the hi-tech systems of today and the future. Only an overview is intended, because it is not within the scope of this article to discuss the many variations of the basic failure mechanisms. Since the trend in integrated circuits has been and will continue toward more complex devices, ramifications for the future are also discussed. The inherent failure mechanisms lead ultimately to failure—failure at the device-manufacturing plant, at the equipment-manufacturing plant, and in the field.

It is useful to describe failure rates in terms of historical bathtub curves (Fig. 1). The regions in this curve include infant mortality, constant or steady-state life, and wear-out. Considering that semiconductor devices usually do not exhibit wear-out within their useful lifetime, the regions of interest for the failure mechanisms discussed are infant mortality and steady-state life. Generally, the more important region is infant mortality, because the failure rates are high (although dropping) and the cost impact of failures is most significant. In gen-

Fig. 1 Bathtub curve of failure rate versus time over the life of an integrated-circuit population

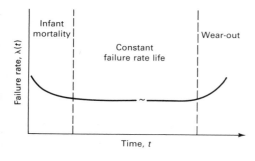

Fig. 2 The reliability model

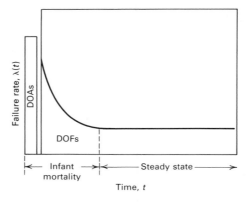

eral, the infant mortality period lasts about a year. It is usually after that period that a steady-state or near-steady-state failure rate occurs.

Infant mortality can be further defined by Fig. 2, in which infant mortality consists of dead-on-arrivals (DOAs) and device-operating failures (DOFs). Dead-on-arrivals are initial failures that, for example, are found at an equipment-manufacturing location. Device-operating failures are the early failures in the field. Dead-on-arrivals are sometimes greater in number than DOFs and are most often due to handling problems, but they may also be due to inherent failure mechanisms in the devices. Conversely, DOFs are most often caused by inherent mechanisms. The total of DOA and DOF percentages constitutes the early failure or infant mortality population. Table 1 lists estimates of infant mortality DOF percentages.

In the early stages of infant mortality, the failures are generally due to device-manufacturing defects, such as oxide pinholes, weak bonds, and other mechanical problems. However, also included are failures due to failure mechanisms inherent in even the best of manufactured devices. Infant mortality ap-

Table 1 Estimates of infant mortality DOF percentages

Device type	Typical range of cumulative percent DOF (1 year)
Transistors	0.02–0.04
Diodes	0.01–0.02
Bipolar TTL integrated circuits(a)	0.05–0.07
Linear integrated circuits	0.10–0.18
Digital CMOS integrated circuits(a)	0.05–0.07
MOS memory integrated circuits	0.07–0.40

(a) Small- or medium-scale integration

pears to have a limited dependence on temperature. The thermal-activation energy, based on data from many sources, appears to be in the range of 0.3 to 0.5 eV, centering around 0.4 eV. As an introduction to some of the time-dependent failure mechanisms, an overview is given in Table 2 of the relevant factors, accelerating factors, and associated activation energies.

Device failures are sometimes categorized as package, chip, and use failures. For example, package failures can be caused by corroded leads and leaking seals, while chip failures can result from dielectric breakdown of metal-oxide semiconductor (MOS) devices or surface inversion. Use failures can result from electrostatic discharge. In truth, package, chip, and use failures are interactive in many cases. Table 3 shows a breakdown of failure causes observed in hermetic and nonhermetic (plastic) packages. These data show the same general failure causes for hermetic and plastic devices. The difference is in the incidence. The relative importance of the failure mechanisms depends on the device. For example, large-scale integration (LSI) devices have more chip-related failures than small-scale integration (SSI) devices, MOS has more oxide failures than bipolar, and low-power high-voltage devices have more corrosion failures.

Table 2 Time-dependent failure mechanisms in silicon semiconductor devices

Device association	Process	Relevant factors	Accelerating factors	Apparent activation energy
Metallization	Corrosion galvanic, electrolytic	Contamination H, V, T	H, V, T	Strong H effect $E_A = 0.4 - 0.6$ eV (for electrolysis)
	Electromigration	T, J, A, gradients of T and J, grain size	T, J	$E_A = 0.5 - 0.77$ eV J^2 to J^4
	Stress cracking	Topology metallurgy	None	None
Bonds and other mechanical interfaces	Intermetallic growth	T, impurities, bond strength	T	Aluminum-gold $E_A = 0.77$ eV
	Fatigue	Temperature cycling, bond strength	T extremes in cycling	...
Silicon oxide and silicon-silicon oxide interface	Dielectric breakdown	E, T	E	$E_A = 0.3$ eV
	Surface charge accumulation	Mobile ions, V, T	T	$E_A = 1.1$ eV
	Charge injection	E, T, Q_{SS}	E, T	$E_A = 1.3$ eV (slow trapping)
Latch-up	Junction breakdown	V, T, Frequency design	None	...

H, humidity; V, voltage; T, temperature; J, current density; E, electric field; A, area; Q_{SS}, surface-state oxide charges

Table 3 Normalized distribution of malfunctions for MOS devices

Defect category	PMOS, %(a)	NMOS, %(b)	CMOS, %(c)
Hermetic packages			
Surface	35	42	52
Bulk	1	3	1
Oxide	25	26	18
Diffusion	12	5	3
Metallization	27	24	26
Nonhermetic packages			
Surface	61	70	74
Bulk	1	2	...
Oxide	12	12	10
Diffusion	8	4	1
Metallization	18	12	15

(a) PMOS, positive metal-oxide semiconductor. (b) NMOS, negative metal-oxide semiconductor. (c) CMOS, complementary metal-oxide semiconductor
Source: Ref 1

Techniques of Failure Analysis

This section will overview the techniques and procedures of failure analysis in integrated circuits. Because there are many different failure mechanisms and because new ones appear as technology advances, there is no prescribed methodology for failure analysis. While some of the more well-known failure mechanisms are readily apparent upon opening the integrated-circuit package, there are subtle failure mechanisms that involve months of probing and experimentation to identify correctly. The principal barrier to identifying a device failure mechanism correctly is its complexity. Modern LSI integrated circuits can contain over one million transistors and can perform a staggering variety of electrical functions. Even so, there are a few techniques that help tilt the balance slightly in favor of a successful failure analysis.

The first is testing. The device must be tested completely before the package is opened to identify the portion of the circuit that is failing and to identify and sort out failure mechanisms induced later during the analysis procedure. The chief tools of testing are computer-driven test sets that can be programmed to check every electrical parameter of the circuit and to record the specific device response in comparison with the normal range of each parameter. Thus, programming computer-driven test sets and in-

Fig. 3 Sectional view of an integrated-circuit chip encapsulated in a plastic package

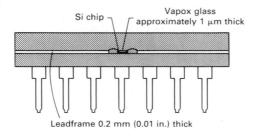

Fig. 4 Acid percolator for selectively dissolving a plastic package

Source: Ref 2

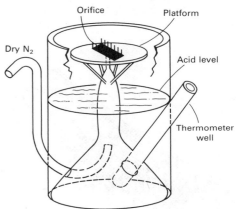

terpreting their test logs are a large part of the failure analysis effort. However, additional efforts are usually necessary to confirm the test set findings. These are carried out on the laboratory bench using power supplies, pulse generators, and oscilloscopes.

To pursue the failure mechanism further, the next step is to expose the chip in the package without damaging it or its electrical connection to the outside pins. This process is termed decapping. In the case of ceramic packages, the top is pried off if it is metal or ground off if it is ceramic. The latter process is analogous to opening a shipping carton with a chain saw in such a way that the egg within is not damaged. It requires practice. If the chip is encapsulated in plastic, the challenge is even greater because the plastic also holds the connecting wires and pins together. Figure 3 shows an illustration representing the chip in the package with its connecting leads to the outside world. Simply dissolving the plastic yields only the chip without any means of operating it. A useful device was invented that is an adaptation of the coffee percolator (Fig. 4). Dry sulfuric acid is kept at a temperature above 250 °C (480 °F) in a modified glass beaker to expel any moisture. The device to be opened is placed upside down over an orifice at the top. Dry nitrogen is then introduced into the inner cone, and it forces small amounts of the etchant to impinge against the surface of the inverted plastic package. This process is allowed to continue until the chip is exposed. If the technique is done properly, the exposed chip will function electrically when the

process is completed. Figure 5 shows a plastic package before and after the percolator has been used to expose the chip.

After decapping, a visual inspection is in order. For modern circuits, magnifications of 800 to 1000 × are needed with bright-field and dark-field capabilities. Polarized light is also helpful. Defects that are obvious at this point are usually photographed, then examined using scanning electron microscopy (SEM). If no defects are visible, more sophisticated techniques are implemented.

At this point, it is helpful if the initial testing has indicated that the device conducts current in excess of the specified normal current. Such excess current can indicate electrical shorts due to metallization corrosion, oxide breakdown due to defects, or electrostatic discharge (these are discussed in the sections "Metallization Failures," "Silicon Oxide Failures," and "Electrostatic/Electrical Overstress Failures," respectively, in this article). If such excess device current is large enough (10 to 20 mA), it is possible to view the defect using a sensitive infrared microscope that can see the excess infrared radiation. Another technique that is also helpful under such circumstances is the use of cholesteric liquid crystal (Ref 3, 4). Figure 6

Fig. 5 Plastic package before (a) and after (b) use of the acid percolator

Source: Ref 2

(a)

(b)

Fig. 6 Schematic representation of the use of cholesteric liquid crystal

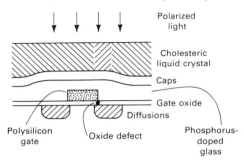

circuit element is best found by a technique called electron beam voltage contrast (Ref 5, 6). In a scanning electron microscope, the image of the surface is produced by collecting the secondary or backscattered electrons, which are those scattered from the surface of the sample. If parts of the sample surface are biased positively, fewer secondary electrons reach the collector because some have been attracted back to the surface of the sample. Thus, a positively biased element of the sample will appear darker than normal. Conversely, a negatively biased circuit element will appear brighter than normal, and an element of the circuit being dynamically switched will appear striped like the old-fashioned barber pole or candy cane. Figure 7 shows an example of an integrated-circuit chip being viewed in a scanning electron microscope using voltage con-

trast. This technique has been extended to the analysis of memory circuits (Ref 7) as well as additional schemes, such as phase-dependent voltage contrast (Ref 8), and useful clocking modifications, such as pulse stretching (Ref 9). A very effective technique using an image processor in conjunction with voltage contrast SEM has been developed that uses powerful image-processing techniques to compare the voltage contrast images of a good device and a faulty device (Ref 10). The difference between these two images is displayed and immediately gives an image of suspect areas on the faulty circuit.

A very useful means of examining the integrity of *p-n* junctions was developed that uses a focused laser beam to scan an area of a biased semiconductor circuit (Ref 11, 12). The helium-neon laser beam is focused through a microscope and penetrates several microns into the surface of the silicon integrated circuit. The laser beam is absorbed by the generation of electron-hole pairs. In the self-field of a normal *p-n* junction, the electrons and holes are separated and create a current through the device. Figure 8 illustrates how the device current is formed. The amplified device current is used to enhance an image of the scanned area obtained from the reflected light. Figure 9 shows an optical schematic of a laser scanner. In recent years, the technique has been enhanced and developed and has proven useful in detecting latch-up in LSI complementary metal-oxide semiconductor (CMOS) circuits (Ref 13, 14).

No general overview of failure analysis techniques is complete without some reference to

shows a schematic illustration. The integrated circuit is coated with a thin layer of cholesteric liquid crystal chosen because its transition temperature is about 40 °C (105 °F). The circuit is operated under a microscope using a polarized light. In the region of an electrical short between circuit elements, the excess heat generated will raise the local temperature of the liquid crystal above the transition temperature. The resultant change in the molecular structure of the liquid crystal above the electrical short causes a localized change in the index of refraction. In polarized light, this can be viewed as a bubble or, more often, as a small region of a different color from its surroundings, thus delineating the location of the circuit defect.

In many cases, there is no excess device current and therefore no hot spot to detect. Here the failure may be due to open metallization or *p-n* junctions that have been shorted for one reason or another. In this case, the defective

Fig. 7 SEM micrograph of an integrated circuit using voltage contrast

Fig. 8 Illustration of device current formed by electron-hole pairs at a *p-n* junction under proper bias

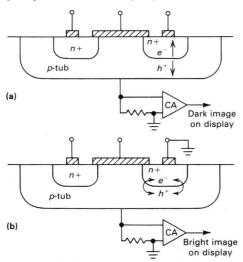

(a)

(b)

Fig. 9 Optical schematic of a laser scanner

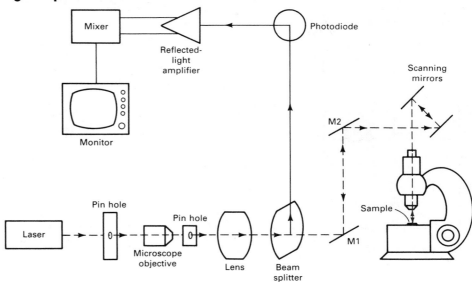

investigation of the defect area after it has been located. Much of this work is done by lapping and staining. Recently, advanced techniques have been developed that permit one to edge lap the circuit to a specific area within 2 μm of the desired location, giving a sectional view of the defective area (Ref 15). Reference 15 is also a standard reference on etching and staining as a means of decorating and identifying various circuit elements after lapping.

Metallization Failures

Corrosion is a common problem because the accelerating factors of electric field, ionic contamination, and moisture are often present. It can affect every metal part of the device. Table 4 lists some examples (a notable exception is polysilicon MOS gate metallization). Susceptibility is a function of size and spacing of chip elements. As the spacing between circuit elements decreases, the electric field increases, thereby accelerating the corrosion processes. Plastic packages are obviously more susceptible because of the relatively easy entry of moisture. Corrosion of aluminum metallization is a major failure mechanism in such packages. On the other hand, devices in hermetic packages are not immune to corrosion due to sealed-in moisture, leaks, and so on.

The two types of corrosion cells of interest are electrolytic and galvanic (Fig. 10). In the case of galvanic corrosion, the electrodes are two dissimilar metals connected in an electrolyte, and the driving force is the difference in the electrochemical potentials. In electrolytic corrosion, the electrodes are similar, and the driving force is the applied potential between them. One example of galvanic corrosion is that in which aluminum bond pads are corroded away around the gold ball in an aluminum-gold

wire bond system. Another example concerns the atmospheric corrosion of gold-plated Kovar leads (Fig. 11). In the presence of moisture, corrosion of the Kovar begins at pinholes in the gold plating. Examples of electrolytic corrosion (discussed below) include anodic corrosion of aluminum in the presence of chloride and cathodic corrosion in the presence of phosphorus.

Corrosion, like all chemical reactions, is considered from two aspects: thermodynamics and kinetics. Thermodynamic considerations yield relative corrosion tendencies. However, these tendencies may have little relation to the rate at which corrosion proceeds, that is, the reaction kinetics.

For the reaction $A + B \rightarrow C + D$, thermodynamic considerations involve the Nernst equation:

$$E = E° - \frac{RT}{2.3\, nF} \log \frac{A_C\, A_D}{A_A\, A_B} \qquad \text{(Eq 1)}$$

where E is the cell electromotive force (emf), $E°$ is the standard oxidation potential, A's are activities, R is the gas constant (8.314 J/mol · K), T is the absolute temperature, n is the number of chemical equivalents, and F is the Faraday constant (96 500 C/equivalent). The quantity $RT/2.3F$ equals 0.059 V at 25 °C (75 °F). For electrolytic corrosion, if the potential on the metal line exceeds E, the reaction will occur; if it is less than E, the reaction is thermodynamically impossible. An illustration of such a condition is an integrated injection logic device metallized with gold and using a relatively low supply voltage of 0.65 V. The reaction that takes place in this case is:

$$Au + 4Cl^- \rightleftharpoons AuCl_4^- + 3e^-$$

$$\tfrac{3}{2}H_2 \rightleftharpoons 3H^+ + 3e^- \qquad \text{(Eq 2)}$$

Table 4 Integrated-circuit materials subject to corrosion

Material	Type of corrosion
Aluminum, gold, nickel, chromium	Chip metallization
Kovar	Leadframe metallization
Silver, tin/lead	Leadframe plating metallization

These are combined to give:

$$Au + 4Cl^- + 3H^+ \rightleftharpoons AuCl_4^- + \tfrac{3}{2}H_2 \qquad \text{(Eq 3)}$$

Using Eq 1:

$$E = E° - \frac{0.059}{3} \log \frac{[AuCl_4^-]}{[H_+]^3\,[Cl^-]^4}$$

If, for example, $[Cl^-] = 1\ M$, $[H_+] = 10^{-7}\ M$, $[Au] = 1\ M$, $[H_2] = 1\ M$, and $[(AuCl_4)^-] = 10^{-5}\ M$, then:

$$E = 1.0 - 0.31 \approx 0.7\ V$$

Therefore, the device working at a maximum voltage of 0.65 V is safe. A lifetime in excess of 1000 h was observed when a bipolar device was operated in salt water (Ref 17).

Kinetics are dominant in most corrosion processes. The important factors are passivity and environment. Aluminum, for example, is a very reactive metal. On thermodynamic grounds, it might be expected to corrode readily. However, the formation in air of a thin, stable oxide effectively passivates it. If kept clean and free of ionic contamination, aluminum films will survive for long periods. The

Fig. 10 Commonly observed corrosion cells

(a) Galvanic corrosion caused by dissimilar metals.
(b) Electrolytic corrosion caused by applied bias

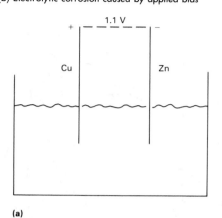

(a)

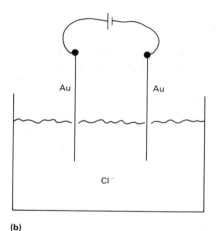

(b)

Fig. 11 Chloride ion pitting of Kovar lead material

Source: Ref 16

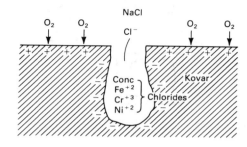

This is followed by a process of chlorine adsorption on the hydrated oxide surface, leading to the formation of a water-soluble aluminum chloride salt:

$$Al(OH)_2^+ + Cl^- \rightarrow Al(OH)_2Cl \qquad (Eq\ 5)$$

If the aluminum metallization is under positive bias, reaction will tend to take place at thinned regions where the electric field is higher. This leads to the dissolution of the aluminum oxide in localized regions. After the surface oxide is dissolved, the chlorine reacts with the underlying aluminum:

$$Al + 4Cl^- \rightarrow Al(Cl)_4^- + 3e^- \qquad (Eq\ 6)$$

The $Al(Cl)_4^-$ will then react with the available water:

$$2AlCl_4^- + 6H_2O \rightarrow 2Al(OH)_3$$
$$+ 6H^+ + 8Cl^- \qquad (Eq\ 7)$$

Chlorine is liberated in this last process and is free to continue the corrosion process. With electrical bias, the chloride ions migrate to the anode, thus enhancing the corrosion of the positively biased element. Figure 12 shows an example of voltage-specific corrosion. There has been some concern about the possible corrosion effects of the addition of copper to aluminum. However, aluminum-copper devices have been shown to have excellent reliability under severe test conditions (Ref 20). The copper is not incorporated in the aluminum oxide, and in the absence of chloride, the corrosion properties of aluminum-copper are similar to those of pure aluminum.

Phosphorus, as contained in phosphosilicate passivation glasses, is an inherent component in NMOS integrated circuits, serving as an Na^+ ion getter. Aluminum corrosion will occur if the phosphorus concentration is above a certain threshold. Cathodic corrosion is the more common form of aluminum corrosion in the pres-

important environmental factors are humidity, the aluminum microstructure, and the substrate surface. Another factor is polarization, that is, the change in potentials on metal lines as a result of circuit bias.

Aluminum Metallization Corrosion. Aluminum is susceptible to corrosion attack by a variety of materials and environments. Because of the specific nature of aluminum corrosion and the variable environmental and contamination factors, few generalizations can be made. However, two frequent causes of corrosion are chlorine and phosphorus.

In the presence of ionic contamination, the passive aluminum oxide skin may be breached, allowing rapid corrosion to occur due to the reaction between the exposed aluminum and moisture in the atmosphere. Chloride contamination is particularly effective. The corrosion process is described in Ref 18 and 19. It begins with the ionization of the hydrated aluminum oxide surface according to the following reaction:

$$Al(OH)_3 \rightleftharpoons Al(OH)_2^+ + OH^- \qquad (Eq\ 4)$$

ence of phosphorus and moisture. The reactions at the anode and cathode are (Ref 21):

Anode	
$2H_2O \rightarrow 4H^+ + O_2 + 4e^-$	(electrolysis of water)
Cathode	
$2H^+ + 2e^- \rightarrow H_2$	(hydrogen ion reduction)
$2Al + 6H^+ \rightarrow 2Al^{3+} + 3H_2$	(aluminum ion formation)
$2Al^{3+} + 3H_2O \rightarrow 2Al(OH)_3 + 3H_2$	(aluminum corrosion)

Under the applied voltage, hydrogen ions migrate to the cathode, where they react with aluminum. The phosphorus reacts with moisture in the package to form strongly acidic compounds. These compounds provide more hydrogen ions at the cathode to accelerate the aluminum corrosion. As a result, cathodic corrosion of aluminum takes place at a rate proportional to the phosphorus concentration. Note that the phosphorus is not used up. The corrosion usually proceeds by grain-boundary attack (Fig. 13). Figure 14 shows the effects of phosphorus additions to dielectric layers on aluminum metallization in close proximity.

Gold Metallization Corrosion. Electrolytic corrosion has been observed to some degree in most gold metallization systems (Ti-Pd-Au, Ni-Cr-Au, Ti-W-Au, Ti-Pt-Au). Gold corrosion produces electrical shorts caused by the migration of gold between two biased conductors, resulting in a metal bridge between them. This is commonly referred to as dendritic growth. Dissolution (corrosion) of the gold occurs at the anode, followed by migration of the soluble corrosion product to the cathode, where it is reduced and electrodeposited in the familiar dendritic pattern. The current density increases rapidly near the tip of the dendrite, encouraging preferential growth of the spike. The most likely reactions are (at the anode):

$$Au + 4Cl^- \rightarrow [AuCl_4]^- + 3e^- \qquad (Eq\ 8)$$

It is thought that concentration gradients cause the $[AuCl_4]^-$ to diffuse to the cathode, where the gold is plated out according to:

$$[AuCl_4]^- \rightarrow Au\ (Metal) + 4Cl^- \qquad (Eq\ 9)$$

Figure 15 shows an example of electrolytic gold corrosion. As with aluminum corrosion, the accelerating factors are temperature, humidity, and voltage.

Electromigration is the transport of metal ions through a conductor resulting from the passage of direct current. It is caused by a modification of the normally random diffusion process to a directional one by the charge carrier electron wind. This directional effect causes ions to migrate or diffuse downstream in terms of electron flow by momentum transfer and causes vacancies to move upstream. Tradi-

Fig. 12 Example of voltage-specific corrosion of aluminum

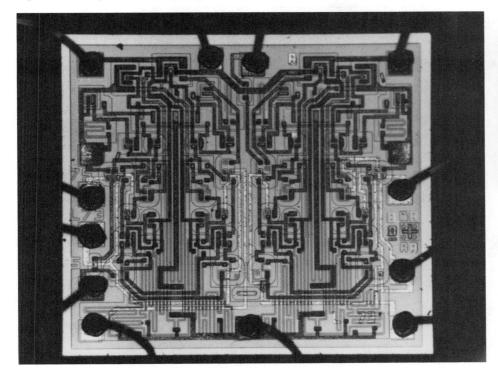

Fig. 13 Example of phosphorus-related corrosion by aluminum metallization

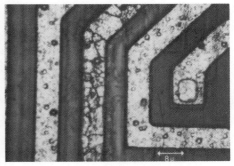

Fig. 14 Effects of phosphorus in the intermediate dielectrics on integrated-circuit reliability

Source: Ref 20

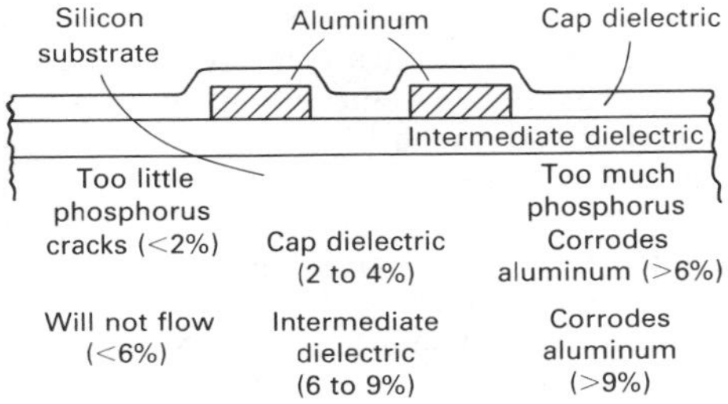

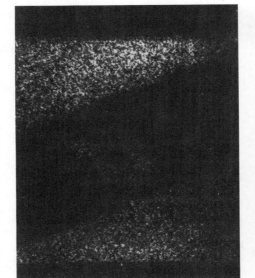

Fig. 15 Example of electrolytic corrosion of gold

(a) Gold x-ray. (b) SEM micrograph. Source: Ref 20

(a)

(b)

2 μm

tionally, failures have been defined as open circuits in the conductor line material (usually aluminum or aluminum alloys). However, growth nodules have also been observed to cause shorts.

The interaction appears basically to be due to nonuniformities in the ion flux. The causes of these nonuniformities have been attributed to temperature gradients or structural effects. Metal films consist of an agglomeration of single-crystal grains in which the grains are oriented randomly (Fig. 16). Grain boundaries provide easy paths for self-diffusion and electromigration as compared to diffusion and electromigration through the lattice. The surface of the metal also provides an easy path for the diffusion. Therefore, grain-boundary and surface diffusion are the more significant factors. Electric current parallel to a temperature gradient leads to accumulation or depletion of metal, depending on the sign of the gradient. Depletion causes voids that may eventually grow to form an open circuit. Accumulation results in the formation of hillocks or whiskers that result in short circuits. Table 5 summarizes the basic factors affecting electromigration failure rates. The dependence of time-to-failure on current density, J, is indicated as being in the range of J^{-2} to J^{-4}, but at current densities above 10^6 A/cm^2, where joule heating starts to become significant, even higher powers of J can result.

Much higher current density would appear to be usable under pulsed direct current (dc) at low duty factors. For long pulses (10^{-3} to 10^3 Hz),

electromigration dominates by vacancy diffusion during the off time. For short pulses ($>10^3$ Hz), failure involves pulse heating leading to thermal fatigue. Current densities up to 10^7 A/cm^2 and above can be used for low duty

cycles. Figure 17 shows data obtained on the effects of pulsed dc on gold film conductors.

Grain size and its homogeneity are important factors affecting electromigration phenomena. Large grain size appears to help by limiting the

Fig. 16 Lattice (D_L), grain-boundary (D_{GB}), and surface (D_S) diffusion

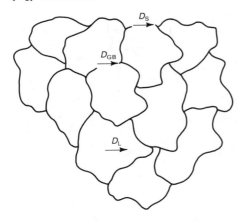

Table 5   Factors affecting electromigration rates

Factor		Effect
Current density . . . . Lifetime $\alpha \dfrac{1}{J N}$		$N = 2$ to 4
Temperature . . . . Lifetime $\alpha \exp\left(\dfrac{E_A}{kT}\right)$		$E_A = 0.5$ to 1.2 eV
Crystal structure . . . . D_S and $D_{GB} \geqslant D_L$		D_S = surface diffusion coefficient D_{GB} = grain-boundary diffusion coefficient D_L = lattice diffusion coefficient

Fig. 17 Electromigration effects of pulsed direct current on gold film conductors

Source: Ref 22

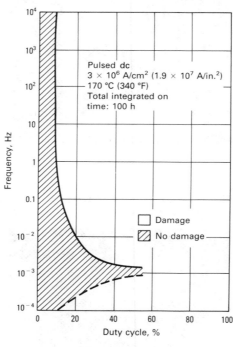

Pulsed dc
3×10^6 A/cm^2 (1.9×10^7 A/in.^{2})
170 °C (340 °F)
Total integrated on
time: 100 h

□ Damage
▨ No damage

Fig. 18 Median time-to-failure of aluminum and an Al-2Cu alloy as a function of current density *J*, temperature *T*, and cross-sectional area *A*

Source: Ref 24

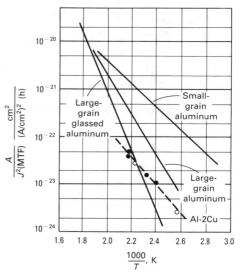

number of diffusion paths; a single-crystal film of aluminum is virtually indestructible (Ref 23). The effect on time-to-failure of some aluminum alloys has been shown to be a function of current density, temperature, and cross-sectional area (Fig. 18). Small-grain crystals have a high density of grain-boundary paths and exhibit a relatively short life at lower temperatures and an activation energy of 0.48 eV. Large-grain crystals without passivation provide electromigration paths mainly at the surface and exhibit moderate lifetimes at the lower temperature and an activation energy 0.8 eV. Large-grain conductors with chemical vapor deposited (CVD) glass passivation exhibit the greatest life—apparently due to reduced grain-boundary and surface diffusion. The activation energy of 1.2 eV approaches that of bulk aluminum (1.48 eV). It is noted, however, that the role of a confining glass is controversial because electromigration can produce cracks in overlying passivation glass layers, particularly for thick narrow stripes.

It has also been found that small amounts of impurities added to pure aluminum films greatly increase the reliability. The effects of small amounts of added copper have been studied by several investigators. The activation energy for electromigration failure as a function of copper concentration has been examined (Ref 25). It was found that the addition of 4 wt% Cu to aluminum increases the activation energy by 0.2 to 0.3 eV over that of pure aluminum films. Figure 18 also shows these and other data for fine-grain aluminum films doped with copper (Ref 25, 26). The activation energy is about 0.6 eV as compared to 0.48 eV for fine-grain pure aluminum film, and the median time-to-failure (MTF) is 40 times larger. In an investigation of Al-2Cu and Al-2Cu-1Si films (Ref 27), it was found that the addition of silicon increased the reliability of the Al-2Cu films even further and prevented junction spiking (see the section "Metallic Interface Failures" in this article).

In 1977, it was reported that greatly increased life was obtained in aluminum sandwich stripes with a thin middle layer of chromium, titanium, or other metal that forms a low-diffusivity intermetallic phase with aluminum upon heating (Ref 28). In another study (Ref 29), improvement in electromigration behavior was reported using an underlying Cr-Cr$_2$O$_3$ material with aluminum-copper metallization. Results showed a ten-fold improvement in electromigration lifetime, an activation energy of 0.7 eV, and a three-fold increase in the current-carrying capability of aluminum-copper. This improvement was related to the structural changes and diffusions that occur between the underlying layer and the aluminum-copper metallization. An order of magnitude improvement has been reported (Ref 30) in electromigration resistance from the use of an underlying layer of chromium (beneath

aluminum-copper) for interconnection lines. Voids formed by electromigration were shown to move, heal, and interact with one another and with obstacles in the metallization as well as with other dynamically occurring structural changes that simultaneously occur. The sum total of these events determines failure due to an open circuit rather than just the simple expansion of a void. Chromium was thought to provide a temporary support of the current, enabling a dynamic healing process to come into play.

The ratio of line width to grain size is another important structural factor influencing electromigration-induced failures. Studies have shown that under conditions of constant current density the lifetime of conductors decreases with decreasing conductor width (Ref 31). This work was extended in another study (Ref 32) to include aluminum-copper, aluminum-copper-silicon, and chromium-silver-chromium conductor types of various widths formed by a variety of processes.

Fig. 19 Width dependence of time to 50% failure, t_{50}, for many lots of aluminum-copper samples

Source: Ref 32

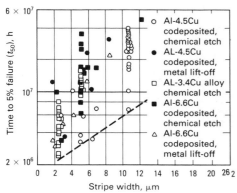

Fig. 20 Width dependence of σ for many lots of aluminum-copper samples

Source: Ref 32

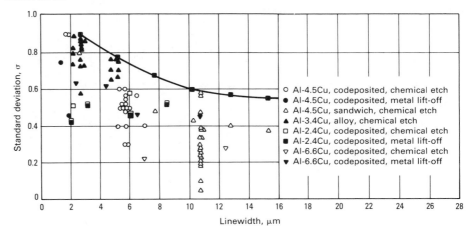

Results were reported in terms of the median lifetime and of σ, the standard deviation, of the population of electromigration-induced failures. Figure 19 shows a plot of median life versus stripe width on different lots of aluminum-copper samples, and Fig. 20 shows a plot of σ versus line width for the same samples. The data in Fig. 19 are all normalized to a current density of 2.0×10^5 A/cm^2 (1.3×10^6 A/in.^{2}).

The dashed line shown in Fig. 19 is the worst-case lower bound for the width dependence of median life, and the solid curve shown in Fig. 20 is the worst-case upper bound for the width dependence of σ. It was postulated that the line-width dependence was due to a grain-size/line-width relationship (Fig. 21). Figure 22 shows schematically the effect of aluminum line width and grain size on median life, σ, and the failure rate. The dashed parts of the curves for small W_n (stripe width/median grain size) show a possible reversal of the trend. The lifetime might be expected to decrease with increasing length of the line because of an increasing probability of a fatal structure defect in longer lines. The combination of long lines (>1 cm, or 0.4 in.) with narrow widths(<3 μm) would appear to present a formidable obstacle to the use of fine-line aluminum for very large scale integration (VLSI) devices. However, the electromigration lifetimes have been determined for a combination of long lines (up to 3 cm, or 1.2 in.) and narrow line widths (down to 1 μm) of evaporated and sputter-source deposited aluminum-copper-silicon films (Ref 33). The lifetimes of the finer-grain sputtered films were significantly shorter than those for the larger grain-evaporated films at smaller line widths (Fig. 23).

Multilayer metallization structures are sometimes used in semiconductor devices, particularly for LSI and VLSI integrated circuits. Electromigration has been investigated in aluminum double-layer systems (Ref 34). A current-density dependence and an activation

Fig. 21 Effects of grain size in metal stripes

(a) Illustration of log normal grain-size distribution. (b) Typical grain structure of a metal stripe. Source: Ref 32

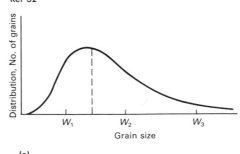

(a)

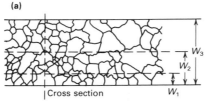

(b)

energy characterization similar to that in single-layer systems were found. In addition, failures were found to be a function of the magnitude and polarity of the electric field applied between adjacent stripes and of the dielectric material used between adjacent stripes.

Integrated-circuit designs that avoid high current (charge-coupled devices, integrated injection logic devices) minimize the risk of electromigration failure. Small, high operating temperature devices, such as microwave transistors and GaAs field-effect transistors (FETs), are the most susceptible. In general, problems with electromigration can be expected to increase as device geometries shrink.

Fig. 22 Effect of stripe width and grain size on t_{50}, σ, and the failure rate

Source: Ref 32

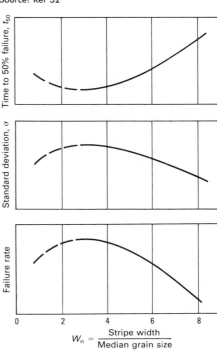

$$W_n = \frac{\text{Stripe width}}{\text{Median grain size}}$$

Polysilicon-Silicide (Polycide) Structures. Polycrystalline silicon (polysilicon) has been used for many years in integrated circuits as an FET gate material as well as an interconnecting metallization. The one disadvantage associated with its use is the fact that its resistivity (30 to 60 Ω/square) is quite high. As interconnection runs became longer and interconnecting metallizations became thinner and

Fig. 23 Effect of line width on electromigration lifetimes normalized to that for 7-μm material

Source: Ref 33

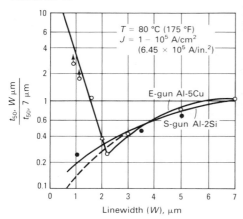

Fig. 24 SEM micrographs of silicon nodules remaining after aluminum has been preferentially etched away

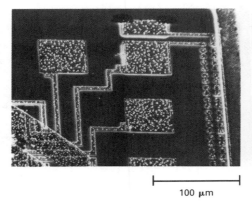

100 μm

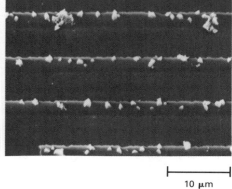

10 μm

narrower, the high resistivity of polysilicon became a distinct disadvantage.

One method of decreasing the resistance of long polysilicon metallization runs used in integrated circuits has been to deposit a metal silicide on top of the polysilicon, making a composite metal silicide/polysilicon (polycide) sandwich. Several metal silicides are used in the semiconductor industry. The more popular seem to be tungsten, titanium, tantalum, and molybdenum. The use of higher and higher current densities in these polycide structures has led to concern about their reliability in terms of resistance of electromigration. A limited amount of work has been done in which polycide structures were stressed along with silicide metallizations (Ref 35). These structures were stressed in an oven ambient of 300 °C (570 °F) and at a current density of 5.3×10^6 A/cm^2 (3.4×10^7 A/in.^{2}). After 2000 h of stressing, the polycide structures showed more damage than the silicide metallization. The polycide structures were composed of TaSi$_2$-polysilicon and WSi$_2$-polysilicon, and the silicide metallization used was TaSi$_2$ and WSi$_2$. The damage was due to electromigration. Failure occurred due to extrusions of aluminum (probably from the aluminum wire bond used for electrical connections). The extrusions resulted from mass transport of aluminum at the polysilicon/silicide interface and from the accumulation of aluminum at the extrusion site, causing blistering and loss of adhesion. The WSi-polysilicon structures were less susceptible to this failure mechanism than the TaSi-polysilicon structures.

Stress Cracking and Voiding. In recent years, there has been greater and greater pressure to reduce the physical size of the topological features of an integrated circuit. A result of this reduction is narrower metallization used as interconnects. A correspondingly critical reduc-

tion has occurred in the diffusion depths used in integrated circuits. In an effort to reduce the chance of aluminum penetration through contact windows and shorting p-n junctions, integrated-circuit manufacturers have employed sputtered metallization using aluminum sources doped with silicon. Thus, the metallization used as interconnections in recent integrated circuits contains small amounts of silicon. Depending on the conditions under which the sputtered deposits were made, silicon in various amounts will precipitate in the form of nodules in the grain boundaries of the aluminum metallization.

In investigating failures in aluminum-silicon metallization, it has been found that nodules of silicon precipitate from the aluminum (Ref 36, 37). In some cases, these nodules have been found to be greater than 1 μm in size. Figure 24 shows an example of silicon nodules. In addition, other impurities, such as nitrogen and hydrogen, have been found in the aluminum lines. The failures have been ascribed to either current crowding by the silicon nodules causing electromigration-like failures or to the increased amount of stress in the aluminum lines due to the presence of the silicon nodules (Ref 37). Silicon nodules, voids, and open metal lines have been observed in 16K dynamic random access memories (DRAMs) that exhibited failures in metal word lines (Ref 38). Voids have been observed in aluminum lines with SiN as the passivation layer (Ref 39). The density and size of the voids are strongly dependent on SiN compressive stress and substrate temperature. The density of metal voids was found to be a high power function of tensile stress in the line, $(\delta/G)^n$, where $6 < n < 9$, δ is the inferred tensile stress, and G is the aluminum shear modulus. The results suggest that a nonsteady-state temperature-dependent creep is responsible.

The cracking phenomena in aluminum films with CVD-deposited SiN as the passivation layer and nitrogen contamination has been investigated (Ref 40). This failure mode was

found to be temperature dependent with a thermal activation energy of 0.35 eV. The conclusion was that nitrogen contamination resulted in a harder, more brittle aluminum metallization that will fail more readily when subjected to strain induced by the silicon nitride passivation layer. It has been reported that aluminum films doped with silicon exhibit substantial increases in bulk resistivity and hardness due to AlN or Al$_2$O$_3$ compound formation at grain-boundary surfaces (Ref 41, 42). It has been stated that at room temperature the thermal-expansion coefficient mismatch between the aluminum-silicon metallization and the silicon dioxide intermediate dielectric will result in the aluminum-silicon metallization being in tensile stress (Ref 43). If the stress is sufficiently high, solid-state self-diffusion will occur even at room temperature, resulting in voids and open metal lines. Figure 25 shows an example of these failures. However, this failure mechanism cannot be accelerated with higher temperatures (Ref 43). It appears that once voids or nodules are formed due to stress, the aluminum metallization may fail due to a creep mechanism.

Fatigue Failure. There are several methods by which integrated-circuit chips are connected to the outside world. In one method, the chip is placed in a recess in a ceramic package, and electrical connections are made to the package leads by means of 0.02-mm (0.001-in.) gold wires that are thermocompression bonded both to the integrated-circuit chip and to the package leads. A variation of this technique has the chip mounted on a metal base, connected to a leadframe by 0.02-mm (0.001-in.) gold wires that are thermocompression bonded, then encapsulated in molded plastic. In both of these packaging configurations, the wire bonds are made with sufficient slack in the wires that there is no strain induced when the temperature of the device is cycled over a range of 25 °C (45 °F) during the normal daily operation of the equipment in which it is incorporated.

Fig. 25 SEM micrographs of cracked aluminum metallization

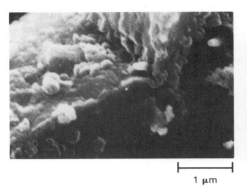

Fig. 26 Two types of packaging

(a) Integrated-circuit chip with gold beam leads bonded circuit-side down to a substrate.
(b) Integrated-circuit chip mounted to a substrate by means of tabs

(a)

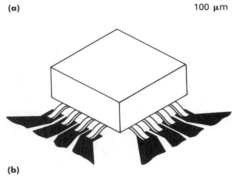

(b)

Fig. 27 Model of a silicone rubber encapsulated integrated-circuit chip

(a) Molded DIP (dual in-line package). (b) RTV (room-temperature vulcanizing) encapsulated device. (c) External RTV neglected. (d) Final model. Source: Ref 47

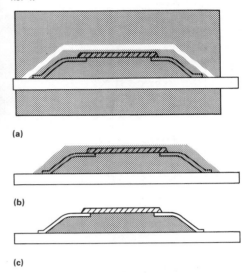

(a)

(b)

(c)

(d)

Fig. 28 Integrated-circuit beam lead showing plastic-strain regions

Source: Ref 47

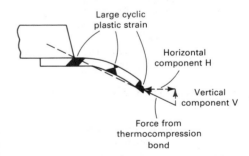

Large ΔT thermal cycling

Large cyclic plastic strain

Horizontal component H

Vertical component V

Force from thermocompression bond

Fig. 29 Analytically derived curves of maximum cyclic strain versus ΔT for beam-leaded SIC

Source: Ref 47

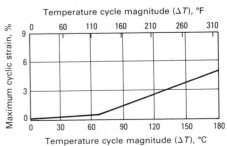

In other types of packaging, the structures are as shown in Fig. 26. Figure 26(a) shows an integrated-circuit chip with gold beam leads. The beam leads are an integral part of the chip metallization fabrication (the chip is shown circuit-side down) and are thermocompression bonded to a ceramic substrate to make connection to the outside world. The gold beam leads are about 0.5 mil thick and 3 mil wide. The free beam length is about 4 mil. Figure 26(b) shows a chip mounted to a ceramic substrate by means of tabs that are 1.3-mil-thick gold-plated copper foil. The tab is bonded to a gold bump on the silicon chip on one end and the ceramic substrate on the other end by thermocompression bonding. The tab is 10 to 20 mils long.

All packaged integrated circuits undergo temperature changes due to climatic changes, equipment start-up, and load variations. Because the integrated-circuit package is composed of several components with different thermal-expansion coefficients, the temperature changes will induce mechanical stresses. Under certain conditions, these stresses lead to beam and wire failures due to fatigue. The model is shown in Fig. 27(d). The integrated circuit was modeled as a flexurally loaded plane-strain plate with the silicone rubber encapsulant as a plane-strain slab (Ref 44). The beams were modeled as an elastic, kinematic strain-hardened plastic material (Ref 45, 46). An investigation (Ref 47) has been conducted that extended a previous analysis of the beams (Ref 48) to include beam shear effects. These studies suggest that as the device (Fig. 27d) undergoes temperature change, the ends of each beam are displaced relative to one another in a manner having a horizontal and vertical component. The horizontal component is due to the different expansion coefficients of the chip and substrate. The vertical component comes from the fact that the expansion coefficient of the silicone rubber is much greater than that of the beams. If the magnitude of the temperature cycle is small enough (about 50 °C, or 90 °F), the cyclic strain is elastic for all points of the beams. Larger temperature excursions give rise to cyclic plastic strains. The plastic strain is believed to occur in three distinct parts of the beam (Fig. 28). The result of this analysis is the analytically derived curve for maximum cyclic strain versus ΔT shown in Fig. 29. An empirical formula (Ref 49) was used to relate Δε, the strain magnitude, to N_f, the cycles to median failure (Ref 47):

$$\Delta\epsilon = MN_f^{-1/2} + \frac{G}{E} N_f^\alpha \qquad \text{(Eq 10)}$$

where $M = 0.4$, $\alpha = -0.02$, and $G/E = 0.0017$.

Equation 10, along with the analytic relationship shown in Fig. 29, yields the analytic prediction in Fig. 30. The empirical data gathered (Ref 47) are included to show the degree of

Fig. 30 Cycles to median failure versus ΔT

Source: Ref 47

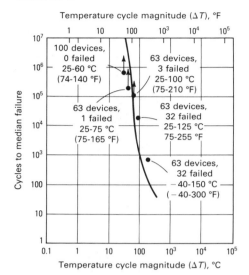

Fig. 31 SEM micrographs of ruptured integrated-circuit beam lead

$\vdash\!\!\!\dashv$ 20 μm

agreement between theory and practice. The analytic results assume a particular device geometry as well as particular encapsulant and gold fatigue properties. Changes in design and material properties have various effects on the analytic curve shown in Fig. 30. Increasing the chip size would shift the entire curve to the left. An increase in beam thickness shifts the curve to the right. Increasing the free beam length increases the cycles to median failure. Lowering the thermal-expansion coefficient, Young's modulus, or the bulk modulus shifts the entire curve to the right. Figure 31 shows SEM micrographs of a semiconductor device beam that had ruptured after 60 cycles from −40 to 130 °C (−40 to 265 °F). The characteristic fatigue fracture is apparent.

Metallic Interface Failures

Metal-Metal Interdiffusion. One failure mechanism due to interdiffusion is that resulting from the improper formation of intermetallic phases in the commonly used gold-aluminum wire bond system. Figure 32 shows an example of such intermetallic formation. Five phases can possibly form from this interdiffusion: $AuAl_2$, $AuAl$, Au_2Al, Au_5Al_2, and Au_4Al. If there is a sufficient supply of gold and aluminum and if the bond sees temperatures above 250 °C (480 °F) for a long enough time, all five phases will be present. Generally, most common phases of the intermetallic formation will be Au_5Al_2, regardless of time or temperature. However, if there is a shortage of gold or aluminum, the bond will contain only aluminum-rich or gold-rich intermetallics after a reasonable time at temperature. In themselves, all of these phases are stronger

Fig. 32 Example of intermetallic formation around a bonded lead

Source: Ref 20

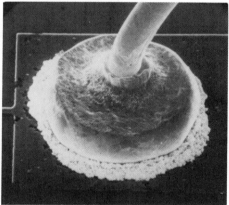

$\vdash\!\dashv$ 10 μm

than aluminum. However, weak or open bonds sometimes occur at temperatures in excess of 300 °C (570 °F). These problems have been related to the formation of Kirkendall voids in intermetallic phases (Fig. 33), which can form as a result of the differing diffusion rates of the aluminum and gold in those phases. An excess vacancy flux is generated on the side from which the faster species diffuses. These vacancies nucleate into voids. Nucleation is fastest in the predominant phase, Au_5Al_2, although it has also been shown to occur in Au_5Al_2 upon continuous aging at temperatures above 300 °C (570 °F). If gold and aluminum are present in sufficient quantities, the formation of intermetallics follows the rate law (Ref 50):

$$K = 5.2 \times 10^{-4} \exp\left[\frac{0.7}{kT}\right] \frac{cm^2}{s} \qquad \text{(Eq 11)}$$

When either the gold or aluminum becomes depleted, the formation continues at a much slower rate. The time for voiding to become continuous in Au_5Al_2 has been measured as a function of temperature (Fig. 34). This graph has served as the basis for design limits for

Fig. 33 The effect of Kirkendall voids on intermetallic formation

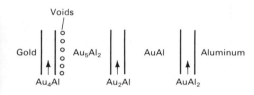

Fig. 34 Time-to-failure caused by continuous voiding in Au_5Al_2 versus temperature

Source: Ref 50

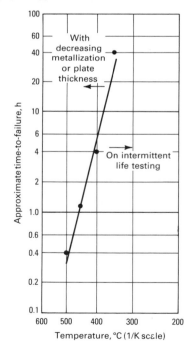

temperature excursions during and after bonding. Thermal cycling was shown to have reduced these times by as much as an order of magnitude. It was postulated that temperature cycling caused stresses due to differences in the thermal expansion between the various intermetallics of gold and aluminum. Microcracks occur to relieve these stresses in the more brittle intermetallics. During subsequent aging, the microcracks serve as sinks for excess vacancies that coalesce to form voids. Upon further cycling, more microcracks will form to nucleate during subsequent aging. In this way, Kirkendall voiding can occur at lower temperatures and shorter times. The solution to this failure mechanism is to keep time/temperature excursions to a relative minimum.

Recently, it has become clear that some impurities interact with the gold and aluminum to impede or accelerate the formation of the intermetallic compounds and, in some cases, to accelerate the formation of Kirkendall voids. Such impurities as bismuth, chromium, cadmium (Ref 51), palladium (Ref 52), nickel, cobalt, iron, and boron (Ref 53) have been shown to precipitate directly ahead of the moving diffusion front by a process similar to zone refining. The precipitated particles act as sinks for vacancies, thereby contributing to the formation of voids. Copper has been shown to have the beneficial effect of slowing the growth rate of the intermetallics without contributing to Kirkendall voiding (Ref 54); silicon has been shown to affect the composition of the gold-aluminum intermetallic phases (Ref 50). At present, there is no evidence showing that silicon contributes directly to the formation of Kirkendall voids.

The advent of the uses of epoxy in molding compounds (Ref 55, 56) or as die attach adhesives (Ref 57) has brought about another mechanism for the formation of Kirkendall voids at low temperatures. The high failure rate in a 200-°C (390-°F) storage test has been attributed to the presence of chlorine impurities and bromine flame-retardant compounds in the epoxy (Ref 55). The large resistance changes in gold wires bonded to aluminum pads have been attributed to voiding caused by the flame retardant in epoxy molding compounds (Ref 56). This failure mechanism has been studied extensively, and the presence of chlorine and bromine in the bond zone has been confirmed as a contributor to low-temperature Kirkendall voiding after aging in epoxy molding compounds at 125, 150, and 180 °C (260, 300, and 355 °F) (Ref 58). Figure 35 shows an example of this type of Kirkendall voiding. An activation energy of 0.8 eV for the gold-aluminum bond degradation by chlorine and bromine was found to agree with the 0.7 eV determined (Ref 50) for the formation of intermetallics with pure gold and aluminum. A mechanism has been proposed in which chlorine and bromine react with halide compounds that remove aluminum from the intermetallic formation by vapor transport (Ref 58).

Interdiffusion at aluminum-silicon contacts is another possible source of device failure. The diffusivity of silicon in thin-film aluminum is much higher than that in bulk samples (Ref 59). The activation energy is

Fig. 35 Photomicrograph of Kirkendall voiding in gold-aluminum bond
Source: Ref 58

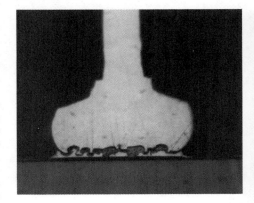

Fig. 36 The diffusivity of silicon in solid aluminum
Source: Ref 59

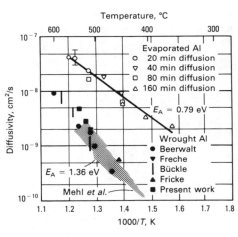

Fig. 37 Alloy penetration pits formed by aluminum-silicon interdiffusion
(a) As-formed. (b) After sintering

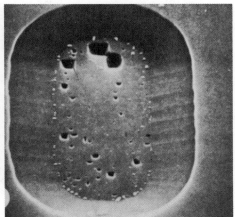

(a)

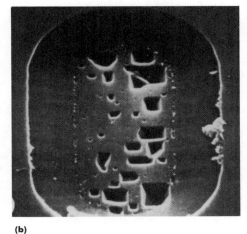

(b)

appreciably lower, presumably due to a much stronger grain-boundary contribution. When an aluminum-silicon metallization is annealed at elevated temperature, interface reactions can occur. As deposited, the silicon in the metallization is in a supersaturated solution that is not thermodynamically stable below 500 °C (930 °F). Silicon precipitation is diffusion controlled. Long anneals or slow cooling gives large precipitation. Fast cooling gives small precipitates and supersaturation (Fig. 36).

Thermal interdiffusion and solubility can result in the shorting of p-n junctions that are underlying or adjacent to ohmic contacts. Shorting occurs when aluminum penetrates into the silicon as spikes by the solid-state dissolution and diffusion of the silicon into the aluminum (Fig. 37a). This phenomenon is particularly risky for very shallow junction diffusions. Figure 37(b) shows alloy penetration pits after sintering for 1 h at 400 °C (750 °F). The etch pits at the silicon-aluminum surface at the

positive contact are caused by the electromigration of dissolved silicon in the aluminum contact. This process results in the shorting of an underlying or adjacent junction when a pit filled with aluminum penetrates the junction. Many manufacturers of integrated circuits use aluminum metallization doped with 1 to 3% Si to avoid this failure mechanism. The addition of the silicon to the aluminum reduces the interdiffusion of silicon in aluminum in the solid state, thereby reducing the aluminum diffusion into silicon to cause shorts. Some adverse effects of silicon inclusion in aluminum metallization have been discussed in the section "Stress Cracking and Voiding" in this article.

Another effect is the degradation of Schottky barrier contacts (Fig. 38). Here the aluminum-doped silicon is p-type, and this thin (100-Å) epitaxial layer raises the barrier height; but subsequent heat treatment tends to precipitate the aluminum atoms onto electrically inactive sites so the barrier drops. Figure 39(a) shows an

Fig. 38 Epitaxial precipitation of silicon in Schottky barrier contacts

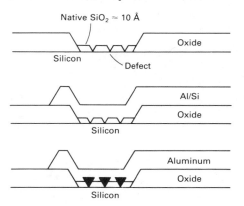

Fig. 39 Epitaxial mesas formed by aluminum-silicon interdiffusion
(a) As-formed. (b) After heat treatment

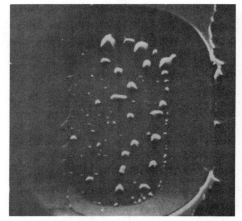

.(a)

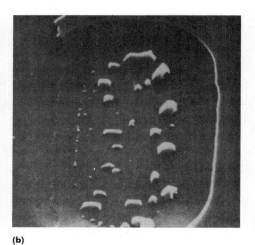

(b)

example of epitaxial mesas formed by this process. Figure 39(b) shows the effects of the heat treatment.

Electromigration at aluminum-silicon contacts is an additional interdiffusion effect. This is not a major problem in power devices, but may become more prevalent as shallow junctions and small contacts come into use. The net effect is that silicon hillocks form at the negative contact, and voids occur at the silicon/aluminum interface at the positive contact. The silicon hillocks due to the precipitation of the electromigrating silicon are resistive and may lead to open-circuit failures. Figure 40 shows an example of these electromigration effects. This effect has been investigated (Ref 60) using the configuration shown in Fig. 41. Mean time-to-failure (open circuit) was monitored at constant current density as a function of contact geometry. A dependence on contact width (Fig. 42) was found where current crowding causes voids to develop at the leading edge of the contact and sweep across the contact interface, causing an open circuit. Results for a constant current density (unspecified, but probably in the 1×10^3-A/cm^2, or 6.45×10^3-A/in.^{2}, range) are shown in Fig. 43. Dependence on contact length, again at a constant current density, is shown in Fig. 44. The dependence was thought to be related to the contact/grain-size ratio. (Small contacts have few aluminum grain boundaries intersecting the silicon interface.) This does not apply to the width dependence, because current crowding breaks up the contact into a series of unit-width contacts. The empirical equation for the time to 50% failure, t_{50}, is (Ref 60):

$$t_{50} = AJ^{-1}W \exp\left[\frac{\alpha}{L} + \frac{E_A}{kT}\right] \qquad (Eq\ 12)$$

where A and α are constants that depend on process conditions; E_A is probably around 0.85 eV for silicon diffusion in aluminum. An investigation of the interdiffusion of aluminum and silicon provided the following relationship

Fig. 40 Example of electromigration effects at contact windows

for the MTF due to this failure mechanism (Ref 61):

$$MTF = \frac{wt}{J^2} 4.396 \times 10^{12} \exp\left[\frac{0.889}{kT}\right] \qquad (Eq\ 13)$$

where w is the film width (in centimeters), t is the film thickness (in centimeters), T is the absolute temperature, k is Boltzmann's constant, and J is the current density. There is good agreement in the value of the activation energy, 0.889 eV. As junction depths have become shallower and contacts smaller, it has become necessary to use a metal barrier, such as tungsten, to eliminate this failure mechanism.

Fig. 41 Illustration of a typical metal-semiconductor contact

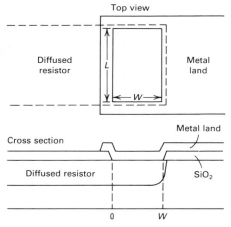

Silicon Oxide Failures

Large-scale integration metal-oxide-silicon devices are characterized by large areas of thin gate oxides. Dielectric breakdown is a major failure mechanism for MOS and CMOS devices.

The trend toward increasing levels of integration of MOS devices has resulted in rapidly decreasing feature sizes in these circuits. For example, although oxide thicknesses for 4K dynamic RAMs were in the range of 1000 Å, they are now in the range of 200 Å for 256K devices. As oxides are further thinned down with scaling for higher levels of integration, dielectric breakdown will become even more of a problem. There are two general categories of dielectric breakdown: defect-related and intrinsic.

Defect-Related Dielectric Breakdown. Figure 45 shows a histogram of breakdown

Fig. 42 Current density of contacts calculated from transmission line model

Source: Ref 60

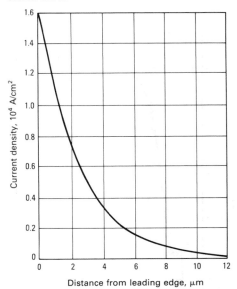

Fig. 43 Dependence of MTF on width of contact for a constant current density J_o

Source: Ref 60

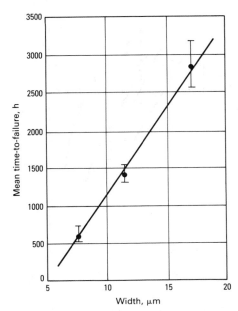

Fig. 44 Dependence of MTF on contact leading edge for a constant $J_o = I/L$

Source: Ref 60

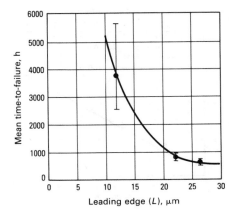

Fig. 45 Distribution of oxide breakdown voltages

Source: Ref 62

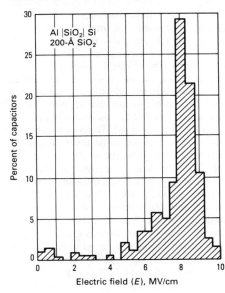

Fig. 46 Oxide breakdown probability as a function of field and time

Source: Ref 67

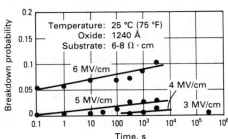

voltages in an Al-SiO$_2$-Si structure in which the SiO$_2$ dielectric is relatively thin. The failures that occur at low voltage are due to defects in the dielectric. Such defects can be cracks, pinholes, included particles, impurity-caused structural anomalies, or regions of thin oxide or lower dielectric strength (Ref 63-65). Oxide breakdown of this nature is a cause of low manufacturing yield and early field failures in MOS and CMOS circuits. For defect-induced breakdown, the thermal-activation energy is of the order of 0.3 or 0.4 eV, depending on whether aluminum or polysilicon is used as one electrode of the structure (Ref 66). While defect-related dielectric breakdown is characterized by a relatively low thermal-activation energy, it is sharply dependent on the voltage stress impressed across the structure (Fig. 46).

The electric-field dependence of dielectric breakdown, both defect-related and intrinsic breakdown, has been investigated by several authors (Ref 68, 69). The phenomenon has been used to investigate the dielectric integrity of thin SiO$_2$ films in the intrinsic-breakdown region and to screen out the low-voltage failures illustrated in Fig. 45, which contribute to infant mortality. The expression for the voltage-acceleration factor, A_v, can be given as:

$$A_v = \exp\left[\frac{C}{t_{OX}}(V_1 - V_2)\right] \qquad \text{(Eq 14)}$$

where t_{OX} is the oxide thickness (in angstroms), C is a constant (in angstroms per volt), V_1 is the voltage applied during stress, and V_2 is the operating voltage. Studies of SiO$_2$ dielectrics by several investigators have yielded values of

C ranging from approximately 400 Å/V (Ref 70, 71) to 1600 Å/V (Ref 68). In terms of the electric field applied across the dielectric, these values correspond to acceleration factors ranging from 55 to 10^7 MV/cm, a troubling discrepancy.

It has recently been postulated that the acceleration factor is temperature dependent (Ref 72). The form of the electric-field acceleration factor has been shown to be $\gamma(T) = B + C/T$, where B and C are constants, and T is the absolute temperature; $\gamma(T)$ is given in inverse megavolts per centimeter, $(MV/cm)^{-1}$. From previous work (Ref 73), $B = -5$ and $C = 0.28/k$, where k is Boltzmann's constant, have been evaluated (Ref 72). The fit is shown in Fig. 47 to extend over a wide temperature range. The actual voltage acceleration is given as $A_v = 10^{\gamma\Delta E}$, where ΔE (in megavolts per centimeter) is the incremented increase of the electric field across the oxide over the nominal electric field.

Intrinsic Dielectric Breakdown. The high-voltage failures shown in Fig. 45 can be represented by a skewed Gaussian distribution and are believed to be due to intrinsic breakdown. For intrinsic breakdown of SiO$_2$, time dependence on electric field is very steep (Ref 66). This dependence is shown in Fig. 48, which illustrates time-to-failure as a function of electric field. Temperature stress is also a factor in intrinsic breakdown. For intrinsic breakdown, the activation energy is approximately 1.4 eV for SiO$_2$ dielectric structures with polysilicon as one electrode.

Early investigation of intrinsic breakdown concerned the effects of sodium contamination

Fig. 47 Dependence of electric-field acceleration parameter on temperature

Source: Ref 72

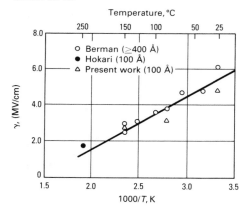
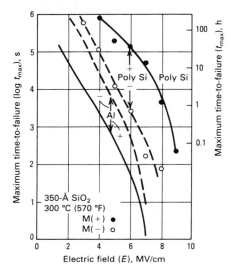

Fig. 48 Maximum time-to-failure as a function of field at 300 °C (570 °F) for polysilicon and aluminum electrodes

Source: Ref 66

Fig. 49 Current through a Na⁺-contaminated 2000-Å SiO₂ film as a function of time

Source: Ref 75

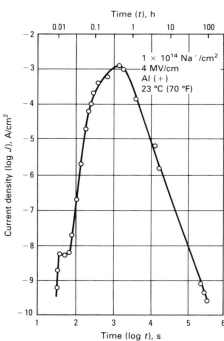

in SiO$_2$ films (Ref 74-76). Sodium-related breakdown was found to be important in time-to-breakdown of MOS capacitors when the gate was positively biased. The exact mechanisms relating sodium contamination and breakdown in SiO$_2$ are not fully understood. However, it has been proposed that mobile sodium ions drifting through the SiO$_2$ film cluster at the Si/SiO$_2$ interface can cause enhanced current injection from the silicon into the silicon oxide. With increased electrical field, the current enhancement leads to a destructive current runaway. It is also possible that clustering could cause local high-field areas where impact ionization might be initiated. The time to dielectric breakdown was found to be inversely proportional to the fourth power of the electric field (Ref 74). The leakage current is also time dependent (Fig. 49). The increasing current at short time is due to the mobile ion current. At longer times, the current is low because the mobile sodium ions have all drifted through the dielectric and clustered at the SiO$_2$/Si interface.

Impact ionization is believed to be related to dielectric breakdown in SiO$_2$ because of the low mobility of holes in SiO$_2$ (Ref 77, 78). Hole mobility values of $\mu = 10^5$ cm^2/V·s at room temperature have been reported (Ref 79). Therefore, after the formation of electron-hole pairs in SiO$_2$, the electrons are swept out rapidly by the electric field. The holes, on the other hand, move much more slowly. The model assumes that this hole cluster leads to an increased local electric field, which in turn leads to more impact ionization leading to breakdown. In general, the effect is described by several characteristics:

- Electrons tunnel into the SiO$_2$ conduction band from the cathode by Fowler-Nordheim tunneling
- These electrons are accelerated by the SiO$_2$ field, and some gain sufficient energy to cause electron-hole pairs by impact ionization

- These secondary electrons are rapidly swept out of the oxide, but the holes, being relatively immobile, remain and increase the field in the oxide in the vicinity of the hole cluster
- The increased field leads to further impact ionization and possibly to dielectric breakdown
- The process is balanced by drift and recombination of holes

An excellent review of the literature on dielectric breakdown in SiO$_2$ dielectrics is given in Ref 80.

The phenomenon of charge trapping in SiO$_2$ dielectrics grown on silicon films has been studied for some time (Ref 81-83). It arises when carriers (electrons or holes) flowing through the oxide become trapped at certain sites. A number of distinct trapping mechanisms have been studied, each giving rise to specific trapping sites. These sites become populated or depopulated depending on the electric field, temperature, trapping cross section, and recombination rate. For each distinct trapping mechanism, a density of trapped charge can be given by:

$$N_t = N_0 \left[1 - \exp\left(-\frac{\sigma Q}{e} \right) \right] \qquad \text{(Eq 15)}$$

where N_t is the area density of trapped charge, N_0 is the density of unfilled traps, σ is the capture cross section, Q is the leakage charge flux, and e is the electron charge. In Fig. 50, the dashed line shows the case of the field without trapped charge, and the solid line represents the effects of a sheet of charge at

Fig. 50 Altered electric field due to trapped charge

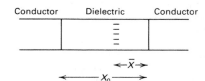

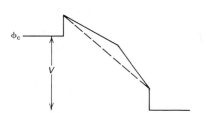

$\bar{X}$. The electric field is reduced to the left of the trapped charge and increased to the right.

A number of investigators (Ref 84-86) have extended investigation of charge trapping to the high electric-field region found in ultrathin oxides. Three or four electron traps and one positive hole trap have been observed. Table 6 gives the trapping parameters found for a 250-Å SiO$_2$ film (Ref 86). The occurrence of a positive hole trapping mechanism is important in understanding the discussion of threshold drift in CMOS transistors given in the section "Charge Injection" in this article.

Table 6 Measured trapping parameters for representative 250-Å oxides

Trapping mechanism	Generation rate G, (Cb^{-1})	Recombination cross section R, (cm^2/Cb)	Centroid $\bar{X}$, Å
1 e^-	3×10^{-11}	<1	125
2 e^-	3×10^{-7}	2.4	25
3 e^-	2×10^{-6}	20	8
4 h^+	Donor trapping		~10

Source: Ref 86

While the spatial distribution of trapped charge cannot be determined, each trapping mechanism can be described in terms of a centroid of charge that can be determined, and the trapped-charge distribution can be thought of in terms of a sheet of charge. Figure 50 shows schematically the location of a sheet of charge at centroid $\bar{X}$ and the modification of the electric field in the oxide due to that sheet of charge. The electric field at any point in the oxide can be thought of as the sum of the electric field due to the applied voltage, V, and an incremental electric field due to the trapped charge (Ref 81):

$$E = \frac{V}{X_0} + \Delta E \qquad \text{(Eq 16)}$$

where X_0 is the thickness of the dielectric, and $\Delta E = \Sigma \Delta Ei$ (the summation of the incremental electric fields arising from the charge trapped by each of the various charging mechanisms). The integrated leakage current is measured during the time-to-breakdown. The value of $E_{\text{crit}} = V/X_0$ in Eq 16 has been estimated, for which $\Delta E = 0$; E_{crit} is the stress field at which breakdown should occur before any leakage current could flow. This value is extrapolated from lower applied fields for which there is leakage current before breakdown. These data were plotted (Fig. 51) and extrapolated to zero accumulated trapped charge. The value of $E_{\text{crit}} = 11.2$ MV/cm was determined (Ref 86).

Silicon Oxide Interface Failures

In this section, integrated-circuit failures that occur when charge accumulates at silicon oxide surfaces are examined. In some cases, the accumulated charge is trapped at interface states; in other cases, there simply arises sufficient charge accumulation to disrupt the electrical function of the integrated circuit. Charge accumulation can occur at the surfaces of field oxides or, in the case of CMOS and NMOS devices, at the surface of gate oxides. In the latter situation, a phenomenon known as threshold drift occurs that is very disruptive to the operation of these devices.

Ionic Contamination. One failure mechanism that often limits the reliability of semicon-

Fig. 51 Charge-to-fail versus applied electric field
Source: Ref 86

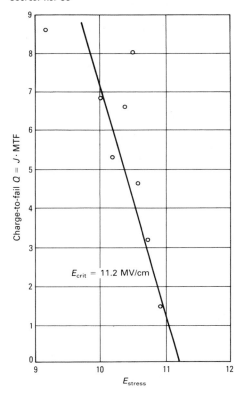

Fig. 51 $E_{\text{crit}} = 11.2$ MV/cm

ductor devices is silicon inversion due to ionic contamination of field oxides. Figure 52 illustrates how this can occur. Here the presence of positive charge in (or on) the field oxide inverts the p-type base material due to induced negative charge, making it an n-type material. Similarly, the presence of negative charges can invert the n-type collector material to p-type. These inversions result in increases in leakage current (shunt) paths, decreases in breakdown voltages, and hence failure of the device. Sodium is probably the worst culprit because it is mobile in SiO$_2$ and will diffuse to the SiO$_2$/Si interface, is a prevalent element, and readily dissociates into a positive ion.

Low surface-dopant concentrations are particularly susceptible—especially at high temperature and humidity, which increase the surface mobility of ions. Silicon inversion can be reduced by using silicon nitride or phosphorus-doped glass passivation. Other solutions, which are not without their penalties, include hermetic encapsulation at low relative humidity (expensive), thick field oxides (step-coverage problems), channel-stops (increased chip size), uniform increase in surface doping (decreases breakdown voltage), and the use of metal field plates over implanted resistors (less area for metal routing). The thermal-activation energy for silicon inversion due to ionic contamination is generally accepted as approximately 1.0 eV.

Fig. 52 Illustration of silicon inversion due to ionic contamination

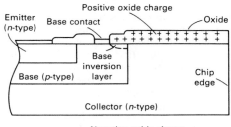
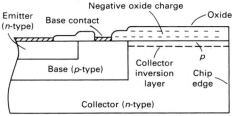

Base and collector inversion layers

Fig. 53 Field-induced junction by mobile surface ions
$-V_g$, gate voltage; V_t, voltage at time t. Source: Ref 87

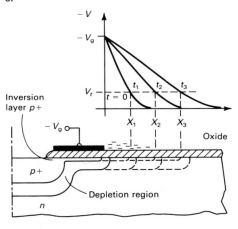

Surface-Charge Accumulation. In addition to the ion drift in bulk oxide described above, there is the possibility of ion movement in lateral electric fields (Ref 87). The drift of mobile ions occurs on the outer surface of the SiO$_2$ insulator or along insulator interfaces. Figure 53 shows the insulator surface condition for a $p+n$ diode. The accumulation of surface charge extends the field plate laterally, causing the inversion layer to extend with time. This can lead to surface leakage and channel formation between circuit elements. The activation energy is estimated at 1 eV. Because the mobility of the ions is strongly dependent on humidity and temperature, the effect is strongly influenced by protective layers. Phosphosilicate glass is widely used to immobilize the ions in MOS structures.

Ion-Induced Threshold Drift. Sodium, introduced during processing into the SiO_2 gate insulator of MOS devices, is easily ionized and can drift readily through the oxide under the influence of an electric field, causing considerable changes in parameters, such as threshold voltage. The closer the ions come to the SiO_2/Si interface, under the influence of the gate voltage, the larger the effect is on the silicon. The ions retain their charge even when they approach very close to the interface.

Equation 17 expresses the change in the threshold voltage (Ref 88):

$$\Delta V_t = \frac{Q_0}{C_0} \qquad \text{(Eq 17)}$$

where Q_0 and C_0 are the total mobile charge and oxide capacitance per unit oxide area, respectively. Equation 17 provides a qualitative illustration of the sensitivity of MOS structures to sodium ion drift. In a layer 1000 Å thick, the migration of 10^{11} ions/cm^2 could result in a threshold voltage shift of 0.5 V. In a smaller (10^{-5}-cm^2) gate area configuration, only 10^6 ions would be required to produce this shift. The forward drift activation energy has been measured as 1.1 eV. The recovery may be faster with an activation energy of about 0.75 eV. Phosphosilicate glass is widely used to immobilize the ions. The use of an O_2-HCl oxidizing ambient has also been reported (Ref 88). This acts to reduce sodium problems by unblocking the silicon surface to sodium.

Interface Traps. Two other types of electrically active sites in the Si/SiO_2 interface region cause performance and reliability problems in MOS devices (Ref 89). These are surface-state oxide charges (Q_{SS}) and interface states (N_{SS}).

Surface-state oxide charge is intrinsic and is generally considered to be located very near the interface. Its density is independent of band bending or applied bias. It is immobile, and its charge state cannot be changed by varying gate bias. Its polarity is always positive. Its magnitude depends on silicon-surface orientation. It is practically independent of impurity type and concentration in the silicon and is independent of SiO_2 thickness for a specific preparation condition. The value of Q_{SS} is related to oxidation conditions. Control is a function of oxidation conditions, such as temperature, wet or dry ambients, and silicon substrate orientation. The density of Q_{SS} is two or three times larger on a (111) surface than on a (100) surface (Ref 89) and is an inverse function of oxidation temperature. Oxidation in dry oxygen is preferred for minimum fixed charge.

Interface states, N_{SS}, are related to the termination of the silicon lattice at the Si/SiO_2 interface and have been described as centers that have energy levels in the silicon band gap and can exchange charge with the silicon (Ref 89). The traps are most likely caused by defects at the Si/SiO_2 interface or distorted or broken silicon-oxygen bonds. Interface traps can be either charged or discharged by varying gate voltages. Interface-trap density has the same dependence on orientation as fixed charge. They most likely occur at the Si/SiO_2 interface. Interface-trap density can be reduced by annealing in hydrogen.

The values Q_{SS} and N_{SS} can change with time during device life, resulting in stability problems. One effect that can change the densities is slow trapping. Temperature-bias aging accelerates the effect, producing an irreversible change in the densities. However, evidence exists that the slow trapping effect is negligible in many practical cases (Ref 89). Another way in which Q_{SS} and N_{SS} densities can change at low temperatures is transfer of hydrogen in or out of the oxide, for example, from moisture in the device package. Protective layers, such as silicon nitride, can provide protection. However, it has been reported (Ref 90) that n-channel MOS transistors, encapsulated with plasma-deposited silicon nitride, exhibited drift in saturation-mode threshold when operated with high drain voltage. The effect was attributed to hydrogen, present in the nitride, diffusing to the Si/SiO_2 interface, creating interface states near the drain diffusion.

Charge Injection. Electrons traversing the gate channel of an n-channel device can be scattered into the gate oxide if they have sufficient energy (Ref 91) (Fig. 54). In saturation, there is a large field in the drain-depletion region. Channel-current electrons are accelerated by the field between the source and drain. For sufficiently high source-drain fields, the electrons are accelerated to energies in excess of the normal energy of free electrons in a lattice. They are termed hot electrons. Impact ionizing and phonon scattering direct some of these electrons to the Si/SiO_2 interface. If their energy is sufficient to surmount the interface barrier, they will be injected into the oxide, and the trapped charge will alter the threshold voltage. Hot-electron injection is increased with reduced channel lengths and oxide thicknesses, shallower diffusions, and increased diffusion doping. All of these will occur with device scaling. The effect is dependent on the effective electron energy or temperature, and the oxide trapping efficiency. The injection current, I_E, is given by:

$$I_E \propto \exp\left(-\frac{E_B}{kT_E}\right) \qquad \text{(Eq 18)}$$

where the interface barrier potential E_B is 3.1 eV, and T_E is the effective electron temperature, which is related to the electric field E_{SI} in the channel by the relationship:

$$kT_E = \frac{q(E_{SI})^2\lambda}{2E_R} \qquad \text{(Eq 19)}$$

where λ is electron free path and E_R is the energy loss per phonon collision.

Fig. 54 MOS transistor in saturation

V_{GS}, voltage of gate referenced to source; V_{DS}, voltage of drain referenced to source; V_{DSAT}, drain saturation voltage; I_B, substrate current; $-V_{SB}$, voltage of substrate referenced to source

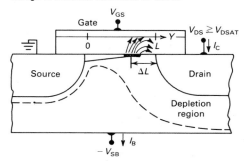

The value E_{SI} increases with increasing channel doping. It increases with decreasing channel length, insulator thickness, source/drain junction depth, and temperature. The value E_{SI} is greatest near the drain, and this is where most electron injection occurs. The trapping efficiency of the charge in the oxide is thought to be related to the density and location of trap sites and the electric field of conductive metallization over the oxide. The injected electrons travel through the oxide and occupy electron trapping levels in the gate oxide (see the section "Silicon Oxide Failures" in this article) or surface interface states discussed above. This in turn causes a shift in the threshold voltage of the transistor; that is, the voltage needed to turn the MOS transistor on or off is shifted from its design value. Continued charge injection and trapping leads to the inability to operate the transistor completely and to circuit failure. A comprehensive treatment of the current knowledge on threshold shift in CMOS circuits is provided in Ref 92.

Silicon *p-n* Junction Failures

Alpha Particle Induced Failures. Ceramic packaging materials contain minute quantities of uranium and thorium. These compounds have been found by quantitative analysis to exist in most materials used in semiconductor packaging. Table 7 lists the results of quantitative analysis of various materials (Ref 93). As shown in Table 7, these impurities undergo radioactive decay, emitting α particles in the process. Those α particles that pass through the active circuit of an NMOS memory will cause it to fail momentarily in a random fashion. Even the lowest level of 0.04 α/cm^2·h shown in Table 7 would be enough to produce a high failure rate for some dynamic memories. The failures that occur are called soft errors. The term soft error refers to a random single-bit failure not related to a physically defective device. Replacement of the device before the failure would not necessarily have prevented the failure, nor does the occurrence of this type

Table 7 Results of analysis of packaging materials

Material	Uranium, ppm(a)	Thorium, ppm(a)	α particles, cm²·h(b)	Zirconium, %(c)
Alumina - A	2.5	0.6	0.6	1
Alumina - B	...	...	0.3	...
Alumina - C	...	...	0.5	...
Glass - A	12	6	29	17.5
Glass - B	2.5	3	5.2	3
Glass - C	17	6	45	25
Glass - D	12	6	18	6
Glass - E	...	...	32	20
Epoxy	...	...	0.01	...
Silicone	...	...	0.01	...
Gold-plated lids	...	...	0.04–1.0(d)	...

(a) Mass spectrographic analysis. (b) α scintillation counting. (c) Emission spectrographic analysis. (d) Activity varied widely.
Source: Ref 93

Fig. 55 Effects of α particles on NMOS dynamic memory storage cell

'0' and '1' are logic states

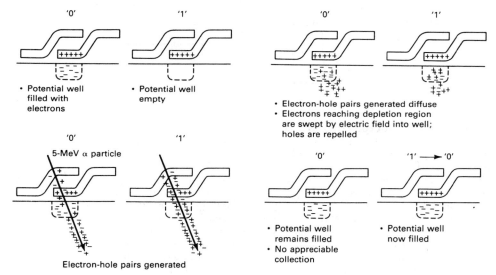

Fig. 56 Error rate versus α flux for two test devices of different critical charges

Source: Ref 93

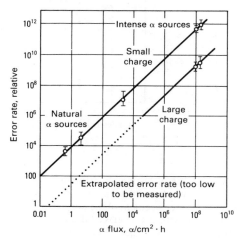

Fig. 57 Error rate versus critical charge for devices from several manufacturers

Source: Ref 93

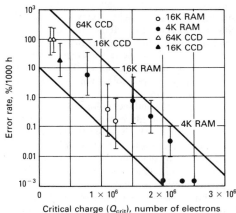

of failure increase the probability of another occurrence with that bit. To illustrate the seriousness of the problem, a typical soft-error rate may be of the order of 1 bit per 1000 h in a system containing 1000 16K memory devices. This rate corresponds to a device failure rate of one per 10^6 h or 1000 failures in time (FITs, one FIT being equal to one failure in 10^9 device hours). Furthermore, this high rate does not improve with time, because the radioactive uranium and thorium compounds have effectively infinite half-lives.

In dynamic RAMs, data are stored as the presence or absence of charge in potential wells in p-type silicon under polysilicon gate electrodes. A refresh cycle is needed to maintain the charge. For example, zero may be represented by a stored charge in the range, depending on the design, of several hundred thousand electrons to a few million. A one may be represented by an empty potential well. A factor called the critical charge, Q_{crit}, differentiates between a one and a zero in terms of the

number of stored electrons. Electron-hole pairs are generated by the α particles and are collected by depletion layers such that electrons end up in the potential well. If collection of electrons exceeds Q_{crit}, then a one can be changed into a zero. Collection does not change a zero into a one. Figure 55 shows the stages of the production of a soft error. The error rate is directly proportional to the α flux, as shown by experimentation (Ref 93). Figure 56 shows the error rate to be directly proportional to α flux for several decades of source intensity. In addition to the flux dependence, the actual soft-error rate of a specific device will depend on such factors as target area, collection efficiency, critical charge, and geometry of the cell. Obviously, these factors are all interrelated as a function of the device design and technology and package type and composition.

Critical charge has been called the most important factor in determining device sensitivity to soft errors. As an example of Q_{crit} in a commercial dynamic RAM, the value for a 16K

design is about 10^6 electrons. Soft-error rates versus Q_{crit} have been measured (Ref 93) for a number of dynamic RAMs and charge-coupled devices (CCDs) from several manufacturers (Fig. 57). In other experiments (Ref 93), it was shown that the critical charge stored can be varied by changing cell geometry and thus storage capacitance or by changing voltages (Fig. 58).

The practical prospects for characterization of a specific device for α-particle soft errors have been accomplished by accelerated testing. For example, consider a system with 1000 RAMs, and assume that one system error per 6 weeks can be tolerated (many cannot tolerate such a high rate). One error per 1000 RAMs per 6 weeks means an error rate per RAM of 0.1% per 1000 h. Verification of this maximum allowable device error rate would require mil-

Fig. 58 Effects of critical charge on soft-error rate

Source: Ref 93

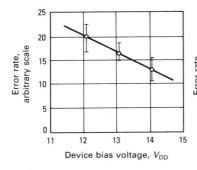

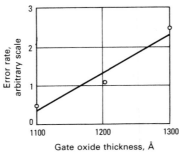

Table 8 Bits needed for error detection and correction

Bits/word	EDAC bits	Overhead, %
8	5	63
16	6	38
32	7	22
64	8	13

Fig. 60 Growth of junction leakage as a function of time

Source: Ref 96

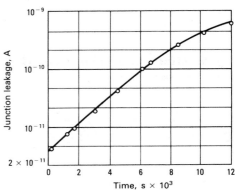

lions of hours of testing—a difficult prospect. The present alternative, accelerated testing, involves the replacement of the normal alumina device lids with a hot glass lid and a foil of thorium. This technique (Ref 93) accelerates the device exposure time by a factor of 10^6.

There are some prospects for reduction of the α flux at the device surface by cleaning up the materials. Sealing glasses, which have high fluxes, are easiest because the most radioactive constituent of the glass can be eliminated. The selection of the proper alumina from those presently available might achieve a four-fold reduction in error rate. It is estimated that a factor of ten reduction in α flux may eventually be achieved. For hermetic packages, an eventual lower limit of the order of 0.001 to 0.01 $\alpha/cm^2 \cdot h$ may be possible. This is near the limit of measurability.

Various design and technology changes have been proposed to reduce α-particle sensitivity. The design of a 64K dynamic RAM has been reported that has a soft error of 100 FITs, which was achieved by designing α immunity into the circuit (Ref 94). One possible method of improvement involves shielding of the chip surface with a material that will stop the particles. A promising technique was recently reported (Ref 95). Use of a thick (2-mil minimum) RTV rubber coating over the chip surface of 4K static RAMs, packaged in a side-brazed multilayer ceramic package was found to shield the chip effectively from α radiation. A soft-error rate of less than 0.1% per 1000 h was observed with this coating, and the coating did not introduce any undesirable effects.

As device trends toward higher density continue, smaller values of Q_{crit} and changes in collection efficiency will be encountered. On the other hand, there will certainly be lower values of α flux in next-generation packages.

Nevertheless, the error rate is still likely to be high, and error detection and correction (EDAC) will be necessary. This is accomplished by reserving some bits for reconstruction of the data following an error. In a DRAM, reconstruction is done when refreshing. Table 8 shows that EDAC becomes more efficient as word length increases.

Fig. 59 Decay diagram for Kr⁸⁵

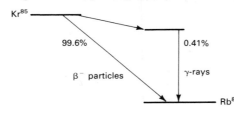

Radioactivity-Induced Failures. There is a second type of failure induced by radioactivity in NMOS memory devices. This failure mode arises in NMOS dynamic memory devices that have residues of radioactive gas entrapped in hermetically sealed packages during hermeticity testing (Ref 96).

The procedure for hermeticity testing is first to pump a vacuum in a chamber containing the devices, then backfill the chamber with nitrogen containing Kr^{85} gas. If there is a faulty seal in a device package, Kr^{85} will be forced into the package and will be detected after the device is removed from the chamber using a scintillation counter. Damaging levels of β radiation can impinge on the device if excessive radioactive gas is forced into the package and not properly detected. The activity of the failed devices ranged from 0.1 to 1.0 MCi of Kr^{85} (Ref 96).

The Kr^{85} decay scheme is shown in Fig. 59. The half-life is 10.7 years. These are two decay paths. Most (99.6%) of the atoms decay by the release of β particles, and the remaining 0.4% decay by a path that emits γ-rays.

There are two kinds of failure mechanisms. The first is a *p-n* junction leakage-current increase due to radiation-induced trapped charge. This mode of failure was discussed in the section "Surface-Charge Accumulation" in this article. The increase in leakage current due to surface charge is shown in Fig. 60 to be exponential with time for a large area *p-n* junction. Measurements on smaller area junctions show much less charge, indicating that degradation occurs at the perimeter surface and not in the bulk.

The second failure mechanism is due to positive charge trapping in the gate oxides of the FET transistors that comprise the NMOS dynamic memory. The β radiation generates electron-hole pairs in the gate oxide of the NMOS FET transistor. The positive charge is trapped near the SiO_2/Si interface (Ref 97). The presence of this charge will have two effects. First, there will be an increase in leakage through the transistor when it is turned off; that is, the FET transistor cannot be turned off completely. Thus, certain charged electrical nodes will not maintain their charge long enough for the device to function. Second, any positive voltage impressed on the gate of the FET must overcome the existing positive charge in the oxide before turning the transistor fully on. This causes the device to slow down.

The effects of entrapped Kr^{85} gas on device reliability have been estimated (Ref 98). A model of the β influence at an oxide has been developed, an electron-hole pair generation rate provided (the change of trapped charge), and the above-mentioned threshold voltage shift predicted. The model can be expressed as:

$$\Delta V_{tx} = -\alpha P^1 t_{ox}^n S\gamma \left[1 - \exp\left(\frac{-t}{\gamma}\right) \right] \qquad \text{(Eq 20)}$$

where ΔV_{tx} is the threshold voltage shift, S is the activity of the Kr^{85} gas measured in Curies, αP^1 and n are constants, t is thickness of oxide, and $\gamma = 9.4 \times 10^4$ h. The following constants have been experimentally determined (Ref 98):

Field and gate oxide	Gate oxide only
$\alpha P^1 = 2.9 \times 10^2$	$\alpha P^1 = 3.7 \times 10^{-4}$
$n = 1.6$	$n = 2.2$

These values show that if a maximum voltage shift of -50 mV is assumed for a 500-Å oxide and if a 10-year lifetime is required, then the maximum allowable level of entrapped Kr[85] gas is 2.7 nCi (Ref 98).

The dose received by a chip in a hermetic package has been computed as a function of the package activity (Ref 99). This relationship is shown in Fig. 61. This graph shows that if a circuit fails at a dose of 10 krad (most NMOS RAMs fail in the range of 0.7 to 5 krad) (Ref 100) and its required lifetime is 10^5 h, then the activity in the package just after leak testing should not exceed 5 nCi. This is in reasonable agreement with the work reported in Ref 98 and confirms the use of the experimentally determined constants given above.

Latch-Up in CMOS. While not an intrinsic failure mechanism, that is, a failure mechanism that arises from an intrinsic defect in integrated-circuit processing, latch-up in CMOS integrated circuits is potentially a very destructive phenomenon. It occurs because of the presence of parasitic pn-pn connections that arise between p-type and n-type MOS transistors in CMOS circuits. These pn-pn parasitic transistors are normally off. However, under certain conditions, they can be switched on, creating a low-resistance path between V_{DD} and V_{SS}. The resulting high current destroys the device, usually by melting the metallization.

A complete discussion of the pn-pn device is given in Ref 101. An abbreviated description is provided in conjunction with Fig. 62, which shows a four-layer pn-pn structure. Enclosed in the dotted line is a pnp bipolar transistor, with the remainder of the structure comprising an npn bipolar transistor. If this structure is forward biased, layer $p1$ is the emitter of the pnp and $n2$ the emitter of the npn. The common reverse-biased collector junction is J_2. The collector current of the pnp drives the base of the npn and vice-versa. In this situation, a little current can be magnified immensely.

Current-voltage curves of this device are shown in Fig. 63. There are two distinct states. The device is off for voltages below V_{BO}, the breakover voltage. At V_{BO}, the device jumps through a negative resistance region to the on state characterized by low resistance. After the device switches on, the voltage collapses, and there is a large increase in current. The pnp device can be triggered by temperature increase, voltage increase, optical excitation, a rapid voltage change, or injected base current into $n1$ or $p2$.

In CMOS circuits, the term latch-up describes the very undesirable breakover to the on state of parasitic pn-pn device. Figure 64 shows the cross section of a CMOS inverter. It shows the parasitic vertical npn and the lateral pnp parasitic transistors. The presence of a latch-up

Fig. 61 The time to accumulate a special dose versus the initial activity of the Kr[85] in the air space above the chip
Source: Ref 99

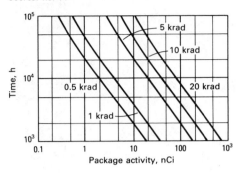

Fig. 62 The *pn-pn* device with schematic representation

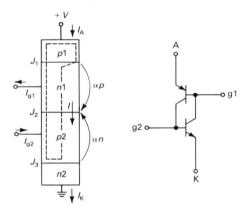

path from the p-channel source at V_{DD} through the n-tub, p-tub, and n-channel source to V_{SS} is shown schematically. The equivalent circuit is shown in Fig. 65. Note the resistances R_A and R_K shunting the base-emitter junctions. If the values of these were zero, the emitter junctions could never become forward biased, and latch-up could not be initiated.

Latch-up on a CMOS circuit can be started if an output (Fig. 66) is brought momentarily or accidentally above V_{DD}. In this situation, holes are injected into the n-tub, are collected at the p-tub, and flow out of the V_{SS} contact. The hole current causes a voltage drop between the p-tub contact and the nearby $n+$ source, which forward biases it, causing the $n+$ source to inject electrons into the p-tub. If this process continues under the right conditions for a sufficiently long time, latch-up occurs.

Avoiding latch-up in CMOS circuits is largely a matter of design lay out rules and careful process control (Ref 102). Process changes involve reducing the substrate resistance by doping it with gold or reducing the parasitic transistor gain by using a heavily doped $p+$ layer at the bottom of the p-tub (Ref

Fig. 63 Current-voltage characteristics of the *pn-pn* device

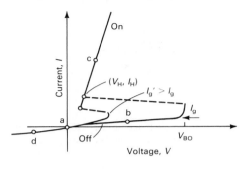

Fig. 64 Cross section of a CMOS inverter with parasitic devices

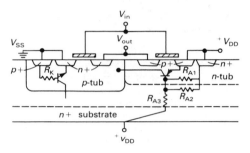

Fig. 65 Equivalent circuit of parasitic devices

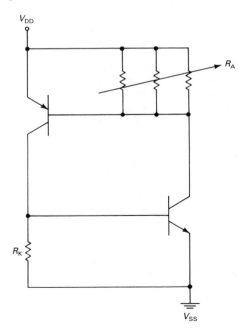

103, 104). The most effective latch-up control seems to be the use of guard bands consisting of p-tubs connected to V_{SS} in order to collect holes and to prevent them from starting latch-up. Tub

Fig. 66 Charge transfer in a CMOS inverter during latch-up

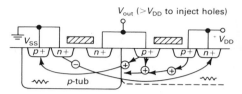

isolation is another technique. However, both tub isolation and guard bands consume space on the circuit. The key is to find the right combination of device structure and circuit layout that will result in cost effective latch-up free CMOS devices.

Electrostatic/Electrical Overstress Failures

While not an intrinsic failure mechanism, electrostatic discharge (ESD) was recognized as a significant cause of integrated circuit device failure in the early 1970s. During the last 15 years, the ESD phenomenon has been identified as a contributor to device failures and therefore has increased costs in all stages of device and equipment production, assembly, test, installation, and field use. This failure mode and its analysis and prevention have become even more significant in recent times as newer devices have incorporated smaller geometries and new technologies.

Early studies of ESD damage to integrated circuits indicate two types of failure mechanisms (Ref 105, 106). The first mechanism is that of dielectric breakdown. This mechanism has been described in detail in the section "Silicon Oxide Failures" in this article. This mechanism is common in the gate oxides of MOS devices. This type of failure can be prevented if a protection circuit is built for each input of an MOS device that limits the voltage that can be developed across the gate oxide, thereby preventing the oxide dielectric breakdown.

The second mechanism is the degradation or breakdown of the p-n junction. Junction degradation or breakdown is frequently observed in p-n junctions connected with input terminals. An excess reverse current flows through a p-n junction during the discharge. If the power through the junction is large enough, the localized junction temperature is increased, and the more lightly doped side becomes intrinsic (Ref 107). A more highly conducting path is formed, resulting in current concentration and increased localized heating. The result is thermal runaway that will melt a small volume of silicon through the junction if sufficient pulse energy is available. The molten silicon will recrystallize upon cooling into a low-resistance path through the junction. Figure 67 shows an example of p-n junction degradation.

Fig. 67 I-V characteristic of device input circuitry
(a) Damaged input after electrostatic discharge. (b) Normal input characteristic

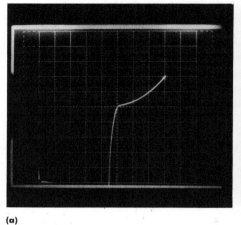

(a)

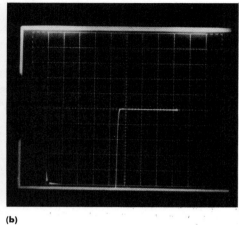

(b)

A model has been proposed for determining the threshold failure level of p-n junctions (Ref 108). The model uses linear heat flow theory and relates the power dissipated per unit area to the thermal characteristics of the semiconductor to determine the local temperature rise and the threshold power density for failure. Failure is assumed to occur when the temperature reaches the melting point of the semiconductor. The Wunsch-Bell relationship is:

$$\frac{P}{A} = \sqrt{\pi\,k\rho\,C_p}\,[T_m - T_i]t^{-1/2} \qquad \text{(Eq 21)}$$

where P is the power, A is the junction area, k is the thermal conductivity, ρ is the density of silicon, C_p is the specific heat of silicon, T_m is the failure temperature, T_i is the initial temperature, and t is time.

The Wunsch-Bell model has been applied to the failure of the junctions of npn transistors by charging a capacitance to a given voltage and then discharging it through a series resistor through the p-n junction (Ref 109). In this case, the average power through the junction is approximated by:

$$P_V = \frac{V_D I_P}{5} + \frac{R_B I_P^2}{10} \qquad \text{(Eq 22)}$$

where I_p is the initial peak value of the current, V_D is the voltage drop across the junction, and R_B is the internal resistance of the semiconductor bulk. The area is given as the cross-sectional area of the current path. The data exhibited good agreement with the Wunsch-Bell model (Fig. 68), showing that a capacitive discharge through p-n junction could be modeled and the limiting power density causing a junction degradation could be predicted.

The two main sources of static electricity that are responsible for failures of integrated circuits have been recognized (Ref 109). These are the discharge of a human body through an inte-

Fig. 68 Experimental data points superimposed on Wunsch-Bell curve
Source: Ref 109

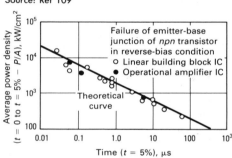

Fig. 69 Human body equivalent circuit
C_b, body capacitance; R_b, body resistance; R_d, device resistance; R_c, resistance to ground

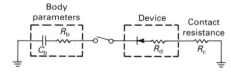

grated circuit and the discharge of a device through one of its own pins. The human body model is shown in Fig. 69. The values of the equivalent circuit have been determined based on the human body as an ESD source (Ref 110, 111). It is generally felt that a value of the body capacitance, C_b, of 100 to 250 pF and body resistance of 1000 to 2000 Ω are appropriate parameters. It is possible to develop body potentials (thousands of volts) that far exceed the potential that will be damaging to devices. With a potential of only 2000 V, the human body stores approximately 0.4 μJ of energy. With the human body equivalent circuit shown

Fig. 70 Charged device model

R_c, resistance to ground; R_d, resistance in device; L_d, inductance of device; C_d, capacitance of device

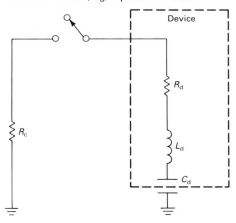

Fig. 71 Device capacitance as a function of orientation and device size

(a) Leads up. (b) Leads down. (c) 45° tilt. Source: Ref 112

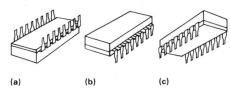

(a)   (b)   (c)

Device capacitance, pF

IC	16 pin (plastic)	18 pin (plastic)	24 pin (plastic)	24 pin (ceramic side-brazed)	40 pin
		Dual inline packages			
a	2.9	3.6	7.1	2.8	5.2
b	2.0	2.3	3.9	3.6	6.6
c	1.4	1.6	2.0	2.0	2.8

IC	24 pin (plastic)	24 pin (ceramic)	68 pin (plastic)	68 pin (ceramic)
		Chip carriers		
a	2.9	4.4	8.0	12.6
b	2.4	4.3	9.1	9.9
c	1.0	0.9	2.4	2.0

in Fig. 69, this energy is released with time constants in tenths of microseconds, providing average powers up to several watts, which are sufficient to rupture gate oxides and melt silicon. In some extreme cases, metallization interconnects are melted, and the chip surface is cratered (Ref 112).

The charged device model is shown in Fig. 70. A value of 1 pF was assigned to the device capacitance C_d (Ref 109). Later investigators (Ref 112, 113) have measured device capacitance up to 52 pF. These data (Fig. 71) show device capacitance as a function of device size and orientation. The average power developed in the two models was compared (Fig. 72), and it was shown that in the charged device model the average power dissipated in the integrated circuit is much greater even though the stored

Fig. 72 Comparison of average powers developed with the human body model and charged device model

DUT, device under test. Source: Ref 112

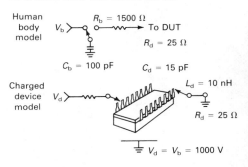

Model	Stored energy, μJ	5τ, ns	P_{ave}, W
Human body	50	750	~2
Charged device	7.5	~10	>750

energy is less (Ref 112). With device capacitance in the 1- to 12-pF range, a device can store up to several microjoules of energy. With low resistance to ground and a leadframe inductance of 10 nH, this energy results in nanosecond discharge pulses with an average power range of several hundred watts—enough to destroy metal. Figure 73 shows damage to an I^2L device that was charged to 1000 V, then discharged to ground. In an MOS device that has been charged to 1000 V, the grounding of one pin results in potential differences across gate oxides sufficient to cause oxide breakdown. Figure 74 shows an example.

Several investigators have determined the damage threshold for devices of different technology. The results of one investigation using the human body model are shown in Table 9. Electrostatic discharge is a cause of failure in all technologies and devices. The difference is in the degree of susceptibility. Metal-oxide semiconductor devices are the most susceptible due to breakdown of the thin gate oxide. In addition to oxide breakdown, junction leakage may be increased to MOS devices by ESD stress, causing a circuit to fail specifications even though it is still functionally operational (Ref 112). In addition to these susceptibilities in NMOS circuits, several authors have demonstrated an ESD-induced electrothermomigration of aluminum metallization through contacts into diffused areas of NMOS LSI devices (Ref 114, 115). In many cases, under the action of an applied electric field the aluminum has diffused through the silicon substrate and penetrated a nearby junction, thus shortening two p-n junctions together. Prevention of this failure mode is best achieved by good design practices, that is, keeping aluminum windows sufficiently far away from p-n junctions.

Many models have been proposed for device ESD protection techniques. For MOS, such techniques are input resistors, input diodes,

Fig. 73 Charged device breakdown of bipolar devices

(a) 400×. (b) 1600×

(a)   ⊢   10 μm

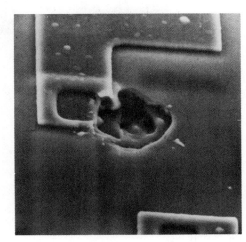

(b)   ⊢   10 μm

Table 9 Susceptibility ranges of various devices exposed to electrostatic discharge

Device type	Range of ESD susceptibility, V
MOSFET	100-200
JFET (junction field-effect transistor)	140-10 000
CMOS	250-2000
Schottky diodes, TTL	300-2500
Bipolar transistors	380-7000
ECL (emitter-controlled logic) (For hybrid use, PC board level)	500- ...
SCR (silicon control rectifier)	680-1000

Source: Ref 108

resistor-diode combinations, field plate diode, thick-oxide MOS transistor, punch-through diode, spark-gap device, and the gated punch-through device. Of these, the most effective

Fig. 74 Charged device breakdown of MOS oxide

(a) 3000×. (b) 10 000×

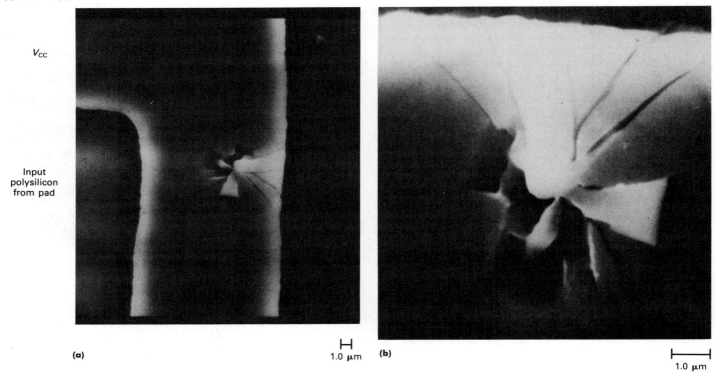

V_{CC}

Input
polysilicon
from pad

(a)

1.0 μm (b)

1.0 μm

Fig. 75 Gated punch-through protection network

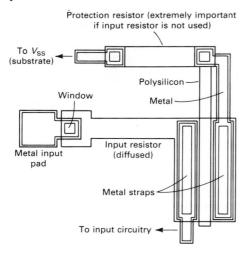

Protection resistor (extremely important if input resistor is not used)

To V_{SS} (substrate)

Polysilicon

Metal

Window

Metal input pad

Input resistor (diffused)

Metal straps

To input circuitry

appears to be the gated punch-through device (Fig. 75). A technique for bipolar linear devices is the phantom-emitter transistor structure illustrated in Fig. 76. This adds a second emitter diffusion shorted to the base contact. Also, there is a deliberate separation of base contact from the normal emitter diffusion. In normal operation, the diffusion contributes nothing. But under ESD stress, it completes a lower breakdown voltage path between the buried collector and the base contact. This transient

Fig. 76 Phantom-emitter transistor structure

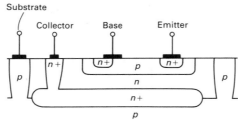

Substrate

Collector   Base   Emitter

$n+$ $n+$ p $n+$

p p

n

$n+$

p

clamping action, with the help of the extrinsic base resistance, limits the ESD pulse current through the vulnerable emitter sidewall, effectively protecting the regular emitter from damage.

Efforts have also been made to improve ESD protection of integrated-circuit devices by optimizing the design parameters of output buffers and input contact window metallization. It has been shown that considerable improvement in ESD threshold voltages could be achieved by changing the spacings between elements of the input-protection device (Ref 116).

Plastic-Package Failures

Since the latter 1970s, there has been an increasing emphasis on the use of molded plastics as an integrated-circuit packaging material. Figure 77 shows a cutaway view of a

Fig. 77 Cutaway view of a plastic-encapsulated integrated circuit

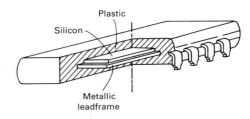

Plastic

Silicon

Metallic leadframe

plastic package. Those failure mechanisms directly associated with plastic packages will be discussed. Some of these failure mechanisms arise from the failure of the package to provide the necessary protection to the integrated circuit, and others arise from direct adverse interactions between the package and the integrated circuit within it. Adverse interactions between ceramic packages and integrated circuits were discussed in the section "Alpha Particle Induced Failures" in this article.

Moisture-Induced Failures. One of the principal failure mechanisms is associated with the penetration of the plastic package by moisture. The penetration can occur at the epoxy/leadframe interface, cracks in the epoxy itself, or through the bulk plastic directly. The moisture either carries impurities into the package from the manufacturing assembly process or reacts with impurities in the plastic or on the integrated-circuit surface. In all cases, the re-

Fig. 78 Plastic-package model of chlorine-induced bond-pad corrosion

The migration path of Cl⁻ ions is shown. Source: Ref 123

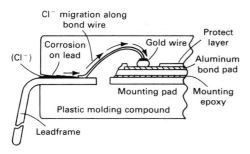

Fig. 79 Modeling of plastic-package stress

Stress maximum is at chip corners. Source: Ref 130

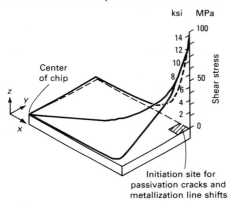

Fig. 80 Crack initiation at a chip corner due to stress

Source: Ref 130

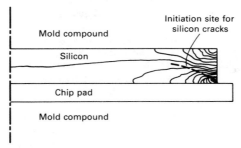

Fig. 81 Relationship of principal and shear stresses in integrated-circuit packages

Source: Ref 130

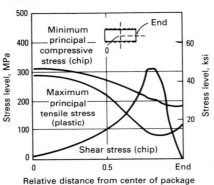

sult is the formation of chemical products that are corrosive to the aluminum metallization on the integrated circuit.

An early recognition of this effect was the discovery of aluminum corrosion due to the interaction of moisture with excessive amounts of chlorine and phosphorus in the intermediate oxide insulation layer in an integrated circuit. The results were both galvanic and electrolytic corrosion of the aluminum (Ref 117). In this study, phosphorus was the chief contributor to corrosion. The effects of other specific impurities, such as sodium, potassium, and chlorine, on the aluminum metallization have been studied (Ref 118). Similar corrosion failures were observed due to chlorine contaminants in the plastic packaging material (Ref 119). More recently, the influence of various encapsulants, including epoxy novolacs, epoxy anhydride, silicone formulations, and phenolics, has been investigated (Ref 120). A wide variety of corrosion behavior was found due to the ionic-impurity content of the plastics and the presence of other impurities resulting from device processing steps.

Corrosion failures due to the migration of corrosive impurities along epoxy/leadframe in-terfaces have been reported by several authors (Ref 121-123). These impurities are inherent in the solder-fluxing process used to solder the components into a printed circuit board. Figure 78 shows how these contaminants can migrate along the leads of a poorly encapsulated device and attack the aluminum bond pad. In the case of plastic-encapsulated integrated circuits operated at higher temperatures, it was found that polymer outgassing occurs at several different temperature ranges (Ref 124). Chlorine was shown to be the dominant product of the outgassing process. The effects of this outgassing on integrated circuits was reported separately (Ref 125). The failure mechanism was that of bond separation due to the Kirkendall effect and was described in the section "Metallic Interface Failures" in this article.

One further failure mechanism due to the penetration of moisture through a plastic encapsulant has been found to be the formation of parasitic MOS transistors due to mobile charge accumulation on the surface of the silicon integrated circuit (Ref 126). This failure mechanism was discussed in the section "Silicon Oxide Interface Failures" in this article.

Stress-Induced Failures. The stress exerted on the integrated-circuit chip by the plastic package has been shown to induce a number of failure mechanisms. Several investigators (Ref 127, 128) have observed failures due to the moisture penetration of stress-induced cracks and defects in the passivation layer of the integrated circuit. The deformation of aluminum metallization in plastic packages as a result of stress has been observed (Ref 129). Computer studies have been conducted of the stresses occurring in plastic-encapsulated integrated-circuit packages (Ref 130). This modeling uses a technique called finite-element analysis, which allows the characterization of the stresses and deformations in the package for various stresses and loads. The model indicates a stress maximum at the corner of the chips (Fig. 79). If severe enough, the stress at the corner can cause the silicon chip to crack (Fig. 80). This type of fracture has been observed earlier (Ref 131). This situation is worse for long, narrow chips that occupy a large fraction of the area of the package. This modeling also shows that the stress concentrations on the chip can be reduced by the proper choice of materials and processing conditions. Another important result of this analysis has been the relationship of principal and shear stresses in the plastic integrated-circuit package. This is shown in Fig. 81. The results of this analysis have been verified using diffused strain-gage experiments that permitted quantitative stress measurements and the mapping of stress across the area of large silicon chips (Ref 132).

ACKNOWLEDGMENT

The authors wish to acknowledge the direct contributions of many colleagues at Bell Laboratories, and those others who contributed to the literature. We wish to acknowledge the support and encouragement of J.N. McGinn, Y. Nakada, and J.C. North. Also, we wish to thank Ms. Kathy Leachman for her efforts in typing and editing the manuscript.

REFERENCES

1. "LSI/Microprocessor Reliability Prediction Model Development," RADC-TR-79-97, Rome Air Development Center, Rome, NY, March 1979
2. N. Carthage and V. Hauser, Selective Decapsulation of Epoxy Encapsulated Integrated Circuits, *AT&T Technol. Tech. Dig.*, Vol 68, 1982, p 7
3. J. Hiatt, A Method of Detecting Hot Spots on Semiconductors Using Liquid Crystals, in *19th Annual Proceedings of the Reliability Physics Symposium*, 1981, p 130-133
4. G.J. West, A Simple Technique for Analysis of ESD Failures of Dynamic RAMs Using Liquid Crystals, in *20th Annual Proceedings of the Reliability Physics Symposium*, 1982, p 185-187
5. R.E. Ripplinger, Voltage Contrast in a Dynamic Mode: Its Construction and Application, *Scan. Elec. Microsc.*, No. 1, 1977, p 211-216

6. J. Bart, Scanning Electron Microscopy for Complex Microcircuit Analysis, in *16th Annual Proceedings of the Reliability Physics Symposium*, 1978, p 108-111

7. J.R. Beall *et al.*, SEM Techniques for the Analyses of Memory Circuits, in *18th Annual Proceedings of the Reliability Physics Symposium*, 1980, p 65-72

8. D. Younken, Phase Dependent Voltage Contrast, in *19th Annual Proceedings of the Reliability Physics Symposium*, 1981, p 264-268

9. J.B. Bindell and J.N. McGinn, Voltage Contrast SEM Observations with Microprocessor Controlled Device Timing, in *18th Annual Proceedings of the Reliability Physics Symposium*, 1980, p 55-58

10. T.C. May, G.L. Scott, E.S. Meieran, P. Winer, and V.R. Rao, Dynamic Fault Imaging of VLSI Random Logic Devices, in *22nd Annual Proceedings of the Reliability Physics Symposium*, 1984, p 95-108

11. C.N. Potter and D.E. Sawyer, Optical Scanning Techniques for Semiconductor Device Screening and Identification of Surface and Junction Phenomena, *Phys. Fail. Elec.*, Vol 5, 1966, p 37

12. D.E. Sawyer and D.W. Berning, Thermal Mapping of Transistors with a Laser Scanner, *Proc. IEEE*, Vol 64, 1976, p 1634-1635

13. D.J. Burns and J.M. Kendall, Imaging Latch-Up Sites in LSI CMOS with a Laser Photoscanner, in *21st Annual Proceedings of the Reliability Physics Symposium*, 1983, p 118

14. F.J. Henley, M.H. Chi, and W.G. Oldham, CMOS Latch-Up Characterization using a Laser Scanner, in *21st Annual Proceedings of the Reliability Physics Symposium*, 1983, p 122-129

15. T. Mills and E.W. Sponheimer, Precision VLSI Cross-Sectioning and Staining, in *20th Annual Proceedings of the Reliability Physics Symposium*, 1982, p 214-220

16. S.C. Kolesar, Principles of Corrosion, in *12th Annual Proceedings of the Reliability Physics Symposium*, 1974, p 155

17. F.W. Hewlett, Jr. and R.A. Pederson, The Reliability of Integrated Injection Logic Circuits for the Bell System, in *14th Annual Proceedings of the Reliability Physics Symposium*, 1976, p 5

18. Z.A. Foroules and M.J. Thubrikar, On the Kinetics of the Breakdown of Passivity of Preanodized Aluminum by Chlorine Ions, *J. Electrochem. Soc.*, Vol 122, 1975, p 1296

19. W.M. Paulson and R.P. Lorrigan, The Effect of Impurities on the Corrosion of Aluminum Metallization, in *14th Annual Proceedings of the Reliability Physics Symposium*, 1976, p 42

20. A.T. English and C.M. Melliar-Smith, Reliability and Failure Mechanisms of Electronic Materials, *Ann. Rev. Mater. Sci.*, Vol 8, 1978, p 459

21. W.M. Paulson and R.W. Kirk, The Effects of Phosphorus-Doped Passivation Glass on the Corrosion of Aluminum, in *12th Annual Proceedings of the Reliability Physics Symposium*, 1974, p 172

22. A.T. English *et al.*, Electromigration in Conductor Stripes Under Pulsed DC Powering, *Appl. Phys. Lett.*, Vol 21, 1972, p 397

23. F.M. d'Heurle and I. Ames, Electromigration in Single Crystal Aluminum Films, *Appl. Phys. Lett.*, Vol 16, 1970, p 80

24. J.R. Black, Physics of Electromigration, in *12th Annual Proceedings of the Reliability Physics Symposium*, 1974, p 142

25. F.M. d'Heurle, N.G. Ainslie, A. Gangulee, and M.C. Shine, Activation Energy for Electromigration Failure in Aluminum Films Containing Copper, *J. Vac. Sci. Technol.*, Vol 9, 1972, p 289

26. E. Hall, "Silicon RF Power Transistor Metallization," ECOM-0164F, U.S. Army Electronics Command, Ft. Monmouth, NJ, Oct 1971

27. P.B. Ghate, Aluminum Alloy Metallization for Integrated Circuits, *Thin Solid Films*, Vol 83, 1981, p 195

28. J.K. Howard *et al.*, Intermetallic Compounds of Al and Transition Metals: Effects of Electromigration in 1-2μ Wide Lines, *1977 Electrochemical Society Fall Meeting*, Ext. Abstract No. 178, p 480-481

29. J.S. Jaspal and H.M. Dabal, A Three-Fold Increase in Current Carrying Capability of Al-Cu Metallurgy by Predepositing a Suitable Underlay Material, in *19th Annual Proceedings of the Reliability Physics Symposium*, 1981, p 238

30. E. Levine and J. Kitchner, Electromigration Induced Damage and Structure Change In CR-AL/Cu and Al/Cu Interconnection Lines, in *22nd Annual Proceedings of the Reliability Physics Symposium*, 1984, p 242

31. B.N. Agarwala *et al.*, Dependence of Electromigration Induced Failure Time on Length and Width of Aluminum Thin-Film Conductors, *J. Appl. Phys.*, Vol 41 (No. 10), 1970, p 3954

32. G.A. Scoggan *et al.*, Width Dependence of Electromigration Life in Al-Cu, Al-Cu-Si and Ag Conductors, in *13th Annual Proceedings of the Reliability Physics Symposium*, 1975, p 151

33. S. Vaidya *et al.*, Electromigration Resistance of Fine-Line Al for VLSI Applications, in *18th Annual Proceedings of the Reliability Physics Symposium*, 1980, p 165

34. T. Wada, H. Higuchi, and T. Ajiki, New Phenomenon of Electromigration in Double-Layer Metallization, in *22nd Annual Proceedings of the Reliability Physics Symposium*, 1983, p 203

35. J.R. Lloyd, M.J. Sullivan, G.S. Hopper, J.T. Coffin, E.T. Severn, and J.L. Iozwiak, Electromigration Failure in Thin Film Silicides and Polysilicon/Silicide (Polycide) Structures, in *21st Annual Proceedings of the Reliability Physics Symposium*, 1983, p 198-202

36. S.B. Herschbein, P.A. Zupal, and J.M. Curry, Effects of Silicon Inclusions on the Reliability of Sputtered Aluminum Silicon Metallization, in *22nd Annual Proceedings of the Reliability Physics Symposium*, 1984, p 134

37. J. Curry, G. Fitzgibbon, Y. Guan, R. Muollo, G. Nelson, and A. Thomas, New Failure Mechanisms in Sputtered Aluminum-Silicon Films, in *22nd Annual Proceedings of the Reliability Physics Symposium*, 1984, p 6

38. S.T. O'Donnell, J.W. Bartling, and G. Hill, Silicon Inclusions in Aluminum Interconnects, in *22nd Annual Proceedings of the Reliability Physics Symposium*, 1984, p 9

39. J.T. Yue, W.P. Funsten, and R.V. Taylor, Stress Induced Voids in Aluminum Interconnects During IC Processing, in *23rd Annual Proceedings of the Reliability Physics Symposium*, 1985, p 126

40. J. Klema, R. Pyle, and E. Domanque, Reliability Implications of Nitrogen Contamination During Deposition of Sputtered Aluminum/Silicon Metal Films, in *22nd Annual Proceedings of the Reliability Physics Symposium*, 1984, p 1

41. L.D. Hartsough and D.R. Denison, "Aluminum-Silicon Sputter Deposition," Technical Report No. 79.01, Perkin-Elmer Corp., Norwalk, CT, 1979

42. L.D. Hartsough and P.S. McLeod, U.S. Patent 4,125,446, Nov 14, 1978

43. T. Turner and K. Wendel, The Influence of Stress on Aluminum Conductor Life, in *23rd Annual Proceedings of the Reliability Physics Symposium*, 1985, p 142

44. S. Timoshenko, *Strength of Materials—Part II*, Van Nostrand Reinhold, 1958

45. W. Prager, The Theory of Plasticity: A Survey of Recent Achievements, *Proc. Inst. Mech. Eng.*, Vol 169, 1955, p 41-57

46. R.T. Shield and H. Ziegler, On Prager's Hardening Rule, *Z. Ange. Math. Phys.*, Vol 9A, 1958, p 260-276

47. J.L. Dais and F.L. Howland, Fatigue Failure of Encapsulated Gold-Beam Lead and Tab Devices, *Trans. IEEE Compon. Hybrids Mfg. Technol.*, Vol CHMT-1, 1978, p 158-166

48. J. Dais, The Mechanics of Gold Beam Leads During Thermocompression Bonding, *Trans. IEEE, Parts, Hybrids, Packages*, Vol PHP-12 (No. 3), 1976, p 241

49. S.S. Manson, *Thermal Stress and Low Cycle Fatigue*, McGraw-Hill, 1966

50. E. Philofsky, Design Limits When Using Gold-Aluminum Bonds, in *9th Annual Proceedings of the Reliability Physics Symposium*, 1971, p 114

51. J.L. Newsome, F.G. Oswald, and W.R. Rodrigues de Miranda, Metallurgical Aspects of Aluminum Wire Bonds to Gold Metallization, in *Proceedings of the International Society of Hybrid Microelectronics*, 1976, p 63-67

52. D.W. Bushmire, Resistance Increase in Gold Aluminum Interconnections with Time and Temperature, *Trans. IEEE Parts, Hybrids, Packages*, Vol PHP-13, 1977, p 152-156

53. C.W. Horsting, Purple Plague and Gold Purity, in *10th Annual Proceedings of the Reliability Physics Symposium*, 1972, p 155

54. J.B. Prather, S.D. Robertson, and J.W. Slemmons, Gold Thick Film Conductors for Aluminum Wire Bonding, in *International Microelectronics Conference* (IMC 1974 West), Feb 1974

55. R.E. Thomas, V. Winchell, K. James, and T. Scharr, Plastic Outgassing Induced Wire Bond Failure, in *Proceedings of the 27th Electronics Components Conference*, 1977, p 182-187

56. R.C. Blish II and L. Parobek, Wire Bond Integrity Test Chip, in *21st Annual Proceedings of the Reliability Physics Symposium*, 1983, p 142-147

57. P.V. Plunkett and J.F. Dalporto, Low Temperature Void Formation in Gold-Aluminum Contacts, in *Proceedings of the 32nd Electronics Components Conference*, 1982, p 421-427

58. R.J. Gale, Epoxy Degradation Induced Au-Al Intermetallics Void Formation in Plastic Encapsulated MOS Memories, in *22nd Annual Proceedings of the Reliability Physics Symposium*, 1984, p 37-47

59. J.O. McCaldin and H. Sankur, Diffusivity and Solubility of Si in the Al Metallization of Integrated Circuits, *Appl. Phys. Lett.*, Vol 19, 1971, p 524

60. G.S. Prokop and R.R. Joseph, Electromigration Failure at Aluminum-Silicon Contacts, *J. Appl. Phys.*, Vol 43, 1972, p 2595

61. J.R. Black, Electromigration of Al-Si Alloy Films, in *16th Annual Proceedings of the Reliability Physics Symposium*, 1978, p 233-239

62. A.T. English and C.M. Melliar-Smith, Reliability and Failure Mechanisms of Electronic Materials, *Ann. Rev. Mater. Sci.*, Vol 8, 1978, p 459

63. G.H. Johnson and M. Stitch, Microcircuit Accelerated Testing Reveals Life Limiting Failure Modes, in *15th Annual Proceedings of the Reliability Physics Symposium*, 1977, p 179

64. P. Brambilla, F. Fantine, P. Malberti, and G. Matlana, CMOS Reliability, *Microelec. Rel.*, Vol 21, 1981, p 191

65. D.G. Edwards, Testing for MOS IC Failure Modes, *Trans. IEEE Rel.*, Vol 31, 1982, p 9

66. C.M. Osburn and E. Bassous, Improved Dielectric Reliability of SiO_2 Films with Polycrystalline Silicon Electrodes, *J. Electrochem. Soc.*, Vol 122, 1975, p 89

67. C.R. Barrett and R.C. Smith, Failure Modes and Reliability of Dynamics RAMs, in *14th IEEE Computer Society International Conference*, 1977

68. D.L. Crook, Method for Determining Reliability Screens for Time Dependent Dielectric Breakdown, in *17th Annual Proceedings of the Reliability Physics Symposium*, 1979, p 1

69. E.S. Anolick and G.R. Nelson, Low Field Time Dependent Dielectric Integrity, in *17th Annual Proceedings of the Reliability Physics Symposium*, 1979, p 8

70. V. Hokari, T. Baba, and N. Kawamura, Reliability of Thin SiO_2 Films Showing Intrinsic Dielectric Integrity, *IEDM Tech. Dig.*, 1982, p 46

71. J. Eachus, J. Klema, and S. Walker, Monitored Burn-In of MOS 64K Dynamic RAMs, *Semicon. Int.*, Vol 7, 1984, p 104

72. J.W. McPherson and D.A. Baglee, Acceleration Factors For Thin Gate Oxide Stressing, in *23rd Annual Proceedings of the Reliability Physics Symposium*, 1985, p 1

73. A. Berman, Time Zero Dielectric Reliability Test by a Ramp Method, in *19th Annual Proceedings of the Reliability Physics Symposium*, 1981, p 204

74. S.I. Raider, Time Dependent Breakdown of Silicon Dioxide Films, *Appl. Phys. Lett.*, Vol 23 (No. 1), 1973, p 34

75. C.M. Osburn and S.I. Raider, The Effect of Mobile Sodium Ions on Field Enhancement Breakdown in SiO_2 Films on Silicon, *J. Electrochem. Soc.*, Vol 120, 1973, p 1369

76. C.M. Osburn and D.W. Ormond, Sodium-Induced Barrier Height Lowering and Dielectric Breakdown in SiO_2 Films on Silicon, *J. Electrochem. Soc.*, Vol 121, 1974, p 1195

77. T.I. DiStefano, Impact Ionization Model For Dielectric Instability and Breakdown, *Appl. Phys. Lett.*, Vol 25 (No. 12), 1974, p 685

78. N. Klein, Current Runaway in Insulators Affected by Impact Ionization and Recombination, *J. Appl. Phys.*, Vol 47 (No. 10), 1976, p 4364

79. R.C. Hughes, Hole Mobility and Transport in Thin SiO_2 Films, *Appl. Phys. Lett.*, Vol 26 (No. 8), 1975, p 436

80. P. Solomon, Breakdown in Silicon Oxide—A Review, *J. Vac. Sci. Technol.*, Vol 14 (No. 15), 1977, p 1122

81. D.R. Young, Characterization of Electron Traps in SiO_2 as Influenced by Processing Parameters, *J. Appl. Phys.*, Vol 52, 1981, p 4090

82. D.R. Young, E.A. Irene, D.J. DeMaria, and R.F. DeKeersmaecker, Electron Trapping in SiO_2 at 295 and 77 degrees K., *J. Appl. Phys.*, Vol 50, 1979, p 6366

83. M. Istumi, Positive and Negative Charging of Thermally Grown SiO_2 Induced by Fowler-Nordheim Emission, *J. Appl. Phys.*, Vol 52, 1981, p 3491

84. C. Jeng, T. Ranganath, C. Huang, H. Jones, and T. Chang, High Field Generation of Electron Traps and Charge Trapping in Ultra Thin SiO_2, *IEDM Tech. Dig.*, 1981, p 388

85. M. Liang and C. Hu, Electron Trapping in Very Thin Thermal Silicon Dioxides, *IEDM Tech. Dig.*, 1981, p 396

86. W.K. Meyer and D.L. Crook, Model for Oxide Wearout Due to Charge Trapping, in *21st Annual Proceedings of the Reliability Physics Symposium*, 1983, p 242

87. W.H. Schroen, Process Testing for Reliability Control, in *16th Annual Proceedings of the Reliability Physics Symposium*, 1978, p 81

88. R.J. Kriegler, Ion Instabilities in MOS Structures, in *12th Annual Proceedings of the Reliability Physics Symposium*, 1974, p 250

89. E.H. Nicollian, Interface Instabilities, in *12th Annual Proceedings of the Reliability Physics Symposium*, 1974, p 267

90. R.C. Sun *et al.*, Effects of Silicon Nitride Encapsulation on MOS Device Stability, in *18th Annual Proceedings of the Reliability Physics Symposium*, 1980, p 244

91. B. Euzant, Hot Electron Efficiency in IGFET Structures, in *15th Annual Proceedings of the Reliability Physics Symposium*, 1977, p 1

92. N. Strojadinovic, S. Dimitryev, and S. Myalkovic, Effects of High Field Stress on Threshold Voltage of CMOS Transistors, *Microelec. Rel.*, Vol 25, 1985, p 275-279

93. T.C. May and M.H. Woods, A New Physical Mechanism for Soft Errors in Dynamic Memories, in *16th Annual Proceedings of the Reliability Physics Symposium*, 1978, p 33

94. R.J. McPartland *et al.*, Alpha-Particle Induced Soft Errors and 64K Dynamic RAM Design Interaction, in *18th Annual Proceedings of the Reliability Physics Symposium*, 1980, p 261

95. M. White *et al.*, The Use of Silicone RTV for Alpha-Particle Protection on Silicon Integrated Circuits, in *19th Annual Proceedings of the Reliability Physics Symposium*, 1981, p 43

96. J.L. Boyle, R.C. McIntyre, R.F. Youtz, and J.T. Nelson, Latent β-Radiation Damage in Hermetically Sealed NMOS Devices, *19th Annual Proceedings of the Reliability Physics Symposium*, 1981, p 34-37

97. E.H. Snow, A.S. Grove, and D.J. Fitzgerald, Effects of Ionizing Radiation on Oxidized Silicon Surfaces and Planar Devices, *Proc. IEEE*, Vol 55, 1967, p 1168

98. D. Fisch, T. Mozdzen, and G. Roberts, The Effects of Entrapped Krypton 85 Gas on Device Reliability, in *21st Annual Proceedings of the Reliability Physics Symposium*, 1983, p 102-105

99. G.A. Sai-Halasz and J.M. Aitken, Projected Life-Times of Circuits in Krypton 85 Contaminated Packages, *IEEE Elec. Dev. Lett.*, Vol EDL-3 (No. 4), 1982, p 106-108

100. D.M. Long, State of the Art Review: Hardness of MOS and Bipolar Integrated Circuits, *IEEE Trans. Nucl. Sci.*, Vol NS-27, 1980, p 674

101. S.M. Sze, *Physics of Semiconductor Devices*, 2nd ed., John Wiley & Sons, 1981, p 190-242

102. B.L. Gregory and B.D. Shafer, Latch-Up in CMOS Integrated Circuits, *Trans. IEEE Nucl. Sci.*, Vol NS-20, 1973, p 293

103. A. Ochon, W. Dawes, and D. Estreich, Latch-Up Control in CMOS Integrated Circuits, *Trans. IEEE Nucl. Sci.*, Vol NS-26, 1979, p 5065

104. D.B. Estreich, A. Ochon, and R.W. Dultton, An Analysis of Latch-Up Prevention in CMOS IC's Using an Epitaxial-Buried Layer Process, *IEDM Tech. Dig.*, 1978, p 230

105. R.L. Minear and G.A. Dodson, Effects of Electrostatic Discharge on Linear Bipolar Integrated Circuits, in *15th Annual Proceedings of the Reliability Physics Symposium*, 1977, p 138

106. A.C. Trigonis, Electrostatic Discharge in Microcircuits, in *Proceedings on the Annual Reliability and Maintainability Symposium*, 1976, p 162

107. W.B. Smith, D.H. Pontius, and P.P. Budenstein, Second Breakdown and Damage in Junction Devices, *Trans. IEEE Elec. Dev.*, Vol ED-20 (No. 8), 1973, p 731-744

108. D.C. Wunsch and R.R. Bell, Determination of Threshold Failure Levels of Semiconductor Diodes and Transistors Due to Pulse Voltages, *Trans. IEEE Nucl. Sci.*, Vol NS-15, 1968, p 244-259

109. T.S. Speakman, A Model for the Failure of Bipolar Silicon Integrated Circuits Subjected to Electrostatic Discharge, in *12th Annual Proceedings of the Reliability Physics Symposium*, 1974, p 60

110. E.R. Greeman and J.R. Beall, Control of Electrostatic Discharge Damage to Semiconductors, in *12th Annual Proceedings of the Reliability Physics Symposium*, 1974, p 304-312

111. L.J. Gallace and H.L. Pujol, The Evaluation of CMOS Static-charge Protection Networks and Failure Mechanisms Associated with Overstress Conditions as Related to Device Life, in *15th Annual Proceedings of the Reliability Physics Symposium*, 1977, p 149-157

112. B.A. Unger, Electrostatic Discharge Failures of Semiconductor Devices, in *19th Annual Proceedings of the Reliability Physics Symposium*, 1981, p 193-199

113. P.R. Bossard, R.G. Chemelli, and B.A. Unger, ESD Damage from Turboelectrically Charged IC Pins, in *Electrical Overstress/Electrostatic Discharge Symposium Proceedings*, 1980

114. L.F. DeChiaro, Electro-Thermomigration in NMOS LSI Devices, in *19th Annual Proceedings of the Reliability Physics Symposium*, 1981, p 223-229

115. T. Brooks, P. Czahor, J. Kita, and M. Obeng, ESD Failure Analysis of NMOS VLSI Chips, in *International Symposium for Testing and Failure Analysis*, 1983, p 152-156

116. L.F. DeChiaro, Device ESD Susceptibility Testing and Design Hardening, in *Electrical Overstress/Electrostatic Discharge Symposium Proceedings*, 1984, p 179-188

117. R.C. Olberg and J.L. Bozarth, Factors Contributing to the Corrosion of Aluminum Metal on Semiconductor Device Packages in Plastics, *Microelec. Rel.*, Vol 15, 1976, p 601-611

118. W.M. Paulson and R.P. Lorrigan, The Effects of Impurities on the Corrosion of Aluminum Metallization, in *14th Annual Proceedings of the Reliability Physics Symposium*, 1976, p 42

119. R.W. Lawson and J.C. Harrison, Plastic Encapsulation of Semiconductors, in *Plastics in Telecommunications Conference Proceedings*, 1974

120. H.M. Berg and W.M. Paulson, Aluminum Corrosion in Plastic Encapsulated Devices, in *Proceedings of the Symposium on Plastic Encapsulated and Polymer Sealed Semiconductor Devices for Army Equipment*, May 1978

121. S.P. Sim and R.W. Lawson, The Influence of Plastic Encapsulants and Passivation Layers on the Corrosion of Thin Aluminum Films Subject to Humidity Stress, in *17th Annual Proceedings of the Reliability Physics Symposium*, 1979, p 103-112

122. P.R. Engle, T. Corbett, and W. Baerg, A New Failure Mechanism of Bond Pad Corrosion in Plastic Encapsulated IC Under Temperature Humidity and Bias Stress, in *Proceedings of the 33rd Electronic Components Conference*, 1983, p 245-252

123. L. Gallace and M. Rosenfield, Reliability of Plastic Encapsulated Integrated Circuits in Moisture Environments, *Qual. Rel. Int.*, Vol 1, 1985, p 105-117

124. R.M. Lum and L.G. Feinstein, Investigation of the Molecular Processes Controlling Corrosion Failure Mechanisms in Plastic Encapsulated Semiconductor Devices, *Microelec. Rel.*, Vol 21, 1981, p 15-31

125. L.G. Feinstein, Do Flame Retardants Affect the Reliability of Molded Plastic Packages, *Microelec. Rel.*, Vol 21, 1981, p 533-541

126. Y. Wakashima, H. Inayoshi, K. Nishi, and N. Nishida, A Study of Parasitic MOS Formation Mechanism in Plastic Encapsulated MOS Devices, in *14th Annual Proceedings of the Reliability Physics Symposium*, 1976, p 223-227

127. H. Inayoshi, K. Nishi, S. Okikawa, and Y. Wakashima, Moisture-Induced Aluminum Corrosion and Stress on the Chip in Plastic Encapsulated LSIs, in *17th Annual Proceedings of the Reliability Physics Symposium*, 1979, p 113-117

128. S. Okikawa, M. Sakimoto, M. Tanaka, T. Sato, T. Toya, and Y. Hare, Stress Analysis of Passivation Film Crack for Plastic Molded LSI Caused by Thermal Stress, in *International Symposium for Testing and Failure Analysis*, 1983, p 275-280

129. M. Isagawa *et al.*, Deformation of Al Metallization in Plastic Encapsulated Semiconductor Devices Caused by Thermal Shock, in *18th Annual Proceedings of the Reliability Physics Symposium*, 1980, p 171-177

130. S. Groothius, W. Schroen, and M. Murtuza, Computer Aided Stress Modeling for Optimizing Plastic Package Reliability, in *23rd Annual Proceedings of the Reliability Physics Symposium*, 1985, p 184-191

131. T.C. Taylor and F.L. Yuan, Thermal Stress and Fracture in Shear-Constrained Semiconductor Device Structures, *Trans. IEEE Elec. Dev.*, Vol ED-9, 1962, p 303-309

132. J.L. Spencer *et al.*, New Quantitative Measurements of IC Stress Introduced by Plastic Package, in *19th Annual Proceedings of the Reliability Physics Symposium*, 1981, p 74-81

Metric Conversion Guide

This Section is intended as a guide for expressing weights and measures in the Système International d'Unités (SI). The purpose of SI units, developed and maintained by the General Conference of Weights and Measures, is to provide a basis for world-wide standardization of units and measure. For more information on metric conversions, the reader should consult the following references:

- "Standard for Metric Practice," E 380, *Annual Book of ASTM Standards,* 1985, ASTM, 1916 Race Street, Philadelphia, PA 19103
- "Metric Practice," ANSI/IEEE 268–1982, American National Standards Institute, 1430 Broadway, New York, NY 10018
- *Metric Practice Guide—Units and Conversion Factors for the Steel Industry,* 1978, American Iron and Steel Institute, 1000 16th Street NW, Washington, DC 20036
- *The International System of Units,* SP 330, 1981, National Bureau of Standards. Order from Superintendent of Documents, U.S. Government Printing Office, Washington, DC 20402
- *Metric Editorial Guide,* 4th ed. (revised), 1985, American National Metric Council, 1010 Vermont Avenue NW, Suite 320, Washington, DC 20005-4960
- *ASME Orientation and Guide for Use of SI (Metric) Units,* ASME Guide SI 1, 9th ed., 1982, The American Society of Mechanical Engineers, 345 East 47th Street, New York, NY 10017

Base, supplementary, and derived SI units

Measure	Unit	Symbol	Measure	Unit	Symbol
Base units			Entropy	joule per kelvin	J/K
			Force	newton	N
Amount of substance	mole	mol	Frequency	hertz	Hz
Electric current	ampere	A	Heat capacity	joule per kelvin	J/K
Length	meter	m	Heat flux density	watt per square meter	W/m^2
Luminous intensity	candela	cd	Illuminance	lux	lx
Mass	kilogram	kg	Inductance	henry	H
Thermodynamic temperature	kelvin	K	Irradiance	watt per square meter	W/m^2
Time	second	s	Luminance	candela per square meter	cd/m^2
			Luminous flux	lumen	lm
			Magnetic field strength	ampere per meter	A/m
Supplementary units			Magnetic flux	weber	Wb
			Magnetic flux density	tesla	T
Plane angle	radian	rad	Molar energy	joule per mole	J/mol
Solid angle	steradian	sr	Molar entropy	joule per mole kelvin	$J/mol \cdot K$
			Molar heat capacity	joule per mole kelvin	$J/mol \cdot K$
			Moment of force	newton meter	$N \cdot m$
Derived units			Permeability	henry per meter	H/m
			Permittivity	farad per meter	F/m
Absorbed dose	gray	Gy	Power, radiant flux	watt	W
Acceleration	meter per second squared	m/s^2	Pressure, stress	pascal	Pa
Activity (of radionuclides)	becquerel	Bq	Quantity of electricity, electric		
Angular acceleration	radian per second squared	rad/s^2	charge	coulomb	C
Angular velocity	radian per second	rad/s	Radiance	watt per square meter steradian	$W/m^2 \cdot sr$
Area	square meter	m^2			
Capacitance	farad	F	Radiant intensity	watt per steradian	W/sr
Concentration (of amount of			Specific heat capacity	joule per kilogram kelvin	$J/kg \cdot K$
substance)	mole per cubic meter	mol/m^3	Specific energy	joule per kilogram	J/kg
Conductance	siemens	S	Specific entropy	joule per kilogram kelvin	$J/kg \cdot K$
Current density	ampere per square meter	A/m^2	Specific volume	cubic meter per kilogram	m^3/kg
Density, mass	kilogram per cubic meter	kg/m^3	Surface tension	newton per meter	N/m
Electric charge density	coulomb per cubic meter	C/m^3	Thermal conductivity	watt per meter kelvin	$W/m \cdot K$
Electric field strength	volt per meter	V/m	Velocity	meter per second	m/s
Electric flux density	coulomb per square meter	C/m^2	Viscosity, dynamic	pascal second	$Pa \cdot s$
Electric potential, potential			Viscosity, kinematic	square meter per second	m^2/s
difference, electromotive force	volt	V	Volume	cubic meter	m^3
Electric resistance	ohm	Ω	Wavenumber	1 per meter	1/m
Energy, work, quantity of heat	joule	J			
Energy density	joule per cubic meter	J/m^3			

Conversion factors

To convert from	to	multiply by
Angle		
degree	rad	1.745 329 E − 02
Area		
in.²	mm²	6.451 600 E + 02
in.²	cm²	6.451 600 E + 00
in.²	m²	6.451 600 E − 04
ft²	m²	9.290 304 E − 02
Bending moment or torque		
lbf · in.	N · m	1.129 848 E − 01
lbf · ft	N · m	1.355 818 E + 00
kgf · m	N · m	9.806 650 E + 00
ozf · in.	N · m	7.061 552 E − 03
Bending moment or torque per unit length		
lbf · in./in.	N · m/m	4.448 222 E + 00
lbf · ft/in.	N · m/m	5.337 866 E + 01
Current density		
A/in.²	A/cm²	1.550 003 E − 01
A/in.²	A/mm²	1.550 003 E − 03
A/ft²	A/m²	1.076 400 E + 01
Electricity and magnetism		
gauss	T	1.000 000 E − 04
maxwell	μWb	1.000 000 E − 02
mho	S	1.000 000 E + 00
Oersted	A/m	7.957 700 E + 01
Ω · cm	Ω · m	1.000 000 E − 02
Ω circular-mil/ft	μΩ · m	1.662 426 E − 03
Energy (impact, other)		
ft · lbf	J	1.355 818 E + 00
Btu (thermochemical)	J	1.054 350 E + 03
cal (thermochemical)	J	4.184 000 E + 00
kW · h	J	3.600 000 E + 06
W · h	J	3.600 000 E + 03
Flow rate		
ft³/h	L/min	4.719 475 E − 01
ft³/min	L/min	2.831 000 E + 01
gal/h	L/min	6.309 020 E − 02
gal/min	L/min	3.785 412 E + 00
Force		
lbf	N	4.448 222 E + 00
kip (1000 lbf)	N	4.448 222 E + 03
tonf	kN	8.896 443 E + 00
kgf	N	9.806 650 E + 00
Force per unit length		
lbf/ft	N/m	1.459 390 E + 01
lbf/in.	N/m	1.751 268 E + 02
Fracture toughness		
ksi $\sqrt{\text{in.}}$	MPa$\sqrt{\text{m}}$	1.098 800 E + 00
Heat content		
Btu/lb	kJ/kg	2.326 000 E + 00
cal/g	kJ/kg	4.186 800 E + 00

To convert from	to	multiply by
Heat input		
J/in.	J/m	3.937 008 E + 01
kJ/in.	kJ/m	3.937 008 E + 01
Length		
Å	nm	1.000 000 E − 01
μin.	μm	2.540 000 E − 02
mil	μm	2.540 000 E + 01
in.	mm	2.540 000 E + 01
in.	cm	2.540 000 E + 00
ft	m	3.048 000 E − 01
yd	m	9.144 000 E − 01
mile	km	1.609 300 E + 00
Mass		
oz	kg	2.834 952 E − 02
lb	kg	4.535 924 E − 01
ton (short, 2000 lb)	kg	9.071 847 E + 02
ton (short, 2000 lb)	kg × 10³(a)	9.071 847 E − 01
ton (long, 2240 lb)	kg	1.016 047 E + 03
Mass per unit area		
oz/in.²	kg/m²	4.395 000 E + 01
oz/ft²	kg/m²	3.051 517 E − 01
oz/yd²	kg/m²	3.390 575 E − 02
lb/ft²	kg/m²	4.882 428 E + 00
Mass per unit length		
lb/ft	kg/m	1.488 164 E + 00
lb/in.	kg/m	1.785 797 E + 01
Mass per unit time		
lb/h	kg/s	1.259 979 E − 04
lb/min	kg/s	7.559 873 E − 03
lb/s	kg/s	4.535 924 E − 01
Mass per unit volume (includes density)		
g/cm³	kg/m³	1.000 000 E + 03
lb/ft³	g/cm³	1.601 846 E − 02
lb/ft³	kg/m³	1.601 846 E + 01
lb/in.³	g/cm³	2.767 990 E + 01
lb/in.³	kg/m³	2.767 990 E + 04
Power		
Btu/s	kW	1.055 056 E + 00
Btu/min	kW	1.758 426 E − 02
Btu/h	W	2.928 751 E − 01
erg/s	W	1.000 000 E − 07
ft · lbf/s	W	1.355 818 E + 00
ft · lbf/min	W	2.259 697 E − 02
ft · lbf/h	W	3.766 161 E − 04
hp (550 ft · lbf/s)	kW	7.456 999 E − 01
hp (electric)	kW	7.460 000 E − 01
Power density		
W/in.²	W/m²	1.550 003 E + 03
Pressure (fluid)		
atm (standard)	Pa	1.013 250 E + 05
bar	Pa	1.000 000 E + 05
in. Hg (32 °F)	Pa	3.386 380 E + 03
in. Hg (60 °F)	Pa	3.376 850 E + 03
lbf/in.² (psi)	Pa	6.894 757 E + 03
torr (mm Hg, 0 °C)	Pa	1.333 220 E + 02

To convert from	to	multiply by
Specific heat		
Btu/lb · °F	J/kg · K	4.186 800 E + 03
cal/g · °C	J/kg · K	4.186 800 E + 03
Stress (force per unit area)		
tonf/in.² (tsi)	MPa	1.378 951 E + 01
kgf/mm²	MPa	9.806 650 E + 00
ksi	MPa	6.894 757 E + 00
lbf/in.² (psi)	MPa	6.894 757 E − 03
MN/m²	MPa	1.000 000 E + 00
Temperature		
°F	°C	5/9 · (°F − 32)
°R	°K	5/9
Temperature interval		
°F	°C	5/9
Thermal conductivity		
Btu · in./s · ft² · °F	W/m · K	5.192 204 E + 02
Btu/ft · h · °F	W/m · K	1.730 735 E + 00
Btu · in./h · ft² · °F	W/m · K	1.442 279 E − 01
cal/cm · s · °C	W/m · K	4.184 000 E + 02
Thermal expansion		
in./in. · °C	m/m · K	1.000 000 E + 00
in./in. · °F	m/m · K	1.800 000 E + 00
Velocity		
ft/h	m/s	8.466 667 E − 05
ft/min	m/s	5.080 000 E − 03
ft/s	m/s	3.048 000 E − 01
in./s	m/s	2.540 000 E − 02
km/h	m/s	2.777 778 E − 01
mph	km/h	1.609 344 E + 00
Velocity of rotation		
rev/min (rpm)	rad/s	1.047 164 E − 01
rev/s	rad/s	6.283 185 E + 00
Viscosity		
poise	Pa · s	1.000 000 E − 01
stokes	m²/s	1.000 000 E − 04
ft²/s	m²/s	9.290 304 E − 02
in.²/s	mm²/s	6.451 600 E + 02
Volume		
in.³	m³	1.638 706 E − 05
ft³	m³	2.831 685 E − 02
fluid oz	m³	2.957 353 E − 05
gal (U.S. liquid)	m³	3.785 412 E − 03
Volume per unit time		
ft³/min	m³/s	4.719 474 E − 04
ft³/s	m³/s	2.831 685 E − 02
in.³/min	m³/s	2.731 177 E − 07
Wavelength		
Å	nm	1.000 000 E − 01

(a) kg × 10³ = 1 metric ton

SI prefixes—names and symbols

Exponential expression	Multiplication factor	Prefix	Symbol
10^{18}	1 000 000 000 000 000 000	exa	E
10^{15}	1 000 000 000 000 000	peta	P
10^{12}	1 000 000 000 000	tera	T
10^9	1 000 000 000	giga	G
10^6	1 000 000	mega	M
10^3	1 000	kilo	k
10^2	100	hecto(a)	h
10^1	10	deka(a)	da
10^0	1	BASE UNIT	
10^{-1}	0.1	deci(a)	d
10^{-2}	0.01	centi(a)	c
10^{-3}	0.001	milli	m
10^{-6}	0.000 001	micro	μ
10^{-9}	0.000 000 001	nano	n
10^{-12}	0.000 000 000 001	pico	p
10^{-15}	0.000 000 000 000 001	femto	f
10^{-18}	0.000 000 000 000 000 001	atto	a

(a) Nonpreferred. Prefixes should be selected in steps of 10^3 so that the resultant number before the prefix is between 0.1 and 1000. These prefixes should not be used for units of linear measurement, but may be used for higher order units. For example, the linear measurement, decimeter, is nonpreferred, but square decimeter is acceptable.

Abbreviations and Symbols

a crystal lattice length along the *a* axis; crack length

A ampere

A area; ratio of the alternating stress amplitude to the mean stress

Å angstrom

AAR Association of American Railroads

ac alternating current

Ac_cm in hypereutectoid steel, temperature at which cementite completes solution in austenite

Ac_1 temperature at which austenite begins to form on heating

Ac_3 temperature at which transformation of ferrite to austenite is completed on heating

ADTT average daily truck traffic

Ae_cm, Ae_1, Ae_3 equilibrium transformation temperatures in steel

AES Auger electron spectroscopy

AFS American Foundrymen's Society

AIME American Institute of Mining, Metallurgical and Petroleum Engineers

AISI American Iron and Steel Institute

AMS Aerospace Material Specification (of SAE)

ANSI American National Standards Institute

AOD argon-oxygen decarburization (or decarburized)

API American Petroleum Institute

Ar_cm temperature at which cementite begins to precipitate from austenite on cooling

Ar′ temperature at start of transformation of austenite to pearlite during cooling

Ar_1 temperature at which transformation to ferrite or to ferrite plus cementite is completed on cooling

Ar_3 temperature at which transformation of austenite to ferrite begins on cooling

ASM American Society for Metals

ASME American Society of Mechanical Engineers

ASTM American Society for Testing and Materials

at.% atomic percent

atm atmosphere (pressure)

b Burgers vector

b crystal lattice length along the *b* axis

B thickness

bal balance or remainder

bcc body-centered cubic

Btu British thermal unit

c crystal lattice length along the *c* axis

C coulomb

cal calorie

CCD charge-coupled device

CCT continuous cooling transformation (diagram)

CE carbon equivalent

CERT constant extension rate test

CFTA Committee of Foundry Technical Associations

cm centimeter

CMOS complementary metal oxide semiconductor

cpm cycles per minute

cps cycles per second

CRT cathode ray tube

CT compact tension (test specimen)

CTOD crack-tip opening displacement

CVD chemical vapor deposition

CVN Charpy V-notch (impact test or specimen)

d day

d used in mathematical expressions involving a derivative (denotes rate of change); depth; diameter

da/dN fatigue crack growth rate

DBTT ductile-brittle transition temperature

dc direct current

diam diameter

DIB diiodobutane

DIP dual in-line package

dm decimeter

DMW dissimilar-metal weld

DNB departure from nucleate boiling

DOA dead on arrival

DOF device operating failure

DPH diamond pyramid hardness (Vickers hardness)

DRAM dynamic random access memory

DT dynamic tear (test)

DWT drop-weight test

DWTT drop-weight tear test

e electron; natural log base, 2.71828

E energy; modulus of elasticity

ECT equicohesive temperature

EDAC error detection and correction

EDM electric discharge machining

EDS energy-dispersive spectroscopy

EDX energy-dispersive x-ray (analysis)

EPFM elastic-plastic fracture mechanics

EPMA electron probe microanalysis

Eq equation

ESD electrostatic discharge

ESR electroslag remelted

et al. and others

ETP electrolytic tough pitch (copper)

eV electron volt

f coefficient of friction

F force

fcc face-centered cubic

FET field-effect transistor

FHWA Federal Highway Administration

Fig. figure

ft foot

g gram

gal gallon

GPa gigapascal

h hour

HAZ heat-affected zone

HB Brinell hardness

hcp hexagonal close-packed

HIC hydrogen-induced cracking

HK Knoop hardness

hp horsepower

HR Rockwell hardness (requires scale designation, such as HRC for Rockwell C hardness)

HSLA high-strength low-alloy

HV Vickers hardness (diamond pyramid hardness)

Hz hertz

IACS International Annealed Copper Standard

ID inner diameter

in. inch

ISO International Organization for Standardization

J joule

J crack growth energy release rate (fracture mechanics); current density

K Kelvin

K stress-intensity factor

ΔK stress-intensity factor range

K_c plane-stress fracture toughness

K_{Ic} plane-strain fracture toughness

K_{Id} dynamic fracture toughness

K_f fatigue notch factor

K_{Iscc} threshold stress intensity for stress-corrosion cracking

K_t stress-concentration factor

K_{th} threshold stress-intensity factor

keV kiloelectron volt

kg kilogram

kgf kilogram force

km kilometer

kPa kilopascal

ksi kips (1000 lb) per square inch

kV kilovolt

L length

L liter

L_{50} median life of a rolling-element bearing

LAMMA laser microprobe mass analysis

lb pound

lbf pound force

LEFM linear elastic fracture mechanics

LME liquid metal embrittlement

LMIE liquid metal induced embrittlement

LMP Larson-Miller parameter

ln natural logarithm (base *e*)

log common logarithm (base 10)

LSI large-scale integration

m meter

M molar solution

M_f temperature at which martensite formation finishes during cooling

M_s temperature at which martensite starts to form from austenite on cooling

max maximum

MDPR mean depth-of-penetration rate

mg milligram

Mg megagram

min minimum; minute

MJ megajoule

mL milliliter

mm millimeter

mol% mole percent

MPa megapascal

mph miles per hour

MTF mean time-to-failure

n strain-hardening exponent

N newton

N fatigue life (number of cycles); normal solution

NACE National Association of Corrosion Engineers

NASA National Aeronautics and Space Administration

NBS National Bureau of Standards

ND normal direction (of a sheet)

NDE nondestructive evaluation

NDT nil-ductility transition; nondestructive testing

NDTT nil-ductility transition temperature

N(E) electron energy distribution

NLEFM nonlinear elastic fracture mechanics

nm nanometer

No. number

NTSB National Transportation and Safety Board

OD outside diameter

OECD Organization for Economic Cooperation and Development

oz ounce

p page

P applied load

Pa pascal

PCB printed circuit board

pH negative logarithm of hydrogen-ion activity

PH precipitation hardenable

P/M powder metallurgy

ppm parts per million

psi pounds per square inch

psig pounds per square inch, gage

PTFE polytetrafluoroethylene

PTH plated through-holes

PVC polyvinyl chloride

PVD physical vapor deposition

q fatigue notch sensitivity

R radius of curvature; ratio of the minimum stress to the maximum stress

RAD ratio-analysis diagram

RAM random access memory

RD rolling direction (of a sheet)

Ref reference

RH reheater

rms root mean square

rpm revolutions per minute

RT_{NDT} reference nil-ductility temperature

RTV room-temperature vulcanizing (rubber)

s second

S siemens

S strain energy; stress

S_a alternating stress amplitude

S_C carbon saturation

S_m mean stress

S_{max} maximum stress

S_{min} minimum stress

S_r range of stress

SAE Society of Automotive Engineers

SATT shear-area transition temperature

SCC stress-corrosion cracking

SCE saturated calomel electrode

SCFM subcritical fracture mechanics

SEM scanning electron microscopy

sfm surface feet per minute

SH superheater

SI Système International d'Unités

SIMS secondary ion mass spectroscopy

SMIE solid metal induced embrittlement

SMYS specified minimum yield stress

SSC sulfide-stress cracking

SSI small-scale integration

t thickness; time

T temperature

T_m, T_M melting temperature

TC total carbon (content)

tcp topologically close-packed

TD transverse direction (of a sheet)

TEG triethylene glycol

TEM transmission electron microscopy

TIR total indicator reading

TL transmission line

TMA thermomechanical analysis

TTU through-transmission ultrasonics

UNS Unified Numbering System (ASTM-SAE)

UTS ultimate tensile strength

UV ultraviolet

V volt

V_{ab} volume rate of abrasive wear per unit length of sliding

VAR vacuum arc remelting

VLSI very large-scale integration

vol volume

vol% volume percent

W watt

W width

WDS wavelength-dispersive spectroscopy

WOL wedge-opening loading

wt% weight percent

XPS x-ray photoelectron spectrosocpy

yr year

Z atomic number

° angular measure; degree

°C degree Celsius (centigrade)

°F degree Fahrenheit

°dH degree of hardness (for water)

⇆ direction of reaction

÷ divided by

= equals

≈ approximately equals

≠ not equal to

≡ identical with

> greater than

≫ much greater than

≥ greater than or equal to

∞ infinity

∝ is proportional to; varies as

∫ integral of

< less than

≪ much less than

≤ less than or equal to

± maximum deviation

− minus; negative ion charge

× diameters (magnification); multiplied by

· multiplied by

Ω ohm

/ per

% percent

+ plus; positive ion charge

√ square root of

~ approximately; similar to

↑ substance passes off as a gas

α angle

Δ change in quantity; an increment; a range

ϵ strain

μin. microinch

μm micron (micrometer)

ν Poisson's ratio

π pi (3.141592)

ρ density

σ stress

σ_{YS} yield stress

Greek Alphabet

A, α	alpha	I, ι	iota	P, ρ	rho
B, β	beta	K, κ	kappa	Σ, σ	sigma
Γ, γ	gamma	Λ, λ	lambda	T, τ	tau
Δ, δ	delta	M, μ	mu	Υ, υ	upsilon
E, ϵ	epsilon	N, ν	nu	Φ, ϕ	phi
Z, ζ	zeta	Ξ, ξ	xi	X, χ	chi
H, η	eta	O, o	omicron	Ψ, ψ	psi
Θ, θ	theta	Π, π	pi	Ω, ω	omega

Index